# THE COSMIC PERSPECTIVE

Astronauts get a unique opportunity to experience a cosmic perspective. Here, astronaut John Grunsfeld has a CD of *The Cosmic Perspective* floating in front of him while he orbits Earth during the Space Shuttle's final servicing mission to the Hubble Space Telescope (May 2009).

Ninth Edition

# THE COSMIC PERSPECTIVE

**JEFFREY BENNETT**
University of Colorado at Boulder

**MEGAN DONAHUE**
Michigan State University

**NICHOLAS SCHNEIDER**
University of Colorado at Boulder

**MARK VOIT**
Michigan State University

Pearson

Director Physical Science Portfolio: Jeanne Zalesky
Physics & Astronomy Portfolio Analyst:
Ian Desrosiers
Content Producer: Mary Tindle
Managing Producer: Kristen Flathman
Product Portfolio Editorial Assistant: Frances Lai
Rich Media Content Producer: Ali Candlin
Full-Service Vendor: Lifland et al., Bookmakers
Copyeditor: Lifland et al., Bookmakers
Compositor: Pearson CSC
Design Manager: Maria Guglielmo
Interior and Cover Designer: Tamara Newnam
Illustrators: Rolin Graphics; Troutt Visuals
Rights & Permissions Project Manager: Ben Ferrini
Rights & Permissions Management: Matthew Perry
Photo Researcher: Matthew Perry
Manufacturing Buyer: Stacey Weinberger
Director of Field Marketing: Tim Galligan
Director of Product Marketing: Allison Rona
Field Marketing Manager: Yez Alayan
Product Marketing Manager: Elizabeth Bell

Cover Photo Credit:
Milky Way over Ahu Tongariki—Anne Dirkse/Getty Images



**Library of Congress Cataloguing-in-Publication Data**

Names: Bennett, Jeffrey O., author. | Donahue, M. (Megan), 1962- author. | Schneider, Nicholas, author. | Voit, Mark, author.
Title: The cosmic perspective / Jeffrey Bennett (University of Colorado at Boulder), Megan Donahue (Michigan State University), Nicholas Schneider (University of Colorado at Boulder), Mark Voit (Michigan State University).
Description: Ninth edition. | New York, NY : Pearson, [2020] | Includes index.
Identifiers: LCCN 2018048551 | ISBN 9780134874364
Subjects: LCSH: Astronomy—Textbooks.
Classification: LCC QB43.3 .C68 2020 | DDC 520—dc23
LC record available at https://lccn.loc.gov/2018048551

Pearson

ISBN-10: 0-13-524460-9 (High School Binding)
ISBN-13: 978-0-13-524460-9 (High School Binding)

www.PearsonSchool.com/Advanced

# DEDICATION

To all who have ever wondered about the mysteries of the universe. We hope this book will answer some of your questions—and that it will also raise new questions in your mind that will keep you curious and interested in the ongoing human adventure of astronomy. And, especially, to Michaela, Emily, Sebastian, Grant, Nathan, Brooke, and Angela. The study of the universe begins at birth, and we hope that you will grow up in a world with far less poverty, hatred, and war so that all people will have the opportunity to contemplate the mysteries of the universe into which they are born.

# Brief Contents

# Explore Modern Astronomy and Its Connections to Our Lives

***The Cosmic Perspective*** provides a thoroughly engaging and up-to-date introduction to astronomy for anyone who is curious about the universe. As respected teachers and active researchers, the authors present astronomy using a coherent narrative and a thematic approach that engages students immediately and guides them through connecting ideas. The **Ninth Edition** features major scientific updates, new content that focuses on the possibility of life in the universe and recent discoveries, and an enhanced focus on cultural diversity among scientists and ethics across science and astronomy. **Mastering Astronomy** includes a wealth of author-created resources for students to use before, during, and after class.

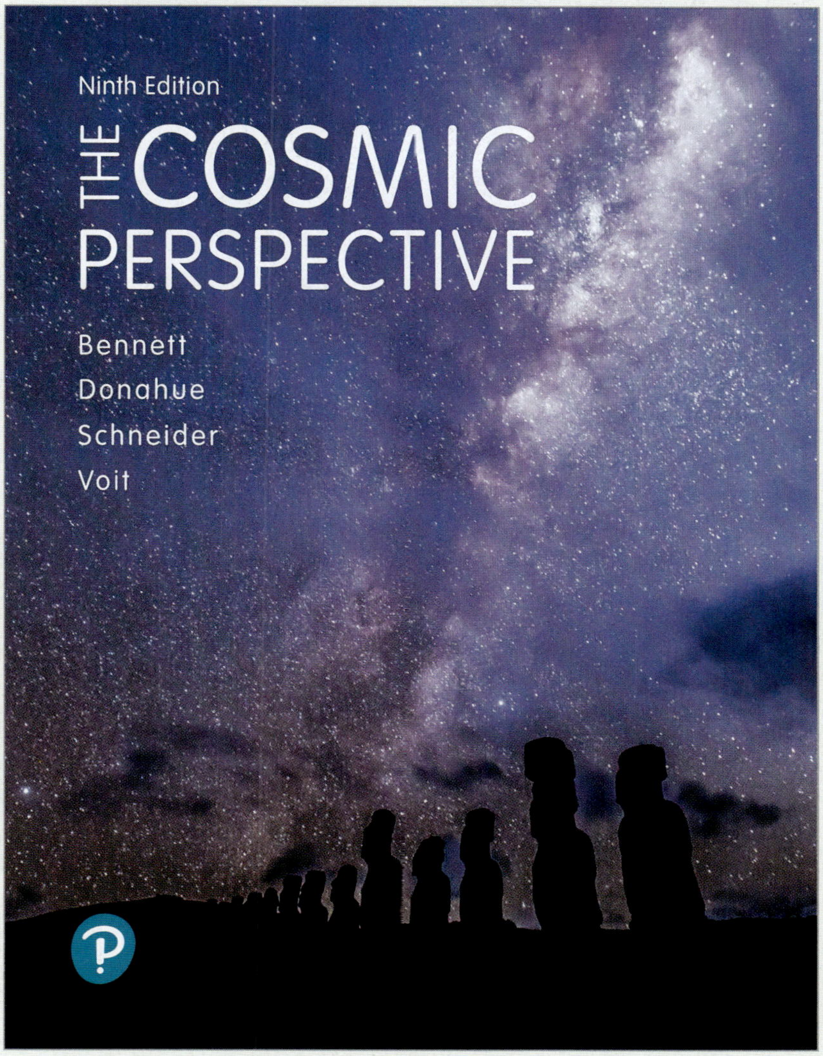


Pearson

# Fully Updated Science Engages Students

P. 311

Chapter 11 has been updated with the latest discoveries from the *Juno* and *Cassini* missions, as well as new understanding of Jupiter's weather, Saturn's rings, and more.

Other updates include new material on the detection of gravitational waves (Chapters S3 and 18), new insights into extrasolar planets (Chapter 13), new discoveries about early life on Earth (Chapter 24), and much more.

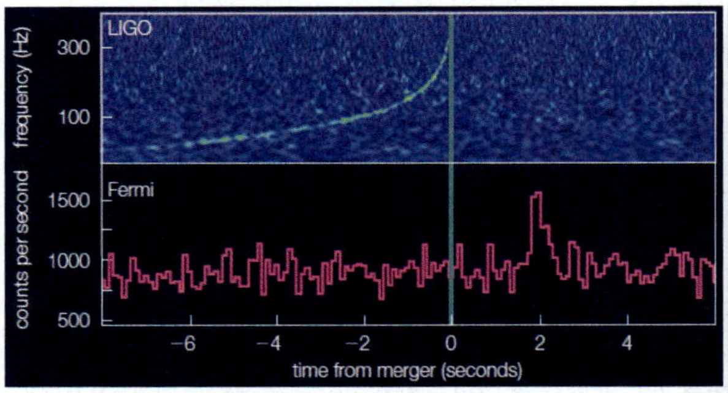


a Top: Detection on August 17, 2017 of the gravitational wave signal from a merger of two neutron stars. The light-blue track standing out from the background noise (darker speckles) shows the steady climb of the signal's frequency from below 100 Hz through more than 300 Hz just before the merger. The vertical green line shows the moment of the merger event. The signal's peak frequency shows that the neutron stars were orbiting one another more than 300 times per second just before the merger. Bottom: Detection by the Fermi observatory of a gamma-ray burst from the merger two seconds after the end of the gravitational-wave signal.

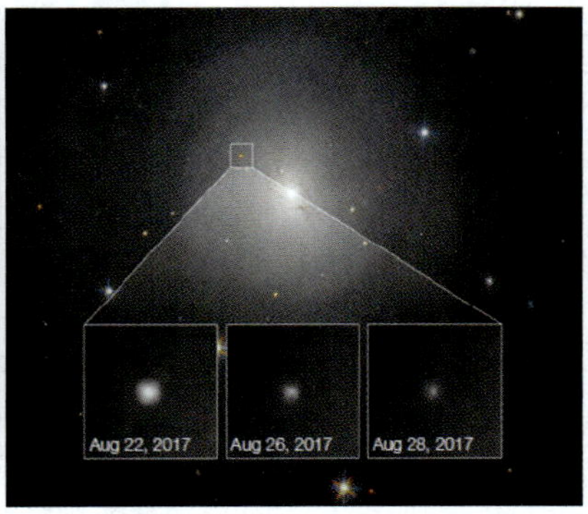


b Visible light from the aftermath of the neutron-star merger shown in part a, as seen by the Hubble Space Telescope. Information from the LIGO signal allowed astronomers to find this afterglow in the galaxy NGC 4993, about 130 million light-years away. The inset frames show the afterglow fading during the two weeks following the gravitational-wave detection.

**FIGURE 18.18** Observed signals from the first detection of a merger of two neutron stars.

P. 572

# Revised End-of-Chapter Exercises Deepen Student Understanding

## Inclusive Astronomy

*Use these questions to reflect on participation in science.*

33. *Group Discussion: Ancestral Astronomy.* No matter what your background, you had ancestors who watched the sky and observed how celestial objects move through it.
**a.** Working independently, choose a particular branch of your ancestry to explore, then gather historical information dating back as far as possible (ideally at least several centuries) about how and why your ancestors made use of their observations of the sky.
**b.** Gather in small groups (two to four students) and take turns sharing what you learned through your research. Be clear about the ancestral group you have chosen and the time period your research covers.
**c.** Make a list of the major uses of astronomical knowledge that your group members found, categorizing the uses as practical, ceremonial/religious, or other. Which uses were most common? Do you see any noticeable differences among the cultures?
**d.** Discuss how cultural or geographical factors may have influenced the astronomical knowledge of these different ancestral groups.

P. 82

**NEW!** Each chapter now has **Inclusive Astronomy** exercises designed to spur student discussion about topics such as why astronomy belongs to everyone and the ways in which the astronomical community is working to address historical inequities.

**The Process of Science,** a major theme integrated throughout the main text, is reinforced with a set of short-answer questions at the end of each chapter, as well as a suite of tutorials available for assignment in Mastering Astronomy.

## The Process of Science

*These questions may be answered individually in short-essay form or discussed in groups, except where identified as group-only.*

34. *What Makes It Science?* Choose a single idea in the modern view of the cosmos as discussed in Chapter 1, such as "The universe is expanding," "The universe began with a Big Bang," "We are made from elements manufactured by stars," or "The Sun orbits the center of the Milky Way Galaxy once every 230 million years."
**a.** Describe how this idea reflects each of the three hallmarks of science, discussing how it is based on observations, how our understanding of it depends on a model, and how that model is testable.
**b.** Describe a hypothetical observation that, if it were actually made, might cause us to call the idea into question. Then briefly discuss whether you think that, overall, the idea is likely or unlikely to hold up to future observations.
35. *The Importance of Ancient Astronomy.* Why was astronomy important to people in ancient times? Discuss both the practical importance of astronomy and the importance it may have had for religious, ceremonial, or philosophical traditions. Which of those roles (practical or religious/ceremonial/philosophical) do you think was more important in leading to the development of modern astronomy? Defend your opinion.
36. *The Impact of Science.* The modern world is filled with ideas, knowledge, and technology that developed through science and application of the scientific method. Discuss some of these things and how they affect our lives. Which of these impacts do you think are positive? Which are negative? Overall, do you think science has benefited the human race? Defend your opinion.

P. 82

See the newly reorganized end-of-chapter exercise sets for additional problem types designed to help your students review key concepts, check their understanding, and learn to think critically. All exercises are also assignable through Mastering Astronomy.

# Mastering Astronomy's Study Area Helps Students Come Prepared to Class . . .

**The Study Area** features self-study Reading, Concept, and Visual quizzes for each chapter, many videos and interactive figures, a set of self-guided tutorials covering key concepts, a media workbook, World Wide Telescope tours, and much more — PLUS access to a full etext of The Cosmic Perspective.

**NEW!** Nearly 100 new videos about key concepts and figures in the text, all written and most narrated by the authors to ensure consistency of terminology and pedagogy. Most videos include embedded pause-and-predict questions that allow students to check their understanding as they watch. Students can use these videos to help prepare for lectures, while instructors will find the same videos with assignable tutorials in the instructor-accessible Item Library.

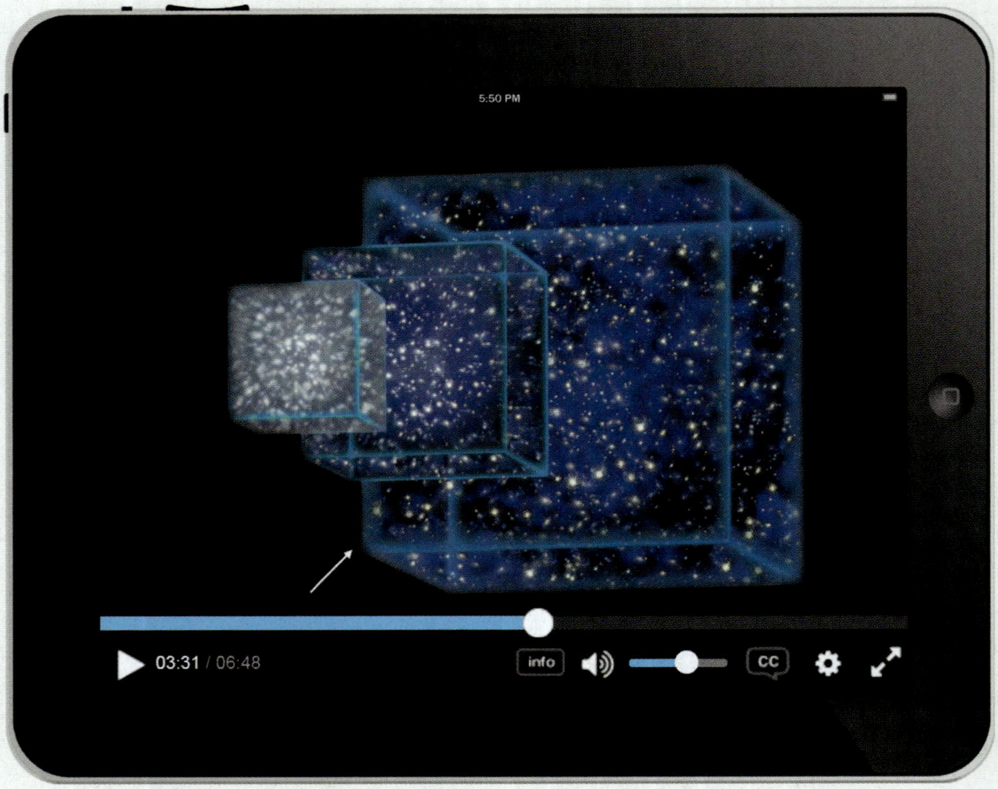

# . . . While Instructors Can Access a Large Library of Homework and Test Questions

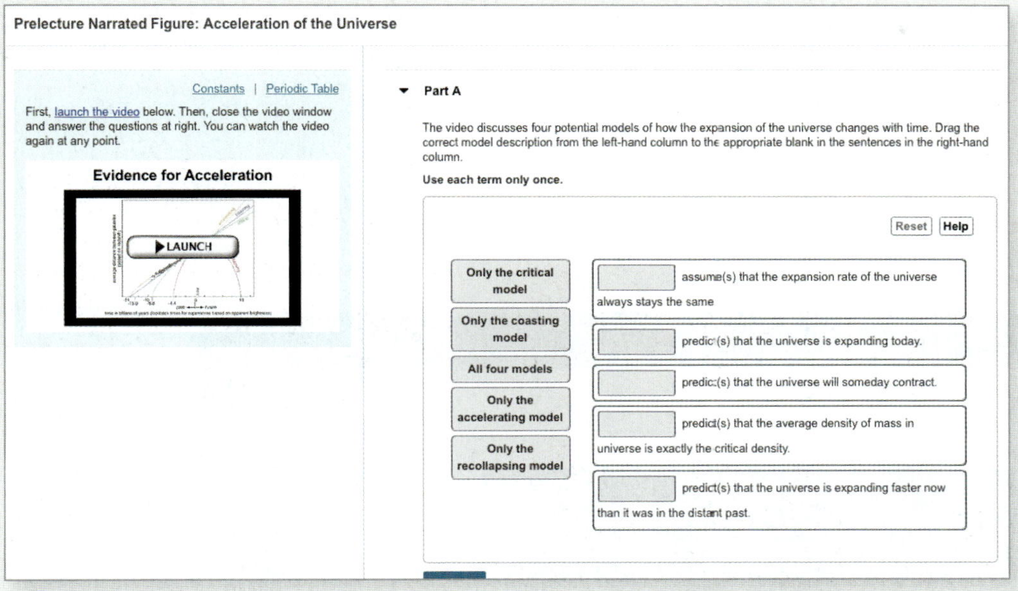


**The Item Library** features more than 250 assignable tutorials—all written or co-written by the textbook authors—including new tutorials based on all of the videos and interactive figures as well as updated tutorials on key concepts, process of science, vocabulary, and much more.

Many of the **assignable tutorials** use ranking or sorting tasks, which research shows to be particularly effective in building conceptual understanding.

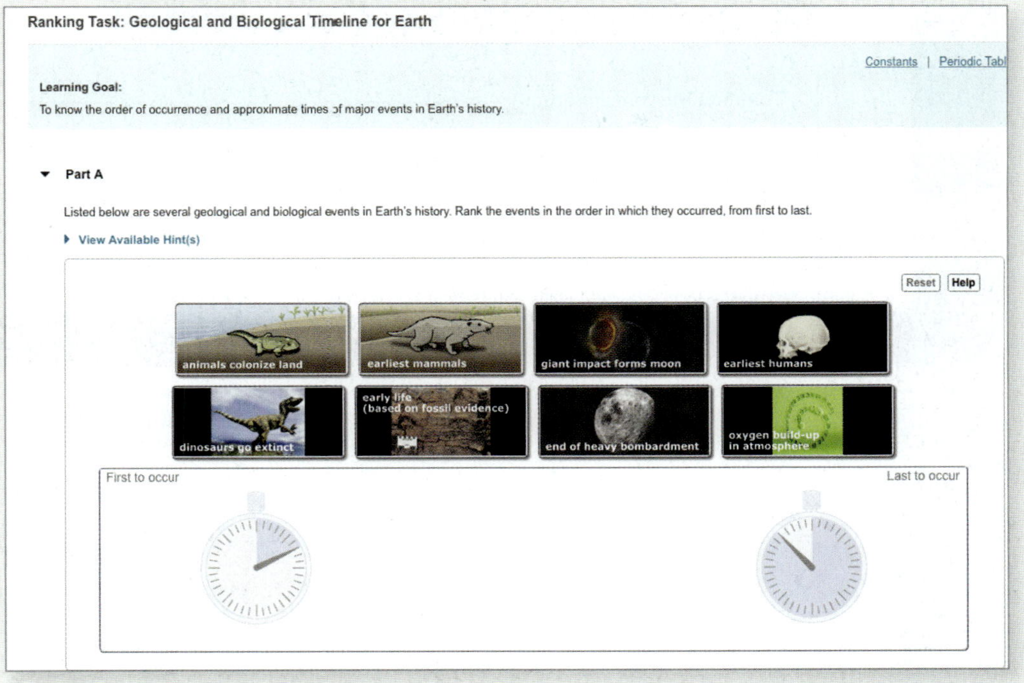


**The Item Library** also includes all end-of-chapter exercises from the book, individual questions from the three self-study quizzes (Reading, Concept, and Visual) for each chapter, and a large test bank.

# Reach Every Student with Pearson eText

**Pearson eText, optimized for mobile,** seamlessly integrates videos and other rich media with the text and gives students access to their textbook anytime, anywhere.

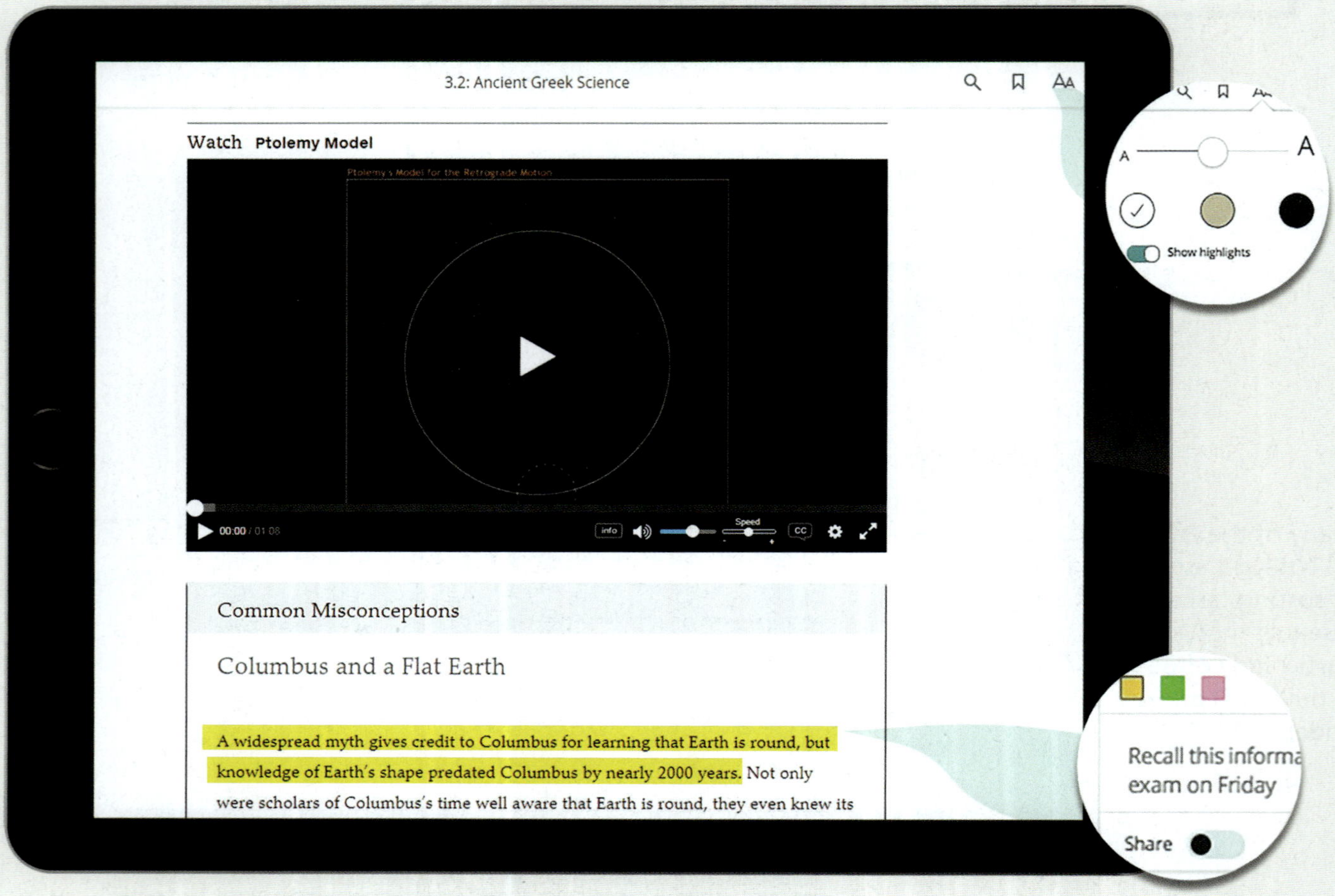

# Engage Students Before and During Class with Dynamic Study Modules and Learning Catalytics

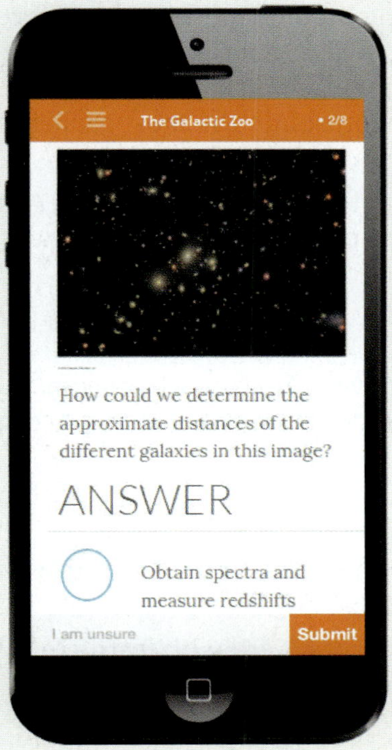


**NEW! Dynamic Study Modules in Mastering Astronomy** help students study effectively—and at their own pace—by keeping them motivated and engaged. The assignable modules rely on the latest research in cognitive science, using methods—such as adaptivity, gamification, and intermittent rewards—to stimulate learning and improve retention.

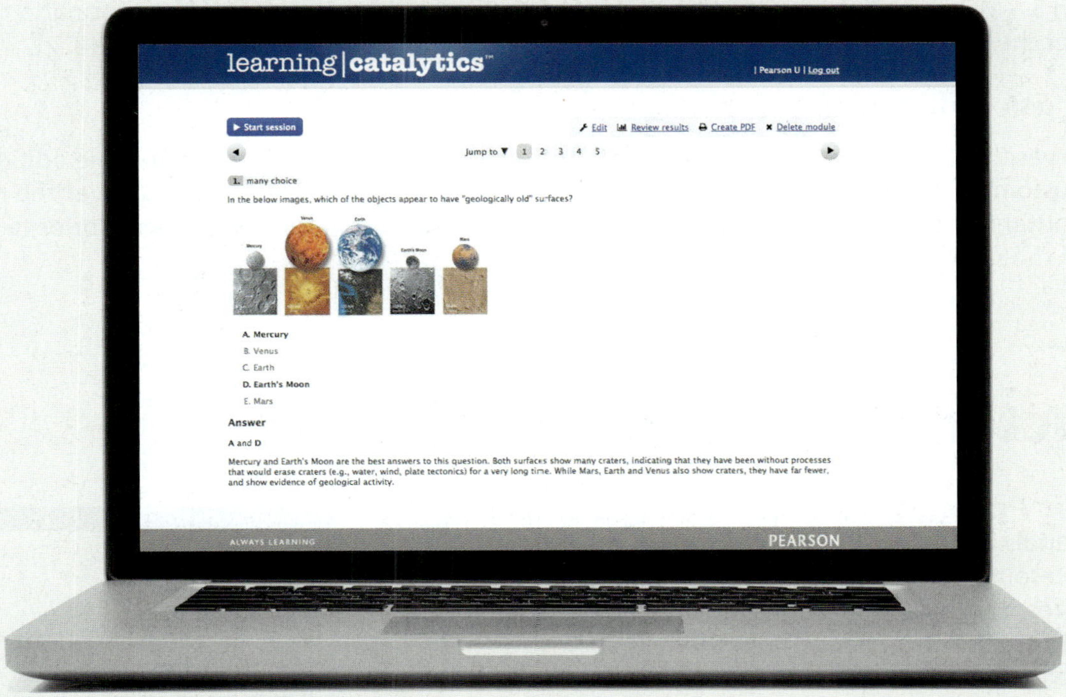


With **Learning Catalytics**, you'll hear from every student when it matters most. You pose a variety of questions that help students recall ideas, apply concepts, and develop critical-thinking skills. Your students respond using their own smartphones, tablets, or laptops. You can monitor responses with real-time analytics and find out what your students do—and don't—understand. Then you can adjust your teaching accordingly, and even facilitate peer-to-peer learning, helping students stay motivated and engaged.

# Teaching and Learning Materials/ Mastering Astronomy™ Access

Mastering Astronomy is the most advanced, most widely used online astronomy tutorial and homework program available. Mastering Astronomy, with Pearson eText, utilizes the Socratic method to coach students through problem-solving techniques, offering hints and simpler questions on request to help students learn, not just practice. Tutorials guide students through the most challenging topics and help them make connections.

Upon textbook purchase, students and teachers are granted access to Mastering Astronomy with Pearson eText. High school teachers can obtain preview or adoption access to Mastering Astronomy in one of the following ways:

### Preview Access

• Teachers can request preview access online by visiting www.PearsonSchool.com/Access_Request. Select Science, choose Initial Access, and complete the form under Option 2. Preview Access information will be sent to the teacher via e-mail.

### Adoption Access

• With the purchase of this program, a Pearson Adoption Access Card with Instructor Manual will be delivered with your textbook purchase. (ISBN: 978-0-13-354087-1)

• Ask your sales representative for a Pearson Adoption Access Card with Instructor Manual. (ISBN: 978-0-13-354087-1)

OR

• Visit PearsonSchool.com/Access_Request, select Science, choose initial Access, and complete the form under Option 3—MyLab/Mastering Class Adoption Access. Teacher and Student access information will be sent to the teacher via e-mail.

### Students, ask your teacher for access.

Pearson reserves the right to change and/or update technology platforms, including possible edition updates to customers during the term of access. This will allow Pearson to continue to deliver the most up-to-date content and technology to customers. Customer will be notified of any change prior to the beginning of the new school year.

# Instructor support you can rely on

***The Cosmic Perspective*** includes a full suite of instructor support materials in the Instructor Resources area in Mastering Astronomy. Resources include lecture presentations, images, reading quizzes, and clicker questions in PowerPoint; labeled and unlabeled JPEGs of all images from the text; an instructor's guide for each chapter; and a test bank.

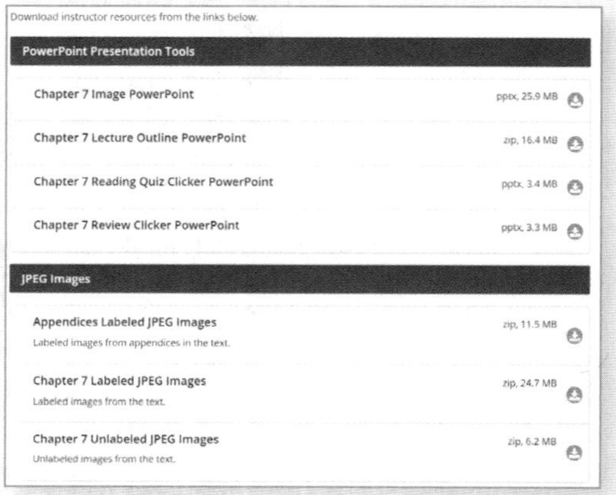

# Detailed Contents

# PART II
# KEY CONCEPTS FOR ASTRONOMY



# PART III
# LEARNING FROM OTHER WORLDS

## PART VII
## LIFE ON EARTH AND BEYOND

# Preface

We humans have gazed into the sky for countless generations. We have wondered how our lives are connected to the Sun, Moon, planets, and stars that adorn the heavens. Today, through the science of astronomy, we know that these connections go far deeper than our ancestors ever imagined. This book tells the story of modern astronomy and the new perspective, *The Cosmic Perspective*, that astronomy gives us on ourselves and our planet.

## Who Is This Book For?

*The Cosmic Perspective* provides a comprehensive survey of modern astronomy suitable for anyone who is curious about the universe, regardless of prior background in astronomy or physics. However, it is designed primarily to serve as a textbook for college courses in introductory astronomy. *The Cosmic Perspective* contains enough material for a full-year introductory astronomy sequence but can be flexibly used for shorter courses as well.

Instructors of shorter courses may also wish to consider several available variations of this textbook. We offer two volumes containing selected chapters from this book: *The Solar System*, which consists of Chapters 1–14 (including S1) and 24, and *Stars, Galaxies, and Cosmology*, which consists of Chapters 1–6 (including S1), S2–S4, and 14–24. Those teaching one-term general survey courses may wish to consider *The Essential Cosmic Perspective*, which covers a smaller set of topics and is tailored to meet the needs of comprehensive one-term survey courses in astronomy, and *The Cosmic Perspective Fundamentals*, which is even shorter and covers only the most fundamental topics in astronomy. All of these options are also available with Mastering™ Astronomy.

## New to This Edition

The underlying philosophy, goals, and structure of *The Cosmic Perspective* remain the same as in past editions, but we have thoroughly updated the text and made a number of other improvements. Here, briefly, is a list of the significant changes you'll find in the ninth edition:

- **Major Chapter-Level Changes:** We have made numerous significant changes to both update the science and improve the pedagogical flow in this edition. The full list is too long to put here, but major changes include the following:
  - In **Chapter 2**, we have reworked the section on eclipses with a new set of art pieces and revised pedagogy to reflect the fact that many students heard about or witnessed the 2017 eclipse.
  - **Chapter 5** has a new Common Misconception box on "Light Paths, Lasers, and Shadows."
  - In **Chapter 6**, we have made three significant updates: a focus on major new and planned observatories, including the James Webb Space Telescope; a new subsection on the role of "big data" in astronomy, using LSST as an example; and an expanded discussion of multi-messenger astronomy, such as the gravitational-wave observatory LIGO.
  - **Chapter 9** has numerous scientific updates based on recent planetary missions, especially in Section 9.4 on Mars, where we discuss how recent evidence from the *Curiosity* rover is providing a new view of past periods of liquid water on Mars. We also discuss recent reanalysis of the cause of dark streaks and gullies on crater walls.
  - **Chapter 10** has similar updates for new data, including the addition of a new Learning Goal in Section 10.4 on Mars, to reflect a deeper discussion of the history of the Martian climate. Section 10.6 on global warming has also been significantly updated, with greater emphasis on expected consequences of the warming.
  - In **Chapter 11**, we have revamped the discussion of Jupiter's weather to include new results from the *Juno* mission. We have also made important scientific updates to the information on Saturn's moons and rings based on results from the final stages of the *Cassini* mission.
  - **Chapter 12** includes updated data and images from the *Dawn*, *Rosetta*, and *New Horizons* missions, along with discussion of the possibility of an undiscovered "Planet 9" and a new Special Topic box on 'Oumuamua, the first confirmed object with origin beyond our solar system to pass through our solar system.
  - **Chapter 13** covers the fast-evolving topic of extrasolar planets and hence has numerous scientific updates and new figures.
  - In **Chapter S3**, we have rewritten the section on gravitational waves to include their recent direct detection.
  - **Chapter 14** includes new discussion and a new figure about the Sun's influence on Earth's climate, and how we can rule out a changing Sun as a cause of recent global warming.
  - In **Chapter 18**, we have almost completely rewritten Section 18.4—changing from the two learning goals in the prior edition to three learning goals—to include the detection of gravitational waves from neutron star and black hole mergers.

- **Chapters 20 and 21** have been updated in light of new research on galactic evolution, some of which is based on the work of two of the authors of this book (Donahue and Voit). Chapter 20 also incorporates updates in describing the cosmic distance scale, including honoring Henrietta Levitt by referring to her period-luminosity relation as *Leavitt's law*.
- In **Chapter 24**, the first section has significant changes to incorporate newly discovered evidence for early life on Earth. The second section has been reworked to update the discussion of searching for life on Mars, and its second learning goal has been reworked to cover more than just the moons of Jupiter and Saturn.

- **Fully Updated Science:** Astronomy is a fast-moving field, and numerous new developments have occurred since the prior edition was published. In addition to the major chapter-level changes described above, we have made many other scientific updates to reflect the latest results from both ground-based and space-based observatories and from spacecraft missions within the solar system.
- **Revamped Exercise Sets:** We have reorganized the end-of-chapter exercise sets in order to place greater emphasis on questions designed to promote discussion and group work.
- **New Feature—Inclusive Astronomy:** The astronomical community is engaged in broad and wide-ranging conversations about inclusion and the persistent lack of diversity in the fields of astronomy and other sciences. To provide sample openings for discussions of inclusion in the classroom, we have (1) added a new set of exercises in every chapter under the heading "Inclusive Astronomy," written to initiate student discussions about topics centered on inclusiveness in astronomy; (2) added a similar set of additional exercises that you can find in the set of Group Activities available in the Study Area of Mastering Astronomy; and (3) replaced many of the chapter-opening epigraphs in order to include a more diverse group of individuals.
- **New Content in Mastering Astronomy:** *The Cosmic Perspective* is much more than a textbook; it is a complete "learning package" that combines the textbook with deeply integrated, interactive media developed to support every chapter of our book. We continually update the material on the Mastering Astronomy website, and for this edition we call your attention to nearly 100 new "prelecture videos," all written by (and most narrated by) the authors, designed to help students understand key concepts. Students can watch the videos at any time in the Study Area, while instructors can find assignable tutorials based on the videos in the instructor-accessible Item Library. In addition to the new videos and their corresponding tutorials, you will find many other new tutorials in the Item Library, as well as a fully updated set of reading, concept, and visual quizzes for each chapter, available in both the Study Area and the assignable Item Library. These resources should be especially valuable to instructors who wish to offer assignments designed to ensure that students are prepared before class and to those using "flipped classroom" strategies.

## The Pedagogical Approach of *The Cosmic Perspective*

*The Cosmic Perspective* offers a broad survey of modern understanding of the cosmos and of how we have built that understanding. Such a survey can be presented in a number of different ways. We have chosen to build *The Cosmic Perspective* around a set of key themes designed to engage student interest and a set of pedagogical principles designed to ensure that all material comes across as clearly as possible to students.

### Themes

Most students enrolled in introductory astronomy courses have little connection to astronomy when their course begins, and many have little understanding of how science actually works. The success of these students therefore depends on getting them engaged in the subject matter. To help achieve this, we have chosen to focus on the following five themes, which are interwoven throughout the book.

*Theme 1: We are a part of the universe and can therefore learn about our origins by studying the universe.* This is the overarching theme of *The Cosmic Perspective*, as we continually emphasize that learning about the universe helps us understand ourselves. Studying the intimate connections between human life and the cosmos gives students a reason to care about astronomy and also deepens their appreciation of the unique and fragile nature of our planet and its life.

*Theme 2: The universe is comprehensible through scientific principles that anyone can understand.* The universe is comprehensible because the same physical laws appear to be at work in every aspect, on every scale, and in every age of the universe. Moreover, while professional scientists generally have discovered the laws, anyone can understand their fundamental features. Students can learn enough in one or two terms of astronomy to comprehend the basic reasons for many phenomena that they see around them—phenomena ranging from seasonal changes and phases of the Moon to the most esoteric astronomical images that appear in the news.

*Theme 3: Science is not a body of facts but rather a process through which we seek to understand the world around us.* Many students assume that science is just a laundry list of facts. The long history of astronomy can show them that science is a process through which we learn about our universe—a process that is not always a straight line to the truth. That is why our ideas about the cosmos sometimes change as we learn more, as they did dramatically when we first recognized that Earth is a planet going around the Sun rather than the center of the universe. In this book, we continually emphasize the nature of science so that students can understand how and why modern theories have gained acceptance and why these theories may change in the future.

*Theme 4: Astronomy belongs to everyone.* Astronomy has played a significant role throughout history in virtually every culture, and the modern science of astronomy owes a debt to these early and largely unsung astronomers. We therefore strive throughout the book to make sure that students understand that astronomical knowledge belongs to everyone, that people of all backgrounds have made and continue to make contributions to astronomical understanding, and that everyone should have the opportunity to study astronomy. Moreover, we seek to motivate students enough to ensure that they will remain engaged in the ongoing human adventure of astronomical discovery throughout their lives, no matter whether they choose to do that only by following the news media or by entering careers relating to astronomy.

*Theme 5: Astronomy affects each of us personally with the new perspectives it offers.* We all conduct the daily business of our lives with reference to some "world view"—a set of personal beliefs about our place and purpose in the universe, which we have developed through a combination of schooling, religious training, and personal thought. This world view shapes our beliefs and many of our actions. Although astronomy does not mandate a particular set of beliefs, it does provide perspectives on the architecture of the universe that can influence how we view ourselves and our world, and these perspectives can potentially affect our behavior. For example, someone who believes Earth to be at the center of the universe might treat our planet quite differently from someone who views it as a tiny and fragile world in the vast cosmos. In many respects, the role of astronomy in shaping world views may represent the deepest connection between the universe and the everyday lives of humans.

## Pedagogical Principles

No matter how an astronomy course is taught, it is very important to present material according to well-established pedagogical principles. The following list briefly summarizes the major pedagogical principles that we apply throughout this book.*

- *Stay focused on the big picture.* Astronomy is filled with interesting facts and details, but they are meaningless unless they fit into a big-picture view of the universe. We therefore take care to stay focused on the big picture (essentially the themes discussed above) at all times. A major benefit of this approach is that although students may forget individual facts and details after the course is over, the big-picture framework should stay with them for life.
- *Always provide context first.* We all learn new material more easily when we understand why we are learning it. In essence, this is simply the idea that it is easier to get somewhere when you know where you are going. We therefore begin the book (Chapter 1) with a broad overview of modern understanding of the cosmos, so that students know what they will be studying in the rest of the book. We maintain this "context first" approach throughout the book by always telling students what they will be learning, and why, before diving into the details.
- *Make the material relevant.* It's human nature to be more interested in subjects that seem relevant to our lives. Fortunately, astronomy is filled with ideas that touch each of us personally. For example, the study of our solar system helps us better understand and appreciate our planet Earth, and the study of stars and galaxies helps us learn how we have come to exist. By emphasizing our personal connections to the cosmos, we make the material more meaningful, inspiring students to put in the effort necessary to learn it.
- *Emphasize conceptual understanding over "stamp collecting" of facts.* If we are not careful, astronomy can appear to be an overwhelming collection of facts that are easily forgotten when the course ends. We therefore emphasize a few key conceptual ideas, which we use over and over again. For example, the laws of conservation of energy and conservation of angular momentum (introduced in Section 4.3) reappear throughout the book, and the wide variety of features found on the terrestrial planets are described in terms of just a few basic geological processes. Research shows that, long after the course is over, students are far more likely to retain such conceptual learning than individual facts or details.
- *Proceed from the more familiar and concrete to the less familiar and abstract.* It's well known that children learn best by starting with concrete ideas and then generalizing to abstractions later. The same is true for many adults. We therefore always try to "build bridges to the familiar"— that is, to begin with concrete or familiar ideas and then gradually draw more general principles from them.
- *Use plain language.* Surveys have found that the number of new terms in many introductory astronomy books is larger than the number of words taught in many first-year courses on a foreign language. In essence, this means the books are teaching astronomy in what looks to students like a foreign language! Clearly, it is much easier for students to understand key astronomical concepts if they are explained in plain English without resorting to unnecessary jargon. We have gone to great lengths to eliminate jargon or, at minimum, to replace standard jargon with terms that are easier to remember in the context of the subject matter.
- *Recognize and address student misconceptions.* Students do not arrive as blank slates. Most students enter our courses not only lacking the knowledge we hope to teach but also holding misconceptions about astronomical ideas. Therefore, to teach correct ideas, we must help students recognize the paradoxes in their prior misconceptions. We address this issue in a number of ways, the most obvious being the presence of many Common Misconceptions boxes. These

*More detail on these pedagogical principles can be found in the Instructor Guide and in the book *On Teaching Science* by Jeffrey Bennett (Big Kid Science, 2014).

summarize commonly held misconceptions and explain why they cannot be correct.

## The Organizational Structure of *The Cosmic Perspective*

*The Cosmic Perspective* is organized into seven broad topical areas (the seven parts in the table of contents), each corresponding to a set of chapters along with related content in Mastering Astronomy. Note that the above themes and pedagogical principles are woven into this structure at every level.

### Part Structure

The seven parts of *The Cosmic Perspective* each approach their set of chapters in a distinctive way designed to help maintain the focus on the five themes discussed earlier. Here, we summarize the philosophy and content of each part. Note that each part concludes with a two-page Cosmic Context spread designed to tie the part content together into a coherent whole.

#### Part I: Developing Perspective (Chapters 1–3, S1)

*Guiding Philosophy: Introduce the big picture, the process of science, and the historical context of astronomy.*

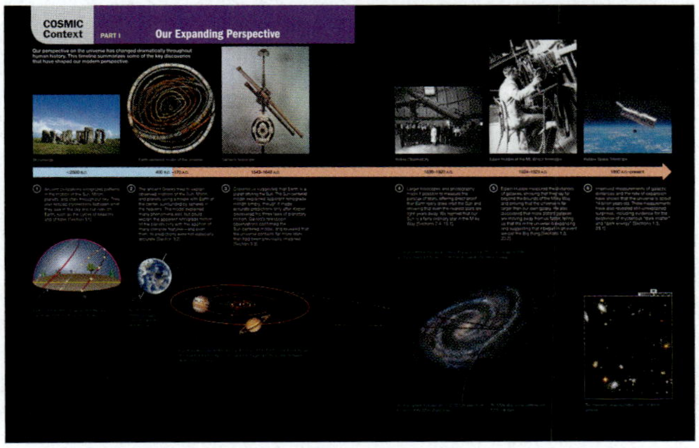


**The Cosmic Context figure for Part I appears on pp. 108–109.**

The basic goal of these chapters is to give students a big-picture overview and context for the rest of the book, as well as to help them develop an appreciation for the process of science and how science has developed through history. Chapter 1 outlines our modern understanding of the cosmos, including the scale of space and time, so that students gain perspective on the entire universe before diving into its details. Chapter 2 introduces basic sky phenomena, including seasons and phases of the Moon, and provides perspective on how phenomena we experience every day are tied to the broader cosmos. Chapter 3 discusses the nature of science, offering a historical perspective on the development of science and giving students perspective on how science works and how it differs from nonscience. The supplementary (optional) Chapter S1 goes into more detail about the sky, including celestial timekeeping and navigation.

#### Part II: Key Concepts for Astronomy (Chapters 4–6)

*Guiding Philosophy: Connect the physics of the cosmos to everyday experiences.*

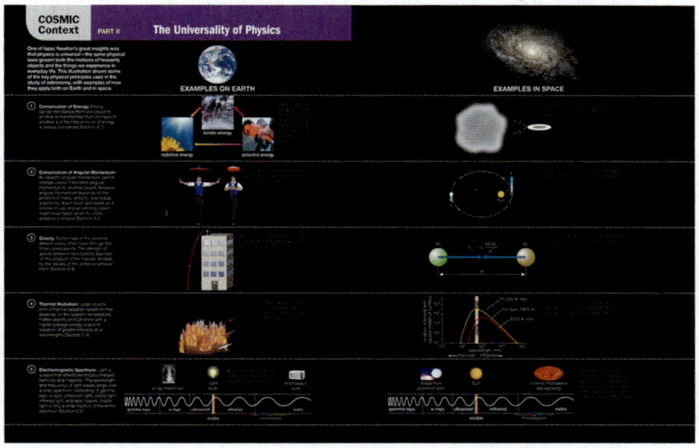


**The Cosmic Context figure for Part II appears on pp. 188–189.**

These chapters lay the groundwork for understanding astronomy through what is sometimes called the "universality of physics"—the idea that a few key principles governing matter, energy, light, and motion explain both the phenomena of our daily lives and the mysteries of the cosmos. Each chapter begins with a section on science in everyday life in which we remind students how much they already know about scientific phenomena from their everyday experiences. We then build on this everyday knowledge to help students learn the formal principles of physics needed for the rest of their study of astronomy. Chapter 4 covers the laws of motion, the crucial conservation laws of angular momentum and energy, and the universal law of gravitation. Chapter 5 deals with the nature of light and matter, the formation of spectra, and the Doppler effect. Chapter 6 covers telescopes and astronomical observing techniques.

#### Part III: Learning from Other Worlds (Chapters 7–13)

*Guiding Philosophy: We learn about our own world and existence by studying about other planets in our solar system and beyond.*

*Note:* Part III is essentially independent of Parts IV through VII and can be covered either before or after them.

**The Cosmic Context figure for Part III appears on pp. 400–401.**

This set of chapters begins in Chapter 7 with a broad overview of the solar system, including an 11-page tour that highlights some of the most important and interesting features of the Sun and each of the planets in our solar system. In the remaining chapters of this part, we seek to explain these features through a true *comparative planetology* approach, in which the discussion emphasizes the *processes* that shape the planets rather than the "stamp collecting" of facts about them. Chapter 8 uses the concrete features of the solar system presented in Chapter 7 to build student understanding of the current theory of solar system formation. Chapters 9 and 10 focus on the terrestrial planets, covering key ideas of geology and atmospheres, respectively. In both chapters, we start with examples from our own planet Earth to help students understand the types of features that are found throughout the terrestrial worlds and the fundamental processes that explain how these features came to be. We then complete each of these chapters by summarizing how the various processes have played out on each individual world. Chapter 11 covers the jovian planets and their moons and rings. Chapter 12 discusses small bodies in the solar system, including asteroids, comets, and dwarf planets. It also covers cosmic collisions, including the impact linked to the extinction of the dinosaurs and views on how seriously we should take the ongoing impact threat. Finally, Chapter 13 turns to the exciting topic of other planetary systems.

## Part IV: A Deeper Look at Nature (Chapters S2–S4)

*Guiding Philosophy: Ideas of relativity and quantum mechanics are accessible to anyone.*

*Note:* These chapters are labeled "supplementary" because coverage of them is optional. Covering them will give your students a deeper understanding of the topics that follow on stars, galaxies, and cosmology, but the later chapters are self-contained so that they may be studied without having read Part IV at all.

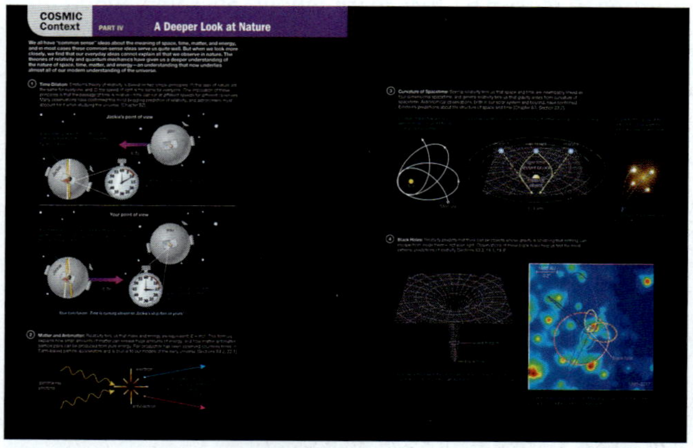


**The Cosmic Context figure for Part IV appears on pp. 466–467.**

Nearly all students have at least heard of things like the prohibition on faster-than-light travel, curvature of spacetime, and the uncertainty principle. But few (if any) students enter an introductory astronomy course with any idea of what these things mean, and they are naturally curious about them. Moreover, a basic understanding of the ideas of relativity and quantum mechanics makes it possible to gain a much deeper appreciation of many of the most important and interesting topics in modern astronomy, including black holes, gravitational lensing, and the overall geometry of the universe. The three chapters of Part IV cover special relativity (Chapter S2), general relativity (Chapter S3), and key astronomical ideas of quantum mechanics (Chapter S4). The main thrust throughout is to demystify relativity and quantum mechanics by convincing students that they are capable of understanding the key ideas despite the reputation of these subjects for being hard or counterintuitive.

## Part V: Stars (Chapters 14–18)

*Guiding Philosophy: We are intimately connected to the stars.*

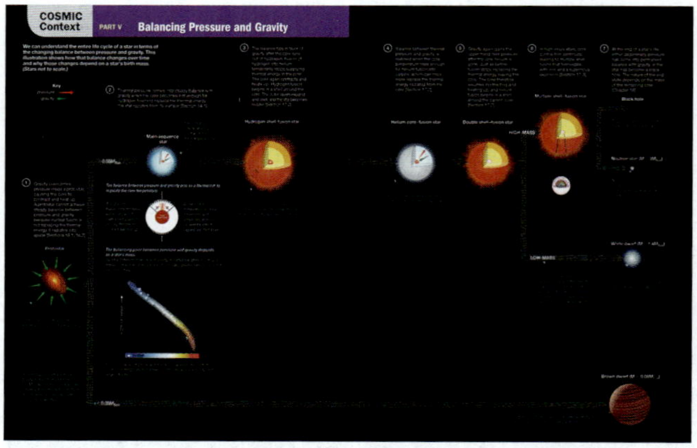


**The Cosmic Context figure for Part V appears on pp. 578–579.**

These are our chapters on stars and stellar life cycles. Chapter 14 covers the Sun in depth so that it can serve as a concrete model for building an understanding of other stars. Chapter 15 describes the general properties of other stars, how we measure these properties, and how we classify stars with the H-R diagram. Chapter 16 covers star birth, and the rest of stellar evolution is discussed in Chapter 17. Chapter 18 focuses on the end points of stellar evolution: white dwarfs, neutron stars, and black holes.

## Part VI: Galaxies and Beyond (Chapters 19–23)

*Guiding Philosophy: Present galaxy evolution and cosmology together as intimately related topics.*

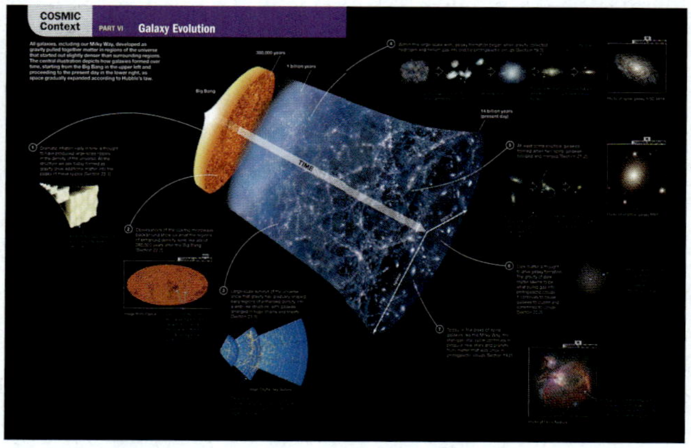


**The Cosmic Context figure for Part VI appears on pp. 696–697.**

These chapters cover galaxies and cosmology. Chapter 19 presents the Milky Way as a paradigm for galaxies in much the same way that Chapter 14 uses the Sun as a paradigm for stars. Chapter 20 describes the properties of galaxies and shows how the quest to measure galactic distances led to Hubble's law and laid the foundation for modern cosmology. Chapter 21 discusses how the current state of knowledge regarding galaxy evolution has emerged from our ability to look back through time. Chapter 22 then presents the Big Bang theory and the evidence supporting it, setting the stage for Chapter 23, which explores dark matter and its role in galaxy formation, as well as dark energy and its implications for the fate of the universe.

### Part VII: Life on Earth and Beyond (Chapter 24)

*Guiding Philosophy: The study of life on Earth helps us understand the search for life in the universe.*

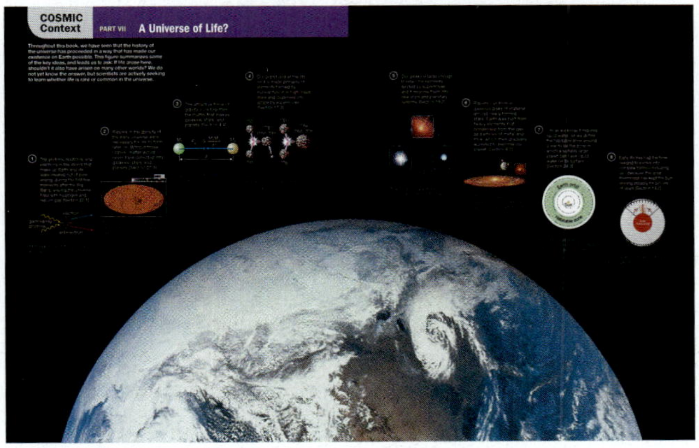


**The Cosmic Context figure for Part VII appears on pp. 728–729.**

This part consists of a single chapter. It may be considered optional, to be used as time allows. Those who wish to teach a more detailed course on astrobiology may wish to consider the text *Life in the Universe*, by Bennett and Shostak.

## Chapter Structure

Each chapter is carefully structured to ensure that students understand the goals up front, learn the details, and pull all the ideas together at the end. Note the following key structural elements of each chapter:

- **Chapter Learning Goals:** Each chapter opens with a page offering an enticing image and a brief overview of the chapter, including a list of the section titles and associated learning goals. The learning goals are presented as key questions designed to help students both to understand what they will be learning about and to stay focused on these key goals as they work through the chapter.
- **Introduction and Epigraph:** The main chapter text begins with a one- to three-paragraph introduction to the chapter material and an inspirational quotation relevant to the chapter.
- **Section Structure:** Chapters are divided into numbered sections, each addressing one key aspect of the chapter material. Each section begins with a short introduction that leads into a set of learning goals relevant to the section—the same learning goals listed at the beginning of the chapter.
- **The Big Picture:** Every chapter narrative ends with this feature, designed to help students put what they have learned in the chapter into the context of the overall goal of gaining a broader perspective on ourselves, our planet, and prospects for life beyond Earth. The final entry in this section is always entitled "My Cosmic Perspective"; it aims to help students see a personal connection between themselves and the chapter content, with the goal of encouraging them to think more critically about the meaning of all that they learn in their astronomy course.
- **Chapter Summary:** The end-of-chapter summary offers a concise review of the learning goal questions, helping to reinforce student understanding of key concepts from the chapter. Thumbnail figures are included to remind students of key illustrations and photos in the chapter.
- **End-of-Chapter Exercises:** Each chapter includes an extensive set of exercises that can be used for study, discussion, or assignment. All of the end-of-chapter exercises are organized into the following subsets:
  - **Visual Skills Check:** This set of questions is designed to help students build their skills at interpreting the many types of visual information used in astronomy.
  - **Chapter Review Questions:** These questions are ones that students should be able to answer from the reading alone.
  - **Does It Make Sense?** (or similar title): Each of these short statements is to be critically evaluated by students, to explain why it does or does not make sense. These exercises are generally easy once students understand a particular concept, but difficult otherwise; this makes these questions an excellent probe of comprehension.
  - **Quick Quiz:** The short multiple-choice quiz allows students to check their basic understanding. Note that, for further self-testing, every chapter also has a Reading, Concept, and Visual quiz available on the Mastering Astronomy website.
  - **Inclusive Astronomy:** These questions are designed to stimulate discussion about participation in science, and in particular about the ideas that (1) astronomy belongs to everyone; (2) all cultures have made contributions to astronomical understanding; (3) opportunities for women and minorities have historically been limited; and (4) the scientific community can take active steps to provide more equitable opportunities for the future.
  - **Process of Science Questions:** These questions, which can be used for discussion or essays, are intended to help students think about how science progresses over time. This set always concludes with an activity designed for group work, in order to promote collaborative learning in class.
  - **Investigate Further:** The remaining questions are designed for home assignment and are intended

to go beyond the earlier review questions. These questions are separated into two groups: Short-Answer/Essay questions, which focus on conceptual interpretation and sometimes on outside research or experiment, and Quantitative Problems, which require some mathematics and are usually based on topics covered in the Mathematical Insight boxes.

## Additional Pedagogical Features

You'll find a number of other features designed to increase student understanding, both within individual chapters and at the end of the book, including the following:

- **Think About It:** This feature, which appears throughout the book in the form of short questions integrated into the narrative, gives students the opportunity to reflect on important new concepts. It also serves as an excellent starting point for classroom discussions.
- **See It for Yourself:** This feature also occurs throughout the book, integrated into the narrative; it gives students the opportunity to conduct simple observations or experiments that will help them understand key concepts.
- **Common Misconceptions:** These boxes address popularly held but incorrect ideas related to the chapter material.
- **Special Topic Boxes:** These boxes address supplementary discussion topics related to the chapter material but not prerequisite to the continuing discussion.
- **Extraordinary Claims Boxes:** Carl Sagan made famous the statement "extraordinary claims require extraordinary evidence." These boxes provide students with examples of extraordinary claims about the universe and how they were either supported or debunked as scientists collected more evidence.
- **Mathematical Insight Boxes:** These boxes contain most of the mathematics used in the book and can be covered or skipped depending on the level of mathematics that you wish to include in your course. The Mathematical Insights use a three-step problem-solving strategy—Understand, Solve, and Explain—that gives students a consistent and explicit structure for solving quantitative homework problems.
- **Annotated Figures:** Key figures in each chapter use the research-proven technique of annotation—the placement on the figure of carefully crafted text (in blue) to guide students in interpreting graphs, following process figures, and translating between different representations.
- **Cosmic Context Two-Page Figures:** These two-page spreads provide visual summaries of key processes and concepts.
- **Wavelength/Observatory Icons:** For astronomical images, simple icons indicate whether the image is a photo, artist's impression, or computer simulation; whether a photo came from ground-based or space-based observations; and the wavelength band used to take the photo.
- **Video Icons:** These icons point to videos available in the Study Area of Mastering Astronomy that are relevant to the topic at hand. Tutorial assessments based on these videos are available for assignment in the instructor Item Library.
- **Cross-References:** When a concept is covered in greater detail elsewhere in the book, a cross-reference to the relevant section is included in brackets (e.g., [**Section 5.2**]).
- **Glossary:** A detailed glossary makes it easy for students to look up important terms.
- **Appendixes:** The appendixes contain a number of useful references and tables, including key constants (Appendix A), key formulas (Appendix B), key mathematical skills (Appendix C), and numerous data tables and star charts (Appendixes D–I).

## Mastering Astronomy

What is the single most important factor in student success in astronomy? Both research and common sense reveal the same answer: study time. No matter how good the teacher or how good the textbook, students learn only when they spend adequate time learning and studying on their own. Unfortunately, limitations on resources for grading have prevented most instructors from assigning much homework despite its obvious benefits to student learning. And limitations on help and office hours have made it difficult for students to make sure they use self-study time effectively. That, in a nutshell, is why we created Mastering Astronomy. For students, it provides adaptive learning designed to coach them individually—responding to their errors with specific, targeted feedback and giving optional hints for those who need additional guidance. For professors, Mastering Astronomy provides unprecedented ability to automatically monitor and record students' step-by-step work and evaluate the effectiveness of assignments and exams.

Note that nearly all the content available at the Mastering Astronomy site for *The Cosmic Perspective* has been written by the textbook authors. This means that students can count on consistency between the textbook and web resources, with both emphasizing the same concepts and using the same terminology and the same pedagogical approaches. This type of consistency ensures that students can study in the most efficient possible way.

All students registered for Mastering Astronomy receive full access to the Study Area, which includes three self-study multiple-choice quizzes for each chapter; a large set of prelecture videos, narrated figures, interactive figures, and math review videos; a set of interactive self-guided tutorials that go into depth on topics that some students find particularly challenging; a downloadable set of group activities; and much more.

Instructors have access to many additional resources, including a large Item Library of assignable material that features more than 250 author-written tutorials, all of the end-of-chapter exercises, all the questions from the self-study quizzes in the Study Area, and a test bank. Instructors also have access to the author-written Instructor Guide and teaching resources including PowerPoint® Lecture Outlines, a complete set of high-resolution JPEGs of all images from the book, and PRS-enabled clicker quizzes based on the book and book-specific interactive media.

## Supplements for *The Cosmic Perspective*

*The Cosmic Perspective* is much more than just a textbook. It is a complete package of teaching, learning, and assessment resources designed to help both teachers and students. In addition to Mastering Astronomy (described above), the following supplements are available with this book:

| Name of Supplement | Instructor or Student Supplement | Description |
|---|---|---|
| *SkyGazer 5.0* (Access code card ISBN 0-321-76518-5, CD ISBN 0-321-89843-2) | Student Supplement | Based on Voyager IV, one of the world's most popular planetarium programs, SkyGazer 5.0 makes it easy for students to learn constellations and explore the wonders of the sky through interactive exercises and demonstrations. Accompanying activities are available in *LoPresto's Astronomy Media Workbook*, Seventh Edition. Both SkyGazer and LoPresto's workbook are available for download. |
| *Starry Night™ College* (ISBN 0-321-71295-1) | Student Supplement | Now available as an additional option with *The Cosmic Perspective*, Starry Night™ College has been acclaimed as the world's most realistic desktop planetarium software. This special version has an easy-to-use point-and-click interface and is available as an additional bundle. The *Starry Night Activity Workbook*, consisting of thirty-five worksheets for homework or lab, based on Starry Night Planetarium software, is available for download in the Mastering Astronomy Study Area or with a Starry Night College access code. |
| *Astronomy Active Learning In-Class Tutorials* (ISBN 0-805-38296-8) by Marvin L. De Jong | Student Supplement | This workbook provides fifty 20-minute in-class tutorial activities to choose from. Designed for use in large lecture classes, these activities are also suitable for labs. The short, structured activities are designed for students to complete on their own or in peer-learning groups. Each activity targets specific learning objectives such as understanding Newton's laws, understanding Mars's retrograde motion, tracking stars on the H-R diagram, or comparing the properties of planets. |
| *Lecture Tutorials for Introductory Astronomy* (ISBN 0-321-82046-0) by Ed Prather, Tim Slater, Jeff Adams, and Gina Brissenden | Student Supplement | These forty-four lecture tutorials are designed to engage students in critical reasoning and spark classroom discussion. |
| *Sky and Telescope* (ISBN 0-321-70620-X) | Student Supplement | This supplement, which includes nine articles with an assessment insert covering general review, Process of Science, Scale of the Universe, and Our Place in the Universe, is available for bundling. |
| *Observation Exercises in Astronomy* (ISBN 0-321-63812-3) by Lauren Jones | Student Supplement | This workbook includes fifteen observation activities that can be used with a number of different planetarium software packages. |
| *Astronomy Lab: A Concept Oriented Approach* (ISBN 0-321-86177-9) by Nate McCrady and Emily Rice | Student Supplement | This modular collection of forty conceptually oriented introductory astronomy labs, housed in the Pearson Custom Library, allows for easy creation of a customized lab manual. |
| *Instructor Resources* (ISBN 0-135-17325-1) | Instructor Supplement | This comprehensive collection of instructor resources includes high-resolution JPEGs of all images from the book; Interactive Figures and Photos™ based on figures in the text; additional applets and animations to illustrate key concepts; PowerPoint® Lecture Outlines that incorporate figures, photos, checkpoint questions, and multimedia; and PRS-enabled clicker quizzes based on the book and book-specific interactive media, to make preparing for lectures quick and easy. These resources are located in the Mastering Astronomy Instructor Resource Area. |
| *Instructor Guide* (ISBN 0-135-17324-4) | Instructor Supplement | The Instructor Guide contains a detailed overview of the text, sample syllabi for courses of different emphasis and duration, suggested teaching strategies, answers or discussion points for all Think About It and See It for Yourself questions in the text, solutions to all end-of-chapter problems, and a detailed reference guide summarizing media resources available for every chapter and section of the book. |
| *Test Bank* (ISBN 0-135-17326-4) | Instructor Supplement | Available in both Word and TestGen formats in the Instructor Resource Center and Mastering Astronomy, the Test Bank contains a broad set of multiple-choice, true/false, and free-response questions for each chapter. The Test Bank is also assignable through Mastering Astronomy. |

## Acknowledgments

Our textbook carries only four author names, but in fact it is the result of hard work by a long list of committed individuals. We could not possibly list everyone who has helped, but we would like to call attention to a few people who have played particularly important roles. First, we thank our past and present editors at Addison-Wesley and Pearson, including Bill Poole, Robin Heyden, Sami Iwata, Ben Roberts, Adam Black, Nancy Whilton, and Jeanne Zalesky, along with many others at Pearson who have contributed to making our books possible. Special thanks to our production teams, especially Sally Lifland, and our art and design team.

We've also been fortunate to have an outstanding group of reviewers, whose extensive comments and suggestions helped us shape the book. We thank all those who have reviewed drafts of the book in various stages, including the following individuals (listed with their affiliations at the time they contributed their reviews): Eric Agol, *University of Washington*; Marilyn Akins, *Broome Community College*; Caspar Amman, *NCAR*; Christopher M. Anderson, *University of Wisconsin*; John Anderson, *University of North Florida*; Peter S. Anderson, *Oakland Community College*; Nahum Arav, *Virginia Technical University*; Phil Armitage, *University of Colorado*; Keith Ashman; Thomas Ayres, *University of Colorado*; Simon P. Balm, *Santa Monica College*; Reba Bandyopadhyay, *University of Florida*; Nadine Barlow, *Northern Arizona University*; Cecilia Barnbaum, *Valdosta State University*; John Beaver, *University of Wisconsin at Fox Valley*; Peter A. Becker, *George Mason University*; Timothy C. Beers, *National Optical Astronomy Observatory*; Jim Bell, *Arizona State University*; Priscilla J. Benson, *Wellesley College*; Zach Berta-Thompson, *University of Colorado*; Rick Binzel, *Massachusetts Institute of Technology*; Philip Blanco, *Grossmont College*; Jeff R. Bodart, *Chipola College*; Howard Bond, *Space Telescope Science Institute*; Bernard W. Bopp, *University of Toledo*; Sukanta Bose, *Washington State University*; David Brain, *University of Colorado*; David Branch, *University of Oklahoma*; John C. Brandt, *University of New Mexico*; James E. Brau, *University of Oregon*; Jean P. Brodie, *UCO/Lick Observatory, University of California, Santa Cruz*; Erik Brogt, *University of Canterbury*; James Brooks, *Florida State University*; Daniel Bruton, *Stephen F. Austin State University*; Thomas Burbine, *Moraine Valley Community College*; Debra Burris, *University of Central Arkansas*; Scott Calvin, *Sarah Lawrence College*; Amy Campbell, *Louisiana State University*; Humberto Campins, *University of Central Florida*; Robin Canup, *Southwest Research Institute*; Eugene R. Capriotti, *Michigan State University*; Eric Carlson, *Wake Forest University*; David A. Cebula, *Pacific University*; Supriya Chakrabarti, *University of Massachusetts, Lowell*; Clark Chapman, *Southwest Research Institute*; Kwang-Ping Cheng, *California State University, Fullerton*; Dipak Chowdhury, *Indiana University—Purdue University Fort Wayne*; Chris Churchill, *New Mexico State University*; Kelly Cline, *Carroll College*; Josh Colwell, *University of Central Florida*; James Cooney, *University of Central Florida*; Anita B. Corn, *Colorado School of Mines*; Philip E. Corn, *Red Rocks Community College*; Kelli Corrado, *Montgomery County Community College*; Peter Cottrell, *University of Canterbury*; John Cowan, *University of Oklahoma*; Kevin Crosby, *Carthage College*; Christopher Crow, *Indiana University—Purdue University Fort Wayne*; Manfred Cuntz, *University of Texas at Arlington*; Charles Danforth, *University of Colorado*; Christopher De Vries, *California State University, Stanislaus*; John M. Dickey, *University of Minnesota*; Mark Dickinson, *National Optical Astronomy Observatory*; Matthias Dietrich, *Worcester State University*; Jim Dove, *Metropolitan State College of Denver*; Doug Duncan, *University of Colorado*; Bryan Dunne, *University of Illinois, Urbana-Champaign*; Suzan Edwards, *Smith College*; Robert Egler, *North Carolina State University at Raleigh*; Paul Eskridge, *Minnesota State University*; Dan Fabrycky, *University of Chicago*; David Falk, *Los Angeles Valley College*; Timothy Farris, *Vanderbilt University*; Harry Ferguson, *Space Telescope Science Institute*; Robert A. Fesen, *Dartmouth College*; Tom Fleming, *University of Arizona*; Douglas Franklin, *Western Illinois University*; Sidney Freudenstein, *Metropolitan State College of Denver*; Martin Gaskell, *University of Nebraska*; Richard Gelderman, *Western Kentucky University*; Harold A. Geller, *George Mason University*; Donna Gifford, *Pima Community College*; Mitch Gillam, *Marion L. Steele High School*; Bernard Gilroy, *The Hun School of Princeton*; Owen Gingerich, *Harvard–Smithsonian* (Historical Accuracy Reviewer); David Graff, *U.S. Merchant Marine Academy*; Richard Gray, *Appalachian State University*; Kevin Grazier, *Jet Propulsion Laboratory*; Robert Greeney, *Holyoke Community College*; Henry Greenside, *Duke University*; Alan Greer, *Gonzaga University*; John Griffith, *Lin-Benton Community College*; David Griffiths, *Oregon State University*; David Grinspoon, *Planetary Science Institute*; John Gris, *University of Delaware*; Bruce Gronich, *University of Texas at El Paso*; Thomasana Hail, *Parkland University*; Andrew Hamilton, *University of Colorado*; Jim Hamm, *Big Bend Community College*; Paul Harderson, *Planetary Science Institute*; Charles Hartley, *Hartwick College*; J. Hasbun, *University of West Georgia*; Joe Heafner, *Catawba Valley Community College*; Todd Henry, *Georgia State University*; David Herrick, *Maysville Community College*; Dennis Hibbert, *Everett Community College*; Scott Hildreth, *Chabot College*; Tracy Hodge, *Berea College*; Lynnette Hoerner, *Red Rocks Community College*; Mark Hollabaugh, *Normandale Community College*; Richard Holland, *Southern Illinois University, Carbondale*; Seth Hornstein, *University of Colorado*; Joseph Howard, *Salisbury University*; James Christopher Hunt, *Prince George's Community College*; Richard Ignace, *University of Wisconsin*; James Imamura, *University of Oregon*; Douglas R. Ingram, *Texas Christian University*; Assad Istephan, *Madonna University*; Bruce Jakosky, *University of Colorado*; Adam G. Jensen, *University of Colorado*; Dave Jewitt, *University of California, Los Angeles*; Adam Johnston, *Weber State University*; Lauren Jones, *Gettysburg College*; Kishor T. Kapale, *Western Illinois University*; William Keel, *University of Alabama*; Julia Kennefick, *University of Arkansas*; Steve Kipp, *University of Minnesota, Mankato*; Kurtis Koll, *Cameron University*; Ichishiro Konno, *University of Texas at San Antonio*; John Kormendy, *University of Texas at Austin*; Eric Korpela, *University of California, Berkeley*; Arthur Kosowsky, *University of Pittsburgh*; Julia Kregenow, *Penn State University*; Kevin Krisciunas, *Texas A&M*; Emily Lakdawalla, *The Planetary Society*; David Lamp, *Texas Technical University*; Ted La Rosa, *Kennesaw State*

*University*; Kenneth Lanzetta, *Stony Brook University*; Kristine Larsen, *Central Connecticut State University*; Ana Marie Larson, *University of Washington*; Stephen Lattanzio, *Orange Coast College*; Chris Laws, *University of Washington*; Larry Lebofsky, *University of Arizona*; Patrick Lestrade, *Mississippi State University*; Nancy Levenson, *University of Kentucky*; Hal Levison, *Southwest Research Institute*; David M. Lind, *Florida State University*; Mario Livio, *Space Telescope Science Institute*; Abraham Loeb, *Harvard University*; Michael LoPresto, *Henry Ford Community College*; William R. Luebke, *Modesto Junior College*; Ihor Luhach, *Valencia Community College*; Darrell Jack MacConnell, *Community College of Baltimore City*; Marie Machacek, *Massachusetts Institute of Technology*; Loris Magnani, *University of Georgia*; Steven Majewski, *University of Virginia*; J. McKim Malville, *University of Colorado*; Geoff Marcy, *University of California, Berkeley*; Mark Marley, *Ames Research Center*; Linda Martel, *University of Hawaii*; Phil Matheson, *Salt Lake Community College*; John Mattox, *Fayetteville State University*; Marles McCurdy, *Tarrant County College*; Stacy McGaugh, *Case Western University*; Kevin McLin, *University of Colorado*; Michael Mendillo, *Boston University*; Steven Merriman, *Moraine Valley Community College*; Barry Metz, *Delaware County Community College*; William Millar, *Grand Rapids Community College*; Dinah Moche, *Queensborough Community College of City University, New York*; Steve Mojzsis, *University of Colorado*; Irina Mullins, *Houston Community College–Central*; Stephen Murray, *University of California, Santa Cruz*; Zdzislaw E. Musielak, *University of Texas at Arlington*; Charles Nelson, *Drake University*; Gerald H. Newsom, *Ohio State University*; Francis Nimmo, *University of California, Santa Cruz*; Tyler Nordgren, *University of Redlands*; Lauren Novatne, *Reedley College*; Brian Oetiker, *Sam Houston State University*; Richard Olenick, *University of Dallas*; John P. Oliver, *University of Florida*; Rachel Osten, *Space Telescope Science Institute*; Stacy Palen, *Weber State University*; Russell L. Palma, *Sam Houston State University*; Bob Pappalardo, *Jet Propulsion Laboratory*; Mark Pecaut, *Rockhurst University*; Jon Pedicino, *College of the Redwoods*; Bryan Penprase, *Pomona College*; Eric S. Perlman, *Florida Institute of Technology*; Peggy Perozzo, *Mary Baldwin College*; Greg Perugini, *Burlington County College*; Charles Peterson, *University of Missouri, Columbia*; Cynthia W. Peterson, *University of Connecticut*; Jorge Piekarewicz, *Florida State University*; Lawrence Pinsky, *University of Houston*; Stephanie Plante, *Grossmont College*; Jascha Polet, *California State Polytechnic University, Pomona*; Matthew Price, *Oregon State University*; Harrison B. Prosper, *Florida State University*; Monica Ramirez, *Aims College, Colorado*; Christina Reeves-Shull, *Richland College*; Todd M. Rigg, *City College of San Francisco*; Elizabeth Roettger, *DePaul University*; Roy Rubins, *University of Texas at Arlington*; April Russell, *Siena College*; Carl Rutledge, *East Central University*; Bob Sackett, *Saddleback College*; Rex Saffer, *Villanova University*; John Safko, *University of South Carolina*; James A. Scarborough, *Delta State University*; Britt Scharringhausen, *Ithaca College*; Ann Schmiedekamp, *Pennsylvania State University, Abington*; Joslyn Schoemer, *Denver Museum of Nature and Science*; James Schombert, *University of Oregon*; Gregory Seab, *University of New Orleans*; Bennett Seidenstein, *Arundel High School*; Larry Sessions, *Metropolitan State College of Denver*; Michael Shara, *American Museum of Natural History*; Anwar Shiekh, *Colorado Mesa University*; Seth Shostak, *SETI Institute*; Ralph Siegel, *Montgomery College, Germantown Campus*; Philip I. Siemens, *Oregon State University*; Caroline Simpson, *Florida International University*; Paul Sipiera, *William Harper Rainey College*; Earl F. Skelton, *George Washington University*; Evan Skillman, *University of Minnesota*; Michael Skrutskie, *University of Virginia*; Mark H. Slovak, *Louisiana State University*; Norma Small-Warren, *Howard University*; Jessica Smay, *San Jose City College*; Dale Smith, *Bowling Green State University*; Brad Snowder, *Western Washington University*; Brent Sorenson, *Southern Utah University*; James R. Sowell, *Georgia Technical University*; Kelli Spangler, *Montgomery County Community College*; John Spencer, *Southwest Research Institute*; Darryl Stanford, *City College of San Francisco*; George R. Stanley, *San Antonio College*; Bob Stein, *Michigan State University*; Peter Stein, *Bloomsburg University of Pennsylvania*; Adriane Steinacker, *University of California, Santa Cruz*; Glen Stewart, *University of Colorado*; John Stolar, *West Chester University*; Irina Struganova, *Valencia Community College*; Jack Sulentic, *University of Alabama*; C. Sean Sutton, *Mount Holyoke College*; Beverley A. P. Taylor, *Miami University*; Brett Taylor, *Radford University*; Jeff Taylor, *University of Hawaii*; Donald M. Terndrup, *Ohio State University*; Dave Tholen, *University of Hawaii*; Nick Thomas, *University of Bern*; Frank Timmes, *Arizona State University*; David Trott, *Metro State College*; David Vakil, *El Camino College*; Trina Van Ausdal, *Salt Lake Community College*; Dimitri Veras, *Cambridge University*; Licia Verde, *Institute of Cosmological Studies, Barcelona*; Nicole Vogt, *New Mexico State University*; Darryl Walke, *Rariton Valley Community College*; Fred Walter, *State University of New York, Stony Brook*; Marina Walther-Antonio, *Mayo Clinic*; James Webb, *Florida International University*; John Weiss, *Carleton College*; Mark Whittle, *University of Virginia*; Paul J. Wiita, *The College of New Jersey*; Francis Wilkin, *Union College*; Lisa M. Will, *Mesa Community College*; Jonathan Williams, *University of Hawaii*; Terry Willis, *Chesapeake College*; Grant Wilson, *University of Massachusetts, Amherst*; Jeremy Wood, *Hazard Community College*; J. Wayne Wooten, *Pensacola Junior College*; Guy Worthey, *Washington State University, Pullman*; Jason Wright, *Penn State University*; Scott Yager, *Brevard College*; Don Yeomans, *Jet Propulsion Laboratory*; Andrew Young, *Casper College*; Arthur Young, *San Diego State University*; Tim Young, *University of North Dakota*; Min S. Yun, *University of Massachusetts, Amherst*; Dennis Zaritsky, *University of Arizona*; Robert L. Zimmerman, *University of Oregon*.

Finally, we thank the many people who have greatly influenced our outlook on education and our perspective on the universe over the years, including Tom Ayres, Fran Bagenal, Forrest Boley, Robert A. Brown, George Dulk, Erica Ellingson, Katy Garmany, Jeff Goldstein, David Grinspoon, Don Hunten, Joan Marsh, Catherine McCord, Dick McCray, Dee Mook, Cherilynn Morrow, Charlie Pellerin, Carl Sagan, J. Michael Shull, John Spencer, and John Stocke.

*Jeff Bennett, Megan Donahue,*
*Nick Schneider, Mark Voit*

# About the Authors

## Jeffrey Bennett

*Jeffrey Bennett*, a recipient of the American Institute of Physics Science Communication Award, holds a B.A. in biophysics (UC San Diego) and an M.S. and Ph.D. in astrophysics (University of Colorado). He specializes in science and math education and has taught at every level from preschool through graduate school. Career highlights include serving 2 years as a visiting senior scientist at NASA headquarters, where he developed programs to build stronger links between research and education, proposing and helping to develop the Voyage scale model solar system on the National Mall (Washington, DC) and developing the free app *Totality by Big Kid Science* to help people learn about total solar eclipses. He is the lead author of textbooks in astronomy, astrobiology, mathematics, and statistics and of critically acclaimed books for the public including *Beyond UFOs* (Princeton University Press), *Math for Life* (Big Kid Science), *What Is Relativity?* (Columbia University Press), *On Teaching Science* (Big Kid Science), and *A Global Warming Primer* (Big Kid Science). He is also the author of six science picture books for children, titled *Max Goes to the Moon, Max Goes to Mars, Max Goes to Jupiter, Max Goes to the Space Station, The Wizard Who Saved the World*, and *I, Humanity*; all six have been launched to the International Space Station and read aloud by astronauts for NASA's Story Time From Space program. His personal website is www.jeffreybennett.com.

## Megan Donahue

*Megan Donahue* is a full professor in the Department of Physics and Astronomy at Michigan State University (MSU), a Fellow of the American Physical Society and of the American Association for the Advancement of Science, and President of the American Astronomical Society (2018–2020). Her research focuses on using x-ray, UV, infrared, and visible light to study galaxies and clusters of galaxies: their contents—dark matter, hot gas, galaxies, active galactic nuclei—and what they reveal about the contents of the universe and how galaxies form and evolve. She grew up on a farm in Nebraska and received an S.B. in physics from MIT, where she began her research career as an x-ray astronomer. She has a Ph.D. in astrophysics from the University of Colorado. Her Ph.D. thesis on theory and optical observations of intergalactic and intracluster gas won the 1993 Robert Trumpler Award from the Astronomical Society for the Pacific for an outstanding astrophysics doctoral dissertation in North America. She continued postdoctoral research as a Carnegie Fellow at Carnegie Observatories in Pasadena, California, and later as an STScI Institute Fellow at Space Telescope. Megan was a staff astronomer at the Space Telescope Science Institute until 2003, when she joined the MSU faculty. She is also actively involved in advising national and international astronomical facilities and NASA, including planning future NASA missions. Megan is married to Mark Voit, and they collaborate on many projects, including this textbook, over 70 peer-reviewed astrophysics papers, and the nurturing of their children, Michaela, Sebastian, and Angela. Megan has run three full marathons, including Boston. These days she runs trails with friends, orienteers, and plays piano and bass guitar for fun and no profit.

## Nicholas Schneider

*Nicholas Schneider* is a full professor in the Department of Astrophysical and Planetary Sciences at the University of Colorado and a researcher in the Laboratory for Atmospheric and Space Physics. He received his B.A. in physics and astronomy from Dartmouth College in 1979 and his Ph.D. in planetary science from the University of Arizona in 1988. His research interests include planetary atmospheres and planetary astronomy. One research focus is the odd case of Jupiter's moon Io. Another is the mystery of Mars's lost atmosphere, which he is helping to answer by leading the Imaging UV Spectrograph team on NASA's *MAVEN* mission now orbiting Mars. Nick enjoys teaching at all levels and is active in efforts to improve undergraduate astronomy education. Over his career he has received the National Science Foundation's Presidential Young Investigator Award, the Boulder Faculty Assembly's Teaching Excellence Award, and NASA's Exceptional Scientific Achievement Medal. Off the job, Nick enjoys exploring the outdoors with his family and figuring out how things work.

## Mark Voit

*Mark Voit* is a full professor in the Department of Physics and Astronomy and Associate Dean for Undergraduate Studies at Michigan State University. He earned his A.B. in astrophysical sciences at Princeton University and his Ph.D. in astrophysics at the University of Colorado in 1990. He continued his studies at the California Institute of Technology, where he was a research fellow in theoretical astrophysics, and then moved on to Johns Hopkins University as a Hubble Fellow. Before going to Michigan State, Mark worked in the Office of Public Outreach at the Space Telescope, where he developed museum exhibitions about the Hubble Space Telescope and helped design NASA's award-winning HubbleSite. His research interests range from interstellar processes in our own galaxy to the clustering of galaxies in the early universe, and he is a Fellow of the American Association for the Advancement of Science. He is married to coauthor Megan Donahue and cooks terrific meals for her and their three children. Mark likes getting outdoors whenever possible and particularly enjoys running, mountain biking, canoeing, orienteering, and adventure racing. He is also author of the popular book *Hubble Space Telescope: New Views of the Universe*.

# How to Succeed in Your Astronomy Course

| If Your Course Is | Times for Reading the Assigned Text (per week) | Times for Homework Assignments (per week) | Times for Review and Test Preparation (average per week) | Total Study Time (per week) |
|---|---|---|---|---|
| 3 credits | 2 to 4 hours | 2 to 3 hours | 2 hours | 6 to 9 hours |
| 4 credits | 3 to 5 hours | 2 to 4 hours | 3 hours | 8 to 12 hours |
| 5 credits | 3 to 5 hours | 3 to 6 hours | 4 hours | 10 to 15 hours |

## The Key to Success: Study Time

The single most important key to success in any college course is to spend enough time studying. A general rule of thumb for college classes is that you should expect to study about 2 to 3 hours per week *outside* of class for each unit of credit. For example, based on this rule of thumb, a student taking 15 credit hours should expect to spend 30 to 45 hours each week studying outside of class. Combined with time in class, this works out to a total of 45 to 60 hours spent on academic work—not much more than the time a typical job requires, and you get to choose your own hours. Of course, if you are working or have family obligations while you attend school, you will need to budget your time carefully.

The table above gives rough guidelines for how you might divide your study time. If you find that you are spending fewer hours than these guidelines suggest, you can probably improve your grade by studying longer. If you are spending more hours than these guidelines suggest, you may be studying inefficiently; in that case, you should talk to your instructor about how to study more effectively.

## Using This Book

Each chapter in this book is designed to make it easy for you to study effectively and efficiently. To get the most out of each chapter, you might wish to use the following study plan.

- A textbook is not a novel, and you'll learn best by reading the elements of this text in the following order:
  1. Start by reading the Learning Goals and the introductory paragraphs at the beginning of the chapter so that you'll know what you are trying to learn.
  2. Get an overview of key concepts by studying the illustrations and their captions and annotations. The illustrations highlight most major concepts, so this "illustrations first" strategy gives you an opportunity to survey the concepts before you read about them in depth. You will find the two-page Cosmic Context figures especially useful.
  3. Read the chapter narrative, trying the Think About It questions and the See It for Yourself activities as you go along, but save the boxed features (e.g., Common Misconceptions, Special Topics) to read later. As you read, make notes on the pages to remind yourself of ideas you'll want to review later. Take notes as you read, but avoid using a highlight pen (or a highlighting tool if you are using an e-book), which makes it too easy to highlight mindlessly.
  4. After reading the chapter once, go back through and read the boxed features.
  5. Review the Chapter Summary, ideally by trying to answer the Learning Goal questions for yourself before reading the given answers.
- After completing the reading as outlined above, test your understanding with the end-of-chapter exercises. A good way to begin is to make sure you can answer all of the Review and Quick Quiz Questions; if you don't know an answer, look back through the chapter until you figure it out.
- Further build your understanding by making use of the videos, quizzes, and other resources available at Mastering Astronomy. These resources have been developed specifically to help you learn the most important ideas in your course, and they have been extensively tested to make sure they are effective. They really do work, and the only way you'll gain their benefits is by going to the website and using them.

## General Strategies for Studying

- Budget your time effectively. Studying 1 or 2 hours each day is more effective, and far less painful, than studying all night before homework is due or before exams. *Note*: Research shows that it can be helpful to create a "personal contract" for your study time (or for any other personal commitment), in which you specify rewards you'll give yourself for success and penalties you'll assess for failings.
- Engage your brain. Learning is an active process, not a passive experience. Whether you are reading, listening to a lecture, or working on assignments, always make sure that your mind is actively engaged. If you find your mind drifting or find yourself falling asleep, make a conscious effort to revive yourself, or take a break if necessary.
- Don't miss class, and come prepared. Listening to lectures and participating in class activities and discussions is much more effective than reading someone else's notes or watching a video later. Active participation will help you retain what you are learning. Also, be sure to complete any assigned reading *before* the class in which it will be discussed. This is crucial, since

class lectures and discussions are designed to reinforce key ideas from the reading.

- Take advantage of resources offered by your professor, whether it be email, office hours, review sessions, online chats, or other opportunities to talk to and get to know your professor. Most professors will go out of their way to help you learn in any way that they can.
- Start your homework early. The more time you allow yourself, the easier it is to get help if you need it. If a concept gives you trouble, do additional reading or studying beyond what has been assigned. And if you still have trouble, ask for help: You surely can find friends, peers, or teachers who will be glad to help you learn.
- Working together with friends can be valuable in helping you understand difficult concepts. However, be sure that you learn *with* your friends and do not become dependent on them.
- Don't try to multitask. Research shows that human beings simply are not good at multitasking: When we attempt it, we do more poorly at all of the individual tasks. And in case you think you are an exception, research has also shown that those people who believe they are best at multitasking are often the worst! So when it is time to study, turn off your electronic devices, find a quiet spot, and concentrate on your work. (If you *must* use a device to study, as with an e-book or online homework, turn off email, text, and other alerts so that they will not interrupt your concentration; some apps will do this for you.)

## Preparing for Exams

- Rework problems and other assignments; try additional questions, including the online quizzes available at Mastering Astronomy, to be sure you understand the concepts. Study your performance on assignments, quizzes, or exams from earlier in the term.
- Study your notes from classes, and reread relevant sections in your textbook. Pay attention to what your instructor expects you to know for an exam.
- Study individually *before* joining a study group with friends. Study groups are effective only if every individual comes prepared to contribute.
- Don't stay up too late before an exam. Don't eat a big meal within an hour of the exam (thinking is more difficult when blood is being diverted to the digestive system).
- Try to relax before and during the exam. If you have studied effectively, you are capable of doing well. Staying relaxed will help you think clearly.

## Presenting Homework and Writing Assignments

All work that you turn in should be of *collegiate quality:* neat and easy to read, well organized, and demonstrating mastery of the subject matter. Future employers and teachers will expect this quality of work. Moreover, although submitting homework of collegiate quality requires "extra" effort, it serves two important purposes directly related to learning:

1. The effort you expend in clearly explaining your work solidifies your learning. Writing (or typing) triggers different areas of your brain than reading, listening, or speaking. As a result, writing something down will reinforce your learning of a concept, even when you think you already understand it.
2. By making your work clear and self-contained (that is, making it a document that you can read without referring to the questions in the text), you will have a much more useful study guide when you review for a quiz or exam.

The following guidelines will help ensure that your assignments meet the standards of collegiate quality:

- Always use proper grammar, proper sentence and paragraph structure, and proper spelling. Do not use texting shorthand, and don't become over-reliant on spell checkers, which may miss "too two three mistakes, to."
- All answers and other writing should be fully self-contained. A good test is to imagine that a friend is reading your work and to ask yourself whether the friend would understand exactly what you are trying to say. It is also helpful to read your work out loud to yourself, making sure that it sounds clear and coherent.
- In problems that require calculation:

    1. Be sure to *show your work* clearly so that both you and your instructor can follow the process you used to obtain an answer. Also, use standard mathematical symbols, rather than "calculator-ese." For example, show multiplication with the × symbol (not with an asterisk), and write $10^5$, not 10^5 or 10E5.
    2. *Word problems should have word answers*. That is, after you have completed any necessary calculations, make sure that any problem stated in words is answered with one or more *complete sentences* that describe the point of the problem and the meaning of your solution.
    3. *Units are crucial.* If your answer has units, be sure they are stated clearly. For example, if you are asked to calculate a distance, be sure you state whether your answer is in miles, kilometers, or some other distance unit.
    4. Express your word answers in a way that would be *meaningful* to most people. For example, most people would find it more meaningful if you expressed a result of 720 hours as 1 month. Similarly, if a precise calculation yields an answer of 9,745,600 years, it may be more meaningfully expressed in words as "nearly 10 million years."

- Include illustrations whenever they help explain your answer, and make sure your illustrations are neat and clear. For example, if you graph by hand, use a ruler to make straight lines. If you use software to make illustrations, be careful not to make them overly cluttered with unnecessary features.
- If you study with friends, be sure that you turn in your own work stated in your own words—you should avoid anything that might give even the *appearance* of possible academic dishonesty.

# Foreword
# The Meaning of *The Cosmic Perspective*

© Neil deGrasse Tyson

**by Neil deGrasse Tyson**

*Astrophysicist Neil deGrasse Tyson is the Frederick P. Rose Director of New York City's Hayden Planetarium at the American Museum of Natural History. He has written numerous books and articles, has hosted the PBS series* NOVA scienceNOW *and the globally popular* Cosmos: A Spacetime Odyssey, *and was named one of the "Time 100"—Time Magazine's list of the 100 most influential people in the world. He contributed this essay about the meaning of "The Cosmic Perspective," abridged from his 100th essay written for* Natural History *magazine.*

*Of all the sciences cultivated by mankind, Astronomy is acknowledged to be, and undoubtedly is, the most sublime, the most interesting, and the most useful. For, by knowledge derived from this science, not only the bulk of the Earth is discovered . . . ; but our very faculties are enlarged with the grandeur of the ideas it conveys, our minds exalted above [their] low contracted prejudices.*

—*James Ferguson,* Astronomy Explained Upon Sir Isaac Newton's Principles, and Made Easy To Those Who Have Not Studied Mathematics (1757)

Long before anyone knew that the universe had a beginning, before we knew that the nearest large galaxy lies two and a half million light-years from Earth, before we knew how stars work or whether atoms exist, James Ferguson's enthusiastic introduction to his favorite science rang true.

But who gets to think that way? Who gets to celebrate this cosmic view of life? Not the migrant farm worker. Not the sweatshop worker. Certainly not the homeless person rummaging through the trash for food. You need the luxury of time not spent on mere survival. You need to live in a nation whose government values the search to understand humanity's place in the universe. You need a society in which intellectual pursuit can take you to the frontiers of discovery, and in which news of your discoveries can be routinely disseminated.

When I pause and reflect on our expanding universe, with its galaxies hurtling away from one another, embedded with the ever-stretching, four-dimensional fabric of space and time, sometimes I forget that uncounted people walk this Earth without food or shelter, and that children are disproportionately represented among them.

When I pore over the data that establish the mysterious presence of dark matter and dark energy throughout the universe, sometimes I forget that every day—every twenty-four-hour rotation of Earth—people are killing and being killed. In the name of someone's ideology.

When I track the orbits of asteroids, comets, and planets, each one a pirouetting dancer in a cosmic ballet choreographed by the forces of gravity, sometimes I forget that too many people act in wanton disregard for the delicate interplay of Earth's atmosphere, oceans, and land, with consequences that our children and our children's children will witness and pay for with their health and well-being.

And sometimes I forget that powerful people rarely do all they can to help those who cannot help themselves.

I occasionally forget those things because, however big the world is—in our hearts, our minds, and our outsize atlases—the universe is even bigger. A depressing thought to some, but a liberating thought to me.

Consider an adult who tends to the traumas of a child: a broken toy, a scraped knee, a schoolyard bully. Adults know that kids have no clue what constitutes a genuine problem, because inexperience greatly limits their childhood perspective.

As grown-ups, dare we admit to ourselves that we, too, have a collective immaturity of view? Dare we admit that our thoughts and behaviors spring from a belief that the world revolves around us? Part the curtains of society's racial, ethnic, religious, national, and cultural conflicts, and you find the human ego turning the knobs and pulling the levers.

Now imagine a world in which everyone, but especially people with power and influence, holds an expanded view of our place in the cosmos. With that perspective, our problems would shrink—or never arise at all—and we could celebrate our earthly differences while shunning the behavior of our predecessors who slaughtered each other because of them.

■■■

Back in February 2000, the newly rebuilt Hayden Planetarium featured a space show called "Passport to the Universe," which took visitors on a virtual zoom from New York City to the edge of the cosmos. En route the audience saw Earth, then the solar system, then the 100 billion stars of the Milky Way galaxy shrink to barely visible dots on the planetarium dome.

I soon received a letter from an Ivy League professor of psychology who wanted to administer a questionnaire to visitors, assessing the depth of their depression after viewing the show. Our show, he wrote, elicited the most dramatic feelings of smallness he had ever experienced.

How could that be? Every time I see the show, I feel alive and spirited and connected. I also feel large, knowing that the goings-on within the three-pound human brain are what enabled us to figure out our place in the universe.

Allow me to suggest that it's the professor, not I, who has misread nature. His ego was too big to begin with, inflated by delusions of significance and fed by cultural assumptions that human beings are more important than everything else in the universe.

In all fairness to the fellow, powerful forces in society leave most of us susceptible. As was I . . . until the day I learned in biology class that more bacteria live and work in one centimeter of my colon than the number of people who have ever existed in the world. That kind of information makes you think twice about who—or what—is actually in charge.

From that day on, I began to think of people not as the masters of space and time but as participants in a great cosmic chain of being, with a direct genetic link across species both living and extinct, extending back nearly 4 billion years to the earliest single-celled organisms on Earth.

■■■

Need more ego softeners? Simple comparisons of quantity, size, and scale do the job well.

Take water. It's simple, common, and vital. There are more molecules of water in an eight-ounce cup of the stuff than there are cups of water in all the world's oceans. Every cup that passes through a single person and eventually rejoins the world's water supply holds enough molecules to mix 1,500 of them into every other cup of water in the world. No way around it: some of the water you just drank passed through the kidneys of Socrates, Genghis Khan, and Joan of Arc.

How about air? Also vital. A single breathful draws in more air molecules than there are breathfuls of air in Earth's entire atmosphere. That means some of the air you just breathed passed through the lungs of Napoleon, Beethoven, Lincoln, and Billy the Kid.

Time to get cosmic. There are more stars in the universe than grains of sand on any beach, more stars than seconds have passed since Earth formed, more stars than words and sounds ever uttered by all the humans who ever lived.

Want a sweeping view of the past? Our unfolding cosmic perspective takes you there. Light takes time to reach Earth's observatories from the depths of space, and so you see objects and phenomena not as they are but as they once were. That means the universe acts like a giant time machine: the farther away you look, the further back in time you see—back almost to the beginning of time itself. Within that horizon of reckoning, cosmic evolution unfolds continuously, in full view.

Want to know what we're made of? Again, the cosmic perspective offers a bigger answer than you might expect. The chemical elements of the universe are forged in the fires of high-mass stars that end their lives in stupendous explosions, enriching their host galaxies with the chemical arsenal of life as we know it. We are not simply in the universe. The universe is in us. Yes, we are stardust.

■■■

Again and again across the centuries, cosmic discoveries have demoted our self-image. Earth was once assumed to be astronomically unique, until astronomers learned that Earth is just another planet orbiting the Sun. Then we presumed the Sun was unique, until we learned that the countless stars of the night sky are suns themselves. Then we presumed our galaxy, the Milky Way, was the entire known universe, until we established that the countless fuzzy things in the sky are other galaxies, dotting the landscape of our known universe.

The cosmic perspective flows from fundamental knowledge. But it's more than just what you know. It's also about having the wisdom and insight to apply that knowledge to assessing our place in the universe. And its attributes are clear:

- The cosmic perspective comes from the frontiers of science, yet is not solely the provenance of the scientist. It belongs to everyone.
- The cosmic perspective is humble.
- The cosmic perspective is spiritual—even redemptive—but is not religious.
- The cosmic perspective enables us to grasp, in the same thought, the large and the small.
- The cosmic perspective opens our minds to extraordinary ideas but does not leave them so open that our brains spill out, making us susceptible to believing anything we're told.
- The cosmic perspective opens our eyes to the universe, not as a benevolent cradle designed to nurture life but as a cold, lonely, hazardous place.
- The cosmic perspective shows Earth to be a mote, but a precious mote and, for the moment, the only home we have.
- The cosmic perspective finds beauty in the images of planets, moons, stars, and nebulae but also celebrates the laws of physics that shape them.

- The cosmic perspective enables us to see beyond our circumstances, allowing us to transcend the primal search for food, shelter, and sex.
- The cosmic perspective reminds us that in space, where there is no air, a flag will not wave—an indication that perhaps flag waving and space exploration do not mix.
- The cosmic perspective not only embraces our genetic kinship with all life on Earth but also values our chemical kinship with any yet-to-be discovered life in the universe, as well as our atomic kinship with the universe itself.

■■■

At least once a week, if not once a day, we might each ponder what cosmic truths lie undiscovered before us, perhaps awaiting the arrival of a clever thinker, an ingenious experiment, or an innovative space mission to reveal them. We might further ponder how those discoveries may one day transform life on Earth.

Absent such curiosity, we are no different from the provincial farmer who expresses no need to venture beyond the county line, because his forty acres meet all his needs. Yet if all our predecessors had felt that way, the farmer would instead be a cave dweller, chasing down his dinner with a stick and a rock.

During our brief stay on planet Earth, we owe ourselves and our descendants the opportunity to explore—in part because it's fun to do. But there's a far nobler reason. The day our knowledge of the cosmos ceases to expand, we risk regressing to the childish view that the universe figuratively and literally revolves around us. In that bleak world, arms-bearing, resource-hungry people and nations would be prone to act on their "low contracted prejudices." And that would be the last gasp of human enlightenment—until the rise of a visionary new culture that could once again embrace the cosmic perspective.

# 1 A Modern View of the Universe

▲ **About the photo:** This Hubble Space Telescope photo shows thousands of galaxies in a region of the sky so small you could cover it with a grain of sand held at arm's length.

## LEARNING GOALS

**1.1 The Scale of the Universe**
- What is our place in the universe?
- How big is the universe?

**1.2 The History of the Universe**
- How did we come to be?
- How do our lifetimes compare to the age of the universe?

**1.3 Spaceship Earth**
- How is Earth moving through space?
- How do galaxies move within the universe?

**1.4 The Human Adventure of Astronomy**
- How has the study of astronomy affected human history?

*It suddenly struck me that that tiny pea, pretty and blue, was the Earth. I put up my thumb and shut one eye, and my thumb blotted out the planet Earth. I didn't feel like a giant. I felt very, very small.*

*—Neil Armstrong on looking back at the Earth from the Moon, July 1969*

**Chapter 1 Overview**

Far from city lights on a clear night, you can gaze upward at a sky filled with stars. Lie back and watch for a few hours, and you will observe the stars marching steadily across the sky. Confronted by the seemingly infinite heavens, you might wonder how Earth and the universe came to be. If you do, you will be sharing an experience common to humans around the world and in thousands of generations past.

Modern science offers answers to many of our fundamental questions about the universe and our place within it. We now know the basic content and scale of the universe. We know the ages of Earth and the universe. And, although much remains to be discovered, we are rapidly learning how the simple ingredients of the early universe developed into the incredible diversity of life on Earth—and, perhaps, life on other worlds as well.

In this first chapter, we will survey the scale, history, and motion of the universe. This "big picture" perspective on our universe will provide a base on which you'll be able to build a deeper understanding in the rest of the book.

# 1.1 The Scale of the Universe

For most of human history, our ancestors imagined Earth to be stationary at the center of a relatively small universe. This idea made sense at a time when understanding was built upon everyday experience. After all, we cannot feel the constant motion of Earth as it rotates on its axis and orbits the Sun, and if you observe the sky you'll see that the Sun, Moon, planets, and stars all appear to revolve around us each day. Nevertheless, we now know that Earth is a planet orbiting a rather average star in a rather typical galaxy in a vast universe.

The historical path to this knowledge was long and complex. In later chapters, we'll see that the ancient belief in an Earth-centered (or *geocentric*) universe changed only when people were confronted by strong evidence to the contrary, and we'll explore how the method of learning that we call *science* enabled us to acquire this evidence. First, however, it's useful to have a general picture of the universe as we know it today.

## What is our place in the universe?

Take a look at the remarkable photo that opens this chapter (on page 1). This photo, taken by the Hubble Space Telescope, shows a piece of the sky so small that you could block your view of it with a grain of sand held at arm's length. Yet it encompasses an almost unimaginable expanse of both space and time. Nearly every object within it is a *galaxy* filled with billions of stars, and some of the smaller smudges are galaxies so far away that their light has taken billions of years to reach us. Let's begin our study of astronomy by exploring what a photo like this one tells us about our own place in the universe.

**Our Cosmic Address** The galaxies that we see in the Hubble Space Telescope photo are just one of several key levels of structure in our universe, all illustrated as our "cosmic address" in **FIGURE 1.1**.

Earth is a *planet* in our **solar system**, which consists of the Sun, the planets and their moons, and countless smaller objects that include rocky *asteroids* and icy *comets*. Keep in mind that our Sun is a *star*, just like the stars we see in our night sky.

Our solar system belongs to the huge, disk-shaped collection of stars called the **Milky Way Galaxy**. A **galaxy** is a great island of stars in space, all held together by gravity and orbiting a common center. The Milky Way is a relatively large galaxy, containing more than 100 billion stars, and we think that most of these stars are orbited by planets. Our solar system is located a little over halfway from the galactic center to the edge of the galactic disk.

Billions of other galaxies are scattered throughout space. Some galaxies are fairly isolated, but most are found in groups. Our Milky Way, for example, is one of the two largest among more than 70 galaxies (most relatively small) in the **Local Group**. Groups of galaxies with many more large members are often called **galaxy clusters**.

On a very large scale, galaxies and galaxy clusters appear to be arranged in giant chains and sheets with huge voids between them; the background of Figure 1.1 represents this large-scale structure. The regions in which galaxies and galaxy clusters are most tightly packed are called **superclusters**, which are essentially clusters of galaxy clusters. Our Local Group is located in the outskirts of the Local Supercluster (also called *Laniakea*, Hawaiian for "immense heaven").

Together, all these structures make up our **universe**. In other words, the universe is the sum total of all matter and energy, encompassing the superclusters and voids and everything within them.

**Think about it** Some people think that our tiny physical size in the vast universe makes us insignificant. Others think that our ability to learn about the wonders of the universe gives us significance despite our small size. What do *you* think?

**Astronomical Distance Measurements** The labels in Figure 1.1 give an approximate size for each structure in kilometers (recall that 1 kilometer ≈ 0.6 mile), but many distances in astronomy are so large that kilometers are not the most convenient unit. Instead, we often use two other units:

- One **astronomical unit (AU)** is Earth's average distance from the Sun, which is about 150 million kilometers (93 million miles). We commonly describe distances within our solar system in AU.
- One **light-year (ly)** is the distance that light can travel in 1 year, which is about 10 trillion kilometers (6 trillion miles). We generally use light-years to describe the distances of stars and galaxies.

Be sure to note that a light-year is a unit of *distance*, not of time. Light travels at the speed of light, which is 300,000 kilometers per second. We therefore say that one *light-second* is about 300,000 kilometers, because that is the distance light travels in one second. Similarly, one

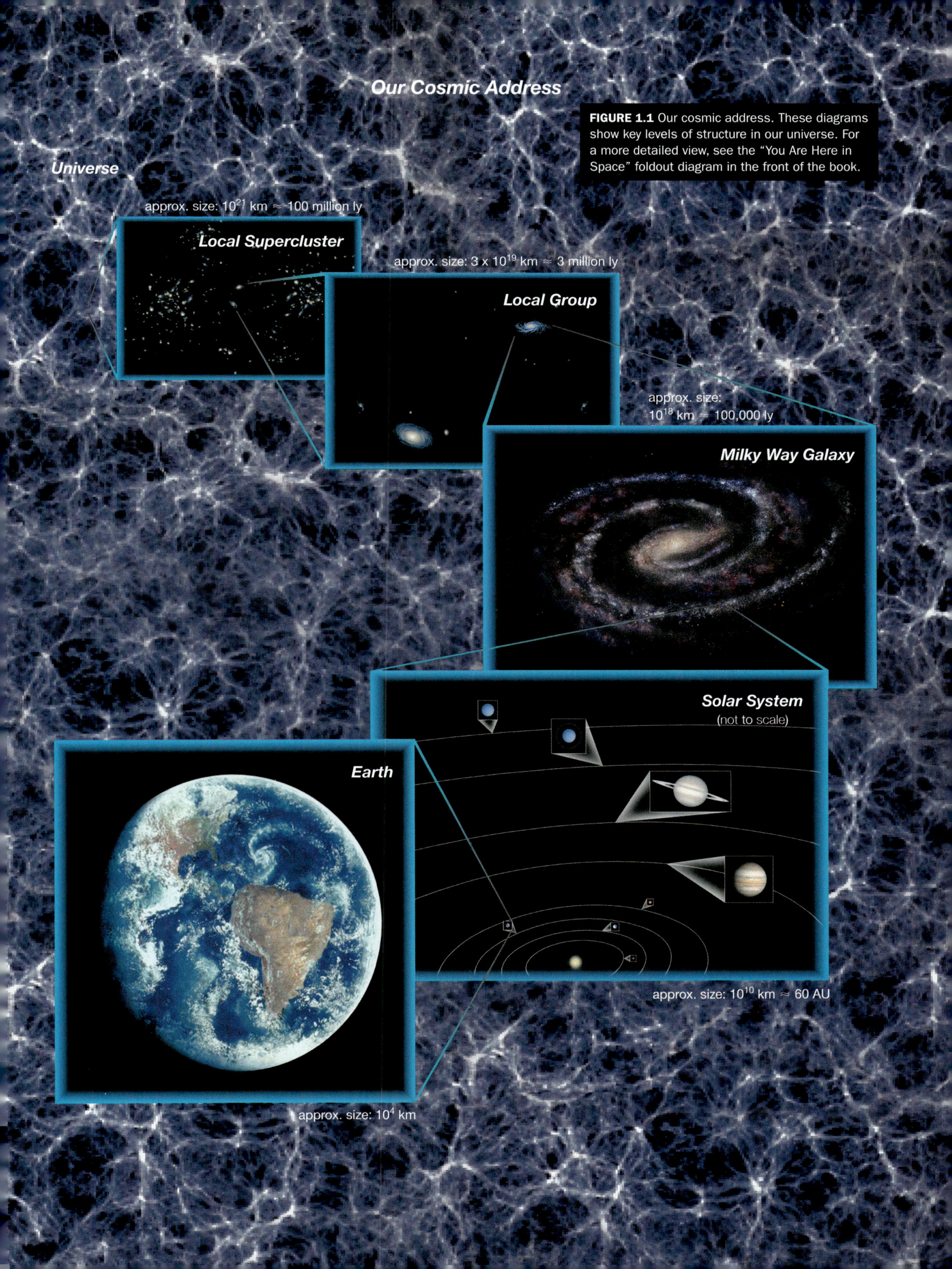


**FIGURE 1.1** Our cosmic address. These diagrams show key levels of structure in our universe. For a more detailed view, see the "You Are Here in Space" foldout diagram in the front of the book.

light-minute is the distance that light travels in one minute, one light-hour is the distance that light travels in one hour, and so on. Mathematical Insight 1.1 (page 6) shows that light travels about 10 trillion kilometers in one year, so that distance represents a light-year.

**Looking Back in Time** The speed of light is extremely fast by earthly standards. It is so fast that if you could make light go in circles, it could circle Earth nearly eight times in a single second. Nevertheless, even light takes time to travel the vast distances in space. Light takes a little more than 1 second to reach Earth from the Moon, and about 8 minutes to reach Earth from the Sun. Stars are so far away that their light takes years to reach us, which is why we measure their distances in light-years.

Consider Sirius, the brightest star in the night sky, which is located about 8 light-years away. Because it takes light 8 years to travel this distance, we see Sirius not as it is today, but rather as it was 8 years ago. The effect is more dramatic at greater distances. The Orion Nebula (**FIGURE 1.2**) is a giant cloud in which stars and planets are forming. It is located about 1350 light-years from Earth, which means we see it as it looked about 1350 years ago. If any major events have occurred in the Orion Nebula since that time, we cannot yet know about them because the light from these events has not yet reached us.

The general idea that light takes time to travel through space leads to a remarkable fact:

> **The farther away we look in distance, the further back we look in time.**

The Andromeda Galaxy (**FIGURE 1.3**) is about 2.5 million light-years away, which means we see it as it looked about 2.5 million years ago. We see more distant galaxies as they were even further in the past. Some of the galaxies in the Hubble Space Telescope photo that opens the chapter are more than 12 billion light-years away, meaning we see them as they were more than 12 billion years ago.

**See it for yourself** The central region of the Andromeda Galaxy is faintly visible to the naked eye and easy to see with binoculars. Use a star chart to find it in the night sky and remember that you are seeing light that spent 2.5 million years in space before reaching your eyes. If students on a planet in the Andromeda Galaxy were looking at the Milky Way, what would they see? Could they know that we exist here on Earth?

It's also amazing to realize that any "snapshot" of a distant galaxy is a picture of both space and time. For

## BASIC ASTRONOMICAL DEFINITIONS

### Astronomical Objects

**star** A large, glowing ball of gas that generates heat and light through nuclear fusion in its core. Our Sun is a star.

**planet** A moderately large object that orbits a star and shines primarily by reflecting light from its star. According to the current definition, an object can be considered a planet only if it (1) orbits a star, (2) is large enough for its own gravity to make it round, and (3) has cleared most other objects from its orbital path. An object that meets the first two criteria but has not cleared its orbital path, like Pluto, is designated a **dwarf planet**.

**moon** (or **satellite**) An object that orbits a planet. The term *satellite* is also used more generally to refer to any object orbiting another object.

**asteroid** A relatively small and rocky object that orbits a star.

**comet** A relatively small and ice-rich object that orbits a star.

**small solar system body** An asteroid, comet, or other object that orbits a star but is too small to qualify as a planet or dwarf planet.

### Collections of Astronomical Objects

**solar system** The Sun and all the material that orbits it, including planets, dwarf planets, and small solar system bodies. Although the term *solar system* technically refers only to our own star system (*solar* means "of the Sun"), it is often applied to other star systems as well.

**star system** A star (sometimes more than one star) and any planets and other materials that orbit it.

**galaxy** A great island of stars in space, all held together by gravity and orbiting a common center, with a total mass equivalent to millions, billions, or even trillions of stars.

**cluster of galaxies** (or **group of galaxies**) A collection of galaxies bound together by gravity. Small collections (up to a few dozen galaxies) are generally called *groups*, while larger collections are called *clusters*.

**supercluster** A gigantic region of space in which many groups and clusters of galaxies are packed more closely together than elsewhere in the universe.

**universe** (or **cosmos**) The sum total of all matter and energy—that is, all galaxies and everything between them.

**observable universe** The portion of the entire universe that can be seen from Earth, at least in principle. The observable universe is probably only a tiny portion of the entire universe.

### Astronomical Distance Units

**astronomical unit (AU)** The average distance between Earth and the Sun, which is about 150 million kilometers. More technically, 1 AU is the length of the semimajor axis of Earth's orbit.

**light-year** The distance that light can travel in 1 year, which is about 10 trillion kilometers (more precisely, 9.46 trillion km).

### Terms Relating to Motion

**rotation** The spinning of an object around its axis. For example, Earth rotates once each day around its axis, which is an imaginary line connecting the North and South Poles.

**orbit** (or **revolution**) The orbital motion of one object around another due to gravity. For example, Earth orbits the Sun once each year.

**expansion (of the universe)** The increase in the average distance between galaxies as time progresses.

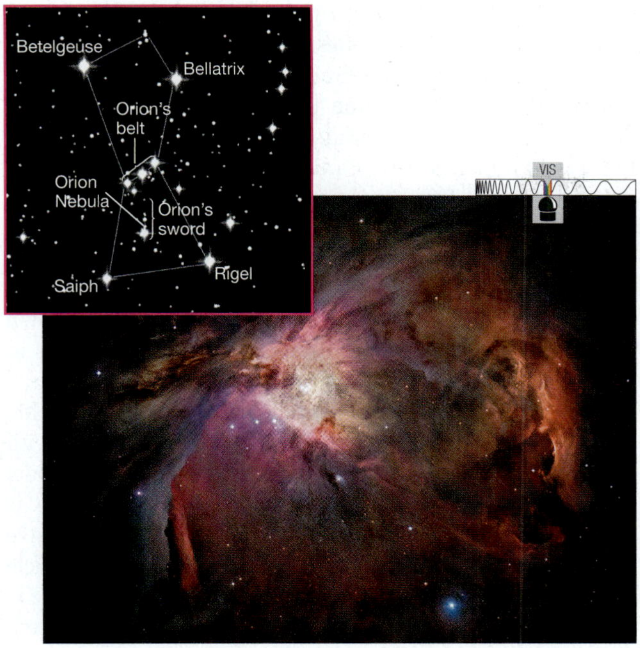


**FIGURE 1.2** The Orion Nebula, located about 1350 light-years away. The inset shows its location in the constellation Orion.

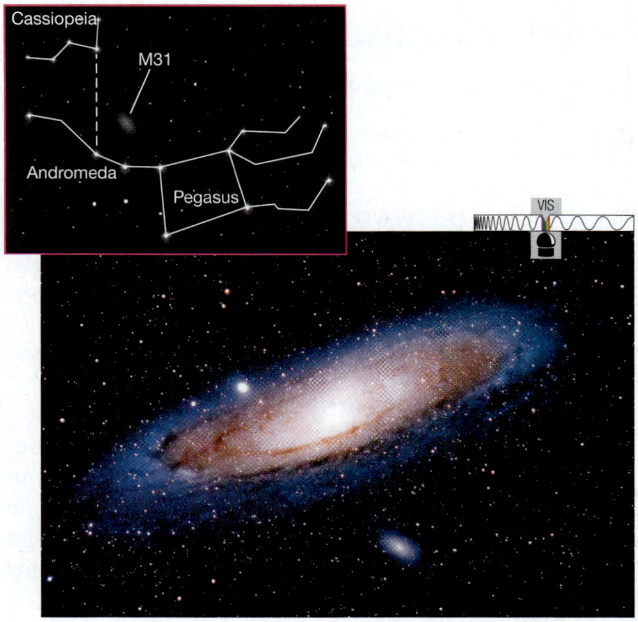


**FIGURE 1.3** The Andromeda Galaxy (also known as M31). When we look at this galaxy, we see light that has been traveling through space for 2.5 million years. The inset shows the galaxy's location in the constellation Andromeda.

example, because the Andromeda Galaxy is about 100,000 light-years in diameter, the light we currently see from the far side of the galaxy must have left on its journey to us some 100,000 years before the light we see from the near side. Figure 1.3 therefore shows different parts of the galaxy spread over a time period of 100,000 years. When we study the universe, it is impossible to separate space and time.

**The Observable Universe** As we'll discuss in Section 1.2, the measured age of the universe is about 14 billion years. This fact, combined with the fact that looking deep into space means looking far back in time, places a limit on the portion of the universe that we can see, even in principle.

**FIGURE 1.4** shows the idea. If we look at a galaxy that is 7 billion light-years away, we see it as it looked 7 billion years ago*—which means we see it as it was when the universe was half its current age. If we look at a galaxy that is 12 billion light-years away (like the most distant ones in the Hubble Space Telescope photo), we see it as it was 12 billion years ago, when the universe was only 2 billion years old. And if we tried to look beyond 14 billion light-years, we'd be looking to a time more than 14 billion years ago—which is before the universe existed and therefore means that there is nothing to see. This distance of 14 billion light-years therefore marks the boundary (or *horizon*) of our **observable universe**—the portion of the entire universe that we can potentially observe. Note that this fact does not put any limit on the size of the *entire* universe, which we assume to be far larger than our observable universe. We simply cannot see or study anything beyond the bounds of our observable universe, because the light from such distances has not yet had time to reach us in a 14-billion-year old universe.

*As we'll discuss in Chapter 20, distances to faraway galaxies in an expanding universe can be described in more than one way; distances like those given here are based on the time it has taken a galaxy's light to reach us (called the *lookback time*).

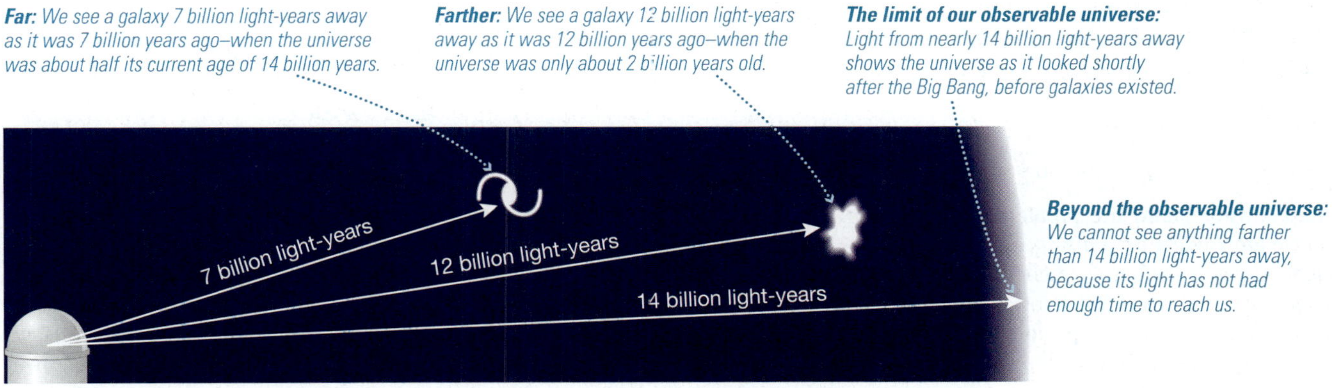


**FIGURE 1.4** The farther away we look in space, the further back we look in time. The age of the universe therefore puts a limit on the size of the *observable* universe—the portion of the entire universe that we can observe, at least in principle.

## COMMON MISCONCEPTIONS

### The Meaning of a Light-Year

Maybe you've heard people say things like "It will take me light-years to finish this homework!" But that statement doesn't make sense, because a light-year is a unit of *distance*, not time. If you are unsure whether the term *light-year* is being used correctly, try testing the statement by using the fact that 1 light-year is about 10 trillion kilometers, or 6 trillion miles. The statement then reads "It will take me 6 trillion miles to finish this homework," which clearly does not make sense.

## How big is the universe?

Figure 1.1 put numbers on the sizes of different structures in the universe, but these numbers have little meaning for most people—after all, they are literally astronomical. To help you develop a greater appreciation of our modern view of the universe, we'll discuss a few ways of putting these numbers into perspective.

**The Scale of the Solar System** One of the best ways to develop perspective on cosmic sizes and distances is to imagine our solar system shrunk down to a scale that would allow you to walk through it. The Voyage scale model solar system (**FIGURE 1.5**) makes such a walk possible by showing sizes and distances in the solar system at *one ten-billionth* of their actual values.

**FIGURE 1.6a** shows the Sun and planets at their actual sizes (but not distances) on the Voyage scale. The model Sun is about the size of a large grapefruit, Jupiter is about the size of a marble, and Earth is about the size of the ball point in a pen. You can immediately see some key facts about our solar system. For example, the Sun is far larger than any of the planets; in mass, the Sun outweighs all the planets combined by a factor of nearly 1000. The planets also vary considerably in size: The storm on Jupiter known as the Great Red Spot (visible near Jupiter's lower left in the painting) could swallow up the entire Earth.

The scale of the solar system is even more remarkable when you combine the sizes shown in Figure 1.6a with the distances illustrated by the map of the Voyage model in **FIGURE 1.6b**. For example, the ball-point-size Earth is located about 15 meters (49 feet) from the grapefruit-size Sun, which means you can picture Earth's orbit as a circle of radius 15 meters around a grapefruit.

Perhaps the most striking feature of our solar system when we view it to scale is its emptiness. The Voyage model shows the planets along a straight path, so we'd need to draw each planet's orbit around the model Sun to show the full extent of our planetary system. Fitting all these orbits would require an area measuring more than a

## MATHEMATICAL INSIGHT 1.1

Problem Solving Part 1

### How Far Is a Light-Year? An Introduction to Astronomical Problem Solving

We can develop greater insight into astronomical ideas by applying mathematics. The key to using mathematics is to approach problems in a clear and organized way. One simple approach uses the following three steps:

**Step 1 Understand the problem:** Ask yourself what the solution will look like (for example, what units will it have? will it be big or small?) and what information you need to solve the problem. Draw a diagram or think of a simpler analogous problem to help you decide how to solve it.

**Step 2 Solve the problem:** Carry out the necessary calculations.

**Step 3 Explain your result:** Be sure that your answer makes sense, and consider what you've learned by solving the problem.

You can remember this process as "Understand, Solve, and Explain," or U-S-E for short. You may not always need to write out the three steps explicitly, but they may help if you are stuck.

**EXAMPLE:** How far is a light-year?

**SOLUTION:** Let's use the three-step process.

**Step 1 Understand the problem:** The question asks how *far*, so we are looking for a *distance*. In this case, the definition of a light-year tells us that we are looking for the *distance that light can travel in 1 year*. We know that light travels at the speed of light, so we are looking for an equation that gives us distance from speed. If you don't remember this equation, just think of a simpler but analogous problem, such as "If you drive at 50 kilometers per hour, how far will you travel in 2 hours?" You'll realize that you simply multiply the speed by the time: $\text{distance} = \text{speed} \times \text{time}$. In this case, the speed is the speed of light, or 300,000 km/s, and the time is 1 year.

**Step 2 Solve the problem:** From Step 1, our equation is that 1 light-year is the speed of light times 1 year. To make the units consistent, we convert 1 year to seconds by remembering that there are 60 seconds in 1 minute, 60 minutes in 1 hour, 24 hours in 1 day, and 365 days in 1 year. (See Appendix C.3 to review unit conversions.) We now carry out the calculations:

$$\begin{aligned} 1 \text{ light-year} &= (\text{speed of light}) \times (1 \text{ yr}) \\ &= \left(300{,}000 \frac{\text{km}}{\text{s}}\right) \times \left(1 \cancel{\text{yr}} \times \frac{365 \cancel{\text{days}}}{1 \cancel{\text{yr}}} \right. \\ &\qquad \left. \times \frac{24 \cancel{\text{hr}}}{1 \cancel{\text{day}}} \times \frac{60 \cancel{\text{min}}}{1 \cancel{\text{hr}}} \times \frac{60 \cancel{\text{s}}}{1 \cancel{\text{min}}} \right) \\ &= 9{,}460{,}000{,}000{,}000 \text{ km } (9.46 \text{ trillion km}) \end{aligned}$$

**Step 3 Explain your result:** In sentence form, our answer is "One light-year is about 9.46 trillion kilometers." This answer makes sense: It has the expected units of distance (kilometers) and it is a long way, which we expect for the distance that light can travel in a year. We say "about" in the answer because we know it is not exact. For example, a year is not exactly 365 days long. In fact, for most purposes, we can approximate the answer further as "One light-year is about 10 trillion kilometers."

**FIGURE 1.5** This photo shows the pedestals housing the Sun (the gold sphere on the nearest pedestal) and the inner planets in the Voyage scale model solar system (Washington, D.C.). The model planets are encased in the sidewalk-facing disks visible at about eye level on the planet pedestals. To the left is the National Air and Space Museum.

kilometer on a side—an area equivalent to more than 300 football fields arranged in a grid. Spread over this large area, only the grapefruit-size Sun, the planets, and a few moons would be big enough to see. The rest of it would look virtually empty (that's why we call it *space!*).

Seeing our solar system to scale also helps put space exploration into perspective. The Moon, the only other world on which humans have ever stepped (**FIGURE 1.7**), lies only about 4 centimeters ($1\frac{1}{2}$ inches) from Earth in the Voyage model. On this scale, the palm of your hand can cover the entire region of the universe in which humans have so far traveled. The trip to Mars is more than 150 times as far as the trip to the Moon, even when Mars is on the same side of its orbit as Earth. And while you can walk from Earth to Pluto in a few minutes on the Voyage scale, the *New Horizons* spacecraft, which flew past Pluto in 2015, took more than 9 years to make the real journey, despite traveling at a speed nearly 100 times that of a commercial jet.

**Distances to the Stars** If you visit the Voyage model in Washington, D.C., you need to walk only about 600 meters

**a** The scaled sizes (but not distances) of the Sun, the planets, and the two largest known dwarf planets.

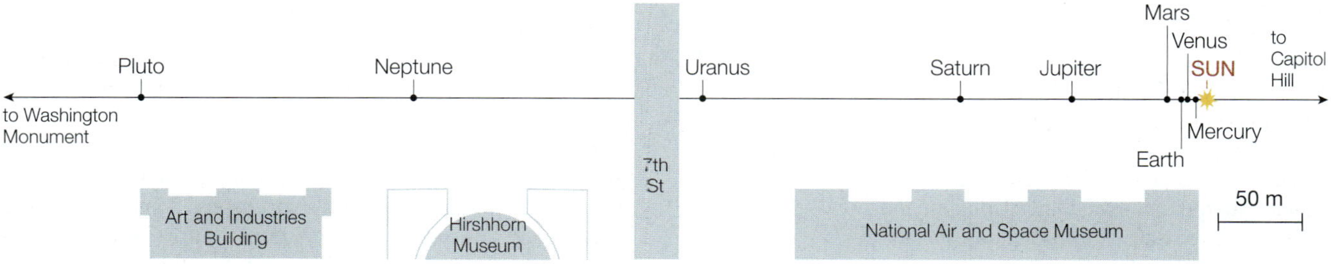


**b** Locations of major objects in the Voyage model (Washington, D.C.).

**FIGURE 1.6** The Voyage scale model represents sizes and distances in the solar system at *one ten-billionth* of their actual values. Planets are lined up in the model, but in reality each planet orbits the Sun independently and a perfect alignment never occurs.

**FIGURE 1.7** This famous photograph from the first Moon landing (*Apollo 11* in July 1969) shows astronaut Buzz Aldrin, with Neil Armstrong reflected in his visor. Armstrong was the first to step onto the Moon's surface, saying, "That's one small step for a man, one giant leap for mankind." (When asked why this photo became so iconic, Aldrin replied, "Location, location, location!")

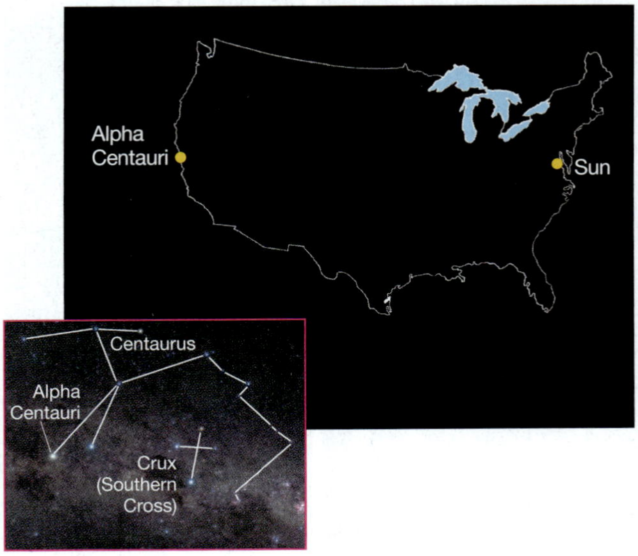


**FIGURE 1.8** On the same 1-to-10 billion scale on which you can walk from the Sun to Pluto in just a few minutes, you'd need to cross the United States to reach Alpha Centauri, the nearest other star system. The inset shows the location and appearance of Alpha Centauri among the constellations.

to go from the Sun to Pluto. How much farther would you have to walk to reach the next star on this scale?

Amazingly, you would need to walk to California. If this answer seems hard to believe, you can check it for yourself. A light-year is about 10 trillion kilometers, which becomes 1000 kilometers on the 1-to-10-billion scale (because 10 trillion $\div$ 10 billion $=$ 1000). The nearest star system to our own, a three-star system called Alpha Centauri (**FIGURE 1.8**), is about 4.4 light-years away. That distance is about 4400 kilometers (2700 miles) on the 1-to-10-billion scale, or roughly equivalent to the distance across the United States.

The tremendous distances to the stars give us some perspective on the technological challenge of astronomy. For example, because the largest star of the Alpha Centauri system is roughly the same size and brightness as our Sun, viewing it in the night sky is somewhat like being in Washington, D.C., and seeing a very bright grapefruit in San Francisco (neglecting the problems introduced by the curvature of Earth). It may seem remarkable that we can see the star at all, but the blackness of the night sky allows the naked eye to see it as a faint dot of light. It looks much brighter through powerful telescopes, but we still cannot see features of the star's surface.

Now, consider the difficulty of detecting *planets* orbiting nearby stars, which is equivalent to looking from Washington, D.C., and trying to find ball points or marbles orbiting grapefruits in California or beyond. When you consider this challenge, it is all the more remarkable to realize that we now have technology capable of finding such planets [**Section 13.1**].

## SPECIAL TOPIC How Many Planets Are There in Our Solar System?

Prior to 2006, children were taught that Pluto was one of nine planets in our solar system. But we now call Pluto a *dwarf planet*, leaving the solar system with only eight planets. Why the change?

When Pluto was discovered in 1930, it was assumed to be similar to other planets. But by the mid-1990s, it had become clear that Pluto is part of a large group of ice-rich objects that share its region of the solar system (called the *Kuiper belt*, which we'll discuss in Chapter 12). Still, as long as Pluto was the largest known of these objects, most astronomers were content to leave the planetary status quo. Change was forced by the 2005 discovery of an object called Eris. Because Eris is slightly larger in mass than Pluto, astronomers could no longer avoid the question of what objects should count as planets.

Official decisions on astronomical names and definitions rest with the International Astronomical Union (IAU), an organization made up of professional astronomers from around the world. In 2006, an IAU vote defined "planet" in a way that left out Pluto and Eris (see Basic Astronomical Definitions on page 4), but added the "dwarf planet" category to accommodate them. These definitions still spark some controversy, but perhaps are best thought of as an example of the difference between the fuzzy boundaries of nature and the human preference for categories. After all, the argument about whether Pluto is a planet or a dwarf planet is not so different from the argument you may hear over whether a particular waterway is a creek, stream, or river.

A more recent question about the number of planets in our solar system has arisen from calculations based on the orbits of the ice-rich objects of the outer solar system. Some astronomers argue that these orbits show patterns indicating the gravitational tug of an undiscovered "planet nine." If such an object is actually found, astronomers may be forced to revisit the definition of "planet" once again.

The vast distances to the stars also offer a sobering lesson about interstellar travel. Although science fiction shows like *Star Trek* and *Star Wars* make such travel look easy, the reality is far different. Consider the *Voyager 2* spacecraft. Launched in 1977, *Voyager 2* flew by Jupiter in 1979, Saturn in 1981, Uranus in 1986, and Neptune in 1989. It is now bound for the stars at a speed of close to 50,000 kilometers per hour—about 100 times as fast as a speeding bullet. But even at this speed, *Voyager 2* would take about 100,000 years to reach Alpha Centauri if it were headed in that direction (which it's not). Convenient interstellar travel remains well beyond our present technology.

**The Size of the Milky Way Galaxy** The vast separation between our solar system and Alpha Centauri is typical of the separations between star systems in our region of the Milky Way Galaxy. We therefore cannot use the 1-to-10-billion scale for thinking about distances beyond the nearest stars, because more distant stars would not fit on Earth with this scale. To visualize the galaxy, let's reduce our scale by another factor of 1 billion (making it a scale of 1 to $10^{19}$).

On this new scale, each light-year becomes 1 millimeter, and the 100,000-light-year diameter of the Milky Way Galaxy becomes 100 meters, or about the length of a football field. Visualize a football field with a scale model of our galaxy centered over midfield (**FIGURE 1.9**). Our entire solar system is a microscopic dot located around the 20-yard line. The 4.4-light-year separation between our solar system and Alpha Centauri becomes just 4.4 millimeters on this scale—smaller than the width of your little finger. If you stood at the position of our solar system in this model, millions of star systems would lie within reach of your arms.

Another way to put the galaxy into perspective is to consider its number of stars—more than 100 billion. Imagine that tonight you are having difficulty falling asleep (perhaps because you are contemplating the scale of the universe). Instead of counting sheep, you decide to count stars. If you are able to count about one star each second, how long would it take you to count 100 billion stars in the Milky Way? Clearly, the answer is 100 billion $(10^{11})$ seconds, but how long is that? Amazingly, 100 billion seconds is more than 3000 years. (You can confirm this by dividing 100 billion by the number of seconds in 1 year.) You would need thousands of years just to *count* the stars in the Milky Way Galaxy, and this assumes you never take a break—no sleeping, no eating, and absolutely no dying!

## MATHEMATICAL INSIGHT 1.2

### The Scale of Space and Time

Scientific Notation, Parts 1 to 3

Making a scale model usually requires nothing more than division. For example, in a 1-to-20 architectural scale model, a building that is actually 6 meters tall will be only $6 \div 20 = 0.3$ meter tall. The idea is the same for astronomical scaling, except that we usually divide by such large numbers that it's easier to work in *scientific notation*—that is, with the aid of powers of 10. (See Appendixes C.1 and C.2 to review these concepts.)

**EXAMPLE 1:** How big is the Sun on a 1-to-10-billion scale?

**SOLUTION:**

**Step 1 Understand:** We are looking for the scaled *size* of the Sun, so we simply need to divide its actual radius by 10 billion, or $10^{10}$. Appendix E.1 gives the Sun's radius as 695,000 km, or $6.95 \times 10^5$ km in scientific notation.

**Step 2 Solve:** We carry out the division:

$$\text{scaled radius} = \frac{\text{actual radius}}{10^{10}} = \frac{6.95 \times 10^5\ \text{km}}{10^{10}} = 6.95 \times 10^{(5-10)}\ \text{km} = 6.95 \times 10^{-5}\ \text{km}$$

Notice that we used the rule that dividing powers of 10 means subtracting their exponents [**Appendix C.1**].

**Step 3 Explain:** We have found an answer, but because most of us don't have a good sense of what $10^{-5}$ kilometer looks like, the answer will be more meaningful if we convert it to centimeters (recalling that $1\ \text{km} = 10^3\ \text{m}$ and $1\ \text{m} = 10^2\ \text{cm}$):

$$6.95 \times 10^{-5}\ \cancel{\text{km}} \times \frac{10^3\ \cancel{\text{m}}}{1\ \cancel{\text{km}}} \times \frac{10^2\ \text{cm}}{1\ \cancel{\text{m}}} = 6.95\ \text{cm}$$

On the 1-to-10-billion scale, the Sun's radius is about 7 centimeters, which is a diameter of about 14 centimeters—about the size of a large grapefruit.

**EXAMPLE 2:** What scale allows the 100,000-light-year diameter of the Milky Way Galaxy to fit on a 100-meter-long football field?

**SOLUTION:**

**Step 1 Understand:** We want to know *how many times larger* the actual diameter of the galaxy is than 100 meters, so we'll divide the actual diameter by 100 meters. To carry out the division, we'll need both numbers in the same units. We can put the galaxy's diameter in meters by using the fact that a light-year is about $10^{13}$ kilometers (see Mathematical Insight 1.1) and a kilometer is $10^3$ meters; because we are working with powers of 10, we'll write the galaxy's 100,000-light-year diameter as $10^5$ ly.

**Step 2 Solve:** We now convert the units and carry out the division:

$$\frac{\text{galaxy diameter}}{\text{football field diameter}} = \frac{10^5\ \cancel{\text{ly}} \times \dfrac{10^{13}\ \cancel{\text{km}}}{1\ \cancel{\text{ly}}} \times \dfrac{10^3\ \cancel{\text{m}}}{1\ \cancel{\text{km}}}}{10^2\ \cancel{\text{m}}} = 10^{(5+13+3-2)} = 10^{19}$$

Note that the answer has no units, because it simply tells us how many times larger one thing is than the other.

**Step 3 Explain:** We've found that we need a scale of 1 to $10^{19}$ to make the galaxy fit on a football field.

**FIGURE 1.9** This painting shows the Milky Way Galaxy on a scale where its diameter is the length of a football field. On this scale, stars are microscopic and the distance between our solar system and Alpha Centauri is only 4.4 millimeters. There are so many stars in our galaxy that it would take thousands of years just to count them out loud.

**Think about it** Contemplate the vast number of stars in our galaxy, and consider that each star is a potential sun for a system of planets. How does this perspective affect your thoughts about the possibilities for finding life—or intelligent life—beyond Earth? Explain.

**The Observable Universe** As incredible as the scale of our galaxy may seem, the Milky Way is only one of more than 100 billion large galaxies (and many more small galaxies) in the observable universe. Just as it would take thousands of years to count the stars in the Milky Way, it would take thousands of years to count all these galaxies.

Think for a moment about the total number of stars in all these galaxies. If we assume 100 billion galaxies and 100 billion stars per galaxy, the total number of stars in the observable universe is roughly 100 billion × 100 billion, or 10,000,000,000,000,000,000,000 $(10^{22})$. How big is this number? Visit a beach. Run your hands through the fine-grained sand. Imagine counting each tiny grain of sand as it slips through your fingers. Then imagine counting every grain of sand on the beach and continuing to count *every* grain of dry sand on *every* beach on Earth (see Mathematical Insight 1.3). If you could actually complete this task, you would find that the number of grains of sand is comparable to the number of stars in the observable universe (**FIGURE 1.10**).

**Think about it** Study the foldout in the front of this book, which illustrates the ideas covered in this section in greater detail. Overall, how does visualizing Earth to scale affect your perspective on our planet and on human existence? Explain.

## MATHEMATICAL INSIGHT 1.3 Order of Magnitude Estimation

In astronomy, numbers are often so large that an estimate can be useful even if it's good only to about the nearest power of 10. For example, when we multiplied 100 billion stars per galaxy by 100 billion galaxies to estimate that there are about $10^{22}$ stars in the observable universe, we knew that the "ballpark" nature of these numbers means the actual number of stars could easily be anywhere from about $10^{21}$ to $10^{23}$. Estimates good to about the nearest power of 10 are called **order of magnitude estimates**.

**EXAMPLE:** Verify the claim that the number of grains of (dry) sand on all the beaches on Earth is comparable to the number of stars in the observable universe.

**SOLUTION:**

**Step 1 Understand:** To verify the claim, we need to estimate the number of grains of sand and see if it is close to our estimate of $10^{22}$ stars. We can estimate the total number of sand grains by dividing the *total volume* of sand on Earth's beaches by the *average volume* of an individual sand grain. Volume is equal to length times width times depth, so the total volume is the total length of sandy beach on Earth multiplied by the typical width and depth of dry sand. That is,

$$\text{total sand grains} = \frac{\text{total volume of beach sand}}{\text{average volume of 1 sand grain}} = \frac{\text{beach length} \times \text{beach width} \times \text{beach depth}}{\text{average volume of 1 sand grain}}$$

We now need numbers to put into the equation. We can estimate the average volume of an individual sand grain by measuring out a small volume of sand, counting the number of grains in this volume, and then dividing the volume by the number of grains. If you do this, you'll find that a reasonable order of magnitude estimate is one-tenth of a cubic millimeter, or $10^{-10}\ \text{m}^3$, per sand grain. We can estimate beach width and depth from experience or photos of beaches. Typical widths are about 20 to 50 meters and typical sand depth is about 2 to 5 meters, so we can make the numbers easy by assuming that the product of beach width times depth is about 100 square meters, or $10^2\ \text{m}^2$. The total length of sandy beach on Earth is more difficult to estimate, but you can look online and find that it is less than about 1 million kilometers, or $10^9$ m.

**Step 2 Solve:** We already have our equation and all the numbers we need, so we just put them in; note that we group beach width and depth together, since we estimated them together in Step 1:

$$\begin{aligned}\text{total sand grains} &= \frac{\text{beach length} \times (\text{beach width} \times \text{beach depth})}{\text{average volume of 1 sand grain}} \\ &= \frac{10^9\ \cancel{\text{m}} \times 10^2\ \cancel{\text{m}^2}}{10^{-10}\ \cancel{\text{m}^3}} \\ &= 10^{[9+2-(-10)]} = 10^{21}\end{aligned}$$

**Step 3 Explain:** Our order of magnitude estimate for the total number of grains of dry sand on all the beaches on Earth is $10^{21}$, which is within a factor of 10 of the estimated $10^{22}$ stars in the observable universe. Because both numbers could easily be off by a factor of 10 or more, we cannot say with certainty that one is larger than the other, but the numbers are clearly comparable.

**FIGURE 1.10** The number of stars in the observable universe is comparable to the number of grains of dry sand on all the beaches on Earth.

## COMMON MISCONCEPTIONS

### Confusing Very Different Things

Most people are familiar with the terms *solar system* and *galaxy*, but few realize how incredibly different they are. Our solar system is a single star system, while our galaxy is a collection of more than 100 billion star systems—so many that it would take thousands of years just to count them. Moreover, if you look at the sizes in Figure 1.1, you'll see that our galaxy is about 100 million times larger in diameter than our solar system. So be careful; numerically speaking, mixing up *solar system* and *galaxy* is a gigantic mistake!

## 1.2 The History of the Universe

Our universe is vast not only in space, but also in time. In this section, we will briefly discuss the history of the universe as we understand it today.

Before we begin, you may wonder how we can claim to know anything about what the universe was like in the distant past. We'll devote much of this textbook to understanding how science enables us to do this, but you already know part of the answer: Because looking farther into space means looking further back in time, we can actually *see* parts of the universe as they were long ago, simply by looking far enough away. In other words, telescopes are somewhat like time machines, enabling us to observe the history of the universe.

### How did we come to be?

**FIGURE 1.11** (pages 12–13) summarizes the history of the universe according to modern science. Let's start at the upper left of the figure, and discuss the key events and what they mean.

**The Big Bang, Expansion, and the Age of the Universe** Telescopic observations of distant galaxies show that the entire universe is *expanding*, meaning that average distances between galaxies are increasing with time. This fact implies that galaxies must have been closer together in the past, and if we go back far enough, we must reach the point at which the expansion began. We call this beginning the **Big Bang**, and scientists use the observed rate of expansion to calculate that it occurred about 14 billion years ago. The three cubes in the upper left portion of Figure 1.11 represent the expansion of a small piece of the universe through time.

The universe as a whole has continued to expand ever since the Big Bang, but on smaller size scales the force of gravity has drawn matter together. Structures such as galaxies and galaxy clusters occupy regions where gravity has won out against the overall expansion. That is, while the universe as a whole continues to expand, individual galaxies and galaxy clusters (and objects within them such as stars and planets) do *not* expand. This idea is also illustrated by the three cubes in Figure 1.11. Notice that as the cube as a whole grew larger, the matter within it clumped into galaxies and galaxy clusters. Most galaxies, including our own Milky Way, formed within a few billion years after the Big Bang.

**Stellar Lives and Galactic Recycling** Within galaxies like the Milky Way, gravity drives the collapse of clouds of gas and dust to form stars and planets. Stars are not living organisms, but they nonetheless go through "life cycles." A star is born when gravity compresses the material in a cloud to the point at which the center becomes dense enough and hot enough to generate energy by **nuclear fusion**, the process in which lightweight atomic nuclei smash together and stick (or fuse) to make heavier nuclei. The star "lives" as long as it can shine with energy from fusion, and "dies" when it exhausts its usable fuel.

In its final death throes, a star blows much of its contents back out into space. The most massive stars die in titanic explosions called *supernovae*. The returned matter mixes with other matter floating between the stars in the galaxy, eventually becoming part of new clouds of gas and dust from which new generations of stars can be born. Galaxies therefore function as cosmic recycling plants, recycling material expelled from dying stars into new generations of stars and planets. This cycle is illustrated in the lower right of Figure 1.11. Our own solar system is a product of many generations of such recycling.

**Star Stuff** The recycling of stellar material is connected to our existence in an even deeper way. By studying stars of different ages, we have learned that the early universe contained only the simplest chemical elements: hydrogen and helium (and a trace of lithium). We and Earth are made primarily of other elements, such as carbon, nitrogen, oxygen, and iron. Where did these other elements come from? Evidence shows that they were manufactured by stars, some through the nuclear fusion that makes stars shine, and most others through nuclear reactions accompanying the explosions that end stellar lives.

By the time our solar system formed, about $4\frac{1}{2}$ billion years ago, earlier generations of stars had already converted up to 2% of our galaxy's original hydrogen and helium into heavier elements. Therefore, the cloud that gave birth to our

# FIGURE 1.11 Our Cosmic Origins

**Throughout this book we will see that human life is intimately connected with the development of the universe as a whole. This illustration presents an overview of our cosmic origins, showing some of the crucial steps that made our existence possible.**

(1) **Birth of the Universe:** The expansion of the universe began with the hot and dense Big Bang. The cubes show how one region of the universe has expanded with time. The universe continues to expand, but on smaller scales gravity has pulled matter together to make galaxies.

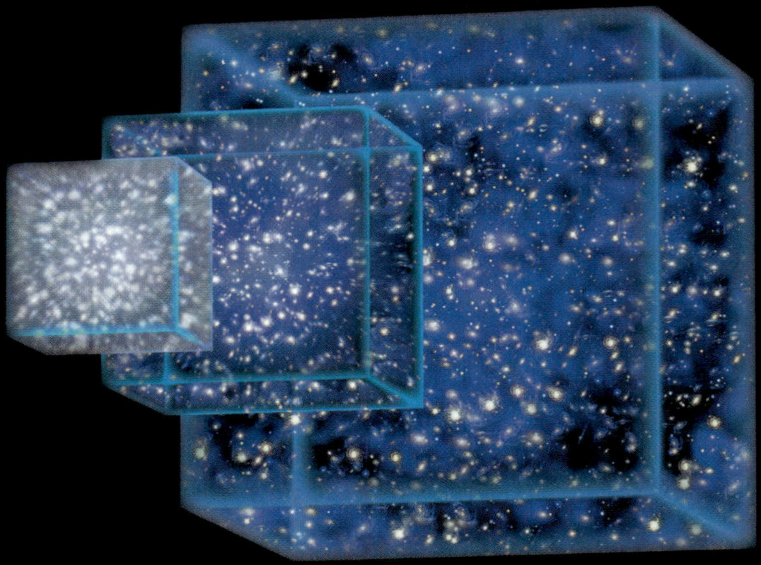

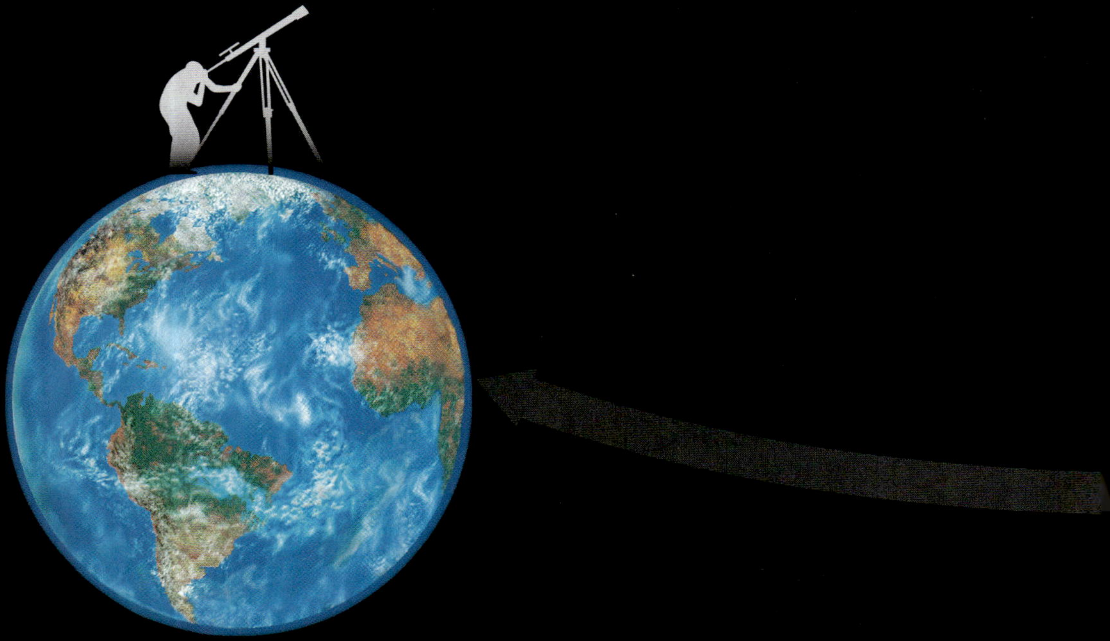

(4) **Earth and Life:** By the time our solar system was born, 4½ billion years ago, about 2% of the original hydrogen and helium had been converted into heavier elements. We are therefore "star stuff," because we and our planet are made from elements manufactured in stars that lived and died long ago.

② **Galaxies as Cosmic Recycling Plants:** The early universe contained only two chemical elements: hydrogen and helium. All other elements were made by stars and recycled from one stellar generation to the next within galaxies like our Milky Way.

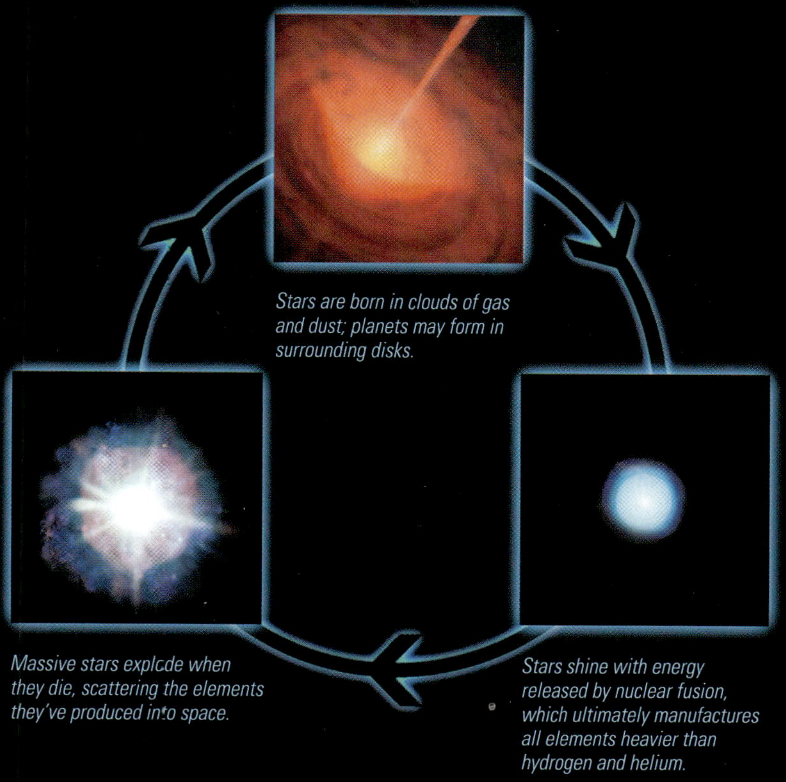


③ **Life Cycles of Stars:** Many generations of stars have lived and died in the Milky Way.

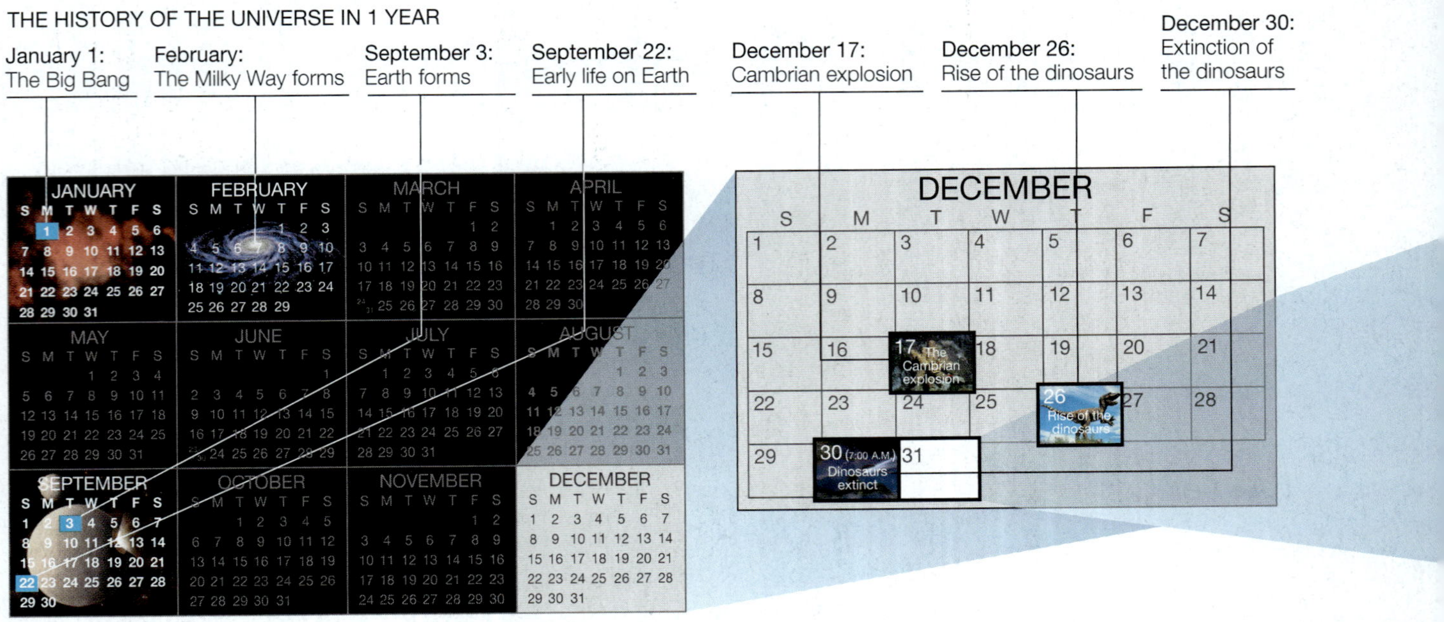


**FIGURE 1.12** The cosmic calendar compresses the 14-billion-year history of the universe into 1 year, so each month represents a little more than 1 billion years. Adapted from the cosmic calendar created by Carl Sagan. (For a more detailed version, see the "You Are Here in Time" foldout diagram in the front of the book.)

solar system was made of roughly 98% hydrogen and helium and 2% other elements. This 2% may sound small, but it was more than enough to make the small rocky planets of our solar system, including Earth. On Earth, some of these elements became the raw ingredients of life, which ultimately blossomed into the great diversity of life on Earth today.

In summary, most of the material from which we and our planet are made was created inside stars that lived and died before the birth of our Sun. As astronomer Carl Sagan (1934–1996) said, we are "star stuff."

### How do our lifetimes compare to the age of the universe?

We can put the 14-billion-year age of the universe into perspective by imagining this time compressed into a single year, so each month represents a little more than 1 billion years. On this *cosmic calendar*, the Big Bang occurred at the first instant of January 1 and the present is the stroke of midnight on December 31 (**FIGURE 1.12**).

On this time scale, the Milky Way Galaxy probably formed in February. Many generations of stars lived and died in the subsequent cosmic months, enriching the galaxy with the "star stuff" from which we and our planet are made.

Our solar system and our planet did not form until early September on this scale, or $4\frac{1}{2}$ billion years ago in real time. By late September, life on Earth was flourishing. However, for most of Earth's history, living organisms remained relatively primitive and microscopic. On the scale of the cosmic calendar, recognizable animals became prominent only in mid-December. Early dinosaurs appeared on the day after Christmas. Then, in a cosmic instant, the dinosaurs disappeared forever—probably because of the impact of an asteroid or a comet [**Section 12.5**]. In real time the death of the dinosaurs occurred some 65 million years ago, but on the cosmic calendar it was only yesterday. With the dinosaurs gone, small furry mammals inherited Earth. Some 60 million years later, or around 9 p.m. on December 31 of the cosmic calendar, early hominids (human ancestors) began to walk upright.

Perhaps the most astonishing fact about the cosmic calendar is that the entire history of human civilization falls into just the last half-minute. The ancient Egyptians built the pyramids only about 11 seconds ago on this scale. About 1 second ago, Kepler and Galileo provided the key evidence that led us to understand that Earth orbits the Sun rather than vice versa. The average college student was born about 0.05 second ago, around 11:59:59.95 p.m. on the cosmic calendar. On the scale of cosmic time, the human species is the youngest of infants, and a human lifetime is a mere blink of an eye.

**Think about it** Study the more detailed cosmic calendar found on the foldout in the front of this book. How does an understanding of the scale of time affect your view of human civilization? Explain.

## 1.3 Spaceship Earth

Wherever you are as you read this book, you probably have the feeling that you're "just sitting here." But, in fact, all of us are moving through space on what noted inventor and philosopher R. Buckminster Fuller (1895–1983) described as *spaceship Earth*.

### How is Earth moving through space?

Let's explore the major motions we are all undergoing with our spaceship Earth.

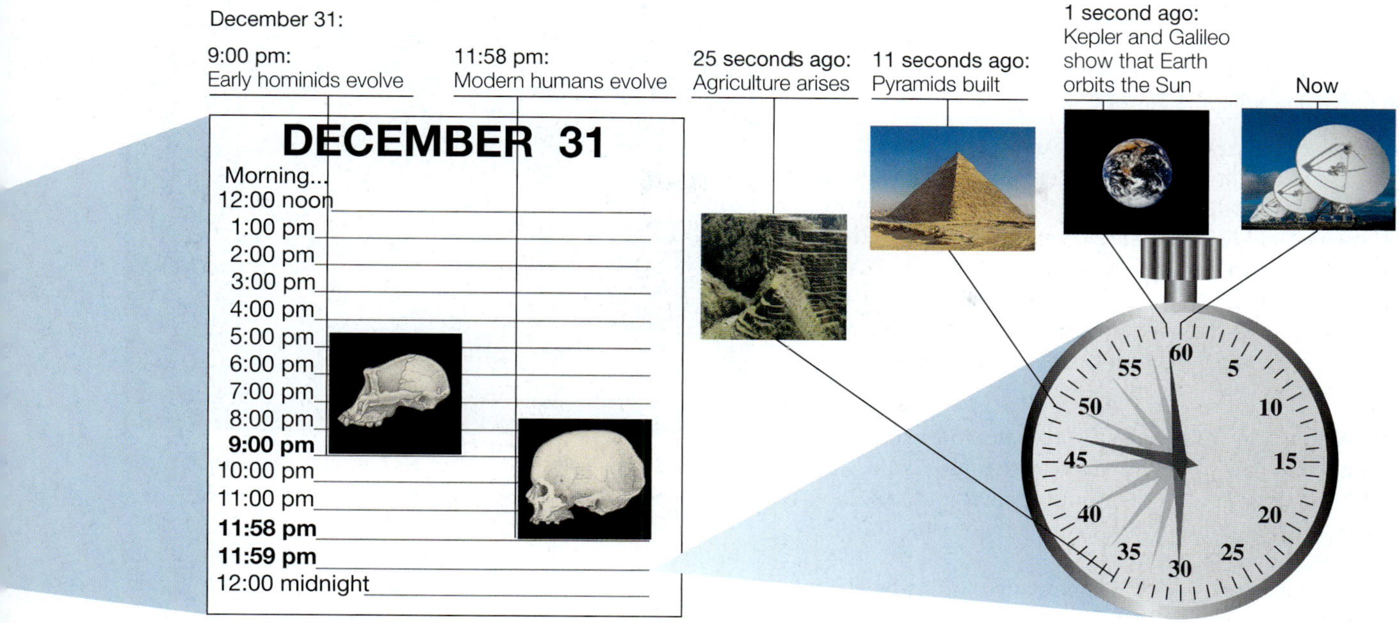


## Earth's Rotation and Orbit

**Rotation and Orbit** The most basic motions of Earth are its daily **rotation** (spin) and its yearly **orbit** (or *revolution*) around the Sun.

Earth rotates once each day around its axis (**FIGURE 1.13**), which is the imaginary line connecting the North Pole to the South Pole. Earth rotates from west to east—counterclockwise as viewed from above the North Pole—which is why the Sun and stars appear to rise in the east and set in the west each day. Although the physical effects of rotation are so subtle that our ancestors assumed the heavens revolved around us, the rotation speed is substantial: Unless you live quite far north or south, you are whirling around Earth's axis at a speed of more than 1000 kilometers per hour (600 miles per hour)—faster than most airplanes travel.

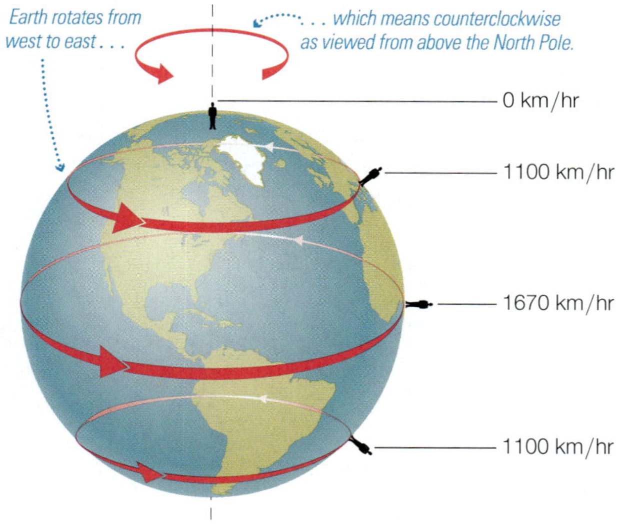


**FIGURE 1.13** As Earth rotates, your speed around Earth's axis depends on your location: The closer you are to the equator, the faster you travel with rotation.

At the same time as it is rotating, Earth also orbits the Sun, completing one orbit each year (**FIGURE 1.14**). Earth's orbital distance varies slightly over the course of each year, but as we discussed earlier, the average distance is one astronomical unit (AU), which is about 150 million kilometers. Again, even though we don't feel this motion, the speed is impressive: We are racing around the Sun at a speed in excess of 100,000 kilometers per hour (60,000 miles per hour), which is faster than any spacecraft yet launched.

As you study Figure 1.14, notice that Earth's orbital path defines a flat plane that we call the **ecliptic plane**. Earth's axis is tilted by $23\frac{1}{2}°$ from a line *perpendicular* to the ecliptic plane. This **axis tilt** happens to be oriented so that the axis points almost directly at a star called *Polaris*, or the *North Star*. Keep in mind that the idea of axis tilt makes sense only in relation to the ecliptic plane. That is, the idea of "tilt" by itself has no meaning in space, where there is no absolute up or down. In space, "up" and "down" mean only "away from the center of Earth" (or another planet) and "toward the center of Earth," respectively.

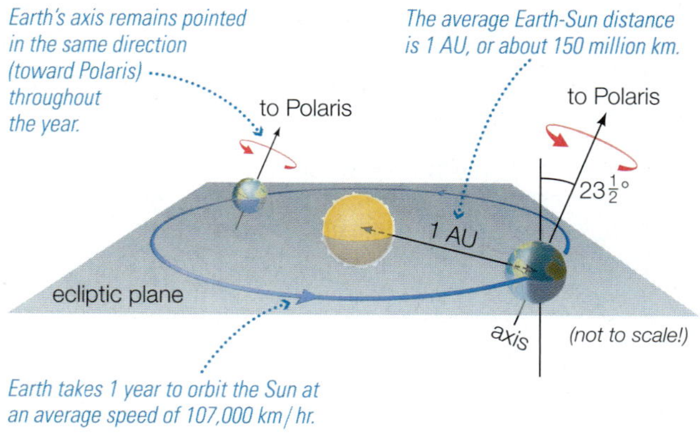


**FIGURE 1.14** This diagram shows key characteristics of Earth's daily rotation and yearly orbit, both of which are counterclockwise as viewed from above the North Pole.

**Think about it** If there is no up or down in space, why do you think that most globes and maps have the North Pole on top? Would it be equally correct to have the South Pole on top or to turn a globe sideways? Explain.

Notice also that Earth orbits the Sun in the same direction that it rotates on its axis: counterclockwise as viewed from above the North Pole. This is not a coincidence but a consequence of the way our planet was born. As we'll discuss in Chapter 8, strong evidence indicates that Earth and the other planets were born in a spinning disk of gas that surrounded our Sun as it formed, and Earth rotates and orbits in the same direction that the disk was spinning.

**Motion Within the Milky Way Galaxy** Rotation and orbit are only a small part of the travels of spaceship Earth. Our entire solar system is on a great journey within the Milky Way Galaxy. There are two major components to this motion, both shown in **FIGURE 1.15**.

First, our solar system is moving relative to nearby stars in our *local solar neighborhood*, the region of the Sun and nearby stars. The small box in Figure 1.15 shows that stars in our local solar neighborhood (like the stars of any other small region of the galaxy) move essentially at random relative to one another. The speeds are quite fast: On average, our Sun is moving relative to nearby stars at a speed of about 70,000 kilometers per hour (40,000 miles per hour), almost three times as fast as the International Space Station orbits Earth. Given these high speeds, you might wonder why we don't see stars racing around our sky. The answer lies in their vast distances from us. You've probably noticed that a distant airplane appears to move through your sky more slowly than

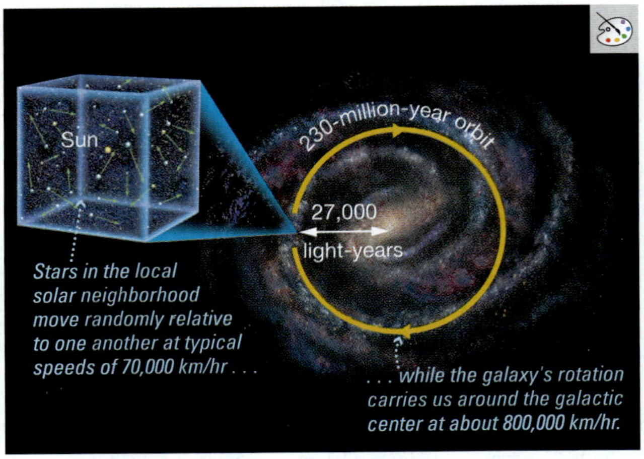


**FIGURE 1.15** This painting illustrates the motion of the Sun both within the local solar neighborhood and around the center of the galaxy.

## MATHEMATICAL INSIGHT 1.4

### Speeds of Rotation and Orbit

Problem Solving Part 3

Building upon prior Mathematical Insights, we will now see how simple formulas—such as the formula for the circumference of a circle—expand the range of astronomical problems we can solve.

**EXAMPLE 1:** How fast is a person on Earth's equator moving with Earth's rotation?

**SOLUTION:**

**Step 1 Understand:** The question *how fast* tells us we are looking for a *speed*. If you remember that highway speeds are posted in miles (or kilometers) per hour, you'll realize that speed is a distance (such as miles) divided by a time (such as hours). In this case, the distance is Earth's equatorial circumference, because that is how far a person at the equator travels with each rotation (see Figure 1.13); we'll therefore use the formula for the circumference of a circle, $C = 2 \times \pi \times \text{radius}$. The time is 24 hours, because that is how long each rotation takes.

**Step 2 Solve:** From Appendix E.1, Earth's equatorial radius is 6378 km, so its circumference is $2 \times \pi \times 6378\text{ km} = 40{,}074\text{ km}$. We divide this distance by the time of 24 hours:

$$\text{rotation speed at equator} = \frac{\text{equatorial circumference}}{\text{length of day}} = \frac{40{,}074\text{ km}}{24\text{ hr}} = 1670\,\frac{\text{km}}{\text{hr}}$$

**Step 3 Explain:** A person at the equator is moving with Earth's rotation at a speed of about 1670 kilometers per hour, which is a little over 1000 miles per hour, or about twice the flying speed of a commercial jet.

**EXAMPLE 2:** How fast is Earth orbiting the Sun?

**SOLUTION:**

**Step 1 Understand:** We are again asked *how fast* and therefore need to divide a distance by a time. In this case, the distance is the circumference of Earth's orbit, and the time is the 1 year that Earth takes to complete each orbit.

**Step 2 Solve:** Earth's average distance from the Sun is 1 AU, or about 150 million ($1.5 \times 10^8$) km, so the orbit circumference is about $2 \times \pi \times 1.5 \times 10^8\text{ km} \approx 9.40 \times 10^8\text{ km}$. The orbital speed is this distance divided by the time of 1 year, which we convert to hours so that we end up with units of km/hr:

$$\text{orbital speed} = \frac{\text{orbital circumference}}{1\text{ yr}} = \frac{9.40 \times 10^8\text{ km}}{1\,\cancel{\text{yr}} \times \frac{365\,\cancel{\text{days}}}{\cancel{\text{yr}}} \times \frac{24\text{ hr}}{\cancel{\text{day}}}} \approx 107{,}000\,\frac{\text{km}}{\text{hr}}$$

**Step 3 Explain:** Earth orbits the Sun at an average speed of about 107,000 km/hr (66,000 mi/hr). Most "speeding bullets" travel between about 500 and 1000 km/hr, so Earth's orbital speed is more than 100 times that of a speeding bullet.

**FIGURE 1.16** This painting shows an edge-on view of the Milky Way Galaxy. Study of galactic rotation shows that although most visible stars lie in the central bulge or thin disk, most of the mass lies in the halo that surrounds and encompasses the disk. Because this mass emits no light that we have detected, we call it *dark matter*.

one flying close overhead. Stars are so far away that even at speeds of 70,000 kilometers per hour, their motions would be noticeable to the naked eye only if we watched them for thousands of years. That is why the patterns in the constellations seem to remain fixed. Nevertheless, in 10,000 years the constellations will be noticeably different from those we see today. In 500,000 years they will be unrecognizable. If you could watch a time-lapse movie made over millions of years, you *would* see stars racing across our sky.

**Think about it** Despite the chaos of motion in the local solar neighborhood over millions and billions of years, collisions between star systems are extremely rare. Explain why. (*Hint*: Consider the sizes of star systems, such as the solar system, relative to the distances between them.)

The second motion shown in Figure 1.15 is much more organized. If you look closely at leaves floating in a stream, their motions relative to one another might appear random, just like the motions of stars in the local solar neighborhood. As you widen your view, you see that all the leaves are being carried in the same general direction by the downstream current. In the same way, as we widen our view beyond the local solar neighborhood, the seemingly random motions of its stars give way to a simpler and even faster motion: rotation of the Milky Way Galaxy. Our solar system, located about 27,000 light-years from the galactic center, completes one orbit of the galaxy in about 230 million years. Even if you could watch from outside our galaxy, this motion would be unnoticeable to your naked eye. However, if you calculate the speed of our solar system as we orbit the center of the galaxy, you will find that it is close to 800,000 kilometers (500,000 miles) per hour.

Careful study of the galaxy's rotation reveals one of the greatest mysteries in science. Stars at different distances from the galactic center orbit at different speeds, and we can learn how mass is distributed in the galaxy by measuring these speeds. Such studies indicate that the stars in the disk of the galaxy represent only the "tip of the iceberg" compared to the mass of the entire galaxy (**FIGURE 1.16**). Most of the mass of the galaxy seems to be located outside the visible disk (occupying the galactic *halo* that surrounds the disk), but the matter that makes up this mass is completely invisible to our telescopes. We therefore know very little about the nature of this matter, which we refer to as **dark matter** (because of the lack of light from it). Studies of other galaxies indicate that they also are made mostly of dark matter, which means this mysterious matter significantly outweighs the ordinary matter that makes up planets and stars; this also means that dark matter must be the dominant source of gravity that has led to the formation of galaxies, clusters, and superclusters. We know even less about the mysterious **dark energy** that astronomers first recognized when they discovered that the expansion of the universe is actually getting faster with time, and that scientists have since found to make up the majority of the total energy content of the universe. We'll discuss the mysteries of dark matter and dark energy in Chapter 23.

## How do galaxies move within the universe?

The billions of galaxies in the universe also move relative to one another. Within the Local Group (see Figure 1.1), some of the galaxies move toward us, some move away from us, and numerous small galaxies (including the Large and Small Magellanic Clouds) apparently orbit our Milky Way Galaxy. Again, the speeds are enormous by earthly standards. For example, the Milky Way and Andromeda galaxies are moving toward each other at about 300,000 kilometers (180,000 miles) per hour. Despite this high speed, we needn't worry about a collision anytime soon. Even if the Milky Way and Andromeda Galaxies are approaching each other head-on, it will be billions of years before any collision begins.

When we look outside the Local Group, however, we find two astonishing facts first discovered by Edwin Hubble (1889–1953), for whom the Hubble Space Telescope was named:

1. Virtually every galaxy outside the Local Group is moving *away* from us.
2. The more distant the galaxy, the faster it appears to be racing away.

These facts might make it sound as if we suffered from a cosmic case of chicken pox, but there is a much more natural explanation: *The entire universe is expanding.* We'll save the details for later in the book, but you can understand the basic idea by thinking about a raisin cake baking in an oven.

**The Raisin Cake Analogy** Imagine that you make a raisin cake in which the distance between adjacent raisins is 1 centimeter. You place the cake into the oven, where it expands as it bakes. After 1 hour, you remove the cake, which has expanded so that the distance between adjacent raisins has increased to 3 centimeters (**FIGURE 1.17**). The expansion of the cake seems fairly obvious. But what would you see if you lived *in* the cake, as we live in the universe?

Pick any raisin (it doesn't matter which one) and call it the Local Raisin. Figure 1.17 shows one possible choice, with three nearby raisins also labeled. The accompanying table summarizes what you would see if you lived within the Local Raisin. Notice, for example, that Raisin 1 starts out at a distance of 1 centimeter before baking and ends up at a distance of 3 centimeters after baking, which means it moves a distance of 2 centimeters farther away from the Local Raisin during the hour of baking. Hence, its speed as seen from the Local Raisin is 2 centimeters per hour. Raisin 2 moves from a distance of 2 centimeters before baking to a distance of 6 centimeters after baking, which means it moves a distance of 4 centimeters farther away from the Local Raisin during the hour. Hence, its speed is 4 centimeters per hour, or twice the speed of Raisin 1. Generalizing, the fact that the cake is expanding means that all the raisins are moving away from the Local Raisin, with more distant raisins moving away faster.

**Think about it** Suppose a raisin started out 10 centimeters from the Local Raisin. How far away would it be after 1 hour, and how fast would it be moving away from the Local Raisin?

Hubble's discovery that galaxies are moving in much the same way as the raisins in the cake, with most moving away from us and more distant ones moving away faster, implies that the universe is expanding much like the raisin cake. If you now imagine the Local Raisin as representing our Local Group of galaxies and the other raisins as representing more distant galaxies or clusters of galaxies, you have a basic picture of the expansion of the universe. Like the expanding dough between the raisins in the cake, *space* itself is growing between galaxies. More distant galaxies move away from us faster because they are carried along with this expansion like the raisins in the expanding cake. You can also now see how observations of expansion allow us to measure the age of the universe: The faster the rate of expansion, the more quickly the galaxies reached their current positions, and therefore the younger the universe must be. It is by precisely measuring the expansion rate that astronomers have learned that the universe is approximately 14 billion years old.

**The Real Universe** There's at least one important distinction between the raisin cake and the universe: A cake has a center and edges, but we do not think the same is true of the entire universe. Anyone living in any galaxy in an expanding universe sees just what we see—other galaxies moving away, with more distant ones moving away faster. Because the view from each point in the universe is about the same, no place can claim to be more "central" than any other place.

It's also important to realize that, unlike the case with a raisin cake, we can't actually *see* galaxies moving apart with time—the distances are too vast for any motion to be noticeable on the time scale of a human life. Instead, we measure the speeds of galaxies by spreading their light into spectra and observing what we call *Doppler shifts* [**Section 5.4**]. This illustrates how modern astronomy depends both on careful observations and on using current understanding of the laws of nature to explain what we see.

**Motion Summary FIGURE 1.18** summarizes the motions we have discussed. As we have seen, we are never truly sitting still. We spin around Earth's axis at more than 1000 kilometers per hour, while our planet orbits the Sun at more than 100,000 kilometers per hour. Our solar system moves among the stars of the local solar neighborhood at a typical speed of 70,000 kilometers per hour, while also orbiting the center of the Milky Way Galaxy at a speed of about 800,000 kilometers per hour. Our galaxy moves among the other galaxies of the

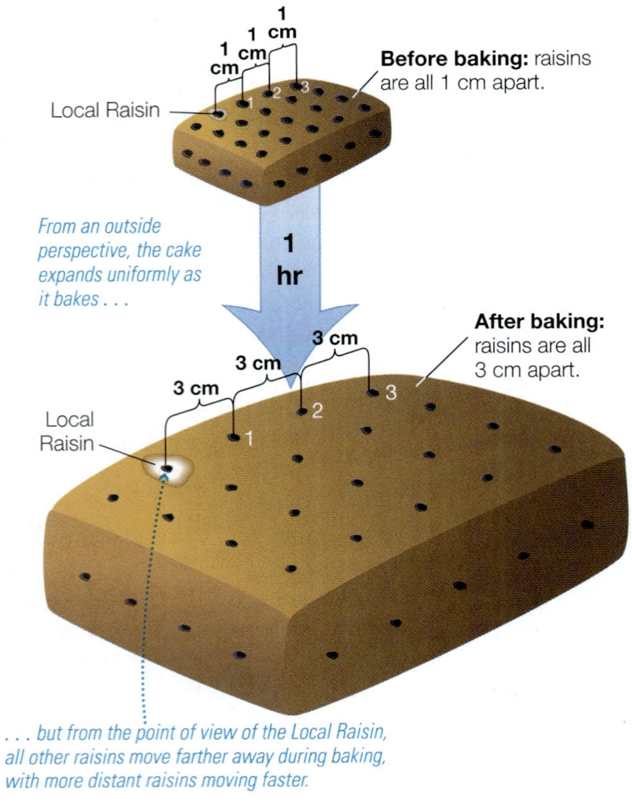


**Distances and Speeds as Seen from the Local Raisin**

| Raisin Number | Distance Before Baking | Distance After Baking (1 hour later) | Speed |
|---|---|---|---|
| 1 | 1 cm | 3 cm | 2 cm/hr |
| 2 | 2 cm | 6 cm | 4 cm/hr |
| 3 | 3 cm | 9 cm | 6 cm/hr |
| ⋮ | ⋮ | ⋮ | ⋮ |

**FIGURE 1.17** An expanding raisin cake offers an analogy to the expanding universe. Someone living in one of the raisins inside the cake could figure out that the cake is expanding by noticing that all other raisins are moving away, with more distant raisins moving away faster. In the same way, we know that we live in an expanding universe because all galaxies outside our Local Group are moving away from us, with more distant ones moving faster.

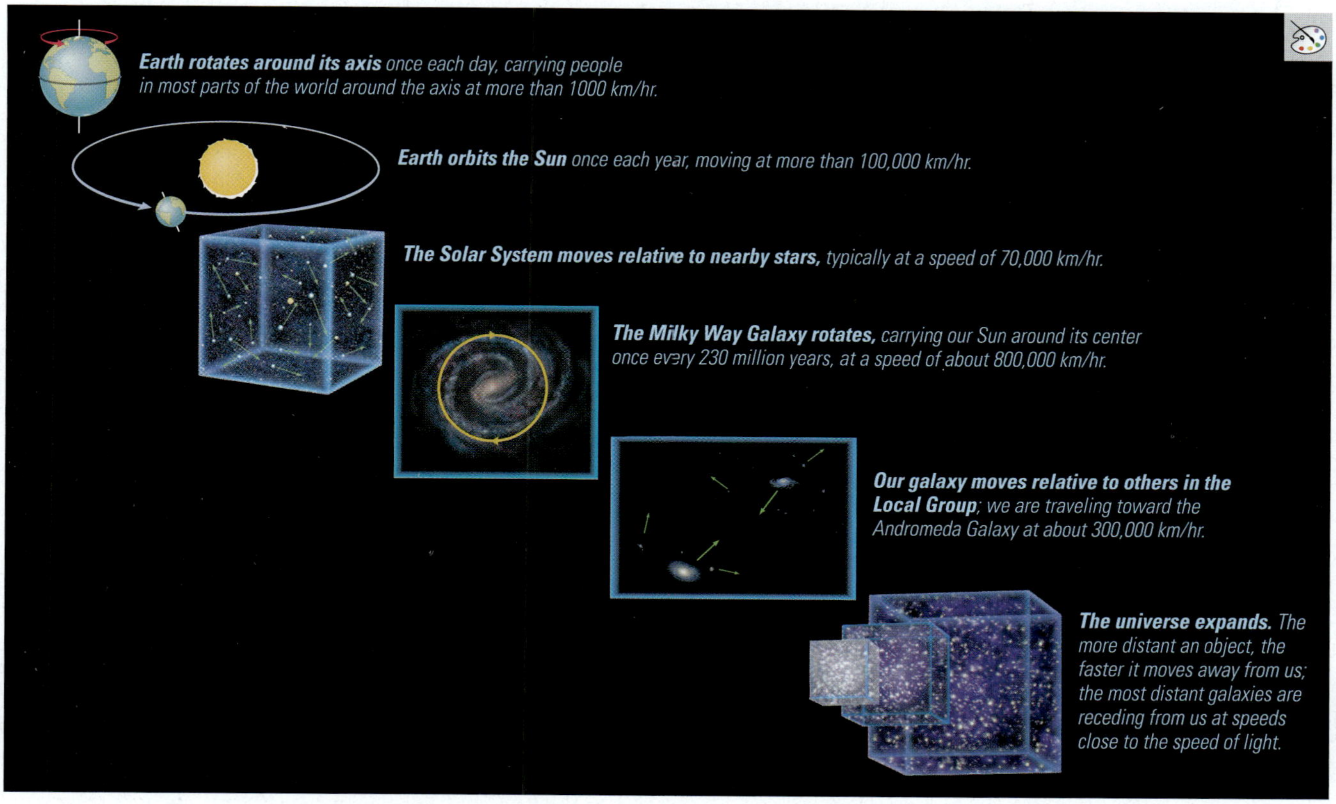


**FIGURE 1.18** This figure summarizes the basic motions of Earth in the universe, along with their associated speeds.

Local Group, while all other galaxies move away from us at speeds that grow greater with distance in our expanding universe. Spaceship Earth is carrying us on a remarkable journey.

## 1.4 The Human Adventure of Astronomy

In relatively few pages, we've laid out a fairly complete overview of modern scientific ideas about the universe. But our goal in this book is not simply for you to be able to recite these ideas. Rather, it is to help you understand the evidence that supports them and the extraordinary story of how they developed.

### How has the study of astronomy affected human history?

Astronomy is a human adventure in the sense that it affects everyone—even those who have never looked at the sky—because the history of astronomy has been so deeply intertwined with the development of civilization. Revolutions in astronomy have gone hand in hand with the revolutions in science and technology that have shaped modern life.

Witness the repercussions of the *Copernican revolution*, which showed us that Earth is not the center of the universe but rather just one planet orbiting the Sun. This revolution, which we will discuss further in Chapter 3, began when Copernicus published his idea of a Sun-centered solar system in 1543. Three later figures—Tycho Brahe, Johannes Kepler, and Galileo—provided the key evidence that eventually led to wide acceptance of the Copernican idea. The revolution culminated with Isaac Newton's uncovering of the laws of motion and gravity. Newton's work, in turn, became the foundation of physics that helped fuel the industrial revolution.

More recently, the development of space travel and the computer revolution have helped fuel tremendous progress in astronomy. We've sent probes to all the planets in our solar system, and many of our most powerful observatories reside in space. On the ground, computer design and control have led to tremendous growth in the size and power of telescopes.

Many of these efforts, and the achievements they spawned, led to profound social change. The most famous example is the fate of Galileo, whom the Vatican put under house arrest in 1633 for his claims that Earth orbits the Sun. Although the Church soon recognized that Galileo was right, he was formally vindicated only in 1992 with a statement by Pope John Paul II. In the meantime, his case spurred great debate in religious circles and profoundly influenced both theological and scientific thinking.

As you progress through this book, keep the context of the human adventure in mind. You will then be learning not just about astronomy, but also about one of the great forces that has shaped our modern world.

These forces will continue to play a role in our future. What will it mean to us when we learn the nature of dark matter and dark energy? How will our view of Earth change when we learn whether life is common or rare in the universe? Only time may answer these questions, but the chapters ahead will give you the foundation you need to understand how we changed from a primitive people looking at patterns in the night sky to a civilization capable of asking deep questions about our existence.

# The BIG Picture PUTTING CHAPTER 1 INTO CONTEXT

*In this first chapter, we developed a broad overview of our place in the universe. As we consider the universe in more depth in the rest of the book, remember the following "big picture" ideas:*

- Earth is not the center of the universe but instead is a planet orbiting a rather ordinary star in the Milky Way Galaxy. The Milky Way Galaxy, in turn, is one of billions of galaxies in our observable universe.
- Cosmic distances are literally astronomical, but we can put them in perspective with the aid of scale models and other scaling techniques. When you think about these enormous scales, don't forget that every star is a sun and every planet is a unique world.
- We are "star stuff." The atoms from which we are made began as hydrogen and helium in the Big Bang and were later fused into heavier elements by massive stars. Stellar deaths released these atoms into space, where our galaxy recycled them into new stars and planets. Our solar system formed from such recycled matter some $4\frac{1}{2}$ billion years ago.
- We are latecomers on the scale of cosmic time. The universe was already more than half its current age when our solar system formed, and it took billions of years more before humans arrived on the scene.
- All of us are being carried through the cosmos on spaceship Earth. Although we cannot feel this motion in our everyday lives, the associated speeds are surprisingly high. Learning about the motions of spaceship Earth gives us a new perspective on the cosmos and helps us understand its nature and history.

**MY COSMIC PERSPECTIVE** The science of astronomy affects all of us on many levels. In particular, it helps us understand how we as humans fit into the universe as a whole, and the history of astronomy has been deeply intertwined with the development of civilization.

# Summary of Key Concepts

## 1.1 The Scale of the Universe

- **What is our place in the universe?** Earth is a planet orbiting the Sun. Our Sun is one of more than 100 billion stars in the **Milky Way Galaxy**. Our galaxy is one of more than 70 galaxies in the **Local Group**. The Local Group is one small part of the **Local Supercluster**, which is one small part of the **universe**.

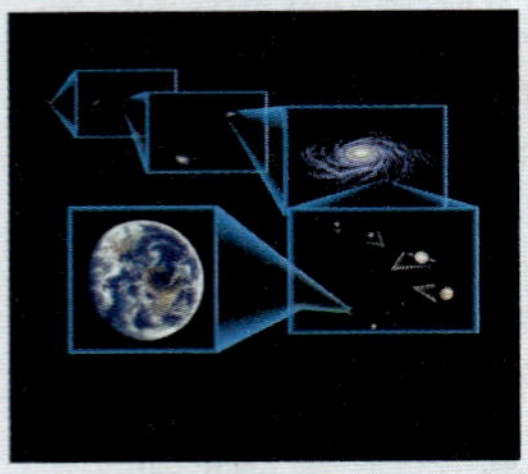

- **How big is the universe?** If we imagine our Sun as a large grapefruit, Earth is a ball point that orbits 15 meters away; the nearest stars are thousands of kilometers away on the same scale. Our galaxy contains more than 100 billion stars—so many that it would take thousands of years just to count them out loud. The **observable universe** contains more than 100 billion galaxies, and the total number of stars is comparable to the number of grains of dry sand on all the beaches on Earth.

## 1.2 The History of the Universe

- **How did we come to be?** The universe began in the **Big Bang** and has been expanding ever since, except in localized regions where gravity has caused matter to collapse into galaxies and stars. The Big Bang essentially produced only two chemical elements: hydrogen and helium. The rest have been produced by stars and recycled within galaxies from one generation of stars to the next, which is why we are "star stuff."

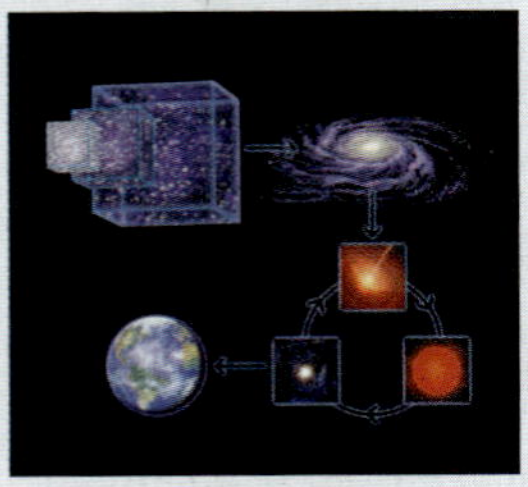

- **How do our lifetimes compare to the age of the universe?** On a cosmic calendar that compresses the history of the universe into 1 year, human civilization is just a few seconds old, and a human lifetime lasts only a fraction of a second.

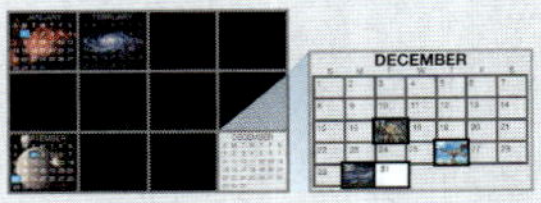


## 1.3 Spaceship Earth

- **How is Earth moving through space?** Earth **rotates** on its axis once each day and **orbits** the Sun once each year. At the same time, we move with our Sun in random directions relative to other stars in our local solar neighborhood, while the galaxy's rotation carries us around the center of the galaxy every 230 million years.

- **How do galaxies move within the universe?** Galaxies move essentially at random within the Local Group, but all galaxies beyond the Local Group are moving away from us. More distant galaxies are moving faster, which tells us that we live in an expanding universe.

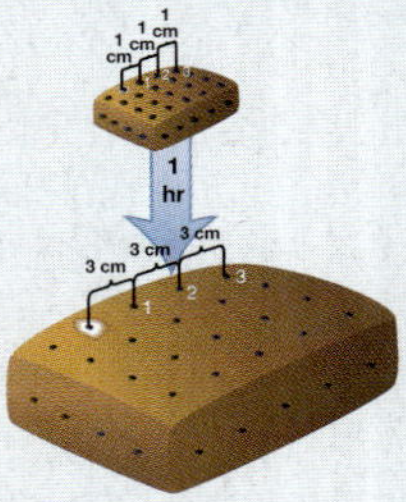


## 1.4 The Human Adventure of Astronomy

- **How has the study of astronomy affected human history?** Throughout history, astronomy has developed hand in hand with social and technological development. Astronomy thereby touches all of us and is a human adventure that all can enjoy.

# Visual Skills Check

*Use the following questions to check your understanding of some of the many types of visual information used in astronomy. For additional practice, try the Chapter 1 Visual Quiz in the Study Area at www.MasteringAstronomy.com.*

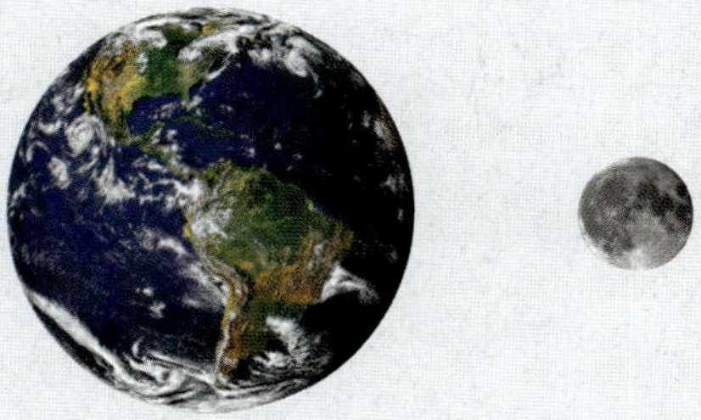

**Useful Data:**

Earth-Sun distance = 150,000,000 km
Diameter of Sun = 1,400,000 km
Earth-Moon distance = 384,000 km
Diameter of Earth = 12,800 km

*The figure above shows the sizes of Earth and the Moon to scale; the scale used is 1 cm = 4000 km. Using what you've learned about astronomical scale in this chapter, answer the following questions. Hint: If you are unsure of the answers, you can calculate them using the data given above.*

1. If you wanted to show the *distance* between Earth and the Moon on the same scale, about how far apart would you need to place the two photos?
   a. 10 centimeters (about the width of your hand)
   b. 1 meter (about the length of your arm)
   c. 100 meters (about the length of a football field)
   d. 1 kilometer (a little more than a half mile)
2. Suppose you wanted to show the Sun on the same scale. About how big would it need to be?
   a. 3.5 centimeters in diameter (the size of a golf ball)
   b. 35 centimeters in diameter (a little bigger than a basketball)
   c. 3.5 meters in diameter (about 11½ feet across)
   d. 3.5 kilometers in diameter (the size of a small town)
3. About how far away from Earth would the Sun be located on this scale?
   a. 3.75 meters (about 12 feet)
   b. 37.5 meters (about the height of a 12-story building)
   c. 375 meters (about the length of four football fields)
   d. 37.5 kilometers (the size of a large city)
4. Could you use the same scale to represent the distances to nearby stars? Why or why not?

# Exercises and Problems

*For instructor-assigned homework and other learning materials, go to www.MasteringAstronomy.com.*

## Chapter Review Questions

### Short-Answer Questions Based on the Reading

1. Briefly describe the major levels of structure (such as planet, star, galaxy) in the universe.
2. Define *astronomical unit* and *light-year*.
3. Explain the statement "*The farther away we look in distance, the further back we look in time.*"
4. What do we mean by the *observable universe*? Is it the same thing as the entire universe?
5. Using techniques described in the chapter, put the following into perspective: the size of our solar system; the distance to nearby stars; the size and number of stars in the Milky Way Galaxy; the number of stars in the observable universe.
6. What do we mean when we say that the universe is *expanding*, and how does expansion lead to the idea of the *Big Bang* and our current estimate of the age of the universe?
7. In what sense are we "star stuff"?
8. Use the cosmic calendar to describe how the human race fits into the scale of time.
9. Briefly explain Earth's daily rotation and annual orbit, defining the terms *ecliptic plane* and *axis tilt*.
10. Briefly describe our solar system's location and motion within the Milky Way Galaxy.
11. Why do scientists suspect that most of our galaxy's mass consists of *dark matter*? Briefly describe the mystery of dark matter and *dark energy*.
12. What key observations lead us to conclude that the universe is expanding? Use the raisin cake model to explain how these observations imply expansion.

### Does It Make Sense?

*Decide whether or not each of the following statements makes sense (or is clearly true or false). Explain clearly; not all of these have definitive answers, so your explanation is more important than your chosen answer.*

*Example:* I walked east from our base camp at the North Pole.

*Solution:* The statement does not make sense because east has no meaning at the North Pole—all directions are south from the North Pole.

13. Our solar system is bigger than some galaxies.
14. The universe is billions of light-years in age.
15. It will take me light-years to complete this homework assignment!
16. Someday we may build spaceships capable of traveling a light-year in only a decade.
17. Astronomers recently discovered a moon that does not orbit a planet.
18. NASA will soon launch a spaceship that will photograph our Milky Way Galaxy from beyond its halo.
19. The observable universe is larger today than it was a few billion years ago.

20. Photographs of distant galaxies show them as they were when they were much younger than they are today.
21. At a nearby park, I built a scale model of our solar system in which I used a basketball to represent Earth.
22. Because nearly all galaxies are moving away from us, we must be located near the center of the universe.

## Quick Quiz

*Choose the best answer to each of the following. For additional practice, try the Chapter 1 Reading and Concept Quizzes in the Study Area at www.MasteringAstronomy.com.*

23. Which of the following correctly lists our "cosmic address" from small to large? (a) Earth, solar system, Milky Way Galaxy, Local Group, Local Supercluster, universe (b) Earth, solar system, Local Group, Local Supercluster, Milky Way Galaxy, universe (c) Earth, Milky Way Galaxy, solar system, Local Group, Local Supercluster, universe
24. An astronomical unit is (a) any planet's average distance from the Sun. (b) Earth's average distance from the Sun. (c) any large astronomical distance.
25. The star Betelgeuse is about 600 light-years away. If it explodes tonight, (a) we'll know because it will be brighter than the full Moon in the sky. (b) we'll know because debris from the explosion will rain down on us from space. (c) we won't know about it until about 600 years from now.
26. If we represent the solar system on a scale that allows us to walk from the Sun to Pluto in a few minutes, then (a) the planets are the size of basketballs and the nearest stars are a few miles away. (b) the planets are marble-size or smaller and the nearest stars are thousands of miles away. (c) the planets are microscopic and the stars are light-years away.
27. The total number of stars in the observable universe is roughly equivalent to (a) the number of grains of sand on all the beaches on Earth. (b) the number of grains of sand on Miami Beach. (c) infinity.
28. When we say the universe is *expanding*, we mean that (a) everything in the universe is growing in size. (b) the average distance between galaxies is growing with time. (c) the universe is getting older.
29. If stars existed but galaxies did not, (a) we would probably still exist anyway. (b) we would not exist because life on Earth depends on the light of galaxies. (c) we would not exist because we are made of material that was recycled in galaxies.
30. Could we see a galaxy that is 50 billion light-years away? (a) Yes, if we had a big enough telescope. (b) No, because it would be beyond the bounds of our observable universe. (c) No, because a galaxy could not possibly be that far away.
31. The age of our solar system is about (a) one-third of the age of the universe. (b) three-fourths of the age of the universe. (c) two billion years less than the age of the universe.
32. The fact that nearly all galaxies are moving away from us, with more distant ones moving faster, helped us to conclude that (a) the universe is expanding. (b) galaxies repel each other like magnets. (c) our galaxy lies near the center of the universe.

## Inclusive Astronomy

*Use these questions to reflect on participation in science.*

33. *Group Discussion: What does a scientist look like?* The purpose of this exercise is to help you identify preconceptions that you or others may have about science and scientists.

**a.** Working independently, make a simple sketch of a professional scientist and write down five words that describe the scientist in your sketch. Then join with a group of two to three other students to share your sketches and word lists.
**b.** Make a list of all the words the group wrote down, then rank them in order of how often they were used.
**c.** Discuss any similarities, differences, or patterns you notice among the scientists described by the group members.
**d.** Discuss whether you and the other members of your group feel as if you have much in common with professional scientists.
**e.** Discuss how your beliefs about what you have (or do not have) in common with scientists might affect your approach to scientific thinking.

## The Process of Science

*These questions may be answered individually in short-essay form or discussed in groups, except where identified as group-only.*

34. *Earth as a Planet.* For most of human history, scholars assumed Earth was the center of the universe. Today, we know that our Sun is just one star in a vast universe. How did science make it possible for us to learn these facts about Earth?
35. *Thinking About Scale.* One key to success in science is finding simple ways to evaluate new ideas, and making a simple scale model is often helpful. Suppose someone tells you that the reason it is warmer during the day than at night is that the day side of Earth is closer to the Sun than the night side. Evaluate this idea by thinking about the size of Earth and its distance from the Sun in a scale model of the solar system.
36. *Looking for Evidence.* In this first chapter, we have discussed the scientific story of the universe but have not yet discussed most of the evidence that backs it up. Choose one idea presented in this chapter—such as the idea that there are billions of galaxies in the universe, or that the universe was born in the Big Bang, or that the galaxy contains more dark matter than ordinary matter—and briefly discuss the type of evidence you would want to see before accepting the idea. (*Hint*: It's okay to look ahead in the book to see the evidence presented in later chapters.)
37. *A Human Adventure.* Astronomical discoveries clearly are important to science, but are they also important to our personal lives? Defend your opinion.
38. *Infant Species.* In the last few tenths of a second before midnight on December 31 of the cosmic calendar, we have developed an incredible civilization and learned a great deal about the universe, but we also have developed technology with which we could destroy ourselves. The midnight bell is striking, and the choice for the future is ours. How far into the next cosmic year do you think our civilization will survive? Defend your opinion.
39. *Group Activity: Counting the Milky Way's Stars.* Work as a group to answer each part. *Note*: This activity works particularly well in groups of four students, with each student taking on one of the following roles: *Scribe*—takes notes on the group's activities; *Proposer*—suggests tentative explanations to the group; *Skeptic*—points out weaknesses in proposed explanations; *Moderator*—leads group discussion and makes sure everyone contributes.

**a.** Work together to estimate the number of stars in the Milky Way from just these two facts: (1) the number of stars within 12 light-years of the Sun, which you can count in Appendix F; (2)

the total volume of the Milky Way's disk (100,000 light-years in diameter and 1000 light-years thick), which is about 1 billion times the volume of the region of your star count.

**b.** Discuss how your value from part a compares to the value given in this chapter. Make a list of possible reasons why your technique may have underestimated or overestimated the actual number.

## Investigate Further

### Short-Answer/Essay Questions

40. *Alien Technology.* Some people believe that Earth is regularly visited by aliens who travel here from other star systems. For this to be true, how much more advanced than our own technology would the alien space travel technology have to be? Write one to two paragraphs to give a sense of the technological difference. (*Hint*: The ideas of scale in this chapter can help you contrast the distance the aliens would have to travel with the distances we currently are capable of traveling.)
41. *Raisin Cake Universe.* Suppose that all the raisins in a cake are 1 centimeter apart before baking and 4 centimeters apart after baking.
**a.** Draw diagrams to represent the cake before and after baking.
**b.** Identify one raisin as the Local Raisin on your diagrams. Construct a table showing the distances and speeds of other raisins as seen from the Local Raisin.
**c.** Briefly explain how your expanding cake is similar to the expansion of the universe.
42. *Scaling the Local Group of Galaxies.* Both the Milky Way Galaxy and the Andromeda Galaxy (M31) have a diameter of about 100,000 light-years. The distance between the two galaxies is about 2.5 million light-years.
**a.** Using a scale on which 1 centimeter represents 100,000 light-years, draw a sketch showing both galaxies and the distance between them to scale.
**b.** How does the separation between galaxies compare to the separation between stars? Based on your answer, discuss the likelihood of galactic collisions in comparison to the likelihood of stellar collisions.
43. *The Hubble eXtreme Deep Field.* The photo that opens this chapter is called the Hubble eXtreme Deep Field. Find this photo on the Hubble Space Telescope website. Learn how it was taken, what it shows, and what we've learned from it. Write a short summary of your findings.
44. *The Cosmic Perspective.* Write a short essay describing how the ideas presented in this chapter affect your perspectives on your own life and on human civilization.

### Quantitative Problems

*Be sure to show all calculations clearly and state your final answers in complete sentences.*

45. *Distances by Light.* Just as a light-year is the distance that light can travel in 1 year, we define a light-second as the distance that light can travel in 1 second, a light-minute as the distance that light can travel in 1 minute, and so on. Calculate the distance in both kilometers and miles represented by each of the following:
**a.** 1 light-second. **b.** 1 light-minute.
**c.** 1 light-hour. **d.** 1 light-day.
46. *Spacecraft Communication.* We use radio waves, which travel at the speed of light, to communicate with robotic spacecraft. How long does it take a message to travel from Earth to a spacecraft at
**a.** Mars at its closest to Earth (about 56 million km)?
**b.** Mars at its farthest from Earth (about 400 million km)?
**c.** Pluto at its average distance from Earth (about 5.9 billion km)?
47. *Saturn vs. the Milky Way.* Photos of Saturn and photos of galaxies can look so similar that children often think the photos show similar objects. In reality, a galaxy is far larger than any planet. About how many times larger is the diameter of the Milky Way Galaxy than the diameter of Saturn's rings? (Data: Saturn's rings are about 270,000 km in diameter; the Milky Way is 100,000 light-years in diameter.)
48. *Galaxy Scale.* Consider the 1-to-$10^{19}$ scale, on which the disk of the Milky Way Galaxy fits on a football field. On this scale, how far is it from the Sun to Alpha Centauri (real distance: 4.4 light-years)? How big is the Sun itself on this scale? Compare the Sun's size on this scale to the actual size of a typical atom (about $10^{-10}$ m in diameter).
49. *Universal Scale.* Suppose we wanted to make a scale model of the Local Group of galaxies in which the Milky Way Galaxy was the size of a marble (about 1 cm in diameter).
**a.** How far from the Milky Way Galaxy would the Andromeda Galaxy be on this scale?
**b.** How far would the Sun be from Alpha Centauri on this scale?
**c.** How far would it be from the Milky Way Galaxy to the most distant galaxies in the observable universe on this scale?
50. *Driving Trips.* Imagine that you could drive your car at a constant speed of 100 km/hr (62 mi/hr), even across oceans and in space. (In reality, the law of gravity would make driving through space at a constant speed all but impossible.) How long would it take to drive
**a.** around Earth's equator? **b.** from the Sun to Earth?
**c.** from the Sun to Pluto? **d.** to Alpha Centauri?
51. *Faster Trip.* Suppose you wanted to reach Alpha Centauri in 100 years.
**a.** How fast would you have to go, in km/hr?
**b.** How many times faster is the speed you found in part a than the speed of our fastest current spacecraft (around 50,000 km/hr)?
52. *Galactic Rotation Speed.* We are located about 27,000 light-years from the galactic center and we orbit the center about once every 230 million years. How fast are we traveling around the galaxy, in km/hr?
53. *Earth Rotation Speed.* Mathematical Insight 1.4 shows how to find Earth's equatorial rotation speed. To find the rotation speed at any other latitude, you need the following fact: The radial distance from Earth's axis at any latitude is equal to the equatorial radius times the *cosine* of the latitude. Use this fact to find the rotation speed at the following latitudes. (*Hint*: When using the cosine (cos) function, be sure your calculator is set to recognize angles in degree mode, not in radian or gradient mode.)
**a.** 30°N **b.** 60°N **c.** your latitude
54. *Order of Magnitude Estimate.* Mathematical Insight 1.3 defines order of magnitude estimates, and in the text we estimated there are about $10^{22}$ stars in the observable universe by assuming that 100 billion large galaxies each have about 100 billion stars. Now, assume that there are also 1 trillion small galaxies, but that each has only 10 million stars on average. How will this affect the estimate of the total number of stars in the observable universe? Explain your answer without calculating.

# 2 Discovering the Universe for Yourself

▲ **About the photo:** This time-exposure photograph shows star paths at Arches National Park, Utah.

## LEARNING GOALS

### 2.1 Patterns in the Night Sky

- What does the universe look like from Earth?
- Why do stars rise and set?
- Why do the constellations we see depend on latitude and time of year?

### 2.2 The Reason for Seasons

- What causes the seasons?
- How does the orientation of Earth's axis change with time?

### 2.3 The Moon, Our Constant Companion

- Why do we see phases of the Moon?
- What causes eclipses?

### 2.4 The Ancient Mystery of the Planets

- Why was planetary motion so hard to explain?
- Why did the ancient Greeks reject the real explanation for planetary motion?

*We had the sky, up there, all speckled with stars, and we used to lay on our backs and look up at them, and discuss about whether they was made, or only just happened.*

—*Mark Twain*, Huckleberry Finn

FIGURE 2.1 This photo shows the Milky Way over Haleakala crater on the island of Maui, Hawaii. The bright spot just below (and slightly left of) the center of the band is the planet Jupiter.

**Chapter 2 Overview**

This is an exciting time in the history of astronomy. New and powerful telescopes are scanning the depths of the universe. Sophisticated space probes are exploring our solar system. Rapid advances in computing technology are allowing scientists to analyze the vast amount of new data and to model the processes that occur in planets, stars, galaxies, and the universe.

One goal of this book is to help *you* share in the ongoing adventure of astronomical discovery. One of the best ways to become a part of this adventure is to do what other humans have done for thousands of generations: Go outside, observe the sky around you, and contemplate the awe-inspiring universe of which you are a part. In this chapter, we'll discuss a few key ideas that will help you understand what you see in the sky.

## 2.1 Patterns in the Night Sky

Today we take for granted that we live on a small planet orbiting an ordinary star in one of many galaxies in the universe. But this fact is not obvious from a casual glance at the night sky, and we've learned about our place in the cosmos only through a long history of careful observations. In this section, we'll discuss major features of the night sky and how we understand them in light of our current knowledge of the universe.

### What does the universe look like from Earth?

Shortly after sunset, as daylight fades to darkness, the sky appears to slowly fill with stars. On clear, moonless nights far from city lights, more than 2000 stars may be visible to your naked eye, along with the whitish band of light that we call the *Milky Way* (**FIGURE 2.1**). As you look at the stars, your mind may group them into patterns that look like familiar shapes or objects. If you observe the sky night after night or year after year, you will recognize the same patterns of stars. These patterns have not changed noticeably in the past few thousand years.

**Constellations** People of nearly every culture gave names to patterns they saw in the sky. We usually refer to such patterns as constellations, but to astronomers the term has a more precise meaning: A **constellation** is a *region* of the sky with well-defined borders; the familiar patterns of stars merely help us locate the constellations. Just as every spot of land in the continental United States is part of some state, every point in the sky belongs to some constellation. **FIGURE 2.2** shows the borders of the constellation Orion and several of its neighbors.

The names and borders of the 88 official constellations (Appendix H) were chosen in 1928 by members of the International Astronomical Union (IAU). Most of the IAU members lived in Europe or the United States, so they chose names familiar in the western world. That is why the official names for constellations visible in the Northern

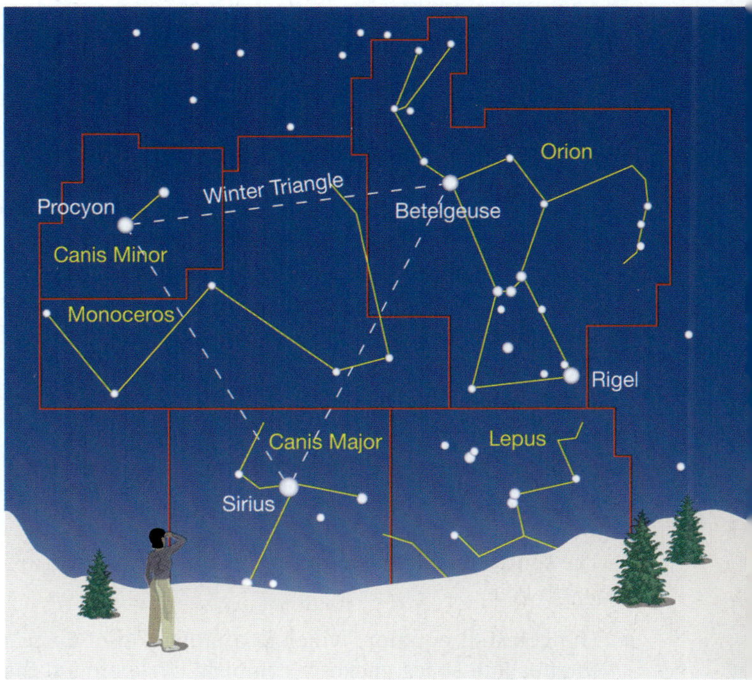


FIGURE 2.2 Red lines mark official borders of several constellations near Orion. Yellow lines connect recognizable patterns of stars. Sirius, Procyon, and Betelgeuse form the *Winter Triangle*, which spans several constellations. This view shows how it appears (looking south) on winter evenings from the Northern Hemisphere.

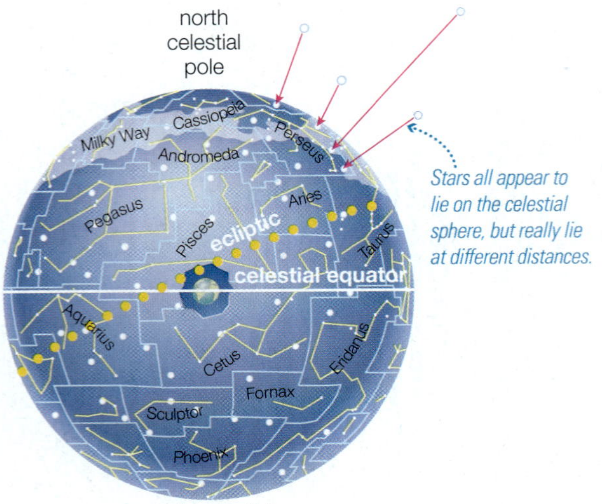


**FIGURE 2.3** The stars and constellations appear to lie on a celestial sphere that surrounds Earth. This is an illusion created by our lack of depth perception in space, but it is useful for mapping the sky.

Hemisphere can be traced back to civilizations of the ancient Middle East, while Southern Hemisphere constellations carry names that originated with 17th-century European explorers.

Recognizing the patterns of just 20 or so constellations is enough to make the sky seem as familiar as your own neighborhood. The best way to learn the constellations is to go out and view them, guided by a few visits to a planetarium, star charts (Appendix I), or sky-viewing apps.

**The Celestial Sphere** The stars in a particular constellation appear to lie close to one another but may be quite far apart in reality, because they may lie at very different distances from Earth. This illusion occurs because we lack depth perception when we look into space, a consequence of the fact that the stars are so far away [**Section 1.1**]. The ancient Greeks mistook this illusion for reality, imagining the stars and constellations to lie on a great **celestial sphere** that surrounds Earth (**FIGURE 2.3**).

We now know that Earth seems to be in the center of the celestial sphere only because it is where we are located as we look into space. Nevertheless, the celestial sphere is a useful illusion, because it allows us to map the sky as seen from Earth. For reference, we identify two special points and two special circles on the celestial sphere (**FIGURE 2.4**).

- The **north celestial pole** is the point directly over Earth's North Pole.
- The **south celestial pole** is the point directly over Earth's South Pole.
- The **celestial equator**, which is a projection of Earth's equator into space, makes a complete circle around the celestial sphere.
- The **ecliptic** is the path the Sun follows as it appears to circle around the celestial sphere once each year. It crosses the celestial equator at a $23\frac{1}{2}°$ angle, because that is the tilt of Earth's axis.

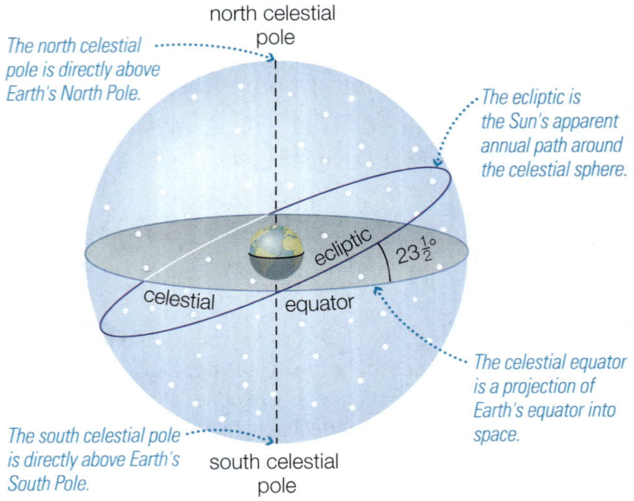


**FIGURE 2.4** This schematic diagram shows key features of the celestial sphere.

**The Milky Way** The band of light that we call the *Milky Way* circles all the way around the celestial sphere, passing through more than a dozen constellations. The widest and brightest parts of the Milky Way are most easily seen from the Southern Hemisphere, which probably explains why the Aborigines of Australia gave names to patterns within the Milky Way in the same way other cultures named patterns of stars. Our Milky Way Galaxy gets its name from this band of light, and the two "Milky Ways" are closely related: *The Milky Way in the night sky traces our galaxy's disk of stars—the galactic plane—as it appears from our location within the* Milky Way Galaxy.

**FIGURE 2.5** shows the idea. Our galaxy is shaped like a thin pancake with a bulge in the middle. We view the universe from our location a little more than halfway out from the center of this "pancake." In all directions that we look within the pancake, we see the many stars and vast interstellar clouds that make up the Milky Way in the night sky; that is why the band of light makes a full circle around our sky. The Milky Way appears somewhat wider in the direction of the constellation Sagittarius, because that is the direction in which we are looking toward the galaxy's central bulge. We have a clear view to the distant universe only when we look *away* from the galactic plane, along directions that have relatively few stars and clouds to block our view.

The dark lanes that run down the center of the Milky Way contain the densest clouds, obscuring our view of stars behind them. In most directions, these clouds prevent us from seeing more than a few thousand light-years into our galaxy's disk. As a result, much of our own galaxy remained hidden from view until just a few decades ago, when new technologies allowed us to peer through the clouds by observing forms of light that are invisible to our eyes (such as radio waves, infrared light, and x-rays [**Section 5.2**]).

**Think about it** Consider a distant galaxy located in the same direction from Earth as the center of our own galaxy (but much farther away). Could we see it with a visible-light telescope? Why or why not?

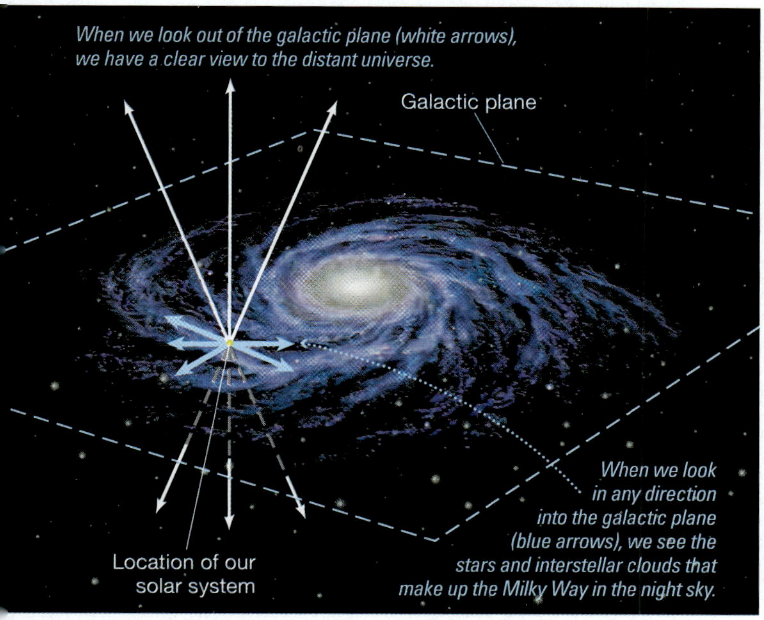


**FIGURE 2.5** This painting shows how our galaxy's structure affects our view from Earth.

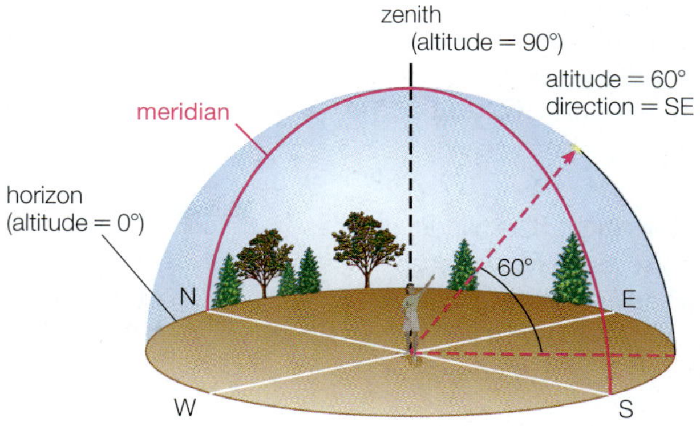


**FIGURE 2.6** From any place on Earth, the local sky looks like a dome (hemisphere). This diagram shows key reference points in the local sky. It also shows how we can describe any position in the local sky by its altitude and direction.

**The Local Sky** The celestial sphere provides a useful way of thinking about the appearance of the universe from Earth. But it is not what we actually see when we go outside. Instead, your **local sky**—the sky as seen from wherever you happen to be standing—appears to take the shape of a hemisphere or dome, which explains why people of many ancient cultures imagined that we lived on a flat Earth under a great dome encompassing the world. The dome shape arises from the fact that we see only half of the celestial sphere at any particular moment from any particular location, while the ground blocks the other half from view.

**FIGURE 2.6** shows key reference features of the local sky. The boundary between Earth and sky defines the **horizon**. The point directly overhead is the **zenith**. The **meridian** is an imaginary half circle stretching from the horizon due south, through the zenith, to the horizon due north.

We can pinpoint the position of any object in the local sky by stating its **direction** along the horizon (sometimes stated as *azimuth*, which is degrees clockwise from due north) and its **altitude** above the horizon. For example, Figure 2.6 shows a person pointing to a star located in the direction of southeast at an altitude of 60°. Note that the zenith has altitude 90° but no direction, because it is straight overhead.

**Angular Sizes and Distances** Our lack of depth perception on the celestial sphere means we have no way to judge the true sizes or separations of the objects we see in the sky. However, we can describe the *angular* sizes or separations of objects without knowing how far away they are.

The **angular size** of an object is the angle it appears to span in your field of view. For example, the angular sizes of the Sun and Moon are each about $\frac{1}{2}°$ (**FIGURE 2.7a**). Note that angular size does not by itself tell us an object's true size, because angular size also depends on distance. The Sun is about 400 times as large in diameter as the Moon, but it has the same angular size in our sky because it is also about 400 times as far away.

The **angular distance** between a pair of objects in the sky is the angle that appears to separate them. For example, the angular distance between the "pointer stars" at the end of the Big Dipper's bowl is about 5° and the angular length of the Southern Cross is about 6° (**FIGURE 2.7b**). You can use your outstretched hand to make rough estimates of angles in the sky (**FIGURE 2.7c**).

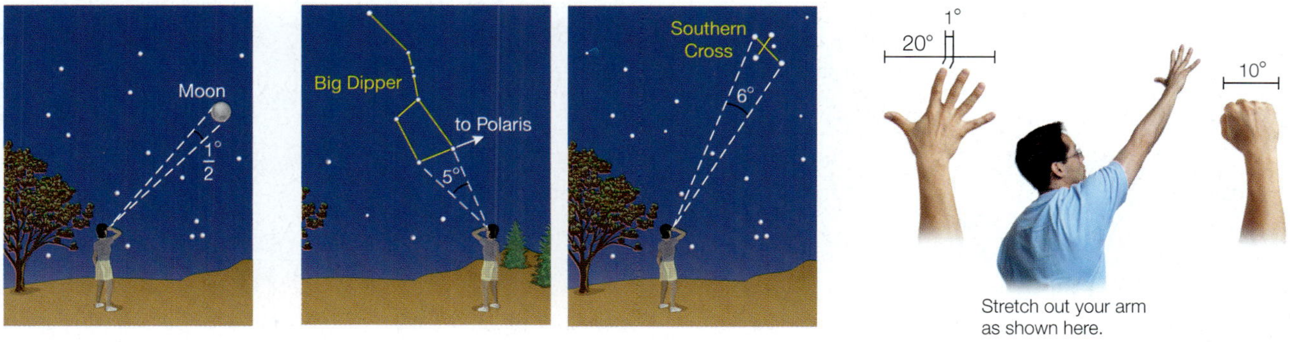


**a** The angular sizes of the Sun and the Moon are about 1/2°.

**b** The angular distance between the "pointer stars" of the Big Dipper is about 5°, and the angular length of the Southern Cross is about 6°.

**c** You can estimate angular sizes or distances with your outstretched hand.

**FIGURE 2.7** We measure *angular sizes* or *angular distances*, rather than actual sizes or distances, when we look at objects in the sky.

For greater precision, we subdivide each degree into 60 **arcminutes** (symbolized by ′) and each arcminute into 60 **arcseconds** (symbolized by ″) as shown in **FIGURE 2.8**. For example, we read 35° 27′15″ as "35 degrees, 27 arcminutes, 15 arcseconds."

**Think about it** Children often try to describe the sizes of objects in the sky (such as the Moon or an airplane) in inches or miles, or by holding their fingers apart and saying "it was THIS big." Can we really describe objects in the sky in this way? Why or why not?

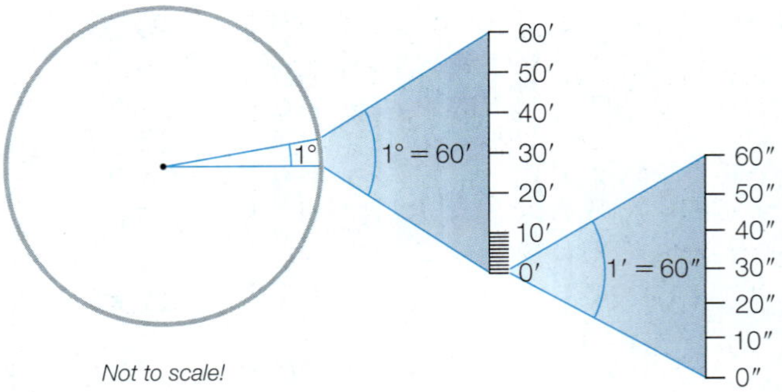


**FIGURE 2.8** We subdivide each degree into 60 arcminutes and each arcminute into 60 arcseconds.

## MATHEMATICAL INSIGHT 2.1 Angular Size, Physical Size, and Distance

If you hold a coin in front of one eye, it can block your entire field of view. But as you move it farther away, it appears to get smaller and it blocks less of your view (**FIGURE 1a**). The coin's true size obviously does not change; what changes is its *angular size*—the angle that it covers as seen from your eye.

**FIGURE 1b** shows how angular size depends on physical size and distance. As long as an object's angular size is relatively small (less than a few degrees), its physical size (diameter) is similar to that of a small piece of a circle with a radius equal to the object's distance from your eye. The object's angular size (in degrees) is therefore the *same fraction* of the full 360° circle as its physical size is of the circle's full circumference (given by the formula $2\pi \times distance$). That is,

$$\frac{\text{angular size}}{360^\circ} = \frac{\text{physical size}}{2\pi \times \text{distance}}$$

Multiplying both sides by 360°, we find

$$\text{angular size} = \text{physical size} \times \frac{360^\circ}{2\pi \times \text{distance}}$$

This formula is often called the **small-angle formula**, because it is valid when the angular size is small.

**EXAMPLE 1:** The two headlights on a car are separated by 1.5 meters. What is their angular separation when the car is 500 meters away?

**SOLUTION:**

**Step 1 Understand:** We can use the small-angle formula by thinking of the "separation" between the two lights as a "size." That is, if we set the physical size to the actual separation of 1.5 meters, the small-angle formula will tell us the angular separation.

**Step 2 Solve:** We simply plug in the given values and solve:

$$\begin{aligned}\text{angular separation} &= \text{physical separation} \times \frac{360^\circ}{2\pi \times \text{distance}} \\ &= 1.5\ \cancel{\text{m}} \times \frac{360^\circ}{2\pi \times 500\ \cancel{\text{m}}} \approx 0.17^\circ\end{aligned}$$

**Step 3 Explain:** We have found that the angular separation of the two headlights is 0.17°. This small angle will be easier to interpret if we convert it to arcminutes. There are 60 arcminutes in 1°, so 0.17° is equivalent to $0.17 \times 60 = 10.2$ arcminutes. In other words, the angular separation of the headlights is about 10 arcminutes, or about a third of the 30 arcminute (0.5°) angular diameter of the Moon.

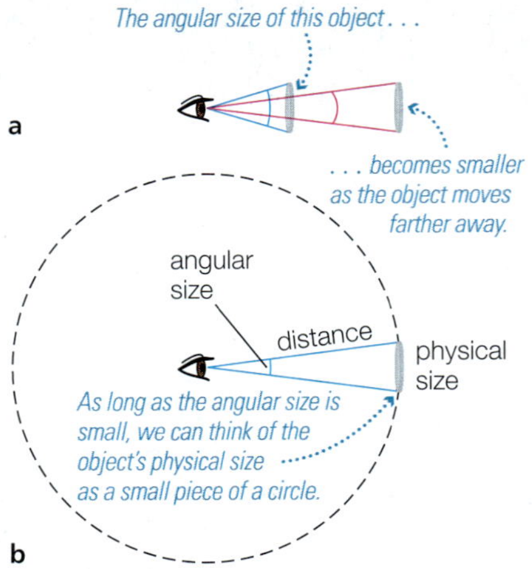


**FIGURE 1** Angular size depends on physical size and distance.

**EXAMPLE 2:** Estimate the Moon's actual diameter from its angular diameter of about 0.5° and its distance of about 380,000 km.

**SOLUTION:**

**Step 1 Understand:** We are seeking to find a physical size (diameter) from an angular size and distance. We therefore need to solve the small-angle formula for the *physical size*, which we do by switching its left and right sides and multiplying both sides by $(2\pi \times distance)/360°$:

$$\text{physical size} = \text{angular size} \times \frac{2\pi \times \text{distance}}{360^\circ}$$

**Step 2 Solve:** We now plug in the given values of the Moon's angular size and distance:

$$\text{physical size} = 0.5^\circ \times \frac{2\pi \times 380{,}000\text{ km}}{360^\circ} \approx 3300\text{ km}$$

**Step 3 Explain:** We have used the Moon's approximate angular size and distance to find that its diameter is about 3300 kilometers. We could find a more exact value (3476 km) by using more precise values for the angular diameter and distance.

## Why do stars rise and set?

If you spend a few hours out under a starry sky, you'll notice that the universe seems to be circling around us, with stars moving gradually across the sky from east to west. Many ancient people took this appearance at face value, concluding that we lie at the center of a universe that rotates around us each day. Today we know that the ancients had it backward: It is Earth that rotates daily, not the rest of the universe.

We can picture the movement of the sky by imagining the celestial sphere rotating around Earth (**FIGURE 2.9**). From this perspective you can see how the universe seems to turn around us: Every object on the celestial sphere appears to make a simple daily circle around Earth. However, the motion can look a little more complex in the local sky, because the horizon cuts the celestial sphere in half. **FIGURE 2.10** shows the idea for a typical Northern Hemisphere location (latitude 40°N). If you study the figure carefully, you'll notice the following key facts about the paths of various stars through the local sky:

- Stars near the north celestial pole are **circumpolar**, meaning that they remain perpetually above the horizon, circling (counterclockwise) around the north celestial pole each day.
- Stars near the south celestial pole never rise above the horizon at all.
- All other stars have daily circles that are partly above the horizon and partly below it, which means they appear to rise in the east and set in the west.

The time-exposure photograph that opens this chapter (page 24) shows a part of the daily paths of stars. Paths of circumpolar stars are visible within the arch; notice that the complete daily circles for these stars are above the horizon, although the photo shows only a portion of each circle. The north celestial pole lies at the center of these circles. The circles grow larger for stars farther from the north celestial pole. If they are large enough, the circles cross the horizon, so that the stars rise in the east and set in the west. The same ideas apply in the Southern Hemisphere, except that circumpolar stars are those near the south celestial pole and they circle clockwise rather than counterclockwise.

**Think about it** Do distant galaxies also rise and set like the stars in our sky? Why or why not?

## Why do the constellations we see depend on latitude and time of year?

If you stay in one place, the basic patterns of motion in the sky will stay the same from one night to the next. However, if you travel far north or south, you'll see a different set of constellations than you see at home. And even if you stay in one place, you'll see different constellations at different times of year. Let's explore why.

### COMMON MISCONCEPTIONS

**The Moon Illusion**

You've probably noticed that the full moon appears to be larger when it is near the horizon than when it is high in your sky. However, this apparent size change is an illusion: If you compare the Moon's angular size to that of a small object (such as a small button) held at arm's length, you'll see that it remains essentially the same throughout the night. The reason is that the Moon's angular size depends on its true size and distance, and while the latter varies over the course of the Moon's monthly orbit, it does not change enough to cause a noticeable effect on a single night. The Moon illusion clearly occurs within the human brain, though its precise cause is still hotly debated. Interestingly, you may be able to make the illusion go away by viewing the Moon upside down between your legs.

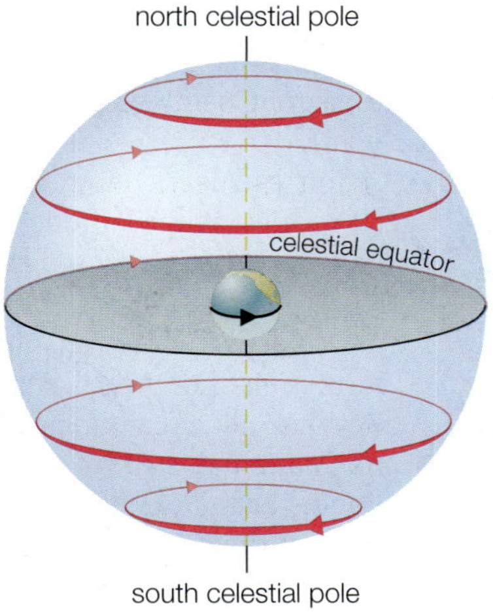


**FIGURE 2.9** Earth rotates from west to east (black arrow), making the celestial sphere *appear* to rotate around us from east to west (red arrows).

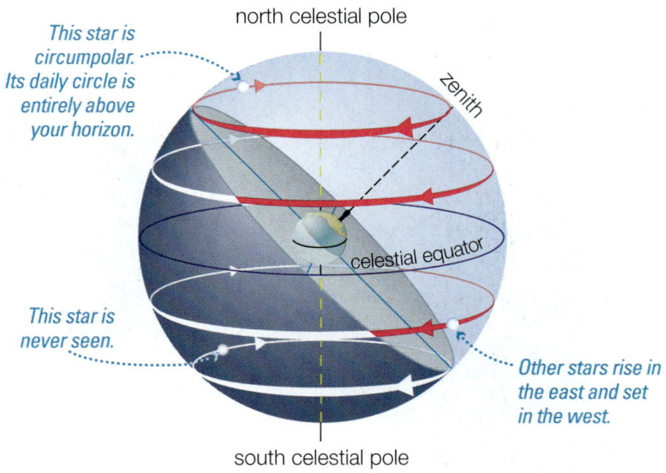


**FIGURE 2.10** The local sky for a location at latitude 40°N. The horizon slices through the celestial sphere at an angle to the celestial equator, causing the daily circles of stars to appear tilted in the local sky. Note: It may be easier to follow the star paths in the local sky if you rotate the page so that the zenith points up.

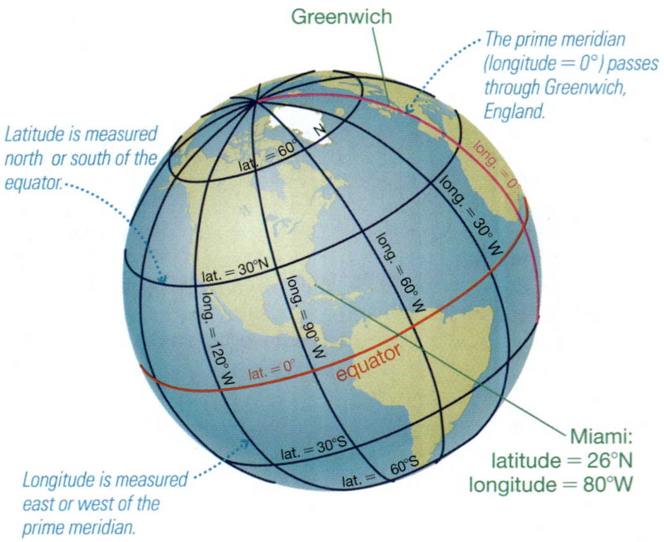


**a** We can locate any place on Earth's surface by its latitude and longitude.

**b** The entrance to the Old Royal Greenwich Observatory, near London. The line emerging from the door marks the prime meridian.

**FIGURE 2.11** Definitions of latitude and longitude.

## COMMON MISCONCEPTIONS

### Stars in the Daytime

Stars may appear to vanish in the daytime and "come out" at night, but in reality the stars are always present. The reason you don't see stars in the daytime is that their dim light is overwhelmed by the bright daytime sky. You *can* see bright stars in the daytime with the aid of a telescope, or if you are fortunate enough to observe a total eclipse of the Sun. Astronauts can also see stars in the daytime. Above Earth's atmosphere, where there is no air to scatter sunlight, the Sun is a bright disk against a dark sky filled with stars. (However, the Sun is so bright that astronauts must block its light if they wish to see the stars.)

**Variation with Latitude** **Latitude** measures north-south position on Earth, and **longitude** measures east-west position (**FIGURE 2.11**). Latitude is defined to be 0° at the equator, increasing to 90°N at the North Pole and 90°S at the South Pole. By international treaty, longitude is defined to be 0° along the **prime meridian**, which passes through Greenwich, England. Stating a latitude and a longitude pinpoints a location on Earth. For example, Miami lies at about 26°N latitude and 80°W longitude.

Latitude affects the constellations we see because it affects the locations of the horizon and zenith relative to the celestial sphere. **FIGURE 2.12** shows how this works for the latitudes of the North Pole (90°N) and Sydney, Australia (34°S). Note that although the sky varies with latitude, it does *not* vary with longitude. For example, Charleston (South Carolina) and San Diego (California) are at about the same latitude, so people in both cities see the same set of constellations at night.

You can learn more about how the sky varies with latitude by studying diagrams like those in Figures 2.10 and 2.12. For example, at the North Pole, you can see only objects that lie on the northern half of the celestial sphere, and they are all circumpolar. That is why the Sun remains above the horizon for 6 months at the North Pole: The Sun lies north of the celestial equator for half of each year (see Figure 2.3), so during these 6 months it circles the sky at the North Pole just like a circumpolar star.

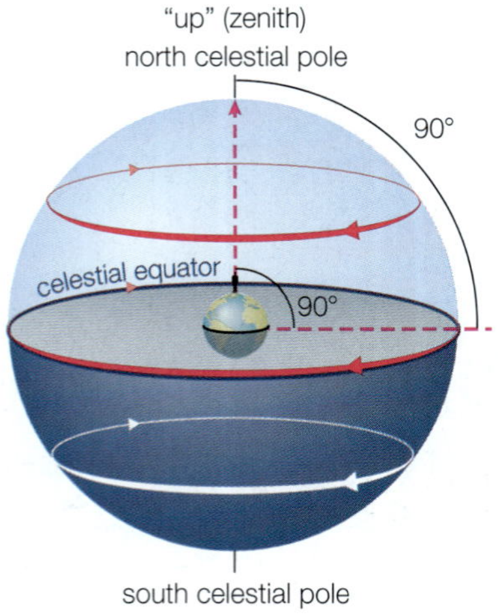


**a** The local sky at the North Pole (latitude 90°N).

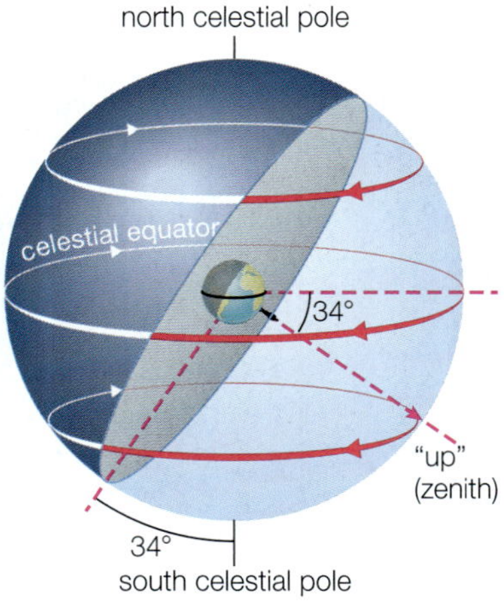


**b** The local sky at latitude 34°S.

**FIGURE 2.12** The sky varies with latitude. Notice that the altitude of the celestial pole that is visible in your sky is always equal to your latitude.

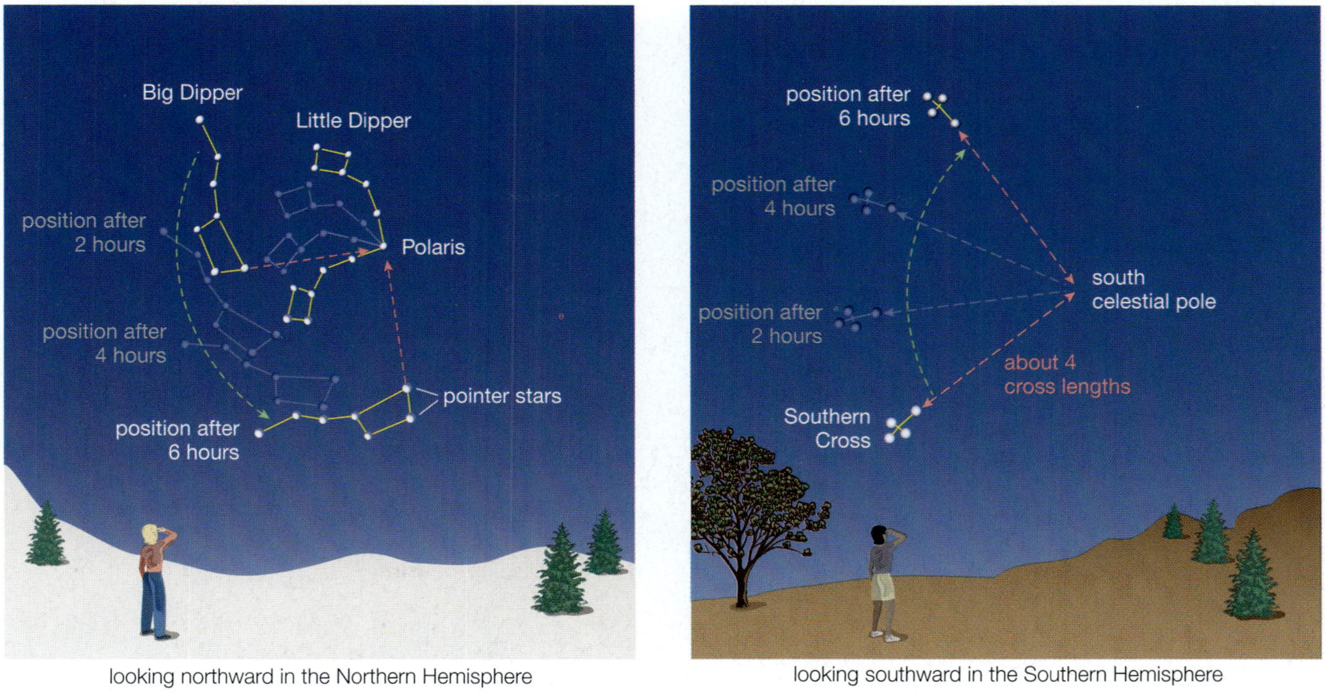


**a** The pointer stars of the Big Dipper point to the North Star, Polaris, which lies within 1° of the north celestial pole. The sky appears to turn *counterclockwise* around the north celestial pole.

**b** The Southern Cross points to the south celestial pole, which is not marked by any bright star. The sky appears to turn *clockwise* around the south celestial pole.

**FIGURE 2.13** You can determine your latitude by measuring the altitude of the celestial pole in your sky.

The diagrams also show a fact that is very important to navigation:

**The altitude of the celestial pole in your sky is equal to your latitude.**

For example, if you see the north celestial pole at an altitude of 40° above your north horizon, your latitude is 40°N. Similarly, if you see the south celestial pole at an altitude of 34° above your south horizon, your latitude is 34°S. You can therefore determine your latitude simply by finding the celestial pole in your sky (Figure 2.13). Finding the north celestial pole is fairly easy, because it lies very close to the star Polaris, also known as the North Star (**FIGURE 2.13a**). In the Southern Hemisphere, you can find the south celestial pole with the aid of the Southern Cross (**FIGURE 2.13b**). We'll discuss celestial navigation and how the sky varies with latitude in more detail in Chapter S1.

**See it for yourself** What is *your* latitude? Use Figure 2.13 to find the celestial pole in your sky, and estimate its altitude with your hand as shown in Figure 2.7c. Is its altitude what you expect?

**Variation with Time of Year** The night sky changes throughout the year because of Earth's changing position in its orbit around the Sun. **FIGURE 2.14** shows how this works. From our vantage point on Earth, the annual orbit of Earth around the Sun makes the Sun *appear* to move steadily eastward along the ecliptic, with the stars of different constellations in the background at different times of year. The constellations along the ecliptic make up what we call the **zodiac**; tradition places 12 constellations along the zodiac, but the official borders include a 13th constellation, Ophiuchus.

The Sun's apparent location along the ecliptic determines which constellations we see at night. For example, Figure 2.14 shows that the Sun appears to be in Leo in late August. We therefore cannot see Leo at this time (because it is in our daytime sky), but we can see Aquarius all night long because of its location opposite Leo on the celestial sphere. Six months later, in February, we see Leo at night while Aquarius is above the horizon only in the daytime.

**See it for yourself** Based on Figure 2.14 and today's date, in what constellation does the Sun currently appear? What constellation of the zodiac will be on your meridian at midnight? What constellation of the zodiac will you see in the west shortly after sunset? Go outside at night to confirm your answers to the last two questions.

## COMMON MISCONCEPTIONS

### What Makes the North Star Special?

Most people are aware that the North Star, Polaris, is a special star. Contrary to a relatively common belief, however, it is *not* the brightest star in the sky. More than 50 other stars are just as bright or brighter. Polaris is special not because of its brightness, but because it is so close to the north celestial pole and therefore very useful in navigation.

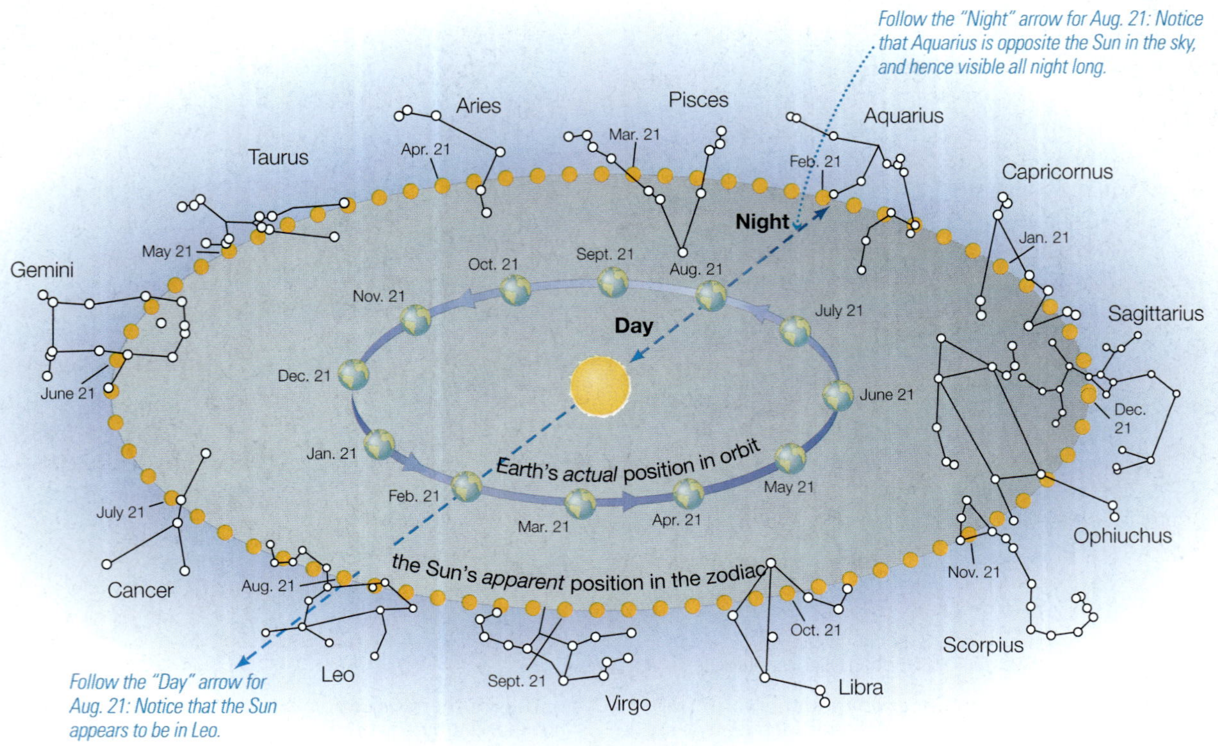


**FIGURE 2.14** The Sun appears to move steadily eastward along the ecliptic as Earth orbits the Sun, so we see the Sun against the background of different zodiac constellations at different times of year. For example, on August 21 the Sun appears to be in Leo, because it is between us and the much more distant stars that make up Leo.

**The Reason for Seasons**

## 2.2 The Reason for Seasons

We have seen how Earth's rotation makes the sky appear to circle us daily and how the night sky changes as Earth orbits the Sun each year. The combination of Earth's rotation and orbit also leads to the progression of the seasons.

### What causes the seasons?

You know that we have seasonal changes, such as longer and warmer days in summer and shorter and cooler days in winter. But why do the seasons occur? The answer is that the tilt of Earth's axis causes sunlight to fall differently on Earth at different times of year.

**FIGURE 2.15** (pages 34–35) illustrates the key ideas. Step 1 illustrates the tilt of Earth's axis, which remains pointed in the same direction in space (toward Polaris) throughout the year. As a result, the orientation of the axis *relative to the Sun* changes over the course of each orbit: The Northern Hemisphere is tipped toward the Sun in June and away from the Sun in December, while the reverse is true for the Southern Hemisphere. That is why the two hemispheres experience opposite seasons. The rest of the figure shows how the changing angle of sunlight on the two hemispheres leads directly to seasons.

Step 2 shows Earth in June, when the axis tilt causes sunlight to strike the Northern Hemisphere at a steeper angle and the Southern Hemisphere at a shallower angle. The steeper sunlight angle makes it summer in the Northern Hemisphere for two reasons. First, as shown in the zoom-out, the steeper angle means more concentrated sunlight, which tends to make it warmer. Second, if you visualize what happens as Earth rotates each day, you'll see that the steeper angle also means the Sun follows a longer and higher path through the sky, giving the Northern Hemisphere more hours of daylight during which it is warmed by the Sun. The opposite is true for the Southern Hemisphere at this time: The shallower sunlight angle makes it winter there because sunlight is less concentrated and the Sun follows a shorter, lower path through the sky.

The sunlight angle gradually changes as Earth orbits the Sun. At the opposite side of Earth's orbit, Step 4 shows that it has become winter for the Northern Hemisphere and summer for the Southern Hemisphere. In between these

### COMMON MISCONCEPTIONS

#### The Cause of Seasons

Many people guess that seasons are caused by variations in Earth's distance from the Sun. But if this were true, the whole Earth would have summer or winter at the same time, and it doesn't: The seasons are opposite in the Northern and Southern Hemispheres. In fact, Earth's slightly varying orbital distance has virtually no effect on the weather. The real cause of the seasons is Earth's axis tilt, which causes the two hemispheres to take turns being tipped toward the Sun over the course of each year.

two extremes, Step 3 shows that both hemispheres are illuminated equally in March and September. It is therefore spring for the hemisphere that is on the way from winter to summer, and fall for the hemisphere on the way from summer to winter.

Notice that the seasons on Earth are caused only by the axis tilt and *not* by changes in Earth's distance from the Sun. Although Earth's orbital distance varies over the course of each year, the variation is fairly small: Earth is only about 3% farther from the Sun at its farthest point (which is in July) than at its nearest (in January). The difference in the strength of sunlight due to this small change in distance is overwhelmed by the effects caused by the axis tilt. If Earth did not have an axis tilt, we would not have seasons.

**Think about it** Jupiter has a very small axis tilt (about 3°). Saturn has an axis tilt of about 27°. Both planets have nearly circular orbits around the Sun. Do you expect Jupiter to have seasons? Do you expect Saturn to have seasons? Explain.

**Solstices and Equinoxes** To help us mark the changing seasons, we define four special moments in the year, each of which corresponds to one of the four special positions in Earth's orbit shown in Figure 2.15.

- The **June solstice**, called the *summer solstice* in the Northern Hemisphere, occurs around June 21 and is the moment when the Northern Hemisphere is tipped most directly toward the Sun and receives the most direct sunlight.
- The **December solstice**, called the *winter solstice* in the Northern Hemisphere, occurs around December 21 and is the moment when the Northern Hemisphere receives the least direct sunlight.
- The **March equinox**, called the *spring* (or *vernal*) *equinox* in the Northern Hemisphere, occurs around March 21 and is the moment when the Northern Hemisphere goes from being tipped slightly away from the Sun to being tipped slightly toward the Sun.
- The **September equinox**, called the *fall* (or *autumnal*) *equinox* in the Northern Hemisphere, occurs around September 22 and is the moment when the Northern Hemisphere first starts to be tipped away from the Sun.

The exact dates and times of the solstices and equinoxes can vary by up to a couple of days from the dates given above, depending on where we are in the leap year cycle. In fact, our modern calendar includes leap years (usually adding one day—February 29—every fourth year) specifically to keep the solstices and equinoxes around the same dates [**Section S1.1**].

We can mark the dates of the equinoxes and solstices by observing changes in the Sun's path through our sky (**FIGURE 2.16**). The equinoxes occur on the only two days of the year on which the Sun rises precisely due east and sets precisely due west; these are also the two days when the Sun is above and below the horizon for equal times of 12 hours (*equinox* means "equal night"). The June solstice occurs on the day on which the Sun follows its longest and highest path through the Northern Hemisphere sky (and its shortest and lowest path through the Southern Hemisphere sky). It is therefore the day on which the Sun rises and sets farthest to the north of due east and due west; it is also the day on which the Northern Hemisphere has its longest hours of daylight and the Sun rises highest in the midday sky. The opposite is true on the day of the December solstice, when the Sun rises and sets farthest to the south and the Northern Hemisphere has its shortest hours of daylight and lowest midday Sun. **FIGURE 2.17** shows how the Sun's position in the sky varies over the course of the year.

**First Days of Seasons** We usually say that each equinox and solstice marks the first day of a season. For example, the day of the June solstice is usually called the "first day of summer" in the Northern Hemisphere. Notice, however, that the Northern Hemisphere has its *maximum* tilt toward the Sun at this time. You might then wonder why we consider the solstice to be the beginning rather than the midpoint of summer.

The choice is somewhat arbitrary, but it makes sense in at least two ways. First, it was much easier for ancient people to identify the days on which the Sun reached extreme positions in the sky—such as when it reached its highest point on the summer solstice—than other days in between. Second, we usually think of the seasons in terms of weather, and the warmest summer weather tends to come 1 to 2 months after the solstice. To understand why, think about what happens when you heat a pot of cold soup. Even though you may have the stove turned on high from the start, it takes a while for the soup to warm up. In the same way, it takes some time for sunlight to heat the ground and oceans from the cold of winter to the warmth of summer. "Midsummer" in terms of weather therefore comes in late July and early August, which makes the June solstice a pretty good choice for the "first day of summer." Similar logic applies to the starting times for spring, fall, and winter.

**Seasons Around the World** The seasons have different characteristics in different parts of the world. High latitudes have more extreme seasons. For example, Vermont has much longer summer days and much longer winter nights than Florida. At the Arctic Circle (latitude 66½°), the Sun remains above the horizon all day long on the June solstice (**FIGURE 2.18**) and never rises on the December solstice (although bending of light by the atmosphere makes the Sun *appear* to be about a half-degree higher than it really is). The most extreme cases occur at the North and South Poles, where the Sun remains above the horizon for 6 months in summer and below the horizon for 6 months in winter.

## COMMON MISCONCEPTIONS

### High Noon

When is the Sun directly overhead in your sky? Many people answer "at noon." It's true that the Sun reaches its *highest* point each day when it crosses the meridian, giving us the term "high noon" (though the meridian crossing is rarely at precisely 12:00). However, unless you live in the Tropics (between latitudes 23.5°S and 23.5°N), the Sun is *never* directly overhead. In fact, any time you can see the Sun as you walk around, you can be sure it is not at your zenith. Unless you are lying down, seeing an object at the zenith requires tilting your head back into a very uncomfortable position.

**COSMIC Context**

# FIGURE 2.15 The Cause of Seasons

**Earth's seasons are caused by the tilt of its rotation axis, which is why the seasons are opposite in the two hemispheres. The seasons do *not* depend on Earth's distance from the Sun, which varies only slightly throughout the year.**

① **Axis Tilt:** Earth's axis points in the same direction throughout the year, which causes changes in Earth's orientation *relative to the Sun*.

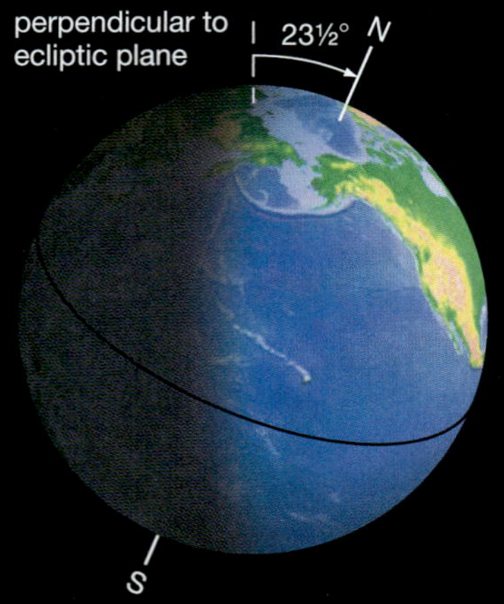


② **Northern Summer/Southern Winter:** In June, sunlight falls more directly on the Northern Hemisphere, which makes it summer there because solar energy is more concentrated and the Sun follows a longer and higher path through the sky. The Southern Hemisphere receives less direct sunlight, making it winter.

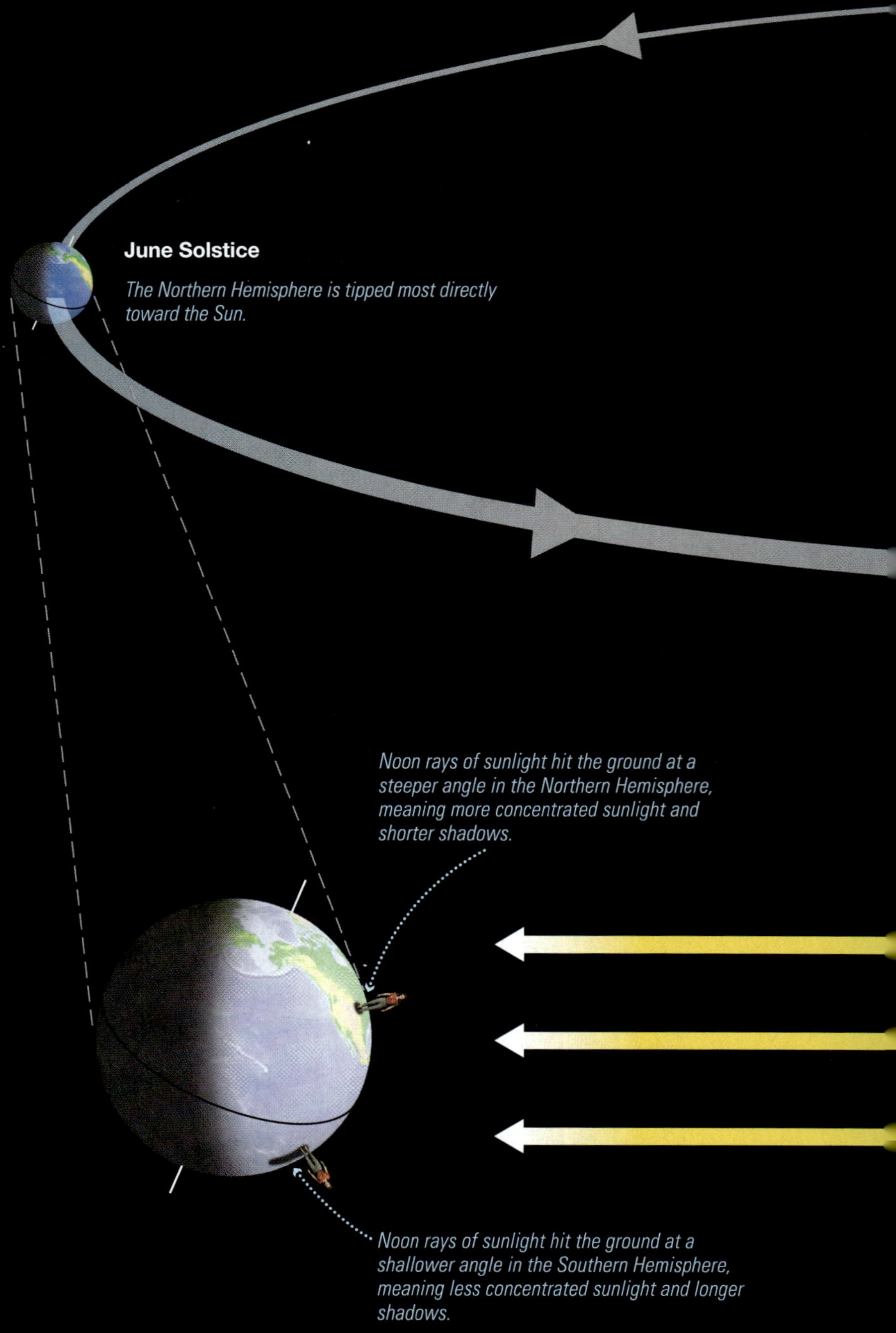


*Interpreting the Diagram*

*To interpret the seasons diagram properly, keep in mind:*

*1. Earth's size relative to its orbit would be microscopic on this scale, meaning that both hemispheres are at essentially the same distance from the Sun.*

*2. The diagram is a side view of Earth's orbit. A top-down view (below) shows that Earth orbits in a nearly perfect circle and comes closest to the Sun in January.*

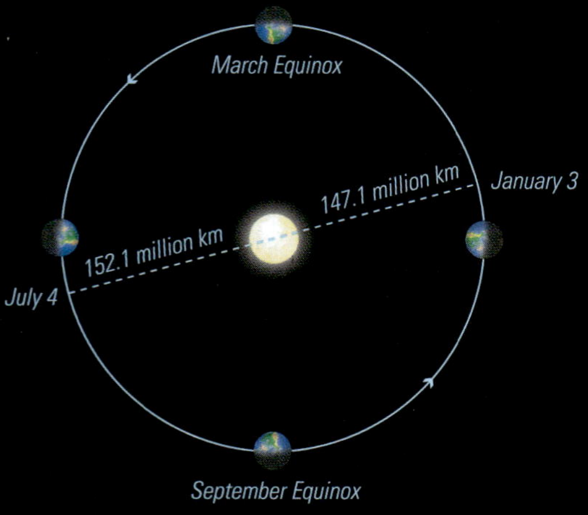

3 **Spring/Fall:** Spring and fall begin when sunlight falls equally on both hemispheres, which happens twice a year: In March, when spring begins in the Northern Hemisphere and fall in the Southern Hemisphere; and in September, when fall begins in the Northern Hemisphere and spring in the Southern Hemisphere.

4 **Northern Winter/Southern Summer:** In December, sunlight falls less directly on the Northern Hemisphere, which makes it winter because solar energy is less concentrated and the Sun follows a shorter and lower path through the sky. The Southern Hemisphere receives more direct sunlight, making it summer.

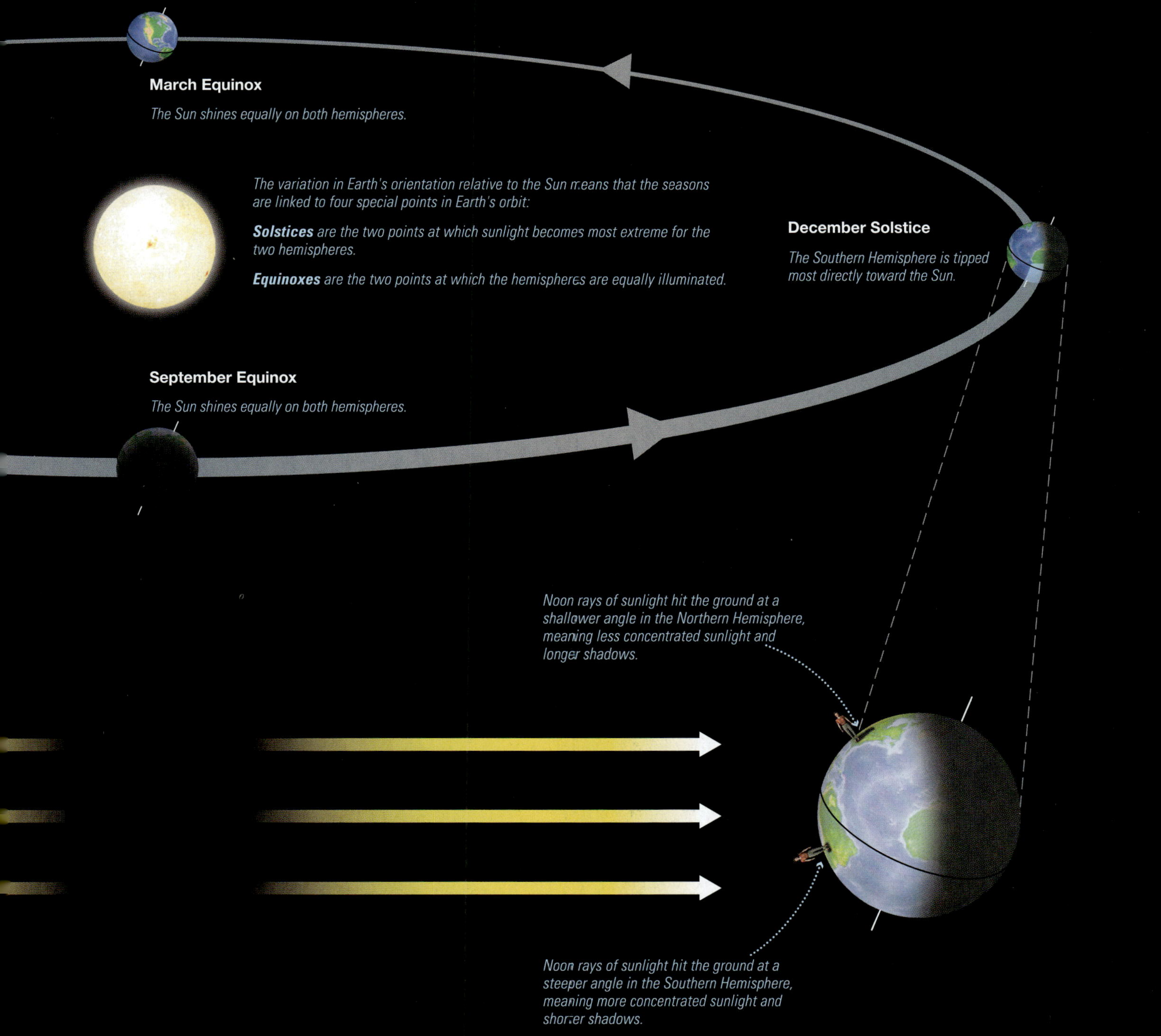

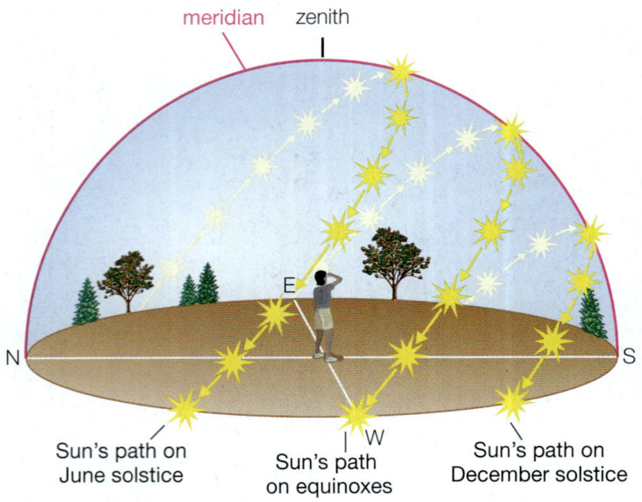


**FIGURE 2.16** This diagram shows the Sun's path on the solstices and equinoxes for a Northern Hemisphere sky (latitude 40°N). The precise paths are different for other latitudes; for example, at latitude 40°S, the paths look similar except tilted to the north rather than to the south. Notice that the Sun rises exactly due east and sets exactly due west only on the equinoxes.

**FIGURE 2.17** This composite photograph shows images of the Sun taken at 7- to 11-day intervals over the course of a year, each at the same time ("mean solar time" [Section S1.1]) and from the same spot (in Carefree, Arizona); the photo looks eastward, so north is to the left and south is to the right. Because this location is in the Northern Hemisphere, the Sun images that are high and to the north represent times near the June solstice and the images that are low and south represent times near the December solstice. The "figure 8" shape (called an *analemma*) arises from a combination of Earth's axis tilt and Earth's varying speed as it orbits the Sun (see the Special Topic, page 92).

Seasons also differ in equatorial regions, because the equator gets its most direct sunlight on the two equinoxes and its least direct sunlight on the solstices. As a result, instead of the four seasons experienced at higher latitudes, equatorial regions generally have rainy and dry seasons, with the rainy seasons coming when the Sun is higher in the sky.

**Why Southern Hemisphere Seasons Are Milder (on Earth)** We've seen that the seasons are caused by Earth's axis tilt, not by Earth's slightly varying distance from the Sun. Still, we might expect the varying orbital distance to play at least some role. For example, the Northern Hemisphere has winter when Earth is closer to the Sun and summer when Earth is farther away (see the lower left diagram in Figure 2.15), so we might expect the Northern Hemisphere to have more moderate seasons than the Southern Hemisphere. In fact, weather records show that the opposite is true: Northern Hemisphere seasons are slightly more extreme than those of the Southern Hemisphere.

The main reason for this surprising fact becomes clear when you look at a map of Earth (**FIGURE 2.19**). Most of Earth's land lies in the Northern Hemisphere, with far more ocean in the Southern Hemisphere. As you'll notice at any

**FIGURE 2.18** This sequence of photos shows the progression of the Sun around the horizon on the summer solstice at the Arctic Circle. Notice that the Sun skims the northern horizon at midnight, then gradually rises higher, reaching its highest point when it is due south at noon.

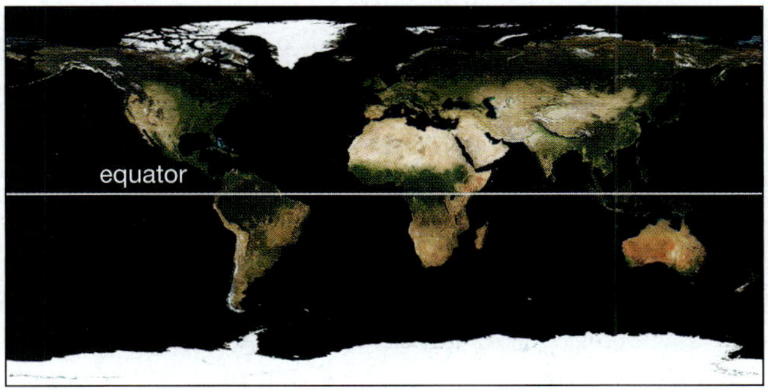


**FIGURE 2.19** Most land lies in the Northern Hemisphere while most ocean lies in the Southern Hemisphere. The climate-moderating effects of water make Southern Hemisphere seasons less extreme than Northern Hemisphere seasons.

beach, lake, or pool, water takes longer to heat or cool than soil or rock (largely because sunlight heats bodies of water to a depth of many meters while heating only the very top layer of land). The water temperature therefore remains fairly steady both day and night, while the ground can heat up and cool down dramatically. The Southern Hemisphere's larger amount of ocean moderates its climate. The Northern Hemisphere, with more land and less ocean, heats up and cools down more easily, which is why it has the more extreme seasons.

Although distance from the Sun plays no role in *Earth's* seasons, it *can* affect seasons on planets with more elliptical orbits. For example, Mars has about the same axis tilt as Earth and therefore has similar seasonal patterns, but its much greater variation in distance from the Sun makes these patterns more extreme in its southern hemisphere than in its northern hemisphere (see Figure 10.24).

## How does the orientation of Earth's axis change with time?

Our calendar keeps the solstices and equinoxes around the same dates each year, but the constellations associated with them change gradually over time. The reason is **precession**, a gradual wobble that alters the orientation of Earth's axis in space.

Precession occurs with many rotating objects. You can see it easily by spinning a top (**FIGURE 2.20a**). As the top spins rapidly, you'll notice that its axis also sweeps out a circle at a slower rate. We say that the top's axis *precesses*. Earth's axis precesses in much the same way, but far more slowly (**FIGURE 2.20b**). Each cycle of Earth's precession takes about 26,000 years. This gradually changes the direction in which the axis points in space.

**Think about it** Was Polaris the North Star in ancient times? Explain.

Note that precession does not change the *amount* of the axis tilt (which stays close to $23\frac{1}{2}°$) and therefore does not affect the pattern of the seasons. However, it changes the points in Earth's orbit at which the solstices and equinoxes occur, and therefore changes the constellations that we see at those times. For example, a couple thousand years ago the June solstice occurred when the Sun appeared in the constellation Cancer, but it now occurs when the Sun appears in Gemini. This explains something you can see on any world map: The latitude at which the Sun is directly overhead on the June solstice $\left(23\frac{1}{2}°\text{N}\right)$ is called the *Tropic of Cancer,* telling us that it was named back when the Sun appeared in Cancer on this solstice.

Precession is caused by gravity's effect on a tilted, rotating object. You have probably seen how gravity affects a top. If you try to balance a nonspinning top on its point,

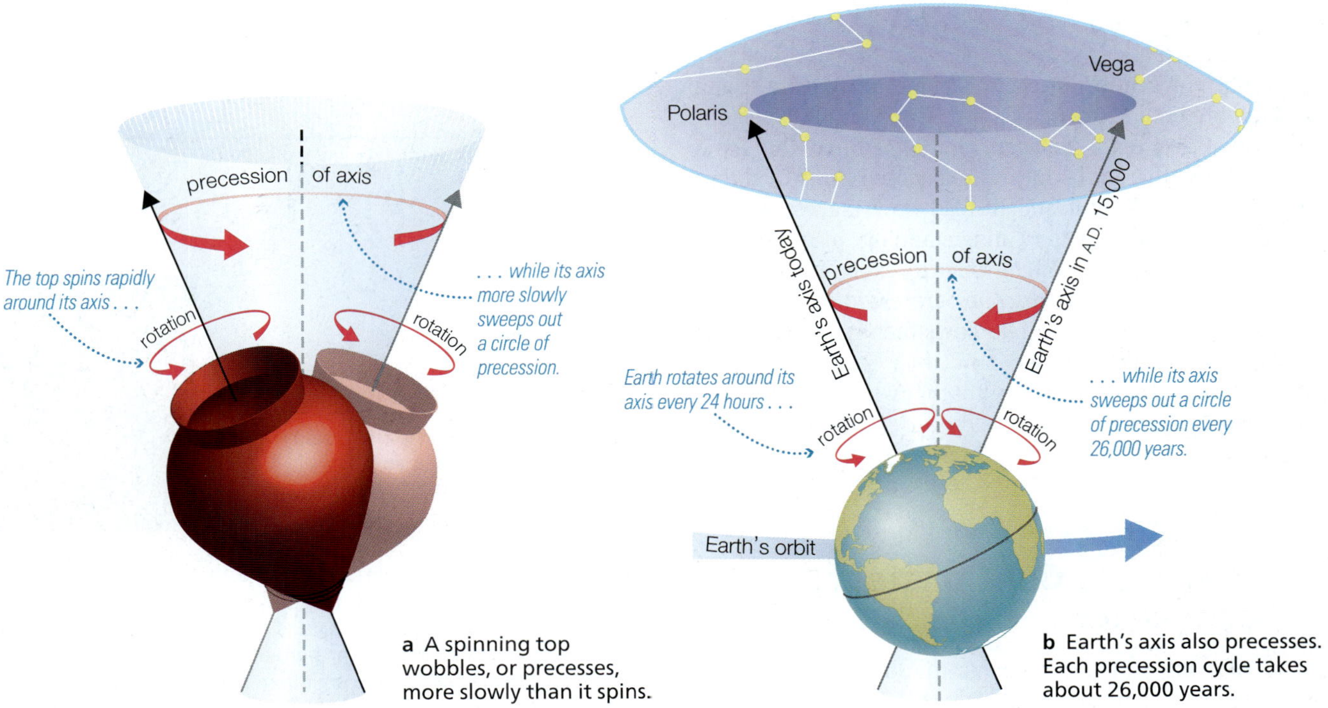


**FIGURE 2.20** Precession affects the orientation of a spinning object's axis but not the amount of its tilt.

## COMMON MISCONCEPTIONS

### Sun Signs

You probably know your astrological "Sun sign." When astrology began a few thousand years ago, your Sun sign was supposed to represent the constellation in which the Sun appeared on your birth date. However, because of precession, this is no longer the case for most people. For example, if your birthday is March 21, your Sun sign is Aries even though the Sun now appears in Pisces on that date. The problem is that astrological Sun signs are based on the positions of the Sun among the stars as they were almost 2000 years ago. Because Earth's axis has moved about 1/13 of the way through its 26,000-year precession cycle since that time, astrological Sun signs are off by nearly a month from the actual positions of the Sun among the constellations today.

it will fall over almost immediately. This happens because a top will inevitably lean a little to one side. No matter how slight this lean, gravity will quickly tip the non-spinning top over. However, if you spin the top rapidly, it does not fall over so easily. The spinning top stays upright because rotating objects tend to keep spinning around the same rotation axis (a consequence of the *law of conservation of angular momentum* [**Section 4.3**]). This tendency prevents gravity from immediately pulling the spinning top over, since falling over would mean a change in the spin axis from near-vertical to horizontal. Instead, gravity succeeds only in making the axis trace circles of precession. As friction slows the top's spin, the circles of precession get wider and wider, and ultimately the top falls over. If there were no friction to slow its spin, the top would spin and precess forever.

The spinning (rotating) Earth precesses because of gravitational tugs from the Sun and Moon. Earth is not quite a perfect sphere, because it bulges at its equator. Because the equator is tilted $23\frac{1}{2}°$ to the ecliptic plane, the gravitational attractions of the Sun and Moon try to pull the equatorial bulge into the ecliptic plane, effectively trying to "straighten out" Earth's axis tilt. However, like the spinning top, Earth tends to keep rotating around the same axis. Gravity therefore does not succeed in straightening out Earth's axis tilt and instead only makes the axis precess. To gain a better understanding of precession and how it works, you might wish to experiment with a simple toy gyroscope. *Gyroscopes* are essentially rotating wheels mounted in a way that allows them to move freely, which makes it easy to see how their spin rate affects their motion. (The fact that gyroscopes tend to keep the same rotation axis makes them very useful in aircraft and spacecraft navigation.)

## 2.3 The Moon, Our Constant Companion

Aside from the Sun, the Moon is the brightest and most noticeable object in our sky. The Moon is our constant companion in space, traveling with us as we orbit the Sun.

**Moon Phases, Parts 1 and 2**

### Why do we see phases of the Moon?

As the Moon orbits Earth, it returns to the same position relative to the Sun in our sky (such as along the Earth-Sun line) about every $29\frac{1}{2}$ days. This time period marks the cycle of **lunar phases**, in which the Moon's appearance in our sky changes as its position relative to the Sun changes. This $29\frac{1}{2}$-day period is also the origin of the word *month* (think "moonth").

**See it for yourself** Like the Sun, the Moon appears to move gradually eastward through the constellations of the zodiac. However, while the Sun takes a year for each circuit, the Moon takes only about a month, which means it moves at a rate of about 360° per month, or $\frac{1}{2}°$—its own angular size—each hour. If the Moon is visible tonight, go out and note its location relative to a few bright stars. Then go out again a couple hours later. Can you notice the Moon's change in position relative to the stars?

**Understanding Phases** The first step in understanding phases is to recognize that sunlight essentially comes at both Earth and the Moon from the same direction. You can see why by studying **FIGURE 2.21**, which shows the Moon's

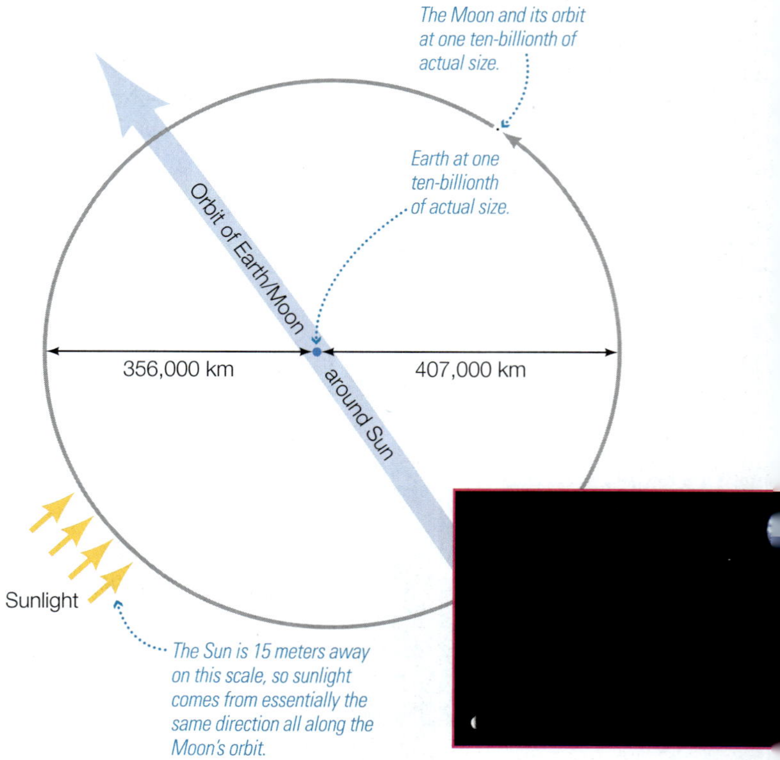


**FIGURE 2.21** The Moon's orbit on the 1-to-10-billion scale introduced in Chapter 1 (see Figure 1.6); black labels indicate the Moon's actual distances when it is nearest and farthest from Earth. The orbit is so small compared to the distance to the Sun that sunlight strikes the entire orbit from the same direction. You can see this in the inset photo, which shows the Moon and Earth photographed from Mars by the *Mars Reconnaissance Orbiter*.

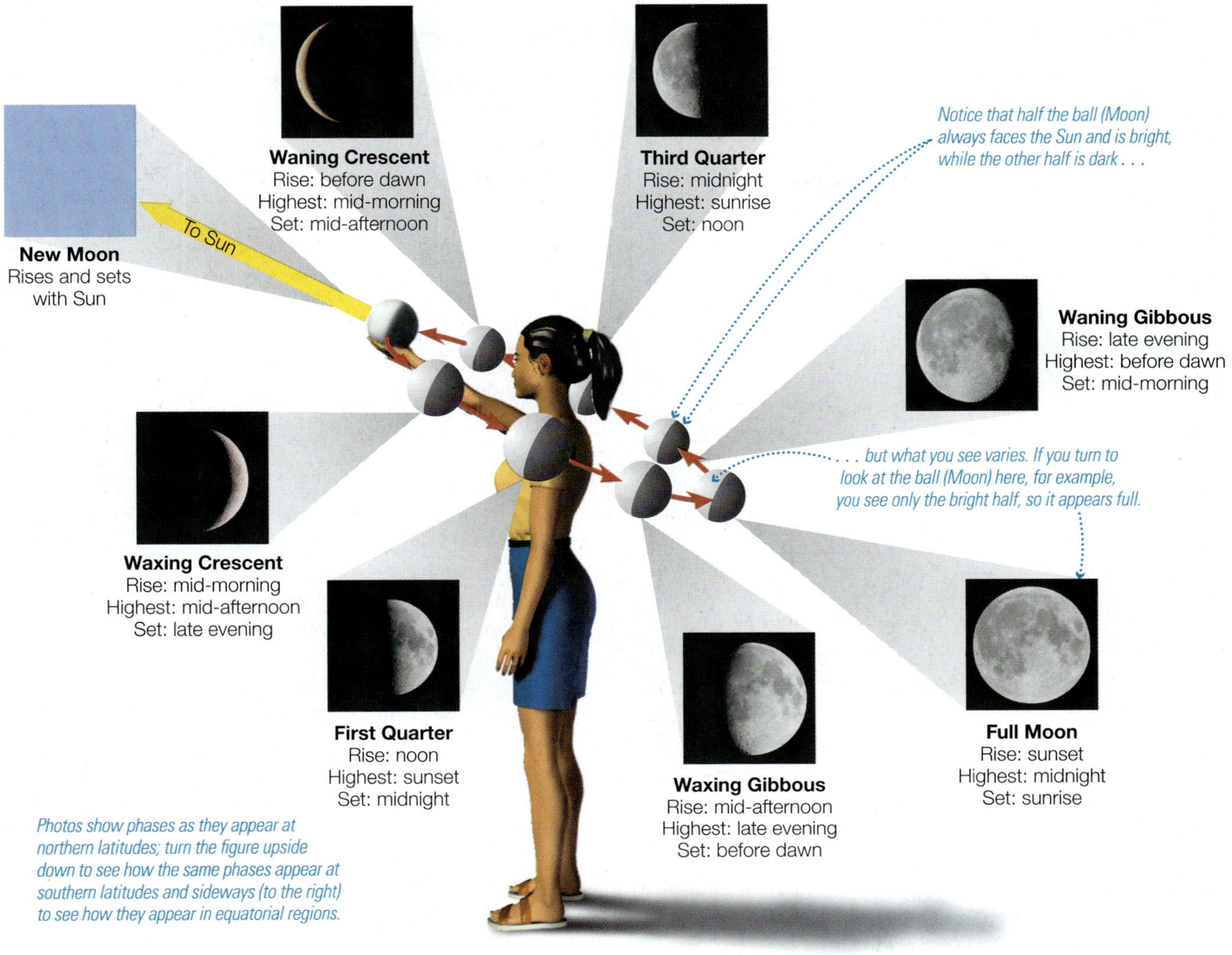


**FIGURE 2.22** In this demonstration, your head represents Earth and the ball represents the Moon as it orbits Earth. The half of the ball (Moon) facing the Sun is always illuminated while the other half is dark, but from your perspective in the middle, you see the ball (Moon) go through the phases shown in the photos. The labels indicate the approximate times of day at which each phase rises, reaches its highest point in the sky, and sets (exact times vary with location, time of year, and orbital details).

orbit on the same scale we used for the model solar system in Chapter 1. Recall that the Sun is located 15 meters away from Earth and the Moon on this scale, which is far enough that the Sun would seem to be in almost precisely the same direction no matter whether you looked at it from Earth or from the Moon.

You can now understand the lunar phases with the simple demonstration illustrated in **FIGURE 2.22**. Take a ball outside on a sunny day. (If it's dark or cloudy, you can use a flashlight instead of the Sun; put the flashlight on a table a few meters away and shine it toward you.) Hold the ball at arm's length to represent the Moon while your head represents Earth. Slowly spin counterclockwise so that the ball goes around you the way the Moon orbits Earth. (If you live in the Southern Hemisphere, spin clockwise because you view the sky "upside down" compared to the Northern Hemisphere.) As you turn, you'll see the ball go through phases just like the Moon's. If you think about what's happening, you'll realize that the phases of the ball result from just two basic facts:

1. Half the ball always faces the Sun (or flashlight) and therefore is bright, while the other half faces away from the Sun and is dark.
2. As you look at the ball at different positions in its "orbit" around your head, you see different combinations of its bright and dark faces.

For example, when you hold the ball directly opposite the Sun, you see only the bright portion of the ball, which represents the "full" phase. When you hold the ball at its "first-quarter" position, half the face you see is dark and the other half is bright.

We see lunar phases for the same reason. Half the Moon is always illuminated by the Sun, but the amount of this illuminated half that we see from Earth depends on the Moon's position in its orbit. The photographs in Figure 2.22 show how the phases look. (The new moon photo shows blue sky, because a new moon is nearly in line with the Sun and therefore hidden from view in the bright daytime sky.)

## COMMON MISCONCEPTIONS

### Shadows and the Moon

Many people guess that the Moon's phases are caused by Earth's shadow falling on its surface, but this is not the case. As we've seen, the Moon's phases are caused by the fact that we see different portions of its day and night sides at different times as it orbits around Earth. The only time Earth's shadow falls on the Moon is during the relatively rare event of a lunar eclipse.

## COMMON MISCONCEPTIONS

### The "Dark Side" of the Moon

Some people refer to the far side of the Moon—meaning the side that we never see from Earth—as the *dark side*. But this is not correct, because the far side is not always dark. For example, during new moon the far side faces the Sun and hence is completely sunlit. In fact, because the Moon rotates with a period of approximately one month (the same time it takes to orbit Earth), points on both the near and the far side have two weeks of daylight alternating with two weeks of darkness. The only time the far side is completely dark is at full moon, when it faces away from both the Sun and Earth.

The Moon's phase also determines the times of day at which we see it in the sky. For example, the full moon must rise around sunset, because it occurs when the Moon is opposite the Sun in the sky. It therefore reaches its highest point in the sky at midnight and sets around sunrise. Similarly, a first-quarter moon must rise around noon, reach its highest point around sunset, and set around midnight, because it occurs when the Moon is about 90° east of the Sun in our sky.

**Think about it** Suppose you go outside in the morning and notice that the visible face of the Moon is half light and half dark. Is this a first-quarter or third-quarter moon? How do you know?

Notice that the phases from new to full are said to be *waxing*, which means "increasing." Phases from full to new are *waning*, or "decreasing." Also notice that no phase is called a "half moon." Instead, we see half the Moon's face at first-quarter and third-quarter phases; these phases mark the times when the Moon is one quarter or three quarters of the way through its monthly cycle (which begins at new moon). The phases just before and after new moon are called *crescent*, while those just before and after full moon are called *gibbous* (pronounced with a hard *g* as in "gift"). A gibbous moon is essentially the opposite of a crescent moon—a crescent moon has a small sliver of light while a gibbous moon has a small sliver of dark. The term *gibbous* literally means "hump-backed," so you can see how the gibbous moon got its name.

**The Moon's Synchronous Rotation** Although we see many *phases* of the Moon, we do not see many *faces*. From Earth we always see (nearly*) the same face of the Moon. This happens because the Moon rotates on its axis in the same amount of time it takes to orbit Earth, a trait called **synchronous rotation**. A simple demonstration shows the idea. Place a ball on a table to represent Earth while you represent the Moon (**FIGURE 2.23**). The only way you can face the ball at all times is by completing exactly one rotation while you complete one orbit. Note that the Moon's synchronous rotation is *not* a coincidence; it is a consequence of Earth's gravity affecting the Moon in much the same way the Moon's gravity causes tides on Earth [**Section 4.5**].

**The View from the Moon** A good way to solidify your understanding of the lunar phases is to imagine that you live on the side of the Moon that faces Earth. For example, what would you see if you looked at Earth when people on Earth saw a new moon? By remembering that a new moon occurs when the Moon is between the Sun and Earth, you'll realize that from the Moon you'd be looking at Earth's

*Because the Moon's orbital speed varies (in accord with Kepler's second law [**Section 3.3**], while its rotation rate is steady, the visible face appears to wobble slightly back and forth as the Moon orbits Earth. This effect, called *libration*, allows us to see a total of about 59% of the Moon's surface over the course of a month, even though we see only 50% of the Moon at any single time.

**a** If you do not rotate while walking around the model, you will not always face it.

**b** You will face the model at all times only if you rotate exactly once during each orbit.

**FIGURE 2.23** The fact that we always see the same face of the Moon means that the Moon must rotate once in the same amount of time it takes to orbit Earth once. You can see why by walking around a model of Earth while imagining that you are the Moon.

## COMMON MISCONCEPTIONS

### Moon in the Daytime and Stars on the Moon

Night is so closely associated with the Moon in traditions and stories that many people mistakenly believe that the Moon is visible only in the nighttime sky. In fact, the Moon is above the horizon as often in the daytime as at night, though it is easily visible only when its light is not drowned out by sunlight. For example, a first-quarter moon is easy to spot in the late afternoon as it rises through the eastern sky, and a third-quarter moon is visible in the morning as it heads toward the western horizon.

Another misconception appears in illustrations that show a star in the dark portion of the crescent moon. The star in the dark portion appears to be in front of the Moon, which is impossible because the Moon is much closer to us than is any star.

**FIGURE 2.24** These illustrations show the two basic types of eclipse. The relative sizes of the Moon and Earth are shown to scale, but the relative distance between them is about 50 times that shown here, the Sun is actually about 100 times larger in diameter than Earth, and the Earth-Sun distance is about 400 times the Earth-Moon distance.

daytime side and hence would see a *full earth*. Similarly, at full moon you would be facing the night side of Earth and would see a *new earth*. In general, you'd always see Earth in a phase opposite the phase of the Moon seen by people on Earth at the same time. Moreover, because the Moon always shows nearly the same face to Earth, Earth would appear to hang nearly stationary in your sky as it went through its cycle of phases.

**Think about it** About how long would each day and night last if you lived on the Moon? Explain.

Thinking about the view from the Moon clarifies another interesting feature of the lunar phases: The dark portion of the lunar face is not *totally* dark. Just as we can see at night by the light of the Moon, if you were in the dark area of the Moon during crescent phase your moonscape would be illuminated by a nearly full (gibbous) Earth. In fact, because Earth is much larger than the Moon, the illumination would be much greater than what the full moon provides on Earth. In other words, sunlight reflected by Earth faintly illuminates the "dark" portion of the Moon's face. We call this illumination the *ashen light*, or *earthshine*, and it enables us to see the outline of the full face of the Moon even when the Moon is not full.

**Eclipses Part 1**

## What causes eclipses?

The lunar phases repeat "moonthly," but sometimes the Moon gets involved in more dramatic events: eclipses. There are two basic types of eclipse (**FIGURE 2.24**):

- A **lunar eclipse** occurs when Earth comes directly between the Sun and Moon, so that Earth's shadow falls on the Moon.
- A **solar eclipse** occurs when the Moon comes directly between the Sun and Earth, so that the Moon's shadow falls on Earth.

Note that, because Earth is much larger than the Moon, Earth's shadow can cover the entire Moon during a lunar eclipse. Therefore, a lunar eclipse can be seen by anyone on the night side of Earth when it occurs. In contrast, the Moon's shadow can cover only a small portion of Earth at any one moment, so you must be located within the relatively narrow pathway through which the shadow moves to see a solar eclipse. That is why we tend to see lunar eclipses more often than solar eclipses, even though both types occur about equally often.

**Shadow Regions** To understand what we actually see during eclipses, we need to look at the three-dimensional geometry of eclipse shadows. **FIGURE 2.25** shows where light from opposite edges of the Sun would fall if we could place a

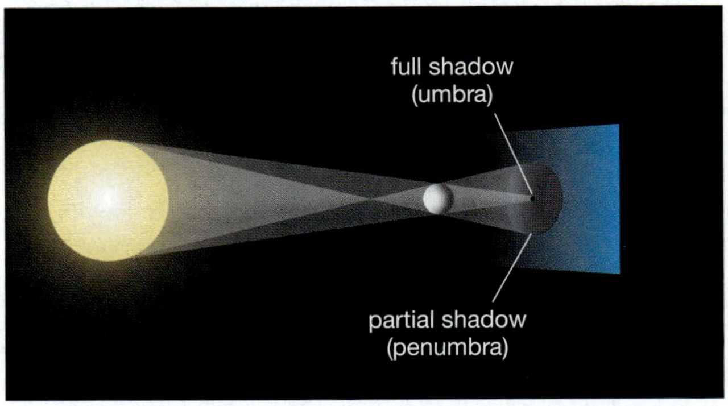


**FIGURE 2.25** This diagram shows the two distinct, cone-shaped regions of the shadow cast on an imaginary screen by a world like Earth or the Moon in sunlight. If you looked back from within the full shadow, you would see the Sun fully blocked from view; if you looked from within the partial shadow, the Sun would be only partly blocked from view.

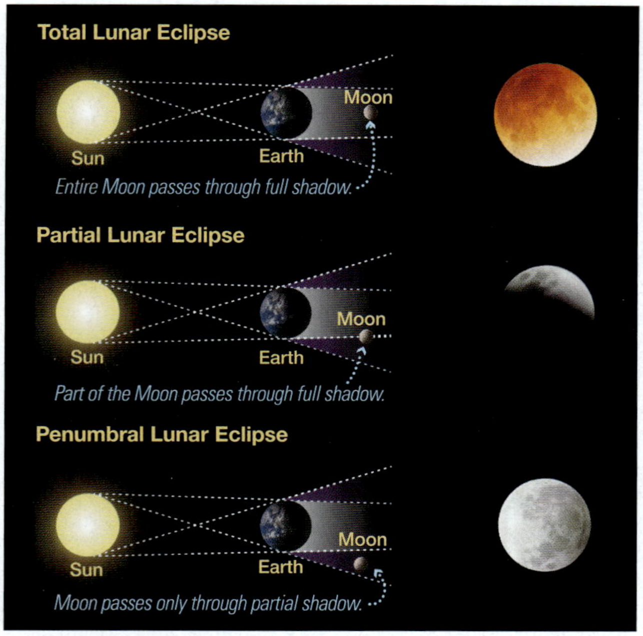


**FIGURE 2.26** The three types of lunar eclipse. The Moon is actually much dimmer during totality than during the partial eclipse, but it is shown here as the eye tends to notice it due to its coloration.

**FIGURE 2.27** This multiple-exposure photograph shows the progression (left to right) of Earth's shadow across the Moon during a total lunar eclipse; totality began (far right) just before the Moon set in the west. Notice the clear curvature of Earth's shadow and the redness of the full moon during totality. The photo was taken in Tenerife, Canary Islands (Spain).

giant screen behind a world like Earth or the Moon. Notice that the shadow consists of two distinct regions: a central **full shadow**, or *umbra*, in which sunlight is fully blocked, and a surrounding **partial shadow**, or *penumbra*, in which light from only part of the Sun is blocked. We can use these ideas to explore the details of lunar and solar eclipses.

**Lunar Eclipses** A lunar eclipse begins at the moment when the Moon's orbit first carries it into Earth's partial shadow. After that, we will see one of three types of lunar eclipse (**FIGURE 2.26**). If the Sun, Earth, and Moon are nearly perfectly aligned, the Moon passes through Earth's full shadow and we see a **total lunar eclipse**. If the alignment is somewhat less perfect, only part of the full moon passes through the full shadow (with the rest in the partial shadow) and we see a **partial lunar eclipse**. If the Moon passes *only* through Earth's partial shadow (penumbra), we see a **penumbral lunar eclipse**. Penumbral eclipses are the most common, but they are the least visually impressive because the full moon darkens only slightly.

Total lunar eclipses can be quite beautiful (**FIGURE 2.27**). Earth's full shadow gradually progresses across the face of the Moon, and the clear curvature of this shadow demonstrates that our world is round. **Totality** begins when the Moon is entirely engulfed in the full shadow and typically lasts about an hour, after which we see the shadow gradually move off the Moon. The Moon becomes dark and eerily red during totality, for reasons you can understand by considering the view of an observer on the eclipsed Moon. This observer would see Earth's night side surrounded by the reddish glow of all the sunrises and sunsets occurring on Earth at that moment, which means that this reddish light illuminates the Moon during totality.

**Solar Eclipses** We can also see three types of solar eclipse (**FIGURE 2.28**). If a solar eclipse occurs when the Moon is in a part of its orbit where it is relatively close to Earth (see Figure 2.21), the Moon's full shadow can cover a small area of Earth's surface (up to about 270 kilometers in diameter). Within this area you will see a **total solar eclipse**. If the eclipse occurs when the Moon is in a part of its orbit that puts it farther from Earth, the full shadow may not reach Earth's surface, leading to an **annular eclipse**—a ring of sunlight surrounding the Moon—in the small region of Earth directly behind the full shadow. In either case, the region of totality or annularity will be surrounded by a much larger region (typically about 7000 kilometers in diameter) that falls within the Moon's partial shadow. Here you will see a **partial solar eclipse**, in which only part of the Sun is blocked from view. (Some solar eclipses are only partial, meaning that no locations on Earth see a total or annular eclipse, because the full shadow passes above or below our planet.) The combination of Earth's rotation and the Moon's orbital motion causes the Moon's shadow to race across the face of Earth at a typical speed of about 1700 kilometers per hour. As a result, the full shadow traces a narrow path across Earth, and totality never lasts more than a few minutes in any particular place.

A total solar eclipse is a spectacular sight. It begins when the disk of the Moon first appears to touch the Sun. Over the next hour or so, the Moon appears to take a larger and larger "bite" out of the Sun. As totality approaches, the sky darkens and temperatures fall. Birds head back to their nests, and crickets begin their nighttime chirping. During the few minutes of totality, the Moon completely blocks the visible disk of the Sun, allowing the faint *corona* to be

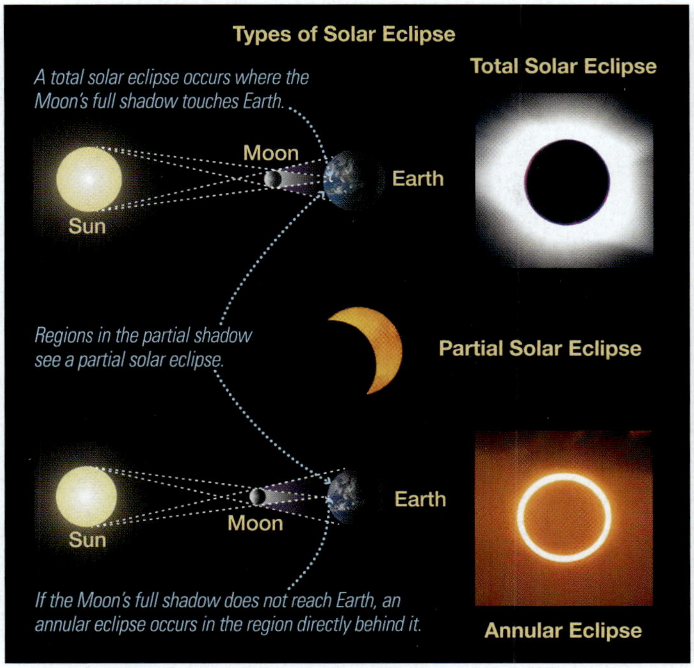


**a** The three types of solar eclipse. (A partial solar eclipse may also be seen in cases when the Moon's full shadow passes above or below Earth.)

**b** This photo from the Deep Space Climate Observatory shows the Moon's full shadow and the darker regions of the partial shadow (arrow) on Earth during a total solar eclipse.

**FIGURE 2.28** During a solar eclipse, the Moon's shadow moves rapidly across the face of Earth.

**FIGURE 2.29** This multiple-exposure photograph shows the progression of the 2017 total solar eclipse over Green River Lake, Wyoming. Totality (central image) lasted about 2 minutes.

seen (**FIGURE 2.29**). The surrounding sky takes on a twilight glow, and planets and bright stars become visible in the daytime. As totality ends, the Sun slowly emerges from behind the Moon over the next couple of hours. However, because your eyes have adapted to the darkness, totality appears to end far more abruptly than it began.

**Eclipses Part 2**

**Conditions for Eclipses** How often do eclipses occur? If you look at a simple Moon phase diagram (such as Figure 2.22), it might seem as if the Sun, Earth, and Moon would line up with every new and full moon, in which case we'd have both a lunar and a solar eclipse every month. But we don't, and the reason is that the Moon's orbit is slightly inclined (by about 5°) to the ecliptic plane (the plane of Earth's orbit around the Sun).

To visualize this inclination, it's helpful to imagine the ecliptic plane as the surface of a pond, as shown in **FIGURE 2.30**. The Moon's orbital inclination means that the Moon spends most of its time either above or below the pond surface (representing the ecliptic plane). The Moon crosses *through* this surface only twice during each orbit—once coming out and once going back in—at the two points called the **nodes** of the Moon's orbit.

Notice that the nodes are aligned approximately the same way (diagonally in Figure 2.30) throughout the year, which means they lie along a nearly straight line with the Sun and Earth about twice each year. Eclipses can occur only during these periods, called **eclipse seasons**, which

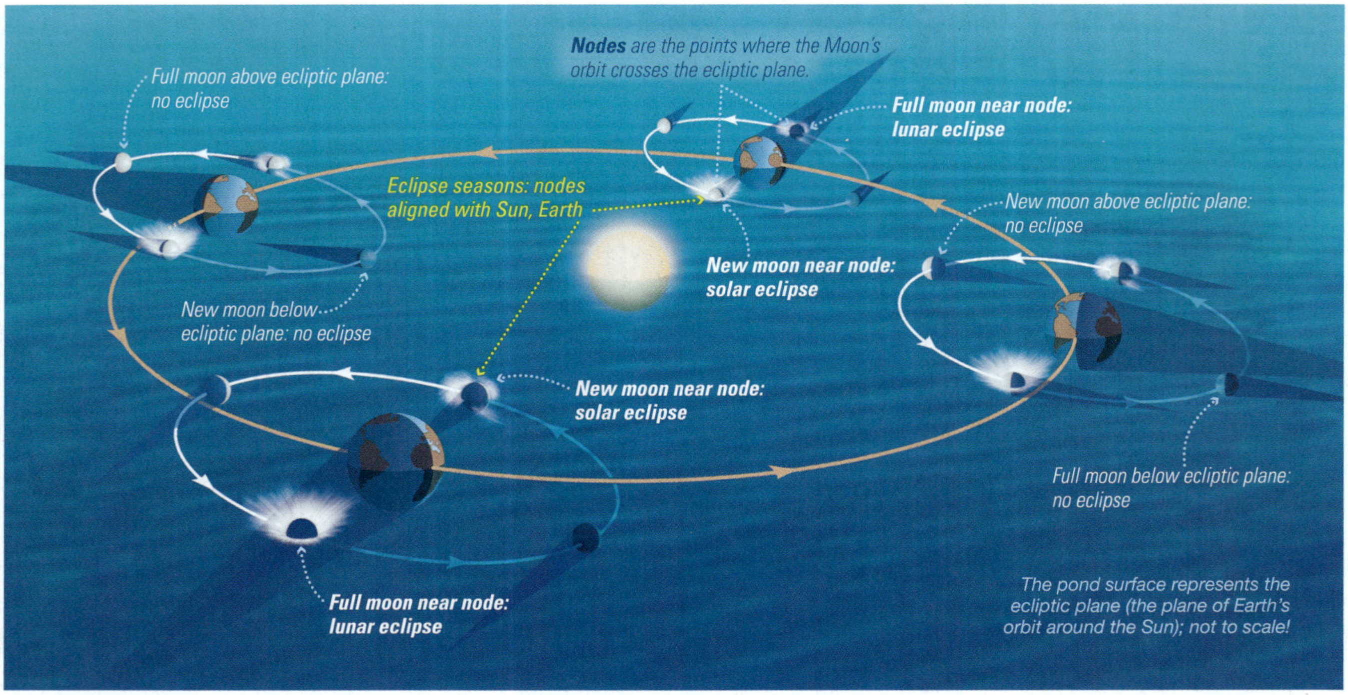


**FIGURE 2.30** This illustration represents the ecliptic plane as the surface of a pond. The Moon's orbit is tilted by about 5° to the ecliptic plane, so the Moon spends half of each orbit above the plane (the pond surface) and half below it. Eclipses occur only during *eclipse seasons*, when the nodes are closely aligned with the Sun and Earth *and* when the lunar phase is either new (for a solar eclipse) or full (for a lunar eclipse).

each last about five weeks (on average). In other words, eclipses can occur only when

1. the phase of the Moon is full (for a lunar eclipse) or new (for a solar eclipse) *and*
2. the new or full moon occurs when the Moon is very close to a node, which means it is during an eclipse season.

Because an eclipse season lasts a few days longer than a cycle of phases, there is always a lunar eclipse (at full moon) and a solar eclipse (at new moon) during each eclipse season. (A second lunar or solar eclipse can occasionally also occur during a single eclipse season.)

**Think about it** Suppose that, today, some parts of the world saw a partial solar eclipse. Is it possible that you would see a lunar eclipse tonight? In two weeks? In two months? Explain.

**Predicting Eclipses** Few phenomena have so inspired and humbled humans throughout the ages as eclipses. For many cultures, eclipses were mystical events associated with fate or the gods, and countless stories and legends surround them. One legend holds that the Greek philosopher Thales (c. 624–546 B.C.) successfully predicted the year (but presumably not the precise time) that a total eclipse of the Sun would be visible in the area where he lived, which is now

## SPECIAL TOPIC Does the Moon Influence Human Behavior?

From myths of werewolves to stories of romance under the full moon, human culture is filled with claims that the Moon influences our behavior. Can we say anything scientific about such claims?

The Moon clearly has important influences on Earth, perhaps most notably through its role in creating tides [**Section 4.5**]. Although the Moon's tidal force cannot directly affect objects as small as people, the ocean tides have indirect effects. For example, fishermen, boaters, and surfers all adjust at least some of their activities to the cycle of the tides.

Another potential influence might come from the lunar phases. Physiological patterns in many species appear to follow the lunar phases; for example, some crabs and turtles lay eggs only at full moon. No human trait is so closely linked to lunar phases, but the average human menstrual cycle is so close in length to a lunar month that it is difficult to believe the similarity is mere coincidence.

Nevertheless, aside from the physiological cycles and the influence of tides on people who live near the oceans, claims that the lunar phase affects human behavior are difficult to verify scientifically. For example, although it is possible that the full moon brings out certain behaviors, it may also simply be that some behaviors are easier to engage in when the sky is bright. A beautiful full moon may bring out your desire to walk on the beach under the moonlight, but there is no scientific evidence to suggest that the full moon would affect you the same way if you were confined to a deep cave.

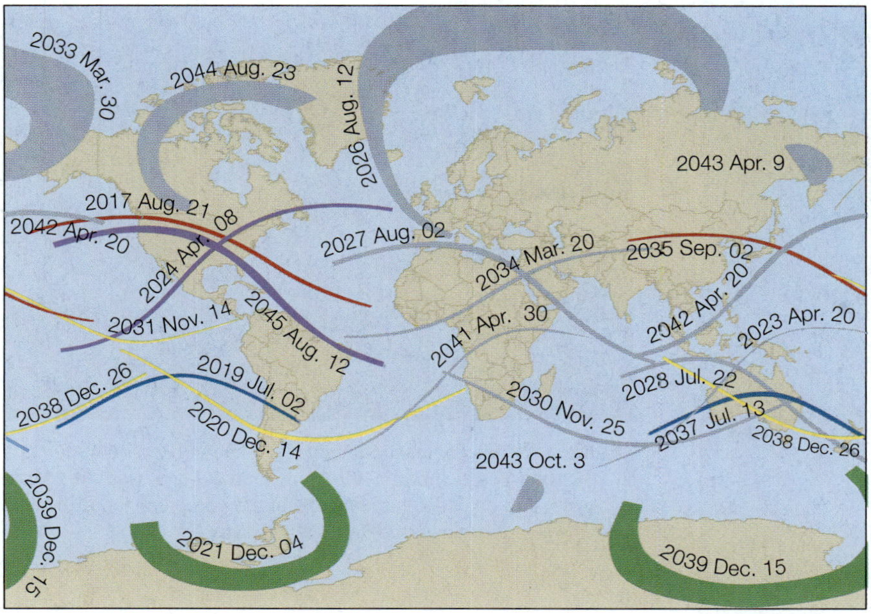


**FIGURE 2.31** This map shows the paths of totality for total solar eclipses from 2017 through 2045. Selected paths representing eclipses of successive saros cycles are shown in color. For example, the paths in red for the 2017 and 2035 eclipses are separated by the 18-year, 11-day saros period. Eclipse predictions by Fred Espenak, NASA GSFC.
Note: Download the free app "Totality by Big Kid Science" (created by one of the textbook authors) to learn more about upcoming total solar eclipses, including the next one in the United States on April 8, 2024.

part of Turkey. Moreover, the eclipse is said to have occurred as two opposing armies (the Medes and the Lydians) were massing for battle, and it so frightened them that they put down their weapons, signed a treaty, and returned home.

Much of the mystery of eclipses probably stems from the relative difficulty of predicting them. Look again at Figure 2.30, focusing on the two eclipse seasons. If this figure told the whole story, eclipse seasons would always occur 6 months apart and predicting eclipses would be easy. For example, if the eclipse seasons always occurred in January and July, eclipses would always occur on the dates of new and full moons in those months. Actual eclipse prediction is more difficult than this because of something the figure does not show: The nodes slowly move around the Moon's orbit (often called "precession of the nodes," a process that has a period of 18.6 years), causing the eclipse seasons to occur slightly *less* than 6 months apart (about 173 days apart).

**TABLE 2.1 Lunar Eclipses 2019–2022***

| Date | Type | Where You Can See It |
|---|---|---|
| Jan. 21, 2019 | total | Pacific, Americas, Europe, Africa |
| July 16, 2019 | partial | S. America, Europe, Africa, Asia, Australia |
| Jan. 10, 2020 | penumbral | Americas, Europe, Africa |
| June 5, 2020 | penumbral | Asia, Australia, Pacific, Americas |
| July 5, 2020 | penumbral | Americas, Europe, Africa |
| Nov. 30, 2020 | penumbral | Asia, Australia, Pacific, Americas |
| May 26, 2021 | total | Australia, Pacific, western Americas |
| Nov. 19, 2021 | partial | Asia, Australia, Pacific, Americas |
| May 16, 2022 | total | Americas, Europe, Africa |
| Nov. 8, 2022 | total | Asia, Australia, Pacific, Americas |

* Dates are based on Universal Time and hence are those in Greenwich, England, at the time of the eclipse; check a news source for the local time and date. Eclipse predictions by Fred Espenak, NASA GSFC.

The combination of the changing dates of eclipse seasons and the $29\frac{1}{2}$-day cycle of lunar phases makes eclipses recur in a cycle of about 18 years, $11\frac{1}{3}$ days, called the **saros cycle**. Astronomers in many ancient cultures identified the saros cycle and used it to make eclipse predictions. For example, in the Middle East the Babylonians achieved remarkable success at predicting eclipses more than 2500 years ago, and the Mayans achieved similar success in Central America; in fact, the Mayan calendar includes a cycle (the *sacred round*) of 260 days—almost exactly $1\frac{1}{2}$ times the 173.32 days between successive eclipse seasons.

However, while the saros cycle allows you to predict *when* an eclipse will occur, the approximately $\frac{1}{3}$ day in the cycle length means that the locations *where* an eclipse will be visible shift about $\frac{1}{3}$ of the way around the world with each cycle. This and other subtleties of eclipses (such as whether a solar eclipse is total or annular, which depends on the Moon's orbital distance at the time of the eclipse) make exact eclipse prediction very difficult, and no ancient culture achieved the ability to predict eclipses in every detail.

Today, we can predict eclipses because we know the precise details of the orbits of Earth and the Moon. **TABLE 2.1** lists upcoming lunar eclipses; notice that, as we expect, eclipses generally come a little less than 6 months apart. (Two lunar eclipses sometimes occur in the same eclipse season and therefore come just a month apart, like those for June 5 and July 5, 2020.) **FIGURE 2.31** shows paths of totality for recent and upcoming total solar eclipses (but not for partial or annular eclipses), using color coding to show selected pairs of eclipses that repeat with the saros cycle.

## 2.4 The Ancient Mystery of the Planets

We've now covered the appearance and motion of the stars, Sun, and Moon in the sky. That leaves us with the planets to discuss. As you'll soon see, planetary motion posed an ancient mystery that played a critical role in the development of modern civilization.

**FIGURE 2.32** This photograph shows a grouping in our sky of all five planets that are easily visible to the naked eye. It was taken near Chatsworth, New Jersey, on April 23, 2002. The next such close grouping of these five planets in our sky will occur in September 2040.

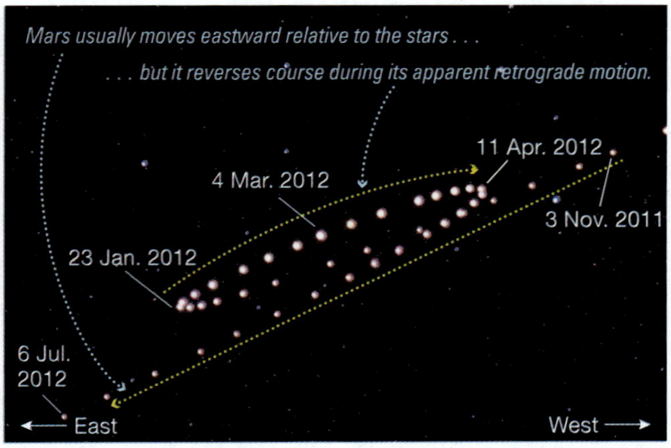


**FIGURE 2.33** This composite of images (taken at 5- to 7-day intervals in 2011 and 2012) shows a retrograde loop of Mars. Note that Mars is biggest and brightest in the middle of the retrograde loop, because that is where it is closest to Earth in its orbit.

Five planets are easy to find with the naked eye: Mercury, Venus, Mars, Jupiter, and Saturn. Mercury is visible infrequently, and only just after sunset or just before sunrise because it is so close to the Sun. Venus often shines brightly in the early evening in the west or before dawn in the east. If you see a very bright "star" in the early evening or predawn sky, it is probably Venus. Jupiter, when it is visible at night, is the brightest object in the sky besides the Moon and Venus. Mars is often recognizable by its reddish color, though you should check a star chart to make sure you aren't looking at a bright red star. Saturn is also easy to see with the naked eye, but because many stars are just as bright as Saturn, it helps to know where to look. (It also helps to know that planets tend not to twinkle as much as stars.) Sometimes several planets may appear close together in the sky, offering a particularly beautiful sight (**FIGURE 2.32**).

**See it for yourself** Using astronomical software or the Internet, find out what planets are visible tonight and where to look for them, then go out and try to find them. Are they easy or difficult to identify?

## Why was planetary motion so hard to explain?

Over the course of a single night, planets behave like all other objects in the sky: Earth's rotation makes them appear to rise in the east and set in the west. But if you continue to watch the planets night after night, you will notice that their movements among the constellations are quite complex. Instead of moving steadily eastward relative to the stars, like the Sun and Moon, the planets vary substantially in both speed and brightness; in fact, the word *planet* comes from a Greek term meaning "wandering star." Moreover, while the planets *usually* move eastward through the constellations, they occasionally reverse course, moving westward through the zodiac (**FIGURE 2.33**). These periods of **apparent retrograde motion** (*retrograde* means "backward") last from a few weeks to a few months, depending on the planet.

For ancient people who believed in an Earth-centered universe, apparent retrograde motion was very difficult to explain. After all, what could make planets sometimes turn around and go backward if everything moves in circles around Earth? The ancient Greeks came up with some very clever ways to explain it, but their explanations (which we'll study in Chapter 3) were quite complex.

In contrast, apparent retrograde motion has a simple explanation in a Sun-centered solar system. You can demonstrate it for yourself with the help of a friend (**FIGURE 2.34a**). Pick a spot in an open area to represent the Sun. You can represent Earth by walking counterclockwise around the Sun, while your friend represents a more distant planet (such as Mars or Jupiter) by walking in the same direction around the Sun at a greater distance. Your friend should walk more slowly than you, because more distant planets orbit the Sun more slowly. As you walk, watch how your friend appears to move relative to buildings or trees in the distance. Although both of you always walk the same way around the Sun, your friend will appear to move backward against the background during the part of your "orbit" in which you catch up to and pass him or her. **FIGURE 2.34b** shows how the same idea applies to Mars. Note that Mars never actually changes direction; it only *appears* to go backward as Earth passes Mars in its orbit. (To understand the apparent retrograde motions of Mercury and Venus, which are closer to the Sun than is Earth, simply switch places with your friend and repeat the demonstration.)

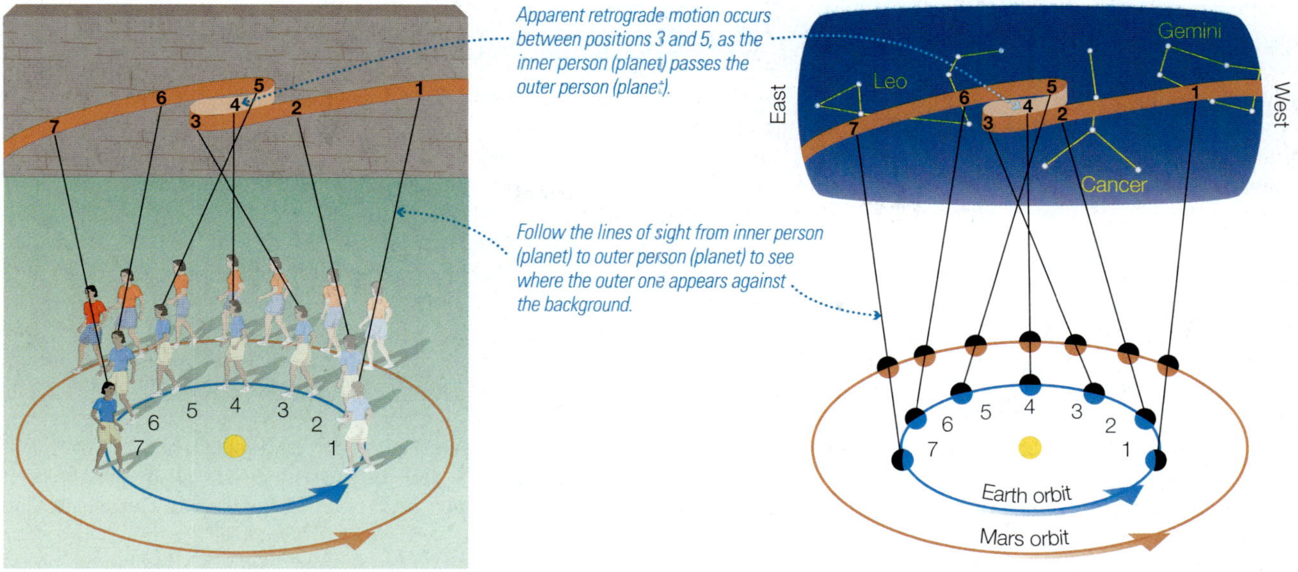


**a** The retrograde motion demonstration: Watch how your friend (in red) usually appears to move forward against the background of the building in the distance but appears to move backward as you (in blue) catch up to and pass her in your "orbit."

**b** This diagram shows the same idea applied to a planet. Follow the lines of sight from Earth to Mars in numerical order. Notice that Mars appears to move westward relative to the distant stars (from point 3 to point 5) as Earth passes it by in its orbit.

**FIGURE 2.34** Apparent retrograde motion—the occasional "backward" motion of the planets relative to the stars—has a simple explanation in a Sun-centered solar system.

**Stellar Parallax**

## Why did the ancient Greeks reject the real explanation for planetary motion?

If the apparent retrograde motion of the planets is so readily explained by recognizing that Earth orbits the Sun, why wasn't this idea accepted in ancient times? In fact, the idea that Earth goes around the Sun was suggested as early as 260 B.C. by the Greek astronomer Aristarchus (see the Special Topic, page 48). Nevertheless, Aristarchus's contemporaries rejected his idea, and the Sun-centered solar system did not gain wide acceptance until almost 2000 years later.

Although there were many reasons the Greeks were reluctant to abandon the idea of an Earth-centered universe, one of the most important was their inability to detect what we call **stellar parallax**. Extend your arm and hold up one finger. If you keep your finger still and alternately close your left eye and right eye, your finger will appear to jump back and forth against the background. This apparent shifting, called *parallax,* occurs because your two eyes view your finger from opposite sides of your nose. If you move your finger closer to your face, the parallax increases. If you look at a distant tree or flagpole instead of your finger, you may not notice any parallax at all. In other words, parallax depends on distance, with nearer objects exhibiting greater parallax than more distant objects.

If you now imagine that your two eyes represent Earth at opposite sides of its orbit around the Sun and that the tip of your finger represents a relatively nearby star, you have the idea of stellar parallax. Because we view the stars from different places in our orbit at different times of year, nearby stars should *appear* to shift back and forth against the background of more distant stars (**FIGURE 2.35**).

Because the Greeks believed that all stars lie on the same celestial sphere, they expected to see stellar parallax in a

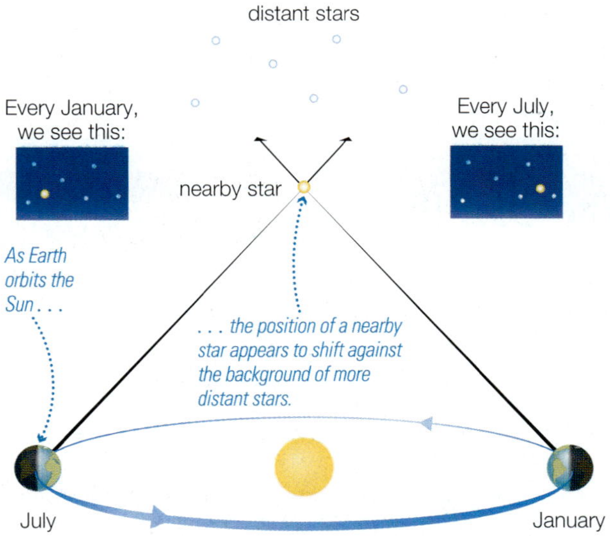


**FIGURE 2.35** Stellar parallax is an apparent shift in the position of a nearby star as we look at it from different places in Earth's orbit. This figure is greatly exaggerated; in reality, the amount of shift is far too small to detect with the naked eye.

## SPECIAL TOPIC Who First Proposed a Sun-Centered Solar System?

You've probably heard of Copernicus, whose work in the 16th century started the revolution that ultimately overturned the ancient belief in an Earth-centered universe [**Section 3.3**]. However, the idea that Earth goes around the Sun was proposed much earlier by the Greek scientist Aristarchus of Samos (c. 310–230 B.C.).

Little of Aristarchus's work survives to the present day, so we cannot know what motivated him to suggest an idea so contrary to the prevailing view of an Earth-centered universe. However, it's likely that he was motivated by the fact that a Sun-centered system offers a much more natural explanation for the apparent retrograde motion of the planets. To account for the lack of detectable stellar parallax, Aristarchus suggested that the stars were extremely far away.

Aristarchus further strengthened his argument by estimating the sizes of the Moon and the Sun. By observing the shadow of Earth on the Moon during a lunar eclipse, he estimated the Moon's diameter to be about one-third of Earth's diameter—only slightly more than the actual value. He then used a geometric argument, based on measuring the angle between the Moon and the Sun at first- and third-quarter phases, to conclude that the Sun must be larger than Earth. (Aristarchus's measurements were imprecise, so he estimated the Sun's diameter to be about 7 times Earth's rather than the correct value of about 100 times.) His conclusion that the Sun is larger than Earth may have been another reason he believed that Earth should orbit the Sun, rather than vice versa.

Although Aristarchus was probably the first to suggest that Earth orbits the Sun, his ideas built on the work of earlier scholars. For example, Heracleides (c. 388–315 B.C.) had previously suggested that Earth rotates, which offered Aristarchus a way to explain the daily circling of the sky in a Sun-centered system. Heracleides also suggested that not all heavenly bodies circle Earth: Based on the fact that Mercury and Venus always stay fairly close to the Sun in the sky, he argued that these two planets must orbit the Sun. In suggesting that *all* the planets orbit the Sun, Aristarchus was extending the ideas of Heracleides and others before him.

Aristarchus gained little support among his contemporaries, but his ideas never died, and Copernicus was aware of them when he proposed his own version of the Sun-centered system. Thus, our modern understanding of the universe owes at least some debt to the remarkable vision of a man born more than 2300 years ago.

slightly different way. If Earth orbited the Sun, they reasoned, at different times of year we would be closer to different parts of the celestial sphere and would notice changes in the angular separation of stars. However, no matter how hard they searched, they could find no sign of stellar parallax. They concluded that one of the following must be true:

1. Earth orbits the Sun, but the stars are so far away that stellar parallax is undetectable to the naked eye.
2. There is no stellar parallax because Earth remains stationary at the center of the universe.

Aside from a few notable exceptions, such as Aristarchus, the Greeks rejected the correct answer (the first one) because they could not imagine that the stars could be *that* far away. Today, we can detect stellar parallax with the aid of telescopes, providing direct proof that Earth really does orbit the Sun. Careful measurements of stellar parallax also provide the most reliable means of measuring distances to nearby stars [**Section 15.1**].

**Think about it** How far apart are opposite sides of Earth's orbit? How far away are the nearest stars? Using the 1-to-10-billion scale from Chapter 1, describe the challenge of detecting stellar parallax.

The ancient mystery of the planets drove much of the historical debate over Earth's place in the universe. In many ways, the modern technological society we take for granted today can be traced directly to the scientific revolution that began in the quest to explain the strange wanderings of the planets among the stars in our sky. We will turn our attention to this revolution in the next chapter.

## The BIG Picture PUTTING CHAPTER 2 INTO CONTEXT

*In this chapter, we surveyed the phenomena of our sky. Keep the following "big picture" ideas in mind as you continue your study of astronomy:*

- You can enhance your enjoyment of astronomy by observing the sky. The more you learn about the appearance and apparent motions of the sky, the more you will appreciate what you can see in the universe.
- From our vantage point on Earth, it is convenient to imagine that we are at the center of a great celestial sphere—even though we really are on a planet orbiting a star in a vast universe. We can then understand what we see in the local sky by thinking about how the celestial sphere appears from our latitude.
- Most of the phenomena of the sky are relatively easy to observe and understand. The more complex phenomena—particularly eclipses and apparent retrograde motion of the planets—challenged our ancestors for thousands of years. The desire to understand these phenomena helped drive the development of science and technology.

**MY COSMIC PERSPECTIVE** No matter how abstract or esoteric the study of astronomy may sometimes seem to be, you can always connect it back to your own personal experience of the sky around us.

# Summary of Key Concepts

## 2.1 Patterns in the Night Sky

- **What does the universe look like from Earth?** Stars and other celestial objects appear to lie on a great **celestial sphere** surrounding Earth. We divide the celestial sphere into **constellations** with well-defined borders. From any location on Earth, we see half the celestial sphere at any one time as the dome of our **local sky**, in which the **horizon** is the boundary between Earth and sky, the **zenith** is the point directly overhead, and the **meridian** runs from due south to due north through the zenith.

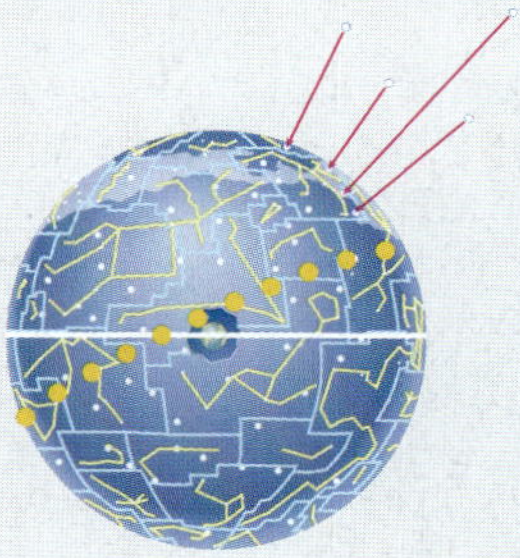

- **Why do stars rise and set?** Earth's rotation makes stars appear to circle around Earth each day. A star whose complete circle lies above our horizon is said to be **circumpolar**. Other stars have circles that cross the horizon, making them rise in the east and set in the west each day.

- **Why do the constellations we see depend on latitude and time of year?** The visible constellations vary with time of year because our night sky lies in different directions in space as we orbit the Sun. The constellations vary with **latitude** because your latitude determines the orientation of your horizon relative to the celestial sphere. The sky does not vary with **longitude**.

## 2.2 The Reason for Seasons

- **What causes the seasons?** The tilt of Earth's axis causes the seasons. The axis points in the same direction throughout the year, so as Earth orbits the Sun, sunlight hits different parts of Earth more directly at different times of year.

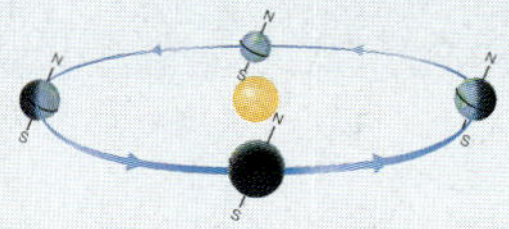

- **How does the orientation of Earth's axis change with time?** Earth's 26,000-year cycle of **precession** changes the orientation of the axis in space, although the tilt remains about $23\frac{1}{2}°$. The changing orientation of the axis does not affect the pattern of seasons, but it changes the identity of the North Star and shifts the locations of the solstices and equinoxes in Earth's orbit.

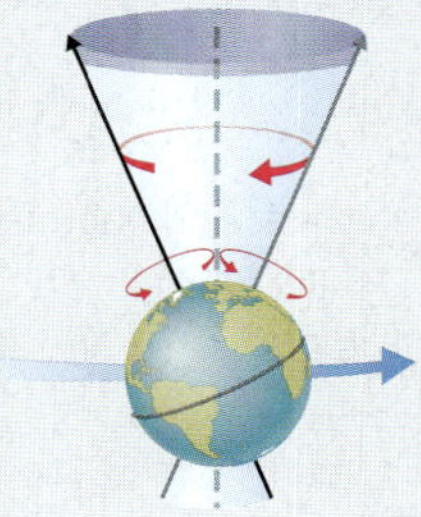

## 2.3 The Moon, Our Constant Companion

- **Why do we see phases of the Moon?** The **phase** of the Moon depends on its position relative to the Sun as it orbits Earth. The half of the Moon facing the Sun is always illuminated while the other half is dark, but from Earth we see varying combinations of the illuminated and dark faces.

- **What causes eclipses?** We see a **lunar eclipse** when Earth's shadow falls on the Moon and a **solar eclipse** when the Moon blocks our view of the Sun. We do not see an eclipse at every new and full moon because the Moon's orbit is slightly inclined to the ecliptic plane.

## 2.4 The Ancient Mystery of the Planets

- **Why was planetary motion so hard to explain?** Planets generally move eastward relative to the stars over the course of the year, but for weeks or months they reverse course during periods of **apparent retrograde motion**. This motion occurs when Earth passes by (or is passed by) another planet in its orbit, but it posed a major mystery to ancient people who assumed Earth to be at the center of the universe.

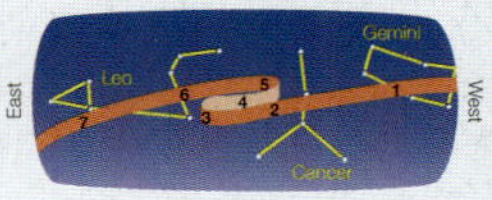


- **Why did the ancient Greeks reject the real explanation for planetary motion?** The Greeks rejected the idea that Earth goes around the Sun in part because they could not detect **stellar parallax**—slight apparent shifts in stellar positions over the course of the year. To most Greeks, it seemed unlikely that the stars could be so far away as to make parallax undetectable to the naked eye, even though that is, in fact, the case.

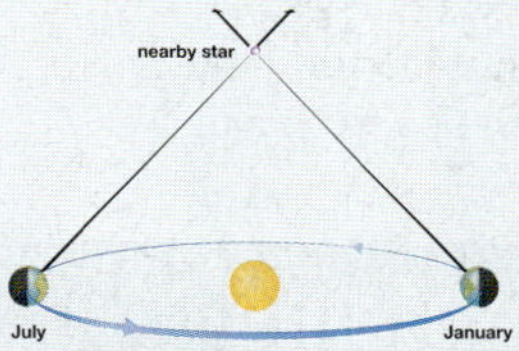

# Visual Skills Check

*Use the following questions to check your understanding of some of the many types of visual information used in astronomy. For additional practice, try the Chapter 2 Visual Quiz in the Study Area at www.MasteringAstronomy.com.*

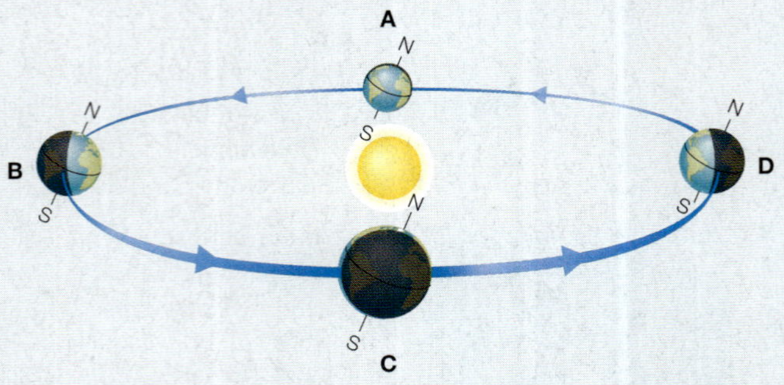


*The figure above is a typical diagram used to describe Earth's seasons.*

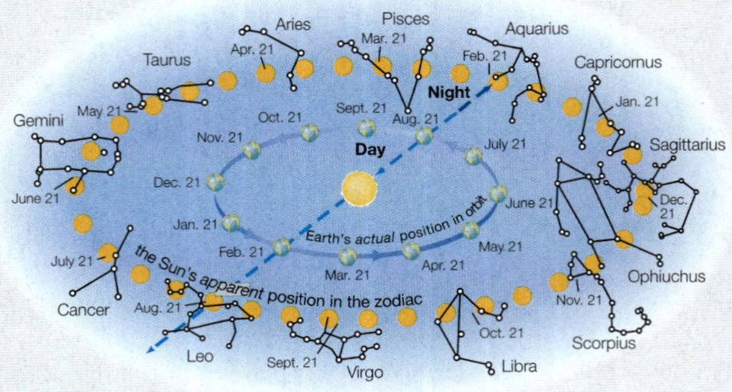


*The figure above (based on Figure 2.14) shows the Sun's path through the constellations of the zodiac.*

1. Which of the four labeled points (A through D) represents the day with the most hours of daylight for the Northern Hemisphere?
2. Which of the four labeled points represents the day with the most hours of daylight for the Southern Hemisphere?
3. Which of the four labeled points represents the beginning of spring for the Southern Hemisphere?
4. The diagram exaggerates the sizes of Earth and the Sun relative to the orbit. If Earth were correctly scaled relative to the orbit in the figure, how big would it be?
   a. about half the size shown b. about 2 millimeters across
   c. about 0.1 millimeter across d. microscopic
5. Given that Earth's actual distance from the Sun varies by less than 3% over the course of a year, why does the diagram look so elliptical?
   a. It correctly shows that Earth is closest to the Sun at points A and C and farthest at points B and D.
   b. The elliptical shape is an effect of perspective, since the diagram shows an almost edge-on view of a nearly circular orbit.
   c. The shape of the diagram is meaningless and is done only for artistic effect.
6. As viewed from Earth, in which zodiac constellation does the Sun appear to be located on April 21?
   a. Leo b. Aquarius
   c. Libra d. Aries
7. If the date is April 21, what zodiac constellation will be visible on your meridian at midnight?
   a. Leo b. Aquarius
   c. Libra d. Aries
8. If the date is April 21, what zodiac constellation will you see setting in the west shortly after sunset?
   a. Scorpius b. Pisces
   c. Taurus d. Virgo

# Exercises and Problems

*For instructor-assigned homework and other learning materials, go to www.MasteringAstronomy.com.*

## Chapter Review Questions

### Short-Answer Questions Based on the Reading

1. What are *constellations*? How did they get their names?
2. Suppose you were making a model of the celestial sphere with a ball. Briefly describe all the things you would need to mark on your celestial sphere.
3. On a clear, dark night, the sky may appear to be "full" of stars. Does this appearance accurately reflect the way stars are distributed in space? Explain.
4. Why does the *local sky* look like a dome? Define *horizon, zenith,* and *meridian.* How do we describe the location of an object in the local sky?
5. Explain why we can measure only *angular sizes* and *angular distances* for objects in the sky. What are *arcminutes* and *arcseconds*?
6. What are *circumpolar stars*? Are more stars circumpolar at the North Pole or in the United States? Explain.
7. What are *latitude* and *longitude*? Does the sky vary with latitude? Does it vary with longitude? Explain.
8. What is the *zodiac,* and why do we see different parts of it at different times of year?
9. Suppose Earth's axis had no tilt. Would we still have seasons? Why or why not?
10. Briefly describe key facts about the solstices and equinoxes.
11. What is *precession*? How does it affect what we see in our sky?
12. Briefly describe the Moon's cycle of *phases.* Can you ever see a full moon at noon? Explain.
13. Why do we always see the same face of the Moon?
14. Why don't we see an *eclipse* at every new and full moon? Describe the conditions needed for a *solar* or *lunar eclipse.*
15. What do we mean by the *apparent retrograde motion* of the planets? Why was this motion difficult for ancient astronomers to explain? How do we explain it today?
16. What is *stellar parallax*? How did an inability to detect it support the ancient belief in an Earth-centered universe?

### Does It Make Sense?

*Decide whether or not each of the following statements makes sense (or is clearly true or false). Explain clearly; not all of these have definitive answers, so your explanation is more important than your chosen answer.*

17. The constellation Orion didn't exist when my grandfather was a child.
18. When I looked into the dark lanes of the Milky Way with my binoculars, I saw a cluster of distant galaxies.
19. Last night the Moon was so big that it stretched for a mile across the sky.
20. I live in the United States, and during a trip to Argentina I saw many constellations that I'd never seen before.
21. Last night I saw Jupiter in the middle of the Big Dipper. (*Hint:* Is the Big Dipper part of the zodiac?)
22. Last night I saw Mars move westward through the sky in its apparent retrograde motion.
23. Although all the known stars rise in the east and set in the west, we might someday discover a star that will rise in the west and set in the east.
24. If Earth's orbit were a perfect circle, we would not have seasons.
25. Because of precession, someday it will be summer everywhere on Earth at the same time.
26. This morning I saw the full moon setting at about the same time the Sun was rising.

### Quick Quiz

*Choose the best answer to each of the following. For additional practice, try the Chapter 2 Reading and Concept Quizzes in the Study Area at www.MasteringAstronomy.com.*

27. Two stars that are in the same constellation (a) must both be part of the same cluster of stars in space. (b) must both have been discovered at about the same time. (c) may actually be very far away from each other.
28. The north celestial pole is 35° above your northern horizon. This tells you that you are at (a) latitude 35°N. (b) longitude 35°E. (c) latitude 35°S.
29. Beijing and Philadelphia have about the same latitude but different longitudes. Therefore, tonight's night sky in these two places will (a) look about the same. (b) have completely different sets of constellations. (c) have partially different sets of constellations.
30. In winter, Earth's axis points toward the star Polaris. In spring, the axis points toward (a) Polaris. (b) Vega. (c) the Sun.
31. When it is summer in Australia, the season in the United States is (a) winter. (b) summer. (c) spring.
32. If the Sun rises precisely due east, (a) you must be located at Earth's equator. (b) it must be the day of either the March or the September equinox. (c) it must be the day of the June solstice.
33. A week after full moon, the Moon's phase is (a) first quarter. (b) third quarter. (c) new.
34. The fact that we always see the same face of the Moon tells us that the Moon (a) does not rotate. (b) rotates with the same period that it orbits Earth. (c) looks the same on both sides.
35. If there is going to be a total lunar eclipse tonight, then you know that (a) the Moon's phase is full. (b) the Moon's phase is new. (c) the Moon is unusually close to Earth.
36. When we see Saturn going through a period of apparent retrograde motion, it means (a) Saturn is temporarily moving backward in its orbit of the Sun. (b) Earth is passing Saturn in its orbit, with both planets on the same side of the Sun. (c) Saturn and Earth must be on opposite sides of the Sun.

### Inclusive Astronomy

*Use these questions to reflect on participation in science.*

37. *Cultural Constellations.* Many cultures have created their own sets of constellations that differ from those used officially in astronomical research. Learn about the constellations of a particular culture of interest to you. How do the patterns of your chosen constellation(s) relate to the patterns of the official constellations listed in Appendix H? Which set of patterns makes more sense to you personally?
38. *Group Discussion: Sharing the Sky.* Astronomers around the world are fond of saying "we all share the same sky." Gather in groups of two to four students to discuss the meaning of this statement.
    **a.** Give each group member a chance to describe the similarities and differences you would expect between the sky you see during a 24-hour period and the sky seen by someone on the opposite side of the world or on the other side of the equator.
    **b.** Discuss whether you think that people making independent studies of the sky in different locations would come to similar conclusions about how the universe works. Would their conclusions be more accurate if they shared perspectives with people from other places? Why or why not?
    **c.** Discuss whether the skies that people saw 2000 years ago differ in any significant ways from the skies we see today.
    **d.** Do you think that people living 2000 years from now will have experiences of the sky that are similar to your own?

### The Process of Science

*These questions may be answered individually in short-essay form or discussed in groups, except where identified as group-only.*

39. *Earth-Centered or Sun-Centered?* Decide whether each of the following phenomena is consistent or inconsistent with a belief in an Earth-centered system. If consistent, describe how. If inconsistent, explain why, and also explain why the inconsistency did not immediately lead people to abandon the Earth-centered model.
    **a.** The daily paths of stars through the sky **b.** Seasons
    **c.** Phases of the Moon **d.** Eclipses
    **e.** Apparent retrograde motion of the planets
40. *Shadow Phases.* Many people incorrectly guess that the phases of the Moon are caused by Earth's shadow falling on the Moon. How would you convince a friend that the phases of the Moon have nothing to do with Earth's shadow? Describe the observations you would use to show that Earth's shadow can't be the cause of phases.
41. *Earth-Centered Language.* Many common phrases reflect the ancient Earth-centered view of our universe. For example, although we now know that day and night arise from Earth's rotation, we speak of "sunrise" or "sunset" as though the Sun were moving around us daily. Identify other common phrases that imply an Earth-centered viewpoint. Do you think this language creates any difficulties in teaching science? Why or why not?
42. *A Flat Earth?* A few relatively famous people have recently made the claim that the Earth is flat. Working in small groups, first find out the basis on which they make these claims, then make a list of key observations that refute the flat Earth claims. Based on what you learn, how can you tell which websites discussing flat Earth claims are scientifically valid and which are not, and what lessons the flat Earth

claims may contribute to the more general issue of "fake news" in our modern society?

43. *Group Activity: Lunar Phases and Time of Day.* Make a copy of the diagram below representing the Moon's orbit as seen from above Earth's North Pole. Note: You may wish to do this activity using the four roles described in Chapter 1, Exercise 39; you may also find it useful to watch the video "Moon Phases, Part 2."

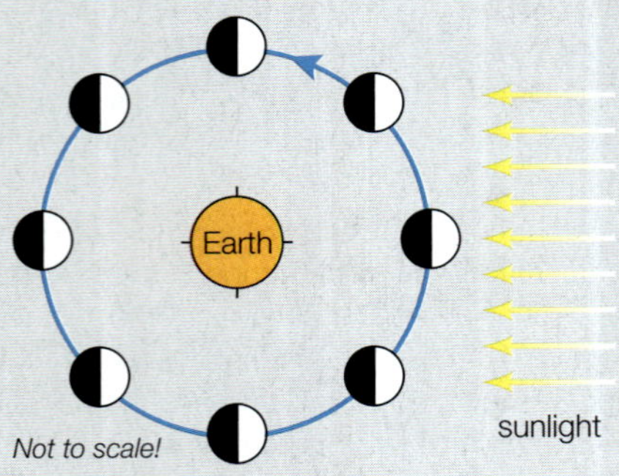


**a.** Label each of the eight Moon positions with the phase that it represents.
**b.** What time of day corresponds to each of the four tick marks on Earth? Label each tick mark accordingly.
**c.** Why doesn't the Moon's phase change during the course of one night? Explain your reasoning.
**d.** At what times of day would a full moon be visible to someone on Earth? Write down when a full moon rises and explain why it appears to rise at that time.
**e.** At what times of day would a third-quarter moon be visible to someone on Earth? Write down when a third-quarter moon sets and explain why it appears to set at that time.
**f.** At what times of day would a waxing crescent moon be visible to someone on Earth? Write down when a waxing crescent moon rises and explain why it appears to rise at that time.

## Investigate Further

### Short-Answer/Essay Questions

44. *New Planet.* A planet in another solar system has a circular orbit and an axis tilt of 35°. Would you expect this planet to have seasons? If so, would you expect them to be more extreme than the seasons on Earth? If not, why not?
45. *Your View of the Sky.*
**a.** What are your latitude and longitude?
**b.** Where does the north (or south) celestial pole appear in your sky?
**c.** Is Polaris a circumpolar star in your sky? Explain.
46. *View from the Moon.* Assume you live on the Moon, near the center of the face that looks toward Earth.
**a.** Suppose you see a full earth in your sky. What phase of the Moon would people on Earth see? Explain.
**b.** Suppose people on Earth see a full moon. What phase would you see for Earth? Explain.
**c.** Suppose people on Earth see a waxing gibbous moon. What phase would you see for Earth? Explain.
**d.** Suppose people on Earth are viewing a total lunar eclipse. What would you see from your home on the Moon? Explain.
47. *View from the Sun.* Suppose you lived on the Sun (and could ignore the heat). Would you still see the Moon go through phases as it orbits Earth? Why or why not?
48. *A Farther Moon.* Suppose the distance to the Moon were twice its actual value. Would it still be possible to have a total solar eclipse? Why or why not?
49. *A Smaller Earth.* Suppose Earth were smaller. Would solar eclipses be any different? If so, how? What about lunar eclipses?
50. *Project: Observing Planetary Motion.* Find out which planets are currently visible in your evening sky. At least once a week, observe the planets and draw a diagram showing the position of each visible planet relative to stars in a zodiac constellation. From week to week, note how the planets are moving relative to the stars. Can you see any of the apparently wandering features of planetary motion? Explain.
51. *Project: Eclipse Trip.* Find details about a future total solar eclipse that you may be able to observe. Create a plan for a trip to see the eclipse, including details of where you will view it, how you will get there, and what you should expect to see.
52. *Project: A Connecticut Yankee.* Read *A Connecticut Yankee in King Arthur's Court* by Mark Twain. What role does an eclipse play in the story? Discuss its significance.

### Quantitative Problems

*Be sure to show all calculations clearly and state your final answers in complete sentences.*

53. *Arcminutes and Arcseconds.* There are 360° in a full circle.
**a.** How many arcminutes are in a full circle?
**b.** How many arcseconds are in a full circle?
**c.** The Moon's angular size is about $\frac{1}{2}$°. What is this in arcminutes? In arcseconds?
54. *Latitude Distance.* Earth's radius is approximately 6370 km.
**a.** What is Earth's circumference?
**b.** What distance is represented by each degree of latitude?
**c.** What distance is represented by each arcminute of latitude?
**d.** Can you give similar answers for the distances represented by a degree or arcminute of longitude? Why or why not?
55. *Angular Conversions I.* The following angles are given in degrees and fractions of degrees. Rewrite them in degrees, arcminutes, and arcseconds.
**a.** 24.3° **b.** 1.59° **c.** 0.1° **d.** 0.01° **e.** 0.001°
56. *Angular Conversions II.* The following angles are given in degrees, arcminutes, and arcseconds. Rewrite them in degrees and fractions of degrees.
**a.** 7°38′42″ **b.** 12′54″ **c.** 1°59′59″ **d.** 1′ **e.** 1″
57. *Sun Diameter.* Use the Sun's approximate distance of 150 million km and angular diameter of about 0.5° to calculate the Sun's physical diameter. Compare your answer to the actual value of 1,390,000 km.
58. *Betelgeuse Diameter.* Estimate the diameter of the supergiant star Betelgeuse from its angular diameter of 0.05 arcsecond and distance of about 600 light-years. Compare your answer to the size of our Sun and the Earth-Sun distance.
59. *Eclipse Conditions.* The Moon's precise equatorial diameter is 3476 km, and its orbital distance from Earth varies between 356,400 and 406,700 km. The Sun's diameter is 1,390,000 km, and its distance from Earth ranges between 147.5 and 152.6 million km.
**a.** Find the Moon's angular size at its minimum and maximum distances from Earth.
**b.** Find the Sun's angular size at its minimum and maximum distances from Earth.
**c.** Based on your answers to parts a and b, is it possible to have a total solar eclipse when the Moon and Sun are both at their maximum distance? Explain.

# 3 The Science of Astronomy

▲ **About the photo:** Astronaut Bruce McCandless orbits Earth as if he were a tiny moon during Space Shuttle mission STS-41-B.

## LEARNING GOALS

**3.1 The Ancient Roots of Science**
- In what ways do all humans use scientific thinking?
- How is modern science rooted in ancient astronomy?

**3.2 Ancient Greek Science**
- Why does modern science trace its roots to the Greeks?
- How did the Greeks explain planetary motion?

**3.3 The Copernican Revolution**
- How did Copernicus, Tycho, and Kepler challenge the Earth-centered model?
- What are Kepler's three laws of planetary motion?
- How did Galileo solidify the Copernican revolution?

**3.4 The Nature of Science**
- How can we distinguish science from nonscience?
- What is a scientific theory?

**3.5 Astrology**
- How is astrology different from astronomy?
- Does astrology have any scientific validity?

*We especially need imagination in science. It is not all mathematics, nor all logic, but is somewhat beauty and poetry.*

*—Maria Mitchell (1818–1889), astronomer and the first woman elected to the American Academy of Arts and Sciences*

**Chapter 3 Overview**

Today we know that Earth is a planet orbiting a rather ordinary star, in a galaxy of more than a hundred billion stars, in an incredibly vast universe. We know that Earth, along with the entire cosmos, is in constant motion. We know that, on the scale of cosmic time, human civilization has existed for only the briefest moment. How did we manage to learn these things?

It wasn't easy. In this chapter, we will trace how modern astronomy grew from its roots in ancient observations, including those of the Greeks. We'll discuss the Copernican revolution, which overturned the ancient belief in an Earth-centered universe and laid the foundation for the rise of our technological civilization. Finally, we'll explore the nature of modern science and how science can be distinguished from nonscience.

## 3.1 The Ancient Roots of Science

The rigorous methods of modern science have proven to be among the most valuable inventions in human history. These methods have enabled us to discover almost everything we now know about nature and the universe, and they also have made our modern technology possible. In this section, we will explore the ancient roots of science, which grew out of experiences common to nearly all people and all cultures.

### In what ways do all humans use scientific thinking?

Scientific thinking comes naturally to us. By about a year of age, a baby notices that objects fall to the ground when she drops them. She lets go of a ball—it falls. She pushes a plate of food from her high chair—it falls, too. She continues to drop all kinds of objects, and they all plummet to Earth. Through her powers of observation, the baby learns about the physical world, finding that things fall when they are unsupported. Eventually, she becomes so certain of this fact that, to her parents' delight, she no longer needs to test it continually.

One day someone gives the baby a helium balloon. She releases it, and to her surprise it rises to the ceiling! Her understanding of nature must be revised. She now knows that the principle "all things fall" does not represent the whole truth, although it still serves her quite well in most situations. It will be years before she learns enough about the atmosphere, the force of gravity, and the concept of density to understand *why* the balloon rises when most other objects fall. For now, she is delighted to observe something new and unexpected.

The baby's experience with falling objects and balloons exemplifies scientific thinking. In essence, science is a way of learning about nature through careful observation and trial-and-error experiments. Rather than thinking differently than other people, modern scientists simply are trained to organize everyday thinking in a way that makes it easier for them to share their discoveries and use their collective wisdom.

**Think about it** Describe a few cases where you have learned by trial and error while cooking, participating in sports, fixing something, or working at a job.

Just as learning to communicate through language, art, or music is a gradual process for a child, the development of science has been a gradual process for humanity. Science in its modern form requires painstaking attention to detail, relentless testing of each piece of information to ensure its reliability, and a willingness to give up old beliefs that are not consistent with observed facts about the physical world. For professional scientists, these demands are the "hard work" part of the job. At heart, professional scientists are like the baby with the balloon, delighted by the unexpected and motivated by those rare moments when they—and all of us—learn something new about the universe.

### How is modern science rooted in ancient astronomy?

Astronomy has been called the oldest of the sciences, because its roots stretch deepest into antiquity. Ancient civilizations did not always practice astronomy in the same ways or for the same reasons that we study it today, but they nonetheless had some amazing achievements. Understanding this ancient astronomy can give us a greater appreciation of how and why science developed through time.

**Practical Benefits of Astronomy** Humans have been making careful observations of the sky for many thousands of years. Part of the reason for this interest in astronomy probably comes from our inherent curiosity as humans, but ancient cultures also discovered that astronomy had practical benefits for timekeeping, keeping track of seasonal changes, and navigation.

One amazing example comes from people of central Africa. Although we do not know exactly when they developed the skill, people in some regions learned to predict rainfall patterns through observations of the Moon. **FIGURE 3.1** shows how the method works. The orientation of the "horns" of a waxing crescent moon (relative to the horizon) varies over the course of the year, primarily because the angle at which the ecliptic intersects the horizon changes during the year. (The orientation also depends on latitude.) In tropical regions in which there are distinct rainy and dry seasons—rather than the four seasons familiar at temperate latitudes—the orientation of the crescent moon can be used to predict how much rainfall should be expected over coming days and weeks.

**Astronomy and Measures of Time** The impact of ancient astronomical observations is still with us in our modern measures of time. The length of our day is the time it takes the Sun to make one full circuit of the sky. The length of a

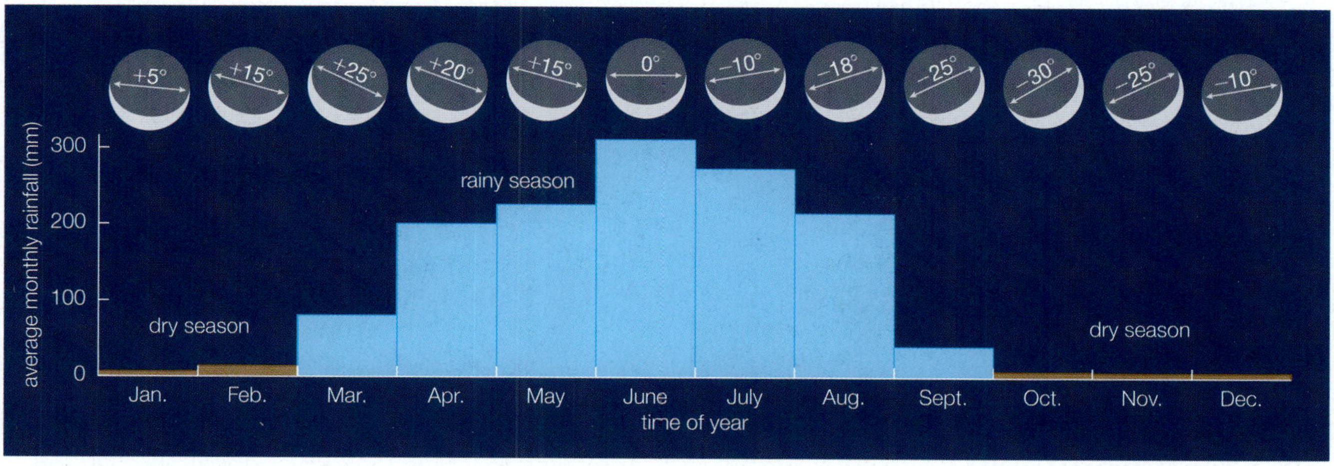


**FIGURE 3.1** In central Nigeria, the orientation of the "horns" of a waxing crescent moon (shown along the top) correlates with the average amount of rainfall at different times of year. Local people could use this fact to predict the weather with reasonable accuracy. (Adapted from *Ancient Astronomers* by Anthony F. Aveni.)

month comes from the Moon's cycle of phases [**Section 2.3**], and our year is based on the cycle of the seasons [**Section 2.2**]. The seven days of the week were named after the seven "planets" of ancient times (**TABLE 3.1**), which were the Sun, the Moon, and the five planets that are easily visible to the naked eye: Mercury, Venus, Mars, Jupiter, Saturn. Note that the ancient definition of *planet* (which meant "wandering star") applied to any object that appeared to wander among the fixed stars. That is why the Sun and Moon were on the list while Earth was not, because we don't see our own planet moving in the sky.

**Think about it** Uranus is faintly visible to the naked eye, but it was not recognized as a planet in ancient times. If Uranus had been brighter, would we now have eight days in a week? Defend your opinion.

Because timekeeping was so important and required precise observations, many ancient cultures built structures or created special devices to help with it. Let's briefly investigate a few of the ways that ancient cultures kept track of time.

**TABLE 3.1 The Seven Days of the Week and the Astronomical Objects They Honor**

*The seven days were originally linked directly to the seven objects. The correspondence is no longer perfect, but the pattern is clear in many languages; some English names come from Germanic gods.*

| Object | Germanic God | English | French | Spanish |
|---|---|---|---|---|
| Sun | — | Sunday | dimanche | domingo |
| Moon | — | Monday | lundi | lunes |
| Mars | Tiw | Tuesday | mardi | martes |
| Mercury | Woden | Wednesday | mercredi | miércoles |
| Jupiter | Thor | Thursday | jeudi | jueves |
| Venus | Fria | Friday | vendredi | viernes |
| Saturn | — | Saturday | samedi | sábado |

**Determining the Time of Day** In the daytime, ancient peoples could tell time by observing the Sun's path through the sky. Many cultures probably used the shadows cast by sticks as simple sundials [**Section S1.3**]. The ancient Egyptians built huge obelisks, often decorated in homage to the Sun, which probably also served as simple clocks (**FIGURE 3.2**).

At night, ancient people could estimate the time from the position and phase of the Moon (see Figure 2.22) or by observing the constellations visible at a particular time (see Figure 2.14). For example, ancient Egyptian star clocks, often found painted on the coffin lids of Egyptian pharaohs, cataloged where particular stars appeared in the sky at various times of night throughout the year. By knowing the date from their calendar and observing the positions of the cataloged stars in the sky, the Egyptians could use the star clocks to estimate the time of night.

We also trace the origins of our modern clock to ancient Egypt. Some 4000 years ago, the Egyptians divided daytime

**FIGURE 3.2** This ancient Egyptian obelisk resides in St. Peter's Square at the Vatican in Rome. It is one of 21 surviving Egyptian obelisks. Shadows cast by the obelisks may have been used to tell time.

a The remains of Stonehenge today.

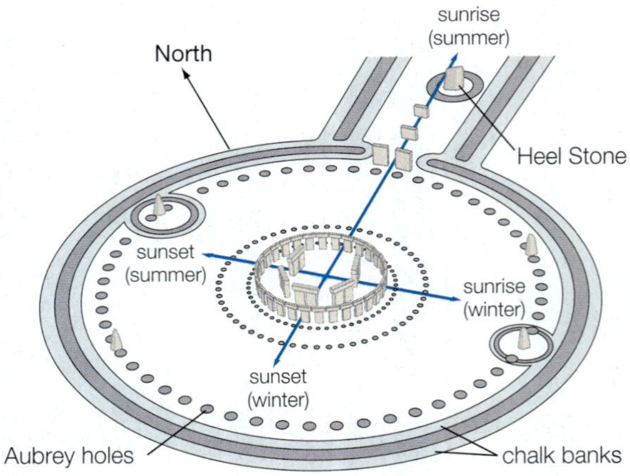


b This sketch shows how archaeologists believe Stonehenge looked upon its completion in about 1550 B.C. Several astronomical alignments are shown as they appear from the center. For example, the Sun rises directly over the Heel Stone on the summer solstice.

**FIGURE 3.3** Stonehenge, in southern England, was built in stages from about 2750 B.C. to about 1550 B.C.

and nighttime into 12 equal parts each, which is how we got our 12 hours each of a.m. and p.m. The abbreviations *a.m.* and *p.m.* stand for the Latin terms *ante meridiem* and *post meridiem*, respectively, which mean "before the middle of the day" and "after the middle of the day."

By about 1500 B.C., Egyptians had abandoned star clocks in favor of clocks that measure time by the flow of water through an opening of a particular size, just as hourglasses measure time by the flow of sand through a narrow neck.* These *water clocks* had the advantage of working even when the sky was cloudy. They eventually became the primary timekeeping instruments for many cultures, including the Greeks, Romans, and Chinese. Water clocks, in turn, were replaced by mechanical clocks in the 17th century and by electronic clocks in the 20th century.

**Marking the Seasons** Many ancient cultures built structures to help them mark the seasons. Stonehenge (**FIGURE 3.3**) is a well-known example that served both as an astronomical device and as a social and religious gathering place. In the Americas, one of the most spectacular structures was the Templo Mayor (**FIGURE 3.4**) in the Aztec city of Tenochtitlán (in modern-day Mexico City), which featured twin temples on a flat-topped pyramid. From the vantage point of a royal observer watching from the opposite side of the plaza, the Sun rose through the notch between the temples on the equinoxes. Before the Conquistadors destroyed it, Spanish visitors reported elaborate rituals at the Templo Mayor, sometimes including human sacrifice, that were held at times determined by astronomical observations. After its destruction, stones from the Templo Mayor were used to build a cathedral in the great plaza of Mexico City.

Many cultures aligned buildings and streets with the cardinal directions (north, south, east, and west), which made it easier to keep track of the rise and set positions of the Sun over the course of the year. This type of alignment is found at such diverse sites as the Egyptian pyramids and the Forbidden City in China and among ceremonial kivas built by the Ancestral Pueblo People of the American southwest (**FIGURE 3.5**). Many modern cities retain this layout, which is why you'll find so many streets that run directly north-south or east-west.

Other structures marked special dates such as the winter or summer solstice. Many such structures can be found around the world, but one of the most amazing is the *Sun Dagger*, made by the Ancestral Pueblo People in Chaco Canyon, New Mexico (**FIGURE 3.6**). Three large slabs of rock in front of a carved spiral produced special patterns of light and shadow at different times of year. For example, a single dagger of sunlight pierced the center of the spiral only at noon on the summer solstice, while two daggers of light bracketed the spiral at the winter solstice.

The Sun Dagger may also have been used to mark a special cycle of the Moon that had ritual significance to the Ancestral Pueblo People. The rise and set positions of the

**FIGURE 3.4** This scale model shows the Templo Mayor and the surrounding plaza as they are thought to have looked.

*Hourglasses using sand were not invented until about the 8th century A.D., long after the advent of water clocks. Natural sand grains vary in size, so making accurate hourglasses required technology for making uniform grains of sand.

**FIGURE 3.5** This large structure, more than 20 meters in diameter, is a kiva in Chaco Canyon, New Mexico. It was built by Ancestral Pueblo People approximately 1000 years ago. Its main axis is aligned almost precisely north-south.

full moon vary in an 18.6-year cycle (the cycle of "precession of the nodes" around the Moon's orbit), so the full moon rises at its most southerly point along the eastern horizon only once every 18.6 years. At this time, known as a "major lunar standstill," a shadow cast by the slabs of rock under the light of the full moon lies tangent to the edge of the spiral in the Sun Dagger; then, 9.3 years later, a similar lunar shadow cuts through the center of the spiral. The major lunar standstill can also be observed with structures at nearby Chimney Rock and in cliff dwellings at Colorado's Mesa Verde National Park.

**Think about it** Review the meaning of the *nodes* of the Moon's orbit and how they precess in an 18.6-year cycle (see Section 2.3). Comment on the sophistication required to have discovered this cycle and built structures to observe phenomena like the major lunar standstill. How is this cycle related to the slightly shorter *saros cycle*?

**FIGURE 3.6** The Sun Dagger. Three large slabs of rock in front of the carved spiral produced patterns of light and shadow that varied throughout the year. Here, we see the single dagger of sunlight that pierced the center of the spiral only at noon on the summer solstice. (Unfortunately, within just 12 years of the site's 1977 discovery, the rocks shifted—probably as a result of erosion of the trail below caused by large numbers of visitors—so the effect no longer occurs.)

**Solar and Lunar Calendars** The tracking of the seasons eventually led to the advent of written calendars. Today, we use a *solar calendar*, meaning a calendar that is synchronized with the seasons so that seasonal events such as the solstices and equinoxes occur on approximately the same dates each year [**Section S1.1**]. However, recall that the length of our month comes from the Moon's $29\frac{1}{2}$-day cycle of phases. Some cultures therefore created *lunar calendars* that aimed to stay synchronized with the lunar cycle, so that the Moon's phase was always the same on the first day of each month.

A basic lunar calendar has 12 months, with some months lasting 29 days and others lasting 30 days; the lengths are chosen to make the average agree with the approximately $29\frac{1}{2}$-day lunar cycle. A 12-month lunar calendar therefore has 354 or 355 days, or about 11 days fewer than a calendar based on the Sun. Such a calendar is still used in the Muslim religion. That is why the month-long fast of Ramadan (the 9th month) begins about 11 days earlier with each subsequent year.

It's possible to keep lunar calendars roughly synchronized with solar calendars by taking advantage of a timing coincidence: 19 years on a solar calendar is almost precisely 235 months on a lunar calendar. As a result, the lunar phases repeat on the same solar dates about every 19 years (a pattern known as the *Metonic cycle*, because it was recognized by the Greek astronomer Meton in 432 B.C.). For example, there was a full moon on December 22, 2018, and there will be a full moon 19 years later, on December 22, 2037. Because an ordinary lunar calendar has only $19 \times 12 = 228$ months in a 19-year period, adding 7 extra months (to make 235) can keep the lunar calendar roughly synchronized to the seasons. The Jewish calendar does this by adding a 13th month in the 3rd, 6th, 8th, 11th, 14th, 17th, and 19th years of each 19-year cycle. This scheme keeps the dates of Jewish holidays within about a 1-month range on a solar calendar, with precise dates repeating every 19 years. It also explains why the date of Easter changes from year to year: The New Testament ties the date of Easter to the Jewish festival of Passover. In a slight modification of the original scheme, most Western Christians now celebrate Easter on the first Sunday after the first full moon after March 21. If the full moon falls on Sunday, Easter is the following Sunday. (Eastern Orthodox churches calculate the date of Easter differently, because they base the date on the Julian rather than the Gregorian calendar [**Section S1.1**].)

**Learning About Ancient Achievements** The study of ancient astronomical achievements is a rich field of research. Many ancient cultures made careful observations of planets and stars, and some left remarkably detailed records. The Chinese, for example, began recording astronomical observations at least 5000 years ago, allowing ancient Chinese astronomers to make many important discoveries. By the 15th century, the Chinese had built a great observatory in Beijing, which still stands today (**FIGURE 3.7**). We can also study written records from ancient Middle Eastern civilizations such as those of Egypt and Babylonia.

**FIGURE 3.7** This photo shows a model of the celestial sphere and other instruments on the roof of the ancient astronomical observatory in Beijing. The observatory was built in the 15th century; the instruments shown here were built later and show a European influence brought by Jesuit missionaries.

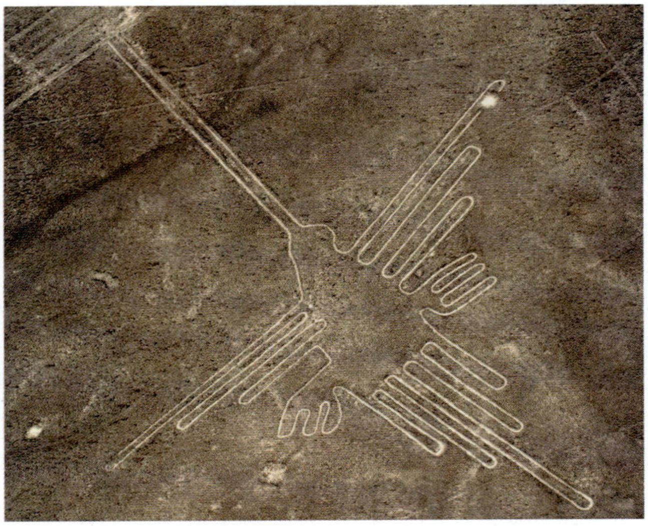

**FIGURE 3.8** Hundreds of lines and patterns are etched in the sand of the Nazca desert in Peru. This aerial photo shows a large figure of a hummingbird.

Other cultures either did not leave clear written records or had records that were lost or destroyed, so we must piece together their astronomical achievements by studying the physical evidence they left behind. This type of study is usually called *archaeoastronomy,* a word that combines archaeology and astronomy.

The cases we've discussed to this point have been fairly straightforward for archaeoastronomers to interpret, but many other cases are more ambiguous. For example, ancient people in what is now Peru etched hundreds of lines and patterns in the sand of the Nazca desert. Many of the lines point to places where the Sun or bright stars rise at particular times of year, but that doesn't prove anything: With hundreds of lines, random chance ensures that many will have astronomical alignments no matter how or why they were made. The patterns, many of which are large figures of animals (**FIGURE 3.8**), have evoked even more debate. Some people think they may be representations of constellations recognized by the people who lived in the region, but we do not know for sure.

**Think about it** Animal figures like that in Figure 3.8 show up clearly only when seen from above. As a result, some UFO enthusiasts argue that the patterns must have been created by aliens. What do you think of this argument? Defend your opinion.

In some cases, scientists studying archaeoastronomy can use other clues to establish the intentions of ancient builders. For example, lodges built by the Pawnee people in Kansas feature strategically placed holes for observing the passage of constellations that figure prominently in Pawnee folklore. The correspondence between the folklore and the structural features provides a strong case for deliberate intent rather than coincidence. Similarly, traditions of the Inca Empire of South America held that its rulers were descendants of the Sun and therefore demanded close watch of the movements of the Sun and stars. This fact supports the idea that astronomical alignments in Inca cities and ceremonial centers, such as the World Heritage Site of Machu Picchu (**FIGURE 3.9**), were deliberate rather than accidental.

A different type of evidence makes a convincing case for the astronomical sophistication of ancient Polynesians, who lived and traveled among the islands of the mid- and South Pacific. Navigation was crucial to their survival because the next island in a journey usually was too distant to be seen. The most esteemed position in Polynesian culture was that of the Navigator, a person who had acquired the knowledge necessary to navigate great distances among the islands. Navigators used detailed knowledge of astronomy for their broad navigational sense, and a deep understanding of wave and swell patterns to locate precise landing points (**FIGURE 3.10**). A Navigator memorized all his knowledge and passed it to the next generation through a well-developed training program. Unfortunately, with the advent of modern navigational technology, many of the skills of the Navigators have been lost.

**FIGURE 3.9** The World Heritage Site of Machu Picchu has structures aligned with sunrise at the winter and summer solstices.

**FIGURE 3.10** A Micronesian stick chart, an instrument used by Polynesian Navigators to represent swell patterns around islands.

## 3.2 Ancient Greek Science

Before a structure such as Stonehenge or the Templo Mayor could be built, careful observations had to be made and repeated over and over to ensure their accuracy. Careful, repeatable observations also underlie modern science. Elements of modern science were therefore present in many early human cultures. If the circumstances of history had been different, almost any culture might have been the first to develop what we consider to be modern science. In the end, however, history takes only one of countless possible paths. The path that led to modern science emerged from the ancient civilizations of the Mediterranean and the Middle East—especially from ancient Greece.

### Why does modern science trace its roots to the Greeks?

Greece gradually rose as a power in the Middle East beginning around 800 B.C. and was well established by about 500 B.C. Its geographical location placed it at a crossroads for travelers, merchants, and armies from northern Africa, Asia, and Europe. Building on the diverse ideas brought forth by the meeting of these many cultures, ancient Greek philosophers soon began their efforts to move human understanding of nature from the mythological to the rational.

**Three Philosophical Innovations** Greek philosophers developed at least three major innovations that helped pave the way for modern science. First, they developed a tradition of trying to understand nature without relying on supernatural explanations and of working communally to debate and challenge each other's ideas. Second, the Greeks used mathematics to give precision to their ideas, which allowed them to explore the implications of new ideas in much greater depth than would have otherwise been possible. Third, while much of their philosophical activity consisted of subtle debates grounded only in thought and was not scientific in the modern sense, the Greeks also saw the power of reasoning from observations. They understood that an explanation could not be right if it disagreed with observed facts.

**Models of Nature** Perhaps the greatest Greek contribution to science came from the way they synthesized all three innovations in creating models of nature, a practice that is central to modern science. Scientific models differ somewhat from the models you may be familiar with in everyday life. In our daily lives, we tend to think of models as miniature physical representations, such as model cars or airplanes. In contrast, a scientific **model** is a conceptual representation created to explain and predict observed phenomena. For example, a scientific model of Earth's climate uses logic and mathematics to represent what we know about how the climate works. Its purpose is to explain and predict climate changes, such as the changes that may occur with global warming. Just as a model airplane does not faithfully represent every aspect of a real airplane, a scientific model may not fully explain all our observations of nature. Nevertheless, even the failings of a scientific model can be useful, because they often point the way toward building a better model.

**From Greece to the Renaissance** The Greeks created models that sought to explain many aspects of nature, including the properties of matter and the principles of motion. For our purposes, the most important of the Greek models was their Earth-centered model of the universe. Before we turn to its details, however, it's worth briefly discussing how ancient Greek philosophy was passed to Europe, where it ultimately grew into the principles of modern science.

Greek philosophy first began to spread widely with the conquests of Alexander the Great (356–323 B.C.). Alexander had a deep interest in science, perhaps in part because Aristotle (see the Special Topic, page 61) had been his personal tutor. Alexander founded the city of Alexandria in Egypt, and his successors founded the renowned Library of Alexandria (**FIGURE 3.11**). Though it is sometimes difficult to distinguish fact from legend in stories of this great Library, there is little doubt that it was once the world's preeminent center of research, housing up to a half million books written on papyrus scrolls. Most were ultimately burned, their contents lost forever.

**Think about it** Estimate the number of books you're likely to read in your lifetime, and compare this number to the half million books that may once have been housed in the Library of Alexandria. Can you think of other ways to put into perspective the loss of ancient wisdom resulting from the destruction of the Library of Alexandria?

The details of the Library's destruction are hazy and subject to disagreement among historians, but the Library appears to have remained an important research center for several hundred years. One account holds that its demise was intertwined with the execution of a woman named Hypatia (A.D. 370–415) in A.D. 415. Hypatia was one of the few prominent female scholars of the ancient world, and some accounts attribute to her important discoveries in mathematics and astronomy. In commemoration of the ancient library, in 2003 Egypt opened a New Library of Alexandria (the *Bibliotheca Alexandrina*), with hopes of again making Alexandria a global center for scientific research.

**a** This painting shows a courtyard of the ancient Library of Alexandria, as it may have looked shortly after its completion.

**b** The New Library of Alexandria in Egypt, which opened in 2003.

**FIGURE 3.11** The ancient Library of Alexandria thrived for centuries, starting some time after about 300 B.C.

The relatively few books from the Library that survive today were preserved primarily thanks to the rise of a new center of intellectual inquiry in Baghdad (in present-day Iraq). As European civilization fell into the period of intellectual decline known as the Dark Ages, scholars of the new religion of Islam sought knowledge of mathematics and astronomy in hopes of better understanding the wisdom of Allah.

Around A.D. 800, the Islamic leader Al-Mamun (A.D. 786–833) established a "House of Wisdom" in Baghdad, where Islamic scholars—often working together with Jews and Christians—translated and thereby saved many ancient Greek works. These scholars were also in frequent contact with Hindu scholars from India, who in turn brought ideas and discoveries from China. Hence, the intellectual center in Baghdad achieved a synthesis of the surviving work of the ancient Greeks and that of the Indians and the Chinese. Using all these ideas as building blocks, scholars in the House of Wisdom developed the mathematics of algebra and many new instruments and techniques for astronomical observation. The latter explains why many official star names come from Arabic; for example, the names of many bright stars begin with *al* (e.g., Aldebaran, Algol), which means "the" in Arabic.

The accumulated knowledge of the Baghdad scholars spread throughout the Byzantine empire (part of the former Roman Empire). When the Byzantine capital of Constantinople (modern-day Istanbul) fell to the Turks in 1453, many Eastern scholars headed west to Europe, carrying with them the knowledge that helped ignite the European Renaissance.

## How did the Greeks explain planetary motion?

The Greek **geocentric model** of the cosmos—so named because it placed a spherical Earth at the center of the universe—developed gradually over a period of several

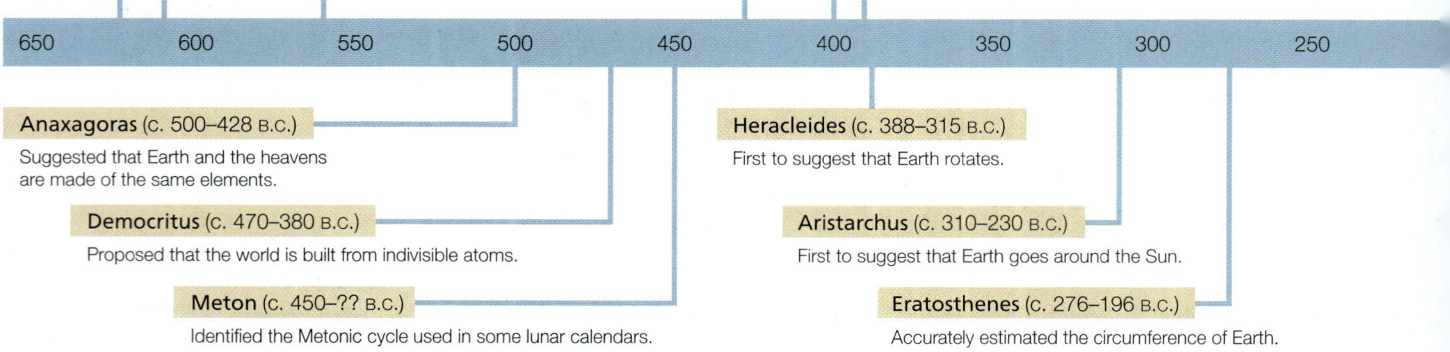

## SPECIAL TOPIC Aristotle

Aristotle (384–322 B.C.) is among the best-known philosophers of the ancient world. Both his parents died when he was a child, and he was raised by a family friend. In his 20s and 30s, he studied under Plato (428–348 B.C.) at Plato's Academy. He later founded his own school, called the Lyceum, where he studied and lectured on virtually every subject. Historical records tell us that his lectures were collected and published in 150 volumes. About 50 of these volumes survive to the present day.

Many of Aristotle's discoveries concerned the nature of plants and animals. He studied more than 500 animal species in detail, dissecting specimens of nearly 50 species, and came up with a strikingly modern classification system. For example, he was the first person to recognize that dolphins should be classified with land mammals rather than with fish. In mathematics, he is known for laying the foundations of mathematical logic. Unfortunately, he was far less successful in physics and astronomy, areas in which many of his claims turned out to be wrong.

Despite his wide-ranging discoveries and writings, Aristotle's philosophies were not particularly influential until many centuries after his death. His books were preserved and valued by Islamic scholars but were unknown in Europe until they were translated into Latin in the 12th and 13th centuries. Aristotle's work gained great influence only after his philosophy was integrated into Christian theology by St. Thomas Aquinas (1225–1274). In the ancient world, Aristotle's greatest influence came indirectly, through his role as the tutor of Alexander the Great.

centuries. Because this model was so important in the history of science, let's briefly trace its development. **FIGURE 3.12** will help you keep track of some of the personalities we will encounter.

**Early Development** We generally trace the origin of Greek science to the philosopher Thales (c. 624–546 B.C.; pronounced "thay-lees"). We encountered Thales earlier because of his legendary prediction of a solar eclipse [**Section 2.3**]. Thales was the first person known to have addressed the question "What is the universe made of?" without resorting to supernatural explanations. His own guess—that the universe fundamentally consists of water and that Earth is a flat disk floating in an infinite ocean—was not widely accepted even in his own time. Nevertheless, just by asking the question he suggested that the world is inherently understandable and thereby inspired others to come up with better models for the structure of the universe.

A more sophisticated idea followed soon after, proposed by a student of Thales named Anaximander (c. 610–546 B.C.). Anaximander suggested that Earth floats in empty space surrounded by a sphere of stars and two separate rings along which the Sun and Moon travel. We therefore credit him with inventing the idea of a celestial sphere [**Section 2.1**]. Interestingly, Anaximander imagined Earth itself to be cylindrical rather than spherical in shape. He probably chose this shape because he knew Earth had to be curved in a north-south direction to explain changes in the constellations with latitude. Because the visible constellations do not change with longitude, he saw no need for curvature in the east-west direction.

We do not know precisely when the Greeks first began to recognize Earth as round, but this idea was taught as early as about 500 B.C. by the famous mathematician Pythagoras (c. 560–480 B.C.). He and his followers may have adopted a spherical Earth in part because they considered a sphere to be geometrically perfect. Later Greeks, including Aristotle, cited observations of Earth's curved shadow on the Moon during lunar eclipses as evidence for a spherical Earth.

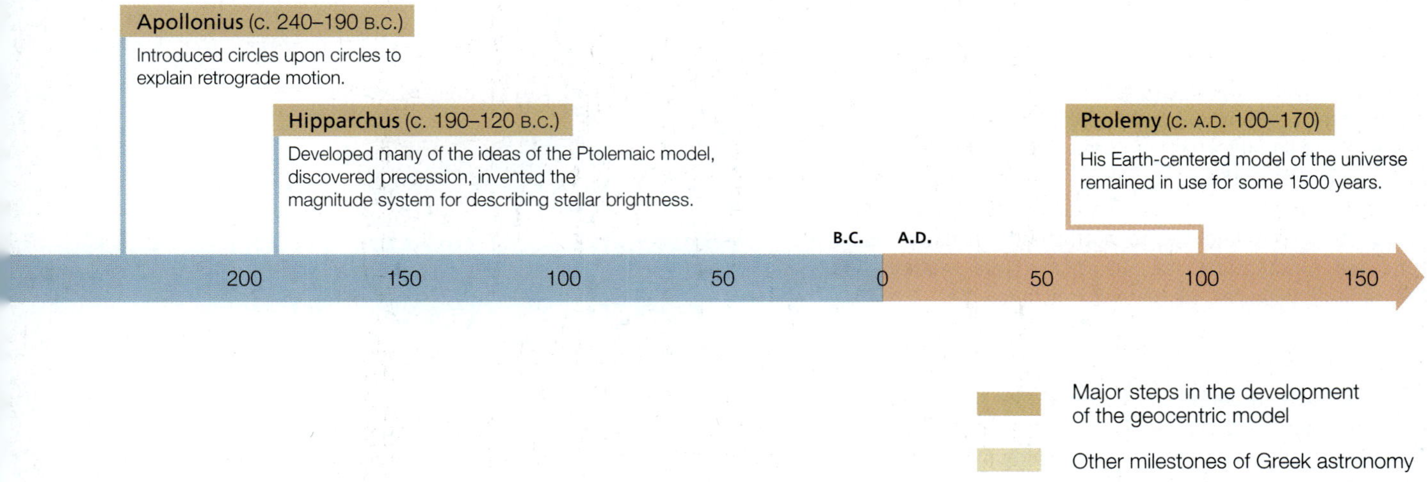


**FIGURE 3.12** Timeline for major Greek figures in the development of astronomy. (All these individuals are discussed in this book, but not necessarily in this chapter.)

**FIGURE 3.13** This model represents the Greek idea of the heavenly spheres (c. 400 B.C.). Earth is a sphere that rests in the center. The Moon, the Sun, and the planets all have their own spheres. The outermost sphere holds the stars.

## COMMON MISCONCEPTIONS

### Columbus and a Flat Earth

A widespread myth gives credit to Columbus for learning that Earth is round, but knowledge of Earth's shape predated Columbus by nearly 2000 years. Not only were scholars of Columbus's time well aware that Earth is round, but they even knew its approximate size: Earth's circumference was first measured in about 240 B.C. by the Greek scientist Eratosthenes. In fact, a likely reason Columbus had so much difficulty finding a sponsor for his voyages was that he tried to argue a point on which he was wrong: He claimed the distance by sea from western Europe to eastern Asia to be much less than scholars knew it to be. When he finally found a patron in Spain and left on his journey, he was so woefully underprepared that the voyage would almost certainly have ended in disaster if the Americas hadn't stood in his way.

The idea of "heavenly perfection" became even more deeply ingrained in Greek philosophy after Plato (428–348 B.C.) asserted that all heavenly objects move in perfect circles at constant speeds and therefore must reside on huge spheres encircling Earth (**FIGURE 3.13**). The Platonic belief in perfection influenced astronomical models for the next 2000 years. Of course, those Greeks who made observations found Plato's model problematic: The apparent retrograde motion of the planets [**Section 2.4**], already well known by that time, clearly showed that planets do *not* move at constant speeds around Earth.

An ingenious solution came from Plato's colleague Eudoxus (c. 400–347 B.C.), who created a model in which the Sun, the Moon, and the planets each had their own spheres nested within several other spheres. Individually, the nested spheres turned in perfect circles. By carefully choosing the sizes, rotation axes, and rotation speeds for the invisible spheres, Eudoxus was able to make them work together in a way that reproduced many of the observed motions of the Sun, Moon, and planets in our sky. Other Greeks refined the model by comparing its predictions to observations and adding more spheres to improve the agreement.

This is how things stood when Aristotle (384–322 B.C.) arrived on the scene. Whether Eudoxus and his followers thought of the nested spheres as real physical objects is not

## SPECIAL TOPIC Eratosthenes Measures Earth

In a remarkable feat, the Greek scientist Eratosthenes accurately estimated the size of Earth in about 240 B.C. He did it by comparing the noon altitude of the Sun on the same day in two locations.

Eratosthenes knew that on the summer solstice, the Sun passed directly overhead in Syene (modern-day Aswan) but came within only 7° of the zenith in Alexandria. He concluded that Alexandria must be 7° of latitude north of Syene (**FIGURE 1**). Because 7° is $\frac{7}{360}$ of a circle, this meant that the north-south distance between the two cities must be $\frac{7}{360}$ of the circumference of Earth.

Eratosthenes estimated the north-south distance between Syene and Alexandria to be 5000 stadia (the *stadium* was a Greek unit of distance). He thereby concluded that

$$\frac{7}{360} \times \text{circumference of Earth} = 5000 \text{ stadia}$$

If you multiply both sides by $\frac{360}{7}$, you'll find that this equation implies that Earth's circumference is about 250,000 stadia.

Based on the actual sizes of Greek stadiums, historians estimate that stadia must have been about $\frac{1}{6}$ km each, making Eratosthenes's estimate about $\frac{250{,}000}{6} = 42{,}000$ kilometers—impressively close to the real value of just over 40,000 kilometers.

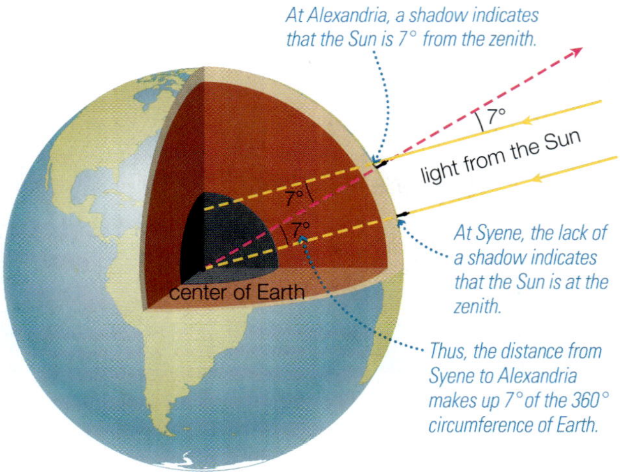


**FIGURE 1** This diagram shows how Eratosthenes concluded that the north-south distance from Syene to Alexandria is $\frac{7}{360}$ of Earth's circumference.

clear, but Aristotle certainly did. In Aristotle's model, all the spheres responsible for celestial motion were transparent and interconnected like the gears of a giant machine. Earth's position at the center was explained as a natural consequence of gravity. Aristotle argued that gravity pulled heavy things toward the center of the universe (and allowed lighter things to float toward the heavens), thereby causing all the dirt, rock, and water of the universe to collect at the center and form the spherical Earth. We now know that Aristotle was wrong about both gravity and Earth's location. However, largely because of his persuasive arguments for an Earth-centered universe, the geocentric view dominated Western thought for almost 2000 years.

**Ptolemy's Synthesis** Greek modeling of the cosmos culminated in the work of Claudius Ptolemy (c. A.D. 100–170; pronounced "TOL-e-mee"). Ptolemy's model still placed Earth at the center of the universe, but it differed in significant ways from the nested spheres of Eudoxus and Aristotle. We refer to Ptolemy's geocentric model as the **Ptolemaic model** to distinguish it from earlier geocentric models.

To explain the apparent retrograde motion of the planets, the Ptolemaic model applied an idea first suggested by Apollonius (c. 240–190 B.C.). This idea held that each planet moved around Earth on a small circle that turned upon a larger circle (**FIGURE 3.14**). (The small circle is sometimes called an *epicycle,* and the larger circle is called a *deferent.*) A planet following this circle-upon-circle motion would trace a loop as seen from Earth, with the backward portion of the loop mimicking apparent retrograde motion.

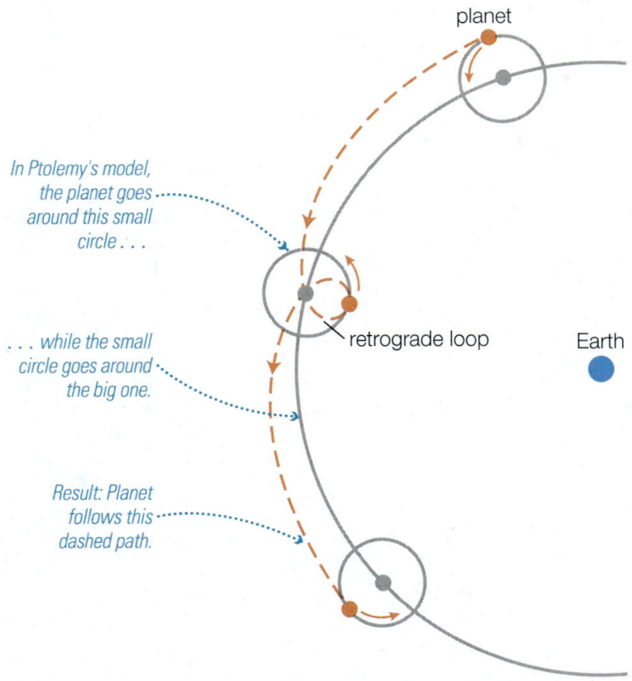


**FIGURE 3.14** This diagram shows how the Ptolemaic model accounted for apparent retrograde motion. Each planet is assumed to move around a small circle that turns upon a larger circle. The resulting path (dashed) includes a loop in which the planet goes backward as seen from Earth.

Ptolemy also relied heavily on the work of Hipparchus (c. 190–120 B.C.), considered one of the greatest of the Greek astronomers. Among his many accomplishments, Hipparchus developed the circle-upon-circle idea of Apollonius into a model that could predict planetary positions. To do this, Hipparchus added several features to the basic idea; for example, he included even smaller circles that moved upon the original set of small circles, and he positioned the large circles slightly off-center from Earth.

Ptolemy's great accomplishment was to adapt and synthesize earlier ideas into a single system that agreed quite well with the astronomical observations available at the time. In the end, he created and published a model that could correctly forecast future planetary positions to within a few degrees of arc, which is about the angular size of your hand held at arm's length against the sky. This was sufficiently accurate to keep the model in use for the next 1500 years. When Ptolemy's book describing the model was translated by Arabic scholars around A.D. 800, they gave it the title *Almagest*, derived from words meaning "the greatest work."

## 3.3 The Copernican Revolution

The Greeks and other ancient peoples developed many important scientific ideas, but what we now think of as science arose during the European Renaissance. Within a half century after the fall of Constantinople, Polish scientist Nicholas Copernicus began the work that ultimately overturned the Earth-centered Ptolemaic model.

### How did Copernicus, Tycho, and Kepler challenge the Earth-centered model?

The ideas introduced by Copernicus fundamentally changed the way we perceive our place in the universe. The story of this dramatic change, known as the **Copernican revolution**, is in many ways the story of the origin of modern science. It is also the story of several key personalities, beginning with Copernicus himself.

**Copernicus** Copernicus was born in Torun, Poland, on February 19, 1473. His family was wealthy and he received an education in mathematics, medicine, and law. He began studying astronomy in his late teens.

By that time, tables of planetary motion based on the Ptolemaic model had become noticeably inaccurate. But few people were willing to undertake the difficult calculations required to revise the tables. The best tables available had been compiled

*Copernicus (1473–1543)*

some two centuries earlier under the guidance of Spanish monarch Alphonso X (1221–1284). Commenting on the tedious nature of the work, the monarch is said to have complained, "If I had been present at the creation, I would have recommended a simpler design for the universe."

In his quest for a better way to predict planetary positions, Copernicus decided to try Aristarchus's Sun-centered idea, first proposed more than 1700 years earlier [**Section 2.4**]. He had read of Aristarchus's work, and recognized the much simpler explanation for apparent retrograde motion offered by a Sun-centered system (see Figure 2.34). But he went far beyond Aristarchus in working out mathematical details of the model. Through this process, Copernicus discovered simple geometric relationships that allowed him to calculate each planet's orbital period around the Sun and its relative distance from the Sun in terms of the Earth-Sun distance (see Mathematical Insight S1.1). The model's success in providing a geometric layout for the solar system convinced him that the Sun-centered idea must be correct.

Copernicus was nevertheless hesitant to publish his work, fearing that his suggestion that Earth moved would be considered absurd. However, he discussed his system with other scholars, including high-ranking officials of the Catholic Church, who urged him to publish a book. Copernicus saw the first printed copy of his book, *De Revolutionibus Orbium Coelestium* ("On the Revolutions of the Heavenly Spheres"), on the day he died—May 24, 1543.

Publication of the book spread the Sun-centered idea widely, and many scholars were drawn to its aesthetic advantages. However, the Copernican model gained relatively few converts over the next 50 years, for a good reason: It didn't work all that well. The primary problem was that while Copernicus had been willing to overturn Earth's central place in the cosmos, he held fast to the ancient belief that heavenly motion must occur in perfect circles. This incorrect assumption forced him to add numerous complexities to his system (including circles on circles much like those used by Ptolemy) to get it to make decent predictions. In the end, his complete model was no more accurate and no less complex than the Ptolemaic model, and few people were willing to throw out thousands of years of tradition for a new model that worked just as poorly as the old one.

**Tycho** Part of the difficulty faced by astronomers who sought to improve either the Ptolemaic or the Copernican model was a lack of quality data. The telescope had not yet been invented, and existing naked-eye observations were not very accurate. Better data were needed, and they were provided by the Danish nobleman Tycho Brahe (1546–1601), usually known simply as Tycho (pronounced "tie-koe").

Tycho was an eccentric genius who lost part of his nose in a sword fight with another student over who was the better mathematician; he designed a replacement nose piece made of silver and gold. In 1563, Tycho decided to observe a widely anticipated alignment of Jupiter and Saturn. To his surprise, the alignment occurred nearly 2 days later than the date Copernicus had predicted. Resolving to improve the state of astronomical prediction, he set about compiling careful observations of stellar and planetary positions in the sky.

*Tycho Brahe (1546–1601)*

Tycho's fame grew after he observed what he called a *nova,* meaning "new star," in 1572. Its lack of observable parallax [**Section 2.4**] led Tycho to conclude that the nova was much farther away than the Moon, a fact that contradicted the ancient Greek belief in unchanging heavens. (Today, we know that Tycho saw a *supernova*—the explosion of a distant star [**Section 17.3**].) In 1577, Tycho made similar observations of a comet and showed that it too lay in the realm of the heavens. Others, including Aristotle, had argued that comets were phenomena of Earth's atmosphere. King Frederick II of Denmark decided to sponsor Tycho's ongoing work, giving him money to build an unparalleled

**FIGURE 3.15** Tycho Brahe in his naked-eye observatory, which worked much like a giant protractor. He could sit and observe a planet through the rectangular hole in the wall as an assistant used a sliding marker to measure the angle on the protractor.

observatory for naked-eye observations (**FIGURE 3.15**). After Frederick II died in 1588, Tycho moved to Prague, where his work was supported by German emperor Rudolf II.

Over a period of three decades, Tycho and his assistants compiled naked-eye observations accurate to within less than 1 arcminute—less than the thickness of a fingernail viewed at arm's length. Despite the quality of his observations, Tycho never succeeded in coming up with a satisfying explanation for planetary motion. He was convinced that the *planets* must orbit the Sun, but his inability to detect stellar parallax [**Section 2.4**] led him to conclude that Earth must remain stationary. He therefore advocated a model in which the Sun orbits Earth while all other planets orbit the Sun. Few people took this model seriously.

**Kepler** Tycho failed to explain the motions of the planets satisfactorily, but he succeeded in finding someone who could: In 1600, he hired the young German astronomer Johannes Kepler (1571–1630). Kepler and Tycho had a strained relationship, but Tycho recognized the talent of his young apprentice. In 1601, as he lay on his deathbed, Tycho begged Kepler to find a system that would make sense of his observations so "that it may not appear I have lived in vain."*

*Johannes Kepler (1571–1630)*

Kepler was deeply religious and believed that understanding the geometry of the heavens would bring him closer to God. Like Copernicus, he believed that planetary orbits should be perfect circles, so he worked diligently to match circular motions to Tycho's observations.

Kepler labored with particular intensity to find an orbit for Mars, which posed the greatest difficulties in matching the data to a circular orbit. After years of calculation, Kepler found a circular orbit that matched all of Tycho's observations of Mars's position along the ecliptic (east-west) to within 2 arcminutes. However, the model did not correctly predict Mars's positions north or south of the ecliptic. Because Kepler sought a physically realistic orbit for Mars, he could not (as Ptolemy and Copernicus had done) tolerate one model for the east-west positions and another for the north-south positions. He attempted to find a unified model with a circular orbit. In doing so, he found that some of his predictions differed from Tycho's observations by as much as 8 arcminutes.

Kepler surely was tempted to attribute these discrepancies to errors by Tycho. After all, 8 arcminutes is barely one-fourth the angular diameter of the full moon. But Kepler trusted Tycho's work. The small discrepancies finally led Kepler to abandon the idea of circular orbits—and to find the correct solution to the ancient riddle of planetary motion. About this event, Kepler wrote:

> *If I had believed that we could ignore these eight minutes [of arc], I would have patched up my hypothesis accordingly. But, since it was not permissible to ignore, those eight minutes pointed the road to a complete reformation in astronomy.*

Kepler's key discovery was that planetary orbits are not circles but instead are a special type of oval called an **ellipse**. You can draw a circle by putting a pencil on the end of a string, tacking the string to a board, and pulling the pencil around (**FIGURE 3.16a**). Drawing an ellipse is similar, except that you must stretch the string around *two* tacks (**FIGURE 3.16b**). The locations of the two tacks are called the **foci** (singular, **focus**) of the ellipse. The long axis of the ellipse is called its *major axis*, each half of which is called a *semimajor axis*; as you'll see shortly, the length of the

*For a particularly moving version of the story of Tycho and Kepler, see Episode 3 of Carl Sagan's *Cosmos* video series.

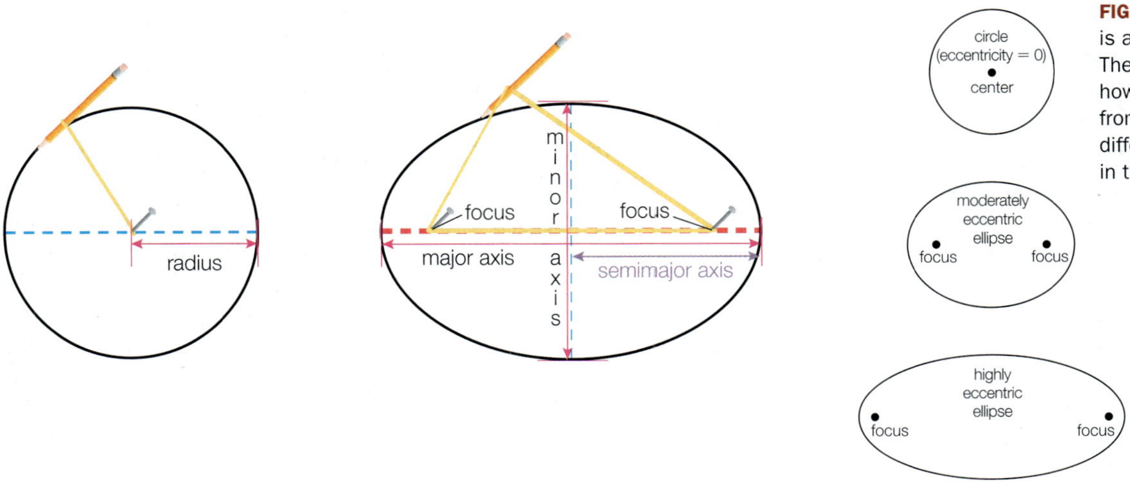


**FIGURE 3.16** An ellipse is a special type of oval. These diagrams show how an ellipse differs from a circle and how different ellipses vary in their eccentricity.

**a** Drawing a circle with a string of fixed length.

**b** Drawing an ellipse with a string of fixed length.

**c** Eccentricity describes how much an ellipse deviates from a perfect circle.

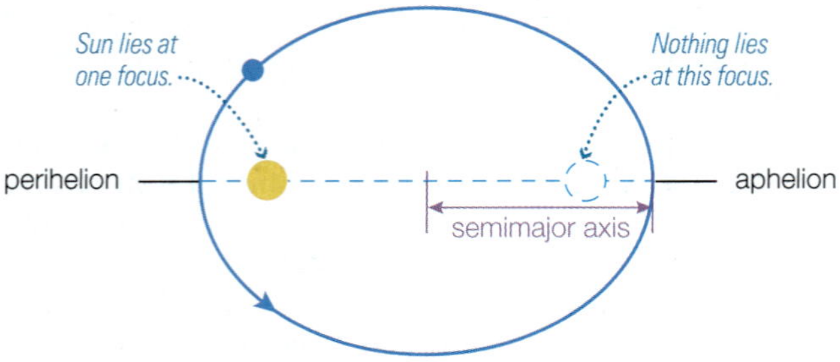


**FIGURE 3.17** Kepler's first law: The orbit of each planet about the Sun is an ellipse with the Sun at one focus. (The eccentricity shown here is exaggerated compared to the actual eccentricities of the planets.)

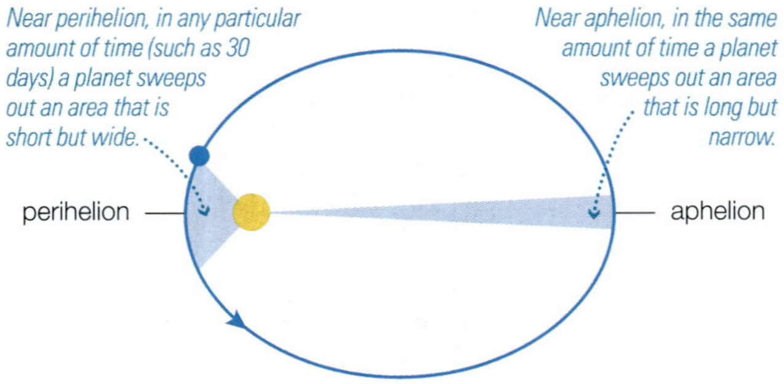


**FIGURE 3.18** Kepler's second law: As a planet moves around its orbit, an imaginary line connecting it to the Sun sweeps out equal areas (the shaded regions) in equal times.

semimajor axis is particularly important in astronomy. The short axis is called the *minor axis*. By altering the distance between the two foci while keeping the length of string the same, you can draw ellipses of varying **eccentricity**, a quantity that describes how much an ellipse is stretched out compared to a perfect circle (**FIGURE 3.16c**). A circle is an ellipse with zero eccentricity, and greater eccentricity means a more elongated ellipse.

Kepler's decision to trust the data over his preconceived beliefs marked an important transition point in the history of science. Once he abandoned perfect circles in favor of ellipses, Kepler soon came up with a model that could predict planetary positions with far greater accuracy than Ptolemy's Earth-centered model. Kepler's model withstood the test of time and became accepted not only as a model of nature but also as a deep, underlying truth about planetary motion.

## What are Kepler's three laws of planetary motion?

Kepler summarized his discoveries with three simple laws that we now call **Kepler's laws of planetary motion**. He published the first two laws in 1609 and the third in 1619.

- **Kepler's first law**: *The orbit of each planet about the Sun is an ellipse with the Sun at one focus* (**FIGURE 3.17**). This law tells us that a planet's distance from the Sun varies during its orbit. Its closest point is called **perihelion** (from the Greek for "near the Sun") and its farthest point is called **aphelion** ("away from the Sun"). The *average* of a planet's perihelion and aphelion distances is the length of its **semimajor axis**. We will refer to this simply as the planet's average distance from the Sun.
- **Kepler's second law**: *A planet moves faster in the part of its orbit nearer the Sun and slower when farther from the Sun, sweeping out equal areas in equal times.* As shown in **FIGURE 3.18**, the "sweeping" refers to an imaginary line connecting the planet to the Sun, and keeping the areas equal means that the planet moves a greater distance (and hence is moving faster) when it is near perihelion than it does in the same amount of time near aphelion.
- **Kepler's third law**: *More distant planets orbit the Sun at slower average speeds, obeying the precise mathematical relationship*

$$p^2 = a^3.$$

The letter $p$ stands for the planet's orbital period in years and $a$ for its average distance from the Sun in astronomical units. **FIGURE 3.19a** shows the $p^2 = a^3$ law graphically. Notice that the square of each planet's orbital period $(p^2)$ is indeed equal to the cube of its average distance from the Sun $(a^3)$. Because Kepler's third law relates orbital distance to orbital time (period), we can use the law to calculate a planet's average orbital speed.* **FIGURE 3.19b** shows the result, confirming that more distant planets orbit the Sun more slowly.

The fact that more distant planets move more slowly led Kepler to suggest that planetary motion might be the result of a force from the Sun. He even speculated about the nature of this force, guessing that it might be related to magnetism. (This idea, shared by Galileo, was first suggested by William Gilbert [1544–1603], an early believer in the Copernican system.) Kepler was right about the existence of a force but wrong in his guess of magnetism. A half century later, Isaac Newton identified the force as gravity [**Section 4.4**].

**Think about it** Suppose a comet has an orbit that brings it quite close to the Sun at its perihelion and beyond Mars at its aphelion, but with an average distance (semimajor axis) of 1 AU. How long would the comet take to complete each orbit of the Sun? Would it spend most of its time close to the Sun, far from the Sun, or somewhere in between? Explain.

## How did Galileo solidify the Copernican revolution?

The success of Kepler's laws in matching Tycho's data provided strong evidence in favor of Copernicus's placement of the Sun at the center of the solar system. Nevertheless, many scientists still voiced objections to the

*To calculate orbital speed from Kepler's third law, remember that speed = distance/time. For a nearly circular planetary orbit, the distance is the orbital circumference, or $2\pi a$, and the time is the orbital period $p$, so the orbital speed is approximately $(2\pi a)/p$. From Kepler's third law, $p = a^{3/2}$. Plugging this value for $p$ into the orbital speed equation, we find that a planet's orbital speed is about $2\pi/\sqrt{a}$; the graph of this equation is the curve in Figure 3.19b.

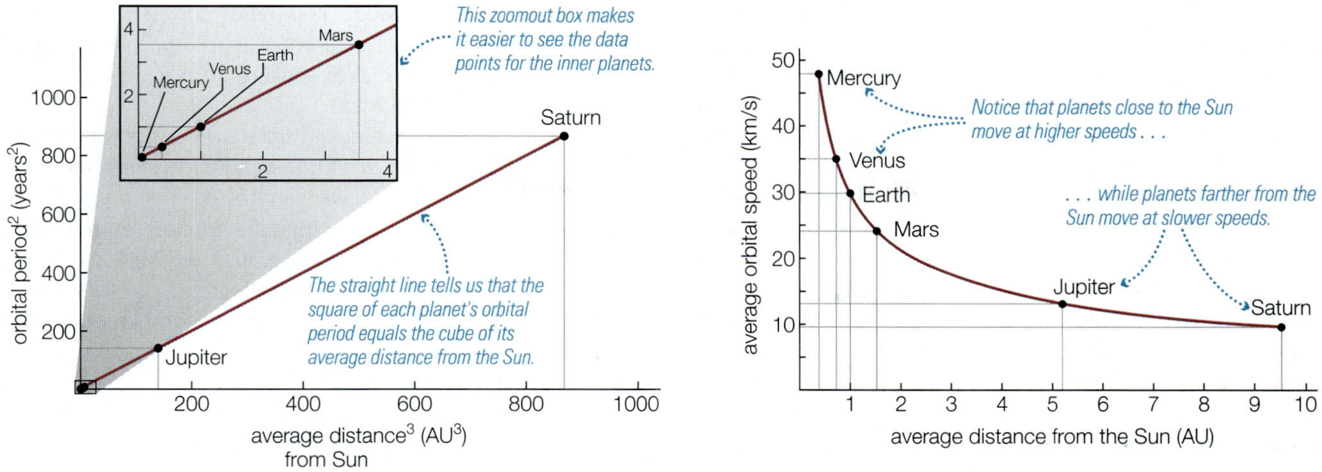


**a** This graph shows that Kepler's third law ($p^2 = a^3$) holds true; the graph shows only the planets known in Kepler's time.

**b** This graph, based on Kepler's third law and modern values of planetary distances, shows that more distant planets orbit the Sun more slowly.

**FIGURE 3.19** Graphs based on Kepler's third law.

Copernican view. There were three basic objections, all rooted in the 2000-year-old beliefs of Aristotle and other ancient Greeks.

- First, Aristotle had held that Earth could not be moving because, if it were, objects such as birds, falling stones, and clouds would be left behind as Earth moved along its way.
- Second, the idea of noncircular orbits contradicted Aristotle's claim that the heavens—the realm of the Sun, Moon, planets, and stars—must be perfect and unchanging.
- Third, no one had detected the stellar parallax that should occur if Earth orbits the Sun [**Section 2.4**].

Galileo Galilei (1564–1642), usually known by only his first name, answered all three objections.

**Galileo's Evidence** Galileo defused the first objection with experiments that almost single-handedly overturned the Aristotelian view of physics. In particular, he used experiments with rolling balls to demonstrate that a moving object remains in motion *unless* a force acts to stop it (an idea now codified in Newton's first law of motion [**Section 4.2**]). This insight explained why objects that share Earth's motion through space—such as birds, falling stones, and clouds—should *stay* with Earth rather than falling behind as Aristotle had argued. This same idea explains why passengers stay with a moving airplane even when they leave their seats.

*Galileo (1564–1642)*

The second objection had already been challenged by Tycho's supernova and comet observations, which demonstrated that the heavens could change. Galileo then shattered the idea of heavenly perfection after he built a telescope in late 1609. (Galileo did not invent the telescope, but his innovations made it much more powerful.) Through his telescope, Galileo saw sunspots on the Sun, which were considered "imperfections" at the time. He also used his telescope to observe that the Moon has mountains and valleys like the "imperfect" Earth by noticing the shadows cast near the dividing line between the light and dark portions of the lunar face (**FIGURE 3.20**). If the heavens were in fact not perfect, then the idea of elliptical orbits (as opposed to "perfect" circles) was not so objectionable.

**FIGURE 3.20** The shadows cast by mountains and crater rims near the dividing line between the light and dark portions of the lunar face show that the Moon's surface is not perfectly smooth.

The third objection—the absence of observable stellar parallax—had been of particular concern to Tycho. Based on his estimates of the distances of stars, Tycho believed that his naked-eye observations were sufficiently precise to detect stellar parallax if Earth did in fact orbit the Sun. Refuting Tycho's argument required showing that the stars were more distant than Tycho had thought and therefore too distant for him to have observed stellar parallax. Although Galileo didn't actually prove this fact, he provided strong evidence in its favor. For example, he saw with his telescope that the Milky Way resolved into countless individual stars. This discovery helped him argue that the stars were far more numerous and more distant than Tycho had believed.

**Sealing the Case** In hindsight, the final nails in the coffin of the Earth-centered model came with two of Galileo's earliest discoveries through the telescope. First, he observed four moons clearly orbiting Jupiter, *not* Earth (**FIGURE 3.21**). By itself, this observation still did not rule out a stationary, central Earth. However, it showed that moons can orbit a moving planet like Jupiter, which overcame some critics' complaints that the Moon could not stay with a moving Earth. Soon thereafter, he observed that Venus goes through phases in a way that makes sense only if it orbits the Sun and not Earth (**FIGURE 3.22**).

With Earth clearly removed from its position at the center of the universe, the scientific debate turned to the question of whether Kepler's laws were the correct model for our solar system. The most convincing evidence came in 1631, when astronomers observed a transit of Mercury across the Sun's face. Kepler's laws had predicted the transit with overwhelmingly better success than any competing model.

**Galileo and the Church** Although we now recognize that Galileo won the day, the story was more complex in his own time, when Catholic Church doctrine still held Earth to be the center of the universe. On June 22, 1633, Galileo was brought before a Church inquisition in Rome and ordered to recant his claim that Earth orbits the Sun. Nearly 70 years old and fearing for his life, Galileo did as ordered and his life was spared. However, legend has it that as he rose from his knees he whispered under his breath, *Eppur si muove*—Italian for

## MATHEMATICAL INSIGHT 3.1 Eccentricity and Planetary Orbits

We describe how much a planet's orbit differs from a perfect circle by stating its orbital eccentricity. There are several equivalent ways to define the eccentricity of an ellipse, but the simplest is shown in **FIGURE 1**. We define $c$ to be the distance from each focus to the center of the ellipse and $a$ to be the length of the semimajor axis. The eccentricity, $e$, is then defined to be

$$e = \frac{c}{a}.$$

Notice that $c = 0$ for a perfect circle, because a circle is an ellipse with both foci *in* the center, so this formula gives an eccentricity of 0 for a perfect circle, just as we expect.

You can find the orbital eccentricities for the planets in tables such as Table E.2 in Appendix E of this book. Once you know the eccentricity, the following formulas allow you to calculate the planet's perihelion and aphelion distances (**FIGURE 2**):

$$\text{perihelion distance} = a(1 - e)$$

$$\text{aphelion distance} = a(1 + e)$$

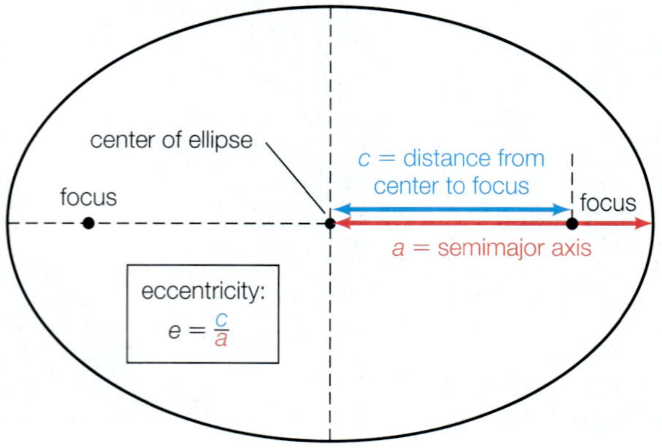


**FIGURE 1**

**EXAMPLE:** What are Earth's perihelion and aphelion distances?

**SOLUTION:**

**Step 1 Understand:** To use the given formulas, we need to know Earth's orbital eccentricity, which Table E.2 gives as $e = 0.017$, and semimajor axis length, which is 1 AU, or $a = 149.6$ million km.

**Step 2 Solve:** We plug these values into the equations:

$$\begin{aligned}\text{Earth's perihelion distance} &= a(1 - e)\\ &= (149.6 \times 10^6\ \text{km})(1 - 0.017)\\ &= 147.1 \times 10^6\ \text{km}\end{aligned}$$

$$\begin{aligned}\text{Earth's aphelion distance} &= a(1 + e)\\ &= (149.6 \times 10^6\ \text{km})(1 + 0.017)\\ &= 152.1 \times 10^6\ \text{km}\end{aligned}$$

**Step 3 Explain:** Earth's perihelion (nearest to the Sun) distance is 147.1 million kilometers and its aphelion (farthest from the Sun) distance is 152.1 million kilometers. In other words, Earth's distance from the Sun varies between 147.1 and 152.1 million kilometers.

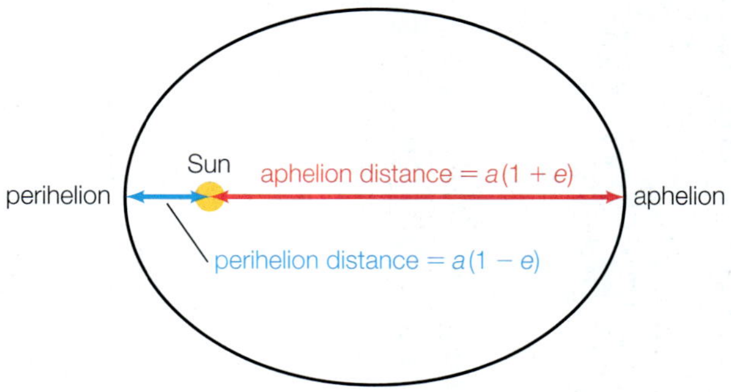


**FIGURE 2**

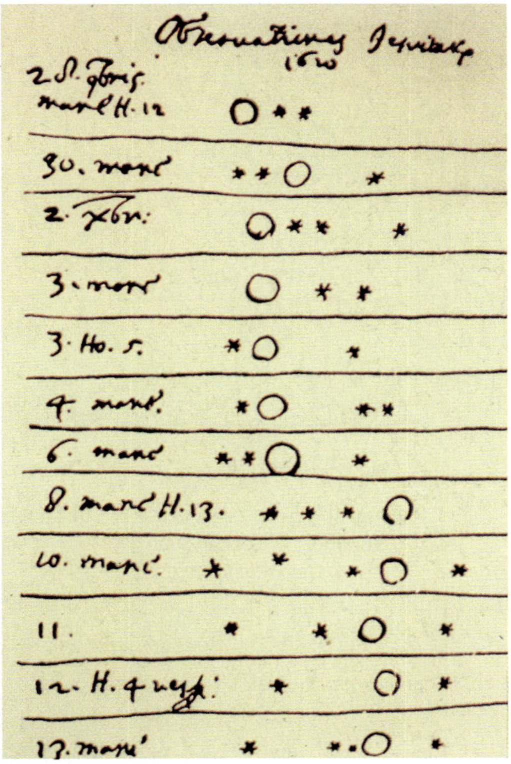

**FIGURE 3.21** A page from Galileo's notebook written in 1610. His sketches show four "stars" near Jupiter (the circle) but in different positions at different times (with one or more sometimes hidden from view). Galileo soon realized that the "stars" were actually moons orbiting Jupiter.

"And yet it moves." (Given the likely consequences if Church officials had heard him say this, most historians doubt the legend; see the Special Topic, page 74.)

The Church did not formally vindicate Galileo until 1992, but Church officials gave up the argument long before that: In 1757, all works backing the idea of a Sun-centered solar system were removed from the Church's index of banned books. Moreover, Catholic scientists have long worked at the forefront of astronomical research, and today's official Church teachings are compatible not only with Earth's planetary status but also with the theories of the Big Bang and the subsequent evolution of the cosmos and of life.

## 3.4 The Nature of Science

The story of how our ancestors gradually figured out the basic architecture of the cosmos exhibits many features of what we now consider "good science." For example, we have seen how models were formulated and tested against observations and modified or replaced when they failed those tests. The story also illustrates some classic mistakes, such as the apparent failure of anyone before Kepler to question the belief that orbits must be circles. The ultimate success of the Copernican revolution led scientists, philosophers, and theologians to reassess the various modes of thinking that played a role in the 2000-year process of discovering Earth's place in the universe. Let's examine how the principles of modern science emerged from the lessons learned in the Copernican revolution.

### How can we distinguish science from nonscience?

It's surprisingly difficult to define the term *science* precisely. The word comes from the Latin *scientia,* meaning "knowledge," but not all knowledge is science. For example, you may know what music you like best, but your musical taste is not a result of scientific study.

**Approaches to Science** One reason science is difficult to define is that not all science works in the same way. For example, you've probably heard that science is supposed to proceed according to something called the "scientific method." As an idealized illustration of this method,

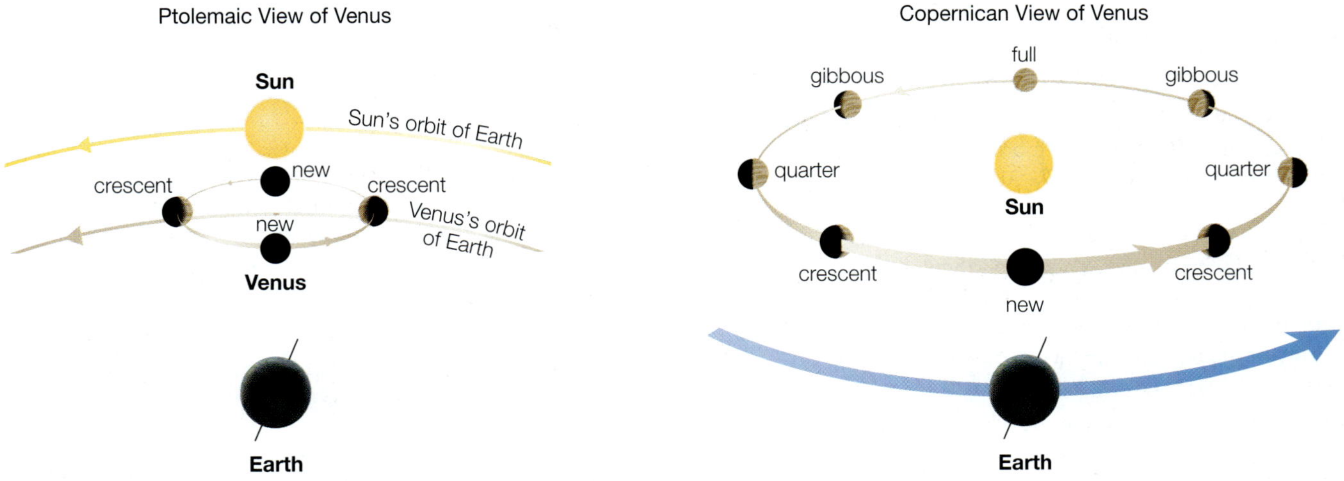


**a** In the Ptolemaic system, Venus orbits Earth, moving around a smaller circle on its larger orbital circle; the center of the smaller circle lies on the Earth-Sun line. If this view were correct, Venus's phases would range only from new to crescent.

**b** In reality, Venus orbits the Sun, so from Earth we can see it in many different phases. This is just what Galileo observed, telling him that Venus orbits the Sun.

**FIGURE 3.22** Galileo's telescopic observations of Venus showed that it orbits the Sun rather than Earth.

consider what you would do if your flashlight suddenly stopped working. You might suspect that the flashlight's batteries had died. This type of tentative explanation, or **hypothesis**, is sometimes called an *educated guess*—in this case, it is "educated" because you already know that flashlights need batteries. Your hypothesis allows you to make a simple prediction: If you replace the batteries with new ones, the flashlight should work. You can test this prediction by replacing the batteries. If the flashlight now works, you've confirmed your hypothesis. If it doesn't, you must revise or discard your hypothesis, perhaps in favor of some other one that you can also test (such as that the bulb no longer works). **FIGURE 3.23** illustrates the basic flow of this process.

The scientific method can be a useful idealization, but real science rarely progresses in such an orderly way. Scientific progress often begins with someone going out and looking at nature in a general way, rather than conducting a careful set of experiments. For example, Galileo wasn't looking for anything in particular when he pointed his telescope at the sky and made his first startling discoveries. Furthermore, scientists are human beings, and their intuition and personal beliefs inevitably influence their work. Copernicus, for example, adopted the idea that Earth orbits the Sun not because he had carefully tested it but because he believed it made more sense than the prevailing view of an Earth-centered universe. While his intuition guided him to the right general idea, he erred in the specifics because he still held Plato's ancient belief that heavenly motion must be in perfect circles.

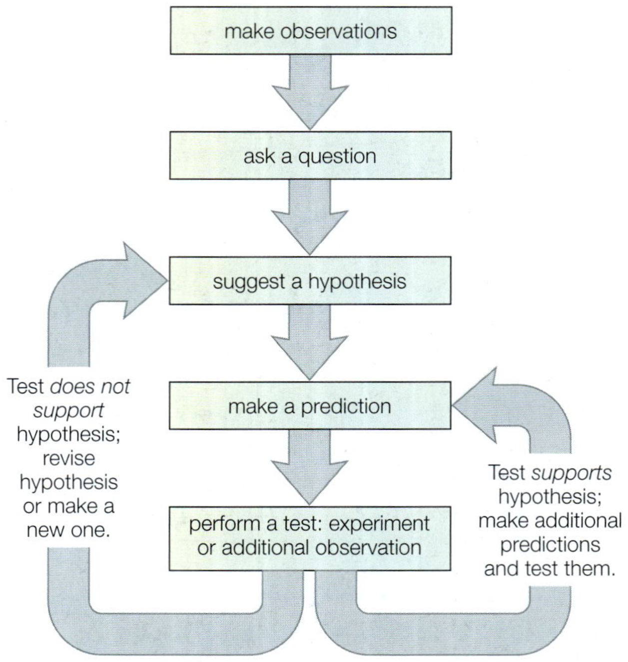


**FIGURE 3.23** This diagram illustrates what we often call the *scientific method*.

Given that the idealized scientific method is an overly simplistic characterization of science, how can we tell what is science and what is not? To answer this question, we must look a little deeper into the distinguishing characteristics of scientific thinking.

## MATHEMATICAL INSIGHT 3.2 Kepler's Third Law

When Kepler discovered his third law, $p^2 = a^3$, he did so only by looking at planet orbits. In fact, it applies much more generally. We'll see its most general form in Mathematical Insight 4.3, but even in its original form we can apply it to any object *if*

1. the object is *orbiting the Sun* or another star of the same mass as the Sun and
2. we measure orbital *periods in years* and *distances in AU.*

**EXAMPLE 1:** What is the orbital period of the dwarf planet (and largest asteroid) Ceres, which orbits the Sun at an average distance (semimajor axis) of 2.77 AU?

**SOLUTION:**

**Step 1 Understand:** We can apply Kepler's third law because both conditions above are met. The first is met because Ceres orbits the Sun. The second is met because we are given the orbital distance in AU, which means Kepler's third law will tell us the orbital period in years.

**Step 2 Solve:** We want the period $p$, so we solve Kepler's third law for $p$ by taking the square root of both sides; we then substitute the given value $a = 2.77$ AU:

$$p^2 = a^3 \Rightarrow p = \sqrt{a^3} = \sqrt{2.77^3} = 4.6.$$

Note that because of the special conditions attached to the use of Kepler's third law in its original form, we do *not* include units when working with it; we know we'll get a period in years as long as we start with a distance in AU.

**Step 3 Explain:** Ceres has an orbital period of 4.6 years, meaning it takes 4.6 years to complete each orbit around the Sun.

**EXAMPLE 2:** A new planet is discovered to be orbiting a star with the same mass as our Sun. The planet orbits the star every 3 months. What is its average distance from its star?

**SOLUTION:**

**Step 1 Understand:** We can use Kepler's third law in its original form if the problem meets the two conditions above. The first condition is met because the planet is orbiting a star with the same mass as our Sun. To meet the second condition, we must convert the orbital period from 3 months to $p = 0.25$ year.

**Step 2 Solve:** We want the distance $a$, so we solve Kepler's third law for $a$ by taking the cube root of both sides; we then substitute the orbital period $p = 0.25$ year:

$$p^2 = a^3 \Rightarrow a = \sqrt[3]{p^2} = \sqrt[3]{0.25^2} = 0.40$$

**Step 3 Explain:** The planet orbits its star at an average distance of 0.4 AU. By comparing this result to the distances of planets in our own solar system given in Table E.2, we find that this planet's average orbital distance is just slightly larger than that of the planet Mercury in our own solar system.

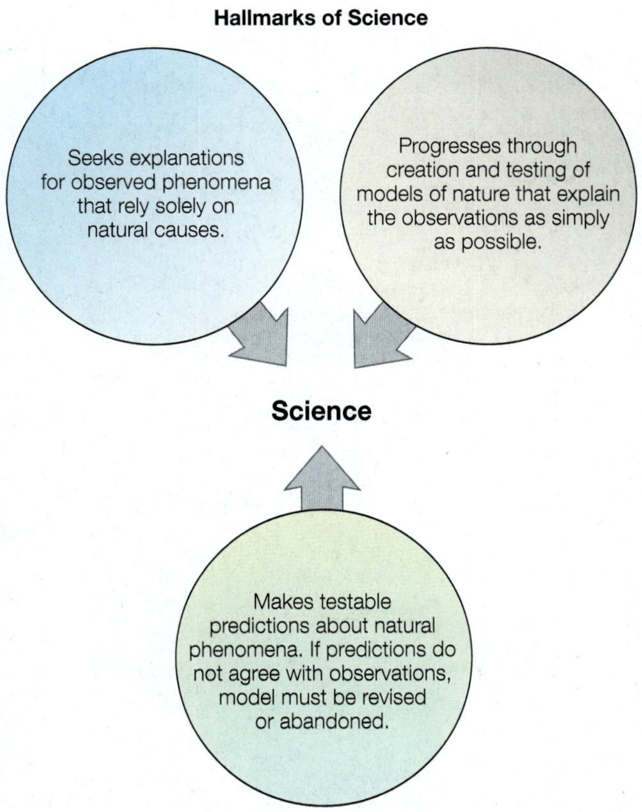


**FIGURE 3.24** Hallmarks of science.

**Hallmarks of Science** One way to define scientific thinking is to list the criteria that scientists use when they judge competing models of nature. Historians and philosophers of science have examined (and continue to examine) this issue in great depth, and different experts express different viewpoints on the details. Nevertheless, everything we now consider to be science shares the following three basic characteristics, which we will refer to as the "hallmarks" of science (**FIGURE 3.24**):

- Modern science seeks explanations for observed phenomena that rely solely on natural causes.
- Science progresses through the creation and testing of models of nature that explain the observations as simply as possible.
- A scientific model must make testable predictions about natural phenomena that will force us to revise or abandon the model if the predictions do not agree with observations.

Each of these hallmarks is evident in the story of the Copernican revolution. The first shows up in the way Tycho's careful measurements of planetary motion motivated Kepler to come up with a better explanation for those motions. The second is evident in the way several competing models were compared and tested, most notably those of Ptolemy, Copernicus, and Kepler. We see the third in the fact that each model could make precise predictions about the future motions of the Sun, Moon, planets, and stars in our sky. Kepler's model gained acceptance because it worked, while the competing models lost favor because their predictions failed to match the observations. **FIGURE 3.25** summarizes the Copernican revolution and how it illustrates the hallmarks of science.

**Occam's Razor** The criterion of simplicity in the second hallmark deserves additional explanation. Remember that Copernicus's original model did *not* match the data noticeably better than Ptolemy's model. If scientists had judged this model solely on the accuracy of its predictions, they might have rejected it immediately. However, many scientists found elements of the Copernican model appealing, such as its simple explanation for apparent retrograde motion. They therefore kept the model alive until Kepler found a way to make it work.

If agreement with data were the sole criterion for judgment, we could imagine a modern-day Ptolemy adding millions or billions of additional circles to the geocentric model in an effort to improve its agreement with observations. A sufficiently complex geocentric model could in principle reproduce the observations with almost perfect accuracy—but it still would not convince us that Earth is the center of the universe. We would still choose the Copernican view over the geocentric view because its predictions would be just as accurate but follow a much simpler model of nature. The idea that scientists should prefer the simpler of two models that agree equally well with observations is called *Occam's razor*, after the medieval scholar William of Occam (1285–1349).

**Verifiable Observations** The third hallmark of science forces us to face the question of what counts as an "observation" against which a prediction can be tested. Consider the claim that aliens are visiting Earth in UFOs. Proponents of this claim say that thousands of eyewitness reports of UFO encounters provide evidence that it is true. But do these personal testimonials count as *scientific* evidence? On the surface, the answer isn't obvious, because all scientific studies involve eyewitness accounts on some level. For example, only a handful of scientists have personally made detailed tests of Einstein's theory of relativity, and it is their personal reports of the results that have convinced other scientists of the theory's validity. However, there's an important difference between personal testimony about a scientific test and a UFO: The first can be verified by anyone, at least in principle, while the second cannot.

Understanding this difference is crucial to understanding what counts as science and what does not. Even though you may never have conducted a test of Einstein's theory of relativity yourself, there's nothing stopping you from doing so. It might require several years of study before you had the necessary background to conduct the test, but you could then confirm the results reported by other scientists. In other words, while you may currently be trusting the eyewitness testimony of scientists, you always have the option of verifying their testimony for yourself.

In contrast, there is no way for you to verify someone's eyewitness account of a UFO. Without hard evidence such as photographs or pieces of the UFO, there is nothing that you could evaluate for yourself, even in principle. (And in those cases where "hard evidence" for UFO sightings has been presented, scientific study has never yet found the

# COSMIC Context

## FIGURE 3.25 The Copernican Revolution

**Ancient Earth-centered models of the universe easily explained the simple motions of the Sun and Moon through our sky, but had difficulty explaining the more complicated motions of the planets. The quest to understand planetary motions ultimately led to a revolution in our thinking about Earth's place in the universe that illustrates the process of science. This figure summarizes the major steps in that process.**

① Night by night, planets usually move from west to east relative to the stars. However, during periods of *apparent retrograde motion,* they reverse direction for a few weeks to months [Section 2.4]. The ancient Greeks knew that any credible model of the solar system had to explain these observations.

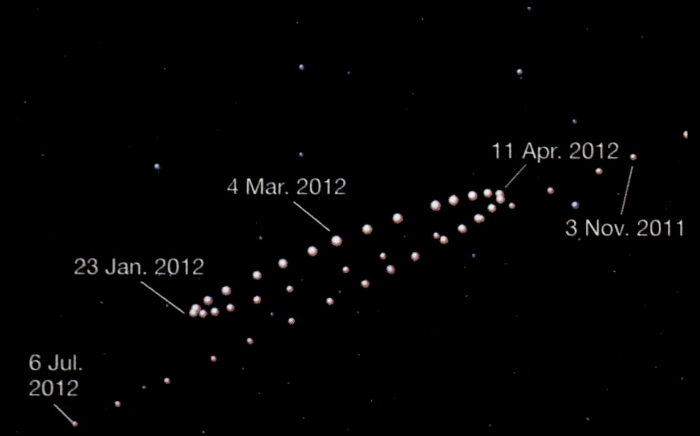


*This composite photo shows the apparent retrograde motion of Mars.*

② Most ancient Greek thinkers assumed that Earth remained fixed at the center of the solar system. To explain retrograde motion, they therefore added a complicated scheme of circles moving upon circles to their Earth-centered model. However, at least some Greeks, such as Aristarchus, preferred a Sun-centered model, which offered a simpler explanation for retrograde motion.

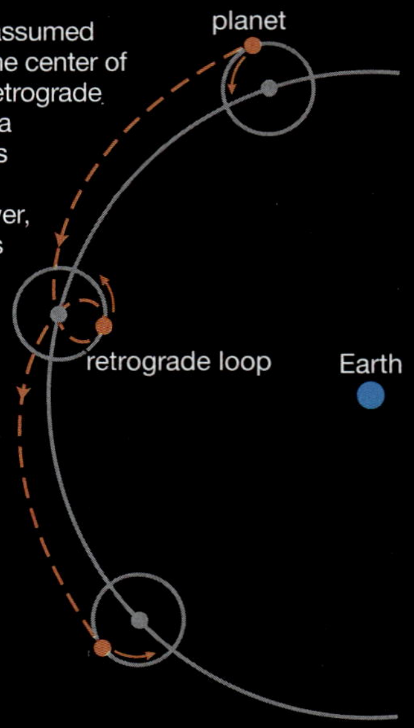


*The Greek geocentric model explained apparent retrograde motion by having planets move around Earth on small circles that turned on larger circles.*

HALLMARK OF SCIENCE **A scientific model must seek explanations for observed phenomena that rely solely on natural causes.** The ancient Greeks used geometry to explain their observations of planetary motion.

***(Left page)***
*A schematic map of the universe from 1539 with Earth at the center and the Sun (Solis) orbiting it between Venus (Veneris) and Mars (Martis).*

***(Right page)***
*A page from Copernicus's* De Revolutionibus, *published in 1543, showing the Sun (Sol) at the center and Earth (Terra) orbiting between Venus and Mars.*

3 By the time of Copernicus (1473–1543), predictions based on the Earth-centered model had become noticeably inaccurate. Hoping for improvement, Copernicus revived the Sun-centered idea. He did not succeed in making substantially better predictions because he retained the ancient belief that planets must move in perfect circles, but he inspired a revolution continued over the next century by Tycho, Kepler, and Galileo.

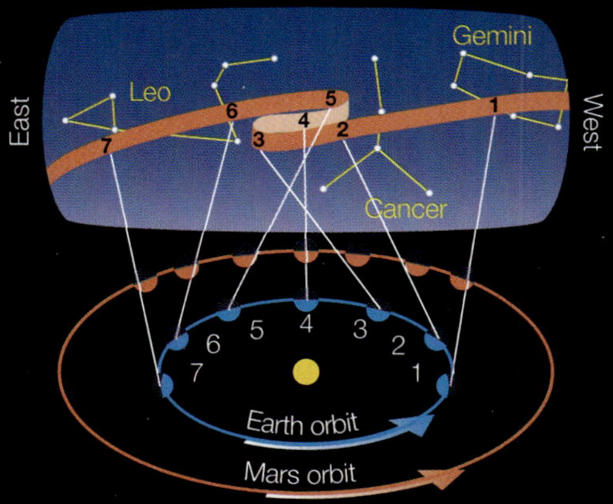


*Apparent retrograde motion is simply explained in a Sun-centered system. Notice how Mars appears to change direction as Earth moves past it.*

HALLMARK OF SCIENCE **Science progresses through creation and testing of models of nature that explain the observations as simply as possible.** Copernicus developed a Sun-centered model in hopes of explaining observations better than the more complicated Earth-centered model.

4 Tycho exposed flaws in both the ancient Greek and Copernican models by observing planetary motions with unprecedented accuracy. His observations led to Kepler's breakthrough insight that planetary orbits are elliptical, not circular, and enabled Kepler to develop his three laws of planetary motion.

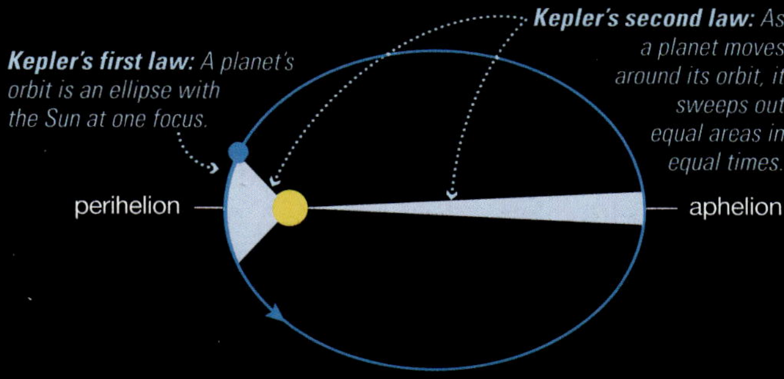


***Kepler's third law:*** *More distant planets orbit at slower average speeds, obeying* $p^2 = a^3$.

HALLMARK OF SCIENCE **A scientific model makes testable predictions about natural phenomena. If predictions do not agree with observations, the model must be revised or abandoned.** Kepler could not make his model agree with observations until he abandoned the belief that planets move in perfect circles.

5 Galileo's experiments and telescopic observations overcame remaining scientific objections to the Sun-centered model. Together, Galileo's discoveries and the success of Kepler's laws in predicting planetary motion overthrew the Earth-centered model once and for all.

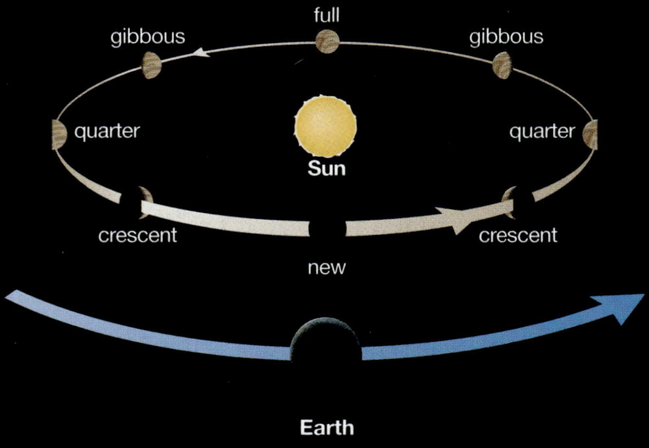


*With his telescope, Galileo saw phases of Venus that are consistent only with the idea that Venus orbits the Sun rather than Earth.*

evidence to be strong enough to support the claim of alien spacecraft [**Section 24.4**].) Moreover, scientific studies of eyewitness testimony show it to be notoriously unreliable, because different eyewitnesses often disagree on what they saw even immediately after an event has occurred. As time passes, memories of the event may change further. In some cases in which memory has been checked against reality, people have reported vivid memories of events that never happened at all. Virtually all of us have experienced this effect: disagreements with a friend about who did what and when. Since both people cannot be right in such cases, at least one person must have a memory that differs from reality.

The demonstrated unreliability of eyewitness testimony explains why it is generally considered insufficient for a conviction in criminal court; at least some other evidence is required. For the same reason, we cannot accept eyewitness testimony by itself as sufficient evidence in science, no matter who reports it or how many people offer similar testimony.

**Science and Pseudoscience** It's important to realize that science is not the only valid way of seeking knowledge. For example, suppose you are shopping for a car, learning to play drums, or pondering the meaning of life. In each case, you might make observations, exercise logic, and test hypotheses. Yet these pursuits clearly are not science, because they are not directed at developing testable explanations for observed natural phenomena. As long as nonscientific searches for knowledge make no claims about how the natural world works, they do not conflict with science.

However, you will often hear claims about the natural world that seem to be based on observational evidence but do not treat evidence in a truly scientific way. Such claims are often called **pseudoscience**, which means "false science." To distinguish real science from pseudoscience, a good first step is to check whether a particular claim exhibits all three hallmarks of science. Consider the example of

## SPECIAL TOPIC And Yet It Moves

The case of Galileo is often portrayed as a simple example of conflict between science and religion, but the reality was much more complex, with deep divisions inside the Church hierarchy. Perhaps the clearest evidence for a more open-minded Church comes from the case of Copernicus, whose revolutionary work was strongly supported by many Church officials. A less-well-known and earlier example concerns Nicholas of Cusa (1401–1464), who published a book arguing for a Sun-centered solar system in 1440. (Copernicus probably was not aware of this work by Nicholas of Cusa.) Nicholas was ordained a priest in the same year that his book was published, and he was later elevated to Cardinal. Clearly, his views caused no problems for Church officials of the time.

Many other scientists received similar support from within the Church. In fact, for most of his life, Galileo counted Cardinals (and even the Pope who later tried him) among his friends. Some historians suspect that Galileo got into trouble less for his views than for the way in which he portrayed them. In 1632—just a year before his famous trial—he published a book in which two fictional characters debated the geocentric and Sun-centered views. He named the character taking the geocentric position Simplicio—essentially "simple-minded"—and someone apparently convinced the Pope that the character was meant to represent him.

If it was personality rather than belief that got Galileo into trouble, he was not the only one. Another early supporter of Copernicus, Giordano Bruno (1548–1600), drew the wrath of the Church after essentially writing that no rational person could disagree with him (not just on the Copernican system but on other matters as well). Bruno was branded a heretic and burned at the stake.

The evidence supporting the idea that Earth rotates and orbits the Sun was quite strong by the time of Galileo's trial in 1633, but it was still indirect. Today, we have much more direct proof that Galileo was correct when he supposedly whispered of Earth, *Eppur si muove*—"And yet it moves."

French physicist Jean Foucault provided the first direct proof of rotation in 1851. Foucault built a large pendulum that he carefully started swinging. Any pendulum tends to swing always in the same plane, but Earth's rotation made Foucault's pendulum appear to twist slowly in a circle. Today, *Foucault pendulums* are a popular attraction at many science centers and museums (**FIGURE 1**). A second direct proof that Earth rotates is provided by the *Coriolis effect*, first described by French physicist Gustave Coriolis (1792–1843). The Coriolis effect [**Section 10.2**], which would not occur if Earth were not rotating, is responsible for things such as the swirling of hurricanes and the fact that missiles that travel great distances on Earth deviate from straight-line paths.

The first direct proof that Earth orbits the Sun came from English astronomer James Bradley (1693–1762). To understand Bradley's proof, imagine that starlight is like rain, falling straight down. If you are standing still, you should hold your umbrella straight over your head, but if you are walking through the rain, you should tilt your umbrella forward, because your motion makes the rain appear to be coming down at an angle. Bradley discovered that observing light from stars requires that telescopes be tilted slightly in the direction of Earth's motion—just like the umbrella. This effect is called the *aberration of starlight*. Stellar parallax also provides direct proof that Earth orbits the Sun, and it was first measured in 1838 by German astronomer Friedrich Bessel.

**FIGURE 1** A Foucault pendulum at the San Diego Natural History Museum.

## COMMON MISCONCEPTIONS

### Eggs on the Equinox

One of the hallmarks of science holds that you needn't take scientific claims on faith. In principle, at least, you can always test them for yourself. Consider the claim, repeated in news reports every year, that the spring equinox is the only day on which you can balance an egg on its end. Many people believe this claim, but you'll be immediately skeptical if you think about the nature of the spring equinox. The equinox is merely a point in time at which sunlight strikes both hemispheres equally (see Figure 2.15). It's difficult to see how sunlight could affect an attempt to balance eggs (especially if the eggs are indoors), and there's nothing special about either Earth's or the Sun's gravity on that day.

More important, you can test this claim directly. It's not easy to balance an egg on its end, but with practice you can do it on any day of the year, not just on the spring equinox. Not all scientific claims are so easy to test for yourself, but the basic lesson should be clear: Before you accept any scientific claim, you should demand at least a reasonable explanation of the evidence that backs it up.

people who claim a psychic ability to "see" the future and use it to make specific, testable predictions. In this sense, "seeing" the future sounds scientific, since we can test it. However, numerous studies have tested the predictions of "seers" and have found that their predictions come true no more often than would be expected by pure chance. If the "seers" were scientific, they would admit that this evidence undercuts their claim of psychic abilities. Instead, they generally make excuses, such as saying that the predictions didn't come true because of "psychic interference." Making testable claims but then ignoring the results of the tests marks the claimed ability to see the future as pseudoscience.

**Objectivity in Science** We generally think of science as being objective, meaning that all people should in principle be able to find the same scientific results. However, there is a difference between the overall objectivity of science and the objectivity of individual scientists.

Science is practiced by human beings, and individual scientists may bring their personal biases and beliefs to their scientific work. For example, most scientists choose their research projects based on personal interests rather than on some objective formula. In extreme cases, scientists have even been known to cheat—either deliberately or subconsciously—to obtain a result they desire. For example, in the late 19th century, astronomer Percival Lowell claimed to see a network of artificial canals in blurry telescopic images of Mars, leading him to conclude that there was a great Martian civilization [**Section 9.4**]. But no such canals exist, so Lowell must have allowed his beliefs about extraterrestrial life to influence the way he interpreted what he saw—in essence, a form of cheating, though almost certainly not intentional.

Bias can occasionally show up even in the thinking of the scientific community as a whole. Some valid ideas may not be considered by any scientist because they fall too far outside the general patterns of thought, or **paradigm**, of the time. Einstein's theory of relativity provides an example. Many scientists in the decades before Einstein had gleaned hints of the theory but did not investigate them, at least in part because they seemed too outlandish.

The beauty of science is that it encourages continued testing by many people. Even if personal biases affect some results, tests by others should eventually uncover the mistakes. Similarly, if a new idea is correct but falls outside the accepted paradigm, sufficient testing and verification of the idea will eventually force a paradigm shift. In that

## SPECIAL TOPIC Logic and Science

In science, we attempt to acquire knowledge through logical reasoning. A logical argument begins with a set of premises and leads to one or more conclusions. There are two basic types of logical argument: deductive and inductive.

In a *deductive argument*, the conclusion follows automatically from the premises, as in this example:

> PREMISE: All planets orbit the Sun in ellipses with the Sun at one focus.
>
> PREMISE: Earth is a planet.
>
> CONCLUSION: Earth orbits the Sun in an ellipse with the Sun at one focus.

Note that the first premise is a general statement that applies to all planets, and the conclusion is a specific statement that applies only to Earth. In other words, we use a deductive argument to *deduce* a specific prediction from a more general theory. If the specific prediction proves to be false, then something must be wrong with the premises from which it was deduced. If it proves true, then we've acquired a piece of evidence in support of the premises.

Now consider the following example of an *inductive argument:*

> PREMISE: Birds fly up but eventually come back down.
>
> PREMISE: People who jump up fall back down.
>
> PREMISE: Rocks thrown up come back down.
>
> PREMISE: Balls thrown up come back down.
>
> CONCLUSION: What goes up must come down.

Notice that the inductive argument begins with specific facts that are used to *generalize* to a broader conclusion. In this case, each premise supports the conclusion, which may explain why the conclusion was thought to be true for thousands of years. However, no amount of additional examples could ever prove the conclusion to be true, and we need only a single counterexample—such as a rocket leaving Earth—to prove the conclusion to be false.

Both types of argument are important in science. We use inductive arguments to build scientific theories, because we *infer* general principles from observations and experiments. We use deductive arguments to make specific predictions from hypotheses and theories, which we can then test. This explains why theories can never be proved true beyond all doubt—they can only be shown to be consistent with ever-larger bodies of evidence. Theories *can* be proved false, however, if they fail to account for observed or experimental facts.

sense, *science ultimately provides a means of bringing people to agreement*, at least on topics that can be subjected to scientific study.

## What is a scientific theory?

The most successful scientific models explain a wide variety of observations in terms of just a few general principles. When a powerful yet simple model makes predictions that survive repeated and varied testing, scientists elevate its status and call it a **theory**. Some famous examples are Isaac Newton's theory of gravity, Charles Darwin's theory of evolution, and Albert Einstein's theory of relativity.

Note that the scientific meaning of the word *theory* is quite different from its everyday meaning, in which we equate a theory more closely with speculation or a hypothesis. For example, someone might say, "I have a new theory about why people enjoy the beach." Without the support of a broad range of evidence that others have tested and confirmed, this "theory" is really only a guess. In contrast, Newton's theory of gravity qualifies as a scientific theory because it uses simple physical principles to explain many observations and experiments. *Theory* is just one of many terms that are used with different meaning in science than in everyday life. **TABLE 3.2** summarizes a few of the most common of these terms.

**TABLE 3.2 Scientific Usage Often Differs from Everyday Usage**

*This table lists some words you will encounter in the media that have a different meaning in science than in everyday life. (Adapted from a table published by Richard Somerville and Susan Joy Hassol in Physics Today.)*

| Term | Everyday Meaning | Scientific Meaning | Example |
|---|---|---|---|
| model | something you build, like a model airplane | a representation of nature, sometimes using mathematics or computer simulations, that is intended to explain or predict observed phenomena | A model of planetary motion can be used to calculate exactly where planets should appear in our sky. |
| hypothesis | a guess or assumption of almost any type | a model that has been proposed to explain some observations but that has not yet been rigorously confirmed | Scientists hypothesize that the Moon was formed by a giant impact, but there is not enough evidence to be fully confident in this model. |
| theory | speculation | a particularly powerful model that has been so extensively tested and verified that we have extremely high confidence in its validity | Einstein's theory of relativity successfully explains a broad range of natural phenomena and has passed a great many tests of its validity. |
| bias | distortion, political motive | tendency toward a particular result | Current techniques for detecting extrasolar planets are biased toward detecting large planets. |
| critical | really important; involving criticism, often negative | right on the edge, near a boundary | A boiling point is a "critical value" because above that temperature, a liquid will boil away. |
| deviation | strangeness or unacceptable behavior | change or difference | The recent deviation in global temperatures from their long-term average implies that something is heating the planet. |
| enhance/ enrich | improve | increase or add more, but not necessarily to make something "better" | "Enhanced colors" means colors that have been brightened. "Enriched with iron" means containing more iron. |
| error | mistake | range of uncertainty | The "margin of error" tells us how closely measured values are likely to reflect true values. |
| feedback | a response | a self-regulating (negative feedback) or self-reinforcing (positive feedback) cycle | Gravity can provide positive feedback to a forming planet: Adding mass leads to stronger gravity, which leads to more added mass, and so on. |
| state (as a noun) | a place or location | a description of current condition | The Sun is in a state of balance, so it shines steadily. |
| uncertainty | ignorance | a range of possible values around some central value | The measured age of our solar system is 4.55 billion years with an uncertainty of 0.02 billion years. |
| values | ethics, monetary value | numbers or quantities | The speed of light has a measured value of 300,000 km/s. |

Despite its success in explaining observed phenomena, a scientific theory can never be proved true beyond all doubt, because future observations may disagree with its predictions. However, anything that qualifies as a scientific theory must be supported by a large, compelling body of evidence.

In this sense, a scientific theory is not at all like a hypothesis or any other type of guess. We are free to change a hypothesis at any time, because it has not yet been carefully tested. In contrast, the results of the many tests that a scientific theory has already passed cannot simply disappear, so if a theory ever fails a new test, a replacement theory must still explain the past results.

Again, the theories of Newton and Einstein offer good examples. A vast body of evidence supports Newton's theory of gravity, but in the late 19th century scientists began to discover cases where its predictions did not perfectly match observations. These discrepancies were explained only when Einstein developed his general theory of relativity in the early 20th century. Still, the many successes of Newton's theory could not be ignored, and Einstein's theory would not have gained acceptance if it had not been able to explain these successes equally well. It did, and that is why we now view Einstein's theory as a broader theory of gravity than Newton's theory. Some scientists today are seeking a theory of gravity that will go beyond Einstein's. If any new theory ever gains acceptance, it will have to match all the successes of Einstein's theory as well as work in new realms where Einstein's theory does not.

**Think about it** When people claim that something is "only a theory," what do you think they mean? Does this meaning of "theory" agree with the definition of a theory in science? Do scientists always use the word *theory* in its "scientific" sense? Explain.

## 3.5 Astrology

We have discussed the development of astronomy and the nature of science in some depth. Now let's talk a little about a subject often confused with the science of astronomy: *astrology*. Although the terms *astrology* and *astronomy* sound very similar, today they describe very different practices. In ancient times, however, astrology and astronomy often went hand in hand, and astrology played an important role in the historical development of astronomy.

### How is astrology different from astronomy?

The basic tenet of astrology is that the apparent positions of the Sun, Moon, and planets among the stars in our sky influence human events. The origins of this idea are easy to understand. After all, the position of the Sun in the sky certainly influences our lives, since it determines the seasons and the times of daylight and darkness, and the Moon's position determines the tides. Because planets also move among the stars, it probably seemed natural to imagine that they might also influence our lives, even if the influences were more subtle.

Ancient astrologers hoped to learn how the positions of the Sun, Moon, and planets influence our lives by charting the skies and seeking correlations with events on Earth. For example, if an earthquake occurred when Saturn was entering the constellation Leo, might Saturn's position have been the cause of the earthquake? If the king became ill when Mars appeared in the constellation Gemini and the first-quarter moon appeared in Scorpio, might another tragedy be in store for the king when this particular alignment of the Moon and Mars next recurred? Surely, the ancient astrologers thought, the patterns of influence would eventually become clear, and they would then be able to forecast human events with the same reliability with which astronomical observations of the Sun could be used to forecast the coming of spring.

Because forecasts of the seasons and forecasts of human events were imagined to be closely related, astrologers and astronomers usually were one and the same in the ancient world. For example, in addition to his books on astronomy, Ptolemy published a treatise on astrology called *Tetrabiblios,* which remains the foundation for much of astrology today. Interestingly, Ptolemy himself recognized that

---

**EXTRAORDINARY CLAIMS** **Earth Orbits the Sun**

In the 21st century, claiming that Earth orbits the Sun will not raise any eyebrows, but it was quite an extraordinary claim in the 3rd century B.C., when Greek astronomer Aristarchus put it forward. To almost everyone else of his time, the idea that the Sun moves while Earth remains stationary seemed like plain common sense. However, Aristarchus was also a mathematician and he used mathematical reasoning to conclude that observations of the sky made more sense if the Sun, and not Earth, was at the center of the solar system (see the Special Topic, page 48).

In this and similar boxes elsewhere in the book, we will look at scientific claims that seemed extraordinary in their time. As astronomer Carl Sagan was fond of saying, "extraordinary claims require extraordinary evidence," and we will discuss how scientific evidence ended up supporting or debunking those claims, or in some cases leaving them still unanswered. Each case will illustrate the self-correcting nature of science: Mistaken ideas are eventually disproved, while a few ideas that once appeared extraordinary end up gaining widespread acceptance.

In the case of Aristarchus, the evidence proving his claim did not become strong enough to convince most other scholars until almost two millennia after his death. Nevertheless, Aristarchus's Sun-centered idea remained alive throughout this time, and apparently influenced Copernicus when he proposed his own, more detailed Sun-centered model. As discussed in this chapter, others including Tycho, Kepler, and Galileo then collected the evidence that ultimately led to widespread acceptance of Aristarchus's extraordinary claim. The case was later sealed after Newton provided a physical understanding of *why* Kepler's laws hold and astronomers collected direct evidence, including measurements of stellar parallax, that proved beyond a shadow of doubt that Earth orbits the Sun.

**Verdict:** Clearly correct.

astrology stood upon a far shakier foundation than astronomy. In the introduction to *Tetrabiblios,* he wrote:

> *[Astronomy], which is first both in order and effectiveness, is that whereby we apprehend the aspects of the movements of sun, moon, and stars in relation to each other and to the earth. . . . I shall now give an account of the second and less sufficient method [of prediction (astrology)] in a proper philosophical way, so that one whose aim is the truth might never compare its perceptions with the sureness of the first, unvarying science. . . .*

Other ancient scientists also recognized that their astrological predictions were far less reliable than their astronomical ones. Nevertheless, confronted with even a slight possibility that astrologers could forecast the future, no king or political leader would dare to be without one. Astrologers held esteemed positions as political advisers in the ancient world and were provided with the resources they needed to continue charting the heavens and human history. Wealthy political leaders' support of astrology made possible much of the development of ancient astronomy.

Throughout the Middle Ages and into the Renaissance, many astronomers continued to practice astrology. For example, Kepler cast numerous *horoscopes*—the predictive charts of astrology—even as he was discovering the laws of planetary motion. However, given Kepler's later descriptions of astrology as "the foolish stepdaughter of astronomy" and "a dreadful superstition," he may have cast the horoscopes solely as a source of much-needed income. Modern-day astrologers also claim Galileo as one of their own, in part for his having cast a horoscope for the Grand Duke of Tuscany. However, while Galileo's astronomical discoveries changed human history, the horoscope was just plain wrong: The Duke died a few weeks after Galileo predicted that he would have a long and fruitful life.

The scientific triumph of Kepler and Galileo in showing Earth to be a planet orbiting the Sun heralded the end of the linkage between astronomy and astrology. Astronomy has since gained status as a successful science that helps us understand our universe, while astrology no longer has any connection to the modern science of astronomy.

## Does astrology have any scientific validity?

Although astronomers gave up on it centuries ago, astrology remains popular with the public. Many people read their daily horoscopes, and some pay significant fees to have personal horoscopes cast by professional astrologers. With so many people giving credence to astrology, is it possible that it has some scientific validity after all?

**Testing Astrology** The validity of astrology can be difficult to assess, because there's no general agreement among astrologers even on such basic things as what astrology is or what it can predict. For example, "Western astrology" is quite different in nature from the astrology practiced in India and China. Some astrologers do not make testable predictions at all; rather, they give vague guidance about how to live one's life. Most daily horoscopes fall into this category. Although your horoscope may seem to ring true at first, a careful read will usually show it to be so vague as to be untestable. A horoscope that says "It is a good day to spend time with friends" may be good advice but doesn't offer much to test.

**See it for yourself** Find a local weather forecast and a horoscope for today. Contrast the nature of their predictions. At the end of the day, you will know if the weather forecast was accurate. Will you be able to say whether the horoscope was accurate? Explain.

Nevertheless, most professional astrologers still earn their livings by casting horoscopes that either predict future events in an individual's life or describe characteristics of the person's personality and life. If the horoscope predicts future events, we can check to see whether the predictions come true. If it describes a person's personality and life, the description can be checked for accuracy. A scientific test of astrology requires evaluating many horoscopes and comparing their accuracy to what would be expected by pure chance. For example, suppose a horoscope states that a person's best friend is female. Because roughly half the population of the United States is female, an astrologer who casts 100 such horoscopes would be expected by pure chance to be right about 50 times. We would be impressed with the predictive ability of the astrologer only if he or she were right much more often than 50 times out of 100.

In hundreds of scientific tests, astrological predictions have never proved to be significantly more accurate than expected from pure chance. Similarly, in tests in which astrologers are asked to cast horoscopes for people they have never met, the horoscopes fail to match actual personality profiles more often than expected by chance. The verdict is clear: The methods of astrology are useless for predicting the past, the present, or the future.

**Examining the Underpinnings of Astrology** In science, observations and experiments are the ultimate judge of any idea. No matter how outlandish an idea might appear, it cannot be dismissed if it successfully meets observational or experimental tests. The idea that Earth rotates and orbits the Sun seemed outlandish for most of human history, yet today it is so strongly supported by the evidence that we consider it a fact. The idea that the positions of the Sun, Moon, and planets among the stars influence our lives might sound outlandish today, but if astrology were to make predictions that came true, adherence to the principles of science would force us to take astrology seriously. However, given that scientific tests of astrology have never found any evidence that its predictive methods work, it is worth looking at its premises to see whether they make sense. Might there be a few kernels of wisdom buried within the lore of astrology?

Let's begin with one of the key premises of astrology: that there is special meaning in the patterns of the stars in the constellations. This idea may have seemed reasonable in ancient times, when the stars were assumed to be fixed on an unchanging celestial sphere, but today we know that the patterns of the stars in the constellations are accidents of the moment. Long ago the constellations did not look the same, and they will also look different in the future [**Section 1.3**]. Moreover, the stars in a constellation don't necessarily have any *physical* association, because two stars that are close together in the sky might lie at vastly different distances (see Figure 2.3). Constellations are only *apparent* associations of stars, with no more physical reality than the water in a desert mirage.

Astrology also places great importance on the positions of the planets among the constellations. Again, this idea might have seemed reasonable in ancient times, when it was thought that the planets truly wandered among the stars. Today we know that the planets only *appear* to wander among the stars, much as your hand might appear to move among distant mountains when you wave it (**FIGURE 3.26**). It is difficult to see how mere appearances could have profound effects on our lives.

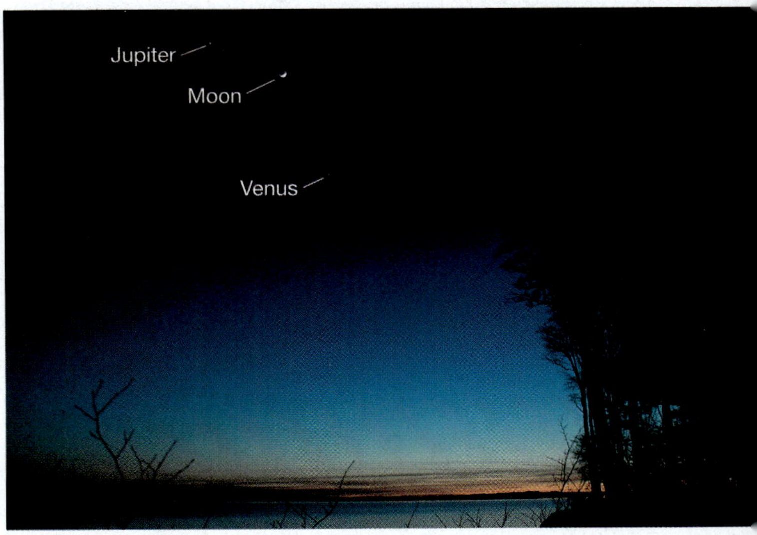


**FIGURE 3.26** This photo shows Jupiter and Venus appearing near the Moon in the night sky. Astrologers place great importance on the relative positions of the Sun, Moon, planets, and stars, but in reality these objects are nowhere near one another in space.

Many other ideas at the heart of astrology are equally suspect. For example, most astrologers claim that a proper horoscope must account for the positions of *all* the planets. Does this mean that all horoscopes cast before the discovery of Neptune in 1846 were invalid? If so, why didn't astrologers notice that something was wrong with their horoscopes and predict the existence of Neptune? (In contrast, astronomers *did* predict its existence; see the Special Topic on page 315.) Most astrologers have included Pluto since its discovery in 1930; does this mean that they should now stop including it, since it has been demoted to dwarf planet, or that they need to include Eris and other dwarf planets, including some that may not yet have been discovered? And why stop with our own solar system; shouldn't horoscopes also depend on the positions of planets orbiting other stars? Given seemingly unanswerable questions like these, there seems little hope that astrology will ever meet its ancient goal of forecasting human events.

## The BIG Picture PUTTING CHAPTER 3 INTO CONTEXT

*In this chapter, we focused on the scientific principles through which we have learned so much about the universe. Key "big picture" concepts from this chapter include the following:*

- The basic ingredients of scientific thinking—careful observation and trial-and-error testing—are a part of everyone's experience. Modern science simply provides a way of organizing this thinking to facilitate the learning and sharing of new knowledge.
- Although our understanding of the universe is growing rapidly today, each new piece of knowledge builds on ideas that came before.
- The Copernican revolution, which overthrew the ancient Greek belief in an Earth-centered universe, unfolded over a period of more than a century. Many of the characteristics of modern science first appeared during this time.
- Science exhibits several key features that distinguish it from nonscience and that in principle allow anyone to come to the same conclusions when studying a scientific question.
- Astronomy and astrology once developed hand in hand, but today they represent very different things.

**MY COSMIC PERSPECTIVE** Modern science, which grew out of the Copernican revolution, affects every one of us both in the way it helps us understand the world (and universe) and in the fact that it has driven the development of virtually all technology.

## Summary of Key Concepts

### 3.1 The Ancient Roots of Science

- **In what ways do all humans use scientific thinking?**
  Scientific thinking relies on the same type of trial-and-error thinking that we use in our everyday lives, but in a carefully organized way.
- **How is modern science rooted in ancient astronomy?**
  Ancient astronomers were accomplished observers who learned to tell the time of day and the time of year, to track cycles of the Moon, and to observe planets and stars. The care and effort that went into these observations helped set the stage for modern science.

## 3.2 Ancient Greek Science

- **Why does modern science trace its roots to the Greeks?** The Greeks developed **models** of nature and emphasized the importance of agreement between the predictions of those models and observations of nature.
- **How did the Greeks explain planetary motion?** The Greek **geocentric model** reached its culmination with the **Ptolemaic model**, which explained apparent retrograde motion by having each planet move on a small circle whose center moves around Earth on a larger circle.

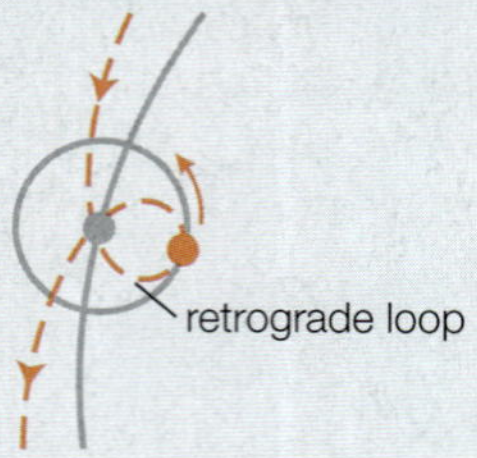


## 3.3 The Copernican Revolution

- **How did Copernicus, Tycho, and Kepler challenge the Earth-centered model?** Copernicus created a Sun-centered model of the solar system designed to replace the Ptolemaic model, but it was no more accurate than Ptolemy's because Copernicus still used perfect circles. Tycho's accurate, naked-eye observations provided the data needed to improve on Copernicus's model. Kepler developed a model of planetary motion that fit Tycho's data.
- **What are Kepler's three laws of planetary motion?** (1) The orbit of each planet is an ellipse with the Sun at one focus. (2) A planet moves faster in the part of its orbit nearer the Sun and slower when farther from the Sun, sweeping out equal areas in equal times. (3) More distant planets orbit the Sun at slower average speeds, obeying the mathematical relationship $p^2 = a^3$.

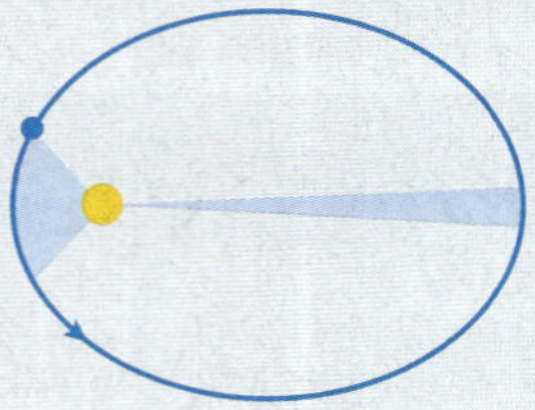

- **How did Galileo solidify the Copernican revolution?** Galileo's experiments and telescopic observations overcame remaining objections to the Copernican idea of Earth as a planet orbiting the Sun. Although not everyone accepted his results immediately, in hindsight we see that Galileo sealed the case for the Sun-centered solar system.

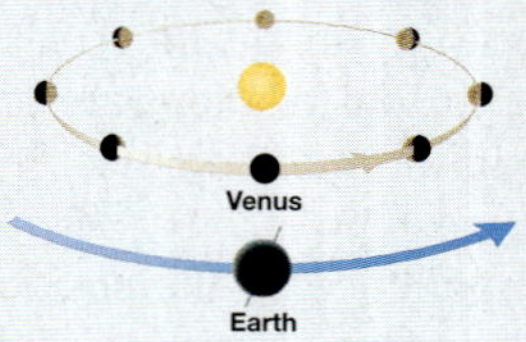


## 3.4 The Nature of Science

- **How can we distinguish science from nonscience?** Science generally exhibits three hallmarks: (1) Modern science seeks explanations for observed phenomena that rely solely on natural causes. (2) Science progresses through the creation and testing of models of nature that explain the observations as simply as possible. (3) A scientific model must make testable predictions about natural phenomena that would force us to revise or abandon the model if the predictions did not agree with observations.
- **What is a scientific theory?** A scientific **theory** is a simple yet powerful model that explains a wide variety of observations using just a few general principles and has been verified by repeated and varied testing.

## 3.5 Astrology

- **How is astrology different from astronomy?** Astronomy is a modern science that has taught us much about the universe. Astrology is a search for hidden influences on human lives based on the apparent positions of planets and stars in the sky; it does not follow the tenets of science.
- **Does astrology have any scientific validity?** Scientific tests have shown that astrological predictions do not prove to be accurate more than we can expect by pure chance, showing that the predictions have no scientific validity.

# Visual Skills Check

*Use the following questions to check your understanding of some of the many types of visual information used in astronomy. For additional practice, try the Chapter 3 Visual Quiz in the Study Area at www.MasteringAstronomy.com.*

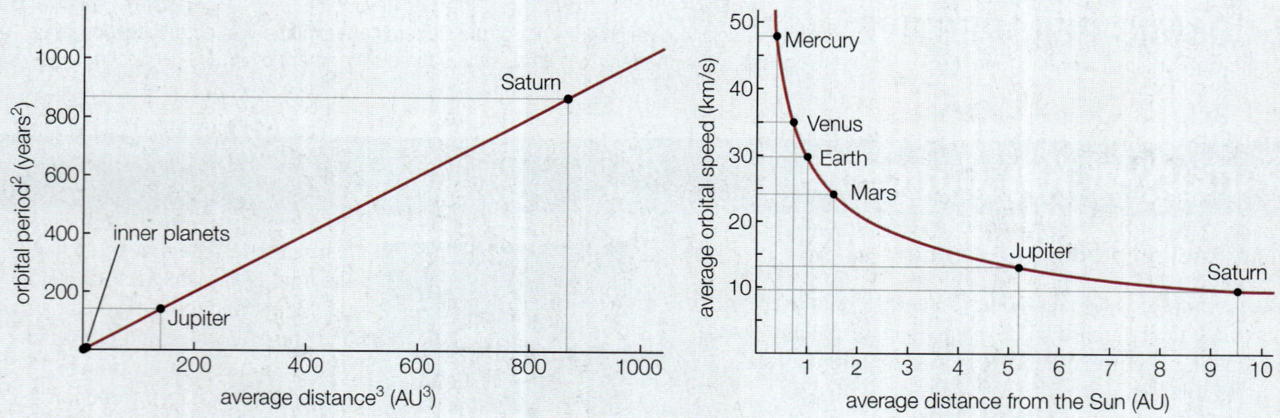


a b

*Study the two graphs above, based on Figure 3.19. Use the information in the graphs to answer the following questions.*

1. Approximately how fast is Jupiter orbiting the Sun?
   a. This cannot be determined from the information provided.
   b. 20 km/s
   c. 10 km/s
   d. a little less than 15 km/s
2. An asteroid with an average orbital distance of 2 AU will orbit the Sun at an average speed that is
   a. a little slower than the orbital speed of Mars.
   b. a little faster than the orbital speed of Mars.
   c. the same as the orbital speed of Mars.
3. Uranus, not shown on graph b, orbits about 19 AU from the Sun. Based on the graph, its approximate orbital speed is between about
   a. 20 and 25 km/s.
   b. 15 and 20 km/s.
   c. 10 and 15 km/s.
   d. 5 and 10 km/s.
4. Kepler's third law is often stated as $p^2 = a^3$. The value $a^3$ for a planet is shown on
   a. the horizontal axis of graph **a**.
   b. the vertical axis of graph **a**.
   c. the horizontal axis of graph **b**.
   d. the vertical axis of graph **b**.
5. On graph **a**, you can see Kepler's third law $(p^2 = a^3)$ from the fact that
   a. the data fall on a straight line.
   b. the axes are labeled with values for $p^2$ and $a^3$.
   c. the planet names are labeled on the graph.
6. Suppose graph **a** showed a planet on the red line directly above a value of 1000 $AU^3$ along the horizontal axis. On the vertical axis, this planet would be at
   a. 1000 $years^2$.
   b. $1000^2$ $years^2$.
   c. $\sqrt{1000}$ $years^2$.
   d. 100 years.
7. How far does the planet in question 6 orbit from the Sun?
   a. 10 AU
   b. 100 AU
   c. 1000 AU
   d. $\sqrt{1000}$ AU

# Exercises and Problems

*For instructor-assigned homework and other learning materials, go to www.MasteringAstronomy.com.*

## Chapter Review Questions

### Short-Answer Questions Based on the Reading

1. In what way is scientific thinking natural to all of us, and how does modern science build upon this everyday type of thinking?
2. Why did ancient peoples study astronomy? Describe an astronomical achievement of at least three ancient cultures.
3. Describe the astronomical origins of our day, week, month, and year.
4. What is a lunar calendar? How can it be kept roughly synchronized with a solar calendar?
5. What do we mean by a *model* in science?
6. Summarize the development of the Greek *geocentric model* through Ptolemy. How did the *Ptolemaic model* account for apparent retrograde motion?
7. What was the *Copernican revolution*, and how did it change the human view of the universe?
8. What is an *ellipse*? Define its *foci*, *semimajor axis*, and *eccentricity*.
9. State and explain the meaning of each of *Kepler's laws of planetary motion*.
10. Describe the three hallmarks of science and how we can see them in the Copernican revolution. What is *Occam's razor*? Why doesn't science accept personal testimony as evidence?
11. What is the difference between a *hypothesis* and a *theory* in science?
12. What is the basic idea behind *astrology*? Explain why this idea seemed reasonable in ancient times but is no longer accepted by scientists.

### Science or Nonscience?

*Decide whether the claim made in each of the following statements could be evaluated scientifically or falls into the realm of nonscience. Explain clearly; not all of these have definitive answers, so your explanation is more important than your chosen answer.*

13. The Yankees are the best baseball team of all time.
14. Several kilometers below its surface, Jupiter's moon Europa has an ocean of liquid water.
15. My house is haunted by ghosts who make the creaking noises I hear each night.
16. There are no lakes or seas on the surface of Mars today.
17. Dogs are smarter than cats.
18. Children born when Jupiter is in the constellation Taurus are more likely to be musicians than other children.
19. Aliens can manipulate time and memory so that they can abduct and perform experiments on people who never realize they were taken.
20. Newton's law of gravity works as well for explaining orbits of planets around other stars as it does for explaining orbits of the planets in our own solar system.
21. God created the laws of motion that were discovered by Newton.
22. A huge fleet of alien spacecraft will land on Earth and introduce an era of peace and prosperity on January 1, 2035.

### Quick Quiz

*Choose the best answer to each of the following. For additional practice, try the Chapter 3 Reading and Concept Quizzes in the Study Area at www.MasteringAstronomy.com.*

23. In the Greek geocentric model, the retrograde motion of a planet occurs when (a) Earth is about to pass the planet in its orbit around the Sun. (b) the planet actually goes backward in its orbit around Earth. (c) the planet is aligned with the Moon in our sky.
24. Which of the following was *not* a major advantage of Copernicus's Sun-centered model over the Ptolemaic model? (a) It

made significantly better predictions of planetary positions in our sky. (b) It offered a more natural explanation for the apparent retrograde motion of planets in our sky. (c) It allowed calculation of the orbital periods and distances of the planets.

25. When we say that a planet has a highly *eccentric* orbit, we mean that (a) it is spiraling in toward the Sun. (b) its orbit is an ellipse with the Sun at one focus. (c) in some parts of its orbit it is much closer to the Sun than in other parts.
26. Earth is closer to the Sun in January than in July. Therefore, in accord with Kepler's second law, (a) Earth travels faster in its orbit around the Sun in July than in January. (b) Earth travels faster in its orbit around the Sun in January than in July. (c) it is summer in January and winter in July.
27. According to Kepler's third law, (a) Mercury travels fastest in the part of its orbit in which it is closest to the Sun. (b) Jupiter orbits the Sun at a faster speed than Saturn. (c) all the planets have nearly circular orbits.
28. Tycho Brahe's contribution to astronomy included (a) inventing the telescope. (b) proving that Earth orbits the Sun. (c) collecting data that enabled Kepler to discover the laws of planetary motion.
29. Galileo's contribution to astronomy included (a) discovering the laws of planetary motion. (b) discovering the law of gravity. (c) making observations and conducting experiments that dispelled scientific objections to the Sun-centered model.
30. Which of the following is *not* true about scientific progress? (a) Science progresses through the creation and testing of models of nature. (b) Science advances only through the scientific method. (c) Science avoids explanations that invoke the supernatural.
31. Which of the following is *not* true about a scientific theory? (a) A theory must explain a wide range of observations or experiments. (b) Even the strongest theories can never be proved true beyond all doubt. (c) A theory is essentially an educated guess.
32. When Einstein's theory of gravity (general relativity) gained acceptance, it demonstrated that Newton's theory had been (a) wrong. (b) incomplete. (c) really only a guess.

## Inclusive Astronomy

*Use these questions to reflect on participation in science.*

33. *Group Discussion: Ancestral Astronomy.* No matter what your background, you had ancestors who watched the sky and observed how celestial objects move through it.
   **a.** Working independently, choose a particular branch of your ancestry to explore, then gather historical information dating back as far as possible (ideally at least several centuries) about how and why your ancestors made use of their observations of the sky.
   **b.** Gather in small groups (two to four students) and take turns sharing what you learned through your research. Be clear about the ancestral group you have chosen and the time period your research covers.
   **c.** Make a list of the major uses of astronomical knowledge that your group members found, categorizing the uses as practical, ceremonial/religious, or other. Which uses were most common? Do you see any noticeable differences among the cultures?
   **d.** Discuss how cultural or geographical factors may have influenced the astronomical knowledge of these different ancestral groups.

## The Process of Science

*These questions may be answered individually in short-essay form or discussed in groups, except where identified as group-only.*

34. *What Makes It Science?* Choose a single idea in the modern view of the cosmos as discussed in Chapter 1, such as "The universe is expanding," "The universe began with a Big Bang," "We are made from elements manufactured by stars," or "The Sun orbits the center of the Milky Way Galaxy once every 230 million years."
   **a.** Describe how this idea reflects each of the three hallmarks of science, discussing how it is based on observations, how our understanding of it depends on a model, and how that model is testable.
   **b.** Describe a hypothetical observation that, if it were actually made, might cause us to call the idea into question. Then briefly discuss whether you think that, overall, the idea is likely or unlikely to hold up to future observations.
35. *The Importance of Ancient Astronomy.* Why was astronomy important to people in ancient times? Discuss both the practical importance of astronomy and the importance it may have had for religious, ceremonial, or philosophical traditions. Which of those roles (practical or religious/ceremonial/philosophical) do you think was more important in leading to the development of modern astronomy? Defend your opinion.
36. *The Impact of Science.* The modern world is filled with ideas, knowledge, and technology that developed through science and application of the scientific method. Discuss some of these things and how they affect our lives. Which of these impacts do you think are positive? Which are negative? Overall, do you think science has benefited the human race? Defend your opinion.
37. *Astronomy and Astrology.* Why do you think astrology remains so popular around the world even though it has failed all scientific tests of its validity? Do you think the popularity of astrology has any positive or negative social consequences? Defend your opinions.
38. *Group Activity: Galileo on Trial.* Conduct a mock trial in which you consider the following three pieces of evidence: (1) observations of mountains and valleys on the Moon; (2) observations of moons orbiting Jupiter; (3) observations of the phases of Venus. Note: This activity works particularly well in groups of four or more students, in which
   - One student takes the role of Galileo, arguing in favor of the idea that Earth orbits the Sun.
   - One student takes the role of Prosecutor, arguing against the idea that Earth orbits the Sun.
   - The remaining students serve as a jury and render a verdict after hearing the arguments and rebuttals from Galileo and the prosecutor.

## Investigate Further

### Short-Answer/Essay Questions

39. *Earth's Shape.* It took thousands of years for humans to deduce that Earth is spherical. For each of the following alternative models of Earth's shape, identify one or more observations that you could make for yourself that would invalidate the model.
   **a.** A flat Earth **b.** A football-shaped Earth
   **c.** A cylindrical Earth, like that proposed by Anaximander

40. *Lunar Calendars.*

**a.** Find the dates of the Jewish festival of Chanukah for this year and the next three years. Based on what you have learned in this chapter, explain why the dates change as they do.

**b.** Find the dates of the Muslim fast for Ramadan for this year and the next three years. Based on what you have learned in this chapter, explain why the dates change as they do.

41. *Copernican Players.* Using a bulleted-list format, make a one-page "executive summary" of the major roles that Copernicus, Tycho, Kepler, and Galileo played in overturning the ancient belief in an Earth-centered universe.

42. *Easter Timing.* Research when different denominations of Christianity celebrate Easter and why they use different dates. Summarize your findings in a one- to two-page report.

43. *Greek Astronomers.* Many ancient Greek scientists had ideas that, in retrospect, seem well ahead of their time. Choose one ancient Greek scientist to study, and write a one- to two-page "scientific biography" of your chosen person.

44. *Influence on History.* Based on what you have learned about the Copernican revolution, write a one- to two-page essay about how you believe it altered the course of human history.

45. *Cultural Astronomy.* Choose a particular culture of interest to you, and research the astronomical knowledge and accomplishments of that culture. Write a two- to three-page summary of your findings.

46. *Astronomical Structures.* Choose an ancient astronomical structure of interest to you (e.g., Stonehenge, Templo Mayor, Pawnee lodges), and research its history. Write a two- to three-page summary of your findings. If possible, also build a scale model of the structure or create detailed diagrams to illustrate how the structure was used.

## Quantitative Problems

*Be sure to show all calculations clearly and state your final answers in complete sentences.*

47. *The Metonic Cycle.* The length of our calendar year is 365.2422 days, and the Moon's monthly cycle of phases averages 29.5306 days in length. By calculating the number of days in each, confirm that 19 solar years is almost precisely equal to 235 cycles of the lunar phases. Show your work clearly; then write a few sentences explaining how this fact can be used to keep a lunar calendar roughly synchronized with a solar calendar.

48. *Chinese Calendar.* The traditional Chinese lunar calendar has 12 months in most years but adds a 13th month to 22 of every 60 years. How many days does this give the Chinese calendar in each 60-year period? How does this compare to the number of days in 60 years on a solar calendar? Based on your answers, explain how this scheme is similar to the scheme used by lunar calendars that follow the Metonic cycle. (*Hint:* You'll need the data given in Problem 47.)

49. *Method of Eratosthenes I.* You are an astronomer on planet Nearth, which orbits a distant star. It has recently been accepted that Nearth is spherical in shape, though no one knows its size. One day, while studying in the library of Alectown, you learn that on the equinox your sun is directly overhead in the city of Nyene, located 1000 kilometers due north of you. On the equinox, you go outside and observe that the altitude of your sun is 80°. What is the circumference of Nearth? (*Hint:* Apply the technique used by Eratosthenes to measure Earth's circumference.)

50. *Method of Eratosthenes II.* You are an astronomer on planet Tirth, which orbits a distant star. It has recently been accepted that Tirth is spherical in shape, though no one knows its size. One day, you learn that on the equinox your sun is directly overhead in the city of Tyene, located 400 kilometers due north of you. On the equinox, you go outside and observe that the altitude of your sun is 86°. What is the circumference of Tirth? (*Hint:* Apply the technique used by Eratosthenes to measure Earth's circumference.)

51. *Mars Orbit.* Find the perihelion and aphelion distances of Mars. (*Hint:* You'll need data from Appendix E.)

52. *Eris Orbit.* The dwarf planet Eris orbits the Sun every 557 years. What is its average distance (semimajor axis) from the Sun? How does its average distance compare to that of Pluto?

53. *New Planet Orbit.* A newly discovered planet orbits a distant star with the same mass as the Sun at an average distance of 112 million kilometers. Its orbital eccentricity is 0.3. Find the planet's orbital period and its nearest and farthest orbital distances from its star.

54. *Halley Orbit.* Halley's Comet orbits the Sun every 76.0 years and has an orbital eccentricity of 0.97.

**a.** Find its average distance from the Sun (semimajor axis).

**b.** Find its perihelion and aphelion distances.

# S1 Celestial Timekeeping and Navigation

SUPPLEMENTARY CHAPTER

▲ **About the photo:** The path of the Sun on the June solstice at the Arctic Circle, from about 11 p.m. to 9 a.m.

## LEARNING GOALS

### S1.1 Astronomical Time Periods

- How do we define the day, month, year, and planetary periods?
- How do we tell the time of day?
- When and why do we have leap years?

### S1.2 Celestial Coordinates and Motion in the Sky

- How do we locate objects on the celestial sphere?
- How do stars move through the local sky?
- How does the Sun move through the local sky?

### S1.3 Principles of Celestial Navigation

- How can you determine your latitude?
- How can you determine your longitude?

*Socrates*: *Shall we make astronomy the next study? What do you say?*

*Glaucon*: *Certainly. A working knowledge of the seasons, months, and years is beneficial to everyone, to commanders as well as to farmers and sailors.*

*Socrates*: *You make me smile, Glaucon. You are so afraid that the public will accuse you of recommending unprofitable studies.*

—*Plato*, Republic

**Chapter S1 Overview**

As the above quote from Plato shows, ancient astronomy served practical needs for timekeeping and navigation. These ancient uses may no longer seem so important in an age when we tell time with digital watches and navigate with the global positioning system (GPS). But knowing the celestial basis of timekeeping and navigation can help us understand the rich history of astronomical discovery, and occasionally still proves useful in its own right. In this chapter, we will explore the apparent motions of the Sun, Moon, and planets in enough detail to learn the basic principles of keeping time and navigating by the stars.

## S1.1 Astronomical Time Periods

Although many people do not realize it, modern clocks and calendars are beautifully synchronized to the rhythms of the heavens. Precision measurements allow us to ensure that our clocks keep pace with the Sun's daily trek across our sky, while our calendar holds the dates of the equinoxes and solstices as steady as possible. In earlier chapters, we saw how this synchronicity took root in ancient observations of the sky. In this section, we will look more closely at basic measures of time and our modern, international system of timekeeping.

### How do we define the day, month, year, and planetary periods?

By now you know that the length of the day is related to Earth's rotation, the length of the month to the cycle of lunar phases, and the length of the year to our orbit around the Sun. However, the relationships are not quite as simple as you might at first guess, because these time periods can be defined in more than one way.

**The Length of the Day** We usually think of a day as the time it takes for Earth to rotate once. But Earth's rotation period is actually about 4 minutes short of 24 hours. What's going on?

Remember that the daily circling of the stars in our sky is an illusion created by Earth's rotation (see Figure 2.9). You can therefore measure Earth's rotation period by measuring how long it takes for any star to go from its highest point in the sky one day to its highest point the next day (**FIGURE S1.1a**). This time period, which we call a **sidereal** (pronounced "sy-DEAR-ee-al") **day**, is about 23 hours 56 minutes (more precisely, $23^{h}\ 56^{m}\ 4.09^{s}$). *Sidereal* means "related to the stars"; note that you'll measure the same time no matter what star you choose. For practical purposes, the sidereal day is Earth's precise rotation period.

Our 24-hour day, which we call a **solar day**, is based on the time it takes for the *Sun* to make one circuit around the local sky. You can measure this time period by measuring how long it takes the Sun to go from its highest point in the sky one day to its highest point the next day (**FIGURE S1.1b**). The solar day is indeed 24 hours on average, although it varies slightly (up to about 25 seconds longer or shorter than 24 hours) over the course of a year.

A simple demonstration shows why the solar day is about 4 minutes longer than the sidereal day. Set an object representing the Sun on a table, and stand a few steps away to represent Earth. Point at the Sun and imagine that you also happen to be pointing toward a distant star that lies in the same direction. If you rotate (counterclockwise) while standing in place, you'll again be pointing at both the Sun and the

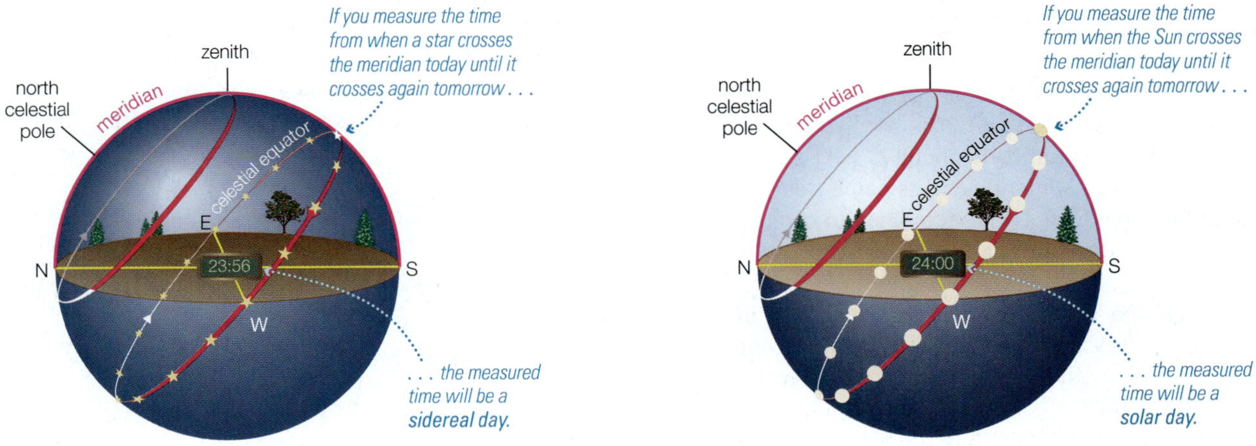


**a** A sidereal day is the time it takes any star to make a circuit of the local sky. It is about 23 hours 56 minutes.

**b** A solar day is the time it takes the Sun to make a circuit of the local sky. Its precise length varies slightly over the course of the year, but the average is 24 hours.

**FIGURE S1.1** Using the sky to measure the length of a day.

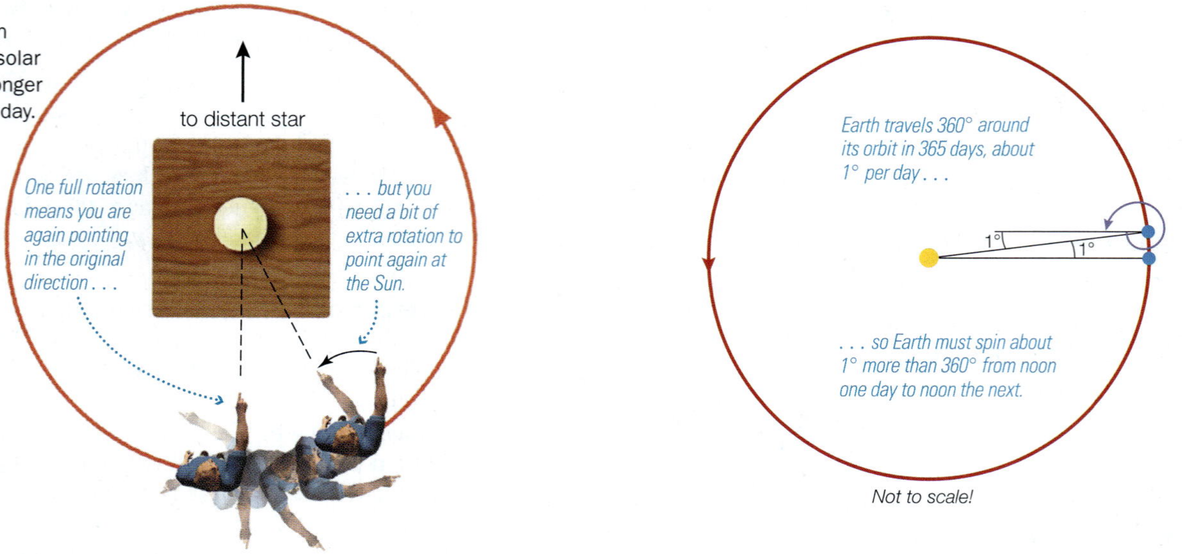


**FIGURE S1.2** A demonstration showing why a solar day is slightly longer than a sidereal day.

**a** One full rotation represents a sidereal day and returns you to pointing in your original direction, but you need to rotate a little extra to return to pointing at the Sun.

**b** Earth travels about 1° per day around its orbit, so a solar day requires about 361° of rotation.

star after one full rotation. However, to show that Earth also orbits the Sun, you should take a couple of steps around the Sun (counterclockwise) as you rotate (**FIGURE S1.2a**). After one full rotation, you will again be pointing in the direction of the distant star, so this represents a sidereal day. But notice that you need to rotate a little extra to point back at the Sun, which is why the solar day is longer than the sidereal day. To figure out how long this extra rotation takes, note that Earth travels about 1° around its orbit each day (**FIGURE S1.2b**). This extra rotation therefore takes about $\frac{1}{360}$ of Earth's rotation period—which is about 4 minutes.

**The Length of the Month** As we discussed in Chapter 2, our month comes from the Moon's $29\frac{1}{2}$-day cycle of phases (think "moonth"). More technically, this $29\frac{1}{2}$-day period is called a **synodic month**. The word *synodic* comes from the Latin *synod*, which means "meeting." A synodic month gets its name from the idea that the Sun and the Moon "meet" in the sky with every new moon.

Just as a solar day is not Earth's true rotation period, a synodic month is not the Moon's true orbital period. Earth's motion around the Sun means that the Moon must complete more than one full orbit of Earth from one new moon to the next (**FIGURE S1.3**). The Moon's true orbital period, or a **sidereal month**, is about $27\frac{1}{3}$ days. Like the sidereal day, the sidereal month gets its name from the fact that it describes how long it takes the Moon to complete an orbit relative to the positions of distant stars.

**The Length of the Year** We can also define a year in two slightly different ways. The time it takes Earth to complete one orbit relative to the stars is called a **sidereal year**. However, our calendar is based on the cycle of the seasons, which we measure as the time from the March equinox one year to the March equinox the next year. This time period, called a **tropical year**, is about 20 minutes shorter than the sidereal year. A 20-minute difference might not seem like much, but it would make a calendar based on the sidereal year get out of sync with the seasons by 1 day every 72 years—a difference that would add up over centuries.

The difference between the sidereal year and the tropical year arises from Earth's 26,000-year cycle of axis precession [**Section 2.2**]. Precession not only changes the orientation of the axis in space but also changes the locations in Earth's orbit at which the seasons occur. Each year, the location of the equinoxes and solstices among the stars shifts about $\frac{1}{26,000}$ of the way around the orbit. If you do the math, you'll find that $\frac{1}{26,000}$ of a year is about 20 minutes, which explains the 20-minute difference between the tropical year and the sidereal year.

**Planetary Periods** Although planetary periods are not used in modern timekeeping, they were important to many ancient cultures. For example, the Mayan calendar was based in part on the apparent motions of Venus. In addition, Copernicus's ability to determine orbital periods of planets with his Sun-centered model played an important role in keeping the model alive long enough for its ultimate acceptance (see Mathematical Insight S1.1).

A planet's **sidereal period** is the time the planet takes to orbit the Sun; again, the name comes from the fact that it is measured relative to distant stars. For example, Jupiter's

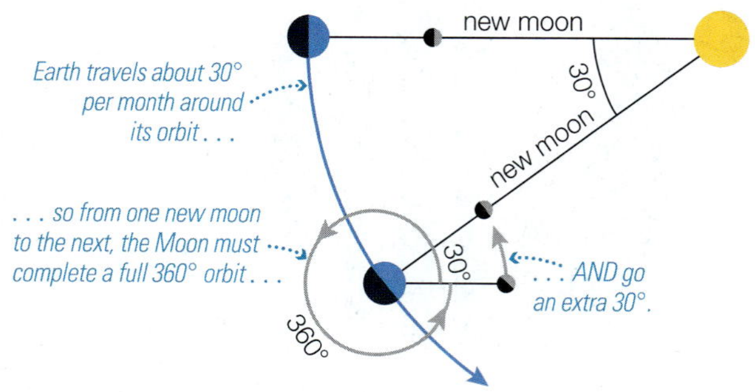


**FIGURE S1.3** The Moon completes one 360° orbit in about $27\frac{1}{3}$ days (a sidereal month), but the time from new moon to new moon is about $29\frac{1}{2}$ days (a synodic month).

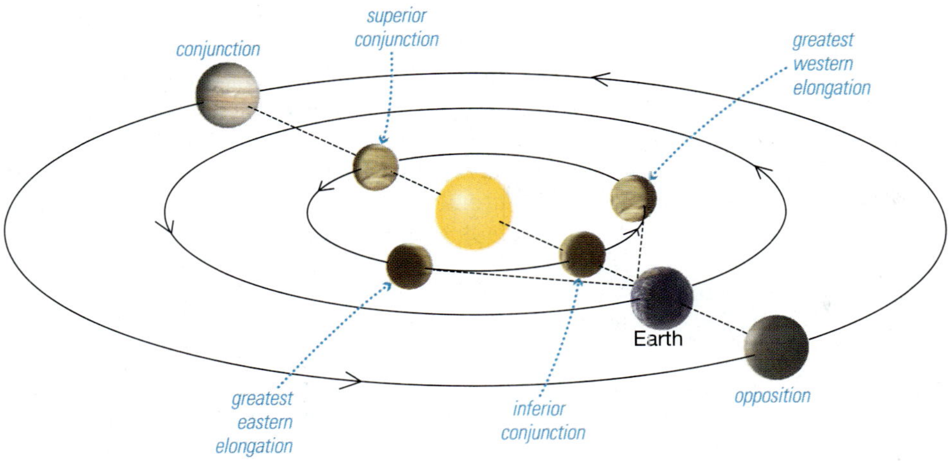


**FIGURE S1.4** This diagram shows important positions of planets relative to Earth and the Sun. For a planet farther than Earth from the Sun (such as Jupiter), conjunction occurs when the planet appears aligned with the Sun in the sky, and opposition occurs when the planet appears on our meridian at midnight. Planets nearer the Sun (such as Venus) have two conjunctions and never get farther from the Sun in our sky than at their greatest elongations. (Adapted from *Advanced Skywatching*, by Burnham et al.)

sidereal period is 11.86 years, meaning it takes a little less than 12 years for Jupiter to complete one circuit through the constellations of the zodiac. Jupiter therefore appears to move through roughly one zodiac constellation each year. If Jupiter is currently in Virgo, it will be in Libra at this time next year and Scorpius the following year, returning to Virgo in about 12 years.

A planet's **synodic period** is the time from when it is lined up with the Sun in our sky once to the next similar alignment. (Again, the term *synodic* refers to the planet's "meeting" the Sun in the sky.) **FIGURE S1.4** shows that the situation is somewhat different for planets nearer the Sun than Earth (that is, Mercury and Venus) than for planets farther away.

Look first at the more distant planet in Figure S1.4. As seen from Earth, this planet will sometimes line up with the Sun in what we call a **conjunction**. At other times, it will appear exactly opposite the Sun in our sky, or at **opposition**. We cannot see the planet during conjunction because it is hidden by the Sun's glare and rises and sets with the Sun in our sky. At opposition, the planet moves through the sky like the full moon, rising at sunset, reaching the meridian at midnight, and setting at dawn. Note that the planet is closest to Earth at opposition and hence appears brightest in our sky at this time.

Now look at the planet that is *nearer* than Earth to the Sun in Figure S1.4. This planet never has an opposition but instead has two conjunctions—an "inferior conjunction" between Earth and the Sun and a "superior conjunction" when the planet appears behind the Sun as seen from Earth. An inner planet also has two points of **greatest elongation**, when it appears farthest from the Sun in our sky. For Venus, these points occur when it appears about 46° east (greatest eastern elongation) or west (greatest western elongation) of the Sun in our sky; it shines brightly at these times in the evening (when it is east of the Sun) or before dawn (when it is west of the Sun). In between, Venus disappears from view for a few weeks with each conjunction. Mercury's pattern is similar, but because it is closer to the Sun, its greatest elongations are only about 28° from the Sun in our sky. That makes Mercury difficult to see, because it is almost always obscured by the glare of the Sun.

**Think about it** Do we ever see Mercury or Venus at midnight? Explain.

As you study Figure S1.4, you might wonder whether Mercury and Venus ever fall directly in front of the Sun at inferior conjunction, creating a mini-eclipse as they block a little of the Sun's light. They do, but only rarely, because their orbital planes are slightly tilted compared to Earth's orbital plane (the ecliptic plane). As a result, Mercury and Venus usually appear slightly above or below the Sun at inferior conjunction. But on rare occasions, we do indeed see Mercury or Venus appear to pass directly across the face of the Sun during inferior conjunction. Such events are called **transits** (**FIGURE S1.5**). Mercury transits occur

**FIGURE S1.5** This photo shows the Venus transit of June 5, 2012 as it appeared in Kansas City at sunset. Venus is the small dot visible near the center right of the Sun's face. The next Venus transit will not occur until 2117.

an average of a dozen times per century; one occurred on Nov. 11, 2019, and the next will occur on Nov. 13, 2032. Venus transits come in pairs 8 years apart, with more than a century between the second of one pair and the first of the next. The most recent transits of Venus occurred in 2004 and 2012, and the next pair will occur in 2117 and 2125.

## How do we tell the time of day?

We base the time of day on the 24-hour solar day, but there are several different ways to define this time.

**Apparent Solar Time** If we base time on the Sun's *actual* position in the local sky, as is the case when we use a sundial (**FIGURE S1.6**, page 90), we are measuring **apparent solar time**. Noon is the precise moment when the Sun is highest in the sky (on the meridian) and the sundial casts its shortest shadow. Before noon, when the Sun is climbing higher in the sky, the apparent solar time is *ante meridiem* ("before the middle of the day"), or a.m. For example, if the Sun will reach the meridian 2 hours from now, the apparent solar time is 10 a.m. After noon, the apparent solar time is *post meridiem* ("after the middle of the day"), or p.m. If the Sun crossed the meridian 3 hours ago, the apparent solar time is 3 p.m. Note that, technically, noon and midnight are *neither* a.m. nor p.m. However, by convention we usually say that noon is 12 p.m. and midnight is 12 a.m.

**Think about it** Is it daytime or nighttime at 12:01 a.m.? 12:01 p.m.? Explain.

**Mean Solar Time** Suppose you set a clock to precisely 12:00 when a sundial shows noon today. If every solar day were precisely 24 hours, your clock would always remain synchronized with the sundial. However, while 24 hours is the *average* length of the solar day, the actual length of the solar day varies throughout the year, so your clock is likely to read a few seconds before or after 12:00 when the sundial reads noon tomorrow, and within a few weeks your

## MATHEMATICAL INSIGHT S1.1 The Copernican Layout of the Solar System

Recall that Copernicus favored the Sun-centered model partly because it allowed him to calculate orbital periods and distances for the planets [**Section 3.3**]. Let's see how.

We cannot directly measure a planet's orbital period, because we look at the planet from different points in our orbit at different times. However, we can measure its *synodic* period from the time between one particular alignment (such as opposition or inferior conjunction) and the next. **FIGURE 1** shows the geometry for a planet *farther than Earth from the Sun* (such as Jupiter), under the assumption of circular orbits (which Copernicus assumed). Note the following key facts:

- The dashed brown curve shows the planet's orbit, which takes a time of one orbital period, $P_{orb}$, to complete.
- The solid brown arrow shows how far the planet travels along its orbit from one opposition to the next. The time between oppositions is defined as its synodic period, $P_{syn}$.
- The dashed blue curve shows Earth's orbit; Earth takes $P_{Earth} = 1$ yr to complete an orbit.
- The solid red curve and extra red arrow show how far Earth goes during the planet's synodic period; it is *more* than one complete orbit because Earth must travel a little "extra" to catch back up with the planet, and the time required for this "extra" distance (the thick red arrow) is the planet's synodic period minus 1 year, or $P_{syn} - 1$ yr.

Now, notice that the angle that the planet sweeps out during its synodic period is equal to the angle that Earth sweeps out as it travels the "extra" distance. Therefore, the *ratio* of the planet's complete orbital period ($P_{orb}$) to its synodic period ($P_{syn}$) must equal the *ratio* of Earth's orbital period (1 yr) to the time required for the "extra" distance (see Appendix C.5 to review ratios). We already found that the time required for this extra distance is $P_{syn} - 1$ yr, so we write

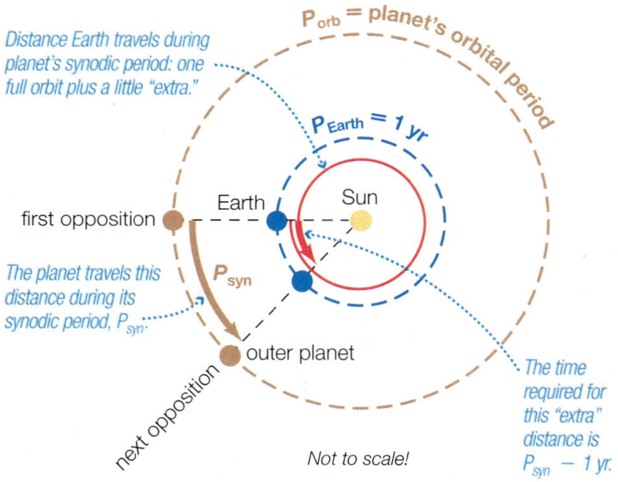


**FIGURE 1**

$$\frac{P_{orb}}{P_{syn}} = \frac{1\text{ yr}}{P_{syn} - 1\text{ yr}}$$

Multiplying both sides by $P_{syn}$ gives us the final equation for a planet farther than Earth from the Sun:

$$\text{outer planets: } P_{orb} = P_{syn} \times \frac{1\text{ yr}}{P_{syn} - 1\text{ yr}}$$

The geometry is slightly different for a planet closer to the Sun (Mercury or Venus). **FIGURE 2** shows that in this case the equal ratios are $1\text{ yr}/P_{syn} = P_{orb}/(P_{syn} - P_{orb})$, leading (with a bit of algebra) to this equation for a planet closer than Earth to the Sun:

$$\text{inner planets: } P_{orb} = P_{syn} \times \frac{1\text{ yr}}{P_{syn} + 1\text{ yr}}$$

clock time may differ from the apparent solar time by several minutes.

Averaging the differences between the time your clock would read and the time a sundial would read defines **mean solar time** (*mean* is another word for *average*). A clock set to mean solar time reads 12:00 each day at the time the Sun crosses the meridian *on average*. The actual mean solar time at which the Sun crosses the meridian varies over the course of the year in a fairly complex way (see the Special Topic, page 94). The result is that, on any given day, a clock set to mean solar time may read anywhere from about 17 minutes before noon to 15 minutes after noon (that is, from 11:43 a.m. to 12:15 p.m.) when a sundial indicates noon.

Although the lack of perfect synchronization with the Sun might at first sound like a drawback, mean solar time is actually more convenient than apparent solar time (the sundial time), at least if you have access to a mechanical or electronic clock. Once set, a reliable mechanical or electronic clock can always tell you the mean solar time. In contrast, measuring apparent solar time requires a sundial, which is useless at night or when it is cloudy.

Like apparent solar time, mean solar time is a *local* measure of time. That is, it varies with longitude because of Earth's west-to-east rotation. For example, clocks in New York are set 3 hours ahead of clocks in Los Angeles. In fact, if clocks were set precisely to local mean solar time, they would vary even over relatively short east-west distances. For example, mean solar clocks in central Los Angeles would be about 2 minutes behind mean solar clocks in Pasadena, because Pasadena is slightly to the east.

**Standard, Daylight, and Universal Time** Clocks displaying mean solar time were once common. But by the late 19th century, particularly in the United States, the growth of railroad travel made mean solar time increasingly problematic. Some states had dozens of different "official" times, usually corresponding to mean solar time in dozens of different cities, and each railroad company made schedules according to its own "railroad time." The many

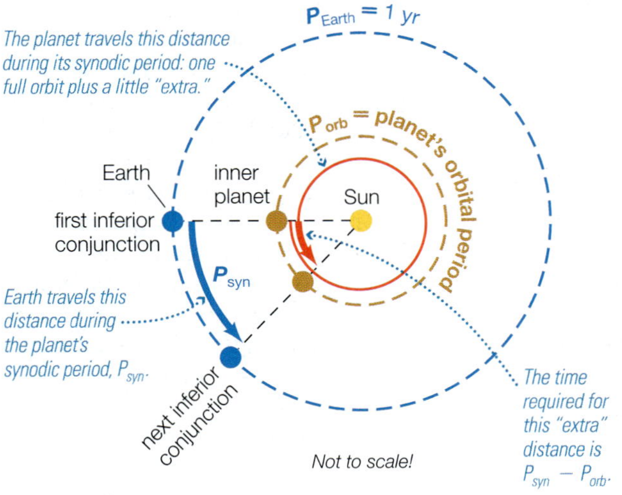


**FIGURE 2**

Copernicus knew the synodic periods of the planets and therefore could use the above equations (in a slightly different form) to calculate true orbital periods. He then used the geometry of planetary alignments to compute distances in terms of the Earth-Sun distance (that is, in AU). His results were quite close to modern values.

**EXAMPLE 1:** Jupiter's synodic period is 398.9 days, or 1.092 years. What is its actual orbital period?

**SOLUTION:** **Step 1 Understand:** We are given Jupiter's synodic period ($P_{syn}$), which is the only value we need to find its orbital period ($P_{orb}$).

**Step 2 Solve:** We use the equation for a planet farther than Earth from the Sun, with $P_{syn} = 1.092$ yr:

$$P_{orb} = P_{syn} \times \frac{1\ \text{yr}}{P_{syn} - 1\ \text{yr}}$$

$$= 1.092\ \text{yr} \times \frac{1\ \text{yr}}{1.092\ \text{yr} - 1\ \text{yr}} = 11.87\ \text{yr}$$

**Step 3 Explain:** Jupiter's orbital period is a little less than 12 years. Notice that, as we expect for a planet farther from the Sun, Jupiter's orbital period is longer than Earth's.

**EXAMPLE 2:** Venus's synodic period is 583.9 days, or 1.599 years. What is its actual orbital period?

**SOLUTION:**

**Step 1 Understand:** We can calculate Venus's orbital period from its given synodic period using the equation for a planet closer to the Sun than Earth.

**Step 2 Solve:** For a planet closer to the Sun and $P_{syn} = 1.599$ yr:

$$P_{orb} = P_{syn} \times \frac{1\ \text{yr}}{P_{syn} + 1\ \text{yr}}$$

$$= 1.599\ \text{yr} \times \frac{1\ \text{yr}}{1.599\ \text{yr} + 1\ \text{yr}} = 0.6152\ \text{yr}$$

**Step 3 Explain:** Venus's orbital period is 0.6152 year, which you can confirm to be equivalent to 224.7 days, or about $7\frac{1}{2}$ months. As we expect, it is shorter than Earth's orbital period of 1 year.

**FIGURE S1.6** A basic sundial consists of a dial, marked by numerals, and a stick, or *gnomon*, that casts a shadow. Here, the shadow is on the Roman numeral I, indicating that the apparent solar time is 1:00 p.m. (The portion of the dial without numerals represents nighttime hours.) Because the Sun's path across the local sky depends on latitude, a particular sundial will be accurate only for a particular latitude.

time systems made it difficult for passengers to follow train schedules.

On November 18, 1883, the railroad companies agreed to a new system that divided the United States into four time zones, setting all clocks within each zone to the same time. That was the birth of **standard time**, which today divides the entire world into time zones (**FIGURE S1.7**). Depending on where you live within a time zone, your standard time may vary somewhat from your mean solar time. In general, the standard time in a particular time zone is the mean solar time in the *center* of the time zone, in which case local mean solar time within a 1-hour-wide time zone never differs by more than about a half-hour from standard time. However, many time zones have unusual shapes to conform to social, economic, and political realities, so larger variations between standard time and mean solar time sometimes occur.

In most parts of the United States, clocks are set to standard time for only part of the year. Between the second Sunday in March and the first Sunday in November, most of the United States changes to **daylight saving time**, which is 1 hour ahead of standard time. Because of the 1-hour advance with daylight saving time, clocks read around 1 p.m. (rather than around noon) when the Sun is on the meridian. Other countries (and some states and localities within the United States) change to daylight saving time on different dates or not at all.

For purposes of navigation and astronomy, it is useful to have a single time for the entire Earth. For historical reasons, this "world" time was chosen to be the mean solar time in Greenwich, England—the place that also defines longitude 0° (see Figure 2.11). Today, this *Greenwich mean time* (*GMT*) is often called **universal time (UT)**. (Outside astronomy, it is more commonly called *universal coordinated time* [UTC]. It is also sometimes called "Zulu time," because Greenwich's time zone is at zero [Z] hours longitude and "Zulu" is the international aviation standard for identifying the letter "Z" in noisy audio conditions.)

## When and why do we have leap years?

Our modern calendar is designed to stay synchronized with the seasons and is therefore based on the tropical year (the time from one March equinox to the next). Getting this synchronization just right was a long process in human history.

The origins of our modern calendar go back to ancient Egypt. By 4200 B.C., the Egyptians were using a calendar that counted 365 days in a year. However, because the length of a year is actually about $365\frac{1}{4}$ days, the Egyptian calendar drifted out of phase with the seasons by about 1 day every 4 years. For example, if the March equinox occurred on March 21 one year, 4 years later it occurred on March 22, 4 years after that on March 23, and so on. Over many centuries, the "March" equinox moved through many different months. To keep the seasons and the calendar synchronized, Julius Caesar decreed the adoption of a new calendar in 46 B.C. This **Julian calendar** introduced the **leap year**: Every fourth year has 366 days (the extra day is Feb. 29), rather than 365, so that the average length of the calendar year is $365\frac{1}{4}$ days.

The Julian calendar originally had the March equinox falling around March 24. If it had been perfectly synchronized with the tropical year, this calendar would have ensured that the March equinox occurred on the same date every 4 years (that is, every leap-year cycle). It didn't work perfectly, however, because the precise length of the tropical year is about 11 minutes short of $365\frac{1}{4}$ days. As a result, the moment of the March equinox advanced by about 11 minutes per year. By the late 16th century, the March equinox was occurring on March 11.

Concerned by this drift in the date of the March equinox, Pope Gregory XIII introduced a new calendar in 1582. This **Gregorian calendar** was much like the Julian calendar, with two important adjustments. First, Pope Gregory decreed that the day in 1582 following October 4 would be October 15. By eliminating the 10 dates from October 5 through October 14, 1582, he pushed the date of the March equinox in 1583 from March 11 to March 21. (He chose March 21 because it was the date of the March equinox in A.D. 325, which was the time

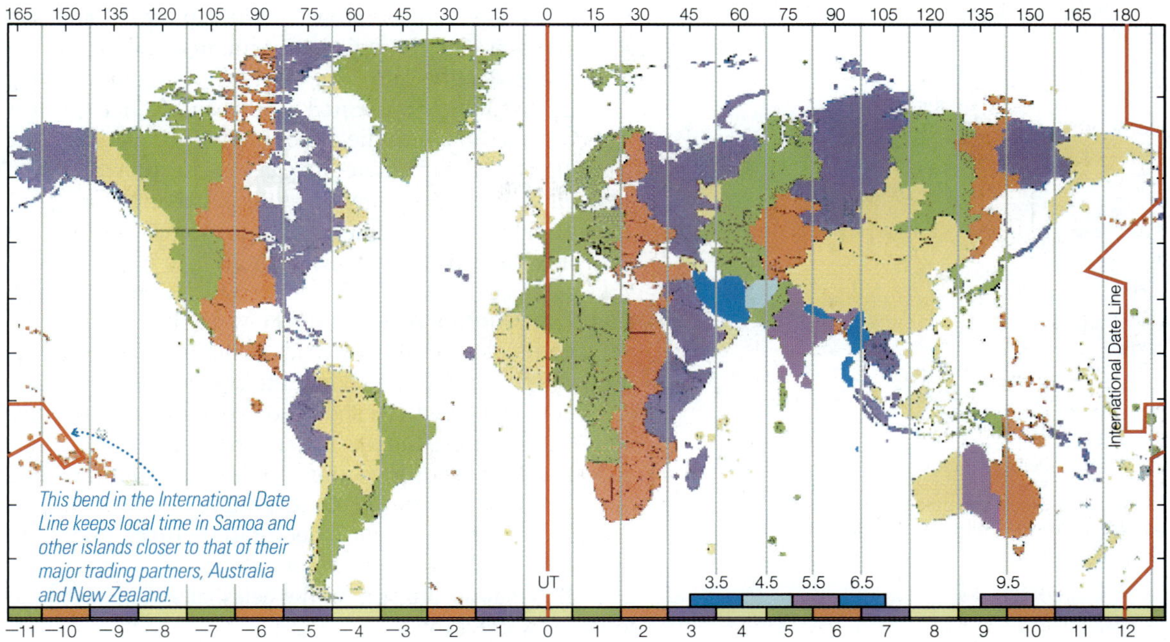


**FIGURE S1.7** Time zones around the world. The numerical scale at the bottom shows hours ahead of (positive numbers) or behind (negative numbers) the time in Greenwich, England; the scale at the top is longitude. The vertical lines show standard time zones as they would be in the absence of political considerations. The color-coded regions show the actual time zones. Note, for example, that all of China uses the same standard time, even though the country is wide enough to span several time zones. Note also that a few countries use time zones centered on a half-hour (the upper set of colored bars), rather than an hour, relative to Greenwich time.

of the Council of Nicaea, the first ecumenical council of the Christian church.) Second, the Gregorian calendar added an exception to the rule of having leap year every 4 years: Leap year is skipped when a century changes (for example, in years 1700, 1800, 1900) *unless* the century year is divisible by 400. For example, 2000 was a leap year because it is divisible by 400 ($2000 \div 400 = 5$), but 2100 will *not* be a leap year. These adjustments make the average length of the Gregorian calendar year almost exactly the same as the actual length of a tropical year, which ensures that the March equinox will occur on March 21 every fourth year for thousands of years to come.

Today, the Gregorian calendar is used worldwide for international communication and commerce. (Many people still use traditional calendars for cultural purposes, such as the Chinese, Islamic, and Jewish calendars.) However, it took time for the Gregorian calendar to come into use in regions not bound to the Catholic Church. For example, the Gregorian calendar was not adopted in England or in the American colonies until 1752, and it was not adopted in China until 1912 or in Russia until 1919.

## S1.2 Celestial Coordinates and Motion in the Sky

We now turn our attention from timekeeping to celestial navigation. First, however, we need to explore the apparent motions of the sky in more detail than we covered in Chapter 2.

### How do we locate objects on the celestial sphere?

Recall from Chapter 2 that the celestial sphere is an illusion, but one that is quite useful when looking at the sky. We can make it even more useful by adding a set of **celestial coordinates** that allow us to describe the precise position of a star (or other object in the sky) on the celestial sphere in much the same way that we use latitude and longitude to locate a city on Earth.

The key starting points for the celestial coordinate system are the north and south celestial poles, the celestial equator, and the ecliptic (**FIGURE S1.8**). To better visualize the celestial sphere, you should make a three-dimensional model with a simple plastic ball. Use a felt-tip pen to mark the north and south celestial poles on your ball, and then add the celestial equator and the ecliptic. Note that the ecliptic crosses the celestial equator on opposite sides of the celestial sphere at an angle of $23\frac{1}{2}°$ (because of the tilt of Earth's axis).

**Equinoxes and Solstices** Recall that the equinoxes and solstices are special moments that occur each year when Earth is at particular positions in its orbit (see Figure 2.15). These positions correspond to the apparent locations of the Sun along the ecliptic shown in Figure S1.8. For example, the March equinox occurs at the moment when the Sun's path along the ecliptic crosses the celestial equator going from south to north, so we also use the term March equinox to refer to this point on the celestial sphere. That is, the term *March equinox* has a dual meaning: It is the *moment* in March when the Sun's path crosses the celestial equator

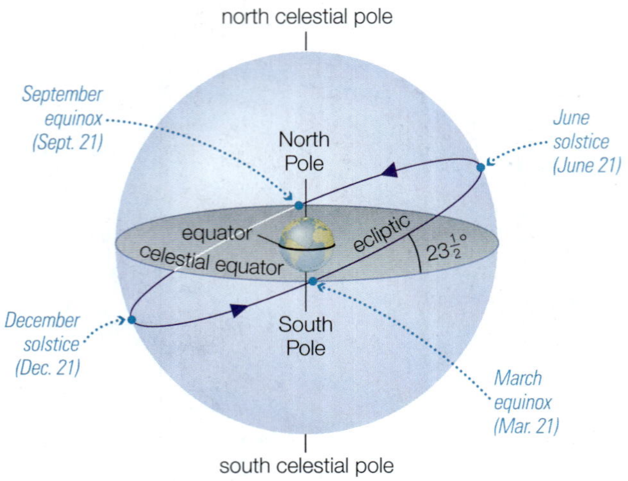


**FIGURE S1.8** Schematic diagram of the celestial sphere without stars. The arrow along the ecliptic indicates the direction in which the Sun appears to move over the course of each year.

and the *point* on the ecliptic at which the Sun appears to be located at that moment. Figure S1.8 shows both the points on the celestial sphere and the approximate dates for each equinox and solstice.

**See it for yourself** Using your plastic ball model of the celestial sphere (which you have already marked with the celestial poles, equator, and ecliptic), mark the locations and approximate dates of the equinoxes and solstices. Based on the dates for these points, approximately where along the ecliptic is the Sun on April 21? On November 21? How do you know?

You can find the locations of the equinoxes and solstices among the constellations with the aid of nearby bright stars (**FIGURE S1.9**). For example, the point marking the March equinox is located in the constellation Pisces and can be found with the aid of the four bright stars that make up the "Great Square of Pegasus." Keep in mind that you can find this point any time it is above the horizon on a clear night, even though the Sun is located at this point only once each year (around March 21).

**Celestial Coordinates** We are now ready to add celestial coordinates to the celestial sphere. Let's begin by reviewing the two other coordinate systems we've used in this book: **FIGURE S1.10a** shows the coordinates of *altitude* and *direction* (or *azimuth**) that we use in the local sky, and **FIGURE S1.10b** shows the coordinates of *latitude* and *longitude* that we use on Earth's surface. Our system of celestial coordinates, called **declination (dec)** and **right ascension (RA)**, is shown in **FIGURE S1.10c**.

Notice that declination on the celestial sphere is similar to latitude on Earth:

- Just as lines of latitude are parallel to Earth's equator, lines of declination are parallel to the celestial equator.
- Just as Earth's equator has lat $= 0°$, the celestial equator has dec $= 0°$.
- Latitude is labeled *north* or *south* relative to the equator, while declination is labeled *positive* or *negative*. For

*Azimuth is usually measured clockwise around the horizon from due north. By this definition, the azimuth of due north is 0°, of due east is 90°, of due south is 180°, and of due west is 270°.

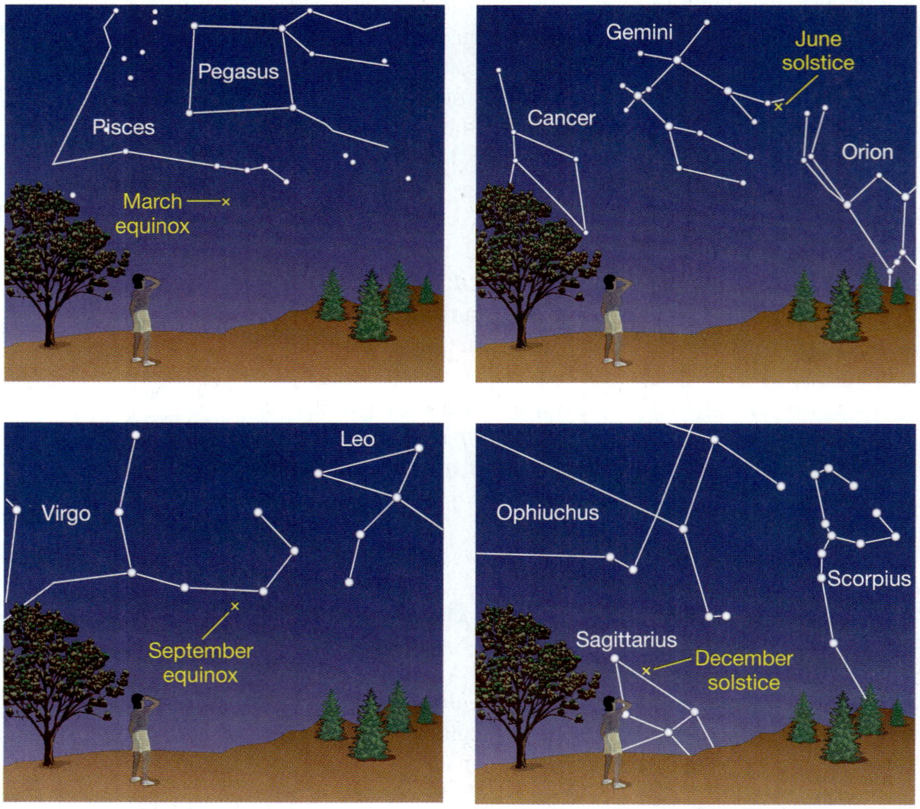


**FIGURE S1.9** These diagrams show the locations among the constellations of the equinoxes and solstices. No bright stars mark any of these points, so you must find them by studying their positions relative to recognizable patterns. The time of day or night at which each point is above the horizon depends on the time of year.

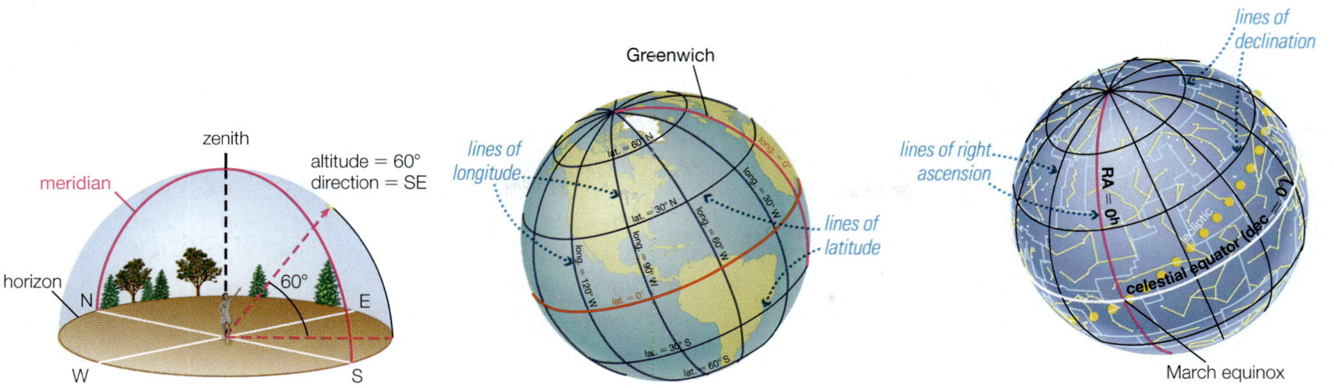


**a** We use altitude and direction to pinpoint locations in the local sky.

**b** We use latitude and longitude to pinpoint locations on Earth.

**c** We use declination and right ascension to pinpoint locations on the celestial sphere.

**FIGURE S1.10** Celestial coordinate systems.

example, the North Pole has lat = 0°N, while the north celestial pole has dec = +90°N; the South Pole has lat = 90°S, while the south celestial pole has dec = −90°.

We find a similar correspondence between right ascension and longitude:

- Just as lines of longitude extend from the North Pole to the South Pole, lines of right ascension extend from the north celestial pole to the south celestial pole.
- Just as there is no natural starting point for longitude, there is no natural starting point for right ascension. By international treaty, longitude zero (the prime meridian) is the line of longitude that runs through Greenwich, England. By convention, right ascension zero is the line of right ascension that runs through the March equinox.
- Longitude is measured in *degrees* east or west of Greenwich, while right ascension is measured in *hours* (and minutes and seconds) east of the March equinox. A full 360° circle around the celestial equator goes through 24 hours of right ascension, so each hour of right ascension represents an angle of $360° \div 24 = 15°$.

As an example of how we use celestial coordinates to locate objects on the celestial sphere, consider the bright star Vega. Its coordinates are dec = +38°44′ and RA = $18^h35^m$ (**FIGURE S1.11**). The positive declination tells us that Vega is 38°44′ *north* of the celestial equator. The right ascension tells us that Vega is 18 hours 35 minutes east of the March equinox. Translating the right ascension from hours to angular degrees, we find that Vega is about 279° east of the March equinox (because 18 hours represents $18 \times 15° = 270°$ and 35 minutes represents $\frac{35}{60} \times 15° \approx 9°$).

**See it for yourself** On your plastic ball model of the celestial sphere, add a scale for right ascension along the celestial equator and add a few circles of declination, such as declination 0°, ±30°, ±60°, and ±90°. Where is Vega on your model?

We can also use the Vega example to see the benefit of measuring right ascension in units of time. All objects with a particular right ascension cross the meridian at the same time. For example, all stars with RA = $0^h$ cross the meridian at the same time the March equinox crosses the meridian, all objects with RA = $1^h$ cross the meridian 1 hour after the March equinox, and so on. Vega's right ascension of $18^h35^m$ tells us that it always crosses the meridian 18 hours 35 minutes after the March equinox crosses the meridian. (This is 18 hours 35 minutes of *sidereal time* later, which is not exactly the same as 18 hours 35 minutes of solar time; see Mathematical Insight S1.2, page 97.) Generalizing, an object's right ascension tells us *how long after* the March equinox the object crosses the meridian.

Note that while we generally think of declination and right ascension as fixed coordinates like latitude and longitude, they are not perfectly constant. Instead, they move slowly relative to distant stars because they are tied to the celestial equator, which moves gradually relative to the constellations with Earth's 26,000-year cycle of axis precession [**Section 2.2**]. (Axis precession does not affect Earth's orbit, so it does not affect the location of the ecliptic among the constellations.) Even over just a few decades,

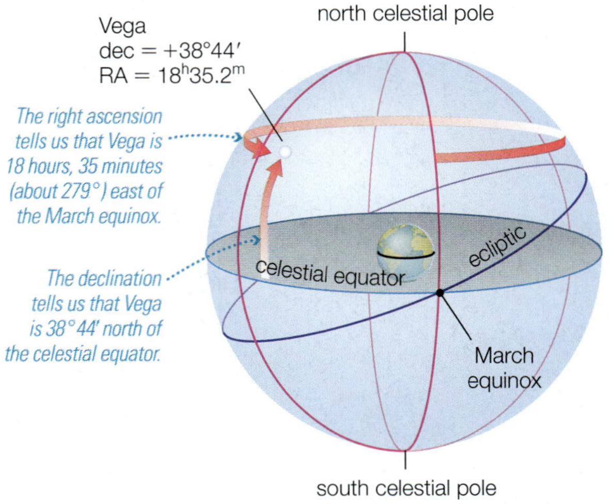


**FIGURE S1.11** This diagram shows how we interpret the celestial coordinates of Vega.

the resulting coordinate changes can be significant enough to make a difference in precise astronomical work, such as aiming a telescope at a particular object. As a result, careful observations require almost continual updating of celestial coordinates. Star catalogs therefore always state the year for which coordinates are given (for example, "epoch 2000"). Astronomical software can automatically calculate day-to-day celestial coordinates for any object in our sky.

**Celestial Coordinates of the Sun** Unlike stars, which remain essentially fixed in the patterns of the constellations on the celestial sphere, the Sun moves gradually along the ecliptic. It takes a year for the Sun to make a full circuit of the ecliptic, which means it moves through all 24 hours of right ascension over the course of the year. The Sun therefore moves approximately one twelfth of the way around the ecliptic each month, meaning that its right ascension changes by about $24 \div 12 = 2$ hours per month. **FIGURE S1.12** shows the ecliptic marked with the Sun's monthly position and a scale of celestial coordinates. From this figure, we can create a table of the Sun's month-by-month celestial coordinates.

**TABLE S1.1** starts from the March equinox, when the Sun has declination 0° and right ascension $0^h$. You can see in

## SPECIAL TOPIC Solar Days and the Analemma

The precise length of a solar day varies from its average of 24 hours for two reasons. The first is Earth's varying orbital speed. Recall that, in accord with Kepler's second law, Earth moves slightly faster—and therefore moves slightly farther along its orbit each day—when it is closer to the Sun in its orbit. The solar day therefore requires more than the average amount of "extra" rotation (see Figure S1.2) during these periods, making these solar days longer than average. Similarly, the solar day requires less than the average amount of "extra" rotation when Earth is farther from the Sun.

The second reason is the tilt of Earth's axis, which makes the ecliptic inclined $23\frac{1}{2}°$ to the celestial equator (see Figure S1.8). Because the length of a solar day depends on the Sun's apparent *eastward* motion along the ecliptic, the inclination would cause solar days to vary in length even if Earth's orbit were perfectly circular. To see why, suppose the Sun appeared to move exactly 1° per day along the ecliptic. Around the times of the solstices, this motion would be entirely eastward, making the solar day slightly longer than average. Around the times of the equinoxes, when the motion along the ecliptic has a significant northward or southward component, the solar day would be slightly shorter than average.

Together, the effects of varying orbital speed and tilt mean the actual length of a solar day can be up to about 25 seconds longer or shorter than the 24-hour average. Because the effects accumulate at particular times of year, the apparent solar time can differ by as much as 17 minutes from the mean solar time. The net result is often depicted visually by an **analemma** (**FIGURE 1**), which looks much like a figure 8. You'll find an analemma printed on many globes, and Figure 2.17 shows a photographic version.

The horizontal scale on the analemma allows you to find the difference between mean and apparent solar time for any date. (The vertical scale shows the Sun's declination.) For example, the dashed line shows that on November 10, a mean solar clock is about 17 minutes "behind the Sun," or behind apparent solar time; this means that if the apparent solar time is 6:00 p.m., the mean solar time is about 5:43 p.m. The annual pattern of variations between mean and apparent solar times is called the **equation of time**. It is often plotted as a graph (**FIGURE 2**), which gives the same results as reading from the analemma.

The discrepancy between mean and apparent solar time also explains why the times of sunrise and sunset don't follow seasonal patterns perfectly. For example, the December solstice (around December 21) has the shortest daylight hours in the Northern Hemisphere, but the earliest sunset occurs around December 7, when the Sun is still well "behind" mean solar time.

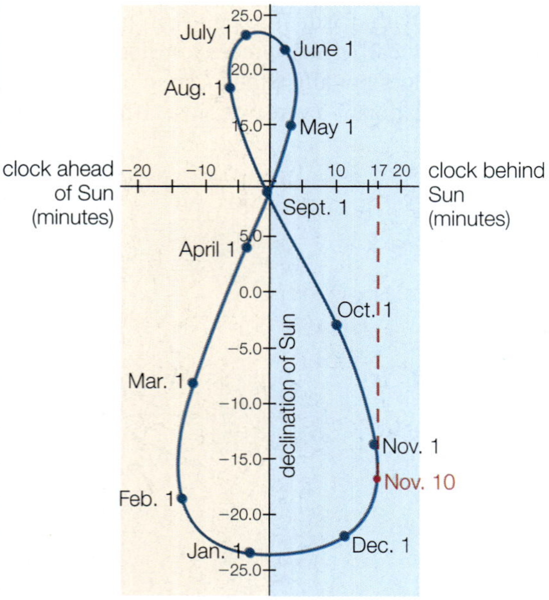


**FIGURE 1** The *analemma* shows the annual pattern of discrepancies between apparent and mean solar time. The dashed red line shows that the maximum discrepancy occurs around November 10.

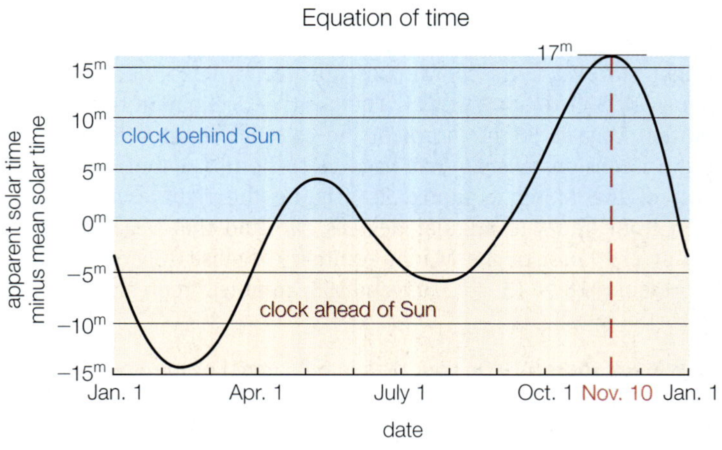


**FIGURE 2** The discrepancies can also be plotted on a graph as the *equation of time*.

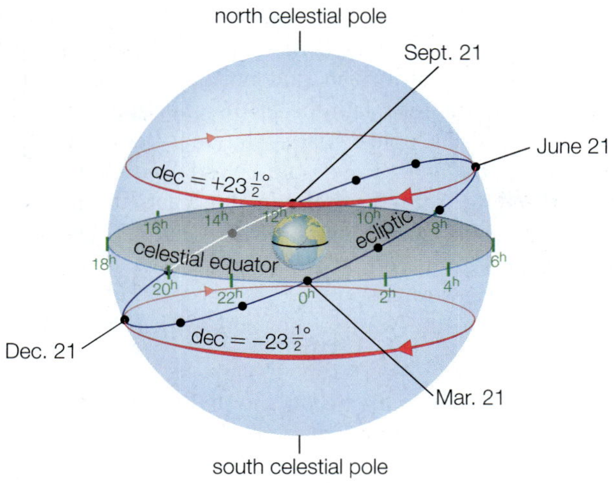


**FIGURE S1.12** We can use this diagram of the celestial sphere to determine the Sun's right ascension and declination at monthly intervals.

**TABLE S1.1 The Sun's Approximate Celestial Coordinates at 1-Month Intervals**

| Approximate Date | RA | Dec |
|---|---|---|
| Mar. 21 (March equinox) | $0^h$ | 0° |
| Apr. 21 | $2^h$ | +12° |
| May 21 | $4^h$ | +20° |
| June 21 (June solstice) | $6^h$ | $+23\frac{1}{2}°$ |
| July 21 | $8^h$ | +20° |
| Aug. 21 | $10^h$ | +12° |
| Sept. 21 (September equinox) | $12^h$ | 0° |
| Oct. 21 | $14^h$ | −12° |
| Nov. 21 | $16^h$ | −20° |
| Dec. 21 (December solstice) | $18^h$ | $-23\frac{1}{2}°$ |
| Jan. 21 | $20^h$ | −20° |
| Feb. 21 | $22^h$ | −12° |

the blue shaded areas of the table that while RA advances steadily through the year, the Sun's declination changes much more slowly around the solstices than around the equinoxes. For example, during the 2 months around the June solstice (that is, between May 21 and July 21), the Sun's declination varies only between +20° and its maximum of $+23\frac{1}{2}°$; a similar idea holds around the December solstice. In contrast, in the two months around the March equinox, the Sun's declination changes by about 24°, from −12° on February 21 to +12° on April 21; again, a similar pattern holds around the September equinox. These facts explain why the number of daylight hours increases rapidly in spring and decreases rapidly in fall, while remaining nearly constant for a couple of months around the solstices.

**See it for yourself** On your plastic ball model of the celestial sphere, add dots along the ecliptic to show the Sun's monthly positions. Based on your model, what are the Sun's approximate celestial coordinates on your birthday?

## How do stars move through the local sky?

Recall that Earth's rotation makes all celestial objects appear to circle around Earth each day (see Figure 2.9), but what we see in the local sky is more complex because we see only half the celestial sphere at one time; the ground blocks our view of the other half. We are now ready to explore the local sky in more depth. As we'll see, the path of any star through your local sky depends only on (1) your latitude and (2) the declination of the star.

**The Sky at the North Pole** Let's begin with the local sky at the North Pole, where the daily paths of stars are easiest to understand. **FIGURE S1.13a** shows your orientation relative to the celestial sphere when you are standing at the North Pole. Your "up" points toward the north celestial

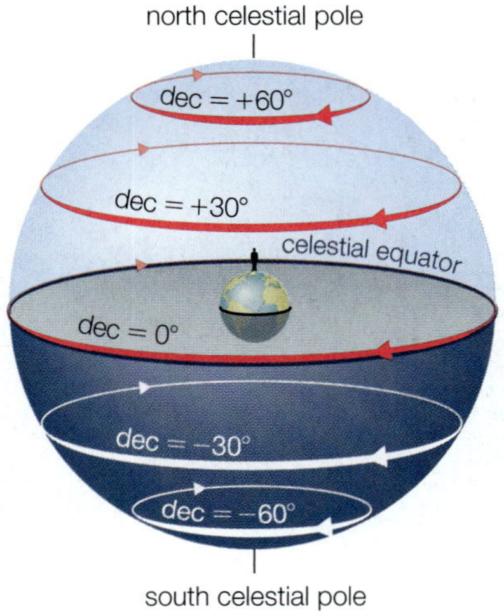


**a** The orientation of the local sky, relative to the celestial sphere, for an observer at the North Pole.

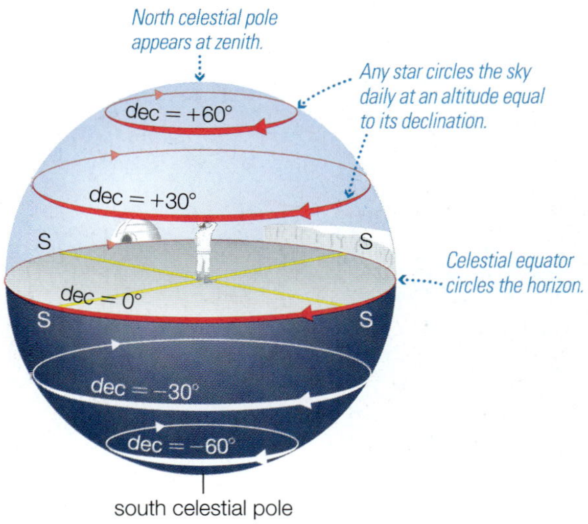


**b** Extending the horizon to the celestial sphere makes it easier to visualize the local sky at the North Pole.

**FIGURE S1.13** The sky at the North Pole.

pole, which therefore marks your zenith. Earth blocks your view of anything south of the celestial equator, which therefore runs along your horizon. To make it easier for you to visualize the local sky, **FIGURE S1.13b** shows your horizon extending to the celestial sphere. The horizon is marked with directions; note that all directions are south from the North Pole, which means we cannot define a meridian (or a longitude) for the North Pole.

Notice that the daily circles of the stars keep them at constant altitudes above or below the North Polar horizon. Moreover, the altitude of any star is equal to its declination. For example, a star with declination +60° circles the sky at an altitude of 60°, and a star with declination −30° remains 30° below your horizon at all times. As a result, all stars north of the celestial equator are circumpolar at the North Pole, meaning that they never fall below the horizon. Stars south of the celestial equator can never be seen at the North Pole. (If you are having difficulty visualizing the star paths, it may help you to watch them as you rotate your plastic ball model of the celestial sphere.)

You should also notice that right ascension does not affect a star's path at all: The path depends only on declination. As we'll see shortly, this rule holds for all latitudes. Right ascension affects only the *time* of day and year at which a star is found in a particular position in your sky.

**The Sky at the Equator** Imagine that you are standing somewhere on Earth's equator (lat = 0°), such as in Ecuador, in Kenya, or on the island of Borneo. **FIGURE S1.14a** shows that "up" points directly away from (perpendicular to) Earth's rotation axis. **FIGURE S1.14b** shows the local sky more clearly by extending the horizon to the celestial sphere and rotating the diagram so that the zenith is up. As it does everywhere except at the poles, the meridian extends from the horizon due south, through the zenith, to the horizon due north.

Look carefully at how the celestial sphere appears to rotate in the local sky. The north celestial pole remains stationary on your horizon due north, with its altitude equal to the equator's latitude of 0° [**Section 2.1**], and the south celestial pole remains stationary on your horizon due south. Exactly half the celestial equator is visible, extending from the horizon due east, through the zenith, to the horizon due west. (The other half lies below the horizon.) As the equatorial sky appears to turn, all star paths rise straight out of the eastern horizon and set straight into the western horizon, with the following features:

- **Stars with dec = 0°** lie *on* the celestial equator and therefore rise due east, cross the meridian at the zenith, and set due west.
- **Stars with dec > 0°** rise north of due east, reach their highest point on the meridian in the north, and set north of due west. Their rise, set, and highest point depend on their declination. For example, a star with dec = +30° rises 30° north of due east, crosses the meridian 30° to the north of the zenith—that is, at an *altitude* of 90° − 30° = 60° in the north—and sets 30° north of due west.
- **Stars with dec < 0°** rise south of due east, reach their highest point on the meridian in the south, and set south of due west. For example, a star with dec = −50° rises 50° south of due east, crosses the meridian 50° to the south of the zenith—that is, at an *altitude* of 90° − 50° = 40° in the south—and sets 50° south of due west.

Because exactly half of any star's daily circle lies above the horizon, every star at the equator is above the horizon for exactly half of each sidereal day, or just under 12 hours (and below the horizon for the other half of the sidereal day).

**Think about it** Are any stars circumpolar at the equator? Are there stars that never rise above the horizon at the equator? Explain.

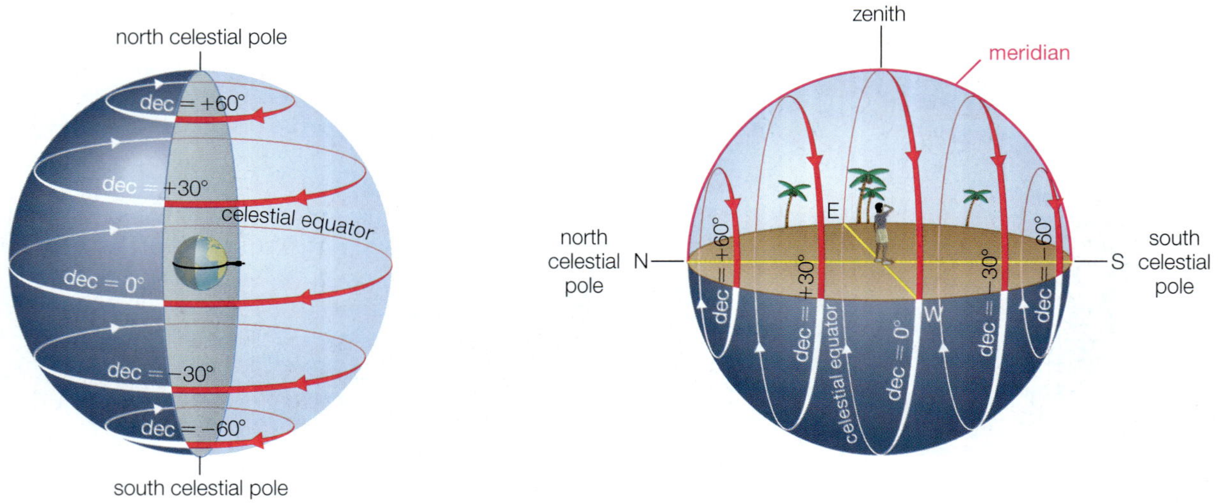


**a** The orientation of the local sky, relative to the celestial sphere, for an observer at Earth's equator.

**b** Extending the horizon and rotating the diagram make it easier to visualize the local sky at the equator.

**FIGURE S1.14** The sky at the equator.

**Skies at Other Latitudes** Star tracks may at first seem more complex at other latitudes, with their mixtures of circumpolar stars and stars that rise and set. However, they are easy to understand if we apply the same basic strategy we've used for the North Pole and equator. Let's consider latitude 40°N, such as in Denver, Indianapolis, Philadelphia, or Beijing. First, as shown in **FIGURE S1.15a**, imagine standing at this latitude on a basic diagram of the rotating celestial sphere. Note that "up" points to a location on the celestial sphere with declination +40°. To make it easier to visualize the local sky, we next extend the horizon and rotate the diagram so that the zenith is up (**FIGURE S1.15b**).

As we expect, the north celestial pole appears 40° above the horizon due north, since its altitude in the local sky is always equal to your latitude. Half the celestial equator is visible, extending from the horizon due east, to the meridian at an altitude of 50° in the south, to the horizon due west. By comparing this diagram to that of the local sky for the equator, you'll notice the following general rule that applies to all latitudes except the poles:

> *The celestial equator always extends from due east on your horizon to due west on your horizon, crossing the meridian at an altitude of 90°* minus *your latitude.*

## MATHEMATICAL INSIGHT S1.2 Time by the Stars

Our everyday clocks are set to solar time, ticking through 24 hours for each day of mean solar time. For astronomical observations, it is also useful to have clocks that tell time by the stars, or **sidereal time**. Just as *solar time* is defined according to the Sun's position relative to the meridian, *sidereal time* is based on the positions of stars relative to the meridian. We define the **hour angle (HA)** of any object on the celestial sphere to be the time *since* it last crossed the meridian (or the higher of its two meridian crossing points for a circumpolar star). For example:

- If a star is crossing the meridian *now*, its hour angle is $0^h$.
- If a star crossed the meridian 3 hours ago, its hour angle is $3^h$.
- If a star will cross the meridian 1 hour from now, its hour angle is $-1^h$ or, equivalently, $23^h$.

By convention, time by the stars is based on the hour angle of the March equinox. That is, the **local sidereal time (LST)** is

$$\text{LST} = \text{HA}_{\text{March equinox}}$$

For example, the local sidereal time is 00:00 when the March equinox is *on* the meridian. Three hours later, when the March equinox is 3 hours west of the meridian, the local sidereal time is 03:00.

Note that, because right ascension tells us how long after the March equinox an object reaches the meridian, the local sidereal time is also equal to the right ascension (RA) of objects currently crossing your meridian. For example, if your local sidereal time is 04:30, stars with RA = $4^h30^m$ are currently crossing your meridian. This idea leads to an important relationship among any object's current hour angle, the current local sidereal time, and the object's right ascension:

$$\text{HA}_{\text{object}} = \text{LST} - \text{RA}_{\text{object}}$$

This formula should make sense: The local sidereal time tells us how long it has been since the March equinox was on the meridian and an object's right ascension tells us how long after the March equinox it crosses the meridian. Therefore, the difference $\text{LST} - \text{RA}_{\text{object}}$ must tell us how long it has been *since* the object crossed the meridian, which is the object's hour angle.

Sidereal time has one important subtlety: Sidereal clocks tick through 24 hours of sidereal time in one sidereal day, which is only about 23 hours 56 minutes of solar time. As a result, a sidereal hour is slightly shorter than a "normal" solar hour, and sidereal clocks gain about 4 minutes per day over solar clocks. Therefore, you cannot easily determine sidereal time from a solar clock. That is why astronomical observatories always have special sidereal clocks in addition to clocks that tell solar time.

**EXAMPLE 1:** Suppose the local solar time is 9:00 p.m. and it is the March equinox (March 21). What is the local sidereal time?

**SOLUTION:**

**Step 1 Understand:** We are asked to find the local sidereal time, which is the hour angle of the March equinox. We therefore need to know the current location of the March equinox in the local sky. The key clue is that it is the day of the March equinox, which is the one day on which the Sun is located in the same position as the March equinox in the sky.

**Step 2 Solve:** We are told that the local solar time is 9:00 p.m., which means that the Sun is 9 hours past the meridian and therefore has an hour angle of 9 hours. Because the March equinox and the Sun are located in the same place on this date, the March equinox also has an hour angle of 9 hours.

**Step 3 Explain:** The hour angle of the March equinox is 9 hours, which means the local sidereal time is LST = 09:00.

**EXAMPLE 2:** Suppose the local sidereal time is LST = 04:00. When will Vega (RA = $18^h35^m$) cross the meridian?

**SOLUTION:**

**Step 1 Understand:** We are given the local sidereal time and Vega's right ascension, so we can use our formula to determine Vega's hour angle, which tells us its current position relative to the meridian.

**Step 2 Solve:** We put the given values into the formula to find Vega's hour angle:

$$\text{HA}_{\text{Vega}} = \text{LST} - \text{RA}_{\text{Vega}} = 4{:}00 - 18{:}35 = -14{:}35$$

**Step 3 Explain:** Vega's hour angle is −14 hours 35 minutes, which means Vega will cross your meridian 14 hours and 35 minutes of sidereal time from now. This also means that Vega crossed your meridian 9 hours and 25 minutes ago (because $14^h35^m + 9^h25^m = 24^h$).

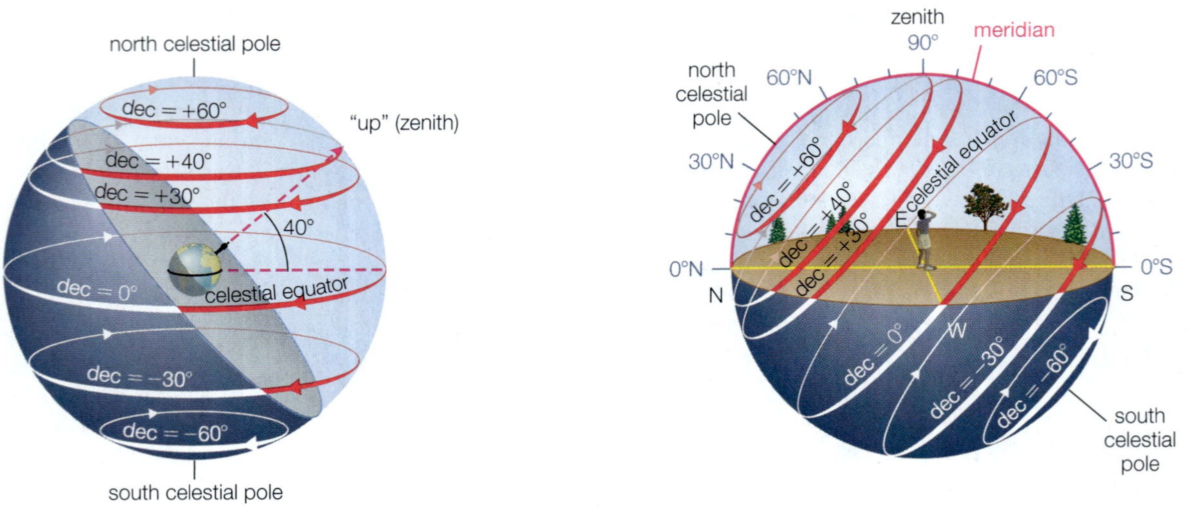


**a** The orientation of the local sky, relative to the celestial sphere, for an observer at latitude 40°N. Because latitude is the angle to Earth's equator, "up" points to the circle on the celestial sphere with declination +40°.

**b** Extending the horizon and rotating the diagram so that the zenith is up make the local sky easier to visualize. The meridian is marked with altitudes and directions.

**FIGURE S1.15** The sky at 40°N latitude.

The celestial equator crosses the meridian south of the zenith for locations in the Northern Hemisphere and north of the zenith for locations in the Southern Hemisphere.

If you study Figure S1.15b carefully, you'll notice the following features of the sky for latitude 40°N:

- **Stars with dec = 0°** lie *on* the celestial equator and therefore follow the celestial equator's path through the local sky. For latitude 40°N, these stars rise due east, cross the meridian at altitude 90° − 40° = 50° in the south, and set due west.
- **Stars with dec > (90° − lat)** are circumpolar. For latitude 40°N, stars with declination greater than 90° − 40° = 50° are circumpolar, because they lie *within* 40° of the north celestial pole; they therefore cross the meridian twice each day.
- **Stars with dec > 0° that are not circumpolar** follow paths parallel to but north of the celestial equator: They rise north of due east and set north of due west, and cross the meridian to the north of the place where the celestial equator crosses it by an amount equal to their declination. For example, because the celestial equator at latitude 40° crosses the meridian at altitude 50° in the south, a star with dec = +30° crosses the meridian at altitude 50° + 30° = 80° in the south. Similarly, a star with dec = +60° crosses the meridian 60° farther north than the celestial equator, which means at altitude 70° in the north (because 50° + 60° = 110°, which means 20° past the zenith, which is 90° − 20° = 70°).
- **Stars with dec < (−90° + lat)** never rise above the horizon. For latitude 40°N, stars with declination less than −90° + 40° = −50° never rise above the horizon, because they lie within 40° of the south celestial pole.
- **Stars with dec < 0° that are sometimes visible** follow paths parallel to but south of the celestial equator: They rise south of due east and set south of due west, and cross the meridian south of the place where the celestial equator crosses it by an amount equal to their declination. For example, a star with dec = −30° crosses the meridian at altitude 50° − 30° = 20° in the south.

Note also that the fraction of any star's daily circle that is above the horizon—and hence the amount of time it is above the horizon each day—depends on its declination. Because exactly half the celestial equator is above the horizon, stars on the celestial equator (dec = 0°) are above the horizon for exactly half of each sidereal day, or about 12 hours. For northern latitudes like 40°N, stars with positive declinations have more than half their daily circles above the horizon and hence are above the horizon for more than 12 hours each day (with the range extending to 24 hours a day for the circumpolar stars). Stars with negative declinations have less than half their daily circles above the horizon and hence are above the horizon for less than 12 hours each day (with the range going to zero for stars that are never above the horizon).

We can apply the same strategy we used in Figure S1.15 to find star paths for other latitudes. **FIGURE S1.16** shows the local sky for latitude 30°S. Note that the south celestial pole is visible to the south and that the celestial equator passes through the northern half of the sky. If you study the diagram carefully, you can see how star tracks depend on declination.

**Think about it** Study Figure S1.16 for latitude 30°S. Describe the path of the celestial equator. Does it obey the 90° *latitude* rule given earlier? Describe how star tracks differ for stars with positive and negative declinations. What declination must a star have to be circumpolar at this latitude?

## How does the Sun move through the local sky?

Just as we've discussed for stars, the Sun's path on any particular day depends only on its declination and your latitude. However, because the Sun's declination changes over the course of the year, the Sun's path also changes.

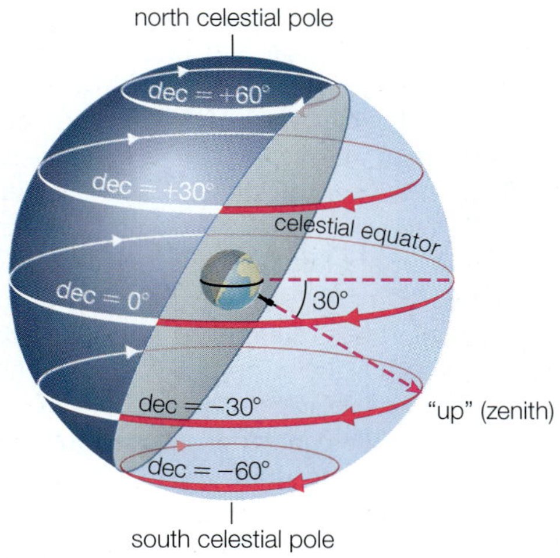


**a** The orientation of the local sky for an observer at latitude 30°S, relative to the celestial sphere. "Up" points to the circle on the celestial sphere with dec = −30°.

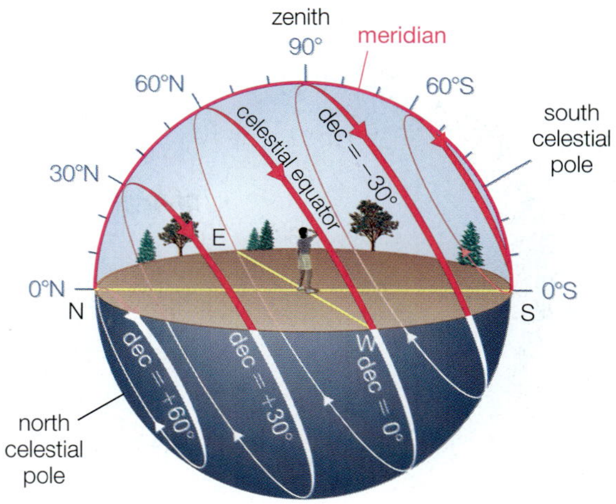


**b** Extending the horizon and rotating the diagram so that the zenith is up make it easier to visualize the local sky. Note that the south celestial pole is visible at altitude 30° in the south, while the celestial equator stretches across the northern half of the sky.

**FIGURE S1.16** The sky at 30°S latitude.

**FIGURE S1.17** shows the Sun's path on the equinoxes and solstices for latitude 40°N. On the equinoxes, the Sun is on the celestial equator (dec = 0°) and therefore follows the celestial equator's path: It rises due east, crosses the meridian at altitude 50° in the south, and sets due west. Like other objects on the celestial equator, it is above the horizon for 12 hours. On the June solstice, the Sun has dec = $+23\frac{1}{2}°$ (see Table S1.1) and therefore rises well north of due east,* reaches an altitude of $50° + 23\frac{1}{2}° = 73\frac{1}{2}°$ when it crosses the meridian in the south, and sets well north of due west. The daylight hours are long because much more than half the Sun's path is above the horizon. On the December solstice, when the Sun has dec = $-23\frac{1}{2}°$, the Sun rises well south of due east, reaches an altitude of only $50° - 23\frac{1}{2}° = 26\frac{1}{2}°$ when it crosses the meridian in the south, and sets well south of due west. The daylight hours are short because much less than half the Sun's path is above the horizon.

We could make a similar diagram to show the Sun's path on various dates for any latitude. However, the $23\frac{1}{2}°$ tilt of Earth's axis makes the Sun's path particularly interesting at the special latitudes shown in **FIGURE S1.18**. Let's investigate.

**The Sun at the North and South Poles** Recall that the celestial equator circles the horizon at the North Pole. **FIGURE S1.19** shows how we use this fact to find the Sun's path in the North Polar sky. Because the Sun appears *on* the celestial equator on the day of the March equinox, the Sun circles the North Polar sky *on the horizon* on March 21, completing a full circle in 24 hours (1 solar day). Over the next 3 months, the Sun continues to circle the horizon each day, circling at gradually higher altitudes as its declination increases. It reaches its highest point on the June solstice, when its declination of $+23\frac{1}{2}°$ means that it circles the North Polar sky at an altitude of $23\frac{1}{2}°$. After the June solstice, the daily circles gradually fall lower over the next 3 months, reaching the horizon on the September equinox. Then, because the Sun's declination is negative for the next 6 months (until the following March equinox), the Sun remains below the North Polar horizon. That is why the North Pole essentially has 6 months of daylight and 6 months of darkness, with an extended twilight that lasts a few weeks beyond the September equinox and an extended dawn that begins a few weeks before the March equinox.

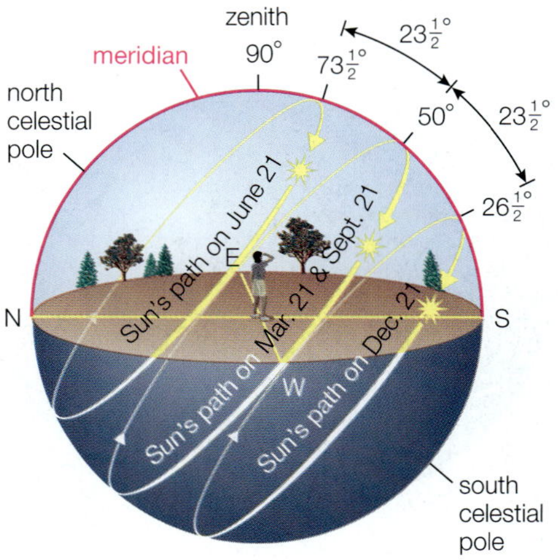


**FIGURE S1.17** The Sun's daily path on the equinoxes and solstices at latitude 40°N.

*Calculating exactly how far north of due east the Sun rises is beyond the scope of this book, but astronomical software and websites can do these calculations for different latitudes.

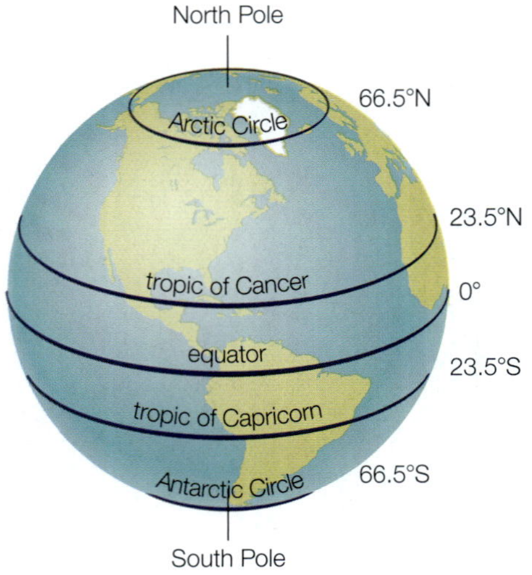


**FIGURE S1.18** Special latitudes defined by the Sun's path through the sky.

The situation is the opposite at the South Pole. Here, the Sun's daily circle first reaches the horizon on the September equinox. The daily circles then rise gradually higher, reaching a maximum altitude of $23\frac{1}{2}°$ on the December solstice (when it is summer in the Antarctic), and then slowly fall back to the horizon on the March equinox. That is, the South Pole has the Sun above the horizon during the 6 months it is below the North Polar horizon.

Two important caveats make the actual view from the Poles slightly different than we've described. First, the atmosphere bends light enough so that when the Sun is near the horizon, it *appears* to be about 1° higher than it really is, which means we can see the Sun even when it is slightly below the horizon. Second, the Sun's angular size of about $\frac{1}{2}°$ means that it does not fall below the horizon at a single moment but instead sets gradually. Together, these effects mean that the Sun *appears* above each polar horizon for several days longer than 6 months each year.

**The Sun at the Equator** At the equator, the celestial equator extends from the horizon due east, through the zenith, to the horizon due west. The Sun therefore follows this path on each equinox, reaching the zenith at local noon (**FIGURE S1.20**). Following the March equinox, the Sun's increasing declination means that, day by day, its path moves gradually northward in the sky. It is farthest north on the June solstice, when it rises $23\frac{1}{2}°$ north of due east, crosses the meridian at altitude $90° - 23\frac{1}{2}° = 66\frac{1}{2}°$ in the north, and sets $23\frac{1}{2}°$ north of due west. Over the next 6 months, it gradually tracks southward until the December solstice, when its path is the mirror image (across the celestial equator) of its June solstice path.

Like all objects in the equatorial sky, the Sun is always above the horizon for half a day and below it for half a day. Moreover, the Sun's track is highest in the sky on the equinoxes and lowest on the solstices. That is why equatorial regions do not have four seasons like temperate regions [**Section 2.2**]. The Sun's path in the equatorial sky also makes it rise and set perpendicular to the horizon every day of the year, making for a more rapid dawn and a briefer twilight than at other latitudes.

**The Sun at the Tropics** The circles of latitude 23.5°N and 23.5°S are called the **tropic of Cancer** and the **tropic of Capricorn**, respectively (see Figure S1.18). The region between these two circles, generally called the *tropics*, represents the parts of Earth where the Sun can sometimes reach the zenith at noon.

**FIGURE S1.21** shows why the tropic of Cancer is special. The celestial equator extends from due east on the horizon to due west on the horizon, crossing the meridian in the south at an altitude of $90° - 23\frac{1}{2}°$ (the latitude) $= 66\frac{1}{2}°$, or $23\frac{1}{2}°$ short of the zenith. Therefore, the Sun reaches the zenith at local noon on the June solstice, when it crosses the meridian $23\frac{1}{2}°$ northward of the celestial equator. The tropic of Cancer marks the northernmost latitude at which the Sun ever reaches the zenith. Similarly, at the tropic of Capricorn, the Sun reaches the zenith at local noon on

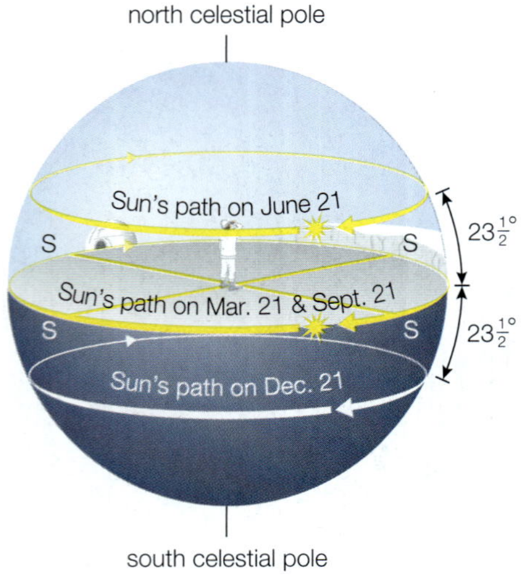


**FIGURE S1.19** Daily path of the Sun on the equinoxes and solstices at the North Pole.

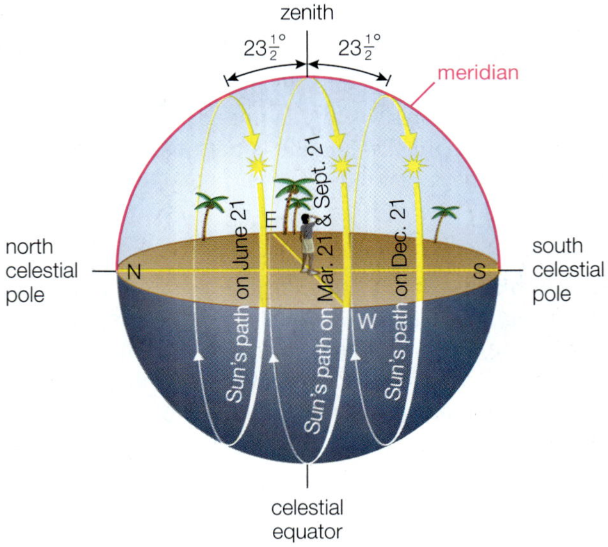


**FIGURE S1.20** Daily path of the Sun on the equinoxes and solstices at the equator.

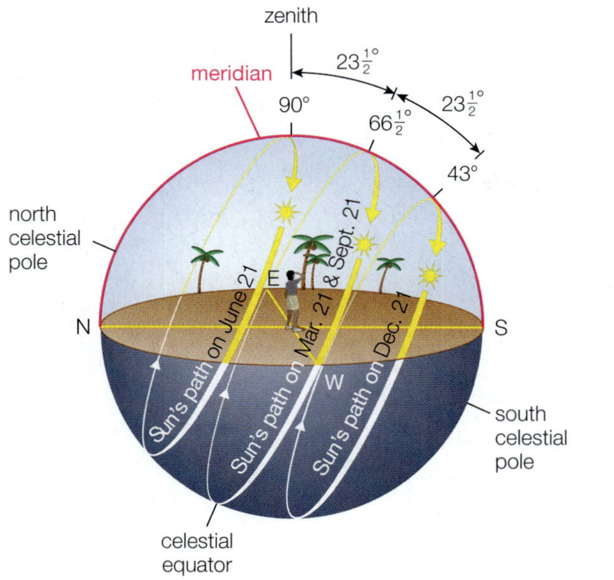


**FIGURE S1.21** Daily path of the Sun on the equinoxes and solstices at the tropic of Cancer.

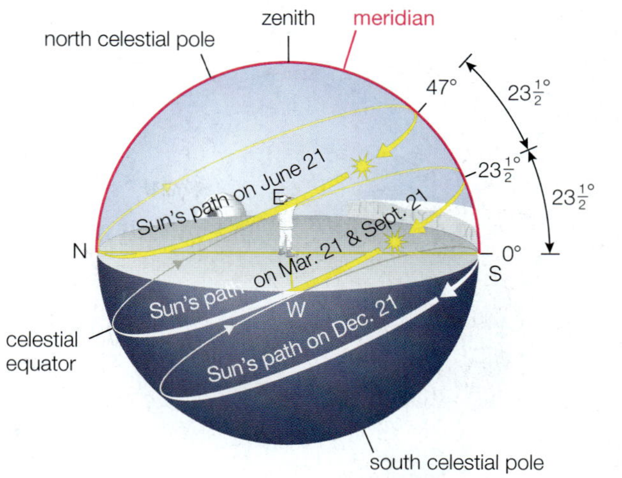


**FIGURE S1.22** Daily path of the Sun on the equinoxes and solstices at the Arctic Circle.

the December solstice (when it is summer for the Southern Hemisphere), making this the southernmost latitude at which the Sun ever reaches the zenith. Between the two tropic circles, the Sun passes through the zenith twice a year; the precise dates vary with latitude.

**The Sun at the Arctic and Antarctic Circles** At the equator, the Sun is above the horizon for 12 hours each day year-round. At latitudes progressively farther from the equator, the daily time that the Sun is above the horizon varies progressively more with the seasons. The special latitudes at which the Sun remains continuously above the horizon for a full day each year are the polar circles: the **Arctic Circle** at latitude 66.5°N and the **Antarctic Circle** at latitude 66.5°S (see Figure S1.18). Poleward of these circles, the length of continuous daylight (or darkness) increases beyond 24 hours, reaching the extreme of 6 months at the North and South Poles.

**FIGURE S1.22** shows why the Arctic Circle is special. The celestial equator extends from due east on the horizon to due west on the horizon, crossing the meridian in the south at an altitude of $90° - 66\frac{1}{2}°$ (the latitude) $= 23\frac{1}{2}°$. As a result, the Sun's path is circumpolar on the June solstice: The Sun skims the northern horizon at midnight, rises through the eastern sky to a noon maximum altitude of 47° in the south (which is the celestial equator's maximum altitude of $23\frac{1}{2}°$ plus the Sun's June solstice declination of $23\frac{1}{2}°$), and then gradually falls through the western sky until it is back on the horizon at midnight (see Figure 2.18). At the Antarctic Circle, the Sun follows the same basic pattern on the December solstice, except that it skims the horizon in the south and rises to a noon maximum altitude of 47° in the north.

Of course, what we *see* is subject to the same caveats we discussed for the North and South Poles: The bending of light by Earth's atmosphere and the Sun's angular size make the Sun *appear* to be slightly above the horizon even when it is slightly below it. As a result, at the Arctic Circle, the Sun seems not to set for several days around the June solstice (rather than for a single day) and appears to peek above the horizon momentarily (rather than not at all) around the December solstice. The same ideas hold for the opposite solstices at the Antarctic Circle.

**Celestial Navigation**

## S1.3 Principles of Celestial Navigation

Imagine that you're on a ship at sea, far from any landmarks. How can you figure out where you are? We now have all the background we need to answer this question.

### How can you determine your latitude?

It's easy to determine your latitude if you can find the north or south celestial pole in your sky, because latitude is equal to the altitude of the celestial pole. In the Northern Hemisphere at night, you can determine your approximate latitude by measuring the altitude of Polaris, which lies within 1° of the north celestial pole. For example, if Polaris has altitude 17°, your latitude is between 16°N and 18°N.

If you want to be more precise, you can determine your latitude from the altitude of *any* star as it crosses your meridian. For example, suppose Vega happens to be crossing your meridian right now and it appears in your southern sky at altitude 78° 44′. Because Vega has dec = +38° 44′ (see Figure S1.11), it crosses your meridian 38° 44′ north of the celestial equator. As shown in **FIGURE S1.23a**, you can conclude that the celestial equator crosses your meridian at an altitude of precisely 40° in the south, and therefore that your latitude is 50°N.

In the daytime, you can find your latitude from the Sun's altitude on your meridian if you know the date and the Sun's declination on that date. For example, suppose the date is March 21 and the Sun crosses your meridian at altitude 70° in the north (**FIGURE S1.23b**). Because the Sun has dec = 0° on March 21, you can conclude that the celestial equator also crosses your meridian in the north at altitude 70°, which means you are at latitude 20°S.

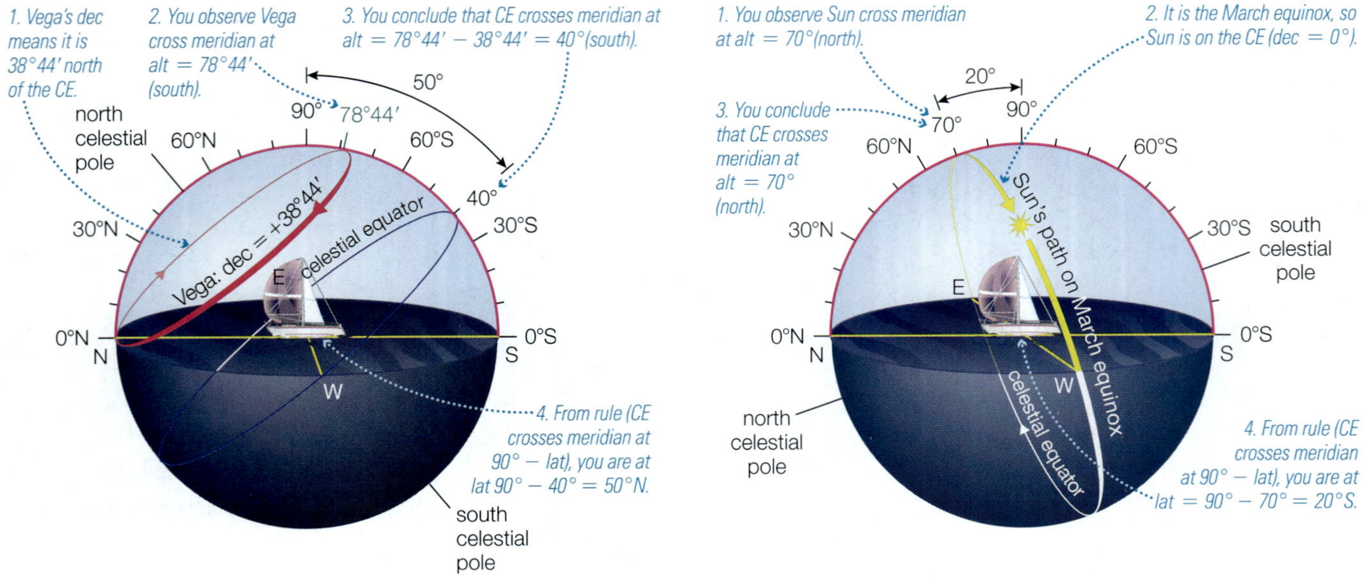


**a** Diagram showing how you can find your latitude by measuring Vega's altitude when it crosses the meridian. You know that you are in the Northern Hemisphere because the CE crosses the meridian in the south.

**b** Diagram showing how you can find your latitude by knowing the date and measuring the Sun's altitude when it crosses the meridian. You know that you are in the Southern Hemisphere because the CE crosses the meridian in the north.

**FIGURE S1.23** Determining latitude from a star and from the Sun. Abbreviations: *dec* for declination, *alt* for altitude, *lat* for latitude, and *CE* for celestial equator.

## How can you determine your longitude?

You can determine your longitude by comparing the current position of an object in your sky with its position as seen from some known longitude. As a simple example (**FIGURE S1.24**), suppose you use a sundial to determine that the apparent solar time is 1:00 p.m., which means the Sun crossed your meridian 1 hour ago. You immediately call a friend in England and learn that it is 3:00 p.m. in Greenwich (or you carry a clock that keeps Greenwich time). You now know that your local time is 2 hours earlier than the local time in Greenwich, which means you are 2 hours west of Greenwich. (An earlier time means that you are *west* of Greenwich, because Earth rotates from west to east.) Each hour corresponds to 15° of longitude, so "2 hours west of Greenwich" means longitude 30°W.

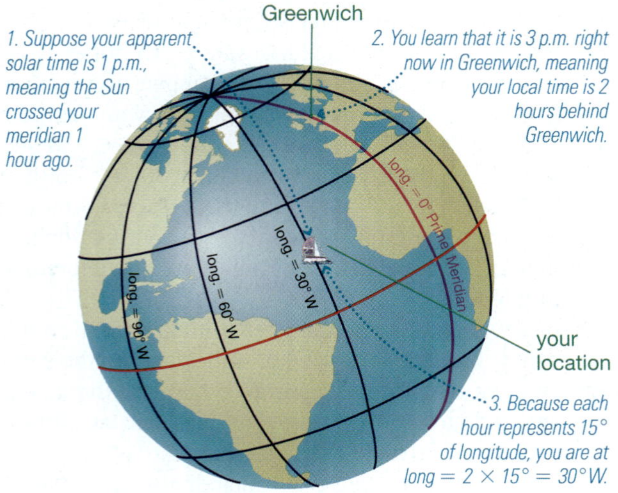


**FIGURE S1.24** You can determine your longitude by comparing your local time to the time in Greenwich.

At night, you can find your longitude by comparing the positions of stars in your local sky and at some known longitude. For example, suppose Vega is on your meridian and a call to your friend reveals that it won't cross the meridian in Greenwich until 6 hours from now. In this case, your local time is 6 hours later than the local time in Greenwich, which means you are 6 hours east of Greenwich, or at longitude 90°E (because $6 \times 15° = 90°$).

**Celestial Navigation in Practice** Although celestial navigation is easy in principle, at least three considerations make it more difficult in practice. First, finding either latitude or longitude requires a tool for measuring angles in the sky. One such device, called an *astrolabe*, was invented by the ancient Greeks and significantly improved by Islamic scholars during the Middle Ages. The astrolabe's faceplate (**FIGURE S1.25a**) could be used to tell time, because it consisted of a rotating star map and horizon plates for specific latitudes. Today you can buy similar rotatable star maps, called *planispheres*. Most astrolabes contained a sighting stick on the back that allowed users to measure the altitudes

### COMMON MISCONCEPTIONS

**Compass Directions**

Most people determine direction with the aid of a compass rather than the stars. However, a compass needle doesn't actually point to true geographic north. Instead, the compass needle responds to Earth's magnetic field and points to *magnetic* north, which can be substantially different from true north. If you want to navigate precisely with a compass, you need a special map that takes into account local variations in Earth's magnetic field. Such maps are available at most camping stores. They are not perfectly reliable, however, because the magnetic field also varies with time. In general, celestial navigation is much more reliable for determining direction than using a compass.

**a** The faceplate of an astrolabe. Many astrolabes had sighting sticks on the back for measuring positions of bright stars.

**b** A copper engraving of Italian explorer Amerigo Vespucci (for whom the Americas were named) using an astrolabe to sight the Southern Cross. The engraving by Philip Galle, from the book *Nova Reperta*, was based on an original by Joannes Stradanus in the early 1580s.

**c** A woodcut of Ptolemy holding a cross-staff (artist unknown).

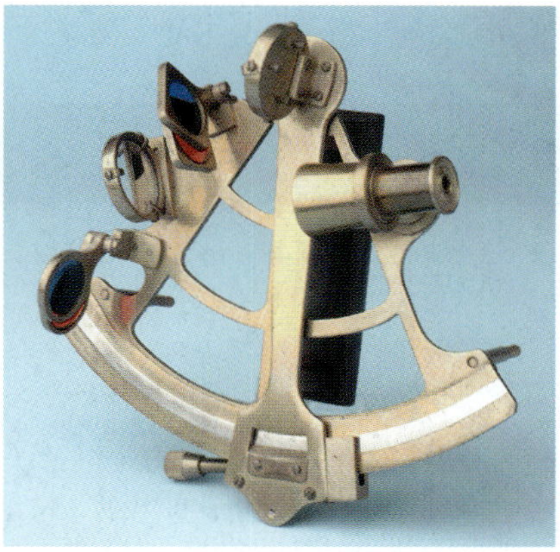

**d** A sextant.

**FIGURE S1.25** Navigational instruments.

of bright stars in the sky. These measurements could then be correlated against special markings under the faceplate (**FIGURE S1.25b**). Astrolabes were effective but difficult and expensive to make. As a result, medieval sailors often measured angles with a simple pair of calibrated perpendicular sticks, called a *cross-staff* or *Jacob's staff* (**FIGURE S1.25c**). A more modern device called a *sextant* allows much more precise angle determinations by incorporating a small telescope for sightings (**FIGURE S1.25d**). Sextants are still used for celestial navigation on many ships. If you want to practice celestial navigation yourself, you can buy an inexpensive plastic sextant at many science-oriented stores.

A second practical consideration is knowing the celestial coordinates of stars and the Sun so that you can determine their paths through the local sky. At night, you can use a table listing the celestial coordinates of bright stars. In addition to knowing the celestial coordinates, you must either know the constellations and bright stars extremely well or carry star charts to help you identify them. For navigating by the Sun in the daytime, you'll need a table listing the Sun's celestial coordinates on each day of the year.

The third practical consideration applies to determining longitude: You need to know the current position of the Sun (or a particular star) in a known location, such as Greenwich, England. Although you could determine this by calling a friend who lives there, it's more practical to carry a clock set to universal time (the time in Greenwich). In the daytime, the clock makes it easy to determine your longitude. If apparent solar time is 1:00 p.m. in your location and the clock tells you that it is 3:00 p.m. in Greenwich, then you are 2 hours west of Greenwich, or at longitude 30°W. The task is more difficult at night, because you must compare the position of a *star* in your sky to its current position in Greenwich. You can do this with the aid of detailed astronomical tables that allow you to determine the current position of any star in the Greenwich sky from the date and the universal time.

Historically, this third consideration created enormous problems for navigation. Before the invention of accurate clocks, sailors could easily determine their latitude but not their longitude. Indeed, most of the European voyages of discovery in the 15th century through the 17th century relied on little more than guesswork about longitude, although some sailors learned complex mathematical techniques for estimating longitude through observations of the lunar phases. More accurate longitude determination, upon which the development of extensive ocean commerce and travel depended, required the invention of a clock that would remain accurate on a ship rocking in the ocean swells. By the early 18th century, solving this problem was considered so important that the British government offered a substantial monetary prize for the solution. John Harrison claimed the prize in 1761, with a clock that lost only 5 seconds during a 9-week voyage to Jamaica.*

**The Global Positioning System** Today, a new type of celestial navigation has supplanted traditional methods.

*The story of the difficulties surrounding the measurement of longitude at sea and how Harrison finally solved the problem is chronicled in *Longitude*, by Dava Sobel (Walker and Company, 1995).

It finds positions relative to satellites of the **global positioning system (GPS)**. In essence, these Earth-orbiting satellites function like artificial stars. The satellites' positions at any moment are known precisely from their orbital characteristics, and they transmit radio signals that can be picked up by GPS receivers in cars, smart phones, and other devices. Your GPS receiver locates three or more of the satellites and then does computations to calculate your position on Earth.

Navigation by GPS is so precise that the ancient practice of celestial navigation is in danger of becoming a lost art. Fortunately, many amateur clubs and societies are keeping the skills of celestial navigation alive.

## The BIG Picture PUTTING CHAPTER S1 INTO CONTEXT

*In this chapter, we built upon concepts from the first three chapters to form a more detailed understanding of celestial timekeeping and navigation. You also learned how to determine paths for the Sun and the stars in the local sky. As you look back at what you've learned, keep in mind the following "big picture" ideas:*

- Our modern systems of timekeeping are rooted in the apparent motions of the Sun through the sky. Although it's easy to forget these roots when you look at a clock or a calendar, the sky was the only guide to time for most of human history.
- The term *celestial navigation* sounds a bit mysterious, but it refers to simple principles that allow you to determine your location on Earth. Even if you're never lost at sea, you may find the basic techniques of celestial navigation useful to orient yourself at night (for example, on your next camping trip).

**MY COSMIC PERSPECTIVE** If you understand the apparent motions of the sky discussed in this chapter and also learn the constellations and bright stars, you'll feel very much "at home" under the stars at night.

## Summary of Key Concepts

### S1.1 Astronomical Time Periods

- **How do we define the day, month, year, and planetary periods?** Each of these is defined in two ways. A **sidereal day** is Earth's rotation period, which is about 4 minutes shorter than the 24-hour **solar day** from noon one day to noon the next day. A **sidereal month** is the Moon's orbital period of about $27\frac{1}{3}$ days; a **synodic month** is the $29\frac{1}{2}$ days required for the Moon's cycle of phases. A **sidereal year** is Earth's orbital period, which is about 20 minutes longer than the **tropical year** from one March equinox to the next. A planet's **sidereal period** is its orbital period, and its **synodic period** is the time from one opposition or conjunction to the next.

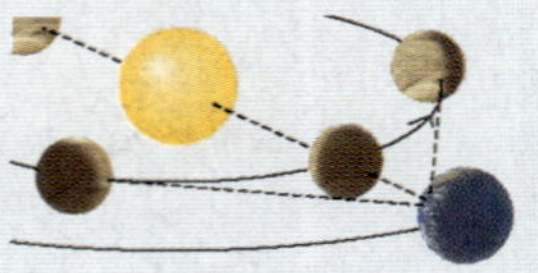

- **How do we tell the time of day?** There are several time measurement systems. **Apparent solar time** is based on the Sun's position in the local sky. **Mean solar time** is also local, but it averages the changes in the Sun's rate of motion over the year. **Standard time** and **daylight saving time** divide the world into time zones. **Universal time** is the mean solar time in Greenwich, England.
- **When and why do we have leap years?** We usually have a **leap year** every 4 years because the length of the year is about $365\frac{1}{4}$ days. However, it is not exactly $365\frac{1}{4}$ days, so our calendar skips a leap year in century years not divisible by 400.

### S1.2 Celestial Coordinates and Motion in the Sky

- **How do we locate objects on the celestial sphere?** **Declination** is given as an angle describing an object's position north or south of the celestial equator. **Right ascension**, usually measured in hours (and minutes and seconds), tells us how far east an object is located relative to the March equinox.

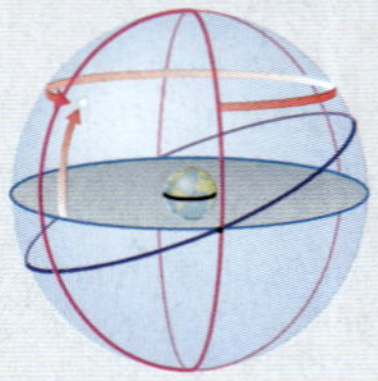

- **How do stars move through the local sky?** A star's path through the local sky depends on its declination and your latitude. Latitude tells you the orientation of your sky relative to the celestial sphere, while declination tells you how a particular star's path compares to the path of the celestial equator through your sky.

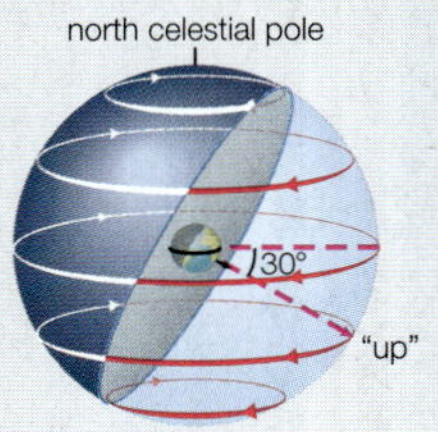


- **How does the Sun move through the local sky?** The Sun's path also depends on its declination and your latitude, but it varies throughout the year because of the Sun's changing declination. The Sun's varying path helps define special latitudes, including the **tropic of Cancer** and **tropic of Capricorn** and the **Arctic Circle** and **Antarctic Circle**.

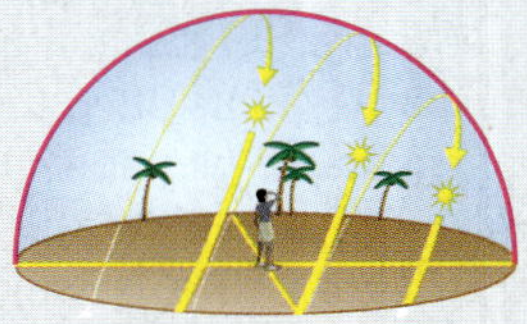

### S1.3 Principles of Celestial Navigation

- **How can you determine your latitude?** You can determine your latitude from the altitude of the celestial pole in your sky or by measuring the altitude and knowing the declination of a star (or the Sun) as it crosses your meridian.

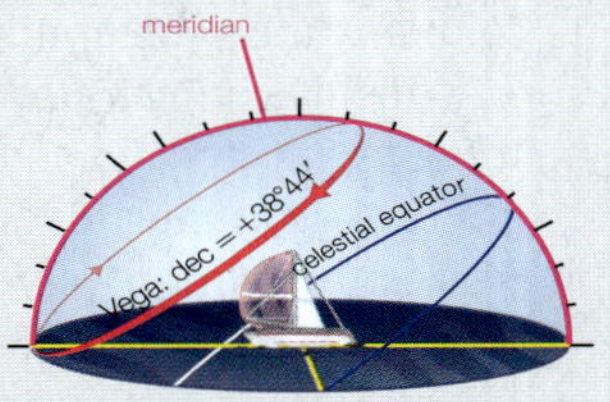


- **How can you determine your longitude?** To determine longitude you must know the position of the Sun or a star in your sky and its position at the same time in the sky of Greenwich, England (or some other specific location). This is most easily done if you have a clock that tells universal time.

# Visual Skills Check

*Use the following questions to check your understanding of some of the many types of visual information used in astronomy. For additional practice, try the Chapter S1 Visual Quiz in the Study Area at www .MasteringAstronomy.com.*

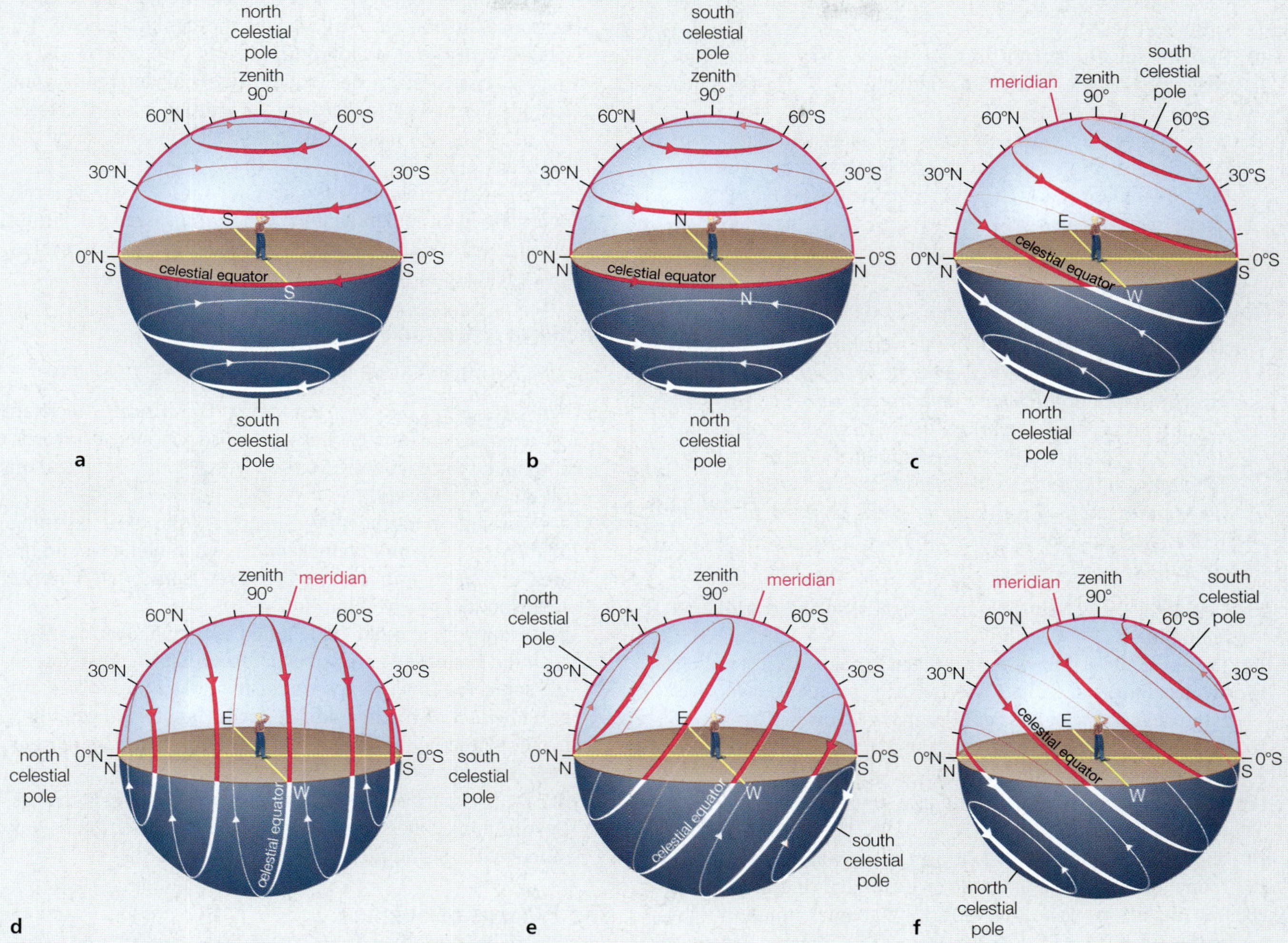


*The six diagrams represent the sky at six different latitudes. Answer the following questions about them.*

1. Which diagram represents the paths of stars at the North Pole?
2. Which diagram represents the paths of stars at the South Pole?
3. Which diagrams represent Southern Hemisphere skies?
4. What latitude is represented in diagram c?
5. Which diagram(s) represent(s) a latitude at which the Sun sometimes passes directly overhead?
6. Which diagram(s) represent(s) a latitude at which the Sun sometimes remains below the horizon during a full 24-hour period?
7. Each diagram shows five star circles. Look at the first circle to the north of the celestial equator on each diagram. Can you characterize the *declination* of stars on this circle? If so, what is it? Can you characterize the *right ascension* of stars on this circle? If so, what is it?

# Exercises and Problems

*For instructor-assigned homework and other learning materials, go to www.MasteringAstronomy.com.*

## Chapter Review Questions

### Short-Answer Questions Based on the Reading

1. Explain the differences between a (a) *sidereal day* and *solar day*, (b) *sidereal month* and *synodic month*, (c) *sidereal year* and *tropical year*, (d) planet's *sidereal period* and *synodic period*.
2. Define *opposition*, *conjunction*, and *greatest elongation* for planets both closer to and farther from the Sun than is Earth.
3. For what planets do we sometimes observe a *transit*? Why?
4. Distinguish among apparent solar time, mean solar time, standard time, daylight saving time, and universal time.
5. Describe the origins of the Julian and Gregorian calendars. Which one do we use today?
6. What do we mean when we say the equinoxes and solstices are points on the celestial sphere? How are these points related to the times of year called the equinoxes and solstices?

7. What are *declination* and *right ascension*? How are they similar to latitude and longitude? How are they different?
8. How and why do the Sun's celestial coordinates change over the course of each year?
9. Suppose you are at the North Pole. Where is the celestial equator? Where is the north celestial pole? Describe the daily motion of the sky. Do the same for the equator and for latitude 40°N.
10. Describe the Sun's path through the local sky on the equinoxes and on the solstices for latitude 40°N. Do the same for the North Pole, South Pole, and equator.
11. What is special about the tropics of Cancer and Capricorn? Describe the Sun's path on the solstices at these latitudes. Do the same for the Arctic and Antarctic Circles.
12. Briefly describe how you can use the Sun or stars to determine your latitude and longitude.

### Does It Make Sense?

*Decide whether or not each of the following statements makes sense (or is clearly true or false). Explain clearly; not all of these have definitive answers, so your explanation is more important than your chosen answer. (Hint: For statements that use coordinates such as altitude, longitude, or declination, be sure to check whether the coordinates are used correctly; for example, it would not make sense to describe a location on Earth's surface by an altitude.)*

13. Last night I saw Venus shining brightly on the meridian at midnight.
14. The apparent solar time was noon, but the Sun was just setting.
15. My mean solar clock said it was 2:00 p.m., but a friend who lives east of here had a mean solar clock that said it was 2:11 P.M.
16. When the standard time is 3:00 p.m. in Baltimore, it is 3:15 p.m. in Washington, D.C.
17. Last night around 8:00 p.m., I saw Jupiter at an altitude of 45° in the south.
18. The latitude of the stars in Orion's belt is about 5°N.
19. Today the Sun is at an altitude of 10° on the celestial sphere.
20. Los Angeles is west of New York by about 3 hours of right ascension.
21. The June solstice is east of the March equinox by 6 hours of right ascension.
22. My UT clock had stopped, but I found my longitude by measuring the altitudes of 14 stars in my local sky.

## Quick Quiz

*Choose the best answer to each of the following. For additional practice, try the Chapter S1 Reading and Concept Quizzes in the Study Area at www.MasteringAstronomy.com.*

23. The time from one March equinox to the next is the (a) sidereal day. (b) tropical year. (c) synodic month.
24. Jupiter is brightest when it is (a) at opposition. (b) at conjunction. (c) closest to the Sun in its orbit.
25. Venus is easiest to see when it is at (a) superior conjunction. (b) inferior conjunction. (c) greatest eastern elongation.
26. In the winter, your wristwatch tells (a) apparent solar time. (b) standard time. (c) universal time.
27. A star located 30° north of the celestial equator has (a) declination = 30°. (b) right ascension = 30°. (c) latitude = 30°.
28. A star's path through your sky depends on your latitude and the star's (a) declination. (b) right ascension. (c) both declination and right ascension.
29. At latitude 50°N, the celestial equator crosses the meridian at altitude (a) 50° in the south. (b) 50° in the north. (c) 40° in the south.
30. At the North Pole on the June solstice, the Sun (a) remains stationary in the sky. (b) reaches the zenith at noon. (c) circles the horizon at altitude $23\frac{1}{2}°$.
31. If you know a star's declination, you can determine your latitude if you also (a) measure its altitude when it crosses the meridian. (b) measure its right ascension. (c) know the universal time.
32. If you measure the Sun's position in your local sky, you can determine your longitude if you also (a) measure its altitude when it crosses the meridian. (b) know its right ascension and declination. (c) know the universal time.

## Inclusive Astronomy

*Use these questions to reflect on participation in science.*

33. *Pair Discussion: Benjamin Banneker.* The African American astronomer Benjamin Banneker was also a farmer, a surveyor, and an author who corresponded with Thomas Jefferson on the subject of slavery.
**a.** Working independently, spend a little time learning about who Benjamin Banneker was, when he lived, and his area of expertise in astronomy. Also, learn what Thomas Jefferson had to say about him.
**b.** Pair up with another student and share what you learned. How did Banneker's area of expertise in astronomy relate to the subject of this chapter and his work as a surveyor?
**c.** Had either of you heard of Banneker before? Do you think he should be better known? Do you think there are reasons why is not more famous? Explain.
**d.** Discuss what Thomas Jefferson had to say about Banneker. Do you think Jefferson's impressions of Banneker made any difference in his attitude toward slavery?

## The Process of Science

*These questions may be answered individually in short-essay form or discussed in groups, except where identified as group-only.*

34. *Transits and the Geocentric Universe.* Ancient people could not observe transits of Mercury or Venus across the Sun, because they lacked instruments for viewing a small dark spot against the Sun. But suppose they *could* have seen transits. How would transit observations have affected the debate over an Earth-centered versus a Sun-centered solar system? Explain.
35. *Geometry and Science.* As discussed in Mathematical Insight S1.1, Copernicus found that a Sun-centered model led him to a simple geometric layout for the solar system, a fact that gave him confidence that his model was on the right track. Did the mathematics actually prove that the Sun-centered model was correct? Use your answer to briefly discuss the role of mathematics in science.
36. *Northern Chauvinism.* Why is the writing on maps and globes usually oriented so that the Northern Hemisphere is at the top, even though there is no up or down in space? How does this relate to the fact that the June solstice has traditionally been called the *summer* solstice? Discuss.

37. *Group Activity: Find Your Way Home.* You and your partners are international spies who have been captured by a criminal gang and flown to a secret compound. You escape . . . but all you have is a watch (set to your previous local time), a star chart, and a world map (marked with longitude and latitude). How do you figure out where you are? How could you use celestial navigation to find your way home? Note: You may wish to do this activity using the four roles described in Chapter 1, Exercise 39.

## Investigate Further

### Short-Answer/Essay Questions

38. *Opposite Rotation.* Suppose Earth rotated in a direction opposite to its orbital direction; that is, suppose it rotated clockwise (as seen from above the North Pole) but orbited counterclockwise. Would the solar day still be longer than the sidereal day? Explain.
39. *No Precession.* Suppose Earth's axis did *not* precess. Would the sidereal year still be different from the tropical year? Explain.
40. *The Sun from Mars.* Mars has an axis tilt of 25.2°, only slightly larger than that of Earth. Compared to that on Earth, is the range of latitudes on Mars for which the Sun can reach the zenith larger or smaller? Is the range of latitudes for which the Sun is circumpolar larger or smaller? Make a sketch of Mars similar to the one for Earth in Figure S1.18.
41. *Fundamentals of Your Local Sky.* Answer each of the following for *your* latitude.
    **a.** Where is the north (or south) celestial pole in your sky?
    **b.** Describe the meridian in your sky, specifying at least three distinct points along it (such as the points at which it meets your horizon and its highest point).
    **c.** Describe the celestial equator in your sky, specifying at least three distinct points along it.
    **d.** Does the Sun ever appear at your zenith? If so, when? If not, why not?
    **e.** What range of declinations makes a star circumpolar in your sky?
    **f.** What is the range of declinations for stars that you can never see in your sky?
42. *Sydney Sky.* Repeat Problem 41 for the local sky in Sydney, Australia (latitude 34°S).
43. *Local Path of the Sun.* Describe the path of the Sun through your local sky for each of the following days:
    **a.** the March and September equinoxes.
    **b.** the June solstice.
    **c.** the December solstice.
    **d.** today. (*Hint*: You can estimate the Sun's RA and dec for today's date from data in Table S1.1.)
44. *Sydney Sun.* Repeat Problem 43 for the local sky in Sydney, Australia (latitude 34°S).
45. *Calendar History.* Investigate the history of the Julian or Gregorian calendar in greater detail. Write a short summary of at least one interesting aspect of the history you learn from your research. (For example, why did Julius Caesar allow one year to have 445 days? How did our months end up with 28, 30, or 31 days?)
46. *Global Positioning System.* Learn more about the global positioning system and its uses. Write a short report summarizing how new uses of GPS may affect our lives over the next 10 years.

### Quantitative Problems

*Be sure to show all calculations clearly and state your final answers in complete sentences.*

47. *Lost at Sea I.* During a vacation, you decide to take a solo boat trip. While contemplating the universe, you lose track of your location. Fortunately, you have some astronomical tables and instruments, as well as a UT clock. You thereby put together the following description of your situation:
    - It is the March equinox.
    - The Sun is on your meridian at altitude 75° in the south.
    - The UT clock reads 22:00.

    **a.** What is your latitude? How do you know?
    **b.** What is your longitude? How do you know?
    **c.** Consult a map. Based on your position, where is the nearest land? Which way should you sail to reach it?
48. *Lost at Sea II.* Repeat Problem 47 for this situation:
    - It is the day of the June solstice.
    - The Sun is on your meridian at altitude $67\frac{1}{2}°$ in the north.
    - The UT clock reads 06:00.
49. *Lost at Sea III.* Repeat Problem 47 for this situation:
    - Your local time is midnight.
    - Polaris appears at altitude 67° in the north.
    - The UT clock reads 01:00.
50. *Lost at Sea IV.* Repeat Problem 47 for this situation:
    - Your local time is 6 a.m.
    - From the position of the Southern Cross, you estimate that the south celestial pole is at altitude 33° in the south.
    - The UT clock reads 11:00.
51. *Orbital and Synodic Periods.* Use each object's given synodic period to find its actual orbital period.
    **a.** Saturn, synodic period = 378.1 days
    **b.** Mercury, synodic period = 115.9 days
    **c.** An asteroid with synodic period = 429 days
52. *Using the Analemma.*
    **a.** It's February 15 and your sundial tells you the apparent solar time is 18 minutes until noon. What is the mean solar time?
    **b.** It's July 1 and your sundial tells you that the apparent solar time is 3:30 p.m. What is the mean solar time?
53. *HA = LST − RA.*
    **a.** It is 4 p.m. on the March equinox. What is the local sidereal time?
    **b.** The local sidereal time is 19:30. When will Vega cross your meridian?
    **c.** You observe a star that has an hour angle of 13 hours ($13^h$) when the local sidereal time is 8:15. What is the star's right ascension?
    **d.** The Orion Nebula has declination of about −5.5° and right ascension of $5^h25^m$. If you are at latitude 40°N and the local sidereal time is 7:00, approximately where does the Orion Nebula appear in your sky?
54. *Meridian Crossings of the Moon and Phobos.* Estimate the time between meridian crossings of the Moon for a person standing on Earth. Repeat your calculation for meridian crossings of the Martian moon Phobos for a person on Mars. Use the Appendixes in the back of the book if necessary.
55. *Mercury's Rotation Period.* Mercury's sidereal day is approximately $\frac{2}{3}$ of its orbital period, or about 58.6 days. Estimate the length of Mercury's solar day. Compare it to Mercury's orbital period of about 88 days.

# COSMIC Context PART I Our Expanding Perspective

**Our perspective on the universe has changed dramatically throughout human history. This timeline summarizes some of the key discoveries that have shaped our modern perspective.**

Stonehenge

Earth-centered model of the universe

Galileo's telescope

| <2500 B.C. | 400 B.C.–170 A.D. | 1543–1648 A.D. |
|---|---|---|

(1) Ancient civilizations recognized patterns in the motion of the Sun, Moon, planets, and stars through our sky. They also noticed connections between what they saw in the sky and our lives on Earth, such as the cycles of seasons and of tides [Section 3.1].

(2) The ancient Greeks tried to explain observed motions of the Sun, Moon, and planets using a model with Earth at the center, surrounded by spheres in the heavens. The model explained many phenomena well, but could explain the apparent retrograde motion of the planets only with the addition of many complex features—and even then, its predictions were not especially accurate [Section 3.2].

(3) Copernicus suggested that Earth is a planet orbiting the Sun. The Sun-centered model explained apparent retrograde motion simply, though it made accurate predictions only after Kepler discovered his three laws of planetary motion. Galileo's telescopic observations confirmed the Sun-centered model, and revealed that the universe contains far more stars than had been previously imagined [Section 3.3].

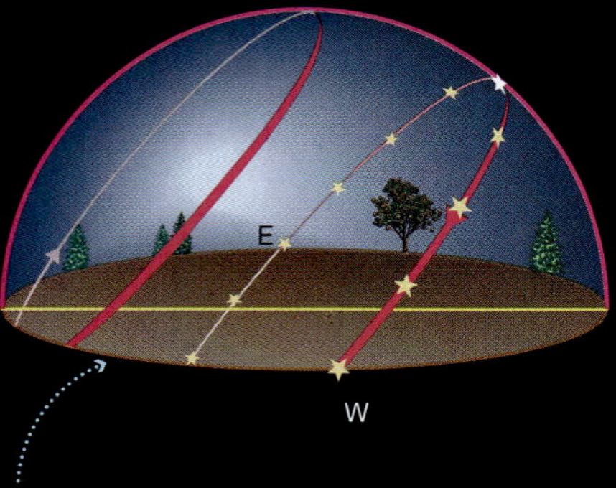


*Earth's rotation around its axis leads to the daily east-to-west motions of objects in the sky.*

*The tilt of Earth's rotation axis leads to seasons as Earth orbits the Sun.*

*Planets are much smaller than the Sun. At a scale of 1 to 10 billion, the Sun is the size of a grapefruit, Earth is the size of a ball point of a pen, and the distance between them is about 15 meters.*

Yerkes Observatory

Edwin Hubble at the Mt. Wilson telescope

Hubble Space Telescope

1838–1920 A.D.

1924–1929 A.D.

1990 A.D.–present

4. Larger telescopes and photography made it possible to measure the parallax of stars, offering direct proof that Earth really does orbit the Sun and showing that even the nearest stars are light-years away. We learned that our Sun is a fairly ordinary star in the Milky Way [Sections 2.4, 15.1].

5. Edwin Hubble measured the distances of galaxies, showing that they lay far beyond the bounds of the Milky Way and proving that the universe is far larger than our own galaxy. He also discovered that more distant galaxies are moving away from us faster, telling us that the entire universe is expanding and suggesting that it began in an event we call the Big Bang [Sections 1.3, 20.2].

6. Improved measurements of galactic distances and the rate of expansion have shown that the universe is about 14 billion years old. These measurements have also revealed still-unexplained surprises, including evidence for the existence of mysterious "dark matter" and "dark energy" [Sections 1.3, 23.1].

*Distances between stars are enormous. At a scale of 1 to 10 billion, you can hold the Sun in your hand, but the nearest stars are thousands of kilometers away.*

*Our solar system is located about 27,000 light-years from the center of the Milky Way Galaxy.*

*The Milky Way Galaxy contains over 100 billion stars.*

*The observable universe contains over 100 billion galaxies.*

# 4 Making Sense of the Universe

## UNDERSTANDING MOTION, ENERGY, AND GRAVITY

▲ **About the photo:** The same laws that govern motion on Earth also govern gargantuan collisions between galaxies.

## LEARNING GOALS

**4.1 Describing Motion: Examples from Daily Life**

- How do we describe motion?
- How is mass different from weight?

**4.2 Newton's Laws of Motion**

- How did Newton change our view of the universe?
- What are Newton's three laws of motion?

**4.3 Conservation Laws in Astronomy**

- Why do objects move at constant velocity if no force acts on them?
- What keeps a planet rotating and orbiting the Sun?
- Where do objects get their energy?

**4.4 The Universal Law of Gravitation**

- What determines the strength of gravity?
- How does Newton's law of gravity extend Kepler's laws?

**4.5 Orbits, Tides, and the Acceleration of Gravity**

- How do gravity and energy allow us to understand orbits?
- How does gravity cause tides?
- Why do all objects fall at the same rate?

*If I have seen farther than others, it is because I have stood on the shoulders of giants.*
—*Isaac Newton*

**Chapter 4 Overview**

The history of the universe is essentially a story about the interplay between matter and energy. This interplay began in the Big Bang and continues today in everything from the microscopic jiggling of atoms to gargantuan collisions of galaxies. Understanding the universe therefore depends on becoming familiar with how matter responds to the ebb and flow of energy.

You might guess that it would be difficult to understand the many interactions that shape the universe, but we now know that just a few physical laws govern the movements of everything from atoms to galaxies. The Copernican revolution spurred the discovery of these laws, and Galileo deduced some of them from his experiments. But it was Sir Isaac Newton who put all the pieces together into a simple system of laws describing both motion and gravity.

In this chapter, we'll discuss Newton's laws of motion, the laws of conservation of angular momentum and of energy, and the universal law of gravitation. Understanding these laws will enable you to make sense of many of the wide-ranging phenomena you will encounter as you study astronomy.

**Newton's Laws of Motion (first 5 minutes)**

## 4.1 Describing Motion: Examples from Daily Life

Think about what happens when you throw a ball to a dog. The ball leaves your hand, traveling in some particular direction at some particular speed. During its flight, the ball is pulled toward Earth by gravity, slowed by air resistance, and pushed by gusts of wind. Despite the complexity of the ball's motion, the dog still catches it.

We humans can perform an even better trick: We have learned how to figure out where the ball will land even before throwing it. In fact, we can use the same basic trick to predict the motions of objects throughout the universe, and we can perform it with such extraordinary precision that we can land a spaceship on target on another world after sending it on a journey of hundreds of millions of kilometers.

Our primary goal in this chapter is to understand how humans have learned to make sense of motion in the universe. We all have experience with motion and a natural intuition as to what motion is, but in science we need to define our ideas and terms precisely. In this section, we'll use examples from everyday life to explore some of the fundamental ideas of motion.

### How do we describe motion?

You are probably familiar with common terms used to describe motion in science, such as *velocity*, *acceleration*, and *momentum*. However, their scientific definitions may differ subtly from those you use in casual conversation. Let's investigate the precise meanings of these terms.

**Speed, Velocity, and Acceleration** A car provides a good illustration of the three basic terms that we use to describe motion:

- The **speed** of the car tells us how far it will go in a certain amount of time. For example, "100 kilometers per hour" (about 60 miles per hour) is a speed, and it tells us that the car will cover a distance of 100 kilometers if it is driven at this speed for an hour.
- The **velocity** of the car tells us both its speed and its direction. For example, "100 kilometers per hour going due north" describes a velocity.
- The car has an **acceleration** if its velocity is changing in any way, whether in speed or direction or both.

Note that while we normally think of *acceleration* as an increase in speed, in science we also say that you are accelerating when you slow down or turn (**FIGURE 4.1**). Slowing represents a negative acceleration, causing your velocity to decrease. Turning means a change in direction—which therefore means a change in velocity—so turning is a form of acceleration even if your speed remains constant.

You can often feel the effects of acceleration. For example, as you speed up in a car, you feel yourself being pushed back into your seat. As you slow down, you feel yourself being pulled forward. As you drive around a curve, you feel yourself being pushed away from the direction of your turn. In contrast, you don't feel such effects when moving at *constant velocity*. That is why you don't feel any sensation of motion when you're traveling in an airplane on a smooth flight.

**The Acceleration of Gravity** One of the most important types of acceleration is the acceleration caused by gravity. In a legendary experiment in which he supposedly

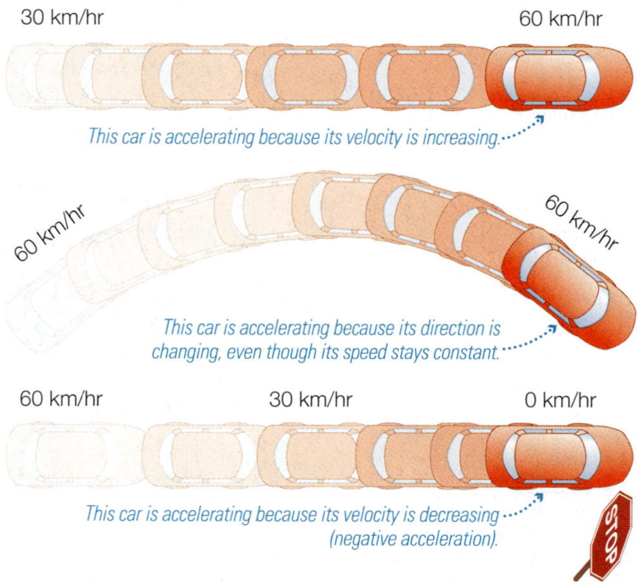


**FIGURE 4.1** Speeding up, turning, and slowing down are all examples of acceleration.

dropped weights from the Leaning Tower of Pisa, Galileo demonstrated that gravity accelerates all objects by the same amount, regardless of their mass. This fact may be surprising because it seems to contradict everyday experience: A feather floats gently to the ground, while a rock plummets. However, air resistance causes this difference in acceleration. If you dropped a feather and a rock on the Moon, where there is no air, both would fall at exactly the same rate. (NASA has an online video showing *Apollo 15* astronaut Dave Scott performing this experiment on the Moon, but with a hammer instead of a rock.)

**See it for yourself** Find a piece of paper and a small rock. Hold both at the same height, and let them go at the same instant. The rock, of course, hits the ground first. Next, crumple the paper into a small ball and repeat the experiment. What happens? Explain how this experiment suggests that gravity accelerates all objects by the same amount.

The acceleration of a falling object is called the **acceleration of gravity**, abbreviated *g*. On Earth, the acceleration of gravity causes falling objects to fall faster by 9.8 meters per second (m/s) or about 10 m/s, with each passing second. For example, suppose you drop a rock from a tall building. At the moment you let it go, its speed is 0 m/s. After 1 second, the rock will be falling downward at about 10 m/s. After 2 seconds, it will be falling at about 20 m/s. In the absence of air resistance, its speed will continue to increase by about 10 m/s each second until it hits the ground (**FIGURE 4.2**). We therefore say that the acceleration of gravity is about 10 *meters per second per second*, or 10 *meters per second squared*, which we write as 10 $m/s^2$ (more precisely, $g = 9.8\ m/s^2$).

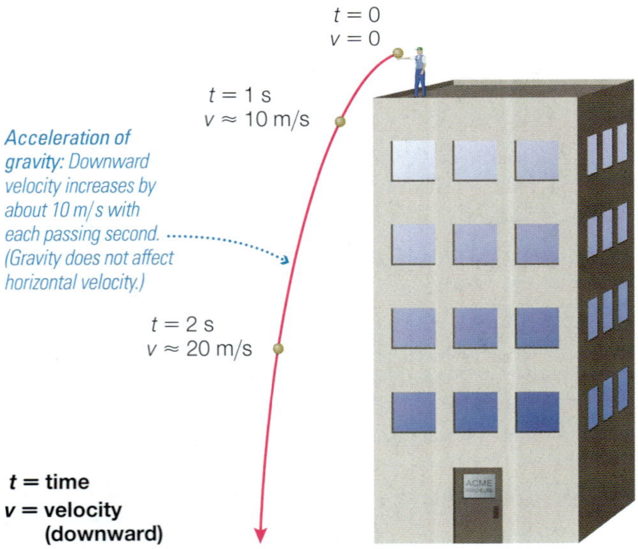


**FIGURE 4.2** On Earth, gravity causes an unsupported object to accelerate downward at about 10 $m/s^2$, which means its downward velocity increases by about 10 m/s with each passing second. (Gravity does not affect horizontal velocity.)

**Momentum and Force** The concepts of speed, velocity, and acceleration describe how an individual object moves, but most of the interesting phenomena we see in the universe result from interactions between objects. We need two additional concepts to describe these interactions:

- An object's **momentum** is the product of its mass and velocity; that is, momentum = mass × velocity.
- The only way to change an object's momentum is to apply a **force** to it.

We can understand these concepts by considering the effects of collisions. Imagine that you're stopped in your car at a red light when a bug flying at a velocity of 30 km/hr due south slams into your windshield. What will happen to your car? Not much, except perhaps a bit of a mess on your windshield. Next, imagine that a 2-ton truck runs the red light and hits you head-on with the same velocity as the bug. Clearly, the truck will cause far more damage. We can understand why by considering the momentum and force in each collision.

Before the collisions, the truck's much greater mass means it has far more momentum than the bug, even though both the truck and the bug are moving with the same velocity. During the collisions, the bug and the truck each transfer some of their momentum to your car. The bug has very little momentum to give to your car, so it does not exert much of a force. In contrast, the truck imparts enough of its momentum to cause a dramatic and sudden change in your car's momentum. You feel this sudden change in momentum as a force, and it can do great damage to you and your car.

The mere presence of a force does not always cause a change in momentum. For example, a moving car is always affected by forces of air resistance and friction with the road—forces that will slow your car if you take your foot off the gas pedal. However, you can maintain a constant velocity, and hence constant momentum, if you step on the gas pedal hard enough to overcome the slowing effects of these forces.

In fact, forces of some kind are always present, such as the force of gravity or the electromagnetic forces acting between atoms. The **net force** (or *overall force*) acting on an object represents the combined effect of all the individual forces put together. There is no net force on your car when you are driving at constant velocity, because the force generated by the engine to turn the wheels precisely offsets the forces of air resistance and road friction. A change in momentum occurs only when the net force is not zero.

Changing an object's momentum means changing its velocity, as long as its mass remains constant. A net force that is not zero therefore causes an object to accelerate. Conversely, whenever an object accelerates, a net force must be causing the acceleration. That is why you feel forces (pushing you forward, backward, or to the side) when you accelerate in your car. We can use the same ideas to understand many astronomical processes. For example, planets are always accelerating as they orbit the Sun, because their direction of travel constantly changes as they go around their orbits. We can therefore conclude that some force must be causing this acceleration. As we'll discuss shortly, Isaac Newton identified this force as gravity.

**Moving in Circles** Think about an ice skater spinning in place (look ahead to Figure 4.10). He isn't going anywhere, so he has no overall velocity and hence no overall momentum. Nevertheless, every part of his body is moving in a circle as he spins, so these parts have momentum even though his overall momentum is zero. Is there a way to describe the total momentum from each part of his body as he spins? Yes—we say that his spin gives him **angular momentum**, which you can also think of as "circling momentum" or "turning momentum." (The term *angular* arises because a complete circle turns through an *angle* of 360°.)

Any object that is either spinning or moving along a curved path has angular momentum, which makes angular momentum very important in astronomy. For example, **FIGURE 4.3** shows that Earth has angular momentum due to its rotation (*rotational angular momentum*) and to its orbit around the Sun (*orbital angular momentum*).

Because angular momentum is a special type of momentum, an object's angular momentum can change only when a special type of force is applied to it. To see why, consider what happens when you try to open a swinging door. Opening the door means making it rotate on its hinges, which means giving the door some angular momentum. Pushing directly on the hinges will have no effect on the door, even if you push with a very strong force. However, even a light force can make the door rotate if you push on the part of the door that is farthest from the hinges. The type of force that can change an object's angular momentum is called a **torque**, which you can think of as a "twisting force." As the door example shows, the amount of torque depends not only on how much force is applied, but also on where it is applied.

Changing a tire offers another familiar example of torque. Turning the bolts on a tire means making them rotate, which requires giving them some angular momentum. A longer wrench allows you to push from farther out than you can with a short wrench, so you can turn the bolts with less force. We will see many more applications of angular momentum in astronomy throughout the rest of the book.

## How is mass different from weight?

In daily life, we usually think of *mass* as something you can measure with a bathroom scale, but technically the scale measures your *weight*, not your mass. The distinction between mass and weight rarely matters when we are talking about objects on Earth, but it is very important in astronomy:

- Your **mass** is the amount of matter in your body.
- Your **weight** (or *apparent weight**) is the *force* that a scale measures when you stand on it; that is, weight depends both on your mass and on the forces (including gravity) acting on your mass.

To understand the difference between mass and weight, imagine standing on a scale in an elevator (**FIGURE 4.4**). Your mass will be the same no matter how the elevator moves, but your weight can vary. When the elevator is stationary or moving at constant velocity, the scale reads your "normal" weight. When the elevator accelerates upward, the floor exerts a greater force than it does when you are at rest. You feel heavier, and the scale verifies your greater weight. When the elevator accelerates downward, the floor and the scale exert a weaker force on you, so the scale registers less weight. Note that the scale shows a weight different from your "normal" weight only when the elevator is *accelerating*, not when it is going up or down at constant speed.

**See it for yourself** Find a small bathroom scale and take it with you on an elevator ride. How does your weight change when the elevator accelerates upward or downward? Does it change when the elevator is moving at constant speed? Explain your observations.

Your mass therefore depends only on the amount of matter in your body and is the same anywhere, but your weight can vary because the forces acting on you can vary. For example, your mass would be the same on the Moon as on Earth, but you would weigh less on the Moon because of its weaker gravity.

**Free-Fall and Weightlessness** Now consider what happens if the elevator cable breaks (see the last frame in Figure 4.4). The elevator and you are suddenly in **free-fall**—falling without any resistance to slow you down. The floor drops away at the same rate that you fall, allowing you to "float" freely above it, and the scale reads zero because you are no longer held to it. In other words, your free-fall has made you **weightless**.

In fact, you are in free-fall whenever there's nothing to *prevent* you from falling. For example, you are in free-fall when you jump off a chair or spring from a diving board or trampoline. Surprising as it may seem, you have therefore experienced weightlessness many times in your life. You can experience it right now simply by jumping off your chair—though your weightlessness lasts for only the very short time until you hit the ground.

**Weightlessness in Space** You've probably seen videos of astronauts floating weightlessly in the International Space Station. But why are they weightless? Many people guess

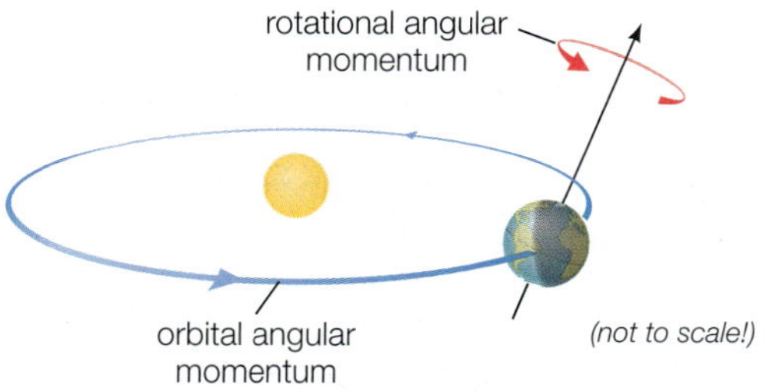


**FIGURE 4.3** Earth has angular momentum due both to its rotation and to its orbit around the Sun.

*Some physics texts distinguish between "true weight," due only to gravity, and "apparent weight," which also depends on other forces (as in an elevator).

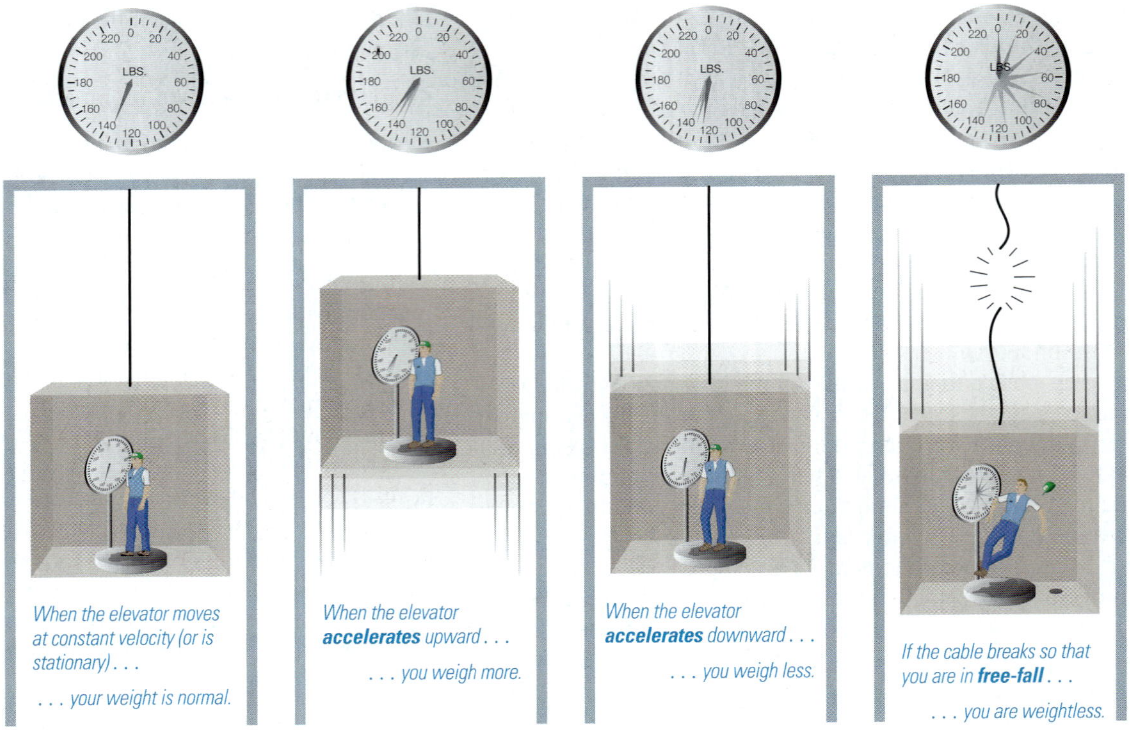


**FIGURE 4.4** Mass is not the same as weight. The man's mass never changes, but his weight is different when the elevator accelerates.

that there's no gravity in space, but that's not true. After all, it is gravity that makes the Space Station orbit Earth. Astronauts are weightless for the same reason that you are weightless when you jump off a chair: They are in free-fall.

Astronauts are weightless the entire time they orbit Earth because they are in a *constant state of free-fall*. To understand this idea, imagine a tower that reaches all the way to the Space Station's orbit, about 350 kilometers above Earth (**FIGURE 4.5**). If you stepped off the tower, you would fall downward, remaining weightless until you hit the ground (or until air resistance had a noticeable effect on you). Now, imagine that instead of stepping off the tower, you ran and jumped out of the tower. You'd still fall to the ground, but because of your forward motion, you'd land a short distance away from the base of the tower.

## COMMON MISCONCEPTIONS

### No Gravity in Space?

If you ask people why astronauts are weightless in space, one of the most common answers is "There is no gravity in space." But you can usually convince people that this answer is wrong by following up with another simple question: Why does the Moon orbit Earth? Most people know that the Moon orbits Earth because of gravity, proving that there is gravity in space. In fact, at the altitude of the Space Station's orbit, the acceleration of gravity is only about 10% less than it is on Earth's surface.

The real reason astronauts are weightless is that they are in a constant state of free-fall. Imagine being an astronaut. You'd have the sensation of free-fall—just as when you jump from a diving board—the entire time you were in orbit. This constant falling sensation makes many astronauts sick to their stomachs when they first experience weightlessness. Fortunately, they quickly get used to the sensation and can then work hard and enjoy the view.

The faster you ran out of the tower, the farther you'd go before landing. If you could somehow run fast enough—about 28,000 km/hr (17,000 mi/hr) at the orbital altitude of the Space Station—a very interesting thing would happen: By the time gravity had pulled you downward as far as the length of the tower, you'd already have moved far enough around Earth that you'd no longer be going down at all. Instead, you'd be just as high above Earth as you'd been all along, but a good portion of the way around the world. In other words, you'd be orbiting Earth.

The Space Station and all other orbiting objects stay in orbit because they are constantly "falling around" Earth. Their constant state of free-fall makes these spacecraft and everything in them weightless.

**Think about it** In the *Hitchhiker's Guide to the Galaxy* books, author Douglas Adams says that the trick to flying is to "throw yourself at the ground and miss." Although this phrase does not really explain flying, which involves lift from air, it describes *orbit* fairly well. Explain.

## 4.2 Newton's Laws of Motion

The complexity of motion in daily life might lead you to guess that the laws governing motion would also be complex. For example, if you watch a falling piece of paper waft lazily to the ground, you'll see it rock back and forth in a seemingly unpredictable pattern. However, the complexity of this motion arises because the paper is affected by a variety of forces, including gravity and the changing forces caused by air currents. If you could analyze the forces individually, you'd find that each force affects the paper's motion in a simple, predictable way. Sir Isaac Newton (1642–1727) discovered the remarkably simple laws that govern motion.

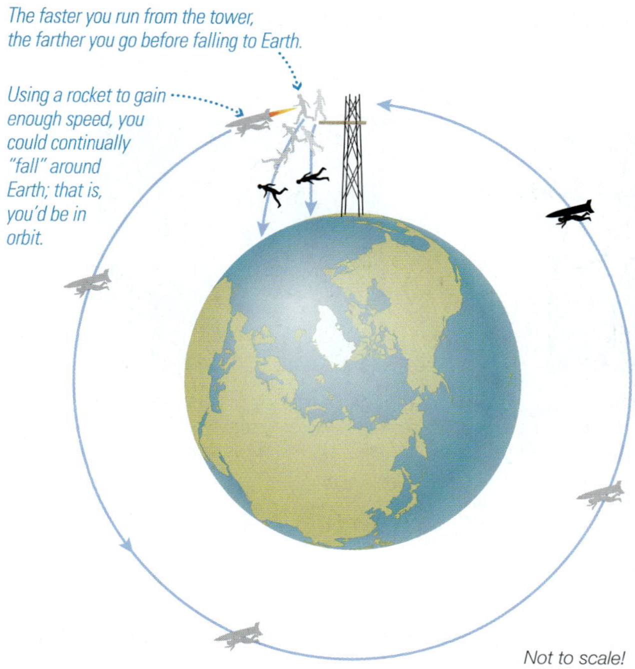


**FIGURE 4.5** This figure explains why astronauts are weightless and float freely in space. If you could leap from a tall tower with enough speed, you could travel forward so fast that you'd orbit Earth. You'd then be in a constant state of free-fall, which means you'd be weightless. *Note:* On the scale shown here, the tower extends far higher than the Space Station's orbit. (Adapted from *Space Station Science* by Marianne Dyson.)

## How did Newton change our view of the universe?

Newton was born in Lincolnshire, England, on Christmas Day in 1642. His father, a farmer who never learned to read or write, died 3 months before his birth. Newton had a difficult childhood and showed few signs of unusual talent. He attended Trinity College at Cambridge, where he earned his keep by performing menial labor, such as cleaning the boots and bathrooms of wealthier students and waiting on their tables.

The plague hit Cambridge shortly after Newton graduated, and he returned home. By his own account, he experienced a moment of inspiration in 1666 when he saw an apple fall to the ground. He suddenly realized that the gravity making the apple fall was the same force that held the Moon in orbit around Earth. In that moment, Newton shattered the remaining vestiges of the Aristotelian view of the world, which for centuries had been accepted as unquestioned truth.

Aristotle had made many claims about the physics of motion, using his ideas to support his belief in an Earth-centered cosmos. He had also maintained that the heavens were totally distinct from Earth, so physical laws on Earth did not apply to heavenly motion. By the time Newton saw the apple fall, the Copernican revolution had displaced Earth from a central position, and Galileo's experiments had shown that the laws of physics were not what Aristotle had believed [**Section 3.3**].

Newton's sudden insight delivered the final blow to Aristotle's view. By recognizing that gravity operates in the heavens as well as on Earth, Newton eliminated Aristotle's distinction between the two realms and brought the heavens and Earth together as one *universe*. This insight also heralded the birth of the modern science of *astrophysics* (although the term wasn't coined until much later), which applies physical laws discovered on Earth to phenomena throughout the cosmos.

Over the next 20 years, Newton's work completely revolutionized mathematics and science. He quantified the laws of motion and gravity, conducted crucial experiments regarding the nature of light, built the first reflecting telescopes, and invented the mathematics of calculus. The compendium of Newton's discoveries is so tremendous that it would take a complete book just to describe them, and many more books to describe their influence on civilization. When Newton died in 1727, at age 84, English poet Alexander Pope composed the following epitaph:

Sir Isaac Newton (1642–1727)

*Nature, and Nature's laws lay hid in the Night.*
*God said, Let Newton be! and all was Light.*

Newton's Laws of Motion (last 1:30)

## What are Newton's three laws of motion?

Newton published the laws of motion and gravity in 1687, in his book *Philosophiae Naturalis Principia Mathematica* ("Mathematical Principles of Natural Philosophy"), usually called *Principia*. He enumerated three laws that apply to all motion, which we now call **Newton's laws of motion**. These laws govern the motion of everything from our daily movements on Earth to the movements of planets, stars, and galaxies throughout the universe. **FIGURE 4.6** summarizes the three laws.

**Newton's First Law** Newton's first law of motion essentially restates Galileo's discovery that objects will remain in motion unless a force acts to stop them:

> **Newton's first law**: *An object moves at constant velocity if there is no net force acting upon it.*

In other words, objects at rest (velocity = 0) tend to remain at rest, and objects in motion tend to remain in motion with no change in either their speed or their direction.

The idea that an object at rest should remain at rest is rather obvious: A car parked on a flat street won't suddenly start moving for no reason. But what if the car is traveling along a flat, straight road? Newton's first law says that the car should keep going at the same speed forever *unless* a force acts to slow it down. You know that the car eventually will come to a stop if you take your foot off the gas pedal, so one or more forces must be stopping the car—in

**Newton's first law of motion:** An object moves at constant velocity unless a net force acts to change its speed or direction.

*Example:* A spaceship needs no fuel to keep moving in space.

**Newton's second law of motion:** Force = mass × acceleration

*Example:* A baseball accelerates as the pitcher applies a force by moving his arm. (Once the ball is released, the force from the pitcher's arm ceases, and the ball's path changes only because of the forces of gravity and air resistance.)

**Newton's third law of motion:** For any force, there is always an equal and opposite reaction force.

*Example:* A rocket is propelled upward by a force equal and opposite to the force with which gas is expelled out its back.

**FIGURE 4.6** Newton's three laws of motion.

this case forces arising from friction and air resistance. If the car were in space, and therefore unaffected by friction or air, it would keep moving forever (though gravity would gradually alter its speed and direction). That is why interplanetary spacecraft need no fuel to keep going after they are launched into space, and why astronomical objects don't need fuel to travel through the universe.

Newton's first law also explains why you don't feel any sensation of motion when you're traveling in an airplane on a smooth flight. As long as the plane is traveling at constant velocity, no net force is acting on it or on you. Therefore, you feel no different from the way you would feel at rest. You can walk around the cabin, play catch with someone, or relax and go to sleep just as though you were "at rest" on the ground.

**Newton's Second Law** Newton's second law of motion tells us what happens to an object when a net force *is* present. We have already seen that a net force will change an object's momentum, accelerating it in the direction of the force. Newton's second law quantifies this relationship and can be written in either of the following two forms:

**Newton's second law:**

$$\textit{force} = \textit{mass} \times \textit{acceleration}\ (F = ma)$$
$$\textit{force} = \text{rate of change in } \textit{momentum}$$

This law explains why you can throw a baseball farther than you can throw a shot in the shot put. The force your arm delivers to both the baseball and the shot equals the product of mass and acceleration. Because the mass of the shot is greater than that of the baseball, the same force from your arm gives the shot a smaller acceleration. Because of its smaller acceleration, the shot leaves your hand with less speed than the baseball and therefore travels a shorter distance before hitting the ground. Astronomically, Newton's second law explains why a large planet such as Jupiter has a greater effect on asteroids and comets than does a small planet such as Earth [**Section 12.2**]. Jupiter exerts a stronger

## MATHEMATICAL INSIGHT 4.1 Units of Force, Mass, and Weight

Newton's second law, $F = ma$, shows that the unit of force is equal to a unit of mass multiplied by a unit of acceleration. Consider a mass of 1 kilogram accelerating at 10 m/s$^2$:

$$\begin{aligned} \text{force} &= \text{mass} \times \text{acceleration} \\ &= 1\ \text{kg} \times 10\ \frac{\text{m}}{\text{s}^2} = 10\ \frac{\text{kg} \times \text{m}}{\text{s}^2} \\ &= 10\ \text{newtons} \end{aligned}$$

We conclude that the standard unit of force, called the **newton**, is equivalent to a *kilogram-meter per second squared*.

We can also use Newton's second law to clarify the difference between mass and weight. Imagine standing on a chair when it is suddenly pulled out from under you. You will immediately begin accelerating downward with the acceleration of gravity, which means the force of gravity acting on you must be your mass times the acceleration of gravity. This force is what physicists call your true weight, and it is the same whether you are falling or standing still:

$$\text{weight} = \text{mass} \times \text{acceleration of gravity}$$

Your apparent weight may differ if forces besides gravity are acting on you at the same time and is zero if you are in free-fall.

Like any force, weight has units of mass times acceleration. Therefore, although we commonly speak of weights in *kilograms*, this usage is not technically correct: Kilograms are a unit of mass, not of force. You may safely ignore this technicality as long as you are dealing with objects on Earth (that are not accelerating). In space or on other planets, the distinction between mass and weight is important and cannot be ignored.

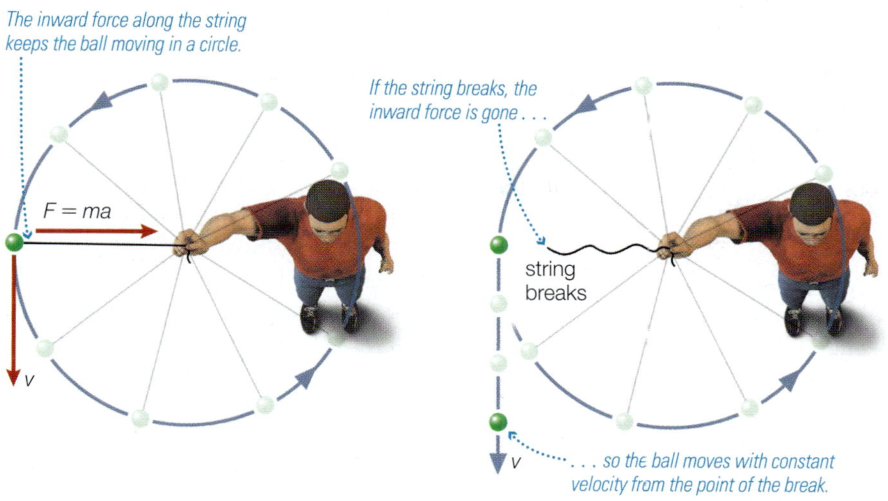

a When you swing a ball on a string, the string exerts a force that pulls the ball inward.

b If the string breaks, the ball flies off in a straight line at constant velocity.

**FIGURE 4.7** Newton's second law of motion tells us that an object going around a curve has an acceleration pointing toward the inside of the curve.

gravitational force on passing asteroids and comets, and therefore sends them scattering with a greater acceleration.

We can also use Newton's second law of motion to understand acceleration around curves. Suppose you swing a ball on a string around your head (**FIGURE 4.7a**). The ball is accelerating even if it has a steady speed, because it is constantly changing direction and therefore has a changing velocity. What makes it accelerate? According to Newton's second law, the taut string must be applying a force to the ball. We can understand this force by thinking about what happens when the string breaks (**FIGURE 4.7b**): With the force gone, the ball flies off in a straight line. Therefore, when the string is intact, the force must be pulling the ball *inward* to keep it from flying off. Because acceleration must be in the same direction as the force, we conclude that the ball has an inward acceleration as it moves around the circle.

The same idea helps us understand the force on a car moving around a curve or a planet orbiting the Sun. In the case of the car, the inward force comes from friction between the tires and the road. The tighter the curve (or the faster the car is going), the greater the force needed to keep the car moving around it. If the inward force due to friction is not great enough, the car skids outward. Similarly, a planet orbiting the Sun always has an acceleration in the direction of the Sun, and gravity is the inward force that causes this acceleration. Indeed, it was Newton's discovery of the precise nature of this acceleration that helped him deduce the law of gravity, which we'll discuss in Section 4.4.

## COMMON MISCONCEPTIONS

### What Makes a Rocket Launch?

If you've ever watched a rocket launch, it's easy to see why many people believe that the rocket "pushes off" the ground. However, the ground has nothing to do with the rocket launch, which is actually explained by Newton's third law of motion. To balance the force driving gas out the back of the rocket, an equal and opposite force must propel the rocket forward. Rockets can be launched horizontally as well as vertically, and a rocket can be "launched" in space (for example, from a space station) with no need for any solid ground.

**Newton's Third Law** Think for a moment about standing still on the ground. Your weight exerts a downward force, so if this force were acting alone, Newton's second law would demand that you accelerate downward. The fact that you are not falling means there must be no *net* force acting on you, which is possible only if the ground is exerting an upward force on you that precisely offsets the downward force you exert on the ground. The fact that the downward force you exert on the ground is offset by an equal and opposite force that pushes upward on you is one example of Newton's third law of motion, which tells us that any force is always paired with an equal and opposite reaction force.

> **Newton's third law**: *For any force, there is always an equal and opposite reaction force.*

This law is very important in astronomy, because it tells us that objects always attract *each other* through gravity. For example, your body always exerts a gravitational force on Earth identical to the force that Earth exerts on you, except that it acts in the opposite direction. Of course, the same force means a much greater acceleration for you than for Earth (because your mass is so much smaller than Earth's), which is why you fall toward Earth when you jump off a chair, rather than Earth falling toward you.

Newton's third law also explains how a rocket works: A rocket engine generates a force that drives hot gas out the back, which creates an equal and opposite force that propels the rocket forward.

## 4.3 Conservation Laws in Astronomy

Newton's laws of motion are easy to state, but they may seem a bit arbitrary. Why, for example, should every force be opposed by an equal and opposite reaction force? In the centuries since Newton first stated his laws, we have learned that they are not arbitrary at all, but instead reflect deeper aspects of nature known as *conservation laws*. In this section, we'll explore three of the most important conservation laws for astronomy: conservation of momentum, of angular momentum, and of energy.

## Why do objects move at constant velocity if no force acts on them?

The first of our conservation laws, the law of **conservation of momentum**, states that as long as there are no external forces, the total momentum of interacting objects cannot change; that is, their total momentum is *conserved*. An individual object can gain or lose momentum only if some other object's momentum changes by a precisely opposite amount.

The law of conservation of momentum is implicit in Newton's laws. To see why, watch a game of pool. Newton's second law tells us that when one pool ball strikes another, it exerts a force that changes the momentum of the second ball. At the same time, Newton's third law tells us that the second ball exerts an equal and opposite force on the first one—which means that the first ball's momentum changes by precisely the same amount as the second ball's momentum, but in the opposite direction. The total combined momentum of the two balls remains the same both before and after the collision (**FIGURE 4.8**). Note that no *external* forces are accelerating the balls.

Rockets offer another good example of conservation of momentum in action. When you fire a rocket engine, the total momentum of the rocket and the hot gases it shoots out the back must stay the same. In other words, the amount of forward momentum the rocket gains is equal to the amount of backward momentum in the gas that shoots out the back. That is why forces between the rocket and the gases are always equal and opposite.

From the perspective of conservation of momentum, Newton's first law makes perfect sense. When no net force acts on an object, there is no way for the object to transfer any momentum to or from any other object. In the absence of a net force, an object's momentum must therefore remain unchanged—which means the object must continue to move exactly as it has been moving.

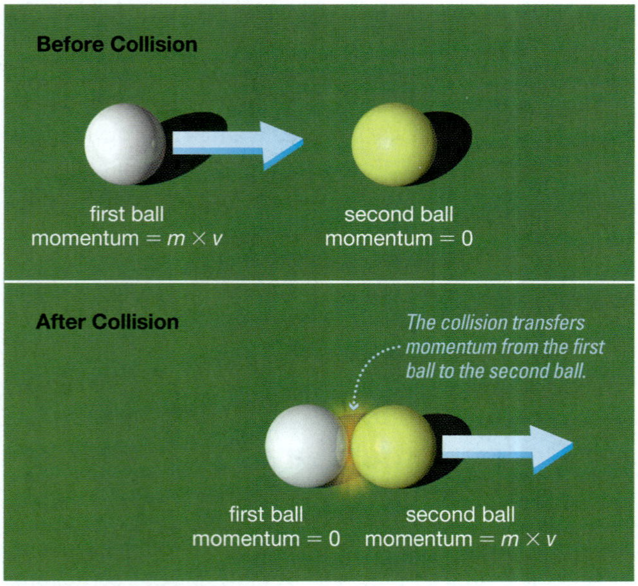


**FIGURE 4.8** Conservation of momentum demonstrated with head-on collision of two balls on a pool table.

According to current understanding of the universe, conservation of momentum is an absolute law that always holds true. For example, it holds even when you jump up into the air. You may wonder, Where do I get the momentum that carries me upward? The answer is that as your legs propel you skyward, they are actually pushing Earth in the other direction, giving Earth's momentum an equal and opposite kick. However, Earth's huge mass renders its acceleration undetectable. During your brief flight, the gravitational force between you and Earth pulls you back down, transferring your momentum back to Earth. The total momentum of you and Earth remains the same at all times.

## What keeps a planet rotating and orbiting the Sun?

Perhaps you've wondered how Earth manages to keep rotating and going around the Sun day after day and year after year. The answer comes from our second conservation law: the law of **conservation of angular momentum**. Recall that rotating or orbiting objects have angular momentum because they are moving in circles or going around curves, and that angular momentum can be changed only by a "twisting force," or *torque*. The law of conservation of angular momentum states that as long as there is no external torque, the total angular momentum of a set of interacting objects cannot change. An individual object can change its angular momentum only by transferring some angular momentum to or from another object. Because astronomical objects can have angular momentum due to both rotation and orbit (see Figure 4.3), let's consider both cases.

**Orbital Angular Momentum** Consider Earth's orbit around the Sun. A simple formula tells us Earth's angular momentum at any point in its orbit:

$$\text{angular momentum} = m \times v \times r$$

where $m$ is Earth's mass, $v$ is its orbital velocity (or, more technically, the component of velocity perpendicular to $r$), and $r$ is the "radius" of the orbit, by which we mean its distance from the Sun (**FIGURE 4.9**). Because there are no objects around to give or take angular momentum from Earth as it orbits the Sun, Earth's orbital angular momentum must always stay the same. This explains two key facts about Earth's orbit:

1. Earth needs no fuel or push of any kind to keep orbiting the Sun—it will keep orbiting as long as nothing comes along to take angular momentum away.
2. Because Earth's angular momentum at any point in its orbit depends on the product of its speed and orbital radius (distance from the Sun), Earth's orbital speed must be faster when it is nearer to the Sun (and the radius is smaller) and slower when it is farther from the Sun (and the radius is larger).

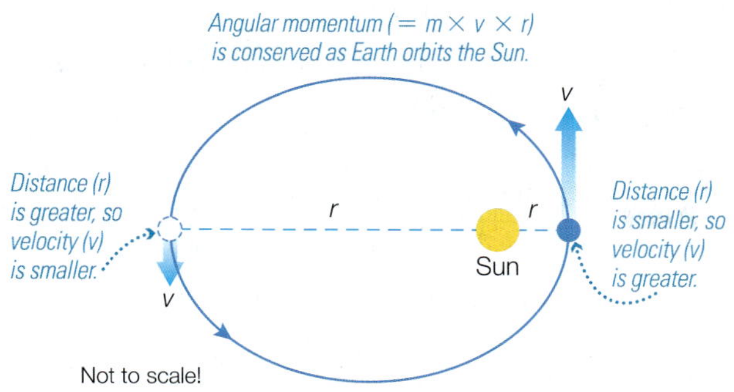


**FIGURE 4.9** Earth's orbital angular momentum stays constant, so Earth moves faster when it is closer to the Sun and slower when it is farther from the Sun.

The second fact is just what Kepler's second law of planetary motion states [**Section 3.3**]. That is, the law of conservation of angular momentum tells us *why* Kepler's law is true.

**Rotational Angular Momentum** The same idea explains why Earth keeps rotating. As long as Earth isn't transferring any of the angular momentum of its rotation to another object, it keeps rotating at the same rate. (In fact, Earth is very gradually transferring some of its rotational angular momentum to the Moon, and as a result Earth's rotation is gradually slowing down; see Section 4.5.)

Conservation of angular momentum also explains why we see so many spinning disks in the universe, such as the disks of galaxies like the Milky Way and disks of material orbiting young stars. The idea is easy to illustrate with an ice skater spinning in place (**FIGURE 4.10**). Because there is so little friction on ice, the angular momentum of the ice skater remains essentially constant. When she pulls in her extended arms, she decreases her radius—which means her velocity of rotation must increase. Stars and galaxies are both born from clouds of gas that start out much larger in size. These clouds almost inevitably have some small net rotation, though it may be imperceptible. Like the spinning skater as she pulls in her arms, they must therefore spin faster as gravity makes them shrink in size. (We'll discuss why the clouds also flatten into disks in Chapter 8.)

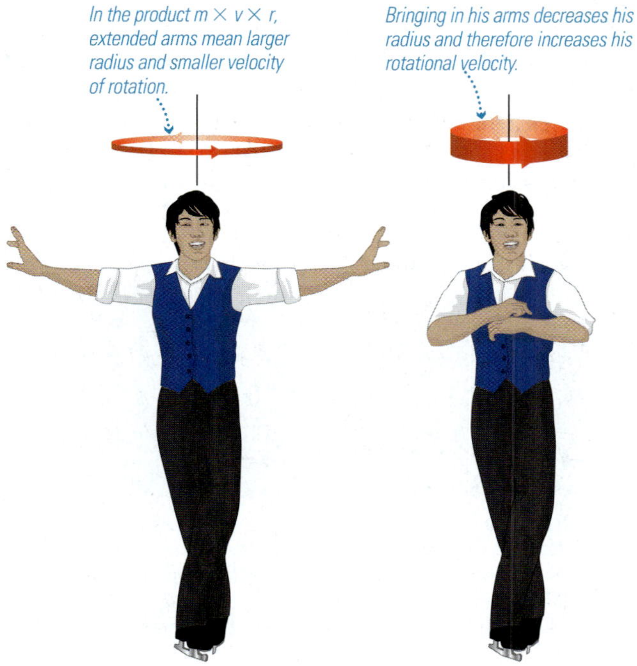


**FIGURE 4.10** A spinning skater conserves angular momentum.

**Think about it** How does conservation of angular momentum explain the spiraling of water going down a drain?

## Where do objects get their energy?

The law of **conservation of energy** tells us that, like momentum and angular momentum, energy cannot appear out of nowhere or disappear into nothingness. Objects can gain or lose energy only by exchanging energy with other objects. Because of this law, the story of the universe is a story of the interplay of energy and matter: All actions involve exchanges of energy or the conversion of energy from one form to another.

Throughout the rest of this book, we'll see numerous cases in which we can understand astronomical processes simply by studying how energy is transformed and exchanged. For example, we'll see that planetary interiors cool with time because they radiate energy into space, and that the Sun became hot because of energy released by the gas that formed it. By applying the laws of conservation of momentum, angular momentum, and energy, we can understand almost every major process that occurs in the universe.

**Basic Types of Energy** Before we can fully understand the law of conservation of energy, we need to know what energy is. In essence, energy is what makes matter move. Because this statement is so broad, we often distinguish between different types of energy. For example, we talk about the energy we get from the food we eat, the energy that makes our cars go, and the energy a light bulb emits. Fortunately, we can classify nearly all types of energy into just three major categories (**FIGURE 4.11**):

- Energy of motion, or **kinetic energy** (*kinetic* comes from a Greek word meaning "motion"). Falling rocks, orbiting planets, and the molecules moving in the air are all examples of objects with kinetic energy. Quantitatively, the kinetic energy of a moving object is $\frac{1}{2}mv^2$, where $m$ is the object's mass and $v$ is its speed.
- Energy carried by light, or **radiative energy** (the word *radiation* is often used as a synonym for *light*). All light carries energy, which is why light can cause changes in matter. For example, light can alter molecules in our eyes—thereby allowing us to see—or warm the surface of a planet.
- Stored energy, or **potential energy**, which might later be converted into kinetic or radiative energy. For example, a rock perched on a ledge has *gravitational* potential energy because it will fall if it slips off the edge, and gasoline contains *chemical* potential energy that can be converted into the kinetic energy of a moving car.

Energy can be converted from one form to another.

**FIGURE 4.11** The three basic categories of energy. Energy can be converted from one form to another, but it can never be created or destroyed, an idea embodied in the law of conservation of energy.

**TABLE 4.1 Energy Comparisons**

| Item | Energy (joules) |
|---|---|
| Energy of sunlight at Earth (per $m^2$ per second) | $1.3 \times 10^3$ |
| Energy from metabolism of a candy bar | $1 \times 10^6$ |
| Energy needed to walk for 1 hour | $1 \times 10^6$ |
| Kinetic energy of a car going 100 km/hr | $1 \times 10^6$ |
| Daily food energy need of average adult | $1 \times 10^7$ |
| Energy released by burning 1 liter of oil | $1.2 \times 10^7$ |
| Thermal energy of parked car | $1 \times 10^8$ |
| Energy released by fission of 1 kilogram of uranium-235 | $5.6 \times 10^{13}$ |
| Energy released by fusion of hydrogen in 1 liter of water | $7 \times 10^{13}$ |
| Energy released by 1-megaton H-bomb | $4 \times 10^{15}$ |
| Energy released by magnitude 8 earthquake | $2.5 \times 10^{16}$ |
| Annual U.S. energy consumption | $10^{20}$ |
| Annual energy generation of Sun | $10^{34}$ |
| Energy released by a supernova | $10^{44}$–$10^{46}$ |

Regardless of which type of energy we are dealing with, we can measure the amount of energy with the same standard units. For Americans, the most familiar units of energy are *Calories*, which are shown on food labels to tell us how much energy our bodies can draw from the food. A typical adult needs about 2500 Calories of energy from food each day. In science, the standard unit of energy is the **joule**. One food Calorie is equivalent to about 4184 joules, so the 2500 Calories used daily by a typical adult is equivalent to about 10 million joules. **TABLE 4.1** compares various energies in joules.

## Thermal Energy—The Kinetic Energy of Many Particles

Although there are only three major categories of energy, we sometimes divide them into various subcategories. In astronomy, the most important subcategory of kinetic energy is **thermal energy**, which represents the collective kinetic energy of the many individual particles (atoms and molecules) moving randomly within a substance like a rock or the air or the gas within a distant star. In such cases, it is much easier to talk about the thermal energy of the object than about the kinetic energies of its billions upon billions of individual particles. Note that all objects contain thermal energy even when they are sitting still, because the particles within them are always jiggling about randomly. These random motions can contain substantial energy: The thermal energy of a parked car due to the random motion of its atoms is much greater than the kinetic energy of the car moving at highway speed.

Thermal energy gets its name because it is related to temperature, but temperature and thermal energy are not quite the same thing. Thermal energy measures the *total* kinetic energy of all the randomly moving particles in a substance, while **temperature** measures the *average* kinetic energy of the particles. For a particular object, a higher temperature simply means that the particles on average have more kinetic energy and hence are moving faster (**FIGURE 4.12**). You're probably familiar with temperatures measured in *Fahrenheit* or *Celsius*, but in science we often use the **Kelvin** temperature scale (**FIGURE 4.13**). The Kelvin scale does not have negative temperatures, because it starts from the coldest possible temperature, known as *absolute zero* (0 K).

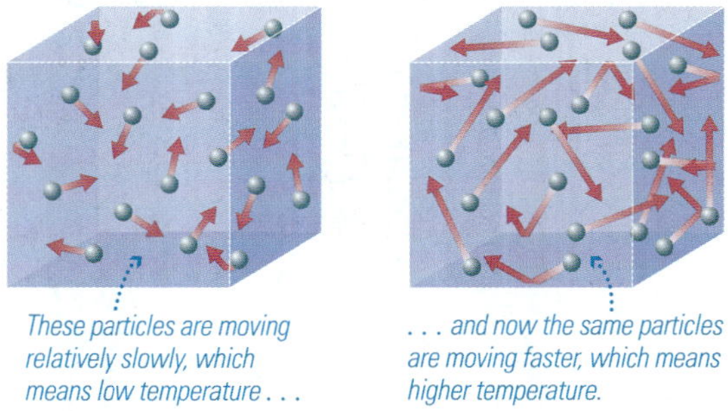


**FIGURE 4.12** Temperature is a measure of the average kinetic energy of the particles (atoms and molecules) in a substance. Longer arrows represent faster speeds.

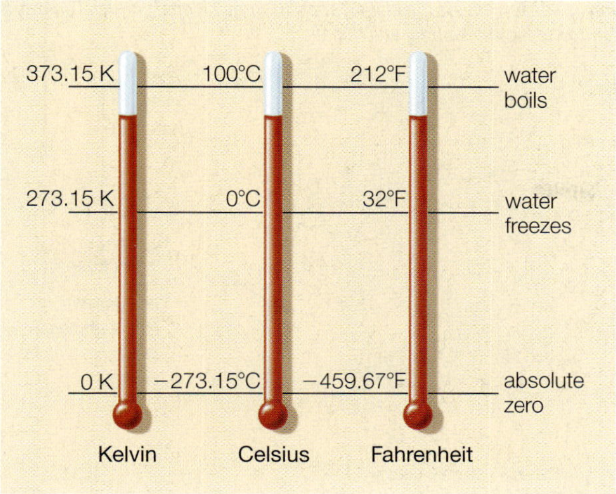


**FIGURE 4.13** Three common temperature scales: Kelvin, Celsius, and Fahrenheit. Scientists generally prefer the Kelvin scale. Note that the degree symbol (°) is not usually used with the Kelvin scale.

Thermal energy depends on temperature, because a higher average kinetic energy for the particles in a substance means a higher total energy. But thermal energy also depends on the number and density of the particles, as you can see by imagining that you quickly thrust your arm in and out of a hot oven and a pot of boiling water (don't try this!). The air in a hot oven is much higher in temperature than the water boiling in a pot (**FIGURE 4.14**). However, the boiling water would scald your arm almost instantly, while you could safely put your arm into the oven air for a few seconds. The reason for this difference is density. In both cases, because the air or water is hotter than your body, molecules striking your skin transfer thermal energy to molecules in your arm. The higher temperature in the oven means that the air molecules strike your skin harder, on average, than the molecules in the boiling water. However, because the *density* of water is so much higher than the density of air (meaning water has far more molecules in the same amount of space), many more molecules strike your skin each second in the water. While each individual molecule that strikes your skin transfers a little less energy in the boiling water than in the oven, the sheer number of molecules hitting you in the water means that more thermal energy is transferred to your arm. That is why the boiling water causes a burn almost instantly.

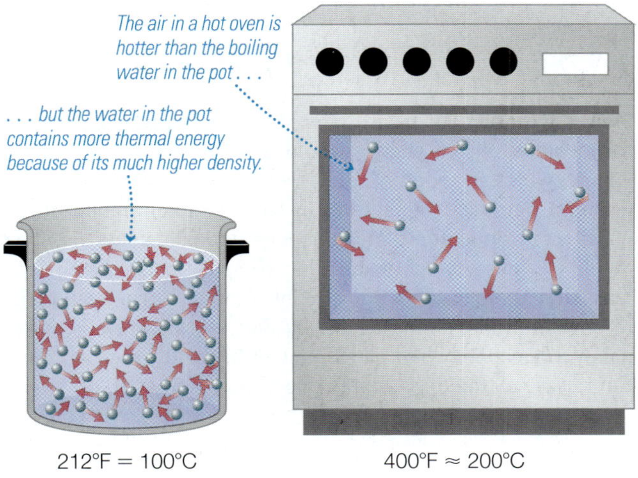


**FIGURE 4.14** Thermal energy depends on both the temperature and the density of particles in a substance.

**Think about it** In air or water that is colder than your body temperature, thermal energy is transferred from you to the surrounding cold air or water. Use this fact to explain why falling into a 32°F (0°C) lake is much more dangerous than standing naked outside on a 32°F day.

The environment in space provides another example of the difference between temperature and heat. Surprisingly, the temperature in low Earth orbit can be several thousand degrees. However, astronauts working outside in Earth orbit are at much greater risk of getting cold than hot.* The reason is the extremely low density: Although the particles striking an astronaut's space suit may be moving quite fast, there are not enough of them to transfer much thermal energy. (You may wonder how the astronauts become cold given that the low density also means the astronauts cannot transfer much of their own thermal energy to the particles in space. It turns out that they lose their body heat by emitting *thermal radiation*, which we will discuss in Section 5.4.)

**Potential Energy in Astronomy** Many types of potential energy are important in astronomy, but two are particularly important: *gravitational potential energy* and the potential energy of mass itself, or *mass-energy*.

An object's **gravitational potential energy** depends on its mass and how far it can fall as a result of gravity. An object has more gravitational potential energy when it is higher and less when it is lower. For example, if you throw a ball up into the air, it has more potential energy when it is high up than when it is near the ground. Because energy must be conserved during the ball's flight, the ball's kinetic energy increases when its gravitational potential energy decreases, and vice versa (**FIGURE 4.15a**). That is why the ball travels fastest (has the most kinetic energy) when it is closest to the ground, where it has the least gravitational potential energy. The higher the ball is, the more gravitational potential energy it has and the slower the ball travels (less kinetic energy). An object very close to Earth's surface has a gravitational potential energy of $mgh$, where $m$ is its mass, $g$ is the acceleration of gravity, and $h$ is its height above the ground.

The same general idea explains how stars become hot (**FIGURE 4.15b**). Before a star forms, its matter is spread out in a large, cold cloud of gas. Most of the individual gas particles are far from the center of this large cloud and

*Note that the situation is the opposite if you are *inside* the Space Station. The low density of space means the Space Station cannot easily shed heat, so it needs cooling systems to prevent the accumulation of heat generated by electronics and human bodies.

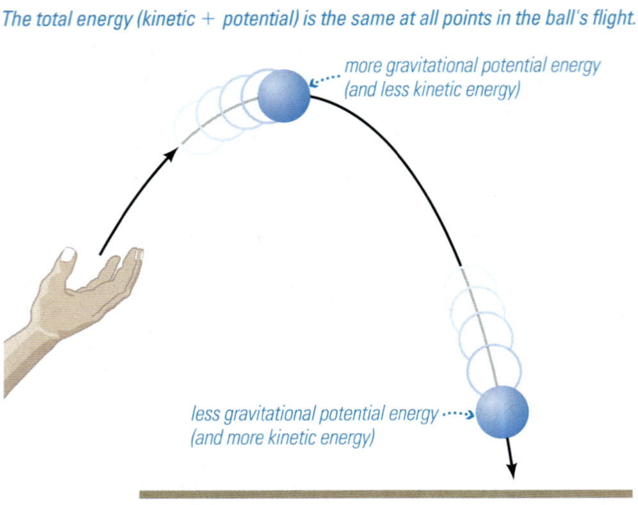


**a** The ball has more gravitational potential energy when it is high up than when it is near the ground.

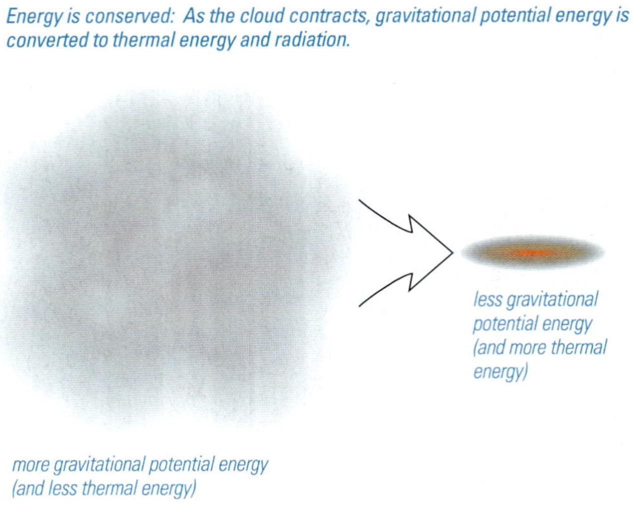


**b** A cloud of interstellar gas contracting because of its own gravity has more gravitational potential energy when it is spread out than when it shrinks in size.

**FIGURE 4.15** Two examples of gravitational potential energy.

therefore have a lot of gravitational potential energy. The particles lose gravitational potential energy as the cloud contracts under its own gravity, and this "lost" potential energy ultimately gets converted into thermal energy, making the center of the cloud hot.

Einstein discovered that mass itself is a form of potential energy, often called **mass-energy**. The amount of potential energy contained in mass is described by Einstein's famous equation

$$E = mc^2$$

where $E$ is the amount of potential energy, $m$ is the mass of the object, and $c$ is the speed of light. This equation tells us that a small amount of mass contains a huge amount of energy. For example, the energy released by a 1-megaton H-bomb, which could destroy a city out to a radius of about 10 kilometers, comes from converting only about 0.1 kilogram of mass (about 3 ounces—a quarter of a can of soda) into energy (**FIGURE 4.16**). The Sun generates energy by converting a tiny fraction of its mass into energy through a similar process of nuclear fusion [**Section 14.2**].

Just as Einstein's formula tells us that mass can be converted into other forms of energy, it also tells us that energy can be transformed into mass. This process is especially important in understanding what we think happened during the early moments in the history of the universe, when

## MATHEMATICAL INSIGHT 4.2 Mass-Energy

It's easy to calculate mass-energies with Einstein's formula $E = mc^2$.

**EXAMPLE:** Suppose a 1-kilogram rock were completely converted to energy. How much energy would it release? Compare this to the energy released by burning 1 liter of oil.

**SOLUTION:**

**Step 1 Understand:** We can compute the total mass-energy of the rock from Einstein's formula and then compare it to the energy released by burning a liter of oil, from Table 4.1.

**Step 2 Solve:** The mass-energy of the rock is

$$\begin{aligned} E = mc^2 &= 1\text{ kg} \times \left(3 \times 10^8 \frac{\text{m}}{\text{s}}\right)^2 \\ &= 1\text{ kg} \times \left(9 \times 10^{16} \frac{\text{m}^2}{\text{s}^2}\right) \\ &= 9 \times 10^{16} \frac{\text{kg} \times \text{m}^2}{\text{s}^2} = 9 \times 10^{16}\text{ joules} \end{aligned}$$

We divide to compare this mass-energy to the energy released by burning 1 liter of oil (12 million joules; see Table 4.1):

$$\frac{9 \times 10^{16}\text{ joules}}{1.2 \times 10^7\text{ joules}} = 7.5 \times 10^9$$

**Step 3 Explain:** We have found that converting a 1-kilogram rock completely to energy would release $9 \times 10^{16}$ joules of energy, which is about 7.5 billion times as much energy as we get from burning 1 liter of oil. In fact, the total amount of oil used by all cars in the United States is approximately 7.5 billion liters per week—which means that complete conversion of the mass of a 1-kilogram rock to energy could yield enough energy to power all the cars in the United States for a week. Unfortunately, no technology available now or in the foreseeable future can release all the mass-energy of a rock.

**FIGURE 4.16** The energy released by this H-bomb comes from converting only about 0.1 kilogram of mass into energy in accordance with the formula $E = mc^2$.

some of the energy of the Big Bang turned into the mass from which all objects, including us, are made [**Section 22.1**]. Scientists also use this idea to search for undiscovered particles of matter, using large machines called *particle accelerators* to create subatomic particles from energy.

**Conservation of Energy** We have seen that energy comes in three basic categories—kinetic, radiative, and potential—and explored several subcategories that are especially important in astronomy: thermal energy, gravitational potential energy, and mass-energy. Now we are ready to return to the question of where objects get their energy. Because energy cannot be created or destroyed, objects always get their energy from other objects. Ultimately, we can always trace an object's energy back to the Big Bang [**Section 1.2**], the beginning of the universe in which all matter and energy is thought to have come into existence.

For example, imagine that you've thrown a baseball. It is moving, so it has kinetic energy. Where did this kinetic energy come from? The baseball got its kinetic energy from the motion of your arm as you threw it. Your arm, in turn, got its kinetic energy from the release of chemical potential energy stored in your muscle tissues. Your muscles got this energy from the chemical potential energy stored in the foods you ate. The energy stored in the foods came from sunlight, which plants convert into chemical potential energy through photosynthesis. The radiative energy of the Sun was generated through the process of nuclear fusion, which releases some of the mass-energy stored in the Sun's supply of hydrogen. The mass-energy stored in the hydrogen came from the birth of the universe in the Big Bang. After you throw the ball, its kinetic energy will ultimately be transferred to molecules in the air or ground. It may be difficult to trace after this point, but it will never disappear.

## 4.4 The Universal Law of Gravitation

Newton's laws of motion describe how objects in the universe move in response to forces. The laws of conservation of momentum, angular momentum, and energy offer an alternative and often simpler way of thinking about what happens when a force causes some change in the motion of one or more objects. However, we cannot fully understand motion unless we also understand the forces that lead to changes in motion. In astronomy, the most important force is gravity, which governs virtually all large-scale motion in the universe.

### What determines the strength of gravity?

Isaac Newton discovered the basic law that describes how gravity works. Newton expressed the force of gravity mathematically with his **universal law of gravitation**. Three simple statements summarize this law:

- Every mass attracts every other mass through the force called *gravity*.
- The strength of the gravitational force attracting any two objects is *directly proportional* to the product of their masses. For example, doubling the mass of *one* object doubles the force of gravity between the two objects.
- The strength of gravity between two objects decreases with the *square* of the distance between their centers. We therefore say that the gravitational force follows an **inverse square law**. For example, doubling the distance between two objects weakens the force of gravity by a factor of $2^2$, or 4.

These three statements tell us everything we need to know about Newton's universal law of gravitation. Mathematically, all three statements can be combined into a single equation, usually written like this:

$$F_g = G\frac{M_1M_2}{d^2}$$

where $F_g$ is the force of gravitational attraction, $M_1$ and $M_2$ are the masses of the two objects, and $d$ is the distance between their centers (**FIGURE 4.17**). The symbol $G$ is a constant

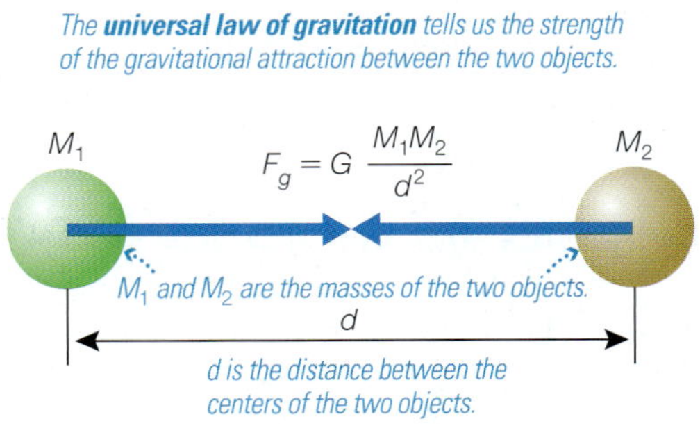


**FIGURE 4.17** The universal law of gravitation is an *inverse square law*, which means that the force of gravity declines with the *square* of the distance $d$ between two objects.

called the **gravitational constant**, and its numerical value has been measured to be $G = 6.67 \times 10^{-11}\ m^3/(kg \times s^2)$.

**Think about it** How does the gravitational force between two objects change if the distance between them triples? If the distance between them drops by half?

## How does Newton's law of gravity extend Kepler's laws?

By the time Newton published *Principia* in 1687, Kepler's laws of planetary motion [**Section 3.3**] had already been known for some 70 years and had proven so successful that there was little doubt about their validity. However, there was great debate among scientists about *why* Kepler's laws hold true.

Newton resolved the debate by showing that Kepler's laws are consequences of the laws of motion and the universal law of gravitation. In particular, with the aid of the mathematics of calculus that he invented, Newton showed that the inverse square law for gravity leads naturally to elliptical orbits for planets orbiting the Sun (with the Sun at one focus), which is Kepler's first law. As we've seen, Kepler's second law (a planet moves faster when it is closer to the Sun) then arises as a consequence of conservation of angular momentum. Kepler's third law (average orbital speed is slower for planets with larger average orbital distance) arises from the fact that gravity weakens with distance from the Sun. Newton also discovered that he could extend Kepler's laws into a more general set of rules about orbiting objects.

Newton's discoveries sealed the triumph of the Copernican revolution. Prior to Newton, it was still possible to see Kepler's model of planetary motion as "just" another model, though it fit the observational data far better than any previous model. By explaining Kepler's laws in terms of basic laws of physics, Newton removed virtually all remaining doubt about the legitimacy of the Sun-centered solar system. By extending the laws to other orbiting objects, he provided us with a way to explain the motions of objects throughout the universe. Let's explore four crucial ways in which Newton extended Kepler's laws.

### Planets Are Not the Only Objects with Elliptical Orbits

Kepler wrote his first two laws for planets orbiting the Sun, but Newton showed that any object going around another object will obey these laws. For example, the orbits of a satellite around Earth, of a moon around a planet, and of an asteroid around the Sun are all ellipses in which the orbiting object moves faster at the nearer points in its orbit and slower at the farther points.

### Ellipses Are Not the Only Possible Orbital Paths

Ellipses (which include circles) are the only possible shapes for **bound orbits**—orbits in which an object goes around another object over and over again. (The term *bound orbit* comes from the idea that gravity creates a *bond* that holds the objects together.) However, Newton discovered that objects can also follow **unbound orbits**—paths that bring an object close to another object just once. For example,

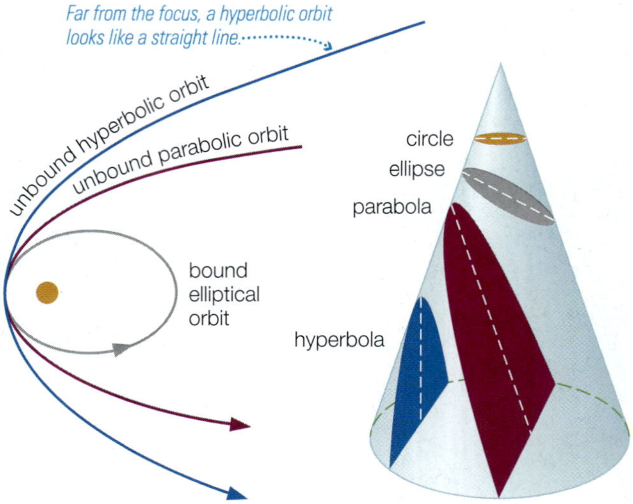


**a** Orbits allowed by the law of gravity.

**b** Ellipses (which include circles), parabolas, and hyperbolas are conic sections, made by slicing a cone at different angles.

**FIGURE 4.18** Newton showed that ellipses are not the only possible orbital paths. Orbits can also be unbound, taking the mathematical shape of either parabolas or hyperbolas.

some comets that enter the inner solar system follow unbound orbits. They come in from afar just once, loop around the Sun, and never return.

More specifically, Newton showed that bound orbits are ellipses, while unbound orbits can be either parabolas or hyperbolas (**FIGURE 4.18a**). Together, these shapes are known in mathematics as the *conic sections*, because they can be made by slicing through a cone at different angles (**FIGURE 4.18b**). Note that objects on unbound orbits still obey the basic principle of Kepler's second law: They move faster when they are closer to the object they are orbiting, and slower when they are farther away.

**Objects Orbit Their Common Center of Mass** We usually think of one object orbiting another object, like a planet orbiting the Sun or the Moon orbiting Earth. However, Newton showed that two objects attracted by gravity actually *both* orbit around their common **center of mass**—the point at which the two objects would balance if they were somehow connected (**FIGURE 4.19**). For example, in a binary star system in which both stars have the same mass, we would see both stars tracing ellipses around a point halfway between them. When one object is more massive than the other, the center of mass lies closer to the more massive object.

The idea that objects orbit their common center of mass holds even for the Sun and planets. However, the Sun is so much more massive than the planets that the center of mass between the Sun and any planet lies either inside or nearly inside the Sun, making it difficult for us to notice the Sun's motion about this center. Nevertheless, with precise measurements we can detect the Sun's slight motion around this center of mass. As we will see in Chapter 13,

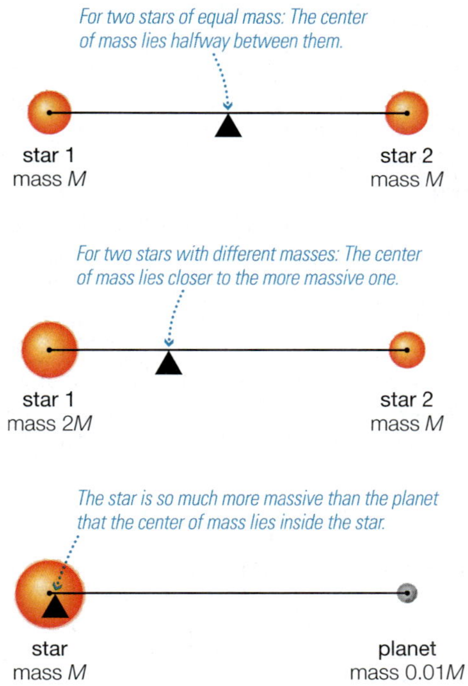


**FIGURE 4.19** Two objects attracted by gravity orbit their common center of mass—the point at which they would balance if they were somehow connected.

**FIGURE 4.20** Newton's version of Kepler's third law shows that when one object orbits a much more massive object, the orbital period depends only on its average orbital distance. The astronaut and the spacecraft share the same orbit and therefore stay together—even as both orbit Earth at a speed of 28,000 km/hr.

astronomers have used this same idea to discover many planets around other stars.

**Orbital Characteristics Tell Us the Masses of Distant Objects** Recall that Kepler's third law is written $p^2 = a^3$, where $p$ is a planet's orbital period in years and $a$ is the planet's average distance from the Sun in AU. Newton found that this statement is a special case of a more general equation that we call **Newton's version of Kepler's third law** (see Mathematical Insight 4.3). This equation allows us to calculate the mass of a distant object if we measure the orbital period and distance of another object orbiting around it; moreover, we can use any units, not just years and AU. For example, we can calculate the mass of the Sun from Earth's orbital period (1 year) and its average distance (1 AU); we can calculate Jupiter's mass from the orbital period and average distance of one of its moons; and we can determine the masses of distant stars if they are members of binary star systems, in which two stars orbit each other. In fact, Newton's version of Kepler's third law is the primary means by which we determine masses throughout the universe.

Newton's version of Kepler's third law also explains another important characteristic of orbital motion. It shows that the orbital period of a *small* object orbiting a much more massive object depends only on its orbital distance, not on its mass. That is why an astronaut does not need a tether to stay close to the spacecraft during a space walk (**FIGURE 4.20**). The spacecraft and the astronaut are both much smaller in mass than Earth, so they stay together because they have the same orbital distance and hence the same orbital period.

## 4.5 Orbits, Tides, and the Acceleration of Gravity

Newton's universal law of gravitation has applications that go far beyond explaining Kepler's laws. In this final section, we'll explore three important concepts that we can understand with the help of the universal law of gravitation: orbits, tides, and the acceleration of gravity.

### How do gravity and energy allow us to understand orbits?

The law of gravitation explains Kepler's laws of planetary motion, which describe the simple and stable orbits of the planets, and Newton's extensions of Kepler's laws explain other stable orbits, such as the orbit of a satellite around Earth or of a moon around a planet. But orbits do not always stay the same. For example, you've probably heard of satellites crashing to Earth from orbit, proving that orbits can sometimes change dramatically. To understand how and why orbits sometimes change, we need to consider the role of energy in orbits.

**Orbital Energy** A planet orbiting the Sun has both kinetic energy (because it is moving around the Sun) and gravitational potential energy (because it would fall toward the Sun if it stopped orbiting). The amount of kinetic energy depends on orbital speed, and the amount of gravitational potential energy depends on orbital distance. Because the planet's distance and speed both vary as it orbits the Sun, its gravitational potential energy and kinetic energy also vary (**FIGURE 4.21**). However, the planet's total **orbital energy**—the

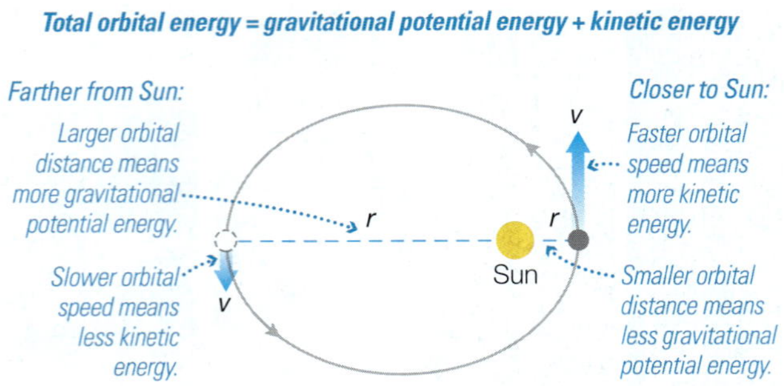


**FIGURE 4.21** The total orbital energy of a planet stays constant throughout its orbit, because its gravitational potential energy increases when its kinetic energy decreases, and vice versa.

sum of its kinetic and gravitational potential energies—stays the same. This fact is a consequence of the law of conservation of energy. As long as no other object causes the planet to gain or lose orbital energy, its orbital energy cannot change and its orbit must remain the same.

Generalizing from planets to other objects leads to an important idea about motion throughout the cosmos: *Orbits cannot change spontaneously.* Left undisturbed, planets would forever keep the same orbits around the Sun, moons would keep the same orbits around their planets, and stars would keep the same orbits in their galaxies.

**Gravitational Encounters** Although orbits cannot change spontaneously, they can change through exchanges of energy. One way that two objects can exchange orbital

## MATHEMATICAL INSIGHT 4.3 Newton's Version of Kepler's Third Law

Newton's version of Kepler's third law relates the orbital periods, distances, and masses of any pair of orbiting objects. Mathematically, we write it as follows:

$$p^2 = \frac{4\pi^2}{G(M_1 + M_2)}a^3$$

where $M_1$ and $M_2$ are the object masses, $p$ is their orbital period, and $a$ is the average distance between their centers. The term $4\pi^2$ is simply a number ($4\pi^2 \approx 4 \times 3.14^2 = 39.44$); $G$ is the gravitational constant, which is measured experimentally.

If we measure the orbital period and distance of one object orbiting another, we can use Newton's equation to calculate the sum $M_1 + M_2$ of the object masses. If one object is much more massive than the other, we essentially learn its mass. For example, when we apply the law to a planet orbiting the Sun, the sum $M_{Sun} + M_{planet}$ is pretty much just $M_{Sun}$ because the Sun is so much more massive than any planet. We can therefore use any planet's orbital period and distance from the Sun to calculate the mass of the Sun.

**EXAMPLE 1:** Earth orbits the Sun in 1 year at an average distance of 150 million kilometers (1 AU). Calculate the Sun's mass.

**SOLUTION:**

**Step 1 Understand:** We will use Newton's version of Kepler's third law. For Earth's orbit around the Sun, this law takes the form

$$(p_{Earth})^2 = \frac{4\pi^2}{G(M_{Sun} + M_{Earth})}(a_{Earth})^3$$

The Sun is much more massive than Earth, so the sum $M_{Sun} + M_{Earth}$ is approximately the Sun's mass alone, $M_{Sun}$. We therefore rewrite the equation as

$$(p_{Earth})^2 \approx \frac{4\pi^2}{G \times M_{Sun}}(a_{Earth})^3$$

**Step 2 Solve:** We know Earth's orbital period ($p_{Earth}$) and average distance ($a_{Earth}$), so the above equation contains only one unknown: $M_{Sun}$. To solve for this unknown, we multiply both sides by $M_{Sun}$ and divide both sides by $(p_{Earth})^2$:

$$M_{Sun} \approx \frac{4\pi^2(a_{Earth})^3}{G(p_{Earth})^2}$$

We now plug in the values $p_{Earth} = 1$ yr, which is the same as $3.15 \times 10^7$ s; $a_{Earth} \approx 150$ million km, or $1.5 \times 10^{11}$ m; and the measured value $G = 6.67 \times 10^{-11}\ m^3/(kg \times s^2)$:

$$M_{Sun} \approx \frac{4\pi^2}{\left(6.67 \times 10^{-11} \frac{m^3}{kg \times s^2}\right)} \frac{(1.5 \times 10^{11}\ m)^3}{(3.15 \times 10^7\ s)^2}$$

$$= 2 \times 10^{30}\ kg$$

**Step 3 Explain:** Simply by knowing Earth's orbital period and average distance, along with the gravitational constant, $G$, we were able to use Newton's version of Kepler's third law to find that the Sun's mass is about $2 \times 10^{30}$ kilograms. (Note: $G$ was not measured until more than 100 years after Newton published *Principia*, so Newton was not able to calculate masses in absolute units.)

**EXAMPLE 2:** A *geosynchronous satellite* orbits Earth with the same period as that of Earth's rotation: 1 sidereal day, or about 23 hours, 56 minutes, 4 seconds [**Section S1.1**]. Calculate the orbital distance of a geosynchronous satellite.

**SOLUTION:**

**Step 1 Understand:** A satellite is much less massive than Earth ($M_{Earth} + M_{satellite} \approx M_{Earth}$), so we can use Newton's version of Kepler's third law in the following form:

$$(p_{satellite})^2 \approx \frac{4\pi^2}{G \times M_{Earth}}(a_{satellite})^3$$

**Step 2 Solve:** We solve for the satellite's distance, $a_{satellite}$, by multiplying both sides of the equation by $(G \times M_{Earth})/4\pi^2$ and then taking the cube root of both sides:

$$a_{satellite} \approx \sqrt[3]{\frac{G \times M_{Earth}}{4\pi^2}(p_{satellite})^2}$$

If you now plug in the given value $p_{satellite} = 1$ sidereal day $\approx 86{,}164$ s, along with Earth's mass and $G$, you will find that $a_{satellite} \approx 42{,}000$ km.

**Step 3 Explain:** We have found that a geosynchronous satellite orbits at a distance of 42,000 kilometers above the *center* of Earth, which is about 35,600 kilometers above Earth's surface.

energy is through a **gravitational encounter**, in which they pass near enough that each can feel the effects of the other's gravity. For example, in the rare cases in which a comet happens to pass near a planet, the comet's orbit can change dramatically. **FIGURE 4.22** shows a comet headed toward the Sun on an unbound orbit. The comet's close passage by Jupiter allows the comet and Jupiter to exchange energy. In this case, the comet loses so much orbital energy that its orbit changes from unbound to bound and elliptical. Jupiter gains exactly as much energy as the comet loses, but the effect on Jupiter is unnoticeable because of its much greater mass.

Spacecraft engineers can use the same basic idea in reverse. For example, on its way to Pluto, the *New Horizons* spacecraft was deliberately sent past Jupiter on a path that allowed it to gain orbital energy at Jupiter's expense. This extra orbital energy boosted the spacecraft's speed; without this boost, it would have needed four extra years to reach Pluto. The effect of the tiny spacecraft on Jupiter was negligible.

A similar dynamic can occur naturally and may explain why most comets orbit far from the Sun. Comets probably once orbited in the same region of the solar system as the large outer planets [**Section 12.3**]. Gravitational encounters with the planets then caused some of these comets to be "kicked out" into much more distant orbits around the Sun; some may have been ejected from the solar system completely.

**Atmospheric Drag** Friction can cause objects to lose orbital energy. A satellite in low-Earth orbit (a few hundred kilometers above Earth's surface) experiences a bit of drag from Earth's thin upper atmosphere. This drag gradually causes the satellite to lose orbital energy until it finally plummets to Earth. The satellite's lost orbital energy is converted to thermal energy in the atmosphere, which is why a falling satellite usually burns up.

Friction may also help explain why the outer planets have so many small moons. These moons may once have orbited the Sun independently, and their orbits could not have changed spontaneously. However, the outer planets probably once were surrounded by clouds of gas [**Section 8.2**], and friction would have slowed objects passing through this gas. Some of these small objects may have lost just enough energy to friction to allow them to be "captured" as moons. Mars may have captured its two small moons in a similar way.

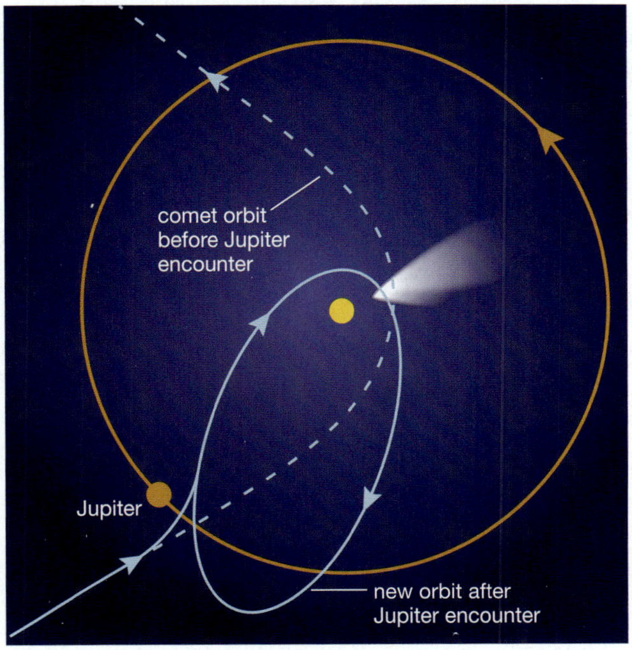


**FIGURE 4.22** This diagram shows a comet in an unbound orbit of the Sun that happens to pass near Jupiter. The comet loses orbital energy to Jupiter, changing its unbound orbit to a bound orbit around the Sun.

**Escape Velocity** An object that gains orbital energy moves into an orbit with a higher average altitude. For example, if we want to boost the orbital altitude of a spacecraft, we can give it more orbital energy by firing a rocket. The chemical potential energy released by the rocket fuel is converted to orbital energy for the spacecraft.

If we give a spacecraft enough orbital energy, it may end up in an unbound orbit that allows it to *escape* Earth completely (**FIGURE 4.23**). For example, when we send a space probe to Mars, we must use a large rocket that gives the probe enough energy to leave Earth orbit. Although it would probably make more sense to say that the probe achieves "escape energy," we instead say that it achieves **escape velocity**. The escape velocity from Earth's surface is about 40,000 km/hr, or 11 km/s; this is the minimum velocity required to escape Earth's gravity for a spacecraft that starts near the surface.

Note that escape velocity does not depend on the mass of the escaping object—*any* object must travel at a velocity of 11 km/s to escape from Earth, whether it is an individual atom or molecule escaping from the atmosphere, a spacecraft being launched into deep space, or a rock blasted into the sky by a large impact. Escape velocity *does* depend on

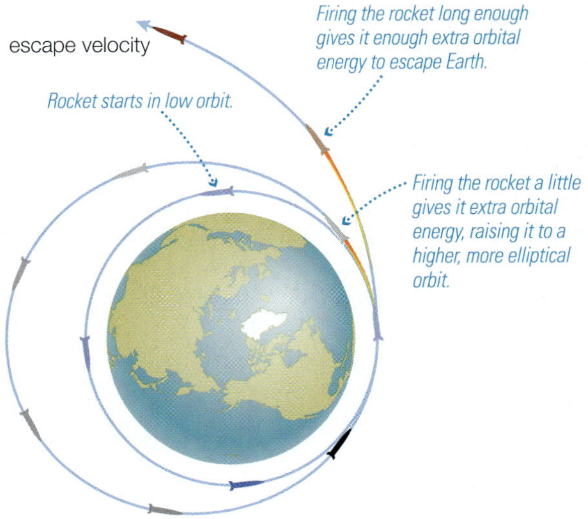


**FIGURE 4.23** An object with escape velocity has enough orbital energy to escape Earth completely.

whether you start from the surface or from someplace high above the surface. Because gravity weakens with distance, it takes less energy—and hence a lower velocity—to escape from a point high above Earth than from Earth's surface.

## How does gravity cause tides?

If you've spent time near an ocean, you've probably observed the rising and falling of the tides. In most places, tides rise and fall twice each day. We can understand the basic cause of tides by examining the gravitational attraction between Earth and the Moon. We'll then see how the same ideas explain many other phenomena that we can observe throughout the universe, including the synchronous rotation of our own Moon and many other worlds.

**The Moon's Tidal Force** Gravity attracts Earth and the Moon toward each other (with the Moon staying in orbit as it "falls around" Earth), but it affects different parts of Earth slightly differently: Because the strength of gravity declines with distance, the gravitational attraction of each part of Earth to the Moon becomes weaker as we go from the side of Earth facing the Moon to the side facing away from the Moon. This difference in attraction creates a "stretching force," or **tidal force**, that stretches the entire Earth to create two tidal bulges, one facing the Moon and one opposite the Moon (**FIGURE 4.24**). If you are still unclear about why there are *two* tidal bulges, think about a rubber band: If you pull on a rubber band, it will stretch in both directions relative to its center, even if you pull on only one side (while holding the other side still). In the same way, Earth stretches on both sides even though the Moon is tugging harder on only one side.

### COMMON MISCONCEPTIONS

#### The Origin of Tides

Many people believe that tides arise because the Moon pulls Earth's oceans toward it. But if that were the whole story, there would be a bulge only on the side of Earth facing the Moon, and hence only one high tide each day. The correct explanation for tides must account for why Earth has *two* tidal bulges.

Only one explanation works: Earth must be stretching from its center in both directions (toward and away from the Moon). This stretching force, or tidal force, arises from the *difference* in the force of gravity attracting different parts of Earth to the Moon. In fact, stretching due to tides affects many objects, not just Earth. Many moons are stretched into slightly oblong shapes by tidal forces caused by their parent planets, and mutual tidal forces stretch close binary stars into teardrop shapes. In regions where gravity is extremely strong, such as near a black hole, tides can have even more dramatic effects.

### MATHEMATICAL INSIGHT 4.4 Escape Velocity

A simple formula allows us to calculate the escape velocity from any planet, moon, or star:

$$v_{\text{escape}} = \sqrt{2 \times G \times \frac{M}{R}}$$

where $M$ is the object's mass, $R$ is the starting distance above the object's center, and $G$ is the gravitational constant. If you use this formula to calculate the escape velocity from an object's *surface*, replace $R$ with the object's radius.

**EXAMPLE 1:** Calculate the escape velocity from the Moon's surface. Compare it to the 11 km/s escape velocity from Earth.

**SOLUTION:**

**Step 1 Understand:** We use the above formula; because we seek the escape velocity from the Moon's *surface*, we use the Moon's radius as $R$. From Appendix E, the Moon's mass and radius are $M_{\text{Moon}} = 7.4 \times 10^{22}$ kg and $R_{\text{Moon}} = 1.7 \times 10^6$ m.

**Step 2 Solve:** We substitute the Moon's mass and radius into the escape velocity formula:

$$\begin{aligned} v_{\text{escape}} &= \sqrt{2 \times G \times \frac{M_{\text{Moon}}}{R_{\text{Moon}}}} \\ &= \sqrt{2 \times \left(6.67 \times 10^{-11} \frac{\text{m}^3}{\text{kg} \times \text{s}^2}\right) \times \frac{(7.4 \times 10^{22}\ \text{kg})}{(1.7 \times 10^6\ \text{m})}} \\ &\approx 2400\ \text{m/s} = 2.4\ \text{km/s} \end{aligned}$$

**Step 3 Explain:** Escape velocity from the Moon's surface is 2.4 km/s, which is less than one-fourth the escape velocity (11 km/s) from Earth's surface.

**EXAMPLE 2:** Suppose a future space station orbits Earth in geosynchronous orbit, 42,000 kilometers above the center of Earth (see Mathematical Insight 4.3). At what velocity must a spacecraft be launched from the station to escape Earth?

**SOLUTION:**

**Step 1 Understand:** We seek the escape velocity from a satellite orbiting 42,000 kilometers above the center of Earth, so we use the escape velocity formula with the mass of Earth ($M_{\text{Earth}} = 6.0 \times 10^{24}$ kg) and $R$ set to the satellite's distance ($R = 42{,}000\ \text{km} = 4.2 \times 10^7$ m).

**Step 2 Solve:** With the above values, we find

$$\begin{aligned} v_{\text{escape}} &= \sqrt{2 \times G \times \frac{M_{\text{Earth}}}{R_{\text{orbit}}}} \\ &= \sqrt{2 \times \left(6.67 \times 10^{-11} \frac{\text{m}^3}{\text{kg} \times \text{s}^2}\right) \times \frac{(6.0 \times 10^{24}\ \text{kg})}{(4.2 \times 10^7\ \text{m})}} \\ &\approx 4400\ \text{m/s} = 4.4\ \text{km/s} \end{aligned}$$

**Step 3 Explain:** The escape velocity from geosynchronous orbit is 4.4 km/s—considerably lower than the 11 km/s escape velocity from Earth's surface. It would therefore require substantially less fuel to launch a spacecraft from the space station than from Earth, which is why some people propose building future spacecraft at future space stations.

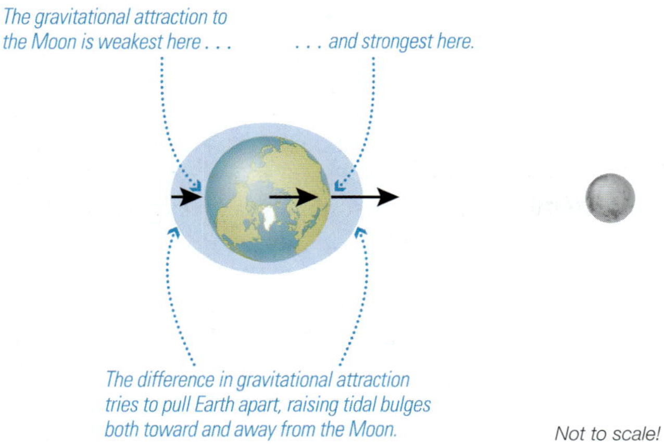


**FIGURE 4.24** Tides are created by the difference in the force of attraction between the Moon and different parts of Earth. The two daily high tides occur as a location on Earth rotates through the two tidal bulges. (The diagram greatly exaggerates the tidal bulges, which raise the oceans only about 2 meters and the land only about a centimeter.)

Tides affect both land and ocean, but we generally notice only the ocean tides because water flows much more readily than land. Earth's rotation carries any location through each of the two bulges each day, creating two high tides. Low tides occur when the location is at the points halfway between the two tidal bulges. Note that, because of its orbital motion around Earth, the Moon reaches its highest point in the sky at any location about every 24 hours 50 minutes, rather than every 24 hours. As a result, the tidal cycle of two high tides and two low tides takes about 24 hours 50 minutes, so each high tide occurs about 12 hours 25 minutes after the previous one.

The height and timing of ocean tides vary considerably from place to place on Earth, depending on factors such as latitude, the orientation of the coastline (such as whether it is north-facing or west-facing), and the depth and shape of any channel through which the rising tide must flow. For example, while the tide rises gradually in most locations, the incoming tide near the famous abbey on Mont-Saint-Michel, France, moves much faster than a person can swim (**FIGURE 4.25**). In centuries past, the Mont was an island twice a day at high tide but was connected to the mainland at low tide. Many pilgrims drowned when they were caught unprepared by the tide rushing in. Another unusual tidal pattern occurs in coastal states along the northern shore of the Gulf of Mexico, where topography and other factors combine to make only one noticeable high tide and low tide each day.

**The Tidal Effect of the Sun** The Sun also exerts a tidal force on Earth, causing Earth to stretch along the Sun-Earth line. You might at first guess that the Sun's tidal force would be more than the Moon's, since the Sun's mass is more than a million times that of the Moon. Indeed, the *gravitational* force between Earth and the Sun is much greater than that between Earth and the Moon, which is why Earth orbits the Sun. However, the much greater distance to the Sun (than to the Moon) means that the *difference* in the Sun's pull on the near and far sides of Earth is relatively small.

The overall tidal force caused by the Sun is a little less than half that caused by the Moon (**FIGURE 4.26**). When the tidal forces of the Sun and the Moon work together, as is the case at both new moon and full moon, we get the especially pronounced *spring tides* (so named because the water tends to "spring up" from Earth). When the tidal forces of the Sun and the Moon counteract each other, as is the case at first- and third-quarter moons, we get the relatively small tides known as *neap tides*.

**Think about it** Explain why any tidal effects on Earth caused by the other planets would be unnoticeably small.

**Tidal Friction** So far, we have talked as if Earth rotated smoothly through the tidal bulges. But because tidal forces stretch Earth itself, the process creates friction, called **tidal friction**. **FIGURE 4.27** shows the effects of this friction. In essence, the Moon's gravity tries to keep the tidal bulges on

**FIGURE 4.25** Photographs of high and low tide at the abbey of Mont-Saint-Michel, France, one of the world's most popular tourist destinations. Before a causeway was built (visible at the far left), the Mont was accessible by land only at low tide. At high tide, it became an island. (You can find great videos online of the tide rushing in and out at Mont-Saint-Michel.)

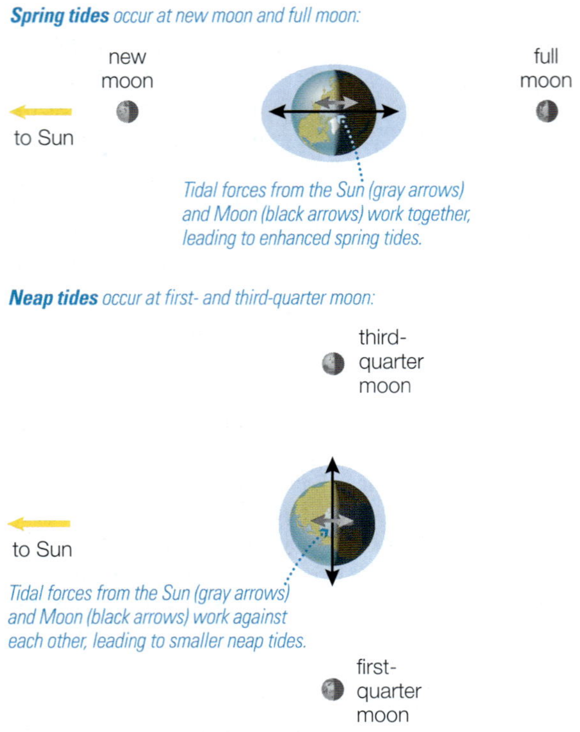


**FIGURE 4.26** The Sun exerts a tidal force on Earth less than half as strong as that from the Moon. When the tidal forces from the Sun and Moon work together at new moon and full moon, we get enhanced *spring tides*. When they work against each other at first- and third-quarter moons, we get smaller *neap tides*.

the Earth-Moon line, while Earth's rotation tries to pull the bulges around with it. The resulting "compromise" keeps the bulges just ahead of the Earth-Moon line at all times.

The slight misalignment of the tidal bulges with the Earth-Moon line causes two important effects. First, the Moon's gravity always pulls back on the bulges, slowing Earth's rotation. Second, the gravity of the bulges pulls the Moon slightly ahead in its orbit, adding orbital energy that causes the Moon to move farther from Earth. These effects are barely noticeable on human time scales—for example, tidal friction increases the length of a day by only about 1 second every 50,000 years*—but they add up over billions of years. Early in Earth's history, a day may have been only 5 or 6 hours long and the Moon may have been one-tenth or less its current distance from Earth. These changes also provide a great example of conservation of angular momentum and energy: The Moon's growing orbit gains the angular momentum and energy that Earth loses as its rotation slows.

**The Moon's Synchronous Rotation** Recall that the Moon always shows (nearly) the same face to Earth, a trait called **synchronous rotation** (see Figure 2.23). Synchronous rotation may seem like an extraordinary coincidence, but it is a natural consequence of tidal friction.

Because Earth is more massive than the Moon, Earth's tidal force has a greater effect on the Moon than the Moon's tidal force has on Earth. This tidal force gives the Moon two tidal bulges along the Earth-Moon line, much like the two tidal bulges that the Moon creates on Earth. (The Moon's tidal bulges are not visible but can be measured in terms of excess mass along the Earth-Moon line.) If the Moon rotated relative to its tidal bulges in the same way as Earth, the resulting tidal friction would cause the Moon's rotation to slow down. This is exactly what we think happened long ago.

The Moon probably once rotated much faster than it does today. As a result, it *did* rotate relative to its tidal bulges, and its rotation gradually slowed. Once the Moon's rotation slowed to the point at which the Moon and its bulges rotated at the same rate—that is, synchronously with the orbital period—there was no further source for tidal friction. The Moon's synchronous rotation was therefore a natural outcome of Earth's tidal effects on the Moon.

**Tidal Effects on Other Worlds** Tidal forces and tidal friction affect many worlds. Synchronous rotation is especially common. For example, Jupiter's four large moons (Io, Europa, Ganymede, and Callisto) keep nearly the same face toward Jupiter at all times, as do many other moons. Pluto and its moon Charon *both* rotate synchronously: Like two dancers, they always keep the same face toward each other. Many binary star systems also rotate in this way. Some moons and planets exhibit variations on synchronous rotation. For example, Mercury rotates exactly three times for every two orbits of the Sun. This pattern ensures that Mercury's tidal bulge always aligns with the Sun at perihelion, where the Sun exerts its strongest tidal force.

*This effect is overwhelmed on short time scales by other effects due to slight changes in Earth's internal mass distribution; these changes can alter Earth's rotation period by a second or more per year, which is why "leap seconds" are occasionally added to or subtracted from the year.

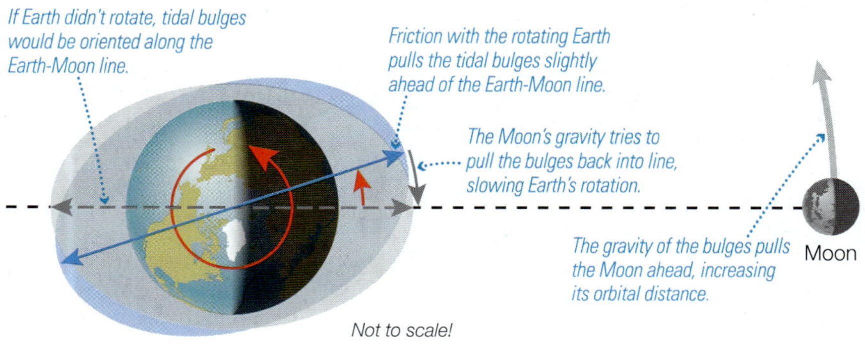


**FIGURE 4.27** Earth's rotation pulls its tidal bulges slightly ahead of the Earth-Moon line, leading to gravitational effects that gradually slow Earth's rotation and increase the Moon's orbital energy and distance.

Tidal forces play other roles in the cosmos as well. They can alter the shapes of objects by stretching them along the line of tidal bulges. In Chapter 11, we'll see how tidal forces also lead to the astonishing volcanic activity of Jupiter's moon Io and probably to a subsurface ocean on its moon Europa. As you study astronomy, you'll encounter many more cases where tides and tidal friction play important roles.

## Why do all objects fall at the same rate?

We will discuss many more applications of the universal law of gravitation in this book, but for now let's look at just one more: Galileo's discovery that the acceleration of a falling object is independent of its mass.

If you drop a rock, the force acting on the rock is the force of gravity. The two masses involved are the mass of Earth and the mass of the rock, which we'll denote $M_{Earth}$ and $M_{rock}$, respectively. The distance is the distance from the *center of Earth* to the center of the rock. If the rock isn't too far above Earth's surface, this distance is approximately the radius of Earth, $R_{Earth}$ (about 6400 kilometers), so the force of gravity acting on the rock is

$$F_g = G\frac{M_{Earth}\,M_{rock}}{d^2} \approx G\frac{M_{Earth}M_{rock}}{(R_{Earth})^2}$$

According to Newton's second law of motion ($F = ma$), this force is equal to the product of the rock's mass and acceleration. That is,

$$G\frac{M_{Earth}\cancel{M_{rock}}}{(R_{Earth})^2} = \cancel{M_{rock}}a_{rock}$$

Note that $M_{rock}$ "cancels" because it appears on both sides of the equation (as a multiplier), giving Galileo's result that the acceleration of the rock—or of any falling object—does not depend on the object's mass.

The fact that objects of different masses fall with the same acceleration struck Newton as an astounding coincidence, even though his own equations showed it to be so. For the next 240 years, this seemingly odd coincidence remained just that—a coincidence—in the minds of most scientists. However, in 1915, Einstein showed that it is not a coincidence at all. Rather, it reveals something deeper about the nature of gravity and of the universe. Einstein described the new insights in his *general theory of relativity* (the topic of Chapter S3).

## MATHEMATICAL INSIGHT 4.5 The Acceleration of Gravity

We've seen that the acceleration of a falling rock near Earth's surface is

$$a_{rock} = G \times \frac{M_{Earth}}{(R_{Earth})^2}$$

Because this formula applies to *any* falling object on Earth, it is the *acceleration of gravity, g.* Calculating $g$ is easy. Simply plug in Earth's mass ($6.0 \times 10^{24}$ kg) and radius ($6.4 \times 10^6$ m):

$$g_{Earth} = G \times \frac{M_{Earth}}{(R_{Earth})^2} = \left(6.67 \times 10^{-11}\frac{m^3}{kg \times s^2}\right) \times \frac{6.0 \times 10^{24}\ kg}{(6.4 \times 10^6\ m)^2} = 9.8\ \frac{m}{s^2}$$

**EXAMPLE 1:** What is the acceleration of gravity on the Moon?

**SOLUTION:**

**Step 1 Understand:** We want the acceleration of gravity on the Moon's surface, so we use the above formula with the Moon's mass ($7.4 \times 10^{22}$ kg) and radius ($1.7 \times 10^6$ m).

**Step 2 Solve:** The formula becomes

$$g_{Moon} = G \times \frac{M_{Moon}}{(R_{Moon})^2} = \left(6.67 \times 10^{-11}\frac{m^3}{kg \times s^2}\right) \times \frac{7.4 \times 10^{22}\ kg}{(1.7 \times 10^6\ m)^2} = 1.7\ \frac{m}{s^2}$$

**Step 3 Explain:** The acceleration of gravity on the Moon is $1.7\ m/s^2$, or about one-sixth that on Earth, so objects on the Moon weigh about one-sixth of what they weigh on Earth.

**EXAMPLE 2:** The Space Station orbits at an altitude of roughly 350 kilometers above Earth's surface. What is the acceleration of gravity at this altitude?

**SOLUTION:**

**Step 1 Understand:** Because the Space Station is significantly above Earth's surface, we cannot use the approximation $d \approx R_{Earth}$ that we used in the text. Instead, we must go back to Newton's second law and set the gravitational force on the Space Station equal to its mass times acceleration. The acceleration in this equation is the acceleration of gravity at the Space Station's altitude.

**Step 2 Solve:** We write Newton's second law with the force being the force of gravity acting between Earth and the Space Station, which we set equal to the Space Station's mass times its acceleration:

$$G \times \frac{M_{Earth}M_{station}}{d^2} = M_{station} \times a_{station}$$

You should confirm that when we solve this equation for the acceleration of gravity, we find

$$a_{station} = G \times \frac{M_{Earth}}{d^2}$$

In this case, the distance $d$ is the 6400-kilometer radius of Earth *plus* the 350-kilometer altitude of the Station, or $d = 6750$ km $= 6.75 \times 10^6$ m. The gravitational acceleration is

$$a_{station} = G \times \frac{M_{Earth}}{d^2} = \left(6.67 \times 10^{-11}\frac{m^3}{kg \times s^2}\right) \times \frac{6.0 \times 10^{24}\ kg}{(6.75 \times 10^6\ m)^2} = 8.8\ \frac{m}{s^2}$$

**Step 3 Explain:** The acceleration of gravity in low Earth orbit is $8.8\ m/s^2$, which is only about 10% less than the $9.8\ m/s^2$ acceleration of gravity at Earth's surface. We see again that lack of gravity cannot be the reason astronauts are weightless in orbit; rather, they are weightless because they are in free-fall.

# The BIG Picture PUTTING CHAPTER 4 INTO CONTEXT

*We've covered a lot of ground in this chapter, from the scientific terminology of motion to the overarching principles that govern motion throughout the universe. Be sure you grasp the following "big picture" ideas:*

- Understanding the universe requires understanding motion. Motion may seem complex, but it can be described simply using Newton's three laws of motion.
- Today, we know that Newton's laws of motion stem from deeper physical principles, including the laws of conservation of momentum, of angular momentum, and of energy. These principles enable us to understand a wide range of astronomical phenomena.
- Newton also discovered the universal law of gravitation, which explains how gravity holds planets in their orbits and much more—including how satellites can reach and stay in orbit, the nature of tides, and why the Moon rotates synchronously around Earth.
- Newton's discoveries showed that the same physical laws we observe on Earth apply throughout the universe. The universality of physics opens up the entire cosmos as a possible realm of human study.

**MY COSMIC PERSPECTIVE** Although the physical laws discussed in this chapter have been presented in the context of astronomy, they also apply on Earth and explain much of what happens to us in our daily lives.

# Summary of Key Concepts

## 4.1 Describing Motion: Examples from Daily Life

- **How do we describe motion?** **Speed** is the rate at which an object is moving. **Velocity** is speed in a certain direction. **Acceleration** is a change in velocity, meaning a change in either speed or direction. **Momentum** is mass × velocity. A **force** can change an object's momentum, causing it to accelerate.
- **How is mass different from weight?** An object's **mass** is the same no matter where it is located, but its **weight** varies with the strength of gravity or other forces acting on the object. An object becomes **weightless** when it is in **free-fall**, even though its mass is unchanged.

## 4.2 Newton's Laws of Motion

- **How did Newton change our view of the universe?** Newton showed that the same physical laws that operate on Earth also operate in the heavens, making it possible to learn about the universe by studying physical laws on Earth.
- **What are Newton's three laws of motion?** (1) An object moves at constant velocity if there is no net force acting upon it. (2) Force = mass × acceleration ($F = ma$). (3) For any force, there is always an equal and opposite reaction force.

## 4.3 Conservation Laws in Astronomy

- **Why do objects move at constant velocity if no force acts on them?** **Conservation of momentum** means that an object's momentum cannot change unless the object transfers momentum to or from other objects. When no force is present, no momentum can be transferred so an object must maintain its speed and direction.
- **What keeps a planet rotating and orbiting the Sun?** **Conservation of angular momentum** means that a planet's rotation and orbit cannot change unless the planet transfers angular momentum to another object. The planets in our solar system do not exchange substantial angular momentum with each other or anything else, so their orbits and rotation rates remain fairly steady.

- **Where do objects get their energy?** Energy is always conserved—it can be neither created nor destroyed. Objects received whatever energy they now have from exchanges of energy with other objects. Energy comes in three basic categories—**kinetic**, **radiative**, and **potential**.

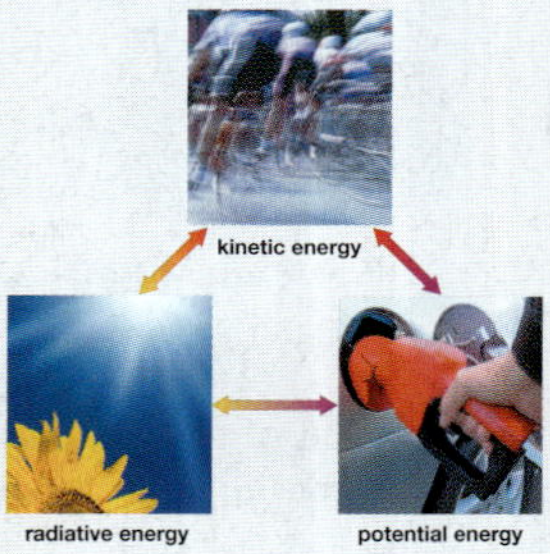


## 4.4 The Universal Law of Gravitation

- **What determines the strength of gravity?** The **universal law of gravitation** states that every object attracts every other object with a gravitational force that is proportional to the product of the objects' masses and declines with the square of the distance between their centers:

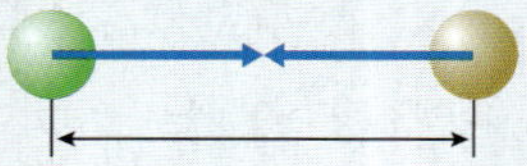

$$F_g = G\frac{M_1M_2}{d^2}$$

- **How does Newton's law of gravity extend Kepler's laws?** (1) Newton showed that any object going around another object will obey Kepler's first two laws. (2) He showed

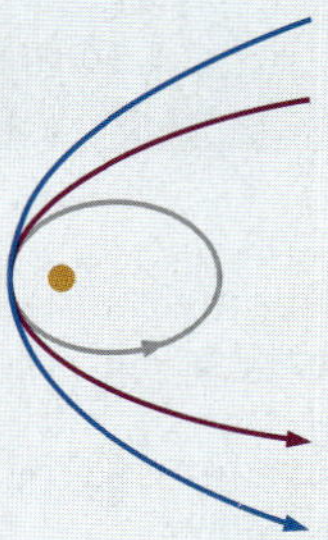

that elliptical **bound orbits** are not the only possible orbital shape—orbits can also be **unbound** in the shape of parabolas or hyperbolas. (3) He showed that two objects actually orbit their common **center of mass**. (4) **Newton's version of Kepler's third law** allows us to calculate the masses of orbiting objects from their orbital periods and distances.

## 4.5 Orbits, Tides, and the Acceleration of Gravity

- **How do gravity and energy allow us to understand orbits?** Gravity determines orbits, and an object cannot change its orbit unless it gains or loses **orbital energy**—the sum of its kinetic and gravitational potential energies—through energy transfer with other objects. If an object gains enough orbital energy, it may achieve **escape velocity** and leave the gravitational influence of the object it was orbiting.

- **How does gravity cause tides?** The Moon's gravity creates a **tidal force** that stretches Earth along the Earth-Moon line, causing Earth to bulge both toward and away from the Moon. Earth's rotation carries us through the two bulges each day, giving us two daily high tides and two daily low tides. Tidal forces also lead to **tidal friction**, which is gradually slowing Earth's rotation and explains the **synchronous rotation** of the Moon.

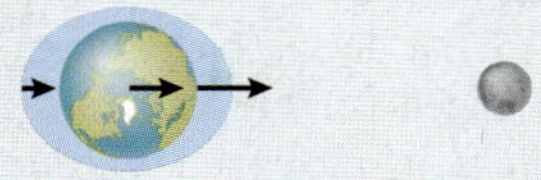

- **Why do all objects fall at the same rate?** Newton's equations show that the acceleration of gravity is independent of the mass of a falling object, so all objects fall at the same rate.

# Visual Skills Check

*Use the following questions to check your understanding of some of the many types of visual information used in astronomy. For additional practice, try the Chapter 4 Visual Quiz in the Study Area at www.MasteringAstronomy.com.*

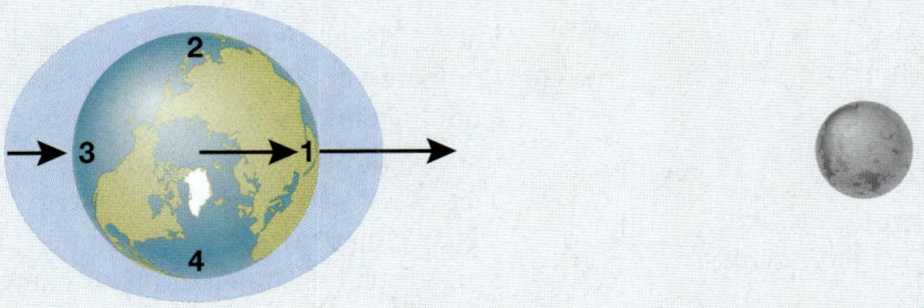


*The figure above, based on Figure 4.24, shows how the Moon causes tides on Earth. Note that the North Pole is in the center of the diagram, so the numbers 1 through 4 label points along Earth's equator.*

1. What do the three black arrows represent?
   a. the tidal force Earth exerts on the Moon
   b. the Moon's gravitational force at different points on Earth
   c. the direction in which Earth's water is flowing
   d. Earth's orbital motion
2. Where is it high tide?
   a. point 1 only
   b. point 2 only
   c. points 1 and 3
   d. points 2 and 4
3. Where is it low tide?
   a. point 1 only
   b. point 2 only
   c. points 1 and 3
   d. points 2 and 4
4. What time is it at point 1?
   a. noon
   b. midnight
   c. 6 a.m.
   d. cannot be determined from the information in the figure
5. The light blue ellipse represents tidal bulges. In what way are these bulges drawn inaccurately?
   a. There should be only one bulge rather than two.
   b. They should be aligned with the Sun rather than the Moon.
   c. They should be much smaller compared to Earth.
   d. They should be more pointy in shape.

# Exercises and Problems

*For instructor-assigned homework and other learning materials, go to www.MasteringAstronomy.com.*

## Chapter Review Questions

### Short-Answer Questions Based on the Reading

1. Define *speed, velocity*, and *acceleration*. What are the units of acceleration? What is the *acceleration of gravity*?
2. Define *momentum* and *force*. What do we mean when we say that momentum can be changed only by a *net force*?
3. What is *free-fall*, and why does it make you *weightless*? Briefly describe why astronauts are weightless in the Space Station.
4. State *Newton's three laws of motion*. For each law, give an example of its application.
5. Describe the laws of conservation of momentum, of angular momentum, and of energy. Give an example of how each is important in astronomy.
6. Define *kinetic energy, radiative energy*, and *potential energy*, with at least two examples of each.
7. Define and distinguish *temperature* and *thermal energy*.
8. What is *mass-energy*? Explain the formula $E = mc^2$.
9. Summarize the *universal law of gravitation* both in words and with an equation.
10. What is the difference between a *bound* and an *unbound orbit*? What orbital shapes are possible?
11. What do we need to know if we want to measure an object's mass with *Newton's version of Kepler's third law*? Explain.
12. Explain why orbits cannot change spontaneously, and how a *gravitational encounter* can cause a change. How can an object achieve *escape velocity*?
13. Explain how the Moon creates tides on Earth. Why do we have two high and low tides each day? How do the tides vary with the phase of the Moon?
14. What is *tidal friction*? What effects does it have on Earth? How does it explain the Moon's synchronous rotation?

### Does It Make Sense?

*Decide whether or not each of the following statements makes sense (or is clearly true or false). Explain clearly; not all of these have definitive answers, so your explanation is more important than your chosen answer.*

15. I've never been to space, so I've never experienced weightlessness.
16. Suppose you could enter a vacuum chamber (a chamber with no air in it) on Earth. Inside this chamber, a feather would fall at the same rate as a rock.
17. If an astronaut goes on a space walk outside the Space Station, she will quickly float away from the station unless she has a tether holding her to the station.
18. I used Newton's version of Kepler's third law to calculate Saturn's mass from orbital characteristics of its moon Titan.
19. If the Sun were magically replaced with a giant rock that had precisely the same mass, Earth's orbit would not change.
20. The fact that the Moon rotates once in precisely the time it takes to orbit Earth once is such an astonishing coincidence that scientists probably never will be able to explain it.
21. Venus has no oceans, so it could not have tides even if it had a moon (which it doesn't).
22. If an asteroid passed by Earth at just the right distance, Earth's gravity would capture it and make it our second moon.
23. When I drive my car at 30 miles per hour, it has more kinetic energy than it does at 10 miles per hour.
24. Someday soon, scientists are likely to build an engine that produces more energy than it consumes.

### Quick Quiz

*Choose the best answer to each of the following. For additional practice, try the Chapter 4 Reading and Concept Quizzes in the Study Area at www.MasteringAstronomy.com.*

25. A car is accelerating when it is (a) traveling on a straight, flat road at 50 miles per hour. (b) traveling on a straight uphill road at 30 miles per hour. (c) going around a circular track at a steady 100 miles per hour.
26. Compared to their values on Earth, on another planet your (a) mass and weight would both be the same. (b) mass would be the same but your weight would be different. (c) weight would be the same but your mass would be different.
27. Which person is weightless? (a) A child in the air as she plays on a trampoline (b) A scuba diver exploring a deep-sea wreck (c) An astronaut on the Moon
28. Consider the statement "There's no gravity in space." This statement is (a) completely false. (b) false if you are close to a planet or moon, but true in between the planets. (c) completely true.
29. To make a rocket turn left, you need to (a) fire an engine that shoots out gas to the left. (b) fire an engine that shoots out gas to the right. (c) spin the rocket clockwise.
30. Compared to its angular momentum when it is farthest from the Sun, Earth's angular momentum when it is nearest to the Sun is (a) greater. (b) less. (c) the same.
31. The gravitational potential energy of a contracting interstellar cloud (a) stays the same at all times. (b) gradually transforms into other forms of energy. (c) gradually grows larger.
32. If Earth were twice as far from the Sun, the force of gravity attracting Earth to the Sun would be (a) twice as strong. (b) half as strong. (c) one quarter as strong.
33. According to the universal law of gravitation, what would happen to Earth if the Sun were somehow replaced by a black hole of the same mass? (a) Earth would be quickly sucked into the black hole. (b) Earth would slowly spiral into the black hole. (c) Earth's orbit would not change.
34. If the Moon were closer to Earth, high tides would (a) be higher than they are now. (b) be lower than they are now. (c) occur three or more times a day rather than twice a day.

## Inclusive Astronomy

*Use these questions to reflect on participation in science.*

35. *The Dual Power of Knowledge.* Einstein's discovery that energy and mass are equivalent has led to technological developments that are both beneficial and dangerous. List as many of these developments as you can, categorizing

them as positive, negative, or neutral. Overall, do you think the human race would be better or worse off if we had never discovered that mass is a form of energy? Defend your opinion.

36. *Group Discussion: Hidden Figures.* The 2017 movie *Hidden Figures* (inspired by the 2016 book by Margot Lee Shetterly) popularized the contributions of Katherine Johnson, Dorothy Vaughan, and Mary Jackson to America's space program.
    **a.** Form groups of three students. Within your group, decide who will research each of the three "hidden figures" listed above, learning about her scientific specialty, how this specialty relates to human understanding of the laws of motion and gravity, and how her work contributed to putting astronauts in space.
    **b.** After completing your individual research, report to the other members of your group on what you have learned.
    **c.** Discuss: In what sense were these individuals pioneers? Is there more than one way in which they can be considered pioneers?

## The Process of Science

*These questions may be answered individually in short-essay form or discussed in groups, except where identified as group-only.*

37. *Testing Gravity.* Scientists are continually trying to learn whether our current understanding of gravity is complete or must be modified. Describe how the observed motion of spacecraft headed out of our solar system (such as the *Voyager* spacecraft) can be used to test the accuracy of our current theory of gravity.
38. *How Does the Table Know?* Thinking deeply about seemingly simple observations sometimes reveals underlying truths that we might otherwise miss. For example, think about holding a golf ball in one hand and a bowling ball in the other. To keep them motionless you must actively adjust the tension in your arm muscles so that each arm exerts a different upward force that exactly balances the weight of each ball. Now, think about what happens when you set the balls on a table. Somehow, the table also exerts exactly the right amount of upward force to keep the balls motionless, even though their weights are very different. How does a table "know" to make the same type of adjustment that you make when you hold the balls motionless in your hands? (*Hint:* Think about the origin of the force pushing upward on the objects.)
39. *Perpetual Motion Machines.* Every so often, someone claims to have built a machine that can generate energy perpetually from nothing. Why isn't this possible according to the known laws of nature? Why do you think claims of perpetual motion machines sometimes receive substantial media attention?
40. *Group Activity: Your Ultimate Energy Source.* Work as a group to answer each part. Note: You may wish to do this activity using the four roles described in Chapter 1, Exercise 39.
    **a.** According to the law of conservation of energy, the energy your body is using right now had to come from somewhere else. Work in groups to make a list going backwards in time describing how the energy you are using right now has proceeded through time.
    **b.** For each item on the group's list, identify the energy as kinetic energy, gravitational potential energy, chemical potential energy, electrical potential energy, mass-energy, or radiative energy.

## Investigate Further

### Short-Answer/Essay Questions

41. *Weightlessness.* Astronauts are weightless when in orbit in the Space Station. Are they also weightless during launch to the station? How about during their return to Earth? Explain.
42. *Units of Acceleration.*
    **a.** If you drop a rock from a very tall building, how fast will it be going after 4 seconds?
    **b.** As you sled down a steep, slick street, you accelerate at a rate of 4 meters per second squared. How fast will you be going after 5 seconds?
    **c.** You are driving along the highway at a speed of 60 miles per hour when you slam on the brakes. If your acceleration is at an average rate of $-20$ miles per hour per second, how long will it take to come to a stop?
43. *Gravitational Potential Energy.* For each of the following, which object has more gravitational potential energy, and how do you know?
    **a.** A bowling ball perched on a cliff ledge or a baseball perched on the same ledge
    **b.** A diver on a 10-meter platform or a diver on a 3-meter diving board
    **c.** A 100-kilogram satellite orbiting Jupiter or a 100-kilogram satellite orbiting Earth (Assume both satellites orbit at the same distance from their planet's center.)
44. *Einstein's Famous Formula.*
    **a.** What is the meaning of the formula $E = mc^2$? Be sure to define each variable.
    **b.** How does this formula explain the generation of energy by the Sun?
    **c.** How does this formula explain the destructive power of nuclear bombs?
45. *The Gravitational Law.*
    **a.** How does quadrupling the distance between two objects affect the gravitational force between them?
    **b.** Suppose the Sun were somehow replaced by a star with twice as much mass. What would happen to the gravitational force between Earth and the Sun?
    **c.** Suppose Earth were moved to one-third of its current distance from the Sun. What would happen to the gravitational force between Earth and the Sun?
46. *Allowable Orbits?*
    **a.** Suppose the Sun were replaced by a star with twice as much mass. Could Earth's orbit stay the same? Why or why not?
    **b.** Suppose Earth doubled in mass (but the Sun stayed the same as it is now). Could Earth's orbit stay the same? Why or why not?
47. *Head-to-Foot Tides.* You and Earth attract each other gravitationally, so you should also be subject to a tidal force resulting from the difference between the gravitational attraction felt by your feet and that felt by your head (at least when you are standing). Explain why you can't feel this tidal force.
48. *Synchronous Rotation.* Suppose the Moon had rotated *more slowly* when it formed than it does now. Would it still have ended up in synchronous rotation? Why or why not?
49. *Geostationary Orbit.* A satellite in *geostationary* orbit appears to remain stationary in the sky as seen from any particular location on Earth.
    **a.** Briefly explain why a geostationary satellite must orbit Earth in 1 *sidereal* day, rather than 1 solar day.
    **b.** Explain why a geostationary satellite must be in orbit around Earth's equator, rather than in some other orbit (such as around the poles).

**c.** Home satellite dishes (such as those used for television) receive signals from communication satellites. Explain why these satellites must be in geostationary orbit.

50. *Space Station Laws.* Visit NASA's website for the International Space Station. Choose two photos that illustrate some facet of Newton's laws of motion or gravity, then briefly explain how these laws can help you understand what each photo shows.

51. *Elevator to Orbit.* Some people have proposed building a giant elevator from Earth's surface to the altitude of geosynchronous orbit. The top of the elevator would then have the same orbital distance and period as any satellite in geosynchronous orbit.
    **a.** Suppose you were to let go of an object at the top of the elevator. Would the object fall? Would it orbit Earth? Explain.
    **b.** Briefly explain why (not counting the huge costs for construction) the elevator would make it much cheaper and easier to put satellites in orbit or to launch spacecraft into deep space.
    **c.** Research ideas for actually building a space elevator. Do you think it is feasible? Do you think it would be a good idea? Explain.

## Quantitative Problems

*Be sure to show all calculations clearly and state your final answers in complete sentences.*

52. *Energy Comparisons.* Use the data in Table 4.1.
    **a.** Compare the energy of a 1-megaton H-bomb to the energy released by a major earthquake.
    **b.** If the United States obtained all its energy from oil, how much oil would be needed each year?
    **c.** Compare the Sun's annual energy output to the energy released by a supernova.

53. *Moving Candy Bar.* Table 4.1 shows that metabolizing a candy bar releases about $10^6$ joules. How fast must the candy bar travel to have the same $10^6$ joules in the form of kinetic energy? Assume the candy bar's mass is 0.2 kg and use the formula kinetic energy $= \frac{1}{2}mv^2$, where $m$ is the object's mass and $v$ is its velocity. Is your answer a faster or slower speed than you expected?

54. *Spontaneous Human Combustion.* Suppose that all the mass in your body were suddenly converted into energy according to the formula $E = mc^2$. How much energy would be released? Compare this to the energy released by a 1-megaton H-bomb (see Table 4.1). What effect would your disappearance have on your surroundings?

55. *Fusion Power.* No one has yet succeeded in creating a commercially viable way to produce energy through nuclear fusion. However, suppose we could build fusion power plants using the hydrogen in water as a fuel. Based on the data in Table 4.1, how much water would we need each minute to meet U.S. energy needs? Could such a reactor power the entire United States with the water flowing from your kitchen faucet? Explain. (*Hint:* Use the annual U.S. energy consumption to find the energy consumption per minute, and then divide by the energy yield from fusing 1 liter of water to figure out how many liters would be needed each minute.)

56. *Understanding Newton's Version of Kepler's Third Law.* Find the orbital period for the planet in each case. (*Hint:* The calculations for this problem are so simple that you will not need a calculator.)
    **a.** A planet with twice Earth's mass orbiting at a distance of 1 AU from a star with the same mass as the Sun
    **b.** A planet with Earth's mass orbiting at a distance of 1 AU from a star with four times the Sun's mass

57. *Using Newton's Version of Kepler's Third Law.*
    **a.** Calculate Earth's approximate mass from the fact that the Moon orbits Earth every 27.3 days at an average distance of 384,000 kilometers.
    **b.** Calculate Jupiter's approximate mass from the fact that its moon Io orbits every 42.5 hours at an average distance of 422,000 kilometers.
    **c.** Calculate the orbital distance for a planet that has an orbital period of 63 days around a star with the mass of the Sun.
    **d.** Calculate the *combined* mass of Pluto and its moon Charon from the fact that Charon orbits Pluto every 6.4 days with a semimajor axis of 19,700 kilometers.
    **e.** Calculate the orbital period of a spacecraft in an orbit 300 kilometers above Earth's surface.
    **f.** Calculate the approximate mass of the Milky Way Galaxy from the fact that the Sun orbits the galactic center every 230 million years at a distance of 27,000 light-years. (As discussed in Chapter 19, this calculation tells us only the mass of the galaxy *within* the Sun's orbit.)

58. *Escape Velocity.* Calculate the escape velocity from each of the following, using data as needed from Appendix E.
    **a.** The surface of Mars
    **b.** The surface of Mars's moon Phobos
    **c.** The surface (cloud tops) of Jupiter
    **d.** Our solar system, starting from Earth's orbit (*Hint:* The mass of our solar system is approximately the same as the mass of the Sun.)
    **e.** Our solar system, starting from Saturn's orbit

59. *Weights on Other Worlds.* Calculate the acceleration of gravity on the surface of each of the following worlds, using data as needed from Appendix E. How much would *you* weigh, in pounds, on each of these worlds?
    **a.** Mars
    **b.** Venus
    **c.** Jupiter's moon Europa
    **d.** Mars's moon Phobos

60. *Gees.* Acceleration is sometimes measured in *gees*, or multiples of the acceleration of gravity: 1 gee (1$g$) means $1 \times g$, or $9.8\ m/s^2$; 2 gees (2$g$) means $2 \times g$, or $2 \times 9.8\ m/s^2 = 19.6\ m/s^2$; and so on. Suppose you experience 6 gees of acceleration in a rocket.
    **a.** What is your acceleration in meters per second squared?
    **b.** You will feel a compression force from the acceleration. How does this force compare to your normal weight?
    **c.** Do you think you could survive this acceleration for long? Explain.

61. *Extra Moon.* Suppose Earth had a second moon, called Swisscheese, with an average orbital distance double the Moon's and a mass about the same as the Moon's.
    **a.** Would Swisscheese's orbital period be longer or shorter than the Moon's? Explain.
    **b.** Use Kepler's third law to find the approximate orbital period of Swisscheese. (*Hint*: You can solve this problem with Kepler's original version of his third law by using the ratios of the orbital periods and distances of the two moons.)
    **c.** Describe how you would expect tides on Earth to differ because of the presence of this second moon. Consider cases in which the two moons are (i) on the same side of Earth, (ii) on opposite sides of Earth, and (iii) 90° apart in their orbits.

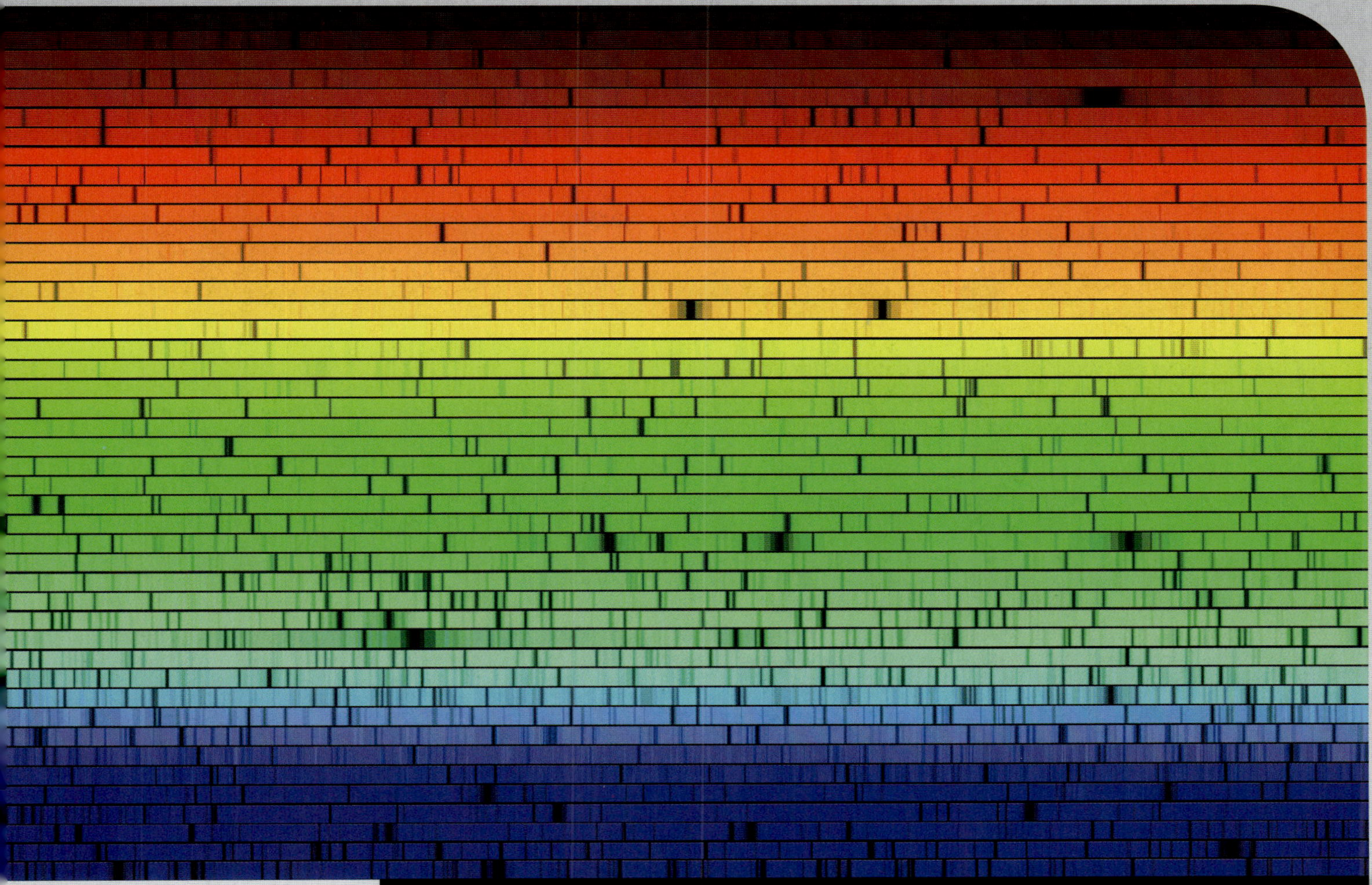

# 5 Light and Matter

## READING MESSAGES FROM THE COSMOS

▲ **About the photo:** The visible light spectrum of the Sun.

## LEARNING GOALS

**5.1 Light in Everyday Life**
- How do we experience light?
- How do light and matter interact?

**5.2 Properties of Light**
- What is light?
- What is the electromagnetic spectrum?

**5.3 Properties of Matter**
- What is the structure of matter?
- What are the phases of matter?
- How is energy stored in atoms?

**5.4 Learning from Light**
- What are the three basic types of spectra?
- How does light tell us what things are made of?
- How does light tell us the temperatures of planets and stars?
- How does light tell us the speed of a distant object?

*May the warp be the white light of morning,*
*May the weft be the red light of evening,*
*May the fringes be the falling rain,*
*May the border be the standing rainbow.*
*Thus weave for us a garment of brightness.*
—*Song of the Sky Loom (Native American)*

**Chapter 5 Overview**

Ancient observers could discern only the most basic features of the light that they saw, such as color and brightness, but we now know that light carries far more information. Today, we can analyze the light of distant objects to learn what they are made of, how hot they are, how fast they are moving, and much more. Light is truly the cosmic messenger, bringing the stories of distant objects to Earth.

In this chapter, we will focus our attention on learning how to read the messages carried by light. We'll begin with a brief look at the basic interactions of light and matter that create those messages, and then study the properties of light and matter. With that background, we'll be ready to explore how a spectrum forms, so that we can understand how light can encode so much information about distant objects.

## 5.1 Light in Everyday Life

What do you see as you look around you? You may be tempted to list nearby objects, but all you're really seeing is *light* that has interacted with those objects. Through intuition and experience, you're able to interpret the colors and patterns of the light and turn them into information about the objects and substances that surround you.

Astronomers study the universe in much the same way. Telescopes collect the light of distant objects, and we use the light to extract information about those objects. The more we understand light and its interactions with matter, the more information we can extract. As a first step in developing this understanding, let's take a closer look at our everyday experience with light.

### How do we experience light?

You can tell that light is a form of energy even without opening your eyes. Outside on a hot, sunny day you can feel your skin warm as it absorbs sunlight. Because greater warmth means more molecular motion, sunlight must be transferring its energy to the molecules in your skin. The energy that light carries is called *radiative energy;* recall that it is one of the three basic categories of energy, along with kinetic and potential energy (see Figure 4.11).

**Energy and Power** Recall that we measure energy in units of *joules* [**Section 4.3**]. With light, however, we are usually more interested in the *rate* at which light transfers energy than in the total amount of energy it carries. After all, we cannot hold light in our hands in the same way that we can hold a rock, which has both thermal energy and gravitational potential energy. The rate of energy flow is called **power**, which we measure in units called **watts**. A power of 1 watt means an energy flow of 1 joule per second:

$$1 \text{ watt} = 1 \text{ joule/s}$$

For example, a 100-watt light bulb requires 100 joules of energy (which you buy from the electric company) for each second it is turned on. Interestingly, the power requirement of an average human—about 10 million joules per day—is close to that of a 100-watt light bulb.

**Light and Color** Everyday experience tells us that light comes in different forms that we call *colors*. You've probably seen a prism split light into the rainbow of light called a **spectrum** (**FIGURE 5.1**), in which the basic colors are red, orange, yellow, green, blue, and violet. We see *white* when these colors are mixed in roughly equal proportions. Light from the Sun or a light bulb is often called *white light,* because it contains all the colors of the rainbow. *Black* is what we perceive when there is no light and hence no color.

The wide variety of all possible colors comes from mixtures of just a few colors in varying proportions. Your television takes advantage of this fact to simulate a huge range of colors by combining only red, green, and blue light; these three colors are often called the *primary colors of vision,* because they are the colors directly detected by cells in your eyes. Colors tend to look different on paper, so artists generally work with an alternative set of primary colors: red, yellow, and blue. If you do graphic design or book printing, you may use the CMYK process, in which the four colors cyan, magenta, yellow, and black are mixed to produce the great variety of colors.

**See it for yourself** Try splashing a tiny amount of water (only a few drops!) onto your TV screen and look closely at the droplets. Relate what you see to the fact that televisions combine red, green, and blue to make all their colors.

You can produce a spectrum with either a prism or a **diffraction grating**, which is a piece of plastic or glass etched with many closely spaced lines. If you have a DVD

**FIGURE 5.1** When we pass white light through a prism, it disperses into a rainbow of color that we call a *spectrum*.

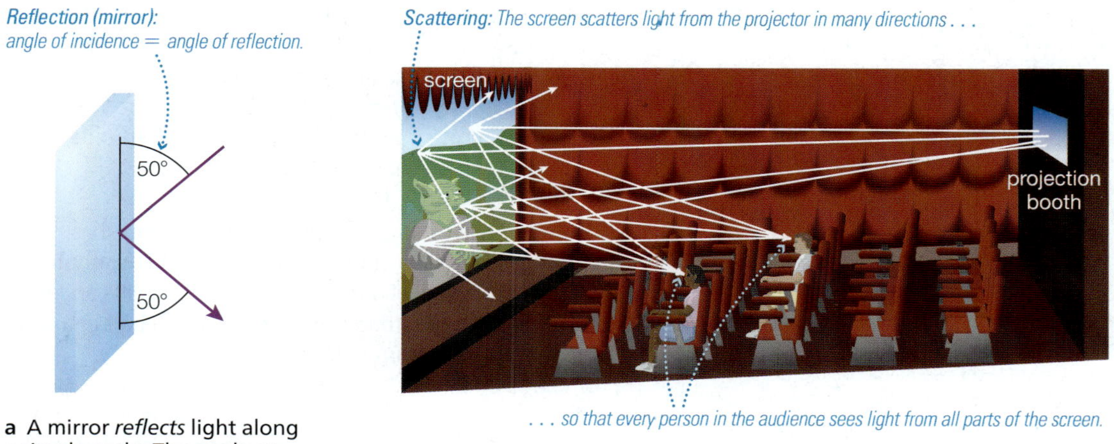


**a** A mirror *reflects* light along a simple path: The angle at which the light strikes the mirror is the same angle at which it is reflected.

**b** A movie screen *scatters* light in many different directions, so that each member of the audience can watch the movie. The pages in a book do the same thing, which is why you can read them from different angles and distances.

**FIGURE 5.2** Reflection and scattering.

(or similar disc) handy, you can make a spectrum for yourself. The disc is etched with many closely spaced circles that act like a diffraction grating, which is why you see rainbows of color when you hold it up to light.

## How do light and matter interact?

Light can interact with matter in four basic ways, all of which are familiar in everyday life:

- **Emission**: A light bulb *emits* visible light; the energy of the light comes from electrical potential energy supplied to the light bulb.
- **Absorption**: When you place your hand near an incandescent light bulb, your hand *absorbs* some of the light, and this absorbed energy warms your hand.
- **Transmission**: Some forms of matter, such as glass or air, *transmit* light, allowing it to pass through.
- **Reflection/scattering**: Light can bounce off matter, leading to what we call *reflection* when the bouncing is all in the same general direction or *scattering* when the bouncing is more random (**FIGURE 5.2**).

Materials that transmit light are said to be *transparent*, and materials that absorb light are called *opaque*. Many materials are neither perfectly transparent nor perfectly opaque. For example, dark sunglasses and clear eyeglasses are both partially transparent, but the dark glasses absorb more and transmit less light. Materials often interact differently with different colors of light. For example, red glass transmits red light but absorbs other colors, while a green lawn reflects (scatters) green light but absorbs all other colors.

Let's put these ideas together to understand what happens when you walk into a room and turn on the light switch (**FIGURE 5.3**). The light bulb begins to emit white light, which is a mix of all the colors in the spectrum. Some

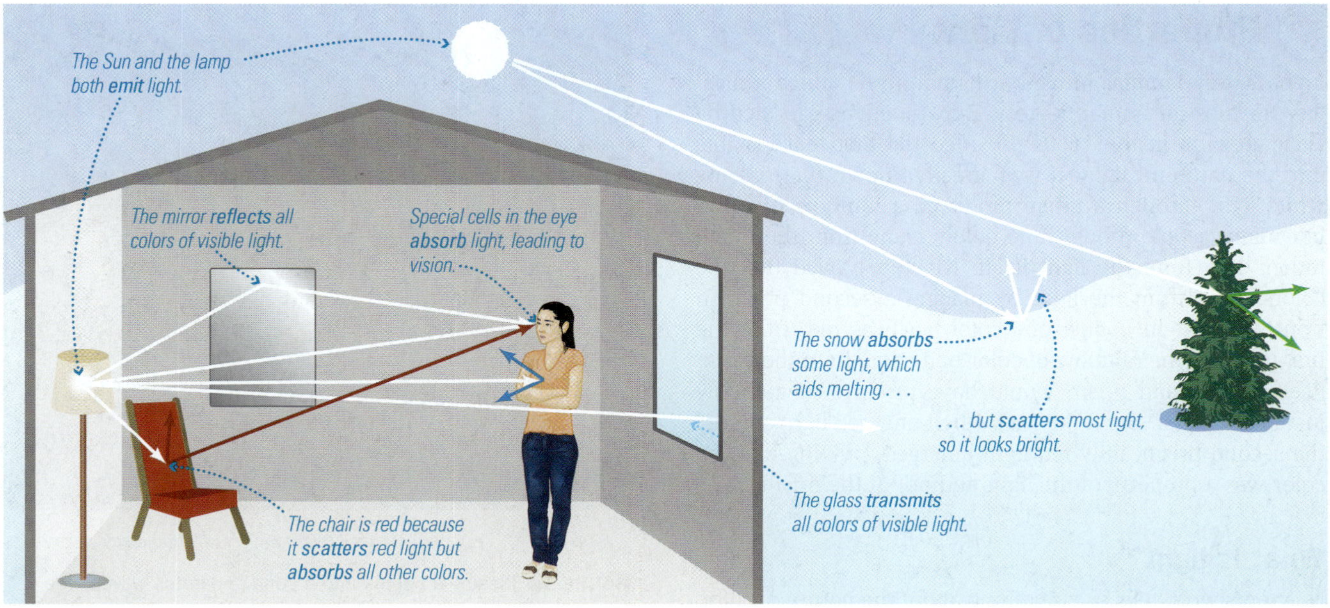


**FIGURE 5.3** This diagram shows examples of the four basic interactions between light and matter: emission, absorption, transmission, and reflection (or scattering).

## COMMON MISCONCEPTIONS

### Light Paths, Lasers, and Shadows

We often think of light as traveling in straight-line "rays," much as laser light appears to do. However, light goes perfectly straight only when there is no matter for it to interact with, which is *not* the general case on Earth. For example, have you ever wondered how daylight comes in through a window even when the Sun is not shining directly in? The answer is that air molecules scatter some of the visible light coming from the Sun, bouncing it in through your windows. The same idea explains why our daytime sky is bright and why shadows are not pitch black; in space, where there is no air to scatter light, shadows are pitch black and so is the sky. In fact, even laser beams are visible to our eyes only because some of their light is scattered toward us by air, which means that you would not see laser light in space unless it was aimed directly at you—a fact that most science fiction movies get wrong!

of this light exits the room, transmitted through the windows. The rest of the light strikes the surfaces of objects inside the room, and the material properties of each object determine the colors it absorbs or reflects. The light coming from each object therefore carries an enormous amount of information about the object's location, shape and structure, and composition. You acquire this information when light enters your eyes, where special cells in your retina absorb it and send signals to your brain. Your brain interprets the messages that light carries, recognizing materials and objects in the process we call *vision*.

All the information that light brings us from the cosmos was encoded by the same four basic interactions between light and matter common to our everyday experience. However, our eyes perceive only a tiny fraction of all the information contained in light. Modern instruments can break light into a much wider variety of colors and can analyze those colors in far greater detail. In order to understand how to decode that information, we need to examine the nature of light and matter more closely.

## 5.2 Properties of Light

Light is very familiar to us, but its nature remained a mystery for most of human history. Experiments performed by Isaac Newton in the 1660s provided the first real insights into the nature of light. It was already known that passing white light through a prism produced a rainbow of color, but many people thought the colors came from the prism rather than from the light itself. Newton proved that the colors came from the light by placing a second prism in front of the light of just one color, such as red, from the first prism. If the rainbow of color had come from the prism itself, the second prism would have produced a rainbow just like the first. But it did not: When only red light entered the second prism, only red light emerged, proving that the color was a property of the light and not of the prism.

### What is light?

Newton's work tells us something about the nature of color, but it still does not tell us exactly what light *is*. Newton himself guessed that light was made up of countless tiny particles. However, other scientists soon conducted experiments demonstrating that light behaves like waves. Thus began one of the most important debates in scientific history: Is light a wave or a particle? To understand this question, and our modern answer to it, we must first understand the differences between particles and waves.

**Particles and Waves in Everyday Life** Marbles, baseballs, and individual atoms and molecules are all examples of *particles*. A particle of matter can sit still or it can move from one place to another. If you throw a baseball at a wall, it obviously travels from your hand to the wall.

In contrast, think about what happens when you toss a pebble into a pond, creating a set of outward moving ripples, or *waves* (**FIGURE 5.4**). These waves consist of *peaks*, where the water is higher than average, and *troughs*, where the water is lower than average. If you watch as the waves pass by a floating leaf, you'll see the leaf rise up with each peak and drop down with each trough, but the leaf itself will *not* travel across the pond's surface with the wave. We conclude that even though the waves are moving outward, the particles (molecules) that make up the water are moving primarily up and down (along with a bit of sloshing back and forth). That is, the waves carry *energy* outward from the place where the pebble landed but do not carry matter along with them. In essence, a particle is a *thing*, while a wave is a *pattern* revealed by its interaction with particles.

Let's focus on three basic properties of waves: wavelength, frequency, and speed.* **Wavelength** is the distance from one peak to the next (or one trough to the next). **Frequency** is the number of peaks passing by any point each second. For example, if the leaf bobs up and down three times each second, then three peaks must be passing

*There is also a fourth wave property, *amplitude*, defined as half the height from trough to peak. Amplitude is related to the brightness of light.

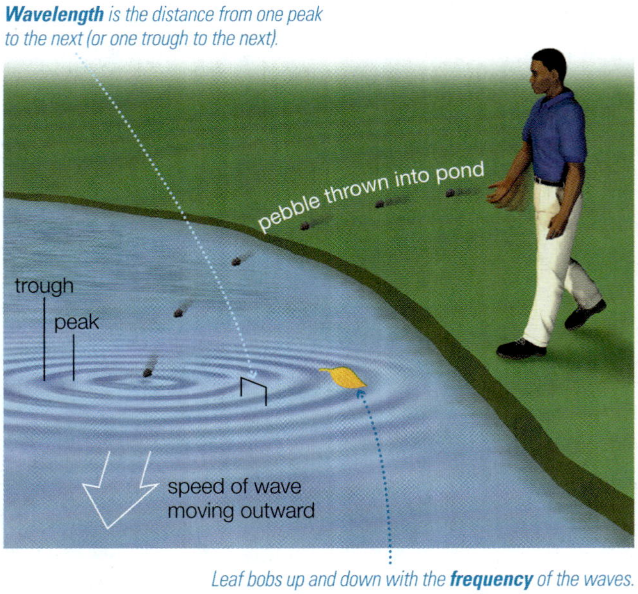


**FIGURE 5.4** Tossing a pebble into a pond generates waves. The waves carry energy outward, but matter, such as a floating leaf and the molecules of the water, only bobs up and down (with a bit of sloshing back and forth) as the waves pass by.

by it each second, which means the waves have a frequency of three **cycles per second**. "Cycles per second" are often called **hertz (Hz)**, so we can also describe this frequency as 3 Hz. The **speed** of the waves tells us how fast their peaks travel across the pond. Because the waves carry energy, the speed essentially tells us how fast the energy travels from one place to another.

A simple formula relates the wavelength, frequency, and speed of any wave. Suppose a wave has a wavelength of 1 centimeter and a frequency of 3 hertz. The wavelength tells us that each time a peak passes by, the wave peak has traveled 1 centimeter. The frequency tells us that three peaks pass by each second. The speed of the wave must therefore be 3 centimeters per second. If you try a few more similar examples, you'll find the general rule

$$\text{wavelength} \times \text{frequency} = \text{speed}$$

**Light as an Electromagnetic Wave** You've probably heard that light is a wave, but it isn't quite like the waves we see in everyday life. More familiar waves always move through some form of matter. For example, the waves on the pond move through the water, causing particles (molecules) of water to vibrate up and down and slosh back and forth, while sound waves move through air, causing air molecules to vibrate back and forth. The vibrations of matter allow the waves to transmit energy from one place to another, even though particles of matter do not travel along with the waves. In contrast to these everyday examples of waves, we do not see anything move up and down when light travels through space. So what, exactly, is "waving" when a light wave passes by?

The answer is what scientists call electric and magnetic fields. The concept of a **field** is a bit abstract, but it is used to describe the strength of force that a particle would experience at any point in space. For example, Earth creates a *gravitational field* that describes the strength of gravity at any distance from Earth, which means that the strength of the field declines with the square of the distance from Earth's center [**Section 4.4**]. Electricity and magnetism also create forces, so their strength in different places can be described in terms of *electric fields* and *magnetic fields*.

Light waves are traveling vibrations of both electric and magnetic fields, so we say that light is an **electromagnetic wave**. Just as the ripples on a pond will cause a leaf to bob up and down, the vibrations of the electric field in an electromagnetic wave will cause any charged particle, such as an electron, to bob up and down. If you could set up electrons in a row, they would wriggle like a snake as light passed by (**FIGURE 5.5a**). The distance between peaks in this row of electrons would tell us the wavelength of the light wave, while the number of times each electron bobbed up and down would tell us the frequency (**FIGURE 5.5b**).

All light travels through empty space at the same speed—the **speed of light** (represented by the letter $c$)—which is about 300,000 kilometers per second. Because the speed of any wave is its wavelength times its frequency, we find a very important relationship between wavelength and frequency for light: *The longer the wavelength, the lower the frequency, and the shorter the wavelength, the higher the frequency*. For example, light waves with a wavelength of 1 centimeter must have half the frequency of light waves with a wavelength of $\frac{1}{2}$ centimeter and one-fourth the frequency of light waves with a wavelength of $\frac{1}{4}$ centimeter (**FIGURE 5.6**).

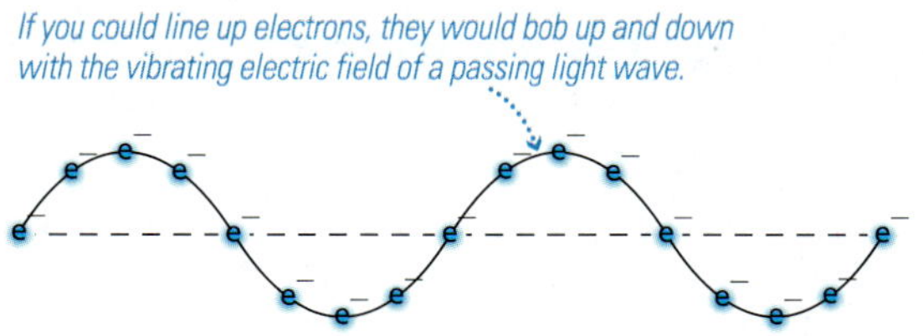


**a** Electrons move when light passes by, showing that light carries a vibrating electric field.

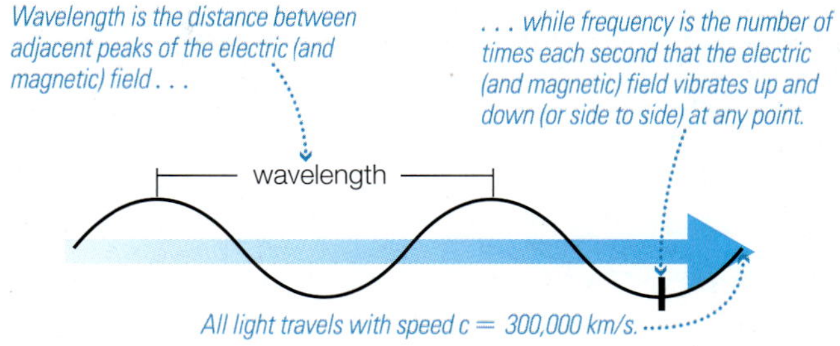


**b** The vibrations of the electric field determine the wavelength and frequency of a light wave. Light also has a magnetic field (not shown) that vibrates perpendicular to the direction of the electric field vibrations.

**FIGURE 5.5** Light is an electromagnetic wave.

**Photons: "Particles" of Light** Waves and particles appear distinctly different in everyday life. For example, no one would confuse the ripples on a pond with a baseball. However, experiments have shown that light behaves as *both* a wave and a particle. We say that light comes in individual "pieces," called **photons**, that have properties of both particles and waves. Like baseballs, photons of light can be counted individually and can hit a wall one at a time. Like waves, each photon is characterized by a wavelength and

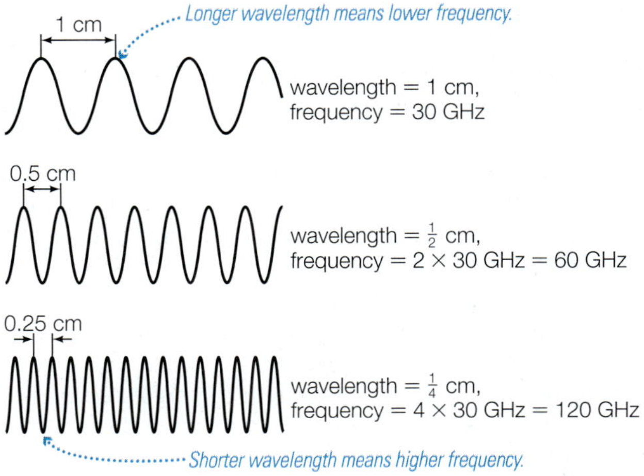


**FIGURE 5.6** Because all light travels through space at the same speed, light of longer wavelength must have lower frequency, and shorter wavelength light has higher frequency. (GHz stands for gigahertz, or $10^9$ Hz.)

## COMMON MISCONCEPTIONS

### Is Radiation Dangerous?

Many people associate the word *radiation* with danger. However, the word *radiate* simply means "to spread out from a center" (note the similarity between *radiation* and *radius* [of a circle]), and *radiation* is just a term to describe energy carried through space. Energy carried by particles of matter, such as protons or neutrons, is called *particle radiation*. Energy carried by light is called *electromagnetic radiation*. In general, radiation is dangerous only if it has high energies that allow it to penetrate and cause damage to body tissues. Lower-energy forms of radiation, such as radio waves, are usually harmless, and visible light radiation from the Sun is necessary to life on Earth. So while some forms of radiation are dangerous, others are harmless or beneficial.

a frequency. The idea that light can be both a wave and a particle may seem quite strange, but it is fundamental to our modern understanding of physics. (We will discuss some of the implications of this *wave-particle duality* in Chapter S4.)

Just as a moving baseball carries a specific amount of kinetic energy, each photon of light carries a specific amount of radiative energy. The shorter the wavelength of the light (or, equivalently, the higher its frequency), the higher the energy of the photons.

To sum up, our modern understanding maintains that (1) light is both a particle and a wave, an idea we describe by saying that light consists of individual photons characterized by wavelength, frequency, and energy, and (2) the wavelength, frequency, and energy of light are simply related because all photons travel through space at the same speed—the speed of light.

**Think about it** Suppose that each of the three waves shown in Figure 5.6 represents a photon of light. Which one has the most energy? Which one has the least energy? Explain.

**The Electromagnetic Spectrum**

## What is the electromagnetic spectrum?

Newton's experiments proved that white light is a mix of all the colors in the rainbow. Later scientists found that just as there are sounds our ears cannot hear (such as the sound of a dog whistle), there is light "beyond the rainbow" that our eyes cannot see. In fact, the light that we can see is only a tiny part of the complete spectrum of light, usually called the **electromagnetic spectrum**; light itself is often called **electromagnetic radiation**. **FIGURE 5.7** shows the way the electromagnetic spectrum is commonly divided into regions according to wavelength (or, equivalently, frequency or energy). Keep in mind that despite the different names, everything in the electromagnetic spectrum represents a form of light and therefore consists of photons that travel through space at the speed of light.

The light that our eyes can see, which we call **visible light**, is found near the middle of the spectrum, with wavelengths ranging from about 400 nanometers at the blue or violet end of the rainbow to about 700 nanometers at the red end. (A nanometer [nm] is a billionth of a meter.) Light with wavelengths somewhat longer than those of red light is called **infrared**, because it lies beyond the red end of the rainbow. **Radio waves** are the longest-wavelength light. The region near the border between infrared and radio waves, where wavelengths range from micrometers to centimeters, is often called **microwaves**. In astronomy, microwaves are sometimes divided further: Wavelengths from about one to a few millimeters are called *millimeter waves,* while wavelengths of tenths of a millimeter are called *submillimeter waves*.

On the other side of the spectrum, light with wavelengths somewhat shorter than those of blue light is called **ultraviolet**, because it lies beyond the blue (or violet) end of the rainbow. Light with even shorter wavelengths is called **x-rays**, and the shortest-wavelength light is called **gamma rays**. Notice that visible light is an extremely small part of the entire electromagnetic spectrum: The reddest red that our eyes can see has only about twice the wavelength of the bluest blue, but the radio waves from your favorite radio station are a billion times longer than the x-rays used in a doctor's office.

The various energies of light explain many familiar effects in everyday life. Radio waves carry so little energy that they have no noticeable effect on our bodies, but they can make electrons move up and down in an antenna, making them useful for radio communication. Molecules moving in a warm object emit infrared light, which is why we sometimes associate infrared light with heat. Receptors in our eyes respond to visible-light photons, making vision possible. Ultraviolet photons carry enough energy to damage skin cells, causing sunburn or cancer. X-ray photons have enough energy to penetrate through skin and muscle but can be blocked by bones or teeth, which is why they can be used to make images of bone or tooth structures.

## COMMON MISCONCEPTIONS

### Can You Hear Radio Waves or See an X-Ray?

Most people associate the term *radio* with sound, but radio waves are a form of *light* with wavelengths too long for our eyes to see. Radio stations encode sounds (such as voices and music) as electrical signals and broadcast the signals as radio waves. What we call "a radio" in daily life is an electronic device that receives these radio waves and decodes them to re-create the sounds played at the radio station. Televisions, cell phones, and other wireless devices also work by encoding and decoding information in the form of light called radio waves.

X-rays are also a form of light, with wavelengths far too short for our eyes to see. In a medical office, a special machine works somewhat like the flash on an ordinary camera but emits x-rays instead of visible light. This machine flashes the x-rays at you, and a detector placed behind you records the x-rays transmitted through your body. You never see the x-rays themselves—you see only the image recorded by the detector. Incidentally, this means that Superman-like "x-ray vision" would be pretty worthless, even if it were possible: People, walls, and other ordinary objects do not emit any x-rays of their own, so there'd be nothing to see with your x-ray vision.

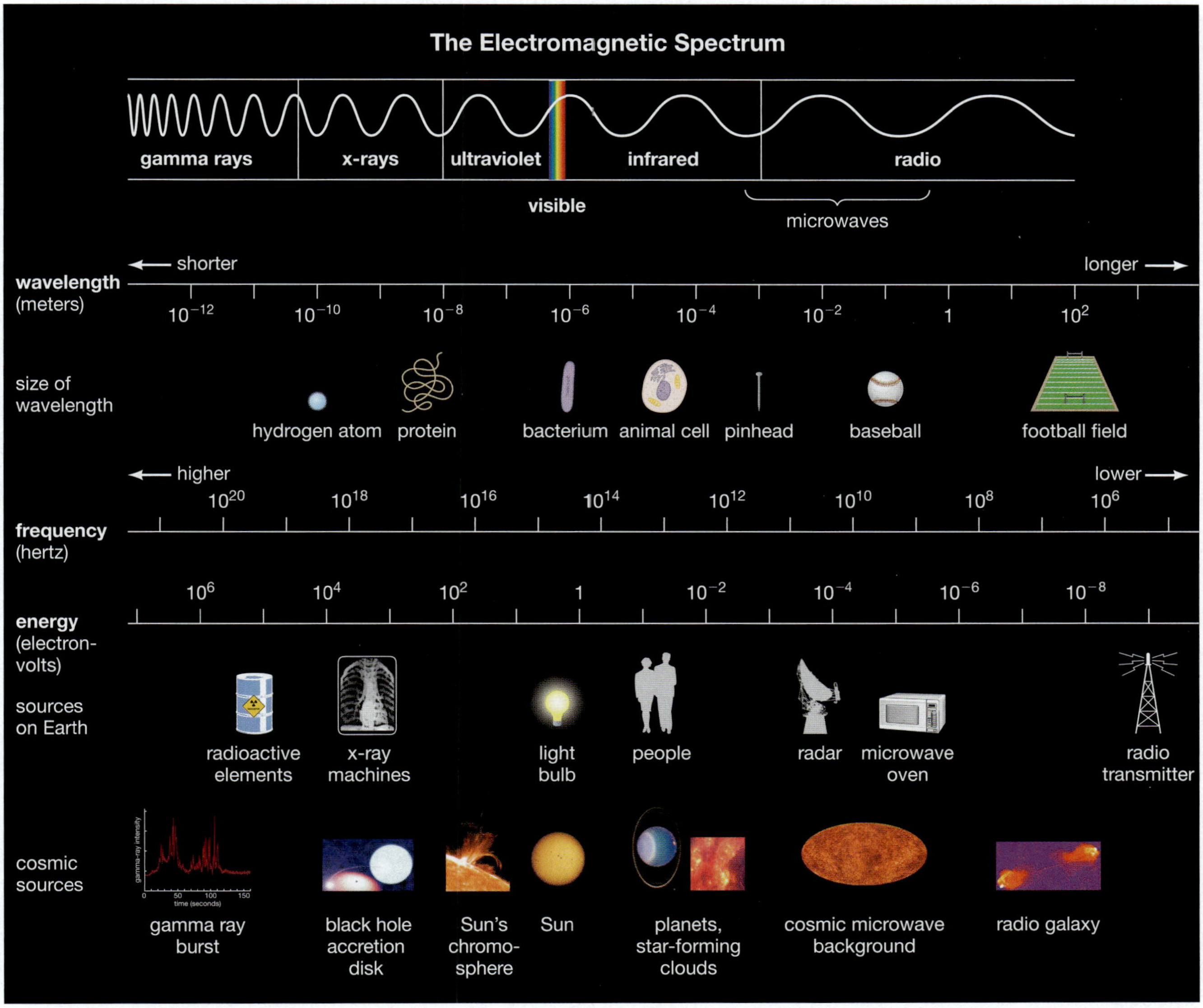


**FIGURE 5.7** The electromagnetic spectrum. Notice that wavelength increases as we go from gamma rays to radio waves, while frequency and energy increase in the opposite direction. (Energy is given in units of *electron-volts,* eV: $1 \text{ eV} = 1.60 \times 10^{-19}$ joule.)

Just as different colors of visible light may be absorbed or reflected differently by the objects we see (see Figure 5.3), the various portions of the electromagnetic spectrum may interact with matter in very different ways. For example, a brick wall is opaque to visible light but transmits radio waves, which is why radios and cell phones work inside buildings. Similarly, glass that is transparent to visible light may be opaque to ultraviolet light. In general, certain types of matter tend to interact more strongly with certain types of light, so each type of light carries different information about distant objects in the universe. That is why astronomers seek to observe light of all wavelengths [**Section 6.4**].

## 5.3 Properties of Matter

Light carries information about matter across the universe, but we are usually more interested in the matter the light is coming from—such as planets, stars, and galaxies—than we are in the light itself. We must therefore explore the nature of matter if we are to decode the messages carried by light.

### What is the structure of matter?

Like the nature of light, the nature of matter remained mysterious for most of human history. Nevertheless, ancient philosophers came up with some ideas that are still with us today.

The ancient Greek philosopher Democritus (c. 470–380 B.C.) wondered what would happen if we could break a piece of matter, such as a rock, into ever smaller pieces. He claimed that the rock would eventually break into particles so small that nothing smaller could be possible. He called these particles *atoms*, a Greek term meaning "indivisible." Building on the beliefs of earlier Greek philosophers, Democritus assumed that all materials were composed from four basic *elements:* fire, water, earth, and air. He proposed that the properties of different elements could be explained by the physical characteristics of their atoms. For example, Democritus suggested that atoms of water were smooth and round, so water flowed and had no fixed shape, while burns were painful because atoms of fire were thorny. He imagined atoms of earth to be rough and jagged, so they could fit together like pieces of a three-dimensional jigsaw puzzle, and he used this idea to suggest that the universe began as a chaotic mix of atoms that slowly clumped together to form our world.

Although Democritus was wrong in his specifics, he was on the right track. All ordinary matter is indeed composed of **atoms**, and the properties of ordinary matter depend on the physical characteristics of its atoms. However, by modern definition, atoms are *not* indivisible because they are composed of even smaller particles.

Atoms come in different types, and each type corresponds to a different chemical **element**. Scientists have identified more than 100 chemical elements, and fire, water, earth, and air are *not* among them. Some of the most familiar

## MATHEMATICAL INSIGHT 5.1 Wavelength, Frequency, and Energy

The relationship *wavelength* × *frequency* = *speed* holds for any wave. For light, which travels (in a vacuum) at speed $c = 3 \times 10^8$ m/s, this relationship becomes

$$\lambda \times f = c$$

where $\lambda$ (the Greek letter *lambda*) stands for wavelength and $f$ stands for frequency. Note that, because $c$ is a constant, frequency must go up when wavelength goes down, and vice versa.

The radiative energy ($E$) carried by a photon of light is given by

$$E = h \times f$$

where $h$ is *Planck's constant* ($h = 6.626 \times 10^{-34}$ joule × s). Energy therefore increases with frequency.

**EXAMPLE 1:** A radio station at 93.3 FM broadcasts radio waves with a frequency of 93.3 megahertz (MHz). What is their wavelength?

**SOLUTION:**

**Step 1 Understand:** Radio waves are a form of light, so they obey the relationship $\lambda \times f = c$. We are given the frequency ($f$) and know the speed of light ($c$), so we can simply solve for the wavelength ($\lambda$). There is one subtlety: The units of frequency, called hertz or "cycles per second," are really just "per second," or 1/s; the reason is that "cycles" is just a descriptive term, with no units itself. So we write 93.3 megahertz as $93.3 \times 10^6$ 1/s.

**Step 2 Solve:** We solve for wavelength by dividing both sides of $\lambda \times f = c$ by $f$, which gives $\lambda = c/f$. We now plug in the speed of light and the frequency to find

$$\lambda = \frac{c}{f} = \frac{3 \times 10^8 \frac{\text{m}}{\cancel{\text{s}}}}{93.3 \times 10^6 \frac{1}{\cancel{\text{s}}}} = 3.2\ \text{m}$$

**Step 3 Explain:** Radio waves with a frequency of 93.3 MHz have a wavelength of 3.2 meters. That is why radio antennas need to be fairly long to transmit or receive these waves.

**EXAMPLE 2:** The middle of the visible spectrum is green light with a wavelength of about 550 nanometers. What is its frequency?

**SOLUTION:**

**Step 1 Understand:** All light obeys the relation $\lambda \times f = c$. In this case we are given the wavelength, so we simply solve the equation for the frequency.

**Step 2 Solve:** Dividing both sides of the equation $\lambda \times f = c$ by $\lambda$ gives $f = c/\lambda$. We plug in the speed of light and the wavelength ($\lambda = 550 \times 10^{-9}$ m) to find

$$f = \frac{c}{\lambda} = \frac{3 \times 10^8 \frac{\cancel{\text{m}}}{\text{s}}}{550 \times 10^{-9}\ \cancel{\text{m}}} = 5.45 \times 10^{14} \frac{1}{\text{s}}$$

**Step 3 Explain:** Green visible light has a frequency of about $5.5 \times 10^{14}$ 1/s, which is $5.5 \times 10^{14}$ Hz, or 550 trillion Hz. This high frequency is one reason the wave properties of light are not obvious in everyday life.

**EXAMPLE 3:** What is the energy of a visible-light photon with a wavelength of 550 nanometers?

**SOLUTION:**

**Step 1 Understand:** The energy of a photon is $E = h \times f$. We are given the photon's wavelength rather than frequency, so we use the fact that $f = c/\lambda$ to write

$$E = h \times f = h \times \frac{c}{\lambda}$$

**Step 2 Solve:** We plug in the wavelength and Planck's constant to find

$$\begin{aligned} E &= h \times \frac{c}{\lambda} \\ &= (6.626 \times 10^{-34}\ \text{joule} \times \cancel{\text{s}}) \times \frac{3 \times 10^8 \frac{\cancel{\text{m}}}{\cancel{\text{s}}}}{550 \times 10^{-9}\ \cancel{\text{m}}} \\ &= 3.6 \times 10^{-19}\ \text{joule} \end{aligned}$$

**Step 3 Explain:** The energy of a single visible-light photon is about $3.6 \times 10^{-19}$ joule. Note that this is barely a *billion-trillionth* of the 100 joules of energy needed each second by a 100-watt light bulb.

chemical elements are hydrogen, helium, carbon, oxygen, silicon, iron, gold, silver, lead, and uranium. Appendix D gives the periodic table of all the elements.

**Atomic Structure** Atoms are made of particles that we call **protons**, **neutrons**, and **electrons** (**FIGURE 5.8**). Protons and neutrons are found in the tiny **nucleus** at the center of the atom. The rest of the atom's volume contains the electrons that surround the nucleus. Although the nucleus is very small compared to the atom as a whole, it contains most of the atom's mass, because protons and neutrons are each about 2000 times as massive as an electron. Note that atoms are incredibly small: Millions could fit end to end across the period at the end of this sentence. The number of atoms in a single drop of water (typically, $10^{22}$ to $10^{23}$ atoms) may exceed the number of stars in the observable universe.

The properties of an atom depend mainly on the **electrical charge** in its nucleus. Electrical charge is a fundamental physical property that describes how strongly an object will interact with electromagnetic fields; total electrical charge is always conserved, just as energy is always conserved. We define the electrical charge of a proton as the basic unit of positive charge, which we write as +1. An electron has an electrical charge that is precisely opposite that of a proton, so we say it has negative charge (−1). Neutrons are electrically neutral, meaning that they have no charge.

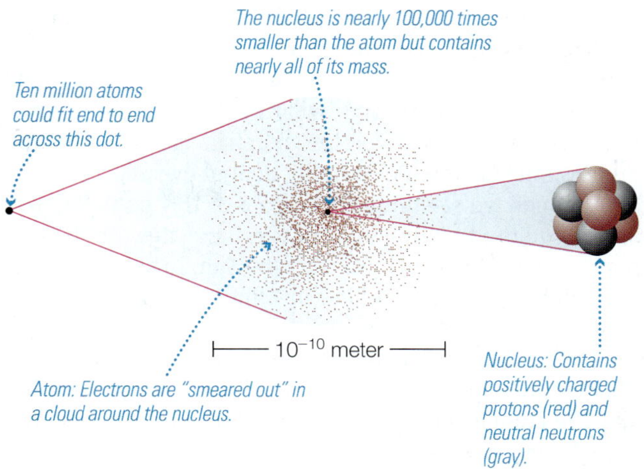


**FIGURE 5.8** The structure of a typical atom. Note that atoms are extremely tiny: The atom shown in the middle is magnified to about 1 billion times its actual size, and the nucleus on the right is magnified to about 100 trillion times its actual size.

## SPECIAL TOPIC What Do Polarized Sunglasses Have to Do with Astronomy?

If you go to the store to buy a pair of sunglasses, you'll face a dizzying array of choices. Sunglasses come in different styles and different tints and with different efficiencies in blocking ultraviolet and infrared light. Most of these choices should make sense to you (well, perhaps not all of the styles), but one option may not be familiar: The labels on some sunglasses say that they are "polarized." What does this mean? The term comes from a property of light, called *polarization*, that has to do with the direction in which a light wave vibrates and how those vibrations change when light bounces off or passes through matter. Polarization is important not only to sunglasses but also to astronomy.

To explore this idea, think about how waves move on a string when you shake one end of it. The string vibrates either up and down or back and forth while the wave itself moves along it in a direction perpendicular to the direction of vibration. Light waves move in a similar way, with the electric and magnetic fields vibrating either up and down or side to side compared to the direction of travel. For example, the wave shown in Figure 5.5 is moving to the right on the page while its electric field vibrates up and down on the page.

The direction of vibration affects the way light interacts with matter. As Figure 5.5 indicates, an electric field that vibrates up and down will make electrons move up and down as the wave passes by. That is, the direction in which the electric field vibrates determines the direction in which charged particles vibrate as the wave passes by. Because the direction of wave vibration matters, we give it a name: the *polarization* of the wave. An individual wave moving toward you can be polarized with its vibrations either up and down or side to side or some combination of those two.

Each light wave (or, more technically, each individual photon) has a particular direction of polarization, although our eyes do not detect it. If all the waves taken together have no preferential direction of vibration, we say that the light is *unpolarized*. However, some physical processes produce waves with a particular direction of polarization, which is where your sunglasses and astronomy come in.

When light reflects off a flat horizontal surface like the ground or a lake, all the reflected light tends to have its electric field vibrating horizontally. (Light with other directions of vibration is absorbed or transmitted.) In other words, the reflected light is horizontally polarized. Polarized sunglasses are designed to block light with horizontal polarization, which is often the cause of "glare." Of course, the polarized glasses work only if you are wearing them horizontally; if you turn a pair of polarized sunglasses so that the two lenses are no longer horizontal to the ground, they will not block glare effectively.

In astronomy, we aren't worried about glare from distant objects, but if we learn that a light source is producing polarized light, this can tell us something about the nature of the source. For example, light that passes through clouds of interstellar dust tends to be polarized, telling us that the dust grains in the cloud must be preferentially absorbing light with electric fields vibrating in a particular direction. More detailed analysis has taught us that the polarization arises because the microscopic dust grains have an elongated shape and tend to be aligned in the same way as a result of magnetic fields within the clouds. Polarization arises in many other astronomical contexts as well, including the study of the leftover radiation from the Big Bang. Although polarization has provided important insights into many astronomical processes, its analysis can be fairly technical, so we will not discuss it much further in this book.

Oppositely charged particles attract and similarly charged particles repel. The attraction between the positively charged protons in the nucleus and the negatively charged electrons that surround it is what holds an atom together. Ordinary atoms have identical numbers of electrons and protons, making them electrically neutral overall.*

Although we can think of electrons as tiny particles, they are not quite like tiny grains of sand and they don't orbit the nucleus the way planets orbit the Sun. Instead, the electrons in an atom form a kind of "smeared out" cloud that surrounds the nucleus and gives the atom its apparent size. The electrons aren't really cloudy, but it is impossible to pinpoint their positions in the atom. The electrons therefore give the atom a size far larger than that of its nucleus even though they represent only a tiny portion of the atom's mass. If you imagine an atom on a scale that makes its nucleus the size of your fist, its electron cloud would be many kilometers wide.

**Atomic Terminology** You've probably learned the basic terminology of atoms in past science classes, but let's review it just to be sure. **FIGURE 5.9** summarizes the key terminology we will use in this book.

Each different chemical element contains a different number of protons in its nucleus. This number is its **atomic number**. For example, a hydrogen nucleus contains just one proton, so its atomic number is 1. A helium nucleus contains two protons, so its atomic number is 2. The *combined* number of protons and neutrons in an atom is called its **atomic mass number**. The atomic mass number of ordinary hydrogen is 1 because its nucleus is just a single proton. Helium usually has two neutrons in addition to its two protons, giving it an atomic mass number of 4. Carbon usually has six protons and six neutrons, giving it an atomic mass number of 12.

*You may wonder why electrical repulsion doesn't cause the positively charged protons in a nucleus to fly apart from one another. The answer is that an even stronger force, called the *strong force*, overcomes electrical repulsion and holds the nucleus together [**Section S4.2**].

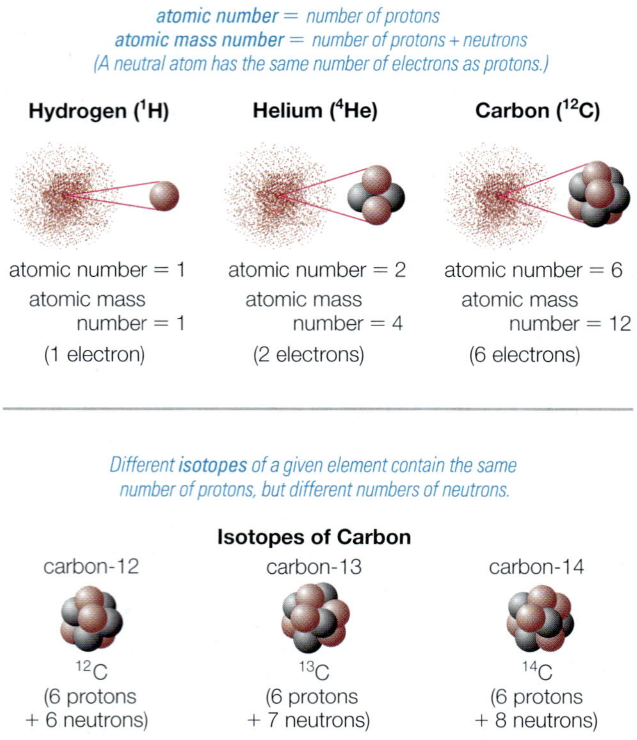


**FIGURE 5.9** Terminology of atoms.

## COMMON MISCONCEPTIONS

### The Illusion of Solidity

Bang your hand on a table. Although the table feels solid, it is made almost entirely of empty space! Nearly all the mass of the table is contained in the nuclei of its atoms. But the volume of each atom is more than a trillion times the volume of its nucleus, so the nuclei of adjacent atoms are nowhere near to touching one another. The solidity of the table comes from a combination of electrical interactions between the charged particles in its atoms and the strange quantum laws governing the behavior of electrons. If we could somehow pack all the table's nuclei together, the table's mass would fit into a microscopic speck. Although we cannot pack matter together in this way, nature can and does—in *neutron stars*, which we will study in Chapter 18.

Every atom of a given element contains exactly the same number of protons, but the number of neutrons can vary. For example, all carbon atoms have six protons, but they may have six, seven, or eight neutrons. Versions of an element with different numbers of neutrons are called **isotopes** of that element. Isotopes are named with their element name and atomic mass number. For example, the most common isotope of carbon has six protons and six neutrons, giving it atomic mass number $6 + 6 = 12$, so we call it *carbon-12*. Other isotopes of carbon are carbon-13 (six protons and seven neutrons) and carbon-14 (six protons and eight neutrons). We sometimes write the atomic mass number as a superscript to the left of the element symbol: $^{12}C$, $^{13}C$, $^{14}C$. We read $^{12}C$ as "carbon-12."

**Think about it** The symbol $^{4}He$ represents helium with an atomic mass number of 4. $^{4}He$ is the most common form of helium, containing two protons and two neutrons. What does the symbol $^{3}He$ represent?

**Molecules** The number of different material substances is far greater than the number of chemical elements because atoms can combine to form **molecules**. Some molecules consist of two or more atoms of the same element. For example, we breathe $O_2$, oxygen molecules made of two oxygen atoms. Other molecules, such as the water molecule, are made up of atoms of two or more different elements. (Molecules with two or more types of atom are often called *compounds*.) The symbol $H_2O$ tells us that a water molecule contains two hydrogen atoms and one oxygen atom. The chemical properties of a molecule are different from those of its individual atoms. For example, molecular oxygen ($O_2$) behaves very differently from atomic oxygen (O), and water behaves very differently from pure hydrogen or pure oxygen.

## What are the phases of matter?

Interactions between light and matter depend on the physical state of the matter, which we usually describe by the matter's **phase**. For example, molecules of $H_2O$ can exist in three familiar phases: as **solid** ice, as **liquid** water, and as the **gas** we call water vapor. But how can the *same* molecules ($H_2O$) look and act so different in different phases?

You are probably familiar with the idea of a **chemical bond**, the name we give to the interactions between electrons that hold the atoms in a molecule together. For example, we say that chemical bonds hold the hydrogen and oxygen atoms together in a molecule of $H_2O$. Similar but much weaker interactions among electrons hold together the many water molecules in a block of ice or a pool of water. We can think of the interactions that keep neighboring atoms or molecules close together as other types of bonds, with the phases of solid, liquid, and gas differing in the strength of the bonds between neighboring atoms and molecules. Phase changes occur when one type of bond is broken and replaced by another. Changes in either pressure or temperature (or both) can cause phase changes, but it's easier to think first about temperature: As a substance is heated, the average kinetic energy of its particles increases, enabling the particles to break the bonds holding them to their neighbors.

**Phase Changes in Water** Water is the only familiar substance that we see in all three phases (solid, liquid, gas) in everyday life, so let's consider what happens to water as an example of how phase changes occur as a substance heats up.

At low temperatures, water molecules have a relatively low average kinetic energy, allowing them to be tightly bound to their neighbors in the *solid* structure of ice. As long as the temperature remains below freezing, the water molecules in ice remain rigidly held together. However, the molecules within this rigid structure are always vibrating, and higher temperature means greater vibrations. If we start with ice at a very low temperature, the molecular vibrations grow gradually stronger as the temperature rises toward the melting point, which is 0°C at ordinary (sea level) atmospheric pressure.

The melting point is the temperature at which the molecules have enough energy to break the solid bonds of ice. The molecules can then move much more freely among one another, allowing the water to flow as a *liquid*. However, the molecules in liquid water are not completely free of one another, as we can tell from the fact that droplets of water can stay intact. Adjacent molecules in liquid water must therefore still be held together by the same type of bond, though it is much looser than the bond that holds them together in solid ice.

If we continue to heat the water, the increasing kinetic energy of the molecules will ultimately break the bonds between neighboring molecules altogether. The molecules will then be able to move freely, and freely moving particles constitute a *gas*. Above the boiling point (100°C at sea level), all the bonds between adjacent molecules are broken, so the water can exist only as a gas.

We see ice melting into liquid water and liquid water boiling into gas so often that it's tempting to think that's the end of the story. However, a little thought should convince you that the reality has to be more complex. For example, you know that Earth's atmosphere contains water vapor that condenses to form clouds and rain. But Earth's surface temperature is well below the boiling point of water, so how is it that our atmosphere can contain water in the gas phase?

The answer lies in the fact that temperature is a measure of the *average* kinetic energy of the particles in a substance [**Section 4.3**]. Individual particles may have energies substantially lower or higher than the average. Even at the low temperatures at which most water molecules are bound together as ice or liquid, a few molecules will always have enough energy to break free of their neighbors and enter the gas phase. In other words, some gas (water vapor) is always present along with solid ice or liquid water. The process by which molecules break free is often called **vaporization**, because the escaped molecules enter the gas (or vapor) phase. More technically, vaporization from a solid is called *sublimation*, while vaporization from a liquid is called *evaporation*. Higher temperatures lead to higher rates of sublimation or evaporation.

**Think about it** Based on what you've learned about phase changes, how would you expect global warming (which is raising Earth's average surface temperature) to affect the total amount of cloud cover on Earth? Explain.

**Molecular Dissociation and Ionization** Above the boiling point, all the water will have entered the gas phase. What happens if we continue to raise the temperature?

The molecules in a gas move freely, but they often collide with one another. As the temperature rises, the molecules move faster and the collisions become more violent. At high enough temperatures, the collisions become so violent that they can break the chemical bonds holding individual water molecules together. The molecules then split into pieces, a process we call **molecular dissociation**. (In the case of water, molecular dissociation usually frees one hydrogen atom and leaves a negatively charged molecule that consists of one hydrogen atom and one oxygen atom [OH]; at even higher temperatures, the OH dissociates into individual atoms.)

### COMMON MISCONCEPTIONS

#### One Phase at a Time?

In daily life, we usually think of $H_2O$ as being in just one phase at a time—that is, as solid ice, liquid water, or gaseous water vapor—with the phase depending on the temperature. In reality, two or even all three phases can exist at the same time. In particular, some sublimation *always* occurs over solid ice, and some evaporation *always* occurs over liquid water. You can tell that evaporation always occurs, because an uncovered glass of water will gradually empty as the liquid evaporates into gas. You can see sublimation by observing the snow pack after a winter storm: Even if the snow doesn't melt into liquid, it gradually disappears as the ice sublimates into water vapor.

At still higher temperatures, collisions can break the bonds holding electrons around the nuclei of individual atoms, allowing the electrons to go free. The loss of one or more negatively charged electrons leaves the remaining atom with a net positive charge. Charged atoms (whether positive or negative) are called **ions**, and the process of stripping electrons from atoms is called **ionization**. At temperatures of several thousand degrees, the process of ionization turns what once was water into a hot gas consisting of freely moving electrons and positively charged ions of hydrogen and oxygen. This type of hot gas, in which atoms have become ionized, is called a **plasma**. Because a plasma contains many charged particles, its interactions with light are different from those of a gas consisting of neutral atoms, which is one reason plasma is sometimes referred to as "the fourth phase of matter." However, because the electrons and ions are not bound to one another, it is also legitimate to call a plasma a gas. That is why we sometimes say that the Sun is made of hot gas and sometimes say that it is made of plasma; both statements are correct.

The degree of ionization in a plasma depends on its temperature and composition. A neutral hydrogen atom contains only one electron, so hydrogen can be ionized only once; the remaining hydrogen ion, designated $H^+$, is simply a proton. Oxygen, with atomic number 8, has eight electrons when it is neutral, so it can be ionized multiple times. *Singly ionized* oxygen is missing one electron, so it has a charge of +1 and is designated $O^+$. *Doubly ionized* oxygen, or $O^{+2}$, is missing two electrons; *triply ionized* oxygen, or $O^{+3}$, is missing three electrons; and so on. At temperatures of several million degrees, oxygen can be *fully ionized*, in which case all eight electrons are stripped away and the remaining ion has a charge of +8.

**FIGURE 5.10** summarizes the changes that occur as we heat water from ice to a fully ionized plasma. Other chemical substances go through similar phase changes, but changes from one phase to another occur at different temperatures for different substances.

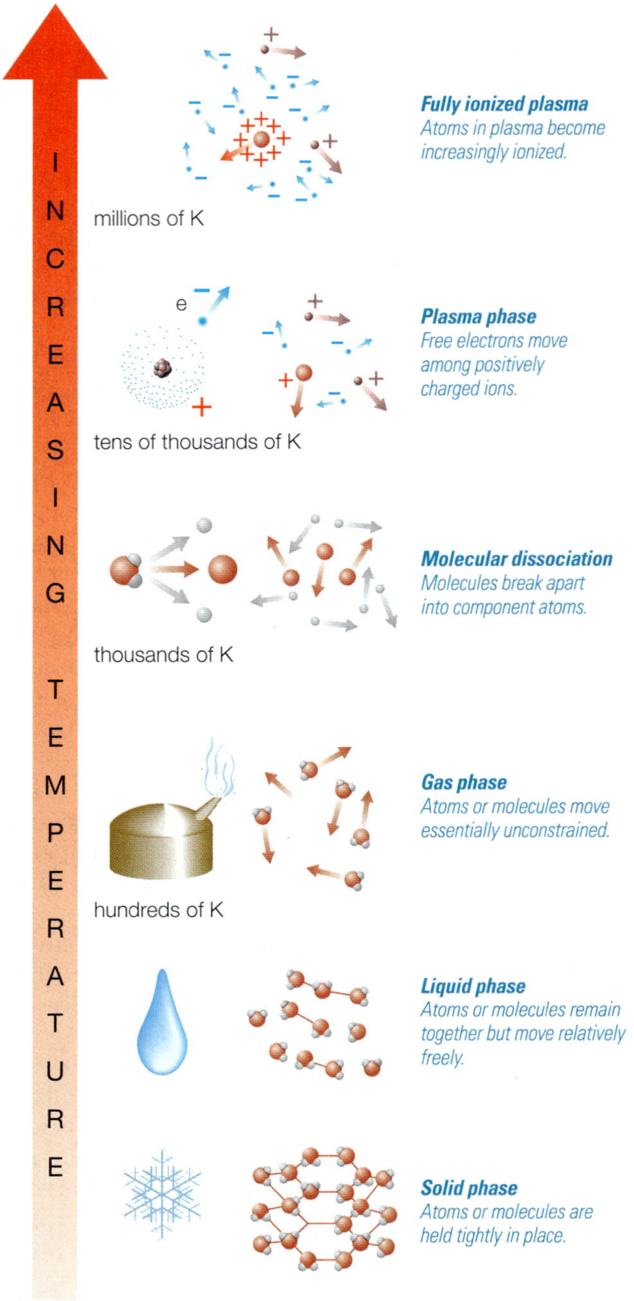


**FIGURE 5.10** The general progression of changes with temperature for water.

**Phases and Pressure** Temperature is the primary factor determining the phase of a substance and the ways in which light interacts with it, but pressure also plays a role. You're undoubtedly familiar with the idea of pressure in an everyday sense: For example, you can put more pressure on your arm by squeezing it. In science, we use a more precise definition: **Pressure** is the *force per unit area* pushing on an object's surface. You feel more pressure when you squeeze your arm because squeezing increases the force on each square centimeter of your arm's surface. Similarly, piling rocks on a table increases the weight (force) on the table, which therefore increases the pressure on the surface of the table; if the pressure becomes too great, the table breaks.

Pressure can affect phases in a variety of ways. For example, deep inside Earth, the pressure is so high that Earth's inner metal core remains solid, even though the temperature is high enough that the metal would melt into liquid under less extreme pressure conditions. On a planetary surface, the *atmospheric pressure* [**Section 10.1**] created by the weight of gas above can determine whether water is stable in liquid form.

Remember that liquid water is always evaporating (or ice sublimating) at a low level, because a few molecules randomly get enough energy to break the bonds holding them to their neighbors. On Earth, enough liquid water has evaporated from the oceans to make water vapor an important ingredient of our atmosphere. Some of these atmospheric water vapor molecules collide with the ocean surface, where they can "stick" and rejoin the ocean—essentially the opposite of evaporation (**FIGURE 5.11**). The greater the pressure from water vapor molecules in our atmosphere,* the higher the rate at which water molecules return to the ocean.

*This pressure is the *vapor pressure* of water in the atmosphere. We can also measure vapor pressure for other atmospheric constituents, and the total gas pressure is the sum of all the individual vapor pressures.

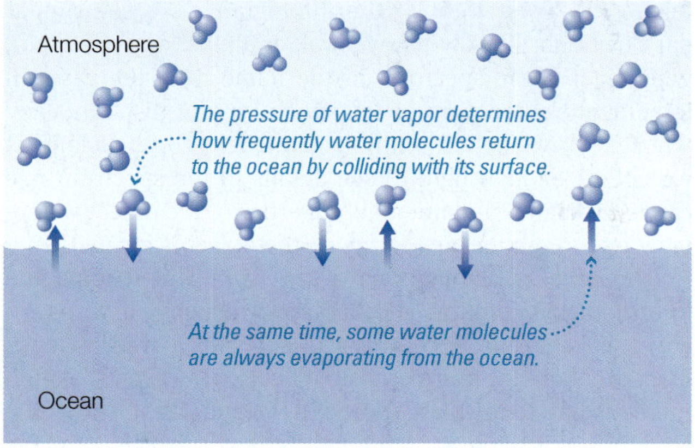


**FIGURE 5.11** Evaporation of water molecules from the ocean is balanced in part by molecules of water vapor in Earth's atmosphere returning to the ocean. The rate at which these molecules return is directly related to the pressure created by water vapor in the atmosphere.

This direct return of water vapor molecules from the atmosphere helps keep the total amount of water in Earth's oceans fairly stable.* On the Moon, where the lack of atmosphere means no pressure from water vapor at all, liquid water would evaporate quite quickly (as long as the temperature were high enough that it did not freeze first). The same is true on Mars, because the atmosphere lacks enough water vapor to balance the rate of evaporation.

High pressure can also cause gases to dissolve in liquid water. For example, sodas are made by mixing water with high-pressure carbon dioxide gas. Because of the high pressure, many more carbon dioxide molecules enter the water than are released, so the water becomes "carbonated"—that is, it has a lot of dissolved carbon dioxide. When you open a bottle of carbonated water, exposing it to air with ordinary pressure, the dissolved carbon dioxide quickly bubbles up and escapes.

## How is energy stored in atoms?

We are now ready to return to the primary goal of this chapter: understanding how we learn about distant objects by studying their light. To produce light, objects must somehow transform energy contained in their matter into the vibrations of electric and magnetic fields that we call light. We therefore need to focus on the charged particles within atoms, particularly the electrons, because only particles that have charge can interact with light.

Atoms contain energy in three different ways. First, by virtue of their mass, they possess mass-energy in the amount $mc^2$. Second, they possess kinetic energy by virtue of their motion. Third, they contain *electrical potential energy* that depends on the arrangement of their electrons around their nuclei. To interpret the messages carried by light, we must understand how electrons store and release this electrical potential energy.

**Energy Levels in Atoms** The energy stored by electrons in atoms has a strange but important property: The electrons can have only particular amounts of energy, and not other energies in between. As an analogy, suppose you're washing windows on a building. If you use an adjustable platform to reach high windows, you can stop the platform at any height above the ground. But if you use a ladder, you can stand only at *particular* heights—the heights of the rungs of the ladder—and not at other heights in between. The possible energies of electrons in atoms are like the possible heights on a ladder. Only a few particular energies are possible; energies between these special few are not possible. The possible energies are known as the **energy levels** of an atom.

**FIGURE 5.12** shows the energy levels of hydrogen, the simplest of all elements. The energy levels are labeled on the left in numerical order and on the right in units of *electron-volts*, or *eV* for short ($1 \text{ eV} = 1.60 \times 10^{-19}$ joule). The lowest possible energy level—called level 1 or the *ground state*—is defined to have an energy of 0 eV. Each of the higher energy levels (sometimes called *excited states*) is labeled with the extra energy of an electron in that level compared to an electron in the ground state.

**Energy Level Transitions** An electron can rise from a low energy level to a higher one or fall from a high level to a lower one. Such changes are called **energy level transitions**. Because energy must be conserved, energy level transitions can occur only when an electron gains or loses the specific amount of energy separating two levels. For example, an electron in level 1 can rise to level 2 only if it gains 10.2 eV of energy. If you try to give the electron 5 eV of energy, it won't accept it because that is not enough energy to reach level 2. Similarly, if you try to give it 11 eV, it won't accept it because that is too much for level 2 but not enough to reach level 3. Once in level 2, the electron can

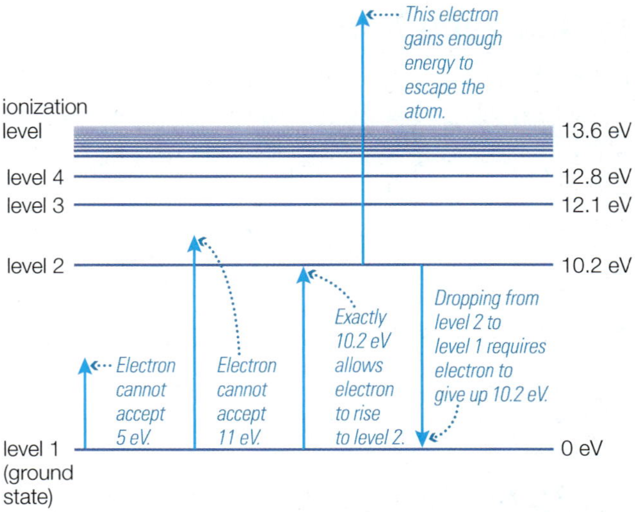


**FIGURE 5.12** Energy levels for the electron in a hydrogen atom. The electron can change energy levels only if it gains or loses the amount of energy separating the levels. If the electron gains enough energy to reach the ionization level, it can escape from the atom, leaving behind a positively charged ion. (The many levels between level 4 and the ionization level are not labeled.)

*Rain and snow also contribute, of course; however, even if Earth's temperature rose enough that raindrops and snowflakes could no longer form, only a small fraction of Earth's ocean water would evaporate before the return rate of water vapor molecules balanced the evaporation rate.

return to level 1 by giving up 10.2 eV of energy. Figure 5.12 shows several examples of allowed and disallowed energy level transitions.

Notice that the amount of energy separating the various levels gets smaller at higher levels. For example, it takes more energy to raise the electron from level 1 to level 2 than from level 2 to level 3, which in turn takes more energy than the transition from level 3 to level 4. If the electron gains enough energy to reach the *ionization level,* it escapes the atom completely, thereby ionizing the atom. Any excess energy beyond the amount needed for ionization becomes kinetic energy of the free-moving electron.

**Think about it** Are there any circumstances under which an electron in a hydrogen atom can lose 2.6 eV of energy? Explain.

**Quantum Physics** If you think about it, the idea that electrons in atoms are restricted to particular energy levels is quite bizarre. It is as if you had a car that could go around a track only at particular speeds and not at speeds in between. How strange it would seem if your car suddenly changed its speed from 5 kilometers per hour to 20 kilometers per hour without first passing through a speed of 10 kilometers per hour! The electron's energy levels in an atom are said to be *quantized,* and the study of the energy levels of electrons (and other particles) is called *quantum physics* (or *quantum mechanics*). We will explore some of the astonishing implications of quantum physics in Chapter S4.

Electrons have quantized energy levels in all atoms, not just in hydrogen. Moreover, the allowed energy levels differ from element to element and from one ion of an element to another ion of the same element. Even molecules have quantized energy levels. As we will see shortly, the different energy levels of different atoms and molecules allow light to carry "fingerprints" that can tell us the chemical composition of distant objects.

## 5.4 Learning from Light

The photograph that opens this chapter (page 137) shows the Sun's visible-light spectrum in great detail, with the rainbow of color stretching in horizontal rows from the upper left to the lower right of the photograph. We see similar dark or bright lines when we look at almost any spectrum, whether it is the spectrum of the flame from the gas grill in someone's backyard or the spectrum of a distant galaxy whose light we collect with a gigantic telescope. As long as we collect enough light to see details in the spectrum, we can learn many fundamental properties of the object we are viewing, no matter how far away the object is located.

The process of obtaining a spectrum and reading the information it contains is called **spectroscopy**. If you project a spectrum produced by a prism onto a wall, it looks like a rainbow (at least for visible light). However, it's often more useful to display spectra as graphs that show the amount, or **intensity**, of the light at each wavelength. For example, consider the spectrum in **FIGURE 5.13**, which plots the intensity of light from an astronomical object at wavelengths ranging from the ultraviolet on the left to the infrared on the right. At wavelengths where a lot of light is coming from the object, the intensity is high, while at wavelengths where there is little light, the intensity is low.*

Our goal in this section is to learn how to interpret astronomical spectra like the one in Figure 5.13. The bumps and wiggles in that spectrum arise from several different processes, making it a good case study. We'll consider these processes one at a time, then return to interpret the full spectrum at the end of this section.

**Three Types of Spectra**

### What are the three basic types of spectra?

Laboratory studies show that spectra come in three basic types† (**FIGURE 5.14**):

1. The spectrum of a traditional, or incandescent, light bulb (which contains a heated wire filament) is a rainbow of color. Because the rainbow spans a broad range of wavelengths without interruption, we call it a **continuous spectrum**.

*More technically, intensity is proportional to the total amount of *energy* transmitted by the light at each wavelength.

†The rules that specify the conditions producing each type are often called *Kirchhoff's laws*.

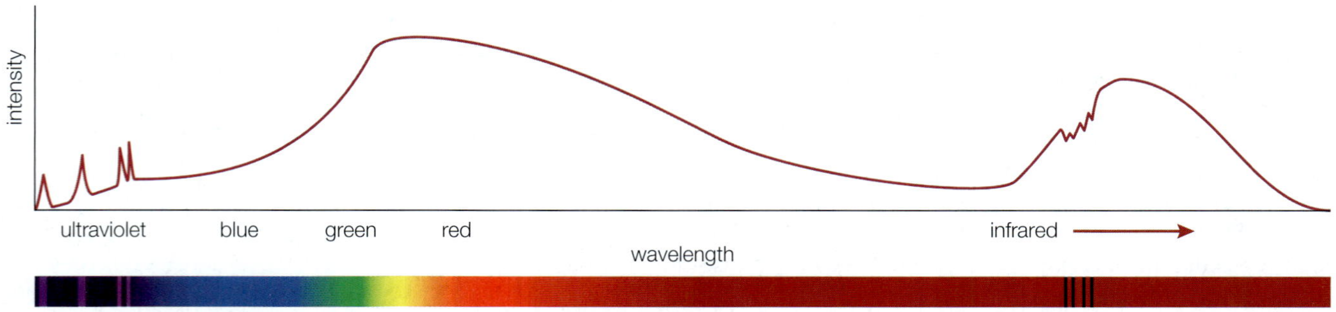


**FIGURE 5.13** A schematic spectrum obtained from the light of a distant object. The "rainbow" at the bottom shows how the light would appear if viewed with a prism or diffraction grating; of course, our eyes cannot see the ultraviolet or infrared light. The graph shows the corresponding intensity of the light at each wavelength. The intensity is high where the rainbow is bright and low where it is dim (such as in places where the rainbow shows dark lines).

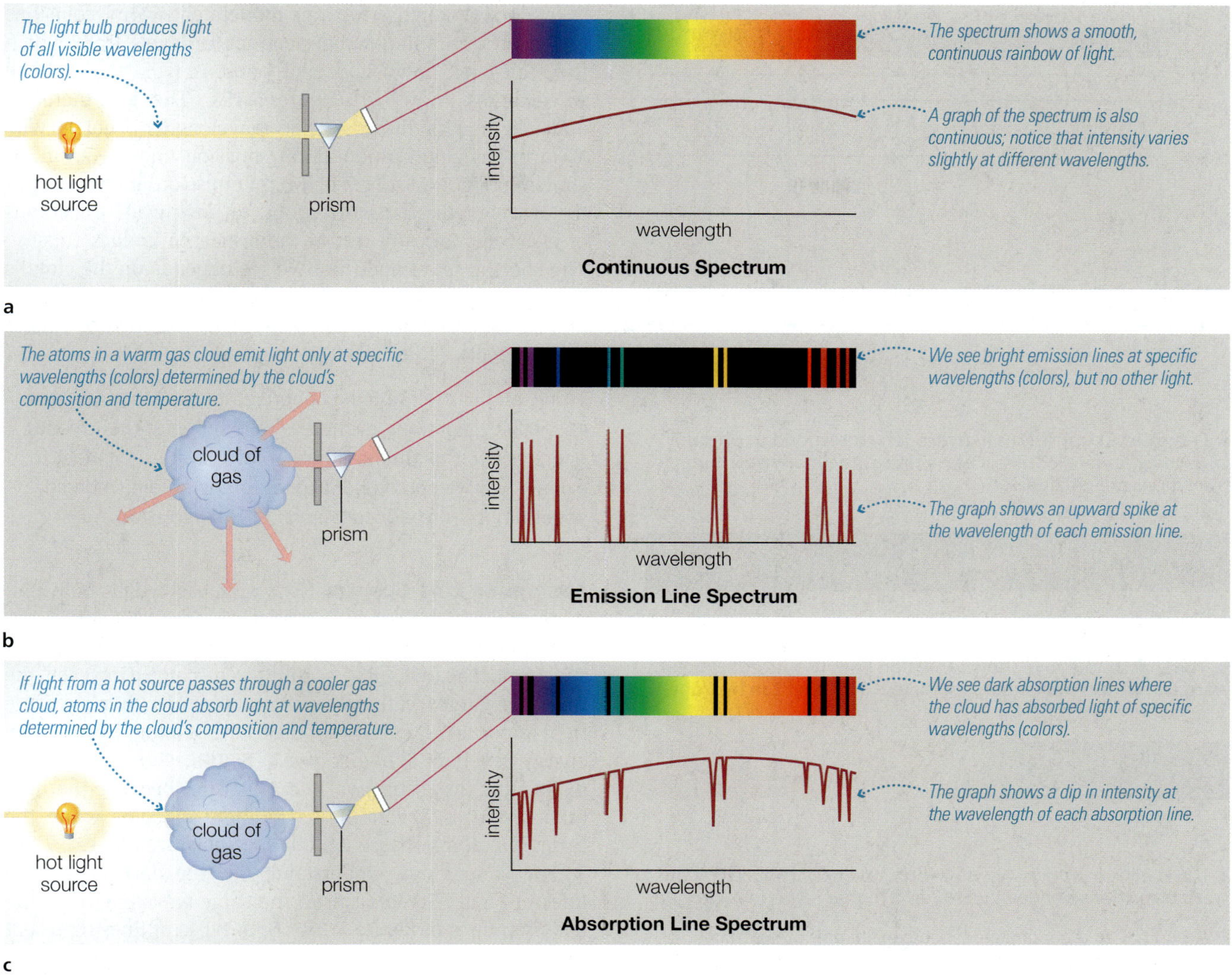


**FIGURE 5.14** These diagrams show examples of the conditions under which we see the three basic types of spectra.

2. A thin or low-density cloud of gas emits light only at specific wavelengths that depend on its composition and temperature. The spectrum therefore consists of bright **emission lines** against a black background and is called an **emission line spectrum**.

3. If the cloud of gas lies between us and a light bulb (and the cloud is cooler than the light bulb or other light source), we still see most of the continuous spectrum of the light bulb. However, the cloud absorbs light of specific wavelengths, so the spectrum shows dark **absorption lines** over the background rainbow, making it what we call an **absorption line spectrum**.

Note that when the spectra are shown as graphs, absorption lines appear as dips on a background of relatively high-intensity light while emission lines look like spikes on a background with little or no intensity.

We can apply these ideas to the solar spectrum that opens this chapter. The many dark absorption lines over a background rainbow of color tell us that we are looking at a hot light source through a cooler gas, much like the situation in Figure 5.14c. For the solar spectrum, the hot light source is the hot interior of the Sun, while the "cloud" is the relatively cool and low-density gas that makes up the Sun's visible surface, or *photosphere* [**Section 14.1**].

## How does light tell us what things are made of?

We have just seen *how* different viewing conditions lead to different types of spectra, so now it is time to discuss *why*. Let's start with emission and absorption line spectra, in which the lines form as a direct consequence of the fact that each type of atom, ion, or molecule possesses a unique set of energy levels.

**Emission Line Spectra** The atoms in any cloud of gas are continually colliding with one another, exchanging energy in each collision. Most of the collisions simply send the atoms flying off in new directions. However, a few of the collisions transfer the right amount of energy to bump an electron from a low energy level to a higher energy level.

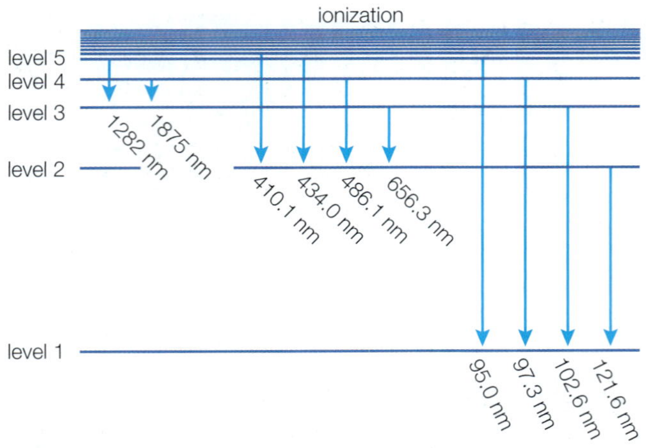


**a** Energy level transitions in hydrogen correspond to photons with specific wavelengths. Only a few of the many possible transitions are labeled.

**b** The spectrum above shows emission lines produced by downward transitions between higher levels and level 2 in hydrogen.

**c** The spectrum above shows absorption lines produced by upward transitions between level 2 and higher levels in hydrogen.

**FIGURE 5.15** An atom emits or absorbs light only at specific wavelengths that correspond to changes in the atom's energy as an electron undergoes transitions between its allowed energy levels.

Electrons can't stay in higher energy levels for long. They always fall back down to the ground state, level 1, usually in a tiny fraction of a second. The energy the electron loses when it falls to a lower energy level must go somewhere, and often it goes into *emitting* a photon of light. The emitted photon must have the same amount of energy that the electron loses, which means that it has a specific wavelength and frequency. **FIGURE 5.15a** shows the energy levels in hydrogen that we saw in Figure 5.12, but it is also labeled with the wavelengths of the photons emitted by various downward transitions of an electron from a higher energy level to a lower one. For example, the transition from level 2 to level 1 emits an ultraviolet photon of wavelength 121.6 nm, and the transition from level 3 to level 2 emits a red visible-light photon of wavelength 656.3 nm.*

As long as the gas remains moderately warm, collisions are always bumping some electrons to levels from which they fall back down and emit photons with some of the wavelengths shown in Figure 5.15a. The gas therefore emits light with these specific wavelengths. That is why a warm gas cloud produces an emission line spectrum, as shown in **FIGURE 5.15b**. The bright emission lines appear at the wavelengths that correspond to downward transitions of electrons, and the rest of the spectrum is dark (black). The specific set of lines that we see depends on the cloud's temperature as well as its composition: At higher temperatures, electrons are more likely to be bumped to higher energy levels.

**Think about it** If nothing continues to heat the hydrogen gas, all the electrons eventually will end up at the lowest energy level (the ground state, or level 1). Use this fact to explain why we should *not* expect to see an emission line spectrum from a very cold cloud of hydrogen gas.

**Absorption Line Spectra** Now, suppose a light bulb illuminates the hydrogen gas from behind (as in Figure 5.14c). The light bulb emits light of all wavelengths, producing a spectrum that looks like a rainbow of color. However, the hydrogen atoms can absorb those photons that have the right amount of energy to raise an electron from a low energy level to a higher one.* **FIGURE 5.15c** shows the result. It is an absorption line spectrum, because the light bulb produces a continuous rainbow of color while the hydrogen atoms absorb light at specific wavelengths.

You should now understand why the dark absorption lines in Figure 5.15c occur at the same wavelengths as the emission lines in Figure 5.15b: Both types of lines represent the same energy level transitions, except in opposite directions. For example, electrons moving downward from level 3 to level 2 in hydrogen can emit photons of wavelength 656.3 nm (producing an emission line at this wavelength), while electrons absorbing photons with this wavelength can rise up from level 2 to level 3 (producing an absorption line at this wavelength).

**Chemical Fingerprints** The fact that hydrogen emits and absorbs light at specific wavelengths makes it possible to detect its presence in distant objects. For example, imagine that you look through a telescope at an interstellar gas cloud, and its spectrum looks like that shown in Figure 5.15b. Because only hydrogen produces this particular set of lines, you can conclude that the cloud is made of hydrogen. In essence, the spectrum contains a "fingerprint" left by hydrogen atoms.

Real interstellar clouds are not made solely of hydrogen. However, the other chemical constituents in the cloud leave fingerprints on the spectrum in much the same way. Every type of atom has its own unique spectral

*Astronomers call transitions between level 1 and other levels the *Lyman* series of transitions. The transition between level 1 and level 2 is Lyman $\alpha$, between level 1 and level 3 is Lyman $\beta$, and so on. Similarly, transitions between level 2 and higher levels are called *Balmer* transitions. Other sets of transitions also have names.

*Of course, the electrons quickly fall back down, which means they can emit photons of the same wavelength they absorbed. However, these photons are emitted in random directions, so we still see absorption lines because photons that were originally coming toward us have been redirected away from our line of sight.

**FIGURE 5.16** Visible-light emission line spectra for helium, sodium, and neon. The patterns and wavelengths of lines are different for each element, giving each a unique spectral fingerprint.

fingerprint, because it has its own unique set of energy levels. **FIGURE 5.16** shows emission line spectra for helium, sodium, and neon. Moreover, different ions (atoms with missing or extra electrons) also produce different fingerprints (**FIGURE 5.17**). For example, the wavelengths of lines produced by doubly ionized neon ($Ne^{+2}$) are different from those of singly ionized neon ($Ne^{+}$), which in turn are different from those of neutral neon (Ne). These differences can help us determine the temperature of a hot gas or plasma, because more highly charged ions will be present at higher temperatures; this fact enables us to use spectra to measure the surface temperatures of stars [**Section 15.1**].

Molecules also produce spectral fingerprints. Like atoms, molecules can produce spectral lines when their electrons change energy levels. But molecules can also produce spectral lines in two other ways. Because they are made of two or more atoms bound together, molecules can vibrate and rotate (**FIGURE 5.18a**). Vibration and rotation also require energy, and the possible energies of rotation and vibration in molecules are quantized much like electron energy levels in atoms. A molecule can absorb or emit a photon when it changes its rate of vibration or rotation. The energy changes in molecules are usually smaller than those in atoms and therefore produce lower-energy photons, and the energy levels also tend to be bunched more closely together than in atoms. Molecules therefore produce spectra with many sets of tightly bunched lines, called **molecular bands** (**FIGURE 5.18b**), that are usually found in the infrared portion of the electromagnetic spectrum.

Over the past century, scientists have conducted laboratory experiments to identify the spectral lines of every chemical element and of many ions and molecules. As a result, when we see lines in the spectrum of a distant object, we can determine what chemicals produced them. For example, if we see spectral lines of hydrogen, helium, and carbon in the spectrum of a distant star, we know that all three elements are present in the star. More detailed analysis even allows us to determine the relative proportions of the various elements. That is how we have learned the chemical compositions of objects throughout the universe.

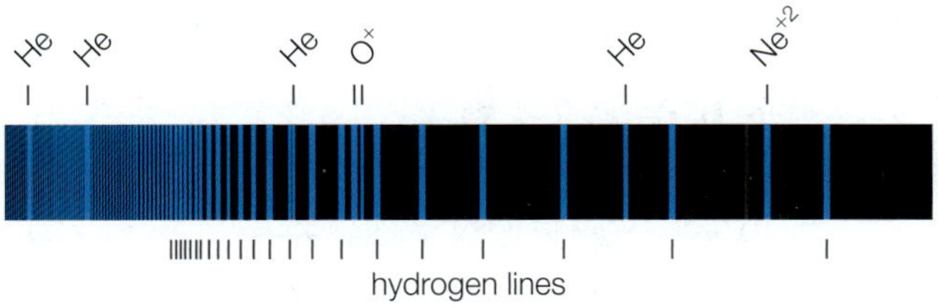


**FIGURE 5.17** The emission line spectrum of the Orion Nebula in a portion of the ultraviolet (about 350–400 nm). The lines are labeled with the chemical elements or ions that produce them (He = helium; O = oxygen; Ne = neon). The many hydrogen lines are all transitions from high levels to level 2.

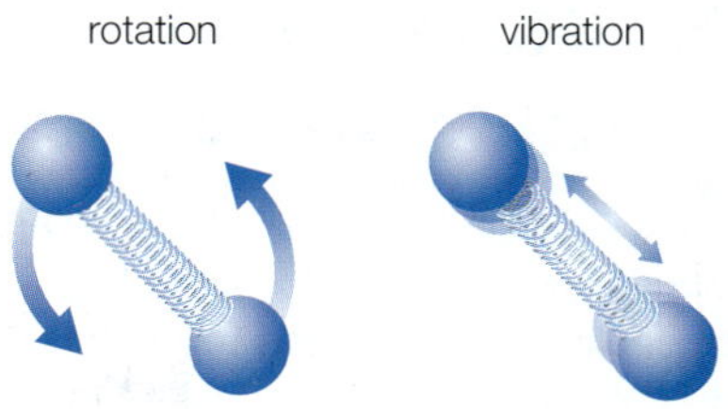


**a** We can think of a two-atom molecule as two balls connected by a spring. Although this model is simplistic, it illustrates how molecules can rotate and vibrate. The rotations and vibrations can have only particular amounts of energy and therefore produce unique spectral fingerprints.

**b** This spectrum of molecular hydrogen ($H_2$) consists of lines bunched into broad molecular bands.

**FIGURE 5.18** Like atoms and ions, molecules emit or absorb light at specific wavelengths.

**Reflected Light Spectra** Some astronomical objects, such as planets and moons, reflect some of the light that falls on them. Reflected light also leaves a mark in spectra that can reveal information about the object, though not with the same level of detail as spectral lines. To understand why, consider the spectrum you would see from a red shirt on a sunny day. The red shirt absorbs blue light and reflects red light, so its visible spectrum will look like the spectrum of sunlight but with blue light missing. Because the shirt itself is too cool in temperature to emit visible light, the missing blue light must be telling you something about the dye in the shirt. In a similar way, the surface materials of a planet determine how much light of different colors is reflected or absorbed. The reflected light gives the planet its color, while the absorbed light heats the surface and helps determine its temperature. Careful study of which colors are absorbed and which are reflected can tell you something about the types of minerals on the surface.

## How does light tell us the temperatures of planets and stars?

We next turn our attention to continuous spectra. Although continuous spectra can be produced in more than one way, light bulbs, planets, and stars produce a particular kind of continuous spectrum that can help us determine their temperatures.

## EXTRAORDINARY CLAIMS We Can Never Learn the Composition of Stars

The limits of science have long been a topic of great debate among both scientists and philosophers. One of the most famous claims about such limits was made in 1835, when the prominent French philosopher Auguste Comte declared that science could never allow us to learn the composition of stars. His rationale was based primarily on the idea that while we could observe things like the sizes of distant objects, we could not learn what they were made of without collecting physical samples to study here on Earth.

As you know from this chapter, Comte's extraordinary claim was proven wrong a few decades later with the advent of spectroscopy. Less than a century after his claim, Cecilia Payne-Gaposchkin used spectroscopy and an understanding of quantum mechanics to unlock the secret to the composition of stars [**Section 15.1**]. Moreover, even if spectroscopy had not been discovered, Comte's claim might still have been proven wrong, because it also presumed that we would never travel out into space to collect samples of distant objects. While stars besides the Sun remain far beyond our reach, sample collection within our solar system is now well within our abilities. It's also worth noting that the possibility of space travel was already being considered seriously by Comte's time. For example, in 1593, Johannes Kepler wrote a letter to Galileo in which he made an extraordinary claim of his own: "Provide ships or sails adapted to the heavenly breezes, and there will be some who will not fear even that void . . . ." Of course, that didn't stop many people from claiming that space travel would prove impossible, some of them holding to that view almost right up until we achieved it.

There may well be limits to what we can learn through science, but the history of science teaches us to be wary of claims about what we may or may not do or learn in the future. Some questions that seem unanswerable today, such as what (if anything) came before the Big Bang, may remain unanswerable tomorrow, but science may open doors to questions that we cannot yet envision.

**Verdict:** Rejected.

**Thermal Radiation: Every Body Does It** In a cloud of gas that produces a simple emission or absorption line spectrum, the individual atoms or molecules are essentially independent of one another. Most photons pass easily through such a gas, except those that cause energy level transitions in the atoms or molecules of the gas. However, the atoms and molecules within most of the objects we encounter in everyday life—such as rocks, light bulb filaments, and people—cannot be considered independent and therefore have much more complex sets of energy levels. These objects tend to absorb light across a broad range of wavelengths, which means that light cannot easily pass through them and light emitted inside them cannot easily escape. The same is true of almost any large or dense object, including planets and stars.

In order to understand the spectra of such objects, let's consider an idealized case in which an object absorbs all photons that strike it and does not allow photons inside it to escape easily. Photons tend to bounce around randomly inside such an object, constantly exchanging energy with its atoms or molecules. By the time the photons finally escape the object, their radiative energies have become randomized so that they are spread over a wide range of wavelengths. The wide wavelength range of the photons explains why the spectrum of light from such an object is smooth, or *continuous,* like a pure rainbow without any absorption or emission lines.

Most important, the spectrum from such an object depends on only one thing: the object's *temperature.* To understand why, remember that temperature represents the average kinetic energy of the atoms or molecules in an object [**Section 4.3**]. Because the randomly bouncing photons interact so many times with those atoms or molecules, they end up with energies that match the kinetic energies of the object's atoms or molecules—which means the photon energies depend only on the object's temperature, regardless of what the object is made of. The temperature dependence of this light explains why we call it **thermal radiation** (sometimes known as *blackbody radiation*) and why its spectrum is called a **thermal radiation spectrum**.

No real object emits a perfect thermal radiation spectrum, but almost all familiar objects—including the Sun, the planets, rocks, and even you—emit light that approximates thermal radiation. **FIGURE 5.19** shows graphs of the idealized thermal radiation spectra of three stars and a human, each with its temperature given on the Kelvin scale (see Figure 4.13). Be sure to notice that these spectra show the intensity of light *per unit surface area,* not the total amount of light emitted by the object. For example, a very large 3000 K star can emit more total light than a small 15,000 K star, even though the hotter star emits much more light per unit area.

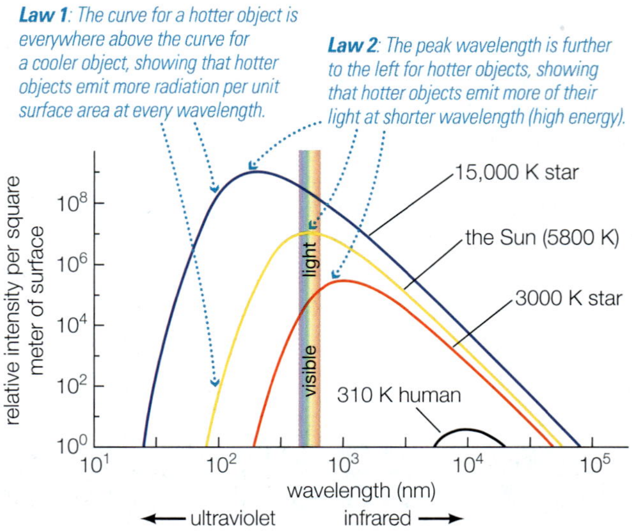


**FIGURE 5.19** Graphs of idealized thermal radiation spectra demonstrate the two laws of thermal radiation: (1) Each square meter of a hotter object's surface emits more light at all wavelengths; (2) hotter objects emit photons with a higher average energy. Notice that the graph uses power-of-10 scales on both axes so that we can see all the curves even though the differences between them are quite large.

## Thermal Radiation Laws

**The Two Laws of Thermal Radiation** If you compare the spectra in Figure 5.19, you'll see that they obey two laws of thermal radiation:

- **Law 1** (the Stefan-Boltzmann law): *Each square meter of a hotter object's surface emits more light at all wavelengths.* For example, each square meter on the surface of the 15,000 K star emits a lot more light at every wavelength than each square meter of the 3000 K star, and the hotter star emits light at some ultraviolet wavelengths that the cooler star does not emit at all.
- **Law 2** (Wien's [pronounced "veen's"] law): *Hotter objects emit photons with a higher average energy,* which means a shorter average wavelength. That is why the peaks of the spectra are at shorter wavelengths for hotter objects. For example, the peak for the 15,000 K star is in ultraviolet light, the peak for the 5800 K Sun is in visible light, and the peak for the 3000 K star is in the infrared.

You can see these laws in action with a fireplace poker (**FIGURE 5.20**). While the poker is still relatively cool, it emits only infrared light, which we cannot see. As it gets hot (above about 1500 K), it begins to glow with visible light, and it glows more brightly as it gets hotter, demonstrating the first law. Its color demonstrates the second law. At first it glows "red hot," because red light has the longest wavelengths of visible light. As it gets even hotter, the average wavelength of the emitted photons moves toward the blue (short-wavelength) end of the visible spectrum. The mix of colors emitted at this higher temperature makes the poker look white to your eyes, which is why "white hot" is hotter than "red hot."

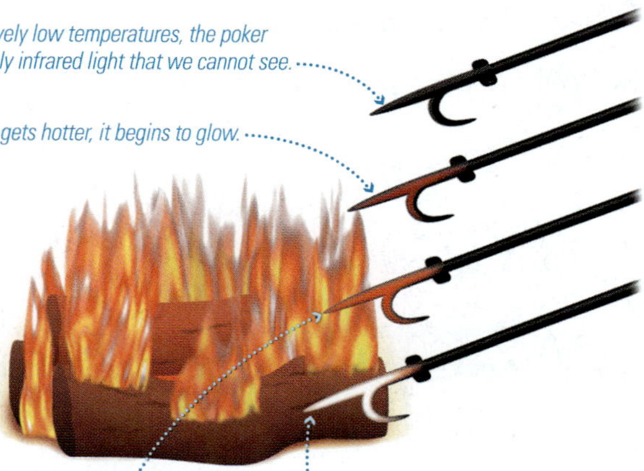


**FIGURE 5.20** A fireplace poker shows the two laws of thermal radiation in action.

**See it for yourself** Find an incandescent light that has a dimmer switch. What happens to the temperature of the bulb (which you can check by placing your hand near it) as you turn the switch up? How does the light change color? Explain how these observations demonstrate the two laws of thermal radiation.

Because thermal radiation spectra depend only on temperature, we can use them to measure the temperatures of distant objects. In many cases we can estimate temperatures simply from the object's color. Notice that while hotter objects emit more light at *all* wavelengths, the biggest difference appears at the shortest wavelengths. At human body

## MATHEMATICAL INSIGHT 5.2 Laws of Thermal Radiation

The two laws of thermal radiation have simple formulas.

Law 1 (*Stefan-Boltzmann law*):

$$\text{emitted power (per square meter of surface)} = \sigma T^4$$

where $\sigma$ (Greek letter *sigma*) is a constant with a measured value of $\sigma = 5.7 \times 10^{-8}\ \text{watt}/(\text{m}^2 \times \text{K}^4)$ and $T$ is on the Kelvin scale (K).

$$\text{Law 2 (}\textit{Wien's law}\text{): } \lambda_{max} \approx \frac{2{,}900{,}000}{T\ \text{(Kelvin scale)}}\ \text{nm}$$

where $\lambda_{max}$ (read as "lambda-max") is the wavelength (in nanometers) of maximum intensity, which is the peak of a thermal radiation spectrum.

**EXAMPLE:** Find the emitted power per square meter and the wavelength of peak intensity for a 10,000 K object that emits thermal radiation.

**SOLUTION:**

**Step 1 Understand:** We can calculate the emitted power per square meter from Law 1 and the wavelength of maximum intensity from Law 2.

**Step 2 Solve:** We plug the object's temperature ($T = 10{,}000$ K) into Law 1 to find the emitted power per square meter:

$$\sigma T^4 = 5.7 \times 10^{-8} \frac{\text{watt}}{\text{m}^2 \times \text{K}^4} \times (10{,}000\ \text{K})^4$$
$$= 5.7 \times 10^{8}\ \text{watt}/\text{m}^2$$

We find the wavelength of maximum intensity with Law 2:

$$\lambda_{max} \approx \frac{2{,}900{,}000}{10{,}000\ \text{(Kelvin scale)}}\ \text{nm} = 290\ \text{nm}$$

**Step 3 Explain:** A 10,000 K object emits 570 million watts per square meter of surface. Its wavelength of maximum intensity is 290 nm, which is in the ultraviolet. Note that we can learn about astronomical objects by using these facts in reverse. For example, if an object's thermal radiation spectrum peaks at a wavelength of 290 nm, its surface temperature must be about 10,000 K. We can then divide its total emitted power by the power it emits per square meter of surface to determine its surface area, from which we can calculate its radius.

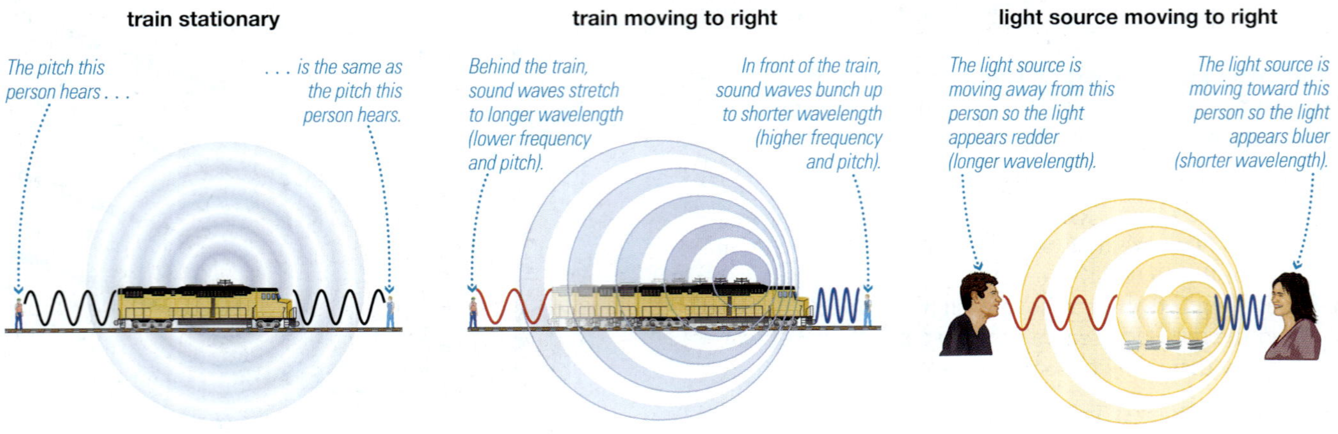


**a** The whistle sounds the same no matter where you stand near a stationary train.

**b** For a moving train, the sound you hear depends on whether the train is moving toward you or away from you.

**c** We get the same basic effect from a moving light source (although the shifts are usually too small to notice with our eyes).

**FIGURE 5.21** The Doppler effect. Each circle represents the crests of sound (or light) waves going in all directions from the source. For example, the circles from the train might represent waves emitted 0.001 second apart.

temperature of about 310 K, people emit mostly in the infrared and emit no visible light at all—which explains why we don't glow in the dark! A relatively cool star, with a 3000 K surface temperature, emits mostly red light. That is why some bright stars in our sky, such as Betelgeuse (in Orion) and Antares (in Scorpius), appear reddish in color. The Sun's 5800 K surface emits most strongly in green light (around 500 nm), but the Sun looks yellow or white to our eyes because it also emits other colors throughout the visible spectrum. Hotter stars emit mostly in the ultraviolet but appear blue-white in color because our eyes cannot see their ultraviolet light. If an object were heated to a temperature of millions of degrees, it would radiate mostly x-rays. Some astronomical objects are indeed hot enough to emit x-rays, such as the Sun's *corona* and hot *accretion disks* around black holes.

## How does light tell us the speed of a distant object?

There is still more that we can learn from light. In particular, we can learn about the motion of distant objects (relative to us) from changes in their spectra caused by the **Doppler effect**.

**The Doppler Effect** You've probably noticed the Doppler effect on the *sound* of a train whistle near train tracks. If the train is stationary, the pitch of its whistle sounds the same no matter where you stand (**FIGURE 5.21a**). But if the train is moving, the pitch sounds higher when the train is coming toward you and lower when it's moving away from you. Just as the train passes by, you can hear the dramatic change from high to low pitch—a sort of "weeeeeeee-ooooooooh" sound. To understand why, we have to think about what happens to the sound waves coming from the train (**FIGURE 5.21b**). When the train is moving toward you, each pulse of a sound wave is emitted a little closer to you. The result is that waves are bunched up between you and the train, giving them a shorter wavelength and higher frequency (pitch). After the train passes you by, each pulse comes from farther away, stretching out the wavelengths and giving the sound a lower frequency.

The Doppler effect causes similar shifts in the wavelengths of light (**FIGURE 5.21c**). If an object is moving toward us, the light waves bunch up between us and the object, so its entire spectrum is shifted to shorter wavelengths. Because shorter wavelengths of visible light are bluer, the Doppler shift of an object coming toward us is called a **blueshift**. If an object is moving away from us, its light is shifted to longer wavelengths. We call this a **redshift** because longer wavelengths of visible light are redder. For convenience, astronomers use the terms *blueshift* and *redshift* even when they aren't talking about visible light.

Spectral lines provide the reference points we use to identify and measure Doppler shifts (**FIGURE 5.22**). For example, suppose we recognize the pattern of hydrogen lines in the spectrum of a distant object. We know the **rest wavelengths** of the hydrogen lines—that is, their wavelengths in stationary clouds of hydrogen gas—from laboratory experiments in which a tube of hydrogen gas is heated so that the wavelengths

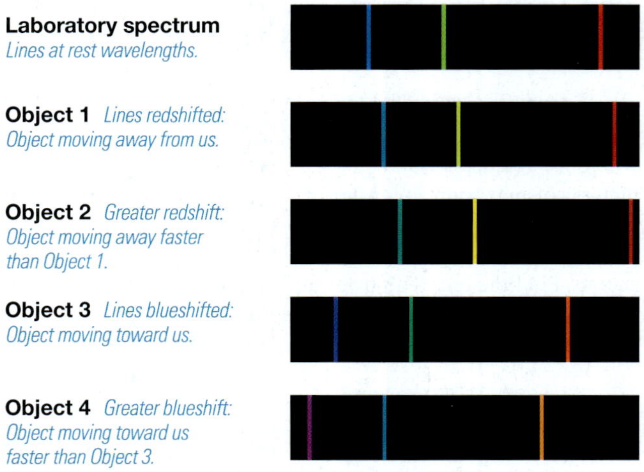


**FIGURE 5.22** Spectral lines provide the crucial reference points for measuring Doppler shifts.

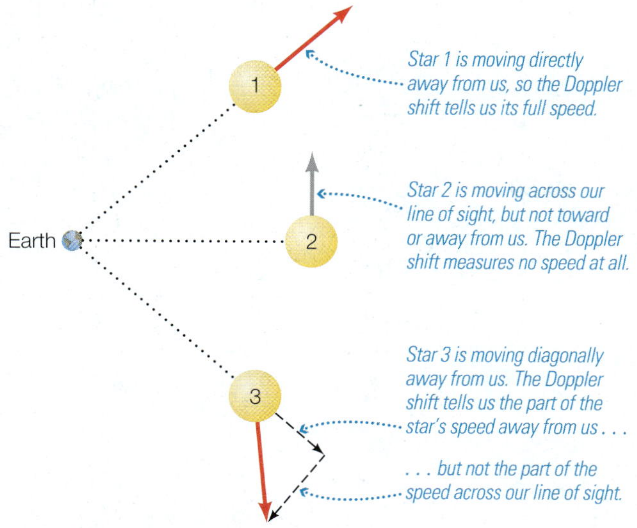


**FIGURE 5.23** The Doppler shift tells us only the portion of an object's speed that is directed toward or away from us. It does not give us any information about how fast an object is moving across our line of sight.

of the spectral lines can be measured. If the hydrogen lines from the object appear at longer wavelengths, then we know they are redshifted and the object is moving away from us. The larger the shift, the faster the object is moving. If the lines appear at shorter wavelengths, then we know they are blueshifted and the object is moving toward us.

**Think about it** Suppose the hydrogen emission line with a rest wavelength of 121.6 nm (the transition from level 2 to level 1) appears at a wavelength of 120.5 nm in the spectrum of a particular star. Given that these wavelengths are in the ultraviolet, is the shifted wavelength closer to or farther from blue visible light? Why, then, do we say that this spectral line is *blueshifted*?

**Components of Motion** It's important to note that a Doppler shift tells us only the part of an object's full motion that is directed toward or away from us (the object's *radial* component of motion). Doppler shifts do not give us any information about how fast an object is moving across our line of sight (the object's *tangential* component of motion). For example, consider three stars all moving at the same speed, with one moving directly away from us, one moving across our line of sight, and one moving diagonally away from us (**FIGURE 5.23**). The Doppler shift will tell us the full speed of only the first star. It will not indicate any speed for the second star, because none of this star's motion is directed toward or away from us. For the third star, the Doppler shift will tell us only the part of the star's velocity that is directed away from us. To measure how fast an object is moving across our line of sight, we must observe it long enough to notice how its position gradually shifts across our sky.

**Rotation Rates** The Doppler effect not only tells us how fast a distant object is moving toward or away from us but also can reveal information about motion *within* the object. For example, suppose we look at spectral lines of a *rotating* planet or star (**FIGURE 5.24**). As the object rotates, light from the part of the object rotating toward us will be blueshifted, light from the part rotating away from us will be redshifted, and light from the center of the object won't be shifted at all. The net effect, if we look at the whole object at once, is to make each spectral line appear *wider* than it would if the object were not rotating. The faster the object is rotating, the broader in wavelength the spectral lines become. We can therefore determine the rotation rate of a distant object by measuring the width of its spectral lines.

**Interpreting a Spectrum**

**Putting It All Together FIGURE 5.25** shows the same spectrum we began with in Figure 5.13, but this time with labels indicating the processes responsible for its various features. The thermal emission peaks in the infrared, corresponding

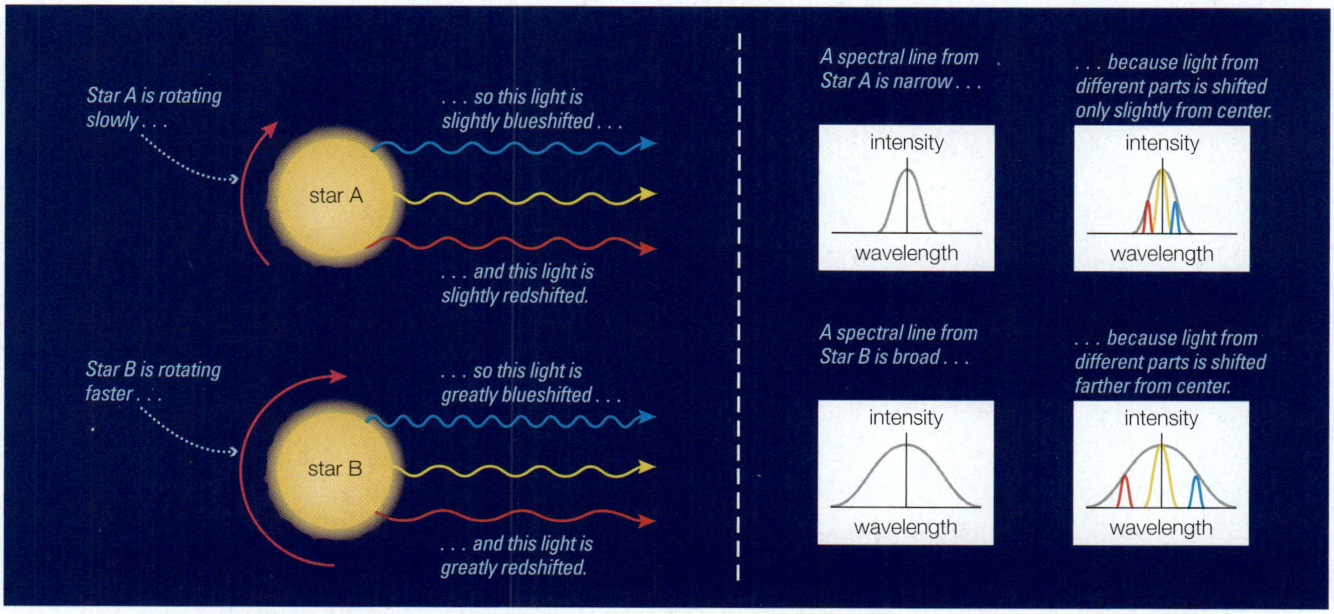


**FIGURE 5.24** This diagram shows how the Doppler effect can tell us the rotation rate even of stars that appear as points of light to our telescopes. Rotation spreads the light of any spectral line over a range of wavelengths, so faster-rotating stars have broader spectral lines.

**COSMIC Context**

# FIGURE 5.25 Interpreting a Spectrum

**An astronomical spectrum contains an enormous amount of information. This figure shows a schematic spectrum of Mars. It is the same spectrum shown in Figure 5.13, but this time describing what we can learn from it.**

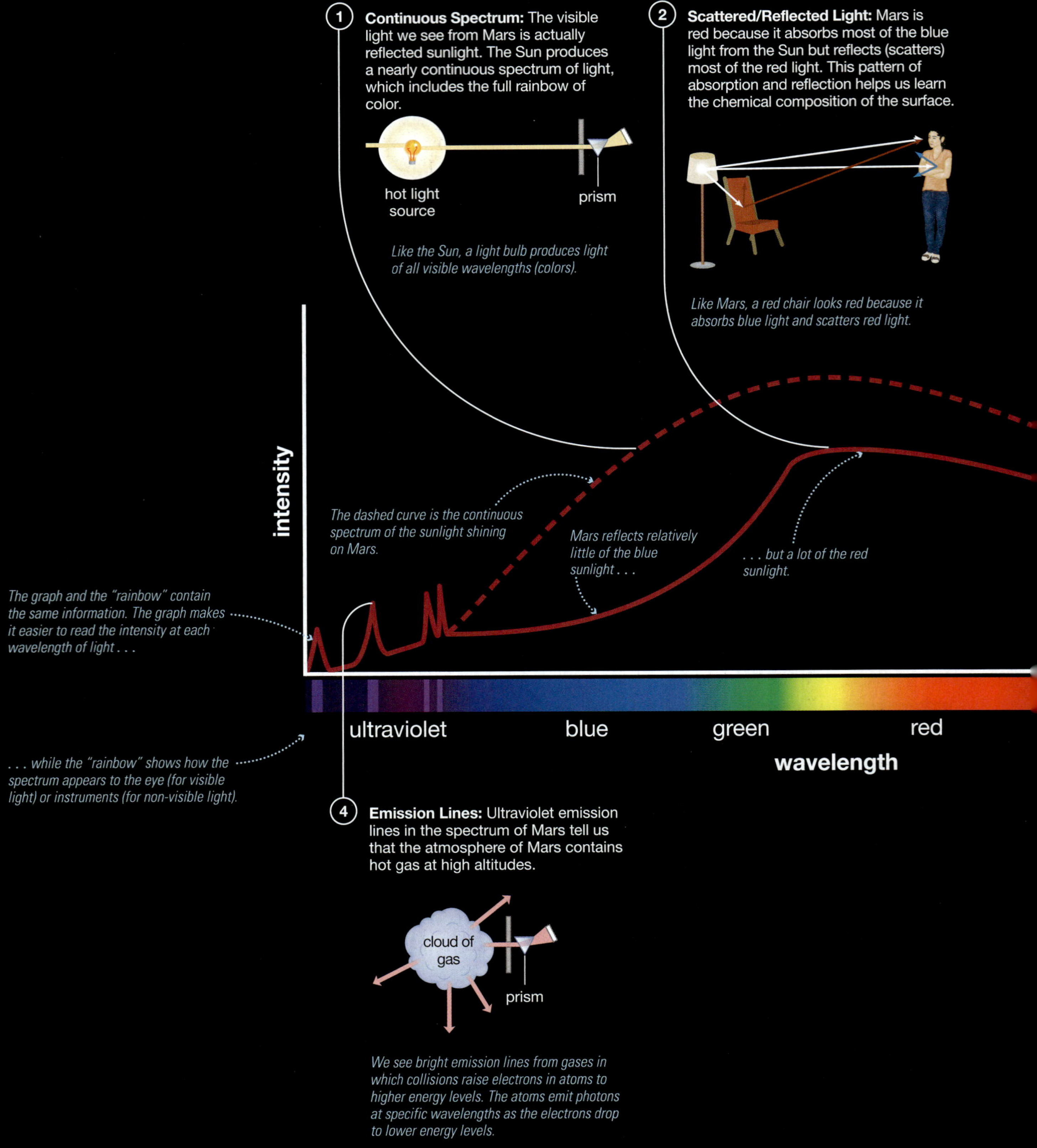

**3** **Thermal Radiation:** Objects emit a continuous spectrum of thermal radiation that peaks at a wavelength determined by temperature. Thermal radiation from Mars produces a broad hump in the infrared, with a peak indicating a surface temperature of about 225 K.

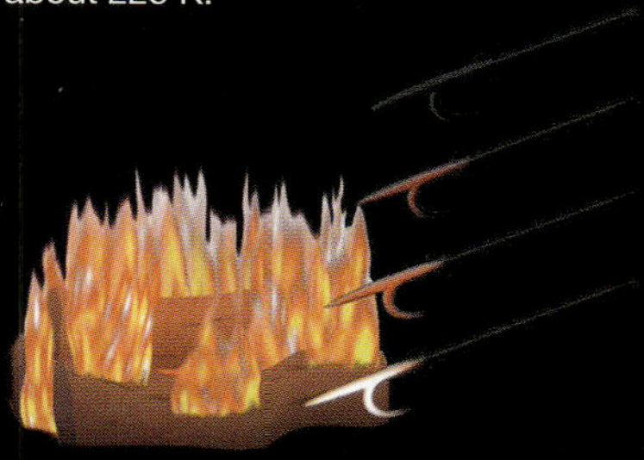

*All objects—whether a fireplace poker, planet, or star—emit thermal radiation. The hotter the object, (1) the more total light (per unit area), and (2) the higher the average energy (shorter average wavelength) of the emitted photons.*

*Mars's thermal radiation peaks in the infrared because it is much cooler than the Sun, which peaks in visible light.*

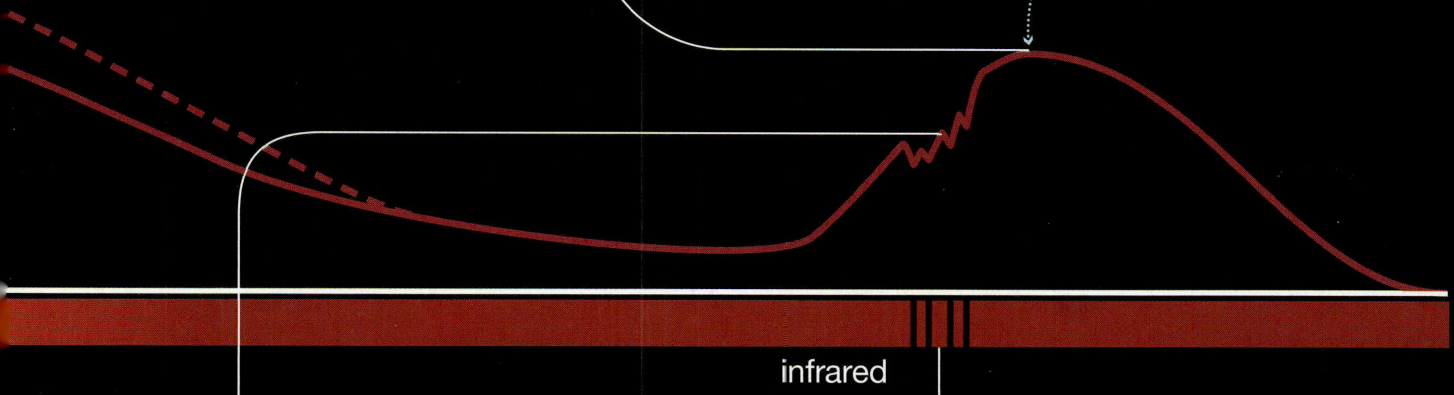


**5** **Absorption Lines:** These absorption lines reveal the presence of carbon dioxide in Mars's atmosphere.

*When light from a hot source passes through a cooler gas, the gas absorbs light at specific wavelengths that raise electrons to higher energy levels. Every different element, ion, and molecule has unique energy levels and hence its own spectral "fingerprint."*

**6** **Doppler Effect:** The wavelengths of the spectral lines from Mars are slightly shifted by an amount that depends on the velocity of Mars toward or away from us as it moves in its orbit around the Sun.

*A Doppler shift toward the red side of the spectrum tells us the object is moving away from us. A shift toward the blue side of the spectrum tells us the object is moving toward us. For planets and stars, Doppler shifts are far too small to be detected by eye.*

to a surface temperature of about 225 K, well below the 273 K freezing point of water. The absorption bands in the infrared come mainly from carbon dioxide, indicating a carbon dioxide atmosphere. The emission lines in the ultraviolet come from hot gas in a high, thin layer of the object's atmosphere. The reflected light looks like the Sun's 5800 K thermal radiation except that much of the blue light is missing, so the object must be reflecting sunlight and must look red in color. Perhaps by now you have guessed that this figure represents the spectrum of the planet Mars.

## MATHEMATICAL INSIGHT 5.3 The Doppler Shift

We can calculate an object's radial (toward or away from us) velocity from its Doppler shift. For speeds that are small compared to the speed of light (less than a few percent of $c$), the formula is

$$\frac{v_{rad}}{c} = \frac{\lambda_{shift} - \lambda_{rest}}{\lambda_{rest}}$$

where $v_{rad}$ is the radial velocity of the object, $\lambda_{rest}$ is the rest wavelength of a particular spectral line, and $\lambda_{shift}$ is the shifted wavelength of the same line. A positive answer means the object is redshifted and moving away from us; a negative answer means it is blueshifted and moving toward us.

**EXAMPLE:** One of the visible lines of hydrogen has a rest wavelength of 656.285 nm, but it appears in the spectrum of the star Vega at 656.255 nm. How is Vega moving relative to us?

**SOLUTION:**

**Step 1 Understand:** We can calculate the radial velocity from the given formula. Note that the line's wavelength in Vega's spectrum is slightly *shorter* than its rest wavelength, which means it is blueshifted and Vega's radial motion is *toward* us.

**Step 2 Solve:** We plug in the rest wavelength ($\lambda_{rest} = 656.285$ nm) and the wavelength in Vega's spectrum ($\lambda_{shift} = 656.255$ nm):

$$\frac{v_{rad}}{c} = \frac{\lambda_{shift} - \lambda_{rest}}{\lambda_{rest}}$$

$$= \frac{656.255\text{ nm} - 656.285\text{ nm}}{656.285\text{ nm}}$$

$$= -4.5712 \times 10^{-5}$$

**Step 3 Explain:** We have found Vega's radial velocity as a fraction of the speed of light; it is negative because Vega is moving toward us. To convert to a velocity in km/s, we multiply by the speed of light:

$$v_{rad} = -4.5712 \times 10^{-5} \times c$$

$$= -4.5712 \times 10^{-5} \times (3 \times 10^5\text{ km/s})$$

$$= -13.7\text{ km/s}$$

Vega is moving *toward* us at 13.7 km/s. This speed is typical of stars in our neighborhood of the galaxy.

## The BIG Picture PUTTING CHAPTER 5 INTO CONTEXT

*This chapter was devoted to one essential purpose: understanding how we learn about the universe by observing the light of distant objects. "Big picture" ideas that will help you keep your understanding in perspective include the following:*

- Light and matter interact in ways that allow matter to leave "fingerprints" on light. We can therefore learn a great deal about the objects we observe by carefully analyzing their light. Most of what we know about the universe comes from information that we receive from light.
- The visible light that our eyes can see is only a small portion of the complete electromagnetic spectrum. Different portions of the spectrum contain different pieces of the story of a distant object, so it is important to study all forms of light.
- There is far more to light than meets the eye. By dispersing the light of a distant object into a spectrum, we can determine the object's composition, surface temperature, motion toward or away from us, and more.

**MY COSMIC PERSPECTIVE** The methods of learning from light for astronomy are also used to learn about many things on Earth. For example, we use light to learn about the atmosphere and global warming, to identify toxic chemicals in water, and to perform medical scans that help diagnose injuries and disease.

## Summary of Key Concepts

### 5.1 Light in Everyday Life

- **How do we experience light?** Light carries radiative energy that it can exchange with matter. **Power** is the *rate* of energy transfer, measured in **watts**: 1 watt = 1 joule/s. The colors of light contain a great deal of information about the matter with which it has interacted.
- **How do light and matter interact?** Matter can emit, absorb, transmit, or reflect (or scatter) light.

### 5.2 Properties of Light

- **What is light?** Light is an **electromagnetic wave**, but it also comes in individual "pieces" called **photons**. Each photon has a precise wavelength, frequency, and energy: The shorter the wavelength, the higher the frequency and energy.
- **What is the electromagnetic spectrum?** In order of decreasing wavelength (increasing frequency and energy),

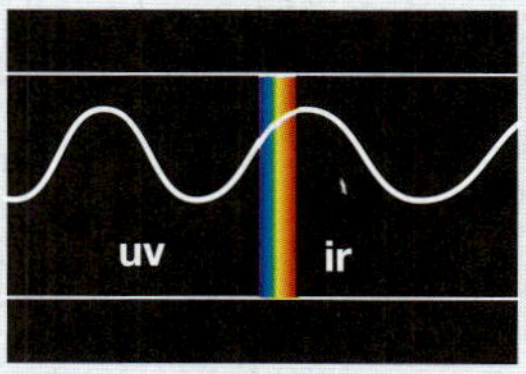


the forms of light are radio waves, microwaves, infrared, visible light, ultraviolet, x-rays, and gamma rays.

## 5.3 Properties of Matter

- **What is the structure of matter?** Ordinary matter is made of **atoms**, which are made of **protons**, **neutrons**, and **electrons**. Atoms of different **chemical elements** have different numbers of protons. **Isotopes** of a particular chemical element all have the same number of protons but different numbers of neutrons. **Molecules** are made from two or more atoms.

- **What are the phases of matter?** The appearance of matter depends on its **phase**: **solid**, **liquid**, or **gas**. Some gas always **vaporizes** from the solid or liquid phases; solids *sublimate* into gas and liquids *evaporate* into gas. At very high temperatures, **molecular dissociation** breaks up molecules and **ionization** strips electrons from atoms; an ionized gas is called a **plasma**.
- **How is energy stored in atoms?** Electrons can exist at particular **energy levels** within an atom. **Energy level transitions**, in which an electron moves from one energy level to another, can occur only when the electron gains or loses just the right amount of energy.

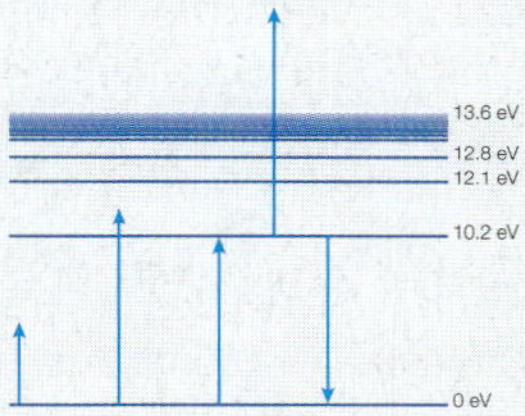


## 5.4 Learning from Light

- **What are the three basic types of spectra?** There are three basic types of spectra: a **continuous spectrum**, which looks like a rainbow of light; an **absorption line spectrum**, in which specific colors are missing from the rainbow; and an **emission line spectrum**, in which we see light only with specific colors against a black background.

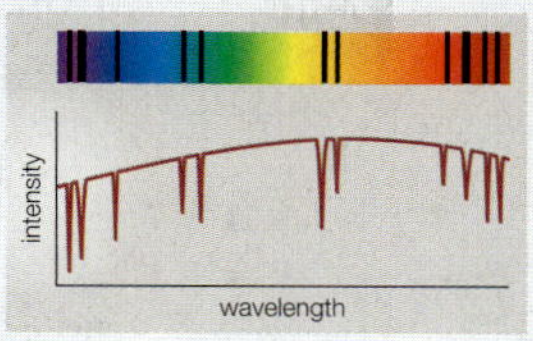


- **How does light tell us what things are made of?** **Emission lines** or **absorption lines** occur only at specific wavelengths that correspond to particular energy level transitions in atoms or molecules. Every kind of atom, ion, and molecule produces a unique set of spectral lines, so we can determine composition by identifying these lines.
- **How does light tell us the temperatures of planets and stars?** Objects such as planets and stars produce **thermal radiation** spectra, the most common type of continuous spectra. We can determine temperature from these spectra because hotter objects emit more total radiation per unit area and emit photons with a higher average energy.

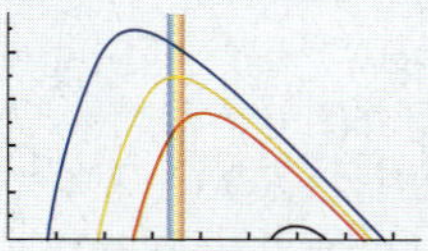

- **How does light tell us the speed of a distant object?** The **Doppler effect** tells us how fast an object is moving toward or away from us. Spectral lines are shifted to shorter wavelengths (a **blueshift**) in objects moving toward us and to longer wavelengths (a **redshift**) in objects moving away from us.

# Visual Skills Check

*Use the following questions to check your understanding of some of the many types of visual information used in astronomy. For additional practice, try the Chapter 5 Visual Quiz in the Study Area at www.MasteringAstronomy.com.*

*The graph above is a schematic spectrum of the planet Mars; it is the same spectrum shown in Figure 5.13. Keeping in mind that Mars reflects visible sunlight and emits infrared light, refer to the numbered features of the graph and answer the following questions.*

1. Which of the six numbered features represents emission lines?
2. Which of the six numbered features represents absorption lines?
3. Which portion(s) of the spectrum represent(s) reflected sunlight?
   a. 1 only  b. 2, 3, and 4  c. 3 and 6  d. the entire spectrum
4. What does the wavelength of the peak labeled 6 tell us about Mars?
   a. its color  b. its surface temperature
   c. its chemical composition  d. its orbital speed
5. What feature(s) of this spectrum indicate(s) that Mars appears red in color?
   a. the wavelength of the peak labeled 3
   b. the wavelength of the peak labeled 6
   c. the fact that the intensity of region 4 is higher than that of region 2
   d. the fact that the peak labeled 3 is higher than the peak labeled 6

# Exercises and Problems

*For instructor-assigned homework and other learning materials, go to www.MasteringAstronomy.com.*

## Chapter Review Questions

### Short-Answer Questions Based on the Reading

1. What is the difference between *energy* and *power*? What units do we use to measure power?
2. What are the four major ways light and matter can interact? Give an example of each from everyday life.
3. Why do we say that light is an *electromagnetic wave*? Describe the relationship among *wavelength, frequency*, and *speed* for light.
4. What is a *photon*? In what way is a photon like a particle? In what way is it like a wave?
5. List the different forms of light in order from lowest to highest energy. Is the order the same from lowest to highest frequency? From shortest to longest wavelength? Explain.
6. Briefly describe the structure and size of an atom. How big is the *nucleus* in comparison to the entire atom?
7. Define *atomic number* and *atomic mass number*. Under what conditions are two atoms different *isotopes* of the same element? What is a *molecule*?
8. What is *electrical charge*? Will an electron and a proton attract or repel each other? How about two electrons? Explain.
9. Describe the phase changes of water as you heat it, starting from its solid phase, ice. What happens at very high temperatures? What is a *plasma*?
10. Describe the *energy levels* that we find for electrons in atoms. Under what circumstances can *energy level transitions* occur?
11. How do we convert a spectrum shown as a band of light (like a rainbow) into a graph of the spectrum?
12. Describe the conditions that lead to each of the three basic types of spectra. Which type is the Sun's visible-light spectrum, and why?
13. How can we use *emission* or *absorption lines* to determine the chemical composition of a distant object?
14. Describe two ways in which the *thermal radiation spectrum* of an 8000 K star would differ from that of a 4000 K star.
15. Describe the *Doppler effect* for light and what we can learn from it. What does it mean to say that radio waves are *blueshifted*? Why does the Doppler effect widen the spectral lines of rotating objects?
16. Describe each of the key features of the spectrum in Figure 5.25 and explain what it tells us about the object.

### Does It Make Sense?

*Decide whether or not each of the following statements makes sense (or is clearly true or false). Explain clearly; not all of these have definitive answers, so your explanation is more important than your chosen answer.*

17. The walls of my room are transparent to radio waves.
18. Because of their higher frequencies, x-rays must travel through space faster than radio waves.
19. If you could see infrared light, you would see a glow from the backs of your eyelids when you closed your eyes.
20. If you had x-ray vision, you could read this entire book without turning any pages.
21. Two isotopes of the element rubidium differ in their number of protons.
22. A "white hot" object is hotter than a "red hot" object.
23. If the Sun's surface became much hotter (while the Sun's size remained the same), the Sun would emit more ultraviolet light but less visible light than it currently emits.
24. If you could view a spectrum of light reflecting off a blue sweatshirt, you'd find the entire rainbow of color (looking the same as a spectrum of white light).
25. Galaxies that show redshifts must be red in color.
26. If a distant galaxy has a substantial redshift (as viewed from Earth), then anyone living in that galaxy would see a substantial redshift in a spectrum of the Milky Way Galaxy.

### Quick Quiz

*Choose the best answer to each of the following. For additional practice, try the Chapter 5 Reading and Concept Quizzes in the Study Area at www.MasteringAstronomy.com.*

27. Why is a sunflower yellow? (a) It emits yellow light. (b) It absorbs yellow light. (c) It reflects yellow light.
28. Compared to red light, blue light has higher frequency and (a) higher energy and shorter wavelength. (b) higher energy and longer wavelength. (c) lower energy and shorter wavelength.
29. Radio waves are (a) a form of sound. (b) a form of light. (c) a type of spectrum.
30. Compared to an atom as a whole, an atomic nucleus is (a) very tiny but has most of the mass. (b) quite large and has most of the mass. (c) very tiny and has very little mass.
31. Some nitrogen atoms have seven neutrons and some have eight neutrons; these two forms of nitrogen are (a) ions of each other. (b) phases of each other. (c) isotopes of each other.
32. Ionization is the process by which (a) electrons escape from atoms. (b) liquid material enters the gas phase. (c) molecules break apart into individual atoms.
33. If you heat a rock until it glows, its spectrum will be (a) a thermal radiation spectrum. (b) an absorption line spectrum. (c) an emission line spectrum.
34. The set of spectral lines that we see in a star's spectrum depends on the star's (a) interior temperature. (b) chemical composition. (c) rotation rate.
35. Compared to the Sun, a star whose spectrum peaks in the infrared is (a) cooler. (b) hotter. (c) larger.
36. A spectral line that appears at a wavelength of 321 nm in the laboratory appears at a wavelength of 328 nm in the spectrum of a distant object. We say that the object's spectrum is (a) redshifted. (b) blueshifted. (c) whiteshifted.

## Inclusive Astronomy

*Use these questions to reflect on participation in science.*

37. *Pair Discussion: The Curies.* The married couple Marie and Pierre Curie were pioneers in the study of matter.
    **a.** Working independently, learn about the Nobel prizes the Curies won, the reasons for those prizes, and the differing roles Marie and Pierre assumed in carrying out their

research. Also, find the overall fraction of Nobel prizes in science that have gone to women.

**b.** Pair up with another student, one of you talking about Marie Curie and the other about Pierre Curie. Take turns describing what each scientist did to help earn their Nobel prizes. Then discuss whether you think they both deserved the Nobel prizes they received, and how well you think they worked together and supported each other.

**c.** Discuss the fraction of Nobel prizes that have gone to women. Does it seem surprising to you? Why or why not?

## The Process of Science

*These questions may be answered individually in short-essay form or discussed in groups, except where identified as group-only.*

38. *Newton's Prisms.* Review the brief discussion in this chapter of how Newton proved that the colors seen when light passes through a prism come from the light itself rather than from the prism. Suppose you wanted to test Newton's findings. Assuming you had two prisms and a white screen, describe how you would arrange the prisms to duplicate Newton's discovery.
39. *Elements in Space.* Astronomers claim that objects throughout the universe are made of the same chemical elements that exist here on Earth. Given that most of these objects are so far away that we can never hope to visit them, why are astronomers so confident that these objects are made from the same set of chemical elements, rather than some completely different types of materials?
40. *The Changing Limitations of Science.* Review the box "Extraordinary Claims: We Can Never Learn the Composition of Stars," which gives an example of how new discoveries can change the apparent limitations of science. Today, other questions seem beyond the reach of science, such as the question of how life began on Earth. Do you think such questions will ever be answerable through science? Defend your opinion.
41. *Democritus and the Path of History.* Besides his belief in atoms, Democritus held several other strikingly modern notions. For example, he maintained that the Moon was a world with mountains and valleys and that the Milky Way was composed of countless individual stars—ideas that weren't generally accepted until the time of Galileo, more than 2000 years later. Unfortunately, we know of Democritus's work only secondhand, because none of the 72 books he is said to have written survived the destruction of the Library of Alexandria. Do you think history might have been different if the work of Democritus had not been lost? Defend your opinion.
42. *Group Activity: Light Around You.* Working in groups, look carefully at all the ways in which light and matter are interacting in the room around you to answer the following questions; explain all your answers clearly. Note: You may wish to do this activity using the four roles described in Chapter 1, Exercise 39.
    **a.** What is emitting light?
    **b.** What is absorbing light?
    **c.** What is responsible for the colors you see?
    **d.** What would the room look like if you observed it with an infrared camera? With an ultraviolet camera? With an x-ray camera?
    **e.** Are there any radio waves in the room? How do you know?

## Investigate Further

### Short-Answer/Essay Questions

43. *Atomic Terminology Practice I.*
    **a.** The most common form of iron has 26 protons and 30 neutrons. State its atomic number, atomic mass number, and number of electrons (if it is neutral).
    **b.** Consider the following three atoms: Atom 1 has 7 protons and 8 neutrons; atom 2 has 8 protons and 7 neutrons; atom 3 has 8 protons and 8 neutrons. Which two are *isotopes* of the same element?
    **c.** Oxygen has atomic number 8. How many times must an oxygen atom be ionized to create an $O^{+5}$ ion? How many electrons are in an $O^{+5}$ ion?
44. *Atomic Terminology Practice II.*
    **a.** What are the atomic number and atomic mass number of a fluorine atom with 9 protons and 10 neutrons? If we could add a proton to this fluorine nucleus, would the result still be fluorine? What if we added a neutron to the fluorine nucleus? Explain.
    **b.** The most common isotope of gold has atomic number 79 and atomic mass number 197. How many protons and neutrons does the gold nucleus contain? If the isotope is electrically neutral, how many electrons does it have? If it is triply ionized, how many electrons does it have?
    **c.** Uranium has atomic number 92. Its most common isotope is $^{238}U$, but the form used in nuclear bombs and nuclear power plants is $^{235}U$. How many neutrons are in each of these two isotopes of uranium?
45. *The Fourth Phase of Matter.*
    **a.** Explain why nearly all the matter in the Sun is in the plasma phase.
    **b.** Based on your answer to part a, explain why plasma is the most common phase of matter in the universe.
    **c.** If plasma is the most common phase of matter in the universe, why is it so rare on Earth?
46. *Energy Level Transitions.* The diagram below shows five transitions (labeled with capital letters A through E) representing an electron moving between energy levels in a hydrogen atom. Explain your answers clearly.

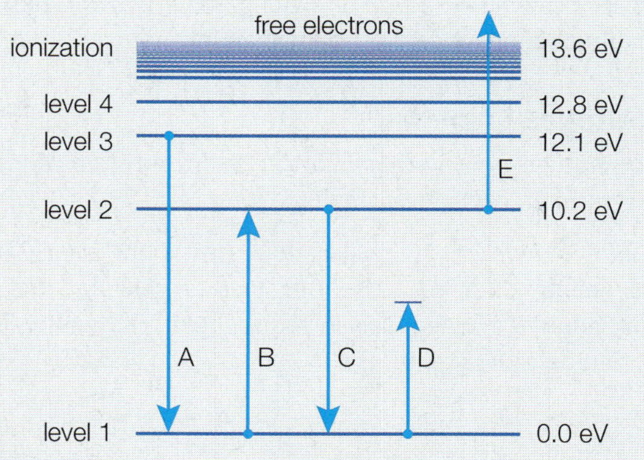


    **a.** Which transition could represent an atom that *absorbs* a photon with 10.2 eV of energy?
    **b.** Which transition could represent an atom that *emits* a photon with 10.2 eV of energy?
    **c.** Which transition represents an electron that is breaking free of the atom?
    **d.** Which transition, as shown, is *not* possible?

**e.** Does transition A represent emission or absorption of light, and how would the wavelength of the emitted or absorbed photon compare to that of the photon involved in transition C?

47. *Spectral Summary.* Clearly explain how studying an object's spectrum can allow us to determine each of the following properties of the object.
**a.** The object's surface chemical composition
**b.** The object's surface temperature
**c.** Whether the object is a low-density cloud of gas or something more substantial
**d.** Whether the object has a hot upper atmosphere
**e.** Whether the object is reflecting blue light from a star
**f.** The speed at which the object is moving toward or away from us
**g.** The object's rotation rate

48. *Orion Nebula.* Much of the Orion Nebula looks like a glowing cloud of gas. What type of spectrum would you expect to see from the glowing parts of the nebula? Why?

49. *The Doppler Effect.* In hydrogen, the transition from level 2 to level 1 has a rest wavelength of 121.6 nm. Suppose you see this line at a wavelength of 120.5 nm in Star A, 121.2 nm in Star B, 121.9 nm in Star C, and 122.9 nm in Star D. Which stars are coming toward us? Which are moving away? Which star is moving fastest relative to us? Explain your answers without doing any calculations.

50. *Your Microwave Oven.* A microwave oven emits microwaves that have just the right wavelength to cause energy level changes in water molecules. Use this fact to explain how a microwave oven cooks your food. Why doesn't a microwave oven make a plastic dish get hot? Why do some clay dishes get hot in the microwave? Why do dishes that aren't themselves heated by the microwave oven sometimes still get hot when you heat food on them? (*Note:* It's not a good idea to put empty dishes in a microwave.)

51. *Light Bulbs.* Look up typical spectra of traditional incandescent light bulbs, compact fluorescent light (CFL) bulbs, and LED bulbs. (Or, if you have a diffraction grating [which you can buy for about $1], observe the spectra for yourself.) How do the spectra of the different types of light bulbs compare? Use what you find to explain why CFLs and LEDs can be much more energy efficient than incandescent bulbs.

## Quantitative Problems

*Be sure to show all calculations clearly and state your final answers in complete sentences.*

52. *Human Wattage.* A typical adult uses about 2500 Calories of energy each day. Use this fact to calculate the typical adult's average *power* requirement, in watts. (*Hint:* 1 Calorie = 4184 joules.)

53. *Electric Bill.* Your electric utility bill probably shows your energy use for the month in units of *kilowatt-hours*. A kilowatt-hour is defined as the energy used in 1 hour at a rate of 1 kilowatt (1000 watts); that is, 1 kilowatt-hour = 1 kilowatt × 1 hour. Use this fact to convert 1 kilowatt-hour into joules. If your bill says you used 900 kilowatt-hours, how much energy did you use in joules?

54. *Radio Station.* What is the wavelength of a radio photon from an AM radio station that broadcasts at 1120 kilohertz? What is its energy?

55. *UV Photon.* What is the energy (in joules) of an ultraviolet photon with wavelength 120 nm? What is its frequency?

56. *X-Ray Photon.* What is the wavelength of an x-ray photon with energy 10 keV (10,000 eV)? What is its frequency? (1 eV = $1.60 \times 10^{-19}$ joule.)

57. *How Many Photons?* Suppose that all the energy from a 100-watt light bulb came in the form of photons with wavelength 600 nm. (Note: In reality, light bulbs distribute energy over many wavelengths; see Exercise 51.)
**a.** Calculate the energy of a *single* photon with wavelength 600 nm.
**b.** How many 600-nm photons would have to be emitted each second to account for all the light from this 100-watt light bulb?
**c.** Based on your answer to part b, explain why we don't notice the particle nature of light in our everyday lives.

58. *Thermal Radiation Laws.*
**a.** Find the emitted power per square meter and wavelength of peak intensity for a 3000 K object that emits thermal radiation.
**b.** Find the emitted power per square meter and wavelength of peak intensity for a 50,000 K object that emits thermal radiation.

59. *Hotter Sun.* Suppose the surface temperature of the Sun were about 12,000 K, rather than 6000 K.
**a.** How much more thermal radiation would the Sun emit?
**b.** What would happen to the Sun's wavelength of peak emission?
**c.** Do you think it would still be possible for life to exist on Earth? Explain.

60. *Taking the Sun's Temperature.* The Sun radiates a total power of about $4 \times 10^{26}$ watts into space. The Sun's radius is about $7 \times 10^{8}$ meters.
**a.** Calculate the average power radiated by each square meter of the Sun's surface. (*Hint:* The formula for the surface area of a sphere is $A = 4\pi r^2$.)
**b.** Using your answer from part a and the Stefan-Boltzmann law, calculate the average surface temperature of the Sun. (*Note:* The temperature calculated this way is called the Sun's *effective temperature*.)

61. *Doppler Calculations.* In hydrogen, the transition from level 2 to level 1 has a rest wavelength of 121.6 nm. Find the speed and direction (toward or away from us) of a star in which this line appears at wavelength
**a.** 120.5 nm. **b.** 121.2 nm. **c.** 121.9 nm. **d.** 122.9 nm.

# 6 Telescopes

## PORTALS OF DISCOVERY

▲ **About the photo:** The Atacama Large Millimeter/submillimeter Array (ALMA) is located in a high desert (altitude 5000 meters) in Chile.

## LEARNING GOALS

### 6.1 Eyes and Cameras: Everyday Light Sensors
- How do eyes and cameras work?

### 6.2 Telescopes: Giant Eyes
- What are the two most important properties of a telescope?
- What are the two basic designs of telescopes?
- What do astronomers do with telescopes?

### 6.3 Telescopes and the Atmosphere
- How does Earth's atmosphere affect ground-based observations?
- Why do we put telescopes into space?

### 6.4 Telescopes Across the Spectrum
- How do we observe invisible light?
- How can multiple telescopes work together?

*All of this has been discovered and observed these last days thanks to the telescope that I have [built], after having been enlightened by divine grace.*
—*Galileo*

**Chapter 6 Overview**

We are in the midst of a revolution in human understanding of the universe, sparked in large part by advances in telescope technology. New technologies are fueling the construction of larger and more powerful telescopes, increasing our ability to collect and process vast amounts of data, and in some cases opening windows to the universe beyond those offered by light alone.

Because telescopes are the portals through which we study the universe, understanding them can help us understand both the triumphs and the limitations of modern astronomy. In this chapter, we explore the basic principles by which telescopes work and some of the technological advances behind the current revolution in astronomy.

## 6.1 Eyes and Cameras: Everyday Light Sensors

We learn about the world around us by observing with our five senses (touch, taste, smell, hearing, sight) and using our brains to analyze and interpret the data that our senses record. The science of astronomy progresses similarly. We collect data about the universe, and then we analyze and interpret the data. Within our solar system, we can analyze some matter directly, such as samples of Earth's surface, meteorites, and surfaces and atmospheres of worlds visited by spacecraft. Nearly all other data about the universe come to us in the form of light, which we collect with telescopes and record with cameras and other instruments. Because telescopes function much like giant eyes, we begin this chapter by examining the principles of eyes and cameras, our everyday light sensors.

### How do eyes and cameras work?

Eyes and cameras work similarly, so let's begin with eyes. The eye is a remarkably complex organ, but its basic components are a *pupil*, a *lens*,* and a *retina* (**FIGURE 6.1**). The pupil controls how much light enters the eye; it dilates (opens wider) in low light and constricts in bright light. The lens bends light to form an image on the retina. The retina contains light-sensitive cells (called *cones* and *rods*) that, when triggered by light, send signals to the brain via the optic nerve.

**See it for yourself** Use a mirror to compare the opening size of your pupils under normal lighting and right after looking at a bright light. What do you notice? Why do eye doctors dilate your pupils during eye exams?

*The lens actually works together with the cornea (the clear part of the eye in front of the pupil), but for simplicity we will consider their combined effects as the effects of the lens.

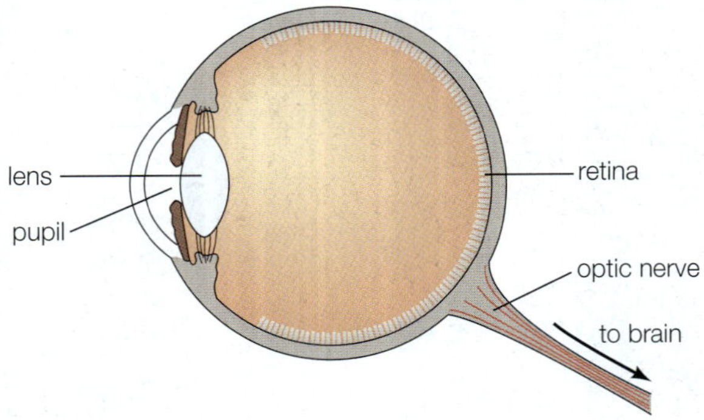


**FIGURE 6.1** A simplified diagram of the human eye.

**Bending Light** The lens of the eye creates an image by bending light in much the same way as a simple glass lens. You can understand why light bends by imagining a light wave coming toward you from far away. The peaks and troughs of the electric and magnetic fields are perpendicular to the light wave's direction of travel, as shown for the approaching light wave in **FIGURE 6.2**. The wave slows down when it hits glass or your eye because light travels more slowly through denser matter than through air. For light coming in at an angle (as in Figure 6.2), this slowing affects the side of the wave nearest the surface first, allowing the far side to catch up. The result is bending (more technically known as *refraction*)—a change in the direction in which the light is traveling. **FIGURE 6.3** shows an example of how Earth's atmosphere bends light from space, distorting the Sun's image at sunset.

**Image Formation** We can visualize the bending of light by drawing simple *rays*, with each ray (drawn as an arrow) representing light coming from a single direction. Light rays that enter the lens farther from the center are bent more, and rays that pass directly through the center are not bent at all. In this way, parallel rays of light, such as those from a distant star, converge to a point called the **focus** (or **focal**

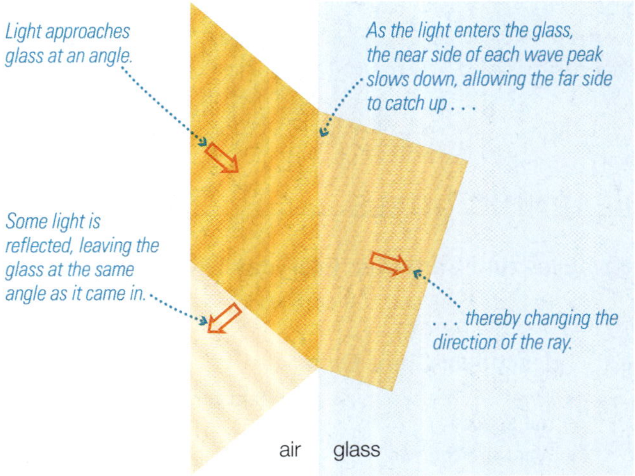


**FIGURE 6.2** Light that hits glass at an angle bends as it enters the glass, a phenomenon called *refraction*. The wide yellow ribbons in this figure represent light waves. The darker bands on those ribbons (perpendicular to the direction in which the light travels) represent the positions of wave peaks.

FIGURE 6.3 Earth's atmosphere also bends light. The Sun looks squashed at sunset because light from the lower portion of the Sun passes through more atmosphere and therefore bends slightly more than light from the upper portion.

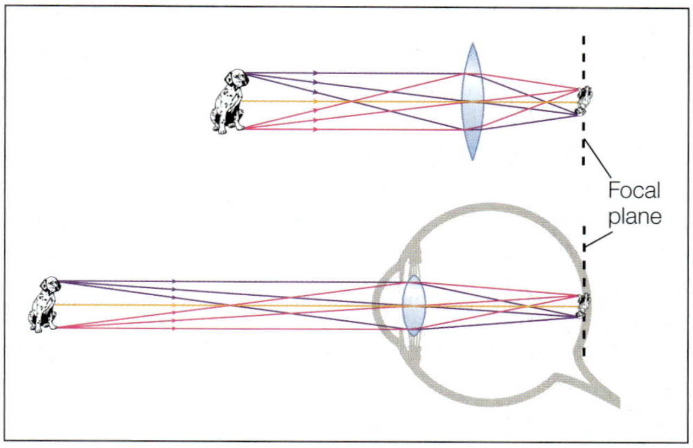


FIGURE 6.5 Light from different parts of an object focuses at different points to make an (upside-down) image of the object.

**point**). FIGURE 6.4 shows the idea for both a glass lens and an eye. The fact that parallel rays of light converge to a sharp focus explains why distant stars appear as *points* of light to our eyes or on photographs.

Light rays that are *not* parallel, such as those from a nearby object, enter a lens from different directions. These rays do not all converge at the focus, but they still follow precise rules as they bend at the lens; we will not discuss these rules in this book, but some of them are illustrated by the ray paths in FIGURE 6.5. The result is the bending of rays to form an **image** of the original object. The place where the image appears in focus is called the **focal plane** of the lens. In an eye with perfect vision, the focal plane is on the retina. (The retina actually is curved, rather than a flat plane, but we will ignore this detail.) Note that the image formed by a lens is upside down. In other words, our eyes actually form upside-down images, which are then flipped right-side-up by our brains.

**Recording Images** If we want to keep an image or study it in detail, it's useful to record it with a camera (FIGURE 6.6). The basic operation of a camera is quite similar to that of an eye. The camera has a small opening for light to enter, much like the pupil of the eye. The camera lens bends the light, bringing it to a focus on a **detector** that makes a permanent record of the image. Today, detectors are nearly always electronic, but older cameras used photographic film. Cameras also have a *shutter* that is analogous to an eyelid: Light can reach the detector only when the shutter is open. We can use the shutter to control the **exposure time** of an image, the amount of time during which light collects on the detector. A longer exposure time means that more light reaches the detector, allowing the detector to record details that might be too faint to be seen in shorter exposures.

Modern detectors use electronic chips that are physically divided into grids of *picture elements,* or **pixels** for short. When a photon of light strikes a pixel, it causes a bit of electric charge to accumulate. Each subsequent photon striking the same pixel adds to this accumulated electric charge. After an exposure is complete, a computer measures the total electric charge in each pixel, thereby determining how many photons have struck each one. The overall image is stored on a memory chip as an array of numbers representing the results from each pixel. Most consumer camera chips now have 10 million or more pixels, and professional cameras can have significantly more.

No detector is perfect, so a variety of trade-offs must be made when images are recorded. For example, a longer exposure can reveal fainter details, but it may also cause

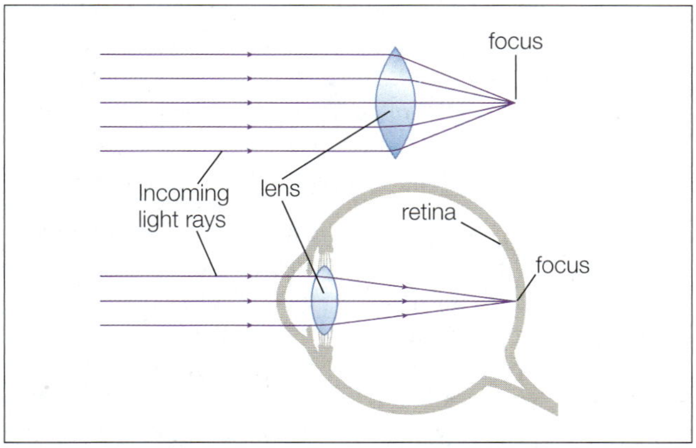


FIGURE 6.4 A glass lens bends parallel rays of light to a point called the *focus* of the lens. In an eye with perfect vision, rays of light are bent to a focus on the retina.

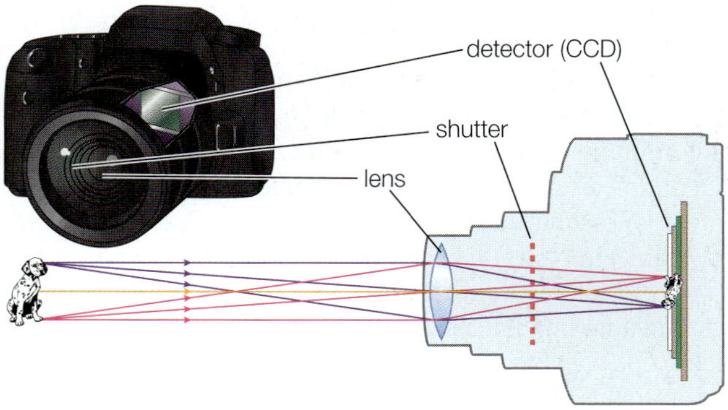


FIGURE 6.6 A camera works much like an eye. When the shutter is open, light passes through the lens to form an image on the detector (which may be film or an electronic device).

bright regions of the image to become overexposed, meaning that so many photons collect that they can no longer be counted accurately. Nevertheless, today's detectors are able to record light over a wide range of brightness levels much more accurately than the photographic film of the past, and the technology continues to improve.

The increasing sophistication of detector technology has provided major payoffs in astronomy. For example, equipping an "old" telescope with new detectors can vastly increase its power. Similarly, obtaining spectra of distant galaxies used to be very time-consuming and labor-intensive, but today astronomers can automate the process with detectors that can simultaneously record hundreds or even thousands of spectra.

**See it for yourself** Examine a digital camera. Where is its lens? Where is its detector? Can you control its exposure time manually? How many pixels does its detector have?

**Image Processing** The photographs you see in most media and in science today are not the original images recorded by cameras. Instead, these images have been combined and manipulated through techniques of *image processing*. Some of the images we see today have been manipulated to change what the camera actually recorded, something that's easily done with almost any photo software. In science, image processing is often used to bring out details that might otherwise remain hidden. For example, image processing can be used to sharpen or adjust colors, to correct over- or underexposure, or to remove artifacts or defects from an image. We will not discuss image processing much in this book, but you should be aware of its tremendous power both to increase the amount of information we can glean from images and, when misused, to distort what we see.

## 6.2 Telescopes: Giant Eyes

Telescopes are essentially giant eyes that can collect far more light than our own eyes. By combining this light-collecting capacity with cameras and other instruments that can record and analyze light in detail, modern telescopes have become extremely powerful scientific instruments.

### What are the two most important properties of a telescope?

Let's begin by investigating the two most fundamental properties of any telescope: its *light-collecting area* and its *angular resolution*.

A telescope's **light-collecting area** tells us how much total light it can collect at one time. Telescopes are generally round, so we usually characterize a telescope's size by the *diameter* of its light-collecting area. For example, a "10-meter telescope" has a light-collecting area that is 10 meters in diameter. Note that, because area is proportional to the *square* of diameter, a relatively small increase in diameter can mean a big increase in light-collecting area. A 10-meter telescope has five times the diameter of a 2-meter telescope, so its light-collecting area is $5^2 = 25$ times as great. The 10-meter telescope has more than 1000 times the diameter of the pupil of your eye, which means it can collect more light than $1000^2 = 1$ million human eyes.

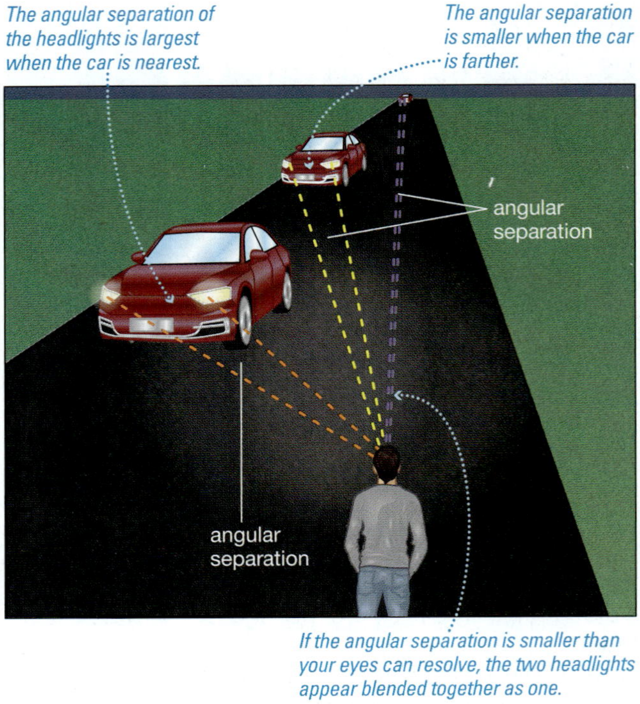


**FIGURE 6.7** Angular separation depends on distance: For a particular physical separation between two objects—such as two headlights on a car or two stars in a binary star system—the angular separation is smaller when the objects are farther away. If the angular separation is smaller than the angular resolution of your eyes or your telescope, the two objects will appear blended together as one.

**Angular resolution** is the *smallest* angle over which we can tell that two dots—or two stars—are distinct. The human eye has an angular resolution of about 1 arcminute $(\frac{1}{60}°)$, meaning that two stars can appear distinct only if they have at least this much angular separation in the sky. If the stars are separated by less than 1 arcminute, our eyes will not be able to distinguish them individually and they will look like a single star. The angular separation between two points of light depends both on their actual separation and on their distance from us [**Section 2.1**]; **FIGURE 6.7** shows the idea.

**See it for yourself** Poke two pin holes fairly close together in a dark sheet of paper. Have a friend hold a flashlight behind the paper and slowly back away until you see the two points of light blend together into one. How does the distance at which the points blend together change if you change the separation of the two holes? Bonus: Measure the separation of the holes and the distance at which the light blends together; then use the small angle formula (Mathematical Insight 2.1) to calculate the angular resolution of your eyes.

Large telescopes can have amazing angular resolution. For example, the 2.4-meter Hubble Space Telescope has an angular resolution of about 0.05 arcsecond (for visible light), which would allow you to read this book from a distance of almost 1 kilometer! Larger telescopes can have even better (smaller) angular resolution, though Earth's atmosphere prevents most ground-based telescopes from achieving their theoretical limits.

The ultimate limit to a telescope's resolving power comes from the properties of light. Because light is an

**FIGURE 6.8** This computer-generated image represents interference between overlapping sets of ripples on a pond. (The colors are for visual effect only.) Where peaks or troughs meet, the effects add to make the water rise extra high or fall extra low. Where peak meets trough, the effects cancel to make the water surface flat. Light waves also exhibit interference.

electromagnetic wave [**Section 5.2**], beams of light can interfere with one another like overlapping sets of ripples on a pond (**FIGURE 6.8**). This *interference* limits a telescope's angular resolution even when all other conditions are perfect. That is why even a high-quality telescope in space cannot have perfect angular resolution (**FIGURE 6.9**).

The angular resolution that a telescope could achieve if it were limited only by the interference of light waves is called its **diffraction limit.** (*Diffraction* is a technical term for the effects of interference that limit telescope resolution.) The diffraction limit depends on both the diameter of the telescope's primary mirror and the wavelength of the light being observed (see Mathematical Insight 6.2). For any particular wavelength of light, a larger telescope has a smaller diffraction limit, meaning it can achieve a better (smaller) angular resolution. For any particular telescope, the diffraction limit is larger (poorer angular resolution) for longer-wavelength light. That is why, for example, a radio telescope must be far larger than a visible-light telescope to achieve the same angular resolution.

## COMMON MISCONCEPTIONS

### Magnification and Telescopes

Many people guess that magnification is the most important function of a telescope. However, even though telescopes—much like telephoto camera lenses or binoculars—can magnify images, the amount of magnification a telescope can provide is *not* one of its crucial properties. No matter how much a telescope image is magnified, you cannot see details if the telescope does not collect enough light to show them or if they are smaller than the angular resolution of the telescope. Magnifying an image too much just makes it look blurry, which is why a telescope's light-collecting area and angular resolution are much more important than its magnification.

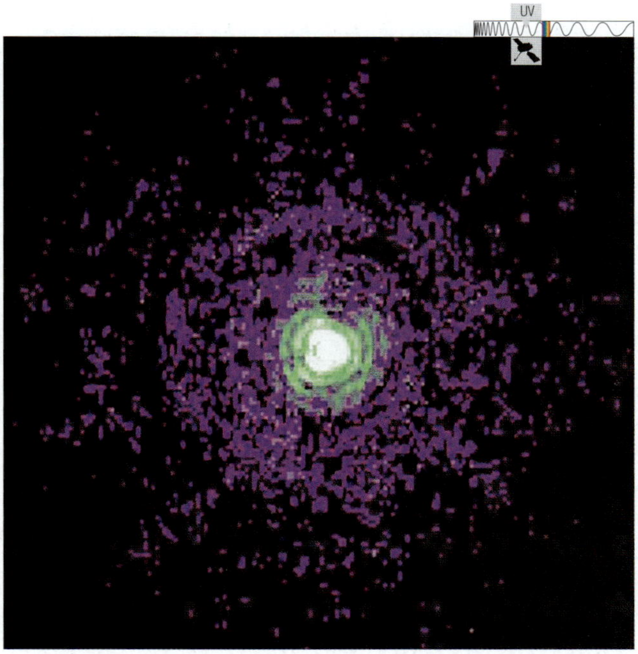


**FIGURE 6.9** When examined in detail, a Hubble Space Telescope image of a star has rings (represented as green and purple in the figure) resulting from the wave properties of light. With higher angular resolution, the rings would be smaller.

## What are the two basic designs of telescopes?

Telescopes come in two basic designs: *refracting* and *reflecting*. A **refracting telescope** operates much like an eye, using transparent glass lenses to collect and focus light (**FIGURE 6.10**). The earliest telescopes, including Galileo's, were refracting telescopes. The world's largest refracting telescope, completed

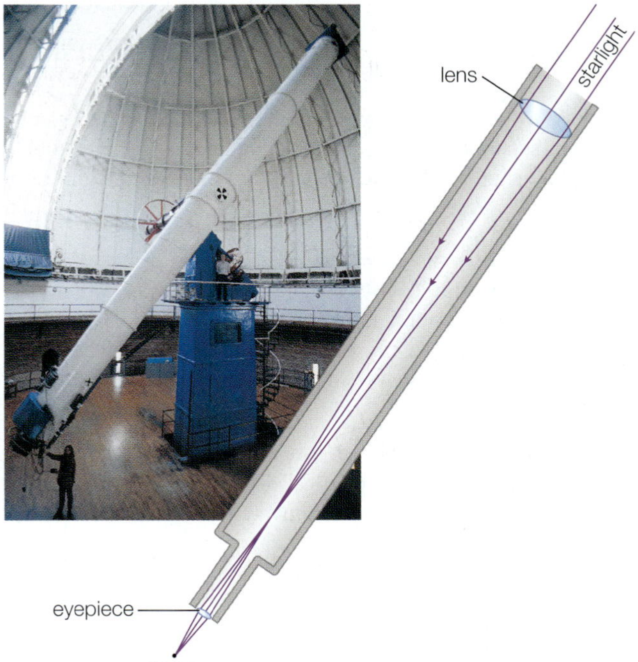


**FIGURE 6.10** A refracting telescope collects light with a large transparent lens (diagram). The photo shows the 1-meter refractor at the University of Chicago's Yerkes Observatory, the world's largest refracting telescope.

in 1897, has a lens that is 1 meter (40 inches) in diameter and a telescope tube that is 19.5 meters (64 feet) long.

A **reflecting telescope** uses a precisely curved *primary mirror* to gather light (**FIGURE 6.11**). This mirror reflects the gathered light to a *secondary mirror* that lies in front of it. The secondary mirror then reflects the light to a focus at a place where the eye or instruments can observe it—sometimes through a hole in the primary mirror and sometimes through the side of the telescope (often with the aid of additional small mirrors). The fact that the secondary mirror prevents some light from reaching the primary mirror might seem like a drawback, but in practice it is not a problem because only a small fraction of the incoming light is blocked.

Nearly all telescopes used in current astronomical research are reflectors, mainly for two practical reasons. First, because light passes *through* the lens of a refracting telescope, lenses must be made from clear, high-quality glass with precisely shaped surfaces on both sides. In contrast, only the reflecting surface of a mirror must be precisely shaped; the quality of the underlying glass is not a factor. Second, large glass lenses are extremely heavy and can be held in place only by their edges. Because the large lens is at the top of a refracting telescope, it is difficult to stabilize refracting telescopes and to prevent large lenses from deforming. The primary mirror of a reflecting telescope is mounted at the bottom, where its weight presents a far less serious problem. (Other problematic features of lenses include the fact that glass is not as transparent to infrared and ultraviolet light, and that a glass lens brings different colors of light into focus at slightly different places, though this problem—called *chromatic aberration*—can be minimized by using combinations of lenses.)

For a long time, the main factor limiting the size of reflecting telescopes was the sheer weight of the glass

## MATHEMATICAL INSIGHT 6.1 Angular Resolution

We often want to know whether a telescope can resolve (see as distinct) two points, such as two stars, based on their physical separation and distance. From Mathematical Insight 2.1, the angular separation of two points is given by

$$\text{angular separation} = \text{physical separation} \times \frac{360^\circ}{2\pi \times \text{distance}}$$

This formula gives an answer in degrees. Because there are 3600 arcseconds in 1 degree (see Figure 2.8), we can rewrite the formula to give a result in arcseconds by multiplying the right side by $\frac{3600''}{1^\circ}$ (recall that $''$ is the symbol for arcseconds). The right side will then read (physical separation)/(distance) times the numbers $360^\circ \times \frac{3600''}{1^\circ} \div 2\pi$, which a calculator shows to be approximately 206,265″. Therefore, the formula becomes

$$\text{angular separation} = 206{,}265'' \times \frac{\text{physical separation}}{\text{distance}}$$

**EXAMPLE 1:** A binary star system is 20 light-years away and its two stars are separated by 200 million kilometers. Can the Hubble Space Telescope resolve the two stars? Assume an angular resolution of 0.05 arcsecond.

**SOLUTION:**

**Step 1 Understand:** The telescope can resolve the two stars if their angular separation is larger than the angular resolution of 0.05 arcsecond. We can calculate the angular separation of the two stars because we know their distance and physical separation.

**Step 2 Solve:** Before we can use the angular separation formula, we must have the physical separation and distance in the same units. It's easiest to convert the light-years to kilometers from the fact that 1 light-year $\approx 10^{13}$ km (see Mathematical Insight 1.1); the 20-light-year distance becomes about $20 \times 10^{13}$ km $= 2 \times 10^{14}$ km. Writing the physical separation in scientific notation as $2 \times 10^8$ km, we find

$$\begin{aligned}\text{angular separation} &= 206{,}265'' \times \frac{\text{physical separation}}{\text{distance}} \\ &= 206{,}265'' \times \frac{2 \times 10^8 \ \cancel{\text{km}}}{2 \times 10^{14} \ \cancel{\text{km}}} = 0.2''\end{aligned}$$

**Step 3 Explain:** The angular separation of the two stars is 0.2 arcsecond. Because this is larger than the telescope's angular resolution of 0.05 arcsecond, the two stars can be distinguished and studied individually.

**EXAMPLE 2:** If you looked at this book with a telescope that had Hubble's angular resolution of 0.05 arcsecond, how far away could you place the book and still be able to read it?

**SOLUTION:**

**Step 1 Understand:** We can read the book if we can resolve its individual letters, so answering this question hinges on determining the relevant physical and angular separations. The letters in this book are about 2 millimeters tall, so one way to think about the question is to ask how closely spaced a set of dots would have to be to look like the letters in this book. If you do some test cases, you'll find that letters made 10 dots tall (and 10 dots wide) are clearly identifiable, so 2-mm-tall letters would have to be composed of dots separated by 0.2 millimeter. We can use this value as the physical separation of the dots. The dots will be resolved if their angular separation is greater than or equal to the telescope's resolution of 0.05 arcsecond, so we use this value as the angular separation. We can then use the angular separation formula to calculate the distance at which the book could be read through the telescope.

**Step 2 Solve:** We solve the angular separation formula for distance:

$$\text{distance} = 206{,}265'' \times \frac{\text{physical separation}}{\text{angular separation}}$$

We substitute 0.05 arcsecond for the angular separation and 0.2 millimeter for the physical separation:

$$\text{distance} = 206{,}265'' \times \frac{0.2 \text{ mm}}{0.05''} \approx 825{,}000 \text{ mm}$$

**Step 3 Explain:** The distance of 825,000 millimeters, or 825 meters, is the distance at which the angular separation of the dots composing the letters would equal the angular resolution of the telescope. Therefore, the book would be readable at a distance of up to 825 meters, or a little less than 1 kilometer.

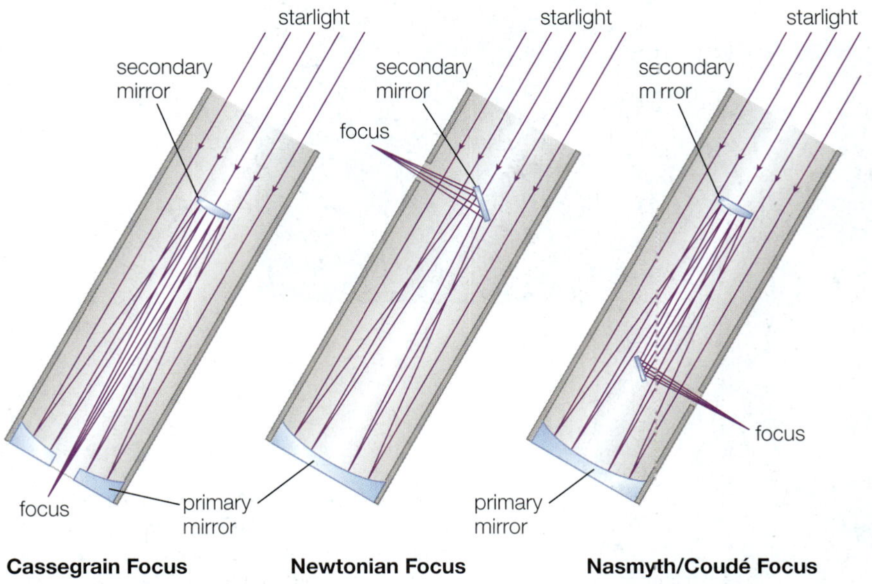


**a** Three variations on the basic design of a reflecting telescope. In all cases, a reflecting telescope collects light with a precisely curved primary mirror that reflects light back upward to the secondary mirror. In the *Cassegrain* design, the secondary mirror reflects the light through a hole in the primary mirror, so that the light can be observed with cameras or instruments beneath the telescope. In the *Newtonian* design, the secondary mirror reflects the light out to the side of the telescope. In the *Nasmyth* and *Coudé* designs, a third mirror is used to reflect light out the side but lower down than in the Newtonian design.

**b** The Gemini North telescope, located on the summit of Mauna Kea, Hawaii, is a reflecting telescope with the Cassegrain design. The primary mirror, visible at the bottom of the large lattice tube, is 8 meters in diameter. The secondary mirror, located in the smaller central lattice, reflects light back down through the hole visible in the center of the primary mirror.

**FIGURE 6.11** Reflecting telescopes.

## MATHEMATICAL INSIGHT 6.2 The Diffraction Limit

A simple formula gives the diffraction limit of a telescope in arcseconds:

$$\text{diffraction limit} \approx 2.5 \times 10^{5\prime\prime} \times \frac{\text{wavelength of light}}{\text{diameter of telescope}}$$

**EXAMPLE 1:** What is the diffraction limit of the 2.4-meter Hubble Space Telescope for visible light with a wavelength of 500 nanometers?

**SOLUTION:**

**Step 1 Understand:** We are given the wavelength of light and the telescope diameter, so we have all the information we need.

**Step 2 Solve:** We plug in the wavelength ($500 \text{ nm} = 500 \times 10^{-9} \text{ m}$) and Hubble's diameter (2.4 m):

$$\begin{aligned}\text{diffraction limit} &\approx 2.5 \times 10^{5\prime\prime} \times \frac{\text{wavelength}}{\text{telescope diameter}}\\ &= 2.5 \times 10^{5\prime\prime} \times \frac{500 \times 10^{-9}\ \cancel{\text{m}}}{2.4\ \cancel{\text{m}}} = 0.05^{\prime\prime}\end{aligned}$$

**Step 3 Explain:** The Hubble Space Telescope has a diffraction-limited angular resolution of 0.05 arcsecond for visible light with a wavelength of 500 nanometers. Therefore, it can in principle resolve objects separated by more than 0.05 arcsecond, while objects separated by less will be blurred together.

**EXAMPLE 2:** How large a telescope would you need to achieve a diffraction limit of 0.001 arcsecond for visible light (wavelength 500 nm)?

**SOLUTION:**

**Step 1 Understand:** We are given the diffraction limit and wavelength, so we simply need to solve the formula for telescope diameter. You should confirm that it becomes

$$\text{telescope diameter} \approx 2.5 \times 10^{5\prime\prime} \times \frac{\text{wavelength}}{\text{diffraction limit}}$$

**Step 2 Solve:** We substitute the given values:

$$\text{telescope diameter} \approx 2.5 \times 10^{5\prime\prime} \times \frac{500 \times 10^{-9}\text{ m}}{0.001^{\prime\prime}} = 125\text{ m}$$

**Step 3 Explain:** A telescope would need a diameter of 125 meters—longer than a football field—to achieve an angular resolution of 0.001 arcsecond for visible light. Note that this would be 50 times the diameter of the Hubble Space Telescope and give an angular resolution 50 times better.

**FIGURE 6.12** The primary mirror of one of the Keck telescopes (Mauna Kea, Hawaii), with a person standing near the bottom center for scale. It is made up of 36 hexagonal mirrors arranged in a honeycomb pattern. The inset shows the two Keck telescopes from above.

needed for their primary mirrors. Recent technological innovations have made it possible to build lighter-weight mirrors, such as the 8-meter mirror shown in Figure 6.11b, or to make many small mirrors work together as one large one, as in the twin 10-meter Keck telescopes (**FIGURE 6.12**).

These new mirror-building technologies are fueling a revolution in the building of large telescopes. Before the 1990s, the 5-meter Hale telescope on Mount Palomar (outside San Diego) reigned for more than 40 years as the most powerful telescope in the world. Today, it does not even make the top-10 list for telescope size (**TABLE 6.1**). Several other large telescopes are currently in planning or under construction, including the Large Synoptic Survey Telescope (8.4 meters, but with a very wide field of view), the Giant Magellan Telescope (effective size of 21 meters), the Thirty Meter Telescope (30 meters), and the European Extremely Large Telescope (39 meters).

## What do astronomers do with telescopes?

Every astronomical observation is unique, and astronomers use many different kinds of instruments and detectors to extract the information contained in the light collected by a telescope. Nevertheless, most observations fall into one of three basic categories: *imaging,* which yields photographs (images) of astronomical objects; *spectroscopy,* in which astronomers obtain and study spectra; and *time monitoring*, which tracks how an object changes with time.* Let's look at each category in a little more detail, and at how astronomers work with the data they receive.

*Some astronomers include a fourth general category called *photometry,* which is the accurate measurement of light intensity from a particular object at a particular time. We do not list this as a separate category because today's detectors can generally perform photometry at the same time that they are being used for imaging, spectroscopy, or time monitoring.

**TABLE 6.1 Largest Optical (Visible-Light) Telescopes**

| Size | Name | Location | Opened* |
|---|---|---|---|
| 10.4 m | Gran Telescopio Canarias | Canary Islands | 2007 |
| 10.2 m | South African Large Telescope | South Africa | 2005 |
| 10 m | Keck I and Keck II | Mauna Kea, HI | 1993/1996 |
| 10 m | Hobby-Eberly | Mt. Locke, TX | 1997 |
| 2 × 8.4 m | Large Binocular Telescope | Mt. Graham, AZ | 2005 |
| 4 × 8.2 m | Very Large Telescope | Cerro Paranal, Chile | 1998/1999/2000/2001 |
| 8.4 m | Large Synoptic Survey Telescope | Cerro Pachón, Chile | 2020 (planned) |
| 8.3 m | Subaru | Mauna Kea, HI | 1999 |
| 8 m | Gemini North and South | Mauna Kea, HI (North)/Cerro Pachón, Chile (South) | 1999/2002 |
| 6.5 m | Magellan I and II | Las Campanas, Chile | 2000/2002 |
| 6.5 m | MMT | Mt. Hopkins, AZ | 2000 |

*The year of "first light," when the telescope began operating.

**Imaging** At its most basic, an imaging instrument is simply a camera. Astronomers often place *filters* in front of a camera to allow only particular colors or wavelengths of light to pass through. In fact, most of the richly hued astronomical images that you see are made by combining images recorded through different filters (**FIGURE 6.13**).

Today, many astronomical images are made from invisible light—light that our eyes cannot see but that can be captured by specialized detectors. You can understand the idea by thinking about x-rays at a doctor's office. When the doctor "takes an x-ray" of your arm, he or she uses a machine that sends x-rays through your arm. The x-rays that pass through are recorded with an x-ray-sensitive detector. Astronomical images work in much the same way. For example, **FIGURE 6.14** shows an x-ray image from the Chandra X-Ray Observatory (which is in space); the telescope collected x-rays and the image was recorded with an x-ray-sensitive detector. In other words, what we see in Figure 6.14 is not the x-rays themselves, but a picture that shows where x-rays hit the detector.

Images made with invisible light cannot have any natural color, because "color" is a property only of visible light. However, we can use color-coding to help us interpret them. For example, the colors in Figure 6.14 correspond to x-rays of different energy. In other cases, images may be color-coded according to the intensity of the light or to physical properties of the objects in the image.

**Think about it** Medical images from CT scans and MRIs are usually displayed in color, even though neither type of imaging uses visible light. What do you think the colors mean in CT scans and MRIs? How are the colors useful to doctors?

**Spectroscopy** Instruments called **spectrographs** use diffraction gratings (or other devices) to separate the various colors of light into spectra, which are then recorded with a detector (**FIGURE 6.15**).

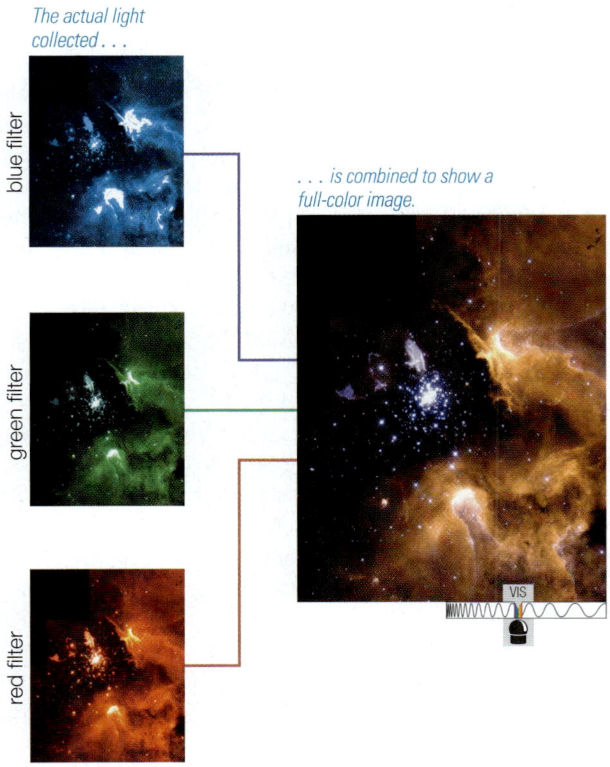


**FIGURE 6.13** Astronomical images are usually made by combining several images taken through different filters. Here, we see how three separate images from the Hubble Space Telescope—each taken with a different filter—were combined to make the image at right. The image shows stars and gas in a star-forming region called NGC 3603, located about 20,000 light-years away.

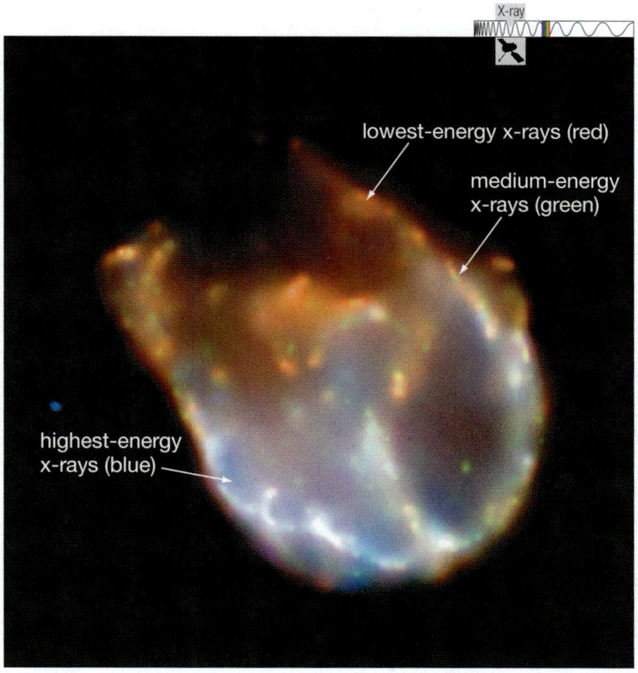


**FIGURE 6.14** X-rays are invisible, but we can color-code the information recorded by an x-ray detector to make an image of the x-ray light from an object. This image, from NASA's Chandra X-Ray Observatory, shows x-ray emission from the debris of a stellar explosion (a supernova remnant named N132D). Different colors represent x-rays of different energy.

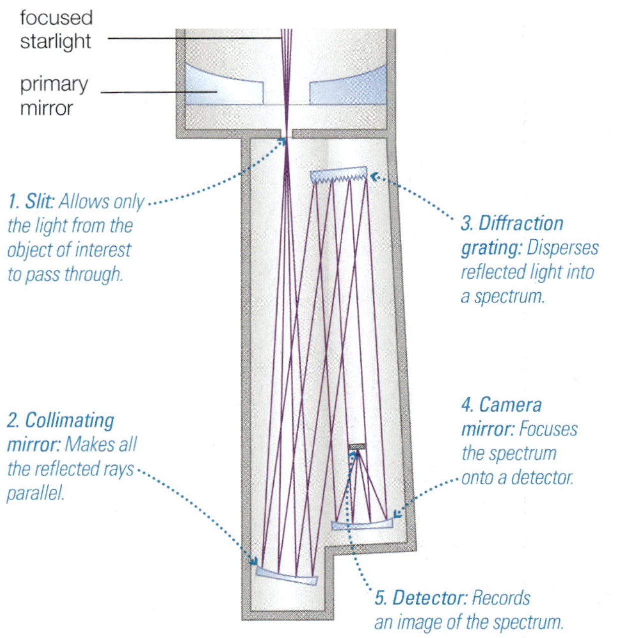


**FIGURE 6.15** The basic design of a spectrograph. In this diagram, the spectrograph is attached to the bottom of a reflecting telescope, with light entering the spectrograph through a hole in the primary mirror. A narrow slit (or small hole) at the entrance to the spectrograph allows only light from the object of interest to pass through.

As we discussed in Chapter 5, a spectrum can reveal a wealth of information about an object, including its chemical composition, temperature, and motion. However, just as the amount of information we can glean from an image depends on the angular resolution, the information we can glean from a spectrum depends on the **spectral resolution:** The higher the spectral resolution, the more detail we can see (**FIGURE 6.16**).

In principle, astronomers would always like the highest possible spectral resolution. However, higher spectral resolution comes at a price. A telescope collects only so much light in a given amount of time, and the spectral resolution depends on how widely the spectrograph spreads out this light. The more the light is spread out, the more total light we need in order for the spectrograph to record it successfully. Making a spectrum of an object therefore requires a longer exposure time than making an image, and high-resolution spectra require longer exposures than low-resolution spectra.

**Time Monitoring** Many astronomical objects vary with time. For example, some stars undergo sudden outbursts, and most stars (including our Sun) vary in brightness as starspots (or sunspots) cover more or less of their surfaces. Some objects vary periodically; for example, small, periodic changes in a star's brightness can reveal the presence of an orbiting planet [**Section 13.1**]. Time monitoring allows us to carefully study such variations.

For a slowly varying object, time monitoring may be as simple as comparing images or spectra obtained at different times. For more rapidly varying sources, time monitoring may require instruments that make rapid multiple exposures, in some cases recording the arrival time of every individual photon.

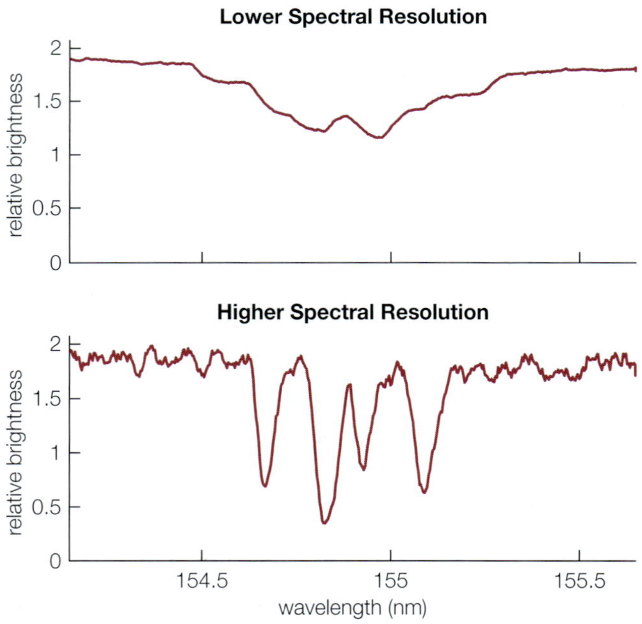


**FIGURE 6.16** These two ultraviolet spectra show the same object in the same wavelength band. However, we see far more detail with higher spectral resolution, including individual spectral lines that appear merged together at lower spectral resolution. (The spectrum shows absorption lines created when interstellar gas absorbs light from a more distant star.)

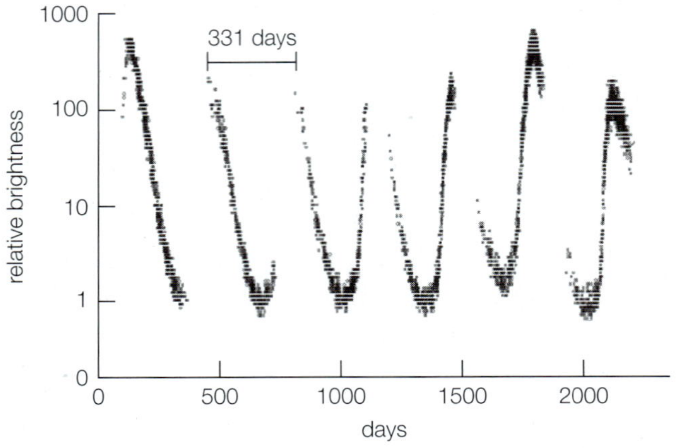


**FIGURE 6.17** This graph shows a light curve for the variable star Mira (in the constellation Cetus), with data spanning several years. Centuries of observation show that Mira's brightness varies with an average period of 331 days.

The results of time monitoring are often shown as **light curves:** graphs that show how an object's intensity varies with time. **FIGURE 6.17** shows a light curve for the star Mira. Notice that Mira's light output varies by more than a factor of 100 as it rises and falls with a period of a little less than one year. Astronomers' ability to conduct time-monitoring observations is expected to undergo a revolution with the Large Synoptic Survey Telescope, which will use the world's largest digital camera (3200 megapixels) to scan the entire sky visible from its location in Chile every three nights; the telescope is currently under construction and expected to begin science operations in 2021.

**Working with Astronomical Data** Many people imagine astronomers spending most of their time in late-night observing sessions, but in fact, most professional astronomers spend far more time working with the vast amounts of astronomical data that are now pouring in from observatories around the world and in space. Indeed, while some astronomers still spend many nights at telescopes, most of the world's largest telescopes are operated only by professional staff, and many telescopes—including all space telescopes—are operated remotely or even autonomously.

The data quantities are truly mind boggling. For example, consider the Large Synoptic Survey Telescope, or LSST (**FIGURE 6.18**). This telescope is unique in having a very wide field of view, allowing each image it takes of the sky to cover an angular area equivalent to that of nearly 50 full moons. It is also equipped with a 3.2-*gigapixel* camera—which has nearly 1000 times as many pixels as a typical phone camera—that will take a 15-second exposure every 20 seconds. As a result, during its planned 10-year survey of the sky, the LSST will produce approximately 200 *petabytes* of data (1 petabyte = $10^{15}$ bytes), which is roughly equivalent to nearly 100 billion typical photos posted on the web. This is far more data than people could possibly analyze on their own, which means that many astronomers are now becoming "data miners," developing computational and artificial intelligence methods for finding important discoveries in the vast data trove.

**FIGURE 6.18** This rendering shows what the Large Synoptic Survey Telescope (LSST) will look like when it opens in 2020. It is located at an altitude of 2700 meters on a mountain called Cerro Pachón in Chile.

More generally, you can think of the job of astronomers today as consisting of three major roles: planning the observations to be obtained with telescopes, analyzing the data received from telescopes, and developing theoretical models to explain the astronomical observations. If you become an astronomer, you will probably focus most of your effort on just one of these roles, but you will work closely and collaboratively with others who focus on the other roles. That is why astronomy today is rarely an individual pursuit, but rather one in which large teams of scientists from all over the world work together to advance human understanding of the universe.

## 6.3 Telescopes and the Atmosphere

Even now in the space age, the vast majority of astronomical observatories are located on Earth's surface, primarily because ground-based telescopes are much less expensive to build, operate, and maintain than telescopes in space. Nevertheless, Earth is far from ideal as an observing site. In this section, we'll explore some of the problems that Earth's atmosphere poses for astronomical observations and learn why, despite the higher costs, dozens of telescopes have been lofted into Earth orbit or beyond.

### How does Earth's atmosphere affect ground-based observations?

For visible-light telescopes, daylight and weather are the most obvious problems with observing from the ground. Our daytime sky is bright because the atmosphere scatters sunlight, and this brightness drowns out the dim light of most astronomical objects. That is why most astronomical observations are practical only at night. Even then, we can observe only when the sky is clear rather than cloudy. (The atmosphere does not scatter most radio waves, so radio telescopes can operate day and night and under cloudy skies.)

The constraints of daylight and weather affect the timing of observations, but by themselves do not hinder observations on clear nights. However, our atmosphere creates three other problems that inevitably affect astronomical observations: the scattering of human-made light, the blurring of images by atmospheric motion, and the fact that most forms of light cannot reach the ground at all.

**Light Pollution** Just as our atmosphere scatters sunlight in the daytime, it also scatters the bright lights of cities at night, creating what astronomers call **light pollution** (**FIGURE 6.19**). Light pollution explains why you cannot see as many stars from a big city as you can from an unpopulated area, and it can seriously hinder astronomical observations.

Light pollution has become an increasing problem as cities have grown, encroaching into areas that were once remote enough to be chosen as sites for major observatories. For example, the 2.5-meter telescope at Mount Wilson, the world's largest when it was built in 1917, would be much more useful today if it weren't located so close to the lights of what was once the small town of Los Angeles. Similar but less severe light pollution affects many other telescopes, including those on Mount Palomar near San Diego and on Kitt Peak near Tucson. Fortunately, many communities are working to reduce light pollution, with benefits not only to astronomers but to everyone who enjoys our ancient connections with the night sky.

**Atmospheric Blurring** The ever-changing motion, or *turbulence,* of air in the atmosphere bends light in continually

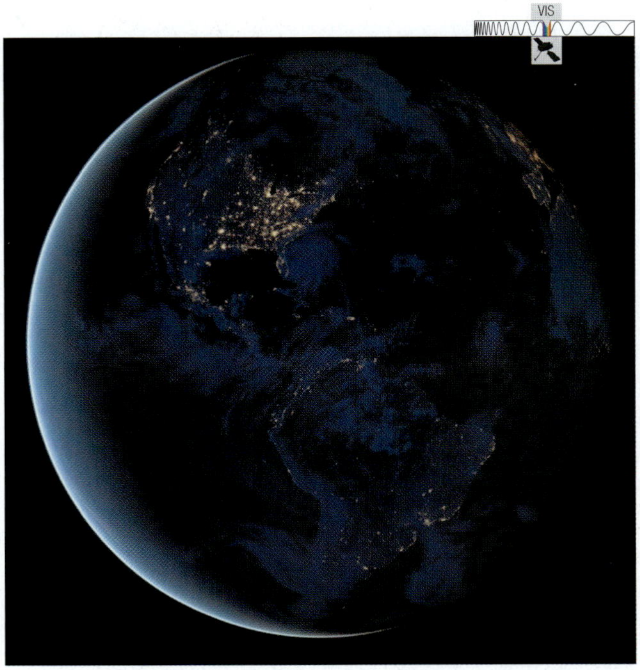


**FIGURE 6.19** Earth at night. It's pretty, but to astronomers it's light pollution. This image, a composite made from hundreds of satellite photos, shows the bright lights of cities around the world as they appear from Earth orbit at night.

shifting patterns. As a result, our view of things outside Earth's atmosphere appears to jiggle around, in much the same way as your view of things outside the water when you look up from the bottom of a swimming pool. This jiggling causes the familiar twinkling of stars, which may be beautiful to the naked eye but blurs astronomical images.

**See it for yourself** Put a coin in a cup of water and stir the water gently so that the coin appears to move around while actually remaining stationary on the bottom. How is what you see similar to the twinkling of stars?

In most cases, the blurring of images by turbulence limits the angular resolution of ground-based telescopes to no better than about 0.5 arcsecond, even if a telescope's diffraction limit is much smaller than that. However, a remarkable technology called **adaptive optics** can dramatically reduce this blurring. The technology works like this: Turbulence causes rays of light from a star to dance around as they reach a telescope; adaptive optics essentially makes the telescope's mirrors do an opposite dance, canceling out the atmospheric distortions (**FIGURE 6.20**). The shape of a mirror (often the secondary or even a third or fourth mirror) is changed slightly many times each second to compensate for the rapidly changing atmospheric distortions. A computer calculates the necessary changes by monitoring distortions in the image of a bright star near the object under study. If there is no bright star near the object of interest, the observatory may shine a laser into the sky to create an *artificial star* (a point of light in Earth's atmosphere) that it can monitor for distortions.

**Locating Ground-Based Observatories** Astronomers can partially mitigate effects of weather, light pollution, and atmospheric blurring by choosing observing sites that are *dark* (limiting light pollution), *dry* (limiting rain and clouds), *calm* (limiting turbulence), and *high* (placing them above at least part of the atmosphere). A handful of sites around the world meet these criteria particularly well and are therefore home to many of the world's most powerful telescopes. Three particularly important sites are the 4300-meter (14,000-foot) summit of Mauna Kea on the Big Island of Hawaii (**FIGURE 6.21**), a 2400-meter-high site on the island of La Palma in Spain's Canary Islands, and, for the southern hemisphere, the 2600-meter-high Paranal Observatory site in Chile.

**FIGURE 6.21** Observatories on the summit of Mauna Kea in Hawaii. Mauna Kea meets all the key criteria for an observing site: It is far from big-city lights, high in altitude, and in an area where the air tends to be calm and dry.

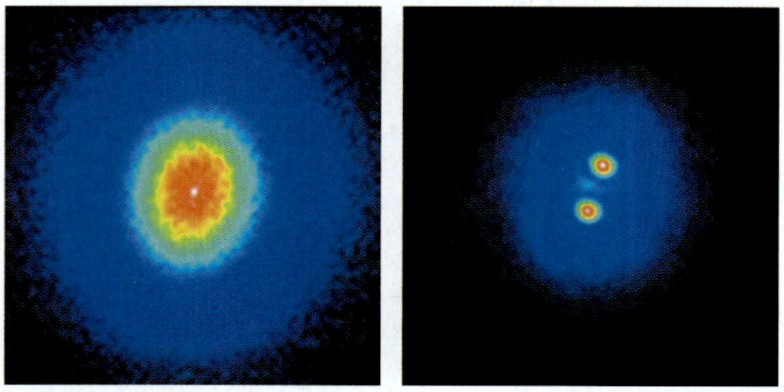

a Atmospheric blurring makes this ground-based image of a double star look like that of a single star.

b When the same telescope is used with adaptive optics, the two stars can be clearly distinguished. The angular separation between the two stars is 0.28 arcsecond.

**FIGURE 6.20**
The technology of adaptive optics can enable a ground-based telescope to overcome most of the blurring caused by Earth's atmosphere. Both images were taken in near-infrared light with the Canada-France-Hawaii telescope. The colors represent infrared brightness, with the brightest light shown in white (center of each star image) and the faintest light in blue to black.

## Why do we put telescopes into space?

The ultimate solution to the problems faced by ground-based observatories is to put telescopes into space, where they are unaffected by the atmosphere. That is one reason why the Hubble Space Telescope (**FIGURE 6.22**) was built and why it has been so successful despite the relatively small size (2.4 meters) of its primary mirror. However,

### COMMON MISCONCEPTIONS

#### Twinkle, Twinkle, Little Star

Twinkling, or apparent variation in the brightness and color of stars, is not intrinsic to the stars. Instead, just as light is bent by water in a swimming pool, starlight is bent by Earth's atmosphere. Air turbulence causes twinkling because it continually changes how the starlight is bent. Hence, stars tend to twinkle more on windy nights and at times when they are near the horizon (and therefore are viewed through a thicker layer of atmosphere). Above the atmosphere, in space, stars do not twinkle at all.

A related misconception holds that planets don't twinkle in our sky. They actually do, but not as much as stars (though they shimmer noticeably in telescopes). The reason is that planets have a measurable angular size in our sky, so the effects of turbulence on any one ray of light are compensated for by the effects of turbulence on others.

## COMMON MISCONCEPTIONS

### Closer to the Stars?

Many people mistakenly believe that space telescopes are advantageous because their locations above Earth put them closer to the stars. You can see why this is wrong by thinking about scale. On the scale of the Voyage model solar system (see Section 1.1), the Hubble Space Telescope is so close to the surface of the millimeter-diameter Earth that you would need a microscope to resolve its altitude, while the nearest stars are thousands of kilometers away. The distances to the stars are effectively the same whether a telescope is on the ground or in space. The real advantages of space telescopes all arise from their being above Earth's atmosphere and the observational problems it presents.

**FIGURE 6.22** The Hubble Space Telescope orbits Earth. Its position above the atmosphere allows it an undistorted view of space. Hubble can observe infrared and ultraviolet light as well as visible light.

there is another and even more important reason for putting observatories in space: *Our atmosphere prevents most forms of light from reaching the ground at all.*

If we studied only visible light, we'd be missing much of the story that light brings to us from the cosmos. Planets are relatively cool and emit primarily infrared light. The hot upper layers of stars like the Sun emit ultraviolet light and x-rays. Some violent cosmic events produce bursts of gamma rays. In fact, most astronomical objects emit light over a broad range of wavelengths. If we want to understand the universe, we must observe light all across the electromagnetic spectrum.

**FIGURE 6.23** shows the approximate depths to which different forms of light penetrate Earth's atmosphere. Only radio waves, visible light (and the very longest wavelengths of ultraviolet light), and small parts of the infrared spectrum can be observed from the ground. In addition, the atmosphere itself

## SPECIAL TOPIC Would You Like Your Own Telescope?

Just a couple decades ago, a decent personal telescope would have set you back a few thousand dollars and taken weeks of practice to learn to use. Today, you can get a good-quality telescope for a few hundred dollars, and built-in computer drives can make it easy to use.

Before you consider buying a telescope, you should understand what a personal telescope can and cannot do. A telescope will allow you to look for yourself at light that has traveled vast distances through space to reach your eyes. This can be a rewarding experience, but the images in your telescope will *not* look like the beautiful photographs in this book, which were obtained with much larger telescopes and sophisticated cameras. In addition, while your telescope can in principle let you see many distant objects, including star clusters, nebulae, and galaxies, it won't allow you to find anything unless you first set it up properly. Even computer-driven telescopes (sometimes called "go to" telescopes) typically take 15 minutes to a half-hour to set up for each use, and longer when you are first learning.

If your goal is just to see the Moon and a few other objects with relatively little effort, you may want to skip the telescope in favor of a good pair of binoculars, which is usually less expensive. Binoculars are generally described by two numbers, such as $7 \times 35$ or $12 \times 50$. The first number is the magnification; for example, "$7 \times$" means that objects will look seven times closer through the binoculars than to your eye. The second number is the diameter of each lens in millimeters. As with telescopes, larger lenses mean more light and better views. However, larger lenses also tend to be heavier and more difficult to hold steady, which means you may need a tripod.

If you decide to get a telescope, the first rule to remember is that magnification is *not* the key factor, and telescopes advertised only by their magnification (such as "650 power") are rarely high quality. Instead, focus on three factors when choosing your telescope:

1. *The light-collecting area* (also called *aperture*). Most personal telescopes are reflectors, so a "6-inch" telescope has a primary mirror that is 6 inches in diameter.
2. *Optical quality*. A poorly made telescope won't do you much good. If you cannot do side-by-side comparisons, stick with a major telescope manufacturer (such as Meade, Celestron, or Orion).
3. *Portability*. A large, bulky telescope can be great if you plan to keep it on a deck, but it will be difficult to carry on camping trips. Depending on how you plan to use your telescope, you'll need to make trade-offs between size and portability.

Most important, remember that a telescope is an investment that you will keep for many years. As with any investment, learn all you can before you settle on a particular model. Read buyers' guides and reviews of telescopes from sources such as *Astronomy*, *Sky and Telescope*, or the Astronomical Society of the Pacific. Talk to knowledgeable salespeople at stores that specialize in telescopes. And find a nearby astronomy club that holds observing sessions at which you can try out some telescopes and learn from experienced telescope users.

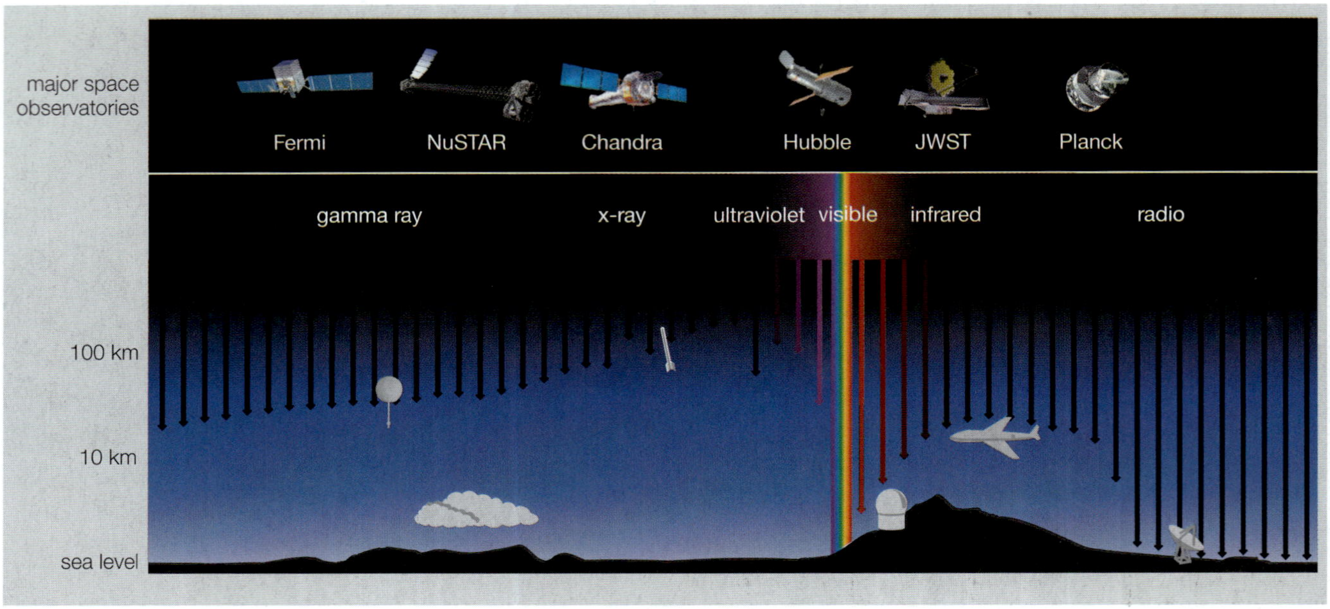


**FIGURE 6.23** This diagram shows the approximate depths to which different wavelengths of light penetrate Earth's atmosphere. Note that most of the electromagnetic spectrum can be observed only from very high altitudes or from space. Major space observatories for different wavelengths are also shown.

glows at many infrared wavelengths, generating a background glare for most infrared observations from the ground. As a result, without space-based observatories, we'd be unable to study light from much of the electromagnetic spectrum.

The Hubble Space Telescope is the most famous observatory in space—and it is used to observe infrared and ultraviolet light in addition to visible light—but there are many other space observatories. Most of these observe parts of the electromagnetic spectrum that do not reach the ground. **TABLE 6.2** lists some of the most important space telescopes. Astronomers are particularly excited about the successor mission to the Hubble Space Telescope, called the James Webb Space Telescope, which is optimized for infrared observations that will allow it to study the highly redshifted light of the most distant galaxies in the universe. Because we see distant objects as they were long ago, this will enable us to observe

**TABLE 6.2 Selected Major Space Observatories**

| Name | Operating Years | Lead Space Agency | Special Features |
|---|---|---|---|
| James Webb Space Telescope (JWST) | 2021* | NASA | Optimized for infrared observations |
| X-ray Astronomy Recovery Mission (XARM) | 2021* | JAXA*** | X-ray imaging and spectroscopy |
| Spektr-RG/eROSITA | 2018 | Russia/Germany | X-ray all-sky survey |
| GAIA | 2013– | ESA** | Precise measurement of stellar distances and positions |
| Nuclear Spectroscopic Telescope Array (NuSTAR) | 2012– | NASA | Imaging of high-energy x-rays |
| Kepler/K2 | 2009–2018 | NASA | Transit search for extrasolar Earth-like planets |
| Planck | 2009–2013 | ESA | Study of the cosmic microwave background |
| Herschel | 2009–2013 | ESA | Far-infrared imaging and spectroscopy |
| Fermi Gamma-Ray Space Telescope | 2008– | NASA | Gamma-ray imaging, spectroscopy, and timing |
| Swift | 2004– | NASA | Study of gamma-ray bursts |
| Galaxy Evolution Explorer (GALEX) | 2003–2012 | NASA | Ultraviolet observations of galaxies |
| Spitzer Space Telescope | 2003– | NASA | Infrared observations of the cosmos |
| Chandra X-Ray Observatory | 1999– | NASA | X-ray imaging and spectroscopy |
| XMM-Newton | 1999– | ESA | X-ray spectroscopy |
| Hubble Space Telescope | 1990– | NASA | Optical, infrared, and ultraviolet observations |

* Scheduled **European Space Agency ***Japan Aerospace Exploration Agency

**FIGURE 6.24** This photo shows a full-size model of the James Webb Space Telescope, with the sunshield at the bottom, on display in Austin, Texas, during the South by Southwest festival.

galaxies as they first began to form in the early universe. The telescope, scheduled for launch in 2021, will be kept far from Earth's heat by being put in an orbit around the Sun at a greater distance than Earth, while an attached sunshield will prevent sunlight from heating the telescope (**FIGURE 6.24**).

**Think about it** Find the current status of the James Webb Space Telescope. Has it now launched successfully? If so, describe its early results. Also list at least three of its key science objectives.

## 6.4 Telescopes Across the Spectrum

As we've discussed, astronomers today study light from across the electromagnetic spectrum, sometimes with observatories on the ground and sometimes with observatories in space. While the basic idea behind all telescopes is the same—light is collected by a primary mirror (sometimes more than one) and ultimately focused on cameras or other instruments—different wavelengths of light pose different challenges for telescope design. In this section, we'll explore how these challenges are met.

### How do we observe invisible light?

Some portions of the infrared and ultraviolet lie near enough to visible wavelengths that the light behaves similarly to visible light. This light can therefore be focused by visible-light telescopes, which is why the Hubble Space Telescope can be used to study infrared and ultraviolet light in addition to visible light. But other wavelengths require different telescope designs. Let's investigate, going in order of decreasing wavelength.

**Radio Telescopes** A specialized kind of radio telescope is now the most common type of telescope in the world: the satellite dish, which is a small radio telescope designed to collect radio waves from a satellite in Earth orbit. Just by looking at a satellite dish, you can see that it operates by the same basic principles as a reflecting telescope (**FIGURE 6.25**).

**FIGURE 6.25** A satellite TV dish is essentially a small radio telescope.

The metal dish is the primary mirror, shaped to bring the radio waves to a focus in front of the dish; that's where you see the receiver, located where a secondary mirror would be in a visible-light telescope. The receiver collects the radio waves reflected by the primary mirror and sends them to the television (or other device).

The differences between satellite dishes and astronomical radio telescopes are in where they look in the sky and their sizes. Communication satellites have geostationary orbits, which means they orbit above Earth's equator in exactly the same amount of time Earth takes to rotate (see Mathematical Insight 4.3), so that a dish aimed at a particular satellite can always point to the same spot in the local sky. In contrast, astronomical radio telescopes point toward cosmic radio sources that, like the Sun and stars, rise and set with Earth's rotation. Astronomical radio telescopes are also larger than satellite dishes, both because they need a large light-collecting area to detect the faint radio waves from cosmic sources and because they are used to make images and therefore require decent angular resolution. (Angular resolution is unimportant for satellite dishes, because they are not used to make images of the satellites in space; radio and television signals are encoded in the radio waves themselves, so the dish needs only to collect the radio waves and send them to a decoding device like a television.)

The long wavelengths of radio waves mean that very large telescopes are necessary to achieve reasonable angular resolution. The world's largest single telescope is China's Five-hundred-meter Aperture Spherical Telescope, or FAST (**FIGURE 6.26**). Completed in 2016, FAST surpassed the previous record holder, Puerto Rico's 305-meter Arecibo radio telescope. Despite their large sizes, these telescopes

**FIGURE 6.26** China's Five-hundred-meter Aperture Spherical Telescope (FAST) is a radio dish with a diameter of 500 meters. It is nestled in a natural valley in southwest Guizhou province.

have an angular resolution of only about 1 arcminute at commonly observed radio wavelengths—about 1000 times worse than the visible-light resolution of the Hubble Space Telescope. Fortunately, through an amazing technique that we'll discuss shortly (interferometry), radio telescopes can work together to achieve much better angular resolution.

If you look again at Figure 6.23, you'll see that radio waves are the only form of light besides visible light that we can observe easily from the ground. Moreover, because the atmosphere does not distort radio waves the way it distorts visible light, there's no inherent advantage to observing from space. However, "radio-wave pollution" is an even more serious impediment to radio astronomy than light pollution is to visible-light astronomy. Humans use many portions of the radio spectrum so heavily that radio signals from cosmic sources are almost completely drowned out. Astronomers hope someday to put radio telescopes into deep space or on the far side of the Moon, where the Moon itself would block out any radio interference from Earth. In addition, because radio telescopes can be made to work together, putting them into space in principle can allow them to be spread out over a much greater distance.

**Infrared Telescopes** Most of the infrared portion of the spectrum is close enough in wavelength to visible light to behave quite similarly, so infrared telescopes generally look much the same as visible-light telescopes. As you can see in Figure 6.23, a few portions of the infrared spectrum can be observed from the tops of high mountains, such as Mauna Kea; the higher you go in the atmosphere, the more infrared light becomes accessible. NASA's airborne observatory called SOFIA (Stratospheric Observatory For Infrared Astronomy) carries a 2.5-meter infrared telescope that looks out through a large hole cut in the body of a Boeing 747 airplane (**FIGURE 6.27**).

One problem with infrared telescopes, particularly at longer wavelengths, is that Earth and even the telescope itself emit infrared thermal radiation [**Section 5.4**] that can interfere with observations of the same wavelengths from the cosmos. The only solution to this problem is to put the telescopes into space, so they get away from Earth's heat,

**FIGURE 6.27** This photograph shows NASA's airborne observatory, SOFIA, with its 2.5-meter infrared telescope. NASA's Airborne Astronomy Ambassadors program provides an opportunity for educators to fly aboard SOFIA during its science missions.

and to keep the telescopes cool, so they emit less infrared light. As noted earlier, the James Webb Space Telescope (see Figure 6.24) will be kept cool by its orbit and sunshield. NASA's Spitzer Space Telescope (**FIGURE 6.28**), launched in 2003, was cooled with liquid helium to just a few degrees above absolute zero.

**Ultraviolet Telescopes** Like infrared light, much of the ultraviolet spectrum is close enough in wavelength to visible light to behave similarly, so in principle it can be collected and focused by visible-light mirrors. However, Earth's atmosphere almost completely absorbs ultraviolet light, making most ultraviolet observations impossible from the ground. (Very-short-wavelength ultraviolet light, sometimes called *extreme ultraviolet,* behaves like x-rays, which we'll discuss below.)

At present, the Hubble Space Telescope is the only major space observatory capable of ultraviolet observations, and no major new ones are near completion. Astronomers hope

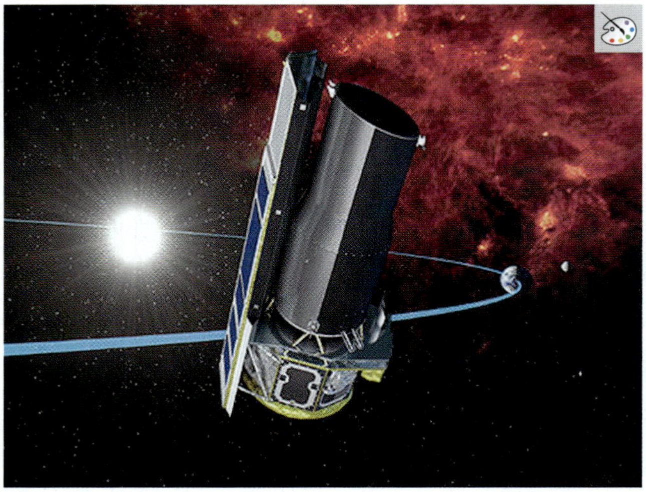

**FIGURE 6.28** This painting shows the Spitzer Space Telescope. The background is an artistic rendition of infrared emission from a star-forming cloud.

a Artist's illustration of the Chandra X-Ray Observatory, which orbits Earth.

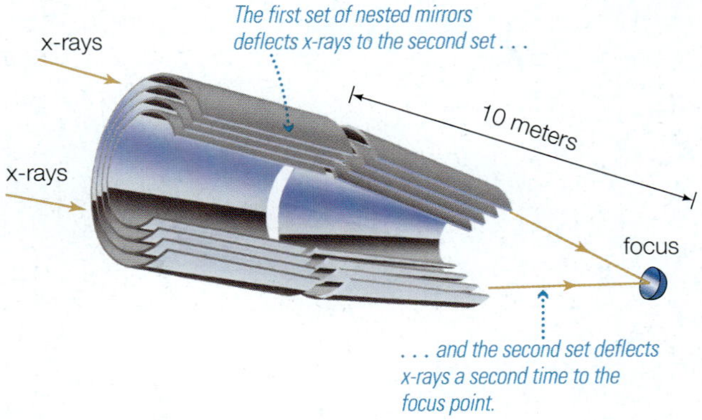


b This diagram shows the arrangement of Chandra's nested, cylindrical x-ray mirrors. Each mirror is 0.8 meter long and between 0.6 and 1.2 meters in diameter.

**FIGURE 6.29** The Chandra X-Ray Observatory focuses x-rays that enter the front of the telescope by deflecting them twice so that they end up focused at the back of the telescope.

that Hubble will continue to operate for many years, but no further servicing missions are planned (the last occurred in 2009), so Hubble will continue to observe only as long as its machinery continues to work and NASA continues to fund it.

**X-Ray Telescopes** No cosmic x-rays reach the ground, so x-ray telescopes must be placed in space. X-rays also pose another challenge: They have sufficient energy to penetrate many materials, including living tissue and ordinary mirrors. While this property makes x-rays useful to medical doctors, it creates headaches for astronomers.

Trying to focus x-rays is somewhat like trying to focus a stream of bullets. If the bullets are fired directly at a metal sheet, they will puncture or damage the sheet. However, if the metal sheet is angled so that the bullets barely graze its surface, it will slightly deflect the bullets. Specially designed mirrors can deflect x-rays in much the same way. Such mirrors are called **grazing incidence mirrors** because x-rays merely graze their surfaces as the rays are deflected toward the focal plane. X-ray telescopes, such as NASA's Chandra X-Ray Observatory and the NuSTAR mission, generally consist of several nested grazing incidence mirrors (**FIGURE 6.29**).

**See it for yourself** If you look straight down at your desktop, you probably cannot see your reflection. But if you glance along the desktop surface (or another smooth surface, such as that of a book), you should see reflections of objects *in front* of you. Explain how these reflections represent *grazing incidence* for visible light.

Chandra offers the best angular resolution of any x-ray telescope yet built, but a European x-ray telescope called XMM-Newton has a larger light-collecting area. Astronomers therefore use the two observatories in the way best suited to their science goals. For example, Chandra is better for making images of x-ray sources, while XMM-Newton's larger light-collecting area allows it to obtain more detailed x-ray spectra. The most recent x-ray telescope in space, NuSTAR, is optimized for imaging higher-energy x-rays than Chandra or XMM.

**Gamma-Ray Telescopes** Gamma rays can penetrate even grazing incidence mirrors and therefore cannot be focused in the traditional sense. Gamma-ray observatories, such as the Fermi Gamma-Ray Space Telescope (launched in 2008), use massive detectors to capture photons and determine the direction they came from. For example, the Large Area Telescope on Fermi weighs 3 tons (**FIGURE 6.30**).

**Looking Beyond Light** Light is not the only form of information that travels through the universe, and astronomers have begun to build and use telescopes designed to observe at least three other types of "cosmic messengers." First, there's an extremely lightweight type of subatomic particle known as the *neutrino* [**Sections S4.2, 14.2**] that is produced by nuclear reactions, including nuclear fusion in the Sun and the reactions that accompany the explosions of distant stars. Astronomers have already had some

**FIGURE 6.30** This artist's rendering shows the Fermi Gamma-Ray Telescope operating in space.

**FIGURE 6.31** This aerial view shows the Laser Interferometer Gravitational-wave Observatory (LIGO) in Hanford, Washington, which works in conjunction with a similar detector in Livingston, Louisiana, to detect gravitational waves. The two LIGO detectors also work with a third similar observatory, called Virgo, which is located in Italy.

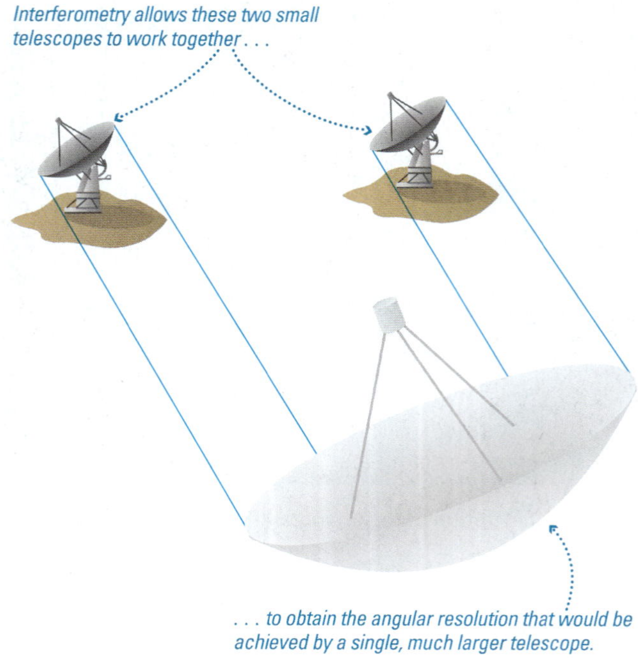


**FIGURE 6.32** This diagram shows the basic idea behind interferometry: Smaller telescopes work together to obtain the angular resolution of a much larger telescope. Note that interferometry improves angular resolution but does not affect the total light-collecting area, which is simply the sum of the light-collecting areas of the individual telescopes.

success with "neutrino telescopes"—typically located in deep mines or under water or ice—which have provided valuable insights about the Sun and stellar explosions. Second, Earth is continually bombarded by very-high-energy subatomic particles from space known as *cosmic rays* [**Section 19.2**]. We still know relatively little about the origin of cosmic rays, but astronomers are now using both satellites and ground-based detectors to catch and study them. Third, Einstein's general theory of relativity predicts the existence of something called *gravitational waves* [**Section S3.4**], which are different in nature from light but travel at the speed of light. The first gravitational wave observatories are now up and running (**FIGURE 6.31**), and they have already made numerous detections of gravitational waves from the mergers of orbiting pairs of neutron stars and black holes [**Section 18.4**].

## How can multiple telescopes work together?

Individual telescopes always face limits on their capabilities. Even in space, the diffraction limit places a fundamental constraint on the angular resolution of a telescope of any particular size. In addition, while astronomers would always like larger telescopes, the current state of technology and budgetary considerations place practical limits on telescope size.

These constraints ultimately limit the amount of light that we can collect with telescopes. Even if we put a group of telescopes together, there's no getting around the fact that their total light-collecting area is simply the sum of their individual areas. However, remember that the two key properties of a telescope are light-collecting area and angular resolution. Amazingly, there is a way to make the angular resolution of a group of telescopes far better than that of any individual telescope.

In the 1950s, radio astronomers developed an ingenious technique for improving the angular resolution of radio telescopes: They learned to link two or more individual telescopes to achieve the angular resolution of a much larger telescope (**FIGURE 6.32**). This technique is called **interferometry** because it works by taking advantage of the wavelike properties of light that cause interference (see Figure 6.8). The procedure relies on precisely timing when radio waves reach each dish and using computers to analyze the resulting interference patterns.

The Karl G. Jansky Very Large Array (JVLA) near Socorro, New Mexico, consists of 27 individual radio dishes that can be moved along railroad tracks laid down in the shape of a Y (**FIGURE 6.33**). The light-gathering capability of the 27 dishes is simply equal to their combined area, equivalent to that of a single telescope 130 meters across. But the JVLA's angular resolution is equivalent to that of a much larger telescope; when the 27 dishes are spaced as widely as possible, the JVLA can achieve an angular resolution that otherwise would require a single radio telescope with a diameter of almost 40 kilometers. Astronomers can achieve even higher angular resolution by linking radio telescopes around the world.

Interferometry is more difficult for shorter-wavelength (higher-frequency) light, but astronomers have achieved many successes. One spectacular example is the Atacama Large Millimeter/submillimeter Array (ALMA), located on a 5000-meter-high plateau in Chile, which combines light from 66 individual telescopes working at millimeter and submillimeter wavelengths (see the chapter-opening photo, page 165). Perhaps even more impressively, the Event Horizon Telescope project is linking ALMA and other radio and submillimeter telescopes around the world with the goal of

**FIGURE 6.33** The Karl G. Jansky Very Large Array (JVLA) in New Mexico consists of 27 telescopes that can be moved along train tracks. The telescopes work together through interferometry and can achieve an angular resolution equivalent to that of a single radio telescope almost 40 kilometers across.

achieving the angular resolution needed to obtain an image of the Milky Way Galaxy's central black hole [**Section 19.4**].

Interferometry is also being used at infrared and visible wavelengths. Indeed, telescopes are now often built in pairs (such as the Keck and Magellan telescope pairs) or with more than one telescope on a common mount (such as the Large Binocular Telescope) so that they can be used for infrared and visible-light interferometry. Astronomers are testing technologies that may allow interferometry to be extended all the way to x-rays. Someday, astronomers may use telescopes in space or on the Moon as giant interferometers, offering views of distant objects that may be as detailed in comparison to Hubble Space Telescope images as Hubble's images are in comparison to those of the naked eye.

## The BIG Picture PUTTING CHAPTER 6 INTO CONTEXT

*In this chapter, we've focused on the technological side of astronomy: the telescopes that we use to learn about the universe. Keep in mind the following "big picture" ideas as you continue to learn about astronomy:*

- Technology drives astronomical discovery. Every time we build a bigger telescope, develop a more sensitive detector, or open up a new wavelength region to study, we learn more about the universe.
- Telescopes work much like giant eyes, enabling us to see the universe in great detail. New technologies for making larger telescopes, along with advances in adaptive optics and interferometry, are making ground-based telescopes more powerful than ever.
- For the ultimate in observing the universe, space is the place! Telescopes in space allow us to detect light from across the entire spectrum while avoiding the distortion caused by Earth's atmosphere.

**MY COSMIC PERSPECTIVE** The modern world is filled with beautiful telescopic images from astronomy, used in everything from artwork to advertisements. Understanding how telescopes work will give you a greater appreciation for the images you see.

# Summary of Key Concepts

## 6.1 Eyes and Cameras: Everyday Light Sensors

- **How do eyes and cameras work?** Your eye brings rays of light to a **focus** (or **focal point**) on your retina. Glass lenses work similarly, so distant objects form an **image** that is in focus on the **focal plane.** A camera has a detector at the focal plane, which can make a permanent record of an image.

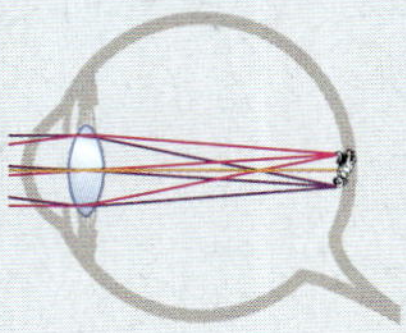

## 6.2 Telescopes: Giant Eyes

- **What are the two most important properties of a telescope?** A telescope's most important properties are its **light-collecting area,** which determines how much light it gathers, and its **angular resolution,** which determines how much detail we can see in its images.
- **What are the two basic designs of telescopes?** A **refracting telescope** forms an image by bending light through a lens. A **reflecting telescope** forms an image by focusing light with mirrors.

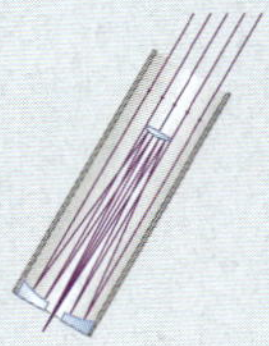

- **What do astronomers do with telescopes?** The three primary uses of telescopes are **imaging** to create pictures of distant objects, **spectroscopy** to study the spectra of distant objects, and **time monitoring** to study how a distant object's brightness changes with time.

## 6.3 Telescopes and the Atmosphere

- **How does Earth's atmosphere affect ground-based observations?** Earth's atmosphere limits visible-light observations to nighttime and clear weather. **Light pollution** can lessen the quality of observations, and atmospheric turbulence makes stars twinkle, blurring their images. The technology of **adaptive optics** can overcome some of the blurring due to turbulence.
- **Why do we put telescopes into space?** Telescopes in space are above Earth's atmosphere and the problems it causes for observations. Most important, telescopes in space can observe all wavelengths of light, while telescopes on the ground can observe only visible light, radio waves, and small portions of the infrared.

## 6.4 Telescopes Across the Spectrum

- **How do we observe invisible light?** Telescopes for other than visible light often use variations on the basic design of a reflecting telescope. Radio telescopes use large metal dishes as their primary mirrors. Infrared telescopes are sometimes cooled to very low temperature. X-ray telescopes use **grazing incidence** reflections rather than direct reflections.

- **How can multiple telescopes work together?** The technique of **interferometry** links multiple telescopes in a way that allows them to obtain the angular resolution of a much larger telescope.

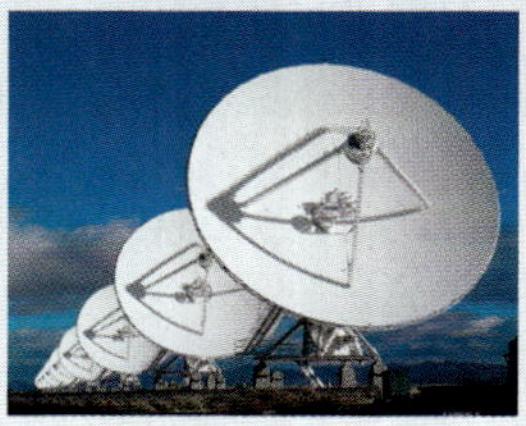

# Visual Skills Check

*Use the following questions to check your understanding of some of the many types of visual information used in astronomy. For additional practice, try the Chapter 6 Visual Quiz in the Study Area at www.MasteringAstronomy.com.*

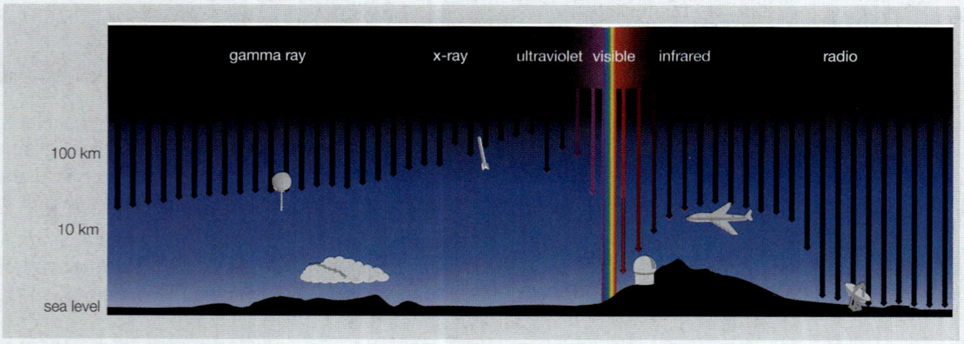


*The figure repeats the portion of Figure 6.23 showing the approximate depths to which different wavelengths of light penetrate Earth's atmosphere. Use this figure to answer the following questions.*

1. Only very small amounts of infrared and ultraviolet light can penetrate all the way to the ground. Based on the diagram, which statement is true?
   a. A small percentage of the incoming light at every infrared and ultraviolet wavelength reaches the ground, while the remaining light at the same wavelengths does not reach the ground.
   b. Most infrared and ultraviolet wavelengths do not reach the ground at all; the only wavelengths that do are the ones closest to the visible portion of the spectrum.
   c. Most infrared and ultraviolet wavelengths do not reach the ground at all; the only wavelengths that do are the ones closest to the radio and x-ray portions of the spectrum.
2. (Choose all that apply.) Observatories on mountaintops can detect
   a. visible light
   b. x-rays
   c. a small portion of the infrared spectrum
   d. very-long-wavelength infrared light
   e. radio waves
3. (Choose all that apply.) An observatory in space could in principle detect
   a. visible light
   b. x-rays
   c. infrared light
   d. ultraviolet light
   e. radio waves
   f. gamma rays
4. What kind of light can be detected from an airplane but not from the ground?
   a. most infrared light
   b. only the shortest-wavelength infrared light
   c. radio waves
   d. x-rays

# Exercises and Problems

*For instructor-assigned homework and other learning materials, go to www.MasteringAstronomy.com.*

## Chapter Review Questions

### Short-Answer Questions Based on the Reading

1. How does your eye focus light? How is a glass lens similar? What do we mean by the *focal plane* of a lens?
2. How does a camera record light? How are images affected by *exposure time*? What are *pixels*?
3. What are the two key properties of a telescope, and why is each important?
4. What is the *diffraction limit*, and how does it depend on a telescope's size and the wavelength of light being observed?
5. How do *reflecting telescopes* differ from *refracting telescopes*? Which type is more commonly used by professional astronomers, and why?
6. What are the three basic categories of astronomical observation, and how is each conducted?
7. What do we mean when we speak of images made from *invisible* light, such as x-ray or infrared images? What do the colors in these images mean?
8. What do we mean by *spectral resolution*? Why is higher spectral resolution more difficult to achieve?
9. List at least three ways in which Earth's atmosphere can hinder astronomical observations. What problem can *adaptive optics* help with?
10. Describe how deeply each portion of the electromagnetic spectrum penetrates Earth's atmosphere. Based on your answers, why are space telescopes so important?
11. Briefly describe how telescopes for invisible wavelengths differ from those for visible light. Besides light, what other types of "cosmic messengers" can astronomers now observe?
12. What is *interferometry*, and how can it improve astronomical observations?

### Does It Make Sense?

*Decide whether or not each of the following statements makes sense (or is clearly true or false). Explain clearly; not all of these have definitive answers, so your explanation is more important than your chosen answer.*

13. The image was blurry because the detector was not placed at the focal plane.
14. I wanted to see faint details in the Andromeda Galaxy, so I photographed it with a very short exposure time.
15. I have a reflecting telescope in which the secondary mirror is bigger than the primary mirror.
16. The photograph shows what appear to be just two distinct stars, but each of those stars is actually a binary star system.
17. My 14-inch telescope has a lower diffraction limit than most large professional telescopes.
18. Now that I've bought a spectrograph, I can use my home telescope for spectroscopy as well as imaging.
19. If you lived on the Moon, you'd never see stars twinkle.
20. New technologies will soon allow astronomers to use x-ray telescopes on Earth's surface.
21. Thanks to adaptive optics, telescopes on the ground can now make ultraviolet images of the cosmos.
22. Thanks to interferometry, a properly spaced set of 10-meter radio telescopes can achieve the angular resolution of a single 100-kilometer radio telescope.

### Quick Quiz

*Choose the best answer to each of the following. For additional practice, try the Chapter 6 Reading and Concept Quizzes in the Study Area at www.MasteringAstronomy.com.*

23. How much greater is the light-collecting area of a 6-meter telescope than that of a 3-meter telescope? (a) two times (b) four times (c) six times
24. Suppose you look at two stars that are separated in the sky by 0.1 arcsecond, using a telescope with an angular resolution of 0.5 arcsecond. What will you see? (a) two distinct stars (b) one point of light that is the blurred image of both stars (c) nothing at all
25. The diffraction limit is a limit on a telescope's (a) size. (b) angular resolution. (c) spectral resolution.
26. The Hubble Space Telescope obtains higher-resolution images than most ground-based telescopes because it is (a) larger. (b) closer to the stars. (c) above Earth's atmosphere.
27. What does it mean if you see the color red in an x-ray image from the Chandra X-Ray Observatory? (a) The object is red in color. (b) The red parts are hotter than the blue parts. (c) It depends; the colors are chosen arbitrarily to represent something about the x-rays recorded by the telescope.
28. The twinkling of stars is caused by (a) variations in stellar brightness with time. (b) light pollution. (c) motion of air in our atmosphere.
29. To achieve the same angular resolution as a visible-light telescope, a radio telescope would need to be (a) much larger. (b) slightly larger. (c) in space.
30. Where should you put a telescope designed for ultraviolet observations? (a) in Earth orbit (b) on an airplane (c) on a high mountaintop
31. Which technology can allow a single ground-based telescope to achieve images as sharp as those from the Hubble Space Telescope? (a) adaptive optics (b) grazing incidence mirrors (c) interferometry
32. Interferometry uses two or more telescopes to achieve (a) a light-collecting area equivalent to that of a much larger telescope. (b) an angular resolution equivalent to that of a much larger telescope. (c) both the light-collecting area and the angular resolution of a much larger telescope.

## Inclusive Astronomy

*Use these questions to reflect on participation in science.*

33. *A Telescope Near You.* For either your current residence, your birthplace, or your home town, find the location of the nearest ground-based, visible-light telescope with a mirror larger than about 2 meters in diameter. What is this telescope used for? Is it open to the public? Discuss whether or how you think this telescope may benefit your community.
34. *Pair Discussion: The Herschels.* William and Caroline Herschel—brother and sister—were pioneers in using large telescopes to study the night sky.
    **a.** Working independently, find out when the Herschels lived, the kinds of telescopes they had at their disposal, and the major accomplishments credited to each of them.
    **b.** Pair up with another student, and decide which one of you will focus on William and which one on Caroline. Then take a turn describing the major accomplishments of the sibling you focus on.
    **c.** Discuss the relationship between William and Caroline Herschel. Were they equal partners? Did they have similar opportunities to make major astronomical discoveries? Why or why not?

## The Process of Science

*These questions may be answered individually in short-essay form or discussed in groups, except where identified as group-only.*

35. *Technology-Driven Science.* Choose one technology described in this chapter and summarize how its development (or improvement) has allowed us to learn more about the universe. Then discuss how we might expect this technology to improve further during the next few decades, and identify at least one currently unanswered question about the universe that will likely be answered with the help of the improved technology.
36. *Research Funding.* Technological innovation clearly drives scientific discovery in astronomy, but the reverse is also true. For example, Newton made his discoveries in part because he wanted to explain the motions of the planets, but his discoveries led to new technologies that have had far-reaching effects on our civilization. Congress often must decide between funding programs with purely scientific purposes (basic research) and programs designed to develop new technologies. If you were a member of Congress, how would you allocate spending between basic research and technology? Why?
37. *LIGO.* Learn about how the Laser Interferometer Gravitational-wave Observatory (LIGO) detects gravitational waves. How small are the length changes it can measure in its arms? Why do astronomers need more than one detector to be confident of its discoveries? Briefly summarize how LIGO works and its latest discoveries.
38. *Group Activity: Which Telescope?* You represent a research group that wishes to observe matter around a black hole; assume that the matter is emitting photons at all wavelengths. You have the opportunity to make your observations with one of the following four telescopes:
    - A radio telescope, 300 meters in diameter, located in Puerto Rico
    - An infrared telescope, 2 meters in diameter, on a spacecraft in orbit around Earth and observing at a wavelength of micrometers ($2 \times 10^{-6}$ m)
    - An infrared telescope, 10 meters in diameter, equipped with adaptive optics, located on Mauna Kea in Hawaii and observing at a wavelength of micrometers ($10^{-5}$ m)
    - An x-ray telescope, 2 meters in diameter, located at the South Pole

    Discuss the pros and cons of each telescope for this observing task, then come to a group consensus in which you rank the four choices from best to worst. Compare your rankings to those of other groups in your class. Note: You may wish to do this activity using the four roles described in Chapter 1, Exercise 39.

## Investigate Further

### Short-Answer/Essay Questions

39. *Image Resolution.* What happens if you take a photograph and blow it up to a larger size? Does it contain more detail than it did before? Explain clearly, and relate your answer to the concepts of magnification and angular resolution in astronomical observations.
40. *Filters.* What would an American flag look like if you viewed it through a filter that transmits only red light? What would it look like through a filter that transmits only blue light?

41. *Type of Observation.* For each of the following, decide what type of observation (imaging, spectroscopy, timing) you would need to make. Explain clearly.
**a.** Studying how a star's hot upper atmosphere changes with time
**b.** Learning the composition of a distant star
**c.** Determining how fast a distant galaxy is moving away from Earth

42. *Telescope Location.* Considering the pros and cons of different locations for ground-based observatories, do you think the place where you live is a good location for a visible-light telescope? Why or why not?

43. *Technology Choice.* Suppose you were building a space-based observatory consisting of five individual telescopes. Which would be the better way to use these telescopes: as five individual telescopes with adaptive optics or as five telescopes linked together for interferometry (without adaptive optics)? Explain your reasoning clearly.

44. *Major Ground-Based Observatories.* Take a virtual tour of one of the world's major astronomical observatories. Write a short report on why the observatory is so useful.

45. *Space Observatory.* Visit the website of a major space observatory, either existing or under development. Write a short report about the observatory, including its purpose, its orbit, and how it operates.

46. *Really Big Telescopes.* Learn about one of the projects to build a very large telescope, such as the Giant Magellan Telescope, the Thirty Meter Telescope, or the European Extremely Large Telescope. Write a short report about the telescope's current status and potential capabilities.

47. *A Lunar Observatory.* Some people have proposed building an astronomical observatory on the Moon. Research the pros and cons of this idea, considering it both in isolation and in the context of plans for a human return to the Moon. If it were up to you, would you recommend that Congress begin funding a program to develop such an observatory? Defend your opinion.

48. *Project: Twinkling Stars.* Using a star chart, identify a few bright stars that should be visible in the early evening. On a clear night, observe each of these stars for a few minutes. Note the date and time, and for each star record the following information: approximate altitude and direction in your sky, brightness compared to other stars, color, and how much the star twinkles compared to other stars. Study your record. Can you draw any conclusions about how brightness and position in your sky affect twinkling? Explain.

49. *Project: Personal Telescope Review.* Find three telescopes that you could buy for under $1000 and evaluate each on the following criteria: light-collecting area, angular resolution, construction quality, and portability. Give each telescope a rating of 1 to 4 stars (4 is best), and state which one you would recommend for purchase.

## Quantitative Problems

*Be sure to show all calculations clearly and state your final answers in complete sentences.*

50. *Light-Collecting Area.*
**a.** How much greater is the light-collecting area of one of the 10-meter Keck telescopes than that of the 5-meter Hale telescope?
**b.** Suppose astronomers built a 100-meter telescope. How much greater would its light-collecting area be than that of the 10-meter Keck telescope?

51. *Close Binary System.* Suppose that two stars in a binary star system are separated by a distance of 100 million kilometers and are located at a distance of 100 light-years from Earth. What is the angular separation of the two stars? Give your answer in both degrees and arcseconds. Can the Hubble Space Telescope resolve the two stars?

52. *Finding Planets.* Suppose you were looking at our own solar system from a distance of 10 light-years.
**a.** What angular resolution would you need to see the Sun and Jupiter as distinct points of light?
**b.** What angular resolution would you need to see the Sun and Earth as distinct points of light?
**c.** How do the angular resolutions you found in parts a and b compare to the angular resolution of the Hubble Space Telescope? Comment on the challenge of making images of planets around other stars.

53. *Diffraction Limit of the Eye.*
**a.** Calculate the diffraction limit of the human eye, assuming a wide-open pupil so that your eye acts like a lens with a diameter of 0.8 centimeter, for visible light of 500-nanometer wavelength. How does this compare to the diffraction limit of a 10-meter telescope?
**b.** Now remember that humans have two eyes that are approximately 7 centimeters apart. Estimate the diffraction limit for human vision, assuming that your two eyes act as an "optical interferometer" that gives you the angular resolution of a single "eye" that is 7 centimeters in diameter.

54. *The Size of Radio Telescopes.* What is the diffraction limit of a 100-meter radio telescope observing radio waves with a wavelength of 21 centimeters? Compare this to the diffraction limit of the Hubble Space Telescope for visible light. Use your results to explain why, to be useful, radio telescopes must be much larger than optical telescopes.

55. *Your Satellite Dish.* Suppose you have a satellite dish that is 0.5 meter in diameter and you want to use it as a radio telescope. What is the diffraction limit on the angular resolution of your dish, assuming that you want to observe radio waves with a wavelength of 21 centimeters? Would it be very useful as an astronomical radio telescope?

56. *Hubble's Field of View.* Large telescopes often have small fields of view. For example, the advanced camera of the Hubble Space Telescope (HST) has a field of view that is roughly square and about 0.06° on a side.
**a.** Calculate the angular area of the HST's field of view in square degrees.
**b.** The angular area of the entire sky is about 41,250 square degrees. How many pictures would the HST have to take with its camera to obtain a complete picture of the entire sky?

57. *Hubble Sky Survey?* In Problem 56, you found out how many pictures the HST would need to take to photograph the entire sky. If you assume that it would take 1 hour to produce each picture, how many years would it take the HST to obtain photos of the entire sky? Use your answer to explain why astronomers would like to have more than one large telescope in space.

58. *Visible-Light Interferometry.* Technological advances are now making it possible to link visible-light telescopes so that they can achieve the same angular resolution as a single telescope over 300 meters in size. What is the angular resolution (diffraction limit) of such a system of telescopes for observations at a wavelength of 500 nanometers?

# COSMIC Context

## PART II The Universality of Physics

**One of Isaac Newton's great insights was that physics is universal—the same physical laws govern both the motions of heavenly objects and the things we experience in everyday life. This illustration shows some of the key physical principles used in the study of astronomy, with examples of how they apply both on Earth and in space.**

### EXAMPLES ON EARTH

① **Conservation of Energy:** Energy can be transferred from one object to another or transformed from one type to another, but the total amount of energy is always conserved [Section 4.3].

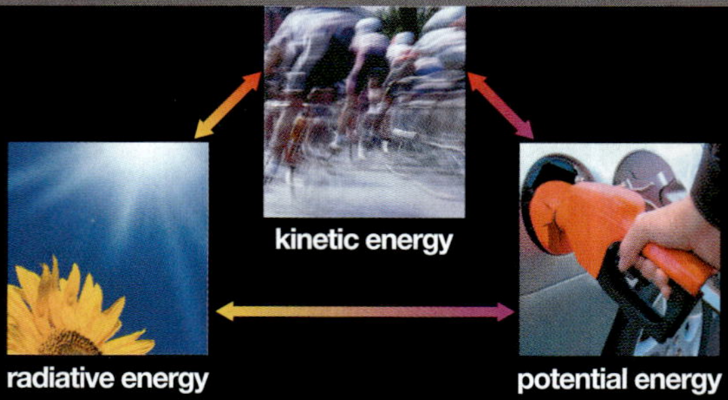


*Plants transform the energy of sunlight into food containing chemical potential energy, which our bodies can convert into energy of motion.*

② **Conservation of Angular Momentum:** An object's angular momentum cannot change unless it transfers angular momentum to another object. Because angular momentum depends on the product of mass, velocity, and radius, a spinning object must spin faster as it shrinks in size and an orbiting object must move faster when its orbital distance is smaller [Section 4.3].

*Conservation of angular momentum explains why a skater spins faster as he pulls in his arms.*

③ **Gravity:** Every mass in the universe attracts every other mass through the force called gravity. The strength of gravity between two objects depends on the product of the masses divided by the square of the distance between them [Section 4.4].

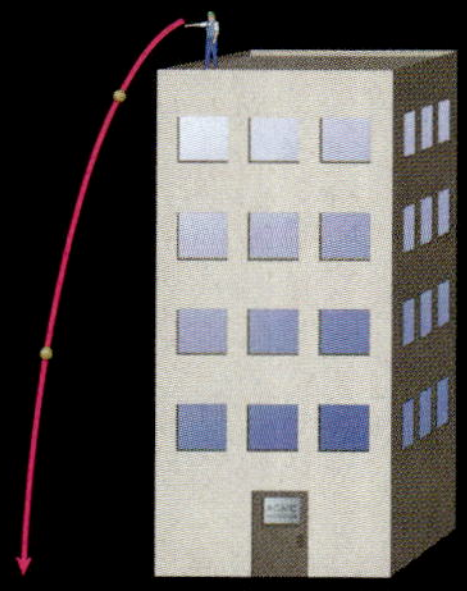

*The force of gravity between a ball and Earth attracts both together, explaining why the ball accelerates as it falls.*

④ **Thermal Radiation:** Large objects emit a thermal radiation spectrum that depends on the object's temperature. Hotter objects emit photons with a higher average energy and emit radiation of greater intensity at all wavelengths [Section 5.4].

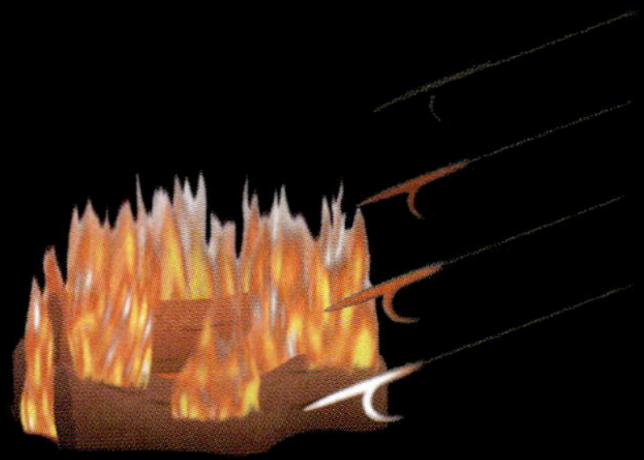

*The glow you see from a hot fireplace poker is thermal radiation in the form of visible light.*

⑤ **Electromagnetic Spectrum:** Light is a wave that affects electrically charged particles and magnets. The wavelength and frequency of light waves range over a wide spectrum, consisting of gamma rays, x-rays, ultraviolet light, visible light, infrared light, and radio waves. Visible light is only a small fraction of the entire spectrum [Section 5.2].

*We encounter many different kinds of electromagnetic radiation in our everyday lives.*

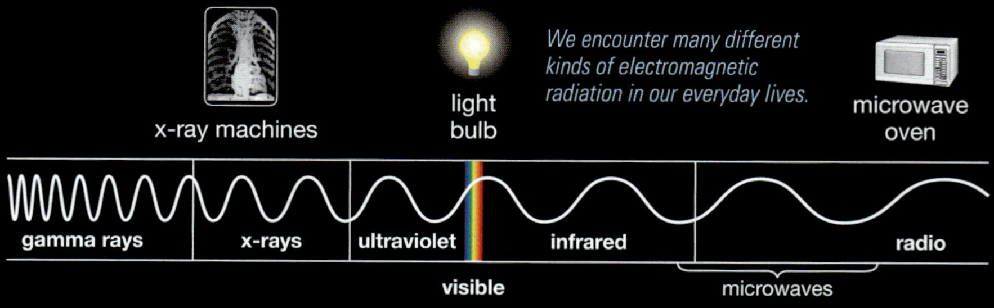

## EXAMPLES IN SPACE

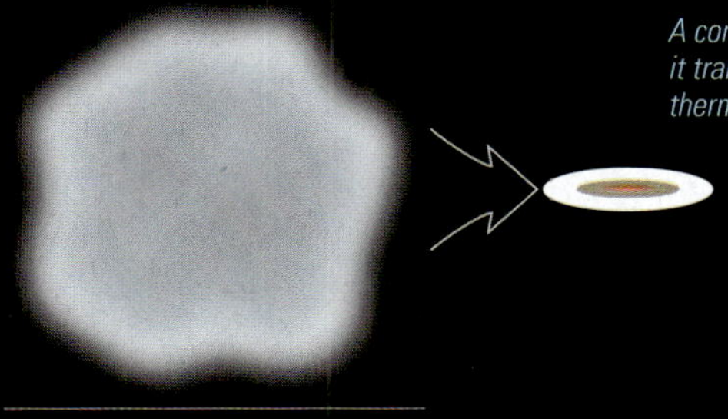

*A contracting gas cloud in space heats up because it transforms gravitational potential energy into thermal energy.*

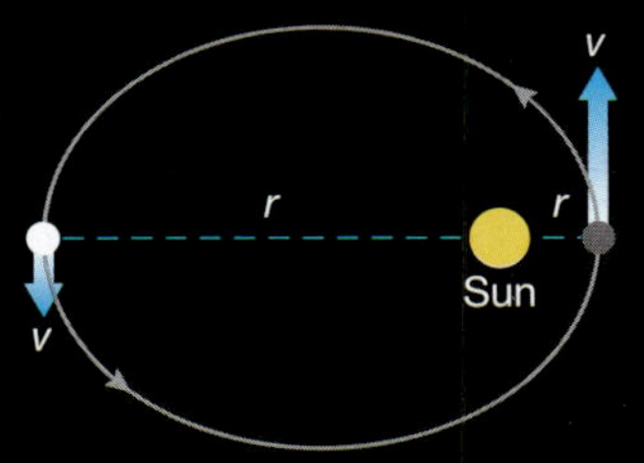


*Conservation of angular momentum also explains why a planet's orbital speed increases when it is closer to the Sun.*

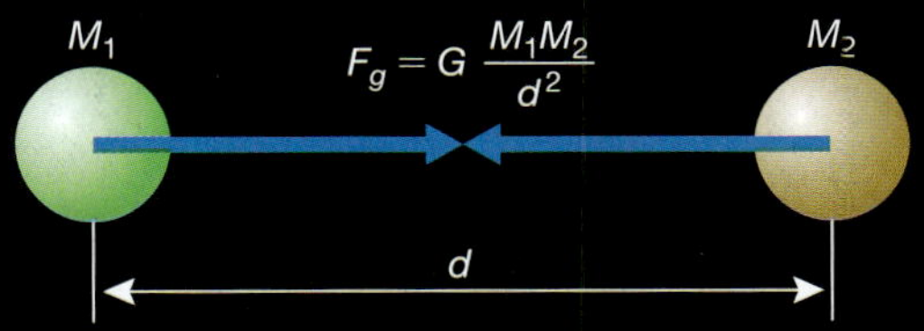


*Gravity also operates in space—its attractive force can act across great distances to pull objects closer together or to hold them in orbit.*

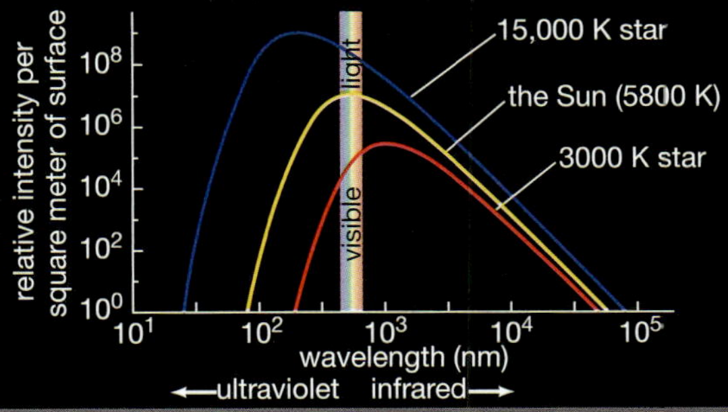


*Sunlight is also a visible form of thermal radiation. The Sun is much brighter and whiter than a fireplace poker because its surface is much hotter.*

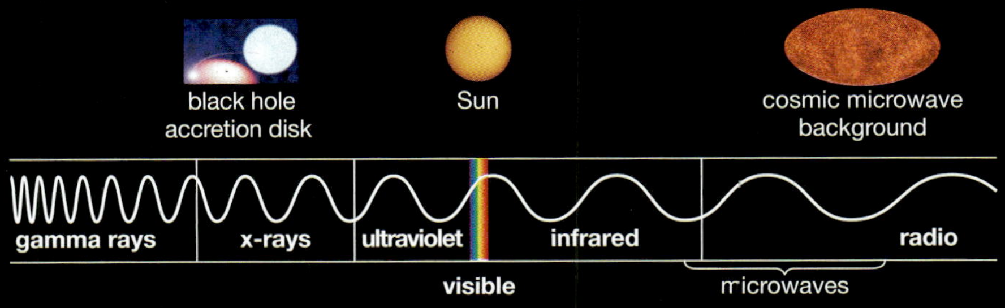


*Many different forms of electromagnetic radiation are present in space. We therefore need to observe light of many different wavelengths to get a complete picture of the universe.*

# 7 Our Planetary System

▲ **About the photo:** Earth, photographed from the outskirts of our solar system by the *Voyager* spacecraft. The "sunbeam" surrounding Earth is an artifact of light scattering in the camera.

## LEARNING GOALS

### 7.1 Studying the Solar System

- What does the solar system look like?
- What can we learn by comparing the planets to one another?

### 7.2 Patterns in the Solar System

- What features of our solar system provide clues to how it formed?

### 7.3 Spacecraft Exploration of the Solar System

- How do robotic spacecraft work?

*I do not know whether humankind is alone in this vast universe. But I do know that we should cherish our existence on this precious speck of matter.*

— *Ban Ki-moon, 8th Secretary General of the United Nations*

**Chapter 7 Overview**

Our ancestors long ago recognized the motions of the planets through the sky, but it has been only a few hundred years since we learned that Earth is also a planet that orbits the Sun. Even then, we knew little about the other planets until the advent of large telescopes. More recently, the dawn of space exploration has brought us far greater understanding of other worlds. We've lived in this solar system all along, but only now are we getting to know it.

In this chapter, we'll explore our solar system like newcomers to the neighborhood. We'll begin by discussing what we hope to learn by studying the solar system, and in the process take a brief tour of major features of the Sun and planets. We'll also explore the major patterns we observe in the solar system—patterns that we will explain in subsequent chapters. Finally, we'll discuss the use of spacecraft to explore the solar system, examining how we are coming to learn so much more about our neighbors.

## 7.1 Studying the Solar System

Galileo's telescopic observations began a new era in astronomy in which the Sun, Moon, and planets could be studied for the first time as *worlds,* rather than as mere lights in the sky. Since that time, we have studied these worlds in different ways. Sometimes we study them individually—for example, when we map the geography of Mars or the atmospheric structure of Jupiter. Other times we compare the worlds to one another, seeking to understand their similarities and differences. This latter approach is called **comparative planetology**. Note that astronomers use the term *planetology* broadly to include moons, asteroids, and comets as well as planets.

We will use the comparative planetology approach for most of our study of the solar system in this book. Before we can compare the planets, however, we must have a general idea of the nature of our solar system and of the characteristics of individual worlds.

### What does the solar system look like?

The first step in getting to know our solar system is to visualize what it looks like as a whole. Imagine having the perspective of an alien spacecraft making its first scientific survey of our solar system. What would we see as we viewed the solar system from beyond the orbits of the planets?

Without a telescope, the answer would be "not much." The Sun and planets are so small compared to the distances between them [**Section 1.1**] that from the outskirts of our solar system, the planets would be only pinpoints of light and even the Sun would be just a small bright dot in the sky. But if we magnify the sizes of the planets by about a thousand times compared to their distances from the Sun and show their orbital paths, we get the central picture in **FIGURE 7.1**.

The ten pages that follow Figure 7.1 offer a brief tour through our solar system, beginning at the Sun, continuing to each of the planets, and concluding with dwarf planets such as Pluto, Ceres, and Eris. The tour highlights a few of the most important features of each world we visit—just enough information so that you'll be ready for the comparative study we'll undertake in later chapters. The side of each page shows the objects to scale, using the 1-to-10-billion scale introduced in Chapter 1. The map along the bottom of each page shows the locations of the Sun and each of the planets in the Voyage scale model solar system (see Figures 1.5 and 1.6), so that you can see their relative distances from the Sun. **TABLE 7.1** follows the tour and summarizes key data.

As you study Figure 7.1, the tour pages, and Table 7.1, you'll quickly see that our solar system is *not* a random collection of worlds, but a system that exhibits many clear patterns. For example, Figure 7.1 shows that all the planets orbit the Sun in the same direction and in nearly the same plane, and the tour pages show that the planets fall into two distinct groups. In science, the existence of patterns like these demands an explanation, and in Chapter 8 we will study the modern theory that explains them quite well. First, however, we need to investigate these patterns in greater detail.

**Think about it** As you read the tour pages (pages 194–203), identify one characteristic of each object that you find particularly interesting and would like to know more about. In addition, try to answer the following questions as you read: (1) Are all the planets made of the same materials? (2) Which planets are "Earth-like" with solid surfaces? (3) How would you organize the planets into groups with common characteristics?

### What can we learn by comparing the planets to one another?

The essence of comparative planetology lies in the idea that we can learn more about an individual world, including our own Earth, by studying it in the context of other objects in our solar system. It is much like learning more about a person by getting to know his or her family, friends, and culture.

(*continued on page 205*)

**COSMIC Context** FIGURE 7.1

# The Solar System

**The solar system's layout and composition offer four major clues to how it formed. The main illustration below shows the orbits of planets in the solar system from a perspective beyond Neptune, with the planets themselves magnified by about a thousand times relative to their orbits. (The Sun is not shown on the same scale as the planets; it would fill the page if it were.)**

1. **Large bodies in the solar system have orderly motions.** All planets have nearly circular orbits going in the same direction in nearly the same plane. Most large moons orbit their planets in this same direction, which is also the direction of the Sun's rotation.

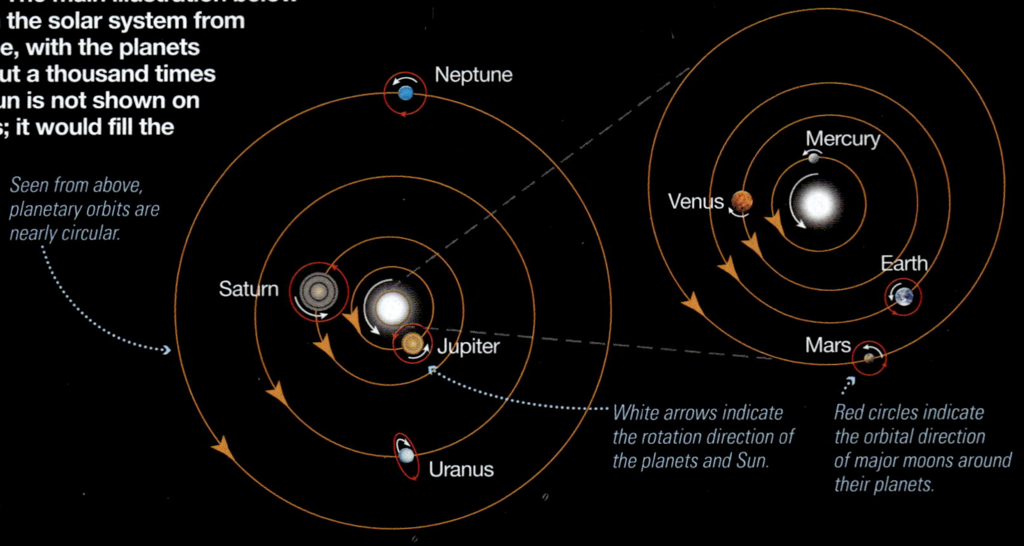


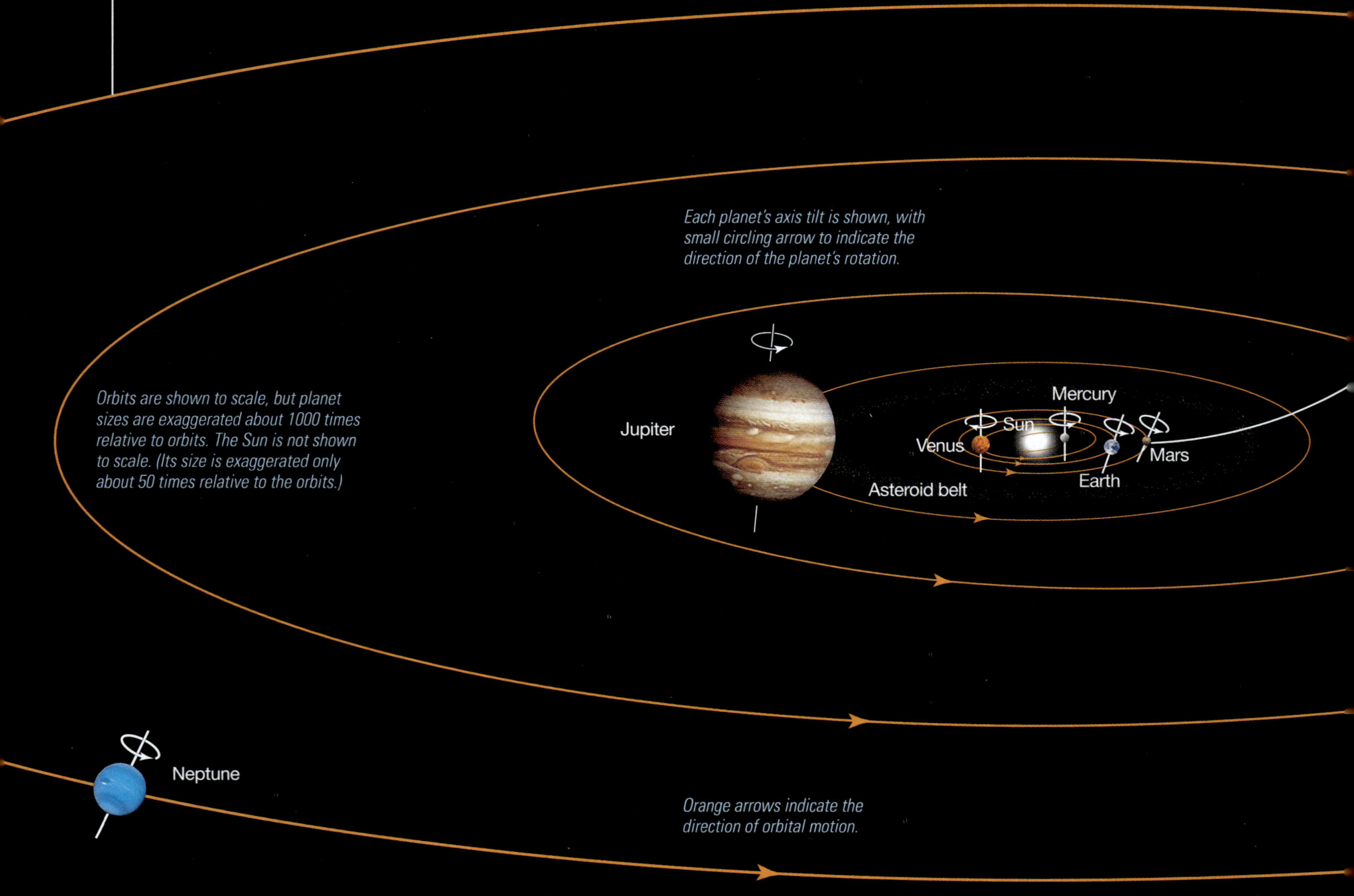

(2) **Planets fall into two major categories:** Small, rocky terrestrial planets and large, hydrogen-rich jovian planets.

***Terrestrial Planets:***
- *small in mass and size*
- *close to the Sun*
- *made of metal and rock*
- *few moons and no rings*

***Jovian Planets:***
- *large mass and size*
- *far from the Sun*
- *made of H, He, and hydrogen compounds*
- *rings and many moons*

(3) **Swarms of asteroids and comets populate the solar system.** Vast numbers of rocky asteroids and icy comets are found throughout the solar system, but are concentrated in three distinct regions.

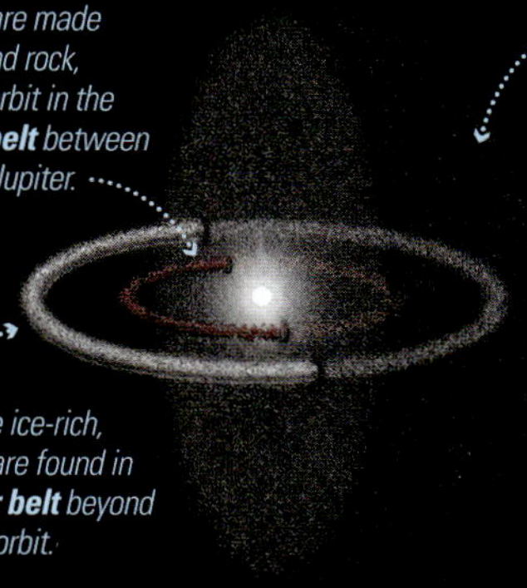

*Asteroids are made of metal and rock, and most orbit in the **asteroid belt** between Mars and Jupiter.*

*Even more comets orbit the Sun in the distant, spherical region called the **Oort cloud**, and only a rare few ever plunge into the inner solar system.*

*Comets are ice-rich, and many are found in the **Kuiper belt** beyond Neptune's orbit.*

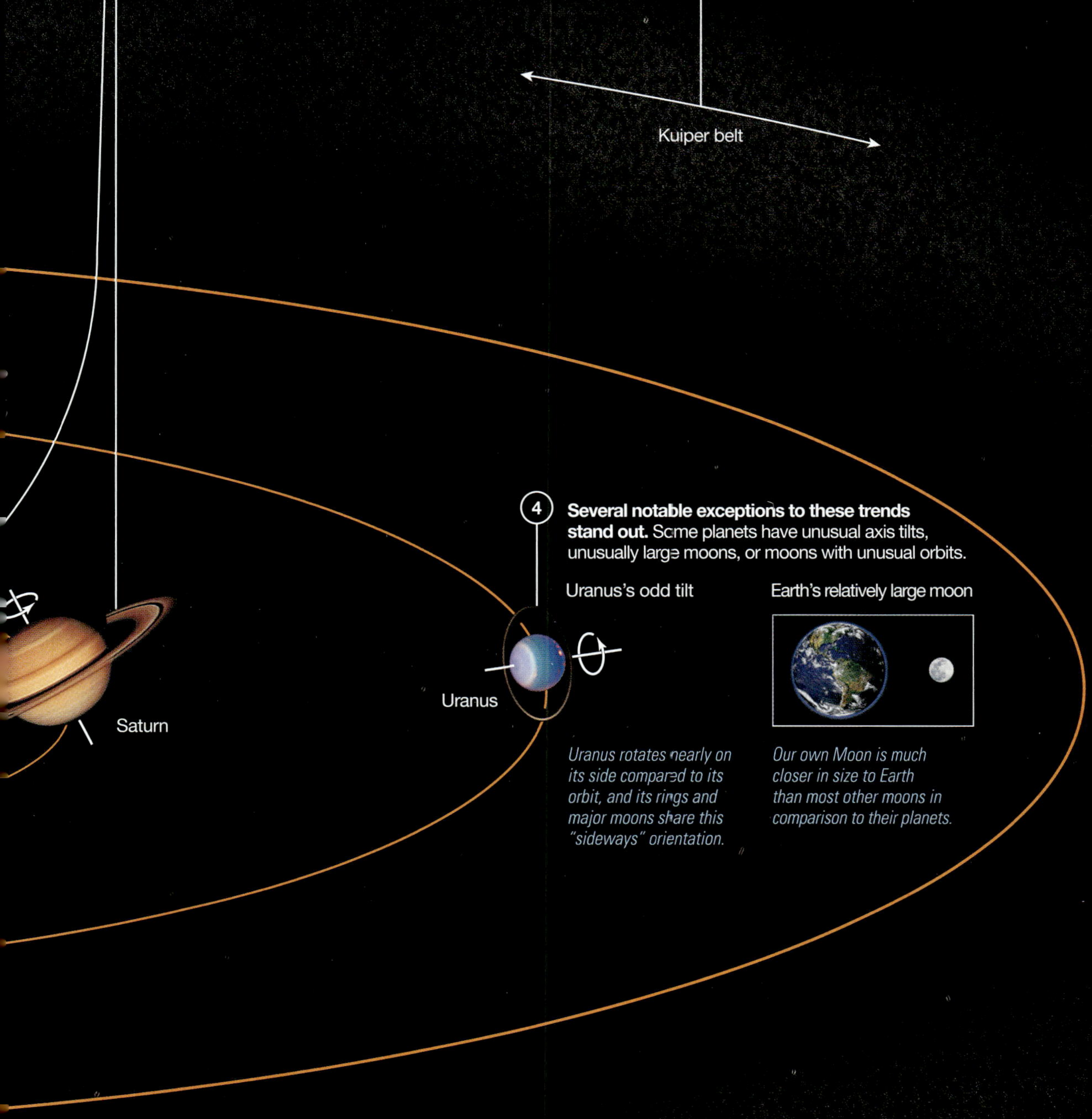


(4) **Several notable exceptions to these trends stand out.** Some planets have unusual axis tilts, unusually large moons, or moons with unusual orbits.

Uranus's odd tilt

*Uranus rotates nearly on its side compared to its orbit, and its rings and major moons share this "sideways" orientation.*

Earth's relatively large moon

*Our own Moon is much closer in size to Earth than most other moons in comparison to their planets.*

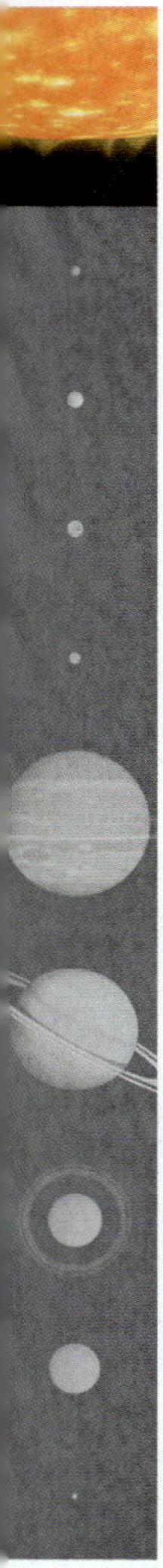

**FIGURE 7.2** The Sun contains more than 99.8% of the total mass in our solar system.

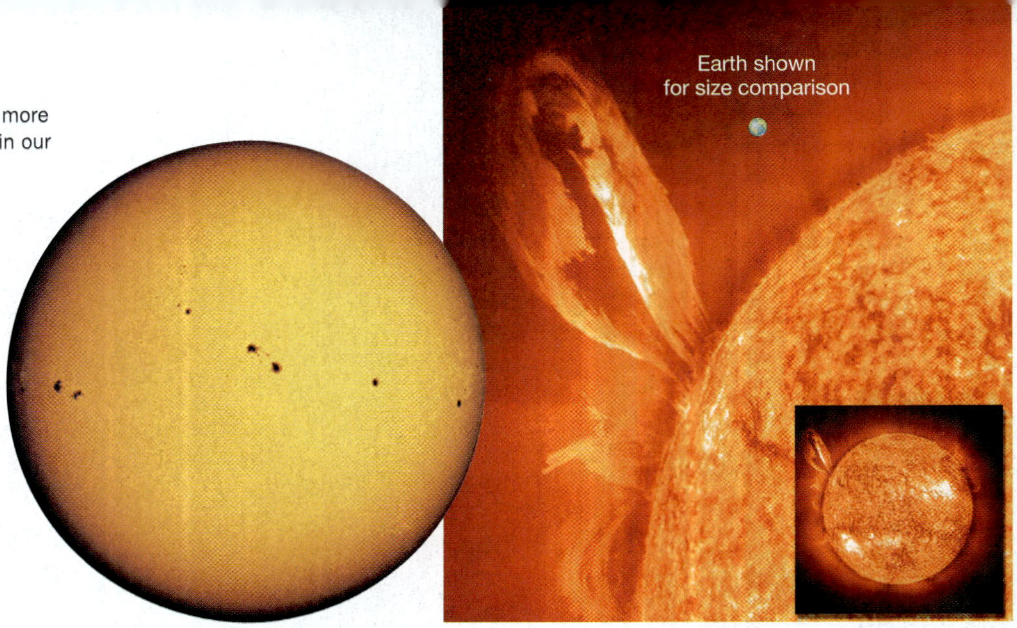


**a** A visible-light photograph of the Sun's surface. The dark splotches are sunspots—each large enough to swallow several Earths.

**b** This ultraviolet photograph, from the *SOHO* spacecraft, shows a huge streamer of hot gas on the Sun.

## The Sun

- Radius: 696,000 km $= 108R_{\text{Earth}}$
- Mass: $333{,}000M_{\text{Earth}}$
- Composition (by mass): 98% hydrogen and helium, 2% other elements

The Sun is by far the largest and brightest object in our solar system. It contains more than 99.8% of the solar system's total mass, making it nearly a thousand times as massive as everything else in the solar system combined.

The Sun's surface looks solid in photographs (**FIGURE 7.2**), but it is actually a roiling sea of hot (about 5800 K, or 5500°C or 10,000°F) hydrogen and helium gas. The surface is speckled with *sunspots* that appear dark in photographs only because they are slightly cooler than their surroundings. Solar storms sometimes send streamers of hot gas soaring far above the surface.

The Sun is gaseous throughout, and the temperature and pressure both increase with depth. The source of the Sun's energy lies deep in its core, where the temperatures and pressures are so high that the Sun is a nuclear fusion power plant. Each second, fusion transforms about 600 million tons of the Sun's hydrogen into 596 million tons of helium. The "missing" 4 million tons becomes energy in accord with Einstein's famous formula, $E = mc^2$ [**Section 4.3**]. Despite losing 4 million tons of mass each second, the Sun contains so much hydrogen that it has already shone steadily for almost 5 billion years and will continue to shine for another 5 billion years.

The Sun is the most influential object in our solar system. Its gravity governs the orbits of the planets. Its heat is the primary influence on the temperatures of planetary surfaces and atmospheres. It is the source of virtually all the light in our solar system—planets and moons shine by virtue of the sunlight they reflect. In addition, charged particles flowing outward from the Sun make up the *solar wind* that interacts with planetary magnetic fields and influences planetary atmospheres. Nevertheless, we can understand almost all the present characteristics of the planets without knowing much more about the Sun than we have just discussed. We'll therefore save more detailed study of the Sun for Chapter 14, where we will study it as our prototype for understanding other stars.

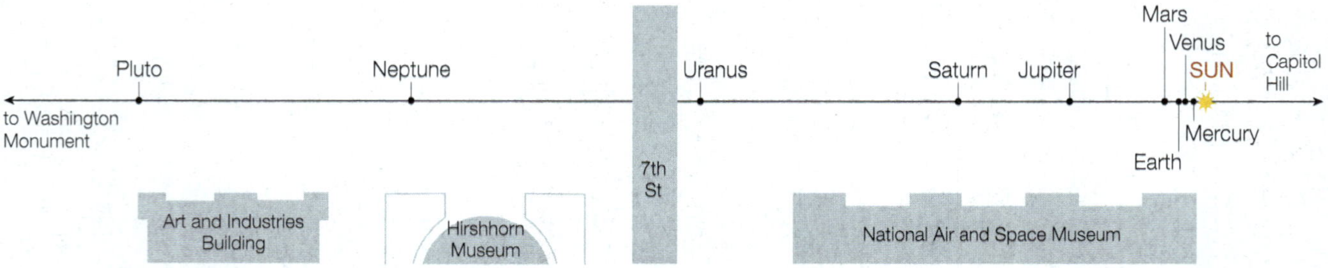


The Voyage scale model solar system represents sizes and distances in our solar system at one ten-billionth of their actual values (see Figure 1.6). The strip along the side of the page shows the sizes of the Sun and planets on this scale, and the map above shows their locations in the Voyage model on the National Mall in Washington, D.C. The Sun is about the size of a large grapefruit on this scale.

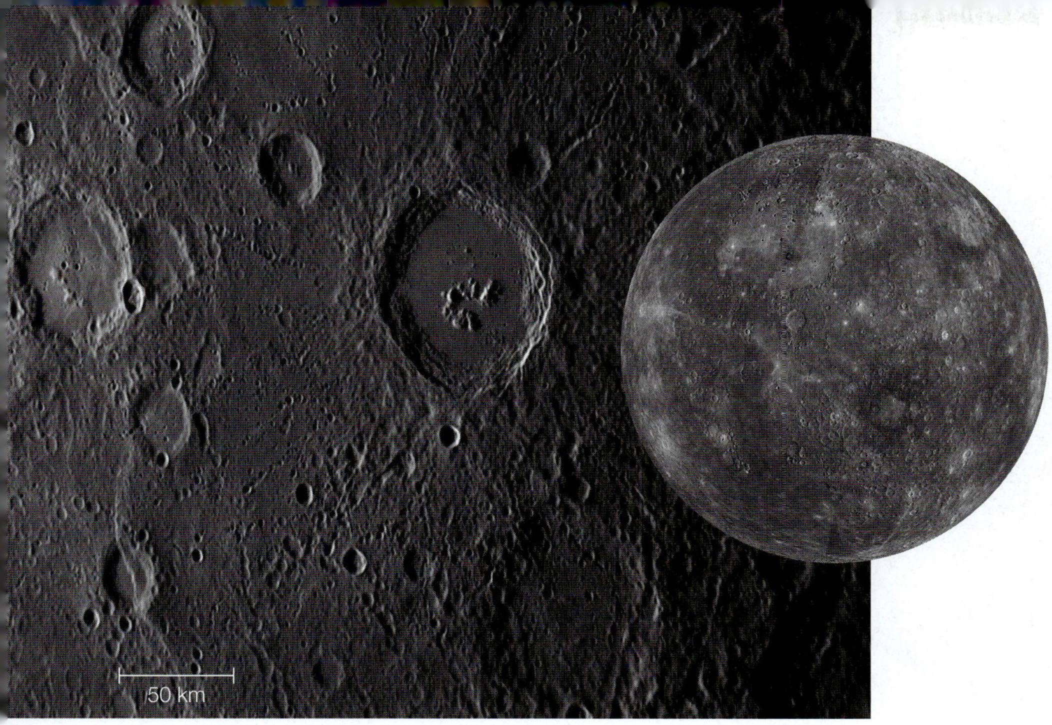


**FIGURE 7.3** The left image shows that Mercury's surface is heavily cratered but also has smooth volcanic plains and long, steep cliffs. The inset shows a global composite. (Images from the *MESSENGER* spacecraft.)

## Mercury

- Average distance from the Sun: 0.39 AU
- Radius: 2440 km = $0.38R_{\text{Earth}}$
- Mass: $0.055M_{\text{Earth}}$
- Average density: 5.43 g/cm$^3$
- Composition: rocks, metals
- Average surface temperature: 700 K (day), 100 K (night)
- Moons: 0

Mercury is the innermost planet of our solar system, and the smallest of the eight planets. It is a desolate, cratered world with no active volcanoes, no wind, no rain, and no life. Because there is virtually no air to scatter sunlight or color the sky, you could see stars even in the daytime if you stood on Mercury with your back toward the Sun.

You might expect Mercury to be very hot because of its closeness to the Sun, but in fact it is a world of both hot and cold extremes. Tidal forces from the Sun have forced Mercury into an unusual rotation pattern [**Section 4.5**]: Its 58.6-day rotation period means it rotates exactly three times for every two of its 87.9-day orbits of the Sun. This combination of rotation and orbit gives Mercury days and nights that last about 3 Earth months each. Daytime temperatures reach 425°C, nearly as hot as hot coals. At night or in shadow, the temperature falls below −150°C, far colder than Antarctica in winter.

Mercury's surface is heavily cratered (**FIGURE 7.3**), much like the surface of our Moon. But it also shows evidence of past geological activity, such as plains created by ancient lava flows and tall, steep cliffs that run hundreds of kilometers in length. These cliffs may be wrinkles from an episode of "planetary shrinking" early in Mercury's history. Mercury's high density (calculated from its mass and volume) indicates that it has a very large iron core, perhaps because it once suffered a huge impact that blasted its outer layers away.

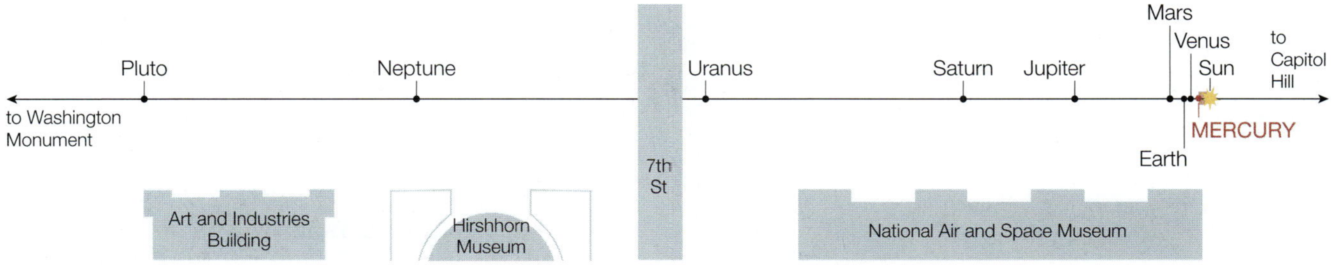

**FIGURE 7.4** The image above shows an artistic rendition of the surface of Venus as scientists think it would appear to our eyes. The surface topography is based on data from NASA's *Magellan* spacecraft. The inset (left) shows the full disk of Venus photographed by NASA's *Pioneer Venus Orbiter* with cameras sensitive to ultraviolet light. (Image above from the Voyage scale model solar system, developed by the Challenger Center for Space Science Education, the Smithsonian Institution, and NASA. Image by David P. Anderson, Southern Methodist University © 2001.)

## Venus

- Average distance from the Sun: 0.72 AU
- Radius: 6051 km $= 0.95R_{\text{Earth}}$
- Mass: $0.82M_{\text{Earth}}$
- Average density: 5.24 g/cm$^3$
- Composition: rocks, metals
- Average surface temperature: 740 K
- Moons: 0

Venus, the second planet from the Sun, is nearly identical in size to Earth. Before the era of spacecraft visits, Venus stood out largely for its strange rotation: It rotates on its axis very slowly and in the opposite direction of Earth, so days and nights are very long and the Sun rises in the west and sets in the east instead of rising in the east and setting in the west. Its surface is completely hidden from view by dense clouds, so we knew little about it until a few decades ago, when spacecraft began to map Venus with cloud-penetrating radar, discovering mountains, valleys, craters, and extensive evidence of past volcanic activity (**FIGURE 7.4**). Because we knew so little about it, some science fiction writers used its Earth-like size, thick atmosphere, and closer distance to the Sun to speculate that it might be a lush, tropical paradise—a "sister planet" to Earth.

The reality is far different. We now know that an extreme *greenhouse effect* bakes Venus's surface to an incredible 470°C (about 880°F), trapping heat so effectively that nighttime offers no relief. Day and night, Venus is hotter than a pizza oven, and the thick atmosphere bears down on the surface with a pressure equivalent to that nearly a kilometer (0.6 mile) beneath the ocean's surface on Earth. Far from being a beautiful sister planet to Earth, Venus resembles a traditional view of hell.

The fact that Venus and Earth are so similar in size and composition but so different in surface conditions suggests that Venus could teach us important lessons. In particular, Venus's greenhouse effect is caused by carbon dioxide, the same gas that is primarily responsible for global warming on Earth. Perhaps further study of Venus may help us better understand and solve some of the problems we face here at home.

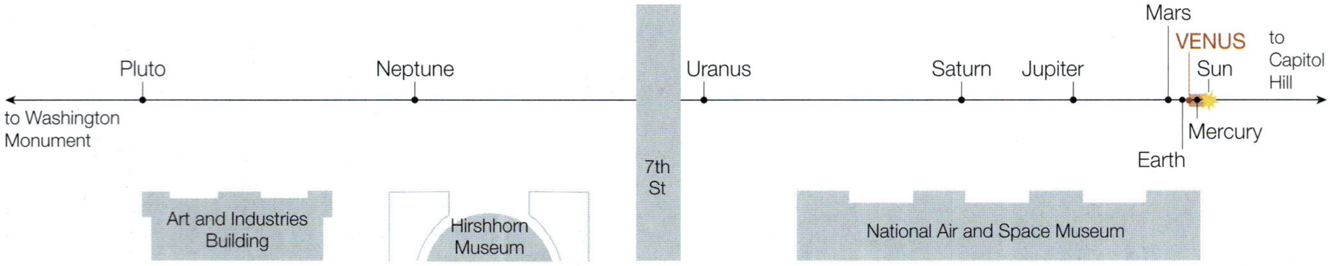

a This image (left), computer generated from satellite data, shows the striking contrast between the day and night hemispheres of Earth. The day side reveals little evidence of human presence, but at night our presence is revealed by the lights of human activity. (From the Voyage scale model solar system, developed by the Challenger Center for Space Science Education, the Smithsonian Institution, and NASA. Image created by ARC Science Simulations © 2001.)

b Earth and the Moon, shown to scale. The Moon is about 1/4 as large as Earth in diameter, while its mass is about 1/80 of Earth's mass. To show the distance between Earth and Moon on the same scale, you'd need to hold these two photographs about 1 meter (3 feet) apart.

**FIGURE 7.5** Earth, our home planet.

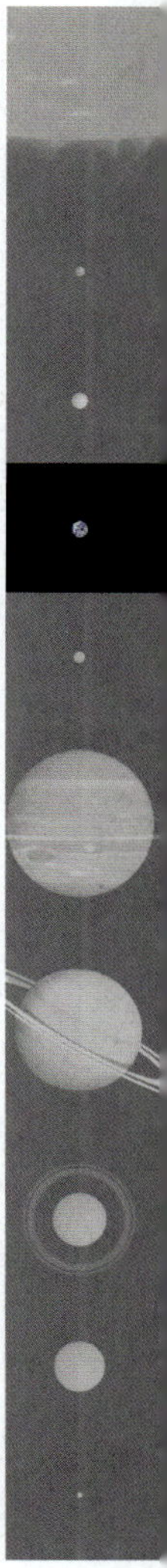

## Earth

- Average distance from the Sun: 1.00 AU
- Radius: 6378 km = $1R_{\text{Earth}}$
- Mass: $1.00M_{\text{Earth}}$
- Average density: 5.52 g/cm$^3$
- Composition: rocks, metals
- Average surface temperature: 290 K
- Moons: 1

Beyond Venus, we next encounter our home planet, Earth, the only known oasis of life in our solar system. Earth is also the only planet in our solar system with oxygen to breathe, ozone to shield the surface from deadly solar ultraviolet radiation, and abundant surface water to nurture life. Temperatures are pleasant because Earth's atmosphere contains just enough carbon dioxide and water vapor to maintain a moderate greenhouse effect.

Despite Earth's small size, its beauty is striking (**FIGURE 7.5a**). Blue oceans cover nearly three-fourths of the surface, broken by the continental land masses and scattered islands. The polar caps are white with snow and ice, and white clouds are scattered above the surface. At night, the glow of artificial lights reveals the presence of an intelligent civilization.

Earth is the first planet on our tour with a moon. The Moon is surprisingly large compared with Earth (**FIGURE 7.5b**); although it is not the largest moon in the solar system, almost all other moons are much smaller relative to the planets they orbit. As we'll discuss in Chapter 8, the leading hypothesis holds that the Moon formed as a result of a giant impact early in Earth's history.

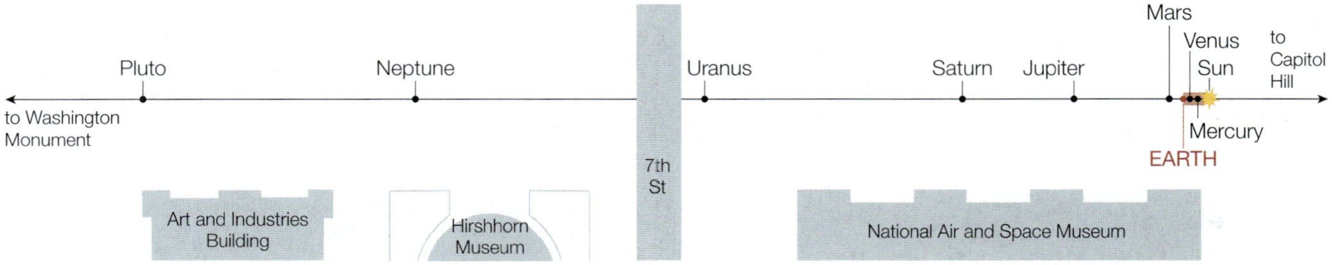

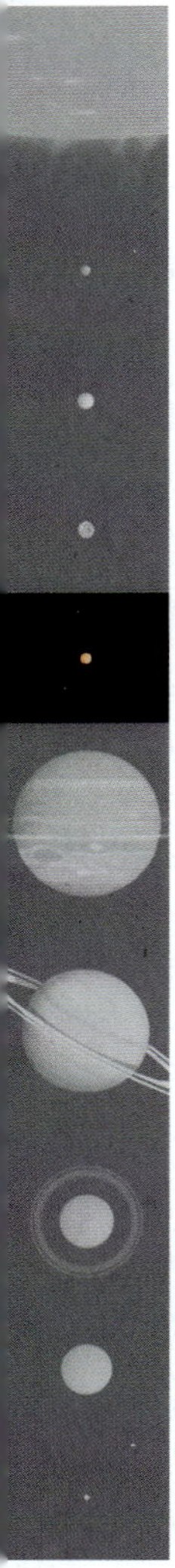

**FIGURE 7.6** The image above shows a self-portrait of NASA's *Curiosity* rover on the floor of Gale Crater, assembled from dozens of separate images taken by the camera on the robot arm. Here, *Curiosity* is on the flank of its primary destination, Mount Sharp, with dark sand dunes visible to the left. The inset shows a close-up of the disk of Mars photographed by the *Viking* orbiter; the horizontal "gash" across the center is the giant canyon Valles Marineris.

## Mars

- Average distance from the Sun: 1.52 AU
- Radius: 3397 km $= 0.53R_{\text{Earth}}$
- Mass: $0.11M_{\text{Earth}}$
- Average density: 3.93 g/cm
- Composition: rocks, metals
- Average surface temperature: 220 K
- Moons: 2 (very small)

The next planet on our tour is Mars, the last of the four inner planets of our solar system (**FIGURE 7.6**). Mars is larger than Mercury and the Moon but only about half Earth's size in diameter; its mass is about 10% that of Earth. Mars has two tiny moons, Phobos and Deimos, which may once have been asteroids that were captured into Martian orbit early in the solar system's history.

Mars is a world of wonders, with ancient volcanoes that dwarf the largest mountains on Earth, a great canyon that runs nearly one-fifth of the way around the planet, and polar caps made of frozen carbon dioxide ("dry ice") and water. Although Mars is frozen today, the presence of dried-up riverbeds, rock-strewn floodplains, and minerals that form in water offers clear evidence that Mars had at least some warm and wet periods in the past. Major flows of liquid water probably ceased at least 3 billion years ago, but some liquid water could persist underground, perhaps flowing to the surface on occasion.

Mars's surface looks almost Earth-like, but you wouldn't want to visit without a space suit. The air pressure is far less than that on top of Mount Everest, the temperature is usually well below freezing, the trace amounts of oxygen would not be nearly enough to breathe, and the lack of atmospheric ozone would leave you exposed to deadly ultraviolet radiation from the Sun.

More than a dozen spacecraft have flown past, orbited, or landed on Mars, and plans are in the works for more. We may even send humans to Mars within the next few decades. By overturning rocks in ancient riverbeds or chipping away at ice in the polar caps, explorers will help us learn whether Mars has ever been home to life.

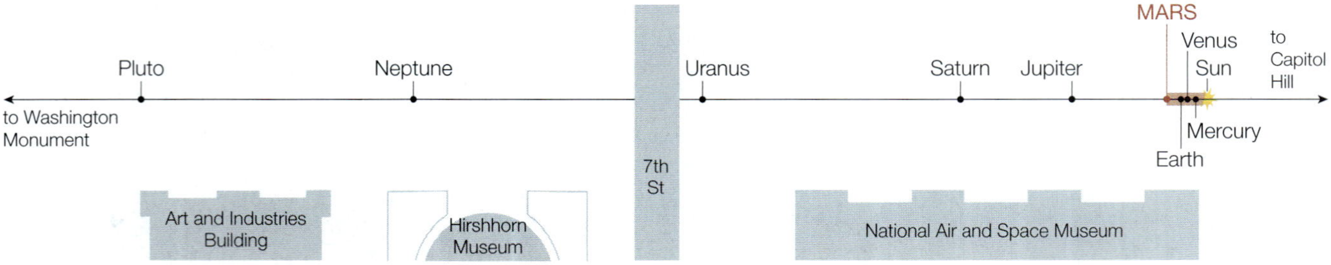

**FIGURE 7.7** This image shows what it would look like to be orbiting near Jupiter's moon Io as Jupiter came into view. Notice the Great Red Spot to the left of Jupiter's center. The extraordinarily dark rings discovered during the *Voyager* missions are exaggerated to make them visible. This computer visualization was created using data from NASA's *Voyager* and *Galileo* missions. (From the Voyage scale model solar system, developed by the Challenger Center for Space Science Education, the Smithsonian Institution, and NASA. Image created by ARC Science Simulations © 2001.)

## Jupiter

- Average distance from the Sun: 5.20 AU
- Radius 71,492 km $= 11.2R_{\text{Earth}}$
- Mass: $318M_{\text{Earth}}$
- Average density: 1.33 g/cm$^3$
- Composition: mostly hydrogen and helium
- Cloud-top temperature: 125 K
- Moons: at least 79

To reach the orbit of Jupiter from Mars, we must traverse a distance that is more than double the total distance from the Sun to Mars, passing through the *asteroid belt* along the way. Upon our arrival, we find a planet much larger than any we have seen so far (**FIGURE 7.7**).

Jupiter is so different from the planets of the inner solar system that we must adopt an entirely new mental image of the term *planet*. Its mass is more than 300 times that of Earth, and its volume is more than 1000 times that of Earth. Its most famous feature—a long-lived storm called the Great Red Spot—is itself large enough to swallow two or three Earths. Like the Sun, Jupiter is made primarily of hydrogen and helium and has no solid surface. If we plunged deep into Jupiter, the increasing gas pressure would crush us long before we reached its core.

Jupiter reigns over dozens of moons and a thin set of rings (too faint to be seen in most photographs). Most of the moons are very small, but four are large enough that we'd call them planets or dwarf planets if they orbited the Sun independently. These four moons—Io, Europa, Ganymede, and Callisto—are often called the *Galilean moons* (because Galileo discovered them), and they display varied and interesting geology. Io is the most volcanically active world in the solar system. Europa has an icy crust that may hide a subsurface ocean of liquid water, making it a promising place to search for life. Ganymede and Callisto may also have subsurface oceans, and their surfaces have many features that remain mysterious.

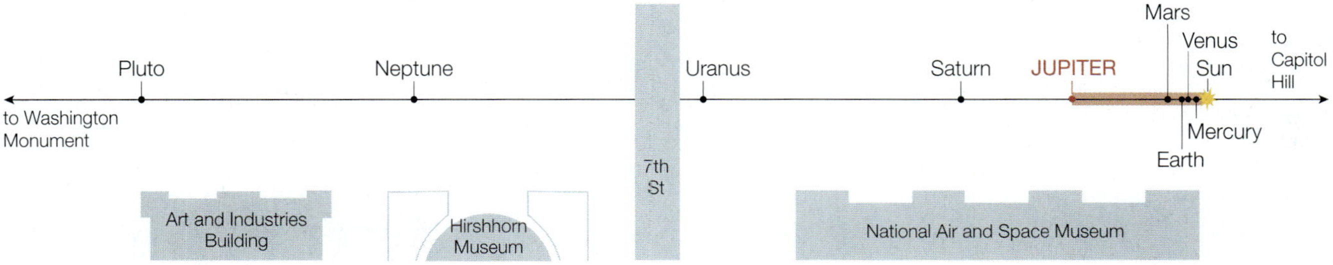

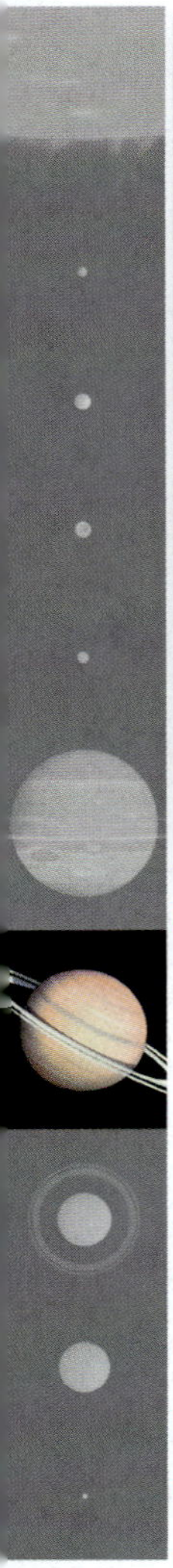

**FIGURE 7.8** *Cassini*'s view of Saturn. We see the shadow of the rings on the upper right portion of Saturn's sunlit face, and the rings become lost in Saturn's shadow on the night side. The inset shows an infrared view of Saturn's largest moon, Titan, which is shrouded in a thick, cloudy atmosphere.

## Saturn

- Average distance from the Sun: 9.54 AU
- Radius: 60,268 km = $9.4R_{\text{Earth}}$
- Mass: $95.2M_{\text{Earth}}$
- Average density: 0.70 g/cm$^3$
- Composition: mostly hydrogen and helium
- Cloud-top temperature: 95 K
- Moons: at least 62

The journey from Jupiter to Saturn is a long one: Saturn orbits nearly twice as far as Jupiter from the Sun. Saturn, the second-largest planet in our solar system, is only slightly smaller than Jupiter in diameter, but its lower density makes it considerably less massive (about one-third of Jupiter's mass). Like Jupiter, Saturn is made mostly of hydrogen and helium and has no solid surface.

Saturn is famous for its spectacular rings (**FIGURE 7.8**). Although all four of the giant outer planets have rings, only Saturn's can be seen easily. The rings look solid from a distance, but in reality they are made of countless small particles, each of which orbits Saturn like a tiny moon. These particles of rock and ice range in size from dust grains to city blocks.

Saturn also has numerous moons, including at least two that are geologically active today: Enceladus, which has ice fountains spraying out from its southern hemisphere, and Titan, the only moon in the solar system with a thick atmosphere. Saturn and its moons are so far from the Sun that Titan's surface temperature is a frigid −180°C, making it far too cold for liquid water. Studies by the *Cassini* spacecraft and its *Huygens* probe, which landed on Titan in 2005, revealed an erosion-carved surface with riverbeds and lakes—but the features are shaped by extremely cold liquid methane or ethane rather than liquid water.

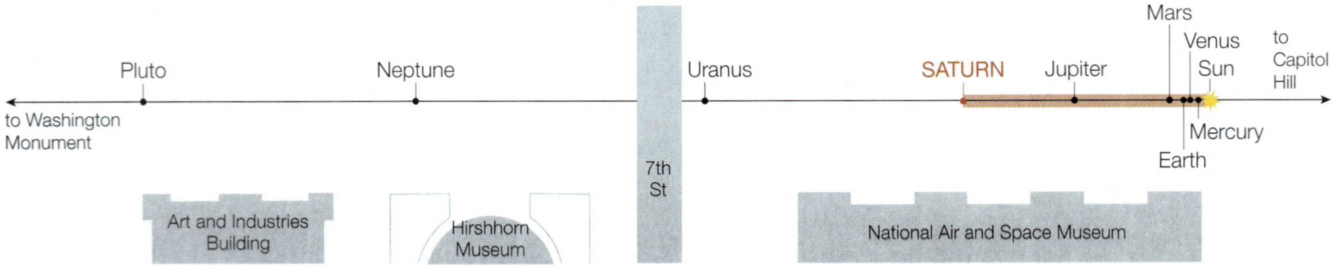

**FIGURE 7.9** This image shows a view of Uranus from high above its moon Ariel. The ring system is shown, although it would actually be too dark to see from this vantage point. This computer simulation is based on data from NASA's *Voyager 2* mission. (From the Voyage scale model solar system, developed by the Challenger Center for Space Science Education, the Smithsonian Institution, and NASA. Image created by ARC Science Simulations © 2001.)

## Uranus

- Average distance from the Sun: 19.2 AU
- Radius: 25,559 km $= 4.0R_{\text{Earth}}$
- Mass: $14.5M_{\text{Earth}}$
- Average density: 1.32 g/cm$^3$
- Composition: hydrogen, helium, hydrogen compounds
- Cloud-top temperature: 60 K
- Moons: at least 27

It's another long journey to the next stop on our tour, as Uranus lies twice as far as Saturn from the Sun. Uranus (normally pronounced "YUR-uh-nus") is much smaller than either Jupiter or Saturn but much larger than Earth. It is made largely of hydrogen, helium, and *hydrogen compounds* such as water ($H_2O$), ammonia ($NH_3$), and methane ($CH_4$). Methane gas gives Uranus its pale blue-green color (**FIGURE 7.9**). Like the other giants of the outer solar system, Uranus lacks a solid surface. More than two dozen moons orbit Uranus, along with a set of rings somewhat similar to those of Saturn but much darker and more difficult to see.

The entire Uranus system—planet, rings, and moon orbits—is tipped on its side compared to the rest of the planets. This extreme axis tilt may be the result of a cataclysmic collision that Uranus suffered as it was forming, and it gives Uranus the most extreme seasonal variations of any planet in our solar system. If you lived on a platform floating in Uranus's atmosphere near its north pole, you'd have continuous daylight for half of each orbit, or 42 years. Then, after a very gradual sunset, you'd enter into a 42-year-long night.

Only one spacecraft has visited Uranus: *Voyager 2*, which flew past all four of the giant outer planets before heading out of the solar system. Much of our current understanding of Uranus comes from that mission, though powerful new telescopes are also capable of studying it. Scientists hope it will not be too long before we can send another spacecraft to study Uranus and its rings and moons in greater detail.

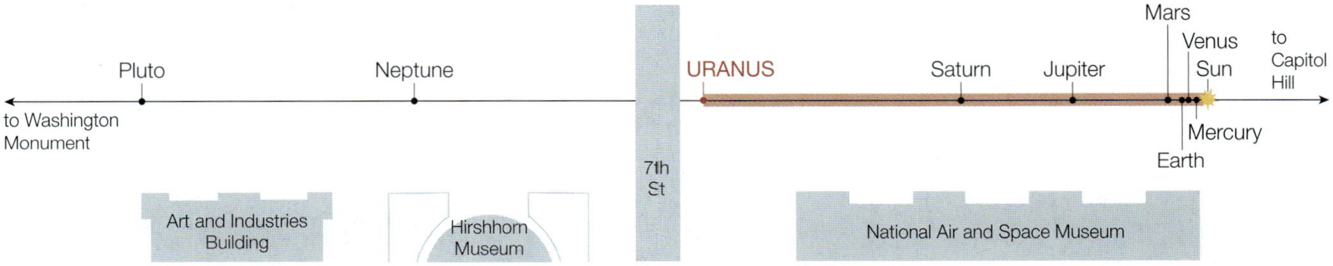

**FIGURE 7.10** This image shows what it would look like to be orbiting Neptune's moon Triton as Neptune itself came into view. The dark rings are exaggerated to make them visible in this computer simulation using data from NASA's *Voyager 2* mission. (From the Voyage scale model solar system, developed by the Challenger Center for Space Science Education, the Smithsonian Institution, and NASA. Image created by ARC Science Simulations © 2001.)

## Neptune

- Average distance from the Sun: 30.1 AU
- Radius: 24,764 km = $3.9R_{\text{Earth}}$
- Mass: $17.1M_{\text{Earth}}$
- Average density: 1.64 g/cm$^3$
- Composition: hydrogen, helium, hydrogen compounds
- Cloud-top temperature: 60 K
- Moons: at least 14

The journey from the orbit of Uranus to the orbit of Neptune is the longest yet in our tour, calling attention to the vast emptiness of the outer solar system. Nevertheless, Neptune looks nearly like a twin of Uranus, although it is more strikingly blue (**FIGURE 7.10**). It is slightly smaller than Uranus in size, but a higher density makes it slightly more massive even though the two planets have very similar compositions. Like Uranus, Neptune has been visited only by the *Voyager 2* spacecraft.

Neptune has rings and numerous moons. Its largest moon, Triton, is larger than Pluto and is one of the most fascinating moons in the solar system. Triton's icy surface has features that appear to be somewhat like geysers, although they spew nitrogen gas rather than water into the sky. Even more surprisingly, Triton is the only large moon in the solar system that orbits its planet "backward"—that is, in a direction opposite to the direction in which Neptune rotates. This backward orbit makes it a near certainty that Triton once orbited the Sun independently before somehow being captured into Neptune's orbit.

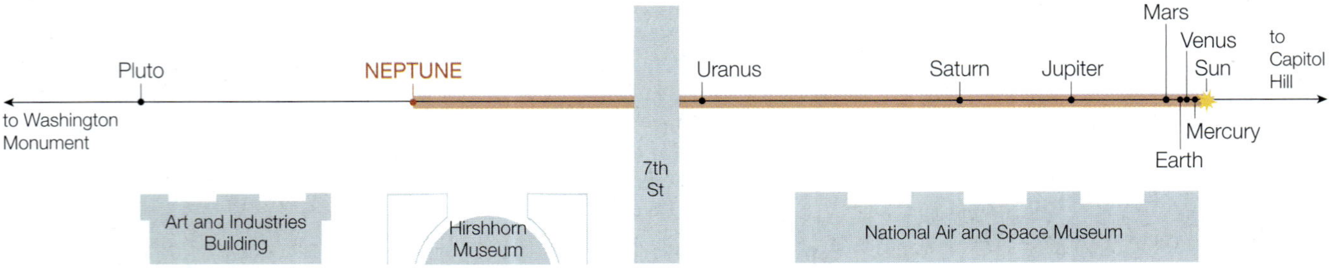

**FIGURE 7.11** Pluto and its largest moon, Charon, photographed by the *New Horizons* spacecraft. Although this image is a composite, it approximates how the spacecraft saw the pair as it approached them in July 2015.

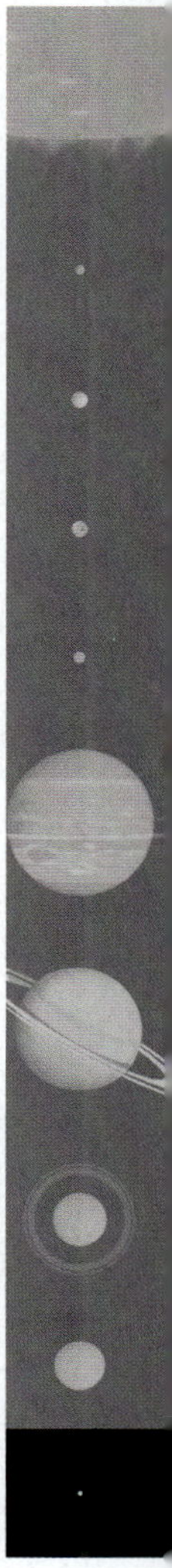

## Dwarf planets: Pluto and more

*Pluto Data:*

- Average distance from the Sun: 39.5 AU
- Radius: 1185 km = $0.19R_{\text{Earth}}$
- Mass: $0.0022M_{\text{Earth}}$
- Average density: 1.9 g/cm$^3$
- Composition: ices, rock
- Average surface temperature: 44 K
- Moons: 5

We conclude our tour at Pluto (**FIGURE 7.11**), which reigned for 75 years as the "ninth planet" in our solar system. However, the 2005 discovery of the slightly more massive Eris, and the fact that dozens of other objects are not much smaller than Pluto and Eris, led scientists to reconsider the definition of "planet." The result was that we now refer to Pluto and Eris as *dwarf planets,* too small to qualify as official planets but large enough to be round in shape. Several other solar system objects also qualify as dwarf planets, including Ceres, the largest asteroid of the asteroid belt.

Pluto and Eris belong to a vast collection of thousands of icy objects that orbit the Sun beyond Neptune, making up what we call the *Kuiper belt*. As you can see in Figure 7.1, the Kuiper belt is much like the asteroid belt, except it is farther from the Sun and composed of comet-like objects rather than rocky asteroids.

Pluto's characteristics help us to think about what it would be like to visit this distant realm. Pluto's average distance from the Sun lies as far beyond Neptune as Neptune lies beyond Uranus, making Pluto extremely cold and dimly lit even in daytime. From Pluto, the Sun is little more than a bright light among the stars. Pluto's largest moon, Charon, is locked together with it in synchronous rotation [**Section 4.5**], so Charon dominates the sky on one side of Pluto but is never seen from the other side.

The great distances and small sizes of Pluto and other dwarf planets make them difficult to study, but two recent spacecraft missions have taught us a lot: the *New Horizons* mission, which flew past Pluto in 2015, and the *Dawn* mission, which began orbiting Ceres in the same year.

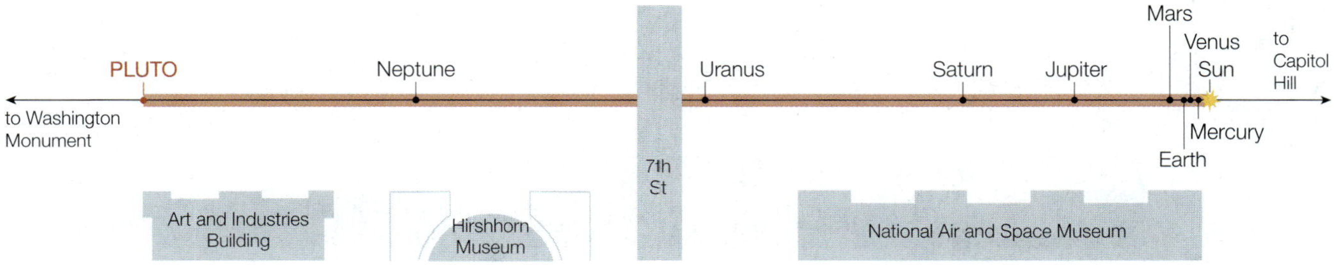

**TABLE 7.1 The Planetary Data**[a]

| Photo | Planet | Relative Size | Average Distance from Sun (AU) | Average Equatorial Radius (km) | Mass (Earth = 1) | Average Density ($g/cm^3$) | Orbital Period | Rotation Period | Axis Tilt | Average Surface (or Cloud-Top) Temperature[b] | Composition | Known Moons (2018) | Rings? |
|---|---|---|---|---|---|---|---|---|---|---|---|---|---|
|  | Mercury | · | 0.387 | 2440 | 0.055 | 5.43 | 87.9 days | 58.6 days | 0.0° | 700 K (day) 100 K (night) | Rocks, metals | 0 | No |
|  | Venus | • | 0.723 | 6051 | 0.82 | 5.24 | 225 days | 243 days | 177.3° | 740 K | Rocks, metals | 0 | No |
|  | Earth | • | 1.00 | 6378 | 1.00 | 5.52 | 1.00 year | 23.93 hours | 23.5° | 290 K | Rocks, metals | 1 | No |
|  | Mars | · | 1.52 | 3397 | 0.11 | 3.93 | 1.88 years | 24.6 hours | 25.2° | 220 K | Rocks, metals | 2 | No |
|  | Jupiter | ● | 5.20 | 71,492 | 318 | 1.33 | 11.9 years | 9.93 hours | 3.1° | 125 K | H, He, hydrogen compounds[c] | 79 | Yes |
|  | Saturn | ● | 9.54 | 60,268 | 95.2 | 0.70 | 29.5 years | 10.6 hours | 26.7° | 95 K | H, He, hydrogen compounds[c] | 62 | Yes |
| 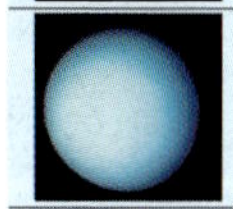 | Uranus | ● | 19.2 | 25,559 | 14.5 | 1.32 | 83.8 years | 17.2 hours | 97.9° | 60 K | H, He, hydrogen compounds[c] | 27 | Yes |
|  | Neptune | ● | 30.1 | 24,764 | 17.1 | 1.64 | 165 years | 16.1 hours | 29.6° | 60 K | H, He, hydrogen compounds[c] | 14 | Yes |
|  | Pluto | · | 39.5 | 1185 | 0.0022 | 1.9 | 248 years | 6.39 days | 112.5° | 44 K | Ices, rock | 5 | No |
| 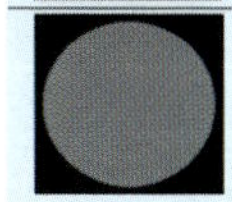 | Eris | · | 67.7 | 1168 | 0.0028 | 2.3 | 557 years | 1.08 days | 78° | 43 K | Ices, rock | 1 | No |

[a] Including the dwarf planets Pluto and Eris; Appendix E gives a more complete list of planetary properties

[b] Surface temperatures for all objects except Jupiter, Saturn, Uranus, and Neptune, for which cloud-top temperatures are listed

[c] Includes water ($H_2O$), methane ($CH_4$), and ammonia ($NH_3$)

(*continued from page 191*)

While we still can learn much by studying planets individually, the comparative planetology approach has demonstrated its value in at least three key ways:

- Comparative study has revealed similarities and differences among the planets that have helped guide the development of our theory of solar system formation, thereby giving us a better understanding of how we came to exist here on Earth.
- Comparative study has given us new insights into the physical processes that have shaped Earth and other worlds—insights that can help us better understand and manage our own planet.
- Comparative study has allowed us to apply lessons from our solar system to the study of the many planetary systems now known around other stars. These lessons help us understand both the general principles that govern planetary systems and the specific circumstances under which Earth-like planets—and possibly life—might exist elsewhere.

The comparative planetology approach should also benefit you as a student by helping you stay focused on *processes* rather than on a collection of facts. We now know so many individual facts about the worlds of our solar system and others that even planetary scientists have trouble keeping track of them all. By concentrating on the processes that shape planets, you'll gain a deeper understanding of how planets, including Earth, actually work.

## 7.2 Patterns in the Solar System

One of our major goals in studying the solar system as a whole is to understand how it formed. In this section, we'll explore key patterns that must be explained by a theory of solar system formation.

**Four Features of the Solar System**

### What features of our solar system provide clues to how it formed?

We have already seen that our solar system is not a random collection of worlds, but rather a family of worlds exhibiting many traits that would be difficult to attribute to coincidence. We could make a long list of such traits, but it is easier to develop a scientific theory by focusing on the more general structure of our solar system. For our purposes, four major features stand out, each corresponding to one of the numbered steps in Figure 7.1:

1. **Patterns of motion among large bodies.** The Sun, planets, and large moons generally orbit and rotate in a very organized way.
2. **Two major types of planets.** The eight planets* divide clearly into two groups: the small, rocky planets that are close together and close to the Sun, and the large, gas-rich planets that are farther apart and farther from the Sun.
3. **Asteroids and comets.** Between and beyond the planets, vast numbers of asteroids and comets orbit the Sun; some are large enough to qualify as dwarf planets. The locations, orbits, and compositions of these asteroids and comets follow distinct patterns.
4. **Exceptions to the rules.** The generally orderly solar system also has some notable exceptions. For example, among the inner planets only Earth has a large moon, and Uranus is tipped on its side. A successful theory must make allowances for such exceptions even as it explains the general rules.

Because these four features are so important to our study of the solar system, let's investigate them in a little more detail.

**Feature 1: Patterns of Motion Among Large Bodies** If you look back at Figure 7.1, you'll notice several clear patterns of motion among the large bodies of our solar system. (In this context, a "body" is simply an individual object such as the Sun, a planet, or a moon.) For example:

- All planetary orbits are nearly circular and lie nearly in the same plane.
- All planets orbit the Sun in the same direction: counterclockwise as viewed from high above Earth's North Pole.
- Most planets rotate in the same direction in which they orbit, with fairly small axis tilts. The Sun also rotates in this direction.
- Most of the solar system's large moons exhibit similar properties in their orbits around their planets, such as orbiting in their planet's equatorial plane in the same direction as the planet rotates.

We consider these orderly patterns together as the first major feature of our solar system. As we'll see in Chapter 8, our theory of solar system formation explains these patterns as consequences of processes that occurred during the early stages of the birth of our solar system.

**Feature 2: Two Types of Planets** Our brief planetary tour showed that the four inner planets are quite different from the four outer planets. We say that these two groups represent two distinct planetary classes: *terrestrial* and *jovian.*

The **terrestrial planets** (*terrestrial* means "Earth-like") are the four planets of the inner solar system: Mercury, Venus, Earth, and Mars. These planets are relatively small and dense, with rocky surfaces and an abundance of metals in their cores. They have few moons, if any, and no rings. We count our Moon as a fifth terrestrial world, because its history has been shaped by the same processes that have shaped the terrestrial planets.

---

*As this book goes to press, a search is on for a possible ninth planet orbiting far beyond Pluto. The search has been spurred by careful study of the orbits of small bodies in this region, which suggest the possible presence of an object of unknown nature with a mass 10 or more times that of Earth.

**TABLE 7.2 Comparison of Terrestrial and Jovian Planets**

| Terrestrial Planets | Jovian Planets |
|---|---|
| Smaller size and mass | Larger size and mass |
| Higher average density | Lower average density |
| Made mostly of rocks and metals | Made mostly of hydrogen, helium, and hydrogen compounds |
| Solid surface | No solid surface |
| Few (if any) moons and no rings | Rings and many moons |
| Closer to the Sun (and closer together), with warmer surfaces | Farther from the Sun (and farther apart), with cool temperatures at cloud tops |

The **jovian planets** (*jovian* means "Jupiter-like") are the four large planets of the outer solar system: Jupiter, Saturn, Uranus, and Neptune. The jovian planets are much larger in size and lower in average density than the terrestrial planets, and they have rings and many moons. They lack solid surfaces and are made mostly of hydrogen, helium, and **hydrogen compounds**—compounds containing hydrogen, such as water ($H_2O$), ammonia ($NH_3$), and methane ($CH_4$). Because these substances are gases under earthly conditions, the jovian planets are sometimes called "gas giants." **TABLE 7.2** contrasts the general traits of the terrestrial and jovian planets.

**Feature 3: Asteroids and Comets** The third major feature of our solar system is the existence of vast numbers of small objects orbiting the Sun. These objects fall into two major groups: asteroids and comets.

**Asteroids** are rocky bodies that orbit the Sun much like planets, but they are much smaller (**FIGURE 7.12**). Most known asteroids are found within the **asteroid belt** between the orbits of Mars and Jupiter (see Figure 7.1).

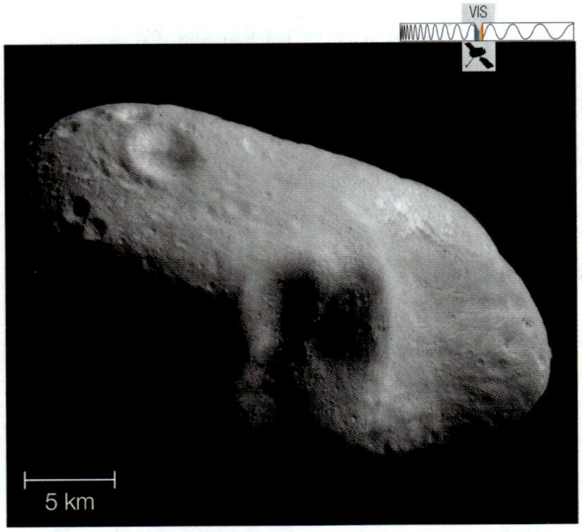


**FIGURE 7.12** The asteroid Eros (photographed from the *NEAR* spacecraft). Its appearance is probably typical of most asteroids. Eros is about 40 kilometers in length. Like other small objects in the solar system, it is not spherical.

**FIGURE 7.13** Comet McNaught over Patagonia, Argentina, in 2007. The fuzzy patches above the comet tail are the Magellanic Clouds, satellite galaxies of the Milky Way.

**Comets** are also small objects that orbit the Sun, but they are made largely of ices (such as water ice, ammonia ice, and methane ice) mixed with rock. You are probably familiar with the occasional appearance of comets in the inner solar system, where they may become visible to the naked eye with long, beautiful tails (**FIGURE 7.13**). These visitors, which may delight sky watchers for a few weeks or months, are actually quite rare among comets. The vast majority of comets never visit the inner solar system. Instead, they orbit the Sun in one of the two distinct regions shown as Feature 3 in Figure 7.1. The first is a donut-shaped region beyond the orbit of Neptune that we call the **Kuiper belt** (*Kuiper* rhymes with *piper*). The Kuiper belt contains at least 100,000 icy objects that are more than 100 kilometers in diameter, of which Pluto and Eris are the largest known. The second cometary region, called the **Oort cloud** (*Oort* rhymes with *court*), is much farther from the Sun and may contain a trillion comets (most just a few kilometers across). These comets have orbits randomly inclined to the ecliptic plane, giving the Oort cloud a roughly spherical shape.

**Feature 4: Exceptions to the Rules** The fourth key feature of our solar system is that there are a few notable exceptions to the general rules. For example, while most of the planets rotate in the same direction as they orbit, Uranus rotates nearly on its side, and Venus rotates "backward" (clockwise as viewed from high above Earth's North Pole). Similarly, while most large moons orbit their planets in the same direction as their planets rotate, many small moons have much more unusual orbits.

One of the most interesting exceptions concerns our own Moon. While the other terrestrial planets have either no moons (Mercury and Venus) or very tiny moons (Mars), Earth has one of the largest moons in the solar system.

**Summary** Now that you have read through the tour of our solar system and the description of its four major features, review them again in Figure 7.1. You should now see clearly that these features hold key clues to the origin of our solar system—the main topic of the Chapter 8.

## 7.3 Spacecraft Exploration of the Solar System

How have we learned so much about the solar system? Much of our knowledge comes from telescopic observations, using both ground-based telescopes and telescopes in Earth orbit such as the Hubble Space Telescope. In one case—our Moon—we have learned a lot by sending astronauts to explore the terrain and bring back rocks for laboratory study. In a few other cases, we have studied samples of distant worlds that have come to us as meteorites. But most of the data fueling the recent revolution in our understanding of the solar system have come from robotic spacecraft. To date, we have sent robotic spacecraft to all the terrestrial and jovian planets, as well as to many moons, asteroids, and comets. In this section, we'll briefly investigate how we use robotic spacecraft to explore the solar system.

### How do robotic spacecraft work?

The spacecraft we send to explore the planets are robots designed for scientific study. All spacecraft have computers used to control their major components, power sources such as solar cells, propulsion systems, and scientific instruments used to study their targets. Robotic spacecraft operate primarily with preprogrammed instructions, but also carry radios that allow them to communicate with controllers on Earth. Most robotic spacecraft make one-way trips, never physically returning to Earth but sending their data back from space in the same way we send radio and television signals.

Broadly speaking, the robotic missions to other worlds fall into four major categories:

- **Flyby.** A spacecraft on a flyby goes past a world just once and then continues on its way.

### SPECIAL TOPIC How Did We Learn the Scale of the Solar System?

This chapter presents the layout of the solar system as we know it today, when we have precise measurements of planetary sizes and distances. But how did we learn the scale of the solar system?

By the middle of the 17th century, Kepler's laws [**Section 3.3**] had provided planetary distances in astronomical units (AU), or distances *relative* to the Earth-Sun distance, but no one yet knew the value of the AU in absolute units like miles or kilometers. A number of 17th-century astronomers proposed ideas for measuring the Earth-Sun distance, but none were practical. Then, in 1716, Edmond Halley (best known for the comet named after him) hit upon the idea that would ultimately solve the problem: He realized that during a planetary *transit*, when a planet appears to pass across the face of the Sun [**Section S1.1**], observers in different locations on Earth would see the planet trace slightly different paths across the Sun. Comparison of these paths could allow calculation of the planet's distance—which would in turn allow determination of the AU—through the simple geometry shown in **FIGURE 1**.

Only Mercury and Venus can produce transits visible from Earth. Halley realized that although Mercury transits occur more often, the measurements would be easier with Venus because its closer distance to Earth means greater separation between the paths in Figure 1. Unfortunately, Venus transits are rare, occurring in pairs 8 years apart about every 120 years. Halley did not live to see a Venus transit, but later astronomers followed his plan, mounting expeditions to observe transits in 1761 and 1769.

The transit observations turned out to be quite difficult in practice, partly as a result of the challenge that long expeditions posed at that time and partly because getting the geometry right required very precise timing of the beginning of the transit. Astronomers discovered that this timing was more complicated than Halley had guessed because of optical effects that occur during a transit. Nevertheless, astronomers studied the data from the 1761 and 1769 transits for many decades, and by the middle of the 19th century the value of the AU had been pinned down to within about 5% of its modern value of 149.6 million kilometers. The next Venus transits occurred in 1874 and 1882. Photography had been invented by then, making observations more reliable, so in principle those transits could have allowed refinement of the AU. However, by that time photography and better telescopes had also made it possible to observe parallax of planets against stars, and by 1877 such observations had given us the value of the AU to within 0.2% of its modern value.

Nowadays, we measure the distance to Venus very precisely by bouncing radio waves off its surface with radar, a technique known as *radar ranging* (see **Figure 20.22**). Because we know the speed of light, measuring the time it takes for the radio waves to make the round trip from Earth to Venus tells us the precise distance. We then use this distance and Venus's known distance in AU to calculate the actual value of the AU. Once we know the value of the AU, we can determine the actual distances of all the planets from the Sun, and we can determine their actual sizes from their angular sizes and distances. Indeed, we now know the layout of the solar system so well that we can launch spacecraft from Earth and send them to precise places on or around distant worlds.

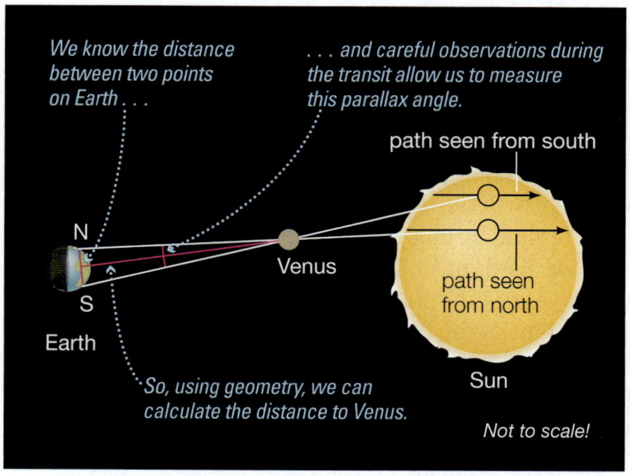


**FIGURE 1** During a transit of Venus, observers at different places on Earth will see it trace slightly different paths across the Sun. The precise geometry of these events therefore allows computation of Venus's true distance, which in turn allows computation of the AU distance. (Adapted from *Sky and Telescope*.)

- **Orbiter.** An orbiter is a spacecraft that orbits the world it is visiting, allowing longer-term study.
- **Lander or probe.** These spacecraft are designed to land on a planet's surface or probe a planet's atmosphere by flying through it. Some landers carry rovers to explore wider regions.
- **Sample return mission.** A sample return mission makes a round trip to return a sample of the world it has studied to Earth.

The choice of spacecraft type depends on both scientific objectives and cost.

**Flybys** Flybys tend to be cheaper than other missions because they are generally less expensive to launch into space. Launch costs depend largely on weight, and onboard fuel is a significant part of the weight of a spacecraft heading to another planet. Once a spacecraft is on its way, the lack of friction or air drag in space means that it can maintain its orbital trajectory through the solar system without using any fuel at all. Fuel is needed only when the spacecraft must change from one trajectory (orbit) to another.

Moreover, some flybys gain more "bang for the buck" by visiting multiple planets. For example, *Voyager 2* flew past Jupiter, Saturn, Uranus, and Neptune before continuing on its way out of our solar system (**FIGURE 7.14**). This trajectory allowed additional fuel savings by permitting use of the gravity of each planet along the spacecraft's path to help boost it onward to the next planet. This technique, known as a *gravitational slingshot,* can not only bend the spacecraft's path but also speed the spacecraft up by essentially stealing a tiny bit of the planet's orbital energy, though the effect on the planet is unnoticeable.

**Think about it** Study the *Voyager 2* trajectory in Figure 7.14. Given that Saturn orbits the Sun every 29 years, Uranus orbits the Sun every 84 years, and Neptune orbits the Sun every 165 years, would it be possible to send another flyby mission to all four jovian planets if we launched it now? Explain.

Although a flyby offers only a relatively short period of close-up study, it can provide valuable scientific information. Spacecraft on flybys generally carry small telescopes, cameras, and spectrographs. Because these instruments are brought relatively close (typically within thousands of kilometers or less) to other worlds, they can obtain much higher resolution images and spectra than the largest telescopes on Earth or in Earth orbit. In addition, flybys sometimes give us information that would be very difficult to obtain from Earth. For example, *Voyager 2* helped us discover Jupiter's rings and learn about the rings of Saturn, Uranus, and Neptune through views in which the rings were backlit by the Sun. Such views are possible only from beyond each planet's orbit.

Spacecraft on flybys may also carry instruments to measure local magnetic field strength or to sample interplanetary dust. The gravitational effects of the planets and their moons on the spacecraft itself provide information about object masses and densities. Like the backlit views of the rings, these types of data cannot be gathered from Earth.

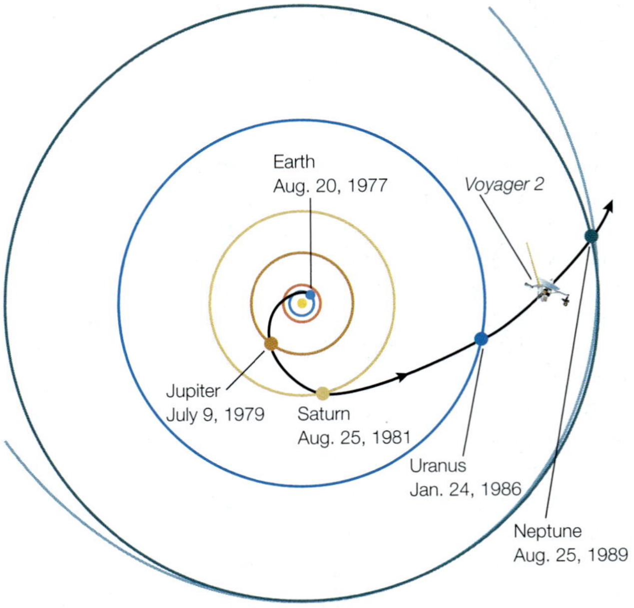


**FIGURE 7.14** The trajectory of *Voyager 2,* which made flybys of the four jovian planets in our solar system.

Indeed, most of what we know about the masses and compositions of moons comes from data gathered by spacecraft that have flown past them.

**Orbiters** An orbiter can study another world for a much longer period of time than a flyby. Like the spacecraft used for flybys, orbiters often carry cameras, spectrographs, and instruments for measuring the strength of magnetic fields. Some missions also carry radar, which can be used to make precise altitude measurements of surface features. Radar has proven especially valuable for the study of Venus and Titan, because it provides our only way of "seeing" through their thick, cloudy atmospheres.

An orbiter is generally more expensive than a flyby for an equivalent weight of scientific instruments, primarily because it must carry added fuel to change from an interplanetary trajectory to a path that puts it into orbit around another world. Careful planning can minimize the added expense. For example, recent Mars orbiters have saved on fuel costs by carrying only enough fuel to enter highly elliptical orbits around Mars. The spacecraft then settled into the smaller, more circular orbits needed for scientific observations by skimming the Martian atmosphere at the low point of every elliptical orbit. Atmospheric drag slowed the spacecraft with each orbit and, over several months, circularized the spacecraft orbit. (This technique is sometimes called *aerobraking.*) We have sent orbiters to the Moon, to the planets Venus, Mars, Jupiter, and Saturn, and to two asteroids and a comet.

**Landers and Probes** The most "up close and personal" study of other worlds comes from spacecraft that send probes into the atmospheres or landers to the surfaces. For example, in 1995, the *Galileo* spacecraft dropped a probe into Jupiter's atmosphere [**Section 11.1**]. The probe collected temperature, pressure, composition, and radiation measurements for about an hour as it descended; it was then destroyed by the heat and pressure of Jupiter's interior.

On planets with solid surfaces, a lander can offer close-up surface views, local weather monitoring, and the ability to carry out automated experiments. Landers have successfully reached the surfaces of the Moon, the planets Venus and Mars, Saturn's moon Titan, asteroids, and a comet. Several Mars landers have included rovers to explore wider areas of the surface, including the *Curiosity* rover that began exploring Mars in 2012. Because of its weight, *Curiosity*'s landing required a particularly spectacular feat of engineering (**FIGURE 7.15**). The spacecraft carrying the lander first used a parachute to slow it down in the Martian atmosphere and then fired rockets that slowed it to a halt about 7 meters above the surface. Finally, a "sky crane" lowered the rover to the surface.

**Sample Return Missions** Although probes and landers can carry out experiments on surface rock or atmospheric samples, the experiments must be designed in advance and the needed equipment must fit inside the spacecraft. One way around these limitations is to design missions in which samples from other worlds can be scooped up and returned to Earth for more detailed study. To date, the only sample return missions have been to the Moon and to asteroids. Many scientists are working toward a sample return mission to Mars; they hope to launch such a mission within the next decade or so. A slight variation on the theme of a sample return mission was the *Stardust* mission, which collected comet dust on a flyby and returned it to Earth in 2006.

**Combination Spacecraft** Many missions combine more than one type of spacecraft. For example, the *Galileo* mission to Jupiter included an orbiter that studied Jupiter and its moons as well as the probe that entered Jupiter's atmosphere. The *Cassini* spacecraft included flybys of Venus, Earth, and Jupiter during its 7-year trip to Saturn. The spacecraft itself was an orbiter that studied Saturn and its moons from 2004 to 2017, but it also carried the *Huygens* probe, which descended through the atmosphere and landed on the surface of Saturn's moon Titan in 2005.

**Exploration—Past, Present, and Future** Over the past several decades, studies using both telescopes on Earth and robotic spacecraft have allowed us to learn the general characteristics of all the major planets and moons in our solar system as well as the general characteristics of asteroids and comets. Telescopes will continue to play an important role in future observations, but for detailed study we will probably continue to depend on spacecraft.

**TABLE 7.3** lists some significant robotic missions that are either currently in operation or far enough along in the planning process that they are likely to be in the news as you study astronomy. The coming years promise many new discoveries as these missions achieve their scientific goals. Over the longer term, all the world's major space agencies have hopes of launching additional and diverse missions to answer many more questions about the nature of our solar system and its numerous worlds.

**See it for yourself** It can be easy with a book full of planetary images to forget that these are real objects, many of which you can see in the night sky. Search the web for "planets tonight" and then go out and see if you can find any of the planets in tonight's sky. Which planets can you see? Why can't you see the others?

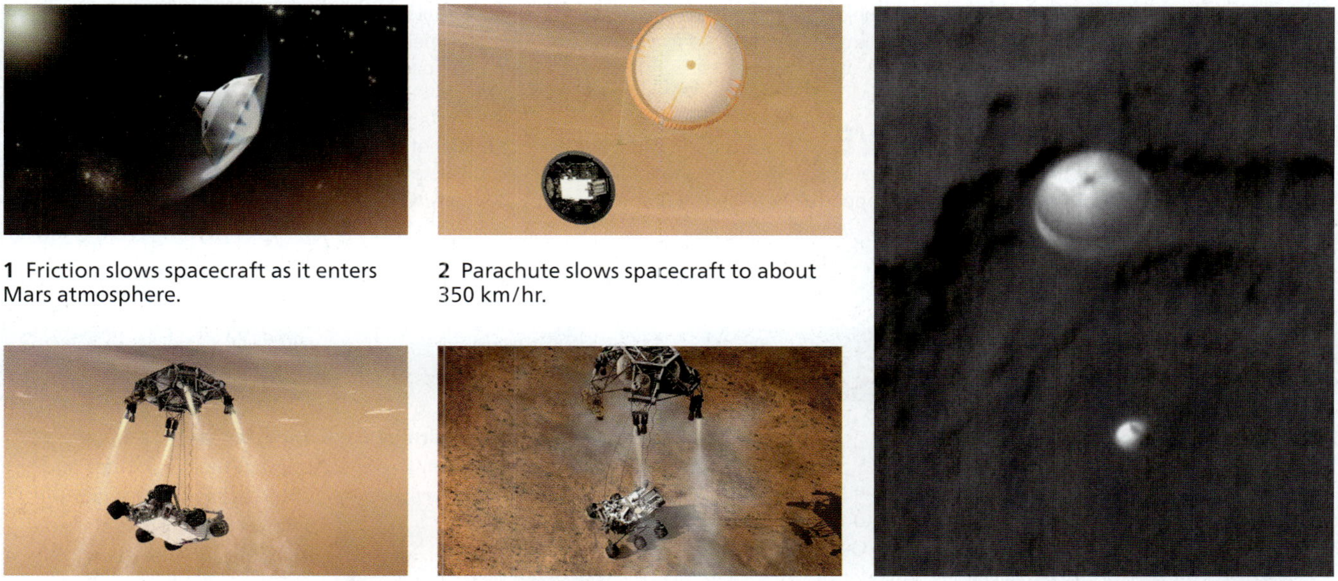

1 Friction slows spacecraft as it enters Mars atmosphere.

2 Parachute slows spacecraft to about 350 km/hr.

3 Rockets slow spacecraft to halt; "sky crane" tether lowers rover to surface.

4 Tether released, the rocket heads off to crash a safe distance away.

As it flew overhead, the *Mars Reconnaissance Orbiter* took this photo of the spacecraft with its parachute deployed.

**FIGURE 7.15** An artist's conception of the landing sequence that brought the *Curiosity* rover to Mars, along with a photo of its descent taken from orbit.

**TABLE 7.3 Current and Upcoming Robotic Missions to Other Worlds (as of 2018)**

| | |
|---|---|
| Mercury | *BepiColombo* is a joint ESA-JAXA* mission with two orbiters to study Mercury's surface, interior, and magnetosphere, scheduled to arrive at Mercury in 2025. |
| Venus | JAXA's *Akatsuki* completed its study of Venus's atmosphere in 2018; no other Venus missions are currently funded. |
| Moon | The United States, China, Japan, India, and Russia all have current or planned robotic missions to explore the Moon. Several privately funded lunar missions are also in development. |
| Mars | Operating missions include the *Opportunity* and *Curiosity* rovers, the *Mars Insight* lander, and orbiters *MAVEN*, *ExoMars*, *Mars Odyssey*, *Mars Reconnaissance Orbiter,* and India's *Mangalyaan.* Missions scheduled to launch in 2020 include NASA's *Mars 2020* rover, ESA's *ExoMars* rover, a Chinese orbiter and lander, a United Arab Emirates orbiter, and a JAXA sample return from Phobos. |
| Asteroids | *Dawn* visited asteroid Vesta and continues to study the dwarf planet Ceres. *Hayabusa-2* is scheduled to return a sample from asteroid 162173 Ryugu in 2020. *OSIRIS-REx* is scheduled to return a sample from asteroid 101955 Bennu in 2023. The *Lucy* mission to Trojan asteroids that share Jupiter's orbit is scheduled to launch in 2021 and arrive in 2027. The *Psyche* mission, named for the metal asteroid it will visit, is scheduled to launch in 2022 and arrive in 2026. |
| Jovian planets | *Voyagers 1* and *2* both continue to operate long after flybys of the jovian planets; they are now well beyond 100 AU from the Sun. The *Juno* orbiter is studying Jupiter's magnetic and gravitational fields to learn about its deep interior and its aurora and clouds. ESA's *Jupiter Icy Moons Explorer* is scheduled to launch in 2022. No new missions are currently scheduled for Saturn, Uranus, or Neptune. |
| Pluto and comets | *New Horizons* flew past Pluto in 2015 and past Kuiper belt comet MU69 on New Year's Day, 2019. |

*ESA = European Space Agency; JAXA = Japan Aerospace Exploration Agency

## The BIG Picture PUTTING CHAPTER 7 INTO PERSPECTIVE

*This chapter introduced the major features of our solar system and discussed some important patterns and trends that provide clues to its formation. As you continue your study of the solar system, keep in mind the following "big picture" ideas:*

- Our solar system is not a random collection of objects moving in random directions. Rather, it is highly organized, with clear patterns of motion and with most objects falling into just a few basic categories.
- Each planet has its own unique and interesting features. Becoming familiar with the planets is an important first step in understanding the root causes of their similarities and differences.
- Much of what we now know about the solar system comes from spacecraft exploration. Choosing the type of mission to send to a planet involves many considerations, from the scientific to the purely political. Many missions are currently under way, offering us hope of learning much more in the near future.

**MY COSMIC PERSPECTIVE** Until just a few decades ago, the planets were never much more than dim lights in the night sky, but we've now come to view them as *worlds*. We should feel privileged to live at such a time, because no generation before or after us will ever have had the same opportunity to be the first to map out the many worlds of our solar system.

# Summary of Key Concepts

## 7.1 Studying the Solar System

- **What does the solar system look like?** The planets are tiny compared to the distances between them. Our solar system consists of the Sun, the planets and their moons, and vast numbers of asteroids and comets. Each world has its own unique character, but there are many clear patterns among the worlds.

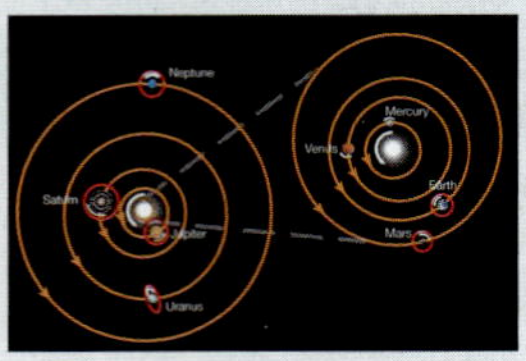


- **What can we learn by comparing the planets to one another?** Comparative studies reveal the similarities and differences that give clues to solar system formation and highlight the underlying processes that give each planet its unique appearance.

## 7.2 Patterns in the Solar System

- **What features of our solar system provide clues to how it formed?** Four major features provide clues: (1) The Sun, planets, and large moons generally rotate and orbit in a very organized way. (2) The planets divide clearly into two groups: **terrestrial** and **jovian**. (3) The solar system contains huge numbers of **asteroids** and **comets**. (4) There are some notable exceptions to these general patterns.

## 7.3 Spacecraft Exploration of the Solar System

- **How do robotic spacecraft work?** Spacecraft can be categorized as flyby, orbiter, lander or probe, or sample return mission. In all cases, robotic spacecraft carry their own propulsion, power, and communication systems, and can operate under preprogrammed control or with updated instructions from ground controllers.

# Visual Skills Check

*Use the following questions to check your understanding of some of the many types of visual information used in astronomy. For additional practice, try the Chapter 7 Visual Quiz in the Study Area at www.MasteringAstronomy.com.*

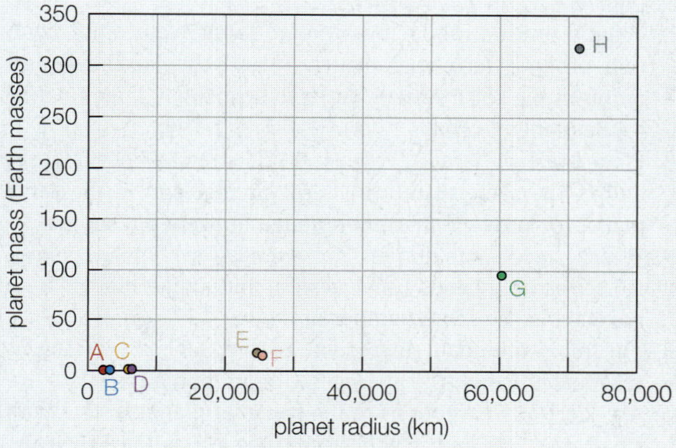


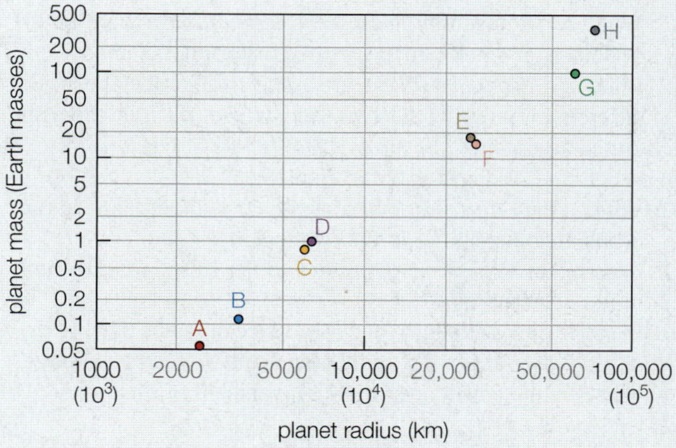


*The plots above show the masses of the eight major planets on the vertical axis and their radii on the horizontal axis. The plot on the left shows the information on a linear scale, meaning that each tick mark indicates an increase by the same amount. The plot on the right shows the same information plotted on an exponential scale, meaning that each tick mark represents another factor-of-ten increase. Before proceeding, convince yourself that the points on each plot are the same.*

1. Based on the information given in Table 7.1, which dots on each plot correspond to which planets? Which correspond to the terrestrial planets, and which to the jovian planets?
2. Notice how the eight planets group roughly into pairs on the graphs. Which planets are in each pair?
3. Which statement most accurately describes the relationship between the largest and smallest planets?
   a. The largest planet is 6000 times as wide (in diameter) and 30 times as massive as the smallest.
   b. The largest planet is 6000 times as wide (in diameter) and 6000 times as massive as the smallest.
   c. The largest planet is 30 times as wide (in diameter) and 30 times as massive as the smallest.
   d. The largest planet is 30 times as wide (in diameter) and 6000 times as massive as the smallest.
4. Answer each of the following questions to compare the two plots.
   a. Which plot, if either, better shows mass and radius information for all the planets?
   b. Which plot, if either, better emphasizes the differences between Jupiter and Saturn?
   c. Which plot, if either, could more easily be extended to show a planet with twice Jupiter's mass or radius?

# Exercises and Problems

*For instructor-assigned homework and other learning materials, go to www.MasteringAstronomy.com.*

## Chapter Review Questions

### Short-Answer Questions Based on the Reading

1. What do we mean by *comparative planetology*? Does it apply only to planets?
2. What would the solar system look like to your naked eye if you could view it from beyond the orbit of Neptune?
3. Briefly describe the overall layout of the solar system as it is shown in Figure 7.1. What are the four major features of our solar system that provide clues to how it formed?
4. For each of the objects in the solar system tour (pages 194–203), describe at least two features that you find interesting.
5. Briefly describe the patterns of motion that we observe among the planets and moons of our solar system.
6. What are the basic differences between the *terrestrial* and *jovian* planets? Which planets fall into each group?
7. What do we mean by *hydrogen compounds*? In what kinds of planets or small bodies are they major ingredients?
8. What are *asteroids*? What are *comets*? Describe the basic differences between the two, and where we find them in our solar system.
9. What kind of object is Pluto? Explain.
10. What is the *Kuiper belt*? What is the *Oort cloud*? How do the orbits of comets differ in the two regions?
11. Describe at least two "exceptions to the rules" that we find in our solar system.
12. Describe and distinguish between space missions that are *flybys, orbiters, landers* or *probes*, and *sample return missions*. What are the advantages and disadvantages of each type?

### Does It Make Sense?

*Decide whether or not each of the following statements makes sense (or is clearly true or false). Explain clearly; not all of these have definitive answers, so your explanation is more important than your chosen answer.*

13. Uranus orbits the Sun in a direction opposite that of all the other planets.
14. If Pluto were as large as the planet Mercury, we would classify it as a terrestrial planet.
15. Comets in the Kuiper belt and Oort cloud have long, beautiful tails that we can see when we look through telescopes.
16. Our Moon is about the same size as moons of the other terrestrial planets.
17. The mass of the Sun compared to the mass of all the planets combined is like the mass of an elephant compared to the mass of a cat.
18. On average, Venus is the hottest planet in the solar system—even hotter than Mercury.
19. The weather conditions on Mars today are much different than they were in the distant past.
20. Moons cannot have atmospheres, active volcanoes, or liquid water.
21. Saturn is the only planet in the solar system with rings.
22. We could probably learn more about Mars by sending a new spacecraft on a flyby than by any other method of studying the planet.

### Quick Quiz

*Choose the best answer to each of the following. For additional practice, try the Chapter 7 Reading and Concept Quizzes in the Study Area at www.MasteringAstronomy.com.*

23. The largest terrestrial planet and jovian planet are, respectively, (a) Venus and Jupiter. (b) Earth and Jupiter. (c) Earth and Saturn.
24. Which terrestrial planets have had volcanic activity at some point in their histories? (a) only Earth (b) Earth and Mars (c) all of them
25. How many of our solar system's large moons orbit their planets in the same direction as their planet rotates? (a) very few (b) about half (c) most
26. Which of the following three kinds of objects resides closest to the Sun on average? (a) comets (b) asteroids (c) jovian planets
27. What's unusual about our Moon? (a) It's the only moon that orbits a terrestrial planet. (b) It's by far the largest moon in the solar system. (c) It's surprisingly large relative to the planet it orbits.
28. Planetary orbits are (a) very eccentric (stretched-out) ellipses and in the same plane. (b) fairly circular and in the same plane. (c) fairly circular but oriented in every direction.
29. Which have more moons on average? (a) jovian planets (b) terrestrial planets (c) Terrestrial and jovian planets both have about the same number of moons.
30. The most abundant ingredient of the Sun and Jupiter is (a) ionized metal. (b) hydrogen. (c) ammonia.
31. Are there any exceptions to the rule that planets rotate with small axis tilts and in the same direction as they orbit the Sun? (a) No (b) Venus is the only exception. (c) Venus and Uranus are exceptions.
32. The *Cassini* spacecraft (a) flew past Pluto. (b) landed on Mars. (c) orbited Saturn.

## Inclusive Astronomy

*Use these questions to reflect on participation in science.*

33. *Group Discussion: Who Named the Planets?* The names of the five planets that can be seen with the naked eye have been in use in the western world for thousands of years. However, beginning with the discoveries of Uranus in 1781 and Neptune in 1846, scientists needed to find names for planets and other newly discovered objects.
**a.** Working in small groups, find out how Uranus and Neptune came to have their current names in English. Do you think these name choices made sense? Do you think the process that led to these names was "fair"? Why or why not?
**b.** Each member of your group should learn about the names of Uranus and Neptune in two other languages, at least one of which is non-European. Discuss the similarities and differences you find among the names in different languages.
**c.** Find out what names were considered for Pluto after its discovery in 1930, who suggested the name "Pluto," and how it came to be official. Are you surprised about who suggested the name?

**d.** Find out how newly discovered objects (such as moons and asteroids) and surface features (such as those recently identified on Mars, Ceres, and Pluto) get their official names today. Do you think this process is appropriate? Make a list of other possibilities that might be considered (for example, allowing the discoverers to choose names or choosing through public, online competitions). If your group were in charge, would you change the current naming process in any way?

## The Process of Science

*These questions may be answered individually in short-essay form or discussed in groups, except where identified as group-only.*

34. *Why Wait?* To explore a planet, we often send first a flyby, then an orbiter, then a probe or a lander. There's no doubt that probes and landers give the most close-up detail, so why don't we send this type of mission first? For the planet of your choice, based just on the information in this chapter, give an example of why such a strategy might cause a mission to provide incomplete information about the planet or to fail outright.
35. *Planetary Priorities.* Suppose you were in charge of developing and prioritizing future planetary missions for NASA. What would you choose as your first priority for a new mission, and why?
36. *Group Activity: Comparative Approach.* Work as a group to answer each part. Note: You may wish to do this activity using the four roles described in Chapter 1, Exercise 39.
**a.** This chapter advocates learning about how planets work by comparing the planets in general, as opposed to studying the individual planets in great depth. Discuss how this approach compares to the way you have been taught about the planets in the past (for example, in grade school or high school).
**b.** Make a list of advantages and any disadvantages of the comparative approach for planetary science.
**c.** Make a list of a few other fields (such as other sciences or social sciences) in which you think a comparative approach is useful. Describe why in each case.

## Investigate Further

### Short-Answer/Essay Questions

37. *Planetary Tour.* Based on the brief planetary tour in this chapter, which planet besides Earth do you think is the most interesting, and why? Defend your opinion clearly in two or three paragraphs.
38. *Patterns of Motion.* In one or two paragraphs, explain why the existence of orderly patterns of motion in our solar system suggests that the Sun and the planets all formed at one time from one cloud of gas, rather than as individual objects at different times.
39. *Solar System Trends.* Answer the following based on the data in Table 7.1.
**a.** Describe the relationship between distance from the Sun and surface temperature. Explain the reason behind the general trend, as well as any notable exceptions to the trend.
**b.** Describe in general how the columns for density, composition, and distance from the Sun support the classification of planets into the terrestrial and jovian categories, with Pluto fitting neither category.
**c.** Describe the trend in orbital periods, and explain it in terms of Kepler's third law.
40. *Comparing Planetary Conditions.* Use both Table 7.1 and Appendix E to answer each of the following.
**a.** Which column of data would you use to find out which planet has the shortest days? Are there any notable differences in the length of a day for the different types of planets? Explain.
**b.** Which column of data would you use to find out which planets should have seasons? Explain.
**c.** Which column tells you how much a planet's orbit deviates from a perfect circle? For each planet, use that column to decide whether you would expect its average surface temperature to vary over the course of its orbit, and why.
41. *Escape Velocity.* Use Appendix E to explore how escape velocity is related to mass and radius. Is the trend what you expect based on what you learned about escape velocity in Chapter 4? Explain.
42. *Current Spacecraft Mission.* Visit the website for a current or upcoming robotic mission of your choice. Write a one- to two-page summary of the mission's basic design, goals, and current status.
43. *Mars Missions.* Visit the website for NASA's Mars Exploration Program. Write a one- to two-page summary of the plans for future exploration of Mars.

### Quantitative Problems

*Be sure to show all calculations clearly and state your final answers in complete sentences.*

44. *Size Comparisons.* How many Earths could fit inside Jupiter (assuming you could fill up all the volume)? How many Jupiters could fit inside the Sun? The equation for the volume of a sphere is $V = \frac{4}{3}\pi r^3$.
45. *Asteroid Orbit.* Ceres, the largest asteroid, has an orbital semimajor axis of 2.77 AU. Use Kepler's third law to find its orbital period. Compare your answer with the value in Table 7.1, and name the planets that orbit just inside and outside Ceres's orbit.
46. *Density Classification.* Calculate the density of a hypothetical planet in our solar system with a mass of $5.97 \times 10^{25}$ kilograms and a radius of 12,800 kilometers; give your answer in units of grams per cubic centimeter. Based just on its density, would we consider it the largest terrestrial planet or the smallest jovian planet? Explain.
47. *Comparative Weight.* Suppose you weigh 100 pounds. Calculate how much you would weigh on each of the other planets in our solar system. Assume you can stand either on the surface or in an airplane in the planet's atmosphere. (*Hint:* Recall from Chapter 4 that weight is mass times the acceleration of gravity; the "surface gravity" column in Appendix E tells you how the acceleration of gravity on other planets compares to that on Earth.)
48. *New Horizons Speed.* On its trajectory to Pluto, the *New Horizons* spacecraft traveled a total distance of about 34 AU over a period of about $9\frac{1}{2}$ years.
**a.** About how fast was it traveling on average? Give your answer in AU per year and kilometers per hour.
**b.** If it continued at this speed, about how long would *New Horizons* take to reach the nearest stars, if it were heading in the correct direction?
49. *Planetary Parallax.* Suppose observers at Earth's North Pole and South Pole use a transit of the Sun by Venus to discover that the angular size of Earth as viewed from Venus would be 62.8 arcseconds. Earth's radius is 6378 kilometers. Estimate the distance between Venus and Earth in kilometers and AU. Compare your answer with information from the chapter.

# 8 Formation of the Solar System

▲ **About the photo:** This is not an artist's conception! It is a real image of a disk in which planets are forming, taken by the Atacama Large Millimeter/submillimeter Array (ALMA); see Figure 8.4b for more details.

## LEARNING GOALS

**8.1 The Search for Origins**
- How did we arrive at a theory of solar system formation?
- Where did the solar system come from?

**8.2 Explaining the Major Features of the Solar System**
- What caused the orderly patterns of motion?
- Why are there two major types of planets?
- Where did asteroids and comets come from?
- How do we explain "exceptions to the rules"?

**8.3 The Age of the Solar System**
- How do we measure the age of a rock?
- How do we know the age of the solar system?

*The evolution of the world may be compared to a display of fireworks that has just ended: some few red wisps, ashes and smoke. Standing on a cooled cinder, we see the slow fading of the suns, and we try to recall the vanished brilliance of the origin of the worlds.*

—*G. Lemaître (1894–1966), astronomer and Catholic priest*

**Chapter 8 Overview**

How did Earth come to be? How old is it? Is it unique? Our ancestors could do little more than guess at the answers to these questions, but today we are able to address them scientifically. As we'll discuss in this chapter, careful study of the major features of our solar system has enabled scientists to put together a detailed theory of how Earth and our solar system were born.

Our theory of solar system formation is important not only because it helps us understand our cosmic origins, but also because it holds the key to understanding the nature of planets. If the planets in our solar system all formed together, then their differences must be attributable to physical processes that occurred during the birth and subsequent evolution of the solar system. Our study of the solar system's birth will therefore form the basis for our comparative study of the planets in subsequent chapters. It will also help us extend these ideas to the myriad of other planetary systems now known to exist, a topic we will study in Chapter 13.

## 8.1 The Search for Origins

The development of any scientific theory is an interplay between observations and attempts to explain those observations [**Section 3.4**]. By the end of the 17th century, the Copernican revolution [**Section 3.3**] and Newton's discovery of the universal law of gravitation [**Section 4.4**] had given us a basic understanding of the layout and motion of the planets and moons in our solar system. It was only natural that scientists would begin to speculate about how this system came to be.

### How did we arrive at a theory of solar system formation?

Recall that a hypothesis can rise to the status of a scientific theory only if it offers a detailed physical model that explains a broad range of observed facts. For our solar system, the most important facts to explain are the four major features discussed in Chapter 7. If a hypothesis fails to explain even one of the four features, then it cannot be correct. If it successfully explains all four, then we might reasonably assume it is on the right track. We therefore arrive at the following four criteria for the success of a solar system formation theory:

1. It must explain the patterns of motion discussed in Chapter 7.
2. It must explain why planets fall into two major categories: small, rocky *terrestrial* planets near the Sun and large, hydrogen-rich *jovian* planets farther out.
3. It must explain the existence of huge numbers of asteroids and comets and why these objects reside primarily in the regions we call the *asteroid belt*, the *Kuiper belt*, and the *Oort cloud*.
4. It must explain the general patterns while at the same time making allowances for exceptions to the general rules, such as the odd axis tilt of Uranus and the existence of Earth's large Moon.

**From Hypothesis to Theory** We generally trace the origins of our modern theory of solar system formation to around 1755, when German philosopher Immanuel Kant proposed that our solar system formed from the gravitational collapse of an interstellar cloud of gas. About 40 years later, French mathematician Pierre-Simon Laplace put forth the same idea independently. Because an interstellar cloud is usually called a *nebula* (Latin for "cloud"), this idea became known as the *nebular hypothesis*.

The nebular hypothesis remained popular throughout the 19th century. By the early 20th century, however, scientists had found a few aspects of our solar system that the hypothesis did not seem to explain well—at least in its original form as described by Kant and Laplace. While some scientists sought to modify the nebular hypothesis, others looked for different ways to explain how the solar system might have formed.

During much of the first half of the 20th century, the nebular hypothesis faced stiff competition from a hypothesis proposing that the planets represent debris from a near-collision between the Sun and another star. According to this *close encounter hypothesis,* the planets formed from blobs of gas that had been gravitationally pulled out of the Sun during the near-collision.

Today, the close encounter hypothesis has been discarded. It began to lose favor when calculations showed that it could not account for either the observed orbital motions of the planets or the neat division of the planets into two major categories (terrestrial and jovian). Moreover, the close encounter hypothesis required a highly improbable event: a near-collision between our Sun and another star. Given the vast separation between star systems in our region of the galaxy, the chance of such an encounter is so small that it would be difficult to imagine it happening even once in order to form our solar system. It certainly could not account for the many other planetary systems that we have discovered in recent years.

While the close encounter hypothesis was losing favor, new discoveries about the physics of planet formation led to modifications of the nebular hypothesis. Using more sophisticated models of the processes that occur in a collapsing cloud of gas, scientists found that the nebular hypothesis offered natural explanations for all four general features of our solar system. Indeed, so much evidence accumulated in favor of the nebular hypothesis that it achieved the status of a scientific *theory* [**Section 3.4**]—the **nebular theory** of our solar system's birth.

**Putting the Theory to the Test** Recall that in science, we can never be sure that a theory is fully complete, because it may always be challenged by new observational or experimental tests. In the case of a theory that claims to explain the origin of *our* solar system, one critical set of tests involves its ability to predict and explain the characteristics of *other* solar systems. As we'll discuss in Chapter 13, studies of other planetary systems have turned up a few surprises that have forced us to consider modifications to the nebular theory. For example, the fact that some systems have jovian-size planets very close to their stars has led scientists to conclude that planets can sometimes migrate inward or outward from the orbits on which they first form. Nevertheless, the mere existence of these other systems means the nebular theory has passed its most important test: Because it claims that planets are a natural outgrowth of the star formation process, it predicts that other planetary systems ought to be common, a prediction that has now been borne out by observations. As a result, the nebular theory today stands on stronger ground than ever. We'll therefore devote the rest of this chapter to understanding the basic theory and how it explains the major features of our solar system.

## Where did the solar system come from?

The nebular theory begins with the idea that our solar system was born from the gravitational collapse of an interstellar cloud of gas, called the **solar nebula**, that collapsed under its own gravity. As we'll discuss in more detail in the next section, this cloud gave birth to the Sun at its center and the planets in a spinning disk that formed around the young Sun.

Where did the gas that made up the solar nebula come from? According to modern science, it was the product of billions of years of galactic recycling that occurred before the Sun and planets were born. Recall that the universe as a whole is thought to have been born in the Big Bang [**Section 1.2**], which essentially produced only two chemical elements: hydrogen and helium. Heavier elements were produced later, some through the nuclear fusion that makes stars shine, and most others through nuclear reactions accompanying the explosions that end stellar lives. The heavy elements then mixed with other interstellar gas that formed new generations of stars (**FIGURE 8.1**).

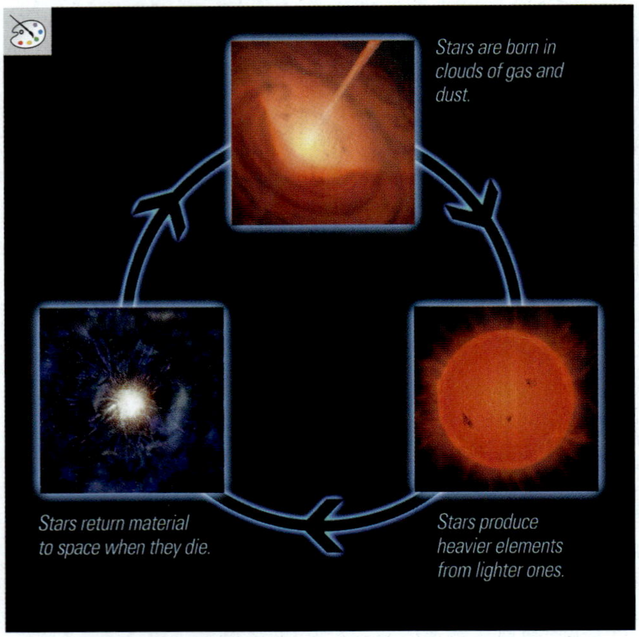


**FIGURE 8.1** This figure, which is a portion of Figure 1.11, summarizes the galactic recycling process.

Although this process of creating heavy elements in stars and recycling them within the galaxy has probably gone on for most of the 14-billion-year history of our universe, only a small fraction of the original hydrogen and helium has been converted into heavy elements. By studying the composition of the Sun, other stars of the same age, and interstellar gas clouds, we have learned that the gas that made up the solar nebula contained (by mass) about 98% hydrogen and helium and 2% all other elements combined. The Sun and planets were born from this gas, and Earth and the other terrestrial worlds were made primarily from the heavier elements mixed within it. As we discussed in Chapter 1, we are "star stuff" because we and our planet are made of elements created by stars that lived and died long ago.

**Think about it** Could a solar system like ours have formed with the first generation of stars after the Big Bang? Explain.

Strong observational evidence supports this scenario. Spectroscopy shows that old stars have a smaller proportion of heavy elements than younger ones, just as we would expect if they were born at a time before many heavy elements had been manufactured. Moreover, visible and infrared telescopes allow us to study stars that are in the process of formation today. **FIGURE 8.2** shows the Orion

**FIGURE 8.2** The Orion Nebula, an interstellar cloud in which new star systems are forming. The insets show close-ups of six young stars that are surrounded by disks of material in which planets could presumably form. The nebula will ultimately give birth to thousands of stars and planetary systems over the next few million years. These images are from the Hubble Space Telescope.

Nebula, in which many stars are in various stages of formation. Just as our scenario predicts, the forming stars are embedded within gas clouds like our solar nebula, and the characteristics of these clouds match what we expect if they are collapsing due to gravity.

## 8.2 Explaining the Major Features of the Solar System

We are now ready to look at the nebular theory in somewhat more detail. In the process, we'll see how it successfully accounts for all four major features of our solar system.

### What caused the orderly patterns of motion?

The solar nebula probably began as a large and roughly spherical cloud of very cold, low-density gas. Initially, this gas was probably so spread out—perhaps over a region a few light-years in diameter—that gravity alone may not have been strong enough to pull it together and start its collapse. Instead, the collapse may have been triggered by a cataclysmic event, such as the impact of a shock wave from the explosion of a nearby star (a supernova).

Once the collapse started, gravity enabled it to continue. Recall that the strength of gravity follows an inverse square law with distance [**Section 4.4**]. The mass of the cloud remained the same as it shrank, so the strength of gravity increased as the diameter of the cloud decreased.

Because gravity pulls inward in all directions, you might at first guess that the solar nebula would have remained spherical as it shrank. Indeed, the idea that gravity pulls in all directions explains why the Sun and the planets are spherical. However, other physical laws also apply, and these explain how orderly motions arose in the solar nebula.

**Heating, Spinning, and Flattening** The details of how a star-forming cloud collapses under gravity can be surprisingly complex [**Section 16.1**], but for our purposes here we need only focus on three key processes that altered the solar nebula's density, temperature, and shape as it shrank (**FIGURE 8.3**):

- **Heating.** The temperature of the solar nebula increased as it collapsed. Such heating represents energy conservation in action [**Section 4.3**]. As the cloud shrank, its gravitational potential energy was converted to the kinetic energy of individual gas particles falling inward. These particles crashed into one another, converting the kinetic energy of their inward fall to the random motions of thermal energy (see Figure 4.15b). The Sun formed in the center, where temperatures and densities were highest.

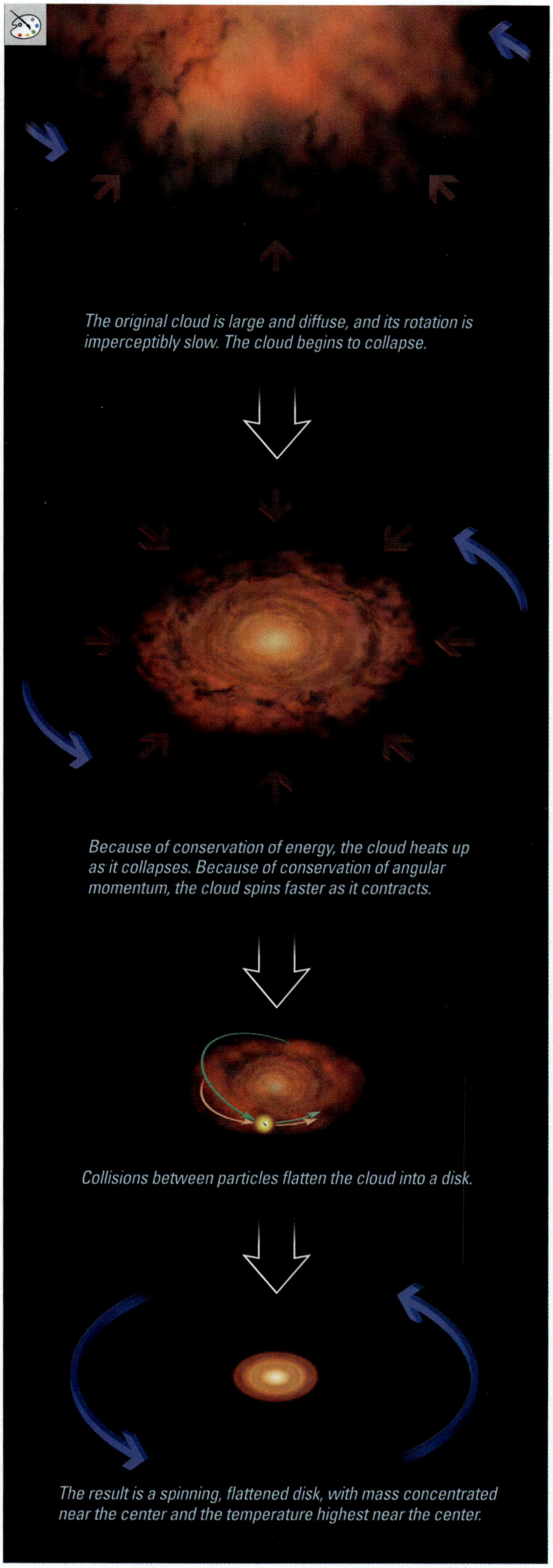


**FIGURE 8.3** This sequence of illustrations shows how the gravitational collapse of a large cloud of gas causes it to become a spinning disk of matter. The hot, dense central bulge becomes a star, while planets can form in the surrounding disk.

- **Spinning.** Like an ice skater pulling in his arms as he spins, the solar nebula rotated faster and faster as it shrank in radius. This increase in rotation rate represents conservation of angular momentum in action [**Section 4.3**]. The rotation of the cloud may have been imperceptibly slow before its collapse began, but the cloud's shrinkage made fast rotation inevitable. The rapid rotation helped ensure that not all the material in the solar nebula collapsed into the center: The greater the angular momentum of a rotating cloud, the more spread out it will be.
- **Flattening.** The solar nebula flattened into a disk. This flattening is a natural consequence of collisions between particles in a spinning cloud. A cloud may start with any size or shape, and different clumps of gas within the cloud may be moving in random directions at random speeds. These clumps collide and merge as the cloud collapses, and each new clump has the average velocity of the clumps that formed it. The random motions of the original cloud therefore become more orderly as the cloud collapses, changing the cloud's original lumpy shape into a rotating, flattened disk. Similarly, collisions between clumps of material in highly elliptical orbits reduce their eccentricities, making the orbits more circular.

The formation of the spinning disk explains the orderly motions of our solar system today. The planets all orbit the Sun in nearly the same plane because they formed in the flat disk. The direction in which the disk was spinning became the direction of the Sun's rotation and the orbits of the planets. Computer models show that planets would have tended to rotate in this same direction as they formed—which is why most planets rotate the same way—though the small sizes of planets compared to the entire disk allowed some exceptions to arise. The fact that collisions in the disk tended to make orbits more circular explains why the planets in our solar system have nearly circular orbits.

**See it for yourself** You can demonstrate the development of orderly motion by sprinkling pepper into a bowl of water and stirring it quickly in random directions. The water molecules continually collide with one another, so the motion of the pepper grains will tend to settle down into a slow rotation representing the average of the original random velocities. Try the experiment several times, stirring the water differently each time. Do the random motions ever cancel out exactly, resulting in no rotation at all? Describe what occurs, and explain how it is similar to what took place in the solar nebula.

**Testing the Model** The same processes should affect other collapsing gas clouds, so we can test our model by searching for disks around other forming stars. Observational evidence does indeed support our model of spinning, heating, and flattening.

The heating that occurs in a collapsing cloud of gas means the gas should emit thermal radiation [**Section 5.4**], primarily in the infrared. We've detected infrared radiation from many nebulae where star systems appear to be forming. More direct evidence comes from flattened, spinning disks around other stars (**FIGURE 8.4**), some of which appear to be ejecting jets of material perpendicular to their disks [**Section 16.2**]. These jets are thought to result from the flow of material from the disk onto the forming star, and they may influence the solar system formation processes.

Other support for the model comes from computer simulations of the formation process. A simulation begins with a set of conditions observed in interstellar clouds. Then, with the aid of a computer, we apply the laws of physics to predict the changes that should occur over time. Computer simulations successfully reproduce most of the general characteristics of motion in our solar system, giving support to the nebular theory.

Additional evidence that our ideas about the formation of flattened disks are correct comes from many other structures in the universe. We expect flattening to occur anywhere orbiting particles can collide, which explains why we find so many cases of flat disks, including the disks of spiral galaxies like the Milky Way, the disks of planetary rings, and the *accretion disks* that surround neutron stars and black holes in close binary star systems [**Section 18.3**].

## Why are there two major types of planets?

The planets began to form after the solar nebula had collapsed into a flattened disk of perhaps 200 AU diameter (about twice the present-day diameter of Pluto's orbit). The churning and mixing of gas in the solar nebula should have ensured that the nebula had the same composition throughout. How, then, did the terrestrial planets end up so different in composition from the jovian planets? The key clue comes from their locations: Terrestrial planets formed in the warm inner regions of the swirling disk, while jovian planets formed in the colder outer regions.

**Condensation: Sowing the Seeds of Planets** In the center of the collapsing solar nebula, gravity drew together enough material to form the Sun. In the surrounding disk, however, the gaseous material was too spread out for gravity alone to clump it together. Instead, material had to begin clumping in some other way and then grow in size until gravity could start pulling it together into planets. In essence, planet formation required the presence of "seeds"—solid bits of matter from which gravity could ultimately build planets.

The basic process of seed formation was probably much like the formation of snowflakes in clouds on Earth: When the temperature is low enough, some atoms or molecules in a gas may bond and solidify. The general process in which solid (or liquid) particles form in a gas is called **condensation**—we say that the particles *condense* out of the gas. (Pressures in the solar nebula were generally too low to allow the condensation of liquid droplets.) These particles start out microscopic in size, but they can grow larger with time.

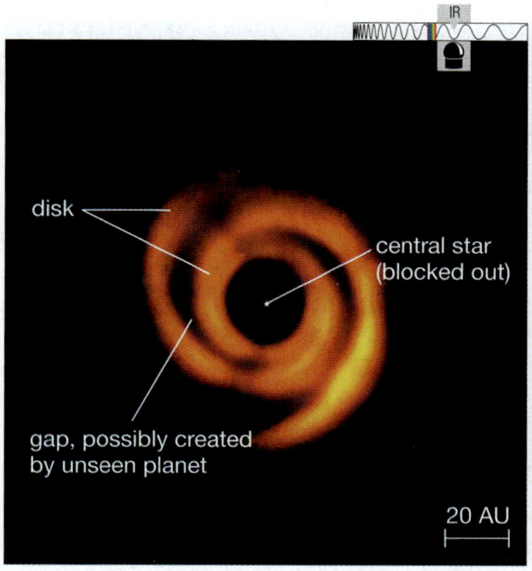


**a** This image from the Very Large Telescope (VLT) in Chile shows a flattened, spinning disk around the star HD 135344B. The spiral pattern may indicate the presence of one or more jovian planets forming in the disk. The central star was blocked out during the observations.

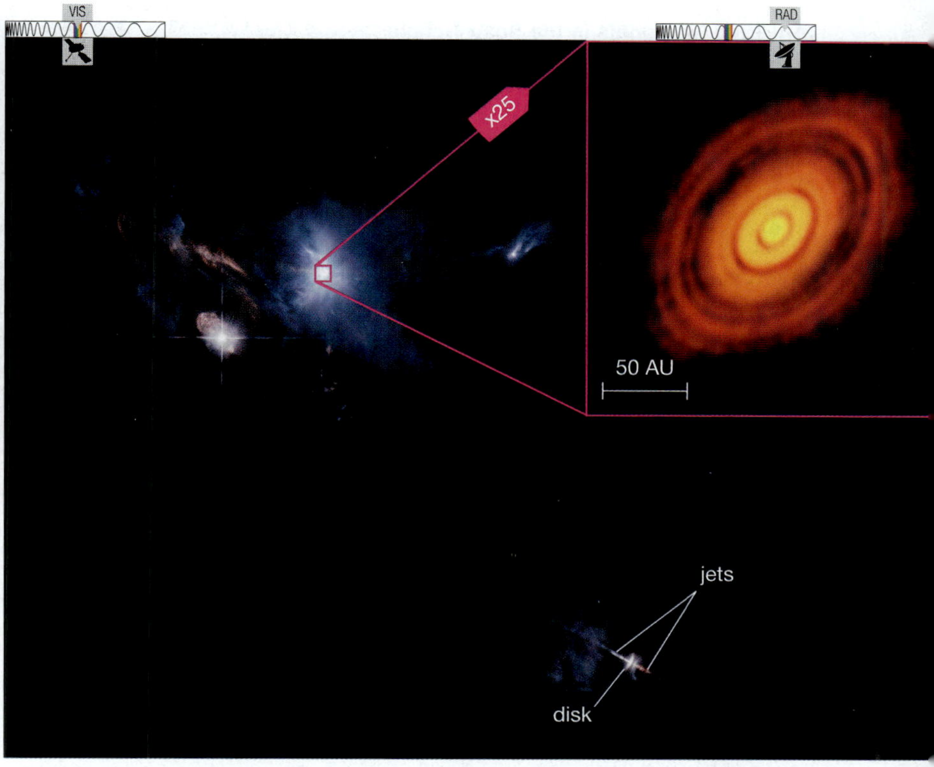


**b** The inset, from the Atacama Large Millimeter/submillimeter Array (ALMA), shows a disk around a star named HL Tauri; the concentric gaps in the disk are almost certainly regions being cleared as planets form. The disk diameter is about three times that of Neptune's orbit around the Sun. The background image, from the Hubble Space Telescope, shows the star-forming region in which this disk is located. Another disk, seen edge-on with jets extending outward, appears at lower right.

**FIGURE 8.4** These images show flattened, spinning disks of material around other stars.

Different materials condense at different temperatures. As summarized in **TABLE 8.1**, the ingredients of the solar nebula fell into four major categories:

- **Hydrogen and helium gas (98% of the solar nebula).** These gases never condense in interstellar space.
- **Hydrogen compounds (1.4% of the solar nebula).** Materials such as water ($H_2O$), methane ($CH_4$), and ammonia ($NH_3$) can solidify into **ices** at low temperatures (below about 150 K under the low pressure of the solar nebula).
- **Rock (0.4% of the solar nebula).** Rocky material is gaseous at high temperatures, but condenses into solid form at temperatures below 500 K to 1300 K, depending on the type of rock.
- **Metal (0.2% of the solar nebula).** Metals such as iron, nickel, and aluminum are also gaseous at high temperatures, but condense into solid form at temperatures below 1000 K to 1600 K, depending on the metal.

Because hydrogen and helium gas made up about 98% of the solar nebula's mass and did not condense, the vast majority of the nebula remained gaseous at all times. However, other materials could condense wherever the temperature allowed (**FIGURE 8.5**). Close to the Sun, it was too hot for any material to condense. A little farther out (near Mercury's current orbit), it was cool enough for metals and some types of rock to condense into tiny solid particles, but other types of rock and all the hydrogen compounds remained gaseous. More types of rock could condense at the distances from the Sun where Venus, Earth, and Mars would form. In the region where the asteroid belt would eventually be located, temperatures were low enough to allow dark carbon-rich minerals to condense, along with minerals containing small amounts of water. (A *mineral* is a component of rock with a particular chemical composition and structure.) It was cold enough for hydrogen compounds to condense into ices only beyond the **frost line**, which lay between the present-day orbits of Mars and Jupiter.

**Think about it** Consider a region of the solar nebula in which the temperature was about 1300 K. What fraction of the material in this region was gaseous? What were the solid particles made of? Answer the same questions for a region with a temperature of 100 K. Would the 100 K region be closer to or farther from the Sun? Explain.

The frost line marked the key transition between the warm inner regions of the solar system where terrestrial planets formed and the cool outer regions where jovian planets formed. Inside the frost line, only metal and rock could condense into solid "seeds." Beyond the frost line, the solid seeds were built of ice along with metal and rock. Moreover, because hydrogen compounds were nearly three

**TABLE 8.1 Materials in the Solar Nebula**

*A summary of the four types of materials present in the solar nebula. The squares represent the relative proportions of each type (by mass).*

| | Examples | Can Condense at Temperatures Below | Relative Abundance (by mass) |
|---|---|---|---|
| **Hydrogen and Helium Gas**  | hydrogen, helium | do not condense in nebula | ■ 98% |
| **Hydrogen Compounds**  | water ($H_2O$), methane ($CH_4$), ammonia ($NH_3$) | 150 K | ■ 1.4% |
| **Rock**  | various minerals | 500–1300 K | ▪ 0.4% |
| **Metal**  | iron, nickel, aluminum | 1000–1600 K | ▪ 0.2% |

times as abundant in the nebula as metal and rock combined (see Table 8.1), the total amount of solid material was far greater beyond the frost line than within it. The stage was set for the birth of two types of planets: planets born from seeds of metal and rock in the inner solar system and planets born from seeds of ice (as well as metal and rock) in the outer solar system.

**Building the Terrestrial Planets** From this point, the story of the inner solar system seems fairly clear. The solid seeds of metal and rock in the inner solar system ultimately grew into the terrestrial planets we see today, but these planets ended up relatively small in size because rock and metal made up such a small amount of the material in the solar nebula.

The process by which small "seeds" grew into planets is called **accretion** (**FIGURE 8.6**). Accretion began with the microscopic solid particles that condensed from the gas of the solar nebula. These particles orbited the forming Sun with the same orderly, circular paths as the gas from which they condensed. Individual particles therefore moved at nearly the same speed as neighboring particles, so "collisions" were more like gentle touches. Although the particles were far too small to attract each other gravitationally at this point, they were able to stick together through electrostatic forces—the same "static electricity" that makes hair stick to a comb. Small particles thereby began to combine into larger ones. As the particles grew in mass,

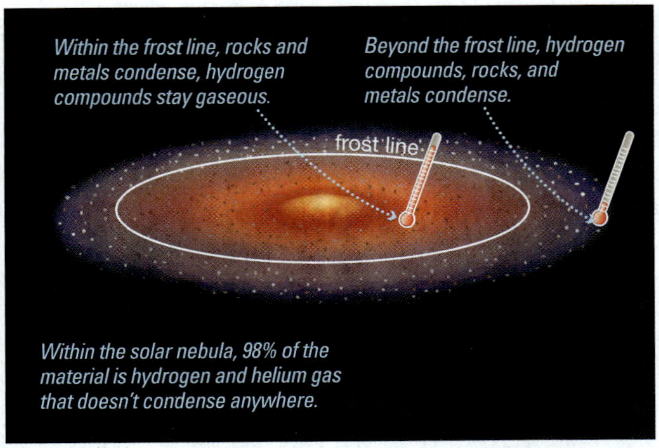


**FIGURE 8.5** Temperature differences in the solar nebula led to different kinds of condensed materials at different distances from the Sun, sowing the seeds for two kinds of planets.

they began to attract each other through gravity, accelerating their growth into boulders large enough to count as **planetesimals**, which means "pieces of planets." Some models suggest that this stage of accretion may have been enhanced by pebble-sized particles gathering together due to friction with the solar nebula gases.

The planetesimals grew rapidly at first. As they grew larger, they had both more surface area to make contact with other planetesimals and more gravity to attract them. Some planetesimals probably grew to hundreds of kilometers in size in only a few million years—a long time in human terms, but only about *one-thousandth* of the present age of the solar system. However, once the planetesimals reached these relatively large sizes, further growth became more difficult.

Gravitational encounters [**Section 4.5**] between planetesimals tended to alter their orbits, particularly those of the smaller ones. With different orbits crossing each other, collisions between planetesimals tended to occur at higher speeds and hence became more destructive. Such collisions tended to shatter planetesimals rather than help them grow. Only the largest planetesimals avoided being shattered and could grow into terrestrial planets.

Computer simulations support this model of the accretion process. Observational evidence comes from **meteorites**,

## COMMON MISCONCEPTIONS

### Solar Gravity and the Density of Planets

Perhaps because we are familiar with how Earth's gravity causes dense rocks to fall downward through less dense water, some students guess that the separation of material in the solar nebula might have been caused by the Sun's gravity pulling denser rocky and metallic materials inward while lighter gases moved outward because gravity couldn't hold them. But this was not the case in the solar nebula, which began with all the ingredients orbiting the Sun together under the influence of the Sun's gravity. The orbit of a particle or a planet does *not* depend on its size or density, so the Sun's gravity was not the cause of the different kinds of planets. Rather, the different temperatures in the solar nebula were the cause.

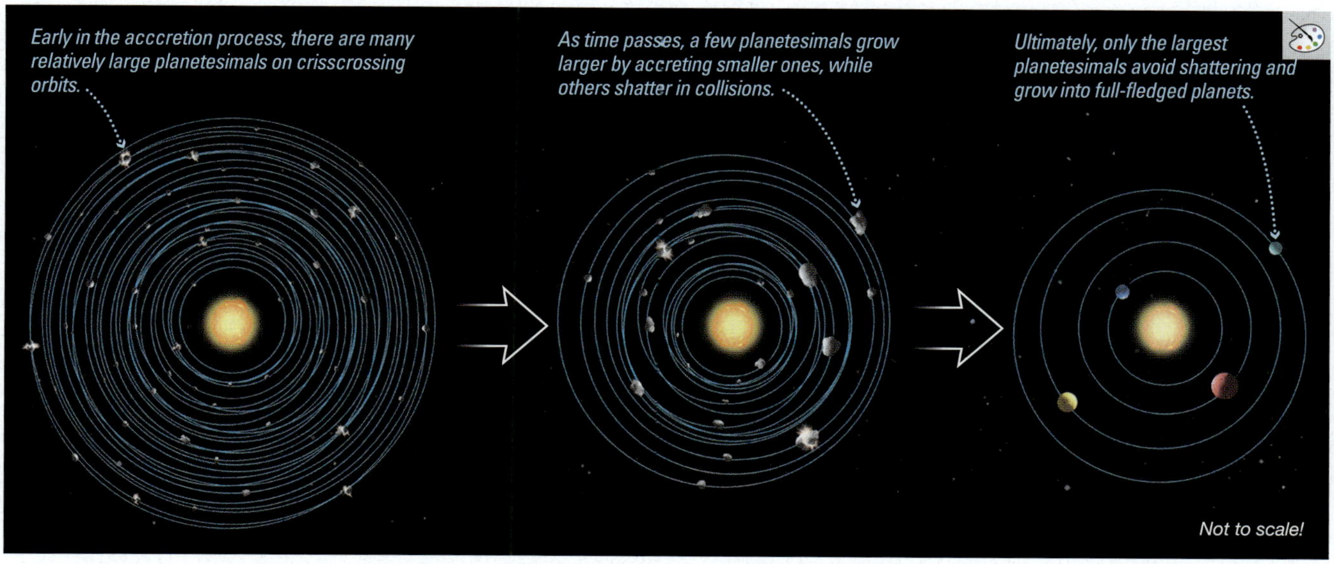


**FIGURE 8.6** These diagrams show how planetesimals gradually accrete into terrestrial planets.

rocks that have fallen to Earth from space. Meteorites that appear to be surviving fragments from the period of condensation contain metallic grains embedded in rocky minerals (**FIGURE 8.7**), just as we would expect if metal and rock condensed in the inner solar system. Meteorites thought to come from the outskirts of the asteroid belt contain abundant carbon-rich materials, and some contain water—again as we would expect for material that condensed in that region.

**Making the Jovian Planets** Accretion should have occurred similarly in the outer solar system, but condensation of ices meant both that there was more solid material and that this material contained ice in addition to metal and rock. The solid objects that reside in the outer solar system today, such as comets and the moons of the jovian planets, still show this ice-rich composition. However, the growth of icy planetesimals cannot be the whole story of jovian planet formation, because the jovian planets contain large amounts of hydrogen and helium gas.

**FIGURE 8.7** Shiny flakes of metal are clearly visible in this slice through a meteorite (a few centimeters across), mixed in among the rocky material. Such metallic flakes are just what we would expect to find if condensation really occurred in the solar nebula as described by the nebular theory.

The leading model for jovian planet formation holds that the largest ice-rich planetesimals became sufficiently massive for their gravity to capture some of the hydrogen and helium gas that made up the vast majority of the surrounding solar nebula. This added gas made their gravity even stronger, allowing them to capture even more gas. Ultimately, the jovian planets accreted so much gas that they bore little resemblance to the icy seeds from which they grew.

This model also explains most of the large moons of the jovian planets. The same processes of heating, spinning, and flattening that made the disk of the solar nebula should also have affected the gas drawn by gravity to the young jovian planets. Each jovian planet came to be surrounded by its own disk of gas, spinning in the same direction as the planet rotated (**FIGURE 8.8**). Moons that accreted from ice-rich planetesimals within these disks ended up with nearly circular orbits going in the same direction as their planet's rotation and lying close to their planet's equatorial plane.

**Clearing the Nebula** The vast majority of the hydrogen and helium gas in the solar nebula never became part of any planet. So what happened to it? Apparently, it was cleared away by a combination of high-energy radiation (ultraviolet and x-rays) from the young Sun and the **solar wind**—a stream of charged particles (such as protons and electrons) continually blown outward in all directions from the Sun. Observations show that stars tend to have much stronger winds and emit much more high-energy radiation when they are young, so the young Sun should have had a strong enough combination of radiation and wind to clear the solar system of its remaining gas.

The clearing of the gas sealed the compositional fate of the planets. If the gas had remained longer, it might have

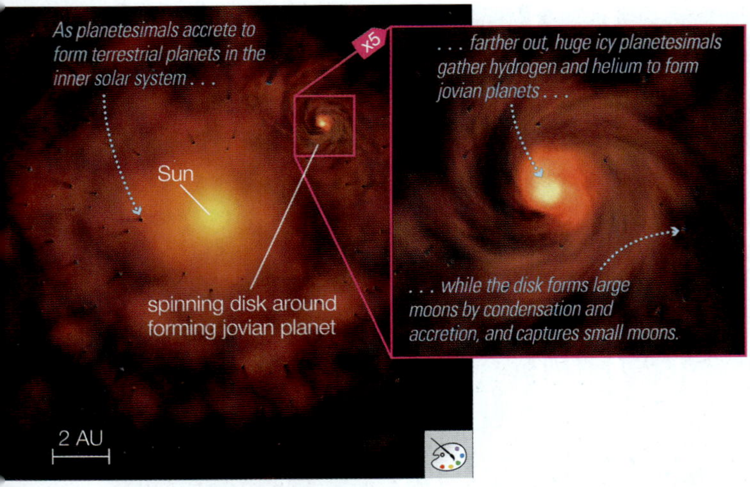


**FIGURE 8.8** The young jovian planets were surrounded by disks of gas, much like the disk of the entire solar nebula but smaller in size. According to the leading model, the planets grew as large, ice-rich planetesimals captured hydrogen and helium gas from the solar nebula. This painting shows the gas and planetesimals surrounding one jovian planet in the larger solar nebula.

continued to cool until hydrogen compounds could have condensed into ices even in the inner solar system. In that case, the terrestrial planets might have accreted abundant ice, and perhaps hydrogen and helium gas as well, changing their basic nature. At the other extreme, if the gas had been blown out earlier, the raw materials of the planets might have been swept away before the planets could fully form. Although these extreme scenarios did not occur in our solar system, they may sometimes occur around other stars. Planet formation may also sometimes be interrupted when radiation from hot neighboring stars drives away material in a solar nebula.

The clearing of the nebula also helps explain what was once considered a surprising aspect of the Sun's rotation. According to the law of conservation of angular momentum, the spinning disk of the solar nebula should have spun fastest near its center, where most of the mass became concentrated. We therefore expect the young Sun to have rotated very fast. But the Sun rotates quite slowly today, with each full rotation taking about a month. If the young Sun really did rotate fast, as the nebular theory seems to demand, how did its rotation slow down?

Angular momentum cannot simply disappear, but it can be transferred from one object to another—and then the other object can be pushed away. A spinning skater can slow his spin by grabbing his partner and then pushing her away. The Sun probably lost angular momentum in a similar way.

The young Sun's rapid rotation would have generated a magnetic field far stronger than that of the Sun today, which in turn would have led to the strong solar wind and to strong surface activity (such as large sunspots and frequent solar flares [**Section 14.3**]) that would explain the young Sun's intense emission of ultraviolet and x-ray radiation. This high-energy radiation ionized gas in the solar nebula, creating many charged particles, while the magnetic field swept through the nebula with the Sun's rapid rotation. Because charged particles and magnetic fields tend to move together, the magnetic field dragged the charged particles along faster than the rest of the nebula, effectively slowing the Sun's rotation and transferring some of the Sun's angular momentum to the nebula. When the nebula was cleared into interstellar space, the gas carried the angular momentum away, leaving the Sun with the greatly diminished angular momentum and slow rotation that we see today (**FIGURE 8.9**).

Although we cannot prove that the young Sun really did lose angular momentum in this way, support for the idea comes from observations of other stars. When we look at young stars in interstellar clouds, we find that nearly all of them rotate rapidly and have strong magnetic fields and strong winds [**Section 16.2**]. Older stars almost invariably rotate slowly, like our Sun. This suggests that nearly all stars have their original rapid rotations slowed by transferring angular momentum to charged particles in their disks—particles that are later swept away—just as we suspect happened with the Sun.

## Where did asteroids and comets come from?

The process of planet formation also explains the origin of the many asteroids and comets in our solar system (including those large enough to qualify as dwarf planets): They are "leftovers" from the era of planet formation. Asteroids are the rocky leftover planetesimals of the inner solar system, while comets are the icy leftover planetesimals of the outer solar system. We'll see in Chapter 12 why most asteroids ended up grouped in the asteroid belt while most comets ended up in either the Kuiper belt or the Oort cloud.

Evidence that asteroids and comets are leftover planetesimals comes from analysis of meteorites, spacecraft visits to comets and asteroids, and theoretical models of solar system formation. In fact, the nebular theory allowed scientists to predict the existence of comets in the Kuiper belt decades before any of them were discovered.

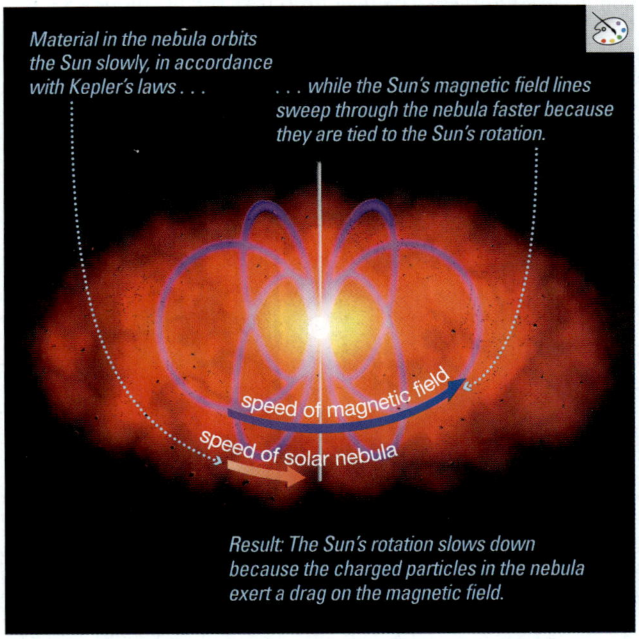


**FIGURE 8.9** The young Sun should have rotated rapidly, but today the Sun rotates quite slowly. As shown in this diagram, the Sun probably lost angular momentum because of drag between slow-moving charged particles in the solar nebula and the Sun's rotating magnetic field (represented by purple loops).

The asteroids and comets that exist today probably represent only a small fraction of the leftover planetesimals that roamed the young solar system. The rest are now gone. Some of these "lost" planetesimals may have been flung into deep space by gravitational encounters, but many others must have collided with the planets. When impacts occur on solid worlds, they leave behind **impact craters** as scars. Impacts have thereby transformed planetary landscapes and, in the case of Earth, altered the course of evolution. For example, an impact is thought to have been responsible for the death of the dinosaurs [**Section 12.5**].

Although impacts occasionally still occur, the vast majority of these collisions occurred in the first few hundred million years of our solar system's history, during the period we call the **heavy bombardment**. Every world in our solar system must have been pelted by impacts during the heavy bombardment (**FIGURE 8.10**), and most of the craters on the Moon and other worlds date from this period.

These early impacts, including many that occurred before the planets finished forming, probably played a key role in making our existence possible. The metal and rock planetesimals that built the terrestrial planets should not have contained water or other hydrogen compounds, because it was too hot for these compounds to condense in our region of the solar nebula. How, then, did Earth come to have the water that makes up our oceans and the gases that first formed our atmosphere? The likely answer is that water, along with other hydrogen compounds, was brought to Earth and other terrestrial planets by the impacts of water-bearing planetesimals that formed farther from the Sun. Recent evidence favors the idea that these planetesimals came from the outer portion of the asteroid belt, where they should have contained small amounts of ice, though some scientists still suspect that ice-rich comets contributed. Either way, the water we drink and the air we breathe probably accreted beyond the orbit of Mars.

**FIGURE 8.10** Around 4 billion years ago, Earth, its Moon, and the other planets were heavily bombarded by leftover planetesimals. This painting shows the young Earth and Moon, with an impact in progress on Earth.

## How do we explain "exceptions to the rules"?

We have seen how the nebular theory accounts for the first three major features of our solar system. For the fourth, the exceptions to the rules, the nebular theory suggests that most of them arose from collisions or close gravitational encounters.

**Captured Moons** We have explained the orbits of most large jovian planet moons by their formation in a disk that swirled around the forming planet. But how do we explain moons with less orderly orbits, such as those that go in the "wrong" direction (opposite their planet's rotation) or that have large inclinations to their planet's equator? These moons are probably leftover planetesimals that originally orbited the Sun but were then captured into planetary orbit.

It's not easy for a planet to capture a moon. An object cannot switch from an unbound orbit (for example, an asteroid whizzing by Jupiter) to a bound orbit (for example, a moon orbiting Jupiter) unless it somehow loses orbital energy [**Section 4.5**]. For the jovian planets, captures probably occurred when passing planetesimals lost energy to friction in the extended and relatively dense gas that surrounded these planets as they formed. The planetesimals would have been slowed by friction with the gas, just as artificial satellites in low orbits are slowed by drag from Earth's atmosphere. If friction reduced a passing planetesimal's orbital energy enough, the planetesimal could have become an orbiting moon. Because of the random nature of the capture process, captured moons would not necessarily orbit in the same direction as their planet or in its equatorial plane. Computer models suggest that this capture process would have worked only on objects up to a few kilometers in size. Most of the small moons of the jovian planets are a few kilometers across, supporting the idea that they were captured in this way. Mars may have similarly captured its two small moons, Phobos and Deimos, at a time when the planet had a much more extended atmosphere than it does today (**FIGURE 8.11**).

**Giant Impacts** Capture processes cannot explain our own Moon, because it is much too large to have been captured by a small planet like Earth. We can also rule out the possibility that our Moon formed simultaneously with Earth, because if both had formed together, they would have accreted from

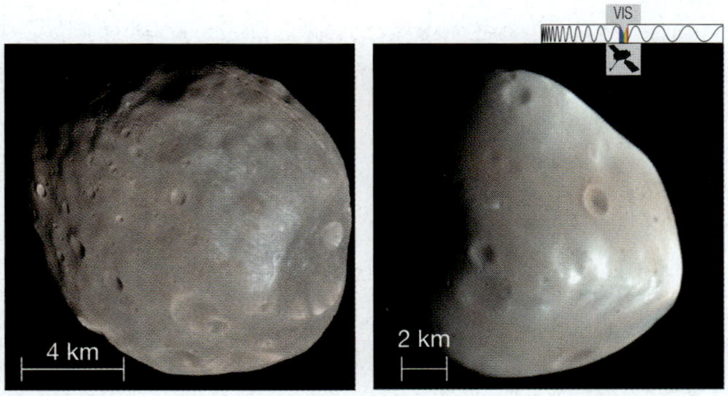


a Phobos    b Deimos

**FIGURE 8.11** The two moons of Mars may be captured asteroids. Phobos is only about 13 kilometers across, and Deimos is only about 8 kilometers across—making each of these two moons small enough to fit within the boundaries of a typical large city. (Images from the *Mars Reconnaissance Orbiter*.)

planetesimals of the same type and would therefore have approximately the same composition and density. But this is not the case: The Moon's density is considerably lower than Earth's, indicating that it has a very different average composition. So how did we get our Moon? Today, the leading hypothesis suggests that it formed as the result of a **giant impact** between Earth and a huge planetesimal.

According to models, a few leftover planetesimals may have been as large as Mars. If one of these Mars-size objects struck a young planet, the blow might have tilted the planet's axis, changed the planet's rotation rate, or completely shattered the planet. The giant impact hypothesis holds that a Mars-size object hit Earth at a speed and angle that blasted Earth's outer layers into space. According to computer simulations, this material could have collected into orbit around our planet, and accretion within this ring of debris could have formed the Moon (**FIGURE 8.12**).

Strong support for the giant impact hypothesis comes from two features of the Moon's composition. First, the Moon's overall composition is quite similar to that of Earth's outer layers—just as we would expect if it were made from material blasted away from those layers. Second, the Moon has a much smaller proportion of easily vaporized ingredients (such as water) than Earth. This fact supports the hypothesis because the heat of the impact would have vaporized these ingredients. As gases, they would not have participated in the subsequent accretion of the Moon.

Giant impacts may also have affected other worlds.* For example, Pluto's moon Charon shows signs of having formed in a giant impact similar to the one thought to have formed our Moon. Mercury has a metal core extending to 85% of its radius—the largest of any planet—which may be the result of a giant impact that blasted away its rocky outer layers. Giant impacts could have been responsible for tilting the axes of many planets (including Earth) and perhaps for tipping Uranus on its side. Venus's slow and backward rotation could also be the result of a giant impact, though some scientists suspect it is a consequence of processes attributable to Venus's thick atmosphere.

Although we may never be able to definitively confirm that a particular giant impact really occurred, the fact that the early solar system must have been filled with leftover planetesimals means we should *expect* there to have been at least a few giant impacts. We conclude that the nebular theory successfully accounts for the fact that there are a few notable exceptions to the rules.

**The Formation of the Solar System**

**Putting It All Together** **FIGURE 8.13** summarizes the nebular theory, showing the processes thought to have led to the four major features of the solar system. Computer models suggest that the entire sequence shown in the figure could have taken place in less than about 50 million years, or only about 1% of the current age of the solar system.

The fact that the nebular theory accounts for the *major* features of our solar system does not mean that it explains *everything* about it. Scientists are still actively working to understand details of planetary formation, both for our own solar system and for others. These details must be enormously complex because of the sheer number of interacting particles, planetesimals, and planets; they also depend on the behavior of the central star, which determines how and when excess nebular gas is cleared. Computer models indicate that the gravitational interactions of planets and small bodies can in some cases nudge orbits enough to cause significant changes to planetary locations. In our solar system, these models suggest that such changes may have moved the jovian planets in a way that led to the formation of the Kuiper belt [**Section 12.4**]. And as we'll discuss in Chapter 13, these types of interactions

*The term *giant impact* is relative to the object being hit; for example, a much smaller object can be "giant" relative to Pluto than relative to Uranus.

**FIGURE 8.12** Artist's conception of the giant impact hypothesis for the formation of our Moon. The fact that ejected material came mostly from Earth's outer rocky layers explains why the Moon contains very little metal. The impact must have occurred more than 4.4 billion years ago, since that is the age of the oldest Moon rocks. As shown, the Moon formed quite close to a rapidly rotating Earth, but over billions of years tidal forces have slowed Earth's rotation and moved the Moon's orbit outward (see Figure 4.27).

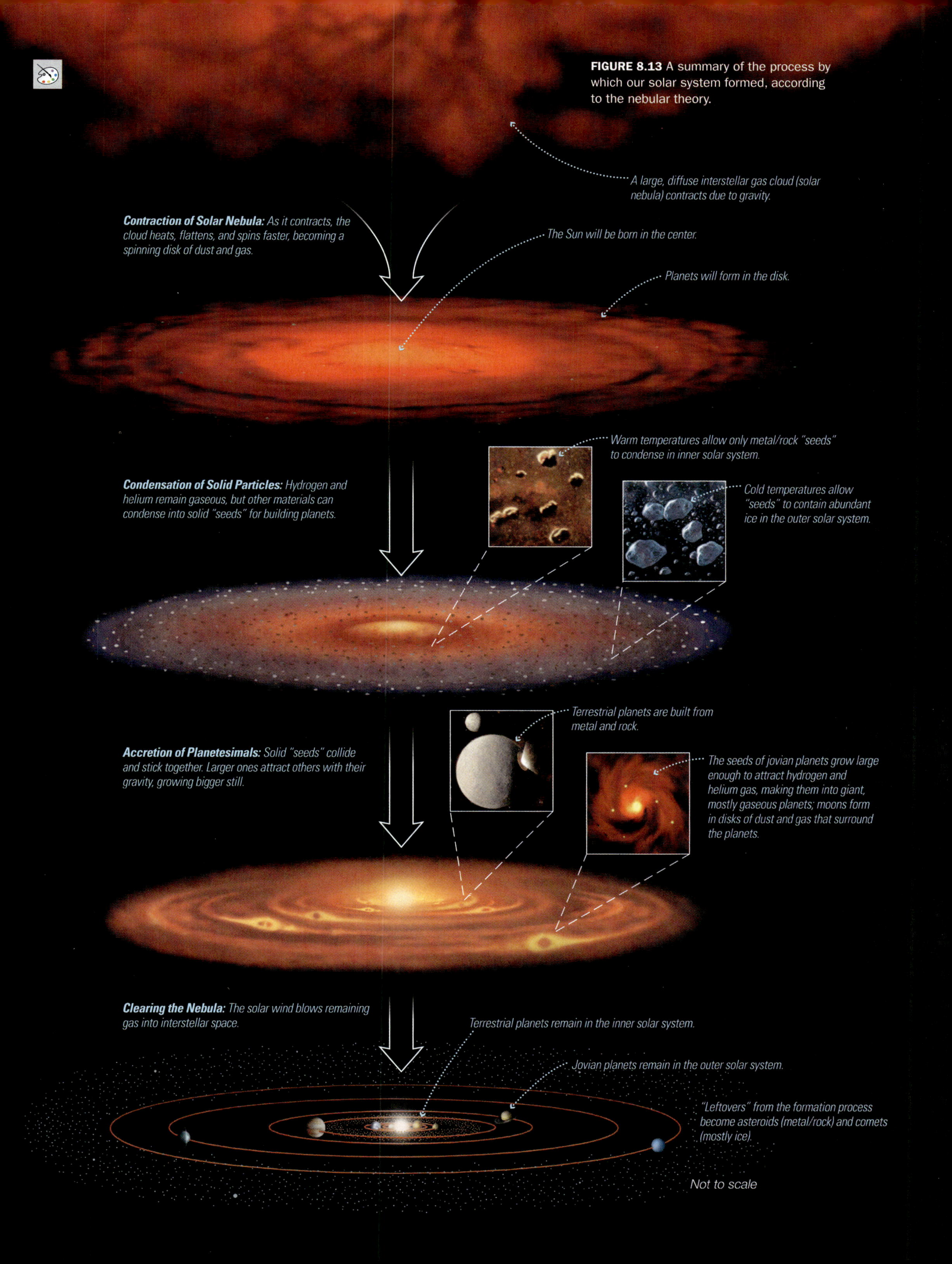


**FIGURE 8.13** A summary of the process by which our solar system formed, according to the nebular theory.

## EXTRAORDINARY CLAIMS A Giant Impact Made Our Moon

In this chapter, you have seen that Earth's large moon is a notable "exception to the rules" of the nebular theory. But how did this exception arise? By the mid-20th century, four competing models had been suggested. The first held that the Moon formed along with Earth through the process of accretion. The second suggested that the Moon was born as an independent "planet" that was captured into Earth orbit. The third, or *fission* (splitting), model suggested that the young, molten Earth rotated so rapidly that it split into two pieces, with the smaller piece becoming the Moon. And a fourth suggested that the Moon had been formed by a giant impact between a wandering planetesimal and Earth. The first three models had weaknesses from the start: Accretion cannot account for the fact that the Moon's density is significantly lower than Earth's, capture would have been too difficult for a body as big in comparison to Earth as the Moon, and fission would have required both that Earth have been spinning improbably fast and that some mechanism have slowed the Moon so that it did not fly into space after it split off. The fourth model was considered so outrageous at the time (it predated studies of how the planets grew through accretion) that it was completely forgotten by the scientific community.

The *Apollo* missions to the Moon (1969–1972) more conclusively ruled out the first three models, based on study of moon rocks returned to Earth. Fortunately, the *Apollo* moon rocks also helped scientists understand the history of cratering on the Moon, leading to a clearer picture of how many large objects had roamed the early solar system. With this in mind, in 1974, astronomers William Hartmann and Donald Davis proposed the giant impact hypothesis, unaware that a similar idea had been suggested previously. They recognized that this idea could explain why the Moon's composition closely matches that of Earth's mantle, and they argued that enough large bodies had roamed the young solar system for such an impact to have been likely. This story is a clear example of how an extraordinary claim that may at first seem completely unreasonable will eventually be reconsidered if enough evidence accumulates for it.

We cannot go back in time to collect direct evidence of a giant impact, so we may never be able to establish with certainty that such an impact really occurred or exactly how it unfolded. Nevertheless, simulations of the event provide strong support for the idea. Many questions still remain to be answered, including the precise nature and size of the impactor, but most scientists now think it very likely that the Moon did indeed form in this way.

**Verdict:** Likely correct, though may never be possible to prove definitively.

probably also explain some of the surprising characteristics we've observed in other planetary systems.

These ideas lead to an interesting question about how much "luck" was involved in our solar system ending up with a planet on which we could evolve. The nebular theory suggests that the early stages of planet formation—including the creation of a spinning disk, condensation within that disk, and the first stages of accretion—were orderly and inevitable. However, the complex interactions that could then occur, and the inherently random nature of giant impacts, suggest that there'd be no guarantee that Earth (or any other planet) would end up with the same orbit and same characteristics if we could turn back the clock and let the solar system form again. For this reason, it's unlikely that any other planetary system is exactly like ours—or that any other world is exactly like Earth—even with the many billions of star systems in our galaxy.

**Think about it** Make a brief list of characteristics of Earth that arose through apparently random processes in our solar system. Overall, how likely do you think it was that a planet like Earth would form? Defend your opinion, and discuss how your answer relates to the possibility of finding Earth-like planets in other star systems.

## 8.3 The Age of the Solar System

The nebular theory seems to explain *how* our solar system was born. But *when* was it born, and how do we know? The answer is that the planets began to form through accretion just over $4\frac{1}{2}$ billion years ago, a fact we learned by determining the age of the oldest rocks in the solar system.

### How do we measure the age of a rock?

The first step in understanding how we've measured the age of our solar system is to understand how we determine the age of an individual rock. A rock is a collection of a great many atoms held together in solid form. The atoms must be older than Earth, having been forged in the Big Bang or in stars that lived long ago. We cannot determine the ages of the individual atoms, because old atoms are indistinguishable from young ones. However, some atoms undergo changes with time that allow us to determine how long they have been held in place within a rock's solid structure. In other words, the age of a rock is the time since its atoms became locked together in their present arrangement, which in most cases means the time *since the rock last solidified.*

Radiometric Dating

**Radiometric Dating** The method by which we measure a rock's age is called **radiometric dating**, and it relies on careful measurement of the rock's proportions of various atoms and isotopes. Recall that each chemical element is uniquely characterized by the number of protons in its nucleus, and that different *isotopes* of the same element differ in their number of neutrons [**Section 5.3**]. The key to radiometric dating lies in the fact that some isotopes are **radioactive**, which is just a fancy way of saying that their nuclei tend to undergo some type of spontaneous change (also called *decay*) with time, such as breaking into two pieces or having a neutron turn into a proton. Decay can change one element into an entirely different one, with different chemical properties.

Consider the radioactive isotope potassium-40 (19 protons and 21 neutrons), which decays when one of its protons turns into a neutron, changing the potassium-40 into argon-40 (18 protons and 22 neutrons). We say that potassium-40 is the *parent isotope*, because it is the original isotope before the decay, and argon-40 is the *daughter isotope* left behind by the decay process. (Potassium-40 also decays by other paths, but we will focus only on decay into argon-40 to keep the discussion simple.)

While the decay of any single nucleus is an instantaneous event, laboratory studies show that a modest amount (millions of atoms or more, which is still only a tiny fraction of a gram) of any radioactive parent isotope will gradually transform itself into the daughter isotope at a very steady rate. You don't have to watch for all that long—rarely longer than a few years at most—before you can measure the rate. You can then use this rate to calculate the **half-life**, which is the time it would take for half of the parent nuclei to decay. Every radioactive isotope has its own unique half-life, which may be anywhere from a fraction of a second to many billions of years. Laboratory measurements show that the half-life for the transformation of potassium-40 into argon-40 is 1.25 billion years.

To understand the idea of half-life better, imagine a small piece of rock that contained 1 microgram of potassium-40 and no argon-40 when it solidified long ago. The half-life of 1.25 billion years means that half the original potassium-40 had decayed into argon-40 by the time the rock was 1.25 billion years old, so at that time the rock contained $\frac{1}{2}$ microgram of potassium-40 and $\frac{1}{2}$ microgram of argon-40.

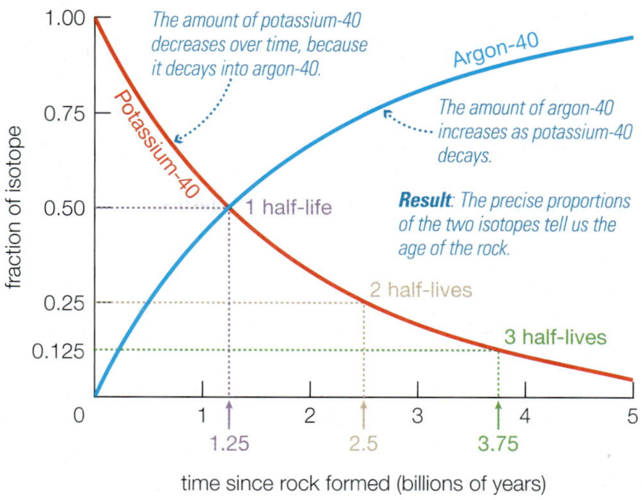


**FIGURE 8.14** Potassium-40 is radioactive, decaying into argon-40 with a half-life of 1.25 billion years. The red curve shows the decreasing amount of potassium-40, and the blue curve shows the increasing amount of argon-40. The remaining amount of potassium-40 drops in half with each successive half-life.

Half of this remaining potassium-40 had then decayed by the end of the next 1.25 billion years, so after 2.5 billion years the rock contained $\frac{1}{4}$ microgram of potassium-40 and $\frac{3}{4}$ microgram of argon-40. After three half-lives, or 3.75 billion years, only $\frac{1}{8}$ microgram of potassium-40 remained, while $\frac{7}{8}$ microgram had become argon-40. **FIGURE 8.14** summarizes the gradual decrease in the amount of potassium-40 and the corresponding rise in the amount of argon-40.

## MATHEMATICAL INSIGHT 8.1 Radiometric Dating

The amount of a radioactive substance decreases by half with each half-life, so we can express the decay process with a simple formula relating the current and original amounts of a radioactive substance in a rock:

$$\frac{\text{current amount}}{\text{original amount}} = \left(\frac{1}{2}\right)^{t/t_{\text{half}}}$$

where $t$ is the time since the rock formed and $t_{\text{half}}$ is the half-life of the radioactive material. We can solve this equation for the age $t$ by taking the base-10 logarithm of both sides and rearranging the terms. The resulting general equation for the age is

$$t = t_{\text{half}} \times \frac{\log_{10}\left(\dfrac{\text{current amount}}{\text{original amount}}\right)}{\log_{10}\left(\dfrac{1}{2}\right)}$$

**EXAMPLE:** You heat and chemically analyze a small sample of a meteorite. Potassium-40 and argon-40 are present in a ratio of approximately 0.85 unit of potassium-40 atoms to 9.15 units of gaseous argon-40 atoms. (The units are unimportant, because only the relative amounts of the parent and daughter materials matter.) How old is the meteorite?

**SOLUTION:**

**Step 1 Understand:** The formula above allows us to find the meteorite's age from the current and original amounts of a parent isotope. We are given that the current amount of the parent isotope potassium-40 is 0.85 unit. We are not told the original amount, but we can figure it out from the fact that the sample now has 9.15 units of the daughter isotope argon-40. As discussed in the text, we can assume that *all* the trapped argon-40 in the meteorite is a decay product of potassium-40, which means that the 9.15 units of argon-40 must originally have been potassium-40. Therefore, the original amount of potassium-40 was $0.85 + 9.15 = 10$ units.

**Step 2 Solve:** We use the formula above, with current amount = 0.85 unit and original amount = 10 units:

$$t = 1.25 \text{ billion yr} \times \frac{\log_{10}\left(\dfrac{0.85}{10}\right)}{\log_{10}\left(\dfrac{1}{2}\right)}$$

$$= 1.25 \text{ billion yr} \times \left(\frac{-1.07}{-0.301}\right)$$

$$= 4.45 \text{ billion yr}$$

**Step 3 Explain:** We have used radiometric dating to determine that the meteorite's age is 4.45 billion years, which means it solidified 4.45 billion years ago. (Note: This example captures the essence of radiometric dating, but real cases generally require more detailed analysis.)

You can now see the essence of radiometric dating. Suppose you find a rock that contains equal numbers of atoms of potassium-40 and argon-40. If you assume that all the argon came from potassium decay (and if the rock shows no evidence of subsequent heating that could have allowed any argon to escape), then it must have taken precisely one half-life for the rock to end up with equal amounts of the two isotopes. You can therefore conclude that the rock is 1.25 billion years old. The only question is whether you are right in assuming that the rock lacked argon-40 when it formed. In this case, knowing a bit of "rock chemistry" helps. Potassium-40 is a natural ingredient of many minerals in rocks, but argon-40 is a gas that does not combine with other elements and did not condense in the solar nebula. If you find argon-40 gas trapped inside minerals, it must have come from radioactive decay of potassium-40.

**Validity of the Method** Radiometric dating is possible with many other radioactive isotopes as well. In many cases, a rock contains more than one radioactive isotope, and agreement between the ages calculated from the different isotopes gives us confidence that we have dated the rock correctly. We can also check results from radiometric dating against those from other methods of measuring or estimating ages. For example, some archaeological artifacts have original dates printed on them, and these dates agree with ages found by radiometric dating. The same agreement is found for wood artifacts that can be dated both by counting tree rings and by the radiometric method.

We can validate the $4\frac{1}{2}$-billion-year radiometric age for the solar system as a whole by comparing it to an age based on detailed study of the Sun. Theoretical models of the Sun, along with observations of other stars, show that stars slowly expand and brighten as they age. The model ages are not nearly as precise as radiometric ages, but they confirm that the Sun is between about 4 and 5 billion years old. Overall, the technique of radiometric dating has been checked in so many ways and relies on such basic scientific principles that there is no serious scientific debate about its validity.

## How do we know the age of the solar system?

Radiometric dating tells us how long it has been since a rock solidified, which is not the same as the age of a planet as a whole. For example, we find rocks of many different ages on Earth. Some rocks are quite young because they formed recently from molten lava; others are much older. The oldest Earth rocks are about 4 billion years old, and some small mineral grains date to almost 4.4 billion years ago, but even these are not as old as Earth itself, because Earth's entire surface has been reshaped through time.

Moon rocks brought back by the *Apollo* astronauts have also been dated, and many are older than the oldest Earth rocks, demonstrating that parts of the Moon's surface have not changed since very early in its history. The oldest Moon rocks contain minerals with a small amount of uranium-238, which decays (in several steps) into lead-206 with a half-life of about 4.5 billion years. Lead and uranium have very different chemical behaviors, and some minerals start with virtually no lead. Laboratory analysis of such minerals in lunar rocks shows that they now contain almost equal proportions of atoms of uranium-238 and lead-206, which means they are about one half-life old. More precise work shows them to be about 4.4 billion years old.

**Think about it** If future scientists examine the same lunar rocks 4.5 billion years from now, what proportions of uranium-238 and lead-206 will they find? Explain.

Although the oldest Moon rocks are older than the oldest Earth rocks, the rocks must still be younger than the Moon and Earth as a whole. In fact, their age tells us that the giant impact thought to have created the Moon must have occurred more than 4.4 billion years ago. But how do we determine when the planets first began to form?

To go all the way back to the origin of the solar system, we must find rocks that have not melted or vaporized since they first condensed in the solar nebula. Meteorites that have fallen to Earth are our source of such rocks. Many meteorites appear to have remained unchanged since they condensed and accreted in the early solar system. Careful analysis of radioactive isotopes in these meteorites shows that the oldest ones formed about 4.56 billion years ago, so this time must mark the beginning of accretion in the solar nebula. Because the planets apparently accreted within about 50 million (0.05 billion) years after that, Earth and the other planets had formed by about 4.5 billion years ago. In other words, the age of our solar system is only about a third of the 14-billion-year age of our universe.

### SPECIAL TOPIC What Started the Collapse of the Solar Nebula?

The solar nebula probably could not have started collapsing on its own, because gravity is weaker when gas is spread out over a large region of space. We therefore suspect that some cataclysmic event triggered the collapse. But what type of event? Although we may never know for sure, radioactive elements and their decay products give us some clues. Radioactive elements are made in the violent stellar explosions called *supernovae,* so their presence in meteorites and on Earth underscores the fact that our solar system is made from the remnants of past generations of stars.

Some meteorites contain the rare isotope xenon-129. Xenon is gaseous even at extremely low temperatures, so it could not have condensed and become trapped in the meteorites as they condensed in the solar nebula. Instead, it must be a product of radioactive decay, and laboratory studies show that its parent isotope is iodine-129. This is a crucial clue, because iodine-129 has a relatively short half-life of 17 million years. We conclude that the xenon-129 we observe in meteorites today is the product of radioactive decay of iodine-129 produced in a supernova that occurred no more than a few tens of millions of years before the collapse of the solar nebula began.

The relatively short time between a supernova and the birth of our solar system suggests that a shock wave from the exploding star may have been the trigger for the nebula's collapse. Once this shock wave got the collapse started, gravity took over and made the rest of the collapse inevitable.

# The BIG Picture PUTTING CHAPTER 8 INTO PERSPECTIVE

*This chapter described the current scientific theory of our solar system's formation, and how this theory explains the major features we observe. As you continue your study of the solar system, keep in mind the following "big picture" ideas:*

- The nebular theory of solar system formation gained wide acceptance because of its success in explaining the major characteristics of our solar system.
- Most of the general features of the solar system were determined by processes that occurred very early in the solar system's history.
- Chance events may have played a large role in determining how individual planets turned out. No one knows how different our solar system might be if the process started over.
- We have learned the age of our solar system—about $4\frac{1}{2}$ billion years—from radiometric dating of the oldest meteorites. This age agrees with ages estimated through a variety of other techniques, making it clear that we are recent arrivals on a very old planet.

**MY COSMIC PERSPECTIVE** Events that happened more than 4 billion years ago may seem distant from our lives, but we wouldn't be here without them; they explain not only how our planet came to exist but also where life-sustaining water comes from and how we came to have the Moon in our sky.

# Summary of Key Concepts

## 8.1 The Search for Origins

- **How did we arrive at a theory of solar system formation?** A successful theory must explain four major features of our solar system: patterns of motion, the existence of two types of planets (terrestrial and jovian), the presence of asteroids and comets, and exceptions to the rules. Developed over a period of more than two centuries, the **nebular theory** explains all four features and also can account for other planetary systems.
- **Where did the solar system come from?** The nebular theory holds that the solar system formed from the gravitational collapse of an interstellar cloud known as the **solar nebula**. The cloud was the product of recycling of gas through many generations of stars within our galaxy. This material consisted of 98% hydrogen and helium and 2% all other elements combined.

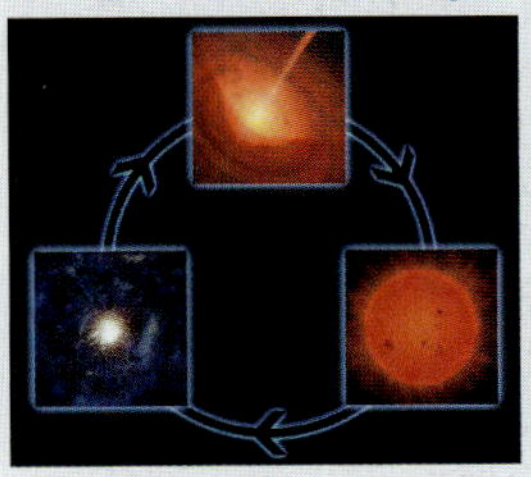

## 8.2 Explaining the Major Features of the Solar System

- **What caused the orderly patterns of motion?** As the solar nebula collapsed under gravity, natural processes caused it to heat up, spin faster, and flatten out as it shrank. The orderly motions we observe today all came from the orderly motion of this spinning disk.

- **Why are there two major types of planets?** The inner regions of the solar nebula were relatively hot, so only metal and rock could condense into tiny solid grains; these grains accreted into larger **planetesimals** that ultimately merged to make the terrestrial planets. Beyond the **frost line**, cooler temperatures also allowed more abundant **hydrogen compounds** to condense into ice, building ice-rich planetesimals; some of these grew large enough for their gravity to draw in hydrogen and helium gas, forming the jovian planets.

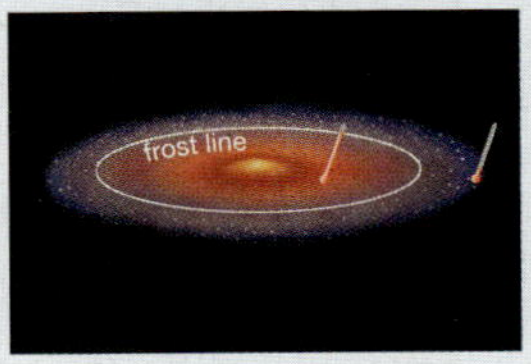


- **Where did asteroids and comets come from?** Asteroids are the rocky leftover planetesimals of the inner solar system, and comets are the ice-rich leftover planetesimals of the outer solar system. These objects still occasionally collide with planets or moons, but the vast majority of impacts occurred during the **heavy bombardment** in the solar system's first few hundred million years.
- **How do we explain "exceptions to the rules"?** Most of the exceptions probably arose from collisions or close encounters with leftover planetesimals. Our Moon is most likely the result of a **giant impact** between a Mars-size planetesimal and the young Earth.

## 8.3 The Age of the Solar System

- **How do we measure the age of a rock? Radiometric dating** is based on carefully measuring the proportions of radioactive isotopes and their decay products within rocks. The ratio of the isotopes changes with time in a steady and predictable way that we characterize by an isotope's **half-life**, the time it takes for half the atoms in a collection to decay.

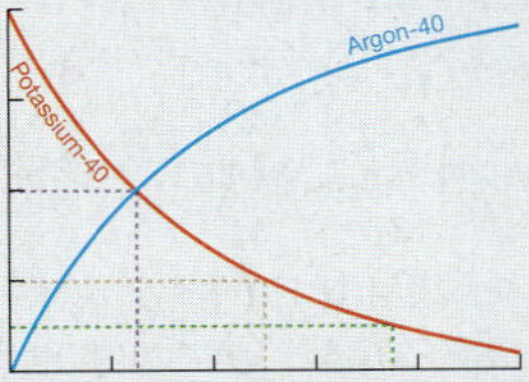


- **How do we know the age of the solar system?** Radiometric dating of the oldest meteorites tells us that accretion began in the solar nebula about 4.56 billion years ago, with the planets forming by about 4.5 billion years ago.

# Visual Skills Check

*Use the following questions to check your understanding of some of the many types of visual information used in astronomy. For additional practice, try the Chapter 8 Visual Quiz in the Study Area at www.MasteringAstronomy.com.*

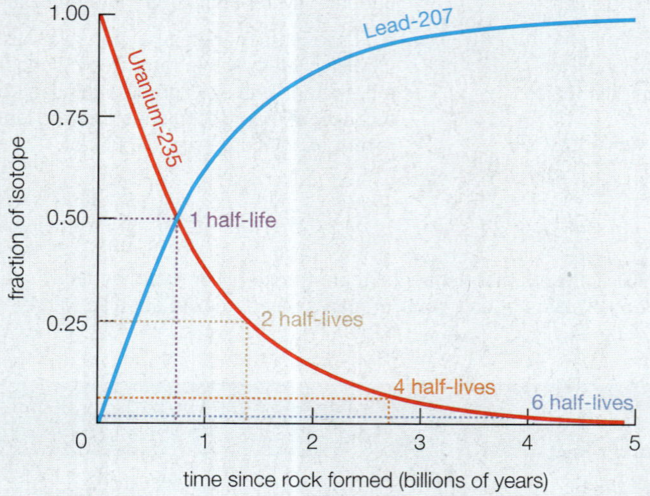


*The graph above, similar to Figure 8.14, shows the radioactive decay of uranium-235 to lead-207. (Uranium-235 also decays to lead-206, but we'll ignore that reaction for the sake of clarity in this example.)*

1. Compare the graph above to Figure 8.14, which shows the decay of potassium-40. Which element is more radioactive (undergoes radioactive decay more quickly)?
   a. uranium-235
   b. potassium-40
   c. Both are equally radioactive.
2. What fraction of the original uranium-235 should be left after 3.5 billion years?
   a. $\frac{1}{2}$ b. $\frac{1}{4}$ c. $\frac{1}{8}$ d. $\frac{1}{32}$ e. $\frac{1}{64}$
3. You find a mysterious rock on the ground and determine that 60% of its uranium-235 has been converted into lead-207. What is the most likely origin of the rock, based on its radiometric age?
   a. It's older than our solar system, so it must have come from another solar system.
   b. It's a meteorite dating back to the formation of the solar system.
   c. It's a volcanic rock nearly a billion years old.
   d. It was just formed this year during the eruption of a nearby volcano.

# Exercises and Problems

*For instructor-assigned homework and other learning materials, go to www.MasteringAstronomy.com.*

## Chapter Review Questions

### Short-Answer Questions Based on the Reading

1. Briefly describe the four major features of our solar system that a formation theory must explain.
2. What is the *nebular theory*, and why is it widely accepted by scientists today?
3. What do we mean by the *solar nebula*? What was it made of, and where did it come from?
4. Describe the three key processes that led the solar nebula to take the form of a spinning disk. What observational evidence supports this scenario?
5. List the approximate condensation temperature and abundance for each of the four categories of materials in the solar nebula. Which ingredients are present in terrestrial planets? In jovian planets? In comets and asteroids?
6. What was the *frost line*? Which ingredients condensed inside and outside the frost line? What role did it play in the formation of two distinct types of planets?
7. Briefly describe the process by which terrestrial planets are thought to have formed.
8. How was the formation of jovian planets similar to that of the terrestrial planets? How was it different? Why did the jovian planets end up with so many moons?
9. What is the *solar wind*, and what roles did it play in the early solar system?
10. How did planet formation lead to the existence of asteroids and comets?
11. What was the *heavy bombardment*, and when did it occur?
12. What is the leading hypothesis for the Moon's formation, and what evidence supports this hypothesis?
13. Describe the technique of *radiometric dating*. What is a *half-life*?
14. How old is the solar system, and how do we know?

### Surprising Discoveries?

*Suppose we found a solar system with the property described. (These are not real discoveries.) In light of what you've learned about the formation of our own solar system, decide whether the discovery should be considered reasonable or surprising. Explain your reasoning.*

15. A solar system has five terrestrial planets in its inner solar system and three jovian planets in its outer solar system.
16. A solar system has four large jovian planets in its inner solar system and seven small terrestrial planets in its outer solar system.
17. A solar system has ten planets that all orbit the star in approximately the same plane. However, five planets orbit in one direction (e.g., counterclockwise), while the other five orbit in the opposite direction (e.g., clockwise).
18. A solar system has 12 planets that all orbit the star in the same direction and in nearly the same plane. The 15 largest moons in this solar system orbit their planets in nearly the same direction and plane as well. However, several smaller moons have highly inclined orbits around their planets.
19. A solar system has six terrestrial planets and four jovian planets. Each of the six terrestrial planets has at least five moons, while the jovian planets have no moons at all.
20. A solar system has four Earth-size terrestrial planets. Each of the four planets has a single moon that is nearly identical in size to Earth's Moon.
21. A solar system has many rocky asteroids and many icy comets. However, most of the comets orbit in the inner solar system, while the asteroids orbit in far-flung regions much like the Kuiper belt and Oort cloud of our solar system.
22. A solar system has several planets similar in composition to the jovian planets of our solar system but similar in mass to the terrestrial planets of our solar system.
23. A solar system has several terrestrial planets and several larger planets made mostly of ice. (*Hint:* What would happen if the solar wind started earlier or later than in our solar system?)
24. Radiometric dating of the oldest meteorites from another solar system shows that they are a billion years younger than rocks from the terrestrial planets of the same system.

## Quick Quiz

*Choose the best answer to each of the following. For additional practice, try the Chapter 8 Reading and Concept Quizzes in the Study Area at www.MasteringAstronomy.com.*

25. How many of the planets orbit the Sun in the same direction as Earth does? (a) a few (b) most (c) all
26. The *nebular theory* holds that (a) our solar system formed from the collapse of an interstellar cloud of gas and dust. (b) each planet formed from the collapse of its own separate nebula. (c) the planets formed as a result of a near-collision between our Sun and another star.
27. The solar nebula was 98% (a) rock and metal. (b) hydrogen compounds. (c) hydrogen and helium.
28. Which of the following did *not* occur during the collapse of the solar nebula? (a) spinning faster (b) heating up (c) concentrating denser materials nearer the Sun
29. What is Jupiter's main ingredient? (a) rock and metal (b) hydrogen compounds (c) hydrogen and helium
30. Which of the following lists the major steps of solar system formation in the correct order? (a) collapse, accretion, condensation (b) collapse, condensation, accretion (c) accretion, condensation, collapse
31. Which of the following is *not* true of the young Sun? (a) It rotated much faster than it does today. (b) It was much brighter and hotter than it is today. (c) It had a much stronger solar wind than it does today.
32. Leftover ice-rich planetesimals are called (a) comets. (b) asteroids. (c) meteorites.
33. The leading hypothesis for the origin of the Moon is that it (a) formed along with Earth. (b) formed from material ejected from Earth in a giant impact. (c) split out of a rapidly rotating Earth.
34. About how old is the solar system? (a) 4.5 million years (b) 4.5 billion years (c) 4.5 trillion years

## Inclusive Astronomy

*Use these questions to reflect on participation in science.*

35. *Group Discussion: Solar Origin Stories.* Ancient cultures understood the immense importance of the Sun and told many stories about its origins. This chapter has told a different origin story, based on the nebular theory, which scientists consider to be supported by a wide body of evidence.
**a.** Working independently, learn about an origin story for the Sun from a culture different from your own, taking the time necessary to understand the value and purpose of that story for the culture that told it.
**b.** Gather in small groups and take turns sharing the origin stories you learned about.
**c.** Make a list of notable similarities and differences among the origin stories your group learned about, then discuss any commonalities you see in the roles the stories played in their cultures.
**d.** Discuss the similarities and differences between the ancient origin stories and the nebular theory. What do you think are the value and purpose of the nebular theory to our own culture?
**e.** Discuss the potential value of the ancient origin stories to our own culture. What do we learn by seeking them out and retelling them?

## The Process of Science

*These questions may be answered individually in short-essay form or discussed in groups, except where identified as group-only.*

36. *Explaining the Past.* Test the nebular theory against each of the three hallmarks of science discussed in Chapter 3. Be as detailed as possible in explaining whether the theory does or does not satisfy these hallmarks. Use your findings to discuss whether it is really possible for science to inform us about how our solar system formed. Defend your opinion.
37. *Theory and Observation.* Discuss the interplay between theory and observation that has led to our modern theory of solar system formation. What role does technology play in allowing us to test this theory?
38. *Unanswered Questions.* As discussed in this chapter, the nebular theory explains many but not all questions about the origin of our solar system. Choose one important but unanswered question about the origin of our solar system and write two or three paragraphs in which you discuss how we might answer this question in the future. Be as specific as possible, focusing on the type of evidence necessary to answer the question and how the evidence could be gathered. What are the benefits of finding answers to this question?
39. *Group Activity: A Cold Solar Nebula.* The excess gas of the solar nebula is presumed to have been cleared away (by the solar wind) at a time when the frost line was located between the orbits of Mars and Jupiter, but study of planets around other stars suggests that gas may be cleared out earlier or later in other solar systems. In this activity, you'll consider how our solar system might have turned out

differently if the excess gas of the solar nebula had not been cleared away until the entire disk of gas had cooled to 50 K. Note: You may wish to do this activity using the four roles described in Chapter 1, Exercise 39.

**a.** Make a list of ingredients that will condense at 50 K.

**b.** Make a list of ways in which the terrestrial planets might have turned out differently under this alternative formation scenario.

**c.** Repeat part b for the jovian planets.

**d.** Discuss the likelihood that your predicted changes would match the actual characteristics of this alternative solar system.

**e.** Come up with additional "what if" scenarios, discussing various ways in which planets might have turned out differently.

## Investigate Further

### Short-Answer/Essay Questions

40. *An Early Solar Wind.* Suppose the solar wind had cleared away the solar nebula before the seeds of the jovian planets could gravitationally draw in hydrogen and helium gas. How would the planets of the outer solar system be different? Would they still have many moons? Explain your answer in a few sentences.
41. *Angular Momentum.* Suppose our solar nebula had begun with much more angular momentum than it did. Do you think planets could still have formed? Why or why not? What if the solar nebula had started with zero angular momentum? Explain your answers in one or two paragraphs.
42. *Two Kinds of Planets.* The jovian planets differ from the terrestrial planets in a variety of ways. Using sentences that members of your family would understand, explain why the jovian planets differ from the terrestrial planets in each of the following: composition, size, density, distance from the Sun, and number of satellites.
43. *Two Kinds of Planetesimals.* Why are there two kinds of "leftovers" from solar system formation? List as many differences between comets and asteroids as you can.
44. *Understanding Radiometric Dating.* Imagine you had the good fortune to find a rocky meteorite in your backyard. Qualitatively, how would you expect its ratio of potassium-40 to argon-40 to be different from that of other rocks in your yard? Explain why, in a few sentences.
45. *Rocks from Other Solar Systems.* Many "leftovers" from planetary formation were likely ejected from our solar system, and the same has presumably happened in other star systems. Given that fact, should we expect to find meteorites that come from other star systems? How rare or common would you expect them to be? (Be sure to consider the distances between stars.) Suppose that we *did* find a meteorite identified as a leftover from another stellar system. What could we learn from it?
46. *Dating the Past.* Radiometric dating is used in many branches of science. For example, it is used to determine ages of fossils that help us learn the history of life on Earth and ages of relics that teach us about the rise of civilization. Research one use of radiometric dating outside of astronomy, writing a short report on how studies are conducted (such as what materials are dated and what radioactive elements are used) and what these studies have concluded.
47. *Lucky to Be Here?* Considering the overall process of solar system formation, do you think formation of a planet like Earth was likely? Could random events in the early history of the solar system have prevented our being here today? What implications do your answers have for the possibility of finding planets with Earth-like conditions around other stars? Defend your opinions.

### Quantitative Problems

*Be sure to show all calculations clearly and state your final answers in complete sentences.*

48. *Radiometric Dating.* You are dating rocks by their proportions of parent isotope potassium-40 (half-life 1.25 billion years) and daughter isotope argon-40. Find the age for each of the following.

    **a.** A rock that contains equal amounts of potassium-40 and argon-40

    **b.** A rock that contains three times as much argon-40 as potassium-40
49. *Lunar Rocks.* You are dating Moon rocks based on their proportions of uranium-238 (half-life of about 4.5 billion years) and its ultimate decay product, lead. Find the age for each of the following.

    **a.** A rock for which you determine that 55% of the original uranium-238 remains, while the other 45% has decayed into lead

    **b.** A rock for which you determine that 63% of the original uranium-238 remains, while the other 37% has decayed into lead
50. *Carbon-14 Dating.* The half-life of carbon-14 is about 5700 years.

    **a.** You find a piece of cloth painted with organic dye. By analyzing the dye, you find that only 77% of the carbon-14 originally in the dye remains. When was the cloth painted?

    **b.** A well-preserved piece of wood found at an archaeological site has 6.2% of the carbon-14 it must have had when it was living. Estimate when the wood was cut.

    **c.** Is carbon-14 useful for establishing Earth's age? Why or why not?
51. *Collapsing Cloud.* The time it takes for a cloud 100,000 AU in radius to collapse in "free-fall" to form a new star is *half* the time it would take for an object to orbit the star on an extremely elliptical orbit with a semimajor axis of 50,000 AU (half the 100,000 AU radius). Use Kepler's third law to find the collapse time, assuming the star has the same mass as the Sun.
52. *Icy Earth.* How massive would Earth have to have been to accrete hydrogen compounds in addition to rock and metal? Assume the same proportions of the ingredients as listed in Table 8.1. What would Earth's mass have to have been for Earth to capture hydrogen and helium gas in the proportion listed in the table?
53. *What Are the Odds?* The fact that all the planets orbit the Sun in the same direction is cited as support for the nebular hypothesis. Imagine that there's a different hypothesis in which planets can be created orbiting the Sun in either direction. Under this hypothesis, what is the probability that eight planets would end up traveling in the same direction? (*Hint:* It's the same probability as that of flipping a coin eight times and getting all heads.)
54. *Spinning Up the Solar Nebula.* The orbital speed of the material in the solar nebula at Pluto's average distance from the Sun was about 5 km/s. What was the orbital speed of this material when it was 40,000 AU from the Sun (before it fell inward with the collapse of the nebula)? Use the law of conservation of angular momentum (see Section 4.3).

# 9 Planetary Geology

## EARTH AND THE OTHER TERRESTRIAL WORLDS

▲ **About the photo:** Volcanic eruptions like this one (Mount Etna, Sicily) play a major role in shaping the surfaces of Earth and other terrestrial worlds.

## LEARNING GOALS

### 9.1 Connecting Planetary Interiors and Surfaces
- What are terrestrial planets like on the inside?
- What causes geological activity?
- Why do some planetary interiors create magnetic fields?

### 9.2 Shaping Planetary Surfaces
- What processes shape planetary surfaces?
- How do impact craters reveal a surface's geological age?
- Why do the terrestrial planets have different geological histories?

### 9.3 Geology of the Moon and Mercury
- What geological processes shaped our Moon?
- What geological processes shaped Mercury?

### 9.4 Geology of Mars
- What geological processes have shaped Mars?
- What geological evidence tells us that water once flowed on Mars?

### 9.5 Geology of Venus
- What geological processes have shaped Venus?
- Does Venus have plate tectonics?

### 9.6 The Unique Geology of Earth
- How is Earth's surface shaped by plate tectonics?
- Was Earth's geology destined from birth?

*Nothing is rich but the inexhaustible wealth of nature. She shows us only surfaces, but she is a million fathoms deep.*

*—Ralph Waldo Emerson (1803–1882)*

**Chapter 9 Overview**

It's easy to take for granted the qualities that make Earth so suitable for human life: a temperature neither boiling nor freezing, abundant water, a protective atmosphere, and a relatively stable environment. But we need look only as far as our neighboring terrestrial worlds to see how fortunate we are. The Moon is airless and barren, and Mercury is much the same. Venus is a searing hothouse, while Mars has an atmosphere so thin and cold that liquid water cannot last on its surface today.

How did the terrestrial worlds come to be so different, and why did Earth alone develop conditions that permit abundant life? We'll explore these questions through careful comparative study of the planets, focusing on the geology of the terrestrial worlds in this chapter and on their atmospheres in Chapter 10. As we will see, the histories of the terrestrial worlds have been determined largely by properties endowed at their births.

## 9.1 Connecting Planetary Interiors and Surfaces

Earth's surface seems solid and steady, but every so often it offers us a reminder that nothing about it is permanent. If you live in Alaska or California, you've probably felt the ground shift beneath you in an earthquake. In Washington State, you may have witnessed the rumblings of Mount St. Helens. In Hawaii, a visit to the active Kilauea volcano will remind you that you are standing on mountains of volcanic rock protruding from the ocean floor.

Volcanoes and earthquakes are not the only processes acting to reshape Earth's surface. They are not even the most dramatic: Far greater change can occur on the rare occasions when an asteroid or a comet slams into Earth. More gradual processes can also have spectacular effects. The Colorado River causes only small changes in the landscape from year to year, but over millions of years its unrelenting flow carved the Grand Canyon. The Rocky Mountains were once twice as tall as they are today; they have been cut down in size through tens of millions of years of erosion by wind, rain, and ice. Entire continents move slowly about, completely rearranging the map of Earth every few hundred million years.

Earth is not alone in having undergone tremendous change since its birth. The surfaces of all five terrestrial worlds—Mercury, Venus, Earth, the Moon, and Mars—must have looked quite similar when they were young, because all are made of rocky material and all were subjected to the impacts of the heavy bombardment [**Section 8.2**]. The great differences in their present-day appearances must therefore be the result of changes that have occurred through time.

**FIGURE 9.1** shows global views of the terrestrial surfaces to scale, as well as sample surface views from orbit. Profound differences are immediately obvious. Mercury and the Moon show the scars of their battering during the heavy bombardment: They are densely covered by craters, except in areas that appear to be volcanic plains. Bizarre bulges and odd volcanoes dot the surface of Venus. Mars has the solar system's largest volcanoes, a huge canyon cutting across its surface, and many features that appear to have been shaped by liquid water but are now dry. Earth has surface features similar to all those on the other terrestrial worlds, and more—including a unique layer of

Heavily cratered Mercury has long steep cliffs (arrow).

Cloud-penetrating radar revealed this twin-peaked volcano on Venus.

A portion of Earth's surface as it appears without clouds.

The Moon's surface is heavily cratered in most places.

Mars has features that look like dry riverbeds; note the impact craters.

**FIGURE 9.1** Global views to scale, along with sample close-ups viewed from orbit, of the five terrestrial worlds. All the images were taken with visible light except those for Venus, which are based on radar data.

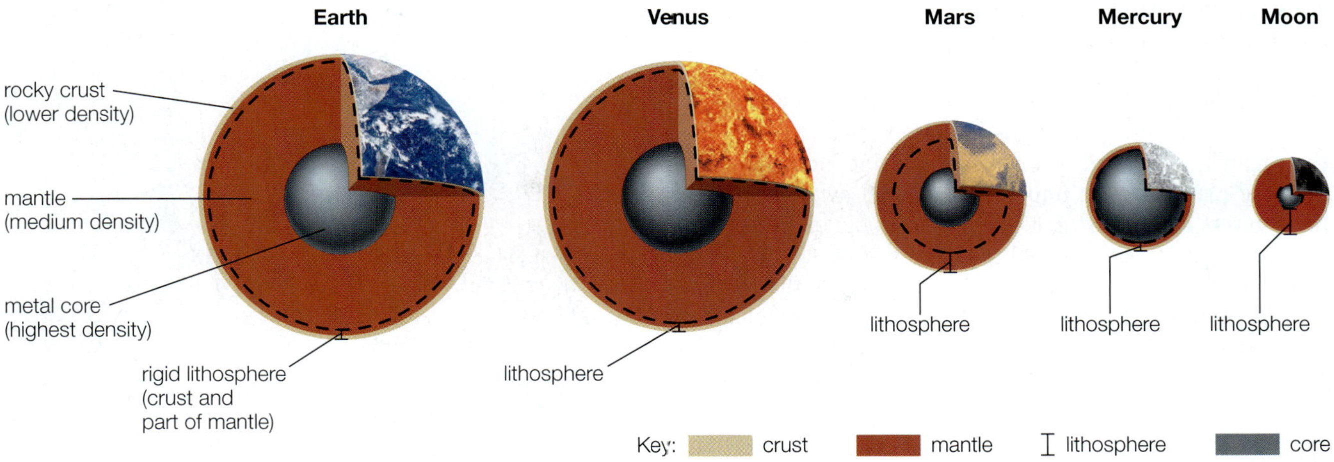


**FIGURE 9.2** Interior structures of the terrestrial worlds, with sizes shown to scale from largest to smallest. Color coding shows the core-mantle-crust layering by density; a dashed circle represents the inner boundary of the lithosphere, defined by the strength of the rock rather than by density. The crust and lithosphere are indicated approximately, as they cannot be drawn to scale on this diagram. Venus and Earth have the thinnest lithospheres, Mercury and the Moon have the thickest, and Mars is in between.

living organisms that covers almost the entire surface of the planet.

Our goal in this chapter is to understand how these differences among the terrestrial worlds came to be. This type of study is called **planetary geology**. Geology is literally the study of Earth (*geo* means "Earth"), so planetary geology is the extension of this science to other worlds. Note that in this usage, the term *planetary* means "planet-like," so it includes moons, dwarf planets, asteroids, and comets.

## What are terrestrial planets like on the inside?

Most geological features are shaped by processes that take place deep beneath the surface, so the first step in our study of planetary geology is to examine planetary interiors. We cannot see inside Earth or any other terrestrial world, but a variety of clues tell us about their internal structures. For Earth, the most direct data come from **seismic waves**, vibrations that travel both through the interior and along the surface after an earthquake (*seismic* comes from the Greek word for "shake"). In much the same way that shaking a gift box offers clues about what's inside, seismic vibrations offer clues about what's inside Earth. We also have seismic data for the Moon, thanks to monitoring stations left on the Moon by the *Apollo* astronauts, and the *Mars Insight* mission, launched in 2018, includes a seismic station for Mars.

We use less direct methods to study the interiors of other worlds. For example, comparing a world's overall average density (found by dividing its total mass by its total volume) to the density of its surface rock tells us how much more dense it must be inside, measurements of a world's gravity from spacecraft can tell us about how its mass is distributed inside it, studies of magnetic fields tell us about the interior layers in which these fields are generated, and volcanic rocks can tell us about interior composition.

**Layering by Density** Our studies have shown that all the terrestrial worlds have layered interiors. We divide these layers by density into three major categories:

- **Core.** The highest-density material, consisting primarily of metals such as nickel and iron, resides in a central core.
- **Mantle.** Rocky material of moderate density—mostly minerals that contain silicon, oxygen, and other elements—forms a thick mantle that surrounds the core.
- **Crust.** The lowest-density rock, which includes the familiar rocks of Earth's surface, forms a thin crust, essentially representing the world's outer skin.

**FIGURE 9.2** shows these layers for the five terrestrial worlds. Although not shown in the figure, Earth's metallic core actually consists of two distinct regions: a solid *inner core* and a molten (liquid) *outer core.** Venus may have a similar core structure, but without seismic data we cannot be sure.

We can understand *why* the interiors are layered by thinking about what happens in a mixture of oil and water: Gravity pulls the denser water to the bottom, driving the less dense oil to the top. This process is called **differentiation**, because it results in layers made of *different* materials. The layered interiors of the terrestrial worlds tell us that they underwent differentiation at some time in the past, which means all these worlds must once have been hot enough inside for their interior rock and metal to melt. Dense metals like iron sank toward the center, driving less dense rocky material toward the surface.

Comparing the terrestrial worlds' interiors provides important clues about their early histories. Models indicate that the relative proportions of metal and rock should have been similar throughout the inner solar system at the time the terrestrial planets formed, which means we should

*Because temperature increases with depth in a planet, it may seem surprising that Earth's inner core is solid while its outer core is molten. The inner core is kept solid by the higher pressure at its greater depth, even though the temperature is also higher.

expect smaller worlds to have correspondingly smaller metal cores. We do indeed see this general pattern in Figure 9.2, but the pattern is not perfect: Mercury's core seems surprisingly big, while the Moon's core seems surprisingly small. These surprises are a major reason scientists suspect that giant impacts affected both worlds [**Section 8.2**]. In Mercury's case, a giant impact that blasted away its outer rocky layers while leaving its core intact could explain why the core is so large compared to the rest of the planet. We can explain the Moon's small core by assuming the Moon formed from debris blasted out of Earth's rocky outer layers (see Figure 8.12): The debris would have contained relatively little high-density metal and therefore would have accreted into an object with a very small metal core.

**Layering by Rock Strength** It's often more useful to categorize interior layers by rock strength instead of density. The idea that rock can vary in strength may seem surprising, since we often think of rock as the very definition of strength. However, like all matter built of atoms, rock is mostly empty space; its apparent solidity arises from electrical bonds between its atoms and molecules [**Section 5.3**]. Although these bonds are strong, they can still break and re-form when subjected to heat or sustained stress, which means that even solid rock can slowly deform and flow over millions and billions of years. The long-term behavior of rock is much like that of Silly Putty, which breaks like a brittle solid when you pull it sharply but deforms and stretches when you pull it slowly (**FIGURE 9.3**). Also like Silly Putty, rock becomes softer and easier to deform when it is warmer.

**See it for yourself** Roll room-temperature Silly Putty into a ball and measure its diameter. Put the ball on a table and gently place a heavy book on top of it. After 5 seconds, measure the height of the squashed ball. Repeat the experiment, but warm the Silly Putty in hot water before you start; repeat again, but cool the Silly Putty in ice water before you start. How does temperature affect the rate of "squashing"? How does the experiment relate to planetary geology?

In terms of rock strength, a planet's outer layer consists of relatively cool and rigid rock, called the **lithosphere** (*lithos* is Greek for "stone"), that essentially "floats" on warmer, softer rock beneath. As shown by the dashed circles in Figure 9.2, the lithosphere encompasses the crust and part of the mantle of each world.

## COMMON MISCONCEPTIONS

### Earth Is Not Full of Molten Lava

Many people guess that Earth is full of molten lava (or *magma*). This misconception may arise partly because we see molten lava emerging from inside Earth when a volcano erupts. However, Earth's mantle and crust are almost entirely solid. The lava that erupts from volcanoes comes from a narrow region of partially molten material beneath the lithosphere. The only part of Earth's interior that is fully molten is the outer core, which lies so deep within the planet that its material never erupts directly to the surface.

**FIGURE 9.3** Silly Putty stretches when pulled slowly but breaks cleanly when pulled rapidly. Rock behaves just the same way, but on a longer time scale.

Notice that lithospheric thickness is closely related to a world's size: Smaller worlds tend to have thicker lithospheres. In fact, while Earth and Venus have lithospheres that extend only a short way into their upper mantles, the smaller worlds (Mars, Mercury, and the Moon) have lithospheres that extend nearly to their cores. The thickness of the lithosphere is very important to geology. A thin lithosphere is brittle and can crack easily. A thick lithosphere is much stronger and inhibits the passage of molten rock from below, making volcanic eruptions and the formation of mountain ranges less likely.

**Why Big Worlds Are Round** The fact that rock can deform and flow also explains why large worlds are spherical while small moons and asteroids are "potato-shaped." The weak gravity of a small object is unable to overcome the rigidity of its rocky material, so the object retains the shape it had when it was born. For a larger world, gravity can overcome the strength of solid rock, slowly deforming and molding it into a spherical shape. Gravity will make any rocky object bigger than about 500 kilometers in diameter into a sphere within about 1 billion years. Larger worlds become spherical more quickly, especially if they are molten (or gaseous) at some point in their history.

## What causes geological activity?

The most interesting aspects of planetary geology are those that cause the surfaces of the terrestrial worlds to change with time. We use the term **geological activity** to describe ongoing changes. Earth is the most geologically active of the terrestrial worlds, with a surface continually being reshaped by volcanic eruptions, earthquakes, erosion, and other geological processes.

Most geological activity is driven by internal heat. For example, volcanoes can erupt only if the interior is hot enough to melt at least some rock into molten lava. But what makes some planetary interiors hotter than others? To find the answer, we must investigate how interiors heat up and cool off. As we'll see, we can ultimately trace a planet's internal heat, and hence its geological activity, back to its size.

**How Interiors Get Hot** A hot interior contains a lot of thermal energy, and the law of conservation of energy tells us that this energy must have come from somewhere [**Section 4.3**]. Although you might first guess that the Sun would be the heat source, this is not the case: Sunlight is the primary heat source for the *surfaces* of the terrestrial worlds, but virtually none of this solar energy penetrates more than a few meters into the ground. Internal heat is a product of the planets themselves, not of the Sun. Three sources of energy explain nearly all the interior heat of the terrestrial worlds (**FIGURE 9.4**):

- **Heat of accretion.** Accretion deposits energy brought in from afar by colliding planetesimals. As a planetesimal approaches a forming planet, its gravitational potential energy is converted to kinetic energy, causing it to accelerate. Upon impact, much of the kinetic energy is converted to heat, adding to the thermal energy of the planet.
- **Heat from differentiation.** When a world undergoes differentiation, the sinking of dense material and rising of less-dense material means that mass moves inward, losing gravitational potential energy. This energy is converted to thermal energy by the friction generated as materials separate by density. The same thing happens when you drop a brick into a pool: As the brick sinks to the bottom, friction with the surrounding water heats the pool—though the amount of heat from a single brick is too small to be noticed.
- **Heat from radioactive decay.** The rock and metal that built the terrestrial worlds contained radioactive isotopes of elements such as uranium, potassium, and thorium. When radioactive nuclei decay, subatomic particles fly off at high speeds, colliding with neighboring atoms and heating them. In essence, this converts some of the mass-energy ($E = mc^2$) of the radioactive nuclei to the thermal energy of the planetary interior.

## SPECIAL TOPIC How Do We Know What's Inside Earth?

Our deepest drills have barely pricked Earth's surface, penetrating much less than 1% of the way into the interior. How, then, can we claim to know what our planet is like on the inside? Much of our information comes from *seismic waves,* vibrations created by earthquakes.

Seismic waves come in two basic types, analogous to the two ways you can generate waves in a Slinky (**FIGURE 1**). Pushing and pulling on one end of a Slinky (while someone holds the other end still) generates a wave in which the Slinky is bunched up in some places and stretched out in others. Waves like this in rock are called P waves. The *P* stands for *primary,* because these waves travel fastest and are the first to arrive after an earthquake, but it is easier to think of *P* as meaning *pressure* or *pushing.* P waves can travel through almost any material—whether solid, liquid, or gas—because molecules can always push on their neighbors no matter how weakly they are bound together. (Sound travels as a pressure wave quite similar to a P wave.)

Shaking a Slinky slightly up and down or side to side generates a different type of wave; in rock, such waves are called S waves. The *S* stands for *secondary* but is easier to remember as meaning *shear* or *side to side.* S waves travel only through solids, because the bonds between neighboring molecules in a liquid or gas are too weak to transmit up-and-down or sideways forces.

The speeds and directions of seismic waves depend on the composition, density, pressure, temperature, and phase (solid or liquid) of the material they pass through. For example, P waves reach the side of the world opposite an earthquake, but S waves do not. This tells us that a liquid layer has stopped the S waves, which is how we know that Earth has a liquid outer core (**FIGURE 2**). More careful analysis of seismic waves, combined with mathematical modeling, has allowed geologists to develop a detailed picture of Earth's interior structure.

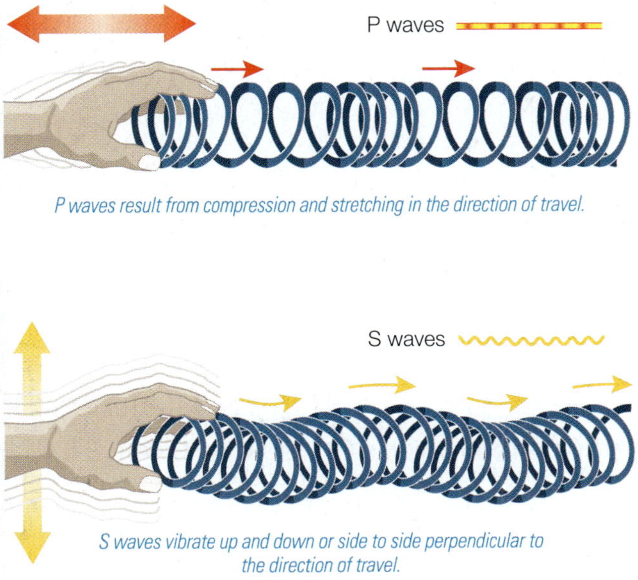


**FIGURE 1** Slinky examples demonstrating P and S waves.

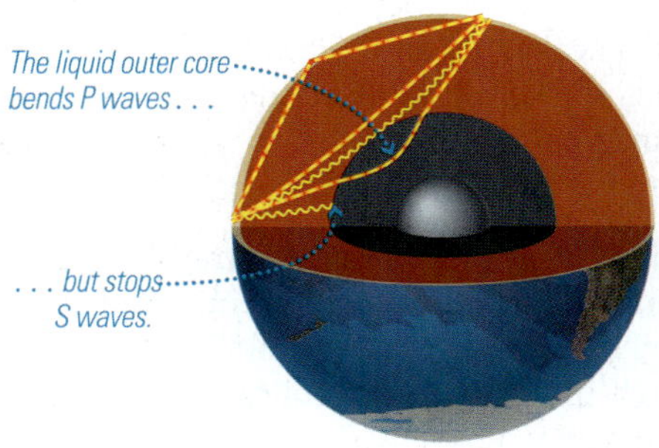


**FIGURE 2** Because S waves do not reach the side of Earth opposite an earthquake, we infer that part of Earth's core is liquid.

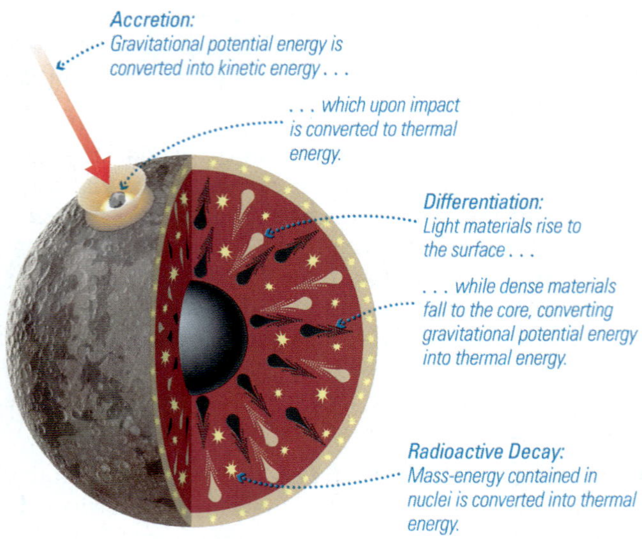


**FIGURE 9.4** The three main heat sources for terrestrial planet interiors are accretion, differentiation, and radioactive decay. Only the last is still a major heat source today.

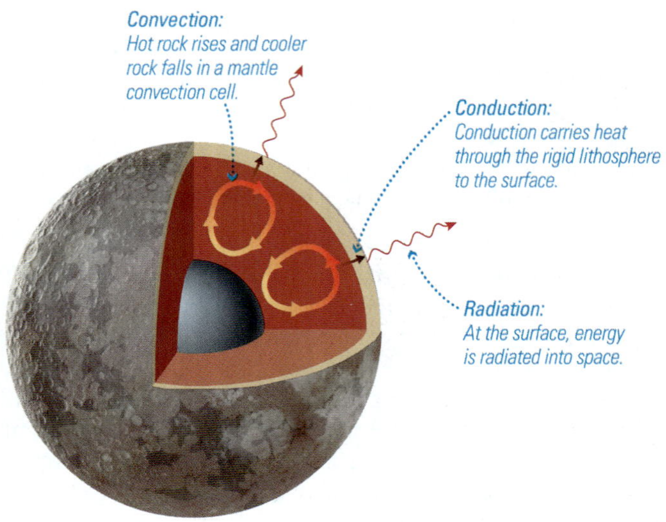


**FIGURE 9.5** The three main cooling processes in planets. Convection can occur only in a planet that is still hot inside.

Note that accretion and differentiation deposited heat into planetary interiors only when the planets were very young. In contrast, radioactive decay provides an ongoing source of heat. Over billions of years, the total amount of heat deposited by radioactive decay has been comparable to or greater than the amount that was deposited initially by accretion and differentiation. However, the rate of radioactive decay has declined with time, so it was an even more significant heat source when the planets were young than it is today.

The combination of the three heat sources explains how the terrestrial interiors ended up with their core-mantle-crust structures. The many violent impacts that occurred during the latter stages of accretion deposited so much energy that the outer layers of the young planets began to melt. This started the process of differentiation, which then released its own additional heat. This heat, along with the substantial heat from early radioactive decay, made the interiors hot enough to melt and differentiate throughout.

**How Interiors Cool Off** Cooling a planetary interior requires transporting heat outward, which also occurs through three basic processes (**FIGURE 9.5**):

- **Convection.** Convection is the process by which hot material expands and rises while cooler material contracts and falls, thereby transporting heat upward; it can occur whenever there is strong heating from below. You can see convection in a pot of soup on a hot burner, and you may be familiar with it in weather: Warm air near the ground tends to rise while cool air above tends to fall.
- **Conduction.** Conduction is the transfer of heat from hot material to cooler material through contact; it is operating when you touch a hot object. Conduction occurs through the microscopic collisions of individual atoms or molecules when two objects are in close contact, because the faster-moving molecules in the hot material tend to transfer some of their energy to the slower-moving molecules of the cooler material.
- **Radiation.** Recall that objects emit *thermal radiation* characteristic of their temperatures [**Section 5.4**]; this radiation (light) carries energy away and therefore cools an object. Planets lose heat to space through radiation; because of their relatively low temperatures, planets radiate primarily in the infrared.

For Earth, convection is the most important heat transfer process in the interior. Hot rock from deep in the mantle gradually rises, slowly cooling as it makes its way upward. By the time it reaches the top of the mantle, the rock has transferred its excess heat to its surroundings, making it cool enough that it begins to fall. This ongoing process creates individual **convection cells** within the mantle, shown as small circles in Figure 9.5; arrows indicate the direction of flow. Keep in mind that mantle convection primarily involves *solid* (not molten) rock, which flows very slowly. At the typical rate of mantle convection on Earth—a few centimeters per year—it would take about 100 million years for a piece of rock to be carried from the base of the mantle to the top.

## COMMON MISCONCEPTIONS

### Pressure and Temperature

Some people guess that Earth's interior should be hot just because the internal pressure is so high. After all, if we compress a gas from low pressure to high pressure, it heats up. But the same is not necessarily true of rock. High pressure compresses rock only slightly, so the compression causes little increase in temperature. Although high pressures and temperatures sometimes go together in planets, they don't have to. In fact, after all the radioactive elements decay (billions of years from now), Earth's deep interior will become quite cool even though the pressure will be the same as it is today. The temperatures inside Earth and the other planets can remain high only if there is a source of heat for the interior, such as accretion, differentiation, or radioactive decay.

Mantle convection stops at the base of the lithosphere, because the rigid rock of the lithosphere does not flow as readily as the rock deeper down. Within the lithosphere, heat continues upward primarily through conduction. (Volcanic eruptions also carry heat through the lithosphere, since hot rock from the interior erupts onto the surface.) The heat then escapes to space through radiation.

**Planetary Size Controls Geological Activity** Just as a hot potato remains hot inside much longer than a hot pea, a large planet can stay hot inside much longer than a small one. You can see why size is the critical factor by picturing a large planet as a smaller planet wrapped in extra layers of rock. The extra rock acts as insulation, so it takes much longer for interior heat to reach the surface.

**See it for yourself** It's easy to demonstrate that larger objects take longer to cool than smaller objects. The next time you are eating something large and hot, cut off a small piece and notice how much more quickly the small piece cools than the rest of it. A similar experiment demonstrates the time it takes a cold object to warm up: Find two ice cubes of the same size, but crack one into small pieces and then compare the rates of melting. Explain your observations.

Size is therefore the primary factor in determining geological activity. The relatively small sizes of the Moon and Mercury allowed their interiors to cool significantly within a billion years or so after they formed. This cooling caused their lithospheres to thicken and confined mantle convection to deeper and deeper layers until it stopped altogether. As a result, the Moon and Mercury are now essentially "dead" geologically, meaning they have little if any heat-driven geological activity.

In contrast, the much larger size of Earth has allowed our planet to stay quite hot inside. Mantle convection keeps interior rock in motion and the heat keeps the lithosphere thin, which is why geological activity can continually reshape the surface. Venus probably remains nearly as active as Earth, thanks to its very similar size. Mars, with a size in between those of the other terrestrial worlds, probably represents an intermediate case: It has cooled significantly during its history, but it probably still retains enough internal heat for at least some geological activity.

**Think about it** In general, which among a group of planets would you expect to have the *thickest* lithosphere: (a) the largest planet, (b) the smallest planet, (c) the planet closest to the Sun, or (d) the planet farthest from the Sun? Why? (Assume the planet has the same composition in all four cases.)

## Why do some planetary interiors create magnetic fields?

Interior heat plays another important role: It can help create a global **magnetic field**. Earth's magnetic field is best known for determining the direction in which a

## MATHEMATICAL INSIGHT 9.1 The Surface Area–to–Volume Ratio

The total amount of heat contained in a planet depends on its volume, but this heat can escape into space only from its *surface*. As heat escapes, more heat flows upward from the interior to replace it until the interior is no hotter than the surface. The time it takes a planet to lose its internal heat is related to the ratio of the *surface area* through which it loses heat to the *volume* that contains heat, or the *surface area–to–volume ratio*:

$$\text{surface area–to–volume ratio} = \frac{\text{surface area}}{\text{volume}}$$

A spherical planet (radius $r$) has surface area $4\pi r^2$ and volume $\frac{4}{3}\pi r^3$, so the ratio becomes

$$\underset{\text{(for a sphere)}}{\text{surface area–to–volume ratio}} = \frac{4\pi r^2}{\frac{4}{3}\pi r^3} = \frac{3}{r}$$

Because $r$ appears in the denominator, we conclude that *larger objects have smaller surface area–to–volume ratios*. Note that this idea holds for objects of any shape, which is why the larger of two objects that start at the same temperature retains heat longer. It also explains why crushed ice cools a drink more quickly than an equal amount of ice cubes: The crushed ice pieces have a greater combined surface area than the ice cubes, which means faster cooling because more ice surface is in direct contact with the surrounding liquid.

**EXAMPLE:** Compare and interpret the surface area–to–volume ratios of the Moon and Earth.

**SOLUTION:**

**Step 1 Understand:** We compare the two surface area–to–volume ratios by dividing the Moon's ratio (which is larger, because the Moon is smaller) by Earth's.

**Step 2 Solve:** We've already found that the surface area–to–volume ratio for a spherical world is $3/r$. Therefore, we find

$$\frac{\text{surface area–to–volume ratio (Moon)}}{\text{surface area–to–volume ratio (Earth)}} = \frac{3/r_{\text{Moon}}}{3/r_{\text{Earth}}} = \frac{r_{\text{Earth}}}{r_{\text{Moon}}}$$

From Appendix E, the radii are $r_{\text{Moon}} = 1738$ km and $r_{\text{Earth}} = 6378$ km. Substituting these values into our equation, we find

$$\frac{\text{surface area–to–volume ratio (Moon)}}{\text{surface area–to–volume ratio (Earth)}} = \frac{6378\text{ km}}{1738\text{ km}} = 3.7$$

**Step 3 Explain:** The Moon's surface area–to–volume ratio is nearly four times as large as Earth's, which means the Moon would cool four times as fast if both worlds started with the same temperature and gained no additional heat. (In fact, Earth's larger size gave it much more internal heat to begin with, which amplifies the difference in heat loss found from the surface area–to–volume ratio alone.)

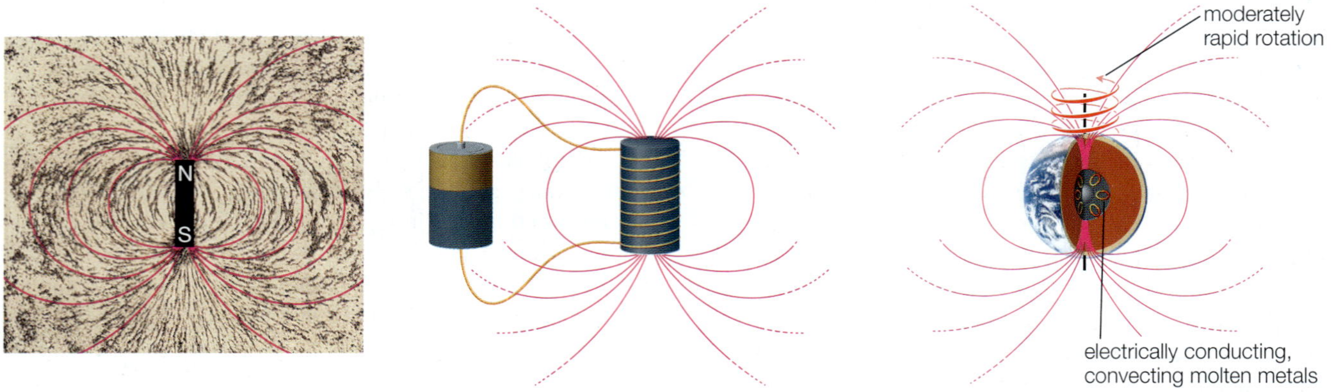


**a** This photo shows how a bar magnet influences iron filings (small black specks) around it. The *magnetic field lines* (red) represent this influence graphically.

**b** A similar magnetic field is created by an electromagnet, which is essentially a wire wrapped around a metal bar and attached to a battery. The field is created by the battery-forced motion of charged particles (electrons) along the wire.

**c** Earth's magnetic field also arises from the motion of charged particles. The charged particles move within Earth's liquid outer core, which is made of electrically conducting, convecting molten metals.

**FIGURE 9.6** Sources of magnetic fields.

compass needle points, but as we'll discuss in Chapter 10, it also creates a *magnetosphere* (see Figure 10.11) that surrounds our planet and diverts the paths of high-energy charged particles coming from the Sun. The magnetic field therefore protects Earth's atmosphere from being stripped away into space by these particles; many scientists suspect that this protection has been crucial to the long-term habitability of Earth, and hence to our own existence.

You are probably familiar with the pattern of the magnetic field created by an iron bar (**FIGURE 9.6a**). Earth's magnetic field is generated by a process more similar to that of an *electromagnet,* in which the magnetic field arises as a battery forces charged particles (electrons) to move along a coiled wire (**FIGURE 9.6b**). Earth does not contain a battery, but charged particles move with the molten metal in its liquid outer core (**FIGURE 9.6c**). Internal heat causes the liquid metal to rise and fall (convection), while Earth's rotation twists and distorts the convection pattern. The result is that electrons in the molten metal move within Earth's outer core in much the same way as in an electromagnet, generating Earth's magnetic field.

We can generalize these ideas to other worlds. There are three basic requirements for a global magnetic field:

1. An interior region of electrically conducting fluid (liquid or gas), such as molten metal
2. Convection in that layer of fluid
3. At least moderately rapid rotation

Earth is the only terrestrial world that meets all three requirements, which is why it is the only terrestrial world with a strong global magnetic field. The Moon has no global magnetic field, presumably because its core has long since cooled and ceased convecting. Mars's core probably still retains some heat, but not enough to drive core convection, which is why it also lacks a global magnetic field today. Venus probably has a molten core layer much like that of Earth, but either its convection or its 243-day rotation period is too slow to generate a magnetic field. Mercury remains an enigma: It possesses a measurable magnetic field despite its small size and slow 59-day rotation. The reason for this may be Mercury's huge metal core, which is apparently still partially molten and convecting.

The same three requirements for a magnetic field also apply to jovian planets and stars. For example, Jupiter's strong magnetic field comes from its rapid rotation and its layer of convecting metallic hydrogen that conducts electricity [**Section 11.1**]. The Sun's magnetic field is generated by the combination of convection of ionized gas (plasma) in its interior and rotation.

**Think about it** Recall that the Sun had a strong solar wind when it was young (which cleared out the remaining gas of the solar nebula [**Section 8.2**]) because its magnetic field was much stronger than it is today. What single factor explains why the young Sun's magnetic field was so strong?

## 9.2 Shaping Planetary Surfaces

We now turn our attention to the surfaces of the terrestrial worlds. These surfaces are remarkably flat. For example, the tallest mountains on Earth rise only about 10 kilometers above sea level, less than 1/600 of Earth's radius—which means they could be represented by grains of sand on a typical globe. Nevertheless, surface features tell us a great deal about the histories of the planets.

**See it for yourself** Find a globe of Earth that shows mountains in raised relief (represents mountains with bumps). Are the mountain heights correctly scaled relative to Earth's size? Explain.

**FIGURE 9.7** Artist's conception of the impact process.

Four Geological Processes

## What processes shape planetary surfaces?

Although we find a huge variety of geological surface features on Earth and the other terrestrial worlds, nearly all of them can be explained by just four major geological processes:

- **Impact cratering:** the creation of bowl-shaped *impact craters* by asteroids or comets striking a planet's surface
- **Volcanism:** the eruption of molten rock, or *lava,* from a planet's interior onto its surface
- **Tectonics:** the disruption of a planet's surface by internal stresses
- **Erosion:** the wearing down or building up of geological features by wind, water, ice, and other phenomena of planetary weather

Let's examine each of these processes in a little more detail.

**Impact Cratering** An impact crater forms when one of the leftover planetesimals from the solar system's formation—meaning an asteroid or comet—slams into a solid surface (**FIGURE 9.7**). An impacting object typically hits the surface at a speed between about 40,000 and 250,000 km/hr (10–70 km/s). At such tremendous speed, the impact releases enough energy to vaporize solid rock and blast out a *crater* (the Greek word for "cup"). Debris from the blast shoots high above the surface and then rains down over a large area. If the impact is large enough, some of the ejected material can escape into space. Small craters far outnumber large ones, demonstrating that many more small planetesimals formed during the birth of the solar system (and more small leftovers still orbit the Sun today) than large ones.

Craters are usually circular, because an impact blasts out material in all directions, regardless of the incoming object's direction. Laboratory experiments suggest that craters are typically about 10 times as wide as the objects that create them and about 10–20% as deep as they are wide. For example, an asteroid 1 kilometer in diameter will blast out a crater about 10 kilometers wide and 1–2 kilometers deep. A large crater may have a central peak, which forms when the center rebounds after impact in much the same way that water rebounds after you drop a pebble into it. **FIGURE 9.8** shows two typical impact craters, one on Earth and one on the Moon.

Details of crater shapes provide clues about geological conditions. For example, Figure 9.9 contrasts three craters

**a** Meteor Crater in Arizona is more than a kilometer across and almost 200 meters deep. It was created around 50,000 years ago by the impact of a metallic asteroid about 50 meters across.

**b** This photo shows a crater, named Tycho, on the Moon. Note the classic shape and central peak.

**FIGURE 9.8** Impact craters.

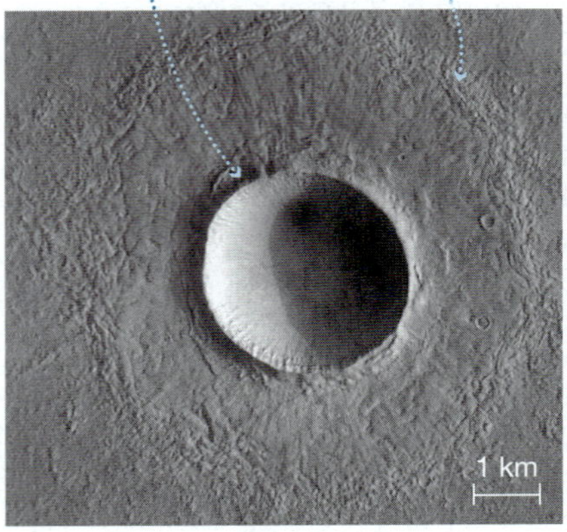


a A crater with a typical bowl shape.

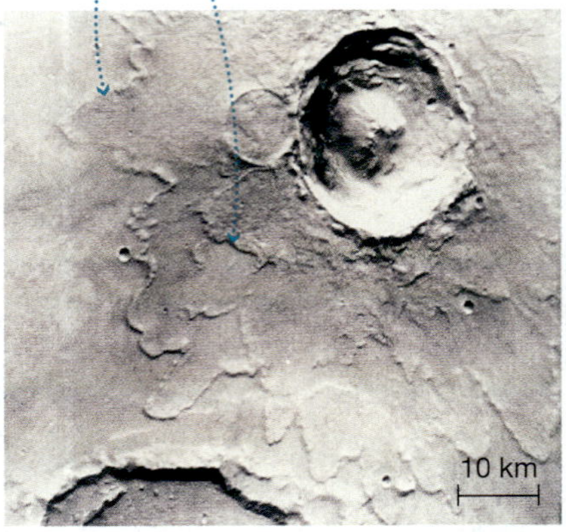


b This crater was probably made by an impact into icy ground.

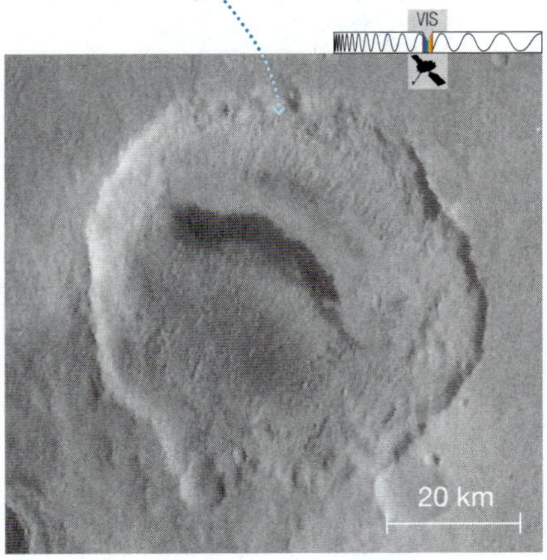


c This crater shows evidence of erosion.

**FIGURE 9.9** Crater shapes on Mars tell us about Martian geology. (Photo a from *Mars Global Surveyor*; photos b and c from *Viking* spacecraft.)

on Mars. The crater in **FIGURE 9.9a** has a simple bowl shape, as we expect from the basic cratering process. The crater in **FIGURE 9.9b** has an extra large bump in its center and appears to be surrounded by mud flows, suggesting that underground water (or ice) melted or vaporized on impact; the muddy debris then flowed across the surface and hardened into the pattern we see today. The crater in **FIGURE 9.9c** shows obvious signs of erosion: It lacks a sharp rim and its floor no longer has a well-defined bowl shape. This suggests that ancient rainfall eroded the crater and that the crater bottom was once a lake. Studies of crater shapes on other worlds provide similar clues to their surface conditions and history.

**FIGURE 9.10** Volcanism. This photo shows the eruption of an active volcano on the flanks of Kilauea on the Big Island of Hawaii. The inset shows the underlying process: Molten rock collects in a chamber (often called a "magma chamber") from which it can erupt upward.

**Volcanism** Volcanism occurs when underground molten rock finds a path to the surface (**FIGURE 9.10**). Molten rock tends to rise for three main reasons. First, it's generally less dense than solid rock, and lower-density materials tend to rise when surrounded by higher-density materials. Second, the solid rock surrounding a chamber of molten rock can squeeze the molten rock, driving it upward under pressure. Third, molten rock often contains trapped gases that expand as it rises, which can make it rise much faster and lead to dramatic eruptions.

The result of an eruption depends on how easily the molten rock, or *lava,* flows across the surface. Lava that is "runny" can flow far before cooling and solidifying, while "thick" lava tends to collect in one place.* Broadly speaking, lava can shape three different types of volcanic features depending on whether it's thick or runny (Figure 9.11):

- The thickest lavas cannot flow far before solidifying and therefore build tall, steep-sided volcanoes (technically called *stratovolcanoes*) (**FIGURE 9.11a**). Examples include Mount Fuji (Japan), Mount Kilimanjaro (Tanzania), and Mount Hood (Oregon). The thick lava can allow particularly great pressure to build up in these volcanoes, making them prone to explosive eruptions that can deposit debris over vast areas.

*The "thickness" of a liquid is described by the technical term *viscosity*, which refers to how much internal friction it has. For example, liquid water has a low viscosity that allows it to flow quickly and easily; honey and molasses flow more slowly because they have higher viscosities.

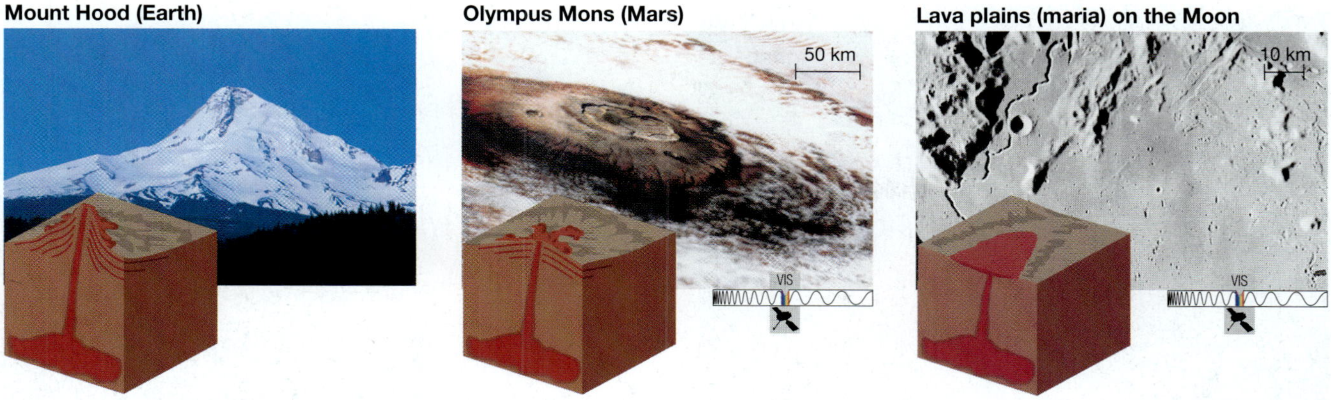


**a** The thickest lavas make steep-sloped volcanoes like Oregon's Mount Hood.

**b** Slightly runnier lava makes shallow-sloped volcanoes, such as Olympus Mons on Mars.

**c** Very runny lava makes flat volcanic plains like these on the Moon. The long, winding channel near the upper left was made by a river of molten lava.

**FIGURE 9.11** The shapes of volcanic structures depend on whether the lava is thick or runny.

- Somewhat runnier lavas can spread some distance before they solidify, creating volcanoes that slope more gradually (technically called *shield volcanoes*). These volcanoes can be very tall but are not very steep. Examples include the mountains of the Hawaiian Islands on Earth and Olympus Mons on Mars (**FIGURE 9.11b**).
- The runniest lavas flow far and flatten out before solidifying, creating vast *volcanic plains* (**FIGURE 9.11c**). Examples include lava plains that make up the Columbia Plateau in Washington and Oregon and vast lava plains that make smooth regions (*maria*) on the Moon.

**See it for yourself** Your kitchen and bathroom probably contain jars, bottles, and tubes with materials that are runny, thick, or in between. By squeezing or pouring these materials, can you match the shapes of the three volcanic features in Figure 9.11?

All the terrestrial worlds (and at least one jovian moon) show evidence of volcanic plains and shallow-sloped (shield) volcanoes, all of which appear to be made of the fairly dense rock type known as *basalt,* which makes up most of Earth's sea floor and volcanic islands such as those of Hawaii. Apparently, basaltic lava is common throughout the solar system. Steep volcanoes are made by lower-density rock that erupts as a thicker lava; this type of lava is common on Earth but rare in the rest of the solar system.

Besides creating volcanic mountains and plains, volcanism can have an even more profound effect. Recall that the terrestrial planets accreted from rocky and metallic planetesimals, with water and other ices brought in only when planetesimals from more distant reaches of the solar system crashed into the growing planets [**Section 8.2**]. The water and gases from these planetesimals became trapped in the interiors of the planets in much the same way that the gas in a carbonated beverage is trapped in a pressurized bottle. Volcanic eruptions can release this gas in a process known as **outgassing**, which occurs both with dramatic volcanic eruptions and with the more gradual escape of gas from volcanic vents (**FIGURE 9.12**). Virtually all the gas that made the atmospheres of Venus, Earth, and Mars—and the water vapor that rained down to form Earth's oceans—originally was released from the planetary interiors by outgassing.

**Tectonics** The term *tectonics* comes from the Greek word *tekton,* for "builder"; notice the same root in the word *architect,* which means "master builder." In geology, tectonics refers to the "building" of surface features by stretching, compression, or other forces acting on the lithosphere. For example, the weight of a volcano can bend or crack the lithosphere beneath it, while a rising plume of hot material can push up on the lithosphere to create a bulge.

Tectonic activity usually goes hand in hand with volcanism, because both require internal heat, and most tectonic activity is a direct or indirect result of mantle convection (**FIGURE 9.13**). The crust can be compressed in places where adjacent convection cells push rock together; this type of compression helped create the Appalachian Mountains of the eastern United States. Cracks and valleys form in places where adjacent convection cells pull the crust apart; the right side of Figure 9.13 shows an example from Mars.

While at least some tectonic activity has occurred on every terrestrial world, it has been particularly important on Earth, where the ongoing stress of mantle convection fractured the lithosphere into more than a dozen pieces, or **plates**. These plates move over, under, and around each other in a process we call **plate tectonics**, which we'll discuss further in Section 9.6. (Because the term *plate tectonics* refers to a process, it is generally considered to be singular rather than plural, despite the *-s* on *tectonics.*) As we discuss other planets, it's important to remember the distinction between *tectonics*, the general geological process, and *plate tectonics*, a specific form of tectonics that may be unique to Earth in our solar system.

**Erosion** The last of our four major geological processes is *erosion,* which refers to the breakdown or transport of surface rock through the action of ice, liquid, or gas.

a The eruption of Mount St. Helens, May 18, 1980.

b More gradual outgassing from a volcanic vent on the island of Vulcano near Sicily.

**FIGURE 9.12** Examples of outgassing, which released the gases that ultimately made the atmospheres of Venus, Earth, and Mars.

Familiar features of erosion include valleys carved by glacial ice, canyons carved by rivers, sand dunes shaped by wind, and river deltas made of sediments carried downstream (**FIGURE 9.14**). Note that while we usually associate erosion with breakdown, sand dunes and river deltas are examples of features *built up* by erosion. Indeed, erosion has built much of Earth's surface rock: Over long periods of time, erosion piled sediments into layers on the floors of oceans and seas, forming **sedimentary rock**. The layered rock of the Grand Canyon is an example, built up long before the Colorado River carved the canyon.

## How do impact craters reveal a surface's geological age?

Notice that impact cratering is the only one of the four processes with an external cause: the random impacts of objects from space. This fact leads to one of the most useful insights in planetary geology: *We can estimate the geological age of any surface region from its number of impact craters, with more craters indicating an older surface.* By "geological age" we mean the age of the surface as it *now* appears: A geologically young surface is dominated by features that have formed relatively recently in the history of the solar system, while a geologically old surface still looks about the same today as it did billions of years ago.

To understand the idea, recall that all the planets were battered by impacts during the heavy bombardment that ended almost 4 billion years ago **[Section 8.2]**, but fewer impacts have occurred since. In places where we see numerous craters, such as on much of the Moon's surface, we must be looking at a surface that has stayed virtually unchanged for billions of years. A surface can have fewer craters only if the scars of ancient impacts have been erased over time by other geological processes, such as volcanic

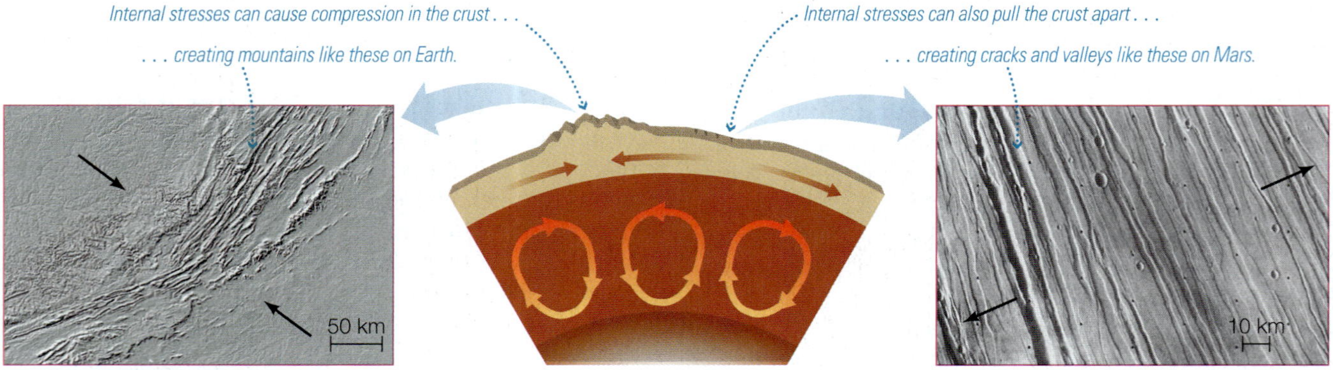


Appalachian Mountains in eastern United States

Ceraunius Valleys on Mars

**FIGURE 9.13** Tectonic forces produce a wide variety of features, including mountain ranges and fractured plains. The Mars image is a visible-light photo from orbit. The Earth image was computer generated from topographical data. (Mars no longer has mantle convection like that shown in the middle figure, but presumably did a few billion years ago before its interior cooled and its lithosphere thickened.)

**a** The Colorado River has been carving the Grand Canyon for millions of years.

**b** Glaciers created Yosemite Valley during ice ages.

**c** Wind erosion wears away rocks and builds up sand dunes.

**d** This river delta is built from sediments worn away by wind and rain and then carried downstream.

**FIGURE 9.14** A few examples of erosion on Earth.

eruptions or erosion. Therefore, surfaces with few craters, such as Earth's surface, must be relatively young.

Careful studies of the Moon have allowed planetary scientists to be more precise about surface ages. The degree of crowding among craters varies greatly from place to place on the Moon (**FIGURE 9.15**). In the **lunar highlands**, craters are so crowded that we see craters on top of other craters. In the **lunar maria**, we see only a few craters on top of generally smooth volcanic plains. Indeed, the maria (Latin for "seas"; singular, *mare*) got their name because they look much like oceans when seen from afar.

Radiometric dating of rocks brought back by the *Apollo* astronauts indicates that those from the lunar highlands are about 4.4 billion years old, telling us that the heavy cratering occurred early in the solar system's history. Rocks from the maria date to 3.0–3.9 billion years ago, telling us that the lava flows that made these volcanic plains had occurred by that time. Because the maria contain only about 3% as many craters as the highlands, we conclude that the heavy bombardment must have ended by about 4 billion years ago, and relatively few impacts have occurred since that time.

In fact, radiometric dating of Moon rocks has allowed scientists to reconstruct the rate of impacts during much of the Moon's history. Because impacts are essentially random events, the same changes in impact rate over time must apply to all the terrestrial worlds. The degree of crater crowding therefore allows us to estimate the geological age of a planetary surface to within a few hundred million years. Although this is nowhere near as accurate as radiometric dating, it has the advantage of requiring nothing more than orbital photos.

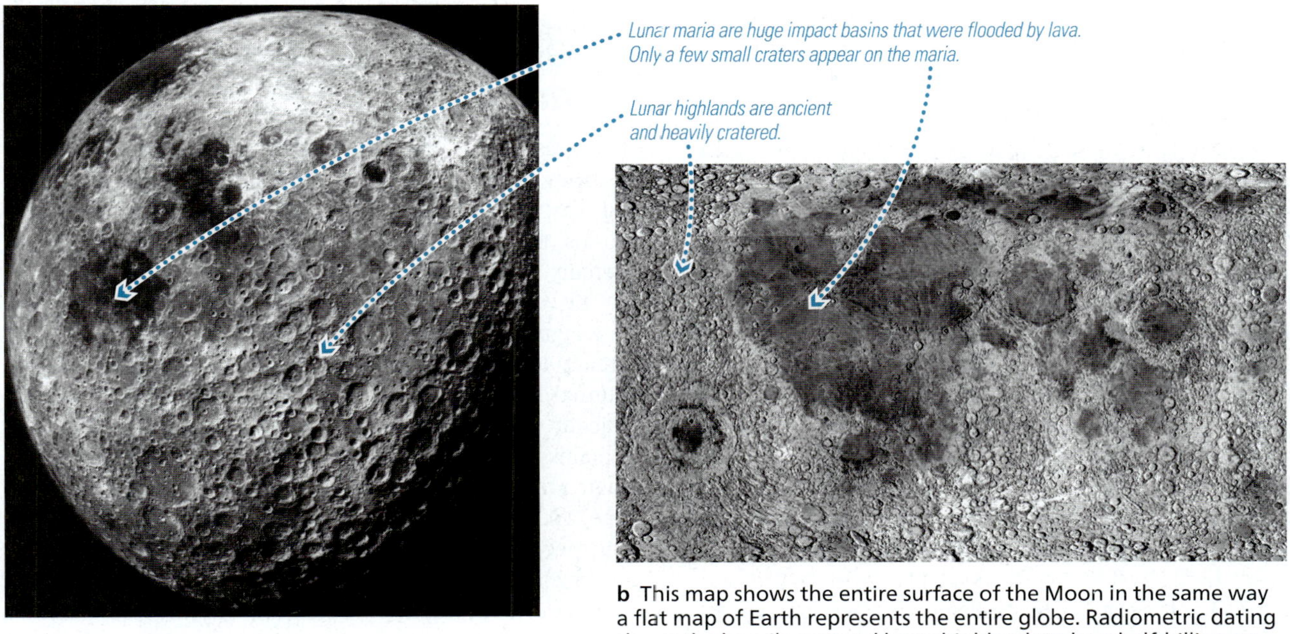


**a** This *Apollo* photograph of the Moon shows that some areas are much more heavily cratered than others. (This view of the Moon is *not* the one we see from Earth.)

**b** This map shows the entire surface of the Moon in the same way a flat map of Earth represents the entire globe. Radiometric dating shows the heavily cratered lunar highlands to be a half-billion or more years older than the darkly colored lunar maria, telling us that the impact rate dropped dramatically after the end of the heavy bombardment.

**FIGURE 9.15** These photos of the Moon show that crater crowding is closely related to surface age.

**Think about it** Mercury is heavily cratered, much like the lunar highlands. Venus has relatively few craters. Which planet has the older surface? Is there any difference between the ages of the two planets as a whole? Explain how the surfaces of Mercury and Venus can have different ages while the planets themselves are the same age.

## Why do the terrestrial planets have different geological histories?

The same four geological processes operate on all solid worlds, but the fact that the different terrestrial worlds show very different geological features tells us that the four processes don't always affect worlds in the same way. For example, Earth has far more features of erosion than any of the other terrestrial worlds. Why do the processes have different effects on different worlds? Ultimately, we can trace the answer back to three fundamental planetary properties: size, distance from the Sun, and rotation rate (**FIGURE 9.16**).

### Planetary Properties Controlling Volcanism and Tectonics

Volcanism and tectonics both require internal heat, which means they depend on planetary size. As shown in Figure 9.16, larger planets have more internal heat and hence more volcanic and tectonic activity.

All the terrestrial worlds probably had some degree of volcanism and tectonics when their interiors were young and hot. As an interior cools, volcanic and tectonic activity subsides. The Moon and Mercury lack ongoing volcanism and tectonics because their small sizes allowed their interiors to cool long ago. Earth has active volcanism and tectonics because it is large enough to still have a hot interior. Venus, nearly the same size as Earth, must still be hot inside and probably also has active volcanism and tectonics. Mars, with its smaller size, has much less volcanism and tectonic activity today than it did in the distant past.

**Planetary Properties Controlling Erosion** Erosion arises from weather phenomena such as wind and rain. Figure 9.16 shows that erosion therefore has links with all three fundamental planetary properties. Planetary size is important because erosion requires an atmosphere, and a terrestrial world can have an atmosphere only if it is large enough to have had significant volcanic outgassing and if its gravity is strong enough to have prevented the gas from escaping to space. Distance from the Sun is important because of its role in temperature: If all else is equal (such as planetary size), the higher temperatures on a world closer to the Sun will make it easier for atmospheric gases to escape into space, while the colder temperatures on a world farther from the Sun may cause atmospheric gases to freeze out. Distance is also important because water erosion is much more effective with liquid water than with water vapor or ice and therefore is strongest when moderate temperature allows water vapor to condense into liquid form. Rotation rate is important because it is the primary driver of winds and other weather [**Section 10.2**]: Faster rotation means stronger winds and storms.

The Moon and Mercury lack significant atmospheres and erosion because they lack outgassing today, and any atmospheric gases they had in the distant past have been lost to space. Mars has limited erosion today; it has only a thin atmosphere because much of the water vapor and carbon dioxide outgassed in its past either escaped into space or lies frozen in its polar caps or beneath the surface. Venus and Earth probably had similar amounts of outgassing, but cooler temperatures on Earth led to condensation and the formation of oceans, allowing wind and weather to drive strong erosion. Most of Venus's gas remained in its atmosphere, making its atmosphere much thicker than Earth's, but Venus has little erosion because its slow rotation rate means that it has very little surface wind and its high temperature means that rain never falls to the surface.

### Planetary Properties Controlling Impact Cratering

Impacts are random events and therefore the *creation* of craters is not "controlled" by fundamental planetary properties. However, because impact craters can be *destroyed* over time, the number of remaining impact craters on a world's surface *is* controlled by fundamental properties. The primary factor is size: Larger worlds have more volcanism and tectonics (and in some cases erosion), processes that tend to cover up or destroy ancient impact craters over time. Mercury and the Moon are heavily cratered today because their small sizes have supported relatively little geological activity, so we can still see most of the impact craters that formed during the heavy bombardment. Mars has had more geological activity and therefore has somewhat fewer impact craters remaining today. The active geologies of Venus and Earth have erased nearly all ancient craters, leaving only those that have formed relatively recently (within the last billion years or so).

# 9.3 Geology of the Moon and Mercury

In the rest of this chapter we will investigate the geological histories of the terrestrial worlds, with the goal of explaining why each ended up as it did. We'll start with the two worlds that have the simplest histories: the Moon and Mercury.

The general appearances of the Moon and Mercury (see Figure 9.1) should make sense based on what you've already learned. Their small sizes mean that most of their internal heat was lost long ago, leaving them without significant ongoing volcanism or tectonics. Small size also explains their lack of significant atmospheres and erosion: Their gravity is too weak to hold gas for long periods of time, and without ongoing volcanism they lack the outgassing needed to replenish gas lost in the past. This overall lack of geological activity means they still retain the scars of ancient impacts, explaining why their surfaces are heavily cratered. Still, both worlds should have had hot interiors early in their histories, which explains why they also show a few features attributable to past volcanic and tectonic activity.

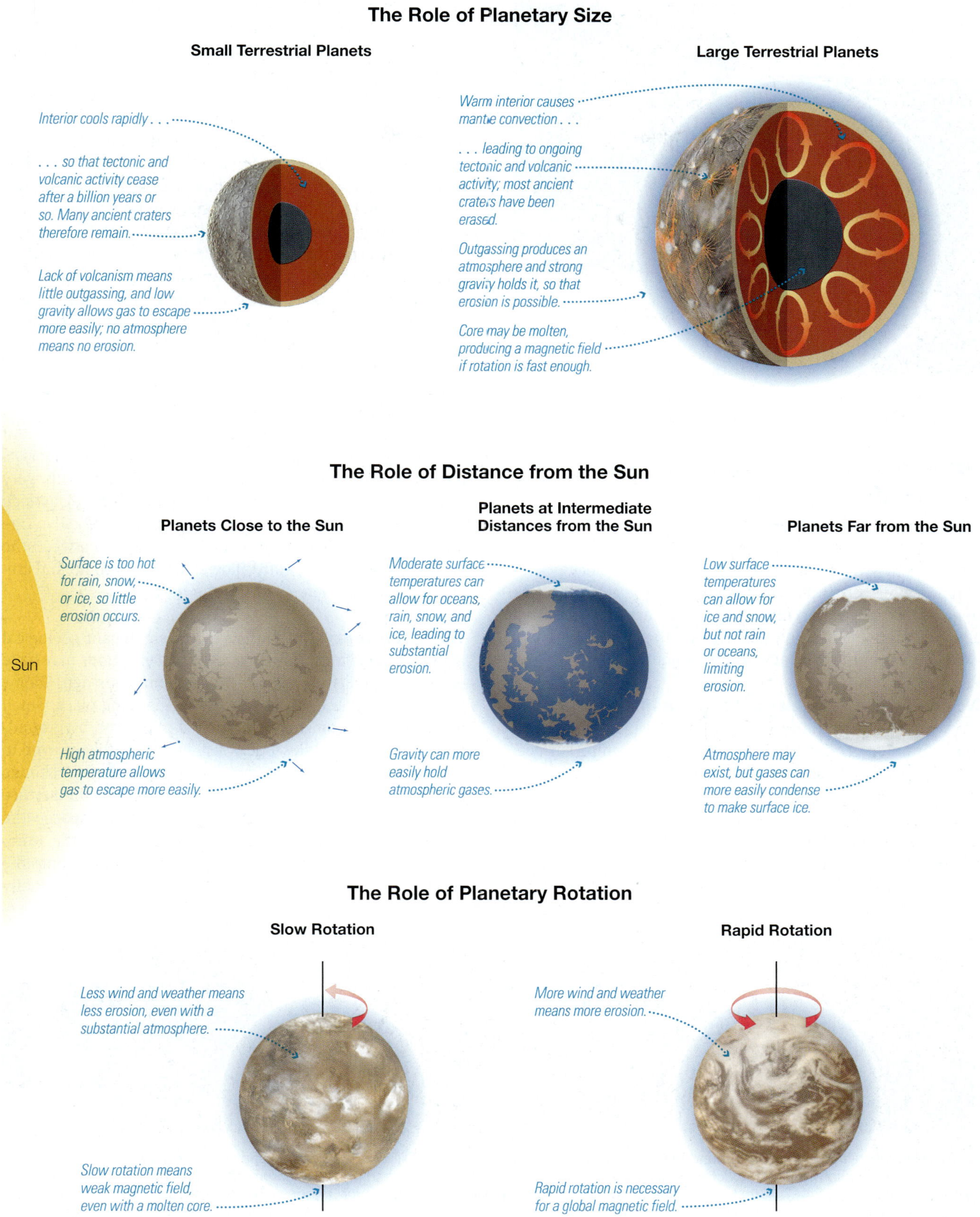


**FIGURE 9.16** A planet's fundamental properties of size, distance from the Sun, and rotation rate are responsible for its geological history. This illustration shows the role of each key property separately, but a planet's overall geological evolution depends on the combination of all these effects.

**FIGURE 9.17** The familiar face of the Moon shows numerous dark smooth maria.

## What geological processes shaped our Moon?

The familiar face of the full moon (**FIGURE 9.17**) reveals some of the surface distinctions we've already discussed. As you saw earlier in Figure 9.15, the bright, heavily cratered regions are lunar highlands and the smooth dark regions are the lunar maria. But why are the maria so smooth, instead of being packed like the lunar highlands with craters from the heavy bombardment?

**Volcanism and Tectonics in the Lunar Maria** **FIGURE 9.18** shows how the maria probably formed. During the heavy bombardment, craters covered the Moon's entire surface. The largest impacts were violent enough to fracture the Moon's lithosphere beneath the huge craters they created. Although there was no molten rock to flood the craters immediately, the ongoing decay of radioactive elements in the Moon's interior ultimately built up enough heat to cause mantle melting between about 3 and 4 billion years ago. Molten rock then welled up through the cracks in the lithosphere, flooding the largest impact craters with lava.

The maria are generally circular because they are flooded craters (and craters are almost always round), and they are dark because the lava consisted of dark, dense, iron-rich rock (basalt). The flat surfaces of the maria tell us that the lunar lava spread easily and far, which means that it must have been among the runniest lava in the solar system. It was so runny that in places it flowed like rivers of molten rock, carving out long, winding channels (see Figure 9.11c). The relative lack of craters within the maria is a consequence of the fact that the lava floods occurred after the heavy bombardment subsided, and relatively few impacts have occurred since that time. The Moon's relatively few tectonic features are also found within the maria. These features, which look much like surface wrinkles, probably were created by small-scale tectonic stresses that arose when the lava that flooded the maria cooled and contracted (**FIGURE 9.19**).

Surprisingly, the Moon's far side looks quite different from the side facing Earth. The far side landscape consists almost entirely of heavily cratered highlands, with very few maria (see Figure 9.15); indeed, the entire far side essentially has a higher altitude than the near side. No one knows what caused this difference in altitude between the two sides, but it apparently allowed lava to well up and create maria much more easily on the near side.

**The Moon Today** The Moon's era of significant geological activity is long gone, though a few small eruptions may have occurred as recently as 50 to 100 million years ago. Today, the Moon is a desolate and nearly unchanging place. Small impacts occur from time to time, as witnessed by ground-based monitoring telescopes, but we are unlikely ever to witness a major one. The only ongoing geological change on the Moon is a very slow "sandblasting" of the surface by *micrometeorites,* sand-size particles from space—the same type of particles that burn up as meteors in the atmospheres of Earth, Venus, and Mars but rain directly onto the surface of the airless Moon. The micrometeorites gradually

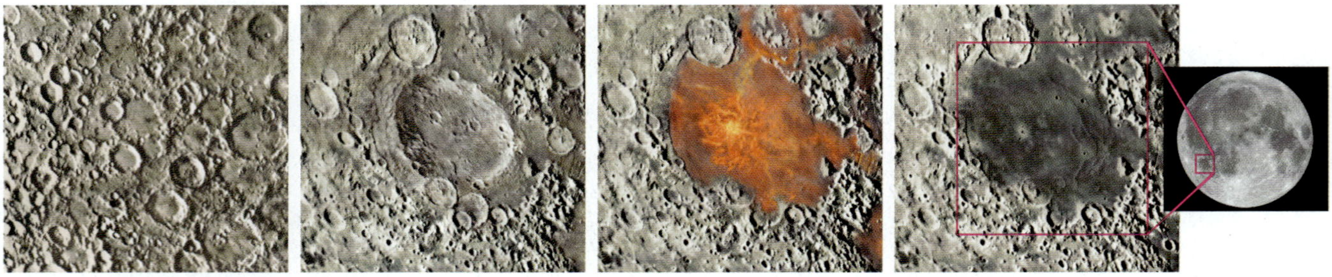

**a** This illustration shows the Mare Humorum region as it probably looked about 4 billion years ago, when it would have been completely covered in craters.

**b** Around that time, a huge impact excavated the crater that would later become Mare Humorum. The impact fractured the Moon's lithosphere and erased the many craters that had existed earlier.

**c** A few hundred million years later, heat from radioactive decay built up enough to melt the Moon's upper mantle. Molten lava welled up through the lithospheric cracks, flooding the impact crater.

**d** This photo shows Mare Humorum as it appears today, and the inset shows its location on the Moon.

**FIGURE 9.18** The lunar maria formed between 3 and 4 billion years ago, when molten lava flooded large craters that had formed hundreds of millions of years earlier. This sequence of diagrams represents the formation of Mare Humorum.

**FIGURE 9.19** The wrinkles within the maria were probably made when the lava that flooded the maria cooled and contracted. (This photo shows Mare Imbrium.)

**a** An astronaut takes the Lunar Roving Vehicle for a spin during the final *Apollo* mission to the Moon (*Apollo 17*, December 1972).

**b** The *Apollo* astronauts left footprints, like this one, in the Moon's powdery "soil." Micrometeorites will eventually erase the footprints, but not for millions of years.

**FIGURE 9.20** The Moon today is geologically dead, but it can still tell us a lot about the history of our solar system.

pulverize the surface rock, which explains why the lunar surface is covered by a thin layer of powdery "soil." You can see this powdery surface in photos from the *Apollo* missions to the Moon (**FIGURE 9.20**). Pulverization by micrometeorites is a very slow process, so the rover tracks and astronaut footprints could remain for millions of years.

Aside from the micrometeorite sandblasting and rare larger impacts, the Moon has been "geologically dead" since the maria formed more than 3 billion years ago. Nevertheless, the Moon remains a prime target of exploration, because study of lunar geology can help us understand both the Moon's history and the history of our solar system. Several robotic missions are currently studying the Moon, and more are planned. In the future, humans may return to the Moon, and lunar resources may someday prove valuable.

**Think about it** It has now been close to 50 years since the last time humans walked on the Moon (in 1972). Do you expect a return of humans to the Moon any time soon? Do you favor such a return? Explain.

## What geological processes shaped Mercury?

Mercury's crater-covered surface looks so much like the Moon's that it's often difficult to tell them apart in photos. Nevertheless, these two worlds have a few important differences.

**Impact Craters and Volcanism on Mercury** Impact craters are visible almost everywhere on Mercury, indicating an ancient surface (**FIGURE 9.21**). However, Mercury's craters are less crowded together in many places than the craters in the most ancient regions of the Moon, suggesting that molten lava later covered up some of the craters that formed on Mercury during the heavy bombardment. Although we have not found evidence of lava flows as large as those that created the lunar maria, the reduced crater crowding and the many smaller lava plains suggest that Mercury had at least as much volcanism as the Moon.

Mercury's largest impact craters, called *basins*, apparently formed from impacts so large and violent that they melted surface rock that then flowed over and filled them. The largest of these, known as *Caloris basin*, spans more than half of Mercury's radius, though it is difficult to see clearly in photographs. The right side of Figure 9.21 shows one of more than a dozen somewhat smaller basins. Many of the basins have few craters within them, indicating that they must have formed at a time when the heavy bombardment was already subsiding. Mercury does not appear to have any features formed by much later lava flows, like those of the lunar maria.

**Tectonic Evidence of Planetary Shrinking** Mercury's most surprising feature is a set of tremendous cliffs that appear to be distributed all over the planet. These cliffs have vertical faces up to 3 or more kilometers high and typically run for hundreds of kilometers across the surface. They probably formed when tectonic forces compressed the crust, causing the surface to crumple. Because crumpling would have shrunk the portions of the surface it affected, Mercury as a whole could not have stayed the same size unless other parts of the surface expanded. However, we find no evidence of large-scale "stretch marks" on Mercury. Can it be that the whole planet simply shrank?

Apparently so. Recall that in addition to being larger than the Moon, Mercury also has a surprisingly large iron core. Mercury therefore gained and retained more internal heat from accretion and differentiation than the Moon, and this heat caused Mercury's core to swell in size. Later, as the core cooled, it contracted by perhaps as much as 20 kilometers in radius (**FIGURE 9.22**). The mantle and lithosphere must have contracted along with the core, generating the

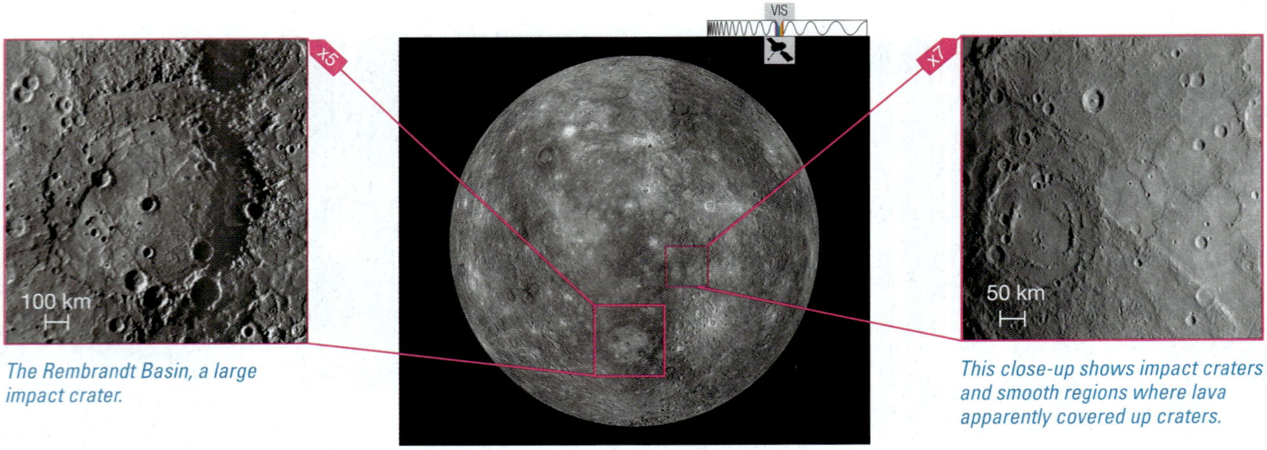


**FIGURE 9.21** Images of Mercury from the *MESSENGER* spacecraft.

tectonic stresses that created the great cliffs. The contraction probably also closed off any remaining volcanic vents, ending Mercury's period of volcanism.

**Mercury Today** Just as Mercury's size and large iron core gave it the heat that led to planetary shrinking, they should also have allowed Mercury to retain internal heat longer than the Moon. Crater counts confirm this idea, suggesting that Mercury remained volcanically and tectonically active (thereby covering up old craters) somewhat longer than the Moon. Nevertheless, Mercury's era of significant volcanic and tectonic activity appears to have ceased within 1 to 2 billion years after its formation. For most practical purposes, Mercury is geologically "dead" today, a fact traceable to its small size.

Although there is no active volcanism on Mercury today, its surface still displays one odd form of activity: Some crater floors appear to be releasing easily vaporized materials from the rock, causing the rock to crumble and make pits nicknamed "hollows" (**FIGURE 9.23**). The release of the vaporized gases leaves behind a light-colored coating, whose composition remains unknown.

Mercury has always seemed an unusual case in our solar system, but the discovery of numerous planets in other solar systems that are rocky and very close to their stars suggests that Mercury may be representative of a class of planets more common elsewhere. Scientists are still learning more about Mercury as they analyze data from the *MESSENGER* mission, which orbited Mercury from 2011 to 2015 (when it crashed to the surface). We should learn even more with the European/Japanese *BepiColombo* mission, which is scheduled to launch in 2018 and then follow a complex path to enter Mercury orbit in 2025.

## 9.4 Geology of Mars

Continuing our tour of the terrestrial worlds in size order from smallest to largest, we now turn to Mars, which has long held a special place in the human imagination. Its bright

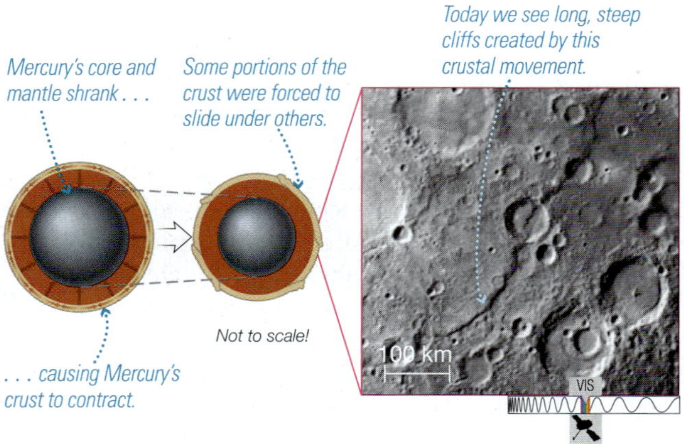


**a** This diagram shows how Mercury's cliffs probably formed as the core shrank and the surface crumpled.

**b** This cliff extends about 100 kilometers in length, and its vertical face is as much as 2 kilometers tall. (Photo from *Mariner 10*.)

**FIGURE 9.22** Long cliffs on Mercury offer evidence that the entire planet shrank early in its history, perhaps by as much as 20 kilometers in radius.

**FIGURE 9.23** The light-colored "hollows" on this crater floor (shown with colors enhanced) are thought to have formed as easy-to-vaporize materials escaped over millions of years, causing the rock to crumble and make hollowed-out pits.

reddish color made it a part of many ancient myths and legends. It inspired even wilder ideas after astronomers noticed some striking similarities to Earth, including a similar axis tilt, a day just slightly longer than 24 hours, the presence of polar caps, and seasonal variations in appearance over the course of the Martian year (about 1.9 Earth years).

Have you ever wondered why people often speak of Martians but rarely of, say, Venusians or Jupiterians? The story goes back to the discoveries of the superficial similarities with Earth (see "Extraordinary Claims: Martians!"), but it got its biggest boost in the 1890s, when Percival Lowell claimed to see complex networks of canals on Mars and argued that the canals were the work of an advanced Martian civilization. Lowell's claims drove rampant speculation about Martians and inspired science fiction fantasies such as H. G. Wells's *The War of the Worlds* (published in 1898). Of course, we now know that no canals actually exist—Lowell's claims appear simply to have been the work of a vivid imagination upon blurry telescopic images (**FIGURE 9.24**)—but Martians have persisted as a cultural meme.

The lack of real Martians may have dampened the enthusiasm of science fiction writers, but scientists find Mars as intriguing as ever. Indeed, although we now know there have never been Martian cities, it's possible that Mars has harbored more primitive life. The search for past or present life on Mars is one of the major goals of Mars exploration, and a first step in this search is understanding Martian geology. To this end we have sent more than two dozen robotic spacecraft to Mars, including more than a half-dozen that have landed on the surface. For the moment, this means that all the known inhabitants of Mars are robots.

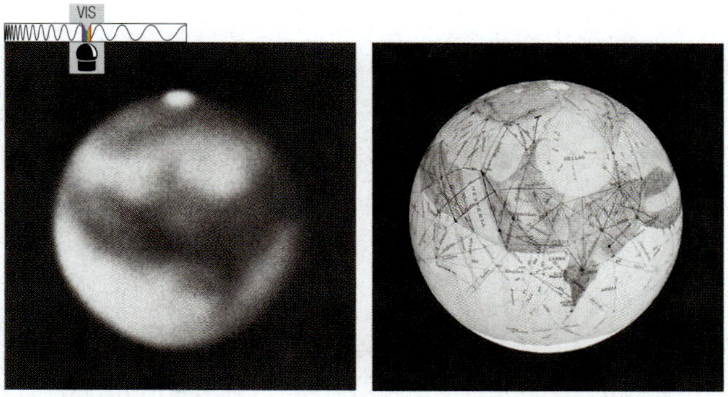


**FIGURE 9.24** The image at left is a telescopic photo of Mars. The image at right is a drawing of Mars made by Percival Lowell. As you can see, Lowell had a vivid imagination. Nevertheless, if you blur your eyes while looking at the photo, you might see how some of the features resemble what Lowell thought he saw.

## What geological processes have shaped Mars?

Based on its intermediate size (see Figure 9.1), we expect Mars to be more geologically active than the Moon or Mercury but less active than Earth or Venus. Observations confirm this basic picture, though Mars's greater distance from the Sun—about 50% farther than Earth—has also played a role in its geological history.

Interestingly, although Mars has only about one-fourth the surface area of Earth, both planets have nearly the same amount of *land* area, because Earth's surface is about three-fourths covered by water. For context, **FIGURE 9.25** shows the full surface of Mars (and the locations of

### EXTRAORDINARY CLAIMS Martians!

The idea that Mars is home to an advanced civilization of Martians certainly qualifies as an extraordinary claim. It also provides a great case study in how carried away people can become with an intriguing idea, even when the evidence is extraordinarily flawed. The story begins with observations of Mars made in the late 18th century by brother and sister astronomers William and Caroline Herschel (best known for discovering the planet Uranus), who discovered the superficial similarities with Earth, which include similar axis tilt, day length, ice caps, and seasonal changes. Taking these resemblances to an unwarranted extreme, by 1784 William Herschel was claiming that Mars was populated by "inhabitants" who enjoyed circumstances "in many ways similar to our own."

The hypothetical Martians got a bigger break about a century later. In 1879, Italian astronomer Giovanni Schiaparelli reported seeing a network of linear features on Mars through his telescope. He named these features *canali*, by which he meant the Italian word for "channels," but it was frequently translated as "canals." Motivated by what sounded like evidence of intelligent life, wealthy American astronomer Percival Lowell built an observatory (still operating in Flagstaff, Arizona) for the study of Mars. He published his first maps of the canals barely a year later, and eventually mapped almost 200 canals (see Figure 9.24). Besides claiming that they were the work of an advanced civilization, he also claimed that they were needed to transport water from the polar caps to drying cities.

Lowell's claims generated great public interest, but his fellow scientists were far more skeptical. They did not see the canals through their own telescopes, and pointed out that Lowell's straight-line canals made no sense, because real canals would follow natural contours of topography (for example, to go around mountains). Nevertheless, much of the public didn't let go of the Martians until 1965, when NASA's *Mariner 4* spacecraft flew past Mars and sent back photos of a barren, cratered surface.

Even then, not everyone gave up. In the late 1970s, NASA's *Viking* orbiters snapped tens of thousands of photographs of Mars, and among them were a few that looked like artifacts of a civilization, including one that resembled a human face. The "face on Mars" became a cottage industry for those who still wished to believe in Martians, but higher-resolution images from later missions confirmed what every scientist already knew: The face and other supposed artifacts of civilization were nothing more than plays of light and shadow, along with the human tendency to see patterns. In the end, the claim of Martian civilization proved to be no more real than the shapes that children see in clouds, or the patterns among stars that led our ancestors to see the mythical gods in the constellations.

**Verdict:** Rejected.

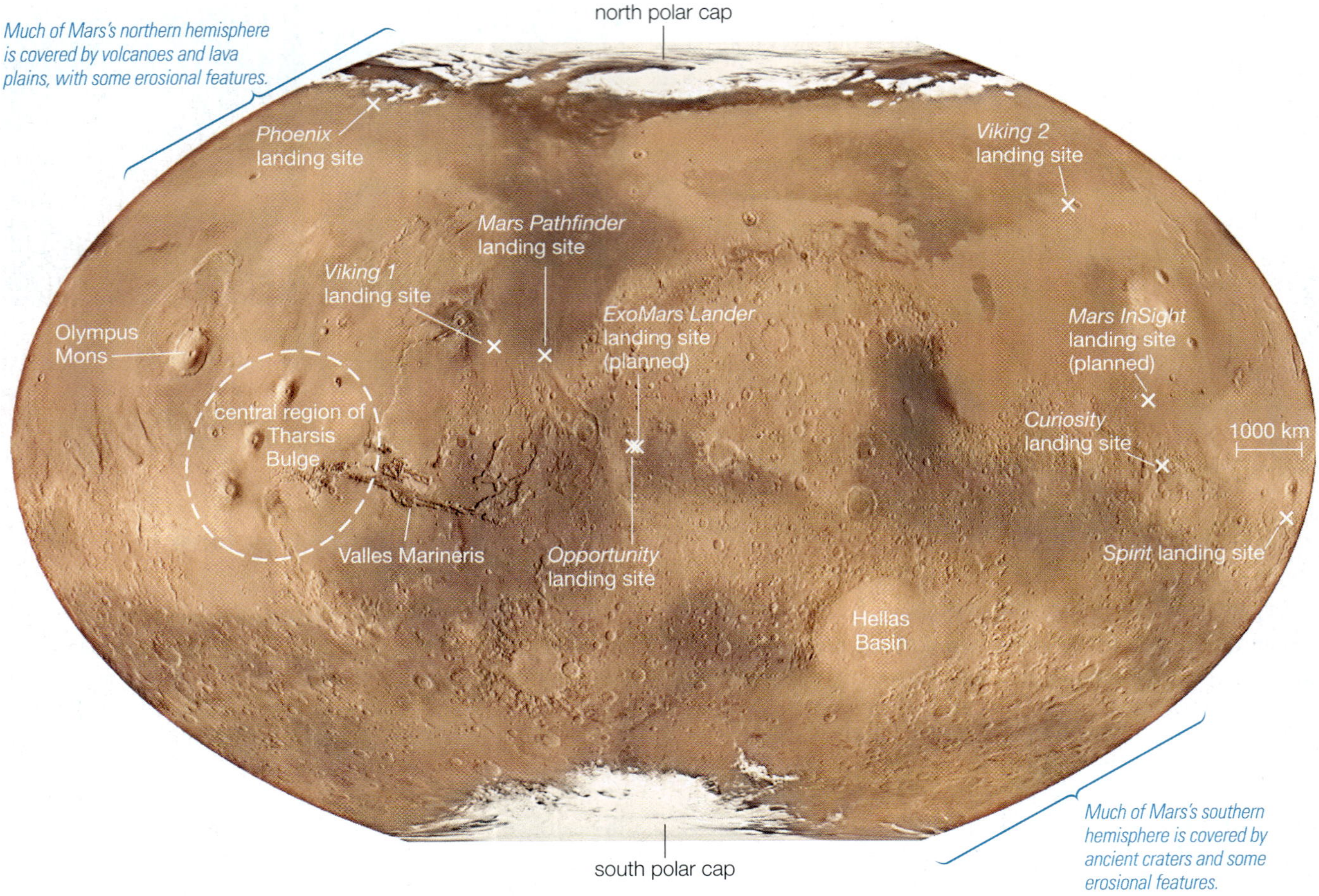


**FIGURE 9.25** This image showing the full surface of Mars is a composite made by combining more than 1000 images with more than 200 million altitude measurements from the *Mars Global Surveyor* mission. Several key geological features are labeled, and landing sites of Mars missions are marked. On the same scale, a map of Earth would be about twice as tall and twice as wide, giving it four times the area, but because Earth's surface is about three-quarters ocean, both planets have nearly the same land area.

particular features we will discuss later), which reveals extensive evidence of all four geological processes.

**Impact Cratering on Mars** Aside from the polar caps, the most striking feature of Figure 9.25 is the dramatic difference in terrain around different parts of Mars. Much of the southern hemisphere has relatively high elevation and is scarred by numerous large impact craters, including the very large crater known as the Hellas Basin. In contrast, the northern plains tend to be below the average Martian surface level and show few impact craters. The differences in cratering tell us that the southern highlands are an older surface than the northern plains, which must have had their early craters erased by other geological processes.

**Think about it** Which fundamental planetary property (size, distance from the Sun, or rotation rate) explains why Earth does not have any terrain that is as heavily cratered as the southern highlands of Mars? Explain.

**Volcanism on Mars** Further study suggests that volcanism was the most important process in erasing craters on the northern plains, although tectonics and erosion also played a part. However, no one knows why volcanism affected the northern plains so much more than the southern highlands or why the two regions differ so much in elevation.

More dramatic evidence of volcanism on Mars comes from several towering but shallow-sloped volcanoes. One of these, Olympus Mons, is the tallest known mountain in the solar system (**FIGURE 9.26**; see also Figure 9.11b). Its base is some 600 kilometers across, large enough to cover an area the size of Arizona. Its peak stands about 26 kilometers above the average Martian surface level, or about three times as high as Mount Everest stands above sea level on Earth. Much of Olympus Mons is rimmed by a cliff that in places is 6 kilometers high.

Olympus Mons and several other large volcanoes are concentrated on or near the continent-size *Tharsis Bulge* (see Figure 9.25). Tharsis is some 4000 kilometers across, and most of it rises several kilometers above the average Martian surface level. It was probably created by a long-lived plume of rising mantle material that bulged the surface upward and provided the molten rock for the eruptions that built the giant volcanoes.

We have not witnessed any ongoing volcanic or tectonic activity on Mars, and we expect Mars to be much less

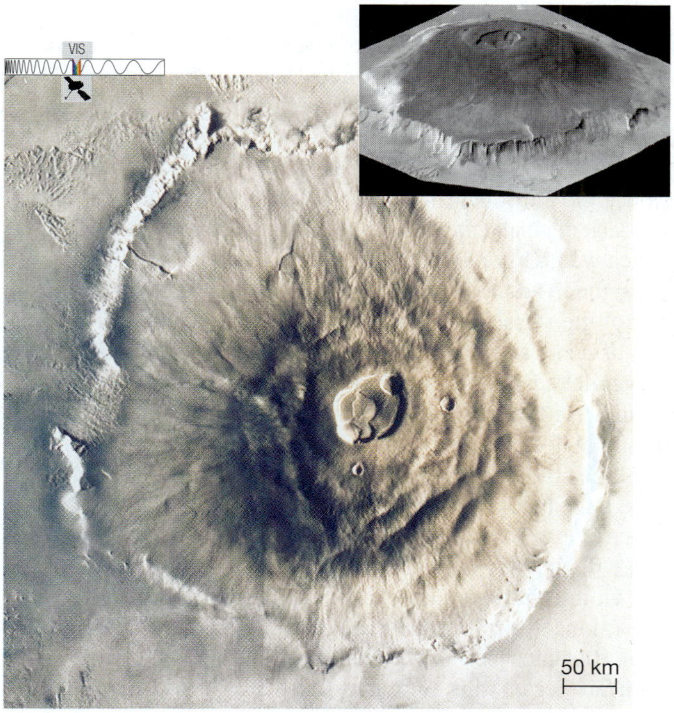


**FIGURE 9.26** Olympus Mons, photographed from orbit; note the tall cliff around its rim and the central volcanic crater from which lava erupted. The inset shows a 3-D perspective on this immense volcano.

volcanically active than Earth, because its smaller size has allowed its interior to cool much more. However, the era of Martian volcanism may not be completely over. Crater counts on the slopes of Martian volcanoes suggest that some lava flows may have occurred as recently as tens of millions of years ago, which is not so long ago in geological terms. In addition, radiometric dating of meteorites that appear to have come from Mars (so-called *Martian meteorites* [**Section 24.2**]) shows some of them to be made of volcanic rock that solidified from molten lava as little as 180 million years ago—also quite recently in the $4\frac{1}{2}$-billion-year history of the solar system. Given this evidence of geologically recent volcanic eruptions, it is likely that Martian volcanoes will erupt again someday. The tentative detection of methane in the Martian atmosphere might also point to ongoing volcanic activity [**Section 24.2**]. Nevertheless, the Martian interior is presumably cooling and its lithosphere thickening. Within a few billion years, Mars will become as geologically dead as the Moon and Mercury.

**Tectonics and Valles Marineris** Mars also has tectonic features, though none on a global scale like the plate tectonics of Earth. The most prominent tectonic feature is the long, deep system of valleys called *Valles Marineris* (**FIGURE 9.27**). Named for the *Mariner 9* spacecraft that first imaged it, Valles Marineris extends almost a fifth of the way along the planet's equator. It is as long as the United States is wide and almost four times as deep as Earth's Grand Canyon. No one knows exactly how Valles Marineris formed. Parts of the canyon are completely enclosed by high cliffs on all sides, so neither flowing lava nor water could have been responsible. However, extensive cracks on its western end run up against the Tharsis Bulge (see Figure 9.25), suggesting a connection between the two features. Perhaps Valles Marineris formed through tectonic stresses accompanying the uplift that created Tharsis, cracking the surface and leaving the tall cliff walls of the valleys.

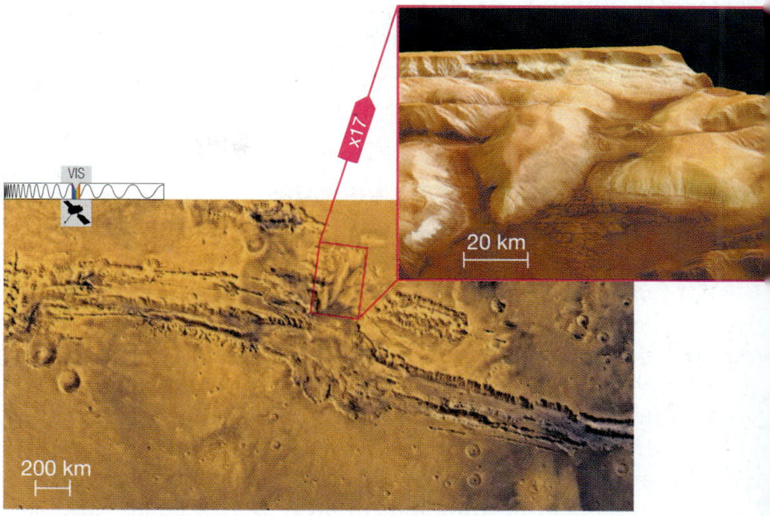


**FIGURE 9.27** Valles Marineris is a huge system of valleys on Mars created in part by tectonic stresses. The inset shows a perspective view looking north across the center of the canyon, obtained by *Mars Express*.

**Erosion on Mars** Impacts, volcanism, and tectonics explain most of the major geological features of Mars, but closer examination shows extensive evidence of erosion by liquid water. For example, **FIGURE 9.28** shows a dry riverbed (another is shown in Figure 9.1) that was almost certainly carved by running water, although we cannot yet say whether the water came from runoff after rainfall, from erosion by water-rich debris flows, or from an underground source. Regardless of the specific mechanism, water is the only substance that could have been liquid under past Martian conditions and that is sufficiently abundant to have

**FIGURE 9.28** This photo, taken by the *Mars Express* spacecraft, shows what appears to be a dried-up meandering riverbed, now filled with dunes of windblown dust. The small impact craters scattered about the site allow scientists to estimate how long ago the water flowed.

created such extensive erosion features. We must therefore look more carefully at the evidence suggesting that Mars once had abundant surface water.

## What geological evidence tells us that water once flowed on Mars?

There are no lakes, rivers, or even puddles of liquid water on the surface of Mars today. We know this not only because we've photographed most of the surface in fairly high resolution but also because the surface conditions do not allow it. In most places and at most times, Mars is so cold that any liquid water would immediately freeze into ice. Even when the temperature rises above freezing, as it often does at midday near the equator, the air pressure is so low that liquid water would quickly evaporate [**Section 5.3**]. In other words, liquid water is *unstable* on Mars today: If you put on a spacesuit and took a cup of water outside your pressurized spaceship, the water would rapidly either freeze or boil away (or some combination of both).

When we combine the clear evidence of water erosion with the fact that liquid water is unstable on Mars today, we conclude that Mars must once have had a very different climate—with warmer temperatures and greater air pressure that would have allowed water to flow and rain to fall—than it does today. Geological evidence indicates that this warmer and wetter period must have ended long ago. For example, notice the impact craters that lie on top of the channels in Figure 9.28. Counts of these craters suggest that no water has flowed through these channels for at least about 3 billion years.

**Orbital Evidence for Ancient Water Flows** Strong evidence that Mars had rain and surface water in the distant past comes from both orbital and surface studies. Figure 9.29 shows just three among thousands of orbital images that provide evidence of water erosion. **FIGURE 9.29a** shows a broad region of the ancient, heavily cratered southern highlands. Notice the indistinct rims of many large craters and the relative lack of small craters. Both facts argue for ancient rainfall, which would have eroded crater rims and erased small craters altogether. **FIGURE 9.29b** shows a three-dimensional perspective of the surface that suggests water once flowed between two ancient crater lakes. **FIGURE 9.29c** shows what looks like a river delta where water flowed into an ancient crater; spectra indicate the presence of clay minerals on the crater floor, presumably deposited by sediments flowing down the river.

Some studies even suggest that the northern plains may once have held a vast ocean (**FIGURE 9.30**), though the evidence is less definitive than the evidence for smaller lakes. The evidence for the ocean comes from features that look like an ancient shoreline. Radar data also suggest that the rock along the proposed shoreline is sedimentary rather than volcanic, just as we would expect if there had once been an ocean; this rock may even contain water ice in its pores and cracks. Additional evidence for an ocean is based on estimates of how much water has been lost to space, a topic we'll discuss in Chapter 10.

**Surface Evidence for Ancient Water Flows** Surface studies further strengthen the case for past water. In 2004, the robotic rovers *Spirit* and *Opportunity* landed on opposite sides of Mars (see Figure 9.25). The twin rovers long outlasted their design lifetime of 3 months, with *Spirit* lasting

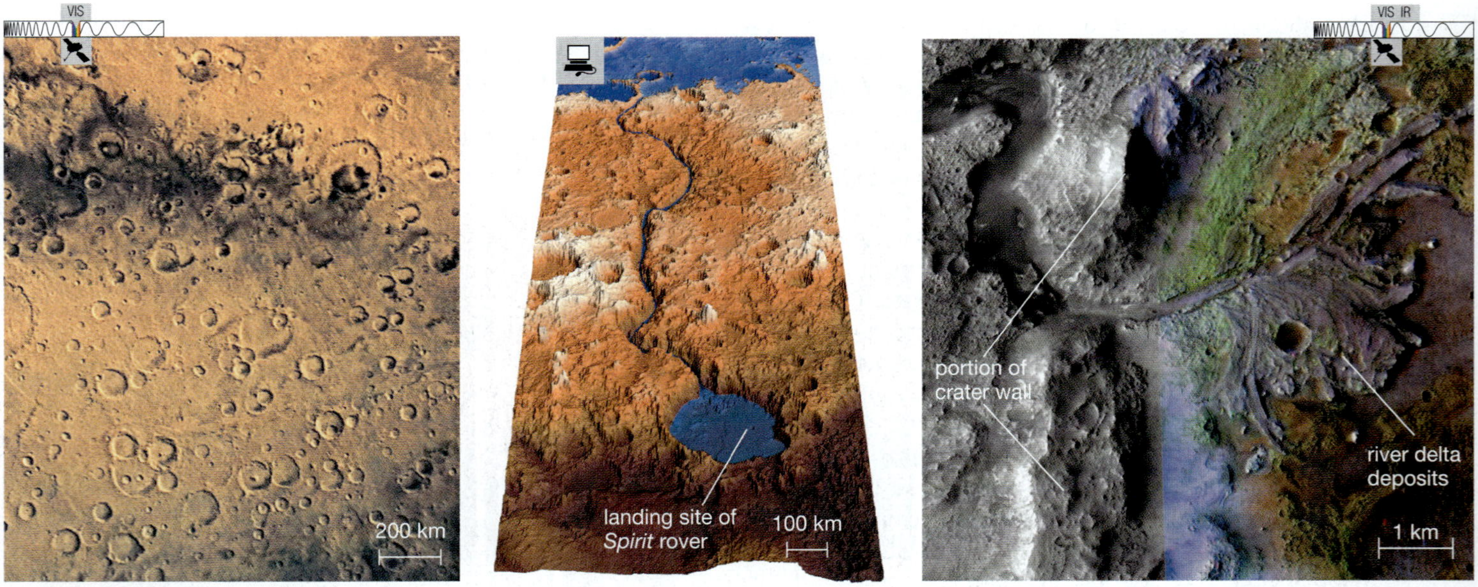


**a** This photo shows a broad region of the southern highlands on Mars. The eroded rims of large craters and the relative lack of small craters suggest erosion by rainfall.

**b** This computer-generated perspective view shows how a Martian valley forms a natural passage between two possible ancient lakes (shaded blue). Vertical relief is exaggerated 14 times to reveal the topography.

**c** Combined visible/infrared image of an ancient river delta that formed where water flowing down a valley emptied into a lake filling a large crater (portions of the crater wall are identified). Clay minerals are identified in green.

**FIGURE 9.29** More evidence of past water on Mars. As of 2018, the regions shown in (b) and (c) were two of three possible landing sites being considered for the *Mars 2020* rover.

**FIGURE 9.30** Mars's northern hemisphere may once have held a vast ocean; this artist's conception shows what it might have looked like some 4 billion years ago.

more than 6 years and *Opportunity* still going as this book was being written, more than 14 years after arrival. The latest rover, *Curiosity* (see Figure 7.6), landed in 2012 at a site called Gale Crater (**FIGURE 9.31**). *Curiosity* carries the most powerful set of scientific instruments ever landed on another world, including cameras, drills, microscopes, rock and soil analyzers, a laser to vaporize rock, and a spectrograph to analyze the vaporized material.

All three rovers have found abundant mineral evidence of past liquid water on the surface. However, the character of the water appears to have differed at different times in Mars's deep past. For example, *Opportunity* discovered tiny spheres—nicknamed "blueberries"—that have apparently eroded out of the underlying rock layers in much the same way as similar spheres found on Earth (**FIGURE 9.32**). The "blueberries" are composed largely of an iron-bearing mineral called *hematite*, while the rock layers in which they were embedded are rich in the sulfur-bearing mineral *jarosite*. Both of these mineral types form in salty, acidic water, suggesting that water like this once percolated through the rocks at *Opportunity*'s landing site.

In contrast, *Curiosity* found evidence that Gale Crater once contained much purer ("drinkable") water (**FIGURE 9.33**), at least at the lower elevations it explored during the first few years of its journey. The dry clay and sedimentary rocks clearly indicate that a lake once resided there, and the clumps of pebbles are so similar in structure and chemical composition to pebbles on Earth that they must have formed in an ancient streambed. *Curiosity* also undertook chemical analysis of some of these low-elevation sedimentary layers, finding that they contain minerals known to form only in relatively pure water, much like that found in most lakes on Earth.

The difference in the purity of the water that apparently once resided at the *Opportunity* and *Curiosity* landing sites suggests that Martian water became saltier and more acidic

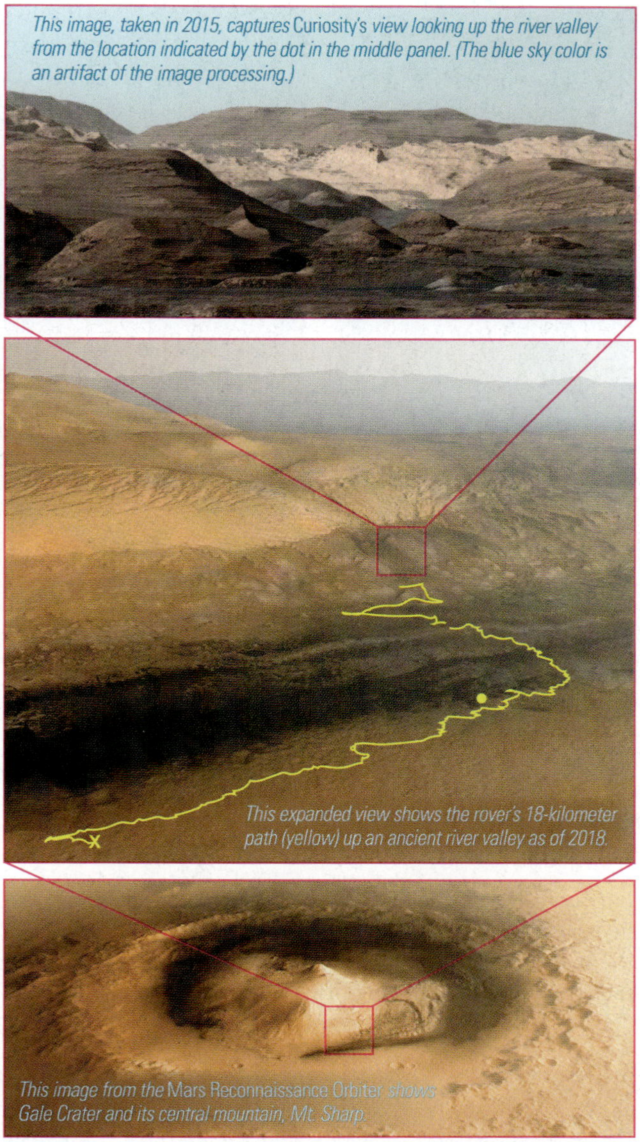


**FIGURE 9.31** The *Curiosity* rover landed in Gale Crater in 2012 at the location marked by the "x" in the middle panel. *Curiosity*'s studies support the idea that, more than 3 billion years ago, a lake filled the crater, allowing layered sedimentary rocks to build up from the crater floor. Later wind erosion exposed much of the crater while leaving the mountain at the center.

over time. Scientists suspect that so far *Curiosity* has been exploring sediments laid down earlier in Martian history, when the water was relatively pure, while *Opportunity* has explored sediments laid down later, after the water became saltier and more acidic. This hypothesis has gained additional support as *Curiosity* has continued its journey along the flanks of the 5-kilometer-tall Mt. Sharp, where preliminary evidence indicates that the higher-elevation sediments formed in saltier and more acidic water.

Orbital studies had already indicated that Mt. Sharp was surrounded by sedimentary rock layers dating to many different times over the past several billion years; indeed, this was the major reason this location was chosen as *Curiosity*'s landing site. *Curiosity*'s on-the-ground studies have now enabled scientists to piece together the following history at this site: A vast lake once filled Gale Crater, though it dried

**Mars (Endurance Crater)** **Earth (Utah)**

**FIGURE 9.32** “Blueberries” on two planets. In both cases the foreground shows hematite “blueberries,” which formed within sedimentary rock layers like those in the background, then eroded out and rolled downhill; the varying tilts of the rock layers hint at changing wind or waves during formation. The background rocks are about twice as far away from the camera in the Earth photo as in the Mars photo (taken by the *Opportunity* rover).

out more than 3 billion years ago. Sediments were deposited on the lake bottom in layers, ultimately filling much of the crater. Even more rock was deposited on top by volcanic eruptions, compressing the sediments into rock. Over the 3 billion years since Mars dried out, winds have eroded away much of the overlying rock, exposing the rock layers visible to *Curiosity*. The oldest layers, which are at the lowest elevations, formed when the water was relatively pure, and the younger and higher layers formed as the water became more acidic and less plentiful. *Curiosity*'s ongoing study of these layers should teach us much more about Mars's geological and climatological history, and perhaps will even shed light on whether Mars has ever been home to life.

### Summary of Evidence for Surface Water on Mars

Together, the orbital and surface evidence makes a very strong case indicating that Mars was once home to abundant surface water, including some that was relatively pure, but only prior to about 3 billion years ago. As we've discussed, this could have been possible only if Mars had both much warmer temperatures and a much greater atmospheric pressure at that time than it has today.

One remaining question is whether any water at all ever still flows on Mars. This is possible in principle, because although liquid water is *unstable* on Mars today, ice could in principle melt and then flow as liquid water for a short time until it refroze or vaporized. For a few years, many scientists thought they had found evidence of such short-lived flows in orbital images showing gullies and dark streaks on many crater walls (**FIGURE 9.34**), which look much like similar features found on Earth where water flows downhill. However, more recent study of these features suggests that they were formed either by the seasonal vaporization of carbon dioxide frozen onto the surface or by dry sand or dust slipping down a slope, as happens on sand dunes.

The bottom line is that if any liquid water ever still flows on Mars, it can be only a minuscule fraction of the water

**a** *Curiosity* drilled into this rock formation of dry clay, finding minerals that should form only in water that was not very acidic or salty.

**b** These clumps of rounded pebbles show a structure nearly identical to that found in typical streambeds on Earth, providing strong evidence that the pebbles were rounded by flowing water before being cemented together into rock.

**c** The even layers of the foreground rocks in this image from *Curiosity* are characteristic of sediment deposited over time in the delta of a river that emptied into a lake.

**FIGURE 9.33** Rock formations photographed by *Curiosity* closely resemble those found among sedimentary rocks on Earth formed in the presence of abundant and fairly pure water.

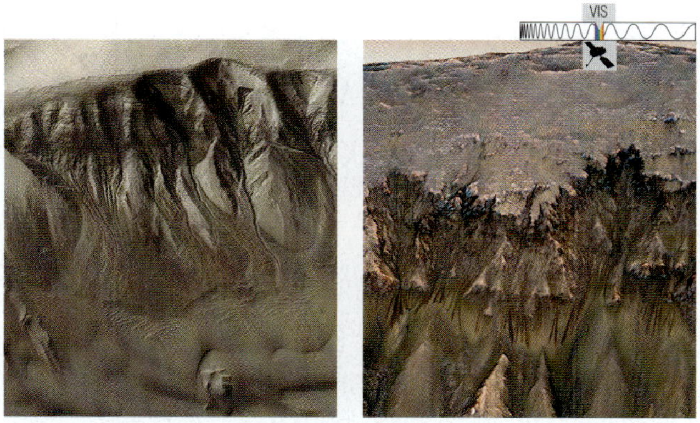


**FIGURE 9.34** Many crater walls on Mars have gullies or dark streaks that look like features formed on Earth by flowing water. However, recent evidence indicates that the Martian features were formed by dry sand or dust, rather than water. Both images are from the *Mars Reconnaissance Orbiter.*

that flowed long ago when riverbeds and lakes existed, at a time when Mars must have been much warmer and must have had a much greater surface pressure than it has today. Ironically, Percival Lowell's supposition that Mars was drying up has turned out to be basically correct, although in a very different way than he imagined. We'll discuss the reasons for Mars's dramatic climate change in Chapter 10.

## 9.5 Geology of Venus

The surface of Venus is searing hot with brutal pressure [**Section 7.1**], making it seem quite unlike the "sister planet" of Earth it is sometimes called. However, beneath the surface, Venus and Earth must be quite similar. The two planets are nearly the same size—Venus is only about 5% smaller than Earth in radius—and their similar densities (see Table 7.1) suggest similar overall compositions. We therefore expect the interiors of Venus and Earth to have the same structure and to retain about the same level of internal heat today. This idea is borne out by the fact that Venus has many geological similarities to Earth, although we also see important differences.

### What geological processes have shaped Venus?

Venus's thick cloud cover prevents us from seeing through to its surface with visible light, but we can study its geological features with *radar mapping*, which bounces radio waves off the surface and uses the reflections to create three-dimensional images. **FIGURE 9.35** shows a global map and selected close-ups made with radar mapping from the *Magellan* spacecraft. Scientists have named most of the major geological features on Venus for goddesses and famous women. Three elevated "continents"

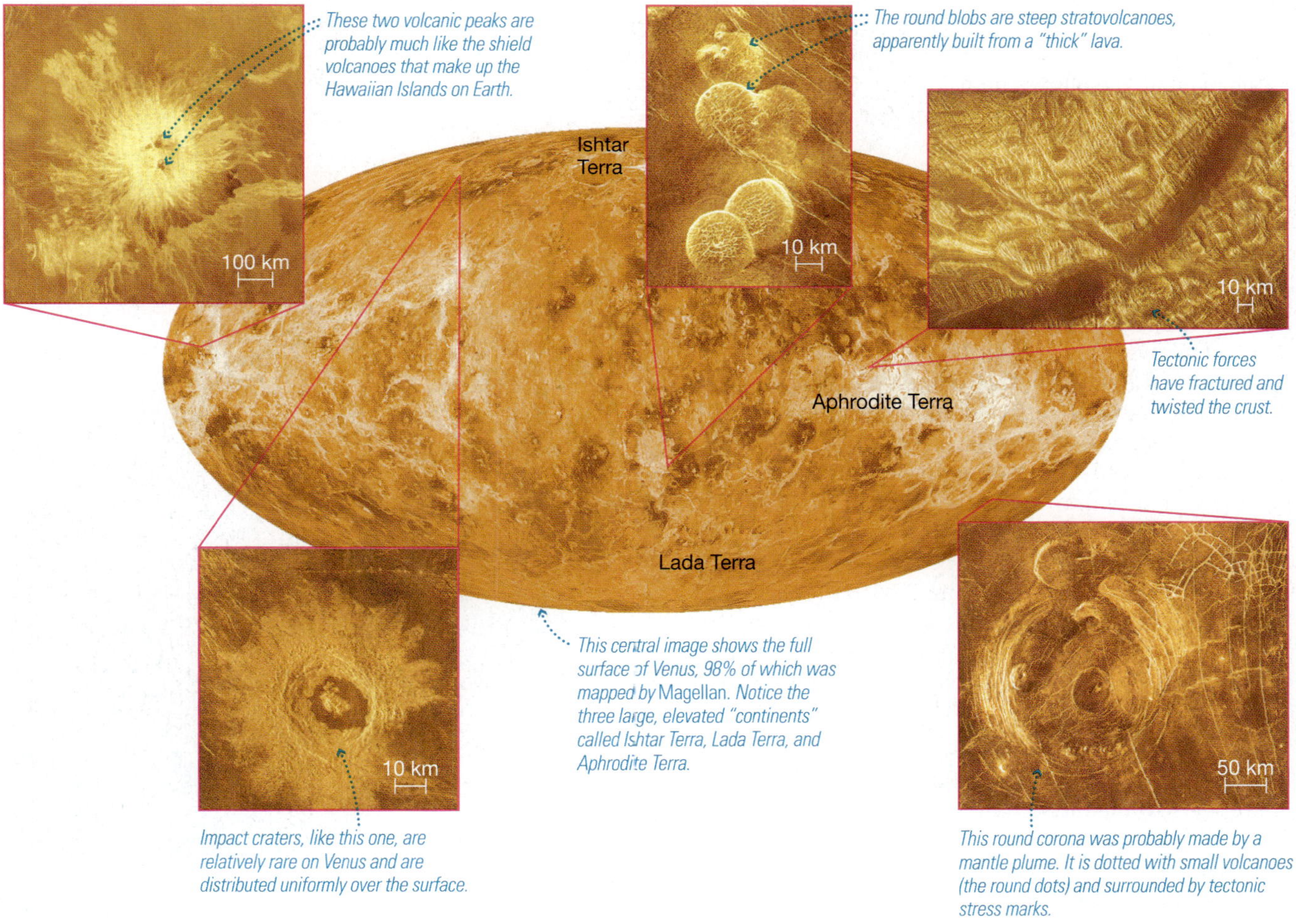


**FIGURE 9.35** The global map and close-ups show Venus's surface as revealed through radar mapping by the *Magellan* spacecraft, which operated from 1990 to 1993 and had resolution good enough to "see" features as small as 100 meters across. Bright regions in the radar images represent rough areas or higher altitudes.

(the three "Terra" in Figure 9.35) are the largest surface features.

**Impact Cratering on Venus** Like Earth, Venus has a relatively small number of impact craters, indicating that its ancient craters were erased by other geological processes. Moreover, while Venus has a few large craters, it lacks the small craters that are most common on other worlds—probably because the small objects that could make such craters burn up completely as they enter Venus's thick atmosphere. But even this thick atmosphere has little effect on objects large enough to make the craters we can see in global views like Figure 9.35.

**Volcanic and Tectonic Features** Venus shows abundant evidence of volcanism with a variety of lava types. Some mountains are shallow-sloped volcanoes, indicating that they were built by eruptions in which the lava was about as runny as that which formed Earth's Hawaiian Islands. But it also has lava plains, which must have formed from a runnier lava, as well as a few steeper-sided volcanoes, indicating eruptions of a thicker lava.

Tectonics has also been very important on Venus, as the entire surface appears to have been extensively contorted and fractured by tectonic forces. Some of these features, including large circular *coronae* (Latin for "crowns") like the one shown in Figure 9.35, provide strong evidence for mantle convection beneath the lithosphere. The coronae were probably pushed upward by hot rising plumes of rock in the mantle. The plumes probably also forced lava to the surface, which would explain why numerous volcanoes are found near coronae.

Venus almost certainly remains geologically active today, since it should still retain nearly as much internal heat as Earth. Although we have not witnessed any volcanic eruptions, at least two lines of evidence suggest recent volcanic activity. First, Venus's clouds contain sulfuric acid. Sulfuric acid is made from sulfur dioxide ($SO_2$) and water, and the sulfur dioxide must have entered the atmosphere through volcanic outgassing. Because sulfur dioxide is gradually removed from the atmosphere by chemical reactions with surface rocks, the existence of sulfuric acid clouds means that volcanic outgassing must have occurred within about the past 100 million years. The second line of evidence narrows the time scale further. The European Space Agency's *Venus Express* spacecraft (which orbited Venus from 2006 until early 2015) detected an infrared spectral feature from rocks on three volcanoes indicating that they erupted within about the past 250,000 years (**FIGURE 9.36**), which is very recent on a planetary time scale.

**Weak Erosion** We might naively expect Venus's thick atmosphere to produce strong erosion, but the view both from orbit and on the surface suggests otherwise. The former Soviet Union sent several landers to Venus in the 1970s and early 1980s. Before the intense surface heat destroyed them, the probes returned images of a bleak volcanic landscape with little evidence of erosion (**FIGURE 9.37**). We can trace the lack of erosion on Venus to two simple facts. First, Venus is far too hot for any type of rain or snow on its surface. Second, as we discussed earlier, Venus's slow rotation—once every 243 days—means it has very little surface wind.

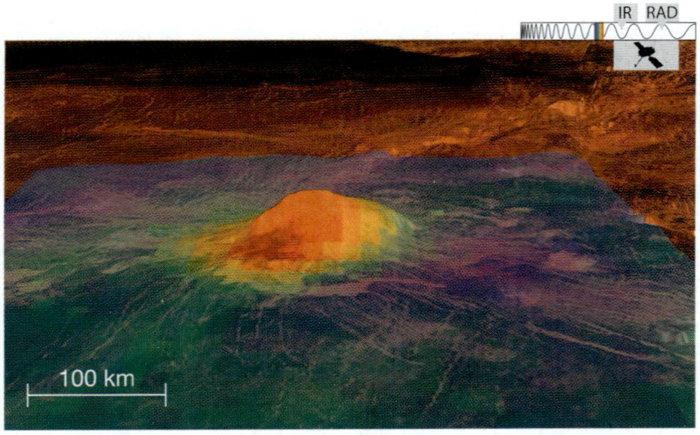


**FIGURE 9.36** This composite image shows a volcano called Idunn Mons on Venus. Surface topography details are from NASA's *Magellan* radar mapper (enlarged about 30 times to make the volcano easier to see), and colors represent infrared data from the *Venus Express* spacecraft. Red indicates relatively new rock that has not been chemically altered by Venus's harsh atmosphere, suggesting that lava flows occurred within the past 250,000 years.

**Think about it** Suppose Venus rotated as fast as Earth. How would this change its relative levels of volcanism, tectonics, and erosion? Explain.

**FIGURE 9.37** This photo from one of the former Soviet Union's *Venera* landers shows Venus's surface; part of the lander is in the foreground, with sky in the background. Many volcanic rocks are visible, hardly affected by erosion despite their presumed age of about 750 million years (the age of the entire surface).

### Does Venus have plate tectonics?

We can easily explain the lack of erosion on Venus, but another "missing feature" of its geology is more surprising: Venus shows no evidence of Earth-like plate tectonics. As we'll discuss in the next section, plate tectonics shapes nearly all of Earth's major geological features, including mid-ocean ridges, deep ocean trenches, and long mountain ranges like the Rockies and Himalayas. Venus lacks any similar features.

Instead, Venus shows evidence of a very different type of global geological change. On Earth, plate tectonics reshapes the surface gradually, so different regions have different ages. In contrast, Venus's relatively few impact craters are distributed fairly uniformly over the entire planet, suggesting that the surface is about the same age everywhere. Crater counts suggest a surface age of about 750 million years, leading us to conclude that the entire surface was somehow "repaved" at that time, erasing all craters that formed earlier. We do not know how much of the repaving was due to tectonic processes and how much to volcanism, but both probably were important. It is even possible that plate tectonics played a role before and during the repaving, only to stop after the repaving episode was over.

The absence of present-day plate tectonics on Venus is a major mystery, though scientists have a few ideas that might explain it. Earth's lithosphere was broken into plates by forces due to the underlying mantle convection. The lack of plate tectonics on Venus therefore suggests either that it has weaker mantle convection or that its lithosphere somehow resists fracturing. The first possibility seems unlikely, because Venus's similarity to Earth in size and density leads us to expect it to have a similar level of mantle convection. Most scientists therefore suspect that Venus's lithosphere resists fracturing into plates because it is thicker and stronger than Earth's.

Even if a thicker and stronger lithosphere explains the lack of plate tectonics on Venus, we are still left with the question of why the lithospheres of Venus and Earth should differ. One possible answer invokes Venus's high surface temperature. Venus is so hot that any water in its crust and mantle has probably been baked out over time. Water tends to soften and lubricate rock, so its loss would have thickened and strengthened Venus's lithosphere. If this hypothesis is correct, then Venus might have had plate tectonics if it had not become so hot in the first place. The mystery of plate tectonics is just one of many reasons scientists would like to explore Venus in greater detail, but as of 2018 no major new missions had been funded.

## 9.6 The Unique Geology of Earth

Most aspects of Earth's geology should make sense in light of what we have found for other terrestrial worlds. Earth's size—largest of the five terrestrial worlds—explains why our planet retains internal heat and remains volcanically and tectonically active. Earth's rampant erosion by water and wind is explained by the combination of our planet's size, distance from the Sun, and rotation rate: Earth is large enough for volcanism and outgassing to have produced an atmosphere, while its distance from the Sun allowed water vapor to condense and fall to the surface as rain, and Earth's moderately rapid rotation drives wind and other weather.

However, Earth's geology stands apart from that of the other terrestrial worlds in one important way: Earth is the only planet shaped primarily by ongoing plate tectonics. The importance of this fact goes far beyond basic geology. As we'll discuss in Chapter 10, plate tectonics plays a crucial role in the stability of Earth's climate. Without this climate stability, it is unlikely that we could be here today.

### How is Earth's surface shaped by plate tectonics?

The term *plate tectonics* refers to the scientific theory that explains much of Earth's surface geology as a result of the slow motion of *plates* that essentially "float" over the mantle, gradually moving over, under, and around each other as convection moves Earth's interior rock. Earth's lithosphere is broken into more than a dozen plates (**FIGURE 9.38**). The plate motions are barely noticeable on human time scales: On average, plates move only a few centimeters per year—about the rate at which your fingernails grow. Nevertheless,

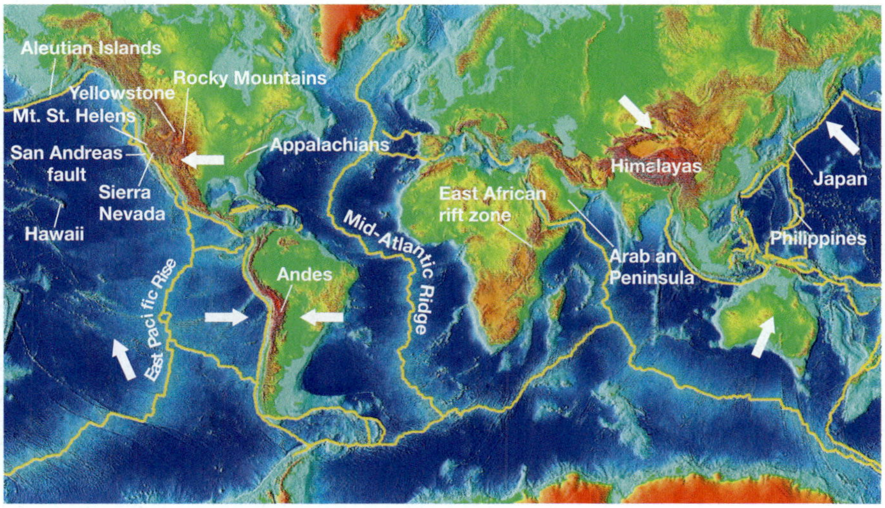


**FIGURE 9.38** This relief map shows known plate boundaries (solid yellow lines), with arrows to represent directions of plate motion. Color represents elevation, progressing from blue (lowest) to red (highest). Labels identify some of the geological features discussed in the text.

over millions of years these motions can completely reshape the surface of our planet.

**The Discovery of Continental Motion** We usually trace the origin of the theory of plate tectonics to a slightly different hypothesis proposed in 1912 by German meteorologist and geologist Alfred Wegener: *continental drift,* the idea that continents gradually drift across the surface of Earth. Wegener got his idea in part from the puzzle-like fit of continents such as South America and Africa (**FIGURE 9.39**). He also noted that similar types of distinctive rocks and rare fossils were found in eastern South America and western Africa, suggesting that these two regions once had been close together.

Despite this evidence, no one at the time knew of a mechanism that could allow the continents to move about. Wegener suggested that Earth's gravity and tidal forces from the Sun and Moon were responsible, but other scientists quickly showed that these forces were too weak to move entire continents. As a result, Wegener's idea of continental drift was widely rejected by geologists for decades after he proposed it.

In the mid-1950s, scientists began to observe geological features that suggested a mechanism for continental motion. In particular, they discovered *mid-ocean ridges* (such as the Mid-Atlantic Ridge shown in Figure 9.38) along which mantle material erupts onto the ocean floor, pushing apart the existing seafloor on either side. This **seafloor spreading** helped explain how the continents could move apart with time. In addition, as more fossil evidence was gathered, it became clear that the continents really were arranged differently in the past, and Wegener's idea of a "continental fit" for Africa and South America ultimately gained acceptance.

Today, geologists can directly measure the slow plate motions by comparing readings taken with the global positioning system (GPS) [**Section S1.3**] on either side of plate boundaries. We also now understand that continental motion is coupled to the underlying mantle convection, driven by the heat released from Earth's interior. Because this idea is quite different from Wegener's original notion of continents plowing through the solid rock beneath them, geologists no longer use the term *continental drift* and instead consider continental motion within the context of plate tectonics.

**Seafloor Crust and Continental Crust** Another key piece of evidence for plate tectonics came with the discovery that Earth's surface has two distinct types of crust, one type found on seafloors and the other on continents (**FIGURE 9.40**). **Seafloor crust** is thinner, denser, and younger than **continental crust**. No other planet shows evidence of such distinct differences in crust from place to place.

Seafloor crust is typically 5–10 kilometers thick and is made primarily of the relatively high-density rock called basalt. Recall that basalt lava is very runny, so it spreads outward when it erupts from volcanoes along mid-ocean ridges. Radiometric dating shows that seafloor crust is quite young—usually less than 200 million years old—indicating that it erupted to the surface relatively recently in geological history. Further evidence of the young age of seafloors comes from studies of impact craters. Large impacts should occur more or less uniformly over Earth's surface, and the oceans are not deep enough to prevent a large asteroid or comet from making a seafloor crater. However, we find far fewer large craters on the seafloor than on the continents, which means that seafloor crust must have been created more recently.

Continental crust is much thicker—typically between 20 and 70 kilometers thick—but it sticks up only slightly higher than seafloor crust because its sheer weight presses it down farther into the mantle below. It is made mostly of rock (such as granite) with lower density than seafloor crust. Continental crust spans a wide range of ages and in some places contains rocks that are up to 4 billion years old.

**The Conveyor Belt of Plate Tectonics** The theory of plate tectonics explains continental motion, seafloor spreading, and the existence of two types of crust as direct results of the way plates move about on Earth. Over millions of years, the movements involved in plate tectonics act like a giant conveyor belt for Earth's lithosphere (**FIGURE 9.41**). Mid-ocean ridges occur at places where mantle material rises upward, creating new seafloor crust and pushing plates apart. The newly formed crust cools and contracts as it spreads sideways from the central ridge, giving seafloor spreading regions their characteristic ridged shape. Along the mid-ocean ridges worldwide, new crust covers an area of about 2 square kilometers every year, enough to replace the entire seafloor within about 200 million years—and

**FIGURE 9.39** The puzzle-like fit of South America and Africa.

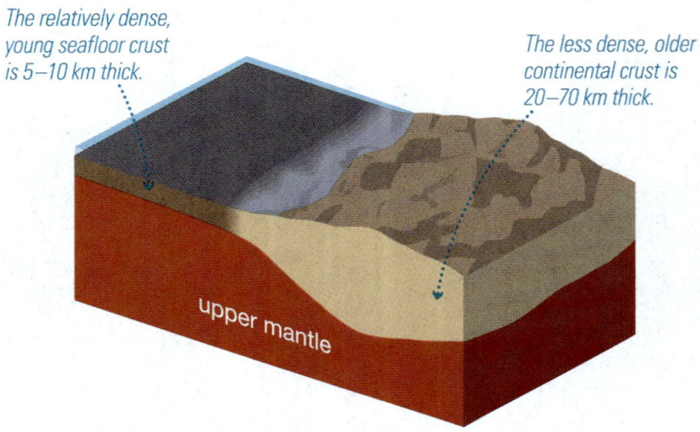


**FIGURE 9.40** Earth has two distinct kinds of crust.

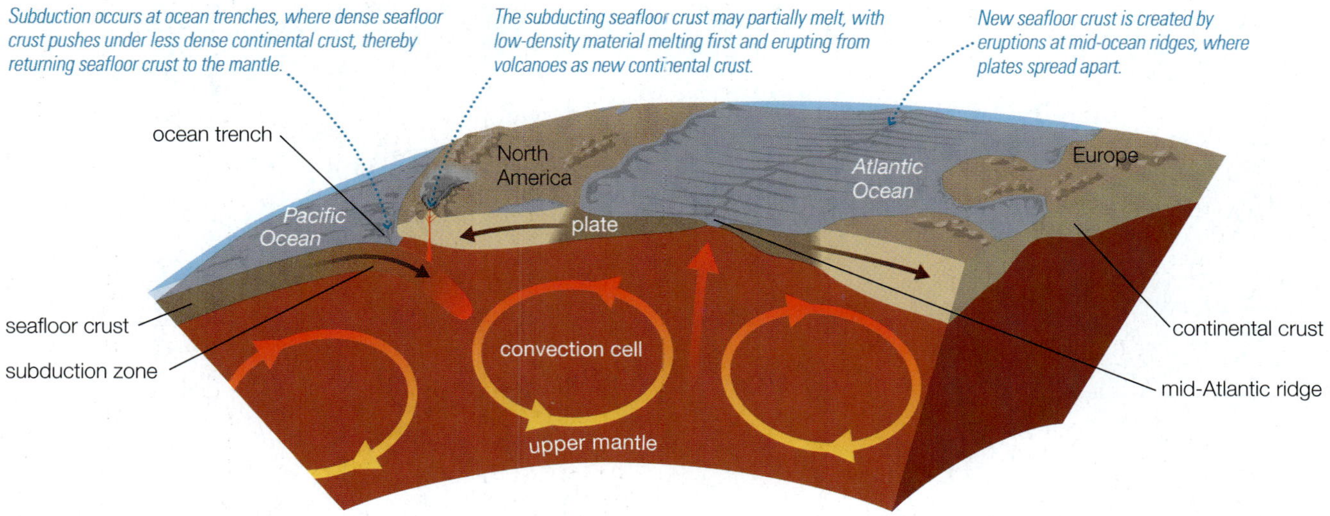


**FIGURE 9.41** Plate tectonics acts like a giant conveyor belt for Earth's lithosphere.

thereby explaining the less-than-200-million-year age of seafloor crust.

Over tens of millions of years, any piece of seafloor crust gradually makes its way across the ocean bottom, then finally gets recycled into the mantle in the process we call **subduction**. Subduction occurs where a seafloor plate meets a continental plate, which is generally somewhat offshore at the edge of a sloping continental shelf. As the dense seafloor crust of one plate pushes under the less-dense continental crust of another plate, it can pull the entire surface downward to form a deep *ocean trench*. At some trenches, the ocean depth is more than 8 kilometers (5 miles).

Beneath a subduction zone, the descending seafloor crust heats up and may begin to melt as it moves deeper into the mantle. If enough melting occurs, the molten rock may erupt upward. This melting tends to occur under the edges of the continents, which is why so many active volcanoes are found along those edges. Moreover, the lowest-density material tends to melt more easily, which is why the continental crust emerging from these landlocked volcanoes is lower in density than seafloor crust. This fact also explains why steep-sided volcanoes, made from low-density but thick lava (see Figure 9.11a), are common on Earth but rare on other worlds: Without plate tectonics to recycle crust, other worlds generally have only volcanic plains or shallow-sloped volcanoes made from higher-density, more runny lava.

The conveyor-like process of plate tectonics is undoubtedly driven by the heat flow from mantle convection, although the precise relationship between the convection cells and the plates remains an active topic of research. For example, it is not clear whether mantle convection simply pushes plates apart at the sites of seafloor spreading or whether the plates are denser than the underlying material and pull themselves down at subduction zones.

**Building Continents** Unlike seafloor crust, continental crust does not get recycled back into the mantle. As a result, present-day continents have been built up over billions of years. However, their histories are not just a simple "piling up" of continental crust. Instead, the continents are continually reshaped by volcanism and stresses associated with plate tectonics, as well as by erosion. **FIGURE 9.42** shows some of the complex geological history of North America, which we can use as an example of how the continents have been built.

The west coast regions of Alaska, British Columbia, Washington, Oregon, and most of California began as numerous volcanic islands in the Pacific. Long ago, plate motion carried these islands from their ocean birthplaces toward the North American mainland, where subduction continues to take the seafloor crust downward into the mantle (see Figure 9.41). Because these islands were made of lower-density continental crust, they remained on the surface and attached themselves to the edge of North America, even as the seafloor crust beneath them slid back into the mantle. Similar processes affect other islands today. Alaska's Aleutian Islands, a string of volcanoes located over a region where one seafloor plate is subducting under another (see the plate boundaries in Figure 9.38), are gradually growing and will someday merge together. Japan and the Philippines represent a later stage in this process; each once contained many small islands that merged into the fewer islands we see today. As these islands continue to grow and merge, they may eventually create a new continent or merge with an existing one.

Other portions of North America have been shaped by erosion. The Great Plains and the Midwest once were ancient seas. Erosion gradually filled these seas with sediment that hardened into sedimentary rock. The Deep South formed from the buildup of sediments that were carried to these regions after eroding from other parts of the continent. Northeastern Canada features some of the oldest continental crust on Earth, worn down over time by erosion.

Long mountain ranges were built by tectonic processes. The Sierra Nevada range in California, which lies over the region where the Pacific plate subducts under North America, formed when partially molten mantle material rose up and pushed the surface rock higher. The original surface rock was sedimentary, but erosion gradually wore it away and left the underlying rock from the mantle exposed; this

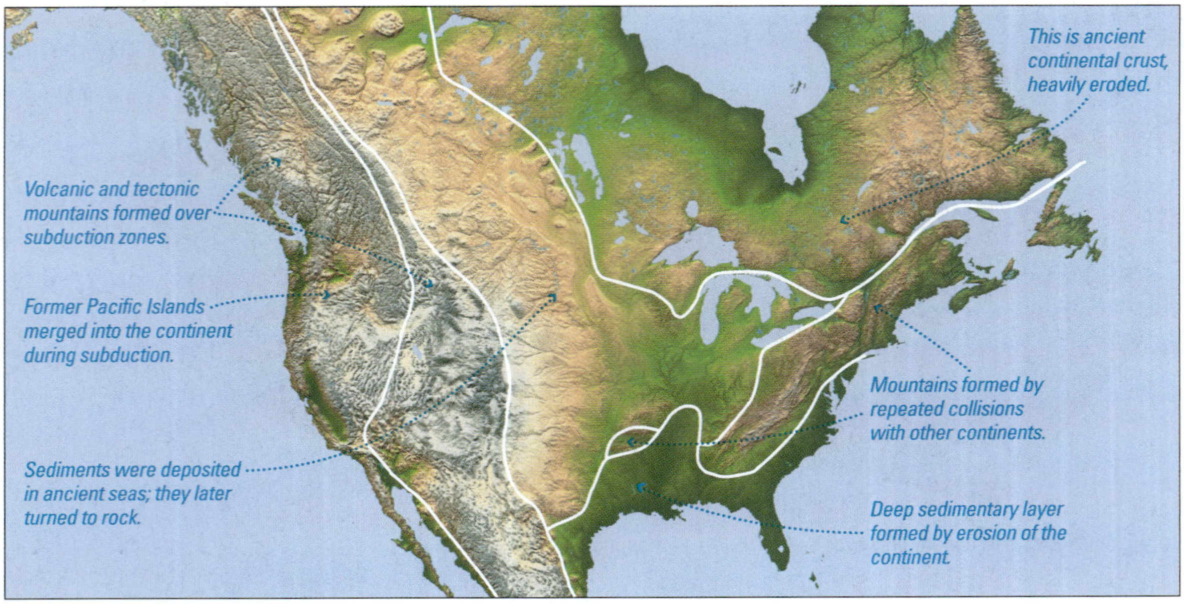


**FIGURE 9.42** The major geological features of North America record the complex history of plate tectonics. Only the basic processes behind the largest features are shown.

rock was mostly of the type called granite, which is why the Sierra Nevada mountains are made largely of granite.

Mountain ranges can also form through collisions of continent-bearing plates. The Himalayas, the tallest mountains on Earth, offer the most famous example of this process (**FIGURE 9.43**). They are still growing taller as the plate carrying the Indian subcontinent rams into the plate carrying Eurasia. Because both colliding plates are made of low-density continental crust, neither plate can subduct under the other. Instead, they push into each other and the resulting compression pushes the land upward to make mountains. The Appalachian range in the eastern United States was built through multiple collisions of continental plates: Over a period of a few hundred million years, North America apparently collided twice with South America and then with western Africa. The Appalachians probably were once as tall as the Himalayas are now, but tens of millions of years of erosion transformed them into the fairly modest mountain range we see today. Similar processes contributed to the formation of the Rocky Mountains in the United States and Canada.

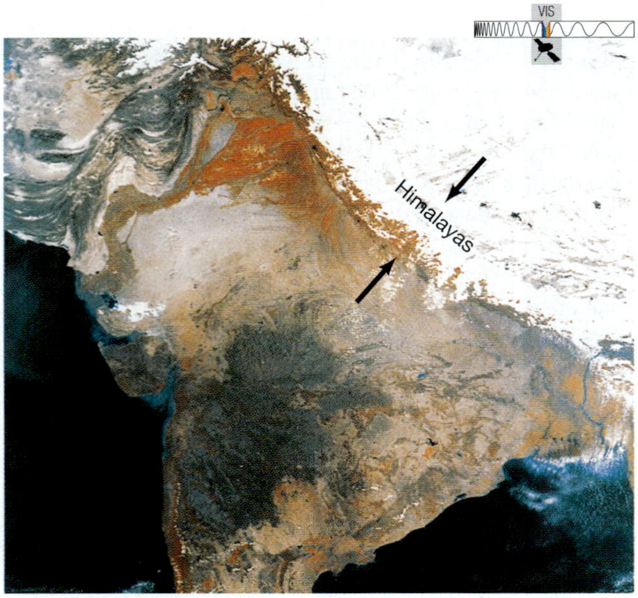


**FIGURE 9.43** This satellite photo shows the Himalayas, which are still slowly growing as the plate carrying India pushes into the Eurasian plate. Arrows indicate directions of plate motion.

**Rifts, Faults, and Earthquakes** So far, we have discussed features formed in places where plates are colliding, either with a seafloor plate subducting under a continental plate or with two continental plates pushing into each other. Different types of features occur in places where plates pull apart or slide sideways relative to each other.

In places where continental plates are pulling apart, the crust thins and can create a large *rift valley*. The East African rift zone is an example (see Figure 9.38). This rift is slowly growing and will eventually tear the African continent apart. At that point, rock rising upward with mantle convection will begin to erupt from the valley floor, creating a new zone of seafloor spreading. A similar process tore the Arabian Peninsula from Africa, creating the Red Sea (**FIGURE 9.44**). The rift valley process is also active in New Mexico's Rio Grande Valley. (The river named the Rio Grande came *after* the valley formed from tectonic processes.)

Places where plates slip sideways relative to each other are marked by what we call **faults**—fractures in the lithosphere. The San Andreas Fault in California marks a line where the Pacific plate is moving northward relative to the continental plate of North America (**FIGURE 9.45**). In about 20 million years, this motion will bring Los Angeles and San Francisco together. The two plates do not slip smoothly against each other; instead, their rough surfaces catch. Stress builds up until it is so great that it forces a rapid and violent shift, causing an earthquake. More generally, earthquakes can occur along any type of plate boundary, because

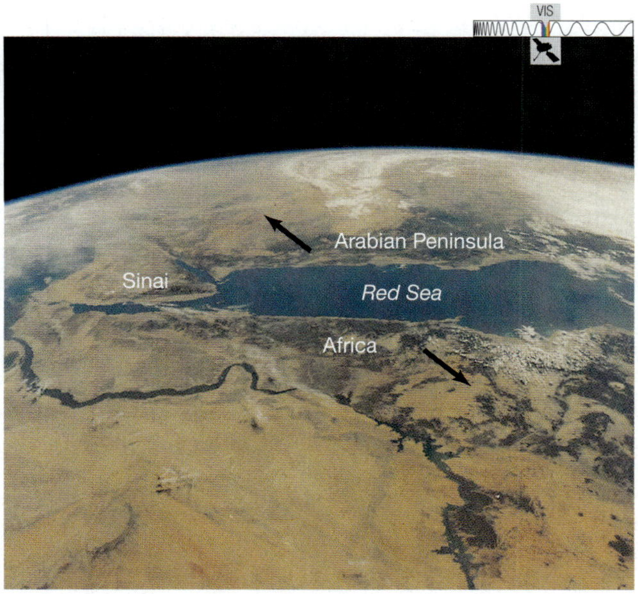


**FIGURE 9.44** When continental plates pull apart, the crust thins and deep rift valleys form. This process tore the Arabian Peninsula from Africa, forming the Red Sea. Arrows indicate directions of plate motion.

**FIGURE 9.45** California's San Andreas Fault marks a boundary where plates are sliding sideways, as shown by the white arrows; asterisks indicate recent earthquakes. The inset shows an aerial photo of the San Andreas Fault (in the Coachella Valley), which runs between the arrows; notice the offset of the land on the two sides of the fault.

plates can become temporarily "stuck" and then suddenly shift when the stress builds to a breaking point.

In contrast to the usual motion of plates, which proceeds at a few centimeters per year, an earthquake can move plates by several *meters* in a few seconds. The movement can raise mountains, level cities, set off destructive tsunamis, and make the whole planet vibrate with seismic waves.

**Think about it** By studying the plate boundaries in Figure 9.38, explain why California, Oregon, and Washington have more earthquakes and volcanoes than the rest of the continental United States. Find the locations of recent earthquakes and volcanic eruptions worldwide. Do the locations fit the pattern you expect?

**Hot Spots** Not all volcanoes occur near plate boundaries. Sometimes, a plume of hot mantle material rises in what we call a **hot spot**. The Hawaiian Islands are the result of a hot spot that has been erupting basaltic lava for tens of millions of years. Plate tectonics has gradually carried the Pacific plate over the hot spot, thereby forming a chain of volcanic islands (**FIGURE 9.46**). Today, most of the lava erupts on the Big Island of Hawaii, giving this island a young, rocky surface. About a million years ago, the Pacific plate lay farther to the southeast (relative to its current location), and the hot spot built the island of Maui. Before that, the hot spot created other islands, including Oahu (3 million years ago), Kauai (5 million years ago), and Midway (27 million years ago). The older islands are more heavily eroded. Midway has eroded so much that it barely

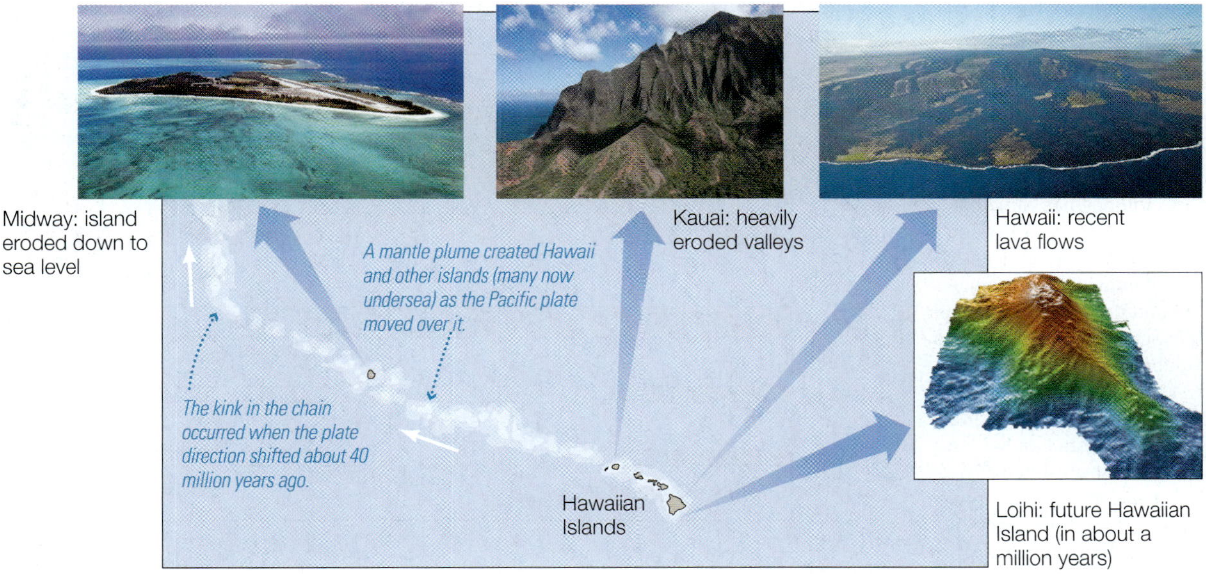


**FIGURE 9.46** The Hawaiian Islands are just the most recent of a very long string of volcanic islands made by a mantle hot spot. The image of Loihi (lower right) was obtained by sonar, as it is still entirely under water.

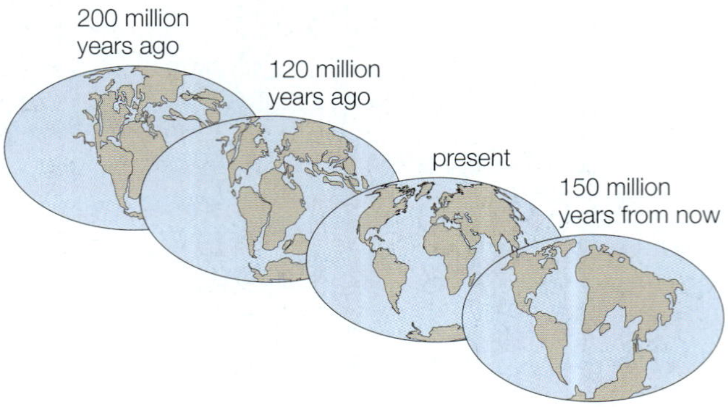


**FIGURE 9.47** Past, present, and future arrangements of Earth's continents. The present continents were all combined into a single "supercontinent" about 200 million years ago.

rises above sea level. If plate tectonics were not moving the plate relative to the hot spot, a single huge volcano would have formed—perhaps looking somewhat like Olympus Mons on Mars (see Figure 9.26).

The movement of the plate over the hot spot continues today, building underwater volcanoes that eventually will rise above sea level to become new Hawaiian Islands. The growth of a future island, named Loihi, is already well under way—prime beach real estate should be available there in a million years or so. Hot spots can also occur beneath continental crust. For example, a hot spot is responsible for the geysers and hot springs of Yellowstone National Park.

**Think about it** Find your hometown or current home on the map in Figure 9.38 or Figure 9.42 (or both). Based on what you have learned about Earth's geology, describe the changes your home has undergone over millions (or billions) of years.

**Plate Tectonics Through Time** We can use the current motions of the plates to project the arrangement of continents millions of years into the past or the future. For example, at a speed of 2 centimeters per year, a plate will travel 2000 kilometers in 100 million years. **FIGURE 9.47** shows two past arrangements of the continents, along with one future arrangement. Note that about 200 million years ago the present-day continents were together in a single "supercontinent," sometimes called *Pangaea* (which means "all lands").

Studies of magnetized rocks (which record the orientation of ancient magnetic fields) and comparisons of fossils found in different places around the world have allowed geologists to map the movement of the continents even further into the past. It seems that, over the past billion years or more, the continents have slammed together, pulled apart, spun around, and changed places on the globe. Central Africa once lay at Earth's South Pole, and Antarctica once was near the equator. The continents continue to move, and their current arrangement is no more permanent than any past arrangement.

## Was Earth's geology destined from birth?

Now that we have completed our geological tour of the terrestrial worlds, let's consider whether fundamental planetary properties shape all geological destinies. If so, then we should be able to predict the geological features of terrestrial worlds we find around other stars. If not, then we still have more to learn before we'll know whether other worlds could have Earth-like geology.

**FIGURE 9.48** summarizes the key trends we've seen among the terrestrial worlds. All the worlds were heavily cratered during the heavy bombardment, but the extent of volcanism and tectonics has depended on planetary size. The interiors

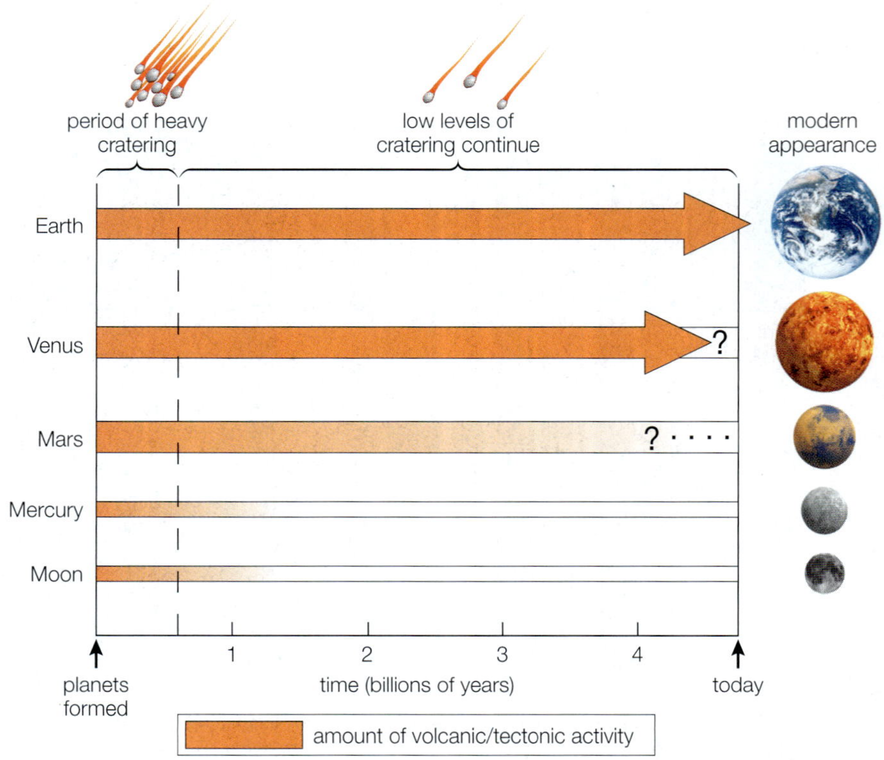


**FIGURE 9.48** This diagram summarizes the geological histories of the terrestrial worlds. The brackets along the top indicate that impact cratering has affected all worlds similarly. The arrows represent volcanic and tectonic activity. A thicker and darker arrow means more volcanic/tectonic activity, and the arrow length tells us how long this activity persisted. Notice this trend's relationship to planetary size: Earth remains active today; Venus has also been active, though we are uncertain whether it remains so; Mars has had an intermediate level of activity and might still have low-level volcanism; Mercury and the Moon have had very little volcanic/tectonic activity. Erosion is not shown, because it has played a significant role only on Earth (ongoing) and Mars (in the past and at low levels today).

of the smallest worlds cooled quickly, so they have not had volcanism or tectonics for billions of years. The interiors of the larger worlds have stayed hot longer, allowing volcanism and tectonics to continue for much longer time periods.

The major remaining question concerns whether fundamental properties can also explain Earth's unique plate tectonics. The lack of plate tectonics on Venus tells us that we cannot attribute plate tectonics solely to size, since Venus is so similar in size to Earth. However, if we are correct in hypothesizing that Venus lacks plate tectonics because its high surface temperature led to loss of water and lithospheric thickening, then a lower surface temperature might have led to plate tectonics on Venus as well as Earth. As we'll discuss in Chapter 10, we can ultimately trace Venus's high surface temperature to its nearness to the Sun. While plenty of uncertainty still remains, it seems likely that the broad geological histories of Earth and the other terrestrial worlds were indeed destined from birth by the fundamental properties of size, distance from the Sun, and rate of rotation.

## The BIG Picture PUTTING CHAPTER 9 INTO PERSPECTIVE

*In this chapter, we have explored the geology of the terrestrial worlds. As you continue your study of the solar system, keep the following "big picture" ideas in mind:*

- The terrestrial worlds all looked much the same when they were born, so their present-day differences are a result of geological processes that occurred in the ensuing $4\frac{1}{2}$ billion years.
- The extent to which different geological processes operate on the different worlds depends largely on their fundamental properties, especially their size.
- A planet's geology is largely destined from its birth—which means we should be able to predict the geology of as-yet-undiscovered planets once we know their fundamental properties.
- Earth has been affected by the same geological processes affecting the other terrestrial worlds. However, erosion is far more important on Earth than on any other terrestrial world, and Earth's unique plate tectonics may be very important to our existence.

**MY COSMIC PERSPECTIVE** Floods, earthquakes, volcanic eruptions, and other "natural disasters" are actually just consequences of the "good fortune" that comes with living on a geologically active planet.

## Summary of Key Concepts

### 9.1 Connecting Planetary Interiors and Surfaces

- **What are terrestrial planets like on the inside?** In order of decreasing density and depth, the interior structure consists of **core**, **mantle**, and **crust**. The crust and part of the mantle together make up the rigid **lithosphere**. In general, a thinner lithosphere allows more geological activity.

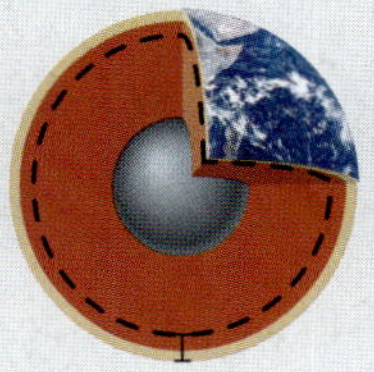

- **What causes geological activity?** Interior heat drives geological activity by causing mantle convection, keeping the lithosphere thin, and keeping the interior partially molten. All the terrestrial interiors were once hot, but larger planets cool slowly, retaining more interior heat and staying geologically active longer.

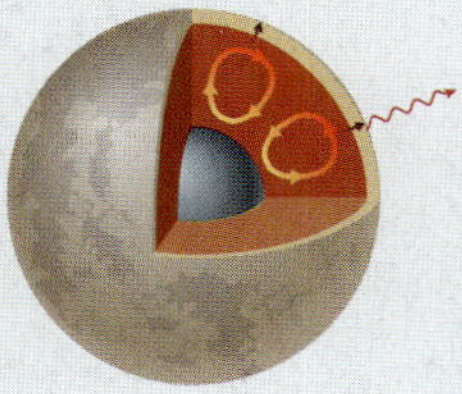

- **Why do some planetary interiors create magnetic fields?** A planetary **magnetic field** requires three things: an interior layer of electrically conducting fluid, convection of that fluid, and rapid rotation. Among the terrestrial planets, only Earth has all three characteristics.

### 9.2 Shaping Planetary Surfaces

- **What processes shape planetary surfaces?** The four major geological processes are **impact cratering**, **volcanism**, **tectonics**, and **erosion**.
- **How do impact craters reveal a surface's geological age?** More craters indicate an older surface. All the terrestrial worlds were battered by impacts when they were young, so those that still have many impact craters must look much the same as they did long ago. Those with fewer impact craters must have had their ancient craters erased by other geological processes.
- **Why do the terrestrial planets have different geological histories?** Fundamental planetary properties, especially size, determine a planet's geological history. Larger worlds have more volcanism and tectonics, and these processes erase more of the world's ancient impact craters. Erosion depends on a planet's size, distance from the Sun, and rotation rate.

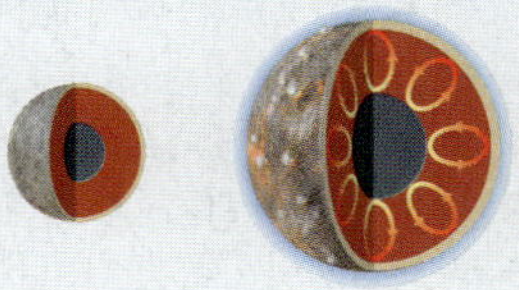

### 9.3 Geology of the Moon and Mercury

- **What geological processes shaped our Moon?** The lunar surface is a combination of extremely ancient, heavily cratered terrain and somewhat younger lava plains called the lunar **maria**. Some small tectonic features are also present. The Moon lacks erosion because it has so little atmosphere.

- **What geological processes shaped Mercury?** Mercury's surface resembles that of the Moon in being shaped by impact cratering and volcanism. It also has tremendous tectonic cliffs that probably formed when the whole planet cooled and contracted in size.

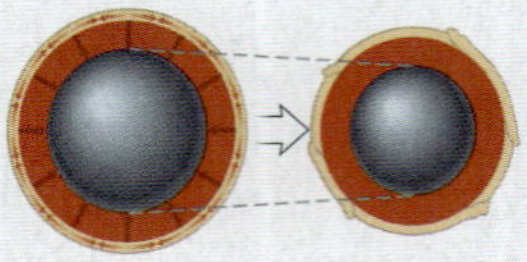

## 9.4 Geology of Mars

- **What geological processes have shaped Mars?** Mars shows evidence of all four geological processes. It has the tallest volcano and the biggest canyon in the solar system, evidence of a period of great volcanic and tectonic activity. It also has abundant craters and evidence of erosion by wind and flowing water.

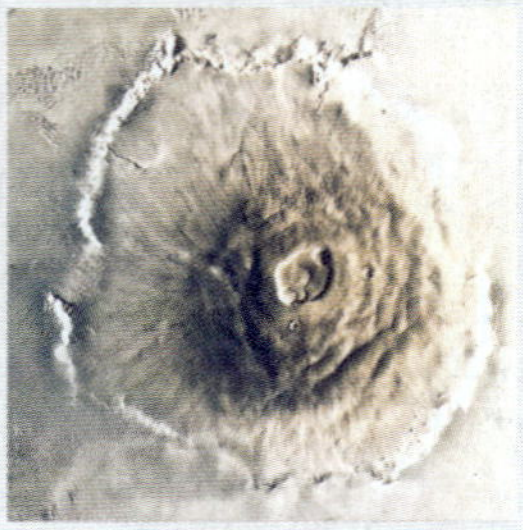

- **What geological evidence tells us that water once flowed on Mars?** Images of dry river channels and eroded craters, along with chemical analysis of Martian rocks, show that water once flowed on Mars. Any periods of rainfall seem to have ended at least 3 billion years ago. Mars still has water ice underground and in its polar caps and could possibly have pockets of underground liquid water.

## 9.5 Geology of Venus

- **What geological processes have shaped Venus?** Venus's surface shows evidence of major volcanic and tectonic activity in the past billion years, as expected for a planet nearly as large as Earth. This activity explains the relative lack of craters. Despite Venus's thick atmosphere, erosion is only a minor factor because of the high temperature (no rainfall) and slow rotation (little wind). Venus almost certainly remains geologically active today.

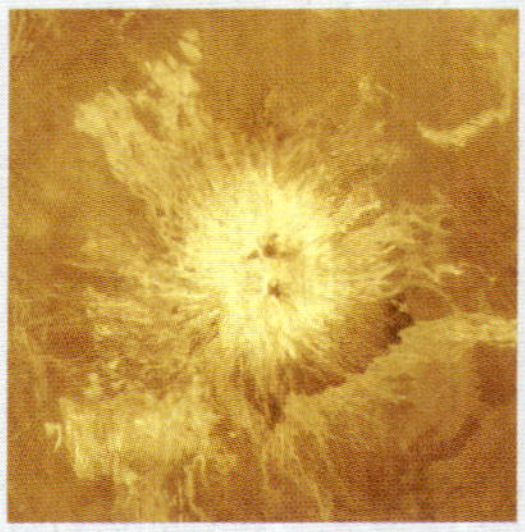

- **Does Venus have plate tectonics?** Venus appears to have undergone planetwide resurfacing, but we do not see evidence of ongoing plate tectonics. The lack of plate tectonics probably means that Venus has a stiffer and stronger lithosphere than Earth, perhaps because the high surface temperature has baked out any water that might have softened the lithospheric rock.

## 9.6 The Unique Geology of Earth

- **How is Earth's surface shaped by plate tectonics?** On Earth, the lithosphere is broken into **plates** that move around in the special type of tectonics that we call **plate tectonics**.

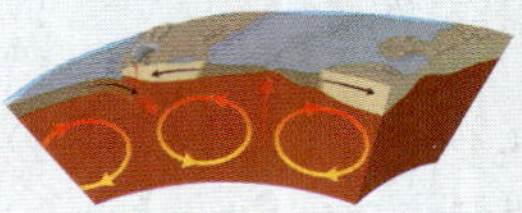

Plate tectonics works much like a giant conveyor belt. New **seafloor crust** emerges from **mid-ocean ridges** and is recycled into the mantle at **subduction zones**, near which lower-density **continental crust** can erupt to build up the continents. Over time, the shifting of plates rearranges the continents on Earth's surface.

- **Was Earth's geology destined from birth?** It seems likely that the geological histories of Earth and the other terrestrial worlds were indeed destined from birth by the fundamental properties of size, distance from the Sun, and rate of rotation.

# Visual Skills Check

*Use the following questions to check your understanding of some of the many types of visual information used in astronomy. For additional practice, try the Chapter 9 Visual Quiz in the Study Area at www.MasteringAstronomy.com.*

*This image from the MESSENGER spacecraft shows evidence of impact cratering, volcanism, and tectonic activity on Mercury. Answer the following questions based on the image. Remember that craters are bowl-shaped and rough-floored when they form, and wipe out any preexisting features in the area. Lava on Mercury appears to be fairly runny and makes flat smooth plains as it spreads out.*

1. Label 1a lies on the rim of a large crater, and label 1b lies on the rim of a smaller one. Which crater must have formed first?
   a. crater 1a
   b. crater 1b
   c. Cannot be determined
2. The region around 2b has far fewer craters than the region around 2c. The crater floor at 2a is also flat and smooth, without many smaller craters on it. Why are regions 2a and 2b so smooth?
   a. Few small craters ever formed in these regions.
   b. Erosion erased craters that once existed in these regions.
   c. Lava flows covered craters that once existed in these regions.

3. A tectonic ridge appears to connect points 3a and 3b, crossing several craters. From its appearance, we can conclude that it must have formed
   a. before the area was cratered
   b. after the area was cratered
   c. at the same time the area was cratered

4. Using your answers to questions 1–3, list the following features in order from oldest to youngest:
   a. the tectonic ridge from 3a to 3b
   b. crater 1a
   c. the smooth floor of crater 1b

# Exercises and Problems

*For instructor-assigned homework and other learning materials, go to www.MasteringAstronomy.com.*

## Chapter Review Questions

### Short-Answer Questions Based on the Reading

1. Describe the core-mantle-crust structures of the terrestrial worlds, and how they imply *differentiation*. What is a *lithosphere*, and how does lithospheric thickness vary among the five terrestrial worlds?
2. Summarize the processes by which planetary interiors get hot and cool off. Why do large planets retain internal heat longer than smaller planets?
3. Why does Earth have a global magnetic field? Why don't the other terrestrial worlds have similarly strong magnetic fields?
4. Define each of the four major geological processes, and give examples of features shaped by each process.
5. What is *outgassing*, and why is it so important to our existence?
6. Why is the Moon so much more heavily cratered than Earth? Explain how crater counts tell us the age of a surface.
7. Summarize the ways in which a terrestrial world's size, distance from the Sun, and rotation rate each affect its relative level of impact cratering, volcanism, tectonics, and erosion.
8. Briefly summarize the geological history of the Moon. How did the lunar maria form?
9. Briefly summarize the geological history of Mercury. How are Mercury's great cliffs thought to have formed?
10. Choose five features on the global map of Mars (Figure 9.25) and explain the nature and likely origin of each.
11. Why isn't liquid water stable on Mars today, and why do we nonetheless think it flowed on Mars in the distant past?
12. Choose at least three major geological features of Venus and explain how we think each one formed.
13. What evidence tells us that Venus was "repaved" about 750 million years ago? What might account for the lack of plate tectonics on Venus?
14. Describe the conveyor-like action of plate tectonics on Earth, and how it explains the differences between seafloor and continental crust.
15. Briefly explain how each of the following geological features of Earth is formed: seafloors, continents, islands, mountain ranges, rift valleys, and faults.
16. To what extent do we think the geologies of the terrestrial worlds were destined from their birth? Explain.

### Surprising Discoveries?

*Suppose we were to make the following discoveries. (These are not real discoveries.) In light of your understanding of planetary geology, decide whether the discovery should be considered reasonable or surprising. (In some cases, both views can be defended.) Explain your answer clearly, ideally in terms of fundamental properties of size, distance from the Sun, and rotation rate.*

17. New photographs reveal sand dunes on Mercury.
18. Seismographs placed on the surface of Mercury record frequent and violent earthquakes.
19. A new orbiter observes a volcanic eruption on Venus.
20. A Venus radar mapper discovers extensive regions of layered sedimentary rocks.
21. Radiometric dating of rocks brought back from one lunar crater shows that the crater was formed only a few tens of millions of years ago.
22. New high-resolution orbital photographs of Mars show many crater bottoms filled with pools of liquid water.
23. Drilling into the slopes of a Martian volcano, a robotic spacecraft discovers liquid water beneath the surface.
24. Clear-cutting in the Amazon rain forest on Earth exposes terrain that is as heavily cratered as the lunar highlands.
25. Seismic studies on Earth reveal a "lost continent" that held great human cities just a few thousand years ago but is now buried deep under water off the western coast of Europe.
26. We find a planet in another solar system that has Earth-like plate tectonics; the planet is the size of the Moon and orbits 1 AU from its star.
27. We find a planet in another solar system that is as large as Earth but as heavily cratered as the Moon.
28. We find a planet in another solar system that has Earth-like seafloor crust and continental crust but apparently lacks plate tectonics or any other kind of crustal motion.

### Quick Quiz

*Choose the best answer to each of the following. For additional practice, try the Chapter 9 Reading and Concept Quizzes in the Study Area at www.MasteringAstronomy.com.*

29. What is the longest-lasting internal heat source responsible for geological activity? (a) accretion (b) radioactive decay (c) sunlight
30. In general, what kind of terrestrial planet would you expect to have the thickest lithosphere? (a) a large planet (b) a small planet (c) a planet far from the Sun
31. The major way in which we learn about Earth's interior is through (a) drilling deep into the crust. (b) studying seismic waves. (c) spacecraft imagery.
32. Which of a planet's fundamental properties has the greatest effect on its level of volcanic and tectonic activity? (a) size (b) distance from the Sun (c) rotation rate
33. What is the name of the rigid outer layer of a planet? (a) crust (b) mantle (c) lithosphere
34. Which describes our understanding of flowing water on Mars? (a) It was never important. (b) It was important once, but is no longer. (c) It is a major process on the Martian surface today.

35. What do we conclude if a planet has few impact craters of any size? (a) The planet was never bombarded by asteroids or comets. (b) Its atmosphere stopped impactors of all sizes. (c) Other geological processes have wiped out craters.
36. How many of the five terrestrial worlds have surfaces being continually reshaped by plate tectonics? (a) one (b) two (c) three or more
37. On how many of the five terrestrial worlds has erosion been an important process? (Be sure that you explain *why* erosion is important on this many worlds and not more.) (a) one (b) two (c) three or more
38. How many of the terrestrial worlds have lava plains or shallow-sloped volcanoes? (a) one (b) three (c) five

## Inclusive Astronomy

*Use these questions to reflect on participation in science.*

39. *Group Discussion: Putting Women in Charge.* On February 23, 2008, NASA arranged work schedules to put an all-woman team in charge of the Mars Exploration Rovers' tactical operations. It was the first time an all-woman team was given full responsibility for a major NASA mission's operations, even for a single day, and it was done in honor of the upcoming Women's History Month.
**a.** Gather in groups of three and discuss why it was considered unusual to have an all-woman team in charge of a NASA mission's operations.
**b.** Discuss societal factors that may discourage women from pursuing careers in space science or space exploration, and how these might be changed to encourage greater inclusion of and participation by women in these fields. Make a list of specific actions you would recommend NASA take.
**c.** Do you think a symbolic event such as the all-woman Mars Rovers' operations team helps to promote long-term inclusion of women in science? Why or why not?

## The Process of Science

*These questions may be answered individually in short-essay form or discussed in groups, except where identified as group-only.*

40. *Mars Attracts.* William Herschel, Giovanni Schiaparelli, and Percival Lowell were all respected astronomers who made important and widely accepted discoveries, yet all three jumped to incorrect conclusions about Mars. Scientifically, where do you think each went wrong, and how seriously? Try to frame your answer in terms of the hallmarks of science discussed in Chapter 3. Do you think their mistakes had any long-term impact on the study of Mars?
41. *What Is Predictable?* Briefly explain why much of a planet's geological history is destined from its birth, and discuss the level of detail that is predictable. For example, was Mars's general level of volcanism predictable? Could we have predicted a mountain as tall as Olympus Mons or a canyon as long as Valles Marineris? Explain.
42. *Worth the Effort?* Politicians often argue over whether planetary missions are worth the expense involved. If you were in Congress, would you support more or fewer missions? Why?
43. *Evidence of Our Civilization.* Discuss how Earth's geological processes would affect the evidence of our current civilization in the distant future, under the assumption that our civilization were to end today. For example, what evidence of our civilization would survive in 100,000 years? In 100 million years? Would future archaeologists or alien visitors be able to know that our civilization had existed? Does thinking about the potential end of our civilization affect the way you think about what we can do to help our civilization survive? Explain.
44. *Unanswered Questions.* Current understanding allows us to paint a broad-brush overview of the geological histories of the terrestrial worlds, but we have much more to learn. Choose one important but unanswered question related to the geology of the terrestrial worlds, and describe why the question is important and how we might answer it in the future. Be as specific as possible, focusing on the type of evidence necessary to answer the question and how the evidence could be gathered.
45. *Group Discussion: Real and Imaginary Planets.* Work in small groups to discuss Figure 9.16. Note: You may wish to do this activity using the four roles described in Chapter 1, Exercise 39.
**a.** Figure 9.16 shows seven "cartoon" worlds. Decide which terrestrial world or worlds in our solar system best match the listed characteristics for each cartoon.
**b.** Study the blue annotations associated with each of the seven cartoon worlds. For each cartoon, find a real planetary image in the chapter that exemplifies the principle described by at least one of its annotations; explain clearly.
**c.** Choose one terrestrial world in our solar system (do not choose Earth), and imagine that *one* of its properties (size, distance, or rotation rate) had been significantly different. How would you expect this world's geological history to have been different?
**d.** Propose two imaginary planets whose size, distance from the Sun, and rotation rates are different from each other and different from those of all of our solar system's terrestrial worlds. Give your planets names and predict their geological histories, clearly explaining how you make your predictions.

## Investigate Further

### Short-Answer/Essay Questions

46. *Dating Planetary Surfaces.* We have discussed two basic techniques for determining the age of a planetary surface: studying the abundance of impact craters and radiometric dating of surface rocks. Which technique seems more reliable? Which technique is more practical? Explain.
47. *Comparative Erosion.* Of Mercury, Venus, the Moon, and Mars, which world has the greatest erosion? Why? Write a paragraph explaining each world's level of erosion, tracing it back to fundamental properties.
48. *Miniature Mars.* Suppose Mars had turned out to be significantly smaller than its current size—say, the size of our Moon. How would this have affected the number of geological features due to each of the four major geological processes? Do you think Mars would be a better or worse candidate for harboring extraterrestrial life? Summarize your answers in two or three paragraphs.
49. *Change in Fundamental Properties.* Choose one property of Earth—either size or distance from the Sun—and suppose that it had been significantly different. Describe how this change might have affected Earth's subsequent geological history and the possibility of our existence today on Earth.
50. *Predictive Geology.* Suppose another star system has a rocky terrestrial planet twice as large as Earth but at the same

distance from its star (which is just like our Sun) and with a similar rotation rate. In one or two paragraphs, describe the type of geology you would expect it to have.

51. *Mystery Planet.* It's the year 2098, and you are designing a robotic mission to a newly discovered planet around a nearby star that is nearly identical to our Sun. The planet is as large in radius as Venus, rotates with the same daily period as Mars, and lies 1.2 AU from its star. Your spacecraft will orbit but not land on the planet.
**a.** Some of your colleagues believe that the planet has no metallic core. How could you support or refute their hypothesis?
**b.** Other colleagues suspect that the planet has no atmosphere, but the instruments designed to study the planet's atmosphere fail because of a software error. However, the spacecraft can still photograph geological features. How could you use the spacecraft's photos of geological features to determine whether a significant atmosphere is (or was) present on this planet?
52. *"Coolest" Surface Photo.* Visit the Astronomy Picture of the Day website, and search for archived images of the terrestrial worlds. Look at many of them, and choose the one you think is the "coolest." Write a short description of what it shows, and explain what you like about it.
53. *Experiment: Planetary Cooling in a Freezer.* Fill two small plastic containers of similar shape but different sizes with cold water and put both into the freezer at the same time. Every hour or so, record the time and estimate the thickness of the "lithosphere" (the frozen layer) in the two tubs. How long does it take the water in each tub to freeze completely? Describe the relevance of your experiment to planetary geology. Extra credit: Plot your results on a graph with time on the $x$-axis and lithospheric thickness on the $y$-axis. What is the ratio of the two freezing times?
54. *Amateur Astronomy: Observing the Moon.* Any amateur telescope has resolution adequate for identifying geological features on the Moon. The light highlands and dark maria should be evident, and shadowing should be visible near the line between night and day. Try to observe the Moon near the first- or third-quarter phase. Sketch or photograph the Moon at low magnification, and then zoom in on a region of interest. Again sketch or photograph your field of view, labeling its features and identifying the geological process that created them. Look for craters, volcanic plains, and tectonic features. Estimate the size of each feature by comparing it to the size of the entire Moon (radius = 1738 km).

## Quantitative Problems

*Be sure to show all calculations clearly and state your final answers in complete sentences.*

55. *Surface Area–to–Volume Ratio.* Compare the surface area–to–volume ratios of **(a)** the Moon and Mars and **(b)** Earth and Venus. In each case, use your answer to discuss differences in the internal heat of the two worlds.
56. *Doubling Your Size.* Just as the surface area–to–volume ratio depends on size, so can other properties. To see how, suppose that your size suddenly doubled—that is, your height, width, and depth all doubled. (For example, if you were 5 feet tall before, you now are 10 feet tall.)
**a.** By what factor has your waist size increased?
**b.** How much more material will be required for your clothes? (*Hint:* Clothes cover the surface *area* of your body.)
**c.** By what factor has your weight increased? (*Hint:* Weight depends on the *volume* of your body.)
**d.** The pressure on your weight-bearing joints depends on how much *weight* is supported by the *surface area* of each joint. How has this pressure changed?
57. *Lunar Footprints.* Assume that the Moon is hit by about 25 million micrometeorite impacts each day (this number comes from observations of meteors in Earth's atmosphere) and that these impacts strike randomly around the Moon's surface. Also assume that it takes about 20 such impacts to destroy a footprint. About how long would it take for one of the footprints left by the *Apollo* astronauts to be erased? (*Hints:* Use the Moon's surface area to determine the impact rate per square centimeter, and estimate the size of a footprint.)
58. *Geological Proportions.* Express the approximate height and width of Olympus Mons (26 km tall and 600 km wide) as percentages of Mars's radius. Repeat for Valles Marineris (7 km deep and 4000 km long). Then compare the height of Mount Everest (9 km tall) and the height and width of the Grand Canyon (1.8 km deep and 450 km long) to Earth's radius, and comment on the relative sizes of geological features on the two planets.
59. *Internal vs. External Heating.* In daylight, Earth's surface absorbs about 400 watts per square meter. Earth's internal radioactivity produces a total of 30 trillion watts that leak out through our planet's entire surface. Calculate the amount of heat from radioactive decay that flows outward through each square meter of Earth's surface (your answer should have units of watts per square meter). Compare quantitatively to solar heating, and comment on why internal heating drives geological activity.
60. *Plate Tectonics.* Typical motion of one plate relative to another is 1 centimeter per year. At this rate, how long would it take for two continents 3000 kilometers apart to collide? What are the global consequences of motion like this?
61. *More Plate Tectonics.* Consider a seafloor spreading zone creating 1 centimeter of new crust over its entire 2000-kilometer length every year. How many square kilometers of surface will be created in 100 million years? What fraction of Earth's surface does this constitute?

# 10 Planetary Atmospheres

## EARTH AND THE OTHER TERRESTRIAL WORLDS

▲ **About the photo:** Viewed from orbit, the atmosphere upon which our lives depend is a remarkably thin layer of gas surrounding our planet.

## LEARNING GOALS

### 10.1 Atmospheric Basics
- What is an atmosphere?
- How does the greenhouse effect warm a planet?
- Why do atmospheric properties vary with altitude?

### 10.2 Weather and Climate
- What creates wind and weather?
- What factors can cause long-term climate change?
- How does a planet gain or lose atmospheric gases?

### 10.3 Atmospheres of the Moon and Mercury
- Do the Moon and Mercury have any atmosphere?

### 10.4 The Atmospheric History of Mars
- What is Mars like today?
- How has Mars's climate differed in the past?
- Why did Mars change?

### 10.5 The Atmospheric History of Venus
- What is Venus like today?
- How did Venus get so hot?

### 10.6 Earth's Unique Atmosphere
- How did Earth's atmosphere end up so different?
- Why does Earth's climate stay relatively stable?
- How is human activity changing our planet?

*Seeing Earth as a small blue ball in the vast dark ocean of space gives you a new perspective on life and what is important. You can see how small we are as compared to the universe and how fragile our lives are.*

— *Anousheh Ansari, first astronaut of Iranian heritage and first female private space explorer*

**Chapter 10 Overview**

Life as we know it would be impossible on Earth without our atmosphere. This thin layer of gas supplies the oxygen we breathe, shields us from harmful ultraviolet and x-ray radiation from the Sun, protects us from continual bombardment by micrometeorites, generates rain-giving clouds, and traps just enough heat to keep Earth habitable.

How did Earth end up with atmospheric conditions that are so favorable to life? In this chapter, we'll explore the answers—and learn why the other terrestrial worlds ended up so different, despite having formed under similar conditions. We'll also discuss why Earth's climate remains relatively stable and how human activity may be threatening that stability, with potential consequences that we are only beginning to understand.

## 10.1 Atmospheric Basics

The atmospheres of the terrestrial worlds are even more varied than their geologies. **FIGURE 10.1** shows global and surface views of each world. **TABLE 10.1** lists general characteristics of their atmospheres. You'll notice vast differences between the worlds.

The Moon and Mercury have so little atmosphere that it's reasonable to call them "airless"; they have no wind or weather of any kind. At the other extreme, Venus is enshrouded by a thick atmosphere composed almost entirely of carbon dioxide, giving it surface conditions so hot and harsh that not even robotic space probes have survived there for long. Mars also has a carbon dioxide atmosphere, but its air is so thin that you'd die within minutes if you ventured outside without a pressurized space suit. Only Earth has the "just right" conditions that allow liquid water on the surface, making it hospitable to life.

Despite these great differences among the terrestrial atmospheres, the same basic processes are at work in all cases. In fact, the same processes are at work on any world with an atmosphere, and in later chapters we'll see how they apply to the atmospheres of the jovian planets, of moons with atmospheres, and even of planets around other stars. Let's begin our comparative study of planetary atmospheres by discussing the basic nature of an atmosphere.

### What is an atmosphere?

An *atmosphere* is a layer of gas that surrounds a world. In most cases, it is a surprisingly thin layer. On Earth, for example, about two-thirds of the air in the atmosphere lies within 10 kilometers of the surface (**FIGURE 10.2**). You could represent this air on a standard globe (to scale) with a layer only as thick as a dollar bill.

Atmospheric air is a mixture of gases that may consist either of individual atoms or of molecules. Temperatures in the terrestrial atmospheres are generally low enough (even on Venus) for atoms to combine into molecules. For example, the air we breathe consists of *molecular* nitrogen ($N_2$) and oxygen ($O_2$), as opposed to individual atoms (N or O). Other common molecules in terrestrial atmospheres include water ($H_2O$) and carbon dioxide ($CO_2$).

**Atmospheric Pressure** Collisions of individual atoms or molecules in an atmosphere create *pressure* [**Section 5.3**] that pushes in all directions. On Earth, for example, the nitrogen and oxygen molecules in the air fly around at average speeds of about 500 meters per second—fast enough to cross your bedroom a hundred times in 1 second. Given that a single breath of air contains more than a billion trillion

**FIGURE 10.1** Views of the terrestrial worlds and their atmospheres from orbit and from the surface. The surface views for Mercury and Venus are artists' conceptions; the others are photos. The global views are visible-light photos taken from spacecraft. (Venus appears in gibbous phase as it was seen by the *Galileo* spacecraft during its Venus flyby en route to Jupiter.)

**TABLE 10.1 Atmospheres of the Terrestrial Worlds**

| World | Composition of Atmosphere | Surface Pressure* | Average Surface Temperature | Winds, Weather Patterns | Clouds, Hazes |
|---|---|---|---|---|---|
| **Mercury** | helium, sodium, oxygen | $10^{-14}$ bar | day: 425°C (797°F)<br>night: −175°C (−283°F) | none: too little atmosphere | none |
| **Venus** | 96% carbon dioxide ($CO_2$)<br>3.5% nitrogen ($N_2$) | 90 bars | 470°C (878°F) | slow winds, no violent storms, acid rain | sulfuric acid clouds |
| **Earth** | 77% nitrogen ($N_2$)<br>21% oxygen ($O_2$)<br>1% argon<br>$H_2O$ (0.4%, but variable)<br>0.04% carbon dioxide ($CO_2$) | 1 bar | 15°C (59°F) | winds, hurricanes, rain, snow | $H_2O$ clouds, pollution |
| **Moon** | helium, sodium, argon | $10^{-14}$ bar | day: 125°C (257°F)<br>night: −175°C (−283°F) | none: too little atmosphere | none |
| **Mars** | 95% carbon dioxide ($CO_2$)<br>2.7% nitrogen ($N_2$)<br>1.6% argon | 0.007 bar | −50°C (−58°F) | winds, dust storms | $H_2O$ and $CO_2$ clouds, dust |

*1 bar = the atmospheric pressure at sea level on Earth.

molecules, you can imagine how frequently molecules collide. On average, each molecule in the air around you will suffer a million collisions in the time it takes to read this paragraph. These collisions create pressure that pushes in all directions, and this pressure holds up the atmosphere so that it does not collapse under its own weight.

A balloon offers a good example of how pressure works in a gas. The air molecules inside a balloon exert pressure, pushing outward as they continually collide with the balloon's inside surface. At the same time, outside air molecules collide with the balloon's outer surface, exerting pressure that by itself would make the balloon collapse. A balloon stays inflated when the inward and outward pressures are balanced (**FIGURE 10.3a**). (We are ignoring the tension in the rubber of the balloon walls.) Imagine that you blow more air into the balloon (**FIGURE 10.3b**). The extra molecules inside mean more collisions with the balloon wall, momentarily making the pressure inside greater than the pressure outside. The balloon therefore expands until the inward and outward pressures are again in balance.

If you heat the balloon (**FIGURE 10.3c**), the gas molecules begin moving faster and collide harder and more frequently with the inside surface, which also momentarily increases the inside pressure until the balloon expands. As it expands, the pressure inside it decreases and the balloon comes back into pressure balance. Conversely, cooling a balloon makes it contract, because the outside pressure momentarily exceeds the inside pressure.

**FIGURE 10.2** Earth's atmosphere, visible in this photograph from the Space Shuttle, is a very thin layer over Earth's surface. Most of the air is in the lowest 10 kilometers of the atmosphere, visible along the edge of the planet.

**See it for yourself** Find an empty plastic bottle with a screwtop that makes a good seal. Warm the air inside by filling the bottle partway with hot water and then shaking and emptying the bottle. Seal the bottle and place it in the refrigerator or freezer for about 15 minutes. What happens, and why? Explain in terms of the individual air molecules.

We can understand **atmospheric pressure** by applying similar principles. Gas in an atmosphere is held down by gravity. The atmosphere above any given altitude therefore has some weight that presses downward, tending to compress the atmosphere beneath it. At the same time, the fast-moving molecules exert pressure in all directions, including upward, which tends to make the atmosphere expand. Planetary atmospheres exist in a perpetual balance between the downward weight of their gases and the upward push of their gas pressure.

The higher you go in an atmosphere, the less the weight of the gas above you, and less weight means less pressure. That is why the pressure decreases as you climb a mountain or ascend in an airplane. You can visualize this

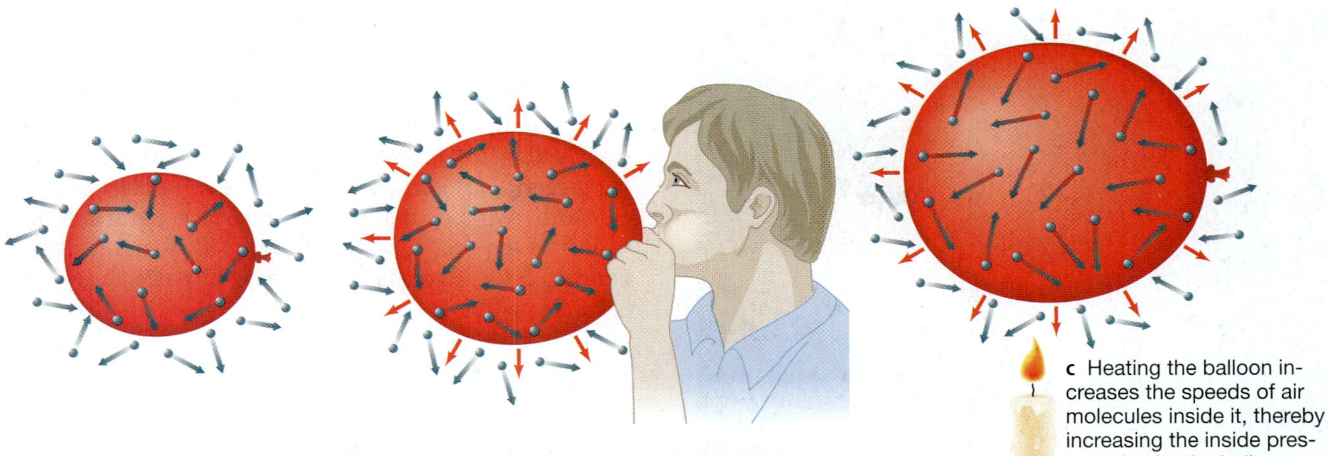

a A balloon stays inflated when the inside and outside pressures are balanced.

b Adding air molecules temporarily increases the pressure inside the balloon, so the balloon expands until the pressure balance is restored.

c Heating the balloon increases the speeds of air molecules inside it, thereby increasing the inside pressure. Again, the balloon expands until the pressure balance is restored.

**FIGURE 10.3** Gas pressure in a balloon depends on both density and temperature.

concept by imagining the atmosphere as a very big stack of pillows (**FIGURE 10.4**). The pillows at the bottom are highly compressed because of the weight of all the pillows above. Going upward, the pillows are less and less compressed because less weight lies on top of them.

In planetary science, we usually measure atmospheric pressure in a unit called the **bar** (as in *barometer*). One bar is roughly equal to Earth's atmospheric pressure at sea level. It is also equivalent to 1.03 kilograms per square centimeter or 14.7 pounds per square inch. In other words, if you gathered up all the air directly above any 1 square inch of Earth's surface, you would find that it weighed about 14.7 pounds.

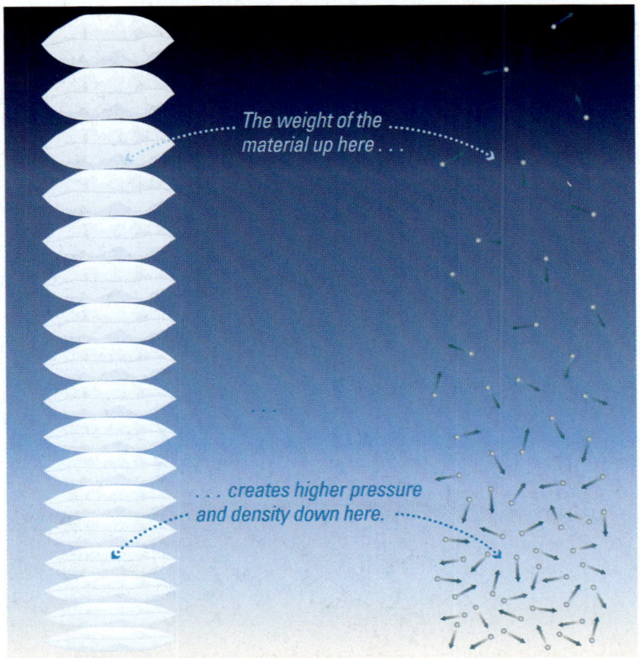


**FIGURE 10.4** Atmospheric density and pressure decrease with altitude for the same reason that a giant stack of pillows would be more compressed at the bottom than at the top. The pressure in each layer is enough to hold up the weight of layers above, counteracting the force of gravity.

**Think about it** Suppose Mars had exactly the same total amount of air above each square inch of surface as Earth. Would the atmospheric pressure be higher, lower, or the same as on Earth? Explain. (*Hint:* Remember that weight depends on the strength of gravity [**Section 4.1**].)

You might wonder why you don't feel Earth's atmospheric weight bearing down on you. After all, if we placed 14.7 pounds of weight on every square inch of your shoulders, you'd certainly feel the downward pressure. You don't notice the atmospheric pressure for two reasons: First, the pressure pushes in all directions, so it pushes upward and inward on you as well as downward. Second, the fluids in your body push outward with an equivalent pressure, so there is no net pressure trying to compress or expand your body. You can tell that the pressure is there, however, because you'll quickly notice any pressure changes. For example, even slight changes in pressure as you go up or down in altitude can cause your ears to "pop." More extreme changes, such as those that affect deep-sea divers when they rise too rapidly, can be deadly. You'd face even greater pressure differences if you visited other planets; as shown in Table 10.1, atmospheric pressure varies by a factor of $10^{16}$ among the terrestrial worlds.

**The "Top" of an Atmosphere** If you study Figure 10.4, you'll see that there is no clear boundary between the atmosphere and space above, because pressure and density decrease gradually with increasing altitude. At some point, the density becomes so low that we can't really think of the gas as "air" anymore. Collisions between atoms or molecules are rare at these altitudes, and the gas is so thin that it would look and feel as if you had already entered space. On Earth, this occurs at an altitude of about 100 kilometers; above that, the sky is black even in the daytime, much like the sky on the Moon. This altitude is often described as "the edge of space."

However, some gas is still present above 100 kilometers. Earth's tenuous upper atmosphere extends for several

**FIGURE 10.5** Low-density gas extends into what we often think of as "space," as you can see from the visible glow of gas around a Space Shuttle's tail in this photograph. Notice the aurora in the background, created by interactions between charged particles and atmospheric gas at altitudes above 100 kilometers, well below the Shuttle. Most of Earth's atmospheric gas lies at altitudes below 10 kilometers.

hundred kilometers above Earth's surface. The International Space Station and many satellites orbit Earth within these outer reaches of the atmosphere (**FIGURE 10.5**). The low-density gas may be barely noticeable under most conditions, but it still exerts drag on orbiting spacecraft. That is why satellites in low-Earth orbit slowly spiral downward, eventually burning up as they reenter the denser layers of the atmosphere [**Section 4.5**].

**How Atmospheres Affect Planets** Earth's atmosphere is obviously important to our existence, but the full range of atmospheric effects is greater than most people realize. To provide context for our study of planetary atmospheres, here is a brief list of key effects of atmospheres that we will study in this chapter:

- As we've seen, atmospheres create pressure that determines whether liquid water can exist on the surface.
- Atmospheres absorb and scatter light. Scattering can make daytime skies bright and absorption can prevent dangerous radiation from reaching the ground.
- Atmospheres can create wind and weather and play a major role in long-term climate change.
- Interactions between atmospheric gases and the solar wind can create a protective *magnetosphere* around planets with strong magnetic fields.
- Atmospheres can make planetary surfaces warmer than they would be otherwise via the **greenhouse effect.**

The greenhouse effect is arguably the most important effect that an atmosphere can have on its planet, and we'll therefore focus our attention on it first.

## How does the greenhouse effect warm a planet?

You've probably heard of the greenhouse effect, because it plays a key role in the problem known as *global warming* [**Section 10.6**]. But you may be surprised to learn that the greenhouse effect is also critical to the existence of life on Earth. Without the greenhouse effect, Earth's surface would be too cold for liquid water to flow and for life to flourish. Let's explore how the greenhouse effect can warm planetary surfaces.

**The Greenhouse Effect** **FIGURE 10.6** shows the basic idea behind the greenhouse effect. The energy that warms a planet comes from sunlight, and in particular from visible light. Some of this visible light is reflected back to space, and the rest is absorbed by the surface. The absorbed energy must ultimately be returned to space, but planetary surfaces are too cool to emit visible light. (Recall that the type of thermal radiation an object emits depends on its temperature [**Section 5.4**].) Instead, planetary surface temperatures are in the range in which they emit mostly infrared light. The greenhouse effect works by temporarily "trapping" some of this infrared light, slowing its return to space.

The greenhouse effect occurs only when an atmosphere contains gases that can absorb the infrared light. Gases that are particularly good at absorbing infrared light are called **greenhouse gases**, and they include water vapor ($H_2O$), carbon dioxide ($CO_2$), and methane ($CH_4$). These gases absorb infrared light effectively because their molecular structures begin rotating or vibrating when they absorb an infrared photon (see Figure 5.18).

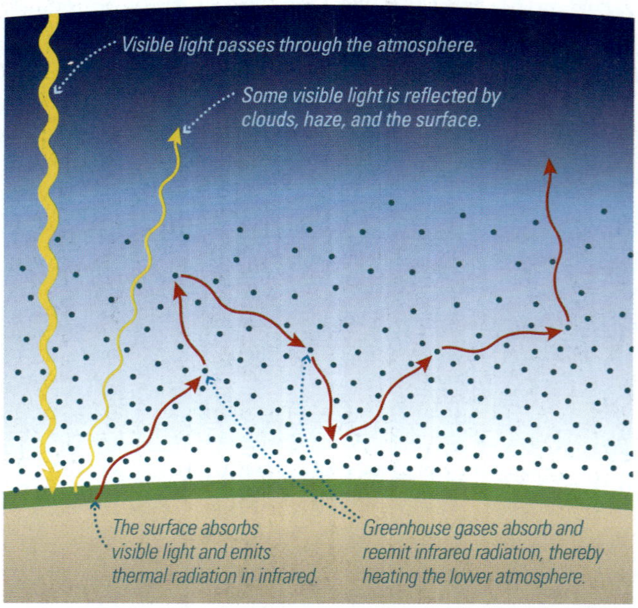


**FIGURE 10.6** This diagram shows how the greenhouse effect makes a planet's surface and lower atmosphere warmer than they would be otherwise. The yellow arrows represent visible light, the red arrows represent infrared light, and the blue dots represent greenhouse gas molecules.

A greenhouse gas molecule that absorbs an infrared photon does not retain this energy for long; instead, it quickly reemits it as another infrared photon, which may head off in any random direction. This photon can then be absorbed by another greenhouse gas molecule, which does the same thing. The net result is that greenhouse gases tend to slow the escape of infrared radiation from the lower atmosphere, while their molecular motions heat the surrounding air. In this way, the greenhouse effect makes the surface and the lower atmosphere warmer than they would be from sunlight alone. The more greenhouse gases present, the greater the degree of surface warming. A blanket offers a good analogy: You stay warmer under a blanket not because the blanket itself provides any heat, but because it slows the escape of your body heat into the cold outside air.

Note that the greenhouse effect by itself does *not* alter a planet's overall energy balance: As long as the strength of the greenhouse effect hasn't changed, the total amount of energy that a planet receives from the Sun will be precisely balanced with the amount of energy it returns to space through reflection and radiation. If it were not, the planet would either heat up (if it received more energy than it returned) or cool down (if it returned more energy than it received). In steady state, all the energy "trapped" by the greenhouse effect still escapes to space, just not as directly as it would otherwise.*

*However, if the greenhouse effect is gaining or losing strength, the energy balance will no longer be perfect. For example, today's current global warming is causing Earth's atmosphere and oceans to gain a combined total of about 250 trillion joules of energy every second.

**Think about it** Clouds on Earth are made of water ($H_2O$), which acts as a very effective greenhouse "gas" even when the water vapor condenses into droplets in clouds. Use this fact to explain why clear nights tend to be colder than cloudy nights.

Incidentally, the name *greenhouse effect* is a bit of a misnomer. The name comes from botanical greenhouses, but greenhouses actually trap heat through a different mechanism than planetary atmospheres: Rather than absorbing infrared radiation, greenhouses stay warm by preventing warm air from rising.

**Greenhouse Warming of the Terrestrial Worlds** We can better appreciate the importance of the greenhouse effect by comparing each planet's average surface temperature—or **global average temperature**—with and without it. Recall that a terrestrial planet's interior heat has very little effect on its surface temperature, so sunlight is the only significant energy source for the surface [**Section 9.1**]. Therefore, without the greenhouse effect, a planet's global average surface temperature would depend on only two things:

- *The planet's distance from the Sun,* which determines the amount of energy received from sunlight. The closer a planet is to the Sun, the greater the intensity of the incoming sunlight.
- *The planet's overall reflectivity,* which determines the relative proportions of incoming sunlight that the planet reflects and absorbs. The higher the reflectivity, the less light absorbed and the cooler the planet.

## MATHEMATICAL INSIGHT 10.1 "No Greenhouse" Temperatures

The "no greenhouse" temperature of a planet depends only on its distance from the Sun, its overall reflectivity, and the Sun's brightness. It can be calculated with the following formula:

$$T_{\text{"no greenhouse"}} = 280\text{ K} \times \sqrt[4]{\frac{(1 - \text{reflectivity})}{d^2}}$$

where $d$ is the distance from the Sun in AU; reflectivity should be stated as a fraction (such as 0.2 rather than 20%). The symbol $\sqrt[4]{\;}$ means the fourth root, or $\frac{1}{4}$ power.

We will not derive the formula here, but you can see how it makes sense. The term (1 − reflectivity) is the proportion of sunlight that the planet absorbs, which is the light that heats its surface. This term is divided by $d^2$ because the amount of energy from sunlight (per unit area) declines with the square of distance from the Sun. The full term $(1 - \text{reflectivity})/d^2$ therefore represents the total amount of energy that the planet absorbs from sunlight (per unit area) each second. This energy warms the surface, which returns the energy to space as thermal radiation. Because the energy emitted by thermal radiation depends on temperature raised to the fourth power (see Mathematical Insight 5.2), calculating the planet's temperature requires taking the fourth root of the absorbed energy. The 280 K in the formula is the temperature of a perfectly black planet with zero reflectivity at 1 AU from the Sun.

**EXAMPLE:** Calculate the "no greenhouse" temperature of Mercury.

**SOLUTION:**

**Step 1 Understand:** We can use the "no greenhouse" formula with Mercury's distance and reflectivity from Table 10.2; its distance is 0.387 AU and we write the reflectivity of 12% as 0.12.

**Step 2 Solve:** We substitute the values to find

$$\begin{aligned} T_{\text{"no greenhouse"}} &= 280\text{ K} \times \sqrt[4]{\frac{(1 - \text{reflectivity})}{d^2}} \\ &= 280\text{ K} \times \sqrt[4]{\frac{(1 - 0.12)}{0.387^2}} = 436\text{ K} \end{aligned}$$

**Step 3 Explain:** Mercury's "no greenhouse" temperature is 436 K. We subtract 273 to convert it to Celsius, finding that 436 K is equivalent to 436 − 273 = 163°C. Note that this agrees with the value in Table 10.2. It also makes sense: Because Mercury has no greenhouse effect, its "no greenhouse" temperature should be roughly midway between its day and night temperatures. The precise halfway point of the day and night temperatures shown in Table 10.2 is 125°C; although this is not an exact match to 163°C, it is close enough for us to be confident that we have the right general idea.

**TABLE 10.2 The Greenhouse Effect on the Terrestrial Worlds**

| World | Average Distance from Sun (AU) | Reflectivity | "No Greenhouse" Average Surface Temperature* | Actual Average Surface Temperature | Greenhouse Warming (actual temperature minus "no greenhouse" temperature) |
|---|---|---|---|---|---|
| Mercury | 0.387 | 12% | 163°C | day: 425°C<br>night: −175°C | 0°C |
| Venus | 0.723 | 75% | −40°C | 470°C | 510°C |
| Earth | 1.00 | 29% | −16°C | 15°C | 31°C |
| Moon | 1.00 | 12% | −2°C | day: 125°C<br>night: −175°C | 0°C |
| Mars | 1.524 | 16% | −56°C | −50°C | 6°C |

*The "no greenhouse" temperature is calculated by assuming no change to the atmosphere other than lack of greenhouse warming. For example, Venus has a lower "no greenhouse" temperature than Earth even though it is closer to the Sun, because the high reflectivity of its bright clouds means that it absorbs less sunlight than Earth.

Note that a planet's reflectivity (sometimes called its *albedo*) depends on its composition and color; darker colors reflect less light. For example, clouds, snow, and ice reflect more than 70% of the light that hits them, absorbing only about 30%, while rocks typically reflect only about 20% of the light that hits them and absorb the other 80%.

**See it for yourself** Find a black shirt and a white shirt of similar thickness. Drape one over each hand and hold your hands near a light bulb. Which hand gets warmer? Explain what's happening using the terms *absorb, reflect,* and *emit.*

Both distance from the Sun and reflectivity have been measured for all the terrestrial worlds. With a little mathematics, these measurements can be used to calculate the "no greenhouse" temperature that each world would have without greenhouse gases (see Mathematical Insight 10.1). **TABLE 10.2** shows the results. The "no greenhouse" temperatures for Mercury and the Moon lie between their actual day and night temperatures, since they have little atmosphere and hence no greenhouse effect. Mars has a weak greenhouse effect that makes its global average temperature only 6°C higher than its "no greenhouse" temperature. Venus is the extreme case, with a greenhouse effect that bakes its surface to a temperature more than 500°C hotter than it would be otherwise.

We can also see why the greenhouse effect is so important to life on Earth. Without the greenhouse effect, our planet's global average temperature would be a chilly −16°C (+3°F), well below the freezing point of water. With it, the global average temperature is about 15°C (59°F), or about 31°C warmer than the "no greenhouse" temperature. This greenhouse warming is even more remarkable when you realize that it is caused by gases, such as water vapor and carbon dioxide, that are only trace constituents of Earth's atmosphere. Most of the atmosphere consists of nitrogen ($N_2$) and oxygen ($O_2$) molecules, which have no effect on infrared light and do not contribute to the greenhouse effect. (Molecules with only two atoms, especially those with two of the same kind of atom, such as $N_2$ and $O_2$, are poor infrared absorbers because they have very few ways to vibrate and rotate.)

**Think about it** Suppose nitrogen and oxygen were greenhouse gases. Assuming our atmosphere still had the same composition and density, how would Earth be different?

## Why do atmospheric properties vary with altitude?

The greenhouse effect can warm a planet's surface and lower atmosphere, but other processes affect the temperature at higher altitudes. The way in which temperature varies with altitude determines what is often called the **atmospheric structure.** Earth's atmospheric structure has four basic layers (**FIGURE 10.7**), each of which affects the planet in a distinct way:

- The **troposphere** is the lowest layer, in which temperature drops with altitude (something you've probably noticed if you've ever climbed a mountain).
- The **stratosphere** begins where the temperature stops dropping and instead begins to rise with altitude. High in the stratosphere, the temperature falls again.*
- The **thermosphere** begins where the temperature again starts to rise at high altitude.
- The **exosphere** is the uppermost region, in which the atmosphere gradually fades away into space.

### Interactions Between Light and Atmospheric Gases

The key to understanding atmospheric structure lies in interactions between atmospheric gases and energy from the Sun. Although visible light dominates the solar spectrum, the Sun also emits significant amounts of ultraviolet light and x-rays. In addition, the planetary surface emits infrared light. Atmospheric gases interact with each of these forms of light in different ways (**FIGURE 10.8**):

- X-rays have enough energy to *ionize* (knock electrons from) almost any atom or molecule. They can therefore be absorbed by virtually all atmospheric gases.

*Technically, the region where the temperature falls again is called the *mesosphere,* but we will not make that distinction in this book.

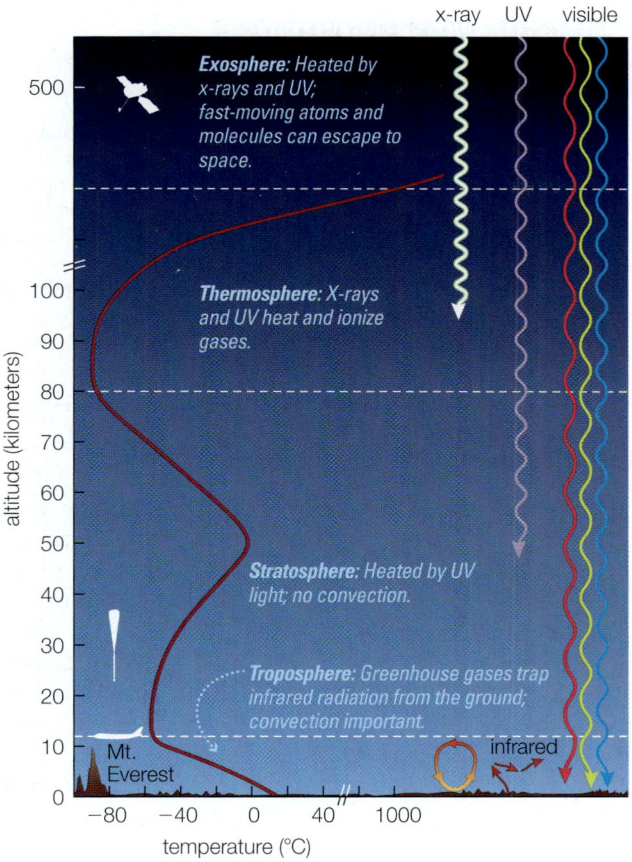


**FIGURE 10.7** Earth's average atmospheric structure. The curve shows average temperature for each altitude. The squiggly arrows to the right represent light of different wavelengths and show where it is typically absorbed.

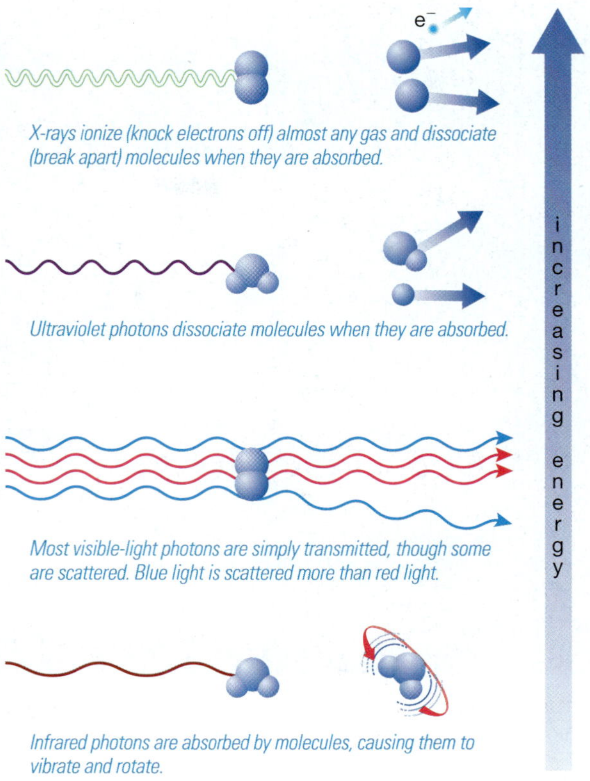


**FIGURE 10.8** The primary effects when light of different energies strikes common atmospheric gases.

- Ultraviolet photons generally do not have enough energy to cause ionization, but they can sometimes break molecules apart. For example, ultraviolet photons can split water ($H_2O$) molecules and are even more likely to be absorbed by weakly bonded molecules, such as ozone ($O_3$), which split apart in the process.
- Visible-light photons generally pass through atmospheric gases without being absorbed, but some are *scattered* so that their direction changes.
- As we've already discussed, infrared photons can be absorbed by greenhouse gases, which are molecules that easily begin rotating and vibrating.

With these ideas in mind, we are ready to examine the reasons for Earth's atmospheric structure, working our way from the ground up.

## COMMON MISCONCEPTIONS

### Temperatures at High Altitude

Many people think that the low temperatures in the mountains are just the result of lower pressures, but Figure 10.7 shows that it's not that simple. The higher temperatures near sea level on Earth are a result of the greenhouse effect, which traps more heat at lower altitudes. If Earth had no greenhouse gases, mountaintops wouldn't be so cold—or, more accurately, sea level wouldn't be so warm.

**Visible Light: Warming the Surface and Coloring the Sky** As we've discussed, most visible sunlight is either absorbed by the surface or reflected to space, but a small amount of the visible light is scattered by atmospheric molecules, and this scattering has two important effects.

First, scattering makes the daytime sky bright, which is why we can't see stars in the daytime. Without scattering, sunlight would travel only in perfectly straight lines, which means we'd see the Sun against an otherwise black sky, just as it appears on the Moon. Scattering also prevents shadows on Earth from being pitch black. On the Moon, shadows receive little scattered sunlight and are extremely cold and dark.

Second, scattering explains why our sky is blue. Visible light consists of all the colors of the rainbow, but not all the colors are scattered equally. Gas molecules scatter blue light (higher energy) much more effectively than red light (lower energy). The difference in scattering is so great that, for practical purposes, we can imagine that only the blue light gets scattered. When the Sun is overhead, this scattered blue light reaches our eyes from all directions and the sky appears blue (**FIGURE 10.9**). At sunset or sunrise, the sunlight must pass through a greater amount of atmosphere on its way to us. Most of the blue light is scattered away, leaving only red light to color the sky.

**Infrared Light and the Troposphere** The surface returns the energy it absorbs by radiating in the infrared. Greenhouse gases absorb this infrared light and warm the troposphere. Because the infrared light comes from the surface, more is absorbed closer to the surface than at higher altitudes, which is why the temperature drops with altitude in

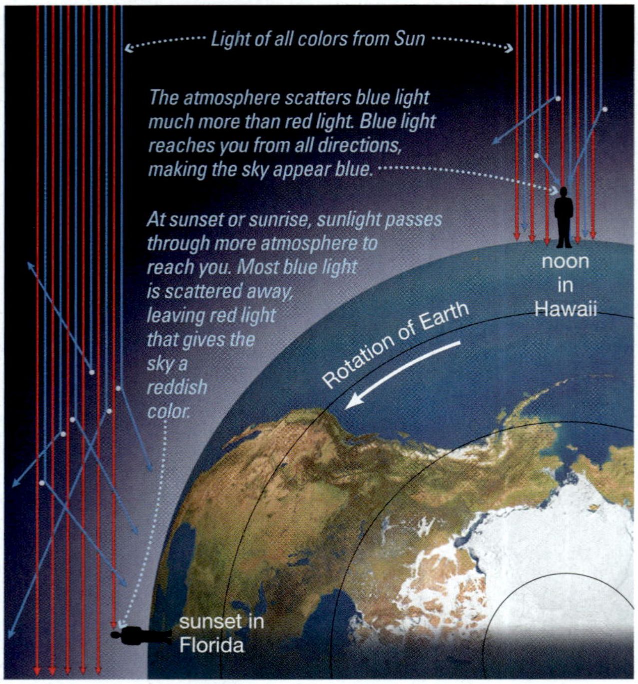


**FIGURE 10.9** This diagram summarizes why the daytime sky is blue and sunsets (and sunrises) are red.

the troposphere. (The relatively small amount of infrared light coming from the Sun does not have a significant effect on the atmosphere.)

The drop in temperature with altitude, combined with the relatively high density of air in the troposphere, explains why the troposphere is the only layer of the atmosphere with storms. The primary cause of storms is the churning of air by convection [**Section 9.1**], in which warm air rises and cool air falls. Recall that convection occurs only when there is strong heating from below; in the troposphere, the heating from the ground can drive convection. In fact, the troposphere gets its name from convection; *tropos* is Greek for "turning."

**Ultraviolet Light and the Stratosphere** Above the troposphere, the air density is too low for greenhouse gases to have much effect, so infrared light from below can travel unhindered through higher layers of the atmosphere and into space. Heating from below therefore has little effect on the stratosphere. Instead, the primary source of heating in the stratosphere is the absorption of solar ultraviolet light by ozone.

Most of this ultraviolet absorption and heating occurs at moderately high altitudes in the stratosphere, which is why temperature tends to *increase* with altitude as we go upward from the base of the stratosphere. This temperature structure prevents convection in the lower stratosphere, because heat cannot rise if the air above is hotter. The lack of convection makes the air relatively stagnant and *stratified* (layered), with layers of warm air overlying cooler air; this stratification explains the name *stratosphere*. The lack of convection also means that the stratosphere has essentially no weather and no rain. Pollutants that reach the stratosphere, including the ozone-destroying chemicals known as chlorofluorocarbons (CFCs), remain there for decades.

## COMMON MISCONCEPTIONS

### Why Is the Sky Blue?

If you ask around, you'll find a wide variety of misconceptions about why the sky is blue. Some people guess that the sky is blue because of light reflecting from the oceans, but that could not explain blue skies over inland areas. Others claim that "air is blue," a vague statement that is clearly wrong: If air molecules emitted blue light, then air would glow blue even in the dark; if they were blue because they reflected blue light and absorbed red light, then no red light could reach us at sunset. The real explanation for the blue sky is light scattering, as shown in Figure 10.9, which also explains our red sunsets.

Note that a planet can have a stratosphere *only* if its atmosphere contains molecules that are particularly good at absorbing ultraviolet photons. Ozone ($O_3$) plays this role on Earth, but the lack of oxygen in the atmospheres of the other terrestrial worlds means that they also lack ozone. As a result, Earth is the only terrestrial world with a stratosphere, at least in our solar system. (The jovian planets have stratospheres due to other ultraviolet-absorbing molecules [**Section 11.1**].)

**X-Rays and the Thermosphere** Because nearly all gases are good x-ray absorbers, x-rays from the Sun are absorbed by the first gases they encounter as they enter the atmosphere. The density of gas in the exosphere is too low for it to absorb significant amounts of these x-rays, so most x-rays are absorbed in the thermosphere. The absorbed energy makes temperatures quite high in the thermosphere (*thermos* is Greek for "hot"), but you wouldn't feel much heat because the density and pressure are so low [**Section 4.3**]. Virtually no x-rays penetrate beneath the thermosphere, which is why x-ray telescopes are useful only on very high-flying balloons, rockets, and spacecraft.

Solar x-rays also ionize a small but important fraction of the thermosphere's gas. The portion of the thermosphere that contains most of the ionized gas is called the *ionosphere*. The ionosphere is very important to radio communication, because it reflects most radio broadcasts back to Earth's surface. Without this reflection, radio communication would work only between locations in sight of each other.

**The Exosphere** The exosphere is the extremely low-density gas that forms the gradual and fuzzy boundary between the atmosphere and space (*exo* means "outermost" or "outside"). The gas density in the exosphere is so low that collisions between atoms or molecules are very rare, although the high temperature means that gas particles move quite rapidly. Lightweight atoms and molecules sometimes reach escape velocity [**Section 4.5**] and fly off into space.

**Comparative Structures of Terrestrial Atmospheres** We can now understand the structures of all the terrestrial atmospheres. The Moon and Mercury have so little gas that they essentially contain only an exosphere and have no structure to speak of. **FIGURE 10.10** contrasts the atmospheric structures of Venus, Earth, and Mars. Notice that

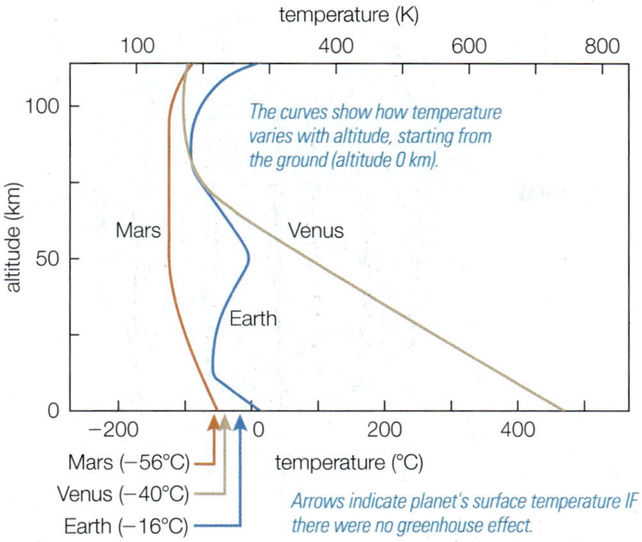


**FIGURE 10.10** Venus, Earth, and Mars all have tropospheres and thermospheres (and exospheres, not shown), but only Earth has middle-atmosphere heating to make a stratosphere. The graph also shows that Venus and Earth are considerably warmer than they would be without the greenhouse effect.

all three planets have a troposphere warmed by the greenhouse effect, and all three have a thermosphere heated by solar x-rays. However, only Earth has the extra "bump" of a stratosphere, because it is the only planet with a layer of ultraviolet-absorbing gas (ozone). Without ozone, the middle altitudes of Earth's atmosphere would be almost as cold as those on Mars.

**Think about it** Would astronauts on the Moon need protection from solar x-rays? What about astronauts on Mars? Explain.

**Magnetospheres and the Solar Wind** There is one other important type of energy coming from the Sun: the low-density flow of subatomic charged particles called the *solar wind* [**Section 8.2**]. The solar wind does not significantly affect atmospheric structure, but it has other important effects.

On the Moon and Mercury, solar wind particles hit the surface, where they can blast atoms free. On Venus and Mars, solar wind particles can strip away atmospheric gas. In contrast, Earth's strong magnetic field creates a **magnetosphere** that acts like a protective bubble surrounding our planet, deflecting most solar wind particles around it (**FIGURE 10.11a**).

The magnetosphere still allows a few solar wind particles to get through, especially near the magnetic poles. Once inside the magnetosphere, these particles move along magnetic field lines, collecting in **charged particle belts** (or *Van Allen belts*, after their discoverer) that encircle our planet. The high energies of the particles in these belts can be hazardous to spacecraft and astronauts passing through them.

Charged particles trapped in the magnetosphere also create the beautiful light of the **aurora** (**FIGURE 10.11b**). Variations in the solar wind can rattle and shake the magnetosphere and give energy to particles trapped there. If a trapped particle gains enough energy, it can follow the magnetic field all the way down to Earth's atmosphere, where it collides with atmospheric atoms and molecules. These collisions cause the atoms and molecules to radiate and produce the moving lights of the aurora. Because the charged particles follow the magnetic field, auroras are most common near the magnetic poles and are best viewed at high latitudes. In the Northern Hemisphere, the aurora is often called the *aurora borealis*, or northern lights. In the Southern Hemisphere, it is called the *aurora australis*, or southern lights. The aurora can also be seen from space,

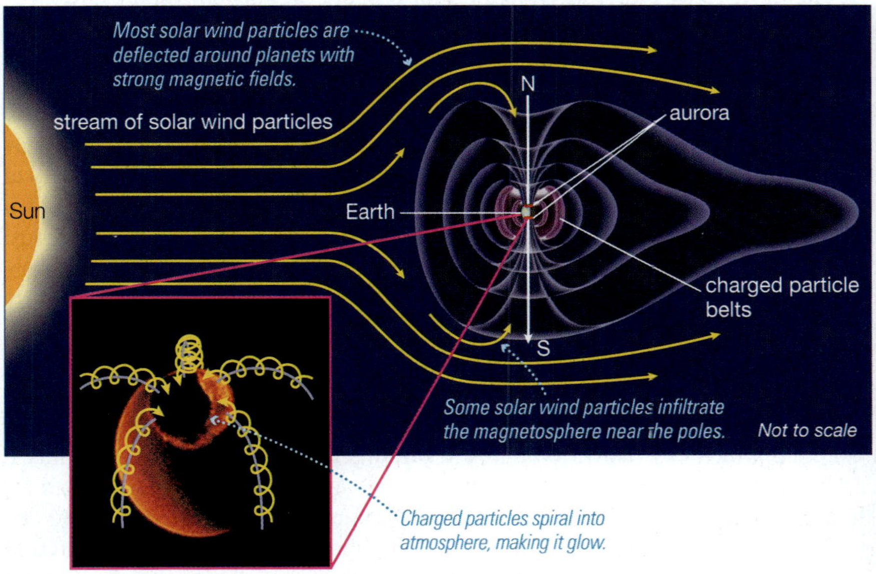


**a** This diagram shows how Earth's magnetosphere deflects solar wind particles. Some particles accumulate in charged particle belts encircling our planet. The inset is a photo of a ring of auroras around the North Pole.

**b** This photograph shows the aurora near Yellowknife, Northwest Territories, Canada. In a video, you would see these lights dancing about in the sky.

**FIGURE 10.11** Earth's magnetosphere acts like a protective bubble, shielding our planet from the charged particles of the solar wind.

where the lights look much like surf in the upper atmosphere (see Figure 10.5). In differing forms, auroras have also been detected on all other planets with significant atmospheres in our solar system, and even on some moons with atmospheres.

## 10.2 Weather and Climate

So far, we've talked mostly about *average* conditions in planetary atmospheres, such as the global average temperature and the average atmospheric structure. However, experience tells us that surface and atmospheric conditions continually change through weather and climate.

Weather and climate are closely related, but they are not quite the same thing. **Weather** refers to the ever-varying combination of winds, clouds, temperature, and pressure that makes some days hotter or cooler, clearer or cloudier, or calmer or stormier than others. **Climate** refers to the *average* of weather over many years. For example, we say that Antarctic deserts have a cold, dry climate, even though it might sometimes rain or snow. Geological records show that climates change naturally, but these changes usually occur gradually over decades, centuries, or millennia.

Weather and climate can be hard to distinguish on a human time scale. For example, a drought lasting a few years may be the result of random weather fluctuations or the beginning of a gradual change to a drier climate. The difficulty in distinguishing between random weather and real climate trends is an important part of the debate about human influences on the climate. We'll return to this topic in Section 10.6; for now, let's focus on understanding weather and climate generally.

### What creates wind and weather?

Wind, rain, and other weather phenomena are all driven by energy in the atmosphere, which means that only planets with atmospheres can have weather. Even then, weather varies dramatically among different worlds. Earth has the most diverse weather of the terrestrial planets, so we'll use it as our example of how weather works.

The complexity of weather makes it difficult to predict, and at best the local weather can be predicted only a week or so in advance (see Special Topic, page 283). Nevertheless, when we look at Earth as a whole, we can identify certain general characteristics of weather.

**Global Wind Patterns** The wind's direction and strength can change rapidly in any particular place, but we find distinctive patterns on a more global scale. For example, winds generally cause storms moving in from the Pacific to hit the West Coast of the United States first and then make their way eastward across the Rocky Mountains and the Great Plains, heading to the East Coast.

**FIGURE 10.12** shows Earth's major **global wind patterns** (or *global circulation*). Notice that the wind direction varies with latitude: Equatorial winds blow from east to west, mid-latitude winds blow from west to east, and high-latitude winds blow like equatorial winds from east to west. Two factors explain this pattern: atmospheric heating and planetary rotation. Let's examine each in turn.

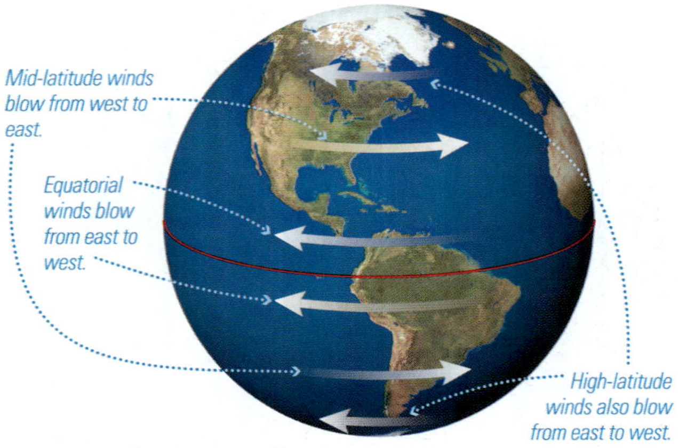


**FIGURE 10.12** Schematic of Earth's global wind patterns. Notice that mid-latitude winds blow from west to east, explaining why storms generally move across the United States from the West Coast toward the East Coast.

**Think about it** Hurricanes usually form over warm, equatorial waters. Based on the wind patterns shown in Figure 10.12, in which direction will a hurricane move after forming off the northeastern coast of South America? Explain why these wind patterns make the Caribbean and the southeastern United States especially vulnerable to hurricanes.

**Atmospheric Heating and Circulation Cells** Atmospheric heating affects global wind patterns because equatorial regions receive more heat from the Sun than polar regions. Warm equatorial air therefore rises upward and flows toward the poles, where cool air descends and flows toward the equator. If Earth's rotation did not influence this process, the result would be two huge **circulation cells** (or *Hadley cells*, after the man who first suggested their existence), one in each hemisphere (**FIGURE 10.13**).

The circulation cells transport heat both from lower to higher altitudes and from the equator to the poles. They therefore make Earth's polar regions much warmer than they would be in the absence of circulation. The same idea applies to different extents on Venus and Mars. On Venus, the dense atmosphere allows the circulation cells to transport so much thermal energy that temperatures are nearly the same at the equator and the poles. On Mars, the circulation cells transport very little heat because the atmosphere is so thin, so the poles remain much colder than the equator.

**Rotation and the Coriolis Effect** Planetary rotation affects global wind patterns through the **Coriolis effect** (named for the French physicist who first explained it), which you can understand by thinking about a spinning merry-go-round (**FIGURE 10.14**). The outer parts of the merry-go-round move at a faster speed than the inner parts, because they have a greater distance to travel around the axis with each rotation. If you sit near the edge and roll a ball toward the center, the ball begins with your relatively high speed around the axis. As it rolls inward, the ball's high speed

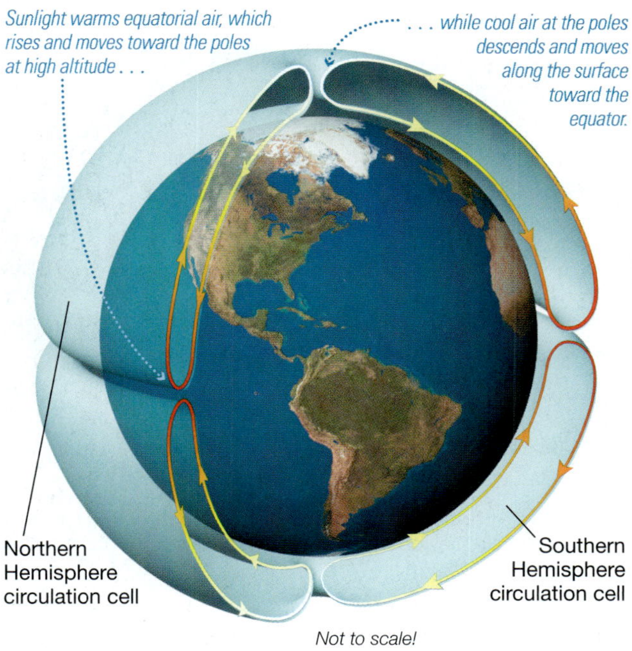


**FIGURE 10.13** Atmospheric heating creates a circulation cell in each hemisphere that carries warm equatorial air toward the poles while cool polar air moves toward the equator. This diagram shows how the circulation cells would work if our planet's rotation didn't affect them. The vertical extent of the circulation cells is greatly exaggerated; in reality, the tops of the cells are only a few kilometers above Earth's surface.

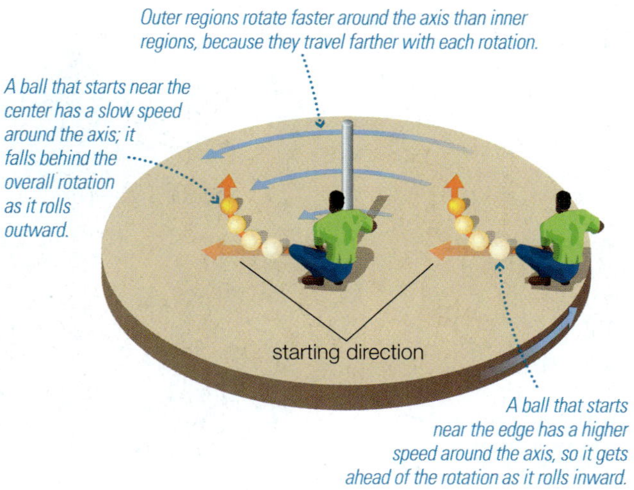


**FIGURE 10.14** The Coriolis effect on a merry-go-round rotating counterclockwise. Notice that the ball's path deviates to the *right* regardless of whether the ball is rolled inward or outward. If the merry-go-round were rotating clockwise, in both cases the ball would veer to the left.

makes it move ahead of the slower-moving inner regions. On a merry-go-round rotating counterclockwise, the ball therefore deviates *to the right* instead of heading straight inward. The Coriolis effect also makes the ball deviate to the right if you roll it outward from a position near the center. In that case, the ball starts with your slower speed around the center and lags behind as it rolls outward to faster-moving regions, again deviating to the right. (If the merry-go-round rotates clockwise rather than counterclockwise, the deviations go to the left instead of the right.)

The Coriolis effect alters the path of air on the rotating Earth in much the same way (**FIGURE 10.15**). Equatorial regions circle around Earth's rotation axis faster than polar regions (see Figure 1.13). Air moving away from the equator therefore has "extra" speed that causes it to move ahead of Earth's rotation to the east, while air moving toward the equator lags behind Earth's rotation to the west. In either case, moving air turns to the *right* in the Northern Hemisphere and to the *left* in the Southern Hemisphere, which explains why storms circulate in opposite directions in the two hemispheres. Uneven heating and cooling of Earth's surface creates regions of slightly higher pressure ("H" on weather maps) or lower pressure ("L" on weather maps) than average. Storms generally occur around low-pressure regions, which draw air inward from surrounding regions; as shown in Figure 10.15a, the Coriolis effect makes the inward-flowing air rotate counterclockwise in the Northern Hemisphere and clockwise in the Southern Hemisphere. This rotation around a low-pressure zone can be quite stable, which is why storms can persist for days or weeks while being carried across the globe.

**Think about it** The Coriolis effect also is important for long-range missiles. Suppose you are in the Northern Hemisphere on Earth and are trying to send a missile to hit a target that lies 5000 kilometers due north of you. Should you aim directly at the target, somewhat to the left, or somewhat to the right? Explain.

The Coriolis effect plays an even more important role in shaping Earth's global wind patterns: It splits each of the two huge circulation cells shown in Figure 10.13 into three smaller circulation cells (**FIGURE 10.16**). You can understand why by considering air flowing southward along the surface from the North Pole. Without rotation, this air would travel 10,000 kilometers due south to the equator. But as Earth rotates, the Coriolis effect diverts this air to the right (westward) well before it reaches the equator, forcing the single large circulation cell to split. The three resulting cells circulate the air somewhat like three interlocking gears.

## COMMON MISCONCEPTIONS

### Toilets in the Southern Hemisphere

A common myth holds that water circulates "backward" in the Southern Hemisphere, so that toilets flush the opposite way, water spirals down sink drains the opposite way, and so on. If you visit an equatorial town, you may even find charlatans who, for a small fee, will demonstrate such changes as they step across the equator. This myth probably arises from the fact that hurricanes really do spiral in opposite directions in the two hemispheres—but it is completely untrue. To understand why, remember that the circulation of hurricanes is due to the Coriolis effect, which is noticeable only when material moves significantly closer to or farther from Earth's rotation axis, which means distances of hundreds of kilometers. It is completely negligible for objects as small as toilets or drains; the direction in these cases is determined by small-scale effects such as the shape of the sink or the initial direction of rotation, which is why you can find water swirling down drains in either direction, no matter where you are located on Earth.

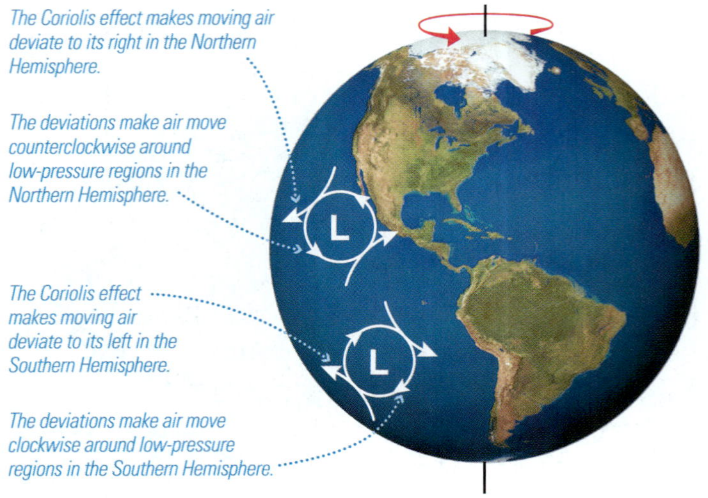


**a** Low-pressure regions ("L") draw in air from surrounding areas, and the Coriolis effect causes this air to circulate counterclockwise in the Northern Hemisphere and clockwise in the Southern Hemisphere.

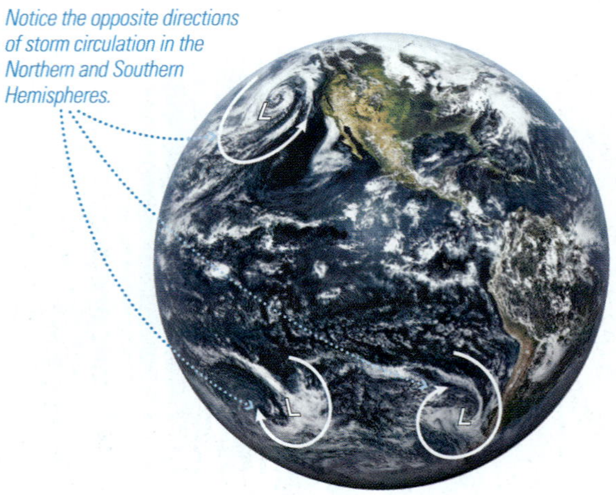


**b** This photograph shows the opposite directions of storm circulation in the two hemispheres.

**FIGURE 10.15** The Coriolis effect works on the rotating Earth much as it does on a merry-go-round, because regions near the equator move at faster speed around the axis than regions near the poles.

These motions explain the global wind directions: Notice that surface air moves toward the equator in the cells near the equator and near the poles, so the Coriolis effect diverts this air into westward winds (see Figure 10.12). In contrast, surface air moves toward the poles in the mid-latitude cells, so the Coriolis effect diverts it into winds that blow eastward. In essence, the Coriolis effect on a rotating planet tends to divert air moving north or south into east-west winds.

The Coriolis effect operates to some extent on all planets. Its strength depends on a planet's size and rotation rate: Larger size and faster rotation both contribute to a stronger Coriolis effect. Among the terrestrial planets, Earth is the only one with a Coriolis effect strong enough to split the two large circulation cells. Venus has a very weak Coriolis effect because of its slow rotation; Mars has a weak Coriolis effect because of its small size. The Coriolis effect is much stronger on the large and fast-rotating jovian planets, and it can therefore split their circulation cells into numerous smaller cells [**Section 11.1**].

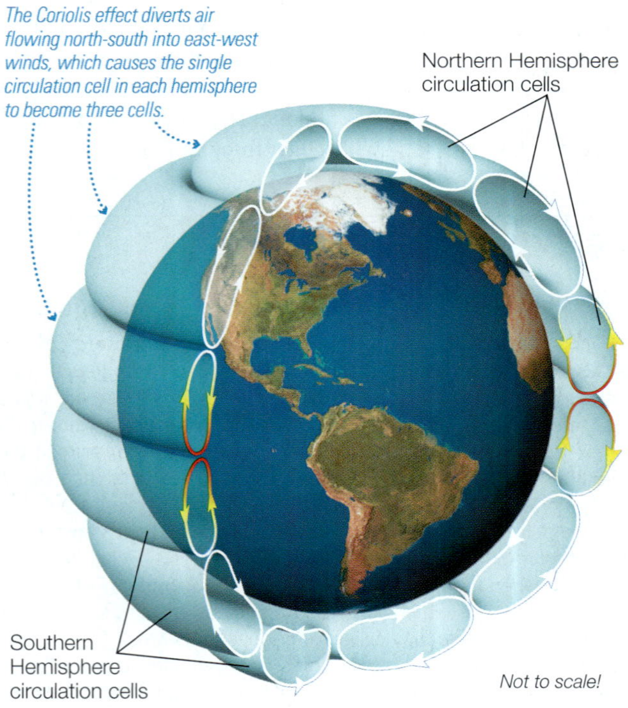


**FIGURE 10.16** On Earth, the Coriolis effect causes each of the two large circulation cells that would be present without rotation (see Figure 10.13) to split into three smaller cells.

**Clouds and Precipitation** In addition to winds, the other key components of weather are rain, snow, and hail, which together are called **precipitation** in weather reports. Precipitation requires clouds. We may think of clouds as imperfections on a sunny day, but they have profound effects on Earth and other planets, despite being made from minor ingredients of the atmosphere (see Table 10.1). Besides being the source of precipitation, clouds can alter a planet's energy balance. Clouds reflect sunlight back to space, thereby reducing the amount of sunlight that warms a planet's surface, but they also tend to be made from greenhouse molecules that contribute to planetary warming.

On Earth, clouds are made from tiny droplets of liquid water or flakes of ice, which you can feel if you walk through a cloud on a mountaintop. Clouds are produced by condensation of water vapor (**FIGURE 10.17**). The water vapor enters the atmosphere through vaporization of surface water or of ice and snow. Convection then carries the water vapor to high, cold regions of the troposphere, where it can condense to form clouds. Clouds can also form as winds blow over mountains, because the mountains force the air high enough to allow condensation. The condensed droplets or ice flakes start out very small, but gradually grow larger. If they get large enough that the upward convection currents can no longer hold them aloft, then they begin to fall toward the surface as rain, snow, or hail.

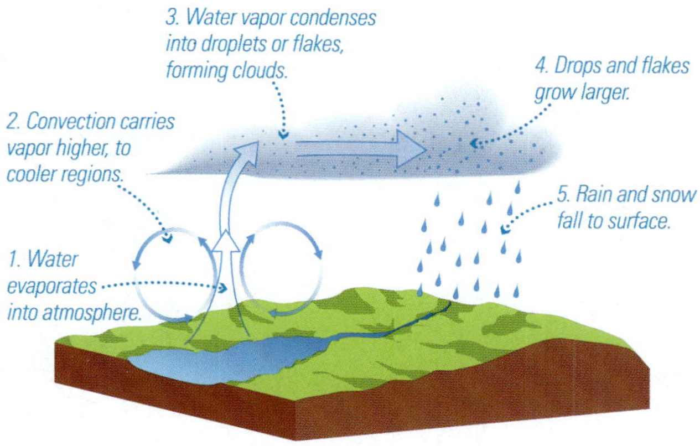


**FIGURE 10.17** The cycle of water on Earth's surface and in the atmosphere.

Stronger convection means more clouds and precipitation. That is why thunderstorms are common on summer afternoons, when the sunlight-warmed surface drives strong convection. The linkage between clouds and convection also explains why Earth has lush jungles near the equator and deserts at latitudes of 20°–30° north or south. Equatorial regions experience high rainfall because they receive more sunlight, which causes more convection. This high rainfall depletes the air of moisture as Earth's circulation cells carry it away from the equator (see Figure 10.16), leaving little moisture to fall as rain at the latitudes of the deserts.

## What factors can cause long-term climate change?

We now turn our attention to climate, which varies much more slowly than weather. Scientists have identified four major factors that can lead to long-term climate change on the terrestrial worlds:

- **Solar brightening.** The Sun has grown gradually brighter with time, increasing the amount of solar energy reaching the planets.
- **Changes in axis tilt.** The tilt of a planet's axis may change over long periods of time.
- **Changes in reflectivity.** An increase in a planet's reflectivity—for example, from increased cloud cover, increased ice cover, or particles released from volcanoes—means a decrease in the amount of sunlight it absorbs, and vice versa.
- **Changes in greenhouse gas abundance.** More greenhouse gases tend to make a planet warmer, and less make it cooler.

**FIGURE 10.18** summarizes these four factors, which we now investigate in a little more detail. Keep in mind that more than one factor may be acting at any given time.

**Solar Brightening** Both theoretical models of the Sun and observations of other Sun-like stars tell us that the Sun has gradually brightened with age [**Section 14.2**]: The Sun today

### SPECIAL TOPIC Weather and Chaos

Scientists today have a very good understanding of the physical laws and mathematical equations that govern the behavior of the atmosphere and oceans. Why, then, do we have so much trouble predicting the weather? To understand the answer, we must look at the nature of scientific prediction.

Prediction always requires knowing two things: (1) the current state of a system, sometimes called its *initial conditions,* and (2) how the system is changing. This is easy for something simple, like a car; if you know where a car is and how fast and in what direction it is traveling, you can predict where it will be a few minutes from now. Weather prediction is more difficult because weather is created by the motions of countless individual atoms and molecules. Scientists therefore attempt to predict the weather by creating a "model world." For example, suppose you overlay a globe of Earth with graph paper and specify the current temperature, pressure, cloud cover, and wind within each square; these are your initial conditions, which you can input into a computer along with a set of equations (physical laws) describing the processes that can change weather from one moment to the next. You can now use your computer model to predict the weather for the next month in New York City. The model might tell you that tomorrow will be warm and sunny, with cooling during the next week and a major storm passing through a month from now.

Now, suppose you run the model again, but you make one minor change in the initial conditions—say, a small change in the wind speed somewhere over Brazil. This slightly different initial condition will not change the weather prediction for tomorrow in New York City, and may only slightly affect the prediction for next week's weather. For next month's weather, however, the two predictions may not agree at all!

The disagreement between the two predictions arises because the laws governing weather can cause tiny changes in initial conditions to be greatly magnified over time. This extreme sensitivity to initial conditions is sometimes called the *butterfly effect:* If initial conditions change by as much as the flap of a butterfly's wings, the resulting long-term prediction may be very different. That is why it is possible to predict the weather with reasonable accuracy a few days in advance, but not a few months in advance.

The butterfly effect is a hallmark of *chaotic systems.* Simple systems are described by linear equations in which, for example, increasing a cause produces a proportional increase in an effect. In contrast, chaotic systems are described by nonlinear equations, which allow for subtler and more intricate interactions. For example, the economy is nonlinear because a rise in interest rates does not automatically produce a corresponding change in consumer spending. Weather is nonlinear because a change in the wind speed in one location does not automatically produce a corresponding change in another location.

Despite their name, chaotic systems are not necessarily random. In fact, many chaotic systems have a kind of underlying order that explains the general features of their behavior even though details at any particular moment remain unpredictable. In a sense, many chaotic systems—like the weather—are "predictably unpredictable."

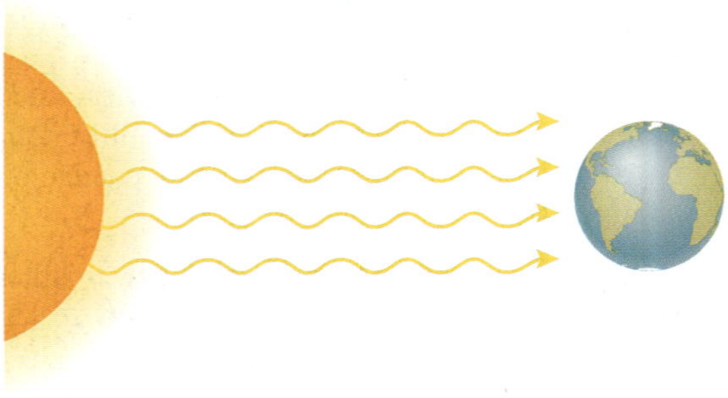

*Solar brightening:* As the Sun brightens with time, the increasing sunlight tends to warm the planets.

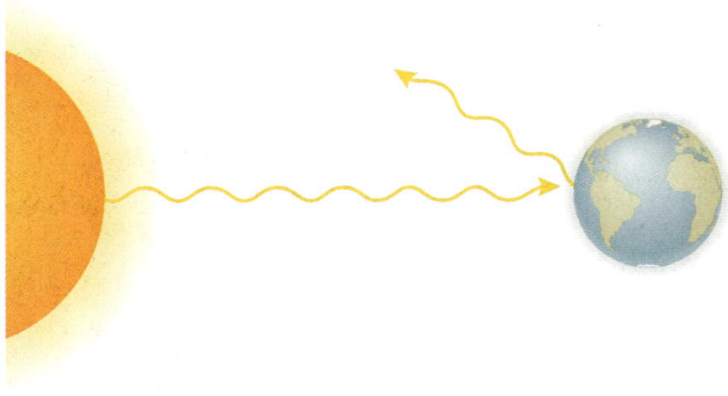

*Changes in reflectivity:* Higher reflectivity tends to cool a planet, while lower reflectivity leads to warming.

*Changes in axis tilt:* Greater tilt makes more extreme seasons, while smaller tilt keeps polar regions colder.

*Changes in greenhouse gas abundance:* An increase in greenhouse gases slows escape of infrared radiation, warming the planet, while a decrease leads to cooling.

**FIGURE 10.18** Four major factors affecting long-term climate change.

is probably about 30% brighter than it was when our solar system was young. This brightening tends to warm climates with time, though we'd expect the effects to be noticeable only over periods of tens of millions of years or longer. Moreover, the brightening is so gradual that other climate change factors can easily overwhelm it. For example, while solar brightening alone would mean that all planets should be warmer today than they were in the past, both Mars and Earth are cooler than they were at times in their early histories.

**Changes in Axis Tilt** Small gravitational tugs from moons, other planets, and the Sun can change a planet's axis tilt over thousands or millions of years. For example, while Earth's current axis tilt is about $23\frac{1}{2}°$, the tilt has varied over tens of thousands of years between about 22° and 25° (**FIGURE 10.19**). These small changes affect the climate by making seasons more or less extreme. Greater tilt means more extreme seasons, with warmer summers and colder winters. The extra summer warmth tends to prevent ice from building up, which reduces the planet's reflectivity and thereby makes the whole planet warmer. Conversely, a smaller tilt means less extreme seasons, which can allow ice to build up and make a planet cooler.

Earth's past periods of smaller axis tilt correlate well with the times of past ice ages (at least over the past few million years), especially when considered along with other small changes in Earth's rotation and orbit. They are therefore thought to be a primary factor in climate changes on Earth. (The cyclical changes in Earth's axis tilt and orbit are often called *Milankovitch cycles,* after the Serbian scientist who suggested their role in climate change.) As we'll discuss shortly, Mars probably experiences much more extreme changes in axis tilt and climate than Earth, and similar climate cycles may occur on some of the icy moons of the outer solar system.

**Changes in Reflectivity** Changes in reflectivity affect climate because they change the proportions of sunlight absorbed and reflected. If a planet reflects more sunlight, it must absorb less, which can lead to planetwide cooling. Microscopic dust particles (called *aerosols*) released by volcanic eruptions can reflect sunlight and cool a planet. Small but measurable planetwide cooling has been detected on Earth following a major volcanic eruption. The cooling can continue for years if the dust particles reach the stratosphere. Changes in reflectivity can also occur as ice caps grow and recede, since ice is highly reflective. More clouds

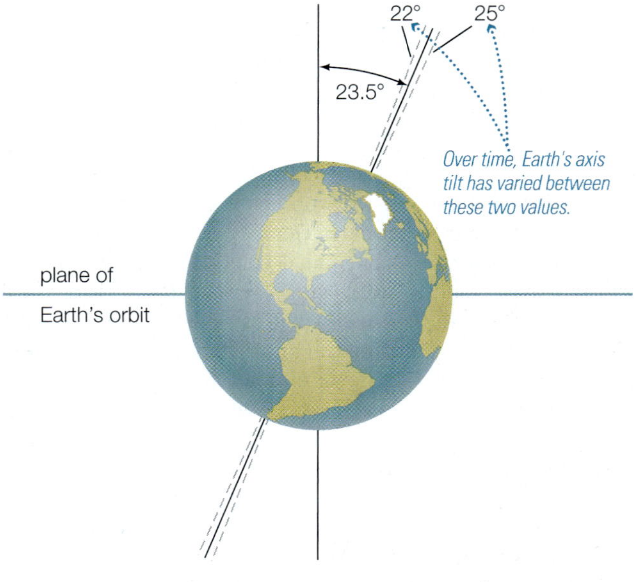


**FIGURE 10.19** Earth's axis tilt is currently about $23\frac{1}{2}°$, but it has varied between about 22° and 25° over tens of thousands of years.

also mean more reflectivity, though this may be offset by the fact that clouds also contribute to greenhouse warming.

Human activity is currently altering Earth's reflectivity, although we are not sure in which direction or by precisely how much. Smog particles can act like volcanic dust, reflecting sunlight before it reaches the ground. Deforestation also increases reflectivity because it removes sunlight-absorbing plants. On the other hand, roads and cities tend to decrease reflectivity, which is why they tend to be hotter than surrounding areas of vegetation, and the loss of Arctic sea ice attributed to global warming decreases reflectivity because liquid water reflects much less sunlight than ice.

**Changes in Greenhouse Gas Abundance** Perhaps the most important factor in long-term climate change is a change in the abundance of greenhouse gases, which strengthens or weakens the greenhouse effect. If the abundance of greenhouse gases increases, the planet generally will warm. If the planet warms enough, increased vaporization of water may add substantial amounts of gas to the planet's atmosphere, leading to an increase in atmospheric pressure. Conversely, if the abundance of greenhouse gases decreases, the planet generally will cool, and atmospheric pressure may decrease as gases freeze.

## How does a planet gain or lose atmospheric gases?

Of the factors that affect planetary climate, changes in greenhouse gas concentrations appear to have had the greatest effect on the long-term climates of Venus, Earth, and Mars. Such changes generally occur as part of more general changes in the abundances of atmospheric gases. We must therefore investigate how atmospheres gain and lose gas.

**Sources of Atmospheric Gas** Terrestrial atmospheres can gain gas in three basic ways (**FIGURE 10.20**):

- **Outgassing.** Volcanic *outgassing* has been the primary source of gases for the atmospheres of Venus, Earth, and Mars. Recall that the terrestrial worlds were built primarily of metal and rock, but impacts of ice-rich planetesimals (from beyond the frost line [**Section 8.2**]) brought in water and gas that became trapped in their interiors during accretion. Studies of volcanic eruptions show that the most common gases released by outgassing are water ($H_2O$), carbon dioxide ($CO_2$), nitrogen ($N_2$), and sulfur-bearing gases ($H_2S$ and $SO_2$).*
- **Vaporization.** After outgassing creates an atmosphere, some atmospheric gases may condense to become surface liquids or ices. The subsequent vaporization of these surface liquids (evaporation) and ices (sublimation) therefore represents a secondary source of atmospheric gas. For example, if a planet warms, the rates of vaporization will increase, adding gas to the atmosphere.
- **Surface ejection.** The tiny impacts of micrometeorites, solar wind particles, and high-energy solar photons can knock individual atoms or molecules free from the surface. This *surface ejection* process explains the small amounts of gas that surround the Moon and Mercury. It is not a source process for planets that already have substantial atmospheres, because the atmospheres prevent small particles and high-energy solar photons from reaching the surface.

**Losses of Atmospheric Gas** Planets can also lose atmospheric gas, and there are four major loss processes (**FIGURE 10.21**). Note that the first two of these processes

*Among these materials, only water existed in modest quantities in the solar nebula. The others were created by chemical reactions that occurred *inside* the planets after gas became trapped.

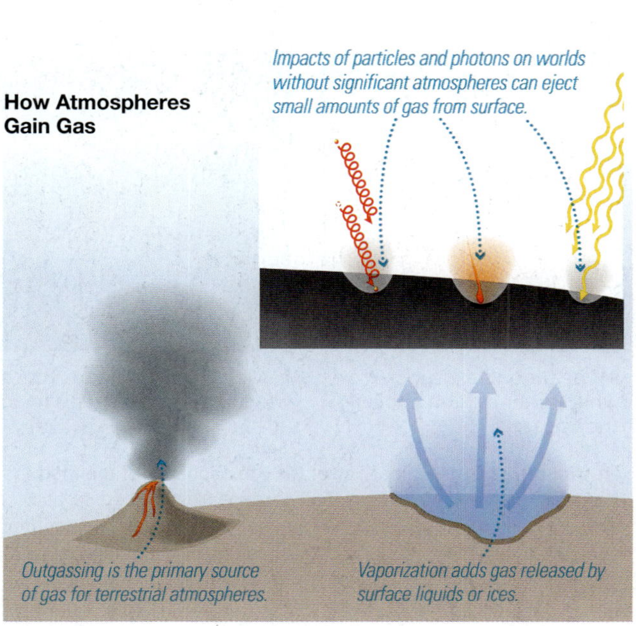


**FIGURE 10.20** Three processes that can provide gas to terrestrial atmospheres.

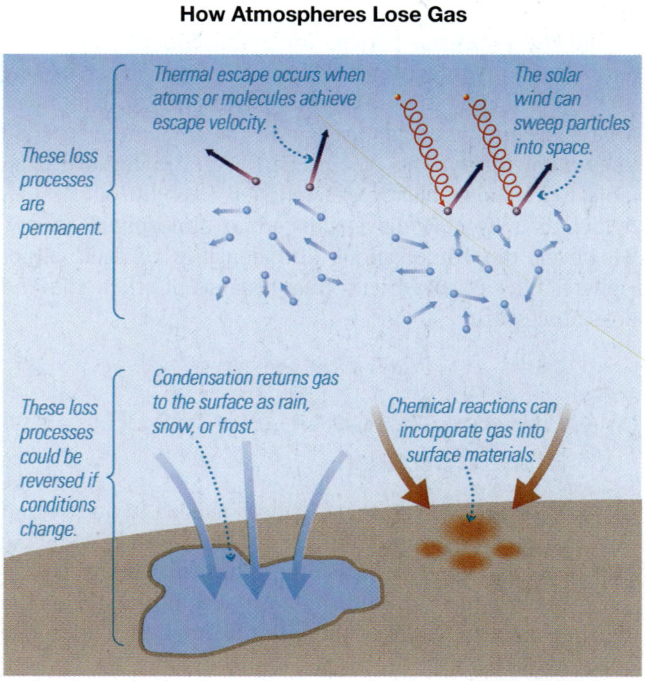


**FIGURE 10.21** Four processes that can remove gas from terrestrial atmospheres.

simply recycle gas from the atmosphere to the planet's surface or interior, while the latter two lead to permanent loss of gas.

- **Condensation.** The condensation of gases that then fall as rain, hail, or snow is essentially the reverse of the release of gas by vaporization. On Mars, for example, it is cold enough for carbon dioxide to condense into dry ice (frozen carbon dioxide), especially at the poles.
- **Chemical reactions.** Some chemical reactions incorporate gas into surface metal or rock. Rusting is a familiar example: Iron rusts when it reacts with oxygen, thereby removing oxygen from the atmosphere and incorporating it into the metal.
- **Solar wind stripping.** For any world without a protective magnetosphere, particles from the solar wind can gradually strip away gas particles into space.
- **Thermal escape.** If an atom or a molecule of gas in a planet's exosphere achieves escape velocity [**Section 4.5**], it will fly off into space. The relative importance of thermal escape on any world depends on its size, distance from the Sun, and atmospheric composition. In general, more thermal escape will occur if a planet is small (so that it has a low escape velocity) or close to the Sun (which makes it hotter, so that atoms and molecules of atmospheric gas are moving faster). Lightweight gases, such as hydrogen and helium, escape more easily than heavier gases, such as carbon dioxide, nitrogen, and oxygen.

**See it for yourself** You can see some atmospheric gain and loss processes in your freezer. Look for loose ice cubes that are gradually shrinking away, or look for the buildup of frost. What process is occurring in each case, and where do these processes occur in the solar system?

All the terrestrial planets are small enough and warm enough for hydrogen and helium to escape. That is why they were unable to hold on to any hydrogen or helium gas that they may have captured from the solar nebula when they were very young. Venus, Earth, and Mars have been able to retain heavier gases released later by outgassing, which is why they have substantial atmospheres today. However, the compositions and densities of their atmospheres have changed with time because of other gain and loss processes.

## 10.3 Atmospheres of the Moon and Mercury

We have now covered the basic ideas needed to understand the atmospheric histories of the terrestrial worlds. In the rest of this chapter, we'll use these to learn how and why each world ended up with its current atmosphere. As we did in Chapter 9, we'll begin with the smallest worlds, the Moon and Mercury.

### Do the Moon and Mercury have any atmosphere?

We usually don't think of the Moon and Mercury as having atmospheres, but they are not totally devoid of gas. However, their gas densities are far too low for sunlight to be scattered or absorbed. The lack of scattering means that, even in broad daylight, you would see a pitch-black sky surrounding the bright Sun. The lack of absorption means that their atmospheres do not have a troposphere, stratosphere, or thermosphere. In essence, the Moon and Mercury have only extremely low-density exospheres, without any other atmospheric layers.

The total amount of gas in the exospheres of the Moon and Mercury is very small. If you could condense the entire atmosphere of either the Moon or Mercury into solid form, you would have so little material that you could almost store it in a dorm room. The low density of the gas means that collisions between atoms or molecules are rare. The gas particles therefore can rise as high as their speeds allow—sometimes even escaping to space. As a result, the exospheres of the Moon and Mercury extend thousands of kilometers above their surfaces (**FIGURE 10.22**).

#### Source and Loss Processes on the Moon and Mercury

The Moon and Mercury may once have had some gas released by volcanic outgassing, but they no longer have volcanic activity, and any gas released in the distant past is long gone. Some of the gas released long ago was probably lost through stripping by the solar wind, but it would have been lost to thermal escape anyway. Mercury cannot hold much of an atmosphere because its small size and high daytime temperature mean that nearly all gas particles eventually achieve escape velocity. The Moon is cooler than Mercury, but its smaller size gives it a lower escape velocity; as a result, gases escape about as easily on both worlds.

The only ongoing source of gas on the Moon and Mercury is the surface ejection that occurs when micrometeorites,

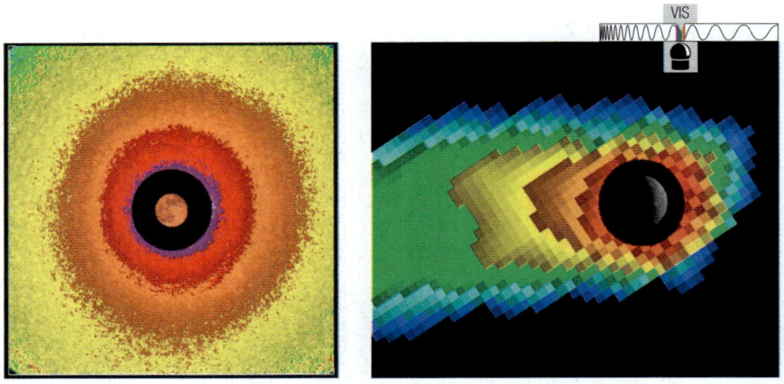


**a** The Moon's exosphere, which extends high above the surface.

**b** Mercury's exosphere, much of which is escaping in this image.

**FIGURE 10.22** These "images" of the atmospheres of the Moon and Mercury—which are essentially exospheres only—are based on data collected with instruments sensitive to emission lines from sodium atoms; the colors represent gas density from highest (red) to lowest (blue). The inset photos of the Moon and Mercury are on the same scale as the images.

solar wind particles, or high-energy solar photons knock free surface atoms and molecules. This gas never accumulates because it is lost as quickly as it is gained. Some of the relatively few particles released from the surface are blasted upward fast enough to achieve escape velocity and therefore escape directly to space. The rest bounce around like tiny rubber balls, arcing hundreds of kilometers into the sky before crashing back down to the surface. Each gas particle typically bounces a few dozen times before being absorbed back into the surface.

**Ice in Polar Craters** There cannot be liquid water on the Moon or Mercury, because the lack of significant atmospheric pressure means that water cannot remain stable in liquid form. Daytime temperatures on both worlds are generally high enough that any water ice would long ago

## MATHEMATICAL INSIGHT 10.2 Thermal Escape from an Atmosphere

Thermal escape depends on the speeds of gas particles (atoms or molecules), because a particle can escape to space only if it reaches or exceeds a world's escape velocity. The fact that gas particles continually collide with one another means that individual particles move at a wide range of speeds. However, these speeds are not totally random; instead, their distribution has a characteristic shape that depends on the temperature and the particle masses. **FIGURE 1** shows this characteristic shape for sodium atoms in the Moon's daytime exosphere ($T = 400$ K). The peak in the figure represents the most common speed, or *peak thermal velocity,* which is given by the following formula:

$$v_{\text{thermal}} = \sqrt{\frac{2kT}{m}}$$

where $m$ is the mass of a single gas particle, $T$ is the temperature on the Kelvin scale, and $k = 1.38 \times 10^{-23}$ joule/K is *Boltzmann's constant.* Note that the peak thermal velocity increases with temperature (because higher temperature means higher average kinetic energy for gas particles) and decreases with particle mass (because at any particular temperature, particles of lighter gases, with smaller $m$, move at faster speeds than particles of heavier gases).

If a gas's peak thermal velocity is greater than the escape velocity, most of the gas particles will quickly escape to space. However, some gas particles can escape even if the peak thermal velocity is much lower. For example, Figure 1 shows that a small fraction of the sodium atoms have speeds exceeding the Moon's escape velocity (about 2.4 km/s), even though the peak thermal velocity (about 0.5 km/s) is well below it. These atoms can escape if they are moving in the right direction (upward) and don't collide with other particles on their way out. (Once these atoms escape, ongoing collisions redistribute the speeds of the remaining atoms, so that the shape of the distribution remains the same and the same small fraction always exceeds escape velocity.) The time it takes for a gas to completely escape depends on how its peak thermal velocity compares to the escape velocity. As a rule of thumb, a peak thermal velocity above about 20% of the escape velocity will allow the gas to be completely lost to space within a few billion years—which is less than the age of our solar system.

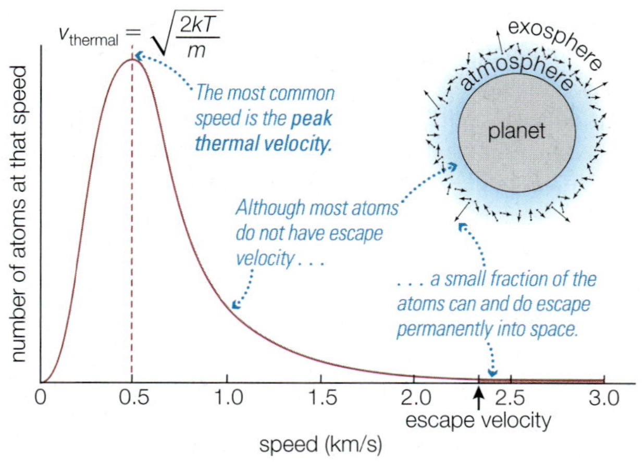


**FIGURE 1** In a gas at a given temperature, different atoms are always moving at different speeds. This plot shows the range of speeds of sodium atoms in the lunar atmosphere.

**EXAMPLE:** Why does the Moon's exosphere contain sodium atoms but virtually no hydrogen gas? Useful data: The Moon's daytime temperature is 400 K and its escape velocity is about 2.4 km/s; the mass of a hydrogen atom is $1.67 \times 10^{-27}$ kg; the mass of a sodium atom is $3.84 \times 10^{-26}$ kg (about 23 times the mass of a hydrogen atom).

**SOLUTION:**

**Step 1 Understand:** The rate of thermal escape depends on how each gas's peak thermal velocity compares to the escape velocity at the Moon's daytime temperature of 400 K. From Figure 1, we know the peak thermal velocity for sodium atoms (about 0.5 km/s). We need to find the peak thermal velocity for hydrogen atoms at the same temperature.

**Step 2 Solve:** We use the given formula and data to calculate the peak thermal velocity for hydrogen atoms at a temperature of 400 K:

$$v_{\text{thermal (H atoms)}} = \sqrt{\frac{2kT}{m_{\text{H atom}}}}$$
$$= \sqrt{\frac{2 \times \left(1.38 \times 10^{-23}\,\frac{\text{joule}}{\text{K}}\right) \times (400\ \text{K})}{1.67 \times 10^{-27}\ \text{kg}}}$$
$$\approx 2600\ \text{m/s} = 2.6\ \text{km/s}$$

**Step 3 Explain:** The peak thermal velocity of the hydrogen atoms is 2.6 km/s, which is slightly *greater* than the Moon's escape velocity of 2.4 km/s. This tells us that hydrogen atoms quickly escape, which is why the Moon cannot retain hydrogen in its atmosphere. In contrast, the peak thermal velocity for sodium atoms (about 0.5 km/s) is only about 20% of the Moon's escape velocity, so their escape rate is slow, and they are continually replenished as micrometeorites, solar wind particles, and high-energy photons eject new sodium atoms from the Moon's surface. That is why the Moon has a thin sodium exosphere.

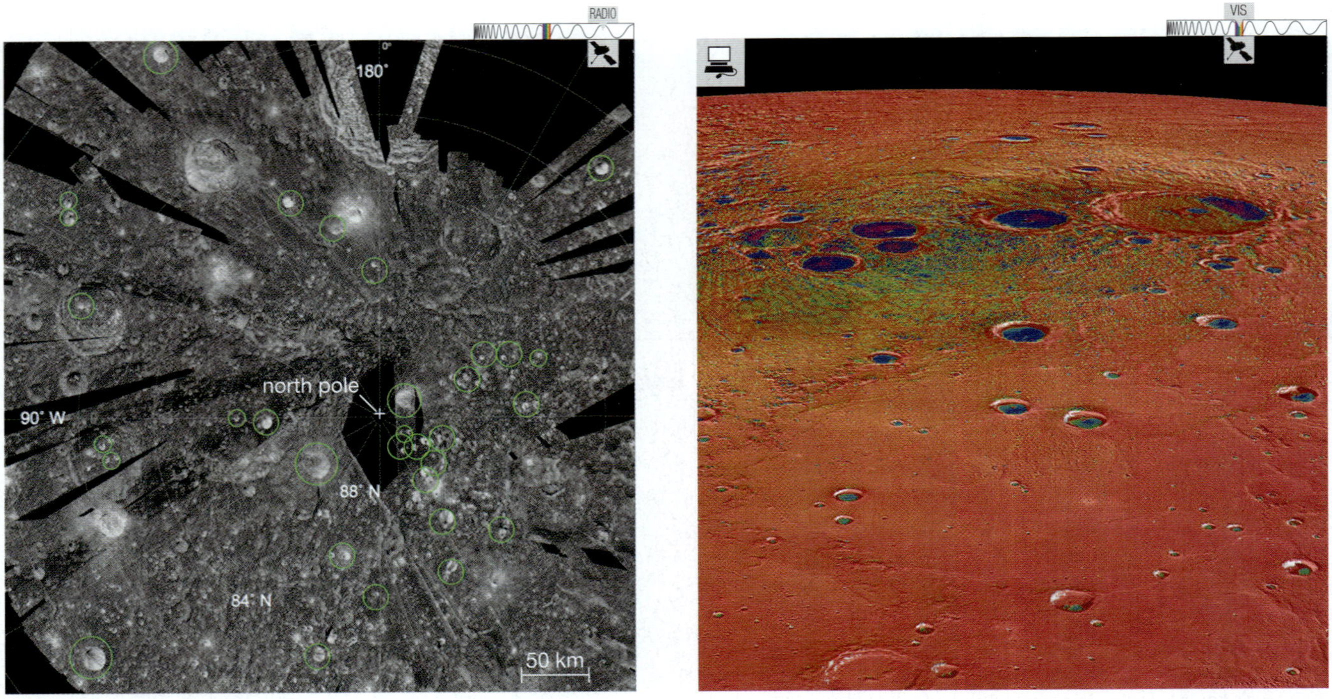


a This radar map shows a region near the Moon's north pole, imaged by a NASA instrument on India's *Chandrayaan-1* spacecraft. The green circles represent craters in which water ice was detected.

b This color-coded image, based on data from the *MESSENGER* spacecraft, represents surface temperatures in a region near Mercury's north pole. The purple regions are perpetually shadowed with temperatures as low as −220°C, and water ice has been detected in these locations.

**FIGURE 10.23** Both the Moon and Mercury have water ice at the bottoms of craters near their poles that are in perpetual shadow.

have vaporized into gas and escaped to space. However, spacecraft observations have confirmed a prediction made decades ago: Both the Moon and Mercury have water ice in permanently shadowed craters near their poles, presumably deposited by comet impacts over millions of years.

For the Moon, the presence of polar ice was first confirmed in 2009, when scientists sent the rocket from the *LCROSS* spacecraft crashing into a crater near the south pole. The debris that splashed upward revealed the presence of water vaporized from ice in the lunar soil. A radar sensor aboard India's *Chandrayaan-1* spacecraft detected ice deposits in similar craters near the Moon's north pole (**FIGURE 10.23a**). More surprisingly, other missions detected small amounts of water mixed into the upper layer of lunar soil over much of the lunar surface; the origin of this water is unknown. Mercury also has ice in its permanently shadowed craters, as first suggested from Earth-based radar measurements and confirmed by multiple instruments on the *MESSENGER* spacecraft (**FIGURE 10.23b**).

## 10.4 The Atmospheric History of Mars

One of our key goals in this chapter is to understand how and why the atmospheres of Venus, Earth, and Mars came to differ so profoundly, despite the fact that all three worlds must have had similar early atmospheres supplied by outgassing. We are now ready to focus on this question as we continue our atmospheric tour with Mars. Recall that Mars is only about 40% larger in radius than Mercury, but its surface reveals a much more fascinating and complex atmospheric history.

### What is Mars like today?

The present-day surface of Mars looks much like deserts or volcanic plains on Earth (for example, see Figure 7.6). However, its thin atmosphere makes Mars quite different. The low atmospheric pressure—less than 1% of that on Earth's surface—explains why liquid water is unstable on the Martian surface [**Section 9.4**] and why visiting astronauts could not survive without pressurized space suits. The atmosphere is made mostly of carbon dioxide, but the total amount of gas is so small that it creates only a weak greenhouse effect. The temperature is usually well below freezing, with a global average of about −50°C (−58°F). The lack of oxygen means that Mars lacks an ozone layer, so much of the Sun's damaging ultraviolet radiation passes unhindered to the surface.

**Martian Seasons** Recall that Mars has an axis tilt similar to that of Earth, which means it undergoes seasonal changes. However, while axis tilt is the only important influence on Earth's seasons, Mars's seasons are also affected by its orbit (**FIGURE 10.24**). Mars's more elliptical orbit puts it significantly closer to the Sun during southern hemisphere summer (and farther from the Sun during southern hemisphere winter), giving its southern hemisphere more extreme

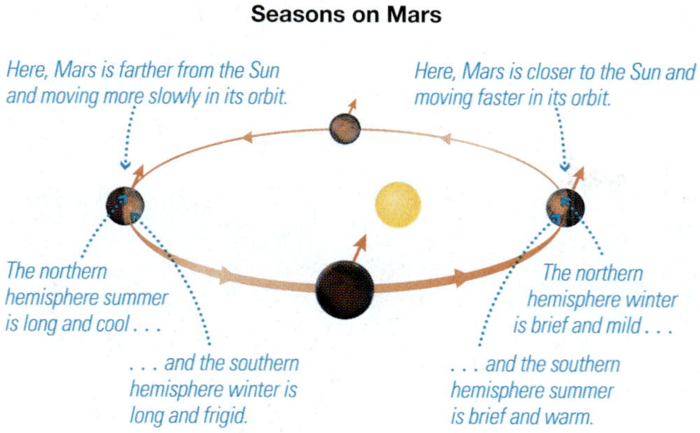


**FIGURE 10.24** The ellipticity of Mars's orbit makes seasons more extreme in the southern hemisphere than in the northern hemisphere.

**FIGURE 10.26** These two ground-based telescope images contrast the appearance of the same face of Mars in the absence (left) and presence (right) of a global dust storm.

seasons—that is, shorter, warmer summers and longer, colder winters—than its northern hemisphere.

Seasonal changes lead to several major features of Martian weather. Temperatures at the winter pole drop so low (about −130°C) that carbon dioxide condenses into "dry ice" at the winter polar cap. Meanwhile, frozen carbon dioxide at the summer pole vaporizes into carbon dioxide gas, and by the peak of summer only a residual cap of water ice remains (**FIGURE 10.25**). The atmospheric pressure therefore increases at the summer pole and decreases at the winter pole. Overall, as much as one-third of the total carbon dioxide of the Martian atmosphere moves seasonally between the north and south polar caps.

**Martian Winds** The strong winds associated with the seasonal cycling of carbon dioxide gas can initiate huge dust storms, particularly when the more extreme summer approaches in the southern hemisphere (**FIGURE 10.26**). At times, the Martian surface becomes almost completely obscured by airborne dust. As the dust settles out, it can change the surface appearance over vast areas (for example, by covering dark regions with brighter dust); such changes fooled astronomers of the past into thinking they were seeing seasonal changes in vegetation.

Martian winds can also spawn *dust devils,* swirling winds that you may have seen over desert sands or dry dirt on Earth. Dust devils look much like miniature tornadoes, but they rise up from the ground rather than coming down from the sky. The air in dust devils is heated from below by the sunlight-warmed ground; it swirls because of the way it interacts with prevailing winds. Dust devils on Mars are especially common during summer in either hemisphere. While many are quite small (**FIGURE 10.27**), some can be far larger than their counterparts on Earth.

Martian winds and dust storms leave Mars with perpetually dusty air, which helps explain the colors of the Martian sky. The air on Mars is so thin that, without suspended dust, the sky would be essentially black even in daytime. However, light scattered by the suspended dust tends to give the sky a yellow-brown color. Different hues can occur as the amount of suspended dust varies, and in the mornings and evenings.

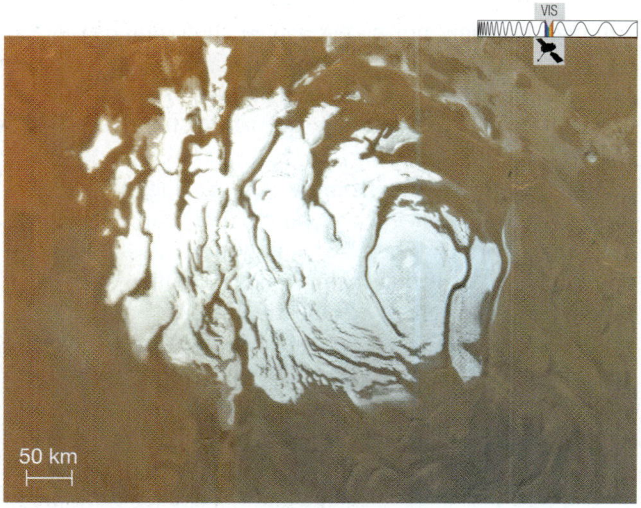


**FIGURE 10.25** This image from the *Mars Global Surveyor* shows the residual south polar cap during summer. A layer of frozen carbon dioxide a few meters thick overlies a much thicker cap of water ice. In winter, the whole area shown in the image is covered in $CO_2$ frost.

**Water Ice on Mars** Although there is no liquid water on Mars's surface today, there is a fair amount of water ice. The polar caps are made mostly of water ice, overlaid with a thin layer (at most a few meters thick) of carbon dioxide ice (see Figure 10.25). Radar instruments on Mars orbiters have found substantial quantities of water frozen in vast layers of dusty ice surrounding both poles and have also discovered icy glaciers at lower latitudes, where they are kept frozen by a protective layer of rocks and dust above the ice. For example, **FIGURE 10.28** shows a dramatic outcrop of exposed ice in a mid-latitude (57°S) cliff; at the top of the cliff, the ice reaches to within a meter or two of the surface. In sum, if all the water ice now known on Mars melted, it could cover the entire planet to an average depth of at least about 10 meters. It is even possible that some liquid water exists underground near sources of volcanic heat, providing a potential home to microscopic life. Recent evidence also suggests that Mars has at least one large underground lake located near its south polar ice cap.

a This image shows a dust devil in action, along with its shadow, as seen from orbit. The length of the shadow tells us that the dust devil extended to an altitude of at least 800 meters.

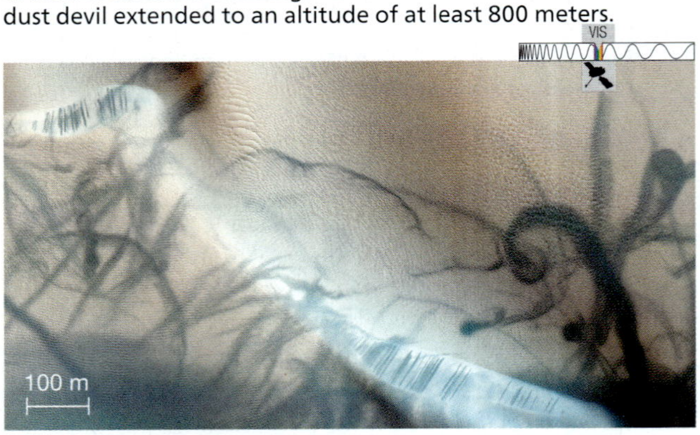


b This orbital image looks down on a set of tracks left behind by dust devils that moved across Martian sand dunes. The dunes generally appear light in color because they are covered by reddish dust deposited by recent dust storms. The underlying sand is darker in color, so we see dark tracks where dust devils have swept the dust away.

**FIGURE 10.27** Dust devils on Mars surface. Both images from the *Mars Reconnaissance Orbiter*.

## How has Mars's climate differed in the past?

Recall from Chapter 9 that we have substantial evidence of past liquid water on Mars. Therefore, the fact that the atmospheric pressure and temperature are too low for stable liquid water today tells us that the Martian climate must have differed in the past. In fact, Mars's climate appears to have undergone at least two types of long-term climate change: (1) changes that recur over time due to a changing axis tilt and (2) an even longer-term change that transformed Mars from a much warmer, wetter planet to the cold desert we see today.

**Mars Climate and Axis Tilt** The climate of Mars does not change much from one year to the next. However, Mars apparently undergoes longer-term cycles of climate change caused by changes in its axis tilt. Theoretical calculations suggest that Mars's axis tilt should vary far more than Earth's: On time scales of hundreds of thousands to millions of years, Mars's axis may swing from tilts as small as 0° to tilts as large as about 60°. This extreme variation arises for two reasons. First, Jupiter's gravity has a greater effect on the axis of Mars than on that of Earth, because Mars's orbit is closer to Jupiter's orbit. Second, Earth's axis is stabilized by the gravity of our relatively large Moon, but Mars's two tiny moons (Phobos and Deimos; see Figure 8.11) are too small to offer any stabilizing influence on its axis.

As we discussed in Section 10.2, such changes in axis tilt should affect both the severity of the seasons and the global average temperature. When Mars's axis tilt is small, the poles may stay in a perpetual deep freeze for tens of thousands of years. With more carbon dioxide frozen at the poles, the atmosphere becomes thinner, lowering the pressure and weakening the greenhouse effect, thereby cooling the entire planet. When the axis is highly tilted, the summer pole becomes much warmer, allowing substantial amounts of water ice to vaporize, along with carbon dioxide, into the atmosphere. The pressure therefore increases, and Mars becomes warmer as the greenhouse effect strengthens—although probably not by enough to allow liquid water to become stable at the surface. The Martian polar regions show layering of dust and ice that probably reflects changes in climate due to the changing axis tilt (**FIGURE 10.29**).

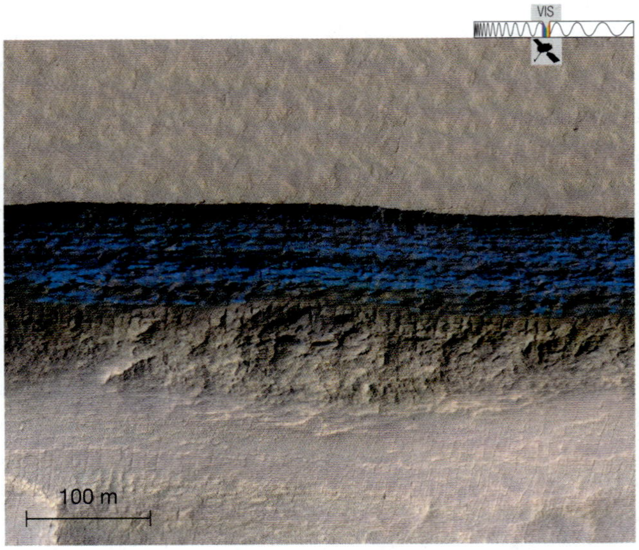


**FIGURE 10.28** This enhanced color image from the *Mars Reconnaissance Orbiter* shows an approximately 80-meter-thick layer of water ice (blue) exposed on the face of a steep cliff on Mars. This particular cliff is located at latitude 57°S, and several similar icy cliffs have been observed at other mid-latitudes, confirming earlier orbital data indicating that water ice is abundant under the Martian surface.

**Longer-Term Climate Change** The geological evidence discussed in Chapter 9 leaves little doubt that Mars had at least some periods when water flowed on its surface and rain could fall, but only before about 3 billion years ago. Because these conditions could have existed only if temperatures were warm enough to keep water from freezing and the atmospheric pressure was high enough for liquid water to be stable, we conclude that Mars once must have had a much thicker atmosphere, with a much stronger greenhouse effect.

The idea that Mars once had a much thicker atmosphere and stronger greenhouse effect makes sense. Calculations suggest that Martian volcanoes should have outgassed enough carbon dioxide to make the atmosphere about 400 times as dense as it is today, and enough water to fill oceans tens or even hundreds of meters deep. Additional

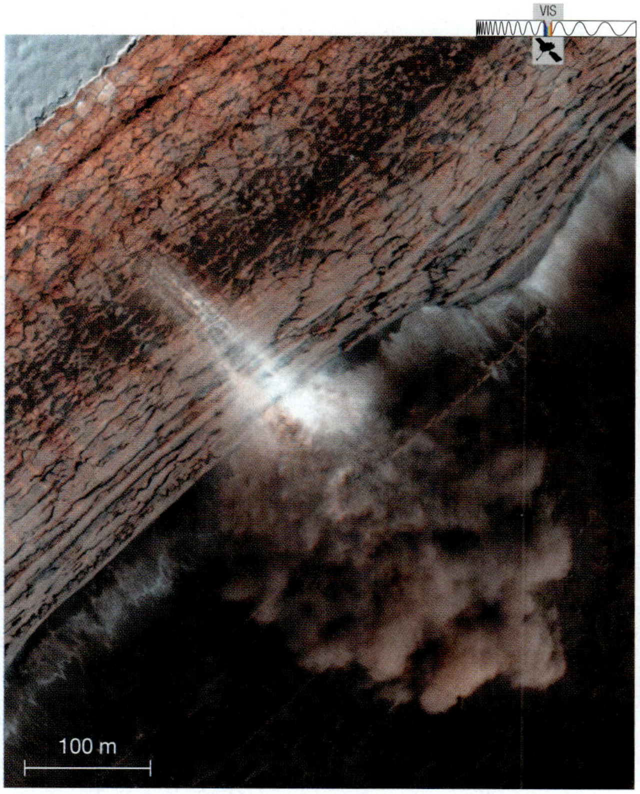


**FIGURE 10.29** The *Mars Reconnaissance Orbiter* captured this image of a landslide in layered terrain in the north polar region. Despite the dark appearance, water makes up the bulk of the material. Layers of dusty ice more than 700 meters thick built up over many cycles of climate change. During the northern spring of 2010, warming conditions apparently weakened the cliff walls and triggered landslides.

evidence that Mars once had much more water comes from measurement of the atmospheric ratio of deuterium to hydrogen, which (for reasons we'll discuss in the next section on Venus) allows us to estimate how much more water a planet had in the past than it does today. Such measurements suggest that Mars once had enough water to cover the entire surface to a depth of more than 130 meters, which is another reason some scientists suspect that Mars once had a northern ocean (see Figure 9.30).

Nevertheless, scientists continue to debate the full extent of the warm and wet periods on Mars. Part of this debate revolves around models of the past and present Martian climate. In particular, models indicate that if *present-day* Mars had as much carbon dioxide and water as we expect it to have had in the distant past, its greenhouse effect would make it warm enough to allow the water to be liquid, and the pressure would be great enough for the water to remain stable. However, because the Sun was dimmer in the distant past (see Figure 10.18), even more greenhouse warming would have been needed to allow for liquid water when Mars was young. The source and strength of this additional greenhouse warming remains a mystery. Some scientists suspect that carbon dioxide ice clouds or atmospheric methane may have created enough additional greenhouse warming to have kept Mars warm and wet for a billion or more years. Others doubt the greenhouse effect could ever have been that strong, in which case Mars may have had only intermittent periods of rainfall, perhaps triggered by the heat of large impacts. In this latter view, ancient lakes, ponds, or oceans would have been completely ice-covered for extended periods.

**Mars Climate Change**

## Why did Mars change?

Regardless of whether Mars was warm and wet only intermittently or for a billion or more years, there's little doubt that Mars underwent major and permanent climate change, turning a world that could sustain liquid water to the cold and dry world we see today. As we've discussed, even limited warm and wet conditions could have occurred only if Mars once had a much denser carbon dioxide atmosphere that allowed for liquid water and possibly oceans, which means the key question is what happened to all that carbon dioxide gas and all that water.

**Loss of Atmospheric Gas** Mars must somehow have lost most of the carbon dioxide gas that once filled its atmosphere. This loss would have weakened the greenhouse effect until the planet essentially froze over. Some of the carbon dioxide condensed and became part of the polar caps. Studies by Mars orbiters and rovers indicate that some may also be chemically bound into **carbonate rocks** (rocks rich in carbon and oxygen). But the bulk of the gas was probably lost to space.

The precise way in which Mars lost its carbon dioxide gas is not fully settled, but the leading hypothesis suggests a close link to a change in Mars's magnetic field (**FIGURE 10.30**). Early in its history, Mars probably had molten, convecting metals in its core, much like Earth today. The combination of this convecting metal with Mars's rotation should have produced a magnetic field and a protective magnetosphere. However, the magnetic field would have weakened as the small planet cooled and core convection ceased, leaving atmospheric gases vulnerable to being stripped into space by solar wind particles. More specifically, the hypothesis suggests that carbon dioxide molecules were dissociated into carbon and oxygen atoms by sunlight or chemical processes, and the resulting atoms were then stripped away by the solar wind.

NASA's *MAVEN* mission, which has been orbiting Mars since 2014, has confirmed that Mars must have lost substantial amounts of carbon dioxide gas over the past few billion years. *MAVEN* measures the rate at which gases escape from Mars's atmosphere today, and the first two panels of **FIGURE 10.31** show that Mars is indeed losing carbon and oxygen atoms that once made up carbon dioxide molecules. Based on the present-day loss rate, scientists can estimate the total carbon dioxide loss over billions of years, and these results confirm that Mars once had far more carbon dioxide gas than it does today.

**Loss of Water** Some water still lies frozen in the polar caps and underground, but most of the water once present on Mars is also probably gone for good. Mars probably lost water in a different way than it lost carbon dioxide.

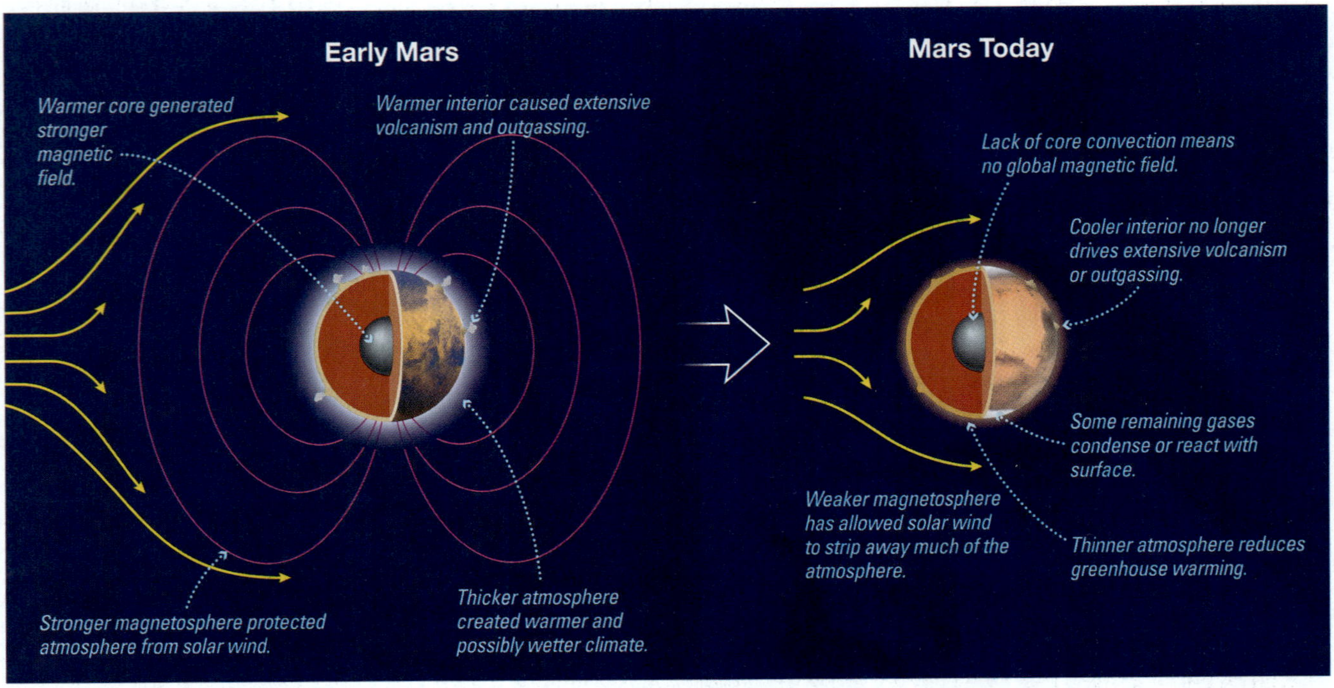


**FIGURE 10.30** Some 3 billion years ago, Mars underwent dramatic climate change, ensuring that rain could never fall again.

Because Mars lacks an ultraviolet-absorbing stratosphere, atmospheric water molecules would have been easily broken apart by ultraviolet photons. The hydrogen atoms that broke away from the water molecules would have been lost rapidly to space through thermal escape; the third panel of Figure 10.31 confirms the loss of hydrogen atoms in this way. Once the hydrogen atoms were lost, the water molecules could not be made whole again. Initially, oxygen from the water molecules would have remained in the atmosphere, but over time this oxygen was lost, too. Some was probably stripped away by the solar wind, and the rest was drawn out of the atmosphere through chemical reactions with surface rock. This process literally rusted the Martian rocks, giving the "red planet" its distinctive tint.

**Think about it** Some people have proposed "terraforming" Mars—that is, making it more Earth-like—by finding a way to release all the carbon dioxide frozen in its polar caps into its atmosphere. If Mars had oceans in the distant past, could this release of gas allow it to have oceans again? Why or why not?

One remaining mystery concerns the evidence we discussed in Chapter 9 indicating that sediments deposited at

**FIGURE 10.31** These ultraviolet images from NASA's *MAVEN* spacecraft show carbon, oxygen, and hydrogen atoms (which came from dissociated carbon dioxide and water molecules) in the Martian atmosphere. The red circle shows the size of Mars; black regions indicate a lack of data. Notice that the atomic gases extend high above the surface, where some of the carbon and oxygen is escaping through solar wind stripping, and hydrogen is being lost to thermal escape. Hydrogen extends the highest because light gases move fastest (see Mathematical Insight 10.2).

earlier times have minerals suggesting that they formed in purer water than sediments deposited later. This mystery may have a simple solution relating to the volcanoes that dot the Martian surface. Volcanic eruptions should have released sulfur-bearing gases, and these gases can dissolve in water and thereby make the water more acidic, which allows more salts to dissolve. Martian water therefore would have become saltier and more acidic with time simply because more sulfur-bearing gases had been released by volcanoes. This scenario is consistent with rover studies of rocks: For example, rocks found at the lower elevations explored by the *Curiosity* rover appear to have formed in purer water than those found at higher elevations or at the *Opportunity* landing site.

**Size as the Critical Factor** In summary, Mars's fate was probably sealed by its relatively small size. It was big enough for volcanism and outgassing to release water and atmospheric gas early in its history, but too small to maintain the internal heat needed to keep this water and gas. As Mars's interior cooled, its volcanoes quieted and released far less gas, while its relatively weak gravity and the loss of its magnetic field allowed existing gas to be stripped away to space. If Mars had been as large as Earth, so that it could still have outgassing and a global magnetic field, it might have a moderate climate today. Mars's distance from the Sun helped seal its fate: Even with its small size, Mars might still have some flowing water if it were significantly closer to the Sun, where the extra warmth could melt the water that remains frozen underground and at the polar caps.

The history of the Martian atmosphere holds important lessons for us on Earth. Mars apparently was once a world with more moderate temperatures and streams, rain, glaciers, lakes, and possibly oceans. It had all the necessities for life as we know it. But this once-hospitable planet turned into a frozen and barren desert at least 3 billion years ago, and it is unlikely that Mars will ever again be warm enough for its frozen water to flow. Any life that may have existed on Mars is either extinct or hidden away in a few choice locations, such as in underground water near not-quite-dormant volcanoes. As we consider the possibility of future climate change on Earth, Mars presents us with an ominous example of how drastically things can change.

## 10.5 The Atmospheric History of Venus

Cloud-covered Venus (**FIGURE 10.32**) presents a stark contrast to Mars, but it is easy to understand why: Its larger size allowed it to retain more interior heat, leading to greater volcanism. The associated outgassing released the vast quantities of carbon dioxide that create Venus's strong greenhouse effect.

Venus becomes more mysterious when we compare it to Earth. Because Venus and Earth are so similar in size, we might naively expect both planets to have had similar atmospheric histories. Clearly, this is not the case. In this section, we'll explore how Venus's atmosphere ended up so different from Earth's, and in the process we'll learn important lessons about the habitability of our own planet.

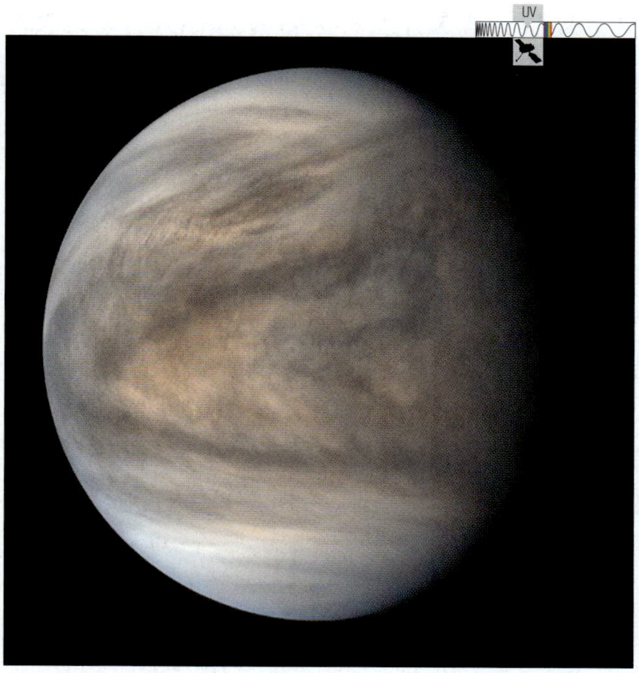


**FIGURE 10.32** This color-coded image shows Venus at ultraviolet wavelengths as seen by the Japanese Space Agency's *Akatsuki* spacecraft. At most wavelengths, clouds completely prevent any view of the surface.

### What is Venus like today?

If you stood on the surface of Venus, you'd feel a searing heat hotter than that of a self-cleaning oven and a tremendous pressure 90 times that on Earth. A deep-sea diver would have to go nearly 1 kilometer (0.6 mile) beneath the ocean surface on Earth to feel comparable pressure. Venus's atmosphere consists almost entirely of carbon dioxide ($CO_2$). It has virtually no molecular oxygen ($O_2$), so you could not breathe the air even if you cooled it to a comfortable temperature.

Moving through the thick air near Venus's surface would feel like a cross between swimming and flying: Its density is about 10% that of water. Looking upward, you'd see a perpetually overcast sky, with only weak sunlight filtering through the thick clouds above. Because the thick atmosphere scatters nearly all the blue light away, the dimly lit sky appears reddish-orange in color.

The weather forecast for the surface of Venus today and every day is dull, dull, dull. Venus's slow rotation (243 Earth days) means a very weak Coriolis effect. As a result, Venus has little wind on its surface and never has hurricane-like storms. The top wind speeds measured by the Soviet Union's *Venera* landers were only about 6 kilometers per hour. No rain falls, because droplets that form and fall from the cool upper atmosphere evaporate long before they reach the ground. The weak Coriolis effect also means that Venus's atmosphere has just two large circulation cells, much like what Earth would have if rotation didn't split its cells (see Figure 10.13). The thick atmosphere makes the circulation so efficient at transporting heat from the equator to the poles that the surface temperature is virtually the same everywhere: The poles are no cooler than the equator, and night is just as searingly hot as day. Moreover, Venus has no seasons because it has virtually no axis tilt,* so temperatures are the same year-round.

The weather is more interesting at high altitudes. Strong convection drives hot air upward; high in the troposphere, where the temperature is 400°C cooler than on the surface (see Figure 10.10), sulfuric acid ($H_2SO_4$) condenses into droplets that create Venus's bright, reflective clouds. The droplets sometimes fall through the upper troposphere as sulfuric acid rain, but they evaporate at least 30 kilometers above the surface. In addition, high-altitude winds circle the planet in just 4 days—much faster than the planet rotates. No one knows why these fast winds blow, but they are responsible for the dynamic cloud patterns visible in **FIGURE 10.33**.

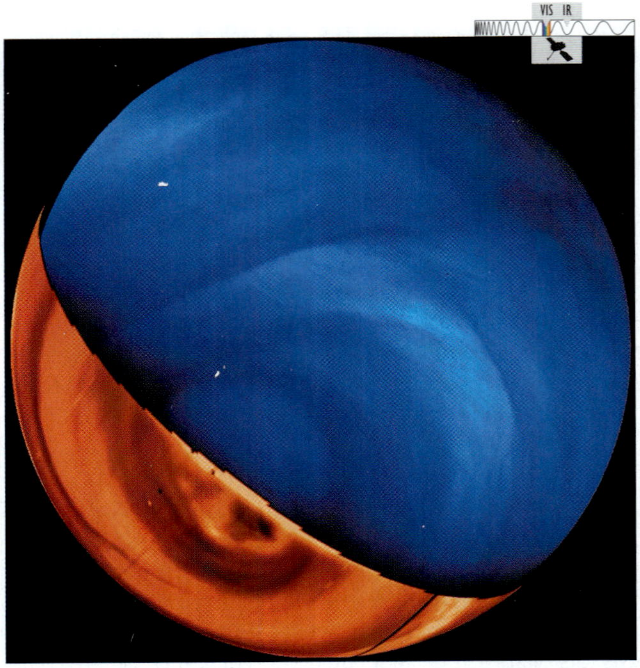


**FIGURE 10.33** This composite image from the *Venus Express* spacecraft combines a visible-wavelength image of the day side (left, shaded red) and an infrared image of the night side (right, shaded blue). Venus's south pole lies at the center.

## How did Venus get so hot?

It's tempting to attribute Venus's high surface temperature solely to the fact that Venus is closer than Earth to the Sun, but Venus would actually be quite cold without its strong greenhouse effect (see Table 10.2), because its bright clouds reflect much more sunlight than Earth. The real question is why Venus has such a strong greenhouse effect.

The simple answer is that Venus has a huge amount of carbon dioxide in its atmosphere—nearly 200,000 times as much as in Earth's atmosphere. However, a deeper question still remains. Given their similar sizes and compositions, we expect Venus and Earth to have had similar levels of volcanic outgassing, and the released gas ought to have had about the same composition on both worlds. Why, then, is Venus's atmosphere so different from Earth's?

### The Fate of Outgassed Water and Carbon Dioxide

We expect that huge amounts of water vapor and carbon dioxide should have been outgassed into the atmospheres of both Venus and Earth. Venus's atmosphere does indeed have an enormous amount of carbon dioxide, but it has virtually no water. Earth's atmosphere has very little of either gas. We conclude that Venus must have somehow lost its outgassed water, while Earth lost both water vapor and carbon dioxide. But where did these gases go?

We can easily account for the missing gases on Earth. The huge amounts of water vapor released into our atmosphere condensed into rain, forming our oceans. In other words, the water is still here, but mostly in liquid rather than gaseous form. The huge amount of carbon dioxide released into our atmosphere is also still here, but in solid form: Carbon dioxide dissolves in water, where it can undergo chemical reactions to make carbonate rocks such as limestone. Earth has about 200,000 times as much carbon dioxide locked up in rocks as in its atmosphere—which means that Earth does indeed have almost as much total carbon dioxide as Venus. Of course, the fact that Earth's carbon dioxide is mostly in rocks rather than in the atmosphere makes all the difference in the world: If this carbon dioxide were in our atmosphere, our planet would be nearly as hot as Venus and certainly uninhabitable.

We are left with the question of what happened to Venus's water. Venus today is incredibly dry. It is far too hot to have any liquid water or ice on its surface; it is even too hot for water to be chemically bound in surface rock, and any water deeper in its crust or mantle was probably baked out long ago. Measurements also show very little water in the atmosphere. Overall, the total amount of water on Venus is about 100,000 times smaller than the total amount on Earth, a fact that explains why Venus retains so much carbon dioxide in its atmosphere: Without oceans, carbon dioxide cannot dissolve or become locked away in carbonate rocks. If it is true that a huge amount of water was outgassed on Venus, the water molecules have somehow disappeared.

The leading hypothesis for the disappearance of Venus's water invokes one of the same processes thought to have removed water from Mars. Ultraviolet light from the Sun broke apart water molecules in Venus's atmosphere. The hydrogen atoms then escaped to space (through thermal escape), ensuring that the water molecules could never reform. The oxygen from the water molecules was lost to a combination of chemical reactions with surface rocks and stripping by the solar wind; Venus's lack of a magnetic field leaves its atmosphere vulnerable to the solar wind.

Acting over billions of years, the breakdown of water molecules and the escape of hydrogen can easily explain the loss of an ocean's worth of water from Venus, and careful study of Venus's atmospheric composition offers evidence that such a loss really occurred. Recall that most hydrogen nuclei contain just a single proton, but a tiny fraction of

*In tables (such as Table 7.1), Venus's axis tilt is usually written as 177.3. This may sound large, but notice that it is nearly 180°—the same tilt as 0° but "upside down." It is written this way because Venus rotates backward compared to its orbit, and backward rotation is equivalent to forward rotation that is upside down.

all hydrogen atoms (about 1 in 6400 measured in Earth's oceans) contain a neutron in addition to the proton, making the isotope of hydrogen that we call *deuterium*. Water molecules that contain one or two atoms of deuterium instead of hydrogen (called *heavy water*) behave chemically just like ordinary water and can be broken apart by ultraviolet light just as easily. However, a deuterium atom is twice as heavy as an ordinary hydrogen atom and therefore does not escape to space as easily when the water molecule is broken apart. If Venus lost a huge amount of hydrogen from water molecules to space, the rare deuterium atoms would have been more likely to remain behind than the ordinary hydrogen atoms. Measurements show that this is the case: The fraction of deuterium among hydrogen atoms is a hundred times higher on Venus than on Earth, suggesting that a substantial amount of water was lost by having its molecules broken apart and its hydrogen lost to space. We cannot determine exactly how much water Venus has lost, but it seems plausible that Venus really did outgas as much water as Earth and then lost virtually all of it.

Of course, Venus could have lost all this water only if it had been in the atmosphere as water vapor, where ultraviolet light could break the molecules apart, rather than in liquid oceans like the water on Earth. Our quest to understand Venus's high temperature therefore leads to one more question: Why didn't Venus, like Earth, end up with oceans to trap its carbon dioxide in carbonate rocks and prevent its water from being lost to space?

**The Runaway Greenhouse Effect** To understand why Venus does not have oceans, we need to consider the role of **feedback processes**—processes in which a change in one property amplifies (positive feedback) or counteracts (negative feedback) the behavior of the rest of the system. You are probably familiar with feedback processes in daily life. For example, if someone brings a microphone too close to a loudspeaker, it picks up and amplifies small sounds from the speaker. These amplified sounds are again picked up by the microphone and further amplified, causing a loud screech. This sound feedback is an example of *positive feedback*, because it automatically amplifies itself. The screech usually leads to a form of *negative feedback:* The embarrassed person holding the microphone moves away from the loudspeaker, thereby stopping the positive sound feedback.

**Think about it** Think of at least one other everyday example each for positive feedback and negative feedback.

With the idea of feedback in mind, let's consider what would happen if we could magically move Earth to the orbit of Venus (**FIGURE 10.34**). The greater intensity of sunlight would almost immediately raise Earth's global average temperature by about 30°C, from its current 15°C to about 45°C (113°F). Although this is still well below the boiling point of water, the higher temperature would lead to increased evaporation of water from the oceans. The higher temperature would also allow the atmosphere to hold more water vapor before the vapor condensed to make rain. The combination of more evaporation and greater atmospheric capacity for water vapor would substantially increase the total amount of water vapor in Earth's atmosphere. Now, remember that water vapor, like carbon dioxide, is a greenhouse gas. The added water vapor would therefore strengthen the greenhouse effect, driving temperatures a little higher. The higher temperatures, in turn, would lead to even more ocean evaporation and more water vapor in the atmosphere, strengthening the greenhouse effect even further. In other words, we'd have a positive feedback process in which each little bit of additional water vapor in the atmosphere would lead to higher temperature and even more water vapor. The process would rapidly spin out of control, resulting in a **runaway greenhouse effect**.

The runaway greenhouse effect would cause Earth to heat up until the oceans were completely evaporated and the carbonate rocks had released all their carbon dioxide back into the atmosphere. By the time the runaway process was complete, temperatures on our "moved Earth" would be even higher than they are on Venus today, thanks to the combined greenhouse effects of carbon dioxide and water vapor in the atmosphere. The water vapor would then gradually disappear, as ultraviolet light broke water molecules apart and the hydrogen escaped to space. In short, moving Earth to Venus's orbit would essentially turn our planet into another Venus.

We have arrived at a simple explanation of why Venus is so much hotter than Earth. Even though Venus is only about 30% closer to the Sun than Earth is, this difference was critical. On Earth, it was cool enough for water to rain down to make oceans. The oceans dissolved carbon dioxide and chemical reactions locked it away in carbonate rocks,

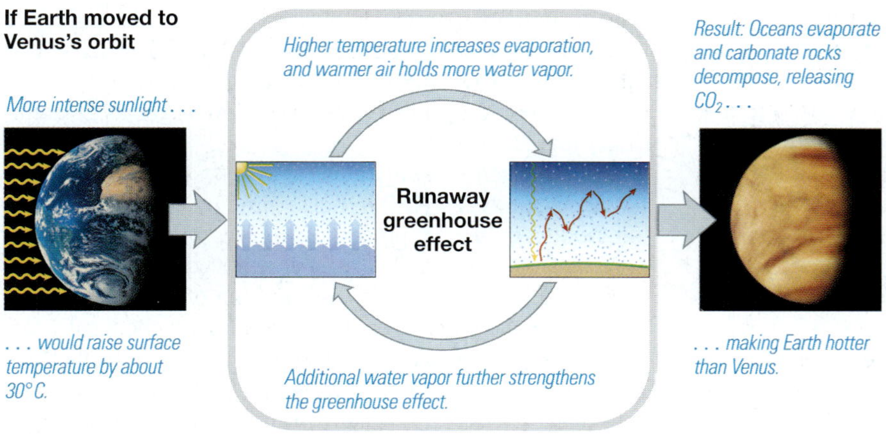


**FIGURE 10.34** This diagram shows how, if Earth were placed at Venus's distance from the Sun, the runaway greenhouse effect would cause the oceans to evaporate completely.

leaving our atmosphere with only enough greenhouse gases to make our planet pleasantly warm. On Venus, the greater intensity of sunlight made it just warm enough that oceans either never formed or soon evaporated, leaving Venus with a thick atmosphere full of greenhouse gases.

The next time you see Venus shining brightly as the morning or evening "star," consider the radically different path it has taken from Earth—and thank your lucky star. If Earth had formed a bit closer to the Sun, or if the Sun had been slightly hotter, our planet might have suffered the same greenhouse-baked fate.

**Think about it** We've seen that moving Earth to Venus's orbit would cause our planet to become Venus-like. If we could somehow move Venus to Earth's orbit, would it become Earth-like? Why or why not?

**A Pleasant Early Venus?** Venus's closeness to the Sun may have sealed its ultimate fate, but it's possible that Venus might have been more moderate in its early history. Recall that the Sun has gradually brightened with age; some 4 billion years ago, the intensity of sunlight shining on the young Venus was not much greater than it is on Earth today. Rain might have fallen, and oceans could have formed. It's even conceivable that life could have arisen on the young Venus.

As the Sun gradually brightened, however, any liquid water or life on Venus was doomed. The runaway greenhouse effect raised the temperature so high that all the water evaporated. In the upper atmosphere, ultraviolet light broke apart the water molecules, and the hydrogen escaped to space. If Venus had oceans in its youth, the water is now gone forever.

We'll probably never know for sure whether Venus ever had oceans. The global "repaving" of Venus's surface by tectonics and volcanism [**Section 9.5**] would long ago have covered up any shorelines or other geological evidence of past oceans, and the high surface temperatures would have baked out any gases that might once have been incorporated into surface rock. If the climate once was pleasant, it's unlikely that any evidence survives to tell the tale.

## 10.6 Earth's Unique Atmosphere

Earth's atmosphere, composed mostly of nitrogen and oxygen, makes our lives possible. It provides our planet with just enough warmth and pressure to enable water to cycle between all three phases (solid ice, liquid water, and gaseous water vapor), it protects us from harmful solar radiation, and it produces the weather patterns that variously bring us days of sunshine, clouds, and rain or snow. In this section, we'll discuss how and why our atmosphere is so hospitable—and how we humans may be altering the very balances upon which we depend for survival.

### How did Earth's atmosphere end up so different?

Most of the major physical differences in the atmospheres of the terrestrial planets should already make sense. For example, the Moon and Mercury lack substantial atmospheres because of their small sizes, while a very strong greenhouse effect causes the high surface temperature on Venus. However, differences in atmospheric composition may seem more surprising, because outgassing should have released the same gases on Venus, Earth, and Mars. How, then, did Earth's atmosphere end up so different? We can break down this general question into four separate questions:

1. Why did Earth retain most of its outgassed water—enough to form vast oceans—while Venus and Mars lost theirs?
2. Why does Earth have so little carbon dioxide ($CO_2$) in its atmosphere compared to Venus, when Earth should have outgassed about as much of it as Venus?
3. Why is Earth's atmosphere composed primarily of nitrogen ($N_2$) and oxygen ($O_2$), when these gases are only trace constituents in the atmospheres of Venus and Mars?
4. Why does Earth have an ultraviolet-absorbing stratosphere, while Venus and Mars do not?

**Water and Carbon Dioxide** We have already answered the first two questions in our discussions of the atmospheres of Mars and Venus. On Mars, some of the outgassed water was lost after solar ultraviolet light broke water vapor molecules apart, and the rest froze and may remain in the polar caps or underground. On Venus, it was too hot for water vapor to condense, so virtually all the water molecules were ultimately broken apart, allowing the hydrogen atoms to escape to space. Earth retained its outgassed water because temperatures were low enough for water vapor to condense into rain and form oceans. Evidence from tiny mineral grains suggests that Earth may have had oceans as early as 4.3–4.4 billion years ago. The oceans, in turn, explain the low level of carbon dioxide in our atmosphere. Most of the carbon dioxide outgassed by volcanism on Earth dissolved in the oceans, where chemical reactions turned it into carbonate rocks. Even today, about 60 times as much carbon dioxide is dissolved in the oceans as is present in the atmosphere, and carbonate rocks contain some 170,000 times as much $CO_2$ as the atmosphere.

**Nitrogen, Oxygen, and Ozone** Turning our attention to the third question, we can easily explain the substantial nitrogen content (77%) of our atmosphere. Nitrogen is the third most common gas released by outgassing, after water vapor and carbon dioxide. Because most of Earth's water ended up in the oceans and most of the carbon dioxide ended up in rocks, our atmosphere was left with nitrogen as its dominant ingredient.

The oxygen content (21%) is a little more mysterious. Molecular oxygen ($O_2$) is not a product of outgassing or any other geological process. Moreover, oxygen is a highly reactive chemical that is easily removed from the atmosphere. Fire, rust, and the discoloration of freshly cut fruits and vegetables are everyday examples of chemical reactions that remove oxygen from the atmosphere (called *oxidation reactions*). Similar reactions between oxygen and surface materials (especially iron-bearing minerals) give

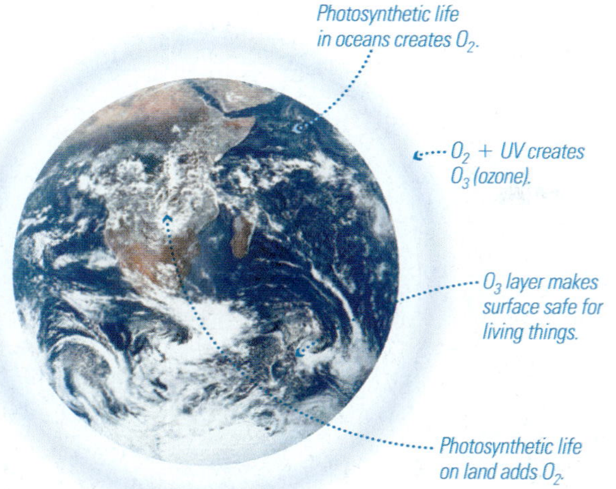


**FIGURE 10.35** The origin of oxygen and ozone in Earth's atmosphere can be traced to life.

## COMMON MISCONCEPTIONS

### Ozone—Good or Bad?

Ozone often generates confusion, because in human terms it is sometimes good and sometimes bad. Near the ground, ozone is one of the main ingredients in urban air pollution, produced as a by-product of automobiles and industry. This is bad for us, because ozone is poisonous to most living creatures. In contrast, the ozone that exists naturally high above the ground, in the stratosphere, serves as a protective shield against the Sun's ultraviolet radiation. This ozone is therefore crucial to our survival, which is why the release of chemicals (such as chlorofluorocarbons, or CFCs) that destroy stratospheric ozone has been a significant environmental problem.

rise to the reddish appearance of much of Earth's rock and clay, including the beautiful reds of Arizona's Grand Canyon. Without continual replenishment, these types of chemical reactions would remove all the oxygen in Earth's atmosphere in just a few million years. We must therefore explain not only how oxygen got into Earth's atmosphere in the first place, but also how it is replenished as chemical reactions remove it.

The answer to the oxygen mystery is *life* (**FIGURE 10.35**). Plants and many microorganisms release oxygen through photosynthesis. Photosynthesis takes in $CO_2$ and, through a complex chain of chemical reactions, releases $O_2$. Because chemical reactions can remove oxygen, it took a long time for oxygen to accumulate in Earth's atmosphere. According to present evidence, it took at least a billion years of photosynthesis before the buildup of atmospheric oxygen began, and Earth's atmosphere probably has had enough oxygen for us to breathe only for the past few hundred million years [**Section 24.1**]. Today, plants and single-celled photosynthetic organisms return oxygen to the atmosphere in approximate balance with the rate at which animals and chemical reactions consume oxygen, keeping the oxygen levels relatively steady.

**Think about it** Suppose that, somehow, all photosynthetic life on Earth died out. What would happen to the oxygen in our atmosphere? Could animals, including us, still survive?

Life and oxygen also explain the presence of Earth's ultraviolet-absorbing stratosphere. In the upper atmosphere, chemical reactions involving solar ultraviolet light transform some of the $O_2$ into molecules of $O_3$, or ozone. The $O_3$ molecule is more weakly bound than the $O_2$ molecule, which allows it to absorb solar ultraviolet energy even better. The absorption of solar energy by ozone heats the upper atmosphere, creating the stratosphere. This ozone layer prevents harmful ultraviolet radiation from reaching the surface. Mars and Venus lack photosynthetic life and therefore have too little $O_2$, and consequently too little ozone, to form a stratosphere.

**Maintaining Balance** We have answered all four basic questions about Earth's atmosphere. However, a deeper look at what we have learned still leaves us with a major mystery. Our oceans exist because Earth has a greenhouse effect that is "just right" to keep them from either freezing or boiling away, and the presence of liquid water allowed life to arise and produce oxygen. But *why* does the amount of carbon dioxide in our atmosphere stay "just right"? For example, why didn't the chemical reactions either remove *all* the carbon dioxide or leave so much of it that the oceans would boil away? The long-term existence of Earth's oceans tells us that our planet has enjoyed remarkable climate stability—presenting a stark contrast to the dramatic climate changes that apparently occurred on Mars and Venus.

## Why does Earth's climate stay relatively stable?

Earth's long-term climate stability has clearly been important to the ongoing evolution of life, and hence to our own relatively recent arrival as a species (see Figure 1.12). Had our planet undergone a runaway greenhouse effect like Venus, life would certainly have been extinguished. If Earth had suffered loss of atmosphere and a global freezing like Mars, any surviving life would have been driven to hide in underground pockets of liquid water.

Earth's climate is not perfectly stable: Our planet has endured numerous ice ages and warm periods in the past. Nevertheless, Earth's temperature has consistently remained in a range in which liquid water could exist and harbor life. This long-term climate stability is even more remarkable when you remember that the Sun has brightened substantially (about 30%) over the past 4 billion years, yet Earth's temperature has managed to stay in nearly the same range throughout this time. Apparently, the strength of the greenhouse effect self-adjusts to keep the climate stable. How does it do this?

**The Carbon Dioxide Cycle** The mechanism by which Earth self-regulates its temperature is called the **carbon dioxide cycle**, or the **$CO_2$ cycle** for short. Let's follow the cycle as illustrated in **FIGURE 10.36**, starting at the top center:

- Atmospheric carbon dioxide dissolves in rainwater, creating a mild acid.
- The mildly acidic rainfall erodes rocks on Earth's continents, and rivers carry the broken-down minerals to the oceans.
- In the oceans, calcium from the broken-down minerals combines with dissolved carbon dioxide and falls to the ocean floor, making carbonate rocks such as limestone.*
- Over millions of years, the conveyor belt of plate tectonics (see Figure 9.41) carries the carbonate rocks to subduction zones, where they are carried downward.
- As they are pushed deeper into the mantle, some of the subducted carbonate rocks melt and release their carbon dioxide, which then outgasses back into the atmosphere through volcanoes.

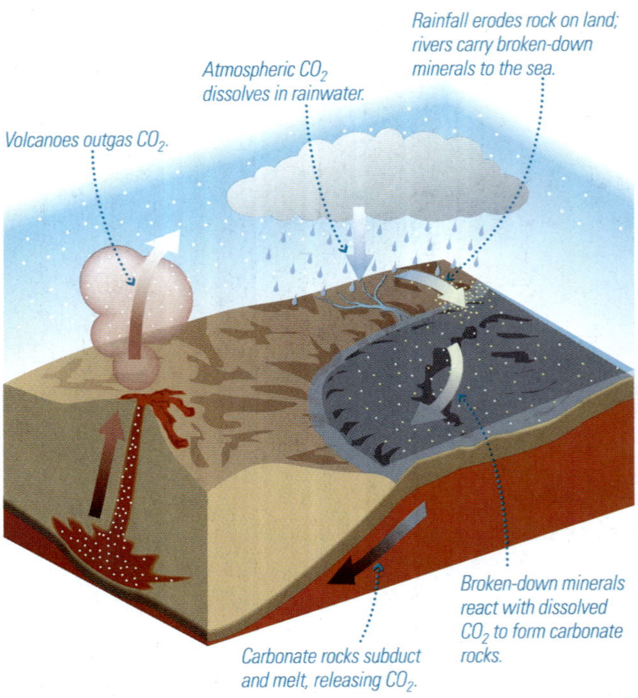


**FIGURE 10.36** This diagram shows how the $CO_2$ cycle continually moves carbon dioxide from the atmosphere to the ocean to rock and back to the atmosphere. Note that plate tectonics (subduction in particular) plays a crucial role in the cycle.

The $CO_2$ cycle acts as a long-term thermostat for Earth, because it has a built-in form of negative feedback that returns Earth's temperature toward "normal" whenever it warms up or cools down (**FIGURE 10.37**). The negative feedback occurs because the overall rate at which carbon dioxide is pulled from the atmosphere is very sensitive to temperature: the higher the temperature, the higher the rate at which carbon dioxide is removed. Note that these cycles occur on geological time scales much longer than human lifetimes, so they are neither the cause of nor the remedy for the rapid changes occurring now.

Consider first what happens if Earth warms up a bit. The warmer temperature means more evaporation and rainfall, ultimately leading more $CO_2$ to dissolve in the oceans and form carbonate rocks. As a result, the atmospheric $CO_2$ concentration eventually drops, leading to a weakened greenhouse effect that cools the planet back down. Similarly, if Earth cools a bit, precipitation decreases and less $CO_2$ is dissolved in rainwater, allowing the $CO_2$ released by volcanism to build up in the atmosphere. The increased $CO_2$ concentration strengthens the greenhouse effect and warms the planet. Overall, the natural thermostat of the carbon dioxide cycle has allowed the greenhouse effect to strengthen or weaken just enough to keep Earth's climate fairly stable, regardless of what other changes have occurred on our planet.

*During the past half billion years or so, the carbonate minerals have been made by shell-forming sea animals, falling to the bottom in the seashells left after the animals die. Without the presence of animals, chemical reactions would do the same thing—and apparently did for most of Earth's history.

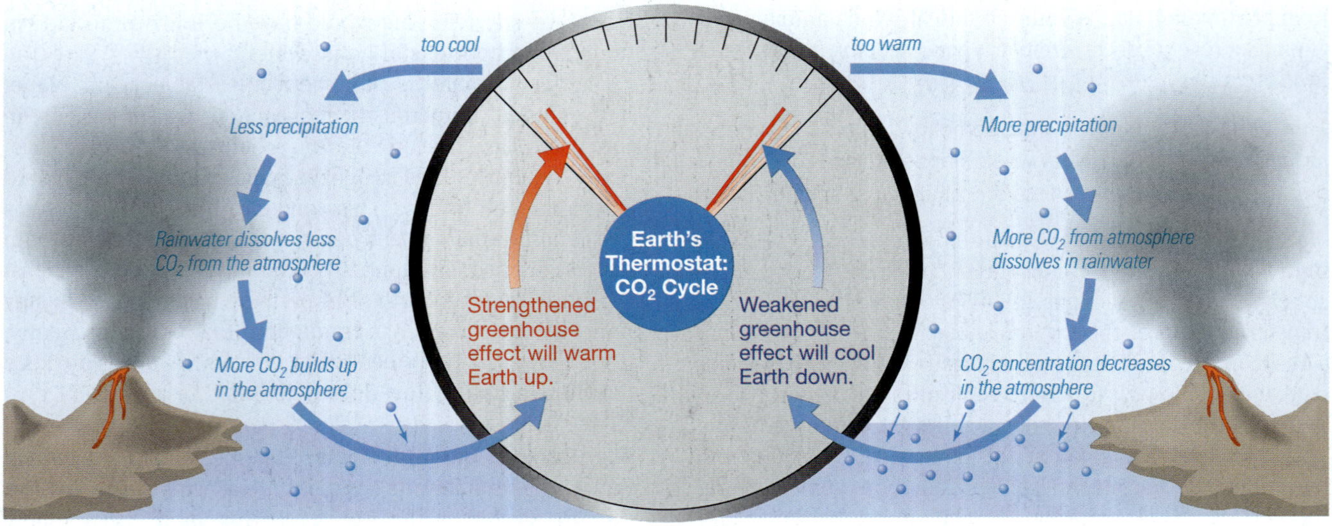


**FIGURE 10.37** The carbon dioxide cycle acts as a thermostat for Earth through negative feedback processes. Cool temperatures cause atmospheric $CO_2$ to increase, and warm temperatures cause atmospheric $CO_2$ to decline.

**Ice Ages and Other Long-Term Climate Change** While Earth's climate has remained stable enough for the oceans to stay at least partly liquid throughout history, significant variations have still occurred. Such variations are possible because the $CO_2$ cycle does not act instantly. When something begins to change the climate, it takes time for the feedback mechanisms of the $CO_2$ cycle to come into play because of their dependence on the gradual actions of plate tectonics and mineral formation in the oceans. Calculations show that it takes hundreds of thousands of years for atmospheric $CO_2$ to stabilize through the $CO_2$ cycle.

**Ice ages** occur when the global average temperature drops by a few degrees. The slightly lower temperatures lead to increased snowfall, which may cover continents with ice down to fairly low latitudes. For example, the northern United States was entirely covered with glaciers during the peak of the most recent ice age, which ended only about 10,000 years ago. The causes of ice ages are complex and not fully understood. Over periods of tens or hundreds of millions of years, the Sun's gradual brightening and the changing arrangement of the continents around the globe have at least in part influenced the climate. During the past few million years—a period too short for solar changes or continental motion to have a significant effect—the ice ages appear to have been strongly influenced by small changes in Earth's axis tilt and other characteristics of Earth's rotation and orbit (the Milankovitch cycles noted earlier).

Geological evidence also points to several particularly long and deep ice ages between about 750 and 580 million years ago. During these periods, glaciers appear to have advanced all the way to the equator. Because ice can reflect up to about 90% of the sunlight hitting it, this increase in global ice would have set up a positive feedback process that would have cooled Earth even further. Geologists suspect that in this way our planet may have entered the periods called **snowball Earth** (**FIGURE 10.38**). We do not know why these episodes occurred or precisely how extreme the cold became. Some models suggest the positive feedback may have driven the global average temperature as low as $-50°C$ ($-58°F$), causing the oceans to freeze to a depth of 1 kilometer or more. Other models suggest the oceans never froze completely, making Earth more of a "slushball" than a snowball. Either way, it seems that Earth became far colder during these periods than in more recent ice ages.

How did Earth recover from a "snowball" phase? The drop in surface temperature would not have affected Earth's interior heat, so volcanic outgassing would have continued to add $CO_2$ to the atmosphere. Oceans covered by ice would have been unable to absorb this $CO_2$ gas, which therefore would have accumulated in the atmosphere and strengthened the greenhouse effect. Eventually (perhaps after as long as 10 million years), the strengthening greenhouse effect would have warmed Earth enough to start melting the ice. The feedback processes that started the snowball Earth episode then moved in reverse. As the ice melted, more sunlight would have been absorbed (because liquid water absorbs more and reflects less sunlight than ice), warming the planet further. In fact, because the $CO_2$ concentration was so high, the warming would have continued well past current temperatures—perhaps taking the global average temperature to higher than 50°C (122°F). In just a few centuries, Earth might have emerged from a "snowball" phase into a "hothouse" phase. Geological evidence supports the occurrence of dramatic increases in temperature at the end of each snowball Earth episode. Earth then recovered over hundreds of thousands of years as the $CO_2$ cycle removed carbon dioxide from the atmosphere.

**Think about it** Suppose Earth did not have plate tectonics. Could the planet ever recover from a "snowball" phase? Explain.

The snowball Earth episodes would have had severe consequences for any life on Earth at the time. Indeed, the end of the snowball Earth episodes roughly coincides with a dramatic increase in the diversity of life on Earth (the *Cambrian explosion* [**Section 24.1**]). Some scientists suspect that the environmental pressures caused by the snowball Earth periods may have led to a burst of evolution. If so, we might not be here today if not for Earth's dramatic climate changes.

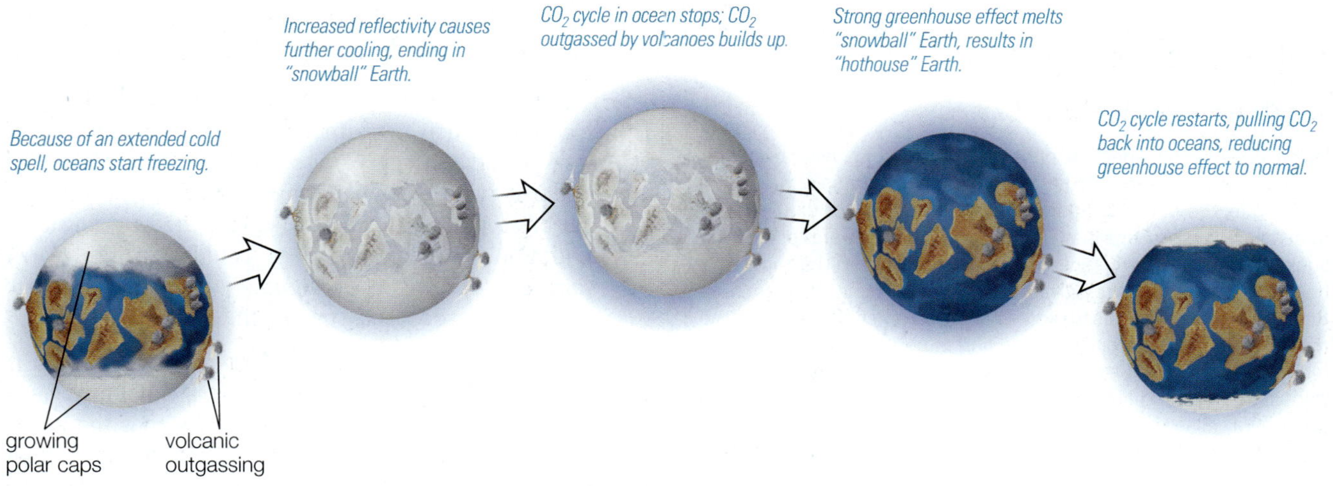


**FIGURE 10.38** The $CO_2$ cycle rescues Earth from a "snowball" phase.

**Earth's Long-Term Future Climate** Despite going through periods of ice ages, snowball Earth episodes, and hothouse phases, Earth's climate has remained suitable for life for some 4 billion years. However, the continuing brightening of the Sun will eventually overheat our planet.

According to some climate models, the warming Sun could cause Earth to begin losing its water as soon as a billion or so years from now. If such models are correct, life on Earth has already completed about 75% of its history on this planet. However, there are enough uncertainties in the models to make it possible that the $CO_2$ cycle will keep the climate steady much longer. Either way, by about 3 to 4 billion years from now, the Sun will have grown so warm that sunlight on Earth will be as intense as it is on Venus today. At that point, the effect will be the same as if we moved Earth to Venus's orbit (see Figure 10.34): a runaway greenhouse effect. Our planet will become a Venus-like hothouse, with temperatures far too high for liquid water to exist and all the $CO_2$ baked out of the rocks released into the atmosphere.

In summary, Earth's habitability will cease sometime between 1 and 4 billion years from now. Although this may seem depressing, remember that a billion years is a very long time—far longer than humans have existed so far. If you want to lose sleep worrying about the future, there are much more immediate threats, including those that we will discuss next.

## How is human activity changing our planet?

We humans are well adapted to the present-day conditions on our planet. The amount of oxygen in our atmosphere, the average temperature of our planet, and the ultraviolet-absorbing ozone layer are just what we need to survive. We have seen that these "ideal" conditions are no accident—they are consequences of our planet's unique geology and biology.

Nevertheless, the stories of the dramatic and permanent climate changes that occurred on Venus and Mars should teach us to take nothing for granted. Our planet may regulate its own climate quite effectively over long time scales, but fossil and geological evidence tells us that substantial and rapid changes in global climate can occur on shorter ones. In some cases, Earth's climate appears to have warmed several degrees Celsius in just decades. Evidence also shows that these past climate changes have had dramatic effects on local climates by raising or lowering sea level as much as tens of meters, altering ocean currents that keep coastlines warm, and transforming rainforests into deserts.

These past climate changes have been due to "natural" causes, such as small changes in Earth's orbit and tilt, major volcanic eruptions, the release of trapped carbon dioxide from oceans, and a variety of other geological processes. Today, however, Earth is undergoing climate change for a new reason: Human activity is rapidly increasing the atmospheric concentration of carbon dioxide and other greenhouse gases. Effects of this increase in greenhouse gas concentration are already apparent: Global average temperatures have risen by about 0.85°C (1.5°F) in the past century (**FIGURE 10.39**). This **global warming** is one of the most important issues of our time.

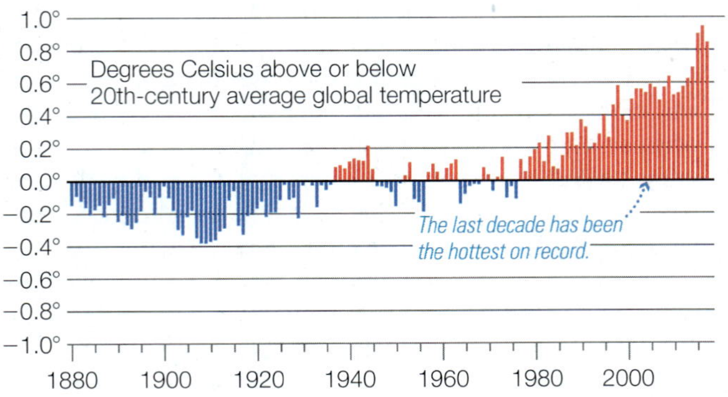


**FIGURE 10.39** Average global temperatures from 1880 through 2017. Notice the clear global warming trend of the past few decades. (Data from the National Climate Data Center.)

## COMMON MISCONCEPTIONS

### The Greenhouse Effect Is Bad

The greenhouse effect is often in the news, usually in discussions about environmental problems, but in itself the greenhouse effect is not a bad thing. In fact, we could not exist without it, because it is responsible for keeping our planet warm enough for liquid water to flow in the oceans and on the surface. The "no greenhouse" temperature of Earth is well below freezing. Why, then, is the greenhouse effect discussed as an environmental problem? The reason is that human activity is adding more greenhouse gases to the atmosphere—and scientists agree that the additional gases are warming Earth's climate. While the greenhouse effect makes Earth livable, it is also responsible for the searing 470°C temperature of Venus, proving that it's possible to have too much of a good thing.

**Global Warming Evidence**

**Global Warming** Global warming has been a hot political issue, both because some people debate its cause and because efforts to slow or stop the warming would require finding new energy sources and making other changes that would dramatically affect the world's economy. Scientifically, however, the case linking global warming with human activity is extremely strong, resting largely on three basic facts:

1. The greenhouse effect is a simple and well-understood scientific model. We can be confident in our understanding of it because it so successfully explains the observed surface temperatures of other planets. Given this basic model, there is no doubt that a rising concentration of greenhouse gases would make our planet warm up more than it would otherwise; the only debate is about how soon and how much.

2. Human activity such as the burning of fossil fuels is clearly increasing the amounts of greenhouse gases in the atmosphere. Observations show that the atmospheric concentration of carbon dioxide is now about 40% higher than it was before the industrial revolution

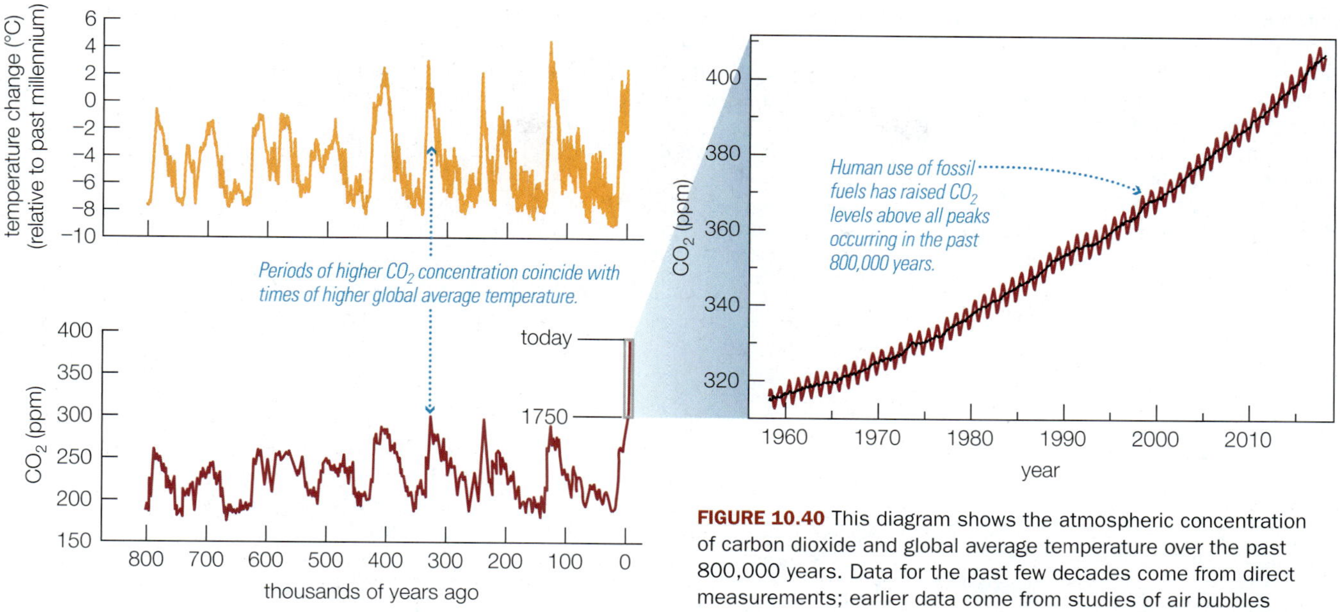


**FIGURE 10.40** This diagram shows the atmospheric concentration of carbon dioxide and global average temperature over the past 800,000 years. Data for the past few decades come from direct measurements; earlier data come from studies of air bubbles trapped in Antarctic ice (ice core samples). The $CO_2$ concentration is measured in parts per million (ppm), which is the number of $CO_2$ molecules among every 1 million air molecules. (Ice core data from the European Project for Ice Coring in Antarctica; inset data from NOAA.)

began or at any other time during the past million years, and it is continuing to rise rapidly (**FIGURE 10.40**). We can be confident that the rise is a result of human activity, because the atmosphere is becoming enriched in molecules of $CO_2$ carrying the distinct ratio of isotopes present in fossil fuels.

3. Climate models that ignore human activity fail to match the observed rise in global temperatures. In contrast, climate models that include the enhanced greenhouse effect from human production of greenhouse gases match the observed temperature trend quite well (**FIGURE 10.41**), providing additional support for the idea that global warming results from human activity.

These facts offer convincing evidence that we humans are now tinkering with the climate in a way that may cause major changes not just in the distant future, but in our own lifetimes. The same models that convince scientists of the reality of human-induced global warming tell us that if current trends in the greenhouse gas concentration continue—that is, if we do nothing to slow our emissions of carbon dioxide and other greenhouse gases—the warming trend will continue to accelerate. By the end of this century, the global average temperature would be 2°C–5°C (4°F–10°F) higher than it is now, giving our children and grandchildren the warmest climate that any generation of *Homo sapiens* has ever experienced.

**Think about it** Based on the rate of rise in the $CO_2$ concentration shown in Figure 10.40, how long will it be until we reach a doubling of the pre-industrial-age concentration of 280 ppm? Discuss the implications of your answer.

**Consequences of Global Warming** A temperature increase of a few degrees might not sound so bad, but small changes in *average* temperature can lead to much more dramatic changes in climate patterns. **FIGURE 10.42** compares recent regional temperatures to the averages from 1951–1980. Notice that almost all regions of the world are now significantly warmer than they were a few decades ago, but some regions warmed much more than others. This is what we mean by "climate change"—the idea that regional effects can be very different from the global average. Models suggest that regional climate changes will be further amplified as the global average temperature continues to increase.

The consequences of global warming are not simply hotter weather. The warming of the atmosphere and oceans means there is more total energy available in the climate

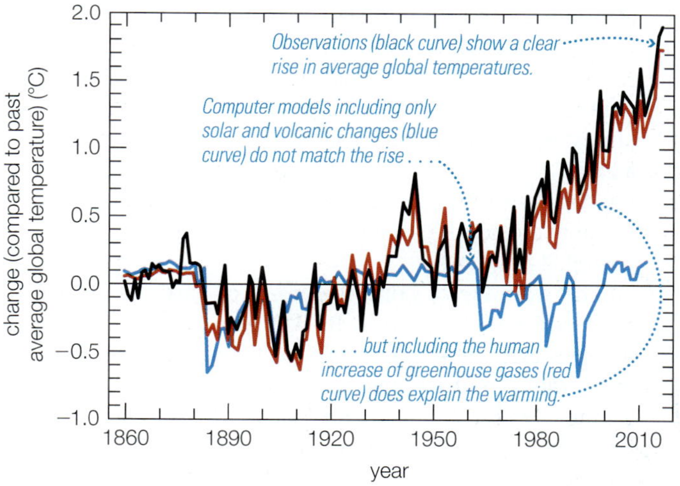


**FIGURE 10.41** This graph compares observed temperature changes (black curve) with the predictions of climate models that include only natural factors such as changes in the brightness of the Sun and effects of volcanoes (blue curve) and models that also include the human contribution to increasing greenhouse gas concentration (red curve). Only the red curve matches the observations well. (The red and blue model curves are averages of many scientists' independent models of global warming, which generally agree with each other within 0.1°C–0.2°C. Data from the Fourth National Climate Assessment.)

**FIGURE 10.42** This map shows how regional average temperatures for the period 2013–2017 compared to the average from 1951–1980. Notice the large regional differences, and the particularly large warming in the north polar region.

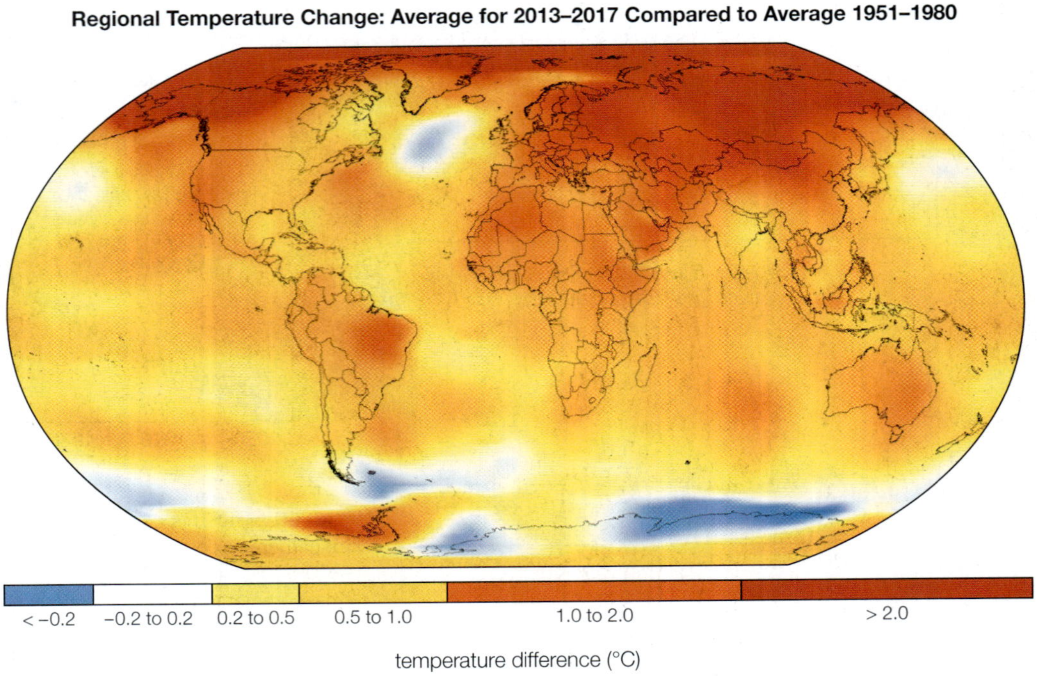


system, a fact that may explain why extreme weather events seem to have increased in frequency and severity in recent decades. For example, the fact that the waters of the Gulf of Mexico are at their warmest temperatures in at least a century may be contributing to the greater strength of hurricanes that have recently blown through the Caribbean. Moreover, warmer temperatures mean more total evaporation both from land and from the oceans, which can contribute both to more severe drought (which can lead to more wildfires) when local conditions are dry and heavier downpours (which can lead to flooding and mudslides) when the rains finally come. The warmer weather also means that spring weather is arriving earlier and fall weather later, which contributes to longer, hotter, and drier summers and higher risk of wildfires. Ironically, the phenomena of greater total energy and more moisture in the atmosphere can also increase the severity of winter storms, leading to the somewhat surprising fact that global warming can cause more severe winter blizzards, a pattern that we may already be seeing in recent winters in the eastern United States.

The warming of polar regions is of particular concern, because it causes ice to melt. This is clearly threatening to the species of these regions (polar bears, which depend on an abundance of ice floes, are already endangered), but it also may create damaging feedbacks. For example, because ice is more reflective than water, the loss of sea ice means an increase in the absorption of energy from the Sun, which can amplify other effects of global warming. Melting sea ice also adds fresh water to the oceans, and some researchers worry that this change in the water's salinity (salt content) could alter major ocean currents, such as the Gulf Stream—a "river" within the ocean that regulates the climate of western Europe and parts of the United States.

Global warming is also expected to cause a rise in sea level. The oceans expand very slightly as they warm, an effect that has already caused sea level to rise about 20 centimeters in the past century and is expected to cause a rise of another 30 centimeters by about 2100. This "thermal expansion" by itself could have devastating effects on coastal communities and low-lying countries such as Bangladesh. It also greatly increases the likelihood of devastating storm surges like those that have accompanied recent hurricanes. Melting ice may cause even greater increases in sea level. While the melting of ice in the Arctic Ocean does not affect sea level—the ice is already floating—melting of landlocked ice does. Such melting appears to be occurring already. For example, the famous "snows" (glaciers) of Mount Kilimanjaro are rapidly retreating and may be gone within the next decade or so. More ominously, recent data suggest a surprisingly rapid change in Greenland's ice sheet, leading some scientists to worry that ice melt could cause sea level to rise by a meter or more—enough to flood much of Florida—by the end of this century. In the longer term, complete melting of the polar ice caps would increase sea level by some 70 meters (more than 200 feet). Although such melting would probably take centuries or millennia, it suggests the disconcerting possibility that future generations will have to send deep-sea divers to explore the underwater ruins of many of our major cities.

The oceans are also suffering in another way. Some of the carbon dioxide released into the atmosphere by human activity (roughly 30% of it) ends up dissolving in the oceans, where it undergoes chemical reactions that make the oceans more acidic. This "ocean acidification" has been tied to the demise of many coral reefs around the world and to less productive fisheries. Acidification, along with the warming temperatures, also reduces the ability of the oceans to absorb carbon dioxide, which could lead to an acceleration of the buildup of carbon dioxide in the atmosphere.

**FIGURE 10.43** (pages 304–305) summarizes the evidence for and consequences of global warming. Fortunately, most scientists believe that we still have time to avert the most serious consequences of global warming, provided that we dramatically and rapidly curtail our greenhouse gas emissions. The most obvious way to reduce these emissions is to improve energy efficiency. Doubling the average gas mileage of cars—which we could do easily with current technology—would immediately cut automobile-related carbon dioxide emissions in half. Other tactics could include replacing fossil fuels with alternative energy sources—such as biofuels or solar, wind, or nuclear energy—or finding ways to capture and bury the carbon dioxide from the fossil fuels that we still use. The key idea to keep in mind is that global warming is a global problem, and significant international cooperation will be required if we hope to solve it.

But there is precedent for success: A few decades ago, we learned that the ozone layer that protects us from ultraviolet radiation was threatened by human-produced chemicals (known as CFCs), which led to international action phasing out those chemicals. As a result, the ozone layer over polar regions has begun to recover from its earlier damage, and we learned that people can be moved to act in the face of a threat to the environment on which we depend for survival.

**Think about it** If you were a political leader, how would *you* deal with the threat of global warming?

## EXTRAORDINARY CLAIMS Human Activity Can Change the Climate

The topic of global warming and associated climate changes is so commonplace in today's news that it's tempting to think of it as a relatively new idea. But the first clear claim that human emissions of carbon dioxide might change the climate dates back to 1896, when Swedish scientist Svante Arrhenius was contemplating possible causes of ice ages, which had recently been identified in the geological record.

The greenhouse effect of gases like carbon dioxide and water vapor had already been discovered, and Arrhenius decided to check whether changes in the carbon dioxide concentration might trigger ice ages. Determining how a change in carbon dioxide concentration will change the temperature requires complex calculations of the many feedbacks involved. Today, scientists do such calculations with detailed computer models, but Arrhenius had only pencil and paper. Nevertheless, after months of tedious calculations, he concluded that a 50% drop in the carbon dioxide concentration would cause Earth's temperature to drop about 5°C—which was indeed enough to bring about an ice age—while doubling it would cause a rise of about 5°C. Moreover, with the help of a colleague, he made an astonishing discovery: Human activities such as the burning of coal were already adding about as much carbon dioxide to the atmosphere each year as volcanoes. This led Arrhenius to make an extraordinary claim: Human activity might eventually raise the carbon dioxide concentration enough to cause significant global warming.

Arrhenius was not particularly concerned by his prediction, because he did not think we could add carbon dioxide fast enough for a doubling to occur in less than several centuries. But the rate of carbon dioxide emissions grew rapidly. In 1958, scientist Charles David Keeling began making the measurements shown in Figure 10.40 (the graph is often called the *Keeling curve* in his honor). These measurements, along with measurements of isotope ratios that show that the added carbon dioxide comes from human activity, leave no doubt that we are causing a substantial rise in the carbon dioxide concentration.

The only question that remained was whether this would really cause global warming, as Arrhenius had predicted. As discussed in the chapter, modern computer models do an excellent job of reproducing past climate, and while there are uncertainties in the precise values, these models generally agree with Arrhenius's prediction that a doubling of the carbon dioxide concentration should cause the global average temperature to rise by a few degrees Celsius. This fact, along with data showing that global warming is already well under way, explains why Arrhenius's extraordinary claim is now well accepted by the vast majority of scientists.

**Verdict:** Strongly supported.

## The BIG Picture PUTTING CHAPTER 10 INTO PERSPECTIVE

*This chapter and the previous chapter have given us a complete "big picture" view of how the terrestrial worlds started out so similar yet ended up so different. As you continue your studies, keep in mind the following important ideas:*

- Atmospheres affect planets in many ways. They absorb and scatter light, distribute heat, and create the weather that can lead to erosion. Perhaps most important, the greenhouse effect allows atmospheres to make a planet warmer than it would be otherwise.
- Atmospheric properties differ widely among the terrestrial worlds, but we can trace these differences to root causes. For example, only the larger worlds have significant atmospheres. The smaller worlds lack the internal heat needed for volcanism and the outgassing that releases atmospheric gases, and they also lack the gravity necessary to retain these gases.
- The histories of Venus and Mars suggest that major climate change is the rule, not the exception. Mars was once warm and wet, but its small size and lack of a magnetic field caused it to lose gas and freeze over 3 billion or more years ago. Venus may once have had oceans, but its proximity to the Sun doomed it to a runaway greenhouse effect.

**MY COSMIC PERSPECTIVE** We are here to talk about these things today only because Earth has managed to be the exception, a planet whose climate has remained relatively stable. We humans are ideally adapted to Earth today, but we have no guarantee that Earth will remain as hospitable in the future, especially as we tinker with the balance of greenhouse gases that has kept our climate stable.

**COSMIC Context**

# FIGURE 10.43 Global Warming

**Scientific studies of global warming apply the same basic approach used in all areas of science: We create models of nature, compare the predictions of those models with observations, and use our comparisons to improve the models. We have found that climate models agree more closely with observations if they include human production of greenhouse gases like carbon dioxide, making scientists confident that human activity is indeed causing global warming.**

(1) The greenhouse effect makes a planetary surface warmer than it would be otherwise because greenhouse gases such as carbon dioxide, methane, and water vapor slow the escape of infrared light radiated by the planet. Scientists have great confidence in models of the greenhouse effect because they successfully predict the surface temperatures of Venus, Earth, and Mars.

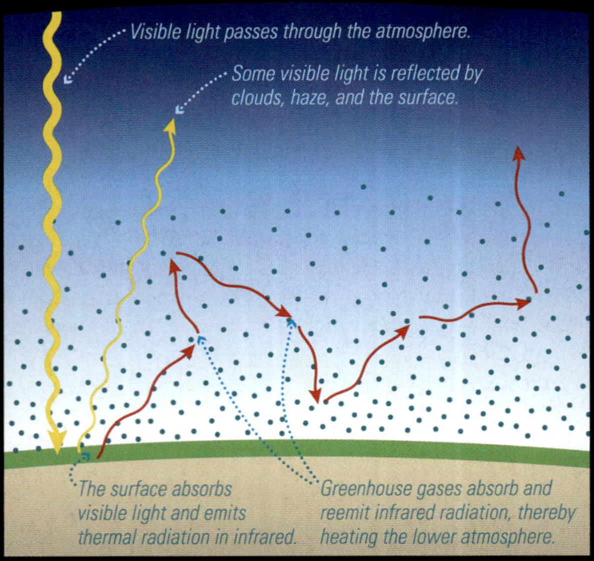


**Global Average Surface Temperature**

| Planet | Temperature Without Greenhouse Effect | Temperature With Greenhouse Effect |
|---|---|---|
| Venus | −40°C | 470°C |
| Earth | −16°C | 15°C |
| Mars | −56°C | −50°C |

*This table shows planetary temperatures as they would be without the greenhouse effect and as they actually are with it. The greenhouse effect makes Earth warm enough for liquid water and Venus hotter than a pizza oven.*

(2) Human activity is adding carbon dioxide and other greenhouse gases to the atmosphere. While the carbon dioxide concentration also varies naturally, its concentration is now much higher than it has been at any time in the previous million years, and it is continuing to rise rapidly.

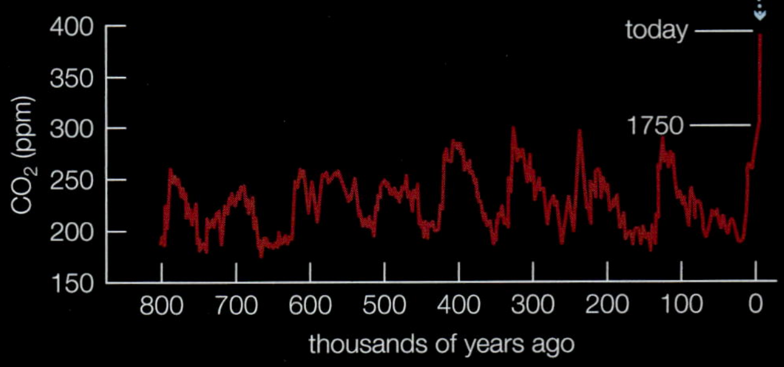

**3** Observations show that Earth's average surface temperature has risen during the last several decades. Computer models of Earth's climate show that an increased greenhouse effect triggered by $CO_2$ from human activities can explain the observed temperature increase.

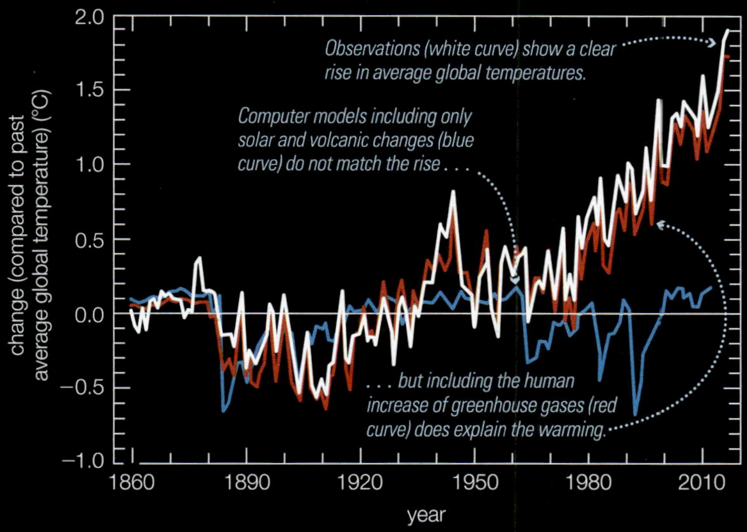


HALLMARK OF SCIENCE **Science progresses through creation and testing of models of nature that explain the observations as simply as possible.** Observations showing a rise in Earth's temperature demand a scientific explanation. Models that include an increased greenhouse effect due to human activity explain those observations better than models without human activity.

**4** Models can also be used to predict the consequences of a continued rise in greenhouse gas concentrations. These models show that, without significant reductions in greenhouse gas emissions, we should expect further increases in global average temperature, rising sea levels, and more intense and destructive weather patterns.

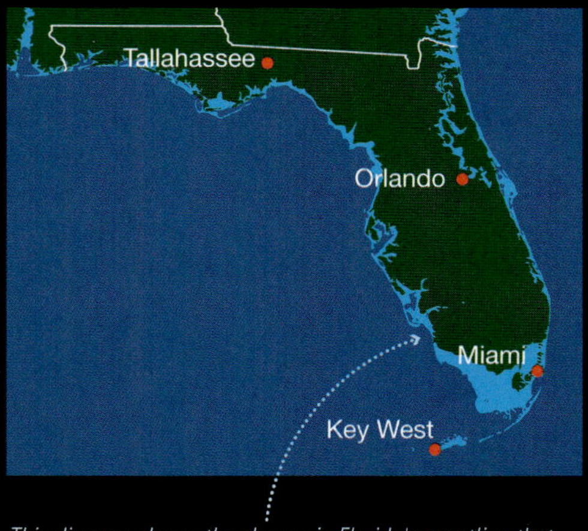


*This diagram shows the change in Florida's coastline that would occur if sea levels rose by 1 meter. Some models predict that this rise could occur within a century. The light blue regions show portions of the existing coastline that would be flooded.*

# Summary of Key Concepts

## 10.1 Atmospheric Basics

- **What is an atmosphere?** An atmosphere is a layer of gas that surrounds a world. It can create pressure, absorb and scatter sunlight, create wind and weather, interact with the solar wind to create a magnetosphere, and cause a **greenhouse effect** that can make a planet's surface warmer than it would be otherwise.

- **How does the greenhouse effect warm a planet?** **Greenhouse gases** such as carbon dioxide, methane, and water vapor absorb infrared light emitted from a planet's surface. The absorbed photons are quickly reemitted, but in random directions. The result acts much like a blanket, slowing the escape of heat from the planet's surface.

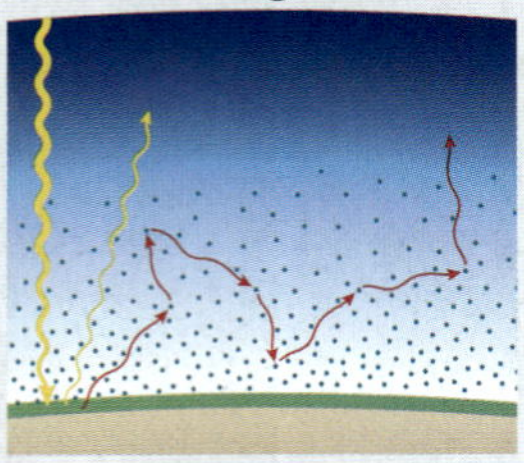

- **Why do atmospheric properties vary with altitude?** **Atmospheric structure** is determined by the way atmospheric gases interact with sunlight. On Earth, the basic structure consists of the **troposphere**, where most greenhouse warming occurs; the **stratosphere**, where **ozone** absorbs ultraviolet light from the Sun; the **thermosphere**, where solar x-rays are absorbed; and the **exosphere**, the extremely low-density outer layer of the atmosphere.

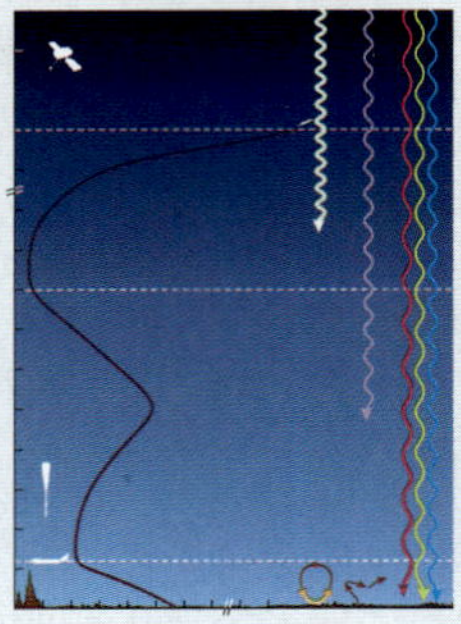

## 10.2 Weather and Climate

- **What creates wind and weather? Global wind patterns** are shaped by atmospheric heating and the **Coriolis effect** caused by a planet's rotation. Convection in the troposphere can lead to the formation of clouds and rain, hail, or snow.

- **What factors can cause long-term climate change?** The four factors that cause climate change are solar brightening (noticeable only over many millions of years), changes in axis tilt, changes in reflectivity, and changes in greenhouse gas abundance.

- **How does a planet gain or lose atmospheric gases?** Three sources of atmospheric gas are outgassing, vaporization of ices and liquids, and—on worlds with little atmosphere—surface ejection by tiny impacts of particles and photons. Four loss processes are condensation, chemical reactions with surface materials, the stripping of gas by the solar wind, and **thermal escape**.

## 10.3 Atmospheres of the Moon and Mercury

- **Do the Moon and Mercury have any atmosphere?** The Moon and Mercury have only very thin exospheres consisting of gas particles released through surface ejection by micrometeorites, solar wind particles, and high-energy solar photons.

## 10.4 The Atmospheric History of Mars

- **What is Mars like today?** Mars is cold and dry, with an atmospheric pressure so low that liquid water is unstable; however, a substantial amount of water is frozen in and near the polar caps. Martian weather is driven largely by seasonal changes that cause carbon dioxide alternately to condense and to vaporize at the poles, creating pole-to-pole winds and sometimes leading to huge dust storms.

- **How has Mars's climate differed in the past?** On time scales of hundreds of thousands to millions of years, Mars undergoes climate change due to large changes in its axis tilt. Mars has undergone even more dramatic climate change over the past 3 to 4 billion years, transforming it from a planet on which liquid water and rainfall were possible to the cold desert we see today.

- **Why did Mars change?** Mars's atmosphere must once have been much thicker, with a stronger greenhouse effect, so change must have occurred due to loss of atmospheric gas. Much of the gas probably was stripped away by the solar wind, which was able to reach the atmosphere as Mars cooled and lost its magnetic field. Water was probably lost as ultraviolet light broke apart water molecules in the atmosphere, and the lightweight hydrogen then escaped to space.

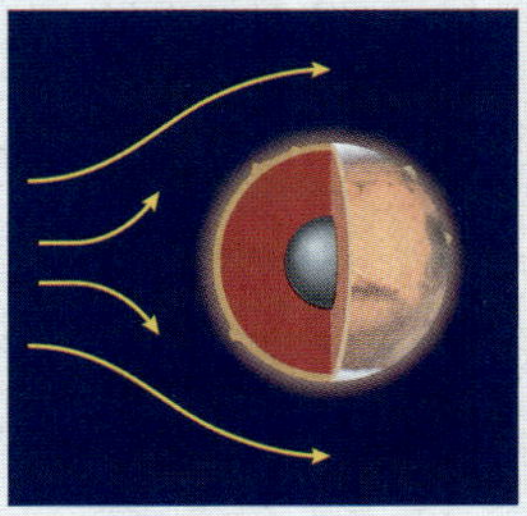

## 10.5 The Atmospheric History of Venus

- **What is Venus like today?** Venus has a thick carbon dioxide atmosphere that creates a strong greenhouse effect, explaining why the planet is so hot. It rotates slowly and therefore has a weak Coriolis effect and weak winds, and it is far too hot for rain to fall. Its atmospheric circulation keeps temperatures about the same day and night, and its lack of axis tilt means no seasonal changes.

- **How did Venus get so hot?** Venus's distance from the Sun ultimately led to a **runaway greenhouse effect**: Venus became too hot to develop liquid oceans like those on Earth. Without oceans to dissolve outgassed carbon dioxide and lock it away in carbonate rocks, all of Venus's carbon dioxide remained in its atmosphere, creating its intense greenhouse effect.

## 10.6 Earth's Unique Atmosphere

- **How did Earth's atmosphere end up so different?** Temperatures on Earth were just right for outgassed water vapor to condense and form oceans. The oceans dissolve carbon dioxide and ultimately lock it away in carbonate rocks, keeping the greenhouse effect moderate. Nitrogen from outgassing remained in the atmosphere. Oxygen and ozone were produced by photosynthesis, which was possible because the moderate conditions allowed the origin and evolution of abundant life.

- **Why does Earth's climate stay relatively stable?** Earth's long-term climate is remarkably stable because of feedback processes that tend to counter any warming or cooling that occurs. The most important feedback process is the **carbon dioxide cycle**, which naturally regulates the strength of the greenhouse effect.

- **How is human activity changing our planet?** Human activity is releasing carbon dioxide and other greenhouse gases into the atmosphere, and scientific evidence confirms that this is causing **global warming**. This warming may have many consequences, including a rise in sea level, an increase in the severity of storms, and dramatic changes in local climates.

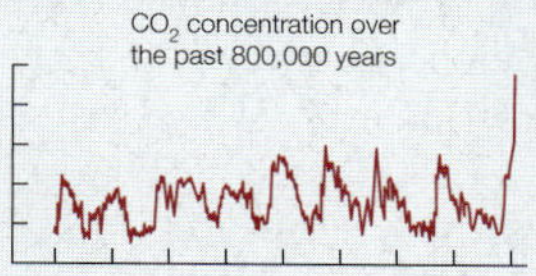


# Visual Skills Check

*Use the following questions to check your understanding of some of the many types of visual information used in astronomy. For additional practice, try the Chapter 10 Visual Quiz in the Study Area at www.MasteringAstronomy.com.*

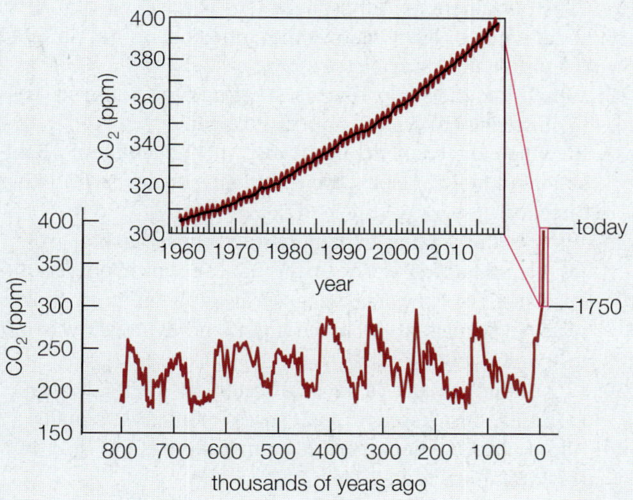


*The graph above, a portion of Figure 10.40, shows the atmospheric concentration of carbon dioxide for the past 800,000 years. Use the information in the graph to answer the following questions. Explain your answers.*

1. Based on the graph, when was the maximum abundance of $CO_2$ over the past 800,000 years?
   a. about 300,000 years ago
   b. about 125,000 years ago
   c. 1750
   d. now
2. Today's $CO_2$ abundance is approximately ___ times the average value over the last 800,000 years.
   a. 18 b. 1.8 c. 1.18 d. 0.18
3. How would you describe the rapid variations in $CO_2$ on the inset graph on the right?
   a. The variations occur randomly every few years.
   b. The variations occur randomly every few months.
   c. The variations occur regularly every few years.
   d. The variations occur regularly every year.
   e. The variations occur regularly every few months.
4. Which physical mechanism is consistent with your answer to question 3?
   a. Major volcanic eruptions every decade or so place additional $CO_2$ in the atmosphere.
   b. Every few years, another huge coal-powered power plant opens up and produces more $CO_2$.
   c. Seasonal changes in plant growth produce a regular yearly variation in $CO_2$ levels, with the Northern Hemisphere (with much more land area) causing almost all the variation.
5. Now study Figure 10.39, which shows global temperatures from 1880. Focus on the period since about 1960, and compare what you notice about the temperatures to what you've learned about carbon dioxide levels in Questions 1–4. Which of the following statements best describes the relationship?
   a. The warmest years correlate exactly with the years of greatest $CO_2$ abundance, indicating that $CO_2$ is the single most important factor controlling global temperature changes.
   b. The warmest years correlate roughly with the years of greatest $CO_2$ abundance, indicating that $CO_2$ is important but other factors also contribute to global temperature changes.
   c. The warmest years are not correlated with years of high $CO_2$ abundance, indicating that factors other than $CO_2$ have the most important effect on global temperature changes.

# Exercises and Problems

*For instructor-assigned homework and other learning materials, go to www.MasteringAstronomy.com.*

## Chapter Review Questions

### Short-Answer Questions Based on the Reading

1. Briefly describe the basic atmospheric characteristics of each of the five terrestrial worlds.
2. Use the balloon analogy to explain the origin of gas pressure. What is *atmospheric pressure*, and why does it decrease with altitude? What is 1 *bar* of pressure?
3. Is there any atmosphere at the orbital altitude of the Space Station? Explain.
4. What is the *greenhouse effect*? Describe how it warms a planet.
5. What factors determine a world's "no greenhouse" surface temperature? Explain the difference between "no greenhouse" and actual temperatures for each of the terrestrial worlds.
6. Describe Earth's basic *atmospheric structure*, from the ground up. How do interactions of sunlight and gases explain the existence of each of the atmospheric layers?
7. Why is the sky blue? Why are sunrises and sunsets red?
8. Why does convection occur in the *troposphere*, leading to active weather, but not in the *stratosphere*?
9. What is ozone? How does the absence of ozone on Venus and Mars explain why these planets lack a stratosphere?
10. What is a *magnetosphere*? Describe its role in protecting any atmosphere from the solar wind and in creating *auroras*.
11. What is the difference between *weather* and *climate*?
12. Describe Earth's *global wind patterns* and the role of *circulation cells*. How does rotation affect these cells?
13. What are clouds made of? How does rain or snow form?
14. Describe each of the four factors that can lead to long-term climate change.
15. Describe each process by which atmospheres gain or lose gas. What factors control *thermal escape*?
16. Why do the Moon and Mercury have so little atmospheric gas?
17. How and why do seasons on Mars differ from seasons on Earth?
18. Describe the leading hypothesis for how Mars lost atmospheric gas. What role does Mars's size play in this process?
19. What do we mean by a *runaway greenhouse effect*? Explain why this process occurred on Venus but not on Earth.
20. Describe four ways in which Earth's atmosphere is unique among the terrestrial worlds, and how each is important to our existence.
21. What is the *carbon dioxide cycle*, and why is it so crucial to life on Earth?
22. Briefly summarize the evidence linking human activity to *global warming*. What are its potential consequences?

### Does It Make Sense?

*Decide whether or not each of the following statements makes sense (or is clearly true or false). Explain clearly; not all these have definitive answers, so your explanation is more important than your chosen answer.*

23. If Earth's atmosphere did not contain molecular nitrogen, x-rays from the Sun would reach the surface.
24. If the molecular oxygen content of Earth's atmosphere increased, our planet would warm up.
25. Earth's oceans must have formed at a time when no greenhouse effect operated on Earth.
26. In the distant past, when Mars had a thicker atmosphere, it also had a stratosphere.
27. If Earth rotated faster, hurricanes would be more common and more severe.
28. Mars would still have seasons even if its orbit around the Sun were perfectly circular rather than elliptical.
29. Mars once may have been warmer than it is today, but it could never have been warmer than Earth.
30. If the solar wind were much stronger, Mercury might develop a carbon dioxide atmosphere.
31. If Earth had as much carbon dioxide in its atmosphere as Venus, our planet would not have oceans.
32. A planet in another solar system has no life but has an Earth-like atmosphere with plentiful oxygen.

### Quick Quiz

*Choose the best answer to each of the following. For additional practice, try the Chapter 10 Reading and Concept Quizzes in the Study Area at www.MasteringAstronomy.com.*

33. Which terrestrial world has the most atmosphere? (a) Venus (b) Earth (c) Mars
34. The greenhouse effect occurs in the (a) troposphere. (b) stratosphere. (c) lithosphere.
35. What kind of light warms the stratosphere? (a) infrared (b) visible (c) ultraviolet
36. Which of the following is a strong greenhouse gas? (a) nitrogen (b) water vapor (c) oxygen
37. In which direction do hurricanes in the Southern Hemisphere rotate? (a) clockwise (b) counterclockwise (c) either direction
38. What is the leading hypothesis for Venus's lack of water? (a) Venus formed closer to the Sun and accreted very little water. (b) Its water is locked away in the crust. (c) Its water molecules were broken apart, and hydrogen was lost to space.
39. What kind of gas is most affected by thermal escape? (a) greenhouse gases (b) light gases (c) all gases equally
40. About what fraction of Earth's atmosphere is $CO_2$? (a) 90% (b) 1% (c) less than 0.1%
41. What causes the release of oxygen into Earth's atmosphere? (a) outgassing (b) vaporization (c) photosynthesis
42. Where is most of the $CO_2$ that has outgassed from Earth's volcanoes? (a) in the atmosphere (b) in space (c) in rocks

## Inclusive Astronomy

*Use these questions to reflect on participation in science.*

43. *Group Discussion: The Global Climate Change Conversation.* Global warming will affect people of all nations, but wealthier nations are often in a better position to study how greenhouse gases affect Earth's climate and to develop technologies and enact policies that minimize those impacts.

**a.** Gather in small groups to discuss and make a list of potential impacts of climate change. In each case, decide if all nations would be affected equally or if effects would be greater for either wealthier or less-developed nations.
**b.** Discuss and make a list of potential steps that could be taken to reduce fossil fuel use and to mitigate climate change and its impacts on how people live. In each case, decide whether the step is likely to be equally difficult for all nations or more difficult for either wealthier or less-developed nations.
**c.** The world's primary forum for sharing scientific assessments of climate change, its impacts and risks, and options for adaptation and mitigation is the Intergovernmental Panel on Climate Change (IPCC), created in 1988 by the United Nations and the World Meteorological Organization. Do you think an organization founded by the United Nations is an appropriate forum for analyzing and sharing scientific findings about climate change? Should the United Nations be making public policy recommendations based on climate-change predictions reported by the IPCC? Defend your opinions to the group.
**d.** Less-developed nations have fewer resources to devote to climate-change research and the development of mitigation technologies, but they will inevitably be affected by the policy decisions made by wealthier nations. To what extent should wealthier nations take this fact into account when developing their own policies? Should they make an active effort to include the voices of less-developed nations in scientific research and/or policy making relating to global warming? Defend your opinions to the group.

## The Process of Science

44. *Unanswered Questions: Mars.* Choose one important but unanswered question about Mars's past, and discuss how we might answer this question in the future. Be as specific as possible, focusing on the type of evidence necessary to answer the question and how the evidence could be gathered. What are the benefits of finding answers to this question?
45. *Humans vs. Robots.* From a scientific standpoint, discuss what (if anything) we could learn about Mars from human missions that we could not learn from robotic missions. Considering only the scientific benefits and the costs, do you think it is worth sending humans to Mars? Are there other potential benefits of a human mission that would make such a mission more worthwhile? Defend your opinions.
46. *Skeptics' Claims on Global Warming.* A small but vocal group of people still dispute that humans are causing global warming. Do some research to find the basis of their claims. Then defend or refute their findings based on your own studies and your understanding of the hallmarks of science discussed in Chapter 3. Note: You may find it useful to visit globalwarmingprimer.com, which features a free online primer (written by one of the authors of this textbook) about global warming and many of the skeptics' claims.
47. *Group Discussion: The Politics of Global Warming.* Polls show that, today, political beliefs strongly influence whether a person accepts the scientific view of global warming presented in this chapter. Use this question to spur discussion of this political divide and how it might be bridged.
**a.** The basic science of global warming is not new; the first scientist to investigate what would happen if we were to double Earth's atmospheric carbon dioxide concentration was Svante Arrhenius, working more than 100 years ago. Each group member should read about his work by doing a search on "Arrhenius and global warming." Discuss why he concluded that an increase in carbon dioxide would cause warming, and whether there was any political dimension to his work.
**b.** Margaret Thatcher, generally considered to be one of the iconic founders of modern conservatism, was among the first global leaders to speak out on global warming. Each group member should read (or view) the speech she delivered to the United Nations on November 8, 1989. Briefly discuss her words and then discuss why so many people now consider global warming to have a political dimension, when it clearly did not (at least in Thatcher's view) at that time.
**c.** Find examples of conservative groups advocating action on global warming today (such as the Climate Leadership Council, the Green Tea Coalition, and RepublicEN.org). Discuss whether you agree with the goals of these groups, and whether learning about them alters your view on whether global warming should be considered a political issue.
**d.** On college campuses, a coalition of campus Republicans and campus Democrats have joined together to form "Students for Carbon Dividends." In Congress, the "Climate Solutions Caucus" allows a member to join only if he or she also brings along a member of the opposing party. Learn about and discuss each of these two groups and their prospects for success with their stated missions.
48. *Group Activity: Moon Colony.* Several nations are considering establishing a permanent human presence on the Moon, and some private companies are also considering it.
**a.** Working independently, research at least one proposed plan for returning humans to the Moon, including estimated costs for the plan.
**b.** Form groups of three to four students and take turns reporting on your findings from part a.
**c.** Considering what you learned from the research, as well as what you've learned in this textbook, make a list of potential *scientific* benefits of sending humans back to the Moon. Then make a separate list of any nonscientific (for example, economic, political, or educational) benefits that might arise from a Moon colony.
**d.** Discuss how the colony might be supplied with food, water, and air. Are there resources on the Moon that could be used to provide any of these, or would they all need to be supplied from Earth?
**e.** Come to a group consensus about whether it is worth the cost of returning to the Moon. If your group is in favor, describe specifics of how you think it should be done; for example, should it be done by individual nations, by an international effort, or by private companies? If your group is against returning humans to the Moon, explain why.

## Investigate Further

### Short-Answer/Essay Questions

49. *Clouds of Venus.* Table 10.2 shows that Venus's surface temperature in the absence of the greenhouse effect is lower than Earth's, even though Venus is closer to the Sun.
**a.** Explain this unexpected result in one or two sentences.
**b.** Suppose Venus had neither clouds nor greenhouse gases. What do you think would happen to the surface temperature of Venus? Why?

**c.** How are clouds and volcanoes linked on Venus? What change in volcanism might result in the disappearance of clouds? Explain.

50. *Atmospheric Structure.* Study Earth's average atmospheric structure (Figure 10.7). Sketch a similar curve for each of the following cases, and explain how and why the structure would be different in each case.

    **a.** Suppose Earth had no greenhouse gases.

    **b.** Suppose the Sun emitted no ultraviolet light.

    **c.** Suppose the Sun had a higher output of x-rays.

51. *Magic Mercury.* Suppose we could magically give Mercury the same atmosphere as Earth. Assuming this magical intervention happened only once, would Mercury be able to keep its new atmosphere? Explain.

52. *A Swiftly Rotating Venus.* Suppose Venus rotated as rapidly as Earth. Briefly explain how and why you would expect it to be different in terms of each of the following: geological processes, atmospheric circulation, magnetic field, and climate history.

53. *Coastal Winds.* During the daytime, heat from the Sun tends to make the air temperature warmer over land near the coast than over the water offshore. At night, land cools off faster than the sea, so temperatures tend to be cooler over land. Use these facts to predict the directions in which winds generally blow during the day and at night in coastal regions. Explain your reasoning in a few sentences. (Diagrams might help.)

54. *Sources and Losses.* Choose one process by which atmospheres can gain gas and one by which they can lose gas. For each process, write a few sentences that describe it and how it depends on each of the following fundamental planetary properties: size, distance from the Sun, and rotation rate.

55. *Two Paths Diverged.* Briefly explain how the different atmospheric properties of Earth and Venus can be explained by the fundamental properties of size and distance from the Sun.

56. *Change in Fundamental Properties.* Choose one property of Earth—either size or distance from the Sun—and suppose that it had been different (for example, size smaller or distance greater). Describe how this change would have affected Earth's subsequent atmospheric history and the possibility of life on Earth.

57. *Feedback Processes in the Atmosphere.* As the Sun gradually brightens in the future, how can the $CO_2$ cycle respond to reduce the warming effect? Which parts of the cycle will be affected? Is this an example of positive or negative feedback?

58. *Earth to Mars.* Section 10.5 discusses what might happen to Earth if it were suddenly moved to the orbit of Venus. What do you think would happen to Earth if it were suddenly moved to the orbital distance of Mars? Write a few sentences explaining your answer.

59. *Terraforming Mars.* Some people have suggested that we might be able to engineer Mars in a way that would cause its climate to warm and its atmosphere to thicken. This type of planet engineering is called *terraforming,* because its objective is to make a planet more Earth-like and easier for humans to live on. Discuss possible ways to terraform Mars. Do any of these ideas seem practical? Do they seem like good ideas? Defend your opinions.

60. *Global Warming Op-Ed.* What, if anything, should we be doing that we are not doing already to alleviate the threat of global warming? Write a one-page editorial summarizing and defending your opinion.

## Quantitative Problems

61. *The Mass of an Atmosphere.* What is the total mass of Earth's atmosphere? You may use the fact that 1 bar is the pressure exerted by 10,000 kilograms pushing down on a square meter in Earth's gravity. Remember that every square meter of Earth experiences this pressure from the atmosphere above it. Alternatively, you may start with the English unit value for pressure of 14.7 pounds per square inch and convert to kilograms for your final answer. Remember that the surface area of a sphere of radius $r$ is $4\pi r^2$.

62. *The Role of Reflectivity.* By assuming 0% and 100% reflectivity (respectively), find the maximum and minimum possible "no greenhouse" temperatures for a planet at 1 AU. What reflectivity would be necessary to keep the average temperature exactly at the freezing point? Compare this value to Earth's actual reflectivity in Table 10.2.

63. *The Cooling Clouds of Venus.* Table 10.2 shows that Venus's temperature in the absence of the greenhouse effect is lower than Earth's, even though Venus is closer to the Sun. What would Venus's "no greenhouse" temperature be if its clouds were more transparent, giving a reflectivity the same as Earth's? What would the actual surface temperature be in this case if the greenhouse effect increased the surface temperature by the same number of degrees that it does today?

64. *Mars's Elliptical Orbit.* Mars's distance from the Sun varies from 1.38 AU to 1.66 AU. How much does this affect its "no greenhouse" surface temperature at different times of year? Comment on how this affects Mars's seasons.

65. *Escape from Venus.*

    **a.** Calculate the escape velocity from Venus's exosphere, which begins about 200 kilometers above the surface. (*Hint:* See Mathematical Insight 4.4.)

    **b.** Calculate and compare the thermal speeds of hydrogen and deuterium atoms at the exospheric temperature of 350 K. The mass of a hydrogen atom is $1.67 \times 10^{-27}$ kilogram, and the mass of a deuterium atom is about twice the mass of a hydrogen atom.

    **c.** In a few sentences, comment on the relevance of these calculations to the question of whether Venus has lost large quantities of water.

# 11 Jovian Planet Systems

▲ **About the photo:** The photo of Saturn was taken by the *Cassini* spacecraft while it was in Saturn's shadow. The small blue dot of light just inside Saturn's rings at the left (about the 10 o'clock position) is Earth, far in the distance.

## LEARNING GOALS

### 11.1 A Different Kind of Planet

- Are jovian planets all alike?
- What are jovian planets like on the inside?
- What is the weather like on jovian planets?
- Do jovian planets have magnetospheres like Earth's?

### 11.2 A Wealth of Worlds: Satellites of Ice and Rock

- What kinds of moons orbit the jovian planets?
- Why are Jupiter's Galilean moons so geologically active?
- What geological activity do we see on Titan and other distant moons?
- Why are small icy moons more geologically active than small rocky planets?

### 11.3 Jovian Planet Rings

- What are Saturn's rings like?
- How do other jovian ring systems compare to Saturn's?
- Why do the jovian planets have rings?

*Do there exist many worlds, or is there but a single world? This is one of the most noble and exalted questions in the study of Nature.*

*—St. Albertus Magnus (1206–1280)*

**Chapter 11 Overview**

In Roman mythology, the namesakes of the jovian planets are rulers among gods: Jupiter is the king of the gods, Saturn is Jupiter's father, Uranus is the lord of the sky, and Neptune rules the sea. However, our ancestors could not have foreseen the true majesty of the four jovian planets. The smallest, Neptune, is large enough to contain the volume of more than 50 Earths. The largest, Jupiter, has a volume some 1400 times that of Earth. These worlds lack solid surfaces and are totally unlike the terrestrial planets. Their many moons and rings only add to their intrigue.

Why should we care about a set of worlds so different from our own? Apart from satisfying our natural curiosity, studies of the jovian worlds help us understand the birth and evolution of planetary systems, which in turn helps us understand our own origins and the types of worlds we may encounter in other planetary systems. In this chapter, we'll focus first on the jovian planets themselves, then on their many moons, and finally on their beautifully complex rings.

## 11.1 A Different Kind of Planet

The great differences between the terrestrial and the jovian planets are a relatively recent discovery in human history. Jupiter and Saturn are easily visible in the night sky, but the naked eye cannot discern any details about their nature. Uranus and Neptune were discovered in 1781 and 1846, respectively (see Special Topic, page 315), so they were unknown even during the Copernican revolution.

Astronomers first recognized the immense sizes of Jupiter and Saturn about 250 years ago. Recall that we need to know both angular size and distance to calculate an object's true size (see Mathematical Insight 2.1). Although Copernicus had figured out the *relative* distances to the known planets in astronomical units (AU), scientists did not establish the absolute scale of the solar system until the 1760s, when they were able to measure the true length of an astronomical unit with data from a transit of Venus (see Special Topic, page 207). Only then could scientists calculate the true sizes of Jupiter and Saturn from their distances and angular sizes (measured through telescopes). Knowing the scale of the solar system also told scientists the distances of orbiting moons, and because the orbital periods of the moons were easy to measure, the orbital distances allowed scientists to use Newton's version of Kepler's third law [**Section 4.4**] to calculate the masses of the jovian planets. Together, the measurements of size and mass revealed the low densities of the jovian planets, proving that these worlds are very different in nature from Earth.

The real revolution in understanding jovian planet systems has come from spacecraft visits to them, which began with the *Pioneer 10* and *Pioneer 11* spacecraft that flew past Jupiter and Saturn in the early 1970s, followed by the *Voyager 1* and *Voyager 2* missions that launched in 1977. Both Voyagers flew past Jupiter and Saturn, and *Voyager 2* continued on to flybys of Uranus and Neptune. More recently, we've sent orbiters to both Jupiter and Saturn. For Jupiter, the orbiters to date have been the *Galileo* spacecraft, which orbited from 1995 to 2003, and the *Juno* mission, which has been in Jupiter orbit since 2016. For Saturn, *Cassini* was in orbit from 2004 until 2017, when the spacecraft was sent plunging into Saturn's atmosphere.

### Are jovian planets all alike?

**FIGURE 11.1** shows a montage of the jovian planets compiled by the *Voyager* spacecraft, with Earth also shown to scale for comparison. The images and the given data reveal several key similarities among the jovian planets. In particular, all four of the jovian planets are immense in size compared to Earth, the largest of the terrestrial planets. All four also have much lower average densities than any of the terrestrial planets, and they all share the fact that their compositions are dominated by hydrogen, helium, and hydrogen compounds, all of which are relatively rare on the terrestrial planets.

The jovian planets also all share rotation rates that are rapid compared to those of the terrestrial worlds, though their lack of solid surfaces can make these rotation rates difficult to measure precisely. For example, we can try to measure their rotation rates by observing their clouds, but this requires care because clouds can appear to move because of winds as well as rotation. As a result, we usually measure jovian planet rotation rates by tracking emissions from charged particles trapped in their magnetic fields. Because these fields are generated deep in the planets' interiors, these measurements tell us how fast their interiors are rotating. The results show that jovian "days" range from about 10 hours on Jupiter and Saturn to 16–17 hours on Uranus and Neptune. Moreover, the large radii of these planets mean that their surface rotation speeds are much greater compared to Earth's than the periods alone would suggest. Interestingly, cloud observations indicate that jovian planet rotation rates vary with latitude: Equatorial regions complete each rotation in less time than polar regions. This variation is possible because these planets are not solid balls like the terrestrial worlds. (The Sun rotates similarly [**Section 14.1**].)

Fast rotation even affects the shapes of the jovian planets (**FIGURE 11.2**). Gravity alone would make these planets into perfect spheres, but rotation makes material bulge outward in the same way you feel yourself flung outward when you ride on a merry-go-round. Material bulges the most near the equator, where speeds around the rotation axis are highest. The size of the equatorial bulge depends on the balance between the inward pull of gravity and the outward push of rotation. The balance tips most strongly toward flattening on Saturn, because of its rapid 10-hour rotation and relatively weak surface gravity, making Saturn about 10% larger in diameter at its equator than from pole to pole. In addition to altering a planet's shape, the equatorial bulge exerts an extra gravitational pull that helps keep moons and rings aligned with the equator.

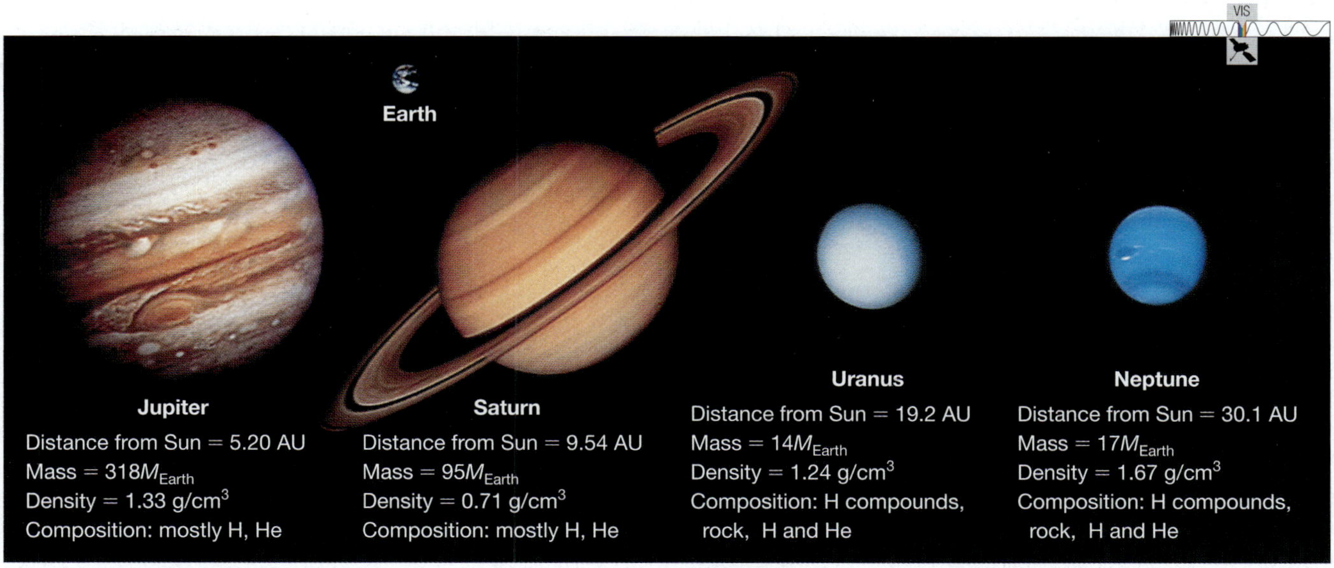


**FIGURE 11.1** Jupiter, Saturn, Uranus, and Neptune, shown to scale with Earth for comparison.

Despite their basic similarities, there are also important differences between the jovian planets, particularly when we contrast Jupiter and Saturn with Uranus and Neptune. The most obvious of these differences is that Jupiter and Saturn are significantly larger and more massive than Uranus and Neptune, but other differences are equally important.

**Composition Differences** Jupiter and Saturn are made mostly of hydrogen and helium, giving them compositions much more similar to that of the Sun than to that of the terrestrial planets. Some people even call Jupiter a "failed star," because it has a starlike composition but lacks the nuclear fusion needed to make it shine. This is a consequence of its size: Although Jupiter is large for a planet, it is about 1/80 as massive as the lowest-mass stars. As a result, its gravity is too weak to compress its interior to the extreme temperatures and densities needed for nuclear fusion. Of course, where some people see a failed star, others see an extremely successful planet.

Uranus and Neptune are much smaller than Jupiter and Saturn, though still much larger than Earth. They also contain significant amounts of hydrogen and helium, but the overall proportions of these materials are much lower than for Jupiter and Saturn. Instead, they are made primarily of hydrogen compounds such as water ($H_2O$), methane ($CH_4$), and ammonia ($NH_3$), along with smaller amounts of metal and rock.

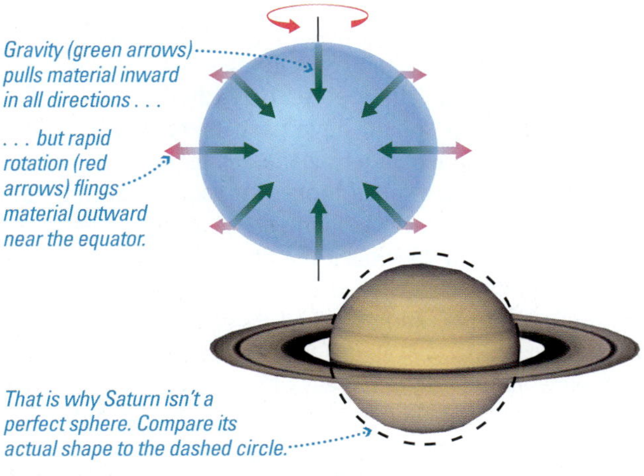


**FIGURE 11.2** Because of their rapid rotation, the jovian planets are not quite spherical. Saturn shows the biggest difference between its actual shape and a perfect sphere.

**Density Differences** Figure 11.1 shows that Saturn is considerably less dense than Uranus or Neptune. This should make sense, because the hydrogen compounds, rock, and metal that make up Uranus and Neptune are normally much more dense than hydrogen or helium gas. However, by the same logic, we'd expect Jupiter to be even less dense than Saturn—but it's not.

**Think about it** Saturn's average density of 0.71 g/cm$^3$ is less than that of water. As a result, it is sometimes said that Saturn could float on a giant ocean. Suppose there really were a gigantic planet with a gigantic ocean and we put Saturn on the ocean's surface. Would it float? If not, what would happen?

We can understand Jupiter's surprisingly high density by thinking about how massive planets are affected by their own gravity. Building a planet of hydrogen and helium is a bit like making one out of fluffy pillows. Imagine assembling a planet pillow by pillow. As each new pillow is added, those on the bottom are compressed more by those above. As the lower layers are forced closer together, their mutual gravitational attraction increases, compressing them even further. At first the stack grows substantially with each pillow, but eventually the growth slows until adding pillows barely increases the height of the stack (**FIGURE 11.3a**).

**See it for yourself** Measure the thickness of your pillow, and then put it at the bottom of a stack of other pillows, folded blankets, or clothing. How much has the stack above the pillow compressed it? Insert your hand between the different layers to feel the pressure differences—and imagine the kind of pressures and compression you'd find in a stack tens of thousands of kilometers tall.

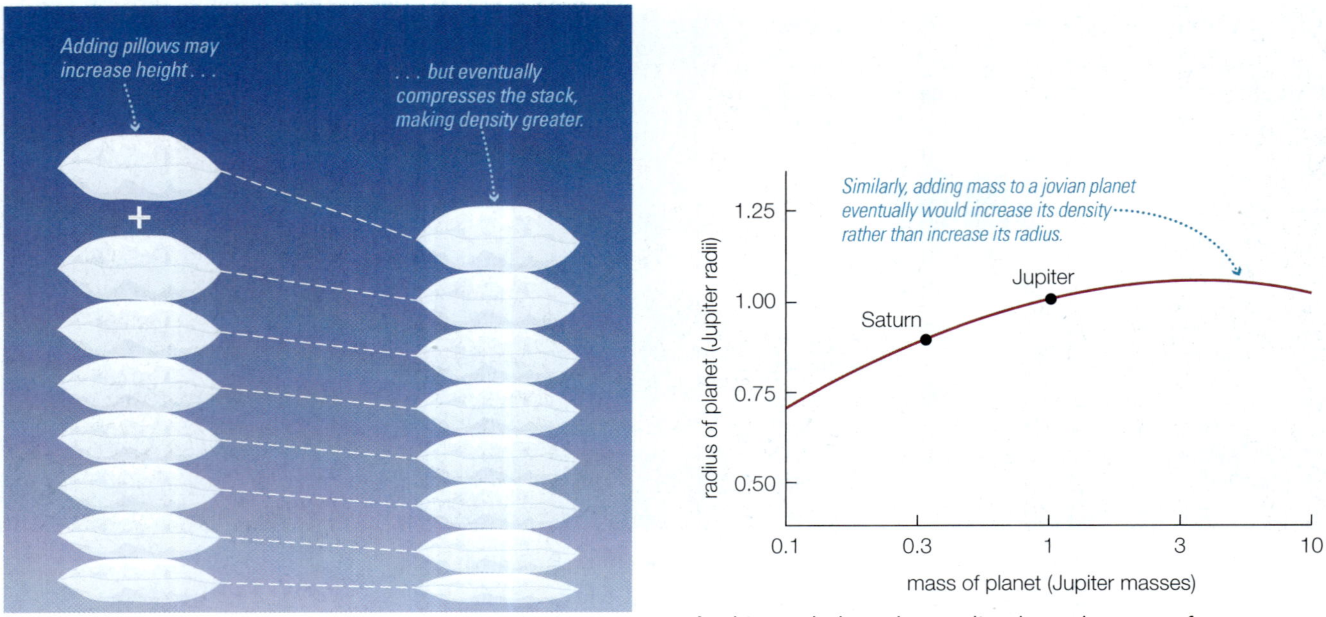


**a** Adding pillows to a stack may increase its height at first but eventually just compresses the stack, making its density greater. Similarly, adding mass to a jovian planet eventually will increase its density rather than increasing its radius.

**b** This graph shows how radius depends on mass for a hydrogen/helium planet. Notice that Jupiter is only slightly larger in radius than Saturn, despite being three times as massive. Gravitational compression of a planet much more massive than Jupiter would actually make it smaller in size.

**FIGURE 11.3** The relationship between mass and radius for a planet made of hydrogen and helium.

This analogy explains why Jupiter is only slightly larger than Saturn in radius even though it is more than three times as massive. The extra mass of Jupiter compresses its interior to a much higher density. More precise calculations show that Jupiter's radius is almost the maximum possible radius for a jovian planet. If much more gas were added to Jupiter, its weight would actually compress the interior enough to make the planet *smaller* rather than larger (**FIGURE 11.3b**). Some extrasolar planets that are larger in mass than Jupiter are therefore smaller in size. In fact, the smallest stars are significantly smaller in radius than Jupiter, even though they are at least 80 times as massive.

**Explaining the Differences** Overall, we've found that Jupiter and Saturn are both much larger and much richer in hydrogen and helium than Uranus and Neptune. This fact already hints at the variety that might be possible among jovian planets, and studies of planets orbiting other stars suggest even greater variety (see Figure 13.16).

The differences between the jovian planets in our solar system can probably be traced to their origins. Recall that, according to the nebular theory, the jovian planets formed in the outer solar system, where it was cold enough for hydrogen compounds to condense into ices [**Section 8.2**]. Because hydrogen compounds were so much more abundant than metal and rock, some of the ice-rich planetesimals of the outer solar system grew to great size. Once these planetesimals became sufficiently massive, their gravity allowed them to draw in the hydrogen and helium gas that surrounded them. Models suggest that all four jovian planets grew from ice-rich planetesimals of about the same mass—roughly 10 times the mass of Earth—but they captured different amounts of hydrogen and helium gas from the surrounding solar nebula.

Jupiter and Saturn captured so much hydrogen and helium gas that these gases now make up the vast majority of their masses. The ice-rich planetesimals from which they grew now represent only about 3% of Jupiter's mass and about 10% of Saturn's mass. Uranus and Neptune pulled in much less gas from the solar nebula, so their masses consist mostly of material from their original ice-rich planetesimals: hydrogen compounds mixed with smaller amounts of rock and metal. In the case of Uranus, its mass of about 14 times that of Earth, combined with the assumption that it formed around an ice-rich planetesimal of about 10 Earth masses, implies that captured hydrogen and helium gas contributed only about a third of its total mass. A similar analysis applies to Neptune, though the fact that it has slightly higher density than Uranus may suggest that it formed around a slightly more massive ice-rich planetesimal.

Why did the different planets capture different amounts of gas? The answer probably lies in their distances from the Sun as they formed. The solid particles that condensed farther from the Sun should have been more widely spread out than those that condensed nearer to the Sun, which means it would have taken longer for them to accrete into large, icy planetesimals. As the nearest jovian planet to the Sun, Jupiter would have been the first to get a planetesimal large enough for its gravity to start drawing in gas, followed by Saturn, Uranus, and Neptune. Because all the planets stopped accreting gas at the same time—when the solar wind blew all the remaining gas into interstellar space—the more distant planets had less time to capture gas and ended up smaller in size. Similar processes, played out to different extents in different planetary systems, may explain much of the wider variety of planetary types that scientists have discovered among planets around other stars.

## What are jovian planets like on the inside?

The jovian planets are often called "gas giants," making it sound as if they were entirely gaseous like air on Earth. Based on their compositions, Jupiter and Saturn may seem to deserve this name; after all, they became giants primarily by capturing so much hydrogen and helium gas. The name may seem less fitting for Uranus and Neptune, since they are made mostly of materials besides pure hydrogen and helium. However, closer inspection shows that the name is a little misleading even for Jupiter and Saturn, because their strong gravity compresses most of the "gas" into forms of matter quite unlike anything we are familiar with in everyday life on Earth. We'll begin our discussion of jovian interiors with Jupiter, then see how the other jovian planets compare.

**Inside Jupiter** Jupiter's lack of a solid surface makes it tempting to think of the planet as "all atmosphere," but you could not fly through Jupiter's interior in the way airplanes fly through air. A spacecraft plunging into Jupiter would soon be destroyed by the increasingly higher temperatures and pressures as it descended. In fact, the *Galileo* spacecraft dropped a scientific probe into Jupiter (in 1995), but the probe survived only to a depth of about 200 kilometers, or about 0.3% of Jupiter's radius. Because we cannot sample it directly, we instead learn about Jupiter's interior through a combination of theoretical modeling and laboratory experiments.

Just as with the terrestrial planets, detailed observations of the strength of any other planet's gravity and of its magnetic field can help us put together a model of the planet's interior structure. We then use the model to predict observable phenomena, such as a planet's average density, and modify the model until the predictions match the data. Today, advanced computer models successfully explain the

### SPECIAL TOPIC How Were Uranus, Neptune, and Pluto Discovered?

The planets Mercury, Venus, Mars, Jupiter, and Saturn are all easily visible to the naked eye and hence were well known to ancient people. In contrast, Uranus, Neptune, and the solar system's many small bodies were discovered relatively recently.

Uranus is actually visible to the naked eye, but it is so faint and moves so slowly in its 84-year orbit of the Sun that ancient people did not recognize it as a planet. It was even recorded as a star on some detailed early sky charts. Uranus was discovered as a planet in 1781 by brother-and-sister astronomers William and Caroline Herschel. William is the more famous of the pair, partly because Caroline's gender made many scientists of the time unwilling to take her seriously, but also because he was nearly 12 years older than his sister and had been the one who drew her into astronomical research. William Herschel originally suggested naming the new planet *Georgium Sidus,* Latin for "George's Star," in honor of his patron, King George III. Fortunately, the idea of "Planet George" never caught on. Instead, many 18th- and 19th-century astronomers referred to the new planet as Herschel. The modern name, Uranus (after the mythological father of Saturn), was first suggested by one of Herschel's contemporaries, astronomer Johann Bode. This name was generally accepted by the middle of the 19th century.

Neptune's discovery more than 60 years later represented an important triumph for Newton's universal law of gravitation and the young science of astrophysics. By that time, careful observations of Uranus had shown its orbit to be slightly inconsistent with that predicted by Newton's law—at least if it were being influenced only by the Sun and the other known planets. In the early 1840s in England, a student named John Adams suggested that the inconsistency could be explained if there were an unseen "eighth planet" orbiting the Sun beyond Uranus. According to the official story, he used Newton's theory to predict the location of the planet but was unable to convince British astronomers to carry out a telescopic search. However, documents from the time suggest his prediction may not have been as precise as the official history claims. Meanwhile, in the summer of 1846, French astronomer Urbain Leverrier independently made very precise calculations. He sent a letter to Johann Galle of the Berlin Observatory, suggesting a search for the eighth planet. On the night of September 23, 1846, Galle pointed his telescope to the position suggested by Leverrier. There, within 1° of its predicted position, he saw the planet Neptune. Hence, Neptune's discovery truly was made by mathematics and physics and was only confirmed with a telescope.

As a side note to this story, Leverrier had such faith in Newton's universal law of gravitation that he also suggested a second unseen planet, this one orbiting closer to the Sun than Mercury does. He got this idea because other astronomers had identified slight discrepancies between Mercury's actual orbit and the orbit predicted by Newton's theory. He assumed that no one had yet seen the planet, which he called Vulcan, because it was so close to the Sun. Leverrier died in 1877, still believing that Vulcan would someday be discovered. But Vulcan does not exist. Mercury's actual orbit does not match the orbit predicted by Newton's law because Newton's theory is not the whole story of gravity. About 40 years after Leverrier's death, Einstein showed that Newton's theory is only an approximation of a broader theory of gravity, known today as Einstein's *general theory of relativity.* Einstein's theory predicts an orbit for Mercury that matches its actual orbit. This match was one of the first key pieces of evidence in favor of Einstein's theory [**Section S3.4**].

Pluto was discovered in 1930 by American astronomer Clyde Tombaugh, and its discovery story at first seemed similar to Neptune's. After accounting for the gravitational tug of Neptune, astronomers found small, lingering discrepancies between observations and predictions for the orbit of Uranus. This seemed to suggest the existence of an undiscovered "ninth planet," and Tombaugh found Pluto just 6° from this planet's predicted position in the sky. However, while initial estimates suggested that Pluto was much larger than Earth, we now know that Pluto is far too small (its mass is only 0.2% that of Earth) to affect the orbit of Uranus. Moreover, the supposed irregularities of Uranus's orbit appear to have been erroneous, caused by the use of incorrect estimates for Neptune's mass; no "ninth planet" is needed to explain the observed orbit. Pluto's discovery near the location predicted for the ninth planet was a coincidence, made possible in part by the fact that many similar objects orbit in the same region of the solar system [**Section 12.4**], and its small size is why we now call it a *dwarf planet.*

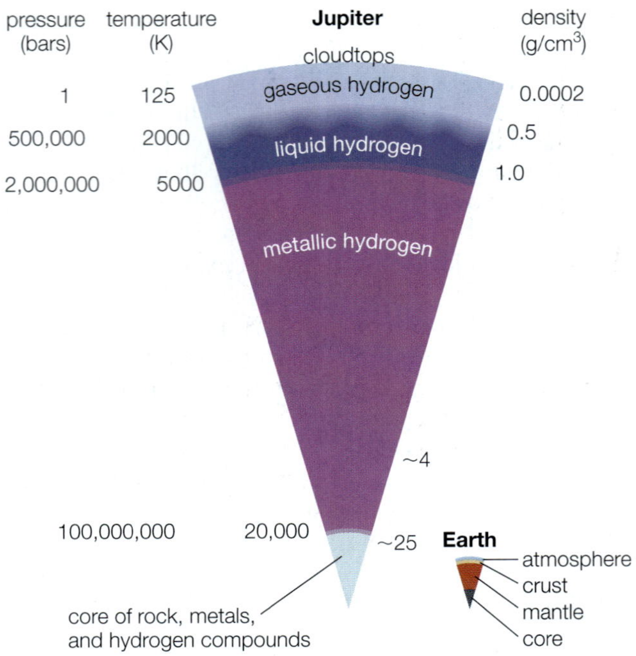


**FIGURE 11.4** Jupiter's interior structure, labeled with the pressure, temperature, and density at various depths. Earth's interior structure is shown to scale for comparison. (The thicknesses of Earth's crust and atmosphere are exaggerated.) Note that Jupiter's core is only slightly larger than Earth but is about 10 times as massive.

observed sizes, densities, atmospheric compositions, and precise shapes of all four jovian planets, giving us confidence the models are reliable. Laboratory studies also provide important data, such as showing us how hydrogen and helium behave under the tremendous temperatures and pressures that exist deep beneath the jovian cloudtops.

This combination of modeling and experiment indicates that Jupiter has fairly distinct interior layers (**FIGURE 11.4**). The layers do not differ much in composition—all except the core are mostly hydrogen and helium. Instead they differ in the *phase* (such as liquid or gas) of their hydrogen. To get a better sense of this layering, imagine plunging head-on into Jupiter in a futuristic spacesuit that could allow you to survive the extreme interior conditions.

Near the cloudtops, you'd find the temperature to be a brisk 125 K (−148°C), the density to be a low 0.0002 $g/cm^3$, and the atmospheric pressure to be about 1 bar (the same as the pressure at sea level on Earth [**Section 10.1**]). As you plunged downward, conditions would quickly become more extreme.

By a depth of 7000 kilometers, about 10% of the way to the center, you'd find the temperature at a scorching 2000 K and the pressure at 500,000 bars. The density here is about 0.5 $g/cm^3$, or about half that of water. Under these conditions, hydrogen acts more like a liquid than a gas, which is why the layer that begins at this depth is labeled *liquid hydrogen* in Figure 11.4. Of course, like the rest of Jupiter, the layer also contains helium and hydrogen compounds.

Continuing to a depth of 14,000 kilometers, about 20% of the way to the center, you'd find the density to have reached about that of water (1.0 $g/cm^3$), the temperature to be near 5000 K (almost as hot as the surface of the Sun), and the pressure at 2 million bars. This extreme pressure forces hydrogen into a compact, metallic form. Just as is the case with everyday metals, electrons are free to move around in *metallic hydrogen,* so it conducts electricity quite well. This layer extends through most of the rest of Jupiter's interior and, as we'll see shortly, it is where Jupiter's magnetic field is generated.

You'd finally reach Jupiter's core at a depth of 60,000 kilometers, about 10,000 kilometers from the center. The core temperature is some 20,000 K and the pressure is about 100 million bars. The core, with a density of about 25 $g/cm^3$, is much denser than any material you'll find on Earth's surface. The core is a mix of hydrogen compounds, rock, and metals, but these materials bear little resemblance to familiar solids or liquids because of the combination of high temperature and extreme pressure. Moreover, while the rock and metal of the terrestrial planets have separated into layers by composition (that is, core, mantle, and crust), Jupiter's core materials are probably all mixed together. The core contains about 10 times as much mass as the entire Earth, but might be only about the same size as Earth because it is compressed to such high density. Alternatively, according to early *Juno* results, the heavy elements of the core might be mixed into a much larger volume of the planet, perhaps extending halfway to the surface. This unexpected result will be tested by additional *Juno* observations.

**Comparing Jovian Interiors** We can extend the ideas from Jupiter to the other jovian planets. Because all four jovian planets have cores of about the same mass, their interiors differ mainly in the hydrogen/helium layers that surround their cores. **FIGURE 11.5** contrasts the four jovian interiors. Remember that while the outer layers are named for the phase of their hydrogen, they also contain helium and hydrogen compounds.

Saturn has the same basic layering as Jupiter, just as we should expect given its similar size and composition, but its lower mass and weaker gravity make the weight of the overlying layers less than on Jupiter. As a result, we must look deeper into Saturn to find each level where pressure changes hydrogen from one phase to another. That is, Saturn has thicker layers of gaseous and liquid hydrogen and a thinner and more deeply buried layer of metallic hydrogen.

Uranus and Neptune have somewhat different layering because their internal pressures never become high enough to form liquid or metallic hydrogen, so they have only a thick layer of gaseous hydrogen surrounding their cores of hydrogen compounds, rock, and metal. This core material may be liquid, making for very odd "oceans" buried deep inside Uranus and Neptune. The cores of Uranus and Neptune are larger in radius than the cores of Jupiter and Saturn, even though they have about the same mass, because they are less compressed by their lighter-weight overlying layers. The less extreme interior conditions also allowed Uranus's and Neptune's cores to differentiate, so hydrogen compounds reside in a layer around a center of rock and metal.

**Internal Heat** Recall that internal heat drives surface geology on the terrestrial worlds. The jovian planets do not have surface geology, because they have no solid surfaces, but internal heat still plays an important role.

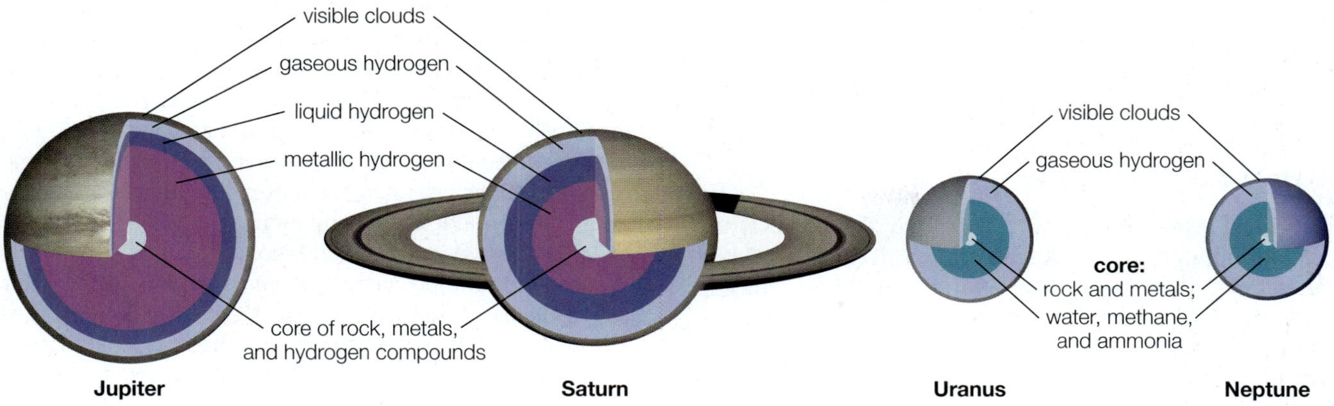


**FIGURE 11.5** These diagrams compare the interior structures of the jovian planets, shown approximately to scale. All four planets have cores of rock, metals, and hydrogen compounds, with masses about 10 times the mass of Earth. They differ primarily in the thicknesses of the hydrogen/helium layers that surround their cores. Notice that the cores of Uranus and Neptune are differentiated into a layer of hydrogen compounds around a center of rock and metals.

Jupiter has a tremendous amount of internal heat, and like any hot object, it gradually loses this heat to space by emitting thermal radiation [**Section 5.4**]. In fact, Jupiter emits almost twice as much energy as it receives from the Sun. This heat contributes significant energy to Jupiter's upper atmosphere. (For comparison, Earth's internal heat contributes only 0.005% as much energy to the surface as sunlight does.)

What explains Jupiter's large emission of heat? Jupiter's large size means it still retains much of the heat it acquired from accretion, differentiation, and radioactivity, but calculations show that the gradual escape of this heat is not enough to explain Jupiter's present-day heat loss. Instead, the leading hypothesis suggests that Jupiter is still slowly contracting. Contraction converts gravitational potential energy to thermal energy, so continued contraction would be an ongoing source of internal heat. Although we have not measured any such contraction, theoretical models suggest that it is probably occurring. Moreover, calculations tell us that contraction could easily explain Jupiter's internal heat, even if the contraction is so gradual that we have little hope of ever measuring it directly.

Regardless of the specific mechanism, Jupiter has undoubtedly lost substantial heat during the 4.5 billion years since its formation. Jupiter's interior must have been much warmer in the distant past, and this heat would have "puffed up" its atmosphere. With a larger size, Jupiter would have reflected more sunlight, so in the distant past Jupiter must have been even more prominent in Earth's sky than it is today.

Saturn also emits nearly twice as much energy as it receives from the Sun, suggesting that it, too, must have some ongoing source of heat. However, Saturn's mass is too small for it to be generating heat by contracting like Jupiter. Instead, Saturn's pressure and its lower interior temperatures may allow helium to condense into liquid form at relatively high levels within the interior. The helium droplets slowly rain down to the deeper interior. This gradual helium rain represents a sort of ongoing *differentiation* [**Section 9.1**], because it means that higher-density material (the liquid helium) is still sinking inside the planet. Spacecraft measurements have confirmed that Saturn's atmosphere has less helium than Jupiter's, just as we would expect if helium has been raining down into Saturn's interior for billions of years.

Neither Uranus nor Neptune has internal conditions that should allow helium rain to form to provide internal heating, and most of their original heat from accretion should have escaped long ago. As a result, Uranus emits virtually no excess internal energy. We might also expect the same to be true of Neptune, but it in fact emits nearly twice as much energy as it receives from the Sun. The only reasonable hypothesis for this internal heat is that Neptune is somehow still contracting, somewhat like Jupiter, thereby converting gravitational potential energy into thermal energy. However, we do not yet know why a planet of Neptune's size would still be contracting more than 4 billion years after its formation.

## What is the weather like on jovian planets?

Jovian atmospheres have dynamic winds and weather, with colorful clouds and enormous storms. These atmospheres are made mostly of hydrogen and helium gas, mixed with small amounts of various hydrogen compounds. The most common compounds are water ($H_2O$), methane ($CH_4$), and ammonia ($NH_3$); this should make sense, because after hydrogen and helium, the three next most common elements in the universe are oxygen, carbon, and nitrogen. Spectroscopy also reveals the presence of small amounts of more complex hydrogen compounds, including acetylene ($C_2H_2$), ethane ($C_2H_6$), propane ($C_3H_8$), and ammonium hydrosulfide ($NH_4SH$). Although all these hydrogen compounds together make up only a minuscule fraction of the jovian planet atmospheres, they are responsible for virtually all aspects of their appearances. Some of these compounds condense to form the clouds that are so prominent in telescope and spacecraft images. Others are responsible for the great variety of colors we see among the jovian planets. Without these compounds in their atmospheres, the jovian planets would be featureless grey balls of gas.

**Think about it** Several gases in Jupiter's atmosphere—including methane, propane, and acetylene—are highly flammable here on Earth. Jupiter has plenty of lightning to provide sparks, so why don't these gases ignite in Jupiter's atmosphere? (*Hint:* What's missing from Jupiter's atmosphere that's necessary for ordinary fire?)

We can understand the jovian atmospheres using the same basic principles that we used with the terrestrial atmospheres. As we did for the jovian planet interiors, let's begin by considering Jupiter.

**Jupiter's Atmosphere** We've learned the basic structure of Jupiter's atmosphere through a combination of telescope and spacecraft observations and data returned from the *Galileo* probe during its 1995 plunge into Jupiter. **FIGURE 11.6** shows that Jupiter, like Earth, has a troposphere, a stratosphere, and a thermosphere [**Section 10.1**]. High above the cloudtops, Jupiter's *thermosphere* consists of very low-density gas heated to about 1000 K by solar x-rays and by energetic particles from Jupiter's magnetosphere. Below the thermosphere (but still above the clouds), we find Jupiter's *stratosphere.* Recall that a planet can have a stratosphere only if it has a gas that can absorb ultraviolet light from the Sun. Ozone plays this role on Earth. Jupiter lacks molecular oxygen and ozone, but has a few minor atmospheric ingredients that absorb solar ultraviolet photons. This absorption gives the stratosphere a peak temperature of around 200 K (−73°C). Chemical reactions driven by the solar ultraviolet photons create a smog-like haze that masks the color and sharpness of the clouds below. Below the stratosphere lies Jupiter's *troposphere,* where the temperature rises with depth because greenhouse gases trap both solar heat and Jupiter's own internal heat.

The fact that temperatures increase with depth in the troposphere leads to strong convection and explains the thick clouds that enshroud Jupiter. Recall that clouds form when a gas condenses to make tiny liquid droplets or solid flakes. Water vapor is the only gas that can condense in Earth's atmosphere, which is why clouds on Earth are made of water droplets or ice flakes that can produce rain or snow. In contrast, Jupiter has three primary cloud layers, each formed by a gas that condenses at a different temperature. You can understand why by imagining that you could watch gas as it rises with convection.

Deep in the troposphere, the temperature is too high for any of the gases that make up clouds to condense, so there are no clouds. When the rising gas reaches the depth marked −100 kilometers in Figure 11.6 (meaning 100 kilometers below the highest cloudtops), the temperature has fallen enough for water vapor to condense to make water clouds, but not cold enough for other gases to condense. As on Earth, these water clouds reflect all the colors of sunlight, making them white, but we rarely see them because they lie so deep within the atmosphere. The rising gas continues to encounter lower and lower temperatures until it reaches an altitude at which ammonium hydrosulfide can condense, making the second cloud layer. These ammonium hydrosulfide clouds reflect brown and red light (though no one knows exactly why), producing many of the dark colors of Jupiter. Finally, the remaining gas reaches an altitude at which it is cold enough for ammonia to condense, making the upper cloud layer. The ammonia clouds reflect all the colors of sunlight, so they are white.

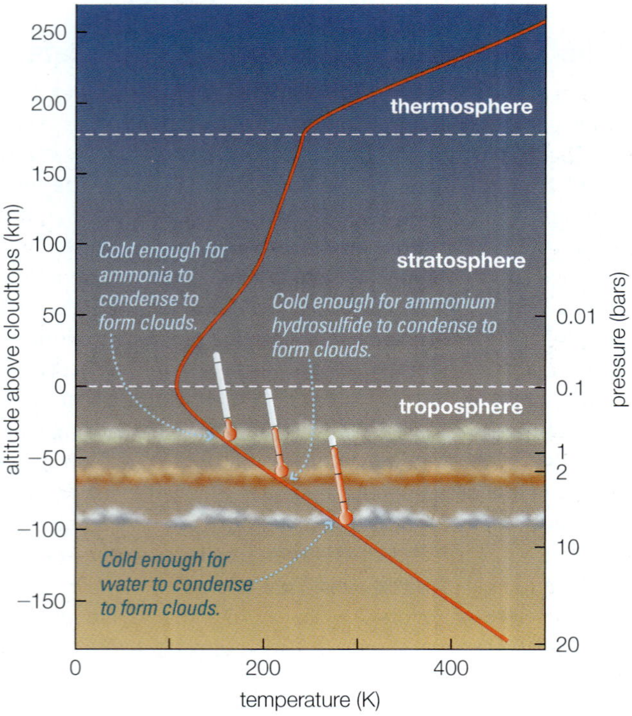


**FIGURE 11.6** This graph shows the temperature structure of Jupiter's atmosphere. Jupiter has at least three distinct cloud layers because different atmospheric gases condense at different temperatures and hence at different altitudes.

**Comparative Atmospheres** Spacecraft and telescope data tell us that the atmospheric structures of the other three jovian planets are quite similar to that of Jupiter, except their atmospheres get progressively cooler with increasing distance from the Sun. These temperature differences lead to some differences in cloud layering, summarized in **FIGURE 11.7**.

Saturn has the same set of three cloud layers as Jupiter, but the lower temperatures mean these layers occur deeper in Saturn's atmosphere. For example, to find the relatively warm temperatures at which water vapor can condense to form water clouds, we must look about 200 kilometers deeper into Saturn than into Jupiter. This fact probably explains Saturn's more subdued colors: Less light penetrates to the depths at which Saturn's clouds are found, and the light they reflect is more obscured by the atmosphere above them. Saturn's cloud layers are also separated by greater vertical distances than Jupiter's, because Saturn's weaker gravity causes less atmospheric compression.

Uranus and Neptune are so cold that any cloud layers similar to those of Jupiter or Saturn would be buried too deep in their atmospheres for us to see. However, the colder high-altitude temperatures of these two worlds (compared to Jupiter and Saturn) allow some of their abundant methane gas to condense into clouds of methane snow. The rest of this methane gas, some of which is high above the clouds, absorbs red light and therefore allows only blue light to penetrate deeper. The methane clouds reflect this blue light upward, giving Uranus and Neptune their distinctive blue colors (**FIGURE 11.8**). Uranus is a lighter blue than Neptune, probably because it has more smog-like haze to

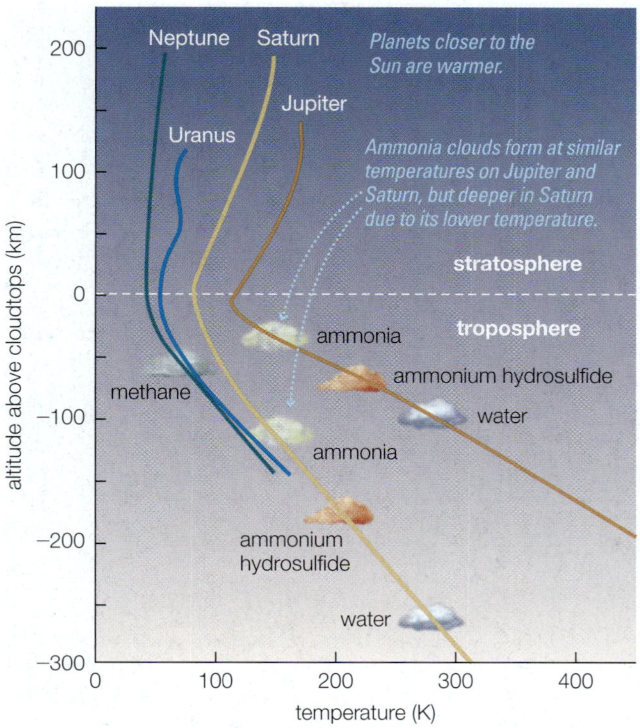


**FIGURE 11.7** The figure contrasts the atmospheric structures and cloud layers of the four jovian planets. Note that the altitude scale is different on this figure than on Figure 11.6, allowing us to see the tropospheres and stratospheres only.

scatter sunlight in its upper atmosphere. The extra haze is probably a result of Uranus's "sideways" axis tilt, which leads to extreme seasons. With one hemisphere remaining sunlit for decades, gases have plenty of time to interact with solar ultraviolet light and to make the chemical ingredients of the haze. Continuous sunlight may also explain why Uranus has a surprisingly hot thermosphere that extends thousands of kilometers above its cloudtops.

**Jupiter's Winds and Storms** We can understand Jupiter's dynamic weather by analogy to Earth [**Section 10.2**]. Recall that on Earth, solar heat causes warm equatorial air to rise and flow toward the poles, while cooler polar air flows toward the equator. By itself, this flow would give Earth two large equator-to-pole *circulation cells* (see Figure 10.13), but Earth's rotation creates a Coriolis effect that splits them into smaller cells (see Figure 10.16). The same basic process occurs on Jupiter, but Jupiter's greater size and faster rotation make its Coriolis effect much stronger. As a result, instead of being split into just three smaller cells, each of Jupiter's circulation cells splits into many alternating bands of rising and falling air. These bands are visible as the stripes of alternating color in photographs of Jupiter.

The alternating colors come from differences in the clouds that form in rising and falling air. As shown in **FIGURE 11.9**, the entire planet is blanketed with the reddish-brown ammonium hydrosulfide clouds that make up Jupiter's middle cloud layer. In contrast, the upper layer of white ammonia clouds forms only in places where rising air carries ammonia gas to altitudes high enough and cold enough for the ammonia to condense. In other words, the white bands (sometimes called *zones*) represent regions of rising air with cold ammonia clouds, while the darker bands (sometimes called *belts*) represent regions of falling air in which we can see down to a lower, warmer cloud level. This distinction between Jupiter's bright and dark bands is analogous to the difference on Earth between cloudy, rainy equatorial regions (regions of generally rising air) and the clear desert skies found at latitudes roughly 20°–30° north and south of the equator (regions of descending air). The infrared photo of Jupiter in Figure 11.9 confirms this idea: Warmer air emits more infrared light, so in the infrared photograph the warmer reddish bands appear brighter than the cooler white bands.

You might wonder why ammonia clouds form only in the rising air and not in the falling air. The answer goes back to the clouds themselves. Jupiter's lower atmosphere has ammonia throughout, so there's always ammonia in the air rising upward. However, once the ammonia condenses to form clouds of ammonia ice, these ice flakes fall back down as ammonia "snow," so the air that continues to rise above the clouds has very little ammonia left in it. When this ammonia-depleted air spills to the north and south and returns downward in the bands of falling air, there is not enough ammonia to form clouds. It's analogous to the way air descending over Earth's desert latitudes is usually depleted of water by precipitation over the equator, leaving it dry.

The alternating bands of rising and falling air shape Jupiter's global wind patterns. As on Earth, the rising and falling air drives slow winds that are directed north or south, but Jupiter's strong Coriolis effect diverts these winds into fast east or west winds. The east-west winds have peak speeds above 400 km/hr, making hurricane winds on Earth seem mild by comparison. The winds are generally strongest at the equator and at the boundaries between bands of rising and falling air. The *Juno* mission has shown that these winds are much more than "skin deep"; they blow even at depths of thousands of kilometers below the cloudtops.

Jupiter's global wind patterns are sometimes interrupted by powerful storms, much as storms can interrupt Earth's

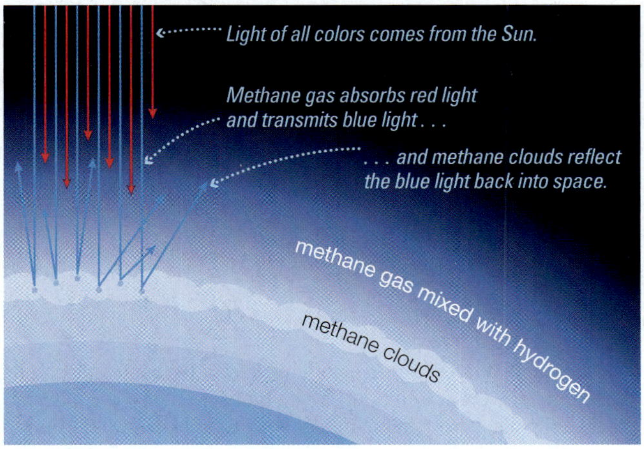


**FIGURE 11.8** Neptune and Uranus look blue because methane gas absorbs red light but transmits blue light. Clouds of methane snowflakes reflect the transmitted blue light back to space.

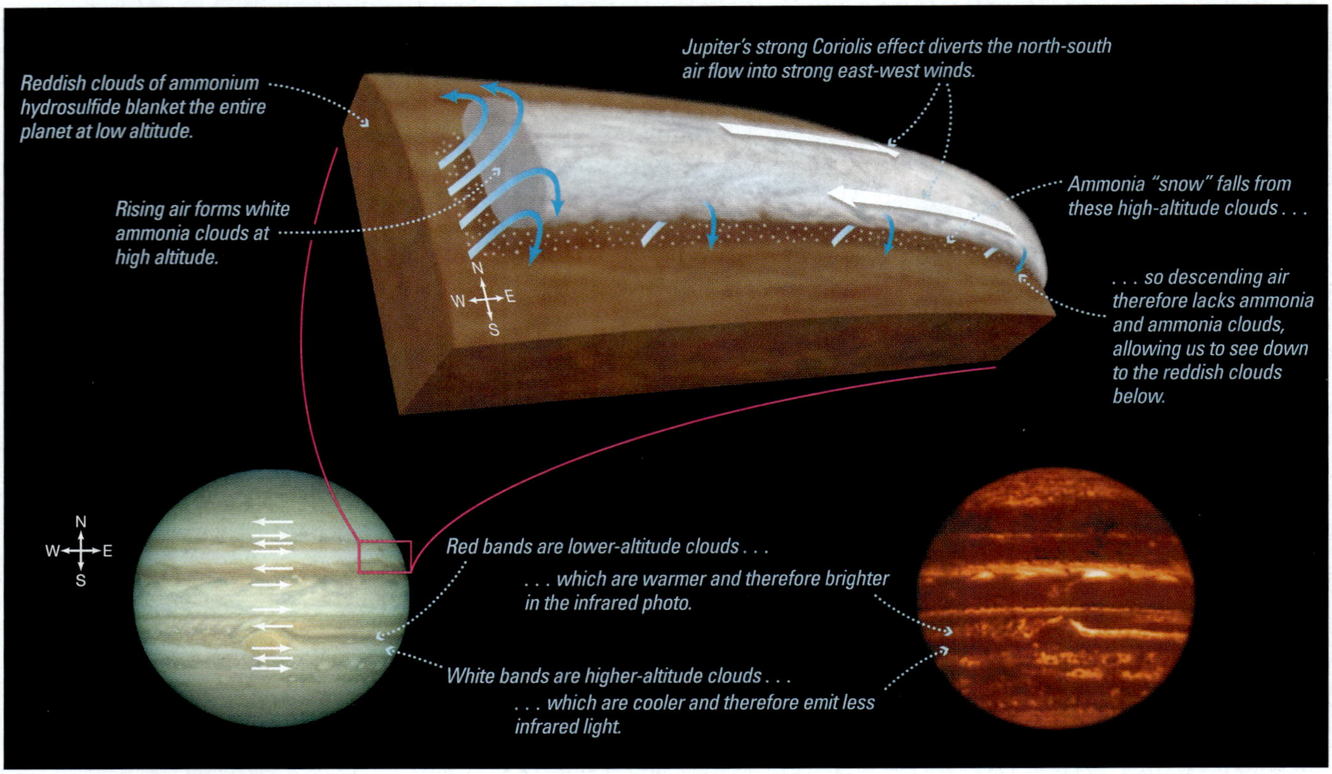


**FIGURE 11.9** Jupiter's bands of color represent alternating regions of rising and falling air: We see white ammonia clouds in regions of rising air, and we see down to the underlying layer of reddish ammonium hydrosulfide clouds in regions of falling air. The white arrows on the visible-light photo indicate wind directions. The infrared photo at the right, taken nearly simultaneously with the visible-light photo, confirms that the red bands represent warmer, deeper clouds and the white bands represent cooler, higher-altitude clouds.

global wind patterns, but Jupiter's storms dwarf those that we see on Earth. Jupiter's most famous feature—the **Great Red Spot**—is a giant storm more than twice as wide as all of planet Earth. It is somewhat like a hurricane, except that its winds circulate around a high-pressure region rather than a low-pressure region (**FIGURE 11.10**). It is also extremely long-lived: Astronomers have seen it throughout the two centuries during which telescopes have been powerful enough to detect it. No one knows why the Great Red Spot has persisted so long, but it might just be a result of Jupiter's structure. On Earth, storms lose energy and dissipate when they pass over land, so perhaps Jupiter's biggest storms can last for centuries simply because no solid surface is present to sap their energy. The bright colors of the Great Red Spot also pose a mystery: We might expect its high-altitude clouds to be white like the high-altitude ammonia clouds elsewhere, but instead, of course, they are red. The colors may be the result of chemicals formed by interactions between the storm's high-altitude gas and solar ultraviolet light, but no one really knows for sure.

Other smaller storms are continually brewing in Jupiter's atmosphere. Cloudtop images show both brown ovals, which are low-pressure storms with their cloudtops deeper in Jupiter's atmosphere, and white ovals, which are high-pressure storms topped with ammonia clouds. *Juno*'s polar orbit also allowed it to discover that the banded patterns visible across most of Jupiter give way to vast numbers of swirling storms near both poles. No one knows what drives these storms or the winds of the Great Red Spot, but scientists hope that ongoing study will help us understand the processes, not least because a better understanding of Jupiter's weather might have applications to our understanding of Earth's. One factor we can rule out is seasonal

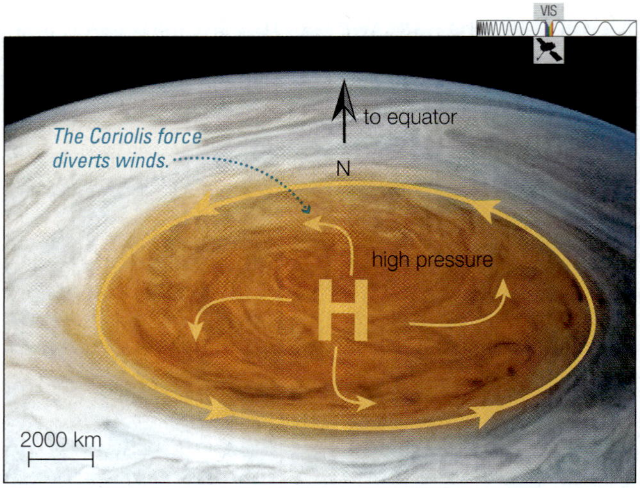


**FIGURE 11.10** This enhanced-color image from the *Juno* spacecraft shows Jupiter's Great Red Spot, a high-pressure storm in Jupiter's southern hemisphere that is large enough to swallow two Earths. The overlaid diagram shows a weather map of the region. The image shape is distorted because of the way the image had to be assembled, given that *Juno* passes only about 13,000 kilometers above Jupiter's cloudtops.

**a** This infrared image from *Juno* looks down over Jupiter's north pole, where eight swirling storms are arranged in a ring around the pole.

**b** Saturn has a banded appearance similar to that of Jupiter (see Figure 11.1). This *Cassini* image also shows a huge whitish storm nearly encircling Saturn's northern hemisphere.

**c** This infrared image of Uranus from the Keck Telescope shows several storms (the bright blotches) and Uranus's thin rings (red).

**d** Neptune's atmosphere, viewed from *Voyager 2*, shows bands and occasional strong storms. The large storm (white arrow) was called the Great Dark Spot.

**FIGURE 11.11** Selected views of weather patterns on the four jovian planets.

changes: Jupiter does not have seasons, because it has no appreciable axis tilt. In fact, Jupiter's polar temperatures are quite similar to its equatorial temperatures, presumably because heat from Jupiter's interior keeps the planet uniformly warm.

**Comparative Winds and Storms** Like Jupiter, the other jovian planets also have dramatic weather patterns (**FIGURE 11.11**). Saturn's are the best studied, thanks to more than a decade of observations from the *Cassini* spacecraft. As on Jupiter, Saturn's rapid rotation creates alternating bands of rising and falling air, along with rapid east-west winds and occasional dramatic storms. In fact, Saturn's winds are even faster than Jupiter's—a surprising finding that scientists have yet to explain. We might expect seasons on Saturn, because it has an axis tilt similar to that of Earth. Some seasonal weather changes have been observed, but Saturn's internal heat keeps temperatures about the same year-round and planetwide.

We know much less about weather on Uranus and Neptune, because each has only had a single spacecraft flyby (*Voyager 2*), and their great distances make them difficult to study telescopically. Nevertheless, we have found some surprises, particularly regarding Uranus. *Voyager 2* saw virtually no clouds or other structure during its 1986 flyby, a fact that scientists attributed to the relatively low amount of internal heat in Uranus. However, more recent observations from the Hubble Space Telescope and ground-based telescopes with adaptive optics have shown raging storms. Perhaps these storms are now brewing because of seasonal change: Thanks to Uranus's extreme axis tilt and 84-year orbit of the Sun [**Section 7.1**], its northern hemisphere began to see sunlight only in 2007, after decades of night.

Neptune's atmosphere shows a banded structure more like that of Jupiter and Saturn. *Voyager 2* also saw a large high-pressure storm, called the *Great Dark Spot,* which looked structurally similar to Jupiter's Great Red Spot. However, the Great Dark Spot did not last long, disappearing from view just 6 years after its discovery. Like Saturn, Neptune has an axis tilt similar to Earth's, but it has relatively little seasonal change because of its internal heat.

## Do jovian planets have magnetospheres like Earth's?

Recall that Earth has a global magnetic field generated by the movements of charged particles in our planet's metallic outer core (see Figure 9.6c). The jovian planets also have global magnetic fields generated by motions of charged particles deep in their interiors. Just as with Earth, these magnetic fields create bubble-like *magnetospheres* that surround the planets and shield them from the solar wind.

**Jupiter's Magnetosphere** Of the four jovian planets, Jupiter has by far the strongest magnetic field; it is some 20,000 times as strong as Earth's magnetic field. As discussed in Chapter 9, a planet can have a global magnetic field if it has (1) an interior region of electrically conducting fluid, (2) convection in that layer of fluid, and (3) at least moderately rapid rotation. In Jupiter's case, the electrically conducting fluid region is its thick layer of metallic hydrogen (see Figure 11.4). The great extent of this region, combined with Jupiter's rapid rotation, explains Jupiter's strong magnetic field. This in turn gives it an enormous magnetosphere, one that begins to deflect the solar wind some 3 million kilometers (about 40 Jupiter radii) in front of the planet (**FIGURE 11.12**). If we could see Jupiter's magnetosphere, it would be larger than the full moon in our sky.

Jupiter's magnetosphere traps far more charged particles than Earth's, largely because it has a source of particles that Earth's lacks. Nearly all the charged particles in Earth's magnetosphere come from the solar wind, but in Jupiter's

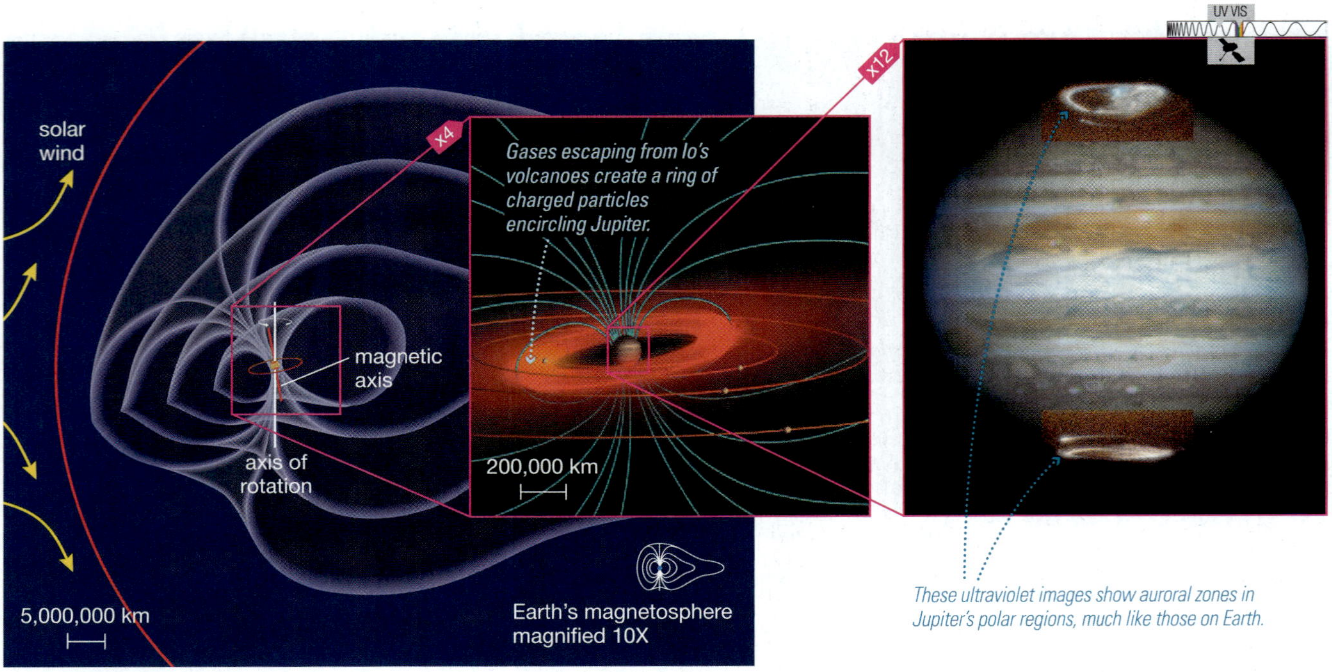


**FIGURE 11.12** Jupiter's strong magnetic field gives it an enormous magnetosphere. Gases escaping from Io feed the donut-shaped torus, and particles entering Jupiter's atmosphere near its magnetic poles contribute to auroras on Jupiter. The image at the right is a composite of ultraviolet images of the polar regions overlaid on a visible image of the whole planet, all taken by the Hubble Space Telescope.

case its volcanically active moon, Io, contributes many additional particles. These particles help create auroras on Jupiter. They also create belts of intense radiation around Jupiter, which can cause damage to orbiting spacecraft.

Jupiter's magnetosphere, in turn, has important effects on Io and the other moons of Jupiter. The charged particles bombard the surfaces of Jupiter's icy moons, with each particle blasting away a few atoms or molecules. This process alters the surface materials and can even generate thin atmospheres, much like the thin atmospheres of Mercury and the Moon [**Section 10.3**]. On Io, bombardment by charged particles leads to the continuous escape of gases that were released by volcanic outgassing. The escaping gases are ionized and feed a donut-shaped charged particle belt, called the *Io torus,* that approximately traces Io's orbit (the red "donut" in Figure 11.12). Scientists hope to learn much more about Jupiter's magnetic field and magnetosphere as the *Juno* mission continues.

**Comparing Jovian Magnetospheres** The other jovian planets also have magnetic fields and magnetospheres, but theirs are much weaker than Jupiter's (although still much stronger than Earth's). The strength of each planet's magnetic field depends primarily on the size of the electrically conducting layer buried in its interior. Saturn's magnetic field is weaker than Jupiter's because it has a thinner layer of electrically conducting metallic hydrogen. Uranus and Neptune, smaller still, have no metallic hydrogen at all. Their relatively weak magnetic fields must be generated in their core "oceans" of hydrogen compounds, rock, and metals.

The size of a planet's magnetosphere depends not only on the magnetic field strength, but also on the pressure of the solar wind against it. The pressure of the solar wind is weaker at greater distances from the Sun, so the magnetospheric "bubbles" surrounding more distant planets are larger than they would be if these planets were closer to the Sun. That is why Uranus and Neptune have moderate-size magnetospheres despite their weak magnetic fields (**FIGURE 11.13**). No other magnetosphere is as full of charged particles as Jupiter's, primarily because no other jovian planet has a satellite like Io. Because trapped particles generate auroras, Jupiter has the brightest auroras, while those of the more distant jovian planets are progressively weaker.

We still have much to learn about the jovian magnetic fields and magnetospheres. For example, we generally expect magnetic fields to be closely aligned with planetary rotation, because the magnetic fields are generated within the rotating interiors of planets. However, while this is the case for Jupiter and Saturn (whose magnetic fields are inclined to their rotation axes by 10° and 0°, respectively), it is not the case for Uranus and Neptune. *Voyager* observations showed that the magnetic field of Uranus is tipped by a whopping 60° relative to its rotation axis, and the magnetic field's center is also significantly offset from the planet's center. Neptune's magnetic field is inclined by 46° to its rotation axis. No one has yet explained these surprising observations.

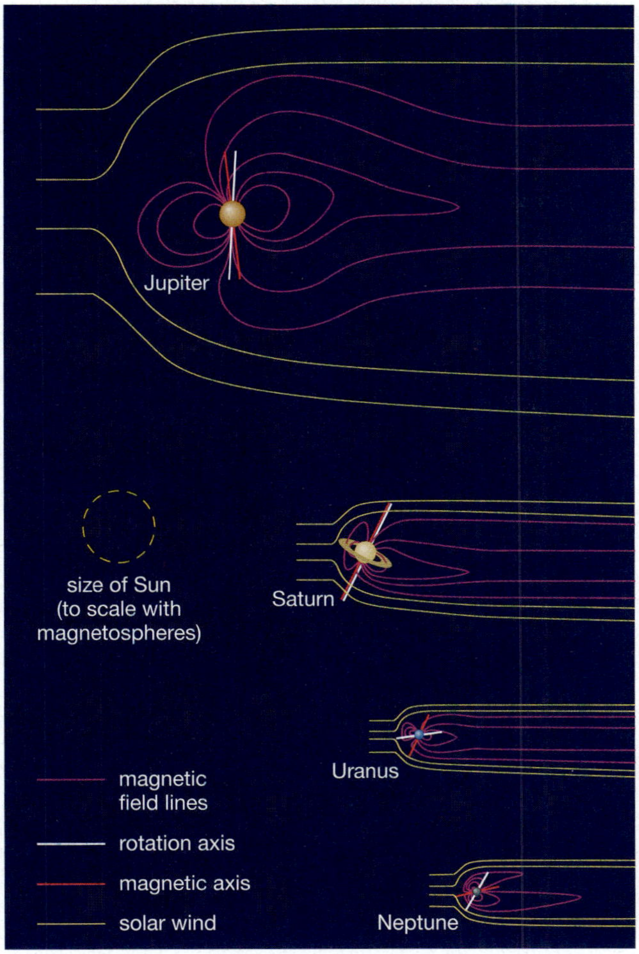


**FIGURE 11.13** Comparison of jovian planet magnetospheres. Note the significant tilts of the magnetic fields of Uranus and Neptune compared to their rotation axes. (Planet sizes exaggerated compared to scale of magnetospheres.)

## 11.2 A Wealth of Worlds: Satellites of Ice and Rock

The jovian planets are majestic and fascinating, but they are only the beginning of our exploration of jovian planet *systems*. Each of the four jovian systems includes numerous moons and a set of rings. The total mass of all the moons and rings put together is minuscule compared to that of any one of the jovian planets, but the remarkable diversity of these satellites makes up for their lack of size.

### What kinds of moons orbit the jovian planets?

We now know of more than 170 moons orbiting the jovian planets. (See Table E.3 in Appendix E for the full list.) Jupiter and Saturn have the most, each with more than 60 known moons. It's helpful to organize these moons into three groups by size: small moons less than about 300 kilometers in diameter, medium-size moons ranging from about 300 to 1500 kilometers in diameter, and large moons more than 1500 kilometers in diameter. These categories are useful because size relates to geological activity. In general, larger moons are more likely to show evidence of past or present geological activity.

**FIGURE 11.14** shows a montage of all the medium-size and large moons. These moons resemble the terrestrial planets in many ways. Each is spherical with a solid surface and its own unique geology. Some possess atmospheres, hot interiors, and even magnetic fields. The two largest—Jupiter's moon Ganymede and Saturn's moon Titan—are larger than the planet Mercury, while four others (Jupiter's moons Io, Europa, and Callisto and Neptune's moon Triton) are larger than the largest known dwarf planets, Pluto and Eris. However, these moons differ from terrestrial worlds in their compositions: Because they formed in the cold outer solar system, most of them contain substantial amounts of ice in addition to metal and rock.

Most of the medium-size and large moons probably formed by accretion within the disks of gas surrounding individual jovian planets [**Section 8.2**]. That explains why their orbits are almost circular and lie close to the equatorial plane of their parent planet, and also why these moons orbit in the same direction in which their planet rotates.

**FIGURE 11.14** The medium-size and large moons of the jovian planets, with sizes shown to scale. The moons are shown in order of distance from their planet, from left (nearest) to right (farthest).

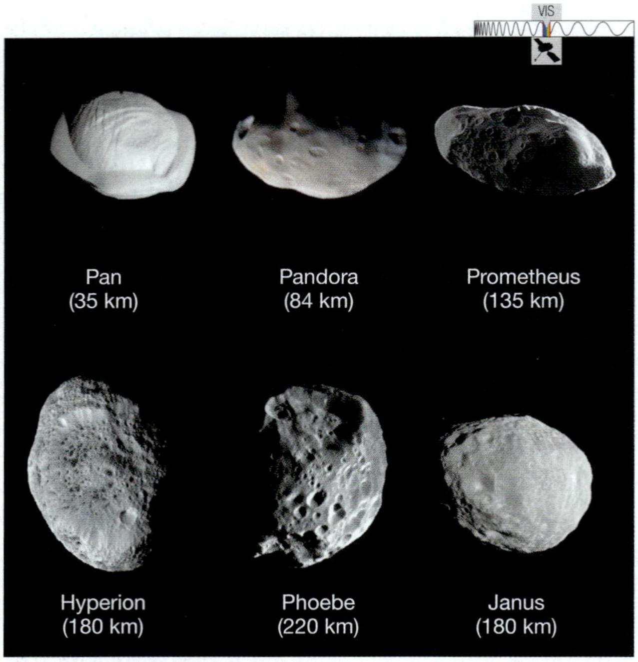


**FIGURE 11.15** These photos from the *Cassini* spacecraft show six of Saturn's smaller moons. All are much smaller than the smallest moons shown in Figure 11.14. The sizes in parentheses represent approximate lengths along their longest axes. Pan orbits within Saturn's thin rings and has collected a ridge of ring dust around its equator.

Nearly all these moons also share an uncanny trait: They always keep the same face turned toward their planet, just as our Moon always shows the same face to Earth. This synchronous rotation arose from the strong tidal forces [**Section 4.5**] exerted by the jovian planets, which caused each moon to end up with equal periods of rotation and orbit regardless of how fast the moon rotated when it formed.

We know much less about the more numerous small moons, many of which are no more than a few kilometers across. Like asteroids, small moons tend to have irregular shapes (**FIGURE 11.15**), because their gravities are too weak to force their rigid material into spheres. In fact, many or most of these small moons probably *are* captured asteroids or comets, which also explains why they do not follow any particular orbital patterns; many even orbit backward relative to their planet's rotation. For the most part, these small moons are probably just chunks of ice and rock held captive by the gravity of a massive jovian planet, with sizes too small to allow for any significant geological activity.

In case you are wondering, scientists originally named Jupiter's moons for the mythical lovers of the Roman god Jupiter. However, Jupiter has so many moons that more recent discoveries are named for more distant mythological relations. Saturn's moon Titan is named for the Greek gods called the Titans, who ruled before Zeus (Jupiter); other moons of Saturn are named for individual Titans. Moons of Uranus all take their names from characters in the works of William Shakespeare and Alexander Pope, while moons of Neptune are all characters related to the sea in Greek and Roman mythology.

## Why are Jupiter's Galilean moons so geologically active?

We now embark on a brief tour of the most interesting medium-size and large moons of the jovian planets. Our first stop is Jupiter, where the four *Galilean moons* (the ones discovered by Galileo [**Section 3.3**]) are each large enough that they would count as planets or dwarf planets if they orbited the Sun (**FIGURE 11.16**).

**Io: The Volcano World** For anyone who thinks of moons as barren, geologically dead places like our own Moon, Io shatters the stereotype. Io is by far the most volcanically active world in our solar system. Large volcanoes pockmark its entire surface (**FIGURE 11.17**), and eruptions are so frequent that they continually repave the surface. Indeed, the surface is so young that it does not have a single impact crater anywhere. Io probably also has tectonic activity, because tectonics and volcanism generally go hand in hand. However, debris from volcanic eruptions has probably buried most tectonic features.

**FIGURE 11.16** This set of photos, taken by the *Galileo* spacecraft, shows global views of the four Galilean moons. Sizes are shown to scale; Io is about the size of Earth's Moon. Colors have been enhanced to show detail.

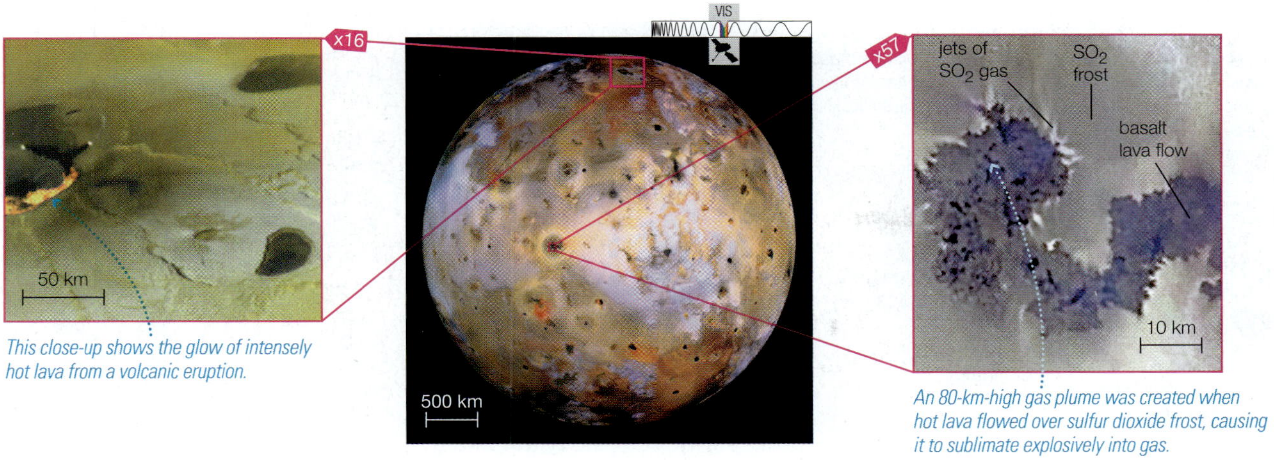


**FIGURE 11.17** Io is the most volcanically active body in the solar system. Most of the black, brown, and red spots on the surface are recently active volcanic features. White and yellow areas are sulfur dioxide and sulfur deposits, respectively, from volcanic gases. (Photographs are from the *Galileo* spacecraft; some colors are slightly enhanced or altered.)

Many of Io's volcanoes are surprisingly similar to those of Earth. Some of the taller volcanoes look like the shallow-sloped volcanoes built from flows of basalt lava that erupts along Earth's mid-ocean ridges. Also as on Earth, Io's eruptions are accompanied by outgassing. The primary gases released from Io's volcanoes are sulfur dioxide ($SO_2$), sulfur, and a hint of sodium. Some of this gas escapes into space, where it supplies ionized gas (plasma) to the Io torus and Jupiter's magnetosphere, and some of it gives Io a very thin atmosphere. Much of the gas condenses and falls back to the surface, where the sulfur gives Io its distinctive red and orange colors and the sulfur dioxide makes a white frost (**FIGURE 11.18**). As hot lava flows across the surface, it can re-vaporize the sulfur dioxide surface ice in much the same way that lava flowing into the ocean vaporizes water on Earth. Io's low gravity and thin atmosphere allow tall plumes of this vaporized sulfur dioxide to rise upward to altitudes of hundreds of kilometers. Over a period of a few months, fallout from these tall volcanic plumes can blanket an area the size of Arizona.

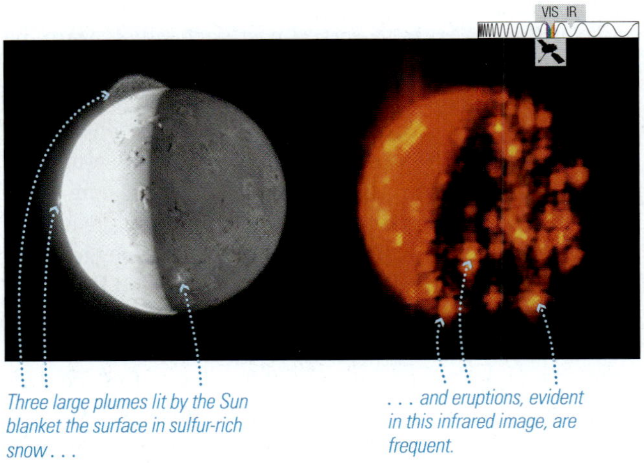


**FIGURE 11.18** Two views of Io's volcanoes: (left) a visible-light image taken by the *New Horizons* spacecraft on its way to Pluto; (right) an infrared image from the *Juno* mission. The images were taken at different times, so the illumination and volcanoes in view are not the same.

Io's active volcanoes tell us that it must be quite hot inside. However, Io is only about the size of our geologically dead Moon, so it should have long ago lost any heat from its birth and it is too small for radioactivity to provide much ongoing heat. How, then, can Io be so hot inside? The only possible answer is that some other ongoing process must be heating Io's interior. Scientists have identified this process and call it **tidal heating**, because it arises from effects of tidal forces exerted by Jupiter.

Just as Earth exerts a tidal force that causes the Moon to keep the same face toward us at all times [**Section 4.5**], a tidal force from Jupiter makes Io keep the same face toward Jupiter as it orbits. But Jupiter's mass makes this tidal force far larger than the tidal force that Earth exerts on the Moon. Moreover, Io's orbit is slightly elliptical, so its orbital speed and distance from Jupiter vary. This variation means that the strength and direction of the tidal force change slightly as Io moves through each orbit, which in turn changes the size and orientation of Io's tidal bulges (**FIGURE 11.19a**). The result is that Io is continuously being flexed in different directions, which generates friction inside it. The flexing heats the interior in the same way that flexing warms Silly Putty. Tidal heating generates tremendous heat on Io—more than 200 times as much heat (per gram of mass) as the radioactive heat driving much of Earth's geology. This heat explains Io's volcanic activity. The energy for tidal heating ultimately comes from Jupiter's rotation, which is gradually slowing, although at a rate too small to be observed.

**See it for yourself** Pry apart the overlapping ends of a paper clip so that you can hold one end in each hand. Flex the ends apart and together until the paper clip breaks. Lightly touch the broken end to your finger or lips—can you feel the warmth produced by flexing? How is this heating similar to the tidal heating of Io?

However, we are still left with a deeper question: Why is Io's orbit slightly elliptical, when almost all other large

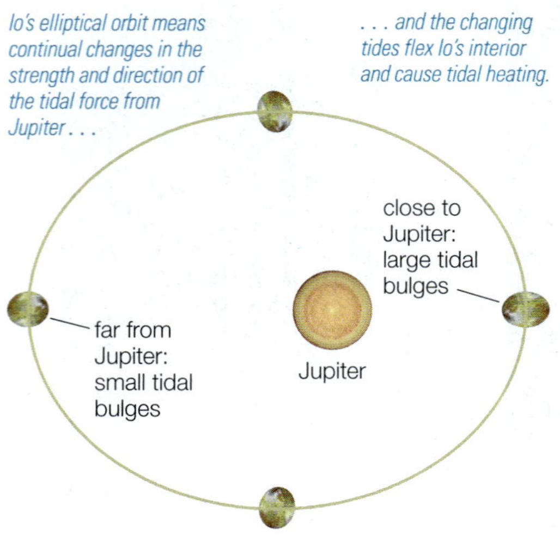


**a** Tidal heating arises because Io's elliptical orbit (exaggerated in this diagram) causes varying tides.

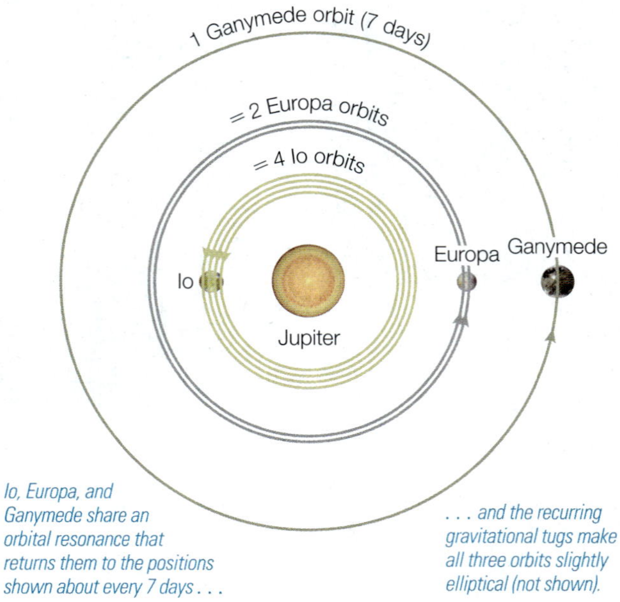


**b** Io's orbit is elliptical because of the orbital resonance Io shares with Europa and Ganymede.

**FIGURE 11.19** These diagrams explain the cause of tidal heating on Io. Tidal heating has a weaker effect on Europa and Ganymede, because they are farther from Jupiter and tidal forces weaken with distance.

satellites have nearly circular orbits? The answer lies in what we call an **orbital resonance**, which means a simple mathematical relationship between the orbital periods of two or more objects. Io, Europa, and Ganymede share an orbital resonance because Io completes exactly four orbits of Jupiter in the same time (about 7 days) that Europa completes two orbits and Ganymede completes one orbit (**FIGURE 11.19b**). The three moons therefore line up periodically, and with each alignment they exert gravitational tugs on one another that are in the same direction and therefore add up over time. This tends to stretch out their orbits, making them slightly elliptical. The effect is much like that of pushing a child on a swing. If timed properly, a series of small pushes can add up to a *resonance* that causes the child to swing quite high; the similar effect for orbits in which the same gravitational tugs repeat with time explains why we call it an *orbital resonance*.* Orbital resonances are quite common, affecting not only the Galilean moons but also planetary rings, the asteroid belt, the Kuiper belt, and the orbits of some of the planets known around other stars.

**Europa: The Water World?** Europa offers a striking contrast to Io. Instead of having active volcanoes dotting its surface, Europa is covered by water ice (see Figure 11.16). Nevertheless, its fractured, frozen surface must hide an interior made hot by the same type of tidal heating that powers Io's volcanoes, though tidal heating is weaker on Europa because it lies farther from Jupiter.

Key evidence about how the hot interior affects the icy surface comes from the fact that Europa has only a handful of impact craters. This fact tells us that some type of ongoing geological activity must have erased the evidence of nearly all past impacts. But what is doing the erasing?

Scientists suspect that the answer is either liquid water rising up from an ocean that lies beneath the icy crust or interior water ice that is just warm enough to undergo convection, so that some of it rises up and flows across the surface. Close-up photos of the surface, combined with the fact that Europa has enough internal heat to melt subsurface ice into liquid water, support the ocean hypothesis. For example, **FIGURE 11.20** shows some of the many double-ridged cracks visible on Europa's surface. As the model at the right shows, these cracks are best explained by assuming that the icy crust can slide on a softer or liquid layer below. The double ridges may form as tidal stresses force parts of the icy crust to scrape past each other, warming and possibly melting the ice along the fault.

Other *Galileo* data provide additional support for the ocean hypothesis. Theoretical models based on *Galileo* measurements of the strength of gravity over different parts of the surface indicate that Europa has a metallic core and a rocky mantle surrounded by enough water to make a layer of ice about 100 kilometers thick. The models suggest that the upper 5 to 25 kilometers of the water should be solidly frozen as an icy crust, but that tidal heating should provide enough warmth to turn the underlying ice into a layer of either liquid water or relatively warm, convecting ice. Evidence favoring the idea that the underlying layer consists of liquid water comes from magnetic field data collected by the *Galileo* spacecraft. Europa is one of only a few moons in the solar system to have a magnetic field, and its magnetic field changes as Jupiter rotates. The simplest way to explain this change is to hypothesize that Europa's magnetic field is created (or *induced*) in response to the rotation of Jupiter's

---

*You may wonder why periodic alignments occur. Like synchronous rotation, they are not a coincidence but rather a consequence of feedback from the tides that the moons raise on Jupiter. The periodic tugs the moons exert on one another actually work to sustain the recurring alignments.

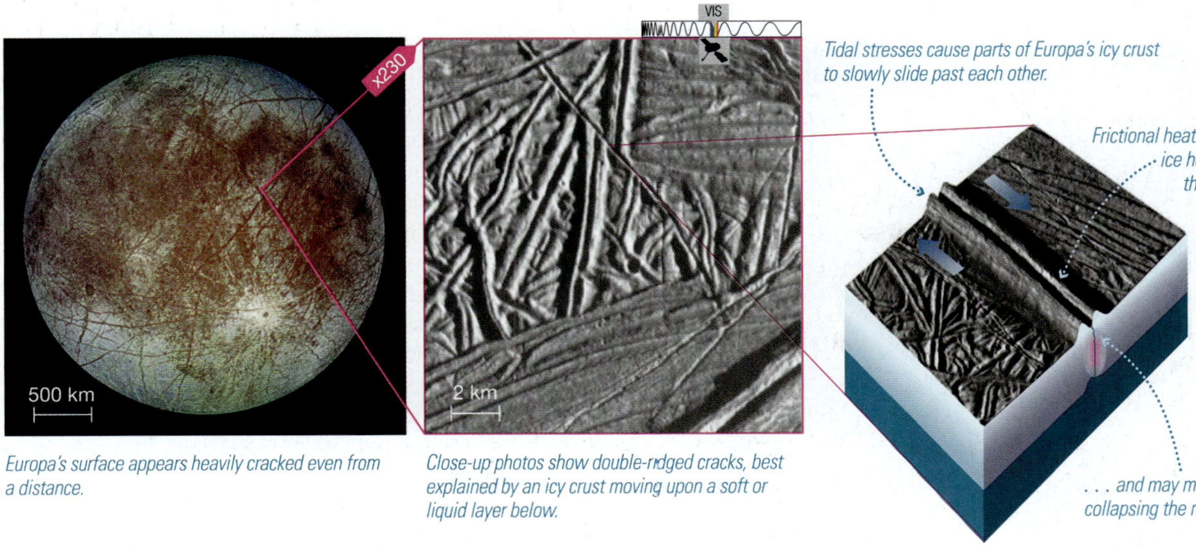


Europa's surface appears heavily cracked even from a distance.

Close-up photos show double-ridged cracks, best explained by an icy crust moving upon a soft or liquid layer below.

**FIGURE 11.20** Europa's icy crust features many double-ridged cracks, which provide evidence for a subsurface ocean. This sequence shows how the cracks may have formed. (Photos are from the *Galileo* spacecraft; colors are enhanced in the global view.)

strong magnetic field. This type of response is possible only if Europa has a *liquid* layer of electrically conducting material. A salty ocean would fit the bill, but convecting ice would not. The data also suggest that Europa's liquid ocean must be global in extent, and that it is about as salty as Earth's oceans. Further support for this idea comes from another of *Galileo's* instruments, which found evidence for salty compounds on Europa's surface—possible seepage from a briny deep.

Taken together, the evidence from surface photos, gravitational measurements, and magnetic fields makes a strong case for a deep ocean of liquid water on Europa (**FIGURE 11.21**). Because Europa must have a hot interior from tidal heating, it seems reasonable to imagine that volcanic vents dot the seafloor and sometimes erupt, creating rising plumes of warm water. These plumes may lead to the formation of subsurface lakes within the icy crust, which could cause cracking of the icy surface above. Surface features that look like jumbled icebergs, such as the region shown on the right in Figure 11.21, may be explained by such cracking. Some astronomers have also reported hints of gas venting from Europa's crust, but the observations haven't been conclusively verified.

If these ideas are correct, then Europa may have a hidden ocean containing more than twice as much liquid water as all of Earth's oceans combined, along with a seafloor dotted with the same types of volcanic vents around which life thrives in Earth's oceans. That is why scientists are so interested in the possibility that Europa might also be home to life—a possibility we'll explore further in Chapter 24.

**Ganymede: King of the Moons** Ganymede is the largest moon in the solar system, and its surface tells of a complex geological history. Like Europa, Ganymede has a surface of water ice. However, while Europa's surface appears relatively

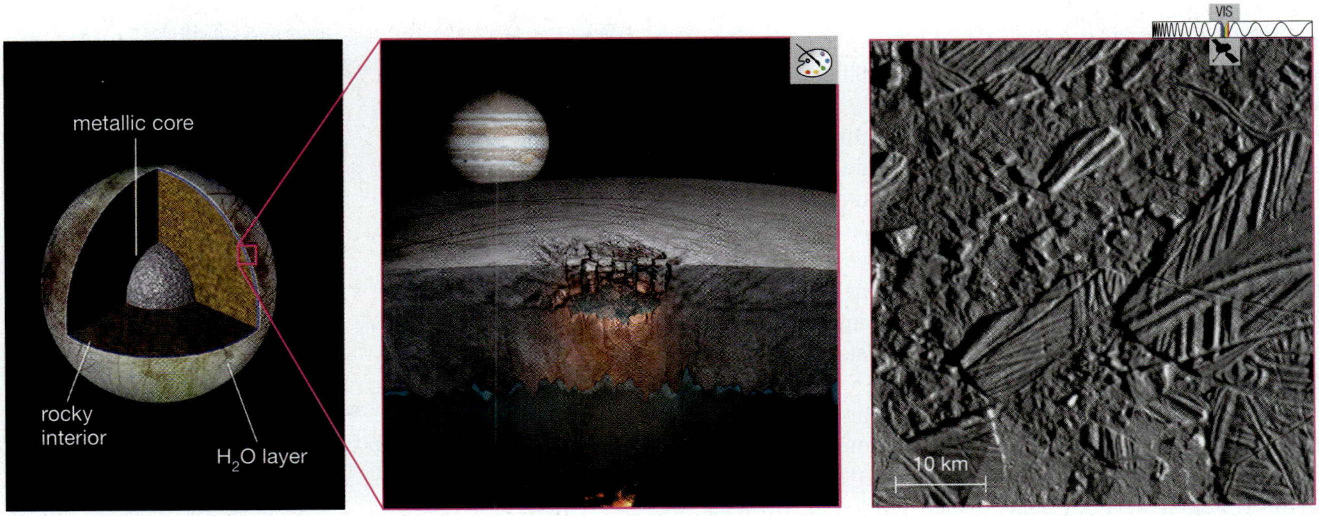


Europa may have a 100-km-thick ocean under an icy crust.

Rising plumes of warm water may sometimes create lakes within the ice, causing the crust above to crack . . .

. . . explaining surface terrain that looks like a jumble of icebergs suspended in a place where liquid or slushy water froze.

**FIGURE 11.21** This diagram shows the leading model of Europa's interior structure, along with a *Galileo* surface photo consistent with this model. (The layer thicknesses are not to scale in the central panel.)

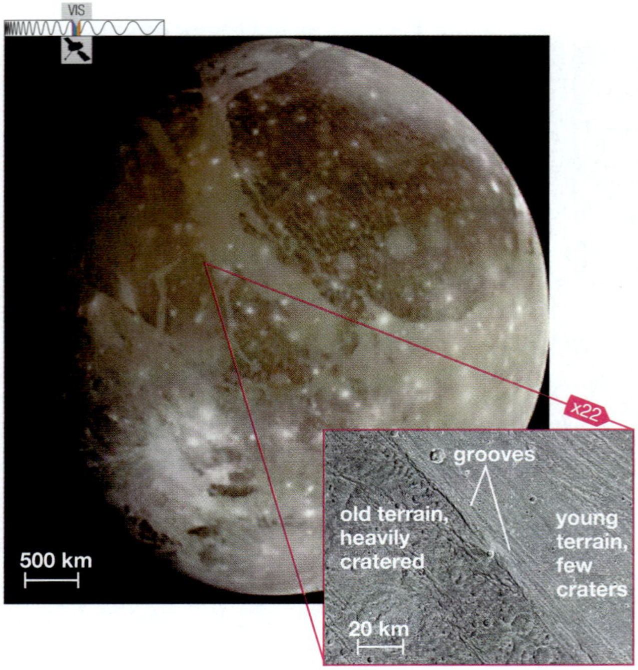


**FIGURE 11.22** Ganymede, the largest moon in the solar system, has both old and young regions on its surface of water ice. Notice that the boundary between the two types of terrain can be quite sharp.

young everywhere, Ganymede's surface appears to have a dual personality (**FIGURE 11.22**). Some regions are dark and densely cratered, suggesting that they look much the same today as they did billions of years ago. Other regions are light-colored with very few craters. In some cases, fairly sharp boundaries separate the two types of terrain.

The young terrain argues for occasional upwelling of liquid water or icy slush to the surface. This material would cover craters before refreezing, explaining why there are so few craters in this terrain. The long grooves are probably made by the eruption of water or slush along a crack in the surface. As the water in the crack freezes, it expands and pushes outward, creating the groove.

If liquid water occasionally wells up to the surface, could it mean that Ganymede has a subsurface ocean like that thought to exist on Europa? As with Europa, the *Galileo* spacecraft found that Ganymede has a magnetic field that varies with Jupiter's rotation, suggesting the presence of a salty ocean beneath the surface. Hubble Space Telescope observations of aurorae around Ganymede also support the case for a subsurface ocean. These aurorae must be caused by Ganymede's magnetic field, and careful study of how the aurorae change with time indicate that the magnetic field varies in a way that makes sense only if Ganymede has a deeply buried (about 150 kilometers beneath the surface) ocean of salty water. Scientists estimate the ocean may be 100 kilometers deep.

Assuming the ocean is real, what heat source keeps it liquid? Because Ganymede is farther from Jupiter than Europa or Io is, its tidal heating is weaker and could not by itself supply enough heat to melt ice today. However, Ganymede's larger size means it should retain more heat from radioactive decay. Perhaps tidal heating and radioactive decay together provide enough heat to make a liquid layer beneath the icy surface.

**Callisto: Last of the Galilean Moons** The outermost Galilean moon, Callisto, looks most like what scientists originally expected for an outer solar system satellite: a heavily cratered iceball (**FIGURE 11.23**). The bright, circular patches on its surface are impact craters. They are probably bright because large impacts blast out "clean" ice from deep underground, and this ice reflects more light than the dirtier ice that has been on the surface longer.

Craters make sense on an old surface like Callisto's, but other features are more difficult to interpret. For example, close-up photos show a dark, powdery substance concentrated in low-lying areas, leaving ridges and crests bright white (small photo in Figure 11.23). The dark powder may be debris left behind when ice vaporizes into gas from Callisto's surface, much as dark material is left behind on a comet's nucleus when more reflective ice vaporizes away [**Section 12.3**].

Despite its relatively large size (the third-largest moon in the solar system), Callisto lacks volcanic and tectonic features. This tells us that it lacks any significant source of internal heat. In fact, gravity measurements by the *Galileo* spacecraft showed that Callisto never underwent differentiation: Dense rock and lighter ice are mixed throughout most of its interior, which means that its interior never warmed significantly. The lack of heat is not surprising: Callisto has no tidal heating because it does not participate in the orbital resonances that affect the other three Galilean moons, and we do not expect its icy interior to contain enough radioactive material to supply much heat through radioactive decay.

Nevertheless, it is possible that Callisto, too, may hide a subsurface ocean. As with Europa and Ganymede, *Galileo* found that Callisto has a varying magnetic field that suggests the presence of a salty interior ocean. Perhaps

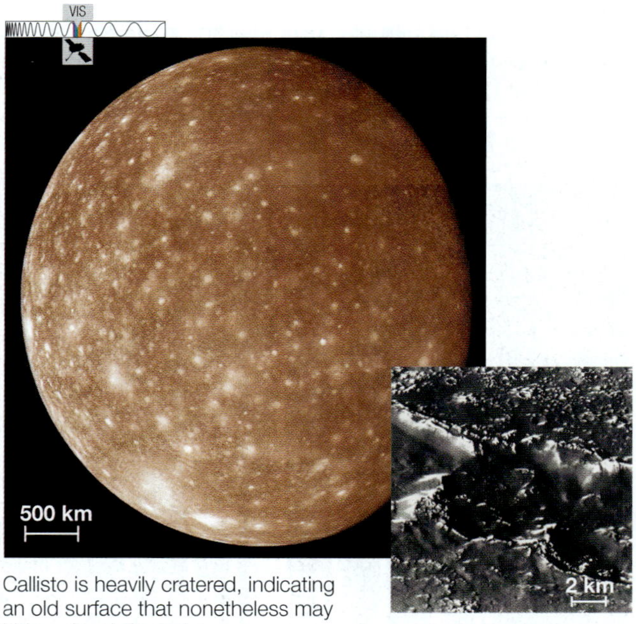


Callisto is heavily cratered, indicating an old surface that nonetheless may hide a deeply buried ocean.

Close-up photo shows a dark powder overlaying the low areas of the surface.

**FIGURE 11.23** Callisto, the outermost of the four Galilean moons, has a heavily cratered, icy surface.

Callisto's surface provides just enough insulation for heat from radioactive decay to melt water beneath the thick, icy crust. If so, we face the intriguing possibility that there could be oceans on three worlds orbiting Jupiter—with far more total ocean than we find here on Earth.

## What geological activity do we see on Titan and other distant moons?

Leaving Jupiter behind, we will continue our tour of moons by discussing the major moons of Saturn, Uranus, and Neptune. We'll go in planetary order, first focusing on moons of Saturn.

**Titan** Saturn's moon Titan is the second-largest moon in the solar system (after Ganymede). It is also unique among the moons of our solar system in having a thick atmosphere—so thick that we cannot see through it with visible light (**FIGURE 11.24**). Titan's color comes from chemicals in its atmosphere, which act much like those that make smog over cities on Earth. The atmosphere is about 95% molecular nitrogen ($N_2$), not that different from the 77% nitrogen content of Earth's atmosphere. However, on Earth the rest of the atmosphere is mostly oxygen, while the rest of Titan's atmosphere consists of argon, methane ($CH_4$), ethane ($C_2H_6$), and other hydrogen compounds.

Titan's atmospheric composition can be understood in terms of our general understanding of atmospheric production and loss processes [**Section 10.2**]. Titan's icy composition supplies methane and ammonia gas through vaporization of surface ices and liquids, and possibly also volcanic eruptions. Solar ultraviolet light breaks down some of those molecules, releasing hydrogen atoms and leaving highly reactive compounds containing carbon and nitrogen. The hydrogen atoms can leave Titan forever by thermal escape, while the remaining molecular fragments can react to make the other ingredients of Titan's atmosphere. For example, the abundant molecular nitrogen is made after ultraviolet light breaks down ammonia ($NH_3$) molecules, and ethane is made from methane.

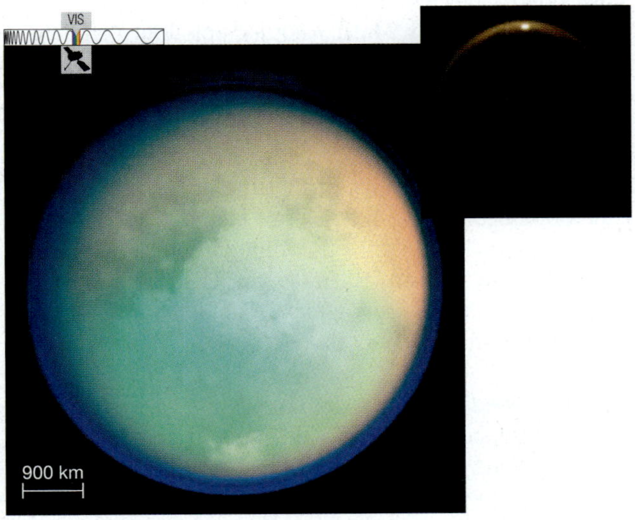


**FIGURE 11.24** Titan, as photographed by the *Cassini* spacecraft, is enshrouded by a thick atmosphere with clouds and haze. *Cassini* was outfitted with filters designed to peer through the atmosphere at specific near-infrared wavelengths of light. The inset shows sunlight reflecting off a lake in Titan's north polar region.

Methane and ethane are both greenhouse gases and therefore give Titan an appreciable greenhouse effect [**Section 10.1**] that makes it warmer than it would be otherwise. Still, because of its great distance from the Sun, its surface temperature is a frigid 93 K (−180°C). The surface pressure on Titan is about 1.5 times the sea-level pressure on Earth, which would be fairly comfortable if not for the lack of oxygen and the cold temperatures.

A moon with a thick atmosphere would be intriguing enough, but we have at least two other reasons for special interest in Titan. First, its complex atmospheric chemistry produces numerous carbon compounds—the chemicals that are the basis of life. Second, although it is far too cold for liquid water to exist on Titan's surface, conditions are right for methane or ethane rain, which creates rivers flowing into lakes and seas. These facts led NASA and the European Space Agency (ESA) to combine forces to explore Titan. In 2005, NASA's *Cassini* "mother ship" released the ESA-built probe called *Huygens* (pronounced "Hoy-guns"), which parachuted to the surface (**FIGURE 11.25**). During its descent, the probe photographed river valleys merging together, flowing down to what looks like an ancient shoreline. On landing, instruments on the probe discovered that the surface has a hard crust but is a bit squishy below, perhaps like sand with liquid mixed in, and photos showed "ice boulders" rounded by erosion. All these results support the idea of a wet climate—but wet with liquid methane rather than liquid water.

Subsequent *Cassini* observations taught us more about Titan. The brighter regions in the central image of Figure 11.25 are icy hills, perhaps made by ice volcanoes. The dark valleys were probably created when methane rain carried down "smog particles" that concentrated on river bottoms. The vast plains into which the valleys appear to empty are low-lying regions, but they do not appear to be liquid. Instead, they are probably covered by smog particles carried down by the rivers and then sculpted into vast dune fields by Titan's global winds. All in all, conditions in Titan's equatorial regions appear to be analogous to those of the desert southwest of the United States, where infrequent rainfall carves valleys and creates vast dry lakes called *playas* where the water evaporates or soaks into the ground.

The polar regions of Titan, revealed by *Cassini* radar, contain numerous lakes of liquid methane or ethane (**FIGURE 11.26**). Images also reveal polar storm clouds and riverbeds leading into the lakes, suggesting that Titan has a methane/ethane cycle resembling the water cycle on Earth. Titan shares Saturn's 27° axis tilt, which gives it long seasons over Saturn's 29.5-year orbit of the Sun. During the long summers at each pole, warm air rises and causes methane rainstorms. Northern summers are apparently wetter, based on the large fraction of lakes located there. This may be due to Saturn's elliptical orbit, which brings it (and Titan) closer to the Sun during northern summer. The more intense northern summer may create stronger circulation and storms with enough rainfall to fill the northern lakes.

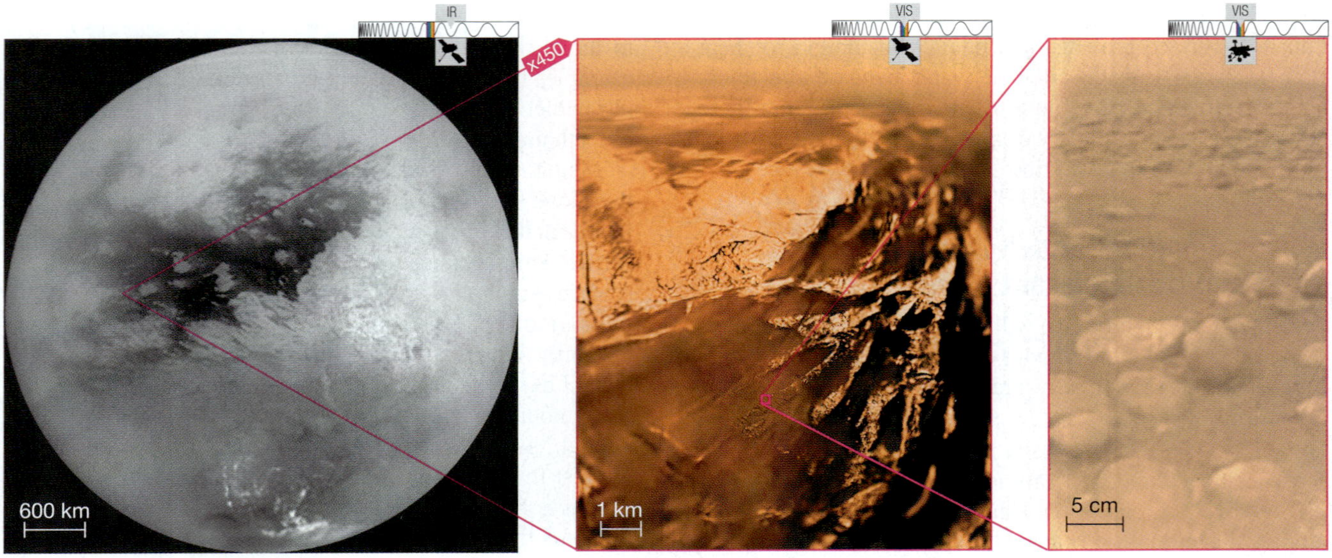


**FIGURE 11.25** This sequence zooms in on the *Huygens* landing site on Titan. Left: A global view taken by the *Cassini* orbiter. Center: An aerial view from the descending probe. Right: A surface view taken by the probe after landing; the "rocks," which are 10 to 20 centimeters across, are presumably made of water ice. Keep in mind that you are looking at the surface of a world more than a billion kilometers away.

Perhaps the most astonishing result from the *Cassini/Huygens* mission is how familiar the landscape looks in this alien environment with these unfamiliar materials. Instead of liquid water, Titan has liquid methane and ethane. Instead of rock, Titan has ice. Instead of molten lava, Titan has a slush of water ice mixed with ammonia. Instead of surface dirt, Titan's surface has smog-like particles that rain out of the sky and accumulate on the ground. Evidently, the similarities between the physical processes that occur on Titan and Earth are far more important in shaping the landscapes than the fact that the two worlds have very different compositions and temperatures.

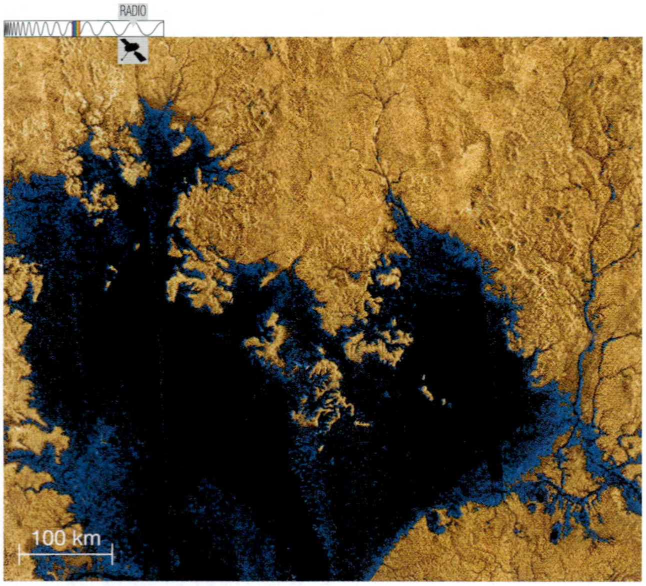


**FIGURE 11.26** Radar image of Ligeia Mare near Titan's north pole, showing lakes of liquid methane and ethane at a temperature of −180°C. Most solid surfaces reflect radar well, and these regions are artificially shaded tan to suggest land. The liquid surfaces reflect radar poorly, and these regions are shaded blue and black to suggest lakes.

**Think about it** What other geological features might you expect on Titan, given its similarities to Earth? How might those features be different, given the differences in temperature and composition?

The abundance of carbon compounds on Titan, along with the similarity of its surface geology to that of Earth, makes it natural to wonder whether Titan could be home to life. Although we cannot rule it out, most scientists think it unlikely that Titan's surface could support life because of the combination of the very low temperatures and the fact that chemical properties of liquid methane and ethane make them seem less suited than liquid water to supporting biology [**Section 24.2**]. Another possibility is subsurface life. *Cassini* observations of small changes in Titan's surface gravity and shape as it orbits Saturn indicate that it probably has a subsurface ocean of liquid water, much like the subsurface ocean on Europa. However, theoretical models of Titan's interior suggest that the bottom of its ocean would be in contact with a layer of high-density ice, rather than with a rocky layer like that thought to exist at the bottom of Europa's ocean. Scientists are unsure whether this difference would have any effect on the potential habitability of Titan's subsurface ocean.

**Saturn's Medium-Size Moons** *Cassini* also flew past many other moons of Saturn, including its six medium-size moons (**FIGURE 11.27**). Each of these shows evidence of a complex geological history and one—Enceladus—remains geologically active today.

Only Mimas, the smallest of these six moons, shows little evidence of past volcanism or tectonics. Of course, like all moons with ancient surfaces, it is heavily cratered, with one huge crater nicknamed Darth Crater because

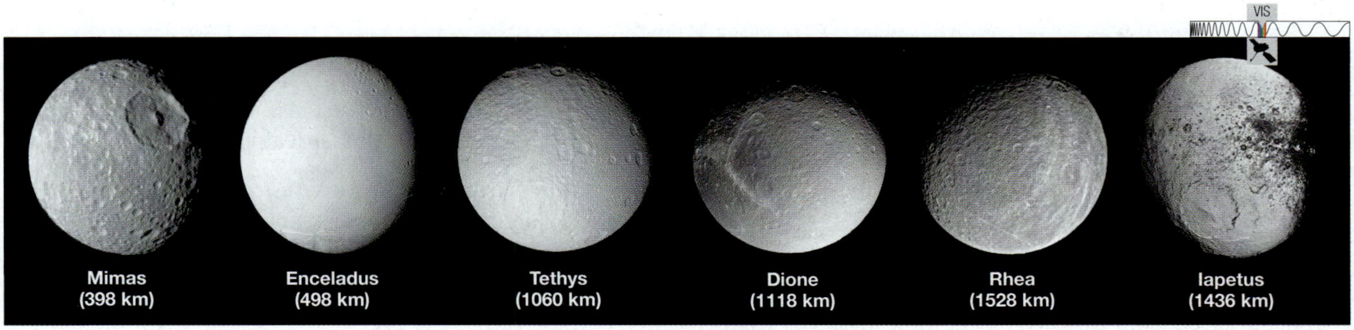


**FIGURE 11.27** Portraits taken by the *Cassini* spacecraft of Saturn's medium-size moons (not to scale).

of Mimas's resemblance to the Death Star in the original *Star Wars* movies. (The crater's official name is Herschel.) The impact that created this crater probably came close to breaking Mimas apart.

Saturn's other medium-size moons all show evidence of volcanism and/or tectonics that must have occurred since the end of the heavy bombardment, though crater counts indicate that the eras of geological activity ended billions of years ago for all except Enceladus. For example, smooth regions visible in the photos in Figure 11.27 probably represent places where icy lava once flowed, and some of the bright streaks (such as the long streaks visible on Dione) are vast sets of tectonic cliffs running parallel to one another.

Iapetus is particularly bizarre (**FIGURE 11.28**). It has an astonishing ridge more than 10 kilometers high that spans nearly half its circumference, curiously aligned along the equator. No one knows its origin, but it is likely the result of tectonic activity. Moreover, much of Iapetus appears coated in dark dust (some visible at the right in the Iapetus image in Figure 11.27) that apparently comes from Phoebe, a dark and dusty moon that orbits Saturn in a backward orbit at a great distance (see Figure 11.15). Small impacts easily eject dust because of Phoebe's low gravity, and the dust forms a ring as it spirals inward toward Iapetus.

This brings us to Enceladus, which provided one of the biggest surprises of the *Cassini* mission. Because it is only about 500 kilometers across—small enough to fit inside the borders of Colorado—we might expect any era of geological activity to have ended long ago. But that is not the case. As shown in **FIGURE 11.29**, its surface has very few impact craters—some regions have none at all—telling us that recent geological activity has erased older craters. Moreover, the strange grooves near its south pole are measurably warmer

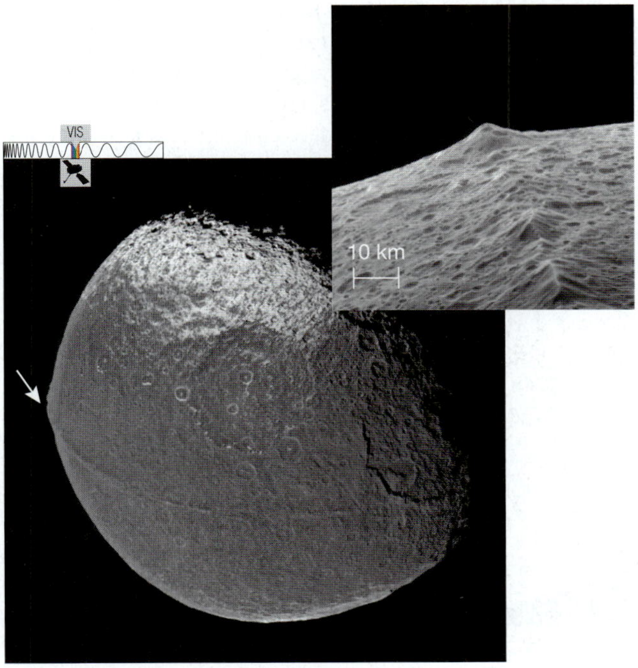


**FIGURE 11.28** Saturn's moon Iapetus has a 10-kilometer equatorial ridge (white arrow) that spans nearly half its circumference. The inset shows a portion of the ridge in perspective.

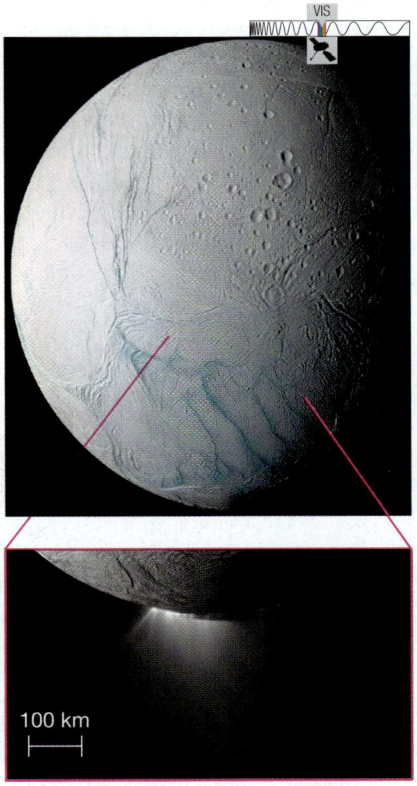


**FIGURE 11.29** *Cassini* photo of Saturn's moon Enceladus. The blue "tiger stripes" near the bottom of the main photo are regions of fresh ice that must have recently emerged from below. The colors are exaggerated; the image is a composite made at near-ultraviolet, visible, and near-infrared wavelengths. The inset shows Enceladus backlit by the Sun, with fountains of ice particles (and water vapor) clearly visible as they spray out of the south polar region.

than the surrounding terrain, and *Cassini* observed this region venting huge clouds of water vapor and ice crystals, some containing salt and rock dust. These fountains must have some subsurface source, and careful measurements of the way Enceladus wobbles on its axis as it orbits Saturn have led scientists to suspect that Enceladus has a global subsurface ocean of liquid water or of a colder water/ammonia mixture. The ocean is estimated to lie 30 to 40 kilometers beneath the moon's surface and may be up to 30 kilometers in depth (see Figure 24.18). This, in turn, makes us wonder about possible life on Enceladus. Internal heat on Enceladus comes from tidal heating through an orbital resonance (with Dione), though scientists were surprised to learn that the heating is enough to make Enceladus active today.

**The Medium-Size Moons of Uranus** Uranus does not have any moons that fall into our large category, but it has five medium-size moons: Miranda, Ariel, Umbriel, Titania, and Oberon (see Figure 11.14). Like other jovian moons, these moons are made largely of ice. Because of Uranus's great distance from the Sun, this ice includes a great deal of ammonia and methane as well as water.

Our only close look at these moons came during the *Voyager* flyby in 1986, which left us with many unanswered questions. For example, Ariel and Umbriel are virtual twins in size, yet Ariel shows evidence of volcanism and tectonics, while the heavily cratered surface of Umbriel suggests a lack of geological activity. Titania and Oberon also are twins in size, but Titania appears to have had much more geological activity than Oberon. No one knows why these two pairs of similar-size moons should vary so greatly in geological activity.

Miranda, the smallest of Uranus's medium-size moons, is the most surprising (**FIGURE 11.30**). We might expect Miranda to be a cratered iceball like Saturn's similar-size moon, Mimas. Instead, *Voyager* images of Miranda show tremendous tectonic features and relatively few craters. Why should Miranda be so much more geologically active than Mimas? Our best guess is that Miranda had an episode of tidal heating billions of years ago, perhaps during a temporary orbital resonance with another moon of Uranus. Mimas apparently never had such an episode of tidal heating and therefore lacks similar features.

**Neptune's Triton: A Captured Moon** The last stop on our tour is Neptune, so distant that only two of its moons (Triton and Nereid) were known to exist before the *Voyager 2* spacecraft visited it. Nereid is Neptune's only medium-size moon, and Triton is its only large moon. Triton is one of the coldest worlds in the solar system, even colder than Pluto because it reflects more of the weak sunlight that reaches the outskirts of the solar system.

Triton may appear to be a typical moon, but it is not. It orbits Neptune "backward" (opposite to Neptune's rotation) and at a high inclination to Neptune's equator. These are telltale signs of a moon that was captured rather than having formed in the disk of gas around its planet. No one knows how a moon as large as Triton could have been captured, but models suggest one possible mechanism: Triton may have once been a member of a binary Kuiper belt object that passed so close to Neptune that Triton lost energy and was captured while its companion gained energy and was flung away at high speed.

Triton's geology is just as surprising as its origin. Triton is smaller than our own Moon, yet its surface shows evidence of relatively recent geological activity (**FIGURE 11.31**). Some regions show evidence of past volcanism, perhaps an icy equivalent to the volcanism that shaped the lunar maria. Other regions show wrinkly ridges (nicknamed "cantaloupe terrain") that appear tectonic in nature; apparently, blobs of ice of different density have risen and fallen, contorting the crust. Triton even has a very thin atmosphere that has left wind streaks on its surface.

What could have generated the heat needed to drive Triton's geological activity? The best guess is tidal heating. Triton probably had a very elliptical orbit and a more rapid rotation when Neptune first captured it. Tidal forces would have circularized its orbit, brought it into synchronous rotation, and perhaps heated its interior enough to cause its geological activity. This idea would also explain why the geological activity has subsided with time.

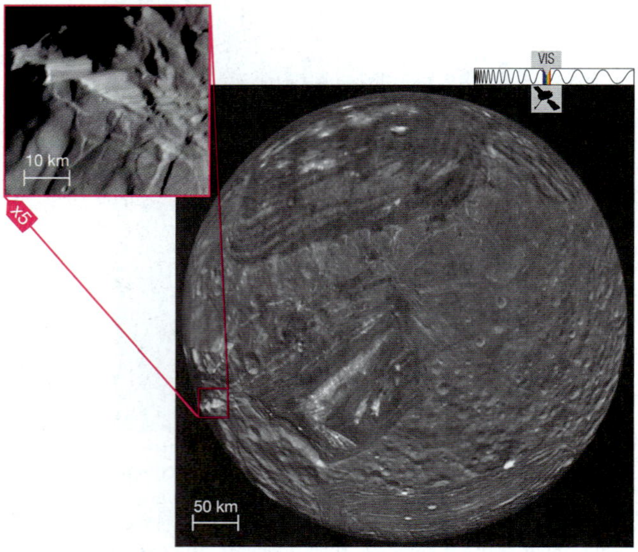


**FIGURE 11.30** The surface of Miranda shows astonishing tectonic activity despite its small size. The cliff walls seen in the inset are higher than those of the Grand Canyon on Earth.

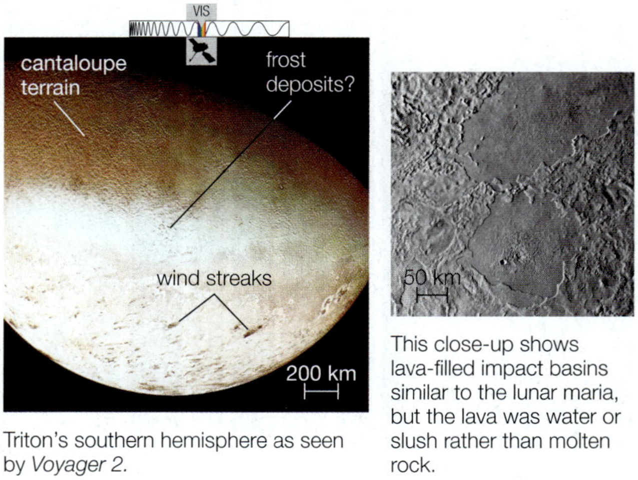


**FIGURE 11.31** Neptune's moon Triton shows evidence of a surprising level of past geological activity.

**Terrestrial Planet Geology**

- Internal heat, primarily from radioactive decay, can cause volcanic and tectonic activity.
- Only large planets retain enough internal heat to stay geologically active today.
- Example: Mars (photo above) probably retains some internal heat. If it had been smaller, like Mercury, it would be geologically "dead" today. If it had been larger, like Earth, it would probably have much more active and ongoing tectonics and volcanism.

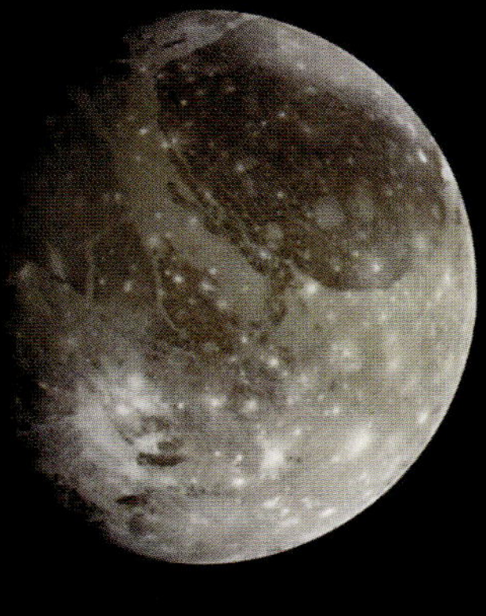

**Jovian Moon Geology**

- Tidal heating can cause tremendous geological activity on moons with elliptical orbits around massive planets.
- Even without tidal heating, icy materials can melt and deform at lower temperatures than rock, increasing the likelihood of geological activity.
- Together, these effects explain why icy moons are much more likely to have ongoing geological activity than rocky terrestrial worlds of the same size.
- Example: Ganymede (photo above) shows evidence of recent geological activity, even though it is similar in size to the geologically dead terrestrial planet Mercury.

**FIGURE 11.32** Jovian moons can be much more geologically active than terrestrial worlds of similar size because of their icy compositions and tidal heating, which is not important on the terrestrial worlds.

### Why are small icy moons more geologically active than small rocky planets?

Based on what we learned when studying the geology of the terrestrial worlds, the active geology of the jovian moons seems out of character with their sizes. Numerous jovian moons remained geologically active far longer than Mercury or our Moon, yet they are no bigger and in many cases are much smaller in size. However, there are two crucial differences between the jovian moons and the terrestrial worlds: icy compositions and tidal heating.

Because they formed far from the Sun, most of the jovian moons contain ices that can melt or deform at far lower temperatures than rock. As a result, they can experience geological activity even when their interiors have cooled to temperatures far below those of rocky worlds. Indeed, except on Io, most of the volcanism that has occurred in the outer solar system is probably "ice volcanism" that produces a lava composed mainly of water, perhaps mixed with methane and ammonia.

The major lesson, then, is that "ice geology" is possible at far lower temperatures than "rock geology." This fact, combined in many cases with tidal heating, explains how the jovian moons have had such interesting geological histories despite their small sizes. In essence, the properties of ice at low temperature are responsible both for the existence of icy moons and for their strong geological activity. **FIGURE 11.32** summarizes the differences between the geology of jovian moons and that of the terrestrial worlds.

## 11.3 Jovian Planet Rings

The jovian planet systems have three major components: the planets themselves, the moons, and their rings. We have already studied the planets and their moons, so we now turn our attention to their amazing rings. All four jovian planets have rings, but we'll begin with Saturn's, since they are by far the most spectacular.

### What are Saturn's rings like?

Saturn's rings have dazzled and puzzled astronomers since Galileo first saw them through his small telescope and suggested that they resembled "ears" on Saturn. You can see the rings through a backyard telescope, but learning their nature requires higher resolution (**FIGURE 11.33**). Earth-based views make the rings appear to be continuous, concentric sheets of material separated by a large gap (called the *Cassini division*). Spacecraft images reveal these "sheets" to be made of many individual rings, each separated from the next by a narrow gap. But even these

**a** This Earth-based telescopic view of Saturn makes the rings look like large, concentric sheets. The dark gap within the rings is called the *Cassini division*.

**b** This image of Saturn's rings from the *Cassini* spacecraft reveals many individual rings separated by narrow gaps.

**c** Artist's conception of particles in a ring system. Particles clump together because of gravity, but small random velocities cause collisions that break them up.

**FIGURE 11.33** Zooming in on Saturn's rings.

appearances are somewhat deceiving. If we could wander into Saturn's rings, we'd find that they are made of countless icy particles ranging in size from dust grains to large boulders, sometimes clumped together by their mutual gravity. All are far too small to be seen individually even from spacecraft like *Cassini* that passed them closely.

**Ring Particle Characteristics** Spectroscopy reveals that Saturn's ring particles are made mostly of relatively reflective water ice. The rings look bright where they contain enough particles to intercept sunlight and scatter it back toward us. We see gaps in places where there are few particles to reflect sunlight.

Each individual ring particle orbits Saturn independently in accord with Kepler's laws, so the rings are much like myriad tiny moons. The individual ring particles are so close together that they collide frequently. In the densest parts of the rings, each particle collides with another every few hours. However, the collisions are fairly gentle: Despite the high orbital speeds of the ring particles, nearby ring particles orbit at nearly the same speed and in the same direction, and therefore touch only gently when they collide.

**Think about it** Which ring particles travel faster: those closer to Saturn or those farther away? Explain why. (*Hint*: Review Kepler's third law.)

The frequent collisions explain why Saturn's rings are one of the thinnest known astronomical structures. They span more than 270,000 kilometers in diameter but are only a few tens of *meters* thick. The rings are so thin that they disappear from view when we see Saturn edge-on, as we do around the equinoxes of its 29.5-year orbit of the Sun.

To understand how collisions keep the rings thin, imagine what would happen to a ring particle on an orbit slightly inclined to the central ring plane. The particle would collide with other particles every time its orbit intersected the ring plane, and its orbital tilt would be reduced with every collision. Before long, these collisions would force the particle to conform to the orbital pattern of the other particles, and any particle that moved away from the narrow ring plane would soon be brought back within it. A similar idea explains why ring particles have almost perfectly circular orbits: Any particle with an elliptical orbit would quickly suffer enough collisions to force it into an orbit matching that of its neighbors.

**Rings and Gaps** Close-up photographs show an astonishing number of rings, gaps, ripples, and other features—as many as 100,000 altogether. Scientists are still struggling to explain all the features, but some general ideas are now clear.

Rings and gaps are caused by particles bunching up at some orbital distances and being forced out at others. This bunching happens when gravity nudges the orbits of ring particles in some particular way. One source of nudging comes from small moons located within the gaps in the rings themselves, sometimes called *gap moons*. The gravity of a gap moon can effectively keep the gap clear of smaller ring particles while creating ripples in the ring edges (**FIGURE 11.34a**). The ripples appear to move in opposite directions on the two sides of the gap, because ring particles on the inner side orbit Saturn slightly faster than the gap moon, while those on the outer side orbit slightly slower than the gap moon. In some cases, two nearby gap moons can force particles between them into a very narrow ring (**FIGURE 11.34b**). In those cases, the gap moons are also called *shepherd moons* because they act like a shepherd, forcing particles into line.

Ring particles also may be nudged by the gravity from larger, more distant moons. For example, a ring particle orbiting about 120,000 kilometers from Saturn's center will circle the planet in exactly half the time it takes the moon Mimas to orbit. Every time Mimas returns to a certain location, the ring particle will also be at its original location and therefore will experience the same gravitational nudge from Mimas. The periodic nudges reinforce one another and clear a gap in the rings—in this case, the large gap visible from Earth (the Cassini division). This type of reinforcement due to repeated gravitational tugs is another example of an *orbital resonance,* much like the orbital resonances that make Io's orbit elliptical (see Figure 11.19). Some resonances create vast numbers of ripples and waves that travel great distances through the rings (**FIGURE 11.35**). Other orbital resonances, caused by moons both within the rings and farther out from Saturn, probably explain most of the intricate structures visible in ring photos.

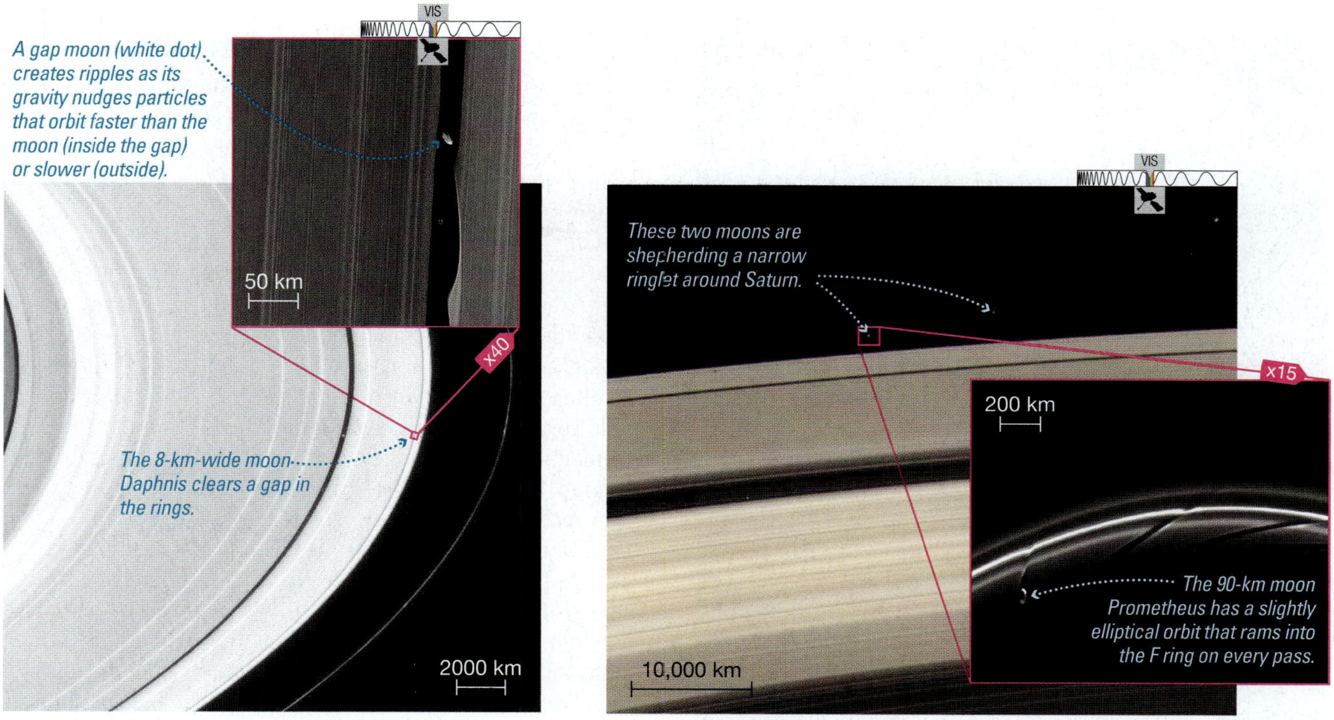


**a** Some small moons create gaps within the rings.

**b** Some moons force particles between them into a very narrow ring, as is the case with the two shepherd moons shown here. The inset shows a close-up of one of them.

**FIGURE 11.34** Small moons within the rings have important effects on ring structure (*Cassini* photos).

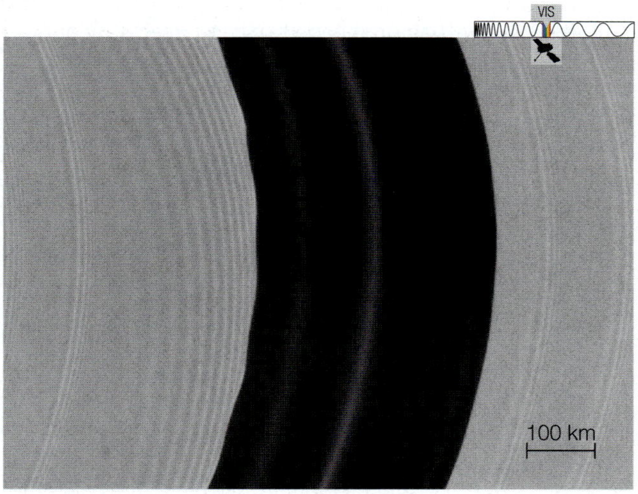


**FIGURE 11.35** This *Cassini* image of a gap in Saturn's rings shows how gravitational tugs from the moon Pan create waves that shape the rings over vast distances.

## How do other jovian ring systems compare to Saturn's?

The rings of Jupiter, Uranus, and Neptune are so much fainter than Saturn's that it took almost four centuries longer to discover them. Ring particles in these three systems are far less numerous, generally smaller, and much darker. Despite these differences, a family portrait of the jovian ring systems shows many similarities (**FIGURE 11.36**). All rings lie in their planet's equatorial plane. Particle orbits are nearly circular, with small orbital tilts relative to the equator. Individual rings and gaps are probably shaped by gap moons, shepherd moons, and orbital resonances.

Uranus's rings were discovered in 1977 during observations of a *stellar occultation*—a star's passage behind Uranus as seen from Earth. During the occultation, the star "blinked" on and off nine times before it disappeared behind Uranus and nine more times as it emerged. Scientists concluded that these nine "blinks" were caused by nine thin rings encircling Uranus. Similar observations of stars passing behind Neptune yielded more confounding results: Rings appeared to be present at some times but not at others. Could Neptune's rings be incomplete or transient?

The *Voyager* spacecraft provided some answers. *Voyager* cameras first discovered thin rings around Jupiter in 1979. After next providing incredible images of Saturn's rings, *Voyager 2* photographed the rings of Uranus as it flew past in 1986. In 1989, *Voyager 2* passed by Neptune and found that it does, in fact, have partial rings—at least when seen from Earth. The space between the ring segments is filled with dust not detectable from Earth. In addition to the dust within the rings of Neptune, *Voyager 2* detected vast dust sheets between the widely separated rings of both Uranus and Neptune.

The differences between the ring systems present us with unsolved mysteries. The larger size, higher reflectivity, and much greater number of particles in Saturn's rings compel us to wonder whether different processes might be at work there. We do not know why Uranus's thin rings have a slight tilt and slightly eccentric orbits, nor do we know why Neptune's rings contain dusty regions that make them appear as partial rings when viewed from Earth. Scientists hope that we may someday learn the answers with new missions to visit Uranus and Neptune, but none are yet under development. Meanwhile, new and powerful ground-based observatories are providing views that in some cases rival those obtained from the *Voyager* spacecraft.

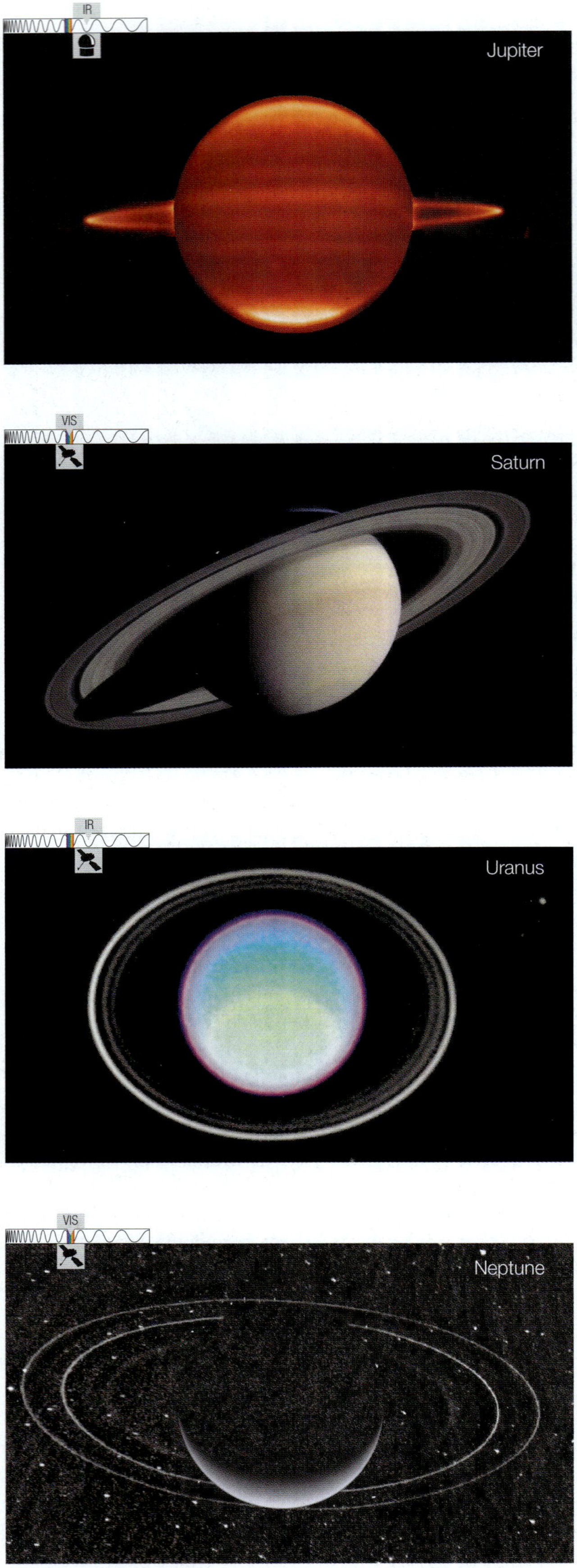


**FIGURE 11.36** Four ring systems (not to scale). The rings differ in appearance and in the composition and sizes of the ring particles. (Jupiter: Keck Telescope, infrared; Saturn: *Cassini*, visible; Uranus: Hubble Space Telescope, infrared; Neptune: *Voyager*, visible.)

## Why do the jovian planets have rings?

The fact that all four jovian planets have ring systems tells us that rings must arise fairly easily around such worlds. To understand how scientists now suspect that this occurs, it's helpful to first discuss why some past ideas no longer seem to work.

One key clue to the origin of ring systems comes from the fact that rings always lie within two to three planetary radii of their planet, which is a region where the tidal forces tugging an object apart become comparable to the gravitational forces holding it together. (This region is often called the *Roche tidal zone.*) Only small objects held together by nongravitational forces—such as the electromagnetic forces that hold solid rock, spacecraft, and human beings together—can avoid being ripped apart in this region. This fact once led many scientists to suspect that Saturn's rings had formed when a large moon's orbit changed and brought it into this region where tidal forces ripped the moon apart. While this hypothesis seemed reasonable when Saturn's rings were the only ones known, it no longer seemed viable once the other jovian ring systems were discovered. The reason is that this hypothesis requires an event that should be extremely rare: Recall that moons don't simply "wander" away from their orbits, so the moon's orbit would have to have been disturbed by a close gravitational encounter with an unusually large asteroid or comet that happened to pass nearby. While such a rare event might have happened once, it seems inconceivable that such an unlikely event could have occurred for all four jovian planets.

Another idea that once seemed reasonable was that ring particles might be leftover bits of rock and ice and dust that condensed in the disks of gas that orbited each jovian planet when it was young. This would explain why all four jovian planets have rings, because tidal forces near each planet would have prevented these particles from accreting into a full-fledged moon. However, we now know that the ring particles cannot be leftovers from the births of the planets, because they could not have survived for billions of years. The larger ring particles are continually being ground down in size, primarily by the impacts of the countless sand-size particles that orbit the Sun—the same types of particles that become meteors in Earth's atmosphere and cause micrometeorite impacts on the Moon [**Section 9.3**]. Millions of years of such tiny impacts would have ground the existing ring particles to dust long ago. The dust particles could not have survived either, because over time they are slightly slowed in their orbits by the pressure of sunlight, and this slowing eventually causes them to spiral into their planet. Ring particles may also be lost in other ways. For example, the thermosphere of Uranus extends up into the ring region, exerting atmospheric drag that also causes ring particles to spiral slowly into the planet. Between the effects of micrometeorite impacts and other processes, *none* of the abundant small particles that now occupy the jovian rings can have been there since the solar system formed more than 4 billion years ago.

We are left with only one reasonable possibility: New particles must be continually supplied to the rings to replace those that are destroyed. These new particles must come

**FIGURE 11.37** This illustration summarizes the origin of rings around the jovian planets.

from a source that lies in each planet's equatorial plane and that consists of objects small enough to avoid being ripped apart by the tidal forces. The most likely source is numerous small "moonlets"—moons the size of gap moons (see Figure 11.34)—that formed in the disks of material orbiting the young jovian planets. As with the ring particles themselves, tiny impacts are gradually grinding away these small moons, but they are large enough to still exist despite $4\frac{1}{2}$ billion years of such sandblasting.

The small moons contribute ring particles in two ways. First, each tiny impact releases particles from a small moon's surface, and these released particles become new, dust-size ring particles. Ongoing impacts ensure that some ring particles are present at all times. Second, occasional larger impacts can shatter a small moon completely, creating a supply of boulder-size ring particles. The frequent tiny impacts then slowly grind these boulders into smaller ring particles. Some of these particles are "recycled" by forming into small clumps, only to come apart again later on; others are ground down to dust and slowly spiral into their planet. In summary, the leading model holds that all ring particles ultimately come from the gradual dismantling of small moons that formed during the birth of the solar system (**FIGURE 11.37**).

The collisions that shatter small moons and generate large ring particles must occur only occasionally and at essentially random times, which means that the numbers and sizes of particles in any particular ring system may vary dramatically over millions and billions of years. Confirmation of this idea came at the end of the *Cassini* mission, when the spacecraft passed between the inner rings and Saturn's atmosphere during its final few orbits before it plunged into Saturn. These passes allowed scientists to determine a more precise mass for the rings (from the gravitational tug on the spacecraft). By considering this mass measurement along with *Cassini's* 12 years of measurements of the rate at which sooty micrometeorites should darken the rings, scientists confirmed that the existing rings could not have remained bright for billions of years, and most likely are only between about 150 and 300 million years old. Therefore, the brilliant spectacle of Saturn's rings appears to be a special treat of our epoch, one that would not have been seen a billion years ago and that might not last long on the time scale of our solar system.

## The BIG Picture PUTTING CHAPTER 11 INTO CONTEXT

*In this chapter, we saw that the jovian planets really are a different kind of planet and, indeed, a different kind of planet system. As you continue your study of the solar system, keep in mind the following "big picture" ideas:*

- The jovian planets dwarf the terrestrial planets. Even some of their moons are as large as terrestrial worlds.
- The jovian planets may lack solid surfaces on which geology can occur, but they are interesting and dynamic worlds with rapid winds, huge storms, strong magnetic fields, and interiors in which common materials behave in unfamiliar ways.
- Despite their relatively small sizes and frigid temperatures, many jovian moons are geologically active by virtue of their icy compositions—a result of their formation in the outer regions of the solar nebula—and tidal heating.
- Ring systems probably owe their existence to small moons formed in the disks of gas that produced the jovian planets billions of years ago. The rings we see today are composed of particles liberated from those moons quite recently.
- Understanding the jovian planet systems forced us to modify many of our earlier ideas about the solar system by adding the concepts of ice geology, tidal heating, and orbital resonances. Each new set of circumstances that we discover offers new opportunities to learn how our universe works.

**MY COSMIC PERSPECTIVE** Jovian moons like Europa, Titan, and Enceladus, once assumed to be geologically dead and uninteresting because of their small sizes and great distances, could turn out to be our nearest neighbors in a biological sense.

# Summary of Key Concepts

## 11.1 A Different Kind of Planet

- **Are jovian planets all alike?** Jupiter and Saturn are made almost entirely of hydrogen and helium, while Uranus and Neptune are made mostly of hydrogen compounds mixed with metals and rock. These differences arose because all four planets started from ice-rich planetesimals of about the same size, but captured different amounts of hydrogen and helium gas from the solar nebula.

- **What are jovian planets like on the inside?** The jovian planets have layered interiors with very high internal temperatures and pressures. All have a core about 10 times as massive as the entire Earth, consisting of hydrogen compounds, metals, and rock. They differ mainly in their surrounding layers of hydrogen and helium, which can take on unusual forms under the extreme internal conditions of the planets.

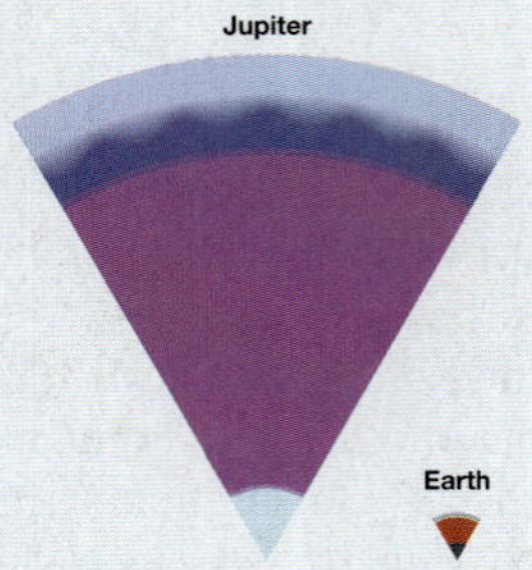


- **What is the weather like on jovian planets?** The jovian planets all have multiple cloud layers that give them distinctive colors, fast winds, and large storms. Some storms, such as Jupiter's **Great Red Spot**, can apparently rage for centuries or longer.

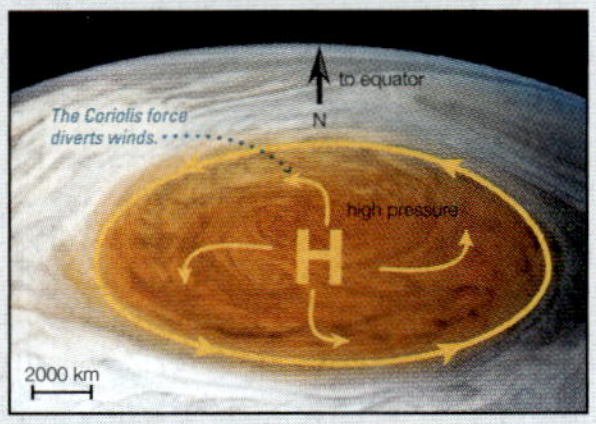


- **Do jovian planets have magnetospheres like Earth's?** Jupiter has a magnetic field 20,000 times as strong as Earth's, which leads to an enormous magnetosphere. Many of the particles in Jupiter's magnetosphere come from volcanic eruptions on Io. Other jovian planets also have magnetic fields and magnetospheres, but they are weaker and smaller than Jupiter's.

## 11.2 A Wealth of Worlds: Satellites of Ice and Rock

- **What kinds of moons orbit the jovian planets?** We can categorize the more than 170 known moons as small, medium-size, or large. Most of the medium-size and large moons probably formed with their planet in the disks of gas that surrounded the jovian planets when they were young. Smaller moons are often captured asteroids or comets.

- **Why are Jupiter's Galilean moons so geologically active?** Io is the most volcanically active object in the solar system, thanks to an interior kept hot by **tidal heating**—which occurs because Io's close orbit is made elliptical by **orbital resonances** with other moons. Europa (and possibly Ganymede) may have a deep, liquid water ocean under its icy crust, also thanks to tidal heating. Callisto is the least geologically active, since it has no orbital resonance or tidal heating, but it may also have a subsurface ocean.

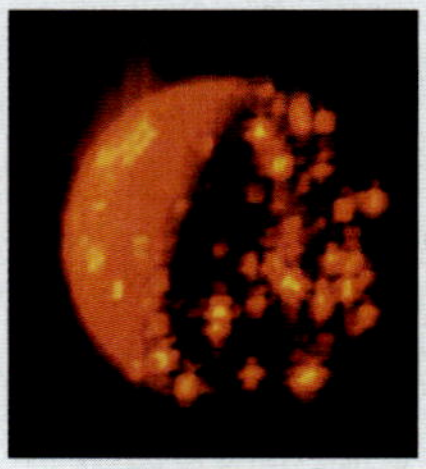

- **What geological activity do we see on Titan and other distant moons?** Many medium-size and large moons show a surprisingly high level of past or present volcanism or tectonics. Titan has a thick atmosphere and ongoing erosion, and Enceladus is also geologically active today. Triton, which apparently was captured by Neptune, also shows signs of recent geological activity.

- **Why are small icy moons more geologically active than small rocky planets?** Ices deform and melt at much lower temperatures than rock, allowing icy volcanism and tectonics at surprisingly low temperatures. In addition, some jovian moons have a heat source—tidal heating—that is not important for the terrestrial worlds.

## 11.3 Jovian Planet Rings

- **What are Saturn's rings like?** Saturn's rings are made up of countless individual particles, each orbiting Saturn independently like a tiny moon. The rings lie in Saturn's equatorial plane, and they are extremely thin. Moons within and beyond the rings create many ringlets and gaps, in part through orbital resonances.

- **How do other jovian ring systems compare to Saturn's?** The other jovian planets have ring systems that are much fainter in photographs. Their ring particles are generally smaller, darker, and less numerous than Saturn's ring particles.

- **Why do the jovian planets have rings?** Ring particles probably come from the dismantling of small moons formed in the disks of gas that surrounded the jovian planets billions of years ago. Small ring particles come from countless tiny impacts on the surfaces of these moons, while larger ones come from impacts that shatter the moons.

# Visual Skills Check

*Use the following questions to check your understanding of some of the many types of visual information used in astronomy. For additional practice, try the Chapter 11 Visual Quiz in the Study Area at www.MasteringAstronomy.com.*

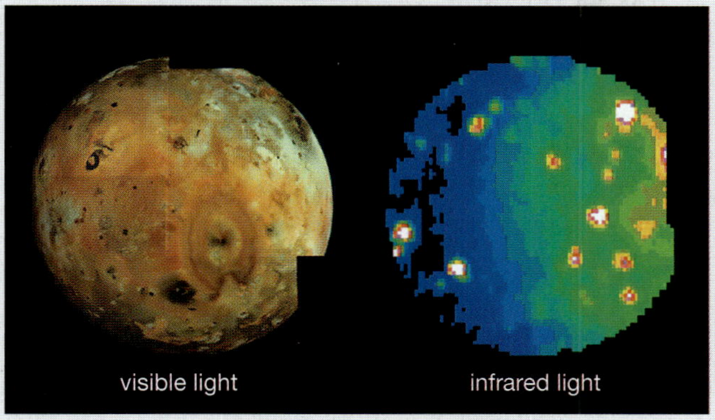


*Left: Approximate colors of Io in visible light; black spots are volcanoes that are active or have recently gone inactive. Right: Infrared thermal emission from Io; bright spots are active volcanoes. (Both images are from* Galileo *data and show the same face of Io, but taken at different times.)*

1. What do the colors in the right image represent?
   a. the actual colors of Io's surface
   b. the colors we would see if we had infrared vision
   c. the intensity of the infrared light
   d. regions of different chemical composition on the surface
2. Which color in the right image represents the highest temperatures?
   a. blue  b. green  c. orange  d. red  e. white
3. The right image was obtained when only part of Io was in sunlight. Based on the colors, which part of the surface was in sunlight?
   a. the left side
   b. the right side
   c. only the peaks of the volcanoes
4. By comparing the two images, what can you conclude about Io's volcanoes?
   a. Every black spot in the visible image has a bright spot in the infrared image, so all of Io's volcanoes were active when the photos were taken.
   b. There are more black spots in the visible image than bright spots in the infrared image, so many of Io's volcanoes were inactive when the photos were taken.
   c. There are more bright spots in the infrared image than black spots in the visible image, so new eruptions must have started after the visible photo was taken.

# Exercises and Problems

*For instructor-assigned homework and other learning materials, go to www.MasteringAstronomy.com.*

## Chapter Review Questions

### Short-Answer Questions Based on the Reading

1. Briefly describe how the differences in composition between the jovian planets can be traced to their formation.
2. Why is Jupiter so much more dense than Saturn? Could a planet be smaller in size than Jupiter but greater in mass?
3. Briefly describe the interior structure of Jupiter and explain why it is layered in this way. How do the interiors of the other jovian planets compare to that of Jupiter?
4. Why does Jupiter have so much internal heat? What generates internal heat on other jovian planets?
5. Briefly describe Jupiter's atmospheric structure and cloud layers. How do the structures and clouds differ on the other jovian planets?
6. How do clouds contribute to Jupiter's colors? Why are Saturn's colors more subdued? Why are Uranus and Neptune blue?
7. Briefly describe Jupiter's weather patterns. What is the *Great Red Spot*?
8. Why does Jupiter have such a strong magnetic field? Describe a few features of Jupiter's magnetosphere and compare it to the magnetospheres of the other jovian planets.
9. Briefly describe how we categorize jovian moons by size. What is the origin of most of the medium-size and large moons? What is the origin of many of the small moons?
10. Describe key features of Jupiter's four Galilean moons, and the role of *tidal heating* and *orbital resonances* in shaping these features.
11. Describe the atmosphere and surface features of Titan. How is Titan's landscape similar to Earth's? How is it different?
12. Summarize the evidence for and some of the mysteries of past or present geological activity on the medium-size moons of Saturn and Uranus.
13. Why do we think Triton is a captured moon? How might its capture be relevant to its geological activity?
14. Briefly explain why icy moons can have active geology at much smaller sizes than rocky worlds.
15. What are planetary rings made of, and how do they differ among the four jovian planets? Briefly describe the effects of gap moons and orbital resonances on ring systems.
16. Explain why we think that ring particles must be replenished over time, and where we think ring particles come from.

## Surprising Discoveries?

*Suppose someone claimed to make the following discoveries. (These are not real discoveries.) In light of what you've learned in this chapter, decide whether each discovery should be considered reasonable or surprising. Explain your reasoning.*

17. Saturn's core is pockmarked with impact craters and dotted with volcanoes erupting lava.
18. Neptune's deep blue color is not due to methane, as previously thought, but instead is due to its surface being covered with an ocean of liquid water.
19. A jovian planet in another star system has a moon as big as Mars.
20. A planet orbiting another star is made primarily of hydrogen and helium and has approximately the same mass as Jupiter but is the same size as Neptune.
21. A previously unknown moon orbits Jupiter outside the orbits of other known moons. It is the smallest of Jupiter's moons but has several large, active volcanoes.
22. A previously unknown moon orbits Neptune in the planet's equatorial plane and in the same direction that Neptune rotates, but it is made almost entirely of metals such as iron and nickel.
23. An icy, medium-size moon orbits a jovian planet in a star system that is only a few hundred million years old. The moon shows evidence of active tectonics.
24. A jovian planet is discovered in a star system that is much older than our solar system. The planet has no moons but has a system of rings as spectacular as the rings of Saturn.
25. Future observations discover rainfall of liquid water on Titan.
26. During a future mission to Uranus, scientists discover that it is orbited by another 20 previously unknown moons.

## Quick Quiz

*Choose the best answer to each of the following. For additional practice, try the Chapter 11 Reading and Concept Quizzes in the Study Area at www.MasteringAstronomy.com.*

27. Which lists the jovian planets in order of increasing distance from the Sun? (a) Jupiter, Saturn, Uranus, Pluto (b) Saturn, Jupiter, Uranus, Neptune (c) Jupiter, Saturn, Uranus, Neptune
28. Why does Neptune appear blue and Jupiter red? (a) Neptune is hotter, which gives bluer thermal emission. (b) Methane in Neptune's atmosphere absorbs red light. (c) Neptune's air molecules scatter blue light, much as Earth's atmosphere does.
29. Why is Jupiter denser than Saturn? (a) It has a larger proportion of rock and metal. (b) It has a larger proportion of hydrogen. (c) Its higher mass and gravity compress its interior.
30. Some jovian planets give off more energy than they receive because of (a) fusion in their cores. (b) tidal heating. (c) ongoing contraction or differentiation.
31. The main ingredients of most satellites of the jovian planets are (a) rock and metal. (b) hydrogen compound ices. (c) hydrogen and helium.
32. Why is Io more volcanically active than our moon? (a) Io is much larger. (b) Io has a higher concentration of radioactive elements. (c) Io has a different internal heat source.
33. What is unusual about Triton? (a) It orbits its planet backward. (b) It does not keep the same face toward its planet. (c) It is the only moon with its own rings.
34. Which moon shows evidence of rainfall and erosion by some liquid substance? (a) Europa (b) Titan (c) Ganymede
35. Saturn's many moons affect its rings through (a) tidal forces. (b) orbital resonances. (c) magnetic field interactions.
36. Saturn's rings (a) have looked basically the same since they formed along with Saturn. (b) were created long ago when tidal forces tore apart a large moon. (c) are continually supplied with new particles by impacts with small moons.

## Inclusive Astronomy

*Use these questions to reflect on participation in science.*

37. *Group Discussion: JUICE and International Collaboration.* The European Space Agency's *JUICE* (for Jupiter Icy Moons Explorer) mission is designed to explore Jupiter's largest icy moons and to assess their potential for hosting life.
**a.** Working independently, find some background on the *JUICE* mission. List its main science objectives and dates currently planned for its launch and its arrival at Jupiter.
**b.** The European Space Agency has 22 member nations that are providing the bulk of the funding for the mission. Should the funding provided by each nation be a factor in determining who gets to analyze the data? For example, should scientists from nations that provided more funding get access to the data before the data are released to other scientists?
**c.** The European Space Agency ultimately plans to share data from the *JUICE* mission widely, but other nations—particularly those in the developing world—may not have the financial resources to support their scientists in analyzing mission data. Should the European Space Agency try to support data analysis by scientists from nations that did not contribute to the mission funding? If so, how? If not, why not?
**d.** Suppose that the mission finds valuable mineral resources on some of these moons. Who should own mineral rights to these resources? Why?

## The Process of Science

*These questions may be answered individually in short-essay form or discussed in groups, except where identified as group-only.*

38. *Europan Ocean.* Scientists strongly suspect that Europa has a subsurface ocean, even though we cannot see through the surface ice. Briefly explain why scientists think the ocean exists. Is this "belief" in a Europan ocean scientific? Explain.
39. *Breaking the Rules.* As discussed in Chapter 9, the geological "rules" for the terrestrial worlds tell us that a world as small as Io should not have any geological activity. However, *Voyager* images of Io's volcanoes proved that the old "rules" had been wrong. Based on your understanding of the nature of science [Section 3.4], do you think this should be seen as a failure in the process of science? Defend your opinion.
40. *Jovian Planet Mission.* We can study terrestrial planets up close by landing on them, but jovian planets have no surfaces to land on. Suppose that you were in charge of planning a long-term mission to "float" in the atmosphere of a jovian planet. Describe the technology you would use and how you would ensure survival for any people assigned to this mission.
41. *Jovian Moon Mission.* Suppose you could choose any one jovian moon to visit. Which one would you pick, and why? What dangers would you face in your visit to this moon? What kinds of scientific instruments would you want to bring along for studies?
42. *Unanswered Questions.* Choose one unanswered question about one of the jovian planets or its moons. Write a few paragraphs discussing the question and the specific types of evidence we would need to answer it.
43. *Group Activity: Comparing Jovian Moons.* Compare the moons of Jupiter, drawing on the data in Appendix E. Note: You may wish to do this activity using the four roles described in Chapter 1, Exercise 39.
**a.** Collect data on Jupiter's four largest moons from Table E.2 in Appendix E and determine which moon has the greatest density.

b. Use Table E.2 to determine what other solar system moon most resembles the moon from part a in mass, radius, and density.

c. Propose a hypothesis about the composition of the moon from part a, based on its resemblance to the moon from part b, and examine potential concerns about the viability of the hypothesis.

d. Use Table E.2 to determine whether there is a trend in density with orbital distance among the major moons of Jupiter; briefly describe any trends.

e. Suggest a hypothesis that accounts for any trend found in part d, and discuss potential concerns about the hypothesis.

f. Develop and describe an experiment that could test the hypotheses in parts c and e.

## Investigate Further

### Short-Answer/Essay Questions

44. *The Importance of Rotation.* Suppose the material that formed Jupiter came together without any rotation so that no "jovian nebula" formed and the planet today wasn't spinning. How else would the jovian system be different? Think of as many effects as you can, and explain each in a sentence.
45. *The Great Red Spot.* Based on the infrared and visible images in Figure 11.9, is Jupiter's Great Red Spot warmer or cooler than nearby clouds? Is it higher or lower in altitude than the nearby clouds? Explain.
46. *Comparing Jovian Planets.* You can do comparative planetology armed only with a telescope and an understanding of gravity.
    **a.** The small moon Amalthea orbits Jupiter at about the distance in kilometers at which Mimas orbits Saturn, yet Mimas takes almost twice as long to orbit. From this observation, what can you conclude about how Jupiter and Saturn differ? Explain.
    **b.** Jupiter and Saturn are not very different in radius. When you combine this information with your answer to part a, what can you conclude? Explain.
47. *Minor Ingredients Matter.* Suppose the jovian planets' atmospheres were composed only of hydrogen and helium, with no hydrogen compounds at all. How would the atmospheres be different in terms of clouds, color, and weather? Explain.
48. *Galilean Moon Formation.* Look up the densities of Jupiter's four Galilean moons in Appendix E, and notice that they follow a trend with distance from Jupiter. Based on what you've learned about condensation in the solar nebula, can you suggest a reason for this trend among the Galilean moons? Next, compare the densities of the moons with the planetary densities in Table E.1. Based on the comparison, do you think it was as hot toward the center of the nebula surrounding Jupiter as it was at the center of the solar nebula? Explain.
49. *Super Jupiter.* Suppose that the solar wind had not cleared the solar nebula until much later in our solar system's history, so that Jupiter accumulated 40 times as much mass as it actually has today.
    **a.** Make a sketch showing how the interior layers of this "Super Jupiter" would compare to the interior layers of the real Jupiter shown in Figure 11.4.
    **b.** How would you expect the cloud layers and colors to differ on Super Jupiter? Explain.
    **c.** How would you expect the wind speeds to be different on Super Jupiter? Explain.
    **d.** How would you expect the magnetic field to be different on Super Jupiter? Explain.
    **e.** Do you think Super Jupiter would have more or less excess heat than the real Jupiter? Explain.
50. *Observing Project: Jupiter's Moons.* Using binoculars or a small telescope, view the moons of Jupiter. Make a sketch of what you see, or take a photograph. Repeat your observations several times (nightly, if possible) over a period of a couple of weeks. Can you determine which moon is which? Can you measure the moons' orbital periods? Can you determine their approximate distances from Jupiter? Explain.
51. *Observing Project: Saturn's Rings.* Using binoculars or a small telescope, view the rings of Saturn. Make a sketch of what you see, or take a photograph. What season is it in Saturn's northern hemisphere? How far do the rings extend above Saturn's atmosphere? Can you identify any gaps in the rings? Describe any other features you notice.

### Quantitative Problems

*Be sure to show all calculations clearly and state your final answers in complete sentences.*

52. *Disappearing Moon.* Io loses about a ton (1000 kilograms) of sulfur dioxide per second to Jupiter's magnetosphere.
    **a.** At this rate, what fraction of its mass would Io lose in 4.5 billion years?
    **b.** Suppose sulfur dioxide currently makes up 1% of Io's mass. When will Io run out of this gas at the current loss rate?
53. *Ring Particle Collisions.* Each ring particle in the densest part of Saturn's rings collides with another about every 5 hours. If a ring particle survived for the age of the solar system, how many collisions would it undergo?
54. *Prometheus and Pandora.* These two moons orbit Saturn at 139,350 and 141,700 kilometers, respectively.
    **a.** Using Newton's version of Kepler's third law, find their two orbital periods. Find the percent difference in their distances from Saturn and in their orbital periods.
    **b.** Consider the two in a race around Saturn: In one Prometheus orbit, how far behind is Pandora (in units of time)? In how many Prometheus orbits will Pandora have fallen behind by one of its own orbital periods? Convert this number of periods back into units of time. This is how often the satellites pass by each other.
55. *Orbital Resonances.* Using the data in Appendix E, identify the orbital resonance relationship between Titan and Hyperion. (*Hint:* If the orbital period of one were 1.5 times the orbital period of the other, we would say that they were in a 3:2 resonance.) Which medium-size moon is in a 2:1 resonance with Enceladus?
56. *Titanic Titan.* What is the ratio of Titan's mass to that of the other satellites of Saturn whose masses are listed in Appendix E? Calculate the strength of gravity on Titan compared to that on Mimas. Comment on how this affects the possibility of an atmosphere on each.
57. *Titan's Evolving Atmosphere.* Titan's exosphere lies nearly 1400 kilometers above its surface. What is the escape velocity from this altitude? What is the thermal speed of a hydrogen atom at the exospheric temperature of about 200 K? Use these answers (and the method of Mathematical Insight 10.2) to comment on whether thermal escape of hydrogen is likely to be important for Titan.
58. *Saturn's Thin Rings.* Saturn's ring system is more than 270,000 kilometers wide and only a few tens of meters thick; let's assume it is 50 meters thick for this problem. Assuming the rings could be shrunk down so that their diameter was the width of a dollar bill (6.6 cm), how thick would the rings be? Compare your answer to the actual thickness of a dollar bill (0.01 cm).

# 12 Asteroids, Comets, and Dwarf Planets

## THEIR NATURE, ORBITS, AND IMPACTS

▲ **About the photo:** Pluto as we've seen it through time. Left: Pluto is circled in this discovery image from 1930. Center: Pluto as revealed in 2010, based on computer image processing of Hubble Space Telescope images. Right: Pluto revealed by the *New Horizons* spacecraft in 2015, showing the same hemisphere as in the Hubble image.

## LEARNING GOALS

### 12.1 Classifying Small Bodies

- What's the difference between an asteroid, a comet, and a dwarf planet?
- What are meteors and meteorites?

### 12.2 Asteroids

- What are asteroids like?
- What do meteorites tell us about asteroids?
- Why is there an asteroid belt?

### 12.3 Comets

- Why do comets grow tails?
- Where do comets come from?

### 12.4 Pluto and the Kuiper Belt

- What is Pluto like?
- Why is there a Kuiper belt?

### 12.5 Cosmic Collisions: Small Bodies Versus the Planets

- Did an impact kill the dinosaurs?
- How great is the impact risk today?
- How do the jovian planets affect impact rates and life on Earth?

*The more clearly we can focus our attention on the wonders and realities of the universe about us, the less taste we shall have for destruction.*

— *Rachel Carson (1907–1964)*

**Chapter 12 Overview**

Asteroids and comets might at first seem insignificant compared to the planets and moons we've discussed so far, but there is strength in numbers, and small bodies probably number in the trillions. The appearance of a comet has more than once altered the course of human history when our ancestors acted on superstitions related to the sighting. More profoundly, asteroids or comets brought water and other chemical ingredients that helped make life on Earth possible, while their occasional impacts scarred our planet with impact craters and sometimes altered the course of biological evolution. Asteroids and comets are also important scientifically: As remnants from the birth of our solar system, they teach us about how our solar system formed.

In this chapter, we will explore the small bodies of our solar system. We'll consider asteroids and the pieces of them that fall to Earth as meteorites, along with comets and dwarf planets like Pluto, Eris, and Ceres. We'll also explore the dramatic effects and ongoing threat of impacts on Earth.

## 12.1 Classifying Small Bodies

So far we have focused our attention on the terrestrial and jovian planets, but as we discussed in Chapters 7 and 8, the Sun is also orbited by vast numbers of small bodies, ranging in size from specks of dust to Pluto and Eris, the largest known of the "dwarf planets." Broadly speaking, all of these objects are leftovers from the era of planet formation, meaning that they formed early in the solar system's history but never became part of a full-fledged planet.

Although we have known of the existence of small objects for a long time, scientific understanding of them has gone through a revolution in the past few decades. This new understanding has helped us build a much clearer picture of our solar system and how it formed, but it has also occasionally created some difficulties in how we define and classify small bodies.

### What's the difference between an asteroid, a comet, and a dwarf planet?

Recall that, today, we use relatively simple definitions of asteroids and comets: Both orbit the Sun and are too small to be considered planets, but asteroids are rocky while comets are ice-rich [**Section 7.2**]. But these definitions have changed over time (and are still subject to some debate), and the dwarf planet category has been used only since 2006, so let's explore these classifications in a little more detail.

**Asteroids** The word *asteroid* means "starlike," but there's really nothing starlike about asteroids; the name is an artifact from the time when all we knew about them was that, like stars, they appeared as points of light in telescopes. This appearance is a result of their small sizes, which make them appear point-like to most telescopes even when they are relatively nearby. Asteroids are generally identified in telescopic images by their motion relative to the stars (**FIGURE 12.1**).

**FIGURE 12.1** Because asteroids orbit the Sun, they move through our sky relative to the stars. In this long-exposure photograph, stars show up as distinct dots, while the motion of an asteroid relative to the stars makes it show up as a short streak.

The first asteroid discovery came in 1801, only about 20 years after the discovery of Uranus. The newly discovered object, Ceres (which later proved to be the largest asteroid), was originally called a "planet." Three more asteroids (Pallas, Juno, and Vesta) were discovered over the next seven years, and as astronomers realized how small these objects were compared to the other planets, they came to be called "minor planets." Interestingly, the alternative term "asteroid" had already been introduced in 1802 (by William Herschel, discoverer of Uranus), but it did not come into wide use until many decades later, after asteroid discoveries began to come at a more rapid pace and it became clear that most of them lie in the *asteroid belt* between Mars and Jupiter (see Figure 7.1).

Hundreds of thousands of asteroids have now been cataloged, and more are discovered almost every night. Newly discovered asteroids first get a provisional name based on the discovery year and month and order of discovery. For example, the first asteroid discovered in January 2021 would be called Asteroid 2021 AA. (When letters run out, numbers are added after the letters.) Once an asteroid has been tracked long enough for its orbit to be calculated from the law of gravity [**Section 4.4**], its discoverer may give it a name, subject to approval by the International Astronomical Union. The earliest discovered asteroids bear the names of mythological figures. More recent discoveries often carry names of scientists, cartoon heroes, pets, rock stars, or textbook authors.

**Comets** For most of human history, comets were familiar only from their occasional presence in the night sky. Every few years, a comet becomes visible to the naked eye, appearing as a fuzzy ball with a long tail. Indeed, the word *comet* comes from the Greek word for "hair," a reference

a Comet Hyakutake.

b Comet Hale-Bopp, photographed over Boulder, Colorado.

**FIGURE 12.2** Brilliant comets can appear at almost any time, as demonstrated by the back-to-back appearances of Comet Hyakutake in 1996 and Comet Hale-Bopp in 1997.

to the appearance of their tails in our sky (**FIGURE 12.2**). Bright comets were hard to miss before the advent of electric lights, and they sometimes had profound effects on human behavior. For example, the ancient Chinese believed that the appearance of a comet led to a time of major and tumultuous change, and comet sightings therefore often sparked such change. Note that while photographs may make comets appear to race across the sky, they do not. If you watch a comet for minutes or hours, it will remain nearly stationary relative to the stars around it in the sky (as they rise and set with Earth's rotation). You'll notice its gradual motion relative to the constellations only over a period of many days, and a comet may remain visible for weeks before it fades from view.

Few ancient cultures made any attempt to explain comets in astronomical terms. In fact, comets were generally thought to be within Earth's atmosphere until 1577, when Tycho Brahe [**Section 3.3**] used observations made from different locations in Europe to prove that a comet lay far beyond the Moon. A century later, Newton correctly deduced that comets orbit the Sun. Then, in a book published in 1705, English scientist Edmond Halley (1656–1742) used Newton's law of gravitation to calculate the orbit of a comet that had been seen in 1682. He showed that it was the same comet that had been observed on numerous prior occasions, which meant that it orbited the Sun every 76 years. Halley predicted that the comet would return in 1758. Although he had died by that time, the comet returned as he predicted and was then named in his honor. Halley's Comet last passed through the inner solar system in 1986, and it will return in 2061.

We now know that the vast majority of comets do not have tails and never come anywhere close to Earth. Instead, they remain in the outer reaches of our solar system, orbiting the Sun far beyond the orbit of Neptune in the two vast reservoirs we call the *Kuiper belt* and the *Oort cloud* (see Figure 7.1, Step 3). The comets that appear in the night sky are the rare ones that have had their orbits changed by the gravitational influences of planets, other comets, or stars passing by in the distance, causing them to venture into the inner solar system. Most of these comets will not return to the inner solar system for thousands of years, if ever. A few happen to pass near enough to a planet to have their orbits changed further, and some (like Halley's) end up on elliptical orbits that periodically bring them close to the Sun.

Today, comets are named for the first observers (up to three) to report a comet discovery to the International Astronomical Union. However, they also carry other designations, which can lead to complex names. For example, the *Rosetta* mission orbited and landed on Comet 67P/Churyumov-Gerasimenko; the "67P" tells us that it is the 67th known periodic comet (meaning it returns again and again), while "Churyumov-Gerasimenko" comes from the names of its two discoverers. The surprise champion of comet discovery is the *SOHO* spacecraft, an orbiting solar observatory. As of 2018, *SOHO* had detected about 3500 "Sun-grazing" comets, most on their last pass by the Sun (**FIGURE 12.3**).

**See it for yourself** Bright comets can be quite photogenic. Search the Internet for comet images taken from Earth. Which is your favorite? Which has the best combination of beauty and scientifically interesting detail?

**Dwarf Planets** Just as the first asteroids were called planets, Pluto too was called a planet after its discovery in 1930. However, as scientists learned more about it, Pluto was recognized as a misfit among the planets because of its small size (a mass only about 0.2% that of Earth), ice-rich composition, and an orbit much more eccentric and more inclined to the ecliptic plane than that of any of the other planets.

Further questions about Pluto's planetary status arose as astronomers better understood the origin of comets. By the 1950s, astronomers realized that many of the comets that visit the inner solar system must be coming from the region of the Kuiper belt, and that Pluto orbits the Sun near the middle of this region. In other words, Pluto began to seem more and more like an unusually large comet. In the 1990s, astronomers began to discover other Pluto-like objects in

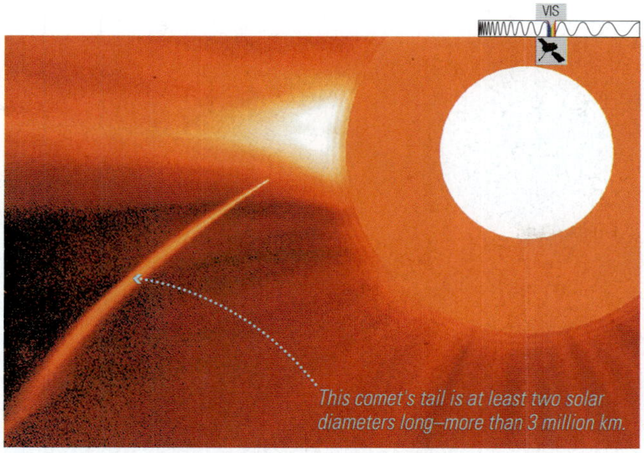


**FIGURE 12.3** Comet SOHO-6's final blaze of glory. The *SOHO* spacecraft observed this "Sun-grazing" comet a few hours before it passed just 50,000 kilometers above the Sun's surface. The comet did not survive its passage, because of the intense solar heating and tidal forces. In this image, the large, orange disk blocks out the Sun; the Sun's size is indicated by the white circle.

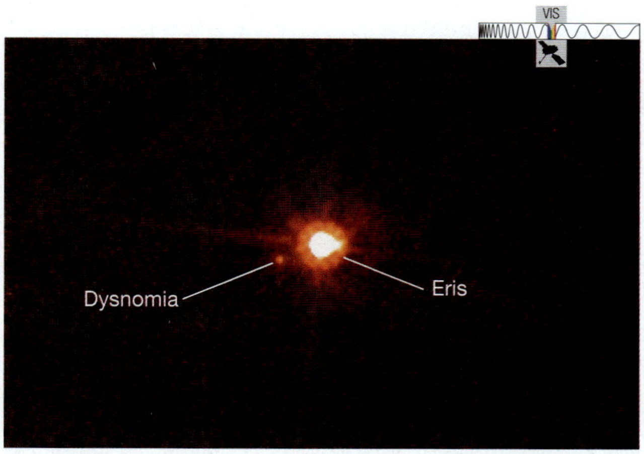


**FIGURE 12.4** Eris and its moon, photographed by the Hubble Space Telescope.

this region, with the only major difference being that these other objects were smaller than Pluto. Apparently, Pluto was not so "unusual" in this part of the solar system.

As time passed, larger and larger objects were found in the Kuiper belt, and in 2005, Caltech astronomer Michael Brown announced the discovery of Eris (**FIGURE 12.4**). Eris is about the same size as Pluto but about 27% larger in mass, and it is orbited by a moon named Dysnomia. Eris is named for a Greek goddess who caused strife and arguments among humans (a commentary on the arguments its discovery caused about the definition of *planet*); Dysnomia was a daughter of Eris and a mythological goddess of lawlessness.

Eris's discovery forced astronomers to confront the question of where to draw the line between planets and non-planets. After all, if Pluto was a planet, then surely the more massive Eris must be one as well. But, in that case, what about some of the other objects that are only slightly smaller than Pluto? Should they also be called planets? Or, given that all these objects have comet-like compositions, should we simply call them all comets?

In 2006, the International Astronomical Union created the *dwarf planet* category to accommodate Pluto, Eris, and other "small bodies" that are large enough to be round (see Special Topic, page 8). Under this definition, two other members of the Kuiper belt—Makemake and Haumea*—have also been designated dwarf planets, as has the asteroid Ceres. However, because the definition depends on roundness and we do not always know the precise shape of a distant object, dozens of other objects may yet join the list. There may also be other dwarf planets that we have not yet discovered in the outer reaches of the solar system, and perhaps even larger objects (such as the hypothesized "Planet 9" [see discussion p. 361]) that may force us to reconsider the definition of *planet* once again. **FIGURE 12.5** compares the sizes of known dwarf planets and several potential dwarf planets to that of Earth.

*Haumea is actually oblong, but it counts as a dwarf planet because it would be round if not for its high rotation rate.

**Think about it** Suppose that we someday discover an object in the Kuiper belt (or in the similar zone of another star system) that is Pluto-like in composition but as large in size as Earth. Under current definitions, the fact that it orbits in the same region as many similar objects would qualify it to be a dwarf planet. Does that seem reasonable for an object the size of Earth? How would *you* classify it? Defend your opinion.

**Fuzzy Boundaries** For most practical purposes, we now have straightforward definitions: Asteroids are rocky leftover planetesimals that orbit the Sun, comets are icy leftover planetesimals, and dwarf planets can be either asteroids or comets that are large enough to be round. However, even these seemingly simple boundaries can be fuzzy. We've already seen that there can be debate on whether an object is "round enough" to count as a dwarf planet. Similarly, asteroids and comets both have a range of compositions that in some cases overlap. For example, some asteroids, including Ceres, appear to contain enough

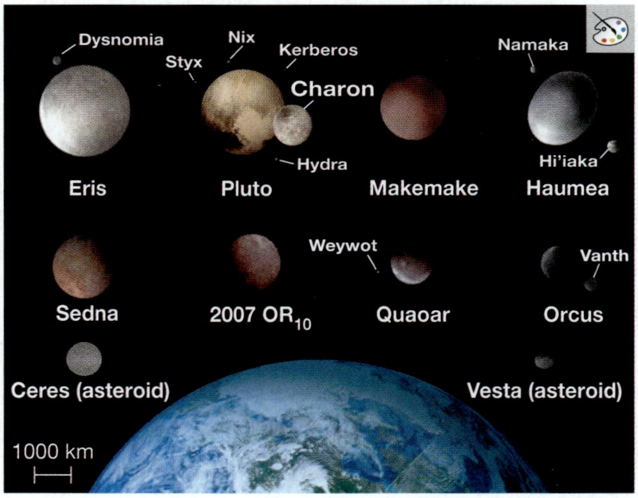


**FIGURE 12.5** The largest known objects of the Kuiper belt (as of 2018) and asteroid belt, along with their known moons, shown to scale with Earth for comparison. All except Pluto, Ceres, and Vesta are paintings based on guesses about their appearances. Pluto, Eris, Makemake, Haumea, and Ceres are officially considered dwarf planets, but some of the others may yet join the list.

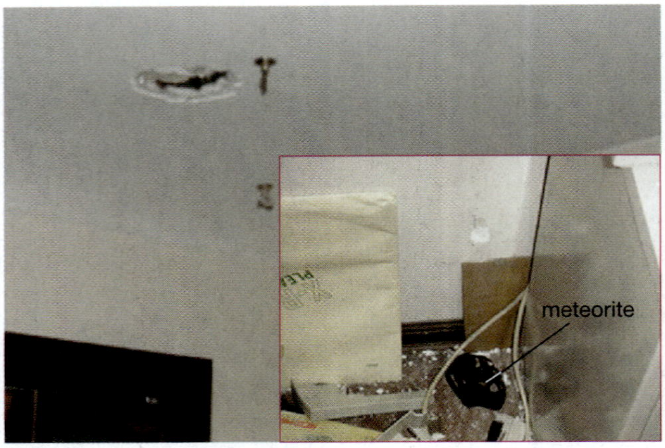


**FIGURE 12.6** On March 26, 2003, a meteorite crashed through the roof of this home in Chicago. The main photo shows the hole in the ceiling; the inset shows the meteorite at rest on the floor. No one was hurt.

**FIGURE 12.7** This large meteorite, called the Ahnighito Meteorite, is located at the American Museum of Natural History in New York. Its dark, pitted surface is a result of its fiery passage through Earth's atmosphere.

ice near their surfaces that they can lose some to vaporization in much the same way as comets.

The most difficult case probably concerns the Kuiper belt, where all the objects, from the smallest boulders to the largest dwarf planets, probably share the same basic composition of ice and rock. In other words, they are all essentially comets of different sizes. That is why we often refer to all of them as *comets* of the Kuiper belt. However, some astronomers object to calling objects "comets" if they never venture into the inner solar system and show tails; therefore, you may also hear these objects referred to as *Kuiper belt objects* (*KBOs*) or *trans-Neptunian objects* (*TNOs*).

## What are meteors and meteorites?

In everyday language, people often use the terms *meteors* and *meteorites* interchangeably. Technically, however, a **meteor** (which means "a thing in the air"; note the similarity to *meteorology*, which is the study of weather) is only a flash of light caused by a particle of dust or rock entering our atmosphere at high speed, not the particle itself. Meteors are sometimes called *shooting stars* or *falling stars*, because some people once thought they really were stars falling from the sky.

The vast majority of the particles that make meteors are no larger than peas and burn up completely before reaching the ground. Only in rare cases is a meteor caused by a chunk of rock large enough to survive the plunge through our atmosphere and leave a **meteorite** (which means "associated with meteors") on the ground. Those cases make unusually bright meteors, called *fireballs*. Observers find a few meteorites each year by following the trajectories of fireballs. Meteorites found in this way can be especially valuable to science, because we can trace their trajectories back to learn exactly where they came from, and because they can be collected before they have much time to be contaminated by terrestrial material.

**Meteorite Falls** People didn't always accept that rocks fall from the sky. Stories of such events arose occasionally in human history, and sometimes even affected it. For example, stories of "fallen stars" influenced the philosophy of the ancient Greek scientist Anaxagoras (see Figure 3.12), who concluded that planets and stars were flaming rocks in the sky. Anaxagoras's assumption that meteorites fell from the heavens made him the first person in history known to believe that the heavens and Earth are made of the same materials, even though his guess about the nature of planets and stars was not quite correct. Many later scientists regarded stories of fallen stars more skeptically. Upon hearing of a meteorite fall in Connecticut, Thomas Jefferson (who was a student of science as well as politics) reportedly said, "It is easier to believe that Yankee professors would lie than that stones would fall from heaven."

Today we know that rocks really do fall from the heavens. More than 1000 meteorite falls have been directly observed, and tens of thousands of meteorites have been found and cataloged. Meteorites are often blasted apart in their fiery descent through our atmosphere, scattering fragments over an area several kilometers across. A direct hit on the head by a meteorite would be fatal, but there are no reliable accounts of human deaths from meteorites, though there have been close calls (**FIGURE 12.6**). Of course, most meteorites fall into the ocean, which covers three-fourths of Earth's surface.

Unless you actually see a meteorite fall, it can be difficult to distinguish a meteorite from an Earth rock. Fortunately, a few clues can help. Meteorites are usually covered with a dark, pitted crust resulting from their fiery passage through the atmosphere (**FIGURE 12.7**). Some have an unusually high metal content, enough to attract a magnet hanging on a string. The ultimate judge of extraterrestrial origin is laboratory analysis: Meteorites often contain elements such as iridium that are very rare in Earth rocks, and even common elements in meteorites tend to have different ratios among their isotopes [**Section 5.3**] than are found in rocks from Earth. If you suspect that you have found a meteorite, many museums will analyze a small chip free of charge.

**Meteorite Origins** The origin of meteorites was long a mystery, but in recent decades we've been able to determine where in our solar system they come from. The most direct evidence comes from the relatively few meteorites whose trajectories have been observed or filmed as they fell to the ground. In every case so far, these meteorites

**FIGURE 12.8** Images of all the asteroids visited by spacecraft as of 2018 shown to scale. Some objects were the destinations of dedicated missions; others were imaged on the way to different targets. The size of Earth's Moon is included for comparison.

clearly originated in the asteroid belt. In other words, the vast majority of meteorites are essentially either very small asteroids or pieces of asteroids.

However, in a few cases, scientists have identified meteorites with compositions that appear to match that of either the Moon or Mars, and careful analysis makes us very confident that these meteorites were indeed chipped off these worlds. This makes sense: Moderately large impacts can blast surface material from terrestrial worlds into interplanetary space, where the rocks can orbit the Sun until they come crashing down on another world. Calculations show that it is not surprising that we should have found a few meteorites from the Moon and Mars in this way. These *lunar meteorites* and *Martian meteorites* therefore represent direct samples from these worlds, which makes them very valuable for scientific study.

## 12.2 Asteroids

We are now ready to turn our attention to studying asteroids and comets in more depth. Scientifically, these small bodies are very important, because many of them remain much as they were when they first formed, some $4\frac{1}{2}$ billion years ago. They therefore have clues to the story of our solar system's birth encoded in their compositions, locations, and numbers. Of course, understanding the clues requires careful study; using small bodies to learn about planetary formation is a bit like picking through the trash in a carpenter's shop to learn how furniture is made. In this section, we'll explore current scientific understanding of asteroids.

### What are asteroids like?

**FIGURE 12.8** shows a montage of the asteroids that had been visited by spacecraft as of 2018. Notice that asteroids come in a wide variety of sizes and shapes, and that their many craters show that they have been battered by impacts over time. Hundreds of thousands of other asteroids have been studied telescopically, allowing us to make clear statements about their general characteristics.

**Asteroid Sizes and Shapes** Ceres, the largest asteroid, is just under 1000 kilometers in diameter, which is a little over a quarter of the Moon's diameter. About a dozen others are large enough that we would call them medium-size moons if they orbited a planet. Smaller asteroids are far more numerous. There are probably more than a million asteroids with diameters greater than 1 kilometer, and many more even smaller in size.

As you can see in Figure 12.8, most asteroids are not spherical. An asteroid's shape depends largely on the strength of its gravity. Only large asteroids have gravity strong enough to have molded them into somewhat spherical shapes, and only Ceres is round enough to be currently counted as a dwarf planet; however, the next two largest asteroids (Pallas and Vesta) are not too far from spherical. The gravity of smaller asteroids is too weak to have reshaped their rocky material, leaving them looking much like potatoes. In some cases, objects that appear to be single asteroids are probably two or more distinct objects held in contact by a weak gravitational attraction, while other small asteroids are little more than weakly bound piles of rubble.

You may wonder how we know so much about asteroid sizes, given that spacecraft have visited only a few and the rest appear as little more than points of light even to powerful telescopes. The answer is that size can be estimated through careful measurements of an asteroid's brightness, which depends on its size, distance, and reflectivity. For example, if two asteroids at the same distance have the same reflectivity, the one that appears brighter must be larger in size. We can determine an asteroid's distance from its position in its orbit. Reflectivity can be measured by comparing the asteroid's visible brightness, which comes from the sunlight it reflects, to its infrared brightness, which depends on the asteroid's temperature and hence tells us how much sunlight it absorbs. Astronomers can then use the reflectivity and distance to calculate the asteroid's size.

**Think about it** Suppose you discover two asteroids of similar size that are equally bright in visible light, but infrared observations tell you that Asteroid 1 is more reflective than Asteroid 2. Which one is farther away? Explain.

Determining asteroid shapes without close-up photographs is more difficult, but we can get some information by monitoring brightness variations as an asteroid rotates: A nonspherical asteroid with a uniformly bright surface will reflect more light when it presents its larger side toward the Sun and our telescopes. In some cases, astronomers have determined shapes by bouncing radar signals off asteroids that have passed close to Earth.

**Asteroid Masses, Densities, and Compositions** The most direct way to measure a distant object's mass is to observe its gravitational effect on another object, and to date this is possible only for the relatively few asteroids visited by spacecraft and for those that have smaller asteroids as tiny orbiting "moons." For example, the asteroid Ida is orbited by Dactyl (both near the upper left in Figure 12.8), which allows us to use Newton's version of Kepler's third law to calculate Ida's mass from Dactyl's orbital period and distance [**Section 4.4**].

Astronomers know precise masses for only a few dozen asteroids, but these cases are important because they allow us to estimate masses of other asteroids and to calculate average density for those with known masses (by dividing by the volumes known from size measurements). Density can offer valuable insights into an asteroid's origin and makeup. For example, the asteroids Eros and Mathilde look similar (aside from their size difference) in Figure 12.8, but their densities reveal significant differences. Eros has a density of 2.4 $g/cm^3$, which is close to the value expected for solid rock of Eros's size. In contrast, Mathilde's density of 1.5 $g/cm^3$ is so low that Mathilde must be a loosely bound "rubble pile," held together by its weak gravity, rather than a solid chunk of rock.

Densities give insight into composition, but we can also learn about composition from spectra; recall that spectra of distant objects contain spectral lines that are essentially "fingerprints" left by the objects' chemical constituents [**Section 5.4**]. Thousands of asteroid spectra have been studied, and the results are consistent with what we expect from our theory of solar system formation: Asteroids are made mostly of metal and rock, because they condensed within the frost line in the solar nebula. Those near the outskirts of the asteroid belt contain larger proportions of dark, carbon-rich material, because this material was able to condense at the relatively cool temperatures found in this region of the solar nebula but not in the regions closer to the Sun; some even contain small amounts of water, telling us that they formed close to the frost line. A few asteroids appear to be made mostly of metals, such as iron, suggesting that they may be fragments of the metal cores of shattered worlds.

**Asteroids Up Close** Our most in-depth studies of asteroids have taken place through spacecraft visits; notice that the labels in Figure 12.8 identify the spacecraft that visited each asteroid. For large asteroids, we've learned much of what we now know from the *Dawn* mission, which spent about 14 months orbiting Vesta in 2011–12, then moved on to Ceres, where it entered orbit in 2015. *Dawn* is expected to remain in Ceres orbit indefinitely, giving the largest asteroid a very tiny moon.

*Dawn* images of Vesta (**FIGURE 12.9**) revealed a battered world that is nearly 25% wider (equatorial) than it is tall (polar). This irregular shape is probably a result of an impact that gouged out a huge impact crater at Vesta's south pole, which means that Vesta was probably once much more spherical in shape. This fact raises the question of whether Vesta should qualify as a dwarf planet, but as of 2018 no official decision had been made.

Vesta's south polar crater is remarkable in many ways. Like many large craters, it has a central mountain formed from the rebound after the impact that made the crater. But rising 23 kilometers from the crater bottom, this mountain is one of the tallest in the solar system. Moreover, the impact excavated so deeply into Vesta's interior that it should have blasted out substantial amounts of rock from the crust and mantle, and measurements from *Dawn* (and from the Hubble Space Telescope) show that rock near the crater bottom has a composition matching that expected deep inside terrestrial worlds. This fact has helped confirm suspicions that Vesta is massive enough to have undergone differentiation, giving it a metallic core, a low-density rocky crust, and a mantle in between. The spectral signature of the crater-bottom rock also matches that of many small asteroids and of many meteorites that have been found on Earth (including the meteorite shown on the right in Figure 12.12b), suggesting that these asteroids and

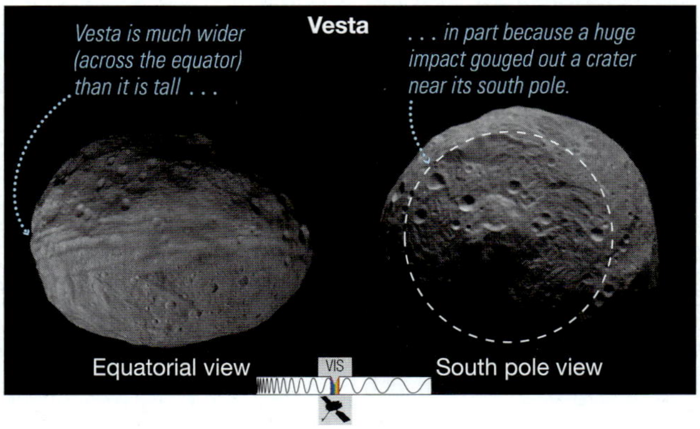


**FIGURE 12.9** Global images of Vesta taken by the *Dawn* spacecraft. The dashed circle is the outline of the huge polar crater.

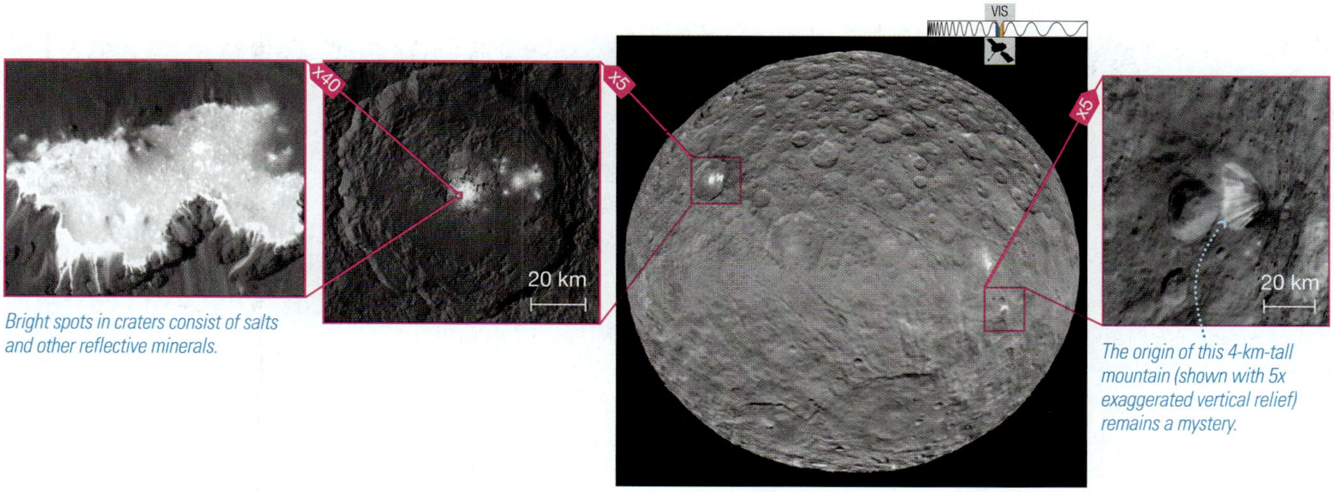


**FIGURE 12.10** The dwarf planet Ceres, imaged by the *Dawn* spacecraft. The global view shows abundant craters and tectonic cracks and faults.

meteorites are pieces of Vesta that were blasted away by the impact that formed the south polar crater.

Vesta still presents numerous mysteries. For example, some regions of the surface show the spectral signature of volcanic rock, and this signature is also present in the asteroids and meteorites that appear to have come from Vesta. However, *Dawn* scientists have not yet identified any ancient volcanoes or lava flows. Perhaps they've been hidden by billions of years of cratering. In addition, Vesta's equator is ringed with ridges that currently defy explanation.

*Dawn*'s studies of Ceres reveal an even more interesting world (**FIGURE 12.10**). As the largest asteroid, Ceres contains nearly as much mass as the rest of the asteroid belt put together. Its low density and location in the outer part of the asteroid belt suggest that water ice makes up a significant fraction of its composition. *Dawn*'s observations confirmed this suggestion but also revealed some surprises. For example, *Dawn* images show bright spots on some craters (left inset in Figure 12.10) that appear to be made of salts and other highly reflective mineral deposits. Scientists suspect that groundwater brought these minerals to the surface, and they were left behind to reflect light after the water vaporized into space. The crater containing the brightest spots sometimes appears to fill with haze or fog. Another surprise is the presence of a single tall mountain (right inset in Figure 12.10); no one yet knows how this mountain formed. The rest of the surface looks more as expected for a dwarf planet: abundant craters and some tectonic fracturing. Spectroscopy reveals the chemical composition of the surface also to be much as expected, with one exception: Ceres has a larger proportion of ammonia-bearing compounds than we would expect under the assumption that it formed inside the frost line.

For future research, scientists are particularly interested in spacecraft that can return samples from asteroid surfaces. This was first achieved by Japan's *Hayabusa* mission, which in 2010 returned a small dust sample from the asteroid Itokawa (**FIGURE 12.11**). The follow-up *Hayabusa-2* reached an asteroid called 162173 Ryugu in 2018 and is expected to return a surface sample to Earth in 2020. NASA's ambitious *OSIRIS-REx* mission, launched in 2016, aims to return a sample in 2023 from the asteroid 101955 Bennu, which is thought to be a carbon-rich asteroid that may be representative of the types of objects that brought to Earth many of the ingredients for oceans, atmosphere, and life.

**Think about it** Find the current status of the *Hayabusa-2* and *OSIRIS-REx* missions. What have we learned from them so far? Explain.

## What do meteorites tell us about asteroids and the early solar system?

We can learn even more about asteroids by studying samples of them, and while spacecraft are only beginning to collect samples directly, we already have tens of thousands of other asteroid samples—the rocks called *meteorites* that fall from the sky.

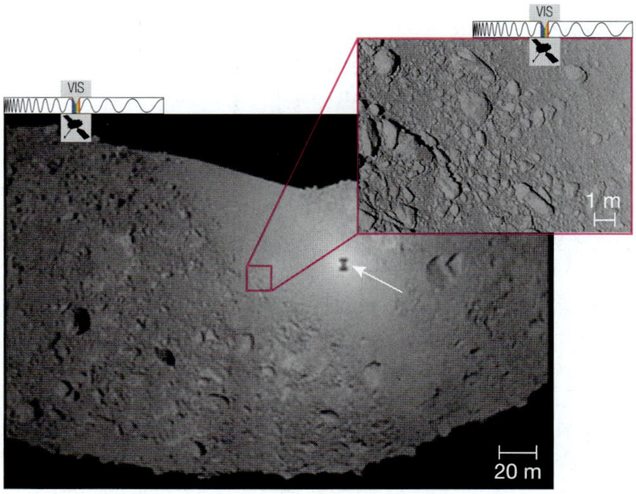


**FIGURE 12.11** The Japanese spacecraft *Hayabusa* (which appropriately means "hawk") scooped up a small sample of the 500-meter-long asteroid Itokawa and returned it to Earth. The main image shows the central portion of the asteroid; the dark spot (arrow) is the shadow of the spacecraft, dominated by its two large solar panels. The inset shows the landing site as the spacecraft descended.

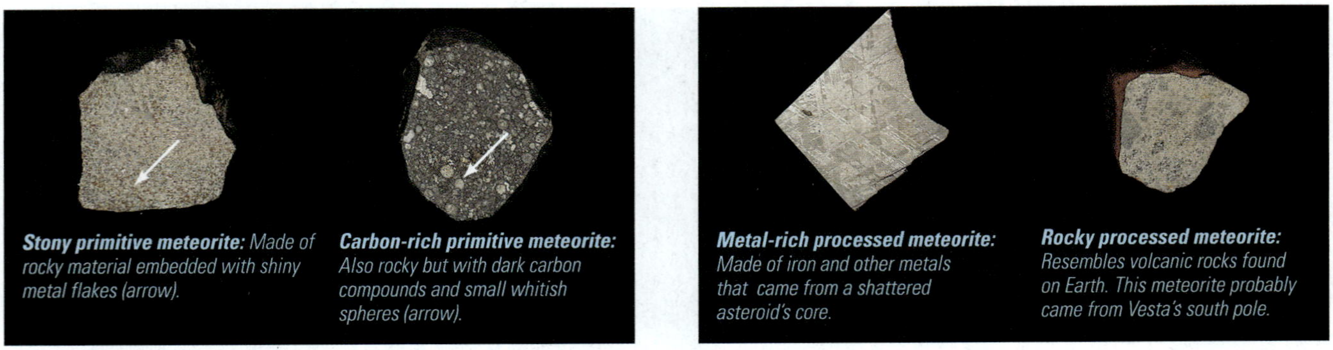


**a** Primitive meteorites.

**b** Processed meteorites.

**FIGURE 12.12** There are two basic types of meteorites: primitive and processed. Each also has two subtypes. They are shown slightly smaller than actual size. (The meteorites have flat faces because they have been sliced with rock saws.)

**Types of Meteorites** Detailed analysis of thousands of meteorites shows that they come in two basic types, primitive and processed,* each of which can be further classified into two subtypes:

- **Primitive meteorites** (**FIGURE 12.12a**) are "primitive" in the sense of being remnants from the birth of our solar system, essentially unchanged since they first accreted in the solar nebula. Radiometric dating [**Section 8.3**] shows them to be the oldest rocks in the solar system. Primitive meteorites come in two major subtypes:
  - *Stony primitive meteorites* are composed of rocky minerals with a small but noticeable fraction of pure metallic flakes mixed in.
  - *Carbon-rich primitive meteorites* are also rocky but contain substantial amounts of carbon compounds and, sometimes, a small amount of water bound to the rock.
- **Processed meteorites** (**FIGURE 12.12b**) apparently once were part of a larger object that "processed" the original material of the solar nebula into another form. Radiometric dating confirms that processed meteorites are slightly younger than primitive meteorites, just as we would expect. The processed meteorites can also be divided into two major subtypes:
  - *Metal-rich processed meteorites* are made mostly of high-density iron and nickel mixed with smaller amounts of other metals. That is, they resemble the terrestrial planet cores in composition.
  - *Rocky processed meteorites* have lower densities and are made of rock with compositions resembling that of terrestrial mantles and crusts. A few have compositions remarkably close to that of the basalts [**Section 9.2**] that erupt from terrestrial volcanoes.

Both types of meteorites teach us important lessons about asteroids and about the birth of our solar system.

*Primitive meteorites are also called *chondrites* because of the roundish features visible in them, known as *chondrules*. Processed meteorites lack these features and hence are called *achondrites* (*a-* means "not" or "without").

**Lessons from Primitive Meteorites** Because primitive meteorites represent samples of early accretion, they hold key clues to when and how the solar system formed. Precise radiometric dating shows primitive meteorites to be 4.56 billion years old, so this must be the time at which accretion in the solar nebula began. The metal flakes in primitive meteorites like the one on the left in Figure 12.12a may represent the tiny particles that first condensed from the gas of the solar nebula. The small roundish features visible in the meteorite on the right in Figure 12.12a may be solidified droplets splashed out from nearby small planetesimals as they accreted. The vast majority of meteorites are primitive, which tells us that even 4½ billion years after the solar system's formation, vast numbers of small leftovers from the era of planet formation still orbit the Sun.

Why do we see differences among primitive meteorites, like those that distinguish the two meteorites in Figure 12.12a? More specifically, why are some primitive meteorites stony in composition while others are carbon-rich, and why do some contain water? Both theory and observation indicate that the answers depend on where the meteorites formed in the solar nebula.

With regard to stony versus carbon-rich primitive meteorites, the nebular theory tells us that most meteorites accreted inside the frost line and are therefore made of metal and rock. However, models tell us that, beyond about 2.5 AU from the Sun, temperatures in the solar nebula were low enough for condensation of carbon compounds. These models suggest that carbon-rich primitive meteorites come from beyond that distance, while stony primitive meteorites come from closer in. Laboratory studies confirm these ideas. The composition of carbon-rich meteorites matches the composition of dark, carbon-rich asteroids found in the outer part of the asteroid belt beyond about 2.5 AU from the Sun. The composition of stony meteorites matches the composition of asteroids in the inner part of the asteroid belt.

The meteorites that contain water take this idea one step further. Both meteorite evidence and computer models suggest the frost line may have been as close in as 2.7 AU, or just a bit beyond the distance at which carbon compounds condensed. Asteroids that formed beyond the frost line contain a small but significant fraction of water (Ceres is an example)—enough to potentially make them the objects that brought the terrestrial planets the water, carbon-rich compounds, and other ingredients necessary for habitability and for life.

**Lessons from Processed Meteorites** The processed meteorites tell a more complex story. Their compositions look similar to those of the cores, mantles, or crusts of the terrestrial worlds. We conclude that these meteorites are fragments of larger asteroids that underwent *differentiation* [**Section 9.1**], in which their interiors melted so that metals sank to the center and rocks rose to the surface. This idea explains why we find two types of processed meteorites.

The metal-rich meteorites must come from large asteroids that underwent differentiation but were subsequently shattered in collisions. In this sense, they represent pieces of a "dissected planet," making them valuable both because we lack the technology to drill for core samples on Earth and because they provide direct proof that large worlds really do undergo differentiation, confirming what we infer from seismic studies of Earth. More generally, the existence of these meteorites attests to the fact that world-shattering collisions must have happened frequently in the early history of the solar system.

**Think about it** Recall that *giant impacts* have been invoked to explain several "exceptions to the rules" in the solar system, including the existence of our relatively large Moon. How does the existence of metal-rich meteorites support the idea that major impacts have occurred? Explain.

Rocky processed meteorites must come from the mantles or crusts of differentiated worlds. Moreover, many of them—including the ones that appear to come from Vesta (see Figure 12.12b)—are so close in composition to volcanic rocks on Earth that they must have been made by lava flows. We conclude that these meteorites are rocks from the surfaces of large asteroids that once were volcanically active, probably chipped off by collisions with smaller asteroids. In addition to Vesta, perhaps as many as a few dozen other large asteroids were geologically active shortly after the formation of the solar system. The fact that the processed meteorites date to only millions of years after the birth of the solar system tells us that this active period was short-lived; the interiors of the geologically active asteroids must have cooled quickly.

**Implications of Asteroid Geology** The lessons we've learned from meteorites, along with our spacecraft visits to asteroids, provide important confirmation of the ideas of geology we discussed for the terrestrial planets and jovian moons. As we expect, small asteroids appear to be geologically "dead," with impact craters as their only features. Many of these asteroids are probably no different in essence from primitive meteorites, except that they are still in orbit rather than having crashed down to Earth.

Processed meteorites and larger asteroids confirm that worlds larger than a few hundred kilometers in diameter contained enough heat for their interiors to melt and undergo differentiation. The rocky processed meteorites demonstrate that at least some of these worlds, including Vesta, were volcanically active in their early histories.

All in all, these studies indicate that dwarf planets like Ceres—as well as "almost dwarf planets" like Vesta—are really planet-like, or perhaps "jovian moon-like," in almost every way. We humans may care about distinguishing planets from dwarf planets from moons, but nature doesn't. All that matters is size and composition, and geology follows accordingly.

## Why is there an asteroid belt?

As noted in Chapter 7, the vast majority of the asteroids orbit the Sun in the *asteroid belt* between the orbits of Mars and Jupiter (**FIGURE 12.13**). These asteroids orbit the Sun in the same direction as the planets, though their orbits tend to be more elliptical and more highly inclined to the ecliptic plane (up to 20°–30°) than those of planets. Note that while science fiction movies often show the asteroid belt as a crowded and hazardous place, it is so large that typical (kilometer-sized) asteroids are millions of kilometers apart on average, and collisions between asteroids are quite rare. Aside from the main asteroid belt, the only other major asteroid groupings are the two sets of *Trojan asteroids*, which share Jupiter's 12-year orbit around the Sun (with one group always staying 60° ahead of Jupiter in its orbit and the other always staying 60° behind); we know relatively little about the Trojan asteroids, but should learn much more with NASA's *Lucy* spacecraft, scheduled to launch in 2021 and then visit several Trojan asteroids during a 12-year mission. In addition, a relatively small number of asteroids have orbits that take them through the inner solar system, where they are probably "impacts waiting to happen"; we'll discuss the dangers posed by these objects in Section 12.5.

Why are asteroids concentrated in the asteroid belt, and why didn't a full-fledged planet form instead? The answer lies with gravitational effects of Jupiter. Virtually all planetesimals that formed inside the orbit of Mars eventually accreted onto one of the inner planets. But planetesimals

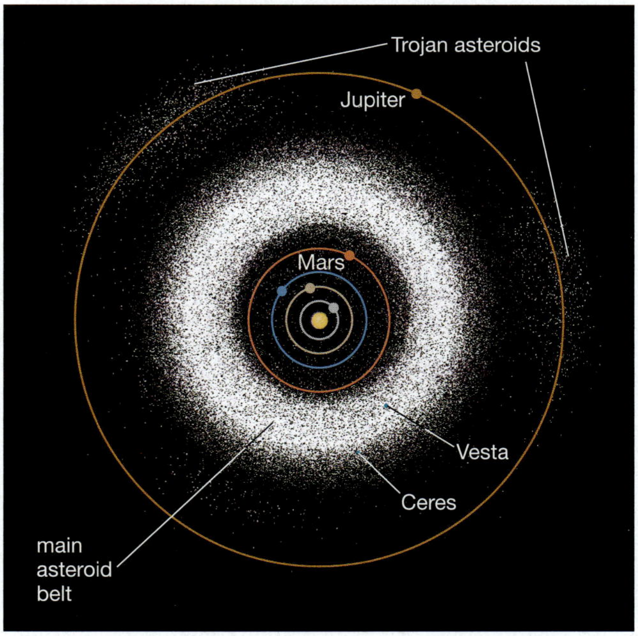


**FIGURE 12.13** This figure shows the positions of more than 150,000 asteroids on a single night, and the locations of Vesta and Ceres when the *Dawn* mission reached each of them. To scale, the asteroids themselves would be much smaller than shown. The asteroids that share Jupiter's orbit, found 60° ahead of and behind Jupiter, are called *Trojan asteroids*.

## COMMON MISCONCEPTIONS

### Dodge Those Asteroids!

Science fiction movies often show brave spacecraft pilots navigating through crowded fields of asteroids, dodging this way and that as they heroically pass through with only a few bumps and bruises. It's great drama, but not very realistic. The asteroid belt looks crowded in Figure 12.13, but in reality it is an enormous region of space. Despite their large numbers, asteroids are so far apart on average that it would take incredibly bad luck to crash into one by accident. Indeed, spacecraft must be carefully guided to fly close enough to an asteroid to take a decent photograph. Future space travelers will have plenty of dangers to worry about, but dodging asteroids is not likely to be one of them.

that formed between Mars and Jupiter were strongly influenced by *orbital resonances* with Jupiter, and only a small fraction of these planetesimals ended up with orbits that have allowed them to remain in the asteroid belt to this day.

Recall that an orbital resonance occurs whenever two objects periodically line up with each other [**Section 11.2**]. In the asteroid belt, an orbital resonance occurs whenever an asteroid has an orbital period that is a simple fraction of Jupiter's orbital period, such as $\frac{1}{2}$, $\frac{1}{4}$, or $\frac{2}{5}$. In those cases, the asteroid experiences repeated tugs from Jupiter that tend to nudge it out of that orbit. For example, an asteroid with an orbital period of 6 years—half of Jupiter's 12-year period—would receive the same gravitational nudge from Jupiter every 12 years and therefore would soon be pushed out of this orbit. The same is true for asteroids with orbital periods of 4 years ($\frac{1}{3}$ of Jupiter's period) and 3 years ($\frac{1}{4}$ of Jupiter's period). Music offers an elegant analogy: When a vocalist sings into an open piano, the strings that produce notes "in resonance" with the voice—not just the note being sung but also notes with a half or a quarter of the note's frequency—are "nudged" and begin to vibrate.

We can see the effect of orbital resonances on a graph showing the numbers of asteroids with various orbital periods (**FIGURE 12.14**). Notice the gaps that indicate a lack of asteroids with periods in resonance with Jupiter, confirming that these orbits have been cleared by the resonances. (The gaps are often called *Kirkwood gaps,* after their discoverer.) Also note that while most orbital resonances result in gaps, some actually gather asteroids. For example, the $\frac{1}{1}$ resonance (where orbital periods are the same as Jupiter's) in Figure 12.14 represents the Trojan asteroids that share Jupiter's orbit.

**Think about it** Why are the gaps due to orbital resonances so easy to see in Figure 12.14 but so hard to see in Figure 12.13? (*Hint:* The top axis in Figure 12.14 is the *average* orbital distance.)

Orbital resonances probably also explain why no planet formed between Mars and Jupiter. Early in the solar system's history, this region probably contained more than enough rocky material to form another terrestrial planet. However, resonances with the young Jupiter disrupted the orbits of planetesimals in this region, sometimes sending them crashing into each other and sometimes kicking them out of the region. Once kicked out, the planetesimals ultimately either crashed into a planet or moon or

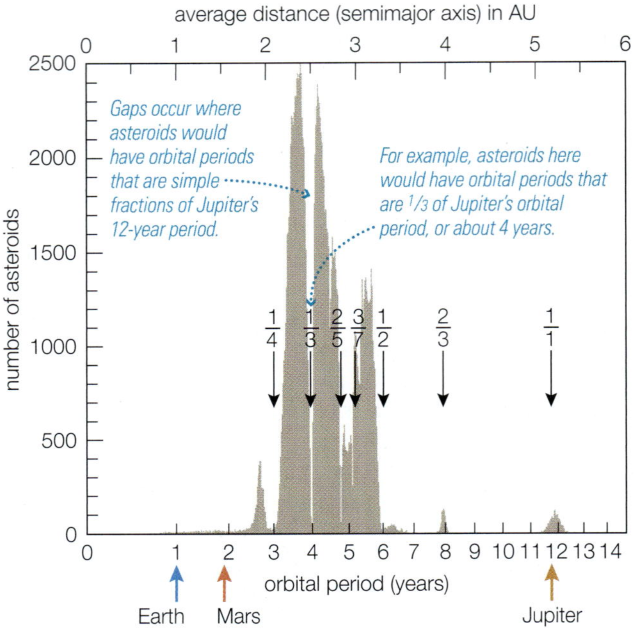


**FIGURE 12.14** This graph shows the numbers of asteroids with various orbital periods, which correspond to different average distances from the Sun (labeled along the top). Notice the gaps created by orbital resonances with Jupiter.

were flung out of the solar system or into the Sun. Over the next $4\frac{1}{2}$ billion years, these ongoing orbital disruptions caused the asteroid belt to lose most of its original mass. In fact, despite their large numbers, the remaining asteroids don't add up to much in total mass. If we could put all the asteroids together and allow gravity to compress them into a sphere, they'd make an object less than 2000 kilometers in diameter—just over half the diameter of our Moon.

The asteroid belt is still undergoing slow change. Jupiter's gravity continues to nudge asteroid orbits, sending asteroids on collision courses with each other and occasionally the planets. A major collision occurs somewhere in the asteroid belt every 100,000 years or so. Over long periods of time, larger asteroids continue to be broken into smaller ones, with each collision also creating numerous dust-size particles. The asteroid belt has been grinding itself down for more than 4 billion years and will continue to do so for as long as the solar system exists.

## 12.3 Comets

We now turn our attention to comets. Although we now know that Pluto and other large objects of the Kuiper belt are essentially large comets, for most of human history the only known comets were the relatively small ones that sometimes enter the inner solar system and grow long tails. Moreover, these small comets are the only ones we've been able to study in great depth. In this section, we'll explore modern understanding of small comets and how we learned that they come from the Kuiper belt and the Oort cloud.

### Why do comets grow tails?

Far from the Sun, small comets must look much like small asteroids, with assorted shapes made possible by the fact that gravity is too weak to compress them into spheres. But

## SPECIAL TOPIC A Visitor from the Stars

Prior to 2017, every object ever observed in our solar system had an orbit consistent with its having always been part of our solar system. For example, all asteroid orbits were consistent with what we expect for leftover planetesimals, and all comets seemed likely to have originated in the Kuiper belt or Oort cloud. But in the fall of 2017, astronomers using the Pan-STARRS survey telescopes in Hawaii discovered an object moving at such a high speed that it clearly had escape velocity from our solar system, suggesting that it was simply "passing through." By tracing its orbit backward, astronomers confirmed that this object—given the name *'Oumuamua* (Hawaiian for "scout")—had entered the solar system at high speed and had not passed near Jupiter or any other object that might have affected its orbit. In other words, 'Oumuamua was the first confirmed object that must have originated around another star.

'Oumuamua is small, perhaps 250 meters in length. It passed within Mercury's orbit without releasing detectable amounts of gas or dust, which means it resembles an asteroid more than a comet. Its brightness varied significantly as it rotated, suggesting either that it is very elongated (perhaps 5–10 times as long as it is wide) or that its surface reflectivity has extreme contrasts. Unfortunately, we'll never learn more about it, because it quickly raced away and became too faint for follow-up observations.

While the discovery of an interstellar visitor like 'Oumuamua was exciting, it was not completely unexpected. Models indicate vast numbers of asteroids and comets have probably been ejected from our own solar system over time, so the same should be true for other solar systems. We are therefore likely to find other visitors, particularly as new survey telescopes begin operations. Perhaps soon, astronomers will discover an interstellar visitor far enough in advance for us to send a spacecraft to explore it up close.

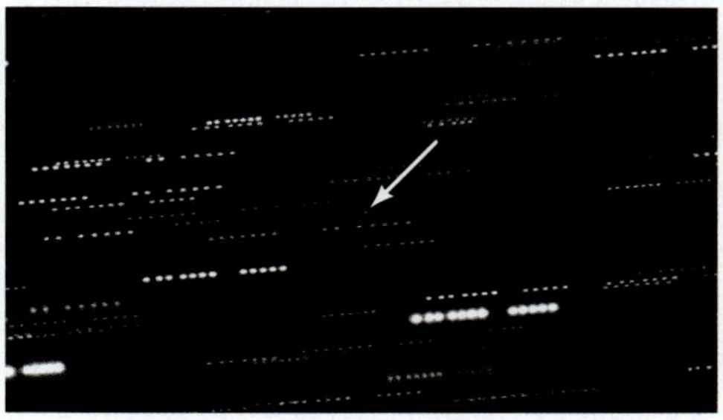

This composite of several individual images shows 'Oumuamua at the center (arrow); the other short streaks of dots are stars that moved during the time the images were taken. The extended trails of other stars relative to 'Oumuamua indicate its high speed moving through our solar system.

appearances change radically for those rare comets that enter the inner solar system. As we'll see, it's the composition of comets, combined with the heat and light of the Sun, that leads to their flashy appearances.

**Comet Composition** We have learned about the composition of comets by studying their spectra. The results are just what we expect for objects that formed in the cold outer solar system: Comets are basically chunks of ice mixed with rocky dust and some more complex chemicals, and hence they are often described as "dirty snowballs."

Spectra confirm the distant origin of comets because they show the presence of compounds that could have condensed only in the cold outer regions of the solar nebula. For example, comet spectra invariably show emission features from hydrogen compounds, including water. Spectra also reveal the presence of carbon dioxide and carbon monoxide, gases that condensed only in the very coldest and most distant regions of the solar nebula. Comets also contain numerous more complex molecules, including some carbon compounds, leading scientists to speculate that comets helped seed Earth with the organic material that made life possible.

**The Flashy Lives of Comets** The "dirty snowball" idea explains how comets grow tails when they are heated by the Sun. To see what happens, let's follow the comet path shown in **FIGURE 12.15**.

Far from the Sun, the comet is completely frozen—in essence, the "dirty snowball" in solid form. For a comet plunging inward, we call this frozen center the **nucleus** of the comet. As the comet accelerates toward the Sun, its surface temperature increases, and ices begin to vaporize into gas that easily escapes the comet's weak gravity. Some of the escaping gas drags dust particles away from the nucleus, and the gas and dust create a huge, dusty atmosphere called a **coma**. The coma grows as the comet continues into the inner solar system, and some of the gas and dust is pushed away from the Sun, forming the comet's tails. Most of the comets that we've seen in the inner solar system have a nucleus no more than about 20 kilometers across. However, the escaping dust and gas can give the coma a radius of tens to hundreds of thousands of kilometers (as big as or bigger than the planet Jupiter), and comet tails can be hundreds of millions of kilometers in length.

Many people mistakenly guess that tails extend behind comets as they travel through their orbits, perhaps because the tails look much like exhaust trails behind rockets. In fact, comet tails generally point away from the Sun, regardless of the direction in which the comet is traveling. Moreover, comets actually have two distinct tails. The **plasma tail** consists of gas that is ionized by ultraviolet light from the Sun and pushed outward by the solar wind; the plasma tail therefore extends almost directly away from the Sun at all times. The **dust tail** consists of dust-size particles that are unaffected by the solar wind and instead are pushed outward by the much weaker pressure of sunlight itself (*radiation pressure*). The dust tail therefore also points generally away from the Sun, but has a slight curve back in the direction the comet came from.

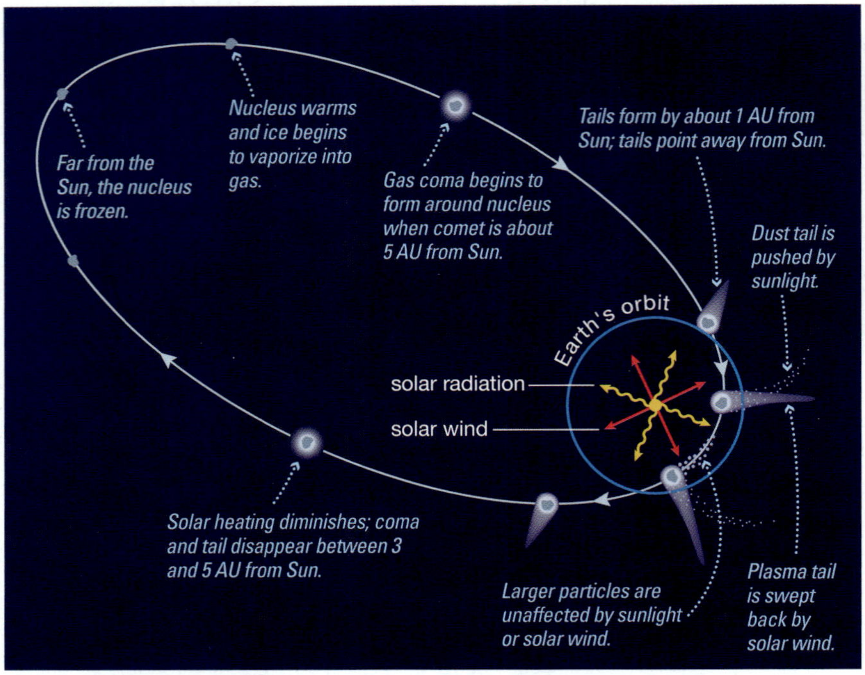


**a** This diagram (not to scale) shows the changes that occur when a comet's orbit takes it on a passage into the inner solar system.

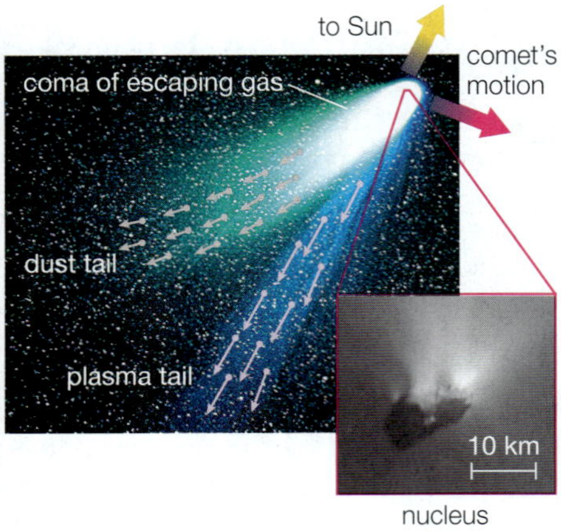


**b** Anatomy of a comet. The larger image is a ground-based photo of Comet Hale-Bopp. The inset shows the nucleus of Halley's Comet photographed by the *Giotto* spacecraft.

**FIGURE 12.15** A comet grows a coma and tail around its nucleus only if it happens to come close to the Sun. Most comets never do this, instead remaining perpetually frozen in the far outer solar system.

After the comet loops around the Sun and begins to head back outward, vaporization declines, the coma dissipates, and the tails disappear. Nothing happens until the comet again comes sunward—in a century, a millennium, a million years, or perhaps never.

Comets that repeatedly visit the inner solar system, like Halley's Comet, cannot last long on the time scale of our solar system. A comet probably loses about 0.1% of its ice on every pass around the Sun, so it could not make more than a few hundred passages before losing most of its original ice. Changes in composition may end the comet's "life" even faster. In a close pass by the Sun, a comet may shed a layer of material a meter thick. Dust that is too heavy to escape accumulates on the surface. This thick, dusty deposit darkens the comet and may eventually block the escape of interior gas, preventing the comet from growing a coma or tails on future passes by the Sun. Several possible fates await a comet after its ices can no longer vaporize into gas and escape. In some cases, the dust layer may disguise the dead comet as an asteroid. In other cases, the comet may come "unglued" and break apart, or even disintegrate along its orbit.

**Spacecraft Missions to Comets** Spacecraft missions have taught us more about comets. Our first clear view of a comet nucleus came from the European Space Agency's *Giotto* spacecraft, which flew past Halley's Comet during its 1986 visit to the inner solar system (see Figure 12.15b). That mission showed the nucleus to be surprisingly dark, reflecting less than 5% of the light that falls on it. Apparently, it doesn't take much rocky or carbon-rich material to darken a comet. Density measurements showed that Halley's nucleus is considerably less dense than water (1 g/cm$^3$), suggesting that the nucleus is part ice and part empty space. The observations also showed that the release of gas can be quite violent, with jets of gas and dust shooting out at speeds of hundreds of meters per second from the interior of the nucleus into space.

In 2004, NASA's *Stardust* mission (**FIGURE 12.16**) used a material called aerogel to capture dust particles from Comet Wild 2 (pronounced "vilt two"). The aerogel was then sealed in a re-entry capsule, which the main spacecraft ejected on a return trajectory to Earth, providing our first direct sample of comet dust. Scientists are still puzzling over the composition of some of the comet dust, which indicates that it must have formed in the inner solar system and somehow become mixed in with other cometary materials that formed in the outer solar system. On July 4, 2005, NASA's

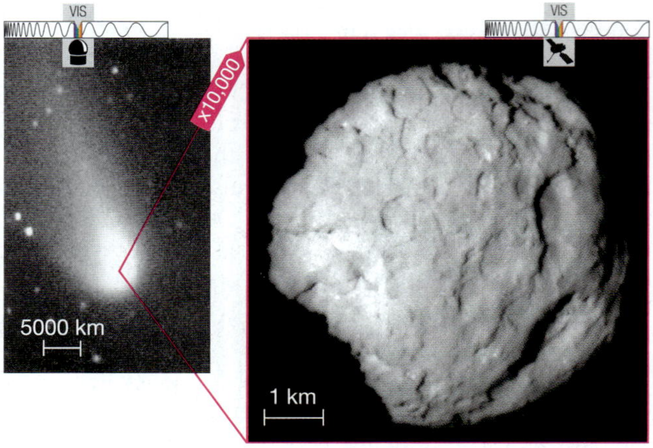


**FIGURE 12.16** Comet Wild 2 as seen from Earth (left) and by the *Stardust* spacecraft. The irregular surface probably shows effects from a combination of impacts and uneven vaporization rates in different regions.

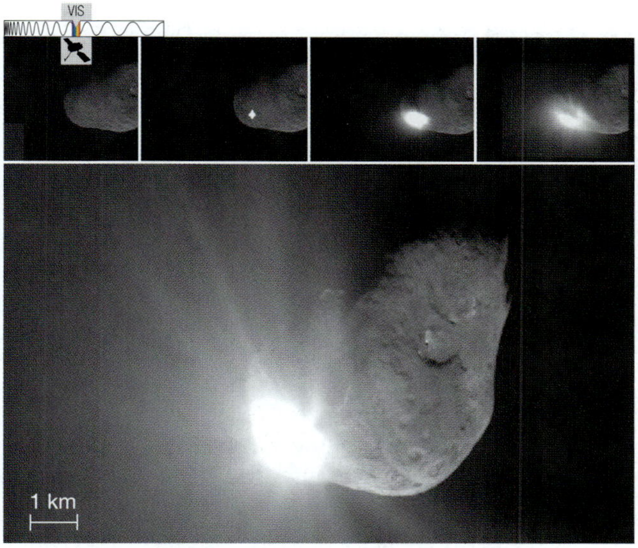


**FIGURE 12.17** The *Deep Impact* spacecraft took this sequence of images (left to right along the top, then below) over a period of 67 seconds as its 370-kg impactor crashed into Comet Tempel 1 on July 4, 2005.

**a** This image (from November 2014) shows jets of material vaporizing into space.

**b** Close-up of a jet shortly before the comet reached perihelion in August 2015.

**FIGURE 12.18** Activity on Comet C-G (67P/Churyumov-Gerasimenko) as imaged by the *Rosetta* spacecraft.

*Deep Impact* spacecraft released a 370-kilogram impactor that slammed into Comet Tempel 1 at 37,000 kilometers per hour (**FIGURE 12.17**). The blast created a plume of hot gas composed of material vaporized from deep within the comet. Spectroscopy showed that this material contained many complex organic molecules. The plume also included a huge amount of dust, telling us that the comet's surface must be covered by dust to a depth of tens of meters.

More recently, from 2014 to 2016, the European Space Agency's *Rosetta* mission spent about two years orbiting Comet 67P/Churyumov-Gerasimenko, or Comet C-G for short. The spacecraft entered orbit when the comet was still almost fully frozen, then stayed with the comet as it approached and passed perihelion (its closest point to the Sun). *Rosetta* thereby was able to observe the changes that occurred as the comet grew its coma and tails (**FIGURE 12.18**). *Rosetta* also sent a small lander, named *Philae*, to the surface of the comet. *Philae* didn't work exactly as planned. It was designed to fire harpoons that would anchor it in place, but instead it bounced twice and tumbled to rest in a shadowed crevice. This location did not provide enough sunlight to recharge the batteries, causing the lander to go silent much sooner than had been hoped. Nevertheless, *Philae* obtained key data during the landing—the first-ever soft landing on a comet—and even came back to life temporarily when the comet's orbital motion turned *Philae*'s landing location back into sunlight. At the end of the mission, the main *Rosetta* spacecraft was also guided to a soft landing on the comet (**FIGURE 12.19**).

The *Rosetta* mission provided a wealth of information on the nature and behavior of comets. One of the most interesting

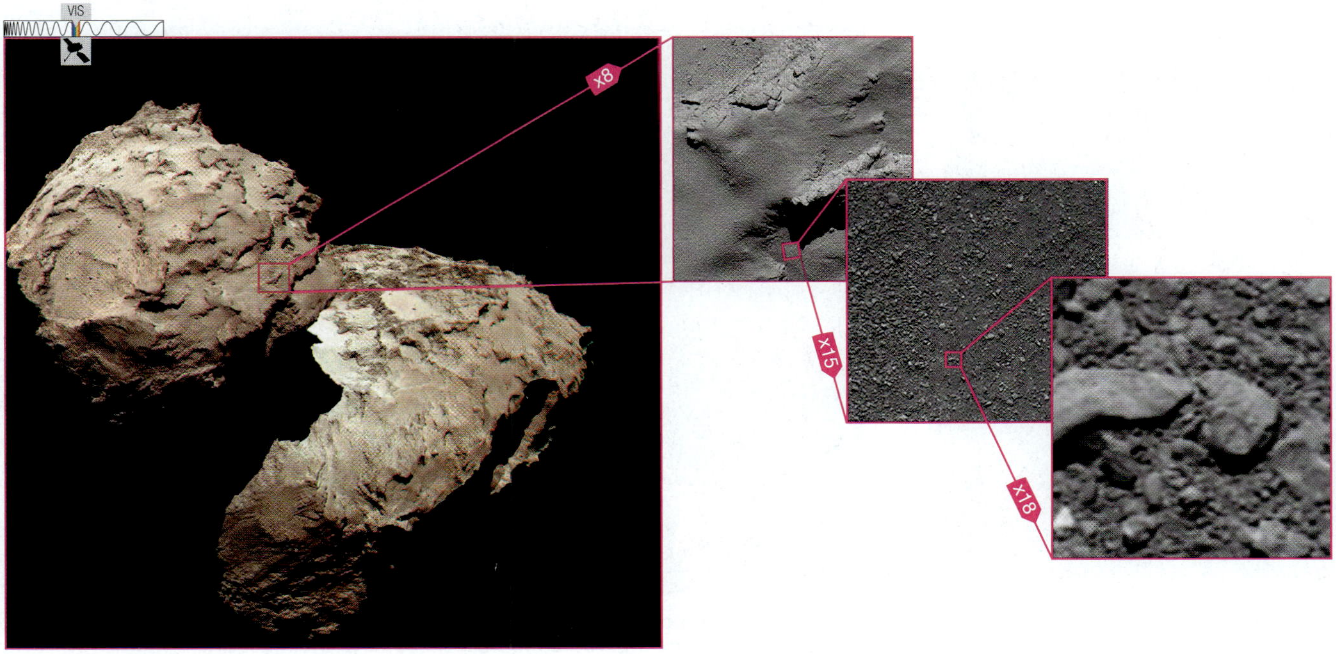


**FIGURE 12.19** The main image shows the 4-kilometer-long nucleus of Comet C-G (67P/Churyumov-Gerasimenko) in enhanced color. The zoom sequence shows increasingly close-up views captured as the *Rosetta* spacecraft descended to the comet's surface. Compared to the surface of a small asteroid (Figure 12.11), the comet is more heavily coated with dust.

results shows that the water ice that drives cometary activity near the Sun is hidden under a substantial crust of dusty material that is composed of rock and carbon-bearing molecules. Virtually no ice is exposed at the surface. This explains the striking features in Figure 12.18b, in which we see a jet of water vapor emanating from a pit or "sinkhole" where the crust has collapsed, exposing the ice at depth.

*Rosetta* also measured the ratio of deuterium to ordinary hydrogen in water vapor streaming from Comet C-G, finding that the ratio is substantially higher than that for water on Earth. Several other comets also have high proportions of deuterium, suggesting that comets were *not* the primary source of the water delivered to the early Earth through impacts. What, then, was the source of Earth's water? Primitive meteorites that contain water have a deuterium-to-hydrogen ratio more similar to Earth's, suggesting that Earth's water came from impacts of asteroids that formed in the outer regions of the asteroid belt (because that is where these meteorites came from). The scientific debate continues, however, because we have not yet measured the deuterium-to-hydrogen ratio in enough comets to fully rule them out as the source of at least some of Earth's water.

**Comet Tails and Meteor Showers** Comets also eject sand- to pebble-size particles that are too big to be affected by either the solar wind or sunlight. These particles essentially form a third, invisible tail that follows the comet around its orbit. They are also the particles responsible for most meteors and meteor showers.

The sand- to pebble-size particles are much too small to be seen themselves, but they enter the atmosphere at such high speeds—up to about 250,000 kilometers per hour (70 km/s)—that they make the surrounding air glow with heat. It is this glow that we see as the brief but brilliant flash of a *meteor*, lasting only until the particle is vaporized by the heat. An estimated 25 million particles of comet dust enter the atmosphere worldwide every day, burning up as meteors and adding hundreds of tons of comet dust to Earth daily.

Comet dust is sprinkled throughout the inner solar system, but the "third tails" of ejected particles are concentrated along the orbits of comets. As a result, while you can typically see a few meteors on any clear night, many more are visible on those nights when our planet is crossing a comet's orbit. You may see dozens of meteors per hour during one of these **meteor showers**, which recur at about the same time each year because the orbiting Earth passes through a particular comet's orbit at the same time each year. The meteors of a meteor shower generally appear to radiate from a particular direction in the sky, for essentially the same reason that snow or heavy rain seems to come from a particular direction in front of a moving car (**FIGURE 12.20**). Because more meteors hit Earth from the front than from behind (just as more snow hits the front windshield of a moving car), meteor showers are best observed in the predawn sky, when part of the sky faces in the direction of Earth's motion. **TABLE 12.1** lists major annual meteor showers and their parent comet, if known. Other planets must have meteor showers too, though only one has been witnessed. After Comet Siding Spring had a close encounter with Mars in October 2014, the *MAVEN* spacecraft detected cometary dust debris throughout the upper atmosphere of Mars.

**a** Meteors appear to radiate from a particular point in the sky for the same reason that we see snow or heavy rain come from a single point in front of a moving car.

**b** This digital composite photo, taken in Australia during the 2001 Leonid meteor shower, shows meteors as streaks of light radiating from the same point in the sky. The large rock is Uluru, also known as Ayers Rock. Note that, by eye, you would not see so many meteors all at once; each meteor would flash across your sky for a few seconds, and even in the best meteor showers you'd likely see one only every few minutes.

**FIGURE 12.20** The geometry of meteor showers.

**TABLE 12.1** **Major Annual Meteor Showers**

| Shower Name | Approximate Date | Associated Comet |
|---|---|---|
| Quadrantids | January 3 | ? |
| Lyrids | April 22 | Thatcher |
| Eta Aquarids | May 5 | Halley |
| Delta Aquarids | July 28 | ? |
| Perseids | August 12 | Swift-Tuttle |
| Orionids | October 22 | Halley |
| Taurids | November 3 | Encke |
| Leonids | November 17 | Tempel-Tuttle |
| Geminids | December 14 | Phaeton |
| Ursids | December 23 | Tuttle |

**See it for yourself** Try to observe the next meteor shower (see Table 12.1); be prepared with a star chart, a marker pen, and a dim (preferably red) flashlight. Each time you see a meteor, record its path on your star chart. Record at least a dozen meteors, and try to determine the *radiant* of the shower—that is, the constellation from which the meteors appear to radiate. Does the meteor shower live up to its name?

## Where do comets come from?

We've stated that comets we see in the inner solar system come from vast reservoirs of comets in the distant outer solar system. But we've never actually seen any small comets at such great distances from the Sun, so you may wonder how we know they are out there. The answer is relatively simple: The comets that we see in the inner solar system must come from *somewhere,* and we can figure out where that somewhere must be by tracing their orbits back. We can then estimate numbers by figuring out how many must reside at great distances to explain the average number that enter the inner solar system each year. It is this type of study that has revealed that comets come from two distinct reservoirs.

**Think about it** Suppose we saw an average of 100 comets entering the inner solar system for the first time each year. If we assume that this has been happening throughout the history of the solar system, what is the minimum number of comets that must reside in a distant reservoir to account for it? Explain.

Most comets that visit the inner solar system do not orbit the Sun in the same direction as the planets, and their elliptical orbits have random orientations. Tracing their orbits back shows that they come from far beyond the orbits of the planets—sometimes nearly a quarter of the distance to the nearest star. These comets must come plunging sunward from the vast spherical region of space that we call the *Oort cloud*. Be sure to note that the Oort cloud is *not* a cloud of gas, but rather a collection of many individual comets. Based on the number of Oort cloud comets that visit the inner solar system, we conclude that the Oort cloud must contain about a trillion ($10^{12}$) comets.

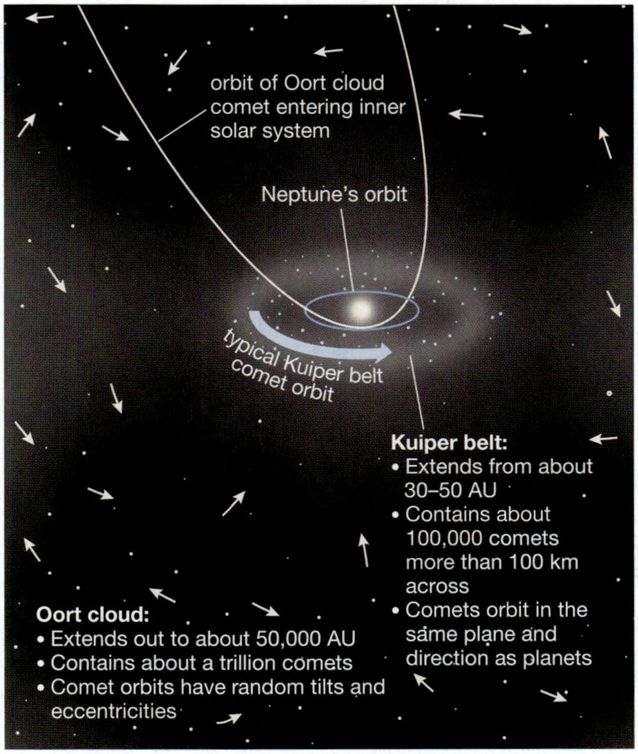


**FIGURE 12.21** The comets we occasionally see in the inner solar system come from two major reservoirs in the outer solar system: the Kuiper belt and the Oort cloud.

A smaller number of the comets that visit the inner solar system have a pattern to their orbits. They travel around the Sun in the same direction and in nearly the same plane as the planets, and their elliptical orbits carry them no more than about twice as far from the Sun as Neptune. These comets must come from the donut-shaped *Kuiper belt* that lies beyond the orbit of Neptune. **FIGURE 12.21** contrasts the general features of the Kuiper belt and the Oort cloud.

How did comets end up in these far-flung regions of the solar system? The only answer that makes scientific sense comes from thinking about what happened to the leftover icy planetesimals that roamed the region in which the jovian planets formed.

The leftover planetesimals that cruised the spaces between Jupiter, Saturn, Uranus, and Neptune were doomed to suffer either a collision or a close gravitational encounter with one of the young jovian planets. Recall that when a small object passes near a large planet, the planet is hardly affected but the small object may be flung off at high speed [**Section 4.5**]. The planetesimals that escaped being swallowed therefore tended to be flung off in all directions. Some may have been cast away at such high speeds that they completely escaped the solar system. The rest ended up on orbits with very large average distances from the Sun, becoming the comets of the Oort cloud. The random directions in which these comets were flung explain why the Oort cloud is roughly spherical in shape. Oort cloud comets are so far from the Sun that they can be nudged by the gravity of nearby stars (and even by the mass of the galaxy as a whole), preventing some of them from ever returning to the region of the planets and sending others plummeting toward the Sun.

Beyond the orbit of Neptune, the icy planetesimals were much less likely to be cast off by gravitational encounters. Instead, they remained in orbits going in the same direction as planetary orbits and concentrated relatively near the ecliptic plane. These are the comets of the Kuiper belt. Kuiper belt comets can be nudged by the gravity of the jovian planets through orbital resonances, sending some on orbits that pass through the inner solar system.

To summarize, the comets of the Kuiper belt seem to have originated farther from the Sun than the comets of the Oort cloud did, even though the Oort cloud comets are now much more distant. The Oort cloud consists of leftover planetesimals that were flung outward after forming between the jovian planets, while the Kuiper belt consists of leftover planetesimals that formed and still remain in the outskirts of the planetary realm.

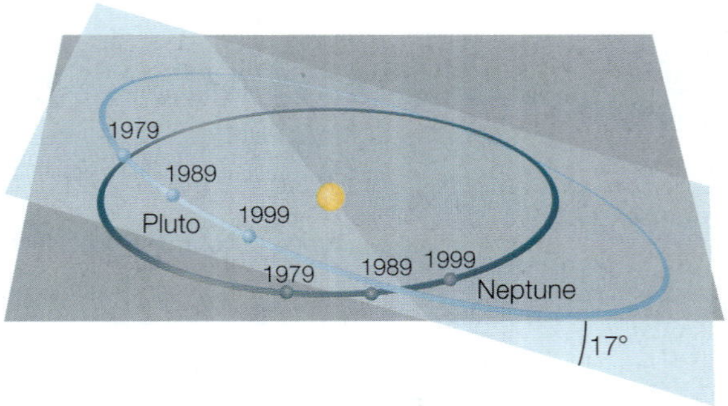


**FIGURE 12.22** Pluto's orbit is significantly elliptical and tilted relative to the ecliptic. Pluto comes closer to the Sun than Neptune does for 20 years in each 248-year orbit, as was the case between 1979 and 1999. There's no danger of a collision, however, thanks to an orbital resonance in which Neptune completes three orbits for every two of Pluto's.

## 12.4 Pluto and the Kuiper Belt

The small comets we have studied in the inner solar system are presumably representative of their frozen cousins in the Kuiper belt and Oort cloud, though our current telescopes cannot detect such small objects at such great distances. In fact, we've never directly detected an object of any size in the Oort cloud. We have, however, detected many moderately large objects in the Kuiper belt. As of 2018, more than 2400 icy objects had been directly observed in the Kuiper belt, allowing scientists to infer that the region contains at least 100,000 objects more than 100 kilometers across. We'll begin our study of the Kuiper belt with Pluto, because we know far more about it than any other object in this region.

### What is Pluto like?

Pluto orbits the Sun at an average distance of nearly 40 AU, and it takes about 248 years to complete a single orbit. Its orbit is much more elliptical and inclined to the ecliptic plane than that of any of the eight planets (**FIGURE 12.22**). In fact, Pluto sometimes comes closer than Neptune to the Sun, although there is no danger of collision: Neptune orbits the Sun precisely three times for every two Pluto orbits, and this stable orbital resonance means that Neptune is always a safe distance away whenever Pluto approaches its orbit.

Pluto has five known moons (**FIGURE 12.23**). The largest, Charon, has more than half the diameter and about $\frac{1}{8}$ the mass of Pluto, and orbits only 20,000 kilometers away. (For comparison, our Moon has a mass $\frac{1}{80}$ of Earth's and orbits 400,000 kilometers away.) Charon is also slightly lower in density than Pluto. These facts have led astronomers to hypothesize that Pluto's moons were created by a *giant impact* similar to the one thought to have formed our Moon [**Section 8.2**]. A large comet crashing into Pluto may have blasted away its low-density outer layers, which then formed a ring around Pluto and eventually re-accreted to make Charon and the smaller moons. Such an impact may also explain why Pluto rotates almost on its side.

**Pluto Before *New Horizons*** Pluto is so far from the Sun that it took astronomers a long time to learn much about it, despite efforts with powerful telescopes. You can understand why with a simple analogy: Trying to see Pluto from Earth is equivalent to looking for a snowball the size of your fist that is about 600 kilometers away—and in very dim light. Nevertheless, astronomers managed to learn a lot even before the 2015 flyby of the *New Horizons* spacecraft.

Much of what we learned came after the 1978 discovery of Charon (other moons were discovered later). Observations of Charon's orbit allowed scientists to calculate Pluto's precise mass by applying Newton's version of Kepler's third law [**Section 4.4**]. The revised mass proved to be key in recognizing how different Pluto really is from

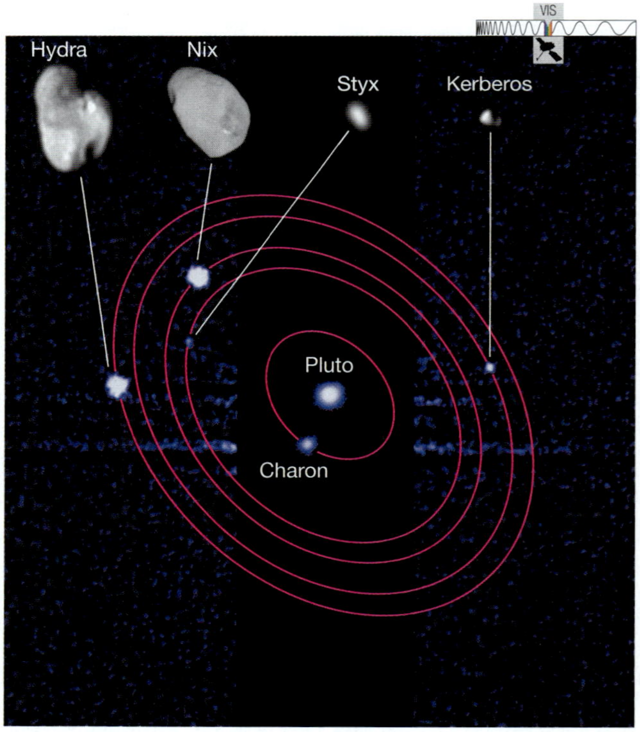


**FIGURE 12.23** The main image shows a "family portrait" of Pluto and its five moons as viewed by the Hubble Space Telescope, along with orbital paths for the moons. The insets show images of the four smaller moons, taken by *New Horizons* during its flyby of Pluto. The images are shown to scale and are magnified 1000 times compared to the scale on which the orbits are shown. (The bluish dots and stripes in the main image come from scattered light within the camera.)

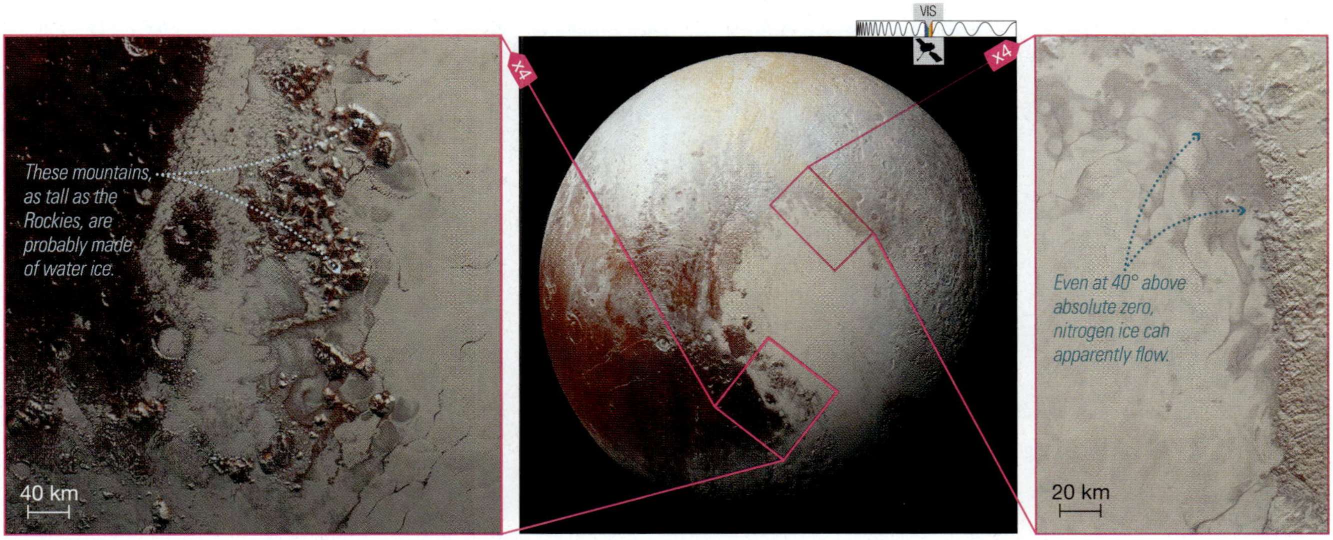


**FIGURE 12.24** *New Horizons* images of Pluto, with enhanced color to identify different regions. The bright, heart-shaped region on the right side of the central image is called Tombaugh Regio, after Pluto's discoverer.

both the terrestrial and the jovian planets. Charon then provided another learning opportunity through some good luck in timing: From 1985 to 1990, Pluto and Charon happened to be aligned in a way that made them eclipse each other every few days as seen from Earth—something that happens only about every 120 years. Detailed analysis of brightness variations during these eclipses allowed the calculation of sizes, masses, and densities for both Pluto and Charon, confirming that both have comet-like compositions of ice and rock. The eclipse data even allowed astronomers to construct rough maps of Pluto's surface markings, and improved telescopic observations, including many with the Hubble Space Telescope, showed a varied surface suggestive of unknown forms of activity.

Pluto is very cold, with an average surface temperature of only 40 K, as we would expect at its great distance from the Sun. Nevertheless, Earth-based observations showed that Pluto has a thin atmosphere of nitrogen, methane, and carbon monoxide, formed by vaporization of surface ices. The amount and composition of atmospheric gas can change as Pluto's distance from the Sun varies along its elliptical orbit—and because of seasonal effects on condensation and vaporization that arise from Pluto's large axis tilt. When atmospheric gas is most abundant, greenhouse gases methane and carbon monoxide make the atmosphere more than 30°C warmer than the icy surface.

The mutual tidal forces acting between Pluto and Charon long ago made them rotate synchronously with each other [**Section 4.5**], which means that Pluto's "day" is the same length as Charon's "month" (orbital period) of 6.4 Earth days. This synchronous rotation also means that Charon is visible from only one side of Pluto, where it would provide a stunning view: From Pluto's surface, Charon would dominate the sky, appearing almost 10 times as large in angular size as our Moon appears from Earth. Moreover, Charon would neither rise nor set but instead hang motionless as it cycled through its phases every 6.4 days. The Sun would appear more than a thousand times fainter than it appears here on Earth and would be no larger in angular size than Jupiter appears in our skies.

**See it for yourself** To get a sense of the visual impact Charon would have in Pluto's sky, imagine that your head is Pluto. To put Charon on the same scale, you'll need a round object about 5 centimeters across, placed about 2 meters from your head. Compare the angular size of your model Charon to the angular size of the Moon in our sky (which is about the width of your pinky fingernail at arm's length).

***New Horizons* at Pluto** The *New Horizons* spacecraft was launched into space in January 2006 at a higher speed than any previous spacecraft, and it used Jupiter's gravity to gain an additional speed boost when it flew past the giant planet just 13 months later. Still, Pluto is so far from the Sun that *New Horizons* had been traveling through space for 9½ years when it flew past Pluto in July 2015 at a speed of about 50,000 kilometers per hour. At closest approach, *New Horizons* was only about 144,000 kilometers from Pluto (and less than 30,000 kilometers from Charon), but the high speed of the flyby meant it had only a few minutes during which to record its highest-resolution images. Scientists spent years planning their operations in order to take full advantage of the short time frame.

Despite the great challenges, the results were spectacular. *New Horizons* images show clear evidence of geological activity on Pluto (**FIGURE 12.24**), some of which must have occurred within the past 100 million years—so recently compared to the age of the solar system that it is likely the activity continues to this day. Vast regions of Pluto lack craters, implying that ancient craters have been erased. The heart-shaped "Tombaugh Regio" shows wide, smooth ice plains that may be flowing glaciers of nitrogen ice. Pluto also has mountains several kilometers tall, comparable in height to the Rocky Mountains on Earth. These cannot be composed of nitrogen ice, which isn't rigid enough to maintain the steep slopes seen. Instead, the mountains must be made of water ice, which is rigid and strong at Pluto's low temperature. Charon also has vast, smooth plains that are the hallmark of recent geological activity, as well as canyons comparable in length and depth to Earth's Grand Canyon (**FIGURE 12.25**).

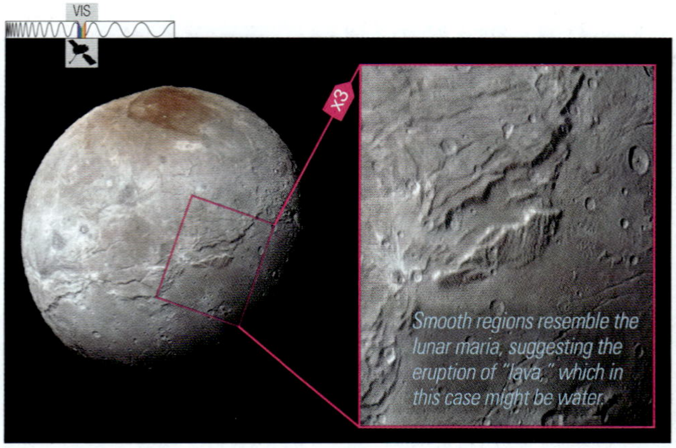


**FIGURE 12.25** *New Horizons* images of Pluto's largest moon, Charon.

**FIGURE 12.26** Pluto's jagged mountains, smooth plains, and hazy atmosphere, seen as *New Horizons* looked back toward the Sun after the encounter.

Scientists are still trying to understand the heat source that drives the geological activity of Pluto and Charon. Neither world is expected to have much heat from radioactive decay, and the tidal lock between Pluto and Charon means that there is no tidal heating. Perhaps the activity occurs because even the frigid 40 K temperature isn't cold enough to make nitrogen ice very rigid, so a small amount of heat from radioactive decay and/or changes in the thickness of the ice over Pluto's elliptical orbit is enough to allow the nitrogen ice to flow. It may even be possible that the internal heat is sufficient to melt subsurface ice, making for yet another icy world with a hidden ocean.

At the end of its brief visit, *New Horizons* looked back to capture a stunning view of Pluto's thin atmosphere above jagged mountains and smooth plains (**FIGURE 12.26**). This atmosphere is composed of gases vaporized from the surface ice. The atmosphere extends thousands of kilometers above Pluto's surface, capped by an exosphere [**Section 10.1**]. The very low gravity in the exosphere allows these gases to escape to space, which means they must be replenished with more vaporization; it's likely that Pluto has lost as much as a kilometer's worth of surface ice in this way over the age of the solar system. The atmosphere was visible when *New Horizons* looked back because it contains enough haze to scatter sunlight forward. The haze probably forms as solar ultraviolet light breaks down methane and nitrogen molecules in the atmosphere, which then react to make longer molecular chains and haze particles (sometimes called *tholins*). These particles probably also explain Pluto's pale brown color: They drift down through the atmosphere and coat the surface; for reasons still unknown, more of them appear to be deposited around Pluto's equator.

Overall, while scientists were surprised by the level of geological activity on Pluto, the findings reinforced what had already been learned elsewhere. In particular, Pluto seems to be a more extreme example of the "ice geology" that allows jovian moons like Europa, Titan, and Enceladus to be active at lower temperatures than are required by the "rock geology" of the terrestrial words (see Figure 11.32). We've also seen that atmospheres can be affected in surprising ways by the differing gases that condense and vaporize at different distances from the Sun: water at Earth, carbon dioxide at Mars, methane at Titan, and nitrogen at Pluto. The general lesson appears to be that worlds can have astonishing similarities in geological activity despite tremendous differences in composition and temperature (a lesson that we might call the "Law of Constant Astonishment"). It will be interesting to see if this lesson applies to worlds not yet explored.

## Why is there a Kuiper belt?

As we've discussed, Pluto shares the Kuiper belt with vast numbers of other objects, including at least one—Eris—that is about the same size as Pluto but more massive, and many others that are not much smaller in size and mass. This abundance of objects makes the Kuiper belt a hot topic for research, as astronomers seek to answer the question of why so many relatively large objects reside so far from the Sun, and what this might tell us about the general nature of planetary systems.

**Clues from Other Kuiper Belt Objects** We don't know nearly as much about any of the other large comets of the Kuiper belt as we do about Pluto, but because they all are thought to have formed in the same distant region of the solar system, we expect them to be similar in nature and composition to Pluto. Evidence to support this idea comes from the fact that several other Kuiper belt comets (including Eris) have known moons, so in those cases we can calculate masses and densities. In every case so far, these results, along with spectral data, support the idea that these objects have comet-like compositions of ice and rock.

**Think about it** By the time you read this book, *New Horizons* should have completed its flyby of an object called MU69, which is much smaller than Pluto (and might have an even smaller moon). Find images from the flyby, and discuss whether they are consistent with what we might expect for a small object in the Kuiper belt.

In addition, there's one other large object (besides Pluto) that we've photographed up close that is almost certainly from the Kuiper belt: Neptune's moon Triton. Recall that Triton's "backward" orbit indicates that it must be a captured object [**Section 11.2**], and it was almost certainly captured from the Kuiper belt. Therefore, *Voyager* 2 images of Triton (see Figure 11.31) probably represent images of a *former* member of the Kuiper belt. Scientifically, Triton suggests at least two important ideas about large comets of the Kuiper belt. First, Triton is larger than both Pluto and Eris (about 15% larger in diameter), suggesting that the Kuiper belt may once have contained more and larger objects than it does today. Second, Triton shows signs of significant past or present geological activity, reinforcing the idea that distant ice-rich worlds can be much more geologically active than we might have guessed from their sizes alone.

The orbits of Kuiper belt comets also provide clues to their nature and origin. Like Pluto, many Kuiper belt comets have stable orbital resonances with Neptune. In fact, hundreds of Kuiper belt comets have the *same* orbital period and average distance from the Sun as Pluto itself (and are nicknamed "Plutinos"). Many other objects have orbits so eccentric that they spend much of their time beyond the main confines of the Kuiper belt. Eris provides an example: It currently lies about twice as far as Pluto from the Sun, but its 557-year orbit is so eccentric that Eris sometimes comes closer than Pluto to the Sun. Eris's orbit also has about twice Pluto's inclination to the ecliptic plane.

One further clue comes from the likelihood that many Kuiper belt comets suffered giant impacts. We've already discussed the likelihood that Pluto's moons formed from a giant impact. Another dwarf planet, Haumea (see Figure 12.5), also shows evidence of having suffered a giant impact; that is the simplest way to explain its very fast rotation (once every 4 hours) and two small moons. In fact, four other known Kuiper belt comets may be fragments from the same impact, because their orbits and spectra closely resemble those of Haumea. This evidence for giant impacts among the relatively small number of Kuiper belt comets that have been studied in any depth suggests that giant impacts must have been very common in the Kuiper belt.

**Origin of the Kuiper Belt** On a basic level, we've already discussed the origin of the Kuiper belt in the context of the nebular theory of solar system formation: It consists of the icy planetesimals that accreted beyond the orbit of Neptune. However, the clues we have discussed from Kuiper belt orbits, compositions, and giant impacts suggest that the full picture is much more complex, and this has led scientists to develop much more sophisticated computer models of what might have occurred as the planets formed.

Current models are works in progress, and there are plenty of mysteries that they cannot yet account for. Nevertheless, a general picture is beginning to emerge, and it adds two important features to our understanding of the birth of the solar system. First, the models suggest that orbital resonances may have played a role not only in shaping the Kuiper belt but also in the formation of the planets themselves. Second, and more intriguingly, the models suggest that some of the planets of our solar system may have undergone *migration*—moving to closer or more distant orbits—after their initial formation.

In particular, the models suggest that neither Uranus nor Neptune could have formed in its current orbit, because the density of the solar nebula at that distance was too low for such a large planet to have accreted. Instead, all four jovian planets probably were much closer together when they formed, and shared the solar system with many other planet-size planetesimals. The models in essence allow scientists to look at what we might have observed in the early solar system, with collisions, close gravitational encounters, and resonances all affecting planetary orbits. Each interaction led to an exchange of orbital energy, with one object moving outward and another moving inward. Overall, the models indicate that Jupiter would have moved somewhat inward from where it formed, while Saturn, Uranus, and Neptune would have moved outward. The planetary migration, in turn, would have led to orbital resonances that stirred the Kuiper belt, ultimately giving it the form it has today. Moreover, the Kuiper belt likely lost more than 99% of its original mass in this process, as most of its objects either collided with a planet or were scattered far off into space.

These new models not only give us new insights into our own system's formation but also hold important lessons for our rapidly growing understanding of other planetary systems. As we'll discuss in Chapter 13, migration seems to have played an even more important role in many other planetary systems. Moreover, when combined with the fact that the timing of nebula clearing may be different in other solar systems, our new understanding suggests that in some cases planets might form with characteristics quite different from those of the terrestrial and jovian planets in our solar system, allowing for a far wider range of possible planetary types.

Finally, it's worth noting that careful study of patterns in the orbits of known Kuiper belt objects has led some astronomers to suggest that an undiscovered object, nicknamed "Planet 9," must have nudged many of these objects into their current orbits. To have done so, Planet 9 would have to be roughly ten times as massive as Earth, which means it should be detectable with current astronomical technology. Extensive searches were under way as of 2018, so if Planet 9 really exists, it could be found fairly soon.

**Think about it** Find the current status of the search for Planet 9. Have any candidate objects been found? If not, do astronomers still suspect the object exists?

## 12.5 Cosmic Collisions: Small Bodies versus the Planets

The chaotic conditions of the early solar system explain why so many small bodies ended up in orbits that could send them crashing into the planets (and into moons and each other). The many impact craters formed during the heavy bombardment [**Section 8.2**], which ended nearly 4 billion years ago, tell us that such collisions were far more common in the past. Nevertheless, plenty of small bodies still roam the solar system, and cosmic collisions still occur on occasion.

Direct proof that large impacts still occur comes from the fact that we have witnessed a few. The Sun is sometimes hit by comets, though only the comets suffer in the process. More dramatically, in 1994 we witnessed a major impact by a comet on Jupiter. The comet, named Shoemaker-Levy 9, or SL9 for short, had already been ripped apart by tidal forces during a previous pass near Jupiter, so it consisted of a string of nuclei rather than a single nucleus (**FIGURE 12.27a**). Comet SL9 was discovered more than a year before it collided with Jupiter, and orbital calculations told astronomers precisely when the collision would occur. When the impacts began, they were observed with nearly every major telescope in existence, as well as by spacecraft that were in position to get a view. Each of the individual nuclei crashed into Jupiter with an energy equivalent to that of a million hydrogen bombs (**FIGURE 12.27b, c**). Comet nuclei barely a kilometer across left scars—some large enough to swallow Earth—that lasted for months before dissipating with Jupiter's strong winds. We've since observed the aftermaths of at least two more impacts on Jupiter (**FIGURE 12.27d**).

**Impact Extinction of the Dinosaurs**

## Did an impact kill the dinosaurs?

There's no doubt that major impacts have occurred on Earth in the past: Geologists have identified more than 150 impact craters on our planet. One of the most recent, Meteor Crater in Arizona (see Figure 9.8a), formed about 50,000 years ago when a metallic asteroid roughly 50 meters across crashed to Earth with the explosive power of a 20-megaton hydrogen bomb. Although the crater is only a bit more than 1 kilometer across, the blast and ejecta probably battered an area covering hundreds of square kilometers. Larger craters presumably formed from impacts that caused more widespread damage. In particular, a growing body of evidence, accumulated over the past three decades, suggests that at least one large impact altered the entire course of evolution.

In 1978, while analyzing geological samples collected in Italy, a scientific team led by father and son Luis and Walter Alvarez made a startling discovery. They found that a thin layer of dark sediments deposited about 65 million years ago—about the time the dinosaurs went extinct—was unusually rich in the element iridium. Iridium is a metal that is rare on Earth's surface (because it sank to Earth's core when our planet underwent differentiation) but common in meteorites. Subsequent studies found the same iridium-rich layer in 65-million-year-old sediments around the world (**FIGURE 12.28**).* The Alvarez team suggested a stunning hypothesis: The extinction of the dinosaurs was caused by the impact of an asteroid or comet.

In fact, the death of the dinosaurs was only a small part of the biological devastation that seems to have occurred 65 million years ago. The fossil record suggests that up to 99% of all living organisms died around that time and that up to 75% of all existing *species* were driven to extinction. This vast loss of life makes the event a clear example of a **mass extinction**—the rapid extinction of a large fraction of all living species. Could it really have been caused by an impact?

*The layer marks what geologists call the *K-T boundary*, because it separates sediments deposited in the Cretaceous and Tertiary periods (the K comes from the German spelling for Cretaceous). The mass extinction that occurred 65 million years ago is therefore called the *K-T event*.

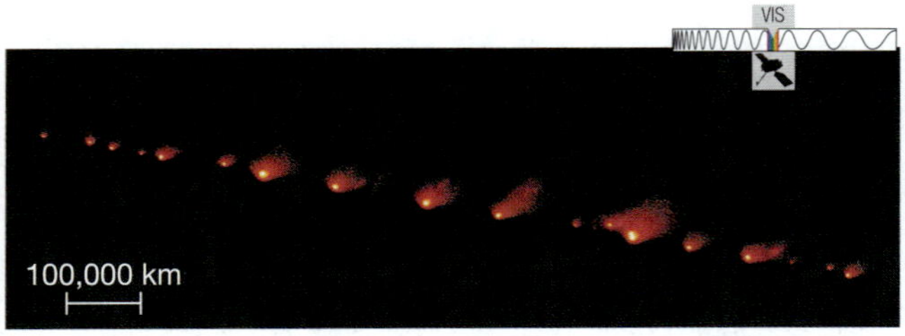


**a** Jupiter's tidal forces ripped Comet SL9 apart, breaking its nucleus into this chain of smaller nuclei.

**b** This painting shows how the SL9 impacts might have looked from the surface of Io. The impacts occurred on Jupiter's night side.

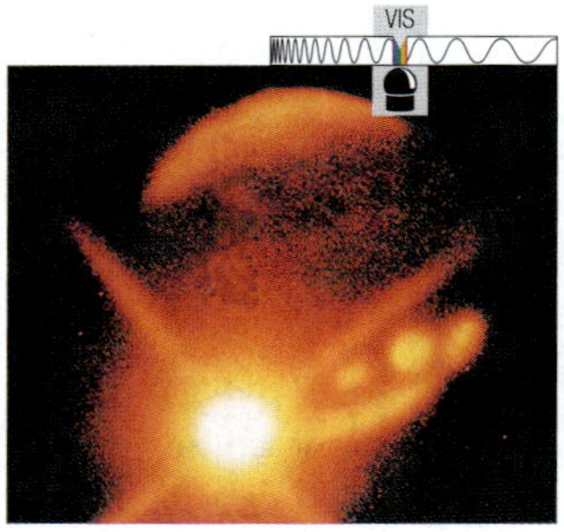


**c** This infrared photo shows the brilliant glow of a rising fireball from the impact of one SL9 nucleus in 1994. Jupiter is the round disk; the impact occurred near the lower left.

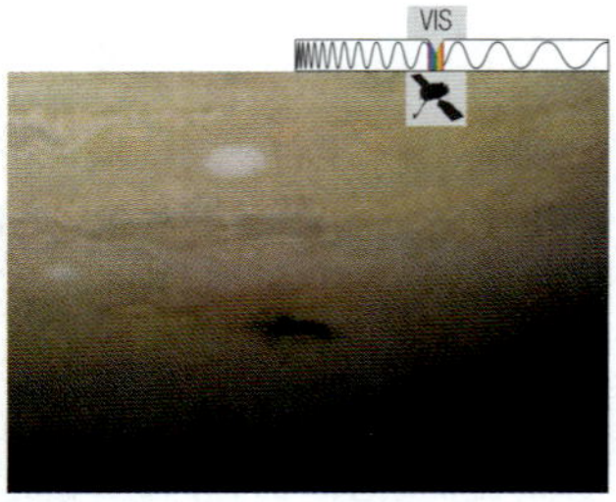


**d** The black spot in this Hubble Space Telescope photo is a scar from the impact of an unknown object that struck Jupiter in July 2009.

**FIGURE 12.27** The impacts of Comet Shoemaker-Levy 9 on Jupiter allowed astronomers their first direct view of a cosmic collision.

**Evidence for the Impact** There's still some scientific debate about whether the impact was the sole cause of the mass extinction or just one of multiple causes, but there's little doubt that a major impact coincided with the death of the dinosaurs. Key evidence comes from further analysis of the sediment layer. Besides being unusually rich in iridium, this layer contains four other unusual features: (1) high abundances of several other metals, including osmium, gold, and platinum; (2) grains of "shocked" quartz, quartz crystals with a distinctive structure that indicates they experienced the high-pressure conditions of an impact; (3) spherical "rock droplets" of a type known to form when

**FIGURE 12.28** Around the world, sedimentary rock layers dating to 65 million years ago share the evidence of the impact of a comet or asteroid. Fossils of dinosaurs and many other species appear only in rocks below the iridium-rich layer.

drops of molten rock cool and solidify in the air; and (4) soot (at some sites) that appears to have been produced by widespread forest fires.

All these features point to an impact. The metal abundances look much like what we commonly find in meteorites rather than what we find elsewhere on Earth's surface. Shocked quartz is also found at other impact sites, including Meteor Crater in Arizona. The rock droplets presumably were made from molten rock splashed into the air by the force and heat of the impact. Some debris would have been blasted so high that it rose above the atmosphere, spreading worldwide before falling back to Earth. On their downward plunge, friction would have heated the debris particles until they became a hot, glowing rain of rock. The soot probably came from vast forest fires ignited by thermal radiation from this impact debris.

Perhaps the most convincing evidence comes from a large impact crater that appears to match the sediment layer in age. The crater, called the Chicxulub Crater after a nearby fishing village, is about 200 kilometers across and is located on the coast of Mexico's Yucatán Peninsula, about half on land and half under water (**FIGURE 12.29**). Its size indicates that it was created by the impact of an asteroid or a comet measuring about 10 kilometers across, which is large enough to account for the iridium and other metals found in the sediment layer.

**The Mass Extinction** If the impact was indeed the cause of the mass extinction, here's how it probably happened. On that fateful day some 65 million years ago, the asteroid or comet slammed into Mexico with the force of a hundred million hydrogen bombs (**FIGURE 12.30**). It apparently hit at a slight angle, sending a shower of red-hot debris across the continent of North America. A huge tsunami sloshed more than 1000 kilometers inland. Much of North American life may have been wiped out almost immediately. Not long after, the hot debris raining around the rest of the world ignited fires that killed many other living organisms.

The longer-term effects were even more severe. Dust and smoke remained in the atmosphere for weeks or months, blocking sunlight and causing temperatures to fall as if Earth were experiencing a harsh global winter. The reduced sunlight would have stopped photosynthesis for up to a year, killing large numbers of species throughout the food chain. Acid rain may have been another by-product, killing vegetation and acidifying lakes around the world. Chemical reactions in the atmosphere probably produced nitrous oxides and other compounds that dissolved in the oceans and killed marine organisms. Recent evidence suggests that a

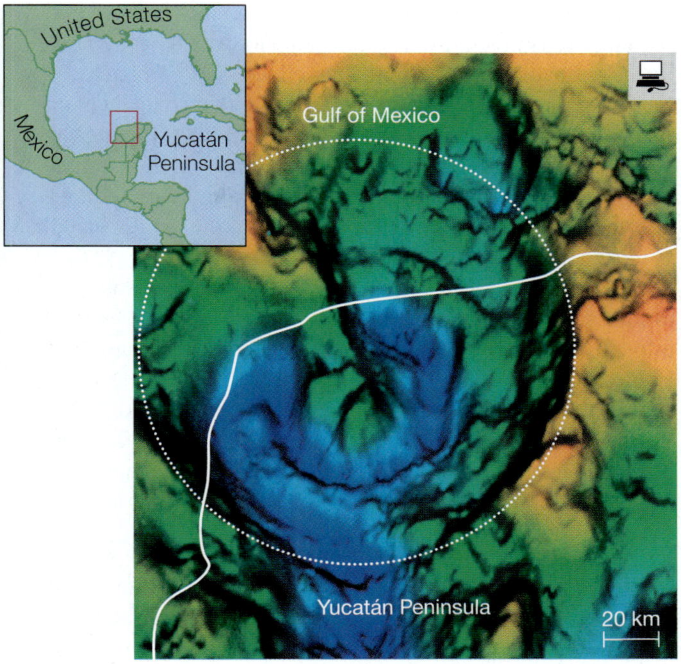


**FIGURE 12.29** This computer-generated image, based on measurements of small local variations in the strength of gravity, shows an impact crater (dashed circle) in the northwest corner of Mexico's Yucatán Peninsula; the inset shows the location.

**FIGURE 12.30** This painting shows an asteroid or comet moments before its impact on Earth, some 65 million years ago. The impact probably caused the extinction of the dinosaurs. If it hadn't occurred, the dinosaurs might still rule Earth today.

period of intense volcanic activity followed the impact—and perhaps was caused by it—and that this activity may have further exacerbated the immediate effects of the impact.

Perhaps the most astonishing fact is not that so many plant and animal species died but that some survived. Among the survivors were a few small mammals. These mammals may have survived in part because they lived in underground burrows and managed to store enough food to outlast the global winter that immediately followed the impact.

The evolutionary impact of the extinctions was profound. For 180 million years, dinosaurs had diversified into a great many species large and small, while mammals (which had arisen at almost the same time as the dinosaurs) had generally remained small and rodent-like. With the dinosaurs gone, mammals became the new kings of the planet. Over the next 65 million years, the mammals rapidly evolved into an assortment of much larger mammals—ultimately including us.

**Controversies and Other Mass Extinctions** There seems little doubt that a major impact coincided with the mass extinction of 65 million years ago, but does it tell the entire story? The jury is still out. The fossil record can be difficult to read when we are trying to understand events that happened over just a few years rather than a few million years. Some scientists suspect that the dinosaurs were already in decline and the impact was only the last straw. Others suggest that major volcanic eruptions also may have played a role.

Was the dinosaur extinction a unique event? Measuring precise extinction rates becomes more difficult as we look to older fossils, but there appear to have been at least four other mass extinctions during the past 500 million years. Indeed, one mass extinction was far worse than the dinosaur event: About 252 million years ago, the "great dying" (also called the "end-Permian" event) killed off the vast majority of all living species. However, neither this nor any of the other mass extinctions have been as closely tied to an impact as the dinosaur extinction, and none show a similar iridium layer. Still, something caused these devastating evolutionary events, and we owe it to ourselves to try to learn what it was so that we can seek ways to avoid a similar fate.

**Think about it** The five identified past mass extinctions each saw Earth lose half or more of its species. Today, species are going extinct at a rate such that we could lose half or more over the next century or two, primarily through habitat loss and hunting. Should this be considered a sixth mass extinction, and if so, how concerned should we be about it? Defend your opinion.

## How great is the impact risk today?

Impacts as large as the one implicated in the demise of the dinosaurs are very rare events. But smaller impacts can still have devastating effects. For example, in 1908, a small asteroid apparently exploded in midair over Tunguska, Siberia, releasing energy equivalent to that of several atomic bombs

### EXTRAORDINARY CLAIMS The Death of the Dinosaurs Was Catastrophic, Not Gradual

By the mid-19th century, geologists recognized that Earth is shaped by both gradual and sudden changes. For example, sediments build up gradually as they are deposited, while volcanic eruptions can spew out vast amounts of rock in a very short time. But which type of process has been more important over the history of Earth? This question became a topic of such great debate among scientists that the two views were given names: *Uniformitarianism* was the view that today's Earth was shaped primarily by slow gradual changes, while *catastrophism* held that the most important processes were dramatic events like global floods and gigantic volcanic eruptions.

A key issue in the debate was Earth's age. Geologists were already aware that major changes had occurred in the past. For example, the commonplace finding of seashells on mountaintops shows that seas can be transformed over time into mountains. If Earth was very old, then there had been plenty of time for such changes to occur gradually, but if Earth was young, catastrophic changes would have been necessary. During the latter half of the 19th century, evidence began to accumulate for a very old Earth, a case that was sealed in the mid-20th century when radiometric dating showed our planet to be more than 4 billion years old. There had clearly been enough time for uniformitarianism to explain most of Earth's features. But could it explain all of them?

The extinction of the dinosaurs posed at least a potential challenge. Dinosaur fossils are abundant in sediment layers dating to more than about 65 million years ago but absent from younger layers, indicating that the dinosaur extinction occurred in a "geological blink of the eye." But that still could have meant a period of a few million years, and most scientists assumed that the extinction occurred through relatively gradual change, such as changes in climate that were detrimental to dinosaur survival. So when Luis and Walter Alvarez proposed that the extinction had been a sudden event due to an asteroid impact, it was an extraordinary claim not only in terms of the dinosaur extinction but also in that it re-introduced the idea of catastrophism as an important part of Earth's history.

The initial evidence, which was from the iridium-rich sediment layer (see Figure 12.28), was strong enough to generate widespread scientific interest in the impact hypothesis, but many scientists (including the Alvarezes themselves) pointed out a major flaw: An impact large enough to deposit so much iridium should have left a noticeable crater, and no such crater was known at the time. That changed with the discovery of the Chicxulub Crater in 1991 (see Figure 12.29). As discussed in this chapter, there is still scientific debate about whether or not the impact was the sole cause of the mass extinction, but at least part of the Alvarezes' extraordinary claim is not subject to any dispute at all: While most of Earth's history can be explained through the gradual processes of uniformitarianism, occasional catastrophic events can and do occur, sometimes with crucial consequences.

**Verdict:** Very likely.

**FIGURE 12.31** This photo shows forests burned and flattened by the 1908 impact over Tunguska, Siberia.

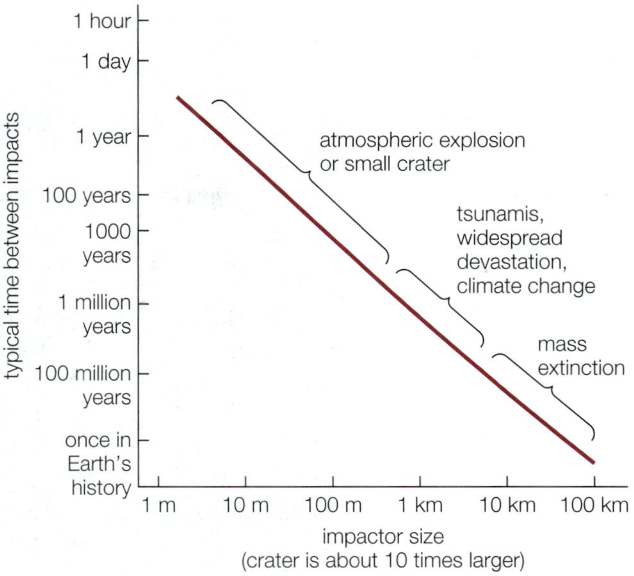


**FIGURE 12.33** This graph shows that larger objects (asteroids or comets) hit Earth less frequently than smaller ones. The labels describe the effects of impacts of different sizes.

(**FIGURE 12.31**). If the asteroid had exploded over a major city instead of Siberia, it would have been the worst natural disaster in human history.

More recently, the people of Chelyabinsk, Russia, got a huge surprise in 2013, when they suddenly saw a brilliant flash of light as a previously undetected asteroid entered the atmosphere above them at a speed of more than 60,000 kilometers per hour (**FIGURE 12.32**). Later estimated to be about 20 meters long with a mass of 10,000 tons, the asteroid streaked across the sky as a giant meteor until friction with the atmosphere made it detonate with the power of a 500-kiloton nuclear bomb. More than a thousand people were injured, mostly by glass shattered by the shock wave. This was in some sense lucky: Had the asteroid's trajectory been directed more vertically toward the ground, the detonation would have occurred at lower altitude, causing much greater damage. The mid-air detonation left no large crater, but meteorites were found scattered across a wide area near Chelyabinsk.

**FIGURE 12.32** This photo shows the meteor trail of the previously undetected 10,000-ton asteroid that detonated in the sky above Chelyabinsk, Russia, on February 15, 2013. The explosion had the power of a 500-kiloton nuclear bomb and caused injuries to more than a thousand people.

More impacts are virtually guaranteed to occur in the future. **FIGURE 12.33** shows how often, on average, we expect Earth to be hit by objects of different sizes, based on geological data from past impacts. The good news is that we are highly unlikely to be hit by an asteroid as large as the one that killed the dinosaurs. Impacts of that size occur tens of millions of years apart on average, which means we face a far greater danger of doing ourselves in than of being done in by a large asteroid or comet. The bad news is that smaller impacts occur much more frequently. Objects a few meters across probably enter Earth's atmosphere every week or so, and objects of the size that caused the Tunguska event probably strike our planet every couple hundred years or so. Until we have more closely monitored the skies for potential impact threats, we cannot discount the possibility of a very damaging impact.

Several efforts to search for potential impact threats are currently under way. Astronomers estimate that there are about 5000 "potentially hazardous asteroids" capable of causing significant regional damage. These estimates further suggest that, as of 2018, about 90% of those larger than 1 kilometer across have been found but only about 30% of those that are smaller but still dangerous have yet been identified. Comets pose a similarly unknown threat; while there are probably far fewer comets that pose an impact threat, we are unlikely to see a comet plunging in from the outer solar system until it is well on its way, so we might have only a few months' or years' warning.

If we were to find an asteroid or a comet on a collision course with Earth, could we do anything about it? Many people have proposed schemes to save Earth by using nuclear

**FIGURE 12.34** The connections among the jovian planets, small bodies, and Earth. The gravity of the jovian planets helped shape both the asteroid belt and the Kuiper belt, and the Oort cloud consists of comets ejected from the jovian planet region by gravitational encounters with these large planets. Ongoing gravitational influences sometimes send asteroids or comets toward Earth.

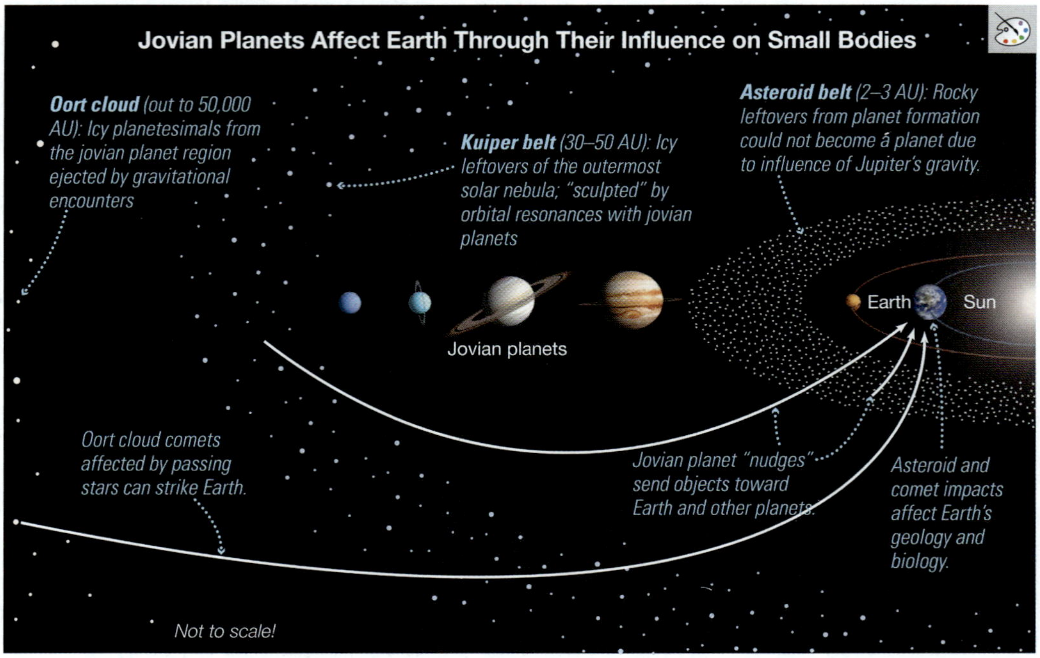


weapons or other means to demolish or divert an incoming asteroid, but no one knows whether current technology is really up to the task. We can only hope that the threat doesn't become a reality before we're ready.

**Think about it** After the Chelyabinsk impact, the following statement circulated widely: "Meteors are nature's way of asking 'How's that space program coming?'" Comment on the meaning of this statement. How much time and money do *you* think we should be spending to counter the impact threat? Defend your opinion.

While near-Earth asteroids may pose real dangers, they may also offer opportunities. These asteroids bring valuable resources tantalizingly close to Earth. Iron-rich asteroids are particularly enticing, because they probably contain many precious metals that have mostly sunk to the core on Earth. It may also be possible to gather fuel and water from asteroids for use in missions to the outer solar system. Both NASA and private companies are working on plans to travel to asteroids and return with materials from them. It may be only a matter of time until we are talking as much about the benefits of asteroids as about their dangers.

## How do the jovian planets affect impact rates and life on Earth?

Ancient people imagined that the mere movement of planets relative to the visible stars in our sky could somehow have an astrological influence on our lives. Scientists no longer give credence to this ancient superstition, but we now know that planets can have a real effect on life on Earth. By shaping the orbits of asteroids and comets, the planets have contributed to cosmic collisions that helped shape our destiny.

The jovian planets, especially Jupiter, have had the greatest effects (**FIGURE 12.34**). As we've discussed, Jupiter disturbed the orbits of rocky planetesimals outside Mars's orbit, preventing a planet from forming and creating the asteroid belt. The jovian planets also ejected icy planetesimals to create the distant Oort cloud of comets, and through migration and orbital resonances they shaped the orbits of the comets in the Kuiper belt. Ultimately, every asteroid or comet that has impacted Earth since the end of the heavy bombardment was in some sense sent our way by the influence of Jupiter or one of the other jovian planets.

We therefore find a deep connection between the jovian planets and the survival of life on Earth. If Jupiter did not exist, the threat from asteroids might be much smaller, since the objects that make up the asteroid belt might instead have become part of a planet. On the other hand, the threat from comets might be much greater: Jupiter probably ejected more comets to the Oort cloud than any other jovian planet, and without Jupiter, those comets might have remained dangerously close to Earth. Of course, even if Jupiter has protected us from impacts, it's not clear whether that has been good or bad for life overall. The dinosaurs appear to have suffered from an impact, but the same impact may have paved the way for our existence. So while some scientists argue that more impacts would have damaged life on Earth, others argue that more impacts might have sped up evolution.

The role of Jupiter has led some scientists to wonder whether we could exist if our solar system had been laid out differently. Could it be that civilizations can arise only in solar systems that happen to have a Jupiter-like planet in a Jupiter-like orbit? No one yet knows the answer to this question. What we do know is that Jupiter and the other jovian planets have had profound effects on life on Earth and will continue to have effects in the future.

## The BIG Picture PUTTING CHAPTER 12 INTO CONTEXT

*In this chapter we concluded the study of our own solar system by focusing on its smallest objects, finding that these objects can have big consequences. Keep in mind the following "big picture" ideas:*

- Asteroids and comets may be small compared to planets, but they are very important scientifically, because they provide evidence that has helped us understand how the solar system formed.
- Pluto, once considered a "misfit" among the planets, is now recognized as just one of many moderately sized objects in the Kuiper belt. In terms of composition, the dwarf planets Pluto and Eris—along with other similar objects—are essentially comets of unusually large size.
- The small bodies are subject to the gravitational whims of the largest. The jovian planets shaped the asteroid belt, the Kuiper belt, and the Oort cloud, and they continue to nudge objects onto collision courses with the planets.
- Collisions not only bring meteorites and leave impact craters but also can profoundly affect life on Earth. An impact probably wiped out the dinosaurs, and future impacts pose a threat that we cannot ignore.

**MY COSMIC PERSPECTIVE** Perhaps more than any other type of astronomical object, the small bodies of the solar system have the potential to touch our lives directly. We can hold pieces of them as meteorites, their impacts pose at least some threat to our survival, their compositions make them potentially useful, and, some 65 million years ago, one of them hit the Earth in an event that may have paved the way for our existence.

## Summary of Key Concepts

### 12.1 Classifying Small Bodies

- **What's the difference between an asteroid, a comet, and a dwarf planet?** In general, an **asteroid** is a rocky leftover planetesimal that orbits the Sun, a **comet** is similar but ice-rich in composition, and a **dwarf planet** can be either an asteroid or a comet that is large enough to be round. However, precise definitions are still debated, and boundaries between categories are sometimes fuzzy.
- **What are meteors and meteorites?** Small pieces of dust or rock continually enter Earth's atmosphere. Technically, a **meteor** is the flash of light that we see as a bit of rock burns up at high speed in the atmosphere, and most meteor particles burn up completely. In the relatively rare cases where the particle is large enough to survive and reach the ground, the rock that we find is called a **meteorite**.

### 12.2 Asteroids

- **What are asteroids like?** Asteroids come in a wide range of sizes and shapes, though small ones are by far the most common. All are rocky in composition, but those at greater distances from the Sun contain more carbon compounds and water. Despite their enormous numbers, the total mass of all asteroids combined is less than that of our Moon.

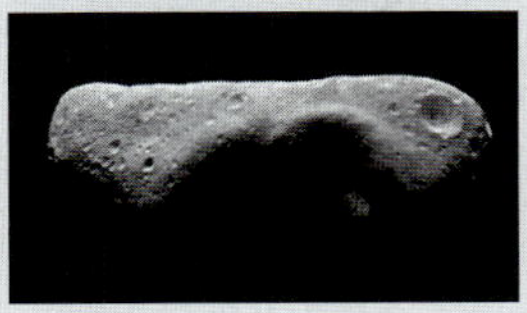

- **What do meteorites tell us about asteroids?** Meteorites represent samples of asteroids. **Primitive meteorites** are essentially unchanged since the birth of the solar system and tell us about the material that accreted to make asteroids and planets. **Processed meteorites** are fragments of larger asteroids that underwent differentiation, telling us that many asteroids had substantial interior heat and volcanism.
- **Why is there an asteroid belt?** Orbital resonances with Jupiter disrupted the orbits of planetesimals in the asteroid belt, preventing them from accreting into a planet. Many were ejected, but some remained and make up the asteroid belt today. Most asteroids in other regions of the inner solar system accreted into one of the planets.

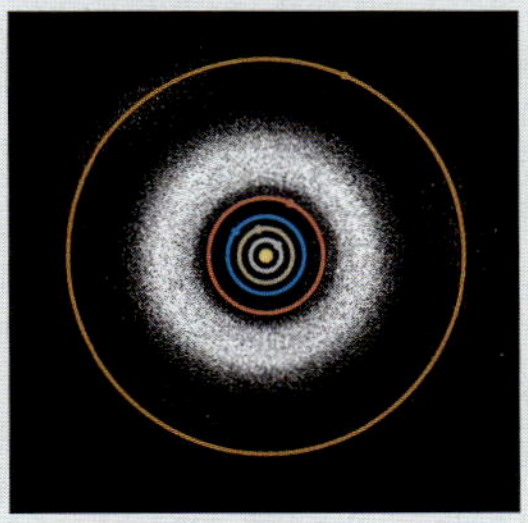

### 12.3 Comets

- **Why do comets grow tails?** The composition of comets makes them much like "dirty snowballs." Far from the Sun, the snowball is frozen as the comet's **nucleus**. As a comet comes inward, ice vaporizes into gas, which along with escaping dust forms a **coma** and two tails: a **plasma tail** of ionized gas and a **dust tail**. Larger particles can also escape, becoming the particles that cause **meteor showers** on Earth.

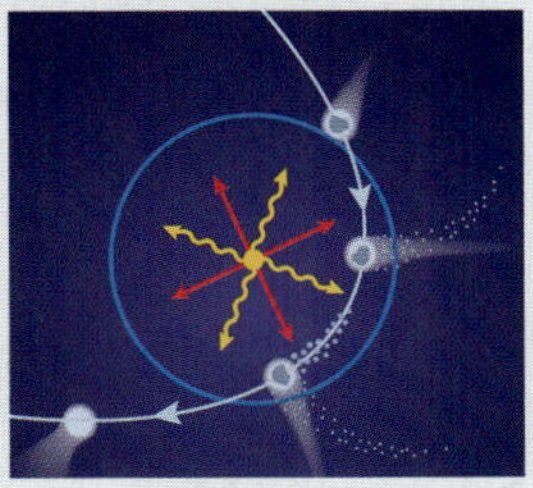

- **Where do comets come from?** Comets come from two reservoirs: the **Kuiper belt** and the **Oort cloud**. The Kuiper belt comets still reside in the region beyond Neptune in which they formed. The Oort cloud comets formed between the jovian planets and were flung out to great distances from the Sun by gravitational encounters with the planets.

## 12.4 Pluto and the Kuiper Belt

- **What is Pluto like?** Pluto is an ice-rich world that orbits the Sun in resonance with Neptune, so that it sometimes comes closer to the Sun than Neptune does, but there is no danger of a collision. It has a large moon, Charon, possibly formed in a giant impact, along with several smaller moons. Pluto has a thin atmosphere, and the *New Horizons* mission revealed a more varied and more active surface than anyone had expected.

- **Why is there a Kuiper belt?** The Kuiper belt consists primarily of leftover planetesimals that formed beyond the orbit of Neptune, but sophisticated models suggest the process was chaotic and violent. The jovian planets probably migrated to their current orbits after forming closer together, while most of the mass of the Kuiper belt was lost through collisions or ejections into deep space.

## 12.5 Cosmic Collisions: Small Bodies Versus the Planets

- **Did an impact kill the dinosaurs?** It may not have been the sole cause, but a major impact clearly coincided with the **mass extinction** in which the dinosaurs died out, about 65 million years ago. Sediments from the time contain iridium and other clear evidence of an impact, and an impact crater of the right age lies near the coast of Mexico.

- **How great is the impact risk today?** Impacts certainly pose a threat, though the probability of a major impact in our lifetimes is fairly low. An impact like the Tunguska event may occur every couple hundred years and would be catastrophic if it occurred in a populated area.

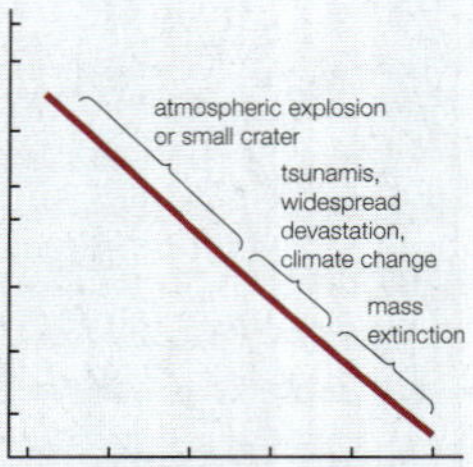


- **How do the jovian planets affect impact rates and life on Earth?** Impacts are always linked in at least some way to the gravitational influences of Jupiter and the other jovian planets. These influences have shaped the asteroid belt, the Kuiper belt, and the Oort cloud, and continue to determine when an object is flung our way.

# Visual Skills Check

*Use the following questions to check your understanding of some of the many types of visual information used in astronomy. For additional practice, try the Chapter 12 Visual Quiz in the Study Area at www.MasteringAstronomy.com.*

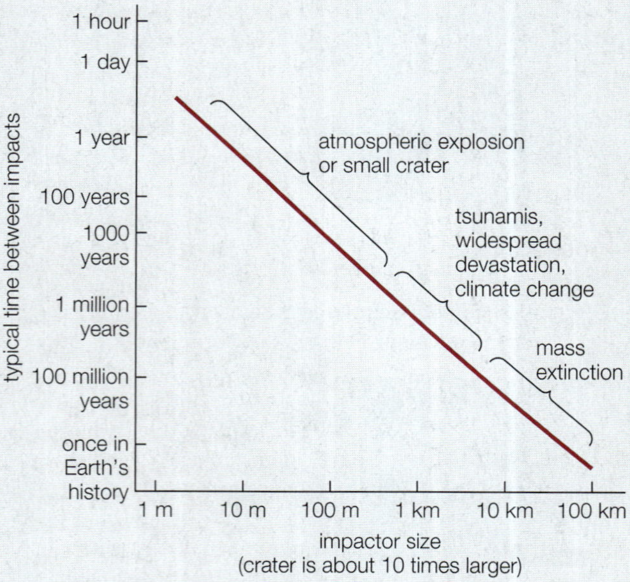


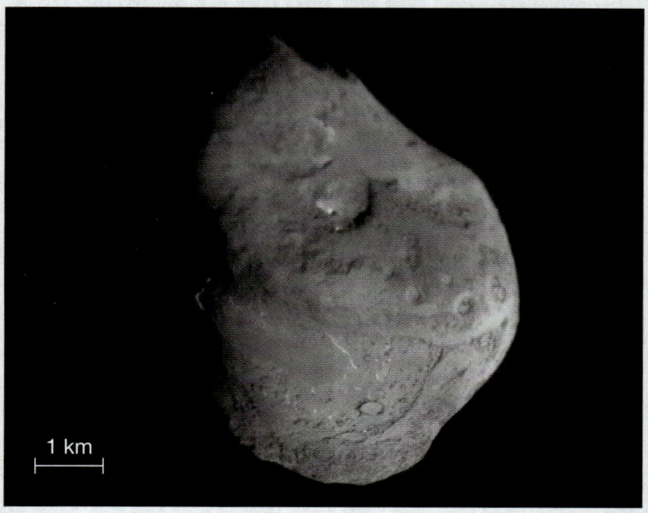

*The graph on the preceding page (Figure 12.33) shows how often impacts occur for objects of different sizes. The photo shows Comet Tempel 1 moments before the Deep Impact spacecraft crashed into it.*

1. Estimate Comet Tempel 1's diameter, using the scale bar in the photo.
2. According to the graph, how frequently do objects the size of Comet Tempel 1 strike Earth?
   a. once in Earth's history
   b. about once every hundred million years
   c. about once every million years
   d. about once every thousand years
3. Consider an object twice the size of Comet Tempel 1. What kind of damage would this object cause if it hit our planet?
   a. mass extinction
   b. widespread devastation and climate change
   c. atmospheric explosion or a small crater
4. Meteor Crater in Arizona is about 1.2 kilometers across. According to the graph, about how big was the object that made this crater? (*Note*: Be sure to read the axis labels carefully.)
   a. 1 meter
   b. 10 meters
   c. 100 meters
   d. 1 kilometer
5. How often do objects big enough to create craters like Meteor Crater impact Earth?
   a. once in Earth's history
   b. about once every ten thousand years
   c. about once every few million years
   d. about once every day, but most burn up in the atmosphere or land in the ocean

# Exercises and Problems

*For instructor-assigned homework and other learning materials, go to www.MasteringAstronomy.com.*

## Chapter Review Questions

### Short-Answer Questions Based on the Reading

1. Briefly define *asteroid, comet, dwarf planet, meteor,* and *meteorite*. How did the discovery of Eris force astronomers to reconsider the definition of *planet*?
2. Briefly describe asteroid sizes, shapes, masses, densities, and compositions. How does the total mass of all asteroids compare to the mass of a terrestrial world?
3. Distinguish between *primitive meteorites* and *processed meteorites* in terms of both composition and origin.
4. What do meteorites and spacecraft observations tell us about the geology of asteroids?
5. Where is the *asteroid belt* located, and why? Explain how orbital resonances with Jupiter affect the asteroid belt.
6. Under what circumstances do we see a comet with a *nucleus*, *coma*, and *tails*? In what direction do the tails point?
7. How are *meteor showers* linked to comets, and why do they recur at about the same time each year?
8. How do we know the *Kuiper belt* and *Oort cloud* exist? Describe each in terms of location, the orbits and number of comets within it, and its likely origin.
9. Briefly describe Pluto and Charon. Why won't Pluto collide with Neptune? How do we think Charon formed?
10. What clues do other Kuiper belt comets offer about the origin of the belt? How is Neptune's moon Triton important to our understanding?
11. Briefly describe how current models explain the origin of the Kuiper belt. What role does planetary *migration* play?
12. Briefly describe the evidence suggesting that an impact caused the *mass extinction* that killed off the dinosaurs. How might the impact have led to the mass extinction?
13. How often should we expect impacts of various sizes on Earth? How serious a threat do we face from these impacts?
14. Briefly summarize the role of the jovian planets in shaping the orbits of small bodies in the solar system and in influencing life on Earth.

### Surprising Discoveries?

*Suppose someone claimed to make the following discoveries. (These are not real discoveries.) Decide whether each discovery should be considered reasonable or surprising. Explain.*

15. A small asteroid that orbits within the asteroid belt has an active volcano.
16. Scientists discover a meteorite that, based on radiometric dating, is 7.9 billion years old.
17. An object that resembles a comet in size and composition is discovered orbiting in the inner solar system.
18. Studies of a large object in the Kuiper belt reveal that it is made almost entirely of rocky (as opposed to icy) material.
19. Astronomers discover a previously unknown comet that will be brightly visible in our night sky about 2 years from now.
20. A mission to Eris finds that it has lakes of liquid water on its surface.
21. Geologists discover a crater from a 5-kilometer object that impacted Earth more than 100 million years ago.
22. Archaeologists learn that the fall of ancient Rome was caused in large part by an asteroid impact in Asia.
23. In another solar system, astronomers discover an object the size of Earth orbiting its star at the distance of the Kuiper belt.
24. Astronomers discover an asteroid with an orbit suggesting that it will impact Earth in the year 2064.

### Quick Quiz

*Choose the best answer to each of the following. For additional practice, try the Chapter 12 Reading and Concept Quizzes in the Study Area at www.MasteringAstronomy.com.*

25. The asteroid belt lies between the orbits of (a) Earth and Mars. (b) Mars and Jupiter. (c) Jupiter and Saturn.
26. Jupiter nudges the asteroids through the influence of (a) tidal forces. (b) orbital resonances. (c) magnetic fields.

27. Can an asteroid be pure metal? (a) No; all asteroids contain rock. (b) Yes; it must have formed where only metal could condense in the solar nebula. (c) Yes; it must be from the core of a shattered asteroid.
28. Did a large terrestrial planet ever form in the region of the asteroid belt? (a) No, because there was never enough mass there. (b) No, because Jupiter prevented one from accreting. (c) Yes, but it was shattered by a giant impact.
29. What does Pluto most resemble? (a) a terrestrial planet (b) a jovian planet (c) a comet
30. How big an object causes a typical "shooting star"? (a) a grain of sand or a small pebble (b) a boulder (c) an object the size of a car
31. Which have the most elliptical and tilted orbits? (a) asteroids (b) Kuiper belt comets (c) Oort cloud comets
32. Which are thought to have formed farthest from the Sun? (a) asteroids (b) Kuiper belt comets (c) Oort cloud comets
33. About how often does a 1-kilometer object strike Earth? (a) every year (b) every million years (c) every billion years
34. What would happen if a 1-kilometer object struck Earth? (a) It would break up in the atmosphere without causing widespread damage. (b) It would cause widespread devastation and climate change. (c) It would cause a mass extinction.

## Inclusive Astronomy

*Use these questions to reflect on participation in science.*

35. *Group Discussion: Who Owns the Asteroids?* Within the next decade or two, it might become profitable for nations or private companies to mine asteroids in pursuit of rare minerals. Several private companies are already laying the groundwork for such mining.
**a.** Gather in small groups (two to four students) and discuss whether nations and/or private companies should be able to claim ownership rights to minerals that they mine from an asteroid and bring to Earth. Make two lists, one of reasons why ownership should be allowed and one of why it shouldn't.
**b.** Debate the pros and cons represented in your two lists and then try to come to a group consensus on whether ownership should be allowed and, if so, what regulations it should be subject to.
**c.** Make a list of ways in which asteroid mining might be used for scientific purposes. If the mining is done for commercial reasons, what obligation should the miners (nations or companies) have to also generate scientific value from their work?

## The Process of Science

*These questions may be answered individually in short-essay form or discussed in groups, except where identified as group-only.*

36. *The Pluto Debate.* Research the decision to demote Pluto to dwarf planet status. In your opinion, is this a good example of the scientific process? Does it exhibit the hallmarks of science described in Chapter 3? Compare your conclusions to opinions you find about the debate, and describe how you think astronomers should handle this or similar debates in the future.
37. *Life-or-Death Astronomy.* In most cases, the study of the solar system has little direct effect on our lives. But the discovery of an asteroid or comet on a collision course with Earth is another matter. How should the standards for verifiable observations described in Chapter 3 apply in this case? Is the potential danger so great that any astronomer with any evidence of an impending impact should spread the word as soon as possible? Or is the potential for panic so great that even higher standards of verification ought to be applied? What kind of review process, if any, would you set in place? Who should be informed of an impact threat, and when?
38. *Group Activity: Assessing Impact Danger.* Work in small groups to assess the risks we face on Earth from meteorite and comet impacts. Note: You may wish to do this activity using the four roles described in Chapter 1, Exercise 39.
**a.** Consider how you might estimate the chance of human civilization being destroyed by an impact during your lifetime. Determine the kinds of information you would need to come up with an estimate, develop a method for making the estimate, and write down your method.
**b.** Analyze Figure 12.33 and determine whether it contains any of the necessary information.
**c.** Apply your group's method to estimate the probability that civilization will be destroyed by an impact during your lifetime, which you can assume to be 100 years for the purpose of this exercise.
**d.** Estimate the probability that an impact will cause "widespread devastation" somewhere on Earth during your lifetime.
**e.** Finding near-Earth asteroids early greatly increases our chances of deflecting them. Given the probabilities you found in parts c and d and considering the damage these events would cause, decide as a group how much money per year should be spent on finding near-Earth asteroids. Explain your reasoning.

## Investigate Further

### Short-Answer/Essay Questions

39. *The Role of Jupiter.* Suppose that Jupiter had never existed. Describe at least three ways in which our solar system would be different, and clearly explain why.
40. *Life Story of an Iron Atom.* Imagine that you are an iron atom in a processed meteorite made mostly of iron. Tell the story of how you got to Earth, beginning when you were part of the gas in the solar nebula 4.6 billion years ago. Include as much detail as possible. Your story should be scientifically accurate but also creative and interesting.
41. *Asteroid Discovery.* You have discovered two new asteroids and named them Albert and Isaac. Both lie at the same distance from Earth and have the same brightness when you look at them through your telescope, but Albert is twice as bright as Isaac at infrared wavelengths. What can you deduce about the relative reflectivities and sizes of the two asteroids? Which would make a better target for a mission to mine metal? Which would make a better target for a mission to obtain a sample of a carbon-rich planetesimal? Explain.
42. *Asteroids vs. Comets.* Contrast the compositions and locations of comets and asteroids, and explain in your own words why they have turned out differently.
43. *Comet Tails.* Describe in your own words why comets have tails. Why do most comets have two distinct visible tails, and why do the tails go in different directions? Why is the third, invisible tail of small pebbles of interest to us on Earth?

44. *Oort Cloud vs. Kuiper Belt.* Explain in your own words how and why there are two different reservoirs of comets. Be sure to discuss where the two groups of comets formed and what kinds of orbits they travel on.
45. *Rise of the Mammals.* Suppose the impact 65 million years ago had not occurred. How do you think our planet would be different? For example, do you think that mammals still would eventually have come to dominate Earth? Would we be here? Defend your opinions.
46. *Project: Dirty Snowballs.* If there is snow where you live or study, make a dirty snowball. (The ice chunks that form behind tires work well.) How much dirt does it take to darken snow? Find out by allowing your dirty snowball to melt in a container and measuring the approximate proportions of water and dirt afterward.

## Quantitative Problems

*Be sure to show all calculations clearly and state your final answers in complete sentences.*

47. *Adding Up Asteroids.* It's estimated that there are a million asteroids 1 kilometer across or larger. If a million asteroids 1 kilometer across were all combined into one object, how big would it be? How many 1-kilometer asteroids would it take to make an object as large as Earth? (*Hint:* You can assume the asteroids are spherical. The expression for the volume of a sphere is $\frac{4}{3}\pi r^3$, where $r$ is the radius.)
48. *Impact Energies.* A relatively small impact crater 20 kilometers in diameter could be made by a comet 2 kilometers in diameter traveling at 30 kilometers per second (30,000 m/s).
**a.** Assume that the comet has a total mass of $4.2 \times 10^{12}$ kilograms. What is its total kinetic energy? (*Hint:* The kinetic energy is equal to $\frac{1}{2}mv^2$, where $m$ is the comet's mass and $v$ is its speed. If you use mass in kilograms and velocity in m/s, the answer for kinetic energy will have units of joules.)
**b.** Convert your answer from part a to an equivalent in megatons of TNT, the unit used for nuclear bombs. Comment on the degree of devastation the impact of such a comet could cause if it struck a populated region on Earth. (*Hint:* One megaton of TNT releases $4.2 \times 10^{15}$ joules of energy.)
49. *The "Near Miss" of Toutatis.* The 5-kilometer asteroid Toutatis passed a mere 1.5 million kilometers from Earth in 2004. Suppose Toutatis were destined to pass *somewhere* within 1.5 million kilometers of Earth. Calculate the probability that this "somewhere" would have meant that it slammed into Earth. Based on your result, do you think it is fair to call the 2004 passage a "near miss"? Explain. (*Hint:* You can calculate the probability by considering an imaginary dartboard of radius 1.5 million kilometers in which the bull's-eye has Earth's radius, 6378 kilometers.)
50. *Room to Roam.* It's estimated that there are a trillion comets in the Oort cloud, which extends out to about 50,000 AU. What is the total volume of the Oort cloud, in cubic AU? How much space does each comet have in cubic AU, on average? Take the cube root of the average volume per comet to find the comets' typical spacing in AU. (*Hints:* For the purpose of this calculation, you can assume the Oort cloud fills the whole sphere out to 50,000 AU. The volume of a sphere is given by $\frac{4}{3}\pi r^3$, where $r$ is the radius.)
51. *Comet Temperatures.* Find the "no greenhouse" temperatures for a comet at distances from the Sun of 50,000 AU (in the Oort cloud), 3 AU, and 1 AU (see Mathematical Insight 10.1). Assume that the comet reflects 3% of the incoming sunlight. At which location will the temperature be high enough for water ice to vaporize (about 150 K)? How do your results explain comet anatomy?
52. *Comet Dust Accumulation.* A few hundred tons of comet dust are added to Earth daily from the millions of meteors that enter our atmosphere. Estimate the time it would take for Earth to get 0.1% heavier at this rate. Is this mass accumulation significant for Earth as a planet? Explain.

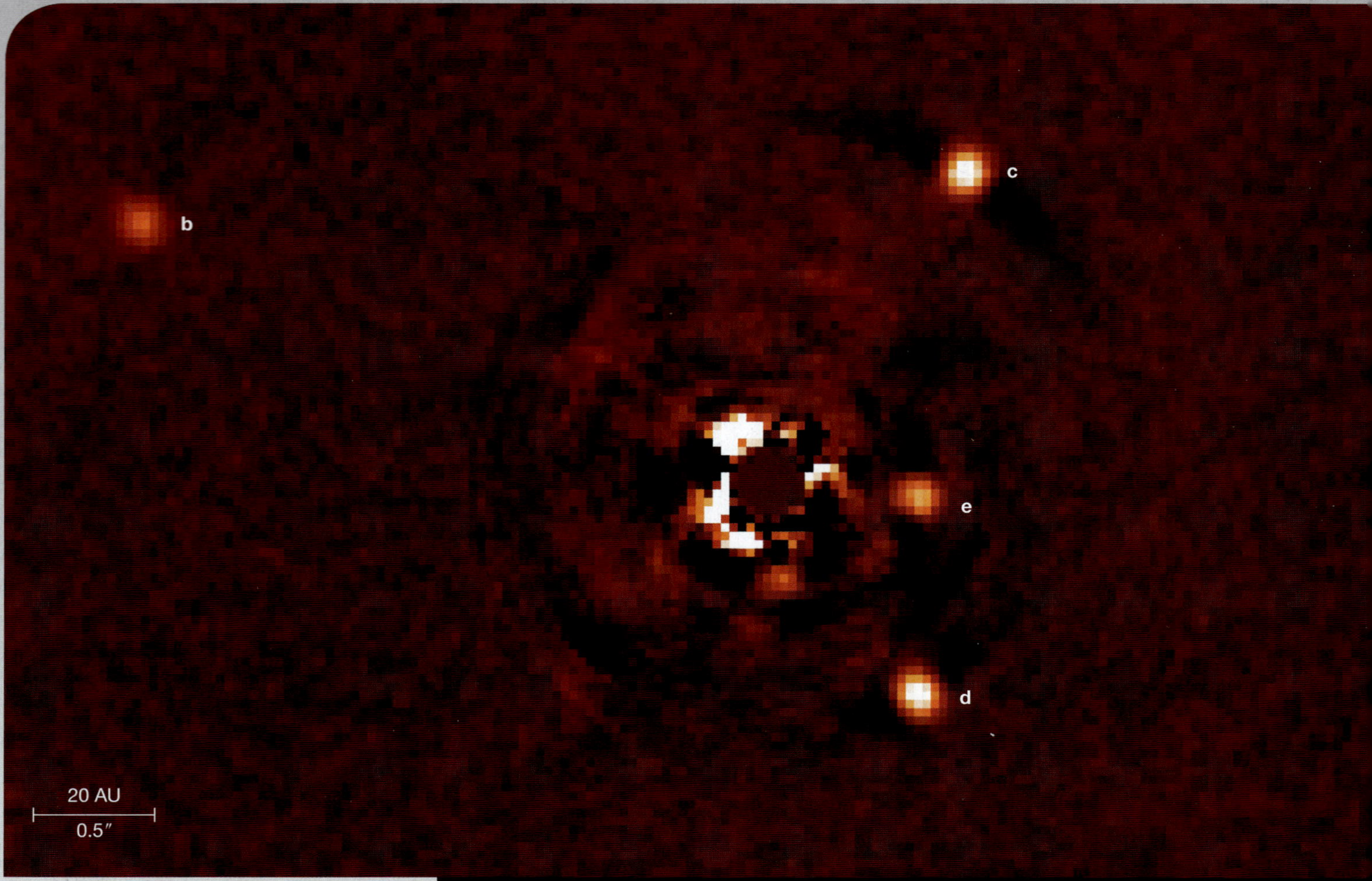


# 13 Other Planetary Systems

## THE NEW SCIENCE OF DISTANT WORLDS

▲ **About the photo:** This infrared image from the Large Binocular Telescope shows direct detection of four planets (marked b, c, d, e) orbiting the star HR 8799. Light from the star itself (center) was mostly blocked out during the exposure, as indicated by the solid red circle.

## LEARNING GOALS

### 13.1 Detecting Planets Around Other Stars

- Why is it so challenging to learn about extrasolar planets?
- How can a star's motion reveal the presence of planets?
- How can changes in a star's brightness reveal the presence of planets?

### 13.2 The Nature of Planets Around Other Stars

- What properties of extrasolar planets can we measure?
- How do extrasolar planets compare to planets in our solar system?

### 13.3 The Formation of Other Solar Systems

- Do we need to modify our theory of solar system formation?
- Are planetary systems like ours common?

### 13.4 The Future of Extrasolar Planetary Science

- How will future observations improve our understanding?

*How vast those Orbs must be, and how inconsiderable this Earth, the Theatre upon which all our mighty Designs, all our Navigations, and all our Wars are transacted, is when compared to them. A very fit consideration, and matter of Reflection, for those Kings and Princes who sacrifice the Lives of so many People, only to flatter their Ambition in being Masters of some pitiful corner of this small Spot.*

—*Christiaan Huygens, c. 1690*

**Chapter 13 Overview**

For more than four centuries after the Copernican revolution taught us that Earth is just one member of our Sun's planetary system, the study of planetary systems remained limited to our own. Then, less than three decades ago, a new scientific revolution began with the first discoveries of planets around other stars. This revolution is now in full bloom, with new discoveries coming at a rapid pace.

The advancing science of extrasolar planets has dramatic implications for our understanding of our place in the universe. The fact that planets are common in the universe makes it seem more likely that we might someday find life elsewhere, perhaps even intelligent life. Moreover, having many more worlds to compare to our own vastly enhances our ability to learn how planets work, which may help us better understand our home planet, Earth. It also allows us to test the nebular theory of solar system formation in new settings. In this chapter, we'll focus our attention on the exciting new science of other planetary systems.

## 13.1 Detecting Planets Around Other Stars

The very idea of planets around other stars, or **extrasolar planets** for short, would have shattered the world views of many people throughout history. After all, cultures of the western world long regarded Earth as the center of the universe, and nearly all ancient cultures imagined the heavens to be a realm distinct from Earth.

The Copernican revolution, which taught us that Earth is a planet orbiting the Sun, opened up the possibility that planets might also orbit other stars. Still, until the 1990s, no extrasolar planets were known. Given that the number of known extrasolar planets is now in the thousands and growing rapidly, you may wonder what took so long.

### Why is it so challenging to learn about extrasolar planets?

We've known for centuries that other stars are distant suns (see Special Topic, page 375), making it natural to suspect that they would have their own planetary systems. The nebular theory of solar system formation, well established decades ago, made such systems seem even more likely, because it predicts that planet formation should be a natural part of the star formation process. However, it took many decades for technology to reach the point at which this prediction could be put to the test.

Detecting extrasolar planets poses a huge technological challenge for two basic reasons. First, as we discussed in Chapter 1, planets are extremely tiny compared to the vast distances between stars. Second, stars are typically *a billion times* brighter than the light reflected by any orbiting planets, so starlight tends to overwhelm any planetary light in photographs. This problem is somewhat lessened—but not eliminated—if we observe in infrared light, because planets emit their own infrared light and stars are usually dimmer in the infrared.

The first step in understanding how scientists are overcoming these challenges is to recognize that there are two general ways of learning about a distant object: *directly*, by obtaining images or spectra of the object, and *indirectly*, by inferring the object's existence or properties without actually seeing the object itself. While scientists have achieved some success in direct detection (as illustrated by the image that opens this chapter), most of our current understanding of extrasolar planets comes from indirect study. There are two major indirect approaches to finding and studying extrasolar planets:

1. Observing the motion of a star to detect the subtle gravitational effects of orbiting planets
2. Observing changes in a star's brightness that occur when one of its planets passes in front of the star as viewed from Earth

The earliest discoveries of extrasolar planets came primarily through the first approach, but we've now found many more planets through the second. We learn even more in cases in which we can combine both indirect approaches, because each can provide different information about a distant planet. Because these two approaches to indirect study are so important to current understanding of extrasolar planets, we will devote the rest of this section to studying them in somewhat more detail.

### How can a star's motion reveal the presence of planets?

Although we usually think of a star as remaining still while planets orbit around it, that is only approximately correct. In reality, all the objects in a star system, including the star itself, orbit the system's "balance point," or *center of mass* [**Section 4.4**]. To understand how this fact allows us to discover extrasolar planets, imagine the viewpoint of extraterrestrial astronomers observing our solar system from afar.

Let's start by considering only the influence of Jupiter (**FIGURE 13.1**). Jupiter is the most massive planet, but the Sun is still about a thousand times as massive. As a result, the center of mass between the Sun and Jupiter lies about one-thousandth of the way from the center of the Sun to the center of Jupiter, putting it just outside the Sun's visible surface. In other words, what we usually think of as Jupiter's 12-year orbit around the Sun is really a 12-year

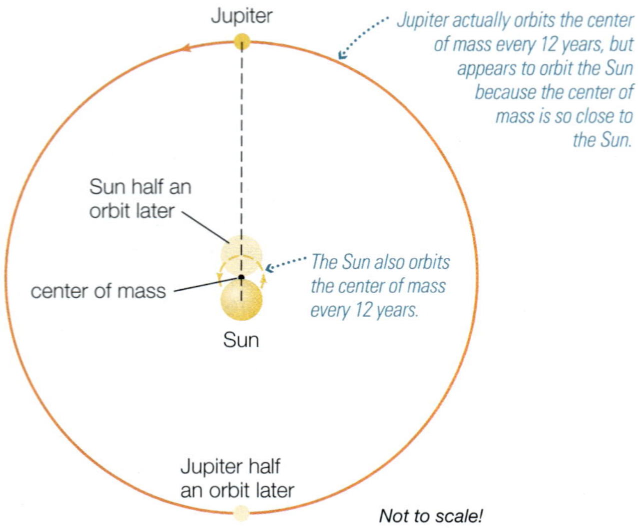


**FIGURE 13.1** This diagram shows how both the Sun and Jupiter orbit around their mutual center of mass, which lies very close to the Sun. The diagram is not to scale; the sizes of the Sun and its orbit are exaggerated about 100 times compared to the size shown for Jupiter's orbit, and Jupiter's size is exaggerated even more.

orbit around the center of mass. Because the Sun and Jupiter are always on opposite sides of the center of mass (that's what makes it a "center"), the Sun must orbit this point with the same 12-year period. The Sun's orbit traces out only a small ellipse with each 12-year period, because the Sun's average orbital distance is barely larger than its own radius; that is why we generally don't notice the Sun's motion. Nevertheless, with sufficiently precise measurements, extraterrestrial astronomers could detect this orbital movement of the Sun and thereby deduce the existence of Jupiter, even without having observed Jupiter itself. They could even determine Jupiter's mass from the orbital characteristics of the Sun as it goes around the center of mass. A more massive planet at the same distance would pull the center of mass farther from the Sun's center, giving the Sun a larger orbit. Because the Sun's period around the center would still be 12 years, the larger orbit would mean a faster orbital speed around the center of mass.

**See it for yourself** To see how a small planet can make a big star wobble, find a pencil and tape a heavy object (such as a set of keys) and a lighter object (perhaps a few coins) to opposite ends. Tie a string (or piece of floss) at the balance point—the center of mass—so that the pencil is horizontal. Then tap the lighter object into "orbit" around the heavier object. What does the heavier object do, and why? How does your setup correspond to a planet orbiting a star? You can experiment further with objects of different weights or shorter pencils; try to explain the differences you see.

Now let's add in the effects of Saturn. Saturn takes 29.5 years to orbit the Sun, so by itself it would cause the Sun to orbit their mutual center of mass every 29.5 years. However, because Saturn's influence is secondary to that of Jupiter, this 29.5-year period appears as a small effect added on top of the Sun's 12-year orbit around its center of mass with Jupiter. In other words, every 12 years the Sun would return to *nearly* the same orbital position around its center of mass with Jupiter, but the precise point of return would move around with Saturn's 29.5-year period. By measuring this motion carefully over many years, an extraterrestrial astronomer could deduce the existence and masses of both Jupiter and Saturn.

The other planets also exert gravitational tugs on the Sun, each contributing a small additional effect to the effects of Jupiter and Saturn (**FIGURE 13.2**). In principle, with sufficiently precise measurements of the Sun's orbital motion made over many decades, an extraterrestrial astronomer could deduce the existence of all the planets of our solar system. If we turn this idea around, it means we can search for planets in other star systems by carefully watching for the tiny orbital motion of a star around the center of mass of its star system.

**The Astrometric Method** Astronomers use two distinct observation methods to search for the gravitational tugs of planets on stars. The first, called the **astrometric method** (*astrometry* means "measurement of the stars"), uses very precise measurements of stellar positions in the sky to look for the slight motion caused by orbiting planets. If a star "wobbles" gradually around its average position (the center of mass), we must be observing the influence of unseen planets.

The astrometric method is essentially the same method that astronomers have long used to identify many binary star systems, in which photos taken at different times allow us to see the motions of two stars orbiting around

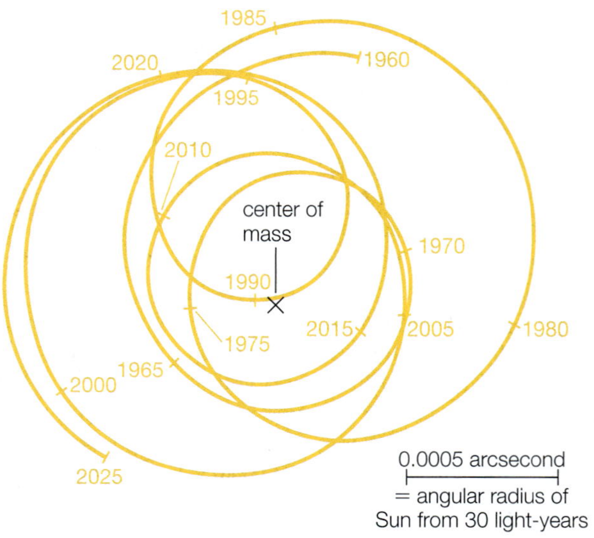


**FIGURE 13.2** This diagram shows the orbital path of the Sun around the center of mass of our solar system as it would appear over a 65-year period (1960–2025) from a distance of 30 light-years away. Notice that the entire range of motion during this period is only about 0.0015 arcsecond, which is almost 100 times smaller than the angular resolution of the Hubble Space Telescope. Nevertheless, if alien astronomers could measure this motion, they could learn of the existence of planets in our solar system.

their center of mass [**Section 15.1**]. However, we face two practical limitations in applying this method to the search for planets.

The first limitation stems from the fact that we are searching for changes in position that are very small even for nearby stars, and these changes become smaller for more distant stars. For example, from a distance of 10 light-years, a Jupiter-size planet in a Jupiter-like orbit (5 AU from a Sun-like star) would cause its star to move slowly over a side-to-side angular distance of only about 0.003 arcsecond—approximately the width of a hair seen from a distance of 5 kilometers. At double the distance (20 light-years) the observed motion would be only half as large (0.0015 arcsecond), and at ten times the distance (100 light-years) it would be only one-tenth as large (0.0003 arcsecond). For this reason, the astrometric method is best suited to searching for relatively massive planets around relatively nearby stars.

To understand the second limitation, consider what would happen if we could move Jupiter into an orbit farther from the Sun. This larger orbit would increase the angular extent of the Sun's side-to-side motion as seen from a distance (because moving Jupiter outward would also move the center of mass outward from the Sun), but Kepler's third law tells us that this move would also increase Jupiter's orbital period. As a result, it would take a much longer time for alien astronomers to recognize Jupiter's effect. For example, if Jupiter moved to Neptune's distance from the Sun, at which the orbital period is 165 years, it might take a century or more of patient observation to be confident that the stellar motion was following an orbital pattern.

As a result of these limitations, the astrometric method has led to few discoveries to date. However, the European Space Agency's *GAIA* mission, launched in 2013, is in the process of obtaining astrometric observations of a billion stars in our galaxy to an accuracy that is in some cases better than 10 microarcseconds, equivalent to the angular width of a human hair viewed from a distance of 2000 kilometers. These observations should ultimately enable *GAIA* to detect thousands of extrasolar planets.

**The Doppler Method** The second method of searching for gravitational tugs due to orbiting planets is the **Doppler method**, which searches for a star's orbital movement around the center of mass by looking for changing Doppler shifts in its spectrum [**Section 5.4**]. Recall that the Doppler effect causes a blueshift when a star is moving toward us and a redshift when it is moving away from us, so alternating blueshifts and redshifts (relative to a star's average Doppler shift) indicate orbital motion around a center of mass (**FIGURE 13.3**).

## SPECIAL TOPIC How Did We Learn That Other Stars Are Suns?

Today we know that stars are other suns—meaning objects that produce enough energy through nuclear fusion to supply light and heat to orbiting planets—but this fact is not obvious from looking at the night sky. After all, the feeble light of stars hardly seems comparable to the majestic light of the Sun. Most ancient observers guessed that stars were much more mundane; typical guesses suggested that they were holes in the celestial sphere or flaming rocks in the sky.

The only way to realize that stars are suns is to know that they are incredibly far away; then, a simple calculation will show that they are actually as bright as or brighter than the Sun [**Section 15.1**]. The first person to make reasonably accurate estimates of the distances to stars was Christiaan Huygens. By *assuming* that other stars were indeed suns, as some earlier astronomers had guessed, Huygens successfully estimated stellar distances. The late Carl Sagan eloquently described the technique:

*Huygens drilled small holes in a brass plate, held the plate up to the Sun and asked himself which hole seemed as bright as he remembered the bright star Sirius to have been the night before. The hole was effectively* $\frac{1}{28{,}000}$ *the apparent size of the Sun. So Sirius, he reasoned, must be 28,000 times farther from us than the Sun, or about half a light-year away. It is hard to remember just how bright a star is many hours after you look at it, but Huygens remembered very well. If he had known that Sirius was intrinsically brighter than the Sun, he would have come up with [a better estimate of] the right answer: Sirius is 8.6 light-years away.**

*Christiaan Huygens (1629–1695)*

Huygens could not actually prove that stars are suns, since his method was based on the assumption that they are. However, his results explained a fact known since ancient times: Stellar parallax is undetectable to the naked eye [**Section 2.4**]. Recall that the lack of detectable parallax led many Greeks to conclude that Earth must be stationary at the center of the universe, but there was also an alternative explanation: Stars are incredibly far away. Even with his original estimate that Sirius was only half a light-year away, Huygens knew that its parallax would be far too small to observe by naked eye or with the telescopes available at the time. Huygens thereby "closed the loop" on the ancient mystery of the nature of stars, showing that their appearance and lack of parallax made perfect sense if they were very distant suns. This new knowledge apparently made a great impression on Huygens, as you can see from the quotation at the beginning of the chapter.

*From *Cosmos,* by Carl Sagan (Random House, 1980). Sagan demonstrates the technique in the *Cosmos* video series, Episode 7.

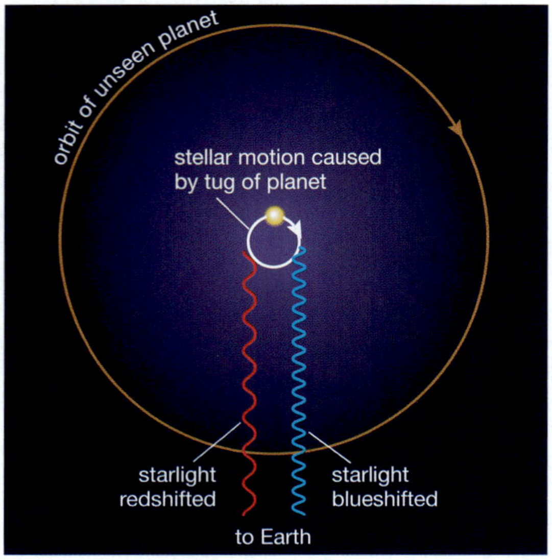


**FIGURE 13.3** The Doppler method for discovering extrasolar planets: The star's Doppler shift alternates toward the blue and toward the red, allowing us to detect its slight motion—caused by an orbiting planet—around the center of mass.

As an example, **FIGURE 13.4** shows data from the first discovery of an extrasolar planet around a Sun-like star: the 1995 discovery of a planet orbiting the star called 51 Pegasi. The 4-day period of the star's motion must be the orbital period of its planet. We therefore know that the planet lies so close to the star that its "year" lasts only 4 Earth days, which means its surface temperature is probably over 1000 K. The Doppler data also allow us to determine the planet's approximate mass, because a more massive planet has a greater gravitational effect on the star (for a given orbital distance) and therefore causes the star to move at higher speed around the system's center of mass. In the case of 51 Pegasi, the data show that the planet has about half the mass of Jupiter. Scientists therefore refer to this planet as a **hot Jupiter**, because it has a Jupiter-like mass but a much higher surface temperature.

Current techniques can measure a star's velocity to a fraction of a meter per second—much slower than walking speed—which corresponds to a Doppler wavelength shift on the order of 0.0000001%. We can therefore find planets that exert a considerably smaller gravitational tug on their stars than the planet orbiting 51 Pegasi does on its star. As of 2018, the Doppler method had been used to detect more than 700 planets, including more than 100 in multiplanet systems.

The Doppler method has three major limitations. First, like the astrometric method, the Doppler method searches for gravitational tugs from orbiting planets and is therefore better for finding massive planets like Jupiter than small planets like Earth. Second, the Doppler method is best suited to identifying these planets when they orbit relatively *close* to their star (in contrast to the astrometric method, which works better for planets farther from their star); close-in planets are easier to detect both because being closer means a stronger gravitational tug and hence a greater velocity for the star (which is easier to measure) and because closer-in planets have shorter orbital periods that allow their orbits to be observed in shorter amounts of time. Together, these first two limitations explain why many of the earliest discoveries of extrasolar planets were of hot Jupiters like the one orbiting 51 Pegasi. The third limitation is that the Doppler method requires a fairly large telescope (and long exposure time) in order to obtain spectra with high enough resolution to reveal very small Doppler shifts, which limits the number of stars that can be studied with this method.

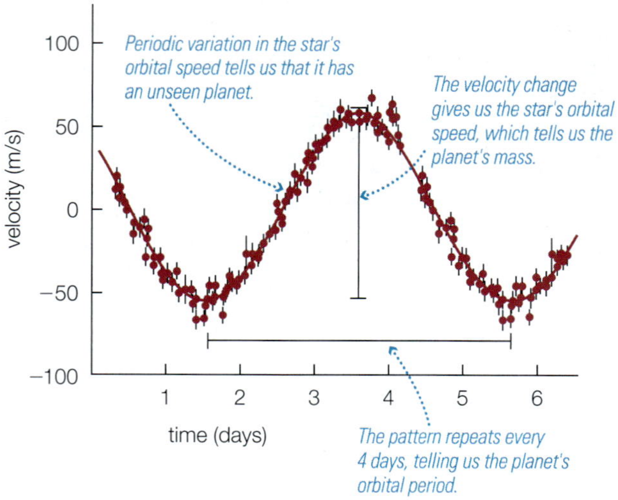


**a** A periodic Doppler shift in the spectrum of the star 51 Pegasi shows the presence of a large planet with an orbital period of about 4 days. Dots are actual data points; bars through dots represent measurement uncertainty.

**b** Artist's conception of the planet orbiting 51 Pegasi, which probably has a mass similar to that of Jupiter but orbits its star at only about one-eighth of Mercury's orbital distance from the Sun. It probably has a surface temperature above 1000 K, making it an example of what we call a hot Jupiter.

**FIGURE 13.4** The discovery of a planet orbiting 51 Pegasi.

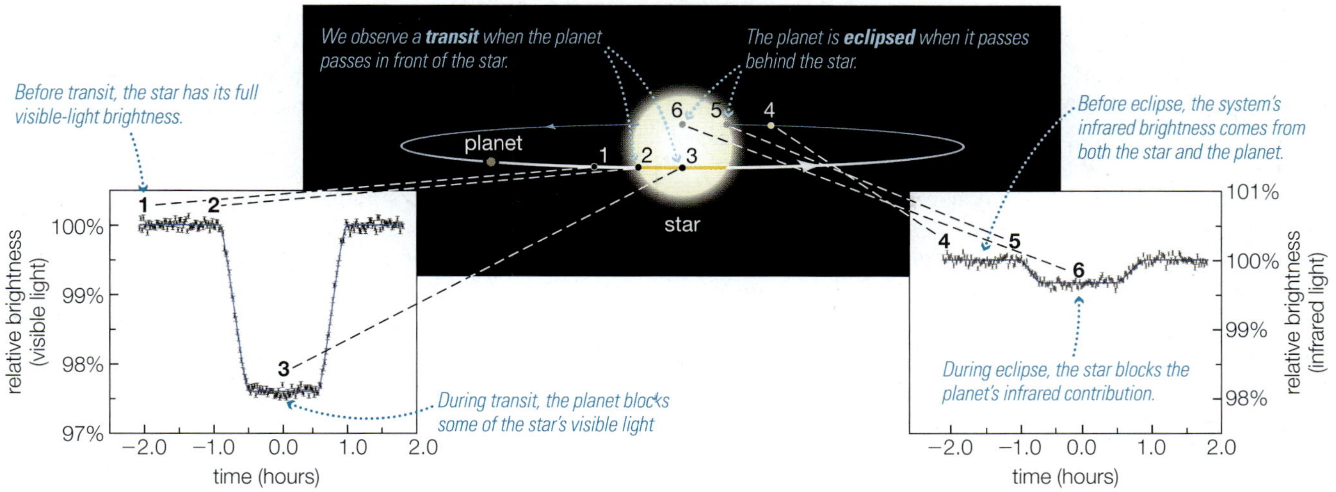


**FIGURE 13.5** This diagram represents transits and eclipses of the planet orbiting the star HD 189733. The graph shows that each transit lasts about 2 hours, during which the star's visible-light brightness dips by about 2.5%. The eclipses are observable in the infrared, because the star blocks the infrared contribution of the planet. The transits and eclipses each occur once during every 2.2-day orbit of the planet.

## How can changes in a star's brightness reveal the presence of planets?

The second general approach to detecting planets indirectly relies on searching for slight changes in a star's brightness caused by orbiting planets. If we were to examine a large sample of stars with planets, a small number of these systems—probably about 1%—would by chance be aligned in such a way that one or more of the star's planets would pass directly between us and the star once each orbit. The result is a **transit**, in which the planet appears to move across the face of the star. We occasionally witness this effect in our own solar system when Mercury or Venus crosses in front of the Sun (see Figure S1.5).

**The Transit Method** Other star systems are so far away that we cannot actually see a planetary dot set against the face of the star as we can see Mercury or Venus against the face of the Sun. Nevertheless, a transiting planet will block a little of its star's light as it passes across the line of sight from us to the star, and the **transit method** searches for these temporary dimmings of a star (**FIGURE 13.5**). The larger the planet, the more dimming it will cause. Some transiting planets also undergo a measurable **eclipse** as the planet goes behind the star; eclipses also cause a small dip in a system's brightness, because during the eclipse we see light only from the star rather than from both the star and the planet. Eclipse observations are more easily accomplished in the infrared, because planets contribute a greater proportion of a system's infrared brightness than visible-light brightness.

**Think about it** Suppose you were an alien astronomer searching for planets going around our Sun. Where would your star system have to be located in order for you to observe transits across the Sun? Which planets in our solar system would be the easiest to discover through transits? Explain.

Modern telescopes can be used to monitor large numbers of stars in search of transits. For example, NASA's *Kepler* mission spent about four years (2009 to 2013) measuring the brightnesses of about 150,000 stars about every 30 minutes in search of dimmings that might indicate transiting planets. Of course, the fact that we see a star dim does not necessarily mean a planet has passed in front of it, because there are many other possible causes for temporary dimming of a star, including intrinsic variations in the star's own brightness. Therefore, to be confident that the transit method has really detected a planet, scientists generally have two requirements: (1) the suspected transit event must be observed to repeat at least three times with a regular period, indicating that the same planet is passing in front of the star repeatedly with each orbit, and (2) follow-up observations with another method, such as the Doppler method, should reveal the same planet.

Transits have been observed with many telescopes, but as of 2018 the vast majority of transit detections had come from the *Kepler* mission, which led to the discovery of more than 2500 confirmed planets,* including dozens that are as small as Earth. The transit method is also the cornerstone of the *TESS* and *CHEOPS* missions that we'll discuss in Section 13.4.

**Advantages and Limitations of the Transit Method** We've already discussed how the most obvious limitation of the transit method—the fact that it can work only for the small fraction of planetary systems in which planet orbits are oriented edge-on to Earth—can be counterbalanced by observing large numbers of stars, so that even a small fraction represents a large number of planets. But the transit

*In addition to these confirmed planets, *Kepler* also found some 2000 additional sets of repeated transits that are likely to be due to planets, but that scientists have so far been unable to confirm by other methods.

method also shares at least two other biases of the other methods we've discussed. First, it can much more easily find planets with shorter orbital periods, simply because observing repeated transits takes longer for planets with longer periods. Second, it can more easily detect larger planets, because they create larger dips in brightness.

The transit method also has several advantages over the astrometric and Doppler methods. Of greatest significance, the transit method can reveal planets far smaller than those that can currently be found with the astrometric or Doppler method. There are two reasons for this advantage. First, although planets as small as (or smaller than) Earth cause only a very small dimming of their star—for example, dimming the brightness of a Sun-like star by less than 0.01%—the *Kepler*, *TESS,* and *CHEOPS* missions all have the sensitivity to detect such tiny changes. Second, in a multiplanet system, the astrometric and Doppler methods can detect the existence of small planets only as small effects added to the much larger gravitational effects of larger planets in the system. In contrast, a small planet leaves a distinct signature with the transit method, because planets with different periods will rarely transit in front of a star at the same time.

Perhaps most remarkably, in some cases transits cause enough stellar dimming to be measurable even with small personal telescopes. A telescope as small as 3 inches in diameter has been used to discover an extrasolar planet, and it's relatively easy to confirm for yourself some of the transits that have already been detected. What was once considered impossible for scientists can now be assigned as homework for students (see Problem 48 at the end of the chapter).

**Summary TABLE 13.1** summarizes the advantages and limitations of the major methods we have discussed for studying extrasolar planets, and Cosmic Context **FIGURE 13.6** (pages 380–381) summarizes how these methods work. Note that while these will likely be the most important methods over the long term, several other strategies have met with some success. More than a dozen planets have been detected by what is sometimes called *microlensing*, an example of gravitational lensing [**Sections S3.4, 23.2**] in which the light of a star is temporarily magnified as another star passes in front of it and bends its light. Careful study of the microlensing event can reveal whether the foreground star has planets, and this method can find planets farther from their stars than those that can currently be found with the Doppler, astrometric, or transit method. The major drawback to this method is that the special alignment necessary for microlensing is a one-time event, which generally means that there is no opportunity for confirmation or follow-up observations. A different strategy takes advantage of the fact that many stars are surrounded by disks of dust. A planet within such a disk can exert small gravitational tugs on

## SPECIAL TOPIC The Names of Extrasolar Planets

The planets in our solar system have familiar names rooted in mythology. Unfortunately, there's not yet a well-accepted scheme for naming extrasolar planets, although the International Astronomical Union has begun a process by which the public can suggest and eventually vote on planet names. In the meantime, astronomers still generally refer to extrasolar planets by the star they orbit, such as "the planet orbiting the star named . . . ." Worse still, the stars themselves often have confusing or even multiple names, reflecting naming schemes used in star catalogs made by different people at different times in history.

A few hundred of the brightest stars in the sky carry names from ancient times. Many of these names are Arabic—such as Betelgeuse, Algol, and Aldebaran—because of the work of the Arab scholars of the Middle Ages [**Section 3.2**]. In the early 17th century, German astronomer Johann Bayer developed a system that gave names to many more stars: Each star is assigned a name based on its constellation and a Greek letter indicating its ranking in brightness within that constellation. For example, the brightest star in the constellation Andromeda is called Alpha Andromedae, the second brightest is Beta Andromedae, and so on. Bayer's system works for only the 24 brightest stars in each constellation, because there are only 24 letters in the Greek alphabet. About a century later, English astronomer John Flamsteed published a more extensive star catalog in which he used numbers once the Greek letters were exhausted; 51 Pegasi gets its name from Flamsteed's catalog. (Flamsteed's numbers are based on position within a constellation rather than brightness.)

As more powerful telescopes made it possible to discover more and fainter stars, astronomers developed many new star catalogs. The names we use today usually come from one of these catalogs. For example, the star HD 209458 appears as star number 209458 in a catalog named for Henry Draper (HD) but compiled primarily by Williamina Fleming, Annie Jump Cannon, and other women astronomers at Harvard [**Section 15.1**]. You may see star names consisting of numbers preceded by other catalog names, including Gliese, Ross, and Wolf; these catalogs are named for the astronomers who compiled them. Because the same star is often listed in several catalogs, a single star can have several different names.

Objects orbiting other stars usually carry the star name plus a letter denoting their order of discovery around that star. If the second object is another star, a capital B is added to the star name; if it's a planet, a lowercase b is added. For example, HD 209458b is the first planet discovered orbiting star number 209458 in the Henry Draper catalog; Upsilon Andromedae d is the third planet discovered orbiting the twentieth brightest star (because upsilon is the twentieth letter in the Greek alphabet) in the constellation Andromeda. Some recently discovered planets have been named for the observing program that discovered them. For example, Kepler 11g is the sixth planet in the eleventh planetary system announced by the *Kepler* mission, and the planets of the TRAPPIST-1 system are named for the TRAnsiting Planets and PlanetesImals Small Telescope (TRAPPIST) that discovered them.

**TABLE 13.1 Major Extrasolar Planet Detection Methods**

| | Description | Key Advantages | Major Limitations | Confirmed Planets (as of mid-2018) |
|---|---|---|---|---|
| **Direct Detection** | Obtain images or spectra of extrasolar planets | ▪ Only method that allows direct study of the planets themselves | ▪ Requires large telescopes and some means of blocking light from star itself | A few |
| **Astrometric Method** | Infer planet existence from small changes in star's position in sky | ▪ Now possible with *GAIA* spacecraft<br>▪ Detects planets in all orbit orientations | ▪ Generally possible only for relatively nearby stars<br>▪ Biased toward finding massive planets that orbit far from their stars<br>▪ May require many years of observation to detect these planets (with large orbital periods) | A few |
| **Doppler Method** | Infer planet existence from star's motion toward/away from us revealed by Doppler shifts in its spectrum | ▪ Possible from ground-based telescopes<br>▪ Detects planets in all orbit orientations except face-on | ▪ Biased toward finding massive planets with close-in orbits<br>▪ Underestimates star's true motion except when system viewed edge-on<br>▪ Requires stellar spectra, which means large telescopes and long observation times | About 700 |
| **Transit Method** | Infer planet existence from slight changes in star's brightness as planet passes in front of (or behind) it | ▪ Allows many stars to be observed at once<br>▪ Can detect very small planets<br>▪ Feasible with small telescopes<br>▪ Can provide atmospheric data during some transits and eclipses | ▪ Possible only for planets with edge-on orbits as viewed from Earth<br>▪ For small planets, requires sensitivity possible only from space observatory | About 3000 |

dust particles to produce gaps, waves, or ripples that may be detectable. Some astronomers are searching for thermal emission attributable to the heat of large impacts on planets in young planetary systems, and others are searching for the special kinds of emission expected from the magnetospheres of jovian planets. As we learn more about extrasolar planets, new search methods are sure to arise.

## 13.2 The Nature of Planets Around Other Stars

The mere existence of planets around other stars has changed our perception of our place in the universe, because it shows that our planetary system is not unique. Scientifically, however, we want to know much more than just that these planets exist. In particular, we'd like to know about the nature of these planets—the topic of this section—and about how their existence fits in with our understanding of how stars and planets are born, which we'll cover in Section 13.3.

### What properties of extrasolar planets can we measure?

Despite the challenge of detecting extrasolar planets, we can learn a surprising amount about them. Depending on the method or methods used, we can determine such planetary characteristics as orbital period and distance, orbital eccentricity, mass, size, density, and even a little bit about a planet's atmospheric composition and temperature.

**Orbital Period and Distance** All three of the major indirect detection methods tell us a planet's orbital period. The astrometric method allows us to observe the star's orbital motion around the system's center of mass, which means we know the star's orbital period; the planet's orbital period must be the same. For the Doppler method, a detected planet's orbital period is simply the time between peaks in the star's velocity curve (see Figure 13.4a), and for the transit method, the orbital period is the time between repeated transits.

Once we know the orbital period, we can determine average orbital distance (semimajor axis) with Newton's version of Kepler's third law [**Section 4.4**]. Recall that for

**COSMIC Context** FIGURE 13.6

# Detecting Extrasolar Planets

**The search for planets around other stars is one of the fastest growing and most exciting areas of astronomy, with known extrasolar planets already numbering well into the thousands. This figure summarizes major techniques that astronomers use to search for and study extrasolar planets.**

(1) **Gravitational Tugs:** We can detect a planet by observing the small orbital motion of its star as both the star and its planet orbit their mutual center of mass. The star's orbital period is the same as that of its planet, and the star's orbital speed depends on the planet's distance and mass. Any additional planets around the star will produce additional features in the star's orbital motion.

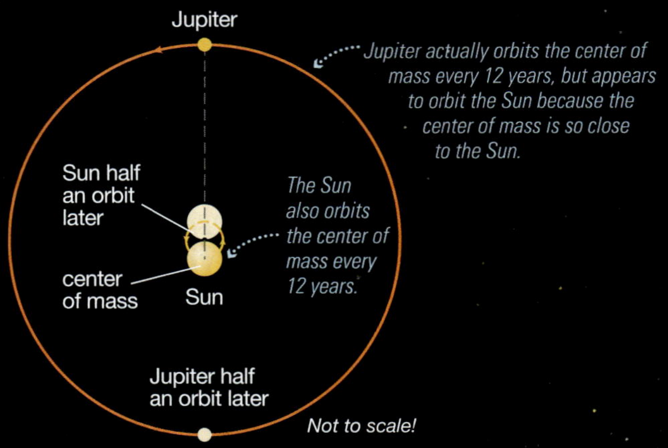

(1a) **The Doppler Method:** As a star moves alternately toward and away from us around the center of mass, we can detect its motion by observing alternating Doppler shifts in the star's spectrum: a blueshift as the star approaches and a redshift as it recedes.

(1b) **The Astrometric Method:** A star's orbit around the center of mass leads to tiny changes in the star's position in the sky. The *GAIA* mission is expected to discover many new planets with this method.

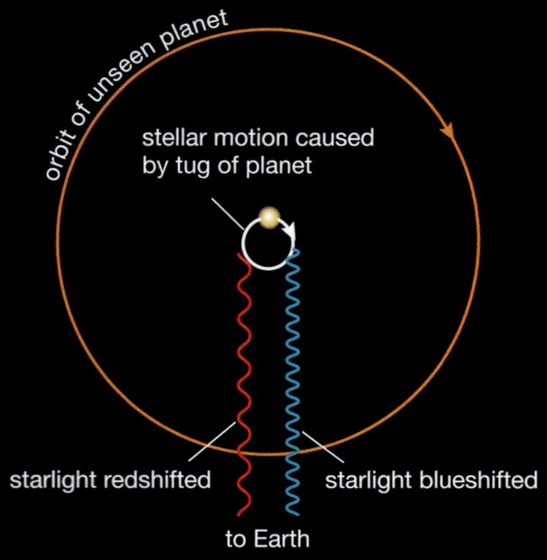


*Current Doppler-shift measurements can detect an orbital velocity as small as 1 meter per second—walking speed.*

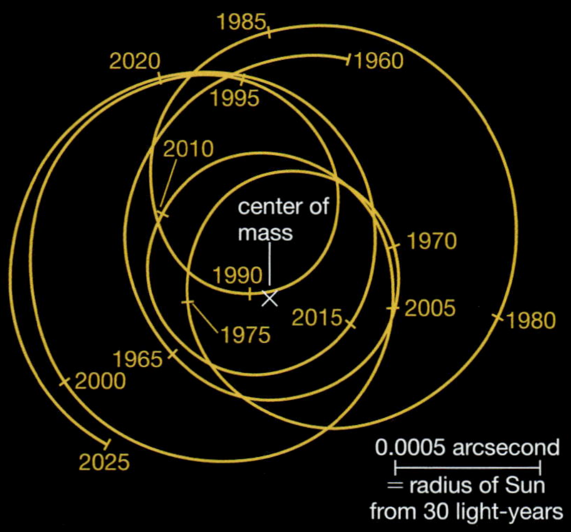


*The change in the Sun's apparent position, if seen from a distance of 10 light-years, would be similar to the angular width of a human hair at a distance of 5 kilometers.*

*Artist's conception of another planetary system, viewed near a ringed jovian planet.*

② **The Transit Method:** If a planet's orbital plane happens to lie along our line of sight, the planet will transit in front of its star once each orbit, causing a dip in the star's visible-light brightness. An *eclipse* may occur half an orbit later, during which the system's infrared brightness will decline because the planet's contribution is blocked by the star.

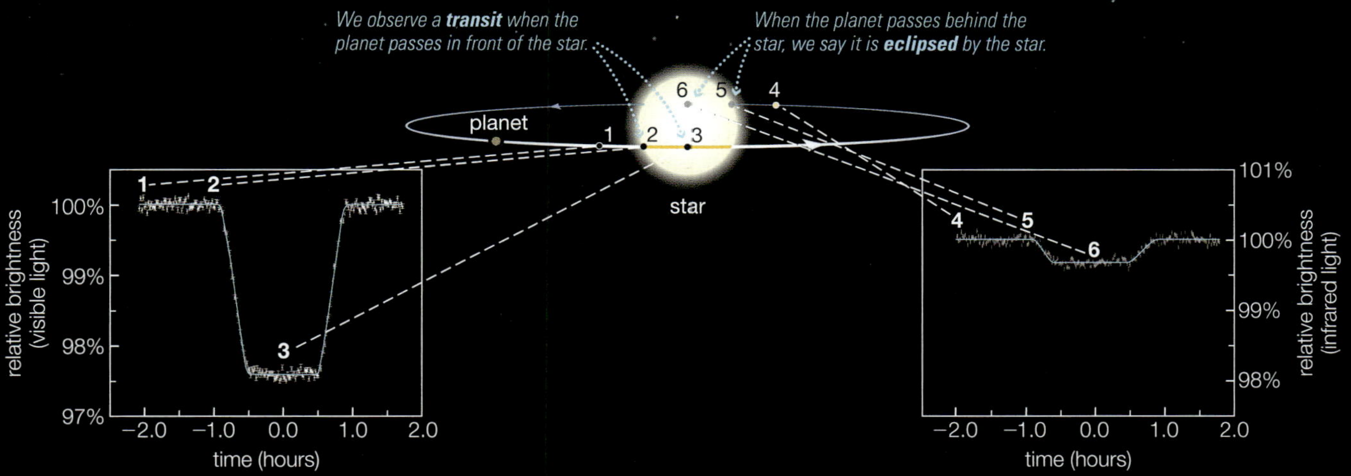


③ **Direct Detection:** In principle, the best way to learn about an extrasolar planet is to observe directly either the visible starlight it reflects or the infrared light it emits. Current technology is capable of direct detection in some cases, but only with very low resolution.

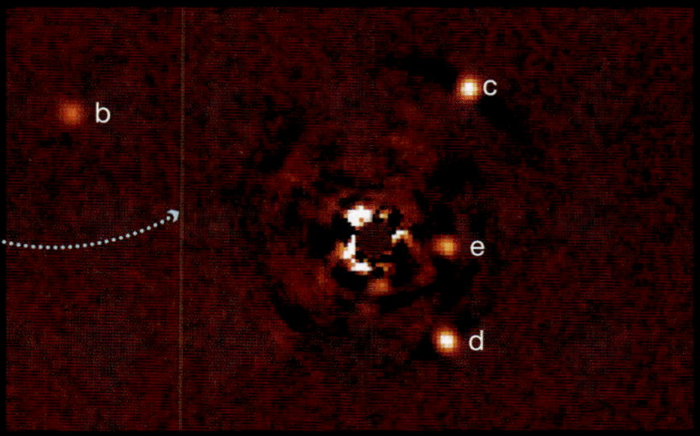


*The Large Binocular Telescope imaged the region around the star HR 8799 in infrared light, discovering four planets labeled **b** through **e** (**a** refers to the star itself).*

a small object like a planet orbiting a much more massive object like a star, this law expresses a relationship between the star's mass, the planet's orbital period, and the planet's average distance. We generally know the masses of the stars with extrasolar planets (through methods we'll discuss in Chapter 15). Therefore, using the star's mass and the planet's orbital period, we can calculate the planet's average orbital distance (see Mathematical Insight 13.1).

**Orbital Eccentricity** All planetary orbits are ellipses, but recall that ellipses vary in eccentricity, which is a measure of how "stretched out" they are (see Figure 3.16). The planets in our solar system all have nearly circular orbits (low eccentricity), which means that their actual distances from the Sun are always relatively close to their average distances. Planets with higher eccentricity swing in close to their star on one side of their orbit and go much farther from their star on the other side.

We can determine eccentricity with both the astrometric and the Doppler method, though most measurements to date come from Doppler data. A planet with a perfectly circular orbit travels at a constant speed around its star, so its velocity curve is perfectly symmetric. Any asymmetry in the Doppler curve tells us that the planet is moving with varying speed and therefore must have a more eccentric elliptical orbit. **FIGURE 13.7** shows Doppler curves for four extrasolar planets; notice that the first three are fairly symmetric, implying low eccentricity, but the fourth planet must have a highly eccentric orbit.

**Planetary Mass** Both the astrometric and the Doppler method measure motions caused by the gravitational tug of a planet, so both can in principle allow us to estimate planetary masses. These methods tell us about planetary masses because, for a given orbital distance, a more massive planet will cause its star to move at higher velocity around the center of mass. (See Mathematical Insight 13.2 for details of the calculations.)

**Think about it** Study the four velocity curves in Figure 13.7. How would each be different if the planet were closer to its star? What if the planet were more massive? Explain.

There is an important caveat for the Doppler method. Recall that Doppler shifts reveal only the part of a star's motion directed toward or away from us (see Figure 5.23). As a result, a planet whose orbit we view face-on (perpendicular to the plane of the orbit) does not cause a Doppler shift in the spectrum of its star, making it impossible to detect such a planet with the Doppler method (**FIGURE 13.8a**). A planet will cause a measureable Doppler shift only if it is orbiting at some other inclination (**FIGURE 13.8b**), and the

## MATHEMATICAL INSIGHT 13.1 Finding Orbital Distances for Extrasolar Planets

Recall that Newton's version of Kepler's third law reads

$$p^2 = \frac{4\pi^2}{G(M_1 + M_2)} a^3$$

In the case of a planet orbiting a star, $p$ is the planet's orbital period, $a$ is its average orbital distance (semimajor axis), and $M_1$ and $M_2$ are the masses of the star and planet, respectively. $G$ is the gravitational constant; $G = 6.67 \times 10^{-11}\ \text{m}^3/(\text{kg} \times \text{s}^2)$. Because a star is so much more massive than a planet, we can approximate $M_{\text{star}} + M_{\text{planet}} \approx M_{\text{star}}$ (see Mathematical Insight 4.3). If we make this approximation, then simple algebra allows us to solve for the average orbital distance $a$:

$$a \approx \sqrt[3]{\frac{GM_{\text{star}}}{4\pi^2} p_{\text{planet}}^2}$$

**EXAMPLE:** Doppler measurements show that the planet orbiting 51 Pegasi has an orbital period of 4.23 days; the star's mass is 1.06 times that of our Sun. What is the planet's average orbital distance?

**SOLUTION:**

**Step 1 Understand:** We are given both the planet's orbital period and the star's mass, so we can use Newton's version of Kepler's third law to find the average orbital distance. To make the units consistent, we convert the given stellar mass to kilograms (using the fact that the Sun's mass is $2 \times 10^{30}$ kg) and the orbital period to seconds:

$$M_{\text{star}} = 1.06 \times M_{\text{Sun}} = 1.06 \times (2 \times 10^{30}\ \text{kg}) = 2.12 \times 10^{30}\ \text{kg}$$

$$p = 4.23\ \text{day} \times \frac{24\ \text{hr}}{1\ \text{day}} \times \frac{3600\ \text{s}}{1\ \text{hr}} = 3.65 \times 10^5\ \text{s}$$

**Step 2 Solve:** We use the above values to find the average distance $a$:

$$a \approx \sqrt[3]{\frac{GM_{\text{star}}}{4\pi^2} p_{\text{planet}}^2}$$

$$= \sqrt[3]{\frac{6.67 \times 10^{-11} \frac{\text{m}^3}{\text{kg} \times \text{s}^2} \times 2.12 \times 10^{30}\ \text{kg}}{4 \times \pi^2} (3.65 \times 10^5\ \text{s})^2}$$

$$= 7.81 \times 10^9\ \text{m}$$

**Step 3 Explain:** The planet orbits its star at an average distance of 7.8 billion meters, or 7.8 million kilometers. It's easier to interpret this number if we convert it to astronomical units (1 AU $\approx 1.50 \times 10^{11}$ meters):

$$a = 7.81 \times 10^9\ \text{m} \times \frac{1\ \text{AU}}{1.50 \times 10^{11}\ \text{m}} = 0.052\ \text{AU}$$

The planet's average orbital distance is 0.052 AU—small even compared to that of Mercury, which orbits the Sun at 0.39 AU. In fact, comparing the planet's 7.8-million-kilometer distance to the size of the star itself (presumably close to the 700,000-kilometer radius of our Sun), we estimate that the planet orbits its star at a distance only a little more than 10 times the star's radius.

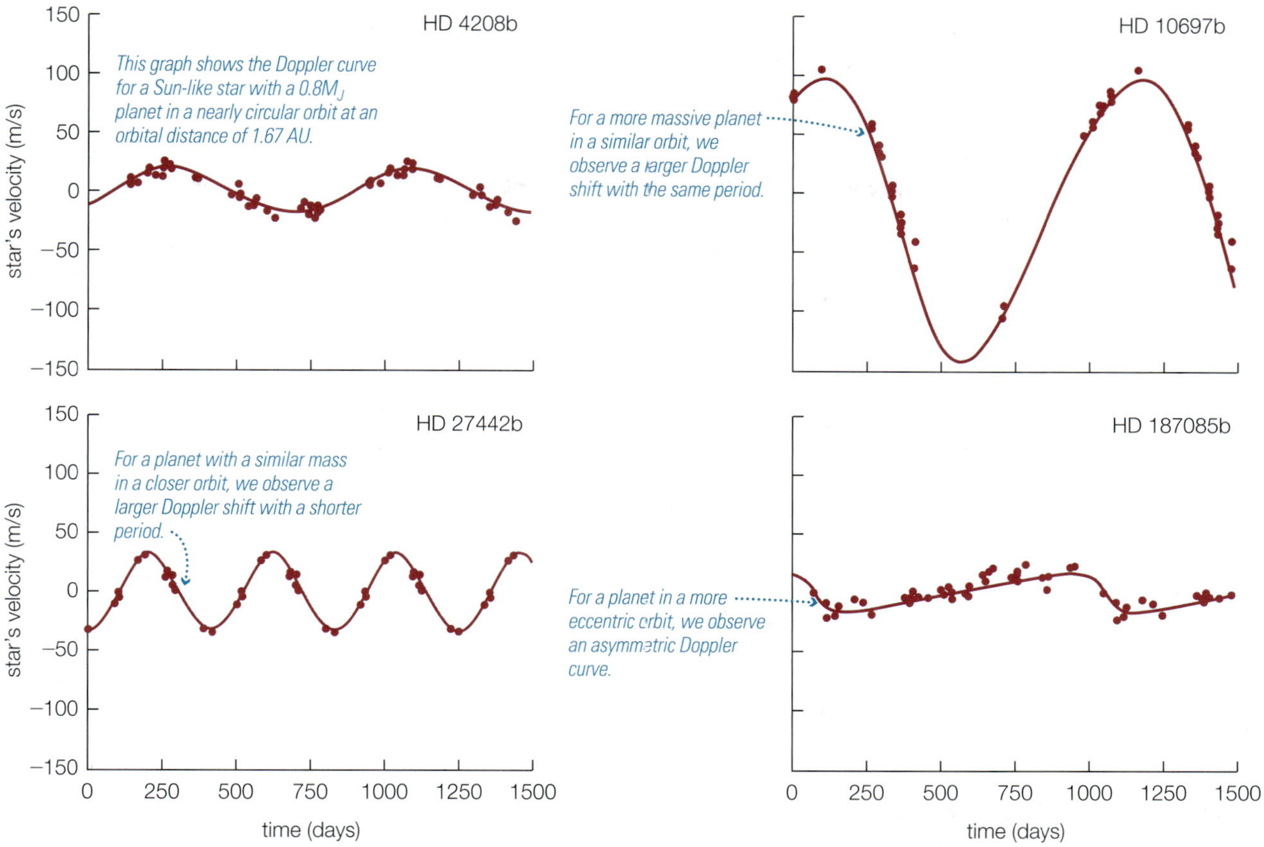


**FIGURE 13.7** Sample data showing velocity curves obtained with the Doppler method for four extrasolar planet systems; dots are data and the curves are best-fits to the data. Shapes of the curves tell us about orbital eccentricity, and magnitudes of the velocity changes tell us about mass.

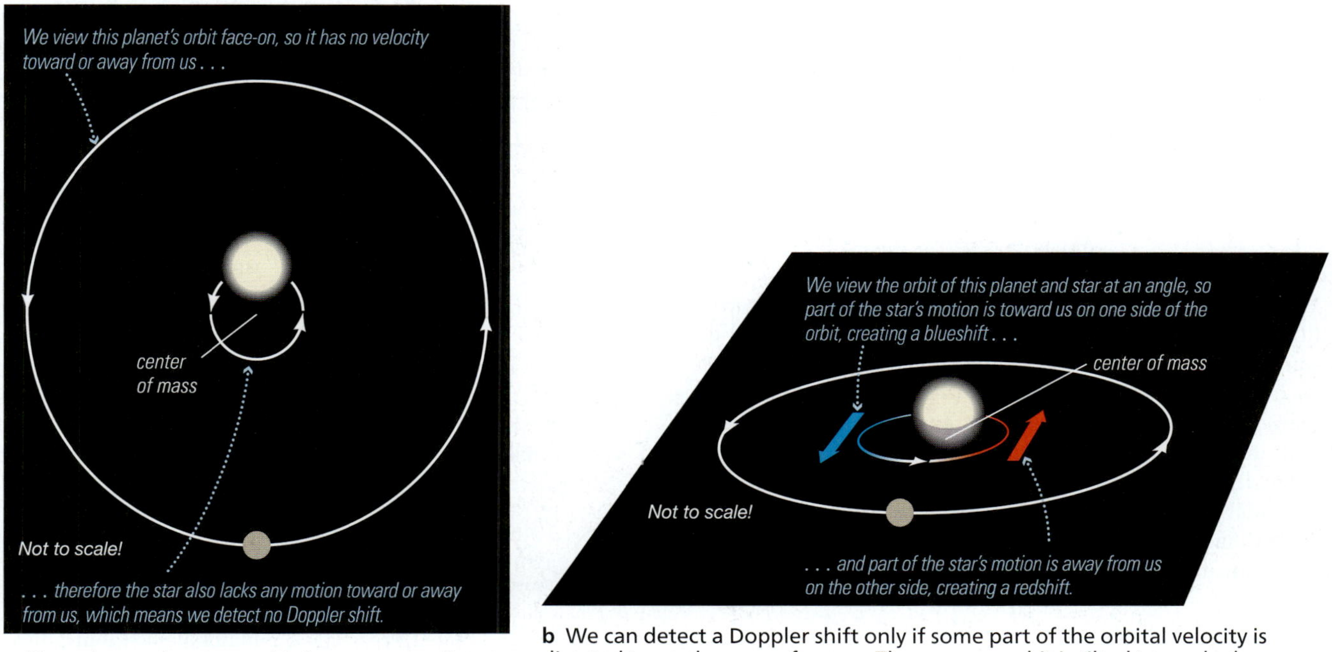


**a** If we view a planetary orbit face-on, we will not detect any Doppler shift at all.

**b** We can detect a Doppler shift only if some part of the orbital velocity is directed toward or away from us. The more an orbit is tilted toward edge-on, the greater the shift we observe.

**FIGURE 13.8** The amount of Doppler shift we observe in a star's spectrum depends on the orientation of the planetary orbit that causes the star's motion.

Doppler shift will tell us the star's *full* velocity only if the orbit is edge-on, in which case we should also be able to detect transits of the planet. In other words, the mass we infer from the Doppler method alone tells us a planet's true mass only if we also observe transits of the same planet. In all other cases, the velocity inferred from Doppler shifts will be less than the full orbital velocity, which means that the mass we calculate will be the planet's *minimum* possible mass (also called a "lower limit" mass). Statistically, however, we expect that the actual planetary mass should be no more than double the minimum mass in at least about 85% of all cases, so masses obtained by the Doppler method provide relatively good estimates for most planets. Moreover, in some cases we can combine Doppler data with other data (such as astrometric data) to determine the planet's precise orbital inclination, in which case we can calculate a more accurate mass.

The transit method cannot by itself tell us masses for single planets, which is one reason scientists often try to follow up transit discoveries with the Doppler method, since the two methods together yield a precise planetary mass. However, the transit method can in some cases reveal planetary masses in multiplanet systems, because the gravitational tug of one planet on another can affect the timing of transits. For example, when another planet is tugging on a transiting planet in the direction in which it is moving, the tug will make the planet move a little faster so that the transit may begin tens of minutes earlier than it would otherwise; it may begin tens of minutes later when the planet is tugging from the opposite direction. Precise timing measurements can therefore allow us to measure slight variations in orbital speed and thereby determine the masses of the planets causing these variations. This capability is particularly important for planets with masses too small to be detectable by the Doppler method with current technology. (For example, this method was used to find the masses of the planets in the TRAPPIST-1 system, discussed below.)

**Planetary Size** A planet's mass provides important information, but to learn about the planet's nature we also need to know the planet's size or radius, since it is possible to have different sizes with the same mass. Transit observations are presently the only means by which we can measure a planet's size or radius. The basic idea is easy to understand: The more of a star's light that a planet blocks during a transit, the larger the planet must be. (See Mathematical Insight 13.3 for details of the calculations.)

## MATHEMATICAL INSIGHT 13.2 Finding Masses of Extrasolar Planets

The Doppler method allows us to find the mass of an extrasolar planet from the law of conservation of momentum [**Section 4.3**]; recall that momentum = mass × velocity. Consider a star with a single planet, both orbiting their common center of mass with the same orbital period. The system as a whole has no momentum relative to this center of mass (which stays in a fixed place between the star and planet), so the planet's momentum must equal the star's momentum (but in the opposite direction). Using $M$ for mass and $v$ for velocity relative to the center of mass, we write

$$M_{star}v_{star} = M_{planet}v_{planet}$$

Solving this equation for $M_{planet}$, we find

$$M_{planet} = \frac{M_{star}v_{star}}{v_{planet}}$$

The Doppler method tells us the star's velocity toward or away from us ($v_{star}$), and as discussed earlier, we generally know the star's mass ($M_{star}$). We can calculate the planet's orbital velocity ($v_{planet}$) from its orbital period $p$ (the time between peaks in the Doppler curve) and its average orbital distance $a$ (calculated with the method in Mathematical Insight 13.1). If we assume a circular orbit, the planet travels a distance $2\pi a$ during each orbit that takes time $p$, so its orbital velocity is $2\pi a_{planet}/p_{planet}$. Substituting this expression for the planet's velocity into the above equation for mass gives

$$M_{planet} = \frac{M_{star}v_{star}\,p_{planet}}{2\pi a_{planet}}$$

Remember that with velocity data from the Doppler method, this formula gives us the *minimum* mass of the planet.

**EXAMPLE:** Estimate the mass of the planet orbiting 51 Pegasi.

**SOLUTION:**

**Step 1 Understand:** From Mathematical Insight 13.1, we know the planet's orbital period ($p = 3.65 \times 10^5$ s) and orbital distance ($a = 7.81 \times 10^9$ m) and the star's mass ($M_{star} = 2.12 \times 10^{30}$ kg). The graph in Figure 13.4a shows that the star's velocity is about $v_{star} = 57$ m/s. We can therefore use the formula above to calculate the planet's mass.

**Step 2 Solve:** We enter the values:

$$\begin{aligned} M_{planet} &= \frac{M_{star}v_{star}\,p_{planet}}{2\pi a_{planet}} \\ &= \frac{(2.12 \times 10^{30}\text{ kg}) \times \left(57\frac{\text{m}}{\text{s}}\right) \times (3.65 \times 10^5\text{ s})}{2\pi \times (7.81 \times 10^9\text{ m})} \\ &\approx 9 \times 10^{26}\text{ kg} \end{aligned}$$

**Step 3 Explain:** The minimum mass of the planet is about $9 \times 10^{26}$ kilograms. This answer will be more meaningful if we convert it to Jupiter masses. From Appendix E, Jupiter's mass is $1.9 \times 10^{27}$ kilograms, so the planet's minimum mass is

$$M_{planet} = 9 \times 10^{26}\text{ kg} \times \frac{1M_{Jupiter}}{1.9 \times 10^{27}\text{ kg}} = 0.47M_{Jupiter}$$

The planet orbiting 51 Pegasi has a minimum mass of just under half Jupiter's mass. Statistically, most planets will have an actual mass less than double the minimum mass, so the planet probably has a mass similar to or a little less than that of Jupiter.

The TRAPPIST-1 planetary system offers a great example. Its central star is a small red star (spectral type is M8 [**Section 15.1**]) with a radius only 12% that of the Sun and a mass only about 84 times that of Jupiter. This low mass is close to the minimum mass for a star and makes the TRAPPIST-1 star quite dim and cool compared to the Sun: Its surface temperature is only about 2550 K (compared to the Sun's 5800 K), and it puts out only about 0.05% as much light as the Sun. Small red stars like TRAPPIST-1 are difficult to study because they are so faint, but they are the most common types of stars in the universe. Scientists are therefore very interested in knowing whether planets are common among such stars, and TRAPPIST-1 is one of the first for which we have good data.

**FIGURE 13.9** shows the brightness of the star TRAPPIST-1 over an 11-day period. Each downward dip represents a transit in which a planet blocks a small fraction of the starlight. Careful study shows that there are seven different planets (represented by the colored dots) that transit the star with periods ranging from 1.5 to 19 days; sometimes more than one planet transits at the same time, causing a larger dip. The panels show a transit of each planet in more detail, along with the planet's size calculated from the depth of the dip. Note that more distant planets have wider dips because they orbit more slowly, which means their transits last longer. When we combine these sizes with the orbital periods and distances, we find that this system's planets are all similar to Earth in size but orbit much closer to their star than Mercury does to our Sun, with orbital distances ranging from 0.01 to 0.06 AU. Nevertheless, because the star is so cool and dim, this puts some of the planets within what we call their star's *habitable zone* [**Section 24.3**], meaning the range of distances at which it might in principle be possible for them to have surface oceans of liquid water.

**Think about it** The properties of the TRAPPIST-1 planets are still being measured and refined. What are the most recent determinations? Have planets been discovered around other small red stars?

**Planetary Density** No single method measures a planet's average density, but we can calculate it if we know a planet's size from the transit method and its mass from the Doppler method. As discussed above, a transiting planet must have an edge-on orbit, which means the Doppler method gives us an exact mass, and the transit data tell us the planet's radius and hence its volume $\left(V = \frac{4}{3}\pi r^3\right)$. We can then calculate its average density simply by dividing the mass by the volume. **FIGURE 13.10** shows this process applied to a planet known as Kepler 10b. Notice that its density of 8.8 $g/cm^3$ is significantly higher than Earth's density of 5.5 $g/cm^3$.

As a lower-density example, consider the planet HD 209458b (the first planet for which scientists obtained both Doppler and transit data); its density of 0.37 $g/cm^3$ is significantly lower than that of Saturn, the lowest-density planet in our solar system. With just these two examples, we see that the variety of densities among extrasolar planets is much greater than that among the planets in our solar system,

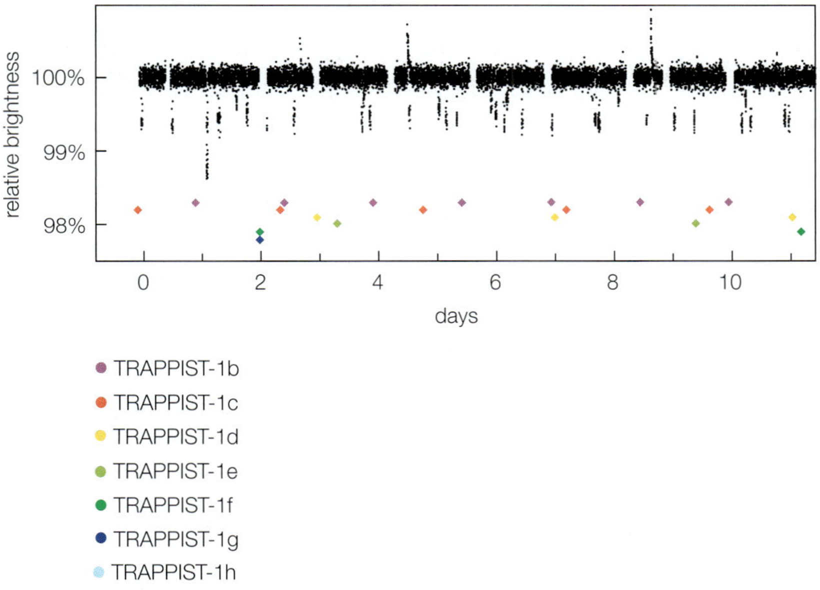


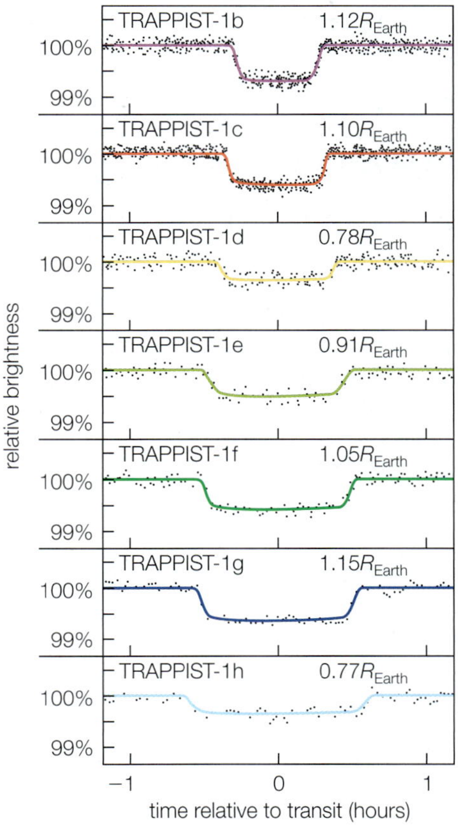


**FIGURE 13.9** The TRAPPIST-1 system. The black dots show the brightness of the central star over an 11-day period. Colored dots indicate transits by seven different planets (identified in the key), sometimes more than one at a time. The panels at right show the dips due to each planet individually and indicate the planet's size in Earth radii. Note that all seven planets have radii within 25% of Earth's radius. (The system's planets are designated b through h; a is the TRAPPIST-1 star.)

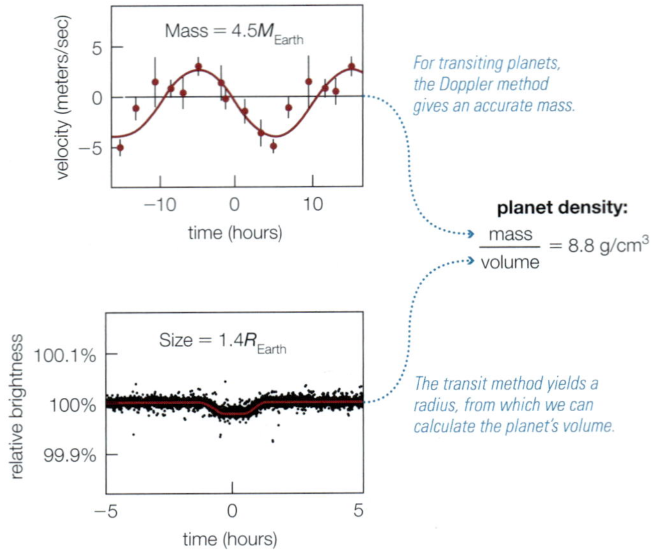


**FIGURE 13.10** This diagram summarizes how we can combine data from the Doppler and transit methods to calculate a planet's average density. Data are for the planet Kepler 10b. (The final density calculation requires converting the mass to grams and the volume to units of cubic centimeters.)

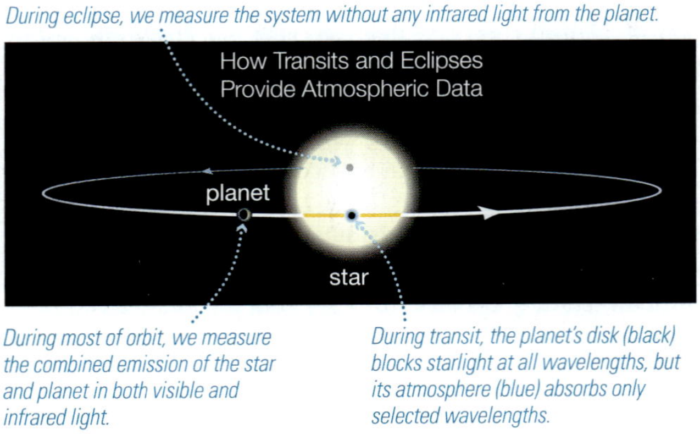


**FIGURE 13.11** This diagram summarizes how transit and eclipse observations can give us information about the atmosphere of the transiting planet, even though we cannot see the light of the planet separately from the light of the star.

suggesting that extrasolar planets come in a wider range of types than just the familiar terrestrial and jovian categories.

**Atmospheric Composition and Temperature** Advancing technology is beginning to allow astronomers to learn about the atmospheres and temperatures of planets. Recall from Chapter 5 that we can learn about atmospheric composition by identifying spectral lines from a planet's atmosphere, and we can learn about temperature from a planet's infrared emission.

Let's first consider atmospheric composition. Direct observations have already provided some data, and are expected to provide much more with the aid of new instruments on ground-based telescopes that can obtain images and spectra of extrasolar planets at the same time. We can also learn about atmospheric composition from transiting systems, as shown in **FIGURE 13.11**: Comparing the system's spectrum taken with the planet in front of its star (the transit) or behind it (the eclipse) to spectra taken at other times can reveal spectral lines caused by gases in the planet's atmosphere. To date, spectra from direct, transit, or eclipse observations have revealed water vapor, methane, hydrogen, sodium, and a handful of other gases in the atmospheres of a few Jupiter-size planets. Astronomers are optimistic that the James Webb Space Telescope will also obtain some spectra of extrasolar planets, particularly those relatively nearby.

Temperature data can also come from both direct observations and transiting systems. For transiting systems, key

## MATHEMATICAL INSIGHT 13.3 Finding Sizes of Extrasolar Planets

We determine planet radii from the fraction of a star's light blocked during a transit. Viewed against the sky, both the star and the planet appear as tiny circular disks. These disks are far too small for our telescopes to resolve, but the fraction of the star's light that is blocked must be equal to the area of the planet's disk ($\pi r_{planet}^2$) divided by the area of the star's disk ($\pi r_{star}^2$). We generally know the approximate radius of the star (from methods we'll discuss in Chapter 15), so the fractional drop in the star's light during a transit is

$$\frac{\text{fraction of}}{\text{light blocked}} = \frac{\text{area of planet's disk}}{\text{area of star's disk}} = \frac{\pi r_{planet}^2}{\pi r_{star}^2} = \frac{r_{planet}^2}{r_{star}^2}$$

Solving for the planet's radius, we find

$$r_{planet} = r_{star} \times \sqrt{\text{fraction of light blocked}}$$

**EXAMPLE:** Figure 13.5 shows a transit of the star HD 189733. The star's radius is about 560,000 kilometers ($0.8R_{Sun}$), and the planet blocks 2.4% of the star's light during a transit. What is the planet's radius?

**SOLUTION:**

**Step 1 Understand:** The star's radius (560,000 km) and the fraction of its light blocked during a transit (2.4% = 0.024) are all we need to calculate the planet's radius.

**Step 2 Solve:** Plugging the numbers into the equation for the planet's radius, we find

$$\begin{aligned} r_{planet} &= r_{star} \times \sqrt{\text{fraction of light blocked}} \\ &= 560{,}000 \text{ km} \times \sqrt{0.024} \\ &\approx 87{,}000 \text{ km} \end{aligned}$$

**Step 3 Explain:** The planet's radius is about 87,000 kilometers, which is about 1.2 times Jupiter's radius of 71,500 kilometers. That is, the planet is about 20% larger than Jupiter in radius.

data come from eclipses. Recall that planets generally emit infrared light, and the amount of infrared emission (per unit area) depends on a planet's temperature. As a planet goes behind its star (the eclipse), the system's infrared brightness will drop because we are no longer seeing the planet's infrared emission. The extent of the drop tells us how much infrared the planet emits, and we can combine knowledge of the amount of infrared emission with the planet's radius (measured by the transits) to calculate an approximate temperature.

We can sometimes learn even more about temperature by monitoring changes in a transiting system's visible-light brightness as a planet presents different phases to Earth. That is, much as we see phases of Venus from Earth (see Figure 3.22b), we see a planet as "new" during a transit and as nearly "full" just before and after an eclipse. *Kepler* successfully measured changes in at least one planet's visible reflected light as the planet went from new to crescent to gibbous to full and back again. With infrared light, we can observe differences in brightness over a planet's day and night sides as it orbits. Moreover, all planets on close-in orbits are expected to have synchronous rotation so that they show the same face to their star at all times, just as the Moon always shows the same face to Earth [**Section 4.5**]. Therefore, we can in principle use infrared observations to create a crude "weather map" of an extrasolar planet. The first success of this technique came with Spitzer Space Telescope observations of HD 189733 and its planet, which showed that the planet reaches a temperature of nearly 1200 K on its day side and 900 K on its night side (**FIGURE 13.12**). Note that the hottest point does not lie exactly at the center of the Sun-facing side, probably because of winds, which models suggest may blow at an incredible 10,000 km/hr.

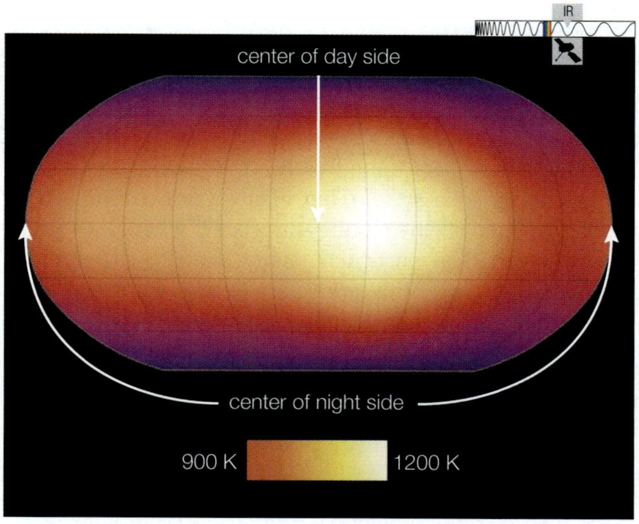


**FIGURE 13.12** The first map of an extrasolar planet's temperatures, derived from infrared observations of HD 189733b made by the Spitzer Space Telescope. Note that this is *not* an actual image, but rather a map constructed from variations in the system's infrared brightness as the planet orbits around its star.

**TABLE 13.2** summarizes how we measure planetary properties.

## How do extrasolar planets compare to planets in our solar system?

The number of known extrasolar planets for which we have measured many key properties is now large enough that we are beginning to gain insight into how these planets compare to the planets of our own solar system. Let's explore the general features of extrasolar planets as we know them today.

**TABLE 13.2 A Summary of How We Measure Properties of Extrasolar Planets**

| | Planetary Property | Method(s) Used | Explanation |
|---|---|---|---|
| Orbital Properties | **period** | Doppler, astrometric, or transit | We directly measure orbital period. |
| | **distance** | Doppler, astrometric, or transit | We calculate orbital distance from orbital period using Newton's version of Kepler's third law (Mathematical Insight 13.1). |
| | **eccentricity** | Doppler or astrometric | Velocity curves and astrometric star positions reveal eccentricity (Figure 13.7). |
| | **inclination** | transit or astrometric | Transits identify edge-on orbits; astrometric data measure any inclination angle. |
| Physical Properties | **mass** | astrometric or Doppler* | We calculate mass based on the amount of stellar motion caused by the planet's gravitational tug (Mathematical Insight 13.2). |
| | **size (radius)** | transit | We calculate size based on the amount of dip in a star's brightness during a transit (Mathematical Insight 13.3). |
| | **density** | transit and Doppler | We calculate density by dividing the mass by the volume (using size from the transit method). See Figure 13.10. |
| | **atmospheric composition and temperature** | transit or direct detection | Transits and eclipses provide data on atmospheric composition and temperature; direct spectroscopy is now also possible in some cases. |

*Transits in some multiplanet systems allow us to determine masses from the mutual gravitational tugs of the planets.

**Orbital Properties** Much as Johannes Kepler first appreciated the true layout of our own solar system [**Section 3.3**], we now have enough data to allow us to step back and see the layout of many other solar systems. A few extrasolar planets have been found with orbits similar to those of planets in our solar system, but many others exhibit orbital properties that seemed quite surprising when they were first discovered. In particular, many planets with Jupiter-like masses or sizes orbit quite close to their stars—in many cases much closer than Mercury orbits the Sun—which is surprising because the jovian planets in our solar system all orbit quite far from the Sun. Moreover, many extrasolar planets have large eccentricities, a clear contrast with the nearly circular orbits of planets in our solar system.

**Think about it** Should we be surprised that we haven't found many planets orbiting as far from their stars as Saturn, Uranus, and Neptune orbit the Sun? Why or why not?

We've also found surprises in multiplanet systems, of which more than 500 are now known. Many of these systems have planets packed much closer to each other than our solar system's planets, often so close that they tug on each other gravitationally in measurable ways. This is the case for the TRAPPIST-1 system (**FIGURE 13.13**), in which the tugs have allowed scientists to determine planetary masses through slight changes in transit timing. Some known multiplanet systems even have orbital resonances among their planets, much like the resonances that occur in our solar system among some jovian moons and in ring systems.

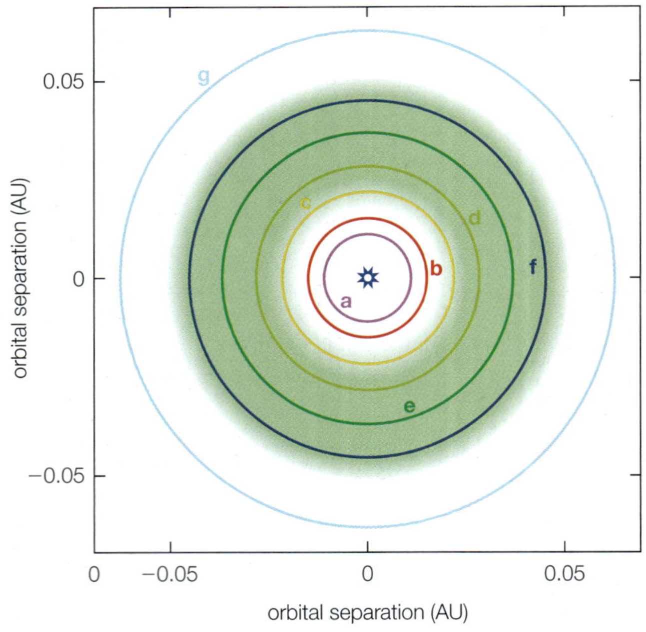


**FIGURE 13.13** A top-down view of planetary orbits in the TRAPPIST-1 system; the orbital colors match the key in Figure 13.9, which also shows that all the planets are similar in size to Earth. Note that all seven planets are much closer to their star than Mercury is to the Sun (0.38 AU). Some are nevertheless in the star's habitable zone (green), which is much smaller and closer to the central star than our Sun's habitable zone because the TRAPPIST-1 star is so much lower in mass and fainter than the Sun [**Section 24.3**].

Keep in mind that these surprising planetary orbits may not be as common as they seem in current data, because these data have all been collected in ways that are biased toward finding planets that orbit close to their stars. Nevertheless, the fact that such surprises exist at all requires explanation, and we'll see in the next section how efforts to come up with an explanation led scientists to realize that there is more to the nebular theory than we might have recognized from our solar system alone.

One other interesting surprise has come from binary star systems. Scientists had not been sure whether planets could form and have stable orbits in binary star systems, but data from the *Kepler* mission have shown that they can, at least in some cases. This means that some worlds have two (or more) "suns" in the sky, much like the fictional planet Tatooine depicted in *Star Wars* (see Figure 24.20).

**Sizes, Masses, and Densities** The *Kepler* mission detected so many planets that scientists can use statistics to estimate the proportions of all stars that have planets of various sizes. **FIGURE 13.14** shows results from these statistics. Two remarkable conclusions are already apparent. First, planets are common: By looking across all size categories, astronomers conclude that at least 70% of all stars harbor at least one planet. Second, many of these planets are quite small, suggesting that Earth-size planets are very common.

Again, keep in mind that these statistics remain incomplete: Because it takes longer to identify planets with larger orbits, the current statistics reflect only planets with relatively short orbital periods (in most cases, periods less than about 50 days for small planets and up to 250 days for large planets). Given that we expect most jovian planets to have long orbital periods, it's likely that there are many more high-mass planets than the current data suggest. The current data may also underestimate the number of small planets, because these planets are so difficult to detect. As

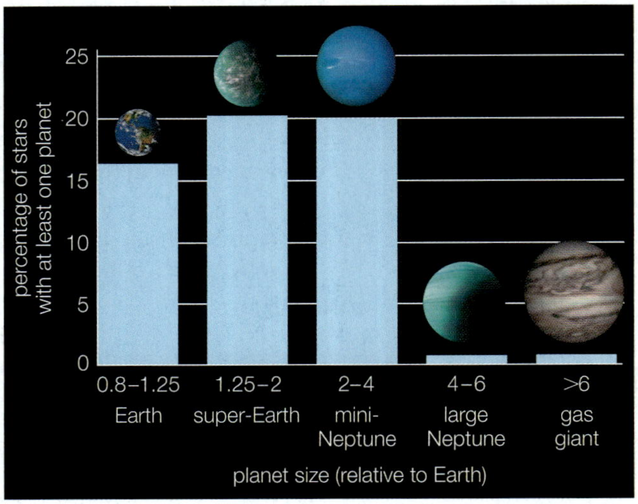


**FIGURE 13.14** This bar chart shows the estimated proportions of all stars that have planets of different size categories, based on *Kepler* results. Because to date *Kepler* data have been analyzed only for planets with relatively short orbital periods, these estimates are nearly certain to increase as additional data are studied.

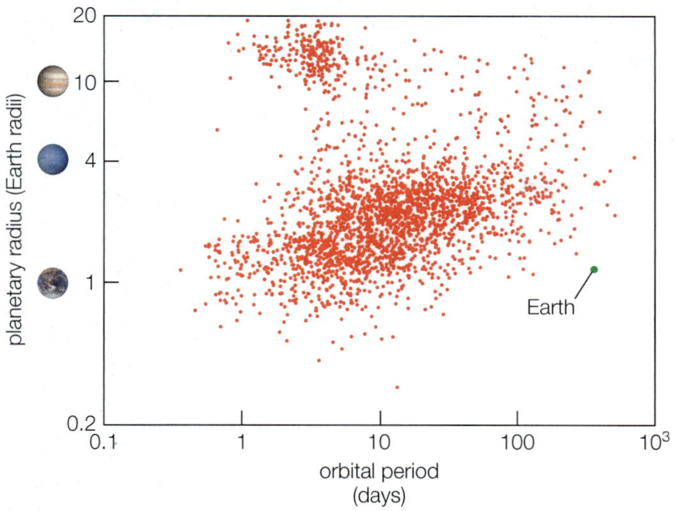


**FIGURE 13.15** This figure shows the orbital periods and sizes of planets identified from *Kepler* data, along with a reference dot for Earth. Note that while many of these planets are similar in size to Earth, the data do not cover a long enough time period to have allowed identification of those with orbital periods as long as Earth's. Many of the Earth-size planets may nevertheless be in their stars' habitable zones, because they orbit stars that are lower in mass than our Sun.

more data are collected and analyzed, the estimated fractions of stars with planets can only increase.

The fact that current statistics are dominated by planets that orbit relatively close to their stars means that most of the planets represented in Figure 13.14 are probably too hot to harbor life. However, **FIGURE 13.15** shows *Kepler* data in a different way: as a graph of planet size versus orbital period. At first glance, the empty space at lower right might seem to suggest that there are few Earth-size planets in Earth-size orbits. But the lack of data points in this region of the graph actually stems from the extreme difficulty of detecting the infrequent, shallow transits such planets make. Accounting statistically for these effects, scientists now estimate that about 20% of stars are likely to have a planet less than twice Earth's size in an orbit within the habitable zone, suggesting that these planets could potentially be homes to life [**Section 24.3**].

In order to determine the general nature of an extrasolar planet, such as whether it is terrestrial or jovian, we need to determine its mass as well as its size, so that we can calculate its average density. Although we currently have both size and mass data for only a limited set of planets, the results have already revealed a surprise: The density range for extrasolar planets is significantly wider than that for the planets in our own solar system. At the extremes, we've identified planets with average densities as great as that of iron and as low as that of Styrofoam, and we've found planets with almost every average density in between.

**The Nature of Extrasolar Planets** We now come to the key question about extrasolar planets: Do they fall into the same terrestrial and jovian categories as the planets in our solar system, or do we find additional types of planets?

We cannot yet know for certain, because we do not yet have direct ways of obtaining spectra with sufficient detail to determine the basic compositions of extrasolar planets. Nevertheless, because we have measured the abundances of different chemical elements among other stars, we know that all planetary systems must start out from gas clouds generally similar in composition to the solar nebula. That is, all star systems are born from gas clouds containing at least about 98% hydrogen and helium, sprinkled with much smaller amounts of ice, rock, and metal (see Table 8.1). Given this fact, knowing a planet's average density gives us great insight into its likely composition, even without being able to observe the planet directly.

More specifically, we can use our understanding of the behavior of different materials to create *models* that will tell us the expected composition of a planet based on its mass and radius, from which we can also calculate its average density. The results are shown in **FIGURE 13.16** for all planets for which both mass (usually from the Doppler method) and radius (from transits) were known as of mid-2015. Be sure you understand the following key features of the figure:

- The horizontal axis shows planetary mass, in units of Earth masses. (The top of the graph shows the equivalent values in Jupiter masses.) Notice that this axis uses a scale that rises by powers of 10 because the masses vary over such a wide range.
- The vertical axis shows planetary radius in units of Earth radii. (The right side of the graph shows the equivalent values in Jupiter radii.)
- Each dot represents one planet for which both mass and radius have been measured. Planets discussed in this chapter are called out by name. Planets of our solar system are marked in green.
- The paintings around the graph show artist conceptions of what representative worlds might look like.
- Average density is easy to calculate from mass and radius, but the different scales used on the two axes make it difficult to read average density directly from the graph. To help with that, the three curves extending from the lower left to the top show three representative average densities.
- The colored regions indicate models representing the expected compositions of planets with the indicated combinations of mass and radius.

As you study Figure 13.16, you'll notice that extrasolar planets show much more variety than the planets of our own solar system. For example, HAT-P-32b has more than twice Jupiter's radius despite having the same mass, giving it an average density of about 0.14 g/cm$^3$—similar to that of Styrofoam. This low average density is probably a result of the fact that this planet orbits only 0.035 AU from its star, putting it more than 10 times closer to its star than Mercury is to the Sun. This close-in orbit gives the planet a very high temperature, which should puff up the planet's atmosphere and may explain why it has such a large size relative to its mass. Near the other extreme, the planet CoRoT-14b is only slightly larger than Jupiter but several times as massive, giving it an average density near 8 g/cm$^3$,

**FIGURE 13.16** Masses and sizes for a sample of extrasolar planets for which both have been measured, compared to those of planets in our solar system. Each dot represents one planet. Dashed lines are lines of constant density for planets of different masses. Colored regions indicate expected planet types based on models of their compositions.

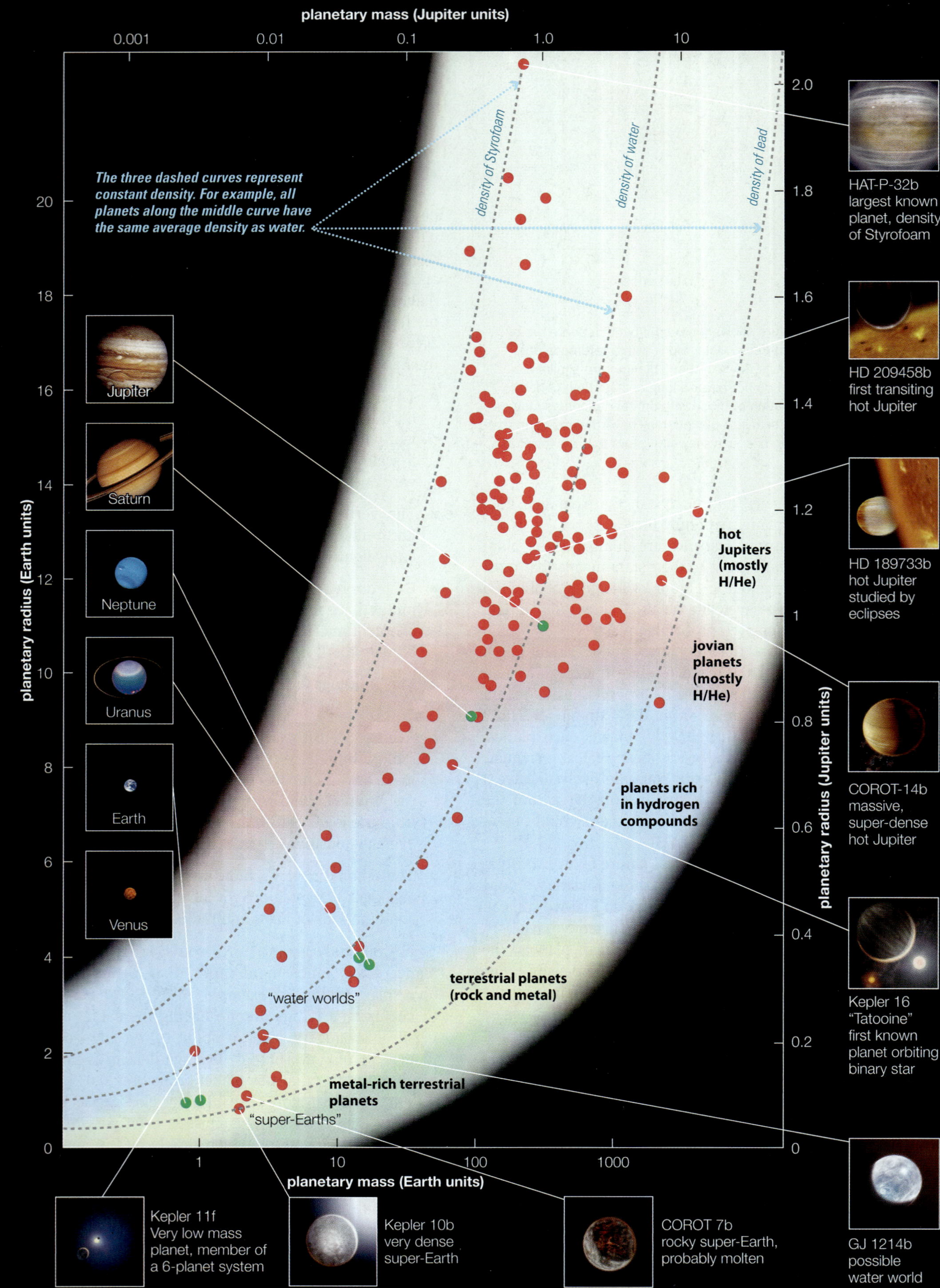

about the same as the density of iron. Although such a high average density might seem surprising, it is not totally unexpected. Recall that jovian planets more massive than Jupiter are expected to have such strong gravity that they can be compressed to smaller sizes and much higher densities (see Figure 11.2b). Despite the wide spread in their densities, both HAT-P-32b and COROT-14b seem clearly to fall into the jovian planet category, being made largely of hydrogen and helium.

We also see many planets that appear to be terrestrial in nature, with compositions of rock and metal. For example, COROT-7b has an average density near 5 $g/cm^3$, comparable to Earth's density. Because it has a mass about 5 times that of Earth, COROT-7b is an example of what is sometimes called a "super-Earth." It orbits very close to its star, so its surface is probably molten. This and the many other known super-Earths are likely to have a rock/metal composition similar to that of the terrestrial worlds in our solar system.

Perhaps the most surprising planets shown in Figure 13.16 are the ones that are not clearly in either the terrestrial or the jovian category. Several planets cluster in the region of Uranus and Neptune, and perhaps share their composition of hydrogen compounds shrouded in an envelope of hydrogen and helium gas. Others (perhaps GJ 1214b) could fit the model for "water worlds," meaning planets likely to be made predominantly of water (perhaps in unusual phases) or of other hydrogen compounds. Alternatively, some of these worlds might be composed of a dense rocky/metallic core and a thick envelope of low-density hydrogen and helium gases.

Finally, it's worth noting that while we've focused here on composition, orbital distance must also play a role in a planet's nature. In particular, because most of the planets known to date orbit fairly close to their stars, they will be much hotter than similar planets in our own solar system. We've already discussed how this could make hot Jupiters have puffed-up atmospheres and terrestrial worlds have molten surfaces. Hot Jupiters might also have very different clouds than we see on the actual Jupiter, such as clouds of mineral flakes instead of clouds of ammonia snow or water droplets. Worlds that might resemble larger versions of Ganymede or Titan in our solar system would be water worlds at closer orbital distances. Even warmer water worlds might become "steam planets" with vast amounts of water vapor in their atmospheres. Also recall that all close-in planets are expected to be tidally locked in synchronous rotation, forever keeping one face toward their star.

The bottom line is that extrasolar planets are abundant and diverse. All the planets in our solar system fall into just two clear types (terrestrial and jovian), but other solar systems have additional types of planets that may defy easy categorization.

**Moons, Rings, and Other Remaining Questions** Although we now have a fair amount of information about planets around other stars, there are many questions that we are not yet able to address. For example, based on what we see in our own solar system, we might expect extrasolar jovian planets to be orbited by rings and many moons, but we do not yet know whether this is the case. Answering the question is important not only to satisfy our curiosity, but also because the answer might have implications for the search for life. For example, astronomers have found numerous cases of extrasolar planets that are jovian in mass but have Earth-like orbits. As we'll discuss in Chapter 24, we don't expect to find life on these jovian planets themselves, but the fact that they orbit within a star's habitable zone makes it conceivable that they could have moons that might be hospitable to life.

## 13.3 The Formation of Other Solar Systems

The discovery of extrasolar planets presents us with an opportunity to test our theory of solar system formation. Can our existing theory explain other planetary systems, or do we have to go back to the drawing board?

### Do we need to modify our theory of solar system formation?

As we discussed in Chapter 8, the nebular theory holds that our solar system's planets formed as a natural consequence of processes that accompanied the formation of our Sun. If the theory is correct, then the same processes should accompany the births of other stars, so the nebular theory clearly predicts the existence of other planetary systems. In that sense, the discovery of extrasolar planets means the theory has passed a major test, because its most basic prediction has been verified. Other key details of the theory also seem supported. For example, the nebular theory says that jovian planet formation begins with condensation of solid particles of rock and ice (see Figure 8.13), which then accrete to larger sizes and capture nebular gas. We therefore expect that such planets should form more easily in a nebula with a higher proportion of rock and ice, and in fact more large planets have been found around stars richer in the elements that make these ingredients.

Nevertheless, extrasolar planets have already presented at least two significant challenges to our theory. One concerns the categories of planets: As we saw in Figure 13.16, many extrasolar planets do not fall neatly into either the terrestrial or the jovian category. An even more significant challenge is posed by the orbits of extrasolar planets. According to the nebular theory, jovian planets form as gravity pulls in gas around large, icy planetesimals that accrete in a spinning disk of material around a young star. The theory therefore predicts that jovian planets should form only in the cold outer regions of star systems (because it must be cold for ice to condense), and that these planets should be born with nearly circular orbits (matching the orderly, circular motion of the spinning disk). The many known extrasolar planets that appear jovian in nature but have close-in or highly elliptical orbits present a direct challenge to these ideas.

**Explaining Planetary Orbits** The nature of science demands that we question the validity of a theory whenever it is challenged by any observation or experiment [**Section 3.4**]. If the theory cannot explain the new observations, then we must revise or discard it. The surprising orbits of many known extrasolar planets have indeed caused scientists to reexamine the nebular theory of solar system formation.

Questioning began almost immediately upon the discovery of the first extrasolar planets. These planets were massive and had close-in orbits, making scientists wonder whether something might be fundamentally wrong with the nebular theory. For example, is it possible for jovian planets to form very close to a star? Astronomers addressed this question by studying many possible models of planet formation and reexamining the entire basis of the nebular theory. Several years of such reexamination did not turn up any good reasons to discard the basic theory, or any alternative means of forming jovian planets close to their stars. While we can't completely rule out the possibility that a major flaw has gone undetected, it now seems much more likely that the basic outline of the nebular theory is correct. Scientists therefore suspect that extrasolar jovian planets were indeed born with circular orbits far from their stars, and that those that now have close-in or eccentric orbits underwent some sort of "planetary migration."

How did these planetary migrations occur? You might think that drag within the solar nebula could cause planets to migrate, much as atmospheric drag can cause satellites in low-Earth orbit to lose orbital energy and eventually spiral into the atmosphere. However, calculations show this drag effect to be negligible. A more likely scenario is that waves propagating through a gaseous disk lead to migration (**FIGURE 13.17**). The gravity of a planet orbiting in a disk can create waves that propagate through the disk, causing material to bunch up as the waves pass by. This "bunched up" matter (in the wave peaks) then exerts a gravitational pull on the planet that reduces its orbital energy, causing the planet to migrate inward toward its star.

This type of migration is not thought to have played a significant role in our own solar system, because the nebular gas was cleared out before it could have much effect. However, planets may form earlier in some other planetary systems, or the nebular gas may be cleared out later, allowing time for jovian planets to migrate substantially inward. In a few cases, the planets may form so early that they end up spiraling all the way into their stars. Indeed, astronomers have noted that some stars have an unusual assortment of elements in their outer layers, suggesting that they may have swallowed planets (including migrating jovian planets and possibly terrestrial planets shepherded inward along with the jovian planets). These ideas are not just hypothetical: Several hot Jupiters orbit so close to their stars that it seems likely that tidal forces will send them on death spirals into their stars on time scales of just millions of years.

**FIGURE 13.17** This figure shows a simulation of waves created by a planet embedded in a disk of material surrounding its star.

Migration may also occur after the nebula has cleared, as long as small planetesimals are still abundant. Astronomers suspect that this type of migration affected the jovian planets in our own solar system. Recall that the Oort cloud is thought to consist of comets that were ejected outward by gravitational encounters with the jovian planets, especially Jupiter [**Section 12.3**]. In that case, the law of conservation of energy demands that Jupiter must have migrated inward, losing the same amount of orbital energy that the comets gained. Jupiter's migration, in turn, may have led to resonances that affected the orbits of the other jovian planets and the objects of the Kuiper belt [**Section 12.4**]. It's not known if this kind of migration is common outside our solar system.

Related mechanisms may explain the surprisingly high orbital eccentricities of many extrasolar planets. For example, planetary migration increases the chances that planets will influence each other gravitationally. In some cases, planets may pass close enough for a gravitational encounter [**Section 4.5**] in which one planet gains enough energy to escape from the star system entirely (becoming an "orphan planet" [**Section 24.3**]) while the other is flung inward into a highly elliptical orbit. In other cases, continuing gravitational tugs may lead to orbital resonances much like those that cause the orbits of Jupiter's moons Io, Europa, and Ganymede to be more elliptical than they would be otherwise (see Figure 11.19b). Simulations show that gravitational influences between planets can drive them into orbits that are very eccentric, highly tilted, or even backward. Resonances can also in some cases prevent planetary collisions, much as the resonance of Neptune and Pluto prevents them from colliding (see Figure 12.22); this is the case for the planets in the TRAPPIST-1 system.

**Explaining Planetary Types** Assuming we are correct about the role of planetary migration in explaining the surprising orbits of hot Jupiters, then the basic tenets of the nebular theory seem to hold. That is, we expect rocky terrestrial worlds to form in the inner regions of solar systems and hydrogen-rich jovian planets to form in the outer regions, although migration may later cause some jovian planets to spiral inward (and likely alter the orbits of smaller planets as well). The remaining mystery, then, is why other systems seem to have planetary types that don't fall neatly into the terrestrial and jovian categories that we identify in our solar system.

Scientists still cannot fully explain the wide range of extrasolar planet properties, but we can envision possible explanations that seem to make sense and that would not fundamentally alter the nebular theory. For example, hydrogen-rich extrasolar planets vary in density by a factor of 100 (see Figure 13.16)—a range far greater than the density range we observe in our solar system—but it seems reasonable to think that much of this range is attributable to increases in temperature caused by some jovian planets being very close to their stars. As we've discussed, heating may puff up their atmospheres to large sizes and low densities, though models cannot yet account for the full density range. Higher densities may be explained by planets having higher masses than their counterparts in our solar system, which compresses them to smaller sizes due to higher gravity (see Figure 11.2).

Similarly, the lack of water world planets in our solar system may not be as mysterious as it seems. Water worlds may be similar to Uranus and Neptune, though in some cases much smaller. These worlds may be much like the ice-rich planetesimals that seeded the formation of jovian planets in our solar system. In that case, perhaps whether water world planets exist depends on when a star clears its nebular gas, halting the epoch of planet formation. In our solar system, this did not occur until the ice-rich planetesimals pulled in vast quantities of hydrogen and helium gas from the solar nebula. In other systems, perhaps an early solar wind blasted out the hydrogen and helium gas before it could be captured.

Super-Earths pose a different mystery: How did these planets gather so much rocky material, especially so close to their stars, given that rocky material represents such a small proportion of the material from which planetary systems are born? The answer is not yet known, though perhaps we should not be too surprised. After all, even though rocky material comprised less than 2% of the solar nebula, this still in principle was enough to build planets much larger than Earth. Perhaps we only need a better understanding of the factors that determine the efficiency with which rocky material can be incorporated into planets. We may similarly need a better understanding of the conditions under which super-Earths or water worlds can capture hydrogen and helium gas, since such capture could lead to planets whose true nature is hiding under enormous envelopes of gas. With hundreds of different planets discovered every year, other surprising planetary types are likely to be found in the future.

### Are planetary systems like ours common?

The validation of the nebular theory leads us to perhaps the most profound question still to be addressed in our study of extrasolar planets: What can we conclude from the fact that we have not yet discovered other planetary systems that are quite like ours? Does this mean that our solar system is of a rare type, or does it simply mean that we have not yet acquired enough data to see how common systems like ours really are?

This is a profound question because of its implications for the way we view our place in the universe. If planetary systems like ours are common, then it seems reasonable to imagine that Earth-like planets—and perhaps life and civilizations—might also be common. But if our solar system is a rarity or even unique, then Earth might be the lone inhabited planet in our galaxy or even the universe.

As we've discussed, current evidence from both transit detections and Doppler discoveries already strongly supports the idea that planetary systems are common. We've also discussed statistical evidence indicating that as many as 20% of stars may be orbited by an approximately Earth-size planet in the habitable zone. However, we do not yet know the actual nature of these worlds, and despite the large number of known planetary systems, it is still too soon to say whether any of these planetary systems are genuinely "like ours." The bottom line is that we do not yet know whether planetary systems like ours are common, but more definitive answers should be coming over the next decade or two.

**Think about it** Not long before most of today's college students were born, the only known planets were those of our own solar system. Today, the evidence suggests that many or most stars have planets. For the galaxy as a whole, that's a change in the estimated number of planets from fewer than 10 to more than 100 billion. Do you think this change should alter our perspective on our place in the universe? Defend your opinion.

## 13.4 The Future of Extrasolar Planetary Science

We have entered a new era in planetary science, one in which our understanding of planetary processes can be based on far more planets than just those of our own solar system. Although our current knowledge of extrasolar planets and their planetary systems is still quite limited, ingenious new observing techniques, dedicated observatories, and ambitious space telescope programs should broaden our understanding dramatically in the coming years and decades. In this section, we'll focus on the more dramatic improvements expected in coming years.

### How will future observations improve our understanding?

As noted earlier, the deepest outstanding questions concern how common solar systems like ours may prove to be and, more specifically, whether other Earth-like planets exist. Keep in mind that there is an important distinction between Earth-*size* and Earth-*like*. The *Kepler* mission found many planets that are Earth-size, but this does not necessarily mean that they are Earth-like in the sense of having features like continents, oceans, plate tectonics, or life.

Learning whether other planets are Earth-like will probably occur in two steps. First, we'll need to find more planets that are Earth-size and have orbits in their star's habitable zone. Second, we'll need some type of direct imaging or spectroscopy to learn whether these planets actually have oceans and atmospheres that might be conducive to life.

**Think about it** How do you think the discovery of other Earth-like planets would change our view of our place in the universe? Defend your opinions.

Scientists are developing a wide range of techniques for trying to carry out such observations, including many that rely on ground-based observatories. However, over the next decade or so, it's likely that the greatest advances will come primarily in three ways: with the ongoing *GAIA* mission, with new small missions following on *Kepler*'s transiting planet success, and with efforts to obtain direct imaging and spectroscopy of extrasolar planets. Let's briefly look at each.

***GAIA*** Section 13.1 briefly discussed the European Space Agency's *GAIA* mission. *GAIA* is a remarkable space observatory designed primarily to make precise measurements of the positions of more than one billion stars in the Milky Way Galaxy. These astrometric measurements enable scientists to calculate precise distances to these stars, thereby facilitating the creation of a three-dimensional map of the galaxy. To make these measurements, *GAIA* actually has two telescopes, each with a collecting area of nearly one square meter, designed to work together to give extremely precise measurements of stellar positions.

Because precise measurements of stellar positions allow scientists to detect changes in those positions, *GAIA* is also capable of discovering planets via the astrometric method, as well as some via the transit method. In fact, mission scientists estimate that they may be able to detect as many as 20,000 extrasolar planets during the life of the mission, which is expected to continue through at least the end of 2020. Patience is necessary, as the astrometric method is most sensitive to planets far from their host stars, which have long orbital periods. As of 2018, the *GAIA* team had verified its ability to detect known planets via the transit method, but scientists were still processing its data in search of astrometric discoveries.

***TESS* and *CHEOPS*** Building on the *Kepler* mission's success, NASA and the European Space Agency (ESA) each have small space telescopes dedicated to observing transits of extrasolar planets. The NASA mission, called the *Transiting Exoplanet Survey Satellite*, or *TESS* (**FIGURE 13.18**), is designed to monitor hundreds of thousands of the brightest and nearest stars over the whole sky in search of transit events. Scientists hope this strategy will enable them to discover thousands of planets orbiting stars that are near enough for follow-up observations with either existing or future telescopes. If humanity ever travels to the stars, it is possible that the first planet we visit will be one discovered by *TESS*.

**FIGURE 13.18** Artist's conception of the *TESS* spacecraft.

The European mission, *CHEOPS* (*CHaracterising ExO-Planet Satellite*), is not designed to discover new planets itself, but instead to make better transit measurements for planets identified by other telescopes, including *TESS*. *CHEOPS* will also target stars around which planets have been discovered with the Doppler method, since a small percentage of the planets are likely to also exhibit transits. If transits are seen, the *CHEOPS* data will tell us that the Doppler-measured masses are exact and also allow us to learn the planets' masses and densities.

**Think about it** *TESS* was launched in 2018, and *CHEOPS* was due for launch in early 2019, shortly after this book went to press. Find the current status of these missions, and briefly describe their important results to date.

**Direct Detection** The indirect planet-hunting methods have started a revolution in planetary science by demonstrating that our solar system is just one of many planetary systems. But it can be very difficult to learn more than planets' most basic properties from indirect measurements. To learn more about their nature, it would be ideal to observe the planets themselves, obtaining images of their surfaces or spectra of their atmospheres.

The glare from stars makes direct detection of planets extremely difficult, especially given how close together the stars and planets lie when observed from Earth. To date, astronomers have found only a few planets by direct detection. One confirmed direct detection made with infrared light is of a jovian planet orbiting the star Beta Pictoris (**FIGURE 13.19a**). This planet's existence was actually first suspected from study of ripples in the dust disk surrounding the star, and then later confirmed through imaging, which has also revealed the planet's orbital motion around its star. **FIGURE 13.19b** shows another confirmed detection with infrared light, revealing at least four planets orbiting the star HR 8799. Astronomers are confident the planets are real because subsequent images have shown changes in their positions due to their orbits around the star.

Scientists are rapidly developing new observing capabilities that should allow for more direct observations in the future from large, ground-based observatories. Scientists are also optimistic that the James Webb Space Telescope, a large infrared space telescope scheduled to launch in 2020, will be able to make some direct observations.

Nevertheless, the quest to find Earth-like worlds is now in a bittersweet period. Thanks to recent discoveries and rapidly advancing technology, astronomers have begun to design advanced space observatories that should be able to obtain images and spectra of Earth-size planets around other stars, with enough resolution to allow us to determine whether they are Earth-like and perhaps even

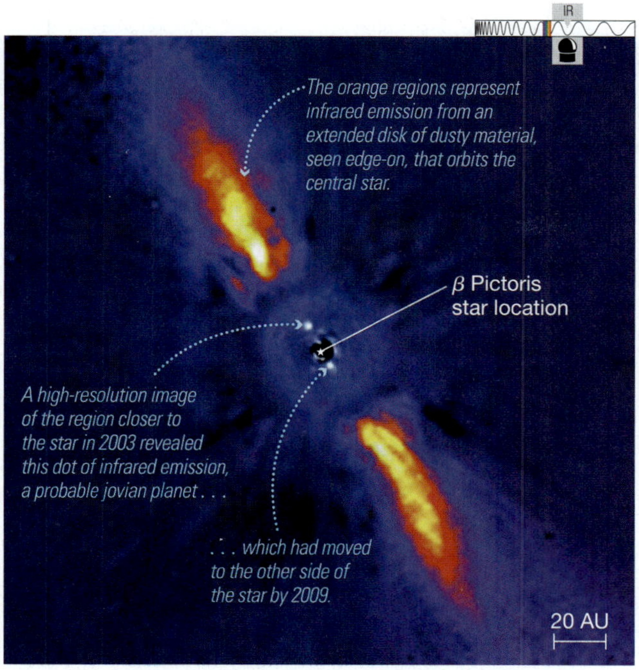


**a** This infrared image composite from the European Southern Observatory Very Large Telescope shows a large debris disk orbiting the star Beta Pictoris and a probable jovian planet that has formed in the disk. Images were taken with the star itself blocked; the star's position has been added digitally.

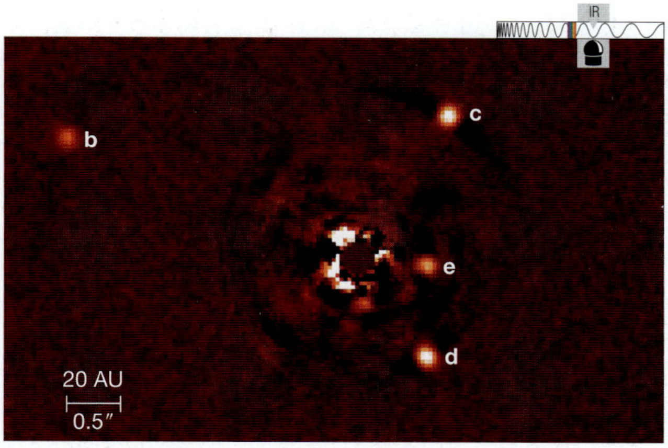


**b** This infrared image from the Large Binocular Telescope shows direct detection of a four-planet system (planets marked b, c, d, e) orbiting the star HR 8799. We know they are planets because they have moved slightly since their discovery. Light from the star itself (center) was mostly blocked out during the exposure, with its remaining light subtracted as much as possible. These planets are much larger, brighter, and farther from their star than jovian planets in our solar system.

**FIGURE 13.19** Direct detections of extrasolar planets.

to detect spectral signatures that would indicate the presence of life. Some designs propose to use a free-flying "star shade" that would maintain a position directly in front of the telescope, where it could be adjusted to block the light of a central star so that the telescope could observe orbiting planets without the interference of starlight. However, while some of these missions could in principle be flown at almost any time, budgetary constraints have pushed them at least 10 to 20 years into the future. Answers to age-old questions lie tantalizingly close, but they will not be obtained until we commit the funds to pay for them.

**Think about it** Suppose you were a member of the United States Congress. How much would you be willing to spend to build a space observatory capable of obtaining images and spectra of Earth-like planets around other stars? Defend your opinion.

## The BIG Picture PUTTING CHAPTER 13 INTO PERSPECTIVE

*In this chapter, we have explored one of the newest areas of astronomy—the study of solar systems beyond our own. As you continue your studies, please keep in mind the following important ideas:*

- In a period of barely two decades, we have gone from knowing of no other planets around other stars to knowing that many or most stars have one or more planets. As a result, there is no longer any question that planets are common in the universe.
- The discovery of other planetary systems represents a striking confirmation of a key prediction of the nebular theory of solar system formation. Nevertheless, the precise characteristics of other planets and planetary systems pose challenges to details of the theory that scientists are still investigating.
- While we have already identified thousands of extrasolar planets, nearly all of these have been found with indirect methods. These methods allow us to determine many properties of the planets, but we will need direct images or spectra to learn about them in much more detail.

**MY COSMIC PERSPECTIVE** Humanity's science fiction future is much less hypothetical now that we know that planets are everywhere and Earth-size planets in habitable zones ex st in our quadrant of the galaxy. Should we start thinking differently about our future?

# Summary of Key Concepts

## 13.1 Detecting Planets Around Other Stars

- **Why is it so challenging to learn about extrasolar planets?** The great challenge stems from the great distances to other stars, the small sizes of planets in comparison, and the vast difference in brightness between stars and planets. Today most planets are found indirectly either by (1) looking for the subtle gravitational tugs of planets on stars or (2) looking for changes in a star's brightness as a planet passes in front of it.
- **How can a star's motion reveal the presence of planets?** We can look for a planet's gravitational effect on its star

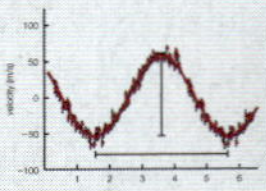

using the **astrometric method**, which looks for small shifts in stellar position, or the **Doppler method**, which looks for the back-and-forth motion of stars revealed by Doppler shifts.
- **How can changes in a star's brightness reveal the presence of planets?** A small fraction of all planetary systems are by chance aligned in such a way that their

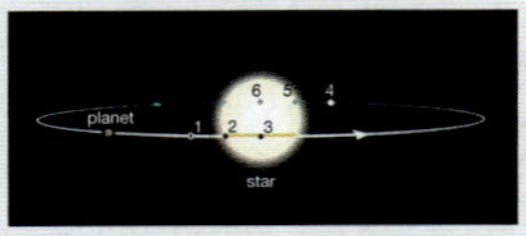


planets pass in front of their star as seen from Earth, creating a **transit** in which the star dims slightly. The planet may also pass behind the star in an **eclipse** on the other side of the orbit, potentially revealing even more information about the planet.

## 13.2 The Nature of Planets Around Other Stars

- **What properties of extrasolar planets can we measure?** All detection methods allow us to determine a planet's orbital period and distance from its star. The astrometric and Doppler methods can provide masses (or minimum masses), while the transit method can provide sizes. When the transit and Doppler methods are used together, we can determine average density. In some cases, transits (and eclipses) can provide other data, including limited data about atmospheric composition and temperature.
- **How do extrasolar planets compare to planets in our solar system?** The known extrasolar planets have a much wider

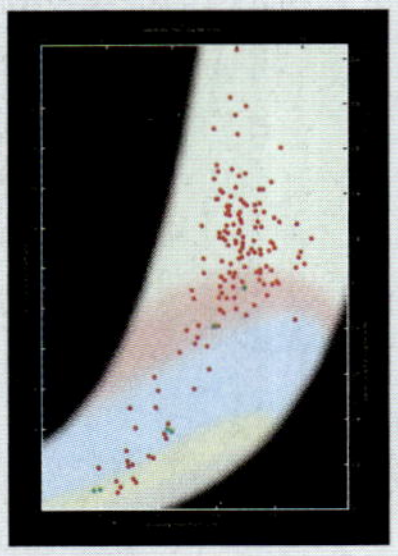

range of properties than the planets in our solar system. Many orbit much closer to their stars and with more eccentric orbital paths; even some jovian planets, called *hot Jupiters*, are found close to their stars. We have also observed properties indicating planetary types, such as water worlds, that do not fall neatly into the traditional terrestrial and jovian categories.

## 13.3 The Formation of Other Solar Systems

- **Do we need to modify our theory of solar system formation?** Our basic theory seems sound, but we have had to modify it to allow for planetary migration and a

wider range of planetary types than we find in our solar system. Many mysteries remain, but they are unlikely to require major change to the nebular theory of solar system formation.
- **Are planetary systems like ours common?** Current evidence suggests that most stars have planets, and at least some are Earth-size and in their stars' habitable zones. Nevertheless, we don't yet have enough data to know for certain whether planetary systems like ours—and planets like Earth—are common.

## 13.4 The Future of Extrasolar Planetary Science

- **How will future observations improve our understanding?**

In the future, scientists hope to learn whether solar systems with layouts like ours are rare or common, and whether Earth-like planets are common. Future observations by the *GAIA*, *TESS*, and *CHEOPS* missions should help answer these questions, but ultimate answers will probably require direct detection with space observatories of the future, perhaps including the James Webb Space Telescope.

## Visual Skills Check

*Use the following questions to check your understanding of some of the many types of visual information used in astronomy. For additional practice, try the Chapter 13 Visual Quiz in the Study Area at www.MasteringAstronomy.com.*

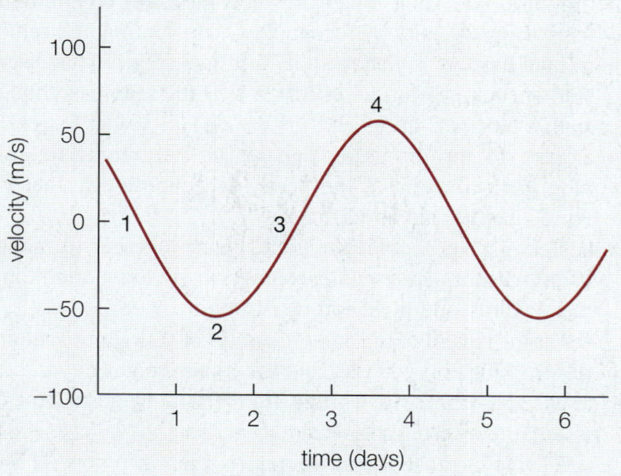


*This plot, based on Figure 13.4a, shows the periodic variations in the Doppler shift of a star caused by a planet orbiting around it. Positive velocities mean the star is moving away from Earth, and negative velocities mean the star is moving toward Earth. (You can assume that the orbit appears edge-on from Earth.) Answer the following questions based on the information in the graph.*

1. How long does it take the star and planet to complete one orbit around their center of mass?
2. What maximum velocity does the star attain?
3. Match the *star's* position at points 1, 2, 3, and 4 in the plot with the descriptions below.
   a. headed straight toward Earth
   b. headed straight away from Earth
   c. closest to Earth
   d. farthest from Earth
4. Match the *planet's* position at points 1, 2, 3, and 4 in the plot with the descriptions in question 3.
5. What would be the most significant change in the plot if the planet were more massive?
   a. It would not change, because it describes the motion of the star, not the planet.
   b. The peaks and valleys would get larger (greater positive and negative velocities) because of larger gravitational tugs.
   c. The peaks and valleys would get closer together (shorter period) because of larger gravitational tugs.

## Exercises and Problems

*For instructor-assigned homework and other study tools, go to www.MasteringAstronomy.com.*

### Chapter Review Questions

#### Short-Answer Questions Based on the Reading

1. Why are *extrasolar planets* hard to detect directly?
2. What are the two major approaches to detecting extrasolar planets indirectly?
3. How can gravitational tugs from orbiting planets affect the motion of a star? Explain how alien astronomers could deduce the existence of planets in our solar system by observing the Sun's motion.
4. Briefly describe the *astrometric method*. What is the *GAIA* mission?
5. Briefly describe the *Doppler method*. Summarize the evidence that the planet orbiting 51 Pegasi is a *hot Jupiter*.
6. How does the *transit method* work? What was the *Kepler* mission?
7. Briefly summarize the planetary properties we can in principle measure with current detection methods.
8. Why does the Doppler method generally allow us to determine only *minimum* planetary masses? In what cases can we be confident that we know precise masses? Explain.
9. How does the transit method tell us planetary size, and in what cases can we also learn mass and density?
10. How do the orbits of known extrasolar planets differ from those of planets in our solar system? Why are these orbits surprising?
11. Summarize the current state of knowledge about extrasolar planet masses and sizes. Based on the evidence, is it likely that smaller planets or larger planets are more common?
12. Summarize the key features shown in Figure 13.16, and briefly describe the nature of planets that would fit each of the model regions shown on the graph.
13. What is planetary migration, and how may it account for the surprising orbits of many extrasolar planets?
14. How can scientists account for the fact that extrasolar planets seem to come in a wider range of types than the planets of our own solar system?
15. Overall, does the nebular theory seem adequate for describing the origins of other planetary systems? Explain.
16. Based on current evidence, how common are planetary systems?
17. Briefly describe how *TESS*, *CHEOPS*, and direct observations should improve our understanding of extrasolar planets in coming years.

#### Does It Make Sense?

*Decide whether or not each of the following statements makes sense (or is clearly true or false). Explain clearly; not all of these have definitive answers, so your explanation is more important than your chosen answer.*

18. An extraterrestrial astronomer surveying our solar system with the Doppler method could discover the existence of Jupiter with just a few days of observation.
19. The fact that we have not yet discovered an Earth-size extrasolar planet in an Earth-like orbit tells us that such planets must be very rare.

20. Within the next few years, astronomers expect to confirm all the planet detections made with the astrometric and Doppler methods by observing transits of these same planets.
21. The infrared brightness of a star system decreases when a planet goes into eclipse.
22. Some extrasolar planets are likely to be made mostly of water.
23. Some extrasolar planets are likely to be made mostly of gold.
24. Current evidence suggests that there could be 100 billion or more planets in the Milky Way Galaxy.
25. It's the year 2025: The *TESS* mission has announced the discovery of numerous planets with Neptune-like orbits around their stars.
26. It's the year 2025: Astronomers have successfully obtained a high-resolution image of an Earth-size extrasolar planet, showing that it has oceans and continents.
27. It's the year 2040: Scientists announce that our first spacecraft to reach an extrasolar planet is now orbiting a planet around a star located near the center of the Milky Way Galaxy.

## Quick Quiz

*Choose the best answer to each of the following. For additional practice, try the Chapter 13 Reading and Concept Quizzes in the Study Area at www.MasteringAstronomy.com.*

28. Which method could detect a planet in an orbit that is face-on to Earth? (a) Doppler method (b) transit method (c) astrometric method
29. Which detection method or methods measure the gravitational tug of a planet on its star, allowing us to estimate planetary mass? (a) the transit method only (b) the Doppler method only (c) the astrometric and Doppler methods
30. Which one of the following can the transit method tell us about a planet? (a) its mass (b) its size (c) the eccentricity of its orbit
31. To determine a planet's average density, we can use (a) the transit method alone. (b) the astrometric and Doppler methods together. (c) the transit and Doppler methods together.
32. Based on the model types shown in Figure 13.16, a planet made almost entirely of hydrogen compounds would be considered a (a) terrestrial planet. (b) jovian planet. (c) water world.
33. Look at the dot for Jupiter in Figure 13.16, then at the red dot directly to Jupiter's left. Compared to the density of Jupiter, the density of the planet represented by that dot is (a) higher. (b) lower. (c) the same.
34. The term "super-Earth" refers to a planet that is (a) the size of Earth but has more water. (b) larger than Earth but on a close-in orbit that makes it much hotter than Earth. (c) similar in composition to Earth but larger in size.
35. What's the best explanation for the location of hot Jupiters? (a) They formed closer to their stars than Jupiter did. (b) They formed farther out, like Jupiter, but then migrated inward. (c) The strong gravity of their stars pulled them in close.
36. Based on computer models, when is planetary migration most likely to occur in a planetary system? (a) early in its history, when there is still a gaseous disk around the star (b) shortly after a stellar wind clears the gaseous disk away (c) late in its history, when asteroids and comets occasionally collide with planets
37. Based on current data, planetary systems appear to be (a) extremely rare. (b) present around about 10% of all stars. (c) present around at least half of all stars.

## Inclusive Astronomy

*Use these questions to reflect on participation in science.*

38. *Group Discussion: Astronomy in Color.* In 2015, a diverse group of scientists, including some international leaders in the study of extrasolar planets, started a blog called Astronomy in Color. Their blog states that they are "committed to increasing diversity by recognizing, confronting and removing the barriers to racial equity and inclusion." This discussion encourages you to reflect on why the scientists thought such a blog was necessary.
    **a.** Prior to the discussion, go to the Astronomy in Color blog, look at some of the posts, and choose one that you would like to read and discuss.
    **b.** Gather in groups of three and work together to make a list of reasons some scientists are trying to draw attention to racial equity and inclusion in astronomy.
    **c.** For the benefit of the other members of your group, summarize the blog post you read and the issues it highlighted.
    **d.** Discuss what you learned from the blog posts. Are the issues they raised connected to broader issues in society? Are some of those issues unique to science in general or to astronomy in particular?
    **e.** What advice would your group give to the astronomical community that might help it to become more inclusive?

## The Process of Science

*These questions may be answered individually in short-essay form or discussed in groups, except where identified as group-only.*

39. *When Is a Theory Wrong?* As discussed in this chapter, in its original form the nebular theory of solar system formation does not explain the orbits of many known extrasolar planets, but it can explain them with modifications, such as allowing for planetary migration. Does this mean the theory was wrong or only incomplete before the modifications were made? Explain. Be sure to look back at the discussion in Chapter 3 of the nature of science and scientific theories.
40. *Refuting the Theory?* Consider the following three hypothetical observations: (1) the discovery of a lone planet that is small and dense like a terrestrial planet but has a Jupiter-like orbit; (2) the discovery of a planetary system in which three terrestrial planets orbit the star beyond the orbital distance of two jovian planets; (3) the discovery that a majority of planetary systems have their jovian planets located nearer to their star than 1 AU and their terrestrial planets located beyond 5 AU. Each of these observations would challenge our current theory of solar system formation, but would any of them shake the very foundations of the theory? Explain clearly for each of the three hypothetical observations.
41. *Unanswered Questions.* As discussed in this chapter, we are only just beginning to learn about extrasolar planets. Briefly describe one important but unanswered question related to the study of planets around other stars. Then write 2–3 paragraphs in which you discuss how we might answer this question in the future. Be as specific as possible, focusing on the type of evidence necessary to answer the question and how the evidence could be gathered. What are the benefits of finding answers to this question?
42. *Group Activity: Time to Move On.* A common theme in science fiction is "leaving home" to find a new planet for humans to live on. Now that we know about thousands of

planets, we can start imagining how to choose one. Note: You may wish to do this activity using the four roles described in Chapter 1, Exercise 39.

**a.** Make a list of characteristics that you would look for in a planet that might make a good home.

**b.** Examine the planets in Figure 13.16. Does this graph give enough information to allow you to determine which planets might make good homes, or poor ones? If not, what's missing?

**c.** Suppose you also knew the orbital distance for each of the planets in Figure 13.16. Would that make it easier to find potential good homes? Why or why not?

## Investigate Further

### Short-Answer/Essay Questions

43. *Explaining the Doppler Method.* Explain how the Doppler method works in terms an elementary school child would understand. It may help to use an analogy to explain the difficulty of direct detection and the general phenomenon of the Doppler shift.

44. *Comparing Methods.* What are the strengths and limitations of the Doppler and transit methods? What kinds of planets are easiest to detect with each method? Are there certain planets that each method cannot detect, even if the planets are very large? Explain. What advantages are gained if a planet can be detected by both methods?

45. *No Hot Jupiters Here.* How do we think hot Jupiters formed? Why didn't one form in our solar system?

46. *Low-Density Planets.* Only one planet in our solar system has a density less than 1 g/cm$^3$, but many extrasolar planets do. Explain why in a few sentences. (*Hint:* Consider the densities of the jovian planets in our solar system, given in Figure 11.1.)

47. *A Year on HD 189733b.* Imagine you were hovering in the upper atmosphere (in a suitable spacecraft) of the planet that orbits the star HD 189733, for which you can find data in Figures 13.5, 13.12, and 13.16. What would you see, and how would the view be different from the view you would have while floating in Jupiter's atmosphere? Consider factors like local conditions, clouds, how the planet's star would appear, and orbital motion.

48. *Project: Detecting an Extrasolar Planet for Yourself.* Most colleges and many amateur astronomers have the equipment necessary to detect known extrasolar planets using the transit method. All that's required is a telescope 10 or more inches in diameter, a CCD camera system, and a computer system for data analysis. The basic method is to take exposures of a few minutes' duration over a period of several hours around the times of predicted transit, and then compare the brightness of the star being transited to that of other stars in the same CCD frame. For complete instructions, see the Study Area of Mastering Astronomy.

### Quantitative Problems

*Be sure to show all calculations clearly and state your final answers in complete sentences.*

49. *Lost in the Glare.* This exercise helps you consider how much harder it would be for an alien astronomer to detect the light from planets in our solar system than to detect the light from the Sun itself.

**a.** Calculate the fraction of the total emitted sunlight that reaches Earth. (*Hint:* Imagine a sphere around the Sun the size of Earth's orbit [area $= 4\pi a^2$], then calculate the fraction of that area taken up by the disk of Earth [area $= \pi r^2_{\text{Earth}}$].)

**b.** Earth reflects 29% of the Sun's light. Based on this fact and your answer to part a, find the ratio of the total amount of light emitted by the Sun to the amount reflected by Earth. What does this tell you about the difficulty of detecting a planet like Earth around another star? How would a "star shade" like that discussed in Section 13.4 help?

**c.** Would detecting Jupiter be easier or harder than detecting Earth? Comment on whether you think Jupiter's larger size or greater distance has a stronger effect on its detectability. You may neglect any difference in reflectivity between Earth and Jupiter.

50. *Transit of TrES-1.* The planet TrES-1, orbiting a distant star, has been detected by both the transit and the Doppler method, so we can calculate its density and get an idea of what kind of planet it is.

**a.** Using the method of Mathematical Insight 13.3, calculate the radius of the transiting planet. The planetary transits block 2% of the star's light. The star TrES-1 has a radius of about 85% of our Sun's radius.

**b.** The mass of the planet is approximately 0.75 times the mass of Jupiter, and Jupiter's mass is about $1.9 \times 10^{27}$ kilograms. Calculate the average density of the planet. Give your answer in grams per cubic centimeter. Compare this density to the average densities of Saturn (0.7 g/cm$^3$) and Earth (5.5 g/cm$^3$). Is the planet likely to be terrestrial or jovian in nature? Explain. (*Hint:* To find the volume of the planet, use the formula for the volume of a sphere: $V = \frac{4}{3}\pi r^3$. Be careful with unit conversions.)

51. *Planet Around 51 Pegasi.* The star 51 Pegasi has about the same mass and luminosity as our Sun and is orbited by a planet with an orbital period of 4.23 days and mass estimated to be 0.6 times the mass of Jupiter.

**a.** Use Kepler's third law to calculate the planet's average distance (semimajor axis) from its star. (*Hint:* Because the mass of 51 Pegasi is about the same as the mass of our Sun, you can use Kepler's third law in its original form, $p^2 = a^3$ [**Section 3.3**]. Be sure to convert the period into years before using this equation.)

**b.** Suppose the planet reflects 15% of the incoming sunlight. Using Mathematical Insight 10.1, calculate its "no greenhouse" average temperature. How does this temperature compare to that of Earth?

**c.** Repeat part b, but assume that the planet is covered in bright clouds that reflect 80% of the incoming sunlight.

**d.** Based on your answers to parts b and c, do you think it is likely that the conditions on this planet could be conducive to life? Explain.

52. *Identical Planets?* Imagine two planets orbiting a star with orbits edge-on to the Earth. The peak Doppler shift for each is 50 m/s, but one has a period of 3 days and the other has a period of 300 days. Calculate the two minimum masses and say which, if either, is larger. (*Hint:* See Mathematical Insight 13.2.)

53. *Finding Orbit Sizes.* The Doppler method allows us to find a planet's semimajor axis using just the orbital period and the star's mass (Mathematical Insight 13.1).

**a.** Imagine that a new planet is discovered orbiting a $2M_{\text{Sun}}$ star with a period of 5 days. What is its semimajor axis?

**b.** Another planet is discovered orbiting a $0.5M_{\text{Sun}}$ star with a period of 100 days. What is its semimajor axis?

**COSMIC Context**

PART III

# Learning from Other Worlds

**Comparing the worlds in the solar system has taught us important lessons about Earth and why it is so suitable for life. This illustration summarizes some of the major lessons we've learned by studying other worlds both in our own solar system and beyond it.**

(1) Comparing the terrestrial worlds shows that a planet's size and distance from the Sun are the primary factors that determine how it evolves through time [Chapters 9, 10].

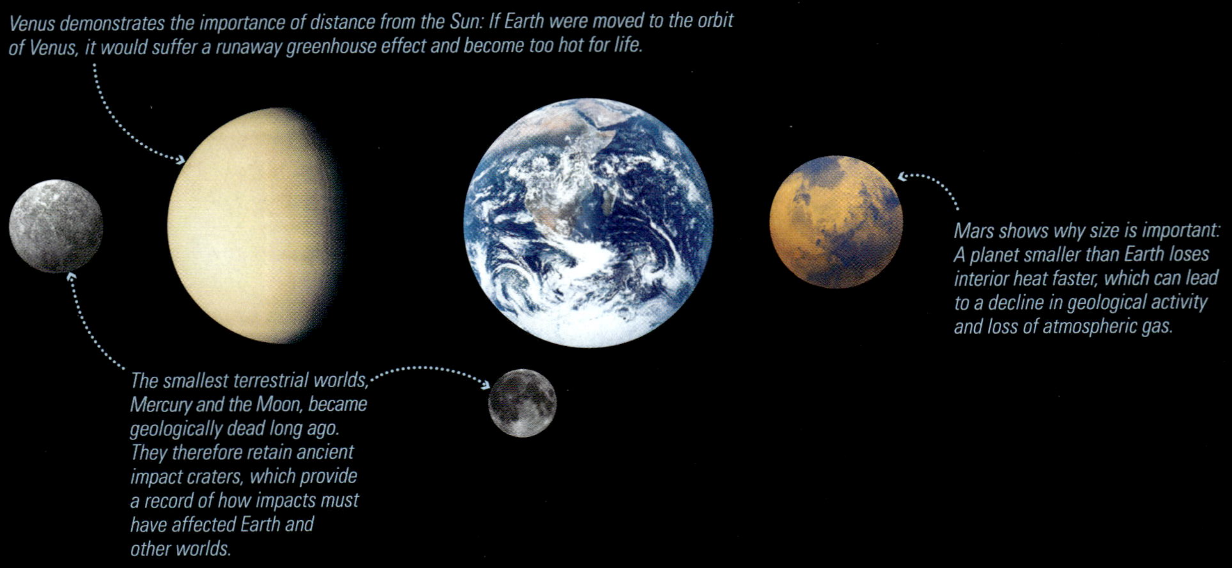

(2) Jovian planets are gas-rich and far more massive than Earth. They and their ice-rich moons have opened our eyes to the diversity of processes that shape worlds [Chapter 11].

*The strong gravity of the jovian planets has shaped the asteroid and Kuiper belts, and flung comets into the distant Oort cloud, ultimately determining how frequently asteroids and comets strike Earth.*

*Our Moon led us to expect all small objects to be geologically dead . . .*

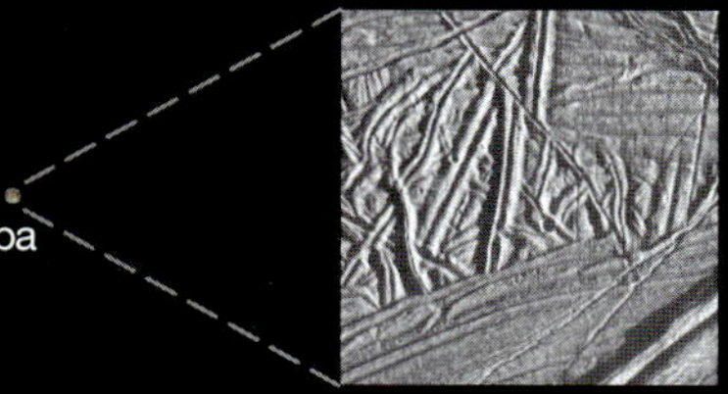


*. . . but Europa—along with Io, Titan, and other moons—proved that tidal heating or icy composition can lead to geological activity, in some cases with subsurface oceans and perhaps even life.*

(3) Asteroids and comets may be small bodies in the solar system, but they have played major roles in the development of life on Earth [Chapter 12].

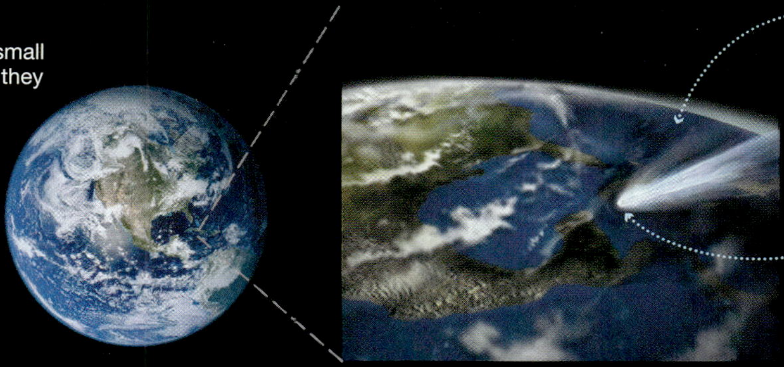

*Comets or water-rich asteroids from the outer asteroid belt brought Earth the ingredients of its oceans and atmosphere.*

*Impacts of comets and asteroids have altered the course of life on Earth and may do so again.*

(4) The discovery of planets around other stars has shown that planetary systems are common. Studies of other solar systems are teaching us new lessons about how planets form and about the likelihood of finding other Earth-like worlds [Chapter 13].

*Rapid advances in extrasolar planet detection have allowed us to find some planets as small as Earth, with some of these planets orbiting within the habitable zones of their stars.*

# S2 Space and Time

▲ **About the photo:** Albert Einstein, shown here in his Princeton office in 1954 (age 75), is best known for developing the theory of relativity, which he published in two parts in 1905 (the special theory) and 1915 (the general theory). This theory forms the basis of our modern understanding of space, time, and gravity.

## LEARNING GOALS

### S2.1 Einstein's Revolution
- What are the major ideas of special relativity?
- What is relative about relativity?
- What is absolute about relativity?

### S2.2 Relative Motion
- What's surprising about the absoluteness of the speed of light?
- Why can't we reach the speed of light?

### S2.3 The Reality of Space and Time
- How does relativity affect our view of time and space?
- Do the effects predicted by relativity really occur?

### S2.4 Toward a New Common Sense
- How can we make sense of relativity?
- How does special relativity offer us a ticket to the stars?

*Henceforth space by itself, and time by itself, are doomed to fade away into mere shadows, and only a kind of union of the two will preserve an independent reality.*

*— Hermann Minkowski, 1908*

**Chapter S2 Overview**

The universe consists of matter and energy moving through space with the passage of time. Up to this point in the book, we have discussed the concepts of space and time as though they were absolute and distinct—just as they seem in everyday life. But what if this appearance is deceiving?

More than a century ago, Albert Einstein discovered that space and time are actually intertwined in a manner that we now understand through Einstein's *theory of relativity.* Contrary to popular myth, this theory is *not* particularly difficult to understand, though it requires us to think in new ways. Moreover, because space and time are such fundamental concepts, understanding relativity is very important to understanding astronomy and the universe. We will therefore devote this chapter and the next to understanding the key ideas of relativity.

## S2.1 Einstein's Revolution

Albert Einstein was born on March 14, 1879, in Ulm, Germany. As a toddler, he was so slow in learning to speak that relatives feared he might have brain damage. Even when he began to show promise, it was in unconventional ways. Young Albert dropped out of high school at the suggestion of a teacher who told him that he would "never amount to anything." Nevertheless, he was admitted to college in Switzerland, in large part because of his proficiency in mathematics.

Einstein hoped to teach after his college graduation, but his Jewish heritage and lack of Swiss citizenship made it difficult for him to find a teaching job. In 1901, he instead took a job with the Swiss patent office, and devoted his free time to the study of unsolved questions in physics.

From the standpoint of history, Einstein burst onto the scene in 1905, when he earned his Ph.D. and published five papers in the *German Yearbook of Physics*. Three of the papers solved three of the greatest mysteries in physics at the time. The first paper dealt with something called the *photoelectric effect,* and essentially presented the first concrete evidence of the wave-particle duality of light that we discussed in Chapter 5. The second paper explained why suspended particles in water jiggle even after the water has been still for a long time (an effect known as *Brownian motion*). Einstein's analysis led to the first direct measurements of molecular motion and molecular and atomic sizes. The third paper introduced the special theory of relativity, which forever changed our understanding of space and time.

Einstein returned to Germany in 1913, where he completed work on his general theory of relativity. After some of its central predictions were confirmed by observation in 1919 [**Section S3.4**], Einstein's fame began to extend beyond the world of science. He received the Nobel Prize in Physics in 1921 and was by this time a household name.

Einstein was a visiting professor at the California Institute of Technology when Hitler came to power in Germany. Recognizing Hitler's evil, Einstein decided to remain in the United States. He was a committed pacifist but nevertheless played an important role in convincing President Franklin Roosevelt to start the Manhattan Project to build an atomic bomb, because he feared that Hitler would otherwise develop an atomic bomb first. He expressed great dismay when the bomb was used against Japan, and he spent much of the rest of his life arguing for a worldwide agreement to ban the further manufacture and use of nuclear weapons. He also worked to promote human rights and equality for men and women of all races and nationalities. Einstein died in Princeton, New Jersey, on April 18, 1955, but remains one of the most famous scientists of all time.

### What are the major ideas of special relativity?

We often talk about the *theory of relativity* as though it were a single theory, and in some sense it is. However, Einstein actually developed his theory in two parts. The **special theory of relativity**, published in 1905, shows that space and time are intertwined but does not deal with the effects of gravity. The **general theory of relativity**, published in 1915, offers a surprising new view of gravity—a view that we will use to help us understand topics such as the expansion and fate of the universe and the strange objects known as *black holes.* In essence, the special theory is "special" because it deals only with the special case in which we ignore the role of gravity, while the general theory is "general" because it includes gravity. We will focus on the special theory of relativity in this chapter and on the general theory in Chapter S3.

Einstein developed his theories with the aid of "thought experiments"—experiments that could be carried out in principle but that would be very difficult in practice. Thought experiments allow us to think through the consequences of simple ideas or statements. Let's start with a thought experiment designed to show you why the special theory of relativity is so surprising.

Imagine that, with the aid of a long tape measure, you carefully measure the distance you walk from home to work and find it to be 5.0 kilometers. You wouldn't expect any argument about this distance. For example, if a friend drives her car along the same route and measures the distance with her car's odometer, she ought to get the same measurement of 5.0 kilometers (assuming her odometer is accurate). Likewise, you would expect agreement about how much time it takes you to walk this distance. Suppose your friend continues driving and you call her just as you leave your house at 8:00 A.M. and again just as you arrive at work at 8:45 A.M. You'd certainly be surprised if she argued that your walk took an amount of time other than 45 minutes.

Our example illustrates that distances and times appear absolute and distinct in our daily lives. We expect everyone to agree on the distance between two points, such as the locations of home and work, and the time between two events, such as leaving home and arriving at work. However, Einstein showed that these expectations are not strictly correct: With extremely precise measurements, the distance you measure between home and work will be *different* from the distance measured by a friend in a car, and you and your friend will also disagree about the time it takes you to walk to work. At ordinary speeds, the differences will be so small as to be unnoticeable. But if your friend could drive at a speed close to the speed of light, the differences would be substantial.

Disagreements about distances and times only mark the beginning of the astonishing ideas contained in the special theory of relativity. In the rest of this chapter, we will see how this theory leads to each of the following ideas:

- No information can travel faster than the speed of light (in a vacuum), and no material object can even reach the speed of light.
- If you observe anyone or anything moving by you at a speed close to the speed of light, you will conclude that time runs more slowly for that person or moving object. That is, a person moving by you ages more slowly than you, a clock moving by you ticks more slowly than your clock, a computer moving by you runs more slowly than your similar computer, and so on.
- If you observe two events to occur simultaneously, such as flashes of light in two different places at the same time, a person moving by you at a speed close to the speed of light may not agree that the two events were simultaneous.
- If you carefully measure the size of something moving by you at a speed close to the speed of light, you will find that its length (in the direction of its motion) is shorter than it would be if the object were not moving.
- If you could measure the mass of something moving by you at a speed close to the speed of light, you would find its mass to be greater than the mass it would have if it were stationary. As we will see, Einstein's famous equation, $E = mc^2$, follows from this fact.

Although these ideas of special relativity may sound like science fiction or fantasy, a vast body of observational and experimental evidence supports their reality. They also follow logically from a few simple premises. If you keep an open mind and think deeply as you read this chapter, you'll soon understand the basic ideas of relativity.

## What is relative about relativity?

Contrary to a common belief, Einstein's theory does *not* in any way tell us that "everything is relative." Rather, the theory takes its name from the idea that *motion* is always relative. You can see why motion must be relative with another simple thought experiment.

Imagine a supersonic airplane that flies at a speed of 1670 km/hr from Nairobi, Kenya, to Quito, Ecuador. How fast is the plane going? At first, this question sounds trivial—we have just said that the plane is going 1670 km/hr.

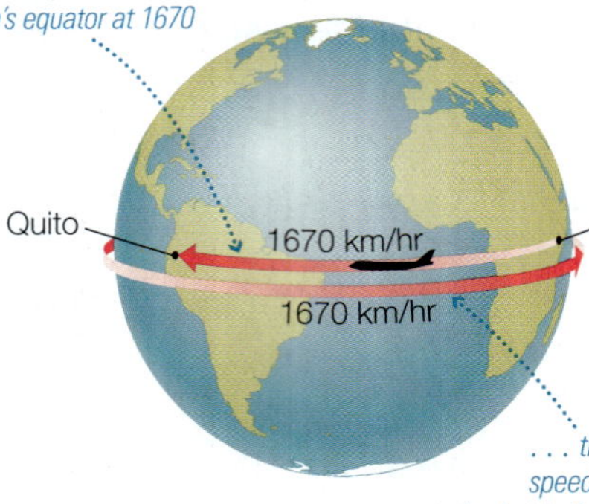


**FIGURE S2.1** A plane flying at 1670 km/hr from Nairobi to Quito (westward) travels precisely opposite Earth's eastward rotation. Viewed from the Moon, the plane appears to remain stationary while Earth rotates underneath it.

But wait. Nairobi and Quito are both nearly on Earth's equator, and the equatorial speed of Earth's rotation is the same 1670 km/hr speed at which the plane is flying (see Figure 1.13), but in the opposite direction (**FIGURE S2.1**). Viewed from the Moon, the plane would therefore appear to stay put *while Earth rotated beneath it.* When the flight began, you would see the plane lift off the ground in Nairobi. The plane would then remain stationary while Earth's rotation carried Nairobi away from it and Quito toward it. When Quito finally reached the plane's position, the plane would drop back down to the ground.

We have two alternative viewpoints about the plane's flight. People on Earth say that the plane is traveling westward across the surface of the Earth. Observers on the Moon say that the plane is stationary while Earth rotates eastward beneath it. Both viewpoints are equally valid.

In fact, there are many other equally valid viewpoints about the plane's flight. Observers looking at the solar system as a whole would see the plane moving at a speed of more than 100,000 km/hr, because that is Earth's speed in its orbit around the Sun. Observers living in another galaxy would see the plane moving at about 800,000 km/ hr with the rotation of the Milky Way Galaxy. The only thing all these observers would agree on is that the plane is traveling at 1670 km/hr *relative to* the surface of the Earth.

The airplane example shows that questions like "Who is really moving?" and "How fast are you going?" have no absolute answers. Einstein's *theory of relativity* tells us that measurements of motion, as well as measurements of time and space, make sense only when we describe whom or what they are being measured relative to.

**Think about it** Suppose you are running on a treadmill and the readout says you are going 8 miles per hour. What is the 8 miles per hour measured relative to? How fast are you going relative to the ground? How fast would an observer on the Moon see you going? Describe a few other possible viewpoints on your speed.

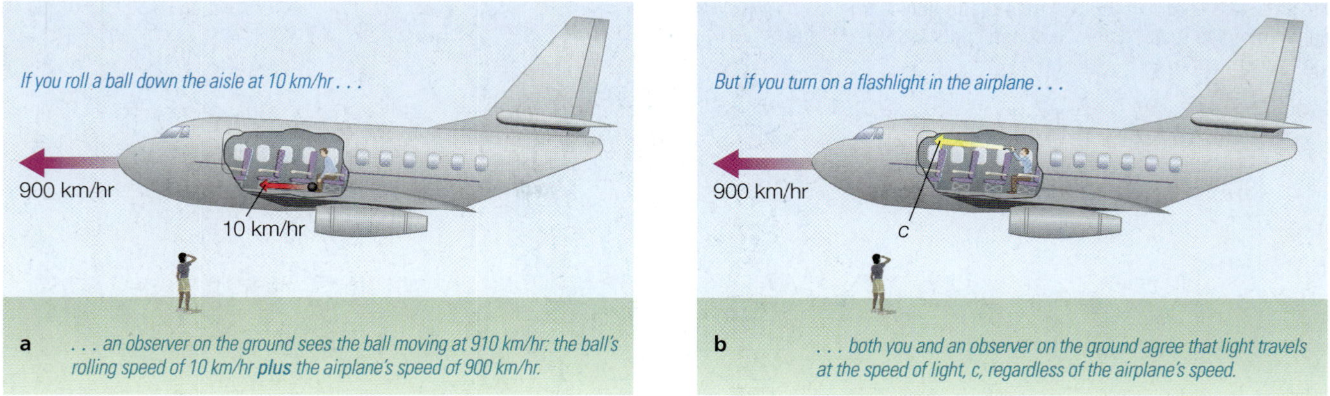


**FIGURE S2.2** Unlike the speed of material objects, the speed of light is the same in all reference frames.

## What is absolute about relativity?

Although the theory of relativity tells us that motion is always relative, the theory's foundations actually rest on the idea that two things in the universe are absolute:

1. The laws of nature are the same for everyone.
2. The speed of light is the same for everyone.

**Understanding the Two Absolutes** The first of the two absolutes, that the laws of nature are the same for everyone, is probably not surprising. If you're on an airplane with the shades drawn during a very smooth flight, you won't feel any sensation of motion, and you would get the same results from any experiments you performed on the airplane that someone else would get performing those experiments on the ground. These equivalent results demonstrate that the laws of nature are the same in the moving airplane and on the ground. In the language of relativity, the ground and the airplane represent different **reference frames** (or *frames of reference*); we say that two people share the same reference frame if they are *not* moving relative to each other. The idea that the laws of nature are the same for everyone means that these laws do not depend on your reference frame.

The second absolute of relativity, that the speed of light is the same for everyone, is far more surprising. In general, we expect people in different reference frames to give different answers for the speed of the same moving object. For example, suppose you roll a ball down the aisle of an airplane. The ball rolls slowly in your airplane reference frame, but someone on the ground would see the ball moving with the airplane's speed *plus* the speed at which it rolls down the aisle (**FIGURE S2.2a**). Now, suppose instead that you turn on a flashlight and measure the speed of the emitted light. The idea that the speed of light is the same for everyone means that a person on the ground will measure exactly the same speed for the light beam, even though the light was emitted inside the moving airplane (**FIGURE S2.2b**). In other words, people in different reference frames can disagree about the speeds of material objects, but everyone always agrees about the speed of light, regardless of where the light comes from.

How do we know that everyone always measures the same speed of light? Observations and experiments are the ultimate judge of any scientific claim, and *the absoluteness of the speed of light is an experimentally verified fact.* We'll discuss some of the experimental evidence later in the chapter. For now, our goal is to understand how the ideas of relativity introduced earlier all follow logically from the two absolutes.

**Making Sense of Relativity** Before we dive into the details, it's worth taking a moment to prepare yourself for the type of change in thinking that relativity will require. One reason relativity has a reputation for being difficult to grasp, despite its underlying simplicity, is that most of its ideas and consequences are not obvious in everyday life. They become obvious only when we deal with speeds close to the speed of light or, in the case of general relativity, with gravitational fields far stronger than that of Earth. Because we don't commonly experience such extreme conditions, we have no *common* sense about them. Therefore, it's not really accurate to say that relativity violates common sense, because the theory is perfectly consistent with everything we have come to expect in daily life.

Making sense of relativity requires only that you learn to view your everyday experiences from a new, broader perspective. Fortunately, you've done this before. At a young age, you learned "common sense" meanings for *up* and *down: Up* is above your head, *down* is toward your feet, and things tend to fall down. One day, however, you learned that Earth is round. When you looked at a globe with the Northern Hemisphere on the top, you were immediately confronted with a **paradox**—a situation that *seems* to violate common sense or to contradict itself. Your common sense told you that Australians should fall off the Earth (**FIGURE S2.3a**), but they don't. To resolve this paradox, you were forced to accept that your "common sense" about *up* and *down* was limited: It works fine for a small area, like a basketball court, but must be revised to apply to the entire Earth. You therefore revised your common sense to accept that *up* and *down* are determined relative to the center of Earth (**FIGURE S2.3b**).

**Think about it** Why do you suppose that most maps show the Northern Hemisphere on the top and the Southern Hemisphere on the bottom? If you hung your map upside down and rewrote the words so that they read right side up, would the map be equally valid?

**FIGURE S2.3** Learning that Earth is round helps children revise their "common sense" understanding of *up* and *down*.

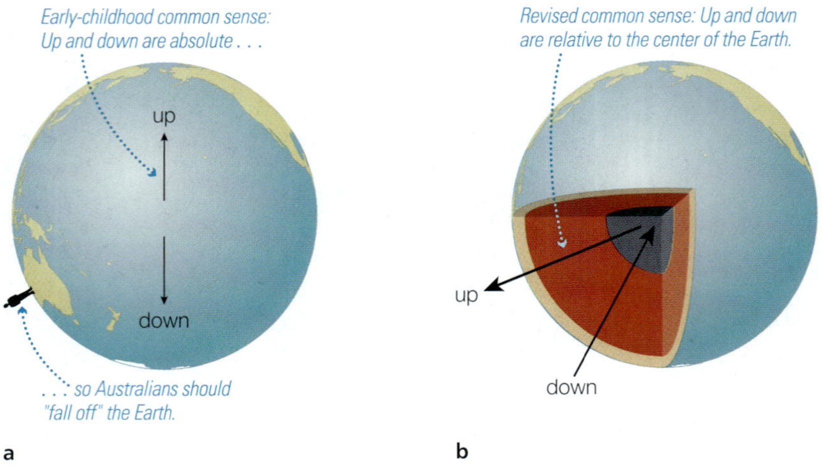


As a teenager, Einstein wondered what the world would look like if he could travel at or beyond the speed of light. He inevitably encountered paradoxes when he thought about this question. Ultimately, he resolved the paradoxes only when he recognized that our common sense ideas about space and time must change if we are to extend them to the realm of very high speeds or very strong gravitational fields. Just as we all once learned a new common sense about up and down, we now must learn a new common sense about space and time.

## S2.2 Relative Motion

We are now ready to explore the theory of relativity by following a series of thought experiments through to their logical conclusions. We will use thought experiments slightly different from the ones Einstein himself used, though the ideas will be essentially the same. Just as Einstein did, we will base all our thought experiments on the assumption that the two absolutes of relativity are true.

### What's surprising about the absoluteness of the speed of light?

The idea that everyone always measures the same speed of light may not sound earth-shattering, but Einstein soon realized that it leads to far-ranging consequences—consequences that force us to let go of many of our intuitive beliefs about how the universe works. We will investigate the implications of the absoluteness of the speed of light by constructing a series of thought experiments about relative motion viewed from different reference frames.

**Thought Experiments at Ordinary Speeds** Because special relativity does not deal with the effects of gravity, we will use thought experiments in which we imagine being in spaceships in deep space, far from any planets or stars and drifting along without engine power. Everything in and around these spaceships is weightless and floats freely, so we call the reference frames of these spaceships **free-float frames** (or *inertial reference frames*). To gain familiarity with the idea, let's begin with thought experiments involving speeds we are familiar with from everyday life.

**Thought Experiment 1** Imagine that you are floating freely in a spaceship (**FIGURE S2.4**). Because you feel no sensation of motion, you perceive yourself to be at rest, or traveling at zero speed. As you look out your window, you see your friend Al in his own spaceship, moving away at a constant speed of 90 km/hr. How does the situation appear to Al?

We can answer the question by logically analyzing the experimental situation. Al has no reason to think he is moving. Like you, he is in a free-float frame, floating freely in a spaceship with the engines off. Al will therefore say that *he* is at rest and that *you* are moving away from him at 90 km/hr.

Both points of view—yours and Al's—are equally valid. You both would find the same results for any experiments performed in your own spaceships, and your research would lead you to exactly the same laws of nature. You could argue endlessly about who is really moving, but your argument would be pointless because all motion is relative.

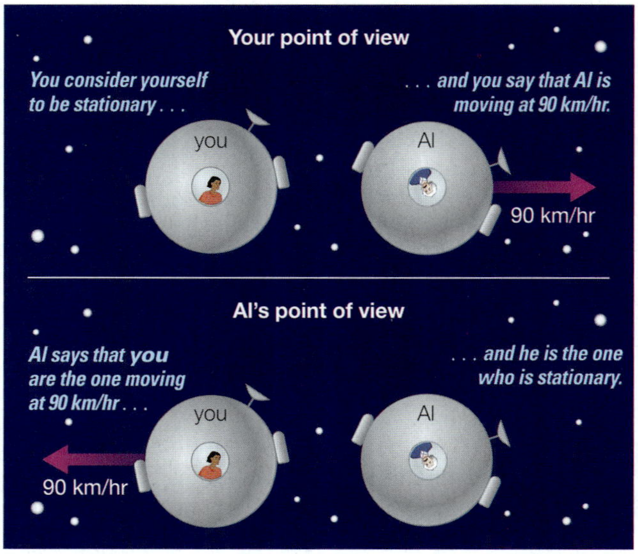


**FIGURE S2.4** Thought Experiment 1.

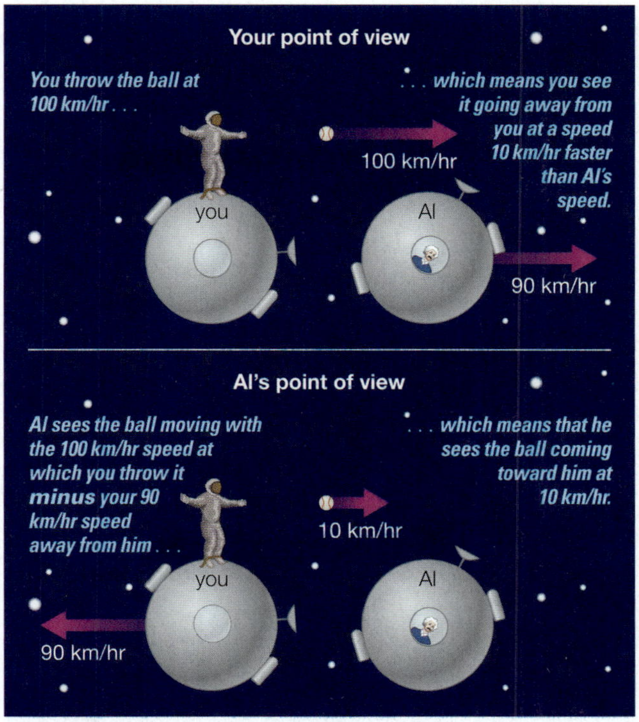


**FIGURE S2.5** Thought Experiment 2.

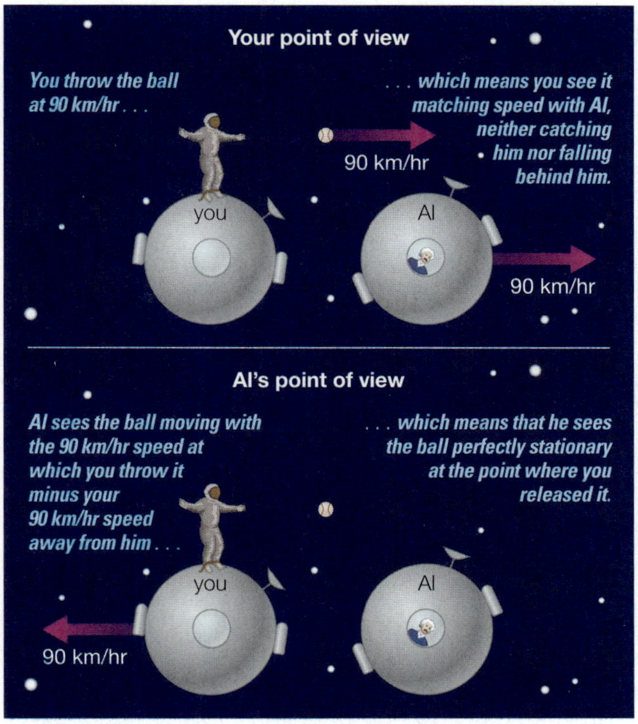


**FIGURE S2.6** Thought Experiment 3.

**Thought Experiment 2** We begin with the same situation as in Thought Experiment 1, but this time you put on your spacesuit and strap yourself to the outside of your spaceship (**FIGURE S2.5**). You happen to have a baseball, which you throw in Al's direction at a speed of 100 km/hr. How fast is the ball moving relative to Al?

From your point of view, Al and the ball are both going in the same direction. Al is going 90 km/hr and the ball is going 100 km/hr, so the ball is going 10 km/hr faster than he is. The ball will therefore overtake and pass him.

From Al's point of view, *he* is stationary and *you* are moving away from him at 90 km/hr. Therefore, he would say the ball was moving even before you threw it, since it would be going with you at your speed of 90 km/hr. Once you release the ball, he sees it moving toward him at 10 km/hr—the ball's speed of 100 km/hr relative to you *minus* the 90 km/hr at which you are moving relative to him. In other words, although you and Al disagree about the speed of the ball, you both agree that the ball will pass him at a *relative* speed of 10 km/hr.

**Thought Experiment 3** This time you throw a baseball in Al's direction at 90 km/hr (**FIGURE S2.6**). From your point of view, the ball is traveling at exactly the same speed as Al. Therefore, you'll see the ball forever chasing him through space, neither catching up with nor falling behind him.

From Al's point of view, the 90 km/hr at which you threw the ball exactly matches your 90 km/hr speed away from him. Therefore, the ball is *stationary* in his reference frame. Think about this for a moment: Before you throw the baseball, Al sees it moving away from him at 90 km/ hr because it is in your hand. At the moment you release the baseball, it suddenly becomes stationary in Al's reference frame, floating in space at a fixed distance from his spaceship. Many hours later, after you have traveled far away, Al will still see the ball floating in the same place. If he wishes, he can put on his spacesuit and go out to retrieve it, or he can just leave it there. From his point of view, it's not going anywhere, and neither is he.

**Think about it** In Thought Experiment 3, suppose you throw the ball in Al's direction at 80 km/hr (rather than 90 km/hr). What will Al see the ball doing? What would he see the ball doing if he were moving *toward* you (rather than away from you) when you threw the ball?

**Thought Experiments at High Speeds** The absoluteness of the speed of light did not come into play in our first three thought experiments because the speeds were small compared to the speed of light. For example, 100 km/hr is less than *one ten-millionth* of the speed of light. Now let's raise the speeds much higher and explore the strange consequences of the absoluteness of the speed of light.

**Thought Experiment 4** Imagine that Al is moving away from you at 90% of the speed of light, or $0.9c$, where $c$ is the speed of light (**FIGURE S2.7**[*]). How does the situation appear to Al?

Other than the much higher speed, this situation is just like that in Thought Experiment 1. Al perceives *himself* to be at rest and sees *you* moving away from him at $0.9c$.

[*]For simplicity, this and most other figures in this chapter ignore effects of length contraction, which would make the moving ships shorter in their direction of motion; only Figure S2.13 shows the length contraction effect.

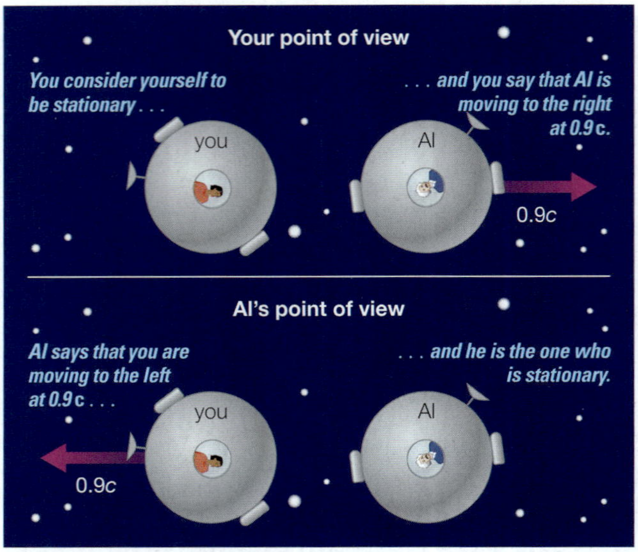


**FIGURE S2.7** Thought Experiment 4.

**Thought Experiment 5** Now, instead of throwing a baseball, you climb out of your spaceship and point a flashlight in Al's direction (**FIGURE S2.8**). How fast is the beam of light moving relative to Al?

From your point of view, Al is going at 90% of the speed of light, or $0.9c$, and the light beam is going in the same direction at the full speed of light, or $c$. Therefore, you see the light beam going $0.1c$ faster than Al.* Nothing should be surprising so far.

From Al's point of view, *he* is stationary and *you* are moving away from him at $0.9c$. Following the "old common sense" used in our earlier thought experiments, we would expect Al to see the light moving toward him at $0.1c$—the light's speed of $c$ minus your speed of $0.9c$. *But this answer is wrong!* Relativity tells us that the speed of light is always the same for everyone. Therefore, Al must see the beam of light coming toward him at $c$, *not* at $0.1c$.

In this case, you and Al no longer agree about his speed *relative* to the speed of the light beam: He'll see the light beam pass by him at the speed of light, $c$, but you'll see it going only $0.1c$ faster than he is going. By our old common sense, this result sounds preposterous. However, we found it by using simple logic, starting with the assumption that the speed of light is the same for everyone. As long as this assumption is true—and, remember, the absoluteness of the speed of light is an experimentally verified fact—our conclusions follow logically.

**Think about it** Suppose Al is moving away from you at a speed of $0.99999c$, which is short of the speed of light by only $0.00001c$, or 3 km/s. At what relative rate would you see a light beam catch up with him? How fast will Al see the light going as it passes?

*You can't actually see a light beam moving forward. When we say that you "see" the beam moving at the speed of light, we really mean that you would find it to be moving at this speed if you made a careful measurement with instruments at rest in your reference frame.

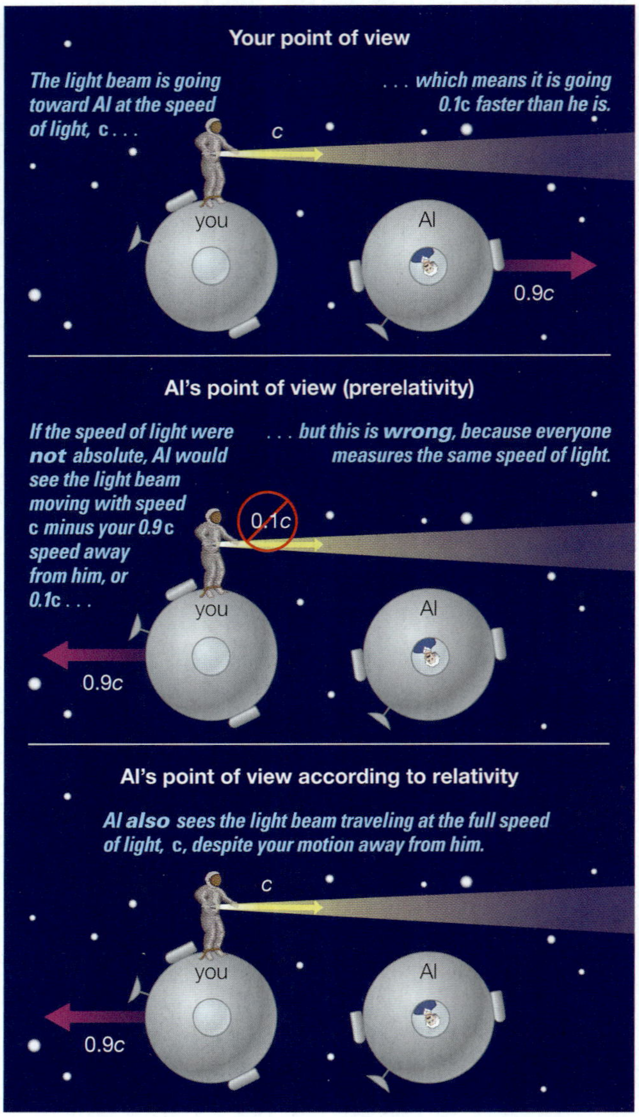


**FIGURE S2.8** Thought Experiment 5.

## Why can't we reach the speed of light?

You might be wondering what Al would see if he were moving away from you at the speed of light or faster, which is essentially the question that Einstein asked when he wondered how the world would look if he could travel at or beyond the speed of light. However, once we accept the absoluteness of the speed of light, it follows that neither Al nor you nor any other material object can ever reach the speed of light, let alone exceed it.

**Thought Experiments (first segment)**

**Thought Experiment 6** You have just built the most incredible rocket imaginable, and you are taking it on a test ride. Soon you are going faster than anyone had ever imagined possible—and then you put the rocket into second gear! You keep going faster and faster and faster. Here is the key question: Are you ever traveling faster than the speed of light?

Before we answer this question, the fact that all motion is relative forces us to answer another question: In what reference frame is your speed being measured? Let's begin with *your* reference frame. Imagine that you turn on your rocket's headlights. Because the speed of light is the same for everyone, you must see the headlight beams traveling at the speed of light—which means they are racing away from your rocket at a speed you'll measure to be 300,000 km/s. The fact that you'll see your headlight beams racing away is true no matter how long you have been firing your rocket engines. In other words, you cannot possibly keep up with your own headlight beams.

The fact that you cannot outrace your headlight beams, combined with the fact that the speed of light is the same for everyone, means that no observer can ever see you reach or exceed the speed of light. Observers in different reference frames will measure your speed differently, but *all* observers will agree on two key points: (1) Your headlight beams are moving out ahead of you, and (2) these light beams are traveling at $c$ = 300,000 km/s. Clearly, if you are being outraced by your headlight beams and if the beams are traveling at the speed of light, you must be traveling *slower* than the speed of light (**FIGURE S2.9**). It does not matter who is measuring your speed. It can be you, someone on Earth, or anyone else in any other reference frame. No one can ever observe you to be traveling as fast as a light beam.

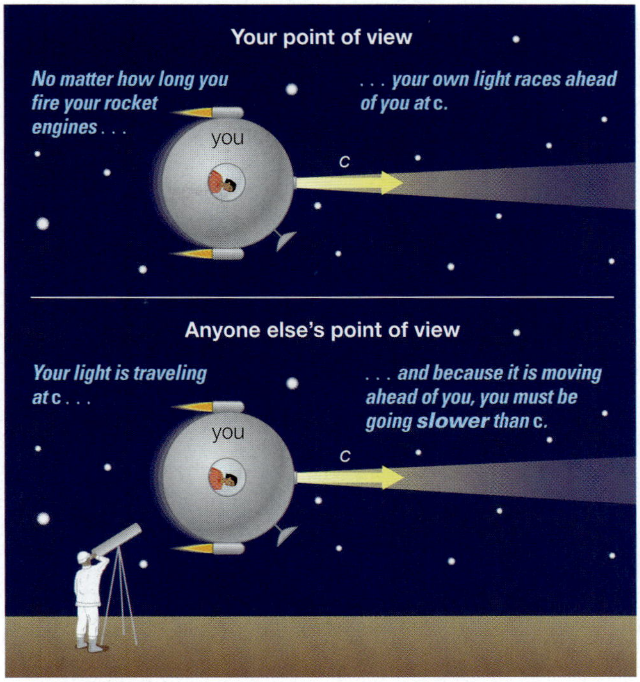


**FIGURE S2.9** Thought Experiment 6.

In case you are still not convinced, let's turn the situation around. Imagine that, as you race by some planet, a person on the planet turns on a light beam. Because the speed of light is absolute, you will see the light beam race past you at $c$ = 300,000 km/s. The person on the planet will also see the light traveling at $c$ = 300,000 km/s and will see the light outrace you. Again, everyone will agree that you are traveling slower than the speed of light.

The same argument applies to any moving object, and it is true with or without headlights. All light travels at the speed of light, including the light that reflects off an object (allowing us to see it) and the infrared light that even cool objects emit as thermal radiation [**Section 5.4**]. As long as the speed of light is absolute, no material object can ever keep up with the light it emits or reflects, which means no material object can reach or exceed the speed of light. Building a spaceship to travel at the speed of light is not a mere technological challenge—it simply cannot be done.

## SPECIAL TOPIC What If Light Can't Catch You?

If you're like most students learning about relativity for the first time, you're probably already looking for loopholes in the logic of our thought experiments. For example, confronted with Thought Experiment 6, you might be tempted to ask, "What happens if you're traveling away from some planet faster than light, so the light from the planet can't catch you?" While it's surely true that light couldn't catch you if you were going faster than the speed of light, this also makes the question moot: If you can't see light from the planet, there is no way for you to know that the planet is there, so you couldn't actually make any measurement that would show your speed to be greater than the speed of light.

In fact, what relativity really tells us about the speed of light is that it is a limit on the speed at which *information* can be transmitted. There are numerous circumstances in the universe in which an object may *seem* to be exceeding the speed of light. However, these circumstances do not provide any means of sending information or objects at speeds faster than the speed of light.

As an example, imagine that you held a laser light and swept it across the sky between two stars that are separated by an angular distance of 90° and are each 10 light-years away from Earth. About 10 years from now, you'd see your laser light first make a dot (an extremely dim one!) on the surface of one star and a few seconds later make a dot on the surface of the second star. That is, your laser dot would seem to have traveled the many light-years from the first star to the second star in just a few seconds—which means at a speed far in excess of the speed of light. However, while the laser dot can carry information from you to each of the individual stars, it clearly cannot be transmitting any information from one star to the other. (After all, you held the laser, not someone at either star.) As a result, there's no violation of relativity.

Other examples of things that *seem* to move faster than the speed of light arise frequently in the world of quantum mechanics (see Chapter S4). According to quantum principles, measuring a particle in one place can (in certain specific circumstances) affect a particle in another place *instantaneously*—even if the particle is many light-years away. In fact, this process has been observed in laboratories over short distances. This instantaneous effect of one particle on another may at first seem to violate relativity, but it does not. The built-in randomness of quantum mechanics prevents this technique from being used to transmit useful information. Moreover, to confirm that the second particle really was affected, we would have to receive a signal carrying information about the particle—and that signal can travel no faster than the speed of light.

## S2.3 The Reality of Space and Time

In the first section of this chapter, we listed five key ideas of special relativity. Each of these ideas is actually a prediction that arises from the two basic assumptions of the theory—that the laws of nature are the same for everyone and that everyone always measures the same speed of light. So far, we have shown only how relativity predicts that no material object can reach or exceed the speed of light. Let's continue down the path we've started on, so that we can see why our old conceptions of time and space must be revised.

### How does relativity affect our view of time and space?

The special theory of relativity tells us that measurements of time, length, and mass can all be different in different reference frames. It also tells us that different observers can disagree about whether two events are simultaneous and about the speeds of different objects. Strange as these ideas may sound, we can continue to use simple thought experiments to show why they follow directly from the premises of relativity. We'll begin by considering how relativity affects our view of time, then move on to other effects.

**Time Differs in Different Reference Frames** In preparation for our next thought experiment, imagine that you are on a moving train tossing a ball straight up so that it bounces straight back down from the ceiling. How do the path and speed of the ball appear to an observer along the tracks outside the train? (We'll ignore the effects of gravity on the ball's speed.)

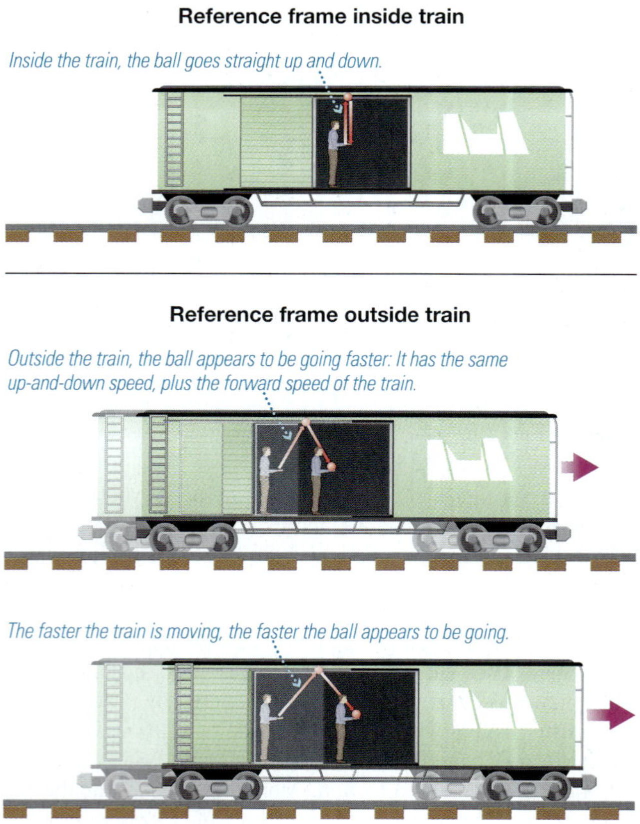


**FIGURE S2.10** A ball tossed straight up and down *inside* a moving train appears to an outside observer to follow a slanted path. Thus, the outside observer sees the ball traveling at a faster total speed than does the person inside the train.

As shown in **FIGURE S2.10**, the outside observer sees the ball going forward with the train at the same time that it is going up and down, so the ball's path always slants forward. Because the outside observer sees the ball moving with this forward speed *in addition to* its up-and-down speed, while you see the ball going only up and down, he would measure a *faster* overall speed for the ball than you would. If the train is moving slowly, the outside observer sees the ball's path slant forward only slightly and would say that the ball's overall speed was only slightly faster than you would report. If the train is moving rapidly, the ball's path leans much farther forward, and the ball appears to be going considerably faster to the outside observer than to you.

**Thought Experiments (second segment)**

**Thought Experiment 7** Inside his spaceship, Al has a laser on his floor that is pointed up to a mirror on his ceiling (**FIGURE S2.11**). He momentarily flashes the laser light and uses a very accurate clock to time how long it takes the light to travel from the floor to the ceiling and back. As Al zips by you at a speed close to the speed of light, you observe his experiment and, using your own very accurate clock, you also time the laser light's trip from Al's floor to his ceiling and back.

Because Al, the laser, and the mirror are all moving from your point of view, the light's path looks slanted as it goes from floor to ceiling and back, just as the ball tossed up and down in the train followed a slanted path from the viewpoint of an outside observer. Therefore, from your point of view, the light travels a *longer* path in going from the floor

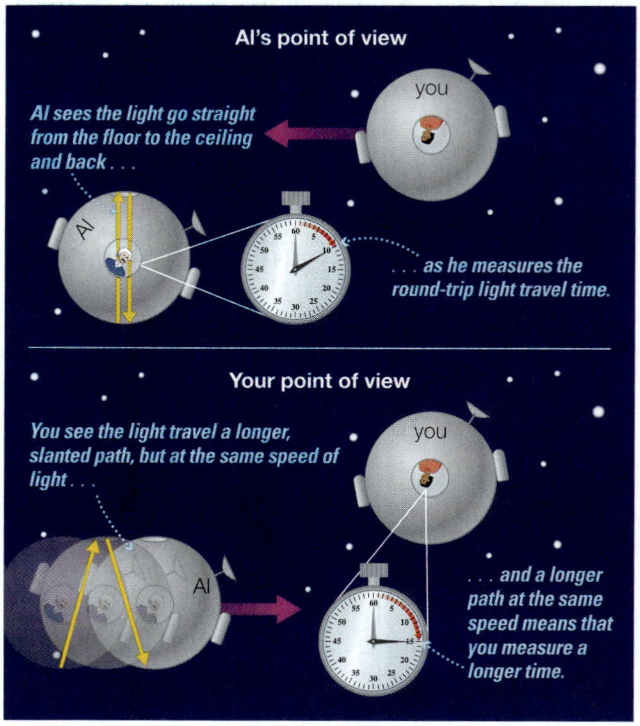


**FIGURE S2.11** Thought Experiment 7.

to the ceiling than it does from Al's point of view, just as the ball in the train took a longer path from the outside observer's point of view.

By our old common sense, this would be no big deal. You and Al would agree on how long it takes the light to go from the floor to the ceiling and back, just as you and an outside observer would agree on how long it takes you to toss a ball up and down on a train. You would explain the light's longer path by saying that the light is moving faster relative to you than it is moving relative to Al, just as the outside observer sees the tossed ball moving faster because of the forward motion of the train. However, because we are dealing with *light,* you and Al must both measure its speed to be the *same*—the speed of light, or 300,000 km/s—even though you see its path slanted forward with the movement of the spaceship. This fact has an astonishing implication.

Remember that, at any given speed, including the speed of light, traveling a longer distance must take a longer time. Because both you and Al see the light traveling at the same speed, but you see it traveling a longer distance, *your clock must record more time* than Al's as the light travels from floor to ceiling and back. And because you can see Al's clock as well as your own, you must see his clock running at a slower rate than yours in order for it to end up reading less elapsed time after the light completes its round trip. In other words, you will see Al's clock running more slowly than your own. Moreover, it doesn't matter what kind of "clock" you and Al use to measure the time for the light trip. *Anything* that can measure time will be going slower in Al's reference frame than in yours, including mechanical clocks, electrical clocks, heartbeats, and biochemical reactions. Our astonishing conclusion: From your point of view, *time itself* is running slower for Al.

How much slower is time running for Al? It depends on his speed relative to you. If he is moving slowly compared to the speed of light, you will scarcely be able to detect the slant of the light path, and your clock and Al's clock will tick at nearly the same rate. The faster he is moving, the more slanted the light path will appear to you and the greater the difference between the rate of his clock and that of yours. Generalizing, we reach the following conclusion:

> *From your point of view, time runs more slowly in the reference frame of anyone moving relative to you. The faster the other reference frame is moving, the more slowly time passes within it.*

This effect is called **time dilation** because it tells us that time is *dilated,* or expanded, in a moving reference frame.

**The Relativity of Simultaneity** Our old common sense also tells us that everyone must agree on whether two events happen at the same time or one event happens before another. For example, if you see two apples—one red and one green—fall from two different trees and hit the ground at the same time, you expect everyone else to agree that they landed at the same time (assuming you've accounted for any difference in the light travel times from the two trees). If you saw the green apple land before the red apple, you would be very surprised if someone else said that the red apple hit the ground first. Well, prepare yourself to be surprised, because our next thought experiment will show that observers in different reference frames will not necessarily agree about the order or simultaneity of events that occur in different places.

Before we go on, note that observers in different reference frames must agree about the order of events that occur in the *same* place. For example, suppose you grab a cookie and eat it. In your reference frame, both events (picking up the cookie and eating it) occur in the same place, so it could not possibly be the case that someone else would see you eat the cookie before you pick it up.

**Thought Experiments (third segment)**

**Thought Experiment 8** Al has a brand-new, extra-long spaceship, and he is coming toward you at a speed of 90% of the speed of light, or 0.9$c$ (**FIGURE S2.12a**). He is in the center of his spaceship, which is totally dark except for a flashing green light at its front end and a flashing red light at its back end. Suppose that you see the green and red lights flash at *exactly the same time,* with the flashes occurring at the instant that Al happens to pass you. During the very short time that the light flashes are traveling toward you, Al's forward motion carries him toward the point where you saw the green flash occur. The green flash will therefore reach him before the red flash, which means

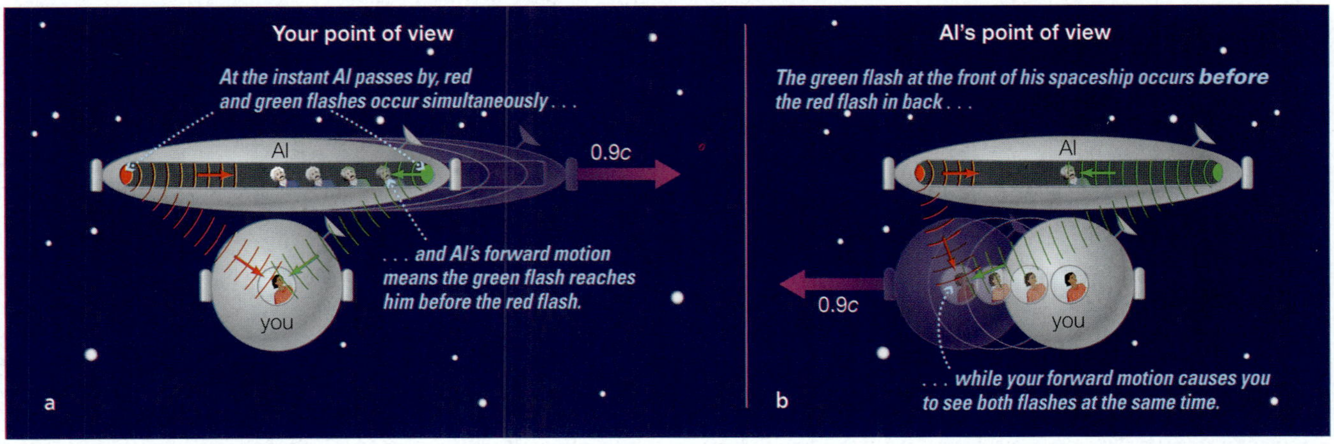


**FIGURE S2.12** Thought Experiment 8.

## MATHEMATICAL INSIGHT S2.1 The Time Dilation Formula

We can find a formula for time dilation by looking at the paths that you and Al see light take from the floor to the ceiling in Figure S2.11. We can use these paths to construct a right triangle as follows (**FIGURE 1**):

- The hypotenuse is the path that *you* see the light take. Let's use $t$ to represent the time *you* measure as the light travels this path. Because *distance* = *speed* × *time* and the light travels at the speed of light, the length of this path is $c \times t$.
- The base of the triangle is the distance that Al's spaceship moves forward during the time $t$. If we use $v$ for the spaceship's speed relative to you, this distance is $v \times t$.
- The vertical side is the path that Al sees the light take. We know that Al measures time differently than you, so we'll use $t'$ ("$t$-prime") to represent the time *he* measures. Therefore, the length of this path is $c \times t'$.

We can now find the time on Al's clock ($t'$) in terms of the time on your clock ($t$) by applying the Pythagorean theorem ($a^2 + b^2 = c^2$, where $a$ and $b$ are side lengths and $c$ is the hypotenuse length) with the side lengths in Figure 1:

$$(ct')^2 + (vt)^2 = (ct)^2$$

Expand the squares:

$$c^2t'^2 + v^2t^2 = c^2t^2$$

Subtract $v^2t^2$ from both sides:

$$c^2t'^2 = c^2t^2 - v^2t^2 = (c^2 - v^2)t^2$$

Divide both sides by $c^2t^2$:

$$\frac{t'^2}{t^2} = \frac{c^2 - v^2}{c^2}$$

Simplify:

$$\frac{t'^2}{t^2} = \frac{c^2}{c^2} - \frac{v^2}{c^2} = 1 - \left(\frac{v}{c}\right)^2$$

Take the square root of both sides:

$$\frac{t'}{t} = \sqrt{1 - \left(\frac{v}{c}\right)^2} \quad \text{or} \quad t' = t\sqrt{1 - \left(\frac{v}{c}\right)^2}$$

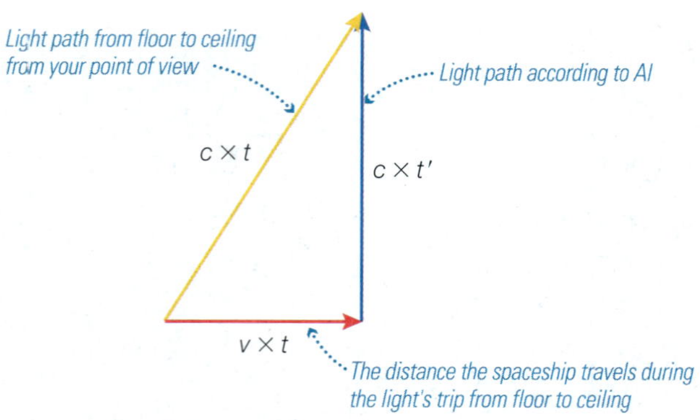


**FIGURE 1** The setup for finding the time dilation formula.

We have found the *time dilation formula,* which tells us the *ratio* of time in a moving reference frame to time in a reference frame at rest. This ratio is graphed against speed in **FIGURE 2**. Note that $t'/t = 1$ at speeds that are small compared to the speed of light, meaning that clocks in both reference frames tick at about the same rate. As $v$ approaches $c$, the amount of time passing in the moving reference frame gets smaller and smaller compared to the time in the reference frame at rest.

**EXAMPLE:** Suppose Al is moving past you at a speed of $0.9c$. While 1 hour passes for you, how much time passes for Al?

**SOLUTION:**

**Step 1 Understand:** To use the time dilation formula, we must decide whose time is $t$ and whose is $t'$. Because Al is moving past you, you expect to see his time running slowly, which means *less* than 1 hour should pass for Al while 1 hour passes for you. Notice that the square root in the time dilation formula always has a value less than 1 (because $v$ is always less than $c$), which means that $t'$ always ends up less than $t$. Therefore, we want $t'$ to represent Al's time and $t$ to represent yours.

**Step 2 Solve:** Al's speed of $v = 0.9c$ means that $v/c = 0.9$, so we use the time dilation formula with this value and your time $t = 1$ hour:

$$\begin{aligned} t' &= t\sqrt{1 - \left(\frac{v}{c}\right)^2} \\ &= (1\text{ hr})\sqrt{1 - (0.9)^2} \\ &= (1\text{ hr})\sqrt{1 - 0.81} \\ &= (1\text{ hr})\sqrt{0.19} \approx 0.44\text{ hr} \end{aligned}$$

**Step 3 Explain:** We have found that only 0.44 hour, or about 26 minutes, passes for Al while 1 hour passes for you. In other words, you will see his time running less than half as fast as yours.

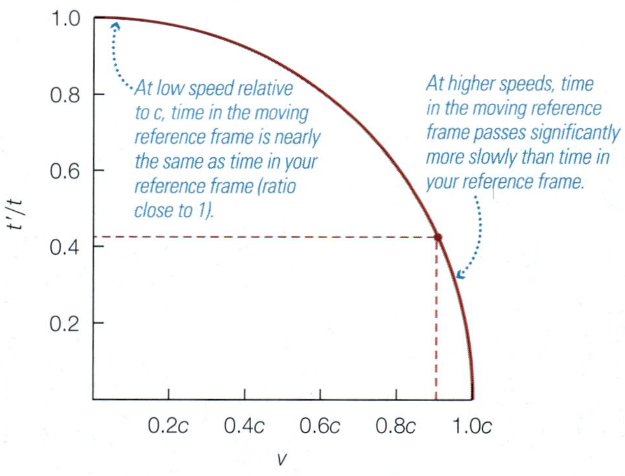


**FIGURE 2** The time dilation factor graphed against speed.

you'll see him illuminated first by green light and then by red light. So far, nothing should be surprising: You see the two flashes simultaneously because you are stationary, but the green flash reaches Al before the red flash because of his forward motion. But let's think about what happens from Al's point of view, in which he is the one who is stationary and you are the one who is moving.

Remember that motion cannot affect the order of events that occur in the same place. Therefore, no one can argue with the fact that you saw the green and red flashes at the same time, and no one can disagree with the fact that the green light reached Al before the red light. However, because Al considers himself to be stationary in the center of his spaceship, the only way that the green light can reach him before the red light is if *the green flash occurred first*. In other words, he'll see the situation as shown in **FIGURE S2.12b**: He'll say that the green flash occurred before the red flash, and that the reason they both reached you at the same time was because your motion was moving you in the direction of the red flash.

Let's summarize the situation: You both agree on the indisputable reality of what each of you saw—that is, you saw the two flashes at the same time, while Al saw the green flash before the red flash. However, you say this reality occurs because the flashes really did occur at the same time and Al was moving toward the green flash, while Al says it occurs because the green flash really occurred first while you were moving away from it. Who is right?

All motion is relative, so both of you are equally correct. In other words, the flashes that are simultaneous in your reference frame occur at different times in Al's reference frame. That is what we mean when we say that the order or simultaneity of events depends on your frame of reference.

**Effects of Motion on Length and Mass** The fact that time is different in different reference frames implies that lengths (or distances) and masses are also affected by motion, although the reasons are a bit subtler. The following two thought experiments use the idea of time dilation to help us understand the effects on length and mass.

**Thought Experiment 9** Al is back in his original spaceship, coming toward you at high speed. As usual, both you and he agree on your *relative* speed. You disagree only about who is stationary and who is moving. Now imagine that Al tries to measure the length of your spaceship as he passes by you. He can do this by measuring the time it takes him to pass from one end of your spaceship to the other; recall that *length* = *speed* × *time*.

In your reference frame, you say that Al is moving by you and therefore you see Al's time running slowly. Because you agree on his relative speed but his clocks record less time than yours as he passes from one end of your spaceship to the other, he must measure the length of your spaceship to be *shorter* than you measure it to be. Now, remember that your viewpoint and Al's viewpoint are equally valid. The fact that the laws of nature are the same for everyone means that if he measures your spaceship to be shorter than it is at rest in your reference frame, you must also measure his to be shorter than it would be

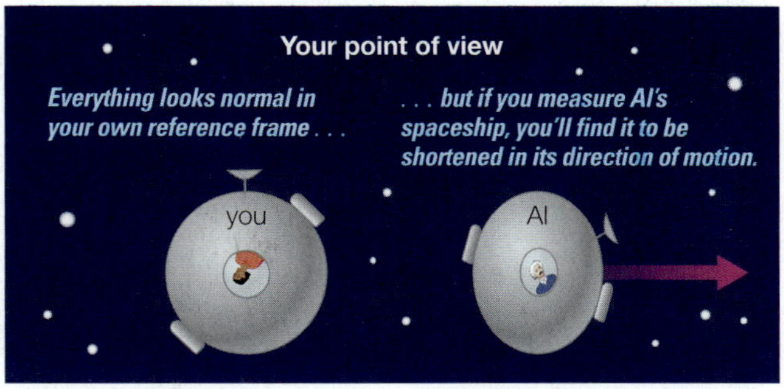


**FIGURE S2.13** People and objects moving relative to you are contracted in their direction of motion. The diagram shows only your point of view, but Al's view would be equivalent: He would see himself at "normal" size and you contracted in your direction of motion.

at rest, an effect called **length contraction**. **FIGURE S2.13** shows that lengths are affected only in the direction of motion. His spaceship is shorter from your point of view, but its height and depth are unaffected. Generalizing, we reach the following conclusion.

> *From your point of view, the lengths of objects moving by you (or the distances between objects moving by you) are shorter in their direction of motion than they would be if the objects were at rest. The faster the objects are moving, the shorter the lengths.*

**Thought Experiment 10** To see how motion affects mass, imagine that Al has an identical twin brother with an identical spaceship, and suppose that his brother is at rest in your reference frame while Al is moving by you at high speed. At the instant Al passes by, you give both Al and his brother identical pushes (**FIGURE S2.14**). That is, you push them both with the same force for the same amount of time. By our old common sense, we'd expect both of them to accelerate by an identical amount, such as gaining 1 km/s of speed relative to you. But we must instead apply our new common sense, which tells us something different.

Remember that, because Al is moving relative to you and his brother, you'll see Al's time running more slowly than yours and his brother's—which means that he experiences the force of your push for a *shorter* time than does his brother. For example, if you and Al's brother measure the duration of the push to be 1 microsecond, Al's clock will show the duration to be less than 1 microsecond. Moreover, because Al feels the force of your push for a shorter time, the push must have a *smaller* effect on Al's velocity than it has on his brother's velocity. In other words, you'll find that your push has less effect on Al than on his brother, despite the fact that you gave them identical pushes. According to Newton's laws of motion [**Section 4.2**], the same push can have a smaller effect on Al's velocity only if his mass is *greater* than his brother's mass. This effect is sometimes called **mass increase***:

*Newton's second law can be stated as "force = mass × acceleration" or as "force = rate of change in momentum." (Recall that momentum = mass × velocity.) For more advanced work, most physicists prefer the second form, and they therefore prefer to think of the effect in relativity as *momentum increase* rather than mass increase. The two viewpoints are equivalent for the level of discussion in this book.

**FIGURE S2.14** Thought Experiment 10.

*From your point of view, objects moving by you have greater mass than they have at rest. The faster an object is moving, the greater the increase in its mass.*

Mass increase provides another way of understanding why no material object can reach the speed of light. The faster an object is moving relative to you, the greater the mass you'll find it to have. Therefore, at higher and higher speeds, the same force will have less and less effect on an object's velocity. As the object's speed approaches the speed of light, you will find its mass to be heading toward infinity. No force can accelerate an infinite mass, so the object can never gain that last little bit of speed needed to push it to the speed of light.

**Velocity Addition** We have just one more important effect to discuss: As the next thought experiment shows, you and Al will disagree about the speed of a material object moving relative to both of you. The only speed you will agree on is the speed of light.

**Thought Experiment 11** Al is moving toward you at $0.9c$ (**FIGURE S2.15**). Your friend Jackie jumps into her spaceship and starts heading in Al's direction at $0.8c$ (from your point of view). How fast will Al see her approaching him?

By our old common sense, Al should see Jackie coming toward him at $0.9c + 0.8c = 1.7c$. We know this answer is wrong, however, because $1.7c$ is faster than the speed of light. Al must see Jackie coming toward him at a speed less than $c$, but she will be coming faster than the speed of $0.9c$ at which Al sees you coming. That is, Al will conclude that Jackie's speed is somewhere between $0.9c$ and $c$. (In this case, the speed turns out to be $0.988c$; see Mathematical Insight S2.2.)

## Do the effects predicted by relativity really occur?

The clear logic of our thought experiments has shown that all the major predictions of special relativity follow directly from the absoluteness of the speed of light and from the fact that the laws of nature are the same for everyone. However, logic alone is not good enough in science; our conclusions remain tentative until they pass observational or experimental tests. Does relativity pass the test?

**The Absoluteness of the Speed of Light** The first thing we might wish to test is the surprising premise of relativity: the absoluteness of the speed of light. In principle, we can test this premise by measuring the speed of light coming from many different objects and going in many different directions and verifying that the speed is always the same.

The speed of light was first measured in the late 17th century, but experimental evidence for the *absoluteness* of the speed of light did not come until 1887, when A. A. Michelson and E. W. Morley performed their now-famous *Michelson-Morley experiment*. This experiment showed that the speed of light is not affected by Earth's motion around the Sun. Countless subsequent experiments have verified and extended the results of the Michelson-Morley experiment. The speed of light coming from opposite sides of the rotating Sun and from the orbiting stars

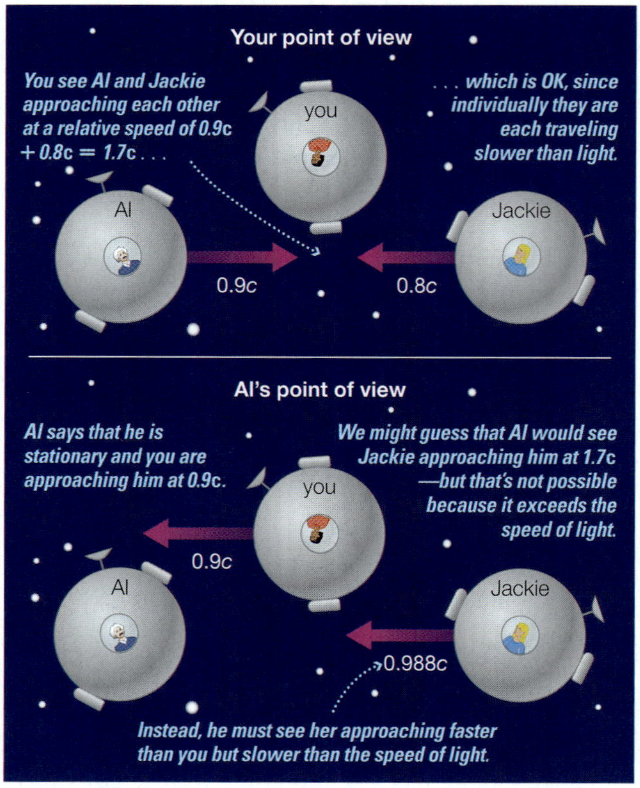


**FIGURE S2.15** Thought Experiment 11.

in binary systems has also been measured, with the same result: The speed of light is always the same.

**Experimental Tests of Special Relativity** Although *we* cannot yet travel at speeds at which the effects of relativity should be obvious, tiny subatomic particles can reach such speeds, thereby allowing us to test the precise predictions of the formulas of special relativity.

In machines called *particle accelerators,* physicists accelerate subatomic particles to speeds near the speed of light and study what happens when the particles collide. The colliding particles have a great deal of kinetic energy, and the collisions convert some of this kinetic energy into mass-energy that emerges as a shower of newly produced particles. Many of these particles have very short lifetimes, at the end of which they decay (change) into other particles. For example, a particle called the $\pi^+$ ("pi plus") meson has a lifetime of about 18 nanoseconds (billionths of a second) when produced at rest. But $\pi^+$ mesons produced at speeds close to the speed of light in particle accelerators last much longer than 18 nanoseconds—and the amount longer is always precisely the amount predicted by the time dilation formula.

## MATHEMATICAL INSIGHT S2.2 Formulas of Special Relativity

We found a formula for time dilation in Mathematical Insight S2.1. Although we will not go through the derivations, it is possible to find similar formulas for length contraction and mass increase. The three formulas are

$$\underset{\text{(moving frame)}}{\text{time}} = \underset{\text{(rest frame)}}{\text{time}} \times \sqrt{1-\left(\frac{v}{c}\right)^2}$$

$$\underset{\text{(moving frame)}}{\text{length}} = (\text{rest length}) \times \sqrt{1-\left(\frac{v}{c}\right)^2}$$

$$\text{moving mass} = \frac{(\text{rest mass})}{\sqrt{1-\left(\frac{v}{c}\right)^2}}$$

There's also a simple formula for velocity addition. Suppose you see Al moving at speed $v_1$ and Al sees a second object moving relative to him at speed $v_2$. By our old common sense, you would see the second object moving at speed $v_1 + v_2$. However, the speed you actually see is

$$\text{speed of second object} = \frac{v_1 + v_2}{1 + \left(\frac{v_1}{c} \times \frac{v_2}{c}\right)}$$

**EXAMPLE 1:** Al is moving by you at 0.99*c* in a spaceship that is 100 meters long at rest. How long is it as it moves by you?

**SOLUTION:**

**Step 1 Understand:** This is a length contraction problem. Because you see Al moving, his spaceship's length should be *shorter* than its rest length of 100 meters.

**Step 2 Solve:** We use the length contraction formula with Al's speed of 0.99*c*, or $v/c = 0.99$, and rest length = 100 m:

$$\underset{\text{(moving frame)}}{\text{length}} = (\text{rest length}) \times \sqrt{1-\left(\frac{v}{c}\right)^2}$$
$$= (100\text{ m}) \times \sqrt{1-(0.99)^2} = 14\text{ m}$$

**Step 3 Explain:** Al's spaceship is 100 meters long at rest, but only 14 meters long when he is moving by you at 99% of the speed of light.

**EXAMPLE 2:** A "super fly" has a rest mass of 1 gram but is capable of flying at 0.9999*c*. What is its mass at that speed?

**SOLUTION:**

**Step 1 Understand:** This time we are asked about mass, so we need the mass increase formula and expect the moving mass to be greater than the rest mass.

**Step 2 Solve:** We use the mass increase formula with the fly's rest mass = 1 g and $v/c = 0.9999$:

$$\text{moving mass} = \frac{(\text{rest mass})}{\sqrt{1-\left(\frac{v}{c}\right)^2}}$$
$$= \frac{1\text{ g}}{\sqrt{1-(0.9999)^2}} = 70.7\text{ g}$$

**Step 3 Explain:** At a speed of 0.9999*c*, or 99.99% of the speed of light, the mass of a fly that is 1 gram at rest becomes 70.7 grams, or more than 70 times its rest mass.

**EXAMPLE 3:** Al is moving toward you at 0.9*c*. Your friend Jackie jumps into her spaceship and, from your point of view, goes in Al's direction at 0.8*c* (see Figure S2.15). How fast will Al see Jackie approaching?

**SOLUTION:**

**Step 1 Understand:** According to Al, your speed is $v_1 = 0.9c$. Jackie's speed *relative to you* is $v_2 = 0.8c$ in the same direction. We therefore need the velocity addition formula to find Jackie's speed relative to Al.

**Step 2 Solve:** We substitute the given values $v_1 = 0.9c$ and $v_2 = 0.8c$ into the velocity addition formula:

$$\underset{\text{(relative to Al)}}{\text{Jackie's speed}} = \frac{v_1 + v_2}{1 + \left(\frac{v_1}{c} \times \frac{v_2}{c}\right)}$$
$$= \frac{0.9c + 0.8c}{1 + (0.9 \times 0.8)}$$
$$= \frac{1.7c}{1.72} = 0.988c$$

**Step 3 Explain:** Al sees Jackie moving toward him at 0.988*c*, or almost 99% of the speed of light. Notice that, as we expect, Jackie's speed relative to Al is faster than yours but slower than the speed of light.

The same experiments allow the mass increase formula to be checked as well. The amount of energy released when high-speed particles collide depends on the particle masses and speeds. Just as relativity predicts, these masses are greater at high speed than they are at low speed—again by amounts that can be precisely predicted by Einstein's formulas.

Particle accelerators even offer experimental evidence that nothing can reach the speed of light. It is relatively easy to get particles traveling at 99% of the speed of light in particle accelerators. However, no matter how much more energy is put into the accelerators, the particle speeds get only fractionally closer to the speed of light. Some particles have been accelerated to speeds within 0.00001% of the speed of light, but none have ever reached the speed of light.

Although the effects of relativity are obvious only at very high speeds, modern techniques of measuring time are so precise that effects can be measured even at ordinary speeds. For example, experiments conducted in 2010 at the National Institute of Standards and Technology verified the predicted amount of time dilation at speeds of less than 10 meters per second (36 kilometers per hour). All in all, special relativity is one of the best-tested theories in physics, and it has passed every experimental test to date with flying colors.

**A Great Conspiracy?** Perhaps you're thinking, "I still don't believe it." After all, how can you know that the scientists who report the experimental evidence are telling the truth? Perhaps physicists are making up the whole thing as part of a great conspiracy designed to confuse everyone else so that they can take over the world!

What you need is evidence that you can see for yourself. So . . . how about nuclear energy? Einstein's famous formula, $E = mc^2$, which explains the energy released in nuclear reactions, is a direct consequence of the special theory of relativity—and one that you can derive for yourself with a bit of algebra (Mathematical Insight S2.3). Every time you see film of an atomic bomb, or use electrical power from a nuclear power plant, or feel the energy of sunlight that the Sun generated through nuclear fusion, you are really experiencing direct experimental evidence of relativity.

Another test you can do yourself is to look through a telescope at a binary star system. If the speed of light were *not* absolute, the speed at which light from each star comes toward Earth would depend on the star's velocity toward us in the binary orbit (**FIGURE S2.16**). Imagine, for example, that one star is currently moving directly away from us in its orbit. If we added and subtracted speeds according to our old common sense, light from the star at this point would approach us at speed $c - v$. Some time later, when the same star was moving toward us in its orbit, its light would approach us at speed $c + v$. This light would therefore tend to catch up with the light that the star emitted from the other side of its orbit. If the orbital speed and distance were just right, we might see the same star on both sides of the orbit at once! More generally, because the light from each star would come toward us at a different speed from each point in its orbit, we would see each star in multiple positions in its orbit simultaneously, making each star appear as a short line of light rather than as a point—*if* the speed of light were not absolute. Therefore, the fact that we

## SPECIAL TOPIC Measuring the Speed of Light

It is difficult to measure the speed of light, because light travels so fast. If you stand a short distance from a mirror and turn on a light, the reflection seems to appear instantaneously. Such observations led Aristotle (384–322 B.C.) to conclude that light travels at infinite speed, a view that many scientists still held as recently as the late 17th century.

One way to make the measurement easier is to place a mirror at increasingly greater distances. If the speed of light truly were infinite, the reflection would always appear instantaneously. However, if it takes time for the light to travel to and from the mirror, you should eventually find a delay between the time you turn on the light and the time you see the reflection. Galileo tried a version of this experiment using the distance between two tall hills, but he was unable to detect any delay (instead of using a mirror, he stationed an assistant on the distant hill to signal back when he saw the light). He concluded that the speed of light, if not infinite, was too fast to be measured between hills on Earth with the technology of his day.

Galileo's idea was extended to space in 1675 by the Danish astronomer Olaus Roemer. Using the four largest moons of Jupiter as his "mirrors," Roemer successfully measured a delay in the time it took for light from the Sun to be reflected and thereby made the first measurement of the speed of light. His technique worked because, by that time, the orbital periods of Jupiter's large moons were well known, so it was possible to predict the precise moment at which Jupiter would eclipse each of its moons. Roemer learned that the eclipses occurred earlier than expected when Earth was closer to Jupiter and later than expected when Earth was farther from Jupiter. He realized that the eclipses actually were occurring at the predicted times, but that the light was taking longer to reach us when Earth was farther from Jupiter. His observations proved that the speed of light is finite and allowed him to estimate its precise value. Using estimates of the Sun-Earth and Sun-Jupiter distances that were available in 1675, Roemer calculated the speed of light to be 227,000 km/s. Redoing his calculations using the presently known values of these distances yields the correct value of 300,000 km/s.

As technology advanced, it became possible to measure light travel time between mirrors at much closer distances. In 1849 and 1850, the French physicists Hippolyte Fizeau and Léon Foucault (also famous for the Foucault pendulum) performed a series of experiments using rotating mirrors to measure the speed of light much more precisely. Modern devices for measuring the speed of light take advantage of the wave properties of light, particularly the fact that light waves can interfere with one another. Such devices, called *interferometers,* were refined by A. A. Michelson and used in the Michelson-Morley experiment. Details of how this experiment worked can be found in many physics texts.

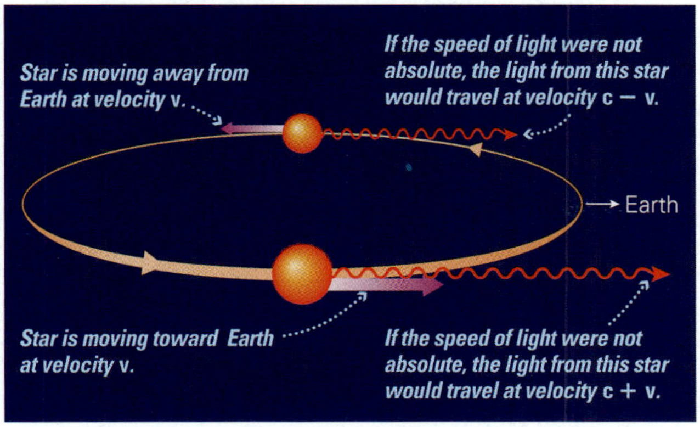


**FIGURE S2.16** If the speed of light were *not* absolute, the speed at which light from each star in a binary system comes toward Earth would depend on the star's velocity toward us in the binary orbit.

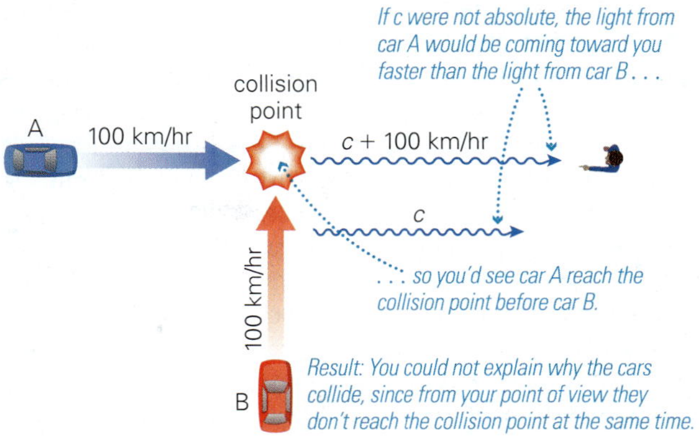


**FIGURE S2.17** If the speed of light were *not* absolute, the light from a car coming toward you would approach you faster than the light from a car going across your line of sight.

always see distinct stars in binary systems demonstrates that the speed of light *is* absolute.*

Finally, you can explore the paradoxes that would occur if the speed of light were not absolute. For example, imagine that two cars, both traveling at about 100 km/hr, collide at an intersection (**FIGURE S2.17**). You witness the collision from far down one street. If the speed of light were *not* absolute, the light from the car that was coming toward you would have a speed of $c + 100$ km/hr, while the light from the other car would approach you at a speed of only $c$. You therefore would see the car coming toward you reach the intersection slightly *before* the other car and thus would see events unfold differently than the passengers in the car or eyewitnesses in other locations. On Earth, this difference would be scarcely noticeable, because 100 km/hr is only about one-millionth the speed of light. However, imagine that you had a super telescope and watched such a collision on a planet located 1 million light-years from Earth:

*This conclusion would not follow if light waves were carried by a medium in the same way that sound waves are carried by air. Scientists in the 19th century believed that such a medium, which they called the *ether*, permeated all of space. The Michelson-Morley experiment ruled out the existence of such a medium, so we are left with the conclusion that $c$ is absolute.

## MATHEMATICAL INSIGHT S2.3 Deriving $E = mc^2$

We can derive the formula $E = mc^2$ from the mass increase formula, which we previously wrote as

$$\text{moving mass} = \frac{\text{(rest mass)}}{\sqrt{1 - \left(\frac{v}{c}\right)^2}}$$

Let's call the moving mass $m$ and the rest mass $m_0$, and use the fact that dividing by a square root is the same as raising to the $-\frac{1}{2}$ power:

$$\text{mass increase formula: } m = m_0\left(1 - \frac{v^2}{c^2}\right)^{-\frac{1}{2}}$$

To continue, we need the following mathematical approximation, which you can verify by using a calculator to check that it holds for small values of $x$ (such as $x = 0.05$ or $x = 0.001$):

$$(1 + x)^{-\frac{1}{2}} \approx 1 - \frac{1}{2}x \text{ (for } x \text{ small compared to 1)}$$

Now, look again at the mass increase formula. The term in parentheses on the right side has the form $(1 + x)^{-\frac{1}{2}}$, as long as we identify $x$ as $-v^2/c^2$. For speeds that are small compared to the speed of light, we can therefore rewrite this term by substituting $x = -v^2/c^2$ into the above approximation:

$$\left(1 + \left[-\frac{v^2}{c^2}\right]\right)^{-\frac{1}{2}} \approx 1 - \frac{1}{2}\left(-\frac{v^2}{c^2}\right)$$

Simplifying, this becomes

$$\left(1 - \frac{v^2}{c^2}\right)^{-\frac{1}{2}} \approx 1 + \frac{1v^2}{2c^2}$$

We can use this result to rewrite the mass increase formula by replacing the term that appears next to $m_0$ with the approximation:

$$m \approx m_0\left(1 + \frac{1v^2}{2c^2}\right)$$

We now expand the right side by multiplying through by $m_0$:

$$m \approx m_0 + \frac{1m_0v^2}{2c^2}$$

Finally, we multiply both sides by $c^2$, so the formula becomes

$$mc^2 \approx m_0c^2 + \frac{1}{2}m_0v^2$$

You may recognize the last term on the right as the *kinetic energy* [**Section 4.3**] of an object with mass $m_0$. Because the other two terms also have units of mass multiplied by speed squared, they also must represent some kind of energy. Einstein recognized that the term on the left represents the *total* energy of a moving object. He then noticed that, even if the speed is *zero* ($v = 0$) so that there is *no* kinetic energy, the equation states that the total energy is $m_0c^2$. That is, a nonmoving object still contains energy by virtue of its mass in an amount equal to its rest mass times the speed of light squared. This is the famous formula $E = mc^2$, which we now see as a part of Einstein's theory of relativity.

Because the light from the first car (the one coming toward you) is coming at a speed one-millionth of the speed of light faster than the light from the second car, over a distance of 1 million light-years it would end up reaching you a full *year* ahead of the light from the second car. In other words, you would see the first car reach the intersection a year before the second car. This poses a paradox: From the viewpoint of the passengers in the cars, they have collided, yet you saw one car reach the collision point long before the other car had even started its journey! If we live in a universe in which images show what really does happen, then the only way to avoid the paradox is to assume that we should not have added the car's speed to the speed of light; that is, we must assume that the speed of light is absolute.

**Think about it** The preceding paradox presents us with two possibilities. If the speed of light is *not* absolute, different people can witness the same events in very different ways. If the speed of light *is* absolute, measurements of time and space are relative. Einstein preferred the latter. Do you? Explain.

Of course, like any scientific theory, the theory of relativity can never be *proved* beyond all doubt. However, it is supported by a tremendous body of evidence, some of which you can see for yourself. This evidence is real and cannot be made to disappear. If anyone ever comes up with an alternative theory, the new theory will have to explain the many experimental results that seem to support relativity so well.

## S2.4 Toward a New Common Sense

We've used thought experiments to show that our old common sense doesn't work, and we've discussed how actual experiments verify the ideas of our thought experiments. But we haven't yet figured out what new common sense should replace the old. Fortunately, while it may take you many years to become completely comfortable with the ideas of relativity, it's relatively easy to see what you need to do to make yourself comfortable.

### How can we make sense of relativity?

Perhaps surprisingly, another thought experiment that may at first make everything seem even more bizarre will help us understand what is really going on, and thereby lead us toward a new common sense.

**Thought Experiment 12** Suppose Al is moving by you at a speed close to the speed of light. From our earlier thought experiments, we know that you'll measure his time as running slowly, his length as having contracted, and his mass as having increased. But what would *he* say?

From Al's point of view, he's not going anywhere—*you* are moving by him at high speed. Because the laws of nature are the same for everyone, he must reach exactly the same conclusions from his point of view that you reach from your point of view. That is, he'll say that *your* time is running slowly, *your* length has contracted, and *your* mass has increased!

Now we have what seems to be a serious argument on our hands. Imagine that you are looking into Al's spaceship with a super telescope. You can clearly see that his time is running slowly because everything he does is in slow motion. You send him a radio message saying, "Hi, Al! Why are you doing everything in slow motion?" Because the radio message travels at the absolute speed of light, Al has no trouble receiving your message,* and he responds with his own radio message back to you.

If you listen to his response, you'll hear it in slow motion—"Hheeeellllloooo tthhheeerrr"—verifying that his time is running slowly. However, if you record the entire message and use your computer to speed up his voice so that it sounds normal, you'll hear Al say, "I'm not moving in slow motion—*you are!*"

You can argue back and forth all you want, but it will get you nowhere. Then you come up with a brilliant idea. You hook up a video camera to your telescope and record a movie showing that Al's clock is moving more slowly than yours. You put your movie into a very fast rocket and shoot it off toward him. You figure that when the movie arrives and he watches it, he'll have visual proof that you are right and that he really is moving in slow motion.

Unfortunately, before you can declare victory in the argument, you learn that Al had the same brilliant idea: A rocket arrives with a movie that he made. As you watch, you see what appears to be clear proof that Al is right—his movie shows that *you* are in slow motion! Apparently, Al really is seeing your time run slowly, just as you are seeing his time run slowly.

How can this be? Think back to our earlier discussion of up and down, and imagine an American child and an Australian child talking by phone. The Australian says, "Isn't the Moon beautiful up in the sky right now?" The American replies, "What are you talking about? The Moon isn't up right now!" According to childhood common sense about up and down, the two children appear to be contradicting each other and could argue endlessly. However, once they realize that up and down are measured *relative* to the center of the Earth, they realize that the argument stems only from incorrect definitions of *up* and *down* (**FIGURE S2.18**). They are both talking about the same Moon in the same place, and their differing claims arise only from the relative nature of "up" and "down."

In much the same way, the argument between you and Al arises because you are using the old common sense in which we think of space and time as absolutes and expect the speed of light to be relative. The theory of relativity tells us that we have it backward. The speed of light is the absolute, and time and space are relative.

By the new common sense, the fact that you and Al disagree about whose time is running slowly is no more surprising than the fact that the two children disagree about whether the Moon is up or down. The disagreement is meaningless because it relies on inadequate definitions of time and space. *Results* are what count, and every experiment that you perform will agree with every experiment

*Of course, because of the Doppler effect [**Section 5.4**], he'll find that the radio waves are blueshifted from the frequency at which you send them as you come toward him, and redshifted as you move away from him. You'll find the same for his radio broadcasts. But as long as you both tune your radios to the correct Doppler-shifted frequencies, you'll have no problem hearing the broadcast messages.

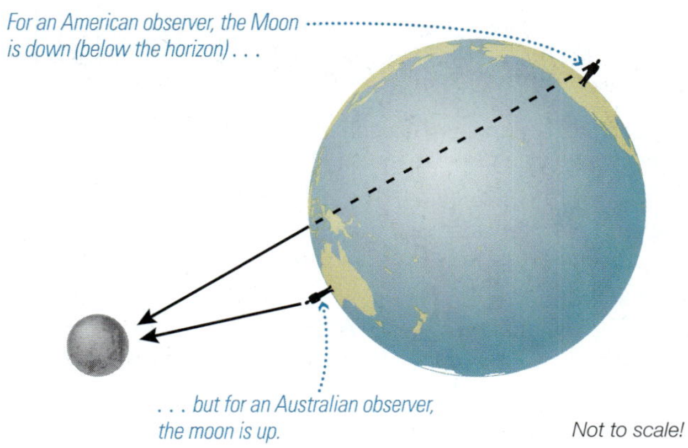


**FIGURE S2.18** The Moon is up for the Australian observer but down for the American observer.

that Al performs. You are both experiencing the same laws of nature, albeit in ways different from those our old common sense would have suggested.

## How does special relativity offer us a ticket to the stars?

One consequence of accepting the special theory of relativity is the fact that we'll never be able to travel to distant stars at a speed faster than the speed of light. However, while this fact might at first make distant stars seem forever out of reach, time dilation and length contraction actually offer a "ticket to the stars"—if we can ever build spaceships capable of traveling at speeds close to the speed of light.

Suppose you take a trip to the star Vega, about 25 light-years away, in a spaceship that travels at a speed very close to the speed of light—say, at $0.999c$. From the point of view of people on Earth, the trip to Vega must take you just over 25 years, since you are traveling at nearly the speed of light for a distance of 25 light-years. The return trip will take another 25 years. If you leave in the year 2050, you will arrive at Vega in the year 2075 and return to Earth in 2100.

However, from your point of view on the spaceship, you can legitimately claim that you and your spaceship never go anywhere; it is Earth and Vega that make the trip. That is, from your point of view, you remain stationary while Earth rushes away from you and Vega rushes toward you at $0.999c$. You'll therefore find the distance from Earth to Vega contracted from its rest length of 25 light-years; with the length contraction formula, the contracted distance turns out to be just over 1 light-year.* Therefore, because Vega is coming toward you at $0.999c$ and has only 1 light-year to travel to reach you from your point of view, you'll be at Vega in only about 1 year. Your return trip to Earth will also take about 1 year, so the round-trip time is only about 2 years from your point of view (**FIGURE S2.19**). If you leave at age 50, you'll return as a 52-year-old.

Although it sounds like a contradiction by our old common sense, both points of view are correct. If you leave in 2050 at age 50, you'll return to Earth at age 52—but in the year 2100. That is, while you will have aged only 2 years, all your surviving friends and family will be 50 years older than when you left.

You could make even longer trips within your lifetime with a sufficiently fast spaceship. For example, the Andromeda Galaxy is about 2.5 million light-years away, so the round trip to any star in the Andromeda Galaxy would take at least 5 million years from the point of view of observers on Earth. However, if you could travel at a speed within 50 parts in 1 trillion of the speed of light (that is, $c - 5 \times 10^{-11}c$ ), the trip would take only about 50 years from your point of view. You could leave Earth at age 30 and return at age 80—but you would return to an Earth on which your friends, your family, and everything you knew had been gone for 5 million years.

Thus, in terms of time, relativity offers only a one-way ticket to the stars. You can go a long distance and return to the *place* that you left, but you cannot return to the epoch from which you left.

*Given that you see the Earth-Vega distance contracted, shouldn't you also claim that it is *Earth's* time running slowly, rather than yours? This question underlies the so-called *twin paradox*, in which you make the trip while your twin sister stays home on Earth (see Chapter S3). The resolution comes from the fact that, because you must turn around at Vega, you effectively change reference frames relative to Earth at least once during your trip. A careful analysis of the changing reference frames is beyond the scope of this book, but it turns out that you do indeed measure less total time than your stay-at-home twin.

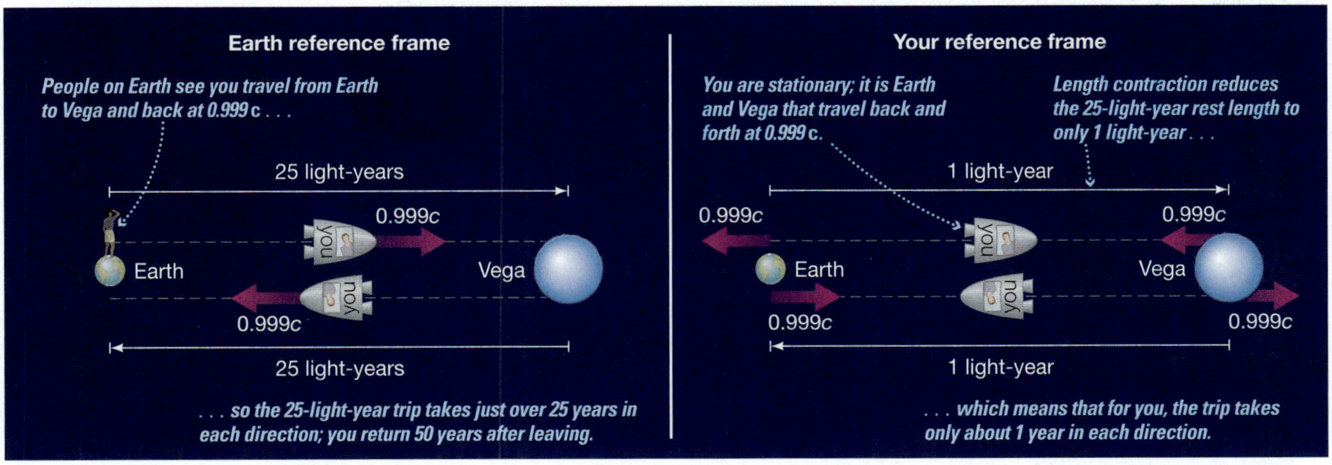


**FIGURE S2.19** A person who travels round-trip at high speed to a distant star will age less than people back home on Earth.

# The BIG Picture PUTTING CHAPTER S2 INTO CONTEXT

*In this chapter, we have studied Einstein's special theory of relativity, learning that space and time are intertwined in remarkable ways that are quite different from what we might expect from everyday experience. As you look back on our new viewpoint about space and time, keep in mind the following "big picture" ideas:*

- The ideas of relativity derive from two simple ideas: The laws of nature are the same for everyone, and the speed of light is absolute. The thought experiments in this chapter showed the consequences of these facts.
- Thought experiments are useful, but the ultimate judge of any theory is observation and experiment. The theory of relativity has been extensively tested and verified.
- Although the ideas of relativity may sound strange at first, you can understand them easily and logically if you allow yourself to develop a "new common sense" that incorporates them.

**MY COSMIC PERSPECTIVE** Most of us would like to believe that we understand reality, and relativity shows us that the reality of space and time is quite different from what we might expect from everyday life. Only by studying relativity can we learn the true nature of the time and space in which we live our lives.

# Summary of Key Concepts

## S2.1 Einstein's Revolution

- **What are the major ideas of special relativity?** Special relativity tells us that different observers can measure time, distance, and mass differently, even though everyone always agrees on the speed of light. It also tells us that no material object can reach or exceed the speed of light (in a vacuum) and that $E = mc^2$.
- **What is relative about relativity?** The theory of relativity is based on the idea that all *motion* is relative. That is, there is no correct answer to the question of who or what is really moving in the universe, so motion can be described only for one object relative to another.

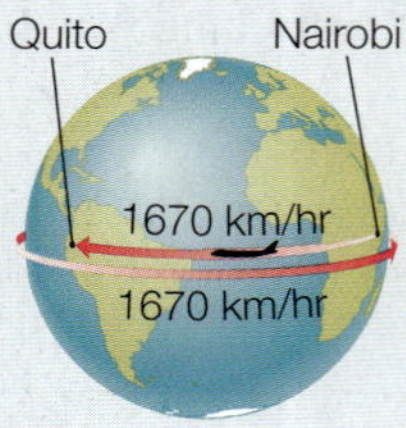


- **What is absolute about relativity?** All the predictions of the special theory of relativity follow from two assumptions, both of which have been experimentally verified: (1) The laws of nature are the same for everyone, and (2) the speed of light is the same for everyone.

## S2.2 Relative Motion

- **What's surprising about the absoluteness of the speed of light?** In our everyday lives, we expect velocities to add simply; for example, an observer on the ground should see a ball in an airplane traveling at the airplane's speed plus the ball's speed. However, this is not true for light, which everyone always measures as traveling at the same speed.

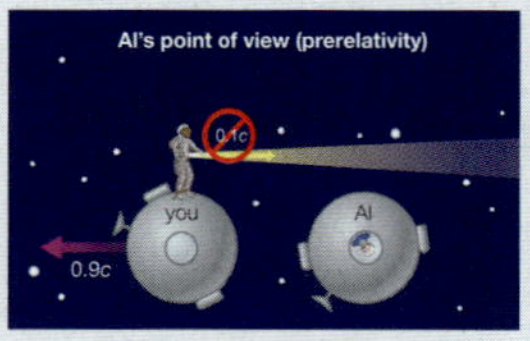


- **Why can't we reach the speed of light?** Light always travels at the same speed, so your own light (the light that you emit or reflect) is always moving ahead of you at the speed of light. All other observers will also see your light moving at the speed of light—and because it is moving ahead of you, the observers will always conclude that you are moving slower than the speed of light.

## S2.3 The Reality of Space and Time

- **How does relativity affect our view of time and space?** If you observe an object moving by you at high speed, you'll find that its time is running more slowly than yours, its length is shorter than its length when at rest, and its mass is greater than its mass when at rest. Moreover, observers in different reference frames may disagree about whether two events are simultaneous and about the speeds of different objects.

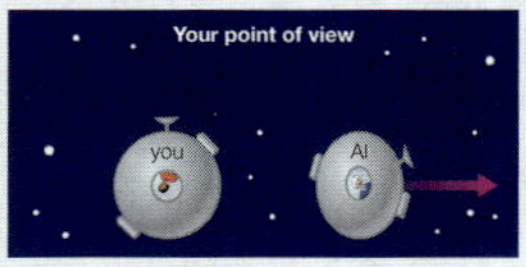


- **Do the effects predicted by relativity really occur?** Experiments with light confirm that its speed is always the same. Experiments with subatomic particles in particle accelerators confirm the predictions of time dilation and mass increase at speeds close to the speed of light, and time dilation has also been verified at relatively low speeds. Nuclear power plants and nuclear bombs release energy in accordance with the formula $E = mc^2$, which is also a prediction of special relativity.

## S2.4 Toward a New Common Sense

- **How can we make sense of relativity?** A person moving by you at high speed will see exactly the same effects on you as you see on her; for example, she'll see your time running slowly while you see her time running slowly. Although this might sound contradictory, it simply tells us that time and space must be relative in much the same way that up and down are relative on Earth.

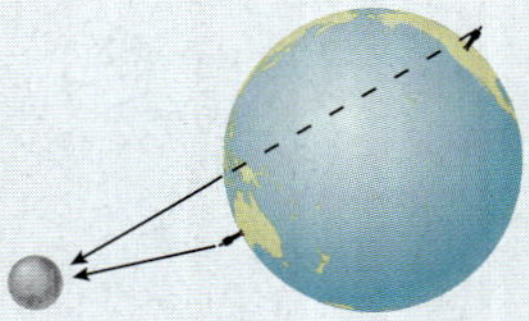

- **How does special relativity offer us a ticket to the stars?** Although the theory tells us that journeys to the stars will always take many years from the point of view of Earth, it also tells us that the time for the passengers will be much shorter if they travel at speeds close enough to the speed of light. Thus, the passengers may be able to make very distant journeys within their lifetimes, even though their friends back on Earth will not be there to greet them when they return.

# Exercises and Problems

*For instructor-assigned homework and other learning materials, go to www.MasteringAstronomy.com.*

## Chapter Review Questions

### Short-Answer Questions Based on the Reading

1. What is the *theory of relativity*? How does *special relativity* differ from *general relativity*?
2. List five major predictions of the special theory of relativity.
3. According to the theory of relativity, what are the two absolutes in the universe? Which one is more surprising, and why?
4. What is a *paradox*? How can a paradox lead us to a deeper understanding of an issue?
5. What do we mean by a *frame of reference*? What is a *free-float frame*?
6. Suppose you see a friend moving by you at some constant speed. Explain why your friend can equally well say that she is stationary and you are moving by her.
7. Construct your own variation on Thought Experiment 6 to prove that no material object can be observed to travel at or above the speed of light.
8. What is *time dilation*? Explain how and why your measurements of time will differ from those of someone moving by you.
9. Explain why observers in different reference frames will not necessarily agree about the order of two events that occur in different places.
10. What is *length contraction*? How will your measurements of the size of a spaceship moving by you differ from your measurements of the same spaceship when it is at rest in your reference frame?
11. What is *mass increase*? How does the mass of an object moving by you compare to its rest mass?
12. Construct your own variation on Thought Experiment 11 to show why velocities must add differently than we would expect according to our "old common sense."
13. Briefly describe experimental tests of special relativity.
14. Why do tests of $E = mc^2$ also test special relativity? Describe several tests of $E = mc^2$ that you can see for yourself.
15. If you watch a friend moving by you at a speed close to that of light, you'll say that her time is running slowly, her length is contracted, and her mass is greater than her rest mass. How will she perceive her own time, length, and mass? Why? How will she perceive *your* time, length, and mass?
16. Suppose you could take a trip to a distant star at a speed very close to the speed of light. How would relativity make it possible for you to make this trip in a reasonably short time? What would you find when you returned home?

### Does It Make Sense?

*Decide whether the statement makes sense (or is clearly true) or does not make sense (or is clearly false). Explain clearly; not all of these have definitive answers, so your explanation is more important than your chosen answer.*

17. Einstein proved that everything is relative.
18. An object moving by you at very high speed will appear to have a higher density than it has at rest.
19. You and a friend agree that you each popped a peanut into your mouth at precisely the same instant, but someone moving past you at high speed may observe that you ate your peanut first.
20. You and a friend agree that you each popped a peanut into your mouth at precisely the same instant, but someone moving past you at high speed may observe that you ate cashews rather than peanuts.
21. Relativity is "only a theory," and we have no way to know whether any of its predictions would really occur at speeds close to the speed of light.
22. The detonation of a nuclear bomb is a test of the special theory of relativity.
23. If you could travel away from Earth at a speed close to the speed of light, you would feel uncomfortably heavy because of your increased mass.
24. Future technology should allow us to build rockets capable of accelerating to speeds much faster than the speed of light.
25. If you see someone's time running slowly in a different reference frame, that person must see your time running fast.
26. If you had a sufficiently fast spaceship, you could leave today, make a round trip to a star 500 light-years away, and return home to Earth in the year 2100.

### Quick Quiz

*Choose the best answer to each of the following. Explain your reasoning with one or more complete sentences. For additional practice, try the Chapter S2 Reading and Concept Quizzes in the Study Area at www.MasteringAstronomy.com.*

27. Which of the following is *not* relative in the special theory of relativity? (a) motion (b) time (c) the speed of light
28. Which of the following must be true of a person who shares your *reference frame*? (a) The person must be sitting right next to you. (b) The person must be the same size as you. (c) The person must not be moving relative to you.
29. Which of the following best describes why your rocket could never reach the speed of light? (a) The absoluteness of the speed of light means you could never keep up with the light coming from you and your rocket. (b) Our technology has not advanced enough to make faster-than-light travel possible. (c) A rocket that could reach the speed of light would have to be bigger than the entire Earth.
30. Carla is traveling past you at a speed close to the speed of light. According to *you*, how much time passes for Carla while 1 minute passes for you? (a) 1 minute (b) less than 1 minute (c) more than 1 minute
31. Carla is traveling past you at a speed close to the speed of light. According to *her*, how much time passes for you while 1 minute passes for her? (a) 1 minute (b) less than 1 minute (c) more than 1 minute
32. A subatomic particle that normally decays in 1 microsecond is created in a particle accelerator and is traveling at close to the speed of light. If you measure its lifetime in that case, you'll find that it is (a) less than 1 microsecond. (b) more than 1 microsecond. (c) 1 microsecond.
33. What does the famous formula $E = mc^2$ have to do with special relativity? (a) Nothing; it comes from a different theory. (b) It is one of the two starting assumptions of special relativity. (c) It is a direct consequence of the theory.

34. What provides the strongest evidence that everyone always measures the same speed of light? (a) precise measurements of the speed of light under different circumstances (b) thought experiments showing that we would encounter paradoxes if it were not true (c) the fact that Einstein said so and he was very smart
35. If you observe people moving by you at very high speed, you will say that their time runs slowly, their lengths are contracted in the direction of motion, and their masses are increased from their rest masses. What will they say about you? (a) Your time runs slowly, your length is contracted in the direction of motion, and your mass is increased from your rest mass. (b) Your time runs fast, your length is expanded in the direction of motion, and your mass is decreased from your rest mass. (c) Your time runs fast, but your length and mass are unaffected.
36. Suppose you had a spaceship so fast that you could make a round-trip journey of 1 million light-years (in Earth's reference frame) in just 50 years of ship time. If you left in the year 2030, you would return to Earth (a) in the year 2080. (b) in the year 2130. (c) a million years from now.

## Inclusive Astronomy

*Use these questions to reflect on participation in science.*

37. *Pair Discussion: Mileva Maric.* We generally credit Einstein for coming up with the theory of relativity, but in fact Mileva Maric certainly played a role.
**a.** Working independently, find out who Mileva Maric was, how she was related to Einstein, and what evidence indicates that she was a serious contributor to the ideas of relativity.
**b.** Pair up with another student to discuss what each of you learned and the significance of Maric's work. Do you think she should be better known for her contributions to relativity?
**c.** Do you think Mileva Maric and Albert Einstein had equal opportunities to contribute to progress in science? Why or why not?
**d.** Does knowing the role of Mileva Maric change your view of Albert Einstein in any way? Why or why not?

## The Process of Science

*These questions may be answered individually in short-essay form or discussed in groups, except where identified as group-only.*

38. *A Paradigm Shift.* Chapter 3 introduced the idea of a paradigm shift, in which general patterns of scientific thought undergo a major or surprising change. How did the development of relativity theory represent such a shift? Did it mean throwing out old ideas or only modifying them? Defend your opinion.
39. *Thought Experiments.* Science demands that hypotheses be tested by real experiments or observations, yet Einstein developed his theory of relativity largely through the use of thought experiments. How, then, did relativity still come to be accepted as a scientific theory? Explain.
40. *Common Sense.* Discuss the meaning of the term *common sense.* How do we develop common sense? Can you think of other examples, besides the example of the meanings of *up* and *down,* of situations in which you've had to change your common sense? Do you think that the theory of relativity contradicts common sense? Why or why not?
41. *Group Discussion: Breaking the Light Barrier.* People long said that no one would travel faster than the speed of sound, but that barrier was broken. Is it possible that we'll someday find a way to travel faster than the speed of light? Discuss in small groups and come to a group consensus for or against this possibility. Note: You may wish to do this activity using the four roles described in Chapter 1, Exercise 39.

## Investigate Further

### Short-Answer/Essay Questions

42. *Stationary Bike.* Suppose you are riding on a stationary bike and the speedometer says you are going 30 km/hr. What does this number mean? What does it tell you about the idea of relative motion?
43. *Walking Trip.* Suppose you go for a walk down the street. In your own reference frame, you are going quite slowly, probably just a few kilometers per hour. About how fast would you appear to be going according to an astronaut on the Moon? Explain.
44. *Relative Motion Practice I.* In all the following, assume that you and your friends are in free-float reference frames.
**a.** Bob is coming toward you at a speed of 75 km/hr. You throw a baseball in his direction at 75 km/hr. What does he see the ball doing?
**b.** Marie is traveling away from you at a speed of 120 km/hr. She throws a baseball at 100 km/hr (according to her) in your direction. What do you see the ball doing?
**c.** José is traveling away from you at 99% of the speed of light when he turns on a flashlight and points it in your direction. How fast will the beam of light be going when it reaches you?
45. *Relative Motion Practice II.* In all the following, assume that you and your friends are in free-float reference frames.
**a.** Carol is going away from you at 75 km/hr, and Sam is going away from you in the opposite direction at 90 km/hr. According to Carol, how fast is Sam going?
**b.** Consider again the situation in part a. Suppose you throw a baseball in Sam's direction at a speed of 120 km/hr. What does Sam see the ball doing? What does Carol see the ball doing?
**c.** Cameron is traveling toward you at 99.9999% of the speed of light when he turns on a flashlight and points it in your direction. How fast will the beam of light be going when it reaches you?
46. *Moving Spaceship.* Suppose you are watching a spaceship go past you at a speed close to the speed of light.
**a.** How do clocks on the spaceship run, compared to your own clocks?
**b.** If you could measure the length, width, and height of the spaceship as it passed by, how would these measurements compare to your measurements of the size of the spaceship if it were stationary?
**c.** If you could measure the mass of the spaceship, how would this value compare to its rest mass?
**d.** How would a passenger on the spaceship view your time, size, and mass?
47. *Relativity of Simultaneity.* Consider the situation in Thought Experiment 8, about the green and red flashes of light at opposite ends of Al's spaceship. Suppose your friend Jackie is traveling in a spaceship in the opposite direction from Al. Further imagine that she is also precisely aligned with you and Al at the instant the two flashes of light occur (in your reference frame).

**a.** According to Jackie, is Al illuminated first by the green flash or the red flash? Explain.

**b.** According to you, which flash illuminates Jackie first? Why?

**c.** According to Jackie, which flash occurs first? Explain. How does Jackie's view of the order of the flashes compare to your view and to Al's view?

48. *Photon Philosophy.* Extend the ideas of time dilation and length contraction to think about how the universe would look if you were a photon traveling at the speed of light. Do you think there's any point in thinking about how a photon "perceives" the universe? If so, discuss any resulting philosophical implications. If not, explain why not.
49. *Ticket to the Stars.* Suppose that we someday acquire the technology to travel among the stars at speeds near the speed of light. Imagine that many people make journeys to many places. Discuss some of the complications that would arise from people aging at different rates depending on their travels.

## Quantitative Problems

*Be sure to show all calculations clearly and state your final answers in complete sentences.*

50. *Time Dilation.*

**a.** A student, after learning about the theory of relativity, decides to try to prolong his life by spending the rest of his life traveling in a car at 55 miles per hour (89 km/hr). If he does this while 70 years pass in his house, how much time will pass in the car? (*Hint:* If you are unable to find a difference, be sure to explain why.)

**b.** A more clever student decides to prolong her life by cruising among the stars at a speed of $0.95c$. How much time will pass on her spacecraft during a period in which 70 years pass on Earth? Will *she* feel as if her life span has been extended? Explain.

**c.** Suppose you stay home on Earth while your twin sister takes a trip to a distant star and back in a spaceship that travels at 99% of the speed of light. If both of you are 25 years old when she leaves and you are 45 years old when she returns, how old is your sister when she gets back?

51. *Length Contraction.*

**a.** Marta has a spaceship that measures 50 meters from end to end when at rest. If she flies past you at $0.75c$, how long will you measure it to be?

**b.** The star Sirius is 8.6 light-years from Earth (in our Earth-based reference frame). Suppose you traveled from Earth to Sirius at $0.92c$. During your trip, how far would you measure the distance from Earth to Sirius to be?

52. *Mass Increase.*

**a.** A spaceship has a rest mass of 500,000 tons. If you could measure its mass when it was traveling at half the speed of light, what would the value be?

**b.** A fly has a mass of 1 gram at rest. How fast would it have to be traveling to have a mass of 3000 kilograms (about that of a large SUV)?

53. *Time Dilation with Subatomic Particles.* A $\pi^+$ meson produced at rest has a lifetime of 18 nanoseconds ($1.8 \times 10^{-8}$ s). Suppose a $\pi^+$ meson is produced in a particle accelerator at a speed of $0.998c$. How long will scientists see the particle last before it decays? Briefly explain how an experiment like this helps verify the special theory of relativity.
54. *Time Dilation on the Space Station.* The International Space Station orbits Earth at a speed of about 30,000 km/hr. While 1 hour passes on Earth, how much time passes on the Station? Assume that both the Station and Earth are in free-float frames, although in reality they are not. (*Hint:* Start by converting the Station's speed to km/s.)
55. *Super Baseball.* For the game of interstellar baseball, teams have used robotic enhancements to allow superhuman feats. Consider a pitcher who can throw a baseball at 80% of the speed of light and is practicing in a spaceship traveling away from Earth at 95% of the speed of light (on the way to the team's next game). Assuming that he throws the baseball in the same direction the spacecraft is traveling, how fast will those of us on Earth measure the baseball to be going?
56. *Racing a Light Beam I.* A long time ago, in a galaxy far, far away, there was a civilization whose inhabitants hosted an Olympic competition every 4 of their years. Unfortunately, the competition had become tarnished by the use of illegal substances designed to aid performance, and many great athletes were disqualified and stripped of their medals. One famed sprinter, named Jo, finally decided that it was pointless to continue racing humans. Instead, he held a press conference to announce that he would race a beam of light! Sponsors lined up, crowds gathered, and the event was sold to a pay-per-view audience. Everything was set. The starting gun was fired. Jo raced out of the starting block and shattered the old world record, running 100 meters in 8.7 seconds. The light beam, represented by a flashlight turned on at precisely the right moment, of course emerged from its "block" at the speed of light. How long did it take the light beam to cover the 100-meter distance? What can you say about the outcome of this race?
57. *Racing a Light Beam II.* Following his humiliation in the first race against the light beam (Problem 56), Jo went into hiding for the next 2 years. By that time, most people had forgotten about both him and the money they had wasted on the pay-per-view event. However, Jo was secretly in training during this time. He worked out hard and tested new performance-enhancing substances. One day, he emerged from hiding and called another press conference. "I'm ready for a rematch," he announced. Sponsors were few this time and spectators scarce in the huge Olympic stadium where Jo and the flashlight lined up at the starting line. But those who were there will never forget what they saw, although it all happened very quickly. Jo blasted out of the starting block at 99.9% of the speed of light. The light beam, emitted from the flashlight, took off at the speed of light. The light beam won again—but barely! After the race, TV commentators searched for Jo, but he seemed to be hiding again. Finally, they found him in a corner of the locker room, sulking under a towel. "What's wrong? You did great!" said the commentators. Jo looked back sadly, saying, "Two years of training and experiments, for nothing!" Let's investigate what happened.

**a.** As seen by spectators in the grandstand, how much faster than Jo is the light beam?

**b.** As seen by Jo, how much faster is the light beam than he is? Explain your answer clearly.

**c.** Using your results from parts a and b, explain why Jo can say that he was beaten just as badly as before, while the spectators can think he gave the light beam a good race.

**d.** Although Jo was disappointed by his performance against the light beam, he did experience one pleasant surprise: The 100-meter course seemed short to him. In Jo's reference frame during the race, how long was the 100-meter course?

# S3 Spacetime and Gravity

▲ **About the photo:** The many long arcs in this photo of the galaxy cluster Abell 2218 are distorted images of galaxies that lie behind the cluster. The galaxy images are distorted because the cluster's gravity curves the space through which their light passes—just as predicted by Einstein's general theory of relativity.

## LEARNING GOALS

### S3.1 Einstein's Second Revolution
- What are the major ideas of general relativity?
- What is the fundamental assumption of general relativity?

### S3.2 Understanding Spacetime
- What is spacetime?
- What is curved spacetime?

### S3.3 A New View of Gravity
- What is gravity?
- What is a black hole?
- How does gravity affect time?

### S3.4 Testing General Relativity
- How do we test the predictions of the general theory of relativity?
- What are gravitational waves?

### S3.5 Hyperspace, Wormholes, and Warp Drive
- Where does science end and science fiction begin?

### S3.6 The Last Word
- How has relativity changed our view of space and time?

*The eternal mystery of the world is its comprehensibility. The fact that it is comprehensible is a miracle.*
—*Albert Einstein (1879–1955)*

**Chapter S3 Overview**

What is gravity? Newton's law of gravity allows us to calculate orbits throughout the universe, but the deeper mystery is how one object can exert a gravitational force on another that may be a great distance away. This mystery was finally solved by Einstein with his general theory of relativity, which explains gravity as a consequence of the structure of space and time.

In this chapter, we will investigate Einstein's revolutionary view of gravity. As we will see, Einstein's theory is crucial to understanding many astronomical phenomena, ranging from the peculiar orbit of Mercury to the bizarre properties of black holes.

## S3.1 Einstein's Second Revolution

Imagine that you and everyone around you believe Earth to be flat. As a wealthy patron of the sciences, you decide to sponsor an expedition to the far reaches of the world. You select two fearless explorers and give them careful instructions. Each is to journey along a perfectly straight path, but they are to travel in opposite directions. You provide each with a caravan for land-based travel and boats for water crossings, and you tell each to turn back only after discovering "something extraordinary."

Some time later, the two explorers return. You ask, "Did you discover something extraordinary?" To your surprise, they answer in unison, "Yes, but we both discovered the same thing: We ran into each other, despite having traveled in opposite directions along perfectly straight paths."

Although this outcome would be extraordinarily surprising if you truly believed Earth to be flat, we are not really surprised because we know that Earth is round (**FIGURE S3.1a**). In a sense, the explorers followed the *straightest possible paths,* but these "straight" lines follow the curved surface of Earth.

Now let's consider a somewhat more modern scenario. You are floating freely in a spaceship somewhere out in space. Hoping to learn more about space in your vicinity, you launch two small probes along straight paths in opposite directions. Each probe is equipped with a camera that transmits pictures back to your spaceship. Imagine that, to your astonishment, the probes one day transmit pictures of each other! That is, although you launched them in opposite directions and neither has ever fired its engines, the probes have somehow met. This might at first sound surprising, but in fact this situation arises quite naturally with orbiting objects. If you launch two probes in opposite directions from a space station, they will meet as they orbit Earth (**FIGURE S3.1b**).

Since the time of Newton, we've generally explained curved paths such as those of the two probes as an effect caused by the force of gravity. However, by analogy with the explorers journeying in opposite directions on Earth, might we instead conclude that the probes meet because *space* is somehow curved? Strange as this idea may sound, it lies at the heart of Einstein's second revolution—a revolutionary view of gravity contained in his *general theory of relativity,* published in 1915.

### What are the major ideas of general relativity?

Our study of the special theory of relativity has already shown that space and time are inextricably linked. More specifically, we say that the three dimensions of space and the one dimension of time together form a *four*-dimensional combination called **spacetime**. When Einstein extended the special theory to the general case that includes gravity, he discovered that matter shapes the "fabric" of spacetime in a manner somewhat analogous to the way heavy weights distort a taut rubber sheet or trampoline (**FIGURE S3.2**). The analogy is not perfect—for example, we cannot place weights "upon" spacetime because all matter exists *within* spacetime—but it is still useful in discussing the principles of general relativity.

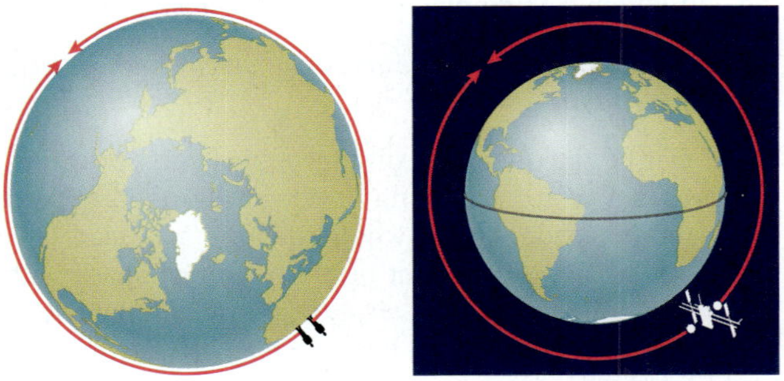

**a** Travelers going in opposite directions along paths that are as straight as possible will meet as they go around Earth, a fact that we attribute to the curvature of Earth's surface.

**b** Two space probes launched in opposite directions in Earth orbit will meet as they orbit Earth, a fact that we usually attribute to the mysterious force of gravity.

**FIGURE S3.1** Travelers on Earth's surface and orbiting objects follow similar-shaped paths, but we usually explain these paths in very different ways.

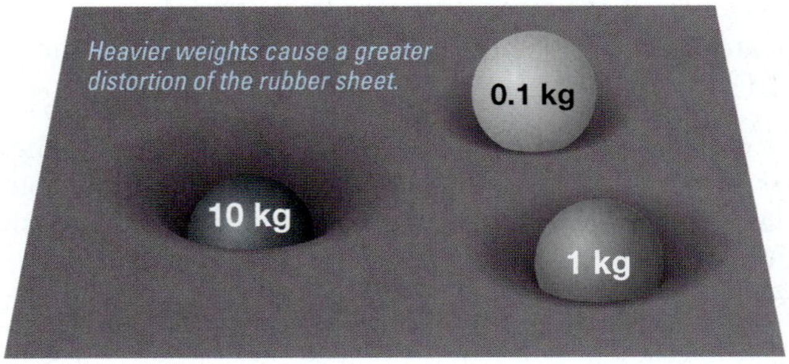


**FIGURE S3.2** A rubber sheet analogy to spacetime: Matter distorts the "fabric" of four-dimensional spacetime in a manner analogous to the way heavy weights distort a taut, two-dimensional rubber sheet. The greater the mass, the greater the distortion of spacetime.

It is difficult to overstate the significance of general relativity to our understanding of the universe. In particular, the following ideas all come directly from this theory:

- Gravity arises from distortions of spacetime. The presence of mass causes the distortions, and the resulting distortions determine how other objects move through spacetime.
- Time runs slowly in gravitational fields. The stronger the gravity, the more slowly time runs.
- *Black holes* can exist in spacetime, and falling into a black hole means leaving the observable universe.
- It is possible for the universe to have a finite volume without having a center or boundaries.
- Large masses that undergo rapid changes in motion or structure emit *gravitational waves* that travel at the speed of light.

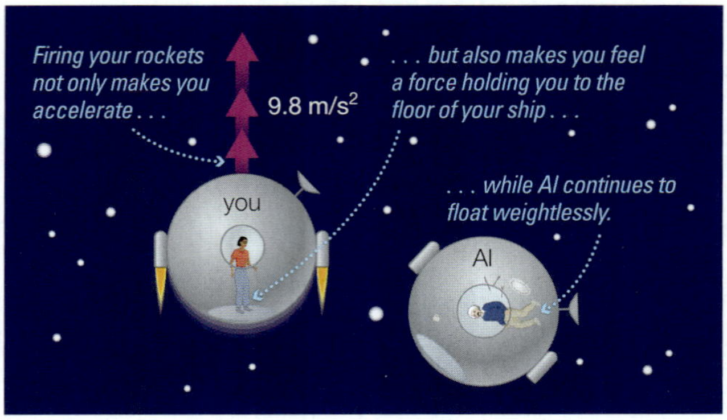


**FIGURE S3.3** In deep space with no engines firing, you and Al would both float freely in your spaceships. But when you fire your engines, you feel a force that presses you against the floor of your spaceship, and Al sees you accelerate away. You might wish to claim that *he* is accelerating away from you, but how, then, do you explain the force that you feel and the fact that Al remains weightless?

## What is the fundamental assumption of general relativity?

Recall that special relativity begins with two key assumptions—that the laws of nature are the same for everyone and that everyone always measures the same speed of light—from which we can derive all the surprising consequences discussed in Chapter S2. General relativity also has a starting assumption that leads to all of its astounding predictions. We can understand the origin of this assumption by thinking more deeply about the idea of relative motion.

**A Thought Experiment with Acceleration** Imagine that you and Al are both floating weightlessly in your spaceships; that is, you are in free-float reference frames. As we saw in Chapter S2, there is no absolute answer to the question "Who is really moving?" because you will both experience the laws of nature in exactly the same way. Now, suppose you decide to fire your rocket engines with enough thrust to give you an acceleration of $1g$ ($= 9.8\ \text{m/s}^2$), which is the acceleration of gravity on Earth [**Section 4.1**]. Al keeps his engines off and therefore sees you flying off with ever-growing speed, so he sends you a radio message saying, "Good-bye, have a nice trip!"

**Think about it** Suppose you start from rest in Al's reference frame and he sees you accelerate at $1g$. Approximately how fast will Al see you going after 1 second? After 10 seconds? After a minute?

If *all* motion is relative, you should be free to claim that you are still stationary and that it is Al who is receding into the distance at ever-faster speeds. You might therefore wish to reply, "Thanks, but I'm not going anywhere. You're the one accelerating into the distance." However, this situation has a new element that was not present when we dealt with constant velocities: Because your rocket engines are firing, you feel a force inside your spaceship that means you will no longer be weightless (**FIGURE S3.3**). In fact, with an acceleration of $1g$, you'll be able to walk on your spaceship floor with your normal Earth weight. Al may therefore respond back, "Oh, yeah? If you're not going anywhere, why are you stuck to the floor of your spaceship, and why do you have your engines turned on? And if I'm accelerating as you claim, why am I weightless?"

You must admit that Al is asking good questions. It certainly *looks* as if you really are the one who is accelerating while Al remains stationary. In other words, it seems that motion is no longer relative when we introduce acceleration. This idea did not sit well with Einstein, because he believed that *all* motion should be relative, regardless of whether the motion was at constant velocity or included an acceleration. Einstein therefore needed some way to explain the force you feel due to firing your rocket engines without necessarily assuming that you are accelerating through space. (Einstein did not actually use spaceships in his thought experiments, but the basic ideas are the same.)

**The Equivalence Principle** In 1907, Einstein hit upon what he later called "the happiest thought of my life." His revelation consisted of the idea that, whenever you feel weight (as opposed to weightlessness), you can attribute it to effects of either acceleration or gravity. This idea is called the **equivalence principle**. Stated more precisely, it says,

> *The effects of gravity are exactly equivalent to the effects of acceleration.**

To clarify the meaning of the equivalence principle, imagine that you are sitting inside with doors closed and window shades down when your room is magically removed from Earth and sent hurtling through space with an acceleration of $1g$ (**FIGURE S3.4**). According to the equivalence principle, you would have no way of knowing that you'd left Earth. Any experiment you performed, such as dropping balls of different weights, would yield the same results you'd get on Earth.

---

*Technically, this equivalence holds only within small regions of space. Over larger regions, the gravity of a massive object varies in ways that would not occur because of acceleration; for example, such variation gives rise to *tidal forces* that do not arise from acceleration.

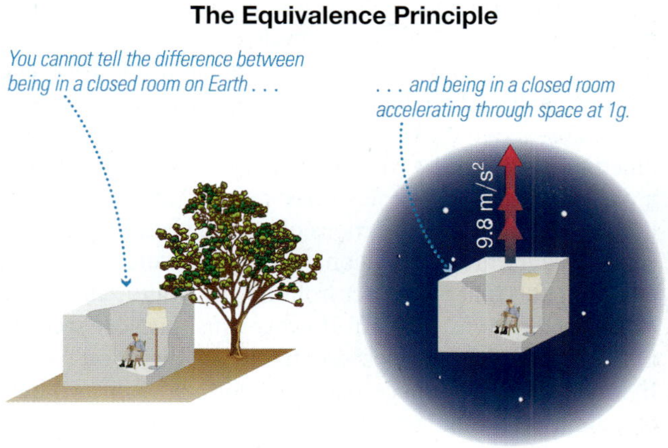


**FIGURE S3.4** The equivalence principle states that the effects of gravity are exactly equivalent to the effects of acceleration. Therefore, you cannot tell the difference between being in a closed room on Earth and being in a closed room accelerating through space at $1g$.

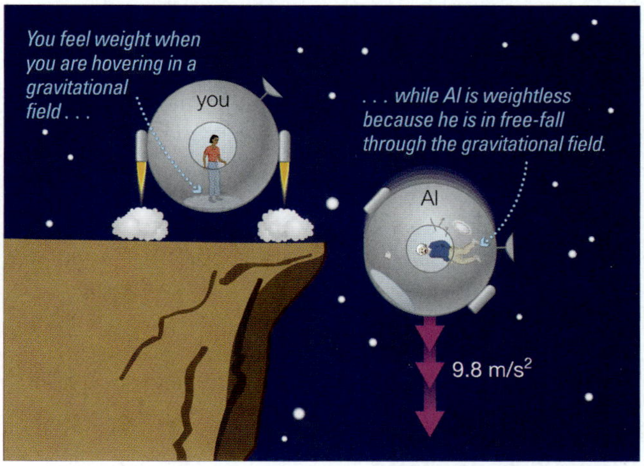


**FIGURE S3.5** According to the equivalence principle, you can claim to be stationary in a gravitational field, using your engines to prevent you from falling. You feel weight due to gravity, while Al is weightless because he is in free-fall.

The equivalence principle gives you a way to answer Al's questions. He asked how you can be stationary when your engines are firing and you feel weight. You can now answer that space is filled with a gravitational field pointing "downward" toward the floor of your spaceship. You are stationary because your rocket engines *prevent* you from falling through this gravitational field, and the gravity also explains your weight. How do you answer Al's second question, about his own weightlessness? Easy: Because he is not using his engines, he and his spaceship are falling through the gravitational field that fills space around you—and anyone in free-fall feels weightless [**Section 4.1**]. In essence, you are claiming that the situation is much as it would be if you were hovering over a cliff while Al had fallen over the edge (**FIGURE S3.5**). To summarize, you respond, "Sorry, Al, but I still say that you have it backward. I'm using my engines to prevent my spaceship from falling, and I feel weight because of *gravity*. You're weightless because you're in free-fall. I hope you won't be hurt by hitting whatever lies at the bottom of this gravitational field!"

The equivalence principle is the starting point for general relativity. As our thought experiment has shown, it allows us to treat *all* motion as relative, just as Einstein thought should be the case. Moreover, as we'll see in the rest of this chapter, the equivalence principle will also lead us to fundamental changes in the way we think about gravity and the universe.

## SPECIAL TOPIC Einstein's Leap

Given that the similarities in the effects of gravity and of acceleration were well known to scientists as far back as the time of Newton, you may wonder why the equivalence principle came as a surprise. The answer is that, before Einstein, the similarities were generally attributed to coincidence. It was as if scientists imagined that nature was showing them two boxes, one labeled "effects of gravity" and the other labeled "effects of acceleration." They shook, weighed, and kicked the boxes but could never find any obvious differences between them. They concluded, "What a strange coincidence! The boxes seem the same from the outside even though they contain different things." Einstein's revelation was, in essence, to look at the boxes and say that it is not a coincidence at all. The boxes appear the same from the outside because they contain the same thing.

In many ways, Einstein's assertion of the equivalence principle represented a leap of faith, although it was a faith he would willingly test through scientific experiment. He proposed the equivalence principle because he thought the universe would make more sense if it were true, not because of any compelling observational or experimental evidence for it at the time. This leap of faith sent him on a path far ahead of his scientific colleagues.

From a historical viewpoint, special relativity was a "theory waiting to happen" because it was needed to explain two significant problems left over from the 19th century: the perplexing constancy of the speed of light, demonstrated in the Michelson-Morley experiment [**Section S2.3**], and some seeming peculiarities of the laws of electromagnetism. Indeed, several other scientists were very close to discovering the ideas of special relativity when Einstein published the theory in 1905, and *someone* was bound to come up with special relativity around that time.

General relativity, in contrast, was a *tour de force* by Einstein. He recognized that unsolved problems remained after completing the theory of special relativity, and he alone took the leap of faith required to accept the equivalence principle. Without Einstein, general relativity probably would have remained undiscovered for at least a couple of decades beyond 1915, the year he completed and published the theory.

## (S3.2) Understanding Spacetime

It's easy to *say* that you can attribute your weight to the effect of gravity or the effect of acceleration, but the two effects tend to *look* very different. A person standing on the surface of Earth appears to be motionless, while an astronaut accelerating through space continually gains speed. How can gravity and acceleration produce such similar effects when they look so different? According to the general theory of relativity, the answer is that they look different only because we're not seeing the whole picture. Instead of looking just at the three dimensions of space, we must learn to "look" at the *four* dimensions of spacetime.

### What is spacetime?

The first step in understanding spacetime is understanding what we mean when we say, for example, that something is two-dimensional or three-dimensional. The concept of **dimension** describes the number of independent directions in which movement is possible (**FIGURE S3.6**). A **point** has zero dimensions. If you were a "geometric prisoner" confined to a point, you'd have no place to go. Sweeping a point back and forth along one direction generates a **line**. The line is one-dimensional because only one direction of motion is possible (going backward is considered the same as going forward by a negative distance). Sweeping a line back and forth generates a two-dimensional **plane**. The two directions of possible motion are, say, lengthwise and widthwise. Any other direction is just a combination of these two. If we sweep a plane up and down, it fills **three-dimensional space**, with the three independent directions of length, width, and depth.

We live in three-dimensional space and cannot visualize any direction that is distinct from length, width, and depth (and combinations thereof). However, just because we cannot *see* "other" directions doesn't mean they don't exist. If we could sweep space back and forth in some "other" direction, we would generate a **four-dimensional space**.

Although we cannot visualize a four-dimensional space, we are able to describe it mathematically. In algebra, we do one-dimensional problems with the single variable $x$, two-dimensional problems with the variables $x$ and $y$, and three-dimensional problems with the variables $x$, $y$, and $z$. A four-dimensional problem simply requires adding a fourth variable, as in $x$, $y$, $z$, and $w$. We could continue to five dimensions, six dimensions, and so forth. Any space with more than three dimensions is called a **hyperspace**, which means "beyond space."

### MATHEMATICAL INSIGHT S3.1 Spacetime Geometry

Spacetime geometry is more complex than the simplest possible four-dimensional geometry, because time enters the equations in a slightly different way than the three spatial dimensions. We can gain insight into the nature of spacetime geometry by thinking about how we define distance in ordinary geometry.

Consider two points in a plane separated by amounts $x = 3$ along the horizontal axis and $y = 4$ along the vertical axis. As shown on the left side of **FIGURE 1**, the *distance* between the two points is $\sqrt{x^2 + y^2}$, which is 5 in this case. Now, consider what happens if we rotate the coordinate axes so that both points lie along the $x$-axis (right side of Figure 1). This rotation changes the $x$ and $y$ separations of the two points, but the distance $\sqrt{x^2 + y^2}$ between the points is still the same. This fact should not be surprising: *Distance* is a real, physical quantity, while the $x$ and $y$ separations depend on the particular coordinate system being used.

*The x- and y-coordinates of two points can be different in different coordinate systems . . .*

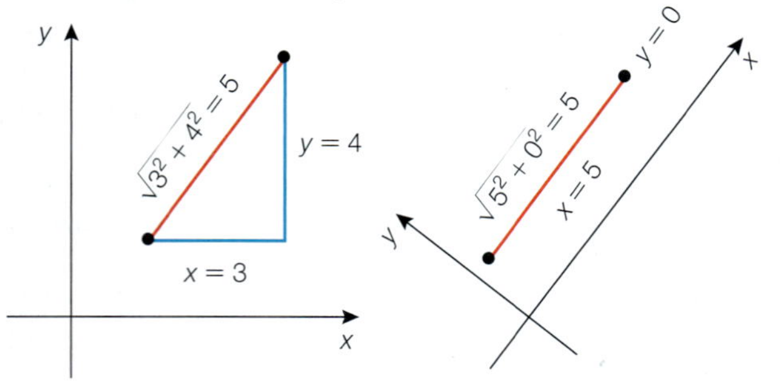


*. . . but the distance between them (red line) is the same either way.*

**FIGURE 1** The distance between two points in a plane is the same regardless of how we set up a coordinate system.

The same idea holds if we add a $z$-axis (perpendicular to the $x$-$y$ plane) to make a three-dimensional coordinate system. Different observers using different coordinate systems can disagree about the $x$, $y$, and $z$ separations, but they will always agree on the distance $\sqrt{x^2 + y^2 + z^2}$.

Spacetime has a fourth axis, which we will call the $t$-axis, for time. Extending our results from two and three dimensions, we might expect that different observers would always agree on a four-dimensional "distance" that we could write as $\sqrt{x^2 + y^2 + z^2 + t^2}$. However, this is not quite the case. Instead, it turns out that different observers will agree on the value of the quantity $\sqrt{x^2 + y^2 + z^2 - t^2}$, which is called the **interval**. (Technically, the interval formula should use $ct$ rather than $t$ so that all the terms have units of length.) That is, different observers can disagree about the values of $x$, $y$, $z$, and $t$ separating two events, but all will agree on the interval between the two events.

The *minus sign* that goes with the time dimension in the interval formula makes the geometry of spacetime surprisingly complex. For example, the three-dimensional distance between two points can be zero only if the two points are in the same place, but the interval between two events can be zero even if they are in different places in spacetime, as long as $x^2 + y^2 + z^2 = t^2$. For example, the interval is zero between any two events connected by a light path on a spacetime diagram. If you study general relativity further, you will see many more examples of how this strange geometry comes into play.

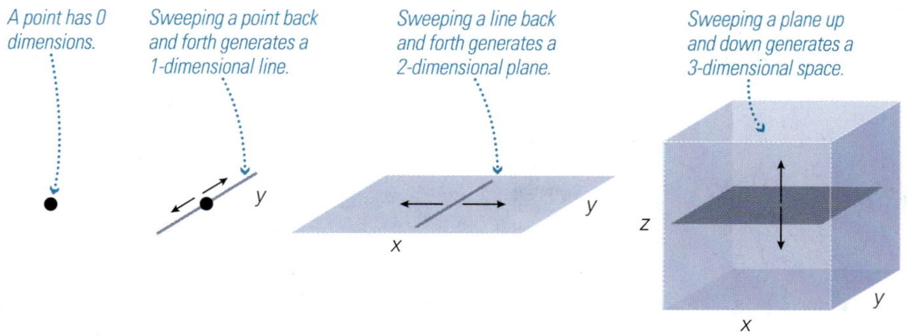


**FIGURE S3.6** An object's number of dimensions is the number of independent directions in which movement is possible within the object. It is zero for a point, one for a line, two for a plane, and three for space.

**Spacetime** Spacetime is a four-dimensional space in which the four directions of possible motion are length, width, depth, and *time.* Note that time is not "the" fourth dimension; it is simply one of the four. (However, time differs in an important way from the other three dimensions. See Mathematical Insight S3.1.)

We cannot picture all four dimensions of spacetime at once, but we can imagine what things would look like if we could. In addition to occupying the three spatial dimensions of spacetime that we ordinarily see, every object would be stretched out through time. Objects that we see as three-dimensional in our ordinary lives would appear as four-dimensional objects in spacetime. If we could see in four dimensions, we could look through time just as easily as we look to our left or right. If we looked at a person, we could see every event in that person's life. If we wondered what really happened during some historical event, we could simply look to find the answer.

**Think about it** Try to imagine how you would look in four dimensions. How would your body, stretched through time, appear? Imagine that you bumped into someone on the bus yesterday. What would this event look like in spacetime?

This spacetime view of objects provides a new way of understanding why different observers can disagree about measurements of time and distance. Because we can't visualize four dimensions, we'll use a three-dimensional analogy. Suppose you give the same book to many different people and ask each person to measure the book's dimensions. Everyone will get the same results, agreeing on the three-dimensional structure of the book. Now, suppose instead that you show each person only a two-dimensional picture of the book rather than the book itself. The pictures may look very different, even though they show the same book in all cases (**FIGURE S3.7**). If the people believed that the two-dimensional pictures reflected reality, they might argue endlessly about what the book really looks like.

In our ordinary lives, we perceive only three dimensions, and we assume that this perception reflects reality. But spacetime is actually four-dimensional. Just as different people can see different two-dimensional pictures of the same three-dimensional book, different observers can see different three-dimensional "pictures" of the same spacetime reality. These different "pictures" are the differing perceptions of time and space of observers in different reference frames. That is why different observers can get different results when they measure time, length, or mass, even though they are all looking at the same spacetime reality. In the words of a famous textbook on relativity,

> *Space is different for different observers.*
> *Time is different for different observers.*
> *Spacetime is the same for everyone.**

**Spacetime Diagrams** Suppose you drive your car along a straight road from home to work as shown in **FIGURE S3.8a**. At 8:00 a.m., you leave your house and accelerate to 60 km/hr. You maintain this speed until you come to a stop sign, where you decelerate to a stop. After the light turns green, you accelerate again to 60 km/hr, which you maintain until you slow to a stop when you reach work at 8:10. What does your trip look like in spacetime?

---

*From E. F. Taylor and J. A. Wheeler, *Spacetime Physics*, 2nd ed. (Freeman, 1992).

**a** A book has an unambiguous three-dimensional shape.

**b** Two-dimensional pictures of the book can look very different.

**FIGURE S3.7** Two-dimensional views of a three-dimensional object can appear different, even though the object has only a single real shape. In a similar way, observers in different reference frames may measure space and time differently (because they perceive only three dimensions at once) even though they are all observing the same four-dimensional spacetime reality.

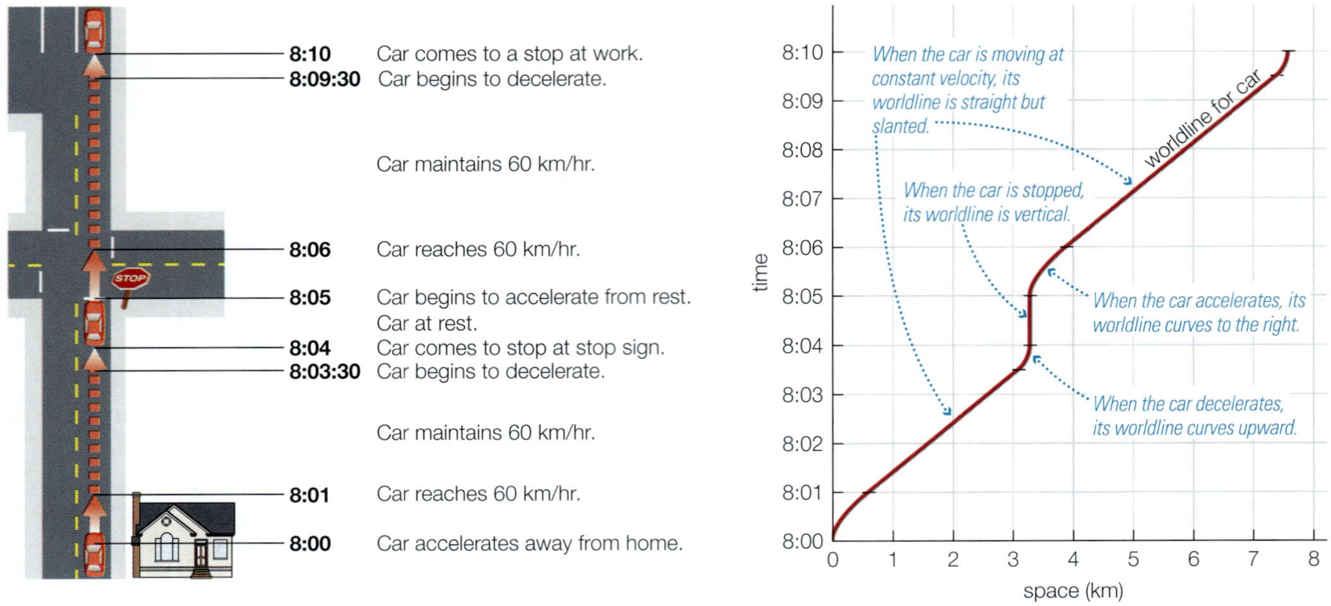


**a** This diagram shows the events that occur during a 10-minute car trip from home to work on a straight road.

**b** We make a spacetime diagram for the trip by putting space (in this case, the car's distance from home) on the horizontal axis and time on the vertical axis.

**FIGURE S3.8** A spacetime diagram allows us to represent one dimension of space and the dimension of time on a single graph.

If we could see all four dimensions of spacetime, we'd see all three dimensions of your car and your trip stretched out through the 10 minutes of time taken for your trip. We can't visualize all four dimensions at once, but in this case we have a special situation: Your trip progressed along only one dimension of space because you took a straight road. Therefore, we can represent your trip in spacetime by drawing a graph showing your path through one dimension of space on the horizontal axis and your path through time on the vertical axis (**FIGURE S3.8b**). This type of graph is called a **spacetime diagram**.

The car's path through four-dimensional spacetime is called its **worldline**. Any particular point along a worldline represents a particular **event**—that is, a specific place and time. For example, the lowest point on the worldline in Figure S3.8b represents the event of leaving your house, which occurs at a place 0 kilometers from home and a time of 8:00 a.m. As you study Figure S3.8b, you'll notice three important properties of worldlines:

1. The worldline of an object at rest is vertical (that is, parallel to the *time* axis). The object is going nowhere in space, but it still moves through time.
2. The worldline of an object moving at constant velocity is *straight* but slanted. The more slanted the worldline, the faster the object is moving.
3. The worldline of an accelerating object is *curved*. If the object's speed is increasing, its worldline curves toward the horizontal. If its speed is decreasing, its worldline gradually becomes more vertical.

In Figure S3.8b, we use units of minutes for time and kilometers for distance. In relativity, it is usually easier to work with spacetime diagrams in which we use units related to the speed of light, such as *seconds* for time and *light-seconds* for distance, so that light beams make 45° lines (because light travels 1 light-second of distance with each second of time). For example, suppose you are sitting still in your chair, so that your worldline is vertical (**FIGURE S3.9a**). If at some particular time you flash a laser beam pointed to your right, the worldline of the light goes diagonally to the right. If you flash the laser to the left a few seconds later, its worldline goes diagonally to the left. Worldlines for several other objects are shown in **FIGURE S3.9b**.

**Think about it** Explain why, in Figure S3.9, the worldlines of all the objects we see in our everyday lives would be nearly vertical.

We can use spacetime diagrams to clarify the relativity of time and space. Suppose you see Al moving past you in a spaceship at $0.9c$. **FIGURE S3.10a** shows the spacetime diagram from your point of view: You are at rest and therefore have a vertical worldline, while Al is moving and has a slanted worldline. Of course, Al claims that *you* are moving by him and therefore would draw the spacetime diagram shown in **FIGURE S3.10b**, in which his worldline is vertical and yours is slanted. Special relativity tells us that you would see Al's time running slowly, while he would see *your* time running slowly (along with effects on length and mass). We know there is no contradiction here, but simply a problem with our old common sense about space and time.

From a four-dimensional perspective, the problem is that the large angle between your worldline and Al's means that neither of you is looking at the other "straight-on" in spacetime. That is, you and Al are both looking at the same

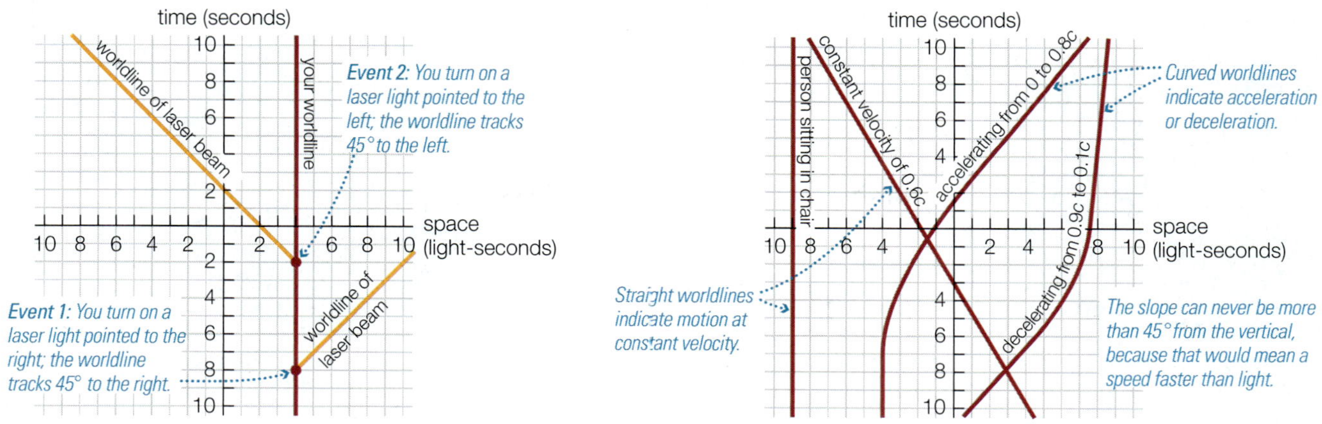


**a** Light follows 45° lines on a spacetime diagram that uses units of seconds for time and light-seconds for space.

**b** This spacetime diagram shows several sample worldlines. Objects at rest have vertical worldlines, objects moving at constant velocity have straight but slanted worldlines, and accelerating objects have curved worldlines.

**FIGURE S3.9** Spacetime diagrams marked with units of seconds for time and light-seconds for space.

four-dimensional reality but from different three-dimensional perspectives. It's not surprising that, like two people looking at each other cross-eyed, if you see Al's time running slowly, he sees the same thing when he looks at you.

## What is curved spacetime?

So far, we've been viewing spacetime diagrams drawn on the flat pages of this book. However, as we discussed in the beginning of this chapter, spacetime can be curved. What do we mean by curved spacetime?

It's easy to visualize the curvature of a *two*-dimensional surface, such as the surface of a bent sheet of paper or the surface of Earth, because we can see it curving *through* the third dimension of space. Note that these surfaces are two-dimensional despite their curvature, because they still allow only two independent directions of travel; for example, two independent directions of travel on Earth's surface are north-south (changes in latitude) and east-west (changes in longitude). Unfortunately, we cannot visualize the curvature of three-dimensional space in a similar way, let alone that of four-dimensional spacetime, because we'd need extra dimensions for them to curve through. Nevertheless, we can determine whether space or spacetime is curved by identifying the rules of geometry that apply. Because we cannot visualize curved space or spacetime, we'll use two-dimensional surfaces in an analogy.

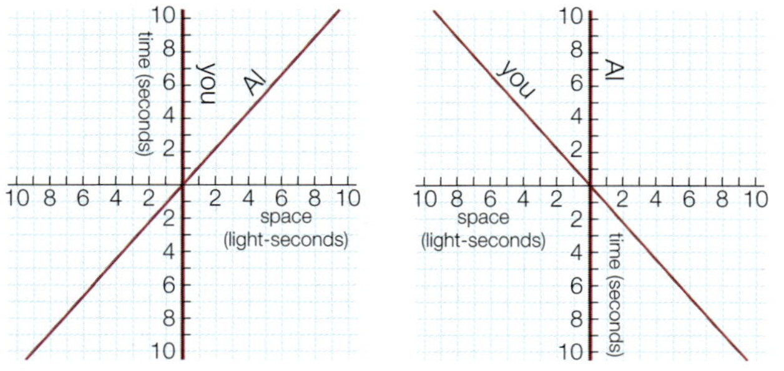


**a** The spacetime diagram from your point of view.

**b** The spacetime diagram from Al's point of view.

**FIGURE S3.10** Spacetime diagrams for the situation in which Al is moving by you at 0.9*c*.

**Three Basic Types of Geometry** Consider Earth's curved, two-dimensional surface. Because Earth's surface is curved everywhere, there really is no such thing as a "straight" line on Earth. For example, if you take a piece of string and lay it across a globe, it will inevitably curve around the surface. This leads us immediately to one fundamental difference between the rules of geometry on Earth's surface and the more familiar rules of geometry in a flat plane: While the shortest distance between two points in a flat plane is always a straight line, the same cannot be true on Earth's surface, because there is no such thing as a straight line.

What rule gives us the shortest distance on Earth's surface? If you experiment by measuring pieces of string stretched in different ways between two points on a globe, you'll find that the shortest and *straightest possible* path between two points on Earth's surface is a piece of a **great circle**—a circle whose center is at the center of Earth (**FIGURE S3.11a**). For example, the equator is a great circle, and any "line" of longitude is part of a great circle. Note that circles of latitude (besides the equator) are *not* great circles because their centers are *not* at the center of Earth. Therefore, if you are seeking the shortest and straightest route between two cities, you must follow a *great-circle route*. For example, Philadelphia and Beijing are both at about 40°N latitude, but the shortest route between them does *not* follow the circle of 40°N latitude. Instead, it follows a great-circle route that extends far to the north (**FIGURE S3.11b**).

**See it for yourself** Find a globe and locate New Orleans and Katmandu (Nepal). Explain why the shortest route between these two cities goes almost directly over the North Pole. Why do you think airplanes try to follow great-circle routes as closely as possible?

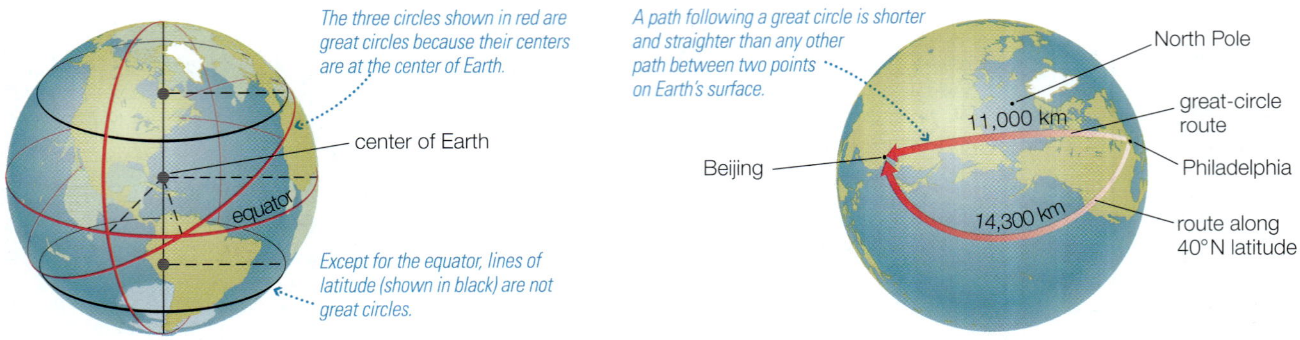


**a** A great circle is any circle on the surface of Earth that has its center at the center of Earth.

**b** The shortest and straightest possible path between two points on Earth is always a piece of a great circle.

**FIGURE S3.11** The straightest possible path between any two points on a sphere must be a segment of a great circle.

Other familiar rules of geometry in a flat plane also are different on Earth's curved surface. **FIGURE S3.12a** shows the straight-line rule and several other geometrical rules on a flat surface, and **FIGURE S3.12b** shows how these rules differ for a spherical surface like that of Earth; in the latter case, we must draw "lines" as portions of great circles, because those are the shortest and straightest possible paths. Notice, for example, that lines that are anywhere parallel on a flat plane stay parallel forever, while lines that start out parallel on a sphere eventually converge (just as lines of longitude start out parallel at Earth's equator but all converge at the North and South Poles). Similarly, the sum of the angles in a triangle is always 180° in a flat plane but is greater than 180° on the surface of a sphere, and the circumference of a circle is $2\pi r$ in a flat plane but is *less* than $2\pi r$ on a spherical surface.

Generalizing these ideas to more than two dimensions, we say that space, or spacetime, has a **flat geometry** if the rules of geometry for a flat plane hold. (Flat geometry is also known as *Euclidean geometry*, after the Greek mathematician Euclid [c. 325–270 B.C.].) For example, if the circumference of a circle in space really is $2\pi r$, then space has a flat geometry. However, if the circumference of a circle in space turns out to be less than $2\pi r$, we say that space has a **spherical geometry** because the rules are those that hold on the surface of a sphere.

Flat and spherical geometries are two of three general types of geometry. The third is called **saddle-shaped geometry** (also called *hyperbolic geometry*) because its rules are most easily visualized on a two-dimensional surface shaped like a saddle (**FIGURE S3.12c**). In this case, lines that start out parallel eventually diverge, the sum of the angles in a triangle is *less* than 180°, and the circumference of a circle is *greater* than $2\pi r$.

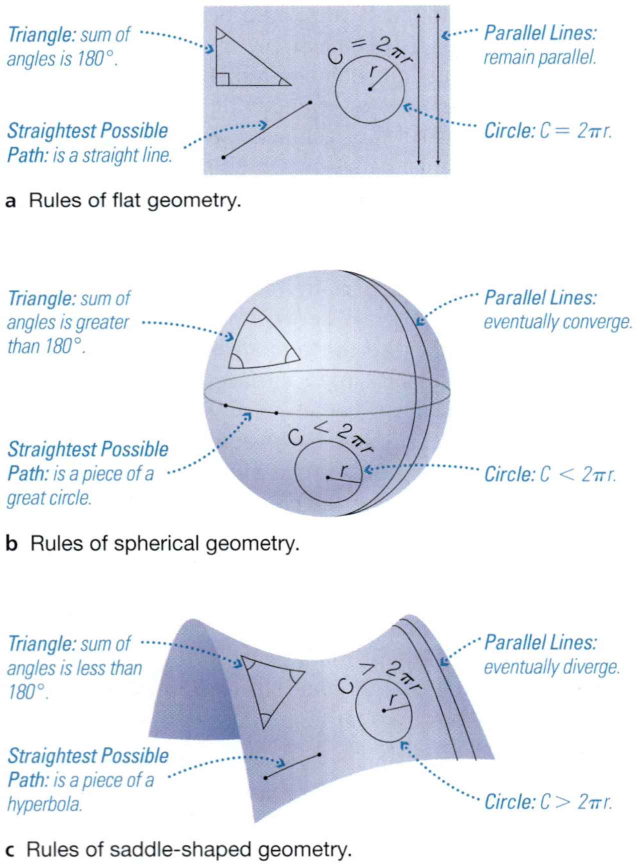


**FIGURE S3.12** These diagrams contrast three basic types of geometry.

**The Geometry of the Universe** The actual geometry of spacetime turns out to be a mixture of all three general types. Earth's surface again provides a good analogy. When we view only small portions of Earth's surface, some regions appear flat while others are curved with hills and valleys. However, when we expand our view to the entire Earth, it's clear that the overall geometry of Earth's surface is like that of the surface of a sphere. In a similar way, but in two more dimensions, the four-dimensional spacetime of our universe obeys different geometrical rules in different regions but presumably has some overall shape. This overall shape must be one of the three general types of geometry shown in Figure S3.12: flat, spherical, or saddle shaped. This fact explains how it is possible for the universe to have no center and no edges, an idea we first discussed in Chapter 1.

In geometry, a plane is infinite in extent, which means it has no center or edges. An idealized, saddle-shaped (hyperbolic) surface is also infinite in extent. Therefore, if the universe has either a flat or a saddle-shaped geometry overall, then spacetime is infinite and the universe has no center and no edges.

In contrast, if the overall geometry of the universe is spherical, then spacetime is finite, much like the surface of Earth. However, it would still have no center or edges. Just as you can sail or fly around Earth's surface endlessly, you could fly through the universe forever and never encounter an edge. And just as the *surface* of Earth has no center—New York is no more "central" than Beijing or any other place on Earth's surface—there would be no center to the universe. (Of course, the three-dimensional Earth *does* have a center, but this center is not part of the two-dimensional surface of Earth and therefore plays no role in our analogy.)

**"Straight" Lines in Curved Spacetime** The rules of geometry give us a way to determine the geometry of any localized region of spacetime. In particular, we can learn the geometry of spacetime by observing the paths of objects that are following the *straightest possible* path between two points in spacetime. For example, if the straightest possible path is truly straight, we know that spacetime is flat in that region. If the straightest possible path is curved, then the shape of the curve must be telling us the shape of spacetime. However, given that we can visualize neither the time part of spacetime nor the curvature of spacetime, how can we know whether an object is traveling on the straightest possible path?

Einstein used the equivalence principle to provide the answer. According to the equivalence principle, we can attribute a feeling of weight either to experiencing a force generated by acceleration or to being in a gravitational field. Similarly, any time we feel weightless, we may attribute it either to being in free-fall or to traveling at constant velocity far from any gravitational fields. Because traveling at constant velocity means traveling in a straight line, Einstein reasoned that objects experiencing weightlessness for *any* reason must be traveling in a "straight" line—that is, along a line that is the straightest possible path between two points in spacetime. In other words,

> *If you are floating freely, then your worldline is following the straightest possible path through spacetime. If you feel weight, then you are not on the straightest possible path.*

This fact provides us with a remarkable way to examine the geometry of spacetime. Recall that any orbit is a free-fall trajectory [**Section 4.1**]. The Space Station is always free-falling toward Earth, but its forward velocity always moves it ahead just enough to "miss" hitting the ground. Earth is constantly free-falling toward the Sun, but our planet's orbital speed keeps us going around and around. According to the equivalence principle, all orbits must therefore represent paths of objects that are following the straightest possible path through spacetime. That is, the shapes and speeds of orbits reveal the geometry of spacetime—a fact that leads us to an entirely new view of gravity.

**Think about it** Suppose you are standing on a scale in your bathroom. Is your worldline following the straightest possible path through spacetime? Explain.

## S3.3 A New View of Gravity

Newton's law of gravity claims that every mass exerts a gravitational attraction on every other mass, no matter how far away they are from each other. However, on close examination, this idea of "action at a distance" is rather mysterious. For example, how does Earth feel the Sun's attraction and know to orbit it? Newton himself was troubled by this idea. A few years after publishing his law of gravity in 1687, Newton wrote,

> *That one body may act upon another at a distance through a vacuum, . . . and force may be conveyed from one to another, is to me so great an absurdity, that I believe no man, who has . . . a competent faculty in thinking, can ever fall into it.**

Nevertheless, for more than 230 years after Newton published his gravitational law, no one found any better way to explain gravity's mysterious "action at a distance." Einstein changed all that when he realized that the equivalence principle allowed him to explain the action of gravity without requiring any long-distance force.

### What is gravity?

Einstein's general theory of relativity removes the idea of "action at a distance" by stating that Earth feels *no* force tugging on it in its orbit, and therefore follows the straightest possible path through spacetime. That is, the fact that Earth goes around the Sun tells us that spacetime itself is curved near the Sun. In other words,

> *What we perceive as gravity arises from the curvature of spacetime.*

**Rubber Sheet Analogy** We cannot actually picture the curvature of spacetime, but a two-dimensional analogy can help us understand the idea. We represent spacetime by an analogy with a stretched rubber sheet. To make the analogy work, we have to ignore any effects of friction on the rubber sheet, because there is no friction in space.

**FIGURE S3.13a** shows a flat rubber sheet representing spacetime in a region where it has a flat geometry. Notice that the radial distances between the circles shown on the sheet are the same, and all the circles have circumferences that follow the flat geometry formula of $2\pi r$. If you rolled a marble across this frictionless sheet, it would roll in a straight line at constant speed. This fact essentially illustrates Newton's first law of motion, which says that objects move at constant velocity when they are not affected by gravity or any other forces.

**FIGURE S3.13b** shows what happens to spacetime around the Sun. We represent the Sun with a heavy mass on the rubber sheet, which causes the sheet to curve and form a bowl-like depression. The circles that were evenly spaced on the flat sheet now become more widely separated (with circumferences increasingly less than $2\pi r$) near the bottom

*Letter from Newton, 1692–1693, as quoted in J. A. Wheeler, *A Journey into Gravity and Spacetime* (Scientific American Library, 1990, p. 2).

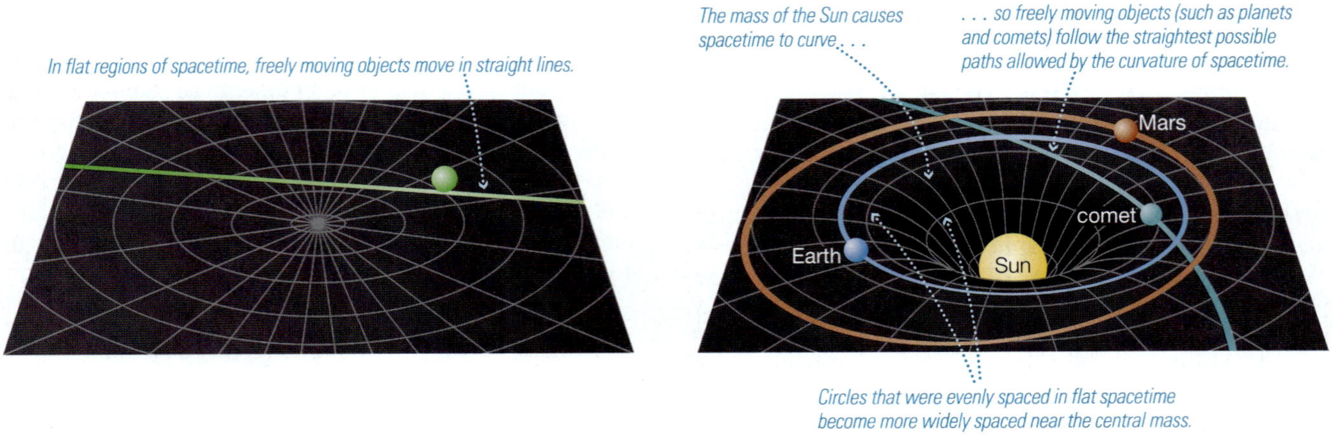


**a** On a flat rubber sheet, evenly spaced circles all have circumference $2\pi r$.

**b** The Sun curves spacetime much like the way a heavy weight curves a rubber sheet.

**FIGURE S3.13** According to general relativity, planets orbit the Sun for much the same reason that you can make a marble go around in a salad bowl: Each planet is going as straight as it can, but the curvature of spacetime causes its path through space to go round and round.

of the bowl, showing that gravity becomes stronger and the curvature of spacetime becomes greater as we approach the Sun's surface. (Notice that the curvature does *not* continue to increase with depth inside the Sun; in fact, the strength of gravity actually weakens near the Sun's center.) If you rolled marbles on this rubber sheet, they could not go in straight lines, because the sheet itself is curved. Instead, the marbles would follow the *straightest possible* paths given the curvature of the sheet. A particular marble's path would depend on the speed and direction with which you rolled it. You'd find that marbles rolled relatively slowly and close to the center would follow circular or elliptical "orbits" around the center of the bowl, while marbles rolled from farther away or at higher speeds could loop around the center on unbound parabolic or hyperbolic paths.

By analogy, general relativity tells us that, depending on their speed and direction, planets or other objects moving freely in space can follow circular, elliptical, or unbound parabolic or hyperbolic orbits—the same orbital shapes that Newton's universal law of gravitation allows [**Section 4.4**]. However, the explanation for these orbits is now quite different from that in Newton's view of gravity. Rather than orbiting because of a mysterious force exerted on them by the distant Sun, the planets orbit because they follow the *straightest possible paths* allowed by the shape of spacetime around them. The central mass of the Sun is not grabbing them, communicating with them, or doing anything else to influence their motion. Instead, it is simply dictating the shape of spacetime around it. In other words,

> *A mass like the Sun causes spacetime to curve, and the curvature of spacetime determines the paths of freely moving masses like the planets.*

**Weightlessness in Space** This idea gives us a new way to explain the weightlessness of astronauts in space. Just as the Sun curves spacetime into a "bowl shape" (but in four dimensions) that makes the straightest possible paths of the planets go round and round, Earth curves spacetime in a way that makes orbiting spacecraft go round and round. In other words, spacecraft orbit Earth because, as long as their engines are off and they are unaffected by atmospheric drag, circular or elliptical orbits are the straightest possible paths they can follow through spacetime in Earth's vicinity. That is, instead of having to invoke the idea of free-fall caused by a gravitational attraction to Earth, we can explain the weightlessness of astronauts in the Space Station simply by recognizing that they are following the straightest possible paths through spacetime.

The same idea holds true for any other orbital trajectory. For example, if we launched a human mission to Mars, we would need to give the spaceship escape velocity from Earth. In the rubber sheet analogy, this means launching it with enough speed so that it can escape from the bowl-shaped region around Earth, like a marble shot fast enough to roll out of the bowl and onto the flatter region far away from it. Except when their rockets were firing, the astronauts would still be weightless throughout the trip because they would be following the straightest possible path. Firing the engines, either to accelerate away from Earth or to decelerate near Mars, would make the spaceship deviate from the straightest possible path, so the astronauts would feel weight during those portions of their journey.

**Limitations of the Analogy** The rubber sheet analogy is useful for understanding how mass affects spacetime, but it also has limitations because it is a two-dimensional representation of a four-dimensional reality. In particular, the analogy has three important limitations that you should keep in mind whenever you use it:

- The rubber sheet is supposed to represent the universe, but it makes no sense to think of placing a mass like the Sun "upon" the universe. Instead, we should think of the masses as being *within* the rubber sheet.
- The rubber sheet allows us to picture only two of the three dimensions of space. For example, it allows us to show that different planets orbit at different distances from the Sun and that some have more highly elliptical orbits than others, but it does not allow us to show the fact that the planets do not all orbit the Sun in precisely the same plane.

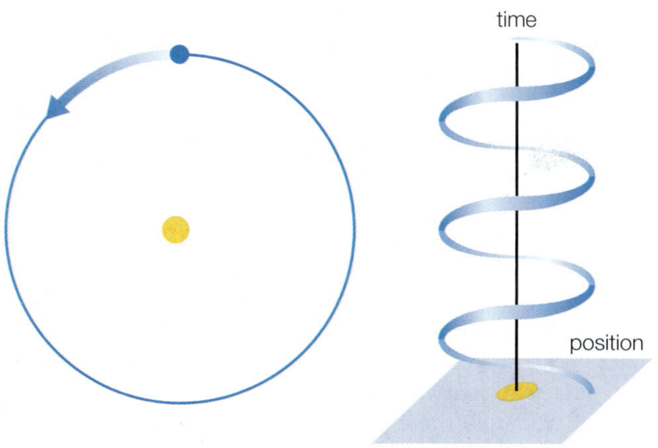


**a** If we ignore time, Earth appears to return to the same point with each orbit of the Sun.

**b** If we include a time axis, we see that Earth never returns to the same point in spacetime because it always moves forward in time.

**FIGURE S3.14** Earth's path through spacetime.

- The rubber sheet analogy does not show the *time* part of spacetime. Bound orbits on the sheet or in space appear to return to the same point with each circuit of the Sun. However, objects cannot return to the same point in spacetime, because they always move forward through time. For example, with each orbit of the Sun, Earth returns to the same place in space (relative to the Sun) but to a time that is a year later (**FIGURE S3.14**).

## What is a black hole?

Greater curvature of spacetime means stronger gravity, and the rubber sheet analogy suggests two basic ways to increase the strength of gravity. First, a larger mass causes greater curvature at any particular distance away from it. For example, the Sun curves spacetime more than any planet, and Earth curves spacetime more than the Moon. Note that this idea is consistent with Newton's law of gravity, which says that increasing the mass of an object increases the gravitational attraction at all distances.

The second way to increase the curvature of spacetime around an object is to leave its mass alone but increase its density by making it smaller in size. For example, suppose we could compress the Sun into a type of "dead" star called a *white dwarf* [**Section 18.1**]. Because its total mass would still be the same, there would be no effect on the curvature of spacetime far from the Sun. However, spacetime would be much more curved near the compressed Sun's surface, reflecting the fact that gravity is much stronger on the surface of a compressed white dwarf than on the surface of the Sun. Again, the idea that the surface gravity on an object of a particular mass grows stronger as the object shrinks in radius is consistent with Newton's law of gravity.

Now, imagine that we could continue to compress the Sun to smaller and smaller sizes. Far from the Sun, this compression would have no effect at all, because the same total mass would still be causing the curvature of spacetime. Near the Sun's surface, however, spacetime would become increasingly curved as we shrank the Sun in size. In fact, if we shrank the Sun enough, we could eventually curve spacetime so much that it would in essence become a bottomless pit—a hole in the observable universe. This is what we call a **black hole** (**FIGURE S3.15**). Note that Newton's view of gravity does not really have any analog to a black hole, because it does not envision the possibility of holes in the universe.

To summarize, a black hole is a place where spacetime is so curved that nothing that falls into it can ever escape. The boundary that marks the "point of no return" is called the **event horizon**, because events that occur within this boundary can have no influence on our observable universe. The idea of black holes is so bizarre that for decades after Einstein published his general theory of relativity, most scientists did not think they could really exist. However, we now have very strong evidence suggesting that black holes are in fact quite common. We'll discuss the nature of black holes in more detail in Chapter 18, and we'll discuss evidence for their existence in both Chapter 18 and Chapter 21.

## How does gravity affect time?

Given that gravity arises from the curvature of spacetime, you should not be surprised to learn that gravity affects time as well as space. We can learn about the effects of gravity on time by considering the effects of accelerated motion and then invoking the equivalence principle.

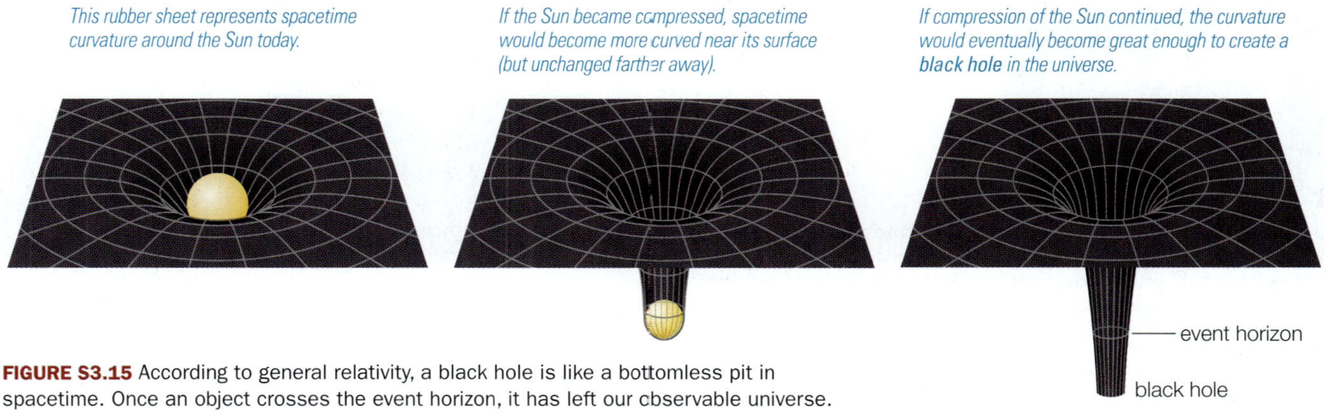


**FIGURE S3.15** According to general relativity, a black hole is like a bottomless pit in spacetime. Once an object crosses the event horizon, it has left our observable universe.

Imagine that you and Al are floating weightlessly at opposite ends of a spaceship. You both have watches that flash brightly each second, which you synchronized beforehand. Because you are both floating freely with no relative motion between you, you are both in the same reference frame. Therefore, you will see each other's watches flashing at the same rate.

Now suppose you fire the spaceship engines so that the spaceship begins to accelerate, with you at the front and Al at the back. When the ship begins accelerating, you and Al will no longer be weightless. The acceleration introduces an even more important change into the situation, which we can understand by imagining the view of someone floating weightlessly outside the spaceship. Remember that observers moving at different relative speeds are in different reference frames. When the spaceship is accelerating, its speed is constantly increasing relative to the outside observer, which means that both you and Al are constantly changing reference frames. Moreover, the flashes from your watches take a bit of time to travel the length of the spaceship. By the time a particular flash from Al's watch reaches you (or a flash from your watch reaches Al), both your reference frames are different from what they were at the time the flash was emitted.

Because you are in the *front* of the accelerating spaceship, your changing reference frames are always carrying you *away* from the point at which each of Al's flashes is emitted. Therefore, the light from each of his flashes will take a little *longer* to reach you than it would if the ship were not accelerating. As a result, instead of seeing Al's flashes 1 second apart, you'll see them coming a little *more* than 1 second apart. That is, you'll see Al's watch flashing more slowly than yours (**FIGURE S3.16a**). You will therefore conclude that time is running more slowly at the back end of the spaceship.

From Al's point of view at the *back* of the accelerating spaceship, his changing reference frames are always carrying him *toward* the point at which each of your flashes is emitted. Therefore, the light from each of your flashes will take a little *less* time to reach him than it would if the ship were not accelerating, so he'll see them coming a little *less* than 1 second apart. He will see your watch flashing *faster* than his and conclude that time is running *fast* at the front end of the spaceship. Note that you and Al agree: Time is running more slowly at the back end of the spaceship and faster at the front end. The greater the acceleration of the spaceship, the greater the difference in the rate at which time passes at the two ends of the spaceship.

Now we apply the equivalence principle, which tells us that we should get the same results for a spaceship at rest in a gravitational field as we do for a spaceship accelerating through space. This means that if the spaceship were at rest on a planet, time would also have to be running more slowly at the bottom of the spaceship than at the top (**FIGURE S3.16b**). That is, time must run more slowly at lower altitudes than at higher altitudes in a gravitational field. This effect is known as **gravitational time dilation**.

The stronger the gravity—and hence the greater the curvature of spacetime—the greater the effect of gravitational time dilation. On an object with relatively weak gravity, like Earth, time passes only slightly more slowly than it does in deep space; the difference is barely detectable even with our most sensitive atomic clocks. However, time runs noticeably more slowly on the surface of the Sun than on Earth, and more slowly on the surface of a white dwarf star than on the Sun. Perhaps you've already guessed that the extreme case is a black hole: To anyone watching from a distance, time comes to a stop at the event horizon. If you could observe clocks placed at varying distances from the black hole, you'd see that clocks nearer the event horizon run more slowly and clocks *at* the event horizon show time to be frozen.

**Think about it** Where would you age more slowly: on Earth or on the Moon? Would you expect the difference to be significant? Explain.

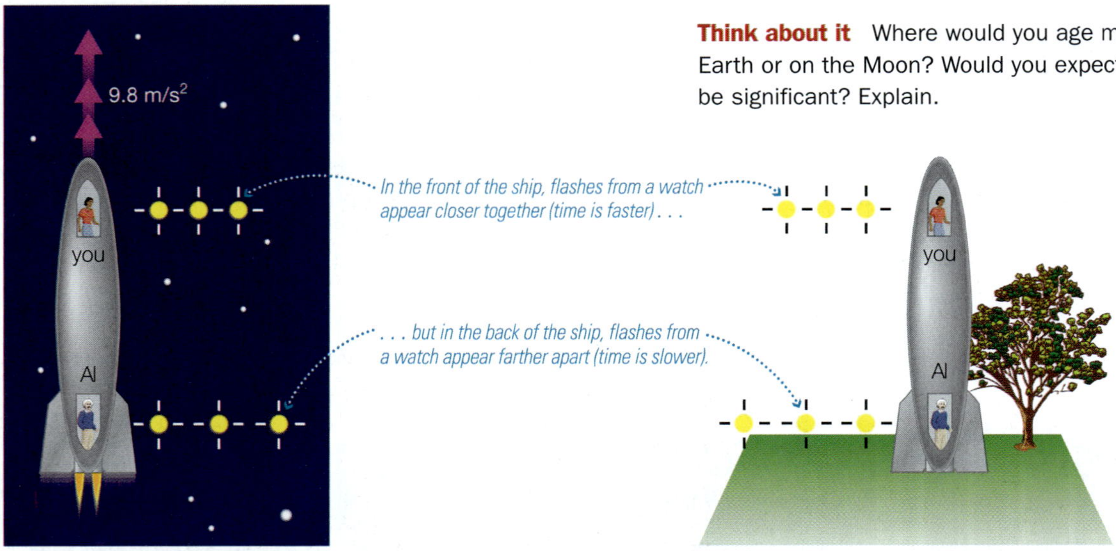


**a** In an accelerating spaceship (but not in one at constant velocity), time must run faster at the front end and more slowly at the back end. The yellow dots represent the flashes from the watches, and the spacing between the dots represents the time between the flashes.

**b** By the equivalence principle, time must also run more slowly at lower altitudes in a gravitational field.

**FIGURE S3.16** Gravity causes time to run more slowly at lower altitudes than at higher altitudes, an effect called *gravitational time dilation*. (Note that the effect occurs even in a uniform gravitational field; that is, it does not depend on the additional fact that gravity tends to weaken at higher altitudes.)

## S3.4 Testing General Relativity

Starting from the principle of equivalence, we've used logic and analogies to develop the ideas of general relativity. However, as always, we should not accept these logical conclusions unless they withstand observational and experimental tests.

### How do we test the predictions of the general theory of relativity?

Like the predictions of special relativity, those of general relativity have faced many tests and have passed in every case to date. Let's examine some of the most important tests of general relativity.

**Mercury's Peculiar Orbit** The first observational test passed by the theory of general relativity concerned the orbit of the planet Mercury. Newton's law of gravity predicts that Mercury's orbit should precess slowly around the Sun because of the gravitational influences of other planets (**FIGURE S3.17**). Careful observations of Mercury's orbit during the 19th century showed that it does indeed precess, but careful calculations made with Newton's law of gravity could not completely account for the observed precession. Although the discrepancy was small, further observations verified that it was real.

Einstein was aware of this discrepancy and, from the time he first thought of the equivalence principle in 1907, he hoped he would be able to explain it. When he finally succeeded in November 1915, he was so excited that he was unable to work for the next three days. He later called the moment of this success the high point of his scientific life.

In essence, Einstein showed that the discrepancy arose because Newton's law of gravity assumes that time is absolute and space is flat. In reality, time runs more slowly and space is more curved on the part of Mercury's orbit that is nearer the Sun. The equations of general relativity take this distortion of spacetime into account, providing a predicted orbit for Mercury that precisely matches its observed orbit.

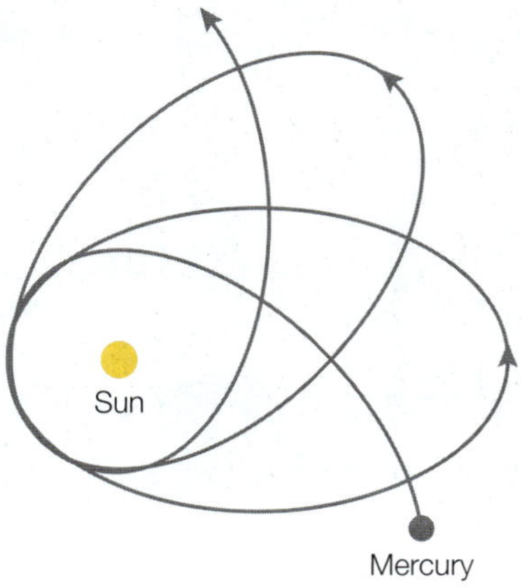


**FIGURE S3.17** Mercury's orbit slowly precesses around the Sun.


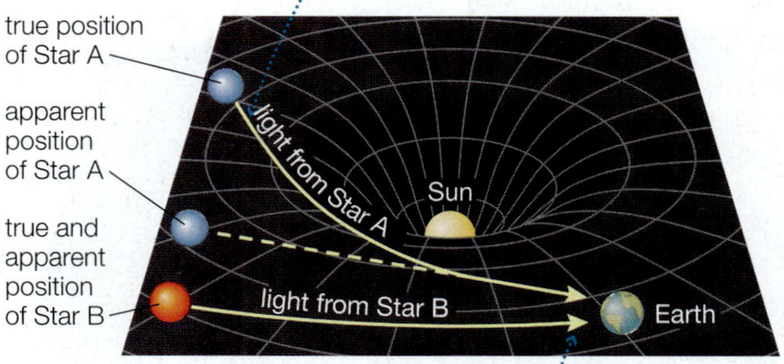


**FIGURE S3.18** When we see starlight that passes near the Sun during a total eclipse, the curvature of spacetime causes a shift in the star's apparent position.

**Think about it** Suppose the perihelion of Mercury's orbit were even closer to the Sun than it actually is. Would you expect the discrepancy between the actual orbit and the orbit predicted by Newton's laws to be greater than or less than it actually is? Explain.

**Gravitational Lensing** We can test Einstein's claim that space is curved by observing the trajectories of light rays moving through the universe. Because light always travels at the same speed, which means it never accelerates or decelerates, light must always follow the straightest possible path. If space itself is curved, then light paths will appear curved as well.

Suppose we could carefully measure the angular separation between two stars during the daytime just when the light from one of the stars passed near the Sun. The curvature of space near the Sun should cause the light beam passing closer to the Sun to curve more than the light beam from the other star (**FIGURE S3.18**). Therefore, the apparent angular separation of the two stars should be smaller than their true angular separation (which we would know from nighttime measurements). This effect was first observed in 1919, when astronomers traveled to South America and Africa to measure stellar positions near the Sun during a total solar eclipse. The results agreed with Einstein's predictions, and the media attention drawn by the eclipse expeditions brought Einstein worldwide fame.

Even more dramatic effects occur when a distant star or galaxy, as seen from Earth, lies directly behind another object with a strong gravitational field (**FIGURE S3.19**). The mass of the intervening object curves spacetime in its vicinity, altering the trajectories of light beams passing nearby. Different light paths can curve so much that they end up converging at Earth, grossly distorting the appearance of the star or galaxy. Depending on the precise four-dimensional geometry of spacetime between us and the observed star or galaxy, the image we see may be magnified or distorted into arcs, rings, or multiple images of the same object (**FIGURE S3.20**). This type of distortion is called **gravitational lensing**, analogous to the lensing of light when it is bent by a glass lens.

Gravitational lensing can be used to search for objects within the Milky Way that are too small and dim to be seen by their own light, such as dim stars, planets, or even black

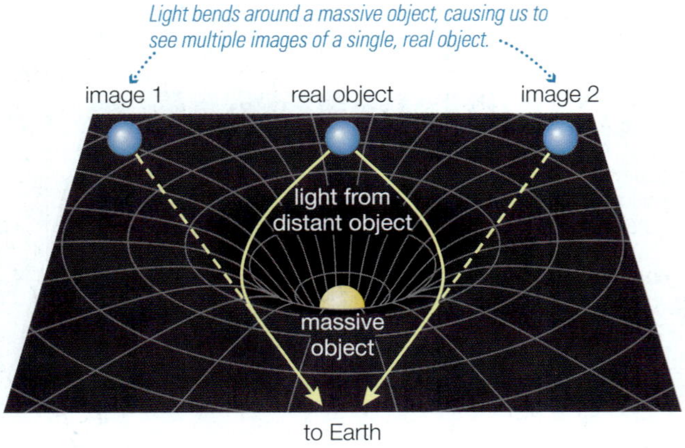


**FIGURE S3.19** Gravitational lensing can create distorted or multiple images of a distant object whose light passes by a massive object on its way to Earth.

holes. If such an object happens to drift across our line of sight to a brighter but more distant star, the small object's gravity will focus more of the star's light directly toward Earth. The distant star will therefore appear brighter than usual for days or weeks as the lensing object passes in front of it, in what is often called a *microlensing* event (**FIGURE S3.21**). We cannot see the object that causes the microlensing, but the duration and strength of the event reveal its mass. Over the past couple of decades, astronomers have undertaken several large observational programs to search for microlensing events. These studies have helped place constraints on the number of small, dim objects that inhabit our galaxy, and in some cases have even been used to identify extrasolar planets [**Section 13.1**]. We'll see more examples of gravitational lensing in Chapter 23.

**Gravitational Time Dilation** We can test the prediction of gravitational time dilation by comparing clocks located in places with different gravitational field strengths. Even in Earth's weak gravity, experiments demonstrate that clocks at low altitude tick more slowly than identical clocks at higher altitude; recent experiments at the National Institute of Standards and Technology successfully measured gravitational time dilation over a distance of only 1 meter. Although the effect would add up to only a few billionths of a second over a human lifetime, the differences agree precisely with the predictions of general relativity. In fact, the global positioning system (GPS) takes these effects into account; if it didn't, it would be far less accurate in locating positions on Earth.

Surprisingly, it's even easier to compare the passage of time on Earth with the passage of time on the surface of the Sun and other stars. Because stellar gases emit and absorb *spectral lines* with particular frequencies [**Section 5.4**], they serve as natural atomic clocks. Suppose that, in a laboratory on Earth, we find that a particular type of gas emits a spectral line with a frequency of 500 trillion cycles per second. If this same gas is present on the Sun, it will also emit a spectral line with a frequency of 500 trillion cycles per second.

However, general relativity claims that time should be running slightly more slowly on the Sun than on Earth. That is, 1 second on the Sun lasts *longer* than 1 second on Earth, or, equivalently, a second on Earth is shorter than a second on the Sun. During 1 second on Earth, we therefore see *fewer* than 500 trillion cycles from the gas on the Sun. Because lower frequency means longer, or redder, wavelengths, the spectral lines from the Sun ought to be *redshifted*. This redshift has nothing to do with the Doppler shifts that we see from moving objects [**Section 5.4**]. Instead, it is a **gravitational redshift**, caused by the fact that time runs more slowly in gravitational fields. Gravitational redshifts have been measured for spectral lines from the Sun and from many other stars. The results agree with the predictions of general relativity, confirming that time slows down in stronger gravitational fields.

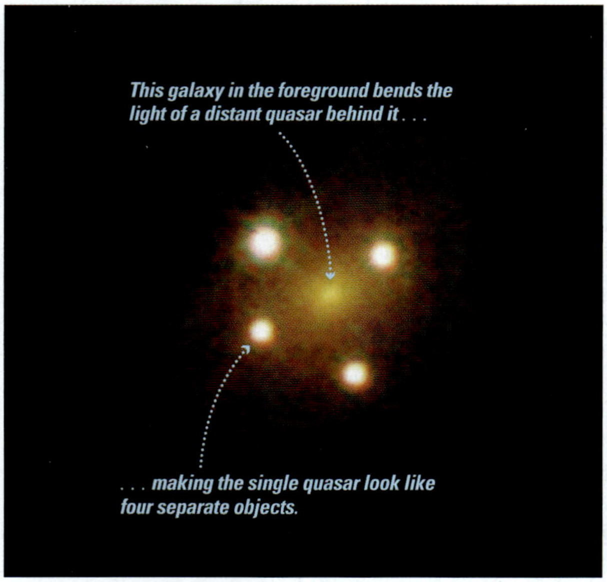


**a** This Hubble Space Telescope image shows an *Einstein cross*, in which light from a single distant quasar has been bent so that it reaches us along four different paths, creating four distinct images of the single object.

**b** The blue ring in this Hubble Space Telescope image, called an *Einstein ring*, is an example of the gravitational lensing that occurs when one galaxy lies almost directly behind another.

**FIGURE S3.20** Examples of gravitational lensing.

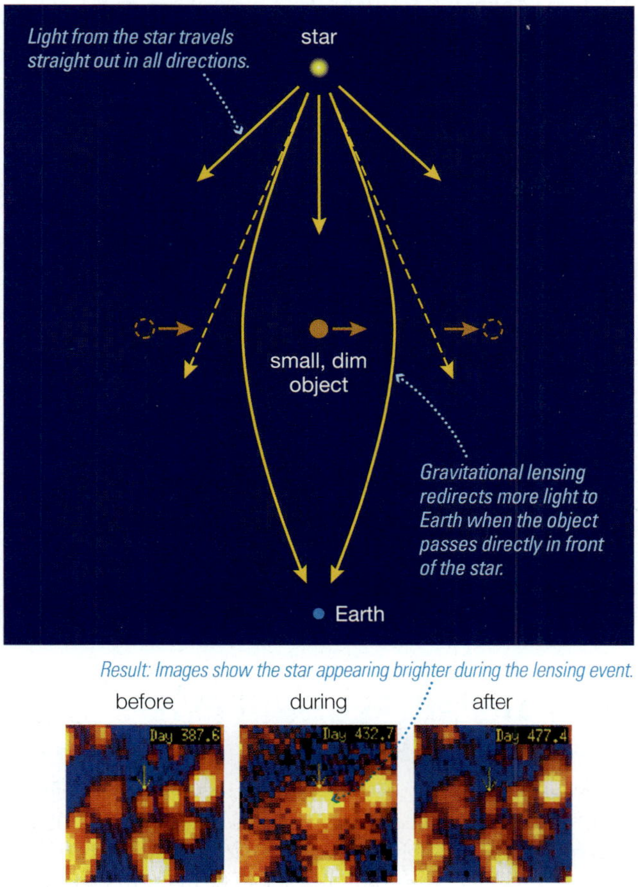


**FIGURE S3.21** When a small, dim object (such as a dim star, a planet, or a black hole) passes in front of a brighter, more distant star, gravitational lensing temporarily makes the star appear brighter. Because of the small objects involved, this type of gravitational lensing is often called *microlensing*.

**Gravitational Waves**

## What are gravitational waves?

According to general relativity, any major change in the curvature of space in one place should propagate outward through space like ripples on a pond. For example, the effect of a star suddenly imploding or exploding should be rather like the effect of dropping a rock into a pond, and two massive objects orbiting each other should generate ripples of curvature in space rather like those caused by a blade turning in water. Einstein called these ripples **gravitational waves**. General relativity predicts that, like light waves, gravitational waves should travel at the speed of light and carry energy but no mass. However, Einstein found that gravitational waves would carry so little energy that he doubted we'd ever be able to detect them or notice their effects. On this point, he proved overly pessimistic, both because the types of objects that emit the strongest gravitational waves had not yet been discovered and because he underestimated our technological advances.

To understand the evidence for gravitational waves, consider a binary system consisting of two of the highly compressed stellar corpses known as *neutron stars* [**Section 18.2**]. Neutron stars typically have masses greater than that of the Sun packed into radii of only about 10 kilometers, which means they can orbit each other much more closely than could two ordinary stars. General relativity predicts that a pair of rapidly orbiting neutron stars should generate gravitational waves that carry away orbital energy, causing the orbit to decay (**FIGURE S3.22**). The process is gradual at first, but the strength and frequency of the gravitational waves should increase as the orbit decays further, so that the process accelerates and the two objects ultimately merge together. The merger should produce a particularly strong pulse of gravitational waves. A similar process should occur with other closely orbiting massive objects, such as two orbiting black holes.

The first binary neutron star was discovered in 1974, and observations soon showed the orbits in this system to be decaying at precisely the rate predicted by general relativity (**FIGURE S3.23**). Other similar systems were subsequently discovered, and they also show orbital decay that matches what general relativity predicts under the assumption that energy is being carried away by gravitational waves. This match between observation and theory gave scientists great confidence that gravitational waves really exist, but the evidence was still indirect.

To achieve direct detection of gravitational waves, scientists built what is commonly called LIGO, short for the Laser Interferometer Gravitational-Wave Observatory (see

*The ripples represent gravitational waves carrying energy away from the system.*

*The loss of orbital energy means the orbit must decay . . .*

*. . . until the two objects ultimately merge, producing a particularly strong pulse of gravitational waves.*

**FIGURE S3.22** This sequence uses a rubber sheet analogy to illustrate the gravitational waves produced by a binary system with two neutron stars. The gravitational waves carry away orbital energy, causing the orbit to decay until the two neutron stars collide and merge together. The time required for this sequence to play out (from left to right) would typically be a few tens of millions of years.

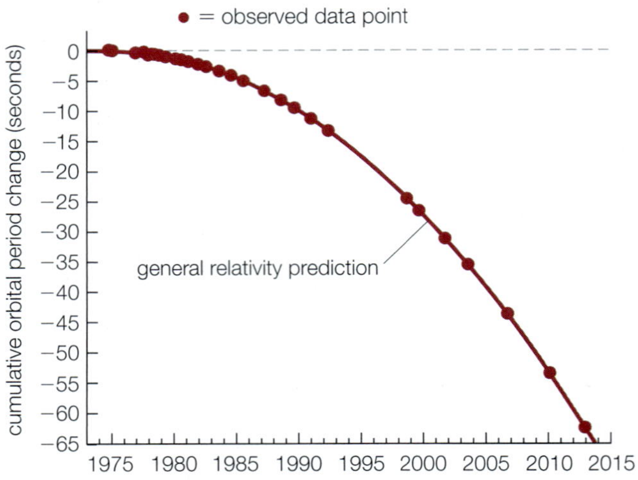


**FIGURE S3.23** This graph shows data since 1974 for the first binary neutron star to be discovered (known as the Hulse-Taylor system after its discoverers). The dots are observed data points, and the red curve shows what general relativity predicts under the assumption that the system emits gravitational waves. The close match provided the first strong evidence for the existence of gravitational waves. (Data courtesy of Joel Weisberg and Yuping Huang.)

Figure 6.31). General relativity tells us that the distortions of space carried by gravitational waves should alternately compress and expand objects as they pass by. LIGO searches for such compression and expansion with laser beams that travel back and forth along a pair of 4-kilometer-long arms arranged in an L-shape (**FIGURE S3.24**). The device is called an *interferometer,* because it uses interference patterns created by the two beams to measure changes in the light-travel times caused by changes in the arm lengths. LIGO is so sensitive that it can measure changes in the arm lengths that are far smaller than the size of a proton! To be confident that such small changes are not just a local artifact, scientists must record the same signal at more than one location, so LIGO began operations with two detectors (one in Washington and one in Louisiana).

LIGO first detected gravitational waves just weeks after it became fully operational in 2015 (see Figure 18.20). Based on the frequency and strength of the gravitational waves, scientists concluded that this pulse came from the merger of two black holes. Moreover, just as would be expected for a signal traveling at the speed of light, one LIGO detector recorded the signal slightly before the other. Several more detections soon followed, and in 2017 LIGO was joined by a third detector (called Virgo), located in Italy. Shortly thereafter, scientists announced the first detection of a neutron star merger of the type shown in Figure S3.22. Having three detectors allowed scientists to pinpoint the location of the neutron star merger closely enough that they could conduct follow-up observations in light across the spectrum from radio waves through gamma rays. These early successes have given astronomers great hope that gravitational waves will provide an important new way for us to study the universe.

**Think about it** How many detections of gravitational waves has LIGO recorded to date? Briefly describe at least three important results from gravitational wave astronomy.

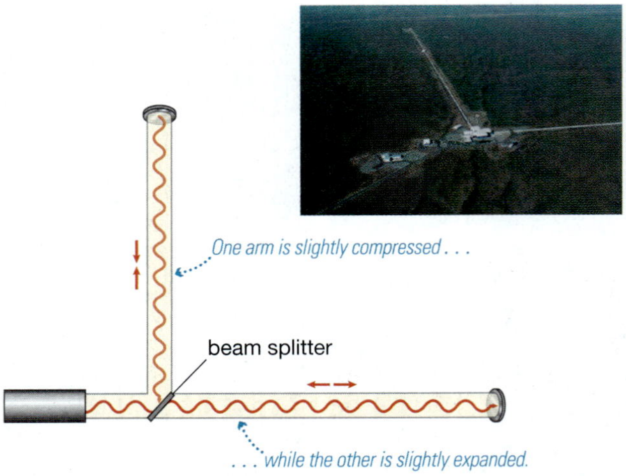


**FIGURE S3.24** This figure shows the arrangement of the LIGO detector arms. The timing of lasers is used to look for extremely tiny changes in the arm lengths as gravitational waves pass by. The inset shows the LIGO detector in Hanford, Washington (also see Figure 6.31).

## S3.5 Hyperspace, Wormholes, and Warp Drive

If you're a fan of *Star Trek, Star Wars,* or other science fiction, you're familiar with spaceships bounding about the galaxy with seemingly little regard for Einstein's prohibition on traveling faster than the speed of light. In fact, these stories do not necessarily have to violate the precepts of relativity as long as they exploit potential "loopholes" in the known laws of nature. General relativity *might* provide the necessary loopholes.

### Where does science end and science fiction begin?

Let's begin with an analogy. Suppose you want to take a trip between Brazil and Indonesia, which happen to lie on exact opposite sides of Earth's surface (**FIGURE S3.25**). Ordinarily, we are restricted to traveling along Earth's surface by car, boat, or plane, and the most direct trip would cover a distance of about 20,000 kilometers. However, suppose you could somehow drill a hole through the center of Earth

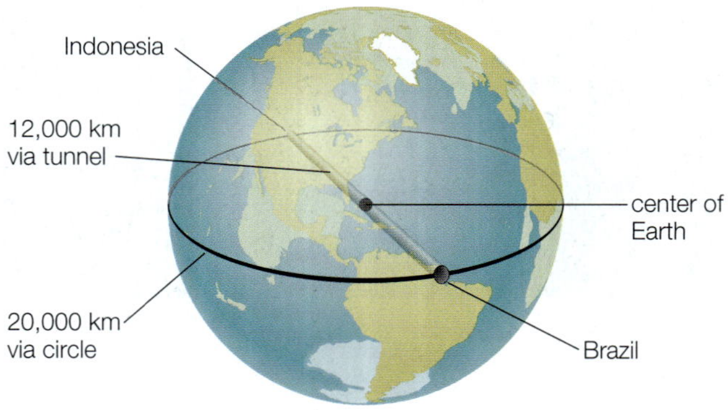


**FIGURE S3.25** If you could take a shortcut *through* Earth, the trip from Brazil to Indonesia would be shorter than is possible on Earth's surface.

and fly through the hole from Brazil to Indonesia. In that case, the trip would be only about 12,000 kilometers. You could thereby fly between Brazil and Indonesia in much less time than we would expect if we thought you could travel only along the surface.

Now consider a trip from Earth to the star Vega, about 25 light-years away. From the point of view of someone who stays home on Earth, this trip must take at least 25 years in each direction. However, suppose space happens to be curved in such a way that Earth and Vega are much closer together as viewed from a multidimensional *hyperspace*, just as Brazil and Indonesia are closer together if we go through Earth than if we must stay on its surface. Further, suppose there is a tunnel through hyperspace, often called

## SPECIAL TOPIC The Twin Paradox

Imagine twin sisters, one of whom stays home on Earth while the other takes a high-speed trip to a distant star and back. In Chapter S2, we said that the twin who takes the trip will age less than the twin who stays home. But shouldn't the traveling twin be allowed to claim that she remained stationary while *Earth* made a trip away from her and back? And in that case, shouldn't she conclude that the twin on Earth would age less? This question underlies the so-called *twin paradox*. It can be analyzed in several different ways, but we will use an approach that offers some insights into the nature of spacetime.

Suppose you and Al are floating weightlessly next to each other with synchronized watches. While you remain weightless, Al uses his engines to accelerate a short distance away from you, decelerate to a stop a bit farther away, and then turn around and return. From your point of view, Al's motion means that you'll see his watch ticking more slowly than yours. Therefore, upon his return, you expect to find that less time has passed for Al than for you. But how does Al view the situation?

The two of you can argue endlessly about who is really moving, but one fact is obvious to both of you: During the trip, you remained weightless while Al felt *weight* holding him to the spaceship floor. Al can account for his weight in either of two ways. First, he can agree with you that he was the one who accelerated, in which case he'll agree that his watch ran more slowly than yours because time runs more slowly in an accelerating spaceship. Alternatively, he can claim that he felt weight because his engines counteracted a gravitational field in which he was stationary while you fell freely. Note, however, that he'll still agree that his watch ran more slowly than yours, because time also runs slowly in gravitational fields. No matter how you or Al look at it, the result is the same: Less time passes for Al.

The left side of **FIGURE 1** shows a spacetime diagram for this experiment. You and Al both moved between the same two events in spacetime: the start and end points of Al's trip. However, your path between the two events is shorter than Al's. Because we have already concluded that less time passes for Al, we are led to a remarkable insight about the passage of time: *Between any two events in spacetime, more time passes on the shorter (and hence straighter) path.* The maximum amount of time you can record between two events in spacetime occurs if you follow the straightest possible path—that is, the path on which you are weightless.

The subtlety arises if Al chooses to claim that he is at rest and attributes his weight to gravity. In that case, he might be tempted to draw the spacetime diagram on the right in Figure 1, on which *he* appears to have the shorter path through spacetime. The rule that more time passes on shorter paths would then seem to imply that *your* watch should have recorded less time than Al's. But there is a problem with Al's diagram: If he wishes to assert that he felt gravity, he must also acknowledge that this gravity curves spacetime in his vicinity. This curvature means that his spacetime diagram must be distorted when he draws it on a flat piece of paper.

Al's problem is analogous to that of planning a flight from Philadelphia to Beijing with a flat map. As we saw in Figure S3.11b, the shortest and straightest path is a great-circle route that passes near the North Pole; this path is much shorter than a route following constant latitude. The fact that a flat map makes the constant latitude route appear shorter (**FIGURE 2**) does not change this reality. A flat map of Earth distorts the paths because the actual geometry of Earth's surface is spherical.

Just as distortions on a map do not change the actual distances between cities, the way we choose to draw a spacetime diagram does not alter the reality of spacetime. The solution to the twin paradox is that the twins do not have identical situations. The twin who turns around at the distant star has a more strongly curved worldline than the stay-at-home twin, which is why the traveler does indeed age less during the journey.

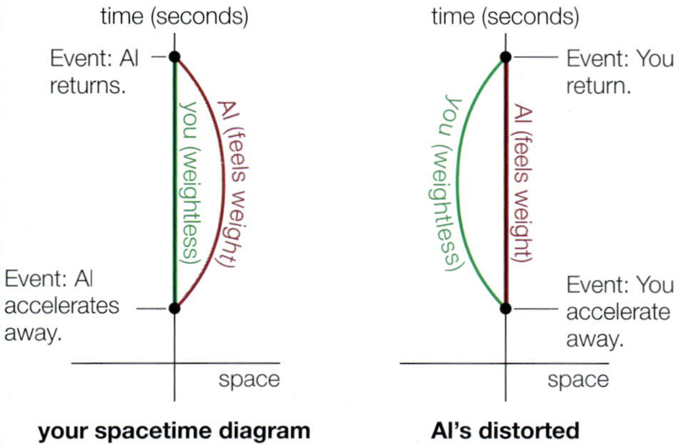


**FIGURE 1** A person floating weightlessly must be following the straightest possible path through spacetime (left). Because this is not the case in Al's diagram (right), his diagram must be distorted.

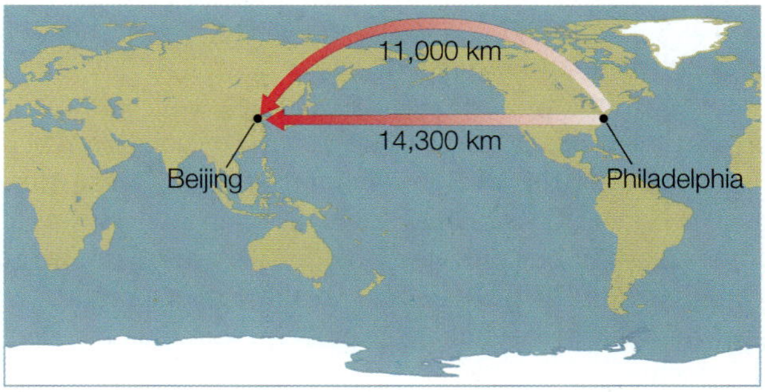


**FIGURE 2** This flat map shows the same two paths on Earth shown in Figure S3.11b. However, the distortion involved in making the map flat means that what looks like a straight line is not really as straight or as short as possible.

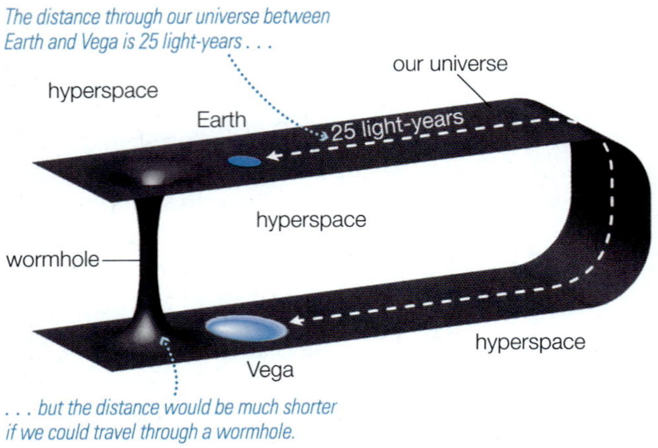


**FIGURE S3.26** The curved sheet represents our universe, in which a trip from Earth to Vega covers a distance of 25 light-years. This trip could be much shorter if a wormhole existed that created a shortcut through hyperspace. (Adapted from a drawing by Caltech physicist Kip Thorne.)

a **wormhole**, through which we can travel (**FIGURE S3.26**). If the tunnel is short—say, just a few kilometers in length—then a spaceship would need to travel only a few kilometers through the wormhole to go from Earth to Vega. The trip might then take only a few minutes in each direction! Relativity is not violated because the spaceship has not exceeded the speed of light. It has simply taken a shortcut through hyperspace.

If no wormhole is available, perhaps we might discover a way to "jump" through hyperspace and return to the universe anywhere we please. Such hyperspace jumps are the fictional devices used for space travel in the *Star Wars* movies. Alternatively, we might discover a way to warp spacetime to our own specifications, thereby allowing us to make widely separated points in space momentarily touch in hyperspace. This fictional device is the basic premise behind *warp drive* in the *Star Trek* series.

Do wormholes really exist and, if so, could we really travel through them? Is it possible that we might someday discover a way to jump into hyperspace or create a warp drive? Our current understanding of physics is insufficient to answer these questions definitively. For the time being, the known laws of physics do not prohibit any of these exotic forms of travel. These loopholes are therefore ideal for science fiction writers, because they might allow rapid travel among distant parts of the universe without violating the established laws of relativity.

However, many scientists believe we will eventually find that these exotic forms of travel are *not* possible. Their primary objection is that wormholes seem to make time travel possible. If you could jump through hyperspace to another place in our universe, couldn't you also jump back to another *time*? If you used a trip through hyperspace to travel into the past, could you prevent your parents from ever meeting?

The paradoxes we encounter when we think about time travel are severe and seem to have no resolution. Most scientists therefore believe that time travel will prove to be impossible, even though we don't yet know of any laws of physics that prohibit it. In the words of physicist Stephen Hawking (1942–2018), time travel should be prohibited "to keep the world safe for historians."

If time travel is not possible, it is much more difficult to see how shortcuts through hyperspace could be allowed. Nevertheless, neither time travel nor travel through hyperspace can yet be ruled out in the same way that we can rule out the possibility of accelerating to a speed greater than the speed of light. Until we learn otherwise, the world remains safe for science fiction writers who choose their fictional space travel techniques with care, avoiding any conflicts with relativity and other known laws of nature.

## S3.6 The Last Word

Having now spent two chapters discussing Einstein's special and general theories of relativity, we are ready to stand back and look at what these theories mean to our understanding of the universe.

### How has relativity changed our view of space and time?

In our daily lives, space and time seem separate and distinct, and for most of human history people simply assumed that this appearance must reflect reality. Thanks to Einstein, we now know otherwise: Space and time are intertwined in ways that earlier generations never even began to consider. As Hermann Minkowski said in 1908, "Henceforth space by itself, and time by itself, are doomed to fade away into mere shadows, and only a kind of union of the two will preserve an independent reality."

In the more than 100 years since Einstein first published his special theory of relativity, both the special and the general theory have been subjected to extensive and precise testing. So far, both theories have passed every test. While it remains possible that future tests will force modifications to the theories of relativity, it seems inconceivable that new discoveries could change their underlying message about the intertwined nature of spacetime, with its implications for our understanding of everything from the passage of time to black holes to the overall geometry of the universe.

The astonishing transformation that relativity brings to our view of space and time has also caused philosophers to debate its implications for concepts such as fate and free will. These debates are fascinating and well worth exploring, but remember that they do not affect the data from which the scientific theories are built. In other words, while there are many different viewpoints on the philosophical implications of relativity, only the scientific aspects of the theory are confirmed by data.

With that in mind, let's turn to Einstein himself for our last word on relativity. Here is what he said about a month before his death, on April 18, 1955:

> *Death signifies nothing. . . . the distinction between past, present, and future is only a stubbornly persistent illusion.**

*This quotation was found with the aid of Alice Calaprice, author of *The Quotable Einstein* (Princeton University Press, 1996).

# The BIG Picture PUTTING CHAPTER S3 INTO CONTEXT

*Just as the Copernican revolution overthrew the ancient belief in an Earth-centered universe, Einstein's revolution overthrew the common belief that space and time are distinct and absolute. We have explored Einstein's revolution in some detail in the past two chapters. Keep in mind the following "big picture" ideas:*

- We live in four-dimensional spacetime. Disagreements among different observers about measurements of time and space occur because the different observers are looking at a single four-dimensional reality from different three-dimensional perspectives.
- Gravity arises from the curvature of spacetime. Once we recognize this fact, the orbits of planets, moons, and all other objects can be understood as natural consequences of the curvature, rather than results of a mysterious "force" acting over great distances.
- Although the predictions of relativity may seem quite bizarre, they have been well verified by many observations and experiments.
- Some questions remain well beyond our current understanding. In particular, we do not yet know whether travel through hyperspace might be possible, in which case some of the imaginative ideas of science fiction might become reality.

**MY COSMIC PERSPECTIVE** Relativity tells us there is a sort of permanence to spacetime, in that every event—including everything you do at every moment of your life—is a part of spacetime. In that sense, you are leaving an indelible mark on the universe. Perhaps if everyone understood that, we'd all be more careful to make sure the mark we leave is one that we are proud of.

# Summary of Key Concepts

## S3.1 Einstein's Second Revolution

- **What are the major ideas of general relativity?** General relativity tells us that gravity arises from the curvature of **spacetime** and that the curvature arises from the presence of masses. This idea leads us to a view of gravity in which time runs more slowly in gravitational fields, black holes can exist in spacetime, and the universe has no center or edges. It also predicts the existence of gravitational waves propagating through space.
- **What is the fundamental assumption of general relativity?** The starting point for general relativity is the **equivalence principle**, which states that the effects of gravity are exactly equivalent to the effects of acceleration.

## S3.2 Understanding Spacetime

- **What is spacetime?** Spacetime is the four-dimensional combination of space and time that forms the "fabric" of our universe.
- **What is curved spacetime?** Spacetime can be curved much like a rubber sheet but in more dimensions. We can recognize spacetime curvature from the rules of geometry. The three possible geometries are a **flat geometry**, in which the ordinary laws of flat (Euclidean) geometry apply; a **spherical geometry**, in which lines that start out parallel tend to converge; and a **saddle-shaped geometry**, in which lines that start out parallel tend to diverge.

## S3.3 A New View of Gravity

- **What is gravity?** Gravity arises from the curvature of spacetime. Mass causes spacetime to curve, and the curvature of spacetime determines the paths of freely moving masses.

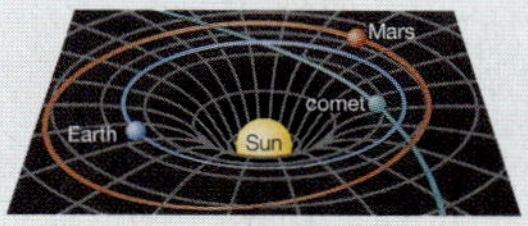


- **What is a black hole?** A **black hole** is a place where spacetime is curved so much that it essentially forms a bottomless pit, making it like a hole in spacetime.

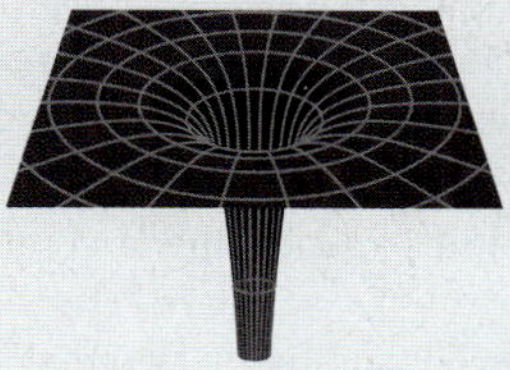

- **How does gravity affect time?** Time runs more slowly in places where gravity is stronger, an effect called **gravitational time dilation**.

## S3.4 Testing General Relativity

- **How do we test the predictions of the general theory of relativity?** Observations of the precession of Mercury's orbit match the precession predicted by Einstein's theory. Observations of stars during eclipses and photos of **gravitational lensing** provide spectacular confirmation of the idea that light can travel curved paths through space. **Gravitational redshifts** observed in the light of objects with strong gravity confirm the slowing of time predicted by general relativity, a prediction that has also been confirmed with clocks at different altitudes on Earth.

- **What are gravitational waves?** General relativity predicts, and observations confirm, that accelerating masses produce **gravitational waves** that travel at the speed of light. Indirect evidence for gravitational waves comes from observing the orbital decay of binary neutron stars. Direct detection has come from the instrument called LIGO, which has detected pulses of gravitational waves produced when orbiting neutron stars or black holes ultimately merge.

## S3.5 Hyperspace, Wormholes, and Warp Drive

- **Where does science end and science fiction begin?** No known physical laws prevent hyperspace, **wormholes**, or warp drive from offering "loopholes" that could allow us to get from one place to another in less time than we could by traveling through ordinary space. However, if any one of them prove to be real, then cause and effect might not be absolute, a proposition troubling to many scientists.

## S3.6 The Last Word

- **How has relativity changed our view of space and time?** Prior to Einstein, space and time were viewed as separate and distinct. We now know that they are deeply intertwined as *spacetime* and that understanding spacetime is crucial to understanding many aspects of astronomy, including black holes and the overall geometry of the universe.

# Exercises and Problems

*For instructor-assigned homework and other learning materials, go to www.MasteringAstronomy.com.*

## Chapter Review Questions

### Short-Answer Questions Based on the Reading

1. What do we mean by the *straightest possible path* on Earth's surface?
2. List five major ideas that come directly from the general theory of relativity.
3. What is the *equivalence principle*? Give an example that clarifies its meaning.
4. What do we mean by *dimension*? Describe a *point*, a *line*, a *plane*, a *three-dimensional space*, and a *four-dimensional space*. What does *hyperspace* mean?
5. Explain the meaning of the statement "Space is different for different observers. Time is different for different observers. Spacetime is the same for everyone."
6. What is a *spacetime diagram*? Define *worldline* and *event*. What is the significance of whether a worldline is vertical, slanted, or curved?
7. How do rules of geometry differ depending on whether the geometry is *flat*, *spherical*, or *saddle shaped*?
8. Explain how the idea of spacetime geometry means that the universe has no center and no edges.
9. How can you tell whether you are following the straightest possible path through spacetime?
10. According to general relativity, what is gravity and why does Earth orbit the Sun? Describe both the use and the limitations of the rubber sheet analogy for picturing gravity.
11. What is a *black hole*? What do we mean by the *event horizon* of a black hole?
12. What is *gravitational time dilation*? What determines how much time is slowed in a gravitational field?
13. Briefly describe several observational tests that support general relativity, including Mercury's orbit, examples of *gravitational lensing*, and measurements of *gravitational redshift*.
14. What are *gravitational waves?* Describe both indirect and direct evidence that they exist.
15. What is the current evidence regarding the possibility of travel through hyperspace or wormholes or by means of warp drive?

### Does It Make Sense?

*Decide whether the statement makes sense (or is clearly true) or does not make sense (or is clearly false). Explain clearly; not all of these have definitive answers, so your explanation is more important than your chosen answer.*

16. The equivalence principle tells us that experiments performed on a spaceship accelerating through space at 1g will give the same results as experiments performed on Earth.
17. The equivalence principle tells us that there's no difference at all between a planet and a human-made spaceship.
18. A person moving by you at high speed will measure time and space differently than you will, but you will both agree that there is just a single reality in spacetime.
19. With a sufficiently powerful telescope, we could search for black holes by looking for funnel-shaped objects in space.
20. Time runs slightly more slowly on the surface of the Sun than it does here on Earth.
21. Telescopes sometimes see multiple images of a single object, just as we would expect from the general theory of relativity.
22. When I walk in circles, I am causing curvature of spacetime.
23. Although special relativity deals only with relativity of motion, general relativity tells us that *everything* is relative.
24. The shortest distance between two points is always a straight line.
25. General relativity tells us that it is impossible to travel through hyperspace or to use anything like *Star Trek*'s "warp drive."

### Quick Quiz

*Choose the best answer to each of the following. Explain your reasoning with one or more complete sentences. For additional practice, try the Chapter S3 Reading and Concept Quizzes in the Study Area at www.MasteringAstronomy.com.*

26. Spacetime is (a) another name for gravity. (b) the combination of time and the three dimensions of space. (c) a curved rubber sheet.
27. The equivalence principle tells us that effects of these two things are indistinguishable: (a) space and time. (b) gravity and acceleration. (c) gravity and curvature of spacetime.

28. The surface of Earth has how many dimensions? (a) one (b) two (c) three
29. If two lines begin parallel but later diverge, the geometry is (a) flat. (b) spherical. (c) saddle shaped.
30. You know that you are following the straightest possible path through spacetime if (a) you are standing still. (b) you are moving directly from one place to another. (c) you are weightless.
31. According to general relativity, Earth goes around the Sun rather than flying straight off into space because (a) gravity creates an invisible bond holding Earth and the Sun together. (b) Earth is going as straight as possible, but the shape of spacetime makes this path go round and round. (c) in its own reference frame, Earth can consider itself to be stationary.
32. On the surface of which of the following objects would time run most slowly? (a) the Sun (b) an object with the same mass as the Sun but twice the radius (c) an object with the same mass as the Sun but only half the radius
33. Gravitational lensing occurs because (a) gravity causes light to slow down. (b) gravity curves space and light always follows the straightest possible path through space. (c) the effects of gravity are indistinguishable from the effects of glass lenses.
34. The first evidence that gravitational waves really exist came from (a) seeing these waves with x-ray telescopes in space. (b) observing the orbital decay of two orbiting neutron stars. (c) Einstein's own calculations with his general theory of relativity.
35. If wormholes are real, which of the following best describes what one is? (a) a place where it is possible to travel faster than light (b) a shortcut between distant parts of the universe (c) a black hole with a wormlike shape

## Inclusive Astronomy

*Use these questions to reflect on participation in science.*

36. *Your Permanent Record.* Relativity is often thought of as being "just for scientists." However, many philosophers have concluded that it has important implications for all of us. For example, consider the fact that, according to our description of spacetime, you exist in spacetime as a "solid" object stretching through time. This can be taken to imply that you cannot erase anything you've ever said or done from spacetime; in a sense, everything you ever do becomes part of some kind of four-dimensional "permanent record." Should this idea affect the way we behave as human beings? If so, how do you think human interactions might change if everyone learned about relativity? Can you think of any other ways in which knowing about relativity might change your life and the lives of others around you?
37. *Group Discussion: International Gravitational Wave Collaboration.* The LIGO detectors in the United States were the first to detect gravitational waves, but other nations have since joined in the effort.

    **a.** Working independently, find out which nations currently have or are building gravitational wave detectors, where those detectors are located, and how much they cost to build. Also, learn what you can about how scientists use multiple detections of a gravitational-wave signal to narrow down where it came from, and how separating detectors widely on Earth can aid in this process.

    **b.** Gather in groups to share what you've learned. Then make a list of reasons several nations working collaboratively can discover much more about gravitational waves than just one nation working alone.

    **c.** The international collaboration in gravitational wave detection raises issues of how the data should be shared for the purposes of publishing scientific papers. Consider three alternative sharing models: (1) Data are available only to the scientists directly involved in the data collection; (2) data are available to all scientists from the nations that paid for the detectors; (3) data are immediately made publicly available to any scientist from any nation. Make a list of the pros and cons of each approach, and try to reach a group consensus on the best way to share the data.

## The Process of Science

*These questions may be answered individually in short-essay form or discussed in groups, except where identified as group-only.*

38. *Hallmarks of Science.* Chapter 3 discusses both the idealized scientific method and a set of hallmarks by which you can distinguish science from nonscience. Did Einstein follow the idealized scientific method in his work on general relativity? Does general relativity satisfy the hallmarks of science? Explain.
39. *Why Keep Testing?* General relativity has been extensively tested in many different ways and has passed every test with flying colors. Yet scientists continue to test it with sensitive measurements and attempts to detect gravitational waves. Why do scientists keep testing a theory that works so well? What would happen if it failed some future test? Explain.
40. *Group Activity: Einstein's Quote.* Discuss the quote from Einstein at the very end of the chapter, and debate whether the distinction among past, present, and future is real or an illusion. Note: You may wish to do this activity using the four roles described in Chapter 1, Exercise 39.

## Investigate Further

### Short-Answer/Essay Questions

41. *Northward Journey.* Suppose two people start out traveling due north from the same latitude, one near Denver and one near San Francisco. Are their paths parallel when they start? Will their paths stay parallel, or will they meet? If they will meet, where will they meet? Summarize your answers with a one- or two-paragraph description of the people's journey, and explain what the results tell us about the geometry of Earth.
42. *Alternative Geometries.* Find three everyday objects that obey the rules of each of the three basic types of geometry—*flat, spherical,* and *saddle shaped.* Describe each object and explain how you know which geometry it has.
43. *Funnels in Space?* Many people have seen rubber sheet diagrams that show black holes looking like funnels, as in Figure S3.15, and therefore they assume that black holes really are funnel shaped. Imagine that you are talking to a person who believes this misconception to be true. Write one or two paragraphs in words the person would understand, explaining how the misconception arose and what a black hole would really look like if we could see one. (For example, what shape would it be?)
44. *Black Hole Sun.* Suppose the Sun were magically replaced by a black hole of precisely the same mass.

    **a.** How would the curvature of spacetime change in the region where the Sun used to be located?

    **b.** How would the curvature of spacetime change in the region of Earth's orbit?

    **c.** What effect would the change have on Earth's orbit?

45. *Galileo and the Equivalence Principle.* Galileo demonstrated that all objects near Earth's surface should fall with the same acceleration, regardless of their mass. According to general relativity, why shouldn't the mass of a falling object affect its rate of fall? Explain in one or two paragraphs.
46. *Movie Science Fiction.* Choose a popular science fiction movie that depicts interstellar travel and study it closely as you watch it. What aspects of the movie are consistent with relativity and other laws of physics? What aspects of the movie violate the laws of relativity or other laws of physics? Write a critical review of the movie and its treatment of science.
47. *Person of the Century.* At the end of the 20th century, *Time* magazine chose Einstein as the "Person of the Century." Find out why, and write a short essay in which you either defend or oppose the choice.
48. *Relativity and Fate.* Read the novel *Slaughterhouse Five* by Kurt Vonnegut. How does he connect relativity with ideas of fate? Do you agree with his assessment of fate? Why or why not?
49. *Research: The Eötvös Experiment.* Galileo's result that all objects fall to Earth with the same acceleration (neglecting air resistance) is very important to general relativity—if it were not true, general relativity would be in serious trouble. Describe the experiments of Baron Roland von Eötvös, who tested Galileo's conclusions in the late 19th century. How did the results of these experiments influence Einstein as he worked on general relativity? Write a short summary.
50. *Research: Wormholes.* Some scientists have thought seriously about wormholes and their consequences. Find and read a popular article or book about wormholes. Write a short summary of your reading, and discuss your opinion of the implications of wormholes raised in it.

## Quantitative Problems

*Be sure to show all calculations clearly and state your final answers in complete sentences.*

51. *Worldlines at Low Speed.* Make a spacetime diagram and draw a worldline for each of the following situations. Explain your drawings.
    **a.** a person sitting still in a chair
    **b.** a person driving by at a constant velocity of 50 km/hr
    **c.** a person driving by at a constant velocity of 100 km/hr
    **d.** a person accelerating from rest to a speed of 50 km/hr
    **e.** a person decelerating from 50 km/hr to a stop
52. *Worldlines at High Speed.* Make a spacetime diagram on which the time axis is marked in seconds and the space axis is marked in light-seconds. Assume you are floating weightlessly and therefore consider yourself at rest.
    **a.** Draw your own worldline.
    **b.** Draw a worldline for Sebastian, who is moving to your right at $0.5c$.
    **c.** Draw a worldline for Michaela, who is moving to your left at $0.7c$.
    **d.** Draw a worldline for Angela, who is traveling away from you at a speed of 100,000 km/s.
53. *Highly Sloped Worldline.* Make a spacetime diagram on which the time axis is marked in seconds and the space axis is marked in light-seconds. Draw a worldline with a slope of 30° (from the horizontal). At what speed would an object have to be traveling to have this worldline? Can any object have this worldline? Explain.
54. *Triangle on Earth.* Draw a simple sketch of Earth (a plain sphere will do), with a triangle on it that connects the following three points: (1) the equator at longitude 0°, (2) the equator at longitude 90°W, and (3) the North Pole. What is the sum of the angles in this triangle? Explain how you know the angles, and what the sum tells you about the geometry of Earth's surface.
55. *Long Trips at Constant Acceleration: Earth Time.* Suppose you stay on Earth and watch a spaceship leave on a long trip at a constant acceleration of $1g$.
    **a.** At an acceleration of $1g$, approximately how long will it take before you see the spaceship traveling away from Earth at *half* the speed of light? Explain. (Use $g = 9.8\ \text{m/s}^2$.)
    **b.** Describe how you will see its speed change as it continues to accelerate. Will it keep gaining speed at a rate of 9.8 m/s each second? Why or why not?
    **c.** Suppose the ship travels to a star that is 500 light-years away. From your perspective on Earth, *approximately* how long will this trip take? Explain.
56. *Long Trips at Constant Acceleration: Spaceship Time.* Consider again the spaceship from Problem 55 on a long trip with a constant acceleration of $1g$. As long as the ship is gone from Earth for many years, the amount of time that passes on the spaceship during the trip turns out to be approximately

$$t_{\text{ship}} = \frac{2c}{g}\ln\left(\frac{g \times D}{c^2}\right)$$

where $D$ is the distance to the destination and ln stands for the natural logarithm (which you can calculate with the "ln" key on most scientific calculators). If $D$ is in meters, $g = 9.8\ \text{m/s}^2$, and $c = 3 \times 10^8\ \text{m/s}$, the answer will be in units of seconds. Use the formula to determine how much time will pass on the ship during its trip to a star that is 500 light-years away. Compare this to the amount of time that will pass on Earth. (*Hint:* Be sure you convert the distance from light-years to meters and your answer from seconds to years.)
57. *Trip to the Center of the Galaxy.* Use the same scenario as in Problem 56, but this time suppose the ship travels to the center of the Milky Way Galaxy, about 27,000 light-years away. How much time will pass on the ship? Compare this to the amount of time that will pass on Earth.
58. *Trip to Another Galaxy.* The Andromeda Galaxy is about 2.5 million light-years away. Suppose you had a spaceship that could constantly accelerate at $1g$. Could you go to the Andromeda Galaxy and come back within your lifetime? Explain. What would you find when you returned to Earth? (*Hint:* You'll need the formula from Problem 56.)
59. *Gravitational Time Dilation on Earth.* For a relatively weak gravitational field, such as that of a planet or an ordinary star, the following formula tells us the fractional amount of gravitational time dilation at a distance $r$ from the center of an object of mass $M_{\text{object}}$:

$$\frac{1}{c^2} \times \frac{GM_{\text{object}}}{r}$$

$[G = 6.67 \times 10^{-11}\ \text{m}^3/(\text{kg} \times \text{s}^2);\ c = 3 \times 10^8\ \text{m/s}.]$ For example, while 1 hour passes in deep space far from the object, the amount of time that passes at a distance $r$ is 1 hour multiplied by the factor above. (This formula does *not* apply to strong gravitational fields, like those near black holes.) Calculate the amount of time that passes on Earth's surface while 1 hour passes in deep space.
60. *Gravitational Time Dilation on the Sun.* Use the formula given in Problem 59 to calculate the percentage by which time runs slower on the surface of the Sun than in deep space. Based on your answer, approximately how much of a gravitational redshift should you expect for a spectral line with a rest wavelength of 121.6 nm?

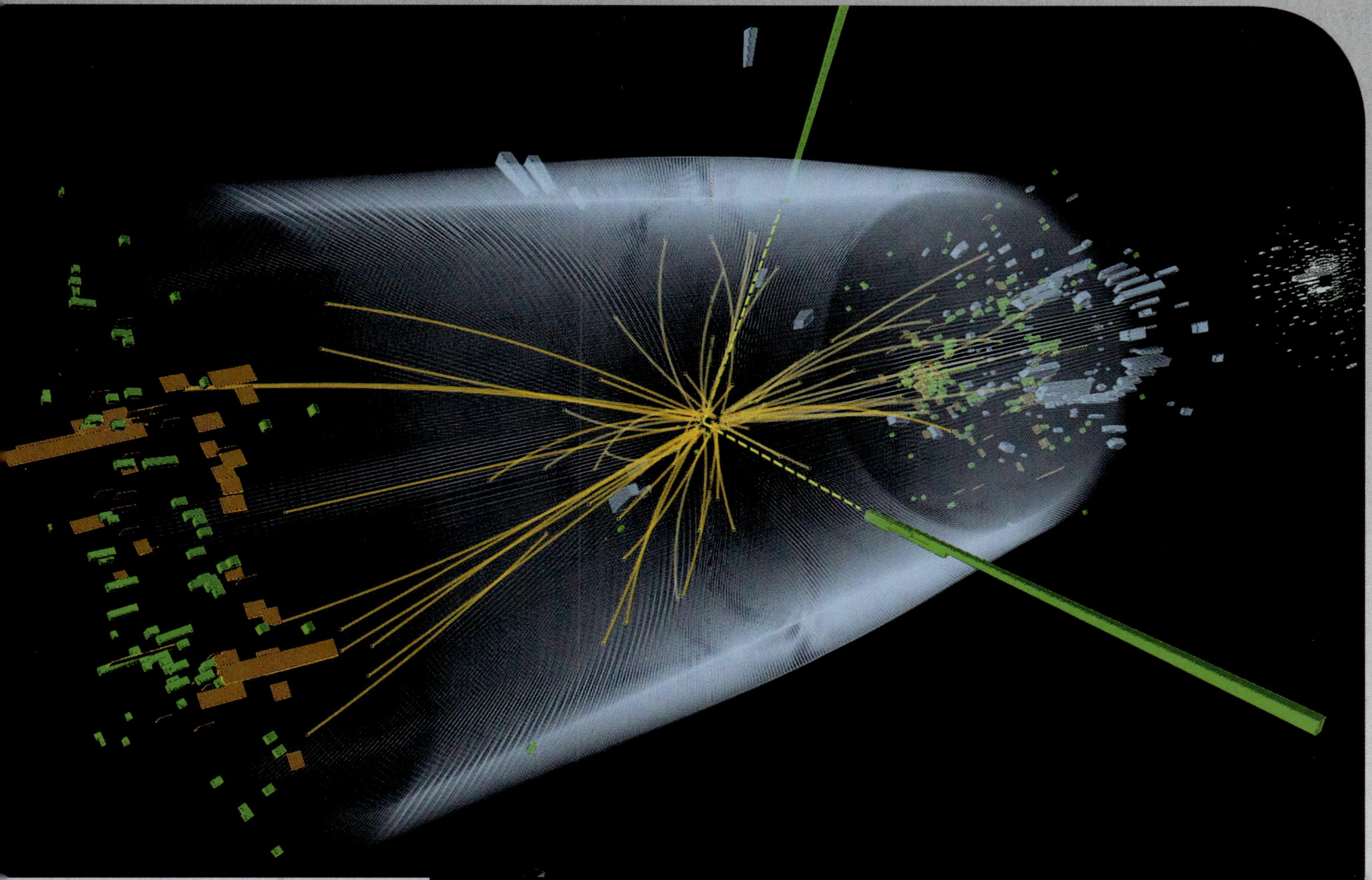

# S4 Building Blocks of the Universe

▲ **About the photo:** Particle tracks recorded at the Large Hadron Collider show detection of the Higgs boson.

## LEARNING GOALS

**S4.1 The Quantum Revolution**

- How has the quantum revolution changed our world?

**S4.2 Fundamental Particles and Forces**

- What are the basic properties of subatomic particles?
- What are the fundamental building blocks of nature?
- What are the fundamental forces in nature?

**S4.3 Uncertainty and Exclusion in the Quantum Realm**

- What is the uncertainty principle?
- What is the exclusion principle?

**S4.4 Key Quantum Effects in Astronomy**

- How do the quantum laws affect special types of stars?
- How is quantum tunneling crucial to life on Earth?
- How empty is empty space?
- Do black holes last forever?

*There is a theory which states that if ever anyone discovers exactly what the Universe is for and why it is here, it will instantly disappear and be replaced by something even more bizarre and inexplicable.*

*There is another which states that this has already happened.*

—*Douglas Adams, from* The Restaurant at the End of the Universe

**Chapter S4 Overview**

The microscopic realm of atoms and subatomic particles may seem far removed from the vast realm of planets, stars, and galaxies. But because everything is made from small particles, we cannot fully understand the universe unless we understand the tiny building blocks of the cosmos.

In this chapter, we will look more deeply into the building blocks of nature, investigating current knowledge of the fundamental particles and forces that make up the universe. We will see that the laws of the microscopic world play a crucial role in diverse processes such as nuclear fusion in the Sun and the collapse of a star into a black hole.

## S4.1 The Quantum Revolution

Around the same time that Einstein was discovering the principles of relativity, he and others were also investigating the behavior of matter and energy. Their discoveries in this area were no less astonishing. In 1905, the same year he published his special theory of relativity, Einstein showed that light behaves like particles (photons) in addition to behaving like waves [**Section 5.2**]. In 1911, British physicist Ernest Rutherford (1871–1937) discovered that atoms consist mostly of empty space, raising the question of how matter can ever feel solid. In 1913, Danish physicist Niels Bohr (1885–1962) suggested that electrons in atoms can have only particular energies; that is, electron energies are *quantized* [**Section 5.3**]. For this reason, the realm of the very small is called the *quantum realm,* and the science of this realm is called **quantum mechanics**.

### How has the quantum revolution changed our world?

Other scientists soon built on the work of Einstein, Rutherford, and Bohr. By the mid-1920s, our ideas about the structure and nature of atoms and subatomic particles were undergoing a total revolution. The repercussions of this *quantum revolution* continue to reverberate today. They have forced us to reexamine our "common sense" about the nature of matter and energy. They have also driven a technological revolution, because the laws of quantum mechanics make modern electronics possible. Most important, at least from an astronomical point of view, the combination of new ideas and new technology enables us to look ever deeper into the heart of matter and energy—the ultimate building blocks of the universe. In this chapter, we will investigate the following key ideas of the quantum revolution:

- Atoms are built from two categories of particles, called *quarks* and *leptons,* which in turn belong to an even more general category called *fermions.* Photons belong to an entirely distinct category of particles called *bosons.*
- *Antimatter* is real and is readily produced in the laboratory. When a particle and its antiparticle meet, the result is mutual annihilation and the release of energy.
- Four basic forces govern all interactions between particles: gravity, electromagnetism, the strong force, and the weak force. These four forces are themselves manifestations of a smaller number of truly fundamental forces, perhaps even of a single unified force.
- Our everyday experience tells us that particles and waves are different, but the quantum laws show that *all* tiny particles exhibit the same *wave-particle duality* that Einstein demonstrated for photons.
- Quantum laws have astronomical consequences. For example, a quantum effect called *degeneracy pressure* can prevent the core of a dying star from collapsing, *quantum tunneling* helps make nuclear fusion possible in the Sun, and phantom-like *virtual particles* may be important to the ultimate fate of black holes and the universe.

No matter how bizarre the quantum laws may seem, they lead to concrete predictions that can be tested experimentally and observationally. Some of these tests require sophisticated technological equipment found only in advanced physics laboratories. Others require billion-dollar particle accelerators. But some are performed every day, right before your eyes. Every time you see a ray of sunlight, you are seeing the product of nuclear reactions made possible by the quantum laws. Every time you turn on a modern electronic device, the laws of quantum mechanics are being put to work for your benefit. Indeed, almost every aspect of our "information society"—including cell phones, computers, and the Internet—has been made possible by the science of quantum mechanics.

## S4.2 Fundamental Particles and Forces

More than 2400 years ago, Democritus proposed that matter was made from particles that he called *atoms* [**Section 5.3**], which meant "indivisible," because he believed that atoms were **fundamental particles** that could not be divided into smaller units. By this definition, the particles we now call atoms are not truly fundamental. By the 1930s, we knew that atoms are made of protons, neutrons, and electrons. Later, as scientists explored more extreme conditions, they found that protons and neutrons are made from even smaller building blocks, and they discovered a variety of other particles that had previously been unknown.

## What are the basic properties of subatomic particles?

Physicists explore the properties of particles under extreme conditions with the aid of *particle accelerators* (sometimes called *atom smashers*). Today, the most powerful accelerator is the Large Hadron Collider on the border between Switzerland and France (**FIGURE S4.1**). Particle accelerators use strong electric fields (along with large magnets) to accelerate familiar particles such as electrons or protons to very high speeds—often extremely close to the speed of light—which leads to a highly concentrated energy release when the particles collide with one another or with a stationary target. Some of this energy spontaneously turns into mass (which is possible because $E = mc^2$ tells us both that matter can be changed into energy and that energy can be changed into matter), producing a shower of particles of many different types.

Careful study of the results from experiments in particle accelerators has led scientists to realize that each distinct particle type has a unique behavior determined by just a few basic properties. The most important properties are mass, electrical charge, and something called *spin*. While mass and electrical charge are familiar from everyday life, spin is evident only in the quantum realm.

Recall that a spinning ice skater or a spinning baseball has rotational angular momentum [**Section 4.3**]. By analogy, subatomic particles have **spin angular momentum**—or **spin**, for short—as they "spin" on their axes. Note, however, that subatomic particles do not really look like tiny spinning balls; rather, *spin* is simply a term used to describe the angular momentum that inherently belongs to a particle. For example, just as all electrons have exactly the same mass and electric charge, all electrons have exactly the same amount of spin. More generally, every particle of a particular type has precisely the same amount of mass, charge, and spin.

Just as electrical charge comes only in distinct amounts—such as +1 for protons, −1 for electrons, and 0 (neutral) for neutrons—spin also comes in distinct amounts. In the units used in quantum mechanics,* values of spin can be either integers (0, 1, 2, . . . ) or half-integers ($\frac{1}{2}, \frac{3}{2}, \frac{5}{2}, \ldots$). Particles with half-integer spins—which include electrons, protons, and neutrons—can be oriented in two ways, usually called *spin up* and *spin down*. These two orientations, which correspond to the two opposite senses of rotation (clockwise and counterclockwise), are often represented by arrows (**FIGURE S4.2**). Keep in mind that denoting a particle's spin by an arrow on a small sphere is a representation of convenience and that subatomic particles are *not* tiny spinning balls.

FIGURE S4.1 Aerial photograph of the Large Hadron Collider. The large circle traces the path of the main particle acceleration ring, which lies underground and has a circumference of 27 kilometers.

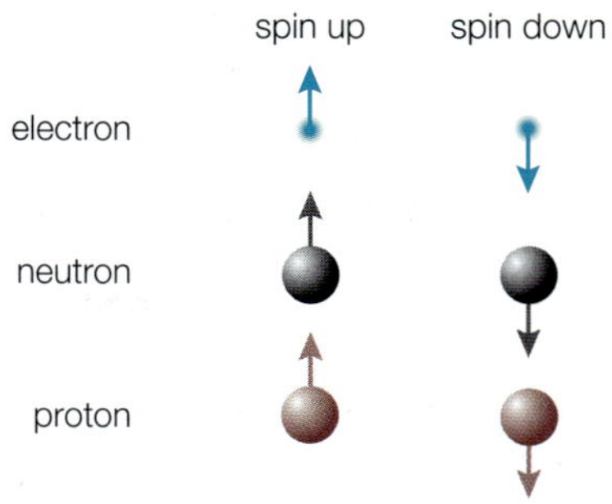


FIGURE S4.2 The two possible states of the spin of an electron, neutron, or proton are spin up and spin down, represented here by arrows.

## What are the fundamental building blocks of nature?

By the 1960s, scientists had discovered dozens of different particle types, each with a unique set of basic properties. Physicist Murray Gell-Mann made sense of this great variety by proposing a scheme in which all particles could be built from just a few fundamental components, and his proposal soon blossomed into what physicists now call the **standard model**. The standard model has proved so successful that it has been used to predict the existence of new particles that were later discovered in particle accelerators.

In the standard model, the most fundamental distinction between particle types depends on spin. Particles with half-integer spins are known as **fermions**, named for Enrico Fermi (1901–1954); particles with integer spins are known as **bosons**, named for Satyendra Bose (1894–1974). The most familiar fermions are electrons, neutrons, and protons. The most familiar bosons are photons, though the *Higgs boson* is arguably more famous. The Higgs boson, named for the physicist (Peter Higgs) who predicted its existence in the 1960s, plays a key role in the standard model. In a spectacular success for the standard model, it was first detected in 2012 at the Large Hadron Collider (see the chapter-opening photo).

**Think about it** What countries run the Large Hadron Collider? Briefly describe its most recent major discovery.

The fermions are further subdivided into two categories known as **quarks** and **leptons**. Both are thought to be fundamental particles that cannot be divided into smaller

*Physicists measure the angular momentum of subatomic particles in units of Planck's constant divided by $2\pi$.

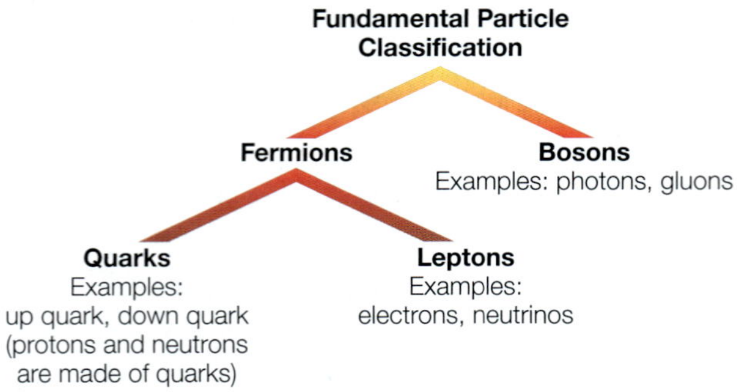


**FIGURE S4.3** Classification of fundamental particles in the standard model.

pieces, but there are important differences between them. The technical distinction surrounds the way they interact with one another: Quarks interact with one another via the *strong force* (which we'll discuss in more detail shortly) while leptons do not. However, a more practical distinction comes in how the particles are found in nature. Quarks are never seen by themselves, but instead are found in combinations that make larger subatomic particles, including protons and neutrons. In contrast, leptons exist as isolated particles; electrons are the most familiar type of lepton. **FIGURE S4.3** summarizes the basic classification scheme for fundamental particles.

**Particles Made from Quarks** Protons and neutrons are the only particles made from quarks that we find in the matter we encounter in everyday life. Protons and neutrons each contain two different types of quarks: the **up quark**, which has an electric charge of $+\frac{2}{3}$, and the **down quark**, which has an electric charge of $-\frac{1}{3}$. (Charge values are in the same units that give $+1$ for protons and $-1$ for electrons.) Two up quarks and one down quark form a proton, giving it an overall charge of $+\frac{2}{3}+\frac{2}{3}-\frac{1}{3}=+1$. One up quark and two down quarks form a neutron, making it neutral: $+\frac{2}{3}-\frac{1}{3}-\frac{1}{3}=0$ (**FIGURE S4.4**).

Although protons and neutrons are the most familiar particles made from quarks, many other particles have been found through particle accelerator experiments and by detectors that capture the high-energy particles from space known as *cosmic rays*. These other particles are generally short-lived, decaying a fraction of a second after they are produced in collisions. Nevertheless, scientists can observe

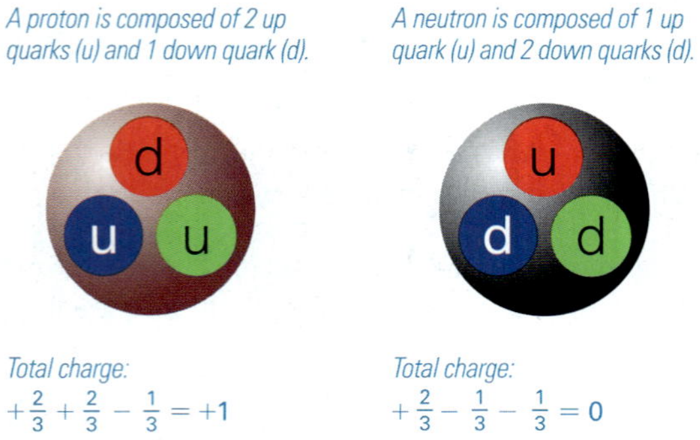


**FIGURE S4.4** Protons and neutrons are made from quarks.

**TABLE S4.1 Fundamental Fermions**

| The Quarks | The Leptons |
|---|---|
| Up | Electron |
| Down | Electron neutrino |
| Strange | Muon |
| Charmed | Muon neutrino |
| Top | Tauon |
| Bottom | Tau neutrino |

their behavior during their short lives using detectors designed to record the tracks that the particles leave as they pass through, decay, or undergo other collisions. The photo that opens this chapter shows an example of particle tracks (for the Higgs boson). Careful analysis of such tracks for particles made from quarks allows scientists to determine the quark composition.

The experiments have revealed several important facts that are now incorporated into the standard model. For example, while scientists have found many particles composed of either two or three quarks, no single quark has ever been detected in isolation. Moreover, despite the fact that individual quarks have charges in units of thirds, the total charge of a particle made from quarks is always $-1$, 0, or $+1$. Although we won't make much use of the distinction in this book, particles built from two quarks are known as *mesons* while particles built from three quarks are called *baryons*.*

In addition, experiments have shown that the up and down quarks that make up protons and neutrons are only two of six different types (or "flavors") of quarks. The other four types of quarks have the rather exotic names *strange, charmed, top,* and *bottom*. The first column of **TABLE S4.1** lists the six types of quarks.

**Leptons** The electron is the only type of lepton that we find in the matter of everyday life, but the standard model predicts the existence of six lepton types to match up with the six quark types. Table S4.1 also lists the six leptons. All six predicted lepton types have been identified experimentally.

Note that three of the leptons bear the name *neutrino*, which means "little neutral one." This name reflects the fact that all three types of neutrinos are electrically neutral and extremely lightweight—far less massive than electrons. No one has yet succeeded in measuring the precise masses of neutrinos, but these particles are extremely common, outnumbering protons, neutrons, and electrons combined by a factor of roughly a billion. Nevertheless, neutrinos are so lightweight that they are thought to make up only a small fraction of the total mass of the universe. In later chapters, we will see that neutrinos are very important in several astronomical processes, including nuclear fusion and the explosive deaths of stars as supernovas.

*Another term, *hadrons* (which appears in the name of the Large Hadron Collider), is used for any particle built from quarks, which means that mesons and baryons are subsets of the hadrons.

**Think about it** In high school, you were probably told that the fundamental building blocks of atoms are protons, neutrons, and electrons. In what sense is this statement true? In what sense would it be better to say that the fundamental building blocks of atoms are quarks and leptons?

**Antimatter** Science fiction fans will be familiar with **antimatter**, popularized as fuel for starship engines in the *Star Trek* series. But antimatter itself is not science fiction; it really exists. In fact, every quark and every lepton has a corresponding *antiquark* and *antilepton.* The antiparticle is an exact opposite of its corresponding ordinary particle. For example, an *antielectron* (also called a *positron*) is identical to an ordinary electron except that it has a positive charge (+1) instead of a negative charge (−1). Note that while particles and antiparticles are opposites in most respects, they are *not* opposite in mass; as far as we know, there is no such thing as negative mass, so the mass of an antiparticle is precisely the same as the mass of its corresponding particle.

When a particle and its corresponding antiparticle meet, the result is mutual **annihilation** (**FIGURE S4.5a**). The combined mass of the particle and the antiparticle turns completely into energy in accord with $E = mc^2$. Because the matter in our universe is predominantly of the ordinary type, antimatter generally does not last very long. Whenever an antiparticle is produced, it quickly meets an ordinary particle, and the two annihilate each other to make energy.

This process also works in reverse. When conditions are right, pure energy can turn into a particle-antiparticle pair. For example, whenever an electron "pops" into existence, an antielectron also pops into existence with it (**FIGURE S4.5b**). This process of **pair production** conserves charge (and satisfies other conservation laws as well). It happens routinely in particle accelerators here on Earth and on a much grander scale in outer space. In fact, during the first few moments after the Big Bang, the universe's energy fields were so intense that particle-antiparticle pairs popped rapidly in and out of existence at virtually every point in space.

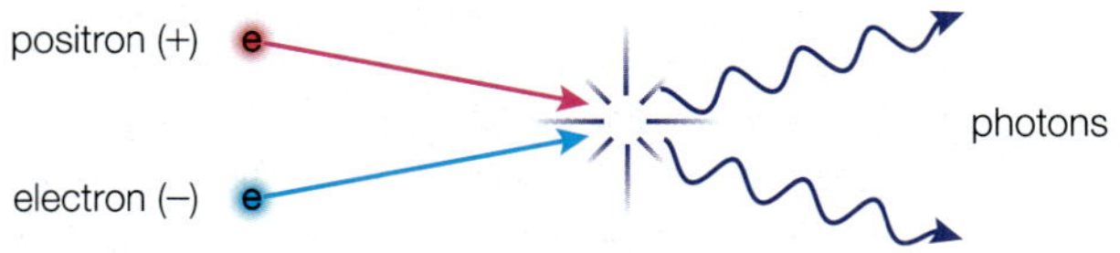


**a** An electron and a positron (antielectron) collide, resulting in annihilation in which all their energy emerges as a pair of photons.

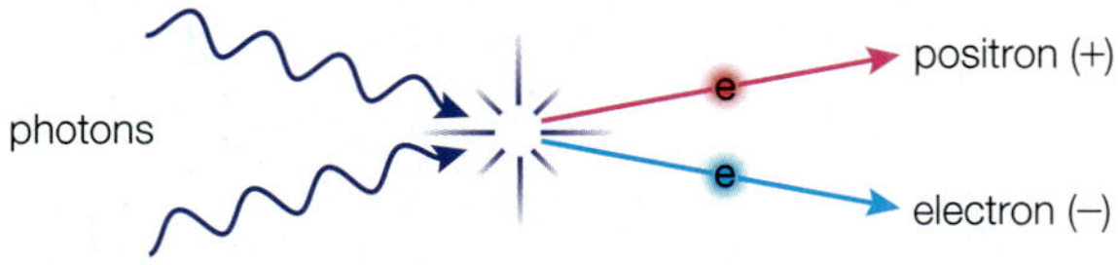


**b** The energy of two photons combines to create an electron and a positron.

**FIGURE S4.5** Every particle of matter has a corresponding antiparticle. When a particle meets its antiparticle, the result is mutual annihilation. Pair production creates a particle and an antiparticle from energy.

When we include antiparticles, the total number of types of quarks and leptons really is twice that shown in Table S4.1. That is, there are 12 types of quarks: the six quarks listed and their six corresponding antiquarks (for example, the up quark and the anti-up quark). Similarly, there are 12 types of leptons: the six listed and their six corresponding antileptons. The net total of 24 different fundamental particles seems quite complex.* As we'll see shortly, this complexity leads many scientists to believe that simplifying principles of particle physics still await discovery.

## What are the fundamental forces in nature?

Without forces, the universe would be infinitely boring, a uniform sea of fundamental particles drifting aimlessly about. Forces supply the means through which particles interact and exchange momentum, attracting or repelling one another depending on their properties. For example, particles with mass interact with one another via the force of *gravity,* and particles with charge interact with one another via the *electromagnetic* force.

**The Four Forces** According to present understanding, only four fundamental forces operate under ordinary conditions in the universe today. **Gravity** and **electromagnetism** are the most familiar of the four. The other two known forces are called the **strong force** and the **weak force**.

The strong and weak forces act only on extremely short-distance scales—so short that these forces can be felt only *within* atomic nuclei. You can see why the strong force must exist by remembering that the nuclei of all elements except hydrogen (which has just a single proton) contain more than one proton. Protons are positively charged, so the electromagnetic force pushes them apart. If the electromagnetic force were unopposed, atomic nuclei would fly apart. The strong force, so named because it is strong enough to overcome electromagnetic repulsion, is what holds nuclei together. The weak force, also very important in nuclear reactions, is a bit more subtle. All particles made from quarks respond to the strong force, but neutrinos, for example, feel only the weak force (and gravity).

According to the standard model, each force is transmitted by an **exchange particle** that transfers momentum between two interacting particles. For example, photons are the exchange particle for the electromagnetic force. To understand the idea, think about how we see a star. Motions of electrons in the star create photons. The photons then cross light-years of space to our eyes, where they generate an electromagnetic disturbance that we see as starlight. In other words, the photons have carried the electromagnetic force from the particles in the star to particles in our eyes.

*Moreover, each of the six quarks and six antiquarks is believed to come in three distinct varieties, called *colors.* (The term *color* is not meant to be literal. Rather, it describes a property of quarks that we cannot visualize.) For example, the up quark comes in three colors, often called red, green, and blue.

**TABLE S4.2 The Four Forces**

| Force | Relative Strength Within Nucleus* | Relative Strength Beyond Nucleus | Exchange Particles | Major Role |
|---|---|---|---|---|
| Strong | 100 | 0 | Gluons | Holding nuclei together |
| Electromagnetic | 1 | 1 | Photons | Chemistry and biology |
| Weak | $10^{-5}$ | 0 | Weak bosons | Nuclear reactions |
| Gravity | $10^{-43}$ | $10^{-43}$ | Gravitons | Large-scale structure |

*The relative values for the strong and weak forces are rough approximations.

In a similar way, exchange particles called *gravitons* are thought to carry the gravitational force through the universe.* *Gluons,* which get their name because they act like glue to bind nuclei together, carry the strong force. The weak force is carried by particles known as *weak bosons,* which come in three types known as $W^+$, $W^-$, and Z. Note that all of the exchange particles are bosons.

**Relative Strength of the Four Forces** TABLE S4.2 lists the four forces and their exchange particles. It also has two columns that describe the relative strengths of the four forces. One column shows the relative strengths of the forces within an atomic nucleus, and the next shows relative strengths outside nuclei. The values in these columns represent the relative attraction or repulsion that would be felt between two particles affected by the forces. For example, consider the relative strengths indicated for the electromagnetic force and gravity. The numbers in the table mean that if we took two protons and put them a certain distance apart, the electromagnetic force of repulsion between them would be $10^{43}$ times as strong as the gravitational force of attraction between them. That is why two protons repel each other through their electrical charge rather than attracting each other through gravity.

The relative strengths explain why the different forces tend to be important in different ways. For example, notice that within the nucleus, the strong force is roughly 100 times as strong as the electromagnetic force. That is why the strong force can hold nuclei together despite the electromagnetic repulsion between protons. The zero relative

*General relativity tells us that gravitational waves transmit changes in spacetime curvature through the universe [**Section S3.4**], so the correspondence between *gravitons* and gravitational waves is analogous to the correspondence between photons and electromagnetic waves. Gravitons have not yet been detected.

## EXTRAORDINARY CLAIMS Faster-Than-Light Neutrinos

Many scientists are trying to learn more about neutrinos (see Table S4.1), which are so lightweight that no one has yet succeeded in learning their mass. Neutrinos interact so weakly with other matter that they can fly through Earth's interior as if it were empty space. In 2011, an Italian research group was measuring the time it took neutrinos produced at the Large Hadron Collider in Switzerland to travel straight through the ground to a detector in Italy when they made an extraordinary claim: The neutrinos were making the trip at a speed about 0.002% faster than the speed of light.

If this measurement was correct, it was a momentous discovery. As discussed in Chapter S2, the absoluteness of the speed of light has been verified over and over again, and clearly implies that nothing can travel through space faster than light (see Figure S2.9). If the neutrinos really were getting from Switzerland to Italy faster than light could, then either our conclusions about the speed of light were somehow incorrect or perhaps neutrinos had some ability to exploit potential "loopholes" (such as wormholes or hyperspace [**Section S3.5**]) that might allow them to get from one place to another without traveling through all the space in between.

The extraordinary nature of the claim made most scientists skeptical from the start. Moreover, other experiments and observations had never shown any hint of faster-than-light neutrino travel. The most stringent test to date was an astronomical one: In 1987, scientists detected neutrinos coming from Supernova 1987A [**Section 17.3**], and their speed was within about 20 parts per billion of the speed of light—just as expected for particles that are nearly massless and emitted with extremely high energies.

The Italian researchers themselves recognized that their extraordinary claim would require extraordinary evidence for anyone to believe it, and they devoted intense effort to reanalyzing every aspect of their experiment. In the end, they discovered a loose fiber optic cable that had caused their timing system to be inaccurate. Once they fixed this and accounted for the errors it had introduced, the faster-than-light measurement went away. Over the following year, several independent teams (using different experiments at the same lab) all found neutrino speeds consistent with speeds slower than light.

This story holds important lessons about how science works. Although the claim ultimately fell apart, the fact that the researchers published it openly gave the entire scientific community a chance to scrutinize their work and to help them understand their results. Because the team was open to the possibility that they had made a mistake, they eventually found the wiring fault, and because the experiment could be duplicated elsewhere, other teams could confirm that neutrinos indeed travel only as fast as expected by Einstein's theory. So while the team received a fair amount of criticism for publishing prematurely, the story ended up as an example of science working well and rooting out the errors of a flawed claim.

**Verdict:** Rejected.

strength for the strong and weak forces outside the nucleus expresses the fact that these forces vanish beyond very short distances, leaving only gravity and electromagnetism to govern the interactions we notice in our daily lives.

In fact, if you remember that a typical atom (with its electrons) is 100,000 times as large (in diameter) as an atomic nucleus [**Section 5.3**], you'll realize that the strong and weak forces cannot even affect interactions between atoms. Gravity is almost equally ineffective between atoms, because it is so incredibly weak compared to electromagnetism. That leaves electromagnetism as the only force that affects the interactions of atoms—which means it is the one and only force that can group electrons and nuclei into atoms, atoms into molecules, and molecules into living cells. Remarkable as it may seem, the force of electromagnetism governs every aspect of chemistry and biology, and in essence is responsible for the existence of life itself.

Given the extreme weakness of gravity in comparison to the other forces, you might wonder how it can be important at all. The answer is that for very massive objects, gravity essentially wins a game of elimination. As we've already seen, the strong and weak forces cannot operate on large distance scales. The electromagnetic force loses importance for massive objects because it effectively cancels itself out. Like gravity, the electromagnetic force can act on large distance scales; in fact, its strength declines with distance in the same way that the gravitational force declines with distance (according to an inverse square law). However, the electromagnetic force differs from gravity in a very important way: It can be either attractive or repulsive depending on the charges of the particles, while gravity always attracts. Large objects always contain virtually equal numbers of protons and electrons, because if they didn't, the strength of the electromagnetic force would quickly either drive out the excess charge or draw in oppositely charged particles to balance it. The charge balance means that large objects are electrically neutral overall, and therefore unresponsive to the electromagnetic force. With the other three forces out of the picture, gravity is the only force left to act between massive objects, and it grows stronger as an object's mass grows larger. That is why gravity dominates the structure of the universe on large scales, despite its weakness among individual particles.

**The Quest for Simplicity** The standard model has four forces that mediate interactions between particles built from six types of quarks and six types of leptons, plus their corresponding antiparticles. It explains many experimental and observational results quite successfully. However, many scientists think that this model is too complicated. They seek an even more basic theory of matter that reduces the number of forces and fundamental particles.

Theoretical work in the 1970s, verified experimentally in the 1980s, showed that the electromagnetic and weak forces are really just two different aspects of a single force, called the *electroweak force*. Many scientists hope that future discoveries will show three or even all four forces to be different aspects of a single, unified force governing *all* interactions in nature. As we will see in Chapter 22, these *unified theories* might be necessary to understanding what happened during the first fraction of a second after the Big Bang.

## S4.3 Uncertainty and Exclusion in the Quantum Realm

We often use the word *particles* as if we were talking about objects that you might hold in your hand, but subatomic particles can exhibit behavior that seems very strange to us. In particular, like photons of light, particles of matter can act both as particles and as waves. In other words, all subatomic particles—whether particles of matter like protons and electrons or exchange particles like photons—exhibit what we call **wave-particle duality**.

The fact that matter can sometimes act like tiny baseballs and other times spread out like water waves explains why even professional physicists have difficulty forming a mental picture of the subatomic world. Nevertheless, the science of quantum mechanics allows us to model the properties of matter with great accuracy, which is why we have been able to use it to build computers and other modern electronic devices. In this section, we will discuss two fundamental laws of quantum mechanics: the *uncertainty principle* and the *exclusion principle*.

### What is the uncertainty principle?

The **uncertainty principle** was first described by Werner Heisenberg (1901–1976) in 1927. Here is one way of stating it:

> ***The Uncertainty Principle****: The more we know about a particle's location, the less we can know about its momentum, and conversely, the more we know about its momentum, the less we can know about its location.*

We can illuminate the meaning of the uncertainty principle by considering how we might measure the trajectories of a ball and an electron. In the case of a ball, we could photograph it with a blinking strobe light. The resulting photograph would show us both where the ball was and where it was going at each moment in time (**FIGURE S4.6**). In scientific terms, knowing the path of the ball means measuring both its *location* and its *velocity* at each instant or, equivalently, its location and its *momentum*. (Recall that momentum is mass times velocity.)

**FIGURE S4.6** A photograph of a ball taken with a blinking strobe light allows us to determine both where the ball was and where it was going at each moment in time.

Now, imagine trying to observe an electron in the same way. We will detect the electron only if it manages to scatter some of the photons of light streaming by it. However, while most photon wavelengths are tiny compared to a ball, they are quite large compared to an electron. The precision with which we can pinpoint the electron's location depends on the photon wavelength. If we use visible light with a wavelength of 500 nanometers, we can measure the electron's location only to within 500 nanometers—which is about *5000 times* the size of a typical atom. That is, if we see a flash of 500-nanometer light from a row of 5000 atoms, we cannot even know which atom contains the electron that caused the flash!

To locate the electron more precisely, we must use shorter-wavelength light, such as ultraviolet light or x-rays, but this creates a problem for measuring its momentum. To determine the electron's momentum we must track its path, which means we must observe a series of flashes from the electron's interactions with photons. However, each photon's energy delivers a "kick" that disturbs the electron and thereby changes the momentum we are trying to measure. The shorter the wavelength of the photon—which means the higher its energy—the more it alters the electron's momentum.

It is almost as if nature were playing a perverse trick on us. Locating the electron precisely requires hitting it with a short-wavelength photon, but the high energy of this photon prevents us from determining the electron's momentum. Conversely, measuring the electron's momentum requires hitting it with low-energy photons that will not disturb it much. But because low-energy light has long wavelengths, we'll no longer have a very good idea of where the electron is located.

**Think about it** Colloquially, we often express the uncertainty principle by saying that we can't know both where a particle is and where it is going. How does this statement relate to the more precise statement that we can't know both the particle's location and its momentum? In what ways is the colloquial statement accurate, and in what ways is it an oversimplification?

The uncertainty principle applies to *all* particles, not just to electrons. In fact, it even applies to macroscopic objects, meaning objects large enough to be seen without a microscope (*macro* comes from the Greek word for "large"). For example, consider what happens when we observe a baseball with visible light with a 500-nanometer wavelength. Just as with the electron, we can locate any part of the baseball only to within 500 nanometers. However, an uncertainty of 500 nanometers (about 0.00002 inch) is negligible compared to the size of the baseball. Moreover, the energy of visible light is so small compared to the energy of the baseball (including its mass-energy) that it has no noticeable effect on the baseball's momentum. That is why Newton's laws work perfectly well when we deal with the motions of baseballs, cars, and planets; they fail us only in the microscopic quantum realm, where we must deal with the implications of the uncertainty principle.

## SPECIAL TOPIC A String Theory of Everything?

During the past 400 years, the history of physics has been marked by discoveries that have revealed nature to be ever simpler. For example, Newton showed that we need only one set of physical laws for both Earth and the heavens, Einstein showed that gravity could be viewed more simply than as a mysterious "action at a distance," and the quantum laws show that forces can be understood through the interactions of exchange particles. Perhaps as a result of this history, most physicists today suspect that nature is even simpler than we now understand it to be in terms of the standard model of fundamental particles and forces discussed in this chapter. Physicists therefore seek a "theory of everything," in which just one or a few equations would describe every interaction that can occur in the universe.

Many approaches have been tried, but in recent decades many physicists have focused their efforts to find a theory of everything on an approach known as *string theory.* String theory is not really a single theory, but rather a family of theories that share a common idea—that all the particles and forces of the standard model arise from tiny structures called *strings.* The term *string* comes from an analogy to little pieces of string that can be shaped into loops or that can vibrate like strings on a violin. However, the strings of string theory are hypothesized to exist in ten or more dimensions. Aside from the three ordinary dimensions of space and the one dimension of time, the remaining six or more dimensions are thought to be "folded up" in a way that does not allow them to be detected on macroscopic scales. Some variations on string theory include multidimensional analogies to membranes, called *branes*; an idea called *M-theory* unites different versions of string theory.

Despite all the work that has gone into string theory, scientists do not yet know whether the basic tenets of the theory are correct. The problem is that, at least so far, most of the predictions of string theory would be observable only at energies in excess of what current particle accelerators can achieve. Therefore, by our usual definitions in science, string theory should really be called "string hypothesis," because it does not meet the high standard of verification required for a scientific theory. (String theory does make some predictions that are consistent with current knowledge of the universe, giving theorists hope that it is on the right track.)

Indeed, in its current form, string theory is really more of a mathematical theory than a scientific one. That is, scientists and mathematicians are developing new theorems as they work out the mathematics of multidimensional strings and branes. Their hope is that as they learn more about the mathematics, they will eventually be able to use it to make testable predictions. Meanwhile, the mathematical results give tantalizing hints of a new view of nature, one that could potentially be as revolutionary as any past revolution in physics.

Nevertheless, it's important to keep in mind that "string theory" has not yet passed the rigorous testing needed to be considered a valid model of the underlying reality of nature. Only through further observations and experiments will we learn whether strings lie at the heart of nature or are simply a dead-end idea that must be abandoned in the ongoing quest to discover a "theory of everything" that may reveal an underlying simplicity in nature.

**Wave-Particle Duality** Our thought experiment about measuring an electron's position and momentum with photons might at first seem to suggest that the electron somehow "hides" its precise path from us. However, the uncertainty principle runs deeper than this—it implies that the electron *does not even have* a precise path. From this point of view, the concepts of location and momentum do not exist independently for electrons in the way they appear to exist for objects in everyday life. Instead of imagining the electron to be following some complex but hidden path, we need to think of the electron as being "smeared out" over some volume of space [**Section 5.3**]. This "smearing out" of electrons and other particles holds the essence of the idea of wave-particle duality.

If we choose to regard the electron as a particle, we are imagining that we can locate it precisely by hitting it with a short-wavelength photon. In that case, we can measure the electron's precise location at each instant in time but can never predict where it will be at the next instant. In other words, we can see the electron in one place at one moment and in another place at another moment, but we'll have no idea how it passed through the regions in between. If we made many such measurements, we could make a map of an atom in which lots of dots represented electron locations at different times, but we could not draw the electron's actual path between the dots. In fact, a mathematical description of quantum mechanics allows us to calculate the *probability* that we'll find the electron in any particular place at any particular time. **FIGURE S4.7** shows the probability patterns for the electron in several energy levels of hydrogen. The brighter regions in the figure show where the electron has a higher probability of being found at any particular instant.

Alternatively, we can choose to regard the electron as a wave. We can measure the momentum of a wave, such as that of a ripple moving across a pond, but we cannot say that a wave has a single precise location (**FIGURE S4.8**). Instead, the wave is spread out over some region of the water. In the same way, regarding an electron as a wave means that we see it spread out over some region of space; that is, the electron is viewed as a three-dimensional wave surrounding an atom. Note that the wave viewpoint also leads us to conclude that the electron is "smeared out" over some volume of space.

The fact that electrons exhibit both particle and wave properties demonstrates that our common sense about the macroscopic world does not translate well to the quantum world. It also explains what we mean when we say that electrons and other subatomic particles exhibit wave-particle duality. We call them *particles* only for convenience. Like a photon of light, each particle has a wavelength.* When the wavelength of the particle is small, we can locate the particle fairly precisely, but its momentum is highly uncertain. When the wavelength of the particle is large, its momentum becomes well defined, but its location grows fuzzy.

*Electrons usually have very small wavelengths and therefore can be used to locate other particles with high precision. This is the principle behind *electron microscopes*, which use short-wavelength electrons rather than light. Electron microscopes can achieve resolutions of less than 0.1 nanometer, far better than the roughly 500-nanometer resolution of visible-light microscopes.

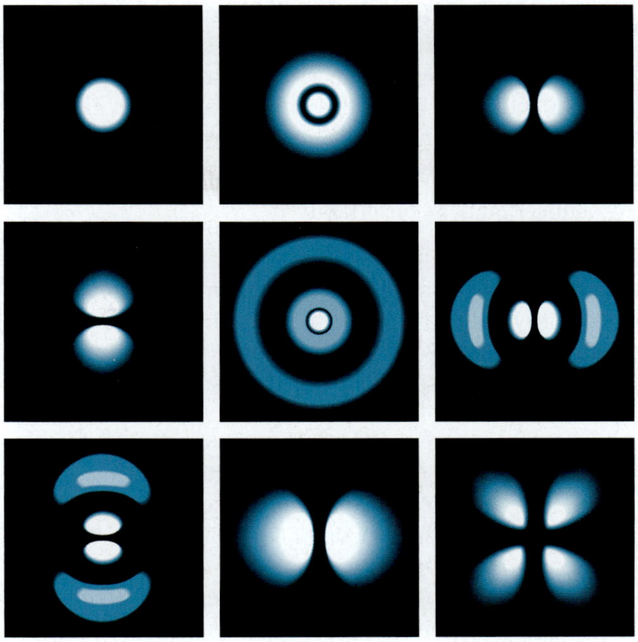

**FIGURE S4.7** If we regard an electron as a particle, we can map out its location within the atom at different times. These diagrams show calculated probabilities for finding the electron at various places in a hydrogen atom; brighter regions are those where the probability of finding the electron is higher. The nucleus, which consists of a single proton, is in the center in each case. The nine patterns shown here represent the probability patterns for an electron in nine different energy levels.

**Quantifying the Uncertainty Principle** We can quantify the uncertainty principle with a simple mathematical statement:

$$\begin{matrix}\text{uncertainty} \\ \text{in location}\end{matrix} \times \begin{matrix}\text{uncertainty} \\ \text{in momentum}\end{matrix} \approx \text{Planck's constant}$$

Planck's constant is a fundamental constant in nature, rather like the gravitational constant ($G$) in Newton's law of gravity and the speed of light ($c$). Its numerical value is $6.626 \times 10^{-34}$ joule $\times$ s. (We encountered Planck's constant earlier in the formula for the energy of a photon [**Section 5.2**].)

The mathematical statement of the uncertainty principle quantifies what we have already learned: Because the product of the uncertainties is roughly constant, when one uncertainty (either location or momentum) goes down, the other must go up. For example, if we determine the location of an electron with a particular amount of uncertainty (such as within 500 nanometers), we can use the formula to calculate the amount of uncertainty in the electron's momentum. The numerical value of Planck's constant is quite small, which explains why uncertainties are not noticeable for macroscopic objects.

**FIGURE S4.8** A wave has a well-defined momentum (represented by the arrow) but not a single precise location. If we regard an electron as a wave, we can imagine it as a three-dimensional wave occupying some volume around an atom, but we cannot pinpoint its location.

A second way of writing the uncertainty principle is mathematically equivalent but leads to additional insights. Instead of expressing the uncertainty principle in terms of location and momentum, this alternative version expresses it in terms of the amount of *energy* that a particle has and *when* it has this energy. This version reads

$$\begin{matrix}\text{uncertainty} \\ \text{in energy}\end{matrix} \times \begin{matrix}\text{uncertainty} \\ \text{in time}\end{matrix} \approx \text{Planck's constant}$$

An amusing way to gain further appreciation for the uncertainty principle in both its forms is to imagine a game of "quantum baseball." Suppose a quantum pitcher is pitching an electron that you must try to hit with your quantum bat. You'll find it extremely difficult. With the first version of the uncertainty principle, the problem is that you'll never know both where the electron is and where it is headed next. You might see it right in front of you, but because you don't know in which direction it's going, you can't know whether to swing straight, up, down, or sideways. With the second version of the uncertainty principle, your problem is that you might know the electron's energy, which tells you how hard you need to swing, but you'll never know *when* to swing. Either way, your likelihood of hitting the electron is completely a matter of random chance, governed by probabilities that can be calculated with the equations of quantum mechanics.

**Think about it** Explain why real baseball players don't have the problems that would arise with quantum baseball.

## What is the exclusion principle?

The second fundamental law of quantum mechanics, called the **exclusion principle**, was first described by Wolfgang Pauli (1900–1958) in 1925. In its simplest sense, the exclusion principle says that two particles cannot be in the same place at the same time. A more complete understanding of the exclusion principle requires investigating the properties of particles a little more deeply.

**The Quantum State of a Particle** Scientists use the term *state* to describe the current conditions of an object. For example, if you are relaxing in a chair, a scientist might say that you are in a "state of rest." A more precise description of your state in the chair might be something like "Your current state is a velocity of zero (at rest), a heart rate of 65 beats per minute, a breathing rate of 12 breaths per minute, a body temperature of 37°C, a metabolic rate of 200 Calories per hour," and so on.

**Think about it** Suppose you have the following information about the current state of a friend: velocity 5 km/hr, heart rate 160 beats per minute, metabolic rate 1200 Calories per hour. Which of the following is your friend most likely doing: (a) walking slowly as she reads a book, (b) driving in her car, (c) riding a bicycle downhill, or (d) swimming at a hard pace? Explain.

Completely describing a person's state would be quite complicated. Fortunately, it's much easier to describe the state of a subatomic particle, called its **quantum state**. In general, the complete quantum state specifies the particle's location, momentum, orbital angular momentum, and spin to the extent that the uncertainty principle allows. Like the energy of an electron in an atom, each property that describes a particle's quantum state is *quantized,* meaning that it can take on only particular values and not other values in between.

**Statement and Meaning of the Exclusion Principle** Earlier, we described the exclusion principle in simple

### SPECIAL TOPIC Does God Play Dice?

Suppose we knew all the laws of nature and, at some particular moment in time, what every single particle in the universe was doing. Could we then predict the future of the universe for all time?

Until the 20th century, most philosophers would have answered "yes." In fact, many philosophers concluded that God was like a watchmaker: God simply started up the universe, and the future was forever after determined. The idea that everything in the universe is predictable from its initial state is called *determinism,* and a universe that runs predictably, like a watch, is called a *deterministic universe.*

The discovery of the uncertainty principle shattered the idea of a deterministic universe because it tells us that, at best, we can make statements only about the *probability* of the precise future location of a subatomic particle. Because everything is made of subatomic particles, the uncertainty principle implies a degree of randomness built into the universe.

The idea that nature is governed by probability rather than certainty unsettled many people, including Einstein. Although he was well aware that the theories of quantum physics had survived many experimental tests, Einstein believed that the theories were incomplete and that scientists would one day discover a deeper level of nature at which uncertainty would be removed. To summarize his philosophical objections to uncertainty, Einstein said, "God does not play dice."

Einstein did more than simply object on philosophical grounds. He also proposed a number of thought experiments in which he showed that the uncertainty principle implies paradoxical results. Claiming that such paradoxes made no sense, he argued that the uncertainty principle must not be correct. In the decades since Einstein's death in 1955, advances in technology have made it possible to perform some of Einstein's quantum thought experiments. The results have proved to be just as strange as Einstein expected them to be, even as they have confirmed the uncertainty principle.

What can we make of an idea, such as the uncertainty principle, that seems to violate common sense at the same time that it survives every experimental test? Under the tenets of science, experiment is the ultimate judge of theory, and we must accept the results despite philosophical objections. In a sense, Einstein's objection that "God does not play dice" reflected his beliefs about how the universe *should* behave. Niels Bohr argued instead that nature need not fit our preconceptions, and he famously replied to Einstein by saying, "Stop telling God what to do."

terms by saying that two particles cannot be in the same place at the same time. Now it is time to state the exclusion principle more precisely. A key point is that the exclusion principle applies only to particles whose spin qualifies them as fermions. For example, the exclusion principle applies to protons, neutrons, and electrons because they are all fermions, but it does not apply to photons because they are bosons. With that in mind, we can state the exclusion principle more precisely as follows:

**The Exclusion Principle:** *Two fermions of the same type cannot occupy the same quantum state at the same time.*

The exclusion principle has many important implications. One of the most important is in chemistry, where it dictates how electrons occupy the various energy levels in atoms. For example, an electron occupying the lowest energy level in an atom necessarily has a particular amount of orbital angular momentum and a restricted range of locations. The electron's energy level fully determines its quantum state, except for its spin. Because electrons have only two possible spin states, up or down, only two electrons can occupy the lowest energy level. If you tried to put a third electron into the lowest energy level, it would have the same spin—and hence the same quantum state—as one of the two electrons already there. The exclusion principle won't allow that, so the third electron must go into a higher energy level (**FIGURE S4.9**). If you take a course in chemistry, you'll learn how a similar analysis of higher energy levels explains the chemical properties of all the elements, including their arrangement in the periodic table of the elements (see Appendix D).

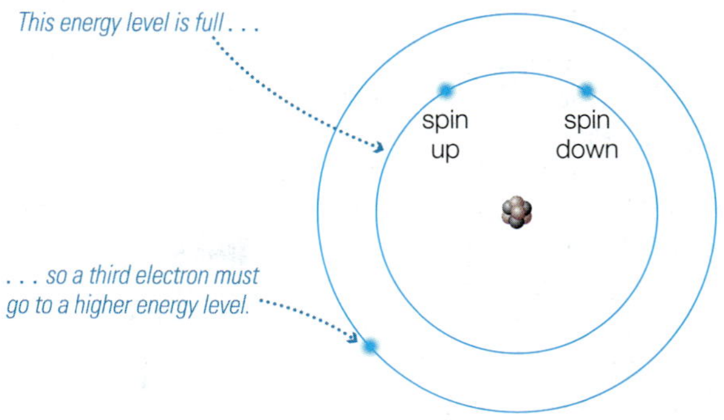


**FIGURE S4.9** The exclusion principle tells us that only two electrons, one with spin up and the other with spin down, can share a single energy level in an atom.

The uncertainty principle and the exclusion principle together determine the sizes of atoms and of everything made of atoms, including your own body. The uncertainty principle ensures that electrons cannot be packed into infinitesimally tiny spaces. If you tried to confine an electron in too small a space in an atom, the uncertainty principle tells us that its momentum would become so large that the electromagnetic attraction of the nucleus could no longer retain it. The exclusion principle ensures that each electron has to have its own space. Together, the two fundamental laws of quantum mechanics explain why matter seems solid even though atoms are almost entirely empty space. In the words of physicist Richard Feynman (1918–1988), "It is the fact that electrons cannot all get on top of each other that makes tables and everything else solid."

## MATHEMATICAL INSIGHT S4.1 Electron Waves in Atoms

In the text, we've said that an electron in an atom is "smeared out" over some volume of space. In fact, the physics is much more precise than this vague statement implies. If we choose to view the electron as a wave, an electron in an atom can be regarded as a *standing wave.*

You are probably familiar with standing waves on a string that is anchored in place at its two ends, such as a violin string. Such waves are called standing waves because each point on the string vibrates up and down but the wave does not appear to move along the length of the string. Moreover, because the string is anchored at both ends, only wave patterns with an integer or half-integer number of wavelengths along the string are possible. Other patterns, such as three-fourths of a wavelength along the string, are not possible because they would mean that one end of the string was broken away from its anchor point (**FIGURE 1**).

An electron viewed as a standing wave is anchored by the electromagnetic force holding it in the atom. As is true for waves on a string, only particular wave patterns are possible; however, wave patterns in an atom are more complex because they are three-dimensional. The allowed wave patterns for an electron in an atom can be calculated with the famous *Schrödinger equation,* developed by Erwin Schrödinger in 1926. These wave patterns correspond directly to the allowed energies of the electron in the atom. That is, the Schrödinger equation enables scientists to predict the allowed energies in different atoms. The fact that the Schrödinger equation successfully predicts the energy levels that are measured in the laboratory is one of the great triumphs of quantum mechanics.

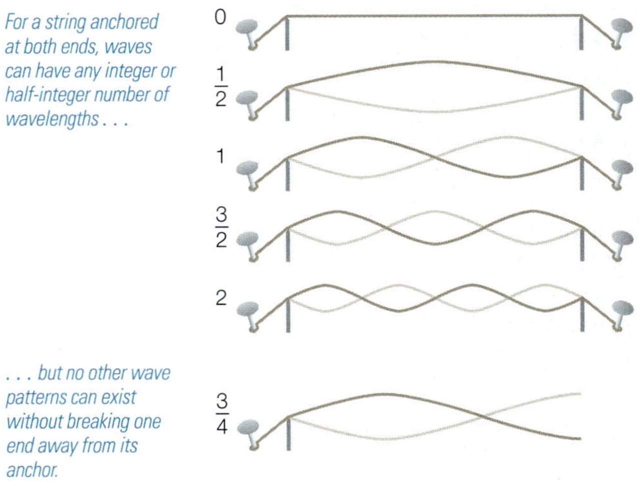


**FIGURE 1**

The exclusion and uncertainty principles also govern the behavior of tightly grouped protons, neutrons, and all other kinds of fermions. Just as these principles dictate the sizes of atoms, they also determine the sizes of nuclei, because they limit how closely protons and neutrons can pack together. As we will see shortly, these quantum effects can even influence the lives of stars.

Before we turn to astronomical implications of the quantum laws, it's worth noting that the fact that photons and other bosons do *not* obey the exclusion principle has other important implications. For example, laser beams are so intense because many photons *can* be in the same quantum state at the same time. In addition, it is possible under special conditions for two or more fermions to act together like a single boson. Such conditions lead to some amazing behaviors, including *superconductivity,* in which electricity flows without any resistance, and *superfluidity,* in which extremely cold liquids flow with no resistance (viscosity) at all.

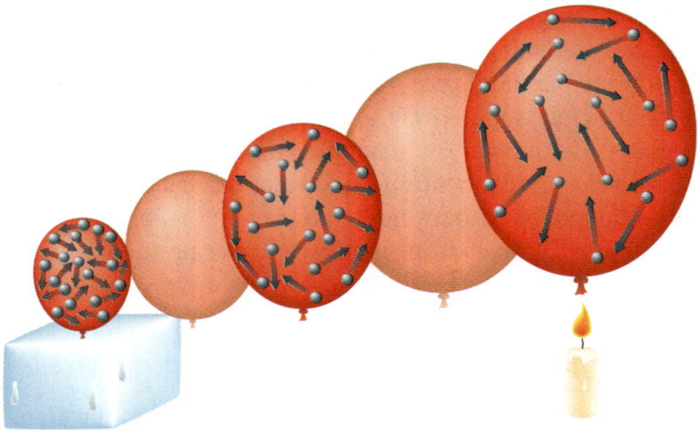


**FIGURE S4.10** Thermal pressure is the familiar type of pressure that keeps a balloon inflated. (The dots in this diagram represent molecules, and the lengths of the arrows represent their speeds.)

## S4.4 Key Quantum Effects in Astronomy

The uncertainty principle and the exclusion principle have clear consequences in the subatomic realm. Amazingly, this microscopic behavior also produces important effects on much larger scales. In fact, we cannot fully understand how stars are born, shine brightly throughout their lives, and die unless we understand the implications of quantum laws. In this section, we investigate four quantum effects of great importance in astronomy: degeneracy pressure, quantum tunneling, virtual particles, and the evaporation of black holes.

Degeneracy Pressure

### How do the quantum laws affect special types of stars?

Under ordinary conditions in gases, pressure and temperature are closely related. For example, suppose we inflate a balloon, filling it with air molecules. The individual molecules zip around inside the balloon, continually bouncing off its walls (**FIGURE S4.10**). The force of these molecules striking the walls of the balloon creates **thermal pressure**, which keeps the balloon inflated. If we cool the balloon by, say, putting it into a freezer, the molecules slow down. Slowing the molecules reduces the force with which they strike the walls, temporarily reducing the thermal pressure and causing the balloon to shrink.* Heating the balloon speeds up the molecules, temporarily raising the thermal pressure and causing the balloon to expand. Note that thermal pressure gets its name because it depends on temperature.

Thermal pressure is the dominant type of pressure at the low to moderate densities that we experience in everyday life. However, quantum effects produce an entirely different type of pressure under conditions of extremely high density, one that does not depend on temperature at all.

Consider what happens when we compress a *plasma,* a mixture of positively charged ions and free electrons [**Section 5.3**]. At first, the energy we expend in compressing the plasma makes the electrons and protons move faster and faster, increasing the pressure and the temperature. Now suppose we let the plasma cool off for a while. As the plasma cools, its pressure drops, enabling us to compress it further. Continuing this process of cooling and compression, we can in principle squeeze the plasma down to a very dense state. However, we cannot continue this process indefinitely. According to the exclusion principle, no two electrons in the plasma can have exactly the same position, momentum, and spin. That is, just as in an atom, all the electrons can't get on top of one another at once. The compression must stop at some point, no matter how cold the plasma becomes. This resistance to compression that stems from the exclusion principle is called **degeneracy pressure**.

The following analogy might help you visualize how degeneracy pressure works. Imagine a small number of people moving from chair to chair in an auditorium filled with folding chairs. Each person can move freely about and sit in any empty chair. The chairs represent available quantum states, the people represent electrons darting from place to place, and their motions represent thermal pressure as the electrons move from one quantum state to the next. The exclusion principle corresponds to the rule that two people cannot sit in the same chair at the same time. As long as the chairs greatly outnumber the people, two people will rarely fight over the same chair (**FIGURE S4.11a**).

Suppose ushers begin removing chairs from the front of the auditorium (compression), gradually forcing people to move to the back. Soon, everyone has to crowd toward the back of the auditorium, where the number of remaining chairs is just slightly larger than the number of people. Now when a person moves to a particular chair, there's a good chance that it's already occupied. Ultimately, when the number of people equals the number of chairs, people

*For a balloon, the shrinking (or expanding) occurs only until pressure balance is restored (meaning that the internal pressure is balanced with the combination of the external air pressure and the pressure created by the surface tension in the balloon itself). At that point, the balloon stabilizes in size.

**a** When there are many more available quantum states (chairs) than electrons (people), an electron is unlikely to try to enter the same state as another electron. The only pressure comes from the temperature-related motion of the electrons, which is thermal pressure.

**b** When the number of electrons (people) approaches the number of available quantum states (chairs), finding an available state requires that the electrons move faster than they would otherwise. This extra motion creates degeneracy pressure.

**FIGURE S4.11** An auditorium analogy to explain degeneracy pressure. Chairs represent available quantum states, and people who must keep moving from chair to chair represent electrons.

can still move from place to place, but only if they swap seats with somebody else (**FIGURE S4.11b**). The ushers can't take away any more chairs, and the compression must stop. In other words, all the available states are filled.

The uncertainty principle also influences degeneracy pressure, though in a way that does not perfectly fit this analogy. In a highly compressed plasma, the available space for each electron is very small, which in essence means that each electron's position is precisely defined. According to the uncertainty principle, its momentum must then be extremely uncertain. For the many electrons in the plasma, the great uncertainty in momentum means they must be moving very fast on average. This requirement that highly compressed electrons have to move quickly holds even if the object is very cold.* This quantum-mechanical trade-off between position and momentum is at the root of degeneracy pressure. If you want to compress lots of electrons into a tiny space, you need to exert an enormous force to rein in their momentum.

Degeneracy pressure caused by the crowding of electrons, or **electron degeneracy pressure**, affects the lives of stars in several different ways. In some cases, it can prevent a collapsing cloud of gas from becoming a star in the first place, creating what is called a *brown dwarf* [**Section 16.3**]. In stars like the Sun, it determines how they begin burning helium near the ends of their lives [**Section 17.2**]. When stars die, most leave behind an extremely dense stellar corpse called a *white dwarf,* which is also supported by electron degeneracy pressure [**Section 18.1**].

Not all stars meet this fate, because electron degeneracy pressure cannot grow infinitely strong. Under extreme compression, the average speed of the electrons begins to approach the speed of light. Nothing can go faster than the speed of light, so there is a limit to how much degeneracy pressure the electrons can exert. Once a dying star reaches that limit, electron degeneracy pressure cannot prevent it from shrinking further. The star then collapses until it becomes a ball of neutrons, called a *neutron star* [**Section 18.2**]. Neutron stars support themselves through **neutron degeneracy pressure**, which is just like electron degeneracy pressure except that it is caused by neutrons and occurs at much higher densities. Neutron degeneracy pressure comes into play only at much higher densities because neutrons have much greater mass than electrons. A neutron moving at close to the speed of light possesses over 1800 times as much momentum as an electron moving at the same speed. Therefore, the positions of neutrons can be over 1800 times as precise, enabling neutrons to occupy a much smaller volume of space.

Neutron degeneracy pressure cannot grow infinitely strong either. It begins to fail when the speed of the neutrons approaches the speed of light. At that point, gravity can make an object shrink further, and according to our present understanding, nothing can stop the collapse of an object once its gravity overcomes neutron degeneracy pressure. Such an object collapses indefinitely, becoming a *black hole* [**Section S3.3**].

*In this sense, an object is *cold* if there's no way to get heat from it. Consider a plasma in which all the available momentum states are filled up to a certain level. To extract heat from the plasma, you'd have to slow down some of its particles. That means moving them to lower-momentum states, which are already taken. The exclusion principle thereby prevents any energy from escaping, so the plasma is cold even though the electrons may be moving at high speeds.

## How is quantum tunneling crucial to life on Earth?

The next quantum effect we'll investigate arises from the uncertainty principle and has important implications not only in astronomy but also for our very existence and for modern technology. Let's start with an analogy.

Imagine that, as the unfortunate result of a case of mistaken identity, you're sitting on a bench in a locked jail cell (**FIGURE S4.12a**). Another bench is on the other side of the cell wall. If you could magically transport yourself from the bench on the inside to the bench on the outside, you'd be free.

Alas, no such magic ever occurs for humans. But what if you were an electron? In that case, the uncertainty principle would prevent us from predicting your precise location. At best, we could state only the probability of your being in various locations. While the probability that you remain in your cell might be very high, there is always some small probability that you will be found outside your cell, despite the existence of the wall. Therefore, you might suddenly find yourself free, thanks to the uncertainty principle (**FIGURE S4.12b**). Although electrons don't get put into jails, an analogous process in which electrons or other subatomic

**a** *A person is confined to the bench inside the jail cell, even though it would take no more energy to sit on the outside bench . . .*

**b** *. . . but an electron can move to the other side of the wall and become free through the process of quantum tunneling.*

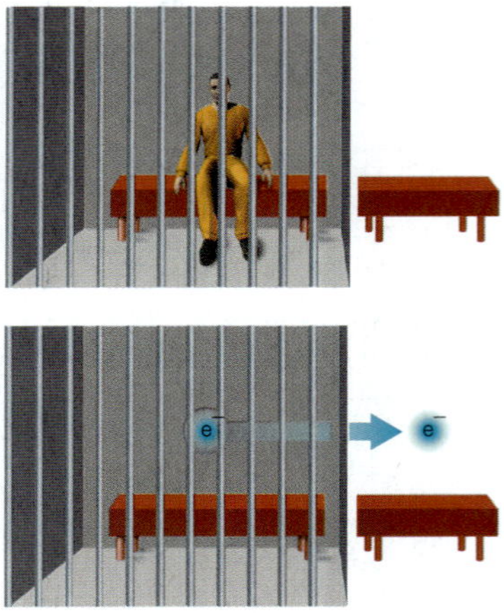


**FIGURE S4.12** A jail cell analogy to explain quantum tunneling.

particles "magically" go through wall-like barriers really does happen. We call it **quantum tunneling**.

We can gain a deeper understanding of quantum tunneling by thinking in terms of the *energy* needed to cross a barrier. If you are sitting in a jail cell, the barrier is the wall of the cell. The reason you cannot escape the cell is that you don't have enough energy to crash through the wall. Just as the cell wall keeps you imprisoned, a barrier of electromagnetic repulsion can imprison an electron that does not have enough energy to crash through it. However, recall that we can write the uncertainty principle in the following form:

$$\begin{matrix}\text{uncertainty} \\ \text{in energy}\end{matrix} \times \begin{matrix}\text{uncertainty} \\ \text{in time}\end{matrix} \approx \text{Planck's constant}$$

Because of the uncertainty inherent in energy, there is always *some* chance that the electron will have more energy than we think at a particular moment, allowing it to cross the barrier that lies in its way. From this point of view, quantum tunneling comes about because of uncertainty in energy rather than uncertainty in location.

The two points of view on quantum tunneling are equivalent, but the latter viewpoint illustrates a rather remarkable "loophole" in the law of conservation of energy. To cross the barrier, the particle must briefly gain some excess energy. Thanks to the uncertainty principle, this "stolen" energy need not come from anywhere as long as the particle returns it within a time period *shorter* than the uncertainty in time. In that case, we cannot be certain that any energy was ever missing! It's like stealing a dollar and putting it right back before anyone notices, so that no harm is done—except that the particle uses the stolen energy to cross the barrier before returning it.

This tale of phantom quantum energy thefts may at first sound utterly ridiculous, but the process of quantum tunneling is readily observed and extremely important. The microchips used in all modern computers and many other modern electronic devices work because of quantum tunneling by electrons. We can control the rate of quantum tunneling, and hence the electric current, by adjusting the "height" of the energy barrier. The higher the energy barrier, the less likely it is that particles will tunnel through it and the lower the electric current. This control over the electric current is critical to modern electronics.

Even more amazing, our universe would look much different were it not for quantum tunneling. The nuclear fusion reactions that power stars occur when atomic nuclei smash together so hard that they stick. However, nuclei tend to repel each other, because they are positively charged (they contain only positive protons and neutral neutrons) and like charges repel. This repulsion creates an electromagnetic barrier that prevents nuclear fusion under most conditions. Even at the high temperatures found deep in the cores of stars like our Sun, atomic nuclei don't have enough energy to crash through the electromagnetic barrier. Instead, they rely on quantum tunneling to "sneak" through the barrier into the region where the strong force dominates. In other words, quantum tunneling is what makes nuclear fusion possible in stars like our Sun—which means that life on Earth could not exist without it. (Quantum tunneling is not as crucial to fusion in stars that have core temperatures much higher than that of the Sun.)

## How empty is empty space?

The same "loophole" in the law of conservation of energy that allows particles to tunnel through otherwise impenetrable barriers also means that space can never be truly empty. According to the uncertainty principle, particles can "pop" into existence from nowhere—their mass made from stolen energy—as long as they "pop" back out of existence before anyone can verify that they ever existed. A somewhat fanciful analogy will help you understand this concept.

Imagine that the law of conservation of energy is enforced by a "great cosmic accountant." (In reality, of course, the law is enforced naturally.) Further, imagine that the cosmic accountant keeps a storehouse of energy in a large bank vault and ensures that any time something borrows some energy, it returns the energy in a precisely equal amount. A particle that pops into existence is like a bank robber who steals some energy from the vault. If the cosmic accountant catches the particle holding its stolen energy, someone will have to pay for the theft. However, the particle won't be caught as long as it pops back out of existence quickly enough, returning the stolen energy before the cosmic accountant notices that anything was ever missing. The length of time the particle has in which to pop back out of existence and return the stolen energy is a time short enough that the uncertainty principle prevents anyone from knowing that energy is missing.

Particles that pop in and out of existence before anyone can possibly detect them are called **virtual particles**. Modern theories of the universe propose that empty space—what we call a *vacuum*—actually "bubbles" with virtual particles that pop rapidly in and out of existence.

**Think about it** Imagine that you write a check for $100, but you have no money in your checking account. Your check is not necessarily doomed to bounce—as long as you deposit the needed $100 before the check clears. Explain how this $100 of "virtual money" is similar in concept to virtual particles.

You might wonder why a vacuum would seem so empty if it's actually teeming with virtual particles. The reason is that there's usually no way to extract the stolen energy from those virtual particles before they pop back out of existence. Consequently, the real particles that we can detect with our instruments cannot gain any energy by interacting with the virtual particles of the vacuum, making these vacuum particles essentially undetectable. However, the fact that we cannot detect these particles doesn't mean that they are unimportant; theory predicts that virtual particles should exert some measurable effects on real particles, and such effects have indeed been observed in specially designed experiments. For example, theory predicts that placing certain metal surfaces very close together should change the number of virtual photons in the vacuum between them, which should cause a change in the force acting between the metal surfaces. This change in force has been measured, and results agree with the theoretical predictions.

Moreover, the combined energy of the teeming sea of virtual particles in the vacuum, known as the **vacuum energy** (or *zero point energy*), may have a major impact on the universe as a whole. When we apply Einstein's general theory of relativity to the quantum idea of virtual particles, we find that the energy associated with the virtual particles in a vacuum can oppose the gravitational force associated with mass. In essence, the vacuum energy can act as a repulsive force, driving the objects in space farther apart even while the gravity of matter tries to pull them together. That is, the vacuum energy could in principle exert enough repulsion to cause the expansion of the universe to accelerate with time, and strong evidence suggests that the expansion rate is indeed increasing [**Section 23.4**]. However, calculations based on current theories of particle physics predict that if virtual particles were responsible for acceleration, the acceleration would be far greater (by many orders of magnitude) than the acceleration we observe. As a result, the question of whether vacuum energy can explain the acceleration of the expansion remains unanswered.

## Do black holes last forever?

Virtual particles are also thought to limit the lifetimes of black holes. We can understand why by extending our cosmic accountant analogy to the case of a virtual electron popping into existence near a black hole. Because electric charge must be conserved (there's no loophole for this conservation law), the virtual electron cannot pop into existence alone. Instead, it must be accompanied by a virtual positron (an antielectron) so that the total electric charge of the two particles is zero. Like all virtual particles, the virtual electron-positron pair essentially comes into existence with stolen energy, so to remain undetected the two particles must return the stolen energy before our imaginary cosmic accountant catches them. They return the energy by quickly annihilating each other (**FIGURE S4.13a**).

However, suppose a virtual electron-positron pair pops into existence very close to, but outside of, the event horizon of a black hole. The electron and positron are supposed to annihilate each other quickly, but a terrible problem arises: One of the particles crosses the event horizon during its brief virtual existence (**FIGURE S4.13b**). From the perspective of our cosmic accountant, this virtual particle

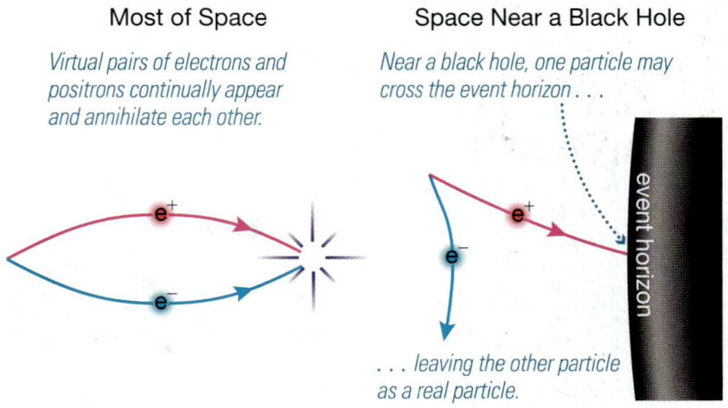


**a** Pairs of virtual electrons and positrons can pop into existence, as long as they annihilate each other before they can be detected.

**b** Near the event horizon of a black hole, this same process leads to the creation of real particles, not just virtual ones. The energy to make these particles comes at the expense of the black hole, which loses as much mass as the new particles gain.

**FIGURE S4.13** Virtual particles may become real particles near a black hole.

was never accounted for in the first place, so there's no problem with the fact that it will never be seen again. But the other particle is suddenly caught like a deer in headlights. Without its virtual mate, it has no way to annihilate itself and is caught red-handed by the cosmic accountant. Because the particle is now holding real rather than virtual energy, the cosmic accountant demands that someone or something pay for the stolen energy. The particle itself cannot pay, because it now has the energy by virtue of its existence and has no way to give it back once its virtual partner is gone. The cosmic accountant therefore makes the black hole pay for the energy by giving up a little bit of its gravitational potential energy.

If we strip away the fanciful imagery of a cosmic accountant, the end result is the creation of *real* particles, not virtual ones, just outside the event horizon of a black hole. Nothing is escaping from inside the black hole. Rather, these real particles are created from the gravitational potential energy of the black hole. Around the black hole, these *real* electrons and positrons annihilate each other, producing real photons that are radiated into space. To an outside observer, the black hole would therefore appear to be radiating, even though nothing ever escapes from inside it. This effect was first predicted by Stephen Hawking in the 1970s and is therefore called **Hawking radiation**. As we've just seen, the ultimate source of Hawking radiation is the gravitational potential energy, and hence the mass, of the black hole. The continual emission of Hawking radiation must therefore cause the black hole to shrink slowly in mass, or *evaporate*, over very long periods of time.

The idea that black holes can evaporate remains untested, but if it is true, it may have profound implications for both the origin and the fate of the universe. Some scientists speculate that black holes of all sizes might have been created during the Big Bang. If so, some of the smaller ones should be evaporating by now, emitting bursts of gamma rays during their final moments of existence. The fact that we have not yet detected any such evaporation sets limits on the number and size of small black holes that might have formed in the Big Bang.

At the other end of time, if the universe continues to expand forever, black holes may be the last large masses left after all the stars have died. In that case, the slow evaporation caused by Hawking radiation will mean that even black holes cannot last forever, and the universe eventually will contain nothing but a fog of photons and subatomic particles, separated from one another by incredible distances as the universe continues to grow in size.

## The BIG Picture PUTTING CHAPTER S4 INTO CONTEXT

*In this chapter, we have studied the quantum revolution and its astronomical consequences. As you look back, keep in mind the following "big picture" ideas:*

- The quantum revolution can be considered the third great revolution in our understanding of the universe. The first was the Copernican revolution, which demolished the ancient belief in an Earth-centered universe. The second was Einstein's discovery of relativity, which radically revised our ideas about space, time, and gravity. The quantum revolution has changed our ideas about the fundamental nature of matter and energy.
- Strange as the laws of quantum mechanics may seem, they can be tested and confirmed through observation and experiment. Quantum laws, like relativity, now stand on very solid ground. So far, they have passed every experimental test devised for them.
- The tiny quantum realm may seem remote from the large scales we are accustomed to in astronomy, but it is exceedingly important. The laws of quantum mechanics are necessary for understanding many astronomical processes, including nuclear fusion in the Sun, the degeneracy pressure that supports stellar corpses, and the possible evaporation of black holes.

**MY COSMIC PERSPECTIVE** The strange laws of the quantum realm are not just scientific curiosities; in fact, they are used in virtually every electronic device, and therefore have a direct and daily impact on our lives.

## Summary of Key Concepts

### S4.1 The Quantum Revolution

- **How has the quantum revolution changed our world?** Quantum mechanics has revolutionized our understanding of particles and forces and made possible the development of modern electronic devices such as computers.

### S4.2 Fundamental Particles and Forces

- **What are the basic properties of subatomic particles?**

The three most important properties of a particle are its *mass*, *charge*, and **spin**.

- **What are the fundamental building blocks of nature?** The **standard model** divides particles into two classes by spin:

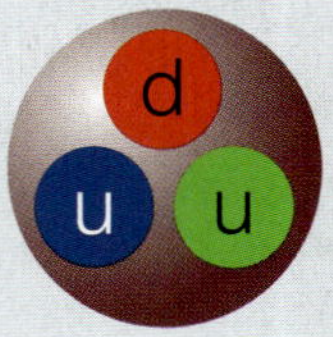


**fermions**, which include protons, neutrons, and electrons, and **bosons**, which include photons. The fermions are further subdivided into **quarks**, which make up protons and neutrons (as well as many other particles), and **leptons**, which include electrons and neutrinos. There are a total of six different types of quarks and six different types of leptons. Each type of particle of matter also has a corresponding particle of **antimatter**.

- **What are the fundamental forces in nature?** The four fundamental forces are **gravity**, **electromagnetism**, the **weak force**, and the **strong force**. Each force is transmitted through space by one or more **exchange particles**; for example, photons are the exchange particle for electromagnetism. Evidence suggests that the four forces are manifestations of a smaller number of truly fundamental forces.

### S4.3 Uncertainty and Exclusion in the Quantum Realm

- **What is the uncertainty principle?** The **uncertainty principle** tells us that we cannot simultaneously know the precise values of an object's position and momentum—or,

equivalently, its energy and the precise time during which it has this energy.

- **What is the exclusion principle?** The **exclusion principle** tells us that two fermions of the same type cannot occupy the same quantum state at the same time. This principle does not apply to bosons.

### S4.4 Key Quantum Effects in Astronomy

- **How do the quantum laws affect special types of stars?** **Degeneracy pressure** is a type of pressure that can occur even in the absence of heat. It arises from the combination of the exclusion principle and the uncertainty principle. It

is the dominant form of pressure in the astronomical objects known as brown dwarfs, white dwarfs, and neutron stars.

- **How is quantum tunneling crucial to life on Earth?** **Quantum tunneling** is a process in which subatomic particles can "tunnel" from one place to another even when they don't actually have enough energy to overcome an energy barrier between the two places. Without this process, fusion in the Sun would not be possible and life on Earth could not exist. It is also important to many other astronomical processes and to modern technology.

- **How empty is empty space?** According to the uncertainty principle, a vacuum cannot be completely empty but must instead be filled with unobservable **virtual particles** that are constantly popping in and out of existence. While virtual particles generally do not affect the particles that we can observe, the **vacuum energy** associated with them may oppose gravity, and in principle could cause the expansion of the universe to accelerate.
- **Do black holes last forever?** No. According to current theory, isolated black holes can gradually evaporate, emitting **Hawking radiation** in the process. Although the theoretical basis of this evaporation seems solid, it has not yet been observed in nature.

# Exercises and Problems

*For instructor-assigned homework and other learning materials, go to www.MasteringAstronomy.com.*

## Chapter Review Questions

### Short-Answer Questions Based on the Reading

1. What do we mean by the *quantum* realm? List five major ideas that come from the laws of *quantum mechanics*.
2. What is *spin*? What are the two basic categories of particles based on spin?
3. List the six *quarks* and six *leptons* in the standard model. Describe the quark composition of a proton and of a neutron.
4. What is *antimatter*? What happens when a particle and its antiparticle meet? Why is antimatter always produced along with matter in *pair production*?
5. List the four fundamental forces in nature, and name the *exchange particles* for each.
6. Describe the relative strengths of the four forces. Why does gravity dominate on large scales, even though it is by far the weakest of the four forces?
7. Why do many scientists believe that the standard model will eventually be replaced by an even simpler model of nature?
8. What is the *uncertainty principle*? How is it related to the idea of *wave-particle duality*?
9. Describe two ways of quantifying the uncertainty principle, and give an example showing the meaning of each.
10. What do we mean by the *quantum state* of a particle?
11. What is the *exclusion principle*? What types of particles obey it? Briefly explain how the exclusion principle determines how electrons fill energy levels in atoms.
12. What is *degeneracy pressure*? How does it differ from *thermal pressure*? How is it important in astronomy?
13. What is *quantum tunneling*? How is it important to modern electronics? How is it important to nuclear fusion in the Sun?
14. What do we mean by *virtual particles*? How might they help explain the observed acceleration of the expansion of the universe? How might they lead to gradual evaporation of black holes through *Hawking radiation*?

### Does It Make Sense?

*Decide whether the statement makes sense (or is clearly true) or does not make sense (or is clearly false). Explain clearly; not all of these have definitive answers, so your explanation is more important than your chosen answer.*

15. Although there are six known types of quarks, ordinary atoms contain only two of these types.
16. If you put a quark and a lepton close together, they'll annihilate each other.
17. There's no such thing as antimatter, except in science fiction.
18. Some particle accelerators have been known to build up a huge electrical charge because of the electrons produced inside them.
19. According to the uncertainty principle, we can never be certain whether one theory is really better than another.
20. The exclusion principle describes the cases in which the uncertainty principle is excluded from being true.
21. No known astronomical objects exhibit any type of degeneracy pressure.
22. Although we speak of four fundamental forces—gravity, electromagnetism, strong force, and weak force—it is likely that these forces are different manifestations of a smaller number of truly fundamental forces.
23. Imagine that, somewhere in deep space, you met a person made entirely of antimatter. Shaking that person's hand would be very dangerous.
24. Someday, we may detect radiation coming from an evaporating black hole.

### Quick Quiz

*Choose the best answer to each of the following. Explain your reasoning with one or more complete sentences. For additional practice, try the Chapter S4 Reading and Concept Quizzes in the Study Area at www.MasteringAstronomy.com.*

25. The fundamental particles of matter are (a) atoms and molecules. (b) quarks and leptons. (c) electrons, protons, and neutrons.
26. When an electron is produced from energy in a particle accelerator, the following also always happens: (a) An antielectron (positron) is produced. (b) A quark is produced. (c) The electron disintegrates into a bunch of quarks.
27. Within an atomic nucleus, the strongest of the four forces is (a) gravity. (b) electromagnetism. (c) the strong force.
28. Across a distance of 1 millimeter, the strongest force acting between two protons is (a) gravity. (b) electromagnetism. (c) the strong force.
29. If we measure a subatomic particle's position very precisely, then (a) its momentum will be highly uncertain. (b) its spin will be highly uncertain. (c) we violate the uncertainty principle.
30. In addition to photons, which of the following also behave sometimes as waves and sometimes as particles? (a) fermions (b) bosons (c) all other subatomic particles
31. Which of the following is not allowed by the exclusion principle? (a) knowing both the precise position and the precise

momentum of an electron (b) having two electrons in the same quantum state (c) having an electron with a spin that is neither up nor down

32. The strength of degeneracy pressure in an object (such as a white dwarf) depends on (a) its temperature. (b) its density. (c) both its temperature and its density.
33. In which of the following objects does degeneracy pressure play the most important role? (a) a neutron star (b) the Sun (c) a star 10 times as massive as the Sun
34. According to modern theories based on quantum mechanics, empty space (a vacuum) is (a) truly empty. (b) bubbling with virtual particles. (c) filled with tiny black holes.

## Inclusive Astronomy

*Use these questions to reflect on participation in science.*

35. *Group Discussion: Large Collaborations.* The world's largest particle accelerators are built and operated by teams of thousands of scientists. As a result, a single scientific paper describing research results from such a project may list hundreds of scientists as co-authors of the paper.
**a.** Scientists' career prospects depend heavily on the number of papers they publish and the significance of their contributions to those papers. Discuss why scientists who participate in a large collaboration may face difficulty in getting recognized for their contributions.
**b.** Often, the job prospects of junior scientists who are members of such collaborations depend on recommendation letters written by senior members of the collaboration. List some of the reasons those letters might be biased.
**c.** List and discuss some ways in which the scientific community can account for the biases that might be present in those recommendation letters.
**d.** Try to come to a group consensus on whether the potential biases you've discussed are likely to hurt the goal of making the scientific community more representative of the community at large. If you conclude that they are, decide what might be done to minimize the problem.

## The Process of Science

*These questions may answered individually in short-essay form or discussed in groups, except where identified as group-only.*

36. *Hallmarks of Science.* This chapter discussed many ideas that may seem quite bizarre, from the existence of quarks to the uncertainty principle. Choose one of these ideas and evaluate it in terms of the hallmarks of science from Chapter 3. Is the idea science? Defend your opinion.
37. *Common Sense vs. Experiment.* Even the most highly trained physicists find the results of quantum mechanics to be strange and counter to their everyday common sense, but the predictions of quantum mechanics have passed every experimental test yet posed for them. Does this difficulty in reconciling common sense with experiment or theory pose any problems for science? Defend your opinion.
38. *The Meaning of the Uncertainty Principle.* When first hearing about it, many people assume that the uncertainty principle means that we cannot *measure* the position and momentum of a particle precisely. According to current understanding, it really tells us that the particle *does not have* a precise position and momentum in the sense that we would expect from everyday life. How do these two viewpoints differ? Discuss the different philosophical consequences of these two viewpoints.
39. *Funding Big Science.* Large particle accelerators cost billions of dollars, more even than large telescopes in space. If you were a member of Congress and the government could afford either a new accelerator or a new large space observatory, which would you choose? Why?
40. *Group Activity: New Particle Discovery.* Form a small group, and imagine that your group is in charge of an important particle physics experiment. A junior scientist comes to your group reporting that she or he has discovered a new particle. Make a list of questions that, as responsible scientists, your group should ask and answer before announcing this claimed discovery to the world. Note: You may wish to do this activity using the four roles described in Chapter 1, Exercise 39.

## Investigate Further

### Short-Answer/Essay Questions

41. *The Strong Force.* The strong force is the force that holds the protons and neutrons in the nucleus together. Based on the fact that most atomic nuclei are stable, briefly explain how you can conclude that the strong force must be even stronger than the electromagnetic force, at least over very short distances.
42. *Chemistry and Biology.* All chemical and biological reactions involve the creation and breaking of chemical bonds, which are bonds created by interactions between the electrons of one atom and the electrons of others. Given this fact, explain why the electromagnetic force governs all chemical and biological reactions. Also explain why the strong force, the weak force, and gravity play no role in these reactions.
43. *Gravity.* In one or two paragraphs, explain both (1) what we mean when we say that gravity is the weakest of the four forces and (2) why it nevertheless dominates the universe on large scales.
44. *Quantum Tunneling and Life.* In one or two paragraphs, explain the role of quantum tunneling in creating the elements from which we are made. (*Hint:* Recall that we are *star stuff* in the sense that the elements of our bodies were produced by nuclear fusion inside stars.)
45. *Antimatter Engines.* In the *Star Trek* series, starships are powered by matter-antimatter annihilation. Explain why we should expect matter-antimatter annihilation to be the most efficient possible source of power. What practical problems would we face in developing matter-antimatter engines?
46. *Exchange of Information.* Suppose an electron is moving up and down at some place located far in the distance. How could we learn that the electron is moving? How long after the movement would we know that it had occurred? Does this idea have any practical importance in astronomy? Explain. (*Hint:* Remember that the photon is the exchange particle for the electromagnetic force.)
47. *Nonquantum Baseball.* In your own words, explain why a baseball player does not have to take into account the uncertainty principle when up at bat.
48. *The Electron.* Based on what you have learned in this chapter, how would you explain an electron to a friend? Be sure that you explain why physicists don't think of electrons as tiny, negatively charged, spinning balls. Write your explanation in one or two paragraphs.

49. *Movie Review.* Several recent movies deal with concepts covered in this chapter, including the Oscar-winning *Theory of Everything* and the award-winning documentary *Particle Fever*, about the discovery of the Higgs boson. Choose one of these movies and write a one-page critical review, focusing on its entertainment and its educational value.

50. *Research: Quantum Computing.* Learn how computer scientists hope to harness quantum effects to build computers much more powerful than any existing today. Briefly summarize the ideas, and write an essay stating your opinion concerning the benefits and drawbacks of developing this technology.

51. *Research: Beyond the Standard Model.* Research some of the ideas that physicists are considering as possible improvements on the standard model. Choose one such idea, and write a short essay describing its potential effect on physics if it is correct, and how the idea may be tested.

## Quantitative Problems

*Be sure to show all calculations clearly and state your final answers in complete sentences.*

52. *Gravity and the EM Force.* In this problem, we compare the strength of gravity to the strength of the electromagnetic (EM) force for two interacting electrons. Because both electrons are negatively charged, they *repel* each other through the EM force. Because electrons have mass, they *attract* each other through gravity. Let's see which effect will dominate. You will need the following information for this problem:
   - The force law for gravitation is

$$F_g = G \times \frac{M_1 M_2}{d^2} \qquad \left(G = 6.67 \times 10^{-11} \frac{\text{N} \times \text{m}^2}{\text{kg}^2}\right)$$

   where $M_1$ and $M_2$ are the masses of the two objects, $d$ is the distance between them, and $G$ is the gravitational constant. (N is the abbreviation for *newton*, the metric unit of force.)
   - The force law for electromagnetism is

$$F_{\text{EM}} = k\frac{q_1 q_2}{d^2} \qquad \left(k = 9.0 \times 10^9 \frac{\text{N} \times \text{m}^2}{\text{Coul}^2}\right)$$

   where $q_1$ and $q_2$ are the *charges* of the two objects (in *coulombs*, the standard unit of charge), $d$ is the distance between them, and $k$ is a constant. (Coul is an abbreviation for *coulomb*.)
   - The *mass* of an electron is $9.10 \times 10^{-31}$ kg.
   - The *charge* of an electron is $-1.6 \times 10^{-19}$ Coul.

   **a.** Calculate the gravitational force, in newtons, that attracts the two electrons if a distance of $10^{-10}$ m (about the diameter of an atom) separates them.
   **b.** Calculate the electromagnetic force, in newtons, that repels the two electrons at the same distance.
   **c.** How many times stronger is the electromagnetic repulsion than the gravitational attraction between the two electrons?

53. *Large-Scale Gravity.* Suppose Earth and the Sun each had an excess charge equivalent to the charge of just one electron. Using the data from Problem 52, compare the gravitational force between Earth and the Sun to the electromagnetic force between them in that case. What does your answer tell you about why gravity dominates on large scales? (*Hint:* See Appendix E for the data you'll need about Earth and the Sun.)

54. *Evaporation of Black Holes.* The time it takes for a black hole to evaporate through the process of Hawking radiation can be calculated using the following formula, in which $M$ is the mass of the black hole in kilograms and $t$ is the lifetime of the black hole in seconds:

$$t = 10{,}240\pi^2 = \frac{G^2 M^3}{hc^4}$$

$$\left(h = 6.63 \times 10^{-34} \frac{\text{kg} \times \text{m}^2}{\text{s}}, G = 6.67 \times 10^{-11} \frac{\text{m}^3}{\text{kg} \times \text{s}^2}\right)$$

   Without doing any calculations, explain how this formula implies that lower-mass black holes have much shorter lifetimes than more massive ones and that the evaporation process accelerates as a black hole loses mass.

55. *Solar Mass Black Holes.* Use the formula from Problem 54 to calculate the lifetime of a black hole with the mass of the Sun ($M_{\text{Sun}} = 2.0 \times 10^{30}$ kg). How does your answer compare to the current age of the universe?

56. *Long-Lived Black Holes.* Some scientists speculate that in the far distant future, the universe will consist only of gigantic black holes and scattered subatomic particles. The largest black holes that conceivably might form would have a mass of about a trillion ($10^{12}$) Suns. Using the formula from Problem 54, calculate the lifetime of such a giant black hole. How does your answer compare to the current age of the universe? (*Hint:* Your calculator may not be able to handle the large numbers involved in this problem, in which case you will need to rearrange the numbers so that you can calculate the powers of 10 without your calculator.)

57. *Mini–Black Holes.* Some scientists speculate that black holes of many different masses might have been formed during the early moments of the Big Bang. Some of these black holes might be mini–black holes, much smaller in mass than those that can be formed by the crush of gravity in today's universe. Use the formula from Problem 54 to calculate the lifetime of a mini–black hole with the mass of Earth (about $6 \times 10^{24}$ kg). Compare this to the current age of the universe.

58. *Black Holes Evaporating Today.* Starting with the formula from Problem 54, calculate the mass of a mini–black hole that was formed in the Big Bang and would be completing its evaporation today. Compare your answer to the mass of Earth. Assume that the universe is 14 billion years old.

59. *Your Quantum Uncertainty.* Suppose you are running at a speed of about 10 km/hr, but there is an uncertainty of 0.5 km/hr in your precise speed. Given your mass, you can calculate your momentum and the uncertainty of that momentum. What is the corresponding quantum limit to the measurement of your position? Is this significant? Why or why not? (*Hint:* You'll need the first form of the uncertainty principle given on page 453; you can use the value of Planck's constant, *h*, from Problem 54.)

60. *An Electron's Quantum Uncertainty.* You are conducting an experiment in which you can measure the location of individual electron collisions to within $10^{-10}$ m. What is the theoretical limit to which you can simultaneously measure the momentum of those collisions? What is the uncertainty in the electron's speed? (The electron has a rest mass of $9.1 \times 10^{-31}$ kg. )

# A Deeper Look at Nature

**We all have "common sense" ideas about the meaning of space, time, matter, and energy, and in most cases these common-sense ideas serve us quite well. But when we look more closely, we find that our everyday ideas cannot explain all that we observe in nature. The theories of relativity and quantum mechanics have given us a deeper understanding of the nature of space, time, matter, and energy—an understanding that now underlies almost all of our modern understanding of the universe.**

1 **Time Dilation:** Einstein's theory of relativity is based on two simple principles: (1) the laws of nature are the same for everyone, and (2) the speed of light is the same for everyone. One implication of these principles is that the passage of time is relative—time can run at different speeds for different observers. Many observations have confirmed this mind-boggling prediction of relativity, and astronomers must account for it when studying the universe [Chapter S2].

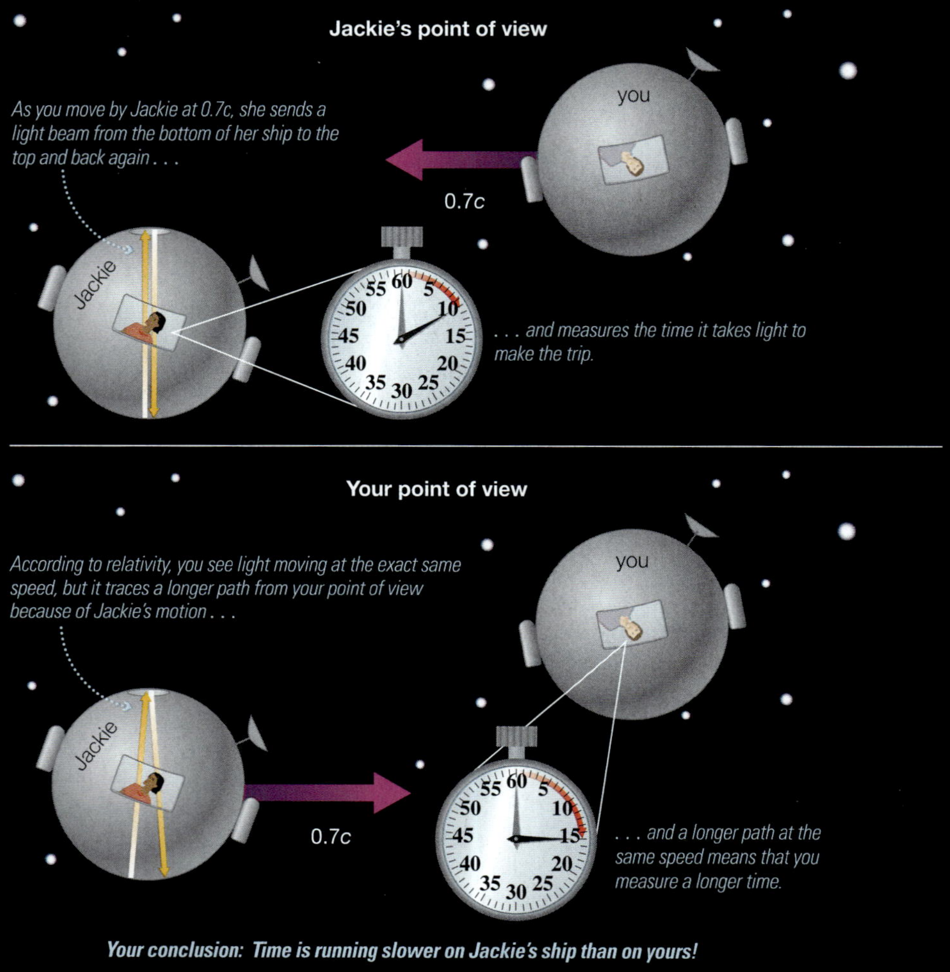


2 **Matter and Antimatter:** Relativity tells us that mass and energy are equivalent: $E = mc^2$. This formula explains how small amounts of matter can release huge amounts of energy, and how matter-antimatter particle pairs can be produced from pure energy. Pair production has been observed countless times in Earth-based particle accelerators and is crucial to our models of the early universe [Sections S4.2, 22.1].

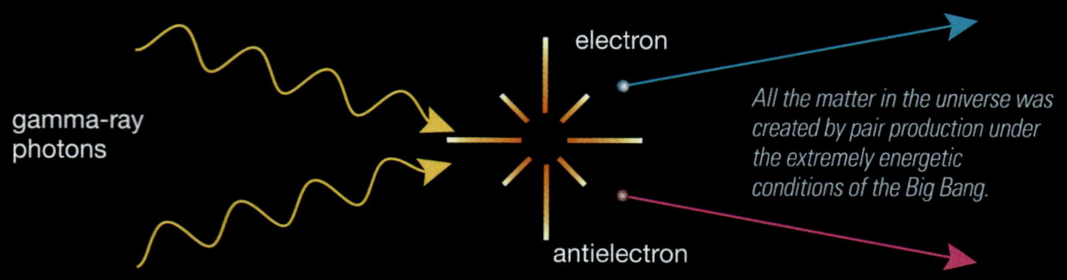

**3** **Curvature of Spacetime:** Special relativity tells us that space and time are inextricably linked as four-dimensional spacetime, and general relativity tells us that gravity arises from curvature of spacetime. Astronomical observations, both in our solar system and beyond, have confirmed Einstein's predictions about the structure of space and time [Chapter S3, Section 23.2].

*Curvature of spacetime precisely explains the precession of Mercury's elliptical orbit around the Sun.*

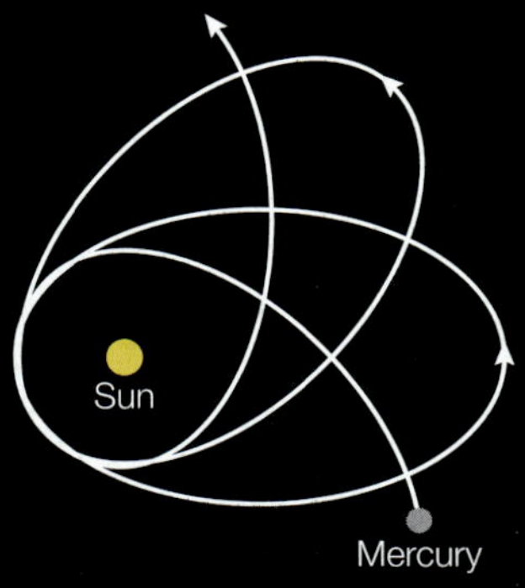


*Spacetime curvature also explains the bending of light known as gravitational lensing.*

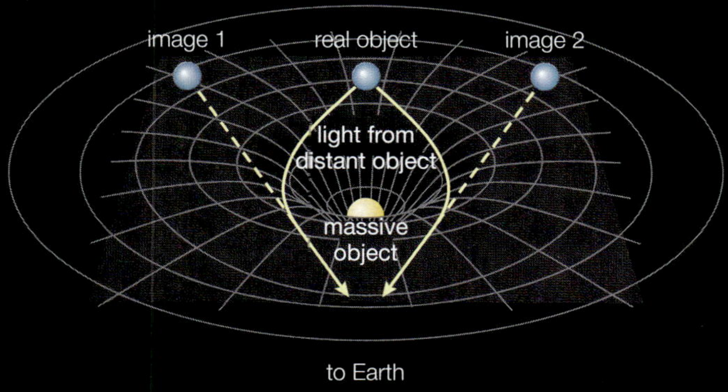


*The gravity of the galaxy at the center of this image bends the light from a single object behind it . . .*

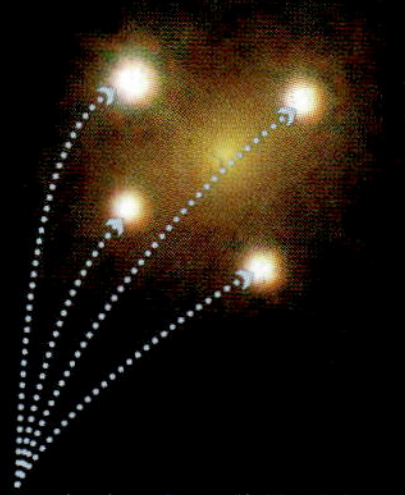

*. . . producing four distinct images of the object.*

**4** **Black Holes:** Relativity predicts that there can be objects whose gravity is so strong that nothing can escape from inside them— not even light. Observations of these *black holes* help us test the most extreme predictions of relativity [Sections S3.3, 18.3, 19.4].

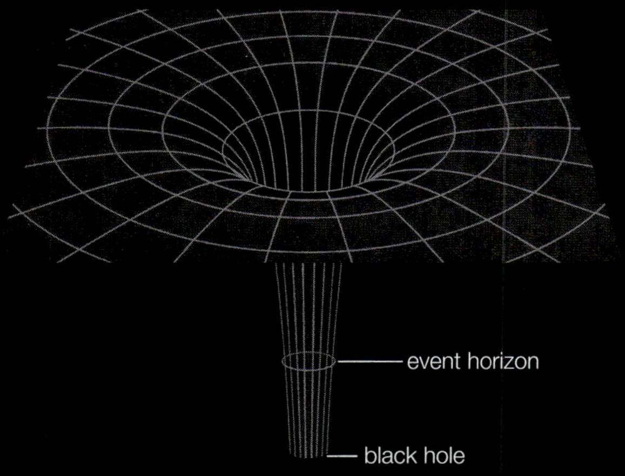


*Squeezing a star down to the size of a small city would make the curvature of spacetime (gravity) around it so extreme that the star would turn into a black hole.*

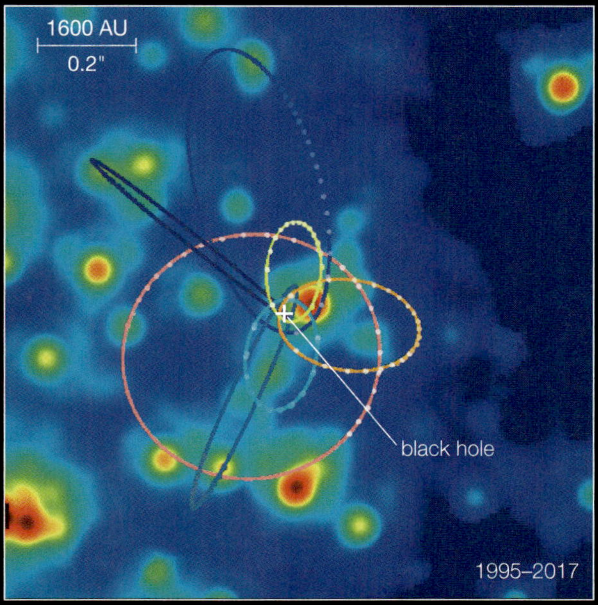


*The orbits of stars at the center of the Milky Way Galaxy indicate that it contains a black hole about 4 million times as massive as our Sun.*

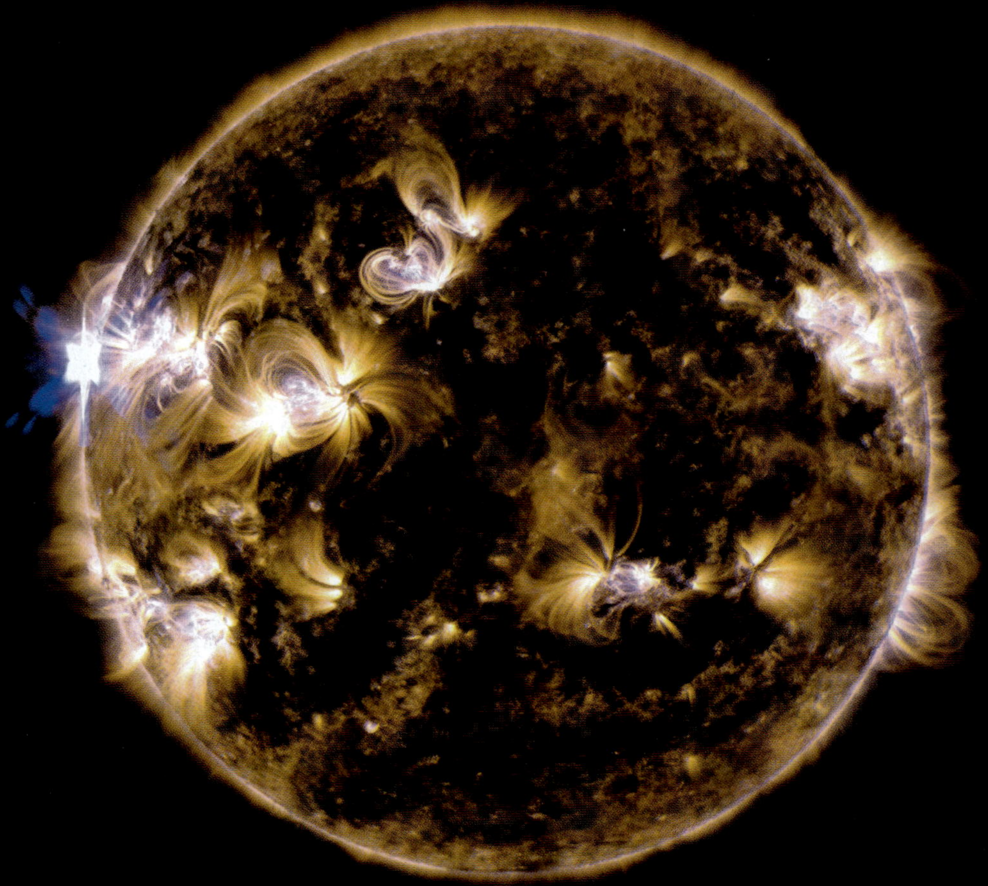

# 14 Our Star

▲ **About the photo:** NASA's *Solar Dynamics Observatory* captured this x-ray image of our Sun, with a solar flare in progress (bright spot at far left just above center). Colors correspond to the strength of the x-ray emission, with the brightest emission in blue and purple.

## LEARNING GOALS

### 14.1 A Closer Look at the Sun
- Why does the Sun shine?
- What is the Sun's structure?

### 14.2 Nuclear Fusion in the Sun
- How does nuclear fusion occur in the Sun?
- How does the energy from fusion get out of the Sun?
- How do we know what is happening inside the Sun?

### 14.3 The Sun-Earth Connection
- What causes solar activity?
- How does solar activity vary with time?

*Give me the splendid silent sun with*
*all his beams full-dazzling.*

—*Walt Whitman (1819–1892), from* Leaves of Grass

**Chapter 14 Overview**

Astronomy today encompasses the study of the entire universe, but the root of the word *astronomy* comes from the Greek word for "star." Although we have learned a lot about the universe up to this point in the book, only now do we turn our attention to the study of the stars, the namesakes of astronomy.

When we think of stars, we usually think of the beautiful points of light visible on a clear night. But the nearest and most easily studied star is visible only in the daytime—our Sun. In this chapter, we will study the Sun in some detail. We will see how the Sun generates the energy that supports life on Earth. Equally important, we will study our Sun as a star so that it can serve as an introduction to subsequent chapters in which we will study stars throughout the universe.

## 14.1 A Closer Look at the Sun

Most people know that the Sun is a star—a giant ball of hot gas that generates light and shines it brightly in all directions. However, scientists realized this fact only recently in human history (see Special Topic, p. 375). In this section, we'll consider the question of what makes the Sun shine, and then take an imaginary plunge into the Sun that will get you better acquainted with its general features.

### Why does the Sun shine?

The Sun's energy is vital to human existence. Ancient peoples certainly recognized that fact. Some worshipped the Sun as a god. Others created mythologies to explain its daily rise and set. But no one who lived before the 20th century knew the energy source for the Sun's light and heat.

**Ancient Ideas** Ancient thinkers often imagined the Sun to be some type of fire, perhaps a lump of burning coal or wood. It was a reasonable suggestion for the times, because people did not know the size or distance of the Sun and therefore did not realize how incredible its energy output really is. Nor did they know how long Earth had existed, so they had no way to know that the Sun had been shining for more than 4 billion years.

The situation began to change in the mid-19th century, after the Sun's distance and size had been measured with reasonable accuracy. Scientists could then calculate the true energy output of the Sun, and this quickly ruled out coal, wood, or any other type of chemical burning: There is simply no way that chemical processes can account for the Sun's huge energy output.

**Gravitational Contraction** In the late 19th century, astronomers came up with an idea that seemed more plausible, at least at first. They suggested that the Sun generates energy by slowly contracting in size, a process called **gravitational contraction** (or *Kelvin-Helmholtz contraction*, after the scientists who suggested it). Recall that a shrinking gas cloud heats up because the gravitational potential energy of gas particles far from the center of the cloud is converted into thermal energy as the gas moves inward (see Figure 4.15b). A gradually shrinking Sun would always have some gas moving inward, converting gravitational potential energy into thermal energy. This thermal energy would keep the inside of the Sun hot.

Because of its large mass, the Sun would need to contract only very slightly each year to maintain its temperature—so slightly that the contraction would have been unnoticeable to 19th-century astronomers. Calculations showed that gravitational contraction could have kept the Sun shining steadily for up to about 25 million years. For a while, some astronomers thought that this idea had solved the ancient mystery of how the Sun shines. However, geologists pointed out a fatal flaw: Studies of rocks and fossils had already suggested that Earth was far older than 25 million years, which meant that gravitational contraction could not account for the Sun's long-term energy generation.

**Einstein's Breakthrough** With both chemical processes and gravitational contraction ruled out as possible explanations for why the Sun shines, scientists were at a loss. There was no known way that an object the size of the Sun could generate so much energy for billions of years. A completely new type of explanation was needed, and it came with Einstein's publication of his special theory of relativity in 1905.

Einstein's theory included his famous equation, $E = mc^2$ [**Sections 4.3, S2.3**], which tells us that mass itself contains an enormous amount of potential energy. Calculations showed that the Sun's mass contained more than enough energy to account for billions of years of sunshine, if the Sun could somehow convert some of its mass into thermal energy. It took a few decades for scientists to work out the details, but by the end of the 1930s we had learned that the Sun converts mass into energy through the process of *nuclear fusion* [**Section 1.2**].

**How Fusion Started** Nuclear fusion requires extremely high temperatures and densities (for reasons we will discuss in the next section). In the Sun, these conditions are found deep in the core. But how did the Sun become hot enough for fusion to begin in the first place?

The answer invokes the mechanism of gravitational contraction. Recall that our Sun was born about $4\frac{1}{2}$ billion years ago from a collapsing cloud of interstellar gas [**Section 8.2**]. The contraction of the cloud released gravitational potential energy, raising the interior temperature and pressure. This process continued until the core finally became hot enough to sustain nuclear fusion, because only then did the Sun produce enough energy to give it the stability that it has today.

**The Stable Sun** The Sun continues to shine steadily today because it has achieved two kinds of balance that keep its size and energy output stable. The first kind of balance, called **gravitational equilibrium** (or *hydrostatic*

**FIGURE 14.1** An acrobat stack is in gravitational equilibrium: The lowest person supports the most weight and feels the greatest pressure, and the overlying weight and underlying pressure decrease for those higher up.

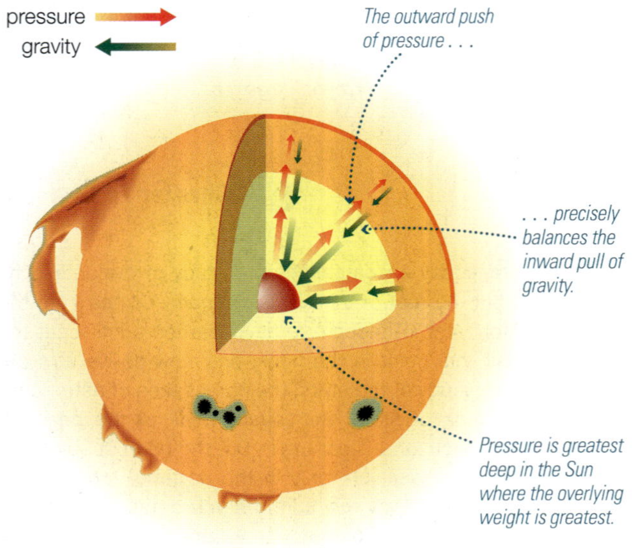


**FIGURE 14.2** Gravitational equilibrium in the Sun: At each point inside, the pressure pushing outward balances the weight of the overlying layers.

The second kind of balance is **energy balance** between the rate at which fusion releases energy in the Sun's core and the rate at which the Sun's surface radiates this energy into space (**FIGURE 14.3**). Energy balance is important because without it, the balance between pressure and gravity would not remain steady. If fusion in the core did not replace the energy radiated from the surface, thereby keeping the total thermal energy content constant, then gravitational contraction would cause the Sun to shrink and force its core temperature to rise.

*equilibrium*), is between the outward push of internal gas pressure and the inward pull of gravity. A stack of acrobats provides a simple example of gravitational equilibrium (**FIGURE 14.1**). The bottom person supports the weight of everyone above, so he must push upward with enough pressure to support all this weight. At each higher level, the overlying weight is less, so it's a little easier for each additional person to hold up the rest of the stack.

Gravitational equilibrium works much the same way in the Sun, except the outward push against gravity comes from internal gas pressure. The Sun's internal pressure precisely balances gravity at every point within it, thereby keeping the Sun stable in size (**FIGURE 14.2**). Because the weight of overlying layers is greater as we look deeper into the Sun, the pressure must increase with depth. Deep in the Sun's core, the pressure makes the gas hot and dense enough to sustain nuclear fusion. The energy released by fusion, in turn, heats the gas and maintains the pressure that keeps the Sun in balance against the inward pull of gravity.

**Think about it** Earth's atmosphere is also in gravitational equilibrium, with the weight of upper layers supported by the pressure in lower layers. Use this idea to explain why the air gets thinner at higher altitudes.

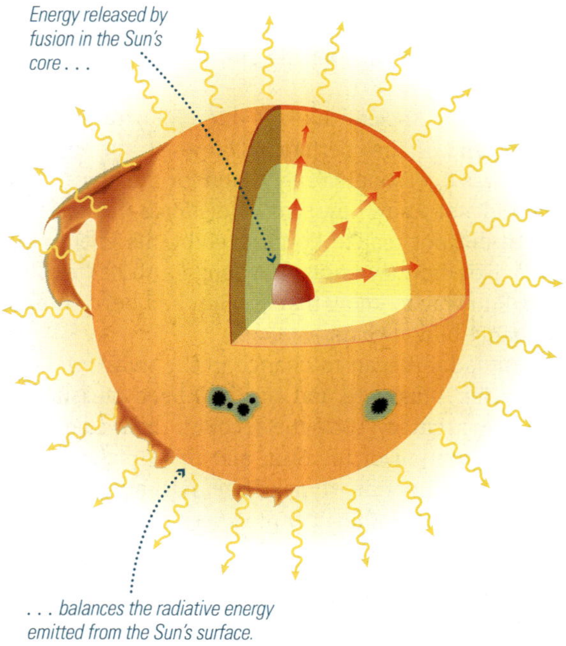


**FIGURE 14.3** Energy balance in the Sun: Fusion supplies energy in the core at the same rate the Sun radiates energy from its surface.

In summary, the answer to the question "Why does the Sun shine?" is that about $4\frac{1}{2}$ billion years ago *gravitational contraction* made the Sun hot enough to sustain nuclear fusion in its core. Ever since, energy liberated by fusion has maintained *gravitational equilibrium* and *energy balance* within the Sun, keeping it shining steadily. The Sun was born with enough nuclear fuel to last about 10 billion years, which means it is now only about halfway through this 10-billion-year lifetime. About 5 billion years from now, when the Sun finally exhausts its nuclear fuel, gravitational contraction will begin once again. As we will see in later chapters, some of the most important and spectacular processes in astronomy arise from the changes that occur as the crush of gravity begins to overcome a star's internal sources of pressure.

## What is the Sun's structure?

The Sun is essentially a giant ball of hot gas or, more technically, *plasma*—a gas in which atoms are ionized because of the high temperature [**Section 5.3**]. The differing temperatures and densities of the plasma at different depths give the Sun the layered structure shown in **FIGURE 14.4**. To make sense of what you see in the figure, let's imagine that you have a spaceship that can somehow withstand the immense heat and pressure as you take an imaginary journey from Earth to the center of the Sun.

**Basic Properties of the Sun** As you begin your journey from Earth, the Sun appears as a whitish ball of glowing gas. Just as astronomers have done in real life, you can use simple observations to determine basic properties of the Sun. Spectroscopy [**Section 5.4**] tells you that the Sun is made almost entirely of hydrogen and helium. From the Sun's angular size and distance (see Mathematical Insight 2.1), you can determine that its radius is just under 700,000 kilometers, or more than 100 times the radius of Earth. Even **sunspots**, visible splotches that appear darker than the surrounding surface, can be larger in size than Earth (**FIGURE 14.5**).

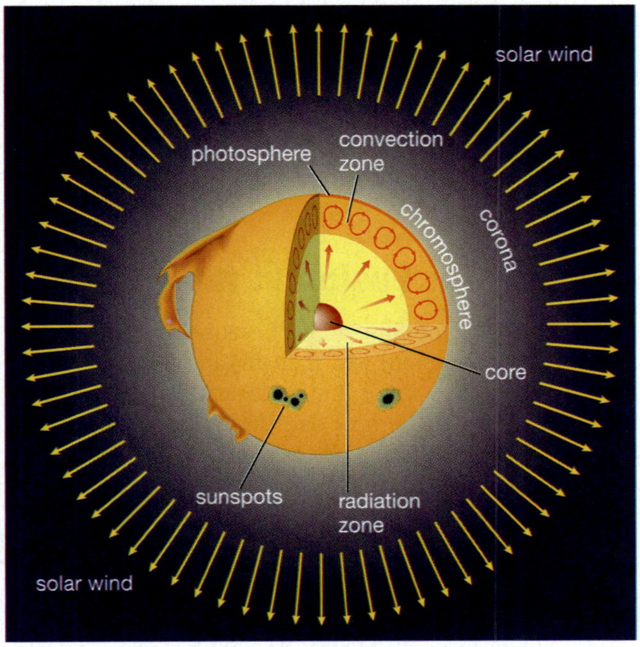


**FIGURE 14.4** The basic structure of the Sun.

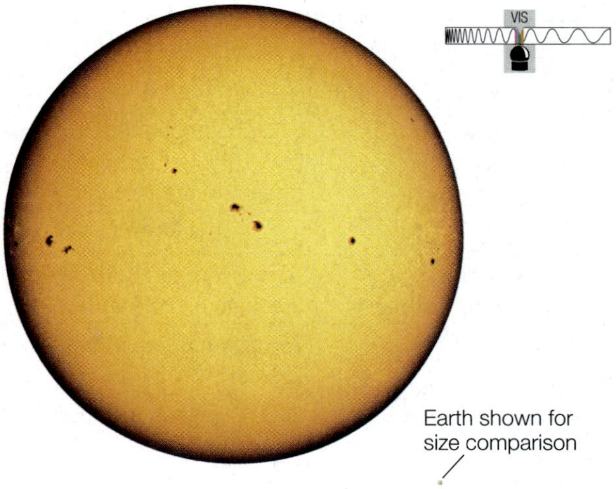


**FIGURE 14.5** This photo of the visible surface of the Sun shows several dark sunspots, each large enough to swallow our entire planet.

You can measure the Sun's mass using Newton's version of Kepler's third law [**Section 4.4**]. It is about $2 \times 10^{30}$ kilograms, which is about 300,000 times the mass of Earth and nearly 1000 times the mass of all the planets in our solar system put together. You can observe the Sun's rotation rate by tracking the motion of sunspots or by measuring Doppler shifts on opposite sides of the Sun. Unlike a spinning ball, the entire Sun does *not* rotate at the same rate: The solar equator completes one rotation in about 25 days, and the rotation period increases with latitude to about 30 days near the solar poles.

**Think about it** As a brief review, describe how astronomers use Newton's version of Kepler's third law to determine the mass of the Sun. What two properties of Earth's orbit do we need to know in order to apply this law? (*Hint*: See Section 4.4.)

The Sun releases an enormous amount of radiative energy into space. Recall that, in science, we measure energy in units of joules [**Section 4.3**]. We define **power** as the *rate* at which energy is used or released [**Section 5.1**]. The standard unit of power is the *watt*, defined as 1 joule of energy per second; that is, 1 watt = 1 joule/s. For example, a 100-watt light bulb requires 100 joules of energy for every second it is left turned on. The Sun's total power output, or **luminosity**, is an incredible $3.8 \times 10^{26}$ watts. If we could somehow capture and store just 1 second's worth of the Sun's luminosity, it would be enough to meet current human energy demands for roughly the next 500,000 years. **TABLE 14.1** summarizes the basic properties of the Sun.

Of course, only a tiny fraction of the Sun's total energy output reaches Earth, since it goes in all directions into space. Most of this energy is radiated in the form of visible and infrared light, but after you've left the protection of Earth's atmosphere, you encounter significant amounts of more dangerous types of solar radiation, including ultraviolet and x-rays.

**TABLE 14.1 Basic Properties of the Sun**

| | |
|---|---|
| Radius ($R_{Sun}$) | 696,000 km (about 109 times the radius of Earth) |
| Mass ($M_{Sun}$) | $2 \times 10^{30}$ kg (about 300,000 times the mass of Earth) |
| Luminosity ($L_{Sun}$) | $3.8 \times 10^{26}$ watts |
| Composition (by percentage of mass) | 70% hydrogen, 28% helium, 2% heavier elements |
| Rotation rate | 25 days (equator) to 30 days (poles) |
| Surface temperature | 5800 K (average); 4000 K (sunspots) |
| Core temperature | 15 million K |

**The Sun's Atmosphere** Even at great distances from the Sun, you and your spacecraft can feel slight effects from the **solar wind**—the stream of charged particles continually blown outward in all directions from the Sun. The solar wind helps shape the magnetospheres of planets [**Sections 10.1** and **11.1**] and blows back the material that forms the plasma tails of comets [**Section 12.3**].

As you approach the Sun more closely, you begin to encounter the low-density gas that represents what we usually think of as the Sun's atmosphere. The outermost layer of this atmosphere, called the **corona**, extends several million kilometers above the visible surface of the Sun. The temperature of the corona is astonishingly high—about 1 million K—explaining why this region emits most of the Sun's x-rays. However, the corona's density is so low that your spaceship absorbs relatively little heat despite the million-degree temperature [**Section 4.3**].

Nearer the surface, the temperature suddenly drops to about 10,000 K in the **chromosphere**, the middle layer of the solar atmosphere and the region that radiates most of the Sun's ultraviolet light. Then you plunge through the lowest layer of the atmosphere, or **photosphere**, which is the visible surface of the Sun. Although the photosphere looks like a well-defined surface from Earth, it consists of gas far less dense than Earth's atmosphere. The temperature of the photosphere averages just under 6000 K, and its surface seethes and churns like a pot of boiling water. The photosphere is also where you'll find sunspots, regions of intense magnetic fields that would cause your compass needle to swing about wildly.

### COMMON MISCONCEPTIONS

#### The Sun Is Not on Fire

We often say that the Sun is "burning," a term that conjures up images of a giant bonfire in the sky. However, the Sun does not burn in the same sense as a fire burns on Earth. Fires generate light through chemical changes that consume oxygen and produce a flame. The glow of the Sun has more in common with the glow of embers left over after the flames have burned out. Much like hot embers, the Sun's surface shines because it is hot enough to emit thermal radiation that includes visible light [**Section 5.4**].

Hot embers quickly stop glowing as they cool, but the Sun keeps shining because its surface is kept hot by the energy rising from its core. Because this energy is generated by nuclear fusion, scientists sometimes say that it is the result of "nuclear burning"—a term intended to suggest nuclear changes in much the same way that "chemical burning" suggests chemical changes. While it may be reasonable to say that the Sun undergoes nuclear burning in its core, it is not accurate to speak of any kind of burning on the Sun's surface, where light is produced primarily by thermal radiation.

**The Sun's Interior** Up to this point in your journey, you may have seen Earth and the stars when you looked back. But blazing light engulfs you as you slip beneath the photosphere. You are inside the Sun, and incredible turbulence tosses your spacecraft about. If you can hold steady long enough to see what is going on around you, you'll notice spouts of hot gas rising upward, surrounded by cooler gas cascading down from above. You are in the **convection zone**, where energy generated in the solar core travels upward, transported by the rising of hot gas and falling of cool gas called *convection* [**Section 9.1**]. The photosphere above you is the top of the convection zone, and convection is the cause of the Sun's seething, churning appearance.

About a third of the way down to the center, the turbulence of the convection zone gives way to the calmer plasma of the **radiation zone**, where energy moves outward primarily in the form of photons of light. The temperature rises to almost 10 million K, and your spacecraft is bathed in x-rays trillions of times more intense than the visible light at the solar surface.

No real spacecraft could survive, but your imaginary one keeps plunging straight down to the solar **core**. There you finally find the source of the Sun's energy: nuclear fusion transforming hydrogen into helium. At the Sun's center, the temperature is about 15 million K, the density is more than 100 times that of water, and the pressure is 200 billion times that on Earth's surface. The energy produced in the core today will take a few hundred thousand years to reach the surface.

With your journey complete, it's time to turn around and head back home. We'll continue this chapter by studying fusion in the solar core and then tracing the flow of the energy generated by fusion as it moves outward through the Sun.

## 14.2 Nuclear Fusion in the Sun

We've seen that the Sun shines because of energy generated by nuclear fusion, and that this fusion occurs under the extreme temperatures and densities found deep in the Sun's core. But exactly how does fusion occur and release energy? And how can we claim to know about something taking place out of sight in the Sun's interior?

Before we begin to answer these questions, it's important to realize that the nuclear reactions that generate energy in the Sun are very different from those used to generate energy in human-built nuclear reactors on Earth. Our nuclear power plants generate energy by splitting large nuclei—such as those of uranium or plutonium—into smaller ones.

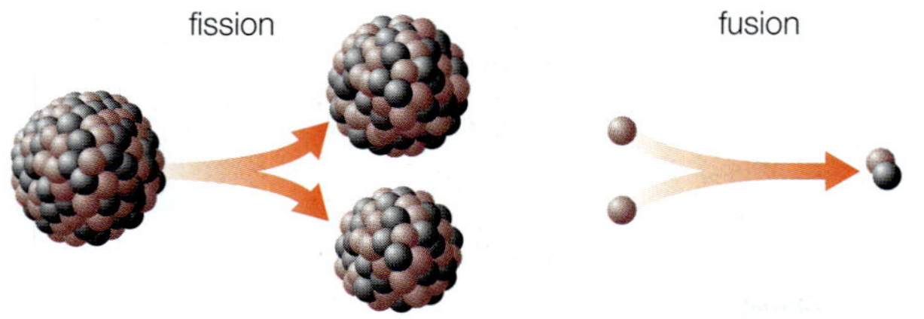


**FIGURE 14.6** Nuclear fission splits a nucleus into smaller nuclei, while nuclear fusion combines smaller nuclei into a larger nucleus.

The process of splitting an atomic nucleus is called **nuclear fission**. In contrast, the Sun makes energy by combining, or fusing, two or more small nuclei into a larger one. That is why we call the process **nuclear fusion**. **FIGURE 14.6** summarizes the difference between fission and fusion.

## How does nuclear fusion occur in the Sun?

Fusion occurs within the Sun because the 15 million K plasma in the solar core is like a "soup" of hot gas full of positively charged atomic nuclei (and negatively charged electrons) whizzing about at extremely high speeds. At any one time, some of these nuclei are on high-speed collision courses with each other. In most cases, electromagnetic forces deflect the nuclei, preventing collisions, because positive charges repel one another. However, if nuclei collide with sufficient energy, they can stick together (fuse) to form a heavier nucleus.

Sticking positively charged nuclei together is not easy. The **strong force**, which binds protons and neutrons together in atomic nuclei, is the only force in nature that can overcome the electromagnetic repulsion between two positively charged nuclei [**Section S4.2**]. In contrast to gravitational and electromagnetic forces, which drop off gradually as the distances between particles increase (by an inverse square law [**Section 4.4**]), the strong force is more like glue or Velcro: It overpowers the electromagnetic force over very small distances but is insignificant when the distances between particles exceed the typical sizes of atomic nuclei. The key to nuclear fusion is pushing the positively charged nuclei close enough together for the strong force to overcome electromagnetic repulsion (**FIGURE 14.7**).

The high pressures and temperatures in the solar core are sufficient for fusion of hydrogen nuclei into helium nuclei. The high temperature is important because the nuclei must collide at very high speeds if they are to come close enough together to fuse. (Quantum tunneling is also important to this process [**Section S4.4**].) The higher the temperature, the more energetic the collisions, making fusion reactions more likely. The high pressure of the overlying layers is necessary because without it, the hot plasma of the solar core would simply explode into space, shutting off the nuclear reactions.

**Think about it** The Sun generates energy by fusing hydrogen into helium, but as we'll see in later chapters, some stars fuse helium or even heavier elements. Do temperatures need to be higher or lower for the fusion of heavier elements? Why? (*Hint*: How does the positive charge of a nucleus affect the difficulty of fusing it to another nucleus?)

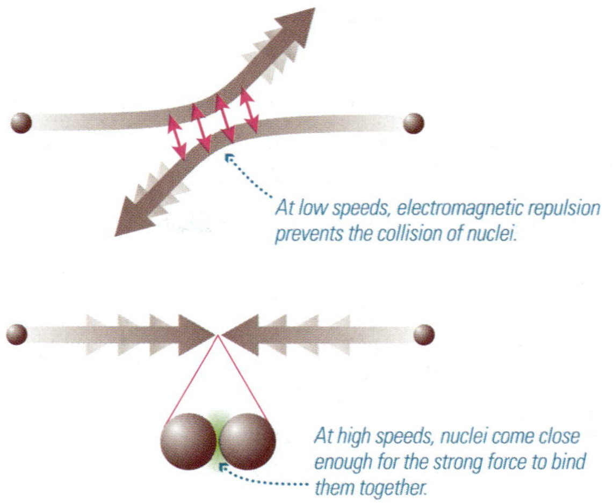


**FIGURE 14.7** Positively charged nuclei can fuse only if a high-speed collision brings them close enough for the strong force to come into play.

**The Proton-Proton Chain** Let's investigate the fusion process in the Sun in a little more detail. Recall that hydrogen nuclei are simply individual protons, while the most common form of helium consists of two protons and two neutrons (see Figure 5.9). The overall hydrogen fusion reaction therefore transforms four individual protons into a helium nucleus containing two protons and two neutrons:

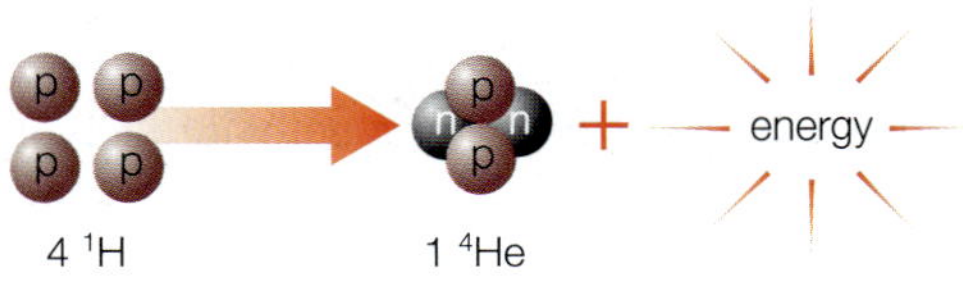


This overall reaction actually proceeds through several steps involving just two nuclei at a time. The sequence of steps that occurs in the Sun is called the **proton-proton chain**, because it begins with collisions between individual protons (hydrogen nuclei). **FIGURE 14.8** illustrates the steps in the proton-proton chain:

***Step 1***: Two protons fuse to form a nucleus consisting of one proton and one neutron, which is the isotope of hydrogen known as *deuterium*. Note that this step converts a proton into a neutron, reducing the total nuclear charge from +2 for the two fusing protons to +1 for the resulting deuterium nucleus. The lost positive charge is carried off by a *positron* (antielectron), the antimatter version of an electron with a positive rather than a negative charge [**Section S4.2**]. A **neutrino**—a subatomic particle with a very tiny mass—is also produced in this step.* (The positron won't last long, because it soon meets up with an ordinary electron, resulting in the creation of two

*Producing a neutrino is necessary because of a law called *conservation of lepton number*: The number of leptons (e.g., electrons or neutrinos [**Chapter S4**]) must be the same before and after the reaction. The lepton number is zero before the reaction because there are no leptons. Among the reaction products, the positron (antielectron) has lepton number −1 because it is antimatter, and the neutrino has lepton number +1. Thus, the total lepton number remains zero.

**Hydrogen Fusion by the Proton-Proton Chain**

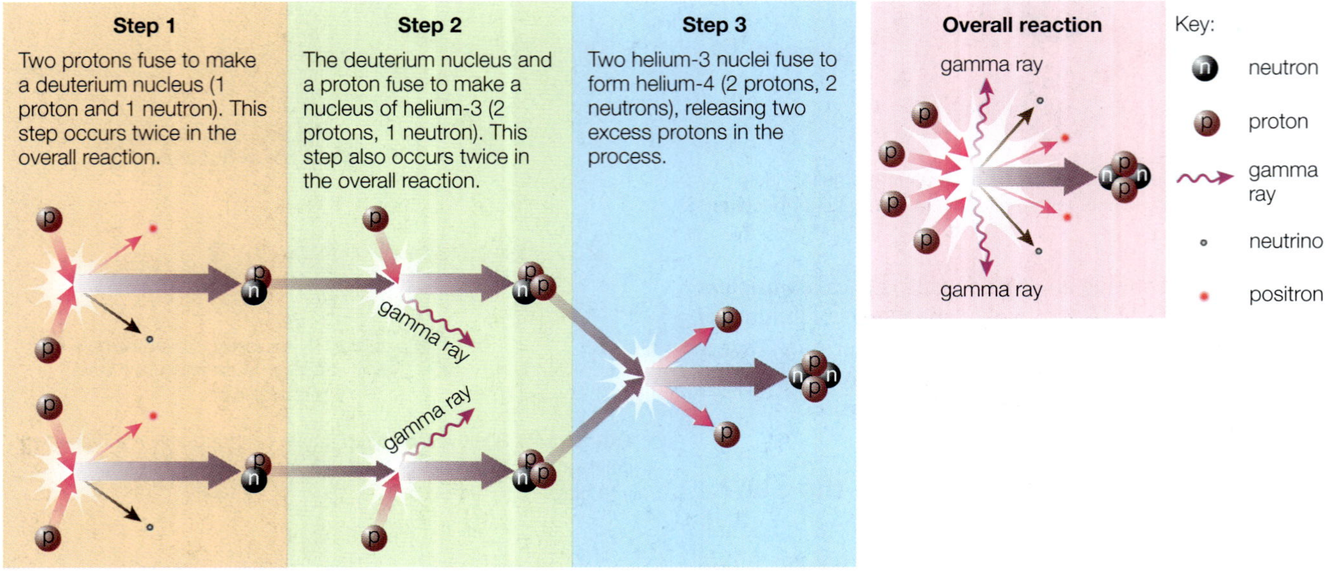


**FIGURE 14.8** In the Sun, four hydrogen nuclei (protons) fuse into one helium-4 nucleus by way of the proton-proton chain. Gamma rays and subatomic particles known as neutrinos and positrons carry off the energy released in the reaction.

gamma-ray photons through matter-antimatter annihilation.) This step must occur twice in the overall reaction, since the reaction requires a total of four protons.

***Step 2***: A fair number of deuterium nuclei are always present along with the protons and other nuclei in the solar core, since Step 1 occurs so frequently in the Sun (about $10^{38}$ times per second). Step 2 occurs when one of these deuterium nuclei collides and fuses with a proton. The result is a nucleus of helium-3, a rare form of helium with two protons and one neutron, along with a gamma-ray photon. This step also occurs twice in the overall reaction.

***Step 3***: The third and final step of the proton-proton chain requires the addition of another neutron to the helium-3, thereby making normal helium-4. This final step can proceed in several different ways, but the most common is through a collision of two helium-3 nuclei. Each of these helium-3 nuclei resulted from a prior, separate occurrence of Step 2 somewhere in the solar core. The final result is a normal helium-4 nucleus and two protons.

Notice that a total of six protons enter the reaction during Steps 1 and 2, with two coming back out in Step 3. The overall reaction therefore combines four protons to make one helium nucleus. The gamma rays and subatomic particles (neutrinos and positrons) carry off the energy released in the reaction.

Fusion of hydrogen into helium generates energy because a helium nucleus has a mass slightly less (by about 0.7%) than the combined mass of four hydrogen nuclei (see Mathematical Insight 14.1). That is, when four hydrogen nuclei fuse into a helium nucleus, a little bit of mass disappears. The disappearing mass becomes energy in accord with Einstein's formula $E = mc^2$. About 98% of the energy emerges as kinetic energy of the resulting helium nuclei and radiative energy of the gamma rays. As we will see, this energy slowly percolates to the solar surface, eventually emerging as the sunlight that bathes Earth. Neutrinos carry off the other 2% of the energy. Overall, fusion in the Sun converts about 600 million tons of hydrogen into 596 million tons of helium every second, which means that 4 million tons of matter is turned into energy each second. Although this sounds like a lot, it is such a small fraction of the Sun's total mass that it does not affect the overall mass of the Sun in any measurable way.

**The Solar Thermostat**

**The Solar Thermostat** Nuclear fusion is the source of all the energy the Sun releases into space. If the fusion rate varied, so would the Sun's energy output, and large variations in the Sun's luminosity would almost surely be lethal to life on Earth. Fortunately, the Sun fuses hydrogen at a steady rate, thanks to a natural feedback process that acts as a thermostat for the Sun's interior. To see how it works, let's examine what would happen if a small change were to occur in the core temperature (**FIGURE 14.9**).

Suppose the Sun's core temperature rose very slightly. The rate of nuclear fusion is extremely sensitive to temperature, so a slight temperature increase would cause the fusion rate to soar as protons in the core collided more frequently and with more energy. Because energy moves slowly through the Sun's interior, this extra energy would be bottled up in the core, temporarily forcing the Sun out of energy balance and raising the core pressure. The push of this pressure would temporarily exceed the pull of gravity, causing the core to expand and cool. This cooling, in turn, would cause the fusion rate to drop back down until the core returned to its original size and temperature, restoring both gravitational equilibrium and energy balance.

A slight drop in the Sun's core temperature would trigger an opposite chain of events. The reduced core temperature

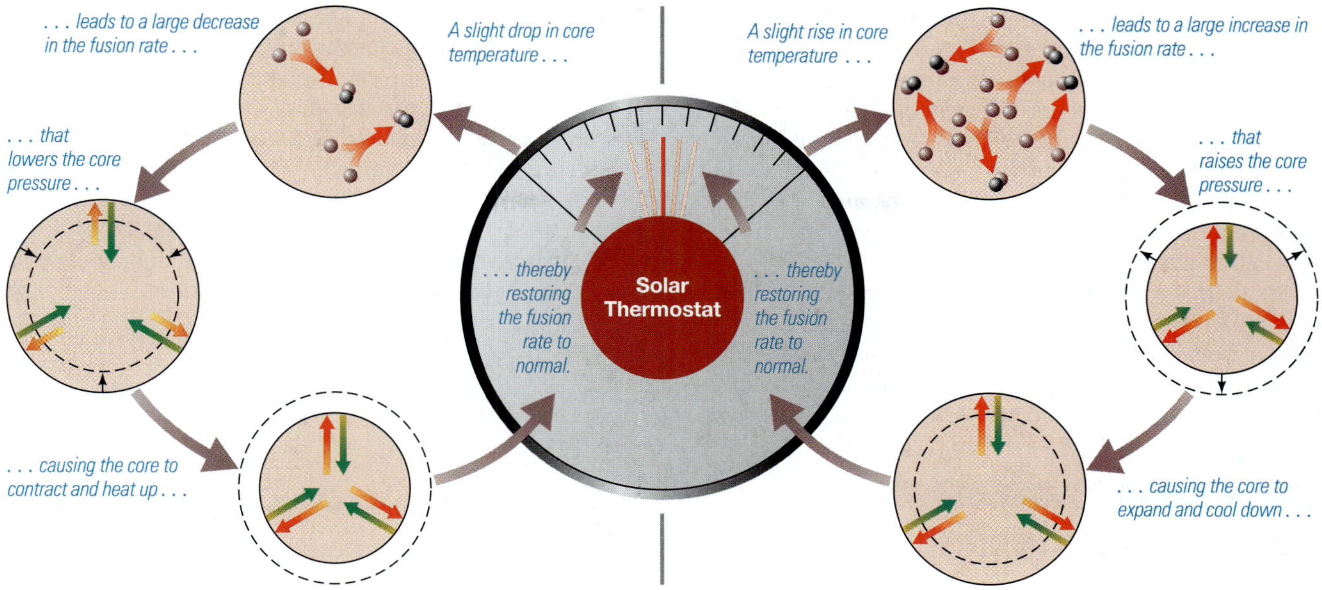


**FIGURE 14.9** The solar thermostat. Gravitational equilibrium regulates the Sun's core temperature. Everything is in balance if the amount of energy leaving the core equals the amount of energy produced by fusion. A rise in core temperature triggers a chain of events that causes the core to expand, lowering its temperature to the original value. A decrease in core temperature triggers the opposite chain of events, also restoring the original core temperature.

would lead to a decrease in the rate of nuclear fusion, causing a drop in pressure and contraction of the core. As the core shrank, its temperature would rise until the fusion rate returned to normal and restored the core to its original size and temperature.

**The Gradually Brightening Sun** While the processes involved in gravitational equilibrium prevent erratic changes in the fusion rate, they also ensure that the fusion rate gradually rises over billions of years. This rise in the fusion rate explains a fact we first encountered when we discussed climate change factors in Chapter 10: the gradual brightening of the Sun with time.

Remember that each fusion reaction converts *four* hydrogen nuclei into *one* helium nucleus. The total number of *independent particles* in the solar core therefore gradually decreases with time. This gradual reduction in the number of particles causes the solar core to shrink. The slow shrinkage, in turn, gradually increases the core temperature and fusion rate, keeping the core pressure high enough to counteract the stronger compression of gravity. Theoretical models indicate that the Sun's core temperature should have increased enough to raise its fusion rate and luminosity by about 30% since the Sun was born $4\frac{1}{2}$ billion years ago.

## How does the energy from fusion get out of the Sun?

The solar thermostat balances the Sun's fusion rate so that the amount of nuclear energy generated in the core equals the amount of energy radiated from the surface as sunlight. However, the journey of solar energy from the core to the photosphere takes hundreds of thousands of years.

Most of the energy released by fusion starts its journey out of the solar core in the form of photons. Although photons travel at the speed of light, the path they take through the Sun's interior zigzags so much that it takes them a very long time to make any outward progress. Deep in the solar interior, the plasma is so dense that a photon can travel only a fraction of a millimeter in any one direction before it interacts with an electron. Each time a photon "collides" with an electron, the photon gets deflected into a new random direction. The photon therefore bounces around the dense interior in a haphazard way (sometimes called a *random walk*) and only very gradually works its way outward from the Sun's center (**FIGURE 14.10**). The technical term for this slow outward migration of photons is **radiative diffusion**; to *diffuse* means to "spread out" and *radiative* refers to the photons of light, or radiation.

**See it for yourself** Radiative diffusion is just one type of diffusion. Another is the diffusion of dye through a glass of water. Try placing a concentrated spot of dye at one point in a still glass of water and observe what happens. The dye starts to spread throughout the entire glass because each individual dye molecule begins a random walk as it bounces among the water molecules. Can you think of any other examples of diffusion in the world around you?

**FIGURE 14.10** A photon in the solar interior bounces randomly among electrons, slowly working its way outward.

Energy released by fusion moves outward through the Sun's radiation zone (see Figure 14.4) primarily by way of these randomly bouncing photons. At the top of the radiation zone, where the temperature has dropped to about 2 million K, the solar plasma absorbs photons more readily, rather than just bouncing them around. This absorption creates the conditions needed for convection [**Section 9.1**], so above this level we find the Sun's convection zone.

Recall that convection occurs because hot gas is less dense than cool gas. Like a hot-air balloon, a hot bubble of solar plasma rises upward through the cooler plasma above it. Meanwhile, cooler plasma from above slides around the rising bubble and sinks to lower layers, where it is heated. The rising of hot plasma and sinking of cool plasma form a cycle that transports energy outward from the base of the convection zone to the photosphere (**FIGURE 14.11a**). There, the density of the gas becomes so low that photons can escape to space, which is why we see the photosphere as the "surface" of the Sun.

The convecting gas gives the photosphere a mottled appearance (**FIGURE 14.11b**): We see bright blobs where hot gas is welling up from below, and the sinking of cooler gas creates the darker borders around those blobs.* If we

*The blobs are formally called *granules*, and the photosphere's mottled appearance is sometimes referred to as *solar granulation*.

## MATHEMATICAL INSIGHT 14.1 Mass-Energy Conversion in Hydrogen Fusion

Hydrogen fusion releases energy because the four hydrogen nuclei (protons) that go into the overall reaction have a slightly greater mass than the helium nucleus that comes out. The mass of a proton is $1.6726 \times 10^{-27}$ kg, so four protons have a mass of $4 \times 1.6726 \times 10^{-27}\text{ kg} = 6.690 \times 10^{-27}$ kg. The mass of a helium-4 nucleus is $6.643 \times 10^{-27}$ kg. Therefore, the mass that "disappears" to become energy is

$$(6.690 \times 10^{-27}\text{ kg}) - (6.643 \times 10^{-27}\text{ kg}) = 0.047 \times 10^{-27}\text{ kg}$$

We find the fractional loss of mass by dividing this lost mass by the original mass of the four protons:

$$\begin{matrix}\text{fractional mass loss} \\ \text{in hydrogen fusion}\end{matrix} = \frac{0.047 \times 10^{-27}\text{ kg}}{6.69 \times 10^{-27}\text{ kg}} = 0.007$$

That is, a fraction 0.007, or 0.7%, of the original hydrogen mass is converted into energy according to Einstein's equation $E = mc^2$.

**EXAMPLE 1:** How much hydrogen is converted to helium each second in the Sun?

**SOLUTION:**

**Step 1 Understand:** The Sun's luminosity is $3.8 \times 10^{26}$ watts (see Table 14.1), which means the Sun produces $3.8 \times 10^{26}$ joules of energy each second. We can use $E = mc^2$ to calculate the total amount of mass needed to produce this energy, and the fact that 0.7% of the hydrogen mass becomes energy to calculate how much hydrogen fuses each second.

**Step 2 Solve:** We start by solving Einstein's equation for the mass, $m$:

$$E = mc^2 \Rightarrow m = \frac{E}{c^2}$$

We plug in $3.8 \times 10^{26}$ joules for the energy ($E$) that the Sun produces each second and $c = 3 \times 10^8$ m/s; to keep the units consistent, recall that 1 joule = 1 kg × m²/s²:

$$m = \frac{E}{c^2} = \frac{3.8 \times 10^{26} \dfrac{\text{kg} \times \text{m}^2}{\text{s}^2}}{\left(3 \times 10^8 \dfrac{\text{m}}{\text{s}}\right)^2} = 4.2 \times 10^9\text{ kg}$$

We've found that the Sun converts 4.2 billion kg of mass to energy each second. Because only 0.7% (=0.007) of the hydrogen mass becomes energy, the total mass of the hydrogen that undergoes fusion each second is

$$\text{mass of hydrogen fused} = \frac{4.2 \times 10^9\text{ kg}}{0.007} = 6.0 \times 10^{11}\text{ kg}$$

**Step 3 Explain:** The Sun fuses 600 billion kilograms of hydrogen each second, converting about 4 billion kilograms of this mass into energy; the rest, about 596 billion kilograms, becomes helium.

**EXAMPLE 2:** How many helium nuclei are created by fusion each second in the Sun?

**SOLUTION:**

**Step 1 Understand:** One way to approach this problem is to divide the total mass lost per second by the mass loss that occurs with each fusion reaction of four hydrogen nuclei into one helium nucleus. The result will tell us how many times this reaction occurs per second.

**Step 2 Solve:** From Example 1, the Sun converts $4.2 \times 10^9$ kilograms of mass into energy each second, and we learned earlier that each individual fusion reaction converts $0.047 \times 10^{-27}$ kilogram of mass into energy. We divide to find the number of fusion reactions per second:

$$\begin{matrix}\text{number of} \\ \text{fusion reactions} \\ \text{(per second)}\end{matrix} = \frac{\text{total mass lost through fusion (per second)}}{\text{mass lost in each fusion reaction}}$$

$$= \frac{4.2 \times 10^9\text{ kg}}{0.047 \times 10^{-27}\text{ kg}}$$

$$= 8.9 \times 10^{37}$$

**Step 3 Explain:** The result, $8.9 \times 10^{37}$, is just a little less than $10^{38}$. In other words, the overall hydrogen fusion reaction occurs in the Sun nearly $10^{38}$ times each second.

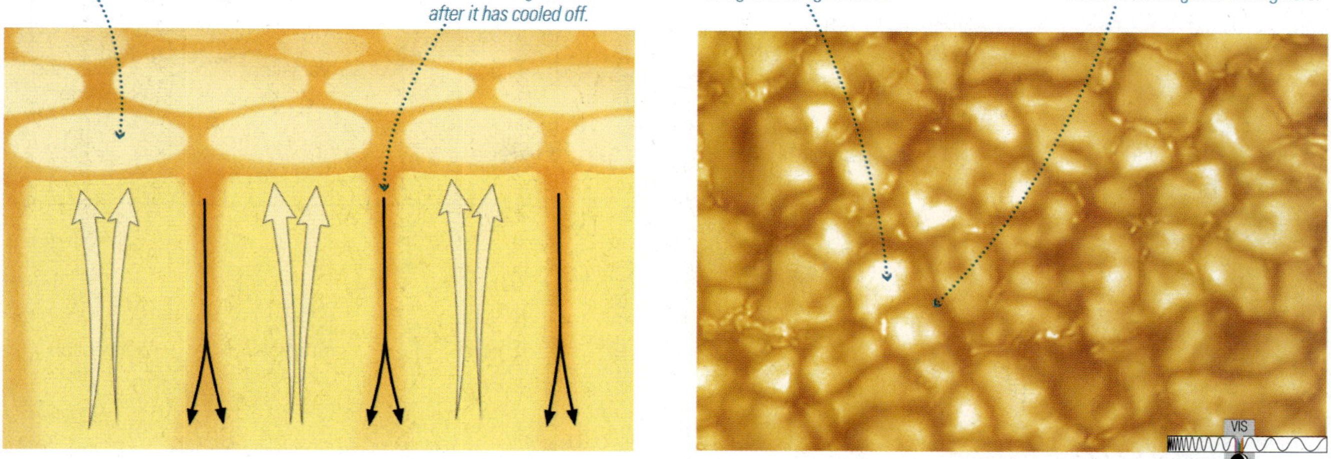


**a** This diagram shows convection beneath the Sun's surface. Hot gas (light yellow arrows) rises while cooler gas (black arrows) descends around it.

**b** This image shows the mottled appearance of the Sun's photosphere. The bright spots, each about 1000 kilometers across, correspond to the rising plumes of hot gas in part a.

**FIGURE 14.11** The Sun's photosphere churns with rising hot gas and falling cool gas as a result of underlying convection.

watched a movie of the photosphere, we'd see its surface bubbling much like a pot of boiling water, in which bubbles rise up and burst on the surface and then are replaced by new bubbles. Each hot blob in the photosphere lasts only a few minutes before being replaced by others bubbling upward. The average temperature of the gas in the photosphere is about 5800 K, but it is hotter in the centers of the blobs and cooler in the darker edges.

To summarize, energy produced by fusion in the Sun's core works its way slowly through the radiation zone through random bounces of photons, then gets carried upward by convection in the convection zone. The photosphere lies at the top of the convection zone and marks the place where the density of gas has become low enough that photons can escape to space. The energy produced hundreds of thousands of years earlier in the solar core finally emerges from the Sun as thermal radiation [**Section 5.4**] produced by the 5800 K gas of the photosphere. Once in space, the photons travel away at the speed of light, bathing the planets in sunlight.

## How do we know what is happening inside the Sun?

We cannot see inside the Sun, so you may wonder how we can claim to know so much about what goes on beneath its surface. We can study the Sun's interior in three different ways: through mathematical models of the Sun, observations of solar vibrations, and observations of solar neutrinos.

**Mathematical Models** The primary way we learn about the interior of the Sun (and other stars) is by creating *mathematical models* that use the laws of physics to predict internal conditions. A basic model starts with the Sun's observed composition and mass and then solves equations that describe gravitational equilibrium and energy balance. With the aid of a computer, we can use the model to calculate the Sun's temperature, pressure, and density at any depth. We can then predict the rate of nuclear fusion in the solar core by combining these calculations with knowledge about nuclear fusion gathered in laboratories on Earth.

If a model is a good description of the Sun's interior, it should correctly predict the radius, surface temperature, luminosity, age, and other observable properties of the Sun. Current models predict these properties quite accurately, giving us confidence that we really do understand what is going on inside the Sun.

**Solar Vibrations** A second way to learn about the inside of the Sun is to observe vibrations of the Sun's surface that are somewhat similar to the vibrations that earthquakes cause on Earth. These vibrations arise from the movement of gas within the Sun, which generates waves of pressure that travel through the Sun like sound waves moving through air. We can observe these vibrations on the Sun's surface by looking for Doppler shifts [**Section 5.4**]. Light from portions of the surface that are rising toward us is slightly blueshifted, while light from portions that are falling away from us is slightly redshifted. The vibrations are relatively small but measurable (**FIGURE 14.12**).

We can deduce a great deal about the solar interior by carefully analyzing these vibrations. (By analogy to seismology on Earth, this type of study of the Sun is called *helioseismology—helios* means "sun.") Results to date confirm that our mathematical models of the solar interior are on the right track (**FIGURE 14.13**), while also providing data that help us to improve the models further.

**Solar Neutrinos** A third way to study the Sun's interior is to observe the neutrinos produced by fusion (see Figure 14.8). Don't panic, but about a *thousand trillion* of these solar neutrinos will zip through your body as you read this sentence—but they will do no damage at all. The reason is

**FIGURE 14.12** This image shows vibrations on the Sun's surface that have been measured from Doppler shifts. Shades of orange show how quickly each spot on the Sun's surface is moving toward or away from us at a particular moment. Dark shades (negative velocities) represent motion toward us; light shades (positive velocities) represent motion away from us. The large-scale change in color from left to right reflects the Sun's rotation, and the small-scale ripples reflect the surface vibrations.

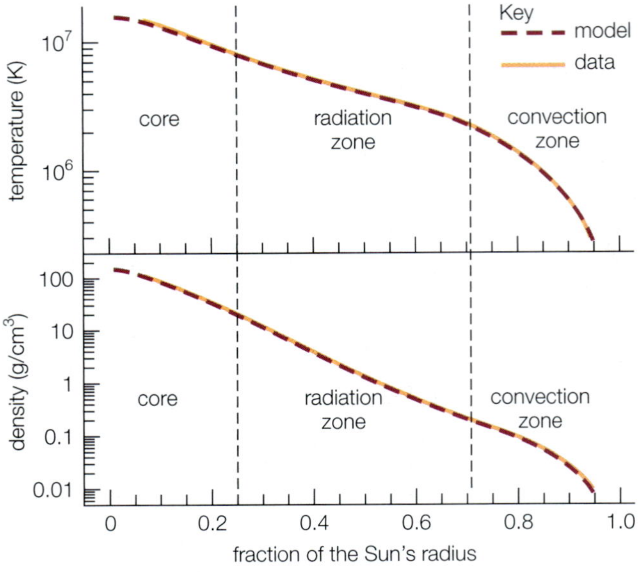


**FIGURE 14.13** These graphs show the close agreement between predictions about the Sun's interior structure based on mathematical models and actual data obtained from observations of the Sun's surface vibrations. The agreement gives us confidence that the models are a good representation of the Sun's interior.

that neutrinos interact with other matter only through the *weak force* [**Section S4.2**] and gravity, not through the electromagnetic force. As a result, they can pass through almost anything. For example, an inch of lead will stop an x-ray, but stopping an average neutrino would require a slab of lead more than a light-year thick!

In principle, neutrinos give us a direct way to study fusion in the Sun's core, because nearly all of them pass straight through the solar interior into space. Traveling at nearly the speed of light, they reach us just minutes after they were produced. In practice, their elusiveness also makes neutrinos dauntingly difficult to detect. Nevertheless, neutrinos *do* occasionally interact with matter, and

## MATHEMATICAL INSIGHT 14.2 Pressure in the Sun: The Ideal Gas Law

The pressure that resists gravity inside the Sun comes from the thermal motions of gas particles. The pressure ($P$) depends on both the gas temperature ($T$) and the *number density* ($n$)—the number of particles contained in each cubic centimeter of gas. The **ideal gas law** expresses this relationship:

$$P = nkT$$

where $k = 1.38 \times 10^{-23}$ joule/K is *Boltzmann's constant*. This law applies to all gases consisting of simple, freely flying particles, like those in the Sun. With $T$ in Kelvin and $n$ in particles per cubic centimeter ($\text{cm}^{-3}$), the ideal gas law gives pressure in units of joules per cubic centimeter ($\text{J/cm}^3$); these units of energy per unit volume are equivalent to units of force per unit area.

**EXAMPLE 1:** The Sun's core contains about $10^{26}$ particles per cubic centimeter at a temperature of 15 million K. Compare the Sun's core pressure to that of Earth's atmosphere at sea level, where the density is about $2.4 \times 10^{19}$ particles per cubic centimeter and the temperature is about 300 K.

**SOLUTION:**

**Step 1 Understand:** We can apply the ideal gas law to both the Sun and Earth's atmosphere, because both essentially consist of freely moving particles. We can compare the two pressures by dividing the larger (the Sun's core) by the smaller (Earth's atmosphere).

**Step 2 Solve:** We divide the two pressures, finding each with the ideal gas law:

$$\frac{P_{\text{Sun(core)}}}{P_{\text{Earth(atmos)}}} = \frac{n_{\text{Sun}}kT_{\text{Sun}}}{n_{\text{Earth}}kT_{\text{Earth}}} = \frac{10^{26}\ \text{cm}^{-3} \times 1.5 \times 10^{7}\ \text{K}}{2.4 \times 10^{19}\ \text{cm}^{-3} \times 300\ \text{K}}$$
$$= 2 \times 10^{11}$$

**Step 3 Explain:** The Sun's core pressure is about 200 billion ($2 \times 10^{11}$) times the atmospheric pressure on Earth.

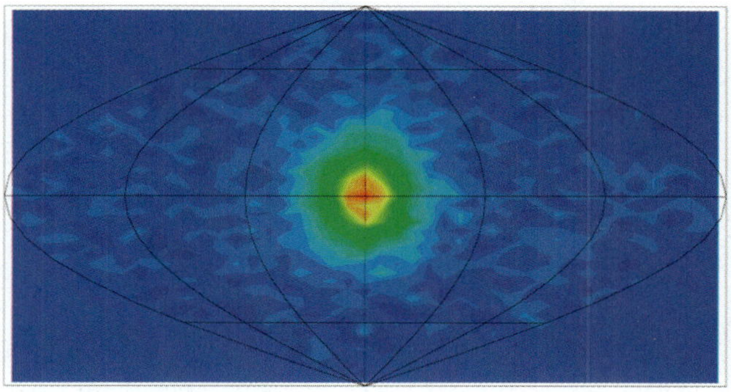

**FIGURE 14.14** This image shows neutrinos detected by the Super-Kamiokande neutrino observatory in Japan. The grid of black lines shows the extent of the entire sky, with the Sun's position at the center, making it clear that the Sun is the source of these neutrinos. The neutrinos are produced by fusion and therefore represent direct proof that fusion really is responsible for the Sun's energy output.

**FIGURE 14.15** This photo shows the inside of the main vessel of the Borexino neutrino detector, located in a mine 1400 meters underground at Italy's Gran Sasso National Lab. During operations, the vessel is filled with 300 tons of a fluid that emits tiny flashes of light when a neutrino is captured; these flashes are recorded by the detectors visible on the inside of the sphere. Borexino is one of the detectors confirming that the number of neutrinos coming from the Sun agrees with the predictions of solar fusion models.

it is possible to capture a few solar neutrinos with a large enough detector. To distinguish neutrino captures from reactions caused by other particles, scientists usually place neutrino detectors deep underground or under the Antarctic ice. The overlying rock or ice blocks most other particles, but neutrinos pass right through.

The fact that we detect solar neutrinos provides direct proof that nuclear fusion is responsible for the Sun's energy (**FIGURE 14.14**). However, attempts to detect solar neutrinos, which began in the 1960s, initially found only about one-third of the number expected based on the Sun's energy output. This disagreement between model predictions and observations came to be called the *solar neutrino problem*. It was solved in the early 2000s. The solution is this: Neutrinos come in three distinct types, called *electron neutrinos, muon neutrinos*, and *tau neutrinos* [**Section S4.2**]. Early solar neutrino detectors could detect only electron neutrinos, which didn't seem like a problem at first, because fusion should produce only this type. However, we now know that neutrinos can change among the three types while passing through matter, so that by the time solar neutrinos reach our detectors, only about one-third of them are still electron neutrinos. We can be confident in this solution, because today's neutrino detectors can detect all three types and confirm that the total number of solar neutrinos matches the predictions (**FIGURE 14.15**).

---

**EXAMPLE 2:** How would the Sun's core pressure change if the Sun fused all its core hydrogen into helium without shrinking and without its core temperature changing? Use your answer to explain what should happen as fusion occurs in the real Sun, in which the core *can* shrink and heat up. To simplify the problem, assume that the Sun's core begins with pure hydrogen that is fully ionized, so that there are two particles for every hydrogen nucleus (the proton plus one electron that must also be present for charge balance), and ends as pure helium with three particles for each helium nucleus (the nucleus plus two electrons).

**SOLUTION:**

**Step 1 Understand:** The core pressure will change if either the temperature or the number density of particles changes. We are assuming that the temperature does not change and that the core does not shrink, so pressure changes are due only to the declining number density. We can therefore find the pressure change simply by finding how fusion changes the number of particles in the core.

**Step 2 Solve:** Recall that hydrogen fusion converts four hydrogen nuclei into one helium nucleus, which means converting eight particles (two each for the four hydrogen nuclei) into only three particles (the helium nucleus and two electrons); therefore, the particle density after fusion is $\frac{3}{8}$ of its original value. If the temperature and volume don't change, the core pressure after all the hydrogen fuses into helium is also $\frac{3}{8}$ of its original value.

**Step 3 Explain:** If the core volume stayed constant, the core pressure at the end of the fusion process would decrease to $\frac{3}{8}$ of its initial value. This decrease in core pressure would tip the balance of gravitational equilibrium in favor of gravity, which is why the core shrinks as fusion progresses in the Sun. This shrinkage decreases the core volume, allowing the number density of particles to stay approximately constant. The gradual shrinkage also produces a slight but gradual rise in core temperature and therefore in the fusion rate, which is why the Sun's luminosity gradually increases with time.

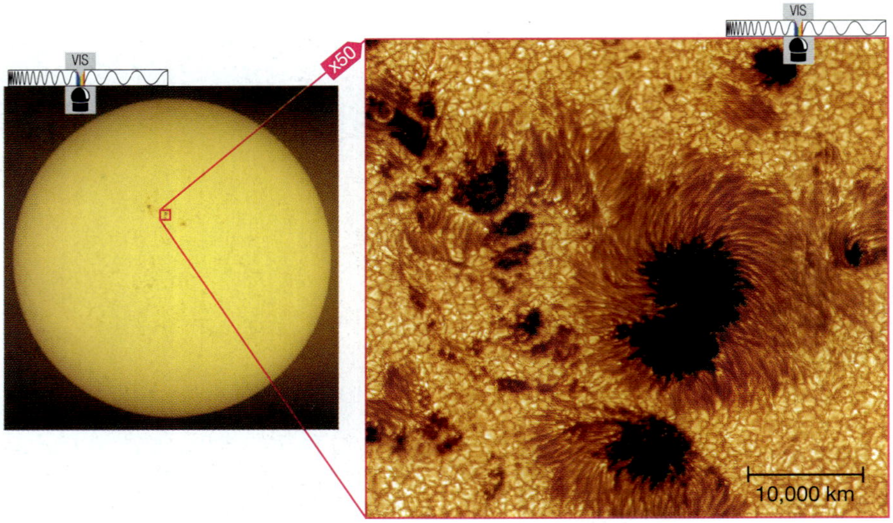


a This close-up view of the Sun's surface shows two large sunspots and several smaller ones. Each of the big sunspots is roughly as large as Earth.

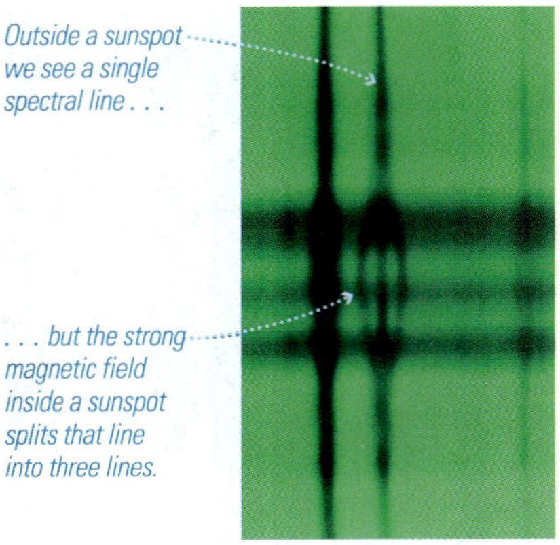


b Very strong magnetic fields split the absorption lines in spectra of sunspot regions. The dark vertical bands are absorption lines in a spectrum of the Sun. Notice that these lines split where they cross the dark horizontal bands corresponding to sunspots.

**FIGURE 14.16** Sunspots are regions of strong magnetic fields.

## 14.3 The Sun-Earth Connection

Energy liberated by nuclear fusion in the Sun's core eventually reaches the solar surface, where it helps create a wide variety of phenomena that we can observe from Earth. Sunspots are the most obvious of these phenomena. Because sunspots and other features of the Sun's surface change with time, they constitute what we call *solar weather*, or **solar activity**. The "storms" associated with solar weather are not just of academic interest. Sometimes they affect our day-to-day life on Earth. In this section, we'll explore solar activity and its far-reaching effects.

### What causes solar activity?

Most of the Sun's surface churns constantly with rising and falling gas and looks like the close-up photo shown in Figure 14.11b. However, larger features sometimes appear, including sunspots, the huge explosions known as *solar flares*, and gigantic loops of hot gas extending high into the Sun's corona. All of these features are created by magnetic fields, which form and change easily in the convecting plasma in the outer layers of the Sun.

**Sunspots and Magnetic Fields** Sunspots are the most striking features of the solar surface (**FIGURE 14.16a**). If you could look directly at a sunspot without damaging your eyes, it would be blindingly bright. Sunspots appear dark in photographs only because they are *less* bright than the surrounding photosphere. They are less bright because they are cooler: The temperature of the plasma in sunspots is about 4000 K, significantly cooler than the 5800 K plasma that surrounds them.

You may wonder how sunspots can be so much cooler than their surroundings. Gas usually flows easily, so you might expect the hotter gas around a sunspot to mix with the cooler gas within it, quickly warming the sunspot. The fact that sunspots stay relatively cool means that something must prevent hot plasma from entering them, and that something turns out to be magnetic fields.

Detailed observations of the Sun's spectral lines reveal sunspots to be regions with strong magnetic fields. These magnetic fields can alter the energy levels in atoms and ions, causing some spectral lines to split into two or more closely spaced lines (**FIGURE 14.16b**). Wherever we see this effect (called the *Zeeman effect*), magnetic fields must be present. Scientists can therefore use this effect to map magnetic fields on the Sun.

To understand how sunspots stay cooler than their surroundings, we must investigate the nature of magnetic fields in a little more depth. Magnetic fields are invisible, but we can represent them by drawing **magnetic field lines** (**FIGURE 14.17**). These lines represent the directions in which compass needles would point if we placed them within the magnetic field. The lines are closer together where the field is stronger and farther apart where the field is weaker. Because these imaginary field lines are easier to visualize than the magnetic field itself, we usually discuss magnetic fields by talking about how the field lines would appear. Charged particles, such as the ions and electrons in the solar plasma, cannot easily move perpendicular to the field lines, so they instead follow spiraling paths along them.

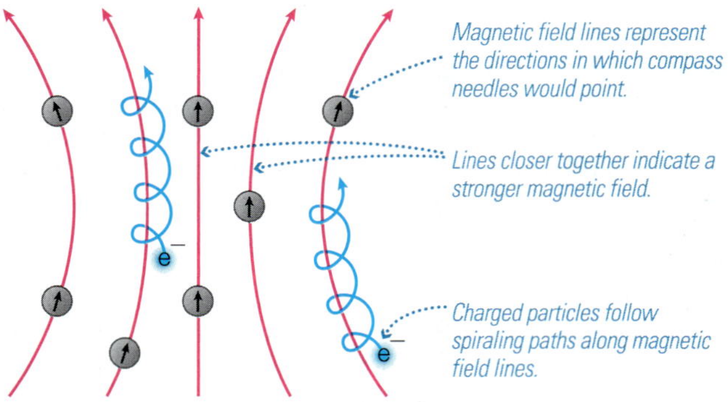


**FIGURE 14.17** We draw magnetic field lines (red) to represent invisible magnetic fields. These lines also represent the directions along which charged particles tend to move.

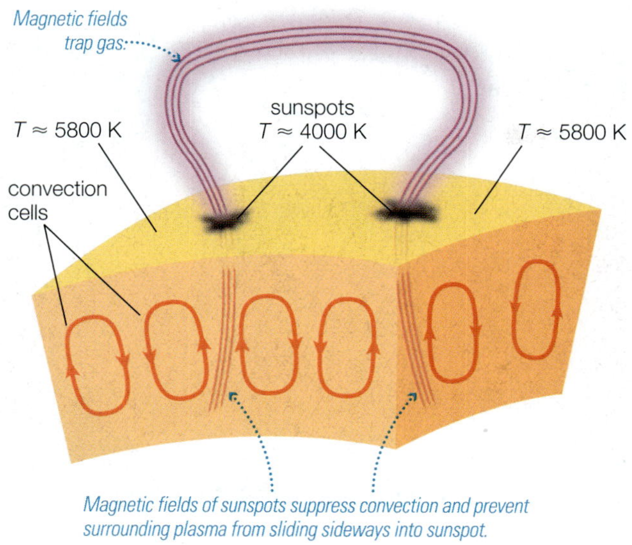


**a** Pairs of sunspots are connected by tightly wound magnetic field lines.

**b** This x-ray photo (from NASA's *TRACE* mission) shows hot gas trapped within looped magnetic field lines.

**FIGURE 14.18** Strong magnetic fields keep sunspots cooler than the surrounding photosphere, and magnetic loops can arc from the sunspots to great heights above the Sun's surface.

Solar magnetic field lines act somewhat like elastic bands, twisted into contortions and knots by turbulent motions in the solar atmosphere. Sunspots occur where tightly wound magnetic fields poke nearly straight out from the solar interior (**FIGURE 14.18a**). These tight magnetic field lines suppress convection within the sunspot and prevent surrounding plasma from entering the sunspot. With hot plasma unable to enter the region, the sunspot plasma becomes cooler than that of the rest of the photosphere. Individual sunspots typically last up to a few weeks, dissolving when their magnetic fields weaken and allow hotter plasma to flow in.

Sunspots tend to occur in pairs, connected by a loop of magnetic field lines that can arc high above the Sun's surface (**FIGURE 14.18b**). Gas in the Sun's chromosphere and corona becomes trapped in these giant loops, called **solar prominences**. Some prominences rise to heights of more than 100,000 kilometers above the Sun's surface (**FIGURE 14.19**). Individual prominences can last for days or even weeks.

**Solar Storms** The magnetic fields winding through sunspots and prominences sometimes undergo dramatic and sudden change, producing short-lived but intense storms on the Sun. The most dramatic of these storms are **solar flares**, which emit bursts of ultraviolet light and x-rays along with charged particles moving at nearly the speed of light (**FIGURE 14.20**).

Flares generally occur in the vicinity of sunspots, which is why we think they are created by changes in magnetic fields. The leading model suggests that solar flares occur when magnetic field lines become so twisted and knotted that they can no longer bear the tension, causing them to snap suddenly and reorganize themselves into a less twisted configuration. The energy released in the process heats the nearby plasma to 100 million K over the next few minutes or hours, generating the intense radiation and high-speed particles that we see from solar flares.

**Heating of the Chromosphere and Corona** As we've seen, many of the most dramatic weather patterns and storms on the Sun involve the very hot gas of the Sun's chromosphere and corona. But why is this gas so hot in the first place?

Recall that temperatures gradually decline as we move outward from the Sun's core to its photosphere. We might expect the decline to continue above the photosphere, but

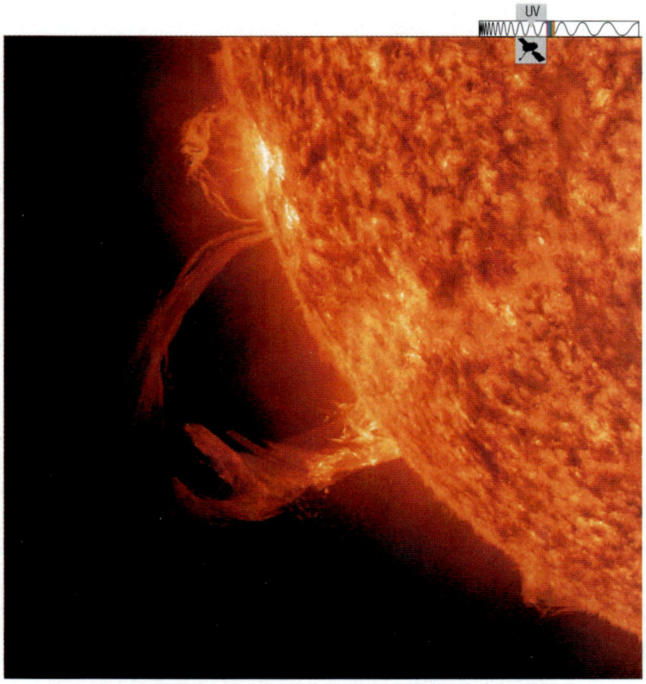


**FIGURE 14.19** A gigantic solar prominence erupts from the solar surface in this ultraviolet-light image (from the *Solar Dynamics Observatory* spacecraft). The height of the prominence is more than 20 times the diameter of Earth.

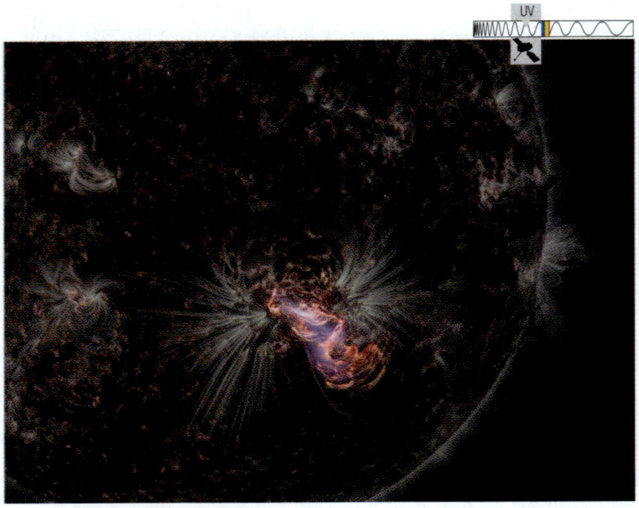


**FIGURE 14.20** This ultraviolet image (from *Solar Dynamics Observatory*) shows a solar flare erupting from the Sun's surface.

instead it reverses, making the chromosphere and corona much hotter than the Sun's surface. Some aspects of this heating remain a mystery today, but we have at least a general explanation: Strong magnetic fields carry energy upward from the churning solar surface to the chromosphere and corona. More specifically, the rising and falling of gas in the convection zone probably shakes tightly wound magnetic field lines beneath the solar surface, transmitting energy upward until it is ultimately deposited as heat. The same magnetic fields that keep sunspots cool therefore make the overlying plasma of the chromosphere and corona hot.

Observations support this model. The gas density in the chromosphere and corona is so low that we cannot see visible light from these layers (except during a total eclipse when the bright photosphere is blocked from view; see Figure 2.29). However, the roughly 10,000 K plasma of the chromosphere emits strongly in the ultraviolet and the 1 million K plasma of the corona is the source of virtually all x-rays from the Sun. X-ray images show that bright spots in the corona tend to be directly above sunspots in the photosphere (**FIGURE 14.21**), indicating that both are created by the same magnetic fields.

**The Solar Wind** Notice that some regions of the corona barely show up in x-ray images; these regions, called **coronal holes**, are nearly devoid of hot coronal gas. More detailed analyses show that the magnetic field lines in coronal holes project out into space like broken rubber bands, allowing particles spiraling along them to escape the Sun altogether. These particles streaming outward from the corona are the source of the solar wind.

The solar wind also gives us something tangible to study. In the same way that meteorites provide us with samples of asteroids we've never visited, solar wind particles captured by satellites provide us with a sample of material from the Sun. Analysis of these solar particles has reassuringly verified that the Sun is made mostly of hydrogen, just as we conclude from studying the Sun's spectrum.

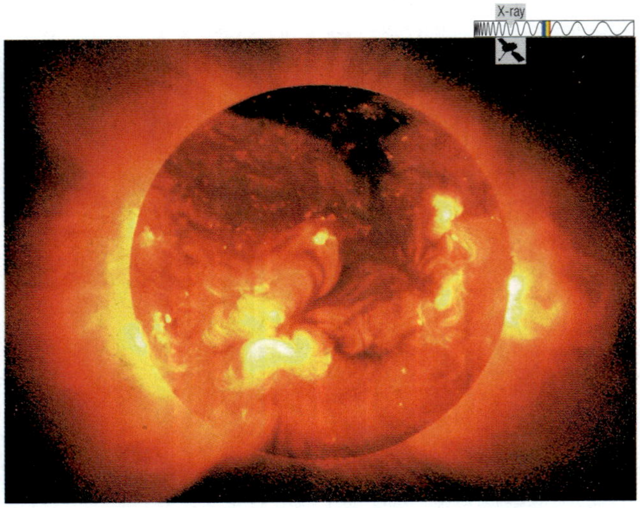


**FIGURE 14.21** An x-ray image of the Sun reveals the 1 million K gas of the corona. Brighter regions (yellow) correspond to areas of stronger x-ray emission. The darker regions (such as near the north pole at the top of this photo) are *coronal holes* from which the solar wind escapes. (From the *Yohkoh Space Observatory*.)

**Effects of Solar Activity on Earth** Flares and other solar storms sometimes eject large numbers of highly energetic charged particles from the Sun's corona. These particles travel outward from the Sun in huge bubbles called **coronal mass ejections** (**FIGURE 14.22**). The bubbles have strong magnetic fields and can reach Earth in a couple of days if they happen to be aimed in our direction. Once a coronal mass ejection reaches Earth, it can create a *geomagnetic storm* in Earth's magnetosphere. On the positive side, these storms can lead to unusually strong auroras (see Figure 10.11) that can be visible throughout much of the United States. On the negative side, they can hamper radio communications, disrupt electrical power delivery, and damage the electronic components in orbiting satellites. In some cases, coronal mass ejections have shut down power systems and destroyed spacecraft electronics.

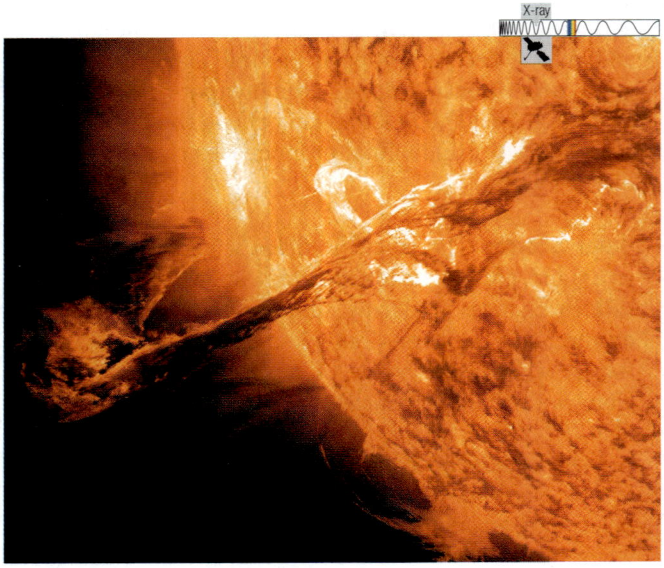


**FIGURE 14.22** This x-ray image from the *Solar Dynamics Observatory* spacecraft shows a solar eruption that led to a coronal mass ejection on August 31, 2012.

Satellites in low Earth orbit are particularly vulnerable during periods of strong solar activity, when the increase in solar x-rays and energetic particles heats Earth's upper atmosphere, causing it to expand. The density of the gas surrounding low-flying satellites therefore rises, exerting drag that saps their energy and angular momentum. If this drag proceeds unchecked, the satellites ultimately plummet back to Earth.

## How does solar activity vary with time?

Solar weather is just as unpredictable as weather on Earth. Individual sunspots can appear or disappear at almost any time, and we have no way to know that a solar storm is coming until we observe it through our telescopes. However, long-term observations have revealed overall patterns in solar activity that make sunspots and solar storms more common at some times than at others. They've also helped us understand the origin of the Sun's strong magnetic fields, which are responsible for all the solar activity we've discussed.

**The Sunspot Cycle** The most notable pattern in solar activity is the **sunspot cycle**—a cycle in which the average number of sunspots on the Sun gradually rises and falls (Figure 14.23). At the time of *solar maximum*, when sunspots are most numerous, we may see dozens of sunspots on the Sun at one time. In contrast, we may see few if any sunspots at the time of *solar minimum*. The frequency of prominences, flares, and coronal mass ejections also follows the sunspot cycle, with these events being most common at solar maximum and least common at solar minimum.

**Think about it** Find out where we are in the sunspot cycle today; are we near minimum or maximum? Based on your answer, should we expect more or fewer solar flares and coronal mass ejections over the coming year? Explain.

Notice that the sunspot cycle varies both in duration and in the peak number of sunspots from one period to the next (**FIGURE 14.23a**). The length of time between maximums averages 11 years, but we have observed it to be as short as 7 years and as long as 15 years. The locations of sunspots on the Sun also vary with the sunspot cycle (**FIGURE 14.23b**). As a cycle begins at solar minimum, sunspots form primarily at mid-latitudes (30° to 40°) on the Sun. The sunspots tend to form at lower latitudes as the cycle progresses, appearing very close to the solar equator as the next solar minimum approaches. Then the sunspots of the next cycle begin to form near mid-latitudes again.

A less obvious feature of the sunspot cycle is that something peculiar happens to the Sun's magnetic field at each solar maximum: The Sun's entire magnetic field starts to flip, turning magnetic north into magnetic south and vice versa. We know this because the magnetic field lines connecting pairs of sunspots (see Figure 14.18a) on the same side of the solar equator all tend to point in the same direction throughout an 11-year cycle. For example, all compass needles might point from the easternmost sunspot to the westernmost sunspot in each sunspot pair north of the solar equator. However, by the time the cycle ends at solar minimum, the magnetic field has reversed: In the subsequent solar cycle, the field lines connecting pairs of sunspots point in the opposite direction. The Sun's complete magnetic cycle (sometimes called the *solar cycle*) therefore averages 22 years, since it takes two 11-year sunspot cycles before the magnetic field is back the way it started.

**Longer-Term Changes in the Sunspot Cycle** Figure 14.23 shows how the sunspot cycle has varied in length and intensity during the past century, the time during which we've had the most complete records of numbers of sunspots. However, astronomers have observed the Sun telescopically

**a** This graph shows how the number of sunspots on the Sun changes with time. The vertical axis shows the percentage of the Sun's surface covered by sunspots. The cycle has a period of approximately 11 years.

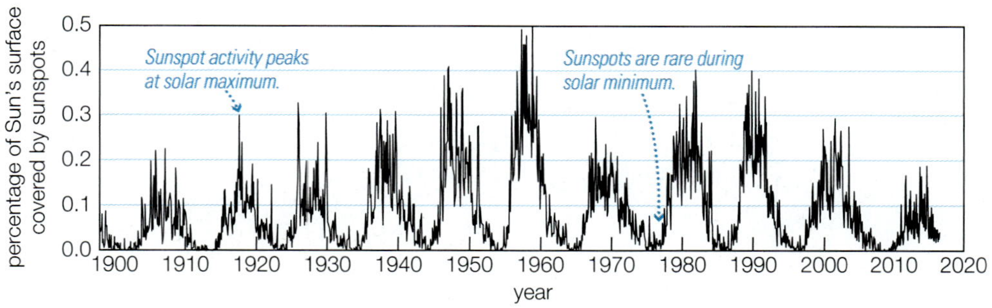


**b** This graph shows how the latitudes at which sunspot groups appear tend to shift during a single sunspot cycle.

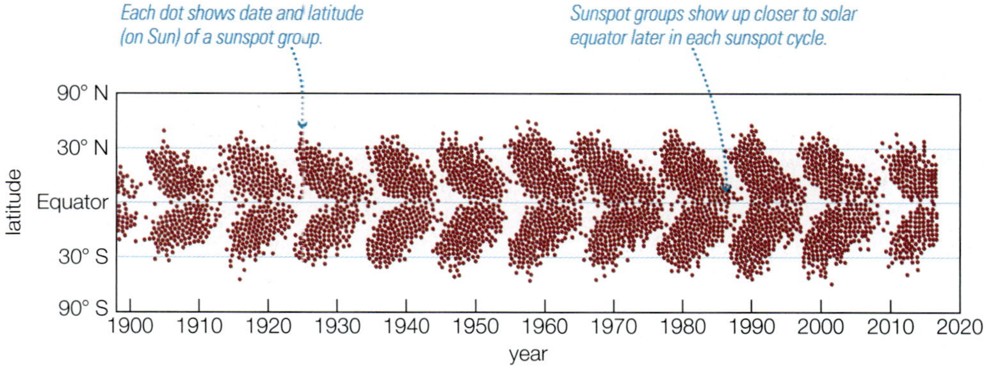


**FIGURE 14.23** Sunspot cycle since about 1900.

**FIGURE 14.24** This graph reconstructs the sunspot cycle over the past 400 years, based on available data from telescopic observations.

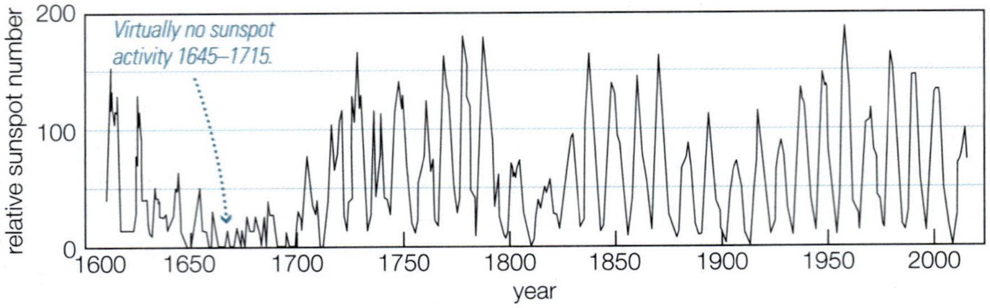


for nearly 400 years, and these longer-term observations suggest that the sunspot cycle can change even more dramatically (**FIGURE 14.24**). For example, astronomers observed virtually no sunspots between the years 1645 and 1715, a period sometimes called the *Maunder minimum* (after husband and wife E. W. Maunder and A. R. Maunder, who identified it in historical sunspot records).

Is it possible that the Maunder minimum is part of a longer-term cycle of solar activity, one lasting much longer than 11 or 22 years? Some scientists have hypothesized that such cycles might exist, but little evidence has been found to back up these claims. Of course, searching for long-term variations is difficult, because we have very limited observational data predating the invention of the telescope.

The search for long-term cycles therefore relies on less direct evidence. For example, we can make educated guesses about past solar activity from historical descriptions of solar eclipses: When the Sun is more active, the corona tends to have longer and brighter "streamers" visible to the naked eye. We can also gauge solar activity further in the past (a few thousand years) by studying the amount of radioactive carbon-14 in tree rings. This amount varies because carbon-14 is produced in the atmosphere by interactions with high-energy *cosmic rays* [**Section 19.2**] coming from beyond our own solar system. During periods of high solar activity, the solar wind tends to grow stronger, shielding Earth from some of these cosmic rays and reducing the production of carbon-14. Because trees steadily incorporate atmospheric carbon into their rings (through their respiration of carbon dioxide), we can estimate the level of solar activity during each year of a tree's life by measuring the level of carbon-14 in the corresponding ring. These data have not yet turned up any clear evidence of longer-term cycles of solar activity, but the search goes on.

**The Cause of the Sunspot Cycle** The precise reasons for the sunspot cycle are not fully understood, but the leading model ties it to a combination of convection and the Sun's rotation. Convection is thought to dredge up weak magnetic fields generated in the solar interior, amplifying them as they rise. The Sun's rotation—faster at its equator than near its poles—then stretches and shapes these fields.

**Think about it** Suppose you take a photograph of the Sun and notice two sunspots: one near the equator and one directly north of it at higher latitude. If you looked again at the Sun in a few days, would you still find one sunspot directly north of the other? Why or why not? Explain how you could use this type of observation to learn how the Sun rotates.

Imagine what happens to magnetic field lines that start out running along the Sun's surface from south to north (**FIGURE 14.25**). At the equator the lines circle the Sun every 25 days, but at higher latitudes they lag behind. As a

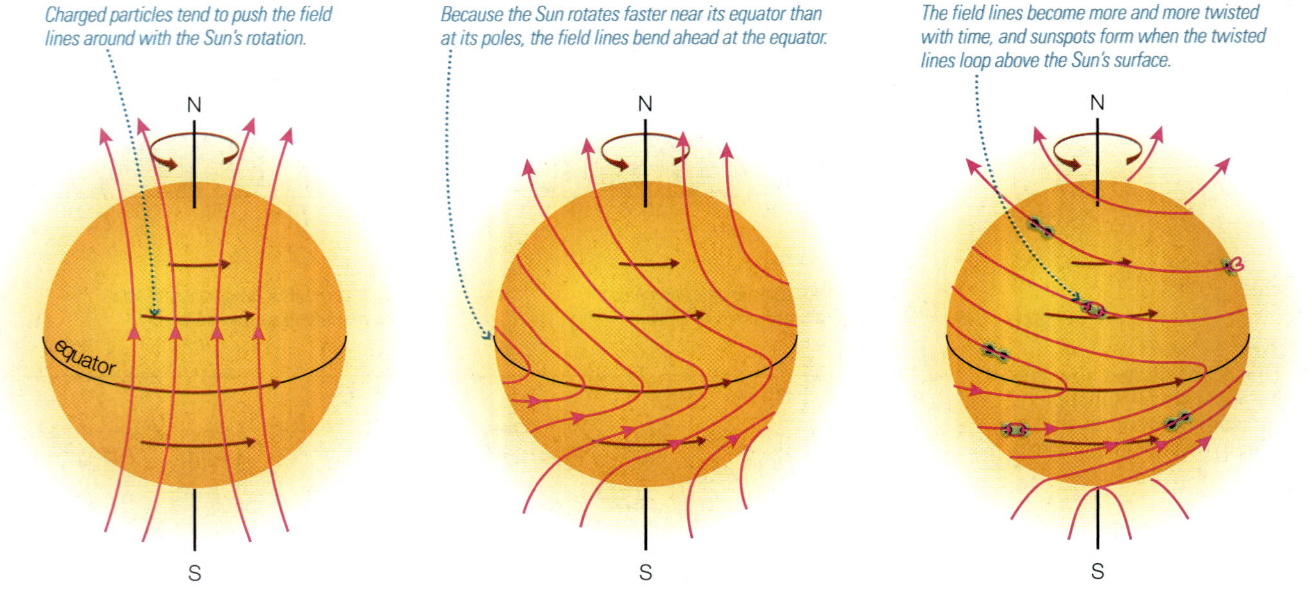


**FIGURE 14.25** The Sun rotates more quickly at its equator than it does near its poles. Because gas circles the Sun faster at the equator, it drags the Sun's north-south magnetic field lines into a more twisted configuration. The magnetic field lines linking pairs of sunspots, depicted here as dark blobs, trace out the directions of these stretched and distorted field lines.

result, the lines gradually get wound more and more tightly around the Sun. This process, operating at all times over the entire Sun, explains the origin of the contorted field lines that generate sunspots and other solar activity.

The detailed behavior of these magnetic fields is quite complex, so scientists attempt to study it with sophisticated computer models. Using these models, scientists have successfully replicated many features of the sunspot cycle, including changes in the number and latitude of sunspots and the magnetic field reversals that occur about every 11 years. However, much remains mysterious, including why the period of the sunspot cycle varies and why solar activity is different from one cycle to the next.

Over extremely long time periods—hundreds of millions to billions of years—these theoretical models predict a gradual lessening of solar activity. Recall that, according to our theory of solar system formation, the Sun must have rotated much faster when it was young [**Section 8.2**], and a faster rotation rate should have meant much more activity. Observations of other stars that are similar to the Sun but rotate faster confirm that these stars are much more active. We find evidence for many more "starspots" on these stars than sunspots on the Sun, and their relatively bright ultraviolet and x-ray emissions suggest that they have brighter chromospheres and coronas—just as we would expect if they are more active than the Sun.

**The Sunspot Cycle and Earth's Climate** Despite the changes that occur during the sunspot cycle, the Sun's total output of energy barely changes at all—the largest measured changes have been less than 0.1% of the Sun's average luminosity. However, the ultraviolet and x-ray output of the Sun, which comes from the magnetically heated gas of the chromosphere and corona, can vary much more significantly. Could any of these changes affect the weather or climate on Earth?

Some data suggest connections between solar activity and Earth's climate. For example, the Maunder minimum from 1645 to 1715, when solar activity seems to have virtually ceased (see Figure 14.24), was a time of exceptionally low temperatures in Europe and North America known as the *Little Ice Age*. However, no one knows whether the low solar activity caused these low temperatures or whether this was just a coincidence. Similarly, some researchers have claimed that certain weather phenomena, such as drought cycles or frequencies of storms, are correlated with the 11- or 22-year cycle of solar activity. A few scientists have even claimed that changes in the Sun may be responsible for Earth's recent global warming, but the Sun's energy output has actually remained nearly steady while the temperature has continued to rise (**FIGURE 14.26**), leaving little doubt that the warming is a result of human activity [**Section 10.6**].

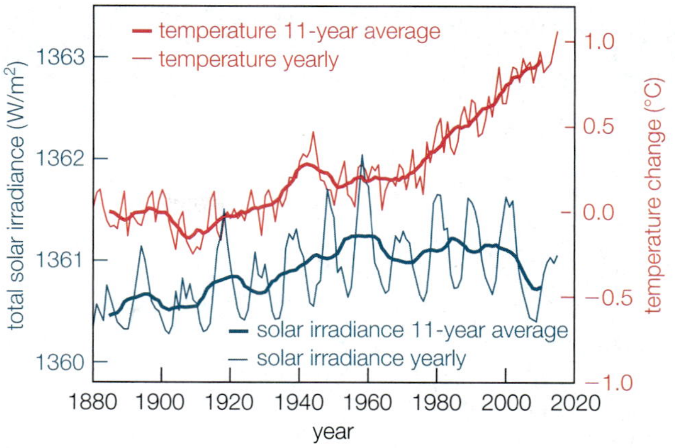


**FIGURE 14.26** This graph compares the global average temperature and the amount of sunlight reaching Earth (the solar irradiance, measured in watts per square meter) since 1880. The light red and blue curves show annual changes while the dark curves show 11-year averages, which smooth out the effects of the 11-year solar cycle. Notice that, for recent decades, the amount of sunlight has remained approximately constant while Earth has warmed, ruling out the Sun as the cause of recent global warming.

## The BIG Picture PUTTING CHAPTER 14 INTO CONTEXT

*In this chapter, we examined our Sun, the nearest star. When you look back at this chapter, make sure you understand these "big picture" ideas:*

- The ancient riddle of why the Sun shines has been solved. The Sun shines with energy generated by fusion of hydrogen into helium in the Sun's core. After a journey through the solar interior lasting several hundred thousand years and an 8-minute journey through space, a small fraction of this energy reaches Earth and supplies sunlight and heat.
- The Sun shines steadily thanks to the balance between pressure and gravity (gravitational equilibrium) and the balance between energy production in the core and energy release at the surface (energy balance). These two kinds of balance create a natural thermostat that regulates the Sun's fusion rate, keeping the Sun shining steadily and allowing life to flourish on Earth.
- The Sun's atmosphere displays its own version of weather and climate, governed by solar magnetic fields. Some solar weather, such as coronal mass ejections, clearly affects Earth's magnetosphere. Other claimed connections between solar activity and Earth's climate are not supported by observational evidence.
- The Sun is important not only as our source of light and heat, but also because it is the only star near enough for us to study in great detail. In the coming chapters, we will use what we've learned about the Sun to help us understand other stars.

**MY COSMIC PERSPECTIVE** The Sun affects us more directly than any other astronomical object. It is the source of all our heat and light, and its "solar weather" can even affect our use of technology.

# Summary of Key Concepts

## 14.1 A Closer Look at the Sun

- **Why does the Sun shine?** The Sun began to shine about $4\frac{1}{2}$ billion years ago when **gravitational contraction** made its core hot enough to sustain nuclear fusion. It has shined steadily ever since because of two types of balance: (1) **gravitational equilibrium**, a balance between the outward push of pressure and the inward pull of gravity, and (2) **energy balance** between the energy released by fusion in the core and the energy radiated into space from the Sun's surface.

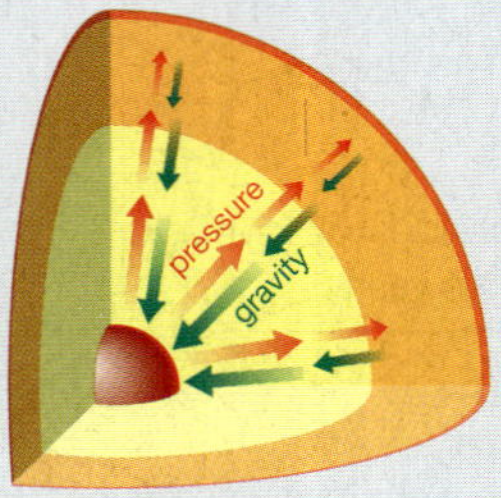


- **What is the Sun's structure?** The Sun's interior layers, from the inside out, are the **core**, the **radiation zone**, and the **convection zone**. Atop the convection zone lies the **photosphere**, the surface layer from which photons can freely escape into space. Above the photosphere are the warmer **chromosphere** and the very hot **corona**.

## 14.2 Nuclear Fusion in the Sun

- **How does nuclear fusion occur in the Sun?** The core's extreme temperature and density are just right for fusion of hydrogen into helium, which occurs via the **proton-proton chain**. Because the fusion rate is so sensitive to temperature, gravitational equilibrium acts as a thermostat that keeps the fusion rate steady.

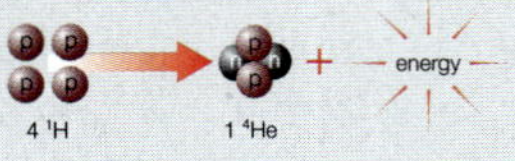


- **How does the energy from fusion get out of the Sun?** Energy moves through the deepest layers of the Sun—the core and the radiation zone—through **radiative diffusion**, in which photons bounce randomly among gas particles. After energy emerges from the radiation zone, convection carries it the rest of the way to the photosphere, where it is radiated into space as sunlight. Energy produced in the core takes hundreds of thousands of years to reach the photosphere.

- **How do we know what is happening inside the Sun?** We can construct theoretical models of the solar interior using known laws of physics and then check the models against observations of the Sun's size, surface temperature, and energy output. We also use studies of solar vibrations and solar **neutrinos**.

## 14.3 The Sun-Earth Connection

- **What causes solar activity? Sunspots** and other changing features of the Sun constitute **solar activity**, which is caused by strong magnetic fields that contort and sometimes snap, creating phenomena that include **flares**, **prominences**, and **coronal mass ejections**. The magnetic fields also carry energy upward, depositing the heat that explains the high temperatures of the chromosphere and corona.

- **How does solar activity vary with time?** The **sunspot cycle**, or the variation in the number of sunspots on the Sun's surface, has an average period of 11 years. The magnetic field flip-flops every 11 years or so, resulting in a 22-year magnetic cycle. The number of sunspots can vary dramatically from one cycle to the next, and sometimes sunspots seem to be absent altogether. The sunspot cycle and other solar activity are tied to the Sun's ever-changing magnetic field, which is created by the combination of convection and the Sun's rotation pattern (faster at the equator than at the poles).

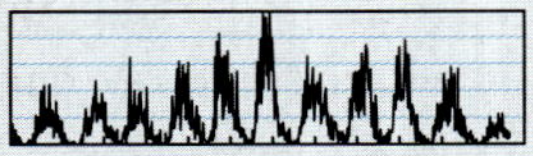

# Visual Skills Check

*Use the following questions to check your understanding of some of the many types of visual information used in astronomy. For additional practice, try the Chapter 14 Visual Quiz in the Study Area at www.MasteringAstronomy.com.*

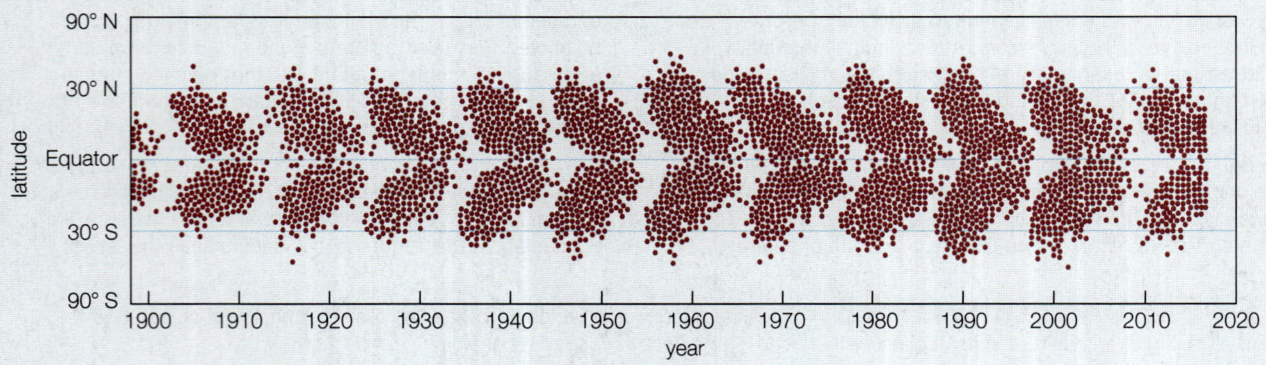

*Figure 14.23b, repeated above, shows the latitudes at which sunspots appeared on the surface of the Sun during the last century. Answer the following questions, using the information provided in the figure.*

1. Which of the following years had the least sunspot activity?
   a. 1930
   b. 1949
   c. 1961
   d. 1987
2. What is the approximate range in latitude over which sunspots appear?
3. According to the figure, how do the positions of sunspots appear to change during one sunspot cycle? Do they get closer to or farther from the equator with time?

# Exercises and Problems

*For instructor-assigned homework and other learning materials, go to www.MasteringAstronomy.com.*

## Chapter Review Questions

### Short-Answer Questions Based on the Reading

1. Briefly describe how *gravitational contraction* generates energy. When was it important in the Sun's history?
2. What two forces are balanced in *gravitational equilibrium*? What does it mean for the Sun to be in *energy balance*?
3. State the Sun's luminosity, mass, radius, and average surface temperature, and put the numbers into a perspective that makes them meaningful.
4. Briefly describe the distinguishing features of each of the layers of the Sun shown in Figure 14.4.
5. What is the difference between nuclear *fission* and nuclear *fusion*? Which one is used in nuclear power plants? Which one takes place in the Sun?
6. Why does nuclear fusion require high temperatures and pressures?
7. What is the overall nuclear fusion reaction in the Sun? Briefly describe the *proton-proton chain*.
8. Describe how a natural "solar thermostat" keeps the core fusion rate steady in the Sun.
9. Why has the Sun gradually brightened with time?
10. Describe how energy generated by fusion makes its way to the Sun's surface. How long does it take?
11. How do mathematical models help us learn about the Sun, and what gives us confidence in the models?
12. What are *neutrinos*? What was the *solar neutrino problem*, and how was it solved?
13. What do we mean by *solar activity*? Describe *sunspots, prominences, flares*, and *coronal mass ejections*.
14. Describe the appearance and temperature of the Sun's photosphere. Why does the surface look mottled? How are sunspots different from the surrounding photosphere?
15. How do magnetic fields keep sunspots cooler than the surrounding plasma? Explain.
16. Why are the chromosphere and corona best viewed with ultraviolet and x-ray telescopes, respectively? Briefly explain how these regions are heated.
17. What is the *sunspot cycle*? Why is it sometimes described as an 11-year cycle and sometimes as a 22-year cycle? Are there longer-term changes in solar activity?
18. Describe the leading model for the sunspot cycle. Does the sunspot cycle influence Earth's climate? Explain.

### Does It Make Sense?

*Decide whether the statement makes sense (or is clearly true) or does not make sense (or is clearly false). Explain clearly; not all of these have definitive answers, so your explanation is more important than your chosen answer.*

19. Before Einstein, gravitational contraction appeared to be a perfectly plausible mechanism for solar energy generation.
20. A temperature rise in the Sun's core is nothing to worry about, because conditions in the core will soon return to normal.
21. If fusion in the solar core ceased today, worldwide panic would break out tomorrow as the Sun began to grow dimmer.
22. Astronomers have recently photographed magnetic fields churning deep beneath the solar photosphere.
23. I wear a lead vest to protect myself from solar neutrinos.
24. If there are few sunspots this year, we should expect many more in about 5 years.
25. News of a solar flare caused concern among businesses involved in communication and electrical power generation.
26. By observing solar neutrinos, we can learn about nuclear fusion deep in the Sun's core.
27. If the Sun's magnetic field somehow disappeared, there would be no more sunspots on the Sun.
28. Scientists are currently building an infrared telescope designed to observe fusion reactions in the Sun's core.

### Quick Quiz

*Choose the best answer to each of the following. For additional practice, try the Chapter 14 Reading and Concept Quizzes in the Study Area at www.MasteringAstronomy.com.*

29. Which of these groups of particles has the greatest mass? (a) a helium nucleus with two protons and two neutrons (b) four electrons (c) four individual protons
30. X-ray images of the Sun generally show the (a) photosphere. (b) chromosphere. (c) corona.
31. Which of these layers of the Sun is coolest? (a) core (b) radiation zone (c) photosphere
32. Scientists estimate the central temperature of the Sun using (a) probes that measure changes in Earth's atmosphere. (b) mathematical models of the Sun. (c) laboratories that create miniature versions of the Sun.
33. Why do sunspots appear darker than their surroundings? (a) They are cooler than their surroundings. (b) They block some of the sunlight from the photosphere. (c) They do not emit any light.
34. At the center of the Sun, fusion converts hydrogen into (a) plasma. (b) radiation and elements like carbon and nitrogen. (c) helium, energy, and neutrinos.
35. Solar energy leaves the core of the Sun in the form of (a) photons. (b) rising hot gas. (c) sound waves.
36. The fact that we observe neutrinos from the Sun provides direct evidence of (a) fusion in the Sun's core. (b) convection in the Sun's interior. (c) the existence of the solar wind.

37. What causes the cycle of solar activity? (a) changes in the Sun's fusion rate (b) changes in the organization of the Sun's magnetic field (c) changes in the speed of the solar wind
38. Which of these things poses the greatest hazard to communication satellites? (a) photons from the Sun (b) solar magnetic fields (c) particles from the Sun

## Inclusive Astronomy

*Use these questions to reflect on participation in science.*

39. *Group Discussion: Arthur Walker, Scientist and Mentor.* Arthur B. C. Walker Jr., an African-American scientist who spent much of his career at Stanford University, was a pioneer in developing the technology for taking ultraviolet and x-ray images of the Sun.
    **a.** Working independently, learn basic details of Dr. Walker's career, including when and where he earned his Ph.D., the name and occupation of his first graduate student, the purpose of the investigative commission he was appointed to in 1986, and the purpose of the award that now bears his name.
    **b.** Gather in small groups to discuss whether Dr. Walker was likely to have faced discrimination in his efforts to become a scientist. Be sure to consider the time period when he was earning his Ph.D.
    **c.** Within the group, share what you've learned about Dr. Walker's first graduate student, including what was posthumously revealed about the student's sexual orientation. Discuss why such a person might have considered Dr. Walker a particularly suitable mentor.
    **d.** Discuss what you learned about the 1986 commission. In addition to scientific and technical knowledge, what other skills do people on such a commission need to have?
    **e.** What is the purpose of the award that bears Dr. Walker's name, and who has won it to date? Does your group think this award is an appropriate way to honor him? Why or why not?

## The Process of Science

*These questions may be answered individually in short-essay form or discussed in groups, except where identified as group-only.*

40. *Inside the Sun.* Scientists claim to know what is going on inside the Sun, even though we cannot observe the solar interior directly. What is the basis for these claims, and how do they align with the hallmarks of science outlined in Section 3.4?
41. *The Solar Neutrino Problem.* Early solar neutrino experiments detected only about a third of the number of neutrinos predicted by the theory of fusion in the Sun. Why didn't scientists abandon their fusion models at this point, and what ultimately led them to the investigations that resolved the discrepancies? Explain how their decision-making process exemplified the hallmarks of science from Chapter 3.
42. *The Role of the Sun.* Briefly discuss how the Sun affects us here on Earth. Be sure to consider not only factors such as its light and warmth but also how the study of the Sun has led us to new understandings in science and to technological developments. Overall, how important has solar research been to our lives?
43. *The Sun and Global Warming.* Some people still claim that part or all of the global warming observed over the past century may be due to changes in the Sun, but the data in Figure 14.26 seem to rule that out (as do several other lines of evidence beyond those discussed in this text). Discuss how the people in your life talk about global warming, what information sources they view as credible, and whether they consult scientific data when making up their minds about the causes of global warming.
44. *Group Activity: The Sun's Future.* Working in small groups, discuss the following questions in light of what you have learned about how the Sun came to shine steadily with energy from fusion in its core. Note: You may wish to do this activity using the four roles described in Chapter 1, Exercise 39.
    **a.** Will the Sun's core temperature go up or down after the core runs out of hydrogen for fusion? Why?
    **b.** If your group thinks the temperature will go up, decide what would ultimately stop the temperature from rising forever. If you think the temperature will go down, decide what could stop it from falling to absolute zero.
    **c.** Propose and describe an experiment or a set of observations that could test whether your answers to parts a and b are correct.

## Investigate Further

### Short-Answer/Essay Questions

45. *The End of Fusion I.* Describe what would happen in the Sun if fusion reactions suddenly ceased.
46. *The End of Fusion II.* If fusion reactions in the Sun were to suddenly cease, would we be able to tell? If so, how?
47. *A Really Strong Force.* How would the interior temperature of the Sun be different if the strong force that binds nuclei together were 10 times as strong?
48. *Covered with Sunspots.* Describe what the Sun would look like from Earth if the entire photosphere were the same temperature as a sunspot.
49. *Inside the Sun.* Describe how scientists determine what the interior of the Sun is like. Why haven't we sent a probe into the Sun to measure what is happening there?
50. *Solar Energy Output.* Observations over the past century show that the Sun's visible-light output varies by less than 1%, but its x-ray output can vary by a factor of 10 or more. Explain why changes in x-ray output can be so much larger than those in the output of visible light.
51. *An Angry Sun.* A *Time* magazine cover once suggested that an "angry Sun" was becoming more active as human activity changed Earth's climate. It's certainly possible for the Sun to become more active at the same time that humans are affecting Earth, but is it possible that the Sun could be responding to human activity? Why or why not?
52. *Research: Current Solar Weather.* Daily information about solar activity is available at NASA's website spaceweather.com. Where are we in the sunspot cycle right now? When is the next solar maximum or minimum expected? Have there been any major solar storms in the past few months? If so, did they have any significant effects on Earth? Summarize your findings in a one- to two-page report.
53. *Research: Solar Observatories in Space.* Visit NASA's website for the Sun-Earth connection and explore some of the current and planned space missions designed to observe the Sun. Choose one mission to study in greater depth, and write a one- to two-page report on the status and goals of the mission and what it has taught or will teach us about the Sun.
54. *Research: Nuclear Power.* There are two basic ways to generate energy from atomic nuclei: through nuclear fission

(splitting nuclei) and through nuclear fusion (combining nuclei). All current nuclear reactors are based on fission, but fusion would have many advantages if we could develop the technology. Research some of the advantages of fusion and some of the obstacles to developing fusion power. Do you think fusion power will be a reality in your lifetime? Explain.

## Quantitative Problems

*Be sure to show all calculations clearly and state your final answers in complete sentences.*

55. *Chemical Burning and the Sun.* When an object burns in a fire, the amount of energy released through this chemical burning is typically about $10^8$ joules per kilogram of mass burned. Use this fact to estimate how long the Sun would last if its energy source were a huge fire releasing chemical energy (rather than its actual source, fusion energy).
56. *The Lifetime of the Sun.* The Sun's chemical composition was about 70% hydrogen when it formed, and about 13% of this hydrogen was available for eventual fusion in the core. (The rest remains in layers of the Sun where the temperature is currently too low for fusion.)
**a.** Use these data and the Sun's mass to calculate the total mass of hydrogen available for fusion over the lifetime of the Sun.
**b.** The Sun fuses about 600 billion kilograms of hydrogen each second. Based on your result from part a, calculate how long the Sun's initial supply of hydrogen can last. Give your answer in both seconds and years.
**c.** Given that our solar system is now about 4.6 billion years old, when will we need to worry about the Sun running out of hydrogen for fusion?
57. *Solar Power Collectors.* This problem leads you through the calculation and discussion of how much solar power can in principle be collected by solar cells on Earth.
**a.** Imagine a giant sphere with a radius of 1 AU surrounding the Sun. What is the surface area of this sphere, in square meters? (*Hint*: The formula for the surface area of a sphere is $4\pi r^2$.)
**b.** Because this imaginary giant sphere surrounds the Sun, the Sun's entire luminosity of $3.8 \times 10^{26}$ watts must pass through it. Calculate the power passing through each square meter of this imaginary sphere in *watts per square meter*. Explain why this number represents the *maximum* power per square meter that a solar collector in Earth orbit can collect.
**c.** List several reasons why the average power per square meter collected by a solar collector on the ground will always be less than what you found in part b.
**d.** Suppose you want to put a solar collector on your roof. If you want to optimize the amount of power you can collect, how should you orient the collector? (*Hint*: The optimal orientation depends on both your latitude and the time of year and day.)
58. *Solar Power for the United States.* Total annual U.S. energy consumption is about $2 \times 10^{20}$ joules.
**a.** What is the average power requirement for the United States, in watts? (*Hint*: 1 watt = 1 joule/s.)
**b.** The best available solar cells can generate an average (day and night) power of about 200 watts/m$^2$. What total area would we need to cover with solar cells to supply all the power needed for the United States? Give your answer in square kilometers and compare it with the total surface area of the continental United States.
59. *The Color of the Sun.* Use Wien's law (see Mathematical Insight 5.2) and the Sun's average surface temperature of about 5800 K to calculate the wavelength of peak thermal emission from the Sun. What color does this wavelength correspond to in the visible-light spectrum? Why do you think the Sun appears white or yellow to our eyes?
60. *The Color of a Sunspot.* Use Wien's law (see Mathematical Insight 5.2) and a typical sunspot temperature of 4000 K to calculate the wavelength of peak thermal emission from a sunspot. What color does this wavelength correspond to in the visible-light spectrum? How does this color compare with the overall color of the Sun?
61. *Solar Mass Loss.* Estimate how much mass the Sun loses through fusion reactions during its 10-billion-year life. You can simplify the problem by assuming the Sun's energy output remains constant. Compare the amount of mass lost with Earth's mass.
62. *Pressure of the Photosphere.* The gas pressure of the photosphere changes substantially from its upper levels to its lower levels. Near the top of the photosphere, the temperature is about 4500 K and there are about $1.6 \times 10^{16}$ gas particles per cubic centimeter. In the middle, the temperature is about 5800 K and there are about $1.0 \times 10^{17}$ gas particles per cubic centimeter. At the bottom of the photosphere, the temperature is about 7000 K and there are about $1.5 \times 10^{17}$ gas particles per cubic centimeter. Use the ideal gas law (Mathematical Insight 14.2) to compare the pressures of each of these layers; explain the reason for the trend that you find. How do these gas pressures compare with Earth's atmospheric pressure at sea level?
63. *Tire Pressure.* Air pressure at sea level is about 15 pounds per square inch. The recommended air pressure in your car tires is about 30 pounds per square inch. Use the ideal gas law to determine how the density of gas particles inside your tires compares with that of gas particles in the outside air. How does this explain what happens when a tire springs a leak? How is the tire's response to a leak similar to the response of the Sun's core as its number of independent particles gradually declines with time? How is the tire's response different?
64. *Personal Energy Content.* The average power of a human body (for living and metabolism) is about 100 watts. Suppose your body could run on fusion power and could convert 0.7% of its mass into energy. How much energy would be available through fusion? For how long could your body then operate on fusion power?

# 15 Surveying the Stars

▲ **About the photo:** This single image captures many types of stars, as well as some of the gas between them. The bright foreground star at the bottom, surrounded by glowing orange gas, is Antares, a supergiant star that marks the "heart" of the scorpion in the constellation Scorpius. To the right of Antares is a much more distant cluster of stars—the globular cluster M4. Figure 15.20 shows the properties of this cluster's stars.

## LEARNING GOALS

### 15.1 Properties of Stars

- How do we measure stellar luminosities?
- How do we measure stellar temperatures?
- How do we measure stellar masses?

### 15.2 Patterns Among Stars

- What is a Hertzsprung-Russell diagram?
- What is the significance of the main sequence?
- What are giants, supergiants, and white dwarfs?
- Why do the properties of some stars vary?

### 15.3 Star Clusters

- What are the two types of star clusters?
- How do we measure the age of a star cluster?

*Classifying the stars has helped materially in all studies of the structure of the universe.*

*—Annie Jump Cannon (1863–1941)*

**Chapter 15 Overview**

On a clear dark night, a few thousand stars are visible to the naked eye. Many more become visible through binoculars, and a powerful telescope reveals so many stars that we could never hope to count them. Like each individual person, each individual star is unique. Like all humans, all stars have much in common.

Today, we know that stars are born from clouds of interstellar gas, shine brilliantly by nuclear fusion for millions to billions of years, and then die, sometimes in dramatic ways. In this chapter, we'll discuss how we study and categorize stars and how we have come to realize that stars, like people, change over their lifetimes.

## 15.1 Properties of Stars

Imagine that an alien spaceship flies by Earth on a simple but short mission: The visitors have just 1 minute to learn everything they can about the human race. In 60 seconds, they will see next to nothing of any individual person's life. Instead, they will obtain a collective "snapshot" of humanity showing people from all stages of life engaged in their daily activities. From this snapshot alone, they must piece together their entire understanding of human beings and their lives, from birth to death.

We face a similar problem when we look at the stars. Compared with stellar lifetimes of millions or billions of years, the few hundred years humans have spent studying stars with telescopes is rather like the aliens' 1-minute glimpse of humanity. We see only a brief moment in any star's life, and our collective snapshot of the heavens consists of such frozen moments for billions of stars. From this snapshot, we have learned to reconstruct the life cycles of stars.

We now know that all stars have a lot in common with the Sun. They all form in great clouds of gas and dust and begin their lives with roughly the same chemical composition as the Sun: about three-quarters hydrogen and one-quarter helium (by mass), with no more than about 2% consisting of elements heavier than helium. Nevertheless, stars are not all the same; they differ in size, age, brightness, and temperature. We'll devote most of this and the next two chapters to understanding how and why stars differ. First, however, let's explore how we measure three of the most fundamental properties of stars: luminosity, surface temperature, and mass.

### How do we measure stellar luminosities?

If you go outside on any clear night, you'll immediately see that stars differ in brightness. Some stars are so bright that we can use them to identify constellations [**Section 2.1**]. Others are so dim that our naked eyes cannot see them at all. However, these differences in brightness do not by themselves tell us how much light the stars are generating, because brightness also depends on distance. For example, the stars Procyon and Betelgeuse, which make up two of the three corners of the Winter Triangle (see Figure 2.2), appear about equally bright in our sky. However, Betelgeuse actually emits about 15,000 times as much light as Procyon. It has about the same brightness in our sky because it is much farther away.

**See it for yourself** Until the 20th century, people classified stars primarily by their brightness and location in our sky. On the next clear night, find a favorite constellation and visually rank the stars by brightness. Then look to see how that constellation is represented on the star charts in Appendix I. Why do the star charts use different size dots for different stars? Do the brightness rankings on the star charts agree with what you see?

Because two similar-looking stars can be generating very different amounts of light, we need to distinguish clearly between a star's brightness in our sky and the actual amount of light that it emits into space (**FIGURE 15.1**):

- When we talk about how bright stars look in our sky, we are talking about **apparent brightness**. More specifically, we define the apparent brightness of any star in our sky as the amount of power (energy per second) reaching us *per unit area*.
- When we talk about how bright stars are in an absolute sense, regardless of their distance, we are talking about **luminosity**—the total amount of power that a star emits into space.

You can understand the difference between apparent brightness and luminosity by thinking about a standard light bulb. The bulb always puts out the same amount of light, so its luminosity doesn't vary. However, its apparent brightness depends on your distance from the bulb: It will look quite bright if you stand very close to it, but quite dim if you are far away.

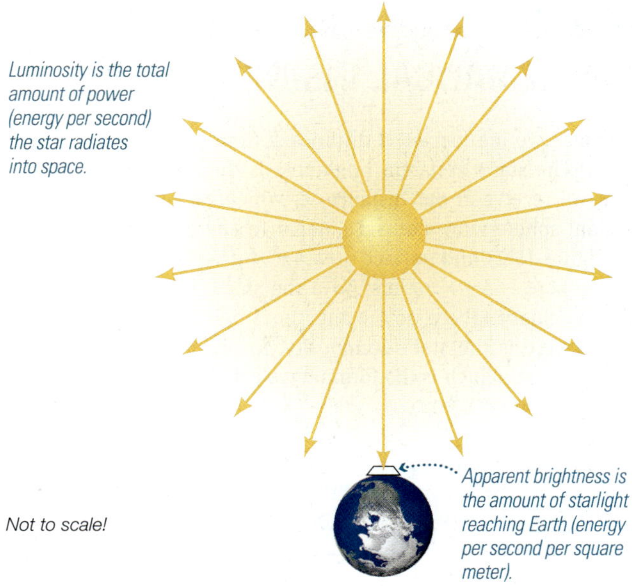


**FIGURE 15.1** Luminosity is a measure of power, and apparent brightness is a measure of power per unit area.

**The Inverse Square Law for Light** The apparent brightness of a star or any other light source obeys an *inverse square law* with distance, much like the inverse square law that describes the force of gravity [**Section 4.4**]. For example, if we viewed the Sun from twice Earth's distance, it would appear dimmer by a factor of $2^2 = 4$. If we viewed it from 10 times Earth's distance, it would appear $10^2 = 100$ times dimmer.

**FIGURE 15.2** shows why apparent brightness follows an inverse square law. The same total amount of light must pass through each imaginary sphere surrounding the star. If we focus on the light passing through the small square on the sphere located at 1 AU, we see that the same amount of light must pass through *four* squares of the same size on the sphere located at 2 AU. Each square on the sphere at 2 AU therefore receives only $\frac{1}{2^2} = \frac{1}{4}$ as much light as the square on the sphere at 1 AU. Similarly, the same amount of light passes through *nine* squares of the same size on the sphere located at 3 AU, so each of these squares receives only $\frac{1}{3^2} = \frac{1}{9}$ as much light as the square on the sphere at 1 AU. Generalizing, the amount of light received per unit area decreases with increasing distance by the square of the distance, making for an inverse square law.

This inverse square law leads to a very simple and important formula relating the apparent brightness, luminosity, and distance of any light source. We will call it the **inverse square law for light**:

$$\text{apparent brightness} = \frac{\text{luminosity}}{4\pi \times \text{distance}^2}$$

Because the standard units of luminosity are watts [**Section 14.1**], the units of apparent brightness are *watts per square meter*. (The $4\pi$ in the formula above comes from the fact that the surface area of a sphere is given by $4\pi \times \text{radius}^2$.)

In principle, we can always determine a star's apparent brightness by carefully measuring the amount of light per square meter we receive from the star. We can then use the inverse square law to calculate a star's luminosity if we can first measure its distance, or to calculate a star's distance if we somehow know its luminosity.

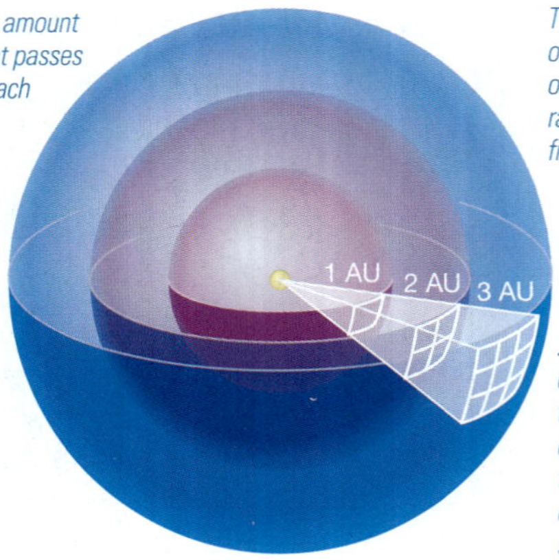


**FIGURE 15.2** The inverse square law for light: The apparent brightness of a star declines with the square of its distance.

**Think about it** Suppose Star A is four times as luminous as Star B. How will their apparent brightnesses compare if they are both the same distance from Earth? How will their apparent brightnesses compare if Star A is twice as far from Earth as Star B? Explain.

**Measuring Apparent Brightness and Calculating Luminosity** We can measure a star's apparent brightness by using a detector that records how much energy strikes its light-sensitive surface each second. For example, such a detector would record an apparent brightness of $2.7 \times 10^{-8}$ watt per square meter from Alpha Centauri A (the brightest of the three stars in the Alpha Centauri system). The only difficulties we face in measuring apparent brightness are making sure the detector is properly calibrated and, for ground-based telescopes, taking into account the absorption of light by Earth's atmosphere.

## MATHEMATICAL INSIGHT 15.1 The Inverse Square Law for Light

Suppose we are located a distance $d$ from a star with luminosity $L$. The star's apparent brightness is the power per unit area that we receive at our distance, $d$, which we find by imagining a giant sphere with radius $d$ (similar to any of the three spheres in Figure 15.2) and surface area $4\pi \times d^2$. (The surface area of any sphere is $4\pi \times \text{radius}^2$.) All the star's light passes through the imaginary sphere, so the apparent brightness at any point on this sphere is the star's luminosity, $L$, divided by the sphere's surface area, which is the formula we call the inverse square law for light:

$$\text{apparent brightness} = \frac{\text{star's luminosity}}{\text{surface area of imaginary sphere}} = \frac{L}{4\pi \times d^2}$$

**EXAMPLE:** What is the Sun's apparent brightness in our sky?

**SOLUTION:**

**Step 1 Understand:** The Sun's apparent brightness is the power per unit area that we receive in the form of sunlight. We find this power with the inverse square law for light, using the Sun's luminosity and Earth's distance from the Sun; for unit consistency, we put the Earth-Sun distance in meters.

**Step 2 Solve:** The Sun's luminosity is $L_{Sun} = 3.8 \times 10^{26}$ watts, and Earth's distance from the Sun is $d = 1.5 \times 10^{11}$ meters. The Sun's apparent brightness is therefore

$$\frac{L}{4\pi \times d^2} = \frac{3.8 \times 10^{26}\ \text{watts}}{4\pi \times (1.5 \times 10^{11}\ \text{m})^2} = 1.3 \times 10^3\ \text{watts/m}^2$$

**Step 3 Explain:** The Sun's apparent brightness is about 1300 watts per square meter at Earth's distance.

No detector can record light of all wavelengths, so any measurement of apparent brightness applies only to some small range of the complete spectrum. For example, if we use a detector that is sensitive only to visible light, then our measurement of a star's apparent brightness allows us to calculate only the star's *visible-light luminosity*. Similarly, when we observe a star with a spaceborne x-ray telescope, we measure only the apparent brightness in x-rays and can calculate only the star's *x-ray luminosity*. We will use the terms **total luminosity** (also called *bolometric luminosity*) and **total apparent brightness** to describe the luminosity and apparent brightness we would measure *if* we could detect photons across the entire electromagnetic spectrum.

Note that the inverse square law for light works perfectly only if the starlight follows an uninterrupted path to Earth. In reality, the light of most stars passes through at least some clouds containing interstellar dust on its way to Earth, and this dust can absorb or scatter some of the star's light [**Section 16.1**]. Today, thanks largely to our modern scheme of stellar classification, we can usually measure the effect of interstellar dust and account for it when we apply the inverse square law for light. A century ago, before astronomers knew of the existence of interstellar dust, they often overestimated stellar distances because they did not realize that the dust was making stars appear less bright than they would without the dust.

**Measuring Distance Through Stellar Parallax** The most direct way to measure a star's distance is with *stellar parallax*, the small annual shifts in a star's apparent position caused by Earth's motion around the Sun [**Section 2.4**]. Recall that you can observe parallax of your finger by holding it at arm's length and looking at it with first one eye closed and then the other. Astronomers measure stellar parallax by comparing observations of a nearby star made 6 months apart (**FIGURE 15.3**). The nearby star appears to shift against the background of more distant stars because we are observing it from two opposite points of Earth's orbit.

We can calculate a star's distance if we know the precise amount of the star's annual shift due to parallax. This means measuring the angle $p$ in Figure 15.3, which we call the star's *parallax angle*. Notice that this angle would be smaller if the star were farther away, meaning that more distant stars have smaller parallax angles. Real stellar parallax angles are very small: Even the nearest stars have parallax angles smaller than 1 arcsecond—well below the approximately 1 arcminute angular resolution of the naked eye [**Section 6.2**]—which explains why the ancient Greeks were never able to measure parallax.

By definition, the distance to an object with a parallax angle of 1 arcsecond is 1 **parsec (pc)**. (The word *parsec* comes from combining the words *parallax* and *arcsecond*.) If we use units of arcseconds for the parallax angle, $p$, a simple formula allows us to calculate distances in parsecs:

$$d \text{ (in parsecs)} = \frac{1}{p \text{ (in arcseconds)}}$$

For example, the distance to a star with a parallax angle of $\frac{1}{2}$ arcsecond is 2 parsecs, the distance to a star with a parallax angle of $\frac{1}{10}$ arcsecond is 10 parsecs, and the distance to a star with a parallax angle of $\frac{1}{100}$ arcsecond is 100 parsecs.

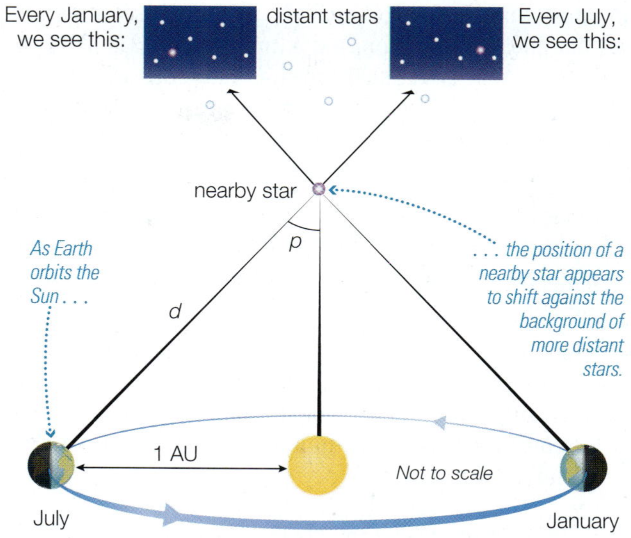


**FIGURE 15.3** Parallax makes the apparent position of a nearby star shift back and forth with respect to distant stars over the course of each year. The angle $p$, called the *parallax angle*, represents half the total parallax shift each year. If we measure $p$ in arcseconds, the distance $d$ to the star in parsecs is $1/p$. The angle in this figure is greatly exaggerated: All stars have parallax angles of less than 1 arcsecond.

Astronomers often state distances in parsecs, kiloparsecs (1000 parsecs), or megaparsecs (1 million parsecs). However, with a bit of geometry, it's possible to show that 1 parsec is equivalent to 3.26 light-years (see Mathematical Insight 15.2). We can therefore modify the above formula slightly to give distances in light-years:

$$d \text{ (in light-years)} = 3.26 \times \frac{1}{p \text{ (in arcseconds)}}$$

In this book, we'll generally state distances in light-years rather than parsecs.

Measuring parallax was the first reliable technique astronomers developed for finding distances to stars, and it remains the only technique that tells us stellar distances without any assumptions about the nature of stars. If we know a star's distance from parallax, we can calculate its luminosity with the inverse square law for light. In fact, parallax measurements are the key to nearly all other distance measurements in the universe, because parallax measurements of the distances of nearby stars are a key link in a chain of techniques that allow measurement of much greater distances [**Section 20.2**]. As of 2018, astronomers using the European *GAIA* spacecraft had made parallax measurements for more than 1 billion stars, out to distances of tens of thousands of light-years.

**The Luminosity Range of Stars** Now that we have discussed how we determine stellar luminosities, it's time to take a quick look at the results. We usually state stellar luminosities in comparison to the Sun's luminosity, $L_{\text{Sun}}$. For example, Proxima Centauri, the nearest of the three stars in the Alpha Centauri system and hence the nearest star besides our Sun, has only about 0.0006 times the luminosity of the Sun, or $0.0006L_{Sun}$. Betelgeuse, the bright left-shoulder star of Orion, has a luminosity of $120{,}000L_{Sun}$, meaning that it is 120,000 times as luminous as the Sun. Overall, studies of the luminosities of many stars have taught us two particularly important lessons:

- Stars have a wide range of luminosities, with our Sun somewhere in the middle. The dimmest stars have luminosities $\frac{1}{10{,}000}$ times that of the Sun ($10^{-4}L_{Sun}$), while the brightest stars are about 1 million times as luminous as the Sun ($10^{6}L_{Sun}$).
- Low-luminosity stars are far more common than high-luminosity stars. For example, even though our Sun is roughly in the middle of the overall range of stellar luminosities, it is more luminous than the vast majority of stars in our galaxy.

**The Magnitude System** Many astronomy resources (including star charts) describe the apparent brightnesses and luminosities of stars in an alternative way: They use the ancient **magnitude system** devised by the Greek astronomer Hipparchus (c. 190–120 B.C.). Although we won't make much use of this system in this book, you might want to become familiar with it in case you encounter it elsewhere.

Hipparchus designated the brightest stars in the sky as "first magnitude," the next brightest "second magnitude," and so on. The faintest visible stars were magnitude 6. Today we call this type of designation an **apparent magnitude** because it describes how bright a star *appears* in the sky. Notice that the magnitude scale runs backward: A larger apparent magnitude means a dimmer apparent brightness. For example, a star of magnitude 4 is dimmer in the sky than a star of magnitude 1.

In modern times, the magnitude system has been extended and more precisely defined. Each difference of 5 magnitudes is defined to represent a factor of exactly 100 in brightness. For example, a magnitude 1 star is 100 times as bright as a magnitude 6 star, and a magnitude 3 star is 100 times as bright as a magnitude 8 star. As a result of this precise definition, stars can have fractional apparent magnitudes and a few bright stars have apparent magnitudes *less than* 1—which means *brighter* than magnitude 1. For example, the brightest star in the night sky, Sirius, has an apparent magnitude of $-1.46$. Appendix F gives apparent magnitudes and actual luminosities for both the nearest stars and the brightest stars visible in the sky.

The modern magnitude system also defines **absolute magnitude** as a way of describing stellar luminosity. A star's absolute magnitude is the apparent magnitude it would have *if* it were at a distance of 10 parsecs (32.6 light-years) from Earth. For example, the Sun's absolute magnitude is about 4.8, meaning that the Sun would have an apparent

## MATHEMATICAL INSIGHT 15.2 The Parallax Formula

We can derive the formula relating a star's distance and parallax angle by studying Figure 15.3. The parallax angle, $p$, is part of a right triangle, and from trigonometry, the *sine* of angle $p$ is the length of the side opposite this angle divided by the length of the hypotenuse. Because the side opposite $p$ is the Earth-Sun distance of 1 AU and the hypotenuse is the distance $d$ to the object, we find

$$\sin p = \frac{\text{length of opposite side}}{\text{length of hypotenuse}} = \frac{1\text{ AU}}{d}$$

Solving for $d$ gives the formula

$$d = \frac{1\text{ AU}}{\sin p}$$

By definition, a parallax angle of 1 arcsecond ($1''$) gives a distance $d = 1$ parsec; putting these numbers in the parallax formula, we find that

$$1\text{ parsec} = \frac{1\text{ AU}}{\sin 1''} = \frac{1\text{ AU}}{4.84814 \times 10^{-6}} = 206{,}265\text{ AU}$$

(You can use a calculator to find $\sin 1''$, which is the same as $\sin[(\frac{1}{3600})^\circ]$.) That is, 1 parsec = 206,265 AU. Because 1 AU = 149.6 million km and 1 light-year = $9.46 \times 10^{12}$ km, we also find that 1 parsec = $3.09 \times 10^{13}$ km = 3.26 light-years.

We need one more fact from geometry to derive the parallax formula given in the text. As long as the parallax angle, $p$, is small, $\sin p$ is proportional to $p$. For example, $\sin 2''$ is twice as large as $\sin 1''$, and $\sin (\frac{1}{2})''$ is half as large as $\sin 1''$. (You can verify these examples with your calculator.) If we use $(\frac{1}{2})''$ instead of $1''$ for the parallax angle in the formula above, we get a distance of 2 parsecs instead of 1 parsec. Similarly, if we use a parallax angle of $(\frac{1}{10})''$, we get a distance of 10 parsecs. Generalizing, we find

$$d\text{ (in parsecs)} = \frac{1}{p\text{ (in arcseconds)}}$$

**EXAMPLE:** Sirius, the brightest star in our night sky, has a measured parallax angle of $0.379''$. Find its distance in parsecs and light-years.

**SOLUTION:**

**Step 1 Understand:** We are given the parallax angle for Sirius in arcseconds, so we use the parallax formula to find its distance. Because the parallax angle is between $0.1''$ and $1''$, we expect the answer to be a distance between 1 and 10 parsecs.

**Step 2 Solve:** When we substitute the parallax angle of $0.379''$ into the formula, the distance to Sirius in parsecs is

$$d\text{ (in parsecs)} = \frac{1}{0.379} = 2.64\text{ parsecs}$$

Because 1 parsec = 3.26 light-years, this distance is equivalent to

$$2.64\text{ parsecs} \times 3.26\frac{\text{light-years}}{\text{parsec}} = 8.60\text{ light-years}$$

**Step 3 Explain:** From its measured parallax angle, the distance to Sirius is 2.64 parsecs, or 8.60 light-years.

magnitude of 4.8 *if* it were 10 parsecs away from us—bright enough to be visible but not conspicuous on a dark night.

## How do we measure stellar temperatures?

A second fundamental property of a star is its surface temperature. You might wonder why we emphasize *surface* temperature rather than interior temperature. The answer is that only surface temperature is directly measurable; interior temperatures are inferred from mathematical models of stellar interiors [**Section 14.2**]. Whenever you hear astronomers speak of the "temperature" of a star, you can be pretty sure they mean surface temperature unless they state otherwise.

Measuring a star's surface temperature is somewhat easier than measuring its luminosity, because the star's distance doesn't affect the measurement. Instead, we determine surface temperature from either the star's color or its spectrum.

**Color and Temperature** Take a careful look at **FIGURE 15.4**. Notice that stars come in almost every color of the rainbow. Simply looking at the colors tells us something about the surface temperatures of the stars. For example, a red star is cooler than a yellow star, which in turn is cooler than a blue star.

Stars come in different colors because they emit thermal radiation [**Section 5.4**]. Recall that a thermal radiation spectrum depends only on the (surface) temperature of the object that emits it (see Figure 5.19). For example, the Sun's 5800 K surface temperature causes it to emit most strongly in the middle of the visible portion of the spectrum, which is why the Sun looks yellow or white. A cooler star, such as Betelgeuse (surface temperature 3650 K), looks red because it emits much more red light than blue light. A hotter star, such as Sirius (surface temperature 9400 K), emits a little more blue light than red light and therefore has a slightly blue color.

Astronomers can measure surface temperature fairly precisely by comparing a star's apparent brightness in two different colors of light. For example, by comparing the amounts of blue light and red light coming from Sirius, astronomers can measure how much more blue light it emits than red light. Because thermal radiation spectra have a distinctive shape (again, see Figure 5.19), this difference between blue and red light output allows astronomers to calculate surface temperature.

**Spectral Type and Temperature** A star's spectral lines provide a second way to measure its surface temperature. In fact, because interstellar dust can affect the apparent colors of stars, temperatures determined from spectral lines are generally more accurate than temperatures determined from colors alone. Stars displaying spectral lines of highly ionized elements must be fairly hot, because it takes a high

## MATHEMATICAL INSIGHT 15.3 The Modern Magnitude System

The modern magnitude system is defined so that a difference of 5 magnitudes corresponds to a factor of exactly 100 in brightness, which means that a difference of 1 magnitude corresponds to a factor of $(100)^{1/5} \approx 2.512$ in brightness. Therefore, if we have two stars, Star 1 and Star 2, with apparent magnitudes $m_1$ and $m_2$, the ratio of their apparent brightnesses is

$$\frac{\text{apparent brightness of Star 1}}{\text{apparent brightness of Star 2}} = (100^{1/5})^{m_2 - m_1}$$

If we replace the apparent magnitudes with absolute magnitudes (designated $M$ instead of $m$), the same formula allows us to calculate the ratio of stellar luminosities:

$$\frac{\text{luminosity of Star 1}}{\text{luminosity of Star 2}} = (100^{1/5})^{M_2 - M_1}$$

**EXAMPLE 1:** The human eye can see stars down to about magnitude 5, while large telescopes can detect objects as faint as magnitude 30 (or fainter in some cases). How much more sensitive are such telescopes than the human eye?

**SOLUTION:**

**Understand:** We imagine that our eye sees "Star 1" with magnitude 5 and the telescope detects "Star 2" with magnitude 30. We can then use the first formula above to find the apparent brightness ratio.

**Solve:** Substituting $m_1 = 5$ and $m_2 = 30$ into the formula, we find

$$\frac{\text{apparent brightness of Star 1}}{\text{apparent brightness of Star 2}} = (100^{1/5})^{m_2 - m_1}$$

$$= (100^{1/5})^{30-5}$$

$$= (100^{1/5})^{25} = 100^5 = 10^{10}$$

**Explain:** The magnitude 5 star is $10^{10}$, or 10 billion, times as bright as the magnitude 30 star, so the telescope is 10 billion times as sensitive as the human eye.

**EXAMPLE 2:** The Sun has an absolute magnitude of about 4.8. Polaris, the North Star, has an absolute magnitude of −3.6. How much more luminous is Polaris than the Sun?

**SOLUTION:**

**Understand:** Luminosity and absolute magnitude are two different ways of describing a star's total power output, and the second formula above gives the relationship between them. We can therefore use that formula with Polaris as Star 1 and the Sun as Star 2.

**Solve:** Substituting $M_1 = -3.6$ for Polaris and $M_2 = 4.8$ for the Sun, we find

$$\frac{\text{luminosity of Polaris}}{\text{luminosity of Sun}} = (100^{1/5})^{M_2 - M_1} = (100^{1/5})^{4.8-(-3.6)}$$

$$= (100^{1/5})^{8.4} = 100^{1.7} \approx 2500$$

**Explain:** Based on the given absolute magnitudes, Polaris is about 2500 times as luminous as the Sun.

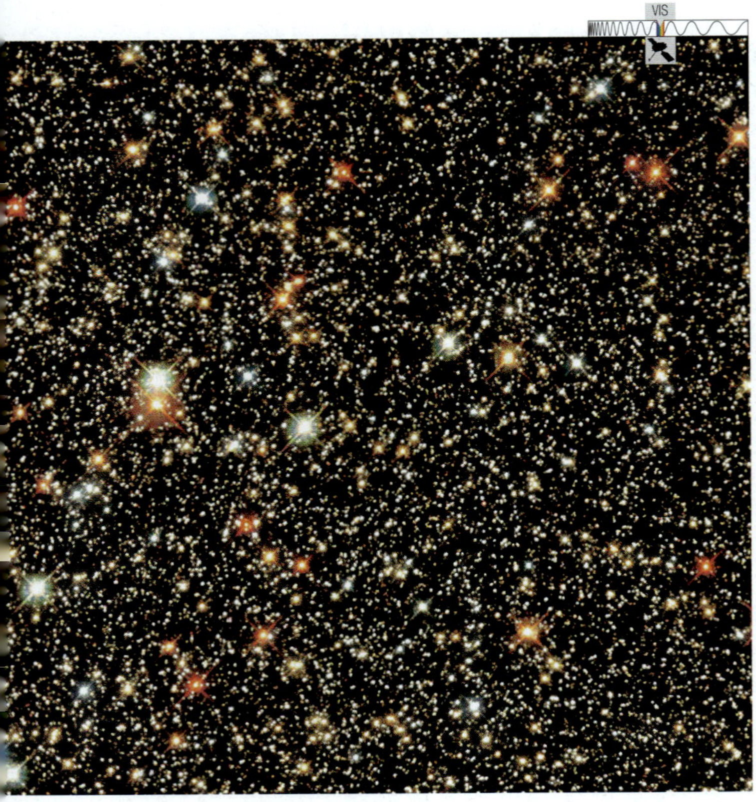


**FIGURE 15.4** This Hubble Space Telescope photo shows a wide variety of stars that differ in color and brightness. Most of the stars in this photo lie about 2000 light-years from the center of our galaxy. We can see these stars despite the fact that they are near the galactic center because of a gap (known as *Baade's window*) in the dusty clouds that obscure our view of most other such stars.

temperature to ionize atoms. Stars displaying spectral lines of molecules must be relatively cool, because molecules break apart into individual atoms unless they are at relatively cool temperatures. The types of spectral lines present in a star's spectrum therefore provide a direct measure of the star's surface temperature.

Astronomers classify stars according to surface temperature by assigning a **spectral type** determined from the spectral lines present in a star's spectrum. The hottest stars, with the bluest colors, are called spectral type O, followed in order of declining surface temperature by spectral types B, A, F, G, K, and M. **TABLE 15.1** summarizes the characteristics of each spectral type. (The sequence of spectral types is sometimes extended beyond type M to include spectral types L, T, and Y, representing starlike objects—usually *brown dwarfs* [**Section 16.3**]—that are cooler than stars of spectral type M.)

**Think about it** The traditional mnemonic for memorizing the OBAFGKM sequence was "Oh, Be A Fine Girl, Kiss Me!" However, since this may sound inappropriate to an enlightened 21st-century ear, try to invent a new mnemonic of your own.

Each spectral type is subdivided into numbered subcategories (such as B0, B1, . . . , B9). The larger the number, the cooler the star. For example, the Sun is designated spectral type G2, which means it is slightly hotter than a G3 star but cooler than a G1 star.

The range of surface temperatures for stars is much narrower than the range of luminosities. The coolest stars, of spectral type M, have surface temperatures below 3700 K. The hottest stars, of spectral type O, have surface temperatures that can exceed 40,000 K. Cool, red stars are much more common than hot, blue stars.

## COMMON MISCONCEPTIONS

### Photos of Stars

Astronomical photographs convey a great deal of information, but they also contain artifacts that are not real. For example, the spikes visible around bright stars are an artifact created by the way starlight interacts with the supports holding the secondary mirror in a telescope [**Section 6.2**], and stars have different sizes in photographs only because bright stars are overexposed, causing their light to spill over a larger region of an image than the light of dimmer stars. Overexposure also explains why the centers of globular clusters and galaxies usually look like big blobs in photographs: The central regions of these objects contain many more stars than the outskirts, and the combined light of so many stars tends to get overexposed, making a blended patch of light. The artifacts can sometimes be useful: The size issue makes it easy to identify the brighter stars, and the spikes generally occur only with point sources of light like stars, which means you can use them to distinguish stars (which have the spikes) from distant galaxies (which don't) in photos like the Hubble eXtreme Deep Field on page 1 of this book.

**History of the Spectral Sequence** You may wonder why the spectral types follow the peculiar order of OBAFGKM. The answer lies in the history of stellar spectroscopy.

Astronomical research never paid well, and many astronomers of the 19th century were able to do research only because of family wealth. One such astronomer was

**TABLE 15.1 The Spectral Sequence**

| Spectral Type | Example(s) | Temperature Range |
|---|---|---|
| O | Stars of Orion's Belt | >33,000 K |
| B | Rigel | 33,000 K–10,000 K |
| A | Sirius | 10,000 K–7500 K |
| F | Polaris | 7500 K–6000 K |
| G | Sun, Alpha Centauri A | 6000 K–5200 K |
| K | Arcturus | 5200 K–3700 K |
| M | Betelgeuse, Proxima Centauri | <3700 K |

Henry Draper (1837–1882), an early pioneer of stellar spectroscopy. After Draper died in 1882, his widow made a series of large donations to Harvard College Observatory for the purpose of building on his work. The observatory director, Edward Pickering (1846–1919), used the gifts to improve the facilities and to hire numerous assistants, whom he called "computers." Pickering added money of his own, as did other wealthy donors.

Many of Pickering's hired computers were women who had studied physics or astronomy at women's colleges such as Wellesley and Radcliffe. Women had few opportunities to advance in science at the time. Harvard, for example, did not allow women to enroll as students and would not hire them as faculty. Pickering's project of studying and classifying stellar spectra provided plenty of work and opportunity for his "computers," and many of the Harvard Observatory women ended up among the most prominent astronomers of the late 19th and early 20th century.

One of the first computers was Williamina Fleming (1857–1911). Following Pickering's suggestion, Fleming classified stellar spectra according to the strength of their hydrogen lines: type A for the strongest hydrogen lines, type B for slightly weaker hydrogen lines, and so on, to type O for stars with the weakest hydrogen lines. Pickering published Fleming's classifications of more than 10,000 stars in 1890.

As more stellar spectra were obtained and the spectra were studied in greater detail, it became clear that the classification scheme based solely on hydrogen lines was inadequate. Ultimately, the task of finding a better classification scheme fell to Annie Jump Cannon (1863–1941), who joined Pickering's team in 1896 (**FIGURE 15.5**).

**FIGURE 15.5** Women astronomers pose with Edward Pickering at Harvard College Observatory in 1913. Annie Jump Cannon is fifth from the left in the back row.

Building on the work of Fleming and another "computer," Antonia Maury (1866–1952), Cannon soon realized that the spectral classes fell into a natural order—but not the alphabetical order determined by hydrogen lines alone. Moreover, she found that some of the original classes overlapped others and could be eliminated. Cannon discovered that the natural sequence consisted of just a few of Fleming and Pickering's original classes in the order OBAFGKM; she also added the subdivisions by number.

Cannon became so adept that she could properly classify a stellar spectrum with little more than a momentary

| Key Absorption Line Features | Brightest Wavelength (Color) | Typical Spectrum |
|---|---|---|
| Lines of ionized helium, weak hydrogen lines | <89 nm (ultraviolet) | O |
| Lines of neutral helium, moderate hydrogen lines | 89–290 nm (ultraviolet)* | B |
| Very strong hydrogen lines | 290–390 nm (violet)* | A |
| Moderate hydrogen lines, moderate lines of ionized calcium | 390–480 nm (blue)* | F |
| Weak hydrogen lines, strong lines of ionized calcium | 480–560 nm (yellow) | G |
| Lines of neutral and singly ionized metals, some molecules | 560–780 nm (red) | K |
| Strong molecular lines | >780 nm (infrared) | M |

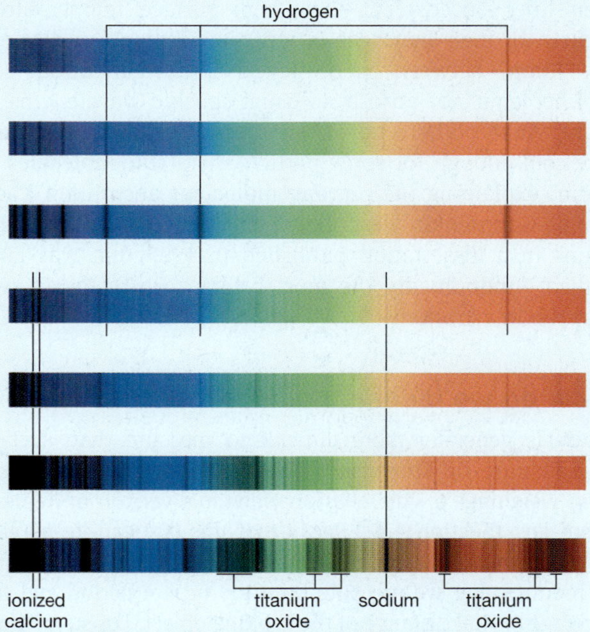


*All stars above 6000 K look more or less white to the human eye because they emit plenty of radiation at all visible wavelengths.

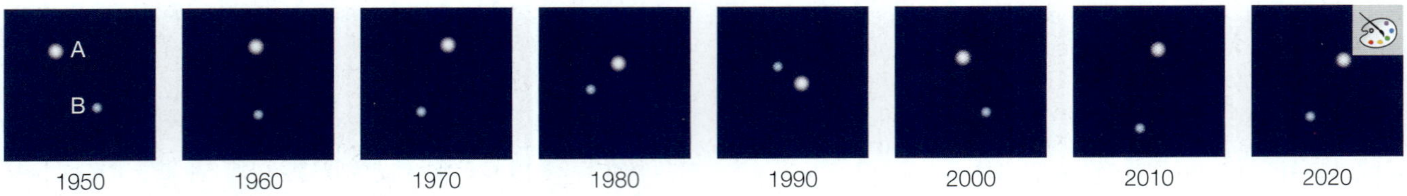


**FIGURE 15.6** Each frame represents the relative positions of Sirius A and Sirius B at 10-year intervals from 1950 to 2020. The back-and-forth "wobble" of Sirius A allowed astronomers to infer the existence of Sirius B even before the two stars could be resolved in telescopic photos. The average orbital separation of the binary system is about 20 AU.

glance. In the course of her career, she personally classified more than 400,000 stars. She became the first woman ever awarded an honorary degree by Oxford University, and in 1923 the League of Women Voters named her one of the 12 greatest living American women.

The astronomical community adopted Cannon's system of stellar classification in 1910. However, no one at that time knew *why* spectra followed the OBAFGKM sequence. Many astronomers guessed, incorrectly, that the different sets of spectral lines indicated different compositions for the stars. The correct answer—that all stars are made primarily of hydrogen and helium and that a star's surface temperature determines the strength of its spectral lines—was discovered at Harvard Observatory by Cecilia Payne-Gaposchkin (1900–1979).

Cecilia Payne-Gaposchkin

Relying on insights from what was then the newly developing science of quantum mechanics, Payne-Gaposchkin showed that the differences in spectral lines from star to star merely reflect changes in the ionization level of the emitting atoms. For example, O stars have weak hydrogen lines because, at their high surface temperatures, nearly all their hydrogen is ionized. Without an electron to "jump" between energy levels, ionized hydrogen can neither emit nor absorb its usual specific wavelengths of light. At the other end of the spectral sequence, M stars are cool enough for some particularly stable molecules to form, explaining their strong molecular absorption lines. Payne-Gaposchkin described her work and her conclusions in a dissertation published in 1925 that was later called "undoubtedly the most brilliant Ph.D. thesis ever written in astronomy."

## How do we measure stellar masses?

Mass is generally more difficult to measure than surface temperature or luminosity. The most dependable method for "weighing" a star relies on Newton's version of Kepler's third law [**Section 4.4**]. Recall that this law can be applied only when we can observe one object orbiting another, and it requires that we measure both the orbital period and the average orbital distance of the orbiting object. For stars, these requirements generally mean that we can apply the law to measure masses only in **binary star systems**—systems in which two stars continually orbit each other. Before we consider how we determine the orbital periods and distances needed to use Newton's version of Kepler's third law, let's look briefly at the different types of binary star systems that we can observe.

**Types of Binary Star Systems** Surveys show that about half of all stars orbit a companion star of some kind and are therefore members of binary star systems. These star systems fall into three classes.

- A **visual binary** is a pair of stars that we can see distinctly (with a telescope) as the stars orbit each other. Sometimes we observe a star slowly shifting position in the sky as if it was a member of a visual binary, but its companion is too dim to be seen. For example, slow shifts in the position of Sirius, the brightest star in the night sky, revealed it to be a binary star long before its companion was discovered (**FIGURE 15.6**).
- A **spectroscopic binary** is identified through observations of Doppler shifts in its spectral lines [**Section 5.4**]. If one star is orbiting another, it periodically moves toward us and away from us in its orbit, which means its spectral lines will show alternating blueshifts and redshifts (**FIGURE 15.7**). Sometimes we see two sets of lines shifting back and forth—one set from each of the two stars in the system (a *double-lined* spectroscopic binary). Other times we see a set of shifting lines from only one star because its companion is too dim to be detected (a *single-lined* spectroscopic binary).

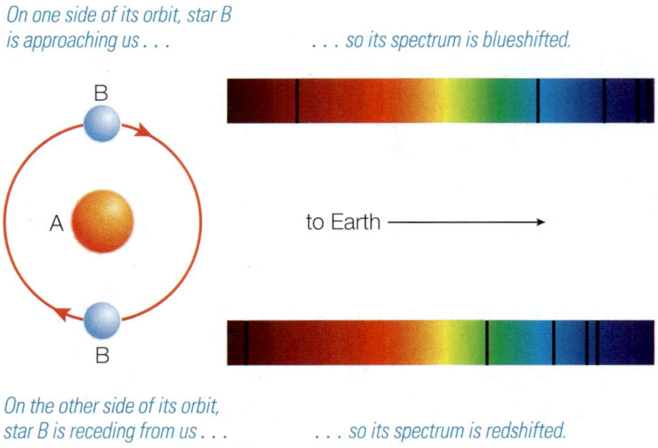


**FIGURE 15.7** The spectral lines of a star in a binary system are alternately blueshifted as it comes toward us in its orbit and redshifted as it moves away from us.

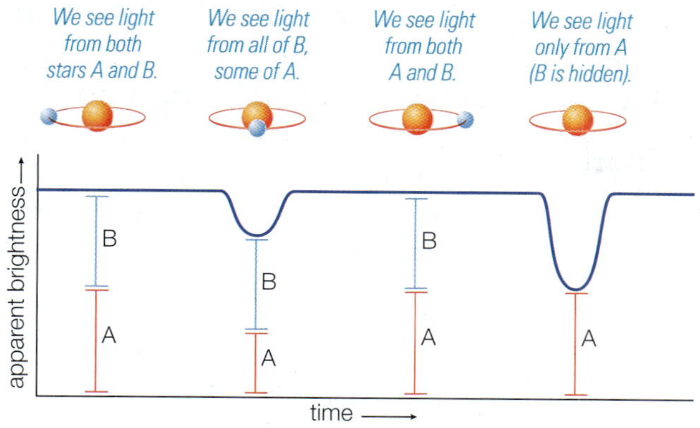


**FIGURE 15.8** The apparent brightness of an eclipsing binary system drops when either star eclipses the other.

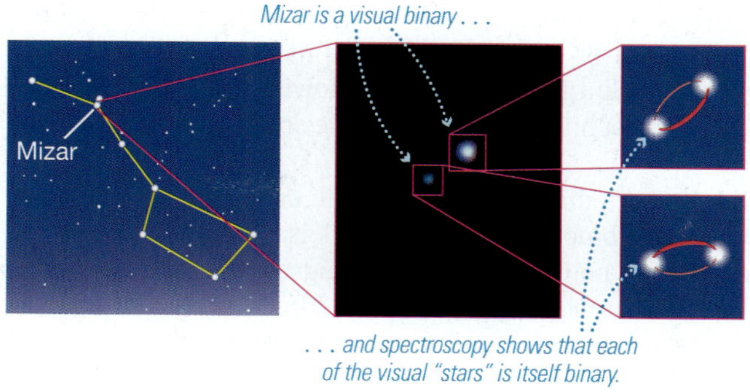


**FIGURE 15.9** Mizar looks like one star to the naked eye but is actually a system of four stars. Through a telescope, Mizar appears to be a visual binary made up of two stars, Mizar A and Mizar B, that gradually change positions, indicating that they orbit each other every few thousand years. Moreover, each of these two "stars" is itself a spectroscopic binary, making for a total of four stars.

- An **eclipsing binary** is a pair of stars that orbit in the plane of our line of sight (**FIGURE 15.8**). When neither star is eclipsed, we see the combined light of both stars. When one star eclipses the other, the apparent brightness of the system drops because some of the light is blocked from our view. A *light curve*, or graph of apparent brightness against time, reveals the pattern of the eclipses. The most famous example of an eclipsing binary is Algol (Arabic for "the ghoul"), the "demon star" in the constellation Perseus. Algol's brightness drops to only a third of its usual level for a few hours about every 3 days as the brighter of its two stars is eclipsed by its dimmer companion.

Note that these three ways of identifying binaries are essentially the same as the three major methods used to detect extrasolar planets [**Section 13.1**]: Observing a visual binary means watching changes in position and hence is equivalent to the *astrometric method*, looking at spectral changes in spectroscopic binaries is equivalent to the *Doppler method*, and eclipsing binaries are essentially undergoing both *transits* and *eclipses*. The only major difference is that these changes are much easier to observe for binary systems, because the influence of a second star is far greater than the influence of a planet.

Some star systems combine two or more of these binary types. For example, telescopic observations reveal Mizar (the second star in the handle of the Big Dipper) to be a visual binary. Spectroscopy then shows that each of the two stars in the visual binary is itself a spectroscopic binary (**FIGURE 15.9**).

**Measuring Masses and Radii in Binary Systems** Even for a binary system, we can apply Newton's version of Kepler's third law to calculate mass only if we can measure both the orbital period and the separation of the two stars. We can directly measure orbital period for all three binary types, but determining the average separation of the stars in a binary system is usually much more difficult. In rare cases, we can measure the separation directly; otherwise, we can calculate the separation only if we know the actual orbital speeds of the stars from their Doppler shifts. Unfortunately, a Doppler shift tells us only the portion of a star's velocity that is directed toward us or away from us (see Figure 5.23). Because orbiting stars generally do not move directly along our line of sight, their actual velocities can be significantly greater than those we measure through the Doppler effect.

The exceptions are eclipsing binary stars. Because these stars orbit in the plane of our line of sight, their Doppler shifts can tell us their true orbital velocities.* Eclipsing binaries are therefore particularly important to the study of stellar masses. As an added bonus, eclipsing binaries allow us to measure stellar radii directly. Because we know how fast the stars are moving across our line of sight as one eclipses the other, we can determine their radii by timing how long each eclipse lasts, in much the same way that we use transits (and eclipses) to measure the radii of extrasolar planets [**Section 13.2**].

Through careful observations of eclipsing binaries and other binary star systems, astronomers have established the masses of many different kinds of stars. The overall range extends from as little as 0.08 times the mass of the Sun ($0.08M_{\text{Sun}}$) to at least 150 times the mass of the Sun ($150M_{\text{Sun}}$). We'll discuss the reasons for this mass range in Chapter 16.

## 15.2 Patterns Among Stars

We have seen that stars come in a wide range of luminosities, surface temperatures, and masses. But are these characteristics randomly distributed among stars, or can we find patterns that might tell us something about stellar lives?

Before reading any further, take another look at Figure 15.4 and think about how you would classify these stars. Almost all of them are at nearly the same distance from Earth, so we can compare their true luminosities by looking at their apparent brightnesses in the photograph. If you look closely, you might notice a couple of important patterns:

- Most of the very brightest stars are reddish in color.
- If you ignore those relatively few bright red stars, there's a general trend to the luminosities and colors among all

*In other binaries, we can calculate an actual orbital velocity from the velocity obtained by the Doppler effect if we also know the system's orbital inclination. Although this is currently possible only in relatively rare cases, the *GAIA* mission should vastly increase the number of cases in which we can combine astrometric and Doppler data.

the rest of the stars: The brighter ones are white with a little bit of blue tint, the more modest ones are similar to our Sun in color with a yellowish-white tint, and the dimmest ones are barely visible specks of red.

Keeping in mind that colors tell us about surface temperature—blue is hotter and red is cooler—you can see that these patterns must be telling us about relationships between surface temperature and luminosity.

Danish astronomer Ejnar Hertzsprung (1873–1967) and American astronomer Henry Norris Russell (1877–1957) recognized these relationships in the first decade of the 20th century. Building on the work of Annie Jump Cannon and others, Hertzsprung and Russell independently decided to make graphs of stellar properties by plotting stellar luminosities on one axis and spectral types on the other. These graphs revealed previously unsuspected patterns among the properties of stars and ultimately unlocked the secrets of stellar life cycles.

The H-R Diagram

## What is a Hertzsprung-Russell diagram?

Graphs of the type made by Hertzsprung and Russell are now called **Hertzsprung-Russell (H-R) diagrams**. These diagrams quickly became one of the most important tools in astronomical research, and they remain central to the study of stars.

**Basics of the H-R Diagram** **FIGURE 15.10** (pp. 502–503) shows how we construct an H-R diagram, with a complete diagram on the right-hand page. Note the following about an H-R diagram:

- The horizontal axis represents stellar surface temperature, which, as we've discussed, corresponds to spectral type. Temperature *decreases* from left to right because Hertzsprung and Russell based their diagrams on the spectral sequence OBAFGKM.

## MATHEMATICAL INSIGHT 15.4 Measuring Stellar Masses

We measure masses in binary star systems by applying Newton's version of Kepler's third law (see Mathematical Insight 4.3). This requires knowing the system's orbital period $p$, which is usually easy to measure, and the semimajor axis $a$, which can be calculated if we know orbital velocities from Doppler-shift measurements. For a binary system in which one star traces a circle of radius $a$ around its companion, the star's orbital velocity is

$$v = \frac{\text{distance traveled in one orbit}}{\text{period of one orbit}} = \frac{2\pi a}{p}$$

Solving for $a$, we find

$$a = \frac{pv}{2\pi}$$

Once we know both $p$ and $a$, Newton's version of Kepler's third law allows us to calculate the *sum* of the masses of the two stars $(M_1 + M_2)$. We can calculate individual masses by comparing the orbital velocities of the two stars around the system's center of mass.

**EXAMPLE:** The spectral lines of two stars in an eclipsing binary system with a circular orbit shift back and forth with a period of 2 years ($p = 6.3 \times 10^7$ seconds). The lines of one star (Star 1) shift twice as far as the lines of the other (Star 2). The Doppler shift indicates an orbital speed of $v = 100{,}000$ m/s for Star 1 relative to Star 2. What are the masses of the two stars?

**SOLUTION:**

**Step 1 Understand:** We can find the sum of the masses with Newton's version of Kepler's third law, which reads

$$p^2 = \frac{4\pi^2}{G(M_1 + M_2)} a^3$$

We rearrange this equation to find the masses:

$$M_1 + M_2 = \frac{4\pi^2}{G} \times \frac{a^3}{p^2}$$

We are given the system's orbital period $p$, and because the orbit is circular, we can find the semimajor axis $a$ as described above from the velocity $v$ of Star 1 relative to Star 2. Once we calculate the sum of the masses $(M_1 + M_2)$, we can determine the individual masses of the stars as follows: Because the lines of Star 1 shift twice as far as those of Star 2, we know that Star 1 moves twice as fast as Star 2, and hence that Star 1 is half as massive as Star 2.

**Step 2 Solve:** We find the semimajor axis $a$ of the system from the system's orbital velocity $v$:

$$a = \frac{pv}{2\pi} = \frac{(6.3 \times 10^7 \text{ s}) \times (100{,}000 \text{ m/s})}{2\pi} = 1.0 \times 10^{12} \text{ m}$$

We now use this value, the value of the gravitational constant $G$, and the given orbital period ($p = 6.3 \times 10^7$ s) to find the sum of the masses:

$$M_1 + M_2 = \frac{4\pi^2}{\left(6.67 \times 10^{-11} \frac{\text{m}^3}{\text{kg} \times \text{s}^2}\right)} \times \frac{(1.0 \times 10^{12} \text{ m})^3}{(6.3 \times 10^7 \text{ s})^2} = 1.5 \times 10^{32} \text{ kg}$$

The two stars have a combined mass of $1.5 \times 10^{32}$ kg, and we know from the Doppler shifts that Star 2 is twice as massive as Star 1. We therefore conclude that Star 2 has a mass of $1.0 \times 10^{32}$ kg and Star 1 has a mass of $0.5 \times 10^{32}$ kg.

**Step 3 Explain:** The masses will be more meaningful if we convert them from kilograms to solar masses, which we do by dividing by the Sun's mass of $2 \times 10^{30}$ kg. Doing so, we find that this binary system consists of one star of mass $50M_{\text{Sun}}$ and another star of mass $25M_{\text{Sun}}$.

- The vertical axis represents stellar luminosity, in units of the Sun's luminosity ($L_{Sun}$). Stellar luminosities span a wide range, so we keep the graph compact by making each tick mark represent a luminosity 10 times as large as that of the prior tick mark.

Each location on the H-R diagram represents a unique combination of spectral type and luminosity. For example, the dot representing the Sun in Figure 15.10 corresponds to the Sun's spectral type, G2, and its luminosity, $1L_{Sun}$. Because luminosity increases upward on the diagram and surface temperature increases leftward, stars near the upper left are hot and luminous. Similarly, stars near the upper right are cool and luminous, stars near the lower right are cool and dim, and stars near the lower left are hot and dim.

**Think about it** Explain how the colors of the stars in Figure 15.10 help indicate stellar surface temperature. Do these colors tell us anything about *interior* temperatures? Why or why not?

The H-R diagram also provides information about stellar radii, because a star's luminosity depends on both its surface temperature and its surface area or radius (see Mathematical Insight 15.5). If two stars have the same surface temperature, one can be more luminous than the other only if it is larger in size. Stellar radii therefore must increase as we go from the high-temperature, low-luminosity corner on the lower left of the H-R diagram to the low-temperature, high-luminosity corner on the upper right. Notice the diagonal lines that represent different stellar radii in Figure 15.10.

**Patterns in the H-R Diagram** Stars do not fall randomly throughout an H-R diagram like Figure 15.10 but instead cluster into four major groups:

- Most stars fall somewhere along the **main sequence**, the prominent streak running from the upper left to the lower right on the H-R diagram. Notice that our Sun is one of these *main-sequence stars*.
- The stars in the upper right are called **supergiants** because they are very large in addition to being very bright.
- Just below the supergiants are the **giants**, which are somewhat smaller in radius and lower in luminosity (but still much larger and brighter than main-sequence stars of the same spectral type).
- The stars near the lower left are small in radius and appear white in color because of their high temperatures. We call these stars **white dwarfs**.

**Luminosity Classes** In addition to the four major groups we've just listed, stars sometimes fall into "in-between" categories. For more precise work, astronomers therefore assign each star to a **luminosity class**, designated with a Roman numeral from I to V. The luminosity class describes the region of the H-R diagram in which the star falls; so despite the name, a star's luminosity class is more closely related to its size than to its luminosity. The basic luminosity classes are I for supergiants, III for giants, and V for

## MATHEMATICAL INSIGHT 15.5 Calculating Stellar Radii

Although we can rarely measure stellar radii directly, we can calculate radii using the laws of thermal radiation. As given in Mathematical Insight 5.2, the amount (power) of thermal radiation emitted by a star of surface temperature $T$ (on the Kelvin scale) is

$$\text{emitted power (per square meter of surface)} = \sigma T^4$$

where the constant $\sigma = 5.7 \times 10^{-8}\ \text{watt}/(\text{m}^2 \times \text{K}^4)$.

The luminosity $L$ of a star is its power per unit area multiplied by its total surface area, and a star of radius $r$ has surface area $4\pi r^2$. That is,

$$L = 4\pi r^2 \times \sigma T^4$$

With a bit of algebra, we can solve this formula for the star's radius $r$:

$$r = \sqrt{\frac{L}{4\pi\sigma T^4}}$$

**EXAMPLE:** The red supergiant star Betelgeuse has a luminosity of $120{,}000L_{Sun}$ and a surface temperature of about 3650 K. What is its radius?

**SOLUTION:**

**Step 1 Understand:** We are given Betelgeuse's luminosity $L$ and surface temperature $T$, so we can use the above formula to find its radius. To make the units consistent with the units given for $\sigma$, we must convert the luminosity to watts.

**Step 2 Solve:** Remembering that $L_{Sun} = 3.8 \times 10^{26}$ watts, we find

$$L_{Bet} = 120{,}000 \times L_{Sun} = 120{,}000 \times 3.8 \times 10^{26}\ \text{watts}$$
$$= 4.6 \times 10^{31}\ \text{watts}$$

Now we can use our formula to calculate radius:

$$r = \sqrt{\frac{L}{4\pi\sigma T^4}}$$
$$= \sqrt{\frac{4.6 \times 10^{31}\ \text{watts}}{4\pi \times \left(5.7 \times 10^{-8}\ \frac{\text{watt}}{\text{m}^2 \times \text{K}^4}\right) \times (3650\ \text{K})^4}}$$
$$= \sqrt{\frac{4.6 \times 10^{31}\ \text{watts}}{1.3 \times 10^{8}\ \frac{\text{watts}}{\text{m}^2}}} = 5.9 \times 10^{11}\ \text{m}$$

**Step 3 Explain:** Betelgeuse has a radius of about 590 billion meters, or 590 million kilometers. This is almost four times the Earth-Sun distance (1 AU ≈ 150 million km), which means that the orbits of all the inner planets of our solar system could fit easily inside Betelgeuse.

**COSMIC Context**

# FIGURE 15.10 Reading an H-R Diagram

**Hertzsprung-Russell (H-R) diagrams are very important tools in astronomy because they reveal key relationships among the properties of stars. An H-R diagram is made by plotting stars according to their surface temperatures and luminosities. This figure shows a step-by-step approach to building an H-R diagram.**

① **An H-R Diagram Is a Graph:** A star's position along the horizontal axis indicates its surface temperature, which is closely related to its color and spectral type. Its position along the vertical axis indicates its luminosity.

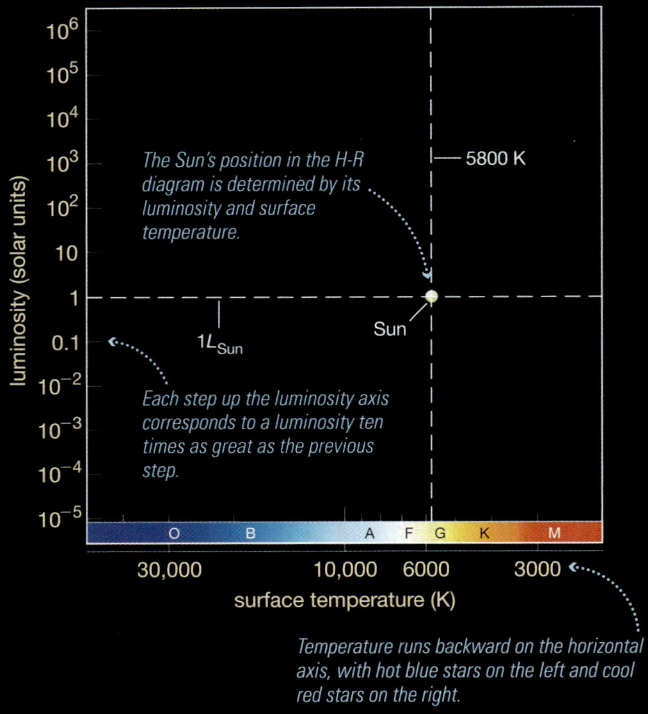


② **Main Sequence:** Our Sun falls along the main sequence, a line of stars extending from the upper left of the diagram to the lower right. Most stars are main-sequence stars, which shine by fusing hydrogen into helium in their cores.

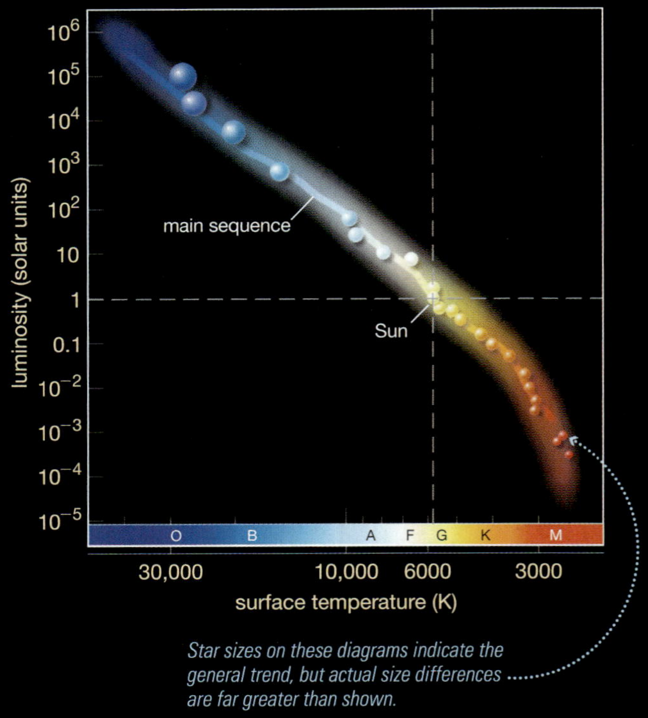


③ **Giants and Supergiants:** Stars in the upper right of an H-R diagram are more luminous than main-sequence stars of the same surface temperature. They must therefore be very large in radius, which is why they are known as *giants* and *supergiants*.

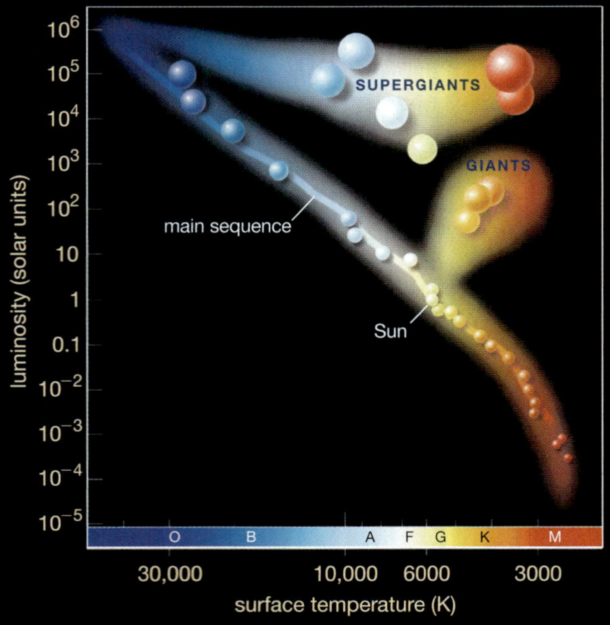


④ **White Dwarfs:** Stars in the lower left have high surface temperatures, dim luminosities, and small radii. These stars are known as *white dwarfs.*

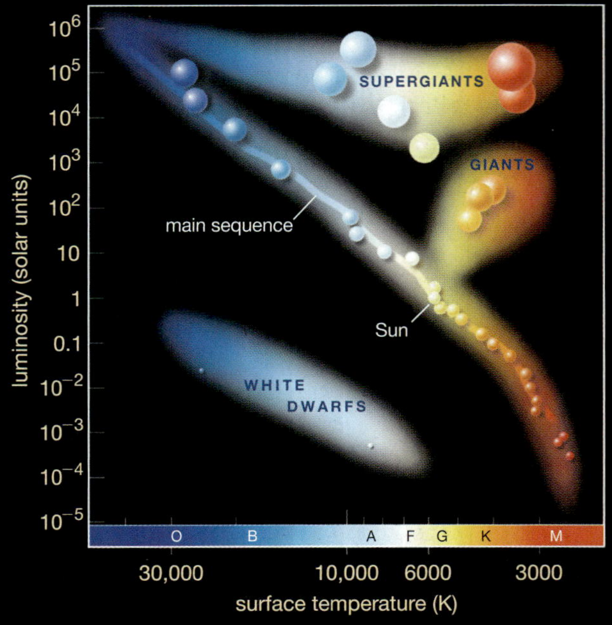

(5) **Masses on the Main Sequence:** Stellar masses (purple labels) decrease from the upper left to the lower right on the main sequence.

(6) **Lifetimes on the Main Sequence:** Stellar lifetimes (green labels) increase from the upper left to lower right on the main sequence: High-mass stars live shorter lives because their high luminosities mean they consume their nuclear fuel more quickly.

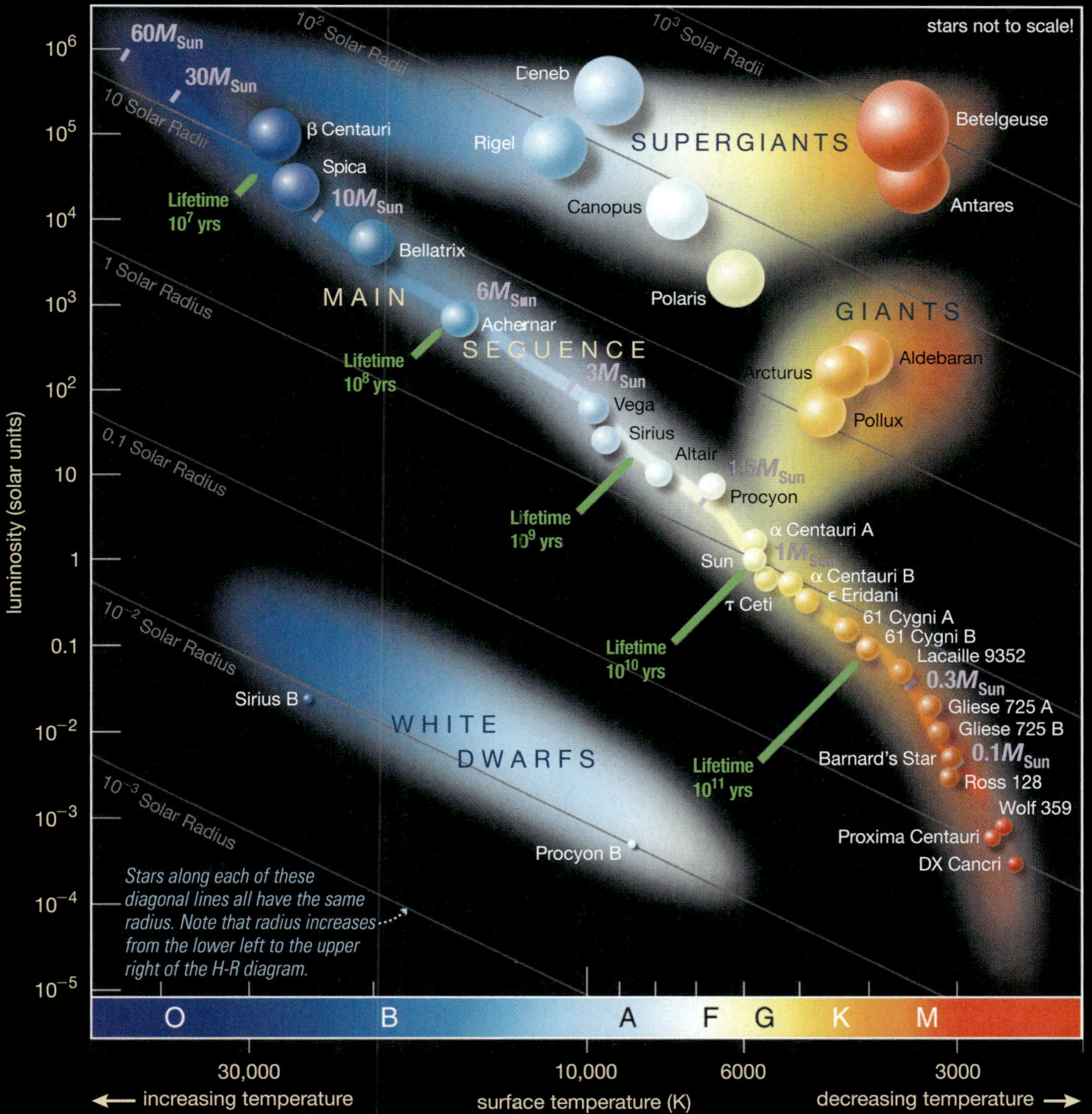

**TABLE 15.2 Stellar Luminosity Classes**

| Class | Description |
|---|---|
| I | Supergiants |
| II | Bright giants |
| III | Giants |
| IV | Subgiants |
| V | Main-sequence stars |

main-sequence stars. Luminosity classes II and IV are intermediate to the others. For example, luminosity class IV represents stars with radii larger than those of main-sequence stars but not quite large enough to qualify them as giants. **TABLE 15.2** summarizes the luminosity classes. White dwarfs fall outside this classification system and are often assigned the luminosity class "wd."

**Complete Stellar Classification** We have now described two different ways of categorizing stars:

- A star's *spectral type*, designated by one of the letters OBAFGKM, tells us its surface temperature and color. O stars are the hottest and bluest, while M stars are the coolest and reddest.
- A star's *luminosity class*, designated by a Roman numeral, is based on its luminosity but also tells us about the star's radius. Luminosity class I stars have the largest radii, with radii decreasing to luminosity class V.

We use both spectral type and luminosity class to fully classify a star. For example, the complete classification of our Sun is G2 V. The G2 spectral type means it is yellow-white in color, and the luminosity class V means it is a hydrogen-fusing, main-sequence star. Betelgeuse is M2 I, making it a red supergiant. Proxima Centauri is M5 V—similar in color and surface temperature to Betelgeuse, but far dimmer because of its much smaller size.

**Think about it** By studying Figure 15.10, determine the approximate spectral type, luminosity class, and radius of the following stars: Bellatrix, Vega, Antares, Pollux, and Proxima Centauri.

## What is the significance of the main sequence?

Most stars, including our Sun, have properties that place them on the main sequence of the H-R diagram. You can see in Figure 15.10 that high-luminosity main-sequence stars have hot surfaces and low-luminosity main-sequence stars have cooler surfaces. This relationship between luminosity and surface temperature comes about because a star's position along the main sequence is closely related to its mass. All stars along the main sequence are fusing hydrogen into helium in their cores, just like the Sun, and mass determines both surface temperature and luminosity because it is the key factor in the star's rate of hydrogen fusion.

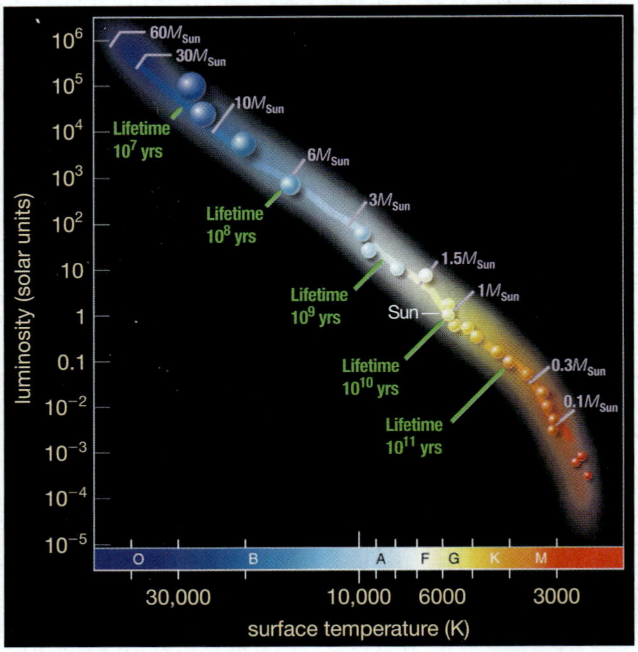


**FIGURE 15.11** The main sequence from Figure 15.10 is isolated here so that you can more easily see how masses and lifetimes vary along it. Notice that more massive hydrogen-fusing stars are brighter and hotter but have shorter lifetimes. (Stellar masses are given in units of solar masses: $1M_{Sun} = 2 \times 10^{30}$ kg.)

**Masses Along the Main Sequence** If you look along the main sequence in Figure 15.10, you'll notice purple labels indicating stellar masses and green labels indicating stellar lifetimes. To make the relationships easier to see, **FIGURE 15.11** repeats the same data but shows only the main sequence rather than the entire H-R diagram. Let's focus first on mass.

Notice that *stellar masses decrease downward along the main sequence*. At the upper end of the main sequence, the hot, luminous O stars can have masses as high as 150 or more times that of the Sun. On the lower end, cool, dim M stars may have as little as 0.08 times the mass of the Sun ($0.08M_{Sun}$). Many more stars fall on the lower end of the main sequence than on the upper end, which tells us that low-mass stars are much more common than high-mass stars.

The orderly arrangement of stellar masses along the main sequence tells us that *mass* is the most important attribute of a hydrogen-fusing star. The reason is that mass determines the balancing point at which the energy released by hydrogen fusion in the core equals the energy lost from the star's surface. The great range of stellar luminosities on the H-R diagram shows that the point of energy balance is very sensitive to mass. For example, a $10M_{Sun}$ star on the main sequence is about 10,000 times as luminous as the Sun.

The relationship between mass and surface temperature is a little more subtle. In general, a very luminous star must have a very large mass or a very high surface temperature, or some combination of both. The most massive main-sequence stars are many thousands of times as luminous as the Sun but only about 10 times the size of the Sun in radius. Their surfaces must be significantly hotter than the Sun's surface to account for their high luminosities. Main-sequence stars more massive than the Sun therefore have higher surface temperatures than the Sun, and those less massive than the Sun have lower surface temperatures. That is why the main sequence slices diagonally from the upper left to the lower right on the H-R diagram.

The fact that mass, surface temperature, and luminosity are all related means that we can estimate a main-sequence star's mass just by knowing its spectral type. For example, any hydrogen-fusing main-sequence star that has the same spectral type as the Sun (G2) must have about the same mass and luminosity as the Sun. Similarly, any main-sequence star of spectral type B1 must have about the same mass and luminosity as Spica (see Figure 15.10). Note that only main-sequence stars follow this simple relationship between mass, temperature, and luminosity; it does *not* hold for giants, supergiants, or white dwarfs.

**Lifetimes Along the Main Sequence** A star is born with a limited supply of core hydrogen and therefore can remain as a hydrogen-fusing main-sequence star for only a limited time—the star's **main-sequence lifetime**. Because stars spend the vast majority of their lives as main-sequence stars, we sometimes refer to the main-sequence lifetime as simply "the lifetime." Like masses, stellar lifetimes vary in an orderly way as we move up the main sequence: Massive stars near the upper end of the main sequence have *shorter* lives than less massive stars near the lower end (see Figure 15.11).

Why do more massive stars have shorter lives? A star's lifetime depends on both its mass and its luminosity. Its mass determines how much hydrogen fuel the star initially contains in its core. Its luminosity determines how rapidly the star uses up its fuel. Massive stars start their lives with a larger supply of hydrogen, but they fuse this hydrogen into helium so rapidly that they end up with shorter lives. For example, a 10-solar-mass star ($10M_{\text{Sun}}$) is born with 10 times as much hydrogen as the Sun. However, its luminosity of $10{,}000L_{\text{Sun}}$ means that it consumes this hydrogen at a rate 10,000 times as fast as the rate in the Sun. With 10 times as much hydrogen being consumed at 10,000 times the rate, the lifetime of a 10-solar-mass star would be only about $\frac{10}{10{,}000} = \frac{1}{1000}$ as long as the Sun's lifetime, or about 10 billion $\div$ 1000 = 10 million years. (Its actual lifetime is a little longer than this, because it can use more of its core hydrogen for fusion than can the Sun.) The most massive stars live even shorter times, in some cases just a few million years.

Cosmically speaking, the lifetimes of very massive stars are remarkably short. That is one reason massive stars are so rare: Most of the massive stars that have ever been born are long since dead. A second reason is that higher-mass stars are born in smaller numbers to begin with [**Section 16.3**]. Indeed, the fact that massive stars exist at all at the present time tells us that stars must form continuously in our galaxy. The massive, bright O stars in our galaxy today formed only recently and will die long before they have a chance to complete even one orbit around the center of the galaxy.

On the other end of the scale, a 0.3-solar-mass main-sequence star emits a luminosity just 0.01 times that of the Sun and consequently lives about $\frac{0.3}{0.01} = 30$ times as long as the Sun, or about 300 billion years. In a universe that is now about 14 billion years old, even the most ancient of these small, dim, red stars of spectral type M still survive and will continue to shine faintly for hundreds of billions of years to come.

**Mass: A Star's Most Fundamental Property** Astronomers began classifying stars by spectral type and luminosity class before they understood why stars vary in these properties. Today, we know that the most fundamental property of any star is its *mass* (**FIGURE 15.12**). A star's mass determines both its surface temperature and its luminosity throughout the main-sequence portion of its life, and these properties in turn explain why higher-mass stars have shorter lifetimes.

**Think about it** Which of the stars labeled in Figure 15.10 has the longest lifetime? Explain.

## What are giants, supergiants, and white dwarfs?

Main-sequence stars fuse hydrogen into helium in their cores, but what about the other classes of stars on the H-R diagram? These other classes all represent stars that have

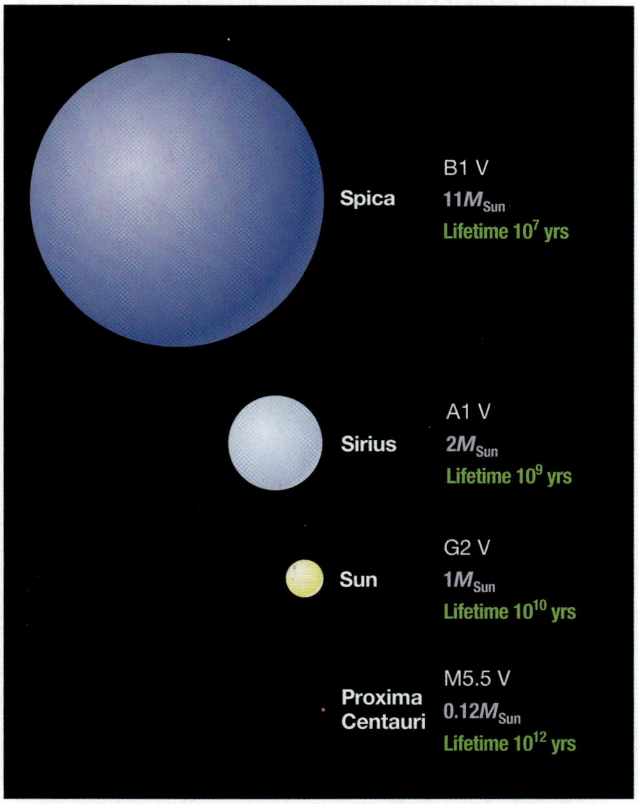


**FIGURE 15.12** Four main-sequence stars shown to scale. The mass of a main-sequence star determines its fundamental properties of luminosity, surface temperature, radius, and lifetime. More massive main-sequence stars are hotter and brighter than less massive ones but have shorter lifetimes.

exhausted the supply of hydrogen in their central cores, so they can no longer generate energy in the same way as our Sun.

**Giants and Supergiants** The bright red stars in Figure 15.4 are giants and supergiants whose properties place them to the upper right of the main sequence in an H-R diagram. The fact that these stars are cooler but much more luminous than the Sun tells us that they must be much larger in radius than the Sun. Remember that a star's surface temperature determines the amount of light it emits per unit surface area [**Section 5.4**]: Hotter stars emit much more light per unit surface area than cooler stars. For example, a blue star emits far more total light than a red star of the same size. A star that is red and cool can be bright only if it has a very large surface area, which means it must be enormous in size.

As we'll discuss in Chapter 17, we now know that giants and supergiants are stars nearing the ends of their lives. They have already exhausted the supply of hydrogen fuel in their central cores and are in essence facing an energy crisis as they try to stave off the inevitable crushing force of gravity. This crisis causes these stars to release fusion energy at a furious rate, which accounts for their high luminosities and causes their surfaces to expand to enormous size (**FIGURE 15.13**). For example, Arcturus and Aldebaran (the eye of the bull in the constellation Taurus) are giant stars more than 10 times as large in radius as our Sun. Betelgeuse, the left shoulder in the constellation Orion, is an enormous supergiant with a radius roughly 900 times that of the Sun, equivalent to almost four times the Earth-Sun distance.

Because giants and supergiants are so bright, we can see them even if they are not especially close to us. Many of the brightest stars in our sky are giants or supergiants, often identifiable by their reddish colors. Overall, however, giants and supergiants are considerably rarer than main-sequence stars. In our snapshot of the heavens, we catch most stars in the act of hydrogen fusion and relatively few in a later stage of life.

**White Dwarfs** Giants and supergiants eventually run out of fuel entirely. A giant with a mass similar to that of our Sun ultimately ejects its outer layers, leaving behind a "dead" core in which all nuclear fusion has ceased. White dwarfs are these remaining embers of former giants. They are hot because they are essentially exposed stellar cores, but they become dim over time because they lack an energy source and radiate only their leftover heat into space. A typical white dwarf is no larger in size than Earth but has a mass similar to that of our Sun. Clearly, white dwarfs must be made of matter compressed to an extremely high density, unlike anything found on Earth. We'll discuss the nature of white dwarfs and other stellar corpses in Chapter 18.

## Why do the properties of some stars vary?

Not all stars shine steadily like our Sun. Any star that varies significantly in brightness with time is called a *variable star*. Certain types of variable stars cannot achieve balance between the power welling up from the core and the power being radiated from the surface. Sometimes the upper layers of such a star are too opaque to allow much energy to escape, so pressure builds up beneath the photosphere and the star expands in size. This expansion puffs up the outer layers until they become transparent enough for the trapped energy to escape. The underlying pressure then drops, allowing the star to contract until the trapping of energy resumes.

In a futile quest for a steady equilibrium, the atmosphere of such a **pulsating variable star** alternately expands and contracts, causing the star to rise and fall in luminosity. **FIGURE 15.14** shows a typical light curve for a pulsating variable star, with the star's brightness graphed against time. Any pulsating variable star has its own particular period between peaks and valleys in luminosity, which we can discover easily from its light curve. These periods can range from as short as several hours to as long as several years.

Most pulsating variable stars inhabit a strip (called the *instability strip*) on the H-R diagram that lies between the main sequence and the red giants (**FIGURE 15.15**). A special

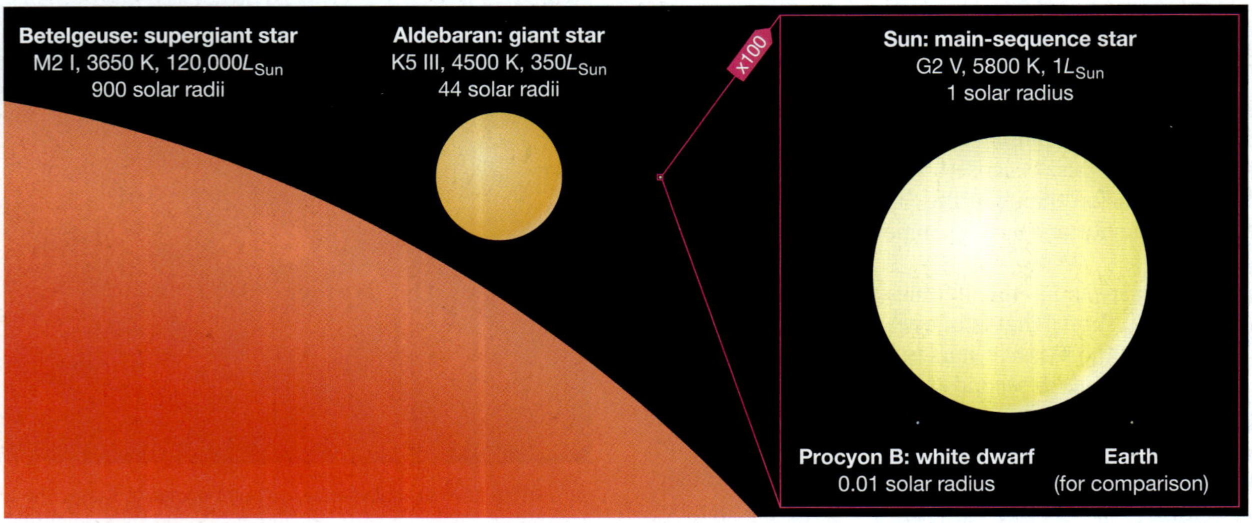


**FIGURE 15.13** The relative sizes of stars. A supergiant like Betelgeuse would fill the inner solar system and extend more than 80% of the way to Jupiter's orbit. A giant like Aldebaran would fill the inner half of Mercury's orbit. The Sun is a hundred times as large in radius as a white dwarf, which is roughly the same size as Earth.

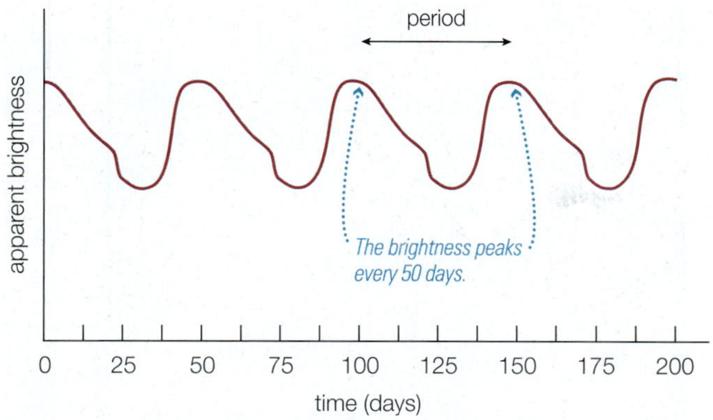


**FIGURE 15.14** A typical light curve for a pulsating variable star. This particular star is a Cepheid variable star with a pulsation period of about 50 days.

category of very luminous pulsating variable stars lies in the upper portion of this strip. They are known as *Cepheid variable stars*, and they are significant both because they are so bright and because their pulsation periods turn out to be closely related to their luminosities. As a result, Cepheids have played a key role in helping us establish the distances to many galaxies beyond the Milky Way, thereby revealing the overall scale of the cosmos. We will discuss the cosmic distance scale further in Chapter 20.

## 15.3 Star Clusters

All stars are born from giant clouds of gas. Because a single interstellar cloud can contain enough material to form many stars, stars usually form in groups [**Section 16.1**]. In our snapshot of the heavens, many stars still congregate in the groups in which they formed. These groups are known as star clusters, and they are extremely useful to astronomers for two key reasons:

1. All the stars in a cluster lie at about the same distance from Earth.
2. All the stars in a cluster formed at about the same time (within a few million years of one another).

Astronomers can therefore use star clusters to compare the properties of stars that all have similar ages.

### What are the two types of star clusters?

Star clusters come in two basic types: modest-size **open clusters** and densely packed **globular clusters**. The two types differ not only in how densely they are packed with stars but also in their locations and ages.

Recall that most of the stars, gas, and dust in the Milky Way Galaxy, including our Sun, lie in the relatively flat *galactic disk;* the region above and below the disk is called the *halo* of the galaxy (see Figure 1.16). Open clusters are always found in the disk of the galaxy and tend to be young. They can contain up to several thousand stars and typically are about 30 light-years across. The most famous open cluster, the *Pleiades* (**FIGURE 15.16**), is a prominent clump of stars in the constellation Taurus. It is sometimes

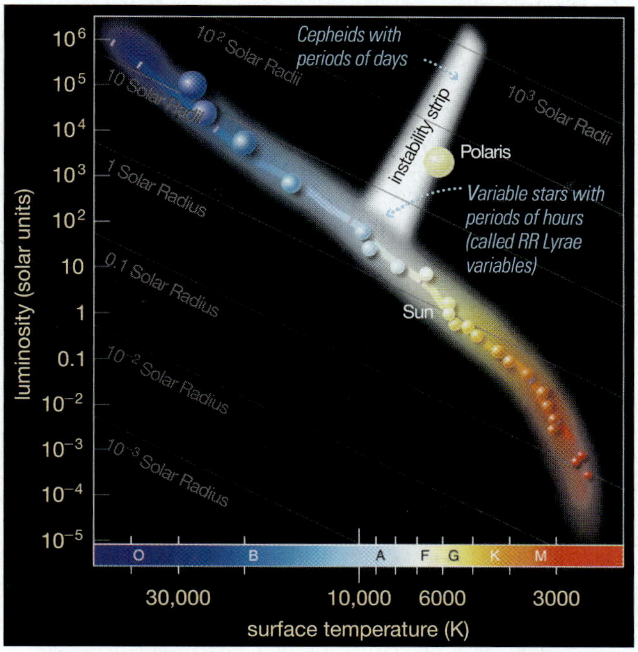


**FIGURE 15.15** An H-R diagram with the instability strip highlighted. Notice that Polaris, the North Star, is a Cepheid variable star.

called the *Seven Sisters*, although only six of the cluster's several thousand stars are easily visible to the naked eye. Other cultures have other names for this beautiful group of stars. In Japan it is called *Subaru*, which is why the logo for Subaru automobiles is a diagram of the Pleiades.

In contrast, most globular clusters are found in the halo, and their stars are among the oldest in the universe. A globular cluster can contain more than a million stars concentrated in a ball typically 60 to 150 light-years across. Its central region can have 10,000 stars packed into a space just a few light-years across (**FIGURE 15.17**). The view from a planet in a globular cluster would be marvelous, with thousands of stars lying closer to that planet than Alpha Centauri is to the Sun.

**FIGURE 15.16** A photo of the Pleiades, a nearby open cluster of stars. The most prominent stars in this open cluster are of spectral type B, indicating that the stars of the Pleiades are no more than 100 million years old, relatively young for a star cluster. The region shown is about 11 light-years across.

**FIGURE 15.17** The globular cluster M80 is more than 12 billion years old. The prominent reddish stars in this Hubble Space Telescope photo are red giant stars nearing the ends of their lives. The central region pictured here is about 15 light-years across.

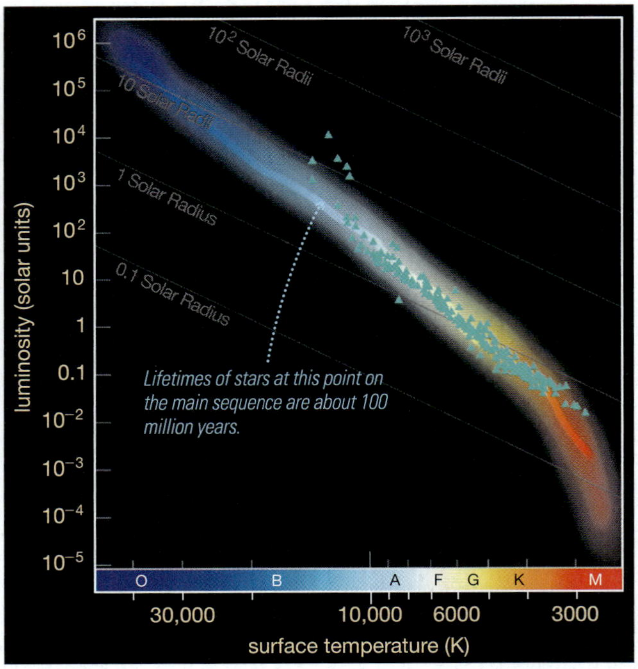


**FIGURE 15.18** An H-R diagram for the stars of the Pleiades. Triangles represent individual stars. The Pleiades cluster is missing its upper main-sequence stars, indicating that these stars have already ended their hydrogen-fusing lives. The main-sequence turnoff point at about spectral type B6 tells us that the Pleiades are approximately 100 million years old.

Because a globular cluster's stars nestle so closely together, they engage in an intricate and complex dance choreographed by gravity. Some stars zoom from the cluster's core to its outskirts and back again at speeds approaching escape velocity from the cluster, while others orbit the dense core more closely. When two stars pass especially close to each other, the gravitational pull between them deflects their trajectories, altering their speeds and sending them careening off in new directions. Occasionally, a close encounter boosts one star's velocity enough to eject it from the cluster. Through such ejections, globular clusters gradually lose stars and grow more compact.

## How do we measure the age of a star cluster?

We can determine the age of a star cluster by plotting its stars on an H-R diagram. **FIGURE 15.18** shows the process for the Pleiades. Note that most of the stars in this cluster lie along the main sequence, with one important exception: At the upper end of the main sequence, stars trail away to the right. That is, the hot, short-lived stars of spectral type O are missing from the main sequence. Apparently, the Pleiades cluster is old enough for hydrogen fusion to have ended in the cores of its main-sequence O stars, but young enough that hydrogen fusion is still happening in the cores of some stars of spectral type B that remain on the main sequence.

The precise point on the H-R diagram at which a cluster's stars diverge from the main sequence is called the **main-sequence turnoff point**. For the Pleiades, it occurs around spectral type B6. The main-sequence lifetime of a B6 star is roughly 100 million years, so this must be the age of the Pleiades. Any star in the Pleiades that was born with a main-sequence spectral type hotter than B6 had a lifetime shorter than 100 million years and is no longer found on the main sequence. Over the next few billion years, the B stars in the Pleiades will die out, followed by the A stars and the F stars. If we could make an H-R diagram for the Pleiades every few million years, we would find its main sequence gradually growing shorter.

**FIGURE 15.19** shows this idea by comparing the main sequences of several open clusters. In each case, *the age of the cluster is equal to the lifetimes of stars at its main-sequence turnoff point.* Stars in a particular cluster that

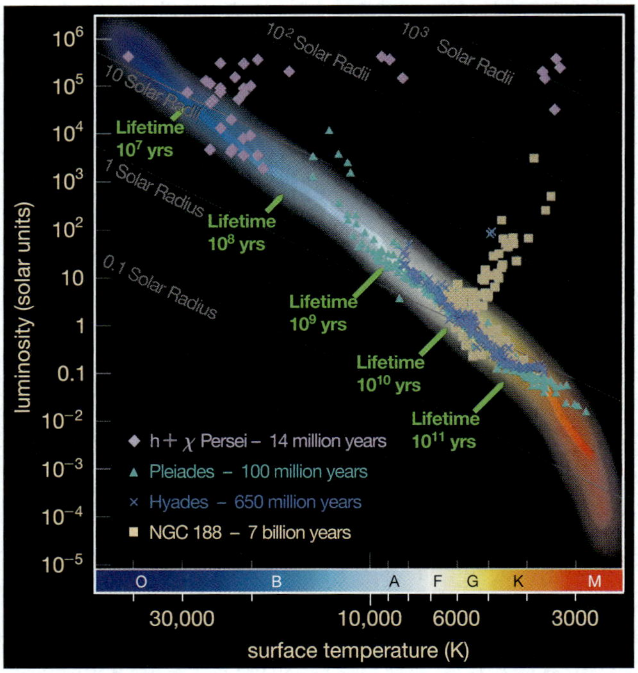


**FIGURE 15.19** This H-R diagram shows stars from the Pleiades and three other clusters. Their differing main-sequence turnoff points indicate very different ages.

once resided above the turnoff point on the main sequence have already exhausted their core supply of hydrogen, while stars below the turnoff point remain on the main sequence.

**Think about it** Suppose a star cluster is precisely 10 billion years old. Where would you expect to find its main-sequence turnoff point? Would you expect this cluster to have any main-sequence stars of spectral type A? Would you expect it to have main-sequence stars of spectral type K? Explain. (*Hint*: What is the lifetime of our Sun?)

The technique of identifying main-sequence turnoff points is our most powerful tool for evaluating the ages of star clusters. We've learned that most open clusters are relatively young, with very few older than about 5 billion years. In contrast, the stars at the main-sequence turnoff points in globular clusters are usually less massive than our Sun (**FIGURE 15.20**). Because stars like our Sun have a lifetime of about 10 billion years and these stars have already died in globular clusters, we conclude that globular cluster stars are older than 10 billion years. More precise studies of the turnoff points in globular clusters, coupled with theoretical calculations of stellar lifetimes, place the ages of these clusters at about 13 billion years, making them the oldest known objects in the galaxy and implying that they began to form within the first billion years of the universe's roughly 14-billion-year history.

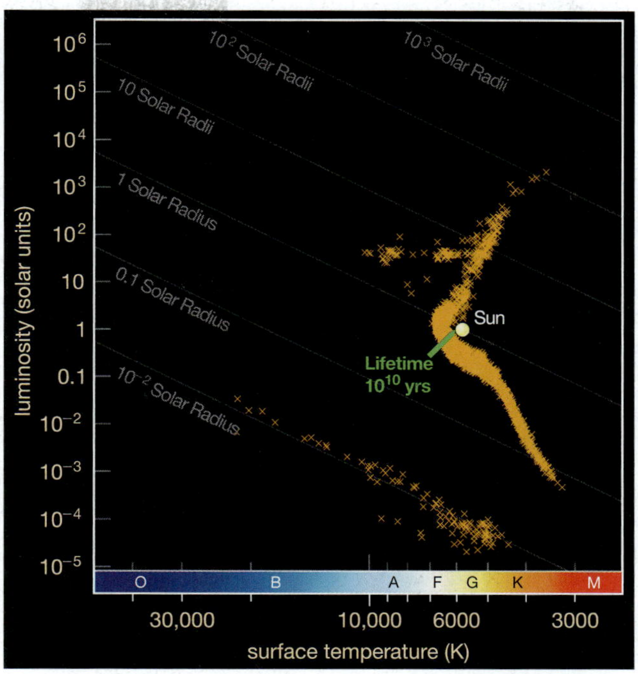


**FIGURE 15.20** This H-R diagram shows stars from the globular cluster M4. The main-sequence turnoff point is in the vicinity of stars like our Sun, indicating an age for this cluster of around 10 billion years. A more technical analysis of this cluster places its age at around 13 billion years.

## The BIG Picture PUTTING CHAPTER 15 INTO CONTEXT

*We have classified the diverse families of stars visible in the night sky. Much of what we know about stars, galaxies, and the universe itself is based on the fundamental properties of stars introduced in this chapter. Make sure you understand the following "big picture" ideas:*

- All stars are made primarily of hydrogen and helium at the time they form. The differences between stars are primarily due to differences in mass and stage of life.
- Stars spend most of their lives as main-sequence stars that fuse hydrogen into helium in their cores. The most massive stars, which are also the hottest and most luminous, live only a few million years. The least massive stars, which are coolest and dimmest, will survive until the universe is many times its present age.
- The key to recognizing the patterns among stars was the H-R diagram, which shows stellar surface temperatures on the horizontal axis and luminosities on the vertical axis. The H-R diagram is one of the most important tools of modern astronomy.
- Much of what we know about the universe comes from studies of star clusters. We can measure a star cluster's age by plotting its stars on an H-R diagram and determining the hydrogen-fusion lifetime of the brightest and most massive stars still on the main sequence.

**MY COSMIC PERSPECTIVE** You may be used to seeing stars on clear dark nights, but you probably never realized how diverse they are in size, color, and age. Next time you're looking up at the stars on a clear dark night, try to find one of the brightest red supergiant stars, such as Betelgeuse or Antares, and notice its redder color.

# Summary of Key Concepts

## 15.1 Properties of Stars

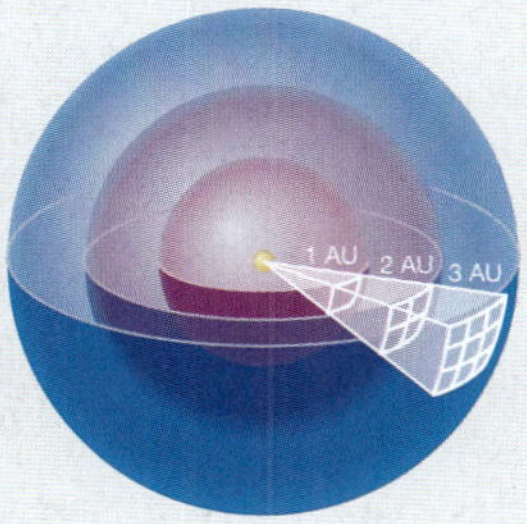


- **How do we measure stellar luminosities?** The **apparent brightness** of a star in our sky depends on both its **luminosity**—the total amount of light it emits into space—and its distance, as expressed by the **inverse square law for light**. We can therefore calculate luminosity from apparent brightness and distance; we can measure the latter through stellar parallax.
- **How do we measure stellar temperatures?** We measure a star's surface temperature from its color or spectrum, and we classify spectra according to the sequence of **spectral types** OBAFGKM, which runs from hottest to coolest. Cool, red stars of spectral type M are much more common than hot, blue stars of spectral type O.

- **How do we measure stellar masses?** We can measure the masses of stars in a **binary star system** using Newton's version of Kepler's third law if we can measure the orbital period and separation of the two stars.

## 15.2 Patterns Among Stars

- **What is a Hertzsprung-Russell diagram?** An **H-R diagram** plots stars according to their surface temperatures (or spectral types) and luminosities. Most stars appear on the H-R diagram in the narrow strip known as the **main sequence**. **Giants** and **supergiants** are to the upper right of the main sequence and **white dwarfs** are to the lower left.

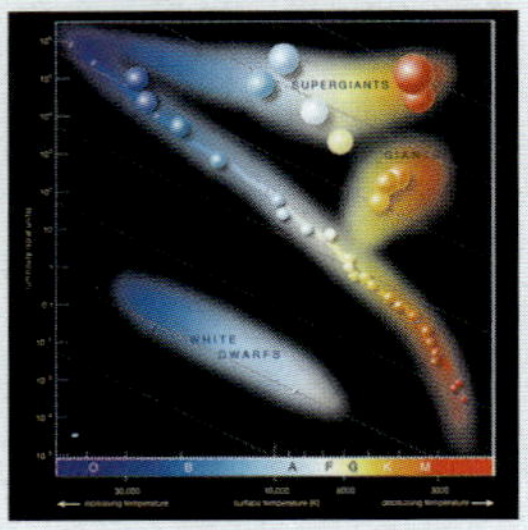


- **What is the significance of the main sequence?** Stars on the main sequence are all fusing hydrogen into helium in their cores. A star's position along the main sequence depends on its mass: High-mass stars are at the upper left and masses become progressively smaller as we move toward the lower right. Lifetimes vary in the opposite way, because higher-mass stars live shorter lives.
- **What are giants, supergiants, and white dwarfs?** Giants and supergiants are stars that have exhausted their core supplies of hydrogen for fusion and are undergoing other forms of fusion at a more rapid rate as they near the ends of their lives. White dwarfs are the exposed cores of stars that have already died, meaning they have no further means of generating energy through fusion.
- **Why do the properties of some stars vary?** Some stars fail to achieve a proper balance between the amount of fusion energy welling up from their cores and the amount of radiative energy emanating from their surfaces. These **pulsating variable stars** therefore alternately expand and contract, periodically rising and falling in luminosity.

## 15.3 Star Clusters

- **What are the two types of star clusters?** **Open clusters** contain up to several thousand stars and are found in the disk of our galaxy. **Globular clusters** contain hundreds of thousands of stars, all closely packed together, and are found mainly in the halo of the galaxy.

- **How do we measure the age of a star cluster?** Because all of a cluster's stars were born at the same time, we can measure a cluster's age by finding the **main-sequence turnoff point** on an H-R diagram of its stars. The cluster's age is equal to the hydrogen-fusion lifetime of the hottest, most luminous stars that remain on the main sequence. Open clusters are much younger than globular clusters, which can be as old as about 13 billion years.

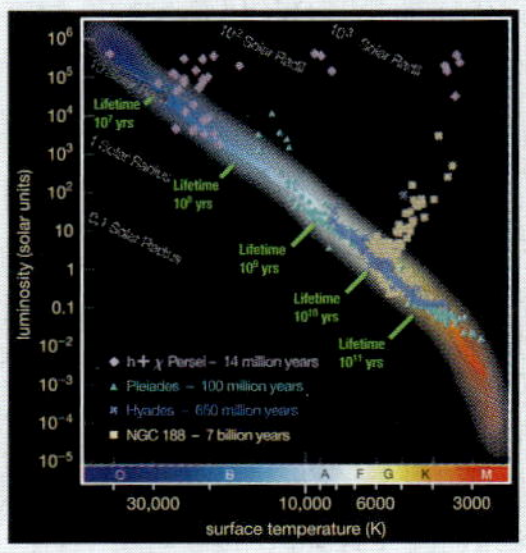


# Visual Skills Check

*Use the following questions to check your understanding of some of the many types of visual information used in astronomy. For additional practice, try the Chapter 15 Visual Quiz in the Study Area at www.MasteringAstronomy.com.*

*The figure to the right, similar to Figure 15.13, uses a zoom-in to compare the sizes of giant and supergiant stars to the sizes of Earth and the Sun.*

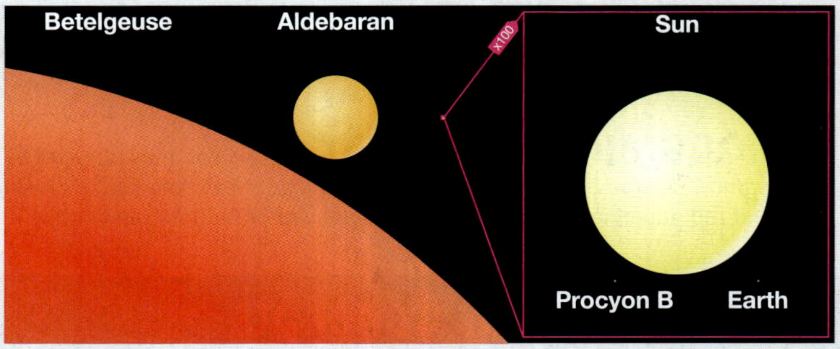


1. Suppose we wanted to represent all of these objects using the 1-to-10-billion scale from Chapter 1, on which the Sun is about the size of a grapefruit. Approximately how large in diameter would the star Aldebaran be on this scale?
   a. 40 centimeters (the size of a typical beach ball)
   b. 6 meters (roughly the size of a dorm room)
   c. 15 meters (roughly the size of a typical house)
   d. 130 meters (slightly larger than a football field)
2. Approximately how large in diameter would the star Betelgeuse be on this same scale?
   a. 40 centimeters (the size of a typical beach ball)
   b. 6 meters (roughly the size of a dorm room)
   c. 130 meters (slightly larger than a football field)
   d. 3 kilometers (the size of a small town)
3. Approximately how large in diameter would the star Procyon B be on this same scale?
   a. 10 centimeters (the size of a large grapefruit)
   b. 1 centimeter (the size of a grape)
   c. 1 millimeter (the size of a grape seed)
   d. 0.1 millimeter (roughly the width of a human hair)

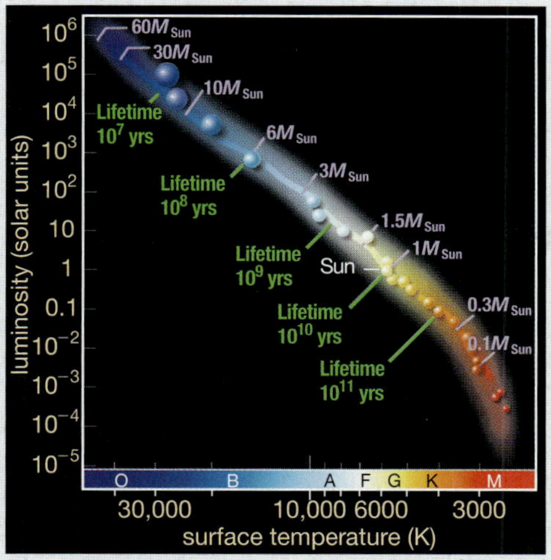


*The H-R diagram to the left is identical to Figure 15.11. Answer the following questions based on the information given in the figure.*

4. What are the approximate luminosity and lifetime of a star whose mass is 10 times that of the Sun?
5. What are the approximate luminosity and lifetime of a star whose mass is 3 times that of the Sun?
6. What are the approximate luminosity and lifetime of a star whose mass is twice that of the Sun?

# Exercises and Problems

*For instructor-assigned homework and other learning materials, go to www.MasteringAstronomy.com.*

## Chapter Review Questions

### Short-Answer Questions Based on the Reading

1. Briefly explain how we can learn about the lives of stars even though their lives are far longer than human lives.
2. In what ways are all stars similar? In what ways do they differ?
3. How is a star's *apparent brightness* related to its *luminosity*? Explain the *inverse square law for light*.
4. How do we use *stellar parallax* to determine a star's distance, and how can we then determine its luminosity?
5. What are *apparent* and *absolute magnitudes*, and how are they related to apparent brightness and luminosity?
6. What do we mean by a star's *spectral type*, and how is spectral type related to surface temperature and color?
7. How was the spectral sequence discovered, and why does it have the order OBAFGKM? Which stars are hottest and coolest in this sequence?
8. What are the three basic types of *binary star systems*? Why are *eclipsing binaries* so important to measuring masses of stars?
9. Draw a sketch of a *Hertzsprung-Russell (H-R) diagram*. Label the *main sequence*, *giants*, *supergiants*, and *white dwarfs*. Where on this diagram do we find stars that are cool and dim? Cool and luminous? Hot and dim? Hot and luminous?
10. What do we mean by a star's *luminosity class*? Briefly explain how we classify stars by spectral type and luminosity class.
11. What is the defining characteristic of a main-sequence star? Briefly explain why massive main-sequence stars are more luminous and have hotter surfaces than less massive main-sequence stars.
12. Which stars have longer lifetimes: massive stars or less massive stars? Explain why.
13. Why is a star's birth mass its most fundamental property?
14. How do giants and supergiants differ from main-sequence stars? What are white dwarfs?
15. How does the luminosity of a *pulsating variable star* change with time?
16. Describe in general terms how *open clusters* and *globular clusters* differ in their numbers of stars, ages, and locations in the galaxy.
17. Explain why H-R diagrams look different for star clusters of different ages. How does the location of the *main-sequence turnoff point* tell us the age of the star cluster?

### Does It Make Sense?

*Decide whether the statement makes sense (or is clearly true) or does not make sense (or is clearly false). Explain clearly; not all of these have definitive answers, so your explanation is more important than your chosen answer.*

18. Two stars that look very different must be made of different kinds of elements.
19. Two stars that have the same apparent brightness in the sky must also have the same luminosity.
20. Sirius looks brighter than Alpha Centauri, but we know that Alpha Centauri is closer because its apparent position in the sky shifts by a larger amount as Earth orbits the Sun.
21. Stars that look red have hotter surfaces than stars that look blue.
22. Some of the stars on the main sequence of the H-R diagram are not converting hydrogen into helium.
23. The smallest, hottest stars are plotted in the lower left-hand portion of the H-R diagram.
24. Stars that begin their lives with the most mass live longer than less massive stars because they have so much more hydrogen fuel.
25. Star clusters with lots of bright, blue stars of spectral type O and B are generally younger than clusters that don't have any such stars.
26. All giants, supergiants, and white dwarfs were once main-sequence stars.
27. Most of the stars in the sky are more massive than the Sun.

### Quick Quiz

*Choose the best answer to each of the following. For additional practice, try the Chapter 15 Reading and Concept Quizzes in the Study Area at www.MasteringAstronomy.com.*

28. If the star Alpha Centauri were moved to a distance 10 times as far from Earth as it is now, its parallax angle would (a) get larger. (b) get smaller. (c) stay the same.
29. What do we need to measure in order to determine a star's luminosity? (a) apparent brightness and mass (b) apparent brightness and temperature (c) apparent brightness and distance
30. What two pieces of information would you need in order to measure the masses of stars in an eclipsing binary system?

(a) the time between eclipses and the average distance between the stars (b) the period of the binary system and its distance from the Sun (c) the velocities of the stars and the Doppler shifts of their absorption lines

31. Which of these stars has the coolest surface temperature? (a) an A star (b) an F star (c) a K star
32. Which of these stars is the most massive? (a) a main-sequence A star (b) a main-sequence G star (c) a main-sequence M star
33. Which of these stars has the longest lifetime? (a) a main-sequence A star (b) a main-sequence G star (c) a main-sequence M star
34. Which of these stars has the largest radius? (a) a supergiant A star (b) a giant K star (c) a supergiant M star
35. Which of these stars has the greatest surface temperature? (a) a $30M_{Sun}$ main-sequence star (b) a supergiant A star (c) a Cepheid variable star
36. Which of these star clusters is youngest? (a) a cluster whose brightest main-sequence stars are white (b) a cluster whose brightest stars are red (c) a cluster containing stars of all colors
37. Which of these star clusters is oldest? (a) a cluster whose brightest main-sequence stars are white (b) a cluster whose brightest main-sequence stars are yellow (c) a cluster containing stars of all colors

## Inclusive Astronomy

*Use these questions to reflect on participation in science.*

38. *Research: Women in Astronomy.* Until fairly recently, men greatly outnumbered women in professional astronomy. Nevertheless, many women made crucial discoveries in astronomy throughout history—including discovering the spectral sequence for stars. Choose a female astronomer from any time period and learn enough about her life and discoveries to write a one- to three-page scientific biography.
39. *Discussion: Annie Jump Cannon's Disability.* Annie Jump Cannon (discussed in Section 15.1) was one of the greatest astronomers of the 20th century, male or female. She also had a significant physical disability: a childhood disease had made her completely deaf by the time she reached adulthood.
**a.** Gather in small groups to discuss the potential impacts of Annie Jump Cannon's deafness on her career, and begin the discussion by making a list of ways in which you think her deafness might have made a career in astronomy more challenging.
**b.** Cannon's greatest contributions came from the fact that she visually classified the spectra of more than 400,000 stars during her lifetime. Discuss whether her deafness might have provided her with any advantages in this work and, if so, list what they might have been.
**c.** When Cannon was considered for membership in the National Academy of Sciences in 1923, one of its members said he could not vote for Cannon because she was deaf. Why do you think someone would have voiced that objection in 1923? Discuss how and why scientists' views about physically disabled people have changed during the past century.

## The Process of Science

*These questions may be answered individually in short-essay form or discussed in groups, except where identified as group-only.*

40. *Classification.* As discussed in the text, Annie Jump Cannon and her colleagues developed our modern system of stellar classification. Why do you think rapid advances in our understanding of stars followed so quickly on the heels of their efforts? What other areas in science have had huge advances in understanding follow directly from improved systems of classification?
41. *Snapshot of the Heavens.* The beginning of the chapter likened the problem of studying the lives of stars to learning about human beings through a 1-minute glance at human life. What could you learn about human life by looking at a single snapshot of a large extended family, including babies, parents, and grandparents? How is the study of such a snapshot similar to what scientists do when they study the lives of stars? How is it different?
42. *Life Spans of Stars.* Scientists estimate the life spans of stars by dividing the total amount of energy available for fusion by the rate at which the stars radiate energy into space. Those calculations predict that the life spans of high-mass stars are shorter than those of low-mass stars. Describe a type of observation that can test this prediction and verify that it is correct.
43. *Group Activity: Comparing Stellar Properties.* The following activity makes use of Tables F.1 and F.2 in Appendix F. Before starting, split into two groups, one of which will gather data from Table F.1 and the other from Table F.2.
**a.** Working in your separate groups, count and record the number of stars of each spectral type (OBAFGKM) and of each luminosity class (I, II, III, IV, V) in your assigned table. Ignore the white dwarfs.
**b.** Working together, compare the counts of spectral types in the two tables. Are there any patterns you can identify? If so, develop a hypothesis that could explain them.
**c.** Are there any patterns to the differences in the counts you found for the luminosity classes in the two tables? If so, develop a hypothesis that could explain them.
**d.** Considering the data from both tables, identify the *nearest* star of each spectral type. Based on your results, discuss whether it is possible that there is an undiscovered star of spectral type O within 200 light-years of the Sun. Explain your reasoning.

## Investigate Further

### Short-Answer/Essay Questions

44. *Stellar Data.* The table below gives basic data for several bright stars; $M_v$ is absolute magnitude and $m_v$ is apparent magnitude. Use these data to answer the following questions. Include a brief explanation with each answer. [*Hint:* Remember that the magnitude scale runs backward, so brighter stars have smaller (or more negative) magnitudes.]

| Star | $M_v$ | $m_v$ | Spectral Type | Luminosity Class |
|---|---|---|---|---|
| Aldebaran | −0.2 | +0.9 | K5 | III |
| Alpha Centauri A | +4.4 | 0.0 | G2 | V |
| Antares | −4.5 | +0.9 | M1 | I |
| Canopus | −3.1 | −0.7 | F0 | II |
| Fomalhaut | +2.0 | +1.2 | A3 | V |
| Regulus | −0.6 | +1.4 | B7 | V |
| Sirius | +1.4 | −1.4 | A1 | V |
| Spica | −3.6 | +0.9 | B1 | V |

a. Which star appears brightest in our sky?
b. Which star appears faintest in our sky?
c. Which star has the greatest luminosity?
d. Which star has the least luminosity?
e. Which star has the highest surface temperature?
f. Which star has the lowest surface temperature?
g. Which star is most similar to the Sun?
h. Which star is a red supergiant?
i. Which star has the largest radius?
k. Which stars have finished fusing hydrogen in their cores?
l. Among the main-sequence stars listed, which one is the most massive?
m. Among the main-sequence stars listed, which one has the longest lifetime?

45. *Interpreting the H-R Diagram.* Using the information in Figure 15.10, describe how Proxima Centauri differs from Sirius.
46. *Parallax from Jupiter.* Suppose you could observe stellar parallax from the orbit of Jupiter. How would it differ from the stellar parallax we observe from Earth? Would it be easier or more difficult to measure stellar distances? Explain.
47. *An Expanding Star.* Describe what would happen to the surface temperature of a star if its radius doubled but there was no change in luminosity.
48. *Colors of Eclipsing Binaries.* Figure 15.8 shows an eclipsing binary system consisting of a small blue star and a larger red star. Explain why the decrease in apparent brightness of the combined system is greater when the blue star is eclipsed than when the red star is eclipsed.
49. *Visual and Spectroscopic Binaries.* Suppose you are observing two binary star systems at the same distance from Earth. Both are spectroscopic binaries consisting of similar types of stars, but only one of these binary systems is a visual binary. Which of these star systems would you expect to have the greater Doppler shifts in its spectra? Explain your reasoning.
50. *Life of a Star Cluster.* Imagine you could watch a star cluster from the time of its birth to an age of 13 billion years. Describe in one or two paragraphs what you would see happening during that time.
51. *Research: GAIA and Parallax.* The European Space Agency's *GAIA* mission, launched in 2013, has measured parallax for more than a billion stars. Learn how this satellite can measure much smaller parallax angles than can be measured from the ground and how those measurements affect our knowledge of the universe. Write a one- to two-page report on your findings.

## Quantitative Problems

*Be sure to show all calculations clearly and state your final answers in complete sentences.*

52. *The Inverse Square Law for Light.* Use Earth's distance of about 150 million kilometers from the Sun, the Sun's apparent brightness in our sky of about 1300 watts/m$^2$, and the inverse square law for light to calculate the apparent brightness that we would measure for the Sun *if* we were located at the following positions.
a. half Earth's distance from the Sun
b. twice Earth's distance from the Sun
c. five times Earth's distance from the Sun
53. *The Luminosity of Alpha Centauri A.* Alpha Centauri A lies at a distance of 4.4 light-years and has an apparent brightness in our night sky of $2.7 \times 10^{-8}$ watt/m$^2$. Recall that 1 light-year $= 9.5 \times 10^{12}$ km $= 9.5 \times 10^{15}$ m.
a. Use the inverse square law for light to calculate the luminosity of Alpha Centauri A.
b. Suppose you have a light bulb that emits 10 watts of visible light. How far away would you have to put the light bulb for it to have the same apparent brightness as Alpha Centauri A in our sky? (*Hint*: Use 10 watts as $L$ in the inverse square law for light, and use the apparent brightness given above for Alpha Centauri A. Then solve for the distance.)
54. *More Practice with the Inverse Square Law for Light.* Use the inverse square law for light to answer each of the following questions.
a. Suppose a star has the same luminosity as our Sun ($3.8 \times 10^{26}$ watts) but is located at a distance of 10 light-years. What is its apparent brightness?
b. Suppose a star has the same apparent brightness as Alpha Centauri A ($2.7 \times 10^{-8}$ watt/m$^2$) but is located at a distance of 200 light-years. What is its luminosity?
c. Suppose a star has a luminosity of $8 \times 10^{26}$ watts and an apparent brightness of $3.5 \times 10^{-12}$ watt/m$^2$. How far away is it? Give your answer in both kilometers and light-years.
d. Suppose a star has a luminosity of $5 \times 10^{29}$ watts and an apparent brightness of $9 \times 10^{15}$ watts/m$^2$. How far away is it? Give your answer in both kilometers and light-years.
55. *Parallax and Distance.* Use the parallax formula to calculate the distance to each of the following stars. Give your answers in both parsecs and light-years.
a. Alpha Centauri: parallax angle of $0.7420''$
b. Procyon: parallax angle of $0.2860''$
56. *The Magnitude System.* Use the definitions of the magnitude system to answer each of the following questions.
a. Which is brighter in our sky: a star with apparent magnitude 2 or a star with apparent magnitude 7? By how much?
b. Which has a greater luminosity: a star with absolute magnitude $-4$ or a star with absolute magnitude $+6$? By how much?
57. *Measuring Stellar Mass.* The spectral lines of two stars in a particular eclipsing binary system shift back and forth with a period of 6 months. The lines of both stars shift by equal amounts, and the amount of the Doppler shift indicates that each star has an orbital speed of 80,000 m/s. What are the masses of the two stars? Assume that each of the two stars traces a circular orbit around their center of mass. (*Hint*: See Mathematical Insight 15.4.)
58. *Calculating Stellar Radii.* Sirius A has a luminosity of $26L_{Sun}$ and a surface temperature of about 9400 K. What is its radius? (*Hint*: See Mathematical Insight 15.5.)
59. *Time Spent as a Red Giant.* The H-R diagram in Figure 15.20 shows a star cluster with a large number of red giants in it.
a. What is the approximate mass of the most massive stars that remain on the main sequence of this star cluster?
b. What is the luminosity of the most luminous stars in the cluster?
c. Compute the ratio of the luminosity from part b to the mass from part a. How does that ratio compare with the Sun's ratio of luminosity to mass?
d. From your answer to part c, estimate the maximum amount of time these very luminous stars can last as red giants.

# 16 Star Birth

▲ **About the photo:** This Herschel Space Observatory image of far-infrared light from the Eagle Nebula shows a region of active star birth within our galaxy.

## LEARNING GOALS

### 16.1 Stellar Nurseries

- Where do stars form?
- Why do stars form?

### 16.2 Stages of Star Birth

- What slows the contraction of a star-forming cloud?
- What is the role of rotation in star birth?
- How does nuclear fusion begin in a newborn star?

### 16.3 Masses of Newborn Stars

- What is the smallest mass a newborn star can have?
- What is the greatest mass a newborn star can have?
- What are the typical masses of newborn stars?

*I can hear the sizzle of newborn stars, and know anything of meaning, of the fierce magic emerging here. I am witness to flexible eternity, the evolving past, and know I will live forever, as dust or breath in the face of stars, in the shifting pattern of winds.*

—*Joy Harjo,* Secrets from the Center of the World

**Chapter 16 Overview**

Our ancestors imagined the stars as fixed and permanent, but this seeming permanence is an illusion. Stars do change with time, and by carefully observing stars and their surroundings, we have unlocked the secrets of their life cycles.

In this chapter, we will see how and why interstellar gas clouds sometimes collapse and give birth to stars, examine the processes that can make star birth surprisingly violent, and discuss why stars can have masses only within a particular range. Ultimately, we will see that the collapse of a gas cloud to make a star marks the beginning of a "battle" between pressure and gravity that will last the rest of the star's life: Gravity inexorably tries to squeeze a star's gas into an ever more compact form, gas pressure tries to resist the crush of gravity, and the outcome hinges on the star's mass.

## 16.1 Stellar Nurseries

The story of star birth begins in the murky depths of interstellar space. Using the technique of main-sequence turnoff point [**Section 15.3**] to determine the ages of star clusters, we have found that the youngest clusters are always associated with dark clouds of gas and dust, indicating that these interstellar clouds are the birthplaces of stars. Theoretical models of star formation confirm this idea, telling us how the physical processes that occur in gas clouds ultimately give birth to stars. In this section, we'll explore the types of clouds that can give birth to stars and learn how gravity overcomes gas pressure during the early stages of star formation.

### Where do stars form?

Star formation is a common process. Counting the number of young stars in our region of the galaxy suggests that an average of two to three stars are born somewhere in the Milky Way Galaxy each year. We see the birthing grounds of those stars with our own eyes whenever we look into the Milky Way on a dark night [**Section 2.1**]. The black patches within the Milky Way are interstellar gas clouds that appear dark because they block our view of the stars behind them. These gas clouds provide the raw material for star formation (**FIGURE 16.1**).

**The Interstellar Medium** We often think of space as "empty," and it is indeed a superb vacuum by earthly standards. However, space is not completely empty, as we find at least some gas and dust everywhere we look. We refer to the gas and dust found in the spaces between stars as the **interstellar medium**.

The gas between the stars is composed mostly of hydrogen and helium. Recall that hydrogen and helium were the only chemical elements produced in the Big Bang [**Section 1.2**]. The very first stars must have been born from clouds made only of hydrogen and helium gas. Since that time, stars have transformed a small fraction of the original hydrogen and helium into heavier elements, as we will discuss in detail in Chapter 17.

We use spectroscopy to measure the abundances of the new elements that stars have added to the interstellar medium. The most straightforward technique is to observe the spectrum of a star whose light has passed through an intervening cloud of interstellar gas. The cloud absorbs some of the star's light, leaving absorption lines in the star's spectrum (see Figure 5.14c). The wavelengths of the lines

**FIGURE 16.1** A star-forming cloud in the constellation Scorpius. The region pictured here is about 50 light-years across.

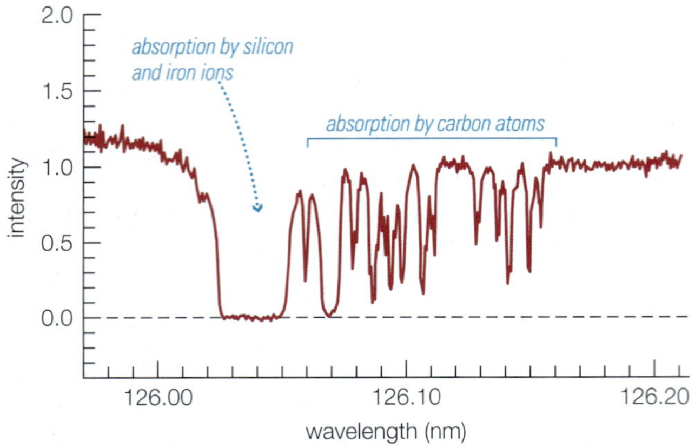


**FIGURE 16.2** This spectrum of a star has absorption lines that allow us to measure the proportions of different elements in an interstellar cloud lying between Earth and the star.

tell us the chemical contents of the cloud, and comparing the amounts of light absorbed by different atoms and molecules allows us to determine the composition of the cloud (**FIGURE 16.2**). Many such measurements have shown that the interstellar medium consists (by mass) of 70% hydrogen, 28% helium, and 2% heavier elements.

Virtually all the gas between stars in the Milky Way Galaxy has approximately the same chemical composition. However, the gas of the interstellar medium can look very different from place to place because of huge differences in temperature and density. Some clouds are extremely hot but low in density, some are cold and relatively dense, and others have conditions in between. We'll discuss the reasons for this variety in Chapter 19, where we'll look more closely at the way in which our galaxy functions as a cosmic recycling plant. Here, we'll focus on the clouds that give birth to stars.

**Star-Forming Clouds** Stars are born in interstellar clouds that are particularly cold and dense. These clouds are usually called **molecular clouds** because they are cold enough and dense enough to allow atoms to combine together into molecules. Typical molecular cloud temperatures are between about 10 and 30 K (recall that 0 K is absolute zero), and the typical average density is about 300 molecules per cubic centimeter—quite high by interstellar standards but still almost a million trillion times less than the density of air at sea level on Earth. The density varies from place to place within a molecular cloud, and high-density regions can be hundreds of times denser than the average.

Molecular hydrogen ($H_2$) is by far the most abundant molecule in these clouds, because hydrogen and helium are the most abundant elements and helium atoms do not combine with other atoms into molecules. Despite its abundance, molecular hydrogen is difficult to detect, because molecular clouds are usually too cold for $H_2$ to produce emission lines that we can study in spectra. As a result, most of what we know about molecular clouds comes from observing spectral lines of molecules that make up only a tiny fraction of a cloud's mass. Carbon monoxide (CO) is the most abundant of these molecules, and it produces radio emission lines that allow us to map the structures of molecular clouds (**FIGURE 16.3**). More than 120 other molecules have also been identified in molecular clouds by their radio emission lines, including water ($H_2O$), ammonia ($NH_3$), and ethyl alcohol ($C_2H_5OH$).

**Interstellar Dust** Not all of the material in a molecular cloud is gaseous. Elements such as carbon, silicon, oxygen, and iron are often found in tiny solid grains of **interstellar dust**. Overall, about half the atoms of elements heavier than helium are found in dust grains. Because these

**a** A visible-light image of the nebula. The dark (horsehead-shaped) region is a molecular cloud.

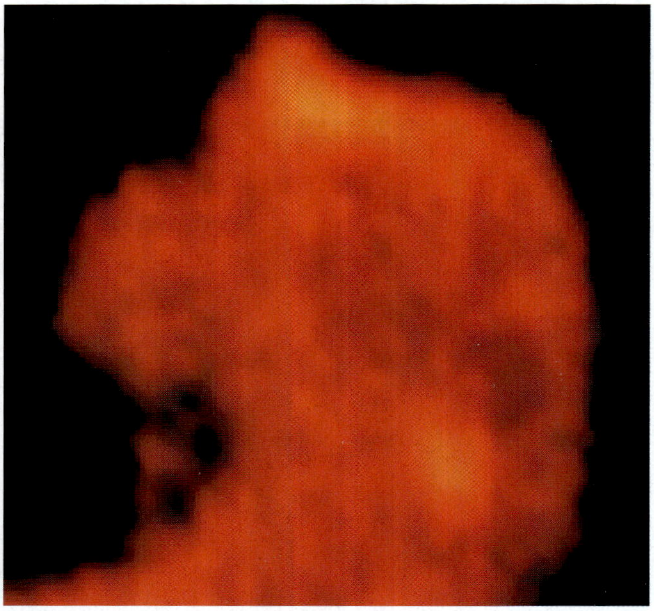

**b** A radio-wave image of the nebula, showing emission from carbon monoxide (CO) molecules.

**FIGURE 16.3** Images of the molecular cloud known as the Horsehead Nebula. Notice that the molecular cloud is dark in visible light but shines with radio emission from carbon monoxide.

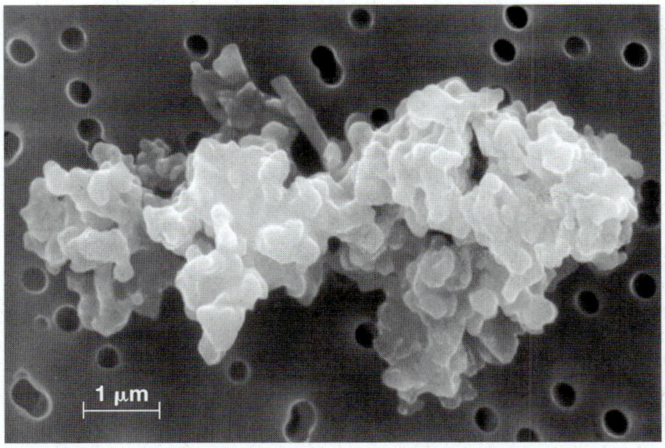


**FIGURE 16.4** This photograph shows a microscopic dust particle from interplanetary space that is suspected to predate the formation of the solar system. Interstellar dust grains are thought to be much like the ones shown in this cluster.

elements make up about 2% of the mass of the interstellar medium, we conclude that interstellar dust grains constitute about 1% of a molecular cloud's mass. Individual dust grains are microscopic in size, usually less than 1 micrometer across, which is smaller than a single cell of bacteria. This small size makes them more like the solid microscopic particles found in smoke than like the sand grains on a beach or the dust particles on the furniture in your house (**FIGURE 16.4**).

Despite the tiny size of these grains, there are so many of them that they profoundly affect how light travels through a molecular cloud. We can see the two most important effects of dust on light in **FIGURE 16.5a**. First, dust grains scatter or absorb virtually all the visible light that enters a molecular cloud, preventing us from seeing stars that lie behind it. That is why regions of the sky with molecular clouds look so black. Second, stars seen near the edges of a molecular cloud appear redder than similar stars outside the cloud, a phenomenon known as **interstellar reddening**.

Interstellar reddening occurs for much the same reason that our Sun appears redder when viewed through smoke or smog: Dust grains block shorter-wavelength (bluer) photons of visible light more easily than longer-wavelength (redder) photons. Near the edges of a molecular cloud, where stars are only partially obscured, the blocking of blue light causes stars to appear redder than their true colors.

**See it for yourself** Next time the Moon rises on a clear night, set aside a few minutes to watch it come up. At first it will look reddish, because moonlight must pass through a large amount of Earth's atmosphere to reach us when the Moon is on the horizon. As the Moon rises, the length of the path of moonlight through the atmosphere decreases, and the Moon's reddening decreases. How many minutes does the rising Moon take to go from red to orange and from orange to yellow?

We can distinguish interstellar reddening from a Doppler shift because reddening doesn't change the wavelengths of a star's spectral lines. As a result, we can determine the amount of reddening by comparing a star's observed color to the color expected for a star of its spectral type [**Section 15.1**]. The amount of reddening tells us how much dusty gas lies between Earth and a distant star, and many such measurements have allowed us to map out the distribution of interstellar dust in our region of the galaxy.

The fact that longer-wavelength light passes more easily through dusty gas is even more helpful when we try to observe stars directly behind a molecular cloud and not just at its edges. Most molecular clouds are thick enough to block all forms of red light from stars directly behind them, but longer-wavelength infrared light can often pass right through (**FIGURE 16.5b**). These clouds appear more and more transparent as we observe light of

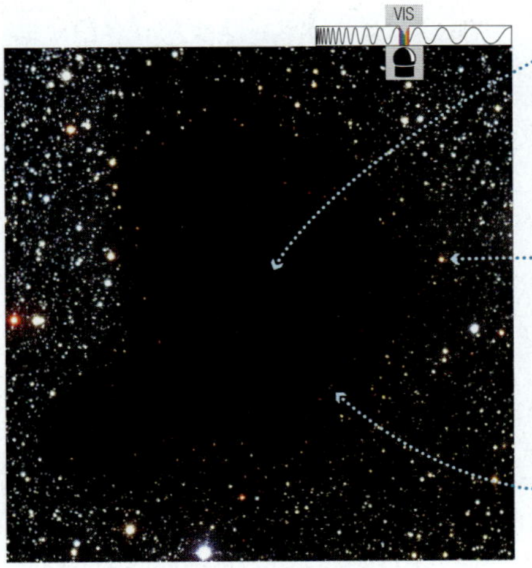


**a** A visible-light image of the dark molecular cloud Barnard 68.

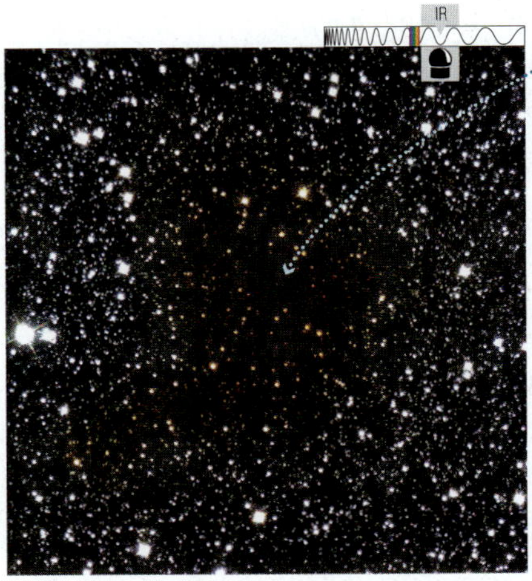


**b** An infrared image of Barnard 68, showing the stars that lie behind the cloud.

**FIGURE 16.5** We can see the effects of interstellar dust on starlight by comparing visible-light and infrared photographs of a molecular cloud. The cloud shown here is about half a light-year across.

**FIGURE 16.6** These Hubble Space Telescope photos show a portion of the Eagle Nebula known as the "pillars of creation," in which new stars are being born. The tallest pillar is about 4 light-years long. The left panel shows a visible-light photo, and the right panel shows an infrared photo. (Figure 16.8 shows emission of longer-wavelength infrared light from a larger portion of the Eagle Nebula.)

progressively longer wavelengths. Infrared observations therefore allow us to see directly through molecular clouds to stars lying at much greater distances. More important, infrared observations reveal new stars embedded within the clouds themselves, caught in the very act of birth (**FIGURE 16.6**).

Much of the radiation produced by young stars within the molecular cloud cannot escape the cloud directly, because dust grains absorb the visible light (and some of the infrared light) from these stars. The absorbed radiative energy heats the dust grains, which may be further heated by radiative energy from newly formed stars just outside the cloud. The dust grains therefore become warm enough to emit thermal radiation in the infrared and microwave bands of the electromagnetic spectrum. Consequently, clouds that appear dark in visible-light photos often glow when observed in long-wavelength infrared light (**FIGURES 16.7** and **16.8**).

**Learning About Star Birth**

## Why do stars form?

Stars form when gravity causes a molecular cloud to contract and the contraction continues until the central object becomes hot enough to sustain nuclear fusion in its core. However, the role of gravity is only one part of the story. For a more complete understanding of star formation, we must also pay attention to how gas pressure resists gravity.

Recall that our Sun remains stable in size because the outward push of gas pressure balances the inward pull of gravity, a balance that we call *gravitational equilibrium* [**Section 14.1**]. The gas pressure is far weaker in the low-density gas of interstellar space, but so is the pull of gravity. In most places within our galaxy, gravity is not strong enough to overcome the internal pressure of interstellar gas,

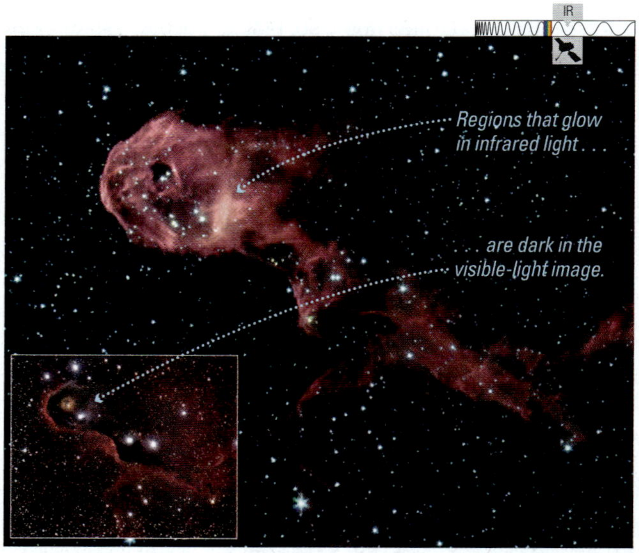


**FIGURE 16.7** Star-forming molecular clouds are colder than stars and therefore emit infrared rather than visible light. This infrared image from the Spitzer Space Telescope shows a star-forming cloud in the constellation Cepheus. The pink color represents infrared light from heated dust grains in the cloud's cold, dense gas. The inset photo shows the same region in visible light. The region in the picture is about 15 light-years across.

**FIGURE 16.8** This infrared image from the Herschel Space Observatory shows the Eagle Nebula; the rectangle indicates the portion of the nebula shown in Figure 16.6. This image combines infrared light over a range of wavelengths, so the colors correspond to different temperatures of the dust grains producing the infrared light. Blue regions contain dust at about 40 K, green regions have cooler dust, and red regions contain dust at about 10 K. The region pictured here is about 50 light-years across.

which is why star formation does not occur everywhere. Molecular clouds are the exception—they are the only places in which gravity can win the battle against pressure and start the formation of stars.

**Gravity versus Pressure** Gravity can create stars only if it can overcome the outward push of the pressure within a gas cloud, which depends on both the density and the temperature of the cloud. To see why, consider gas particles inside a balloon. The outward pressure the gas exerts on the balloon is the collective force that the colliding gas particles apply to each unit of area on the balloon's surface [**Section 5.3**]. The pressure goes up if you increase the density of the gas in the balloon (by blowing more air into it), because the higher density means more particles crashing into each piece of the balloon's surface. The pressure also goes up if you heat the balloon, because the higher temperature means the gas particles move faster and therefore hit the surface both more frequently and with greater force. (These ideas are quantified in the *ideal gas law*; see Mathematical Insight 14.2.)

This type of pressure, which depends on both density and temperature, affects not only balloons and interstellar clouds but virtually all the gases we've encountered so far in this book, including the atmospheres of planets and the plasma throughout the interior of the Sun. However, later in this chapter we will see that other types of pressure (degeneracy pressure, radiation pressure) sometimes come into play, so for clarity we will refer to the temperature-dependent pressure in ordinary gas clouds as **thermal pressure**.

Thermal pressure can resist gravity in most interstellar gas clouds, because their low gas densities keep gravity quite weak. Only in molecular clouds, where gas densities are tens to thousands of times greater than average, is gravity strong enough to overcome thermal pressure. Gravity is stronger in molecular clouds because more mass is packed into each cubic centimeter of volume, but thermal pressure isn't much greater in molecular clouds than in other clouds because of the low temperatures.

Gravity therefore has the upper hand in many molecular clouds. Nevertheless, observations suggest that gravity can form stars more easily if some other force (besides gravity) initiates the compression of a molecular cloud. For example, a collision between two molecular clouds can compress both clouds, driving up the gas density within each of them. The increase in density strengthens the force of gravity within the clouds, tipping the balance between gravity and pressure in favor of gravity and thereby triggering star formation. Similarly, debris from an exploding star can trigger star formation when it collides with and compresses a molecular cloud.

**Preventing a Pressure Buildup** Regions of a molecular cloud in which the gravitational attraction is stronger than the thermal pressure are forced to contract. We have already seen the consequences of *gravitational contraction* in our study of the Sun [**Section 14.1**]. As gravity makes a gas cloud shrink, it converts some of the cloud's gravitational potential energy into thermal energy. If the cloud cannot get rid of that thermal energy as quickly as it is released, then thermal energy builds up inside the cloud. If all this thermal energy remained within the cloud, it would raise the cloud's temperature and thermal pressure, eventually bringing the process of star formation to a halt.

Molecular clouds avoid this fate because they quickly rid themselves of any thermal energy that builds up. Collisions between gas molecules in the cloud transform the thermal energy into photons by exciting the rotational and vibrational energy levels of those molecules, which then produce emission lines in the infrared and radio portions of the spectrum [**Section 5.4**]. As long as the photons produced by these colliding molecules can escape the cloud, the cloud's temperature can remain low. With no significant buildup of thermal energy, gravity continues to dominate over thermal pressure, so all parts of the cloud that remain cool can contract to form stars.

**See it for yourself** You can observe the conversion of gravitational potential energy to thermal energy for yourself with a lump of soft clay. Drop the clay repeatedly onto a hard floor. Each drop converts gravitational potential energy (which becomes kinetic energy during the fall) into thermal energy (released when the clay hits the ground). Can you feel the clay getting warmer after many drops?

**Clustered Star Formation** Observations of star-forming regions show that most stars are born in large clusters. Stars tend to form in clusters because gravity is stronger in a high-mass gas cloud, making it easier for gravity to overcome the outward force due to thermal pressure. Precise calculations show that at the temperatures and densities of typical molecular clouds, gravity can begin to overcome thermal pressure only in clouds with masses greater than a few hundred times the mass of the Sun (see Mathematical Insight 16.1). However, star-forming clouds usually contain thousands of times the mass of the Sun, which raises an important question: How do these clouds resist gravity long enough to grow to such large masses before they begin to form stars?

We are still uncertain of all the processes that enable massive clouds to resist the crush of gravity, but at least two factors probably play important roles. First, observations show that individual gas clumps within molecular clouds move at substantially different speeds, indicating that the overall gas motion is turbulent, much like the motion of water in the wake of a rapidly moving boat. In other words, the internal motions of the cloud would tear it apart if not for the attractive force of gravity. Stars can form in such a cloud only if gravity is strong enough to overcome the turbulent gas motion, and overcoming that motion can require considerably more mass than is needed to overcome thermal pressure alone.

Second, magnetic fields can help the cloud resist gravity. Observations tell us that magnetic fields are important in molecular clouds: Light from stars usually travels through space with its electric and magnetic fields vibrating in random directions, but starlight that has passed through a molecular cloud often has its electric and magnetic fields aligned in particular directions. (This type of alignment of electric and magnetic fields in light is called *polarization*; see the Special Topic on page 145.) Such an alignment can happen if magnetic fields are threading the cloud, causing the cloud's dust grains to line up like iron filings near a bar magnet. Now, recall that magnetic fields tend to force charged particles to move on spiral paths *along* magnetic field lines while preventing the particles from moving perpendicular to the field lines (see Figure 14.17). A magnetic field threading through a molecular cloud therefore tends to prevent charged particles inside the cloud from moving in certain directions. The overall proportion of charged particles in a molecular cloud is relatively small, but these particles exert friction on other (neutral) particles that are moving perpendicular to the field lines. The friction is great enough that a magnetic field can inhibit the movement of all the gas in a cloud (**FIGURE 16.9**). Depending on the strength of the magnetic field, this process can slow or even halt the gravitational collapse of a molecular cloud.

## MATHEMATICAL INSIGHT 16.1 Gravity versus Pressure

A gas cloud can form stars only if the inward force due to gravity is stronger than the outward force due to the thermal pressure inside the cloud. We can therefore determine the minimum mass necessary for a cloud to contract by comparing the strength of these two forces. We will not go through the derivation here, but the resulting equation states that the balance point between thermal pressure and gravity occurs at the following mass:

$$M_{\text{balance}} = 18M_{\text{Sun}}\sqrt{\frac{T^3}{n}}$$

where the gas temperature $T$ is on the Kelvin scale and the number density $n$ is in units of particles per cubic centimeter. Gravity is stronger than thermal pressure in a cloud with a mass greater than the mass $M_{\text{balance}}$, and thermal pressure is stronger than gravity in a cloud with a mass less than this mass. [$M_{\text{balance}}$ is often called the *Jeans mass*, because the formula is based on ideas first worked out by astronomer Sir James Jeans (1877–1946).]

**EXAMPLE 1:** A typical molecular cloud has a temperature of 30 K and an average density of 300 particles per cubic centimeter. What minimum cloud mass is required to form stars?

**SOLUTION:**

**Step 1 Understand:** The cloud can form stars only if it has enough mass for gravity to overpower thermal pressure, which means a mass greater than the $M_{\text{balance}}$ we calculate with the formula above.

**Step 2 Solve:** Using $T = 30$ K and $n = 300$ particles/cm$^3$, we find

$$M_{\text{balance}} = 18M_{\text{Sun}}\sqrt{\frac{30^3}{300}} = 18M_{\text{Sun}}\sqrt{90} \approx 171M_{\text{Sun}}$$

**Step 3 Explain:** A typical molecular cloud can form stars only if its mass is greater than about $171M_{\text{Sum}}$. Because this mass is much larger than the masses of most stars, we conclude that star-forming clouds generally contain enough mass to form many stars.

**EXAMPLE 2:** The density of a star-forming cloud increases as the cloud contracts, and its temperature can remain low as long as the cloud radiates away the thermal energy released by gravitational contraction. Suppose the cloud has reached a density of 300,000 particles/cm$^3$ but still has a temperature of 30 K. What mass is needed for star formation to continue?

**SOLUTION:**

**Step 1 Understand:** As in Example 1, we are looking for a mass greater than $M_{\text{balance}}$, but this time we have a higher density.

**Step 2 Solve:** Using $T = 30$ K and $n = 300{,}000$ particles/cm$^3$, we find

$$M_{\text{balance}} = 18M_{\text{Sun}}\sqrt{\frac{30^3}{300{,}000}} = 18M_{\text{Sun}}\sqrt{0.09} \approx 5.4M_{\text{Sun}}$$

**Step 3 Explain:** When the gas density reaches 300,000 particles per cubic centimeter, the mass required for star formation falls to $5.4M_{\text{Sun}}$. That is why a large star-forming cloud ultimately fragments into pieces that can form individual stars.

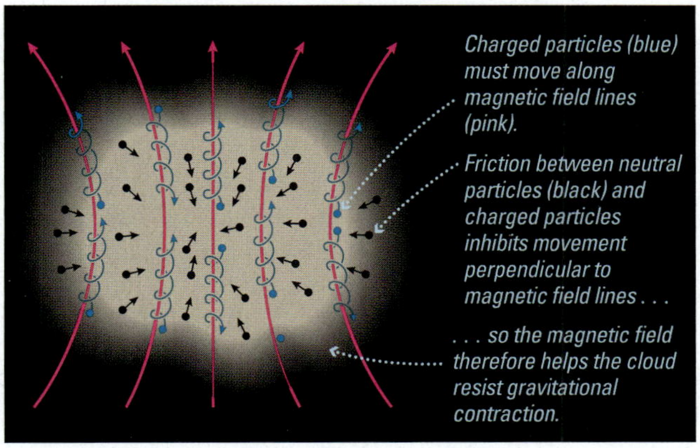


**FIGURE 16.9** Magnetic field lines threading a molecular cloud can hinder its collapse by preventing particles from traveling perpendicular to the field lines.

**Fragmentation of a Molecular Cloud** Gravity in a molecular cloud with a large enough mass is strong enough to surmount all these obstacles, which is why the cloud collapses. But why does a large molecular cloud form many individual stars instead of a single extremely massive star? The answer goes back to the battle between gravity and pressure.

Recall that the force of gravity follows an inverse square law [**Section 4.4**]. The strength of gravity therefore increases as the cloud shrinks in size. If photons emitted by molecules in the cloud can carry away the gravitational potential energy released by contraction (keeping the interior temperature and thermal pressure low), we have a situation in which gravity is strengthening more than thermal pressure. The contraction of the cloud therefore gives gravity a growing advantage in the battle against pressure, which in turn means that a smaller total amount of mass is necessary for gravity to win the battle (**FIGURE 16.10**).

Because molecular clouds are turbulent and lumpy, there are plenty of small, dense clumps within a contracting cloud that become able to shrink on their own. A large molecular cloud therefore splits into numerous individual cloud fragments (sometimes called *molecular cloud cores*). Each of these fragments will go on to become a star system containing either a single star or two or more stars that orbit each other.

**Isolated Star Formation** A molecular cloud does not necessarily need to be very massive to form a star, as long as it is unusually dense and cold. For example, in gas with a density of a few tens of thousands of molecules per cubic centimeter and a temperature of 10 K, gravity can overcome pressure even in a cloud with just a few solar masses of material. Small, isolated molecular clouds with these properties are indeed observed, and they tend to form just one or a few stars at a time (**FIGURE 16.11**). However, it is not yet clear how these small clouds become compressed to such high densities in isolated regions of space.

**The First Generation of Stars** Spectroscopic observations of our Sun and many other nearby stars show that their outer layers have about the same composition that we find in interstellar clouds today: about 70% hydrogen and 28% helium, with the remaining 2% consisting of elements heavier than helium, which astronomers sometimes call *heavy elements*.* These measurements tell us that the composition of gas clouds in our galaxy, and of the stars they produce, has not

*Astronomers also sometimes refer to elements heavier than helium as *metals*—a very different use of the term than in daily life, where *metal* refers to substances like copper, silver, and gold that shine when polished and often are good electrical conductors.

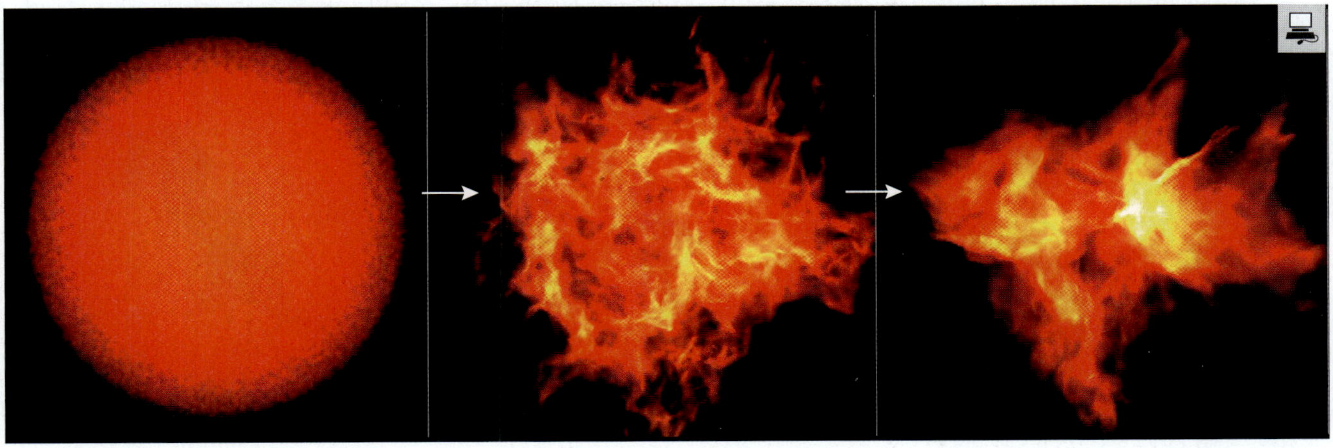

**a** The simulation begins with a turbulent gas cloud 1.2 light-years across and containing $50M_{\text{Sun}}$ of gas.

**b** Random motions in the cloud cause it to become lumpy, with some regions denser than others. If gravity can overcome thermal pressure in these dense regions, they can collapse to form even denser lumps of matter.

**c** The large cloud therefore fragments into many smaller lumps of matter corresponding to the bright yellow regions in this image. Each lump can go on to form one or more new stars.

**FIGURE 16.10** Computer simulation of a fragmenting molecular cloud. Gravity attracts matter to the densest regions of a molecular cloud. If gravity can overcome thermal pressure in these dense regions, they collapse to form even denser knots of gaseous matter. Each of these knots can form one or more new stars.

**FIGURE 16.11** Each of these relatively small molecular clouds, silhouetted against a background of glowing red gas, contains only enough mass to form a few stars. The larger clouds are about 1.4 light-years across.

changed much during the last 5 billion years or so. However, if we go back in time far enough, the composition of interstellar gas must have been different, because virtually all elements heavier than helium have been produced through the lives and deaths of stars [**Section 1.2**]. The abundances of these elements in the oldest known stars confirm this idea: Stars in the oldest globular clusters, with ages greater than 12 billion years, contain less than 0.1% of their mass in the form of elements heavier than helium, just as we would expect for stars that were born before most of the heavy elements had been made [**Section 19.3**]. Extending this idea even further back in time, the very first generation of stars must have been born before any heavy elements had been produced, so they must have formed in clouds of gas containing only hydrogen and helium from the Big Bang.

These first-generation stars cannot have been born in quite the same way that stars are born today, because their birth clouds could not have contained molecules like carbon monoxide to radiate away thermal energy. Given the importance of photon emission from carbon monoxide to the process of molecular cloud contraction, how did this first generation of stars form?

We have not yet found any stars that were born containing only hydrogen and helium, but we can use supercomputer simulations to model how they might have formed (**FIGURE 16.12**). These models show that emission of photons from hydrogen molecules in the birth clouds of these stars could have helped keep them cool, but not as cold as today's molecular clouds. More specifically, the first generation of stars must have been born in clouds that never cooled below a temperature of about 100 K—considerably warmer than the 10–30 K temperatures of molecular clouds today—because temperatures exceeding 100 K are necessary to excite hydrogen molecules.

The relatively high temperatures of these first-generation molecular clouds would have made it more difficult for gravity to overcome pressure, requiring stars to form in relatively massive cloud fragments that would have made predominantly high-mass stars. Such massive stars have very short lifetimes [**Section 15.2**], which would explain why we don't find any first-generation stars in today's universe: They would all have died off long ago. The large masses also help explain why the proportion of heavy elements

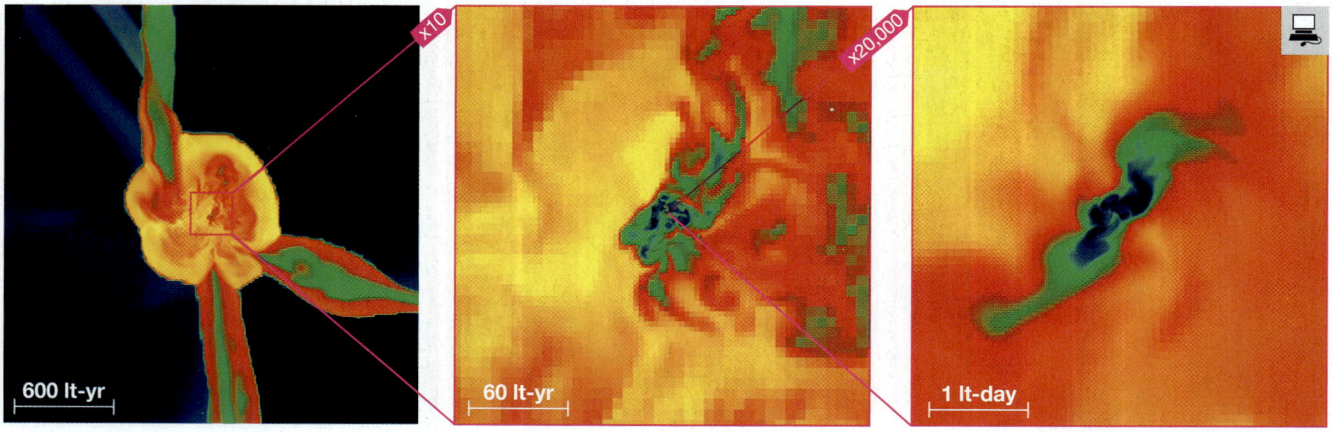


**a** The orange blob is one of the first gas clouds that gravity forms in this simulation of the early universe.

**b** Hydrogen molecules near the center of the blob allow the gas there to cool to 100–300 K.

**c** At the center of the cloud, about 200$M_{Sun}$ of gas collects into a newly forming star. The cloud cannot fragment further, because there are no complex molecules to cool it to lower temperatures.

**FIGURE 16.12** Frames from a supercomputer simulation modeling the formation of one of the first stars in the universe. Color indicates gas temperature: Blue and green represent cool gas (100–300 K), orange and red represent warm gas (500–3000 K), and yellow represents gas at intermediate temperatures (300–500 K).

rose quickly in the young universe. As we'll discuss in Chapter 17, massive stars produce heavy elements and then release those elements into space when they die. The massive first-generation stars would therefore have seeded the interstellar medium with substantial quantities of heavy elements that could be incorporated into all subsequent generations of stars.

## 16.2 Stages of Star Birth

We have seen that gravity can overcome pressure in a molecular cloud, forcing a large cloud to contract and fragment into many smaller clouds that eventually become stars. But how exactly does a contracting cloud become a full-fledged star that shines with energy released by nuclear fusion? Once again, the answer lies in the ongoing battle between the inward pull of gravity and the outward push of thermal pressure. As long as the interior of the cloud can continue to radiate away its thermal energy, the cloud remains cool and the pressure remains too weak to slow the crush of gravity. However, as the cloud becomes smaller and denser, it becomes more difficult for radiation to escape from the interior. The thermal energy becomes trapped, raising the interior temperature enough that thermal pressure begins to slow the cloud's collapse. In this section, we'll investigate what happens from the time a contracting cloud fragment first starts to shrink to the time its internal temperature grows high enough for nuclear fusion.

### What slows the contraction of a star-forming cloud?

A cloud fragment will continue to collapse on itself as long as it remains cold, and it can remain cold as long as photons emitted by molecules within the cloud can carry away the energy released by gravitational contraction. It's easy for photons emitted by molecules to escape from the cloud early in the process of contraction, while the cloud's density is still fairly low, because the photons are unlikely to run into any other molecules after they are emitted. This situation begins to change, however, as continued contraction packs the cloud's molecules closer together.

**Trapping of Thermal Energy** Cloud contraction makes it increasingly difficult for emitted photons to escape, because the growing density of the cloud increases the chances that a photon on its way out will be absorbed, leaving the absorbing molecule in an excited state. Collisions between the excited molecule and other molecules can then change the absorbed photon's energy back into thermal energy, which prevents that energy from escaping the cloud. A similar effect happens with dust grains. For a while, dust grains can help keep the cloud cool by emitting infrared light, but as contraction packs the dust grains closer together, the infrared photons have trouble escaping the cloud.

The central region of the cloud fragment eventually grows dense enough to trap almost all the radiation inside it. When that happens, the inner regions of the contracting cloud can no longer radiate away their heat. The central temperature and pressure begin to rise dramatically, and the rising pressure pushes back against the crush of gravity, slowing the contraction. This change marks the first stage of star formation. The dense center of the cloud fragment is now a **protostar** (sometimes referred to as a *pre-main-sequence star*)—a clump of gas that will become a new star. Protostars look starlike, with surface temperatures and luminosities similar to those of true stars. However, a protostar is not yet a true star because its core is not yet hot enough for nuclear fusion.

**Growth of a Protostar by Gas Infall** A protostar's mass grows with time because a molecular cloud fragment contracts in an "inside-out" fashion. Gravity is strongest near the protostar, where the gas density is greatest. The gas in the outer part of the cloud fragment feels a weaker gravitational pull, so it initially remains behind as the protostar forms. However, because the gas beneath has already contracted to make the protostar, the outer regions of the cloud fragment are left with little pressure support from below. Like something that was resting on a trapdoor that has just opened, the gas from these outer regions begins to rain down onto the protostar, gradually increasing its mass.

This rain of mass onto the protostar continues until the gas surrounding the protostar is gone, either because the gas runs out or because something blows the remaining gas away. The star itself can blow the surrounding gases away through the combined effects of its radiation and stellar wind, much as the Sun's radiation and solar wind are thought to have cleared out excess gas during the formation of our solar system [**Section 8.2**]. Observations show that young stars tend to have strong stellar winds, lending support to the idea that they can blow gas out of the system. Alternatively, the winds of other nearby young stars may blow away some of the gas surrounding a protostar (**FIGURE 16.13**).

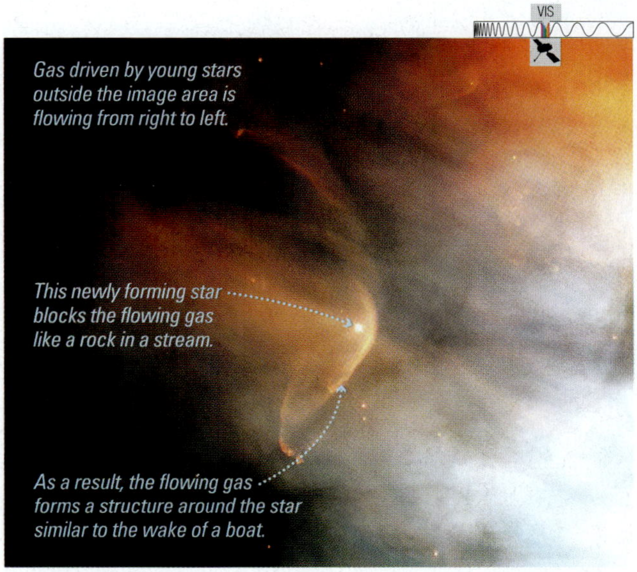


**FIGURE 16.13** The curved structure around this forming star shows that the bright young stars outside the lower right side of the frame are blowing gas past the new star, possibly stripping away gas that has not yet fallen onto it.

## What is the role of rotation in star birth?

So far, our description of star birth may make it sound like the gravitational equivalent of squishing a piece of paper into a tight ball. However, observations of young stars show that their births can be quite violent. For example, many young stars appear to be shooting jets of gas into deep space. These phenomena are consequences of the cloud's rotation, and in particular the law of *conservation of angular momentum* [**Section 4.3**].

**Protostellar Disks** Random motions of gas particles inevitably give a gas cloud some small overall rotation, although it may be imperceptibly slow. However, the law of conservation of angular momentum ensures that, like the rotation of an ice skater pulling in his arms, the cloud's rotation must become much faster as the cloud shrinks in size [**Section 4.3**]. This faster rotation prevents gas from raining directly down onto a protostar, causing it instead to settle into a **protostellar disk** (**FIGURE 16.14**). Notice the similarity to the spinning disk of gas from which the planets of the solar system formed [**Section 8.2**].

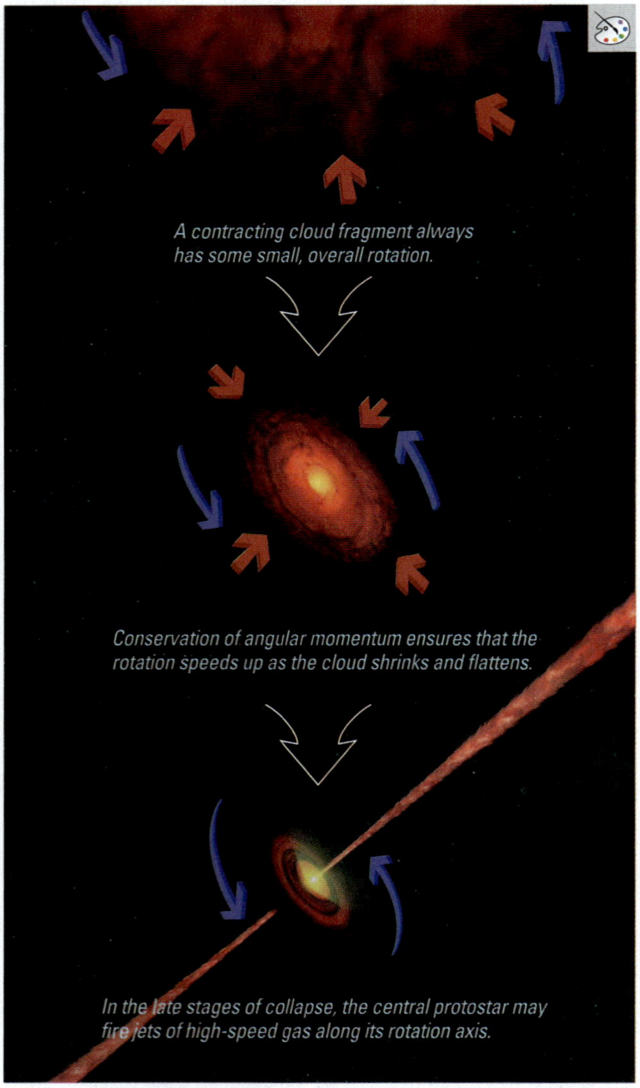


**FIGURE 16.14** Artist's conception of star birth.

Observations of newly forming stars show that they do indeed have disks around them (see Figure 8.4). Sometimes these disks coalesce into planetary systems like our solar system. We do not yet know how often this occurs, but the high fraction of stars with extrasolar planets suggests that it must be quite common [**Section 13.2**]. In the context of star formation, the most important feature of the disk around a protostar is that it helps to transfer angular momentum away from the infalling gas, enabling the protostar to grow more massive.

Gas in a protostellar disk can gradually spiral inward toward the protostar because of friction. Individual gas particles in the disk orbit in accord with Kepler's laws [**Section 3.3**], which means that gas in the inner parts of the disk moves faster than gas in the outer parts. These differences in orbital speed mean that gas in any particular part of the disk "rubs" against slower-moving gas with a slightly larger orbit, generating friction and heat in much the same way that rubbing your palms together makes them warm. The friction slowly causes the orbits of individual gas particles to shrink until the gas particles fall onto the surface of the protostar, thereby increasing the protostar's mass. Because the process in which material falls onto another body is called *accretion* [**Section 8.2**], a disk in which friction causes material to spiral inward is often called an *accretion disk*. A protostellar disk is one example of an accretion disk; we will encounter more examples in Chapter 18.

The protostellar disk is probably also important to slowing the rotation of the protostar itself. The protostar's rapid rotation generates a strong magnetic field. As the magnetic field lines sweep through the protostellar disk, they transfer angular momentum to outlying material, slowing the protostar's rotation (see Figure 8.9). The strong magnetic field may also help to generate a strong **protostellar wind**—an outward flow of particles similar to the solar wind [**Section 14.1**]. The wind can transfer additional angular momentum from the protostar to interstellar space, accounting for observations showing that older stars rotate more slowly than young ones.

**Protostellar Jets** Observations show that many young protostars fire high-speed streams of gas, or **jets**, into interstellar space (**FIGURE 16.15**). We generally see two jets shooting outward in opposite directions along the protostar's rotation axis. Sometimes the jets are lined with glowing blobs of gas, which are presumably clumps of matter swept up as the jets plow into the surrounding interstellar material.*

No one knows exactly how protostars generate these jets, but the fact that they are aligned with the disk's rotation axis indicates that angular momentum plays a large role. The most promising models for explaining the jets rely on magnetic fields to link the angular momentum of the

*These clumps of matter are often called *Herbig-Haro objects* after the two astronomers who discovered them; their nature was unknown at the time, which is why they were given the generic name "objects."

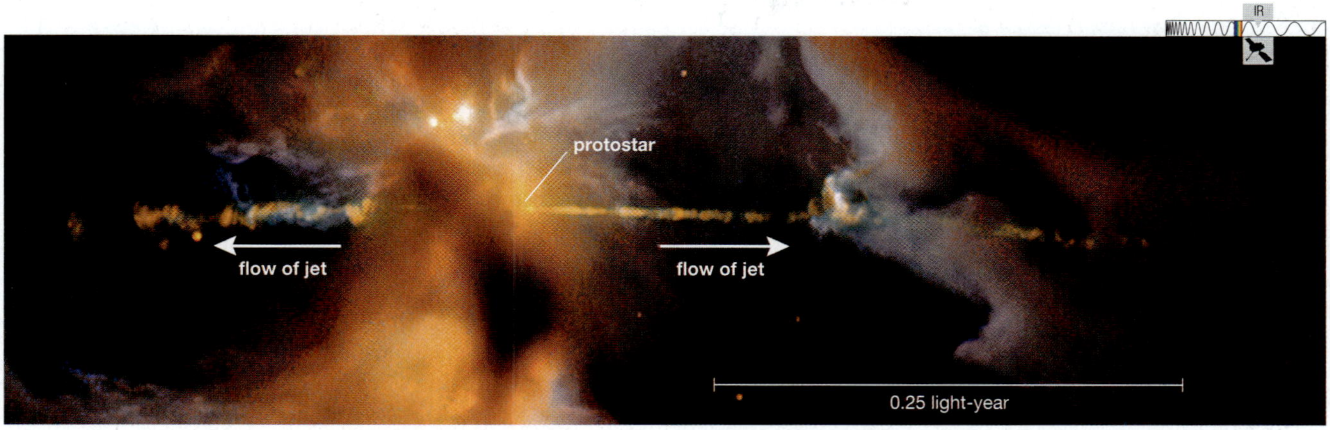


a This Hubble Space Telescope image shows two jets of material being shot in opposite directions by a protostar.

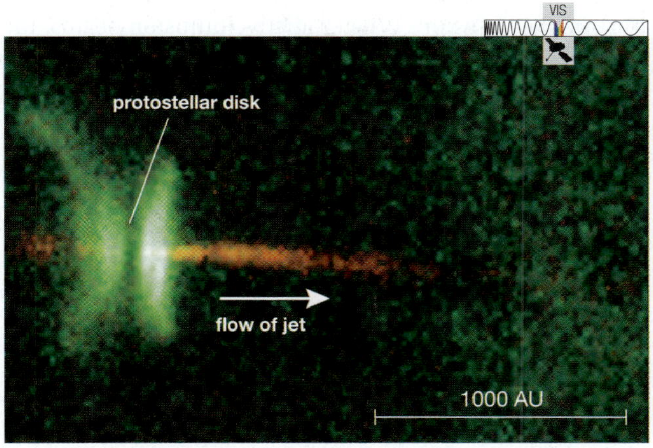


b This Hubble Space Telescope image shows a close-up view of jets (red) and a disk of gas (green) around a protostar. We are seeing the disk nearly edge-on. The top and bottom surfaces of the disk are glowing, but we cannot see the darker middle layers of the disk.

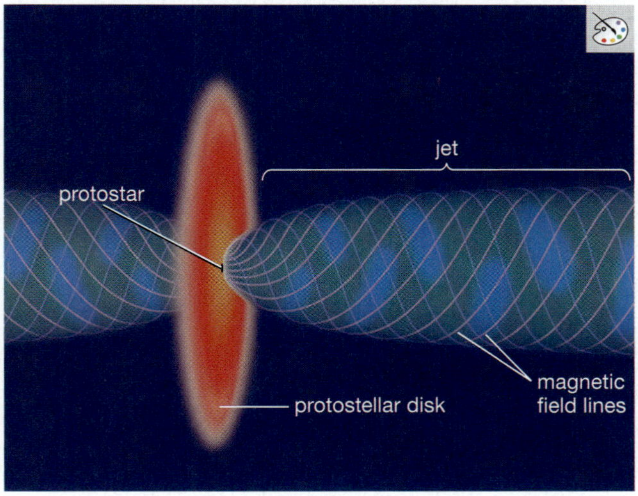


c This schematic drawing illustrates one hypothesis for how protostars create jets, which relies on the magnetic field lines thought to thread the protostellar disk. As the disk spins, it twists the magnetic field lines. Charged particles from the disk's surface can then fly outward along the twisted magnetic field lines.

**FIGURE 16.15** Many protostars fire jets of gas into interstellar space.

protostar's disk to the outflowing gas in the jet. Magnetic field lines passing through the protostellar disk get twisted into a ropelike configuration by the disk's rotation, and this twisted field may help channel jets of charged particles along the rotation axis.

Together, winds and jets can help clear away the cocoon of gas that surrounds a forming star, revealing the protostar within. They also carry away some of the protostar's angular momentum, and the fast-moving material pumps significant amounts of kinetic energy into the surrounding molecular cloud. The disruptive effects of jets are therefore likely to cause some of the turbulent motions we observe in molecular clouds.

**Single Star or Binary?** Angular momentum is also part of the reason so many stars belong to binary star systems. As a molecular cloud contracts, breaks up into fragments, and forms protostars, some of those protostars end up quite close to one another. Gravity can pull two neighboring protostars closer together, but they usually don't crash into each other. Instead, they go into orbit around each other, because each pair of protostars has a certain amount of angular momentum. Pairs with large amounts of angular momentum have large orbits, and those with smaller amounts orbit closer together.

In some cases, gravitational interactions between the binary pair of protostars and other protostars and gas clumps in their vicinity can remove angular momentum from the binary system. If that happens, then their orbit gets smaller, and the two stars can end up quite close to each other. The resulting pair is called a **close binary**. These star systems typically have orbital separations of less than 0.1 AU and orbital periods of only a few days.

## How does nuclear fusion begin in a newborn star?

Once a protostar has accreted a significant amount of mass, its interior grows quite hot. Ultimately, the temperature at the protostar's center becomes high enough for nuclear fusion. However, a protostar with a mass similar to the Sun's must wait millions of years for fusion to begin, because the process of gravitational contraction slows once the star starts trapping its thermal energy inside.

**From Protostar to the Main Sequence** The central temperature of a protostar is typically about 1 million K when its wind and jets blow away the surrounding gas, and its energy comes from gravitational contraction. To ignite fusion, the protostar needs to contract further to boost the central temperature.

Surprisingly, the key factor in allowing the central temperature to rise is radiation of energy from the protostar's surface into space. The energy going into space comes from thermal energy inside the star. Without this loss of thermal energy, the interior pressure would hold gravity at bay, so the protostar would stop contracting and its central temperature would remain fixed. The radiation from the surface allows the star to lose enough thermal energy for gravitational contraction to continue. Calculations show that a contracting protostar retains *half* the thermal energy released by gravitational contraction, which explains the rising temperature in the core.

A protostar becomes a true star when its core temperature exceeds 10 million K, making it hot enough for hydrogen fusion to operate efficiently through the *proton-proton chain* (see Figure 14.8). The ignition of fusion halts the protostar's gravitational contraction and marks what we consider the birth of a star. The newborn star's interior structure stabilizes because the energy produced in the center matches the amount radiated from its surface. The star is now a hydrogen-fusing *main-sequence star* [**Section 15.2**].

The length of time from the formation of a protostar to the birth of a main-sequence star depends on the star's mass. Massive stars do everything faster. The contraction of a high-mass protostar into a main-sequence star of spectral type O or B takes less than a million years. A star like our Sun takes about 30 million years to go from the beginning of the protostellar stage to becoming a main-sequence star. A very-low-mass star of spectral type M may spend more than a hundred million years as a protostar. The most massive stars in a young star cluster therefore can complete a whole life cycle before the smallest stars even begin to fuse hydrogen in their cores.

**The Surface Temperature of a Protostar** Although the core of a protostar heats up steadily as it contracts, the surface temperature—which is the only temperature we can observe directly—changes in a more complex way. Recall that the surface of any starlike object is its photosphere, the layer of gas from which photons can escape into space. When a protostar first begins to contract, its surface gradually heats up as the protostar shrinks. However, once the protostar contracts to the point at which its surface temperature reaches 3000 K, the temperature remains steady for most of the rest of the contraction process.

To understand why the surface temperature holds at 3000 K, we need to consider the ease with which energy flows through a protostar's outer layers. During most of the protostar's contraction, the dominant form of energy transport within its interior is convection. The convecting gas below a protostar's surface therefore rises until it reaches a layer at which the photons can escape into space. The temperature of this layer remains 3000 K even as the protostar contracts, because at greater temperatures collisions can strip electrons from hydrogen atoms. Some of these electrons then bind to neutral hydrogen atoms, making negatively charged hydrogen ions ($H^-$ ions) that are very effective at absorbing visible-light photons. As a result, photons remain trapped within the convecting gas as it rises and cools until it reaches the 3000 K layer, at which point the gas becomes fully neutral and transparent, allowing the photons to escape into space.

The situation in a contracting protostar with a mass similar to the Sun's changes when its interior grows hot enough to strip almost all of the electrons from its atoms. Radiation can then flow more freely through the star's interior, and the main energy transport mechanism through most of the protostar switches from convection to radiative diffusion. When radiative diffusion begins, the increased flow of energy through the protostar allows both the luminosity and the surface temperature of the protostar to increase until fusion ignites in the core. The ignition of fusion brings the star into energy balance, which stabilizes the luminosity and surface temperature and ends the contraction.

**Birth Stages on a Life Track** We can summarize the transitions that occur during star birth with a special type of H-R diagram. Instead of showing luminosities and surface temperatures for many different stars (as on a standard H-R diagram), this special H-R diagram shows part of a **life track** (also called an *evolutionary track*) for a single star in relation to the standard main sequence. Each point along a star's life track represents its surface temperature and luminosity at some point during its life.

**FIGURE 16.16** shows a life track leading to the birth of a $1M_{\text{Sun}}$ star like our Sun. This pre-birth period includes four distinct stages:

***Stage 1—Formation of a Protostar.*** The protostar forms within a collapsing cloud fragment. At first it is concealed by a shroud of dusty molecular gas, which is later cleared away by winds and jets. During this stage, energy moves within the protostar to the surface primarily through convection. The stage ends when the surface temperature reaches about 3000 K, placing the protostar on the right side of the H-R diagram, and the combination of this temperature and a large surface area gives it a luminosity between about $10L_{\text{Sun}}$ and $100L_{\text{Sun}}$.

***Stage 2—Convective Contraction.*** The protostar's surface temperature remains near 3000 K as long as convection remains the dominant mechanism for energy transport. While this condition holds, gravitational contraction leads to a decrease in the protostar's luminosity, because its radius becomes smaller while its surface temperature stays nearly constant. Consequently, the protostar's life track drops almost straight downward on the H-R diagram.

***Stage 3—Radiative Contraction.*** The protostar's surface temperature begins to rise when the primary energy transport mechanism switches from convection

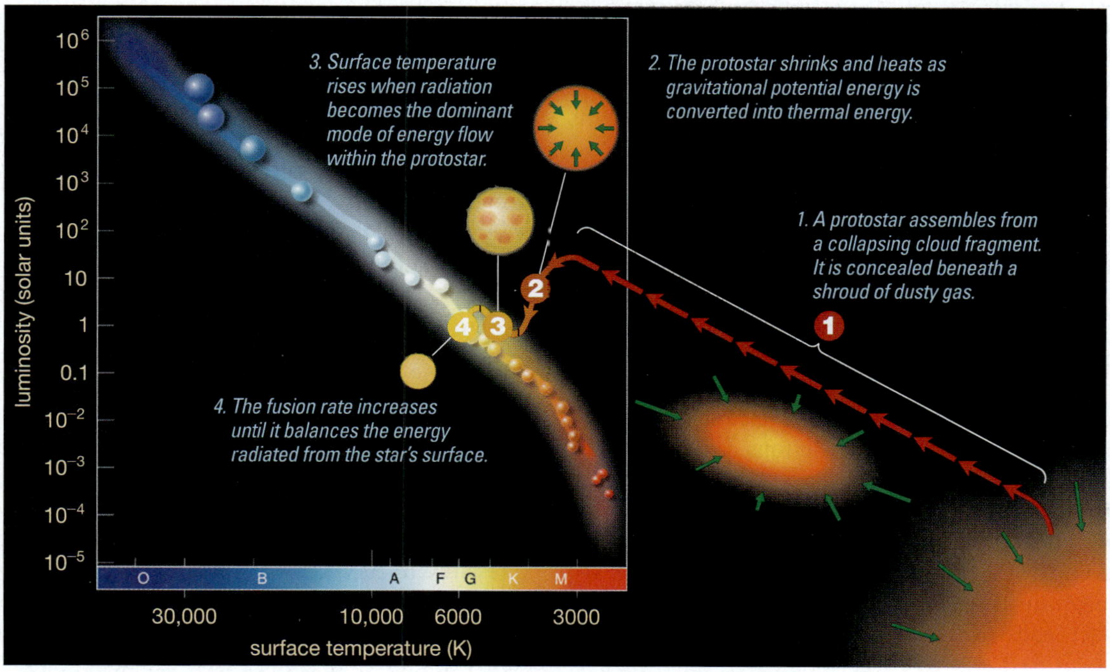


**FIGURE 16.16** The life track of a $1M_{Sun}$ star from protostar to main-sequence star.

to radiative diffusion. This rise in temperature brings a slight rise in luminosity, even though the protostar continues to contract. The life track therefore bends toward higher surface temperature and slightly higher luminosity. During this stage, hydrogen nuclei begin to fuse into helium nuclei, but the energy released is small compared with the amount of energy radiated away. The core temperature and rate of fusion increase gradually for a few tens of millions of years.

***Stage 4—Self-Sustaining Fusion.*** Fusion becomes self-sustaining when the fusion rate becomes high enough to balance the rate at which radiative energy escapes from the surface. At this point, the star settles into its hydrogen-fusing main-sequence life.

Protostars of different masses go through similar stages as they approach the main sequence, but they progress through those stages at different rates. **FIGURE 16.17** illustrates life tracks for several protostars of different masses.

**Think about it** Explain in your own words what we mean by a star's *life track*. Why do we say that Figures 16.16 and 16.17 show only *pre-main-sequence* life tracks? In general terms, predict the appearance on Figure 16.17 of pre-main-sequence life tracks for a $25M_{Sun}$ star and a $0.1M_{Sun}$ star.

## 16.3 Masses of Newborn Stars

In Chapter 15, we saw that the masses of stars span a range from about $0.08M_{Sun}$ at the low end to at least $150M_{Sun}$ at the high end. But why aren't there stars with masses outside this range? The answer once again comes down to the battle between gravity and pressure.

### What is the smallest mass a newborn star can have?

Contracting clouds with masses that are too low never become stars because their central temperatures never climb above the 10 million K threshold needed for efficient nuclear fusion. Instead, a type of pressure called **degeneracy pressure** halts gravitational contraction before hydrogen fusion can begin. Like thermal pressure,

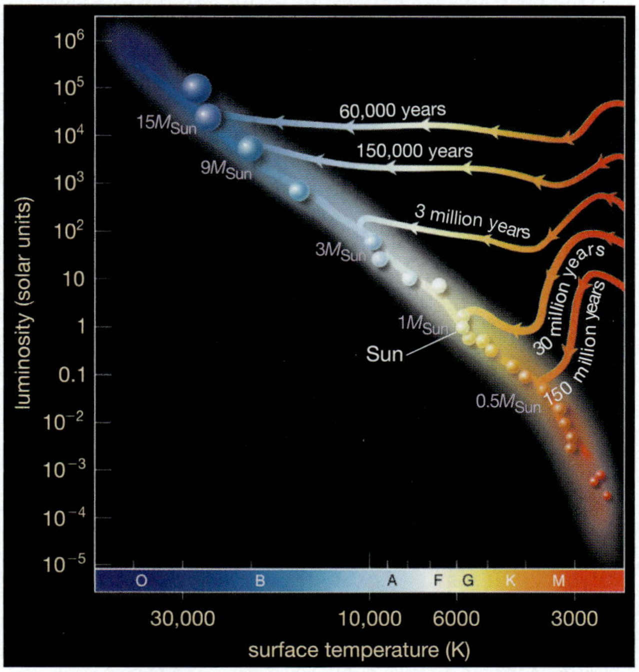


**FIGURE 16.17** The life tracks from protostar to the main sequence for stars of several different masses.

degeneracy pressure pushes outward against the force of gravity, but it differs from thermal pressure in one crucial way: Degeneracy pressure depends only on density and not on temperature. As we'll see, this feature of degeneracy pressure prevents objects with masses below about $0.08M_{Sun}$ from becoming true stars. Note that $0.08M_{Sun}$ is about 80 times the mass of Jupiter.

**The Origin of Degeneracy Pressure** We discussed degeneracy pressure in some detail in Chapter S4. The basic idea is that the laws of quantum mechanics prevent subatomic particles from getting too close together. In much the same way that electrons in atoms face restrictions that allow them to occupy only particular energy levels [**Section 5.3**], quantum laws restrict how closely electrons can be packed in a gas. Under most circumstances, these restrictions have little effect on the motions or locations of the electrons, and hence little effect on the pressure. However, in a protostar with a mass below $0.08M_{Sun}$, the electrons become packed closely enough for these restrictions to matter. We can see how this fact leads to degeneracy pressure with a simple analogy.

Imagine an auditorium in which the laws of quantum mechanics dictate the spacing between chairs, and people represent electrons (see Figure S4.11). As though playing a game of musical chairs, the people are always moving from seat to seat, just as electrons must remain constantly in motion. Most protostars are like auditoriums with many more available chairs than people, so the people (electrons) can easily find chairs as they move about. However, the cores of protostars with masses below $0.08M_{Sun}$ are like much smaller auditoriums with so few chairs that the people (electrons) fill nearly all of them. Because there are virtually no open seats, the people (electrons) can't all squeeze into a smaller section of the auditorium.

This resistance to squeezing is the origin of degeneracy pressure. If the people were really like electrons, quantum laws would also require them to move faster and faster to find open seats as they were squeezed into a smaller section. However, their speeds would have nothing to do with temperature. Degeneracy pressure and the particle motion that goes with it arise *only* because of the restrictions on where the particles can go, which is why temperature does not affect them.

**Brown Dwarfs** Because degeneracy pressure halts the contraction of a protostar with a mass less than $0.08M_{Sun}$ before the release of fusion energy can balance the energy radiated from the protostar's surface, the result is a "failed star" that slowly radiates away its internal thermal energy, gradually cooling with time. Such objects, called **brown dwarfs**, occupy a fuzzy gap between what we call a planet and what we call a star. Because degeneracy pressure does *not* rise and fall with temperature, the gradual cooling of a brown dwarf's interior does not weaken its degeneracy pressure. In the constant battle of any "star" to resist the crush of gravity, brown dwarfs are winners, albeit dim ones. Their degeneracy pressure will not diminish with time, so gravity will never gain the upper hand. Note that, despite their name, brown dwarfs radiate primarily in the infrared and actually look deep red or magenta in color (**FIGURE 16.18**).

Brown dwarfs are far dimmer than normal stars and therefore are extremely difficult to detect, even if they are quite nearby. The existence of brown dwarfs was predicted for decades before the first one was identified in 1995. Many more brown dwarfs have since been discovered (**FIGURE 16.19**).

**FIGURE 16.18** Artist's conception of a brown dwarf, orbited by a planet (to its left) in a system with multiple stars. The reddish color approximates how a brown dwarf would appear to human eyes. The bands are shown because we expect brown dwarfs to look more like giant jovian planets than stars.

**FIGURE 16.19** An infrared image showing brown dwarfs (circled) in the Orion Nebula. They are easier to spot in star-forming regions like this one than elsewhere in our galaxy, because young brown dwarfs still have much of the thermal energy left by the process of gravitational contraction. They therefore emit detectable amounts of infrared light.

Observational studies show that brown dwarfs are not all the same, so astronomers have extended the classification scheme for stars so that it can also be used for brown dwarfs. Recall that we classify ordinary stars using the spectral sequence OBAFGKM, in which spectral type O represents stars with the hottest surface temperatures and spectral type M represents stars with the lowest surface temperatures. To accommodate starlike objects cooler than M stars, astronomers have added three new spectral types, called L, T, and Y. Objects with surface temperatures between 2200 K and about 1400 K are classified as *L dwarfs,* those between 1400 K and 500 K are *T dwarfs*, and those with surfaces cooler than 500 K are *Y dwarfs.* All T and Y dwarfs are too cool to be stars and must therefore be brown dwarfs. L dwarfs seem to include both young brown dwarfs and some true hydrogen-fusing stars at the extreme lower end of the stellar mass range. The only sure way to determine whether an L dwarf is a brown dwarf or a hydrogen-fusing star is to measure its mass.

## What is the greatest mass a newborn star can have?

The maximum mass a star can have is not as well defined as the minimum mass. Observations of newly formed star clusters have identified eclipsing binary systems containing stars with masses greater than $100M_{\text{Sun}}$. These and other observations have long suggested a maximum mass of at least $150M_{\text{Sun}}$ (**FIGURE 16.20**). However, the maximum might be higher. For example, several stars in the Large Magellanic Cloud may have masses as high as $300M_{\text{Sun}}$, though their masses are uncertain because they are not in eclipsing binary systems.

The reason stars have a maximum mass is yet another form of pressure, called **radiation pressure**, which gets its name from the fact that it is caused by light. Photons of light exert a slight amount of pressure when they strike matter, because they have both energy and momentum [**Section 5.2**]. We don't ordinarily notice this pressure, because the force from any light source on Earth is quite small compared with the force of gravity. In space, however, the pressure of sunlight is significant enough that it must be taken into account in designing space missions such as the James Webb Space Telescope [**Section 6.3**]. It can in principle be used to accelerate spacecraft that unfurl large solar sails. In massive stars, some of which are millions of times more luminous than the Sun, nuclear fusion generates so much energy that the radiation bouncing around inside the star exerts a pressure stronger than thermal pressure. Theoretical models indicate that stars with masses above about $100M_{\text{Sun}}$ should generate energy so furiously that gravity cannot resist the force of radiation pressure. Such stars effectively blow away any extra mass by driving their outer layers into space.

In fact, radiation pressure is so powerful for stars with masses above $100M_{\text{Sun}}$ that it's unclear how they can form in the first place. Perhaps they form through the collisions and mergers of several smaller protostars. Such collisions can occur in the centers of particularly dense, young star clusters. Once an extremely massive star forms, it rapidly loses mass as radiation pressure drives its outer layers into space. Models of such stars indicate that they cannot last more than a few hundred thousand years before blowing themselves apart.

**FIGURE 16.20** This Hubble Space Telescope image shows the central region of the star-forming cloud NGC 3603, which has given birth to a cluster of young, massive stars. The cluster contains an eclipsing binary system in which the two stars have measured masses of about $120M_{\text{Sun}}$ and $90M_{\text{Sun}}$, making them two of the highest-mass stars for which we have a reliable mass measurement. The region shown is about 17 light-years across, located about 20,000 light-years away.

## What are the typical masses of newborn stars?

We have discussed the minimum and maximum masses of stars, but how massive is a typical newborn star? Astronomers address this question in two ways. First, they count how many stars of each mass are present in a young star cluster to determine the relative proportions of stars of each spectral type. However, this technique usually underestimates the number of low-mass stars, because these stars are not bright enough to be seen in any but the nearest of star clusters. Astronomers therefore use a second technique to get a handle on the overall number of low-mass stars: They count the number of low-mass stars in the vicinity of the Sun and use this count to estimate how many such stars exist throughout the galaxy.

The results indicate that in a newly formed star cluster, stars with low masses must greatly outnumber stars with high masses (**FIGURE 16.21**). For every star with a mass above $10M_{\text{Sun}}$, there are typically 10 stars with masses between $2M_{\text{Sun}}$ and $10M_{\text{Sun}}$, 50 stars with masses between $0.5M_{\text{Sun}}$ and $2M_{\text{Sun}}$, and a few hundred stars with masses below $0.5M_{\text{Sun}}$. We do not yet fully understand the processes that determine how many stars form in

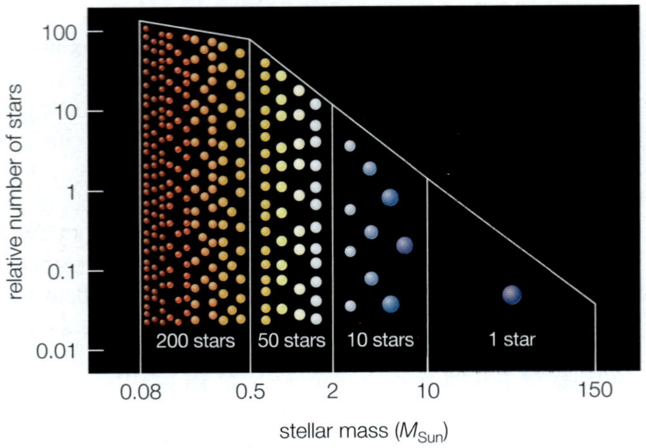


**FIGURE 16.21** Demographics of newborn stars. This schematic diagram shows how many stars of each mass are produced for every star greater than $10M_{Sun}$ in an episode of star formation. Very massive stars ($10M_{Sun}$ to $150M_{Sun}$) are relatively rare; lower-mass stars ($0.08M_{Sun}$ to $2M_{Sun}$) are much more common.

each mass range, but they probably have to do with the turbulent gas motions in the birth cloud and the subsequent pattern of fragmentation as the cloud contracts to form stars.

Notice that, although the Sun lies toward the middle of the overall range of stellar masses, most stars in a new star cluster are less massive than the Sun. High-mass stars are rare, and low-mass stars are extremely common. With the passage of time, the balance tilts even more in favor of the low-mass stars as the high-mass stars die away. We'll see in the next chapter why the more massive stars die younger, as we investigate the life stories of stars.

## The BIG Picture PUTTING CHAPTER 16 INTO CONTEXT

*In this chapter we saw how a star's life begins. As you review the chapter, remember these "big picture" ideas:*

- Star birth occurs when gravity squeezes a cloud of gas to the point that it forms a star with a core hot enough to sustain nuclear fusion. The key to understanding this process—and all other aspects of a star's life—lies in analyzing the ever-present "battle" between the inward pull of gravity and the outward push of pressure.
- Stars form in molecular clouds because these clouds have the relatively high densities and low temperatures needed to give gravity the upper hand in its battle against pressure. A single molecular cloud typically fragments into many pieces as it contracts, leaving behind a star cluster with many more low-mass stars than high-mass stars.
- Stars are born with a wide range of masses, but lower-mass stars are born in far greater numbers than higher-mass stars. The process of star formation also gives birth to brown dwarfs, which are too small to shine with energy generated by nuclear fusion, and planets often form in the disks of material that surround young stars.

**MY COSMIC PERSPECTIVE** Star birth is one of the key steps in the sequence of cosmic events that eventually led to human life. For example, many of the water molecules now in your body first formed when hydrogen and oxygen atoms combined to make water in the molecular cloud that produced the solar nebula.

## Summary of Key Concepts

### 16.1 Stellar Nurseries

- **Where do stars form?** Stars form in cold, relatively dense **molecular clouds**. Solid grains of interstellar dust prevent visible light from passing through these clouds, but we can use infrared observations to see what's going on inside them.

- **Why do stars form?** A star can form in a molecular cloud only when gravity is strong enough to overpower the outward push of **thermal pressure**. Stars tend to form in clusters because gravity can more easily overcome pressure in more massive molecular clouds. A large cloud fragments into many smaller clumps of gas as it contracts, because the advantage of gravity over pressure increases as a clump of gas shrinks in size.

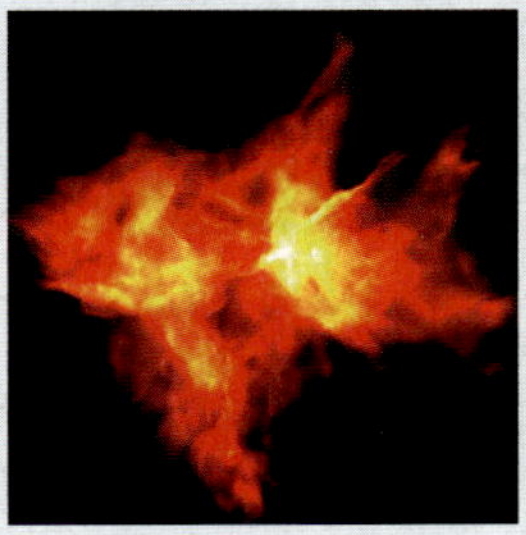

## 16.2 Stages of Star Birth

- **What slows the contraction of a star-forming cloud?** A contracting cloud begins its transformation into a star when its core starts trapping the thermal energy released by gravitational contraction. As pressure begins to push back harder, the contraction slows down and the central part of the cloud becomes a **protostar**. Meanwhile, matter from the surrounding cloud rains down on the protostar, increasing its mass.

- **What is the role of rotation in star birth?** Conservation of angular momentum ensures that a young protostar spins rapidly, and much of the material falling inward toward a protostar ends up in a spinning **protostellar disk;** planets may form in this disk. Friction in the disk can transfer angular momentum away from the inner parts of the disk, allowing gas to accrete more easily onto the protostar. Some protostars drive powerful **jets** outward along the disk's rotation axis. Along with strong **protostellar winds**, these jets can disrupt gas in the surrounding molecular cloud.

- **How does nuclear fusion begin in a newborn star?** Nuclear fusion becomes self-sustaining when a protostar's core temperature rises above 10 million K. In order to reach this temperature, the protostar must keep radiating some of its thermal energy, so that it can continue contracting. During the late phases of star formation, the protostar's luminosity declines and its surface temperature increases, and we can represent these changes with a **life track** on an H-R diagram.

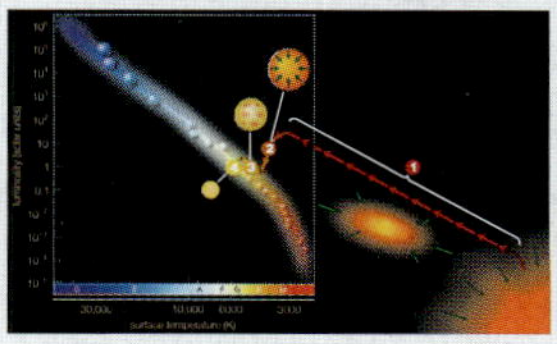

## 16.3 Masses of Newborn Stars

- **What is the smallest mass a newborn star can have?** A quantum mechanical effect known as **degeneracy pressure** halts the contraction of a protostar with a mass less than 0.08 times that of the Sun before its temperature grows high enough to sustain fusion. The minimum mass for a star is therefore $0.08M_{Sun}$. Starlike objects with masses below this limit are called **brown dwarfs**.

- **What is the greatest mass a newborn star can have?** The flood of photons coming from an extremely massive star exerts **radiation pressure** that can drive a star's outer layers into interstellar space. This form of pressure should blow apart stars with masses somewhat above $100M_{Sun}$, though the precise upper limit is uncertain.

- **What are the typical masses of newborn stars?** Low-mass stars are far more numerous than high-mass stars. For every star with a mass above $10M_{Sun}$ in a newborn star cluster, there are typically 10 stars with masses between 2 and $10M_{Sun}$, 50 stars with masses between 0.5 and $2M_{Sun}$, and a few hundred stars with masses below $0.5M_{Sun}$.

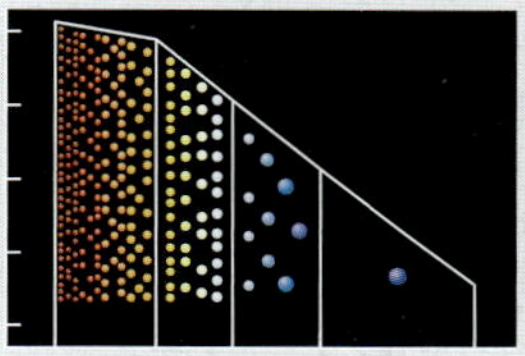

# Visual Skills Check

*Use the following questions to check your understanding of some of the many types of visual information used in astronomy. For additional practice, try the Chapter 16 Visual Quiz in the Study Area at www.MasteringAstronomy.com.*

*This frame from Figure 16.12 is from a supercomputer model of the formation of one of the first stars in the universe. Color indicates gas temperature: Blue and green represent cool gas (100–300 K), orange and red represent warm gas (500–3000 K), and yellow represents gas at intermediate temperatures (300–500 K). Use the information in the figure to answer the following questions.*

1. What is the approximate width of the region inside the black square?
2. What is the approximate diameter of the roundish region with a temperature of 300–500 K?
3. The image shows three different streams of gas that are flowing into the central cloud. Does most of the cool gas in these streams flow down the middle of the stream or along the edges of the stream?
4. Does the cold streaming gas heat up or cool down when it enters the central cloud?

# Exercises and Problems

*For instructor-assigned homework and other learning materials, go to www.MasteringAstronomy.com.*

## Chapter Review Questions

### Short-Answer Questions Based on the Reading

1. What is the *interstellar medium*? What is its chemical composition, and how do we measure it?
2. What is a *molecular cloud*? How does a molecular cloud compare in temperature and density with the rest of the interstellar medium?
3. What is *interstellar dust*? How does it interact with visible light? What are the consequences for our view of the heavens, and how is that view different in infrared light?
4. What features of molecular clouds make the conditions favorable for star formation?
5. What happens to the thermal energy released into molecular clouds as gravity makes them contract? Why doesn't it build up and stop star formation?
6. Why do stars tend to form in clusters? Describe the process by which a single cloud gives birth to an entire cluster of stars.
7. Why do we think the very first stars were much more massive than the Sun?
8. What happens to a contracting cloud when its thermal energy can no longer escape the cloud's interior in the form of photons? How does the trapped thermal energy affect the process of star formation?
9. What is a *protostar*? How does it form? Why does its mass increase with time?
10. What is a *protostellar disk*? Describe how such a disk enables additional matter to accrete onto the protostar.
11. What are *jets*? Why do we think they are related to a protostar's rotation? How do they affect the cloud surrounding the protostar?
12. Describe the final stages a protostar goes through before fusion begins in its core. How are these stages represented on a *life track*?
13. What is *degeneracy pressure*, and how does it differ from *thermal pressure*? Explain why degeneracy pressure can support a stellar core against gravity even when the core becomes very cold.
14. What is the minimum mass for a star, and why can't objects with lower masses be true stars? What is a *brown dwarf*?
15. What is the maximum mass of a star? What kind of pressure limits how massive a star can be?
16. How do the numbers of low-mass stars compare with those of higher-mass stars in new star clusters?

### Does It Make Sense?

*Decide whether the statement makes sense (or is clearly true) or does not make sense (or is clearly false). Explain clearly; not all of these have definitive answers, so your explanation is more important than your chosen answer.*

17. If you want to get a more accurate count of the number of stars in our galaxy, use an infrared telescope to observe them instead of a visible-light telescope.
18. A molecular cloud needs to trap all the energy released by gravitational contraction in order for its center to become hot enough for fusion.
19. Low-mass stars form more easily in clouds that are unusually cold and dense.
20. The current mass of any star is the same as the mass it had when it first became a protostar.
21. The rotation of a protostar always speeds up with time because a protostar is surrounded by a spinning disk.
22. Some of the stars in a star cluster live their entire lives and then die off before many of the cluster's stars initiate fusion.
23. Protostars are generally best observed in ultraviolet light because their surfaces have to get very hot before fusion can begin.
24. Degeneracy pressure exists only in objects that are very cold.
25. If Jupiter were 10 times as massive, we would consider it a brown dwarf, and if it were 100 times as massive, it would be a star.
26. Most of the stars that formed from the same cloud as the Sun have already died off.

### Quick Quiz

*Choose the best answer to each of the following. For additional practice, try the Chapter 16 Reading and Concept Quizzes in the Study Area at www.MasteringAstronomy.com.*

27. Which of these colors of light passes most easily through interstellar clouds? (a) red light (b) green light (c) blue light
28. Molecular clouds stay cool because their molecules emit photons. Which of these molecules produces the largest number of photons in a molecular cloud? (a) molecular hydrogen ($H_2$) (b) carbon monoxide (CO) (c) water ($H_2O$)
29. What happens to a cloud's thermal pressure if its temperature falls while its density rises? (a) Thermal pressure goes up. (b) Thermal pressure goes down. (c) More information is needed to determine what thermal pressure does.
30. What happens within a contracting cloud in which gravity is stronger than pressure and temperature remains constant? (a) It breaks into smaller fragments. (b) Thermal pressure starts to push back more effectively against gravity. (c) It traps all the energy released by gravitational contraction.
31. Why are the very first stars thought to have been much more massive than the Sun? (a) The clouds that made them were much more massive than today's star-forming clouds. (b) The temperatures of the clouds that made them were higher because the clouds consisted entirely of hydrogen and helium. (c) Star-forming clouds were much denser early in time.
32. What slows down the contraction of a star-forming cloud when it makes a protostar? (a) production of fusion energy (b) magnetic fields (c) trapping of thermal energy inside the protostar
33. What effects are jets and magnetic fields thought to have on a protostar? (a) They carry away angular momentum, helping the protostar grow in mass. (b) They carry away thermal energy, helping the protostar cool. (c) They transfer gravitational energy to the core, causing fusion to start.
34. Which kind of pressure prevents the formation of stars of extremely large mass? (a) thermal pressure (b) radiation pressure (c) degeneracy pressure
35. Which kinds of stars are most common in a newly formed star cluster? (a) O stars (b) G stars (c) M stars

## Inclusive Astronomy

*Use these questions to reflect on participation in science.*

36. *Group Discussion: ALMA and International Collaboration.* The world's premier observatory for studying star birth is the Atacama Large Millimeter/submillimeter Array (ALMA), located on a high plateau in the Atacama Desert of Chile (see the photo on page 165). Science agencies from many nations collaborated to build ALMA and shared the $1.4 billion cost.
**a.** There is great demand among astronomers for observing time at ALMA, which can typically execute only about 10% of the observations that astronomers propose for a given year. Discuss whether access to ALMA observing time should be granted to participating countries in proportion to the amount of money they have invested in the project or whether proposals from astronomers in every nation should be treated equally, with acceptance based only on the quality of the proposed investigation. Make a list of the pros and cons of these differing approaches.
**b.** Astronomers from Chile are given 10% of all ALMA observing time, thanks to an agreement negotiated before ALMA was built. This agreement in essence treats Chile's abundance of outstanding sites for astronomical research as a natural resource that entitles the country to compensation, which in this case takes the form of efforts to help it broaden its inclusion in astronomical research. Do you think this policy makes sense? Does it benefit the astronomical community as a whole? Why or why not?
**c.** What general obligations do you think astronomical funding agencies have to local people with ancestral ties to the land on which telescopes are built?

## The Process of Science

*These questions may be answered individually in short-essay form or discussed in groups, except where identified as group-only.*

37. *Judging an Incomplete Model.* The general picture of star formation presented in this chapter is widely accepted, but even the most detailed models for star formation cannot yet explain *all* of the observations made by astronomers. For example, they do not explain why some molecular clouds are many times larger than the minimum mass for star formation, nor do they explain the distribution of stellar birth masses. Discuss the observational evidence that supports the general picture of star formation. Then express your level of confidence in this basic view. Do you find it convincing? Why or why not?
38. *Predicting the Properties of Brown Dwarfs.* Models of star formation predict that objects less massive than $0.08M_{\text{Sun}}$ become brown dwarfs instead of true hydrogen-fusing stars. Once they have formed, these objects must cool because fusion cannot replace the thermal energy lost from their surfaces. How would you then expect the properties of brown dwarfs in an older star cluster to compare with those of brown dwarfs in a younger one? Propose an observing program that could test your hypothesis.
39. *Life in a Molecular Cloud?* As far as we know, molecular clouds are the only places other than planets that contain the kinds of complex molecules needed to support life, including water molecules and many more complex carbon compounds. Do you think it is possible for life to exist in a molecular cloud? What would life have to be like to survive there?
40. *A Star Is "Born."* Our discussion of star formation in this chapter talks about star "birth," even though stars are not really living things like humans, plants, or animals. In what sense is star birth like the birth of a living being? How is it different? Do you think it is appropriate to use the word *birth* in connection with star formation?
41. *Group Activity: Conservation Laws and Star Formation.* Note: You may wish to do this activity using the four roles described in Chapter 1, Exercise 39.
**a.** The amount of thermal energy inside a protostar increases with time, even though the protostar is losing radiative energy from its surface. How can we tell how much radiative energy the protostar is losing and how much thermal energy remains in the star? Which type of energy can we measure and which type do we infer from the law of *conservation of energy*?
**b.** A star-forming cloud flattens into a disk that spins faster as the cloud contracts because of *conservation of angular momentum.* Then material from the disk gradually spirals inward toward the newly forming star at the center, accreting onto its surface and adding to its mass. What would we find if we could measure the orbital speeds of gas particles in the accretion disk? Suppose there were no friction in the disk orbiting the protostar; would this affect the orbital speeds? Overall, how would a lack of friction change the process?

## Investigate Further

### Short-Answer/Essay Questions

42. *Interstellar Dust.* Describe how our view of the night sky would be different if there were no dust grains in the interstellar medium.
43. *Pressure vs. Gravity.* Suppose pressure and gravity are perfectly balanced within a certain molecular cloud. Describe what would happen to that balance if the temperature suddenly dropped. What would happen if the temperature suddenly rose? What would happen if the density suddenly increased without a change in temperature? What would happen if the cloud gained a little bit of mass?
44. *Protostars.* Describe the life story of a protostar from its beginning as part of a molecular cloud to the moment hydrogen fusion begins. Give as many details as possible. How would that life story be different if the protostar formed in a cloud without any angular momentum?
45. *Approach to the Main Sequence.* As a protostar approaches the main sequence, its radius shrinks because it is converting gravitational potential energy into thermal energy to replace the thermal energy it is radiating away. Figure 16.16 shows the stages that a star like the Sun goes through before it becomes a main-sequence star. How do you think the time the star spends as a $100L_{\text{Sun}}$ protostar compares with the time it spends as a $10L_{\text{Sun}}$ protostar? Explain your reasoning.
46. *Understanding Life Tracks.* Compare the life tracks of pre-main-sequence stars in Figure 16.17. Which star's luminosity changes the least as the star approaches the main sequence? Which star's luminosity changes the most? Which star's surface temperature changes the most? Which star's surface temperature changes the least?
47. *Degeneracy Pressure.* Describe how Jupiter would be different if there were no such thing as degeneracy pressure.
48. *Brown Dwarfs.* How are brown dwarfs like jovian planets? In what ways are brown dwarfs like stars?
49. *A Newborn Star Cluster.* Sketch the development of a newborn star cluster on a series of H-R diagrams. Assume that all the stars are initially protostars with a surface temperature

that places them on the extreme right-hand side of the H-R diagram in Figure 16.17. What would the H-R diagram of this star cluster look like after 100,000 years? One million years? Ten million years? One hundred million years?

50. *Light from a Newborn Star Cluster.* Suppose a new star cluster is born with one O star, 10 A stars, 100 G stars, and 1000 M stars. Which stellar type dominates the light output from the cluster? What would the color of this star cluster appear to be if you observed it from a distance so great that you could not make out the individual stars?

51. *Research: Molecules in Space.* More than a hundred different kinds of molecules have been identified in molecular clouds, including some of the molecules necessary for life. Find a list of molecules that have been discovered in space. Choose one that is necessary for life and write a short summary of how it is important to life, how it was discovered in space, and where it has been found in space.

## Quantitative Problems

*Be sure to show all calculations clearly and state your final answers in complete sentences.*

52. *Water in Molecular Clouds.* The water molecules now in your body were once part of a molecular cloud. Only about one-millionth of the mass of a molecular cloud is in the form of water molecules, and the mass density of such a cloud is roughly $10^{-21}$ g/cm$^3$. Estimate the volume of a piece of molecular cloud that has the same amount of water as your body. How does this volume compare with the volume of the entire Earth?

53. *Interstellar Dust Grains.* Dark interstellar gas clouds contain so many dust grains that starlight cannot pass through, even though the dust grains are tiny and the spaces between them are quite large by earthly standards. A typical dust grain has a radius of about $10^{-7}$ meter and a mass of about $10^{-14}$ gram.
**a.** Estimate how many dust particles there are in a cloud containing $1000M_{Sun}$ of dusty gas if 1% of the cloud's mass is in the form of dust grains.
**b.** Estimate the total surface area these grains would cover if you could put them side by side. You can assume that the grains are approximately spherical so each grain covers an area $\pi r^2$, where $r$ is the grain's radius. State your answer in square light-years.
**c.** Estimate the total surface area the cloud covers, assuming that its matter density is like that of a typical molecular cloud, about $10^{-21}$ g/cm$^3$. [*Hint:* First calculate the cloud's volume from its mass and density. Then determine the cloud's radius using the formula for the radius of a sphere, $R = (3 \times \text{volume}/4\pi)^{1/3}$.] State your answer in square light-years.
**d.** Based on your answers to parts b and c, what do you think the chances are that a photon passing through the cloud will hit a dust grain?

54. *An Isolated Star-Forming Cloud.* Isolated molecular clouds can have a temperature as low as 10 K and a particle density as great as 100,000 particles per cubic centimeter. What minimum mass does such a cloud need to form a star?

55. *Masses of the First Stars.* Models of the first star-forming clouds indicate that they had a temperature of roughly 200 K and a particle density of roughly 300,000 particles per cubic centimeter at the time they started trapping their internal thermal energy. Estimate the mass at which thermal pressure balances gravity for these values of pressure and temperature. How does that mass compare with the Sun's mass? What is the estimated lifetime of a star with that mass?

56. *Internal Temperature of the Sun.* The Sun is essentially a gas cloud in which the forces of pressure and gravity balance each other. We can therefore use the equation in Mathematical Insight 16.1 to estimate the interior temperature of the Sun from its mass and particle density.
**a.** What is the average number density of particles within the Sun, given that the average mass per particle is about $10^{-24}$ gram? (*Hint:* The volume of a sphere of radius $r$ is equal to $4\pi r^3/3$.)
**b.** What is the approximate temperature necessary for gas pressure to balance gravity within the Sun, given the average particle density from part a?
**c.** How does your estimate compare with the internal core temperature of the Sun?

57. *Internal Temperature of a Brown Dwarf.* The maximum temperature inside a brown dwarf is the temperature at which thermal pressure would balance gravity. In reality, degeneracy pressure halts the contraction of a brown dwarf before it can reach that maximum value. You can use the formula in Mathematical Insight 16.1 to estimate the maximum interior temperature of a brown dwarf with a mass of $0.05M_{Sun}$ and a radius of $0.1R_{Sun}$.
**a.** What is the average number density of particles inside the brown dwarf, given that the average mass per particle is about $10^{-24}$ gram? (*Hint:* The volume of a sphere of radius $r$ is equal to $4\pi r^3/3$.)
**b.** What is the approximate temperature necessary for gas pressure to balance gravity within the brown dwarf, given the average particle density from part a?
**c.** How does that temperature compare with the $1 \times 10^7$ K needed to sustain hydrogen fusion?
**d.** Explain how conditions inside the brown dwarf would change if you raised its mass to $0.1M_{Sun}$.

58. *Angular Momentum of a Close Binary.* Some close binary star systems have orbital periods as short as several days. Here we will estimate the angular momentum of such a system with two $1M_{Sun}$ stars in a circular orbit having an orbital period of 10 days.
**a.** Determine the average orbital separation of this system using Newton's version of Kepler's third law.
**b.** Determine the average speed of each star with respect to the center of mass of the system.
**c.** The center of mass of this system lies at the midpoint between the two stars because their masses are equal. The distance of each star from the center of mass is therefore equal to half the orbital separation. What is the angular momentum of each star with respect to the center of mass?
**d.** Suppose the Sun had the same angular momentum as these two stars combined. Estimate the speed at which the Sun's surface would be moving around the Sun's rotation axis.
**e.** How does the speed from part d compare with the escape velocity from the Sun's surface? What would happen if the Sun were spinning this fast?

59. *Mass of a Brown Dwarf.* Suppose you observe a binary system containing a main-sequence star and a brown dwarf. The orbital period of the system is 1 year, and the average separation of the system is 1 AU. You then measure the Doppler shifts of the spectral lines from the main-sequence star and the brown dwarf, finding that the orbital speed of the brown dwarf in the system is 20 times that of the main-sequence star. How massive is the brown dwarf?

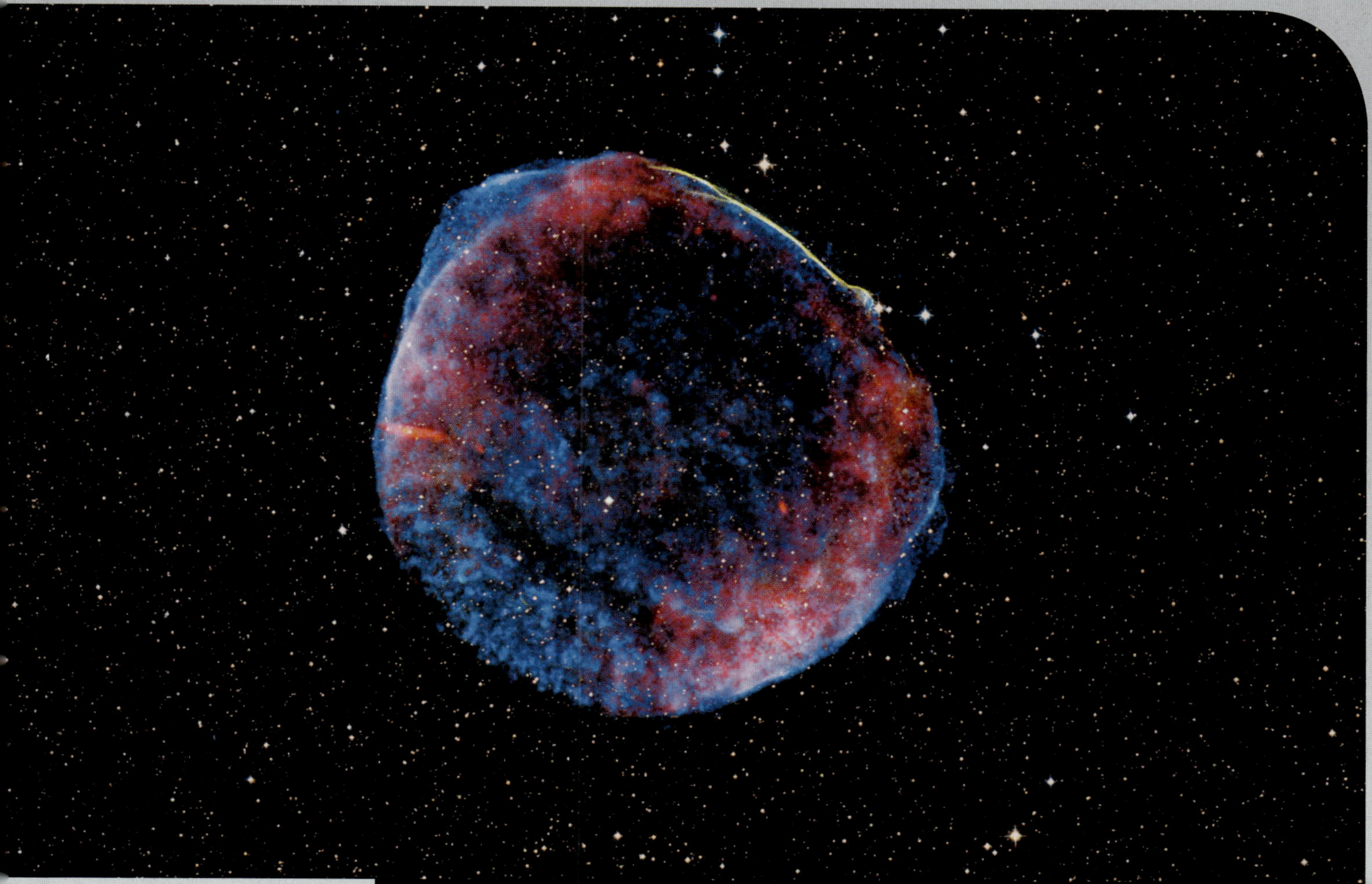

# 17 Star Stuff

▲ **About the photo:** This composite photograph, made from x-ray, radio, and visible-light images, shows a star that has exploded, ejecting newly made elements into interstellar space.

## LEARNING GOALS

**17.1 Lives in the Balance**

- How does a star's mass affect nuclear fusion?

**17.2 Life as a Low-Mass Star**

- What are the life stages of a low-mass star?
- How does a low-mass star die?

**17.3 Life as a High-Mass Star**

- What are the life stages of a high-mass star?
- How do high-mass stars make the elements necessary for life?
- How does a high-mass star die?

**17.4 The Roles of Mass and Mass Exchange**

- How does a star's mass determine its life story?
- How are the lives of stars with close companions different?

*When I look up at the night sky, I'm reminded that yes, we are part of this universe—we are in this universe. But more importantly, the universe is in us. Many people feel small, because they're small and the universe is big. But we should instead feel big, knowing that our atoms came from the stars.*

—*Neil deGrasse Tyson*

**Chapter 17 Overview**

We inhale oxygen with every breath. Iron-bearing hemoglobin in our blood carries this oxygen through our bodies. Chains of carbon and nitrogen form the backbone of the proteins, fats, and carbohydrates in our cells. Calcium strengthens our bones, while sodium and potassium ions communicate signals through our nerves. What does all this biology have to do with astronomy? The profound answer is that life is based on elements created by stars.

We've already discussed in general terms how the elements in our bodies came to exist. Hydrogen and helium were produced in the Big Bang, and stars later created heavier elements that stellar explosions scattered into space. There, in the spaces between the stars, these elements mixed with interstellar gas and became incorporated into subsequent generations of stars.

In this chapter, we will discuss the origins of the elements in greater detail by delving into the lives of stars. As you read, keep in mind that no matter how far removed the stars may seem from our everyday lives, they are connected to us in the most intimate way possible: Without the lives and deaths of stars, none of us would be here. We are truly made from "star stuff."

## 17.1 Lives in the Balance

The story of a star's life is in many ways the story of an extended battle between two opposing forces: gravity and pressure. We studied the early stages of this battle in Chapter 16. We examined the conditions under which gravity can overcome pressure in interstellar gas, causing fragments of a molecular cloud to contract into protostars, and we saw that gravity's advantage over pressure continues until fusion begins in a star's core. Once hydrogen fusion begins, the energy it generates balances the energy the star radiates into space. With energy balanced, the star's internal pressure stabilizes and halts the crush of gravity. The star is then in a state of equilibrium much like that of our Sun, with thermal pressure balancing gravity and fusion energy from the core balancing the flow of radiative energy from the star's surface [**Section 14.2**].

A star can remain in this state of balance for millions to billions of years, but it will eventually exhaust the hydrogen in its core. When that happens, fusion ceases in the core, and gravity regains the upper hand over pressure. The battle between pressure and gravity then grows increasingly dramatic, with a final outcome that depends on the star's mass at birth.

### How does a star's mass affect nuclear fusion?

We discussed the significance of stellar birth masses in Chapter 16. Main-sequence stars with large masses have much greater luminosities than ones with small masses, which means that their cores must release fusion energy at much greater rates.

Stars with large masses have greater fusion rates because they come into energy balance with higher core temperatures than stars of lower mass. All stars approach gravitational equilibrium through gravitational contraction, which converts gravitational potential energy into thermal energy. However, stars of greater mass release larger amounts of gravitational potential energy and therefore heat up more rapidly, achieving the temperatures necessary for hydrogen fusion earlier in the process of contraction. As a result, these stars come into equilibrium with a larger size, a greater luminosity, and a higher core temperature than less massive stars. The rapid increase in fusion rates as core temperature rises [**Section 14.2**] is what allows massive stars to achieve equilibrium with fusion rates far higher than those in lower-mass stars. High-mass stars therefore consume their hydrogen so rapidly that they end up with much shorter lifetimes than low-mass stars, even though they have more hydrogen for fusion. In other words, the mass of a main-sequence star determines both its luminosity and its lifetime because it determines the core temperature and fusion rate at which the star can remain in gravitational equilibrium and energy balance.

A star's mass also determines what happens when the star finally exhausts its core supply of hydrogen. Once the hydrogen is gone, fusion shuts down and the central core can no longer support itself against the crush of gravity. The core contracts, and the star's mass determines whether it eventually becomes hot enough to fuse helium or heavier elements. Even at the end of its life, when a star can no longer generate energy through fusion of any kind, its fate depends on the mass it had at birth. As we will see, relatively low-mass stars like our Sun end up as white dwarfs, while high-mass stars die violently and leave behind either a neutron star or a black hole.

To simplify our discussion of stellar lives, it's useful to divide stars into three basic groups by mass:

- **Low-mass stars** are stars born with less than about 2 solar masses ($2M_{Sun}$) of material.
- **Intermediate-mass stars** have birth masses between about 2 and 8 solar masses.
- **High-mass stars** are those stars born with masses greater than about 8 solar masses.

We will focus primarily on the dramatic differences between the lives of low- and high-mass stars. The life stages of intermediate-mass stars are quite similar to the corresponding stages of high-mass stars until the very ends of their lives, so we include them in our discussion of high-mass stars.

As we discuss the life stories of stars in detail, you might wonder how we can claim to understand what happens

**FIGURE 17.1** This star cluster (NGC 1818) is about 50 million years old. The photo shows both blue main-sequence stars and red supergiants. The supergiants must have had birth masses slightly greater than $7M_{Sun}$, because the most massive main-sequence stars remaining in the cluster are around $7M_{Sun}$.

inside distant stars over time periods of millions to billions of years. As always in science, our understanding comes from comparisons of theoretical models with detailed observations. On the theoretical side, we use mathematical models based on the known laws of physics to predict the interior structures and life cycles of stars. On the observational side, we study stars in star clusters (**FIGURE 17.1**). Recall that we can determine the ages of star clusters by finding their main-sequence turnoff points on H-R diagrams [**Section 15.3**]. By comparing clusters of different ages, we learn what stars of different masses are like at these ages. Occasionally, we even catch a star in its death throes. Our theoretical predictions of the life cycles of stars agree quite well with the observations and confirm the idea that nuclear fusion in stars has produced essentially all the elements heavier than helium. In the remainder of this chapter, we will examine in detail our modern understanding of the life stories of stars and how they manufacture the variety of elements that make our lives possible.

## 17.2 Life as a Low-Mass Star

In the grand hierarchy of stars, our Sun is rather mediocre. We should be thankful for this mediocrity. If the Sun had been a high-mass star, it would have lasted only a few million years, dying before life could have arisen on Earth. Instead, the Sun has shone steadily for about $4\frac{1}{2}$ billion years, providing the light and heat that have allowed life to thrive on our planet. Other low-mass stars have similarly long lives. In this section, we investigate the lives and deaths of low-mass stars like our Sun.

**Lives of Low-Mass Stars, Part 1**

### What are the life stages of a low-mass star?

Our Sun is in the middle of its roughly 10-billion-year life as a hydrogen-fusing, main-sequence star. We therefore expect it to continue to shine steadily for billions of years to come. Eventually, however, the Sun will exhaust its hydrogen and undergo a series of dramatic changes leading up to its death. Let's begin our study of a low-mass star's life stages with a look at what will happen to the Sun.

**Main-Sequence Life: Slow and Steady** As we discussed in Chapter 14, the Sun fuses hydrogen into helium in its core via the *proton-proton chain* [**Section 14.2**]. The Sun shines steadily because *gravitational equilibrium* (see Figure 14.2) and *energy balance* (see Figure 14.3) work together as a self-regulating *solar thermostat* that keeps the Sun's fusion rate and overall luminosity quite steady.

Other low-mass stars generate energy and shine steadily in much the same way as our Sun throughout their main-sequence lives. Models indicate that these stars also have interior structures similar to the Sun's, determined by the way energy moves through them. As in the Sun, energy released by nuclear fusion takes hundreds of thousands of years to travel from the core to the surface, moving outward through a combination of *radiative diffusion* and *convection* [**Section 14.2**]. Radiative diffusion transports energy through the random bounces of photons from one electron to another, and convection transports energy through the rising of hot plasma and the sinking of cool plasma. Once the energy reaches the surface (photosphere), it escapes to space.

The interiors of main-sequence stars differ primarily in the depth of their convection zones, with convection extending deeper in lower-mass stars (**FIGURE 17.2**). Deep inside a star like the Sun, high temperatures allow radiative diffusion

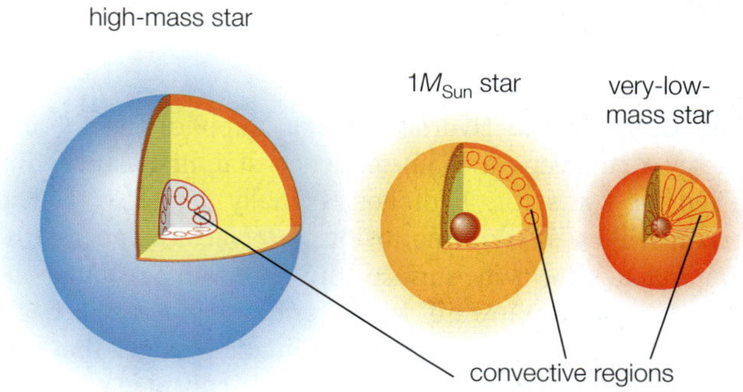


**FIGURE 17.2** Among main-sequence stars, convection zones extend deeper in lower-mass stars. High-mass stars have convective cores but no convection zones near their surfaces.

to carry energy outward at the same rate as fusion produces it in the core. Convection occurs only in the Sun's outer layers, starting at a depth about 70% of the way from the center to the surface, where the cooler temperatures make it more difficult for photons to transport energy. Stars less massive than the Sun have cooler interiors and hence deeper convection zones. In very-low-mass stars, the convection zone extends all the way down to the core. Higher-mass stars have hotter interiors and hence shallower convection zones, and the highest-mass stars have no convection zones at all near their surfaces. However, high-mass stars can have convective cores, because they produce energy so furiously that radiative diffusion cannot transport it out of the core quickly enough. Convection therefore transports energy out of the cores of these high-mass stars, but radiative diffusion takes over throughout the rest of their interiors.

Convection plays a major role in determining whether a star has activity similar to that of the sunspot cycle on our Sun [**Section 14.3**]. Recall that the Sun's activity arises from the twisting and stretching of its magnetic fields by convection and rotation. The most dramatically active stars are very-low-mass stars (spectral type M) that happen to have fast rotation rates in addition to their deep convection zones. The churning interiors of these stars are in a constant state of turmoil, twisting and knotting their magnetic field lines. When these field lines suddenly snap and reconfigure themselves, releasing energy from the magnetic field, spectacular flares can occur. For a few minutes or hours, the flare can produce more radiation in x-rays than the total amount of light coming from the star in infrared and visible light. Life on a planet near one of these **flare stars** might be quite difficult.

Aside from flares and other kinds of surface activity, the long lives of low-mass stars remain relatively uneventful as long as hydrogen fusion continues in the core. During that time, the luminosities of low-mass stars gradually rise for reasons discussed in Chapter 14. Just as in the Sun, fusion in the core of a low-mass star reduces the number of independent particles in the core: Each fusion reaction converts four independent protons into just one independent helium nucleus. As the number of particles drops, the core must shrink and heat up in order to keep pressure in balance with gravity. This slight but continual rise in core temperature slowly raises the fusion rate and therefore the luminosity of the star as it ages. Much more dramatic changes occur when nuclear fusion finally exhausts the star's central supply of hydrogen.

**Red Giant Stage** Hydrogen fusion supplies the thermal energy that keeps a main-sequence star in balance. But when the star's core hydrogen is finally depleted, nuclear fusion will cease. With no fusion to replace the energy the star radiates from its surface, the core will no longer be able to resist the inward pull of gravity, and it will begin to shrink. As an example of the dramatic changes that ultimately occur in all low-mass stars, let's consider what will happen to our own Sun as it passes through its final life stages about 5 billion years from now.

Somewhat surprisingly, the Sun's outer layers will expand outward at this time, even though its core will be shrinking under the crush of gravity. At first, the Sun's life track on an H-R diagram (**FIGURE 17.3**) will move almost horizontally to the right as the Sun grows in size to become a **subgiant**. Then, as the expansion of the outer layers continues, the Sun's luminosity will begin to increase substantially, and its life track will turn upward on the H-R diagram. Over a period of about a billion years—or about 10% as long as its main-sequence lifetime—the Sun will slowly grow in size and luminosity to become a **red giant**. At the peak of its red giant stage, the Sun will be more than 100 times as large in radius and more than 1000 times as luminous as it is today.

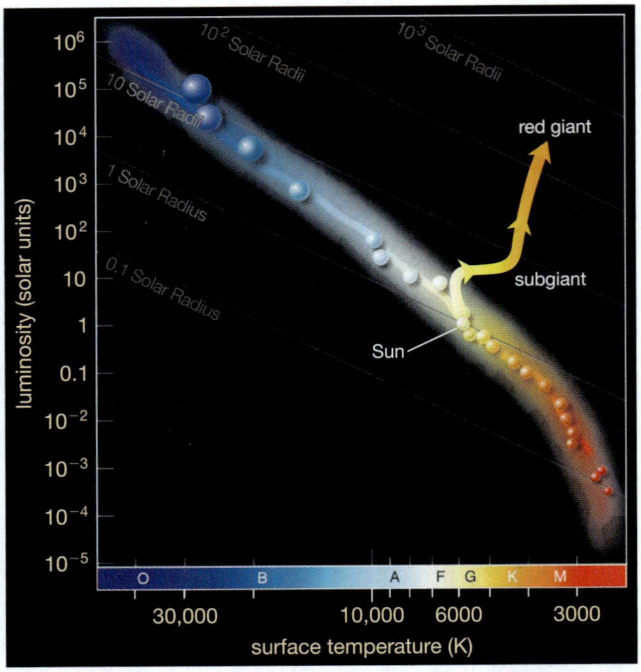


**FIGURE 17.3** The life track of a $1M_{Sun}$ star on an H-R diagram from the end of its main-sequence life until it becomes a red giant.

To understand why the Sun's outer layers will expand even while its core is shrinking, we need to think about the composition of the core at the end of the Sun's main-sequence life. After the core exhausts its hydrogen, it will be made almost entirely of helium, because helium is the product of hydrogen fusion. However, the gas surrounding the core will still contain hydrogen that has never previously undergone fusion. Because gravity shrinks both the *inert* (non-fusing) helium core and the surrounding *shell* of hydrogen, the hydrogen shell will soon become hot enough for **hydrogen shell fusion**—hydrogen fusion in a shell around the core (**FIGURE 17.4**). In fact, the shell will become so hot that hydrogen shell fusion will proceed at a much higher rate than hydrogen core fusion does today. This increase in energy output will cause a buildup of thermal pressure inside the Sun, which will push its surface outward until the luminosity rises to match the elevated fusion rate. That is why the Sun will become a huge red giant as seen from the outside, even while most of its mass remains buried deep in its shrinking core.

The situation will grow more extreme as long as the helium core remains inert. Today, the self-correcting feedback process of the solar thermostat regulates the Sun's fusion rate: A rise in the fusion rate causes the core to inflate and cool until the fusion rate drops back down

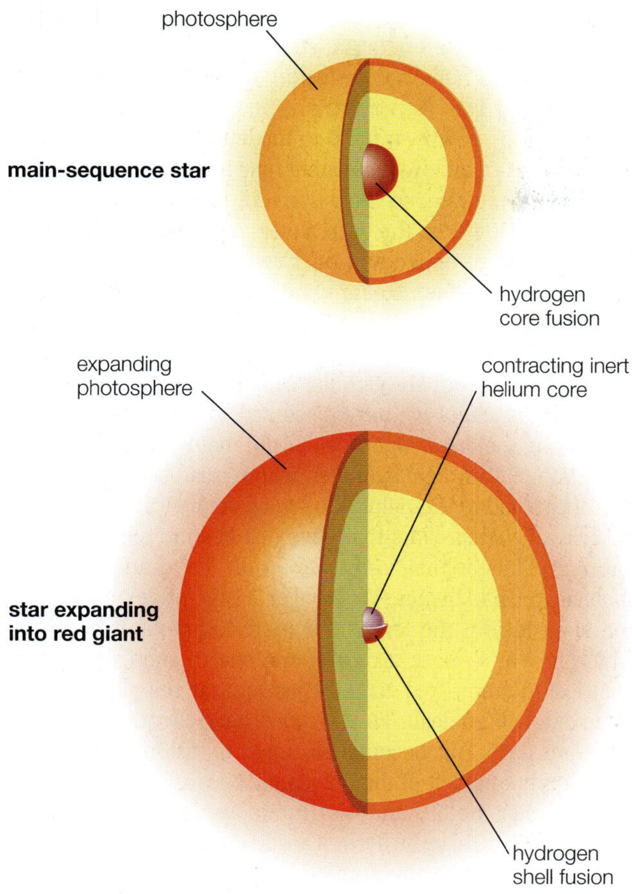


**FIGURE 17.4** After a star ends its main-sequence life, its inert helium core contracts while hydrogen shell fusion begins. The high rate of fusion in the hydrogen shell forces the star's upper layers to expand outward. (Not to scale.)

(see Figure 14.9). In contrast, thermal energy generated in the hydrogen-fusing shell of a red giant cannot do anything to inflate the inert core that lies underneath. Instead, newly produced helium keeps adding to the mass of the helium core, amplifying its gravitational pull and shrinking it further. The hydrogen-fusing shell shrinks along with the core, growing hotter and denser. The fusion rate in the shell consequently rises, feeding even more helium to the core. The star is caught in a vicious circle with a broken thermostat.

The core and shell will continue to contract and heat up—with the Sun as a whole continuing to grow larger and more luminous—until the temperature of the inert helium core reaches about 100 million K. At that point, it will be hot enough for helium nuclei to begin to fuse together, and the Sun will enter the next stage of its life. Meanwhile, the Sun's increasing radius will weaken the pull of gravity at its surface, allowing large amounts of mass to escape via the *solar wind*. Observations of *stellar winds* from red giants show that they carry away much more matter than the solar wind carries away from the Sun today, but at much slower speeds.

We expect other low-mass stars to expand into red giants in much the same way, but like all phases of stellar lives, the process occurs faster for more massive stars and more slowly for less massive stars. In fact, stars with masses much less than that of the Sun have such long main-sequence lifetimes that none of them can yet have reached the red giant stage in a 14-billion-year-old universe. However, theoretical models tell us that when these very-low-mass stars finally reach this point in the distant future, degeneracy pressure will halt the collapse of their inert helium cores before they become hot enough to fuse helium. As a result, the "dead" cores of these stars will become white dwarfs made mostly of helium, or *helium white dwarfs*.

**Think about it** Before you read on, briefly summarize why a star grows larger and brighter after it exhausts its core hydrogen. When does the growth of a red giant finally halt, and why? How would a star's red giant stage be different if the temperature required for helium fusion were around 200 million K, rather than 100 million K? Why?

**Helium Fusion** Recall that fusion occurs only when two nuclei come close enough together for the attractive *strong force* to overcome electromagnetic repulsion [**Section 14.2**]. Helium nuclei have two protons (and two neutrons) and hence a greater positive charge than a hydrogen nucleus with its single proton. The greater charge means that helium nuclei repel one another more strongly than hydrogen nuclei. **Helium fusion** therefore occurs only when nuclei slam into one another at much higher speeds than those needed for hydrogen fusion, which means that helium fusion requires much higher temperatures.

The helium fusion process (often called the *triple alpha reaction* because helium nuclei are sometimes called *alpha particles*) converts three helium nuclei into one carbon nucleus:

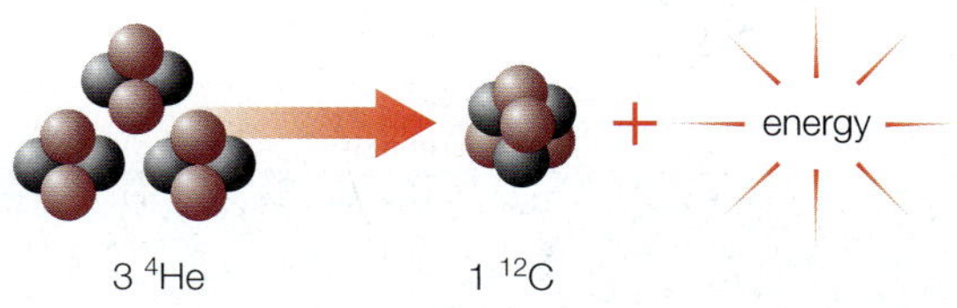


Energy is released because the carbon-12 nucleus has a slightly lower mass than the three helium-4 nuclei, and the lost mass becomes energy in accord with $E = mc^2$.

The ignition of helium fusion in a low-mass star like the Sun has one subtlety. Theoretical models show that the thermal pressure in the inert helium core is too low to counteract gravity. Instead, the pressure fighting against gravity is *degeneracy pressure*—the same type of pressure that supports brown dwarfs [**Section 16.3**]. Because degeneracy pressure does *not* increase with temperature, the onset of helium fusion heats the core rapidly without causing it to expand in size, at least at first. Instead, the rising temperature causes the helium fusion rate to spike drastically in what is called a **helium flash**.

The helium flash releases an enormous amount of energy into the core. In a matter of seconds, the temperature rises so much that thermal pressure again becomes dominant and degeneracy pressure is no longer important. In fact, the thermal pressure soon becomes strong enough to push back against gravity. The core begins to expand and pushes the hydrogen-fusing shell outward, lowering its temperature and its fusion rate. The result is that, even though helium core fusion *and* hydrogen shell fusion

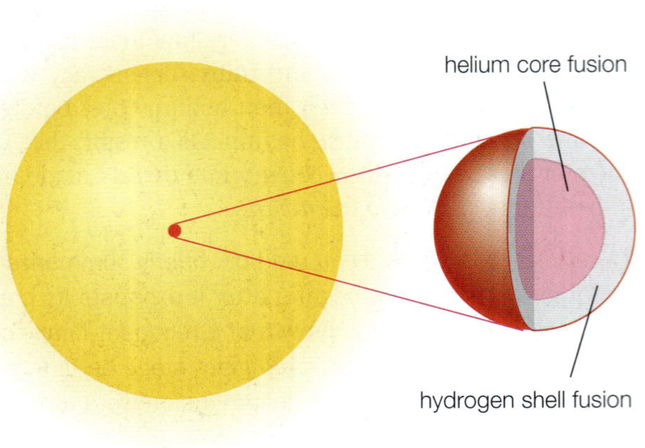


**FIGURE 17.5** Core structure of a helium core-fusion star. Helium fusion causes the core and hydrogen-fusing shell to expand and slightly cool, thereby reducing the overall energy generation rate relative to that occurring during the red giant stage. The outer layers shrink, so a helium-fusing star is smaller than a red giant of the same mass.

are taking place simultaneously in the star (**FIGURE 17.5**), energy production falls from the peak reached during the red giant stage, reducing the star's luminosity and allowing its outer layers to contract somewhat. As the outer layers contract, the star's surface temperature increases, so its color turns back toward yellow from red. After having spent about a billion years expanding into a luminous red giant, the Sun will decline in size and luminosity as it becomes a *helium core-fusion star*. With fusion once again operating in the core, the star regains the same sort of balance it had as a main-sequence star, except now it is helium fusion that keeps the central temperature steady.

Because the helium core-fusion star is now smaller and hotter than it was as a red giant, its life track on the H-R diagram drops downward and to the left (**FIGURE 17.6a**). The helium cores of all low-mass stars fuse helium into carbon at about the same rate, so these stars all have about the same luminosity. However, the outer layers of these stars can have different masses depending on how much mass they have expelled through their stellar winds. Stars that expelled more mass end up with smaller radii and higher surface temperatures and hence are farther to the left on the H-R diagram.

The H-R diagram of a globular cluster shows low-mass stars in all the life stages we have discussed so far (**FIGURE 17.6b**). Stars along the lower right, below the main-sequence turnoff point, are still in their hydrogen-fusing main-sequence stage. Just above and to the right of the main-sequence turnoff point we see *subgiants*—stars that have just begun their expansion into red giants as their cores have shut down and hydrogen shell fusion has begun. The longer a star undergoes hydrogen shell fusion, the larger and more luminous it becomes, which is why we see a continuous line of stars right up to the most luminous red giants. These are the red giants on the verge of a helium flash. The stars that have already undergone a helium flash and become helium core-fusion stars appear below and to the left of the red giants, because they are somewhat smaller, hotter, and less luminous than they were at the moment of helium flash. These helium core-fusion stars all have about the same luminosity but can differ in surface temperature, and so they trace out a horizontal line on the H-R diagram known as the **horizontal branch**.

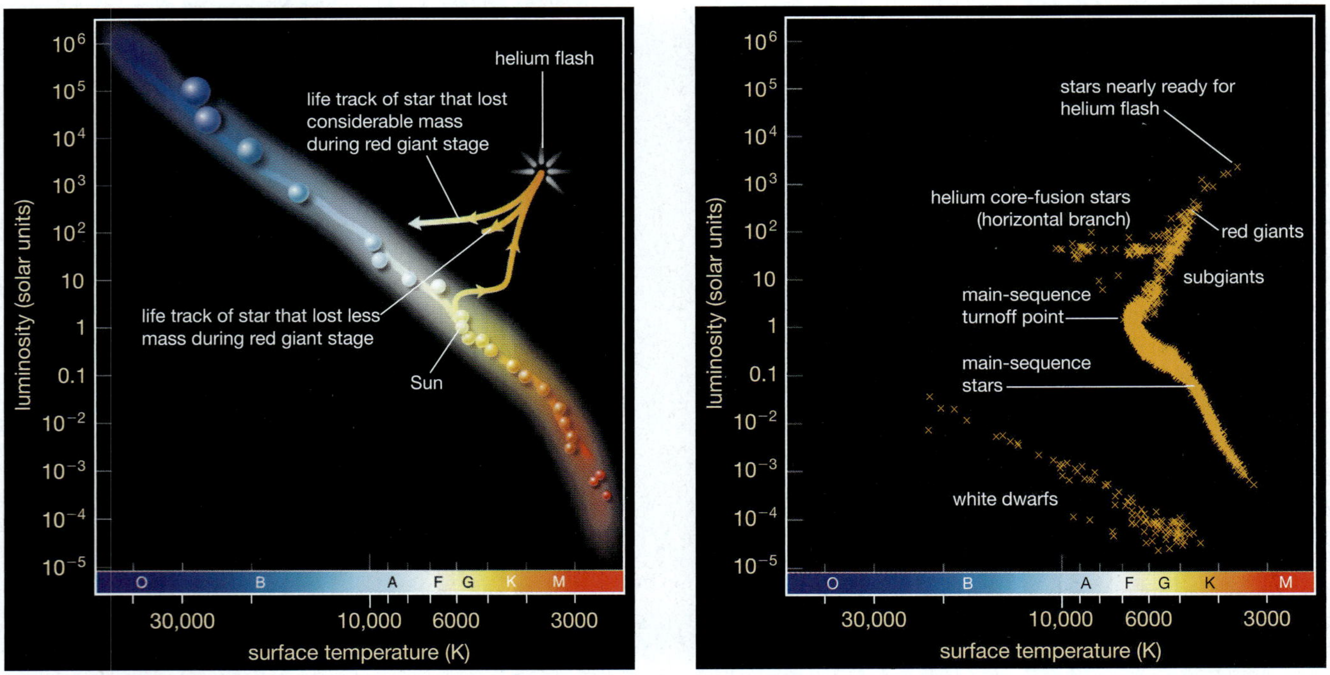


**a** Helium fusion begins with the helium flash, after which the star's surface shrinks and heats, making the star's life track move downward and to the left on the H-R diagram.

**b** An H-R diagram of the globular cluster M4 shows low-mass stars in several different life stages.

**FIGURE 17.6** After a low-mass star exhausts its core hydrogen, hydrogen shell fusion causes it to grow into a red giant. Once its helium core becomes hot enough for fusion, it temporarily settles down as a helium core-fusion star.

Lives of Low-Mass Stars, Part 2

## How does a low-mass star die?

It is only a matter of time until a helium core-fusion star converts all its core helium into carbon. In the Sun, the core helium will run out after about 100 million years—only about 1% as long as the Sun's 10-billion-year main-sequence lifetime. When the core helium is exhausted, fusion will again cease, and the star will once again go out of energy balance. The core, now made of the carbon produced by helium fusion, will begin to shrink once more under the crush of gravity.

**Last Gasps** The exhaustion of core helium will cause the Sun to expand once again, just as it did when it became a red giant. This time, the trigger for the expansion will be helium fusion in a shell around the inert carbon core. Meanwhile, hydrogen fusion will continue in a shell atop the helium layer. The Sun will have become a *double shell–fusion star*. Both shells will contract along with the inert core, driving their temperatures and fusion rates so high that the Sun will expand to an even greater size and luminosity than in its first red giant stage. Theoretical models show that helium fusion inside such a star never reaches equilibrium but instead proceeds in a series of **thermal pulses** during which the fusion rate spikes upward every few thousand years.

The furious fusion rates in the helium and hydrogen shells cannot last for more than a few million years. The Sun's only hope of extending its life at this stage lies with its carbon core, but this is a false hope for a low-mass star like the Sun. Carbon fusion is possible only at temperatures above about 600 million K, and degeneracy pressure will halt the collapse of the Sun's core before it ever gets that hot. With the carbon core unable to undergo fusion and provide a new source of energy, the Sun will have reached the end of its life.

During this final stage, the huge size of a dying star gives it only a very weak grip on its outer layers. As the star's luminosity and radius rise, increasing amounts of matter flow outward with the star's stellar wind. Meanwhile, during each thermal pulse, strong convection dredges up carbon from the core, enriching the surface of the star with carbon. Red giants with carbon-rich photospheres have cool, low-speed stellar winds, and the temperature of the gas in these winds drops with distance from the stellar surface. At a distance where the gas temperature is between about 1000 and 2000 K, some of the gas atoms in these slow-moving winds begin to stick together in microscopic clusters, forming small, solid particles of dust. The carbon-rich dust particles continue their slow drift with the stellar wind into interstellar space, where they become interstellar dust grains [**Section 16.1**]. Ultimately, most of the carbon in your body (and in all life on Earth) was manufactured in red giants and blown into space by their stellar winds.

**Think about it** Suppose the universe contained only low-mass stars. Would elements heavier than carbon exist? Why or why not?

**Planetary Nebula** The Sun's end will be beautiful to those who witness it, as long as they stay far away. Through winds and other processes, the Sun will eject its outer layers into space, creating a huge shell of gas expanding away from the inert carbon core. The exposed core will still be very hot and will therefore emit intense ultraviolet radiation. This radiation will ionize the gas in the expanding shell, making it glow brightly as a **planetary nebula**. **FIGURE 17.7** shows two of many known examples of planetary nebulae around other low-mass stars that have recently died in this way.* Note that, despite their name, planetary nebulae have nothing to do with planets. The name comes from the fact that nearby planetary nebulae look much like planets through small telescopes, appearing as simple disks.

The glow of the Sun's planetary nebula will fade as the exposed core cools and the ejected gas disperses into space. The nebula will disappear within about 100,000 years, leaving the carbon core behind as a *white dwarf*. Recall from Chapter 15 that white dwarfs are small in radius and often quite hot. We can now understand why: They are small in radius because they are the exposed cores of dead stars, supported against the crush of gravity by degeneracy pressure. They are often hot because some of them were only recently the centers of stars and have not yet had time to cool much.

In the ongoing battle between gravity and a star's internal pressure, white dwarfs are in a sort of stalemate. As long as no mass is added to the white dwarf from some other source (such as a companion star in a binary system), neither the strength of gravity nor the strength of the degeneracy pressure that holds gravity at bay will ever change. A white dwarf is therefore little more than a decaying corpse that will cool for the indefinite future, eventually disappearing from view as it becomes too cold to emit any more visible light.

**FIGURE 17.8** summarizes the life stages of a $1M_{Sun}$ star on an H-R diagram, starting from the time it reaches the main sequence and continuing until it produces a planetary nebula and leaves a white dwarf behind. We have already examined the life track to the point at which the star becomes a helium core-fusion star (see Figures 17.3 and 17.6). On this new diagram, we can see what happens after helium fusion ends in the core. The life track again turns upward as the star enters its second red giant phase, this time with energy generated by fusion in shells of both helium and hydrogen. As the star ejects the gases of the planetary nebula, the dashed curve indicates that we shift from plotting the surface temperature of a red giant to plotting the surface temperature of the exposed stellar core left behind. The curve becomes solid again near the lower left, indicating that this core is a hot white dwarf. From that point, the curve continues downward and to the right as the remaining ember cools and fades.

**The Fate of Earth** The death of the Sun will obviously have consequences for Earth, and some of these consequences will begin even before the Sun enters the final stages of its life. The Sun will gradually brighten during its remaining time as a main-sequence star, just as it has been brightening since its birth more than 4 billion years ago [**Section 14.2**]. This ongoing rise in luminosity will be small compared to what will happen during the red giant stage, but it will probably be enough to cause a *runaway greenhouse effect* [**Section 10.5**] on Earth somewhere between about 1 and 4 billion years from now, making Earth's oceans boil away.

*The most spectacular planetary nebulae appear to come from low-mass stars that are more massive than the Sun and also members of close binary star systems.

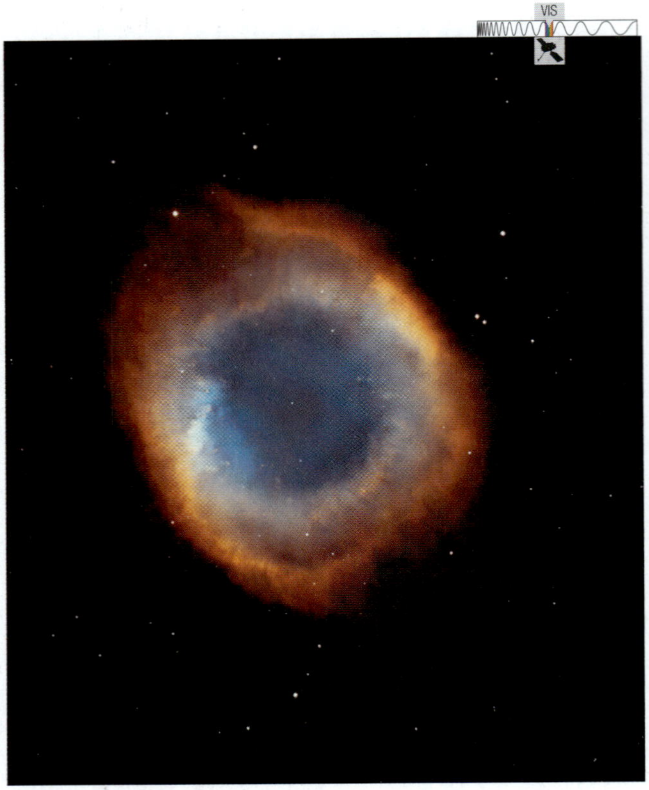


a Helix Nebula. The central white dot is the hot white dwarf.

b The Butterfly Nebula. The hot white dwarf is hidden in the dark ring of dust at the center.

**FIGURE 17.7** Hubble Space Telescope photos of planetary nebulae, which form when low-mass stars in their final death throes cast off their outer layers of gas, leaving behind the hot core that ejected the gas. The hot core ionizes and energizes the shells of gas that surround it. As the gas disperses into space, the hot core remains as a white dwarf.

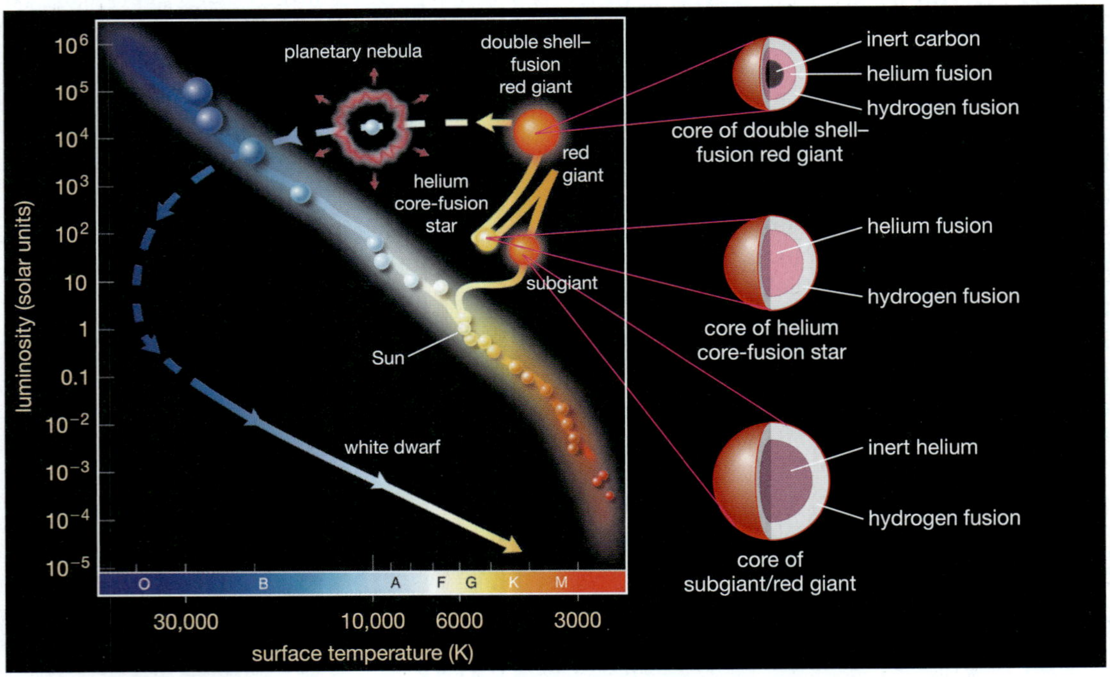


**FIGURE 17.8** The life track of a $1M_{Sun}$ star from the time it first becomes a hydrogen-fusing main-sequence star to the time it dies as a white dwarf. Core structure is shown at key stages.

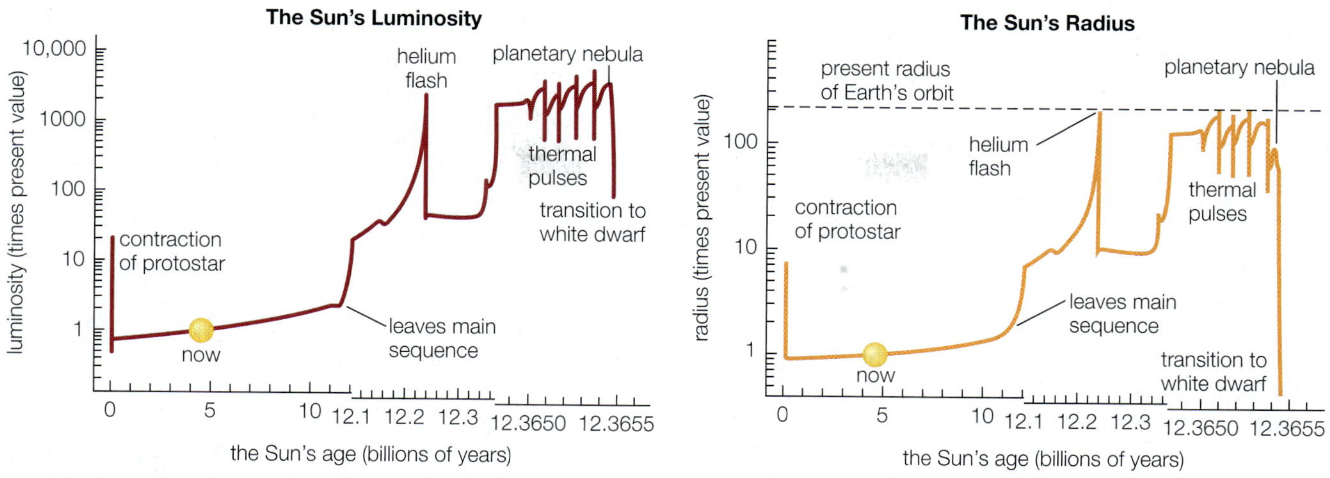


**a** Changes in the Sun's luminosity over time.

**b** Changes in the Sun's radius over time.

**FIGURE 17.9** Evolution of the Sun. These graphs show results from a theoretical model of how the Sun's luminosity (left) and radius (right) should change throughout its life. (This model gives a main-sequence lifetime of about 11 billion years, slightly greater than the more commonly quoted 10-billion-year lifetime.)

Temperatures on Earth will rise even more dramatically when the Sun finally exhausts its core supply of hydrogen, somewhere around the year A.D. 5,000,000,000, and conditions will become even worse as the Sun grows into a red giant over the next several hundred million years. Just before the helium flash, the Sun will be more than 1000 times as luminous as it is today (**FIGURE 17.9a**), and this huge luminosity will heat Earth's surface to more than 1000 K. Clearly, any surviving humans will need to have found a new home. Saturn's moon Titan [**Section 11.2**] might not be a bad choice. Its surface temperature will have risen from well below freezing to about the present temperature of Earth.

The Sun will shrink and cool somewhat after its helium flash turns it into a helium core-fusion star, providing a temporary lull in Earth's incineration. However, this respite will last only 100 million years or so, and then Earth will suffer one final disaster. After exhausting its core helium, the Sun will expand again during its last million years. Its luminosity will soar to thousands of times what it is today, and its radius will grow to nearly the present radius of Earth's orbit—so large that solar prominences might lap at Earth's surface (**FIGURE 17.9b**). Finally, the Sun will eject its outer layers as a planetary nebula that will drift away into interstellar space. If Earth is not destroyed, its charred surface will be cold and dark in the faint, fading light of the white dwarf that the Sun will become.

**Star Stuff**

## 17.3 Life as a High-Mass Star

Human life would be impossible without both low- and high-mass stars. The long lives of low-mass stars allow evolution to proceed for billions of years, but only high-mass stars produce the full array of elements on which life depends.

The early stages of a high-mass star's life are similar to the early stages of the Sun's life, except they proceed much more rapidly. But the late stages of life are quite different for high-mass stars. The cores of low-mass stars never become hot enough to fuse elements heavier than helium. Heavier nuclei contain more positively charged protons and therefore repel each other more strongly than lighter nuclei. As a result, these nuclei can fuse only at extremely high temperatures—temperatures that occur only in the cores of high-mass stars nearing the ends of their lives, when the immense weight of their overlying layers bears down on a core that has already exhausted its hydrogen fuel.

The highest-mass stars proceed to fuse increasingly heavy elements until they have exhausted all possible fusion sources. When fusion finally stops for good, gravity causes the core to implode suddenly. As we will soon see, the implosion of the core causes the star to self-destruct in the titanic explosion we call a *supernova*. The fast-paced life and cataclysmic death of a high-mass star are surely among the great dramas of the universe.

### What are the life stages of a high-mass star?

Like all other stars, a high-mass star forms out of a cloud fragment that gravity forces to contract into a protostar. Just as in a low-mass star, hydrogen fusion begins when the gravitational potential energy released by the contracting protostar makes the core hot enough. However, hydrogen fusion proceeds through a different set of steps inside high-mass stars, which is part of the reason these stars live such brief but brilliant lives.

**Hydrogen Fusion in a High-Mass Star** Recall that a low-mass star like our Sun fuses hydrogen into helium through the *proton-proton chain* (see Figure 14.8). In a high-mass star, gravitational compression makes the hydrogen core hotter than in lower-mass stars. The greater core temperature makes it possible for protons to slam into carbon, oxygen, or nitrogen nuclei as well as into other protons. Although carbon, nitrogen, and oxygen make up less than 2% of the material from which stars form in interstellar space, this

2% is more than enough to be useful in a stellar core. The carbon, nitrogen, and oxygen act as catalysts for hydrogen fusion, enabling it to proceed at a far higher rate than would be possible by the proton-proton chain alone. (A *catalyst* is something that aids the progress of a reaction without being consumed in the reaction.) This faster chain of hydrogen fusion reactions is called the **CNO cycle**, with the letters C, N, and O standing for carbon, nitrogen, and oxygen.

**FIGURE 17.10** shows the six steps of the CNO cycle. Notice that the overall reaction of the CNO cycle is the same as that of the proton-proton chain: Four hydrogen nuclei fuse into one helium-4 nucleus. The amount of energy generated in each reaction cycle therefore is also the same—it is equal to the difference in mass between the four hydrogen nuclei and the one helium nucleus multiplied by $c^2$. However, hydrogen fusion proceeds much more rapidly through the CNO cycle than through the proton-proton chain in the cores of high-mass stars.

The enormous fusion rates in high-mass stars mean there are many more photons bouncing around inside them, as well as many more released into space, and all these photons can exert a significant amount of *radiation pressure*. Recall that radiation pressure ultimately blows apart the highest-mass stars, which is why there is an upper limit to stellar masses [**Section 16.3**]. Near the photosphere of a very-high-mass star, the radiation pressure can drive a wind moving at speeds greater than 1000 kilometers per second, fast enough to cross the United States in about 5 seconds. Moreover, this wind can expel as much as 1/100,000 of a solar mass of gas per year, which may sound small but means the star would send a mass equivalent to that of our Sun hurtling into space in only 100,000 years. Such a wind cannot last long because it would blow away all the mass of even a very massive star in just a few million years.

**Think about it** Did the very first high-mass stars in the history of the universe produce energy through the CNO cycle? Explain.

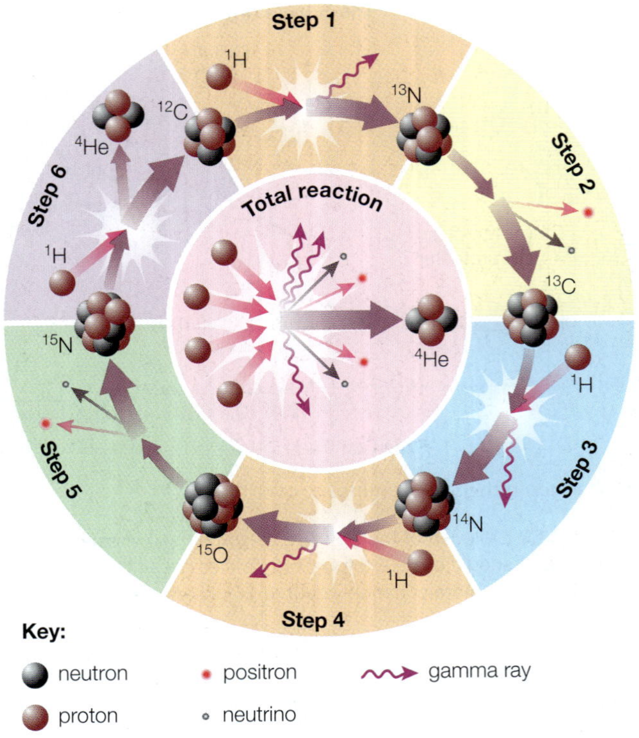


**FIGURE 17.10** This diagram illustrates the six steps of the CNO cycle by which massive stars fuse hydrogen into helium. The overall result is the same as that of the proton-proton chain: Four hydrogen nuclei fuse to make one helium nucleus. The carbon, nitrogen, and oxygen nuclei are catalysts that help the overall cycle proceed but are neither consumed nor created.

**Becoming a Supergiant** With hydrogen fusion proceeding at a fast rate via the CNO cycle, high-mass stars soon begin to run low on core hydrogen fuel. For example, a $25M_{\text{Sun}}$ star can last only a few million years as a main-sequence star. As its core hydrogen runs out, a high-mass star responds much like a low-mass star, but much faster. It develops a hydrogen-fusing shell, and its outer layers begin

## SPECIAL TOPIC How Long Is 5 Billion Years?

The Sun's demise in about 5 billion years might at first seem worrisome, but 5 billion years is a very long time. It is longer than Earth has yet existed, and human time scales pale by comparison. A single human lifetime, if we take it to be about 100 years, is only $2 \times 10^{-8}$, or two hundred-millionths, of 5 billion years. Because $2 \times 10^{-8}$ of a human lifetime is about 1 minute, we can say that a human lifetime compared to the life expectancy of the Sun is roughly the same as 60 heartbeats compared to a human lifetime.

What about human creations? The Egyptian pyramids have often been described as "eternal," but they are slowly eroding because of wind, rain, air pollution, and the impact of tourists. All traces of them will have vanished within a few hundred thousand years. While a few hundred thousand years may seem like a long time, the Sun's remaining lifetime is more than 1000 times as long.

On a more somber note, we can gain perspective on 5 billion years by considering evolutionary time scales. During the past century, our species has acquired sufficient technology and power to destroy human life totally, if we so choose. However, even if we make that unfortunate choice, some species (including many insects) are likely to survive.

Would another intelligent species ever emerge on Earth? We have no way to know, but we can look to the past for guidance. Many species of dinosaurs were biologically quite advanced, if not truly intelligent, when they were suddenly wiped out about 65 million years ago. Some small rodent-like mammals survived, and here we are 65 million years later. We therefore might guess that another intelligent species could evolve some 65 million years after a human extinction. If these beings also destroyed themselves, another species could evolve 65 million years after that, and so on.

Even at 65 million years per shot, Earth would have *nearly 80* more chances for an intelligent species to evolve in 5 billion years (5 billion ÷ 65 million = 77). Perhaps one of those species will not destroy itself, and future generations might move on to other star systems by the time the Sun finally dies. Perhaps this species will be our own.

to expand outward, ultimately turning it into a *supergiant*. At the same time, the core contracts, and this gravitational contraction releases energy that raises the core temperature until it becomes hot enough to fuse helium into carbon. However, there is no helium flash in stars of more than 2 solar masses. Their core temperatures are so high that thermal pressure remains strong, preventing degeneracy pressure from being a factor. Helium fusion therefore ignites gradually, just as hydrogen fusion did at the beginning of the star's life.

A high-mass star fuses helium into carbon so rapidly that it is left with an inert carbon core after just a few hundred thousand years. Once again, the absence of fusion leaves the core without an energy source to fight off the crush of gravity. The inert carbon core shrinks, the crush of gravity intensifies, and the core pressure, temperature, and density all rise. Meanwhile, a helium-fusing shell forms between the inert core and the hydrogen-fusing shell. The star's outer layers swell further.

Up to this point, the life stories of intermediate-mass stars ($2–8M_{Sun}$) and high-mass stars ($>8M_{Sun}$) are very similar, except that all stages proceed more rapidly in high-mass stars. However, degeneracy pressure prevents the cores of intermediate-mass stars from reaching the temperatures required to fuse carbon or oxygen and produce anything heavier. These stars eventually blow away their upper layers and finish their lives as white dwarfs. The rest of a high-mass star's life, on the other hand, is unlike anything that a low- or intermediate-mass star ever experiences.

## How do high-mass stars make the elements necessary for life?

A low-mass star can't make elements heavier than carbon because degeneracy pressure halts the contraction of its inert carbon core before it can get hot enough for fusion. A high-mass star has no such problem. The crush of gravity in a high-mass star keeps its carbon core so hot that degeneracy pressure never comes into play. After helium fusion stops, the gravitational contraction of the carbon core continues until the core reaches the 600 million K required to fuse carbon into heavier elements.

Carbon fusion provides the core with a new source of energy that restores gravitational equilibrium, but only temporarily. In the highest-mass stars, carbon fusion may last only a few hundred years. When the core carbon has been depleted, the core again begins to collapse, shrinking and heating once more until it can fuse a still heavier element. The star is engaged in the final phases of a desperate battle against the ever-strengthening crush of gravity. The star will ultimately lose the battle, but it will be a victory for life in the universe: In the process of its struggle against gravity, the star will produce the heavy elements of which Earth-like planets and living things are made.

**Fusion of Heavier Nuclei** The nuclear reactions in a high-mass star's final stages of life become quite complex, and many different reactions may take place simultaneously. The simplest sequence of fusion stages occurs through successive **helium-capture reactions**—reactions in which a helium nucleus fuses with some other nucleus (**FIGURE 17.11a**). Helium-capture reactions can change carbon into oxygen, oxygen into neon, neon into magnesium, and so on.*

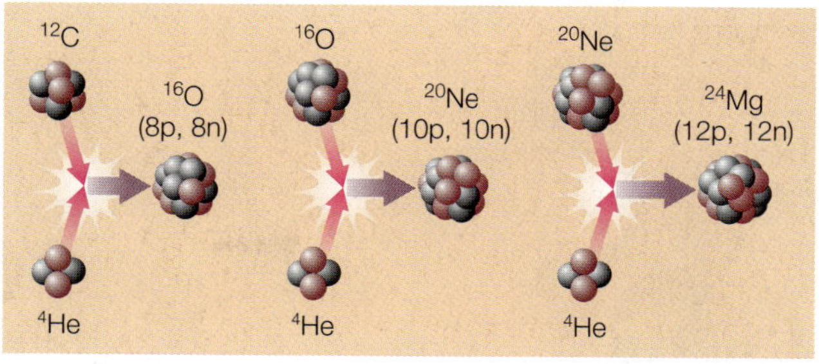


**a** Helium-capture reactions.

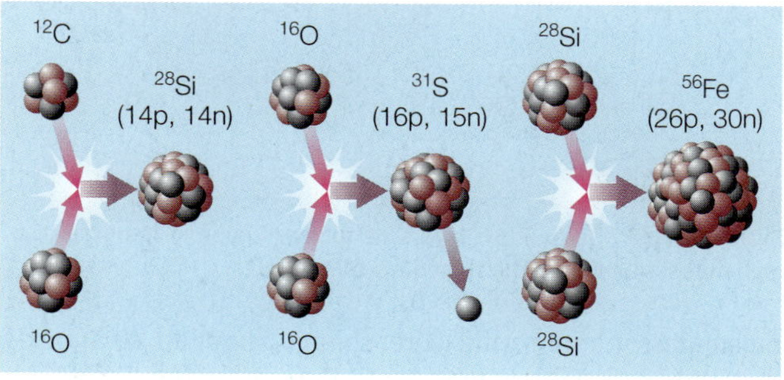


**b** Other reactions. (Note: Fusion of two silicon nuclei first produces nickel-56, which decays rapidly to cobalt-56 and then to iron-56.)

**FIGURE 17.11** A few of the many nuclear reactions that occur in the final stages of a high-mass star's life.

At high enough temperatures, a star's core can fuse heavy nuclei to one another. For example, fusing carbon to oxygen creates silicon, fusing two oxygen nuclei creates sulfur, and fusing two silicon nuclei generates iron (**FIGURE 17.11b**). Some of these heavy-element reactions release free neutrons, which may fuse with heavy nuclei to make still rarer elements. The star is forging the variety of elements that, on Earth at least, became the stuff of life.

Each time the core depletes the elements it is fusing, it shrinks and heats until it becomes hot enough for other fusion reactions. Meanwhile, a new type of shell fusion ignites between the core and the overlying shells. Near the end, the star's central region resembles the inside of an onion, with layer upon layer of shells fusing different elements (**FIGURE 17.12**). During the star's final few days, iron begins to pile up in a silicon-fusing core.

Despite the dramatic events taking place in its interior, the high-mass star's outer appearance changes slowly. As each stage of core fusion ceases, fusion in the surrounding shell intensifies and further inflates the star's outer layers. Each time the core flares up, the outer layers contract somewhat but the star's overall luminosity remains about the same. The result is that the star's life track zigzags across the top of the H-R diagram (**FIGURE 17.13**). In the

*These reactions can still proceed even after the star has used up its initial supply of core helium because there are other reactions that release helium nuclei. For example, when two carbon nuclei fuse together, the reaction can produce a neon nucleus and a helium nucleus.

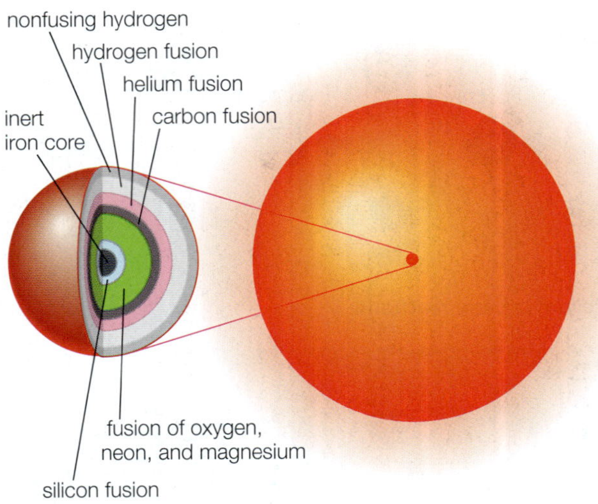


**FIGURE 17.12** The multiple layers of nuclear fusion in the core of a high-mass star during the final days of its life.

most massive stars, the core changes happen so quickly that the outer layers don't have time to respond, and the star progresses steadily toward becoming a red supergiant.

One of these massive, red supergiant stars happens to be the star Betelgeuse, the upper left shoulder of Orion. Its radius is about 900 solar radii, about four times the Sun-Earth distance, and it is a relatively nearby 600 light-years away. We have no way of knowing what stage of nuclear fusion is taking place in Betelgeuse's core as we currently see the star in the sky. It may have thousands of years of nuclear fusion still ahead, or we may be seeing it as iron piles up in its core. If the latter is the case, then sometime in the next few days we will witness one of the most dramatic events that ever occurs in the universe.

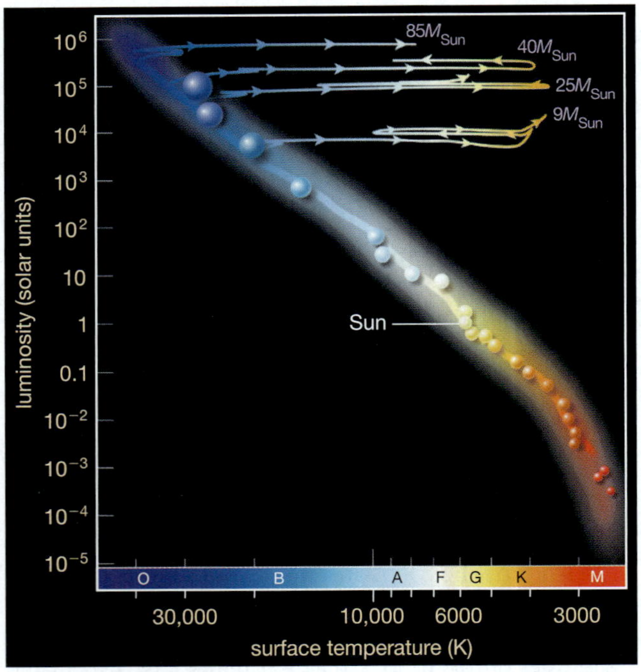


**FIGURE 17.13** Life tracks on the H-R diagram from main-sequence star to red supergiant for a few high-mass stars. Labels on the tracks give the star's mass at the beginning of its main-sequence life. Because of the strong wind from such a star, its mass can be considerably smaller when it leaves the main sequence. (Based on models from A. Maeder and G. Meynet.)

**Iron: Bad News for the Stellar Core** During its final few days, the high-mass core continues to shrink and heat as iron accumulates from nuclear fusion in the surrounding shells. If iron were like the elements in prior stages of nuclear fusion, this core contraction would stop when iron fusion ignited. However, iron is unique among the elements in a very important way: It is the one element from which it is *not* possible to generate any kind of nuclear energy.

To understand why iron is unique, remember that only two basic processes can release nuclear energy: *fusion* of light elements into heavier ones and *fission* of very heavy elements into not-so-heavy ones (see Figure 14.6). Recall that hydrogen fusion converts four protons (hydrogen nuclei) into a helium nucleus that consists of two protons and two neutrons. This fusion reaction generates energy (in accord with $E = mc^2$) because the *mass* of the helium nucleus is less than the combined mass of the four hydrogen nuclei that fused to create it, even though the total number of *nuclear particles* (protons and neutrons combined) does not change.

In other words, fusing hydrogen into helium generates energy because helium has a lower *mass per nuclear particle* than hydrogen. Similarly, fusing three helium-4 nuclei into one carbon-12 nucleus generates energy because carbon has a lower mass per nuclear particle than helium, which means that some mass disappears and becomes energy in this fusion reaction. The decrease in mass per nuclear particle from hydrogen to helium to carbon is part of a general trend shown in **FIGURE 17.14**.

The mass per nuclear particle tends to decrease as we go from light elements to iron, which means that fusion of light nuclei into heavier nuclei generates energy. This trend

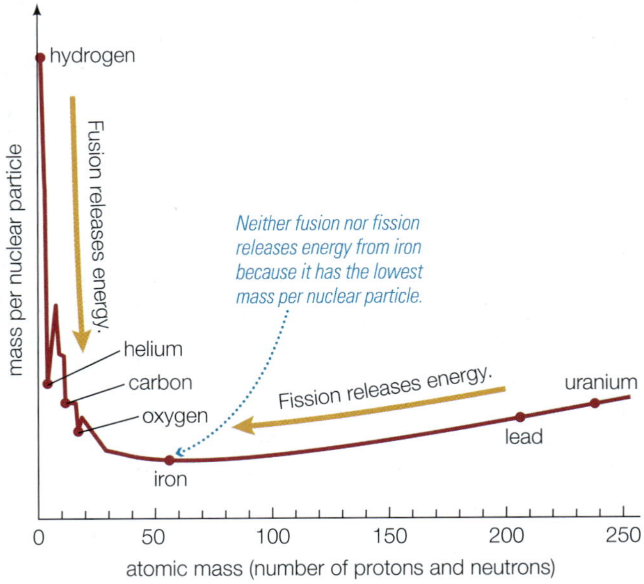


**FIGURE 17.14** Overall, the average mass per nuclear particle declines from hydrogen to iron and then increases. Selected nuclei are labeled to provide reference points. (This graph shows the most general trends only. A more detailed graph would show numerous up-and-down bumps superimposed on the general trends. The vertical scale is arbitrary, but shows the general idea.)

reverses beyond iron: The mass per nuclear particle tends to *increase* as we look to still heavier elements. As a result, elements heavier than iron can generate nuclear energy only through fission into lighter elements. For example, uranium has a greater mass per nuclear particle than lead, so uranium fission (which ultimately leaves lead as a byproduct) must convert some mass into energy.

Iron has the lowest mass per nuclear particle of all nuclei and therefore cannot release energy by either fusion or fission. Once the matter in a stellar core turns to iron, it can generate no further energy. The core's only hope of resisting the crush of gravity lies with degeneracy pressure, but iron keeps piling up until even degeneracy pressure cannot support it. What ensues is the ultimate nuclear waste catastrophe: The star explodes as a supernova, scattering all the newly made elements into interstellar space.

**Think about it** How would the universe be different if hydrogen, rather than iron, had the lowest mass per nuclear particle? Why?

**Evidence for the Origin of Elements** Before we look at how a supernova happens, let's consider the evidence that indicates we actually understand the origin of the elements. We cannot see inside stars, so we cannot directly observe elements being created. However, the signature of nuclear reactions in massive stars is written in the patterns of elemental abundances across the universe.

For example, if massive stars really produce heavy elements (that is, elements heavier than hydrogen and helium) and scatter these elements into space when they die, the total amount of these heavy elements in interstellar gas should gradually increase with time (because additional massive stars have died). We expect stars born recently to contain a greater proportion of heavy elements than stars born in the distant past, because the younger stars formed from interstellar gas that contained more heavy elements. Stellar spectra confirm this prediction: Elements besides hydrogen and helium can make up as little as 0.1% of the total mass of very old stars in globular clusters, while these elements can make up as much as 2–3% of the mass of young stars in open clusters.

We gain even more confidence in our model of element creation when we compare the abundances of various elements in the cosmos. Because helium-capture reactions add two protons (and two neutrons) at a time, we expect nuclei with even numbers of protons to outnumber those with odd numbers of protons that fall between them. Observations confirm that even-numbered nuclei such as carbon, oxygen, and neon have higher abundances than the odd-numbered ones (**FIGURE 17.15**). Similarly, we expect elements heavier than iron to be extremely rare, because they are made primarily by rare fusion reactions (most of which occur shortly before or during a supernova*). Again, observations verify this prediction.

*Nuclei heavier than iron grow by capturing neutrons, which then decay into protons, releasing an electron and an antineutrino with each decay. Because this process requires energy, it happens most rapidly in supernova explosions and in collisions between the neutron stars that supernovae leave behind [**Section 18.2**]. However, it can also happen slowly to heavy nuclei in the helium-fusing zones of giant and supergiant stars.

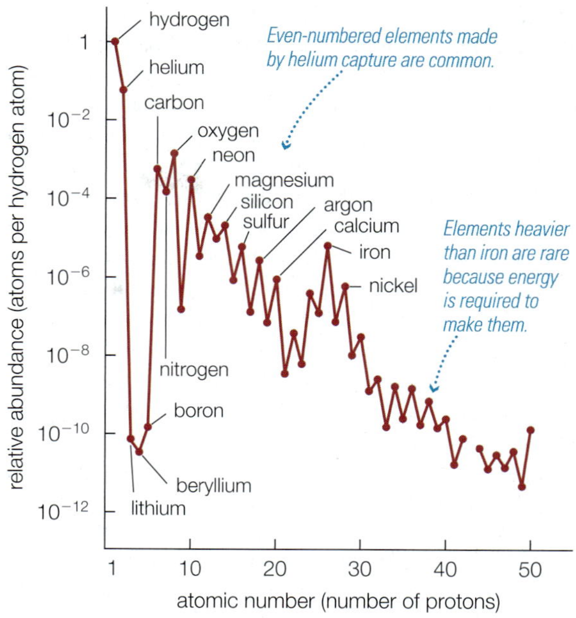


**FIGURE 17.15** The observed abundances of elements in the Milky Way, relative to the abundance of hydrogen (set to 1 in this comparison). For example, the graph shows a nitrogen abundance of about $10^{-4}$, or 1/10,000, which means that there are about 10,000 times as many hydrogen atoms in the galaxy as nitrogen atoms.

## How does a high-mass star die?

Let's return now to our high-mass star, with iron piling up in its core. As we've discussed, it has no hope of generating any energy by fusion of this iron. After shining brilliantly for a few million years, the star will not live to see another day.

**The Supernova Explosion** The degeneracy pressure that briefly supports the inert iron core arises because the laws of quantum mechanics prohibit electrons from getting too close together [**Section S4.4**]. Once gravity pushes the electrons past the quantum mechanical limit, however, they can no longer exist freely. In an instant, the electrons disappear by combining with protons to form neutrons, releasing neutrinos in the process (**FIGURE 17.16**). The degeneracy pressure provided by the electrons instantly vanishes, and gravity has free rein.

In a fraction of a second, an iron core with a mass comparable to that of our Sun and a size larger than that of Earth collapses into a ball of neutrons just a few kilometers across. The collapse halts only because the neutrons have a degeneracy pressure of their own. The entire core then resembles a giant atomic nucleus. If you recall that ordinary atoms are made almost entirely of empty space [**Section 5.3**] and that

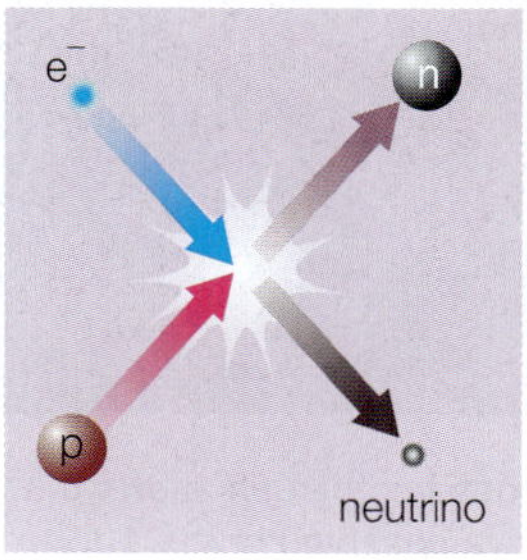


**FIGURE 17.16** During the final, catastrophic collapse of a high-mass stellar core, electrons and protons combine to form neutrons, a process accompanied by the release of neutrinos.

almost all of their mass is in their nuclei, you'll realize that a giant atomic nucleus must have an astoundingly high density.

The gravitational collapse of the core releases an enormous amount of energy—more than 100 times what the Sun will radiate over its entire 10-billion-year lifetime. Where does this energy go? It drives the star's outer layers into space in a titanic explosion called a **supernova**. The ball of neutrons left behind is called a **neutron star**. In some cases, the remaining mass may be so large that gravity also overcomes neutron degeneracy pressure, and the core continues to collapse until it becomes a *black hole* [**Sections S3.3, 18.3**].

Theoretical models of supernovae successfully reproduce the observed energy outputs of real supernovae, but the precise mechanism of the explosion is not yet clear. Two general processes could contribute to the explosion. In the first process, neutron degeneracy pressure halts the gravitational collapse, causing the core to rebound slightly and ram into overlying material that is still falling inward. However, current models of supernovae suggest that rebound also needs to be energized by the neutrinos formed when electrons and protons combine to make neutrons. Although neutrinos rarely interact with anything [**Section 14.2**], so many are produced when the core implodes that they can add enough energy to the shock wave produced by the rebounding core to propel the star's upper layers outward at a speed of 10,000 kilometers per second—fast enough to travel the distance from the Sun to Earth in only about 4 hours.

The heat of the explosion makes the gas shine with dazzling brilliance. For about a week, a supernova blazes as powerfully as 10 billion Suns, rivaling the luminosity of a moderate-size galaxy. The ejected gases slowly cool and fade in brightness over the next several months, continuing to expand outward until they eventually mix with other gases in interstellar space. The scattered debris from the supernova carries with it the variety of elements produced in the star's nuclear furnace, as well as additional elements created when some of the neutrons produced during the core collapse slammed into other nuclei. Millions or billions of years later, this debris may be incorporated into a new generation of stars. We are truly "star stuff," because we and our planet were built from the debris of stars that exploded long ago.

**See it for yourself** To see an effect similar to the core-rebound process in a supernova, find a tennis ball and a basketball. Place the tennis ball directly on top of the basketball and drop them together on a hard floor. How does the speed at which the tennis ball bounces back up compare with the speed at which it fell? How is the response of the tennis ball like the response of the supernova's outer layers to the rebound of the core?

**Historical Supernovae Observations** The study of supernovae owes a great debt to astronomers of many different epochs and cultures. Careful scrutiny of the night skies allowed ancient people to identify several supernovae whose remains can still be seen. The most famous example is the Crab Nebula in the constellation Taurus. The Crab Nebula is a **supernova remnant**—an expanding cloud of debris from a supernova explosion (**FIGURE 17.17**).

**FIGURE 17.17** This Hubble Space Telescope photograph shows the Crab Nebula, the remnant of the supernova observed in A.D. 1054.

A spinning neutron star lies at the center of the Crab Nebula, providing evidence that supernovae really do create neutron stars. Photographs taken years apart show that the nebula is growing larger at a rate of several thousand kilometers per second. Calculating backward from its present size, we can trace the supernova explosion that created it to somewhere near A.D. 1100. Thanks to observations made by ancient astronomers, we can be even more precise.

The official history of the Song Dynasty in China contains a record of a remarkable celestial event, the sudden appearance and gradual dimming of a "guest star" in a location corresponding to the Crab Nebula. The Chinese date of that event corresponds to July 4, 1054, telling us precisely when the Crab supernova became visible from Earth. Descriptions of this particular supernova also appear in Japanese astronomical writings and in an Arabic medical textbook. Some people have claimed that it is even recorded in Native American paintings in the southwestern United States, but these claims now seem doubtful. Curiously, European records do not mention this supernova, even though it would have been clearly visible.

Other historical records of supernovae have allowed us to age-date additional supernova remnants, which in turn allows us to determine the kinds of supernovae that produced the remnants and to assess how frequently stars explode in our region of the Milky Way Galaxy. At least four supernovae have been observed during the past thousand years, appearing as brilliant new stars for a few months in the years 1006, 1054, 1572, and 1604. The supernova of 1006, the brightest of these four, could be seen during the daytime and cast shadows at night.

**Think about it** When Betelgeuse explodes as a supernova, it will be as bright as the full moon in our sky. If our ancestors had seen Betelgeuse explode a few hundred or a few thousand years ago, do you think it could have had any effect on human history? How do you think our modern society would react if we saw Betelgeuse explode tomorrow?

**Modern Supernova Observations** No supernova has been seen in our own galaxy since 1604, but modern astronomers routinely discover supernovae in other galaxies. The nearest of these extragalactic supernovae, and the only one near enough to be visible to the naked eye, burst into view in 1987 (**FIGURE 17.18**). Because it was the first supernova detected that year, it was given the name **Supernova 1987A**. This stellar explosion was located in the *Large Magellanic Cloud,* a small galaxy that orbits the Milky Way and is visible only from Earth's southern latitudes. The Large Magellanic Cloud is about 150,000 light-years away, so the star really exploded some 150,000 years ago.

As the nearest supernova witnessed in four centuries, Supernova 1987A provided a unique opportunity to study a supernova and its debris in detail. Astronomers from all over the world traveled to the Southern Hemisphere to observe it, and several orbiting spacecraft added observations in many different wavelengths of light.

One of the most remarkable findings from Supernova 1987A was a burst of neutrinos, recorded by neutrino detectors in Japan and Ohio. The neutrino data confirmed that the explosion released most of its energy in the form of neutrinos, suggesting that we are correct in believing that the stellar core undergoes sudden collapse to a ball of neutrons. The capture of neutrinos from Supernova 1987A also spurred scientific interest in building more purposefully designed "neutrino telescopes." Many are now operating or are in development, offering a new way to study events in the distant universe.

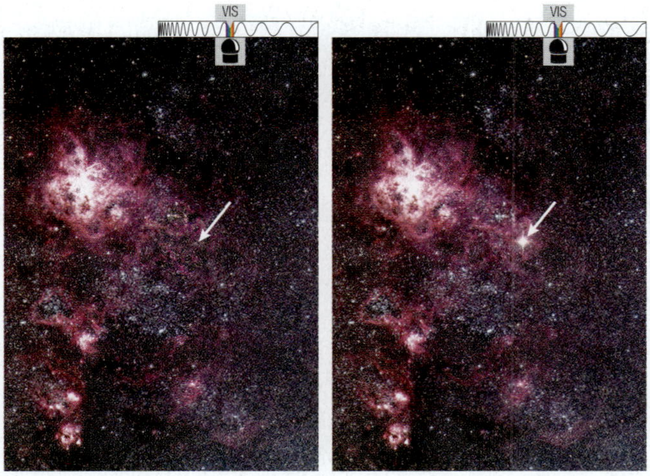


**Before.** The arrow points to the star observed to explode in 1987.

**After.** For a few weeks, the supernova shined brightly enough to be visible to the naked eye and then gradually faded.

**FIGURE 17.18** Before and after photos of the location of Supernova 1987A.

## 17.4 The Roles of Mass and Mass Exchange

Throughout this chapter, we have focused on the key role of birth mass in determining a star's destiny. However, we have so far treated stars as if they all lived in isolation, even though nearly half the stars we see in the sky are actually binary star systems. In this final section, we will first summarize what we have learned about the lives of stars when they are not part of close binary star systems. We will then examine how a star's life story can change if it happens to orbit another star closely enough that mass can sometimes flow from one star to the other.

### How does a star's mass determine its life story?

A star's birth mass determines its life cycle because that mass governs how nuclear fusion progresses in the core. Fusion proceeds relatively slowly in low-mass stars and does not make elements much heavier than carbon. These stars therefore live long lives and die in planetary nebulae, leaving behind white dwarfs composed mostly of carbon. Fusion reactions proceed somewhat faster in the hotter cores of intermediate-mass stars. However, these stars never manage to make iron and also die in planetary nebulae, leaving white dwarfs that can contain elements heavier than carbon. Fusion proceeds most quickly in the very hot cores of high-mass stars, eventually leading to iron production. When too much iron accumulates, the high-mass star explodes as a supernova, leaving behind a neutron star or a black hole.

**FIGURE 17.19** summarizes the stages in the life cycles of stars by focusing on two illustrative cases: a high-mass star with a mass 25 times that of our Sun and a low-mass star with the same mass as our Sun. Remember that all types of stars grow larger and redder when they exhaust their core hydrogen, and during these late stages they make the elements necessary for human existence. Much of the carbon in our bodies was made in low- and intermediate-mass stars and then blown into space by their stellar winds and planetary nebulae. Most of the heavier elements that our bodies rely on were made in high-mass stars and expelled in supernova explosions.

### How are the lives of stars with close companions different?

For the most part, stars in binary systems proceed from birth to death as if they were isolated. However, exceptions can occur in close binary star systems. Algol, the "demon star" in the constellation Perseus, is a good example. It appears as a single star to our eyes and telescopes, but it is actually an eclipsing binary star system [**Section 15.1**] consisting of two stars that orbit each other closely: a $3.7M_{Sun}$ main-sequence star and a $0.8M_{Sun}$ subgiant.

**COSMIC Context**

## FIGURE 17.19 Summary of Stellar Lives

**All stars spend most of their time as main-sequence stars and then change dramatically near the ends of their lives. This figure shows the life stages of a high-mass star and a low-mass star, using the cosmic calendar from Chapter 1 to illustrate the relative lengths of these life stages. On this calendar, the 14-billion-year lifetime of the universe corresponds to a single year. *(Stars not to scale.)***

### LIFE OF A HIGH-MASS STAR ($25M_{Sun}$)

*This high-mass star goes from protostar to supernova in about 6 million years, corresponding to less than 4 hours on the cosmic calendar.*

① **Protostar:** A star system forms when a cloud of interstellar gas collapses under gravity.

② **Blue main-sequence star:** In the core of a high-mass star, four hydrogen nuclei fuse into a single helium nucleus by the series of reactions known as the CNO cycle.

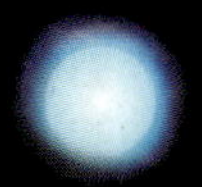

③ **Red supergiant:** After core hydrogen is exhausted, the core shrinks and heats. Hydrogen fusion begins around the inert helium core, causing the star to expand into a red supergiant.

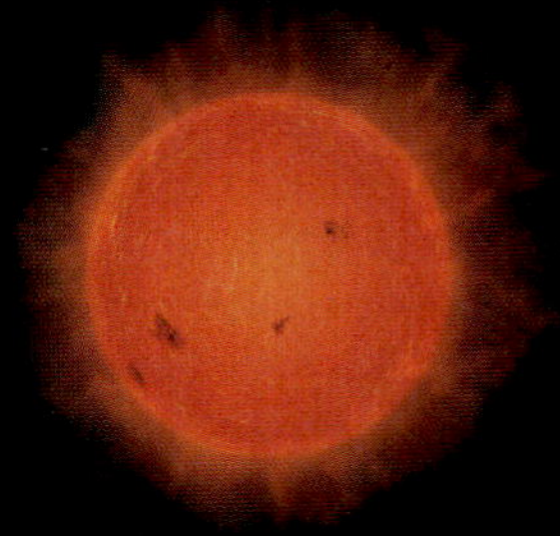

| | ① | ② | ③ |
|---|---|---|---|
| **Actual Length of Stage** | 40,000 years | 5 million years | 100,000 years |
| **Time on Cosmic Calendar** | 12:00:00 AM → 12:01:30 AM | 12:01:30 AM → 3:10:00 AM | 3:10:00 AM → 3:14:00 AM |

*These times correspond to the life stages of a $25M_{Sun}$ star born around midnight on a typical day of the cosmic calendar.*

### LIFE OF A LOW-MASS STAR ($1M_{Sun}$)

*This low-mass star goes from protostar to planetary nebula in about 11.5 billion years, corresponding to 10 months on the cosmic calendar.*

① **Protostar:** A star system forms when a cloud of interstellar gas collapses under gravity.

② **Yellow main-sequence star:** In the core of a low-mass star, four hydrogen nuclei fuse into a single helium nucleus by the series of reactions known as the proton–proton chain.

③ **Red giant star:** After core hydrogen is exhausted, the core shrinks and heats. Hydrogen fusion begins around the inert helium core, causing the star to expand into a red giant.

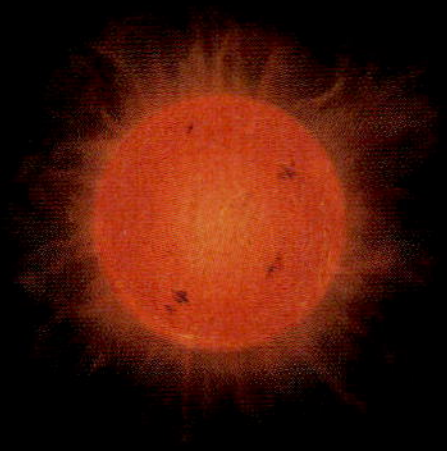

| | ① | ② | ③ |
|---|---|---|---|
| **Actual Length of Stage** | 30 million years | 10 billion years | 1 billion years |
| **Time on Cosmic Calendar** | March 1 → March 2 | March 2 → November 30 | November 30 → December 27 |

*These dates correspond to the life stages of a $1M_{Sun}$ star born in early March on the cosmic calendar.*

4 **Helium core-fusion supergiant:** Helium fusion begins when the core temperature becomes hot enough to fuse helium into carbon. The core then expands, slowing the rate of hydrogen fusion and allowing the star's outer layers to shrink.

1 million years

3:14:00 AM → 3:52:00 AM

5 **Multiple shell–fusion supergiant:** After the core runs out of helium, it shrinks and heats until fusion of heavier elements begins. Late in life, the star fuses many different elements in a series of shells while iron collects in the core.

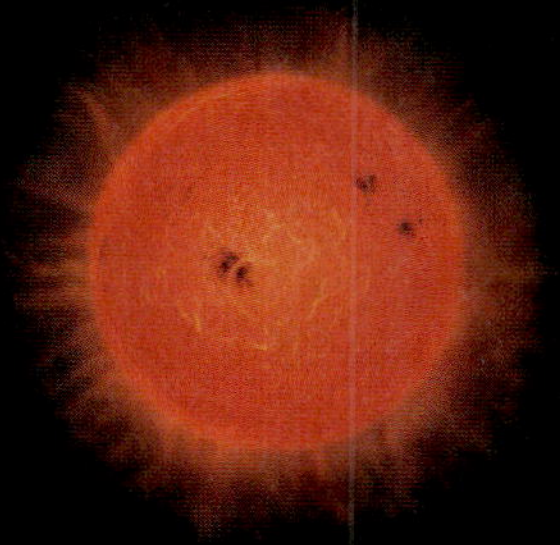

10,000 years

3:52:00 AM → 3:52:23 AM

6 **Supernova:** Iron cannot provide fusion energy, so it accumulates in the core until degeneracy pressure can no longer support it. Then the core collapses, leading to the catastrophic explosion of the star.

a few months

3:52:23 AM

7 **Neutron star or black hole:** The core collapse forms a ball of neutrons, which may remain as a neutron star or collapse further to make a black hole.

indefinite

–

4 **Helium core-fusion star:** Helium fusion begins when the core becomes hot enough to fuse helium into carbon. The core then expands, slowing the rate of hydrogen fusion and allowing the star's outer layers to shrink.

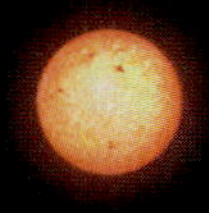

100 million years

December 27 → December 30

5 **Double shell–fusion red giant:** Helium fusion begins around the inert carbon core after the core helium is exhausted. The star then enters its second red giant phase, with fusion in both a hydrogen shell and a helium shell.

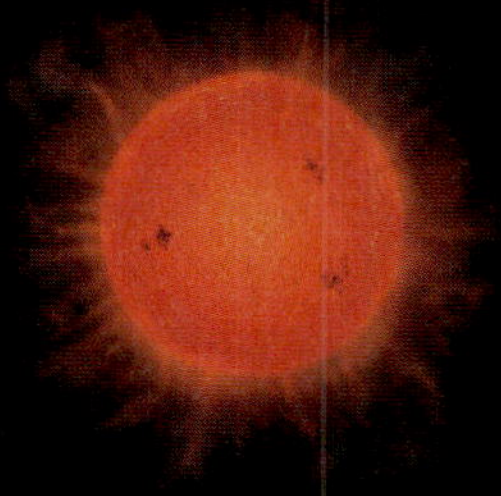

30 million years

December 30 → December 31

6 **Planetary nebula:** The dying star expels its outer layers in a planetary nebula, leaving behind the exposed inert core.

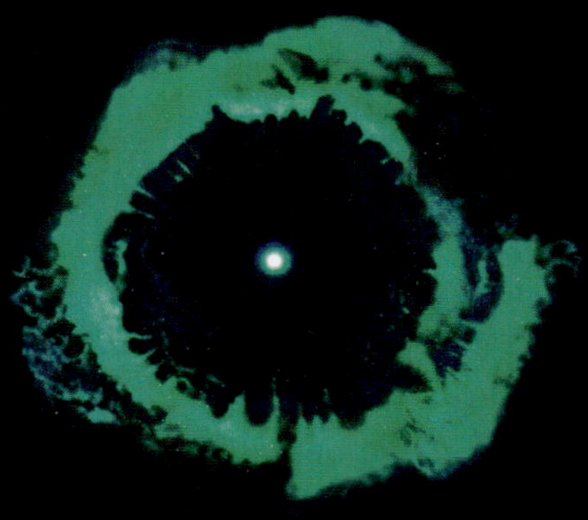

10,000 years

December 31

7 **White dwarf:** The remaining white dwarf is made primarily of carbon and oxygen because the core of the low-mass star never grows hot enough to produce heavier elements.

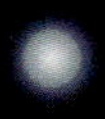

indefinite

–

*The lifetime of this $1M_{Sun}$ star is almost 2000 times as long as that of a $25M_{Sun}$ star.*

A moment's thought reveals that something quite strange is going on. The stars of a binary system are born at the same time and therefore must both be the same age. We know that more massive stars live shorter lives, and therefore the more massive star must exhaust its core hydrogen and become a subgiant before the less massive star does. How, then, can Algol's less massive star be a subgiant while the more massive star is still fusing hydrogen as a main-sequence star?

This *Algol paradox* reveals some of the complications in stellar life cycles that can arise in close binary systems. The two stars in a close binary are near enough to exert significant tidal forces on each other [**Section 4.5**]. The gravity of each star attracts the near side of the other star more strongly than it attracts the far side. The stars therefore stretch into elongated shapes rather than remaining spherical. In addition, the stars become *tidally locked* so that they always show the same face to each other, much as the Moon always shows the same face to Earth.

During the time that both stars are main-sequence stars, the tidal forces have little effect on their lives. However, when the more massive star (which exhausts its core hydrogen sooner) begins to expand into a red giant, gas from its outer layers can spill onto its companion. This **mass exchange** occurs when the giant grows so large that its tidally distorted outer layers succumb to the gravitational attraction of the smaller companion star. The companion then begins to gain mass at the giant's expense.

The solution to the Algol paradox should now be clear (**FIGURE 17.20**). The $0.8M_{Sun}$ subgiant *used to be* much more massive. As the more massive star, it was the first to begin expanding into a red giant. As it expanded, however, it transferred so much of its matter onto its companion that it is now the less massive star.

The future may hold even more interesting events for Algol. The $3.7M_{Sun}$ star is still gaining mass from its subgiant companion. Its life cycle is therefore accelerating as its increasing gravity raises its core hydrogen fusion rate. Millions of years from now, it will exhaust its hydrogen and begin to expand into a red giant itself. At that point, it can begin to transfer mass *back* to its companion. Even more amazing things can happen in other mass-exchange systems, particularly when one of the stars is a white dwarf or a neutron star. But that is a topic for the next chapter.

Algol shortly after its birth. The higher-mass star (left) evolved more quickly than its lower-mass companion (right).

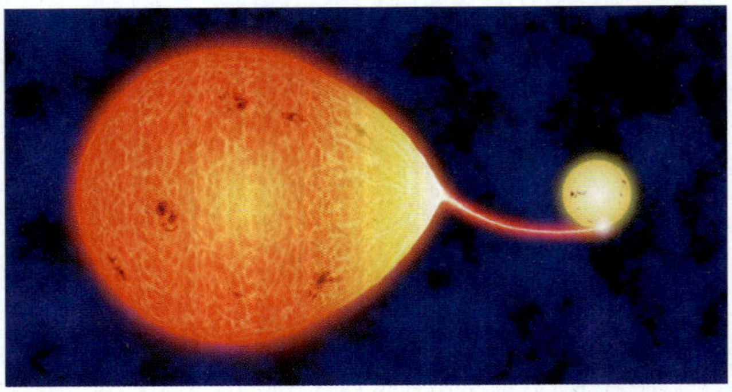

Algol at onset of mass transfer. When the more massive star expanded into a red giant, it began losing some of its mass to its normal, hydrogen core-fusion companion.

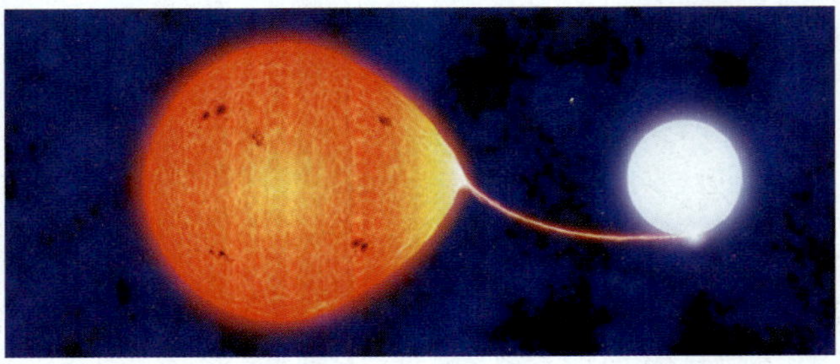

Algol today. As a result of the mass transfer, the red giant has shrunk to a subgiant, and the normal star on the right is now the more massive of the two stars.

**FIGURE 17.20** Artist's conception of the development of the Algol close binary system.

## The BIG Picture PUTTING CHAPTER 17 INTO CONTEXT

*In this chapter, we have seen how the origin of the elements, first discussed in Chapter 1, is intimately linked to the lives and deaths of stars. As you look back over this chapter, remember these "big picture" ideas:*

- Virtually all elements in the universe besides hydrogen and helium were forged in stars. We and our planet are therefore made of stuff produced in stars that lived and died long ago.
- Low-mass stars like our Sun live long lives and die with the ejection of planetary nebulae, leaving behind white dwarfs.
- High-mass stars live fast and die young, exploding dramatically as supernovae and leaving behind neutron stars or black holes.
- Close binary stars can exchange mass, altering the usual course of stellar evolution.

**MY COSMIC PERSPECTIVE** One of the deepest connections we have with the cosmos is through the many elements that make up our bodies, which were formed by nuclear fusion in stars.

# Summary of Key Concepts

## 17.1 Lives in the Balance

- **How does a star's mass affect nuclear fusion?** Stars of greater mass have hotter core temperatures, allowing fusion to proceed more quickly and enabling fusion of heavier elements to take place. A star's *mass* at birth therefore determines almost every aspect of its life and death. To understand the general characteristics of stellar lives, we divide stars into three groups by mass: **low-mass stars**, with masses less than $2M_{Sun}$; **intermediate-mass stars**, with masses between $2M_{Sun}$ and $8M_{Sun}$; and **high-mass stars**, with masses above $8M_{Sun}$.

## 17.2 Life as a Low-Mass Star

- **What are the life stages of a low-mass star?** A low-mass star spends most of its life generating energy by fusing hydrogen in its core via the proton-proton chain. When core hydrogen is exhausted, the core begins to shrink while the star as a whole expands to become a **red giant**, with **hydrogen shell fusion** around an inert helium core. When the core becomes hot enough, a **helium flash** initiates **helium fusion** in the core, which fuses helium into carbon. This phase lasts until core helium is exhausted. Low-mass stars never become hot enough for carbon fusion, so at this point their lives must come to an end.

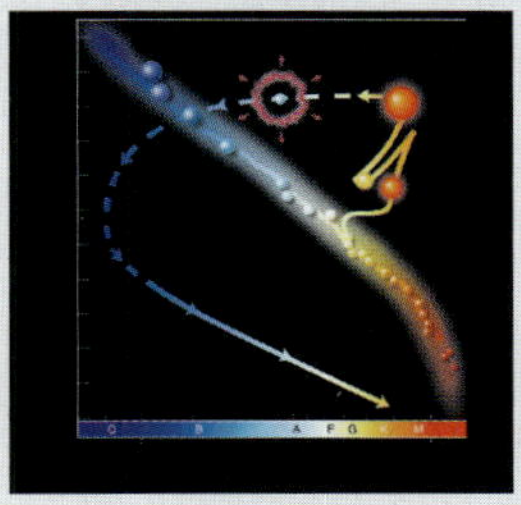

- **How does a low-mass star die?** The core again shrinks after helium core fusion ceases. Helium shell fusion begins around the inert carbon core beneath the hydrogen-fusing shell. The outer layers expand again, making the star into a double shell–fusion star. The star's energy generation never reaches equilibrium during this time; instead, the star experiences a series of **thermal pulses** and ultimately expels its outer layers into space as a **planetary nebula**. The remaining "dead" stellar core is a **white dwarf**.

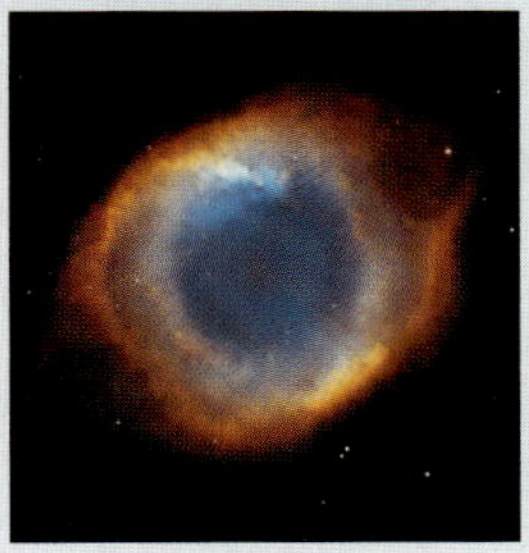

## 17.3 Life as a High-Mass Star

- **What are the life stages of a high-mass star?** A high-mass star lives a much shorter life than a low-mass star, fusing hydrogen into helium via the **CNO cycle**. After exhausting its core hydrogen, a high-mass star begins hydrogen shell fusion and then goes through a series of stages, fusing successively heavier elements as its core temperature rises. While that happens, the star swells in size to become a supergiant.

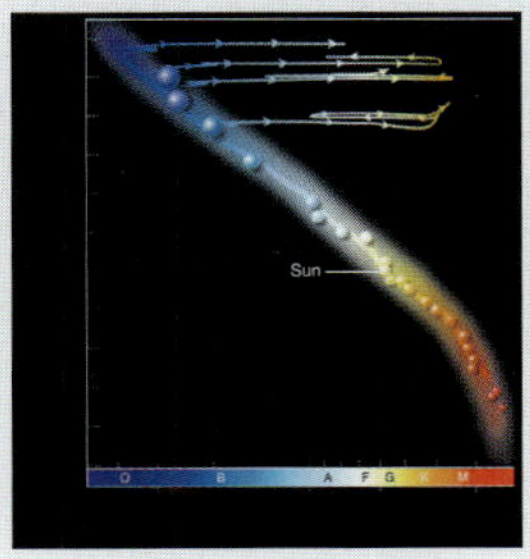


- **How do high-mass stars make the elements necessary for life?** In its final stages of life, a high-mass star's core becomes hot enough to fuse carbon and other heavy elements. The variety of different fusion reactions produces a wide range of elements—including all the elements necessary for life—that are then released into space when the star dies.

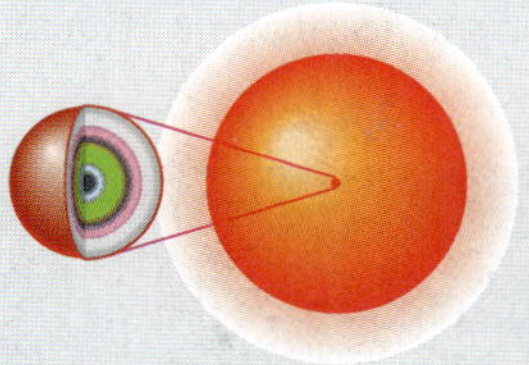

- **How does a high-mass star die?** A high-mass star dies in the cataclysmic explosion called a **supernova**, scattering newly produced elements into space and leaving a neutron star or a black hole behind. The supernova occurs after fusion causes iron to pile up in the core of the high-mass star. Because iron fusion cannot release energy, the core cannot hold off the crush of gravity for long. In the instant that gravity overcomes degeneracy pressure, the core collapses and the star explodes. The expelled gas may be visible for a few thousand years as a **supernova remnant**.

## 17.4 The Roles of Mass and Mass Exchange

- **How does a star's mass determine its life story?** A star's mass determines how it lives its life. Low-mass stars never get hot enough to fuse carbon into heavier elements in their cores, and they end their lives by expelling their outer layers and leaving white dwarfs behind. High-mass stars live short but brilliant lives, ultimately dying in supernova explosions.

- **How are the lives of stars with close companions different?** When one star in a close binary system begins to swell in size at the end of its main-sequence stage, it can start to transfer mass to its companion. This **mass exchange** can change the remaining life histories of both stars.

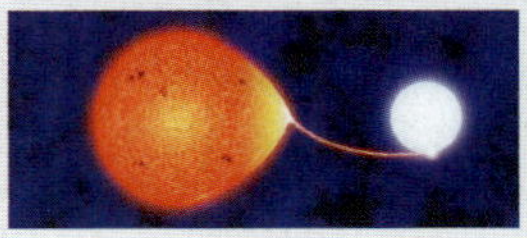

# Visual Skills Check

*Use the following questions to check your understanding of some of the many types of visual information used in astronomy. For additional practice, try the Chapter 17 Visual Quiz in the Study Area at www .MasteringAstronomy.com.*

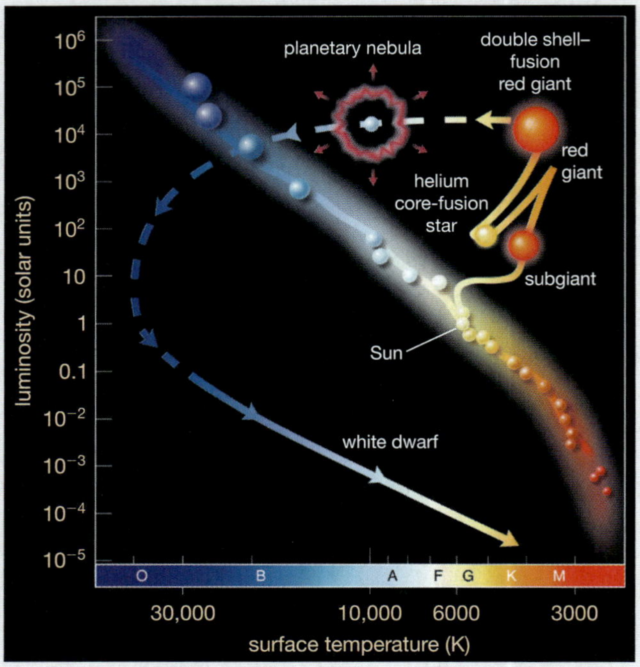


*This figure, similar to the left side of Figure 17.8, shows the future life stages of the Sun on an H-R diagram. Answer the following questions, using the information provided in the figure.*

1. What will the Sun's approximate luminosity be during the subgiant stage?
2. When the Sun is a red giant, what will its approximate surface temperature be?
3. Just before the Sun produces a planetary nebula, what will its approximate luminosity be?
4. When the Sun becomes a white dwarf with a surface temperature similar to its current surface temperature, what will its luminosity be?

# Exercises and Problems

*For instructor-assigned homework and other learning materials, go to www.MasteringAstronomy.com.*

## Chapter Review Questions

### Short-Answer Questions Based on the Reading

1. Why is mass so important to a star's life? How and why do we divide stars into groups by mass?
2. What do all *low-mass stars* have in common? Why do they differ in their levels of surface activity? What are *flare stars*?
3. When a star exhausts its core hydrogen fuel, the core contracts but the star as a whole expands. Why?
4. What is the *helium fusion* reaction, and why does it require much higher temperatures than hydrogen fusion? Why will helium fusion in the Sun begin with a *helium flash*?
5. Why does the H-R diagram of a globular cluster show a *horizontal branch*? What are the characteristics of the stars on the horizontal branch?
6. What happens to a low-mass star after it exhausts its core helium? Why can't it fuse carbon into heavier elements?
7. How do red giants manufacture carbon-rich dust grains, and why are these important to life?
8. What is a *planetary nebula*? What happens to the core of a star after a planetary nebula occurs?
9. What will happen to Earth as the Sun changes in the future?
10. Summarize the stages of life that we see on the Sun's life track in Figure 17.8. Be sure to explain both the changes that occur in the Sun's core with each stage and the changes that are observable from outside the Sun.
11. In broad terms, explain how the life of a *high-mass star* differs from that of a low-mass star. How do *intermediate-mass stars* fit into this picture?
12. Describe some of the nuclear reactions that can occur in high-mass stars after they exhaust their core helium. Why does this continued nuclear fusion occur in high-mass stars but not in low-mass stars?
13. Why can't iron be fused to release energy?
14. Summarize some of the observational evidence supporting our ideas about how heavy elements form in massive stars.
15. What event initiates a *supernova*, and why is a neutron star or a black hole left behind? What observational evidence supports our understanding of supernovae?
16. Why can the lives of close binary stars differ from those of single stars? Describe the *Algol paradox* and its resolution.

### Does It Make Sense?

*Decide whether or not each of the following statements makes sense (or is clearly true or false). Explain clearly; not all of these have definitive answers, so your explanation is more important than your chosen answer.*

17. The iron in my blood came from a star that blew up more than 4 billion years ago.
18. Humanity will eventually have to find another planet to live on, because one day the Sun will blow up as a supernova.

19. I sure am glad hydrogen has a higher mass per nuclear particle than many other elements. If it had the lowest mass per nuclear particle, none of us would be here.
20. I just discovered a $3.5M_{Sun}$ main-sequence star orbiting a $2.5M_{Sun}$ red giant. I'll bet that red giant was more massive than the $3.5M_{Sun}$ star when it was a main-sequence star.
21. If the Sun had been born $4\frac{1}{2}$ billion years ago as a high-mass star rather than as a low-mass star, Jupiter would have Earth-like conditions today, while Earth would be hot like Venus.
22. If you could look inside the Sun today, you'd find that its core contains a much higher proportion of helium and a lower proportion of hydrogen than it did when the Sun was born.
23. Most of the supernova explosions that occur in a star cluster happen during its first 100 million years.
24. Globular clusters generally contain lots of white dwarfs.
25. After hydrogen fusion stops in a low-mass star, its core cools off until the star becomes a red giant.
26. The oxygen in my lungs comes from supernova explosions.

## Quick Quiz

*Choose the best answer to each of the following. For additional practice, try the Chapter 17 Reading and Concept Quizzes in the Study Area at www.MasteringAstronomy.com.*

27. Which of these stars has the hottest core? (a) a white main-sequence star (b) an orange main-sequence star (c) a red main-sequence star
28. Which of these stars has the hottest core? (a) a blue main-sequence star (b) a red supergiant (c) a red main-sequence star
29. Which of these stars does *not* have fusion occurring in its core? (a) a red giant (b) a red main-sequence star (c) a blue main-sequence star
30. After the helium flash in a low-mass star, the star's luminosity (a) goes up. (b) goes down. (c) stays the same.
31. What would stars be like if hydrogen had the smallest mass per nuclear particle? (a) Stars would be brighter. (b) All stars would be red giants. (c) Nuclear fusion would not occur in stars of any mass.
32. What would stars be like if carbon had the smallest mass per nuclear particle? (a) Supernovae would be more common. (b) Supernovae would never occur. (c) High-mass stars would be hotter.
33. What would you be most likely to find if you returned to the solar system in 10 billion years? (a) a neutron star (b) a white dwarf (c) a black hole
34. Which of these stars has the shortest life expectancy? (a) an isolated $1M_{Sun}$ star (b) a $1M_{Sun}$ star in a close binary system with a $0.8M_{Sun}$ star (c) a $1M_{Sun}$ star in a close binary system with a $2M_{Sun}$ star
35. What happens to the core of a high-mass star after it runs out of hydrogen? (a) It shrinks and heats up. (b) It shrinks and cools down. (c) Helium fusion begins right away.
36. Which of these elements can't be made by nuclear fusion in the core of a star? (a) calcium (b) uranium (c) oxygen

## Inclusive Astronomy

*Use these questions to reflect on participation in science.*

37. *Connections to the Stars.* In ancient times, many people believed that our lives were somehow influenced by the patterns of the stars in the sky. Modern science has not found any evidence to support this belief, but instead has found that we have a connection to the stars on a much deeper level: We are "star stuff." Discuss in some detail our real connections to the stars, as established by modern astronomy. Do you think these connections have any philosophical implications for how we view our lives and our civilization? Explain.
38. *Group Discussion: Inclusion of Amateurs.* Observations and careful record keeping by amateur observers of the sky have been enormously beneficial to professional researchers, and particularly to studies of variable and exploding stars.
**a.** Section 17.3 discusses how historical supernova observations, particularly those made by the Chinese, have assisted modern research on supernova remnants. Should we consider those ancient observers to be amateurs or professionals? Is the distinction meaningful? Defend your opinion.
**b.** The American Association of Variable Star Observers (AAVSO) is a large community that has made observations and kept records of stellar brightness variations, including supernova explosions, for more than a century. The vast majority of these observations were made by amateurs, not professionals. Do you think amateur observations can be trusted? Why or why not?
**c.** Imagine you are a professional astronomer. How might you go about verifying that observations made by amateurs are sufficiently accurate to support professional research?
**d.** What do you think of the term *amateur* in this context? What does it mean to you? Do you think it could be biasing your viewpoint on the work of amateur astronomers?

## The Process of Science

*These questions may be answered individually in short-essay form or discussed in groups, except where identified as group-only.*

39. *Predicting the Sun's Future.* Models of stellar evolution make detailed predictions about the fate of the Sun. Describe a piece of evidence that supports each of the following model predictions:
**a.** The Sun cannot continue supplying Earth with light and heat forever.
**b.** The Sun will become a red giant before the end of its life.
**c.** The Sun will leave behind a white dwarf after it dies.
40. *Blue Stragglers.* Notice in Figure 17.6b the small number of stars that appear to be on the main sequence but above the cluster's main-sequence turnoff point. These stars are known as *blue stragglers*, because they are bluer and more massive than the stars we expect to find on the cluster's main sequence.
**a.** Why is the presence of blue stragglers surprising?
**b.** One possible explanation for blue stragglers is that there is something wrong with our theory of stellar evolution, but astronomers instead suspect that a blue straggler forms through the merger of two smaller stars. Why do astronomers prefer the merger explanation? Can you think of a way in which the merger explanation could be tested?
41. *Group Activity: Comparing Models of Stars with Data.* Work as a group to evaluate each of the following predictions. You will find it useful to study the star cluster data in Figures 15.18, 15.19, and 15.20. Note: You may wish to do this activity using the four roles described in Chapter 1, Exercise 39.
**a.** Type O stars have shorter lives than type G stars.
**b.** Type K supergiant stars produce iron before they explode as supernovae.

**c.** Type F stars become much more luminous near the ends of their lives than they were as main-sequence stars.

**d.** Type O stars do not become more luminous near the ends of their lives but do become redder.

**e.** Type M stars should have longer lives than type K stars.

**f.** Stars similar to the Sun reach a maximum size of about 100 solar radii during the red giant stage.

**g.** Type K main-sequence stars will become red giants when their cores run out of hydrogen.

**h.** Some stars become white dwarfs at the ends of their lives.

**i.** White dwarfs cool with time but do not change much in radius.

## Investigate Further

### Short-Answer/Essay Questions

42. *Homes to Civilization?* We do not yet know how many stars have Earth-like planets, nor do we know the likelihood that such planets might harbor advanced civilizations like our own. However, some stars can probably be ruled out as candidates for having planets that support advanced civilizations. For example, given that it took a few billion years for humans to evolve from the first life forms on Earth, it seems unlikely that advanced life would have had time to evolve on a planet around a star that is only a few million years old. For each of the following stars, decide whether you think it is possible that it could have a planet that harbors an advanced civilization. Explain your reasoning in one or two paragraphs.
    **a.** A $10M_{Sun}$ main-sequence star
    **b.** A $1.5M_{Sun}$ main-sequence star
    **c.** A $1.5M_{Sun}$ red giant
    **d.** A $1M_{Sun}$ helium core-fusion star
    **e.** A red supergiant
    **f.** A flare star
43. *Rare Elements.* Lithium, beryllium, and boron are elements with atomic numbers 3, 4, and 5, respectively. Despite their being three of the five simplest elements, Figure 17.15 shows that they are rare compared to many heavier elements. Suggest a reason for their rarity. (*Hint:* Consider the process by which helium fuses into carbon.)
44. *Future Skies.* As a red giant, the Sun will have an angular size in Earth's sky of about 30°. What will sunset and sunrise be like? About how long will they take? Do you think the color of the sky will be different from what it is today? Explain.
45. *Research: Historical Supernovae.* Historical accounts describe supernovae in the years 1006, 1054, 1572, and 1604. Choose one of these supernovae and learn more about historical records of the event. Did the supernova influence human history in any way? Write a two- to three-page summary of your research findings.
46. *Picturing Star Birth and Death.* Photographs of stellar birthplaces (i.e., molecular clouds) and death places (e.g., planetary nebulae and supernova remnants) can be strikingly beautiful, but only a few such photographs are included in this chapter. Search the web for additional images. Look not only for photos taken in visible light, but also for those taken in other wavelengths. Put the photographs you find into a personal online journal, along with a one-paragraph description of what each photograph shows. Include at least 20 images.
47. *Humanity in* A.D. *5,000,000,000.* Do you think it is likely that humanity will survive until the Sun begins to expand into a red giant 5 billion years from now? Why or why not?

### Quantitative Problems

*Be sure to show all calculations clearly and state your final answers in complete sentences.*

48. *Density of a Red Giant.* Near the end of the Sun's life, its radius will extend nearly to the distance of Earth's orbit. Estimate the volume of the Sun at that time, using the formula for the volume of a sphere ($4\pi r^3/3$). Using that result, estimate the average matter density of the Sun at that time. How does that density compare with the density of water ($1 \text{ g/cm}^3$)? How does it compare with the density of Earth's atmosphere at sea level (about $10^{-3} \text{ g/cm}^3$)?
49. *Escape Velocity from a Red Giant.* What is the escape velocity from a red giant with a mass of $1M_{Sun}$ and a radius of $100R_{Sun}$? How does that velocity compare with the escape velocity from the Sun? Describe how your results help account for the fact that red giants have strong stellar winds.
50. *Roasting the Earth.* During its final days as a red giant, the Sun will reach a peak luminosity of about $3000L_{Sun}$. Earth will therefore absorb about 3000 times as much solar energy as it does now, and it will need to radiate 3000 times as much thermal energy to keep its surface temperature in balance. Estimate the temperature Earth's surface will need to attain in order to radiate that much thermal energy. You will need to use the formula for emitted power per unit area in Mathematical Insight 15.5.
51. *Supernova Betelgeuse.* The distance of the red supergiant Betelgeuse is approximately 600 light-years. If it were to explode as a supernova, it would be one of the brightest stars in the sky. Right now, the brightest star other than the Sun is Sirius, with a luminosity of $26L_{Sun}$ and a distance of 8.6 light-years. How much brighter than Sirius would the Betelgeuse supernova be in our sky if it reached a maximum luminosity of $10^{10}L_{Sun}$?
52. *Construction of Elements.* Using the periodic table in Appendix D, determine which elements are made by the following nuclear fusion reactions. (You can assume the total number of protons in the reaction remains constant.)
    **a.** fusion of a carbon nucleus with another carbon nucleus
    **b.** fusion of a carbon nucleus with a neon nucleus
    **c.** fusion of an iron nucleus with a helium nucleus
53. *Expansion of the Crab Nebula.* Observations of the Crab Nebula taken over several decades show that gas blobs that are now 100 arcseconds from the center of the nebula are moving away from the center by about 0.11 arcsecond per year. Use that information to estimate the year in which the explosion ought to have been observed. How does that year compare with the year in which the supernova that produced the nebula was actually observed?
54. *Algol's Orbital Separation.* The Algol binary system consists of a $3.7M_{Sun}$ star and a $0.8M_{Sun}$ star with an orbital period of 2.87 days. Use Newton's version of Kepler's third law to calculate the orbital separation of the system. How does that separation compare with the typical size of a red giant star?
55. *The Speed of Supernova Debris.* In the year 2001, debris moving outward from Supernova 1987A was seen hitting a preexisting ring of gas at a distance of 0.7 light-year around the star that exploded. Calculate the speed of that debris. How does the speed you find compare with the speed of light?

# 18 The Bizarre Stellar Graveyard

▲ **About the photo:** This composite photograph made from x-ray and visible light shows the neutron star amidst the hot gas at the center of the Crab Nebula, a supernova remnant.

## LEARNING GOALS

### 18.1 White Dwarfs
- What is a white dwarf?
- What can happen to a white dwarf in a close binary system?

### 18.2 Neutron Stars
- What is a neutron star?
- How were neutron stars discovered?
- What can happen to a neutron star in a close binary system?

### 18.3 Black Holes: Gravity's Ultimate Victory
- What is a black hole?
- What would it be like to visit a black hole?
- Do black holes really exist?

### 18.4 Extreme Events
- What causes gamma-ray bursts?
- What happens when neutron stars merge?
- What happens when black holes merge?

*Scientists should never claim that something is absolutely true. You should never claim perfect, or total, or 100% because you never ever get there.*

— *Jocelyn Bell, discoverer of neutron stars*

**Chapter 18 Overview**

Welcome to the afterworld of stars, the fascinating domain of white dwarfs, neutron stars, and black holes. To scientists, these dead stars are ideal laboratories for testing the most extreme predictions of general relativity and quantum theory. To most other people, the eccentric behavior of stellar corpses demonstrates that the universe is stranger than they ever imagined.

Dead stars behave in unusual and unexpected ways that challenge our minds and stretch the boundaries of what we believe is possible. Stars that have finished nuclear fusion have only one hope of staving off the crushing power of gravity: the quantum mechanical effect of degeneracy pressure. But even this strange pressure cannot save the most massive stellar cores. In this chapter, we will study the bizarre properties and occasional catastrophes of the eerie inhabitants of the stellar graveyard.

## 18.1 White Dwarfs

In the previous chapter, we saw that stars of different masses leave different types of stellar corpses. Low-mass stars like the Sun leave behind white dwarfs when they die. High-mass stars die in the titanic explosions known as supernovae, leaving behind neutron stars or black holes. Let's begin our study of stellar corpses with white dwarfs.

### What is a white dwarf?

As discussed in Chapters 15 and 17, a **white dwarf** is essentially the exposed core of a star that has died and shed its outer layers in a planetary nebula [**Section 17.2**]. It is quite hot when it first forms, because it was recently the inside of a star, but it slowly cools with time. White dwarfs are similar to the Sun in mass but similar to Earth in radius [**Section 15.2**], which is why they are generally quite dim compared to stars like the Sun. However, the hottest white dwarfs can shine quite brightly in high-energy ultraviolet and x-ray light (**FIGURE 18.1**).

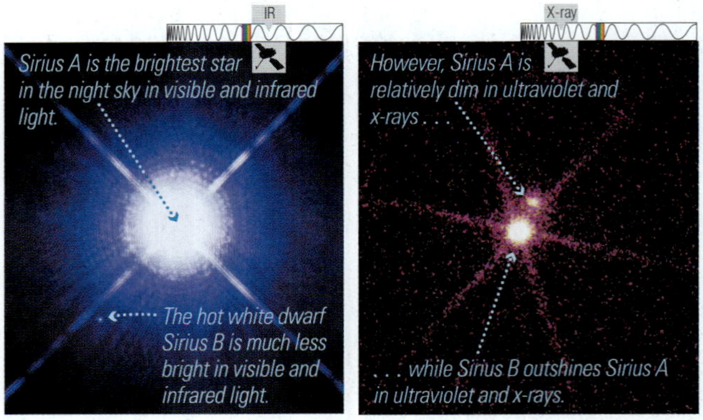


**a** Sirius as seen in infrared light by the Hubble Space Telescope. **b** Sirius as seen by the Chandra X-Ray Telescope.

**FIGURE 18.1** Sirius, the brightest star in the night sky, is actually a binary system consisting of a main-sequence star and a white dwarf. The main-sequence star is much brighter in infrared and visible light, but the hot white dwarf shines more brightly in high-energy light. (The spikes in the images are artifacts of the telescope optics.)

A white dwarf's combination of starlike mass and small size makes gravity very strong near its surface. If gravity were unopposed, it would crush the white dwarf to an even smaller size, so some sort of pressure must be pushing back equally hard to keep the white dwarf stable. Because there is no fusion to maintain heat and pressure inside a white dwarf, the pressure that opposes gravity must come from some other source. As discussed in Chapters S4 and 17, the source is *degeneracy pressure*—a type of pressure that arises when subatomic particles are packed as closely as the laws of quantum mechanics allow. More specifically, the degeneracy pressure in white dwarfs arises from closely packed electrons, so we call it **electron degeneracy pressure**. A white dwarf exists in a state of balance because the outward push of electron degeneracy pressure matches the inward crush of gravity.

**White Dwarf Composition, Density, and Size** Because a white dwarf is the core left over after a star has ceased nuclear fusion, its composition reflects the products of the star's final nuclear fusion stage. The white dwarf left behind by a $1M_{\text{Sun}}$ star like our Sun will be made mostly of carbon, since stars like the Sun fuse helium nuclei to create carbon in their final stage of life. The cores of very-low-mass stars never become hot enough for helium fusion reactions and will end up as helium white dwarfs. Some intermediate-mass stars progress beyond carbon fusion but do not create any iron. These stars leave behind white dwarfs containing large amounts of oxygen or even heavier elements.

Despite its ordinary-sounding composition, a scoop of matter from a white dwarf would be unlike anything ever seen on Earth. A typical white dwarf has the mass of the Sun ($1M_{\text{Sun}}$) compressed into a volume the size of Earth. If you recall that Earth is smaller than a typical sunspot, you'll realize that packing the entire mass of the Sun into the volume of Earth is no small feat. The density of such a white dwarf is so great that a teaspoon of its material would weigh several tons—as much as a small truck—if you could bring it to Earth.

More massive white dwarfs are actually smaller in size than less massive ones. For example, a $1.3M_{\text{Sun}}$ white dwarf is half the diameter of a $1.0M_{\text{Sun}}$ white dwarf (**FIGURE 18.2**). The more massive white dwarf is smaller because its greater gravity can compress its matter to a much greater density. According to the laws of quantum mechanics, the electrons in a white dwarf respond to this compression by moving faster, which makes the degeneracy pressure strong enough to resist the greater force of gravity. The most massive white dwarfs are therefore the smallest.

The fact that more massive white dwarfs are smaller in size also explains why red giants become more luminous as they age [**Section 17.2**]. Degeneracy pressure supports the inert helium core of a low-mass red giant, so this core is essentially a white dwarf buried inside a star. As the hydrogen-fusing shell above it deposits more helium onto the

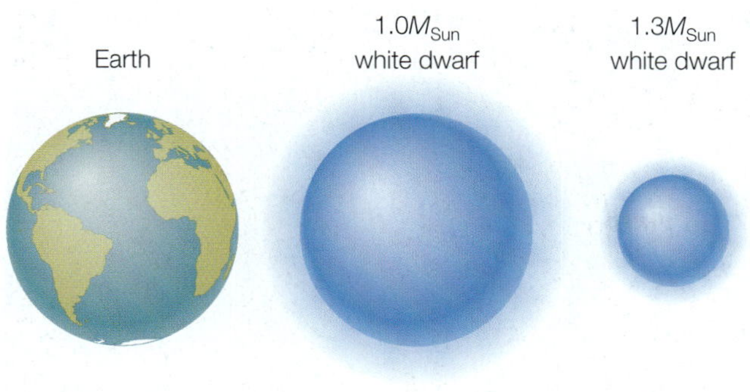


**FIGURE 18.2** Contrary to what you might expect, more massive white dwarfs are actually *smaller* (and therefore denser) than less massive white dwarfs. Earth is shown for scale.

degenerate core, the mass of the core continually increases. Just as more massive white dwarfs must be smaller in size, adding mass to the degenerate core must cause it to contract. The surrounding hydrogen-fusing shell shrinks along with it, and this shrinking makes the shell hotter and increases its fusion rate. Therefore, as long as the core remains inert and fusion occurs in a shell overlying it, the luminosity of the red giant must steadily increase.

**The White Dwarf Limit** The fact that electron speeds are higher in more massive white dwarfs leads to a fundamental limit on the maximum mass of a white dwarf. Theoretical calculations show that electron speeds would reach the speed of light in a white dwarf with a mass of about 1.4 times the mass of the Sun ($1.4M_{Sun}$). Because neither electrons nor anything else can travel faster than the speed of light, no white dwarf can have a mass greater than $1.4M_{Sun}$, a mass known as the **white dwarf limit** (also called the *Chandrasekhar limit*, after its discoverer).

Strong observational evidence supports this theoretical limit on the mass of a white dwarf. Many known white dwarfs are members of binary systems, and hence we can measure their masses [**Section 15.1**]. In every observed case, the white dwarfs have masses below $1.4M_{Sun}$, just as we expect from theory.

## What can happen to a white dwarf in a close binary system?

Left to itself, a single white dwarf will never again shine as brightly as the star it once was. With no source of fuel for fusion, it will simply cool with time. Its size will never change, because its electron degeneracy pressure will forever keep it stable against the crush of gravity. However, the situation can be quite different for a white dwarf in a close binary system.

**Accretion Disks** A white dwarf in a close binary system can gradually gain mass if its companion is a main-sequence or giant star (**FIGURE 18.3**). When a clump of mass first spills over from the companion to the white dwarf, it has some small orbital velocity. The law of conservation of angular momentum dictates that the clump must orbit faster and faster as it falls toward the white dwarf's surface. The infalling matter therefore forms a whirlpool-like disk around the white dwarf. Because the process in which material falls onto another body is called *accretion* [**Sections 8.2, 16.2**], this rapidly rotating disk is called an **accretion disk**.

Just as in the *protostellar disks* that form from material accreting onto protostars [**Section 16.2**], gas particles in a white dwarf's accretion disk move on orbits that obey Kepler's laws, so gas in the inner region of the disk orbits faster than gas in the outer region. The differences in gas particle speeds lead to friction that removes orbital energy from the inner region, and the loss of energy means that gas gradually spirals inward and eventually settles onto the white dwarf. The major differences between protostellar disks and the accretion disks around white dwarfs are in size, orbital speed, and temperature. The stellar mass and small size of a white dwarf combine to make its surface

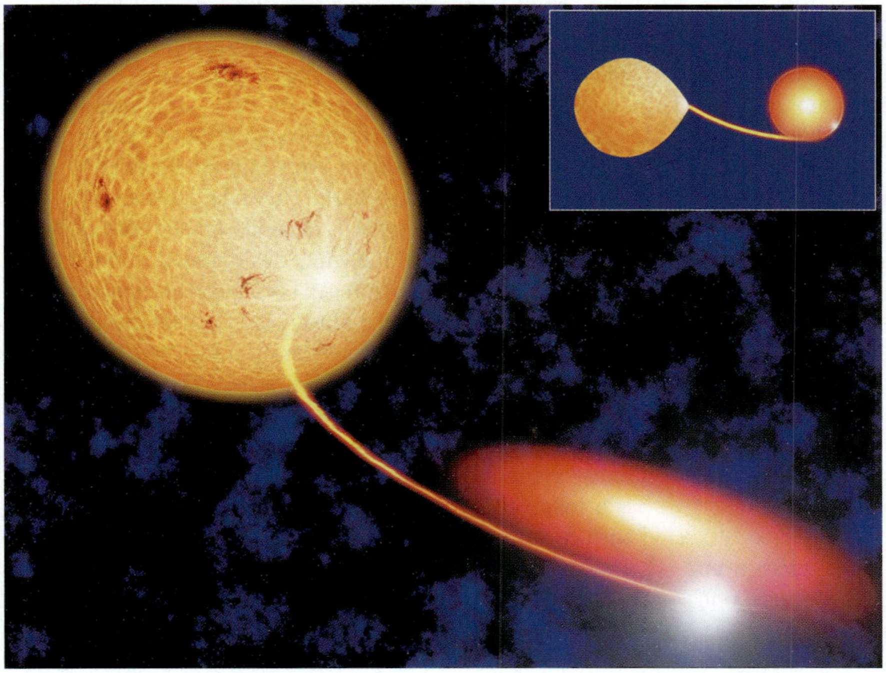

**FIGURE 18.3** This artist's conception shows how mass spilling from a companion star (left) toward a white dwarf (right) forms an accretion disk. The white dwarf itself is in the center of the accretion disk—too small to be seen on this scale. Matter streaming onto the disk creates a hot spot where the stream joins the disk. The inset shows how the system looks from above rather than from the side.

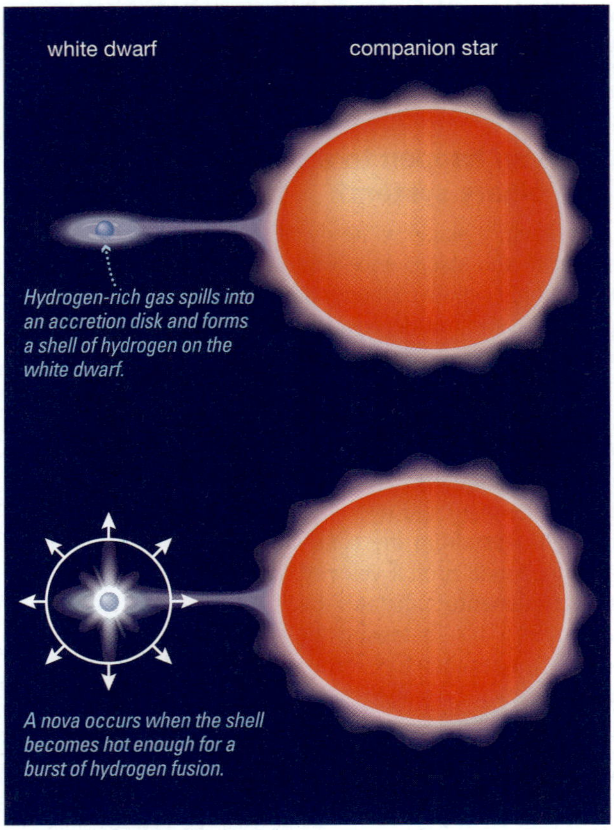


**a** Diagram of the nova process.

**b** Hubble Space Telescope image showing blobs of gas ejected from the nova T Pyxidis. The bright spot at the center of the blobs is the binary star system that generated the nova.

**FIGURE 18.4** A nova occurs when hydrogen fusion ignites on the surface of a white dwarf in a binary star system.

gravity far stronger than that of a protostar, which means the gas in the accretion disk around a white dwarf moves at much higher orbital speeds than the gas in a protostellar disk. Because of the higher orbital speeds, more orbital energy can be turned into heat, making the disk around a white dwarf much hotter than the one around a protostar.

Accretion can provide a "dead" white dwarf with a new energy source as long as its companion keeps feeding matter into the accretion disk. The heat generated by friction can make the accretion disk hot enough to radiate visible and ultraviolet light, and sometimes even x-rays. Although accretion disks around white dwarfs are far too small to be seen directly, we should be able to detect their intense ultraviolet or x-ray radiation.

Searches for this radiation have turned up strong evidence for accretion disks around many white dwarfs. In some cases, the brightness of these systems is highly variable—sudden increases in brightness by a factor of 10 or more may persist for a few days and then fade away, only to repeat a few weeks or months later. Such brightening events (sometimes called *dwarf novae*) probably arise when instabilities in the accretion disk cause some of the matter to fall suddenly onto the white dwarf's surface, with an accompanying conversion of gravitational potential energy into thermal and radiative energy.

**Novae** The hydrogen spilling toward the white dwarf from its companion gradually spirals inward through the accretion disk and eventually falls onto the surface of the white dwarf. The white dwarf's strong gravity compresses this hydrogen gas into a thin surface layer. Both the pressure and the temperature rise as the layer builds up with more accreting gas. When the temperature at the bottom of the layer reaches about 10 million K, hydrogen fusion suddenly ignites.

The white dwarf blazes back to life while fusion proceeds in its hydrogen layer. This thermonuclear flash causes the binary system to shine for a few glorious weeks as a **nova** (**FIGURE 18.4a**). A nova is far less luminous than a supernova, but still can shine as brightly as 100,000 Suns. It generates heat and pressure, ejecting most of the material that has accreted onto the white dwarf. This material expands outward, creating a nova remnant that sometimes remains visible years after the nova explosion (**FIGURE 18.4b**).

Accretion resumes after a nova explosion subsides, so the entire process can repeat itself. The time between successive novae in a particular system depends on the rate at which hydrogen accretes onto the white dwarf's surface and on how highly compressed this hydrogen becomes. The compression of hydrogen is greatest for the most massive white dwarfs, which have the strongest surface gravities. In some cases, novae have been observed to repeat after just a few decades. More commonly, thousands of years pass between nova outbursts.

Notice that, according to our modern definitions, a nova and a supernova are quite different events: A nova is a relatively minor detonation of hydrogen fusion on the surface of a white dwarf in a close binary system, while a supernova is the total explosion of a star. However, the word *nova* simply means "new." Historically, *nova* referred to any star that appeared to the naked eye where none had been visible before. Because supernovae are far more luminous than novae—the light of 10 billion Suns in a supernova versus that of 100,000 Suns in a nova—a very distant supernova can appear as bright in our sky as a nova that is relatively close. Therefore, people could not distinguish between novae and supernovae until modern astronomical methods enabled us to measure their distances.

**White Dwarf Supernovae** Each time a nova occurs, the white dwarf ejects some of its mass. Each time a nova subsides, the white dwarf begins to accrete matter again. Theoretical models cannot yet tell us whether the net result should be a gradual increase or decrease in the white dwarf's mass. If its mass increases, the white dwarf can gradually approach the white dwarf limit. The day it reaches that limit will be the white dwarf's last.

Recall that white dwarfs are made largely of carbon. As a white dwarf's mass closes in on $1.4M_{Sun}$, its temperature rises to the point where carbon fusion can begin. Carbon fusion ignites almost instantly throughout the white dwarf, much like the sudden helium flash in low-mass red giants [**Section 17.2**], but releasing far more energy. The white dwarf explodes completely in what we will call a **white dwarf supernova**.

Alternatively, a white dwarf may reach the white dwarf limit by merging with its binary-star companion. Ordinarily, two objects in orbit around each other will remain in those orbits unless a third object (or surrounding gas) interacts with them. However, two white dwarfs orbiting especially close together emit *gravitational waves* [**Section S3.4**], which carry energy and angular momentum away from the binary system. The two white dwarfs therefore gradually spiral inward toward each other, ultimately merging together. If the total mass of the two white dwarfs exceeds $1.4M_{Sun}$, the result of this merger is a white dwarf supernova.

**Think about it** According to our understanding of novae and white dwarf supernovae, can either of these events ever occur with a white dwarf that is *not* a member of a binary star system? Explain.

The carbon detonation that creates a white dwarf supernova is quite different from the iron catastrophe that leads to a supernova at the end of the life of a high-mass star [**Section 17.3**], which we will call a **massive star supernova**.* Astronomers can distinguish between the two types of supernova by studying their light. Both types shine brilliantly, with peak luminosities about 10 billion times that of the Sun ($10^{10}L_{Sun}$), but the luminosities of white dwarf supernovae fade quickly during the first few weeks and then decline more gradually, while the decline in brightness of a massive star supernova is often more complicated (**FIGURE 18.5**). In addition, spectra of white dwarf supernovae lack hydrogen lines (because white dwarfs contain very little hydrogen), while these lines are prominent in the spectra of most massive star supernovae.

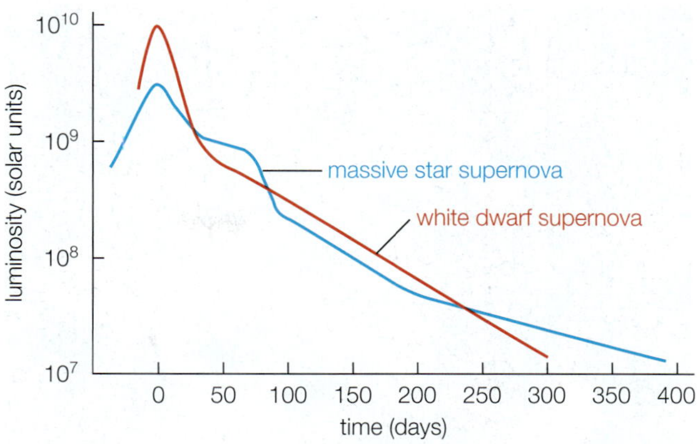


**FIGURE 18.5** The curves on this graph show how the luminosities of two different supernovae fade with time. The white dwarf supernova fades quickly at first and then more gradually a few weeks after the peak, while the massive star supernova fades in a more complicated pattern.

Besides being dramatic, white dwarf supernovae provide one of the primary means by which we measure large distances in the universe. Because all white dwarf supernovae occur in white dwarfs near the $1.4M_{Sun}$ limit, their light curves look quite similar and their maximum luminosities are nearly identical. The same is not true of massive star supernovae, because they occur in stars of many different masses. The fact that all white dwarf supernovae are similar means that once we know the true luminosity of one white dwarf supernova, we essentially know the luminosities of them all. (There is some variation in peak luminosities, but observations have allowed astronomers to identify which white dwarf supernovae are slightly more or less luminous and therefore to correct for this effect.) Whenever we discover a white dwarf supernova in a distant galaxy, we can determine the galaxy's distance by using the inverse square law for light [**Section 15.1**]. In Chapter 20, we'll look more deeply at the role of white dwarf supernovae in helping us establish the cosmic distance scale.

## 18.2 Neutron Stars

White dwarfs with densities of several tons per teaspoon may seem incredible, but neutron stars are stranger still. The possibility that neutron stars might exist was first proposed in the 1930s, but many astronomers thought it preposterous that nature could make anything so bizarre. Nevertheless, a vast amount of evidence now makes it clear that neutron stars really exist.

### What is a neutron star?

A **neutron star** is the ball of neutrons created by the collapse of the iron core in a massive star supernova (**FIGURE 18.6**). Typically just 10 kilometers in radius yet more massive than the Sun, neutron stars are essentially giant atomic nuclei made almost entirely of neutrons and held together by gravity. Like white dwarfs, neutron stars resist the crush of

*Observationally, astronomers classify supernovae as *Type II* if their spectra show hydrogen lines, and *Type I* otherwise. All Type II supernovae are assumed to be massive star supernovae. A Type I supernova can be either a white dwarf supernova or a massive star supernova in which the star blew away all its hydrogen before exploding. Type I supernovae appear in three classes whose light curves differ, called *Type Ia*, *Type Ib*, and *Type Ic*. Only Type Ia supernovae are thought to be white dwarf supernovae.

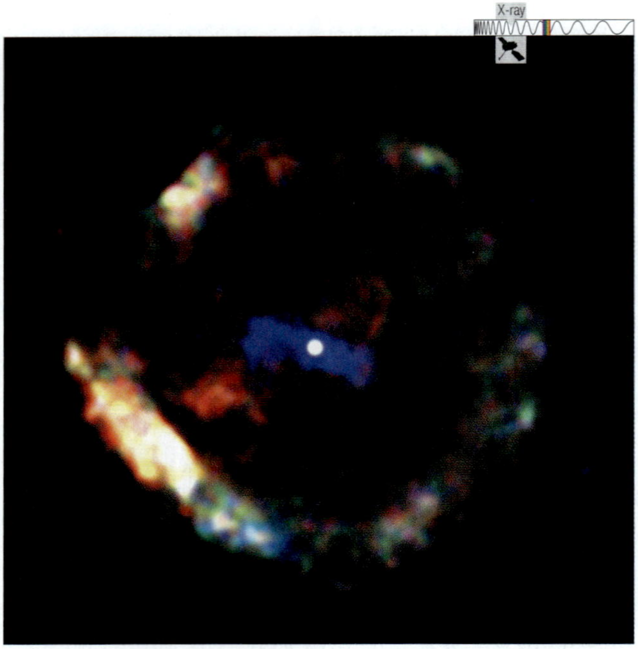


**FIGURE 18.6** This x-ray image from the Chandra X-Ray Observatory shows the supernova remnant G11.2-03, the remains of a supernova observed by Chinese astronomers in A.D. 386. The white dot at the center represents x-rays from the neutron star left behind by the supernova. The different colors correspond to different x-ray wavelength bands. The region pictured is about 23 light-years across.

gravity with the degeneracy pressure that arises when particles are packed as closely as nature allows. In the case of neutron stars, however, it is neutrons rather than electrons that are closely packed, so we say that **neutron degeneracy pressure** supports them against the crush of gravity.

**Neutron Star Gravity** The force of gravity at the surface of a neutron star is awe-inspiring. Escape velocity is about half the speed of light. If you foolishly chose to visit a neutron star's surface, you would be squashed immediately into a microscopically thin puddle of subatomic particles.

Things would be only slightly less troubling if a bit of neutron star could somehow come to visit you. A paper clip with the density of neutron star material would outweigh Mount Everest. If such a paper clip magically appeared in your hand, you could not prevent it from falling. Down it would plunge, passing through Earth like a rock falling through air. It would gain speed until it reached Earth's center, and its momentum would carry it onward until it slowed to a stop on the other side of the planet. Then it would fall back down again. If it came in from space, each plunge of the neutron star material would drill a different hole through the rotating Earth. In the words of astronomer Carl Sagan, the inside of Earth would "look briefly like Swiss cheese" (until the melted rock flowed to fill in the holes) by the time friction finally brought the piece of neutron star to rest at Earth's center.

In the unfortunate event that an *entire* neutron star came to visit you, it would not fall at all. Because it would be only about 10 kilometers in radius, the neutron star would probably fit in your hometown. Remember, however, that it would be 300,000 times as massive as Earth. As a result, the neutron star's immense surface gravity would quickly destroy your hometown and the rest of civilization. By the time the dust settled, the former Earth would have been squashed into a shell no thicker than your thumb on the surface of the neutron star.

**Neutron Star Structure** Theoretical models predict that the state of matter varies with depth in a neutron star. Near the surface, where the pressure is similar to that inside a white dwarf, the neutron star probably has a *crust* composed largely of electrons and positively charged atomic nuclei. Deeper inside, the immense pressure of the overlying layers ensures that individual atomic nuclei have disintegrated and almost all the electrons have combined with protons, which is why the interior of a neutron star is made almost entirely of neutrons.

The properties of neutron-rich matter under the extreme pressure that prevails inside neutron stars are still somewhat uncertain because it is so difficult to re-create those conditions in laboratories on Earth. Observations of neutron stars therefore can provide unique information about how matter behaves under such conditions. Theoretical models predict that matter deep inside a neutron star could be a *superfluid*, meaning that the matter flows without experiencing any friction whatsoever. It may also be *superconducting*, meaning that electricity can pass through it without experiencing any resistance.

## How were neutron stars discovered?

The first observational evidence for neutron stars came in 1967, when a 24-year-old graduate student named Jocelyn Bell discovered a strange source of radio waves. Bell had helped her adviser, Antony Hewish, build a radio telescope ideal for discovering fluctuating sources of radio waves. She was busily trying to interpret the flood of data pouring out of this instrument in October 1967 when she noticed a peculiar signal. After ruling out other possibilities, she concluded that *pulses* of radio waves were arriving from somewhere near the direction of the constellation Cygnus at intervals of precisely 1.337301 seconds (**FIGURE 18.7**).

**Pulsars** The pulses coming from Cygnus were very surprising because no known astronomical object pulsated so regularly. In fact, the pulsations came at such precise intervals that they were nearly as reliable for measuring time as the most precise human-made clocks. For a while, the mysterious source of the radio waves was dubbed "LGM" for Little

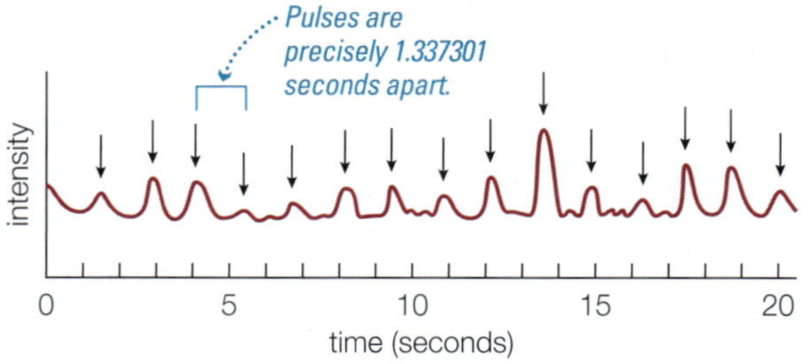


**FIGURE 18.7** About 20 seconds of data from the first pulsar discovered by Jocelyn Bell in 1967.

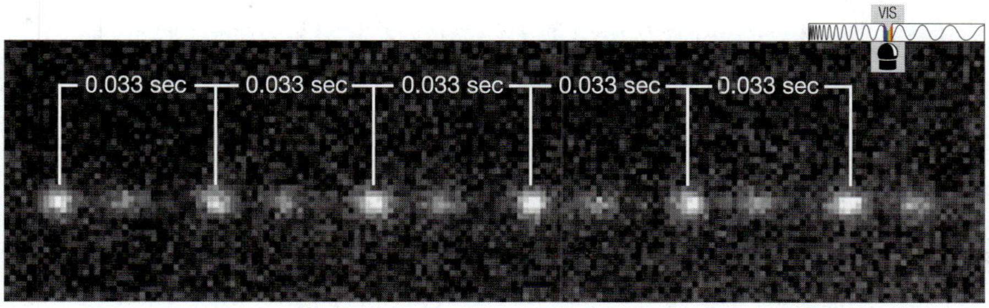


**FIGURE 18.8** This time-lapse image of the pulsar at the center of the Crab Nebula, a supernova remnant, shows its main pulse recurring every 0.033 second. The fainter pulses are thought to come from the pulsar's other lighthouse-like beam. (Photo from the Very Large Telescope of the European Southern Observatory.)

Green Men—only half-jokingly. Today we refer to such rapidly pulsing radio sources as **pulsars**.

The mystery of pulsars was soon solved. By the end of 1968, astronomers had found two smoking guns: Pulsars sat at the centers of two supernova remnants, the Vela Nebula and the Crab Nebula (**FIGURE 18.8**). The pulsars are neutron stars left behind by supernova explosions.

The pulsations arise because the neutron star is spinning rapidly as a result of the conservation of angular momentum [**Section 4.3**]: As an iron core collapses into a neutron star, its rotation rate must increase as it shrinks in size. The collapse also bunches the magnetic field lines running through the core far more tightly, greatly amplifying the strength of the magnetic field. Shortly after its formation in a supernova, a neutron star's magnetic field is a *trillion* times as strong as Earth's. This intense magnetic field directs beams of radiation out along the magnetic poles, although we do not yet know exactly how. If a neutron star's magnetic poles are not aligned with its rotation axis, the beams of radiation sweep round and round (**FIGURE 18.9**). Like lighthouses, neutron stars emit fairly steady beams of light, but we see a pulse of light each time one of those beams sweeps past Earth.

Pulsars are not quite perfect clocks. The continual twirling of a pulsar's magnetic field generates electromagnetic radiation that carries away energy and angular momentum, causing the neutron star's rotation rate to slow gradually. The pulsar in the Crab Nebula, for example, currently spins about 30 times per second. Two thousand years from now, it will spin less than half as fast. Eventually, a pulsar's spin slows

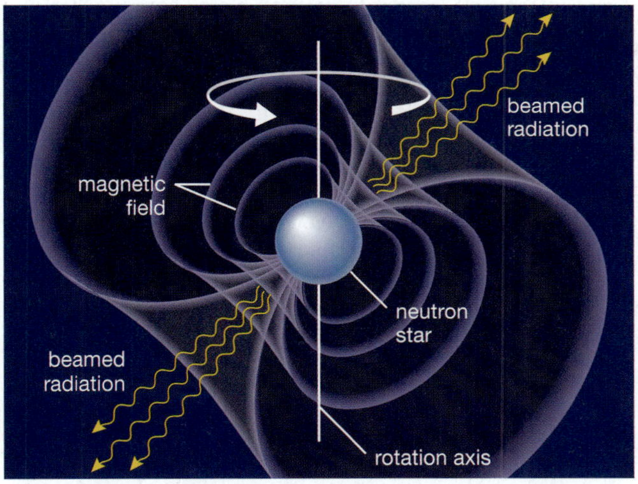


**a** A pulsar is a rotating neutron star that beams radiation along its magnetic axis.

**b** If the magnetic axis is not aligned with the rotation axis, the pulsar's beams sweep through space like lighthouse beams. Each time one of the pulsar's beams sweeps across Earth, we observe a pulse of radiation.

**FIGURE 18.9** Radiation from a rotating neutron star can appear to pulse like the beams from a lighthouse.

so much and its magnetic field becomes so weak that we can no longer detect it. In addition, some spinning neutron stars may be oriented so that their beams do not sweep past our location. We therefore have the following rule: All pulsars are neutron stars, but not all neutron stars are pulsars.

**Think about it** Suppose we observe no pulses of radiation from a neutron star. Is it possible that a civilization in some other star system would see this neutron star as a pulsar? Explain.

We know that pulsars must be neutron stars because no other massive object could spin so fast. A white dwarf, for example, can spin no faster than about once per second without being torn apart, because a faster spin would mean its surface would be rotating faster than the escape velocity. Pulsars have been discovered that rotate as fast as 625 times per second. Only an object as extremely small and dense as a neutron star could spin so fast without breaking apart.

**Pulsar Timing and Pulsar Planets** Several important discoveries have been made by precisely measuring the time intervals between a pulsar's pulses. For example, binary systems in which both objects are neutron stars have been used to test Einstein's general theory of relativity [**Section S3.4**]. Einstein's theory predicts that such systems should radiate *gravitational waves*, and the energy lost in this way should cause the orbits of the two neutron stars to decay. Observations show that such orbital decay does indeed occur at precisely the rate that Einstein's theory predicts (see Figure S3.22), giving scientists confidence that gravitational waves really exist.

Pulsar timing was also used in the very first confirmed discovery of extrasolar planets. In 1992, measurements showed that the pulsar PSR B1257+12 has a slightly varying pulsation rate. Detailed analysis revealed that the changes in the pulsar's pulsation rate could be explained by gravitational tugs of three orbiting planets, with orbital periods of about 25, 67, and 98 days, respectively. This discovery came as an immense surprise to astronomers, because these planets are orbiting the remains of a star that exploded. It seems unlikely that planets could have survived their star's red supergiant stage, let alone the supernova. As a result, astronomers suspect that the planets formed *after* the explosion. For example, if the pulsar once had a stellar companion that came close enough for the pulsar's strong tidal force to rip it apart, the debris from the companion could have formed a disk around the pulsar in which planets could accrete, much as planets accrete in disks of material around ordinary stars.

## What can happen to a neutron star in a close binary system?

Much as with white dwarfs, a neutron star in a close binary system can brilliantly burst back to life as gas overflowing from a companion star creates a hot, swirling accretion disk around it. However, in the neutron star's mighty gravitational field, infalling matter releases an amazing amount of gravitational potential energy. Dropping a brick onto a neutron star would liberate as much energy as an atomic bomb.

**X-Ray Binaries** The huge amount of energy released by infalling matter from the companion star makes a neutron star's accretion disk much hotter and much more luminous than the accretion disk around a white dwarf. The high temperatures in the inner regions of the accretion disk cause it to radiate powerfully in x-rays. Some close binaries with neutron stars emit 100,000 times as much energy in x-rays as our Sun emits in all wavelengths of light combined. Because of this intense x-ray emission, close binaries that contain accreting neutron stars are often called **x-ray binaries**. We have identified hundreds of x-ray binaries in the disk of the Milky Way Galaxy. Note that the systems we see as x-ray binaries consist of one neutron star and one still-living star (main-sequence or giant/supergiant). We'll save discussion of systems with two neutron stars for Section 18.4.

The emission from most x-ray binaries pulsates rapidly as the neutron star spins. However, while other pulsars tend to slow down with time, the pulsation rates of x-ray binaries tend to accelerate, presumably because matter accreting onto the neutron star adds angular momentum (**FIGURE 18.10**). Some of these neutron stars rotate so fast that they pulsate every few thousandths of a second (and are called *millisecond pulsars*).

**X-Ray Bursts** Like accreting white dwarfs that occasionally erupt into novae, accreting neutron stars sporadically erupt with a pronounced spike in luminosity (**FIGURE 18.11**). Because these eruptions release energy primarily in the form of x-rays, we call them **x-ray bursts**, and the systems that produce them are known as **x-ray bursters**. Much like novae, x-ray bursts result from the sudden ignition of nuclear fusion. However, while novae occur when hydrogen fusion ignites on the surface of a white dwarf in a close binary system, x-ray bursts arise from the ignition of helium fusion on the neutron star in a close binary system.

Hydrogen-rich material from the companion star continually spirals through the accretion disk to the neutron star's

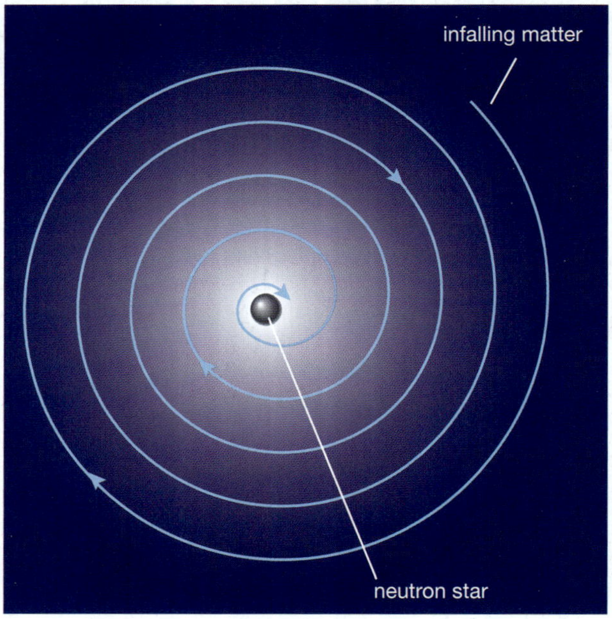


**FIGURE 18.10** Matter accreting onto a neutron star adds angular momentum, increasing the neutron star's rate of spin.

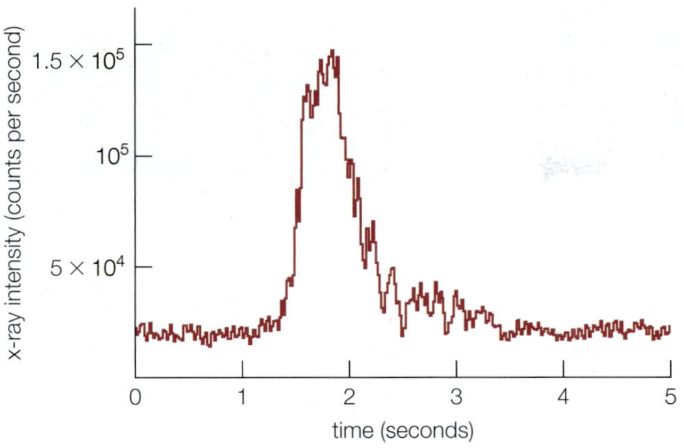


**FIGURE 18.11** Light curve of an x-ray burst. In this particular burst, the x-ray luminosity of the neutron star spiked to over six times its usual brightness in a matter of seconds.

surface, coating the surface with a thin layer of fresh hydrogen gas. Pressures at the bottom of this hydrogen layer are high enough to maintain steady fusion, which produces a layer of helium beneath the hydrogen. The steady fusion keeps the hydrogen layer from becoming more than about 1 meter thick, while the helium produced by fusion gradually adds to the mass of the underlying helium layer.

The temperature of the helium layer slowly climbs as helium accumulates within it. When the temperature reaches about 100 million K, the layer suddenly ignites with helium fusion. The helium fuses rapidly to make carbon and heavier elements, generating a burst of energy that escapes the neutron star in the form of x-rays. During an x-ray burst, fusion reactions in the helium layer and in the overlying hydrogen layer create heavier elements. However, fresh hydrogen from the companion star soon begins to replenish the hydrogen layer, and the entire process is soon ready to repeat. Typical x-ray bursters flare every few hours to every few days. Each burst lasts only a few seconds, but during those seconds the system can radiate 100,000 times as much power as the Sun, all in x-rays. Within a minute after a burst, the x-ray burster cools back down and accretion resumes.

## 18.3 Black Holes: Gravity's Ultimate Victory

The story of stellar corpses would be strange enough if it ended with white dwarfs and neutron stars, but it does not. Sometimes, the gravity in a stellar corpse becomes so strong that nothing can prevent the corpse from collapsing under its own weight. The stellar corpse collapses without end, crushing itself out of existence and forming perhaps the most bizarre type of object in the universe: a *black hole.*

### What is a black hole?

The idea of a black hole was first suggested in the late 18th century by British philosopher John Michell and French physicist Pierre Laplace. It was already known from Newton's laws that the escape velocity from any object depends only on its mass and size. Making an object of a particular mass more compact raises its escape velocity [**Section 4.5**]. Michell and Laplace speculated about objects compact enough to have an escape velocity greater than the speed of light. Because they worked in a time before it was known that light always travels at the same speed, they assumed that light emitted from such an object would behave like a rock thrown upward, eventually slowing to a stop and falling back down.

Einstein's general theory of relativity eventually showed that black holes are considerably stranger than Michell and Laplace thought, but their basic idea was correct. It is indeed possible for an object's gravity to be so strong that not even light can escape. These objects weren't originally called black holes, but physicist John Wheeler started popularizing this term in 1967, long after they were first proposed. The "black" in the name comes from the fact that no light can escape. The "hole" part of the name captures the idea that a black hole is like a hole in the observable universe in the following sense: If you enter a black hole, you leave the region of the universe that we can observe with a telescope and you can never return.

**The Event Horizon** The boundary between the inside of a black hole and the universe outside is called the **event horizon**. The event horizon essentially marks the point of no return for objects entering a black hole: It is the boundary around a black hole at which the escape velocity equals the speed of light. The boundary tends to be spherical because the velocity needed to escape a black hole's gravity depends on the distance to its center, which is the same for every point on the event horizon. Nothing that passes within this boundary can ever escape. The event horizon gets its name from the fact that we have no hope of learning about any events that occur within it.

The structure of space and time near the event horizon is extraordinarily strange. Space and time are not distinct, as we usually think of them, but instead are bound up together as four-dimensional **spacetime** [**Section S3.2**]. Moreover, Einstein's general theory of relativity tells us that what we perceive as gravity arises from *curvature of spacetime.* The concept of curvature of spacetime is challenging to think about because we cannot visualize four dimensions at once, let alone visualize their curvature. However, as discussed in more detail in Chapter S3, we can understand the basic idea with a two-dimensional analogy.

**See it for yourself** Curved space has different rules of geometry than "flat" space, an idea you can see with a plastic ball and a marker. Draw a straight line on the ball going one-quarter of the way around it, then make a right-angle turn and draw a line one-quarter of the way around the ball in your new direction, and finish the triangle with a straight line back to where you started. Measure all three angles. How does the sum of the three angles of the triangle on your sphere compare with the 180° sum you would find for a triangle on a flat plane?

**FIGURE 18.12** uses a rubber sheet to represent two-dimensional slices through spacetime. In this analogy, the sheet is flat—corresponding to weak gravity—in a region far from any mass (Figure 18.12a). The sheet becomes curved near a massive object, corresponding to strong gravity

a A two-dimensional representation of "flat" spacetime. Each pair of circles is separated by the same radial distance.

b The way a mass affects the rubber sheet is similar to the way gravity curves spacetime. The circles become more widely separated—indicating greater curvature—as we move closer to the mass.

c The curvature of spacetime becomes greater and greater as we approach a black hole, and a black hole itself is like a bottomless pit in spacetime.

**FIGURE 18.12** We can use two-dimensional rubber sheets in an analogy to curvature in four-dimensional spacetime.

(Figure 18.12b), and greater curvature means stronger gravity. In this representation, the curvature of spacetime near a black hole is so great that the rubber sheet forms a bottomless pit (Figure 18.12c). Keep in mind that the rubber sheet is only an analogy meant to illustrate the extreme distortion of spacetime near the event horizon. A real black hole is spherical, not funnel-shaped like the rubber sheet.

**The Size of a Black Hole** We usually think of the "size" of a black hole as the radius of its event horizon. As the rubber sheet analogy in Figure 18.12c shows, it is possible to draw a series of concentric circles around a black hole. However, it is not really possible to measure a *radius* for these circles, because their centers lie within the event horizon and hence are not part of our observable universe. We therefore define the radius of a circle around a black hole as the radius it *would* have if geometry were flat (Euclidean), as it is in Figure 18.12a. (That is, we simply define the radius to be the circle's circumference divided by $2\pi$.) The radius of the event horizon is known as the **Schwarzschild radius**, for Karl Schwarzschild (1873–1916). Schwarzschild computed his famous radius from Einstein's general theory of relativity only a month after Einstein published the theory. Moreover, Schwarzschild did this work while serving in the German Army on the Russian front during World War I. Sadly, he died less than a year later of an illness contracted during the war.

The Schwarzschild radius of a black hole depends only on its mass. A black hole with the mass of the Sun has a Schwarzschild radius of about 3 kilometers—only a little smaller than the radius of a neutron star of the same mass. More massive black holes have larger Schwarzschild radii. For example, a black hole with 10 times the mass of the Sun has a Schwarzschild radius of about 30 kilometers.

**Properties of a Black Hole** A collapsing stellar core becomes a black hole at the moment it shrinks to a size smaller than its Schwarzschild radius. At that moment, the core disappears within its own event horizon. The black hole still contains all the mass and exerts the full amount of gravity associated with that mass, but its outward appearance tells us nothing about what fell in. Any information about the object that collapsed to make the black hole is lost. For example, if we found an isolated black hole floating through space, we'd have no way to know whether it was made from the collapse of the iron core of a massive star during a supernova, from the merger of two neutron stars, or from some completely different process. The black hole similarly destroys nearly all information about matter that falls into it at later times. Tossing 1 kilogram of lead into a black hole would have precisely the same effect as tossing 1 kilogram of precious diamonds into it—in both cases, the black hole's mass would increase by 1 kilogram, and there would be no other evidence left to tell us what went in.

Theory tells us that mass is one of only three basic properties of a black hole. The second property is electric charge, but this is relatively unimportant: If a black hole had any positive or negative charge, it would quickly attract oppositely charged particles from its surroundings, making it electrically neutral. The third property of a black hole is its angular momentum. Conservation of angular momentum dictates that black holes should rotate rapidly when they form in the collapse of a rotating star. Much like an ice-skater pulling in his arms, a collapsing stellar core should rotate faster and faster as it shrinks in size.

A black hole's rotation can affect its surroundings. Einstein's general theory of relativity predicts that any massive rotating object should drag neighboring regions of spacetime around in circles. Scientists have successfully measured this effect (known as *frame-dragging*) near our rotating Earth, but it should be far stronger near the event horizon of a black hole. The dragging of spacetime around a rotating black hole would tend to accelerate infalling objects in the direction of rotation, preventing the objects from falling straight downward. This effect can change the shape of a black hole's event horizon, because matter and energy moving along with the black hole's rotation can resist falling into the black hole more easily than matter and energy moving in the opposite direction.

**Singularity and the Limits to Knowledge** What lies inside a black hole? We cannot answer this question observationally, even in principle, because no information can ever emerge from within the event horizon. Nevertheless, we can use our understanding of the laws of physics to predict what should occur inside a black hole. Because nothing can stop the crush of gravity in a black hole, all the matter that forms a black hole should ultimately be crushed to an infinitely tiny and dense point in the black hole's center. We call this point a **singularity**. (In a rapidly rotating black hole, the singularity may stretch into the shape of a ring around the center, rather than being a precise point.)

Unfortunately, the idea of a singularity pushes up against the limits of scientific knowledge today. The problem is that two very successful theories make very different predictions about the nature of a singularity. Einstein's theory of general relativity, which successfully explains how gravity works throughout the universe, predicts that spacetime should grow infinitely curved as it enters the pointlike singularity. Quantum physics, which successfully explains the nature of atoms and the spectra of light, predicts that spacetime should fluctuate chaotically near the singularity. These are clearly different claims, and no overarching theory that can reconcile them has yet been found. We will not fully understand how a singularity behaves until scientists develop a new theory that encompasses both general relativity and quantum mechanics. Some promising candidates for such a theory are currently being pursued, but we so far lack a way to test them experimentally, and the mathematics needed to make predictions is so complex that some of it has not yet been invented. In the meantime, the uncertainty in our current knowledge is a gold mine for science fiction writers, who speculate about using objects like black holes for exotic forms of travel through spacetime [**Section S3.5**].

Voyage to a Black Hole

## What would it be like to visit a black hole?

Imagine that you are a pioneer of the future, making the first visit to a black hole. Your target is a black hole with a mass of $10M_{\text{Sun}}$ and a Schwarzschild radius of 30 kilometers. As your spaceship approaches the black hole, you fire its engines to put the ship on a circular orbit a few thousand kilometers from the event horizon. This orbit will be stable—there is no need to worry about getting "sucked in."

Your first task is to test Einstein's general theory of relativity. This theory predicts that time should run more slowly as the force of gravity grows stronger. It also predicts that light coming out of a strong gravitational field should show a redshift, called a *gravitational redshift,* that is due to gravity rather than to the Doppler effect. You test these predictions with the aid of two identical clocks whose numerals glow with blue light. You keep one clock aboard the ship and push the other one, with a small rocket attached, directly toward the black hole (**FIGURE 18.13**). The small rocket automatically fires its engines just enough so that the clock falls gradually toward the event horizon. Sure enough, the clock on the rocket ticks more slowly as it heads toward the black hole, and its light becomes increasingly redshifted. When the clock reaches a distance of about 10 kilometers above the event horizon, you see it ticking only half as fast as the clock on your spaceship, and its numerals are red instead of blue.

## MATHEMATICAL INSIGHT 18.1 The Schwarzschild Radius

The Schwarzschild radius ($R_S$) of a black hole is

$$\text{Schwarzschild radius} = R_S = \frac{2GM}{c^2}$$

where $M$ is the black hole's mass, $G = 6.67 \times 10^{-11}\ \text{m}^3/(\text{kg} \times \text{s}^2)$ is the gravitational constant, and $c = 3 \times 10^8$ m/s is the speed of light. With a bit of calculation, this formula can also be written as

$$\text{Schwarzschild radius} = R_S = 3.0 \times \frac{M}{M_{\text{Sun}}}\ \text{km}$$

**EXAMPLE 1:** What is the Schwarzschild radius of a black hole with a mass of $10M_{\text{Sun}}$?

**SOLUTION:**

**Step 1 Understand:** We are given the black hole's mass in solar units, so it will be easiest to compute its Schwarzschild radius with the second version of the formula above.

**Step 2 Solve:** We substitute $M = 10M_{\text{Sun}}$ into the formula:

$$R_S = 3 \times \frac{10M_{\text{Sun}}}{M_{\text{Sun}}}\ \text{km} = 30\ \text{km}$$

**Step 3 Explain:** The Schwarzschild radius of a $10M_{\text{Sun}}$ black hole is about 30 kilometers, which is smaller than the radius of most large cities.

**EXAMPLE 2:** As far as we know, black holes in the present-day universe form only when an object's mass exceeds the roughly $3M_{\text{Sun}}$ neutron star limit. However, some scientists speculate that *mini–black holes* might have formed during the Big Bang. Suppose a mini–black hole has the mass of Earth (about $6 \times 10^{24}$ kg). What is its Schwarzschild radius?

**SOLUTION:**

**Step 1 Understand:** This time we are given the black hole's mass in kilograms, so it is easier to find the Schwarzschild radius with the first version of the formula.

**Step 2 Solve:** We set $M = 6 \times 10^{24}$ kg and use the formula with the given values of $G$ and $c$:

$$R_S = 2 \times \frac{\left(6.67 \times 10^{-11}\ \dfrac{\text{m}^3}{\text{kg} \times \text{s}^2}\right) \times (6 \times 10^{24}\ \text{kg})}{\left(3 \times 10^8\ \dfrac{\text{m}}{\text{s}}\right)^2} \approx 0.009\ \text{m}$$

**Step 3 Explain:** The mini–black hole would have a Schwarzschild radius of only 9 millimeters, making it small enough to fit on the tip of your finger. But don't try to hold it—it would weigh as much as Earth!

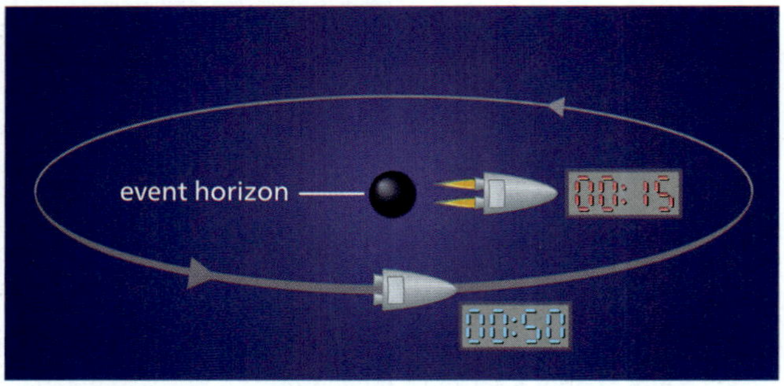


**FIGURE 18.13** Time runs more slowly on the clock nearer to the black hole, and a gravitational redshift makes its glowing blue numerals appear red from your orbiting spaceship.

The rocket has to expend fuel rapidly to keep the clock hovering in the strong gravitational field, and the fuel soon runs out. The clock plunges toward the black hole. From your safe vantage point inside your spaceship, you see the clock ticking more and more slowly as it falls. However, you soon need a radio telescope to "see" it, as the light from the clock face shifts from the red part of the visible spectrum to the infrared, and on into the radio. Finally, its light becomes so redshifted that no conceivable telescope could detect it. Just as the clock vanishes from view, you see that the time on its face has frozen.

Curiosity overwhelms the better judgment of one of your colleagues. He hurriedly climbs into a space suit, grabs the other clock, resets it, and jumps out of the air lock on a trajectory aimed straight for the black hole. Down he falls, clock in hand. He watches the clock, but because he and the clock are traveling together, its time seems to run normally and its numerals stay blue. From his point of view, time seems to neither speed up nor slow down. When his clock reads, say, 00:30, he and the clock pass through the event horizon. There is no barrier, no wall, no hard surface. The event horizon is a mathematical boundary, not a physical one. From his point of view, the clock keeps ticking. He is inside the event horizon, the first human being ever to vanish into a black hole.

Back on the spaceship, you watch in horror as your overly curious friend plunges to his death. Yet, from your point of view, he will *never* cross the event horizon. You'll see time coming to a stop for him and his clock just as he vanishes from view because of the huge gravitational redshift of light. When you return home, you can play a video for the judges at your trial, proving that your friend is still outside the black hole. Strange as it may seem, all this is true according to Einstein's theory. From your point of view, your friend takes *forever* to cross the event horizon (even though he vanishes from view because of his ever-increasing redshift). From his point of view, it is but a moment's plunge before he passes into oblivion.

The truly sad part of this story is that your friend did not live to experience the crossing of the event horizon. The force of gravity grew so quickly as he approached the black hole that it pulled much harder on his feet than on his head, simultaneously stretching him lengthwise and squeezing him from side to side (**FIGURE 18.14**). In essence, your friend was stretched in the same way the oceans are stretched by the tides, except that the *tidal force* near the black hole is trillions of times stronger than the tidal force of the Moon on Earth [**Section 4.5**]. No human could survive it.

If he had thought ahead, your friend might have waited to make his jump until you visited a much larger black hole, like one of the *supermassive black holes* thought to reside in the centers of many galaxies [**Section 21.3**]. A 1 billion solar mass ($10^9 M_{Sun}$) black hole has a Schwarzschild radius of 3 billion kilometers—about the distance from our Sun to Uranus. Although the difficulty of escaping from the event horizon of any black hole is equally great, the larger size of a supermassive black hole makes its tidal forces much weaker and hence nonlethal. Your friend could safely plunge through the event horizon. Unfortunately, anything he saw or learned on his continuing plunge toward oblivion would be known to him alone, because there would be no way for him to send information back to you on the outside.

## COMMON MISCONCEPTIONS

### Black Holes Don't Suck

What would happen if our Sun suddenly became a black hole? For some reason, the idea that Earth and the other planets would inevitably be "sucked in" by the black hole has become part of our popular culture. But it is not true. Although the sudden disappearance of the Sun's light and heat would make our planet cold and dark, Earth's orbit would not change.

Newton's law of gravity tells us that the allowed orbits in a gravitational field are ellipses, hyperbolas, and parabolas [**Section 4.4**]. Note that "sucking" is not on the list! A spaceship would get into trouble only if it came so close to a black hole—within about three times the Schwarzschild radius—that the force of gravity would deviate significantly from what Newton's law predicts. Otherwise, a spaceship passing near a black hole would simply swing around it on an ordinary orbit (ellipse, parabola, or hyperbola). In fact, because most black holes are so small—typical Schwarzschild radii are far smaller than the radius of any star or planet—a black hole is actually one of the most difficult things in the universe to fall into by accident.

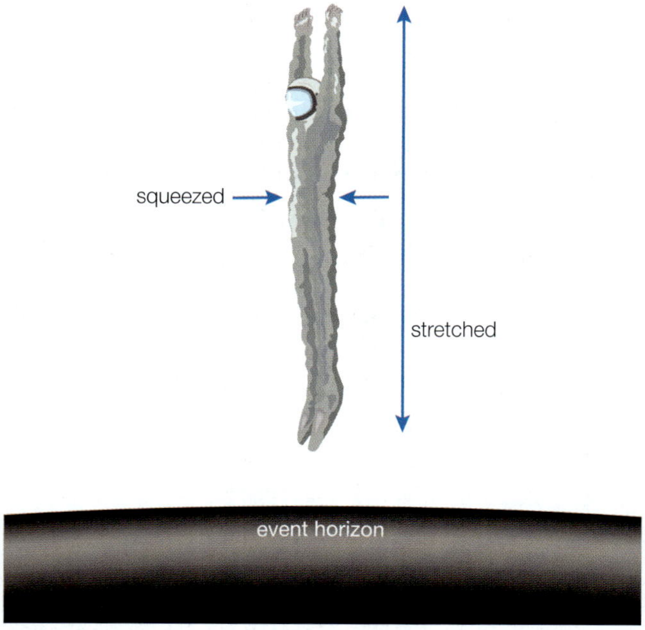


**FIGURE 18.14** Tidal forces would be lethal near a black hole formed by the collapse of a star. The black hole would pull more strongly on the astronaut's feet than on his head, stretching him lengthwise and squeezing him from side to side.

## Do black holes really exist?

As was the case for neutron stars, at first most astronomers who contemplated the idea of black holes thought them too strange to be true. Today, however, both observational evidence and our understanding of physics indicate that black holes are fairly common.

**The Formation of a Black Hole** The idea that black holes ought to exist comes from considering how they might form. Recall that white dwarfs cannot exceed $1.4M_{Sun}$, because gravity overcomes electron degeneracy pressure above that mass. Calculations show that the mass of a neutron star has a similar limit that lies somewhere between about 2 and 3 solar masses. Above this mass, neutron degeneracy pressure cannot hold off the crush of gravity in a collapsing stellar core.

A massive star supernova occurs when the electron degeneracy pressure supporting an iron core succumbs to gravity, causing the core to collapse catastrophically into a ball of neutrons [**Section 17.3**]. That is why most supernovae leave neutron stars behind. However, theoretical models show that very massive stars might not succeed in blowing away all their upper layers. If enough matter falls back onto the neutron core, its mass may rise above the neutron star limit. Likewise, a merger between two neutron stars can produce an object that exceeds the neutron star mass limit.

As soon as the core exceeds the neutron star limit, gravity overcomes the neutron degeneracy pressure and the core collapses once again. This time, no known force can keep the core from collapsing into oblivion as a black hole. Moreover, another consequence of Einstein's theory of relativity makes it highly unlikely that any as-yet-unknown force could intervene and prevent the collapse.

Recall that Einstein's theory tells us that energy is equivalent to mass ($E = mc^2$) [**Section 4.3**], implying that

### EXTRAORDINARY CLAIMS Neutron Stars and Black Holes Are Real

Theoretical calculations predicted the existence of neutron stars and black holes long before their observational discovery, but many astronomers considered these theoretical results too strange to be true.

Subrahmanyan Chandrasekhar, an astrophysicist from India, was only 19 when he completed the calculations showing that there is a white dwarf limit of $1.4M_{Sun}$, and he boldly predicted that a more massive white dwarf would collapse under the force of gravity. He did this work in 1930 while traveling by ship to England, where he hoped to impress the eminent British astrophysicist Sir Arthur Stanley Eddington. However, Eddington ridiculed Chandrasekhar for believing that white dwarfs could collapse. Neutrons had not yet been discovered, little was known about fusion, and no one had any idea what supernovae were. The idea of gravity achieving an ultimate victory seemed nonsensical to Eddington, who speculated that some type of force must prevent gravity from crushing any object.

*Subrahmanyan Chandrasekhar*

*Sir Arthur Stanley Eddington*

A few more radical thinkers took collapsing stars more seriously. A Russian physicist, Lev Davidovich Landau, independently computed the white dwarf limit in 1932. Neutrons were discovered just a few months later, and Landau speculated that stellar corpses above the white dwarf limit might collapse until neutron degeneracy pressure halted the crush of gravity. While most astronomers found the idea of neutron stars to be unacceptably weird, two European scientists who had emigrated to California, Fritz Zwicky and Walter Baade, were not so skeptical. Without knowing of Landau's ideas, they independently concluded that neutron stars were possible. In 1934, they suggested that a supernova might result when a stellar core collapses and forms a neutron star—an extraordinarily insightful guess. By 1938, physicist Robert Oppenheimer, working at Berkeley, was contemplating whether neutron stars had a limiting mass of their own. He and his coworkers concluded that the answer was yes and that neutron degeneracy pressure could not resist the crush of gravity when the mass rose above a few solar masses. Because no known force could keep such a star from collapsing indefinitely, Oppenheimer speculated that gravity would achieve ultimate victory, crushing the star into a black hole.

*Lev Davidovich Landau*

*Robert Oppenheimer*

Astronomers gradually came to accept Chandrasekhar's $1.4M_{Sun}$ white dwarf limit, because observations found no white dwarfs more massive than this. However, most astronomers held to a belief that high-mass stars would inevitably shed enough mass late in life to prevent the formation of a more massive collapsed object. Jocelyn Bell's 1967 discovery of pulsars shattered this belief. Within a few months, Thomas Gold of Cornell University correctly suggested that the pulsars were spinning neutron stars. These discoveries forced astronomers to acknowledge that nature was far stranger than they had expected. The verification that neutron stars really exist made the prospect of the still stranger black holes much less difficult to accept.

*Jocelyn Bell*

Chandrasekhar, who had long since moved to the University of Chicago, was awarded a Nobel Prize in 1983 for his lifelong contributions to astronomy. Landau won a Nobel Prize in 1962 for his work on condensed states of matter. Oppenheimer went on to lead the Manhattan Project, which developed the atomic bomb in 1945. Eddington died in 1944, still convinced that white dwarf stars could not collapse. And in 2018, Jocelyn Bell Burnell won a Breakthrough Prize for the discovery of pulsars.

**Verdict:** Strong evidence supports the reality of neutron stars and black holes.

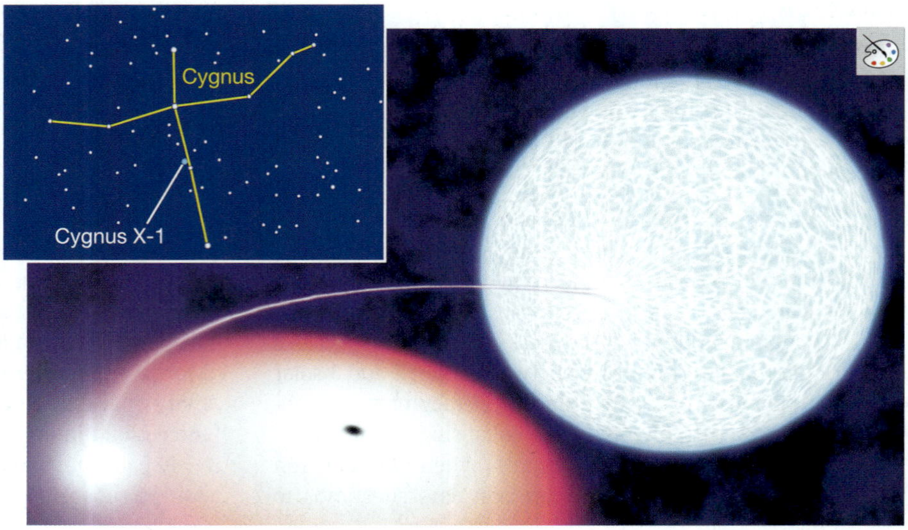


**FIGURE 18.15** Artist's conception of the Cygnus X-1 system, whose name reflects the fact that it is the brightest x-ray source in the constellation Cygnus. The x-rays come from the high-temperature gas in the accretion disk surrounding the black hole.

energy, like mass, must exert some gravitational attraction. The gravity of pure energy usually is negligible, but pure energy becomes a powerful source of gravity in a stellar core collapsing beyond the neutron star limit. Once the core collapses beyond this point, the energy associated with the rapidly rising temperature and pressure acts like additional mass, making the crushing power of gravity even stronger. The more the core collapses, the stronger gravity gets. To the best of our understanding, *nothing* can halt the crush of gravity at this point. The core collapses without end, forming a black hole. Gravity has achieved its ultimate triumph.

**Observational Evidence for Black Holes** The fact that black holes emit no light might make it seem as if they should be impossible to detect. However, a black hole's gravity can influence its surroundings in ways that reveal its presence. Astronomers have discovered many objects that show the telltale signs of an unseen gravitational influence that has a large enough mass to suggest that it is a black hole, and these discoveries provided the first strong observational evidence for black holes.

For black holes formed by supernovae, the key evidence comes from careful studies of x-ray binaries. Recall that the accretion disks around neutron stars in close binary systems can emit strong x-ray radiation, making an x-ray binary. The accretion disk forms because the neutron star's strong gravity pulls mass from the companion star. Because a black hole has even stronger gravity than a neutron star, a black hole in a close binary system should also be surrounded by a hot, x-ray–emitting accretion disk. Some x-ray binaries may therefore contain black holes rather than neutron stars.

One of the most promising black hole candidates is in an x-ray binary called Cygnus X-1 (**FIGURE 18.15**). This system contains an extremely luminous star with an estimated mass of $19M_{\text{Sun}}$. Based on Doppler shifts of its spectral lines, astronomers have concluded that this star orbits a compact, unseen companion with a mass of about $15M_{\text{Sun}}$. Although there is some uncertainty in these mass estimates, the mass of the invisible accreting object clearly exceeds the $3M_{\text{Sun}}$ neutron star limit. It is therefore too massive to be a neutron star, so according to current knowledge it must be a black hole. A few dozen other x-ray binaries offer similarly strong evidence for black holes formed from the collapse of massive stellar cores.

**Think about it** Recall that some x-ray binaries that contain neutron stars emit frequent x-ray bursts and are called *x-ray bursters*. Could an x-ray binary that contains a black hole exhibit the same type of x-ray bursts? Why or why not? (*Hint:* Where do the x-ray bursts occur in an x-ray binary with a neutron star?)

An even greater body of observational evidence suggests the existence of *supermassive black holes*—some with masses millions or billions of times that of our Sun—residing at the centers of many galaxies. As we will discuss in Chapter 21, these black holes must have formed in a different way than the stellar-mass black holes in x-ray binaries did, and they are thought to power some of the most luminous objects in the universe.

In addition, astronomers have recently detected clear evidence of black holes not with light, but with gravitational waves. This line of evidence is the strongest evidence we have for the existence of black holes, and it comes from observations of some of the most extreme events that ever occur in the universe.

## 18.4 Extreme Events

Supernovae that produce neutron stars are rare and enormously energetic events. However, other events can be even more energetic, including supernovae that produce black holes and the mergers of pairs of neutron stars or black holes. To identify such events, astronomers watch for extreme outbursts of energy.

### What causes gamma-ray bursts?

In the early 1960s, the United States began launching a series of top-secret satellites designed to look for gamma rays emitted by nuclear bomb tests. The satellites soon began detecting occasional bursts of gamma rays, typically lasting a few seconds (**FIGURE 18.16**). It took several years for military scientists to become convinced that these **gamma-ray bursts** were coming from space, not from some sinister human activity, and the military then made the

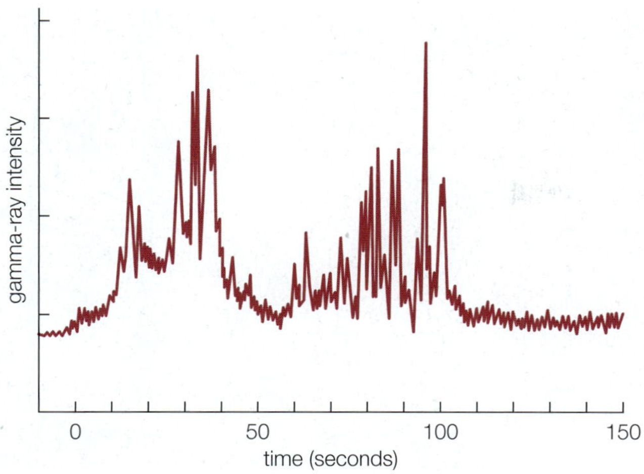


**FIGURE 18.16** The intensity of a gamma-ray burst can fluctuate dramatically over time periods of just a few seconds.

discovery public. We have since learned that gamma-ray bursts represent explosions of almost unimaginable power.

Gamma-ray bursts originating billions of light-years from Earth can produce afterglows of visible light that can be seen through binoculars. To shine so brightly from such great distances, gamma-ray bursts must be by far the most powerful bursts of energy that we observe in the universe. If a gamma-ray burst shined light equally in all directions, like a light bulb, then its total luminosity would briefly exceed the combined luminosity of a million galaxies like our Milky Way. Because such a high luminosity is very difficult to explain, scientists think it's likely that gamma-ray bursts channel their energy into narrow searchlight beams, like pulsars, so only some are visible from Earth. But even in this case, for a brief moment, a gamma-ray burst is far more luminous than anything else in the universe.

What could cause such massive outbursts of energy? At least some gamma-ray bursts come from extremely powerful supernova explosions. A supernova that forms a neutron star does not release enough energy to produce the luminosity of the brightest gamma-ray bursts. However, a supernova that forms a black hole crushes even more matter into an even smaller radius, releasing many times more gravitational potential energy than one that forms a neutron star. This kind of event (sometimes called a *hypernova*) might be powerful enough to explain the most extreme gamma-ray bursts.

The primary evidence linking gamma-ray bursts to exploding stars comes from observing them in other wavelengths of light, including visible light and x-rays. Some gamma-ray telescopes are capable of rapidly pinpointing the location of a gamma-ray burst in the sky, thereby allowing other types of telescopes to be pointed at the gamma-ray source almost as soon as the burst is detected. Observations like these have shown that at least some gamma-ray bursts coincide with powerful supernova explosions (**FIGURE 18.17**). In a number of other cases, the gamma-ray bursts are observed to come from distant galaxies that are actively forming new stars. Those observations also support the idea that gamma-ray bursts are produced in the explosions of extremely massive stars, because such stars are very short-lived and therefore should be found only in places where stars are actively forming.

Supernovae forming black holes probably account for most gamma-ray bursts, but about 30% of all gamma-ray bursts are somewhat different in character. Bursts of this second type last only a few seconds, making them more difficult to catch in action with telescopes for other wavelengths. Nevertheless, recent observations have confirmed what astronomers long suspected might be their cause, which brings us to the next topic.

**Gravitational Waves**

## What happens when neutron stars merge?

Just like two closely orbiting white dwarfs, two neutron stars orbiting close together will emit gravitational waves, ultimately causing the two neutron stars to spiral together and merge (see Figure S3.22). The total energy released during this cataclysmic merger can be even greater than that of a massive star supernova, and it happens in an environment rich in neutrons. As a result, neutron-star mergers can produce a blend of rare elements that differ from those made in a massive star supernova. Most of the neutron-star matter remains gravitationally bound to the system, but some of it is ejected into interstellar space and then reincorporated into new star systems. Models of these mergers indicate that they produce most of the gold, platinum, and rare-earth elements present in the universe (see Appendix D). In other words, the models indicate that the gold in your jewelry was probably made by neutron-star mergers.

Those models remained hypothetical until the LIGO experiment (see Figure 6.31) obtained the first detection of

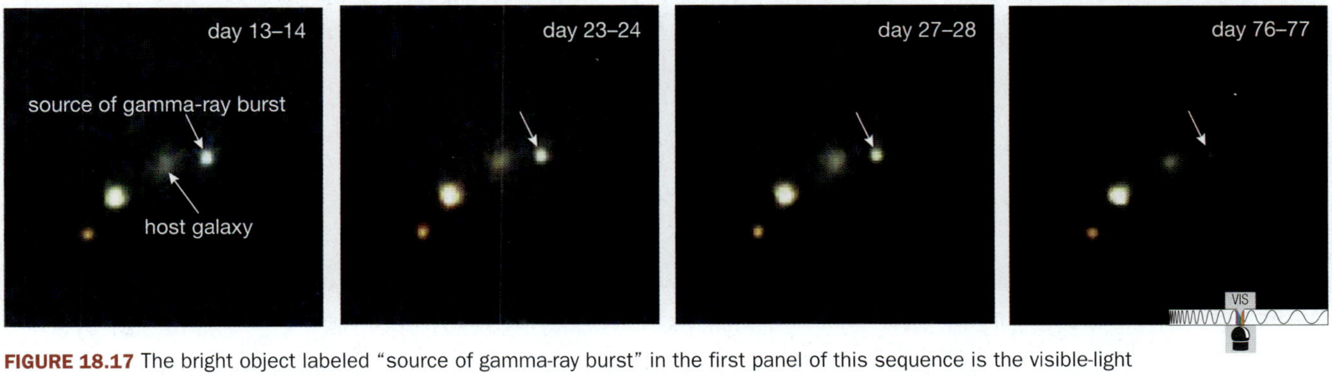


**FIGURE 18.17** The bright object labeled "source of gamma-ray burst" in the first panel of this sequence is the visible-light afterglow of a gamma-ray burst in the outskirts of a distant galaxy (labeled "host galaxy"). Continued monitoring of the light from this object once the initial afterglow faded showed that it had the light curve of a massive star supernova.

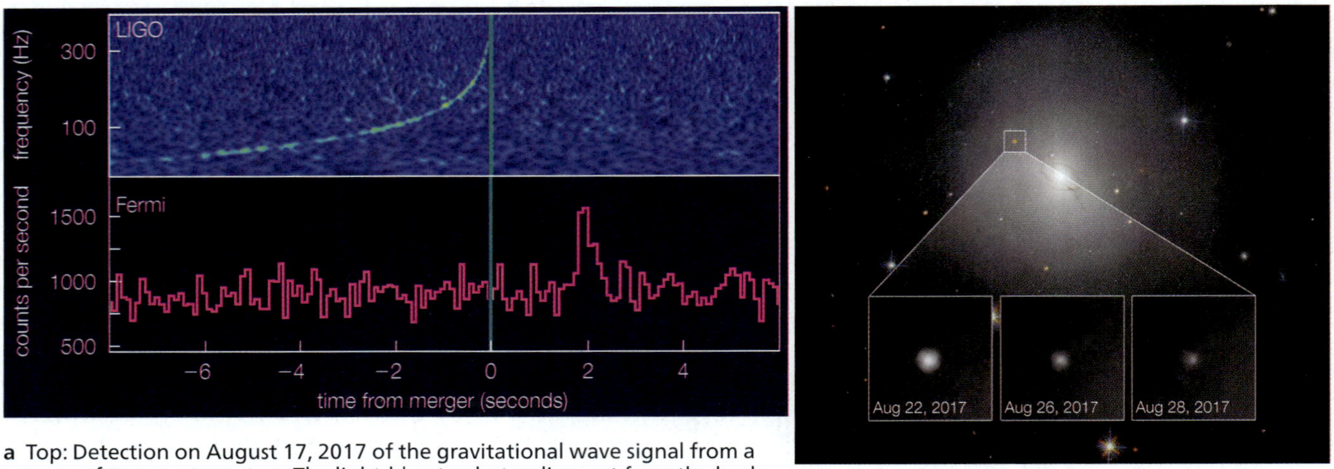


**a** Top: Detection on August 17, 2017 of the gravitational wave signal from a merger of two neutron stars. The light-blue track standing out from the background noise (darker speckles) shows the steady climb of the signal's frequency from below 100 Hz through more than 300 Hz just before the merger. The vertical green line shows the moment of the merger event. The signal's peak frequency shows that the neutron stars were orbiting one another more than 300 times per second just before the merger. Bottom: Detection by the Fermi observatory of a gamma-ray burst from the merger two seconds after the end of the gravitational-wave signal.

**b** Visible light from the aftermath of the neutron-star merger shown in part a, as seen by the Hubble Space Telescope. Information from the LIGO signal allowed astronomers to find this afterglow in the galaxy NGC 4993, about 130 million light-years away. The inset frames show the afterglow fading during the two weeks following the gravitational-wave detection.

**FIGURE 18.18** Observed signals from the first detection of a merger of two neutron stars.

gravitational waves from a merger between two neutron stars on August 17, 2017 (**FIGURE 18.18a**, top panel). A burst of gamma rays, seen by the *Fermi* satellite 2 seconds after the peak of the gravitational-wave burst, confirmed that neutron-star mergers can produce the short type of gamma-ray bursts (Figure 18.18a, bottom panel). Rapid observational follow-up in many other bands of the electromagnetic spectrum found an afterglow (sometimes called a *kilonova*), showing that the merger happened in a galaxy called NGC 4993, located about 130 million light-years from Earth (**FIGURE 18.18b**).

## What happens when black holes merge?

We have seen that white dwarfs and neutron stars in close binary systems can spiral together and merge because of the emission of gravitational waves. Binary black holes can do the same thing, but their greater masses (compared to white dwarfs and neutron stars) lead to much stronger emission of gravitational waves (**FIGURE 18.19**). As with the white dwarf and neutron star systems, the two black holes can ultimately merge together, in this case making a single more massive black hole.

**Think about it** Is the radius of the event horizon of the black hole resulting from the merger of a black-hole binary system larger or smaller than the event-horizon radii of the two original black holes? Explain.

Models predict that gravitational waves from a black-hole binary become particularly intense during the final death spiral preceding a merger. The resulting pulses of gravitational radiation are so powerful that we can detect black-hole mergers that happen billions of light-years away. LIGO first detected such gravitational waves in 2015 (**FIGURE 18.20**), and the close match between those signals and supercomputer models of black-hole mergers represents the strongest evidence we have for the existence of black holes.

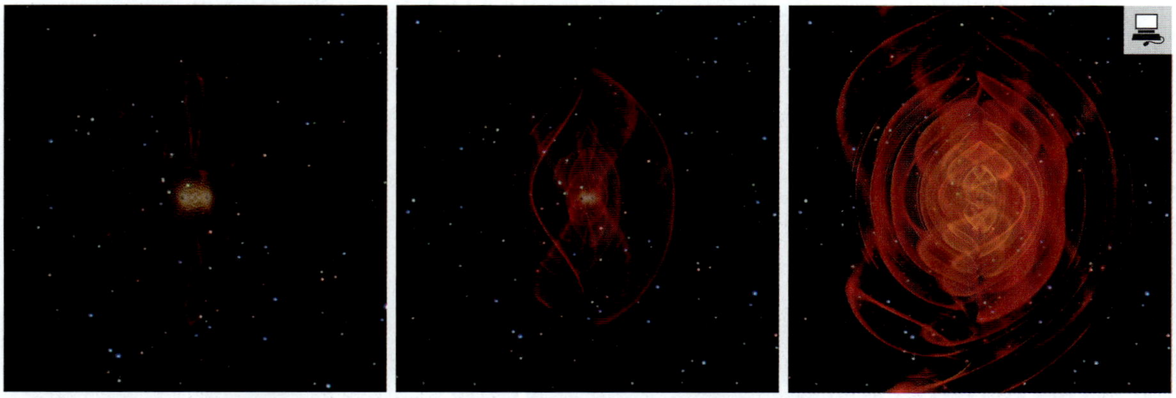

**FIGURE 18.19** These three panels illustrate gravitational waves coming from a black-hole binary. They are based on a supercomputer simulation of the final moments before the black holes merge, with time progressing from left to right. Gravitational waves are depicted in red, orange, and yellow, in order of increasing intensity. The sequence shows that the gravitational-wave intensity from a black-hole binary system rapidly increases as the black holes spiral together.

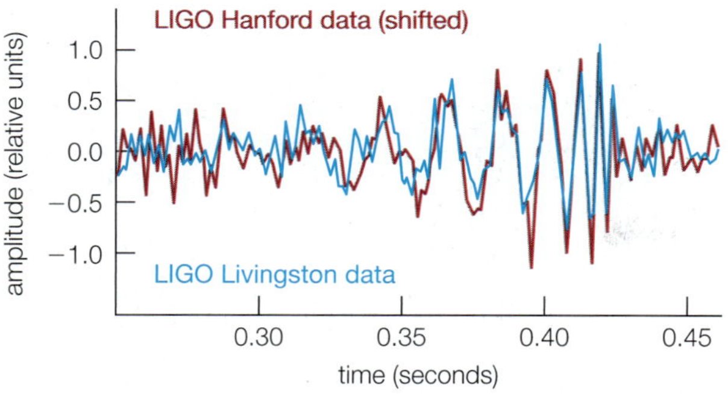


**FIGURE 18.20** Gravitational-wave signal observed on September 14, 2015 by both detectors of the LIGO experiment, which are located in Hanford, Washington and Livingston, Louisiana. In order to measure the signal, the LIGO detectors needed to be sensitive enough to record vibrations (strain) of less than one part in $10^{21}$. Analysis of the signal shows that it came from the merger of two black holes, each about 30 solar masses, at a distance of 1.3 billion light-years.

## The BIG Picture PUTTING CHAPTER 18 INTO CONTEXT

*We have now seen the mind-bending consequences of stellar death. As you think about the unusual objects described in this chapter, try to keep in mind these "big picture" ideas:*

- Despite the strange nature of stellar corpses, clear evidence exists for white dwarfs and neutron stars, and the case for black holes is very strong.
- White dwarfs, neutron stars, and black holes can all have close stellar companions from which they accrete matter. These binary systems produce some of the most spectacular events in the universe, including novae, white dwarf supernovae, and x-ray bursters.
- Black holes are holes in the observable universe that strongly warp space and time around them. The nature of black hole singularities remains beyond the frontier of current scientific understanding.

**MY COSMIC PERSPECTIVE** The idea of a black hole—something with a pull so strong that nothing can escape—is a common metaphor in popular culture, but we humans have little to fear from real black holes. They are at such great distances that their powerful gravitational forces have little chance of ever affecting us.

## Summary of Key Concepts

### 18.1 White Dwarfs

- **What is a white dwarf?** A **white dwarf** is the core left over from a low-mass star, supported against the crush of gravity by **electron degeneracy pressure**. A white dwarf typically has the mass of the Sun compressed into a size no larger than that of Earth.
- **What can happen to a white dwarf in a close binary system?** A white dwarf in a close binary system can acquire hydrogen from its companion through an **accretion disk** in which matter swirls toward the white dwarf's surface. As hydrogen builds up on the white dwarf's surface, it may ignite nuclear fusion and produce a **nova** that, for a few weeks, may shine as brightly as 100,000 Suns. In extreme cases, ongoing accretion or a merger of two white dwarfs may cause a white dwarf's mass to approach the **white dwarf limit** of $1.4M_{Sun}$, at which point the white dwarf will explode as a **white dwarf supernova**.

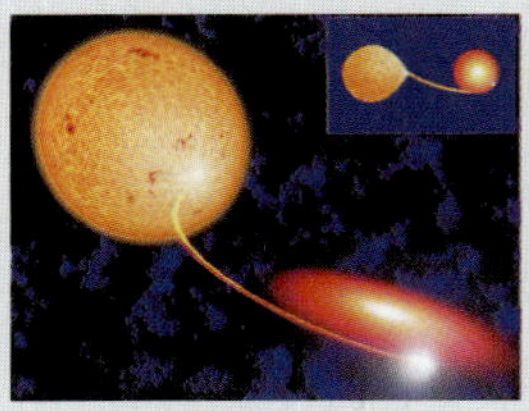

### 18.2 Neutron Stars

- **What is a neutron star?** A **neutron star** is the ball of neutrons created by the collapse of the iron core in a massive star supernova. It resembles a giant atomic nucleus 10 kilometers in radius but more massive than the Sun.
- **How were neutron stars discovered?** Neutron stars spin rapidly when they are born, and their strong magnetic fields can direct beams of radiation that sweep through space as the neutron stars spin. We see such neutron stars as **pulsars**, and these pulsars provided the first direct evidence for the existence of neutron stars.

- **What can happen to a neutron star in a close binary system?** Neutron stars in close binary systems can accrete hydrogen-rich material from their companions, forming dense, hot accretion disks. The hot gas emits strongly in x-rays, so we see these systems as **x-ray binaries**. In some of these systems, helium fusion ignites frequently on the neutron star's surface, resulting in **x-ray bursts**.

### 18.3 Black Holes: Gravity's Ultimate Victory

- **What is a black hole?** A black hole is a place where gravity has crushed matter into oblivion, creating a hole in the universe from which nothing can escape, not even light. The **event horizon** marks the boundary between our observable universe

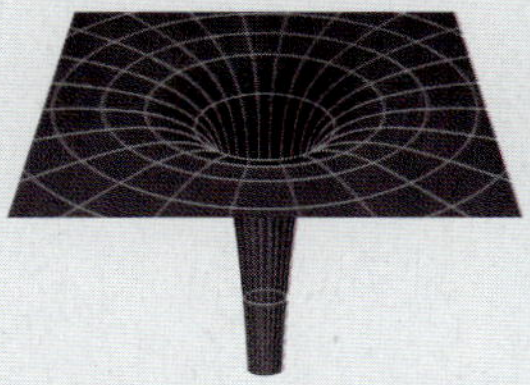

and the inside of the black hole; the size of a black hole is characterized by its **Schwarzschild radius**. A black hole has only three basic properties: its mass, electric charge, and angular momentum.

- **What would it be like to visit a black hole?** You could orbit a black hole just like any other object of the same mass. However, you'd see some strange effects if you watched an object fall toward the black hole: Time would seem to run more slowly for the object, and light from it would appear increasingly redshifted as the object fell closer to the black hole. The object would never appear to pass through the event horizon but instead would disappear from view as its light became so redshifted that no instrument could detect it.

- **Do black holes really exist?** No known force can stop the collapse of a stellar corpse with a mass above the neutron star limit of 2 to 3 solar masses, and theoretical studies of supernovae suggest that such objects should sometimes form. Observational evidence supports this idea: Some x-ray binaries include compact objects far too massive to be neutron stars, making it likely that they are black holes.

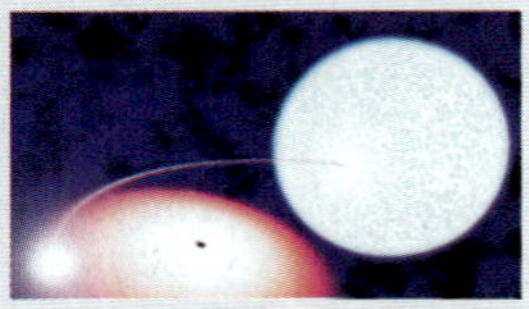

## 18.4 Extreme Events

- **What causes gamma-ray bursts? Gamma-ray bursts** occur in distant galaxies but shine so brightly in the sky that they must be the most powerful explosions we ever observe in the universe. Most gamma-ray bursts appear to come from unusually powerful supernova explosions that may create black holes. Less common short gamma-ray bursts may come from mergers of neutron stars in close binary systems.

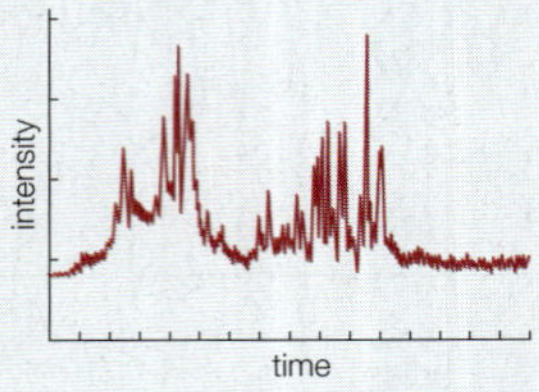


- **What happens when neutron stars merge?** Close binary systems consisting of two neutron stars emit gravitational waves that cause the average orbital distance to decrease with time until the neutron stars merge. Those mergers may be the primary sources of gold, platinum, uranium, and various other rare elements in the universe.

- **What happens when black holes merge?** Close binary systems consisting of two black holes also emit gravitational waves that lead to mergers. Gravitational waves from such systems have been directly detected and represent the strongest evidence we have for the existence of black holes.

# Visual Skills Check

*Use the following questions to check your understanding of some of the many types of visual information used in astronomy. For additional practice, try the Chapter 18 Visual Quiz in the Study Area at www.MasteringAstronomy.com.*

*Figure 18.5, repeated to the right, shows how the luminosities of supernovae change with time. Answer the following questions using the information provided in the figure.*

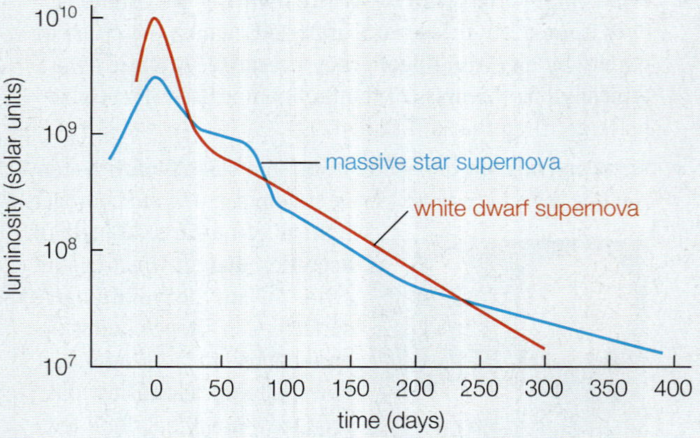


1. At peak brightness, the white dwarf supernova is approximately ______ times as luminous as the massive star supernova at peak brightness.
   a. 1.5
   b. 3
   c. 10
   d. 100
2. The luminosity of the white dwarf supernova 175 days after it reaches peak brightness is about ______ of its luminosity at peak brightness.
   a. 30%
   b. 10%
   c. 3%
   d. 1%
3. Approximately how many days does it take for a white dwarf supernova to decline to 10% of its peak brightness?
   a. 3 days
   b. 30 days
   c. 170 days
   d. 300 days
4. Approximately how many days does it take for a massive star supernova to decline to 10% of its peak brightness?
   a. 10 days
   b. 30 days
   c. 100 days
   d. 300 days
5. Approximately how many days does it take for a massive star supernova to decline to 1% of its peak brightness?
   a. 3 days
   b. 30 days
   c. 170 days
   d. 300 days

# Exercises and Problems

*For instructor-assigned homework and other learning materials, go to www.MasteringAstronomy.com.*

## Chapter Review Questions

### Short-Answer Questions Based on the Reading

1. What is *degeneracy pressure*, and how is it important to the existence of white dwarfs and neutron stars? What is the difference between *electron degeneracy pressure* and *neutron degeneracy pressure*?
2. Describe the mass, size, and density of a typical *white dwarf*. How does the size of a white dwarf depend on its mass?
3. What happens to the electron speeds in a more massive white dwarf, and how does this behavior lead to the *white dwarf limit* for mass?
4. What is an *accretion disk*? Describe how an accretion disk can provide a white dwarf with a new source of energy.
5. What is a *nova*? Describe the process that creates a nova and what a nova looks like.
6. What processes may cause a *white dwarf supernova*? Observationally, how do we distinguish white dwarf and massive star supernovae?
7. Describe the mass, size, and density of a typical *neutron star*. What would happen if a neutron star came to your hometown?
8. How do we know that *pulsars* are neutron stars? Are all neutron stars also pulsars? Explain.
9. Explain how the presence of a neutron star can make a close binary star system appear to us as an *x-ray binary*. Why do some of these systems appear to us as *x-ray bursters*?
10. In what sense is a black hole like a hole in the observable universe? Define the *event horizon* and the *Schwarzschild radius*, and describe the three basic properties of a black hole.
11. What do we mean by the *singularity* of a black hole? How do we know that our current theories are inadequate to explain what happens at the singularity?
12. Suppose you are falling into a black hole. How will you perceive the passage of your own time? How will outside observers see time passing for you? Briefly explain why your trip into a stellar-mass black hole is likely to be lethal.
13. Why do we think that supernovae should sometimes form black holes? What observational evidence supports the existence of black holes?
14. What are *gamma-ray bursts*, and how do we think they are produced?
15. Why can emission of gravitational waves lead to mergers of white dwarfs, neutron stars, and black holes?
16. Describe the event that led to the first detection of gravitational waves. How was it detected?

### Does It Make Sense?

*Decide whether or not each of the following statements makes sense (or is clearly true or false). Explain clearly; not all of these have definitive answers, so your explanation is more important than your chosen answer.*

17. The white dwarf at the center of the Helix Nebula has a mass three times the mass of our Sun.
18. I observed a white dwarf supernova occurring at the location of a single (not binary) white dwarf.
19. If you want to find a pulsar, you should look near the remnant of a supernova described by ancient Chinese astronomers.
20. It's the year 2030, and scientists have just learned that there is a $10M_{\text{Sun}}$ black hole lurking near Pluto's orbit.
21. If your spaceship flew within a few thousand kilometers of a black hole, you and your ship would be rapidly sucked into it.
22. We can detect black holes with x-ray telescopes because matter falling into a black hole emits x-rays after it smashes into the event horizon.
23. From your point of view, an object falling toward a black hole will *never* cross the event horizon.
24. The best way to search for black holes is to look for small black circles in the sky.
25. Gamma-ray bursts are more likely to be observed in galaxies that are rapidly forming new stars than in galaxies containing only old stars.
26. Gravitational waves are best observed with the Hubble Space Telescope.

### Quick Quiz

*Choose the best answer to each of the following. For additional practice, try the Chapter 18 Reading and Concept Quizzes in the Study Area at www.MasteringAstronomy.com.*

27. Which of these objects has the *smallest* radius? (a) a $1.2M_{\text{Sun}}$ white dwarf (b) a $0.6M_{\text{Sun}}$ white dwarf (c) Jupiter
28. What happens if a white dwarf reaches the 1.4 $M_{\text{Sun}}$ limit? (a) It explodes as a white dwarf supernova. (b) It collapses to become a neutron star. (c) It collapses to become a black hole.
29. If we see a *nova*, we know that we are observing (a) a rapidly rotating neutron star. (b) a gamma ray–emitting supernova. (c) a white dwarf in a binary system.
30. A *pulsar* is (a) an unstable high-mass star. (b) an accreting white dwarf. (c) a rapidly rotating neutron star.
31. What would happen if the Sun suddenly became a black hole without changing its mass? (a) The black hole would quickly suck in Earth. (b) Earth would gradually spiral into the black hole. (c) Earth would remain in the same orbit.
32. What makes us think that Cygnus X-1 contains a black hole? (a) We can directly observe that one member of the system emits no light. (b) The unseen object orbited by a luminous star is too massive to be a neutron star. (c) The strong x-ray emission from the system means it must contain a black hole.
33. Viewed from a distance, how would a flashing red light appear as it fell into a black hole? (a) It would appear to flash more quickly. (b) Its flashes would appear bluer. (c) Its flashes would shift to the infrared part of the spectrum.
34. Which of these black holes exerts the *weakest* tidal force on an object near its event horizon? (a) a $10M_{\text{Sun}}$ black hole (b) a $100M_{\text{Sun}}$ black hole (c) a $10^6M_{\text{Sun}}$ black hole
35. Current evidence indicates that most gamma-ray bursts come from (a) supernovae that leave a black hole behind. (b) unusually massive x-ray bursters. (c) the merger of two black holes.
36. Why do some pairs of neutron stars collide and merge? (a) Occasionally a neutron star moving through space will collide head-on with another neutron star. (b) Gravitational waves from close neutron star binary systems carry away orbital energy and angular momentum. (c) Electromagnetic waves from pulsars carry away angular momentum.

## Inclusive Astronomy

*Use these questions to reflect on participation in science.*

37. *Black Holes in Popular Culture.* Expressions such as "it disappeared into a black hole" are now common in popular culture. Give a few other examples of popular expressions in which the term *black hole* is used but is not meant to be taken literally. In what ways are these uses correct in their analogies to real black holes? In what ways are they incorrect? Why do you think such an esoteric scientific idea as that of a black hole has captured the public imagination?
38. *Group Discussion: Graduate Students and Nobel Prizes.* A few astronomers have had the opportunity to do Nobel Prize–winning research while still in graduate school. This discussion asks you to reflect on the reasons some of these people have actually received the Nobel Prize and others have not.
**a.** The Extraordinary Claims box on page 569 briefly discusses Subrahmanyan Chandrasekhar, who won the 1983 Nobel Prize in Physics in part for his discovery in 1930 of the white dwarf limit. He did that work while still a graduate student, and at the time he was ridiculed by the famed astronomer Sir Arthur Stanley Eddington. Chandrasekhar, who was from India, thought that at least some of the ridicule may have been racially motivated. Do you find this possibility plausible? Why or why not? Do you think it could have played a role in the delay of more than 50 years between the time of Chandrasekhar's discovery and his eventual Nobel Prize?
**b.** The Extraordinary Claims box and Section 18.2 discuss the work of Jocelyn Bell Burnell, who discovered the first pulsar in 1967 while working as a graduate student under the mentorship of Antony Hewish. Hewish had led work on the design and construction of the radio telescope that Burnell used in making her discovery. The 1974 Nobel Prize in Physics was awarded based on the pulsar discovery, but it was awarded only to Hewish rather than jointly to him and Burnell. Do you think Burnell should have shared the prize? Why or why not? In what sense could Hewish have been more deserving? Could cultural biases have influenced the Nobel Committee? Do you think similar cultural biases could still be influential today? Why or why not?

## The Process of Science

*These questions may be answered individually in short-essay form or discussed in groups, except where identified as group-only.*

39. *Too Strange to Be True?* Despite strong theoretical arguments for the existence of neutron stars and black holes, many scientists rejected the possibility that such objects could exist until they were confronted with very strong observational evidence. Some people claim that this type of scientific skepticism demonstrates an unwillingness on the part of scientists to give up their deeply held scientific beliefs. Others claim that this type of skepticism is necessary for scientific advancement. What do you think? Defend your opinion.
40. *Gravitational-Wave Detection.* Go to the website for the Laser Interferometer Gravitational Observatory. Look up the detections of gravitational waves that it has made. How many detections have there been so far? What kinds of objects have made the detected waves? What are the approximate distances of those objects? Write a short report on what you learn.
41. *Unanswered Questions.* We have seen in this chapter that theoretical models make numerous predictions about the nature of black holes but leave many questions unanswered. Briefly describe one important but unanswered question related to black holes. If you think it will be possible to answer that question in the future, describe how we would find an answer, being as specific as possible about the evidence necessary to answer the question. If you think the question will never be answered, explain why you think it is impossible to answer.
42. *Group Activity: A Closer Look at Cygnus X-1.* The close binary system Cygnus X-1 contains a supergiant of spectral type O and a black hole with a mass of about $15M_{Sun}$. Consider this system and the artist's conception of it in Figure 18.15 as you work together as a group to do the following. Note: You may wish to do this activity using the four roles described in Chapter 1, Exercise 39.
**a.** Use the H-R diagram in Figure 15.10 to estimate the radius of the supergiant star.
**b.** Use Mathematical Insight 18.1 to estimate the Schwarzschild radius of the black hole. How much larger is the supergiant star than the black hole?
**c.** The supergiant star in Figure 18.15 is drawn to be about 5 centimeters across. Use your result from part b to determine approximately how large (in radius) the black hole would be in this figure if it were drawn to scale. How far off is the figure from this approximate size?
**d.** Discuss whether it is easy or difficult for matter to fall directly into a black hole of this size.
**e.** Notice that the stream of matter from the supergiant star in Figure 18.15 is flowing off to the side of the black hole and onto an accretion disk, rather than going straight into the black hole. Take turns proposing reasons the stream goes off to the side, and then come to a group consensus on the reason for this flow of matter.
**f.** Given your prior results, discuss whether or not the rendering of Figure 18.15 is an "accurate enough" portrayal of the system. What, if anything, would your group do differently if you were asked to make an artist's rendering of the Cygnus X-1 system?

## Investigate Further

### Short-Answer/Essay Questions

*Life Stories of Stars. Write a one- to two-page life story for the scenarios in Problems 43 through 46. Each story should be detailed and scientifically correct but also creative. That is, it should be entertaining and at the same time prove that you understand stellar evolution. Be sure to state whether "you" are a member of a binary system.*

43. You are a white dwarf whose mass is $0.8M_{Sun}$.
44. You are a neutron star whose mass is $1.5M_{Sun}$.
45. You are a black hole whose mass is $10M_{Sun}$.
46. You are a white dwarf in a close binary system that is accreting matter from its companion star.
47. *Census of Stellar Corpses.* Which kind of object do you think is most common in our galaxy: white dwarfs, neutron stars, or black holes? Explain your reasoning.
48. *Fate of an X-Ray Binary.* The x-ray bursts that happen on the surface of an accreting neutron star are not powerful enough to accelerate the exploding material to escape velocity. Predict what will happen in an x-ray binary system in which the companion star eventually feeds more than 3 solar masses of matter into the neutron star's accretion disk.
49. *Why Black Holes Are Safe.* Explain why the principle of conservation of angular momentum makes it very difficult to fall into a black hole.

50. *Surviving the Plunge.* The tidal forces near a black hole with a mass similar to that of a star would tear a person apart before that person could fall through the event horizon. Black hole researchers have pointed out that a fanciful "black hole life preserver" could help counteract those tidal forces. The life preserver would need to have a mass similar to that of an asteroid and would need to be shaped like a flattened hoop placed around the person's waist. In what direction would the gravitational force from the hoop pull on the person's head? In what direction would it pull on the person's feet? Based on your answers, explain in general terms how the gravitational forces from the "life preserver" would help to counteract the black hole's tidal forces.
51. *Research: Black Holes.* Andrew Hamilton, a professor at the University of Colorado, maintains a website with a great deal of information about black holes and what it would be like to visit one. Visit his site and investigate some aspect of black holes that you find to be of particular interest. Write a short report on what you learn.

## Quantitative Problems

*Be sure to show all calculations clearly and state your final answers in complete sentences.*

52. *A Black Hole I?* You've just discovered a new x-ray binary, which we will call Hyp-X1 ("Hyp" for hypothetical). The system Hyp-X1 contains a bright, B2 main-sequence star orbiting an unseen companion. The separation of the stars is estimated to be 20 million kilometers, and the orbital period of the visible star is 4 days.
**a.** Use Newton's version of Kepler's third law to calculate the sum of the masses of the two stars in the system. (*Hint:* See Mathematical Insight 15.4.) Give your answer in both kilograms and solar masses ($M_{Sun} = 2.0 \times 10^{30}$ kg).
**b.** Determine the mass of the unseen companion. Is it a neutron star or a black hole? Explain. (*Hint:* A B2 main-sequence star has a mass of about $10M_{Sun}$.)
53. *A Black Hole II?* You've just discovered another new x-ray binary, which we will call Hyp-X2 ("Hyp" for hypothetical). The system Hyp-X2 contains a bright, G2 main-sequence star orbiting an unseen companion. The separation of the stars is estimated to be 12 million kilometers, and the orbital period of the visible star is 5 days.
**a.** Use Newton's version of Kepler's third law to calculate the sum of the masses of the two stars in the system. (*Hint:* See Mathematical Insight 15.4.) Give your answer in both kilograms and solar masses ($M_{Sun} = 2.0 \times 10^{30}$ kg).
**b.** Determine the mass of the unseen companion. Is it a neutron star or a black hole? Explain. (*Hint:* A G2 main-sequence star has a mass of $1M_{Sun}$.)
54. *White Dwarf Density.* A typical white dwarf has a mass of about $1.0M_{Sun}$ and the radius of Earth (about 6400 kilometers). Calculate the average density of a white dwarf, in *kilograms per cubic centimeter*. How does this compare to the mass of familiar objects?
55. *Neutron Star Density.* A typical neutron star has a mass of about $1.5M_{Sun}$ and a radius of 10 kilometers.
**a.** Calculate the average density of a neutron star, in *kilograms per cubic centimeter*.
**b.** Compare the mass of 1 $cm^3$ of neutron-star material to the mass of Mount Everest ($\approx 5 \times 10^{10}$ kg).
56. *Schwarzschild Radii.* Calculate the Schwarzschild radius (in kilometers) for each of the following.
**a.** A $10^8 M_{Sun}$ black hole in the center of a quasar
**b.** A $5M_{Sun}$ black hole that formed in the supernova of a massive star
**c.** A mini–black hole with the mass of the Moon
**d.** A mini–black hole formed when a super-advanced civilization decides to punish you (unfairly) by squeezing you until you become so small that you disappear inside your own event horizon
57. *The Crab Nebula Pulsar Winds Down.* Theoretical models of the slowing of pulsars predict that the age of a pulsar is approximately equal to $p/2r$, where $p$ is the pulsar's current period and $r$ is the rate at which the period is slowing with time. Observations of the pulsar in the Crab Nebula show that it pulses 30 times a second, so $p = 0.0333$ second, but the time interval between pulses is growing longer by $4.2 \times 10^{-13}$ second with each passing second, so $r = 4.2 \times 10^{-13}$ second per second. Using that information, estimate the age of the Crab Nebula pulsar. How does your estimate compare with the true age of the pulsar, which was born in the supernova observed in A.D. 1054?
58. *A Water Black Hole.* A clump of matter does not need to be extraordinarily dense in order to have an escape velocity greater than the speed of light, as long as its mass is large enough. You can use the formula for the Schwarzschild radius $R_S$ to calculate the volume, $\frac{4}{3}\pi R_S^3$, inside the event horizon of a black hole of mass $M$. What does the mass of a black hole need to be in order for its mass divided by its volume to be equal to the density of water (1 $g/cm^3$)?
59. *Energy of a Supernova.* In a massive star supernova explosion, a stellar core collapses to form a neutron star roughly 10 kilometers in radius. The gravitational potential energy released in such a collapse is approximately equal to $GM^2/r$, where $M$ is the mass of the neutron star, $r$ is its radius, and $G = 6.67 \times 10^{-11}\ m^3/(kg \times s^2)$ is the gravitational constant. Using this formula, estimate the amount of gravitational potential energy released in a massive star supernova explosion. How does it compare with the amount of energy released by the Sun during its entire main-sequence lifetime?
60. *Energy of a Black Hole Merger.* Models indicate that the gravitational-wave signal shown in Figure 18.20 came from the merger of two black holes with masses of $29M_{Sun}$ and $36M_{Sun}$, a merger that resulted in a single black hole with a mass of $62M_{Sun}$. The difference in total mass between the start and finish of the merger corresponds to the amount of energy carried away in the form of gravitational waves. Use Einstein's formula $E = mc^2$ to calculate this amount of energy. Be sure to put the mass in units of kilograms and the speed of light in meters per second, so that the resulting energy has units of joules. Compare your answer to the energy released by a supernova (from Problem 59) and to the energy released by the Sun during its entire main-sequence life.
61. *Challenge Problem: A Neutron Star Comes to Town.* Suppose a neutron star suddenly appeared in your hometown. How thick a layer would Earth form as it wrapped around the neutron star's surface? Assume that the layer formed by Earth has the same average density as the neutron star. (*Hint:* Consider the mass of Earth to be distributed in a spherical *shell* over the surface of the neutron star and then calculate the thickness of such a shell with the same mass as Earth. The volume of a spherical shell is approximately its surface area times its thickness: $V_{shell} = 4\pi r^2 \times$ thickness. Because the shell will be thin, you can assume that its radius is the radius of the neutron star.)

**COSMIC Context** PART V

# Balancing Pressure and Gravity

**We can understand the entire life cycle of a star in terms of the changing balance between pressure and gravity. This illustration shows how that balance changes over time and why those changes depend on a star's birth mass. *(Stars not to scale.)***

**Key**

pressure (red arrow)

gravity (green arrow)

(1) Gravity overcomes pressure inside a protostar, causing the core to contract and heat up. A protostar cannot achieve steady balance between pressure and gravity because nuclear fusion is not replacing the thermal energy it radiates into space [Sections 16.1, 16.2].

**Protostar**

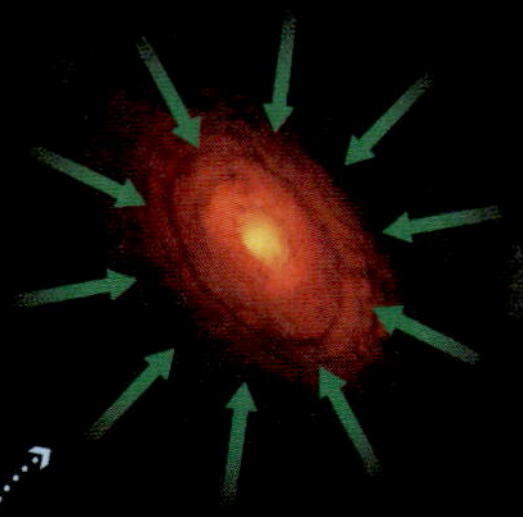

*Contraction converts gravitational potential energy into thermal energy.*

*Degeneracy pressure balances gravity in objects of less than $0.08M_{Sun}$ before their cores become hot enough for steady fusion. These objects never become stars and end up as brown dwarfs.*

$< 0.08M_{Sun}$

$> 0.08M_{Sun}$

(2) Thermal pressure comes into steady balance with gravity when the core becomes hot enough for hydrogen fusion to replace the thermal energy the star radiates from its surface [Section 14.1].

**Main-sequence star**

*Pressure balances gravity at every point within a main-sequence star.*

***The balance between pressure and gravity acts as a thermostat to regulate the core temperature:***

*A drop in core temperature decreases fusion rate, which lowers core pressure, causing the core to contract and heat up.*

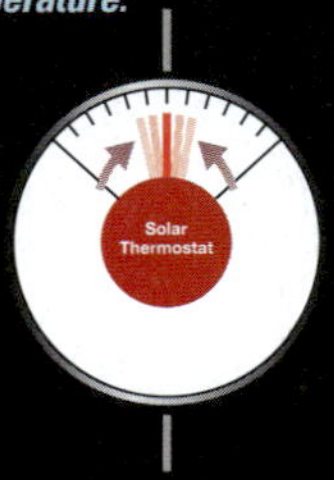


*A rise in core temperature increases fusion rate, which raises core pressure, causing the core to expand and cool down.*

***The balancing point between pressure and gravity depends on a star's mass:***
*Balance between pressure and gravity in high-mass stars results in a higher core temperature, a higher fusion rate, greater luminosity, and a shorter lifetime.*

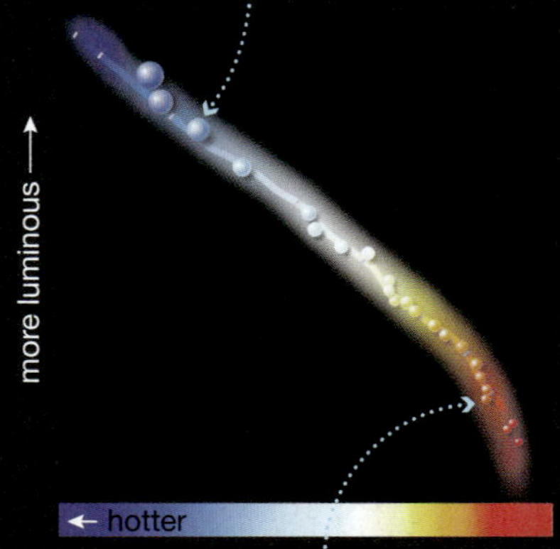


*Balance between pressure and gravity in low-mass stars results in a cooler core temperature, a slower fusion rate, less luminosity, and a longer lifetime.*

(3) The balance tips in favor of gravity after the core runs out of hydrogen. Fusion of hydrogen into helium temporarily stops supplying thermal energy in the core. The core again contracts and heats up. Hydrogen fusion begins in a shell around the core. The outer layers expand and cool, and the star becomes redder [Section 17.2].

**Hydrogen shell-fusion star**

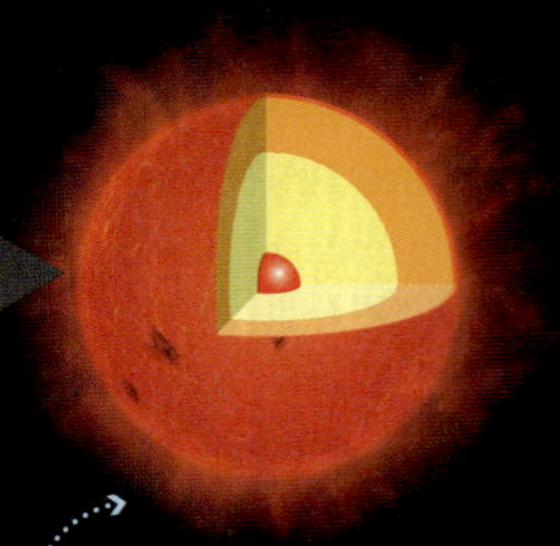

*Luminosity continually rises because core contraction causes the temperature and fusion rate in the hydrogen shell to rise.*

4. Balance between thermal pressure and gravity is restored when the core temperature rises enough for helium fusion into carbon, which can once more replace the thermal energy radiated from the core [Section 17.2].

5. Gravity again gains the upper hand over pressure after the core helium is gone. Just as before, fusion stops replacing the thermal energy leaving the core. The core therefore resumes contracting and heating up, and helium fusion begins in a shell around the carbon core [Section 17.2].

6. In high-mass stars, core contraction continues, leading to multiple shell fusion that terminates with iron and a supernova explosion [Section 17.3].

7. At the end of a star's life, either degeneracy pressure has come into permanent balance with gravity or the star has become a black hole. The nature of the end state depends on the mass of the remaining core [Chapter 18].

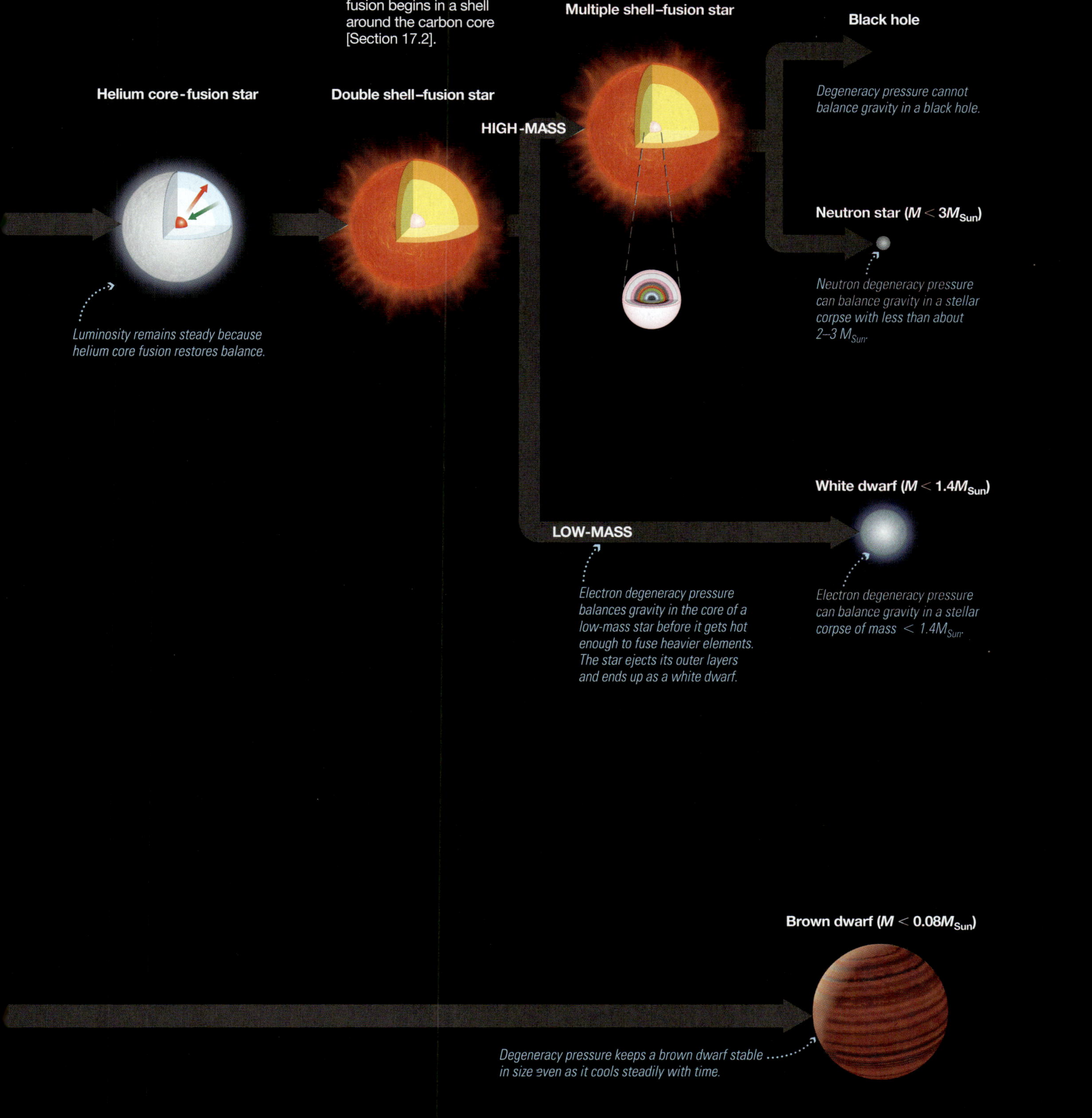

# 19 Our Galaxy

▲ **About the photo:** This photo from the Spitzer Space Telescope shows infrared light from stars and dusty gas clouds within the central 600 light-years of the Milky Way Galaxy.

## LEARNING GOALS

**19.1 The Milky Way Revealed**
- What does our galaxy look like?
- How do stars orbit in our galaxy?

**19.2 Galactic Recycling**
- How is gas recycled in our galaxy?
- Where do stars tend to form in our galaxy?

**19.3 The History of the Milky Way**
- What do halo stars tell us about our galaxy's history?
- How did our galaxy form?

**19.4 The Galactic Center**
- What is the evidence for a black hole at our galaxy's center?

*I still look up at the stars, and it gives me hope, and it gives me energy. I think one of the things we have to think about is, we are all a part of this universe.*

— *Mae Jemison, first African American woman in space*

**Chapter 19 Overview**

In previous chapters, we saw how stars forge new elements and expel them into space, and how new generations of stars and planets are born in interstellar clouds enriched with these stellar by-products. But we have not yet discussed how all these processes interact. You are probably familiar with the idea that all living species on Earth interact with one another and with the land, water, and air to form a large, interconnected ecosystem. In a similar way, but on a much larger scale, our galaxy is a nearly self-contained system that cycles matter from stars into interstellar space and back into stars again. The birth of our solar system and the evolution of life on Earth would not have been possible without this "galactic ecosystem."

In this chapter, we will study our galaxy and the processes that maintain its ecosystem, along with current understanding of galactic history and the evidence for an extremely massive black hole at the galaxy's center. Through it all, we will see that we are not only "star stuff" but "galaxy stuff"—the product of eons of complex recycling and reprocessing of matter and energy in the Milky Way Galaxy.

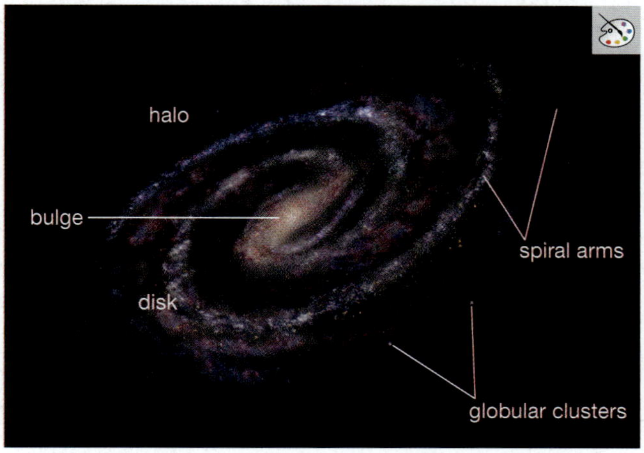


a Artist's conception of the Milky Way viewed from the outside.

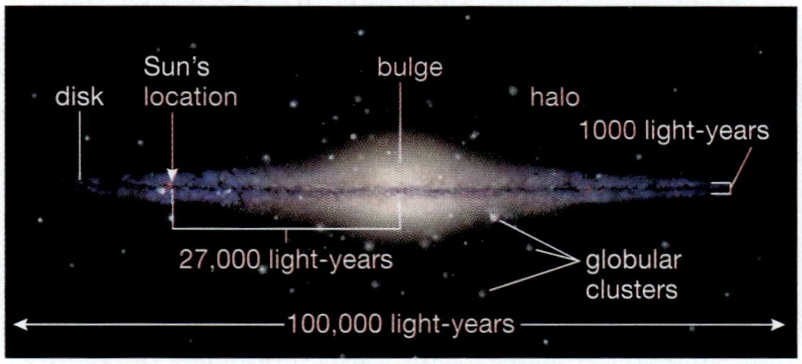


b Edge-on schematic view of the Milky Way.

**FIGURE 19.1** The Milky Way Galaxy.

## 19.1 The Milky Way Revealed

On a dark night, you can see a faint band of light slicing across the sky through several constellations, including Sagittarius, Cygnus, Perseus, and Orion. This band of light looked like a flowing ribbon of milk to the ancient Greeks, so we now call it the *Milky Way* (see Figure 2.1). In the early 17th century, Galileo used his telescope to prove that the light of the Milky Way comes from myriad individual stars. Together these stars make up the kind of stellar system we call a *galaxy,* echoing the Greek word for "milk," *galactos.*

The true size and shape of our Milky Way Galaxy are hard to guess from how it looks in our night sky. Because we live inside the galaxy, trying to determine its structure is somewhat like trying to draw a picture of your house without ever leaving your bedroom. The fact that much of our galaxy's visible light is hidden from our view makes the task even more difficult. Nevertheless, by carefully observing our galaxy and comparing it to others that we see from the outside, we now have a good understanding of the processes that shape it. In this section, we'll begin our exploration of the galaxy by investigating its basic structure and orbital motion.

### What does our galaxy look like?

Our Milky Way Galaxy holds more than 100 billion stars and is just one among some 100 billion galaxies in the observable universe [**Section 1.1**]. Ours is a vast **spiral galaxy**, so named because of the spectacular **spiral arms** illustrated in **FIGURE 19.1a**. If we viewed our galaxy from the side, as shown in **FIGURE 19.1b**, we'd see that the spiral arms are part of a fairly flat **disk** of stars with a bright central **bulge**. The entire disk is surrounded by a dimmer, rounder **halo**. Most of the galaxy's bright stars reside in its disk. The most prominent stars in the halo are found in about 200 *globular clusters* of stars [**Section 15.3**].

The galaxy's disk of stars is about 100,000 light-years in diameter but only about 1000 light-years thick. Our Sun is located in the disk about 27,000 light-years from the galactic center—a little more than halfway out from the center to the edge of the disk. Remember that this distance is incredibly vast [**Section 1.1**]: The few thousand stars visible to the naked eye together fill only a tiny dot in a picture like that in Figure 19.1.

It took us a long time to learn these facts about the Milky Way's size and shape. Recall that the galactic disk is filled with interstellar gas and dust—known collectively as the *interstellar medium* [**Section 16.1**]—that obscures our view of most of our galaxy when we observe just the visible light. This fact long fooled astronomers into believing that we lived near our galaxy's center, a view refuted only around 1920, when astronomer Harlow Shapley showed that the Milky Way's globular clusters are centered on a point tens of thousands of light-years from our Sun (see Special Topic, p. 582).

We now know that the Milky Way is a relatively large galaxy. Within our Local Group of galaxies (see the foldout in the front of the book), only the Andromeda Galaxy is

## COMMON MISCONCEPTIONS

### The Halo of a Galaxy

Astronomers refer to the collection of stars that surround the bulge and disk of our galaxy as the *halo.* However, the word "halo" sometimes causes confusion because it makes people think of the round hoop-like bands of light that can be found encircling the heads of angels and saints in medieval paintings. That's not the shape astronomers have in mind. Our galaxy's halo is roughly spherical and doesn't have a hole in the middle. If you could see it from afar through a very powerful telescope, it would look like a round, faintly glowing ball of light centered on the bulge and encompassing the galactic disk.

comparable in size. The Milky Way's strong gravity influences smaller galaxies in its vicinity. For example, two small galaxies known as the *Large Magellanic Cloud* and the *Small Magellanic Cloud*—both visible to the naked eye from the Southern Hemisphere—orbit the Milky Way at distances of some 150,000 and 200,000 light-years, respectively. Two other small galaxies lie even closer to the Milky Way, but their visible light is obscured from view by the gas and dust in the Milky Way's disk. These two galaxies—known as the Sagittarius Dwarf and Canis Major Dwarf—are each in the process of colliding with the Milky Way's disk. Our galaxy's tidal forces will ultimately rip both of these small galaxies apart.

Keep in mind that while these four nearby galaxies are quite small as galaxies go, they are still vast objects, each containing perhaps 1 billion to a few billion stars. This makes them 1000 or more times the size of typical globular clusters, so they are small only in comparison to an enormous structure like our Milky Way Galaxy.

## How do stars orbit in our galaxy?

A spiral galaxy like ours may look as if it should rotate like a giant pinwheel, but the galaxy is not a solid structure. Each individual star follows its own orbital path around the center of the galaxy. The nature of this orbital path depends primarily on whether the star resides in the disk or in the halo or bulge (**FIGURE 19.2**).

**Orbits of Disk Stars** Stars in the disk orbit the galactic center in roughly circular paths that all go in the same direction in nearly the same plane. However, if you could stand outside the Milky Way and watch it for a few billion years, the disk would resemble a huge merry-go-round. Like horses on a merry-go-round, individual stars bob up and down through the disk as they orbit. The general orbit of a star around the galaxy arises from its gravitational attraction toward the galactic center, while the bobbing arises from the localized pull of gravity within the disk itself. A star that is "too far" above the disk is pulled back into the disk by gravity. Because the density of interstellar gas is too low to slow the star, it flies through the disk until it is "too far" *below* the disk on the other side. Gravity then pulls it back in the other direction. This ongoing process produces the bobbing of the stars.

## SPECIAL TOPIC How Did We Learn the Structure of the Milky Way?

For most of human history, we knew the Milky Way only as an indistinct river of light in the sky. In 1610, Galileo used his telescope to discover that the Milky Way is made up of innumerable stars, but we still did not know the size or shape of our galaxy.

In the late 18th century, British astronomers William and Caroline Herschel (brother and sister) tried to determine the shape of the Milky Way more accurately by counting how many stars lie in each direction. Their approach suggested that the Milky Way's width was five times its thickness. In the early 20th century, Dutch astronomer Jacobus Kapteyn and his colleagues used a more sophisticated star-counting method to gauge the size and shape of the Milky Way. Their results seemed to confirm the Herschels' findings and suggested that the Sun lies very near the center of the galaxy.

Kapteyn's results made astronomers with a sense of history slightly nervous. Only four centuries earlier, before Copernicus challenged the Ptolemaic system, astronomers had believed Earth was the center of the universe. Kapteyn's placement of the Sun near the Milky Way's center seemed to give Earth a central place again. Kapteyn knew that obscuring material, acting like some kind of interstellar fog, could deceive us by hiding the rest of the galaxy, but he found no evidence for such a fog.

While Kapteyn was counting stars, American astronomer Harlow Shapley was studying globular clusters. He found that these clusters appeared to be centered on a location tens of thousands of light-years from the Sun. Shapley concluded that this location marked the true center of our galaxy and that Kapteyn must be wrong.

Today we know that Shapley was right. The Milky Way's interstellar medium is the "fog" that misled Kapteyn. Robert Trumpler, working at California's Lick Observatory in the 1920s, established the existence of this dusty gas by studying open clusters of stars. By assuming that all open clusters had about the same diameter, he estimated their distances from their apparent sizes in the sky, much as you might estimate the distances of cars at night from the apparent separation of their headlights. He found that stars in distant clusters appeared dimmer than expected based on their estimated distance, just as a car's headlights might appear in foggy weather. Trumpler concluded that light-absorbing material fills the spaces between the stars, partially obscuring the distant clusters and making them appear fainter than they would otherwise. We thereby learned that interstellar material had been deceiving earlier astronomers and that the stars visible in the night sky occupy a minuscule portion of the observable universe.

Understanding the effects of interstellar dust allowed astronomers to take it into account in their observations, helping us learn the true size and shape of our galaxy. Then, beginning in the 1950s, careful studies of the motions of stars and gas clouds (made with radio observations of the 21-centimeter line from atomic hydrogen gas) gradually uncovered the detailed structure of the galactic disk, showing us the locations and motions of the spiral arms.

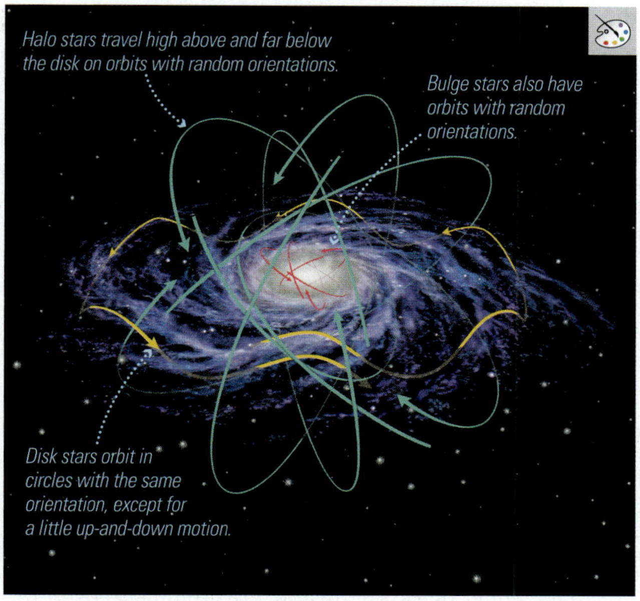


**FIGURE 19.2** Characteristic orbits of disk stars (yellow), bulge stars (red), and halo stars (green) around the galactic center. (The yellow path exaggerates the up-and-down motion of the disk star orbits.)

The up-and-down motions of the disk stars give the disk its thickness of about 1000 light-years—a great distance by human standards, but only about 1% of the disk's 100,000-light-year diameter. In the vicinity of our Sun, each star's orbit takes more than 200 million years, and each up-and-down "bob" takes a few tens of millions of years.

The galaxy's rotation is unlike that of a merry-go-round in one important respect: On a merry-go-round, horses near the edge move much faster than those near the center. But in our galaxy's disk, the orbital velocities of stars near the edge and those near the center are about the same. Stars closer to the center therefore complete each orbit in less time than stars farther out.

**Orbits of Halo Stars** The orbits of stars in the halo are much less organized. Individual halo stars also orbit the galactic center, but the orientations of their paths are relatively random, and neighboring stars may circle the galactic center in opposite directions. Halo stars swoop from high above the disk to far below it and back again, plunging through the disk at velocities so great that the disk's gravity hardly alters their trajectories. Several fast-moving halo stars are currently passing through the disk not too far from our own solar system.

These swooping orbits explain why the halo is much puffier than the disk. Halo stars soar to heights above the disk far greater than the heights achieved by disk stars as they bob up and down. As we'll discuss later, the differences between orbits in the disk and orbits in the bulge and halo provide an important clue to how our galaxy formed.

**Think about it** Is there much danger that a halo star swooping through the disk of the galaxy will someday hit the Sun or Earth? Why or why not? (*Hint*: Consider typical distances between stars on the 1-to-10-billion scale introduced in Chapter 1.)

**Orbits of Bulge Stars** The orbits of stars in the bulge are more difficult to measure than those of disk and halo stars, because the bulge is much farther away from us than the nearest disk and halo stars. Bulge-star orbits were once thought to be similar to halo-star orbits because the bulge is puffier than the disk. However, recent observations have shown that the bulge is more complex, with a range of orbital properties among its stars. Some bulge stars have orbits with random orientations like halo stars, while others orbit in the same general direction as disk stars but with more elongated orbits. These orbital patterns give the bulge a cigar-like shape that, if we could view it from outside the Milky Way, would look like a bar stretching across the center of the galaxy.

**Stellar Orbits and the Mass of the Galaxy** The Sun's orbital path is fairly typical for a disk star. By measuring the speeds of globular clusters relative to the Sun, astronomers have determined that the Sun and its neighbors orbit the center of the Milky Way at a speed of about 220 kilometers per second (about 800,000 km/hr [**Section 1.3**]). Even at this speed, it takes the Sun about 230 million years to

## SPECIAL TOPIC How Do We Determine Stellar Orbits?

Astronomers learned how stars orbit in the Milky Way by measuring the motions of many different stars relative to the Sun. Although these measurements are easy in principle, they can be difficult in practice.

Determining a star's precise motion relative to the Sun requires knowing the star's true velocity through space. However, our primary means of measuring speeds in the universe—the Doppler effect—can tell us only a star's *radial velocity,* the component of its velocity directed toward or away from us (see Figure 5.23). If we want to know the star's true velocity, we must also measure its *tangential velocity,* the component of its velocity directed across our line of sight.

Tangential velocity is difficult to measure because of the vast distances to stars. Over tens of thousands of years, the tangential velocities of stars cause their apparent positions in our sky to change, which alters the shapes of the constellations. These changes are far too small for human eyes to notice. However, we can measure tangential velocities for many stars by comparing telescopic photographs taken years or decades apart. For example, if photographs taken 10 years apart show that a star has moved across our sky by an angle of 1 arcsecond, we know the star is moving at an angular rate of 0.1 arcsecond per year. We can then convert this angular rate of motion—often called the star's *proper motion*—to a tangential velocity if we also know the star's distance. The European Space Agency's *GAIA* mission [**Section 13.4**] has now measured proper motions giving orbital information for more than a billion stars located throughout our galaxy.

complete one orbit around the galactic center. Early dinosaurs were just emerging on Earth when our Sun last visited this side of the galaxy.

The orbital motion of the Sun and other stars gives us a way to determine the mass of the galaxy. Recall that Newton's law of gravity determines how quickly objects orbit one another. This fact, embodied in Newton's version of Kepler's third law [**Section 4.4**], allows us to determine the mass of a relatively large object when we know the period and average distance of a much smaller object in orbit around it.

For example, we can use the Sun's orbital velocity and its distance from the galactic center to determine the mass of our galaxy lying *within* the Sun's orbit (see Mathematical Insight 19.1). To understand why the Sun's orbital motion allows us to calculate only the mass within the Sun's orbit, rather than the total mass of the galaxy, we need to consider the difference between the gravitational effects of mass within the Sun's orbit and those of mass beyond its orbit. Every part of the galaxy exerts gravitational forces on the Sun as it orbits, but the net force from matter outside the Sun's orbit is relatively small because the pulls from opposite sides of the galaxy virtually cancel one another. In contrast, the net gravitational forces from mass within the Sun's orbit all pull the Sun in the same direction—toward the galactic center. The Sun's orbital velocity therefore responds almost exclusively to the gravitational pull of matter inside its orbit. By using the Sun's 27,000-light-year orbital distance and 220-km/s orbital velocity in Newton's version of Kepler's third law, we find that the total amount of mass within the Sun's orbit is about $2 \times 10^{41}$ kg, or about 100 billion times the mass of the Sun.

Similar calculations based on the orbits of more distant stars in the Milky Way have revealed one of the greatest mysteries in astronomy—one that we first encountered in Chapter 1 (see Figure 1.15). Photographs of spiral galaxies make it appear as if most of their mass were concentrated near their centers. However, orbital motions tell us that just the opposite is true. If most of the mass were concentrated near the galaxy's center, the orbital speeds of more distant stars would be slower, just as the orbital speeds of the

## MATHEMATICAL INSIGHT 19.1 Using Stellar Orbits to Measure Galactic Mass

The mass of the galaxy is much greater than the mass of any individual star or gas cloud that orbits within it, so we can use Newton's version of Kepler's third law to calculate the mass of the galaxy *within* an object's orbit. We use the law in the form that is valid when one object is much more massive than the other (see Mathematical Insight 4.3):

$$p^2 = \frac{4\pi^2}{G \times M} \times a^3$$

where $M$ is the mass of the massive object; $p$ and $a$ are the orbital period and semimajor axis of a smaller orbiting object, respectively; and the gravitational constant is $G = 6.67 \times 10^{-11}\ \text{m}^3/(\text{kg} \times \text{s}^2)$. Solving for $M$ gives the equation

$$M = \frac{4\pi^2 \times a^3}{G \times p^2}$$

Because we usually measure orbital *velocities* rather than orbital periods for stars and gas clouds, the above equation is easier to use if we convert it to a form that uses velocity ($v$) rather than period ($p$). In Mathematical Insight 15.4, we found that the orbital speed of an object with a circular orbit is $v = 2\pi a/p$; solving for $p$, we find

$$p = \frac{2\pi a}{v}$$

With this expression for $p$, our equation for $M$ becomes

$$M = \frac{4\pi^2 \times a^3}{G \times \left(\frac{2\pi a}{v}\right)^2}$$

We simplify by expanding the square and canceling like terms:

$$M = \frac{4\pi^2 \times a^3}{G \times \frac{4\pi^2 a^2}{v^2}} = \frac{a \times v^2}{G}$$

Finally, because stars and gas clouds generally have circular orbits around the galactic center, we replace $a$ with the orbital radius $r$. We then write the mass as $M_r$, to remind us that the formula tells us the mass of the galaxy *within* a distance $r$ from the center:

$$M_r = \frac{r \times v^2}{G}$$

We will call this equation the **orbital velocity law**, because it allows us to calculate the galaxy's mass from the orbital velocity (and orbital distance) of a star or a gas cloud.

**EXAMPLE:** Calculate the mass of the Milky Way Galaxy within the Sun's orbit.

**SOLUTION:**

**Step 1 Understand:** We can use the orbital velocity law with the Sun's orbital velocity of 220 km/s and distance of 27,000 light-years. For consistency with the given units of $G$, we convert distances to meters, so $v = 2.2 \times 10^5$ m/s and $r = 2.6 \times 10^{20}$ m (because 1 light-year $= 9.46 \times 10^{15}$ meters).

**Step 2 Solve:** We use the Sun's values for $v$ and $r$ in the orbital velocity law:

$$M_r = \frac{r \times v^2}{G} = \frac{(2.6 \times 10^{20}\ \text{m}) \times \left(2.2 \times 10^5\ \frac{\text{m}}{\text{s}}\right)^2}{6.67 \times 10^{-11}\ \frac{\text{m}^3}{\text{kg} \times \text{s}^2}} = 1.9 \times 10^{41}\ \text{kg}$$

**Step 3 Explain:** The mass of the Milky Way Galaxy *within* the Sun's orbit is about $2 \times 10^{41}$ kg, which is about $10^{11}$, or 100 billion, times the Sun's mass of $2 \times 10^{30}$ kg.

planets decline with distance from the Sun [**Section 4.4**]. Instead, we find that orbital speeds remain about the same out to great distances from the galactic center, telling us that most of the galaxy's mass resides far from the center and is distributed throughout the halo. Because we see few stars and virtually no gas or dust in the halo, we conclude that most of the galaxy's mass must not give off any light that we can detect, and hence we refer to it as *dark matter*. We will discuss dark matter in more detail in Chapter 23.

## 19.2 Galactic Recycling

The Milky Way is our home galaxy, but its importance to our existence runs much deeper. The birth of our solar system could not have occurred without the galactic recycling that takes place within the disk of the galaxy and its interstellar medium.

Galactic recycling makes new generations of stars possible and gradually changes the chemical composition of the interstellar medium. Recall that the early universe contained only the two lightest elements, hydrogen and helium. Virtually all other elements (which astronomers tend to refer to as "heavy elements") have been produced and released into space by stars. These elements mix with the existing interstellar gas, raising its proportion of heavy elements, and the galaxy then recycles this gas into new generations of stars. Today, thanks to more than 10 billion years of galactic recycling, elements heavier than helium constitute about 2% of the galaxy's gaseous content by mass. The remaining 98% still consists of hydrogen (about 71%) and helium (about 27%).

**Think about it** Recall from Chapter 15 that stars in our galaxy's globular clusters are all extremely old, while stars in open clusters are relatively young. Based on this fact, which stars do you think contain a higher proportion of heavy elements: stars in globular clusters or stars in open clusters? Explain.

The galactic recycling process may sound simple enough, but it raises at least one key question. The supernova explosions that scatter most heavy elements into space send debris flying out at speeds of several thousand kilometers per second—much faster than the escape velocity from the galaxy. So how have the chemical riches produced by stars managed to remain in our galaxy? The answer turns out to depend on interactions between the matter expelled by supernovae and the interstellar medium that fills the galactic disk. In this section, we'll examine these interactions in more detail and learn why our existence owes as much to galactic recycling as it does to the manufacturing of heavy elements by stars.

### How is gas recycled in our galaxy?

The galactic recycling process proceeds in several stages, making up what we will call the **star–gas–star cycle**, summarized in **FIGURE 19.3**. We have already discussed the ideas shown in the lower three frames of Figure 19.3: Stars are born when gravity causes the collapse of molecular

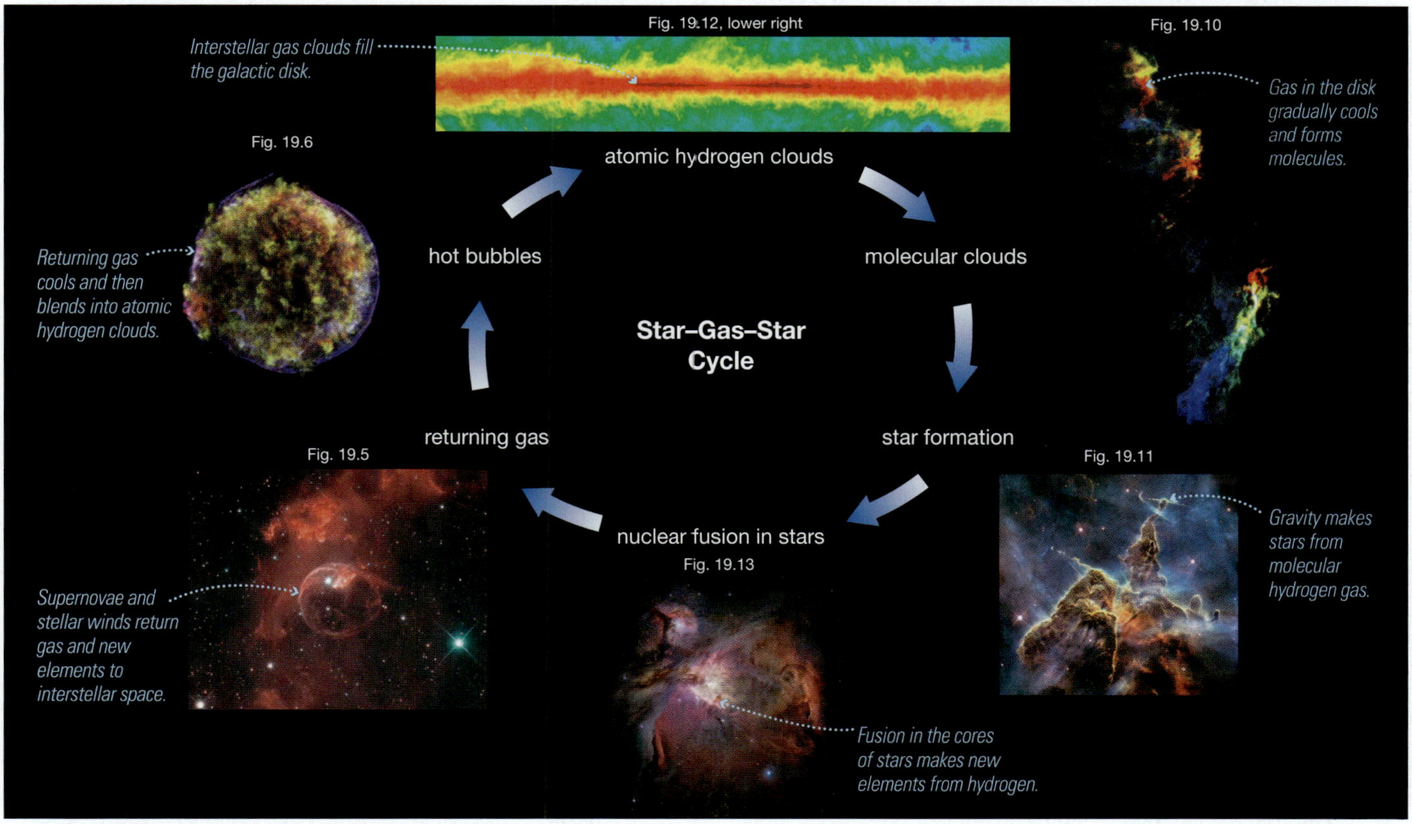


**FIGURE 19.3** A pictorial representation of the star–gas–star cycle. These photos appear individually later in the chapter. Their figure numbers are indicated.

**FIGURE 19.4** A dying low-mass star returns gas to the interstellar medium in a planetary nebula. This Hubble Space Telescope image shows a planetary nebula known as the Cat's Eye Nebula; it is about 1.2 light-years in diameter.

clouds, they shine for millions or billions of years with energy produced by nuclear fusion, and in their deaths they ultimately return much of their material back to the interstellar medium. Let's now complete the loop and look at the remainder of the cycle, starting with the gas ejected by dying stars and finishing back at star birth.

**Gas from Dying Stars** All stars return much of their original mass to interstellar space in two basic ways: through stellar winds that blow throughout their lives and through "death events" of planetary nebulae (for low-mass stars) or supernovae (for high-mass stars). Low-mass stars generally have weak stellar winds while they are on the main sequence. Their winds grow stronger and carry more material into space when they become red giants. By the time a low-mass star like the Sun ends its life with the production of a planetary nebula [**Section 17.2**], it has returned almost half its original mass to the interstellar medium (**FIGURE 19.4**).

High-mass stars lose mass much more dynamically and explosively. The powerful winds from supergiants and massive O and B stars recycle large amounts of matter into the galaxy. At the ends of their lives, these stars explode as supernovae. The high-speed gas ejected into space by supernovae or powerful stellar winds sweeps up surrounding interstellar material, excavating a **bubble** of hot, ionized gas (**FIGURE 19.5**). These hot bubbles are quite common in the disk of the galaxy, but they are not always easy to detect. While some emit strongly in visible light and others are hot enough to emit large amounts of x-rays, many bubbles are evident only through their emission of radio waves.

The bubbles created by supernovae can have even more dramatic effects on the interstellar medium than those created by fast-moving stellar winds. Supernovae generate

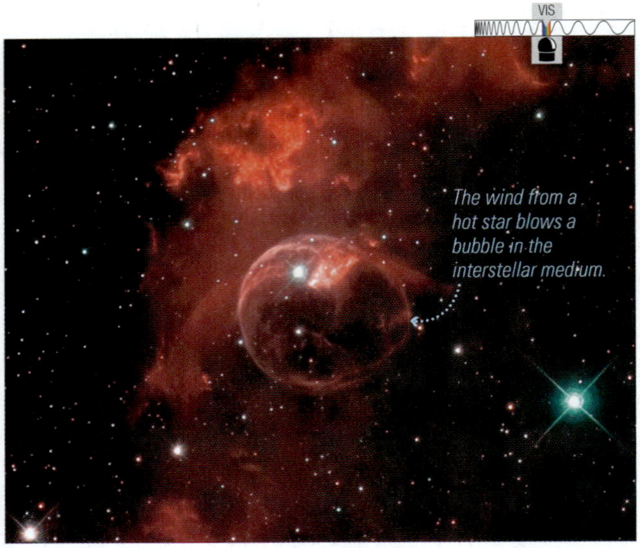


**FIGURE 19.5** This photo shows a bubble of hot, ionized gas blown by the wind from the hot star near its center. Although it looks much like a soap bubble, it is actually an expanding shell of hot gas about 10 light-years in diameter. It glows where gas piles up as the bubble sweeps outward through the interstellar medium.

*shock fronts*—abrupt, high-gas-pressure "walls" that move faster than the speed at which sound waves can travel through interstellar space. A shock front sweeps up surrounding gas as it travels, creating a wall of fast-moving gas on its leading edge. When we observe a *supernova remnant* [**Section 17.3**], we are seeing walls of gas that have been compressed, heated, and ionized by the outward-moving shock front.

**FIGURE 19.6** shows a young supernova remnant whose shocked gas is hot enough to emit x-rays. In contrast,

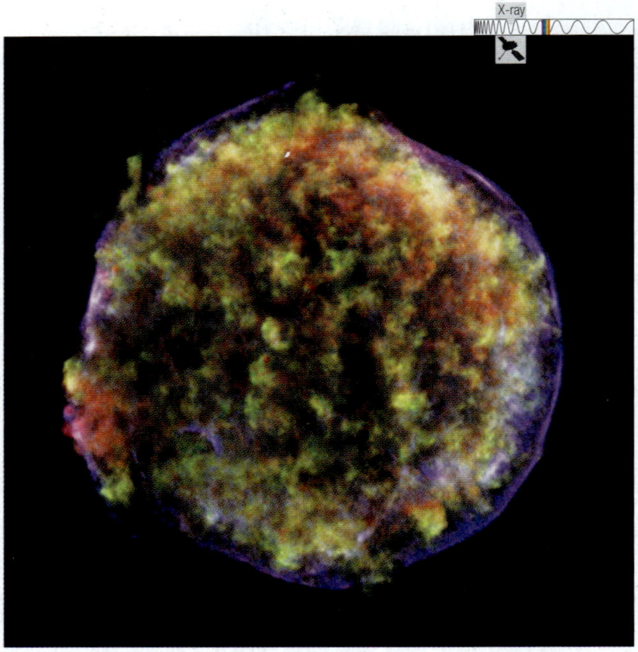


**FIGURE 19.6** This image from the Chandra X-Ray Observatory shows x-ray emission from hot gas in a young supernova remnant—the remnant from the supernova observed by Tycho Brahe in 1572. The most energetic x-rays (blue) come from 20-million-degree gas just behind the expanding shock front. Less energetic x-rays (green and red) come from the 10-million-degree debris ejected by the exploded star. The remnant is about 20 light-years across.

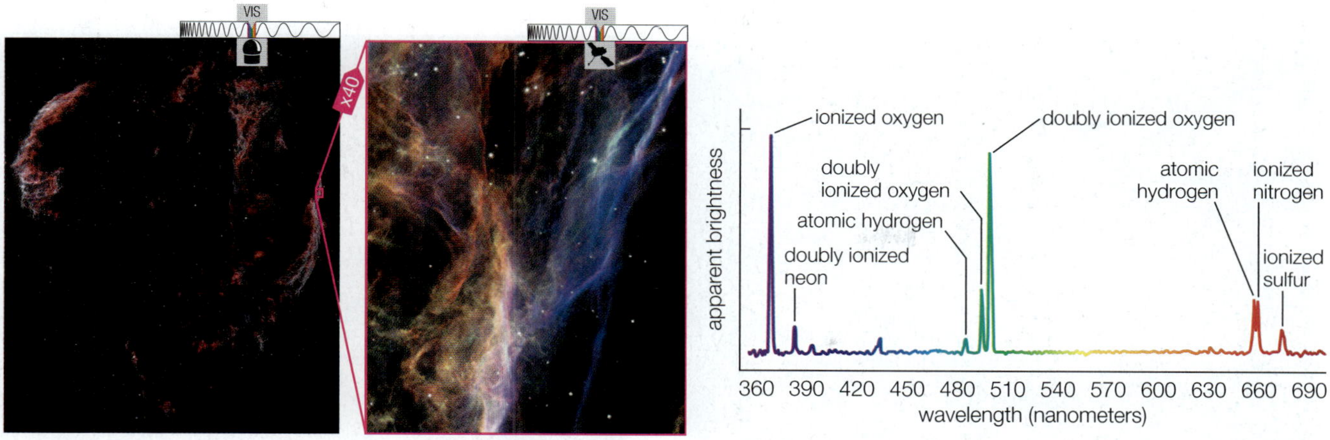


**a** This visible-light image shows the entire supernova remnant, which is about 130 light-years across and spans an angular width in our sky six times that of the full Moon.

**b** This Hubble Space Telescope image shows fine filamentary structure in a small piece of the remnant. The colors come from emission lines of the atoms and ions indicated in part c.

**c** A visible-light spectrum of the Cygnus Loop shows the strong emission lines that account for the distinct colors in the Hubble Space Telescope image.

**FIGURE 19.7** Emission of visible light from an older supernova remnant, the Cygnus Loop.

the older supernova remnant shown in **FIGURE 19.7** is cooler because its shock front has swept up more material and has distributed its energy more widely. Eventually, the shocked gas will radiate away most of its original energy, and the expanding wall of gas will slow to subsonic speeds and merge with the surrounding interstellar medium.

The distinctive shapes of hot bubbles make them fairly easy to spot with x-ray telescopes when they lie at some distance from us. However, the nearest bubble is not so obvious because we're living inside it. X-ray observations show that our local solar neighborhood (the Sun and nearby stars) is surrounded in all directions by hot, x-ray–emitting gas. Beyond this hot gas, which extends to distances of up to a few hundred light-years, lies a region of much cooler gas. We therefore conclude that we live within a *Local Bubble* in which one or more supernovae must have detonated during the past several million years.

In addition to their role as the movers and shakers of the interstellar medium, shock fronts from supernovae can act as subatomic particle accelerators. Some of the electrons in supernova remnants accelerate to nearly the speed of light as they interact with the shock front. These fast electrons emit radio waves (of a type sometimes called *synchrotron radiation*) as they spiral around magnetic field lines threading the supernova remnant (**FIGURE 19.8**).

Supernovae can affect more than just the interstellar medium. They may also affect life by generating **cosmic rays**, which are electrons, protons, and atomic nuclei that zip through interstellar space at close to the speed of light [**Section S4.2**]. Some cosmic rays penetrate Earth's atmosphere and reach Earth's surface, where they can cause genetic mutations in living organisms. On average, about one cosmic-ray particle strikes your body each second. The cosmic-ray bombardment rate is 100 times greater at the high altitudes at which jet planes fly, and even more cosmic rays funnel along magnetic field lines to Earth's magnetic poles.

**Superbubbles and Fountains** The bubble created by a single supernova can grow to a diameter of about a hundred light-years before it slows and merges with surrounding interstellar gas. But in some regions of the Milky Way, we see cavities of hot gas that are more than a thousand light-years across. These huge cavities presumably arise when many individual bubbles combine to form a much larger **superbubble**. Recall that stars tend to form in clusters in which all the stars are about the same age. As a result, the hottest, most massive stars in a cluster can end their lives and explode within a few hundred thousand years of one another. The shock fronts from the individual supernovae soon overlap, combining their energy into one very powerful shock front that we see as a superbubble. Subsequent supernova explosions inside the superbubble add even more energy to the shock front.

In many places in our galactic disk, we see what appear to be elongated superbubbles extending from young clusters of stars to distances of 3000 light-years or more above the disk. These probably are places where superbubbles have grown so large that they cannot be contained within the Milky Way's disk. Once the superbubble breaks out

## COMMON MISCONCEPTIONS

### The Sound of Space

In many science fiction movies, a thunderous sound accompanies the demolition of a spaceship. If the moviemakers wanted to be more realistic, they would silence the explosion. On Earth, we perceive sound when sound waves—which are waves of alternately rising and falling pressure—cause trillions of gas atoms to push our eardrums back and forth. Although sound waves can and do travel through interstellar gas, the extremely low density of this gas means that only a handful of atoms per second would collide with something the size of a human eardrum. It would therefore be impossible for a human ear (or a similar-size microphone) to register any sound, which is why the real sound of space is silence.

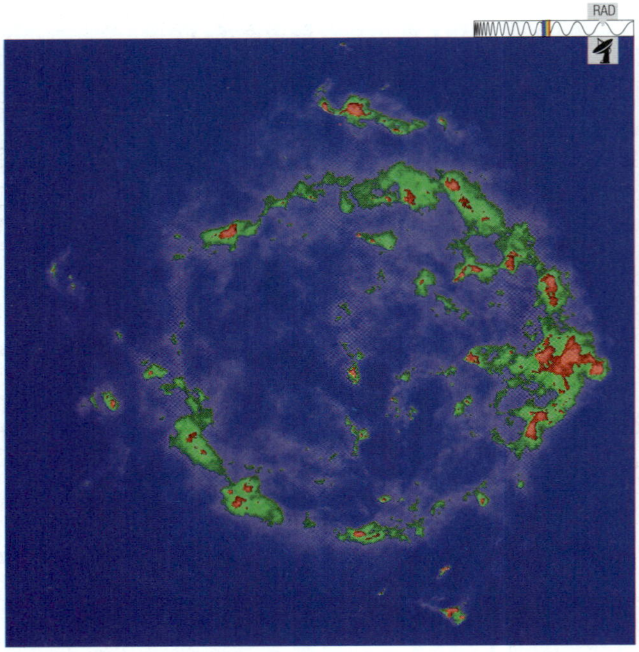


a Radio emission caused by electrons spiraling around magnetic field lines in the young supernova remnant Cassiopeia A. This remnant is about 10 light-years in diameter.

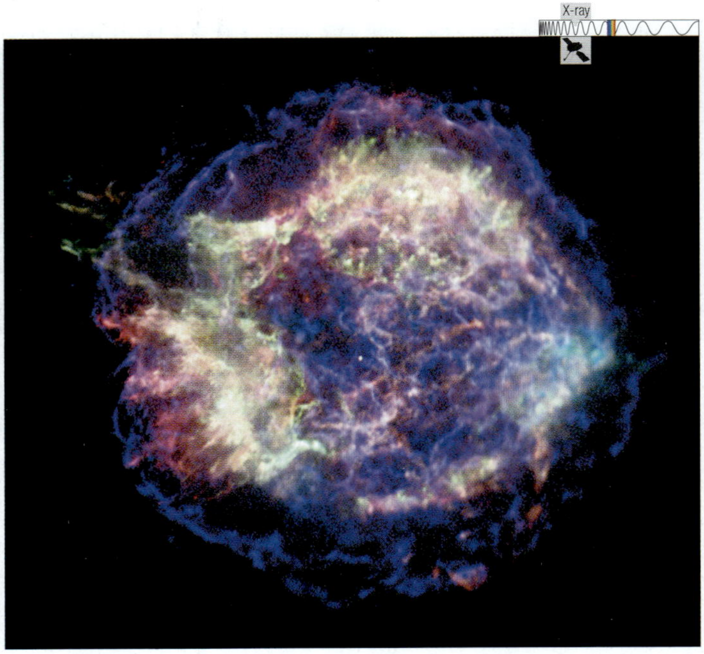


b This image shows x-ray emission from the hot gas of Cassiopeia A, as seen by the Chandra X-Ray Observatory. Red indicates the lowest-energy x-rays and blue indicates the highest-energy x-rays.

**FIGURE 19.8** Supernova remnant Cassiopeia A.

of the disk, where nearly all the Milky Way's gas resides, nothing remains to slow its expansion except gravity. The result is a *blowout* that is in some ways similar to a volcanic eruption, but on a galactic scale: Hot gas erupts from the disk, spreading out as it shoots upward into the galactic halo.

Theoretical modeling of blowouts suggests that they continually cycle gas between the Milky Way's disk and the halo (**FIGURE 19.9**). According to this model, sometimes called the **galactic fountain**, the gravity of the galactic disk slows and eventually halts the rise of the gas from a blowout. Near the top of its trajectory, the ejected gas starts to cool and form clouds. Gravity then causes these clouds to rain back down into the disk, where their contents mix with the gas throughout a large region of the galaxy. If the galactic fountain model is correct, then superbubbles and blowouts play an important role in the galaxy-wide recycling system that has made our existence possible.

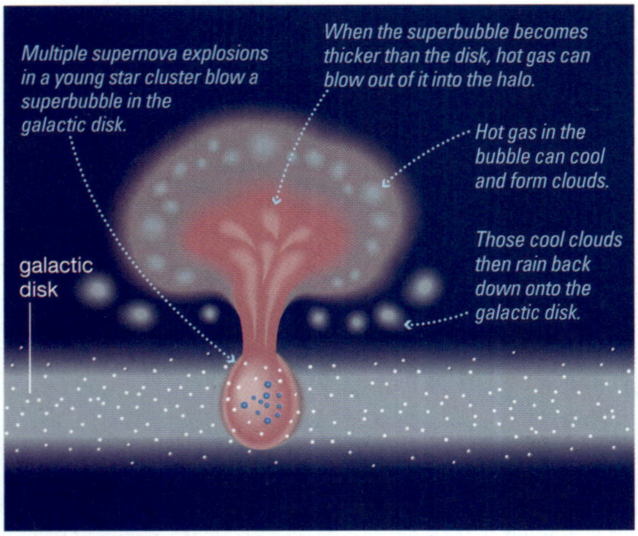


a The galactic fountain model.

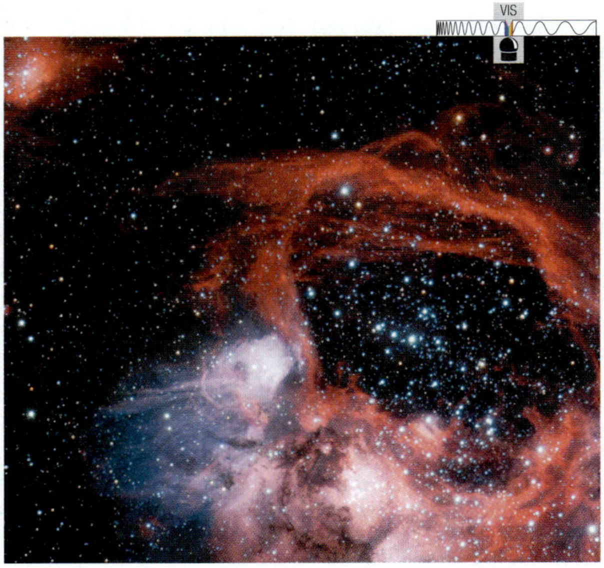


b This image from the European Southern Observatory's Very Large Telescope shows a superbubble of hot gas (red) blowing out from a star-forming region in the Large Magellanic Cloud. The region inside the red bubble is about 250 light-years in diameter.

**FIGURE 19.9** According to the galactic fountain model, superbubbles formed by the shock fronts of many supernovae blow gas out of the galactic disk, and this gas then cools into gas clouds that rain back down on the disk.

The galactic fountain model is plausible but difficult to verify. Observationally, we do indeed see some hot gas high above the galaxy's disk. We also see cooler clouds that appear to be raining down from the halo. However, we can see this "rain" only directly above and below us, making it difficult to be certain that galactic fountains circulate the products of supernovae throughout the Milky Way.

**Cooling and Cloud Formation** The hot, ionized gas in bubbles heated by supernovae is dynamic and widespread but represents a relatively small fraction of the gas in the Milky Way. Most of the gas is much cooler—cool enough that hydrogen atoms remain neutral rather than being ionized. We therefore refer to this gas as **atomic hydrogen gas**, although the hydrogen is mixed with neutral atoms of helium and heavier elements in the usual proportions for the galaxy (by mass about 71% hydrogen, 27% helium, and 2% heavier elements). After the gas that makes bubbles or superbubbles cools, it becomes part of this widespread atomic hydrogen gas in the galaxy.

We can map the distribution of atomic hydrogen gas in the Milky Way with radio observations. Atomic hydrogen emits a spectral line with a wavelength of 21 centimeters, which lies in the radio portion of the electromagnetic spectrum. We detect the radio emission from this **21-centimeter line** coming from all directions, telling us that atomic hydrogen gas is distributed throughout the galactic disk. Based on the overall strength of the 21-centimeter emission, the total amount of atomic hydrogen gas in our galaxy must be about 5 billion solar masses, which is less than 10% of the galaxy's mass in the form of stars.

Atomic hydrogen gas tends to be found in two distinct forms: large, tenuous clouds of warm (10,000 K) atomic hydrogen and smaller, denser clouds of cool (100 K) atomic hydrogen. If you were to take an interstellar voyage across the Milky Way, you would spend the majority of your time cruising through regions of warm atomic hydrogen interspersed with bubbles of hot, ionized gas. Every thousand light-years or so, you would encounter a cooler, denser cloud of atomic hydrogen.

In the warm regions, you would detect about one atom per cubic centimeter and a weak magnetic field (weaker than Earth's magnetic field by a factor of about 100,000). In the cooler clouds, you would find a density of about 100 atoms per cubic centimeter and a stronger magnetic field. As we discussed in Chapter 16, these magnetic fields may provide support to clouds that gravity would otherwise cause to collapse and form stars.

Matter remains in the atomic hydrogen stage of the star–gas–star cycle for millions of years. Gravity slowly draws blobs of this gas together into tighter clumps, which radiate energy more efficiently as they grow denser. The blobs therefore cool and contract, forming clouds of cooler and denser gas. This process takes a much longer time than the other steps in the cycle from star death to star birth, which is why so much of the Milky Way's gas is in the atomic hydrogen stage of the star–gas–star cycle. During this long period of cooling, disturbances produced by supernova explosions occurring in the vicinity of the cooling gas keep it stirred up, which is probably why cold, star-forming clouds are often observed to be turbulent.

Clouds of atomic hydrogen also contain a small but important amount of interstellar dust. Recall that interstellar *dust grains* are tiny solid particles that form in the winds of red giant stars and resemble particles of smoke [**Section 16.1**]. Once formed, dust grains remain in the interstellar medium until they are heated and destroyed by a passing shock front or incorporated into a protostar. Dust grains make up only about 1% of the mass of atomic hydrogen clouds, but they are responsible for the absorption of visible light that prevents us from seeing through the disk of the galaxy.

**From Atomic to Molecular Clouds** As the temperature drops further in the center of a cool cloud of atomic hydrogen, hydrogen atoms combine into molecules (primarily into hydrogen molecules, $H_2$), making a *molecular cloud* [**Section 16.1**]. Recall that molecular clouds are the coldest, densest collections of gas in the interstellar medium, and they are the birthplaces of stars. They often congregate into *giant molecular clouds* that contain up to a million solar masses of gas (**FIGURE 19.10**). The total mass of molecular clouds in the Milky Way is somewhat

**FIGURE 19.10** This image shows the complex structure of a molecular cloud in the constellation Orion. The colored picture was made by measuring Doppler shifts of emission lines from carbon monoxide molecules, and the colors indicate gas motions: Bluer parts are moving toward us and redder parts are moving away from us (relative to the cloud as a whole). This enormous cloud is about 1600 light-years away and several hundred light-years across.

uncertain, but it is probably about the same as the total mass of atomic hydrogen gas—about 5 billion solar masses. Temperatures throughout much of this molecular gas hover only a few degrees above absolute zero.

Molecular clouds are heavy and dense compared to the rest of the interstellar gas and therefore tend to settle toward the central layers of the Milky Way's disk. This tendency creates a phenomenon you can see with your own eyes: the dark lanes running through the luminous band of light in our sky that we call the Milky Way (see Figure 2.1).

**Completing the Cycle** As we discussed in Chapter 16, a large molecular cloud gives birth to a cluster of stars. Once a few stars form in a newborn cluster, their radiation begins to erode the surrounding gas in the molecular cloud. Ultraviolet photons from high-mass stars heat and ionize the gas, and winds and radiation pressure push the ionized gas away. This kind of feedback prevents much of the gas in a molecular cloud from turning into stars.

**FIGURE 19.11** A portion of the Carina Nebula, as seen by the Hubble Space Telescope. The dark blobs of gas are molecular clouds, and stars are currently forming in the densest parts of these clouds. The region pictured here is about 3 light-years across.

The process of molecular cloud erosion is vividly illustrated in the Carina Nebula, a complex of clouds where new stars are forming (**FIGURE 19.11**). The dark, lumpy structures are molecular clouds. Outside the upper edge of the picture, newly formed massive stars glow with ultraviolet radiation. This radiation sears the surface of the molecular clouds, destroying molecules and stripping electrons from atoms. As a result, matter "evaporates" from the molecular clouds and joins the hotter ionized gas encircling them.

We have arrived back where we started in the star–gas–star cycle. The most massive stars now forming in the Carina Nebula will explode within a few million years, filling the region with bubbles of hot gas and newly formed heavy elements. The expanding bubbles will slow and cool as their gas merges with the widespread atomic hydrogen gas in the galaxy. Eventually, this gas will cool enough to coalesce into molecular clouds, forming new stars, new planets, and perhaps even new civilizations.

Despite the recycling of matter from one generation of stars to the next, the star–gas–star cycle cannot go on forever. With each new generation of stars, some of the galaxy's gas becomes permanently locked away in brown dwarfs that never return material to space and in the corpses left behind when stars die (white dwarfs, neutron stars, and black holes). The interstellar medium therefore is slowly running out of gas, and the rate of star formation will gradually taper off until star formation ceases, perhaps some 50 billion years from now.

**Putting It All Together: The Distribution of Gas in the Milky Way** Different regions of the galaxy are in different stages of the star–gas–star cycle. Because the cycle proceeds over such a long period of time compared to a human lifetime, each stage appears to us as a snapshot. We therefore see the interstellar medium in a wide variety of manifestations, ranging from the tenuous million-degree gas of bubbles to the cold, dense gas of molecular clouds. **TABLE 19.1** summarizes the different states of interstellar gas in the galactic disk.

We can see how these different states of gas are arranged by observing our galaxy in different wavelengths of light. **FIGURE 19.12** shows seven views of the entire sky. Each view represents the Milky Way Galaxy in a particular wavelength band, made by photographing the sky in every direction from

**TABLE 19.1 Typical States of Gas in the Interstellar Medium**

| | State of Gas | | |
|---|---|---|---|
| | **Hot Bubbles** | **Atomic Hydrogen Clouds** | **Molecular Clouds** |
| **Primary Constituent** | Ionized hydrogen | Atomic hydrogen | Molecular hydrogen |
| **Approximate Temperature** | 1,000,000 K | 100–10,000 K | 30 K |
| **Approximate Density (atoms per cm$^3$)** | 0.01 | 1–100 | 300 |
| **Description** | Pockets of gas heated by stellar winds or supernovae | The most common form of gas, filling much of the galactic disk | Regions of star formation |

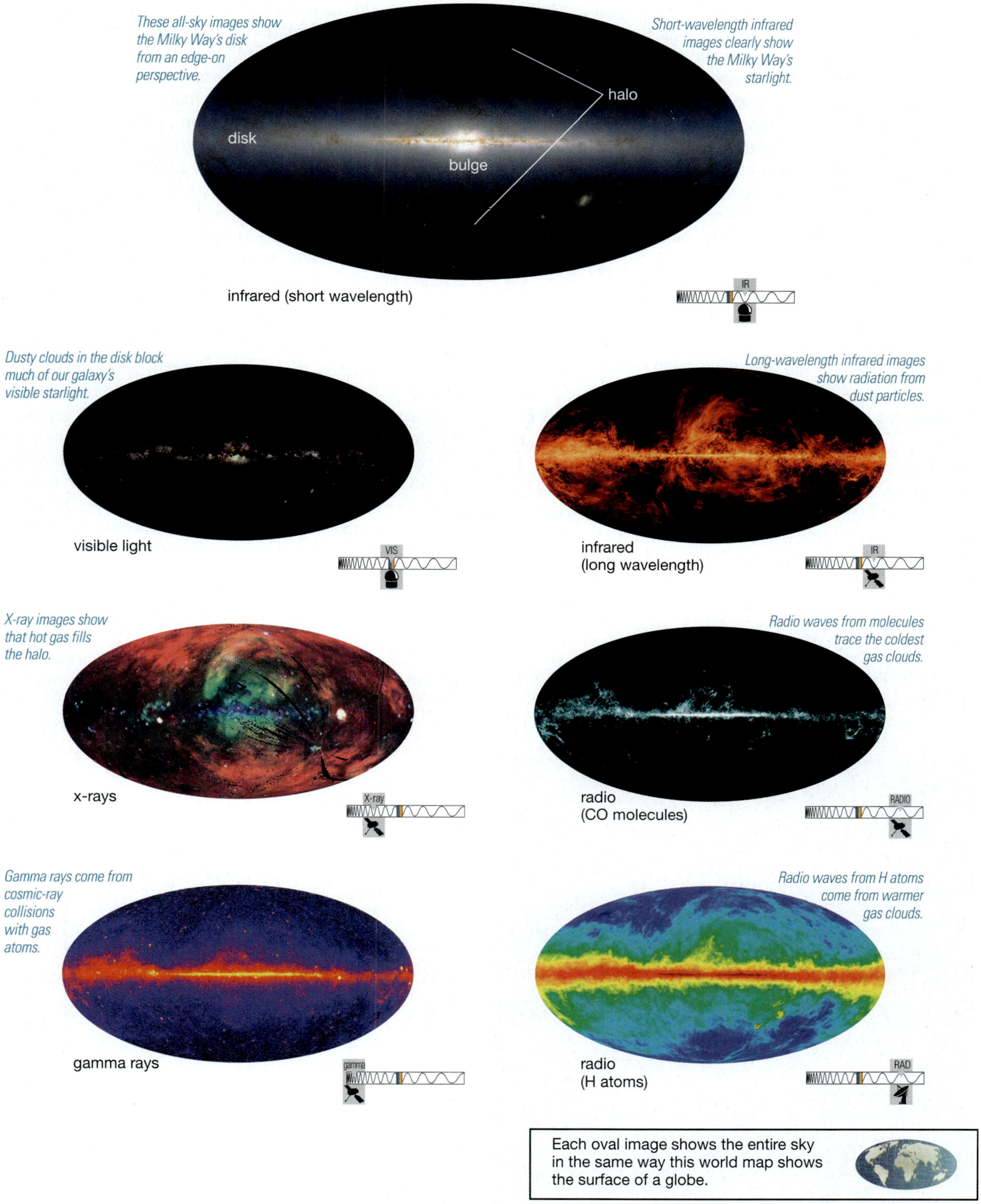


**FIGURE 19.12** All-sky views of the Milky Way Galaxy as it appears in different portions of the spectrum. Observing our galaxy in different wavelengths of light shows us the relationships between its stars, gas clouds, and dust particles. In each image, the brightness corresponds to the intensity of light with that wavelength.

Earth. The infrared image at the top shows what the Milky Way would look like if dusty gas clouds were not blocking our visible-light view. Notice that the dark patches along the disk in the visible-light view correspond to bright regions of long-wavelength infrared emission from dust, as well as to radio emission from molecular and atomic gas clouds. In the x-ray image, the halo is brighter than the disk, because it is filled with x-ray–emitting hot gas. In contrast, the disk is brighter than the halo in the gamma-ray image, because collisions between cosmic rays and the nuclei of gas atoms in cold, dense gas clouds produce most of those gamma rays.

**Think about it** Carefully compare the x-ray image in Figure 19.12 with the radio image of CO (carbon monoxide) molecules. What is the relationship between the dark patches in the x-ray image and the bright patches in the radio image? Explain why this relationship exists.

## Where do stars tend to form in our galaxy?

The star–gas–star cycle has operated continuously since the Milky Way's birth, yet new stars are not spread evenly across the galaxy. Some regions seem much more fertile than others. Regions rich in molecular clouds tend to spawn new stars easily, while gas-poor regions do not. However, molecular clouds are dark and hard to see. Other signatures of star formation are much more obvious. A quick tour of some star-forming galactic environments will help you spot where the action is.

**Star-Forming Regions** Hot, massive stars signal a region of active star formation. Because these stars live fast and die young, they don't get much of a chance to move very far from their birthplaces and generally die while their lower-mass birthmates are still forming.

Regions of active star formation can be extraordinarily beautiful. Near hot stars we often find colorful, wispy blobs of glowing gas known as **ionization nebulae**.* These nebulae glow because ultraviolet photons from the hot stars can ionize the nebula's atoms or raise their electrons to high energy levels, and the atoms emit light as the electrons return to lower energy levels [**Section 5.4**]. The Orion Nebula, about 1350 light-years away in the "sword" of the constellation Orion, is among the most famous (**FIGURE 19.13a**; also see Figure 1.2).

Most of the striking colors in an ionization nebula come from spectral lines produced by particular atomic transitions (**FIGURE 19.13b**). For example, the transition in which an electron in a hydrogen atom falls from energy level 3 to energy level 2 generates a red photon with a wavelength of 656.3 nanometers (see Figure 5.15). Ionization nebulae appear predominantly red in photographs because of all the red photons released by this particular transition. (Jumps from level 2 to level 1 are even more common, but they produce ultraviolet photons that can be studied only with ultraviolet telescopes in space.) Transitions in other elements produce spectral lines of other colors.

*Several other names are sometimes used for ionization nebulae, including *emission nebulae* and *H II regions*; the latter comes from the fact that scientists sometimes use Roman numerals to describe the ionization state, so "H I" is neutral hydrogen and "H II" is ionized hydrogen.

**a** A Hubble Space Telescope photo of the Orion Nebula, an ionization nebula energized by ultraviolet photons from hot stars.

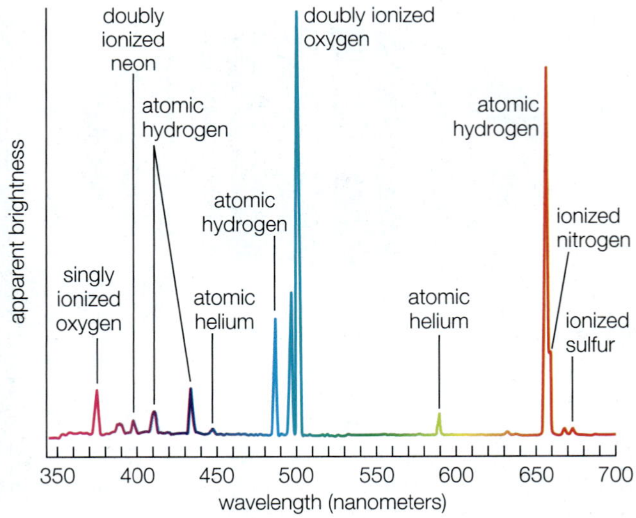


**b** A spectrum of the Orion Nebula. The prominent emission lines reveal the atoms and ions that emit most of the light. Through careful study of these lines, we can determine the nebula's chemical composition.

**FIGURE 19.13** The Orion Nebula.

**FIGURE 19.14** The blue tints in this nebula in the constellation Ophiuchus are produced by reflected light.

**FIGURE 19.15** A photo of the Trifid Nebula and its surroundings. (The region pictured is about 50 light-years across.)

The blue and black tints in some star-forming regions have a different origin. Starlight reflected from dust grains produces the blue colors, because interstellar dust grains scatter blue light much more readily than red light (**FIGURE 19.14**). These *reflection nebulae* are always bluer in color than the stars supplying the light; the effect is similar to the scattering of sunlight in our atmosphere that makes the sky blue [**Section 10.1**]. The black regions of nebulae are dark, dusty gas clouds that block our view of the stars beyond them. **FIGURE 19.15** shows a multicolored nebula characteristic of a hot-star neighborhood.

**Think about it** In Figure 19.15, identify the red ionized regions, the blue reflecting regions, and the dark obscuring regions. Briefly explain the origin of the colors in each region.

## COMMON MISCONCEPTIONS

### What Is a Nebula?

The term *nebula* means "cloud," but in astronomy it can refer to many different kinds of objects, which sometimes leads to misconceptions. Many astronomical objects look "cloudy" through small telescopes, and in past centuries astronomers called any such object a nebula as long as they were sure it wasn't a comet. For example, galaxies were called nebulae because they looked like fuzzy round or spiral blobs.

Using the term *nebula* to refer to a galaxy now sounds somewhat dated, given the enormous differences between these distant star systems and the much smaller clouds of gas that populate the interstellar medium. Nevertheless, some people still refer to spiral galaxies as "spiral nebulae." Today, we generally use the term *nebula* to refer to true interstellar clouds, but be aware that the term is still sometimes used in other contexts.

**Spiral Arms** Taking a broader view of our galaxy, we can see that the spiral arms must be full of newly forming stars because they bear all the hallmarks of star formation. They are home to both molecular clouds and numerous clusters of young, bright, blue stars surrounded by ionization nebulae. Detailed images of other spiral galaxies show these characteristics more clearly (**FIGURE 19.16**). Hot blue stars and ionization nebulae trace out the arms, while the stars between the arms are generally redder and older. We also see enhanced amounts of molecular and atomic gas in the spiral arms, and streaks of interstellar dust often obscure the inner sides of the arms themselves. Spiral arms therefore contain both young stars and the material necessary to make new stars.

At first glance, spiral arms look as if they ought to move with the stars, like the fins of a giant pinwheel in space. However, observations indicate that stars throughout most of the disk orbit at approximately the same speed. Stars near the center of the galaxy—which travel in circles with smaller radii—therefore complete an orbit in much less time than stars far from the center. As a result, if the spiral arms moved along with the stars, the central parts of the arms would wind around the galaxy many more times than the outer parts. After a few galactic rotations, the arms would become wound into a tight coil. We do not see such tightly wound coils of arms in galaxies, so we conclude that spiral arms are more like swirling ripples in a whirlpool than like the fins of a giant pinwheel.

More specifically, spiral arms exhibit unusually high levels of star formation. Theoretical models suggest that the spiral pattern of star formation is caused by disturbances called **spiral density waves** that propagate through

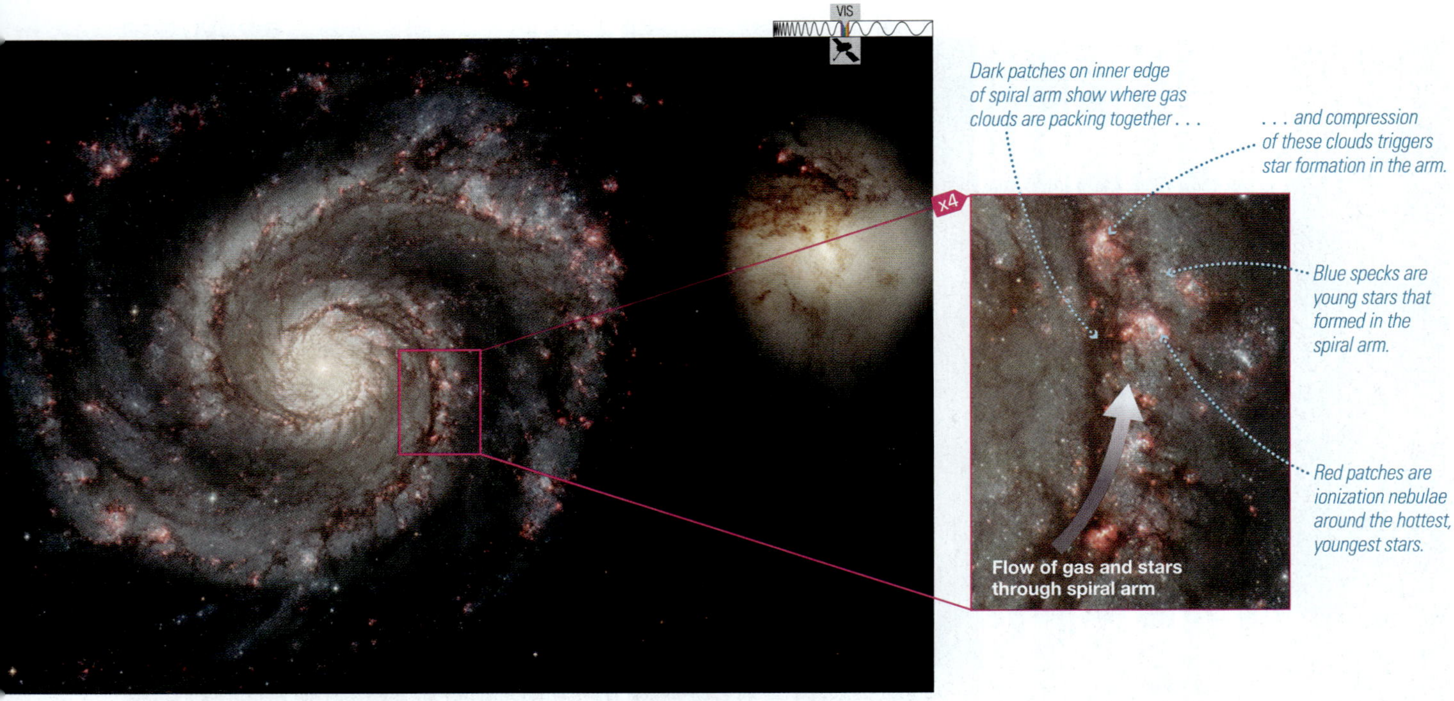


**FIGURE 19.16** This photo from the Hubble Space Telescope shows Galaxy M51's two magnificent spiral arms, along with a smaller galaxy that is currently interacting with one of those arms. Notice that the spiral arms are much bluer in color than the central bulge. Because massive blue stars live for only a few million years, the relative blueness of the spiral arms tells us that stars must be forming more actively within them than elsewhere in the galaxy. (The large image shows a region roughly 90,000 light-years across.)

the gaseous disk of a spiral galaxy like the Milky Way. As a spiral density wave moves through the disk, the changing gravitational forces among stars in the wave alternately cause stars and gas clouds to pack more densely together and then move farther apart. The density changes have little effect on stars, because they always remain too widely separated to collide with each other. However, when a spiral density wave packs gas clouds closer together, gravity within the clouds becomes strong enough to trigger the formation of new star clusters (**FIGURE 19.17**). Supernova explosions from massive stars in these clusters can then compress the surrounding clouds further, triggering even more star formation. As this long-lasting pattern of increased density and star formation propagates through the gas and stars of the galaxy's disk, the disk's rotation gradually stretches it into a spiral arm.

## 19.3 The History of the Milky Way

Having discussed the basic properties of the Milky Way, we are now ready to turn our attention to the history of our galaxy. As we will see, the differences between disk stars and halo stars provide crucial clues about their origins, which have helped astronomers develop a basic model of how our galaxy formed.

### What do halo stars tell us about our galaxy's history?

We have already seen how the disorderly orbits of halo stars differ from the organized orbits of disk stars. Two other key differences distinguish halo stars from disk stars. First, the main-sequence turnoff points for globular clusters [**Section 15.3**] indicate that stars in the halo are generally very old (at least 12 billion years), while disk stars have many different ages. Second, spectra show that halo stars contain smaller proportions of heavy elements than disk stars. Based on these differences, astronomers divide most of the Milky Way's stars into two distinct populations.

1. The **disk population** (sometimes called *Population I*) consists of stars that follow the orderly orbital patterns of the disk. This population includes both young stars and old stars, all of which have heavy-element proportions near 2%, like that of our Sun.
2. The **halo population** (or *Population II*) consists of stars that orbit the center of the galaxy with many different inclinations, so that they cross through the disk during their orbits. Stars in this population are all old and therefore low in mass. Halo stars can have heavy-element proportions as low as 0.02%—meaning that heavy elements are about 100 times rarer in these stars than in the Sun.

Stars in the bulge exhibit a mixture of disk-star and halo-star characteristics, with a wide range of ages and heavy-element proportions.

We can understand why halo stars differ from disk stars by looking at how the Milky Way's gas is distributed. The halo does not contain the cold, dense molecular clouds required for star formation. Instead, the gas is far hotter and much more spread out. Because star-forming molecular clouds are found only in the disk, new stars can be born only in the disk and not in the halo.

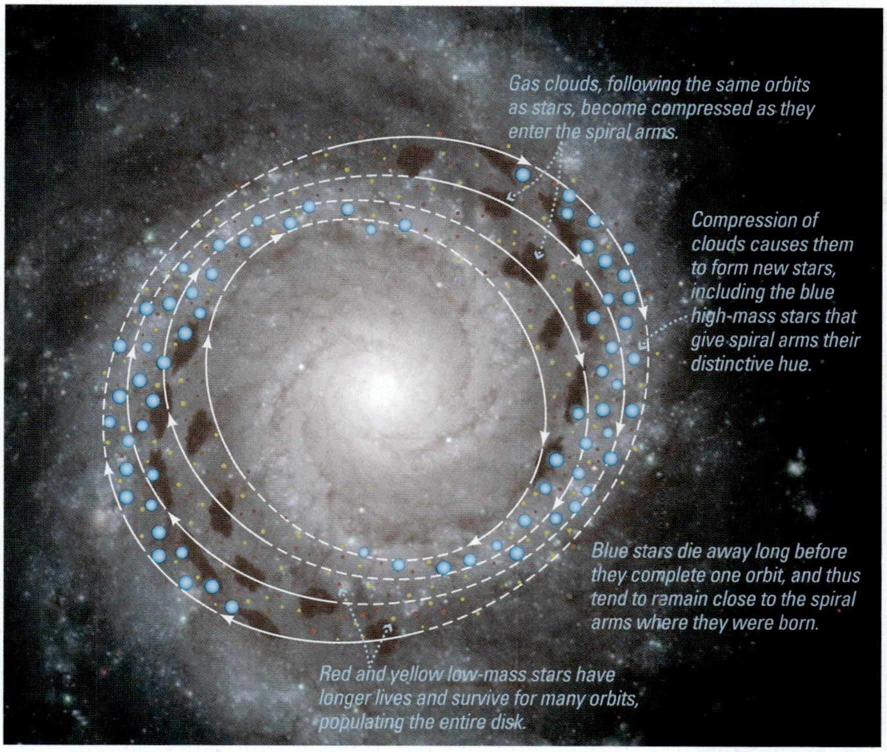


**FIGURE 19.17** Schematic illustration of star formation in a spiral density wave. Stars and gas clouds pass through spiral arms as they orbit the galaxy. Each arm is a self-sustaining pattern because gravity slows stars down as they pass through an arm, causing the stars to be more densely packed in the arms.

The relative lack of heavy elements in halo stars indicates that they must have formed early in the galaxy's history—before many supernova explosions had added heavy elements to star-forming clouds. We therefore conclude that the halo has lacked the gas needed for star formation for a very long time. Apparently, all the Milky Way's cool gas settles into the disk. The only stars that still survive in the halo are long-lived, low-mass stars.

**Think about it** How does the halo of our galaxy resemble the distant future fate of the galactic disk? Explain.

## How did our galaxy form?

Any model of our galaxy's formation must account for the differences between disk stars and halo stars. The simplest such model, shown in **FIGURE 19.18**, has our galaxy beginning as a giant **protogalactic cloud** of hydrogen and helium gas, along with an even larger amount of dark matter.

Early on, the gravity associated with this protogalactic cloud would have drawn in matter from all directions, creating a cloud that was blobby in shape with little or no organized rotation. Gravity would then have caused localized regions within the cloud to contract and fragment, just as in present-day star-forming clouds [**Section 16.1**]. As a result, the orbits of stars that formed early on could have had any orientation, accounting for the randomly oriented orbits of stars in the halo. This model also explains the old ages of halo stars, because it tells us that they were the first stars to form in the galaxy. The low proportions of heavy elements found in halo stars are then simply a consequence of early formation, before multiple generations of stars added heavy elements to star-forming clouds.

As time passed, the remaining gas continued to contract under the force of gravity. The law of conservation of angular momentum ensured that the gas in the collapsing protogalactic cloud (but not the dark matter) ultimately settled into the shape of a flattened, spinning disk, following the same basic processes that occurred on a much smaller scale during the formation of our solar system [**Section 8.2**]. Collisions among gas particles tended to average out their random motions, leading them to acquire orbits in the same direction and in the same plane. Stars that formed within this spinning disk therefore share the organized motion of the disk. The ages of disk stars, which range from newly born to more than 10 billion years, are explained by the fact that the star–gas–star cycle has supported ongoing star formation ever since the disk formed.

Although the model of Figure 19.18 accounts for the basic differences between halo and disk stars, more detailed studies of heavy-element proportions suggest that this model is a bit too simplistic. If the Milky Way had formed from a single protogalactic cloud, it would have steadily accumulated heavy elements during its inward collapse as stars formed and exploded within it. In that case, the outermost stars in the halo would be the oldest and would have the smallest proportions of heavy elements, and the heavy-element proportions would steadily rise as we looked at halo stars orbiting closer to the disk and bulge. But that is not the pattern we observe.

Instead, we find variations in heavy-element proportions, suggesting that the Milky Way's oldest stars formed in relatively small protogalactic clouds, each with a few globular clusters. These clouds later collided and combined to create a single galaxy, the Milky Way (**FIGURE 19.19**). Similar processes may still be under way. Recall that the Sagittarius and Canis Major dwarf galaxies are currently being torn apart as they crash through the Milky Way's disk. A billion years from now, their stars will be indistinguishable

**FIGURE 19.18** This four-picture sequence illustrates a simple schematic model of galaxy formation, showing how a spiral galaxy might develop.

**FIGURE 19.19** This painting shows a model of how the Milky Way's halo may have formed. The characteristics of stars in the Milky Way's halo indicate that several smaller gas clouds, already bearing some stars and globular clusters, merged to form the Milky Way's protogalactic cloud. These stars and star clusters then remained in the halo while the gas settled into the Milky Way's disk.

from halo stars, because they will all be circling the Milky Way on orbits that carry them high above the disk. Evidence indicates that this process has also occurred in the past: Some halo stars move in organized streams that are probably the remnants of dwarf galaxies torn apart long ago by the Milky Way's gravity.

**Think about it** If the preceding scenario is true, then the Milky Way suffered several collisions early in its history. Explain why we should not be surprised that galaxy collisions (or collisions between protogalactic clouds) were rather common in the distant past. (*Hint:* How did the average separations of galaxies in the past compare to their average separations today?)

The distribution of elements also suggests that when the disk finally formed, it began with only about 10% of the amount of heavy elements that it has today. However, the heavy-element concentration must have been greater in the bulge, because the oldest stars of the bulge have heavy-element proportions similar to that of our Sun, even though they are more than 10 billion years old. The rest of the disk gained heavy elements more gradually through the ongoing star–gas–star cycle, which is why young disk stars tend to have higher proportions of heavy elements than older ones.

## 19.4 The Galactic Center

The center of the Milky Way Galaxy lies in the direction of the constellation Sagittarius and does not look particularly special to our unaided eyes. However, if we could remove the interstellar dust that obscures our view, the galaxy's central bulge would be one of the night sky's most spectacular sights. And deep within the bulge, at the very center of the Milky Way, evidence indicates the presence of a gigantic black hole.

The Galactic Center

## What is the evidence for a black hole at our galaxy's center?

Although the Milky Way's clouds of gas and dust prevent us from seeing visible light from the center of the galaxy, we can peer into the heart of our galaxy with radio, infrared, and x-ray telescopes. Within a few hundred light-years of the center, infrared observations reveal swirling clouds of gas and millions of stars. Bright radio emissions trace out the magnetic fields that thread this turbulent region. Then, as we zoom down toward the exact center, we find a source of radio emission called Sagittarius A* (pronounced "Sagittarius A-star"), or Sgr A* for short, which is quite unlike any other radio source in our galaxy.

**See it for yourself** Use the star charts in Appendix I (or a star-finding app) to find the constellation Sagittarius, which is easily visible on summer and fall evenings and looks like a teapot with a handle on the left and a spout on the right. The center of the Milky Way Galaxy is near the tip of the spout. Can you see the Milky Way's faint band of light passing through the constellations Sagittarius, Cygnus, and Cassiopeia?

Several hundred stars crowd the region within about 1 light-year of Sgr A*, and their motions indicate the presence of an extremely massive object. **FIGURE 19.20** shows the orbital paths that have been observed for some of these stars. Applying Newton's version of Kepler's third law to the orbits of these stars tells us that the central object has a mass of about 4 million solar masses, all packed into a region of space just a little larger than our solar system. An object that massive within such a small space is almost certainly a black hole.

Some aspects of Sgr A* remain puzzling. Most suspected black holes are thought to accumulate matter through accretion disks that radiate brightly in x-rays. These include black holes in binary star systems like Cygnus X-1 [**Section 18.3**] and some giant black holes at the centers of other galaxies [**Section 21.3**]. If the black hole at the center of our galaxy had an accretion disk, its x-ray light would easily penetrate the dusty gas of our galaxy and appear fairly bright to our x-ray telescopes. Yet at most times, the x-ray emission from Sgr A* has been relatively faint.

Observations of Sgr A* made in other wavelengths of light are helping us understand this surprising behavior. For example, enormous x-ray flares have been observed coming from the location of the black hole (**FIGURE 19.21**). These sudden changes in x-ray brightness probably come from energy released by comet-size lumps of matter torn apart by the black hole's tidal forces just before disappearing beneath its event horizon. If we continue to observe similar x-ray flares from Sgr A*, then the explanation for its generally low x-ray brightness may be that matter falls into it in big chunks instead of in the smooth, swirling flow of an accretion disk.

**FIGURE 19.22** summarizes these multi-wavelength studies of the galactic center in a zoom-in sequence. Until we better understand our galaxy's central black hole, it is sure to remain a favorite target for observation.

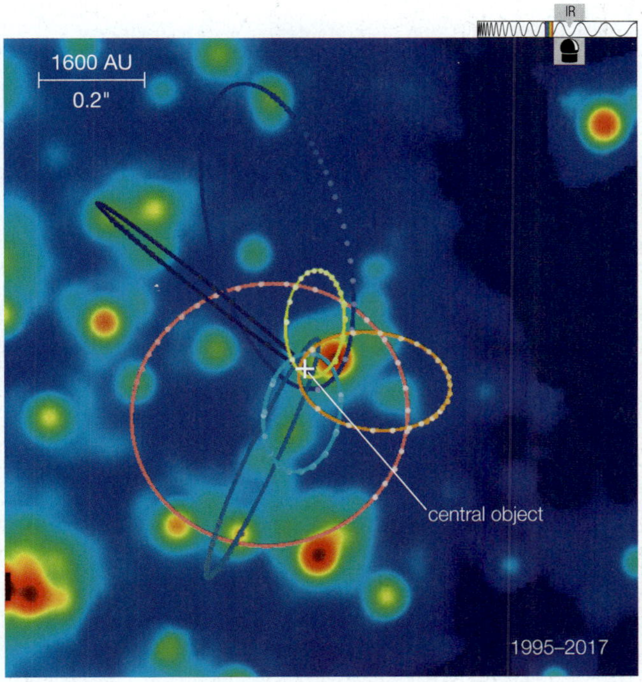


**FIGURE 19.20** Evidence for a black hole at the center of our galaxy. Each set of colored dots shows the positions of a particular star at 1-year intervals as observed with the Keck Telescope. Calculated orbits for these stars are also shown. By applying Newton's version of Kepler's third law to these orbits, we infer that the central object has a mass about 4 million times that of our Sun, packed into a space so small that it is almost certainly a black hole. (The 1600 AU shown on the scale bar is equivalent to about 9 light-days.)

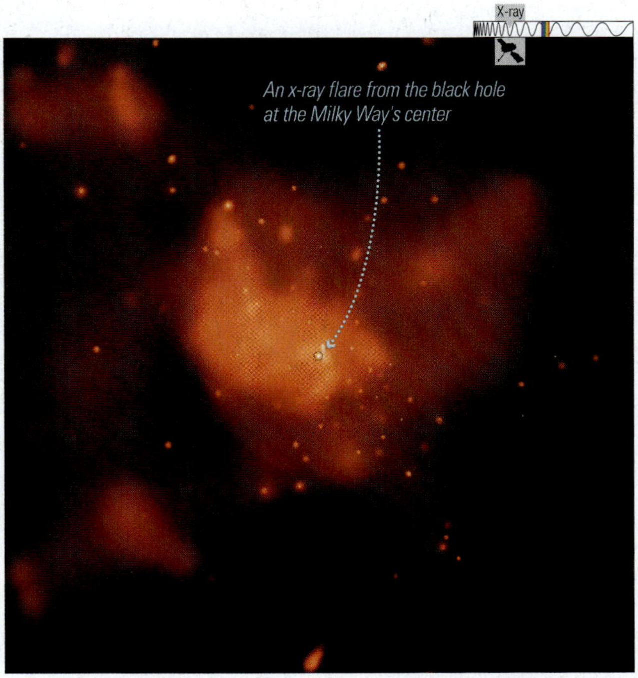


**FIGURE 19.21** This x-ray image from the Chandra X-Ray Observatory shows the central 60 light-years of our galaxy and an x-ray flare from the massive black hole thought to reside there.

## COSMIC Context

# FIGURE 19.22 The Galactic Center

**Zooming into the galactic center in different wavelengths of light reveals strong evidence that the central object, called Sgr A*, is a 4-million-solar-mass black hole.**

1. Dusty gas clouds in our galaxy's disk prevent us from observing visible light from the galactic center.

2. Infrared light passes more easily through dusty clouds than visible light, allowing us to observe the stars (white) and gas clouds (red) at our galaxy's center.

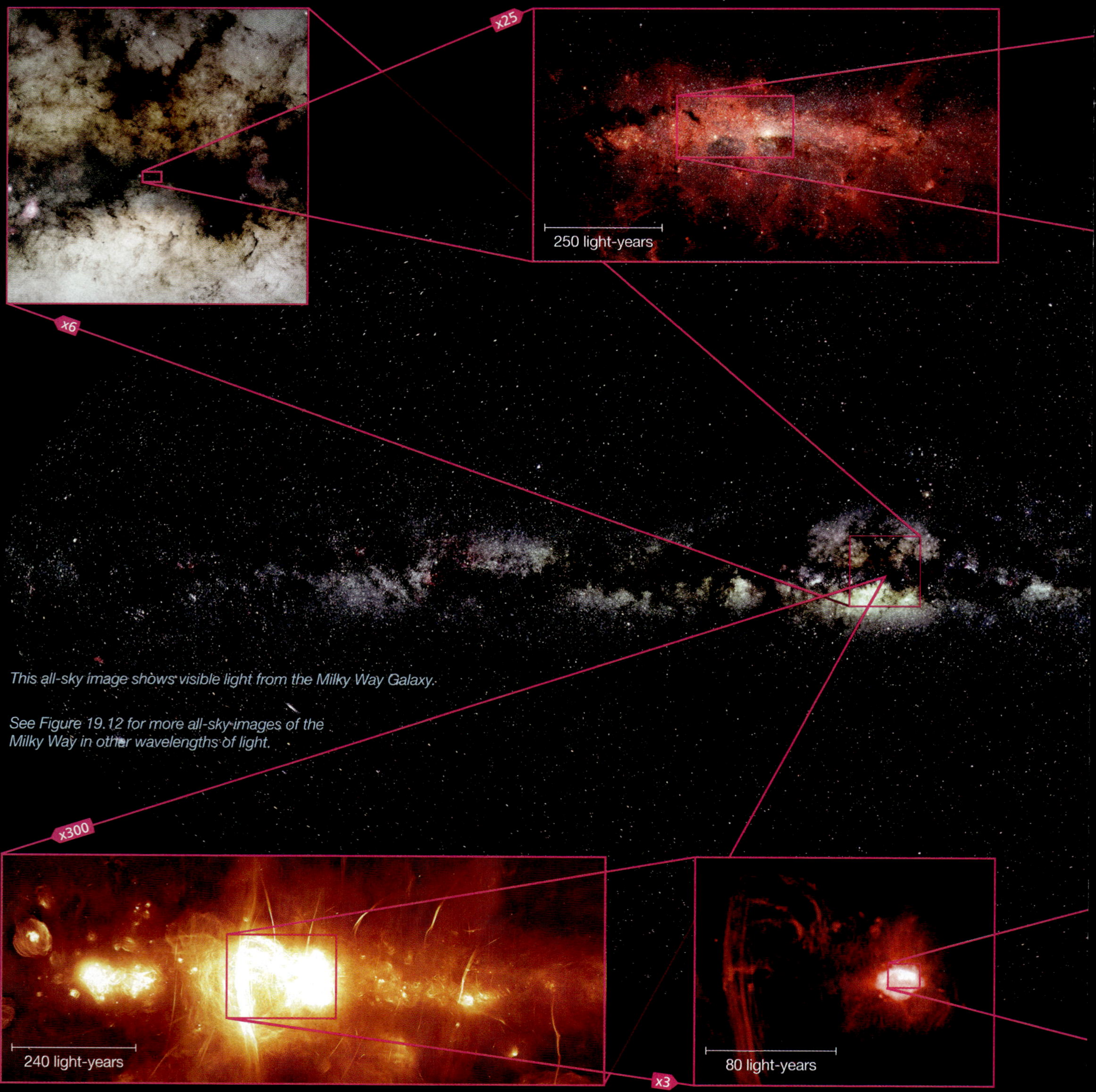


*This all-sky image shows visible light from the Milky Way Galaxy.*

*See Figure 19.12 for more all-sky images of the Milky Way in other wavelengths of light.*

7. Radio waves, which also pass through dusty gas clouds, provide further evidence for a central black hole. Here, we see the strong radio emission from the central few hundred light-years.

8. In this radio-wave image, vast threads of emission trace magnetic field lines near the galactic center.

3 Combining x-ray and infrared observations shows that the galactic center contains hot gas (blue) in addition to its many stars and gas clouds.

*Blue and violet represent x-rays.*
*Yellow represents shorter-wavelength IR light.*
*Red represents longer-wavelength IR light.*

4 Zooming in with the Hubble Space Telescope, we find infrared light from swirling gas clouds within a few light-years of the center.

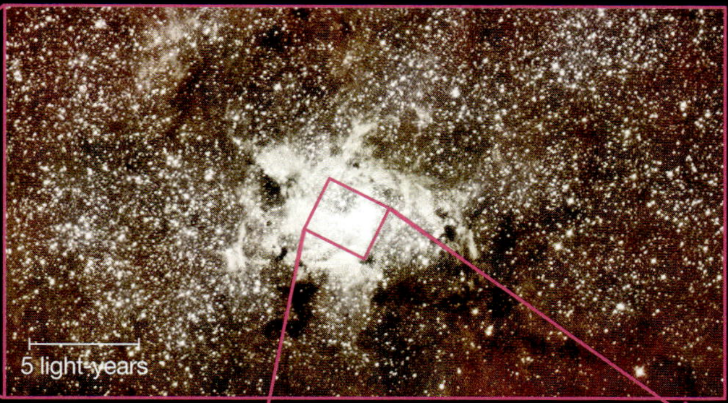


5 Large ground-based telescopes equipped with adaptive optics [Section 6.3] provide sharp views of the stars within a light-year of the galactic center, the location of the object called Sgr A*.

6 Observing the orbits of the innermost stars allows us to calculate the mass of Sgr A*: about 4 million $M_{Sun}$. So much mass in such a small space can only be a black hole.

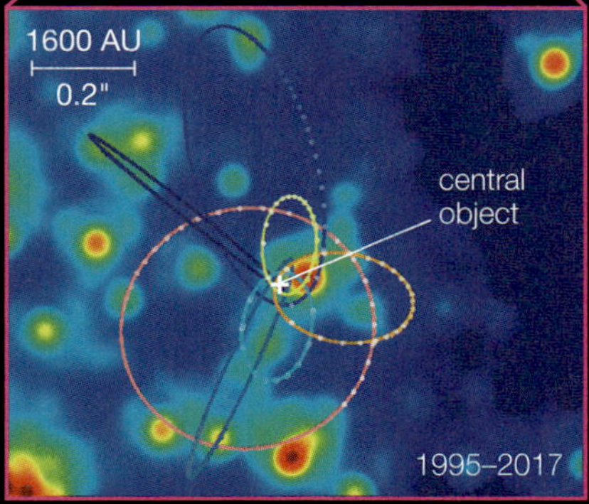

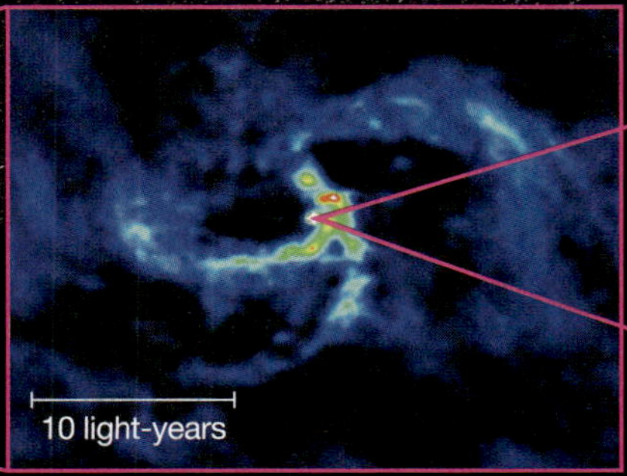


9 The swirling gas clouds seen in the Hubble image (4) are also seen in this radio-wave image, and the exact center is the bright radio source Sgr A*.

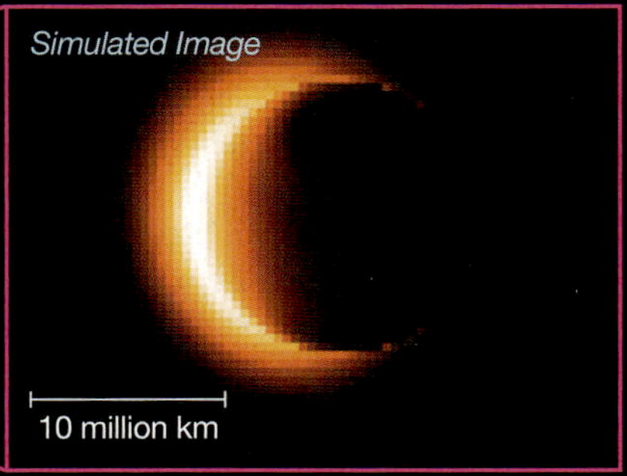


10 Radio telescopes may soon allow astronomers to zoom in to the central black hole's event horizon using the technique of interferometry [Section 6.4].

## The BIG Picture PUTTING CHAPTER 19 INTO CONTEXT

*In this chapter, we have explored the structure, motion, and history of our galaxy, along with the recycling of gas that has made our existence possible. When you review this chapter, pay attention to these "big picture" ideas:*

- The inability of visible light to pass through interstellar gas and dust concealed the true nature of our galaxy until recent times. Modern astronomical instruments reveal the Milky Way Galaxy to be a dynamic system of stars and gas that continually gives birth to new stars and planetary systems.
- Stellar winds and explosions make interstellar space a violent place. Hot gas tears through the atomic hydrogen gas that fills much of the galactic disk, leaving expanding bubbles and fast-moving clouds in its wake. All this violence might seem quite dangerous, but it performs the great service of mixing new heavy elements into the gas of the Milky Way.
- Orbital motions at our galaxy's center indicate that it contains a black hole with 4 million times the mass of our Sun. As we will see in Chapter 21, central black holes appear to be a common feature of galaxies.

**MY COSMIC PERSPECTIVE** Although the elements from which we are made were forged in stars, we could not exist if stars were not organized into galaxies. The Milky Way Galaxy acts as a giant recycling plant, converting gas expelled from each generation of stars into the next generation and allowing some of the heavy elements to solidify into planets like our own.

## Summary of Key Concepts

### 19.1 The Milky Way Revealed

- **What does our galaxy look like?** The Milky Way Galaxy is a **spiral galaxy** consisting of a thin **disk** about 100,000 light-years in diameter with a central **bulge** and a spherical **halo** that surrounds the disk. The disk contains an interstellar medium of gas and dust, while the halo contains only a small amount of hot gas and virtually no cold gas.

- **How do stars orbit in our galaxy?** Stars in the disk all orbit the galactic center in about the same plane and in the same direction. Halo stars also orbit the center of the galaxy, but their orbits are randomly inclined to the disk of the galaxy. Some bulge stars orbit like halo stars, while others orbit more like disk stars. Orbital motions of stars allow us to determine the distribution of mass in our galaxy.

### 19.2 Galactic Recycling

- **How is gas recycled in our galaxy?** Stars are born from the gravitational collapse of gas clumps in molecular clouds. Massive stars explode as supernovae when they die, creating hot **bubbles** in the interstellar medium that contain the new elements made by these stars. This gas cools and mixes into the surrounding interstellar medium, forming clouds of **atomic hydrogen gas**. This gas can cool further to make molecular clouds in which new stars are born, completing the **star–gas–star cycle**.

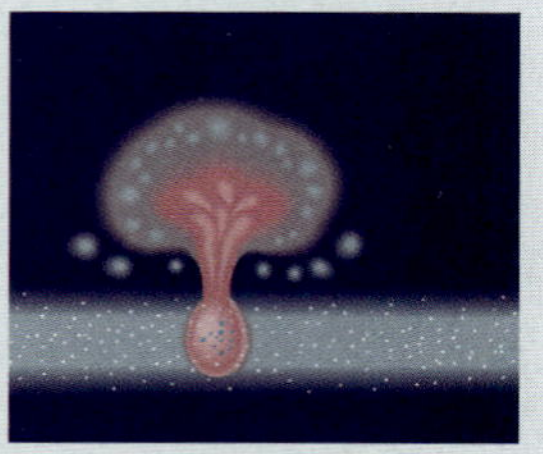

- **Where do stars tend to form in our galaxy?** Active star-forming regions, marked by the presence of hot, massive stars and **ionization nebulae**, are found mostly in **spiral arms**. The spiral arms represent regions where a **spiral density wave** has compressed gas clouds to make star formation more likely.

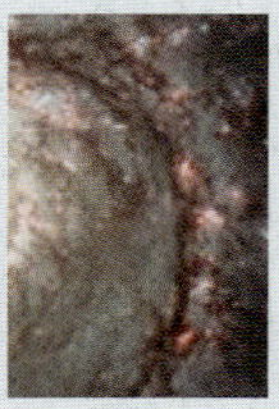

### 19.3 The History of the Milky Way

- **What do halo stars tell us about our galaxy's history?** The stars of the **halo population** are old, low-mass stars that have a much smaller proportion of heavy elements than the stars of the **disk population**. Halo stars therefore must have formed early in the galaxy's history, before the gas settled into a disk.
- **How did our galaxy form?** Halo stars probably formed in several different **protogalactic clouds** of hydrogen and helium gas. Gravity pulled those clouds and stars together to form a single larger cloud. The collapse of this cloud continued until it formed a spinning disk around the galactic center. Stars have formed continuously in the disk since that time, but stars no longer form in the halo.

### 19.4 The Galactic Center

- **What is the evidence for a black hole at our galaxy's center?** Orbits of stars near the center of our galaxy indicate that it contains a black hole about 4 million times as massive as the Sun. The black hole appears to be powering a bright source of radio emission known as Sgr A*.

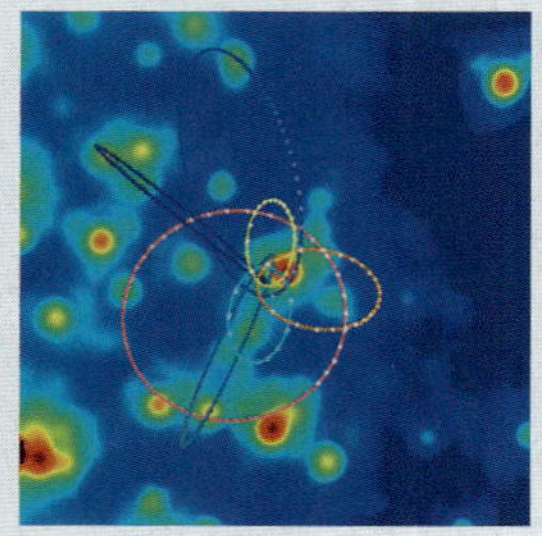

# Visual Skills Check

*Use the following questions to check your understanding of some of the many types of visual information used in astronomy. For additional practice, try the Chapter 19 Visual Quiz in the Study Area at www .MasteringAstronomy.com.*

*The images to the right show what the central region of our galaxy looks like from our viewpoint in different wavelengths of light. Each image shows the same region of the sky. Use the following questions to check your understanding of what these images show and what we can learn by comparing them.*

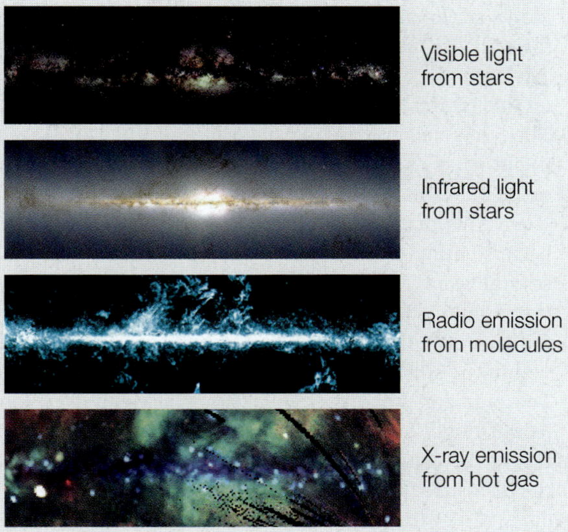


1. The image of radio emission uses different colors to represent different levels of intensity. Which color represents the brightest radio emission? Which color represents the lowest levels of brightness?
2. The image of x-ray emission uses different colors to represent different levels of brightness. The dark blue color represents the least bright x-ray emission. Which color represents the brightest x-ray emission?
3. How do regions showing strong radio emission from molecules look in the visible-light image? Are they bright or are they dark?
4. How do regions showing strong radio emission from molecules look in the infrared-light image? Are they bright or are they dark?
5. Based on a comparison of the radio, infrared, and visible-light images, we can conclude that gas clouds containing molecules ________.
   a. absorb roughly equal amounts of infrared starlight and visible starlight
   b. absorb substantial amounts of visible starlight but no infrared starlight
   c. absorb infrared starlight less effectively than they absorb visible starlight
6. Compare the radio and x-ray images. Can you conclude from this comparison that gas clouds containing molecules absorb x-ray light?

# Exercises and Problems

*For instructor-assigned homework and other learning materials, go to www.MasteringAstronomy.com.*

## Chapter Review Questions

### Short-Answer Questions Based on the Reading

1. Draw simple sketches of our galaxy as it would appear face-on and edge-on. Identify the *disk, bulge, halo,* and *spiral arms,* and indicate the galaxy's approximate dimensions.
2. What are the Large and Small Magellanic Clouds, and the Sagittarius and Canis Major Dwarfs?
3. Describe the basic characteristics of stars' orbits in the disk, halo, and bulge of our galaxy.
4. How can we use orbital properties to learn about the mass of the galaxy? What have we learned?
5. Summarize the *star–gas–star cycle* shown in Figure 19.3.
6. What creates a *bubble* of hot, ionized gas? What happens to the gas in the bubble over time?
7. Why do star clusters make *superbubbles*? What happens to those bubbles when they expand beyond the galactic disk?
8. How does a *galactic fountain* help circulate new elements within the Milky Way?
9. What are *cosmic rays*? Where do they come from?
10. What do we mean by *atomic hydrogen gas*? How common is it, and how do we map its distribution in the galaxy?
11. Briefly summarize the different types of gas present in the disk of the galaxy, and describe how they appear when we view the galaxy in different wavelengths of light.
12. What are *ionization nebulae,* and why are they found near hot, massive stars?
13. How do we know that spiral arms do not rotate like giant pinwheels? What makes spiral arms bright?
14. What triggers star formation within a spiral arm? How do we think spiral arms are maintained?
15. Briefly describe the characteristics that distinguish *disk stars* from *halo stars*.
16. What evidence suggests that the Milky Way formed from the merger of smaller protogalactic clouds?
17. What is Sgr A*? What evidence suggests that it contains a massive black hole?

### Does It Make Sense?

*Decide whether or not each of the following statements makes sense (or is clearly true or false). Explain clearly; not all of these have definitive answers, so your explanation is more important than your chosen answer.*

18. We did not understand the true size and shape of our galaxy until NASA launched satellites into the galactic halo, enabling us to see what the Milky Way looks like from the outside.
19. Planets like Earth probably didn't form around the very first stars because there were so few heavy elements back then.
20. If I could see infrared light, the galactic center would look much more impressive.
21. Many spectacular ionization nebulae are seen throughout the Milky Way's halo.
22. The carbon in my diamond ring was once part of interstellar dust grains.
23. The Sun's velocity around the Milky Way tells us that most of our galaxy's dark matter lies near the center of the galactic disk.
24. We know that a black hole lies at our galaxy's center because numerous stars near it have vanished over the past several years, telling us that they've been sucked in.
25. If we could watch a time-lapse movie of a spiral galaxy over millions of years, we'd see many stars being born and dying within the spiral arms.
26. The star–gas–star cycle will keep the Milky Way looking just as bright in 100 billion years as it looks now.
27. Halo stars orbit the center of our galaxy much faster than the disk stars.

## Quick Quiz

*Choose the best answer to each of the following. For additional practice, try the Chapter 19 Reading and Concept Quizzes in the Study Area at www.MasteringAstronomy.com.*

28. What is the shape of the Milky Way's halo? (a) round like a ball (b) flat like a disk (c) flat like a disk but with a hole in the center
29. Where are most of the Milky Way's globular clusters found? (a) in the disk (b) in the bulge (c) in the halo
30. Why do disk stars bob up and down as they orbit the galaxy? (a) because the gravitational pull of other disk stars always pulls them toward the disk (b) because of friction with the interstellar medium (c) because the halo stars keep knocking them back into the disk
31. How do we determine the Milky Way's mass outside the Sun's orbit? (a) from the Sun's orbital velocity and its distance from the center of our galaxy (b) from the orbits of halo stars near the Sun (c) from the orbits of stars and gas clouds orbiting the galactic center at greater distances than the Sun
32. Which part of the galaxy has gas with the hottest average temperature? (a) the disk (b) the halo (c) the bulge
33. What is the typical percentage (by mass) of elements other than hydrogen and helium in stars that are forming right now in the vicinity of the Sun? (a) 20% (b) 2% (c) 0.02%
34. Which of these forms of radiation passes most easily through the disk of the Milky Way? (a) red light (b) blue light (c) infrared light
35. Where would you be most likely to find an ionization nebula? (a) in the halo (b) in the disk between spiral arms (c) in a spiral arm
36. Which kind of star is most likely to be found in the halo? (a) an O star (b) an A star (c) an M star
37. We measure the mass of the black hole at the galactic center from (a) the orbits of stars in the galactic center. (b) the orbits of gas clouds in the galactic center. (c) the amount of radiation coming from the galactic center.

## Inclusive Astronomy

*Use these questions to reflect on participation in science.*

38. *Group Discussion: Alternative Names for the Milky Way.* Today, we call our galaxy the Milky Way Galaxy, after the band of light that appeared "milky" to the ancient Greeks. But other cultures had other names for that band of light.
**a.** Working independently, learn at least three other names that have been used for the Milky Way and the stories behind those names. Then gather in small groups and share the names and stories that each of you found.
**b.** Discuss and then vote on the name other than "Milky Way" that your group favors. What does your group think are the strengths of that name?
**c.** Discuss the reasons your group's favorite name did *not* become widely used as the name of our galaxy, and make a list of cultural factors that may have favored the name "Milky Way Galaxy." Do you think some of those other names should become better known? Why or why not?

## The Process of Science

*These questions may be answered individually in short-essay form or discussed in groups, except where identified as group-only.*

39. *Discovering the Structure of the Milky Way.* The story of how we came to learn the structure of the Milky Way (see the Special Topic on page 582) provides an excellent example of how science progresses. What features of the Milky Way's appearance in our sky led scientists to conclude that it is much wider than it is thick? Why did they originally believe that the Sun was near the Milky Way's center? What key observations forced scientists to change their views about the location of the Sun within the Milky Way?
40. *Formation of the Milky Way.* Figure 19.18 outlines a basic model that accounts for some but not all of the features of the Milky Way. What observational evidence indicates that the Milky Way's protogalactic cloud contained virtually no elements other than hydrogen and helium? What evidence suggests that the halo stars formed first and disk stars formed later? What features of the Milky Way are *not* explained by this basic model?
41. *Galactic Ecosystem.* We have likened the star–gas–star cycle in our Milky Way to the ecosystem that sustains life on Earth. Here on our planet, water molecules cycle from the sea to the sky to the ground and back to the sea. Our bodies convert atmospheric oxygen molecules into carbon dioxide, and plants convert the carbon dioxide back into oxygen molecules. How are the cycles of matter on Earth similar to the cycles of matter in the galaxy? How do they differ? Do you think the term *ecosystem* is appropriate in discussions of the galaxy?
42. *Galaxy Stuff.* In the chapters on stars, we learned why we are "star stuff." Based on what you've learned in this chapter, explain why we are also "galaxy stuff." Does the fact that the entire galaxy was involved in bringing forth life on Earth change your perspective on Earth or on life in any way? If so, how? If not, why not?
43. *Group Activity: Star Clusters and Milky Way Structure.* Use the all-sky diagram in Appendix H and the information about globular and open clusters below to do the following. Note: You may wish to do this activity using the four roles described in Chapter 1, Exercise 39.

- Globular clusters are found in the following constellations: Aquarius, Aquila, Canes Venatici, Capricornus, Carina, Centaurus, Columba, Coma Berenices, Hercules, Hydra, Lepus, Lynx, Lyra, Musca, Ophiuchus, Pegasus, Puppis, Sagittarius, Scorpius, Sculptor, Serpens, Tucana, and Vela.
- Open clusters with ages of 100 million years or less are found in the following constellations: Canis Major, Carina, Cassiopeia, Crux, Cygnus, Monoceros, Norma, Ophiuchus, Perseus, Puppis, Sagittarius, Scorpius, Scutum, Serpens, Taurus, and Vela.

**a.** Split your group into two teams (e.g., two people each), with one team focusing on globular clusters and the other on open clusters. The members of each team should work together to locate on the all-sky map (Appendix H) the constellations in which their clusters are found and to identify patterns in these locations relative to the Milky Way's disk and relative to the galactic center.

**b.** The full group should compare the locations of the two sets of clusters and describe the differences you've found.

**c.** One member of the group should offer a hypothesis to explain the differences. Others can then suggest alternative hypotheses.

**d.** Discuss the group's set of hypotheses and list possible observations that astronomers could perform to determine which hypothesis is more likely to be correct.

## Investigate Further

### Short-Answer/Essay Questions

44. *Unenriched Stars.* Suppose you discovered a star made purely of hydrogen and helium. How old do you think it would be? Explain your reasoning.
45. *Enrichment of Star Clusters.* The gravitational pull of an isolated globular cluster is weak enough that a single supernova explosion can blow all of the cluster's interstellar gas out of the cluster. How might this fact be related to observations indicating that stars ceased to form in globular clusters long ago? How might it be related to the fact that globular clusters are deficient in elements heavier than hydrogen and helium? Summarize your answers in one or two paragraphs.
46. *High-Velocity Star.* The average speed of stars in the solar neighborhood *relative to the Sun* is about 20 km/s. Suppose you discover a star in the solar neighborhood that is moving relative to the Sun at a much higher speed—say, 200 km/s. What kind of orbit does this star probably have around the Milky Way? In what part of the galaxy does it spend most of its time? Explain.
47. *Future of the Milky Way.* Imagine that you could make a time-lapse movie from outside the Milky Way Galaxy over the next 100 billion years. How would you expect the appearance of our galaxy to change over that time period? Explain.
48. *Orbits at the Galactic Center.* Using the information in Figure 19.20, identify the two stars that reach the highest speeds, based on the positions of the color-coded dots representing the yearly motions of each star. Explain how the orbits of those two stars illustrate Kepler's first two laws.
49. *A Nonspinning Galaxy.* How would the development of the Milky Way Galaxy have been different if it had collapsed from protogalactic clouds that had no net angular momentum? Describe how you think our galaxy would look today, and explain your reasoning.
50. *Gas Distribution in the Milky Way.* Make a sketch of the gas distribution in the plane of the Milky Way, based on the photographs in Figure 19.12. In your sketch, map out where the molecular clouds are and draw the layer of atomic hydrogen gas that surrounds them. Then add the locations of the most prominent bubbles of hot gas. Explain why each of these components of the Milky Way's interstellar medium is found in the location where you have drawn it.
51. *Research: The Galactic Center.* Search for recent images of the center of the Milky Way Galaxy, along with information about the massive black hole thought to reside there. Write a short report, with pictures, giving an update on current knowledge.

### Quantitative Problems

*Be sure to show all calculations clearly and state your final answers in complete sentences.*

52. *Mass of the Milky Way's Halo.* The Large Magellanic Cloud orbits the Milky Way at a distance of roughly 160,000 light-years from the galactic center and a velocity of about 300 km/s. Use these values in the orbital velocity law (Mathematical Insight 19.1) to estimate the Milky Way's mass within 160,000 light-years from the center. (The value you obtain is a fairly rough estimate because the orbit of the Large Magellanic Cloud is not circular.)
53. *Mass of the Central Black Hole.* Suppose you observe a star orbiting the galactic center at a speed of 1000 km/s in a circular orbit with a radius of 20 light-days. Calculate the mass of the object the star is orbiting.
54. *Mass of a Globular Cluster.* Stars in the outskirts of a globular cluster are typically about 50 light-years from the cluster's center, which they orbit at speeds of about 10 km/s. Use these data to calculate the mass of a typical globular cluster.
55. *Mass of Saturn.* The innermost rings of Saturn orbit in a circle with a radius of 67,000 km at a speed of 23.8 km/s. Use the orbital velocity law to compute the mass contained within the orbit of those rings. Compare your answer with the mass of Saturn listed in Appendix E.
56. *Pressure vs. Gravity in Hot Ionized Gas.* Use the formula in Mathematical Insight 16.1 to estimate the mass of a gas cloud in which pressure balances gravity for a gas temperature of $10^6$ K and a number density of 0.01 particle per cubic centimeter. Use your answer to explain why hot gas must cool down before it can collect into star-forming clouds.
57. *Pressure vs. Gravity in Warm Atomic Gas.* Use the formula in Mathematical Insight 16.1 to estimate the mass of a gas cloud in which pressure balances gravity for a gas temperature of $10^4$ K and a number density of 1 particle per cubic centimeter. Compare your answer to the 1 million solar masses in a typical giant molecular cloud. Use your answer to explain why additional cooling is necessary in order to create such clouds.
58. *Pressure vs. Gravity in Cool Atomic Gas.* Use the formula in Mathematical Insight 16.1 to estimate the mass of a gas cloud in which pressure balances gravity for a gas temperature of $10^2$ K and a number density of 100 particles per cubic centimeter. Use your answer to explain why star-forming clouds typically contain enough gas to form several thousand stars.
59. *The Speed of Supernova Debris.* The kinetic energy $E_{kinetic}$ of an amount of mass $M$ traveling at a velocity $v$ is given by the formula $E_{kinetic} = Mv^2/2$. The total kinetic energy of the matter ejected from a supernova explosion is about $10^{44}$ joules. Determine the typical speed at which that matter is ejected from a supernova with a mass of $10M_{Sun}$. Compare that speed with the Sun's orbital speed around our galaxy. Based on your comparison, do you think the galaxy's gravity would be strong enough to retain the supernova debris if there were no interstellar medium to slow it down? Explain.

# 20 Galaxies and the Foundation of Modern Cosmology

▲ **About the photo:** The inner regions of spiral galaxy M66 in the constellation Leo.

## LEARNING GOALS

**20.1 Islands of Stars**
- How are the lives of galaxies connected with the history of the universe?
- What patterns do we find among the properties of galaxies?

**20.2 Measuring Galactic Distances**
- How do we measure the distances to galaxies?
- How did Hubble prove that galaxies lie far beyond the Milky Way?
- What is Hubble's law?

**20.3 The Age of the Universe**
- How does Hubble's law tell us the age of the universe?
- How does expansion affect distance measurements?
- Why does the observable universe have a horizon?

*Those who have learned to walk on the threshold of the unknown worlds, by means of the sciences, may then, with the fair white wings of imagination, hope to soar further into the unexplored amidst which we live.*

— *Ada Lovelace (1815–1852)*

**Chapter 20 Overview**

Barely a century ago, no one knew for certain whether the universe extended beyond the bounds of our own Milky Way. Today, we know that our observable universe contains some 100 billion galaxies, and that these galaxies come in a variety of shapes and sizes. We also know that our universe is expanding, so that nearly all galaxies are being carried away from one another with time.

In this chapter, we'll explore how we have come to know these things, and what they tell us about the age and size of the universe. We'll begin by getting acquainted with the patterns we observe among galaxies. Then we'll discuss how the distances to galaxies are measured and how such measurements led to the discovery of universal expansion. Along the way, we'll see how these discoveries—many made by Edwin Hubble, for whom the Hubble Space Telescope was named—heralded the birth of the modern science of cosmology.

## 20.1 Islands of Stars

Look back at Figure 1.1. Up to this point in the book, we have focused our attention on only the first three levels of structure shown in that figure: planets, star systems, and the Milky Way Galaxy. Now, we are ready to turn our attention to the rest of the universe, which is filled with galaxies from here to its most distant reaches. In this first section, we'll discuss how astronomers categorize these galaxies, which will provide the background we need to explore how galaxies formed, why they differ from one another, and what they can tell us about the universe itself.

### How are the lives of galaxies connected with the history of the universe?

**FIGURE 20.1** shows an amazing Hubble Space Telescope image, known as the Hubble eXtreme Deep Field, obtained by combining images in which the telescope focused on a single tiny patch of the sky for a total of 23 days. The entire field of view is less than one-tenth the angular diameter of the full moon in our sky (which itself is less than the angular width of your fingernail held at arm's length), yet almost every one of the thousands of blobs of light in it is a

**FIGURE 20.1** The Hubble eXtreme Deep Field (XDF) required 23 days of exposure time with the Hubble Space Telescope and shows thousands of galaxies in a typical patch of the sky. To visualize the size of this patch on the celestial sphere, picture on your own thumb the tiny version of the XDF on the thumbnail to the right, then hold your thumb at arm's length.

galaxy—an island of stars bound together by gravity, much like our own Milky Way. Photos like this one allow us to estimate the total number of galaxies in the observable universe, because this patch of sky is quite average and representative of the sky as a whole. We can therefore simply count the number of galaxies in the photo and multiply by the number of such photos it would take to capture the entire sky. The results tell us that the observable universe contains well over 100 billion large galaxies and many more small ones.

As you study Figure 20.1, notice that galaxies come in many sizes, colors, and shapes. Some are large, some small. Some are reddish, some whitish. Some appear round, and some appear flat. We would like to understand *why* galaxies differ in these ways, but it is not easy to learn their histories. Just as with stars, our observations capture only the briefest snapshot of any galaxy's life, leaving us to piece together the life story of a typical galaxy from pictures of different galaxies at various life stages.

The nature of the universe complicates this task because, unlike the case for stars, we do not find galaxies of many different ages in our local neighborhood. Most nearby galaxies appear to be similar in age to our own Milky Way Galaxy, meaning that they formed more than 10 billion years ago. To study young galaxies, we must look far back into the universe's past—which means looking to great distances. Remember that the farther away we look in space, the further back we look in time. So when we observe galaxies that are very far away, we see them as they were when they and the universe were young (see Figure 1.4). Almost all of the galaxies in Figure 20.1 are billions of light-years away, which means we are seeing them as they looked billions of years ago, at an earlier stage in their development.

These relationships between a galaxy's age, its distance, and the age of the universe make it impossible to consider the histories of galaxies without at the same time considering the evolution of the universe as a whole. As a result, the study of galaxies is intimately connected with **cosmology**—the study of the overall structure and evolution of the universe. We will begin our quest to understand the lives of galaxies by learning to recognize patterns among the properties of nearby galaxies, because their details are easier to observe than those of the distant galaxies in Figure 20.1.

## What patterns do we find among the properties of galaxies?

Broadly speaking, astronomers classify galaxies into three types based on their appearance:

- **Spiral galaxies**, such as our own Milky Way, look like flat white disks with yellowish bulges at their centers. The disks are filled with cool gas and dust, interspersed with hotter ionized gas, and usually display beautiful spiral arms.
- **Elliptical galaxies** are redder, more rounded, and often elongated like a football. Compared with spiral galaxies, elliptical galaxies contain very little cool gas and dust, though they often contain very hot, ionized gas.
- **Irregular galaxies** appear neither disklike nor rounded.

The differing colors of galaxies arise from the different kinds of stars that populate them: Spiral and irregular galaxies look white because they contain stars of all different colors and ages, while elliptical galaxies look redder because old, reddish stars produce most of their light. Galaxies also come in a wide range of sizes, from *dwarf galaxies* containing less than 1 million ($10^6$) stars to *giant galaxies* with more than 1 trillion ($10^{12}$) stars.

**Think about it** Take a moment to try to classify the larger galaxies in Figure 20.1. How many appear spiral? Elliptical? Irregular? Do the colors of galaxies seem related to their shapes?

**Spiral Galaxies** Like the Milky Way, other spiral galaxies also have a thin *disk* and a central *bulge* (**FIGURE 20.2**). The bulge merges smoothly into a nearly invisible *halo* that can

**a** M101, a face-on spiral galaxy.

**b** NGC 4414, a spiral galaxy whose disk is somewhat tilted to our line of sight.

**c** NGC 891, a spiral galaxy seen nearly edge-on. Note the central streak of dust associated with the disk.

**FIGURE 20.2** These three photos show spiral galaxies from three different perspectives, ranging from nearly face-on to nearly edge-on. In each case, the region pictured is about 100,000 light-years across. (NGC stands for the *New General Catalog*, a listing of more than 7000 objects published in 1888.)

**FIGURE 20.3** NGC 4594 (the Sombrero Galaxy) is a spiral galaxy with a large bulge and a dusty disk that we see almost edge-on. A much larger but nearly invisible halo surrounds the entire galaxy. The bulge and halo together make up the spheroidal component of the galaxy. This image shows a region of the galaxy about 82,000 light-years across.

**FIGURE 20.4** NGC 1300, a barred spiral galaxy about 110,000 light-years in diameter.

extend to a radius of more than 100,000 light-years. A spiral galaxy's halo is considerably more difficult to see than its bulge and disk because halo stars are generally dim and spread over a large volume of space.

Recall that the disk and the halo of the Milky Way represent two distinct populations of stars [**Section 19.3**]: *Disk stars* orbit in the same plane and include stars of all ages and masses, while *halo stars* have randomly oriented orbits and are all old and low in mass. The Milky Way's bulge stars, on the other hand, exhibit a mixture of disk and halo properties. Other spiral galaxies share these characteristics, but in spiral galaxies with particularly large bulges, the orbits of bulge stars have more in common with the orbits of halo stars than with those of disk stars. We will therefore simplify our discussion of galaxies by dividing their stars into two primary components:

- The **disk component** consists of stars and dusty gas clouds that follow orderly, nearly circular orbits around the galactic center. Galaxy disks always contain a gaseous *interstellar* medium, but the amounts and proportions of molecular, atomic, and ionized gas differ from one galaxy to the next.
- The **halo component** (sometimes called the *spheroidal* or *ellipsoidal* component), which we will take to include both the halo and the central bulge, generally has a rounded or elliptical shape. The halo component generally contains little cool gas and dust, and its stars have orbits with many different inclinations. The disk is embedded within it.

All spiral galaxies have both a disk component and a halo component, but there are variations on the general theme. Bulge sizes can vary dramatically; for example, **FIGURE 20.3**, which shows a spiral galaxy with an unusually large bulge, illustrates the general shape of the halo component. Other spiral galaxies—known as *barred spirals*—appear to have a straight bar of stars cutting across the center, with spiral arms curling away from the ends of the bar (**FIGURE 20.4**). Astronomers suspect that the Milky Way itself is a barred spiral galaxy, because of the elongation of our galaxy's bulge.

Some galaxies have disk and halo components like spiral galaxies but appear to lack spiral arms (**FIGURE 20.5**). These *lenticular galaxies* (*lenticular* means "lens-shaped") are sometimes considered an intermediate class between spirals and ellipticals, because they tend to have less cool gas than normal spirals but more than ellipticals. Most of the large galaxies in the universe (75–85%) are spiral or lenticular.

**Elliptical Galaxies** Elliptical galaxies differ from spiral galaxies primarily in that they have *only* a halo component and lack a significant disk component. That is, elliptical galaxies look much like the bulge and halo of a spiral galaxy without a disk. Elliptical galaxies come in a wide range of sizes (**FIGURE 20.6**). Relatively rare *giant elliptical galaxies* are among the most massive galaxies in the universe, while small *dwarf elliptical galaxies* are much more common.

**FIGURE 20.5** The lenticular galaxy NGC 4866. A few streaks of dusty gas can be seen in this galaxy's disk, but it does not contain any noticeable spiral arms. The region pictured is about 75,000 light-years across.

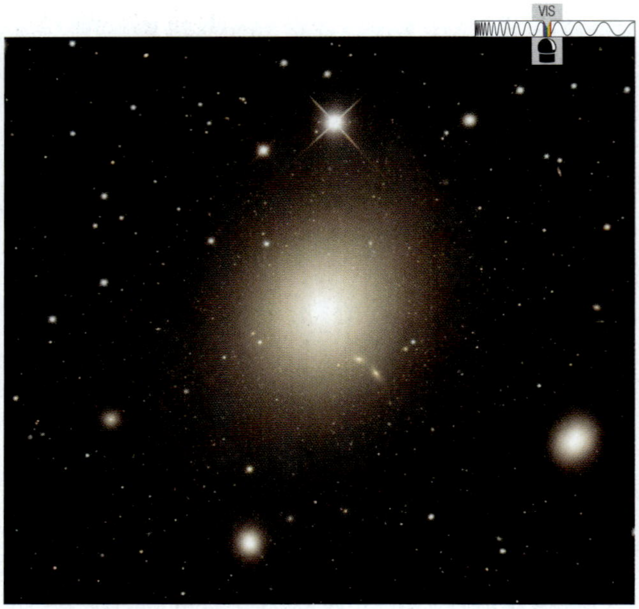


a M87, a giant elliptical galaxy in the Virgo Cluster, is one of the most massive galaxies in the universe. The region shown is more than 300,000 light-years across.

b Leo I is one of at least a dozen dwarf elliptical galaxies in the Local Group. It is only about 2500 light-years across.

**FIGURE 20.6** Elliptical galaxies come in a wide range of sizes.

Elliptical galaxies usually contain very little dust or cool gas. As a result, they generally have little or no ongoing star formation. Elliptical galaxies therefore look red or yellow in color because they lack the hot, young, blue stars found in the disks of spiral galaxies. However, large elliptical galaxies sometimes contain substantial amounts of very hot gas that emits x-rays, much like the gas in the hot bubbles created by supernovae and powerful stellar winds in the Milky Way [**Section 19.2**].

The most numerous galaxies in our Local Group lie at the other end of the galaxy size scale. They are round and diskless like large elliptical galaxies, but they are particularly small and much less bright. Because of these distinct differences, astronomers assign these galaxies to a special subtype (known as *dwarf spheroidal galaxies*). Their lack of brightness makes them almost impossible to detect at distances greater than a few million light-years. Nevertheless, astronomers suspect this galaxy subtype to be the most common in the universe, based on the large proportion of them in the Local Group.

**Irregular Galaxies** Some of the galaxies we see nearby fall into neither of the two major categories. The class of *irregular galaxies* is a miscellaneous class, encompassing small galaxies such as the Magellanic Clouds and larger "peculiar" galaxies that appear to be in disarray (**FIGURE 20.7**). These blobby star systems are usually white and dusty, like the disks of spirals. Their colors tell us that they contain young, massive stars.

Irregular galaxies represent only a small fraction of the nearby large galaxies, but telescopic observations probing deeper into the universe show that irregular shapes are

a The Large Magellanic Cloud, a small companion to the Milky Way. It is about 30,000 light-years across.

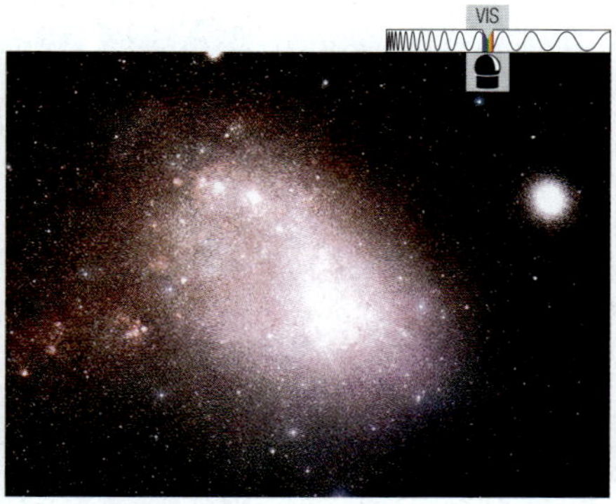


b The Small Magellanic Cloud, a smaller companion to the Milky Way. It is about 18,000 light-years across.

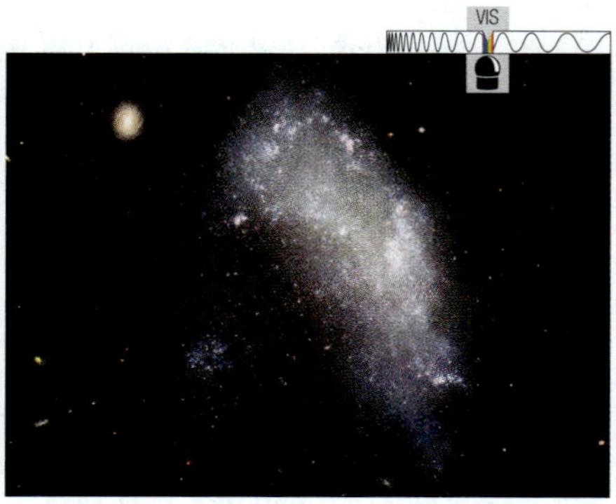


c NGC 1427A, an irregular galaxy with scattered patches of star formation. The region pictured is about 50,000 light-years across.

**FIGURE 20.7** Irregular galaxies.

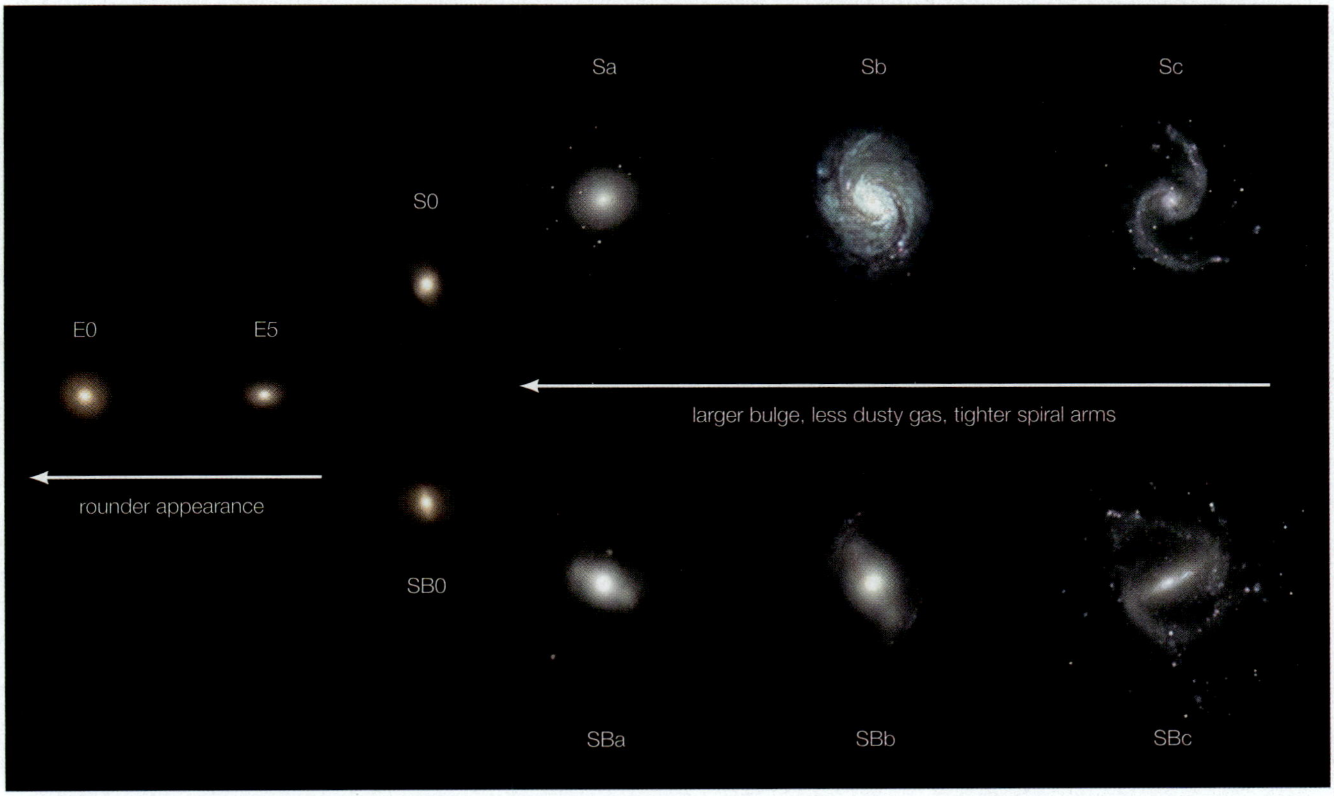


**FIGURE 20.8** This "tuning fork" diagram illustrates Hubble's galaxy classes.

more common among distant galaxies. Because the light of more distant galaxies has taken longer to reach us, these observations tell us that irregular galaxies were more common when the universe was younger.

**Patterns in Galaxy Shapes** Early attempts to understand the relationships among the major galaxy types focused primarily on shape. Astronomer Edwin Hubble devised a system that organized spiral and elliptical galaxies into a diagram shaped like a tuning fork (**FIGURE 20.8**). Elliptical galaxies appear on the "handle" at the left, designated by the letter *E* and a number. The larger the number, the flatter the elliptical galaxy: An E0 galaxy is a sphere, and the numbers increase to the highly elongated type E7. The two prongs of the fork show spiral galaxies, designated by the letter *S* for ordinary spirals and *SB* for barred spirals, followed by a lowercase *a*, *b*, or *c*: The bulge size decreases from a to c, while the amount of dusty gas increases. Galaxies with properties intermediate between those of spirals and ellipticals, such as lenticular galaxies, are designated *S0* or *SB0*. The diagram does not show irregular galaxies, which are designated *Irr*.

Notice that the relative prominence of the halo and disk components changes gradually from one side of the diagram to the other. At the far left are elliptical galaxies with no elongation and no disk component at all. At the far right are spiral galaxies (and barred spiral galaxies) with a large star-forming disk, a small bulge, and a nearly invisible halo. The large but continuous range in properties shows that the main feature determining a galaxy's type is the prominence of its disk component relative to the halo component.

Astronomers once suspected that Hubble's tuning fork represented an evolutionary sequence in which galaxies flattened and spread out as they aged, but we now know that is not the case. Instead, other ways of looking at patterns among galaxies have yielded deeper insights.

**Patterns in Galaxy Color and Luminosity** Astronomers have now measured the properties of millions of galaxies, and can learn more about how galaxies evolve by plotting those properties on a diagram much like the H-R diagram for stars, with galaxy color along the horizontal axis and galaxy luminosity along the vertical axis. **FIGURE 20.9** shows one such diagram. Notice that most galaxies fall into one of two major groups, distinguished primarily by their levels of ongoing star formation:

- Galaxies in the **blue cloud** are blue because they contain numerous hot, young stars that signify active star formation. This star formation is possible because these galaxies tend to be spiral or irregular galaxies with cold, star-forming gas clouds.
- Galaxies in the **red sequence** usually contain only older stars, which is why they are redder in color. Most galaxies in the red sequence are elliptical in shape, and the ones at the top of the diagram are the most luminous galaxies in the universe.

Galaxies that are now members of the red sequence must once have been members of the blue cloud. The reason is simple: A galaxy cannot have old stars now without having experienced an earlier episode of active star formation. We

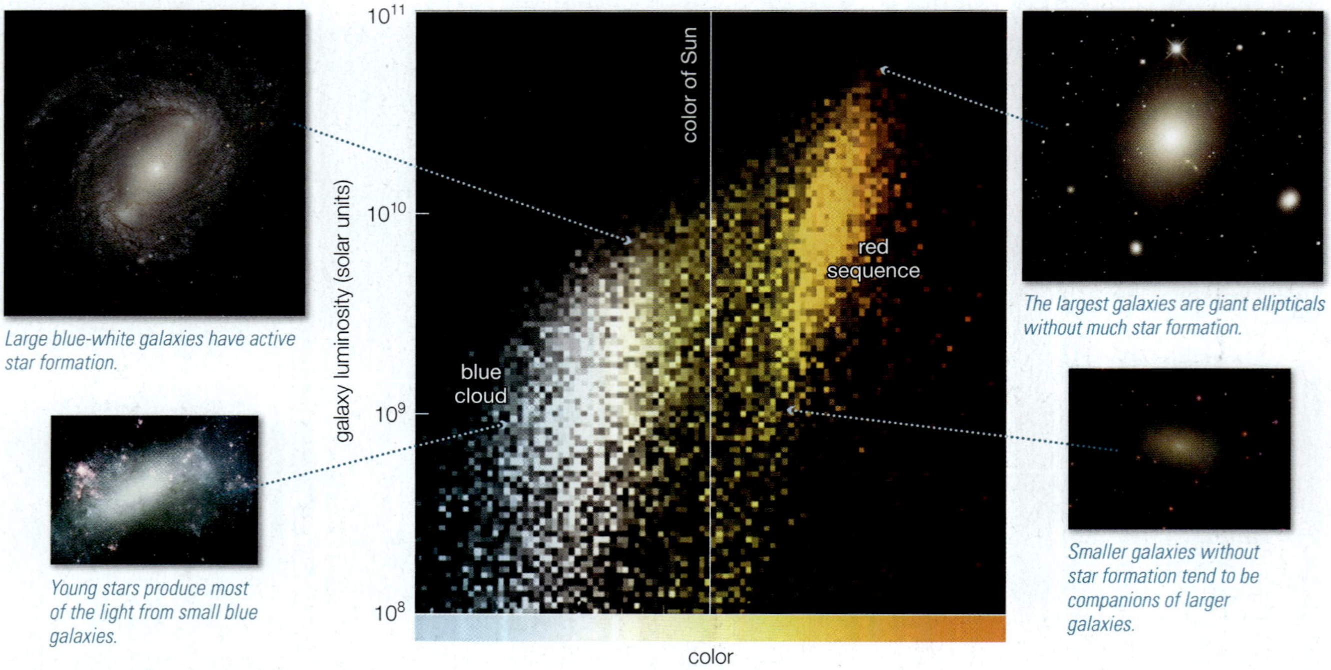


**FIGURE 20.9** A graph plotting galaxy luminosities against colors shows that galaxies fall into two main groups: the *blue cloud*, consisting mostly of spiral and irregular galaxies that are blue to white in color, and the *red sequence*, consisting mostly of elliptical galaxies that are redder in color. The brightness of each small square reflects the number of galaxies with the corresponding color and luminosity. (Data from the Sloan Digital Sky Survey.)

therefore infer that all galaxies start as members of the blue cloud, but some then transition to the red sequence, presumably because they use up or lose all of the cold gas clouds necessary for star formation. The fact that the most luminous galaxies tend to be members of the red sequence suggests that star formation shuts down sooner in larger galaxies than in smaller ones, but that does not appear to be the whole story. Instead, the patterns we see among galaxy groups and clusters suggest that a galaxy's environment (where it resides in relation to neighboring galaxies) also plays an important role.

**Patterns in Galaxy Groups and Clusters** Most of the galaxies in the universe are gravitationally bound together with neighboring galaxies. Spiral galaxies with ongoing star formation are often found in loose collections of up to a few dozen galaxies, called *groups*. Our Local Group is one example (see Figure 1.1). **FIGURE 20.10** shows another galaxy group.

Larger collections are called galaxy *clusters* and may contain hundreds or thousands of individual galaxies concentrated in a region that can be 10 million or more light-years across (**FIGURE 20.11**). Elliptical galaxies in which star formation has ceased make up about half the large galaxies in the central regions of clusters, even though they represent only a small minority (about 15%) of the large galaxies found outside clusters.

These features of groups and clusters indicate that a galaxy's local environment may help determine whether it ends up as a spiral galaxy or an elliptical galaxy. For example, the fact that elliptical galaxies—which generally belong to the red sequence—tend to be found in the crowded central regions of clusters suggests that collisions between blue-cloud spiral galaxies may transform them into red-sequence elliptical galaxies. Also, the most luminous galaxies in the universe tend to be found at the centers of galaxy clusters, suggesting that they may have grown large by dismantling and consuming smaller galaxies that came too close. We'll explore these ideas in greater depth in Chapter 21. First, however, we need to understand how we measure galactic distances, on which our understanding of the expansion and age of the universe is based.

**FIGURE 20.10** Hickson Compact Group 87, a small group of galaxies consisting of a large edge-on spiral galaxy below the center of this photo, two smaller spiral galaxies above it, and an elliptical galaxy to its right. The whole group is about 170,000 light-years in diameter. (The other objects in this photograph are foreground stars in our own galaxy.)

FIGURE 20.11 Central part of the galaxy cluster Abell 1689. Almost every object in this photograph is a galaxy belonging to the cluster. Yellowish elliptical galaxies outnumber the whiter spiral galaxies. The region pictured is about 2 million light-years across. (A few stars from our own galaxy appear in the foreground as white dots centered on cross-shaped spikes.)

## 20.2 Measuring Galactic Distances

To learn more about a galaxy than just its shape and color, we need to know its distance. Only then can we calculate its true size and luminosity, and learn whether other galaxies that appear close to it in the sky are actual neighbors or are located at vastly different distances. Remember that distance is related to age: We see more distant galaxies as they were longer ago in the past, when the universe was younger. As a result, the distances of galaxies are also crucial to our understanding of galaxy evolution and of the size and age of the universe.

In this section, we will explore the techniques astronomers use to measure distances throughout the universe and see how Edwin Hubble applied some of those techniques to prove that we live in an expanding universe full of galaxies. Then, in the final section of this chapter, we'll be ready to explore how Hubble's discoveries have led to our modern measurements of the size and age of the universe.

### How do we measure the distances to galaxies?

Our distance determinations depend on a chain of methods in which each step allows us to measure greater distances. We have already discussed the use of parallax to measure distances to nearby stars [**Section 15.1**]. Because parallax is the apparent shift in a star's position as Earth orbits the Sun, measuring distances by parallax requires knowing the precise Sun-Earth distance, or astronomical unit (AU). That is why the *first* link in the distance-measurement chain is **radar ranging**, in which radio waves transmitted from Earth bounce off a planet or other object and return to Earth (see Special Topic, p. 207). Radio waves travel at the speed of light, so the round-trip travel time for radar signals from Venus told us Venus's distance from Earth. Scientists then used Kepler's laws and a little geometry to calculate the length of an AU. Parallax is then the *second* link in the chain, taking us from distances within the solar system to distances within the Milky Way.

**Standard Candles** The next links in the distance-measurement chain rely on light sources that can serve as **standard candles**, meaning objects whose luminosity we already know. Such objects are useful because we can use their known luminosity and apparent brightness to calculate their distance with the *inverse square law for light* [**Section 15.1**].

**FIGURE 20.12** shows how standard candles work. All of the lights in the illustration are streetlamps of identical luminosity, but the lamps at greater distance appear dimmer (by the square of their distance). We can therefore determine each lamp's distance from its known luminosity. For example, suppose we know that the lamps all emit a luminosity of 1000 watts. Then if one lamp appears only ¼ as bright as another, we know that it must be twice as far away as the other.

Unlike light bulbs, astronomical objects do not come marked with wattage. However, an astronomical object can still serve as a standard candle if we have some way of knowing its true luminosity. Many astronomical objects meet this requirement. For example, any star that is a twin of our Sun—that is, a main-sequence star with spectral type G2—should have about the same luminosity as the Sun. Therefore, if we measure the apparent brightness of a Sun-like star, we can assume it has the Sun's luminosity ($3.8 \times 10^{26}$ watts) and use the inverse square law for light to estimate its distance.

Beyond the distances to which we can measure parallax, we use standard candles for most cosmic distance measurements. These distance measurements always have some uncertainty, because no astronomical object is a perfect standard candle. The challenge of measuring astronomical distances comes down to the challenge of finding the

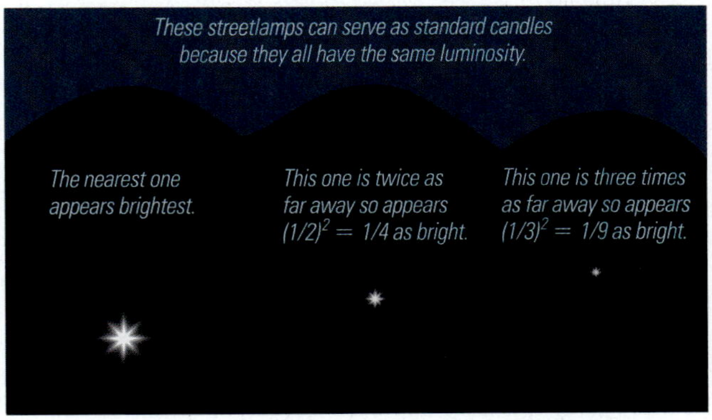


FIGURE 20.12 These streetlights are standard candles because they all put out the same amount of light, allowing us to determine their relative distances from their relative brightnesses.

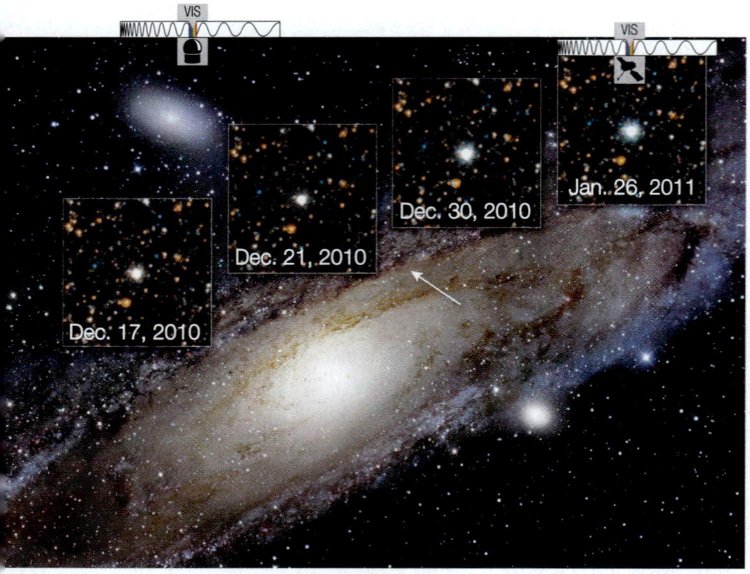


**FIGURE 20.13** A Cepheid variable star in the Andromeda Galaxy, observed by the Hubble Space Telescope. A white arrow in the background image shows the location of this star within the galaxy. The inset images show the star's changing brightness over a period of several weeks. Notice, for example, that it was brighter on December 30, 2010 and January 26, 2011 than it was on December 17, 2010 or December 21, 2010. A more complete set of observations shows that this star pulsates with a period of about 31 days. (This particular star was the first Cepheid that Hubble himself discovered in the Andromeda galaxy, enabling him to estimate the galaxy's distance.)

objects that make the best standard candles. The more confident we are about an object's true luminosity, the more certain we are of its distance.

**Cepheid Variables** The key to measuring distances beyond the Milky Way is to identify objects that can serve as standard candles and that are also bright enough to be detected at great distances. The most useful such objects for relatively nearby galaxies are a special class of extremely luminous stars known as **Cepheid variable stars**, or **Cepheids** for short. Recall that Cepheids are very bright examples of *pulsating variable stars* [**Section 15.2**]: They actually pulsate in size, growing brighter as they grow larger and then dimming as they shrink back in size. A Cepheid's pulsations cause its luminosity (and hence also its apparent brightness) to alternately brighten and dim with a period that can range from a few days to a few months. **FIGURE 20.13** shows four images of a Cepheid in which you can clearly see the changes in brightness. Astronomers can use these changes to plot a light curve (see Figure 15.14) from which we can identify the period of the Cepheid.

Cepheids are valuable as standard candles because a Cepheid's pulsation period is closely related to its luminosity: The longer the period, the more luminous the star (**FIGURE 20.14**). This relation is often called the *Cepheid period-luminosity relation,* but we will call it **Leavitt's law*** because it was first discovered by Henrietta Leavitt in 1912 through careful observations of Cepheids in the Large Magellanic Cloud. The distance to this small galaxy was not yet known at the time, so Leavitt was unable to determine the true luminosities of the Cepheids. However, because all the stars in the Large Magellanic Cloud are at approximately the same distance from us, she was able to discover the relationship by comparing the apparent brightness of these Cepheids. Today, we know the relationship between variability period and luminosity among Cepheid variables quite accurately, because we have measured the distances to nearby Cepheids with parallax and used these distances and the stars' apparent brightnesses to calculate true luminosities.

*The American Astronomical Society has recommended this change in terminology to honor the critical role that Henrietta Levitt's work played in Hubble's subsequent discoveries. For more discussion of this change in terminology, see Exercise 36 at the end of this chapter.

## MATHEMATICAL INSIGHT 20.1 Standard Candles

The inverse square law for light tells us how an object's apparent brightness depends on its luminosity and distance:

$$\text{apparent brightness} = \frac{\text{luminosity}}{4\pi \times (\text{distance})^2}$$

With a little algebra, we can rewrite this law in a form that enables us to calculate distance if we know luminosity and apparent brightness, as we generally do for any object that qualifies as a *standard candle*:

$$\text{distance} = \sqrt{\frac{\text{luminosity}}{4\pi \times (\text{apparent brightness})}}$$

**EXAMPLE:** With a telescope, you measure a star's apparent brightness to be $1.0 \times 10^{-12}$ watt/m$^2$. The star has the same spectral type and luminosity class as the Sun. How far away is it?

**SOLUTION:**

**Step 1 Understand:** We can use the star as a standard candle because its spectral classification tells us that it must have about the luminosity of our Sun, which is $3.8 \times 10^{26}$ watts. We can use this luminosity and the measured apparent brightness to calculate the star's distance with the inverse square law for light.

**Step 2 Solve:** Using the second form of the law above with the given numbers, we find

$$\text{distance} = \sqrt{\frac{3.8 \times 10^{26}\ \text{watts}}{4\pi \times \left(1.0 \times 10^{-12}\ \dfrac{\text{watt}}{\text{m}^2}\right)}} \approx 5.5 \times 10^{18}\ \text{m}$$

**Step 3 Explain:** The star is about $5.5 \times 10^{18}$ meters away. We convert this to light-years by dividing by $9.5 \times 10^{15}$ meters/light-year (see Appendix A); the result is about 580 light-years.

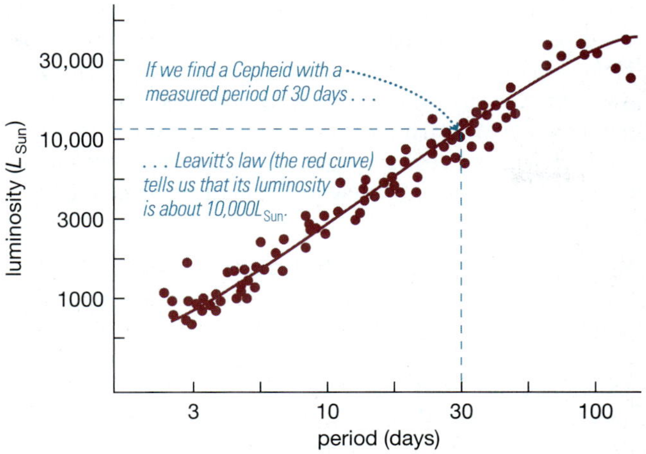


**FIGURE 20.14** The data (red dots) on this graph show that the period of a Cepheid variable star is very closely related to its luminosity, leading to the relationship (red curve) known as *Leavitt's law*. Measuring a Cepheid's period therefore allows us to determine its luminosity and calculate its distance. (Cepheids actually come in two types with two different relations between period and luminosity. The relation here is for Cepheids with heavy-element content similar to that of our Sun, or "Type I Cepheids.")

As an example of how we use Cepheids to measure distance, look at the point marked with dashed lines in Figure 20.14. Notice that this star's period of 30 days corresponds to a luminosity of about 10,000 Suns. In other words, the fact that this Cepheid's brightness peaks every 30 days means it is effectively screaming out, "Hey, everybody, my luminosity is 10,000 times that of the Sun!" More generally, once we measure a Cepheid's period, we can use Leavitt's law to learn its luminosity (within about 10%) and can then use the inverse square law for light to determine its distance.

Cepheids have been used for almost a century to measure distances to nearby galaxies, and as we'll discuss shortly, they played a critical role in Edwin Hubble's discoveries. Today, we can use Cepheids to measure distances to galaxies up to 100 million light-years away. This distance may sound large, but it is still quite small compared with the distances of the galaxies in Figure 20.1. To go farther, we use the distances determined with Cepheids to learn the luminosities of even brighter standard candles.

**Distant Standard Candles** Astronomers have identified several types of objects that can be used as standard candles for measuring very large distances, but the most valuable are *white dwarf supernovae.* Recall that white dwarf supernovae are exploding white dwarf stars that have reached the $1.4M_{Sun}$ limit [**Section 18.1**]. These supernovae all have nearly the same luminosity, probably because they all come from stars of similar mass that explode in nearly the same way. Although white dwarf supernovae are rare in any individual galaxy, numerous examples have been observed within about 50 million light-years of the Milky Way (**FIGURE 20.15**). We can determine the true luminosities of those supernovae by using Cepheids to measure the distances to the galaxies in which they

**FIGURE 20.15** A white dwarf supernova observed in 1994 in the galaxy NGC 4256.

occurred. These measurements not only tell us the true luminosities of white dwarf supernovae, but also confirm that their luminosities are all about the same and therefore that we can use them as standard candles.

Because white dwarf supernovae are so bright—about 10 billion solar luminosities at their peak—we can detect them even when they occur in galaxies billions of light-years away (**FIGURE 20.16**). Although the number of galaxies whose distances we can measure with this technique is relatively small (because it works only for galaxies in which a white dwarf supernova has been seen), these galaxies have allowed us to calibrate yet another technique—one that relies on Hubble's discoveries about the expansion of the universe.

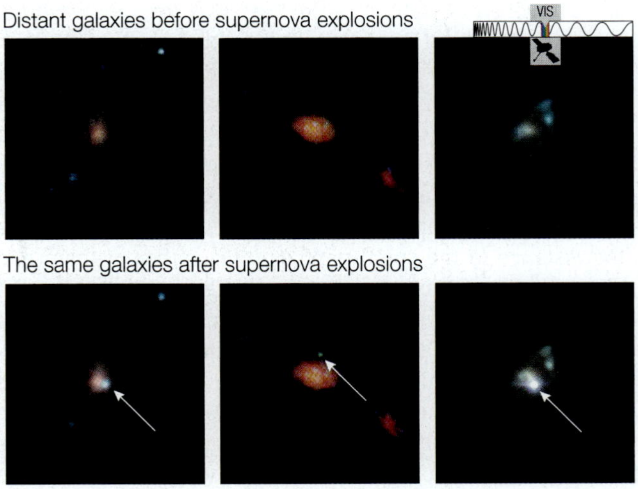


**FIGURE 20.16** White dwarf supernovae. White arrows in the lower images indicate the supernovae, and the upper images show what these galaxies looked like without supernovae. The first two (from left) are supernovae that exploded when the universe was approximately half its current age; the one to the right exploded about 9 billion years ago.

## How did Hubble prove that galaxies lie far beyond the Milky Way?

Now that we've seen how modern astronomers measure distances to galaxies, let's look at how some earlier distance measurements changed our view of the universe and enabled us to measure its size and age—a dramatic example of the process of science in action. We can trace the beginning of this revised understanding back to discoveries made by Edwin Hubble.

**The Great Debate** Less than a century ago, we did not know for certain whether other galaxies existed beyond the Milky Way. That may seem surprising, since large telescopes had already been used to catalog thousands of objects we know as galaxies today. However, the telescopes of the time were not yet powerful enough to resolve individual stars within these galaxies, and astronomers argued about whether the objects were distant galaxies or much nearer clouds of gas. The debate over the nature and location of these objects centered around those that had noticeably spiral shapes, which were known as "spiral nebulae."

The outcome of this debate would have profound implications for our general view of the universe. If the spiral nebulae were nearby gas clouds, then the Milky Way represented the entire universe, with nothing lying beyond it. But if they were separate and distinct islands of stars, as had been suggested by German philosopher Immanuel Kant in 1755, then the universe was vastly larger than the Milky Way alone. Kant's view ultimately proved correct, but guessing correctly is never good enough in science. Science demands hard evidence, and little of relevance to the nature of the spiral nebulae was forthcoming for 150 years after Kant's good guesswork.

In the early 20th century, advances in astronomical technology and data collection began to arm astronomers with more facts, but at first the facts only inflamed the debate about the nature of the spiral nebulae. The issue became so contentious that the National Academy of Sciences sponsored what has since become known in astronomical circles as "The Great Debate." The debate was held in Washington, D.C., on April 26, 1920, and it pitted Harlow Shapley, a rising young astronomer who would the following year become director of the Harvard Observatory, against Heber Curtis of the Lick Observatory. Shapley took the view that the spiral nebulae were gas clouds internal to the Milky Way, while Curtis argued in favor of Kant's islands of stars.

Both Shapley and Curtis knew that the key issue was the distance to the spiral nebulae, but neither of them had direct distance measurements. Shapley argued that the spiral nebulae must be relatively nearby because an apparent nova outburst seen in the Andromeda spiral in 1885 had an apparent brightness similar to that of nova outbursts that were definitely within our own galaxy. He also called attention to observations claiming a rotation period of just 100,000 years for a spiral nebula called M101, which meant that it had to be relatively small and nearby or else the implied rotation speed would have been greater than the speed of light. Curtis countered that many other nova outbursts had been observed in the Andromeda spiral, all much dimmer than the one seen in 1885 (and therefore implying much greater distance), and questioned the accuracy of the measured rotation of M101.

Neither Shapley nor Curtis scored a convincing victory at the debate, and years passed before astronomers learned what had misled Shapley: Andromeda's "nova" of 1885 was actually a *supernova*, explaining its surprising brightness, and reexamination of the M101 data showed that they did not support the claimed rotation rate. Nevertheless, a decisive resolution to the Great Debate came only a few years after the debate itself.

**Hubble and the Andromeda Galaxy** Edwin Hubble was finishing graduate school in 1917 when he was invited to join the staff at the Mount Wilson Observatory in Pasadena, California. It was a prestigious invitation for someone so young, because work was just finishing on Mount Wilson's jewel, a 100-inch telescope that would be the world's largest for the next 30 years (**FIGURE 20.17**). Hubble delayed accepting the invitation in order to serve in World War I, but joined the staff at Mount Wilson in 1919.

Once settled at Mount Wilson, Hubble turned his attention to the study of the spiral nebulae (**FIGURE 20.18**). With the 100-inch telescope, he could see what looked like individual stars in the Andromeda spiral, suggesting that it was indeed a separate galaxy. But his real breakthrough came when he discovered that some of these stars were dimming and brightening with a regular period, marking them as Cepheids, which meant he could determine their luminosities from Leavitt's law, which she had already discovered by that time. Hubble measured the periods of the Cepheids in Andromeda by comparing photos of the galaxy taken on different nights, used Leavitt's law to estimate their luminosities, and then used the inverse square law for light to estimate the distance to the Andromeda Galaxy.

With hindsight, we now know that Hubble underestimated the true distance by about half, because no one yet knew that Cepheids come in two different types that obey two different relationships between period and luminosity.* The version of this relationship originally discovered

*The two types of Cepheids, creatively called Type I and Type II, differ in their heavy-element content. Type I Cepheids have Sun-like heavy-element content and are found in the disk of the galaxy. Type II Cepheids have much lower heavy-element content and are found in the halo of the galaxy. For any particular period of variability, a Type I Cepheid is about four times as luminous as a Type II Cepheid.

**FIGURE 20.17** The 100-inch telescope on Mount Wilson, outside Los Angeles.

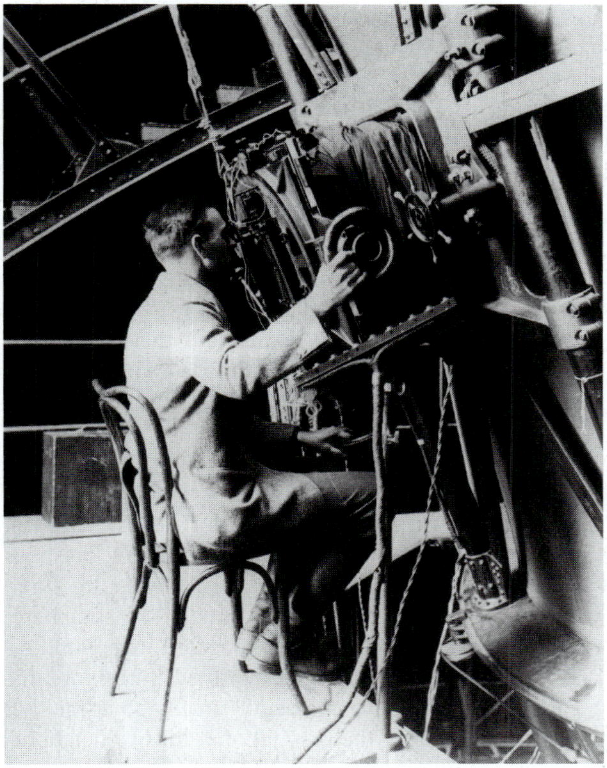

**FIGURE 20.18** Edwin Hubble at the Mount Wilson Observatory.

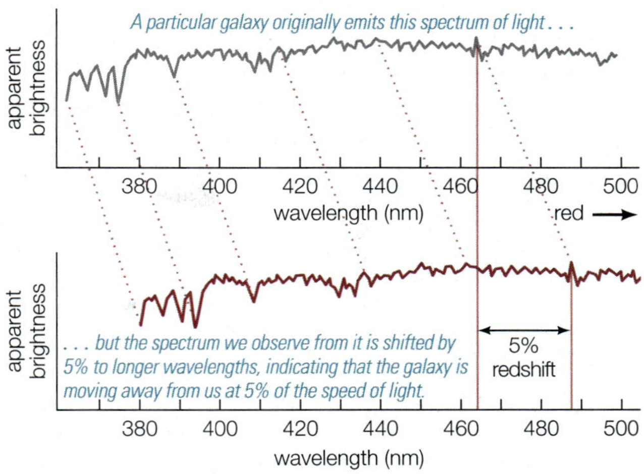


**FIGURE 20.19** Redshifted galaxy spectrum.

by Henrietta Leavitt applied to the dimmer type, but the Cepheids that Hubble was observing were of the brighter type. As a result, he underestimated the actual luminosities of the Andromeda Cepheids, and he therefore underestimated their distances. Nevertheless, Hubble still succeeded in showing that the Cepheids in Andromeda lie far beyond the outer reaches of stars in the Milky Way. The debate was over: Hubble had proved once and for all that the Andromeda "nebula" was in fact a separate galaxy.

This single scientific discovery dramatically changed our view of the universe. Rather than wondering if the Milky Way might be the entire universe, we suddenly knew that it is just one among many galaxies in an enormous universe. The stage was set for an even greater discovery.

## What is Hubble's law?

After Hubble determined the distance to the Andromeda Galaxy, he and his coworkers proceeded to estimate the distances to many more galaxies. Within just a few years, he made one of the most astonishing discoveries in the history of science: that the universe is expanding.

**Distance and Redshift** Astronomers were already aware that, aside from the Andromeda galaxy (and a few other nearby galaxies), nearly all galaxy spectra show a *redshift* (**FIGURE 20.19**). Recall that a redshift in a spectrum is typically caused by the *Doppler effect*. A shift to longer (redder) wavelength indicates that the source of the spectrum is moving away from us, and the amount of the shift depends on the speed of that motion [**Section 5.4**].

### SPECIAL TOPIC Who Discovered the Expanding Universe?

We generally credit Hubble with the discovery of the expanding universe, but like most scientific discoveries, this one was not made in isolation, and it can be difficult to determine exactly how much credit each scientist deserves.

There is no dispute that Hubble and his primary assistant Milton Humason made the distance measurements on which the discovery was based. However, the redshift measurements that Hubble initially used were actually made years earlier by astronomer Vesto Slipher of the Lowell Observatory.

More intriguingly, Belgian priest and astronomer Georges Lemaître (the French pronunciation is approximately "Leh-meht-reh") published a version of what we now call Hubble's law in 1927—two years before Hubble's own publication—and even calculated a value for the rate of universal expansion (Hubble's constant). This fact has led some science historians to suggest that Lemaître deserves more credit than Hubble, but there are at least a couple of caveats. First, Hubble was unaware of Lemaître's paper, which received little notice because it was published in French in an obscure Belgian journal. Second, Lemaître based his work on Hubble's already-published distances to galaxies, along with Slipher's already-published redshifts. Moreover, when Lemaître translated his paper into English for a broader audience in 1931, he left out the parts that contained his discovery of universal expansion.* Perhaps Lemaître was simply being modest, but it seems clear that he was willing to cede credit to Hubble. Either way, there's no doubt that Lemaître was the first person to publish the discovery of the expanding universe, but also no doubt that Hubble made the discovery independently and was the one who made it famous.

*The fact that Lemaître himself left these findings out of his translation came to light only recently, through the efforts of Mario Livio at the Space Telescope Institute.

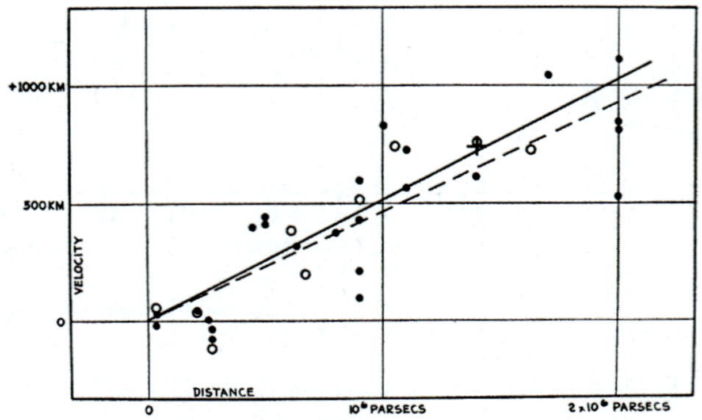


**FIGURE 20.20** Hubble's original velocity-distance diagram. Although Hubble underestimated the true galactic distances, he still discovered the general trend (represented by the straight lines) in which more distant galaxies move away from us at higher speeds. He determined the galaxy speeds by measuring redshifts in the galaxy spectra. (Units on the vertical axis are kilometers per second.)

In the years following his discovery of Cepheids in Andromeda, Hubble sought to estimate more galaxy distances and measure the redshifts of those galaxies. Because even Cepheids were too dim to be seen in most of these galaxies, Hubble needed brighter standard candles for his distance estimates. One of his favorite techniques was to use the brightest object he could see in each galaxy as a standard candle, because he assumed these objects to be very bright stars that would always have about the same luminosity.

In 1929, Hubble announced his conclusion: The more distant a galaxy, the greater its redshift and hence the faster it moves away from us. As we discussed in Chapter 1 (see Figure 1.17), this discovery implies that the entire universe is expanding.

Hubble's original assertion was based on an amazingly small sample of galaxies (**FIGURE 20.20**). Even more incredibly, he had grossly underestimated the luminosities of his standard candles. The "brightest stars" he had been using as standard candles were really entire *clusters* of bright stars. Fortunately, Hubble was both bold and lucky. Subsequent studies of much larger samples of galaxies showed that they are indeed receding from us, but they are even farther away than Hubble thought.

**Hubble's Law** The discovery that galaxies at greater distances are moving away from us at greater speeds can be expressed quantitatively with a simple formula called **Hubble's law**, in which $v$ stands for a galaxy's *recession speed*, meaning the speed at which it is moving away from us, and $d$ stands for its distance:

$$v = H_0 \times d$$

The $H_0$ (pronounced "H-naught") in this equation is a number called **Hubble's constant**.

We usually write Hubble's law in the above form because it expresses the key idea that speeds of galaxies depend on their distances. However, astronomers more often use the law in reverse: They measure a galaxy's speed (from its redshift) and then use Hubble's law to estimate its distance.

Because Hubble's law in principle applies to all distant galaxies for which we can measure a redshift, it is the most useful technique for determining distances to galaxies that are very far away. Nevertheless, we encounter two practical difficulties when we try to use Hubble's law to measure galactic distances:

1. Galaxies do not obey Hubble's law perfectly. Hubble's law gives an exact distance only for a galaxy whose speed is determined solely by the expansion of the universe. In reality, nearly all galaxies experience gravitational tugs from other galaxies, and these tugs alter their speeds from the values predicted by Hubble's law.
2. Even when galaxies obey Hubble's law well, the distances we find with it are only as accurate as our best measurement of Hubble's constant.

The first problem is most serious for nearby galaxies. Within the Local Group, for example, Hubble's law does not work at all: The galaxies in the Local Group are gravitationally bound together with the Milky Way and therefore are *not* moving away from us in accord with Hubble's law. However, Hubble's law works fairly well for more distant galaxies, which have recession speeds so great that any motions caused by the gravitational tugs of neighboring galaxies are tiny in comparison.

The second problem means that, even for distant galaxies, we can know only *relative* distances until we pin down the true value of $H_0$. For example, Hubble's law tells us that a galaxy moving away from us at 20,000 km/s is twice as far away as one moving at 10,000 km/s, but we can determine the actual distances of the two galaxies only if we know $H_0$.

One of the main missions of the Hubble Space Telescope was to measure an accurate value of $H_0$ (**FIGURE 20.21**).

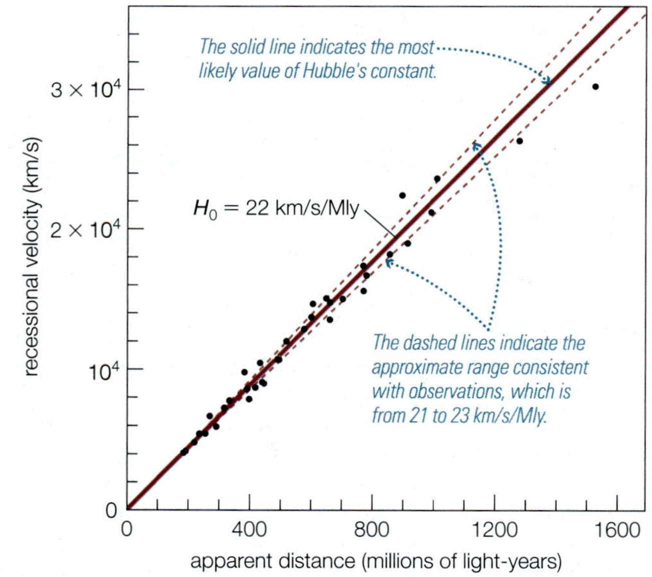


**FIGURE 20.21** The points on this figure show the apparent distances of white dwarf supernovae and the recession velocities of the galaxies in which they exploded. The fact that these points all fall close to a straight line demonstrates that these supernovae are good standard candles.

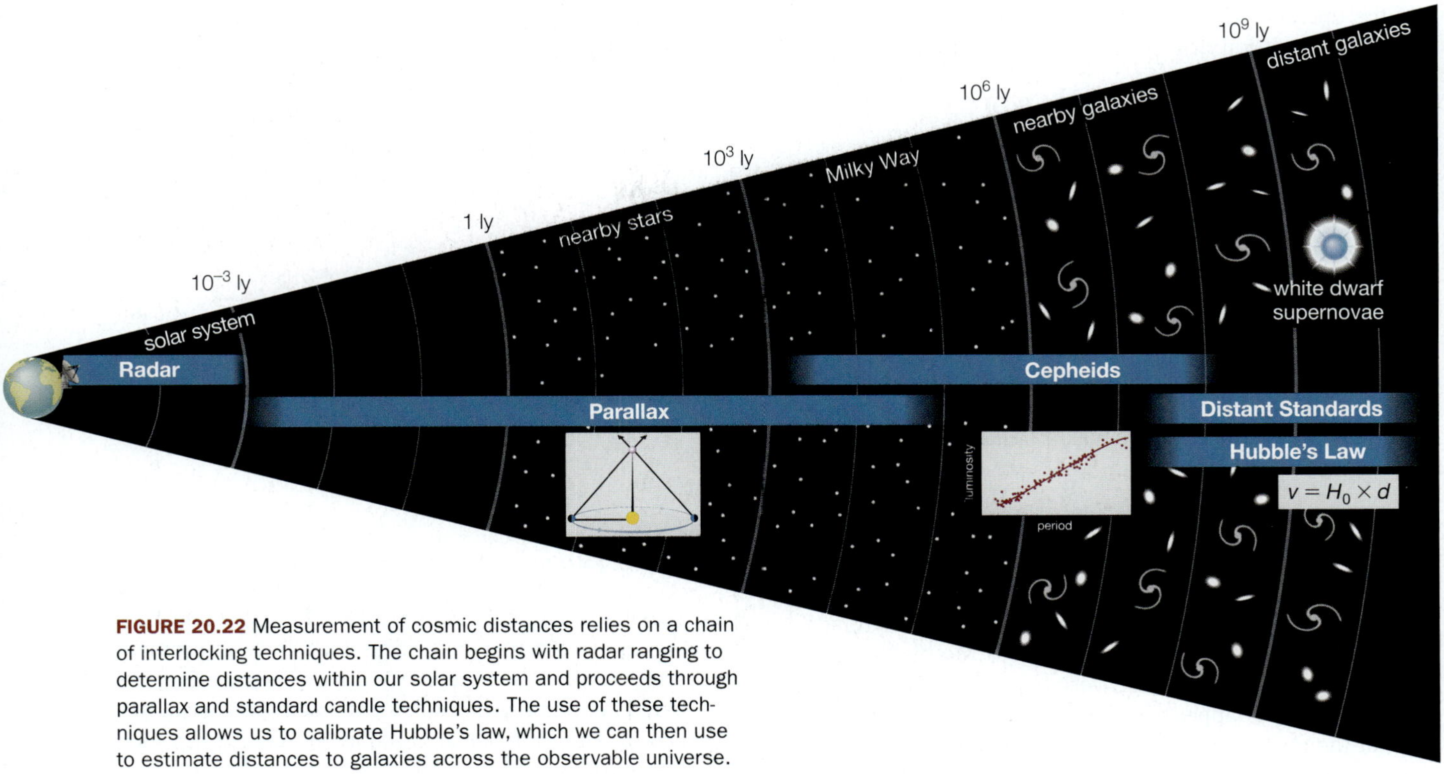


**FIGURE 20.22** Measurement of cosmic distances relies on a chain of interlocking techniques. The chain begins with radar ranging to determine distances within our solar system and proceeds through parallax and standard candle techniques. The use of these techniques allows us to calibrate Hubble's law, which we can then use to estimate distances to galaxies across the observable universe.

Astronomers used the telescope to measure distances to Cepheids in galaxies out to about 100 million light-years, and this in turn allowed them to determine the luminosities of distant standard candles such as white dwarf supernovae. Plotting galactic distances measured with those distant standard candles against the velocities indicated by their redshifts has pinned down the value of $H_0$ to somewhere between 21 and 23 *kilometers per second per million light-years*. That is, a galaxy's speed away from us is between 21 and 23 km/s for every million light-years it is from us. For example, with this range of values for Hubble's constant, Hubble's law predicts that a galaxy located 100 million light-years away would be moving away from us at a speed of between 2100 and 2300 kilometers per second.

**Distance Chain Summary** **FIGURE 20.22** summarizes the techniques we have discussed, showing how they make up a chain of measurements that allows us to determine distances across the universe:

- **Radar ranging.** We measure the Earth-Sun distance by bouncing radio waves off Venus and using some geometry.
- **Parallax.** We measure the distances to nearby stars by observing how their positions appear to change as Earth orbits the Sun. These distances rely on our knowledge of the Earth-Sun distance, determined with radar ranging.
- **Cepheid variables.** Leavitt's law allows us to determine the luminosities of Cepheids from their measured periods, which in turn allows us to calculate their distances using the inverse square law for light. We learn the precise relationship between Cepheid periods and luminosities from relatively nearby Cepheids for which distances have been measured using parallax.
- **Distant standards.** By measuring distances to relatively nearby galaxies with Cepheids, we learn the true luminosities of white dwarf supernovae. We can then use those supernovae as standard candles to measure great distances throughout the universe.
- **Hubble's law.** Distances measured to galaxies with white dwarf supernovae allow us to measure Hubble's constant, $H_0$. Once we know $H_0$, we can use Hubble's law to determine a galaxy's distance from its redshift.

Note that the uncertainty increases with each link in the distance chain. As a result, although we know the Earth-Sun distance at the beginning of the chain extremely accurately, distances to the farthest reaches of the observable universe remain uncertain by about 5%.

## 20.3 The Age of the Universe

Hubble's discoveries dramatically changed not only our conception of the size of the universe, but also our conception of its history. Before scientists became aware of the universe's expansion, their discussions about whether the universe was eternal or had a distinct beginning were purely philosophical, because there was no scientific evidence to go on. Our ignorance about the universe's history tripped up even Einstein, who introduced an additional term into his equations of general relativity in order to get them to fit his assumption that the universe should be static and eternal (see Special Topic, p. 685). After Hubble discovered the expansion of the universe, it became increasingly clear that the universe must have had a beginning, and that, as a result, the portion of it that is observable to us must be finite in size.

## How does Hubble's law tell us the age of the universe?

As we discussed in Chapter 1, galaxies throughout the universe are moving away from one another, and this fact implies that galaxies must have been closer together in the past. Tracing this trend back in time, we reason that all the matter in the observable universe started very close together and that the entire universe came into being at a single moment—the Big Bang. We can determine when that moment occurred if we develop a little deeper understanding of the observed expansion.

**The Cosmological Principle** It may be tempting to think of the expanding universe as a ball of galaxies expanding into a void, but this is not the case. To the best of our knowledge, the universe is not expanding *into* anything. As far as we can tell, there is no edge to the distribution of galaxies in the universe. On very large scales, the distribution of galaxies appears to be relatively smooth, meaning that the overall appearance of the universe around you would look more or less the same no matter where you were located. The idea that the universe has no center or edge, and on large scales looks about the same everywhere, is often called the **Cosmological Principle**. Although we cannot prove this principle to be true, it is consistent with all our observations of the universe [**Section 23.3**].

**The Balloon Analogy** How can the universe be expanding if it's not expanding *into* anything? In Chapter 1, we likened

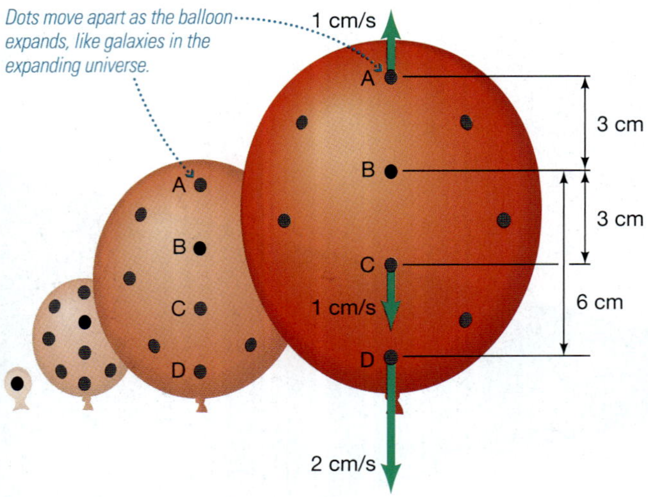


**FIGURE 20.23** As the balloon expands, the dots move apart in the same way that galaxies move apart in our expanding universe.

the expanding universe to a raisin cake baking [**Section 1.3**], but a cake has a center and edges that grow into empty space as it bakes. A better analogy involves something that can expand but that has no center and no edges. The surface of a balloon can fit the bill, as could an infinite surface such as a sheet of rubber that extends to infinity in all directions.

Because it's hard to visualize infinity, let's use the surface of a balloon as our analogy to the expanding universe (**FIGURE 20.23**). Note that this analogy uses the balloon's two-dimensional *surface* to represent all three dimensions of space. The surface of the balloon represents the entire universe; the spaces inside and outside the balloon have no meaning in this analogy. Aside from the reduced number of dimensions, the analogy works well because the balloon's

## MATHEMATICAL INSIGHT 20.2 Redshift

The *redshift* of an object, usually designated by the letter $z$, is defined as the fractional difference between the observed wavelength ($\lambda_{\text{observed}}$) of a line in the object's spectrum and the wavelength the line would have if the object were standing still ($\lambda_{\text{rest}}$):

$$\text{redshift}(z) = \frac{\lambda_{\text{observed}} - \lambda_{\text{rest}}}{\lambda_{\text{rest}}}$$

Because the redshift is a ratio, all lines in a particular object's spectrum should have the same redshift.

For relatively nearby galaxies (with $z$ much less than 1), redshift is simply related to velocity:

$$v = c \times z$$

where $c = 3.0 \times 10^5$ km/s is the speed of light. A more complex formula can be used for cases with larger redshift ($z$ close to or greater than 1), but we will not consider such cases in this book.

**EXAMPLE:** Two visible-light emission lines of hydrogen have rest wavelengths of 656.3 nanometers and 486.1 nanometers, but they appear at 662.9 nanometers and 491.0 nanometers, respectively, in the spectrum of a distant galaxy. What is the redshift of the galaxy, and how fast is it moving away from us?

**SOLUTION:**

**Step 1 Understand:** We are given the observed and rest wavelengths for the spectral lines, so we can calculate the redshift, $z$. As long as this redshift is much less than 1, we can then use the formula $v = c \times z$ to calculate the galaxy's speed.

**Step 2 Solve:** For the first line we have $\lambda_{\text{rest}} = 656.3$ nm and $\lambda_{\text{observed}} = 662.9$ nm, which gives a redshift of

$$z = \frac{662.9\text{ nm} - 656.3\text{ nm}}{656.3\text{ nm}} = 0.010$$

You should confirm for yourself that the second line gives the same redshift. Because this redshift is much less than 1, the galaxy's speed is

$$v = c \times z = (3.0 \times 10^5\text{ km/s}) \times 0.010 = 3000\text{ km/s}$$

**Step 3 Explain:** The galaxy has a redshift of $z = 0.010$, from which we conclude that it is moving away from us at a speed of 3000 km/s, or about 1% of the speed of light.

spherical surface has no center and no edges, just as no city is the center of Earth's surface and no edges exist where you could walk or sail off of Earth. We can represent galaxies with plastic dots attached to the balloon, and we can make our model universe expand by inflating the balloon.

**See it for yourself** Make a one-dimensional model of the expanding universe with a rubber band and some paper clips. Cut the rubber band so that you can stretch it out along a line. Then attach four paper clips along it. Pin down the ends of the rubber band and measure the distances from one paper clip to each of the others. Then unpin one of the ends, stretch the rubber band more, pin it down again, and remeasure the distances. How much have they changed? How do your measurements illustrate Hubble's law?

**Hubble's Constant and the Age of the Universe** We can now see how Hubble's law leads us to an age estimate for the universe. Imagine that some miniature scientists are living on dot B in Figure 20.23. Suppose that, 3 seconds after the balloon begins to expand, they measure the following:

Dot A is 3 cm away and moving at 1 cm per second.

Dot C is 3 cm away and moving at 1 cm per second.

Dot D is 6 cm away and moving at 2 cm per second.

These observations are exactly what Hubble's law would tell them to expect: More distant dots are moving away faster. Moreover, because the miniature scientists have measured distances and speeds for multiple dots, they have the information necessary to discover their own version of Hubble's law. Recall that this law can be written $v = H_0 \times d$. For the balloon, $v$ and $d$ would represent any dot's speed and distance, and "Hubble's law" for the balloon would be the following:

$$v = \left(\frac{1}{3\text{ s}}\right) \times d$$

**Think about it** Confirm that this formula gives the correct values for the speeds of dots C and D, as seen from dot B, 3 seconds after the balloon begins expanding. How fast would a dot located 9 centimeters from dot B move, according to the scientists on dot B?

## MATHEMATICAL INSIGHT 20.3 Understanding Hubble's Law

Hubble's law is one of the most important tools of modern astronomy, and calculations with it require understanding the measured value of Hubble's constant, which in this book we give in units of *kilometers per second per million light-years*, or km/s/Mly for short. [You may also hear astronomers quote Hubble's constant in *kilometers per second per megaparsec* (km/s/Mpc); because 1 parsec = 3.26 light-years, you can convert km/s/Mly to km/s/Mpc simply by multiplying by 3.26.]

What do these strange-sounding units mean? They tell us that if we measure the distance to a distant galaxy in millions of light-years, then Hubble's law gives us the galaxy's velocity in kilometers per second. We can clarify the idea with some simple examples. Because the current best estimate of Hubble's constant puts it between 21 and 23 km/s/Mly, let's use the middle value of 22 km/s/Mly:

- For a galaxy located 100 million light-years away, Hubble's law tells us to expect it to be moving away from us at a speed of

$$v = H_0 \times d = 22\,\frac{\text{km/s}}{\cancel{\text{Mly}}} \times 100\ \cancel{\text{Mly}} = 2200\text{ km/s}$$

- For a galaxy located 110 million light-years away, Hubble's law tells us to expect it to be moving away from us at a speed of

$$v = H_0 \times d = 22\,\frac{\text{km/s}}{\cancel{\text{Mly}}} \times 110\ \cancel{\text{Mly}} = 2420\text{ km/s}$$

  Notice that this is 220 km/s faster than the speed for the galaxy located at a distance of 100 million light-years—just as we should expect, since Hubble's constant tells us that each additional million light-years of distance means an additional 22 km/s of speed.

- For a galaxy located 700 million light-years away, Hubble's law tells us to expect it to be moving away from us at a speed of

$$v = H_0 \times d = 22\,\frac{\text{km/s}}{\cancel{\text{Mly}}} \times 700\ \cancel{\text{Mly}} = 15{,}400\text{ km/s}$$

  Notice that this speed is quite high—more than 5% of the speed of light.

Although the above examples use Hubble's law to calculate velocities, it is much more common to use Hubble's law to calculate distances, especially for galaxies whose distances cannot be estimated with other techniques.

**EXAMPLE:** Estimate the distance to a galaxy whose redshift indicates that it is moving away from us at a speed of 22,000 km/s. Assume that Hubble's constant is $H_0 = 22$ km/s/Mly.

**SOLUTION:**

**Step 1 Understand:** To find a galaxy's distance from its recession speed, we must solve Hubble's law for the distance $d$. Dividing both sides of the law ($v = H_0 \times d$) by $H_0$ gives

$$d = \frac{v}{H_0}$$

With this form, we can find a galaxy's distance simply by plugging in its recession speed and Hubble's constant.

**Step 2 Solve:** Putting in the given values, we find

$$d = \frac{v}{H_0} = \frac{22{,}000\ \cancel{\text{km/s}}}{22\,\frac{\cancel{\text{km/s}}}{\text{Mly}}} = 1000\text{ Mly}$$

**Step 3 Explain:** The galaxy's distance is about 1000 million, or 1 billion, light-years.

## COMMON MISCONCEPTIONS

### What Is the Universe Expanding Into?

When you first learn about the expansion of the universe, it's natural to assume that the universe must be expanding "into" something. The name "Big Bang" does evoke a mental image of a huge explosion that sent matter flying outward into a vast space that was previously empty, but the scientific view of the Big Bang is very different. According to modern science, the Big Bang filled *all of* space with matter and energy.

How can the Big Bang have started "all of space" expanding without its expanding into some empty void? A complete answer requires Einstein's general theory of relativity, which tells us that space can be "curved" in ways we can measure but cannot visualize [**Chapter S3**]. But you can understand the key ideas with the balloon analogy. Remember that the balloon's surface represents all of space; just as the surface has no center and no edges, space has no center and no edges. In the analogy, the expanding universe is like the expanding surface of the balloon, and the Big Bang was the moment when an extremely tiny balloon first started growing in size. As the balloon expands, any two points on the surface get farther apart because the surface itself is stretching, not because they are moving into a previously empty region above the surface. In the same way, galaxies move apart in the expanding universe because the space between them is stretching, not because they move into a space that was previously empty.

If the miniature scientists think of their balloon as a bubble, they might call the number relating distance to velocity—the term $1/(3\text{ s})$ in the preceding formula—the "bubble constant." An especially insightful miniature scientist might flip over the "bubble constant" and find that it is exactly equal to the time elapsed since the balloon started expanding. That is, the "bubble constant" $1/(3\text{ s})$ tells them that the balloon has been expanding for 3 seconds. Perhaps you see where we are heading.

Just as the inverse of the "bubble constant" tells the miniature scientists that their balloon has been expanding for 3 seconds, the inverse of the Hubble constant, or $1/H_0$, tells us something about how long our universe has been expanding. The "bubble constant" for the balloon depends on when it is measured, but it is always equal to 1 divided by the time since the balloon started expanding. Similarly, the Hubble constant actually changes with time but stays roughly equal to 1 divided by the age of the universe. We call it a "constant" because it is the same at all locations in the universe, and because its value does not change noticeably on the time scale of human civilization.

A simple estimate based only on the value of Hubble's constant puts the age of the universe at a little less than 14 billion years. To derive a more precise value, we need to know whether the expansion has been speeding up or slowing down over time, a question we will examine more closely in Chapter 23. If the rate of expansion has slowed with time (because of gravity, for example), then the expansion rate would have been faster in the past. In that case, the universe would have reached its current size faster than we would guess by assuming a constant expansion rate, which means the universe would be *younger* than the age we get from $1/H_0$. If the expansion rate has accelerated with time—and current evidence indicates a small acceleration [**Section 23.4**]—then the universe's age is somewhat more than $1/H_0$. The best available evidence suggests that the universe is very close to 14 billion years old.

## MATHEMATICAL INSIGHT 20.4 Age from Hubble's Constant

The reciprocal of Hubble's constant, or $1/H_0$, tells us the age of the universe *if* the expansion rate has remained constant over time. Let's calculate this age from the measured value of $H_0 \approx 22$ km/s/Mly.

It's easier to find the reciprocal of $H_0$ if we first rewrite $H_0$ in more convenient units. Because a light-year is the speed of light ($c = 3 \times 10^5$ km/s) times 1 year, a million light-years is $10^6 \times c \times 1$ yr. We can therefore rewrite Hubble's constant as

$$H_0 = 22\,\frac{\text{km/s}}{\text{Mly}} = 22\,\frac{\text{km/s}}{10^6 \times (3 \times 10^5\text{ km/s}) \times 1\text{ yr}}$$

Taking the reciprocal, we find

$$\frac{1}{H_0} = \frac{10^6 \times (3 \times 10^5\text{ km/s}) \times 1\text{ yr}}{22\text{ km/s}}$$

$$= \frac{3 \times 10^{11}\text{ yr} \times \cancel{\text{km/s}}}{22\ \cancel{\text{km/s}}} = 1.36 \times 10^{10}\text{ yr}$$

In other words, if the expansion rate has never changed, the value $H_0 = 22$ km/s/Mly implies that the universe is $1.36 \times 10^{10}$, or 13.6 billion, years old.

**EXAMPLE:** How would the estimated age of the universe differ if the measured value of $H_0$ were 44 km/s/Mly rather than 22 km/s/Mly?

**SOLUTION:**

**Step 1 Understand:** Because our estimate for the age of the universe depends on the *reciprocal* of Hubble's constant, we expect a larger value for the constant to mean a *smaller* estimated age.

**Step 2 Solve:** In this example, Hubble's constant is taken to be twice the current value, and the reciprocal of 2 is 1/2. Therefore, our estimated age for the universe would be 1/2 of the 13.6 billion years we found earlier, or 6.8 billion years.

**Step 3 Explain:** If Hubble's constant were twice as large as our best current measurement of it, the age of the universe would be only about half what we find from the current measurements. We can understand this result by noting that Hubble's law, $v = H_0 \times d$, means that the value of $H_0$ is the recession speed of any galaxy divided by its distance. Doubling Hubble's constant would mean doubling galaxy recession speeds, so galaxies would have taken only half as long to reach their current distances.

## How does expansion affect distance measurements?

The expansion of the universe introduces a complication when we try to specify the distance of a faraway galaxy, because its distance is always changing. Here's the problem. Suppose you are trying to determine the distance to a galaxy using observations of a supernova that actually occurred 1 billion years ago. The photons you observe from the supernova and the galaxy it resides in must have traveled a total distance of 1 billion light-years, because they travel at the speed of light and their journey took 1 billion years (**FIGURE 20.24**). However, this 1 billion light-year distance is *smaller* than the current distance to the supernova's galaxy and *larger* than its distance at the time the supernova occurred, because the galaxy's distance has been continually increasing during the time it took for the supernova's light to travel to us. So what should we say if someone asks us the "distance" to the galaxy?

**Lookback Time** The changing distance to the galaxy means that we could potentially choose to answer the question about the galaxy's distance by citing its distance at the time the light left, its distance today, or something in between. Given this choice, most astronomers find it more useful to describe the galaxy's distance by using the fact that its light has taken 1 billion years to reach us. We call this time the galaxy's **lookback time**, because it means that we are seeing the galaxy as it looked 1 billion years ago. Astronomers commonly state large distances in terms of lookback times. For example, when astronomers say that a galaxy is "1 billion light-years away," they usually mean that the galaxy has a lookback time of 1 billion light-years.

**Think about it** Explain why there is no question about meaning when we talk about distances to objects within the Local Group, but distances become difficult to define when we talk about objects much farther away.

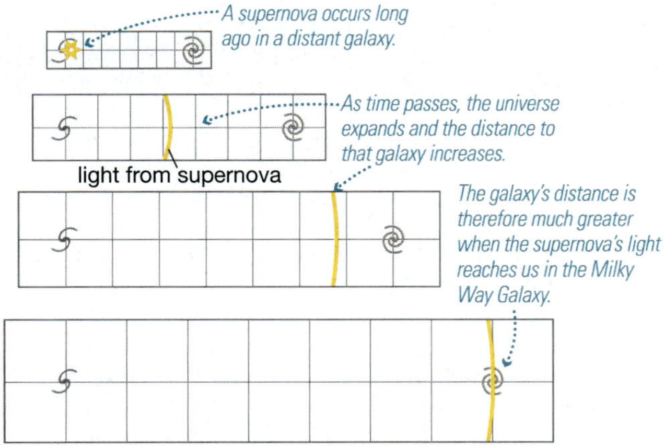


**FIGURE 20.24** Distances to faraway objects keep changing in an expanding universe. This figure shows how the distance to a galaxy changes during the time it takes for the light from a supernova explosion in that galaxy to reach us. Because the distance changes, the most useful way to describe a galaxy's "distance" is to state its *lookback time*—the time it took for the light to make its journey to Earth.

**Cosmological Redshift** The redshifts of distant galaxies call for a similar shift in thinking. One way to interpret a galaxy's redshift is in terms of how fast the galaxy is moving away from us (the $v$ in Hubble's law). However, our balloon analogy suggests an alternative interpretation: We can think of galaxies as remaining essentially stationary—like the dots glued to the balloon—while the space between them grows. From this perspective, the expansion of the universe causes photon wavelengths to become longer (redder) with time in much the same way that wavy lines on a rubber sheet stretch out as the sheet expands (**FIGURE 20.25**).

In essence, we have a choice when we interpret the redshift of a distant galaxy: We can think of the redshift either as being caused by the Doppler effect as the galaxy moves away from us or as being a **cosmological redshift** that arises from photon stretching in an expanding universe. For distant galaxies, astronomers find it more useful to adopt the latter perspective, in which the expansion of space itself is carrying galaxies along for the ride and stretching the wavelengths of photons. A galaxy's redshift therefore tells us how much space has expanded during the time it took for the galaxy's light to travel to us, and its lookback time tells us the distance its light has traveled.

## Why does the observable universe have a horizon?

When we began our discussion of the expanding universe, we stressed that the universe as a whole does not seem to have an edge. However, as we discussed in Chapter 1, the age of the universe limits the size of our *observable* universe (see Figure 1.4): We cannot see any object beyond our **cosmological horizon**, which represents the distance at which lookback times equal the 14-billion-year age of the universe. In other words, the age of the universe fundamentally limits the size of our observable universe.

Note that the general ambiguity in distances makes it difficult to assign a unique distance to the cosmological horizon. That is, while we can unambiguously state that the cosmological horizon lies at a lookback time of 14 billion years in

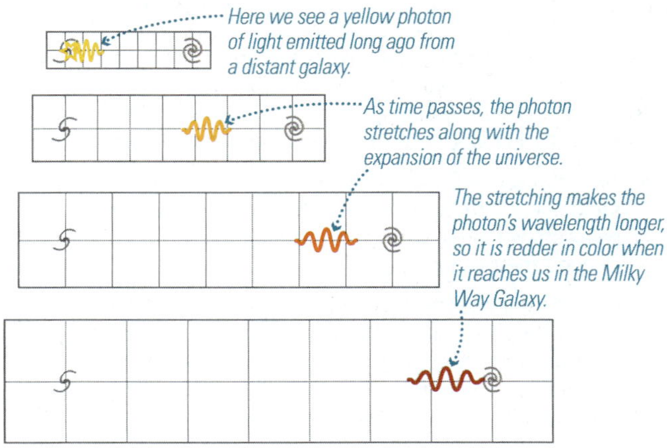


**FIGURE 20.25** As the universe expands, photon wavelengths stretch. This *cosmological redshift* shifts them toward the red side of the spectrum.

all directions, it's not as meaningful to say that the horizon is "14 billion light-years away." In fact, universal expansion has pushed the matter we see near the horizon to a much greater distance during the 14 billion years it has taken for that light to reach us. Calculations based on Einstein's equations of general relativity tell us that this matter is now about 47 billion light-years away. As a result, you may sometimes hear astronomers stating such a distance to the horizon, but it is easier to think in terms of the lookback time of 14 billion years.

Our quest to study galaxies and measure their distances has now brought us to the very limits of the observable universe. The galaxies in Figure 20.1 at the beginning of this chapter extend from relatively nearby almost all the way to the cosmological horizon. In the next chapter, we will explore what these galaxies have been doing during the 14-billion-year history of the universe.

## COMMON MISCONCEPTIONS

### Beyond the Horizon

Perhaps you're thinking there must be something beyond the cosmological horizon. After all, we can see farther and farther with each passing second. The new matter we see had to come from somewhere, didn't it? The problem with this reasoning is that the cosmological horizon, unlike a horizon on Earth, is a boundary in time, not in space.

At any one moment, in whatever direction we look, the cosmological horizon lies at the beginning of time and encompasses a certain volume of the universe. At the next moment, the cosmological horizon still lies at the beginning of time, but it encompasses a slightly larger volume. When we peer into the distant universe, we are looking back in both space and time. We cannot look "past the horizon" because we cannot look back to a time before the universe began.

## MATHEMATICAL INSIGHT 20.5 Cosmological Redshift and the Stretching of Light

The cosmological interpretation of redshift in terms of the stretching of light waves offers us a simple way to measure how the average distances between galaxies change with time. In Mathematical Insight 20.2, we saw that the difference between the observed wavelength of a spectral line from a galaxy ($\lambda_{\text{observed}}$) and the wavelength that line would have if the galaxy were standing still ($\lambda_{\text{rest}}$) can be expressed in terms of the redshift $z = (\lambda_{\text{observed}} - \lambda_{\text{rest}})/\lambda_{\text{rest}}$. Adding the number 1 to both sides of that expression and simplifying the right-hand side leads to the equation

$$1 + z = \frac{\lambda_{\text{observed}}}{\lambda_{\text{rest}}}$$

Because the redshift of a very distant galaxy is cosmological in nature, the ratio $\lambda_{\text{observed}}/\lambda_{\text{rest}}$ for such a galaxy tells us how much the light waves from this galaxy have been stretched since the time the light was emitted. Because this stretching comes from the expansion of the universe itself, it is also proportional to the average distance between galaxies. If we can measure the redshift $z$ of a distant galaxy, we can then determine the factor $1 + z$ by which the distances between galaxies have changed during the time it took for light to reach us from that galaxy.

**EXAMPLE:** A distant galaxy has a measured redshift of $z = 1.5$. By what factor have the average distances between galaxies increased since the light we see today left that distant galaxy?

**SOLUTION:**

**Step 1 Understand:** As discussed above, the relationship between redshift and the expansion of the universe is based on $1 + z$, so we simply need to calculate this factor for the galaxy.

**Step 2 Solve:** The galaxy's redshift is $z = 1.5$, so

$$1 + z = 1 + 1.5 = 2.5$$

**Step 3 Explain:** The result of $1 + z = 2.5$ for this galaxy tells us that the average distances between galaxies today are 2.5 times as large as they were when the light we see left the galaxy.

## The BIG Picture PUTTING CHAPTER 20 INTO CONTEXT

*The picture could hardly get any bigger than it has in this chapter. Looking back through both space and time, we have seen a wide variety of galaxies extending nearly to the limits of the observable universe. As you look back, keep sight of these "big picture" ideas:*

- The universe is filled with galaxies that come in a variety of shapes and sizes. In order to learn the histories of these galaxies, we must account for how the universe itself has evolved through time.
- Much of our current understanding of the structure and evolution of the universe is based on measurements of distances to faraway galaxies. These measurements rely on a carefully constructed chain of techniques, in which each link in the chain builds on the links that come before it.
- It has been less than a century since Hubble first proved that the Milky Way is only one of billions of galaxies in the universe. This discovery, and his subsequent discovery of universal expansion, provided the foundation on which modern cosmology has been built. Measurements of the rate of expansion tell us that our universe was born about 14 billion years ago.
- As the universe expands, it carries the galaxies within it along for the ride. We therefore observe redshifts in the spectra of distant galaxies, and these redshifts allow us to determine how long it has been (the lookback time) since the light from these galaxies left on its way to reach us.

**MY COSMIC PERSPECTIVE** Measurements of galaxy distances by Edwin Hubble nearly 100 years ago led to a profound insight into our origins: The universe is not eternal and unchanging. Instead, it expands with time, starting at a moment about 14 billion years ago.

# Summary of Key Concepts

## 20.1 Islands of Stars

- **How are the lives of galaxies connected with the history of the universe?** Because nearby galaxies tend to be similar in age to the Milky Way, we know that most galaxies formed when the universe was much younger and have aged along with the universe itself. The lives of galaxies are therefore intimately connected with the evolution of the universe, making the study of galaxies a part of **cosmology**—the study of the overall structure and evolution of the universe.

- **What patterns do we find among the properties of galaxies?**

Galaxies come in three major types. **Spiral galaxies** have prominent disks and spiral arms. **Elliptical galaxies** are rounder and redder than spiral galaxies and contain less cool gas and dust. **Irregular galaxies** are neither disklike nor rounded in appearance. Grouping them according to luminosity and color shows that spiral and irregular galaxies tend to belong to a star-forming **blue cloud** in a diagram like an H-R diagram for stars, while elliptical galaxies tend to fall along a non-star-forming **red sequence**. When we look at how galaxies are grouped in space, we find that spiral galaxies tend to reside in relatively small groups while elliptical galaxies are usually found in large clusters.

## 20.2 Measuring Galactic Distances

- **How do we measure the distances to galaxies?** The chain of methods by which we measure galaxy distances begins with **radar ranging** in our own solar system and parallax measurements of distances to nearby stars, then relies on **standard candles** to measure greater distances. Important standard candles include **Cepheid variable stars**, which obey the period-luminosity relationship known as **Leavitt's law**, and **white dwarf supernovae**, which can be seen even at enormous distances.

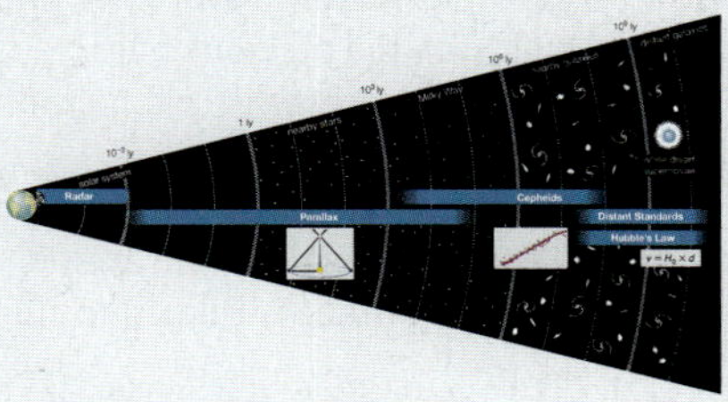

- **How did Hubble prove that galaxies lie far beyond the Milky Way?** Using the largest telescope in the world at the time, Hubble identified individual Cepheids in the Andromeda Galaxy, enabling him to determine their luminosities and hence their distances from Leavitt's law. The results showed conclusively that Andromeda is much too far away to be part of the Milky Way.

- **What is Hubble's law? Hubble's law** tells us that more distant galaxies are moving away faster: $v = H_0 \times d$, where $H_0$ is **Hubble's constant**. It allows us to determine a galaxy's distance from the speed at which it is moving away from us, which we can measure from the redshift of its spectrum.

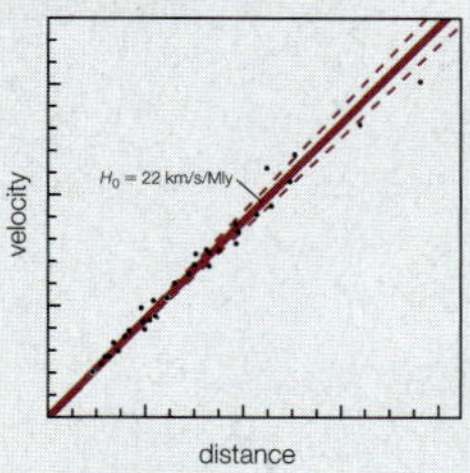


## 20.3 The Age of the Universe

- **How does Hubble's law tell us the age of the universe?** The inverse of Hubble's constant tells us how long it would have taken the universe to reach its present size *if* the expansion rate had never changed. Based on Hubble's constant and estimates of how it has changed with time, we estimate the age of the universe to be about 14 billion years.

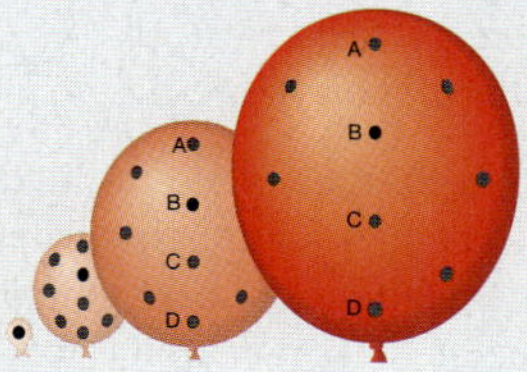

- **How does expansion affect distance measurements?** Distances between galaxies are always changing because of the expansion of the universe. It is therefore best to express the distance to a faraway galaxy in terms of its **lookback time**—the time it has taken for the galaxy's light to reach us. The expansion of the universe during that time stretches the light coming from the galaxy, leading to a **cosmological redshift** directly related to the galaxy's lookback time.

- **Why does the observable universe have a horizon?** In a universe that is 14 billion years old, we cannot possibly see objects with a lookback time of more than 14 billion years, because nothing existed more than 14 billion years ago. Therefore, the observable universe must have a **cosmological horizon** that extends to a lookback time of 14 billion years in all directions.

# Visual Skills Check

*Use the following questions to check your understanding of some of the many types of visual information used in astronomy. For additional practice, try the Chapter 20 Visual Quiz in the Study Area at www.MasteringAstronomy.com.*

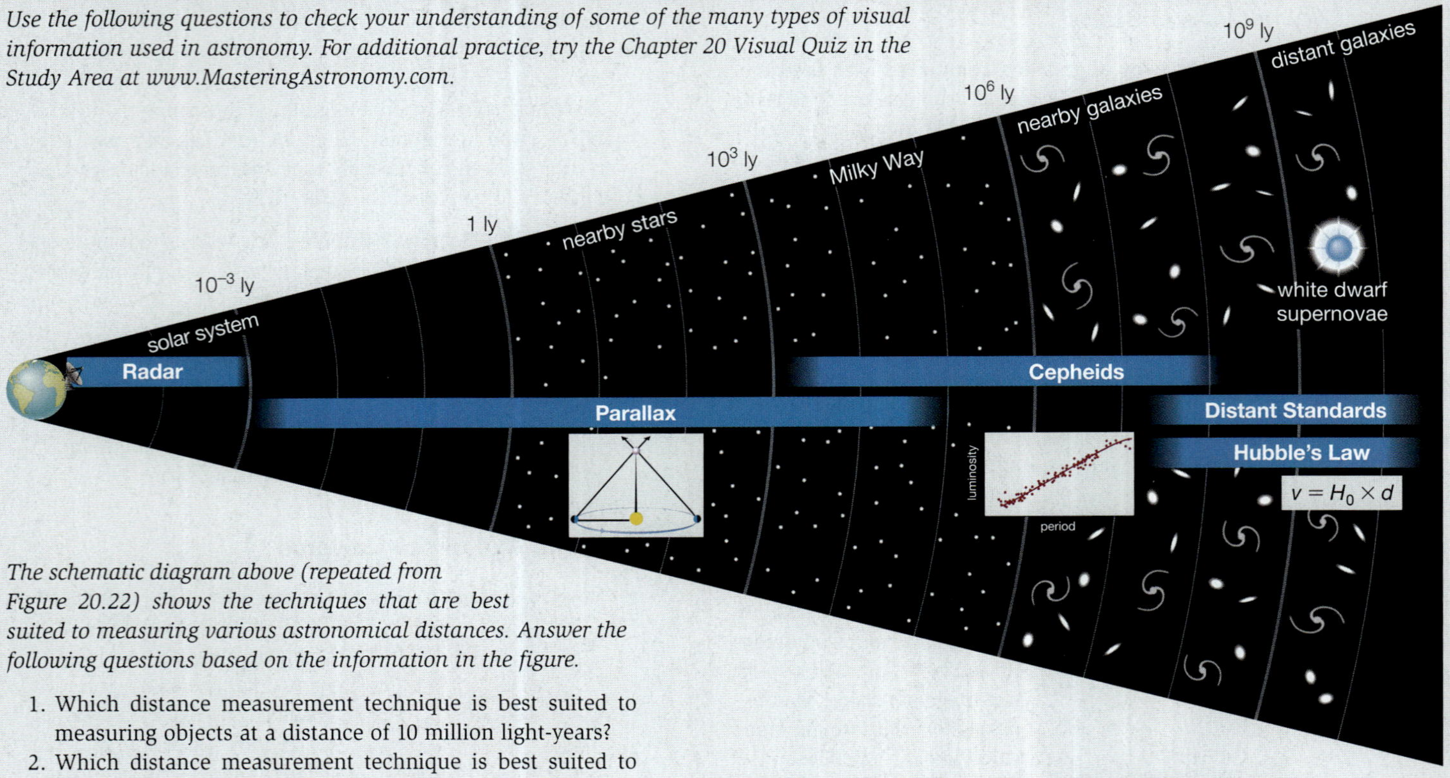


*The schematic diagram above (repeated from Figure 20.22) shows the techniques that are best suited to measuring various astronomical distances. Answer the following questions based on the information in the figure.*

1. Which distance measurement technique is best suited to measuring objects at a distance of 10 million light-years?
2. Which distance measurement technique is best suited to measuring objects at a distance of 10 light-years?
3. What range of astronomical distances can we currently measure using both Cepheids and distant standards (white dwarf supernovae)?
4. What range of astronomical distances can we currently measure using Hubble's law?
5. Which technique is best for measuring the distances of nearby galaxies?
6. Hubble's law requires calibration with distances measured by standard candle techniques. Which standard candle technique is best for measuring the distances of very distant galaxies?

# Exercises and Problems

*For instructor-assigned homework and other study tools, go to www.MasteringAstronomy.com.*

## Chapter Review Questions

### Short-Answer Questions Based on the Reading

1. Why do we need to understand the evolution of the universe in order to understand the lives of galaxies?
2. What are the three major types of galaxies, and how do their appearances differ?
3. Describe the differences between normal spiral galaxies, barred spiral galaxies, and lenticular galaxies.
4. Distinguish between spiral and elliptical galaxies in terms of the presence or absence of a *disk component* and a *halo component*. How does this difference explain the lack of hot, young stars in elliptical galaxies?
5. Where are the *blue cloud* and the *red sequence* on a plot of the relationship between galaxy color and luminosity? On which part of the plot would you tend to find the most luminous galaxies?
6. How do galaxies found in clusters tend to differ in type from isolated galaxies and galaxies found in smaller groups?
7. What do we mean by a *standard candle*? Explain how we can use standard candles to measure distances.
8. What is *Leavitt's law*? Explain how Hubble used it to prove that the Andromeda Galaxy lies beyond the bounds of the Milky Way.
9. What is *Hubble's law*? Explain what we mean when we say that *Hubble's constant* is between 21 and 23 kilometers per second per million light-years.
10. Summarize each of the major links in the distance chain, and explain how each link builds on the previous ones.
11. What is the *Cosmological Principle*, and how is it important to our understanding of the universe?
12. How is the expansion of the surface of an inflating balloon similar to the expansion of the universe? Use the balloon analogy to explain why Hubble's constant is related to the age of the universe.
13. What do we mean by the *lookback time* to a distant galaxy? Briefly explain why lookback times are less ambiguous than distances for discussing objects very far away.
14. What is the *cosmological horizon*, and what determines how far away it lies?
15. What do we mean by a *cosmological redshift*? How does our interpretation of a distant galaxy's redshift differ if we think of it as a cosmological redshift rather than as a Doppler shift?

### Does It Make Sense?

*Decide whether or not each of the following statements makes sense (or is clearly true or false). Explain clearly; not all of these have definitive answers, so your explanation is more important than your chosen answer.*

16. If you want to find a lot of elliptical galaxies, you'll have better luck looking in clusters of galaxies than elsewhere in the universe.
17. Cepheid variables make good standard candles because they all have exactly the same luminosity.
18. If the standard candles you are using are less luminous than you think they are, then the distances you determine from them will be too small.
19. Galaxy A is moving away from me twice as fast as galaxy B. That probably means it's twice as far away.
20. After measuring a galaxy's redshift, I used Hubble's law to estimate its distance.
21. The center of the universe is more crowded with galaxies than any other place in the universe.
22. The lookback time to the Andromeda Galaxy is about 2.5 million light-years.
23. I'd love to live in one of the galaxies near our cosmological horizon, because then I could see the black void into which the universe is expanding.
24. If someone in a galaxy with a lookback time of 4.6 billion years had a superpowerful telescope, that person could see our solar system in the process of its formation.
25. We can't see galaxies beyond the cosmological horizon because their light is too dim.

### Quick Quiz

*Choose the best answer to each of the following. For additional practice, try the Chapter 20 Reading and Concept Quizzes in the Study Area at www.MasteringAstronomy.com.*

26. Which of these galaxies do we see at its oldest age? (a) a galaxy in the Local Group (b) a galaxy observed at a distance of 5 billion light-years (c) a galaxy observed at a distance of 10 billion light-years
27. Which of these galaxies would you most likely find at the *center* of a large cluster of galaxies? (a) a large spiral galaxy (b) a large elliptical galaxy (c) a small irregular galaxy
28. In which of these galaxies would you be least likely to find an ionization nebula heated by hot young stars? (a) a large spiral galaxy (b) a large elliptical galaxy (c) a small irregular galaxy
29. About how many galaxies are there in a typical cluster of galaxies? (a) about 10 (b) a few dozen (c) a few hundred
30. According to Leavitt's law, we can determine the luminosity of a Cepheid in a distant galaxy by (a) measuring its parallax. (b) measuring its period of brightening and dimming. (c) measuring its recession speed.
31. Which kind of object is the best standard candle for measuring distances to extremely distant galaxies? (a) parallax (b) a Cepheid variable star (c) a white dwarf supernova
32. When the ultraviolet light from hot stars in very distant galaxies finally reaches us, it arrives at Earth in the form of (a) x-rays. (b) slightly more energetic ultraviolet light. (c) visible light.
33. Why do virtually all the galaxies in the universe appear to be moving away from our own? (a) We are located near where the Big Bang happened. (b) We are located near the center of the universe. (c) Expansion causes all galaxies to move away from nearly all others.
34. If you observed the redshifts of galaxies at a given distance to be twice as large as they are now, then you would determine a value for Hubble's constant that was (a) twice as large as its current value. (b) equal to its current value. (c) half its current value.
35. What would be your estimate for the age of the universe if you measured Hubble's constant to be 11 kilometers per second per million light-years? (a) 7 billion years (b) 14 billion years (c) 28 billion years

### Inclusive Astronomy

*Use these questions to reflect on participation in science.*

36. *Group Discussion: Leavitt's Law.* This book follows the recommendation of the American Astronomical Society in referring to what was previously called the Cepheid "period-luminosity relation" as *Leavitt's law*.
    **a.** Working independently, do a search on "Leavitt's law" to learn about the history behind this name.
    **b.** Gather into small groups to share what you've learned and to discuss the reasons behind the change from "period-luminosity relation" to "Leavitt's law." Make a list of pros and cons of the change in terminology.
    **c.** Have each group member state which name ("period-luminosity relation" or "Leavitt's law") she or he prefers, and why.
    **d.** Can you come to a group consensus about which term is preferable? If so, prepare a brief summary to explain your choice to others. If not, make a list of reasons for the disagreement.

### The Process of Science

*These questions may be answered individually in short-essay form or discussed in groups, except where identified as group-only.*

37. *The Shapley-Curtis Debate.* Section 20.2 briefly summarizes the 1920 debate between Harlow Shapley and Heber Curtis on the nature of "spiral nebulae." Do additional research to learn about the evidence presented by Shapley and Curtis to support their opposing positions. Which position would you have found more convincing at the time, based only on the evidence presented at the debate? Explain.
38. *Deviations from Hubble's Law.* Suppose you are measuring distances and velocities of galaxies in order to test Hubble's law. You find that 90% of the galaxies have velocities that are within 200 kilometers per second of the predictions of Hubble's law but 10% have velocities that deviate from the predictions by up to 1000 kilometers per second. Propose a hypothesis that would explain these deviations from Hubble's law and outline a set of observations that you could use to test your hypothesis.
39. *Hubble's Revolution.* Just a century ago, astronomers were still not sure whether the Milky Way was the only large collection of stars in the universe. Then Edwin Hubble measured the distance of the Andromeda Galaxy, proving once and for all that our Milky Way was just one among many galaxies in the universe. How was this change in our "cosmic perspective" similar to the Copernican revolution? How was it different?
40. *Group Activity: Counting Galaxies.* The following activity is based on the Hubble eXtreme Deep Field in Figure 20.1. Note: You may wish to do this activity using the four roles described in Chapter 1, Exercise 39.

**a.** Working individually, estimate the number of galaxies in the image.
**b.** Gather into small groups, where each member presents his or her method and estimate to the others. Do the estimates differ? By how much?
**c.** Discuss any significant differences in the methods and the resulting estimates, with the goal of understanding the pros and cons of different estimation methods.
**d.** As a group, come up with an estimation method that incorporates the best features of the team members' individual methods, and apply this method to come up with your group's final estimate for the number of galaxies in the image. Then use the fact that this image shows about 1/30,000,000 of the entire sky to come up with an estimate for the total number of galaxies in the observable universe. Compare your estimate with those of other groups in your class.

# Investigate Further

## Short-Answer/Essay Questions

41. *Deep Fields with Hubble.* The Hubble eXtreme Deep Field in Figure 20.1 was chosen for observation in large part because it is a completely ordinary part of the sky. Why do you think astronomers would want to devote so much precious telescope time to observing totally ordinary regions of the sky in such great detail? Explain.
42. *Supernovae in Other Galaxies.* In which type of galaxy would you be more likely to observe a massive star supernova: a giant elliptical galaxy or a large spiral galaxy? Explain.
43. *Hubble's Galaxy Types.* How would you classify the following galaxies using the system illustrated in Figure 20.8? Justify your answers.
**a.** Galaxy NGC 4594 (Figure 20.3)
**b.** Galaxy M101 (Figure 20.2a)
**c.** Galaxy NGC 4414 (Figure 20.2b)
**d.** Galaxy NGC 1300 (Figure 20.4)
**e.** Galaxy M87 (Figure 20.6a)
44. *Color and Luminosity.* The luminosity of our Milky Way galaxy is about $1.5 \times 10^{10} L_{\text{Sun}}$. Describe how the Milky Way's light output compares with those of the most luminous galaxies in the blue cloud, based on the information in Figure 20.9. Approximately how much more luminous than the Milky Way are the most luminous galaxies in the red sequence?
45. *Distance Measurements.* The techniques astronomers use to measure distances are not so very different from the ones you use every day. Describe how standard candle measurements are similar to the way you estimate the distance to an oncoming car at night.
46. *Cepheids as Standard Candles.* Suppose you are observing Cepheids in a nearby galaxy. You observe one Cepheid with a period of 8 days between peaks in brightness and another with a period of 35 days. Use Leavitt's law (Figure 20.14) to estimate the luminosity of each Cepheid.
47. *Galaxies at Great Distances.* The most distant galaxies that astronomers have observed are much easier to study with infrared light than with visible light. Explain why that is the case.
48. *Universe on a Balloon.* In what ways is the surface of a balloon a good analogy for the universe? In what ways is this analogy limited? Explain why a miniature scientist living in a polka dot on the balloon would observe all other dots to be moving away, with more distant dots moving away faster.
49. *Lookback Time and the Ages of Galaxies.* Suppose you observe a distant galaxy with a lookback time of 10 billion years. What was the maximum possible age of that galaxy when the light we are now observing from it began its journey to Earth? Explain your reasoning. (*Hint:* Assume the galaxy was born less than a billion years after the Big Bang.)
50. *Research: Greatest Lookback Time.* Look for recent discoveries of objects with the largest lookback times. What is the current record for the most distant known object? How was it observed, and what kind of object is it?

## Quantitative Problems

*Be sure to show all calculations clearly and state your final answers in complete sentences.*

51. *Cepheids in M100.* Consider these actual data for three Cepheids in the galaxy M100; the luminosities were determined with Leavitt's law from the observed periods (measured with observations from the Hubble Space Telescope).
    - Cepheid 1: luminosity = $3.9 \times 10^{30}$ watts
      apparent brightness = $9.3 \times 10^{-19}$ watt/m$^2$
    - Cepheid 2: luminosity = $1.2 \times 10^{30}$ watts
      apparent brightness = $3.8 \times 10^{-19}$ watt/m$^2$
    - Cepheid 3: luminosity = $2.5 \times 10^{30}$ watts
      apparent brightness = $8.7 \times 10^{-19}$ watt/m$^2$

    Compute the distance to M100 with data from each of the three Cepheids. Do the three distance computations agree? Based on your results, estimate the uncertainty in the distance you have found.
52. *Redshift and Hubble's Law.* Imagine that you have obtained spectra for three galaxies and found the following measurements for the observed wavelength of a hydrogen emission line that has a rest wavelength of 656.3 nanometers:
    - Galaxy 1: 659.6 nanometers
    - Galaxy 2: 664.7 nanometers
    - Galaxy 3: 679.2 nanometers

    **a.** Calculate the redshift, $z$, for each of these galaxies.
    **b.** From its redshift, calculate the speed at which each of the galaxies is moving away from us. Give your answers both in kilometers per second and as a fraction of the speed of light.
    **c.** Use Hubble's law to estimate the distance to each galaxy. Assume that $H_0 = 22$ km/s/Mly.
53. *Estimating the Universe's Age.* Suppose that Hubble's constant were $H_0 = 33$ km/s/Mly (which is *not* its actual value). What would the approximate age of the universe be in that case?
54. *Hubble's First Attempt.* Edwin Hubble's first attempt to measure the universe's expansion rate was flawed because the standard candles he was using were not properly calibrated. Look at Figure 20.21. Estimate the value of $H_0$ corresponding to the solid line in the figure. Remember that $10^6$ parsecs = 3.26 million light-years. What is the approximate age of the universe indicated by that erroneous value of $H_0$?
55. *Extremely Distant Galaxies.* The most distant galaxies observed to date have a redshift of approximately $z = 10$. How does the wavelength of the light we observe from those galaxies compare with its original wavelength when it was emitted?
56. *Stretching of the Universe.* The most distant white dwarf supernova observed as of 2018 had a redshift of $z = 1.9$. How does the average distance between galaxies now compare with the average distance between galaxies at the time the supernova exploded?

# 21 Galaxy Evolution

▲ **About the photo:** This composite image of both visible light and radio waves from the galaxy Hercules A shows two powerful jets of high-energy plasma emanating from a tiny region around the galaxy's central black hole.

## LEARNING GOALS

**21.1 Looking Back Through Time**
- How do we observe the life histories of galaxies?
- How do we study galaxy formation?

**21.2 The Lives of Galaxies**
- Why do galaxies differ?
- How does gas cycle through galaxies?

**21.3 The Role of Supermassive Black Holes**
- What is the evidence for supermassive black holes at the centers of galaxies?
- Do supermassive black holes regulate galaxy evolution?

**21.4 Gas Beyond the Stars**
- How do we study the gas outside of galaxies?

*I had always seen myself as a star; I wanted to be a galaxy.*

— *Twyla Tharp, American dancer and choreographer*

Chapter 21 Overview

The spectacle of galaxies strewn like beautiful islands across the universe invites us to ponder their origins. What processes create the majestic spiral arms seen in many galaxies, and the elliptical or irregular shapes of others?

If we look closely, we see even more spectacular sights. Some galaxies are engaged in titanic collisions with others, sometimes igniting tremendous bursts of star formation in which new stars are born and massive stars explode 100 times as frequently as in the Milky Way. Other galaxies harbor supermassive black holes encircled by accretion disks that generate extraordinary luminosities. Narrow streams of matter jet at nearly the speed of light from a few of these galaxies into intergalactic space.

The origins of galaxies and the incredible phenomena they exhibit puzzled astronomers for much of the 20th century, but we are beginning to understand how galaxies evolve. In this chapter, we will sift through the fascinating clues that hint at how galaxies formed and developed, pausing now and again to admire the fantastic spectacles that we have uncovered in our quest for understanding.

## 21.1 Looking Back Through Time

In Chapter 20, we saw how the distances to galaxies are measured and how such measurements revealed the expansion of our universe. We also saw how the expansion rate indicates an age for the universe of about 14 billion years. With that understanding, we are now ready to turn our attention to the study of how galaxies form and develop in our expanding universe—a subject known as **galaxy evolution**.

As we study how galaxies evolve, keep in mind that we would not be here if not for the galaxy-wide recycling processes that have gradually transformed the Milky Way's primordial gases into stars and planets [**Section 19.2**]. The same type of star–gas–star cycle has operated for at least some period of time in the history of all galaxies, so the differences we observe among galaxies provide clues to how individual galaxies form and evolve. Throughout this chapter, we will develop a deeper understanding of our cosmic origins by exploring both how galaxies first formed and how they can end up looking so different from one another. We'll begin by considering how astronomers study the lives of galaxies.

### How do we observe the life histories of galaxies?

Observational evidence about the lives of galaxies comes from deep images of the universe, such as the Hubble Space Telescope image we saw in Figure 20.1. Powerful telescopes can be used as time machines to observe the life histories of galaxies, because the farther we look into the universe, the further we look back in time [**Section 1.1**]. The most distant galaxies already had some stars in place more than 13 billion years ago, which means the ages of these stars match those of the oldest stars in our own galaxy. It therefore seems safe to assume that most galaxies began to form at about this time, in which case most galaxies would be roughly the same age today. A galaxy's lookback time is therefore linked directly to its age. A galaxy with a lookback time of 13 billion years in a 14-billion-year-old universe must be less than a billion years old, while nearby galaxies (with lookback times much less than 1 billion years) can be more than 13 billion years old.*

This linkage between lookback time and age gives us a remarkable opportunity: Simply by photographing galaxies at different distances, we can assemble "family albums" of galaxies in different stages of development. Pictures of the most distant galaxies show galaxies in their childhood, and pictures of the nearest show mature galaxies as they are today. **FIGURE 21.1** shows partial family albums for elliptical, spiral, and irregular galaxies, as well as a set of galaxies (left panel) seen when they and the universe were very young. Each individual photograph shows a single galaxy at a single stage in its life, and measuring the redshift of each galaxy allows us to place it on a timeline of the universe. Grouping these photographs by galaxy type then allows us to see how galaxies of a particular type have changed through time.

**Think about it** We've used the word *today* in a very broad sense. For example, a relatively nearby galaxy may be located, say, 20 million light-years away, so we see it as it was 20 million years ago. In what sense is this "today"?

### How do we study galaxy formation?

Although current technology allows us to see some galaxies at a time when the universe was only about 1 billion years old, we cannot yet observe the earliest stages of galaxy formation. Scientists hope to see farther back in time with the James Webb Space Telescope (scheduled for launch in 2021; see Figure 6.24), but we must rely on theoretical models to study the earliest stages of galaxy formation. The most successful models for galaxy formation start from two key assumptions, both of which are backed by strong observational evidence [**Section 22.2**]:

- Hydrogen and helium gas, along with dark matter, filled all of space *almost* uniformly when the universe was very young—say, in the first million years after its birth.
- However, the distribution of matter was not *perfectly* uniform—certain regions of the universe started out ever so slightly denser than others, and these enhanced-density regions served as "seeds" for the formation of galaxies.

*A useful rule of thumb is that a galaxy's approximate age can be no greater than 14 billion years minus its lookback time.

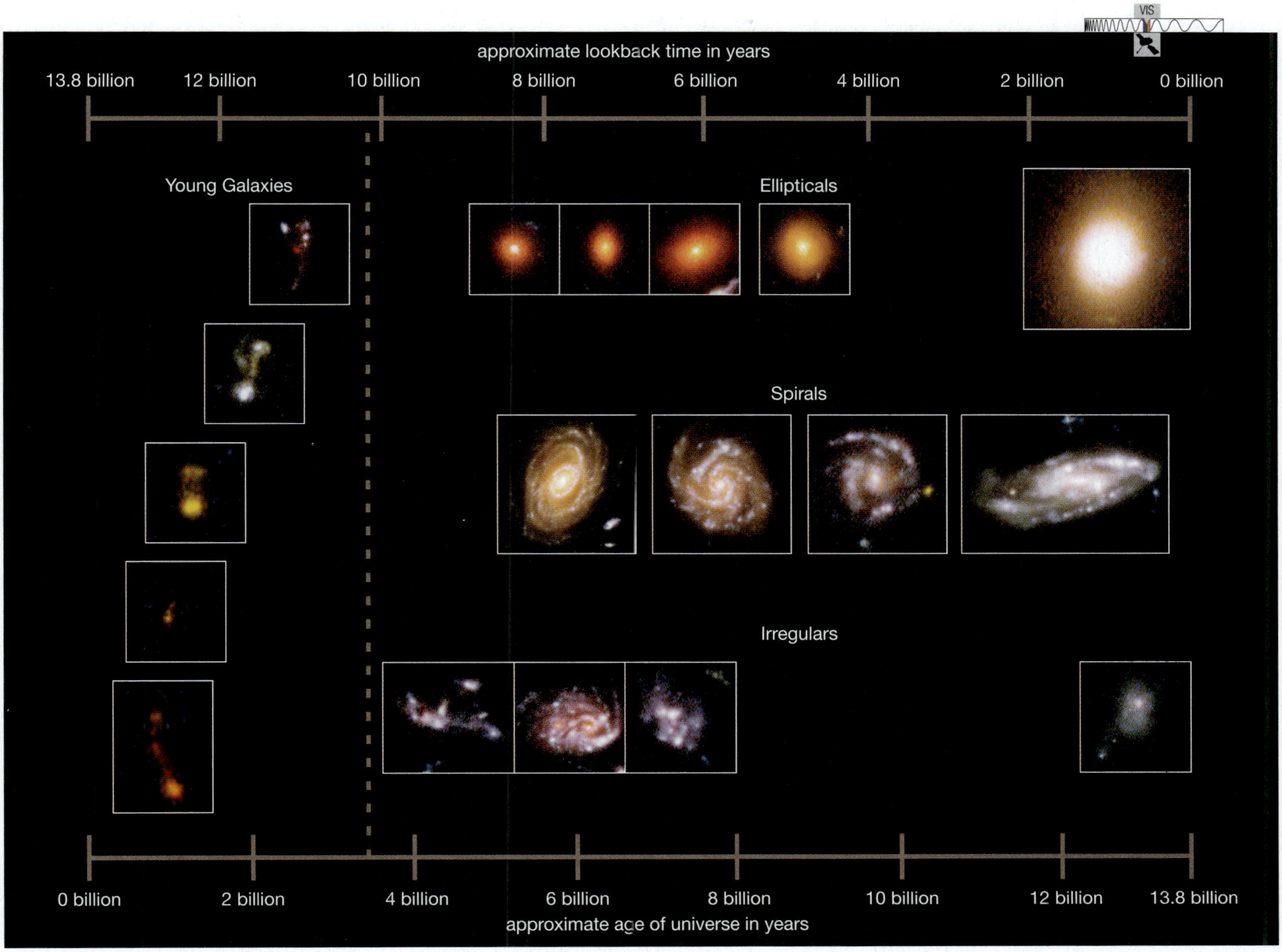


**FIGURE 21.1** Family albums for elliptical, spiral, and irregular galaxies of different ages, plus some very young galaxies shown on the far left. These photos are all zoomed-in images of galaxies from the Hubble eXtreme Deep Field (see Figure 20.1). We see more distant galaxies as they were when they were younger; the approximate age of the universe is indicated along the bottom horizontal axis, with lookback time shown at the top. (The figure assumes an age of 13.8 billion years for the universe, which is the best current estimate.)

Beginning from these assumptions, we can use the laws of physics and computer simulations to model how galaxies formed (**FIGURE 21.2**). The models show that the slightly greater pull of gravity around the enhanced-density seeds would have halted the expansion of these regions within about a billion years, even as the universe as a whole continued to expand [**Section 22.3**]. The material in these regions then contracted into *protogalactic clouds* much like those that eventually formed our Milky Way.

When we discussed the Milky Way, we outlined a basic model for galaxy formation that accounts for the primary features of spiral galaxies [**Section 19.3**]. In that model, the collapsing protogalactic clouds that are destined to form a spiral galaxy initially cool as they contract, because they can radiate away their thermal energy. The first generation of stars then grows from the densest, coldest clumps of gas in those clouds, forming some very massive stars that live and die within just a few million years [**Section 16.1**]. The supernovae of those massive stars seed the rest of the protogalactic cloud system with its first sprinkling of heavy elements and generate shock fronts that heat the surrounding interstellar gas. This heating slows the collapse of the protogalactic clouds and the rate at which stars form within them, allowing time for additional gas to settle into a rotating disk as the protogalactic-cloud system merges to form a galaxy.

This basic model accounts for some of the most obvious features of spiral galaxies. In particular, it explains why the Milky Way and other spiral galaxies have a *disk population* of stars that orbit the galactic center in a fairly flat plane and a *halo population* of stars with more randomly oriented orbits (see Figure 19.2). The halo population consists of stars that were born before the galaxy's rotation became organized, which is why they have randomly oriented orbits around the galactic center. The disk population consists of stars born after the galaxy's gas settled into a rotating disk, which is why they all have similar orbits around the center of the galaxy (see Figure 19.18).

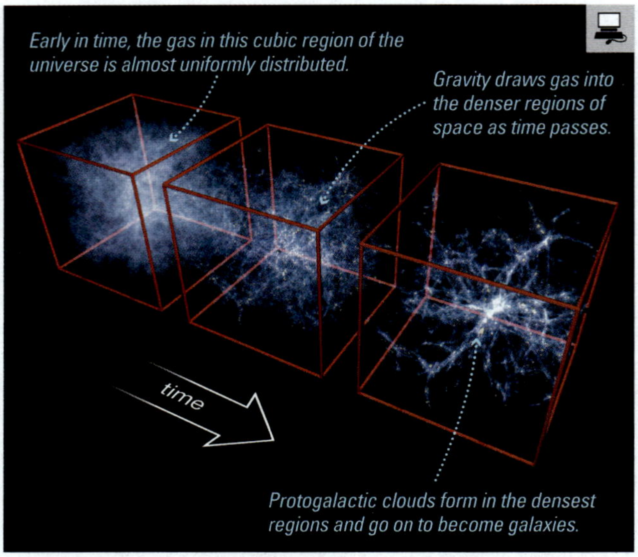


**FIGURE 21.2** A computer simulation of the formation of protogalactic clouds. The simulated region of space is about 500 million light-years wide and goes on to form numerous simulated galaxies.

However, the basic model leaves at least two major questions unanswered. First, it assumes that galaxies formed in regions of slightly enhanced density, but does not tell us where these density enhancements came from. The origin of density enhancements in the early universe is one of the major puzzles in astronomy, and we'll revisit it in Chapters 22 and 23. Second, while the basic model plausibly accounts for the main features of spiral galaxies, it does not explain why some galaxies are elliptical and others irregular. That is the question to which we turn our attention next.

## 21.2 The Lives of Galaxies

The distinct differences between spiral, elliptical, and irregular galaxies clearly tell us that their life stories are not the same. We would like to tell the life story of each type of galaxy from beginning to end as completely as we told the life stories of stars, but many aspects of galaxy evolution remain mysterious and subject to active, ongoing research. Nevertheless, we now know the general outlines of the life stories of galaxies and have some promising ideas about why galaxies come in different types. In this section, we'll examine some of those ideas and show how they fit into the overall picture of galaxy evolution, even though the picture itself is not yet complete.

### Why do galaxies differ?

The early stages in the formation and collapse of a protogalactic-cloud system are thought to have been similar for all galaxies. Therefore, to account for the existence of elliptical galaxies, we must ask why some galaxies do not end up with gas-rich disks, as suggested by our basic model for spiral-galaxy formation. The hypotheses proposed to explain the absence of gas-rich disks in ellipticals fall into two categories: (1) Elliptical galaxies end up looking different from spirals because of different birth conditions, or (2) their beginnings are similar to those of spiral galaxies, but they later become elliptical through interactions with other galaxies.

**Birth Conditions** Hypotheses in the first category trace a galaxy's type back to the characteristics of the protogalactic-cloud system from which it formed. Two factors may play a role:

- **Protogalactic rotation** (**FIGURE 21.3a**). A galaxy's type might be determined in part by the rotation of its protogalactic-cloud system. If the original system had a significant amount of angular momentum, it would have rotated quickly as it collapsed. The galaxy it produced would therefore have tended to form a disk, making it a spiral galaxy. If the protogalactic-cloud system had little or no angular momentum, its gas might not have formed a disk at all, so the resulting galaxy would be elliptical.
- **Protogalactic density** (**FIGURE 21.3b**). A galaxy's type might also be determined in part by the density of the protogalactic clouds from which it formed. Protogalactic clouds with relatively high gas density would have radiated energy more effectively and cooled more quickly, thereby allowing more rapid star formation. If the star formation proceeded fast enough, all the gas could have been turned into stars before any of it had time to settle into a disk. The resulting galaxy would therefore lack a disk, making it an elliptical galaxy. In contrast, lower-density clouds would have formed stars more slowly, leaving plenty of gas to form the disk of a spiral galaxy.

Evidence for the role of birth conditions comes from a few giant elliptical galaxies at very great distances (**FIGURE 21.4**). These galaxies lack blue and white stars, indicating that new stars no longer form within them, even though we are seeing them as they were when the universe was quite young. This finding suggests that all the stars in these elliptical galaxies were born at almost the same time, which is consistent with the idea that all the stars formed before a disk could develop.

**Later Interactions** Differences in birth conditions probably play important roles in the overall story of why some galaxies have gas-rich disks and others do not. However, birth conditions are probably not the whole story, because they ignore one key fact: Galaxies rarely evolve in perfect isolation.

Think back to our scale model solar system in Chapter 1. On a scale on which the Sun was the size of a grapefruit, the nearest star was like another grapefruit a few thousand kilometers away. Because the average distances between stars are so huge compared to the sizes of stars, collisions between stars are extremely rare. However, if we rescale the universe so that our *galaxy* is the size of a grapefruit, the Andromeda Galaxy is like another grapefruit only about 3 meters away, and a few smaller galaxies lie considerably closer. In other words, the average distances between galaxies are not much larger than the sizes of galaxies, meaning that collisions between galaxies are inevitable. Our own

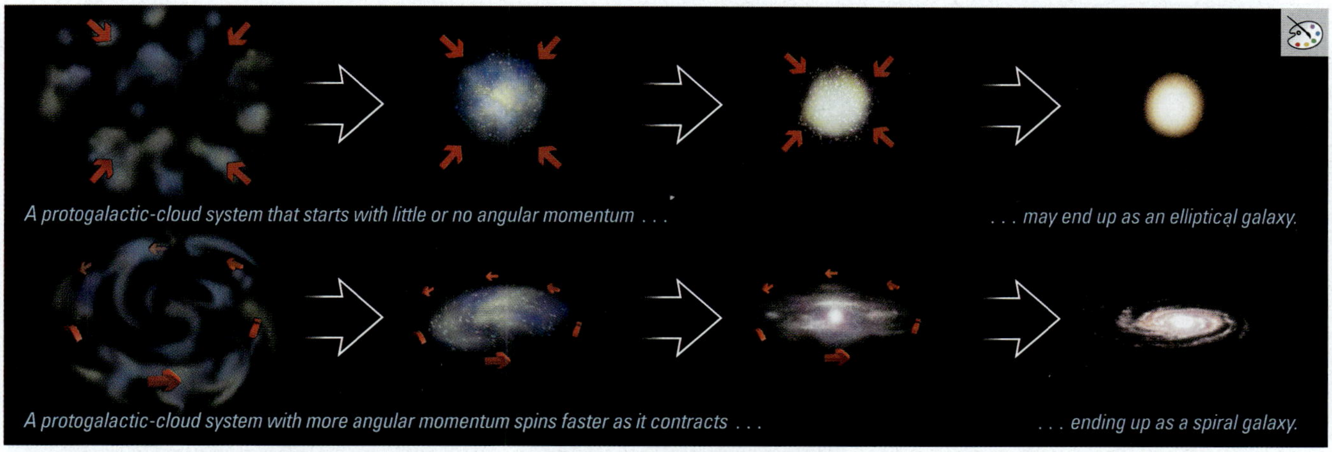


**a** The angular momentum of a galaxy's protogalactic-cloud system may determine whether it ends up spiral or elliptical.

**b** The gas density of a galaxy's protogalactic clouds may determine whether it ends up spiral or elliptical.

**FIGURE 21.3** These diagrams show two ways in which a galaxy's birth conditions may have determined whether it ended up spiral or elliptical.

**FIGURE 21.4** The light we observe from the distant elliptical galaxy called HUDF-JD2 (circled) left that galaxy when the universe was about 800 million years old. Even though it is very young, the galaxy contains about eight times as many stars as the Milky Way, and its color indicates that few new stars are forming within it.

Milky Way Galaxy is not immune. About 80,000 light-years away from us, directly behind the galactic bulge, a small elliptical galaxy (the Sagittarius Dwarf [**Section 19.1**]) is currently crashing through the Milky Way's disk.

Galaxy collisions are spectacular events that unfold over hundreds of millions of years (**FIGURE 21.5**). During our short lifetimes, we can at best see a snapshot of a collision in progress, distorting the shapes of the colliding galaxies. Collisions must have been even more frequent in the distant past, when the universe was smaller and galaxies were closer together. Observations confirm that distorted-looking galaxies—probably galaxy collisions in progress—were more common in the early universe than they are today (**FIGURE 21.6**).

We can learn much more about galactic collisions with the aid of computer simulations that allow us to "watch" collisions that in nature take hundreds of millions of years to unfold. These simulations show that a collision between two spiral galaxies can create an elliptical galaxy (**FIGURE 21.7**). Tremendous tidal forces between the colliding galaxies tear apart the two disks, randomizing the orbits of their stars. Meanwhile, a large fraction of their gas sinks to the center of the collision and rapidly forms new stars. Supernovae and stellar winds eventually blow away the rest of the gas. When the cataclysm finally settles down, the merger of the two spirals has produced a single elliptical galaxy. Little gas is left for a disk, and the orbits of the stars have random orientations.

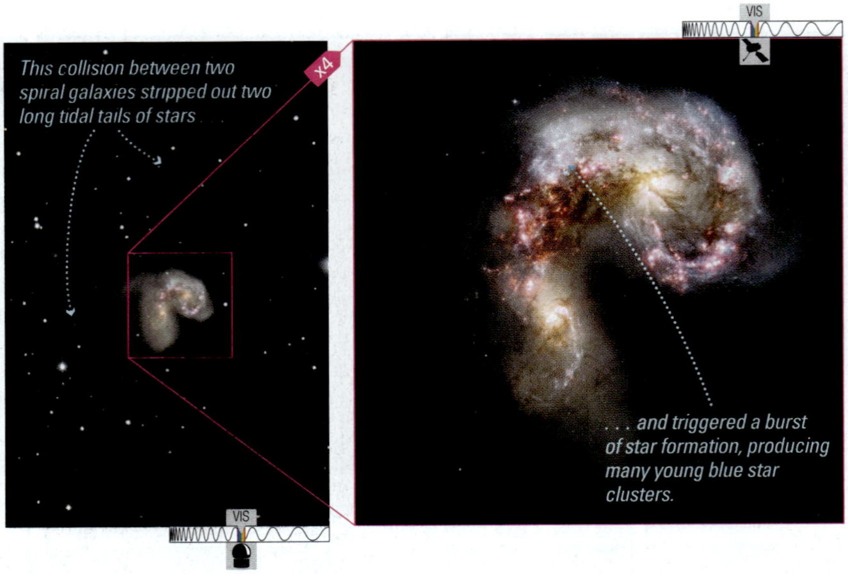


**FIGURE 21.5** A pair of colliding spiral galaxies known as the Antennae (NGC 4038/4039). The image taken from the ground (left) reveals their vast tidal tails, while the close-up from the Hubble Space Telescope shows the burst of star formation at the center of the collision.

**Evidence for Interactions** Observations support the idea that at least some elliptical galaxies result from collisions and mergers. Elliptical galaxies dominate the galaxy populations at the cores of dense clusters of galaxies, where collisions should be most frequent. This fact may mean that any spirals once present became ellipticals through collisions.

Stronger evidence comes from the structural details of elliptical galaxies, which often attest to a violent past. Some elliptical galaxies have stars and gas clouds with orbits suggesting that they are leftover pieces of galaxies that merged in a past collision. One example is the elliptical galaxy NGC 474, shown in **FIGURE 21.8a**. Although it is not a member of a cluster, this galaxy's unusual structure is probably the result of a collision. The shells surrounding this galaxy are made of stars on orbits that are hard to explain if the entire galaxy came from a single protogalactic cloud. The stars that we see in the shells probably plunge back and forth through the central part of the galaxy, each swinging from one side to the other like a pendulum. The shells represent the extreme ends of these plunging orbits, where the stars spend most of their time.

The most decisive evidence that collisions affect the evolution of elliptical galaxies comes from observations of the **central dominant galaxies** found at the centers of many dense clusters. Central dominant galaxies are giant elliptical galaxies that apparently grew to huge sizes by consuming other galaxies through collisions. They frequently contain several tightly bound clumps of stars that probably were the centers of individual galaxies before being swallowed by the giant (**FIGURE 21.8b**). This process of *galactic cannibalism* can create central dominant galaxies more than 10 times as massive as the Milky Way, making them the most massive galaxies in the universe.

Yet another mechanism that can transform spiral galaxies in clusters into ellipticals is gas removal. The central regions of dense galaxy clusters tend to be filled with very hot gas [**Section 23.2**]. When a spiral galaxy cruises through the center of such a cluster, the hot gas exerts drag forces that slow the galaxy's gas but not its stars. The galaxy's stars therefore continue to move freely along their way, but the gas is left behind. If the disk has not yet formed many stars, the galaxy will evolve to look more like an elliptical galaxy as its massive stars die away. Its disk will fade, while its bulge and halo will remain prominent. If the disk has already formed a large number of stars when its gas is stripped, the remaining galaxy will look like a spiral galaxy without its gaseous disk, which makes it a *lenticular galaxy* (see Figure 20.5).

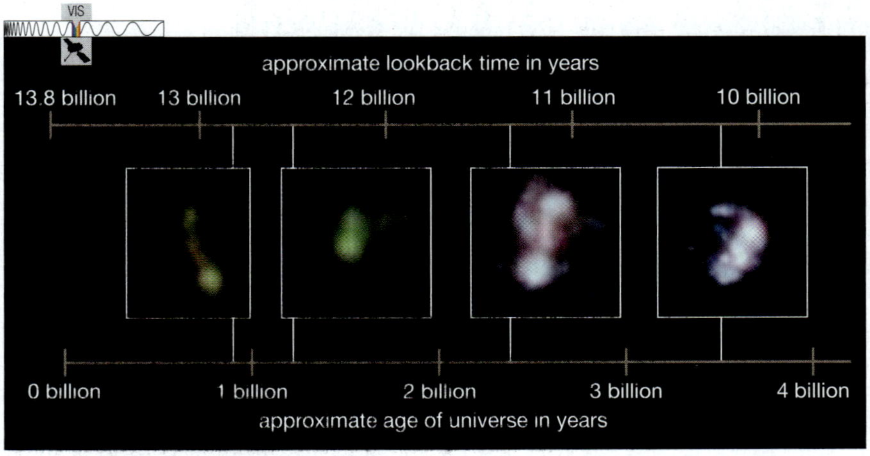


**FIGURE 21.6** These Hubble Space Telescope photographs offer a zoomed-in view of young galaxies from the Hubble eXtreme Deep Field (see Figure 20.1). Notice that these galaxies do not look like either the spiral or the elliptical galaxies that are common in the present-day universe, and instead appear to be undergoing collisions. We infer that galaxy collisions were much more common in the early universe than they are today.

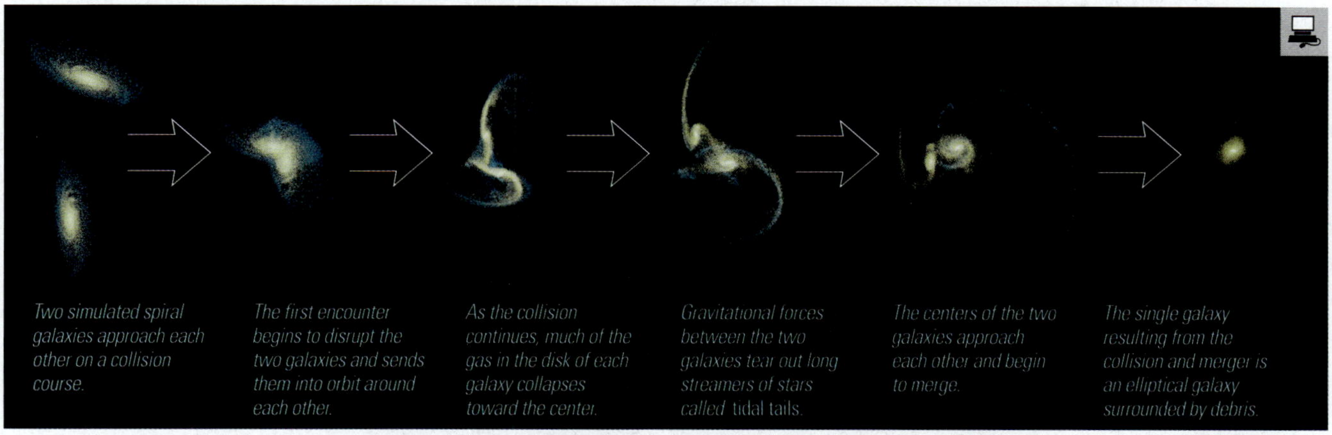


**FIGURE 21.7** Several stages in a supercomputer simulation of a collision between two spiral galaxies that results in an elliptical galaxy. At least some of the elliptical galaxies in the present-day universe formed in this way. The whole sequence spans about 1.5 billion years.

**An Incomplete Answer** Birth conditions and subsequent interactions probably both play important roles in galaxy evolution, but we are not yet certain which is more influential. Nevertheless, when we consider both kinds of mechanisms together, they do seem to account for the basic differences between galaxy types. The birth-condition scenarios may explain why the vast majority of galaxies are either spiral or elliptical in shape. The interaction scenarios may explain why ellipticals are more common in clusters while spirals are more common outside of clusters. Even the relatively small fraction of galaxies that are irregular may be explained by these ideas: At least some irregulars probably are galaxies undergoing some sort of disruptive interaction.

These ideas also appear to explain the patterns of galaxy color and galaxy luminosity originally shown in Figure 20.9 and repeated along with explanations in **FIGURE 21.9**. All galaxies are thought to begin as star-forming systems that are blue-white in color, making them members of the blue cloud. However, some of them later transition to the red sequence. The upper portion of the red sequence forms through mergers that produce large, red elliptical galaxies. The lower portion of the red sequence consists of smaller galaxies that have used up their cool gas, so that star formation has ceased and only cool, red stars are left. Astronomers are conducting observations and planning further ones to test these ideas and complete our understanding of galaxy evolution.

**a** Elliptical galaxy NGC 474 is surrounded by several distinct shells of stars that were likely produced by collisions with other galaxies.

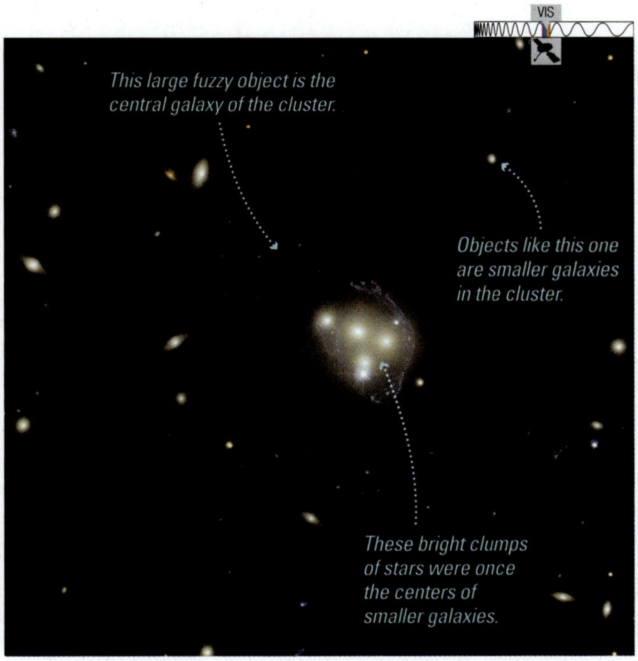


**b** This image shows the central dominant galaxy of the cluster Abell 3827, which has apparently grown by consuming smaller galaxies that have collided with it. Notice that the center of this galaxy contains multiple clumps of stars that probably once were the centers of individual galaxies. (The blue streaks encircling the central galaxy represent light from a much more distant galaxy that has been distorted by the phenomenon of gravitational lensing [Section 23.2].)

**FIGURE 21.8** Evidence for past collisions in elliptical galaxies.

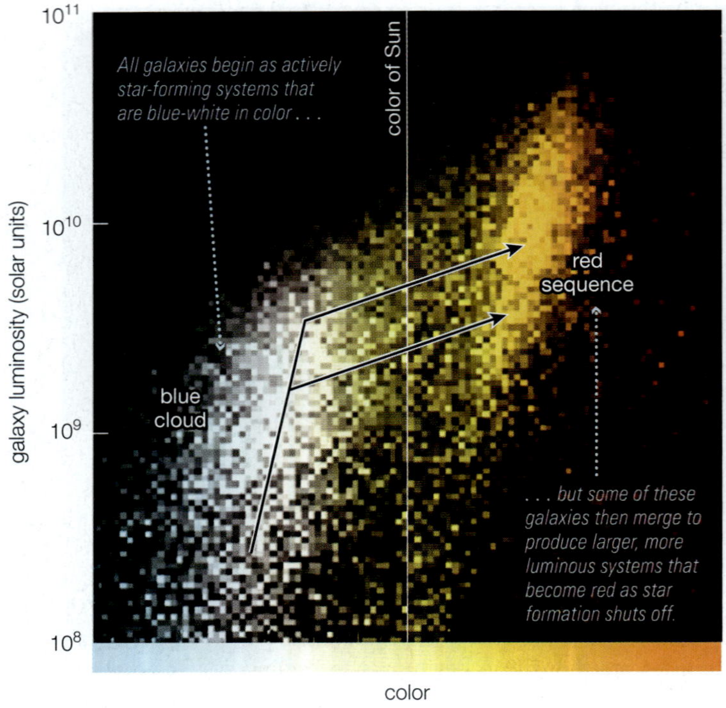


**FIGURE 21.9** This figure shows the data from Figure 20.9 and explains how the patterns arise, according to our current understanding of galaxy evolution. All galaxies are thought to begin as star-forming systems, putting them in the *blue cloud* at the left. Mergers of these galaxies can later produce larger galaxies, some of which cease forming new stars. The black lines show how mergers cause the observed properties of galaxies to change, shifting the galaxies up and to the right onto the *red sequence*.

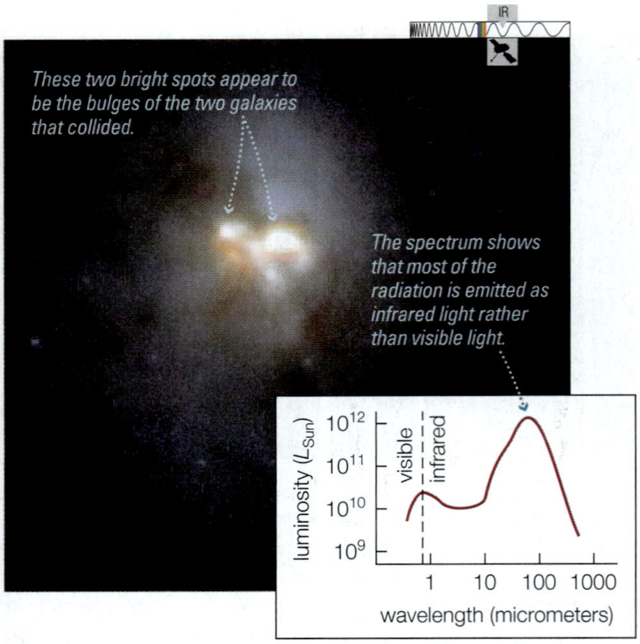


**FIGURE 21.10** Infrared observations of Arp 220, a large starburst galaxy. (The region shown in the photo is about 10,000 light-years across.)

## How does gas cycle through galaxies?

Our discussion so far has assumed that the star–gas–star cycle in spiral galaxies supports ongoing star formation at a fairly steady rate, while elliptical galaxies consist mostly of stars that formed in the distant past. However, observations show that the star-formation rates of galaxies can be quite variable, depending on the amount of cold gas available for star formation at any one time. In particular, a small percentage of galaxies—known as **starburst galaxies**—are undergoing "starbursts" in which new stars form within them at an astonishingly rapid rate. These starbursts demonstrate that a galaxy's star–gas–star cycle is not always steady and depends critically on the rate at which hot gas in a galaxy's halo can cool and resupply the cold clouds needed for star formation.

**Bursts of Star Formation** The Milky Way Galaxy produces an average of about one new star per year, but some starburst galaxies form new stars at rates exceeding 100 stars per year. At this rate, a starburst would consume *all* of a galaxy's available gas in less than 100 million years. Because most galaxies have been around for billions of years, we conclude that a starburst must be a relatively short-term stage in a galaxy's life. After the burst of star formation ends, a starburst galaxy presumably returns to being an "ordinary" spiral, elliptical, or irregular galaxy.

Starburst galaxies are best observed at infrared wavelengths, because they are filled with star-forming molecular clouds that absorb visible and ultraviolet light [**Section 16.1**]. In fact, starburst galaxies can seem almost ordinary to visible-light telescopes, because the impressive energy of their young stars is hidden from view. **FIGURE 21.10** shows an infrared image of an especially luminous starburst galaxy, along with its spectrum from visible through infrared wavelengths. Note that the visible output of this galaxy is about 10 billion solar luminosities ($10^{10}L_{Sun}$), not very different from the total luminosity of the Milky Way. However, the galaxy's infrared output is a trillion times that of our Sun ($10^{12}L_{Sun}$), which makes it 100 times brighter in infrared light than in visible light and tells us that it must be forming stars at about 100 times the rate of the Milky Way.

**Ejection of Gas** The high rate of star formation in a starburst galaxy leads to a high rate of supernova explosions from short-lived, massive stars. Just as in the Milky Way, each supernova generates a shock front that creates a *bubble* of hot gas [**Section 19.2**]. The shock fronts from several nearby supernovae quickly overlap and blend into a much larger *superbubble*. That's usually the end of the story in the Milky Way, but in a starburst galaxy the drama is just beginning. Supernovae continue to explode inside the superbubble, adding to its thermal and kinetic energy. When the superbubble starts to break through the disrupted gaseous disk, it expands even faster. Hot gas then erupts into the galaxy's halo, creating a **galactic wind**.

Galactic winds consist of low-density but extremely hot gas, typically with temperatures of 10–100 million K (**FIGURE 21.11**). They do not emit much visible light, but they generate x-rays. X-ray telescopes in orbit have detected

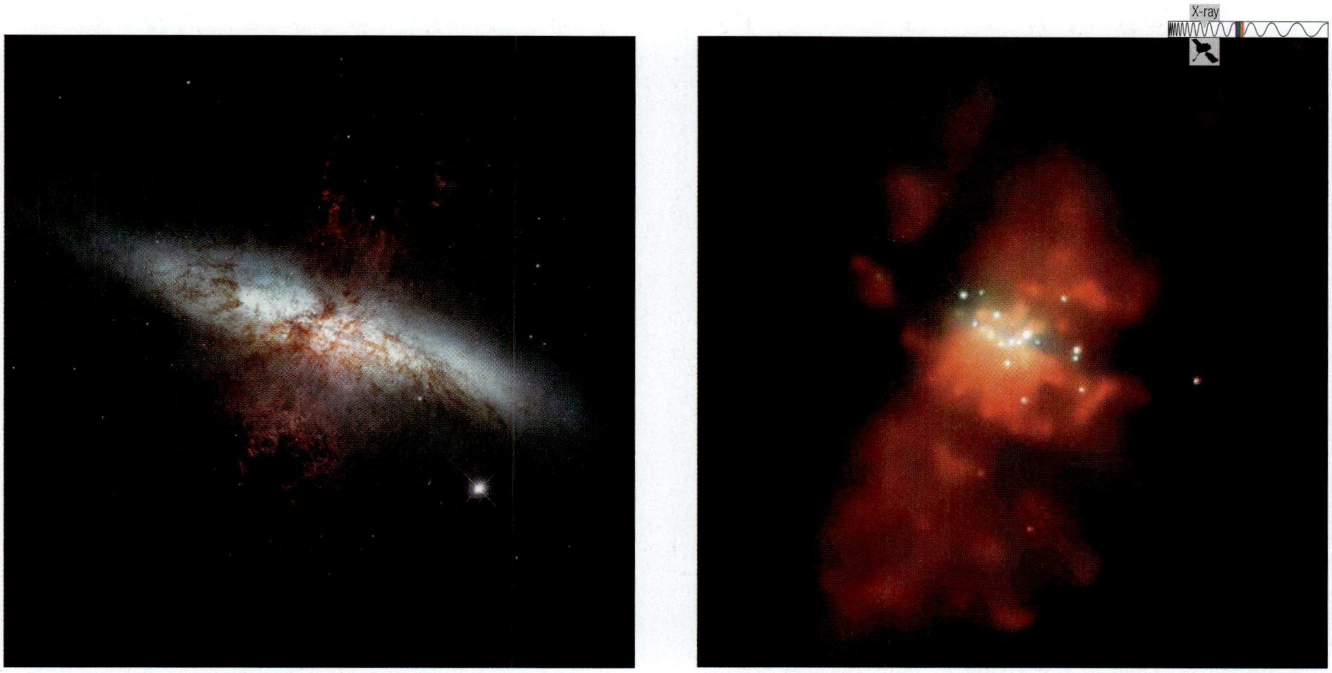


**a** This visible-light photograph (from the Hubble Space Telescope) shows violently disturbed gas (red) blowing out both above and below the disk.

**b** This x-ray image from the Chandra X-Ray Observatory shows the same region as the visible-light photograph in part a. The reddish region represents x-ray emission from very hot gas blowing out of the disk. The bright dots in the galactic disk represent x-ray emission from accretion disks around black holes or neutron stars, probably produced by recent supernovae.

**FIGURE 21.11** Visible-light and x-ray views of a starburst galaxy called M82, showing its galactic wind. Both images show the same region, which is about 16,000 light-years across.

pockets of x-ray emission surrounding the disks of some starburst galaxies, presumably coming from the outflowing galactic wind. Sometimes we also see the glowing remnants of a punctured superbubble extending out into space.

Supernova-driven galactic winds can have a dramatic impact on subsequent star formation, especially when they occur in small galaxies (**FIGURE 21.12**). The winds can blow out of these galaxies on all sides, driving much of the gas beyond the distances of their outermost stars. As a result, star formation in these galaxies may shut down for billions of years, then later resume if some of the gas outside of the galaxy falls back into it. Some of the small elliptical galaxies in the Local Group appear to have experienced multiple starbursts, because their stars seem to fall into a few distinct groups that each have an age billions of years different from the ages of the other groups.

**Think about it** Dwarf galaxies that have undergone bursts of star formation tend to have fewer heavy elements than do large galaxies. Why do you think that is? (*Hint:* What happens to the heavy elements produced by the burst of star formation?)

**Regulation of the Gas Supply** Starbursts and the distinct age patterns we see among stars in small galaxies show that a galaxy's star-formation rate is not always steady. However, averaged over long periods of time, the amount of gas that a galaxy transforms into stars cannot exceed the supply of cold gas available for star formation. The existence of galactic winds therefore shows that a galaxy's star–gas–star cycle is self-limiting. If a burst of star formation heats a large amount of a galaxy's gas and ejects it into the halo, then star formation will slow down until the heated gas cools enough to condense into cold clouds that fall back into the galaxy's disk. This interplay between heating and cooling regulates a galaxy's gas supply, at least in spiral and irregular galaxies, so that star formation happens over a long period of time, rather than all at once.

As a consequence, a galaxy's star–gas–star cycle may extend over a region many times larger than the galaxy's visible disk. Models of galaxy evolution suggest that this large region may actually contain the majority

**FIGURE 21.12** This x-ray/visible composite photo of dwarf starburst galaxy NGC 1569 shows hot gas blowing out in several different directions. X-ray light from this hot galactic wind is shown in blue, and visible light from disturbed hydrogen gas is shown in red.

of a galaxy's gas, in the form of a hot atmosphere that produces a light rain of cold clouds falling back toward the disk. These models are challenging to test, because the hot gas is so spread out that it is difficult to detect the x-ray and ultraviolet radiation that the gas produces. Nevertheless, evidence is building in favor of models in which galactic evolution depends on the cycling of gas in and out of galactic disks. And as we'll see in the next section, central black holes may also play a critical role in this process.

## 21.3 The Role of Supermassive Black Holes

Galaxy collisions and starbursts may be spectacular, but even more awe-inspiring events can occur deep in the centers of galaxies. A small percentage of galaxies emit truly extreme amounts of radiation from a tiny central region, sometimes accompanied by powerful jets of material shooting outward at nearly the speed of light (**FIGURE 21.13**). What could possibly release so much energy from within such a small region of space? In this section, we'll investigate the evidence that has led scientists to conclude that the culprits are **supermassive black holes**, much like the one at the Milky Way's center [**Section 19.4**] but in many cases far more massive.

**FIGURE 21.13** The active galactic nucleus in the elliptical galaxy M87. The bright yellow spot is the active nucleus, and the blue streak is a jet of particles shooting outward from the nucleus at nearly the speed of light.

### What is the evidence for supermassive black holes at the centers of galaxies?

Galaxies with unusually bright centers are called *active galaxies,* and the bright central regions themselves are known as **active galactic nuclei**. The existence of active galactic nuclei was first recognized in the 1940s, when astronomers noticed that about 1% of present-day galaxies—that is, galaxies we see nearby—have unusually large luminosities coming from small regions near their centers.*

The mystery posed by active galactic nuclei became deeper after the discovery in the early 1960s of **quasars**, which we now know to be the most luminous examples of active galactic nuclei (**FIGURE 21.14**). The cosmological redshifts of quasars indicated that they were generally billions of light-years from Earth (**FIGURE 21.15**), which implied luminosities that were in some cases more than 1000 times

*The galaxies that contain these active galactic nuclei are often called *Seyfert galaxies* after astronomer Carl Seyfert, who in 1943 grouped galaxies with active galactic nuclei into a special class.

**FIGURE 21.14** This Hubble Space Telescope photo shows a quasar named 3C 273, the first quasar to be discovered. Its luminosity is more than 1 trillion times that of the Sun. The photo shows a region of space 275,000 light-years across at the distance of this quasar. Spikes of light in the image are artifacts produced by scattering of the quasar's light within the telescope; the quasar lies at the intersection of those spikes. The host galaxy of the quasar is only faintly visible around that intersection because the quasar itself is so much brighter.

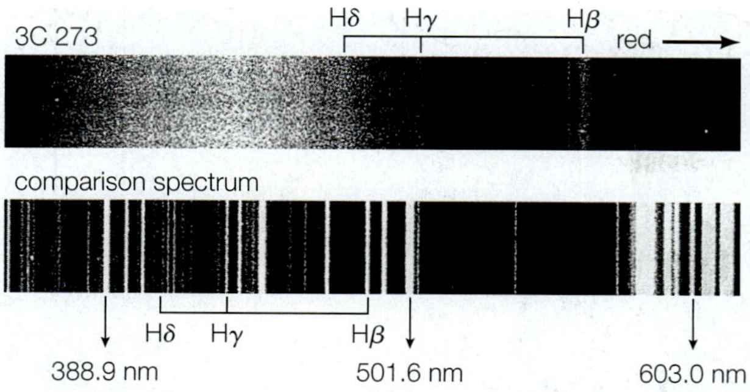


**FIGURE 21.15** These two photographs show a spectrum of the quasar 3C 273 (top) and a comparison spectrum with spectral lines at their rest wavelengths (bottom). The lines labeled H$\beta$, H$\gamma$, and H$\delta$ are hydrogen emission lines. Note their significant redshift in the quasar spectrum relative to the comparison spectrum.

that of the entire Milky Way. These results were so astonishing that, for a while, some scientists feared the distances of quasars had somehow been grossly overestimated. However, observations eventually verified quasar distances and proved that quasars are the bright centers of very distant galaxies.

The fact that most quasars are at very large distances means that quasars were much more common early in the history of the universe than they are now. Apparently, the engines responsible for supplying the awesome luminosities of quasars somehow become dormant as galaxies age. As we will see, this fact helped astronomers realize that activity in the center of a galaxy is linked to the overall evolution of galaxies.

**Sizes of Active Galactic Nuclei** Quasars and other active galactic nuclei puzzled astronomers for decades, not only because of their large luminosities but also because those luminosities are being generated in remarkably small regions of space. Our best visible-light images show these active galactic nuclei to be smaller than 100 light-years across, and higher-resolution radio observations limit their size to no more than about 3 light-years across. Rapid changes in luminosity show them to be even smaller than that.

To understand how variations in luminosity give us clues about an object's size, imagine that you were a master of the universe and you wanted to signal one of your fellow masters a billion light-years away. An active galactic nucleus would make an excellent signal beacon because it is so bright. However, suppose the smallest active galactic nucleus you could find was 1 light-year across. Each time you flashed it on, the photons from the front end would reach your fellow master a full year before the photons from the back end. If you flashed it on and off more than once a year, your signal would be smeared out. Similarly, if you used a source that was 1 light-day across, you could transmit signals that could flash on and off no more than once a day. Using this logic and the fact that the luminosities of active galactic nuclei sometimes double in a matter of hours, we conclude that their light sources must be less than a few light-hours across. In other words, the incredible luminosities of active galactic nuclei and quasars are being generated in a volume of space no bigger than our solar system.

**The Black Hole Hypothesis** The only known way to account for the extreme luminosities of active galactic nuclei is with a model in which the central object is a supermassive black hole surrounded by a swirling accretion disk of very hot gas (**FIGURE 21.16**). This idea is much like the idea we used earlier to explain the emission from x-ray binary star systems [**Section 18.2**]. The gravitational potential energy of matter falling toward a black hole is converted into kinetic energy, and collisions between infalling particles convert kinetic energy into thermal energy. The resulting heat causes this matter to emit the intense radiation we observe. As in x-ray binaries, we expect that the infalling matter swirls through an accretion disk before it disappears within the event horizon of the black hole.

Matter falling into a black hole can generate awesome amounts of energy. As a chunk of matter falls to the event horizon of a black hole, as much as 10–40% of its mass-energy ($E = mc^2$) can be converted into thermal energy and ultimately to radiation (see Mathematical Insight 21.1). (The precise amount within the 10–40% range depends on the black hole's spin.) Accretion by black holes can therefore produce light far more efficiently than nuclear fusion, which converts less than 1% of mass-energy into photons. This mechanism is so efficient that it can account for the enormous luminosity of a quasar if we assume that the black hole swallows the equivalent of about one star per year.

The supermassive black hole model also accounts for the small sizes implied by quasar variability. A black hole accreting a solar mass of matter per year could grow to contain a billion solar masses if accretion continued at that rate for a billion years. But the event horizon of such a massive black hole would have a radius of about 3 light-hours, which would still fit within the orbit of Neptune.

**FIGURE 21.16** Artist's conception of an accretion disk surrounding a supermassive black hole.

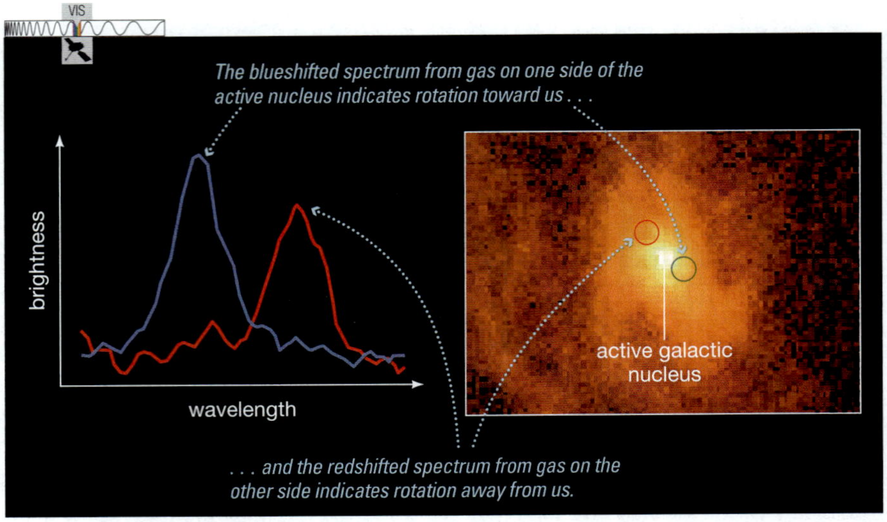


**FIGURE 21.17** This Hubble Space Telescope photo shows gas near the center of the galaxy M87, and the graph shows Doppler shifts of spectra from gas 60 light-years from the center on opposite sides (the circled regions in the photo). The Doppler shifts tell us that gas is orbiting the galactic center at about 800 km/s. From this speed and the orbital velocity law, the central object must have a mass 2–3 billion times that of the Sun.

**Evidence from Orbits** The black hole hypothesis can account for the extreme luminosities of active galactic centers, but how can we determine whether it is correct? Black holes themselves do not emit any light, so we must infer their existence from the ways in which they alter their surroundings. We have already seen how the orbits of stars at the center of the Milky Way indicate the presence of a 4-million-solar-mass black hole [**Section 19.4**]. Similar observations provide evidence of supermassive black holes at the centers of many other galaxies.

In fact, observations indicate that supermassive black holes are quite common, and it is even possible that *all* galaxies contain supermassive black holes at their centers. One prominent example is the giant elliptical galaxy M87 (see Figure 21.13): Images and spectra reveal that gas some 60 light-years from M87's center is orbiting an unseen object at a speed of hundreds of kilometers per second (**FIGURE 21.17**). This high-speed orbital motion indicates that the central object has a mass at least 2–3 billion times that of our Sun (see Mathematical Insight 21.2).

## MATHEMATICAL INSIGHT 21.1 Feeding a Black Hole

The fact that 10–40% of the mass-energy of matter falling into a black hole is radiated away as energy allows us to determine how much mass is accreting onto the black hole in an active galactic nucleus. For example, if 10% of the mass-energy is radiated away, then the amount of energy radiated by an infalling mass $m$ is $E = \frac{1}{10}mc^2$. We simply solve for $m$ to find the mass accreting onto the black hole:

$$\text{accreted mass} = m = 10 \times \frac{E}{c^2}$$

If the percentage of the mass converted to energy is larger, then a *smaller* amount of mass is needed. For example, if 20% of the mass becomes radiated energy, then only half as much mass is needed as under the assumption of 10%.

**EXAMPLE:** Consider a quasar with a luminosity of $10^{40}$ watts. How many solar masses of material must the central black hole consume each year if 10% of the mass-energy is radiated away? What if the quasar radiates away 25% of the mass-energy?

**SOLUTION:**

**Step 1 Understand:** The luminosity of $10^{40}$ watts means that the quasar radiates $10^{40}$ joules of energy each second (because 1 watt = 1 joule/s); the black hole must accrete enough mass to account for this energy. We can therefore use the above formula, remembering that 1 joule = 1 kg $\times$ m$^2$/s$^2$, to calculate how much mass the black hole accretes each second. Multiplying by the number of seconds in a year will tell us the annual accretion amount in kilograms, which we can convert to solar masses ($1M_{\text{Sun}} = 2.0 \times 10^{30}$ kg). The formula assumes that only 10% of the mass-energy is radiated away; if 25% were radiated away, then the amount of mass needed to explain the radiation would be only 10/25 = 0.4 times as much.

**Step 2 Solve:** We use the formula found above with the energy $E = 10^{40}\ \text{kg} \times \text{m}^2/\text{s}^2$ and the speed of light $c = 3 \times 10^8$ m/s:

$$m = 10 \times \frac{E}{c^2} = 10 \times \frac{10^{40}\ \dfrac{\text{kg} \times \text{m}^2}{\text{s}^2}}{\left(3 \times 10^8\ \dfrac{\text{m}}{\text{s}}\right)^2} = 1.1 \times 10^{24}\ \text{kg}$$

This is the mass accreted each second, so we find the annual accretion rate by multiplying by the number of seconds in 1 year:

$$1.1 \times 10^{24}\ \frac{\text{kg}}{\text{s}} \times 60\ \frac{\text{s}}{\text{min}} \times 60\ \frac{\text{min}}{\text{hr}} \times 24\ \frac{\text{hr}}{\text{day}} \times 365\ \frac{\text{day}}{\text{yr}}$$
$$= 3.5 \times 10^{31}\ \frac{\text{kg}}{\text{yr}}$$

We now convert from kilograms to solar masses:

$$\frac{3.5 \times 10^{31}\ \text{kg/yr}}{2.0 \times 10^{30}\ \text{kg}/M_{\text{Sun}}} \approx 17M_{\text{Sun}}/\text{yr}$$

**Step 3 Explain:** To account for the quasar's luminosity, its central supermassive black hole must accrete the equivalent of about 17 Suns per year if it converts 10% of the accreted mass into energy that it radiates away. If it radiates away 25% of the mass-energy, then it must accrete about $0.4 \times 17 \approx 7$ solar masses of material each year.

**a** This composite image of M106 shows emission of visible light (yellow), x-rays (blue), radio waves (purple), and long-wavelength infrared light (red). The radio waves show that jets of energetic plasma are shooting out from the center of the galaxy, and the x-rays show that those jets are heating gas above and below the galaxy's disk.

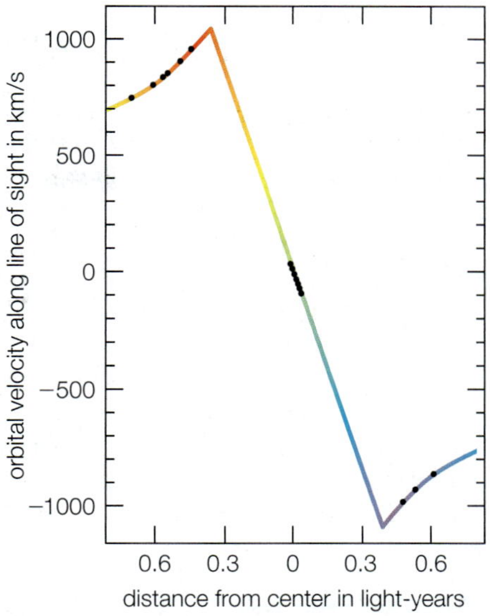


**b** The mass of the central black hole can be calculated (with Newton's version of Kepler's third law) from the orbits of molecular clouds around it. The graph shows cloud velocities measured from their Doppler shifts (dots) and a model curve (color-coded to indicate the sense of the Doppler shift) that assumes a central object with a mass of 36 million Suns; the excellent match shows that this is the mass of the central black hole.

**FIGURE 21.18** The nearby spiral galaxy known as M106 has an active galactic nucleus powered by a supermassive black hole with a mass 36 million times that of our Sun.

A supermassive black hole is the only thing we know of that could be so massive yet so small.

The nearby spiral galaxy M106 (which has an active galactic nucleus) offers even more persuasive evidence. A ring of molecular clouds orbits the nucleus of this galaxy in a circle less than 1 light-year in radius (**FIGURE 21.18**). We can pinpoint these clouds because they amplify the microwave emission lines of water molecules.* The Doppler shifts of these emission lines allow us to determine the orbits of the clouds very precisely. Their orbital motion tells us that the clouds are

*These clouds contain water *masers* that generate beams of microwaves very similar to laser beams. The word *laser* stands for "light amplification by stimulated emission of radiation"; the *m* in *maser* simply substitutes "*m*icrowave" for "light."

## MATHEMATICAL INSIGHT 21.2 Weighing Supermassive Black Holes

We weigh supermassive black holes the same way we weigh almost everything else in the universe: by measuring the velocity $v$ and orbital radius $r$ of the matter circling the central black hole and then applying the orbital velocity law (see Mathematical Insight 19.1) to find the mass $M$ within a distance $r$ of the galactic center:

$$M_r = \frac{r \times v^2}{G}$$

**EXAMPLE:** Doppler shifts show that ionized gas in the nucleus of the active galaxy M87 orbits at a speed of about 800 km/s at a radius of 60 light-years ($5.6 \times 10^{17}$ meters). Calculate the amount of mass that lies within 60 light-years of the galactic center.

**SOLUTION:**

**Step 1 Understand:** We are given both the orbital radius ($r$) and the orbital velocity ($v$) of gas clouds orbiting the center of the galaxy, so we can apply the orbital velocity law to determine the mass that lies within the distance $r$ of the galactic center. For the units to work out properly, we use the orbital radius as given in meters and convert the orbital velocity to meters per second (800 km/s = 800,000 m/s, or $8.0 \times 10^5$ m/s).

**Step 2 Solve:** Substituting the given values into the orbital velocity law, we find

$$\begin{aligned} M_r &= \frac{r \times v^2}{G} \\ &= \frac{(5.6 \times 10^{17}\ \text{m}) \times (8.0 \times 10^5\ \text{m/s})^2}{6.67 \times 10^{-11}\ \text{m}^3/(\text{kg} \times \text{s}^2)} \\ &= 5.4 \times 10^{39}\ \text{kg} \end{aligned}$$

**Step 3 Explain:** The mass that resides within the orbits of the gas clouds is about $5.4 \times 10^{39}$ kg. This mass is easier to interpret if we convert it to solar masses:

$$M_r = 5.4 \times 10^{39}\ \text{kg} \times \left(\frac{1 M_{\text{Sun}}}{2.0 \times 10^{30}\ \text{kg}}\right) = 2.7 \times 10^9 M_{\text{Sun}}$$

The central region of the galaxy has a mass equivalent to that of about 2.7 billion Suns. Presumably, nearly all this mass is in the central black hole.

circling a single invisible object with a mass of 36 million solar masses.

Taken as a whole, the evidence to date strongly favors the hypothesis that supermassive black holes power quasars and other active galactic nuclei. However, we do not yet know how those black holes came to reside at the centers of galaxies. Some scientists have suggested that the black holes formed first out of gas at the centers of protogalactic clouds and that their energy output regulated the growth of the galaxy around them. Other scientists have suggested that clusters of neutron stars resulting from extremely dense starbursts at the centers of young galaxies might have somehow coalesced to form enormous black holes. Whatever the formation process, it must have happened quite early in time, because the first quasars were already active when the universe was less than a billion years old.

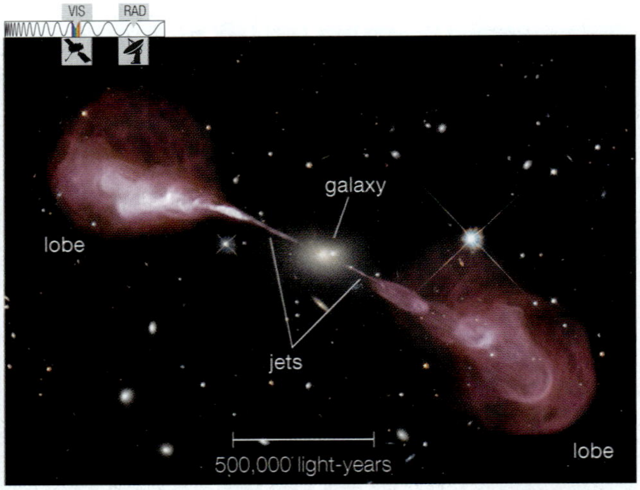


**FIGURE 21.20** This composite image of radio galaxy Hercules A shows visible light in full color and radio waves in red.

## Do supermassive black holes regulate galaxy evolution?

Even though we see most quasars at great distances—when the universe and the galaxies containing the quasars were young—the black holes responsible for the energy output of quasars must still be present at the centers of some of today's galaxies. The difference is that supermassive black holes in the present-day universe tend to be consuming less material and therefore radiate less energy into space. This realization is what led astronomers to search for evidence of supermassive black holes in nearby galaxies. As we've discussed, observations have shown that such black holes exist in many, and perhaps even all, galaxies. More intriguingly, these observations have also revealed the simple relationship shown in **FIGURE 21.19**, in which the mass of a galaxy's central black hole depends directly on the mass of its central bulge. (More specifically, the black hole mass is typically about 0.002 times the mass of the galaxy's bulge.) This relationship appears to hold for galaxies with a wide range of sizes, from small spiral galaxies to giant elliptical galaxies.

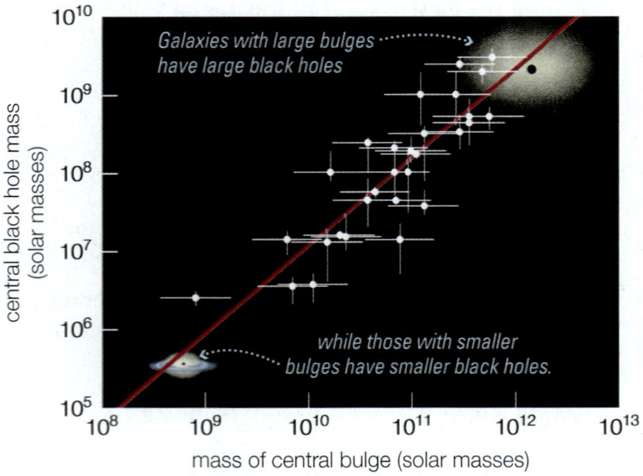


**FIGURE 21.19** The relationship between the mass of a galaxy's bulge and the mass of its supermassive black hole.

The consistency of this relationship is striking and somewhat surprising. After all, a galaxy's bulge occupies an enormous volume of space (typically thousands of light-years across) compared to the central black hole, making it hard to see why the properties of one would be so closely linked to those of the other. However, our studies of starbursts suggest a possible connection: The bulge consists of stars that must have formed in gas clouds, and a central black hole can produce radiation and other forms of activity that can in principle disrupt or eject gas clouds from a galaxy and heat the surrounding gas in much the same way as galactic winds disrupt starburst galaxies. Astronomers do not yet fully understand the link between central black holes and their host galaxies, but mounting evidence indicates that these black holes may play a critical role in regulating galaxy evolution.

**Radio Galaxies** One clue to the relationship between supermassive black holes and galaxy evolution comes from **radio galaxies**, which get their name from their unusually strong radio-wave emission. Modern arrays of radio telescopes (using interferometry [**Section 6.4**]) can resolve the structure of that emission in vivid detail, showing that most of it comes from pairs of huge *radio lobes* on either side of the galaxy (**FIGURE 21.20**).

These radio images show clearly that the plasma in a radio galaxy's lobes has been shot out from the galaxy's center. Apparently, activity around the central supermassive black hole drives two huge jets of matter in opposite directions into space. By comparing radio images taken several years apart, we can track the motions of various plasma blobs in the jets, some of which move at speeds close to the speed of light (**FIGURE 21.21**). Estimates of the amount of energy spewing outward in these jets suggest that they can have a dramatic impact on the hot gas surrounding a galaxy.

**Disruption of the Star–Gas–Star Cycle** The fact that the jets end in radio lobes, rather than continuing outward as narrow streams, tells us that the jets must be running into something, and that something is presumably the hot gas

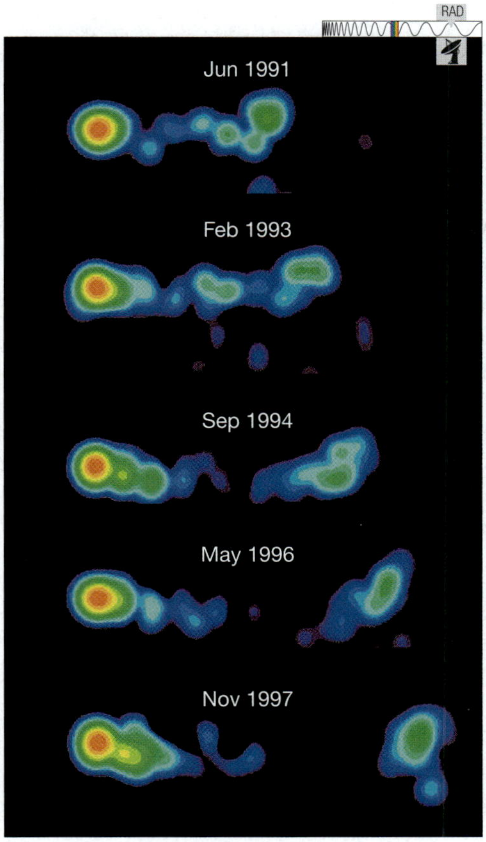


**FIGURE 21.21** These radio images, taken over a period of several years, show a blob of plasma moving at almost the speed of light in a jet extending from the quasar 3C 345; the quasar is on the left in each image.

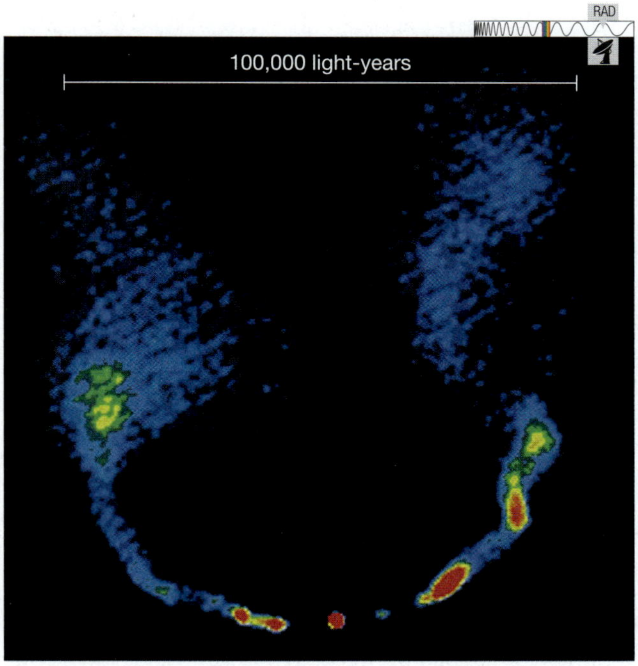


**FIGURE 21.22** Radio galaxy NGC 1265. The lobes are swept back because the galaxy is moving (in the direction of the bottom of the page) relative to the surrounding intergalactic gas.

thought to surround galaxies. In some cases we even see radio galaxies with bent jets and swept-back lobes, indicating that those galaxies are traveling through the surrounding hot intergalactic gas (**FIGURE 21.22**).

The energy of the plasma jets adds further heat to the hot gas around galaxies, which can prevent the hot gas from cooling. Although it is usually very difficult to detect the hot gas surrounding galaxies, the jets sometimes make the gas hot enough to emit detectable x-rays. Observations with x-ray telescopes confirm that the jets are indeed interacting strongly with the surrounding gas and depositing large amounts of heat into it (**FIGURE 21.23**).

**a** A composite image of the radio galaxy Cygnus A showing visible light (yellow), radio waves (red), and x-rays (blue). The bright spot at the center is visible light from the galaxy itself. Two long jets of energetic particles revealed by the radio-wave image are shooting in opposite directions from the active galactic nucleus into the hot gas surrounding the galaxy. The x-ray image of the hot gas shows the disturbances the jets are creating as they burrow through it.

**b** A composite image of the radio galaxy Centaurus A showing visible light (full color), x-rays (blue), and radio waves (orange). The galaxy's unusual appearance — like an elliptical galaxy with prominent dark streaks indicating the presence of cold, dusty gas clouds — suggests that it has experienced a merger within the last few hundred million years. The jets emanating from its active galactic nucleus are heating the surrounding gas, particularly on the lower right, where a (blue) "wake" of hot gas can be seen around the (orange) radio lobe.

**FIGURE 21.23** Examples of galaxies around which jets ejected by an active galactic nucleus are heating the surrounding gas.

**Black Holes and Galaxy Evolution** Studies of both radio galaxies and galactic winds are showing that energetic activity within a galaxy can regulate a galaxy's gas supply at distances that can exceed 10 times the extent of the visible stars. In small galaxies the energy input from starbursts can strongly limit the amount of star formation that happens, but in large galaxies the energy input from black holes may be even more important. Today, the leading hypothesis for why the universe's largest galaxies end up at the top of the red sequence in Figure 21.9 starts by suggesting that a central black hole forms very early in a galaxy's history. Sporadic outbursts of energy fueled by accretion of gas onto that black hole can then periodically disrupt the galaxy's supply of star-forming gas and eventually shut the supply off.

This hypothesis can successfully explain the relationship shown in Figure 21.19 between bulge mass and black hole mass, at least in larger galaxies. Furthermore, recent observations suggest that these energetic outbursts may be the critical factor that prevents the largest galaxies from forming even more stars than they already contain. If this interpretation of the observations is correct, large central black holes are *required* in large galaxies, in order to explain why current rates of star formation within them are relatively low.

## 21.4 Gas Beyond the Stars

Galaxies cannot continue forming stars for billions of years without a continuous supply of cold gas. Most of that gas comes either from recycling of hot gas through a galaxy's halo or from gas clouds falling into a galaxy from even greater distances. The hottest gas can sometimes be seen with x-ray telescopes (see Figures 21.18 and 21.23), but observing cooler gas can be much harder. Radio telescopes show that cool gas in the disks of galaxies can extend well beyond where the stars are, but we would like to know the properties of the gas that is falling into those disks from even greater distances, because that is the gas supply needed to sustain star formation.

**Gas Between Galaxies**

### How do we study the gas outside of galaxies?

Although it is difficult to observe cool gas outside of galaxies directly, careful studies of quasars offer another way to study this gas. The most distant quasars inhabit the outskirts of our observable universe. Photons from some of these quasars began journeying to Earth when the universe was less than 1 billion years old. Along the way, these photons have passed through numerous gas clouds in the spaces between galaxies. The spectra of quasars therefore contain valuable information about the properties of those intergalactic gas clouds, encoded within the many absorption lines present in these spectra.

Recall that atoms tend to absorb light at very specific wavelengths, so every time light from a quasar passes through an intergalactic cloud, some of the atoms in the cloud absorb photons and create absorption lines. Because most of the gas in the universe is hydrogen, the most prominent absorption lines are those created by hydrogen atoms. More importantly for quasar spectra, we see these hydrogen lines at different redshifts for gas clouds at different distances. The reason is cosmological redshift [**Section 20.3**], which means that light from more distant clouds in our expanding universe is more highly redshifted. As a result, studies of absorption lines in quasar spectra can tell us what has been happening to the gas outside of and around galaxies over a time period stretching nearly all the way back to when galaxies started to form (**FIGURE 21.24**).

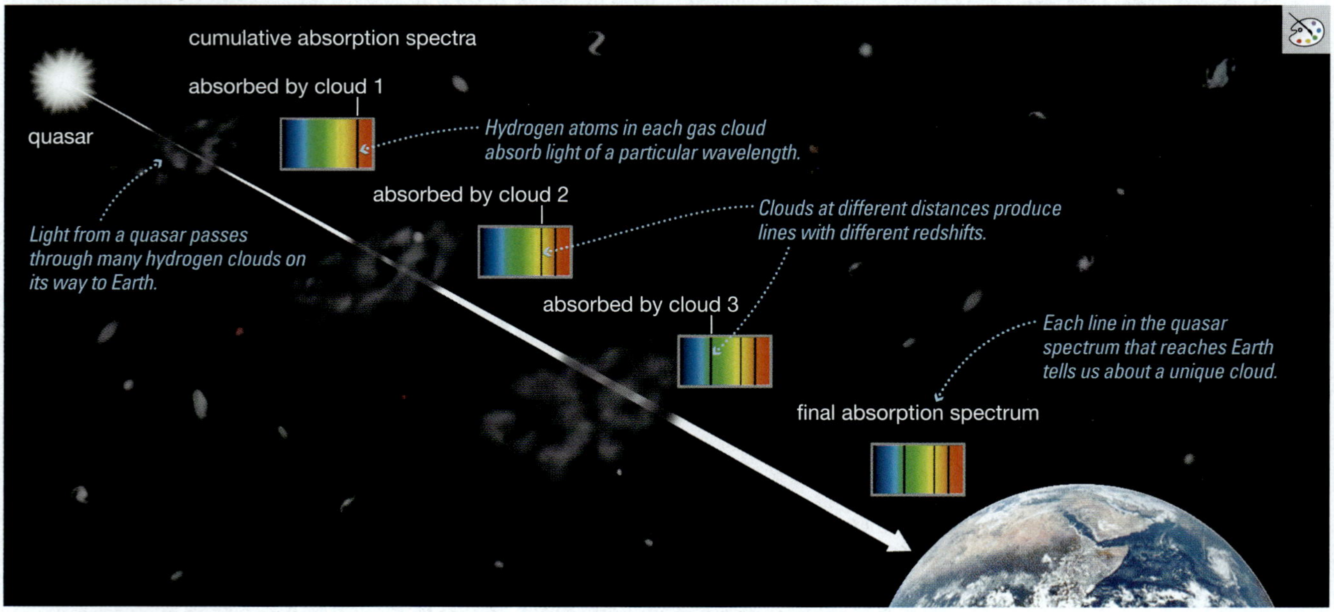


**FIGURE 21.24** This schematic illustration shows how interstellar gas clouds leave their mark on the spectra of quasars. Each hydrogen absorption line in the quasar spectrum tells us about the hydrogen atoms in a unique intergalactic gas cloud between the quasar and Earth. As we study absorption lines in progressively redder parts of the spectrum, we learn about clouds in progressively earlier stages of the universe's development.

Analyses of these absorption lines show that a large majority of the gas in the universe, around 90% of it, lies outside of galaxies. This finding implies that the transformation of cosmic gas into stars within galaxies is remarkably inefficient, probably because of the heat input from galactic winds and energetic outflows from active galactic nuclei. Moreover, the absorption lines in quasar spectra indicate that elements manufactured by stars (that is, elements other than hydrogen and helium) are common in intergalactic clouds within about 1 million light-years of galaxies, suggesting that outflows of gas and energy from galaxies affect large regions around them. As our ability to detect and study intergalactic gas improves, astronomers are hopeful that we will learn much more about both galaxy evolution and the processes through which supermassive black holes regulate star formation in their host galaxies.

## The BIG Picture PUTTING CHAPTER 21 INTO CONTEXT

*We have not yet solved the whole puzzle of galaxy evolution, but in this chapter we have described some of its crucial pieces. As you look back, keep sight of these "big picture" ideas:*

- We can study galaxy evolution by looking back through time: At great distances we see galaxies as they were when the universe was young, and nearby we see mature galaxies as they exist today. These observations, along with theoretical modeling, are helping us understand the lives of galaxies.
- Galaxies probably all begin as systems of protogalactic clouds, but they do not always evolve peacefully. Some galaxies suffer gargantuan collisions with their neighbors, often with dramatic results.
- The tremendous energy outputs of quasars and other active galactic nuclei, including those of radio galaxies, are probably powered by gas accreting onto supermassive black holes. The centers of many present-day galaxies must still contain the supermassive black holes that once enabled them to shine as quasars.

**MY COSMIC PERSPECTIVE** The star–gas–star cycle in the Milky Way is essential to our existence, but observations of the Milky Way alone do not tell us everything we would like to know about how that cycle works. The tools of modern astronomy tell us much more because they show us how that cycle governs star formation in galaxies of many different types and throughout most of cosmic time.

# Summary of Key Concepts

## 21.1 Looking Back Through Time

- **How do we observe the life histories of galaxies?**

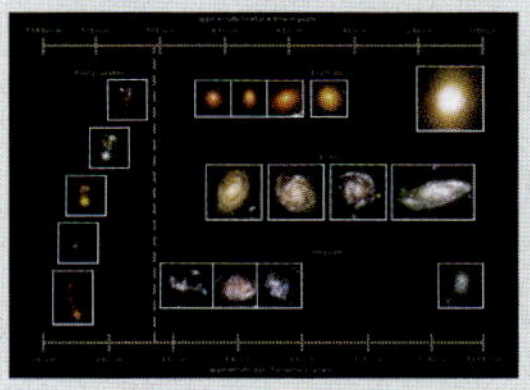

Today's telescopes enable us to observe galaxies of many different ages because they are powerful enough to detect light from objects with lookback times almost as large as the age of the universe. We can therefore assemble "family albums" of galaxies at different distances and lookback times.

- **How do we study galaxy formation?** The most successful models of galaxy formation assume that galaxies formed as gravity pulled together regions of the universe that were ever so slightly denser than their surroundings. Gas collected in protogalactic clouds, and stars began to form as the gas cooled.

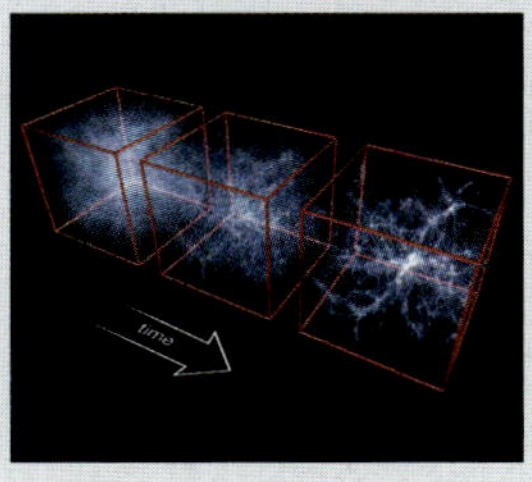


## 21.2 The Lives of Galaxies

- **Why do galaxies differ?** Differences between present-day galaxies arise both from conditions in their protogalactic-cloud systems and from collisions with other galaxies. Slowly rotating or high-density systems of protogalactic clouds may form elliptical rather than spiral galaxies. Ellipticals may also form through the collision and merger of two spiral galaxies.
- **How does gas cycle through galaxies? Starburst galaxies** in which new stars are forming at a very high rate—sometimes more than 100 times the star-formation rate of the Milky Way—show us that star formation is not always steady. A strong starburst can lead to a supernova-driven **galactic wind** that ejects much of a galaxy's gas into its halo, along with large amounts of heat. Such events regulate the star–gas–star cycle in galaxies, because the rate at which galaxies can form stars, when averaged over long periods of time, is limited by the rate at which cooling of gas in the halo feeds cold clouds into a galaxy's disk.

## 21.3 The Role of Supermassive Black Holes

- **What is the evidence for supermassive black holes at the centers of galaxies?** Some galaxies have unusually bright centers known as **active galactic nuclei**; the most luminous of these are known as **quasars**. Quasars are generally found at very great distances, telling us that they were much more common early in the history of the universe. Quasars and other active galactic nuclei are thought to be powered by **supermassive black holes**. As matter falls into a supermassive black hole through an accretion disk, its gravitational potential energy is efficiently transformed into thermal energy and then into light. Observations of orbiting stars and gas clouds in the nuclei of galaxies suggest that *all* galaxies may harbor supermassive black holes at their centers.

- **Do supermassive black holes regulate galaxy evolution?**

The masses we measure for central black holes are closely related to the properties of the galaxy around them, suggesting that the growth of these black holes is closely tied to the process of *galaxy evolution*. Observations of **radio galaxies** show that active galactic nuclei can produce powerful jets of matter that heat the surrounding gas. These outbursts may regulate star formation in galaxies by limiting the rate at which hot gas surrounding the galaxy can cool.

## 21.4 Gas Beyond the Stars

- **How do we study the gas outside of galaxies?** Each cloud of gas through which a quasar's light passes on its long journey to Earth can produce absorption lines in the quasar spectrum. Studies of these absorption lines show that most of the gas in the universe lies outside of galaxies and that galaxies eject many of the heavy elements made by their stars into intergalactic space.

# Visual Skills Check

*Use the following questions to check your understanding of some of the many types of visual information used in astronomy. For additional practice, try the Chapter 21 Visual Quiz in the Study Area at www.MasteringAstronomy.com.*

*In the figure above (repeated from Figure 21.20), the scale bar shows a distance of 500,000 light-years. Use the information in the figure to answer the following questions.*

1. This image combines observations made with visible-light and radio telescopes. Which color in the image represents the radio emission?
2. What is the approximate distance from the far edge of one lobe to the far edge of the other?
3. What is the approximate size of the visible galaxy producing the jets?
4. Suppose we looked at this image without its overlay of radio emission (that is, we looked only at the visible-light image). What features of the image would no longer be visible? Select all that apply.
   a. the background galaxies
   b. the central galaxy
   c. the jets
   d. the lobes

# Exercises and Problems

*For instructor-assigned homework and other study tools, go to www.MasteringAstronomy.com.*

## Chapter Review Questions

### Short-Answer Questions Based on the Reading

1. What do we mean by *galaxy evolution*? How do telescopic observations allow us to study galaxy evolution? How do theoretical models help us study galaxy formation?
2. What are the starting assumptions for models of galaxy formation? Describe how a spiral galaxy is thought to form.
3. Describe two ways in which conditions in a protogalactic-cloud system might lead to the birth of an elliptical rather than a spiral galaxy.
4. What happens when two galaxies collide? How might elliptical galaxies form from collisions of spiral galaxies? What evidence supports the idea that galaxy collisions sometimes occur?
5. Describe some of the consequences of galaxy collisions. Why were collisions more common in the past?
6. What is a *starburst galaxy*? Explain how we know that starbursts must be only temporary stages in a galaxy's life.
7. What is a *galactic wind*? What causes it? How is it similar to a superbubble in the Milky Way, and how is it different?
8. Define *active galactic nuclei* and *quasars*. How can we use variations in luminosity to set limits on the sizes of their emitting regions?
9. What evidence suggests that supermassive black holes really exist?
10. How are the masses of supermassive black holes related to the masses of the bulges of their surrounding galaxies? What does this suggest about the role of supermassive black holes in galaxy evolution?
11. What is a *radio galaxy*? How can radio galaxies affect the gas surrounding them?
12. Briefly describe how studies of absorption lines in quasar spectra allow us to study the gas outside of galaxies. Why is that gas important to galaxy evolution?

### Does It Make Sense?

*Decide whether or not each of the following statements makes sense (or is clearly true or false). Explain clearly; not all of these have definitive answers, so your explanation is more important than your chosen answer.*

13. Galaxies that are more than 10 billion years old are too far away for us to see even with our most powerful telescopes.
14. Elliptical galaxies are more likely to form in denser regions of space.
15. If the Andromeda Galaxy someday collides and merges with the Milky Way, the resulting galaxy may be elliptical.
16. NGC 9645 is a starburst galaxy that has been forming stars at the same furious pace for 10 billion years.
17. The energy from supernova explosions can drive a large proportion of the interstellar gas out of a small galaxy.
18. Astronomers proved that quasar 3C 473 contains a supermassive black hole when they discovered that its center is completely dark.
19. The black hole at the center of our own galaxy may once have powered an active galactic nucleus.
20. Radio galaxies emit only radio waves and no visible light.
21. Activity around a galaxy's central black hole can lead to heating and disruption of gas more than a million light-years from its center.
22. Quasar spectra can tell us about intergalactic gas clouds that might otherwise remain invisible.

### Quick Quiz

*Choose the best answer to each of the following. For additional practice, try the Chapter 21 Reading and Concept Quizzes in the Study Area at www.MasteringAstronomy.com.*

23. When we observe a distant galaxy whose photons have traveled for 10 billion years before reaching Earth, we are seeing that galaxy as it was when the universe was about (a) 10 billion years old. (b) 7 billion years old. (c) 4 billion years old.
24. Which of these statements is a key assumption in our most successful models for galaxy formation? (a) The distribution of matter was perfectly uniform early in time. (b) Some regions of the universe were slightly denser than others. (c) Galaxies formed around supermassive black holes.
25. A collision between two large spiral galaxies is likely to produce (a) a large elliptical galaxy. (b) a large spiral galaxy. (c) one large spiral galaxy and one large elliptical galaxy.
26. A collision and merger of two large elliptical galaxies will eventually produce (a) a large elliptical galaxy. (b) a large spiral galaxy. (c) a large irregular galaxy.
27. Starburst galaxies are especially bright in (a) visible light. (b) ultraviolet light. (c) infrared light.
28. The rate at which supernovae explode in a starburst galaxy that is forming stars 10 times as fast as the Milky Way is (a) about the same as the rate in the Milky Way. (b) about 10 times the rate in the Milky Way. (c) about 0.1 times the rate in the Milky Way.
29. The luminosity of a quasar is generated in a region the size of (a) the Milky Way. (b) a star cluster. (c) the solar system.
30. The primary source of a quasar's energy is (a) chemical energy. (b) nuclear energy. (c) gravitational potential energy.
31. Observations indicate that galaxies with more massive central black holes tend also to have (a) a greater mass of stars in their central bulges. (b) a greater overall luminosity. (c) a more elliptical shape.
32. Which of the following quasars would you expect to have the largest number of hydrogen absorption lines in its spectrum? (a) a quasar with a lookback time of 1 billion years (b) a quasar with a lookback time of 8 billion years (c) a quasar with a lookback time of 13 billion years

## Inclusive Astronomy

*Use these questions to reflect on participation in science.*

33. *Group Discussion: Who Gets to Use JWST?* As this book goes to press, the James Webb Space Telescope (JWST) is scheduled for launch in 2021. It is a joint project of NASA, the European Space Agency, and the Canadian Space Agency. Once in space, JWST will be the world's premier facility for many astronomical observations, including those relating to galaxy evolution. Demand for telescope time will be

enormous, and the people in charge of the proposal process will want to maximize the telescope's scientific productivity while simultaneously minimizing the impact of cultural biases on who gets access.

**a.** Working independently, learn about the process of submitting proposals for observations with JWST. Then gather into small groups to share what you've learned.

**b.** Imagine that your group were in charge of reading the proposals and deciding who should be awarded telescope time. What kinds of people would you consult to determine which proposals had the greatest scientific merit and whether those proposals were feasible, given the technological capabilities of JWST?

**c.** Do you think there is a risk that the judgments of the people you consult might be culturally biased in any way? Make a list of potential cultural biases that could in principle affect their judgments.

**d.** Overall, how would your group structure the process to minimize the effects of cultural bias? Do you think your process would be foolproof, or might there still be biases?

## The Process of Science

*These questions may be answered individually in short-essay form or discussed in groups, except where identified as group-only.*

34. *The Case for Supermassive Black Holes.* The case for supermassive black holes at the centers of galaxies is strong, but it is based on evidence from orbiting objects, because the black holes themselves emit no light. How compelling do you find this evidence? Would you say that astronomers have proved the case for these black holes "beyond a reasonable doubt"? Defend your opinion.
35. *Black Holes and Galaxy Evolution.* Astronomers do not yet know the cause of the relationship between the mass of a galaxy's supermassive black hole and the total mass of stars in its bulge (see Figure 21.19), but this chapter discussed the hypothesis that it arises because energetic outbursts from the black hole periodically disrupt a galaxy's star–gas–star cycle. Suggest possible observations that could be made to test this hypothesis.
36. *Unanswered Questions.* Briefly describe one important but unanswered question related to galaxy evolution. If you think it will be possible to answer that question in the future, describe how we might find an answer, being as specific as possible about the evidence necessary. If you think the question will never be answered, explain why you think it is impossible to answer.
37. *Group Exercise: Testing Galaxy Evolution.* Imagine that astronomers discovered galaxies with the following characteristics; note that these are *not* real discoveries:
    - an extremely large elliptical galaxy with a lookback time of 13 billion years
    - a small irregular galaxy at a distance of 10 million light-years from Earth
    - a large spiral galaxy, 20 million light-years from Earth, consisting entirely of stars made only of hydrogen and helium
    - a small spiral galaxy, 5 million light-years from Earth, with one of the most massive black holes ever discovered at its center
    - a spiral galaxy approximately the size of the Milky Way with a lookback time of 13 billion years

    Work in small groups to decide whether each discovery would be easy or difficult to explain in terms of our current ideas about how galaxies form and evolve. After discussing all of the hypothetical discoveries, rank them in order from easiest to explain to most difficult to explain, and give your reasons. Note: You may wish to do this activity using the four roles described in Chapter 1, Exericise 39.

## Investigate Further

### Short-Answer/Essay Questions

38. *Life Story of a Spiral.* Imagine that you are a spiral galaxy. Describe your life history from birth to the present day. Your story should be detailed and scientifically consistent, but also creative. That is, it should be entertaining while at the same time incorporating current scientific ideas about the formation of spiral galaxies.
39. *Life Story of an Elliptical.* Imagine that you are an elliptical galaxy. Describe your life history from birth to the present. There are several possible scenarios for the formation of elliptical galaxies, so choose one and stick to it. Be creative while also incorporating scientific ideas that demonstrate your understanding.
40. *The Color of an Elliptical Galaxy.* Explain how the color of an old elliptical galaxy has changed during the last 10 billion years. How might you be able to use the galaxy's color to determine when it formed?
41. *Very Early Collisions.* Suppose two protogalactic clouds collide while they are still made primarily of gas, before forming many stars. Could the resulting system end up as a spiral galaxy? Explain your reasoning.
42. *Life in Colliding Galaxies.* Suppose the Milky Way were currently undergoing a collision with another large spiral galaxy. Do you think this collision would affect life on Earth? Why or why not? How would the night sky look if our galaxy were in the midst of such a collision?
43. *A Small-Scale Starburst.* The Large Magellanic Cloud, a small companion galaxy to the Milky Way, is currently undergoing a small-scale starburst. Inspect the picture of the Large Magellanic Cloud in Figure 20.7a for evidence of widespread star formation. Describe the evidence you find.
44. *Orbits Around Supermassive Black Holes.* The data in Figure 21.18b show orbital velocities of gas clouds at distances of 0.4 to 0.7 light-year from the center of the galaxy M106, derived from the Doppler shifts of emission lines. Suppose you observed emission lines from gas clouds orbiting more than 1 light-year from the center. How would you expect the Doppler shifts of those lines to be different, assuming that the gas really is orbiting a supermassive black hole?
45. *Absorption Lines in Quasar Spectra.* Based on your understanding of galaxy evolution, what patterns would you expect to see among the redshifts of absorption lines from elements other than hydrogen? Would you expect those absorption lines to have redshifts similar to those of galaxies along the line of sight to the quasar, or would their redshifts be evenly distributed throughout intergalactic space? Would you expect to see any absorption lines with redshifts greater than that of the quasar? Explain your reasoning.
46. *Research: Greatest Redshift.* As of summer 2018, the most distant quasar known had a redshift of $z = 7.54$, meaning that the wavelengths of its spectral lines are $1 + 7.54 = 8.54$ times their rest wavelengths. Find the current record holder

for the largest redshift. Write a few paragraphs describing the object and its discovery.

## Quantitative Problems

*Be sure to show all calculations clearly and state your final answers in complete sentences.*

47. *Distances Between Galaxies.* If you were to divide the present-day universe into cubes with sides 10 million light-years long, each cube would contain, on average, about one galaxy similar in size to the Milky Way. Now suppose you traveled back in time, to an era when the average distance between galaxies was one quarter of its current value, corresponding to a cosmological redshift of $z = 3$. How many galaxies similar in size to the Milky Way would you expect to find, on average, in cubes of that same size? In order to simplify the problem, assume that the total number of galaxies of each type has not changed. Based on your answer, would you expect collisions to be much more frequent at that time or only moderately more frequent?
48. *A Nearby Starburst.* The galaxy M82, shown in Figure 21.11, is one of the nearest starburst galaxies. Even though it is considerably smaller than the Milky Way, it is currently forming stars more quickly, converting approximately $10M_{\text{Sun}}$ of gas into stars each year. The total amount of gas in its interstellar medium is currently about $10^9 M_{\text{Sun}}$. About how much longer can the starburst in M82 last? Based on your answer, describe how the appearance of M82 will change with time.
49. *Your Last Hurrah.* Suppose you fall into an accretion disk that sweeps you into a supermassive black hole. On your way down, the disk radiates 10% of your mass-energy, $E = mc^2$.
    **a.** What is your mass in kilograms? (Recall that 1 kg = 2.2 lb on Earth.) Calculate how much radiative energy will be produced by the accretion disk as a result of your fall into the black hole.
    **b.** Calculate approximately how long a 100-watt light bulb would have to shine to radiate this same amount of energy.
50. *The Black Hole in M106.* The molecular clouds circling the active nucleus of the galaxy M106 orbit at a speed of about 1000 km/s, with an orbital radius of 0.49 light-year $= 4.8 \times 10^{15}$ meters. Use the orbital velocity law (see Mathematical Insight 21.2) to calculate the mass of the central black hole. Give your answer both in kilograms and in solar masses ($1M_{\text{Sun}} = 2.0 \times 10^{30}$ kg).
51. *The Black Hole in M31.* Measurements of star motions at the center of the Andromeda Galaxy, also known as M31, show that stars about 3 light-years from the center are orbiting at a speed of 400 km/s. These stars are suspected to be orbiting a supermassive black hole. Use the orbital velocity law from Mathematical Insight 21.2 to estimate the mass of this black hole.
52. *Building a Supermassive Black Hole.* The black hole in the galaxy M87 has a mass of about 3 billion solar masses. Let us assume that most of that mass flowed into the black hole through an accretion disk that radiated 10% of the mass-energy passing through it. In that case, what would be the total amount of energy radiated by the accretion disk during the history of the black hole? What would be the average luminosity of the accretion disk, if it continuously radiated that energy over a period of 10 billion years? How does that average luminosity compare with the luminosity of the Milky Way?
53. *Stars and Clouds in Colliding Galaxies.* When two spiral galaxies collide, the stars generally do not run into each other, but the gas clouds do collide, triggering a burst of new star formation.
    **a.** Estimate the probability that our Sun would collide with another star in the Andromeda Galaxy if a collision between the Milky Way and the Andromeda Galaxy were happening at the present time. To simplify the problem, assume that each galaxy has 100 billion stars exactly like the Sun spread evenly over a circular disk with a radius of 100,000 light-years. (*Hint:* First calculate the total area of 100 billion circles with the radius of the Sun and then compare that total area to the area of the galactic disk.)
    **b.** Estimate the probability that a gas cloud in our galaxy would collide with another gas cloud in the Andromeda Galaxy. To simplify the problem, assume that each galaxy contains 100,000 clouds of warm hydrogen gas, that each cloud has a radius of 300 light-years, and that these clouds are spread evenly over a circular disk with a radius of 100,000 light-years. (*Hint:* Use the same method as in part a.)
    **c.** Compare the probabilities you found in parts a and b and use them to explain what you see in Figure 21.5.

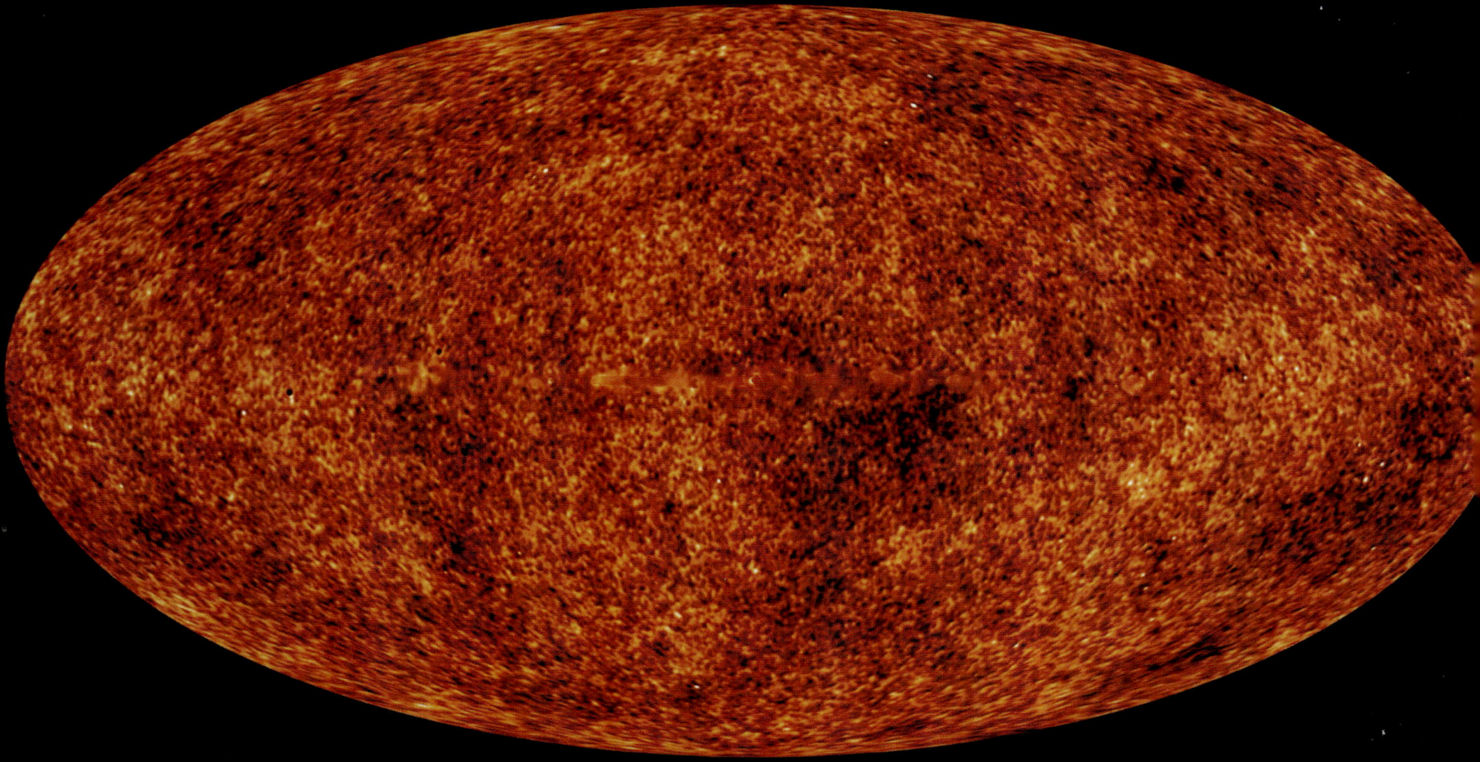

# 22 The Birth of the Universe

▲ **About the photo:** The Big Bang theory is supported by several key lines of evidence, including the existence and characteristics of the cosmic microwave background, shown here in this all-sky image from the *Planck* spacecraft.

## LEARNING GOALS

**22.1 The Big Bang Theory**
- What were conditions like in the early universe?
- How did the early universe change with time?

**22.2 Evidence for the Big Bang**
- How do observations of the cosmic microwave background support the Big Bang theory?
- How do the abundances of elements support the Big Bang theory?

**22.3 The Big Bang and Inflation**
- What key features of the universe are explained by inflation?
- Did inflation really occur?

**22.4 Observing the Big Bang for Yourself**
- Why is the darkness of the night sky evidence for the Big Bang?

*Studying how the universe began, there's something magical about pushing back the frontiers of knowledge. That's something that is almost part of being human, and I'm certain that will continue.*

— *Sally Ride (1951–2012), physicist and first American woman in space*

**Chapter 22 Overview**

Throughout this book, we have studied how the matter produced in the early universe gradually assembled to make galaxies, stars, and planets. However, we have not yet answered one big question: Where did *matter itself* come from?

To answer this question, we must go beyond the most distant galaxies and even beyond what we can see near the horizon of the universe. We must go back not just to the origins of matter and energy but all the way back to the beginning of time itself. As we will see, while many questions about the birth of the universe remain unanswered, we now seem to have some understanding of events that must have unfolded as far back as the first fraction of a second after the Big Bang.

## 22.1 The Big Bang Theory

Is it really possible to study the origin of the entire universe? Not long ago, before we were able to measure the size and age of the observable universe, this topic was considered unfit for scientific study. Scientific attitudes began to change with Edwin Hubble's discovery that the universe is expanding, which led to the insight that all things very likely sprang into being at a single moment in time, in an event that we have come to call the *Big Bang*. Today, powerful telescopes allow us to view how galaxies have changed over the past 14 billion years, and at great distances we see young galaxies still in the process of forming [**Section 21.1**]. These observations confirm that the universe has been gradually aging.

Unfortunately, we cannot see back to the very beginning of time. Light from the most distant galaxies shows us what the universe looked like when it was a few hundred million years old. Light from earlier times is more difficult to observe because we are looking back to a time before stars existed. Ultimately, we face an even more fundamental problem. The universe is filled with a faint glow of radiation that appears to be the remnant heat of the Big Bang. This faint glow is light that has traveled freely through space since the universe was about 380,000 years old, which is when the universe first became transparent to light. Before that time, light could not pass freely through the universe, so there is no possibility of seeing light from earlier times. Just as we must rely on theoretical modeling to determine what the Sun is like on the inside, we must use modeling to investigate what the universe was like during its earliest moments.

### What were conditions like in the early universe?

The scientific theory that predicts what the universe was like early in time is called the **Big Bang theory**. It is based on applying known and tested laws of physics to the idea that everything we see today began as an incredibly hot and dense collection of matter and radiation. The Big Bang theory successfully describes how expansion and cooling of this unimaginably intense mixture of particles and photons could have led to the present universe of stars and galaxies, and it explains several aspects of today's universe with impressive accuracy. Our main goal in this chapter is to understand the evidence supporting the Big Bang theory, but first we must explore what the theory tells us about the early universe.

Observations demonstrate that the universe is cooling with time as it expands, implying that it was hotter and denser in the past. Calculating exactly how hot and dense the universe must have been when it was more compressed is much like calculating how the temperature and density of gas in a balloon change when you squeeze it, except that the conditions become much more extreme. **FIGURE 22.1** shows how the temperature of the universe has changed with time, according to such calculations. Note that for most of the universe's history, even back to times just minutes after the Big Bang, conditions were no more extreme than those found in many places in the universe today, such as in the interiors of stars, and therefore can be understood with the same laws of physics that we've applied throughout most of this book. However, at very early times, temperatures were so high that different processes came into play.

**Particle Creation and Annihilation** The universe was so hot during the first few seconds that photons could transform themselves into matter, and vice versa, in accordance with Einstein's formula, $E = mc^2$ [**Section 4.3**]. Reactions that create and destroy matter are now relatively rare in the

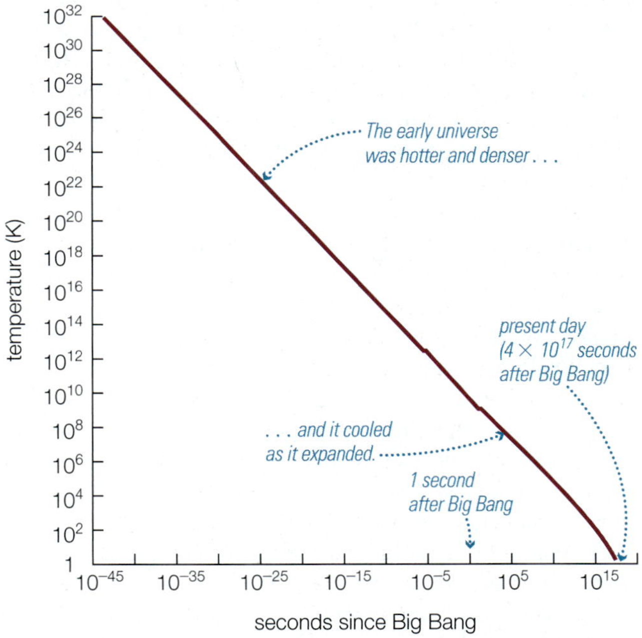


**FIGURE 22.1** This graph shows how the temperature of the universe has cooled with time. Notice that both axis scales use powers of 10, so although most of the graph shows temperatures during the first second of the Big Bang, the far right extends to the present (14 billion years $\approx 4 \times 10^{17}$ s). (The small kinks in the graph at temperatures of $10^{12}$ K and $10^{9}$ K correspond to the moments when the universe became too cold to produce new protons and new electrons, respectively.)

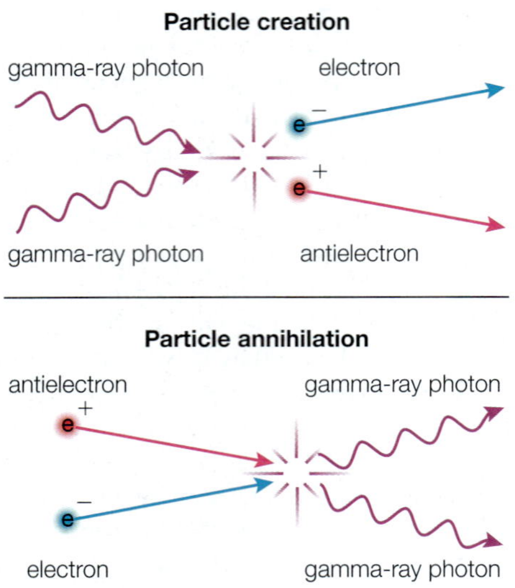


**FIGURE 22.2** Electron-antielectron creation and annihilation. Reactions like these constantly converted photons to particles, and vice versa, in the early universe.

universe at large, but physicists can reproduce many such reactions in particle accelerators such as the Large Hadron Collider (Figure S4.1).

One such reaction is the creation or destruction of an *electron-antielectron pair* (**FIGURE 22.2**). When two photons collide with a total energy greater than twice the mass-energy of an electron (the electron's mass times $c^2$), they can create two brand-new particles: a negatively charged electron and its positively charged twin, the *antielectron* (also known as a *positron*). The electron is a particle of **matter**, and the antielectron is a particle of **antimatter**. The reaction that creates an electron-antielectron pair also runs in reverse. When an electron and an antielectron meet, they *annihilate* each other totally, transforming all their mass-energy back into photon energy. In order to conserve both energy and momentum, an annihilation reaction must produce two photons instead of just one.

Similar reactions can produce or destroy any particle-antiparticle pair, such as a proton and an antiproton or a neutron and an antineutron. The early universe therefore was filled with an extremely hot and dense blend of photons, matter, and antimatter, converting furiously back and forth. Despite all these vigorous reactions, describing conditions in the early universe is straightforward, at least in principle. We simply need to use the laws of physics to calculate the proportions of the various forms of radiation and matter at each moment in the universe's early history. The only difficulty is our incomplete understanding of the laws of physics.

To date, physicists have observed the behavior of matter and energy at temperatures as high as those that existed in the universe just *one ten-billionth* ($10^{-10}$) of a second after the Big Bang, giving us confidence that we actually understand what was happening at that early time. Our understanding of physics under the more extreme conditions that prevailed even earlier is less certain, but we have some ideas about what the universe was like when it was a mere $10^{-38}$ second old, and perhaps a glimmer of what it was like at the age of just $10^{-43}$ second. These tiny fractions of a second may seem insignificant, but the early universe went through some enormous changes during these very small amounts of time.

## Fundamental Forces

**Fundamental Forces** To understand the changes that occurred in the early universe, it helps to think in terms of *forces*. Everything that happens in the universe today is governed by four distinct forces: *gravity, electromagnetism,* the *strong force,* and the *weak force* [**Section S4.2**]. We have already encountered examples of each of these forces in action.

Gravity is the most familiar of the four forces, providing the "glue" that holds planets, stars, and galaxies together. The electromagnetic force, which depends on the electrical charge of a particle instead of its mass, is far stronger than gravity. It is therefore the dominant force between particles in atoms and molecules, responsible for all chemical and biological reactions. However, the existence of both positive and negative electrical charges causes the electromagnetic force to lose out to gravity on large scales, even though both forces decline with distance by an inverse square law. Most large astronomical objects (such as planets and stars) are electrically neutral overall, making the electromagnetic force unimportant on that scale. Gravity therefore becomes the dominant force for such objects, because more mass always means more gravity.

The strong and weak forces operate only over extremely short distances, making them important within atomic nuclei but not on larger scales. The strong force binds protons and neutrons together in atomic nuclei [**Section 14.2**]. The weak force plays a crucial role in nuclear reactions such as fission and fusion, and it is the only force besides gravity that affects weakly interacting particles such as neutrinos.

Although the four forces behave quite differently from one another, current models of fundamental physics predict that they are just different aspects of a smaller number of more fundamental forces (**FIGURE 22.3**). These models

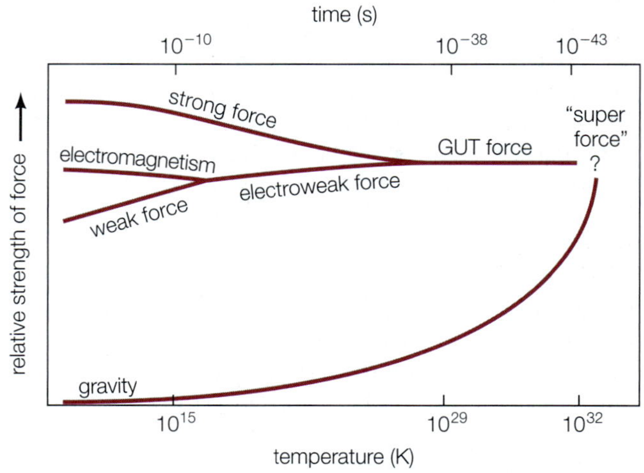


**FIGURE 22.3** The four forces are distinct at low temperatures but may merge at very high temperatures, such as those that prevailed during the first fraction of a second after the Big Bang.

predict that at the high temperatures that prevailed in the early universe, the four forces would not have been as distinct as they are today.

As an analogy, think about ice, liquid water, and water vapor. These three substances are quite different from one another in appearance and behavior, yet they are just different phases of the single substance $H_2O$. Experiments have shown that, in a similar way, the electromagnetic and weak forces lose their separate identities under conditions of very high temperature or energy and merge together into a single **electroweak force**. At even higher temperatures and energies, the electroweak force may merge with the strong force and ultimately with gravity. Models that predict the merger of the electroweak and strong forces are called **grand unified theories**, or **GUTs** for short. The merger of the strong, weak, and electromagnetic forces is therefore often called the *GUT force*. Many physicists suspect that at even higher energies, the GUT force and gravity merge into a single "super force" that governs the behavior of everything. (Current ideas for linking all four forces go by names that include *supersymmetry, superstrings, supergravity*, and "theory of everything.")

If these ideas are correct, then the universe was governed solely by the super force in the first instant after the Big Bang. As the universe expanded and cooled, the super force split into gravity and the GUT force, which then split further into the strong and electroweak forces. Ultimately, all four forces became distinct. As we'll see shortly, these changes in the fundamental forces probably occurred before the universe was one ten-billionth of a second old.

## How did the early universe change with time?

The Big Bang theory uses scientific understanding of particles and forces to reconstruct the history of the universe. Here we will outline this history as a series of *eras*, or time periods. Each era is distinguished from the next by some major change in physical conditions as the universe cools. You'll find it useful to refer to the timeline shown in **FIGURE 22.4** as you read along. Notice that the time scale in Figure 22.4 runs by powers of 10, which means that early eras were very brief even though they appear spread out on the figure. It will take you longer to read this chapter than it took the universe to progress through the first five eras we will discuss, by which point the chemical composition of the early universe had already been determined.

**The Planck Era** The first era after the Big Bang is called the **Planck era**, named for physicist Max Planck. It represents times before the universe was $10^{-43}$ second old. According to the laws of quantum mechanics, there should have been substantial energy fluctuations from point to point during this very early time. Because energy and mass are equivalent, Einstein's theory of general relativity tells us that these energy fluctuations must have generated a rapidly changing gravitational field that would have randomly warped space and time. These fluctuations are predicted to have been so large that our current understanding of physics is inadequate to describe what might have been happening. The main

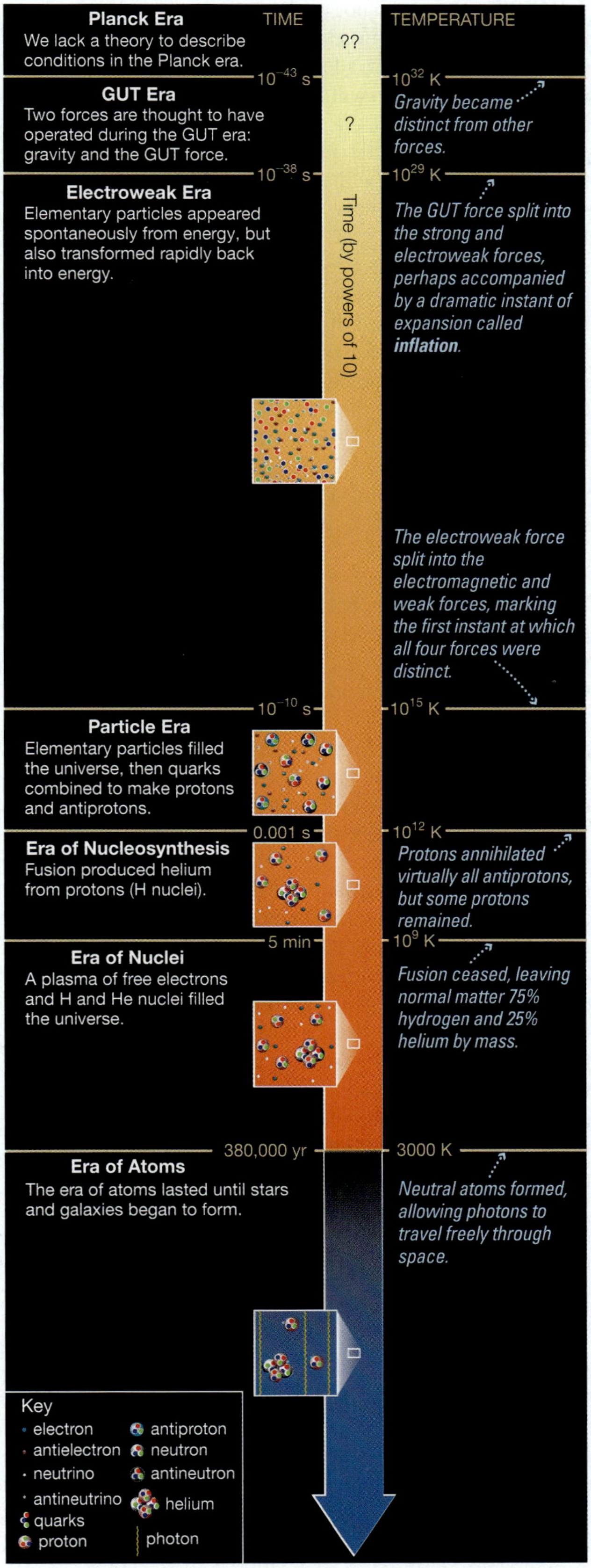


**FIGURE 22.4** A timeline for the eras of the early universe. The only era not shown is the era of galaxies, which began with the birth of stars and galaxies when the universe was a few hundred million years old.

problem is that we do not yet have a theory that links quantum mechanics (our successful theory of the very small) and general relativity (our successful theory of the very big). Perhaps someday we will be able to merge these theories of the very small and the very big into a single "theory of everything" (see Special Topic, p. 454). Until that happens, science cannot describe the universe during the Planck era.

Nevertheless, we have at least some idea of how the Planck era ended. If you look back at Figure 22.3, you'll see that all four forces are thought to merge into the single, unified super force at temperatures above $10^{32}$ K—the temperatures that prevailed during the Planck era. In that case, the Planck era would have been a time of ultimate simplicity, when just a single force operated in nature, and it came to an end when the temperature dropped low enough for gravity to become distinct from the other three forces, which were still merged as the GUT force. By analogy to the way ice crystals form as a liquid cools, we say that gravity "froze out" at the end of the Planck era.

**The GUT Era** The next era is called the **GUT era**, named for the grand unified theories (GUTs) that predict the merger of the strong, weak, and electromagnetic forces into a single GUT force at temperatures above $10^{29}$ K (see Figure 22.3). Although different versions of grand unified theories disagree in many details, they all predict that the GUT era was a time during which two forces—gravity and the GUT force—operated in the universe. It came to an end when the GUT force split into the strong and electroweak forces, which happened when the universe was a mere $10^{-38}$ second old; note that this is less than a trillion-trillion-trillionth of a second.

Our current understanding of physics allows us to say only slightly more about the GUT era than about the Planck era, and none of our ideas about the GUT era have been sufficiently tested to give us great confidence about what occurred during that time. However, if the grand unified theories are correct, the freezing out of the strong and electroweak forces may have released an enormous amount of energy, causing a sudden and dramatic expansion of the universe that we call **inflation**. In a mere $10^{-36}$ second, pieces of the universe the size of an atomic nucleus may have grown to the size of our solar system. Inflation sounds bizarre, but as we will discuss later, it explains several important features of today's universe.

**The Electroweak Era** The splitting of the GUT force marked the beginning of an era during which three distinct forces operated: gravity, the strong force, and the electroweak force. We call this time the **electroweak era**, indicating that the electromagnetic and weak forces were still merged together. Intense radiation continued to fill all of space, as it had since the Planck era, spontaneously producing matter and antimatter particles that almost immediately annihilated each other and turned back into photons.

The universe continued to expand and cool throughout the electroweak era, dropping to a temperature of $10^{15}$ K when it reached an age of $10^{-10}$ second. This temperature is still 100 million times hotter than the temperature in the core of the Sun today, but it was low enough for the electromagnetic and weak forces to freeze out from the electroweak force. After this instant ($10^{-10}$ second), all four forces were forever distinct in the universe.

The end of the electroweak era marks an important transition not only in the physical universe, but also in human understanding of the universe. The theory that unified the weak and electromagnetic forces, which was developed in the 1970s, predicted the emergence of new types of particles (called the W and Z bosons, or *weak bosons*) at temperatures above the $10^{15}$ K temperature that pervaded the universe when it was $10^{-10}$ second old. In 1983, particle-accelerator experiments reached energies equivalent to such high temperatures for the first time. The new particles showed up just as predicted, produced from the extremely high energy in accord with $E = mc^2$. We therefore have direct experimental evidence concerning the conditions in the universe at the end of the electroweak era. We do *not* have any direct experimental evidence of conditions before that time. Our theories concerning the earlier parts of the electroweak era and the GUT era consequently are much more speculative than our theories describing the universe from the end of the electroweak era to the present.

**The Particle Era** As long as the universe was hot enough for the spontaneous creation and annihilation of particles to continue, the total number of particles was roughly in balance with the total number of photons. Once it became too cool for this spontaneous exchange of matter and energy to continue, photons became the dominant form of energy in the universe. We refer to the time between the end of the electroweak era and the moment when spontaneous particle production ceased as the **particle era**, to emphasize the importance of subatomic particles during this period.

During the early parts of the particle era (and during earlier eras), photons turned into all sorts of exotic particles that we no longer find freely existing in the universe today, including *quarks*—the building blocks of protons and neutrons [**Section S4.2**]. By the end of the particle era, all quarks had combined into protons and neutrons, which shared the universe with other particles such as electrons and neutrinos.

The particle era came to an end when the universe reached an age of 1 millisecond (0.001 second), at which point the temperature had fallen to $10^{12}$ K and it was no longer hot enough for protons and antiprotons to be produced spontaneously from photon energy. If the universe had contained equal numbers of protons and antiprotons (or neutrons and antineutrons) at the end of the particle era, all of the pairs would have annihilated each other, creating photons and leaving essentially no matter in the universe. From the obvious fact that the universe contains matter, we conclude that protons must have slightly outnumbered antiprotons at the end of the particle era.

We can estimate the ratio of matter to antimatter by comparing the present numbers of protons and photons in the universe. These two numbers should have been similar in the very early universe, but today photons outnumber protons by about a billion to one. This ratio indicates that for every billion antiprotons in the early universe, there must have been about a billion and one protons. That is, for each 1 billion protons and antiprotons that annihilated each other at the end of the particle era, a single proton was left over. This seemingly slight excess of matter over antimatter accounts for all the ordinary matter in the present-day universe. Some of those protons (and neutrons) left over from when the universe was 0.001 second old are the very ones that make up our bodies.

**The Era of Nucleosynthesis** The eras we have discussed so far all occurred within the first 0.001 second of the universe's existence—less time than it takes you to blink an eye. At this point, the protons and neutrons left over after the annihilation of antimatter began to fuse and make heavier nuclei. However, the temperature of the universe remained so high that gamma rays blasted apart most of those nuclei as fast as they formed. This dance of fusion and demolition marked the **era of nucleosynthesis**, which ended when the universe was about 5 minutes old. By this time, the density in the expanding universe had dropped so much that fusion no longer occurred, even though the temperature was still about a billion Kelvin ($10^9$ K) —much hotter than the temperature of the Sun's core.

When fusion ceased at the end of the era of nucleosynthesis, the chemical content of the universe had become (by mass) about 75% hydrogen and 25% helium, along with trace amounts of deuterium (hydrogen with a neutron) and lithium (the next heaviest element after hydrogen and helium). Except for the small proportion of matter that stars later forged into heavier elements, the chemical composition of the universe remains the same today.

**The Era of Nuclei** After fusion ceased, the universe consisted of a very hot plasma of hydrogen nuclei, helium nuclei, and free electrons. This basic picture held for about the next 380,000 years as the universe continued to expand and cool. The fully ionized nuclei moved independently of electrons (rather than being bound with electrons in neutral atoms) during this period, which we call the **era of nuclei**. Throughout this era, photons bounced rapidly from one electron to the next, just as they do deep inside the Sun today [**Section 14.2**], never managing to travel far between collisions. Any time a nucleus managed to capture an electron to form a complete atom, one of the photons quickly ionized it.

The era of nuclei came to an end when the expanding universe was about 380,000 years old. At this point, the temperature had fallen to about 3000 K—roughly half the temperature of the Sun's surface today. Hydrogen and helium nuclei finally captured electrons for good, forming stable, neutral atoms for the first time. With electrons now bound into atoms, the universe became transparent, as if a thick fog had suddenly lifted. Photons, formerly trapped among the electrons, began to stream freely across the universe. We still see these photons today as the *cosmic microwave background*, which we will discuss shortly.

**The Eras of Atoms and Galaxies** We've already discussed the rest of the universe's history in earlier chapters. The end of the era of nuclei marked the beginning of the **era of atoms**, when the universe consisted of a mixture of neutral atoms and plasma (ions and electrons), along with a large number of photons. Because the density of matter in the universe differed slightly from place to place, gravity slowly drew atoms and plasma into the higher-density regions, which assembled into protogalactic clouds [**Section 21.1**]. Stars then formed in these clouds, and the clouds subsequently merged to form galaxies.

The first full-fledged galaxies had formed by the time the universe was about 1 billion years old, beginning what we call the **era of galaxies**, which continues to this day. Generation after generation of star formation in galaxies steadily builds elements heavier than helium and incorporates them into new star systems. Some of these star systems develop planets, and on at least one of these planets life burst into being a few billion years ago. Now here we are, thinking about it all.

**Early Universe Summary** FIGURE 22.5 summarizes the major ideas from our brief overview of the history of the universe as it is described by the Big Bang theory. In the rest of this chapter, we will discuss the evidence that supports this theory. Before you read on, be sure to study the visual summary presented in Figure 22.5.

## 22.2 Evidence for the Big Bang

Like any scientific theory, the Big Bang theory is a model of nature designed to explain a set of observations. The model was inspired by Edwin Hubble's observation that the universe is expanding, which implies that the universe must have been much denser and hotter in the past. However, like any scientific model, the Big Bang model had to make testable predictions about other observable features of the universe, and it has gained wide scientific acceptance because of two major predictions that have been verified:

- The Big Bang theory predicts that the radiation that began to stream across the universe at the end of the era of nuclei should still be present today. Sure enough, we find that the universe is filled with what we call the **cosmic microwave background**. Its characteristics precisely match what the theory predicts.
- The Big Bang theory predicts that some of the original hydrogen in the universe should have fused into helium during the era of nucleosynthesis. Observations of the actual helium content of the universe closely match the predicted amount of helium.

### How do observations of the cosmic microwave background support the Big Bang theory?

The discovery of the cosmic microwave background was announced in 1965. Arno Penzias and Robert Wilson, two physicists working at Bell Laboratories in New Jersey, were calibrating a sensitive microwave antenna designed for

**COSMIC Context** **FIGURE 22.5**

# The Early Universe

**The Big Bang theory is a scientific model that explains how the present-day universe developed from an extremely hot and dense beginning. This schematic diagram shows how conditions in the early universe changed as the universe expanded and cooled with time.**

(2) As the universe cooled down, it may have undergone a brief period of very rapid expansion known as *inflation* that could account for several key properties of today's universe.

(1) Our expanding universe must have started out much hotter and denser than it is today because the expansion caused matter and energy to cool down and spread out with time.

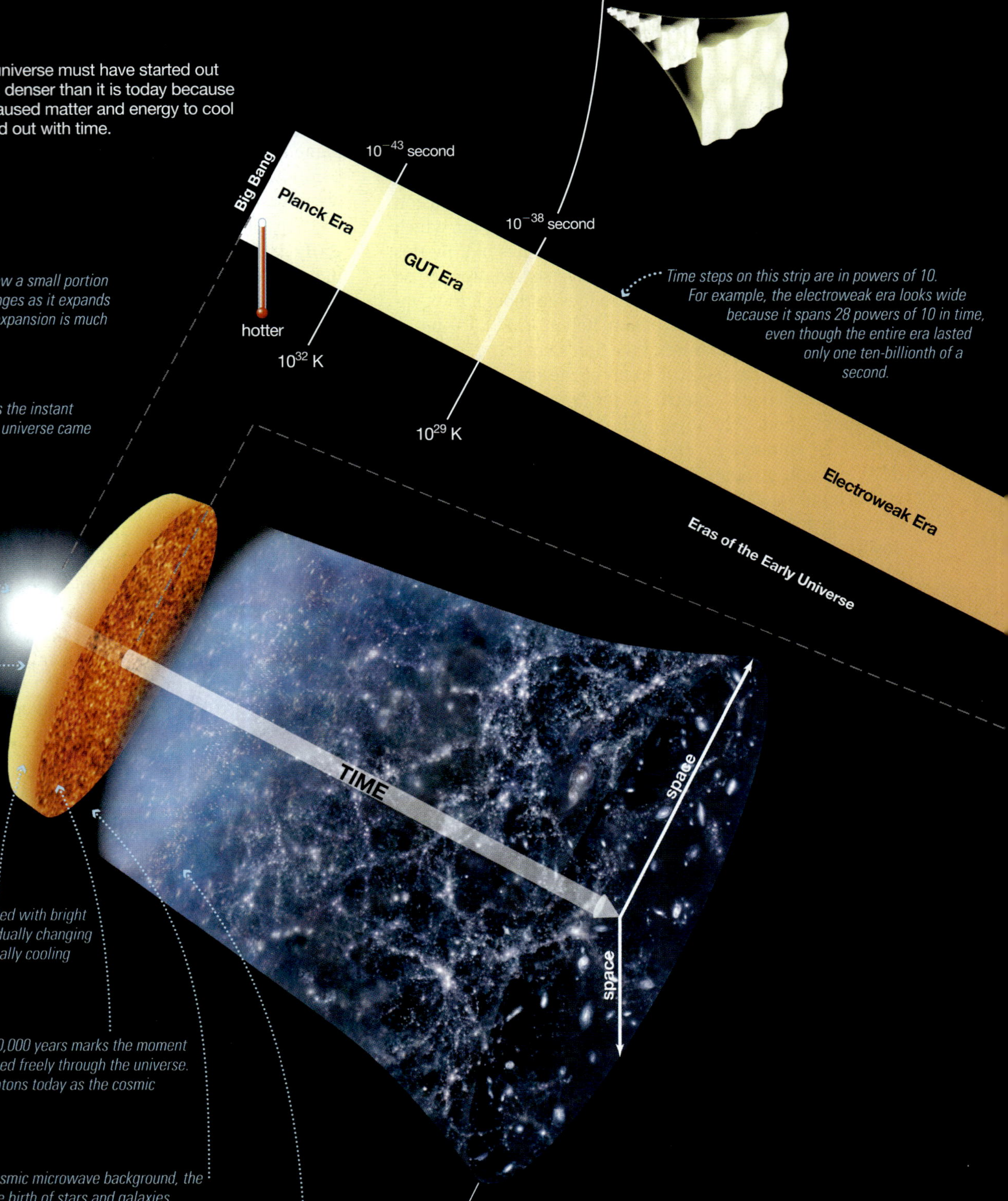


*Time steps on this strip are in powers of 10. For example, the electroweak era looks wide because it spans 28 powers of 10 in time, even though the entire era lasted only one ten-billionth of a second.*

*This illustration depicts how a small portion of the entire universe changes as it expands with time, but the actual expansion is much greater than that shown.*

*This bright spot represents the instant of the Big Bang, when the universe came into existence.*

*This dramatic widening represents inflation—the rapid expansion that may have happened at the end of the GUT era.*

*The early universe was filled with bright light everywhere. The gradually changing color represents the gradually cooling temperature over time.*

*This blotchy surface at 380,000 years marks the moment when photons first streamed freely through the universe. We can still see those photons today as the cosmic microwave background.*

*After the release of the cosmic microwave background, the universe was dark until the birth of stars and galaxies.*

*The era of galaxies was under way by the time the universe was about a billion*

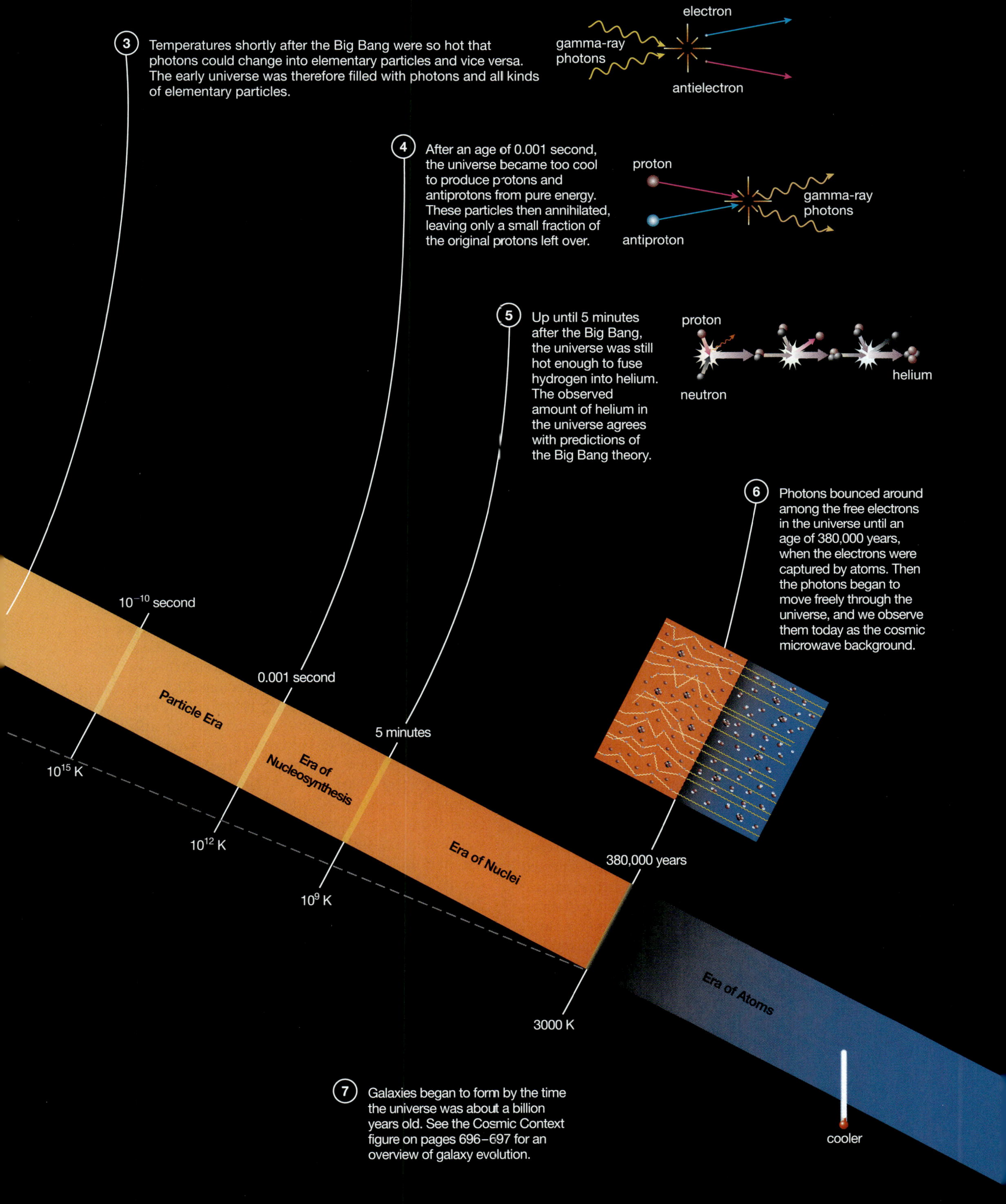

3 Temperatures shortly after the Big Bang were so hot that photons could change into elementary particles and vice versa. The early universe was therefore filled with photons and all kinds of elementary particles.
electron
gamma-ray photons
antielectron
4 After an age of 0.001 second, the universe became too cool to produce p-otons and antiprotons from pure energy. These particles then annihilated, leaving only a small fraction of the original protons left over.
proton
antiproton
gamma-ray photons
5 Up until 5 minutes after the Big Bang, the universe was still hot enough to fuse hydrogen into helium. The observed amount of helium in the universe agrees with predictions of the Big Bang theory.
proton
neutron
helium
6 Photons bounced around among the free electrons in the universe until an age of 380,000 years, when the electrons were captured by atoms. Then the photons began to move freely through the universe, and we observe them today as the cosmic microwave background.
$10^{-10}$ second
Particle Era
0.001 second
Era of Nucleosynthesis
5 minutes
Era of Nuclei
380,000 years
Era of Atoms
$10^{15}$ K
$10^{12}$ K
$10^{9}$ K
3000 K
cooler
7 Galaxies began to form by the time the universe was about a billion years old. See the Cosmic Context figure on pages 696–697 for an overview of galaxy evolution.

**FIGURE 22.6** Arno Penzias and Robert Wilson, discoverers of the cosmic microwave background, with the Bell Labs microwave antenna.

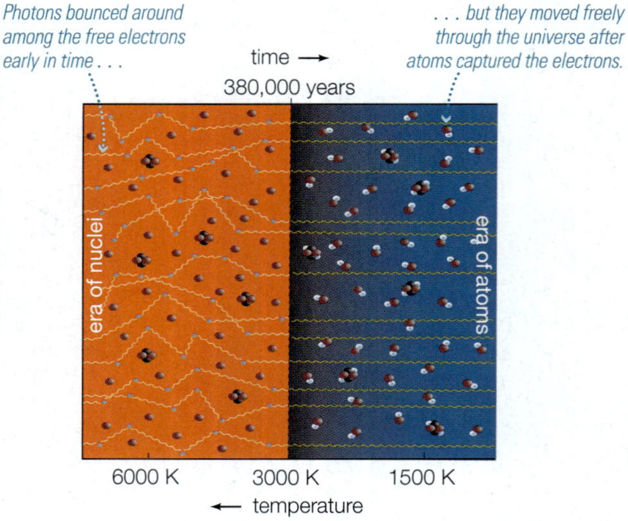


**FIGURE 22.7** Photons (yellow squiggles) frequently collided with free electrons during the era of nuclei and thus could travel freely only after electrons became bound into atoms. This transition was something like the transition from a dense fog to clear air. The photons released at the end of the era of nuclei, when the universe was about 380,000 years old, make up the cosmic microwave background. Precise measurements of these microwaves tell us what the universe was like at this moment in time.

satellite communications (**FIGURE 22.6**). (*Microwaves* fall within the radio portion of the electromagnetic spectrum; see Figure 5.7.) Much to their chagrin, they kept finding unexpected "noise" in every measurement they made. The noise was the same no matter where they pointed the antenna, indicating that it came from all directions in the sky and ruling out any possibility that it came from any particular astronomical object or any place on Earth.

Meanwhile, physicists at nearby Princeton University were busy calculating the expected characteristics of the radiation left over from the heat of the Big Bang, which was first predicted by George Gamow and his colleagues in the 1940s. They concluded that, if the Big Bang had really occurred, this radiation should be permeating the entire universe and should be detectable with a microwave antenna. The Princeton group soon met with Penzias and Wilson to compare notes, and both teams realized that the "noise" from the Bell Labs antenna was the predicted cosmic microwave background—the first strong evidence that the Big Bang had really happened. Penzias and Wilson received the 1978 Nobel Prize in physics for their discovery.

**Origin of the Cosmic Microwave Background** The cosmic microwave background consists of microwave photons that have traveled through space since the end of the era of nuclei, when most of the electrons in the universe joined with nuclei to make neutral atoms, which interact less strongly with photons. With very few free electrons left to block them, most of the photons from that time have traveled unobstructed through the universe ever since (**FIGURE 22.7**). When we observe the cosmic microwave background, we essentially are seeing back to the end of the era of nuclei, when the universe was only 380,000 years old.

**Characteristics of the Cosmic Microwave Background** The Big Bang theory predicts that the cosmic microwave background should have an essentially perfect thermal radiation spectrum [**Section 5.4**], because it came from the heat of the universe itself. Moreover, the theory predicts the approximate wavelength at which this thermal radiation spectrum should peak. As we discussed earlier, the theory tells us that the radiation of the cosmic microwave background broke free when the universe had cooled to a temperature of about 3000 K, similar to the surface temperature of a red giant star. The spectrum of the cosmic microwave background therefore should have originally peaked at a wavelength of about 1000 nanometers, just like the thermal radiation from a red star. Because the universe has since expanded by a factor of about 1000, the wavelengths of these photons should by now have stretched to about 1000 times their original lengths [**Section 20.3**]. We therefore expect the peak wavelength of the cosmic microwave background now to be about a millimeter, squarely in the microwave portion of the spectrum and corresponding to a temperature of a few degrees above absolute zero.

In the early 1990s, a NASA satellite called the *Cosmic Background Explorer (COBE)* was launched to test these ideas about the cosmic microwave background. The results were a stunning success for the Big Bang theory, and earned the 2006 Nobel Prize in physics for *COBE* team leaders George Smoot and John Mather. As shown in **FIGURE 22.8**, the cosmic microwave background does indeed have a perfect thermal radiation spectrum, with a peak corresponding to a temperature of 2.73 K.

**Think about it** Suppose the cosmic microwave background did not really come from the heat of the universe itself but instead came from many individual stars and galaxies. Explain why, in that case, we would not expect it to have a perfect thermal radiation spectrum. How does the spectrum of the cosmic microwave background lend support to the Big Bang theory?

*COBE* and its successor missions, the *Wilkinson Microwave Anisotropy Probe (WMAP)* and the European *Planck*

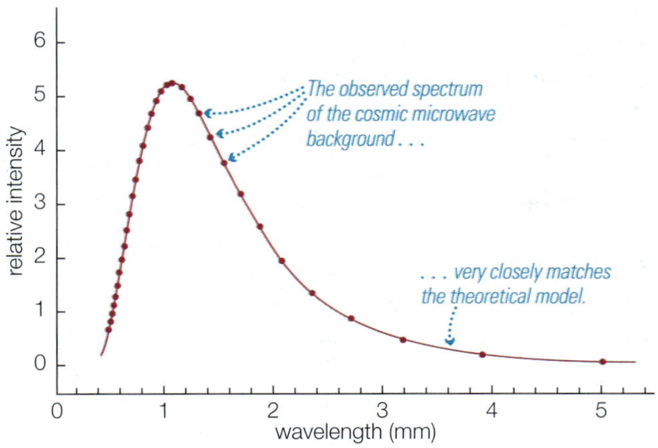


**FIGURE 22.8** This graph shows the spectrum of the cosmic microwave background recorded by NASA's *COBE* satellite. A theoretically calculated thermal radiation spectrum (smooth curve) for a temperature of 2.73 K perfectly fits the data (dots). This excellent fit is important evidence in favor of the Big Bang theory.

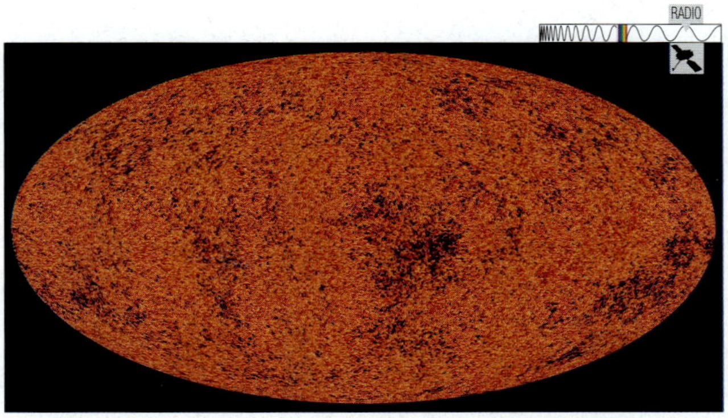


**FIGURE 22.9** This all-sky map shows temperature differences in the cosmic microwave background measured by *Planck*. The background temperature is about 2.73 K everywhere, but the brighter regions of this picture are slightly less than 0.0001 K hotter than the darker regions—indicating that the early universe was very slightly lumpy at the end of the era of nuclei. We are essentially seeing what the universe was like at the surface marked "380,000 years" in Figure 22.5. Gravity later drew matter toward the centers of these lumps, forming the structures we see in the universe today.

satellite, have also mapped the temperature of the cosmic microwave background in all directions (**FIGURE 22.9**). The temperature turns out to be extraordinarily uniform throughout the universe—just as the Big Bang theory predicts it should be—with variations from one place to another of only a few parts in 100,000.* Moreover, these slight variations also represent a predictive success of the Big Bang theory. Recall that our theory of galaxy formation depends on the assumption that the early universe was *not quite* perfectly uniform; some regions of the universe must have started out slightly denser than other regions, so that they could serve as seeds for galaxy formation [**Section 21.1**]. The small variations in the temperature of the cosmic microwave background indicate that the density of the early universe really did differ slightly from place to place.

In fact, detailed observations of these small temperature variations are very important to studies of galaxy evolution, because all large structures in the universe are thought to have formed around the regions of slightly enhanced density [**Section 23.3**]. Measuring the patterns of variations in the cosmic microwave background therefore tells us both about what must have happened at even earlier times to create the variations and about the starting conditions that we should use in models of galaxy evolution.

*Earth's motion (such as our orbit around the Sun and the Sun's orbit around the center of the galaxy) means that we are moving relative to the cosmic microwave background radiation, causing a slight blueshift (about 0.12%) in the direction we're moving and a slight redshift in the opposite direction. Scientists must first subtract these effects before analyzing and making maps of the temperature of the background radiation.

## EXTRAORDINARY CLAIMS The Universe Doesn't Change with Time

Although the Big Bang theory enjoys wide acceptance among scientists today, alternative ideas have been proposed and considered. One of the cleverest alternatives, developed in the late 1940s, was called the *steady state universe hypothesis*. This hypothesis accepted the fact that the universe is expanding but rejected the idea of a Big Bang, instead postulating that the universe is infinitely old and always looks about the same on large scales.

This claim might not have seemed extraordinary before Edwin Hubble discovered that the universe is expanding. However, the claim was made two decades after that discovery and raised the following obvious question: If the universe has been expanding forever, shouldn't every galaxy be infinitely far away from every other galaxy? Proponents of the steady state hypothesis answered by claiming that new galaxies continually form in the gaps that open up as the universe expands, thereby keeping the average distance between galaxies the same at all times. In a sense, the steady state hypothesis said that the creation of the universe is an ongoing and eternal process rather than one that happened all at once with a Big Bang.

Two key discoveries caused the steady state hypothesis to lose favor. First, the 1965 discovery of the cosmic microwave background matched a prediction of the Big Bang theory but was not adequately explained by the steady state hypothesis. Second, a steady state universe should look about the same at all times, but observations made with increasingly powerful telescopes during the last half-century show that galaxies at great distances look younger than nearby galaxies. As a result of these predictive failures, the steady state hypothesis is no longer considered to be viable.

**Verdict:** Rejected.

## How do the abundances of elements support the Big Bang theory?

The Big Bang theory also solves what had previously been another long-standing astronomical problem: the origin of cosmic helium. Everywhere in the universe, about three-quarters of the mass of ordinary matter (not including dark matter) is hydrogen and about one-quarter is helium. The Milky Way's helium fraction is about 28%, and no galaxy has a helium fraction lower than 25%. Although helium is produced by hydrogen fusion in stars, calculations show that this production can account for only a small proportion of the total observed helium. We therefore conclude that the majority of the helium in the universe must already have been present in the protogalactic clouds that preceded the formation of galaxies.

The Big Bang theory makes a specific prediction about the helium abundance. As we discussed earlier, the theory explains the existence of helium as a consequence of fusion that occurred during the era of nucleosynthesis, when the universe itself was hot enough to fuse hydrogen into helium. Combining the current microwave background temperature of 2.73 K with the number of protons we observe in the universe tells us precisely how hot the universe must have been in the distant past, allowing scientists to calculate exactly how much helium should have been made. The result—25% helium—is another impressive success of the Big Bang theory.

**Helium Formation in the Early Universe** To see why 25% of ordinary matter became helium, we need to understand what protons and neutrons were doing during the 5-minute era of nucleosynthesis. Early in this era, when the universe's temperature was $10^{11}$ K, nuclear reactions could convert protons into neutrons, and vice versa. As long as the universe remained hotter than $10^{11}$ K, these reactions kept the numbers of protons and neutrons nearly equal. But as the universe cooled, neutron-proton conversion reactions began to favor protons.

Neutrons are slightly more massive than protons, and therefore reactions that convert protons to neutrons require energy to proceed (in accordance with $E = mc^2$). As the temperature fell below $10^{11}$ K, the energy required for neutron production was no longer readily available, so the rate of these reactions slowed. In contrast, reactions that convert neutrons into protons release energy and therefore are unhindered by cooler temperatures. By the time the temperature of the universe fell to $10^{10}$ K, protons had begun to outnumber neutrons because the conversion reactions ran only in one direction. Neutrons changed into protons, but the protons didn't change back.

For the next few minutes, the universe was still hot and dense enough for nuclear fusion to take place. Protons and neutrons constantly combined to form *deuterium*—the rare form of hydrogen that contains a neutron in addition to a proton in the nucleus—and deuterium nuclei fused to form helium (**FIGURE 22.10**). However, during the early part of the era of nucleosynthesis, the helium nuclei were almost immediately blasted apart by one of the many gamma rays that filled the universe.

Fusion began to create long-lasting helium nuclei when the universe was about 1 minute old and had cooled to a temperature at which it contained few destructive gamma rays. Calculations show that the proton-to-neutron ratio at this time should have been about 7 to 1. Moreover, almost all the available neutrons should have been incorporated into nuclei of helium-4. **FIGURE 22.11** shows that, based on the 7-to-1 ratio of protons to neutrons, the universe should have had a composition of 75% hydrogen and 25% helium by mass at the end of the era of nucleosynthesis. This match between the predicted and observed helium ratios provides strong support for the Big Bang theory.

**Think about it** Briefly explain why it should not be surprising that some galaxies contain a little more than 25% helium, but why it would be very surprising if some galaxies contained less. (*Hint:* Think about how the relative amounts of hydrogen and helium in the universe are affected by fusion in stars.)

**Abundances of Other Light Elements** Why didn't the Big Bang produce heavier elements? By the time stable helium nuclei formed, when the universe was about a minute old,

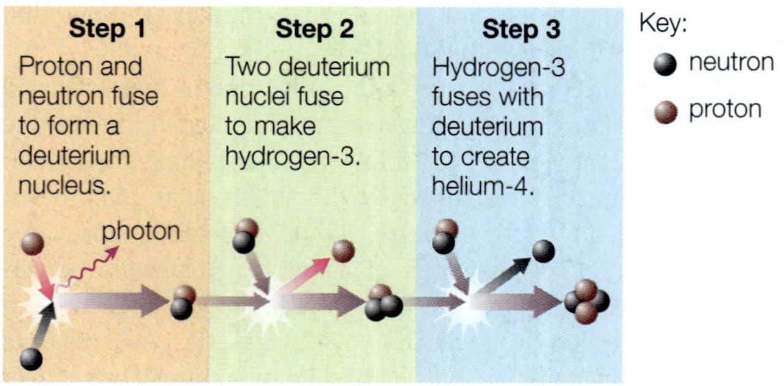


**FIGURE 22.10** During the 5-minute-long era of nucleosynthesis, virtually all the neutrons in the universe fused with protons to form helium-4. This figure illustrates one of several possible reaction pathways.

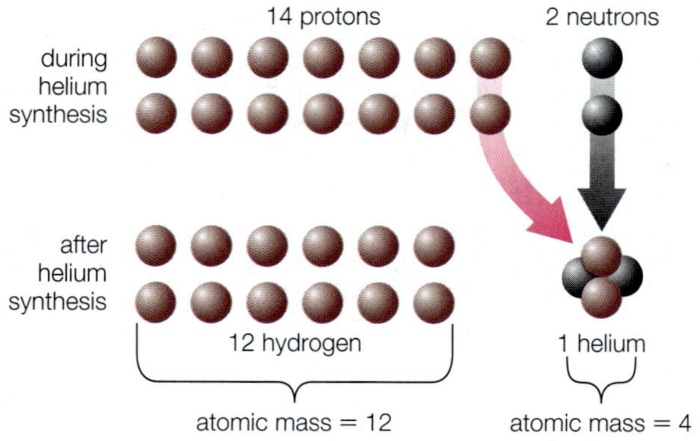


**FIGURE 22.11** Calculations show that protons outnumbered neutrons 7 to 1, which is the same as 14 to 2, during the era of nucleosynthesis. The result was 12 hydrogen nuclei (individual protons) for each helium nucleus. Therefore, the predicted hydrogen-to-helium mass ratio is 12 to 4, which is the same as 75% to 25%, in agreement with the observed abundance of helium.

the temperature and density of the rapidly expanding universe had already dropped too far for a process like carbon production (three helium nuclei fusing into a carbon nucleus [**Section 17.2**]) to occur. Reactions between protons, deuterium nuclei, and helium were still possible, but most of these reactions led nowhere. In particular, fusing two helium-4 nuclei results in a nucleus that is unstable and falls apart in a fraction of a second, as does fusing a proton to a helium-4 nucleus.

A few reactions involving hydrogen-3 (also known as *tritium*) or helium-3 can create long-lasting nuclei. For example, fusing helium-4 and hydrogen-3 produces lithium-7. However, the contributions of these reactions to the overall composition of the universe were minor because hydrogen-3 and helium-3 were so rare. Models of element production in the early universe show that, before the cooling of the universe shut off fusion entirely, such reactions generated only trace amounts of lithium, the next heavier element after helium. Aside from hydrogen, helium, and lithium, all other elements were forged much later in the nuclear furnaces of stars. (Beryllium and boron, which are heavier than lithium but lighter than carbon, were created later when the high-energy particles known as *cosmic rays* broke apart heavier nuclei that formed in stars—see Appendix D.)

## 22.3 The Big Bang and Inflation

When we discussed the eras of the universe earlier in the chapter, we noted that the universe is thought to have undergone a sudden and dramatic expansion, called *inflation*, which may have occurred at the end of the GUT era, when the universe was $10^{-38}$ second old. (In some models of inflation, the dramatic expansion can happen later, up until the end of the electroweak era.) This idea first emerged in 1981, when physicist Alan Guth was considering the consequences of the separation of the strong force

### MATHEMATICAL INSIGHT 22.1 Temperature and Wavelength of Background Radiation

Figure 22.8 shows that the cosmic microwave background has a nearly perfect thermal radiation spectrum for an object at a temperature of 2.73 K. Wien's law (see Mathematical Insight 5.2) therefore tells us that the wavelength of photons at the peak of the spectrum is

$$\lambda_{max} \approx \frac{2{,}900{,}000}{T\ (\text{Kelvin})}\ \text{nm} = \frac{2{,}900{,}000}{2.73}\ \text{nm} = 1.1 \times 10^6\ \text{nm}$$

Because $10^6$ nm = 1 mm, this peak wavelength is equivalent to 1.1 millimeters. But what was the wavelength of the cosmic microwave photons in the past?

From Mathematical Insight 20.5, the universe has grown in size by a factor of $1 + z$ since the time light left objects that we observe to have a redshift $z$. Therefore, we find the peak wavelength of cosmic microwave photons at that time by dividing the current peak wavelength by $1 + z$:

$$\lambda_{max}\ (\text{at redshift } z) \approx \frac{1.1\ \text{mm}}{1 + z}$$

Combining this result with Wien's law and a little algebra, we find a simple formula for the temperature of the universe at any earlier time at which we see objects with redshift $z$:

$$T_{universe}\ (\text{at redshift } z) \approx 2.73\ \text{K} \times (1 + z)$$

**EXAMPLE 1:** Photons first moved freely when the universe had cooled to a temperature of about 3000 K. What was the peak wavelength of the photons at that time?

**SOLUTION:**

**Step 1 Understand:** We can simply use Wien's law relating peak wavelength to temperature.

**Step 2 Solve:** We use the temperature of 3000 K in Wien's law:

$$\lambda_{max} \approx \frac{2{,}900{,}000}{T\ (\text{Kelvin})}\ \text{nm} = \frac{2{,}900{,}000}{3000}\ \text{nm} = 970\ \text{nm}$$

**Step 3 Explain:** The peak wavelength of the photons when they first began to travel freely was about 970 nanometers, which is in the infrared portion of the electromagnetic spectrum fairly close to the wavelength of red visible light (see Figure 5.7).

**EXAMPLE 2:** How much has the expansion of the universe stretched the wavelengths of the background radiation since it began to travel freely through the universe?

**SOLUTION:**

**Step 1 Understand:** We can use the formula that relates the temperature of the background radiation to the cosmological redshift $z$. We are given the 3000 K temperature, so we need to find the stretching factor $(1 + z)$.

**Step 2 Solve:** We divide both sides of the earlier equation by the current temperature of the universe, 2.73 K, to find

$$1 + z = \frac{T_{universe}\ (\text{at redshift } z)}{2.73\ \text{K}}$$

In this case, we are looking for the stretching factor corresponding to the time when the universe had a temperature of 3000 K. Plugging this value into the formula, we find

$$1 + z = \frac{3000\ \text{K}}{2.73\ \text{K}} \approx 1100$$

**Step 3 Explain:** The expansion of the universe has stretched photons by a factor of about 1100 since the time they first began to travel freely across the universe, when the universe was about 380,000 years old. (The answer has no units because it is the *ratio* of the size of the universe now to the size of the universe then.)

from the GUT force that marked the end of the GUT era. Some theories of high-energy physics predict that this separation of forces would have released enormous amounts of energy, and Guth realized that this energy might have caused a short period of inflation. He found that, in a mere $10^{-36}$ second, inflation could have caused the universe to expand by a factor of $10^{30}$. Strange as this idea may sound, it appears to explain several otherwise mysterious features of the present-day universe. Moreover, recent evidence from detailed studies of the cosmic microwave background has provided support for the hypothesis that an early period of inflation really occurred.

## What key features of the universe are explained by inflation?

The Big Bang theory has gained wide acceptance because of the strong evidence from the cosmic microwave background and the abundance of helium in the universe. However, without inflation, the theory leaves several major features of our universe unexplained. The three most pressing questions are the following:

- *Where did the density enhancements that led to galaxies come from?* Recall that successful models of galaxy formation start from the assumption that gravity could collect matter together in regions of the early universe that had slightly enhanced density [**Section 21.1**]. We know from our observations of variations in the cosmic microwave background that such regions of enhanced density were present in the universe at an age of 380,000 years, but we have not yet explained how these density variations came to exist.
- *Why is the large-scale universe nearly uniform?* Although the slight variations in the cosmic microwave background show that the universe is not *perfectly* uniform on large scales, the fact that it is smooth to within a few parts in 100,000 is remarkable enough that we would not expect it to have occurred by pure chance.
- *Why is the geometry of the universe flat?* Einstein's general theory of relativity tells us that the overall geometry of the universe can be curved, like the surface of a balloon or a saddle [**Section S3.2**]. However, observational efforts to measure the large-scale geometry of the universe have not yet detected any curvature. As far as we can tell, the large-scale geometry of the universe is flat, which means it is at the precise balance point between being curved like a balloon and curved like a saddle. This precise balance is another fact that is difficult to attribute to chance.

Taking a scientific approach to the early universe demands that we seek answers to these questions that rely on natural processes, and the hypothesis of inflation provides such answers. That is, if we assume inflation occurred, we find that the density enhancements, large-scale uniformity, and flat geometry are all natural and expected consequences. Note that inflation is a *scientific* hypothesis because it is testable with observations we can perform today, and confidence in the hypothesis is growing because it has passed all the tests it has faced so far.

### Density Enhancements: Giant Quantum Fluctuations

To understand how inflation explains the origin of the density enhancements that led to galaxies, we need to recognize a special feature of energy fields. Laboratory-tested principles of quantum mechanics, especially the uncertainty principle [**Section S4.3**], tell us that on very small scales, the energy fields at any point in space are always fluctuating. The distribution of energy through space is therefore very slightly irregular, even in a complete vacuum. The tiny quantum "ripples" that make up the irregularities can be characterized by a wavelength that corresponds roughly to their size. In principle, quantum ripples in the very early universe could have been the seeds for density enhancements that later grew into galaxies. However, the wavelengths of the original ripples were far too small to explain density enhancements like those we see imprinted on the cosmic microwave background.

Inflation would have dramatically increased the wavelengths of these quantum fluctuations. The rapid growth of the universe during the period of inflation would have stretched tiny ripples from a size smaller than that of an atomic nucleus to the size of our solar system (**FIGURE 22.12**), making them large enough to become the density enhancements from which galaxies and larger structures later formed. If that's the case, then the large-scale structure of today's universe started as tiny quantum fluctuations just before the period of inflation.

### Uniformity: Equalizing Temperatures and Densities

The remarkable uniformity of the cosmic microwave background might at first seem quite natural, but with further thought it becomes difficult to explain. Imagine observing the cosmic microwave background in a certain part of the sky. You are seeing microwaves that have traveled through the universe since the end of the era of nuclei, just 380,000 years after the Big Bang. You are therefore seeing a region

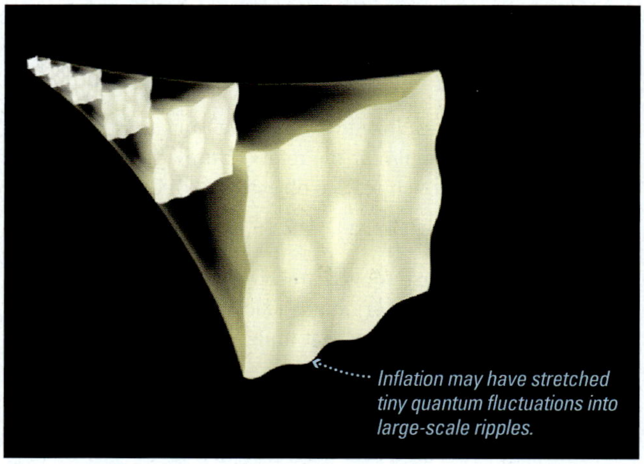


**FIGURE 22.12** During inflation, ripples in spacetime would have stretched by a factor of perhaps $10^{30}$. The peaks of these ripples then would have become the density enhancements that produced all the structure we see in the universe today.

of the universe as it was some 14 billion years ago, when the universe was only 380,000 years old. Now imagine turning around and looking at the background radiation coming from the opposite direction. You are also seeing this region at an age of 380,000 years, and it looks virtually identical in temperature and density. The surprising part is this: The two regions are billions of light-years apart on opposite sides of our observable universe, and we are seeing them as they were when they were only 380,000 years old. They can't possibly have exchanged light or any other information, because a signal traveling at the speed of light has not had time to travel from one region to the other (**FIGURE 22.13**). So how did they come to have the same temperature and density?

The inflation hypothesis answers this question by saying that even though the two regions cannot have had any contact *since* the time of inflation, they were in contact prior to that time. Before the onset of inflation, when the universe was $10^{-38}$ second old, the two regions were less than $10^{-38}$ light-second away from each other. Radiation traveling at the speed of light would therefore have had time to bounce between the two regions, and this exchange of energy equalized their temperatures and densities. Inflation then pushed these equalized regions to much greater distances, far out of contact with each other. Like criminals getting their stories straight before being locked in separate jail cells, the two regions (and all other parts of the observable universe) came to the same temperature and density before inflation spread them far apart.

Because inflation caused different regions of the universe to separate so far in such a short period of time, many people wonder whether it violates Einstein's theories saying that nothing can move faster than the speed of light. It does not, because nothing actually *moves* through space as a result of inflation or the ongoing expansion of the universe. Recall that the expansion of the universe is the expansion of *space itself*. Objects may in some cases be separating from one another at a speed faster than the speed of light, but no matter or radiation is able to travel between them during that time. In essence, inflation opens up a huge gap in space between objects that were once close together. The objects get very far apart, but nothing ever travels between them at a speed that exceeds the speed of light.

**Geometry: Balancing the Universe** The third question asks why the overall geometry of the universe is "flat." To understand this idea, we must consider the geometry of the universe in a little more detail.

Recall that Einstein's general theory of relativity tells us that the presence of matter can curve the structure of spacetime [**Section S3.3**]. We cannot visualize this curvature in all three dimensions of space (or all four dimensions of spacetime), but we can detect its presence by its effects on how light travels through the universe. Although the curvature of the universe can vary from place to place, the universe as a whole must have some overall shape. Almost any shape is possible, but all the possibilities fall into just three general categories (**FIGURE 22.14**). Using analogies to objects that we can see in three dimensions, scientists refer to these three categories of shape as *flat* (or critical), *spherical* (or closed), and *saddle shaped* (or open).

According to general relativity, the overall geometry depends on the average density of matter and energy in the universe, and the geometry can be flat only if the combined density of matter plus energy is precisely equal to a value known as the **critical density**. If the universe's

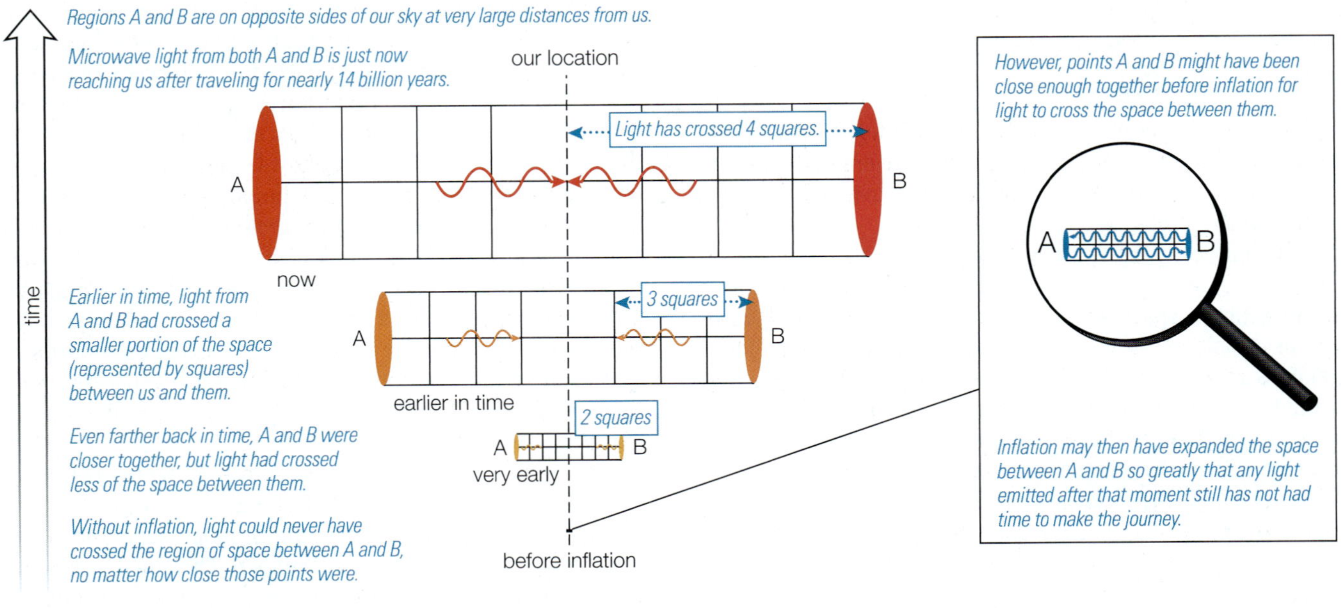


**FIGURE 22.13** This schematic figure shows that, without inflation, regions of space that we observe on opposite sides of the sky would never have been able to exchange light signals, which poses a puzzle: How did the regions come to have the same temperature and density, as indicated by microwave observations? Inflation provides a solution to this puzzle by hypothesizing that A and B were able to exchange light signals until the rapid expansion of inflation pushed them much farther apart.

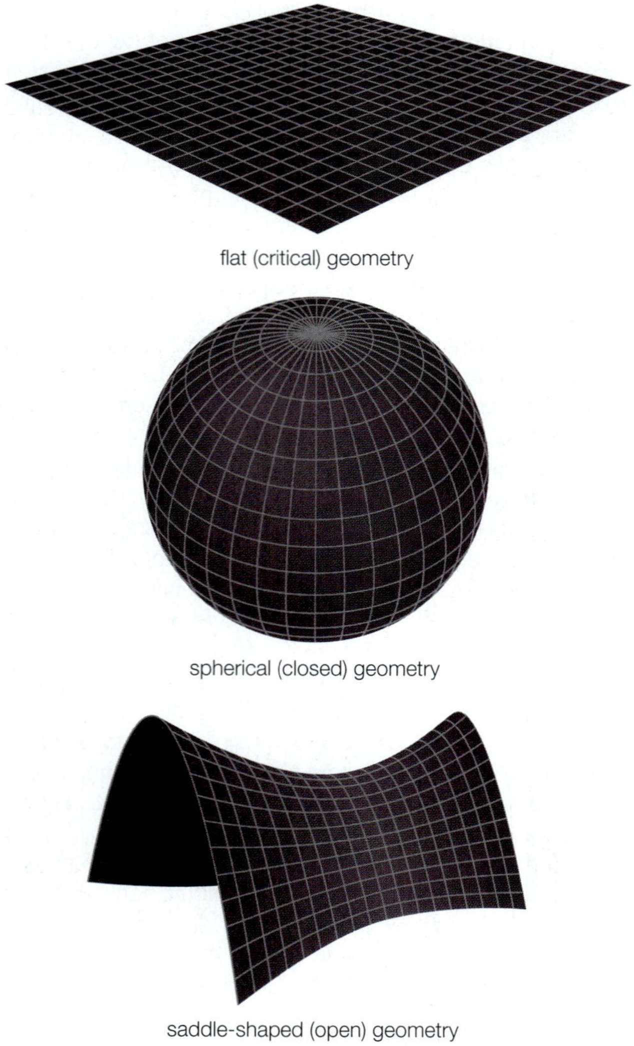


**FIGURE 22.14** The three possible categories of overall geometry for the universe. Keep in mind that the real universe has these "shapes" in more dimensions than we can see.

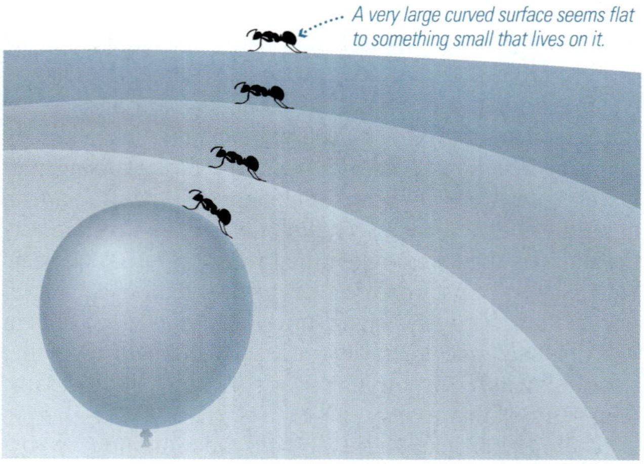


**FIGURE 22.15** As a balloon expands, its surface seems increasingly flat to an ant crawling along it. Inflation is thought to have made the universe seem flat in a similar way.

average density is less than the critical density, then the overall geometry is saddle-shaped. If its average density is greater than the critical density, then the overall geometry is spherical.

Inflation can explain why the overall geometry is so close to being flat. In terms of Einstein's theory, the effect of inflation on spacetime curvature is similar to the flattening of a balloon's surface when you blow into the balloon (**FIGURE 22.15**). The flattening of space caused by inflation would have been so enormous that any curvature the universe might have had previously would be noticeable only on size scales much larger than that of the observable universe. Inflation therefore makes the overall geometry of the universe appear flat, which means that the overall density of matter plus energy must be very close to the critical density.

## Did inflation really occur?

We've seen that inflation offers natural answers to our three key questions about the universe, but did it really happen? We cannot directly observe the universe at the very early time when inflation is thought to have occurred. Nevertheless, we can test the idea of inflation through careful studies of the cosmic microwave background. Scientists are only beginning to make observations that test inflation, but the findings to date are consistent with the idea that an early inflationary episode made the universe uniform and flat while planting the seeds of structure formation.

The strongest tests of inflation to date come from studying the patterns of temperature differences in the cosmic microwave background, and in particular the maps of the cosmic microwave background made by the *WMAP* and *Planck* satellites (see Figure 22.9). Remember that these maps show tiny temperature differences corresponding to density variations in the universe at the end of the era of nuclei, when the universe was about 380,000 years old. However, models of inflation predict that these density enhancements were created much earlier, when inflation caused tiny quantum ripples to expand into seeds of structure. Moreover, detailed models of inflation predict how the density enhancements should have been distributed in the early universe, and this allows scientists to calculate the angular separation that we should observe among the temperature variations in the cosmic microwave background. For example, if the universe is precisely flat, as inflation predicts it should be, then the largest temperature differences in the cosmic microwave background should be separated by an average of about 1°. Careful observations of the temperature variations in the microwave background can therefore tell us about the structure of the universe at that very early time.

The graph in **FIGURE 22.16** shows how temperature differences between patches of sky in the cosmic microwave background depend on the angular separations of these patches on the celestial sphere. The dots represent data from observations by *Planck* and other microwave telescopes, and the red curve shows the inflation-based model that best fits the observations. Notice the excellent fit between the model predictions and the data. Moreover, the same model also makes specific predictions about other characteristics of our universe, such as its composition and

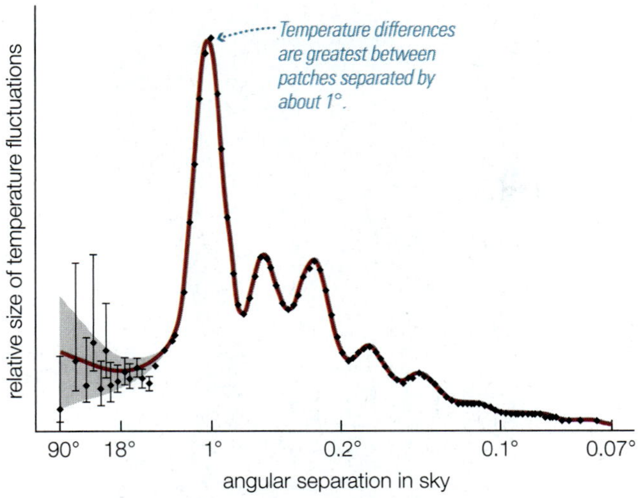


**FIGURE 22.16** All-sky maps of the cosmic microwave background like the one in Figure 22.9 allow scientists to measure temperature differences between different patches of the sky. This graph shows how the typical sizes of those temperature differences depend on the angular separation of the patches of sky. The data points represent actual measurements of the cosmic microwave background, while the red curve is the prediction of a model that relies on inflation to produce slight variations of temperature and density in the universe. Note the close agreement between the data and the model. (Bars indicate the uncertainty range in the data points.)

age, and these also agree well with observations. The general agreement between models and observations provides persuasive evidence in favor of the Big Bang in general, and also inspires confidence in the hypothesis of inflation.

A more direct test of the inflation hypothesis seeks another kind of pattern in the cosmic microwave background. The extreme stretching of space and time that occurred with the instant of inflation should have generated large gravitational waves [**Section S3.4**] that rippled through the universe. Models of the early universe predict that these gravitational waves should produce a distinctive pattern in how the light of the cosmic microwave background is polarized (see Special Topic, p. 145). Scientists are currently gathering more precise polarization data in order to look for this pattern which, if found, would provide strong support for the inflation hypothesis.

**Think about it** Search for news on attempts to observe polarization of the cosmic microwave background. What is the current scientific view on whether any such observations have provided evidence for inflation?

The bottom line is that, all things considered, inflation does a remarkable job of explaining features of our universe that are otherwise unaccounted for in the Big Bang theory. Many astronomers and physicists therefore suspect that some process akin to inflation did affect the early universe, but the details of the interaction between high-energy particle physics and the evolving universe remain unclear. If these details can be worked out successfully, we face an amazing prospect: a breakthrough in our understanding of the very smallest particles, achieved by studying the universe on the largest observable scales.

## 22.4 Observing the Big Bang for Yourself

You might occasionally read an article in a newspaper or a magazine questioning whether the Big Bang really happened. We will never be able to prove with absolute certainty that the Big Bang theory is correct. However, no one has come up with any other model of the universe that so successfully explains so much of what we see. As we have discussed, the Big Bang model makes specific predictions that we have observationally verified, including predictions about the characteristics of the cosmic microwave background and the composition of the universe. It also naturally explains many other features of the universe. So far, we know of nothing that is inconsistent with the Big Bang model.

The Big Bang theory's very success has made it a target for respected scientists, skeptical nonscientists, and crackpots alike. The nature of scientific work requires that we test established wisdom to make sure it is valid. A sound scientific disproof of the Big Bang theory would be a discovery of great importance. However, stories touted in the news media as disproofs of the Big Bang usually turn out to be disagreements over details rather than fundamental problems that threaten to bring down the whole theory. Yet scientists must keep refining the theory and tracking down disagreements, because once in a while a small disagreement blossoms into a full-blown scientific revolution.

You don't need to accept all you have read without question. The next time you are musing on the universe's origins, try an experiment for yourself. Go outside on a clear night, look at the sky, and ask yourself why it is dark.

**See it for yourself** How dark is the night sky where you live? Go outside and observe it on a moonless night. Estimate the total number of stars that are visible to you. How many stars like those do you think it would take to completely cover the entire sky?

### Why is the darkness of the night sky evidence for the Big Bang?

If the universe were infinite, unchanging, and everywhere the same, then the entire night sky would blaze as brightly as the Sun. Johannes Kepler [**Section 3.3**] was one of the first people to reach this conclusion, but we now refer to the idea as **Olbers' paradox** after German astronomer Heinrich Olbers (1758–1840).

To better understand Olbers' paradox, imagine that you were in a dense forest on a flat plain. If you looked in any direction, you'd likely see a tree. If the forest was small, you might be able to see through some gaps in the trees to the open plain, but larger forests would have fewer gaps (**FIGURE 22.17**). An infinite forest would have no gaps at all—a tree trunk would block your view along any line of sight.

The universe is like a forest of stars in this respect. In an unchanging universe with an infinite number of stars, we would see a star in every direction, making every point in the sky as bright as the Sun's surface. Even the presence of obscuring dust would not change this result. The intense starlight would heat the dust over time until it too glowed like the Sun or evaporated away.

**a** In a large forest, a tree will block your view no matter where you look. Similarly, in an unchanging universe with an infinite number of stars, we would expect to see stars in every direction, making the sky bright even at night.

**b** In a small forest with a smaller number of trees, you can see open spaces beyond the trees. Because the night sky is dark, the universe must similarly have spaces in which we see nothing beyond the stars, which means either that the number of stars is finite or that the universe changes in a way that prevents us from seeing an infinite number of them.

**FIGURE 22.17** Olbers' paradox can be understood by thinking of the view through a forest.

There are only two ways out of this dilemma. Either the universe has a finite number of stars, in which case we would not see a star in every direction, or it changes over time in some way that prevents us from seeing an infinite number of stars. For several centuries after Kepler first recognized the dilemma, astronomers leaned toward the first option. Kepler himself preferred to believe that the universe had a finite number of stars because he thought it had to be finite in space, with some kind of dark wall surrounding everything. Astronomers in the early 20th century tended to believe that the universe was infinite in space but that we lived inside a finite collection of stars. They thought of the Milky Way as an island floating in a vast black void. However, subsequent observations showed that galaxies fill all of space more or less uniformly. We are therefore left with the second option: The universe changes over time.

The Big Bang theory resolves Olbers' paradox in a particularly simple way. It tells us that we can see only a finite number of stars because the universe began at a particular moment. While the universe may contain an infinite number of stars, we can see only those that lie within the observable universe, inside our cosmological horizon [**Section 20.3**]. There are other ways in which the universe could change over time and prevent us from seeing an infinite number of stars, so Olbers' paradox does not *prove* that the universe began with a Big Bang. However, we must have some explanation for why the sky is dark at night, and no explanation besides the Big Bang also explains so many other observed properties of the universe so well.

## The BIG Picture PUTTING CHAPTER 22 INTO CONTEXT

*Our "big picture" now extends all the way back to the earliest moments in time. When you think back on this chapter, keep in mind the following ideas:*

- Predicting conditions in the early universe is straightforward, as long as we know how matter and energy behave under such extreme conditions.
- Our current understanding of physics allows us to reconstruct the conditions thought to have prevailed in the universe back to the first $10^{-10}$ second. Our understanding is less certain back to $10^{-38}$ second. Beyond $10^{-43}$ second, we run up against the present limits of human knowledge.
- Although it may sound strange to talk about the universe during its first fraction of a second, our ideas about the Big Bang rest on a solid foundation of observational, experimental, and theoretical evidence. We cannot say with absolute certainty that the Big Bang really happened, but no other model has so successfully explained how our universe came to be as it is.

**MY COSMIC PERSPECTIVE** The Big Bang happened long ago, but our connection to it is intimate. All the protons, neutrons, and electrons in your body were made from the intense energy present during the first few moments of the early universe.

# Summary of Key Concepts

## 22.1 The Big Bang Theory

- **What were conditions like in the early universe?** The early universe was filled with radiation and elementary particles. It was so hot and dense that the energy of radiation could turn into particles of **matter** and **antimatter**, which then collided and turned back into radiation.

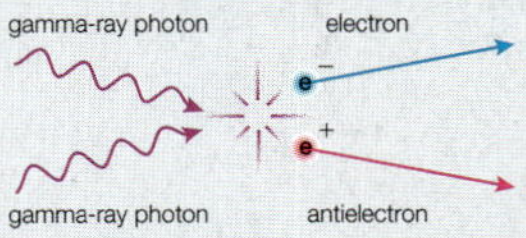


- **How did the early universe change with time?** The universe has progressed through a series of eras, each marked by unique physical conditions. We know little about the **Planck era**, when the four forces may have all behaved as one. Gravity became distinct at the start of the **GUT era**, and electromagnetism and the weak force became distinct at the end of the **electroweak era**. Matter particles annihilated all the antimatter particles by the end of the **particle era**. Fusion of protons and neutrons into helium ceased at the end of the **era of nucleosynthesis**. Hydrogen nuclei captured all the free electrons, forming hydrogen atoms, at the end of the **era of nuclei**. Galaxies began to form at the end of the **era of atoms**. The **era of galaxies** continues to this day.

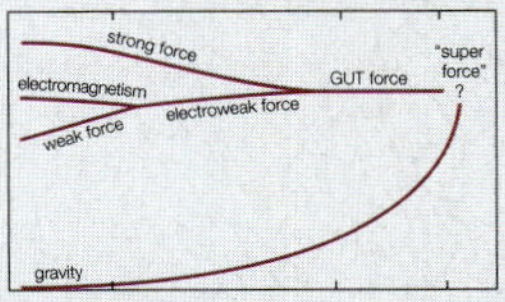


## 22.2 Evidence for the Big Bang

- **How do observations of the cosmic microwave background support the Big Bang theory?** Telescopes that can detect microwaves allow us to observe the **cosmic microwave background**—radiation left over from the Big Bang. Its spectrum matches the characteristics expected of the radiation released at the end of the era of nuclei, confirming a key prediction of the Big Bang theory.

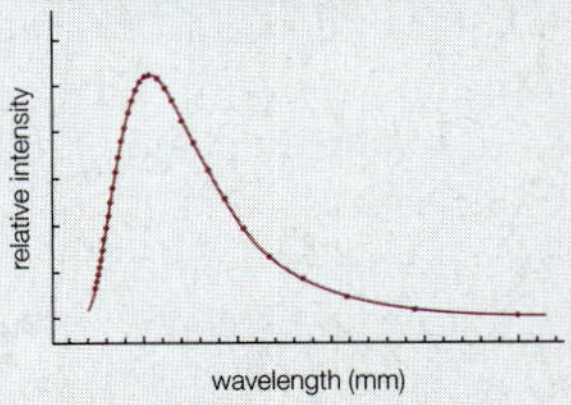


- **How do the abundances of elements support the Big Bang theory?** The Big Bang theory predicts the ratio of protons to neutrons during the era of nucleosynthesis, and from this predicts that the chemical composition of the universe should be about 75% hydrogen and 25% helium (by mass). The prediction matches observations of the cosmic abundances of elements, another spectacular confirmation of the Big Bang theory.

## 22.3 The Big Bang and Inflation

- **What key features of the universe are explained by inflation?** The hypothesis that the universe underwent a rapid and dramatic period of **inflation** successfully explains three key features of the universe that are otherwise mysterious: (1) the density enhancements that led to galaxy formation, (2) the smoothness of the cosmic microwave background, and (3) the "flat" geometry of the observable universe.

- **Did inflation really occur?** We can test the idea of inflation because it makes specific predictions about the patterns we should observe in the cosmic microwave background. Observations made with microwave telescopes so far match those predictions, lending credence to the idea that inflation (or something much like it) really occurred.

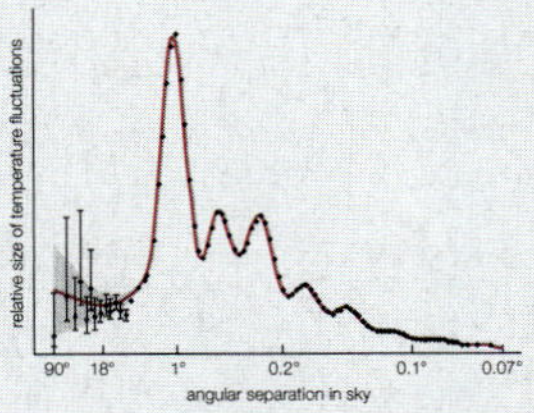


## 22.4 Observing the Big Bang for Yourself

- **Why is the darkness of the night sky evidence for the Big Bang?** **Olbers' paradox** tells us that if the universe were infinite, unchanging, and everywhere the same, the entire night sky would be as bright as the surface of the Sun, and it would not be dark at night. The Big Bang theory solves this paradox by telling us that the night sky is dark because the universe has a finite age, which means we can see only a finite number of stars in the sky.

# Visual Skills Check

*Use the following questions to check your understanding of some of the many types of visual information used in astronomy. For additional practice, try the Chapter 22 Visual Quiz at www.MasteringAstronomy.com.*

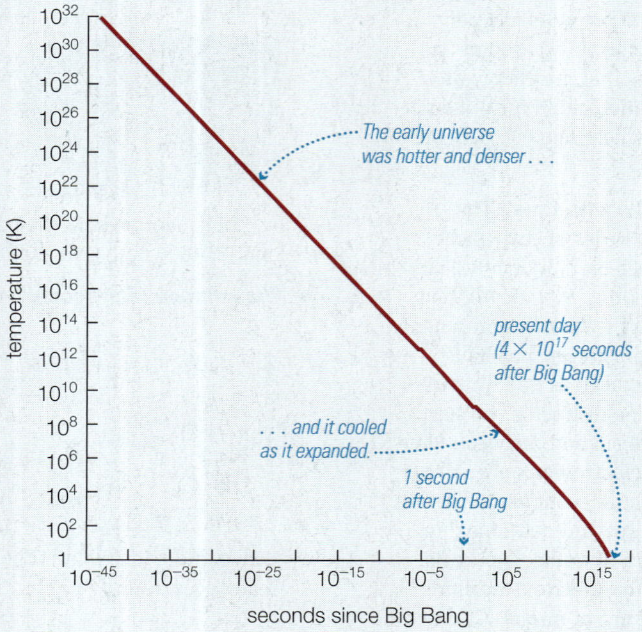


*This graph (repeated from Figure 22.1) shows how the temperature of the universe has declined with time; the graph spans many orders of magnitude for both temperature and time. Answer the following questions based on the information provided in the graph.*

1. What was the approximate temperature of the universe at an age of $10^{15}$ s?
   a. about 1 K
   b. about 100 K
   c. about $10^5$ K
   d. about $10^{15}$ K
2. What was the approximate temperature of the universe at an age of 5 minutes?
   a. about 300 K
   b. about $10^6$ K
   c. about $10^9$ K
   d. about $10^{12}$ K
3. How much cooler is the universe now (at an age of $4 \times 10^{17}$ s) than it was at an age of 1 second?
   a. Its current temperature is one hundred-millionth ($10^{-8}$) the temperature at an age of 1 second.
   b. Its current temperature is one hundred-thousandth ($10^{-5}$) the temperature at an age of 1 second.
   c. Its current temperature is one-hundredth ($10^{-2}$) the temperature at an age of 1 second.
   d. Its current temperature is one ten-billionth ($10^{-10}$) the temperature at an age of 1 second.

# Exercises and Problems

*For instructor-assigned homework and other learning materials, go to www.MasteringAstronomy.com.*

## Chapter Review Questions

### Short-Answer Questions Based on the Reading

1. Explain what we mean by the *Big Bang theory*.
2. What is *antimatter*? How were particle-antiparticle pairs created and destroyed in the early universe?
3. What are the four forces that operate in the universe today? Why do we think there were fewer forces operating in the early universe?
4. Make a list of the major eras in the history of the universe, summarizing the important events thought to have occurred during each era.
5. Why can't our current theories describe the conditions that existed in the universe during the *Planck era*?
6. What are *grand unified theories*? According to these theories, how many forces operated during the *GUT era*? How are these forces related to the four forces that operate today?
7. What do we mean by *inflation*, and when do we think it occurred?
8. Why do we think there was slightly more matter than antimatter in the early universe? What happened to all the antimatter, and when?
9. How long did the *era of nucleosynthesis* last? Explain why this era was so important in determining the chemical composition of the universe.
10. When we observe the *cosmic microwave background*, at what age are we seeing the universe? How long have the photons in the background been traveling through space? Explain.
11. Briefly describe how the cosmic microwave background was discovered. How do the existence and nature of this radiation support the Big Bang theory?
12. How does the chemical abundance of helium in the universe support the Big Bang theory? Explain.
13. Describe three key questions about the universe that are answered by inflation, and explain how inflation answers each of them.
14. What observational evidence supports the hypothesis of inflation? Be sure to explain how observations of the cosmic microwave background can tell us about the universe at the much earlier time when inflation occurred.
15. What is *Olbers' paradox*, and how is it resolved by the Big Bang theory?

### Does It Make Sense?

*Decide whether the statement makes sense (or is clearly true) or does not make sense (or is clearly false). Explain clearly; not all of these have definitive answers, so your explanation is more important than your chosen answer.*

16. According to the Big Bang theory, the early universe had nearly equal amounts of matter and antimatter.
17. According to the Big Bang theory, the cosmic microwave background was created when energetic photons ionized the neutral hydrogen atoms that originally filled the universe.
18. Observed characteristics of the cosmic microwave background can be explained by assuming that it comes from individual stars and galaxies.
19. According to the Big Bang theory, most of the helium in the universe was created by nuclear fusion in the cores of stars.
20. According to the hypothesis of inflation, large-scale structures in the universe may have originated as tiny quantum fluctuations.
21. According to the hypothesis of inflation, the "flat" geometry of the universe most likely arose by chance.
22. Inflation is a nice idea, but there are no known ways to test whether it really happened.
23. In the distant past, what is now the cosmic microwave background consisted primarily of infrared light.
24. The main reason the night sky is dark is that stars are so far away.
25. Patterns in the cosmic microwave background tell us about conditions in the early universe that ultimately led to galaxy formation.

### Quick Quiz

*Choose the best answer to each of the following. For additional practice, try the Chapter 22 Reading and Concept Quizzes in the Study Area at www.MasteringAstronomy.com.*

26. The current temperature of the universe as a whole is (a) absolute zero. (b) a few K. (c) a few thousand K.
27. The charge of an antiproton is (a) positive. (b) negative. (c) neutral.
28. When a proton and an antiproton collide, they (a) repel each other. (b) fuse together. (c) convert into two photons.
29. Which of the following does *not* provide strong evidence for the Big Bang theory? (a) observations of the cosmic microwave background (b) observations of the amount of hydrogen in the universe (c) observations of the ratio of helium to hydrogen in the universe
30. When the universe was 380,000 years old, its thermal radiation spectrum consisted mostly of (a) radio and microwave photons. (b) visible and infrared photons. (c) x-ray and ultraviolet photons.
31. Which of the following does inflation help to explain? (a) the uniformity of the cosmic microwave background (b) the amount of helium in the universe (c) the temperature of the cosmic microwave background
32. Which of the following does inflation help to explain? (a) the origin of hydrogen (b) the origin of galaxies (c) the origin of atomic nuclei
33. Which of these pieces of evidence supports the idea that inflation really happened? (a) the enormous size of the observable universe (b) the large amount of dark matter in the universe (c) the apparently "flat" geometry of the universe
34. What is the earliest time from which we observe light in the universe? (a) a few hundred million years after the Big Bang (b) a few hundred thousand years after the Big Bang (c) a few minutes after the Big Bang
35. Which of the following best explains why the night sky is dark? (a) The universe is not infinite in space. (b) The universe has not always looked the way it looks today. (c) The distribution of matter in the universe is not uniform on very large scales.

## Inclusive Astronomy

*Use these questions to reflect on participation in science.*

36. *Group Discussion: Astronomy and Technological Development.* The stated mission of the *International Astronomical Union* (IAU), an organization of professional astronomers from around the world, is to "promote and safeguard the science of astronomy in all its aspects through international cooperation." As part of this mission, the IAU promotes astronomy as a means of broadly enhancing technological research and education in the developing world.
**a.** Working independently, go to the website of the IAU's *Office of Astronomy for Development* (OAD) and learn about at least two of the activities it is supporting.
**b.** Gather into small groups and tell each other about the OAD's activities. Make a list of impacts that those activities might have on the countries in which they are happening.
**c.** How do you think the IAU should determine whether the OAD's efforts are successful? What processes would you recommend for deciding if or how particular activities should be changed?
**d.** Within your group, briefly debate the relative merits for developing countries of spending money on astronomy versus spending money on sciences with more direct practical applications. Summarize the outcome of your debate.

## The Process of Science

*These questions may answered individually in short-essay form or discussed in groups, except where identified as group-only.*

37. *Unanswered Questions.* Briefly describe one important but unanswered question about the events that happened shortly after the Big Bang. If you think it will be possible to answer that question in the future, describe how we might find an answer, being as specific as possible about the evidence necessary. If you think the question will never be answered, explain why you think it is impossible to answer.
38. *Darkness at Night.* Imagine that you were Johannes Kepler, pondering the darkness of the night sky without any knowledge of the Big Bang or the expanding universe. Come up with a hypothesis for the darkness of the night sky that would have been plausible in Kepler's time. Propose an experiment that scientists might be able to perform today to test that hypothesis.
39. *The Moment of Creation.* The timeline of the universe discussed in this textbook takes us back only to within $10^{-43}$ second of creation, not to the moment of creation itself. Do you think it will *ever* be possible for science to consider the moment of creation, or to answer questions such as *why* the Big Bang happened? Defend your opinions.
40. *Discussion: The Big Bang.* How convincing do you find the evidence for the Big Bang theory? What are its strengths? What does it fail to explain? Do *you* think the Big Bang really happened? Defend your opinion.
41. *Group Activity: Testing the Big Bang Theory.* Listed below are five hypothetical observations that are *not* predicted by the Big Bang theory. For each one, imagine what would happen if it were actually observed—whether it could be explained with the existing Big Bang theory, could be explained with a revision to the Big Bang theory, or would force us to abandon the Big Bang theory. Write down your team's reasoning for each case. Note: You may wish to do this activity using the four roles described in Chapter 1, Exercise 39.
    - Hypothetical observation 1: a star cluster with an age of 15 billion years
    - Hypothetical observation 2: a galaxy with an age of 10 million years
    - Hypothetical observation 3: a galaxy at a distance of 10 billion light-years whose spectrum is blueshifted
    - Hypothetical observation 4: a galaxy containing 90% hydrogen and 10% helium
    - Hypothetical observation 5: evidence for an increase in the cosmic microwave background temperature with time

## Investigate Further

### Short-Answer/Essay Questions

42. *Life Story of a Proton.* Tell the life story of a proton from its formation shortly after the Big Bang to its presence in the nucleus of an oxygen atom you have just inhaled. Your story should be creative and imaginative, but it should also demonstrate your scientific understanding of as many stages in the proton's life as possible. You can draw on material from the entire book, and your story should be three to five pages long.
43. *Creative History of the Universe.* The story of creation as envisioned by the Big Bang theory is quite dramatic, but it is usually told in a fairly straightforward, scientific way. Tell the story more dramatically, in the form of a short story, play, or poem. Be as creative as you wish, but be sure to remain true to the science as it is understood today.
44. *Re-creating the Big Bang.* Particle accelerators on Earth can push particles to extremely high speeds. When these particles collide, the amount of energy associated with the colliding particles is much greater than the mass-energy these particles have when at rest. As a result, these collisions can produce many other particles out of energy. Explain in your own words how the conditions that occur in these accelerators are similar to the conditions that prevailed shortly after the Big Bang. Also, point out some of the differences between what happens in particle accelerators and what happened in the early universe.
45. *Betting on the Big Bang Theory.* If you had $100, how much money would you wager on the proposition that we have a reasonable scientific understanding of what the universe was like when it was 1 minute old? Explain your bet in terms of the scientific evidence presented in this chapter.
46. *"Observing" the Early Universe.* Explain why we will never be able to observe the era of nucleosynthesis through direct detection of the radiation emitted at that time. How do we learn about this era?
47. *Element Production in the Big Bang.* Nucleosynthesis in the early universe was unable to produce more than trace amounts of elements heavier than helium. Using the information in Figure 17.14, which shows the mass per nuclear particle for many different elements, explain why producing elements like lithium (3 protons), boron (4 protons), and beryllium (5 protons) was so difficult.
48. *Evidence for the Big Bang.* Make a list of at least seven observed features of the universe that are satisfactorily explained by the Big Bang theory and the idea of inflation.

## Quantitative Problems

*Be sure to show all calculations clearly and state your final answers in complete sentences.*

49. *Energy from Antimatter.* The total annual U.S. power consumption is about $2 \times 10^{20}$ joules. Suppose you could supply that energy by combining pure matter with pure antimatter. Estimate the total mass of matter-antimatter fuel you would need to supply the United States with energy for 1 year. How does that mass compare with the amount of matter in your car's gas tank? (A gallon of gas has a mass of about 4 kilograms.)
50. *Gravity vs. the Electromagnetic Force.* The amount of electromagnetic force between two charged objects can be computed with an inverse square law similar to Newton's universal law of gravitation; for the electromagnetic force, the law is

$$F = k \times \frac{(\text{charge of object 1}) \times (\text{charge of object 2})}{d^2}$$

In this formula, the charges must be given in units of coulombs (abbreviated C), the distance $d$ between the objects' centers must be in meters, and the constant $k = 9 \times 10^9 \text{ kg} \times \text{m}^3/(\text{C}^2 \times \text{s}^2)$.
**a.** Compute the gravitational force between your body and Earth using Newton's universal law of gravitation (see Section 4.4 or Appendix B).
**b.** Now suppose all the electrons suddenly disappeared from Earth, making it positively charged, and all the protons in your body suddenly changed into neutrons, making you negatively charged. Compute the strength of the electromagnetic force between the electrons in your body and the protons in Earth. Assume that the charge per unit of mass of both you and Earth is $5 \times 10^7$ C/kg.
**c.** Compare the electromagnetic force from part b to the gravitational force from part a. Use that result to explain why gravity is considered weaker than the electromagnetic force.
51. *Background Radiation During Galaxy Formation.* What was the peak wavelength of the background radiation at the time light left the most distant galaxies we can currently see? Assume those galaxies have a cosmological redshift of $z = 10$. What is the temperature corresponding to that peak wavelength?
52. *Expansion Since the Era of Nucleosynthesis.* Compare the peak wavelength of the radiation in the universe at the end of the era of nucleosynthesis with the peak wavelength of the radiation in the universe currently. Assume the temperature at the end of the era of nucleosynthesis was $10^9$ K. How much have the wavelengths of the photons in the universe been stretched since that time?
53. *Temperature of the Universe.* What will the temperature of the cosmic microwave background be when the average distances between galaxies are twice as large as they are today?
54. *Uniformity of the Cosmic Microwave Background.* The temperature of the cosmic microwave background differs by only a few parts in 100,000 across the sky. Compare this level of uniformity to that of the surface of a table that is 1 meter square. How big would the largest bumps on that table be if its surface were smooth to one part in 100,000? Could you see bumps of that size on the table's surface?
55. *Daytime at "Night."* According to Olbers' paradox, the entire sky would be as bright as the surface of a typical star if the universe were infinite in space, unchanging in time, and the same everywhere. However, conditions would not need to be quite that extreme for the "nighttime" sky to be as bright as the daytime sky.
**a.** Using the inverse square law for light from Mathematical Insight 15.1, determine the apparent brightness of the Sun in our sky.
**b.** Using the inverse square law for light, determine the apparent brightness our Sun would have if it were at a distance of 10 billion light-years.
**c.** From your answers to parts a and b, estimate how many stars like the Sun would need to exist at a distance of 10 billion light-years for their total apparent brightness to equal that of our Sun.
**d.** Compare your answer to part c with the estimate of the total number of stars in our observable universe from Mathematical Insight 1.3. Use your answer to explain why the night sky is much darker than the daytime sky. How much larger would the total number of stars need to be for "night" to be as bright as day?

# 23 Dark Matter, Dark Energy, and the Fate of the Universe

▲ **About the photo:** This composite image of the Bullet Cluster of galaxies shows that the x-ray–emitting hot gas (red) is in a different location from the majority of the mass revealed by gravitational lensing (blue).

## LEARNING GOALS

**23.1 Unseen Influences in the Cosmos**
- What do we mean by dark matter and dark energy?

**23.2 Evidence for Dark Matter**
- What is the evidence for dark matter in galaxies?
- What is the evidence for dark matter in clusters of galaxies?
- Does dark matter really exist?
- What might dark matter be made of?

**23.3 Structure Formation**
- What is the role of dark matter in galaxy formation?
- What are the largest structures in the universe?

**23.4 Dark Energy and the Fate of the Universe**
- What is the evidence for an accelerating expansion?
- Why is flat geometry evidence for dark energy?
- What is the fate of the universe?

*The scientific observer of Nature is a kind of mystic seeker in the act of prayer.*
— *Muhammad Iqbal (1877–1938)*

**Chapter 23 Overview**

Over the past several chapters, we have discussed the strong evidence supporting our scientific understanding of the history of the universe. This evidence indicates that our universe was born about 14 billion years ago in the Big Bang and has been expanding ever since. However, in regions that began with slightly enhanced density, gravity took hold and built galaxies, within which some of the hydrogen and helium atoms produced in the Big Bang were assembled into stars. We exist today because galactic recycling has incorporated heavier elements made by early generations of stars into new star systems containing planets like Earth.

Scientists broadly agree on this basic outline of universal history, but many details are not yet well understood and a few more significant mysteries remain. In this chapter, we'll explore two of the greatest remaining mysteries, which concern the nature of the so-called *dark matter* and *dark energy* that appear to make up most of the content of our universe. As we will see, these mysteries likely hold the key not only to a more complete understanding of how the universe has evolved to date, but also to the eventual fate of the universe.

## 23.1 Unseen Influences in the Cosmos

What is the universe made of? You might think this would be a simple question for an astronomer to answer, but based on all the evidence available today, the answer is "We do not yet know."

It might seem incredible that we still do not know the composition of most of the universe, and you might wonder why we should be so clueless. After all, astronomers can measure the chemical composition of distant stars and galaxies from their spectra, so we know that stars and gas clouds are made almost entirely of hydrogen and helium, with small amounts of heavier elements mixed in. But notice the key words "chemical composition." When we say these words, we are talking about the composition of material built from atoms of elements such as hydrogen, helium, carbon, and iron.

While it is true that all familiar objects—including people, planets, and stars—are built from atoms, the same may not be true of the universe as a whole. In fact, we now have good reason to think that the universe is *not* composed primarily of atoms. Instead, observations indicate that the universe consists largely of a mysterious form of mass known as *dark matter* and a mysterious form of energy known as *dark energy*.

### What do we mean by dark matter and dark energy?

It's easy for scientists to talk about dark matter and dark energy, but what do these terms really mean? They are nothing more than names given to unseen influences in the cosmos. In both cases, observational evidence leads us to think that there is something out there, but we do not yet know exactly what the "something" is.

We might naively think that the major source of gravity that holds galaxies together should be the same gas that makes up their stars. However, observations suggest otherwise. By carefully observing gravitational effects on matter that we can see, such as stars or glowing clouds of gas, we've learned that there must be far more matter than meets the eye. Because this matter gives off little or no light, we call it **dark matter**.* In other words, dark matter is simply a name we give to whatever unseen influence is causing the observed gravitational effects. We've already discussed dark matter briefly in Chapters 1 and 19, noting that studies of the Milky Way's rotation suggest that most of our galaxy's mass is distributed throughout its halo while most of the galaxy's light comes from stars and gas clouds in the thin galactic disk (see Figure 1.16).

We infer the existence of the second unseen influence from careful studies of the expansion of the universe. After Edwin Hubble first discovered the expansion, it was generally assumed that gravity must slow the expansion with time. However, evidence collected during the last two decades indicates that the expansion of the universe is actually accelerating, implying that some mysterious force counteracts the effects of gravity on very large scales. **Dark energy** is the name most commonly given to the source of this mysterious force, though you may occasionally hear the same unseen influence attributed to *quintessence* or to a *cosmological constant*. Note that while dark matter really is "dark" compared to ordinary matter (because it gives off no detectable light), there's nothing unusually "dark" about dark energy—after all, we don't expect to see light from the mere presence of a force or a gravitational field.

Before we continue, it's important to think about dark matter and dark energy in the context of science. Strange as these ideas may seem, they have emerged from careful scientific study conducted in accordance with the hallmarks of science discussed in Chapter 3 (see Figure 3.24). Dark matter and dark energy were each proposed to exist because they seem the simplest ways to explain observed motions in the universe. They've each gained credibility because models of the universe that assume their existence make testable predictions and, at least so far, further observations have borne out some of those predictions. Even if we someday conclude that we were wrong to infer the existence of dark matter or dark energy, we will still need alternative explanations for the observations made to date. One way or the other, what we learn as we explore the mysteries of these unseen influences will forever change our view of the universe.

*It does not necessarily appear dark or black to our eyes; indeed, our telescopes see right through it because it does not interact with light. In that sense, a better name might be *transparent matter*.

## 23.2 Evidence for Dark Matter

Scientific evidence for dark matter has been building for decades and is now at the point where dark matter seems almost indispensable to explaining the current structure of the universe. For that reason, we will devote most of this chapter to dark matter and its presumed role as the dominant source of gravity in our universe, saving further discussion of dark energy for the final section of the chapter. In this section, we'll begin our discussion of dark matter by examining the evidence for its existence and what the evidence indicates about its nature.

### What is the evidence for dark matter in galaxies?

Several distinct lines of evidence point to the existence of dark matter, including observations of our own galaxy, of other galaxies, and of clusters of galaxies. Let's start with individual galaxies and then proceed to clusters.

**Dark Matter in the Milky Way** In Chapter 19, we saw how the Sun's motion around the galaxy reveals the total amount of mass within its orbit. We can similarly use the orbital motion of any other star to measure the mass of the Milky Way within that star's orbit. In principle, we could determine the complete distribution of mass in the Milky Way by doing the same thing with the orbits of stars at every different distance from the galactic center.

In practice, interstellar dust obscures our view of disk stars more than a few thousand light-years away from us, making it very difficult to measure stellar velocities. However, radio waves penetrate this dust, and clouds of atomic hydrogen gas emit a spectral line at the radio wavelength of 21 centimeters [**Section 19.2**]. Measuring the Doppler shift of this 21-centimeter line tells us a cloud's velocity toward or away from us. With a little geometry, we can then determine the cloud's orbital speed.

We can summarize the results of these measurements with a diagram that plots the orbital speed of objects in the galaxy against their orbital distances. As a simple example of such a diagram, sometimes called a *rotation curve*, consider a plot of the rotation speed of a merry-go-round in terms of distance from its center. Every object on a merry-go-round goes around the center in the same amount of time (the rotation period of the merry-go-round). But because objects farther from the center move in larger circles, they must move at faster speeds. The speed is proportional to distance from the center, so the graph illustrating the relationship between speed and distance is a steadily rising straight line (**FIGURE 23.1a**).

In contrast, orbital speeds in our solar system *decrease* with distance from the Sun (**FIGURE 23.1b**). This drop-off in speed with distance occurs because virtually all the mass of the solar system is concentrated in the Sun. The gravitational force holding a planet in its orbit therefore decreases with distance from the Sun, and a smaller force means a lower orbital speed. Orbital speeds must drop similarly with distance in any other astronomical system that has its mass concentrated at its center.

**FIGURE 23.1c** shows how orbital speed depends on distance in the Milky Way Galaxy. Each individual dot represents the orbital speed and distance from the galactic center of a particular star or gas cloud, and the curve running through the dots represents a best fit to the data. Notice that the orbital speeds remain approximately constant beyond the inner few thousand light-years, so most of the curve is relatively flat. Because this is so different from the declining orbital speeds found in our solar system, we conclude that most of the Milky Way's mass must *not* be concentrated at its center. Instead, the orbits of progressively more distant gas clouds must encircle more and more mass. The Sun's orbit encompasses about 100 billion solar masses, but a circle twice as large surrounds twice as much mass, and a larger circle surrounds even more mass.

To summarize, orbital speeds in the Milky Way imply that most of our galaxy's mass lies well beyond the orbit of our Sun. A more detailed analysis suggests that most of this mass is distributed throughout the spherical halo that surrounds the disk of our galaxy, extending to distances well beyond those at which we observe globular clusters and other halo stars. Moreover, the total amount of this mass is more than *10 times* the total mass of all the stars in the disk. Because we have detected very little radiation

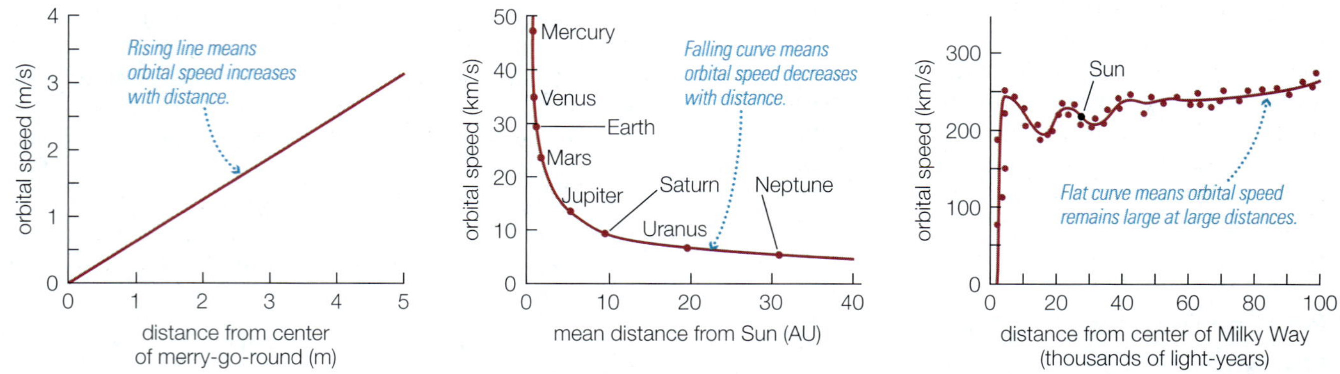


**a** A rotation curve for a merry-go-round is a rising straight line.

**b** The rotation curve for the planets in our solar system.

**c** The rotation curve for the Milky Way Galaxy. Dots represent actual data points for stars or gas clouds.

**FIGURE 23.1** These graphs show how orbital speed depends on distance from the center in three different systems.

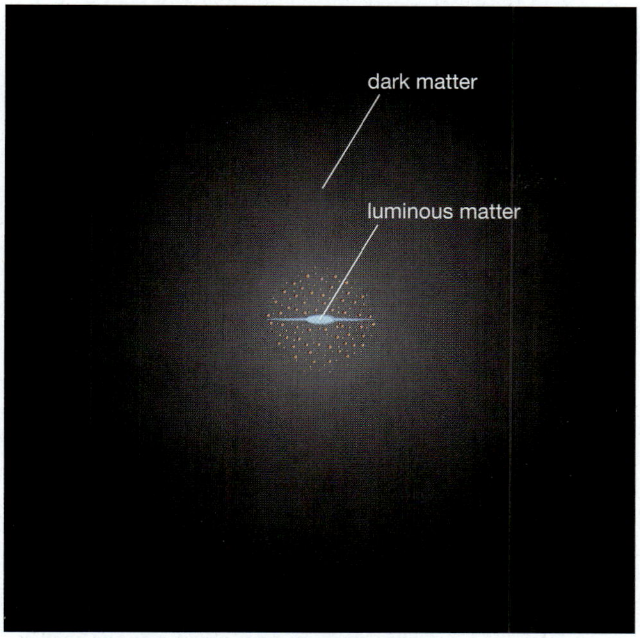


**FIGURE 23.2** The dark matter associated with the Milky Way occupies a much larger volume than the galaxy's luminous matter. The radius of this dark-matter halo may be 10 times as large as that of the galaxy's halo of stars.

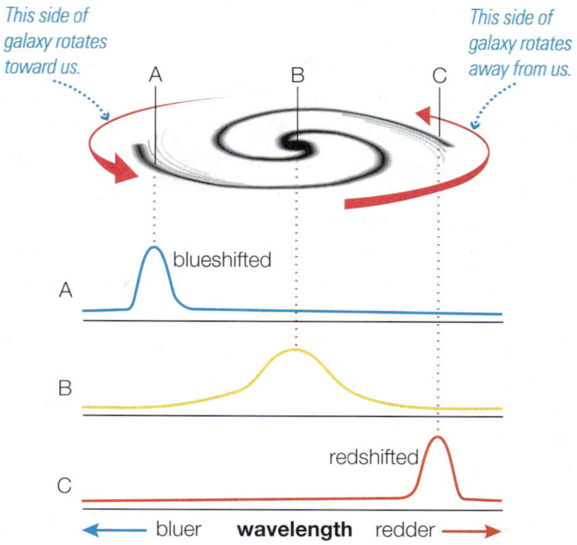


**FIGURE 23.3** Measuring the orbital speeds of gas in a spiral galaxy with the 21-centimeter line of atomic hydrogen. Blueshifted lines on the left side of the disk show how fast that side is moving toward us. Redshifted lines on the right side show how fast that side is moving away from us. (This diagram assumes that we first subtract a galaxy's average redshift, so that the shifts that remain are due to rotation.)

coming from this enormous amount of mass, it qualifies as dark matter. If we are interpreting the evidence correctly, the luminous part of the Milky Way's disk must be rather like the tip of an iceberg, marking only the center of a much larger clump of mass (**FIGURE 23.2**).

**Think about it** Suppose we made a graph of orbital speeds and distances for the moons orbiting Jupiter. Which graph in Figure 23.1 would it most resemble? Why?

**Dark Matter in Other Spiral Galaxies** Other galaxies also seem to contain vast quantities of dark matter. We can determine the amount of dark matter in a galaxy by comparing the galaxy's mass to its luminosity. (More formally, astronomers calculate the galaxy's *mass-to-light ratio;* see Mathematical Insight 23.1.) The procedure is fairly simple in principle. First, we use the galaxy's luminosity to estimate the amount of mass that the galaxy contains in the form of stars. Next, we determine the galaxy's total mass by applying the law of gravity to observations of the orbital velocities of stars and gas clouds. If this total mass is larger than the mass that we can attribute to stars, then we infer that the excess mass must be dark matter.

We can measure a galaxy's luminosity as long as we can determine its distance with one of the techniques discussed in Chapter 20. We simply point a telescope at the galaxy in question, measure its apparent brightness, and calculate its luminosity from its distance and the inverse square law for light [**Section 15.1**]. Measuring the galaxy's total mass requires measuring orbital speeds as far from the galaxy's center as possible. For spiral galaxies, this is usually done with radio observations of the 21-centimeter line from atomic hydrogen gas clouds, because these clouds can generally be found at greater distances from the galactic center than stars. Doppler shifts of the 21-centimeter line tell us how fast a cloud is moving toward us or away from us (**FIGURE 23.3**).

Once we've measured orbital speeds and distances, we can make a graph similar to Figure 23.1c for any spiral galaxy. **FIGURE 23.4** shows a few examples illustrating that, like the Milky Way, most other spiral galaxies have orbital speeds that remain high even at great distances from their centers. Again as in the Milky Way, detailed analysis tells us that most spiral galaxies contain a great deal of matter in extended halos, and that in total they have at least 10 times as much mass in dark matter as they do in stars.

**Dark Matter in Elliptical Galaxies** We must use a different technique to determine masses of elliptical galaxies, because they do not have large well-organized disks in which we can easily measure how the orbital speeds

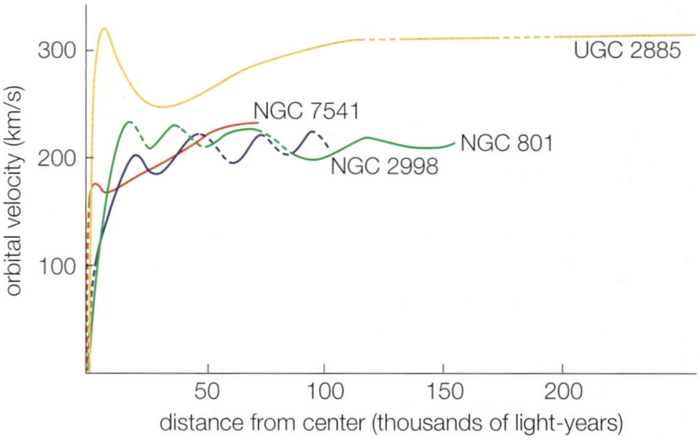


**FIGURE 23.4** Graphs of orbital speed versus distance for four spiral galaxies. In each galaxy, the orbital speeds remain nearly constant over a wide range of distances from the center, indicating that dark matter is common in spiral galaxies.

of stars depend on distance, and they lack atomic hydrogen clouds. However, the orbital speeds of their stars still depend on the amount of mass within their orbits, which allows us to measure mass from the width of an elliptical galaxy's spectral lines. If we look at the galaxy as a whole, its spectral lines come from the combination of all its stars. Because each star has its own orbital speed around the center of the galaxy, each produces its own Doppler shift that contributes to the overall appearance of the galaxy's spectral lines. Some stars are moving toward the center and others away, so their combined effect is to change any spectral line from a narrow line at a particular wavelength to a *broadened* line spanning a range of wavelengths. The greater the broadening of the spectral line, the faster the stars must be moving (**FIGURE 23.5**).

When we compare spectral lines from different regions of elliptical galaxies, we find that the speeds of the stars remain fairly constant as we look farther from the galaxy's center. Just as in spirals, we conclude that most of the matter in elliptical galaxies must lie beyond the distance where the light trails off and hence must be dark matter. The evidence for dark matter is even more convincing for cases in

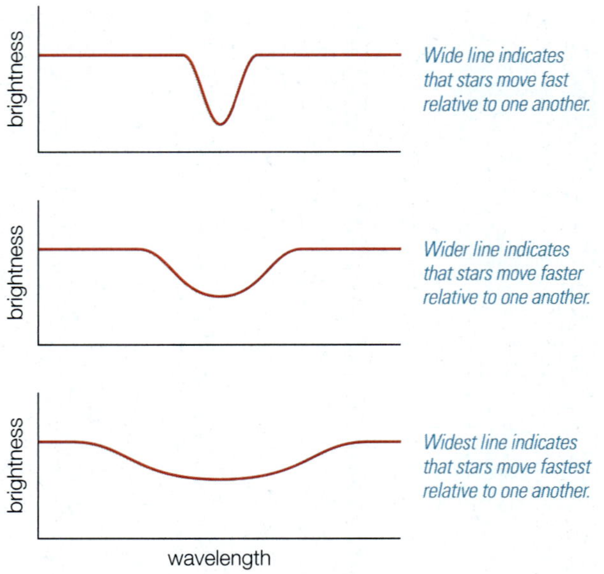


**FIGURE 23.5** The broadening of absorption lines in an elliptical galaxy's spectrum tells us how fast its stars are moving relative to one another.

## MATHEMATICAL INSIGHT 23.1 Mass-to-Light Ratio

An object's mass-to-light ratio ($M/L$) is its total mass in units of *solar masses* divided by its total *visible* luminosity in units of *solar luminosities.* For example, the mass-to-light ratio of the Sun is

$$\frac{M}{L}\text{ for Sun} = \frac{1M_{\text{Sun}}}{1L_{\text{Sun}}} = 1\frac{M_{\text{Sun}}}{L_{\text{Sun}}}$$

We read this answer with its units as "1 solar mass per solar luminosity." The following examples clarify the idea of the mass-to-light ratio and explain what it can tell us about the existence of dark matter.

**EXAMPLE 1:** What is the mass-to-light ratio of a $1M_{\text{Sun}}$ red giant with a luminosity of $100L_{\text{Sun}}$?

**SOLUTION:**

**Step 1 Understand:** Finding a mass-to-light ratio simply requires knowing an object's total mass in solar masses and its total luminosity in solar luminosities. We have been given both.

**Step 2 Solve:** We divide to find the mass-to-light ratio:

$$\frac{M}{L} = \frac{1M_{\text{Sun}}}{100L_{\text{Sun}}} = 0.01\frac{M_{\text{Sun}}}{L_{\text{Sun}}}$$

**Step 3 Explain:** The red giant has a mass-to-light ratio of 0.01 solar mass per solar luminosity. Note that the ratio is *less* than 1 because a red giant puts out *more* light per unit mass than the Sun. More generally, stars *more luminous* than the Sun have mass-to-light ratios *less* than 1 and stars *less luminous* than the Sun have mass-to-light ratios *greater* than 1.

**EXAMPLE 2:** The Milky Way Galaxy contains about 90 billion ($9 \times 10^{10}$) solar masses of material within the Sun's orbit, and the total luminosity of stars within that same region is about 15 billion ($1.5 \times 10^{10}$) solar luminosities. What is the mass-to-light ratio of the matter in our galaxy within the Sun's orbit?

**SOLUTION:**

**Step 1 Understand:** Again, we simply divide the mass of this region by its luminosity, both in solar units.

**Step 2 Solve:** The mass-to-light ratio within the Sun's orbit is

$$\frac{M}{L} = \frac{9 \times 10^{10}M_{\text{Sun}}}{1.5 \times 10^{10}L_{\text{Sun}}} = 6\frac{M_{\text{Sun}}}{L_{\text{Sun}}}$$

**Step 3 Explain:** The mass-to-light ratio of the matter within the Sun's orbit is about 6 solar masses per solar luminosity. This is *greater* than the Sun's ratio of 1 solar mass per solar luminosity, telling us that most matter in this region is *dimmer* per unit mass than our Sun. This is not surprising, because most stars are smaller and dimmer than our Sun.

**EXAMPLE 3:** Observations of orbital speeds in a spiral galaxy indicate that its total mass is $5 \times 10^{11}M_{\text{Sun}}$; its luminosity is $1.5 \times 10^{10}L_{\text{Sun}}$. What is its mass-to-light ratio?

**SOLUTION:**

**Step 1 Understand:** This problem is essentially the same as the others, but with different implications.

**Step 2 Solve:** We divide the galaxy's mass by its luminosity:

$$\frac{M}{L} = \frac{5 \times 10^{11}M_{\text{Sun}}}{1.5 \times 10^{10}L_{\text{Sun}}} = 33\frac{M_{\text{Sun}}}{L_{\text{Sun}}}$$

**Step 3 Explain:** The galaxy has a mass-to-light ratio of 33 solar masses per solar luminosity, which is more than five times the mass-to-light ratio for the matter in the Milky Way Galaxy within the Sun's orbit. We conclude that, on average, the mass in this galaxy is *much less* luminous than the mass found in the inner regions of the Milky Way, suggesting that the galaxy must contain a lot of mass that emits little or no light.

which we can measure the speeds of globular star clusters orbiting at large distances from the center of an elliptical galaxy. These measurements suggest that elliptical galaxies, like spirals, contain 10 or more times as much mass in dark matter as they do in the form of stars.

## What is the evidence for dark matter in clusters of galaxies?

Observations of galaxy clusters suggest that they contain 40 or more times as much mass in dark matter as in stars, which is a greater proportion than is found in individual galaxies. The evidence for dark matter in clusters comes from three different ways of measuring cluster masses: measuring the speeds of galaxies orbiting the center of the cluster, studying the x-ray emission from hot gas between the cluster's galaxies, and observing how the cluster bends light as a *gravitational lens.*

**Orbits of Galaxies in Clusters** The idea of dark matter is not particularly new. In the 1930s, astronomer Fritz Zwicky was already arguing that clusters of galaxies held enormous amounts of this mysterious stuff (**FIGURE 23.6**). Few of his colleagues paid attention, but later observations supported Zwicky's claims.

Zwicky was one of the first astronomers to think of galaxy clusters as huge swarms of galaxies bound together by gravity. It seemed natural to him that galaxies clumped closely together in space should all be orbiting one another, just like the stars in a star cluster. He therefore assumed that he could measure cluster masses by observing galaxy motions and applying Newton's laws of motion and gravitation.

Armed with a spectrograph, Zwicky measured the redshifts of the galaxies in a particular cluster and used these redshifts to calculate the speeds at which the individual galaxies are moving away from us. He determined the *recession speed* of the cluster as a whole—that is, the speed at which the expansion of the universe carries it away from us—by averaging the speeds of its individual galaxies.

**FIGURE 23.6** Fritz Zwicky, discoverer of dark matter in clusters of galaxies. Zwicky had an eccentric personality, but some of his ideas that seemed strange in the 1930s proved correct many decades later.

Once he knew the recession speed for the cluster, Zwicky could subtract this speed from each individual galaxy's speed to determine the speeds of galaxies relative to

## MATHEMATICAL INSIGHT 23.2 Finding Cluster Masses from Galaxy Orbits

Recall that we can use the *orbital velocity law* (see Mathematical Insight 19.1) to calculate the mass, $M_r$, contained *within* a distance $r$ of a galaxy's center:

$$M_r = \frac{r \times v^2}{G}$$

This formula also applies to galaxy clusters if we consider $r$ as the distance from the center of the cluster and assume the galaxies have circular orbits.

**EXAMPLE:** A galaxy cluster has a radius of 6.2 million light-years, and Doppler shifts show that galaxies orbit the cluster center at an average speed of approximately 1350 km/s. Find the cluster's mass.

**SOLUTION:**

**Step 1 Understand:** We can use the orbital velocity law, but to make the units consistent we must convert the radius into meters and the speed into meters per second.

**Step 2 Solve:** You can confirm for yourself that the radius of 6.2 million light-years is equivalent to $5.9 \times 10^{22}$ meters; the speed of 1350 km/s becomes $1.35 \times 10^6$ m/s. Substituting, we find

$$\begin{aligned} M_r &= \frac{r \times v^2}{G} \\ &= \frac{(5.9 \times 10^{22}\ \text{m}) \times (1.35 \times 10^6\ \text{m/s})^2}{6.67 \times 10^{-11}\ \text{m}^3/(\text{kg} \times \text{s}^2)} \\ &= 1.6 \times 10^{45}\ \text{kg} \end{aligned}$$

**Step 3 Explain:** The result is easier to interpret if we convert from kilograms to solar masses ($1M_{\text{Sun}} = 2.0 \times 10^{30}$ kg):

$$M_r = 1.6 \times 10^{45}\ \text{kg} \times \frac{1M_{\text{Sun}}}{2 \times 10^{30}\ \text{kg}} \approx 8.0 \times 10^{14} M_{\text{Sun}}$$

The cluster mass is about 800 trillion solar masses, which is equivalent to about 800 galaxies as large in mass as the Milky Way (including dark matter).

**FIGURE 23.7** A distant cluster of galaxies in both visible light and x-ray light. The visible-light photo shows the individual galaxies. The blue-violet overlay shows the x-ray emission from extremely hot gas in the cluster, with blue representing the hottest gas and violet representing cooler gas. Evidence for dark matter comes both from the observed motions of the visible galaxies and from the temperature of the hot gas. (The region shown is about 8 million light-years across.)

the cluster center. Of course, this method told him only the average *radial* component (the speed toward or away from us) of the actual galaxy velocities [**Section 5.4**], but by averaging over enough individual galaxies, Zwicky could get a good average orbital velocity for the cluster's galaxies as a whole. Once he knew the average orbital velocity of the galaxies, he could use Newton's universal law of gravitation to estimate the cluster's mass (see Mathematical Insight 23.2). Finally, he compared the cluster's mass to its luminosity.

To his surprise, Zwicky found that clusters of galaxies have much greater masses than their luminosities would suggest. That is, when he estimated the total mass of stars necessary to account for the overall luminosity of a cluster, he found that it was far less than the mass he measured by studying galaxy speeds. He concluded that most of the matter within these clusters must not be in the form of stars and instead must be almost entirely dark. Many astronomers disregarded Zwicky's result, believing that he must have done something wrong to arrive at such a strange result. However, more sophisticated measurements ultimately confirmed Zwicky's original finding.

**Hot Gas in Clusters** A second method for measuring a cluster's mass relies on observing x-rays from the hot gas that fills the space between its galaxies (**FIGURE 23.7**). This gas (sometimes called the *intracluster medium*) is so hot—typically tens of millions of Kelvin—that it emits primarily x-rays and does not appear in visible-light images. Measurements show that for large clusters, this hot gas can represent up to seven times the mass found in stars.

The hot gas can tell us about dark matter because its temperature depends on the total mass of the cluster. The gas in most clusters is nearly in a state of *gravitational equilibrium*—that is, the outward gas pressure balances gravity's inward pull [**Section 14.1**]. In this state of balance, the average kinetic energies of the gas particles are determined primarily by the strength of gravity and hence by the amount of mass within the cluster. Because the temperature of a gas reflects the average kinetic energies of its particles, the gas temperatures we measure with x-ray telescopes tell us the average speeds of the x-ray–emitting particles (see Mathematical Insight 23.3). We can then use these particle speeds to determine the cluster's total mass.

The results obtained with this method agree well with the results found by studying the orbital motions of the cluster's galaxies. Even after we account for the mass of the hot gas, we find that the amount of dark matter in clusters of galaxies is at least 40 times the combined mass of the stars in the cluster's galaxies. In other words, the gravity of dark matter seems to be binding the galaxies of a cluster together in much the same way gravity helps bind matter within individual galaxies.

**Think about it** What would happen to a cluster of galaxies if you instantly removed all the dark matter without changing the velocities of the galaxies?

**Gravitational Lensing** The methods of measuring galaxy and cluster masses that we've discussed so far all ultimately rely on Newton's laws, including his universal law of gravitation. But can we trust these laws on such large-size scales? One way to check is to measure masses in a different way. Today, astronomers can do this with observations of *gravitational lensing*.

Gravitational lensing occurs because masses distort spacetime—the "fabric" of the universe [**Section S3.3**]. Massive objects can therefore act as **gravitational lenses**,

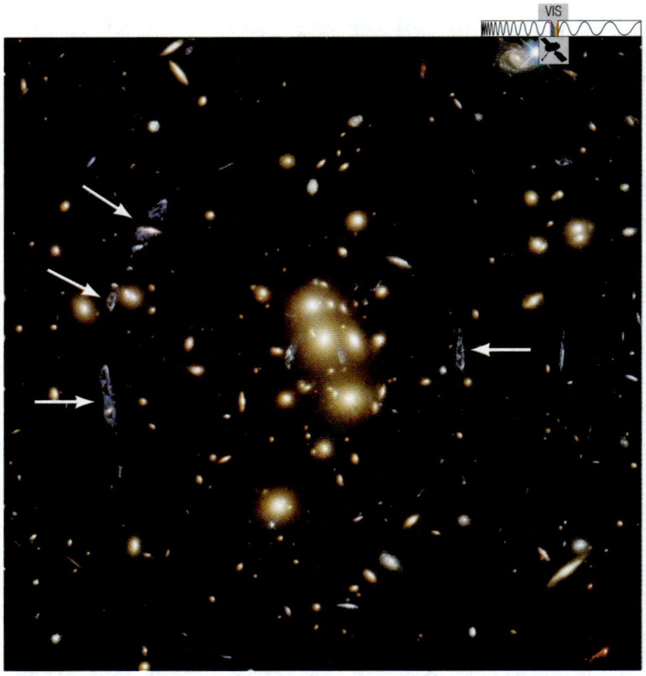


**FIGURE 23.8** This Hubble Space Telescope photo shows a galaxy cluster (known as CL0024+1654) acting as a gravitational lens. The yellow elliptical galaxies are cluster members. The small blue ovals (such as those indicated by the arrows) are multiple images of a single galaxy that lies almost directly behind the cluster's center. (The picture shows a region about 1.4 million light-years across.)

bending light beams that pass nearby. This prediction of Einstein's general theory of relativity was first verified in 1919 during an eclipse of the Sun [**Section S3.4**]. Because the light-bending angle of a gravitational lens depends on the mass of the object doing the bending, we can measure the masses of objects by observing how strongly they distort light paths.

**FIGURE 23.8** shows a striking example of how a cluster of galaxies can act as a gravitational lens. Notice the multiple blue ovals that are found at several positions around the central clump of yellow galaxies. These ovals all represent gravitationally lensed images of a single blue galaxy that lies almost directly behind the center of the cluster, at a much greater distance. We see multiple images because photons from the more distant galaxy do not follow straight paths to Earth. Instead, the cluster's gravity bends the photon paths, allowing light from the galaxy to arrive at Earth from a few slightly different directions (**FIGURE 23.9**). Each alternative path produces a separate, distorted image of the blue galaxy.

Multiple images of a gravitationally lensed galaxy are rare. They occur only when a distant galaxy lies directly behind the lensing cluster. However, single distorted images of gravitationally lensed galaxies are quite common.

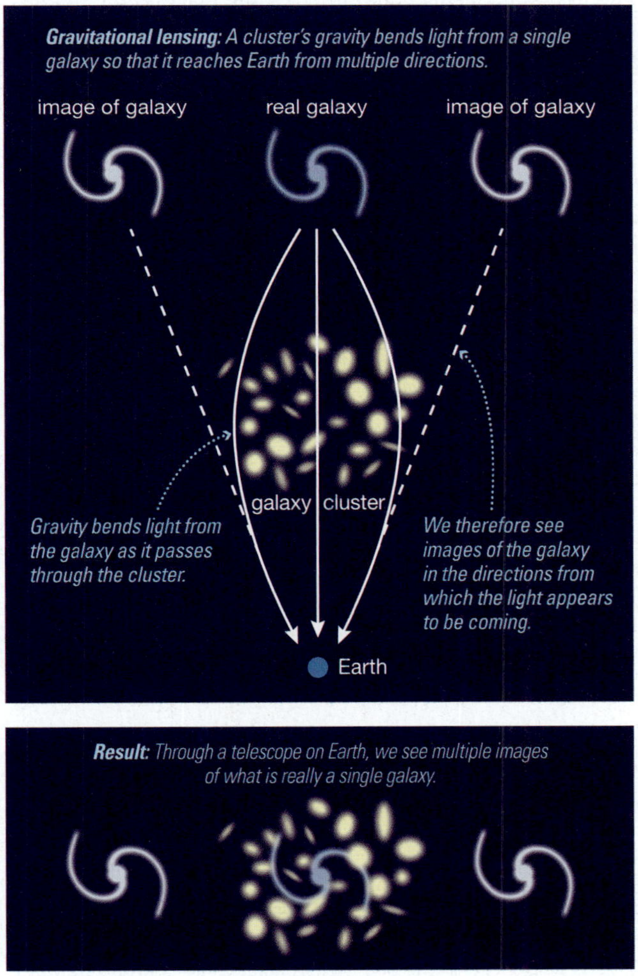


**FIGURE 23.9** A cluster's powerful gravity bends light paths from background galaxies to Earth. If light arrives from several different directions, we see multiple images of the same galaxy.

**FIGURE 23.10** Hubble Space Telescope photo of the cluster Abell 383. The thin, elongated objects are images of background galaxies distorted by the cluster's gravity. By measuring these distortions, astronomers can determine the total amount of mass in the cluster. (The region pictured is about 1 million light-years across.)

**FIGURE 23.10** shows a typical example. This picture shows numerous normal-looking galaxies and several arc-shaped galaxies. The oddly curved galaxies are not members of the cluster, nor are they really curved. They are normal galaxies lying far beyond the cluster whose images have been distorted by the cluster's gravity.

Careful analyses of the distorted images created by clusters enable us to measure cluster masses without using Newton's laws. Instead, Einstein's general theory of relativity tells us how massive these clusters must be to generate the observed distortions. Cluster masses derived in this way generally agree with those derived from galaxy velocities and x-ray temperatures. It is reassuring that the three different methods all indicate that clusters of galaxies hold substantial amounts of dark matter.

## Does dark matter really exist?

Astronomers have made a strong case for the existence of dark matter, but is it possible that there's a completely different explanation for the observations we've discussed? Addressing this question gives us a chance to see how science progresses.

All the evidence for dark matter rests on our understanding of gravity. For individual galaxies, the case for dark matter rests primarily on applying Newton's laws of motion and gravity to observations of the orbital speeds of stars and gas clouds. We've used the same laws to make the case for dark matter in clusters, along with additional evidence based on gravitational lensing predicted by Einstein's general theory of relativity. It therefore seems that one of the following must be true:

1. Dark matter really exists, and we are observing the effects of its gravitational attraction.
2. There is something wrong with our understanding of gravity that is causing us to mistakenly infer the existence of dark matter.

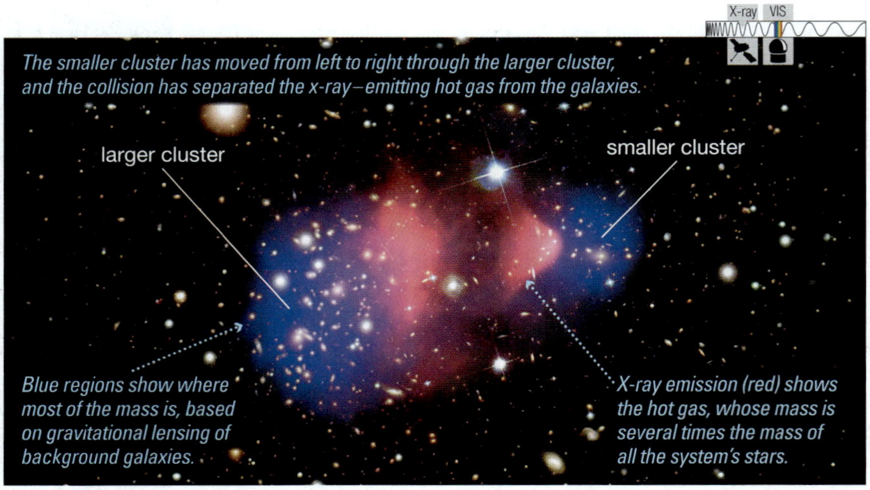


**FIGURE 23.11** Observations of the Bullet Cluster show strong evidence for dark matter. The Bullet Cluster actually consists of two galaxy clusters—the smaller one is emerging from a high-speed collision with the larger one. A map of the system's overall mass (blue) made from gravitational lensing observations does *not* line up with x-ray observations (red) showing the location of the system's hot gas. This fact is difficult to explain without dark matter because the gas contains several times as much mass as all the cluster's stars combined. However, it is easy to explain if dark matter exists: The collision has simply stripped the hot gas away from the dark matter on which it was previously centered.

We cannot completely rule out the second possibility, but most astronomers consider it very unlikely. Newton's laws of motion and gravity are among the most trustworthy tools in science. They have proven to make extremely accurate predictions of orbital motions in the solar system and in binary star systems. We have therefore used them time and again to measure masses of celestial objects from their orbital properties [**Section 4.4**]. They have revealed the general relationships between the masses of stars and their outward appearances and have told us about many things we can't see directly, such as the masses of orbiting neutron stars in x-ray binaries and of black holes in active galactic nuclei. Einstein's general theory of relativity likewise stands on solid ground, having been repeatedly tested and verified to high precision in many observations and experiments. We therefore have good reason to trust our current understanding of gravity.

Moreover, many scientists have made valiant efforts to come up with alternative models of gravity that could account for the observations without invoking dark matter. So far, no one has succeeded in doing so in a way that can also explain the many other observations accounted for by our current theories of gravity. Meanwhile, astronomers keep making observations that are difficult to explain without dark matter. For example, observations of colliding galaxy clusters show that most of the mass detected by gravitational lensing is *not* in the same place as the hot gas, even though the hot gas is several times more massive than the cluster's stars (**FIGURE 23.11**). This finding is at odds with alternative models of gravity, which predict that the hot gas should be doing most of the gravitational lensing.

In essence, our high level of confidence in our current understanding of gravity, combined with observations that

## MATHEMATICAL INSIGHT 23.3 Finding Cluster Masses from Gas Temperature

To find a cluster's mass from the temperature of its hot, x-ray–emitting gas, we need a formula relating the gas temperature to the speeds of individual particles in the gas, which is mostly hydrogen. Although we will not present a derivation here, the following formula applies:

$$v_H = (140 \text{ m/s}) \times \sqrt{T}$$

where $v_H$ is the average orbital speed of the hydrogen nuclei and $T$ is the temperature on the Kelvin scale. Once we find the speeds of the hydrogen nuclei, we can use them in the orbital velocity law to find the cluster mass.

**EXAMPLE:** The galaxy cluster from Mathematical Insight 23.2, with a radius of 6.2 million light-years, is filled with hot gas at a temperature of $9 \times 10^7$ K. Use this temperature to find the cluster's mass.

**SOLUTION:**

**Step 1 Understand:** We can use the formula relating speed and temperature to find the average orbital speed of hydrogen nuclei, which we can then use as the velocity ($v$) in the orbital velocity law. We already know the cluster's radius, which is the only other information we need.

**Step 2 Solve:** Using the given formula and the temperature of $9 \times 10^7$ K, we find that the average orbital speed of the hydrogen nuclei is

$$\begin{aligned} v_H &= (140 \text{ m/s}) \times \sqrt{T} \\ &= (140 \text{ m/s}) \times \sqrt{9 \times 10^7} \\ &= 1.3 \times 10^6 \text{ m/s} \end{aligned}$$

We now find the cluster's mass from the orbital velocity law, using the above value as $v$ and the cluster's radius ($r$ = 6.2 million ly $\approx 5.9 \times 10^{22}$ m):

$$\begin{aligned} M_r &= \frac{r \times v^2}{G} \\ &= \frac{(5.9 \times 10^{22} \text{ m}) \times (1.3 \times 10^6 \text{ m/s})^2}{6.67 \times 10^{-11} \text{ m}^3/(\text{kg} \times \text{s}^2)} \\ &\approx 1.5 \times 10^{45} \text{ kg} \end{aligned}$$

**Step 3 Explain:** The cluster's mass is $1.5 \times 10^{45}$ kilograms, which you can confirm to be about 750 trillion solar masses. This is very close to the 800 trillion solar masses found in Mathematical Insight 23.2 from the galaxy speeds, so the two methods of estimating mass agree well.

seem consistent with dark matter but not with alternative hypotheses, gives us high confidence that dark matter really exists. While we should always keep an open mind about the possibility of future changes in our understanding, we will proceed for now under the assumption that dark matter is real.

**Think about it** Should the fact that we have three different ways of measuring cluster masses give us confidence that we really do understand gravity and that dark matter really does exist? Why or why not?

## What might dark matter be made of?

We have seen strong evidence that dark matter really exists and that it contains far more mass than we observe in the stars and gas found in galaxies and clusters of galaxies. But what exactly is all this dark stuff? There are two basic possibilities:

- It could be made of *ordinary matter* (also called *baryonic matter**), meaning the familiar type of matter built from protons, neutrons, and electrons, but in forms too dark for us to detect with current technology.
- It could consist of one or more types of *exotic matter* (also called *nonbaryonic matter*), meaning particles of matter that are different from those in ordinary atoms and that are dark because they do not interact with light at all.

*Ordinary matter is often called *baryonic* matter, because the protons and neutrons that make up most of its mass belong to a category of particles known as *baryons* (which is the technical term for particles made up of three quarks [**Section S4.2**]).

A first step in distinguishing between the two possibilities is to measure the total amount of matter that is out there. When discussing the matter content of the universe as a whole, astronomers usually focus on density rather than mass. That is, they take the total amount of some type of matter (such as stars, gas, or dark matter) found in a large but typical volume of space and divide by the volume to determine the average density of this type of matter. These densities are then stated as percentages of the *critical density*—the density of mass-energy needed to make the geometry of the universe flat [**Section 22.3**]. The critical density is quite small, equivalent to only $10^{-29}$ gram per cubic centimeter—roughly what we'd find with an average of just a few hydrogen atoms in a volume the size of a closet. Still, the critical density is larger than the average matter density of the observable universe.

The observations of galaxies and clusters that we have discussed indicate that the amount of matter in stars is only about 0.5% of the critical density, while the total amount of matter represents about 30% of the critical density. Clearly, there is a lot of dark matter that needs to be accounted for.

**Ordinary Matter: Not Enough** Could all this dark matter simply be ordinary matter in some hard-to-observe form? After all, matter doesn't necessarily need to be exotic to be dark. Astronomers consider matter to be "dark" as long as it is too dim for us to see at the great distances of the halo of our galaxy or beyond. Your body is dark matter, because our telescopes could not detect you if you were somehow flung into the halo of our galaxy. Planets, the "failed stars" known as brown dwarfs [**Section 16.3**], and even some faint red main-sequence stars also qualify as dark matter, because they are too dim for current telescopes to see in the halo.

### EXTRAORDINARY CLAIMS Most of the Universe's Matter Is Dark

Scientists always take a risk when they publish what they think are groundbreaking results. If their results turn out to be in error, their reputations may suffer. When it came to dark matter, the pioneers in its discovery risked their entire careers. A case in point is Fritz Zwicky and his extraordinary proclamations in the 1930s about dark matter in clusters of galaxies. Most of his colleagues considered him an eccentric who leapt to premature conclusions.

Another pioneer in the discovery of dark matter was Vera Rubin, an astronomer at the Carnegie Institution. Working in the 1960s, she became the first woman to observe under her own name at California's Palomar Observatory, then the largest telescope in the world. (Margaret Burbidge was permitted to observe at Palomar earlier but was required to apply for time under the name of her husband, also an astronomer.) Rubin first saw the gravitational signature of dark matter in spectra that she recorded of stars in the Andromeda Galaxy. She noticed that stars in the outskirts of Andromeda moved at surprisingly high speeds, suggesting a stronger gravitational attraction than the mass of the galaxy's stars alone could explain.

Working with a colleague, Kent Ford, Rubin went on to measure orbital speeds of hydrogen gas clouds in many other spiral galaxies (by studying Doppler shifts in the spectra of hydrogen gas) and discovered that the behavior seen in Andromeda is common. Although Rubin and Ford did not immediately recognize the significance of the results, they were soon arguing that the universe must contain substantial quantities of dark matter.

For a while, many other astronomers had trouble believing the results. Some astronomers suspected that the bright galaxies studied by Rubin and Ford were unusual for some reason. So Rubin and Ford went back to work, obtaining orbital measurements for fainter galaxies. By the 1980s, the evidence that Rubin, Ford, and other astronomers had compiled by measuring rotation curves was so overwhelming that even the critics came around. The orbital speeds could not be explained by the visible mass acting under the known theory of gravity, so either something is amiss in the theory of gravity or vast amounts of dark matter must be present in spiral galaxies. The risks of the pioneers had paid off in a groundbreaking discovery.

**Verdict:** Strongly supported, though there is still some chance that the observations could indicate a problem with our theory of gravity rather than the existence of dark matter.

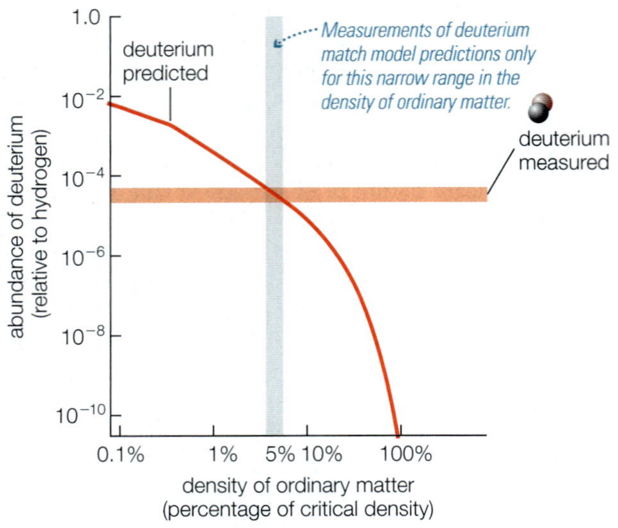


**FIGURE 23.12** This graph shows how the measured abundance of deuterium leads to the conclusion that the density of ordinary matter is about 5% of the critical density. The horizontal band shows the measured abundance of deuterium. The curve shows predictions of the deuterium abundance based on the Big Bang theory, and how it depends on the density of ordinary matter in the universe. Notice that the prediction matches up with the measurements only in the gray vertical strip, which represents a density of about 5% of the critical density.

However, calculations made with the Big Bang model allow scientists to place limits on the total amount of ordinary matter in the universe. Recall that, during the era of nucleosynthesis, protons and neutrons first fused to make deuterium nuclei (nuclei consisting of one proton and one neutron) and these then fused into helium [**Section 22.2**]. The fact that some deuterium nuclei still exist in the universe indicates that this process stopped before all the deuterium nuclei were used up. The amount of deuterium in the universe today therefore tells us about the density of protons and neutrons (ordinary matter) during the era of nucleosynthesis: The higher the density, the more efficiently fusion would have proceeded. A higher density in the early universe would have therefore left less deuterium in the universe today, and a lower density would have left more deuterium.

Calculations based on the observed deuterium abundance indicate that the overall density of ordinary matter in the universe is around 5% of the critical density (**FIGURE 23.12**). Detailed studies of temperature patterns in the cosmic microwave background (see Figures 22.9 and 22.17) give the same result. Because this 5% is much too small to account for the 30% of the critical density consisting of matter, we conclude that most of the dark matter cannot be made from ordinary matter.

**Exotic Matter: The Leading Hypothesis** We are left with the idea that the undiscovered dark matter is made of exotic particles, and probably of a type of exotic particle that has not yet been discovered. Let's begin to explore this possibility by taking another look at a type of exotic particle that we first encountered in connection with nuclear fusion in the Sun: neutrinos [**Section 14.2**]. Neutrinos are dark by nature because they have no electrical charge and cannot emit electromagnetic radiation of any kind. Moreover, they are never bound together with charged particles in the way that neutrons are bound in atomic nuclei, so their presence cannot be revealed by associated light-emitting particles. In fact, neutrinos interact with other forms of matter through only two of the four forces: gravity and the *weak force* [**Sections S4.2, 22.1**]. For this reason, neutrinos are said to be *weakly interacting particles*.

The dark matter in galaxies cannot be made of neutrinos, because these very-low-mass particles travel through the universe at enormous speeds and can easily escape a galaxy's gravitational pull. But what if other weakly interacting particles exist that are similar to neutrinos but considerably heavier? They, too, would evade direct detection, but they would move more slowly, which means that their mutual gravity could hold together a large collection of them. Such hypothetical particles are called **weakly interacting massive particles**, or **WIMPs** for short. Note that they are subatomic particles, so the "massive" in their name is relative—they are massive only in comparison to lightweight particles like neutrinos. Such particles could make up most of the mass of a galaxy or cluster of galaxies, but they would be completely invisible in all wavelengths of light. Most astronomers now consider it likely that WIMPs make up the majority of dark matter, and hence the majority of all matter in the universe.

This hypothesis would also explain why dark matter seems to be distributed throughout spiral galaxy halos rather than concentrated in flattened disks like the visible matter. Recall that galaxies are thought to have formed as gravity pulled together matter in regions of slightly enhanced density in the early universe [**Section 21.1**]. This matter would have consisted mostly of WIMPs mixed with some ordinary hydrogen and helium gas. The ordinary gas could collapse to form a rotating disk because individual gas particles could lose orbital energy: Collisions among many gas particles can convert some of their orbital energy into radiative energy that escapes from the galaxy in the form of photons. In contrast, WIMPs cannot produce photons, and they rarely interact and exchange energy with other particles. As the gas collapsed to form a disk, WIMPs would therefore have remained stuck in orbits far out in the galactic halo—just where most dark matter seems to be located.

**Searching for Dark Matter Particles** The case for the existence of WIMPs seems fairly strong but is still circumstantial. Detecting the particles directly would be much more convincing, and physicists are currently searching for them in two different ways. The first and most direct way is with detectors that can potentially capture WIMPs from space. Because these particles are thought to interact only very weakly, the search requires building large, sensitive detectors deep underground, where they are shielded from other particles from space. As of 2018, these detectors have provided some tantalizing signals, but so far no clear evidence that dark matter particles really exist.

The second way scientists are currently searching for dark matter particles is with particle accelerators, and in particular with the Large Hadron Collider (see Figure S4.1). None of the particles found as of 2018 has the characteristics of a WIMP, but the search is continuing.

**Think about it** Find the latest results in the search for dark matter particles. Have any major claims or discoveries been made? Explain.

## 23.3 Structure Formation

The nature of dark matter remains enigmatic, but we are rapidly learning more about its role in the universe. We've already seen that dark matter appears to be the primary source of gravity in galaxies and galaxy clusters. We therefore suspect that the gravitational attraction of dark matter is what formed structures like galaxies and clusters in the first place.

### What is the role of dark matter in galaxy formation?

Stars, galaxies, and clusters of galaxies are all *gravitationally bound systems*—their gravity is strong enough to hold them together. In most of the gravitationally bound systems we have discussed so far, gravity has completely overwhelmed the expansion of the universe. That is, while the universe as a whole is expanding, space is *not* expanding within our solar system, our galaxy, or our Local Group of galaxies.

Recall that our best model of how galaxies formed, outlined in Section 21.2, envisions them growing from slight density enhancements that were present in the very early universe. During the first few million years after the Big Bang, the universe expanded everywhere. Gradually, the stronger gravity in regions of enhanced density pulled in matter until these regions stopped expanding and became protogalactic clouds, even as the universe as a whole continued to expand. As we've already discussed, dark matter consisting of WIMPs would have remained in the galactic halos because of its inability to radiate away orbital energy, while ordinary gas collapsed to form stars and galactic disks.

**Think about it** State whether each of the following is a gravitationally bound system, and explain why: (a) Earth; (b) a hurricane on Earth; (c) the Orion Nebula; (d) a supernova.

Galaxy clusters probably formed similarly. Early on, all the galaxies that would eventually constitute a cluster were flying apart with the expansion of the universe. The gravity of the dark matter associated with the cluster eventually reversed the trajectories, so that the galaxies ultimately fell back inward and began orbiting each other with random orientations, much like the stars in the halo of our galaxy.

Some clusters apparently have not yet finished forming, because their immense gravity is still drawing in new galaxies. For example, the relatively nearby Virgo Cluster of galaxies (about 60 million light-years away) appears to be tugging on the Milky Way and other galaxies of the Local Group. The evidence comes from careful study of galaxy speeds. Plugging the Virgo Cluster's distance into Hubble's law tells us the speed at which the Milky Way and the Virgo Cluster should be drifting apart as a result of universal expansion [**Section 20.3**]. However, the measured speed is about 220 kilometers per second slower than the speed we predict from Hubble's law alone. We conclude that this 220 kilometers per second discrepancy (sometimes called a *peculiar velocity*) arises because the Virgo Cluster's gravity is pulling us back against the flow of universal expansion. In other words, while the Milky Way and other galaxies of our Local Group are still moving away from the Virgo Cluster with the expansion of the universe, the rate at which we are separating from the cluster is slowing with time.

Many other large clusters of galaxies also appear to be drawing in new members, judging from the velocities of galaxies near the outskirts of those clusters. On even larger scales, clusters themselves seem to be tugging on one another, hinting that they might be parts of even bigger gravitationally bound systems, called **superclusters**, that are still in the early stages of formation (**FIGURE 23.13**). Clearly, the gravity of dark matter is having an enormous and ongoing influence on large structures in the universe.

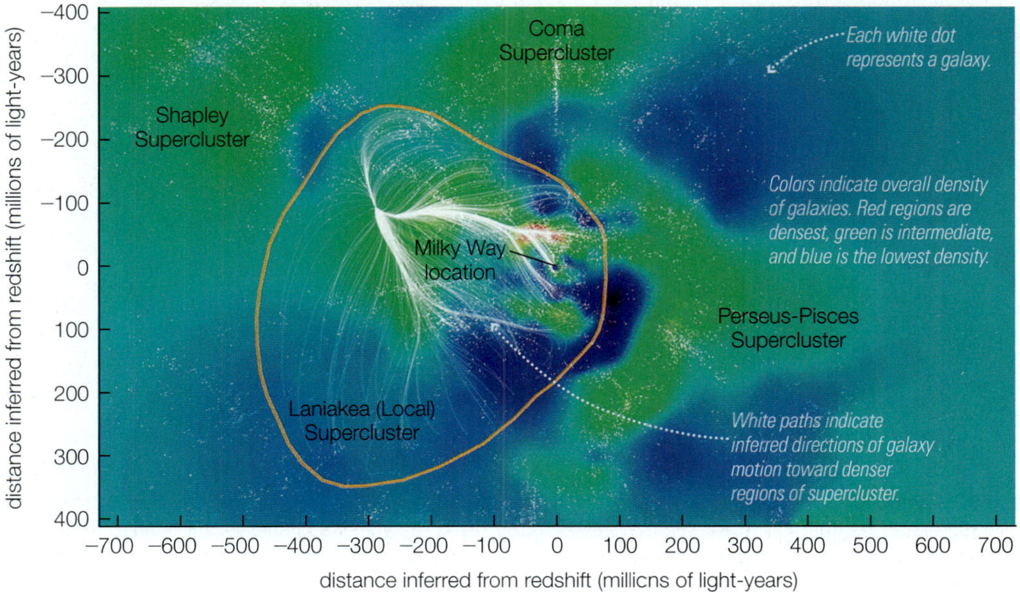


**FIGURE 23.13** This diagram shows how galaxies move within the Local Supercluster (or Laniakea), which is the region within the orange line. White dots represent locations of individual galaxies, with distances from the Milky Way inferred from Hubble's law. Galaxies flow toward denser regions of the supercluster along the white lines, which have been inferred from a combination of observations and modeling.

## What are the largest structures in the universe?

Beyond about 300 million light-years from Earth, deviations from Hubble's law owing to gravitational tugs are insignificant compared with the universal expansion, so Hubble's law becomes our primary method for measuring galaxy distances [**Section 20.2**]. Using this law, astronomers can make maps of the distribution of galaxies in space. Such maps reveal **large-scale structures** much vaster than clusters of galaxies.

**Mapping Large-Scale Structures** Making maps of galaxy locations requires an enormous amount of data. A long-exposure photo showing galaxy positions is not enough, because it does not tell us the galaxy distances. We must also measure the redshift of each individual galaxy so that we can estimate its distance by applying Hubble's law. These measurements once required intensive labor, and up until a couple of decades ago it took years of effort to map the locations of just a few hundred galaxies. However, astronomers have since developed technology that allows them to measure redshifts for hundreds of galaxies during a single night of telescopic observation. As a result, we now have redshift measurements—and hence estimated distances—for millions of distant galaxies.

**FIGURE 23.14** shows the distribution of galaxies in three slices of the universe, each extending farther out in distance than the one before. Our Milky Way Galaxy is located at the vertex at the far left, and each dot represents an entire galaxy of stars. The slice at the left comes from one of the first surveys of large-scale structures, performed at the Harvard-Smithsonian Center for Astrophysics (CfA) in the 1980s. This map, which required years of effort by many astronomers, dramatically revealed the complex structure of our corner of the universe. It showed that galaxies are not scattered randomly through space but are instead arranged in huge chains and sheets that span many millions of light-years. Clusters of galaxies are located at the intersections of these chains. Between these chains and sheets of galaxies lie giant empty regions called **voids**. The other two slices show data from the more recent Sloan Digital Sky Survey, which has measured redshifts for more than a million galaxies spread across about one-fourth of the sky.

Some of the structures in these pictures are amazingly large. The so-called Sloan Great Wall, clearly visible in the center slice, extends more than 1 billion light-years from end to end. Immense structures such as these apparently have not yet collapsed into randomly orbiting, gravitationally bound systems.

The universe may still be growing structures on these very large scales. However, there seems to be a limit to the size of the largest structures. If you look closely at the rightmost slice in Figure 23.14, you'll notice that the overall distribution of galaxies appears nearly uniform on scales larger than about a billion light-years. In other words, on very large scales the universe looks much the same everywhere, in agreement with what we expect from the *Cosmological Principle* [**Section 20.3**].

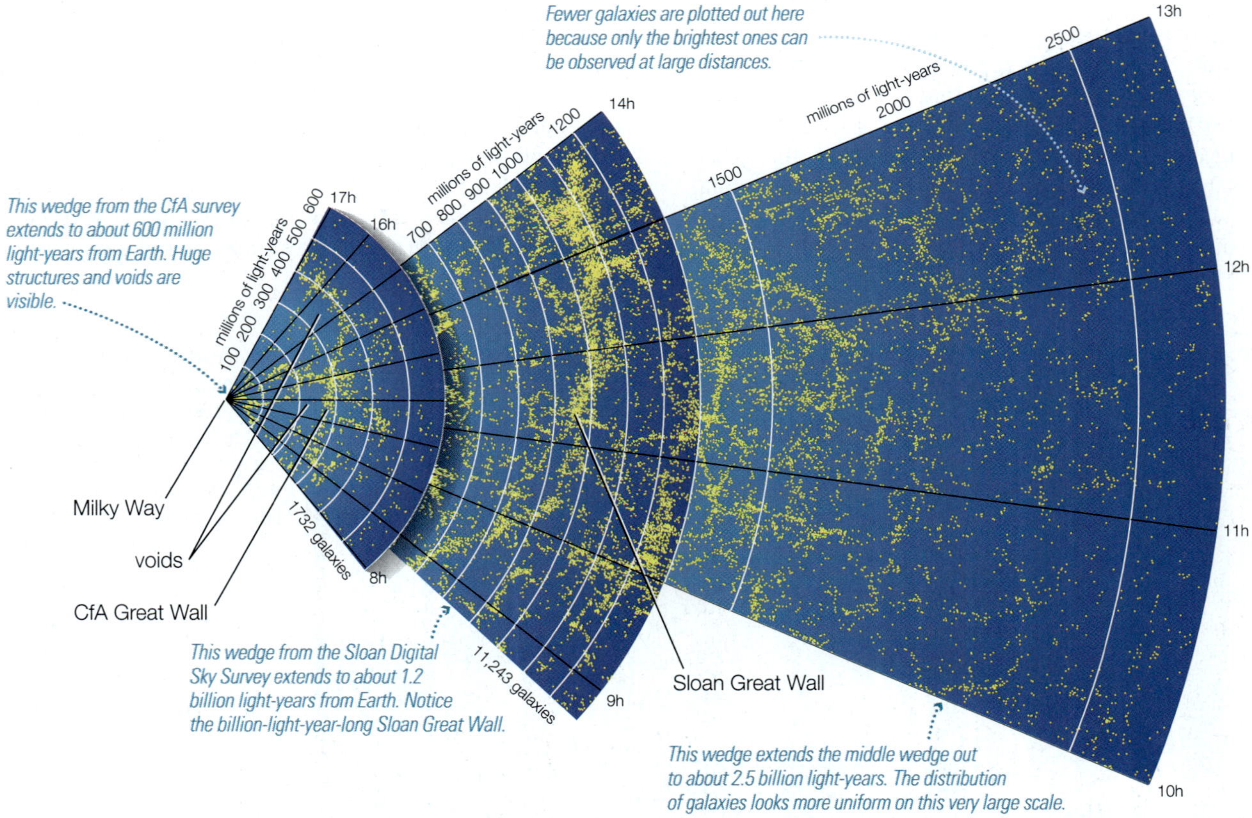


**FIGURE 23.14** Each of these three wedges shows a "slice" of the universe extending outward from our own Milky Way Galaxy. The dots represent galaxies, shown at their measured distances from Earth. We see that galaxies trace out long chains and sheets, surrounded by huge voids containing very few galaxies. (The wedges are shown flat but actually are a few angular degrees in thickness; the CfA wedge at left does not actually line up with the two Sloan wedges.)

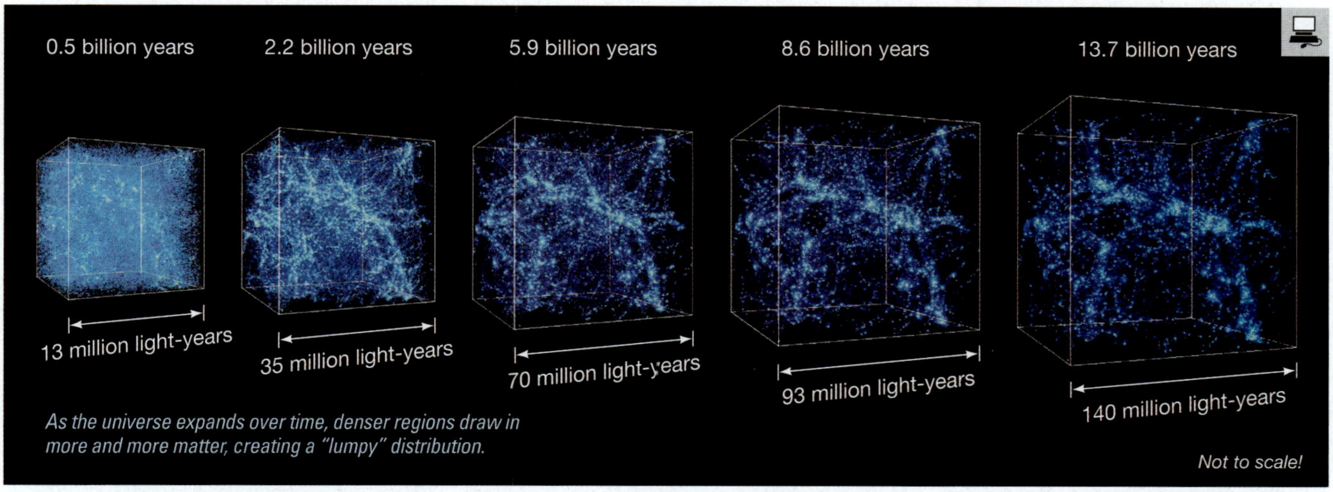


**FIGURE 23.15** Frames from a supercomputer simulation of structure formation. The five boxes depict the development of a cubical region that is now 140 million light-years across. The labels above the boxes give the age of the universe, and the labels below give the size of the box as it expands with time. Notice that the distribution of matter is only slightly lumpy when the universe is young (left frame). Structures grow more pronounced with time as the densest lumps draw in more and more matter.

**The Origin of Large Structures** Why is gravity collecting matter on such enormous scales? Just as we suspect that galaxies formed from regions of slightly enhanced density in the early universe, we suspect that these larger structures were regions of enhanced density. Galaxies, clusters, superclusters, and the Sloan Great Wall probably all started as mildly high-density regions of different sizes. The voids in the distribution of galaxies probably started as mildly low-density regions.

According to this model of structure formation, the structures we see in today's universe should mirror the original distribution of dark matter very early in time. Supercomputer models of structure formation in the universe can now simulate the growth of galaxies, clusters, and larger structures from tiny density enhancements as the universe evolves (**FIGURE 23.15**). Models of extremely large regions reveal how dark matter should be distributed throughout the entire observable universe (**FIGURE 23.16**). The results of these models look remarkably similar to the slices of the universe in Figure 23.14, bolstering our confidence in this scenario. Moreover, the patterns of mass distribution are consistent with the patterns of density enhancements revealed in maps of the cosmic microwave background. Overall, we now seem to have a basic picture of how galaxies and large-scale structures formed in the universe, perhaps starting from quantum fluctuations that occurred when the universe was a tiny fraction of a second old [**Section 22.3**].

## 23.4 Dark Energy and the Fate of the Universe

Over the past few chapters, we have seen that the large-scale development of the universe to date has been governed by two competing processes:

1. the ongoing expansion that began in the Big Bang, which tends to drive galaxies apart from one another, and
2. the gravitational attraction of matter in the universe, which assembles galaxies and larger-scale structures around the density enhancements that emerged from the Big Bang.

These ideas naturally lead us to one of the ultimate questions in astronomy: How will the universe end? After Edwin Hubble discovered the expansion of the universe, astronomers generally assumed that the end would be like one of the two fates in a famous poem by Robert Frost:

*Some say the world will end in fire,*
*Some say in ice.*
*From what I've tasted of desire*
*I hold with those who favor fire.*
*But if it had to perish twice,*
*I think I know enough of hate*
*To say that for destruction ice*
*Is also great*
*And would suffice.*

—Robert Frost (American poet 1874–1963), "Fire and Ice"

If gravity were strong enough, the expansion would someday halt and reverse; the universe would then begin collapsing and heating back up, eventually ending in a fiery and cataclysmic crunch. Alternatively, if the total strength of gravity were too weak, gravity would never slow the expansion enough for it to halt and reverse, leading to an icy end in which the universe would grow ever colder as its galaxies moved ever farther apart. Much to astronomers' surprise, more than two decades of observation now show that the expansion is not slowing at all, but instead is accelerating with time. As we've discussed, this implies the existence of some repulsive force produced by a mysterious form of energy we have come to call *dark energy*.

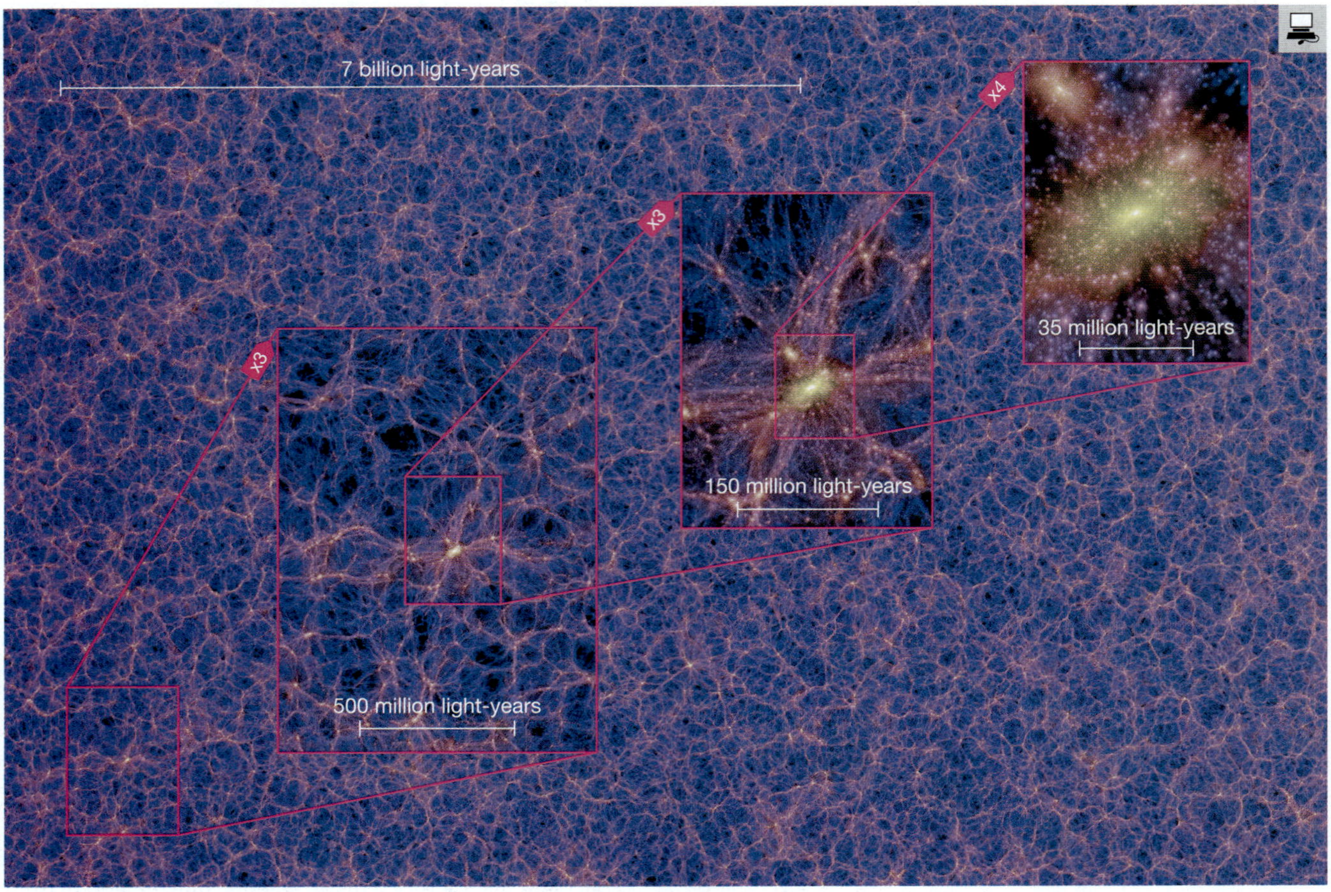


**FIGURE 23.16** These images from an extremely large computer simulation illustrate the structure of dark matter in the universe. The main image shows a region similar in size to our observable universe, and the image sequence zooms in on a massive cluster of galaxies. The images show structure as it would appear if we could see dark matter—the brightest clumps in the image represent the highest densities of dark matter. Notice that the large-scale distribution of dark matter has a uniform web-like pattern.

**Evidence for Acceleration**

## What is the evidence for an accelerating expansion?

Astronomers can learn how the expansion rate of the universe has changed with time by carefully measuring the recession speeds and distances of very distant galaxies. Because those measurements show how fast galaxies were moving when the universe was much younger, careful analysis can tell us whether galaxy speeds have slowed down, held steady, or sped up over time.

**Four Expansion Models** To understand the evidence that has led astronomers to conclude that the expansion is accelerating, let's consider how different amounts of gravity and repulsion cause the expansion rate to change with time, according to the following four general models:

- A **recollapsing universe**. In the case of extremely strong gravitational attraction and no repulsive force, the expansion would continually slow down with time and eventually would stop entirely and then reverse. Galaxies would come crashing back together, and the universe would end in a fiery "Big Crunch." We call this a *recollapsing* universe, because the final state, with all matter collapsed together, would look much like the state in which the universe began in the Big Bang.
- A **critical universe**. In the case of gravitational attraction that was not quite strong enough to reverse the expansion in the absence of a repulsive force, the expansion would decelerate forever, leading to a universe that would never collapse but would expand ever more slowly as time progressed. We call this a *critical* universe, because calculations show that it is what we would expect if the total density of the universe were the critical density and only matter (and not dark energy) contributed to this density.
- A **coasting universe**. In the case of weak gravitational attraction and no repulsive force, galaxies would always move apart at approximately the speeds they have today. We call this a *coasting* universe, because it is what we would find if no forces acted to change the expansion rate, much as a spaceship can coast through space at constant speed if no forces act to slow it down or speed it up.
- An **accelerating universe**. In the case of a repulsive force strong enough to overpower gravity, the expansion would *accelerate* with time, causing galaxies to recede from one another with ever-increasing speed.

**FIGURE 23.17** shows these four models schematically. Notice that, because we can measure the present size and expansion rate of the universe, all four models agree on these values along the horizontal red line that indicates the

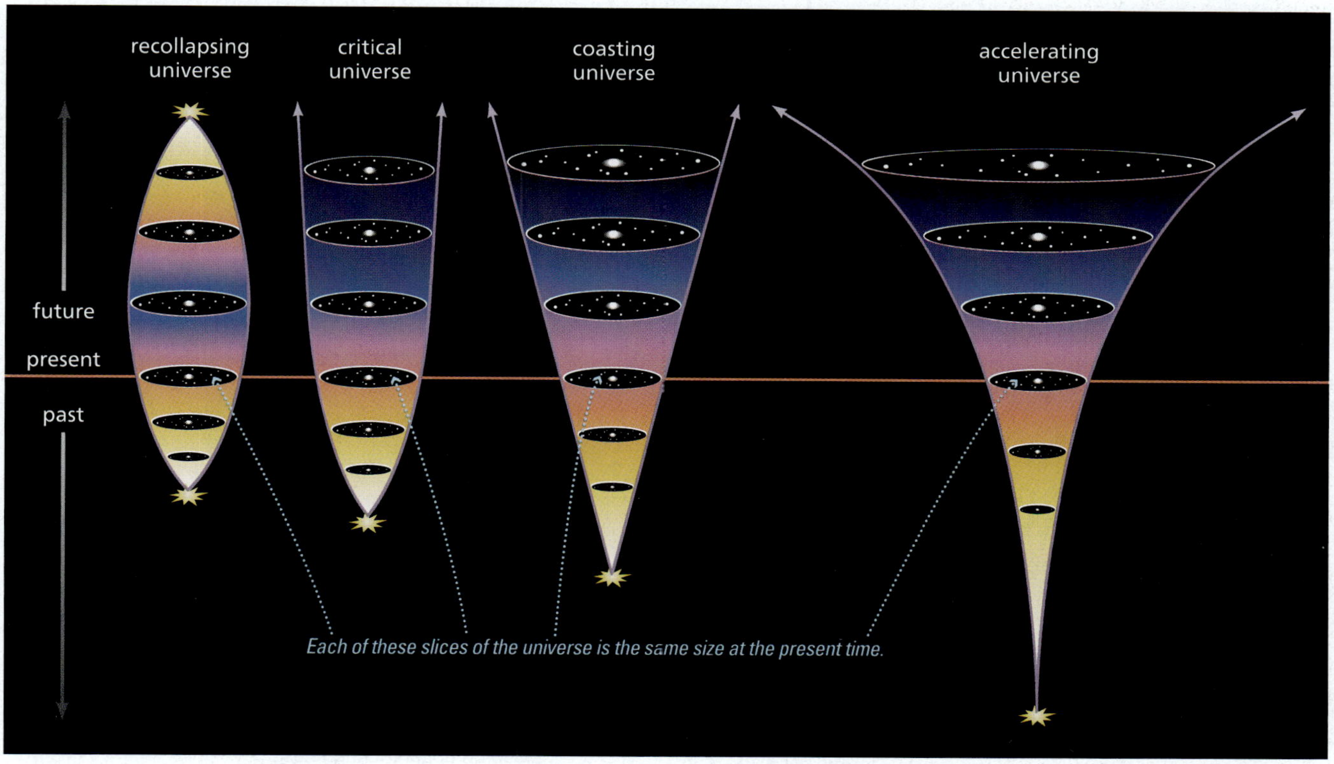


**FIGURE 23.17** Four general models for how the universal expansion rate could in principle change with time. Each diagram shows how the size of a circular slice of the universe changes with time in a particular model. The slices are the same size at the present time, marked by the red line, but the models make different predictions about the sizes of the slices in the past and future.

present time. However, the four models each extend different lengths into the past, which means that each model leads to a different age for the universe today. The coasting model assumes that the expansion rate never changes, and its starting point therefore indicates the age of the universe that we would infer from Hubble's constant alone. The recollapsing and critical models both give younger ages for the universe, because these models assume that galaxy speeds would have been faster in the past, so galaxies would have reached their current distances in less total time. The accelerating model leads to an older age, because it assumes that galaxy speeds would have been slower in the past, so galaxies would have required more time to reach their current distances.

**Evidence for Acceleration** **FIGURE 23.18** shows the key evidence that has led astronomers to favor the accelerating model. The four solid curves show how the four general models predict that the average distance between galaxies should have changed with time. The curves for the accelerating, coasting, and critical universes always continue upward as time increases, because in these cases the universe is always expanding. The steeper the slope, the faster the expansion. In the recollapsing case, the curve begins on an upward slope but eventually turns around and declines as the universe contracts. All the curves pass through the same point and have the same slope at the moment labeled "now," because the current separation between galaxies

## SPECIAL TOPIC Einstein's "Greatest Blunder"

Shortly after Einstein completed his general theory of relativity in 1915, he found that it predicted that the universe could not be standing still: The mutual gravitational attraction of all the matter would make the universe collapse. Because Einstein thought at the time that the universe should be eternal and static, he decided to alter his equations. In essence, he inserted a "fudge factor" called the *cosmological constant* that acted as a repulsive force to counteract the attractive force of gravity.

Had he not been so convinced that the universe should be standing still, Einstein might instead have come up with the correct explanation for why the universe is not collapsing: because it is still expanding from the event of its birth. After Hubble discovered universal expansion, Einstein supposedly called his invention of the cosmological constant "the greatest blunder" of his career.

Now that observations of very distant galaxies (using white dwarf supernovae as standard candles) have shown that the universe's expansion is accelerating, Einstein's idea of a universal repulsive force doesn't seem so far-fetched. In fact, observations to date are consistent with the idea that dark energy has properties virtually identical to those that Einstein originally proposed for the cosmological constant. In particular, the amount of dark energy in each volume of space seems to remain unchanged while the universe expands, as if the vacuum of space itself were constantly rippling with energy—which is just what the cosmological constant does in Einstein's equations. We'll need more measurements to know for sure, but it is beginning to seem that Einstein's greatest blunder may not have been a blunder after all.

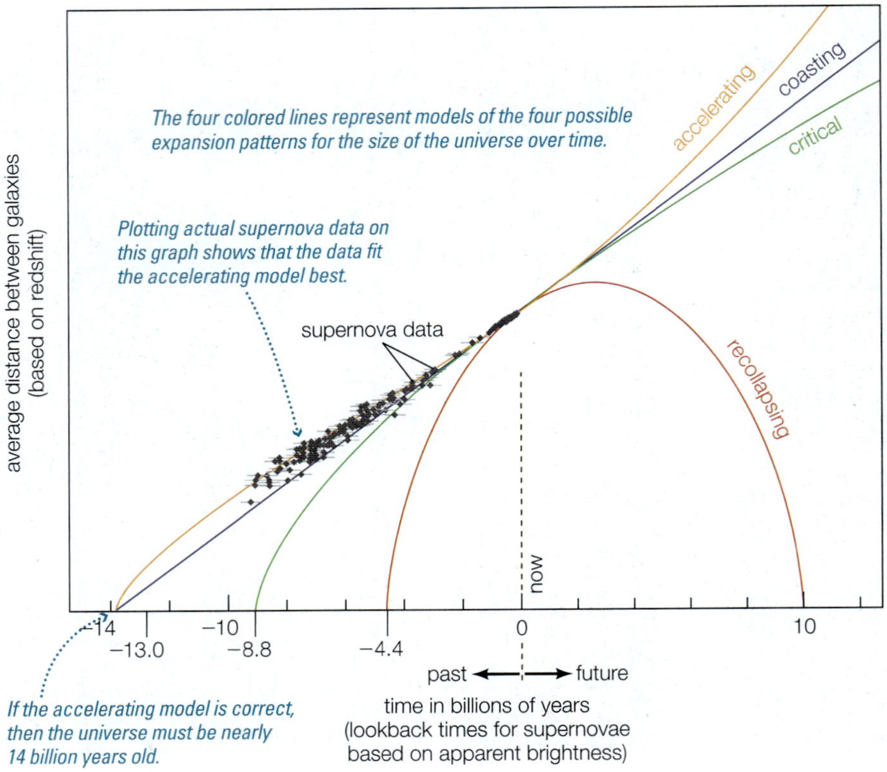


**FIGURE 23.18** Data from white dwarf supernovae are shown, along with four possible models for the expansion of the universe. Each curve shows how the average distance between galaxies changes with time for a particular model. A rising curve means that the universe is expanding, and a falling curve means that the universe is contracting. Notice that the supernova data fit the accelerating universe better than the other models.

and the current expansion rate in each case must agree with observations of the present-day universe.

**See it for yourself** Toss a ball in the air, and observe how it rises and falls. Then make a graph to illustrate your observations, with time on the horizontal axis and height on the vertical axis. Which universe model does your graph most resemble? What is the reason for that resemblance? How would your graph look different if Earth's gravity were not as strong? Would the time for the ball to rise and fall be longer or shorter?

The black dots in Figure 23.18 represent actual data, which are based on distance measurements using white dwarf supernovae as standard candles [**Section 20.2**]. (The horizontal line through each dot indicates the range of uncertainty in the measured lookback time.) Although there is some scatter in the data points, they clearly fit the curve for the accelerating model better than that for any of the other models. In other words, the observations agree best with a model of the universe in which the expansion is accelerating with time.

The discovery of an accelerating expansion, first announced in 1998, came as a great surprise to virtually all astronomers. For several years after the announcement, many astronomers feared that the measurements were being misinterpreted, but additional data have only strengthened the evidence of acceleration. In recognition of the importance of this discovery, three of the leaders of the observing teams—Saul Perlmutter, Brian Schmidt, and Adam Riess—were awarded the 2011 Nobel Prize in physics.

**The Nature of Dark Energy** The acceleration of the expansion clearly implies the existence of some force that acts to push galaxies apart, and the source of this force has been dubbed *dark energy*. Keep in mind, however, that scientists have little idea of what the nature of dark energy might actually be. None of the four known forces in nature could provide a force to oppose gravity, and while some theories of fundamental physics suggest ways in which energy could fit the bill, no known type of energy produces the right amount of acceleration.

Continued observations of distant supernovae have the potential to tell us exactly how large an effect dark energy has had throughout cosmic history and whether the strength of this effect has changed with time. Already there are some intriguing hints. For example, it appears that the expansion was *not* accelerating immediately after the Big Bang, indicating that gravity was strong enough to slow the expansion during the universe's first few billion years, when its matter density was greater. Apparently, it took a few billion years for expansion to spread out the matter and weaken gravity enough for dark energy to cause acceleration. (The curve for the accelerating model in Figure 23.18 shows this scenario.) Interestingly, this type of behavior is consistent with an idea that Einstein once introduced but later disavowed in his general theory of relativity, leading some scientists to suggest that dark energy might successfully be described by a term in Einstein's equations that describe gravity (see Special Topic, p. 685). But even if this idea turns out to be correct, we remain a long way from an actual understanding of dark energy's nature.

## Why is flat geometry evidence for dark energy?

The evidence for the existence of dark energy provided by observations of an accelerating expansion seems quite strong, but it is important to remember that it is based on measurements of distances to white dwarf supernovae. While we have good reason to think that these supernovae make reliable standard candles, having just a single source of evidence would be cause for at least some concern. Fortunately, a second and independent line of evidence also supports the existence of dark energy, and it gives results that are fully consistent with the results indicating an accelerating expansion.

**Flatness and Dark Energy** Recall that Einstein's general theory of relativity tells us that the overall geometry of the universe can take one of three general forms—spherical, flat, or saddle shaped (see Figure 22.15)—and that careful observations of the cosmic microwave background provide strong evidence that the actual geometry is flat (see Figure 22.17). As we discussed in Chapter 22, a flat geometry implies that the total density of *matter plus energy* in the universe must be exactly equal to the critical density.

Now, recall that observations of dark matter indicate that the total density of matter amounts to only about 30% of the critical density. In that case, the remaining 70% of the universe's density must be in the form of energy. Tellingly, the amount of dark energy required to explain the observed acceleration of the expansion also is about 70% of the critical density. The startling conclusion: About 70% of the total mass-energy of the universe takes the form of dark energy.

**Inventory of the Universe** We began this chapter by noting that astronomers today must admit the embarrassing fact that we do not yet know what most of the universe is made of. It appears to be made of things we call dark matter and dark energy, but we do not yet know the true nature of either one. Nevertheless, the observations we have discussed allow us to make quantitative statements about our ignorance. According to the model that best explains the temperature patterns that the *Planck* mission observed in the cosmic microwave background, the total density of matter plus energy in the universe is equal to the critical density, and the universe's mass-energy has the following components, which are summarized as a pie chart in **FIGURE 23.19**:

- Ordinary matter (made up of protons, neutrons, electrons) makes up about 5% of the total mass-energy of the universe. Note that this model prediction agrees with what we find from observations of deuterium in the universe. Some of this matter is in the form of stars (about 0.5% of the universe's mass-energy). Most of the rest consists of gas that lies outside of the luminous parts of galaxies, like the hot gas found in galaxy clusters, but is harder to detect.
- Some form of exotic dark matter—most likely weakly interacting massive particles (WIMPs)—makes up about 27% of the mass-energy of the universe, in close agreement with what we infer from measurements of the masses of clusters of galaxies.
- Dark energy makes up the remaining 68% of the mass-energy of the universe, accounting both for the observed acceleration of the expansion and for the pattern of temperatures in the cosmic microwave background.

**FIGURE 23.20** summarizes the evidence we have discussed for the existence of dark matter and dark energy. We may not yet know what either dark matter or dark energy actually is, but our measurements of how much matter and energy may be out there are becoming quite precise.

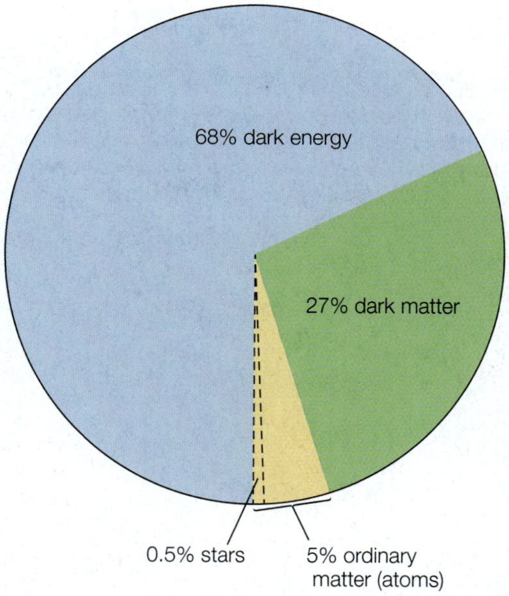


**FIGURE 23.19** This pie chart shows the proportion of each of the major components of matter and energy in the universe, based on current evidence.

**The Age of the Universe** Models that explain the temperature variations in the cosmic microwave background not only give us an inventory of the universe but also make precise predictions about the age of the universe. According to the model that best agrees with the data (the same model used for the inventory above), the age of the universe is about 13.8 billion years, with an uncertainty of about 0.1 billion years (100 million years). That is why, throughout this book, we have said that the universe is "about 14 billion years old." Note that this age is in good agreement with what we infer from Hubble's constant and observed changes in the expansion, and also agrees well with the fact that the oldest stars in the universe appear to be about 13 billion years old.

## What is the fate of the universe?

*Into its own beyond,*
*Till it exhausts itself and lies down cold,*
*Its last star going out.*

—*Rebecca Elson (1960–1999), astronomer and poet, from "Let There Always Be Light"*

We are now ready to return to the question of the fate of the universe. If we think in terms of Robert Frost's poetry at the beginning of this section, the recollapsing universe is the only one of our four possible expansion models that has an end in fire, and the data do not fit that model. Therefore, it seems that the universe is doomed to expand forever, its galaxies receding ever more quickly into an icy, empty future. The end, it would seem, is more likely to be like that described in the poetry excerpt from Rebecca Elson above.

COSMIC Context

FIGURE 23.20 **Dark Matter and Dark Energy**

**Scientists suspect that most of the matter in the universe is *dark matter* we cannot see, and that the expansion of the universe is accelerating because of a *dark energy* we cannot directly detect. Both dark matter and dark energy have been proposed to exist because they bring our models of the universe into better agreement with observations, in accordance with the process of science. This figure presents some of the evidence supporting the existence of dark matter and dark energy.**

① **Dark Matter in Galaxies:** Applying Newton's laws of gravity and motion to the orbital speeds of stars and gas clouds suggests that galaxies contain much more matter than we observe in the form of stars and glowing gas.

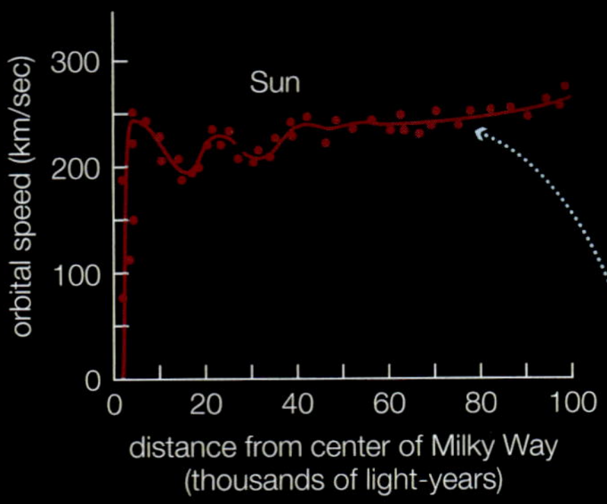


*Orbital speeds of stars and gas clouds remain high even quite far from our galaxy's center . . .*

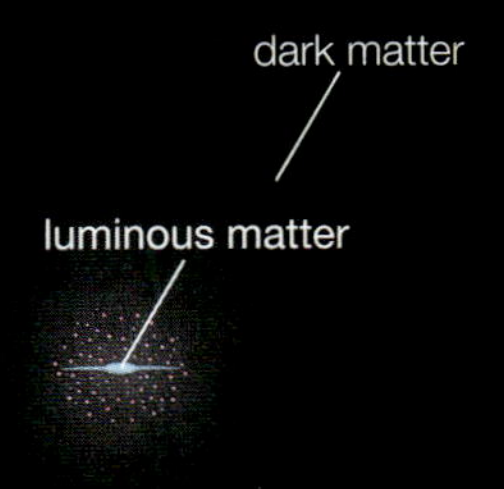


*. . . indicating that the visible portion of our galaxy lies at the center of a much larger volume of dark matter.*

HALLMARK OF SCIENCE **A scientific model must seek explanations for observed phenomena that rely solely on natural causes.** Orbital motions within galaxies demand a natural explanation, which is why scientists proposed the existence of dark matter.

② **Dark Matter in Clusters:** Further evidence for dark matter comes from studying galaxy clusters. Observations of galaxy motions, hot gas, and gravitational lensing all suggest that galaxy clusters contain far more matter than we can directly observe in the form of stars and gas.

*This cluster of galaxies acts as a* **gravitational lens** *to bend light from a single galaxy behind it into the multiple blue shapes in this photo. The amount of bending allows astronomers to calculate the total amount of matter in the cluster.*

HALLMARK OF SCIENCE **Science progresses through creation and testing of models of nature that explain the observations as simply as possible.** Dark matter accounts for our observations of galaxy clusters more simply than alternative hypotheses.

**3 Structure Formation:** If dark matter really is the dominant source of gravity in the universe, then its gravitational force must have been what assembled galaxies and galaxy clusters in the first place. We can test this prediction using supercomputers to model the formation of large-scale structures both with and without dark matter. Models with dark matter provide a better match to what we actually observe in the universe.

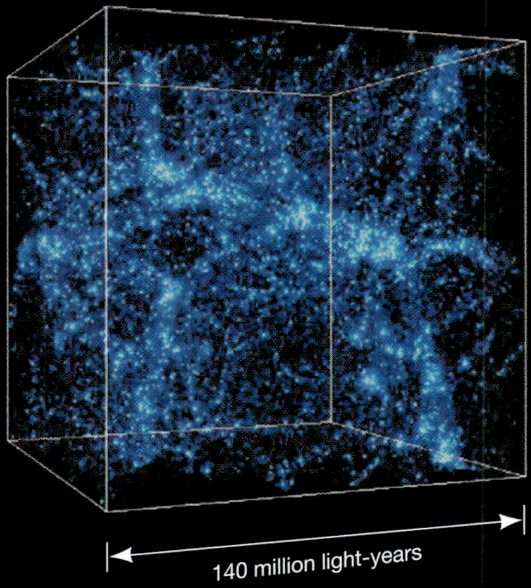


*Supercomputer models in which dark matter is the dominant source of gravity show galaxies organized into strings and sheets similar in size and shape to those we actually observe in the universe.*

**4 Universal Expansion and Dark Energy:** The expansion of a universe consisting primarily of dark matter would slow down over time because of gravity, but observations have shown that the expansion is actually speeding up. Scientists hypothesize that a mysterious *dark energy* is causing the expansion to accelerate. Models that include both dark matter and dark energy agree more closely with observations of distant supernovae and the cosmic microwave background than models containing dark matter alone.

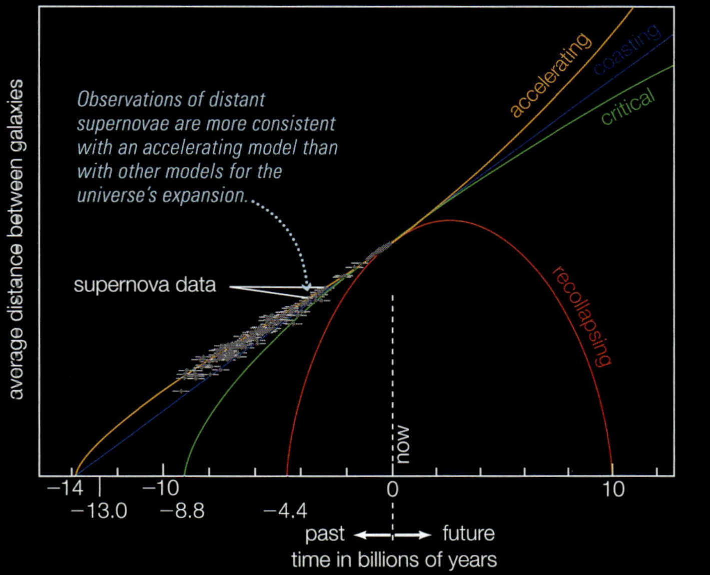


HALLMARK OF SCIENCE **A scientific model makes testable predictions about natural phenomena. If predictions do not agree with observations, the model must be revised or abandoned.** Observations of the universe's expansion have forced us to modify our models of the universe to include dark energy along with dark matter.

**Think about it** Do you think that one of the possible fates (fire or ice) is preferable to the other? Why or why not?

**The Next $10^{100}$ Years** What will happen in the distant future if the universe continues to expand forever? We can use our current understanding of physics to hypothesize about the answer.

First, the answer obviously depends on how much the expansion of the universe accelerates in the future. Some scientists speculate that the repulsive force due to dark energy might strengthen with time. In that case, perhaps in a few tens of billions of years, the growing repulsive force would tear apart our galaxy, our solar system, and even matter itself in a catastrophic event sometimes called the "Big Rip." However, evidence for this type of growing repulsion is very weak.

It seems more likely that the expansion will continue to accelerate more gradually, in which case galaxies and galaxy clusters will remain gravitationally bound far into the future. Galaxies will not always look the same, however, because the star–gas–star cycle [**Section 19.2**] cannot continue forever. With each generation of stars, more mass becomes locked up in planets, brown dwarfs, white dwarfs, neutron stars, and black holes. Eventually, about a trillion years from now, even the longest-lived stars will burn out, and the galaxies will fade into darkness.

At this point, the only new action in the universe will occur on the rare occasions when two objects—such as two brown dwarfs or two white dwarfs—collide within a galaxy. The vast distances separating star systems in galaxies make such collisions extremely rare. For example, the probability of our Sun (or the white dwarf that it will become) colliding with another star is so small that it would be expected to happen only once in a quadrillion ($10^{15}$) years. However, given a long enough period of time, even low-probability events will eventually happen many times. If a star system experiences a collision once in a quadrillion years, it will experience about 100 collisions in 100 quadrillion ($10^{17}$) years. By the time the universe reaches an age of $10^{20}$ years, star systems will have suffered an average of 100,000 collisions each, making a time-lapse history of any galaxy look like a cosmic game of billiards.

These multiple collisions will severely disrupt galaxies. As in any gravitational encounter, some objects lose energy in such collisions and some gain energy. Objects that gain enough energy will be flung into intergalactic space, to be carried away from their home galaxies with the expansion of the universe. Objects that lose energy will eventually fall to the galactic center, leaving behind a single supermassive black hole where a galaxy used to be. The remains of the universe will consist of widely separated black holes with masses as great as a trillion solar masses, and widely scattered planets, brown dwarfs, and stellar corpses. If Earth somehow survives, it will be a frozen chunk of rock in the darkness of the expanding universe, billions of light-years away from any other solid object.

If grand unified theories [**Section 22.1**] are correct, Earth still cannot last forever. These theories predict that protons will eventually fall apart. The predicted lifetime of protons is extremely long: a half-life of at least $10^{33}$ years. However, if protons really do decay, then by the time the universe is $10^{40}$ years old, Earth and all other atomic matter will have disintegrated into radiation and subatomic particles.

The final phase may come through a mechanism proposed by physicist Stephen Hawking. Recall that he predicted that black holes must eventually "evaporate," turning their mass-energy into *Hawking radiation* [**Section S4.4**]. The process is so slow that we do not expect to be able to see it from any existing black holes, but if it really occurs, then black holes in the distant future will disappear in brilliant bursts of radiation. The largest black holes will last the longest, but even trillion-solar-mass black holes will evaporate sometime after the universe reaches an age of $10^{100}$ years. From then on, the universe will consist only of individual photons and subatomic particles, each separated by enormous distances from the others. Nothing new will ever happen, and no events will ever occur that would allow an omniscient observer to distinguish past from future. In a sense, the universe will finally have reached the end of time.

**Forever Is a Long Time** Lest any of this sound depressing, keep in mind that we are talking about incredibly long times. Remember that $10^{11}$ years is already nearly 10 times the current age of the universe (because 14 billion years is the same as $1.4 \times 10^{10}$ years), $10^{12}$ years is another 10 times that, and so on. A time of $10^{100}$ years is so long that we can scarcely describe it, but one way to think about it (thanks to the late Carl Sagan) is to imagine that you wanted to write on a piece of paper a number that consisted of a 1 followed by $10^{100}$ zeros (that is, the number $10^{10^{100}}$). It sounds easy, but a piece of paper large enough to hold all those zeros *would not fit in the observable universe* today. If that still does not alleviate your concerns, you may be glad to know that a few creative thinkers are already speculating about ways in which the universe might avoid an icy fate or undergo rebirth, even after the end of time.

Perhaps of greater significance, speculating about the future of the universe means speculating about forever, and forever leaves us with a very long time in which to make new discoveries. After all, it is only in the past century that we learned that we live in an expanding universe, and only in the past couple of decades that we were surprised to learn that the expansion is accelerating. The universe may yet hold other surprises that might force us to rethink what might happen between now and the end of time.

# The BIG Picture PUTTING CHAPTER 23 INTO CONTEXT

*We have found that there may be much more to the universe than meets the eye. Dark matter too dim for us to see seems to far outweigh the stars, and a mysterious dark energy may be even more prevalent. Together, dark matter and dark energy have probably been the dominant agents of change in the overall history of the universe. Here are some key "big picture" points to remember about this chapter:*

- Dark matter and dark energy sound very similar, but they are each hypothesized to explain different observations. Dark matter is thought to exist because we detect its gravitational influence. Dark energy is a term given to the source of the force that may be accelerating the expansion of the universe.
- Either dark matter exists or we do not understand how gravity operates across galaxy-size distances. There are many reasons to be confident about our understanding of gravity, leading most astronomers to conclude that dark matter is real.
- Dark matter seems to be by far the most abundant form of mass in the universe, and therefore the primary source of the gravity that has formed galaxies and larger-scale structures from tiny density enhancements that existed in the early universe. We still do not know what dark matter is, but we suspect it is largely made up of some type of as-yet-undiscovered subatomic particles.
- The existence of dark energy is supported by evidence from observations both of the expansion rate through time and of temperature variations in the cosmic microwave background. Together, these observations have led to a model of the universe that gives us precise values for the inventory of its contents and its age.
- The fate of the universe seems to depend on whether the expansion of the universe continues forever, and the acceleration of the expansion suggests that it will. Nevertheless, forever is a long time, and only time will tell whether new discoveries will alter our speculations about the distant future.

**MY COSMIC PERSPECTIVE** Dark matter may seem far removed from our lives here on Earth, but our galaxy and the stars within it would not have formed without the assistance of its gravitational attraction.

# Summary of Key Concepts

## 23.1 Unseen Influences in the Cosmos

- **What do we mean by dark matter and dark energy?** Dark matter and dark energy have never been directly observed, but each has been proposed to exist because it seems the simplest way to explain a set of observed motions in the universe. **Dark matter** is the name given to the unseen mass whose gravity governs the observed motions of stars and gas clouds. **Dark energy** is the name given to the form of energy thought to be causing the expansion of the universe to accelerate.

## 23.2 Evidence for Dark Matter

- **What is the evidence for dark matter in galaxies?** The orbital velocities of stars and gas clouds in galaxies do not change much with distance from the center of the galaxy. Applying Newton's laws of gravitation and motion to these orbits leads to the conclusion that the total mass of a galaxy is far larger than the mass of its stars. Because no detectable visible light is coming from this matter, we call it dark matter.

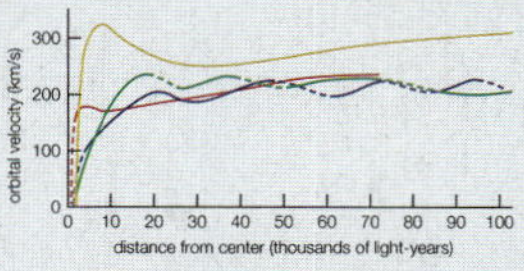


- **What is the evidence for dark matter in clusters of galaxies?** We have three different ways of measuring the amount of dark matter in clusters of galaxies: from galaxy orbits, from the temperature of the hot gas in clusters, and from the **gravitational lensing** predicted by Einstein. All these methods are in agreement, indicating that the total mass of a galaxy cluster is at least 40 times the mass of its stars.

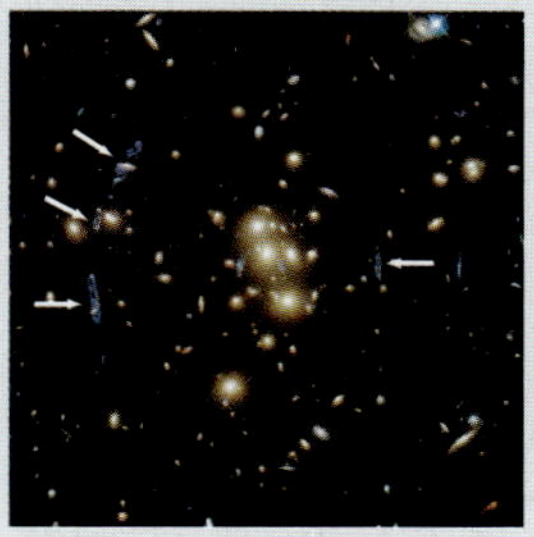

- **Does dark matter really exist?** We infer that dark matter exists from its gravitational influence on the matter we can see, leaving two possibilities: Either dark matter exists or there is something wrong with our understanding of gravity. We cannot rule out the latter possibility, but we have good reason to be confident about our current understanding of gravity and the idea that dark matter is real.
- **What might dark matter be made of?** Some of the dark matter could be ordinary (baryonic) matter in the form of dim stars or planetlike objects, but the amount of deuterium left over from the Big Bang and the patterns in the cosmic microwave background both indicate that ordinary matter adds up to only about one-sixth of the total amount of matter. The rest of the matter is hypothesized to be exotic (nonbaryonic) dark matter consisting of as-yet-undiscovered subatomic particles called **WIMPs**.

## 23.3 Structure Formation

- **What is the role of dark matter in galaxy formation?** Because most of a galaxy's mass is in the form of dark matter, the gravity of that dark matter is probably what formed protogalactic clouds and then galaxies from slight density enhancements in the early universe.

- **What are the largest structures in the universe?** Galaxies appear to be distributed in gigantic chains and sheets that surround great **voids**. These **large-scale structures** trace their origin directly back to regions of slightly enhanced density early in time.

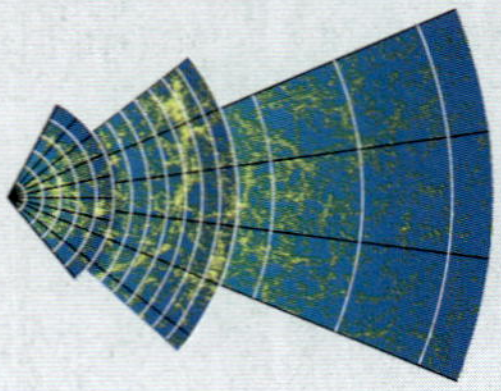

## 23.4 The Fate of the Universe

- **What is the evidence for an accelerating expansion?** Observations of distant supernovae show that the expansion of the universe has been speeding up for the last several billion years. No one knows the nature of the mysterious force causing this acceleration. However, its characteristics are consistent with models in which the force is produced by a form of dark energy that pervades the universe.

- **Why is flat geometry evidence for dark energy?** Observations of the cosmic microwave background also support the existence of dark energy because they demonstrate that the overall geometry of the universe is nearly flat. According to Einstein's general theory of relativity, the universe can be flat only if the total amount of mass-energy it contains is equal to the critical density, but measurements of the total amount of matter show that it represents only about 30% of the critical density. We therefore infer that about 70% of the total mass-energy is in the form of dark energy—the same amount implied by the supernova observations.

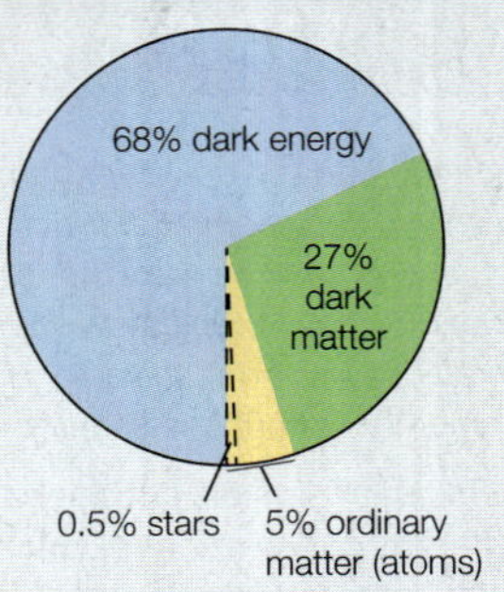


- **What is the fate of the universe?** If dark energy is indeed what's driving the acceleration of the universe's expansion, then we expect the expansion to continue accelerating into the future, as long as the effects of dark energy do not change with time and there are no other factors that affect the fate of the universe.

# Visual Skills Check

*Use the following questions to check your understanding of some of the many types of visual information used in astronomy. For additional practice, try the Chapter 23 Visual Quiz in the Study Area at www .MasteringAstronomy.com.*

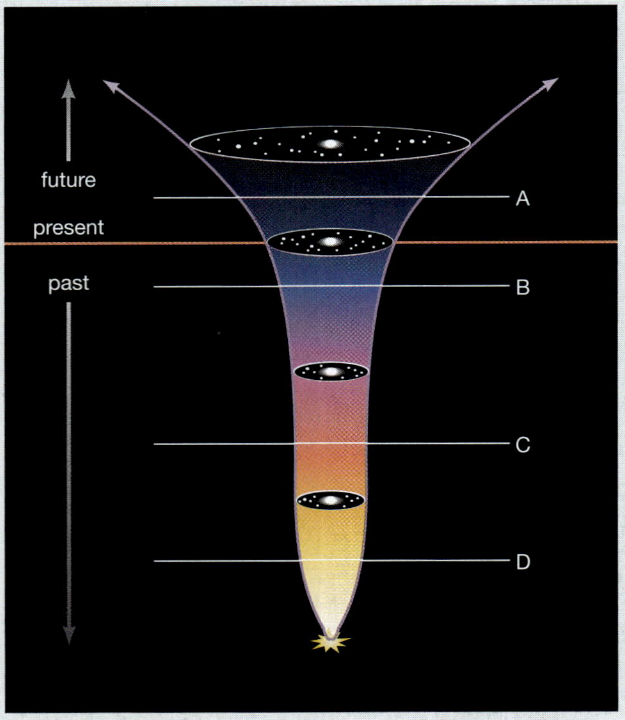


*The schematic figure to the left shows a more complicated expansion history than the four idealized models in Figure 23.17. Answer the following questions, using the information given in this figure.*

1. At time A, is the expansion of the universe accelerating, coasting, or decelerating?
2. At time B, is the expansion of the universe accelerating, coasting, or decelerating?
3. At time C, is the expansion of the universe accelerating, coasting, or decelerating?
4. At time D, is the expansion of the universe accelerating, coasting, or decelerating?

# Exercises and Problems

*For instructor-assigned homework and other learning materials, go to www.MasteringAstronomy.com.*

## Chapter Review Questions

### Short-Answer Questions Based on the Reading

1. Define *dark matter* and *dark energy,* and clearly distinguish between them. What types of observations have led scientists to propose the existence of each of these unseen influences?
2. Describe how orbital speeds in the Milky Way depend on distance from the galactic center. How does this relationship indicate the presence of large amounts of dark matter?
3. How do orbital speeds depend on distance from the galactic center in other spiral galaxies, and what does this tell us about dark matter in spiral galaxies?
4. How do we measure the masses of elliptical galaxies? What do these masses lead us to conclude about dark matter in elliptical galaxies?
5. Briefly describe the three different ways of measuring the mass of a cluster of galaxies. Do the results from the different methods agree? What do they tell us about dark matter in galaxy clusters?
6. What is *gravitational lensing*? Why does it occur? How can we use it to estimate the masses of lensing objects?
7. Briefly explain why the conclusion that dark matter exists rests on the assumption that we understand gravity correctly. Is it possible that our understanding of gravity is not correct? Explain.
8. In what sense is dark matter "dark"? Briefly explain why objects like you, planets, and even dim stars qualify as dark matter.
9. What evidence indicates that most of the matter in the universe cannot be ordinary (baryonic) matter?
10. Explain what we mean when we say that a neutrino is a *weakly interacting particle.* Why can't the dark matter in galaxies be made of neutrinos?
11. What do we mean by *WIMPs*? Why does it seem likely that dark matter consists of these particles, even though we do not yet know what they are?
12. Briefly explain why dark matter is thought to have played a major role in the formation of galaxies and larger structures in the universe. What evidence suggests that larger structures are still forming?
13. What do the *large-scale structures* of the universe look like? Explain why we think these structures reflect the density patterns of the early universe.
14. Describe and compare the four general patterns for the expansion of the universe: *recollapsing, critical, coasting,* and *accelerating*. Observationally, how can we decide which of the four general expansion models best describes the present-day universe?
15. How do observations of distant supernovae provide evidence for dark energy?
16. How do observations of the cosmic microwave background provide evidence for dark energy?
17. Based on current evidence, what is the overall inventory of the mass-energy content of the universe?
18. What implications does the evidence for dark energy have for the fate of the universe?

### Does It Make Sense?

*Decide whether or not each of the following statements makes sense (or is clearly true or false). Explain clearly; not all of these have definitive answers, so your explanation is more important than your chosen answer.*

19. Strange as it may sound, most of both the mass and the energy in the universe may take forms that we are unable to detect directly.
20. A cluster of galaxies is held together by the mutual gravitational attraction of all the stars in the cluster's galaxies.
21. We can estimate the total mass of a cluster of galaxies by studying the distorted images of galaxies whose light passes through the cluster.
22. Clusters of galaxies are the largest structures that we have so far detected in the universe.
23. The primary evidence for an accelerating universe comes from observations of young stars in the Milky Way.
24. There is no doubt remaining among astronomers that the fate of the universe is to expand forever.
25. Dark matter is called "dark" because it blocks light from traveling between the stars.
26. Dark energy is the energy associated with the motion of particles of dark matter.
27. Evidence that the expansion of the universe is accelerating comes from observations showing that the average distance between galaxies is increasing faster now than it was 5 billion years ago.
28. If dark matter consists of WIMPs, then we should be able to observe photons produced by collisions between these particles.

### Quick Quiz

*Choose the best answer to each of the following. For additional practice, try the Chapter 23 Reading and Concept Quizzes in the Study Area at www.MasteringAstronomy.com.*

29. Dark matter is inferred to exist because (a) we see lots of dark patches in the sky. (b) it explains how the expansion of the universe can be accelerating. (c) we can observe its gravitational influence on visible matter.
30. Dark energy has been hypothesized to exist in order to explain (a) observations suggesting that the expansion of the universe is accelerating. (b) the high orbital speeds of stars far from the center of our galaxy. (c) explosions that seem to create giant voids between galaxies.
31. Measurements of how orbital speeds depend on distance from the center of our galaxy tell us that stars in the outskirts of the galaxy (a) orbit the galactic center just as fast as stars closer to the center. (b) rotate rapidly on their axes. (c) travel in straight, flat lines rather than elliptical orbits.
32. Strong evidence for the existence of dark matter comes from observations of (a) our solar system. (b) the center of the Milky Way. (c) clusters of galaxies.
33. A photograph of a cluster of galaxies shows distorted images of galaxies that lie behind it at greater distances. This is an example of what astronomers call (a) dark energy. (b) spiral density waves. (c) gravitational lensing.

34. Based on the observational evidence, is it possible that dark matter doesn't really exist? (a) No, the evidence for dark matter is too strong for us to think it could be in error. (b) Yes, but only if there is something wrong with our current understanding of how gravity should work on large scales. (c) Yes, but only if all the observations themselves are in error.
35. Based on current evidence, which of the following is considered a likely candidate for the majority of the dark matter in galaxies? (a) subatomic particles that we have not yet detected in particle physics experiments (b) swarms of relatively dim red stars (c) supermassive black holes
36. Which region of the early universe was most likely to become a galaxy? (a) a region whose matter density was lower than average (b) a region whose matter density was higher than average (c) a region with an unusual concentration of dark energy
37. The major evidence for the idea that the expansion of the universe is accelerating comes from observations of (a) white dwarf supernovae. (b) the orbital speeds of stars within galaxies. (c) the evolution of quasars.
38. Which of the following possible types of universe would *not* expand forever? (a) a critical universe (b) an accelerating universe (c) a recollapsing universe

## Inclusive Astronomy

*Use these questions to reflect on participation in science.*

39. *Group Discussion: Vera Rubin and the Nobel Prize.* As of 2018, no one had yet received a Nobel Prize for any of the discoveries implying that dark matter exists. Vera Rubin would have seemed a likely candidate for such a prize, for reasons briefly noted in the Extraordinary Claims box on page 679. Unfortunately, however, she died in 2016, and Nobel Prizes can be awarded only to people who are still living.
**a.** Working independently, read at least one article about whether Vera Rubin should have won a Nobel Prize.
**b.** Gather into small groups and discuss the articles that you read. What are the arguments in favor of a Nobel Prize for Rubin? Did you come across anyone arguing that a Nobel Prize would *not* have been appropriate?
**c.** Many eminent female astronomers in the 21st century already view Rubin as a pioneer and role model. Do you think receiving a Nobel Prize would have further increased her positive impact on women's participation in astronomy? Explain your reasoning.
**d.** When considering candidates for the Nobel Prize, would it be appropriate for the Prize Committee to consider potential impacts of the award on diversity and inclusion in science? Make a list of the pros and cons of considering these impacts.

## The Process of Science

*These questions may be answered individually in short-essay form or discussed in groups, except where identified as group-only.*

40. *Dark Matter.* Overall, how convincing do you consider the case for the existence of dark matter? Write a short essay in which you explain what we mean by dark matter, describe the evidence for its existence, and discuss your opinion about the strength of the evidence.
41. *Alternative Gravity.* Suppose someone proposed a new theory of gravity that claimed to explain observations of motion in galaxies and clusters of galaxies without the need for dark matter. Briefly describe at least one other test that you would expect the new theory to be able to pass if it was, in fact, a better theory of gravity than general relativity, which is currently our best explanation of how gravity works.
42. *The Nature of Dark Matter.* Find and study recent reports on the possible nature of dark matter. Write a report or hold a group discussion on the latest ideas about what dark matter is made of.
43. *Dark Energy.* Overall, how convincing do you consider the case for the existence of dark energy? Write a short report or hold a group discussion on what we mean by dark energy, the evidence for its existence, and your opinion about the strength of the evidence.
44. *Group Activity: Dark Matter and Distorted Galaxies.* Study the gravitational lensing shown in Figure 23.9. Note: You may wish to do this activity using the four roles described in Chapter 1, Exercise 39.
**a.** Notice in Figure 23.9 how gravitational lensing causes the image of a background galaxy to shift to a position *farther* from the center of the cluster. Individually, draw simple sketches of how you'd expect the lensed image of a *spherical* galaxy to look. As a group, briefly discuss your sketches. Do they agree? If not, do you think that one prediction is better than the others?
**b.** Make a copy of the diagram below on a large piece of paper. Use a straight edge to make sure that the lines are straight and that they all intersect at the same place; the circle should be close to the point of intersection, as shown. Note that the point at which the lines intersect represents the center of a galaxy cluster, and the circle represents the true position and shape of a spherical galaxy at a much greater distance from Earth.

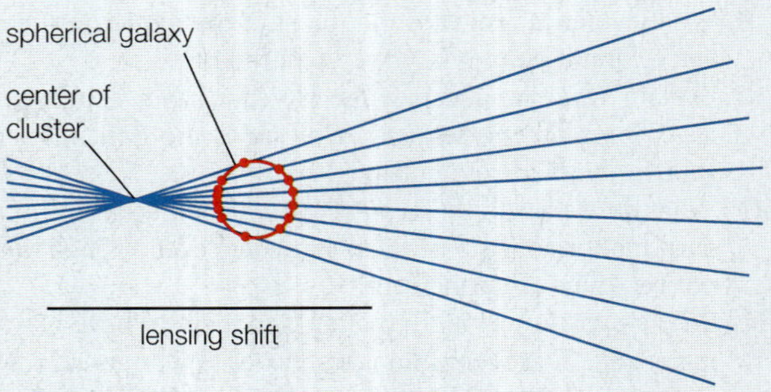


**c.** Now work together to determine the effect of the cluster's lensing shift on the galaxy's image as follows. Step 1: For each dot on the circle, draw another dot farther to the right along the same line, so that the distance between the two dots is equal to the length of the line labeled "lensing shift." Step 2: Connect the new dots to see the shape of the lensed image.
**d.** How do your results from part c compare to the predictions you made in part a? How do the results compare with the lensed galaxy images in Figure 23.10? Discuss possible reasons for any discrepancies you find.

## Investigate Further

### Short-Answer/Essay Questions

45. *The Future Universe.* Based on current evidence concerning the growth of structure in the universe, briefly describe what you would expect large-scale structures in the universe to look like about 10 billion years from now.

46. *Dark Matter and Life.* State and explain at least two reasons one might argue that dark matter is (or was) essential for life to exist on Earth.

47. *Orbital Speed vs. Radius.* For each of the following three hypothetical galaxies, draw a graph showing how orbital speed depends on distance from the galactic center. Make sure the horizontal axis has approximate distances labeled.
**a.** a galaxy with all its mass concentrated at its center
**b.** a galaxy with constant mass density within 20,000 light-years of its center, and zero density beyond that distance
**c.** a galaxy with constant mass density within 20,000 light-years of its center, and beyond that an enclosed mass that increases proportionally with distance from the center

48. *Dark Energy and Supernova Brightness.* When astronomers began measuring the brightnesses and redshifts of distant white dwarf supernovae, they expected to find that the expansion of the universe was slowing down. Instead they found that it was speeding up. Were the distant supernovae brighter or fainter than expected? Explain why. (*Hint:* In Figure 23.18, the position of a supernova on the vertical axis depends on its redshift. Its position on the horizontal axis depends on its brightness—supernovae seen further back in time are not as bright as those seen closer in time.)

49. *What Is Dark Matter?* Describe at least three possible constituents of dark matter. Explain how we would expect each to interact with light, and how we might go about detecting its existence.

50. *Alternative Gravity.* How would gravity have to be different in order for us to be able to explain the rotation curves of galaxies without the need for dark matter? Would gravity at very large distances need to be stronger or weaker than expected? Explain.

51. *Photo Journal: Gravitational Lenses.* Gravitational lensing occurs in numerous astronomical situations. Compile a catalog of examples from the web, with photos of lensed stars, quasars, and galaxies. Give a one-paragraph explanation of what is shown in each photo.

## Quantitative Problems

*Be sure to show all calculations clearly and state your final answers in complete sentences.*

52. *White Dwarf M/L.* What is the mass-to-light ratio of a $1M_{\text{Sun}}$ white dwarf with a luminosity of $0.001L_{\text{Sun}}$?

53. *Supergiant M/L.* What is the mass-to-light ratio of a $30M_{\text{Sun}}$ supergiant star with a luminosity of $300{,}000L_{\text{Sun}}$?

54. *Solar System M/L.* What is the mass-to-light ratio of the solar system?

55. *Mass from Orbital Velocities.* Study the graph of orbital speeds for the spiral galaxy NGC 7541, which is shown in Figure 23.4.
**a.** Use the orbital velocity law to determine the mass (in solar masses) of NGC 7541 enclosed within a radius of 30,000 light-years from its center. (*Hint:* 1 light-year $= 9.461 \times 10^{15}$ m.)
**b.** Use the orbital velocity law to determine the mass of NGC 7541 enclosed within a radius of 60,000 light-years from its center.
**c.** Based on your answers to parts a and b, what can you conclude about the distribution of mass in this galaxy?

56. *Weighing a Cluster.* A cluster of galaxies has a radius of about 5.1 million light-years ($4.8 \times 10^{22}$ m) and an intracluster medium with a temperature of $6 \times 10^{7}$ K. Estimate the mass of the cluster. Give your answer in both kilograms and solar masses. Suppose that the combined luminosity of all the stars in the cluster is $8 \times 10^{12}L_{\text{Sun}}$. What is the cluster's mass-to-light ratio?

57. *Cluster Mass from Hot Gas.* The gas temperature of the Coma Cluster of galaxies is about $9 \times 10^{7}$ K. What is the mass of this cluster within 15 million light-years of the cluster center?

58. *From Newton to Dark Matter.* Show that the equation $M_r = r \times v^2/G$ from Mathematical Insight 23.2 is equivalent to Newton's version of Kepler's third law from Mathematical Insight 4.3. Assume that one mass is much larger than the other mass and that the orbit is circular. (*Hint:* What is the mathematical relationship between period $p$ and orbital velocity $v$ and orbital radius $r$ for a circular orbit?)

59. $10^{100}$ *Years.* Based on current understanding, the final stage in the history of a perpetually expanding universe would come about $10^{100}$ years from now. Such a large number is easy to write but difficult to understand. This problem investigates some of the incredible properties of very large numbers.
**a.** The current age of the universe is around $10^{10}$ years. How much longer is a trillion years than this current age? How much longer is $10^{15}$ years? $10^{20}$ years?
**b.** Suppose protons decay with a half-life of $10^{32}$ years. When will the number of remaining protons be half its current amount? When will it be a quarter of its current amount? How many half-lives will have gone by when the universe reaches an age of $10^{34}$ years? What fraction of the original protons will remain at this time? Is it reasonable to conclude that *all* protons in today's universe will be gone by the time the universe is $10^{40}$ years old? Explain. (*Hint:* See Mathematical Insight 8.1.)

# Galaxy Evolution

**All galaxies, including our Milky Way, developed as gravity pulled together matter in regions of the universe that started out slightly denser than surrounding regions. The central illustration depicts how galaxies formed over time, starting from the Big Bang in the upper left and proceeding to the present day in the lower right, as space gradually expanded according to Hubble's law.**

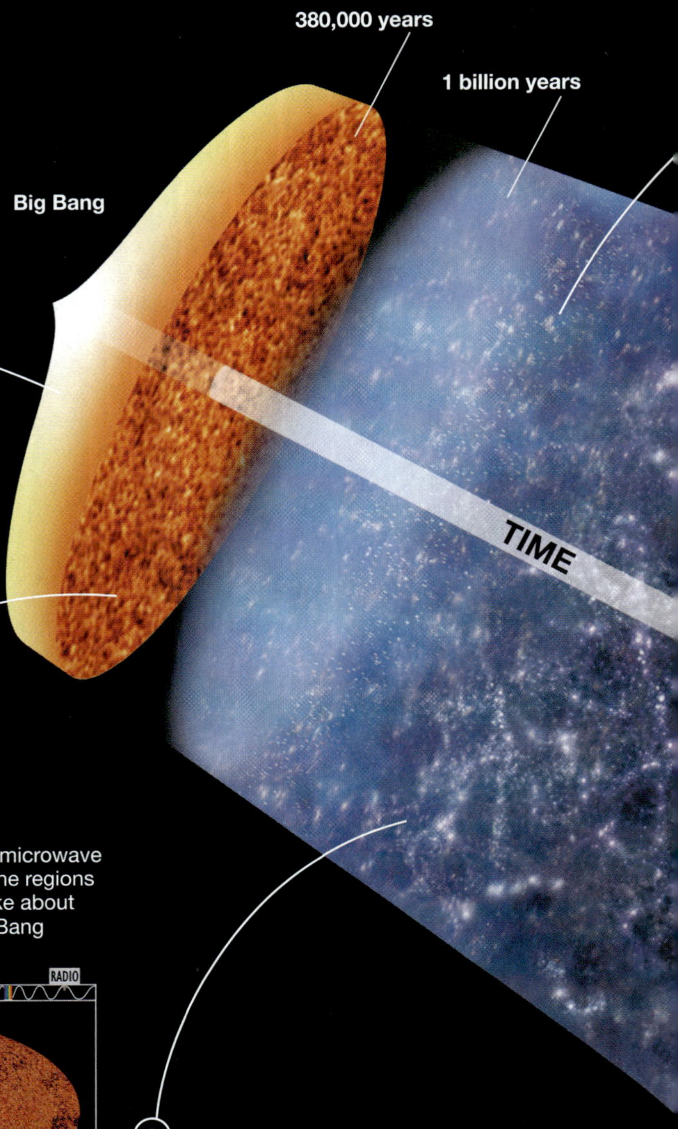


(1) Dramatic inflation early in time is thought to have produced large-scale ripples in the density of the universe. All the structure we see today formed as gravity drew additional matter into the peaks of these ripples [Section 22.3].

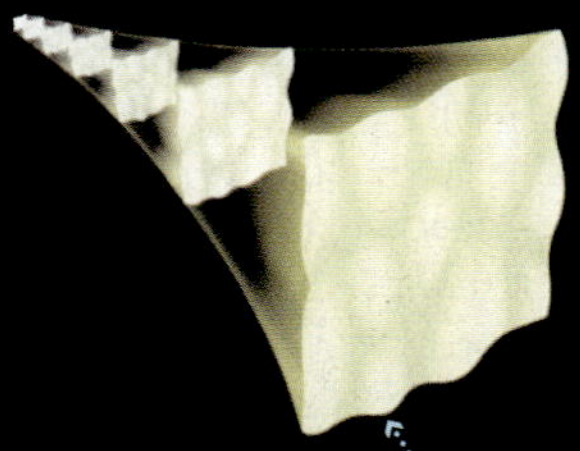

*Inflation may have stretched tiny quantum fluctuations into large-scale ripples.*

(2) Observations of the cosmic microwave background show us what the regions of enhanced density were like about 380,000 years after the Big Bang [Section 22.2].

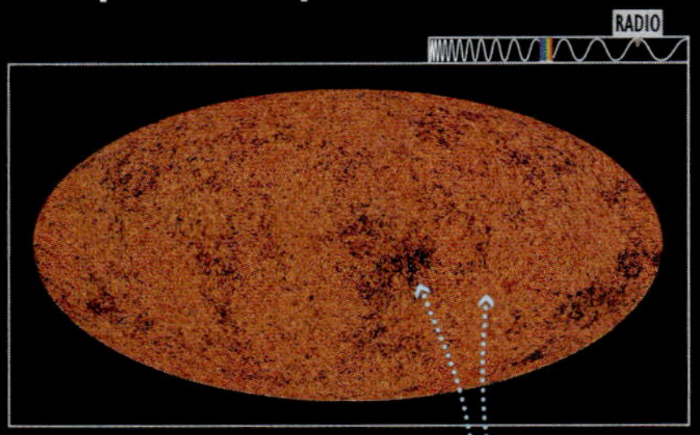


Image from *Planck*

*Variations in the cosmic microwave background show that regions of the universe differed in density by only a few parts in 100,000.*

(3) Large-scale surveys of the universe show that gravity has gradually shaped early regions of enhanced density into a web-like structure, with galaxies arranged in huge chains and sheets [Section 23.3].

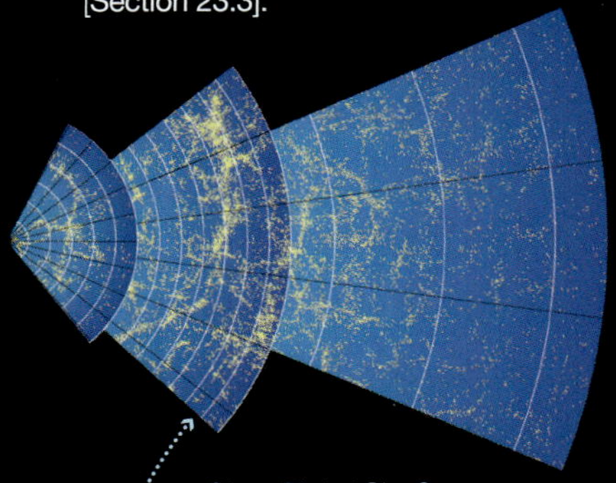

Sloan Digital Sky Survey

*The web-like patterns of structure observed in large-scale galaxy surveys agree with those seen in large-scale computer simulations of structure formation.*

4 Within this large-scale web, galaxy formation began when gravity collected hydrogen and helium gas into blobby protogalactic clouds [Section 19.3].

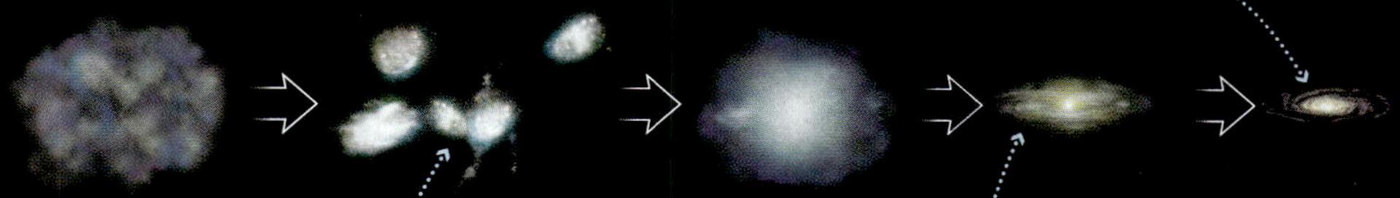

*Stars that form in the disk orbit in orderly circles.*

*Spiral galaxies probably formed through mergers of smaller protogalactic clouds.*

*The gas settled into a spinning disk while stars that had already formed remained in the halo.*

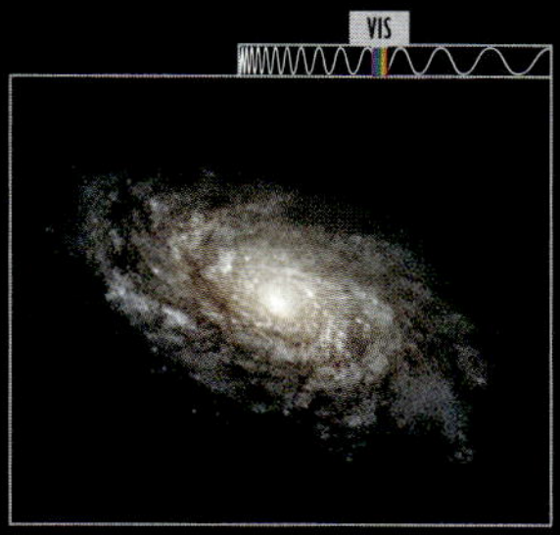


Photo of spiral galaxy NGC 4414

5 At least some elliptical galaxies formed when two spiral galaxies collided and merged [Section 21.2].

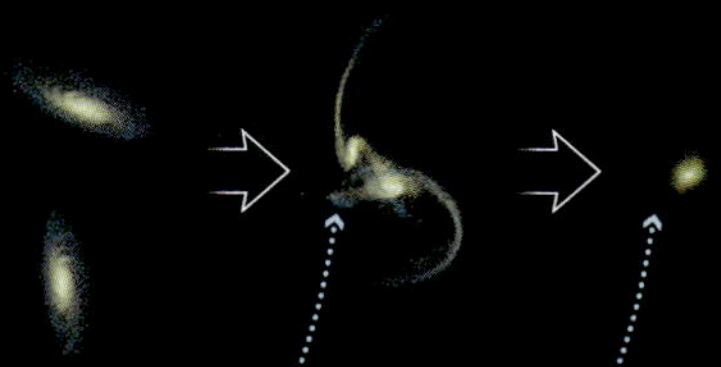

*A collision disrupts the orderly orbits of stars in the disks of spiral galaxies and triggers a starburst.*

*After the starburst, almost no cold gas is left for star formation.*

Photo of elliptical galaxy M87

6 Dark matter is thought to drive galaxy formation. The gravity of dark matter seems to be what pulled gas into protogalactic clouds. It continues to cause galaxies to cluster and sometimes to collide [Section 23.3].

*Measurements indicate a large amount of invisible dark matter surrounds the visible stars in each galaxy.*

7 Today, in the disks of spiral galaxies like the Milky Way, the star–gas–star cycle continues to produce new stars and planets from matter that was once in protogalactic clouds [Section 19.2].

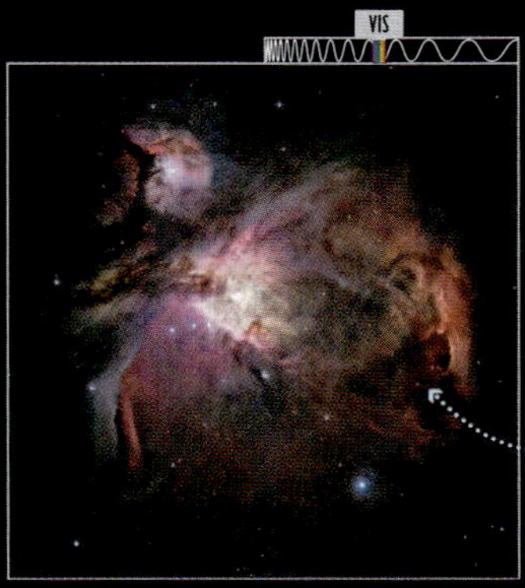


*New stars and planetary systems—some perhaps much like our own—are currently forming in the Orion Nebula, 1500 light-years from Earth.*

Photo of Orion Nebula

# 24 Life in the Universe

▲ **About the photo:** The Allen Telescope Array (Hat Creek, California) is used in the search for extraterrestrial intelligence (SETI).

## LEARNING GOALS

**24.1 Life on Earth**
- When did life arise on Earth?
- How did life arise on Earth?
- What are the necessities of life?

**24.2 Life in the Solar System**
- Could there be life on Mars?
- Could there be life in the outer solar system?

**24.3 Life Around Other Stars**
- What are the requirements for surface habitability?
- What kinds of extrasolar worlds might be habitable?
- How could we detect life on extrasolar planets?

**24.4 The Search for Extraterrestrial Intelligence**
- How many civilizations are out there?
- How does SETI work?

**24.5 Interstellar Travel and Its Implications for Civilization**
- How difficult is interstellar travel?
- Where are the aliens?

*We, this people, on a small and lonely planet*
*Travelling through casual space*
*Past aloof stars, across the way of indifferent suns*
*To a destination where all signs tell us*
*It is possible and imperative that we learn*
*A brave and startling truth.*
—*Maya Angelou, from* A Brave and Startling Truth

Chapter 24 Overview

Throughout this book, we have explored the evidence supporting the modern scientific story of the universe. We have seen very strong evidence that we live in an expanding universe born about 14 billion years ago in the Big Bang, and in which life has been made possible by the recycling of gas by stars within our galaxy. We have seen how the processes that give birth to stars also give birth to planets, and explored how our own planet came to be. But we have not yet explored the most profound question of all: Are we alone?

The universe contains worlds beyond imagination—more than 100 billion star systems in our galaxy alone, and more than 100 billion large galaxies in the observable universe. But we do not yet know whether any world besides Earth has ever been home to life. In this chapter, we will discuss current understanding of the origin of life on Earth, and how we are using this understanding to consider the possibility of finding life elsewhere. In the process, we will see that the search for extraterrestrial life may have astonishing implications for our own future.

## 24.1 Life on Earth

It may seem that aliens are everywhere. Aliens abound in television shows and movies, and it's not hard to find websites claiming alien atrocities or a government conspiracy to hide alien corpses in "Area 51." Most scientists are deeply skeptical of such reports (see Extraordinary Claims, p. 717), but scientific interest in the possibility of alien life is quite real. The scientific search for life in the universe even has its own name: **astrobiology.**

Most research in astrobiology focuses on one of the following three major areas: (1) seeking to understand the origin and evolution of life on Earth, so that we might learn the conditions under which life could arise and evolve elsewhere; (2) searching for worlds, both in our own solar system and beyond, that might have conditions suitable for life; and (3) searching for evidence of actual life on other worlds. In this first section, we'll focus our attention on the first area of astrobiology research.

### When did life arise on Earth?

An important step in learning about the origin and evolution of life on Earth is finding out when life first arose. We cannot draw definitive conclusions from the single example of Earth, but an early origin of life would be consistent with the idea that life can arise easily under the conditions that prevailed on the early Earth, while a late origin of life might suggest that it was difficult for life to arise, in which case life on other worlds seems less likely. The weight of evidence today supports the idea that life arose quickly and easily on Earth.

We learn about the history of life on Earth through the study of **fossils**, relics of organisms that lived and died long ago. Most fossils form when dead organisms fall to the bottom of a sea (or other body of water) and are gradually buried by layers of sediments. The sediments are produced by erosion on land and carried by rivers to the sea. Over millions of years, sediments pile up on the seafloor, and the weight of the upper layers compresses underlying layers into rock. Erosion or tectonic activity can later expose the fossils (**FIGURE 24.1**). In some places, such as the Grand Canyon, the sedimentary layers record hundreds of millions of years of Earth's history (**FIGURE 24.2**). Some fossils are remarkably well preserved (**FIGURE 24.3**), though the vast majority of dead organisms decay completely and leave no fossils behind.

**The Geological Time Scale** The key to reconstructing the history of life is to determine the dates when fossil organisms lived. The *relative* ages of fossils found in different layers are easy to determine: Deeper layers formed earlier and contain more ancient fossils. Radiometric dating [**Section 8.3**] confirms these relative ages and gives us fairly precise absolute ages for fossils. Based on the layering of rocks and fossils, geologists divide Earth's $4\frac{1}{2}$-billion-year history into a set of distinct intervals that make up what we call the **geological time scale. FIGURE 24.4** shows the

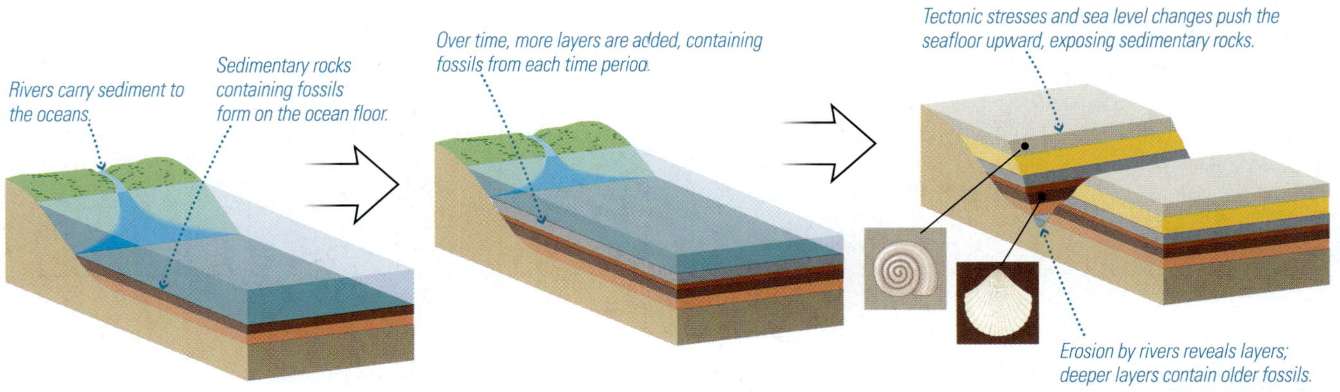


**FIGURE 24.1** Formation of sedimentary rock. Each layer represents a particular time and place in Earth's history and is characterized by fossils of organisms that lived in that time and place.

**FIGURE 24.2** The rock layers of the Grand Canyon record more than 500 million years of Earth's history.

**FIGURE 24.3** Dinosaur fossils preserved in sandstone in Dinosaur National Monument, which straddles Utah and Colorado.

names of some of these intervals on a timeline, along with numerous important events in Earth's history.

**Think about it** Based on Figure 24.4, how does the length of time during which animals and plants have lived on land compare to the length of time during which life has existed? How does the length of time during which humans have existed compare to the length of time since mammals and dinosaurs first arose?

**Fossil Evidence for the Early Origin of Life** You might wonder why the geological time scale shows so much more detail for the last few hundred million years than it does for earlier times. The answer is that fossils become increasingly difficult to find as we look deeper into Earth's history, for three major reasons. First, older rocks are much rarer than younger rocks, because most of Earth's surface is geologically young. Second, even when we find very old rocks, they often have been subject to transformations (caused by heat and pressure) that would have destroyed any fossil evidence they may have contained. Third, nearly all life prior to a few hundred million years ago was microscopic, and microscopic fossils are much more difficult to identify.

Despite these difficulties, geologists have found a few very old rocks that suggest life was already thriving on Earth 3.5 billion years ago, and possibly for hundreds of

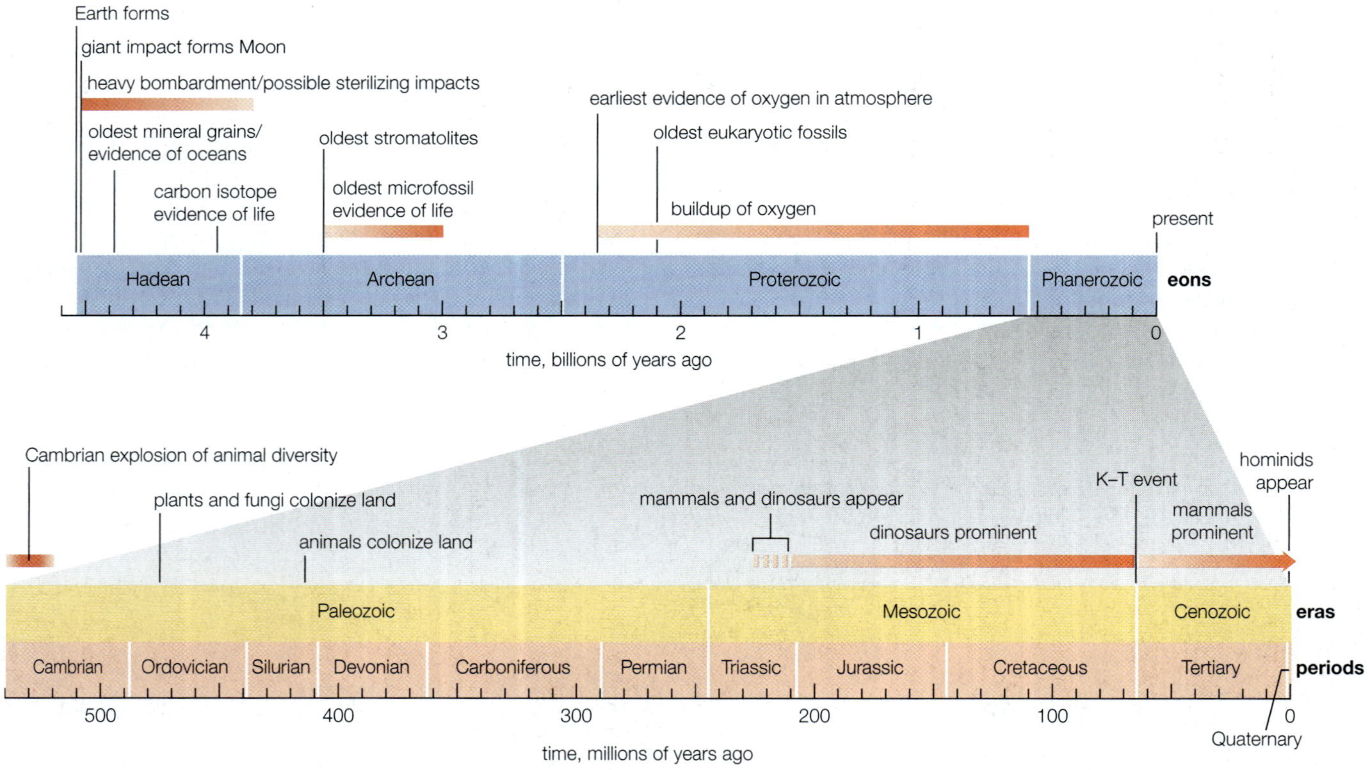


**FIGURE 24.4** Major intervals of the geological time scale. Notice that the lower timeline is an expanded view of the last portion of the upper timeline. The eons, eras, and periods are defined by changes observed in the fossil record. The absolute ages come from radiometric dating. (The *K-T event* is the geological term for the impact linked to the mass extinction of the dinosaurs [**Section 12.5**].)

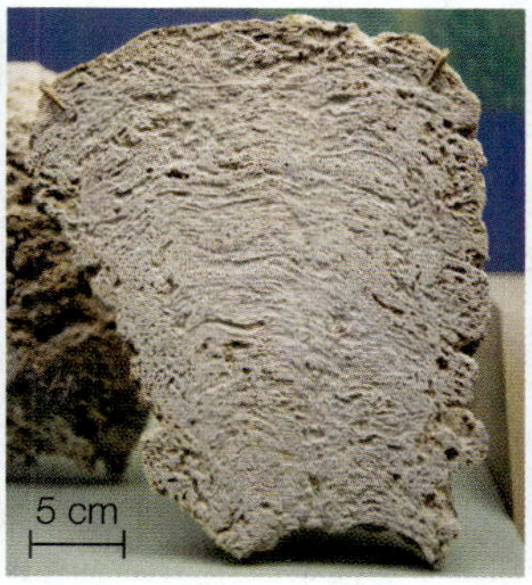

**a** These knee-high mats at Shark Bay, Western Australia, are colonies of microbes known as "living stromatolites."

**b** The banded structure in this section from one of the Shark Bay stromatolites is formed by layers of sediment attaching to the microbes in the mats.

**c** This section of a 3.5-billion-year-old stromatolite (found in the Strelley Pool Formation in Western Australia) shows the same type of structure found in living stromatolites. The black layers are organic deposits that are the remains of ancient microbes. (The ruler is marked in centimeters.)

**FIGURE 24.5** Rocks called *stromatolites* offer evidence of microbial life existing as early as 3.5 billion or more years ago.

millions of years before that. One strong line of evidence comes from rocks called *stromatolites*, which are strikingly similar in size, shape, and interior structure to sections of modern-day, mat-shaped formations known as "living stromatolites" (**FIGURE 24.5**). Living stromatolites are formed by colonies of microbes, and they contain layers of sediment intermixed with microbes of different species. Microbes near the top generate energy through photosynthesis, and those beneath obtain energy from carbon compounds left as waste products by the photosynthetic microbes. The living stromatolites grow in size as sediments are deposited over them, forcing the microbes to migrate upward in order to remain at the depths to which they are adapted. This gradual migration creates the layered structures.

The similarity of structure between the ancient stromatolites and the living stromatolites suggests a similar origin, implying that stromatolites are fossil remnants of early life. There is some controversy about the biological origin of stromatolites, because geological processes of sedimentation can mimic their layering. However, chemical analyses of the structures seen in stromatolites also support a biological origin, making most scientists confident that they offer evidence for the existence of microbial colonies as far back as 3.5 billion years ago; one recent study claims to have found fragments of even older stromatolites, dating to about 3.7 billion years ago, though this claim is still being investigated. Moreover, if the microbes that made the stromatolites are like the microbes in the living mats today, the implication is that at least some of these ancient microbes produced energy by photosynthesis. Because photosynthesis is a fairly sophisticated metabolic process, we presume that it must have taken at least a moderately long time for this process to evolve in living organisms. In other words, the oldest stromatolites suggest that photosynthetic life already existed some 3.5 to 3.7 billion years ago, in which case more primitive life must have existed even earlier.

Other evidence for early life comes from recognizable fossils of microbes, known as *microfossils*. Although it is often difficult to distinguish a true microfossil from a microscopic structure formed by mineral processes, scientists are confident that they have found microfossils dating to nearly 3.5 billion years ago (**FIGURE 24.6**). This is consistent with the ages of the oldest stromatolites, adding strength to the case that life not only existed but was well established on Earth by that time.

It may not even be possible to find stromatolites or microfossils much older than the oldest ones found to date, because older rocks have generally undergone too much transformation to preserve intact fossils. However, another line of evidence, based on study of carbon isotopes, may point to the existence of life even further back in time. Carbon has two stable isotopes: carbon-12, with six protons and six neutrons in its nucleus, and carbon-13, which has one extra neutron (see Figure 5.9). Living organisms incorporate carbon-12 slightly more easily than carbon-13. As a result, the fraction of carbon-13 is always a bit lower in living organisms and fossils than in rock samples that lack fossils. All life and all fossils tested to date (including the microfossil shown in Figure 24.6) show the same characteristic ratio of the two carbon isotopes. This ratio has been found

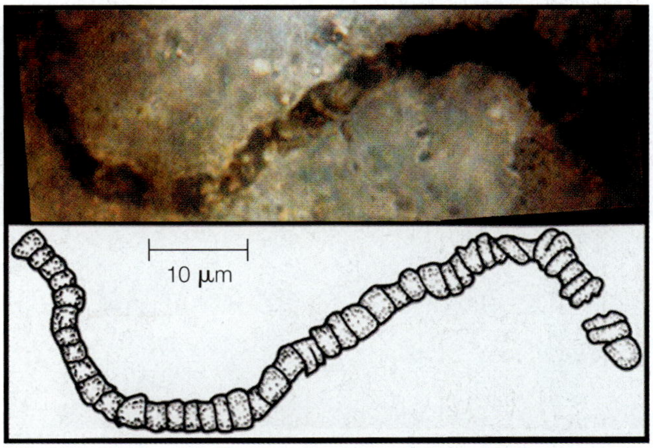


**FIGURE 24.6** This image shows one of many microfossils found in a rock formation in Australia (the "Apex chert"), which radiometric dating shows to be 3.465 billion years old. The structure (shown more clearly in the diagram at bottom), chemical composition, and isotopic composition (the ratio of carbon-13 to carbon-12) all indicate that this is a fossil of ancient, microscopic life.

in at least two of the world's oldest known rock formations. One formation, found on the island of Akilia near Greenland, has a carbon isotope ratio indicative of life in rocks dating to more than 3.85 billion years ago. Another, in Labrador, Canada, has such a ratio in rocks that are about 100 million years older, or about 3.95 billion years old (**FIGURE 24.7**). While we cannot completely rule out alternative explanations, the most likely conclusion is that these rocks contained life when they formed. In that case, life was present on Earth at least 3.95 billion years ago. Moreover, if the carbon isotope evidence really does indicate life at that time, the rarity of such old rocks suggests that life must already have been widespread on Earth, because otherwise extraordinary luck would have been needed for us to have found the evidence.

**How much earlier might life have arisen?** Evidence from ancient mineral grains* indicates that Earth may already have had oceans as early as about 4.3 billion years ago, suggesting at least the possibility that life could have existed at that time. However, recall that the first few hundred million years of the solar system's history was the period of the *heavy bombardment* [**Section 8.2**], during which Earth should have been struck numerous times by large asteroids or comets. The sizes of lunar craters from this time period suggest that some impacts of the heavy bombardment should have released enough energy to completely vaporize the oceans, and perhaps to melt portions of Earth's crust. These impacts are therefore often called "sterilizing impacts," because they could potentially have extinguished any life that already existed. Moreover, even if some life managed to survive these impacts, the surviving microbes would almost certainly have been in deep ocean or underground environments, adding to the difficulty of finding present-day rocks that might preserve evidence of this ancient life.

*This evidence comes from chemical analysis of tiny mineral grains known as zircons. Although these grains are found embedded in much younger sedimentary rocks, radiometric dating shows that some zircons solidified as much as 4.38 billion years ago.

**FIGURE 24.7** This rock outcrop in Labrador, Canada, may contain some of the oldest evidence of life on Earth. Although the graphite structures are not fossilized cells, their carbon isotope ratio indicates that life must have been present when they formed.

**Summary: Life on Earth Arose Quickly** It would have been difficult for life to survive if it arose much more than about 4 billion years ago, and even if it did, finding evidence of such life might be difficult or impossible. When we combine this fact with the stromatolite and microfossil evidence indicating widespread life by 3.5 billion years ago and the carbon isotope evidence that pushes this date back to about 3.95 billion years ago, we are led to a remarkable conclusion: We have evidence of life on Earth going nearly as far back in time as such evidence could exist, and perhaps nearly as far back as life could have taken hold. The implication is that life on Earth arose relatively easily, suggesting that the same might have happened on many other worlds.

## How did life arise on Earth?

The early origin of life is suggestive of the idea that life arose easily, but we would be more confident of this idea if we could learn *how* life arose on Earth. Scientists explore this question primarily by learning about how life has evolved over time, which tells us what early life may have looked like, and then by doing laboratory experiments to learn how life might have started.

**The Theory of Evolution**

**The Theory of Evolution** The fossil record clearly shows that life has gone through great changes over time. If we are to understand how life arose, we must understand what causes these changes so that we can trace life back to its origin. The unifying theory through which scientists understand life on Earth is the **theory of evolution,** first published by Charles Darwin in 1859.

*Evolution* simply means "change with time." Other scientists (some going back to ancient times) had already recognized evidence for evolution in the fossil record, but no one before Darwin successfully explained how species might undergo change. In essence, the fossil record provides strong evidence that evolution *has* occurred, while Darwin's theory of evolution explains *how* it occurs.

Darwin studied relationships between living species (most famously in the Galápagos Islands) as well as fossils of extinct species. Based on the evidence he collected, he put forth a simple model to explain how evolution occurs. As described by biologist Stephen Jay Gould (1941–2002), Darwin built his model from "two undeniable facts and an inescapable conclusion":

- **Fact 1: overproduction and competition for survival.** Any localized population of a species has the potential to produce far more offspring than the local environment can support with resources such as food and shelter. This overproduction leads to a competition for survival among the individuals of the population.
- **Fact 2: individual variation.** Individuals in a population of any species vary in many heritable traits (traits passed from parents to offspring). No two individuals are exactly alike, and some individuals possess traits that make them better able to compete for food and other vital resources.

- **The inescapable conclusion: unequal reproductive success.** In the struggle for survival, those individuals whose traits best enable them to survive and reproduce will, on average, leave the largest number of offspring that in turn survive to reproduce. Therefore, in any local environment, heritable traits that enhance survival and successful reproduction will become progressively more common in succeeding generations.

It is this unequal reproductive success that Darwin called **natural selection**, which is simply the idea that genetic traits that provide a reproductive advantage will naturally win out (be "selected") over less advantageous traits. Natural selection is the primary mechanism of evolution, because it explains how species can change over time as traits that improve adaptation to a local environment are passed down through generations. If enough small individual variations accumulate, natural selection can even give rise to an entirely new species. By studying the variety of species in places like the Galápagos Islands, Darwin backed his model with so much evidence that it quickly gained the status of a scientific *theory* [**Section 3.4**], and ongoing research during the more than 150 years since he published the theory has only given it further support.

**The Mechanism of Evolution** Darwin's theory of evolution by natural selection tells us that species adapt and change by passing hereditary traits from one generation to the next. However, Darwin did not know precisely how these traits were passed down, nor did he know why there is variation among individuals or how new traits can appear in a population. In essence, his theory *predicted* that there must be some mechanism for heredity that would preserve most hereditary information from prior generations but still allow for small changes. Today, we know that this mechanism is embodied in the properties of the molecule called **DNA** (short for "*d*eoxyribo*n*ucleic *a*cid"). In that sense, the discovery of DNA was a tremendous predictive success for Darwin's theory.

Living organisms reproduce by copying DNA and passing these copies on to their descendants. A molecule of DNA consists of two long strands—somewhat like the interlocking strands of a zipper—wound together in the spiral shape known as a *double helix* (**FIGURE 24.8**). The instructions for assembling a living organism are written in the precise order of four chemical bases (abbreviated A, T, G, and C for the first letters of their chemical names) that make up the interlocking portions of the DNA "zipper." These bases pair up in a way that ensures that both strands of a DNA molecule contain the same genetic information. By unwinding and allowing new strands (made from chemicals floating around inside a cell) to form alongside the original ones, a single DNA molecule can give rise to two identical copies of itself. That is how genetic material is copied and passed on to future generations.

Evolution occurs because the transfer of genetic information from one generation to the next is not always perfect. An organism's DNA may occasionally be altered by copying errors or by external influences, such as ultraviolet light from the Sun or exposure to toxic or radioactive chemicals. Any change in an organism's DNA is called a **mutation**. Many mutations are lethal, killing the cell in which the

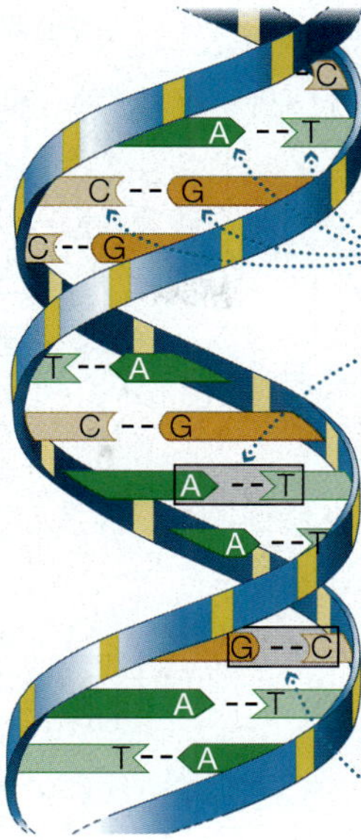


**FIGURE 24.8** This diagram represents a small piece of a DNA molecule, which looks like a zipper twisted into a spiral. Hereditary information is contained in the "teeth" linking the strands. These "teeth" are the DNA bases. Only four DNA bases are used, and they can link up between the two strands only in specific ways: T attaches only to A, and C attaches only to G. (The color coding is arbitrary and is used only to represent different types of chemical groups; in the backbone, blue and yellow represent sugar and phosphate groups, respectively.)

mutation occurs. Some, however, may improve a cell's ability to survive and reproduce, allowing the cell to pass this improvement to its offspring.

Our understanding of the molecular mechanism of natural selection has put the theory of evolution on a stronger foundation than ever. While no theory can ever be proved true beyond all doubt, the theory of evolution is as solid as any theory in science, including the theory of gravity and the theory of atoms. Biologists routinely witness evolution occurring before their eyes among laboratory microorganisms, or over periods of just a few decades among plants and animals subjected to environmental stress. Moreover, the theory of evolution has become the underpinning of virtually all modern biology, medicine, and agriculture. For example, agricultural scientists apply the idea of natural selection to develop pest control strategies that reduce the populations of harmful insects without harming the populations of beneficial ones; medical researchers test new drugs on animals that are genetically similar to humans, because the theory of evolution tells us that genetically similar species should have somewhat similar physiological responses; and biologists study the genetic relationships between organisms by comparing their DNA.

**The First Living Organisms** The basic chemical nature of DNA is virtually identical among all living organisms. This fact, along with other biochemical similarities shared by all living organisms, tells us that all life on Earth today can trace its origins to a common ancestor that lived long ago.

We are unlikely to find fossils of the earliest organisms, but we can learn about early life through careful studies of the DNA of living organisms. Biologists can determine the

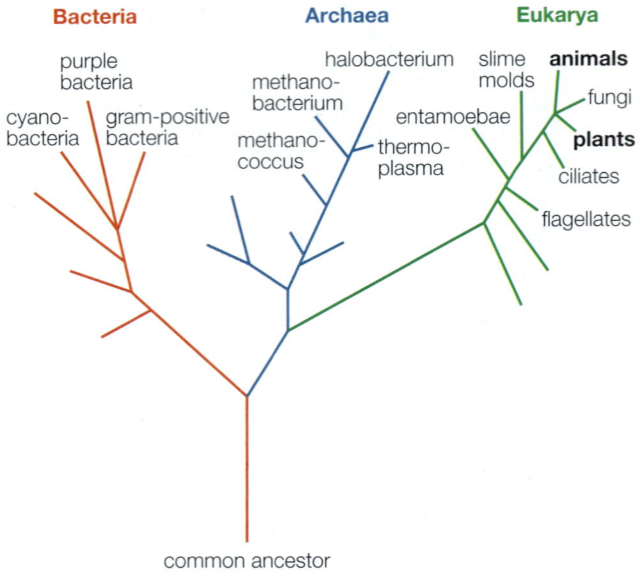


**FIGURE 24.9** The tree of life, showing evolutionary relationships determined by comparison of DNA sequences in different organisms. Just two small branches represent *all* plant and animal species. (Only a few of the many relationships are shown and labeled.)

**FIGURE 24.10** This photograph shows a volcanic vent on the ocean floor that spews out hot, mineral-rich water. DNA studies indicate that the microbes living near these vents are evolutionarily older than most other living organisms, hinting that early life may have arisen in similar environments.

evolutionary relationships among living species by comparing the sequences of bases in their DNA. For example, two organisms whose DNA sequences differ in five places for a particular gene are probably more distantly related than two organisms whose gene sequences differ in only one place. Many such DNA comparisons suggest that all living organisms are related in a way depicted schematically by the "tree of life" in **FIGURE 24.9**. Although details in the structure of this tree remain uncertain (particularly because species can sometimes swap genes), it indicates that life on Earth is divided into three major groupings, or *domains,* called *bacteria, archaea,* and *eukarya,* and that all three domains share a common ancestor. Notice that plants and animals represent only two tiny branches of the domain *eukarya.*

Organisms on branches located closer to the root of the tree of life must contain DNA that is evolutionarily older, suggesting that they more closely resemble the organisms that lived early in Earth's history. Some of the modern-day organisms that appear to be evolutionarily oldest are microbes that live in very hot water around seafloor volcanic vents (**FIGURE 24.10**). Unlike most life at Earth's surface, which depends on sunlight, these organisms get energy from chemical reactions in the hot, mineral-rich water around the volcanic vents.

The idea that early organisms might have lived in such "extreme" conditions may at first seem surprising, but it makes sense when we think about it. The deep ocean environment would have been protected from harmful ultraviolet radiation that bathed Earth's surface before our atmosphere had oxygen or an ozone layer, and may also have offered protection from the effects of impacts. Moreover, the chemical pathways used to extract energy from mineral-rich hot water are simpler than other chemical pathways (such as photosynthesis) used by living organisms to obtain energy. For these reasons, many scientists suspect that life arose near deep-sea vents, or in similar environments that provided chemical energy.

**The Transition from Chemistry to Biology** The theory of evolution explains how the earliest organisms evolved into the great diversity of life on Earth today. But where did the first organisms come from? We may never know for sure, because we are unlikely to find fossils that show the transition from nonlife to life. However, over the past several decades, scientists have conducted laboratory experiments designed to mimic the conditions that existed on the young Earth. These experiments suggest that life could have arisen through natural chemical reactions.

Such experiments were first performed in the 1950s (**FIGURE 24.11**), and they have been refined and improved

**FIGURE 24.11** Stanley Miller poses with a reproduction of the experimental setup he first used in the 1950s to study pathways to the origin of life. (He worked with Harold Urey, so the experiment is called the Miller-Urey experiment.)

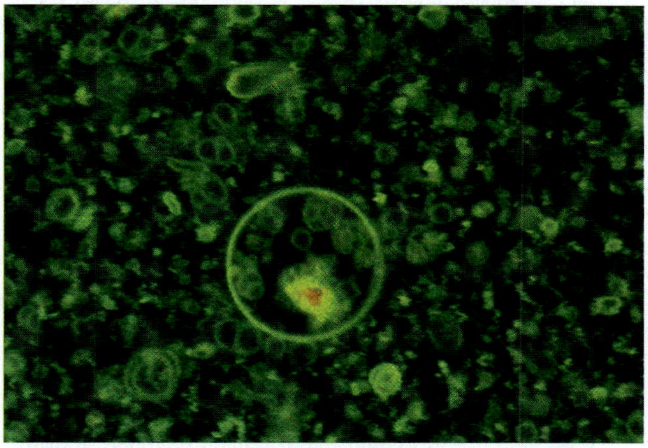

**FIGURE 24.12** This microscopic photo (made with the aid of fluorescent dyes) shows short strands of RNA (red) contained within an enclosed membrane (green circle), both of which formed spontaneously with the aid of clay minerals beneath them.

since that time. In essence, experiments mix chemicals thought to have been present on the early Earth and then "spark" the chemicals with electricity to simulate lightning or other energy sources. The chemical reactions that follow have produced nearly all the major molecules of life, including amino acids and DNA bases. Many of these molecules are also found in meteorites, suggesting that some of the building blocks of life may have arrived from space.

More recent laboratory experiments have shown that naturally occurring clays, which should have been abundant in the early oceans, can catalyze the formation of short strands of RNA, a molecule much like a single strand of DNA. Some of these short RNA molecules can in turn catalyze other reactions. These same clay minerals can also lead to the formation of microscopic enclosed membranes—often called "pre-cells" because of their resemblance to living cells—sometimes with RNA inside them (**FIGURE 24.12**).

Because some RNA molecules (though not those so far produced in the laboratory) are capable of self-replication, we can envision the process summarized in **FIGURE 24.13**. Naturally formed chemical building blocks were catalyzed by clay to make self-replicating RNA molecules, some of which became enclosed in microscopic pre-cells. Pre-cells in which RNA replicated faster and more accurately were more likely to spread, leading to a type of positive feedback that would have encouraged even faster and more accurate replication. With these types of chemical reactions occurring all over Earth, it may have been only a matter of time until some of the pre-cells turned into RNA-based life, which eventually evolved into DNA-based life. We may never know whether life really arose in this way, but unless we are missing some major piece of the puzzle, it seems plausible that life started easily through a natural sequence of chemical processes.

**Could Life Have Migrated to Earth?** Our scenario suggests that life could have arisen naturally here on Earth. However, an alternative possibility is that life arose somewhere else first—perhaps on Venus or Mars—and then migrated to Earth on meteorites. Remember that we have collected meteorites that were blasted by impacts from the surfaces of the Moon and Mars [**Section 12.1**]. Calculations suggest that Venus, Earth, and Mars all should have exchanged many tons of rock, especially in the early days of the solar system when impacts were more common.

The idea that life could travel through space to Earth once seemed outlandish. After all, it's hard to imagine a more forbidding environment than that of space, with no air, no water, and constant bombardment by dangerous radiation from the Sun and stars. However, the presence of complex carbon compounds in meteorites and comets tells us that the building blocks of life can survive in space, and tests have shown that some microbes can survive in space for years. There's even a group of known animal species that can survive some time in space (**FIGURE 24.14**).

In a sense, Earth, Venus, and Mars have been "sneezing" on each other for billions of years. Life could conceivably have originated on any of these three planets and been transported to the others. It's an intriguing thought, but it does not change our basic scenario for the origin of life—it simply moves it from one planet to another.

**A Brief History of Life on Earth** We are now ready to take a quick look at the history of life on Earth, summarized in the geological timeline of Figure 24.4. Earth formed about $4\frac{1}{2}$ billion years ago, and the giant impact thought to have formed the Moon [**Section 8.2**] probably happened soon after. Mineral evidence suggests that Earth had oceans by

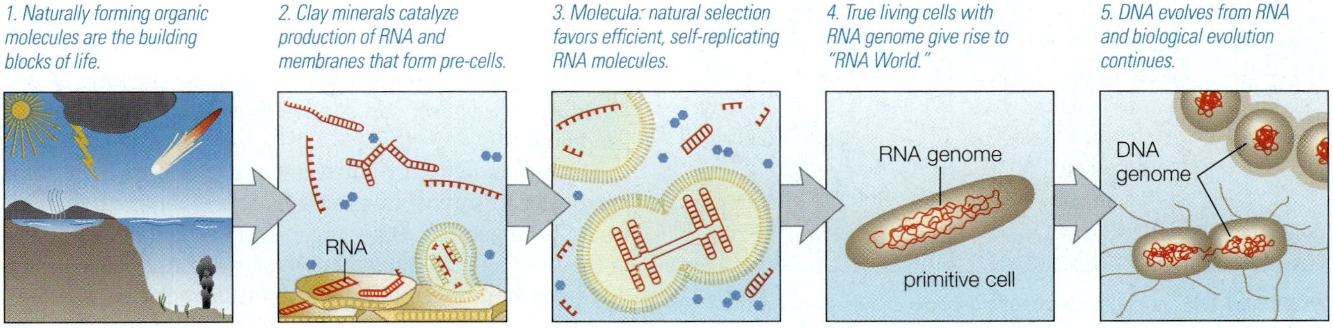


**FIGURE 24.13** A summary of the steps by which chemistry on the early Earth may have led to the origin of life.

**FIGURE 24.14** It almost doesn't look real, but this photograph shows a tiny animal called a tardigrade (also called a "water bear") that is about a millimeter long. Tardigrades can survive an incredible range of "extreme" conditions, including at least some time in the near-vacuum of space.

about 4.3 billion years ago, and those early oceans would have been natural laboratories for chemical reactions that could have led to life. If life formed very early, it may have been disrupted by the heavy bombardment. But once life took hold, evolution rapidly diversified it.

Despite the rapid pace of evolution, the most complex organisms remained single-celled for at least a billion years after life first arose. All of these organisms probably lived in the oceans, because the lack of a protective ozone layer made the land inhospitable. Things began to change only when oxygen started building up in Earth's atmosphere.

Nearly all the oxygen in our atmosphere was originally released through photosynthesis by single-celled organisms known as *cyanobacteria* (**FIGURE 24.15**). Fossil evidence indicates that cyanobacteria were producing oxygen through photosynthesis by at least 2.7 billion years ago, and possibly for hundreds of millions of years before that. However, oxygen did not immediately begin to accumulate in the atmosphere. For hundreds of millions of years, chemical reactions with surface rocks pulled oxygen back out of the atmosphere nearly as fast as the cyanobacteria could produce it. But these tiny organisms were abundant and persistent, and eventually the surface rock was so saturated with oxygen that the rate of oxygen removal slowed down. Oxygen then began to accumulate in the atmosphere, though it may not have reached a level that we could have breathed until just a few hundred million years ago.

**FIGURE 24.15** This photo shows microscopic chains of modern cyanobacteria. The ancestors of these living organisms produced essentially all the oxygen in Earth's atmosphere.

Today, we often think of oxygen as a necessity for life. However, oxygen was probably poisonous to most organisms living before about 2 billion years ago (and remains a poison to many microbes still living today). The rise of atmospheric oxygen therefore caused tremendous evolutionary pressure and may have been a major factor in the evolution of complex plants and animals.

There were undoubtedly many crucial changes as primitive microbes gradually evolved into multicellular organisms and early plants and animals, but the fossil record does not allow us to pinpoint the times at which all these changes occurred. However, we see a dramatic change in the fossil record during the *Cambrian period,* beginning about 542 million years ago. During this period, animal life evolved from tiny and primitive organisms into all the basic body types (phyla) that we find on Earth today. This remarkable diversification occurred in such a short time relative to Earth's history that it is often called the *Cambrian explosion.*

Early dinosaurs and mammals arose some 225 to 250 million years ago, but dinosaurs at first proved more successful and dominated for well over 100 million years. Their sudden demise 65 million years ago [**Section 12.5**] paved the way for the evolution of large mammals—including humans. The earliest humans appeared on the scene only a few million years ago, or after 99.9% of Earth's history to date had already gone by. Our few centuries of industry and technology have come after 99.99999% of Earth's history.

## What are the necessities of life?

Our understanding of the timing and possible origins of life on Earth suggests that life might similarly arise on other worlds, but this would be the case only if they contained the necessities of life. So what are these necessities? While it is possible that life on other worlds could be quite different from life on Earth, it's easiest to begin the search for life by looking for conditions in which organisms from Earth could live.

If we think about ourselves, the requirements for life seem fairly stringent: We need abundant oxygen in an atmosphere that is otherwise not poisonous, we need temperatures in a fairly narrow range, and we need abundant and varied food sources. However, the discovery of life in "extreme" environments—such as in the hot water near undersea volcanic vents—shows that many microbes (often called *extremophiles*) can survive in a much wider range of conditions.

Organisms living in hot water prove that at least some microbes can survive in much higher temperatures than we would have guessed. Other organisms live in other extremes. In the freezing cold but very dry valleys of Antarctica, scientists have found microbes that live *inside* rocks, surviving on tiny droplets of liquid water and energy from sunlight. Life, including tardigrades (see Figure 24.14), has been found in

Antarctic lakes, buried so deep beneath the ice that it has not been exposed to open air or sunlight for millions of years. In many places, microscopic life has been found deep underground in water that fills pores within subterranean rock. We have found life thriving in environments so acidic, alkaline, or salty that humans would be poisoned almost instantly. We have even found microbes that can survive high doses of radiation, making it possible for them to survive for many years in the radiation-filled environment of space.

If we compare all the different forms of life on Earth, we find that life as a whole has only three basic requirements:

1. A source of nutrients (atoms and molecules) from which to build living cells
2. Energy to fuel the activities of life, whether from sunlight, from chemical reactions, or from the heat of Earth itself
3. Liquid water

## SPECIAL TOPIC Evolution and the Schools

You're probably familiar with the public debate about teaching evolution in schools. Much of the controversy comes from misunderstandings on both sides. Scientists sometimes fail to show respect for people's individual religious beliefs, while those opposed to teaching evolution often misunderstand the overwhelming amount of scientific evidence that supports the theory.

The debate centers around the question of what counts as science and what does not, as you can see by considering a common question: Do you believe in creationism or evolution? This question may be philosophically and theologically interesting, but it is poorly phrased from a scientific standpoint: Science is about evidence, not beliefs. Scientific evidence clearly shows that the universe is much older than Earth, that Earth is some $4\frac{1}{2}$ billion years old, that life on Earth has undergone dramatic change since it first arose nearly 4 billion years ago, and that this change has occurred through the mechanism of evolution by natural selection. Nevertheless, all the evidence in the world can never prove a scientific theory to be true beyond all doubt. To take a somewhat extreme case, you could accept the overwhelming evidence for evolution and still believe that God created Earth and the universe a mere 6000 years ago by, for example, assuming that God put all the evidence for evolution in place. Because the evidence would look the same in this case as it would if evolution actually occurred, science cannot say anything about the validity of such a belief.

We can gain deeper insight by examining the theory of evolution using the hallmarks of science from Chapter 3. The first hallmark says that science seeks explanations for observed phenomena that rely on natural causes, and the theory of evolution does exactly that: It explains the fossil record and observed differences between species in terms of natural selection. Note that this hallmark does *not* preclude the possibility of divine intervention; it just says that if divine intervention has occurred, we cannot study it scientifically. The second and third hallmarks remind us that science progresses through the creation and testing of models designed to explain our observations, and that we must modify or discard models that fail our tests. The theory of evolution exhibits these hallmarks because it is continually subject to testing and verification, and it has indeed been modified as new evidence has come to light. For example, Darwin's original theory has been improved and strengthened by modern molecular understanding of DNA and by observations of evolution in microbes (such as bacteria gaining resistance to antibiotics) and in small, localized populations of plants and animals.

The fact that evolution is such a well-established and important scientific theory makes it a key part of the science curriculum for schools, and few people disagree. However, many people have suggested teaching alternative ideas such as *creationism* or *intelligent design* alongside evolution in school curricula. The scientific response is very simple: Scientists take no position on the general school curriculum, but the *science* curriculum should not include these alternative ideas because they are not science, which we can see by again considering the hallmarks of science.

Let's start with *creationism*, the idea that the world was divinely created. In fact, we can't even define this idea clearly enough to test it against the hallmarks of science. Many religions incorporate stories of creation, and these stories are often quite different from one another. For example, many Native American religious beliefs speak of creation in terms that bear little resemblance to the story in Genesis. Even among Christians who claim a literal belief in the Bible, there are differences of interpretation. Some biblical literalists argue that the creation must have occurred in just 6 days, as the first chapter of Genesis seems to say, but others suggest that the term "day" in Genesis does not necessarily mean 24 hours and that Genesis is therefore compatible with a much older Earth.

The related idea of *intelligent design* holds that the complexity of life is too great to have arisen naturally, and therefore there must have been a "designer." Intelligent design can sound much like science, especially because most of its proponents accept both the idea of an old Earth and the idea that some living things have changed through time as indicated by the fossil record. However, the assumption of an intelligent designer does not yield testable predictions, because it posits that the designer is beyond our scientific comprehension; in essence, the idea of intelligent design means that there is no point in continuing to look for natural causes, because a supernatural process made the universe and life. Whether or not this idea is true, it isn't science. As an analogy, consider the collapse of a bridge. If an engineer declares that the collapse was an act of God, he could well be right—but that belief won't help him learn how to design a better bridge. It is the scientific quest for a natural understanding of life—embodied in the theory of evolution—that has led to the discovery of genetics, DNA, and virtually all modern medicine.

So where does this discussion leave us with regard to the question of teaching evolution in schools? Let's start with what our discussion does *not* tell us. It does not tell us that everyone needs to accept the theory of evolution—whether you choose to do so is up to you. Nor does it tell us whether divine intervention has guided evolution—science does not say anything about the existence or role of God. What it *does* tell us is that the theory of evolution is a clear and crucial part of the study of science, and that without it we cannot fully understand modern science and its impact on our world.

These requirements give us a basic road map for the search for life elsewhere. If we want to find life on other worlds, it makes sense to start by searching for worlds that offer these basic necessities.

Interestingly, only the third requirement (liquid water) seems to pose much of a constraint. Molecules that could serve as building blocks for life are present almost everywhere—even on meteorites and comets. Many worlds are large enough to retain internal heat that could provide energy for life, and virtually all worlds have sunlight (or starlight) bathing their surfaces, although the inverse square law for light [**Section 15.1**] means that light provides less energy to worlds farther from their star. Nutrients and energy should therefore be available to some degree on almost every planet and moon. In contrast, liquid water is relatively rare. Therefore, the search for liquid water—or possibly a liquid of some other type—drives the current search for life in our universe.

## 24.2 Life in the Solar System

The liquid water requirement rules out most of the worlds in our solar system as possible homes for life. Mercury and the Moon are barren and dry, Venus is far too hot to have liquid water on its surface (though its clouds contain droplets of water laced with acid), and most of the small bodies of the outer solar system are too cold. The jovian planets may have droplets of liquid water in some of their clouds, but the strong vertical winds on these planets mean any droplets cannot stay liquid for long, making these planets unlikely homes for life. Nevertheless, several worlds *do* show evidence of liquid water or some other liquid mixture that seems to offer at least some possibility of life. The most notable are Mars and a few of the jovian moons, especially Jupiter's moon Europa. In this section, we'll consider the possibility of extraterrestrial life in our own solar system.

### Could there be life on Mars?

When we search for worlds that have the necessities of life as we know them from life on Earth, we are looking for what we call **habitable worlds**. In other words, a habitable world is one that seems to have all the necessities of life.

**Think about it** Is it possible for a world to be habitable but not actually have life? Is it possible for a world to have life but not be habitable? Explain.

Mars seems to have fit the bill for habitability in the past, because strong evidence indicates that liquid water flowed on Mars during its early history [**Section 9.4**]. This fact, along with growing evidence that water persisted for long periods of time on the Martian surface, makes scientists excited about the prospect that we might someday find fossil evidence of past life on Mars.

Perhaps even more intriguing is the possibility that life might still exist on Mars, if it ever took hold in the past. Although Mars lacks surface liquid water today, it still has abundant water ice both on its surface and underground [**Section 10.4**]. It therefore seems likely that at least some pockets of liquid water might still exist underground,

### SPECIAL TOPIC What Is Life?

You may have noticed that while we've been talking about *life*, we haven't actually defined the term. It's surprisingly difficult to draw a clear boundary between life and nonlife, but one way to start is by listing distinguishing features common to all known life. For example, most or all living organisms on Earth appear to share the following six key properties:

1. **Order:** Living organisms are not random collections of molecules but rather have molecules arranged in orderly patterns that form cell structures.
2. **Reproduction:** Living organisms are capable of reproducing.
3. **Growth and development:** Living organisms grow and develop in patterns determined in part by heredity.
4. **Energy utilization:** Living organisms use energy to fuel their many activities.
5. **Response to the environment:** Living organisms actively respond to changes in their surroundings. For example, organisms may alter their chemistry or movements in the presence of a food source.
6. **Evolutionary adaptation:** Life evolves through natural selection, in which organisms pass on traits that make them better adapted to survival in their local environments.

These six properties are all important, but biologists today regard evolution as the most fundamental and unifying of them. Evolutionary adaptation is the only property that can explain the great diversity of life on Earth. Moreover, understanding how evolution works allows us to understand how all the other properties came to be. A simple definition of *life* might therefore be "something that can reproduce and evolve through natural selection."

This definition of life will probably suffice for most practical purposes, but some cases may still challenge it. For example, computer scientists can now write programs (that is, lines of computer code) that can reproduce themselves (that is, create additional sets of identical lines of code). By adding instructions that allow random changes to the programs, computer scientists can even make "artificial life" that evolves on a computer. Should this "artificial life," which consists of nothing but electronic signals processed by computer chips, be considered alive?

The fact that we have such difficulty distinguishing the living from the nonliving on Earth suggests that we should be very cautious about constraining our search for life elsewhere. No matter what definition of life we choose, there's always the possibility that we'll someday encounter something that challenges it. Nevertheless, the properties of reproduction and evolution seem likely to be shared by most if not all life in the universe and therefore provide a useful starting point as we consider how to explore the possibility of extraterrestrial life.

particularly near sources of volcanic heat. Indeed, evidence has recently been found for at least one likely subterranean lake on Mars (near the south polar ice cap), giving a further push to the idea that if life ever got started on Mars, it might still survive in places where underground water exists today.

**Searching for Life on Mars** The most direct way to search for life on Mars is by looking for it with landers and rovers. Unfortunately, a direct search turns out to be more difficult than we first imagined, a lesson learned from the two *Viking* landers that reached Mars in 1976. Each was equipped with a robotic arm for scooping up soil samples, which were fed into several on-board robotically controlled experiments that had been designed to search for evidence of life. For example, one of the experiments mixed scooped-up Martian soil with nutrients from Earth and then looked for evidence that the nutrients were being taken up by microscopic organisms in the soil.

Unfortunately, the *Viking* experiments produced confounding results. While some of them at first seemed suggestive of life, scientists later realized that the results could also be explained by chemical reactions that would occur under the conditions found on Mars. Moreover, one of the experiments looked for evidence of organic molecules* in the soil and found none at all, which seemingly precluded the existence of carbon-based life. Only after more than 30 years had passed since the *Viking* experiments did scientists finally realize that they had missed an important part of the soil chemistry on Mars: The soil contains a salt (known as *perchlorate*) that destroys organic molecules when it is heated as it was in the *Viking* experiments. The *Curiosity* rover has since found strong evidence that the Martian soil *does* contain at least some organic molecules (**FIGURE 24.16**).

**FIGURE 24.16** This is an artist's conception of *Curiosity*'s ChemCam instrument firing its laser at a rock on Mars, so that it can spectroscopically analyze the composition of the vaporized rock. ChemCam is one of the instruments on *Curiosity* used to search for organic molecules on Mars. The inset shows a rock that *Curiosity* first drilled (large hole) and then studied spectroscopically by using its laser to vaporize the dusty debris; the laser made the row of smaller holes.

The main lesson from the *Viking* experiments, then, was that searching for life with robotic experiments requires a deeper understanding of Martian soil chemistry and other issues than we yet have, which is a major reason subsequent missions have sought to help us understand these issues, rather than attempting a direct search for life. In addition, the *Viking* experiments all looked for life essentially by trying to "culture" it (grow it under controlled conditions), and biologists have since learned that most life on Earth cannot be cultured in that way. It now seems that a better approach to searching for life is to look for biochemical markers. The difficulty lies in deciding what markers to look for. For example, on Earth we can search for life in almost any sample by looking for DNA, but Martian life may not use DNA, and even if it does, its DNA might be sufficiently different from the DNA of Earth life that detectors that work on Earth may not work on Mars. The same is true for experiments that might search for various proteins.

For this reason, scientists are particularly interested in a "sample return mission" in which a robotic spacecraft would collect rocks on Mars and bring them back to Earth, where they could be analyzed in many more ways than would be possible with pre-designed robotic experiments sent to Mars. The primary difficulty with such a mission is cost, because it requires not only sending a sophisticated spacecraft to Mars but also sending it with a rocket and enough fuel for the return trip. Still, scientists are optimistic that such a mission might be possible within a decade or so. Indeed, NASA's 2020 Mars rover includes a cache system to grab and store some samples that could later be picked up and returned to Earth by future missions.

**Think about it** Perhaps the best way to search for life on Mars would be by sending humans there, and some plans call for doing so soon enough that the first human being to set foot on Mars may now be a student in elementary school, high school, or college. Do you know anyone who might be that person? Could it be you?

**Methane on Mars** A less direct way to investigate the possibility of life on Mars is to look at atmospheric gases that might provide clues. Scientists are particularly intrigued by the detection of methane ($CH_4$) in the Martian atmosphere. Atmospheric methane is quickly destroyed by ultraviolet light from the Sun, so its presence implies that it is being continually released from the Martian surface.

Claims of methane gas detection were first announced in 2003 and 2004, based on telescopic observations from Earth and the *Mars Express* orbiter at Mars. Those signals were weak enough that many scientists doubted the claims, but

*The term *organic molecule* (or *organic compound*) is generally used for carbon compounds that could *in principle* be building blocks of carbon-based life (because *organic* means "related to life"). Therefore, the presence of organic molecules does not by itself indicate that life is or was present, only that potential building blocks for life are available.

more recent observations suggest the methane is real. The *Curiosity* rover has instruments that allow it to occasionally collect small samples of the Martian air, which it then studies spectroscopically as it shines a laser light through the sample. *Curiosity* had done this more than 30 times as of 2018, usually measuring an atmospheric methane concentration between about 0.4 and 0.7 parts per *billion*, though with occasional spikes to as high as 7 parts per billion. While these are very low concentrations (for comparison, Earth's atmospheric methane concentration is about 1800 parts per billion), they still suggest some ongoing source of methane release. Moreover, the variations measured by *Curiosity* appear to track with the Martian seasons, and Earth-based observations suggest that the methane concentration sometimes has higher spikes than *Curiosity* has detected, with one spike reaching about 45 parts per billion.

What could be producing the Martian methane? We know of at least three possibilities. The least likely is that the methane is produced by chemical reactions that occur when comet dust enters the atmosphere; we cannot completely rule this out, but estimates of the amount of comet dust entering the atmosphere fall far short of what would be needed to explain the methane observations. The other clear possibilities are geological activity or life, both of which might release methane in a way that varies with time. For example, life might emit methane in seasonal or other cyclical patterns, while geological sources might release differing amounts of gas as seasonally varying winds plug or unplug holes from which gases can escape from underground.

Scientists hope to learn much more about methane on Mars from observations by the European Space Agency's *ExoMars Trace Gas Orbiter*, which settled into Martian orbit and began science observations in mid-2018. Meanwhile both the geological and the biological possibilities are relevant to the search for life. If the source is geological, the amount of volcanic heat necessary for methane release would probably also be sufficient to maintain pockets of liquid water underground, and this liquid water could be a potential home for life. And, of course, if the source is biological, it should be only a matter of time until we are able to identify the organisms responsible.

**The Debate over Martian Meteorites** An entirely different approach to the search for life on Mars relies on studies of meteorites whose chemical composition suggests they came from Mars [**Section 12.1**]. Martian meteorites made a major news splash in the 1990s when one of them (called ALH84001) was claimed to contain evidence of life, though subsequent studies showed that the claimed evidence could also be explained through nonbiological processes. However, numerous other Martian meteorites have since been found, and several are old enough (according to radiometric dating) to have been on Mars at a time when liquid water may have existed on the surface. One of these, known as the *Tissint meteor*, fell to Earth 40 miles from the Moroccan town of the same name in 2011. It too has been claimed to contain evidence of Martian life, on the basis of carbon deposits lining fissures in its interior. One hypothesis suggests that these carbon deposits came from fluids that seeped into the rock a half-billion years ago, lining the rock with evidence of past life. However, many experts remain skeptical of these claims and point out that the carbon may also be the result of volcanic action.

The debate over possible evidence of life in Martian meteorites is likely to continue, and it is unclear whether these meteorites could ever provide the extraordinary level of evidence needed to confirm the extraordinary claim that they harbor fossils of life. Nevertheless, the debate has taught us that these meteorites can at minimum provide valuable information about the geological history of Mars, and the findings to date lend strong additional support to the idea that Mars once had water, heat sources, and perhaps organic molecules, all of which strengthen the case for the planet's past habitability.

## Could there be life in the outer solar system?

Beyond Mars, the cold temperatures of the outer solar system make it unlikely that we could ever find surface liquid water. However, as we discussed in Chapters 11 and 12, numerous worlds may have subsurface lakes or oceans of liquid water, and at least one world—Saturn's moon Titan—has surface lakes of ethane and methane.

**Jupiter's Potentially Habitable Moons** Recall that three of Jupiter's moons show evidence of having subsurface oceans: Europa, Ganymede, and Callisto. Of these three, Europa is considered the strongest candidate for potential life. The ice and rock from which Europa formed undoubtedly included the chemical ingredients necessary for life, and Europa's internal heating (primarily due to tidal heating) is strong enough to power volcanic vents on its seafloor. If we are correct in guessing that life on Earth may have arisen near undersea volcanic vents, Europa would seem to have everything needed for life to originate. Indeed, Europa is such a promising candidate for life that while some scientists ask "What are the odds that life could arise on a strange world like Europa?", others ask "What are the chances that all the necessities for life could be available together for 4.5 billion years without life arising?"

The possibility of life on Europa is especially interesting because, given the enormous depth of the suspected ocean (see Figure 11.21), it could in principle include very large creatures. However, the potential energy sources for life on Europa are far more limited than the energy sources for life on Earth, mainly because sunlight could not fuel photosynthesis in the subsurface ocean. As a result, most scientists suspect that any life that might exist on Europa would probably be microscopic and primitive. Energy considerations also explain why scientists consider life to be less likely on Ganymede and Callisto, because these moons would have even less energy for life in their subsurface oceans than Europa.

Searching for life on Europa will probably be difficult, since the ocean is buried beneath several kilometers of ice. However, as we discussed in Chapter 11, study of Europa's icy surface indicates that at least some water occasionally wells up from below and freezes, so it is possible that we could find frozen microbes or other evidence of life through surface spectroscopy from an orbiting spacecraft or direct sampling of the surface by a lander. There is also evidence that water or gas is sometimes vented into space, so orbiting probes could search for evidence of life by sampling this gas. More ambitiously, some scientists are considering the possibility of someday sending a robotic submarine that would melt its way through the ice to search for life in the ocean below.

**Titan and Enceladus** Saturn's moons Titan and Enceladus are both considered promising candidates for life. Let's start with Titan.

Recall that Titan's surface is far too cold for liquid water, but it has lakes and rivers of liquid methane and ethane [**Section 11.2**]. Although many biologists think it unlikely, it is possible that this liquid could support life as water does on Earth. There are also at least slim prospects for water-based life on Titan. The ultraviolet light that hits Titan's atmosphere produces a wide range of carbon compounds that, over billions of years, should have accumulated as a deep layer of sediment on the surface; these sediments may be the material in the dunes found on Titan (**FIGURE 24.17**). Occasional impacts by comets or asteroids would provide enough heat locally to melt any water ice and create pockets of warm water that might persist for a thousand years or so, offering at least a chance for some interesting organic chemistry, if not the development of life.

Perhaps even more promising is *Cassini*'s discovery of several features on Titan's surface that appear to be volcanic cones, each between about 1000 and 1500 meters high. The material erupting from these cold volcanoes would likely include at least some liquid water, probably mixed with ammonia or methane, from reservoirs under Titan's frozen crust. If so, these locations would in essence be volcanic "hot springs" in which temperatures might rise slightly above 0°C, offering a slim chance that life could have originated in such places and gained a foothold. Another possibility is suggested by the recent discovery of bacteria existing in an oily lake on the Caribbean island of Trinidad. The bacteria are found in extremely tiny droplets of water within the lake's sludge, and they break down the oil to supply themselves with the energy necessary to live. Perhaps a watery mixture brought to the surface of Titan by volcanic action might support a similar type of microbial life in its hydrocarbon seas.

One additional possibility for life on Titan is a subsurface ocean. *Cassini* measurements showed small changes in Titan's shape as it orbits Saturn, suggesting that it may have an ocean of liquid water or a colder ammonia/water mixture far beneath its icy crust.

Enceladus seems at least equally promising, since evidence discussed in Chapter 11 supports the idea that it has a global subsurface ocean (**FIGURE 24.18**). Moreover, Enceladus's ice fountains presumably spray out material from this ocean, which becomes frozen as it emerges. This offers the tantalizing possibility that, if the ocean contains life, the icy fountains might spray out flash-frozen life forms. For this reason, scientists hope to send a mission that could sample the ice spray in search of life on Enceladus. (*Cassini* flew through the ice fountains, but was not equipped with instruments to sample the ice, since the existence of the fountains was not known when *Cassini* was built.)

**Beyond Saturn** The discovery of geological activity and of the likely existence of a subsurface ocean on Enceladus has led scientists to realize that potential habitats for life might be more widespread than had previously been imagined. In particular, while we know that Enceladus is tidally heated, scientists still do not fully understand why there is enough heating to melt subsurface ice, suggesting that we might find similar surprises elsewhere in the solar system. For example, recall that *Voyager* observations of Neptune's moon Triton (see Figure 11.31) show evidence of at least some geological activity, so it is possible that Triton could hold subsurface liquid water. Similarly, the surprising geological activity revealed in the *New Horizons* flyby of Pluto [**Section 12.4**] has scientists speculating about subsurface lakes or oceans even on that distant world. We

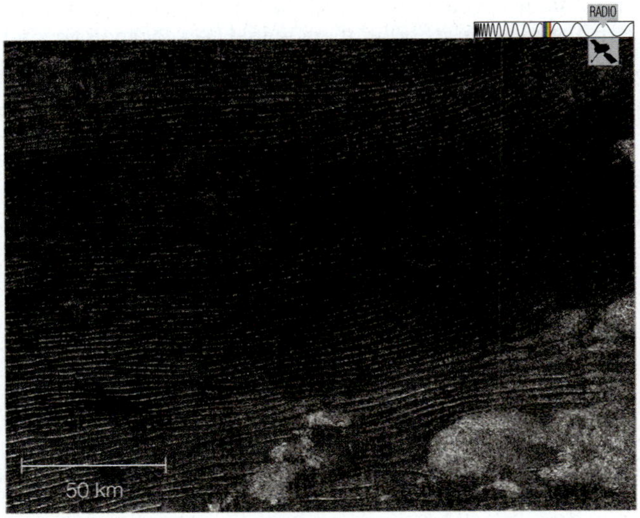


**FIGURE 24.17** Dunes on Titan. The dark streaks in this radar image are thought to be windblown dunes, possibly made of hydrocarbon sediments.

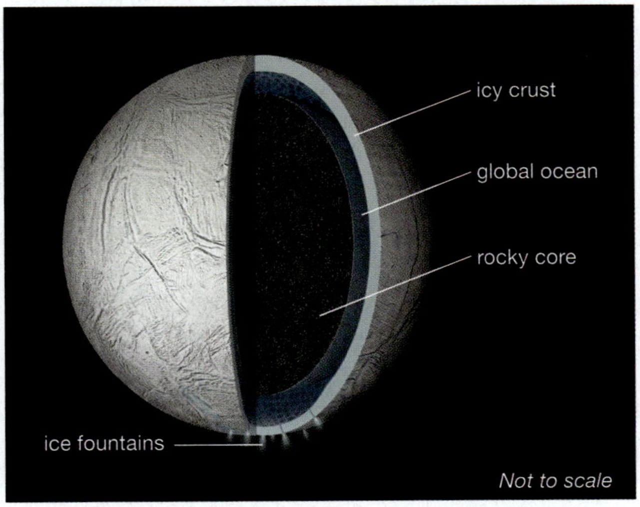


**FIGURE 24.18** Artist's conception of the interior of Enceladus, based on study of the moon's gravity made by the *Cassini* spacecraft. If life exists in the subsurface ocean, we might find evidence of it spraying into space from the ice fountains in the southern hemisphere.

**FIGURE 24.19** The approximate habitable zones around our Sun, a star with one-half the mass of our Sun (spectral type K), and a star with one-tenth the mass of our Sun (spectral type M), shown to scale. The habitable zone becomes increasingly smaller and closer in for stars of lower mass and luminosity.

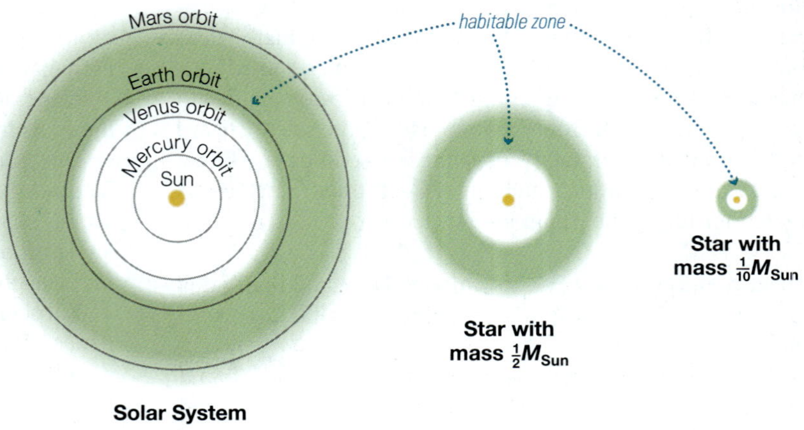


do not yet know whether any world in our solar system besides Earth has ever been home to life, but it is becoming increasingly clear that there are numerous places worth searching for evidence of life.

## 24.3 Life Around Other Stars

Studies of extrasolar planets suggest that our galaxy contains billions of planetary systems [**Section 13.2**], so the prospects for life elsewhere might seem quite good. But numbers alone don't tell the whole story. In this section, we'll consider the prospects for life on worlds orbiting other stars.

### What are the requirements for surface habitability?

When we consider the search for life beyond our solar system, we must distinguish between *surface* life like that on Earth and *subsurface* life like that we envision as a possibility on Mars or Europa. While large telescopes could in principle allow us to discover surface life on extrasolar planets, no foreseeable technology will allow us to find life that is hidden deep underground in other star systems (unless the subsurface life has a noticeable effect on the planet's atmosphere). Therefore, we'll begin our consideration of extrasolar life by focusing on the search for planets with habitable surfaces, meaning surfaces with temperatures and pressures that could allow liquid water to exist.

As we discussed in Chapter 13, current data already suggest that most stars have planets. Indeed, the *Kepler* mission found planets even among stars that were once thought to be poor candidates for having planets, including binary stars and stars with much lower proportions than the Sun of the "heavy elements" (elements besides hydrogen and helium) that can condense to start the accretion process.

Our technology is not yet good enough for us to determine whether extrasolar planets have habitable surfaces, but we can make educated guesses if we know the requirements for long-term surface habitability. If we think back on all the ideas we've discussed in this book, we find that there are four major factors that have made Earth so hospitable to the long-term evolution of life:

1. A distance from the Sun that is great enough to allow water vapor to condense as rain and make oceans, but not so far that all the water freezes
2. Volcanism that released trapped gases from the interior, including water vapor and carbon dioxide, to make the atmosphere and oceans
3. Plate tectonics that support a climate-regulating carbon dioxide cycle
4. A planetary magnetic field that protects the atmosphere from the solar wind

Let's briefly examine each of these factors in more detail.

**The Habitable Zone** We know from our study of Venus that even if a planet seems to be geologically much like Earth, it will be too hot for life if it is too close to its star. In particular, if we moved Earth inward toward the Sun, there is some location between Earth's orbit and the orbit of Venus at which Earth would suffer a runaway greenhouse effect (see Figure 10.34). This location marks the inner boundary of what we call the Sun's **habitable zone**—the range of distances from the Sun at which a planet like Earth could have oceans and surface life. Similarly, if we moved Earth outward in its orbit, there would be a location beyond which Earth would freeze. This location must be somewhere near the orbit of Mars, because we know Mars had liquid water in the past but is now frozen. (In fact, many scientists suspect that if Mars had been larger, it might still be habitable.) The left diagram in **FIGURE 24.19** shows the approximate boundaries of the Sun's habitable zone today.*

We can similarly map the habitable zones of other stars. For ordinary (main-sequence) stars, the location and extent of a habitable zone depend only on the star's mass, because mass determines the star's surface temperature and total energy output (luminosity). Because lower-mass stars are cooler and less luminous, their habitable zones are

*Because the Sun (like other stars) gradually brightens as it ages [**Section 14.2**], the habitable zone gradually moves outward with time. Earth has been in the Sun's habitable zone throughout the past 4 billion years, even as the boundaries have gradually moved outward. The narrower region that has been within the habitable zone for such a long period of time is sometimes called the *continuously habitable zone*.

**FIGURE 24.20** This painting shows a hypothetical planet much like Earth—with the lights of its own civilization visible on the night side—orbiting a close binary star system.

closer in and smaller in total extent. Figure 24.19 compares the habitable zones of two lower-mass stars to that of our Sun. It's worth noting that, in many cases, habitable zones are also possible in binary and multiple-star systems. For example, if two stars in a binary system are widely enough separated, planets can orbit one star without being perturbed much by the other. Alternatively, if two stars orbit closely, there can be a habitable zone that encircles both of them, offering up the prospect of worlds with two suns in the sky (**FIGURE 24.20**).

**Think about it** The habitable zone is sometimes called the "Goldilocks zone." Recall or read the story "Goldilocks and the Three Bears," and then explain how the "Goldilocks zone" got its name.

Note that in the search for life, we generally don't pay much attention to stars that are significantly more massive than the Sun, for two reasons. First, higher-mass stars are much less common than lower-mass stars (see Figure 16.21). Second, recall that higher-mass stars have shorter lifetimes. The example of Earth suggests that life may not take hold for up to a few hundred million years after a planet first forms, so we would expect to find life only around stars with lifetimes of hundreds of millions of years or more. This constraint rules out stars with more than a few times the mass of our Sun.

**Volcanism and Plate Tectonics** Being in a star's habitable zone is not enough to make a world habitable. The Moon, for example, is clearly in our Sun's habitable zone (because it is at the same distance from the Sun as Earth), but it is not habitable. Long-term surface habitability also requires an atmosphere and a relatively stable climate.

Recall that the terrestrial atmospheres came from volcanic outgassing, and that volcanism and tectonics are driven by internal heat [**Section 9.1**]. We therefore conclude that any terrestrial world that is similar to or larger than Venus and Earth in size should retain internal heat for many billions of years, allowing plenty of time for life to arise and flourish.

Most scientists suspect that internal heat and outgassing were enough to produce the conditions that led to an origin of life on Earth, and therefore that the same would be true for other worlds of similar size in habitable zones. However, the fact that Earth has remained habitable for some 4 billion years is due to its long-term climate stability, which we trace to the climate self-regulation that comes from the carbon dioxide cycle (see Figure 10.36). The carbon dioxide cycle, in turn, depends on plate tectonics, suggesting that plate tectonics may be important for long-term habitability. If so, then a key question about habitability around other stars is whether plate tectonics should generally be expected on Earth-size planets in habitable zones. The answer to this question is not yet known, because Venus provides an example of a nearly Earth-size planet that does not show evidence of plate tectonics. But recall that one hypothesis ties Venus's lack of plate tectonics to its high surface temperature [**Section 9.5**], which comes from the runaway greenhouse effect that occurred because Venus is too close to the Sun to be in the habitable zone. If this hypothesis is correct, then Venus might have had plate tectonics if it were in the habitable zone, in which case we might expect plate tectonics on any Earth-size world in a habitable zone. The same might be true of terrestrial worlds larger than Earth, though we do not know.

Overall, internal heat and volcanism appear to be necessary conditions for an origin of life on a planet's surface, while plate tectonics is helpful but may or may not be required for long-term habitability. Still, the great diversity among the billions of expected Earth-like planets argues that we should keep an open mind about other possible ways that long-term habitability might occur.

**Global Magnetic Field** Recall that the leading hypothesis for explaining how Mars changed from being warm and wet to its current frozen condition invokes the loss of its magnetic field (see Figure 10.30). Once the magnetic field was gone, the atmosphere was vulnerable to the solar wind, and over time Mars lost so much atmospheric gas that it could no longer maintain much heat through the greenhouse effect. Earth has been spared a similar fate because its magnetic field protects our atmosphere from solar wind particles. (Venus's lack of magnetic field means it has lost a lot of gas, but it had so much gas to begin with that it still has a thick atmosphere.)

If this hypothesis is correct, then Earth's global magnetic field has been crucial to maintaining conditions for life over a period of some 4 billion years. We might therefore expect a magnetic field to make long-term habitability more likely on other worlds. Recall that the basic requirements for a planet to maintain a global magnetic field are enough internal heat to maintain core convection and fast enough rotation to twist and distort the convection pattern (see Figure 9.6). The first requirement will already be met by any world that is large enough to have ongoing volcanism and tectonics, so only the second requirement—fast enough rotation—is new on our list. However, it probably is not difficult to meet, as many worlds likely either spin relatively fast, like Earth, or have large enough cores to support convection and magnetic fields even with slower rotation rates. Again, the sheer number and diversity of expected terrestrial planets make it seem reasonable to think that some worlds without magnetic fields might maintain enough of their atmospheres to remain habitable.

**A Recipe for Habitability?** Having a habitable surface requires being in the habitable zone, and volcanism, plate tectonics, and a magnetic field all require that a planet be at least moderately large (larger than Mars, perhaps as large as or larger than Earth). The question is whether having this size is *sufficient* to have those or other characteristics that could support long-term habitability, or whether Earth has attributes that will prove to be rare even among worlds that seem superficially similar.

Although we do not know for sure, most planetary scientists suspect that Earth-like size is indeed sufficient. In that case, we would have a simple recipe for long-term surface habitability: Any planet within a star's habitable zone that is similar to Earth in size would be expected to have a habitable surface. If this is correct, then current data suggest there should be billions of worlds with habitable surfaces in our galaxy alone.

## What kinds of extrasolar worlds might be habitable?

One of the key lessons we've learned in the study of extrasolar planets is that planets come in a wider range of types than we find in our own solar system. This fact makes scientists wonder if habitability might also be broader when we consider extrasolar worlds. Here, we'll explore a few of the intriguing possibilities.

**Moons with Habitable Surfaces** While Europa, Enceladus, and other moons may have subsurface oceans of liquid water, no moons in our solar system have liquid water on their surfaces. However, we can envision at least two possible ways in which such moons might exist in other star systems.

One possibility is that they might form as a consequence of giant impacts like that thought to have formed our own Moon. For example, while our Moon has a mass only about $\frac{1}{80}$ that of Earth, we might imagine that a similar giant impact on a super-Earth could lead to the formation of a moon with an Earth-like mass. To understand the second possibility, recall that many other solar systems have jovian-size planets that have apparently migrated inward [**Section 13.3**]. Like the jovian planets of our own solar system, we might expect such planets to have been born with numerous large moons, and it is possible that these moons would remain in orbit as the planet migrates inward. If so, and if the planet ends up within the habitable zone, then the moons might be large enough to have habitable surfaces.

**Super-Earths and Water Worlds in Extended Habitable Zones** The standard way of calculating habitable zones—such as the zones shown in Figure 24.19—assumes rocky planets (like Venus, Earth, and Mars) warmed by the greenhouse effect of water vapor and carbon dioxide in their atmospheres. But is it possible that other types of worlds might have habitable surfaces over a broader zone? Some scientists suspect that this might indeed be the case for worlds somewhat larger than Earth, such as super-Earths and water worlds (see Figure 13.16). Because these planets can be several times as massive as Earth, their atmospheres might retain substantial amounts of hydrogen gas captured from a solar nebula during planet formation. Hydrogen can also act as a greenhouse gas, which means it could keep these planets warm enough to maintain surface liquid water well beyond the boundaries usually assumed for habitable zones. Some estimates suggest that these planets could have habitable surfaces even at distances up to 10 or more times as far from their star as the standard calculations would suggest. For example, a super-Earth or water world with a substantial hydrogen atmosphere might be habitable at Saturn's distance from the Sun, or even beyond. Research into this idea is actively continuing, raising the intriguing possibility that surface habitability may be much more common than we have generally assumed.

**Subsurface Habitability** Just as we found in our own solar system, where we've identified Mars and several jovian moons as possibly having habitable subsurface regions, it is possible that there could be far more worlds with subsurface than surface habitability. These worlds also fall into several categories. First, there are planets, like Mars, that are too small to have the climate stability that would have allowed them to maintain habitable surfaces for billions of years, but that might have had habitable surfaces for a shorter time and might then have subsurface liquid water for far longer. Second, there are worlds that might be similar to Europa, Enceladus, and perhaps even Pluto, with potential zones of subsurface habitability. Beyond that, super-Earths or water worlds located beyond habitable zones might also offer opportunities for finding subsurface liquid water. For example, because we expect super-Earths to contain much more internal heat than Earth, they might have deep subsurface zones of liquid water even if they are far enough from their star to have completely frozen surfaces.

**Orphan Planets** There is another category of potentially habitable worlds that we have not yet discussed: **orphan planets** (sometimes called *rogue planets*), which do not orbit a star. These "planets" pose a definitional challenge since we usually *define* a planet as something that orbits a star, but it's easy to understand where the idea comes from. Recall that one mechanism by which jovian planets may migrate inward in some planetary systems is through close gravitational encounters in which one object loses energy and moves inward while the other object gains energy and is flung outward. Given that migration appears to be quite common in planetary systems, it seems reasonable to expect that many planets have been flung outward into interstellar space in this way. It's also possible that planet-size bodies might form independently (without first being born around a star) from fragments of collapsing interstellar clouds. Orphan planets would be far too dim to be detected directly with current technology, but an orphan world can be detected if it happens to pass in front of a background star, bending the star's light to create a *microlensing* event [**Section 13.1**]. One search for such events has suggested that orphan planets may be twice as numerous as the stars in our galaxy.

You might at first guess that planets without a star would lack energy for life, but plenty of life on Earth lives without sunlight in the deep oceans or underground, and some of this life survives on energy traceable to Earth's internal heat alone. A planet like Earth retains internal heat for billions of years, and therefore might have active volcanism and tectonics—and the prospects of subsurface liquid water—even if it were not orbiting a star. In that case, orphan planets might frequently have subsurface habitable zones.

Some researchers speculate that some orphan planets might even have *surface* habitability. Much as with the idea of extended habitable zones, this might be possible if the planets have a thick enough hydrogen atmosphere. In that case, internal heat leaking outward might be trapped by the greenhouse effect, and in some places this might raise the surface temperature above the freezing point of water. Indeed, with a thick enough atmosphere, even a planet the size of Earth could potentially offer such conditions while floating freely in interstellar space. A key question, then, is whether a thick hydrogen atmosphere is likely around planets similar in size to Earth. No one really knows, but remember that the main reason Earth and the other terrestrial planets did not retain hydrogen from the solar nebula is that their small sizes allowed the lightweight hydrogen to escape. Because the speed of hydrogen atoms depends on the temperature, which would be far lower for planets ejected into interstellar space, it is possible that hydrogen atmospheres might be retained, if they formed in the first place.

**The Bottom Line: Wide-Ranging Possibilities** Overall, there seems to be great potential for habitability throughout the universe, both on planets that might be Earth-like in character and appearance and on a variety of worlds that might have other forms of habitability. Moreover, we have focused primarily on habitability by life that uses water as its liquid medium, but the case of Titan makes us wonder whether other liquids might also be used. Of course, being habitable does not necessarily mean having life. We are therefore ready to turn our attention to the question of how we might actually discover life beyond the solar system.

**Think about it** Considering all the factors we have discussed, do *you* believe that life will prove to be rare or common in the universe? Defend your opinion.

### How could we detect life on extrasolar planets?

At present, we have no way to search for actual life on any of the many extrasolar worlds that may potentially be habitable. None of our indirect methods for learning about these worlds (astrometric, Doppler, transit [**Section 13.1**]) can offer information that would tell us whether they have life, and our direct detection capabilities remain too rudimentary to detect the presence of life. However, we expect our direct observational capabilities to improve dramatically in coming decades, and as a result, scientists are working on strategies that might allow future telescopes to search for life on worlds around other stars. Note that, as we've briefly discussed already, we will probably be limited to searching for evidence of surface life; detecting subsurface life on such distant worlds will be far more difficult, if it is possible at all.

If our telescope technology becomes sufficiently powerful, we can hope to learn about potential life through both images and spectra. For example, large telescopes in space may be able to obtain images with high enough resolution to tell us whether extrasolar planets have continents and oceans, and perhaps even allow us to monitor seasonal changes.

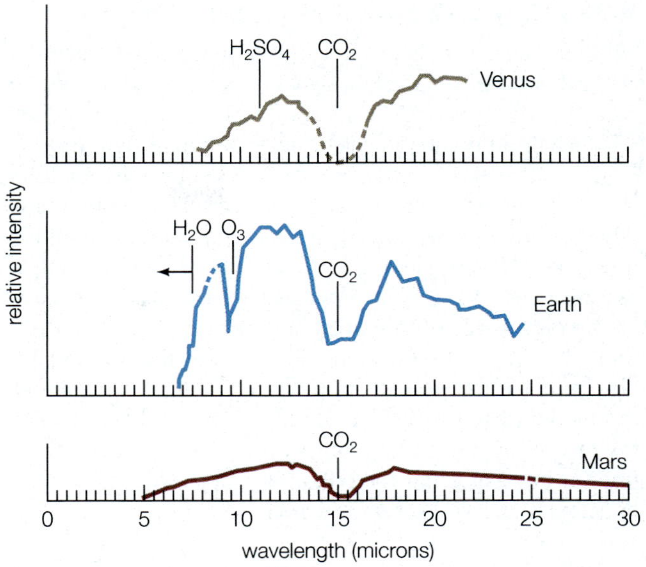


**FIGURE 24.21** The infrared spectra of Venus, Earth, and Mars as they might be seen from afar, showing absorption features that point to the presence of carbon dioxide ($CO_2$), ozone ($O_3$), and sulfuric acid ($H_2SO_4$) in their atmospheres. While carbon dioxide is present in all three spectra, only our own planet has appreciable oxygen (and hence ozone)—a product of photosynthesis. If we could make similar spectral analyses of distant planets, we might detect atmospheric gases that would indicate life.

Spectra should prove even more useful. Moderate-resolution infrared spectra can reveal the presence and abundance of many atmospheric gases (**FIGURE 24.21**), and careful analysis of atmospheric makeup might tell us whether a planet has life. On Earth, for example, the large abundance of oxygen (21% of our atmosphere) is a direct result of photosynthetic life. Abundant oxygen in the atmosphere of a distant world might similarly indicate the presence of life. Other evidence might come from any ratio of atmospheric gases that would naturally change unless it were being actively maintained by the respiration of living organisms. Although we cannot do it yet, it seems likely that within a few decades we will have methods for learning whether distant planets have life.

## 24.4 The Search for Extraterrestrial Intelligence

If life turns out to be common, it's natural to wonder whether other worlds might also have *intelligent* life and civilizations. If such civilizations exist, we might be able to find them simply by listening for signals that they are sending into interstellar space, either in deliberate attempts to contact other civilizations or as a means of communicating among themselves. The search for signals from other civilizations is generally known as the **search for extraterrestrial intelligence**, or **SETI** for short.

## How many civilizations are out there?

SETI efforts have a chance to succeed only if other advanced civilizations are broadcasting signals that we could receive. To judge the chances of SETI success, we'd need to know how many civilizations are broadcasting such signals right now.

Given that we do not even know whether microbial life exists anywhere beyond Earth, we certainly don't know whether other civilizations exist, let alone how many there might be. Nevertheless, for the purposes of planning a search for extraterrestrial intelligence, it is useful to have an organized way of thinking about the number of civilizations that might be out there. To keep our discussion simple, let's consider only the number of potential civilizations in our own galaxy. We can always extend our estimate to the rest of the universe by simply multiplying the result we find for our galaxy by 100 billion, the approximate number of galaxies in the observable universe.

**The Drake Equation**

**The Drake Equation** In 1961, astronomer Frank Drake wrote a simple equation, now known as the **Drake equation**, designed to summarize the factors that would determine the number of civilizations that currently share the Milky Way Galaxy with us and that are capable of interstellar communication. In a form slightly modified from the original,* the Drake equation looks like this:

$$\text{Number of civilizations} = N_{HP} \times f_{life} \times f_{civ} \times f_{now}$$

This equation will make sense once you understand the meaning of each factor:

- $N_{HP}$ is the number of habitable planets in the galaxy; that is, it is the number of planets that could potentially have life.
- $f_{life}$ is the fraction of habitable planets that actually *have* life. For example, $f_{life} = 1$ would mean that all habitable planets have life, and $f_{life} = \frac{1}{1{,}000{,}000}$ would mean that only 1 in a million habitable planets has life. Therefore, the product $N_{HP} \times f_{life}$ tells us the number of life-bearing planets in the galaxy.
- $f_{civ}$ is the fraction of life-bearing planets on which a civilization capable of interstellar communication *has at some time* arisen. For example, $f_{civ} = \frac{1}{1000}$ would mean that such a civilization has existed on 1 out of 1000 planets with life, while the other 999 out of 1000 have not had a species that learned to build radio transmitters, high-powered lasers, or other devices for interstellar conversation. When we multiply this factor by the first two factors to form the product $N_{HP} \times f_{life} \times f_{civ}$, we get the total number of planets on which intelligent beings have evolved and developed a communicating civilization at some time in the galaxy's history.
- $f_{now}$ is the fraction of these civilization-bearing planets that happen to have a civilization *now,* as opposed to, say, millions or billions of years in the past. This factor is important because we can hope to contact only civilizations that are broadcasting signals we could receive at present. (In estimating $f_{now}$, we assume that the light-travel time for signals from other stars has been taken into account.)

*Dr. Drake personally approved using this modified form and still calling it the "Drake equation."

Because the product of the first three factors tells us the total number of civilizations that have *ever* arisen in the galaxy, multiplying by $f_{now}$ tells us how many civilizations we could potentially make contact with today. In other words, the result of the Drake equation should in principle tell us the number of civilizations that we might hope to contact. Unfortunately, we can make a reasonable estimate only for the first term ($N_{HP}$) in the Drake equation, which means we cannot actually calculate its result. Nevertheless, the equation is a useful way of organizing our thinking, as we can see by considering the potential values for each of its factors.

**Think about it** Try the following sample numbers in the Drake equation. Suppose that there are 1000 habitable planets in our galaxy, that 1 in 10 habitable planets has life, that 1 in 4 planets with life has at some point had an intelligent civilization, and that 1 in 5 civilizations that have ever existed is in existence now. How many civilizations would exist at present? Explain.

**The Number of Life-Bearing Planets** Let's begin with the first two factors in the Drake equation, whose product ($N_{HP} \times f_{life}$) tells us the number of life-bearing planets in our galaxy. As we've discussed, the statistics on known extrasolar planets show that Earth-size planets are common, and such planets are probably also common in habitable zones. Overall, it seems reasonable to suppose that a significant fraction of all stars have at least one habitable planet, which means there should be many billions—perhaps more than 100 billion—of habitable planets in our galaxy.

The factor $f_{life}$ presents more difficulty, because we do not yet have a reliable way to estimate the fraction of habitable planets on which life actually arose. Still, we've argued that the rapid appearance of life on Earth and the fact that laboratory experiments suggest possible ways in which life could have arisen both suggest that life should be likely on habitable worlds. In that case, we might expect most or all habitable planets to have life, making the fraction $f_{life}$ close to 1. Of course, until we have solid evidence that life arose anywhere else, such as on Mars, it is also possible that Earth has been very lucky and that $f_{life}$ is so close to zero that life has never arisen on any other planet in our galaxy.

**The Question of Intelligence** Even if life-bearing planets are very common, civilizations capable of interstellar communication might not be. The fraction of life-bearing planets that at some time have such civilizations, $f_{civ}$, depends on at least two things. First, a planet would have to have a species evolve with sufficient intelligence to develop interstellar communication. In other words, the planet needs a species at least as smart as we are. Second, that species would have to develop a civilization with technology at least as advanced as ours.

Although we cannot really be sure, most scientists suspect that only the first requirement is difficult to meet. A fundamental assumption in nearly all science today is that we are not "special" in any particular way. We live on a fairly typical planet orbiting an ordinary star in a normal galaxy, and we assume that living creatures elsewhere—whether they prove to be rare or common—would be subjected to evolutionary pressures similar to those that have operated on Earth. Therefore, if species with intelligence similar to ours have evolved elsewhere, we assume that they would have similar sociological drives that would eventually lead them to develop the technology necessary for interstellar communication.

If this assumption is correct, then the fraction $f_{civ}$ depends primarily on the question of whether sufficient intelligence is rare or common among life-bearing planets. As with the question of life of any kind, the short answer is that we just don't know, but we can get at least some insight by considering what happened on Earth.

Look again at Figure 24.4. While life arose quite quickly on Earth, nearly all life remained microbial until just a few hundred million years ago, and it took nearly 4 billion years for us to arrive on the scene. This slow progress toward intelligence might suggest that producing a civilization is very difficult even when life is present. On the other hand, more than half the stars in the Milky Way are older than our Sun, so if Earth's case is typical, then plenty of planets have existed long enough for intelligence to arise.

Another way to address the question is by considering our level of intelligence in comparison to that of other animals on Earth. We can get a rough measure of intelligence by comparing brain mass to total body mass (a measure

## EXTRAORDINARY CLAIMS Aliens Are Visiting Earth in UFOs

In this chapter, we discuss aliens as a possibility, not a reality. However, opinion polls suggest that up to half the American public believes that aliens are already visiting us. What can science say about this extraordinary claim?

The bulk of the claimed evidence for alien visitation consists of sightings of UFOs—unidentified flying objects. Many thousands of UFOs are reported each year, and no one doubts that unidentified objects are being seen. The question is whether they are alien spacecraft.

Aliens have long been a staple of science fiction, but modern interest in UFOs began with a widely reported sighting in 1947. While flying a private plane near Mount Rainier in Washington State, businessman Kenneth Arnold saw nine mysterious objects streaking across the sky. He told a reporter that the objects "flew erratic, like a saucer if you skip it across the water." (One possible explanation is that he saw meteors skipping across the atmosphere.) He did *not* say that the objects were saucer-shaped, but the reporter wrote of "flying saucers." The story was front-page news throughout America, and "flying saucers" soon invaded popular culture, if not our planet.

The flying saucer reports also interested the U.S. Air Force, largely out of concern that the UFOs might represent new types of aircraft developed by the Soviet Union. For two decades, the Air Force hired teams of academics to study UFO reports. In most cases, these experts were able to specify a plausible identification of the UFO. The explanations included bright stars and planets, aircraft rockets, balloons, birds, meteors, atmospheric phenomena, and the occasional hoax. In a few cases, the investigators could not deduce what was seen, but their overall conclusion was that there was no reason to believe the UFOs were either highly advanced Soviet craft or visitors from other worlds.

Believers discounted the Air Force denials, claiming to have other evidence of alien visitation. So far, none of this evidence has withstood scientific scrutiny. Photographs and film clips are nearly always either fuzzy or obviously faked. Crop circles are easily made by pranksters, and claimed pieces of alien spacecraft turn out to have more mundane origins. Champions of alien visitation generally explain away the lack of clear evidence either as a government cover-up or as a failure of the scientific community to take the subject seriously, but neither explanation is compelling.

It's conceivable that a government might *try* to put the lid on evidence of alien visits, though the motivation for doing so is unclear. The usual explanations are that the public couldn't handle the news or that the government is taking secret advantage of the alien materials to design new military hardware (via "reverse engineering"). However, given that half the population already believes in alien visitors, the shock of discovery would seem unlikely to cause panic. As for reverse engineering extraterrestrial spacecraft, any society that could routinely cross interstellar distances would be far beyond us technologically. Reverse engineering their spaceships is as unlikely as Neandertals constructing personal computers just because a laptop somehow landed in their cave. In addition, while a government might successfully hide evidence for a short time, over decades the lure of talk show fame and riches would surely cause someone to reveal the conspiracy. Moreover, unless the aliens landed only in the United States, the conspiracy would have to include all governments on Earth, which seems highly unlikely given the world's political conditions.

Claims that the scientific community is uninterested also fall apart on scrutiny. Scientists are continually competing with one another to be the first to make a great discovery, and clear evidence of alien visitors would rank high on the all-time list. The fact that few scientists are engaged in such study reflects not a lack of interest, but a lack of evidence worthy of study.

Of course, absence of evidence is not evidence of absence. Most scientists are open to the possibility that we might someday find evidence of alien visits, and given what we've learned about the prospects of life in the universe, it is at least plausible to imagine that such visits might occur. So far, however, we have no hard evidence to support the belief that aliens are already here.

**Verdict:** Not established. While it's conceivable that aliens could be visiting Earth, the claimed evidence is insufficient to warrant confidence in such an extraordinary claim.

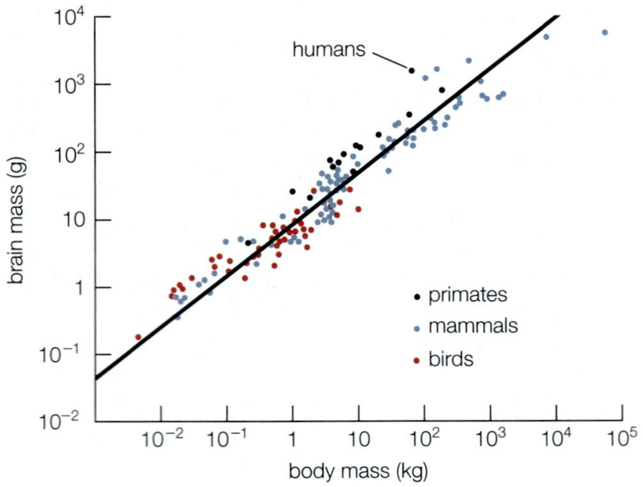


**FIGURE 24.22** This graph shows how brain mass compares to body mass for some mammals (including primates) and birds. The straight line represents an average of the ratio of brain mass to body mass, so animals that fall above the line are smarter than average and animals that fall below the line are less smart. Note that the scale uses powers of 10 on both axes. (Data from Harry J. Jerison.)

sometimes called the *encephalization quotient,* or EQ). **FIGURE 24.22** shows the brain weights for a sampling of birds and mammals plotted against their body weights. There is a clear and expected trend in that heavier animals have heavier brains. By drawing a straight line that fits these data, we can define an average value of brain mass for each body mass. Animals whose brain mass falls above the line are smarter than average, while animals whose brain mass falls below the line are less mentally agile. Keep in mind that it is the *vertical* distance above the line that tells us how much smarter a species is than the average, and that the scale goes in powers of 10 on both axes. If you look closely, you'll see that the data point for humans lies significantly farther above the line than the data point for any other species. By this measure of intelligence, we are far smarter than any other species that has ever existed on Earth.

Some people use this fact to argue that even on a planet with complex life, a species as intelligent as we are would be very rare. They say that even if there is an evolutionary drive toward intelligence in general, it takes extreme luck to reach our level of intelligence. After all, while it's evolutionarily useful to have enough intelligence to capture prey and evade other predators, it's not clear why natural selection would lead to brains big enough to build spacecraft. However, the same data can be used to reach an opposite conclusion. The scatter in the levels of intelligence among different animals tells us that some variation should be expected, and statistical analysis shows that we are not unreasonably far above the average. It might therefore be inevitable that *some* species would develop our level of intelligence on any planet with complex life.

**Technological Lifetimes** For the sake of argument, let's assume that life and intelligence are reasonably likely, so thousands or millions of planets in our galaxy have at some time given birth to a civilization capable of interstellar communication. In that case, the final factor in the Drake equation, $f_{now}$, determines the likelihood of there being someone whom we could contact now. The value of this factor depends on how long civilizations survive.

Consider our own example. In the roughly 12 billion years during which our galaxy has existed, we have been capable of interstellar communication via radio for only about 60 years. Therefore, if we were to destroy ourselves tomorrow, other civilizations could have received signals from us during only 60 years out of the galaxy's 12-billion-year existence, equivalent to 1 part in 200 million of the galaxy's history. If such a short technological lifetime is typical of civilizations, $f_{now}$ would be only $\frac{1}{200{,}000{,}000}$, and some 200 million civilization-bearing planets would need to have existed at one time or another in the Milky Way in order for us to have a good chance of finding another civilization out there now.

However, we'd expect $f_{now}$ to be so small only if we are on the brink of self-destruction—after all, the fraction will grow larger for as long as our civilization survives. That is, if civilizations are at all common, survivability is the key factor in whether any are out there now. If most civilizations self-destruct shortly after achieving the technology for interstellar communication, then we are almost certainly alone in the galaxy at present. But if most survive and thrive for thousands or millions of years, the Milky Way may be brimming with civilizations—most of them far more advanced than our own.

**Think about it** Describe a few reasons a civilization capable of interstellar communication would also be capable of self-destruction. Overall, do you believe our civilization can survive for thousands or millions of years? Defend your opinion.

## How does SETI work?

If there are indeed other civilizations out there, then in principle we ought to be able to make contact with them. Based on our current understanding of physics, it seems likely that even very advanced civilizations would communicate much as we do—by encoding signals in radio waves or other forms of light. To search for alien radio signals, most SETI researchers use large radio telescopes, such as the Allen Telescope Array (see the photo that opens this chapter). A few researchers are studying other parts of the electromagnetic spectrum. For example, some scientists use visible-light telescopes to search for communications encoded as laser pulses. Of course, advanced civilizations may well have invented communication technologies that we cannot even imagine, let alone detect.

A good way to think about our chances of picking up an alien signal is to imagine what aliens would need to do to pick up signals from us. Since about the 1950s, we have been sending relatively high-powered transmissions into space in the form of television broadcasts. Therefore, anyone within about 60 to 70 light-years of Earth could in principle watch our old television shows (perhaps a frightening thought). However, in order to detect our broadcasts, they would need far larger and more sensitive radio telescopes than we have today. If their technology were at the same level as ours,

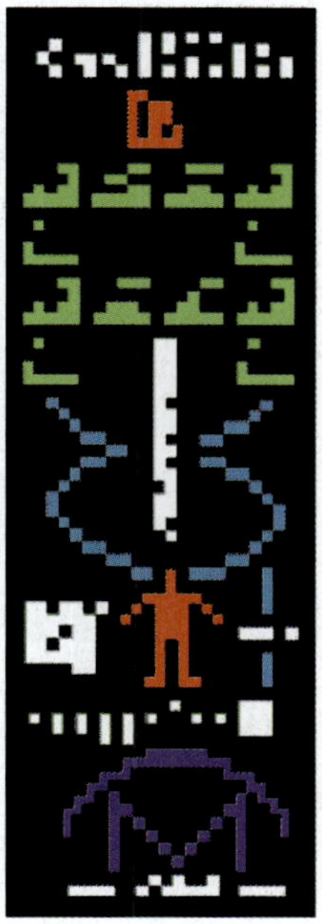

**FIGURE 24.23** In 1974, a short message was broadcast to the globular cluster M13 using the Arecibo radio telescope. The picture shown here was encoded by using two different radio frequencies, one for "on" and one for "off" (the colors used here are arbitrary). To decode the message, the aliens would need to realize that the bits are meant to be arranged in a rectangular grid as shown, but that should not be difficult: The grid has 73 rows and 23 columns, and aliens would presumably know that these are both prime numbers. The picture represents the Arecibo radio dish, our solar system, a human stick figure, and a schematic of DNA and the eight simple molecules used in its construction.

they could receive a signal from us only if we deliberately broadcast an unusually high-powered transmission.

To date, humans have made only a few attempts to broadcast our existence in this way. The most famous was a 3-minute transmission made in 1974 with a planetary radar transmitter on the Arecibo radio telescope, in which a simple pictorial message was beamed toward the globular cluster M13 (**FIGURE 24.23**). This target was chosen in part because it contains a few hundred thousand stars, seemingly offering a good chance that at least one has a civilization around it. However, M13 is about 25,000 light-years from Earth, so it will take some 25,000 years for our signal to get there and another 25,000 years for any response to make its way back to Earth.

Several SETI projects under way or in development would be capable of detecting signals like the one we broadcast from Arecibo if they came from civilizations within a few hundred light-years. These SETI efforts scan millions of radio frequency bands simultaneously. If anyone nearby is deliberately broadcasting on an ongoing basis, we have a good chance of detecting the signals.

**Think about it** SETI efforts are often controversial because of their cost (which is privately funded) and uncertain chance of success, but supporters say the cost is justified because contact with an extraterrestrial intelligence would be such an important discovery. What do *you* think?

## 24.5 Interstellar Travel and Its Implications for Civilization

So far, we have discussed ways of detecting distant civilizations without leaving the comfort of our own planet. Could we ever actually visit other worlds in other star systems? A careful analysis of this question turns out to have profound implications for the future of our civilization. To see why, we first need to consider the prospects for achieving interstellar travel.

### How difficult is interstellar travel?

In many science fiction movies, our descendants race around the galaxy in starships, circumventing nature's prohibition on faster-than-light travel by entering hyperspace, wormholes, or warp drive. Unfortunately, we do not have good reason to think that any of these science fiction technologies are really possible. In that case, we will be limited to speeds slower than the speed of light, and today we cannot even begin to approach that speed. Nevertheless, we have already sent out our first emissaries to the stars, and there's no reason to believe that we won't develop better technologies in the future.

**The Challenge of Interstellar Travel** We have launched five spacecraft that will leave our solar system and eventually travel among the stars: the planetary probes *Pioneer 10, Pioneer 11, Voyager 1, Voyager 2,* and *New Horizons*. These spacecraft are traveling about as fast as anything ever built by humans, but their speeds are still less than $\frac{1}{10,000}$ of the speed of light. It would take each of them some 100,000 years just to reach the next nearest star system (Alpha Centauri), but their trajectories won't take them anywhere near it. Instead, they will simply continue their journey without passing close to any nearby stars, wandering the Milky Way for millions or even billions of years to come. The *Pioneer* and *Voyager* spacecraft carry greetings from Earth, just in case someone comes across one of them someday (**FIGURE 24.24**).

If we want to make interstellar journeys within human lifetimes, we will need starships that can travel at speeds close to the speed of light. We will need entirely new types of engines to reach such high speeds. The energy requirements of interstellar spacecraft may pose an even more daunting challenge. For example, the energy needed to accelerate a single ship the size of *Star Trek*'s *Enterprise* to just half the speed of light would be more than 2000 times the total annual energy use of the world today. Clearly, interstellar travel will require vast new sources of energy. In addition, fast-moving starships will require new types of shielding to protect crew members from instant death. As a starship travels through interstellar gas at near-light speed, ordinary atoms and ions will hit it like a deadly flood of high-energy cosmic rays.

If we succeed in building starships capable of traveling at speeds close to the speed of light, the crews will face significant social challenges. According to the well-tested principles of Einstein's theory of relativity, time will run much more slowly on a spaceship that travels at high speed to the stars than it does here on Earth [**Section S2.4**]. For example, in a ship traveling at an average speed of 99.9% of the speed of light, the 50-light-year round trip to the star Vega would take the travelers only about 2 years—but more than 50 years would pass on Earth while they were gone. The crew would therefore need only 2 years' worth of provisions and would age only 2 years during the voyage, but they would return to a world quite different from the one they left. Family and friends would be older or deceased, new technologies might have made their knowledge and skills obsolete, and many political and social changes might have occurred in their absence. The crew would face a difficult adjustment when they came home to Earth.

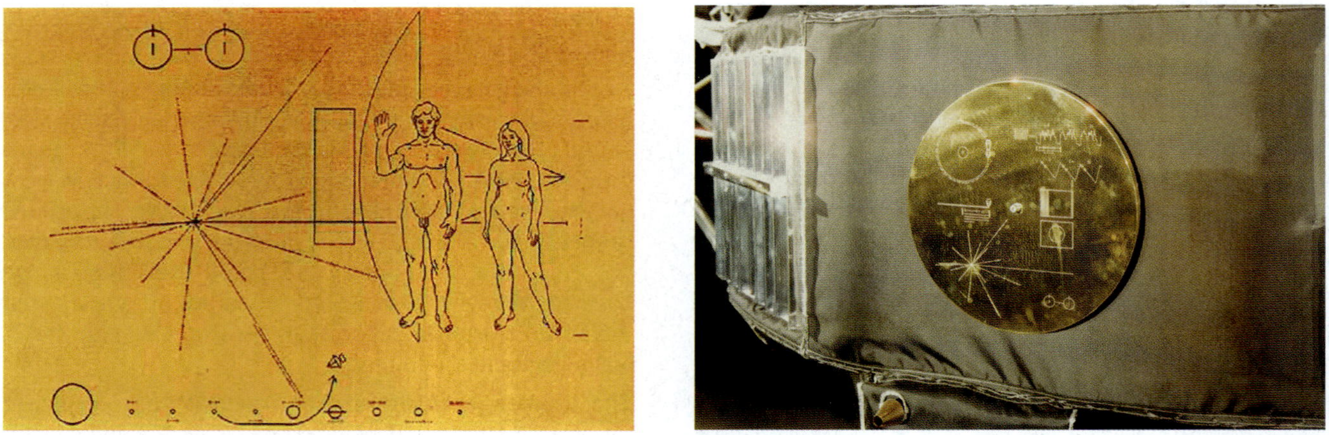

a The *Pioneer* plaque, about the size of an automobile license plate. The human figures are shown in front of a drawing of the spacecraft to give them a sense of scale. The "prickly" graph to their left shows the Sun's position relative to nearby pulsars, and Earth's location around the Sun is shown below. Binary code indicates the pulsar periods; because pulsars slow with time, the periods will allow someone reading the plaque to determine when the spacecraft was launched.

b *Voyagers 1* and *2* carry a phonograph record—a 12-inch gold-plated copper disk containing music, greetings, and images from Earth.

**FIGURE 24.24** Messages aboard the *Pioneer* and *Voyager* spacecraft, which are bound for the stars.

**Starship Design** Despite all the difficulties, some scientists and engineers have already proposed designs that could in principle take us to nearby stars. In the 1960s, a group of scientists proposed *Project Orion,* which envisioned accelerating a spaceship with repeated detonations of relatively small hydrogen bombs. Each explosion would take place a few tens of meters behind the spaceship and would propel the ship forward as the vaporized debris struck a "pusher plate" on the back of the spacecraft (**FIGURE 24.25**). Calculations showed that a spaceship accelerated by the rapid-fire detonation of a million H-bombs could reach Alpha Centauri in just over a century. In principle, we could build an *Orion* spacecraft with existing technology, though it would be very expensive and would require an exception to the international treaty banning nuclear detonations in space.

**FIGURE 24.25** Artist's conception of the *Project Orion* starship, showing one of the small hydrogen-bomb detonations that would propel it. Debris from the detonation strikes the "pusher plate" at the back of the spaceship. The central sections (enclosed in a lattice) hold the bombs, and the front sections house the crew.

A century to the nearest star isn't bad, but it still wouldn't make interstellar travel easy. Unfortunately, no available technology could go much faster. The problem is mass: Making a rocket faster requires more fuel, but adding fuel adds mass and makes it more difficult for the rocket to accelerate. Calculations show that even in the best case, rockets carrying nuclear fuel could achieve speeds no more than a few percent of the speed of light. However, we can envision some possible future technologies that might get around this problem.

One idea suggests powering starships with engines that generate energy through matter-antimatter annihilation. While nuclear fusion converts less than 1% of the mass of atomic nuclei into energy, matter-antimatter annihilation [**Section 22.1**] converts *all* the annihilated mass into energy. Starships with matter-antimatter engines could potentially reach speeds of 90% or more of the speed of light. At these speeds, the slowing of time predicted by relativity becomes noticeable, putting many nearby stars within a few years' journey for the crew members. However, because no natural reservoirs of antimatter exist, we would have to be able to manufacture many tons of antimatter and then store it safely for the trip—capabilities that are far beyond our present means.

An even more speculative and futuristic design, known as an *interstellar ramjet,* would collect interstellar hydrogen with a gigantic scoop, using the collected gas as fuel for its nuclear engines (**FIGURE 24.26**). By collecting fuel along the way, the ship could avoid carrying the weight of fuel on board. However, because the density of interstellar gas is so low, the scoop would need to be enormous. As astronomer Carl Sagan said, we are talking about "spaceships the size of worlds."

**FIGURE 24.26** Artist's conception of a spaceship powered by an interstellar ramjet. The giant scoop in the front (left) would collect interstellar hydrogen for use as fusion fuel.

The bottom line is that while we face enormous obstacles to achieving interstellar travel, there's no reason to think it's impossible. If we can avoid self-destruction and if we continue to explore space, our descendants might well journey to the stars.

## Where are the aliens?

Imagine that we survive long enough to become interstellar travelers and that we begin to colonize habitable planets around nearby stars. As each colony grows, it may send out explorers to other star systems. Even if our starships traveled at relatively low speeds—say, a few percent of the speed of light—we could have dozens of outposts around nearby stars within a few centuries. In 10,000 years, our descendants would be spread among stars within a few hundred light-years of Earth. In a few million years, we could have outposts throughout the Milky Way Galaxy. We would have become a true galactic civilization.

Now, if we take the idea that *we* could develop a galactic civilization within a few million years and combine it with the reasonable (though unproved) idea that civilizations ought to be common, we are led to an astonishing conclusion: Someone else should already have created a galactic civilization.

To see why, let's take some sample numbers. Suppose the overall odds of a civilization arising around a star are about the same as your odds of winning the lottery, or 1 in a million. Using a low estimate of 100 billion stars in the Milky Way Galaxy, this would mean some 100,000 civilizations in our galaxy alone. Moreover, current evidence suggests that stars and planetary systems like our own could have formed for at least 5 billion years before our solar system was even born, in which case the first of these 100,000 civilizations would have arisen at least 5 billion years ago. Others would have arisen, on average, about every 50,000 years. Under these assumptions, we would expect the youngest civilization besides ours to be some 50,000 years ahead of us technologically, and most would be millions or billions of years ahead of us.

We thereby encounter a strange paradox: Plausible arguments suggest that a galactic civilization should already exist, yet we have so far found no evidence of such a civilization. This paradox is often called *Fermi's paradox,* after the Nobel Prize–winning physicist Enrico Fermi. During a 1950 conversation with other scientists about the possibility of extraterrestrial intelligence, Fermi responded to speculations by asking, "So where is everybody?"

This paradox has many possible solutions, but broadly speaking we can group them into three categories:

1. *We are alone.* There is no galactic civilization because civilizations are extremely rare—so rare that we are the first to have arisen on the galactic scene, perhaps even the first in the universe.

2. *Civilizations are common, but no one has colonized the galaxy.* There are at least three possible reasons this might be the case. Perhaps interstellar travel is much harder or more expensive than we have guessed, and civilizations are unable to venture far from their home worlds. Perhaps the desire to explore is unusual, and other societies either never leave their home star systems or stop exploring before they've colonized much of the galaxy. Most ominously, perhaps many civilizations have arisen, but they have all destroyed themselves before achieving the ability to colonize the stars.

3. *There IS a galactic civilization,* but it has not yet revealed its existence to us.

We do not know which, if any, of these explanations is the correct solution to the question "Where are the aliens?" However, each category of solution has astonishing implications for our own species.

Consider the first solution—that we are alone. If this is true, then our civilization is a remarkable achievement. It implies that through all of cosmic evolution, among countless star systems, we are the first matter in our galaxy or the universe ever to know that the rest of the universe exists. Through us, the universe has attained self-awareness. Some philosophers and many religions argue that the ultimate purpose of life is to become truly self-aware. If so, and if we are alone, then the destruction of our civilization and the loss of our scientific knowledge would represent an inglorious end to something that took the universe some 14 billion years to achieve. From this point of view, humanity becomes all the more precious, and the collapse of our civilization would be all the more tragic.

The second category of solutions has much more terrifying implications. If thousands of civilizations before us have all failed to achieve interstellar travel on a large scale, what hope do we have? Unless we somehow think differently than all other civilizations, this solution says that we will never go far in space. Because we have always explored when the opportunity arose, this solution almost inevitably leads to the conclusion that failure will come about because we destroy ourselves. We hope that this answer is wrong.

The third solution is perhaps the most intriguing. It says that we are newcomers on the scene of a galactic civilization that has existed for millions or billions of years before us. Perhaps this civilization is deliberately leaving us alone for the time being and will someday decide the time is right to invite us to join it.

No matter what the answer turns out to be, learning it will surely mark a turning point in the brief history of our species. Moreover, this turning point is likely to be reached within the next few decades or centuries. We already have the ability to destroy our civilization. If we do so, then our fate is sealed. But if we survive long enough to develop technology that can take us to the stars, the possibilities seem almost limitless.

## The BIG Picture PUTTING CHAPTER 24 INTO CONTEXT

*Throughout our study of astronomy, we have taken a "big picture" view of understanding how we fit into the universe. Here, at last, we have returned to Earth and examined the role of our own generation in the big picture of human history. Tens of thousands of past human generations have walked this Earth. Ours is the first generation with the technology to study the far reaches of our universe, to search for life elsewhere, and to travel beyond our home planet. It is up to us to decide whether we will use this technology to advance our species or to destroy it.*

*Imagine for a moment the grand view, a gaze across the centuries and millennia from this moment forward. Picture our descendants living among the stars, having created or joined a great galactic civilization. They will have the privilege of experiencing ideas, worlds, and discoveries far beyond our wildest imagination. Perhaps, in their history lessons, they will learn of our generation—the generation that history placed at the turning point and that managed to steer its way past the dangers of self-destruction and onto the path to the stars.*

## Summary of Key Concepts

### 24.1 Life on Earth

- **When did life arise on Earth?** Fossil evidence puts the origin of life at least 3.5 billion years ago, and carbon isotope evidence pushes this date to about 3.95 billion years ago. Life therefore arose within a few hundred million years after Earth's birth and possibly in a much shorter time.

- **How did life arise on Earth?** Genetic evidence suggests that all life on Earth evolved from a common ancestor, which may have resembled microbes that live today in hot water near undersea volcanic vents. We do not know how this first organism arose, but laboratory experiments suggest that it may have been the result of natural chemical processes on the early Earth. Once life arose, it rapidly diversified and evolved through **natural selection**.

- **What are the necessities of life?** Life on Earth thrives in a wide range of environments, and in general seems to require only three things: a source of nutrients, a source of energy, and liquid water.

### 24.2 Life in the Solar System

- **Could there be life on Mars?** Mars once had conditions that may have been conducive to the origin of life. If life arose, it might still survive in pockets of liquid water underground.

- **Could there be life in the outer solar system?** Numerous worlds in the outer solar system may have subsurface lakes or oceans of liquid water, including Jupiter's moons Europa, Ganymede, and Callisto; Saturn's moons Enceladus and Titan; Neptune's moon Triton; and Pluto. Titan could also have life using methane or ethane as a liquid medium, since it has lakes of those substances on its surface.

### 24.3 Life Around Other Stars

- **What are the requirements for surface habitability?** To have a habitable surface, a planet must reside in its star's **habitable zone**. Earth's long-term habitability has also been made possible by volcanism to create an atmosphere, plate tectonics to help maintain a stable climate, and a magnetic field to protect the atmosphere from the solar wind. It seems likely, but not certain, that other Earth-size worlds in habitable zones would also have habitable conditions on their surfaces.

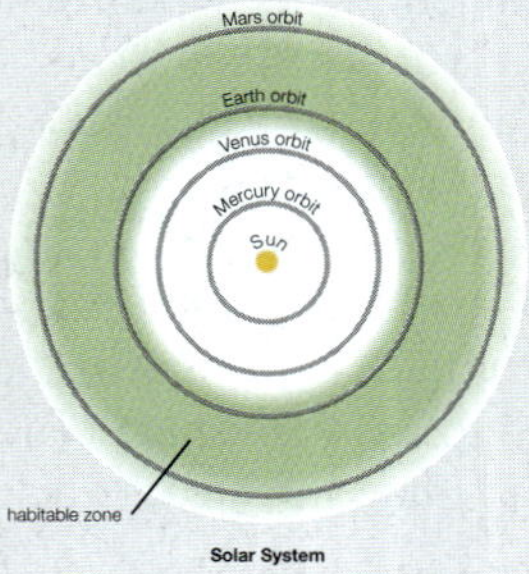


- **What kinds of extrasolar worlds might be habitable?** Surface habitability seems possible for planets or moons similar in size and composition to Earth and located within the habitable zone, and the habitable zone may extend farther for super-Earths or water worlds with thick hydrogen atmospheres. Subsurface habitability may be even more common, since it is possible on any world with enough internal heat to keep water liquid beneath the surface. **Orphan planets**, which do not orbit a star, also offer intriguing possibilities for subsurface life, and possibly even for surface life if they have thick enough atmospheres.

- **How could we detect life on extrasolar planets?** Future telescopes should allow us to obtain crude images or spectra of planets within stellar habitable zones. An image of an extrasolar planet—even if only a few pixels in size—might indicate the presence of continents and oceans or of seasonal changes. Spectroscopic analysis could tell us much more, and might reveal atmospheric gases that would be evidence of life.

## 24.4 The Search for Extraterrestrial Intelligence

- **How many civilizations are out there?** We don't know, but the **Drake equation** gives us a way to organize our thinking about the question. The equation (in a modified form) says that the number of civilizations in the Milky Way Galaxy with which we could potentially communicate is $N_{HP} \times f_{life} \times f_{civ} \times f_{now}$, where $N_{HP}$ is the number of habitable planets in the galaxy, $f_{life}$ is the fraction of habitable planets that actually have life on them, $f_{civ}$ is the fraction of life-bearing planets on which a civilization capable of interstellar communication has at some time arisen, and $f_{now}$ is the fraction of all these civilizations that exist now.
- **How does SETI work?** **SETI**, the **search for extraterrestrial intelligence**, generally involves efforts to detect signals—such as radio or laser communications—coming from civilizations on other worlds.

## 24.5 Interstellar Travel and Its Implications for Civilization

- **How difficult is interstellar travel?** Convenient interstellar travel remains well beyond our capabilities because of the technological requirements for engines, the enormous energy needed to accelerate spacecraft to speeds near the speed of light, and the difficulties of shielding the crew from radiation. Nevertheless, people have proposed ways around all these difficulties, and it seems reasonable to think that we will someday achieve interstellar travel if our civilization survives long enough.

- **Where are the aliens?** A civilization capable of interstellar travel ought to be able to colonize the galaxy in a few million years or less, and the galaxy was around for billions of years before Earth was even born. It therefore seems that some civilization should have colonized the galaxy long ago—yet we have no evidence of other civilizations. Every possible explanation for this surprising fact has astonishing implications for our species and our place in the universe.

# Visual Skills Check

*Use the following questions to check your understanding of some of the many types of visual information used in astronomy. For additional practice, try the Chapter 24 Visual Quiz in the Study Area at www.MasteringAstronomy.com.*

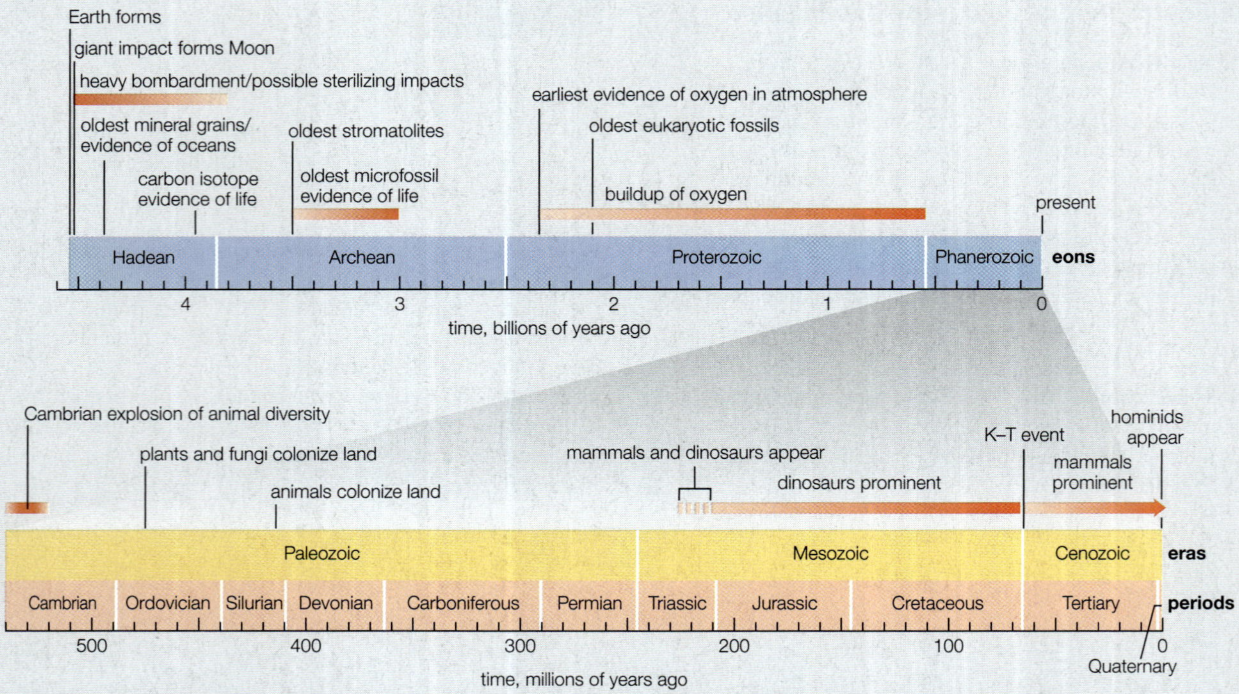


*The figure above, which repeats Figure 24.4, shows the geological time scale. Use this figure to answer the following questions.*

1. List the following events in the order in which they occurred, from first to last.
   a. earliest humans
   b. earliest animals
   c. impact causes extinction of dinosaurs
   d. earliest mammals
   e. earliest plants living on land
   f. first time there is significant oxygen in Earth's atmosphere
   g. first life on Earth
2. List the following time frames in order of how long they lasted, from longest to shortest.
   a. Hadean eon
   b. Proterozoic eon
   c. Paleozoic era
   d. Cretaceous period
3. Which time frame(s) do we live in today? (More than one may apply.)
   a. Quaternary period
   b. Tertiary period
   c. Cenozoic era
   d. Phanerozoic eon
   e. Paleozoic era
4. How long did the Cambrian explosion last?
   a. less than 1 year
   b. about a decade
   c. about 10,000 years
   d. about 20 million years
   e. about 500 million years
5. When did the heavy bombardment end?
   a. about 4.5 billion years ago
   b. between about 4.3 and 4.5 billion years ago
   c. between about 3.8 and 4.0 billion years ago
   d. exactly 3.95 billion years ago
6. How long have mammals been present on Earth?
   a. about 1 million years
   b. about 65 million years
   c. about 225 million years
   d. about 510 million years

# Exercises and Problems

*For instructor-assigned homework and other learning materials, go to www.MasteringAstronomy.com.*

## Chapter Review Questions

### Short-Answer Questions Based on the Reading

1. What is *astrobiology*, and what type of research does it involve?
2. How do we study the history of life on Earth? Describe the *geological time scale* and a few of the major events along it.
3. Summarize the evidence pointing to an early origin of life on Earth. How far back in Earth's history did life exist?
4. Why is the *theory of evolution* so critical to our understanding of the history of life on Earth? Explain how evolution proceeds by *natural selection*, and what happens to DNA that allows species to evolve.
5. How are laboratory experiments helping us study the origin of life on Earth? Explain.
6. Give a brief overview of the history of life on Earth. What evidence points to a common ancestor for all life? How and when did oxygen accumulate in Earth's atmosphere? When did larger animals diversify on Earth?
7. Is it possible that life migrated to Earth from elsewhere? Explain.
8. Describe the range of environments in which life thrives on Earth. What three basic requirements apply to life in all these environments?
9. What is a *habitable world*? Which worlds in our solar system seem potentially habitable, and why?
10. Briefly summarize the current status of the search for life on Mars.
11. What do we mean by a star's *habitable zone*? What key factors have given Earth long-term surface habitability, and do they seem likely on other worlds? Explain.
12. What types of worlds might support surface habitability? What types of worlds might have subsurface habitability? Explain.
13. What is *SETI*? Describe the capabilities of current SETI efforts.
14. What is the *Drake equation*? Define each of its factors, and describe the current state of understanding about the potential values of each factor.
15. Why is interstellar travel so difficult? Describe a few technologies that might someday make it possible.
16. What is *Fermi's paradox*? Describe several potential solutions to the paradox and the implications of each to our civilization.

### Fantasy or Science Fiction?

*Based on our present understanding of science, decide whether each of the following futuristic scenarios is plausible or whether it is unlikely to be possible. Explain clearly; not all of these have definitive answers, so your explanation is more important than your chosen answer.*

17. The first human explorers on Mars discover the ruins of an ancient civilization, including remnants of tall buildings and temples.
18. The first human explorers on Mars drill a hole into a Martian volcano to collect a sample of soil from several meters underground. On analyzing the soil, they discover that it holds living microbes resembling terrestrial bacteria but with a different biochemistry.
19. In 2040, a spacecraft lands on Europa and melts its way through the ice into the Europan ocean. It finds numerous strange, living microbes, along with a few larger organisms that feed on the microbes.
20. It's the year 2075. A giant telescope on the Moon, consisting of hundreds of small telescopes linked together across a distance of 500 kilometers, has just captured a series of images of a planet around a distant star that clearly show seasonal changes in vegetation.
21. A century from now, after completing a careful study of planets around stars within 100 light-years of Earth, astronomers discover that the most diverse life exists on a planet orbiting a young star that formed just 100 million years ago.
22. In 2040, a brilliant teenager working in her garage builds a coal-powered rocket that can travel at half the speed of light.
23. In the year 2750, we receive a signal from a civilization telling us that the *Voyager 2* spacecraft recently crash-landed on its planet, which orbits a nearby star.
24. Crew members of the matter-antimatter spacecraft *Star Apollo*, which left Earth in the year 2165, return to Earth in the year 2450, looking only a few years older than when they left.
25. Aliens from a distant star system invade Earth with the intent to destroy us and occupy our planet, but we successfully fight them off when their technology proves no match for ours.
26. A single great galactic civilization exists. It originated on a single planet long ago but is now made up of beings from many different planets, all assimilated into the galactic culture.

### Quick Quiz

*Choose the best answer to each of the following. For additional practice, try the Chapter 24 Reading and Concept Quizzes in the Study Area at www.Mastering Astronomy.com.*

27. Fossil evidence suggests that life on Earth arose (a) almost immediately after Earth formed. (b) very soon after the end of the heavy bombardment. (c) about a billion years before the rise of the dinosaurs.
28. The theory of evolution is (a) a scientific theory backed by extensive evidence. (b) one of several competing scientific models that all seem equally successful in explaining the nature of life on Earth. (c) essentially just a guess about how life changes through time.
29. Plants and animals are (a) the two major forms of life on Earth. (b) the only organisms that have DNA. (c) just two small branches of the diverse "tree of life" on Earth.
30. Which of the following is a reason early living organisms on Earth could not have survived on land? (a) the lack of an ozone layer (b) the lack of oxygen in the atmosphere (c) the fact that these organisms were single-celled
31. According to current understanding, the key requirement for life is (a) photosynthesis. (b) liquid water. (c) an ozone layer.

32. Which of the following worlds is *not* considered a candidate for harboring life? (a) Europa (b) Mars (c) the Moon
33. How does the habitable zone around a star of spectral type G compare to that around a star of spectral type M? (a) It is larger. (b) It is hotter. (c) It is closer to its star.
34. In the Drake equation, suppose that the term $f_{life}$ was equal to $\frac{1}{2}$. What would this mean? (a) Half the stars in the Milky Way Galaxy have a planet with life. (b) Half of all life-forms in the universe are intelligent. (c) Half of the habitable worlds in the galaxy actually have life, while the other half don't.
35. The amount of energy that would be needed to accelerate a large spaceship to half the speed of light is (a) about 100 times the energy needed to launch the Space Shuttle. (b) more than 2000 times the current annual world energy consumption. (c) more than the amount of energy released by a supernova.
36. According to current scientific understanding, the idea that the Milky Way Galaxy might be home to a civilization millions of years more advanced than ours is (a) a virtual certainty. (b) extremely unlikely. (c) one reasonable answer to Fermi's paradox.

## Inclusive Astronomy

*Use these questions to reflect on participation in science.*

37. *The Turning Point.* The end of this chapter presented the idea that humanity is now at a turning point, and that current generations have a greater responsibility for the future than did any previous generation. Do you agree with this assessment? If so, how can we ensure that we deal with this responsibility wisely, in a way that builds a better future for the entire human race? If not, why not?
38. *Group Discussion: Who Speaks for Earth?* Form small groups to discuss the issues that would arise if SETI were to receive a message from an alien civilization.
    **a.** The message would presumably be received first by a relatively small group of astronomers. Work together to write up a short protocol specifying what these scientists should do upon receiving the message. For example, should they announce the discovery immediately and share the message immediately?
    **b.** Responding to the message would make our existence known to the aliens. Is this a good idea? Make a list of pros and cons of sending a response.
    **c.** Assume that we have decided to send a response. Given that the response will represent the entire human race, who should write it, and what type of information should it offer about our species and our planet?

## The Process of Science

*These questions may be answered individually in short-essay form or discussed in groups, except where identified as group-only.*

39. *The Science of Astrobiology.* The study of astrobiology is sometimes criticized as being the study of something for which we have no evidence, since we do not yet have evidence of life beyond Earth. Is astrobiology a science or speculation? Defend your opinion.
40. *Astrobiology Funding.* Imagine that you were a member of Congress and your job included deciding how much government funding to allocate to research in different areas of science. How would the amount you allotted to the search for life in the universe compare to the amount you allotted to research in other areas of astronomy and planetary science? Why?
41. *Breakthrough Starshot.* The "Breakthrough Starshot" initiative proposes to send a fleet of tiny, robotic spaceships to the Alpha Centauri star system. Learn about how this effort is intended to work, who is funding it, and what its goals are. Overall, do you think it is realistic? Do you think it is worthwhile? Defend your opinions.
42. *The 100-Year Starship.* The "100-Year Starship" is a research project started by NASA and the U.S. Department of Defense to investigate the possibility of sending humans on interstellar trips. Learn about this effort. Based on your findings and the challenges of interstellar flight discussed in this chapter, when (if ever) do you think humans are likely to begin traveling among the stars? Why?
43. *Unanswered Questions.* In a sense, this entire chapter was about one big unanswered question: Are we alone in the universe? But as we attempt to answer that "big" question, we encounter many smaller questions that we might wish to answer along the way. Describe one currently unanswered question about life in the universe that we might be able to answer with new missions or experiments over the next couple of decades. What kinds of evidence will we need to answer the question? How will we know when it is answered?
44. *Group Discussion: Solution to the Fermi Paradox.* Discuss the various possible solutions to the question "Where are the aliens?" Can your group reach a consensus on which possibility is the most likely? If so, write a brief summary of why you chose this solution and what it means for our future. If not, record the number of group members supporting each possibility, and summarize the key points of disagreement.
45. *Group Activity: Habitable Planets?* Work in small groups to rank the four hypothetical planets described below in order from most likely to support life to least likely to support life. Assume the planets are all approximately the same size as Earth, and explain your reasons for each ranking. Note: You may wish to do this activity using the four roles described in Chapter 1, Exercise 39.
    Planet 1: orbits a star of spectral type B (approximately 10 solar masses) in a circular orbit within the habitable zone
    Planet 2: orbits a Sun-like star in a circular orbit at a distance twice Earth's orbital distance from the Sun
    Planet 3: orbits a star with a luminosity one-quarter the Sun's luminosity in a circular orbit at a distance one-half Earth's orbital distance from the Sun
    Planet 4: orbits a Sun-like star in an elliptical orbit, with its orbital distance ranging from 1 AU to 10 AU

## Investigate Further

### Short-Answer/Essay Questions

46. *Artificial Selection.* Suppose you lived hundreds of years ago (before we knew about genetic engineering) and wanted to breed a herd of cows that would provide more milk than the cows in your current herd. How would you have gone about it? How would this process of "artificial selection" be similar to natural selection? How would it be different?
47. *Statistics of One.* Much of the search for life in the universe is based on what we know about life on Earth. Write a short

essay discussing the pros and cons of basing our general understanding of biology on the example of Earth.

48. *Most Likely to Have Life.* Suppose you were asked to vote in a contest to name the world in our solar system (besides Earth) "most likely to have life." Which world would you vote for? Explain and defend your choice in a one-page essay.
49. *Likely Suns.* Study the stellar data for nearby stars given in Appendix F, Table F.1. Which star on the list would you expect to have the largest habitable zone? The second-largest habitable zone? If we rule out multiple-star systems, which star would you expect to have the highest probability of having a habitable planet? Explain your answers.
50. *Is Life Common?* Based on what you have learned in this book, do you think life will ultimately prove to be rare, common, or something in between? Write a one- to two-page essay explaining and defending your opinion.
51. *What's Wrong with This Picture?* Many science fiction stories have imagined the galaxy divided into a series of empires, each having arisen from a different civilization on a different world, that hold one another at bay because they all have about the same level of military technology. Is this a realistic scenario? Explain.
52. *Aliens in the Movies.* Choose a science fiction movie (or television show) that involves an alien species. Do you think aliens like those depicted could really exist? Write a one- to two-page critical review of the movie or show, focusing primarily on the question of whether it portrays the aliens in a scientifically reasonable way.

## Quantitative Problems

*Be sure to show all calculations clearly and state your final answers in complete sentences.*

53. *Nearest Civilization.*
   **a.** Suppose there are 10,000 civilizations in the Milky Way Galaxy. If the civilizations were randomly distributed throughout the disk of the galaxy, about how far (on average) would it be to the nearest civilization? (*Hint:* Start by finding the area of the Milky Way's disk, assuming that it is circular and 100,000 light-years in diameter. Then find the average area per civilization, and use the distance across this area to estimate the distance between civilizations.)
   **b.** Repeat part a, but this time assume that there are only 100 civilizations in the galaxy.
54. *SETI Search.* Suppose there are 10,000 civilizations broadcasting radio signals in the Milky Way Galaxy right now. On average, how many stars would we have to search before we would expect to hear a signal? Assume there are 500 billion stars in the galaxy. How would the answer change if there were only 100 civilizations instead of 10,000?
55. *SETI Signal.* Consider a civilization broadcasting a signal with a power of 10,000 watts. The Arecibo radio telescope, which is about 300 meters in diameter, could detect this signal if it was coming from as far away as 100 light-years. Suppose instead that the signal is being broadcast from the other side of the Milky Way Galaxy, about 70,000 light-years away. How large a radio telescope would we need to detect this signal? (*Hint:* Use the inverse square law for light.)
56. *Cruise Ship Energy.* Suppose we have a spaceship about the size of a typical ocean cruise ship today, which means it has a mass of about 100 million kilograms, and we want to accelerate the ship to a speed of 10% of the speed of light.
   **a.** How much energy would be required? (*Hint:* You can find the answer simply by calculating the kinetic energy of the ship when it reaches its cruising speed; because 10% of the speed of light is still small compared to the speed of light, you can use this formula: kinetic energy $= \frac{1}{2} \times m \times v^2$.)
   **b.** How does your answer compare to total world energy use at present, which is about $5 \times 10^{22}$ joules per year?
   **c.** The typical cost of energy today is roughly 5¢ per 1 million joules. At this price, how much would it cost to generate the energy needed by this spaceship?
57. *Matter-Antimatter Engine.* Consider the spaceship from Problem 56. Suppose you wanted to generate the energy to get it to cruising speed using matter-antimatter annihilation. How much antimatter would you need to produce and take on the ship? (*Hint:* When matter and antimatter meet, they turn all their mass into energy equivalent to $mc^2$.)

# COSMIC Context

## PART VII A Universe of Life?

**Throughout this book, we have seen that the history of the universe has proceeded in a way that has made our existence on Earth possible. This figure summarizes some of the key ideas, and leads us to ask: If life arose here, shouldn't it also have arisen on many other worlds? We do not yet know the answer, but scientists are actively seeking to learn whether life is rare or common in the universe.**

1. The protons, neutrons, and electrons in the atoms that make up Earth and life were created out of pure energy during the first few moments after the Big Bang, leaving the universe filled with hydrogen and helium gas [Section 22.1].

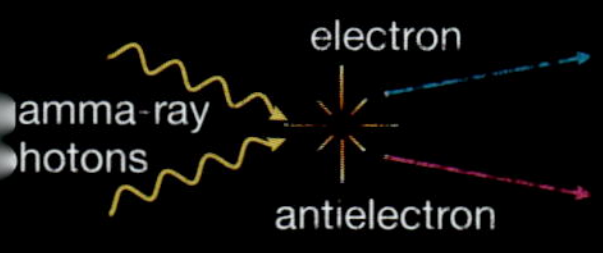


*latter can be created from energy.*
*$= mc^2$.*

2. Ripples in the density of the early universe were necessary for life to form later on. Without those ripples, matter would never have collected into galaxies, stars, and planets [Section 22.3].

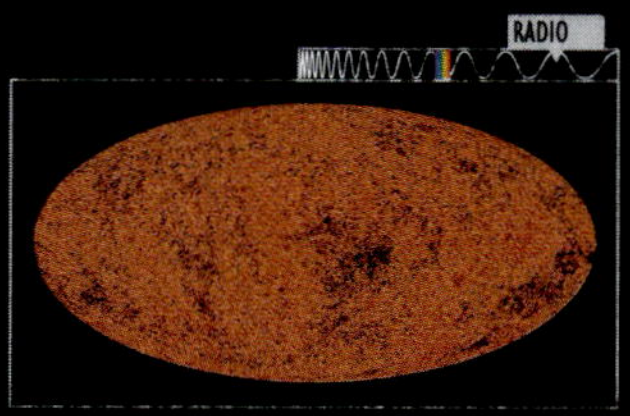


*We observe the seeds of structure formation in the cosmic microwave background.*

3. The attractive force of gravity pulls together the matter that makes galaxies, stars, and planets [Section 4.4].

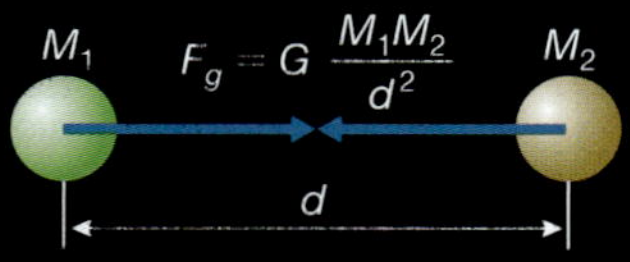


*Every piece of matter in the universe pulls on every other piece.*

4. Our planet and all the life on it is made primarily of elements formed by nuclear fusion in high-mass stars and dispersed into space by supernovae [Section 17.3].

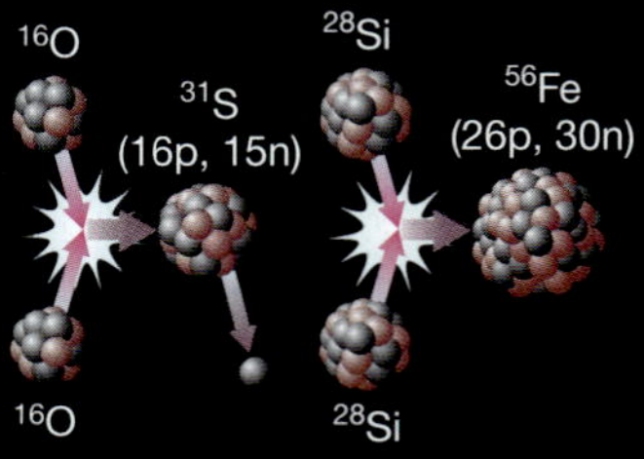


*High-mass stars have cores hot enough to make elements heavier than carbon.*

5 Our galaxy is large enough to retain the elements ejected by supernovae, and it recycles them into new stars and planetary systems [Section 19.2].

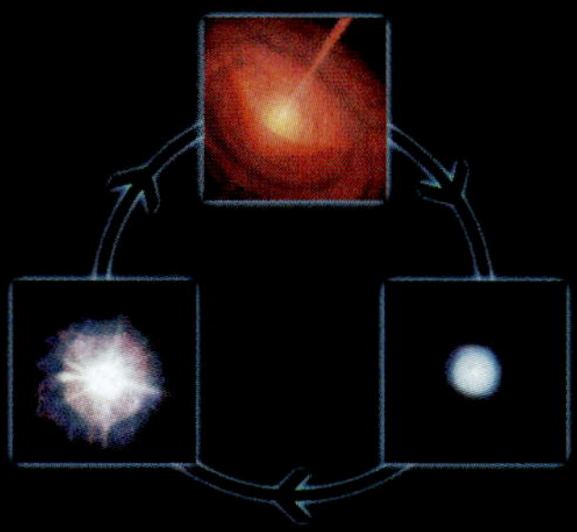

*New elements mix with the interstellar medium, which then forms new stars and planets.*

6 Planets can form in gaseous disks of material around newly formed stars. Earth was built from heavy elements that condensed from the gas as particles of metal and rock, which then gradually accreted to become our planet [Section 8.2].

*Terrestrial planets formed in warm, inner regions of the solar nebula; jovian planets formed in cooler, outer regions.*

7 Life as we know it requires liquid water, so we define the habitable zone around a star to be the zone in which a suitably large planet can have liquid water on its surface [Section 24.3].

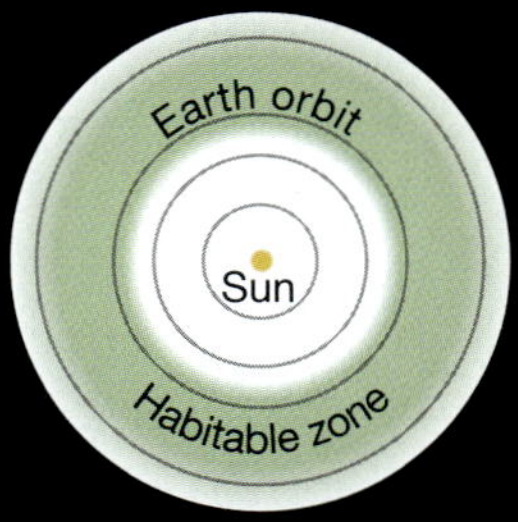


*The Sun's habitable zone (green) occupies a region from beyond the orbit of Venus to near the orbit of Mars.*

8 Early life has had the time needed to evolve into complex forms—including us—because the solar thermostat has kept the Sun shining steadily for billions of years [Section 14.2].

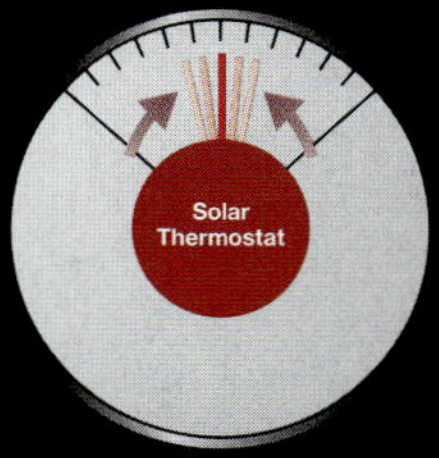


*The solar thermostat keeps the Sun's fusion rate stable.*

# Credits

**Foldout Front** © ESA and the Planck Collaboration, NASA, The Hubble Heritage Team, STScI, AURA, Stephen J. Mojzsis **Back** Russell Shively/Shutterstock, Inc., Colin Keates, © Dorling Kindersley, Courtesy of the Natural History Museum, London, Dave King, © Dorling Kindersley, Courtesy of the Natural History Museum, London, Linda Whitwam, © Dorling Kindersley, Pius Lee/Shutterstock, NASA
**Frontmatter p. ii** NASA **p. xxx** Jeffrey Bennett, Megan Donahue **p. xxxi** Nicholas Schneider, Mark Voit **p. xxxiv** Image courtesy of Neil deGrasse Tyson
**Chapter 1 Opener** NASA, ESA, S. Beckwith (STScI) and the HUDF Team **1.1** (Milky Way Galaxy) Michael Carroll **1.2** NASA, ESA, M. Robberto (Space Telescope Science Institute/ESA) and the Hubble Space Telescope Orion Treasury Project Team **1.3** Jerry Lodriguss/Science Source **1.5** Jeffrey Bennett **1.7** NASA **1.8** (inset) Photo copyright by Fred Espenak, www.AstroPixels.com **1.9** Danny Hooks/Alamy Stock Photo; (Milky Way Galaxy) Michael Carroll **1.10** Jeffrey Bennett **1.12** (second from left to right) Sompol/Shutterstock; NASA; Jon Arnold Images Ltd/Alamy Stock Photo **1.16** NASA
**Chapter 2 Opener** D. Nunuk/Science Source **2.1** Wally Pacholka **2.11b** geog-photos/Alamy Stock Photo **2.17** Frank Zullo/Science Source **2.18** Arnulf Husmo/Getty Images **2.19** Reto Stöckli/NASA Earth Observatory **2.21** (inset) NASA/JPL-Caltech/Univ. of Arizona **2.22** NASA/Bill Dunford **2.26** (top) Brian Kushner/Alamy Stock Photo; (bottom) Dr Ajay Kumar Singh/Shutterstock **2.27** Image courtesy of Itahisa Gonzalez Alvarez **2.28a** (top to bottom) JBCorl/Shutterstock; Sarah M. Golonka/Alamy Stock Photo; Phillip Jones/Stocktrek Images/Alamy Stock Photo **2.28b** Deep Space Climate Observatory (NASA/NOAA) **2.29** Ben Cooper/LaunchPhotography.com **2.32** Gerard Lodriguss/Science Source **2.33** Cenk E. Tezel, Tunç Tezel (TWAN)
**Chapter 3 Opener** NASA **3.2** Karl Kost/Alamy Stock Photo **3.3a** Timo Kohlbacher/Shutterstock **3.4** Willinger, L./SuperStock **3.5** National Park Service **3.6** © Photo by William E. Woolam **3.7** Jeffrey Bennett **3.8** Travelpix/Alamy Stock Photo **3.9** Amy Nichole Harris/Shutterstock **3.10** Walter Maeyer Edwards/National Geographic Creative **3.11a** Sheila Terry/SPL/Science Source **3.11b** Patrick Pihl/Interpix/Alamy Stock Photo **p. 63** World History Archive/SuperStock **p. 64** Ian Woolcock/SuperStock **3.15** Gianni Tortoli/Science Source **p. 65** New York Public Library/Science Source **p. 67** A. Burkatovski/Fine Art Images/SuperStock **3.20** Mike Hill/Alamy Stock Photo **3.21** Galileo Galieli **3.25** (top left) Tunç Tezel; (bottom left) Cosmographia Petri Apiani/Boston Public Library; (bottom right) The Library of Congress Collection **p. 74** Renaud Visage/age fotostock/Alamy Stock Photo **3.26** Kim Christensen/Alamy Stock Photo
**Chapter S1 Opener** Arnulf Husmo/Getty Images **S1.5** John Sleezer/Kansas City Star/MCT via Getty Images **S1.6** Dennis MacDonald/Alamy Stock Photo **S1.25a** Science and Society/SuperStock **S1.25b** Bettmann/Getty Images **S1.25c** Photo Researchers/Science Source **S1.25d** budhi/iStock/Getty Images **p. 108** (top left) Timo Kohlbacher/Shutterstock; (top right) Ian Woolcock/SuperStock **p. 109** (top left to right) Yerkes Observatory; Pictorial Press Ltd/Alamy Stock Photo; NASA; (bottom) NASA, ESA, S. Beckwith (STScI) and the HUDF Team
**Chapter 4 Opener** NASA and The Hubble Heritage Team (STScI) **p. 115** Sir Godfrey Kneller **4.6a** Goddard Institute for Space Studies/NASA **4.6b** Bruce Kluckhohn/Getty Images **4.6c** Goddard Institute for Space Studies/NASA **4.11** (clockwise from top) Stockbyte/Getty Images; Don Hammond/Getty Images; Eric Gevaert/Alamy Stock Photo **4.16** U.S. Department of Energy **4.20** NASA **4.25** (left) Travel Pix Collection/Jon Arnold Images/SuperStock; (right) Photononstop/SuperStock
**Chapter 5 Opener** N. A. Sharp, NOAO/NSO/Kitt Peak FTS/AURA/NSF **Epigraph** This poem from "Songs of the Tewa" by Herbert Joseph Spinden appears courtesy of Sunstone Press, Box 2321, Santa Fe, NM 87504 **5.1** Richard Megna/Fundamental Photographs **5.7** (Sun's chromosphere) TRACE/NASA; (Sun) Solar Dynamics Observatory/NASA; (planets, star-forming clouds) NASA/JPL/STScI; (cosmic microwave background) © ESA and the Planck Collaboration; (radio galaxy) Image courtesy of NRAO/AUI; R. Perley, C. Carilli & J. Dreher
**Chapter 6 Opener** Babak Tafreshi/Getty Images **6.3** Solvin Zankl/Nature Picture Library **6.9** NASA **6.10** Yerkes Observatory Photograph **6.11b** NOAO/AURA/NSF **6.12** Laurie Hatch; (inset) Richard Wainscoat/Alamy Stock Photo **6.13** Wolfgang Brandner (JPL/IPAC), Eva K. Grebel (U. Wash.), You-Hua Chu (UIUC), NASA **6.14** NASA/CXC/NCSU/K. J. Borkowski et al. **6.18** Todd Mason, Mason Productions Inc./LSST Corporation **6.19** NASA Earth Observatory/NOAA NGDC **6.20** Images courtesy of Canada-France-Hawaii Telescope (CFHT) Corporation **6.21** Richard Wainscoat/Alamy Stock Photo **6.22** STS-82 Crew, STScI, NASA **6.23** NASA/JPL-Caltech **6.24** NASA/Chris Gunn **6.25** Spencer Grant/PhotoEdit **6.26** Xinhua/Alamy Stock Photo **6.27** Jim Ross/NASA **6.28** NASA **6.29a** NASA **6.30** NASA **6.31** Handout/MCT/Newscom **6.33** Jon Arnold Images Ltd/Alamy Stock Photo **p. 188** (Earth) NASA; (conservation of energy, clockwise) Stockbyte/Getty Images; Don Hammond/Getty Images; Eric Gevaert/Alamy Stock Photo **p. 189** (cosmic microwave background) © ESA and the Planck Collaboration
**Chapter 7 Opener** NASA/JPL **7.2** NASA Marshal Space Flight Center **7.3** NASA/Johns Hopkins University Applied Physics Laboratory/Carnegie Institution of Washington **7.4** David P. Anderson/Science Source; (global view) NASA/GSFC **7.5a** © ARC Science Simulations **7.5b** NASA Earth Observing System **7.6** NASA/JPL-Caltech/MSSS; (global view) NASA/USGS **7.7** © ARC Science Simulations **7.8** NASA/JPL/Space Science Institute **7.9** © ARC Science Simulations **7.10** © ARC Science Simulations **7.11** NASA/Johns Hopkins University Applied Physics Laboratory/Southwest Research Institute **Table 7.1** NASA Goddard Space Flight Center/APL/SwRI/JPL/Space Science Institute **7.12** NEAR Project, JHU APL, NASA **7.13** Miloslav Druckmuller (Brno University of Technology) **7.15** NASA/JPL-Caltech/Univ. of Arizona
**Chapter 8 Opener** ALMA (ESO/NAOJ/NRAO), ESA/Hubble and NASA. Acknowledgment: Judy Schmidt **8.2** NASA, ESA, M. Robberto (Space Telescope Science Institute/ESA), the Hubble Space Telescope Orion Treasury Project Team and L. Ricci (ESO) **8.4a** ESO, T. Stolker et al. **8.4b** ALMA (ESO/NAOJ/NRAO), ESA/Hubble and NASA. Acknowledgment: Judy Schmidt **8.7** Nick Schneider **8.11** NASA/JPL-Caltech/Univ. of Arizona **8.12** Michael Carroll
**Chapter 9 Opener** Tom Pfeiffer/VolcanoDiscovery/Getty Images **9.1** NASA/JPL/USGS **9.3** Richard Megna/Fundamental Photographs **9.6a** Jules Bucher/Science Source **9.7** Don Davis **9.8a** Brad Snowder, Western Washington University **9.8b** NASA/USGS Lunar Orbiter Digitization Project **9.9** NASA/JPL/MSSS **9.10** Paul Chesley/Getty Images **9.11a** zschnepf/Shutterstock **9.11b** Mariner 9/JPL/NASA **9.11c** NASA Earth Observing System **9.12a** Images & Volcanoes/Science Source

**9.12b** Stocktrek Images/Richard Roscoe/Getty Images **9.13** (left) NASA Earth Observing System; (right) NASA/JPL **9.14a** ShuPhotography/Shutterstock **9.14b** Robert Zehetmayer/Alamy Stock Photo **9.14c** Mytho/Fotolia **9.14d** Florian Werner/AGE Fotostock **9.15a** NASA/Lunar and Planetary Institute **9.15b** © ESA/Space-X (Space Exploration Institute) **9.17** (left) Claudio Divizia/Shutterstock; (right) NASA (image by Lunar Reconnaissance Orbiter)–JMARS **9.18** (inset) Claudio Divizia/Shutterstock **9.19** NASA Earth Observing System **9.20** NASA **9.21** NASA/Johns Hopkins University Applied Physics Laboratory/Carnegie Institution of Washington **9.22b** NASA/JPL **9.23** NASA/Johns Hopkins University Applied Physics Laboratory/Carnegie Institution of Washington **9.24** (left) NASA Earth Observing System; (right) Lowell Observatory **9.25** NASA/Goddard Institute for Space Studies **9.26** NASA/JPL **9.27** NASA/JPL/USGS; (inset) © ESA/Mars Express **9.28** © ESA/DLR/FU Berlin (G. Neukum) **9.29a** EROS Data Center, U.S. Geological Survey **9.29b** R. P. Irwin III and G. A. Franz, National Air and Space Museum, Smithsonian Institution **9.29c** NASA Earth Observing System **9.30** NASA/GSFC **9.31** Seán Doran/NASA/JPL-Caltech/ESA/DLR/FU Berlin/MSSS **9.32** (left) NASA/JPL/Cornell; (right) Jurassic Navajo Sandstone in Northern Arizona. Image taken by Marjoriz A. Chan, University of Utah **9.33** NASA/JPL-Caltech/MSSS and PSI **9.34** (left) NASA/JPL/MSSS; (right) NASA/JPL-Caltech/Univ. of Arizona **9.35** NASA/JPL **9.36** NASA/JPL-Caltech/ESA **9.37** Space Research Institute of the Russian Academy of Sciences/Ted Stryk **9.38** NOAA National Geophysical Data Center **9.43** NASA **9.44** NASA **9.45** NASA **9.46** (clockwise from top left) Ronen Zilberman/Associated Press; Nina B/Shutterstock; Melissa Tse/Getty Images; Alexander Malahoff/NOAA **9.48** NASA **Chapter 10 Opener** ISS Expedition 7 Crew/EOL/NASA **10.1** (left to right) Johns Hopkins University Applied Physics Laboratory/Carnegie Institution of Washington/NASA; David P. Anderson/Science Source; Dmitry Pichugin/Shutterstock; NASA Earth Observing System **10.2** NASA Earth Observing System **10.5** NASA Earth Observing System **10.11a** Dr. L. A. Frank, The University of Iowa/NASA **10.11b** Douglas K. Duncan **10.15b** GSFC/NASA/GOES Project **10.22** Jody K. Wilson, Michael Mendillo, Harlan E. Spence/American Geophysical Union **10.23a** NASA **10.23b** Johns Hopkins University Applied Physics Laboratory/Carnegie Institution of Washington/NASA **10.25** JPL/MSSS/NASA **10.26** NASA/Damian Peach **10.27** NASA/JPL/Univ. of Arizona **10.28** NASA/JPL-Caltech/UA/USGS **10.29** NASA/JPL/Univ. of Arizona **10.31** University of Colorado/NASA **10.32** JAXA/ISAS/DARTS/Damia Bouic **10.33** NASA **10.34** NASA **10.35** NASA **10.42** Map from Goddard Space Flight Center/NASA, accessed on May 5, 2018 **Chapter 11 Opener** NASA/JPL/Space Science Institute **11.1** NASA/JPL **11.9** NASA **11.10** NASA/JPL-Caltech/SwRI/MSSS/Jason Major **11.11a** NASA/JPL-Caltech/SwRI/ASI/INAF/JIRAM **11.11b** NASA/JPL-Caltech/Space Science Institute **11.11c** Lawrence Sromovsky/University of Wisconsin-Madison/W. W. Keck Observatory **11.11d** NASA/JPL **11.12** (left) NASA/JPL/Univ. of Arizona; (right) J. Clarke and G Ballester (University of Michigan), J. Trauger and R. Evans (Jet Propulsion Laboratory) and NASA **11.15** NASA/JPL-Caltech/Space Science Institute **11.16** NASA/JPL/DLR **11.17** NASA/JPL/Univ. of Arizona **11.18** (left) NASA/Johns Hopkins University Applied Physics Laboratory/Southwest Research Institute; (right) NASA/JPL-Caltech/SwRI/ASI/INAF/JIRAM/Roman Tkachenko **11.20** (left) NASA/JPL/DLR; (right) NASA/JPL/ASU **11.21** (center) Britney Schmidt/Dead Pixel FX/Univ. of Texas at Austin; (right) NASA/JPL **11.22** NASA/JPL/DLR **11.23** NASA/JPL/Processed by Kevin M. Gill; (inset) NASA/JPL/ASU **11.24** NASA/JPL/Space Science Institute; (inset) NASA/JPL/Univ. of Arizona/DLR **11.25** (left to right) NASA/JPL/Space Science Institute; ESA/NASA/JPL/Univ. of Arizona; NASA/JPL/ESA/Univ. of Arizona **11.26** NASA/JPL-Caltech/ASI/Cornell **11.27** NASA/JPL-Caltech/Space Science Institute **11.28** NASA/JPL/Space Science Institute **11.29** NASA/JPL-Caltech/Space Science Institute **11.30** NASA/JPL/USGS **11.31** NASA/JPL/USGS **11.32** (left) Image Processing by Jody Swann/Tammy Becker/Alfred McEwen, using the PICS (Planetary Image Cartography System) image processing system developed at the U.S. Geological Survey in Flagstaff, Arizona; (right) NASA/JPL **11.33a** NASA **11.33b** NASA/JPL/Space Science Institute **11.33c** NASA/JPL/University of Colorado **11.34** NASA/JPL-Caltech/Space Science Institute **11.35** NASA **11.36** NASA/JPL/STScI/Space Science Institute **p. 339** NASA/JPL/Univ. of Arizona **Chapter 12 Opener** (left to right) Lowell Observatory Archives; NASA, ESA, and M. Buie/Southwest Research Institute; Johns Hopkins University Applied Physics Laboratory/Southwest Research Institute/NASA **12.1** ESO **12.2a** Image courtesy of Peter Ceravolo **12.2b** Frank Zullo/Science Source **12.3** LASCO/SOHO Consortium/NRL/ESA/NASA **12.4** NASA, ESA, and M. Brown (California Institute of Technology) **12.5** NASA **12.6** Ivan and Colby Navarro **12.7** Jonathan Blair/Getty Images **12.8** ESA/NASA/JAXA/RAS/JHUAPL/UMD/OSIRIS; Montage: Emily Lakdawalla (Planetary Society) & Ted Stryk **12.9** (left) JPL-Caltech/UCLA/MPS/DLR/IDA/NASA; (right) NASA/JPL-Caltech/UCLA/MPS/DLR/IDA **12.10** NASA/JPL-Caltech/UCLA/MPS/DLR/IDA/LPI; (global view) NASA/JPL-Caltech/UCLA/MPS/DLR/IDA/PSI **12.11** Image courtesy ISAS/JAXA and University of Tokyo; (inset) Institute of Space and Astronomical Science (ISAS) **12.12** Robert A. Haag/Nick Schneider **p. 353** ESO/K. Meech et al. **12.15b** NASA/JPL/James W. Young; (inset) © ESA/MPAE, 1986, 1996 **12.16** NASA/K. Meech; (inset) NASA/JPL-Caltech **12.17** NASA/JPL-Caltech/UMD **12.18** © ESA/Rosetta/MPS for OSIRIS Team MPS/UPD/LAM/IAA/SSO/INTA/UPM/DASP/IDA **12.19** © ESA/Rosetta/MPS for OSIRIS Team MPS/UPD/LAM/IAA/SSO/INTA/UPM/DASP/IDA/Ramy El-Maary **12.20b** Vic & Jen Winter, ICSTARS Astronomy **12.23** NASA, ESA, and L. Frattare (STScI); Science Credit: NASA, ESA, and M. Showalter (SETI Institute) **12.24** NASA/Johns Hopkins University Applied Physics Laboratory/Southwest Research Institute **12.25** NASA/Johns Hopkins University Applied Physics Laboratory/Southwest Research Institute **12.26** NASA/Johns Hopkins University Applied Physics Laboratory/Southwest Research Institute **12.27a** H. A. Weaver, T. E. Smith (Space Telescope Science Institute), and NASA **12.27b** Illustration by Joe Bergeron **12.27c** Mount Stromlo and Siding Spring Observatories, ANU/Science Source **12.27d** NASA, ESA, H. Hammel (Space Science Institute, Boulder, Colo.), and the Jupiter Impact Team **12.28** Johnson, Kirk R. **12.29** V. L. Sharpton/USRA-Lunar and Planetary Institute **12.31** Image Asset Management/AGE Fotostock **12.32** AP Video/Associated Press **p. 368** Univ. Maryland, JPL-Caltech, NASA **Chapter 13 Opener** A.-L. Maire/LBTO **p. 375** bilwissedition Ltd. & Co. KG/Alamy Stock Photo **13.6** (bottom right) A.-L. Maire/LBTO **13.9** From Michaël Gillon et al., "Seven temperate terrestrial planets around the nearby ultracool dwarf star TRAPPIST-1," *Nature* volume 542, pages 456–460 (23 February 2017), Figure 1-a, 1-c **13.12** JPL-Caltech/H. Knutson (Harvard-Smithsonian CfA)/NASA **13.13** From Michaël Gillon et al., "Seven temperate terrestrial planets around the nearby ultracool dwarf star TRAPPIST-1," *Nature* volume 542, pages 456–460 (23 February 2017), Figure 1-d **13.16** JPL/STScI/NASA Goddard Space Flight Center **13.18** NASA Goddard Space

Flight Center/Chris Meaney **13.19a** ESO/A.-M. Lagrange et al. **13.19b** A.-L. Maire/LBTO **pp. 400–401** NASA/JPL/ASU
**Chapter S2 Opener** Bettmann/Getty Images
**Chapter S3 Opener** Andrew Fruchter (STScI) et al., WFPC2, HST, NASA **Epigraph** *Journal of the Franklin Institute* by FRANKLIN INSTITUTE (PHILADELPHIA, PA.) Reproduced with permission of PERGAMON **S3.20** ESA/Hubble & NASA **S3.22** NASA/Dana Berry, Sky Works Digital **S3.24** (inset) NASA
**Chapter S4 Opener** CMS Collaboration/CERN/Science Source **Epigraph** Quote from Douglas Adams, *Hitchhiker's Guide to the Galaxy*. New York: Ballantine Books, 1980 **S4.1** ZUMA Press, Inc./Alamy Stock Photo **S4.6** Terry Oakley/Alamy Stock Photo **p. 467** (bottom right) UCLA Galactic Center Group–W. M. Keck Observatory Laser Team
**Chapter 14 Opener** NASA/SDO/AIA **14.1** i4images rm/Alamy Stock Photo **14.5** NASA/SDO and the AIA, EVE, and HMI science teams **14.11b** Swedish 1-m Solar Telescope (SST)/Institute for Solar Physics, Sweden (ISP) **14.12** NASA/SDO/HMI Quick-Look Dopplergram **14.14** Kamioka Observatory, ICRR (Institute for Cosmic Ray Research), The University of Tokyo **14.15** Paolo Lombardi/INFN-MI **14.16a** Swedish 1-m Solar Telescope (SST)/Institute for Solar Physics, Sweden (ISP) **14.16b** NOAO/AURA/NSF **14.18b** NASA/TRACE **14.19** SDO/NASA **14.20** SDO/AIA/NASA **14.21** NASA/ISAS/JAXA/Yohkoh Soft X-ray Telescope **14.22** SDO/AIA/GSFC/NASA
**Chapter 15 Opener** Alan Dyer/Stocktrek Images/Getty Images **15.4** AURA/STScI/Hubble Heritage Team/NASA **15.5 and p. 498** Harvard University Archives/Harvard College Observatory **15.9** David Nash **15.16** Robert Gendler/Stocktrek Images/Science Source **15.17** The Hubble Heritage Team (AURA/STScI/NASA)
**Chapter 16 Opener** ESA/Herschel/PACS/SPIRE/Hill, Motte, HOBYS Key Programme Consortium **Epigraph** From *Secrets from the Center of the World* by Joy Harjo and Stephen Strom. © 1989 The Arizona Board of Regents. Reprinted by permission of the University of Arizona Press **16.1** Stocktrek Images/Getty Images **16.3** ESO/ESA/CSO **16.4** Donald E. Brownlee, University of Washington, Seattle, and Elmar Jessberger, Institut für Planetologie, Münster, Germany **16.5** ESO **16.6** ESA/Hubble Space Telescope/NASA **16.7** JPL-Caltech/W. Reach (SSC/Caltech)/NASA; (inset) Canada-France-Hawaii Telescope/J.-C. Cuillandre/Coelum, and the Digitized Sky Survey/NASA **16.8** ESA/Herschel/PACS/SPIRE/Hill, Motte, HOBYS Key Programme Consortium **16.10** Courtesy of Matthew Bate, University of Exeter **16.11** NASA and The Hubble Heritage Team (STScI/AURA). Acknowledgment: Bo Reipurth (University of Hawaii) (STScI/AURA)/NASA **16.12** Courtesy of Tom Abel, Michael L. Norman and Greg Bryan (2002) **16.13** NASA and the Hubble Heritage Team (STScI/AURA)/NASA **16.15a** NASA and ESA. Acknowledgment: NASA, ESA, the Hubble Heritage (STScI/AURA)/Hubble-Europe (ESA) Collaboration, D. Padgett (GSFC), T. Megeath (University of Toledo), and B. Reipurth (University of Hawaii)/NASA **16.15b** C. Burrows (STScI & ESA), the WFPC 2 Investigation Definition Team, and NASA **16.18** NASA/JPL **16.19** ESO **16.20** NASA, R. Elson and R. Sword (IoA Cambridge)/NASA
**Chapter 17 Opener** NASA, ESA, Zolt Levay (STScI)/NASA **Epigraph** Courtesy of Neil deGrasse Tyson. Used with permission **17.1** NASA, R. Elson and R. Sword (IoA Cambridge)/NASA **17.7a** NASA, NOAO, ESA, the Hubble Helix Nebula Team, M. Meixner (STScI), and T. A. Rector (NRAO)/NASA **17.7b** NASA, ESA, and the Hubble SM4 ERO Team **17.17** NASA, ESA, J. Hester and A. Loll (Arizona State University)/NASA **17.18** ESO **17.19** NASA/JPL
**Chapter 18 Opener** X-ray: NASA/CXC/ASU/J. Hester et al.; optical: NASA/HST/ASU/J. Hester et al./NASA **Epigraph** Professor Dame Jocelyn Bell Burnell, From *Beautiful Minds*, BBC Series 1 **18.1a** NASA, H. E. Bond and E. Nelan (Space Telescope Science Institute, Baltimore, Md.); M. Barstow and M. Burleigh (University of Leicester, U.K.); and J. B. Holberg (University of Arizona)/NASA **18.1b** SAO/CXC/NASA **18.4b** NASA, M. Shara, R. Williams (STScI), R. Gilmozzi (ESO)/NASA **18.6** X-ray: NASA/CXC/Eureka Scientific/M. Roberts et al.; radio: NRAO/AUI/NSF/NASA **18.8** ESO **p. 569** Bettmann/Getty Images; LOC/Science Source; INTERFOTO/Personalities/Alamy Stock Photo; CORBIS/Corbis via Getty Images; PA Images/Alamy Stock Photo **18.17** NASA, Andrew Fruchter (STScI)/NASA **18.18a** LIGO, Virgo, Fermi, Swope, DLT40 **18.18b** NASA and ESA. Acknowledgment: A. J. Levan (U. Warwick), N. R. Tanvir (U. Leicester), and A. Fruchter and O. Fox (STScI)/NASA **18.19** NASA, S. Kulkarni, J. Bloom, P. Price, Caltech-NRAO GRB Collaboration
**Chapter 19 Opener** NASA/JPL-Caltech/S. Stolovy (Spitzer Science Center/Caltech) **19.4** NASA/ESA/STScI **19.5** Takahashi/RC-Optical Gallery/Robert J. Vanderbei **19.6** NASA/CXC/SAO **19.7a** Stocktrek Images/Superstock **19.7b** NASA, ESA, and the Hubble Heritage (STScI/AURA)–ESA/Hubble Collaboration **19.8a** National Radio Astronomy Observatory (NRAO) **19.8b** X-ray: NASA/CXC/UNAM/Ioffe/D. Page, P. Shternin et al.; optical: NASA/STScI **19.9b** ESO/Manu Mejias **19.10** Bally, J., Langer, W. D., Stark, A. A., Wilson, R. W. 1987. Filamentary structure in the Orion molecular cloud. *The Astrophysical Journal* 312, L45–L49 **19.11** NASA, ESA, and M. Livio and the Hubble 20th Anniversary Team (STScI) **19.12** (top to bottom, left to right) Atlas Image courtesy of 2MASS/UMass/IPAC-Caltech/NASA/NSF; Axel Mellinger, A Color All-Sky Panorama Image of the Milky Way, *Publ. Astron. Soc. Pacific* 121, 1180–1187 (2009); ESA/NASA/JPL-Caltech; NASA, Max-Planck-Institut für extraterrestrische Physik (MPE) and S. L. Snowden; ESA/NASA/Planck Collaboration; NASA/DOE/Fermi LAT Collaboration; Mike Carroll **19.13a** NASA, ESA, M. Robberto (Space Telescope Science Institute/ESA) and the Hubble Space Telescope Orion Treasury Project Team **19.14** ESO/Igor Chekalin **19.15** ESO **19.16** NASA, ESA, S. Beckwith (STScI), and The Hubble Heritage Team (STScI/AURA) **19.17** NASA/JPL **19.20** UCLA Galactic Center Group–W. M. Keck Observatory Laser Team **19.21** NASA/Penn State/G. Garmire et al. **19.22** (main image) Axel Mellinger, A Color All-Sky Panorama Image of the Milky Way, *Publ. Astron. Soc. Pacific* 121, 1180–1187 (2009); (clockwise) S. Kulkarni, J. Bloom, P. Price, Caltech–NRAO GRB Collaboration; NASA/JPL-Caltech/S. Stolovy (Spitzer Science Center/Caltech); NASA/JPL-Caltech/ESA/CXC/STScI; European Southern Observatory; ESO/S. Gillessen et al.; UCLA Galactic Center Group–W. M. Keck Observatory Laser Team; European Southern Observatory; European Southern Observatory; x-ray (blue): NASA/CXC/Northwestern/F. Zadeh et al.; millimeter wavelength (green): Nobeyama/M. Tsuboi; radio (red): NRAO/VLA/F. Zadeh et al.; South African Radio Astronomy Observatory (SARAO)
**Chapter 20 Opener** NASA, ESA, Hubble Heritage, S. Van Dyk (JPL/IPAC), R. Chandar (U. Toledo), D. De Martin & R. Gendler **20.1** NASA, ESA, G. Illingworth (UCO/Lick Observatory and the University of California, Santa Cruz), R. Bouwens (UCO/Lick Observatory and Leiden University) and the HUDF09 Team **20.2a** NASA, ESA, CXC, SSC, and STScI **20.2b** Composite Image Data: Subaru Telescope (NAOJ), Hubble Legacy Archive, Michael Joner, David Laney (West Mountain Observatory, BYU); Processing: Robert Gendler **20.2c** NASA, ESA, W. Freedman (U. Chicago) et

al., & the Hubble Heritage Team (AURA/STScI), SDSS; Processing: Judy Schmidt **20.3** NASA and The Hubble Heritage Team (STScI/AURA) **20.4** NASA, ESA, and the Hubble Heritage Team (STScI/AURA). Acknowledgment: P. Knezek (WIYN) **20.5** NASA/ESA Hubble Space Telescope **20.6a** Robert Gendler/Stocktrek Images/Science Source **20.6b** Sloan Digital Sky Survey **20.7a** ESO **20.7b** Royal Observatory, Edinburgh/Anglo-Australian Telescope Board/Science Source **20.7c** NASA, ESA, and The Hubble Heritage Team (STScI/AURA). Acknowledgment: M. Gregg (Univ. Calif.–Davis and Inst. for Geophysics and Planetary Physics, Lawrence Livermore Natl. Lab.) **20.8** (top left) ESO/P. Grosbøl; (left) NASA, ESA, Hubble Space Telescope; (bottom right) NASA/JPL-Caltech/Roma Tre Univ. **20.9** (clockwise from center) Gemini Observatory–GMOS-S Commissioning Team; Robert Gendler/Stocktrek Images/Science Source; Sloan Digital Sky Survey (SDSS) Collaboration, www.sdss.org; Royal Observatory, Edinburgh/Anglo-Australian Telescope Board/Science Source; ESA/Hubble & NASA. Acknowledgment: Judy Schmidt (Geckzilla) **20.10** GMOS-S Commissioning Team, Gemini Observatory **20.11** ESA/JPL-Caltech/STScI/NASA **20.13** NASA, ESA, and the Hubble Heritage Team (STScI/AURA). Acknowledgment: R. Gendler **20.15** High-Z Supernova Search Team, HST, NASA **20.16** (left to right, top to bottom) NASA, ESA, S. Perlmutter (UC Berkeley, LBNL), A. Koekemoer (STScI), M. Postman (STScI), A. Riess (STScI/JHU), J. Nordin (LBNL, UC Berkeley), D. Rubin (Florida State University), and C. McCully (Rutgers University); ESA, S. Faber (UCSC), A. Riess (JHU/STScI), S. Rodney (JHU), and the CANDELS team, NASA; NASA; ESA; A. Riess, STScI; NASA, ESA, S. Perlmutter (UC Berkeley, LBNL), A. Koekemoer (STScI), M. Postman (STScI), A. Riess (STScI/JHU), J. Nordin (LBNL, UC Berkeley), D. Rubin (Florida State University), and C. McCully (Rutgers University); NASA, ESA, S. Faber (UCSC), A. Riess (JHU/STScI), S. Rodney (JHU), and the CANDELS team; NASA; ESA; and A. Riess, STScI **20.17** Courtesy of the Huntington Library and Observatories of the Carnegie Institution of Washington **20.18** Huntington Library/SuperStock

**Chapter 21 Opener** NASA, ESA, S. Baum and C. O'Dea (RIT), R. Perley and W. Cotton (NRAO/AUI/NSF), and the Hubble Heritage Team (STScI/AURA) **21.1** NASA/JPL-Caltech/ESA **21.2** Modification of work by CXC/MPE/V. Springel/Max Planck Institute for Astrophyics, Garching **21.4** NASA/JPL-Caltech/ESA **21.5** (left) NASA, ESA & Ivo Saviane (European Southern Observatory)/Robert Gendler; (right) ESA/Hubble & NASA **21.6** NASA **21.8a** P.-A. Duc (CEA, CFHT), Atlas 3D Collaboration **21.8b** ESO **21.10** R. Thompson (Univ. of Arizona) et al., NICMOS, HST, NASA **21.11a** NASA, ESA, and The Hubble Heritage Team (STScI/AURA) **21.11b** NASA/CXC/SAO/PSU/CMU **21.12** NASA, ESA, HEIC, and The Hubble Heritage Team (STScI/AURA) **21.13** J. A. Biretta et al., Hubble Heritage Team (STScI/AURA), NASA **21.14** NASA/ESA and J. Bahcall (IAS) **21.17** Holland Ford, Space Telescope Science Institute/Johns Hopkins University; Richard Harms, Applied Research Corp.; Zlatan Tsvetanov, Arthur Davidsen, and Gerard Kriss at Johns Hopkins; Ralph Bohlin and George Hartig at Space Telescope Science Institute; Linda Dressel and Ajay K. Kochhar at Applied Research Corp. in Landover, Md.; and Bruce Margon from the University of Washington in Seattle; NASA **21.18a** NASA/CXC/JPL-Caltech/STScI/NSF/NRAO/VLA **21.20** NASA, ESA, S. Baum and C. O'Dea (RIT), R. Perley and W. Cotton (NRAO/AUI/NSF), and the Hubble Heritage Team (STScI/AURA) **21.21** Image courtesy of NRAO/AUI **21.22** Image courtesy of NRAO/AUI and C. O'Dea & F. Owen **21.23a** X-ray: NASA/CXC/SAO; optical: NASA/STScI; radio: NSF/NRAO/AUI/VLA **21.23b** X-ray: NASA/CXC/CfA/R. Kraft et al.; submillimeter: MPIfR/ESO/APEX/A. Weiss et al.; optical: ESO/WFI

**Chapter 22 Opener** © ESA and the Planck Collaboration **22.6** ASSOCIATED PRESS **22.9** © ESA and the Planck Collaboration **22.17a** Greg Thow/Getty Images **22.17b** WDG Photo/Shutterstock

**Chapter 23 Opener** X-ray: NASA/CXC/CfA/M. Markevitch et al.; lensing map: NASA/STScI, ESO WFI, Magellan/U.Arizona/ D. Clowe et al.; optical: NASA/STScI, Magellan/U.Arizona/D. Clowe et al. **23.6** Caltech Image Archives/California Institute of Technology **23.7** X-ray: NASA/CXC/IfA/C. Ma et al.; optical: NASA/STScI/IfA/C. Ma et al. **23.8** ESA, H. Lee & H. Ford (Johns Hopkins U.)/NASA **23.10** NASA, ESA, J. Richard (CRAL) and J.-P. Kneib (LAM). Acknowledgment: Marc Postman (STScI) **23.11** X-ray: NASA/CXC/CfA/M. Markevitch et al.; lensing map: NASA/STScI, ESO WFI, Magellan/Univ. of Arizona/D. Clowe et al.; optical: NASA/STScI, Magellan/Univ. of Arizona/D. Clowe et al. **23.13** R. Brent Tully (U. Hawaii) et al., SDvision, DP, CEA/Saclay **23.15** Simulation and Visualization by Andrey Kravtsov (University of Chicago) and Anatoly Klypin (NMSU) **23.16** Image courtesy of Volker Springel & the Virgo Consortium/Max Planck Institute for Astrophyics, Garching **23.20** (center left) ESA, H. Lee & H. Ford (Johns Hopkins University)/NASA; (center right) Simulation and Visualization by Andrey Kravtsov (University of Chicago) and Anatoly Klypin (NMSU) **p. 696** (Cosmic microwave background) © ESA and the Planck Collaboration **p. 697** (NGC 4414) The Hubble Heritage Team, STScI, AURA, NASA; (M87) ESA/Hubble/RELICS/NASA; (Orion Nebula) ESA, M. Robberto (Space Telescope Science Institute/ESA) and the Hubble Space Telescope Orion Treasury Project Team, NASA

**Chapter 24 Opener** Seth Shostak **Epigraph** Quote from Maya Angelou, *A Brave and Startling Truth.* New York: Random House, © 1995 **24.2** Darlene Cutshall/Shutterstock **24.3** James L. Amos/Science Source **24.5a** Jane Gould/Alamy Stock Photo **24.5b** B Christopher/Alamy Stock Photo **24.5c** NASA JPL/Abigail Allwood **24.6** Courtesy of J. William Schopf **24.7** Dr. Tsuyoshi Komiya, University of Tokyo **24.10** OET/Nautilus Live **24.11** David McNew/Thomson Reuters (Markets) LLC **24.12** Martin M. Hanczyc **24.14** Eye of Science/Science Source **24.15** M I (Spike) Walker/Alamy Stock Photo **24.16** JPL-Caltech/LANL/J.-L. Lacour, CEA/NASA; (inset) JPL-Caltech/MSSS/NASA **24.17** JPL/Space Science Institute/Antoine Lucas/NASA **24.18** JPL-Caltech/NASA **24.20** Michael Carroll **24.23** National Astronomy and Ionosphere Center **24.24a** NASA Ames **24.24b** Space Frontiers/Archive Photos/Getty Images **24.25** Illustration by Joe Bergeron **24.26** Illustration by Joe Bergeron **pp. 728–729** (Global view) NASA; (second from left) © ESA and the Planck Collaboration

# Appendixes

# A Useful Numbers

## Astronomical Distances

1 AU $\approx 1.496 \times 10^8$ km $= 1.496 \times 10^{11}$ m

1 light-year $\approx 9.46 \times 10^{12}$ km $= 9.46 \times 10^{15}$ m

1 parsec (pc) $\approx 3.09 \times 10^{13}$ km $\approx 3.26$ light-years

1 kiloparsec (kpc) $= 1000$ pc $\approx 3.26 \times 10^3$ light-years

1 megaparsec (Mpc) $= 10^6$ pc $\approx 3.26 \times 10^6$ light-years

## Universal Constants

Speed of light: $c = 3.00 \times 10^5$ km/s $= 3 \times 10^8$ m/s

Gravitational constant: $G = 6.67 \times 10^{-11} \dfrac{\text{m}^3}{\text{kg} \times \text{s}^2}$

Planck's constant: $h = 6.63 \times 10^{-34}$ joule $\times$ s

Stefan-Boltzmann constant: $\sigma = 5.67 \times 10^{-8} \dfrac{\text{watt}}{\text{m}^2 \times \text{K}^4}$

Mass of a proton: $m_\text{p} = 1.67 \times 10^{-27}$ kg

Mass of an electron: $m_\text{e} = 9.11 \times 10^{-31}$ kg

## Useful Sun and Earth Reference Values

Mass of the Sun: $1M_\text{Sun} \approx 2 \times 10^{30}$ kg

Radius of the Sun: $1R_\text{Sun} \approx 696{,}000$ km

Luminosity of the Sun: $1L_\text{Sun} \approx 3.8 \times 10^{26}$ watts

Mass of Earth: $1M_\text{Earth} \approx 5.97 \times 10^{24}$ kg

Radius (equatorial) of Earth: $1R_\text{Earth} \approx 6378$ km

Acceleration of gravity on Earth: $g = 9.8$ m/s$^2$

Escape velocity from surface of Earth: $v_\text{escape} = 11.2$ km/s $= 11{,}200$ m/s

## Astronomical Times

1 solar day (average) $= 24^\text{h}$

1 sidereal day $\approx 23^\text{h}56^\text{m}4.09^\text{s}$

1 synodic month (average) $\approx 29.53$ solar days

1 sidereal month (average) $\approx 27.32$ solar days

1 tropical year $\approx 365.242$ solar days

1 sidereal year $\approx 365.256$ solar days

## Energy and Power Units

Basic unit of energy: 1 joule $= 1 \dfrac{\text{kg} \times \text{m}^2}{\text{s}^2}$

Basic unit of power: 1 watt $= 1$ joule/s

Electron-volt: 1 eV $= 1.60 \times 10^{-19}$ joule

# B Useful Formulas

- Universal law of gravitation for the force between objects of mass $M_1$ and $M_2$, with distance $d$ between their centers:

$$F = G\frac{M_1 M_2}{d^2}$$

- Newton's version of Kepler's third law, which applies to any pair of orbiting objects, such as a star and planet, a planet and moon, or two stars in a binary system; $p$ is the orbital period, $a$ is the distance between the centers of the orbiting objects, and $M_1$ and $M_2$ are the object masses:

$$p^2 = \frac{4\pi^2}{G(M_1 + M_2)}a^3$$

- Escape velocity at distance $R$ from center of object of mass $M$:

$$v_{\text{escape}} = \sqrt{\frac{2GM}{R}}$$

- Relationship between a photon's wavelength $(\lambda)$, frequency $(f)$, and the speed of light $(c)$:

$$\lambda \times f = c$$

- Energy of a photon of wavelength $\lambda$ or frequency $f$:

$$E = hf = \frac{hc}{\lambda}$$

- Stefan-Boltzmann law for thermal radiation at temperature $T$ (on the Kelvin scale):

$$\text{emitted power per unit area} = \sigma T^4$$

- Wien's law for the peak wavelength $(\lambda_{\text{max}})$ thermal radiation at temperature $T$ (on the Kelvin scale):

$$\lambda_{\text{max}} = \frac{2{,}900{,}000}{T}\text{ nm}$$

- Doppler shift (radial velocity is positive if the object is moving away from us and negative if it is moving toward us):

$$\frac{\text{radial velocity}}{\text{speed of light}} = \frac{\text{shifted wavelength} - \text{rest wavelength}}{\text{rest wavelength}}$$

- Angular separation $(\alpha)$ of two points with an actual separation $s$, viewed from a distance $d$ (assuming $d$ is much larger than $s$):

$$\alpha = \frac{s}{2\pi d} \times 360^\circ$$

- Inverse square law for light ($d$ is the distance to the object):

$$\text{apparent brightness} = \frac{\text{luminosity}}{4\pi d^2}$$

- Parallax formula (distance $d$ to a star with parallax angle $p$ in arcseconds):

$$d\ (\text{in parsecs}) = \frac{1}{p\ (\text{in arcseconds})}$$

$$d\ (\text{in light-years}) = 3.26 \times \frac{1}{p\ (\text{in arcseconds})}$$

- The orbital velocity law, to find the mass $M_r$ contained within the circular orbit of radius $r$ for an object moving at speed $v$:

$$M_r = \frac{r \times v^2}{G}$$

# C A Few Mathematical Skills

This appendix reviews the following mathematical skills: powers of 10, scientific notation, working with units, the metric system, and finding a ratio. You should refer to this appendix as needed while studying the book.

## C.1 Powers of 10

Powers of 10 indicate how many times to multiply 10 by itself. For example:

$$10^2 = 10 \times 10 = 100$$
$$10^6 = 10 \times 10 \times 10 \times 10 \times 10 \times 10 = 1{,}000{,}000$$

Negative powers are the reciprocals of the corresponding positive powers. For example:

$$10^{-2} = \frac{1}{10^2} = \frac{1}{100} = 0.01$$
$$10^{-6} = \frac{1}{10^6} = \frac{1}{1{,}000{,}000} = 0.000001$$

**TABLE C.1** lists powers of 10 from $10^{-12}$ to $10^{12}$. Note that powers of 10 follow two basic rules:

1. A positive exponent tells how many zeros follow the 1. For example, $10^0$ is a 1 followed by no zeros, and $10^8$ is a 1 followed by eight zeros.
2. A negative exponent tells how many places are to the right of the decimal point, including the 1. For example, $10^{-1} = 0.1$ has one place to the right of the decimal point; $10^{-6} = 0.000001$ has six places to the right of the decimal point.

### Multiplying and Dividing Powers of 10

Multiplying powers of 10 simply requires adding exponents, as the following examples show:

$$10^4 \times 10^7 = \underbrace{10{,}000}_{10^4} \times \underbrace{10{,}000{,}000}_{10^7} = \underbrace{100{,}000{,}000{,}000}_{10^{4+7} = 10^{11}} = 10^{11}$$

$$10^5 \times 10^{-3} = \underbrace{100{,}000}_{10^5} \times \underbrace{0.001}_{10^{-3}} = \underbrace{100}_{10^{5+(-3)} = 10^2} = 10^2$$

**TABLE C.1 Powers of 10**

| Zero and Positive Powers | | | Negative Powers | | |
|---|---|---|---|---|---|
| **Power** | **Value** | **Name** | **Power** | **Value** | **Name** |
| $10^0$ | 1 | One | | | |
| $10^1$ | 10 | Ten | $10^{-1}$ | 0.1 | Tenth |
| $10^2$ | 100 | Hundred | $10^{-2}$ | 0.01 | Hundredth |
| $10^3$ | 1000 | Thousand | $10^{-3}$ | 0.001 | Thousandth |
| $10^4$ | 10,000 | Ten thousand | $10^{-4}$ | 0.0001 | Ten-thousandth |
| $10^5$ | 100,000 | Hundred thousand | $10^{-5}$ | 0.00001 | Hundred-thousandth |
| $10^6$ | 1,000,000 | Million | $10^{-6}$ | 0.000001 | Millionth |
| $10^7$ | 10,000,000 | Ten million | $10^{-7}$ | 0.0000001 | Ten-millionth |
| $10^8$ | 100,000,000 | Hundred million | $10^{-8}$ | 0.00000001 | Hundred-millionth |
| $10^9$ | 1,000,000,000 | Billion | $10^{-9}$ | 0.000000001 | Billionth |
| $10^{10}$ | 10,000,000,000 | Ten billion | $10^{-10}$ | 0.0000000001 | Ten-billionth |
| $10^{11}$ | 100,000,000,000 | Hundred billion | $10^{-11}$ | 0.00000000001 | Hundred-billionth |
| $10^{12}$ | 1,000,000,000,000 | Trillion | $10^{-12}$ | 0.000000000001 | Trillionth |

$$10^{-8} \times 10^{-5} = \underbrace{0.00000001}_{10^{-8}} \times \underbrace{0.00001}_{10^{-5}}$$
$$= \underbrace{0.0000000000001}_{10^{-8+(-5)} = 10^{-13}} = 10^{-13}$$

Dividing powers of 10 requires subtracting exponents, as in the following examples:

$$\frac{10^5}{10^3} = \underbrace{100{,}000}_{10^5} \div \underbrace{1000}_{10^3}$$
$$= \underbrace{100}_{10^{5-3} = 10^2} = 10^2$$

$$\frac{10^3}{10^7} = \underbrace{1000}_{10^3} \div \underbrace{10{,}000{,}000}_{10^7}$$
$$= \underbrace{0.0001}_{10^{3-7} = 10^{-4}} = 10^{-4}$$

$$\frac{10^{-4}}{10^{-6}} = \underbrace{0.0001}_{10^{-4}} \div \underbrace{0.000001}_{10^{-6}}$$
$$= \underbrace{100}_{10^{-4-(-6)} = 10^2} = 10^2$$

## Powers of Powers of 10

We can use the multiplication and division rules to raise powers of 10 to other powers or to take roots. For example:

$$(10^4)^3 = 10^4 \times 10^4 \times 10^4 = 10^{4+4+4} = 10^{12}$$

Note that we can get the same end result by simply multiplying the two powers:

$$(10^4)^3 = 10^{4\times 3} = 10^{12}$$

Because taking a root is the same as raising to a fractional power (e.g., the square root is the same as the $\frac{1}{2}$ power, the cube root is the same as the $\frac{1}{3}$ power, etc.), we can use the same procedure for roots, as in the following example:

$$\sqrt{10^4} = (10^4)^{1/2} = 10^{4\times(1/2)} = 10^2$$

## Adding and Subtracting Powers of 10

Unlike multiplying and dividing powers of 10, there is no shortcut for adding or subtracting powers of 10. The values must be written in longhand notation. For example:

$$10^6 + 10^2 = 1{,}000{,}000 + 100 = 1{,}000{,}100$$
$$10^8 + 10^{-3} = 100{,}000{,}000 + 0.001 = 100{,}000{,}000.001$$
$$10^7 - 10^3 = 10{,}000{,}000 - 1000 = 9{,}999{,}000$$

## Summary

We can summarize our findings using $n$ and $m$ to represent any numbers:

- To *multiply* powers of 10, *add* exponents:

  $10^n \times 10^m = 10^{n+m}$

- To *divide* powers of 10, *subtract* exponents:

  $\dfrac{10^n}{10^m} = 10^{n-m}$

- To *raise* powers of 10 to other powers, multiply exponents: $(10^n)^m = 10^{n\times m}$
- To add or subtract powers of 10, first write them out longhand.

# C.2 Scientific Notation

When we are dealing with large or small numbers, it's generally easier to write them with powers of 10. For example, it's much easier to write the number 6,000,000,000,000 as $6 \times 10^{12}$. This format, in which a number *between* 1 and 10 is multiplied by a power of 10, is called **scientific notation**.

## Converting a Number to Scientific Notation

We can convert numbers written in ordinary notation to scientific notation with a simple two-step process:

1. Move the decimal point to come after the *first* nonzero digit.
2. The number of places the decimal point moves tells you the power of 10; the power is *positive* if the decimal point moves to the left and *negative* if it moves to the right.

   *Examples:*

$$3042 \xrightarrow{\text{decimal needs to move 3 places to left}} 3.042 \times 10^3$$

$$0.00012 \xrightarrow{\text{decimal needs to move 4 places to right}} 1.2 \times 10^{-4}$$

$$226 \times 10^2 \xrightarrow{\text{decimal needs to move 2 places to left}} (2.26 \times 10^2) \times 10^2 = 2.26 \times 10^4$$

## Converting a Number from Scientific Notation

We can convert numbers written in scientific notation to ordinary notation by the reverse process:

1. The power of 10 indicates how many places to move the decimal point; move it to the *right* if the power of 10 is positive and to the *left* if it is negative.

2. If moving the decimal point creates any open places, fill them with zeros.

*Examples:*

$$4.01 \times 10^2 \xrightarrow[\text{2 places to right}]{\text{move decimal}} 401$$

$$3.6 \times 10^6 \xrightarrow[\text{6 places to right}]{\text{move decimal}} 3{,}600{,}000$$

$$5.7 \times 10^{-3} \xrightarrow[\text{3 places to left}]{\text{move decimal}} 0.0057$$

### Multiplying or Dividing Numbers in Scientific Notation

Multiplying or dividing numbers in scientific notation simply requires operating on the powers of 10 and the other parts of the number separately.

*Examples:*

$$\begin{aligned}(6 \times 10^2) \times (4 \times 10^5) &= (6 \times 4) \times (10^2 \times 10^5) \\ &= 24 \times 10^7 = (2.4 \times 10^1) \times 10^7 \\ &= 2.4 \times 10^8\end{aligned}$$

$$\begin{aligned}\frac{4.2 \times 10^{-2}}{8.4 \times 10^{-5}} = \frac{4.2}{8.4} \times \frac{10^{-2}}{10^{-5}} &= 0.5 \times 10^{-2-(-5)} = 0.5 \times 10^3 \\ &= (5 \times 10^{-1}) \times 10^3 = 5 \times 10^2\end{aligned}$$

Note that, in both these examples, we first found an answer in which the number multiplied by a power of 10 was *not* between 1 and 10. We therefore followed the procedure for converting the final answer to scientific notation.

### Addition and Subtraction with Scientific Notation

In general, we must write numbers in ordinary notation before adding or subtracting.

*Examples:*

$$\begin{aligned}(3 \times 10^6) + (5 \times 10^2) &= 3{,}000{,}000 + 500 \\ &= 3{,}000{,}500 = 3.0005 \times 10^6\end{aligned}$$

$$\begin{aligned}(4.6 \times 10^9) - (5 \times 10^8) &= 4{,}600{,}000{,}000 - 500{,}000{,}000 \\ &= 4{,}100{,}000{,}000 = 4.1 \times 10^9\end{aligned}$$

When both numbers have the *same* power of 10, we can factor out the power of 10 first.

*Examples:*

$$\begin{aligned}(7 \times 10^{10}) + (4 \times 10^{10}) &= (7 + 4) \times 10^{10} \\ &= 11 \times 10^{10} = 1.1 \times 10^{11}\end{aligned}$$

$$\begin{aligned}(2.3 \times 10^{-22}) - (1.6 \times 10^{-22}) &= (2.3 - 1.6) \times 10^{-22} \\ &= 0.7 \times 10^{-22} = 7.0 \times 10^{-23}\end{aligned}$$

## C.3 Working with Units

Showing the units of a problem as you solve it usually makes the work much easier and also provides a useful way of checking your work. If an answer does not come out with the units you expect, you probably did something wrong. In general, working with units is very similar to working with numbers, as the following guidelines and examples show.

### Five Guidelines for Working with Units

Before you begin any problem, think ahead and identify the units you expect for the final answer. Then operate on the units along with the numbers as you solve the problem. The following five guidelines may be helpful when you are working with units:

1. Mathematically, it doesn't matter whether a unit is singular (e.g., meter) or plural (e.g., meters); we can use the same abbreviation (e.g., m) for both.

2. You cannot add or subtract numbers unless they have the *same* units. For example, 5 apples + 3 apples = 8 apples, but the expression 5 apples +3 oranges cannot be simplified further.

3. You *can* multiply units, divide units, or raise units to powers. Look for key words that tell you what to do.

   - *Per* suggests division. For example, we write a speed of 100 kilometers per hour as

     $$100\ \frac{\text{km}}{\text{hr}} \quad \text{or} \quad 100\ \frac{\text{km}}{1\ \text{hr}}$$

   - *Of* suggests multiplication. For example, if you launch a 50-kg space probe at a launch cost *of* \$10,000 per kilogram, the total cost is

     $$50\ \cancel{\text{kg}} \times \frac{\$10{,}000}{\cancel{\text{kg}}} = \$500{,}000$$

   - *Square* suggests raising to the second power. For example, we write an area of 75 square meters as 75 $m^2$.

   - *Cube* suggests raising to the third power. For example, we write a volume of 12 cubic centimeters as 12 $cm^3$.

4. Often the number you are given is not in the units you wish to work with. For example, you may be given that the speed of light is 300,000 km/s but need it in units of m/s for a particular problem. To convert the units, simply multiply the given number by a *conversion factor*: a fraction in which the numerator (top of the fraction) and denominator (bottom of the fraction) are equal, so that the value of the fraction is 1; the number in the denominator must have the units that you wish to change. In the case of changing the speed of light from units of km/s to m/s, you need a conversion factor for kilometers to meters. Thus, the conversion factor is

$$\frac{1000\ \text{m}}{1\ \text{km}}$$

Note that this conversion factor is equal to 1, since 1000 meters and 1 kilometer are equal, and that the units to be changed (km) appear in the denominator. We can now convert the speed of light from units of km/s to m/s simply by multiplying by this conversion factor:

$$\underbrace{300{,}000\ \frac{\cancel{\text{km}}}{\text{s}}}_{\text{speed of light in km/s}} \times \underbrace{\frac{1000\ \text{m}}{1\ \cancel{\text{km}}}}_{\text{conversion from km to m}} = \underbrace{3 \times 10^{8}\ \frac{\text{m}}{\text{s}}}_{\text{speed of light in m/s}}$$

Note that the units of km cancel, leaving the answer in units of m/s.

5. It's easier to work with units if you replace division with multiplication by the reciprocal. For example, suppose you want to know how many minutes are represented by 300 seconds. We can find the answer by dividing 300 seconds by 60 seconds per minute:

$$300\ \text{s} \div 60\ \frac{\text{s}}{\text{min}}$$

However, it is easier to see the unit cancellations if we rewrite this expression by replacing the division with multiplication by the reciprocal (this process is easy to remember as "invert and multiply"):

$$300\ \text{s} \div 60\ \frac{\text{s}}{\text{min}} = \underbrace{300\ \cancel{\text{s}} \times \underbrace{\frac{1\ \text{min}}{60\ \cancel{\text{s}}}}_{\text{invert}}}_{\text{and multiply}} = 5\ \text{min}$$

We now see that the units of seconds (s) cancel in the numerator of the first term and the denominator of the second term, leaving the answer in units of minutes.

## More Examples of Working with Units

**Example 1.** How many seconds are there in 1 day?

**Solution:** We can answer the question by setting up a *chain* of unit conversions in which we start with 1 *day* and end up with *seconds*. We use the facts that there are 24 hours per day (24 hr/day), 60 minutes per hour (60 min/hr), and 60 seconds per minute (60 s/min):

$$\underbrace{1\ \cancel{\text{day}}}_{\text{starting value}} \times \underbrace{\frac{24\ \cancel{\text{hr}}}{\cancel{\text{day}}}}_{\text{conversion from day to hr}} \times \underbrace{\frac{60\ \cancel{\text{min}}}{\cancel{\text{hr}}}}_{\text{conversion from hr to min}} \times \underbrace{\frac{60\ \text{s}}{\cancel{\text{min}}}}_{\text{conversion from min to s}} = 86{,}400\ \text{s}$$

Note that all the units cancel except *seconds*, which is what we want for the answer. There are 86,400 seconds in 1 day.

**Example 2.** Convert a distance of $10^8$ cm to km.

**Solution:** The easiest way to make this conversion is in two steps, since we know that there are 100 centimeters per meter (100 cm/m) and 1000 meters per kilometer (1000 m/km):

$$\underbrace{10^{8}\ \text{cm}}_{\text{starting value}} \times \underbrace{\frac{1\ \text{m}}{100\ \text{cm}}}_{\text{conversion from cm to m}} \times \underbrace{\frac{1\ \text{km}}{1000\ \text{m}}}_{\text{conversion from m to km}}$$

$$= 10^{8}\ \cancel{\text{cm}} \times \frac{1\ \cancel{\text{m}}}{10^{2}\ \cancel{\text{cm}}} \times \frac{1\ \text{km}}{10^{3}\ \cancel{\text{m}}} = 10^{3}\ \text{km}$$

Alternatively, if we recognize that the number of kilometers should be smaller than the number of centimeters (because kilometers are larger), we might decide to do this conversion by dividing as follows:

$$10^{8}\ \text{cm} \div \frac{100\ \text{cm}}{\text{m}} \div \frac{1000\ \text{m}}{\text{km}}$$

In this case, before carrying out the calculation, we replace each division with multiplication by the reciprocal:

$$\begin{aligned} 10^{8}\ \text{cm} \div \frac{100\ \text{cm}}{\text{m}} \div \frac{1000\ \text{m}}{\text{km}} &= 10^{8}\ \text{cm} \times \frac{1\ \text{m}}{100\ \text{cm}} \times \frac{1\ \text{km}}{1000\ \text{m}} \\ &= 10^{8}\ \cancel{\text{cm}} \times \frac{1\ \cancel{\text{m}}}{10^{2}\ \cancel{\text{cm}}} \times \frac{1\ \text{km}}{10^{3}\ \cancel{\text{m}}} \\ &= 10^{3}\ \text{km} \end{aligned}$$

Note that we again get the answer that $10^8$ cm is the same as $10^3$ km, or 1000 km.

**Example 3.** Suppose you accelerate at 9.8 $m/s^2$ for 4 seconds, starting from rest. How fast will you be going?

**Solution:** The question asks "how fast?" so we expect to end up with a speed. Therefore, we multiply the acceleration by the amount of time you accelerated:

$$9.8\,\frac{m}{s^2} \times 4\,s = (9.8 \times 4)\,\frac{m \times \not{s}}{s^{\not{2}}} = 39.2\,\frac{m}{s}$$

Note that the units end up as a speed, showing that you will be traveling 39.2 m/s after 4 seconds of acceleration at 9.8 $m/s^2$.

**Example 4.** A reservoir is 2 km long and 3 km wide. Calculate its area, in both square kilometers and square meters.

**Solution:** We find its area by multiplying its length and width:

$$2\text{ km} \times 3\text{ km} = 6\text{ km}^2$$

Next we need to convert this area of 6 $km^2$ to square meters, using the fact that there are 1000 meters per kilometer (1000 m/km). Note that we must square the term 1000 m/km when converting from $km^2$ to $m^2$:

$$6\text{ km}^2 \times \left(1000\,\frac{m}{km}\right)^2 = 6\text{ km}^2 \times 1000^2\,\frac{m^2}{km^2}$$

$$= 6\,\cancel{km^2} \times 1{,}000{,}000\,\frac{m^2}{\cancel{km^2}}$$

$$= 6{,}000{,}000\text{ m}^2$$

The reservoir area is 6 $km^2$, which is the same as 6 million $m^2$.

## C.4 The Metric System (SI)

The modern version of the metric system, known as *Système Internationale d'Unites* (French for "International System of Units") or **SI**, was formally established in 1960. Today, it is the primary measurement system in nearly every country in the world with the exception of the United States. Even in the United States, it is the system of choice for science and international commerce. The basic units of length, mass, and time in the SI are

- the **meter** for length, abbreviated m
- the **kilogram** for mass, abbreviated kg
- the **second** for time, abbreviated s

Multiples of metric units are formed by powers of 10, using a prefix to indicate the power. For example, *kilo* means $10^3$ (1000), so a kilometer is 1000 meters; a microgram is 0.000001 gram, because *micro* means $10^{-6}$, or one millionth. Some of the more common prefixes are listed in **TABLE C.2**.

**TABLE C.2 SI (Metric) Prefixes**

| Small Values | | | Large Values | | |
|---|---|---|---|---|---|
| **Prefix** | **Abbreviation** | **Value** | **Prefix** | **Abbreviation** | **Value** |
| Deci | d | $10^{-1}$ | Deca | da | $10^1$ |
| Centi | c | $10^{-2}$ | Hecto | h | $10^2$ |
| Milli | m | $10^{-3}$ | Kilo | k | $10^3$ |
| Micro | $\mu$ | $10^{-6}$ | Mega | M | $10^6$ |
| Nano | n | $10^{-9}$ | Giga | G | $10^9$ |
| Pico | p | $10^{-12}$ | Tera | T | $10^{12}$ |

### Metric Conversions

**TABLE C.3** lists conversions between metric units and units used commonly in the United States. Note that the conversions between kilograms and pounds are valid only on Earth, because they depend on the strength of gravity.

**TABLE C.3 Metric Conversions**

| To Metric | From Metric |
|---|---|
| 1 inch = 2.540 cm | 1 cm = 0.3937 inch |
| 1 foot = 0.3048 m | 1 m = 3.28 feet |
| 1 yard = 0.9144 m | 1 m = 1.094 yards |
| 1 mile = 1.6093 km | 1 km = 0.6214 mile |
| 1 pound = 0.4536 kg | 1 kg = 2.205 pounds |

**Example 1.** International athletic competitions generally use metric distances. Compare the length of a 100-meter race to that of a 100-yard race.

**Solution:** Table C.3 shows that 1 m = 1.094 yd, so 100 m is 109.4 yd. Note that 100 meters is almost 110 yards; a good "rule of thumb" to remember is that distances in meters are about 10% longer than the corresponding number of yards.

**Example 2.** How many square kilometers are in 1 square mile?

**Solution:** We use the square of the miles-to-kilometers conversion factor:

$$(1\text{ mi}^2) \times \left(\frac{1.6093\text{ km}}{1\text{ mi}}\right)^2 = (1\,\cancel{mi^2}) \times \left(1.6093^2\,\frac{km^2}{\cancel{mi^2}}\right)$$

$$= 2.5898\text{ km}^2$$

Therefore, 1 square mile is 2.5898 square kilometers.

## C.5 Finding a Ratio

Suppose you want to compare two quantities, such as the average density of Earth and the average density of Jupiter. The way we do such a comparison is by dividing, which tells us the *ratio* of the two quantities. In this case, Earth's average density is 5.52 $g/cm^3$ and Jupiter's average density is 1.33 $g/cm^3$ (see Figure 11.1), so the ratio is

$$\frac{\text{average density of Earth}}{\text{average density of Jupiter}} = \frac{5.52 \text{ g/cm}^3}{1.33 \text{ g/cm}^3} = 4.15$$

Notice how the units cancel on both the top and the bottom of the fraction. We can state our result in two equivalent ways:

- The ratio of Earth's average density to Jupiter's average density is 4.15.
- Earth's average density is 4.15 times Jupiter's average density.

Sometimes, the quantities that you want to compare may each involve an equation. In such cases, you could, of course, find the ratio by first calculating each of the two quantities individually and then dividing. However, it is much easier if you first express the ratio as a fraction, putting the equation for one quantity on top and the other on the bottom. Some of the terms in the equation may then cancel out, making any calculations much easier.

**Example 1.** Compare the kinetic energy of a car traveling at 100 km/hr to that of the same car traveling at 50 km/hr.

**Solution:** We do the comparison by finding the ratio of the two kinetic energies, recalling that the formula for kinetic energy is $\frac{1}{2}mv^2$. Since we are not told the mass of the car, you might at first think that we don't have enough information to find the ratio. However, notice what happens when we put the equations for each kinetic energy into the ratio, calling the two speeds $v_1$ and $v_2$:

$$\frac{\text{K.E. car at } v_1}{\text{K.E. car at } v_2} = \frac{\frac{1}{2}m_{\text{car}}\, v_1^2}{\frac{1}{2}m_{\text{car}}\, v_2^2} = \frac{v_1^2}{v_2^2} = \left(\frac{v_1}{v_2}\right)^2$$

All the terms cancel except those with the two speeds, leaving us with a very simple formula for the ratio. Now we put in 100 km/hr for $v_1$ and 50 km/hr for $v_2$:

$$\frac{\text{K.E. car at 100 km/hr}}{\text{K.E. car at 50 km/hr}} = \left(\frac{100 \text{ km/hr}}{50 \text{ km/hr}}\right)^2 = 2^2 = 4$$

The ratio of the car's kinetic energies at 100 km/hr and 50 km/hr is 4. That is, the car has four times as much kinetic energy at 100 km/hr as it has at 50 km/hr.

**Example 2.** Compare the strength of gravity between Earth and the Sun to the strength of gravity between Earth and the Moon.

**Solution:** We do the comparison by taking the ratio of the Earth–Sun gravity to the Earth–Moon gravity. In this case, each quantity is found from the equation of Newton's law of gravity. (See Section 4.4.) Thus, the ratio is

$$\frac{\text{Earth–Sun gravity}}{\text{Earth–Moon gravity}} = \frac{G\dfrac{M_{\text{Earth}}M_{\text{Sun}}}{(d_{\text{Earth–Sun}})^2}}{G\dfrac{M_{\text{Earth}}M_{\text{Moon}}}{(d_{\text{Earth–Moon}})^2}} = \frac{M_{\text{Sun}}}{(d_{\text{Earth–Sun}})^2} \times \frac{(d_{\text{Earth–Moon}})^2}{M_{\text{Moon}}}$$

Note how all but four of the terms cancel; the last step comes from replacing the division with multiplication by the reciprocal (the "invert and multiply" rule for division). We can simplify the work further by rearranging the terms so that we have the masses and distances together:

$$\frac{\text{Earth–Sun gravity}}{\text{Earth–Moon gravity}} = \frac{M_{\text{Sun}}}{M_{\text{Moon}}} \times \frac{(d_{\text{Earth–Moon}})^2}{(d_{\text{Earth–Sun}})^2}$$

Now it is just a matter of looking up the numbers (see Appendix E) and calculating:

$$\frac{\text{Earth–Sun gravity}}{\text{Earth–Moon gravity}} = \frac{1.99 \times 10^{30} \text{ kg}}{7.35 \times 10^{22} \text{ kg}} \times \frac{(384.4 \times 10^3 \text{ km})^2}{(149.6 \times 10^6 \text{ km})^2} = 179$$

In other words, the Earth–Sun gravity is 179 times stronger than the Earth–Moon gravity.

# D The Periodic Table of the Elements

**Key**

| |
|---|
| 20 — Atomic number<br>**Ca** — Element's symbol<br>Calcium — Element's name<br>40.08 — Atomic mass*<br>Colors indicate element origin** |

| 1 | 2 | 3 | 4 | 5 | 6 | 7 | 8 | 9 | 10 | 11 | 12 | 13 | 14 | 15 | 16 | 17 | 18 |
|---|---|---|---|---|---|---|---|---|---|---|---|---|---|---|---|---|---|
| 1<br>**H**<br>Hydrogen<br>1.00794 | | | | | | | | | | | | | | | | | 2<br>**He**<br>Helium<br>4.003 |
| 3<br>**Li**<br>Lithium<br>6.941 | 4<br>**Be**<br>Beryllium<br>9.01218 | | | | | | | | | | | 5<br>**B**<br>Boron<br>10.81 | 6<br>**C**<br>Carbon<br>12.011 | 7<br>**N**<br>Nitrogen<br>14.007 | 8<br>**O**<br>Oxygen<br>15.999 | 9<br>**F**<br>Fluorine<br>18.988 | 10<br>**Ne**<br>Neon<br>20.179 |
| 11<br>**Na**<br>Sodium<br>22.990 | 12<br>**Mg**<br>Magnesium<br>24.305 | | | | | | | | | | | 13<br>**Al**<br>Aluminum<br>26.98 | 14<br>**Si**<br>Silicon<br>28.086 | 15<br>**P**<br>Phosphorus<br>30.974 | 16<br>**S**<br>Sulfur<br>32.06 | 17<br>**Cl**<br>Chlorine<br>35.453 | 18<br>**Ar**<br>Argon<br>39.948 |
| 19<br>**K**<br>Potassium<br>39.098 | 20<br>**Ca**<br>Calcium<br>40.08 | 21<br>**Sc**<br>Scandium<br>44.956 | 22<br>**Ti**<br>Titanium<br>47.88 | 23<br>**V**<br>Vanadium<br>50.94 | 24<br>**Cr**<br>Chromium<br>51.996 | 25<br>**Mn**<br>Manganese<br>54.938 | 26<br>**Fe**<br>Iron<br>55.847 | 27<br>**Co**<br>Cobalt<br>58.9332 | 28<br>**Ni**<br>Nickel<br>58.69 | 29<br>**Cu**<br>Copper<br>63.546 | 30<br>**Zn**<br>Zinc<br>65.39 | 31<br>**Ga**<br>Gallium<br>69.72 | 32<br>**Ge**<br>Germanium<br>72.59 | 33<br>**As**<br>Arsenic<br>74.922 | 34<br>**Se**<br>Selenium<br>78.96 | 35<br>**Br**<br>Bromine<br>79.904 | 36<br>**Kr**<br>Krypton<br>83.80 |
| 37<br>**Rb**<br>Rubidium<br>85.468 | 38<br>**Sr**<br>Strontium<br>87.62 | 39<br>**Y**<br>Yttrium<br>88.9059 | 40<br>**Zr**<br>Zirconium<br>91.224 | 41<br>**Nb**<br>Niobium<br>92.91 | 42<br>**Mo**<br>Molybdenum<br>95.94 | 43<br>**Tc**<br>Technetium<br>(98) | 44<br>**Ru**<br>Ruthenium<br>101.07 | 45<br>**Rh**<br>Rhodium<br>102.906 | 46<br>**Pd**<br>Palladium<br>106.42 | 47<br>**Ag**<br>Silver<br>107.868 | 48<br>**Cd**<br>Cadmium<br>112.41 | 49<br>**In**<br>Indium<br>114.82 | 50<br>**Sn**<br>Tin<br>118.71 | 51<br>**Sb**<br>Antimony<br>121.75 | 52<br>**Te**<br>Tellurium<br>127.60 | 53<br>**I**<br>Iodine<br>126.905 | 54<br>**Xe**<br>Xenon<br>131.29 |
| 55<br>**Cs**<br>Cesium<br>132.91 | 56<br>**Ba**<br>Barium<br>137.34 | | 72<br>**Hf**<br>Hafnium<br>178.49 | 73<br>**Ta**<br>Tantalum<br>180.95 | 74<br>**W**<br>Tungsten<br>183.85 | 75<br>**Re**<br>Rhenium<br>186.207 | 76<br>**Os**<br>Osmium<br>190.2 | 77<br>**Ir**<br>Iridium<br>192.22 | 78<br>**Pt**<br>Platinum<br>195.08 | 79<br>**Au**<br>Gold<br>196.967 | 80<br>**Hg**<br>Mercury<br>200.59 | 81<br>**Tl**<br>Thallium<br>204.383 | 82<br>**Pb**<br>Lead<br>207.2 | 83<br>**Bi**<br>Bismuth<br>208.98 | 84<br>**Po**<br>Polonium<br>(209) | 85<br>**At**<br>Astatine<br>(210) | 86<br>**Rn**<br>Radon<br>(222) |
| 87<br>**Fr**<br>Francium<br>(223) | 88<br>**Ra**<br>Radium<br>226.0254 | | 104<br>**Rf**<br>Rutherfordium<br>(263) | 105<br>**Db**<br>Dubnium<br>(262) | 106<br>**Sg**<br>Seaborgium<br>(266) | 107<br>**Bh**<br>Bohrium<br>(267) | 108<br>**Hs**<br>Hassium<br>(277) | 109<br>**Mt**<br>Meitnerium<br>(268) | 110<br>**Ds**<br>Darmstadtium<br>(281) | 111<br>**Rg**<br>Roentgenium<br>(272) | 112<br>**Cn**<br>Copernicium<br>(285) | 113<br>**Nh**<br>Nihonium<br>(286) | 114<br>**Fl**<br>Flerovium<br>(289) | 115<br>**Mc**<br>Moscovium<br>(288) | 116<br>**Lv**<br>Livermorium<br>(293) | 117<br>**Ts**<br>Tennessine<br>(294) | 118<br>**Og**<br>Oganesson<br>(294) |

**Lanthanide Series**

| | | | | | | | | | | | | | | |
|---|---|---|---|---|---|---|---|---|---|---|---|---|---|---|
| 57<br>**La**<br>Lanthanum<br>138.906 | 58<br>**Ce**<br>Cerium<br>140.12 | 59<br>**Pr**<br>Praseodymium<br>140.908 | 60<br>**Nd**<br>Neodymium<br>144.24 | 61<br>**Pm**<br>Promethium<br>(145) | 62<br>**Sm**<br>Samarium<br>150.36 | 63<br>**Eu**<br>Europium<br>151.96 | 64<br>**Gd**<br>Gadolinium<br>157.25 | 65<br>**Tb**<br>Terbium<br>158.925 | 66<br>**Dy**<br>Dysprosium<br>162.50 | 67<br>**Ho**<br>Holmium<br>164.93 | 68<br>**Er**<br>Erbium<br>167.26 | 69<br>**Tm**<br>Thulium<br>168.934 | 70<br>**Yb**<br>Ytterbium<br>173.04 | 71<br>**Lu**<br>Lutetium<br>174.967 |

**Actinide Series**

| | | | | | | | | | | | | | | |
|---|---|---|---|---|---|---|---|---|---|---|---|---|---|---|
| 89<br>**Ac**<br>Actinium<br>227.028 | 90<br>**Th**<br>Thorium<br>232.038 | 91<br>**Pa**<br>Protactinium<br>231.036 | 92<br>**U**<br>Uranium<br>238.029 | 93<br>**Np**<br>Neptunium<br>237.048 | 94<br>**Pu**<br>Plutonium<br>(244) | 95<br>**Am**<br>Americium<br>(243) | 96<br>**Cm**<br>Curium<br>(247) | 97<br>**Bk**<br>Berkelium<br>(247) | 98<br>**Cf**<br>Californium<br>(251) | 99<br>**Es**<br>Einsteinium<br>(252) | 100<br>**Fm**<br>Fermium<br>(257) | 101<br>**Md**<br>Mendelevium<br>(258) | 102<br>**No**<br>Nobelium<br>(259) | 103<br>**Lr**<br>Lawrencium<br>(260) |

*Atomic masses are fractions because they represent a weighted average of atomic masses of different isotopes—in proportion to the abundance of each isotope on Earth.

**The colors correspond to different element origins as indicated below; for elements with more than one origin, the sizes of the colored regions indicate the approximate percentages of that element with each origin.

**Big Bang:** All of the hydrogen and almost all of the helium in the universe was produced during the first 5 minutes after the Big Bang [Section 22.1].

**Dying Low-Mass Stars:** Most of the lithium, carbon, and nitrogen in the universe was produced by nuclear fusion in low-mass stars and released into space when those stars died [Section 17.2]. Those dying stars can also make relatively rare elements through reactions that add neutrons to nuclei that have more protons than iron (see footnote on page 547). If enough neutrons are added, some of them then decay into protons, raising the atomic number of the nucleus.

**Massive Star Supernovae:** Most of elements from oxygen (O) through rubidium (Rb) were produced by nuclear fusion in massive stars and released into space through massive star supernovae [Section 17.3].

Color coding based on work by Jennifer Johnson.

**White Dwarf Supernovae:** About half of the elements near iron (Fe) on the periodic table were produced by nuclear fusion in white dwarf supernova explosions [Section 18.1].

**Neutron-Star Mergers:** Many of the rarest elements on the periodic table, including gold (Au), platinum (Pt), and uranium (U), are thought to have been produced in mergers between two neutron stars [Section 18.4]. Models of nuclear fusion indicate that they had to be formed in environments extremely rich in neutrons and unlike the proton-rich environments within massive star supernovae.

**Cosmic Rays:** The elements beryllium (Be) and boron (B) are not easily formed by nuclear fusion in stars. Instead, they are made primarily by demolition of elements of greater atomic number, most likely through collision with high-energy cosmic rays [Section 19.2].

**Rare Radioactive Elements:** These elements can be made in supernova explosions or neutron-star mergers but rapidly decay into other elements. Consequently, they are either rarely found or never found in nature.

# E Solar System Data

**TABLE E.1 Sun, Planets, and Dwarf Planets**

| | Physical Properties | | | | | | | Orbital and Rotational Properties | | | | | | |
|---|---|---|---|---|---|---|---|---|---|---|---|---|---|---|
| | Radius[a] | | Mass | | | | | Distance from Sun[b] | | | | | | |
| Name | km | Earth units | kg | Earth units | Average Density ($g/cm^3$) | Surface Gravity (Earth = 1) | Escape Velocity (km/s) | AU | $10^6$ km | Orbital Period (years) | Orbital Inclination[c] (degrees) | Orbital Eccentricity | Rotation Period (Earth days)[d] | Axis Tilt (degrees) |
| **The Sun** | | | | | | | | | | | | | | |
| Sun | 695,000 | 109 | $1.99 \times 10^{30}$ | 333,000 | 1.41 | 27.5 | — | — | — | — | — | — | 25.4 | 7.25 |
| **Planets** | | | | | | | | | | | | | | |
| Mercury | 2440 | 0.382 | $3.30 \times 10^{23}$ | 0.055 | 5.43 | 0.38 | 4.43 | 0.387 | 57.9 | 0.2409 | 7.00 | 0.206 | 58.6 | 0.0 |
| Venus | 6051 | 0.949 | $4.87 \times 10^{24}$ | 0.815 | 5.25 | 0.91 | 10.4 | 0.723 | 108.2 | 0.6152 | 3.39 | 0.007 | −243.0 | 177.3 |
| Earth | 6378 | 1.00 | $5.97 \times 10^{24}$ | 1.00 | 5.52 | 1.00 | 11.2 | 1.00 | 149.6 | 1.0 | 0.00 | 0.017 | 0.9973 | 23.45 |
| Mars | 3397 | 0.533 | $6.42 \times 10^{23}$ | 0.107 | 3.93 | 0.38 | 5.03 | 1.524 | 227.9 | 1.881 | 1.85 | 0.093 | 1.026 | 25.2 |
| Jupiter | 71,492 | 11.19 | $1.90 \times 10^{27}$ | 317.9 | 1.33 | 2.36 | 59.5 | 5.203 | 778.3 | 11.86 | 1.31 | 0.048 | 0.41 | 3.08 |
| Saturn | 60,268 | 9.46 | $5.69 \times 10^{26}$ | 95.18 | 0.70 | 0.92 | 35.5 | 9.54 | 1427 | 29.5 | 2.48 | 0.056 | 0.44 | 26.73 |
| Uranus | 25,559 | 3.98 | $8.66 \times 10^{25}$ | 14.54 | 1.32 | 0.91 | 21.3 | 19.19 | 2870 | 84.01 | 0.77 | 0.046 | −0.72 | 97.92 |
| Neptune | 24,764 | 3.81 | $1.03 \times 10^{26}$ | 17.13 | 1.64 | 1.14 | 23.6 | 30.06 | 4497 | 164.8 | 1.77 | 0.010 | 0.67 | 29.6 |
| **Dwarf Planets (recognized as of 2018[f])** | | | | | | | | | | | | | | |
| Ceres | 473 | 0.007 | $9.4 \times 10^{20}$ | 0.00015 | 2.17 | 0.030 | 0.51 | 2.77 | 442 | 4.60 | 10.59 | 0.79 | 0.38 | 3.0 |
| Pluto | 1187 | 0.186 | $1.30 \times 10^{22}$ | 0.0022 | 1.86 | 0.057 | 1.21 | 39.48 | 5906 | 248.0 | 17.14 | 0.248 | −6.39 | 119.6 |
| Haumea | 852 × 569[g] | 0.183 | $4.01 \times 10^{21}$ | 0.00066 | 1.9–3.3 | 0.045 | 0.84 | 43.13 | 6884 | 283.3 | 28.22 | 0.195 | 0.16 | ? |
| Makemake | 715 | 0.183 | ? | ? | ? | ? | ? | 45.79 | 7408 | 300.9 | 28.96 | 0.159 | 0.32 | ? |
| Eris | 1163 | 0.183 | $1.67 \times 10^{22}$ | 0.0028 | 2.52 | 0.08 | 1.38 | 67.78 | 10,166 | 558 | 44.19 | 0.4412 | ? | ? |

[a]Measured at the equator.
[b]Semimajor axis of the orbit.
[c]With respect to the ecliptic.
[d]A negative sign indicates rotation is backward relative to other planets. The rotation rate is measured relative to the stars (the sidereal rate).
[e]With respect to its orbit.
[f]Under the IAU definition of 2006, five worlds are currently designated "dwarf planets," another six are under consideration, and dozens more may be considered.
[g]Haumea's fast rotation leads to a football-like shape.

TABLE E.2 **Satellites of the Solar System (as of 2018)**[a]

| Planet Satellite | Radius or Dimensions[b] (km) | Distance from Planet ($10^3$ km) | Orbital Period[c] (Earth days) | Mass[d] (kg) | Density[d] ($g/cm^3$) | Notes About the Satellite |
|---|---|---|---|---|---|---|
| **Earth** | | | | | | **Earth** |
| Moon | 1738 | 384.4 | 27.322 | $7.349 \times 10^{22}$ | 3.34 | *Moon:* Probably formed in giant impact. |
| **Mars** | | | | | | **Mars** |
| Phobos | 1.3 × 11 × 9 | 9.38 | 0.319 | $1.3 \times 10^{16}$ | 1.9 | *Phobos, Deimos*: May be captured asteroids. |
| Deimos | 8 × 6 × 5 | 23.5 | 1.263 | $1.8 \times 10^{15}$ | 2.2 | |
| **Jupiter** | | | | | | **Jupiter** |
| Small inner moons (4 moons) | 8–83 | 128–222 | 0.295–0.674 | — | — | *Metis, Adrastea, Amalthea, Thebe*: Small moonlets within and near Jupiter's ring system. |
| Io | 1821 | 421.6 | 1.769 | $8.933 \times 10^{22}$ | 3.57 | *Io*: Most volcanically active object in the solar system. |
| Europa | 1565 | 670.9 | 3.551 | $4.797 \times 10^{22}$ | 2.97 | *Europa*: Possible oceans under icy crust. |
| Ganymede | 2634 | 1070.0 | 7.155 | $1.482 \times 10^{23}$ | 1.94 | *Ganymede*: Largest satellite in solar system; unusual ice geology. |
| Callisto | 2403 | 1883.0 | 16.689 | $1.076 \times 10^{23}$ | 1.86 | *Callisto*: Cratered iceball. |
| Irregular group 1 (10 moons) | 4–85 | 7500–19,000 | 130–457 | — | — | *Themisto, Leda, Himalia, Lysithea, Elara, and others*: Probable captured moons with inclined orbits. |
| Irregular group 2 (61 moons) | 1–30 | 17,000–29,000 | 490–980 | — | — | *Ananke, Carme, Pasiphae, Sinope, and others*: Probable captured moons in inclined backward orbits. |
| **Saturn** | | | | | | **Saturn** |
| Small inner moons (12) | 3–89 | 117–212 | 0.5–1.2 | — | — | *Pan, Atlas, Prometheus, Pandora, Epimetheus, Janus, and others*: Small moonlets within and near Saturn's ring system. |
| Mimas | 199 | 185.52 | 0.942 | $3.70 \times 10^{19}$ | 1.17 | *Mimas, Enceladus, Tethys*: Small and medium-size iceballs, many with interesting geology. |
| Enceladus | 249 | 238.02 | 1.370 | $1.2 \times 10^{20}$ | 1.24 | |
| Tethys | 530 | 294.66 | 1.888 | $6.17 \times 10^{20}$ | 1.26 | |
| Calypso and Telesto | 8–12 | 294.66 | 1.888 | — | — | *Calypso and Telesto*: Small moonlets sharing Tethys's orbit. |
| Dione | 559 | 377.4 | 2.737 | $1.08 \times 10^{21}$ | 1.44 | *Dione*: Medium-size iceball, with interesting geology. |
| Helene and Polydeuces | 2–16 | 377.4 | 2.737 | $1.6 \times 10^{16}$ | — | *Helene and Polydeuces*: Small moonlets sharing Dione's orbit. |
| Rhea | 764 | 527.04 | 4.518 | $2.31 \times 10^{21}$ | 1.33 | *Rhea*: Medium-size iceball, with interesting geology. |
| Titan | 2575 | 1221.85 | 15.945 | $1.35 \times 10^{23}$ | 1.88 | *Titan*: Dense atmosphere shrouds surface; ongoing geological activity. |
| Hyperion | 180 × 140 × 112 | 1481.1 | 21.277 | $2.8 \times 10^{19}$ | — | *Hyperion*: Only satellite known not to rotate synchronously. |
| Iapetus | 718 | 3561.3 | 79.331 | $1.59 \times 10^{21}$ | 1.21 | *Iapetus*: Bright and dark hemispheres show greatest contrast in the solar system. |
| Phoebe | 110 | 12,952 | −550.4 | $1 \times 10^{19}$ | — | *Phoebe*: Very dark; material ejected from Phoebe may coat one side of Iapetus. |
| Irregular groups (37 moons) | 2–16 | 11,300–25,200 | 450–930 | — | — | Probable captured moons with highly inclined and/or backward orbits. |

| | | | | | | |
|---|---|---|---|---|---|---|
| **Uranus** | | | | | | **Uranus** |
| Small inner moons (13 moons) | 5–81 | 49–98 | 0.3–0.9 | — | — | *Cordelia, Ophelia, Bianca, Cressida, Desdemona, Juliet, Portia, Rosalind, Cupid, Belinda, Perdita, Puck, Mab*: Small moonlets within and near Uranus's ring system. |
| Miranda | 236 | 129.8 | 1.413 | $6.6 \times 10^{19}$ | 1.26 | *Miranda, Ariel, Umbriel, Titania, Oberon*: Small and medium-size iceballs, with some interesting geology. |
| Ariel | 579 | 191.2 | 2.520 | $1.35 \times 10^{21}$ | 1.65 | |
| Umbriel | 584.7 | 266.0 | 4.144 | $1.17 \times 10^{21}$ | 1.44 | |
| Titania | 788.9 | 435.8 | 8.706 | $3.25 \times 10^{21}$ | 1.59 | |
| Oberon | 761.4 | 582.6 | 13.463 | $3.01 \times 10^{21}$ | 1.50 | |
| Irregular group (9 moons) | 5–95 | 4280–21,000 | 260–2800 | — | — | *Francisco, Caliban, Stephano, Trinculo, Sycorax, Margaret, Prospero, Setebos, Ferdinand*: Probable captured moons; several in backward orbits. |
| **Neptune** | | | | | | **Neptune** |
| Small inner moons (5 moons) | 29–96 | 48–74 | 0.30–0.55 | — | — | *Naiad, Thalassa, Despina, Galatea, Larissa*: Small moonlets within and near Neptune's ring system. |
| Proteus | 218 × 208 × 201 | 117.6 | 1.121 | $6 \times 10^{19}$ | — | |
| Triton | 1352.6 | 354.59 | -5.875 | $2.14 \times 10^{22}$ | 2.0 | *Triton*: Probable captured Kuiper belt object—largest captured object in solar system. |
| Nereid | 170 | 5588.6 | 360.125 | $3.1 \times 10^{19}$ | — | *Nereid*: Small, icy moon; very little known. |
| Irregulars (6 moons) | 12–27 | 16,600–49,300 | 1880–9750 | — | — | *2002 N1, N2, N3, N4, 2003 N, 2004 N1*: Possible captured moons in inclined or backward orbit. |
| **Pluto** | | | | | | **Pluto** |
| Charon | 606 | 19.6[e] | 6.38 | $1.59 \times 10^{21}$ | 1.702 | *Charon*: More than half the diameter (and about 1/8 the mass) of Pluto; may have formed in giant impact. |
| Styx | 1.8–9.8 | 42.4 | 20.2 | — | — | *Styx, Nix, Kerberos, Hydra*: Small moons, possibly also formed along with Charon after giant impact on Pluto. |
| Nix | 54 × 41 × 36 | 48.7 | 24.9 | — | — | |
| Kerberos | 2.6–14 | 57.8 | 32.2 | — | — | |
| Hydra | 43 × 33[f] | 64.7 | 38.2 | — | — | |
| **Haumea** | | | | | | **Haumea** |
| Namaka | 85 | 39 | 34.7 | — | — | *Namaka, Hi'iaka*: Values are very approximate. |
| Hi'iaka | 160 | 45.5 | 149.1 | — | — | |
| **Makemake** | | | | | | **Makemake** |
| S/2015 (136472) 1 | 80 | — | — | — | — | *S/2015 (136472) 1*: Discovered in 2016, most properties not yet measured. |
| **Eris** | | | | | | **Eris** |
| Dysnomia | 50 | 37.4 | 15.8 | — | — | *Dysnomia*: Approximate properties determined in 2007. |

[a]*Note*: Authorities differ substantially on many of the values in this table.
[b]$a \times b \times c$ values for the dimensions are the approximate lengths of the axes (center to edge) for irregular moons.
[c]Negative sign indicates backward orbit relative to the planet's rotation.
[d]Masses and densities are most accurate for those satellites visited by a spacecraft on a flyby. Masses for the smallest moons have not been measured but can be estimated from the radius and an assumed density.
[e]Distance to system center of mass is $17.5 \times 10^3$ km.
[f]Third dimension not measured.

# F Stellar Data

**TABLE F.1 Stars Within 12 Light-Years**

| Star | Distance (ly) | Spectral Type | | RA h | RA m | Dec deg (°) | Dec arcmin (′) | Luminosity ($L/L_{Sun}$) |
|---|---|---|---|---|---|---|---|---|
| Sun | 0.000016 | G2 | V | — | — | — | — | 1.0 |
| Proxima Centauri | 4.2 | M5.0 | V | 14 | 30 | −62 | 41 | 0.0006 |
| α Centauri A | 4.4 | G2 | V | 14 | 40 | −60 | 50 | 1.6 |
| α Centauri B | 4.4 | K0 | V | 14 | 40 | −60 | 50 | 0.53 |
| Barnard's Star | 6.0 | M4 | V | 17 | 58 | +04 | 42 | 0.005 |
| Wolf 359 | 7.8 | M5.5 | V | 10 | 56 | +07 | 01 | 0.0008 |
| Lalande 21185 | 8.3 | M2 | V | 11 | 03 | +35 | 58 | 0.03 |
| Sirius A | 8.6 | A1 | V | 06 | 45 | −16 | 42 | 26.0 |
| Sirius B | 8.6 | DA2 | White dwarf | 06 | 45 | −16 | 42 | 0.002 |
| BL Ceti | 8.7 | M5.5 | V | 01 | 39 | −17 | 57 | 0.0009 |
| UV Ceti | 8.7 | M6 | V | 01 | 39 | −17 | 57 | 0.0006 |
| Ross 154 | 9.7 | M3.5 | V | 18 | 50 | −23 | 50 | 0.004 |
| Ross 248 | 10.3 | M5.5 | V | 23 | 42 | +44 | 11 | 0.001 |
| ϵ Eridani | 10.5 | K2 | V | 03 | 33 | −09 | 28 | 0.37 |
| Lacaille 9352 | 10.7 | M1.0 | V | 23 | 06 | −35 | 51 | 0.05 |
| Ross 128 | 10.9 | M4 | V | 11 | 48 | +00 | 49 | 0.003 |
| EZ Aquarii A | 11.3 | M5 | V | 22 | 39 | −15 | 18 | 0.0006 |
| EZ Aquarii B | 11.3 | — | — | 22 | 39 | −15 | 18 | 0.0004 |
| EZ Aquarii C | 11.3 | — | — | 22 | 39 | −15 | 18 | 0.0003 |
| 61 Cygni A | 11.4 | K5 | V | 21 | 07 | +38 | 42 | 0.17 |
| 61 Cygni B | 11.4 | K7 | V | 21 | 07 | +38 | 42 | 0.10 |
| Procyon A | 11.4 | F5 | IV–V | 07 | 39 | +05 | 14 | 8.6 |
| Procyon B | 11.4 | DA | White dwarf | 07 | 39 | +05 | 14 | 0.0005 |
| Gliese 725 A | 11.5 | M3 | V | 18 | 43 | +59 | 38 | 0.02 |
| Gliese 725 B | 11.5 | M3.5 | V | 18 | 43 | +59 | 38 | 0.01 |
| GX Andromedae | 11.6 | M1.5 | V | 00 | 18 | +44 | 01 | 0.03 |
| GQ Andromedae | 11.6 | M3.5 | V | 00 | 18 | +44 | 01 | 0.003 |
| ϵ Indi A | 11.8 | K5 | V | 22 | 03 | −56 | 45 | 0.30 |
| ϵ Indi B | 11.8 | T1.0 | Brown dwarf | 22 | 04 | −56 | 46 | — |
| ϵ Indi C | 11.8 | T6.0 | Brown dwarf | 22 | 04 | −56 | 46 | — |
| DX Cancri | 11.8 | M6.0 | V | 08 | 30 | +26 | 47 | 0.0003 |
| τ Ceti | 11.9 | G8.5 | V | 01 | 44 | −15 | 57 | 0.67 |
| GJ 1061 | 12.0 | M5.0 | V | 03 | 36 | −44 | 31 | 0.001 |

*Note:* These data were provided by the RECONS project, courtesy of Dr. Todd Henry (January, 2010). The luminosities are all total (bolometric) luminosities. The DA stellar types are white dwarfs. The coordinates are for the year 2000. The bolometric luminosity of the brown dwarfs is primarily in the infrared and has not been measured accurately yet.

**TABLE F.2 Twenty Brightest Stars**

| | | RA | | Dec | | | | | |
|---|---|---|---|---|---|---|---|---|---|
| **Star** | **Constellation** | **h** | **m** | **deg (°)** | **arcmin (′)** | **Distance (ly)** | **Spectral Type** | **Apparent Magnitude** | **Luminosity ($L/L_{Sun}$)** |
| Sirius | Canis Major | 6 | 45 | −16 | 42 | 8.6 | A1 V | −1.46 | 26 |
| Canopus | Carina | 6 | 24 | −52 | 41 | 313 | F0 Ib–II | −0.72 | 13,000 |
| $\alpha$ Centauri | Centaurus | 14 | 40 | −60 | 50 | 4.4 | G2 V | −0.01 | 1.6 |
| | | | | | | | K0 V | 1.3 | 0.53 |
| Arcturus | Boötes | 14 | 16 | +19 | 11 | 37 | K2 III | −0.06 | 170 |
| Vega | Lyra | 18 | 37 | +38 | 47 | 25 | A0 V | 0.04 | 60 |
| Capella | Auriga | 5 | 17 | +46 | 00 | 42 | G0 III | 0.75 | 70 |
| | | | | | | | G8 III | 0.85 | 77 |
| Rigel | Orion | 5 | 15 | −08 | 12 | 772 | B8 Ia | 0.14 | 70,000 |
| Procyon | Canis Minor | 7 | 39 | +05 | 14 | 11.4 | F5 IV–V | 0.37 | 7.4 |
| Betelgeuse | Orion | 5 | 55 | +07 | 24 | 643 | M2 Iab | 0.41 | 120,000 |
| Achernar | Eridanus | 1 | 38 | −57 | 15 | 144 | B5 V | 0.51 | 3600 |
| Hadar | Centaurus | 14 | 04 | −60 | 22 | 525 | B1 III | 0.63 | 100,000 |
| Altair | Aquila | 19 | 51 | +08 | 52 | 17 | A7 IV–V | 0.77 | 10.5 |
| Acrux | Crux | 12 | 27 | −63 | 06 | 321 | B1 IV | 1.39 | 22,000 |
| | | | | | | | B3 V | 1.9 | 7500 |
| Aldebaran | Taurus | 4 | 36 | +16 | 30 | 65 | K5 III | 0.86 | 350 |
| Spica | Virgo | 13 | 25 | −11 | 09 | 260 | B1 V | 0.91 | 23,000 |
| Antares | Scorpio | 16 | 29 | −26 | 26 | 604 | M1 Ib | 0.92 | 38,000 |
| Pollux | Gemini | 7 | 45 | +28 | 01 | 34 | K0 III | 1.16 | 45 |
| Fomalhaut | Piscis Austrinus | 22 | 58 | −29 | 37 | 25 | A3 V | 1.19 | 18 |
| Deneb | Cygnus | 20 | 41 | +45 | 16 | 2500 | A2 Ia | 1.26 | 170,000 |
| $\beta$ Crucis | Crux | 12 | 48 | −59 | 40 | 352 | B0.5 IV | 1.28 | 37,000 |

*Note:* Three of the stars on this list, Capella, $\alpha$ Centauri, and Acrux, are binary systems with members of comparable brightness. They are counted as single stars because that is how they appear to the naked eye. All the luminosities given are total (bolometric) luminosities. The coordinates are for the year 2000.

# G Galaxy Data

**TABLE G.1 Galaxies of the Local Group**

| Galaxy Name | Distance (millions of ly) | Type[a] | RA h | RA m | Dec deg (°) | Dec arcmin (′) | Luminosity (millions of $L_{Sun}$) |
|---|---|---|---|---|---|---|---|
| Milky Way | — | Sbc | — | — | — | — | 15,000 |
| WLM | 3.0 | Irr | 00 | 02 | −15 | 30 | 50 |
| IC 10 | 2.7 | dIrr | 00 | 20 | +59 | 18 | 160 |
| Cetus | 2.5 | dE | 00 | 26 | −11 | 02 | 0.72 |
| NGC 147 | 2.4 | dE | 00 | 33 | +48 | 30 | 131 |
| And III | 2.5 | dE | 00 | 35 | +36 | 30 | 1.1 |
| NGC 185 | 2.0 | dE | 00 | 39 | +48 | 20 | 120 |
| NGC 205 | 2.7 | E | 00 | 40 | +41 | 41 | 370 |
| And VIII | 2.7 | dE | 00 | 42 | +40 | 37 | 240 |
| M32 | 2.6 | E | 00 | 43 | +40 | 52 | 380 |
| M31 | 2.5 | Sb | 00 | 43 | +41 | 16 | 21,000 |
| And I | 2.6 | dE | 00 | 46 | +38 | 00 | 4.7 |
| SMC | 0.19 | Irr | 00 | 53 | −72 | 50 | 230 |
| And IX | 2.9 | dE | 00 | 52 | +43 | 12 | — |
| Sculptor | 0.26 | dE | 01 | 00 | −33 | 42 | 2.2 |
| LGS 3 | 2.6 | dIrr | 01 | 04 | +21 | 53 | 1.3 |
| IC 1613 | 2.3 | Irr | 01 | 05 | +02 | 08 | 64 |
| And V | 2.9 | dE | 01 | 10 | +47 | 38 | — |
| And II | 1.7 | dE | 01 | 16 | +33 | 26 | 2.4 |
| M33 | 2.7 | Sc | 01 | 34 | +30 | 40 | 2800 |
| Phoenix | 1.5 | dIrr | 01 | 51 | +44 | 27 | 0.9 |
| Fornax | 0.45 | dE | 02 | 40 | −34 | 27 | 15.5 |
| EGB0427 + 63 | 4.3 | dIrr | 04 | 32 | +63 | 36 | 9.1 |
| LMC | 0.16 | Irr | 05 | 24 | −69 | 45 | 1300 |
| Carina | 0.33 | dE | 06 | 42 | −50 | 58 | 0.4 |
| Canis Major | 0.025 | dIrr | 07 | 15 | −28 | 00 | — |
| Leo A | 2.2 | dIrr | 09 | 59 | +30 | 45 | 3.0 |
| Sextans B | 4.4 | dIrr | 10 | 00 | +05 | 20 | 41 |
| NGC 3109 | 4.1 | Irr | 10 | 03 | −26 | 09 | 160 |
| Antlia | 4.0 | dIrr | 10 | 04 | −27 | 19 | 1.7 |
| Leo I | 0.82 | dE | 10 | 08 | +12 | 18 | 4.8 |
| Sextans A | 4.7 | dIrr | 10 | 11 | −04 | 42 | 56 |
| Sextans | 0.28 | dE | 10 | 13 | −01 | 37 | 0.5 |
| Leo II | 0.67 | dE | 11 | 13 | +22 | 09 | 0.6 |
| GR 8 | 5.2 | dIrr | 12 | 59 | +14 | 13 | 3.4 |
| Ursa Minor | 0.22 | dE | 15 | 09 | +67 | 13 | 0.3 |
| Draco | 2.7 | dE | 17 | 20 | +57 | 55 | 0.3 |
| Sagittarius | 0.08 | dE | 18 | 55 | −30 | 29 | 18 |
| SagDIG | 3.5 | dIrr | 19 | 30 | −17 | 41 | 6.8 |
| NGC 6822 | 1.6 | Irr | 19 | 45 | −14 | 48 | 94 |
| DDO 210 | 2.6 | dIrr | 20 | 47 | −12 | 51 | 0.8 |
| IC 5152 | 5.2 | dIrr | 22 | 03 | −51 | 18 | 70 |
| Tucana | 2.9 | dE | 22 | 42 | −64 | 25 | 0.5 |
| UKS2323-326 | 4.3 | dE | 23 | 26 | −32 | 23 | 5.2 |
| And VII | 2.6 | dE | 23 | 38 | +50 | 35 | — |
| Pegasus | 3.1 | dIrr | 23 | 29 | +14 | 45 | 12 |
| And VI | 2.8 | dE | 23 | 52 | +24 | 36 | — |

[a]Types beginning with S are spiral galaxies classified according to Hubble's system (see Chapter 21). Type E galaxies are elliptical or spheroidal. Type Irr galaxies are irregular. The prefix d denotes a dwarf galaxy. This list is based on a list originally published by M. Mateo in 1998 and augmented by discoveries of Local Group galaxies made between 1998 and 2005.

**TABLE G.2 Nearby Galaxies in the Messier Catalog[a, b]**

| Galaxy Name | RA | | Dec | | | | | |
|---|---|---|---|---|---|---|---|---|
| (M / NGC)[c] | h | m | deg (°) | arcmin (′) | $RV_{hel}$[d] | $RV_{gal}$[e] | Type[f] | Nickname |
| M31 / NGC 224 | 00 | 43 | +41 | 16 | −300 ± 4 | −122 | Spiral | Andromeda |
| M32 / NGC 221 | 00 | 43 | +40 | 52 | −145 ± 2 | 32 | Elliptical | |
| M33 / NGC 598 | 01 | 34 | +30 | 40 | −179 ± 3 | −44 | Spiral | Triangulum |
| M49 / NGC 4472 | 12 | 30 | +08 | 00 | 997 ± 7 | 929 | Elliptical/Lenticular/Seyfert | |
| M51 / NGC 5194 | 13 | 30 | +47 | 12 | 463 ± 3 | 550 | Spiral/Interacting | Whirlpool |
| M58 / NGC 4579 | 12 | 38 | +11 | 49 | 1519 ± 6 | 1468 | Spiral/Seyfert | |
| M59 / NGC 4621 | 12 | 42 | +11 | 39 | 410 ± 6 | 361 | Elliptical | |
| M60 / NGC 4649 | 12 | 44 | +11 | 33 | 1117 ± 6 | 1068 | Elliptical | |
| M61 / NGC 4303 | 12 | 22 | +04 | 28 | 1566 ± 2 | 1483 | Spiral/Seyfert | |
| M63 / NGC 5055 | 13 | 16 | +42 | 02 | 504 ± 4 | 570 | Spiral | Sunflower |
| M64 / NGC 4826 | 12 | 57 | +21 | 41 | 408 ± 4 | 400 | Spiral/Seyfert | Black Eye |
| M65 / NGC 3623 | 11 | 19 | +13 | 06 | 807 ± 3 | 723 | Spiral | |
| M66 / NGC 3627 | 11 | 20 | +12 | 59 | 727 ± 3 | 643 | Spiral/Seyfert | |
| M74 / NGC 628 | 01 | 37 | +15 | 47 | 657 ± 1 | 754 | Spiral | |
| M77 / NGC 1068 | 02 | 43 | −00 | 01 | 1137 ± 3 | 1146 | Spiral/Seyfert | |
| M81 / NGC 3031 | 09 | 56 | +69 | 04 | −34 ± 4 | 73 | Spiral/Seyfert | |
| M82 / NGC 3034 | 09 | 56 | +69 | 41 | 203 ± 4 | 312 | Irregular/Starburst | |
| M83 / NGC 5236 | 13 | 37 | −29 | 52 | 516 ± 4 | 385 | Spiral/Starburst | |
| M84 / NGC 4374 | 12 | 25 | +12 | 53 | 1060 ± 6 | 1005 | Elliptical | |
| M85 / NGC 4382 | 12 | 25 | +18 | 11 | 729 ± 2 | 692 | Spiral | |
| M86 / NGC 4406 | 12 | 26 | +12 | 57 | −244 ± 5 | −298 | Elliptical/Lenticular | |
| M87 / NGC 4486 | 12 | 30 | +12 | 23 | 1307 ± 7 | 1254 | Elliptical/Central Dominant/Seyfert | Virgo A |
| M88 / NGC 4501 | 12 | 32 | +14 | 25 | 2281 ± 3 | 2235 | Spiral/Seyfert | |
| M89 / NGC 4552 | 12 | 36 | +12 | 33 | 340 ± 4 | 290 | Elliptical | |
| M90 / NGC 4569 | 12 | 37 | +13 | 10 | −235 ± 4 | −282 | Spiral/Seyfert | |
| M91 / NGC 4548 | 12 | 35 | +14 | 30 | 486 ± 4 | 442 | Spiral/Seyfert | |
| M94 / NGC 4736 | 12 | 51 | +41 | 07 | 308 ± 1 | 360 | Spiral | |
| M95 / NGC 3351 | 10 | 44 | +11 | 42 | 778 ± 4 | 677 | Spiral/Starburst | |
| M96 / NGC 3368 | 10 | 47 | +11 | 49 | 897 ± 4 | 797 | Spiral/Seyfert | |
| M98 / NGC 4192 | 12 | 14 | +14 | 54 | −142 ± 4 | −195 | Spiral/Seyfert | |
| M99 / NGC 4254 | 12 | 19 | +14 | 25 | 2407 ± 3 | 2354 | Spiral | |
| M100 / NGC 4321 | 12 | 23 | +15 | 49 | 1571 ± 1 | 1525 | Spiral | |
| M101 / NGC 5457 | 14 | 03 | +54 | 21 | 241 ± 2 | 360 | Spiral | |
| M104 / NGC 4594 | 12 | 40 | −11 | 37 | 1024 ± 5 | 904 | Spiral/Seyfert | Sombrero |
| M105 / NGC 3379 | 10 | 48 | +12 | 35 | 911 ± 2 | 814 | Elliptical | |
| M106 / NGC 4258 | 12 | 19 | +47 | 18 | 448 ± 3 | 507 | Spiral/Seyfert | |
| M108 / NGC 3556 | 11 | 09 | +55 | 57 | 695 ± 3 | 765 | Spiral | |
| M109 / NGC 3992 | 11 | 55 | +53 | 39 | 1048 ± 4 | 1121 | Spiral | |
| M110 / NGC 205 | 00 | 40 | +41 | 41 | −241 ± 3 | −61 | Elliptical | |

[a]Galaxies identified in the catalog published by Charles Messier in 1781; these galaxies are relatively easy to observe with small telescopes.

[b]Data obtained from NED: NASA/IPAC Extragalactic Database (http://ned.ipac.caltech.edu). The original Messier list of galaxies was obtained from SED, and the list data were updated to 2001 and M102 was dropped.

[c]The galaxies are identified by their Messier number (M followed by a number) and NGC number, which comes from the *New General Catalog* published in 1888.

[d]Radial velocity in kilometers per second, with respect to the Sun (heliocentric). Positive values mean motion away from the Sun; negative values are toward the Sun.

[e]Radial velocity in kilometers per second, with respect to the Milky Way Galaxy, calculated from the $RV_{hel}$ values with a correction for the Sun's motion around the galactic center.

[f]Galaxies are first listed by their primary type (spiral, elliptical, or irregular) and then by any other special categories that apply (see Chapter 21).

**TABLE G.3 Nearby, X-Ray Bright Clusters of Galaxies**

| Cluster Name | Redshift | Distance[a] (billions of ly) | Temperature of Intracluster Medium (millions of K) | Average Orbital Velocity of Galaxies[b] (km/s) | Cluster Mass[c] ($10^{15}M_{Sun}$) |
|---|---|---|---|---|---|
| Abell 2142 | 0.0907 | 1.26 | 101. ± 2 | 1132 ± 110 | 1.6 |
| Abell 2029 | 0.0766 | 1.07 | 100. ± 3 | 1164 ± 98 | 1.5 |
| Abell 401 | 0.0737 | 1.03 | 95.2 ± 5 | 1152 ± 86 | 1.4 |
| Coma | 0.0233 | 0.32 | 95.1 ± 1 | 821 ± 49 | 1.4 |
| Abell 754 | 0.0539 | 0.75 | 93.3 ± 3 | 662 ± 77 | 1.4 |
| Abell 2256 | 0.0589 | 0.82 | 87.0 ± 2 | 1348 ± 86 | 1.4 |
| Abell 399 | 0.0718 | 1.00 | 81.7 ± 7 | 1116 ± 89 | 1.1 |
| Abell 3571 | 0.0395 | 0.55 | 81.1 ± 3 | 1045 ± 109 | 1.1 |
| Abell 478 | 0.0882 | 1.23 | 78.9 ± 2 | 904 ± 281 | 1.1 |
| Abell 3667 | 0.0566 | 0.79 | 78.5 ± 6 | 971 ± 62 | 1.1 |
| Abell 3266 | 0.0599 | 0.84 | 78.2 ± 5 | 1107 ± 82 | 1.1 |
| Abell 1651a | 0.0846 | 1.18 | 73.1 ± 6 | 685 ± 129 | 0.96 |
| Abell 85 | 0.0560 | 0.78 | 70.9 ± 2 | 969 ± 95 | 0.92 |
| Abell 119 | 0.0438 | 0.61 | 65.6 ± 5 | 679 ± 106 | 0.81 |
| Abell 3558 | 0.0480 | 0.67 | 65.3 ± 2 | 977 ± 39 | 0.81 |
| Abell 1795 | 0.0632 | 0.88 | 62.9 ± 2 | 834 ± 85 | 0.77 |
| Abell 2199 | 0.0314 | 0.44 | 52.7 ± 1 | 801 ± 92 | 0.59 |
| Abell 2147 | 0.0353 | 0.49 | 51.1 ± 4 | 821 ± 68 | 0.56 |
| Abell 3562 | 0.0478 | 0.67 | 45.7 ± 8 | 736 ± 49 | 0.48 |
| Abell 496 | 0.0325 | 0.45 | 45.3 ± 1 | 687 ± 89 | 0.47 |
| Centaurus | 0.0103 | 0.14 | 42.2 ± 1 | 863 ± 34 | 0.42 |
| Abell 1367 | 0.0213 | 0.30 | 41.3 ± 2 | 822 ± 69 | 0.41 |
| Hydra | 0.0126 | 0.18 | 38.0 ± 1 | 610 ± 52 | 0.36 |
| C0336 | 0.0349 | 0.49 | 37.4 ± 1 | 650 ± 170 | 0.35 |
| Virgo | 0.0038 | 0.05 | 25.7 ± 0.5 | 632 ± 41 | 0.20 |

*Note:* This table lists the 25 brightest clusters of galaxies in the x-ray sky from a catalog by J. P. Henry (2000).

[a]Cluster distances were computed using a value for Hubble's constant of 21.5 km/s/million light-years.

[b]The average orbital velocities given in this column are the velocity component along our line of sight. This velocity should be multiplied by the square root of 2 to get the average orbital velocity.

[c]This column gives each cluster's mass within the largest radius at which the intracluster gas can be in gravitational equilibrium. Because our estimates of that radius depend on Hubble's constant, these masses are inversely proportional to Hubble's constant, which we have assumed to be 21.5 km/s/million light-years.

# H The 88 Constellations

## Constellation Locations

These two charts each show half of the celestial sphere in projection, so that you can use them to learn the approximate locations of the constellations. The grid lines are marked by the right ascension and declination.

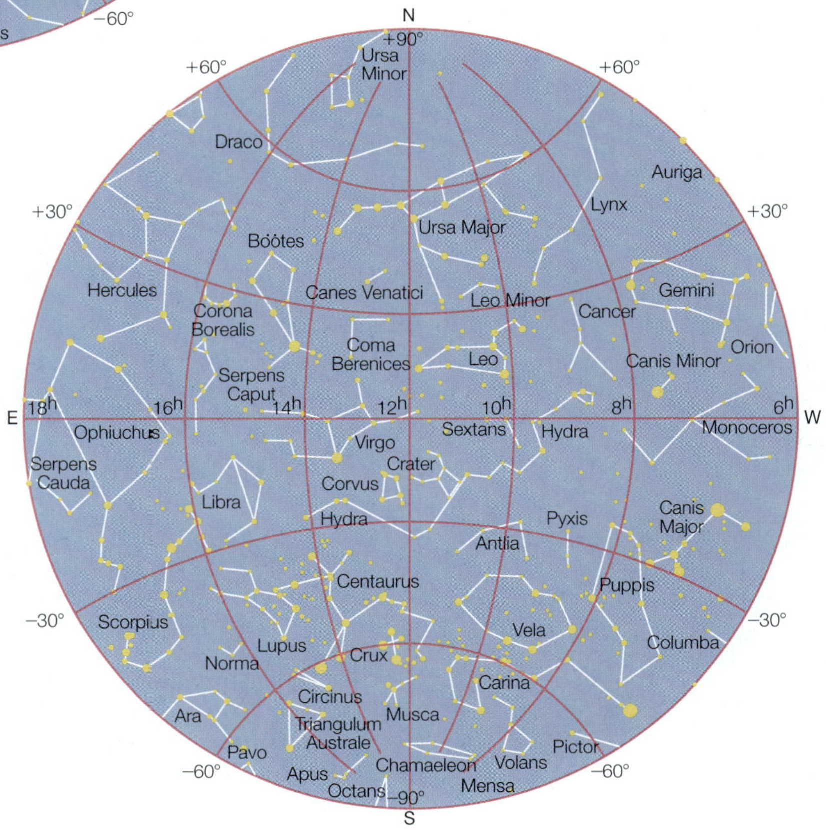

## Constellation Names (English Equivalent in Parentheses)

Andromeda (The Chained Princess)
Antlia (The Air Pump)
Apus (The Bird of Paradise)
Aquarius (The Water Bearer)
Aquila (The Eagle)
Ara (The Altar)
Aries (The Ram)
Auriga (The Charioteer)
Boötes (The Herdsman)
Caelum (The Chisel)
Camelopardalis (The Giraffe)
Cancer (The Crab)
Canes Venatici (The Hunting Dogs)
Canis Major (The Great Dog)
Canis Minor (The Little Dog)
Capricornus (The Sea Goat)
Carina (The Keel)
Cassiopeia (The Queen)
Centaurus (The Centaur)
Cepheus (The King)
Cetus (The Whale)
Chamaeleon (The Chameleon)
Circinus (The Drawing Compass)
Columba (The Dove)
Coma Berenices (Berenice's Hair)
Corona Australis (The Southern Crown)
Corona Borealis (The Northern Crown)
Corvus (The Crow)
Crater (The Cup)
Crux (The Southern Cross)
Cygnus (The Swan)
Delphinus (The Dolphin)
Dorado (The Goldfish)
Draco (The Dragon)
Equuleus (The Little Horse)
Eridanus (The River)
Fornax (The Furnace)
Gemini (The Twins)
Grus (The Crane)
Hercules
Horologium (The Clock)
Hydra (The Sea Serpent)
Hydrus (The Water Snake)
Indus (The Indian)
Lacerta (The Lizard)
Leo (The Lion)
Leo Minor (The Little Lion)
Lepus (The Hare)
Libra (The Scales)
Lupus (The Wolf)
Lynx (The Lynx)
Lyra (The Lyre)
Mensa (The Table)
Microscopium (The Microscope)
Monoceros (The Unicorn)
Musca (The Fly)
Norma (The Level)
Octans (The Octant)
Ophiuchus (The Serpent Bearer)
Orion (The Hunter)
Pavo (The Peacock)
Pegasus (The Winged Horse)
Perseus (The Hero)
Phoenix (The Phoenix)
Pictor (The Painter's Easel)
Pisces (The Fish)
Piscis Austrinus (The Southern Fish)
Puppis (The Stern)
Pyxis (The Compass)
Reticulum (The Reticle)
Sagitta (The Arrow)
Sagittarius (The Archer)
Scorpius (The Scorpion)
Sculptor (The Sculptor)
Scutum (The Shield)
Serpens (The Serpent)
Sextans (The Sextant)
Taurus (The Bull)
Telescopium (The Telescope)
Triangulum (The Triangle)
Triangulum Australe (The Southern Triangle)
Tucana (The Toucan)
Ursa Major (The Great Bear)
Ursa Minor (The Little Bear)
Vela (The Sail)
Virgo (The Virgin)
Volans (The Flying Fish)
Vulpecula (The Fox)

## All-Sky Constellation Map

This map of the entire sky shows the locations of all the constellations, in much the same way that a world map shows all of the countries on Earth. It does not use the usual celestial coordinate system of right ascension and declination, but instead is oriented so that the Milky Way Galaxy's center is at the center of the map and the Milky Way's disk (shown in shades of lighter blue) stretches from left to right across the map.

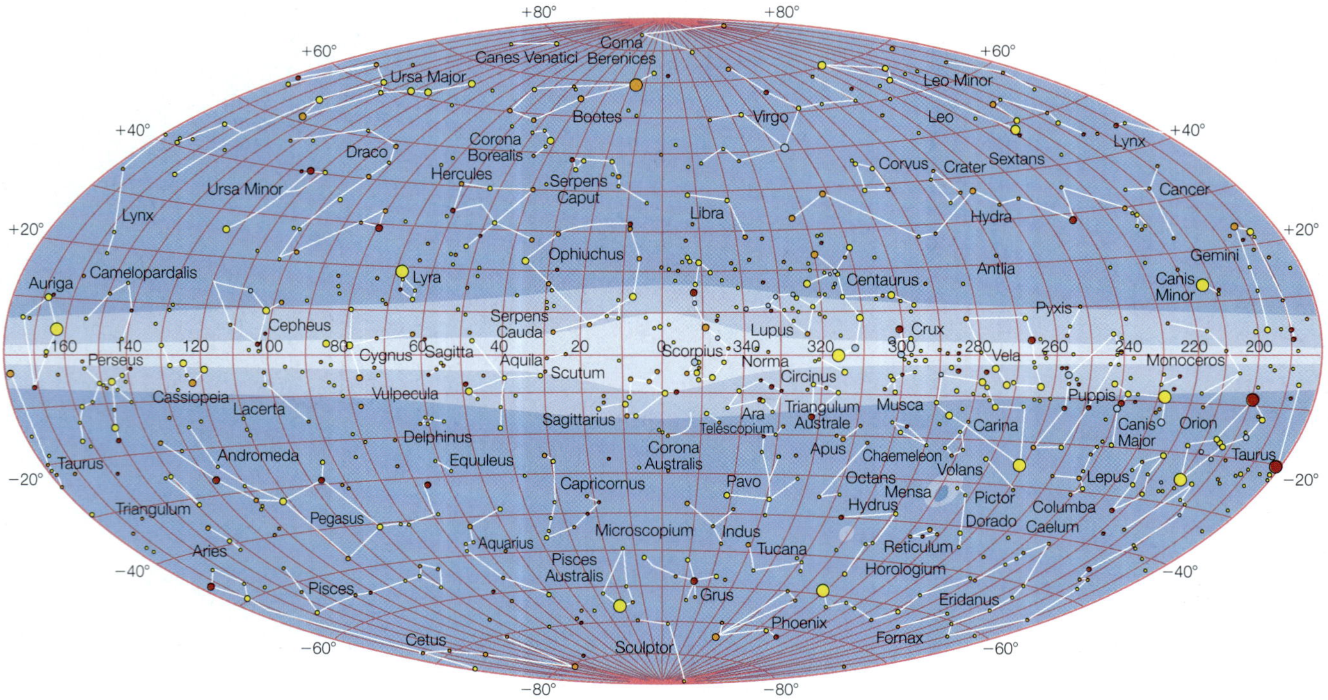

# I Star Charts

## How to use the star charts:

Check the times and dates under each chart to find the best one for you. Take it outdoors within an hour or so of the time listed for your date. Bring a dim flashlight to help you read it.

On each chart, the round outside edge represents the horizon all around you. Compass directions around the horizon are marked in yellow. Turn the chart around so that the edge marked with the direction you're facing (for example, north, southeast) is down. The stars above this horizon now match the stars you are facing. Ignore the rest until you turn to look in a different direction.

The center of the chart represents the sky overhead, so a star plotted on the chart halfway from the edge to the center can be found in the sky halfway from the horizon to straight up.

The charts are drawn for 40°N latitude (for example, Denver, New York, Madrid). If you live far south of there, stars in the southern part of your sky will appear higher than on the chart and stars in the north will be lower. If you live far north of there, the reverse is true.

*See pages A-22 through A-25 for full-size star charts.*

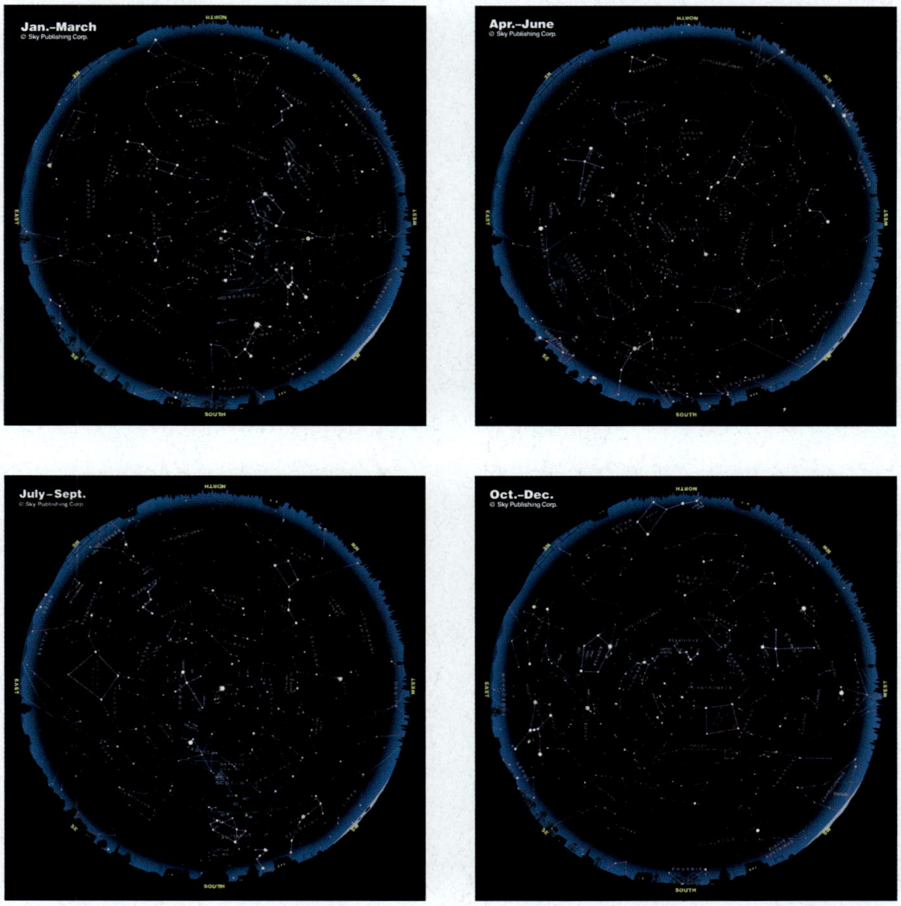


Star charts © 1999 *Sky & Telescope*

© 1999 *Sky & Telescope*

## Use this chart January, February, and March.

| | | |
|---|---|---|
| Early January—1 A.M. | Early February—11 P.M. | Early March—9 P.M. |
| Late January—Midnight | Late February—10 P.M. | Late March—Dusk |

## Use this chart April, May, and June.

| | | |
|---|---|---|
| Early April—3 A.M.* | Early May—1 A.M.* | Early June—11 P.M.* |
| Late April—2 A.M.* | Late May—Midnight* | Late June—Dusk |

*Daylight Saving Time

## Use this chart July, August, and September.

| | | |
|---|---|---|
| Early July—1 A.M.* | Early August—11 P.M.* | Early September—9 P.M.* |
| Late July—Midnight* | Late August—10 P.M.* | Late September—Dusk |

*Daylight Saving Time

## Use this chart October, November, and December.

| | | |
|---|---|---|
| Early October—1 A.M.* | Early November—10 P.M. | Early December—8 P.M. |
| Late October—Midnight* | Late November—9 P.M. | Late December—7 P.M. |

*Daylight Saving Time

# J Key to Icons on Figures

*You'll see the following icons on figures throughout the book. They are used to indicate the wavelength of light shown in each image, and to identify photo-realistic artworks and images made by computer simulations.*

Indicates an artist's representation

Indicates a graphic generated using computer simulations

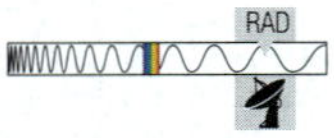

Indicates an image based on data from an observatory on Earth observing radio waves

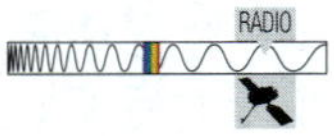

Indicates an image based on data from a spacecraft observing radio waves

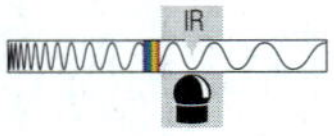

Indicates an image based on data from an observatory on Earth observing infrared light

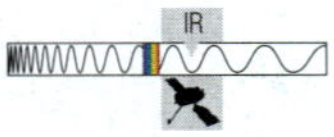

Indicates an image based on data from a spacecraft observing infrared light

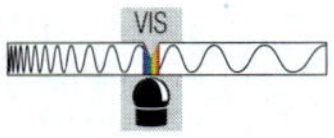

Indicates an image based on data from an observatory on Earth observing visible light

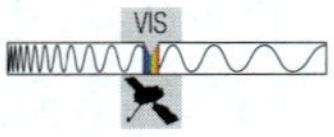

Indicates an image based on data from a spacecraft observing visible light

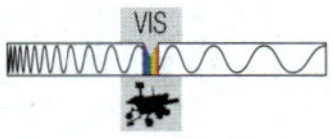

Indicates an image based on data from a rover observing visible light

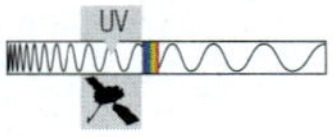

Indicates an image based on data from a spacecraft observing ultraviolet light

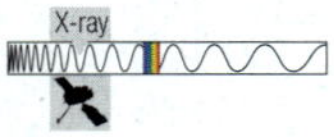

Indicates an image based on data from a spacecraft observing x-rays

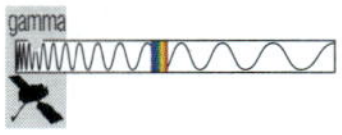

Indicates an image based on data from a spacecraft observing gamma rays

# Glossary

**absolute magnitude** A measure of an object's luminosity; defined to be the apparent magnitude the object would have if it were located exactly 10 parsecs away.

**absolute zero** The coldest possible temperature, which is 0 K = −273.15°C.

**absorption** (of light) The process by which matter absorbs radiative energy.

**absorption line** A dark band (a "line") on an otherwise bright rainbow of light, occurring when light viewed through a diffraction element such as a prism shows a deficit of photons at or near a specific wavelength.

**accelerating universe** A universe in which a repulsive force (*see* cosmological constant) causes the expansion of the universe to accelerate with time. Its galaxies will recede from one another increasingly faster, and it will become cold and dark more quickly than a coasting universe.

**acceleration** The rate at which an object's velocity changes. Its standard units are $m/s^2$.

**acceleration of gravity** The acceleration of a falling object. On Earth, the acceleration of gravity, designated by $g$, is $9.8\ m/s^2$.

**accretion** The process by which small objects gather together to make larger objects.

**accretion disk** A rapidly rotating disk of material that gradually falls inward as it orbits a starlike object (e.g., white dwarf, neutron star, or black hole).

**active galactic nuclei** The unusually luminous centers of some galaxies, thought to be powered by accretion onto supermassive black holes. Quasars are the brightest type of active galactic nuclei; radio galaxies also contain active galactic nuclei.

**active galaxy** A term sometimes used to describe a galaxy that contains an *active galactic nucleus*.

**adaptive optics** A technique in which telescope mirrors flex rapidly to compensate for the bending of starlight caused by atmospheric turbulence.

**Algol paradox** A paradox concerning the binary star Algol, which contains a subgiant star that is less massive than its main-sequence companion.

**altitude** (above horizon) The angular distance between the horizon and an object in the sky.

**amino acids** The building blocks of proteins.

**analemma** The figure-8 path traced by the Sun over the course of a year when viewed at the same place and the same time each day; it represents the discrepancies between apparent and mean solar time.

**Andromeda Galaxy** M31, the Great Galaxy in Andromeda; the nearest large spiral galaxy to the Milky Way.

**angular momentum** Momentum attributable to rotation or revolution. The angular momentum of an object moving in a circle of radius $r$ is the product $m \times v \times r$.

**angular resolution** (of a telescope) The smallest angular separation that two pointlike objects can have and still be seen as distinct points of light (rather than as a single point of light).

**angular size** (or **angular distance**) A measure of the angle formed by extending imaginary lines outward from our eyes to span an object (or the space between two objects).

**annihilation** *See* matter–antimatter annihilation.

**annular solar eclipse** A solar eclipse during which the Moon is directly in front of the Sun but its angular size is not large enough to fully block the Sun; thus, a ring (or *annulus*) of sunlight is still visible around the Moon's disk.

**Antarctic Circle** The circle on Earth with latitude 66.5°S.

**antielectron** The antimatter equivalent of an electron. It is identical to an electron in virtually all respects, except it has a positive rather than a negative electrical charge.

**antimatter** Any particle with the same mass as a particle of ordinary matter but whose other basic properties, such as electrical charge, are precisely opposite.

**aphelion** The point at which an object orbiting the Sun is farthest from the Sun.

**apogee** The point at which an object orbiting Earth is farthest from Earth.

**apparent brightness** The amount of light reaching us *per unit area* from a luminous object; often measured in units of $watts/m^2$.

**apparent magnitude** A measure of the apparent brightness of an object in the sky, based on the ancient system developed by Hipparchus.

**apparent retrograde motion** The apparent motion of a planet, as viewed from Earth, during the period of a few weeks or months when it moves westward relative to the stars in our sky.

**apparent solar time** Time measured by the actual position of the Sun in the local sky, defined so that noon is when the Sun is *on* the meridian.

**arcminute** (or **minute of arc**) 1/60 of 1°.

**arcsecond** (or **second of arc**) 1/60 of an arcminute, or 1/3600 of 1°.

**Arctic Circle** The circle on Earth with latitude 66.5°N.

**asteroid** A relatively small and rocky object that orbits a star; asteroids are officially considered part of a category known as "small solar system bodies."

**asteroid belt** The region of our solar system between the orbits of Mars and Jupiter in which asteroids are heavily concentrated.

**astrobiology** The study of life on Earth and beyond; it emphasizes research into questions of the origin of life, the conditions under which life can survive, and the search for life beyond Earth.

**astrometric method** The detection of extrasolar planets through the side-to-side motion of a star caused by gravitational tugs from the planet.

**astronomical unit (AU)** The average distance (semimajor axis) of Earth from the Sun, which is about 150 million km.

**atmosphere** A layer of gas that surrounds a planet or moon, usually very thin compared to the size of the object.

**atmospheric pressure** The surface pressure resulting from the overlying weight of an atmosphere.

**atmospheric structure** The layering of a planetary atmosphere due to variations in temperature with altitude. For example, Earth's atmospheric structure from the ground up consists of the troposphere, stratosphere, thermosphere, and exosphere.

**atomic hydrogen gas** Gas composed mostly of hydrogen atoms, though in space it is generally mixed with helium and small amounts of other elements as well; it is the most common form of interstellar gas.

**atomic mass number** The combined number of protons and neutrons in an atom.

**atomic number** The number of protons in an atom.

**atoms** The basic units of the chemical elements, consisting of a nucleus made from protons and neutrons, surrounded by a cloud of electrons.

**aurora** Dancing lights in the sky caused by charged particles entering our atmosphere; called the *aurora borealis* in the Northern Hemisphere and the *aurora australis* in the Southern Hemisphere.

**axis tilt** (of a planet in our solar system) The amount by which a planet's axis is tilted with respect to a line perpendicular to the ecliptic plane.

**azimuth** (usually called **direction** in this book) Direction around the horizon from due north, measured clockwise in degrees. For example, the azimuth of due north is 0°, due east is 90°, due south is 180°, and due west is 270°.

**bar** The standard unit of pressure, approximately equal to Earth's atmospheric pressure at sea level.

**barred spiral galaxies** Spiral galaxies that have a straight bar of stars cutting across their centers.

**baryonic matter** Ordinary matter made from atoms (so called because the nuclei of atoms contain protons and neutrons, which are both baryons).

**baryons** Particles, including protons and neutrons, that are made from three quarks.

**basalt** A type of dark, high-density volcanic rock that is rich in iron and magnesium-based silicate minerals; it forms a runny (easily flowing) lava when molten.

**belts** (on a jovian planet) Dark bands of sinking air that encircle a jovian planet at a particular set of latitudes.

**Big Bang** The name given to the event thought to mark the birth of the universe.

**Big Bang theory** The scientific theory of the universe's earliest moments, stating that all the matter in our observable universe came into being at a single moment in time as an extremely hot, dense mixture of subatomic particles and radiation.

**Big Crunch** The name given to the event that would presumably end the universe if gravity ever reverses the universal expansion and the universe someday begins to collapse.

**binary star system** A star system that contains two stars.

**biosphere** The "layer" of life on Earth.

**blackbody radiation** *See* thermal radiation.

**black hole** A bottomless pit in spacetime. Nothing can escape from within a black hole, and we can never again detect or observe an object that falls into a black hole.

**black hole merger** An event in which two black holes merge to form a single more massive black hole, accompanied by the release of gravitational waves.

**blazer** A term sometimes used to describe an active galactic nucleus in which a powerful jet of material happens to be aimed nearly in the direction of Earth.

**BL Lac objects** A class of active galactic nuclei that probably represent the centers of radio galaxies whose jets happen to be pointed directly at us.

**blowout** Ejection of the hot, gaseous contents of a superbubble when it grows so large that it bursts out of the cooler layer of gas filling the galaxy's disk.

**blueshift** A Doppler shift in which spectral features are shifted to shorter wavelengths, observed when an object is moving toward the observer.

**bosons** Particles, such as photons, to which the exclusion principle does not apply.

**bound orbits** Orbits on which an object travels repeatedly around another object; bound orbits are elliptical in shape.

**brown dwarf** An object too small to become an ordinary star because electron degeneracy pressure halts its gravitational collapse before fusion becomes self-sustaining; brown dwarfs have masses less than $0.08M_{Sun}$.

**bubble** (interstellar) An expanding shell of hot, ionized gas driven by stellar winds or supernovae, with very hot and very low density gas inside.

**bulge** (of a spiral galaxy) The central portion of a spiral galaxy that is roughly spherical (or football shaped) and bulges above and below the plane of the galactic disk.

**Cambrian explosion** The dramatic diversification of life on Earth that occurred between about 540 and 500 million years ago.

**carbonate rock** A carbon-rich rock, such as limestone, that forms underwater from chemical reactions between sediments and carbon dioxide. On Earth, most of the outgassed carbon dioxide currently resides in carbonate rocks.

**carbon dioxide cycle ($CO_2$ cycle)** The process that cycles carbon dioxide between Earth's atmosphere and surface rocks.

**carbon stars** Stars whose atmospheres are especially carbon-rich, thought to be near the ends of their lives; carbon stars are the primary sources of carbon in the universe.

**Cassini division** A large, dark gap in Saturn's rings, visible through small telescopes on Earth.

**CCD (charge-coupled device)** A type of electronic light detector that has largely replaced photographic film in astronomical research.

**celestial coordinates** The coordinates of right ascension and declination that fix an object's position on the celestial sphere.

**celestial equator (CE)** The extension of Earth's equator onto the celestial sphere.

**celestial navigation** Navigation on the surface of the Earth accomplished through observations of the Sun and stars.

**celestial sphere** The imaginary sphere on which objects in the sky appear to reside when observed from Earth.

**Celsius** (temperature scale) The temperature scale commonly used in daily activity internationally, defined so that, on Earth's surface, water freezes at 0°C and boils at 100°C.

**center of mass** (of orbiting objects) The point at which two or more orbiting objects would balance if they were somehow connected; it is the point around which the orbiting objects actually orbit.

**central dominant galaxy** A giant elliptical galaxy found at the center of a dense cluster of galaxies, apparently formed by the merger of several individual galaxies.

**Cepheid variable star** (or **Cepheid** for short) A particularly luminous type of pulsating variable star that follows *Leavitt's law* (also called the period–luminosity relation) and hence is very useful for measuring cosmic distances.

**Chandrasekhar limit** *See* white dwarf limit.

**charged particle belts** Zones in which ions and electrons accumulate and encircle a planet.

**chemical element** *See* element (chemical).

**chemical enrichment** The process by which the abundance of heavy elements (heavier than helium) in the interstellar medium gradually increases over time as these elements are produced by stars and released into space.

**chemical potential energy** Potential energy that can be released through chemical reactions; for example, food contains chemical potential energy that your body can convert to other forms of energy.

**chondrites** Another name for primitive meteorites. The name comes from the round chondrules within them. *Achondrites*, meaning "without chondrules," is another name for processed meteorites.

**chromosphere** The layer of the Sun's atmosphere below the corona; most of the Sun's ultraviolet light is emitted from this region, in which the temperature is about 10,000 K.

**circulation cells** (or **Hadley cells**) Large-scale cells (similar to convection cells) in a planet's atmosphere that transport heat between the equator and the poles.

**circumpolar star** A star that always remains above the horizon for a particular latitude.

**climate** The long-term average of weather.

**close binary** A binary star system in which the two stars are very close together.

**closed universe** A universe in which spacetime curves back on itself to the point where its overall shape is analogous to that of the surface of a sphere.

**cluster of galaxies** A collection of a few dozen or more galaxies bound together by gravity; smaller collections of galaxies are simply called *groups*.

**cluster of stars** A group of anywhere from several hundred to a million or so stars; star clusters come in two types—open clusters and globular clusters.

**CNO cycle** The cycle of reactions by which intermediate- and high-mass stars fuse hydrogen into helium.

**coasting universe** A model of the universe in which the universe expands forever with little change in its rate of expansion; in the absence of a repulsive force (*see* cosmological constant), a coasting universe is one in which the actual mass density is *smaller* than the critical density.

**color-coded image** An image that represents information or forms of light in any way that makes an object appear different than it would appear if we looked at its true, visible-light colors. Sometimes called a *false-color image*.

**coma** (of a comet) The dusty atmosphere of a comet, created by sublimation of ices in the nucleus when the comet is near the Sun.

**comet** A relatively small, icy object that orbits a star. Like asteroids, comets are officially considered part of a category known as "small solar system bodies."

**comparative planetology** The study of the solar system by examining and understanding the similarities and differences among worlds.

**compound** (chemical) A substance made from molecules consisting of two or more atoms with different atomic numbers.

**condensates** Solid or liquid particles that condense from a cloud of gas.

**condensation** The formation of solid or liquid particles from a cloud of gas.

**conduction** (of energy) The process by which thermal energy is transferred by direct contact from warm material to cooler material.

**conjunction** (of a planet with the Sun) An event in which a planet and the Sun line up in our sky.

**conservation of angular momentum** (law of) The principle that, in the absence of net torque (twisting force), the total angular momentum of a system remains constant.

**conservation of energy** (law of) The principle that energy (including mass-energy) can be neither created nor destroyed, but can only change from one form to another.

**conservation of momentum** (law of) The principle that, in the absence of net force, the total momentum of a system remains constant.

**constellation** A region of the sky; 88 official constellations cover the celestial sphere.

**continental crust** The thicker lower-density crust that makes up Earth's continents. It is made when remelting of seafloor crust allows lower-density rock to separate and erupt to the surface. Continental crust ranges in age from very young to as old as about 4 billion years (or more).

**continuous spectrum** A spectrum (of light) that spans a broad range of wavelengths without interruption by emission or absorption lines.

**convection** The energy transport process in which warm material expands and rises while cooler material contracts and falls.

**convection cell** An individual small region of convecting material.

**convection zone** (of a star) A region in which energy is transported outward by convection.

**Copernican revolution** The dramatic change, initiated by Copernicus, that occurred when we learned that Earth is a planet orbiting the Sun rather than the center of the universe.

**core of a planet** The dense central region of a planet that has undergone differentiation.

**core of a star** The central region of a star, in which nuclear fusion can occur.

**Coriolis effect** The effect due to rotation that causes air or objects on a rotating surface or planet to deviate from straight-line trajectories.

**corona** (solar) The tenuous uppermost layer of the Sun's atmosphere; most of the Sun's x-rays are emitted from this region, in which the temperature is about 1 million K.

**coronal holes** Regions of the corona that barely show up in x-ray images because they are nearly devoid of hot coronal gas.

**coronal mass ejections** Bursts of charged particles from the Sun's corona that travel outward into space.

**cosmic microwave background** The remnant radiation from the Big Bang, which we detect using radio telescopes sensitive to microwaves (which are short-wavelength radio waves).

**cosmic rays** Particles such as electrons, protons, and atomic nuclei that zip through interstellar space at close to the speed of light.

**cosmological constant** The name given to a term in Einstein's equations of general relativity. If it is not zero, then it represents a repulsive force or a type of energy (sometimes called *dark energy* or *quintessence*) that might cause the expansion of the universe to accelerate with time.

**cosmological horizon** The boundary of our observable universe, which is where the lookback time is equal to the age of the universe. Beyond this boundary in spacetime, we cannot see anything at all.

**Cosmological Principle** The idea that matter is distributed uniformly throughout the universe on very large scales, meaning that the universe has neither a center nor an edge.

**cosmological redshift** The redshift we see from distant galaxies, caused by the fact that expansion of the universe stretches all the photons within it to longer, redder wavelengths.

**cosmology** The study of the overall structure and evolution of the universe.

**cosmos** An alternative name for the universe.

**crescent phase** The phase of the Moon (or of a planet) in which just a small portion (less than half) of the visible face is illuminated by sunlight.

**critical density** The precise average density for the entire universe that marks the dividing line between a recollapsing universe and one that will expand forever.

**critical universe** A model of the universe in which the universe expands more and more slowly as time progresses; in the absence of a repulsive force (*see* cosmological constant), a critical universe is one in which the average mass density *equals* the critical density.

**crust** (of a planet) The low-density surface layer of a planet that has undergone differentiation.

**curvature of spacetime** A change in the geometry of space that is produced in the vicinity of a massive object and is responsible for the force we call gravity. The overall geometry of the universe may also be curved, depending on its overall mass-energy content.

**cycles per second** Units of frequency for a wave; describes the number of peaks (or troughs) of a wave that pass by a given point each second. Equivalent to *hertz*.

**dark energy** Name sometimes given to energy that could be causing the expansion of the universe to accelerate. *See* cosmological constant.

**dark matter** Matter that we infer to exist from its gravitational effects but from which we have not detected any light; dark matter apparently dominates the total mass of the universe.

**daylight saving time** Standard time plus 1 hour, so that the Sun appears on the meridian around 1 p.m. rather than around noon.

**decay** (radioactive) *See* radioactive decay.

**December solstice** Both the point on the celestial sphere where the ecliptic is farthest south of the celestial equator and the moment in time when the Sun appears at that point each year (around December 21).

**declination (dec)** The angular north-south distance between the celestial equator and a location on the celestial sphere; analogous to latitude, but on the celestial sphere.

**deferent** The large circle upon which a planet follows its circle-upon-circle path around Earth in the (Earth-centered) Ptolemaic model of the universe. *See also* epicycle.

**degeneracy pressure** A type of pressure unrelated to an object's temperature, which arises when electrons (electron degeneracy pressure) or neutrons (neutron degeneracy pressure) are packed so tightly that the exclusion and uncertainty principles come into play.

**degenerate object** An object, such as a brown dwarf, white dwarf, or neutron star, in which degeneracy pressure is the primary pressure pushing back against gravity.

**density** (mass) The amount of mass per unit volume of an object. The average density of any object can be found by dividing its mass by its volume. Standard metric units are kilograms per cubic meter, but density is more commonly stated in units of grams per cubic centimeter.

**deuterium** A form of hydrogen in which the nucleus contains a proton and a neutron, rather than only a proton (as is the case for most hydrogen nuclei).

**differential rotation** Rotation in which the equator of an object rotates at a different rate than the poles.

**differentiation** The process by which gravity separates materials according to density, with high-density materials sinking and low-density materials rising.

**diffraction grating** A finely etched surface that can split light into a spectrum.

**diffraction limit** The angular resolution that a telescope could achieve if it were limited only by the interference of light waves; it is smaller (i.e., better angular resolution) for larger telescopes.

**dimension** (mathematical) Measure of the number of independent directions in which movement is possible; for example, the surface of Earth is two-dimensional because only two independent directions of motion are possible (north-south and east-west).

**direction** (in local sky) One of the two coordinates (the other is altitude) needed to pinpoint an object in the local sky. It is the direction, such as north, south, east, or west, in which you must face to see the object. *See also* azimuth.

**disk** (of a galaxy) The portion of a spiral galaxy that looks like a disk and contains an interstellar medium with cool gas and dust; stars of many ages are found in the disk.

**disk population** The stars that orbit within the disk of a spiral galaxy; sometimes called *Population I.*

**DNA (deoxyribonucleic acid)** The molecule that constitutes the genetic material of life on Earth.

**Doppler effect** (or **Doppler shift**) The effect that shifts the wavelengths of spectral features in objects that are moving toward or away from the observer.

**Doppler method** The detection of extrasolar planets through the motion of a star toward and away from the observer caused by gravitational tugs from the planet.

**double shell–fusion star** A star that is fusing helium into carbon in a shell around an inert carbon core and is fusing hydrogen into helium in a shell at the top of the helium layer.

**down quark** One of the two quark types (the other is the up quark) found in ordinary protons and neutrons. It has a charge of $-\frac{1}{3}$.

**Drake equation** An equation that lays out the factors that play a role in determining the number of communicating civilizations in our galaxy.

**dust** (or **dust grains**) Tiny solid flecks of material; in astronomy, we often discuss interplanetary dust (found within a star system) or interstellar dust (found between the stars in a galaxy).

**dust tail** (of a comet) One of two tails seen when a comet passes near the Sun (the other is the *plasma tail*). It is composed of small solid particles pushed away from the Sun by the radiation pressure of sunlight.

**dwarf elliptical galaxy** A small elliptical galaxy with less than about a billion stars.

**dwarf galaxies** Relatively small galaxies, consisting of less than about 10 billion stars.

**dwarf planet** An object that orbits the Sun and is massive enough for its gravity to have made it nearly round in shape, but that does not qualify as an official planet because it has not cleared its orbital neighborhood. The dwarf planets of our solar system include the asteroid Ceres and the Kuiper belt objects Pluto, Eris, Haumea, and Makemake.

**dwarf spheroidal galaxy** A subclass of dwarf elliptical galaxies that are round and diskless like other elliptical galaxies but are particularly small and much less bright. They are the most numerous subtype of galaxy in the Local Group and possibly in the universe.

**Earth-orbiters** (spacecraft) Spacecraft designed to study Earth or the universe from Earth orbit.

**eccentricity** A measure of how much an ellipse deviates from a perfect circle; defined as the center-to-focus distance divided by the length of the semimajor axis.

**eclipse** An event in which one astronomical object casts a shadow on another or crosses our line of sight to the other object.

**eclipse seasons** Periods during which lunar and solar eclipses can occur because the nodes of the Moon's orbit are aligned with Earth and the Sun.

**eclipsing binary** A binary star system in which the two stars happen to be orbiting in the plane of our line of sight, so that each star will periodically eclipse the other.

**ecliptic** The Sun's apparent annual path among the constellations.

**ecliptic plane** The plane of Earth's orbit around the Sun.

**ejecta** (from an impact) Debris ejected by the blast of an impact.

**electrical charge** A fundamental property of matter that is described by its amount and as either positive or negative; more technically, a measure of how a particle responds to the electromagnetic force.

**electromagnetic field** An abstract concept used to describe how a charged particle would affect other charged particles at a distance.

**electromagnetic radiation** Another name for light of all types, from radio waves through gamma rays.

**electromagnetic spectrum** The complete spectrum of light, including radio waves, infrared light, visible light, ultraviolet light, x-rays, and gamma rays.

**electromagnetic wave** A synonym for *light*, which consists of waves of electric and magnetic fields.

**electromagnetism** (or **electromagnetic force**) One of the four fundamental forces; it is the force that dominates atomic and molecular interactions.

**electron degeneracy pressure** Degeneracy pressure exerted by electrons, as in brown dwarfs and white dwarfs.

**electrons** Fundamental particles with negative electric charge; the distribution of electrons in an atom gives the atom its size.

**electron-volt (eV)** A unit of energy equivalent to $1.60 \times 10^{-19}$ joule.

**electroweak era** The era of the universe during which only three forces operated (gravity, strong force, and electroweak force), lasting from $10^{-38}$ second to $10^{-10}$ second after the Big Bang.

**electroweak force** The force that exists at high energies when the electromagnetic force and the weak force exist as a single force.

**element** (chemical) A substance made from individual atoms of a particular atomic number.

**ellipse** A type of oval that happens to be the shape of bound orbits. An ellipse can be drawn by moving a pencil along a string whose ends are tied to two tacks; the locations of the tacks are the *foci* (singular: *focus*) of the ellipse.

**elliptical galaxies** Galaxies that appear rounded in shape, often longer in one direction, like a football. They have no disks and contain little cool gas and dust compared to spiral galaxies, though they often contain hot, ionized gas.

**elongation** (greatest) For Mercury or Venus, the point at which it appears farthest from the Sun in our sky.

**emission** (of light) The process by which matter emits energy in the form of light.

**emission line** A bright band (a "line") of single color, superimposed on a fainter or completely absent rainbow of light, occurring when light viewed through a diffraction element such as a prism shows an excess of photons at or near a specific wavelength.

**emission nebula** *See* ionization nebula.

**energy** Broadly speaking, what can make matter move. The three basic types of energy are kinetic, potential, and radiative.

**energy balance** (in a star) The balance between the rate at which fusion releases energy in the star's core and the rate at which the star's surface radiates this energy into space.

**epicycle** The small circle upon which a planet moves while simultaneously going around a larger circle (the *deferent*) around Earth in the (Earth-centered) Ptolemaic model of the universe.

**equation of time** An equation describing the discrepancies between apparent and mean solar time.

**equinox** *See* March equinox *and* September equinox.

**equivalence principle** The fundamental starting point for general relativity, which states that the effects of gravity are exactly equivalent to the effects of acceleration.

**era of atoms** The era of the universe lasting from about 500,000 years to about 1 billion years after the Big Bang, during which it was cool enough for neutral atoms to form.

**era of galaxies** The present era of the universe, which began with the formation of galaxies when the universe was about 1 billion years old.

**era of nuclei** The era of the universe lasting from about 3 minutes to about 380,000 years after the Big Bang, during which matter in the universe was fully ionized and opaque to light. The cosmic background radiation was released at the end of this era.

**era of nucleosynthesis** The era of the universe lasting from about 0.001 second to about 3 minutes after the Big Bang, by the end of which virtually all of the neutrons and about one-seventh of the protons in the universe had fused into helium.

**erosion** The wearing down or building up of geological features by wind, water, ice, and other phenomena of planetary weather.

**eruption** The process of releasing hot lava onto a planet's surface.

**escape velocity** The speed necessary for an object to completely escape the gravity of a large body such as a moon, planet, or star.

**evaporation** The process by which atoms or molecules escape into the gas phase from a liquid.

**event** Any particular point along a worldline; all observers will agree on the reality of an event but may disagree about its time and location.

**event horizon** The boundary that marks the "point of no return" between a black hole and the outside universe; events that occur within the event horizon can have no influence on our observable universe.

**evolution** (biological) The gradual change in populations of living organisms responsible for transforming life on Earth from its primitive origins to its great diversity today.

**exchange particle** A type of subatomic particle that transmits one of the four fundamental forces; according to the standard model of physics, these particles are always exchanged whenever two objects interact through a force.

**excited state** (of an atom) Any arrangement of electrons in an atom that has more energy than the ground state.

**exclusion principle** The law of quantum mechanics that states that two fermions cannot occupy the same quantum state at the same time.

**exosphere** The hot, outer layer of an atmosphere, where the atmosphere "fades away" to space.

**expansion** (of the universe) The idea that the space between galaxies or clusters of galaxies is growing with time.

**exposure time** The amount of time during which light is collected to make a single image.

**extrasolar planet** A planet orbiting a star other than our Sun.

**extremophiles** Living organisms that are adapted to conditions that are "extreme" by human standards, such as very high or low temperature or a high level of salinity or radiation.

**Fahrenheit** (temperature scale) The temperature scale commonly used in daily activity in the United States; defined so that, on Earth's surface, water freezes at 32°F and boils at 212°F.

**fall equinox** *See* September equinox, which is commonly called the fall equinox by people living in the Northern Hemisphere.

**false-color image** *See* color-coded image.

**fault** (geological) A place where lithospheric plates slip sideways relative to one another.

**feedback processes** Processes in which a small change in some property (such as temperature) leads to changes in other properties that either amplify or diminish the original small change.

**fermions** Particles, such as electrons, neutrons, and protons, that obey the exclusion principle.

**Fermi's paradox** The question posed by Enrico Fermi about extraterrestrial intelligence—"So where is everybody?"—which asks why we have not observed other civilizations even though simple arguments would suggest that some ought to have spread throughout the galaxy by now.

**field** An abstract concept used to describe how a particle would interact with a force. For example, the idea of a gravitational field describes how a particle would react to the local strength of gravity, and the idea of an electromagnetic field describes how a charged particle would respond to forces from other charged particles.

**filter** (for light) A material that transmits only particular wavelengths of light.

**fireball** A particularly bright meteor.

**first-quarter phase** The phase of the Moon that occurs one-quarter of the way through each cycle of phases, in which precisely half of the visible face is illuminated by sunlight.

**fission** *See* nuclear fission.

**flare star** A small, spectral type M star that displays particularly strong flares on its surface.

**flat geometry** (or **Euclidean geometry**) The type of geometry in which the rules of geometry for a flat plane hold, such as that the shortest distance between two points is a straight line and that the sum of the angles in a triangle is 180°.

**flat universe** A universe in which the overall geometry of spacetime is flat (Euclidean), as would be the case if the density of the universe was equal to the critical density.

**flybys** Spacecraft that fly past a target object (such as a planet), usually just once, as opposed to entering a bound orbit of the object.

**focal plane** The place where an image created by a lens or mirror is in focus.

**foci** Plural of *focus*.

**focus** (of a lens or mirror) The point at which rays of light that were initially parallel (such as those from a distant star) converge.

**focus** (of an ellipse) One of two special points within an ellipse that lie along the major axis; these are the points around which we could stretch a pencil and string to draw an ellipse. When one object orbits a second object, the second object lies at one focus of the orbit.

**force** Anything that can cause a change in momentum.

**formation properties** (of planets) In this book, for the purpose of understanding geological processes, four formation properties that planets are defined to be born with: size (mass and radius), distance from the Sun, composition, and rotation rate.

**fossil** Any relic of an organism that lived and died long ago.

**frame of reference** *See* reference frame.

**free-fall** The condition in which an object is falling without resistance; objects are weightless when in free-fall.

**free-float frame** A frame of reference in which all objects are weightless and hence float freely.

**frequency** The rate at which peaks of a wave pass by a point, measured in units of 1/s, often called *cycles per second* or *hertz*.

**frost line** The boundary in the solar nebula beyond which ices could condense; only metals and rocks could condense within the frost line.

**fundamental forces** The four forces that appear to govern all interactions that occur in the universe today: gravity, electromagnetism, the strong force, and the weak force.

**fundamental particles** Subatomic particles that cannot be divided into anything smaller.

**fusion** *See* nuclear fusion.

**galactic cannibalism** The term sometimes used to describe the process by which large galaxies merge with other galaxies in collisions. *Central dominant galaxies* are products of galactic cannibalism.

**galactic fountain** A model for the cycling of gas in the Milky Way Galaxy in which fountains of hot, ionized gas rise from the disk into the halo and then cool and form clouds as they sink back into the disk.

**galactic wind** A wind of low-density but extremely hot gas flowing out from a starburst galaxy, created by the combined energy of many supernovae.

**galaxy** A great island of stars in space, containing millions, billions, or even trillions of stars, all held together by gravity and orbiting a common center.

**galaxy cluster** *See* cluster of galaxies.

**galaxy evolution** The formation and development of galaxies.

**Galilean moons** The four moons of Jupiter that were discovered by Galileo: Io, Europa, Ganymede, and Callisto.

**gamma-ray burst** A sudden burst of gamma rays from deep space. At least some relatively long (many seconds to minutes) gamma-ray bursts come from powerful supernovae (hypernovae), while at least some very short bursts (less than a few seconds) come from neutron star mergers.

**gamma rays** Light with very short wavelengths (and hence high frequencies)—shorter than those of x-rays.

**gap moons** Tiny moons located within a gap in a planet's ring system. The gravity of a gap moon helps clear the gap.

**gas phase** The phase of matter in which atoms or molecules can move essentially independently of one another.

**gas pressure** The force (per unit area) pushing on any object due to surrounding gas. *See also* pressure.

**general theory of relativity** Einstein's generalization of his special theory of relativity so that the theory also applies when we consider effects of gravity or acceleration.

**genetic code** The "language" that living cells use to read the instructions chemically encoded in DNA.

**geocentric model** Any of the ancient Greek models that were used to predict planetary positions under the assumption that Earth lay in the center of the universe.

**geocentric universe** The ancient belief that Earth is the center of the entire universe.

**geological activity** Processes that change a planet's surface long after formation, such as volcanism, tectonics, and erosion.

**geological processes** The processes responsible for a planet's surface features; the four basic geological processes are impact cratering, volcanism, tectonics, and erosion.

**geological time scale** The time scale used by scientists to describe major eras in Earth's past.

**geology** The study of surface features (on a moon, planet, or asteroid) and the processes that create them.

**geostationary satellite** A satellite that appears to stay stationary in the sky as viewed from Earth's surface, because it orbits in the same time it takes Earth to rotate and orbits in Earth's equatorial plane.

**geosynchronous satellite** A satellite that orbits Earth in the same time it takes Earth to rotate (one sidereal day).

**giant galaxies** Galaxies that are unusually large, typically containing a trillion or more stars. Most giant galaxies are elliptical, and many contain multiple nuclei near their centers.

**giant impact** A collision between a forming planet and a very large planetesimal, such as is thought to have formed our Moon.

**giant molecular cloud** A very large cloud of cold, dense interstellar gas, typically containing up to a million solar masses worth of material. *See also* molecular clouds.

**giants (luminosity class III)** Stars that appear just below the supergiants on the H-R diagram because they are somewhat smaller in radius and lower in luminosity.

**gibbous phase** The phase of the Moon (or of a planet) in which more than half but less than all of the visible face is illuminated by sunlight.

**global positioning system (GPS)** A system of navigation by satellites orbiting Earth.

**global warming** An expected increase in Earth's global average temperature caused by human input of carbon dioxide and other greenhouse gases into the atmosphere.

**global wind patterns** (or **global circulation**) Wind patterns that remain fixed on a global scale, determined by the combination of surface heating and the planet's rotation.

**globular cluster** A spherically shaped cluster of up to a million or more stars; globular clusters are found primarily in the halos of galaxies and contain only very old stars.

**gluons** The exchange particles for the strong force.

**grand unified theory (GUT)** A theory that unifies three of the four fundamental forces—the strong force, the weak force, and the electromagnetic force (but not gravity)—in a single model.

**granulation** (on the Sun) The bubbling pattern visible in the photosphere, produced by the underlying convection.

**gravitation** (law of) *See* universal law of gravitation.

**gravitational constant** The experimentally measured constant $G$ that appears in the law of universal gravitation:

$$G = 6.67 \times 10^{-11} \frac{m^3}{kg \times s^2}$$

**gravitational contraction** The process in which gravity causes an object to contract, thereby converting gravitational potential energy into thermal energy.

**gravitational encounter** An encounter in which two (or more) objects pass near enough so that each can feel the effects of the other's gravity and they can therefore exchange energy.

**gravitational equilibrium** A state of balance in which the force of gravity pulling inward is precisely counteracted by pressure pushing outward; also referred to as *hydrostatic equilibrium*.

**gravitational lensing** The magnification or distortion (into arcs, rings, or multiple images) of an image caused by light bending through a gravitational field, as predicted by Einstein's general theory of relativity.

**gravitationally bound system** Any system of objects, such as a star system or a galaxy, that is held together by gravity.

**gravitational potential energy** Energy that an object has by virtue of its position in a gravitational field; an object has more gravitational potential energy when it has a greater distance that it can potentially fall.

**gravitational redshift** A redshift caused by the fact that time runs slowly in gravitational fields.

**gravitational time dilation** The slowing of time that occurs in a gravitational field, as predicted by Einstein's general theory of relativity.

**gravitational waves** Waves, predicted by Einstein's general theory of relativity, that are created by changes in a local gravitational field and cause distortions of space-time as they propogate outward through the universe at the speed of light.

**gravitons** The exchange particles for the force of gravity.

**gravity** One of the four fundamental forces; it is the force that dominates on large scales.

**grazing incidence** (in telescopes) Reflections in which light grazes a mirror surface and is deflected at a small angle; commonly used to focus high-energy ultraviolet light and x-rays.

**great circle** A circle on the surface of a sphere whose center is at the center of the sphere.

**greatest elongation** *See* elongation (greatest).

**Great Red Spot** A large, high-pressure storm on Jupiter.

**greenhouse effect** The process by which greenhouse gases in an atmosphere make a planet's surface temperature warmer than it would be in the absence of an atmosphere.

**greenhouse gases** Gases, such as carbon dioxide, water vapor, and methane, that are particularly good absorbers of infrared light but are transparent to visible light.

**Gregorian calendar** Our modern calendar, introduced by Pope Gregory in 1582.

**ground state** (of an atom) The lowest possible energy state of the electrons in an atom.

**group of galaxies** A few to a few dozen galaxies bound together by gravity. *See also* cluster of galaxies.

**GUT era** The era of the universe during which only two forces operated (gravity and the grand-unified-theory, or GUT, force), lasting from $10^{-43}$ second to $10^{-38}$ second after the Big Bang.

**GUT force** The proposed force that exists at very high energies when the strong force, the weak force, and the electromagnetic force (but not gravity) all act as one.

**H II region** *See* ionization nebula.

**habitable world** A world with environmental conditions under which life could *potentially* arise or survive.

**habitable zone** The region around a star in which planets could potentially have surface temperatures at which liquid water could exist.

**Hadley cells** *See* circulation cells.

**half-life** The time it takes for half of the nuclei in a given quantity of a radioactive substance to decay.

**halo** (of a galaxy) The spherical region surrounding the disk of a spiral galaxy.

**halo component** (of a galaxy) The portion of any galaxy that is spherical (or football-like) in shape and contains very little cool gas; it generally contains only very old stars. For spiral galaxies, the halo component includes both the halo and the bulge (but not the disk); elliptical galaxies have only a halo component.

**halo population** Stars that orbit within the halo component of a galaxy; sometimes called *Population II*. Elliptical galaxies have only a halo population (they lack a disk population), while spiral galaxies have halo population stars in their bulges and halos.

**Hawking radiation** Radiation predicted to arise from the evaporation of black holes.

**heavy bombardment** The period in the first few hundred million years after the solar system formed during which the tail end of planetary accretion created most of the craters found on ancient planetary surfaces.

**heavy elements** In astronomy, generally all elements *except* hydrogen and helium.

**helium-capture reactions** Fusion reactions that fuse a helium nucleus into some other nucleus; such reactions can fuse carbon into oxygen, oxygen into neon, neon into magnesium, and so on.

**helium flash** The event that marks the sudden onset of helium fusion in the previously inert helium core of a low-mass star.

**helium-fusing star** (or **helium core-fusion star**) A star that is currently fusing helium into carbon in its core.

**helium fusion** The fusion of three helium nuclei to form one carbon nucleus; also called the *triple-alpha reaction*.

**hertz (Hz)** The standard unit of frequency for light waves; equivalent to units of 1/s.

**Hertzsprung-Russell (H-R) diagram** A graph plotting individual stars as points, with stellar luminosity on the vertical axis and spectral type (or surface temperature) on the horizontal axis.

**high-mass stars** Stars born with masses above about $8M_{Sun}$; these stars will end their lives by exploding as supernovae.

**horizon** A boundary that divides what we can see from what we cannot see.

**horizontal branch** The horizontal line of stars that represents helium-fusing stars on an H-R diagram for a cluster of stars.

**horoscope** A predictive chart made by an astrologer; in scientific studies, horoscopes have never been found to have any validity as predictive tools.

**hot Jupiter** A class of planet that is Jupiter-like in size but orbits very close to its star, causing it to have a very high surface temperature.

**hot spot** (geological) A place within a plate of the lithosphere where a localized plume of hot mantle material rises.

**hour angle (HA)** The angle or time (measured in hours) since an object was last on the meridian in the local sky; defined to be 0 hours for objects that are on the meridian.

**Hubble's constant** A number that expresses the current rate of expansion of the universe; designated $H_0$, it is usually stated in units of km/s/Mpc. The reciprocal of Hubble's constant is the age the universe would have *if* the expansion rate had never changed.

**Hubble's law** Mathematical expression of the idea that more distant galaxies move away from us faster: $v = H_0 \times d$, where $v$ is a galaxy's speed away from us, $d$ is its distance, and $H_0$ is Hubble's constant.

**hydrogen compounds** Compounds that contain hydrogen and were common in the solar nebula, such as water ($H_2O$), ammonia ($NH_3$), and methane ($CH_4$).

**hydrogen shell fusion** Hydrogen fusion that occurs in a shell surrounding a stellar core.

**hydrosphere** The "layer" of water on Earth consisting of oceans, lakes, rivers, ice caps, and other liquid water and ice.

**hydrostatic equilibrium** *See* gravitational equilibrium.

**hyperbola** The precise mathematical shape of one type of unbound orbit (the other is a parabola) allowed under the force of gravity; at great distances from the attracting object, a hyperbolic path looks like a straight line.

**hypernova** A term sometimes used to describe a supernova (explosion) of a star so massive that it leaves a black hole behind.

**hyperspace** Any space with more than three dimensions.

**hypothesis** A tentative model proposed to explain some set of observed facts, but which has not yet been rigorously tested and confirmed.

**ice ages** Periods of global cooling during which the polar caps, glaciers, and snow cover extend closer to the equator.

**ices** (in solar system theory) Materials that are solid only at low temperatures, such as the hydrogen compounds water, ammonia, and methane.

**ideal gas law** The law relating the pressure, temperature, and number density of particles in an ideal gas.

**image** A picture of an object made by focusing light.

**imaging** (in astronomical research) The process of obtaining pictures of astronomical objects.

**impact** The collision of a small body (such as an asteroid or comet) with a larger object (such as a planet or moon).

**impact basin** A very large impact crater, often filled by a lava flow.

**impact crater** A bowl-shaped depression left by the impact of an object that strikes a planetary surface (as opposed to burning up in the atmosphere).

**impact cratering** The excavation of bowl-shaped depressions (*impact craters*) by asteroids or comets striking a planet's surface.

**impactor** The object responsible for an impact.

**inflation** (of the universe) A sudden and dramatic expansion of the universe thought to have occurred at the end of the GUT era.

**infrared light** Light with wavelengths that fall in the portion of the electromagnetic spectrum between radio waves and visible light.

**inner solar system** Generally, the region of our solar system out to about the orbit of Mars.

**intensity** (of light) A measure of the amount of energy coming from light of specific wavelength in the spectrum of an object.

**interferometry** A telescopic technique in which two or more telescopes are used in tandem to produce much better angular resolution than the telescopes could achieve individually.

**intermediate-mass stars** Stars born with masses between about $2M_{\text{Sun}}$ and $8M_{\text{Sun}}$; these stars end their lives by ejecting a planetary nebula and becoming a white dwarf.

**interstellar cloud** A cloud of gas and dust between the stars.

**interstellar dust grains** Tiny solid flecks of carbon and silicon minerals found in cool interstellar clouds; they resemble particles of smoke and form in the winds of red giant stars.

**interstellar medium** The gas and dust that fills the space between stars in a galaxy.

**interstellar ramjet** A hypothesized type of spaceship that uses a giant scoop to sweep up interstellar gas for use in a nuclear fusion engine.

**interstellar reddening** The change in the color of starlight as it passes through dusty gas. The light appears redder because dust grains absorb and scatter blue light more effectively than red light.

**intracluster medium** Hot, x-ray-emitting gas found between the galaxies within a cluster of galaxies.

**inverse square law** A law followed by any quantity that decreases with the square of the distance between two objects.

**inverse square law for light** The law stating that an object's apparent brightness depends on its actual luminosity and the inverse square of its distance from the observer:

$$\text{apparent brightness} = \frac{\text{luminosity}}{4\pi \times (\text{distance})^2}$$

**inversion** (atmospheric) A local weather condition in which air is colder near the surface than higher up in the troposphere—the opposite of the usual condition, in which the troposphere is warmer at the bottom.

**ionization** The process of stripping one or more electrons from an atom.

**ionization nebula** A colorful, wispy cloud of gas that glows because neighboring hot stars irradiate it with ultraviolet photons that can ionize hydrogen atoms; also called an *emission nebula* or *H II region*.

**ionosphere** A portion of the thermosphere in which ions are particularly common (because of ionization by x-rays from the Sun).

**ions** Atoms with a positive or negative electrical charge.

**Io torus** A donut-shaped charged-particle belt around Jupiter that approximately traces Io's orbit.

**irregular galaxies** Galaxies that look neither spiral nor elliptical.

**isotopes** Forms of an element that have the same number of protons but different numbers of neutrons.

**jets** High-speed streams of gas ejected from an object into space.

**joule** The international unit of energy, equivalent to about 1/4000 of a Calorie.

**jovian nebulae** The clouds of gas that swirled around the jovian planets, from which the moons formed.

**jovian planets** Giant gaseous planets similar in overall composition to Jupiter.

**Julian calendar** The calendar introduced in 46 B.C. by Julius Caesar and used until the Gregorian calendar replaced it.

**June solstice** Both the point on the celestial sphere where the ecliptic is farthest north of the celestial equator and the moment in time when the Sun appears at that point each year (around June 21).

**Kelvin** (temperature scale) The temperature scale most commonly used in science, defined such that absolute zero is 0 K and water freezes at 273.15 K.

**Kepler's first law** Law stating that the orbit of each planet about the Sun is an ellipse with the Sun at one focus.

**Kepler's laws of planetary motion** Three laws discovered by Kepler that describe the motion of the planets around the Sun.

**Kepler's second law** The principle that, as a planet moves around its orbit, it sweeps out equal areas in equal times. This tells us that a planet moves faster when it is closer to the Sun (near perihelion) than when it is farther from the Sun (near aphelion) in its orbit.

**Kepler's third law** The principle that the square of a planet's orbital period is proportional to the cube of its average distance from the Sun (semimajor axis), which tells us that more distant planets move more slowly in their orbits; in its original form, written $p^2 = a^3$. *See also* Newton's version of Kepler's third law.

**kilonova** A name occasionally applied to the observed afterglow of events thought to be due to the merger of two neutron stars.

**kinetic energy** Energy of motion, given by the formula $\frac{1}{2}mv^2$.

**Kirchhoff's laws** A set of rules that summarizes the conditions under which objects produce thermal, absorption line, or emission line spectra. In brief: (1) An opaque object produces thermal radiation. (2) An absorption line spectrum occurs when thermal radiation passes through a thin gas that is cooler than the object emitting the thermal radiation. (3) An emission line spectrum occurs when we view a cloud of gas that is warmer than any background source of light.

**Kirkwood gaps** On a plot of asteroid semimajor axes, regions with few asteroids as a result of orbital resonances with Jupiter.

**K-T event** (or **K-T impact**) The collision of an asteroid or comet 65 million years ago that caused the mass extinction best known for wiping out the dinosaurs. *K* and *T* stand for the geological layers above and below the event.

**Kuiper belt** The comet-rich region of our solar system that resides between about 30 and 100 AU from the Sun. Kuiper belt comets have orbits that lie fairly close to the plane of planetary orbits and travel around the Sun in the same direction as the planets.

**Kuiper belt object** Any object orbiting the Sun within the region of the Kuiper belt, although the term is most often used for relatively large objects. For example, Pluto and Eris are considered large Kuiper belt objects.

**Large Magellanic Cloud** One of two small, irregular galaxies (the other is the Small Magellanic Cloud) located about 150,000 light-years away; it probably orbits the Milky Way Galaxy.

**large-scale structure** (of the universe) Generally, the structure of the universe on size scales larger than that of clusters of galaxies.

**latitude** The angular north-south distance between Earth's equator and a location on Earth's surface.

**leap year** A calendar year with 366 rather than 365 days. Our current calendar (the Gregorian calendar) incorporates a leap year every 4 years (by adding February 29) except in century years that are not divisible by 400.

**Leavitt's law** The relation, discovered by Henrietta Leavitt, that describes how the luminosity of a Cepheid variable star is related to the period between peaks in its brightness: the longer the Cepheid's period, the more luminous the star. Also called the *Cepheid period–luminosity relation.*

**length contraction** The effect in which you observe lengths to be shortened in reference frames moving relative to you.

**lens** (gravitational) *See* gravitational lensing.

**lenticular galaxies** Galaxies that look lens-shaped when seen edge-on, resembling spiral galaxies without arms. They tend to have less cool gas than normal spiral galaxies but more cool gas than elliptical galaxies.

**leptons** Fermions *not* made from quarks, such as electrons and neutrinos.

**life track** A track drawn on an H-R diagram to represent the changes in a star's surface temperature and luminosity during its life; also called an *evolutionary track.*

**light-collecting area** (of a telescope) The area of the primary mirror or lens that collects light in a telescope.

**light curve** A graph of an object's intensity against time.

**light gases** (in solar system theory) Hydrogen and helium, which never condense under solar nebula conditions.

**light pollution** Human-made light that hinders astronomical observations.

**light-year (ly)** The distance that light can travel in 1 year, which is 9.46 trillion km.

**liquid phase** The phase of matter in which atoms or molecules are held together but move relatively freely.

**lithosphere** The relatively rigid outer layer of a planet; it generally encompasses the crust and the upper portion of the mantle.

**Local Bubble** (interstellar) The bubble of hot gas in which our Sun and other nearby stars apparently reside. *See also* bubble (interstellar).

**Local Group** The group of about 40 galaxies to which the Milky Way Galaxy belongs.

**local sidereal time (LST)** Sidereal time for a particular location, defined according to the position of the March equinox in the local sky. More formally, the local sidereal time at any moment is defined to be the hour angle of the March equinox.

**local sky** The sky as viewed from a particular location on Earth (or another solid object). Objects in the local sky are pinpointed by the coordinates of *altitude* and *direction* (or azimuth).

**local solar neighborhood** The portion of the Milky Way Galaxy that is located relatively close (within a few hundred to a couple thousand light-years) to our Sun.

**Local Supercluster** The supercluster of galaxies to which the Local Group belongs.

**longitude** The angular east-west distance between the prime meridian (which passes through Greenwich, England) and a location on Earth's surface.

**lookback time** The amount of time since the light we see from a distant object was emitted. If an object has a lookback time of 400 million years, we are seeing it as it looked 400 million years ago.

**low-mass stars** Stars born with masses less than about $2M_{Sun}$; these stars end their lives by ejecting a planetary nebula and becoming a white dwarf.

**luminosity** The total power output of an object, usually measured in watts or in units of solar luminosities ($L_{Sun} = 3.8 \times 10^{26}$ watts).

**luminosity class** A category describing the region of the H-R diagram in which a star falls. Luminosity class I represents supergiants, III represents giants, and V represents main-sequence stars; luminosity classes II and IV are intermediate to the others.

**lunar eclipse** An event that occurs when the Moon passes through Earth's shadow, which can happen only at full moon. A lunar eclipse may be total, partial, or penumbral.

**lunar maria** The regions of the Moon that look smooth from Earth and actually are impact basins.

**lunar month** *See* synodic month.

**lunar phase** *See* phase (of the Moon or a planet).

**magma** Underground molten rock.

**magnetic braking** The process by which a star's rotation slows as its magnetic field transfers its angular momentum to the surrounding nebula.

**magnetic field** The region surrounding a magnet in which it can affect other magnets or charged particles.

**magnetic field lines** Lines that represent how the needles on a series of compasses would point if they were laid out in a magnetic field.

**magnetosphere** The region surrounding a planet in which charged particles are trapped by the planet's magnetic field.

**magnitude system** A system for describing stellar brightness by using numbers, called *magnitudes*, based on an ancient Greek way of describing the brightnesses of stars in the sky. This system uses *apparent magnitude* to describe a star's apparent brightness and *absolute magnitude* to describe a star's luminosity.

**main sequence** The prominent line of points (representing *main-sequence stars*) running from the upper left to the lower right on an H-R diagram.

**main-sequence lifetime** The length of time for which a star of a particular mass can shine by fusing hydrogen into helium in its core.

**main-sequence stars (luminosity class V)** Stars whose temperature and luminosity place them on the main sequence of the H-R diagram. Main-sequence stars release energy by fusing hydrogen into helium in their cores.

**main-sequence turnoff point** The point on a cluster's H-R diagram where its stars turn off from the main sequence; the age of the cluster is equal to the main-sequence lifetime of stars at the main-sequence turnoff point.

**mantle** (of a planet) The rocky layer that lies between a planet's core and crust.

**March equinox** Both the point in Pisces on the celestial sphere where the ecliptic crosses the celestial equator and the moment in time when the Sun appears at that point each year (around March 21).

**Martian meteorites** Meteorites found on Earth that are thought to have originated on Mars.

**mass** A measure of the amount of matter in an object.

**mass-energy** The potential energy of mass, which has an amount $E = mc^2$.

**mass exchange** (in close binary star systems) The process in which tidal forces cause matter to spill from one star to a companion star in a close binary system.

**mass extinction** An event in which a large fraction of the species living on Earth go extinct, such as the event in which the dinosaurs died out about 65 million years ago.

**mass increase** (in relativity) The effect in which an object moving past you seems to have a mass greater than its rest mass.

**massive star supernova** A supernova that occurs when a massive star dies, initiated by the catastrophic collapse of its iron core; often called a *Type II supernova*.

**mass-to-light ratio** The mass of an object divided by its luminosity, usually stated in units of solar masses per solar luminosity. Objects with high mass-to-light ratios must contain substantial quantities of dark matter.

**matter–antimatter annihilation** An event that occurs when a particle of matter and a particle of antimatter meet and convert all of their mass-energy to photons.

**mean solar time** Time measured by the average position of the Sun in the local sky over the course of the year.

**meridian** A half-circle extending from your horizon (altitude 0°) due south, through your zenith, to your horizon due north.

**metallic hydrogen** Hydrogen that is so compressed that the hydrogen atoms all share electrons and thereby take on properties of metals, such as conducting electricity. It occurs only under very high-pressure conditions, such as those found deep within Jupiter.

**metals** (in solar system theory) Elements, such as nickel, iron, and aluminum, that condense at fairly high temperatures.

**meteor** A flash of light caused when a particle from space burns up in our atmosphere.

**meteorite** A rock from space that lands on Earth.

**meteor shower** A period during which many more meteors than usual can be seen.

**Metonic cycle** The 19-year period, discovered by the Babylonian astronomer Meton, over which the lunar phases occur on the same dates.

**microwaves** Light with wavelengths in the range of micrometers to millimeters. Microwaves are generally considered to be a subset of the radio-wave portion of the electromagnetic spectrum.

**mid-ocean ridges** Long ridges of undersea volcanoes on Earth, along which mantle material erupts onto the ocean floor and pushes apart the existing seafloor on either side. These ridges are essentially the source of new seafloor crust, which then makes its way along the ocean bottom for millions of years before returning to the mantle at a subduction zone.

**Milankovitch cycles** The cyclical changes in Earth's axis tilt and orbit that can change the climate and cause ice ages.

**Milky Way** The name used both for our galaxy and for the band of light we see in the sky when we look into the plane of our galaxy.

**millisecond pulsars** Pulsars with rotation periods of a few thousandths of a second.

**minor planets** An alternative name for *asteroids*.

**model** (scientific) A representation of some aspect of nature that can be used to explain and predict real phenomena without invoking myth, magic, or the supernatural.

**molecular bands** The tightly bunched lines in an object's spectrum that are produced by molecules.

**molecular cloud fragments** (or **molecular cloud cores**) The densest regions of molecular clouds, which usually go on to form stars.

**molecular clouds** Cool, dense interstellar clouds in which the low temperatures allow hydrogen atoms to pair up into hydrogen molecules ($H_2$).

**molecular dissociation** The process by which a molecule splits into its component atoms.

**molecule** Technically, the smallest unit of a chemical element or compound; in this text, the term refers only to combinations of two or more atoms held together by chemical bonds.

**momentum** The product of an object's mass and velocity.

**moon** An object that orbits a planet.

**moonlets** Very small moons that orbit within the ring systems of jovian planets.

**mutations** Errors in the copying process when a living cell replicates itself.

**natural selection** The process by which mutations that make an organism better able to survive get passed on to future generations.

**neap tides** The lower-than-average tides on Earth that occur at first- and third-quarter moon, when the tidal forces from the Sun and Moon oppose each other.

**nebula** A cloud of gas in space, usually one that is glowing.

**nebular capture** The process by which icy planetesimals capture hydrogen and helium gas to form jovian planets.

**nebular theory** The scientific theory that describes how our solar system formed from a cloud of interstellar gas and dust.

**net force** The overall force to which an object responds; the net force is equal to the rate of change in the object's momentum, or equivalently to the object's mass × acceleration.

**neutrino** A type of fundamental particle that has extremely low mass and responds only to the weak force; neutrinos are leptons and come in three types—electron neutrinos, muon neutrinos, and tau neutrinos.

**neutron degeneracy pressure** Degeneracy pressure exerted by neutrons, as in neutron stars.

**neutrons** Particles with no electrical charge found in atomic nuclei, built from three quarks.

**neutron star** The compact corpse of a high-mass star left over after a supernova; it typically has a mass comparable to the mass of the Sun in a volume just a few kilometers in radius.

**neutron star merger** An event in which two neutron stars that were orbiting in a binary system merge together, accompanied by the release of a tremendous amount of energy, some of which is in the form of gravitational waves and some in the form of a gamma-ray burst.

**newton** The standard unit of force in the metric system:

$$1 \text{ newton} = 1 \frac{\text{kg} \times \text{m}}{\text{s}^2}$$

**Newton's first law of motion** Principle that, in the absence of a net force, an object moves with constant velocity.

**Newton's laws of motion** Three basic laws that describe how objects respond to forces.

**Newton's second law of motion** Law stating how a net force affects an object's motion. Specifically, force = rate of change in momentum, or force = mass × acceleration.

**Newton's third law of motion** Principle that, for any force, there is always an equal and opposite reaction force.

**Newton's universal law of gravitation** *See* universal law of gravitation.

**Newton's version of Kepler's third law** A generalization of Kepler's third law used to calculate the masses of orbiting objects from measurements of orbital period and distance; usually written as

$$p^2 = \frac{4\pi^2}{G(M_1 + M_2)} a^3$$

**nodes** (of Moon's orbit) The two points in the Moon's orbit where it crosses the ecliptic plane.

**nonbaryonic matter** Matter that is not part of the normal composition of atoms, such as neutrinos or the hypothetical WIMPs; more technically, particles that are not made from three quarks.

**nonscience** As used in this book, any way of searching for knowledge that makes no claim to follow the scientific method, such as seeking knowledge through intuition, tradition, or faith.

**north celestial pole (NCP)** The point on the celestial sphere directly above Earth's North Pole.

**nova** The dramatic brightening of a star that lasts for a few weeks and then subsides; it occurs when a burst of hydrogen fusion ignites in a shell on the surface of an accreting white dwarf in a binary star system.

**nuclear fission** The process in which a larger nucleus splits into two (or more) smaller particles.

**nuclear fusion** The process in which two (or more) smaller nuclei slam together and make one larger nucleus.

**nucleus** (of a comet) The solid portion of a comet—the only portion that exists when the comet is far from the Sun.

**nucleus** (of an atom) The compact center of an atom made from protons and neutrons.

**observable universe** The portion of the entire universe that, at least in principle, can be seen from Earth.

**Occam's razor** A principle often used in science, holding that scientists should prefer the simpler of two models that agree equally well with observations; named after the medieval scholar William of Occam (1285–1349).

**Olbers' paradox** A paradox pointing out that if the universe were infinite in both age and size (with stars found throughout the universe), then the sky would not be dark at night.

**Oort cloud** A huge, spherical region centered on the Sun, extending perhaps halfway to the nearest stars, in which trillions of comets orbit the Sun with random inclinations, orbital directions, and eccentricities.

**opacity** A measure of how much light a material absorbs compared to how much it transmits; materials with higher opacity absorb more light.

**opaque** A word used to describe a material that absorbs light.

**open cluster** A cluster of up to several thousand stars; open clusters are found only in the disks of galaxies and often contain young stars.

**open universe** A universe in which spacetime has an overall shape analogous to the surface of a saddle.

**opposition** The point at which a planet appears opposite the Sun in our sky.

**optical quality** The ability of a lens, mirror, or telescope to obtain clear and properly focused images.

**orbit** The path followed by a celestial body because of gravity; an orbit may be *bound* (elliptical) or *unbound* (parabolic or hyperbolic).

**orbital energy** The sum of an orbiting object's kinetic and gravitational potential energies.

**orbital resonance** A situation in which one object's orbital period is a simple ratio of another object's period, such as 1/2, 1/4, or 5/3. In such cases, the two objects periodically line up with each other, and the extra gravitational attractions at these times can affect the objects' orbits.

**orbital velocity law** A variation on Newton's version of Kepler's third law that allows us to use a star's orbital speed and distance from the galactic center to determine the total mass of the galaxy contained *within* the star's orbit; mathematically,

$$M_r = \frac{r \times v^2}{G}$$

where $M_r$ is the mass contained within the star's orbit, $r$ is the star's distance from the galactic center, $v$ is the star's orbital velocity, and $G$ is the gravitational constant.

**orbiters** (of other worlds) Spacecraft that go into orbit of another world for long-term study.

**organic molecule** Generally, any molecule containing carbon and of a type associated with carbon-based life, though not necessarily a product of life. Note that we do not usually consider molecules such as carbon dioxide ($CO_2$) and carbonate minerals to be organic, since they are more commonly found independent of life.

**outer solar system** Generally, the region of our solar system beginning at about the orbit of Jupiter.

**outgassing** The process of releasing gases from a planetary interior, usually through volcanic eruptions.

**oxidation** Chemical reactions, often with rocks on the surface of a planet, that remove oxygen from the atmosphere.

**ozone** The molecule $O_3$, which is a particularly good absorber of ultraviolet light.

**ozone depletion** The decline in levels of atmospheric ozone found worldwide on Earth, especially in Antarctica, in recent years.

**ozone hole** A place where the concentration of ozone in the stratosphere is dramatically lower than is the norm.

**pair production** The process in which a concentration of energy spontaneously turns into a particle and its antiparticle.

**parabola** The precise mathematical shape of a special type of unbound orbit allowed under the force of gravity. If an object in a parabolic orbit loses only a tiny amount of energy, it will become bound.

**paradigm** (in science) A general pattern of thought that tends to shape scientific study during a particular time period.

**paradox** A situation that, at least at first, seems to violate common sense or contradict itself. Resolving paradoxes often leads to deeper understanding.

**parallax** The apparent shifting of an object against the background, due to viewing it from different positions. *See also* stellar parallax.

**parallax angle** Half of a star's annual back-and-forth shift due to stellar parallax; it is related to the star's distance according to the formula

$$\text{distance in parsecs} = \frac{1}{p}$$

where $p$ is the parallax angle in arcseconds.

**parsec (pc)** The distance to an object with a parallax angle of 1 arcsecond; approximately equal to 3.26 light-years.

**partial lunar eclipse** A lunar eclipse during which the Moon becomes only partially covered by Earth's umbral shadow.

**partial solar eclipse** A solar eclipse during which the Sun becomes only partially blocked by the disk of the Moon.

**particle accelerator** A machine designed to accelerate subatomic particles to high speeds in order to create new particles or to test fundamental theories of physics.

**particle era** The era of the universe lasting from $10^{-10}$ to 0.001 second after the Big Bang, during which subatomic particles were continually created and destroyed, and ending when matter annihilated antimatter.

**peculiar velocity** (of a galaxy) The component of a galaxy's velocity relative to the Milky Way that deviates from the velocity expected by Hubble's law.

**penumbra** The lighter, outlying regions of a shadow.

**penumbral lunar eclipse** A lunar eclipse during which the Moon passes only within Earth's penumbral shadow and does not fall within the umbra.

**perigee** The point at which an object orbiting Earth is nearest to Earth.

**perihelion** The point at which an object orbiting the Sun is closest to the Sun.

**period–luminosity relation** *See* Leavitt's law.

**phase** (of matter) The state determined by the way in which atoms or molecules are held together; the common phases are solid, liquid, and gas.

**phase** (of the Moon or a planet) The state determined by the portion of the visible face of the Moon (or of a planet) that is illuminated by sunlight. For the Moon, the phases cycle through new, waxing crescent, first-quarter, waxing gibbous, full, waning gibbous, third-quarter, waning crescent, and back to new.

**photon** An individual particle of light, characterized by a wavelength and a frequency.

**photosphere** The visible surface of the Sun, where the temperature averages just under 6000 K.

**pixel** An individual "picture element" in a digital image.

**Planck era** The era of the universe prior to the Planck time.

**Planck's constant** A universal constant, abbreviated $h$, with a value of $h = 6.626 \times 10^{-34}$ joule $\times$ s.

**Planck time** The time when the universe was $10^{-43}$ second old, before which random energy fluctuations were so large that our current theories are powerless to describe what might have been happening.

**planet** A moderately large object that orbits a star and shines primarily by reflecting light from its star. More precisely, according to a definition approved in 2006, a planet is an object that (1) orbits a star (but is itself neither a star nor a moon); (2) is massive enough for its own gravity to give it a nearly round shape; and (3) has cleared the neighborhood around its orbit. Objects that meet the first two criteria but not the third, including Ceres, Pluto, and Eris, are designated *dwarf planets*.

**planetary geology** The extension of the study of Earth's surface and interior to apply to other solid bodies in the solar system, such as terrestrial planets and jovian planet moons.

**planetary migration** A process through which a planet can move from the orbit on which it is born to a different orbit that is closer to or farther from its star.

**planetary nebula** The glowing cloud of gas ejected from a low-mass star at the end of its life.

**planetesimals** The building blocks of planets, formed by accretion in the solar nebula.

**plasma** A gas consisting of ions and electrons.

**plasma tail** (of a comet) One of two tails seen when a comet passes near the Sun (the other is the *dust tail*). It is composed of ionized gas blown away from the Sun by the solar wind.

**plates** (on a planet) Pieces of a lithosphere that apparently float upon the denser mantle below.

**plate tectonics** The geological process in which plates are moved around by stresses in a planet's mantle.

**polarization** (of light) The property of light describing how the electric and magnetic fields of light waves are aligned; light is said to be *polarized* when all of the photons have their electric and magnetic fields aligned in some particular way.

**Population I** *See* disk population.

**Population II** *See* halo population.

**positron** *See* antielectron.

**potential energy** Energy stored for later conversion into kinetic energy; includes gravitational potential energy, electrical potential energy, and chemical potential energy.

**power** The rate of energy usage, usually measured in watts (1 watt = 1 joule/s).

**precession** The gradual wobble of the axis of a rotating object around a vertical line.

**precipitation** Condensed atmospheric gases that fall to the surface in the form of rain, snow, or hail.

**pressure** The force (per unit area) pushing on an object. In astronomy, we are generally interested in pressure applied by surrounding gas (or plasma). Ordinarily, such pressure is related to the temperature of the gas (*see* thermal pressure). In objects such as white dwarfs and neutron stars, pressure may arise from a quantum effect (*see* degeneracy pressure). Light can also exert pressure (*see* radiation pressure).

**primary mirror** The large, light-collecting mirror of a reflecting telescope.

**prime focus** (of a reflecting telescope) The first point at which light focuses after bouncing off the primary mirror; located in front of the primary mirror.

**prime meridian** The meridian of longitude that passes through Greenwich, England; defined to be longitude 0°.

**primitive meteorites** Meteorites that formed at the same time as the solar system itself, about 4.6 billion years ago. Primitive meteorites from the inner asteroid belt are usually stony, and those from the outer belt are usually carbon-rich.

**processed meteorites** Meteorites that apparently once were part of a larger object that "processed" the original material of the solar nebula into another form. Processed meteorites can be rocky if chipped from the surface or mantle, or metallic if blasted from the core.

**proper motion** The motion of an object in the plane of the sky, perpendicular to our line of sight.

**protogalactic cloud** A huge, collapsing cloud of intergalactic gas from which an individual galaxy formed.

**proton–proton chain** The chain of reactions by which low-mass stars (including the Sun) fuse hydrogen into helium.

**protons** Particles with positive electrical charge found in atomic nuclei, built from three quarks.

**protoplanetary disk** A disk of material surrounding a young star (or protostar) that may eventually form planets.

**protostar** A forming star that has not yet reached the point where sustained fusion can occur in its core.

**protostellar disk** A disk of material surrounding a protostar; essentially the same as a protoplanetary disk, but may not necessarily lead to planet formation.

**protostellar wind** The relatively strong wind from a protostar.

**protosun** The central object in the forming solar system that eventually became the Sun.

**pseudoscience** Something that purports to be science or may appear to be scientific but that does not adhere to the testing and verification requirements of the scientific method.

**Ptolemaic model** The geocentric model of the universe developed by Ptolemy in about 150 A.D.

**pulsar** A neutron star from which we observe rapid pulses of radiation as it rotates.

**pulsating variable stars** Stars that grow alternately brighter and dimmer as their outer layers expand and contract in size.

**quantum laws** The laws that describe the behavior of particles on a very small scale; *see also* quantum mechanics.

**quantum mechanics** The branch of physics that deals with the very small, including molecules, atoms, and fundamental particles.

**quantum state** The complete description of the state of a subatomic particle, including its location, momentum, orbital angular momentum, and spin, to the extent allowed by the uncertainty principle.

**quantum tunneling** The process in which, thanks to the uncertainty principle, an electron or other subatomic particle appears on the other side of a barrier that it does not have the energy to overcome in a normal way.

**quarks** The building blocks of protons and neutrons; quarks are one of the two basic types of fermions (leptons are the other).

**quasar** The brightest type of active galactic nucleus.

**radar mapping** Imaging of a planet by bouncing radar waves off its surface, especially important for Venus and Titan, where thick clouds mask the surface.

**radar ranging** A method of measuring distances within the solar system by bouncing radio waves off planets.

**radial motion** The component of an object's motion directed toward or away from us.

**radial velocity** The portion of any object's total velocity that is directed toward or away from us. This part of the velocity is the only part that we can measure with the Doppler effect.

**radiation pressure** Pressure exerted by photons of light.

**radiation zone** (of a star) A region of the interior in which energy is transported primarily by radiative diffusion.

**radiative diffusion** The process by which photons gradually migrate from a hot region (such as the solar core) to a cooler region (such as the solar surface).

**radiative energy** Energy carried by light; the energy of a photon is Planck's constant times its frequency, or $h \times f$.

**radioactive decay** The spontaneous change of an atom into a different element, in which its nucleus breaks apart or a proton turns into an electron. This decay releases heat in a planet's interior.

**radioactive element** (or **radioactive isotope**) A substance whose nucleus tends to fall apart spontaneously.

**radio galaxy** A galaxy that emits unusually large quantities of radio waves; thought to contain an active galactic nucleus powered by a supermassive black hole.

**radio lobes** The huge regions of radio emission found on either side of radio galaxies. The lobes apparently contain plasma ejected by powerful jets from the galactic center.

**radiometric dating** The process of determining the age of a rock (i.e., the time since it solidified) by comparing the present amount of a radioactive substance to the amount of its decay product.

**radio waves** Light with very long wavelengths (and hence low frequencies)—longer than those of infrared light.

**random walk** A type of haphazard movement in which a particle or photon moves through a series of bounces, with each bounce sending it in a random direction.

**recession velocity** (of a galaxy) The speed at which a distant galaxy is moving away from us because of the expansion of the universe.

**recollapsing universe** A model of the universe in which the collective gravity of all its matter eventually halts and reverses the expansion, causing the galaxies to come crashing back together and the universe to end in a fiery Big Crunch.

**red giant** A giant star that is red in color.

**red-giant winds** The relatively dense but slow winds from red giant stars.

**redshift** (Doppler) A Doppler shift in which spectral features are shifted to longer wavelengths, observed when an object is moving away from the observer.

**reference frame** (or **frame of reference**) In the theory of relativity, what two people (or objects) share if they are *not* moving relative to each other.

**reflecting telescope** A telescope that uses mirrors to focus light.

**reflection** (of light) The process by which matter changes the direction of light.

**reflection nebula** A nebula that we see as a result of starlight reflected from interstellar dust grains. Reflection nebulae tend to have blue and black tints.

**refracting telescope** A telescope that uses lenses to focus light.

**resonance** *See* orbital resonance.

**rest wavelength** The wavelength of a spectral feature in the absence of any Doppler shift or gravitational redshift.

**retrograde motion** Motion that is backward compared to the norm. For example, we see Mars in apparent retrograde motion during the periods of time when it moves westward, rather than the more common eastward, relative to the stars.

**revolution** The orbital motion of one object around another.

**right ascension (RA)** The angular east-west distance between the March equinox and a location on the celestial sphere; analogous to longitude, but on the celestial sphere.

**rings** (planetary) The collections of numerous small particles orbiting a planet within its Roche tidal zone.

**Roche tidal zone** The region within two to three planetary radii (of any planet) in which the tidal forces tugging an object apart become comparable to the gravitational forces holding it together; planetary rings are always found within the Roche tidal zone.

**rocks** (in solar system theory) Materials common on the surface of Earth, such as silicon-based minerals, that are solid at temperatures and pressures found on Earth but typically melt or vaporize at temperatures of 500–1300 K.

**rotation** The spinning of an object around its axis.

**rotation curve** A graph that plots rotational (or orbital) velocity against distance from the center for any object or set of objects.

**runaway greenhouse effect** A positive feedback cycle in which heating caused by the greenhouse effect causes more greenhouse gases to enter the atmosphere, which further enhances the greenhouse effect.

**saddle-shaped geometry** (or **hyperbolic geometry**) The type of geometry in which the rules—such as that two lines that begin parallel eventually diverge—are most easily visualized on a saddle-shaped surface.

**Sagittarius Dwarf** A small dwarf elliptical galaxy that is currently passing through the disk of the Milky Way Galaxy.

**saros cycle** The period over which the basic pattern of eclipses repeats, which is about 18 years $11\frac{1}{3}$ days.

**satellite** Any object orbiting another object.

**scattered light** Light that is reflected into random directions.

**Schwarzschild radius** A measure of the size of the event horizon of a black hole.

**science** The search for knowledge that can be used to explain or predict natural phenomena in a way that can be confirmed by rigorous observations or experiments.

**scientific method** An organized approach to explaining observed facts through science.

**scientific theory** A model of some aspect of nature that has been rigorously tested and has passed all tests to date.

**seafloor crust** On Earth, the thin, dense crust of basalt created by seafloor spreading.

**seafloor spreading** On Earth, the creation of new seafloor crust at mid-ocean ridges.

**search for extraterrestrial intelligence (SETI)** The name given to observing projects designed to search for signs of intelligent life beyond Earth.

**secondary mirror** A small mirror in a reflecting telescope, used to reflect light gathered by the primary mirror toward an eyepiece or instrument.

**sedimentary rock** A rock that formed from sediments created and deposited by erosional processes. The sediments tend to build up in distinct layers, or *strata*.

**seismic waves** Earthquake-induced vibrations that propagate through a planet.

**selection effect** (or **selection bias**) A type of bias that arises from the way in which objects of study are selected and that can lead to incorrect conclusions. For example, when you are counting animals in a jungle it is easiest to see brightly colored animals, which could mislead you into thinking that these animals are the most common.

**semimajor axis** Half the distance across the long axis of an ellipse; in this text, it is usually referred to as the *average* distance of an orbiting object, abbreviated $a$ in the formula for Kepler's third law.

**September equinox** Both the point in Virgo on the celestial sphere where the ecliptic crosses the celestial equator and the moment in time when the Sun appears at that point each year (around September 21).

**Seyfert galaxies** The name given to a class of galaxies that are found relatively nearby and that have nuclei much like those of quasars, except that they are less luminous.

**shepherd moons** Tiny moons within a planet's ring system that help force particles into a narrow ring; a variation on *gap moons*.

**shield volcano** A shallow-sloped volcano made from the flow of low-viscosity basaltic lava.

**shock wave** A wave of pressure generated by gas moving faster than the speed of sound.

**sidereal day** The time of 23 hours 56 minutes 4.09 seconds between successive appearances of any particular star on the meridian; essentially, the true rotation period of Earth.

**sidereal month** The time required for the Moon to orbit Earth once (as measured against the stars); about $27\frac{1}{4}$ days.

**sidereal period** (of a planet) A planet's actual orbital period around the Sun.

**sidereal time** Time measured according to the position of stars in the sky rather than the position of the Sun in the sky. *See also* local sidereal time.

**sidereal year** The time required for Earth to complete exactly one orbit as measured against the stars; about 20 minutes longer than the tropical year on which our calendar is based.

**silicate rock** A silicon-rich rock.

**singularity** The place at the center of a black hole where, in principle, gravity crushes all matter to an infinitely tiny and dense point.

**Small Magellanic Cloud** One of two small, irregular galaxies (the other is the Large Magellanic Cloud) located about 150,000 light-years away; it probably orbits the Milky Way Galaxy.

**small solar system body** An asteroid, comet, or other object that orbits a star but is too small to qualify as a planet or dwarf planet.

**snowball Earth** Name given to a hypothesis suggesting that, some 600–700 million years ago, Earth experienced a period in which it became cold enough for glaciers to exist worldwide, even in equatorial regions.

**solar activity** Short-lived phenomena on the Sun, including the emergence and disappearance of individual sunspots, prominences, and flares; sometimes called *solar weather*.

**solar circle** The Sun's orbital path around the galaxy, which has a radius of about 28,000 light-years.

**solar day** 24 hours, which is the average time between appearances of the Sun on the meridian.

**solar eclipse** An event that occurs when the Moon's shadow falls on Earth, which can happen only at new moon. A solar eclipse may be total, partial, or annular.

**solar flares** Huge and sudden releases of energy on the solar surface, probably caused when energy stored in magnetic fields is suddenly released.

**solar luminosity** The luminosity of the Sun, which is approximately $4 \times 10^{26}$ watts.

**solar maximum** The time during each sunspot cycle at which the number of sunspots is the greatest.

**solar minimum** The time during each sunspot cycle at which the number of sunspots is the smallest.

**solar nebula** The piece of interstellar cloud from which our own solar system formed.

**solar neutrino problem** A now-solved problem in which, during the latter decades of the 20th century, there appeared to be disagreement between the predicted and observed number of neutrinos coming from the Sun.

**solar prominences** Vaulted loops of hot gas that rise above the Sun's surface and follow magnetic field lines.

**solar sail** A large, highly reflective (and thin, to minimize mass) piece of material that can "sail" through space using pressure exerted by sunlight.

**solar system** (or **star system**) A star (or sometimes more than one star) and all the objects that orbit it.

**solar thermostat** *See* stellar thermostat; the solar thermostat is the same idea applied to the Sun.

**solar wind** A stream of charged particles ejected from the Sun.

**solid phase** The phase of matter in which atoms or molecules are held rigidly in place.

**solstice** *See* December solstice *and* June solstice.

**sound wave** A wave of alternately rising and falling pressure.

**south celestial pole (SCP)** The point on the celestial sphere directly above Earth's South Pole.

**spacetime** The inseparable, four-dimensional combination of space and time.

**spacetime diagram** A graph that plots a spatial dimension on one axis and time on another axis.

**special theory of relativity** Einstein's theory that describes the relativity of time and space based on the fact that the laws of nature are the same for everyone and that everyone always measures the same speed of light.

**spectral lines** Bright or dark lines that appear in an object's spectrum, which we can see when we pass the object's light through a prismlike device that spreads out the light like a rainbow.

**spectral resolution** The degree of detail that can be seen in a spectrum; the higher the spectral resolution, the more detail we can see.

**spectral type** A way of classifying a star by the lines that appear in its spectrum; it is related to surface temperature. The basic spectral types are designated by letters (OBAFGKM, with O for the hottest stars and M for the coolest) and are subdivided with numbers from 0 through 9.

**spectrograph** An instrument used to record spectra.

**spectroscopic binary** A binary star system whose binary nature is revealed because we detect the spectral lines of one or both stars alternately becoming blueshifted and redshifted as the stars orbit each other.

**spectroscopy** (in astronomical research) The process of obtaining spectra from astronomical objects.

**spectrum** (of light) *See* electromagnetic spectrum.

**speed** The rate at which an object moves. Its units are distance divided by time, such as m/s or km/hr.

**speed of light** The speed at which light travels, which is about 300,000 km/s.

**spherical geometry** The type of geometry in which the rules—such as that lines that begin parallel eventually meet—are those that hold on the surface of a sphere.

**spheroidal component** (of a galaxy) *See* halo component.

**spheroidal galaxy** Another name for an *elliptical galaxy. See also* dwarf spheroidal galaxy.

**spheroidal population** *See* halo population.

**spin** *See* spin angular momentum.

**spin angular momentum** The inherent angular momentum of a fundamental particle; often simply called *spin*.

**spiral arms** The bright, prominent arms, usually in a spiral pattern, found in most spiral galaxies.

**spiral density waves** Gravitationally driven waves of enhanced density that move through a spiral galaxy and are responsible for maintaining its spiral arms.

**spiral galaxies** Galaxies that look like flat white disks with yellowish bulges at their centers. The disks are filled with cool gas and dust, interspersed with hotter ionized gas, and usually display beautiful spiral arms.

**spreading centers** (geological) Places where hot mantle material rises upward between plates and then spreads sideways, creating new seafloor crust.

**spring equinox** *See* March equinox, which is commonly called the spring equinox by people living in the Northern Hemisphere.

**spring tides** The higher-than-average tides on Earth that occur at new and full moon, when the tidal forces from the Sun and Moon both act along the same line.

**standard candle** An object for which we have some means of knowing its true luminosity, so that we can use its apparent brightness to determine its distance with the luminosity–distance formula.

**standard model** (of physics) The current theoretical model that describes the fundamental particles and forces in nature.

**standard time** Time measured according to the internationally recognized time zones.

**star** A large, glowing ball of gas that generates energy through nuclear fusion in its core. The term *star* is sometimes applied to objects that are in the process of becoming true stars (e.g., protostars) and to the remains of stars that have died (e.g., neutron stars).

**starburst galaxy** A galaxy in which stars are forming at an unusually high rate.

**star cluster** *See* cluster of stars.

**star–gas–star cycle** The process of galactic recycling in which stars expel gas into space, where it mixes with the interstellar medium and eventually forms new stars.

**star system** *See* solar system.

**state** (quantum) *See* quantum state.

**steady state theory** A now-discredited hypothesis that held that the universe had no beginning and looks about the same at all times.

**Stefan–Boltzmann constant** A constant that appears in the laws of thermal radiation, with value

$$\sigma = 5.67 \times 10^{-8} \frac{\text{watt}}{\text{m}^2 \times \text{K}^4}$$

**stellar evolution** The formation and development of stars.

**stellar parallax** The apparent shift in the position of a nearby star (relative to distant objects) that occurs as we view the star from different positions in Earth's orbit of the Sun each year.

**stellar thermostat** The regulation of a star's core temperature that comes about when a star is in both energy balance (the rate at which fusion releases energy in the star's core is balanced with the rate at which the star's surface radiates energy into space) and gravitational equilibrium.

**stellar wind** A stream of charged particles ejected from the surface of a star.

**stratosphere** An intermediate-altitude layer of Earth's atmosphere that is warmed by the absorption of ultraviolet light from the Sun.

**stratovolcano** A steep-sided volcano made from viscous lavas that cannot flow very far before solidifying.

**string theory** New ideas, not yet well-tested, that attempt to explain all of physics in a much simpler way than current theories.

**stromatolites** Rocks thought to be fossils of ancient microbial colonies.

**strong force** One of the four fundamental forces; it is the force that holds atomic nuclei together.

**subduction** (of tectonic plates) The process in which one plate slides under another.

**subduction zones** Places where one plate slides under another.

**subgiant (luminosity class IV)** A star that is between being a main-sequence star and being a giant; subgiants have inert helium cores and hydrogen-fusing shells.

**sublimation** The process by which atoms or molecules escape into the gas phase from the solid phase.

**summer solstice** *See* June solstice, which is commonly called the summer solstice by people living in the Northern Hemisphere.

**sunspot cycle** The period of about 11 years over which the number of sunspots on the Sun rises and falls.

**sunspots** Blotches on the surface of the Sun that appear darker than surrounding regions.

**superbubble** Essentially a giant interstellar bubble, formed when the shock waves of many individual bubbles merge to form a single giant shock wave.

**superclusters** The largest known structures in the universe, consisting of many clusters of galaxies, groups of galaxies, and individual galaxies.

**super-Earth** An extrasolar planet whose size and density suggest it has an Earth-like composition, made up largely of rock and metal.

**supergiants (luminosity class I)** The very large and very bright stars that appear at the top of an H-R diagram.

**supermassive black holes** Giant black holes, with masses millions to billions of times that of our Sun, thought to reside in the centers of many galaxies and to power active galactic nuclei.

**supernova** The explosion of a star.

**Supernova 1987A** A supernova witnessed on Earth in 1987; it was the nearest supernova seen in nearly 400 years and helped astronomers refine theories of supernovae.

**supernova remnant** A glowing, expanding cloud of debris from a supernova explosion.

**surface area–to–volume ratio** The ratio defined by an object's surface area divided by its volume; this ratio is larger for smaller objects (and vice versa).

**synchronous rotation** The rotation of an object that always shows the same face to an object that it is orbiting because its rotation period and orbital period are equal.

**synchrotron radiation** A type of radio emission that occurs when electrons moving at nearly the speed of light spiral around magnetic field lines.

**synodic month** (or **lunar month**) The time required for a complete cycle of lunar phases, which averages about $29\frac{1}{2}$ days.

**synodic period** (of a planet) The time between successive alignments of a planet and the Sun in our sky; measured from opposition to opposition for a planet beyond Earth's orbit, or from superior conjunction to superior conjunction for Mercury and Venus.

**tangential motion** The component of an object's motion directed across our line of sight.

**tangential velocity** The portion of any object's total velocity that is directed across (perpendicular to) our line of sight. This part of the velocity cannot be measured with the Doppler effect. It can be measured only by observing the object's gradual motion across our sky.

**tectonics** The disruption of a planet's surface by internal stresses.

**temperature** A measure of the average kinetic energy of particles in a substance.

**terrestrial planets** Rocky planets similar in overall composition to Earth.

**theories of relativity** (special and general) Einstein's theories that describe the nature of space, time, and gravity.

**theory** (in science) *See* scientific theory.

**theory of evolution** The scientific theory, first advanced by Charles Darwin, that explains how evolution occurs through the process of natural selection.

**thermal emitter** An object that produces a thermal radiation spectrum; sometimes called a *blackbody*.

**thermal energy** The collective kinetic energy, as measured by temperature, of the many individual particles moving within a substance.

**thermal escape** The process in which atoms or molecules in a planet's exosphere move fast enough to escape into space.

**thermal pressure** The ordinary pressure in a gas arising from motions of particles that can be attributed to the object's temperature.

**thermal pulses** The predicted upward spikes in the rate of helium fusion, occurring every few thousand years, that occur near the end of a low-mass star's life.

**thermal radiation** The spectrum of radiation produced by an opaque object that depends only on the object's temperature; sometimes called *blackbody radiation*.

**thermosphere** A high, hot, x-ray-absorbing layer of an atmosphere, just below the exosphere.

**third-quarter phase** The phase of the Moon that occurs three-quarters of the way through each cycle of phases, in which precisely half of the visible face is illuminated by sunlight.

**tidal force** A force that occurs when the gravity pulling on one side of an object is larger than that on the other side, causing the object to stretch.

**tidal friction** Friction within an object that is caused by a tidal force.

**tidal heating** A source of internal heating created by tidal friction. It is particularly important for satellites with eccentric orbits, such as Io and Europa.

**time dilation** The effect in which you observe time running more slowly in reference frames moving relative to you.

**time monitoring** (in astronomical research) The process of tracking how the light intensity from an astronomical object varies with time.

**torque** A twisting force that can cause a change in an object's angular momentum.

**total apparent brightness** *See* apparent brightness. The word "total" is sometimes added to make clear that we are talking about light across all wavelengths, not just visible light.

**totality** (eclipse) The portion of a total lunar eclipse during which the Moon is fully within Earth's umbral shadow or a total solar eclipse during which the Sun's disk is fully blocked by the Moon.

**total luminosity** *See* luminosity. The word "total" is sometimes added to make clear that we are talking about light across all wavelengths, not just visible light.

**total lunar eclipse** A lunar eclipse in which the Moon becomes fully covered by Earth's umbral shadow.

**total solar eclipse** A solar eclipse during which the Sun becomes fully blocked by the disk of the Moon.

**transit** An event in which a planet passes in front of a star (or the Sun) as seen from Earth. Only Mercury and Venus can be seen in transit of our Sun. The search for transits of extrasolar planets is an important planet detection strategy.

**transmission** (of light) The process in which light passes through matter without being absorbed.

**transparent** A word used to describe a material that transmits light.

**tree of life** (evolutionary) A diagram that shows relationships between different species as inferred from genetic comparisons.

**triple-alpha reaction** *See* helium fusion.

**Trojan asteroids** Asteroids found within two stable zones that share Jupiter's orbit but lie 60° ahead of and behind Jupiter.

**tropical year** The time from one March equinox to the next, on which our calendar is based.

**Tropic of Cancer** The circle on Earth with latitude 23.5°N, which marks the northernmost latitude at which the Sun ever passes directly overhead (which it does at noon on the June solstice).

**Tropic of Capricorn** The circle on Earth with latitude 23.5°S, which marks the southernmost latitude at which the Sun ever passes directly overhead (which it does at noon on the December solstice).

**tropics** The region on Earth surrounding the equator and extending from the Tropic of Capricorn (latitude 23.5°S) to the Tropic of Cancer (latitude 23.5°N).

**troposphere** The lowest atmospheric layer, in which convection and weather occur.

**turbulence** Rapid and random motion.

**21-cm line** A spectral line from atomic hydrogen with wavelength 21 cm (in the radio portion of the spectrum).

**ultraviolet light** Light with wavelengths that fall in the portion of the electromagnetic spectrum between visible light and x-rays.

**umbra** The dark central region of a shadow.

**unbound orbits** Orbits on which an object comes in toward a large body only once, never to return; unbound orbits may be parabolic or hyperbolic in shape.

**uncertainty principle** The law of quantum mechanics that states that we can never know both a particle's position and its momentum, or both its energy and the time it has the energy, with absolute precision.

**universal law of gravitation** The law expressing the force of gravity ($F_g$) between two objects, given by the formula

$$F_g = G\frac{M_1M_2}{d^2}$$

$$\left(\text{where } G = 6.67 \times 10^{-11} \frac{\text{m}^3}{\text{kg} \times \text{s}^2}\right)$$

**universal time (UT)** Standard time in Greenwich, England (or anywhere on the prime meridian).

**universe** The sum total of all matter and energy.

**up quark** One of the two quark types (the other is the down quark) found in ordinary protons and neutrons. It has a charge of $+\frac{2}{3}$.

**vaporization** The process by which atoms or molecules escape into the gas phase from the liquid or solid phase. More technically, vaporization from a liquid is called *evaporation* and vaporization from a solid is called *sublimation*.

**velocity** The combination of speed and direction of motion; it can be stated as a speed in a particular direction, such as 100 km/hr due north.

**virtual particles** Particles that "pop" in and out of existence so rapidly that, according to the uncertainty principle, they cannot be directly detected.

**viscosity** The thickness of a liquid described in terms of how rapidly it flows; low-viscosity liquids flow quickly (e.g., water), while high-viscosity liquids flow slowly (e.g., molasses).

**visible light** The light our eyes can see, ranging in wavelength from about 400 to 700 nm.

**visual binary** A binary star system in which both stars can be resolved through a telescope.

**voids** Huge volumes of space between superclusters that appear to contain very little matter.

**volatiles** Substances, such as water, carbon dioxide, and methane, that are usually found as gases, liquids, or surface ices on the terrestrial worlds.

**volcanic plains** Vast, relatively smooth areas created by the eruption of very runny lava.

**volcanism** The eruption of molten rock, or lava, from a planet's interior onto its surface.

**waning** (phases) The set of phases in which less and less of the visible face of the Moon is illuminated; the phases that come after full moon but before new moon.

**water world** An extrasolar planet whose size and density suggest it is made largely of water, without a substantial amount of hydrogen or helium gas.

**watt** The standard unit of power in science; defined as 1 watt = 1 joule/s.

**wavelength** The distance between adjacent peaks (or troughs) of a wave.

**waxing** (phases) The set of phases in which more and more of the visible face of the Moon is becoming illuminated; the phases that come after new moon but before full moon.

**weak bosons** The exchange particles for the weak force.

**weak force** One of the four fundamental forces; it is the force that mediates nuclear reactions, and it is the only force besides gravity felt by weakly interacting particles.

**weakly interacting particles** Particles, such as neutrinos and WIMPs, that respond only to the weak force and gravity; that is, they do not feel the strong force or the electromagnetic force.

**weather** The ever-varying combination of winds, clouds, temperature, and pressure in a planet's troposphere.

**weight** The net force that an object applies to its surroundings; in the case of a stationary body on the surface of Earth, it equals mass × acceleration of gravity.

**weightlessness** A weight of zero, as occurs during free-fall.

**white dwarf limit** (or **Chandrasekhar limit**) The maximum possible mass for a white dwarf, which is about $1.4M_{Sun}$.

**white dwarfs** The hot, compact corpses of low-mass stars, typically with a mass similar to that of the Sun compressed to a volume the size of Earth.

**white dwarf supernova** A supernova that occurs when an accreting white dwarf reaches the white dwarf limit, ignites runaway carbon fusion, and explodes like a bomb; often called a *Type Ia supernova*.

**WIMPs** A possible form of dark matter consisting of subatomic particles that are dark because they do not respond to the electromagnetic force; stands for *weakly interacting massive particles*.

**winter solstice** *See* December solstice, which is commonly called the winter solstice by people living in the Northern Hemisphere.

**worldline** A line that represents an object on a spacetime diagram.

**wormholes** The name given to hypothetical tunnels through hyperspace that might connect two distant places in the universe.

**x-ray binary** A binary star system that emits substantial amounts of x-rays, thought to be from an accretion disk around a neutron star or black hole.

**x-ray burster** An object that emits a burst of x-rays every few hours to every few days; each burst lasts a few seconds and is thought to be caused by helium fusion on the surface of an accreting neutron star in a binary system.

**x-ray bursts** Bursts of x-rays coming from sudden ignition of fusion on the surface of an accreting neutron star in an x-ray binary system.

**x-rays** Light with wavelengths that fall in the portion of the electromagnetic spectrum between ultraviolet light and gamma rays.

**Zeeman effect** The splitting of spectral lines by a magnetic field.

**zenith** The point directly overhead, which has an altitude of 90°.

**zodiac** The constellations on the celestial sphere through which the ecliptic passes.

**zones** (on a jovian planet) Bright bands of rising air that encircle a jovian planet at a particular set of latitudes.

# Index

Page references preceded by a "t" refer to tables.
Page references followed by an "n" refer to notes.